SEMI-RIGID CONNECTIONS HANDBOOK

EDITED BY
WAI-FAH CHEN
NORIMITSU KISHI
MASATO KOMURO

J.ROSS PUBLISHING

ISBN: 978-1-932159-99-8

Printed and bound in the U.S.A. Printed on acid-free paper
10 9 8 7 6 5 4 3 2 1

Library of Congress Cataloging-in-Publication Data

Semi-rigid connections handbook / edited by Wai-Fah Chen, Norimitsu Kishi, and Masato Komuro.
p. cm. — (J. Ross Publishing civil & environmental engineering series)
Includes bibliographical references and index.
ISBN 978-1-932159-99-8 (hardcover : alk. paper)
1. Columns, Iron and steel--Specifications. 2. Steel framing (Building) 3. Deformations (Mechanics) 4. Structural dynamics. I. Chen, Wai-Fah, 1936- II. Kishi, Norimitsu, 1949- III. Komuro, Masato, 1969-
TA492.C7S46 2010
624.1'772--dc22
2010035278

Direct all inquiries to J. Ross Publishing, Inc., 5765 N. Andrews Way, Fort Lauderdale, FL 33309.

Phone: (954) 727-9333
Fax: (561) 892-0700
Web: www.jrosspub.com

CIVIL AND ENVIRONMENTAL ENGINEERING SERIES

Wai-Fah Chen, Editor-in-Chief

Semi-rigid Connections Handbook
by Wai-Fah Chen, Norimitsu Kishi, and Masato Komuro

Elastic Beam Calculations Handbook
by Jih-Jiang Chyu

Deepwater Foundations and Pipeline Geomechanics
by William O. McCarron

Sulfur Concrete for the Construction Industry: A Sustainable Development Approach
by Abdel-Mohsen Onsy Mohamed and Maisa El Gamal

Contents

Section IV. Steel-concrete Composite Connections

Section V. Case Study

Preface

The steel framework is one of the most commonly used structural systems in modern construction. The analysis of such structural systems is governed by the assumptions adopted in the modeling of their structural elements, especially those concerning the behavior of beam-to-column connections. Conventional analysis and design of this framework are performed using the assumption of a fully rigid or ideally pinned connection. The assumption of a fully rigid connection implies that no relative rotation of the connection occurs and the end moment of the beam is completely transferred to the column. On the other hand, the pinned connection implies that no restraint for rotation of the connection exists and the connection moment is always zero. The popularity of these idealized models results from the fact that they are simple to use and easy to implement in the analysis and design of steel frameworks. However, as evident from experimental observations, all beam-to-column connections used in current practice possess some stiffness and fall between the two extreme cases of fully rigid and ideally pinned.

In general, the connection that welds the beam directly to the flange of the column is considered fully rigid. However, the connection that fastens the beam to the column with some top and bottom angles and a web plate or with all angles, using either bolts or rivets, (with some angles and/or a plate, bolts or rivets,) displays a nonlinear behavior and lies somewhere between the fully rigid and perfectly pinned conditions, commonly called semi-rigid connections. The nonlinear semi-rigid connection behaviors can be seen clearly in many experimental results. On the other hand, single and double web-angles and header plate connections generally have much less connection stiffness and may be treated as pinned connections. The end-plate connection has connection stiffness close to that of a rigid condition, while the top- and seat-angle connection lies somewhere in between these two extremes and is usually treated as a semi-rigid connection. The recent development of the limit-state approach to design has focused particular attention on the limits of resistance and serviceability. To this end, we need accurate information regarding the behavior of structures throughout the entire range of loading up to ultimate load considering the characteristics of connection stiffness.

Research on the topic of steel frames with semi-rigid connections (Partially Restrained (PR) construction or PR connection) has been conducted over the past 20 years. Despite significant research and development efforts, usage of PR principles has nevertheless been slow in coming to the profession caused in part by the lack of easy access to reliable test data on these connections and also due to the lack of software for practical implementation. With the publication of the 2005 AISC specifications as well as Eurocode 3, practical implementation of the use of PR connections in structural systems is now a real possibility. This handbook presents a simple and comprehensive introduction that will help design practitioners implement these new developments into engineering practice. Beginning with a discussion of the new specifications and classifications of these connections, we go on to show (on the basis of the collected connections database) practical mathematical models for computer implementation and provide case studies on these frames including composite construction. With the help of the user-friendly list of collected data in tabular form with illustrative figures, information on semi-rigid connections is now available in a single publication and may ultimately result in its widespread usage among practitioners.

The scope of the book is indicated by the table of contents with the following features:

- Introduces the 2005 AISC specifications with a cross reference to the counterpart of the new Eurocode 3 on semi-rigid construction
- Includes more than 700 semi-rigid connection test data in tabular form with figures
- Provides connection models for analysis and design with case studies
- Includes the recent development of steel-concrete composite connections with case studies

We are grateful to the following contributors for their advice, help, and original work at various stages during the preparation of the book: S. L. Chan, Y. Goto, J. Y. R. Liew, and E. Lui.

W. F. Chen
N. Kishi
M. Komuro

About the Editors

Dr. Wai-Fah Chen is a well-respected leader in the field of plasticity, structural stability, and structural steel design over the past half century. Having headed the engineering school and department at the University of Hawaii and Purdue University, Chen is a widely cited author for his contributions in the fields of mechanics, materials, and computing. He is the author or coauthor of 20 engineering books and the recipient of several national engineering awards, including the 1990 Shortridge Hardesty Award from the American Society of Civil Engineers and the 2003 Lifetime Achievement Award from the American Institute of Steel Construction. He is a member of the U.S. National Academy of Engineering and an honorary member of the American Society of Civil Engineers.

Dr. Norimitsu Kishi received his Ph.D. in civil engineering from Hokkaido University-Japan in 1977. From March 1985 to March 1986, he was an overseas researcher at Purdue University under Professor W. F. Chen. He is currently professor and director of the Structural Mechanics Laboratory in the Department of Civil Engineering and Architecture at the Muroran Institute of Technology-Japan.

Dr. Masato Komuro graduated with a degree in civil engineering in 1992 from the Muroran Institute of Technology-Japan, where he also obtained his Ph.D. in structural engineering in 2001. From May 2007 to April 2008, he was a visiting researcher in the Department of Civil & Environmental Engineering at the University of Hawaii-Manoa. He is currently assistant professor in the Department of Civil Engineering and Architecture at the Muroran Institute of Technology.

Contributors

S. L. Chan
Department of Civil and Structural Engineering, The Hong Kong Polytechnic University, Hung Hom, Kowloon, Hong Kong

Hong Chen
Offshore Technology Development Pte Ltd, Keppel Offshore & Marine, Singapore

Yoshiaki Goto
Dr. Eng. Professor, Department of Civil Engineering, Nagoya Institute of Technology, Japan

Dennis Lam
Department of Civil Engineering, University of Leeds, U.K.

J. Y. Richard Liew
Department of Civil Engineering, National University of Singapore, Singapore

Y. P. Liu
Department of Civil and Structural Engineering, The Hong Kong Polytechnic University, Hung Hom, Kowloon, Hong Kong

Eric M. Lui
Meredith Professor, Department of Civil and Environmental Engineering, Syracuse University, NY

Zhiling Zhang
Graduate Student, Department of Civil and Environmental Engineering, Syracuse University, NY

This book has free material available for download from the Web Added Value™ resource center at *www.jrosspub.com*

At J. Ross Publishing we are committed to providing today's professional with practical, hands-on tools that enhance the learning experience and give readers an opportunity to apply what they have learned. That is why we offer free ancillary materials available for download on this book and all participating Web Added Value™ publications. These online resources may include interactive versions of material that appears in the book or supplemental templates, worksheets, models, plans, case studies, proposals, spreadsheets, and assessment tools, among other things. Whenever you see the WAV™ symbol in any of our publications, it means bonus materials accompany the book and are available from the Web Added Value™ Download Resource Center at www.jrosspub.com.

Downloads for *Semi-rigid Connections Handbook* consist of updated connection data, the Steel Connection Data Bank (SCDB) program, references, and a link to the semi-rigid website at the Muroran Institute of Technology, Japan.

1

Classification and AISC Specification

Yoshiaki Goto
Dr. Eng. Professor, Department of Civil Engineering, Nagoya Institute of Technology, Japan

1.1 Classification of Connections

1.1.1 General

A beam-to-column connection is generally subject to axial force, shear force, and bending moment for its in-plane behavior. However, the deformation of the connection caused by axial and shear forces are usually small when compared to the deformation caused by bending moment. Consequently, for practical purposes, only the effect of moment on the rotational deformation of connections shown in Figure 1.1 needs to be considered. The in-plane behavior of connections can be represented by the moment-rotation (M–θ_r) curves illustrated in Figure 1.2. All types of actual beam-to-column connections possess some rotational stiffness that falls between the two extreme cases of fully rigid and ideally pinned. Thus, the modeling of connections as semi-rigid is more realistic. However, in engineering practice, some connections can be considered pinned if their stiffness is so small that the connections are incapable of transmitting any significant moment, thus permitting almost free rotation. Similarly, some connections can be considered rigid if their rigidity is so large that no significant slope discontinuity exists between the adjoining members. The assumption of ideally pinned or rigid connections considerably simplifies the design and analysis procedures of framed structures. For that reason, it is beneficial and practical to estimate whether the connec-

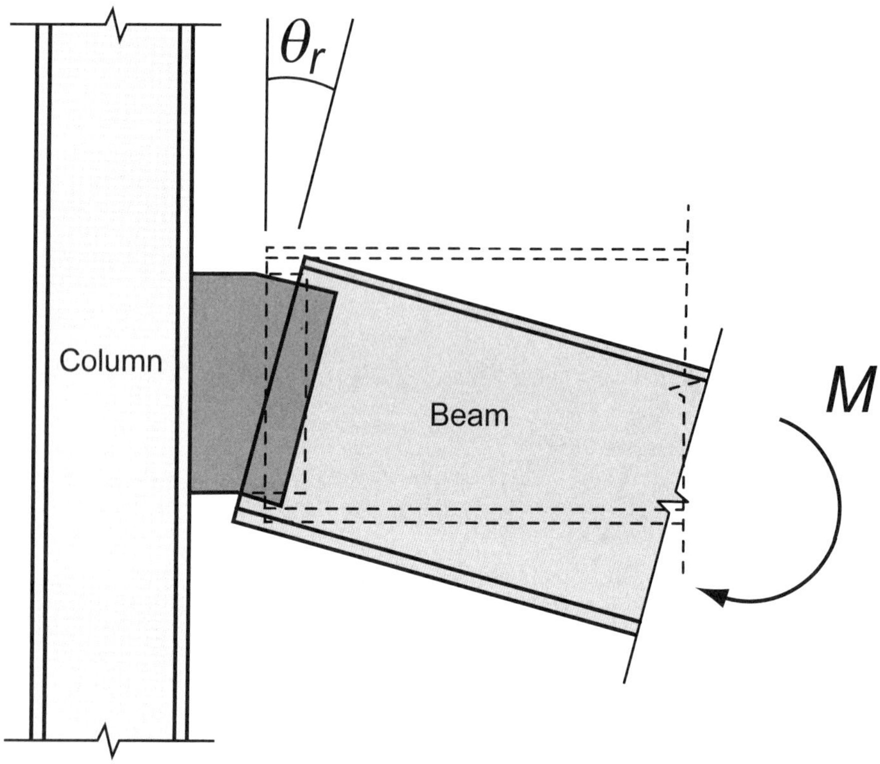

FIGURE 1.1 Rotational deformation of a connection.

tions can be assumed rigid, semi-rigid, or pinned. The classification of connections should be made based on the behavior of the frames at serviceability and ultimate-limit states. Primary index properties that govern the moment-rotation characteristics of connections are stiffness, strength, and rotational capacity. These properties are important factors that are used for the classification. However, the requirement for strength and rotational capacity of connections is not always necessary. This is because the strength and rotational capacity of connections are determined during the design procedure, such that they exceed the calculated moment and rotation demand at the ultimate-limit state of frames.

The classification systems of connections have been presented by Bjorhovde et al. (1990), Goto et al. (1998), Nethercot et al. (1998), Eurocode 3 (CEN, 2005), and ANSI/AISC360-06 (AISC, 2005). The following explains and discusses these existing classification systems.

1.1.2 Eurocode 3 Classification System (CEN 2005)

Eurocode 3 (2005) models connections as simple, semicontinuous, and continuous in structural analysis. In the simple connection model, a connection is assumed not to transmit bending moment and may be modeled as ideally pinned. In the semicontinuous connection model, the behavior of a connection has to be taken into account and modeling of the so-called semi-rigid

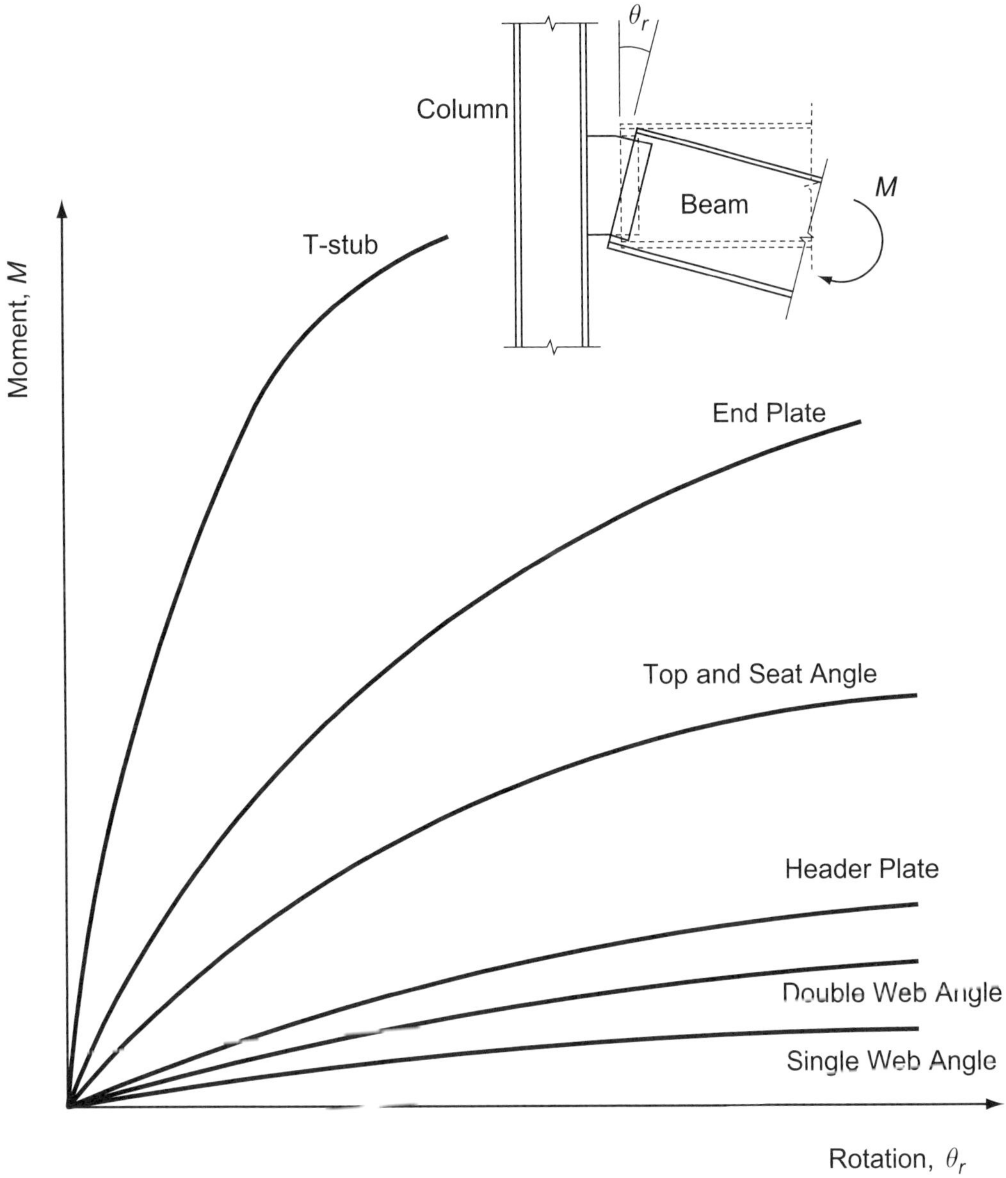

FIGURE 1.2 Moment-rotation curves of semi-rigid connections.

connection is necessary in the structural analysis. In the continuous connection model, the behavior of a connection is assumed to have no effect on the analysis and may be modeled as so-called rigid. The above three types of connection models are classified based on the two independent criteria expressed in terms of rotational stiffness and strength, respectively. The scheme of modeling of connections is shown in Table 1.1, specifically for the case of nonlinear elastic-plastic structural analysis. This is because the types of connection models used in this analysis are determined based on both stiffness and strength of a connection in contrast to

TABLE 1.1 Types of EC3 connection models used in elastic-plastic structural analysis

Type of connection model	Classification by stiffness[1]	Classification by strength[2]
Continuous	Rigid	Full strength
Semi-continuous	Rigid	Partial strength
	Semi-rigid	Full strength Partial strength
Simple	Nominally pinned	Nominally pinned

Remarks: 1) See Table 1.2; 2) See Table 1.3

those used in elastic analysis or rigid-plastic analysis where either stiffness or strength is considered in the determination of connection modeling.

As shown in Table 1.2, connections are classified as rigid, semi-rigid, or nominally pinned based on the initial rotational stiffness R_{ki} of connections. The boundary between rigid and semi-rigid connections is determined, such that the load-carrying capacity of a semi-rigid portal frame is higher than 95 percent of the strength of the corresponding rigid frame. A frame model used to calculate the load-carrying capacity is a pin-based portal frame shown in Figure 1.3. The load carrying capacity is estimated by the Marchant-Rankin formula (Horne and Merchant, 1965). Since the effect of connection stiffness on the frame strength is dependent on whether the frame is braced or unbraced, the boundaries for the classification of connections are differently specified for these two types of frames. These boundaries are expressed in terms of a nondimensional initial connection stiffness parameter k_b defined as $k_b = R_{ki}/(EI_b/L_b)$.

Similar to the classification based on the rotational stiffness, connections are classified as full strength, partial strength or nominally pinned, depending on the ratio between the design mo-

TABLE 1.2 EC3 classification of connections by stiffness

Connections		$k_b = R_{ki}/(EI_b/L_b)$	Other requirement
Rigid	Braced frames	$8 < k_b$	
	Unbraced frames	$25 < k_b$	$K_b/K_c \geq 0.1$
Semi-rigid	Braced frames	$0.5 \leq k_b \leq 8$	
	Unbraced frames	$25 < k_b$	$K_b/K_c < 0.1$
		$0.5 \leq k_b \leq 25$	
Nominally pinned		$k_b < 0.5$	

Remarks:

Braced frames = those where the bracing system reduces the horizontal displacement by at least 80%

R_{ki} = initial rotational stiffness of connections

K_b = the mean value of I_b/L_b for all the beams at the top of that story

K_c = the mean value of I_c/L_c for all the columns in that story

I_b = the second moment of inertia of a beam

I_c = the second moment of inertia of a column

L_b = the span of a beam

L_c= the story height of a column

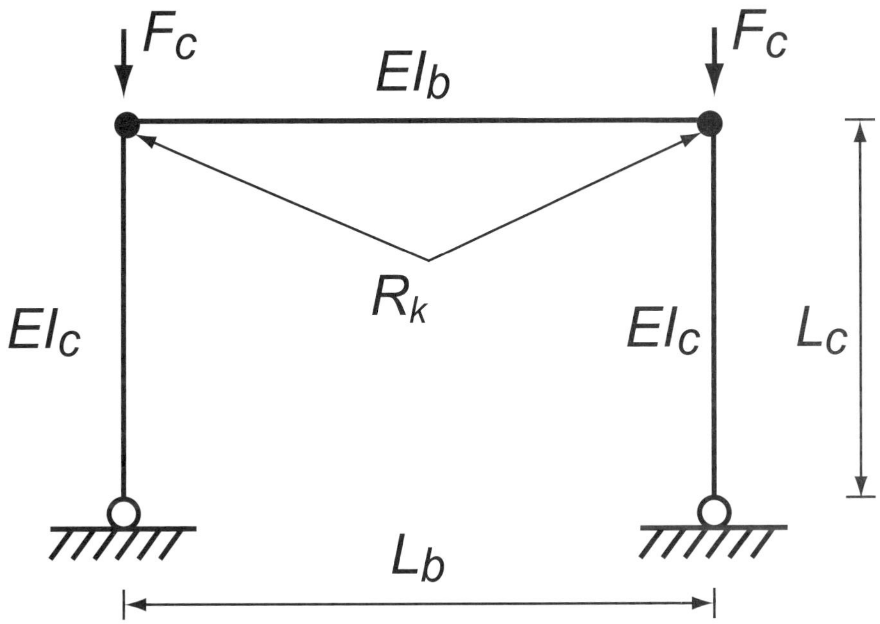

FIGURE 1.3 Pin-based portal frame.

ment resistance of connections $M_{j,Rd}$ and the design plastic moment resistance of the connected members such as beams $M_{b,pl,Rd}$ and columns $M_{c,pl,Rd}$. The boundaries for this classification are different according to whether the connections are located at the top of columns or within the height of columns (Figure 1.4). The classification based on the connection strength is summarized in Table 1.3.

Although the required rotational capacity of connections is not specified in EC3 classification, being different from the classifications by Bjorhovde et al. (1990), Goto et al. (1998) and Nethercot et al. (1998), the connections must be designed such that the deformation of connections imposed by the internal forces and moment is within the deformation capacity of the connections. An almost similar scheme is adopted by the AISC specification (AISC, 2005).

EC3 considers the behavior of frames at the ultimate-limit state in the classification of connections. However, it does not explicitly take into account the behavior at the serviceability limit state. Furthermore, in the evaluation of the load-carrying capacity of frames, EC3 adopts an approximate formula (i.e., the Merchant-Rankine formula). In addition, a single-story portal frame model used in this evaluation may be too simple to reflect the effect of layout and member details of frames in practice.

FIGURE 1.4 Full-strength of connections: (a) top of column; (b) within column height.

TABLE 1.3 EC3 classification of connections by strength

Connection		$M_{j,Rd}$	Other requirements
Full-strength	Top of column	$1.0 \le \left(\frac{M_{j,Rd}}{M_{b,pl,Rd}} \text{ or } \frac{M_{j,Rd}}{M_{c,pl,Rd}} \right)$	
	Within column height	$1.0 \le \left(\frac{M_{j,Rd}}{M_{b,pl,Rd}} \text{ or } \frac{M_{j,Rd}}{2M_{c,pl,Rd}} \right)$	
Partial-strength	Top of column	$0.25 < \left(\frac{M_{j,Rd}}{M_{b,pl,Rd}} \text{ and } \frac{M_{j,Rd}}{M_{c,pl,Rd}} \right) < 1.0$	
	Within column height	$0.25 < \left(\frac{M_{j,Rd}}{M_{b,pl,Rd}} \text{ and } \frac{M_{j,Rd}}{2M_{c,pl,Rd}} \right) < 1.0$	
Nominally pinned	Top of column	$\left(\frac{M_{j,Rd}}{M_{b,pl,Rd}} \text{ or } \frac{M_{j,Rd}}{M_{c,pl,Rd}} \right) \le 0.25$	1. Capable of transmitting internal forces without developing significant moment that might adversely affect the members or the structure as a whole 2. Capable of accepting the resulting rotation under the design load
	Within column height	$\left(\frac{M_{j,Rd}}{M_{b,pl,Rd}} \text{ or } \frac{M_{j,Rd}}{2M_{c,pl,Rd}} \right) \le 0.25$	

Remarks:

$M_{j,Rd}$ = the design moment resistance of a connection

$M_{b,pl,Rd}$ = the design plastic moment resistance of the connected beam

$M_{c,pl,Rd}$ = the design plastic moment resistance of the connected column

A precise elastic-plastic finite-displacement analysis showed that the EC3 boundary between rigid and semi-rigid connections is on the whole restrictive in terms of the ultimate strength of frames (Goto and Miyashita, 1995).

1.1.3 AISC Classification System (AISC 2005)

AISC specification classifies connections as fully restrained (FR), partially restrained (PR), and simple connections, based on the secant stiffness R_{ks} at service load and the strength M_n, respectively, of a connection, as summarized in Table 1.4. This classification is applied to both sway and nonsway frames. Simple and FR connections may be idealized as pinned and rigid, respectively, for the purpose of structural analysis. For structures with PR connections, the connection flexibility must be estimated and included in the structural analysis by the semi-rigid connection model.

The boundary between the fully restrained and partially restrained connections expressed in terms of the secant connection stiffness parameter $R_{ks}L_b/EI_b$ at the service load is determined such that the ratio of the connection moment obtained from semi-rigid analysis to that obtained from rigid analysis is 90 percent. L_b and EI_b are the length and bending rigidity of the connected beam. The connection moment to determine the classification boundary is calculated by using a simple beam with semi-rigid connections at both ends, as shown in Figure 1.5 (Leon, 1994).

TABLE 1.4 AISC classification of connections

<table>
<tr><th colspan="2">Connections</th><th>Stiffness</th><th>Strength</th><th>Ductility</th></tr>
<tr><td colspan="2">Simple</td><td>$R_{ks}L_b/EI_b \le 2$</td><td>• Satisfy shear force demand at strength-limit state
• $M \le 0.02\ M_{pb}$ at 0.02 radians</td><td>• Satisfies rotation demand at the strength-limit state or
• $\theta_u = 0.03$ radians</td></tr>
<tr><td colspan="2">PR</td><td>$2 < R_{ks}L_b/EI_b < 20$</td><td>• Satisfy combined shear force and moment demand at strength-limit state</td><td>• Satisfies rotation demand at the strength-limit state or
• $\theta_u = 0.03$ radians</td></tr>
<tr><td rowspan="2">FR</td><td>With partial-strength of connected beam</td><td rowspan="2">$20 \le R_{ks}L_b/EI_b$</td><td rowspan="2">• Satisfy combined shear force and moment demand at strength-limit state</td><td>• Satisfies rotation demand at the strength-limit state or
• $\theta_u = 0.03$ radians</td></tr>
<tr><td>With full-strength of connected beam</td><td>• No requirement</td></tr>
</table>

Remarks:
R_{ks} = secant stiffness of connections at service load
I_b = second moment of inertia of beam
L_b = span of beam
E = Young's modulus
M_{pb} = plastic moment of beam
θ_u = rotational capacity

As for the flexural strength M_n of connection, an explicit condition in the classification is imposed on the simple connections with no flexural strength. A connection with its moment of $M \le 0.2M_{pb}$ (M_{pb} = fully plastic moment of the connected beam) at the rotation of 0.02 radians is considered to have no flexural strength for design analysis. The strengths of FR and PR connections must be adequate to resist the connection moment demand imposed by the design load. Therefore, it is possible in the AISC specification for an FR connection to have strength less than that of a connected beam. In this case, the semi-rigid action of the connection will appear if the plastification occurs in the connection. This implies that the rigid connection model is only

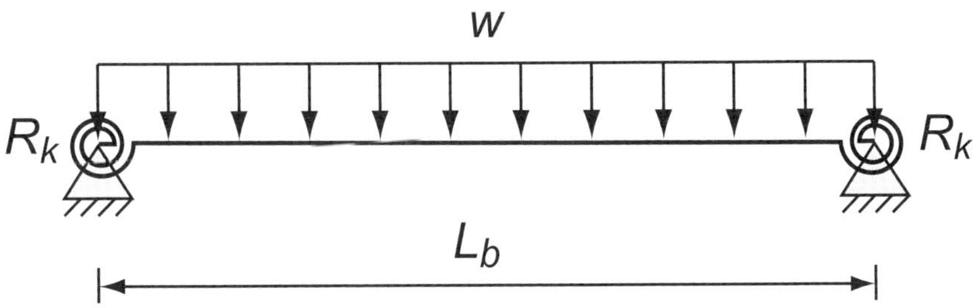

FIGURE 1.5 Simple beam with semi-rigid connections at both ends.

applicable before the plastication of the connection. It is also possible for a PR connection to have strength greater than that of the beam. Considering the above, the strength requirement for FR connections in the AISC is not as strict as that for rigid connections in EC3 where rigid connections must always have strength greater than the full strength of the connected beams or columns.

The rotation capacity θ_u of connections is determined such that the rotation of connections at the strength-limit state of frames will not exceed θ_u. If the strength of a connection is substantially higher than that of the connected beam, the ductility of the structural system is controlled by the connected beam. However, if a connection has marginally higher or partial strength of the connected beam, the connection may experience some deformation before the structural system reaches its strength-limit state. In the above case, the ductility demand for FR, PR, and simple connections is calculated by a structural analysis in which the behavior of the connections is considered. This implies that the ductility required of the connection will depend on the particular application. In the absence of accurate analysis, the rotation capacity of 0.03 radians is considered adequate. The rotation capacity of connections defined by the AISC is explained in 1.2.2.4.

1.1.4 Bjorhovde et al. Classification System

A classification system proposed by Bjorhovde et al. (1990), as shown in Figure 1.6, is expressed in terms of nondimensional parameters of M/M_{pb} and $\theta_r/\{M_{pb}(5d)/EI_b\}$, where d and EI_b are the depth and bending rigidity of the connected beam, respectively, and M_{pb} is the fully plastic moment capacity of the beam. As a reference beam length, $5d$ is assumed. Therefore, the initial stiffness of a connection is expressed in the form $EI_b/(\bar{\alpha}d)$ where $\bar{\alpha}d$ is the equivalent length of the beam that gives the beam the same stiffness as the connection. If it is assumed that L_b is equal to $5d$, the Bjorhovde's classification parameter $\bar{\alpha}$ is related to the EC3 (CEN, 2005) connection stiffness parameter k_b as $k_b = 5/\bar{\alpha}$.

Based on the stiffness and strength of connections, connections are classified as rigid, semi-rigid, and flexible. Flexible connections may be modeled as ideally pinned in structural analysis. In addition, minimum rotational capacity inversely proportional to M/M_{pb} is specified for the respective connections. The Bjorhovde's classification system is intended for the case where prior knowledge concerning the member and structural details is unavailable.

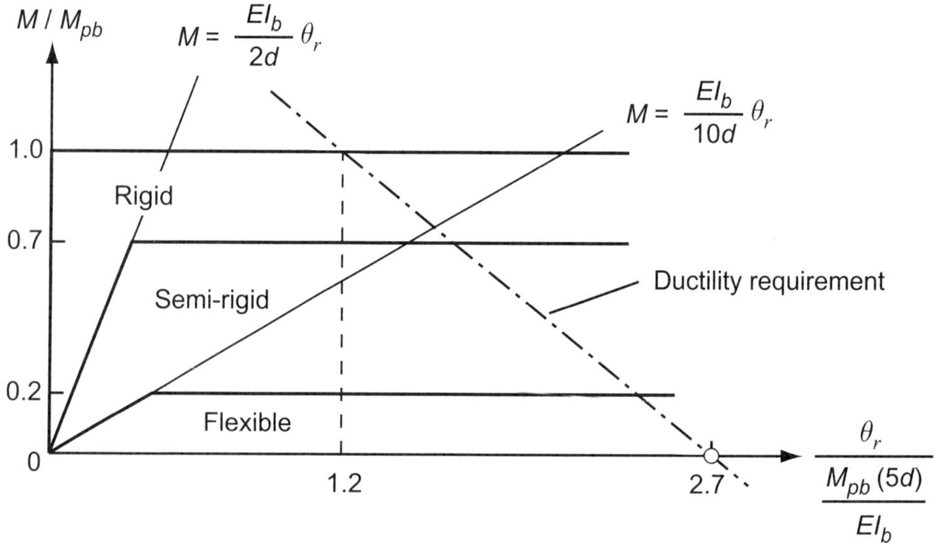

FIGURE 1.6 Classification (Bjorhovde et al., 1990).

However, this classification system is considerably approximate in nature, because it does not consider the overall behavior of frames. In fact, a precise elastic-plastic geometrically nonlinear analysis showed that the behavior of rigid frames in some cases is not assured by the Bjorhovde's condition of rigid connections where the moment capacity of connections is higher than the 70 percent of the fully plastic moment of the connected beam (Goto and Miyashita, 1995).

1.1.5 Nethercot et al. Classification System

In this classification system, a classification of connections is made based on their stiffness, strength, and rotational capacity. The categories used in the classification are fully-connected, partially-connected, pin-connected and non-structural connections. These categories are specified both for serviceability and for ultimate-limit states as illustrated in Figures 1.7 and 1.8, respectively. In the classification, parameter α defined as $\alpha = 8(EI_c/L_c)/(EI_b/L_b)$ is considered for classification. Nethercot's classification of connections considers the stiffness of a column and a beam that a connection is connecting. In the category of fully-connected connections, the behavior of connections is assumed not to affect the behavior of framed structures and may be modeled as so-called rigid. In the category of partially-connected connections, the behavior of connections has to be taken into account and modeling of the so-called semi-rigid connection is necessary in structural analysis. Pin-connected connections are not assumed to transmit bending moment and may be modeled as ideally pinned. The non-structural connections are those that do not satisfy the requirement for strength and rotational capacity. The non-structural connections are not regarded as a structural element.

In the classification at the ultimate-limit state, as shown in Figure 1.8, the stiffness boundary between the fully-connected connection and partially-connected connection is determined such that the ratio of the connection moment obtained from semi-rigid analysis to that obtained from rigid analysis is 95 percent. In addition, the fully-connected connections and partially-connected connections with the stiffness $k_b = R_{ki}/(EI_b/L_b)$ greater than $1/(0.53 - 1/\alpha)$ are required to have a moment capacity at least equal to the connected beam moment capacity. In this case, a required

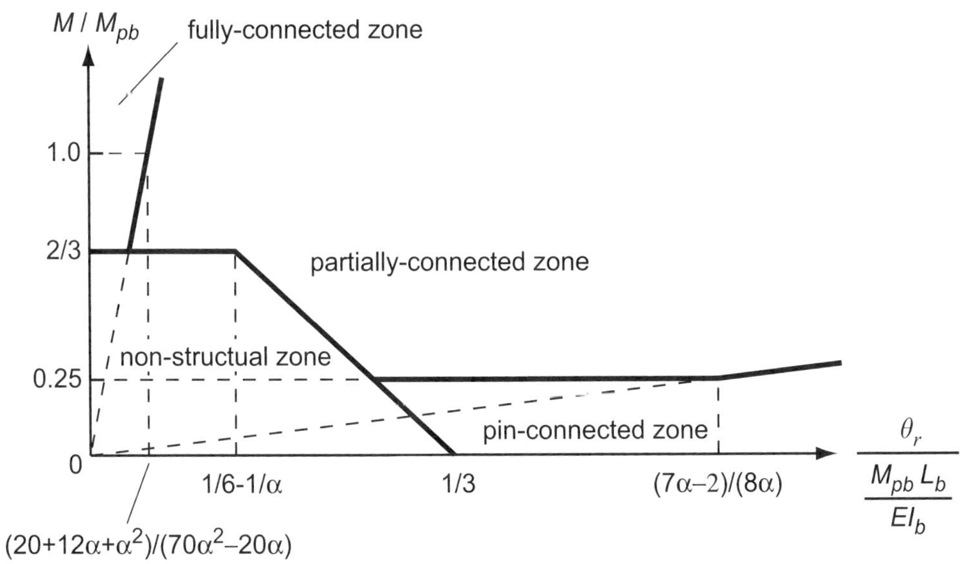

FIGURE 1.7 Classification at serviceability limit state (Nethercot et al., 1998).

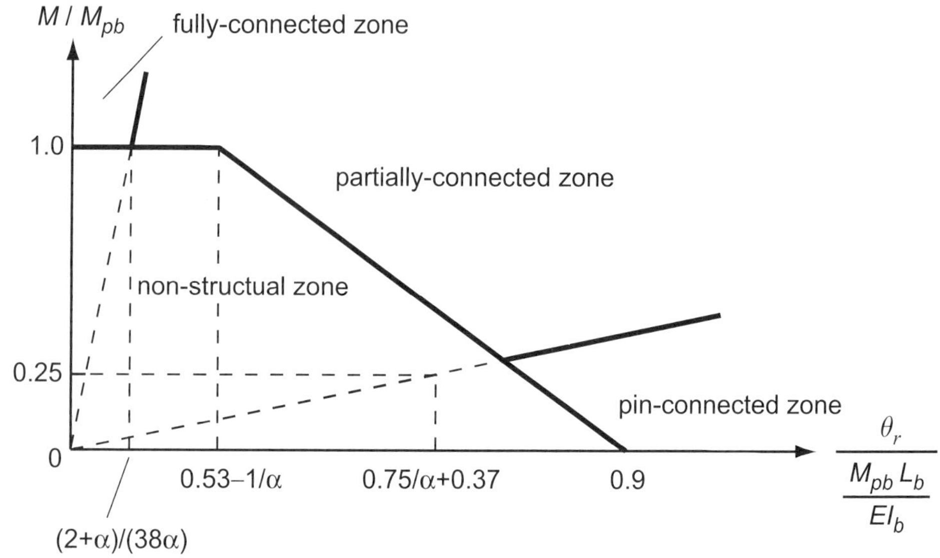

FIGURE 1.8 Classification at ultimate-limit state (Nethercot et al., 1998).

rotational capacity is not specified for connections, considering that the rotation will be mainly concentrated at the beam end.

The stiffness boundary between partially-connected and pin-connected connections at the ultimate-limit state is calculated under the condition that the partially-connected connection moment is 25 percent of rigid connection moment. To obtain the connection moment, a single span subframe under uniformly distributed beam load as illustrated in Figure 1.9 is used. This subframe model may be interpreted as a subassemblage of nonsway frames. Since partially-connected connections and pin-connected connections generally undergo some amount of rotation, the minimum required rotational capacity, which is inversely proportional to M/M_{pb} is specified for the connections with the initial stiffness $k_b = R_{ki}/(EI_b/L_b)$ less than $1/(0.53 - 1/\alpha)$.

In the classification at the serviceability limit state shown in Figure 1.7, the stiffness boundary between the fully-connected and partially-connected connections is determined as $k_b = (70\alpha^2 - 20\alpha)/(20 + 12\alpha + \alpha^2)$ such that the ratio of the beam displacement obtained from rigid frame analysis to that obtained from semi-rigid frame analysis is 90 percent. In addition, the fully-connected connections and partially-connected connections with the stiffness $k_b = R_{ki}/(EI_b/L_b)$ greater than $(2/3)/(1/6 - 1/\alpha)$ are required to have a moment capacity that exceeds at least 2/3 of the connected-beam moment capacity. The stiffness boundary between partially-connected and pin-connected connections is calculated as $k_b = 2\alpha/(7\alpha - 2)$ under the condition that the beam deflection ratio between partially-connected connections and pin-connected connections is 90 percent. This boundary is applied to the range where $\theta_r EI_b/(M_{pb}L_b) \geq (7\alpha - 2)/(2\alpha)$. In the range $\theta_r EI_b/(M_{pb}L_b) \leq (7\alpha - 2)/(2\alpha)$, the strength boundary between partially-connected and pin-connected connections is given at the location of 25 percent of the connected beam moment capacity. Similar to the classification at the ultimate-limit state, the minimum rotational capacity inversely proportional to M/M_{pb} is specified for the partially-connected and pin-connected connections with the initial stiffness $k_b = R_{ki}/(EI_b/L_b)$ less than $2/\{3(1/6 - 1/\alpha)\}$.

Although the rotational capacity and strength of connections are specified for both serviceability and ultimate-limit states in the Nethercot et al. classification system, the required rotational capacity and strength at the ultimate-limit state normally dominates. So, there is little reason to

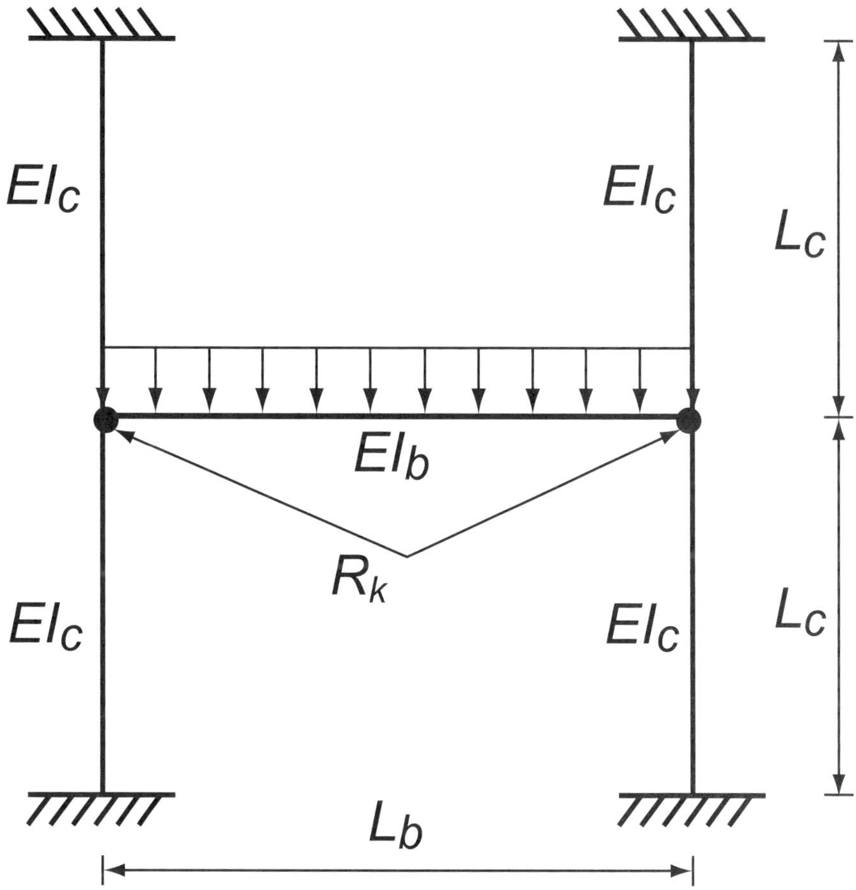

FIGURE 1.9 Single-span subframe under uniformly distributed beam load.

specify these values in the classification at the serviceability limit states. Furthermore, it will be cumbersome to use different connection models in the frame analysis for design, depending on whether the limit state of concern is the ultimate or serviceability limit state.

1.1.6 Goto et al. Classification System

Goto et al. (1998) proposed a more detailed and precise classification system between rigid and semi-rigid connections considering the overall behavior of frames not only at the ultimate-limit state but also at the serviceability limit state. Herein, corrections are made to the mistakes in the original paper (Goto et al., 1998). In addition, a modification is made to improve the original classification scheme.

To take into account the behavior of frames in the classification of connections, several subassemblages are used. These subassemblages approximately represent the behavior of the respective parts of the multistory multibay frames shown in Figure 1.10. The subassemblages are chosen by considering the deformation patterns of the respective parts of the sway and nonsway frames illustrated in Figure 1.11. The subassemblages so chosen are summarized in Figure 1.12 together with loading patterns. In this figure, the notations As ~ Fs and An ~ Fn are used to express the respective subassemblages that represent the parts of the frames in Figure 1.11.

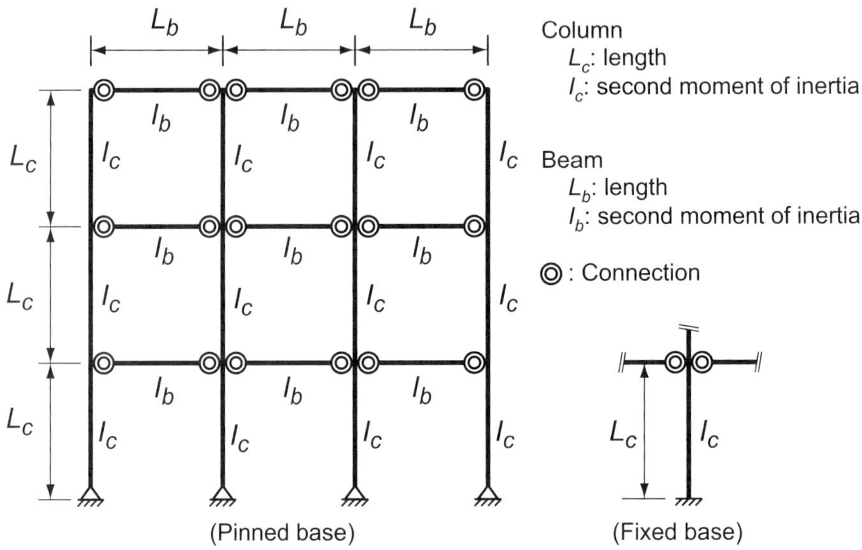

FIGURE 1.10 Multistory and multibay frame.

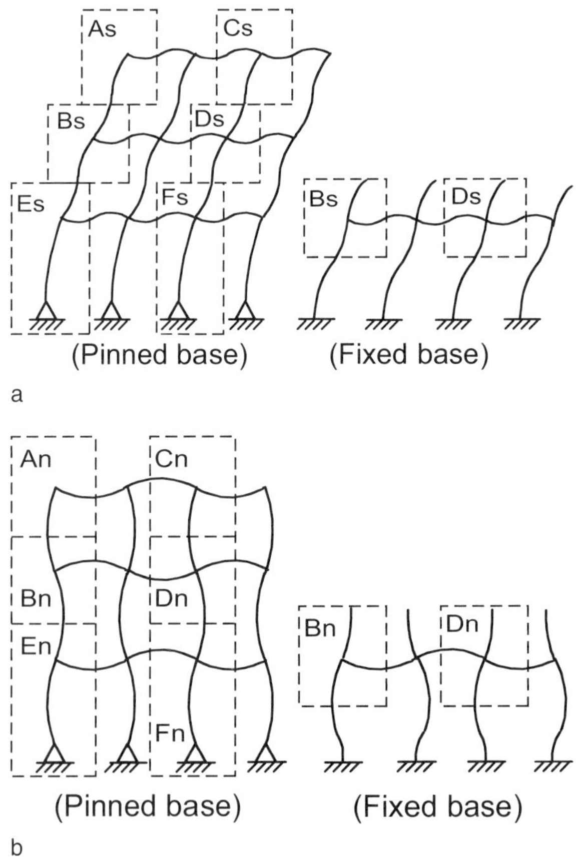

FIGURE 1.11 Deformation pattern of multistory, multibay frames: (a) sway; (b) nonsway.

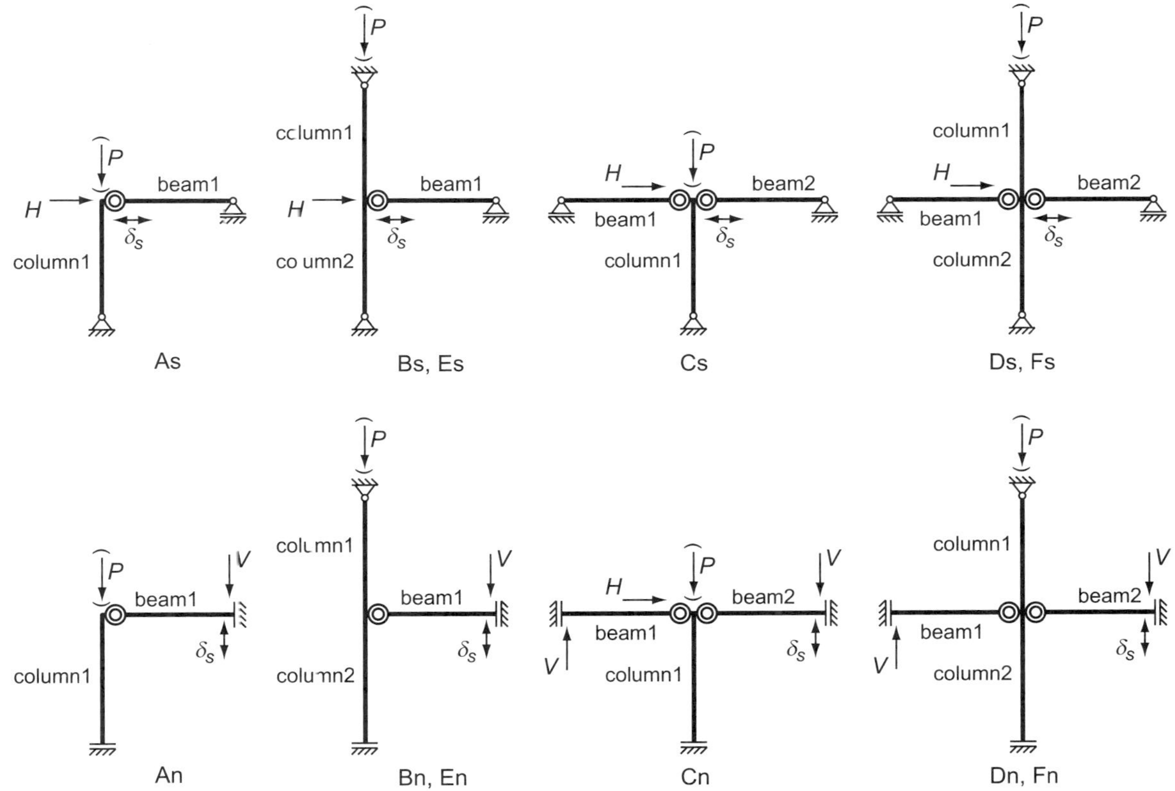

	As, An	Bs, Bn	Cs, Cn	Ds, Dn	Es, En	Fs, Fn
column1	$L_c/2$	$L_c/2$	$L_c/2$	$L_c/2$	$L_c/2$	$L_c/2$
column2	--	$L_c/2$	--	$L_c/2$	L_c	L_c
beam1	$L_b/2$	$L_b/2$	$L_b/2$	$L_b/2$	$L_b/2$	$L_b/2$
beam2	--	--	$L_b/2$	$L_b/2$	--	$L_b/2$

$\left.\begin{matrix} V \\ H \end{matrix}\right\}$ = loads at serviceability limit state

P = load at ultimate limit state

Note : As ~ Fs , An ~ Fn denote the subassemblages of multistory multibay frames shown in Fig. 1.11.

FIGURE 1.12 Subassemblages.

It is assumed that the initial connection stiffness affects the behavior of frames at the serviceability limit state and the boundary value of the initial stiffness between rigid and semi-rigid connections is derived based on the elastic displacement δ_s. This elastic displacement for sway subassemblages is calculated under horizontal force H, while elastic displacement for nonsway subassemblages are calculated under vertical force V. At the serviceability limit state, the effect of vertical compressive force P on the displacement δ_s is small; therefore, the elastic displacement is calculated by the small displacement theory, ignoring the effect of P. The criterion used to classify the connections as rigid is determined based on the tolerance Δ defined by (Equation 1.1):

$$\Delta = (\delta_s - \delta_r)/\delta_r \tag{1.1}$$

where δ_s is a displacement of a subassemblage with semi-rigid connections and δ_r is a displacement of the corresponding rigid subassemblage.

The boundary of the initial stiffness R^b_{ki} between rigid and semi-rigid connections so determined in terms of nondimentionalized initial stiffness parameter $k^b_b = R^b_{ki}/(EI_b/L_b)$ are summarized in Table 1.5, classified according to the respective types of subassemblages. In this table, to compare with EC3 boundary between rigid and semi-rigid connections, numerical results are also shown for $\Delta = 0.05$ where $G = (EI_b/L_b)/(EI_c/L_c) = 1.4$ is assumed similar to the EC3 classification. In Table 1.5, the mistakes in the original paper (Goto et al., 1998) are corrected.

To consider the connection behavior at the ultimate-limit state, the nonlinear moment-rotation relations are assumed to be expressed by the three-parameter model (Kishi et al.,1993) shown in Equation 1.2:

$$m = \bar{m}_u \bar{\theta} \big/ (1 + \bar{\theta}^n)^{1/n} \tag{1.2}$$

TABLE 1.5 Goto et al. (1998) classification for initial stiffness of connections

Subassemblages		$k^b_b = R^b_{ki}/(EI_b/L_b)$	$k^b_b = R^b_{ki}/(EI_b/L_b)$ $G = (EI_b/L_b)/(EI_c/L_c) = 1.4$ $\Delta = 0.05$
Sway	As, Ds, Es	$\dfrac{6}{(1+G)\Delta}$	50.0
	Bs	$\dfrac{6}{(1+G/2)\Delta}$	70.6
	Cs, Fs	$\dfrac{6}{(1+2G)\Delta}$	31.6
Nonsway	An, Dn	$\dfrac{6}{(4G+1)(G+1)\Delta} - \dfrac{2}{G+1}$	6.7
	Bn	$\dfrac{12}{(4G+1)(G+1)\Delta} - \dfrac{4}{G+1}$	17.4
	Cn	$\dfrac{6}{(8G+1)(2G+1)\Delta} - \dfrac{2}{2G+1}$	10.0

Remarks: $G = (EI_b/L_b)/(EI_c/L_c)$, $\Delta = (\delta_s - \delta_r)/\delta_r$

where $m = M/M_{pb}$, $\bar{m}_u = M_u/M_{pb}$, $\bar{\theta} = \theta_r/\theta_0$, $\theta_0 = M_u/R_{ki}$. M, M_u, and M_{pb} are connection moment, ultimate-connection moment capacity, and fully plastic moment of connected beam, respectively.

Equation 1.2 has different shapes as illustrated in Figure 1.13 for $\bar{m}_u = 1.0$ according to the values of n. As can be seen from Equation 1.2, the connection curve is uniquely determined by three parameters. That is, ultimate-moment capacity M_n, initial stiffness R_{ki}, and shape parameter n. The shape parameter n is determined such that the connection model given by Equation 1.2 best fits the experimental data. In the original classification system by Goto et al. (1998), empirical equations (Kishi and Chen, 1993) based on a statistical analysis of the connection data bank by Chen and Kishi (1989) were used in order to estimate the values of n for the respective connection types. This is to reduce the independent parameters of Equation 1.2 to R_{ki} and M_u in view of the simplification in the classification procedure. However, the empirical equations for the estimation of the shape parameter n are considerably approximate and may impair the accuracy of the classification system. Therefore, the boundary moment-rotation curves are herein determined in terms of R_{ki}, M_u, and n, being different from the original classification system (Goto et al., 1998). The boundary values for the initial stiffness R_{ki} between rigid and semi-rigid connections are determined, as shown in Table 1.5, from the behavior of subassemblages at the serviceability limit state. On the other hand, the boundary values for the ultimate-moment capacity M_u and shape parameter n are expressed in terms of interaction equations that are determined by the behavior of subassemblages at the ultimate-limit state.

In order to classify the connections to be rigid from the ultimate behavior of frames, EC3 uses the following criterion, which only considers the ultimate strength of the frames:

$$\frac{F_{ur} - F_{us}}{F_{ur}} \leq 0.05 \tag{1.3}$$

where F_{ur}, F_{us} are ultimate strengths, respectively, of rigid and semi-rigid frames.

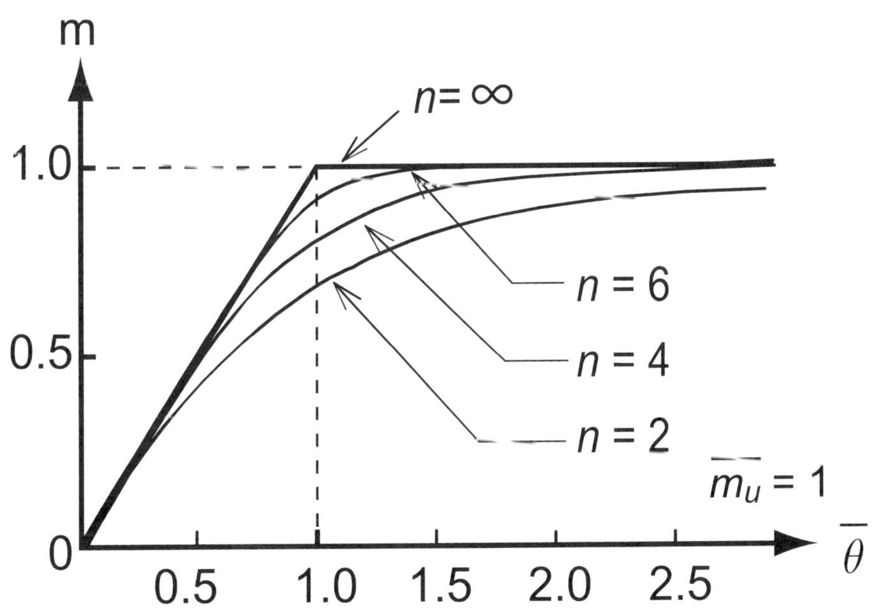

FIGURE 1.13 Three-parameter power model.

The criterion expressed by Equation 1.3, however, may be insufficient because the displacement of frames at the ultimate-limit state is not reflected. Therefore, the following classification criterion is used based on the tolerance Δ_u, which takes into account both strength and displacement at the ultimate-limit state:

$$\sqrt{\left\{\frac{(F_{ur}-F_{us})}{F_{ur}}\right\}^2+\left\{\frac{(d_{us}-d_{ur})}{d_{ur}}\right\}^2} \leq \Delta_u \tag{1.4}$$

where (F_{ur}, d_{ur}) are the coordinate values of the limit point of an equilibrium curve for a rigid subassemblage expressed in terms of force-displacement relation, while (F_{us}, d_{us}) are the coordinate values nearest to (F_{ur}, d_{ur}) of an equilibrium curve for the corresponding semi-rigid subassemblage, as shown in Figure 1.14. It should be noted that (F_{us}, d_{us}) is not necessarily a limit point of the equilibrium curve for the semi-rigid subassemblage. As the tolerance Δ_u, the following value is used:

$$\Delta_u = \sqrt{(0.05)^2+(0.05)^2} \cong 0.07 \tag{1.5}$$

Based on the classification criterion for the ultimate behavior of semi-rigid subassemblages given by Equation 1.4 and Equation 1.5, the boundary values between rigid and semi-rigid connections are determined for the ultimate-moment capacity M_u and shape parameter n expressed in terms of interaction equations. The ultimate behaviors of the subassemblages are calculated by geometrically and materially nonlinear analysis with monotonically increasing the vertical displacement at the loading point of compressive column load P (Figure 1.12). In this analysis, the horizontal load H and the vertical load V are ignored. The analysis method used for this calculation precisely considers the geometrical and material nonlinearities in the structural response.

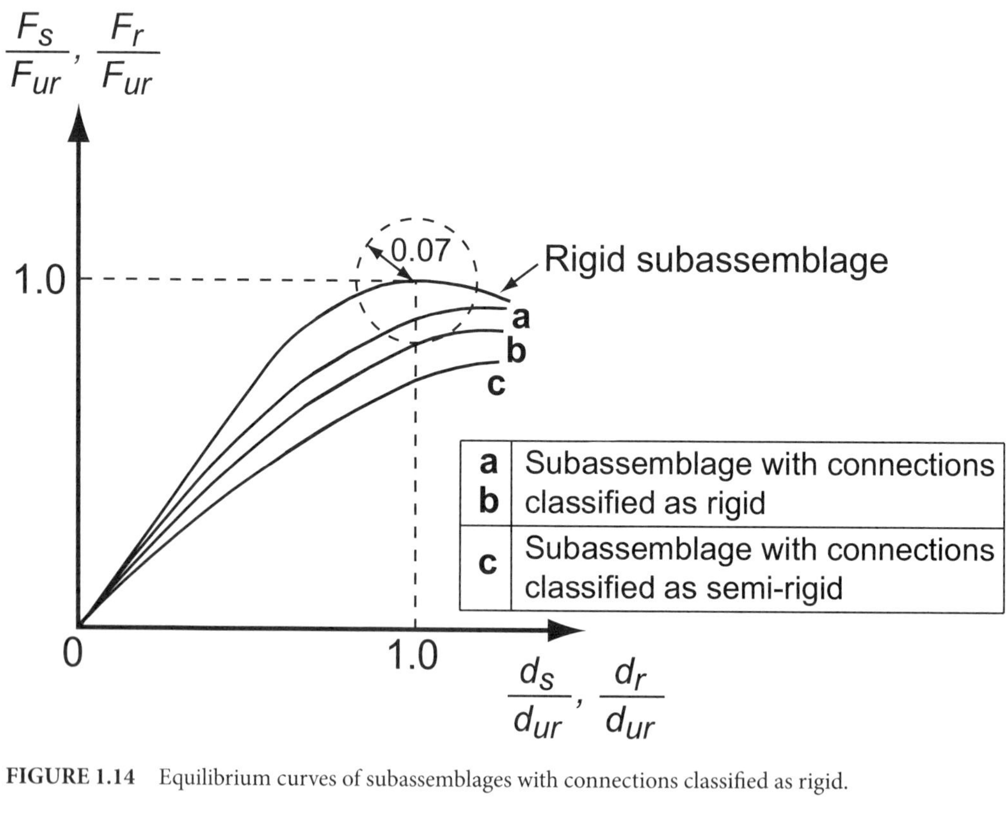

FIGURE 1.14 Equilibrium curves of subassemblages with connections classified as rigid.

Furthermore, initial geometrical imperfections as well as residual stresses are taken into account, following the ECCS recommendation (ECCS, 1984). The stress-strain relation used for the respective subassemblages and initial imperfections are shown in Figure 1.15.

Extensive numerical analyses are carried out on the twelve types of subassemblages (Figure 1.12) with various member sizes to identify the boundary values for the connection moment capacity $\bar{m}_u$ (= M_u/M_{pb}) and shape parameter n, following the criterion given by Equation 1.4 and Equation 1.5. In Equation 1.4, F_{ur} and F_{us} are the ultimate values of P for the respective subassemblages, while d_{ur} and d_{us} correspond to the ultimate-vertical displacement at the loading point

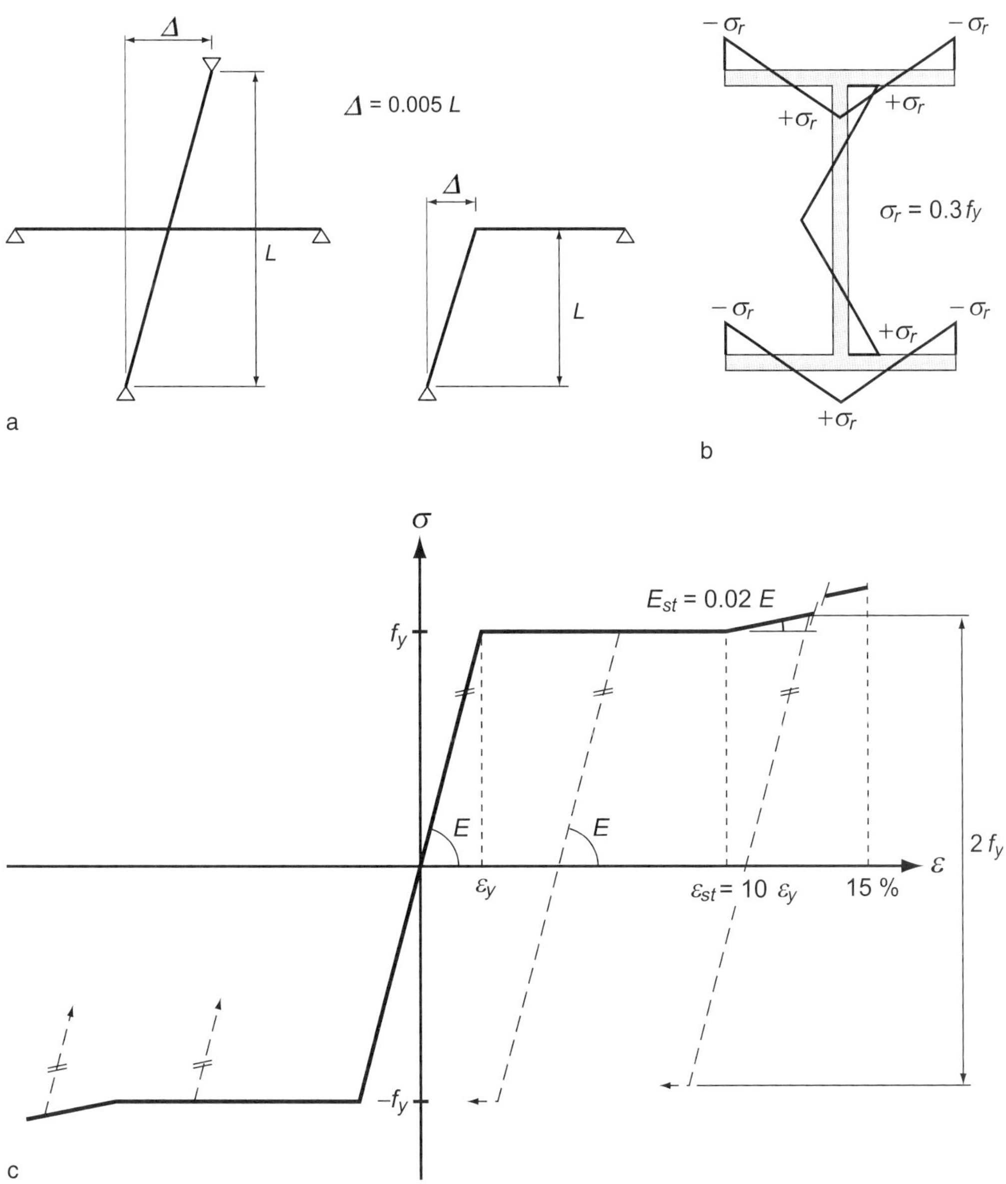

FIGURE 1.15 Imperfections and material properties of subassemblages: (a) initial geometric imperfections; (b) residual stress; (c) stress-strain relation.

TABLE 1.6 Coefficients $c_1 \sim c_{10}$ for interaction equations between boundary moment capacity $\bar{m}_u^b$ and boundary shape parameter n^b

Subassemblage	c_1	c_2	c_3	c_4	c_5	c_6	c_7	c_8	c_9	c_{10}
As	−0.0020	−0.0026	0.0413	0.9931	−0.0231	−0.1252	0.0385	0.0000	0.0554	−0.0036
Bs	−0.2933	0.0263	0.3465	1.3991	−0.2221	−1.0559	0.2122	−0.0041	0.5701	0.0517
Cs	−0.0321	−0.0170	0.0688	1.0498	−0.0352	0.0649	−0.0242	0.0081	−0.0126	0.0095
Ds	0.0242	0.0054	−0.0208	0.9793	0.0112	−0.0470	0.0112	−0.0032	0.0216	−0.0095
Es	−0.2143	0.0522	−0.3322	1.4543	−0.1051	−0.9124	0.1507	−0.0125	0.7557	0.0754
Fs	0.0035	−0.0015	0.0097	0.9887	−0.0207	−0.0196	0.0111	0.0002	0.0472	−0.0005
An	0.0340	0.0101	−0.3594	1.3748	−0.1771	−1.2436	0.2559	−0.0002	0.8971	−0.0097
Bn	0.0050	0.0234	−0.1281	1.0543	−0.2008	−1.1667	0.3115	−0.0030	0.7797	−0.0064
Cn	0.0296	0.0068	−1.3965	3.4374	−0.0161	−2.4585	−0.1592	0.0030	1.5685	−0.0001
Dn	0.0340	0.0101	−0.3594	1.3748	−0.1771	−1.2436	0.2559	−0.0002	0.8971	−0.0097
En	0.0426	−0.0077	0.0327	0.7482	−0.0893	−0.1739	0.1159	0.0018	0.3105	−0.0106
Fn	0.0329	0.0036	0.1224	0.4415	−0.0299	0.0703	0.0974	−0.0016	0.2883	−0.0203

of P. Based on the numerical data obtained for the boundary moment capacity $\bar{m}_u^b = M_u^b / M_{pb}$ and boundary shape parameter n^b, the interaction equations between these two quantities for the respective subassemblages are empirically derived through curve-fitting techniques. All the empirical interaction equations are assumed here to have the following form:

$$n^b = c_1 + c_2 G + c_3 \lambda + c_5 G\lambda + c_6 \lambda \bar{m}_u^b + c_7 G\bar{m}_u^b + c_8 G^2 + c_9 \lambda^2 + c_{10} (\bar{m}_u^b)^2 \tag{1.6}$$

where $c_1 \sim c_{10}$ are curve-fitting coefficients determined by the least square method, according to the types of subassemblages.

G and λ that denote relative stiffness and normalized column slenderness ratio, respectively, are expressed by:

$$G = \frac{(I_b / L_b)}{(I_c / L_c)}, \tag{1.7a}$$

$$\lambda = \left(\frac{L_c}{\pi r}\right)\sqrt{\frac{f_y}{E}} \tag{1.7b}$$

where r is the radius of gyration of member cross section; E is Young's modulus; f_y is yield stress of member material. G and λ are introduced to consider the layout and member details of the subassemblages.

The values of the curve-fitting coefficients $c_1 \sim c_{10}$ applicable to the following parameter range are summarized in Table 1.6:

$$0.7 \le G \le 4.2, \quad 0.2 \le \lambda \le 1.0, \quad 0.8 \le \bar{m}_u \le 1.2 \tag{1.8a-c}$$

The required rotational capacity $\bar{\theta}_u$ is defined as the maximum rotation $\bar{\theta}_{max}$ of the connections that is experienced up to the maximum load carrying capacity of the semi-rigid frames (Goto and Miyashita, 1995). The connections classified as rigid generally exhibit smaller rotation than do the connections that have the boundary moment-rotation characteristics between rigid and semi-rigid connections. This implies that the required rotational capacity calculated based on the boundary moment-rotation characteristics between rigid and semi-rigid connections will provide an upper bound that will be safe for the design of connections classified as rigid. Therefore, the required rotational capacities $\bar{\theta}_u$ for the respective subassemblages are calculated by the nonlinear numerical analyses, assuming that the connection behavior is expressed by the boundary moment-rotation curve between rigid and semi-rigid connection as given by:

$$m = \bar{m}_u^b \bar{\theta} / (1 + \bar{\theta}^{n^n})^{1/n^n} \tag{1.9}$$

where $\bar{\theta} = \theta_r / \theta_0$ and $\theta_0 = M_u^b / R_{ki}^b$.

Based on these numerical results, the required rotational capacities $\bar{\theta}_u = \theta_u / \theta_0$ for the respective subassemblages classified as rigid are approximated by the empirical formulas as shown:

$$\bar{\theta}_u = d_1 + d_2 G + d_3 \lambda + d_5 G\lambda + d_6 \lambda \bar{m}_u^b + d_7 G\bar{m}_u^b + d_8 G^2 + d_9 \lambda^2 + d_{10} (\bar{m}_u^b)^2 \tag{1.10}$$

where $d_1 \sim d_{10}$ are the curve-fitting coefficients shown in Table 1.7. These coefficients are obtained by the least square method for the range given by Equation 1.8.

The validity of the proposed classification system is precisely examined for four types of sway and nonsway frames (Goto et al., 1998).

TABLE 1.7 Coefficients $d_1 \sim d_{10}$ for formulas to predict θ_u

Subassemblage	d_1	d_2	d_3	d_4	d_5	d_6	d_7	d_8	d_9	d_{10}
As	−0.0054	0.0024	−0.0021	1.0006	0.0070	0.0150	−0.0009	−0.0008	−0.0294	0.0035
Bs	−0.0819	0.0455	−0.2690	1.1447	0.0539	−0.0769	−0.0034	−0.0098	−0.0001	0.0485
Cs	0.0674	−0.0280	0.1349	0.9662	−0.0358	−0.1781	0.0348	0.0039	0.0860	−0.0336
Ds	−0.0067	0.0062	−0.0204	0.9981	0.0119	0.0302	−0.0038	−0.0017	−0.0466	0.0065
Es	−0.0746	0.0602	−0.6142	1.1700	0.1798	0.0040	−0.0437	−0.0171	−0.0588	0.0725
Fs	−0.0022	0.0004	−0.0345	1.0047	0.0219	0.0131	−0.0037	−0.0007	−0.0278	0.0033
An	−0.0208	−0.0489	0.1193	0.9820	−0.0830	−0.0559	0.0549	0.0156	0.1246	0.0076
Bn	−0.0093	−0.0359	0.1597	1.0099	−0.0473	−0.0854	0.0116	0.0099	0.0389	−0.0069
Cn	−0.0601	−0.1395	1.8213	1.7711	−0.3256	−5.7450	0.3240	0.0397	−0.5660	−0.0753
Dn	−0.0160	−0.0489	0.0883	0.9737	−0.0785	−0.0104	0.0531	0.0153	0.1319	0.0095
En	−0.0057	−0.0058	0.0292	1.0029	−0.0139	−0.0052	0.0003	0.0023	0.0146	−0.0002
Fn	−0.1933	0.0147	−0.0243	0.6966	−0.0513	1.5172	0.0828	0.0067	−0.0959	0.0714

1.1.7 Comparison of Existing Classification Systems

1.1.7.1 General

Classification of connections is normally made in terms of the index properties such as stiffness, strength, and rotation capacity. Among these index properties, AISC classification (AISC, 2005) adopts only stiffness except for the classification of simple connections where strength requirement is considered in addition to the stiffness. Eurocode 3 (CEN, 2005) adopts both stiffness and strength. The other classification systems (Bjorhovde et al., 1990; Nethercot et al., 1998; Goto et al., 1998) utilize all three indexes. Herein, the respective classification systems are compared with one another in terms of stiffness, strength, and rotation capacity.

1.1.7.2 Classification of Connection Stiffness

The connection stiffness is the most important index in the classification systems. The effect of connection stiffness on the overall behavior of frames is governed by the layout and details of members such as beams and columns. In addition, this effect is different according to whether the frames are sway or nonsway. Normally, the classification based on sway frames tends to yield a more restrictive criterion.

Eurocode 3 (CEN, 2005) specifies different classification criteria for initial connection stiffness according to whether the horizontal displacement is restrained or not. However, the model used to derive a classification criterion is a pin-based portal frame (Figure 1.3). Furthermore, nondimensional connection stiffness parameters are simplified to include only the bending rigidity of the connected beam. This implies that the rigidity of the connected column is ignored.

AISC specification (AISC, 2005) uses the secant connection stiffness at the service load for classification. A simple beam model with semi-rigid connections at both ends (Figure 1.5) is used to derive the criteria for classification. As a result, this classification system ignores the sway behavior of frames as well as the rigidity of columns.

Bjorhovde's classification criterion (Bjorhovde et al., 1990) that uses the initial stiffness of connections is intended for the case where prior knowledge concerning the member and structural details is unavailable. Therefore, this classification system ignores the overall behavior of frames.

Nethercot's classification system (Nethercot et al., 1998) adopts the initial stiffness of connections as a stiffness index. This classification takes into account the bending rigidities of the connected beam and column. The classification criteria are derived both for serviceability limit state and for ultimate-limit state. These criteria are based on a single-span nonsway subframe under uniformly distributed load applied on the beam, as illustrated in Figure 1.9. This subframe model may be interpreted as a subassemblage of nonsway frames. Although Nethercot considers the bending rigidities of the connected beam and column in the classification, the submodel used to derive the classification are too simple to reflect the actual behavior of frames, being similar to the AISC and EC3 classification. Furthermore, it will be cumbersome to use different connection models in the frame analysis for design, depending on whether the limit state of concern is the ultimate or serviceability limit state.

A more detailed classification criterion for the boundary between rigid and semi-rigid connections is proposed by Goto et.al. (1998) in terms of initial connection stiffness. To take into account the behavior of frames in the classification of connections, several subassemblages are used to derive the classification criterion such that the subassemblages will appropriately represent the behavior of the respective parts of the multistory multibay frames with sway or nonsway displacement. As a result, this criterion includes the effect of the bending rigidities of the connected beams and columns.

1.1.7.3 Connection Strength and Rotation Capacity

In AISC specification, the required strength and rotation capacity of connections depend on a particular application. That is, the strength of connections is designed to resist the moment demand under the design load, while the rotational capacity must be larger than the rotational demand of connections at the strength-limit state of frames. Almost similar to AISC specification, connections are designed in EC3 to have a rotational capacity that meets their rotational deformation demand under the design load. The rotational demand has to be calculated by frame analysis except for continuous connections (CEN, 2005) and FR connections (AISC, 2005) that have the strength larger than that of the connected beams. Specifically, in the case of semicontinuous connections (CEN, 2005) and PR connections (AISC, 2005), as well as in the case of FR connections with the partial strength of a connected beam (AISC, 2005), the connection behavior has to be appropriately reflected in the frame analysis. As can be seen from the connection modeling explained above, it will be more appropriate to classify the FR connections with the partial strength of a connected beam (AISC, 2005) as PR connections.

1.2 AISC Specification

1.2.1 Introduction

In the 2005 ANSI/AISC Specification (AISC, 2005), connections are largely classified into two types: Simple and moment connections. A simple connection transmits a negligible moment across the connection. In the analysis of the structure, simple connections may be assumed to allow unrestrained relative rotation between the framing elements being connected. These connections may be modeled as pinned. A moment connection transmits moment across the connection. Two types of moment connections—fully restrained (FR) moment connection and partially restrained (PR) moment connections—are permitted. An FR connection transfers moment with a negligible rotation between the connected members. In the analysis of the structure, the FR connection may be generally assumed to allow no relative rotation and be modeled as rigid. A PR connection transfers moment but the rotation between connected members is not negligible. In the analysis of the structure, the PR connection is modeled as semi-rigid and the moment-rotation response characteristics of the PR connection shall be included in the connection model. For simple and FR connections, the connection proportions are established after the final analysis of the structural design is completed. In contrast, the design of PR connections is inherently iterative because the values of connection proportions must be assumed in order to establish the force-deformation characteristics of the connections that are needed to carry out the structural analysis. Herein, the design of semi-rigid connections is explained and discussed based on the commentary of the 2005 ANSI/AISC specification.

1.2.2 Connection Classification

1.2.2.1 General

A connection is a medium through which forces and moments are transmitted from one member to another. A beam-to-column connection is generally subject to axial force, shear force, bending moment, and torsion. Regarding its in-plane behavior, the effect of torsion can be excluded. Furthermore, the effects of axial and shear forces are usually small compared to that of bending moment. Consequently, for practical purposes, only the effect of moment on the rotational deformation needs to be considered regarding the deformation of connections. As depicted in Figure 1.1, the connection rotates by an amount θ_r when a moment M is applied. The angle θ_r corresponds to the relative rotation between the beam and the column at the connection. Therefore, in the AISC specification, the in-plane behavior of semi-rigid connection is represented by the $M - \theta_r$ curve, referred to as moment-rotation curve. The moment-rotation curve of a connection is assumed to represent the behavior of a region consisting of the column and beam along with the connecting elements. The primary index properties that govern the moment-rotation characteristics of connections adopted in the AISC specification are stiffness, strength, and ductility that will be explained in the next three sections.

1.2.2.2 Connection Stiffness

Most connections exhibit nonlinear behavior at low moment-rotation levels. Therefore, the initial stiffness of a connection R_{ki} inadequately characterizes the connection behavior at service load levels. Furthermore, many connection types do not exhibit a reliable initial stiffness, or they exist only for a small moment-rotation range. The secant stiffness R_{ks} at service loads shown in Figure 1.16 is taken as an index property of connection stiffness for the connection classification. R_{ks} is defined as M_s/θ_s where M_s and θ_s are the moment and rotation, respectively, at service loads.

1.2.2.3 Connection Strength

The strength of a connection is the maximum moment M_n in the moment-rotation relation shown in Figure 1.16. The strength of a connection can be determined based on an ultimate-limit-state model of the connection, or from a physical test. If the moment-rotation response does not exhibit a peak load, then the strength can be taken as the moment at a rotation of 0.02 radians.

1.2.2.4 Connection Ductility

The connection ductility (the rotation capacity) θ_u can be defined as the value of the connection rotation at the point where the resisting strength of connection has dropped to $0.8M_n$ as shown in Figure 1.16. If the connection does not exhibit a $0.2M_n$ drop of the resisting moment beyond 0.03

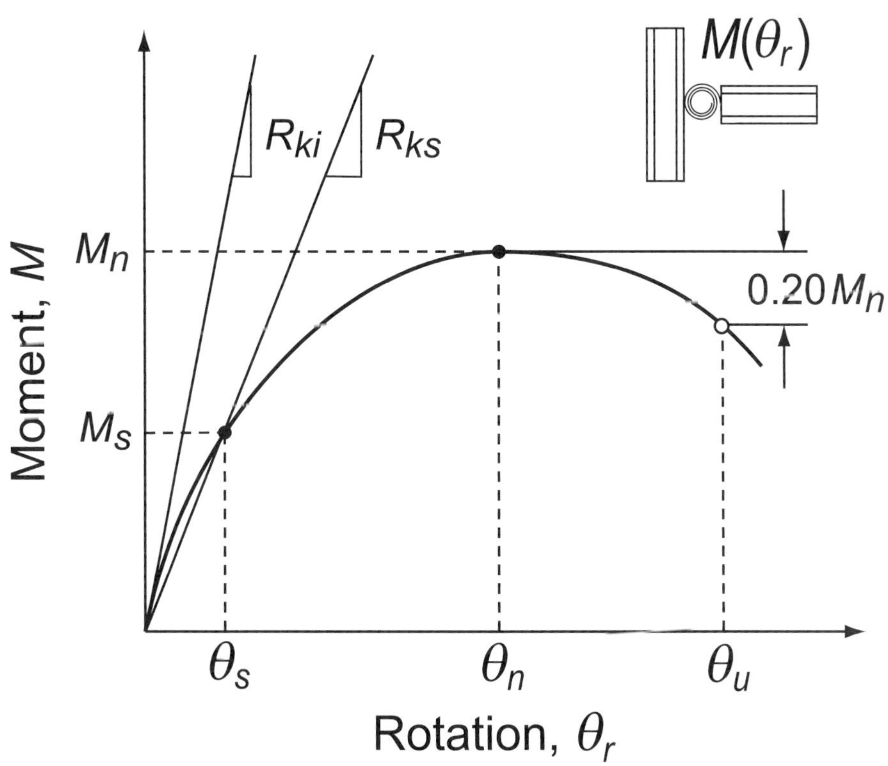

FIGURE 1.16 Definition of stiffness, strength, and ductility characteristics of the moment-rotation response for partially restrained connections.

radians, the connection ductility θ_u is taken as 0.03 radians. The latter criterion is intended to apply to connections where there is no significant loss in strength until large rotations occur.

1.2.2.5 Classification Scheme

The AISC classification scheme for simple, FR, and PR connections is summarized in Table 1.4. The classification boundaries are expressed in terms of the nondimentionalized secant connection stiffness $R_{ks}L_b/EI_b$ at service load where L_b and EI_b are the length and the bending rigidity, respectively, of the connected beam. This classification is applied to both sway and nonsway frames, being different from the Eurocode3 (CEN 2005) where different classification boundaries are proposed according to whether the frame is braced or unbraced.

As for the connection-moment strength M_n, an explicit condition in the classification is imposed only on the simple connections. The connection moment of $M \leq 0.2M_{pb}$ at 0.02 radians is considered to have no flexural strength for design. The strengths of FR and PR connections must be adequate to resist the connection moment and shear force demands imposed by the design load. Therefore, it is possible in the AISC specification for an FR connection to have strength less than that of a connected beam. Furthermore, it is also possible for a PR connection to have strength greater than the strength of the beam. The strength requirement for FR connections is not as strict as that for rigid connections in EC3 where rigid connections must always have strength greater than the full strength of the connected beams or columns. However, if the FR connections have marginally higher or partial strength of the connected beam, it is necessary to calculate the rotational demand of the connections under the factored load by using the semi-rigid modeling of connections. Therefore, in view of the simplification of design analysis procedure, it will be of little use to classify connections with marginally higher or partial strength of the connected beam as FR connections. This connection type should be classified as PR.

The rotation capacity θ_u of connections is required to exceed their demands calculated at the strength-limit state of frames. In the absence of an accurate analysis, a rotation capacity θ_u of 0.03 radians is considered adequate. The above requirement must also be applied to the FR connections with marginally higher or partial strength of the connected beam. However, for the FR connections with strength substantially higher than that of the connected beam, there is no requirement in their rotation capacities. This is because the deformation is controlled by the plastification of the connected beam, and the FR connections exhibit an elastic behavior at the strength-limit state.

Figure 1.17 schematically shows the examples of moment-rotation responses of simple, PR, and FR connections classified according to the AISC scheme.

1.2.3 Structural Analysis and Design for Frames with PR Connections

If the connections in a steel framework are classified as either simple or FR, it is allowed to perform the design analysis for the frame assuming that the connections are either ideally pinned or fully rigid. The assumption of fully rigid connection implies that no relative rotation of connection occurs, while the assumption of pinned connection implies that no restraint for rotation of connection exists and the connection moment is always zero. The assumption of the simple or the FR connection simplifies considerably the analysis and design procedures. The proportions of these connections can be established after the final analysis for the structural design is completed. Simple connections are generally designed under the shear force and must meet the required rotation without introducing moment strength and stiffness that may significantly alter the mode of response. On the other hand, FR connections will be designed for the combined effect of forces resulting from moment and shear and must have sufficient stiffness to transmit moment and maintain the original angle between connected members.

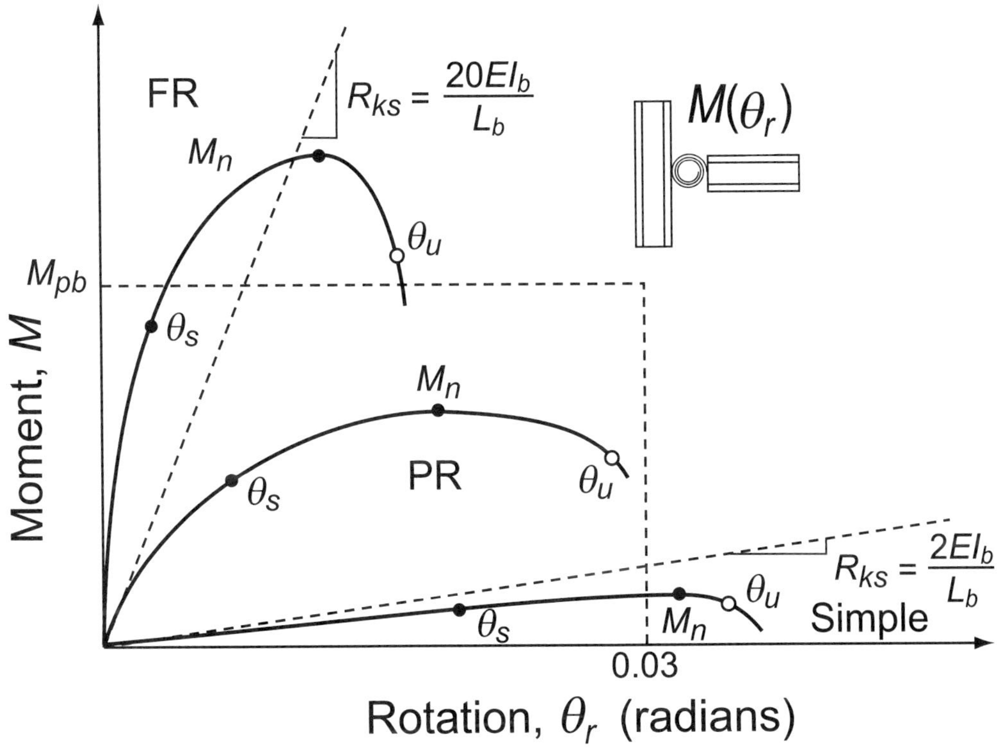

FIGURE 1.17 Classification of moment-rotation response of fully restrained (FR), partially restrained (PR), and simple connections.

When PR connections are used, the relevant moment-rotation relation must be included in the analysis of the structure to determine member forces, connection forces, displacements, and frame stability. Therefore, PR construction first requires that the connection moment-rotation characteristics must be known and second, requires that these characteristics be incorporated into the analysis and member design. In contrast to the design of simple and FR connections, it should be noted that the design of PR connections is inherently iterative because the connection proportions must be assumed in order to establish their moment-rotation characteristics that are used in the structural analysis. The PR connections must transfer the moment similar to the FR connections, whereas the PR connections must have the rotational capacities that meet the rotation demands similar to the simple connections. Therefore, the connections have to be designed under the combined effect of moment and shear such that their rotational capacity satisfies the rotational demand at the strength-limit state.

Typical moment-rotation curves for many PR connections are available from existing connection databases. When utilizing the databases, care must be taken not to extrapolate data to sizes or conditions beyond those used to develop the database since other failure modes may control. For the connections out of the range of the databases, it may be possible to derive the moment-rotation characteristics from tests, component models, or finite element studies. Details of connection databases and connection models are explained in Chapters 4 and 5.

Usually, design of PR construction requires separate analyses for the serviceability and strength-limit states. For the serviceability limit state, an analysis using linear springs with a stiffness given by R_{ks} is sufficient if the resistance demanded of the connection is well below the strength. When subjected to strength load combinations at the ultimate-limit states, a more careful procedure is needed so that the connection characteristics assumed in the analysis are consistent with those of the actual connection response. The response will become nonlinear as the applied moment approaches the connection strength. In particular, the effect of the connection nonlinearity on second-order moment and stability checks need to be considered.

The design requirements for simple, PR, and FR connections are shown in Table 1.4.

References

AISC, *Specification for Structural Steel Buildings*, ANSI/AISC 360-05, American Institute of Steel Construction, Chicago, IL, 2005.

Bijorhovde, R., Colson, A., and Brozzeti, J., Classification system for beam-to-column connections, *Journal of Structural Engineering*, ASCE, 116(11), 3059-3076, 1990.

CEN, *Eurocode 3, Design of Steel Structures*, Part 1-1: General rules and rules for buildings, EN 1993-1-1, Part 1-8: Design of joints, EN 1993-1-8, European Committee for Standardization, Brussels, Belgium, 2005.

Chen, W. F. and Kishi, N., Semi-rigid steel beam-to-column connections: Data base and modeling, *Journal of Structural Engineering*, ASCE, 115(7), 105-119, 1989.

European Commission for Constructional Steelwork (ECCS), *Ultimate-limit state calculation of sway frames with rigid joints*, Systems Publication No.33, Technical Working Group 8.2, Brussels, Belgium, 1984.

Goto, Y. and Miyashita, S., Classification system for rigid and semi-rigid connections, *Journal of Structural Engineering*, ASCE, 124(7), 750-757, 1998.

Goto, Y. and Miyashita, S., Validity of classification systems of semi-rigid connections, *Engineering Structures*, 17(8), 544-553, 1995.

Horne, M. R. and Merchant, W., *The stability of frames*, Pergamon Press, 1965.

Kishi, N., Chen, W. F., Goto, Y., and Matsuoka, K. G., Design aid of semi-rigid connections for frame analysis, *AISC Engineering Journal*, 30(3), 90-107, 1993.

Leon, R. T., Composite semi-rigid construction, *AISC Engineering Journal*, 31(2), 54-67, 1998.

Nethercot, D. A., Ahmed, T. Q., and Li, B., Unified classification system for beam-to-column connections, *Journal of Constructional Steel Research*, 45(1), 39-65, 1998.

2

Effects of Semi-rigid Connections on Structural Members and Frames

Eric M. Lui
Meredith Professor, Department of Civil and Environmental Engineering,
Syracuse University, NY

Zhiling Zhang
Graduate Student, Department of Civil and Environmental Engineering,
Syracuse University, NY

2.1 Introduction

A connection is a medium through which forces and moments are transmitted from one member to another. In a conventional analysis and design of steel frameworks, the frames are often conveniently analyzed and designed under the simplifications that the connections behave either as ideally pinned or fully rigid. The use of the ideally pinned condition implies that no moment will be transmitted between the beam and the column. As far as rotation is concerned, the beam and column that are jointed together by a pin will behave independently. On the other extreme, the use of the fully rigid condition implies that no relative rotation will occur between the adjoining members. The angle between the beam and the column remains virtually unchanged as the frame deforms under the applied loads. Although the use of these idealized joint behaviors drastically simplifies the analysis and design procedures, the predicted response of the frame may not be realistic as most connections used in actual practice transmit some moment and experience some deformation.

According to the American Institute of Steel Construction (AISC, 2005), a connection is considered semi-rigid or partially restrained (PR) if its stiffness R_k falls in the range $2EI_b/L_b < R_k < 20EI_b/L_b$, where EI_b and L_b are the flexural rigidity and length of the adjoining beam, respectively. For connections that are classified as semi-rigid, a more realistic analysis procedure should be used to assess frame behavior. In this chapter, the effect of connection flexibility on column, beam, and frame responses as well as its implication on design will be presented.

2.1.1 Connection Behavior

Although a general set of forces that may be transmitted in a typical beam-to-column connection includes axial force, shearing force, bending moment, and torsion, for simplicity in analysis and design, usually only the effect of in-plane bending moment and the resulting connection rotational deformation is considered. As a result, aside from a few exceptions (Cerfontaine and Jaspart, 2002; Urbonas and Daniunas, 2005, 2006), analytical and experimental work on connections conducted in the past few decades have focused primarily on their flexural response (Frye and Morris, 1976; Lui and Chen, 1987, 1988; Cunningham, 1990). The flexural behavior of connections is best described by their moment-rotation (M-θ_r) curves. A connection moment-rotation curve relates the beam moment M transmitted by the connection to the in-plane rotational deformation θ_r experienced by the connection as shown in the inset of Figure 2.1, in which schematic M-θ_r curves for a variety of commonly-used semi-rigid connections are shown.

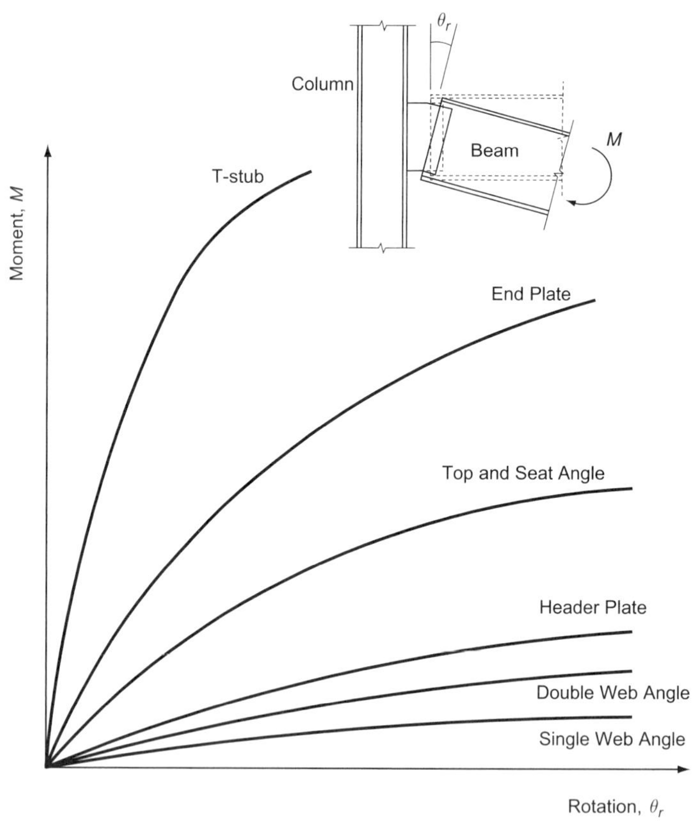

FIGURE 2.1 Moment rotation behavior of commonly used semi-rigid connections.

The single web angle connection represents a flexible connection and the T-stub connection represents a rather stiff connection. Several observations can be made from this figure:

1. All types of connections exhibit M-θ_r responses that fall between the extreme cases of ideally pinned (the horizontal axis) and fully rigid (the vertical axis) conditions.
2. For the same moment, a larger θ_r is experienced by a more flexible connection. Conversely, for a specific value of θ_r, less moment is transmitted by a more flexible connection.
3. The maximum moment that a connection can transmit (herein referred to as the nominal connection moment capacity, M_{cn}) decreases with the more flexible connection.
4. The M-θ_r relationships for all semi-rigid connections are typically nonlinear over virtually the entire loading range.

Generally speaking, linear approximation of the initial portion of the curve is acceptable for a frame designed under a serviceability limit state, but becomes unacceptable if the frame is to be designed under a strength limit state.

When a moment is applied to a connection, the connection rotates according to the curves shown in Figure 2.1. However, if the direction of moment is reversed, the connection will unload and follow a different path that is almost linear and with a slope more or less equal to the initial slope of the M-θ_r curve. This phenomenon is depicted in Figure 2.2.

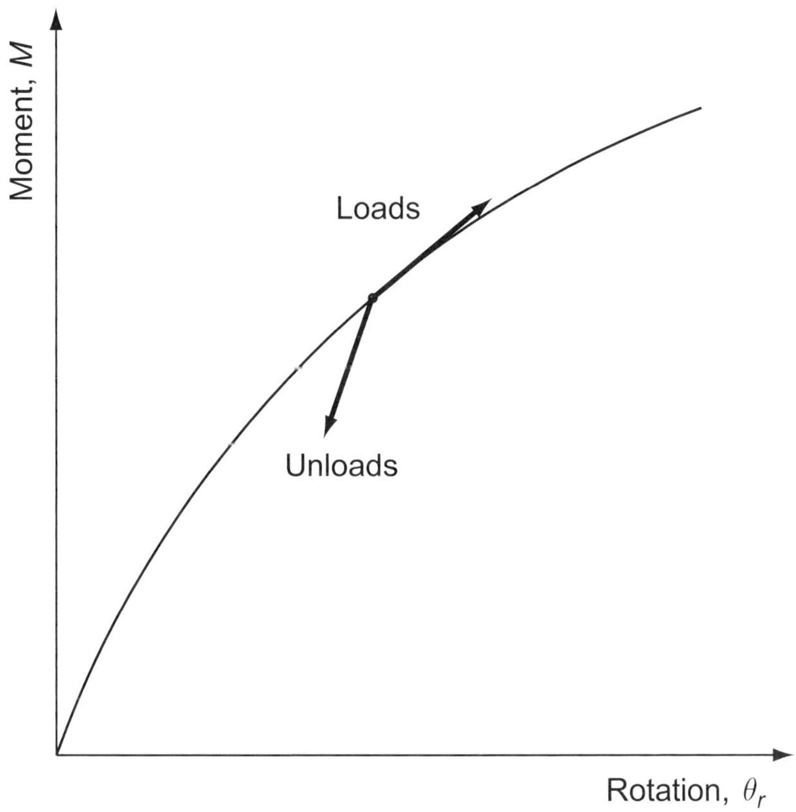

FIGURE 2.2 Loading and unloading behavior of semi-rigid connections.

Because of this behavior, identical connections may not always behave identically. This can best be illustrated by the simple portal frame shown in Figure 2.3. Under the action of the gravity load, the connections at the ends of the beam will experience a moment equal to M_g on the M-θ_r curve (Figure 2.3a). Now, if a lateral force is applied, the leeward connection will continue to load but the windward connection will unload (Figure 2.3b). As a result, the apparent stiffness of the connections under the action of this lateral force will be different. Depending on the magnitude of M_g and the characteristic of the connection M-θ_r curve, the difference in stiffness of the two connections may be quite substantial. In the extreme scenario when the gravity load moment equals the nominal moment capacity of the connections, the leeward connection will behave virtually like a pinned connection, whereas the windward connection will respond like a linear elastic connection. This loading and unloading characteristic of the connection must be properly modeled in order to reliably predict the response of the frame.

Empirical equations describing the moment-rotation behavior of various connection types have been proposed by a number of researchers using parametric or curve-fitting techniques. A summary of these equations can be found in Chen and Lui (1991), Chen et al. (1996), Chen (2000), and in this handbook. Connection behavior needs to be addressed in terms of stiffness, strength, and rotational ductility. Each of these attributes will be examined briefly in Sections 2.1.1.1 through 2.1.1.3 (ASCE, 1997).

2.1.1.1 Connection Stiffness

Although almost all connection M-θ_r curves exhibit nonlinear behavior from practically the onset of loading, it may be desirable for design purposes to employ linear approximations. To account for the loading and unloading behavior of a connection, a minimum of two connection stiffness values need to be identified—the initial stiffness R_{ki} and the ultimate stiffness R_{ku}. For R_{ki}, a somewhat conservative approach is to approximate its value using the secant stiffness obtained as the slope of a line connecting the origin of the M-θ_r curve to the point M_g, as shown in Figure 2.3a. The value R_{ku} can be taken as the slope of the M-θ_r curve at M_{cn} (i.e., the connection stiffness at its nominal moment capacity). Alternatively, R_{ku} can conservatively be taken as zero.

2.1.1.2 Connection Strength

A connection is said to have reached its nominal moment capacity (M_{cn}) if the moment in the connection is at or near the flat portion of the M-θ_r curve. Connection strength, however, is only relevant when compared to the strength of the adjoining beam. If we define M_{pb} as the plastic moment capacity of the beam, a connection is referred to as a full strength connection if $M_{cn}/M_{pb} \geq 1.0$, and partial strength if otherwise. It is important to note that M_{cn} obtained from many of the M-θ_r equations or the connection databases corresponds to only one mode of failure (typically flexural yielding of the connecting elements). Consequently, the equations cannot be used to determine M_{cn} if other failure modes such as bolt shear, bolt tension, component or weld fractures, etc., control the design.

The distinction between partial and full strength connections is important regarding the required ductility. It should be noted that most partially restrained connections are partial strength, resulting in important consequences in design (Leon, 1994).

2.1.1.3 Connection Rotational Ductility

If properly designed, most partial-strength semi-rigid connections exhibit ductile behavior. The inelastic deformation capacity, defined as the ratio of the rotation of the connection at M_{cn} to the rotation of the connection at M_s (the service load moment) i.e., $\mu = \theta_{cn}/\theta_s$ is often greater than 10 in monotonic tests. However, for full-strength semi-rigid connections, the moment introduced in the joint will be capped by M_{pb}, resulting in a shift of deformations to the plastic hinge of the

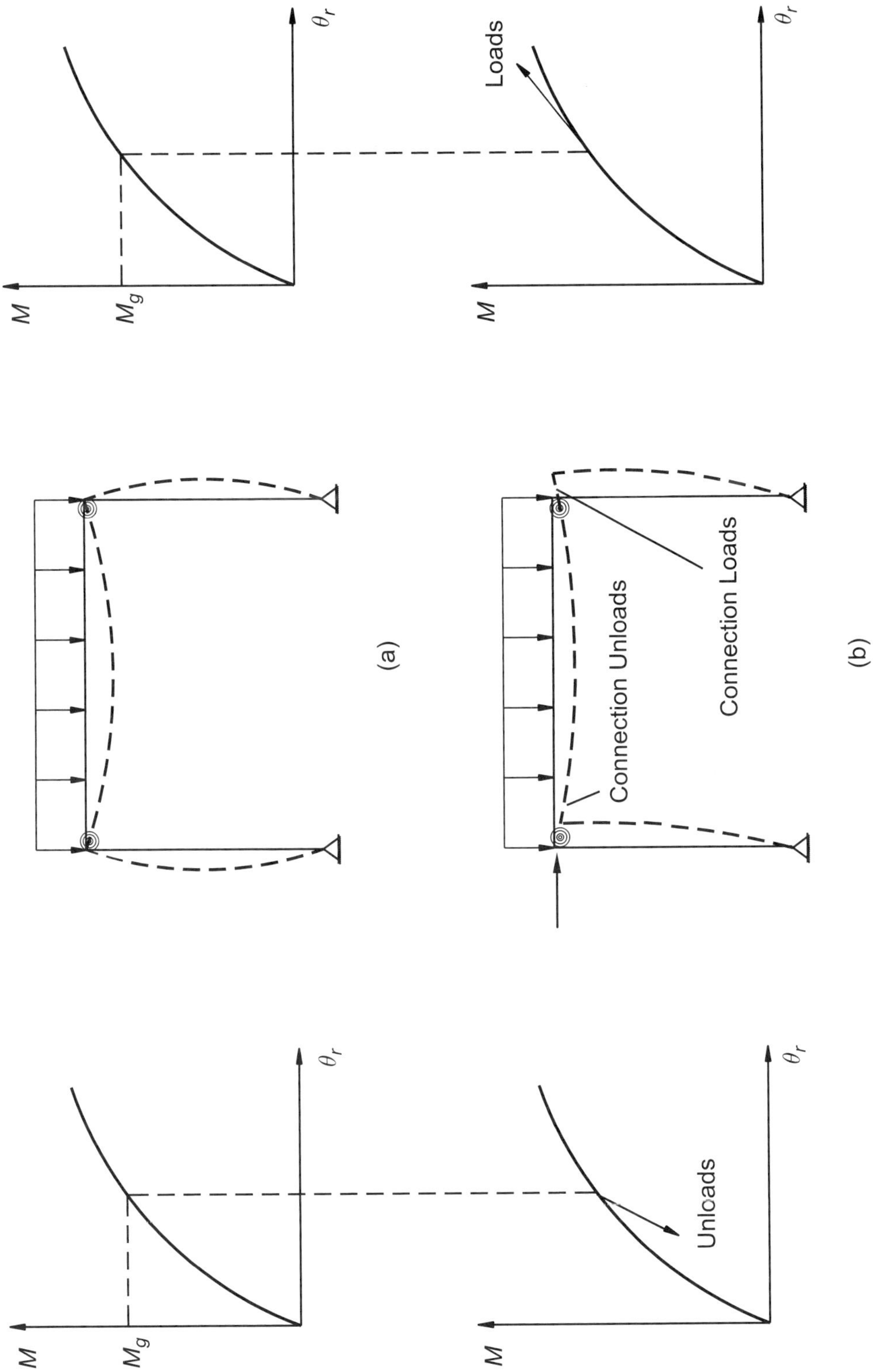

FIGURE 2.3 Semi-rigid frame with connections undergoing loading and unloading.

adjoining beam. The ductility of the connection is therefore less of a concern than if the semi-rigid connection is partial strength. For partial-strength semi-rigid connections, the large ductility demands result in both the need for careful detailing of the connections and the need for a simplified check to insure that θ_{cn} is not exceeded. It should be remembered that ductility demand is highly dependent on load history. The remarks made herein are meant explicitly for monotonic loading cases and not for stability calculations under large seismic loads.

Over the past three decades, extensive studies on the effect of connection flexibility on semi-rigid frames have been conducted and published in various journals. Some of the work has been summarized in book or monograph form. For example, see Chen and Lui (1991), Chen et al., (1996), Chen (2000), and Faella et al. (2000), among others.

2.2 Effects of Semi-rigid Connections on Columns

2.2.1 Column Effective Length Factor

The pin-ended column often serves as a benchmark to which other columns with different end conditions are referenced. Once the compressive strength or critical load of a pin-ended column is known, the strength of the same column with other end conditions can be obtained by using the effective length factor K, defined as:

$$K = \sqrt{\frac{P_e}{P_{cr}}} \tag{2.1}$$

where:

P_{cr} = critical load of the end-restrained column
P_e = Euler load of a pin-ended column having the same length as the end-restrained column
= $\pi^2 EI_c / L_c^2$, where E is the modulus of elasticity, I_c and L_c are the moment of inertia and length of the column, respectively

For a perfectly straight elastic column, the critical load can be obtained by writing the differential equation that conforms to the specific end conditions of the column. The eigenvalue of the characteristic equation that corresponds to the differential equation is the critical load for that column (Timoshenko, 1961; Chen and Lui, 1987). The effective length factor can then be determined from Equation 2.1.

Table 2.1 shows the theoretical and suggested design K values for six idealized conditions in which joint translation and rotation are assumed either fully realized or nonexistent.

A more general case for a perfectly straight column restrained elastically against rotation at ends A and B by rotational springs with spring constants R_{kA} and R_{kB}, and restrained elastically against translation at end B by a translational spring with spring constant T_k, is shown in Figure 2.4.

Using the free body diagrams of the column and the rotational springs shown in the figure, the critical load of this column can be found by first writing moment equilibrium equations of the three free bodies and then setting the determinant of the coefficient matrix of the equilibrium equations to zero:

$$\det\begin{pmatrix} K_1 s_{ii} + R_{kA} & K_1 s_{ij} & -K_2(s_{ii}+s_{ij}) \\ K_1 s_{ij} & K_1 s_{ii} + R_{kB} & -K_2(s_{ii}+s_{ij}) \\ -K_2(s_{ii}+s_{ij}) & -K_2(s_{ii}+s_{ij}) & 2K_3(s_{ii}+s_{ij}) - \dfrac{P}{L_c} + T_k \end{pmatrix} = 0 \tag{2.2}$$

TABLE 2.1 Effective length factors for columns with different boundary conditions

Case	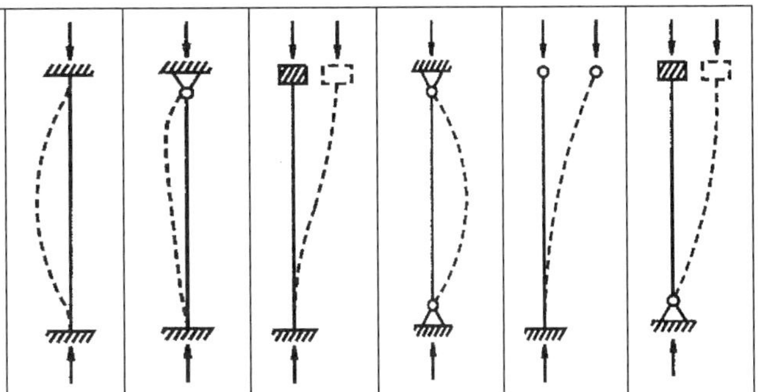					
Support conditions	Fixed Fixed	Pinned Fixed	Guided Fixed	Pinned Pinned	Free Fixed	Guided Pinned
Theoretical K	0.5	0.7	1.0	1.0	2.0	2.0
Recommended K when idealized conditions are not realized	0.65	0.8	1.2	1.0	2.1	2.0

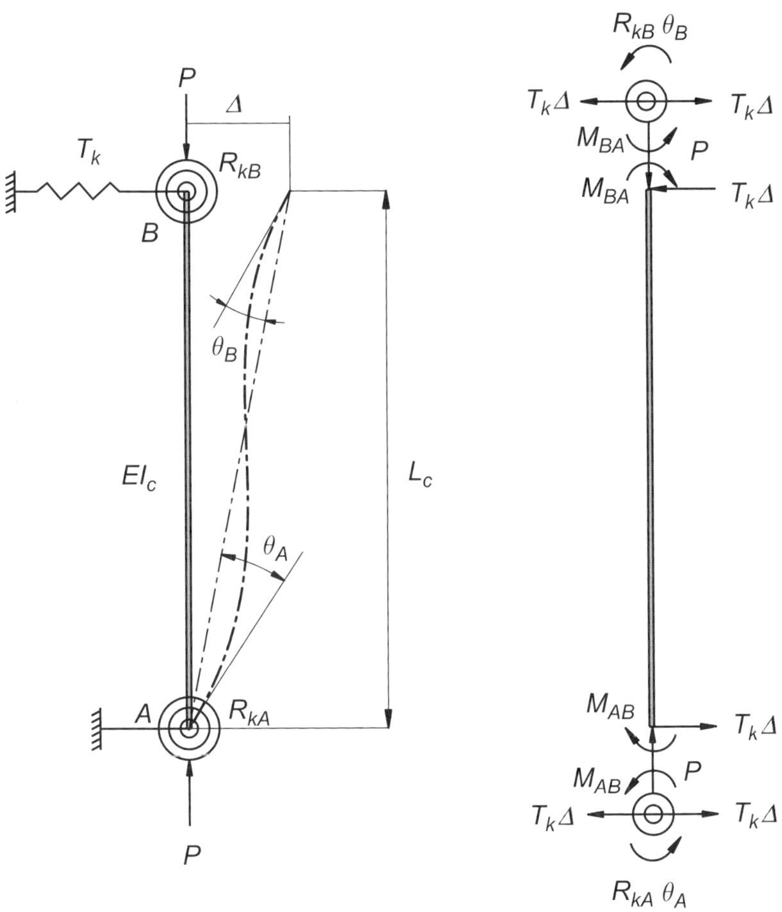

FIGURE 2.4 An end-restrained elastic column.

where

$$K_1 = \frac{EI_c}{L_c}, \quad K_2 = \frac{EI_c}{L_c^2}, \quad K_3 = \frac{EI_c}{L_c^3} \tag{2.3}$$

$$s_{ii} = \frac{kL_c\left(\sin kL_c - kL_c \cos kL_c\right)}{2 - 2\cos kL_c - kL_c \cos kL_c} \tag{2.4}$$

$$s_{ij} = \frac{kL_c\left(kL_c - \sin kL_c\right)}{2 - 2\cos kL_c - kL_c \sin kL_c} \tag{2.5}$$

in which

$$k = \sqrt{\frac{P}{EI_c}} \tag{2.6}$$

The smallest P value that satisfies Equation 2.2 is the critical load for the column of Figure 2.4. The effective length factor can then be obtained from Equation 2.1.

Although the effective length factor discussed above was derived from elastic analysis, studies by Chapius and Galambos (1982) and Lui and Chen (1983) have shown that they will be conservative in most cases for initially-crooked elastic-plastic columns provided that the adjoining beams remain elastic.

In real structures, a column usually exists as part of a frame. The effective length factor for columns as part of a frame has been derived by Julian and Lawrence (1959) and LeMessurier (1972).

By assuming that (1) the girders are rigidly connected to the columns, (2) all columns of a story buckle simultaneously, and at the onset of buckling the rotations of the ends of the girders are equal and opposite for braced frames and equal in magnitude and direction for unbraced frames, and (3) all members are prismatic and the behavior is elastic, the following equations for the effective length factor can be written.

For braced (or sway prevented) frames:

$$\frac{G_A G_B}{4}\left(\frac{\pi}{K}\right)^2 + \left(\frac{G_A + G_B}{2}\right)\left[1 - \frac{\pi/K}{\tan\left(\pi/K\right)}\right] + \frac{2\tan\left(\pi/2K\right)}{\pi/K} = 1 \tag{2.7}$$

For unbraced (or sway permitted) frames:

$$\frac{G_A G_B\left(\pi/K\right)^2 - 36}{6\left(G_A + G_B\right)} - \frac{\pi/K}{\tan\left(\pi/K\right)} = 0 \tag{2.8}$$

where G_A and G_B are the stiffness distribution factors at the Ath and Bth ends of the column respectively. They are defined as:

$$G = \frac{\sum \tau\left(I/L\right)_c}{\sum\left(I/L\right)_b} \tag{2.9}$$

where the subscripts c and b refer to the columns and beams, respectively, and τ is a factor to account for inelasticity in the column given by AISC (2005) as:

$$\tau = \begin{cases} 1.0 & \text{if } \dfrac{P}{P_y} \le 0.5 \\ 4\dfrac{P}{P_y}\left(1 - \dfrac{P}{P_y}\right) & \text{if } \dfrac{P}{P_y} > 0.5 \end{cases} \tag{2.10}$$

where P is the axial force in the column and P_y is the yield load of the cross-section. The summation in the numerator covers all columns that come together at a joint and the summation in the denominator covers all beams that come together at the same joint. If the column is supported on a pinned support, the theoretical G factor evaluated using Equation 2.9 at the supported end is infinity. However, recognizing that an ideally pinned support does not exist in real structures, a G factor of 10 is recommended for use. On the other hand, if the column is fixed at the base, the theoretical G factor is 0. Again, because an ideally fixed support does not exist, a G factor of 1 is recommended for use.

Equations 2.7 and 2.8 form the bases for the alignment charts shown in Figure 2.5 that are used extensively by designers in the United States.

2.2.2 *K* Factor for Columns in Semi-rigid Frames

In using the alignment charts of Figures 2.5a and 2.5b to determine column effective length factors, one must first compute the column end restraint parameters G_A and G_B using Equation 2.9, which assume the beams are rigidly connected to the column in question. If the effect of connection flexibility is to be accounted for, these G factors need to be modified as follows for sway prevented and sway permitted semi-rigid frames.

2.2.2.1 Sway-prevented Frames

For the sway prevented frame, an assumption commonly used is that the restraining beams bend in single curvature with the near end and far end rotations equal but opposite. As a result, the effect of connection flexibility can be incorporated into the G factor computation by replacing the

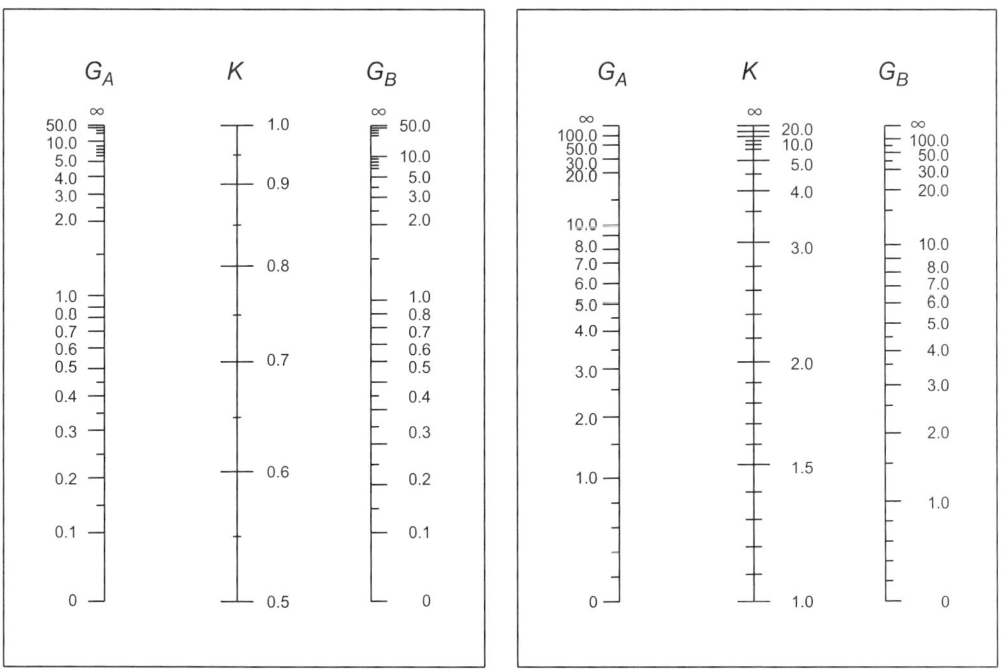

a b

FIGURE 2.5 Alignment charts: (a) sway-prevented (braced) frame; (b) sway-permitted (unbraced) frame.

beam moment of inertia I in the denominator of Equation 2.9 by a modified value I' (Lui, 1995, 1996) given by:

$$I' = 3I\left[\frac{1}{2\left(1+\frac{3}{\bar{R}_k}\right)-\frac{dM_F}{dM_N}}\right] \tag{2.11}$$

where $\bar{R}_k = R_k L/EI$, in which I and L are the moment of inertia and length of the adjoining beam, and dM_F/dM_N is the ratio of the incremental far end to near end connection moments. Because connections at the opposite end of the adjoining beam in a sway-prevented frame normally undergo either loading or unloading simultaneously at incipient instability, one can circumvent the difficulty of having to determine the instantaneous stiffness and the incremental end moments of the loading connections simply by assuming that these connections have zero stiffness. Therefore, only the stiffness of the unloading connections warrants consideration. With this assumption, one can conservatively use:

A. If both the near and far end connections unload:

$$I' = I\left[\frac{1}{\left(1+\frac{2}{\bar{R}_{ki}}\right)}\right] \tag{2.12}$$

where $\bar{R}_{ki} = R_{ki}L/EI$, R_{ki} being the initial connection stiffness (see Section 2.1.1.1)

B. If both the near and far end connections load, $I' = 0$

If we substitute $I = I' = 0$ into Equation 2.9, we obtain $G_A = G_B = \infty$, giving $K = 1$. This condition represents the most severe case for the column in a sway-prevented frame. For simplicity and conservativeness, an effective length factor of unity can always be used for the design of sway-prevented columns.

2.2.2.2 Sway-permitted Frames

An assumption used in developing the alignment chart for the sway-permitted frame subassemblage is that the restraining beams bend in reverse curvature, with the near-end and far-end rotations being equal. With this in mind, one approach to account for the effect of connection flexibility is to replace I in the denominator of Equation 2.9 by I' (Lui, 1995, 1996) given by:

$$I' = I\left[\frac{1}{2\left(1+\frac{3}{\bar{R}_k}\right)-\left(\frac{dM_F}{dM_N}\right)}\right] \tag{2.13}$$

where the definition of $\bar{R}_k$ and dM_F/dM_N are the same as for Equation 2.11. However, it should be noted that in applying the above equation, I' can be positive as well as negative. A positive I' means the beam is imposing restraint on the adjoining column; a negative I' means the beam is destabilizing the adjoining column at incipient instability. If I' is negative, the corresponding G factor will also be negative. Negative G factor is theoretically possible, but beyond the range of the alignment chart shown in Figure 2.5. If negative G factors are encountered, K must be obtained directly by solving Equation 2.8.

It is important to note that for sway permitted semi-rigid frames, the values of dM_F/dM_N will rarely equal one (i.e., the value assumed in the development of the effective length alignment charts). This is because in most cases, a connection at one end of the beam will tend to load while another connection at the other end of the beam will tend to unload at incipient side-sway instability of the structure.

In general, accurate values for R_k and dM_F/dM_N needed for Equation 2.13 are not readily obtainable. Nevertheless, we will discuss three possible approaches for determining these values. In decreasing order of sophistication, they are:

1. Perform an advanced second-order inelastic analysis and obtain R_k and dM_F/dM_N directly. This analysis must include all the significant connection and member nonlinearities. Of course, if one performs such an analysis, the analysis in itself would be verification of the structure's stability, and the application of Equation 2.13 would be unnecessary. Therefore, this procedure is only useful for assessing the ability of Equation 2.13 to predict a frame's stability behavior, given "exact" values for the relevant parameters.
2. Use approximate values for R_k from simplified bilinear or trilinear M-θ_r curves. In this approach, one may assume that the stiffness of the loading connection is equal to R_{ku}, and that of the unloading connection is equal to R_{ki}.
3. Employ a simplified behavioral model in which the connection stiffness and incremental moment are assumed to be zero for all loading connections, and R_{ki} is used for all unloading connections. It should be noted that unless the connections have reached their nominal moment capacities under gravity loads acting on the beam, this approach often leads to an overly conservative design. However, the advantage of applying this approach is that only two connection parameters—R_{ki} and M_{cn}—need to be known, and analytical expressions for these two parameters are available in the literature (for example, Kishi and Chen, 1990; Chen and Lui, 1991; Kishi et al.,1993a, 1993b, 1994).

Because of the ease with which the third approach can be applied to semi-rigid frame analysis, the rest of this chapter presupposes its use. Therefore, the equations that follow are predicated on the assumptions of the third approach. For this situation, the modified beam moment of inertia I' becomes:

A. If the near end connection unloads and the far end connection loads:

$$I' = I\left[\frac{1}{2\left(1+\frac{3}{\overline{R}_k}\right)}\right] \tag{2.14}$$

where $\overline{R}_k = R_{ki}L/EI$, R_{ki} being the initial connection stiffness (see Section 2.1.1.1).

B. If the near end connection loads and the far end connection unloads, $I' = 0$.

For the special case when the connection stiffness are the same at both ends of the beam (i.e., $R_{kA} = R_{kB}$), dM_F/dM_N most likely would be equal to unity (e.g., the connections are identical on each end of the beam, and the total moments M_F and M_N are the same at each of the ends). Upon substituting $dM_F/dM_N = 1$ into Equation 2.13, one obtains:

$$I' = I\left[\frac{1}{1+\frac{6}{\overline{R}_k}}\right] \tag{2.15}$$

However, it should be noted that the condition $R_{kA} = R_{kB}$ is seldom realized because of the loading or unloading characteristic of the connections. Equation 2.13 is a more general equation, and thus it is more appropriate for use in computing effective length factors for columns in semi-rigid frames.

2.2.2.3 Procedure for Evaluating *K* Using the Modified Moment of Inertia Approach

The steps for employing the modified moment of inertia approach for determining effective length factors for columns in semi-rigid frames are as follows:

1. Calculate the modified moment of inertia I' for the beam with semi-rigid connections.
2. Replace I with I' and calculate G factors using Equation 2.9.
3. Use the appropriate alignment chart in Figure 2.5a or 2.5b to obtain a nomograph effective length factor K for the column.
4. Modify the K values obtained in Steps 1-3 for the non-leaning columns using the following equation for sway permitted frames where leaning columns are present:

$$K = \sqrt{\frac{\sum_{all} P}{P} \frac{\tau I_c}{\sum_{non-leaning\ columns} \frac{\tau I_c}{K^2}}} \tag{2.16}$$

In Equation 2.16, τ is given by Equation 2.10, P and I_c are the column axial force and moment of inertia, respectively, and ΣP is the sum of the axial force for all columns in a story, including both the leaning and non-leaning columns. Leaning columns are columns that are pinned at both ends. They carry gravity loads, but do not contribute to the lateral stability of the frame. They are to be designed for $K = 1$. If leaning columns are present in the story, the K values must be calculated only for the non-leaning columns. Non-leaning columns are columns that can provide lateral stability to the frame. All columns that have a predictable amount of rotational end restraint are classified as non-leaning columns.

2.2.2.4 Other Approaches for *K* Factor Computation

Another approach (Lui, 1992, 1995) to evaluate K for columns in a sway-permitted semi-rigid frame is to perform a first-order analysis on the frame modeled as rigid with the following provisions:

1. Replace I with I' computed from Equation 2.13 for all beams with semi-rigid connections.
2. Apply a set of fictitious lateral load ΣH equals to $0.001\Sigma P_u$ at each story, where ΣP_u is the sum of all factored gravity loads acting on that story.
3. Determine the interstory lateral deflection Δ_0 due to ΣH.
4. Evaluate the effective length factor of the column using the equation:

$$K = \sqrt{\left(\frac{\pi^2 E \tau I_c}{P_u L_c^2}\right)\left[\left(\sum_{all} \frac{P_u}{L_c}\right)\left(\frac{1}{5 \sum_{non-leaning\ columns} \eta} + \frac{\Delta_0}{\sum H}\right)\right]} \tag{2.17}$$

where η is a member stiffness index given by:

$$\eta = \left[3 + 4.8\left(\frac{M_1}{M_2}\right) + 4.2\left(\frac{M_1}{M_2}\right)^2\right]\frac{E \tau I_c}{L_c^3} \tag{2.18}$$

in which M_1/M_2 is the ratio of the smaller to larger end moment of the column. The other terms in Equation 2.17 and Equation 2.18 are defined as before.

Other approaches used to evaluate K factor for columns in semi-rigid frames are given by Aristizabal-Ochoa (1994, 1997) and Xu and Liu (2002).

Regardless of the approach used, it should be noted that unless $\overline{R}_k$ is small or G is large, K is not particularly sensitive to changes in the connection stiffness; and when $\overline{R}_k$ is increased beyond 20, little change in K will occur (ASCE, 1997). This means the uncertainties that are often associated with determining connection stiffness will not have a significant influence on the computed K values.

2.2.2.5 Column Strength Design Equations

Once the effective length factor K is computed, the nominal column strength P_n can be evaluated using the following AISC (2005) equations:

For $\frac{KL}{r} \le 4.71\sqrt{\frac{E}{f_y}}$:

$$P_n = \left(0.658^{\frac{f_y}{F_e}}\right) f_y A_g \tag{2.19a}$$

For $\frac{KL}{r} > 4.71\sqrt{\frac{E}{f_y}}$:

$$P_n = 0.877 F_e A_g \tag{2.19b}$$

where KL is the effective length, r is the radius of gyration, E is the modulus of elasticity, f_y is the material yield stress, A_g is the gross cross-sectional area, and:

$$F_e = \frac{\pi^2 E}{\left(\frac{KL}{r}\right)^2} \tag{2.20}$$

Example:

For the semi-rigid portal frame shown in Figure 2.6, knowing $\overline{R}_k = R_k L/EI = 10$ for the connections at B and C, determine the elastic K factor for column AB if:

A = The loading or unloading behavior of the connections is ignored.
B = The loading or unloading behavior of the connections is taken into consideration.

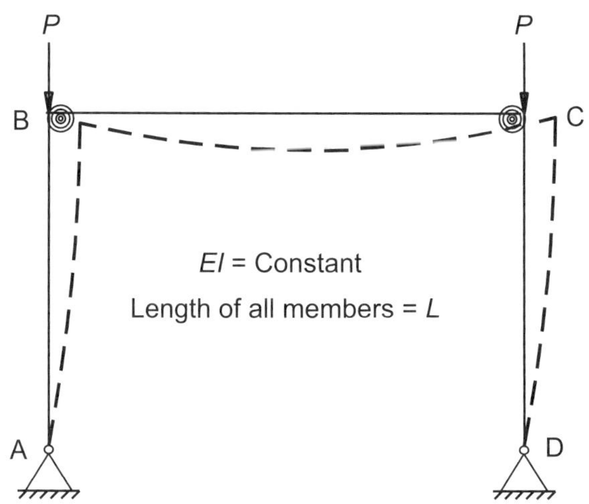

FIGURE 2.6 Example: K-factor calculation.

A. If the loading or unloading behavior of the connections is ignored, both connections *B* and *C* will have identical stiffness that is equal to $10EI/L$.

Using Equation 2.13 with $dM_F/dM_N = 1$ and $\bar{R}_k = 10$, we obtain:

$$I' = \frac{I}{1.6}$$

and from Equation 2.9 with $\tau = 1$, we obtain $G_A = \infty$ and $G_B = 1.6$. Note that the theoretical rather than the recommended G_A value of 10 is used in this example so the result can be directly compared with the theoretical solution.

From the alignment chart of Figure 2.5b, *K* for column *AB* is determined to be 2.5. This calculated value of *K* compares favorably with the theoretical *K* value of 2.52 (Lui and Chen, 1988).

B. If the loading or unloading behavior of the connections is considered, the connection stiffness for connection *B* and connection *C* will be different. If we conservatively assume a connection hinge has formed at *C*, the stiffness R_k for connection *B* (which undergoes unloading) will be $10EI/L$, but R_k for connection *C* is now zero, making column *CD* a leaning column.

Using Equation 2.13 with $dM_F/dM_N=0$ and $\bar{R}_k = 10$, we obtain:

$$I' = \frac{I}{2.6}$$

and from Equation 2.9 with $\tau = 1$, we obtain $G_A = \infty$ and $G_B = 2.6$. Again, the theoretical rather the recommended G_A value of 10 is used so the result can be compared with the theoretical solution directly.

From the alignment chart of Figure 2.5b, *K* is determined to be 2.8. However, because of the presence of the leaning column *CD*, further modification to *K* is necessary.

Using Equation 2.16 with $\tau = 1$ and recognizing that column *AB* is the only non-leaning column of the frame, we obtain:

$$K = \sqrt{\frac{2P}{P}\frac{I}{\left(\frac{I}{2.8^2}\right)}} = 3.96$$

The theoretical *K* value for column *AB* obtained from a system buckling analysis is 3.93. Thus, good agreement is obtained.

In reality, the stiffness for connection *C* when it is undergoing loading will fall between 0 and $10EI/L$, giving an incremental moment ratio dM_F/dM_N in the range of 0 and 1. The actual *K* value for column *AB* will lie anywhere between 2.52 and 3.93. $K = 3.93$ is thus a conservative estimate of the effective length factor for the column.

2.3 Effects of Semi-rigid Connections on Beams

For design purposes, beam cross-sections are classified into compact sections, noncompact sections, and slender compression elements. A compact section is a section that is capable of developing the cross-section plastic moment capacity and sustaining a large rotation at the plastic moment. A noncompact section is a section that may or may not develop the plastic moment capacity. Even if the plastic moment capacity is developed, a noncompact section normally cannot sustain a large rotation at the plastic moment because of, but not limited to, local buckling. A slender compression element is a cross-section component that is susceptible to local buckling under a compressive force.

The bending capacity of a compact beam depends on its laterally unbraced length, the axis of bending, the type and pattern of the applied loads, and the support conditions. If the beam is short or adequately braced when bent about its strong axis or if it is bent about its weak axis, the bending capacity will be equal to the plastic moment capacity of the cross-section. On the other hand, if a slender beam without adequate lateral brace is bent about its strong axis, the bending capacity will be controlled by the elastic critical lateral torsional buckling strength. When an intermediate length beam is bending about its strong axis, the bending capacity will be governed by the inelastic lateral torsional buckling strength. As for the type and pattern of the applied loads, the most severe load case is one in which the beam is subjected to a pair of equal and opposite moments. This is because under this load condition, the internal bending moment along the entire length of the beam is a constant and equal to the applied end moments. Finally, as for the support conditions, the more rotational restraint the support provides to the beam, the less susceptible the beam will be to lateral torsional instability.

2.3.1 Lateral Torsional Buckling Load for Compact Beams

When an ideal I-beam is loaded in the plane of the web, in-plane deformation (i.e., bending about the strong axis) will occur at the commencement of the load and increases as the applied force increases. As the magnitude of the load approaches a certain critical value, out-of-plane deformation (i.e., bending about the weak axis) and twisting (i.e., rotational about the longitudinal axis) of the beam will increase rapidly and the beam is said to have reached an unstable state. This critical load is referred to as the lateral torsional buckling load of the beam. It is obtained as the eigenvalue of the bifurcation problem of the beam (Timoshenko, 1961; Chen and Lui, 1987).

The theoretical critical moment that corresponds to the elastic lateral torsional buckling of a beam for various support and loading conditions is given by Johnson (1976):

$$M_{cr} = C_1 \frac{\pi^2 EI_y}{(KL)^2}\left(C_2 g + C_3 j + \sqrt{(C_2 g + C_3 j)^2 + \frac{C_w}{I_y}\left[1 + \frac{GJ(KL)_2^2}{\pi_2 EC_w}\right]}\right) \quad (2.21)$$

where

C_w = warping constant
E = modulus of elasticity
g = distance from shear center to the point of application of the transverse load, taken positive when the load is below the shear center
I_y = moment of inertia about weak or y-axis
$j = S + \frac{1}{2I_x}\int_A y(x^2 + y^2)dA$, , in which S is the distance from the centroid of the cross section to the shear center, and I_x is the moment of inertia about the strong or x-axis
J = torsional constant
K = effective length factor for the beam
C_1, C_2, C_3 = coefficients which are functions of loading and support conditions (Johnston, 1976)

Beams used in practice usually fail at loads below M_{cr} due to the presence of geometrical and material imperfections. For such beams, lateral deformation and twist commence at the start of loading and increase as the magnitude of the loads increases. The maximum load, referred to as the ultimate load, is obtained as the peak point of the load-deflection curve.

To obtain this ultimate load, a load-deflection approach must be used and recourse to numerical analysis is inevitable. Analytical and experimental studies to determine the ultimate loads of beams are available in the literature (for example, Galambos, 1963; Yoshida and Imoto, 1973; Vinnakota, 1977; Yura et al., 1978; Fukumoto et al., 1980; Fukumoto and Itoh, 1981; Trahair, 1993). In general, the presence of residual stresses and geometric imperfections will decrease the nominal strength of the beam. The effect of residual stresses is to cause early yielding, and depending on the types of loadings and residual stress distributions, the nominal strength will be affected differently. The effect of geometrical imperfections is to aggravate the deformation, and so the beam will reach an unstable state at a lower load. Evidently, this effect is more pronounced for slender beams.

2.3.1.1 Beam Strength Design Equations

Based on a semi-empirical approach that takes into consideration geometrical and material non-linearities, AISC (2005) recommends the use of the following equations to calculate the nominal flexural strength of compact I-shaped beams with lateral unbraced length of L_b bent about the strong axis.

For $L_b \le L_p$:

$$M_n = M_{pb} = f_y Z_x \tag{2.22}$$

For $L_p < L_b \le L_r$:

$$M_n = C_b \left[M_p - (M_p - 0.7 f_y S_x) \left(\frac{L_b - L_p}{L_r - L_p} \right) \right] \le M_{pb} \tag{2.23}$$

For $L_b > L_r$:

$$M_n = C_b S_x \frac{\pi^2 E}{\left(\dfrac{L_b}{r_{ts}} \right)^2} \sqrt{1 + 0.078 \frac{J}{S_x h_o} \left(\frac{L_b}{r_{ts}} \right)^2} \le M_{pb} \tag{2.24}$$

The various terms in the above equations are defined as:

$$L_p = 1.76 r_y \sqrt{\frac{E}{f_y}} \tag{2.25}$$

$$L_r = 1.95 r_{ts} \frac{E}{0.7 f_y} \sqrt{\frac{J}{S_x h_o}} \sqrt{1 + \sqrt{1 + 6.76 \left(\frac{0.7 f_y}{E} \frac{S_x h_o}{J} \right)^2}} \tag{2.26}$$

where

$$r_{ts}^2 = \frac{\sqrt{I_y C_w}}{S_x}, \quad h_o = d - t_f \tag{2.27}$$

in which C_w is the warping constant, E is the modulus of elasticity, f_y is the material yield stress, J is the torsional constant, M_{pb} is the plastic moment capacity of the beam, r_y is the radius of gyration about the weak axis, S_x is the elastic section modulus about the strong axis, Z_x is the plastic section modulus about the strong axis, and C_b is a factor to account for the type and pattern of the applied loads on the beam. It is expressed as:

$$C_b = \frac{12.5 M_{max}}{2.5 M_{max} + 3 M_A + 4 M_B + 3 M_C} \le 3.0 \tag{2.28}$$

in which M_{max}, M_A, M_B and M_C are the absolute value of the maximum, quarter-point, midpoint, and three-quarter point moments in the unbraced segment of the beam, respectively.

Note that the above formulae are applicable only to compact I-shaped sections bent about the strong axis in which lateral torsional instability is the limit state. If the same section is bent about the weak axis, the nominal flexural strength will be the plastic moment capacity about the weak axis.

2.3.2 Beams with Semi-rigid Connections

Like columns, the support conditions of a beam will have an effect on its flexural capacity. Beams that are simply-supported at the ends will have a lower nominal strength than beams that are rigidly supported. In the next two sections, the effect of semi-rigid support condition on beams will be discussed.

2.3.2.1 Analysis of Semi-rigid Beams

For beams with semi-rigid connections, most of the analyses have been focused on using computer-based techniques. However, the analysis of beams with semi-rigid connections can be carried out as rigidly connected beams with the appropriate connection modeling (Cunningham, 1990; Geschwindner, 1991; Chikho and Kirby, 1995). The theory and analysis procedures are discussed by Xu in Chen (2000). The effects of connection flexibility are modeled using rotational springs having stiffness R_k at the ends of the beam. To reflect the relative stiffness of the beam and the rotational end-spring connections, an end-fixity factor is used (Monforton and Wu, 1963; Chen, 2000). It is defined as:

$$\eta = \frac{\alpha}{\phi} = \frac{1}{1 + \dfrac{3}{\bar{R}_k}} \tag{2.29}$$

where α is the member end rotation, ϕ is the combined rotation of the member and the connection due to a unit end moment (see Figure 2.7), $\bar{R}_k = R_k L_b / EI_b$ is defined as before in Equation 2.11. Note that η falls between zero and one, and it defines the stiffness of each end connection relative to the adjoining beam.

Upon the introduction of the end-fixity factor η to account for semi-rigid behavior, the analysis of semi rigid beams can be made through minor modification of the conventional moment distribution method for rigidly connected beams. The steps are outlined in items 1–4:

1. Evaluate end-fixity factors η_i and η_j according to Equation 2.29 at the *i*th and *j*th ends of the beam with semi-rigid connections.

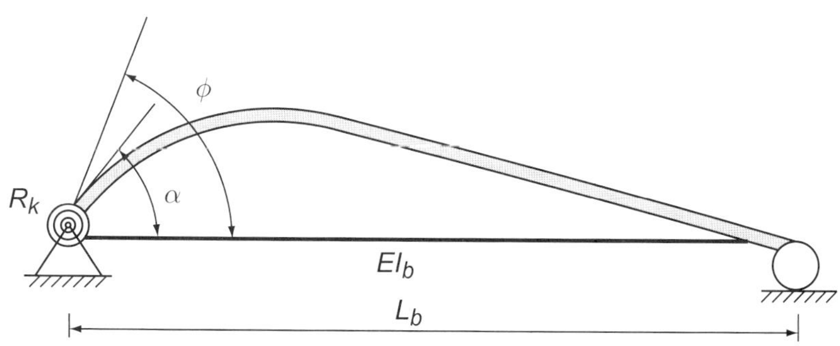

FIGURE 2.7 A beam restrained by a semi-rigid connection.

2. Calculate member stiffness factors K_{ij} and K_{ji} for each semi-rigid beam and obtain moment distribution factor DF_{ij} using the following equations:

$$K_{ij} = \left(\frac{12\eta_i}{4-\eta_i\eta_j}\right)\frac{EI_b}{L_b} \tag{2.30}$$

$$K_{ji} = \left(\frac{12\eta_j}{4-\eta_i\eta_j}\right)\frac{EI_b}{L_b} \tag{2.31}$$

$$DF_{ij} = \frac{K_{ij}}{\sum K_{ij}} \tag{2.32}$$

3. Compute the end reaction moments M_{ij} and M_{ji}.
4. Sum the member end moments at each joint to determine the unbalanced moment, and complete the rest of the analysis by conventional moment distribution method for rigid continuous beams.

2.3.2.2 Design of Semi-rigid Beams

For I-beams, the theoretical critical moment that corresponds to the elastic lateral torsional buckling load is given by Equation 2.21. If pinned supports are used at both ends of the beam and a loading consists of a pair of equal and opposite moments are applied to the beam ends, Equation 2.21 can be reduced to:

$$M_{cr} = \frac{\pi}{L_b}\sqrt{EI_y GJ}\sqrt{1+\frac{\pi^2}{L_b^2}\frac{EC_w}{GJ}} \tag{2.33}$$

By performing a series of second-order inelastic analyses using different values for R_k and L_b, an empirical equation that can be used for estimating the lateral torsional buckling load of a simply supported I-beam with in-plane semi-rigid connections having connection stiffness in the range $0 < \tilde{R}_k < 20$ is given by:

$$M_n = M_{cr} + \frac{3.56\times10^5\,\tilde{R}_k}{L_b^{1.32}} \tag{2.34}$$

In the above equation, M_{cr} is given by Equation 2.33, L_b is the length of the beam in inches (1 inch = 2.54 cm), and $\tilde{R}_k = R_k L_b / EI_y$, in which R_k is the connection stiffness, and I_y is the moment of inertial of the beam about its weak axis. Note the increase in M_n when $\tilde{R}_k$ increases or when L_b decreases. In the event that the applied loads are something other than a pair of equal and opposite moments, Equation 2.34 needs to be multiplied by C_b given in Equation 2.28.

Example:

Consider the I-beam shown in Figure 2.8. If the beam span L_b is 30 feet (or 360 inches or 9.1 m) and the connections at *A* and *B* are semi-rigid connections with an in-plane rotational stiffness of $10EI_y/L_b$, determine the nominal moment capacity of the beam. The section properties are: I_y = 42.6 in^4 (1.77 × 10^3 cm^4), J = 0.769 in^4 (32 cm^4), C_w = 619 in^6 (1.66 × 10^5 cm^6). The material properties are: E = 29,000 ksi (200 MPa), G = 11,200 ksi (77.2 MPa).

Using Equations 2.33 and 2.34, M_n is obtained as:

$$M_n = 970 + \frac{3.56\times10^5(10)}{360^{1.32}} = 2470 \text{ kip-in} \quad (279 \text{ kN-m})$$

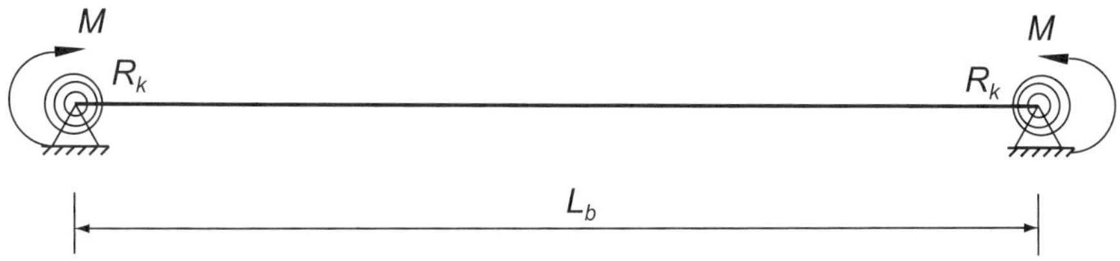

FIGURE 2.8 Analysis of semi-rigid beam.

This value is within 5 percent of M_n = 2580 kip-in (291 kN-m) obtained from a second-order inelastic finite element analysis of the beam.

2.4 Effects of Semi-rigid Connections on Frames

Semi-rigid frames are frames for which the connections joining the beams and columns are neither fully rigid nor perfectly pinned. In reality, all steel frames are semi-rigid in nature as all connections exhibit a certain degree of flexibility under loads. According to AISC (2005), the effect of semi-rigidity needs to be considered when the connection stiffness R_k, compared to the in-plane bending stiffness EI_b/L_b of the beam, falls in the range $2 < R_k L_b / EI_b < 20$. The incorporation of connection flexibility into frame analysis normally entails an iterative-type solution process. This is because most connections exhibit nonlinear response almost from the start of loading. The nonlinear behavior of connections is attributed to a number of factors such as local yielding, local buckling, stress concentration, strain hardening, and bolt slip, etc. Some commonly used connection models have been discussed in this handbook. The selection of the proper model for a specific type of analysis depends on the nature of that analysis. In certain situations, it is possible to model the connections as linear elements with constant stiffness. For instance in a bifurcation analysis, we are interested in locating adjacent equilibrium configurations in the immediate neighborhood of the perfect one. As a result, only the initial stiffness property that corresponds to the instantaneous deformation of the structure is of interest. It is therefore reasonable to model the connections as linear elements since only the initial stiffness of the connections is needed for such analysis. The assumption of linear connection behavior is also acceptable in a serviceability analysis. In the analysis of a structure for serviceability, the usual premise is that deformation is small. For small deformation, the nonlinear behavior of most connections is not significant and so the use of a linear model is justifiable. However, in a strength analysis it is important to use a proper nonlinear connection model.

Over the past decades, a large number of studies have been conducted and published on the influence of connection flexibility on frame responses (for example, Chen and Lui, 1987; Lui and Chen, 1987, 1988; Lui, 1988; Yau and Chan, 1994; Rodrigues et al., 1998; Aristizabal-Ochoa, 2001; van Keulen et al., 2003; Gizejowski et al., 2004; Sekulovic and Nefovska-Danilovic, 2004). In this section, several general approaches by which connection flexibility can be accounted for in frame analysis will be presented. This is followed by a discussion on the design of semi-rigid frames.

2.4.1 Analysis of Semi-rigid Frames

Nonlinearities in flexibly connected frames consist of three types: (1) the nonlinear M-θ_r relationship of the connection, (2) the geometrical nonlinearity of the member (P-δ effect) and of the frame (P-Δ effect), and (3) the material nonlinearity or yielding in the members of the frame.

Thus, depending on the types of nonlinearity and the degree of accuracy required, three different analysis techniques can be employed (Chen and Lui, 1991).

2.4.1.1 Elastic Bifurcation Analysis of Semi-rigid Frames

In a bifurcation or an eigenvalue analysis of semi-rigid frames, the load that corresponds to a state of equilibrium bifurcation of the frame is sought. This load, referred to as the critical load, is the theoretical maximum load that the semi-rigid frame can sustain. To evaluate this critical load, it is necessary to write the equations of equilibrium of the frame with respect to this deformed configuration taking into account the effect of connection flexibility as described below.

2.4.1.1.1 Beam Stiffness Relationship

Consider a beam element subjected to end moments (M_A, M_B) and an axial force (P) with connections, modeled as rotational springs, attached at both ends as shown in Figure 2.9. Because we are only concerned with the instantaneous deformation of the frame in an eigenvalue analysis, it is justifiable to consider only the initial stiffness of the connection. Denoting R_{kiA} and R_{kiB} as the initial connection stiffness of connections A and B, respectively, the relative rotation between the joint and the beam end (i.e., rotational deformation of the connection) can be expressed as:

$$\theta_{rA} = \frac{M_A}{R_{kiA}}, \quad \theta_{rB} = \frac{M_B}{R_{kiB}} \tag{2.35}$$

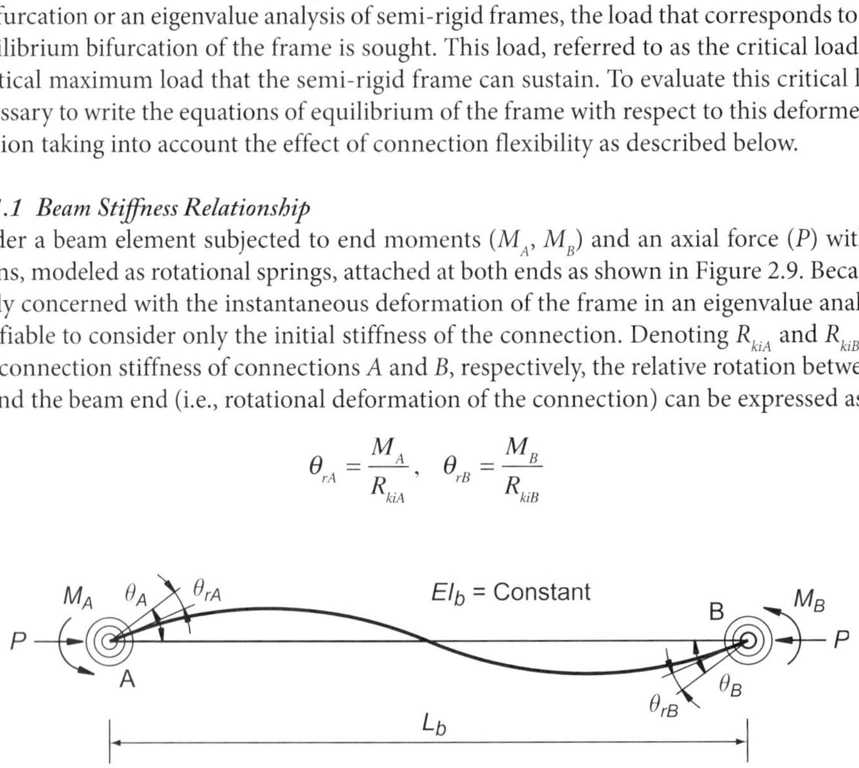

FIGURE 2.9 Beam element with semi-rigid connections.

If we denote θ_A and θ_B as the joint rotations at the Ath and Bth ends of the beam, respectively, the slope-deflection equations for the beam modified for the presence of connections can be expressed as:

$$M_A = \frac{EI_b}{L_b}\left[s_{ii}\left(\theta_A - \frac{M_A}{R_{kiA}}\right) + s_{ij}\left(\theta_B - \frac{M_B}{R_{kiB}}\right)\right] \tag{2.36a}$$

$$M_A = \frac{EI_b}{L_b}\left[s_{ij}\left(\theta_A - \frac{M_A}{R_{kiA}}\right) + s_{jj}\left(\theta_B - \frac{M_B}{R_{kiB}}\right)\right] \tag{2.36b}$$

where EI_b is flexural rigidity and L_b is the length of the beam. s_{ii} and s_{ij} are stability functions, which for a tensile axial force, are given by:

$$s_{ii} = s_{jj} = \frac{(kL_b)^2 \cosh kL_b - kL_b \sinh kL_b}{2 - 2\cosh kL_b + kL_b \sinh kL_b} \tag{2.37a}$$

$$s_{ij} = s_{ji} = \frac{kL_b \sinh kL_b - (kL_b)^2}{2 - 2\cosh kL_b + kL_b \sinh kL_b} \tag{2.37b}$$

and for a compressive axial force, are given by:

$$s_{ii} = s_{jj} = \frac{kL_b \sin kL_b - (kL)^2 \cos kL_b}{2 - 2\cos kL_b - kL_b \sin kL_b} \tag{2.38a}$$

$$s_{ij} = s_{ji} = \frac{(kL_b)^2 - kL_b \sin kL_b}{2 - 2\cos kL_b - kL_b \sin kL_b} \tag{2.38b}$$

in which k is defined in Equation 2.6.

Solving Equations 2.36a and 2.36b for M_A and M_B gives:

$$M_A = \frac{EI_b}{L_b}\left[s_{ii}^{*}\theta_A + s_{ij}^{*}\theta_B\right] \tag{2.39a}$$

$$M_B = \frac{EI_b}{L_b}\left[s_{ji}^{*}\theta_A + s_{jj}^{*}\theta_B\right] \tag{2.39b}$$

where

$$s_{ii}^{*} = \left[s_{ii} + \frac{EI_b s_{ii}^2}{L_b R_{kiB}} - \frac{EI_b s_{ij}^2}{L_b R_{kiB}}\right] / \left[\left(1 + \frac{EI_b s_{ii}}{L_b R_{kiA}}\right)\left(1 + \frac{EI_b s_{ii}}{L_b R_{kiB}}\right) - \left(\frac{EI_b}{L_b}\right)^2 \frac{s_{ij}^2}{R_{kiA} R_{kiB}}\right] \tag{2.40a}$$

$$s_{jj}^{*} = \left[s_{ii} + \frac{EI_b s_{ii}^2}{L_b R_{kiA}} - \frac{EI_b s_{ij}^2}{L_b R_{kiA}}\right] / \left[\left(1 + \frac{EI_b s_{ii}}{L_b R_{kiA}}\right)\left(1 + \frac{EI_b s_{ii}}{L_b R_{kiB}}\right) - \left(\frac{EI_b}{L_b}\right)^2 \frac{s_{ij}^2}{R_{kiA} R_{kiB}}\right] \tag{2.40b}$$

$$s_{ij}^{*} = s_{ji}^{*} = s_{ij} / \left[\left(1 + \frac{EI_b s_{ii}}{L_b R_{kiA}}\right)\left(1 + \frac{EI_b s_{ii}}{L_b R_{kiB}}\right) - \left(\frac{EI_b}{L_b}\right)^2 \frac{s_{ij}^2}{R_{kiA} R_{kiB}}\right]. \tag{2.40c}$$

Equations 2.39a and 2.39b can be transformed into a member stiffness relationship of a six degree-of-freedom plane frame member shown in Figure 2.10 by relating the member end forces r_i, $i = 1$ to 6, to the member basic forces P, M_A, and M_B as:

$$\begin{Bmatrix} r_1 \\ r_2 \\ r_3 \\ r_4 \\ r_5 \\ r_6 \end{Bmatrix} = \begin{bmatrix} 1 & 0 & 0 \\ 0 & \frac{1}{L_b} & \frac{1}{L_b} \\ 0 & 1 & 0 \\ -1 & 0 & 0 \\ 0 & -\frac{1}{L_b} & -\frac{1}{L_b} \\ 0 & 0 & 1 \end{bmatrix} \begin{Bmatrix} P \\ M_A \\ M_B \end{Bmatrix} + \begin{Bmatrix} 0 \\ \frac{P\Delta}{L_b} \\ 0 \\ 0 \\ -\frac{P\Delta}{L_b} \\ 0 \end{Bmatrix} \tag{2.41}$$

where (Δ is the relative joint translation of the member.

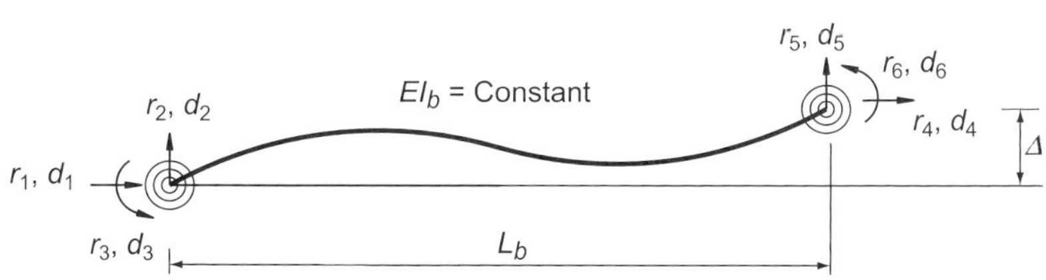

FIGURE 2.10 2-D beam element.

Similarly, the member basic displacements u, θ_A, and θ_B, where u is the axial deformation of the member, and the member end displacements d_i, $i = 1$ to 6, can be related by:

$$\begin{pmatrix} u \\ \theta_A \\ \theta_B \end{pmatrix} = \begin{bmatrix} 1 & 0 & 0 & -1 & 0 & 0 \\ 0 & \dfrac{1}{L_b} & 1 & 0 & -\dfrac{1}{L_b} & 0 \\ 0 & \dfrac{1}{L_b} & 0 & 0 & -\dfrac{1}{L_b} & 1 \end{bmatrix} \begin{pmatrix} d_1 \\ d_2 \\ d_3 \\ d_4 \\ d_5 \\ d_6 \end{pmatrix} \tag{2.42}$$

Equations 2.41 and 2.42 can be combined by realizing that the basic force-displacement relationship is given by:

$$\begin{pmatrix} P \\ M_A \\ M_B \end{pmatrix} = \frac{EI_b}{L_b} \begin{bmatrix} \dfrac{A}{I_b} & 0 & 0 \\ 0 & s_{ii}^* & s_{ij}^* \\ 0 & s_{ji}^* & s_{jj}^* \end{bmatrix} \begin{pmatrix} u \\ \theta_A \\ \theta_B \end{pmatrix} \tag{2.43}$$

Upon substitution of Equation 2.42 into Equation 2.43, and then into Equation 2.41, we obtain:

$$\begin{pmatrix} r_1 \\ r_2 \\ r_3 \\ r_4 \\ r_5 \\ r_6 \end{pmatrix} = \frac{EI_b}{L_b} \begin{bmatrix} \dfrac{A}{I_b} & 0 & 0 & -\dfrac{A}{I_b} & 0 & 0 \\ & \dfrac{\left(s_{ii}^* + 2s_{ij}^* + s_{jj}^*\right) - \left(kL_b\right)^2}{L_b^2} & \dfrac{\left(s_{ii}^* + s_{ij}^*\right)}{L_b} & 0 & \dfrac{-\left(s_{ii}^* + 2s_{ij}^* + s_{jj}^*\right) + \left(kL_b\right)^2}{L_b^2} & \dfrac{\left(s_{ij}^* + s_{jj}^*\right)}{L_b} \\ & & s_{ii}^* & 0 & \dfrac{-\left(s_{ii}^* + s_{ij}^*\right)}{L_b} & s_{ij}^* \\ & & & \dfrac{A}{I_b} & 0 & 0 \\ & sym. & & & \dfrac{\left(s_{ii}^* + 2s_{ij}^* + s_{jj}^*\right) - \left(kL_b\right)^2}{L_b^2} & \dfrac{-\left(s_{ij}^* + s_{jj}^*\right)}{L_b} \\ & & & & & s_{jj}^* \end{bmatrix} \begin{pmatrix} d_1 \\ d_2 \\ d_3 \\ d_4 \\ d_5 \\ d_6 \end{pmatrix} \tag{2.44}$$

Symbolically, Equation 2.44 can be written as:

$$\mathbf{r} = \mathbf{k}_{\mathbf{beam}} \mathbf{d} \tag{2.45}$$

If the axial force in the beam is small, the beam stiffness matrix k_{beam} in Equation 2.45 can be simplified by setting $s_{ii} = s_{jj} = 4$ and $s_{ij} = s_{ji} = 2$ in Equation 2.40a-c, and with $k = \sqrt{\left(P/EI_b\right)} = 0$, we have:

$$k_{beam} = \frac{EI_b}{L_b}\begin{bmatrix} \frac{A}{I_b} & 0 & 0 & -\frac{A}{I_b} & 0 & 0 \\ & \frac{\left(s_{ii}^* + 2s_{ij}^* + s_{jj}^*\right)}{L_b^2} & \frac{\left(s_{ii}^* + s_{ij}^*\right)}{L_b} & 0 & -\frac{\left(s_{ii}^* + 2s_{ij}^* + s_{jj}^*\right)}{L_b^2} & \frac{\left(s_{ij}^* + s_{jj}^*\right)}{L_b} \\ & & s_{ii}^* & 0 & -\frac{\left(s_{ii}^* + s_{ij}^*\right)}{L_b} & s_{ij}^* \\ & & & \frac{A}{I_b} & 0 & 0 \\ & sym. & & & \frac{\left(s_{ii}^* + 2s_{ij}^* + s_{jj}^*\right)}{L_b^2} & -\frac{\left(s_{ij}^* + s_{jj}^*\right)}{L_b} \\ & & & & & s_{jj}^* \end{bmatrix} \tag{2.46}$$

where

$$s_{ii}^* = \left[4 + \frac{12EI_b}{L_b R_{kiB}}\right] \Big/ \left[\left(1 + \frac{4EI_b}{L_b R_{kiA}}\right)\left(1 + \frac{4EI_b}{L_b R_{kiB}}\right) - \left(\frac{EI_b}{L_b}\right)^2 \frac{4}{R_{kiA} R_{kiB}}\right] \tag{2.47a}$$

$$s_{jj}^* = \left[4 + \frac{12EI_b}{L_b R_{kiA}}\right] \Big/ \left[\left(1 + \frac{4EI_b}{L_b R_{kiA}}\right)\left(1 + \frac{4EI_b}{L_b R_{kiB}}\right) - \left(\frac{EI_b}{L_b}\right)^2 \frac{4}{R_{kiA} R_{kiB}}\right] \tag{2.47b}$$

$$s_{ij}^* = s_{ji}^* = 2 \Big/ \left[\left(1 + \frac{4EI_b}{L_b R_{kiA}}\right)\left(1 + \frac{4EI_b}{L_b R_{kiB}}\right) - \left(\frac{EI_b}{L_b}\right)^2 \frac{4}{R_{kiA} R_{kiB}}\right] \tag{2.47c}$$

2.4.1.1.2 Column Stiffness Relationship

Referring to Figure 2.11, the force-displacement relationship of the column taking into account the geometrical nonlinear effects in the absence of in-span loadings can be expressed symbolically as (Chen and Lui, 1991):

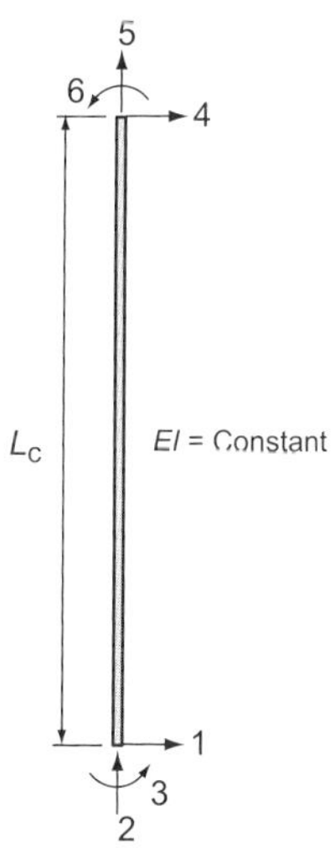

$$\mathbf{r} = \mathbf{k}_{\text{column}}\mathbf{d} \tag{2.48}$$

where

$\mathbf{r} = \{r_1, r_2, r_3, r_4, r_5, r_6\}^T$ is the member end force vector

$\mathbf{d} = \{d_1, d_2, d_3, d_4, d_5, d_6\}^T$ is the member end displacement vector

FIGURE 2.11 2-D column element.

and:

$$
\mathbf{k}_{\text{column}} = \frac{EI_c}{L_c}\begin{bmatrix}
\frac{12}{L_c^2}\phi_1 & 0 & \frac{6}{L_c}\phi_2 & -\frac{12}{L_c^2}\phi_1 & 0 & \frac{6}{L_c}\phi_2 \\
 & \frac{A}{I_c} & 0 & 0 & -\frac{A}{I_c} & 0 \\
 & & 4\phi_3 & -\frac{6}{L_c}\phi_2 & 0 & 2\phi_4 \\
 & & & \frac{12}{L_c^2}\phi_1 & 0 & -\frac{6}{L_c}\phi_2 \\
 & sym. & & & \frac{A}{I_c} & 0 \\
 & & & & & 4\phi_3
\end{bmatrix} \tag{2.49}
$$

is the column stiffness matrix, whereby
For a tensile axial force:

$$\phi_1 = \frac{(kL_c)^3 \sinh kL_c}{12(2 - 2\cosh kL_c + kL_c \sinh kL_c)} \tag{2.50a}$$

$$\phi_2 = \frac{(kL_c)^2 (\cosh kL_c - 1)}{6(2 - 2\cosh kL_c + kL_c \sinh kL_c)} \tag{2.50b}$$

$$\phi_3 = \frac{kL_c (kL_c \cosh kL_c - \sinh kL_c)}{4(2 - 2\cosh kL_c + kL_c \sinh kL_c)} \tag{2.50c}$$

$$\phi_4 = \frac{kL_c (\sinh kL_c - kL_c)}{2(2 - 2\cosh kL_c + kL_c \sinh kL_c)} \tag{2.50d}$$

and for a compressive axial force:

$$\phi_1 = \frac{(kL_c)^3 \sin kL_c}{12(2 - 2\cos kL_c - kL_c \sin kL_c)} \tag{2.51a}$$

$$\phi_2 = \frac{(kL_c)^2 (1 - \cos kL_c)}{6(2 - 2\cos kL_c - kL_c \sin kL_c)} \tag{2.51b}$$

$$\phi_3 = \frac{kL_c (\sin kL_c - kL_c \cos kL_c)}{4(2 - 2\cos kL_c - kL_c \sin kL_c)} \tag{2.51c}$$

$$\phi_4 = \frac{kL_c (kL_c - \sin kL_c)}{2(2 - 2\cos kL_c - kL_c \sin kL_c)} \tag{2.51d}$$

In the above equations, k is defined in Equation 2.6.

If the axial force in the column is zero, it can be shown by applying the L'Hospital rule successively that $\phi_1 = \phi_2 = \phi_3 = \phi_4 = 1$. For a small axial force, the above equations may become numerically unstable. To circumvent this situation and also to avoid the use of different expressions for compressive and tensile forces, a power series expansion (Goto and Chen, 1986) can be used to simplify the expressions to:

$$\phi_1 = \left\{1+\sum_{n=1}^{\infty}\frac{1}{(2n+1)!}\left[(kL_c)^2\right]^n\right\} / \left\{1+\sum_{n=1}^{\infty}\frac{24(n+1)}{(2n+4)!}\left[(kL_c)^2\right]^n\right\} \quad (2.52a)$$

$$\phi_2 = \left\{1+\sum_{n=1}^{\infty}\frac{2}{(2n+2)!}\left[(kL_c)^2\right]^n\right\} / \left\{1+\sum_{n=1}^{\infty}\frac{24(n+1)}{(2n+4)!}\left[(kL_c)^2\right]^n\right\} \quad (2.52b)$$

$$\phi_3 = \left\{1+\sum_{n=1}^{\infty}\frac{6(n+1)}{(2n+3)!}\left[(kL_c)^2\right]^n\right\} / \left\{1+\sum_{n=1}^{\infty}\frac{24(n+1)}{(2n+4)!}\left[(kL_c)^2\right]^n\right\} \quad (2.52c)$$

$$\phi_4 = \left\{1+\sum_{n=1}^{\infty}\frac{6}{(2n+3)!}\left[(kL_c)^2\right]^n\right\} / \left\{1+\sum_{n=1}^{\infty}\frac{24(n+1)}{(2n+4)!}\left[(kL_c)^2\right]^n\right\} \quad (2.52d)$$

where the quantity:

$$(kL_c)^2 = \frac{PL_c^2}{EI_c} \quad (2.53)$$

is positive for a tensile force and negative for a compressive force.

Because the stiffness matrix expressed in Equation 2.49 is a function of the axial force in the member, which depends on the displaced configuration of the structure that is not known in advance, the solution can only be obtained through an iterative process.

The beam stiffness matrix k_{beam} and the column stiffness matrix k_{column} can be assembled to form the structure stiffness matrix K for a given structure. The critical load P_{cr} can be obtained from the characteristic equation:

$$\det\left|\mathbf{K}\right| = 0 \quad (2.54)$$

Example:

To demonstrate the above procedure for the bifurcation analysis of a semi-rigid frame, the pinned-based portal frame shown in Figure 2.12a is used. The frame is subjected to a pair of concentrated loads applied to the columns. The connections at both ends of the beams are assumed to be identical. When the loads attain their critical values, the frame will buckle. Since the frame is unbraced, the preferred buckling mode will be the sway mode as shown in Figure 2.12b. The sway-buckling mode of the frame can be uniquely defined by three displacement variables θ_A, θ_B, and Δ as shown in the figure. By assembling the column and beam element stiffness matrices, the stiffness relationship of the frame can be expressed as:

$$\begin{pmatrix} 0 \\ 0 \\ 0 \end{pmatrix} = \frac{EI}{L}\begin{bmatrix} 4\phi_3 & 2\phi_4 & -\frac{6}{L}\phi_2 \\ & 4\phi_3 + s_{ii}^* + s_{ij}^* & -\frac{6}{L}\phi_2 \\ sym. & & \frac{12}{L^2}\phi_1 \end{bmatrix}\begin{pmatrix} \theta_A \\ \theta_B \\ \Delta \end{pmatrix} \quad (2.55)$$

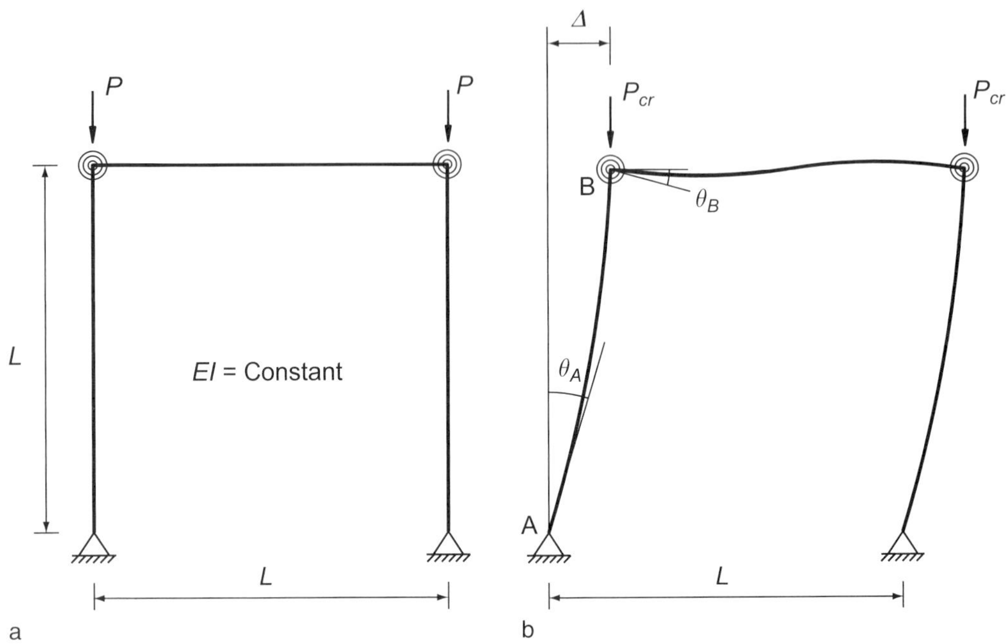

FIGURE 2.12 Buckling analysis of a semi-rigid portal frame.

or symbolically:

$$\mathbf{R} = \mathbf{0} = \mathbf{KD} \tag{2.56}$$

For a nontrivial solution, we must have:

$$\det|\mathbf{K}| = 0 \tag{2.57}$$

The critical loads computed using various $\overline{R}_{ki} = R_{ki}L/EI$ values where R_{ki} is the initial stiffness of the connections are given in Table 2.2. Also shown in the table are the critical load ratios of the semi-rigid frames when compared to a rigid frame of the same dimensions. As can be seen, the change in P_{cr} is rather drastic when $\overline{R}_{ki}$ is small, but it becomes less noticeable as $\overline{R}_{ki}$ increases. When $\overline{R}_{ki} = 20$, the buckling load of the semi-rigid frame reaches 92.3 percent that of the corresponding rigid frame.

2.4.1.2 Load Deflection Analysis of Semi-rigid Frames

In a bifurcation analysis only the magnitude of the critical or bifurcation load is obtained. No information regarding the magnitude of deformation at bifurcation or the behavior of the frame after bifurcation of equilibrium can be obtained. To obtain such information, a load-deflection analysis is required. Figures 2.13a and 2.13b show the post-bifurcation load-lateral deflection behavior of a rectangular semi-rigid portal frame with and without imperfections (Goto et al., 1991). The frame in Figure 2.13a is subjected to a uniformly distributed load applied on the beam, whereas the frame in Figure 2.13b is subjected to a pair of concentrated loads applied on the columns. The magnitude of the concentrated loads is such that they give the same value of vertical load on the frame as does the distributed load. From the figures it can be seen that the behavior of the frame differs quite substantially depending on the type of vertical loads applied. In the figures, the solid line represents the behavior of the perfect frame and the dashed line represents the

TABLE 2.2 Buckling load comparison for a semi-rigid portal frame

$\bar{R}_{ki}$	$P_{cr}L^2/EI$	$P_{cr,\ semi\text{-}rigid}/P_{cr,\ rigid}$
0.5	0.398	0.219
1	0.659	0.362
2	0.976	0.536
4	1.28	0.703
6	1.42	0.78
8	1.51	0.83
10	1.56	0.857
12	1.6	0.879
14	1.63	0.896
16	1.65	0.907
18	1.67	0.918
20	1.68	0.923
∞ (rigid)	1.82	1

behavior of the imperfect frame. Imperfections are induced by applying a small lateral load equal to αP_{cr} to the frame. For the perfect frame subjected to a uniformly distributed load, the post-bifurcation load-deflection curve rises above the bifurcation load and does not experience stability failure until the limit point is reached. On the contrary, for the perfect frame subjected to the concentrated loads, the post-bifurcation load-deflection curve drops drastically after bifurcation indicating that no post-buckling strength is present. The reason for this behavior can be explained in terms of the loading and unloading behavior of the connections.

For the frame subjected to the distributed load, the connections start to load as soon as the load is applied. As the applied load increases, the stiffness of the connections decreases because

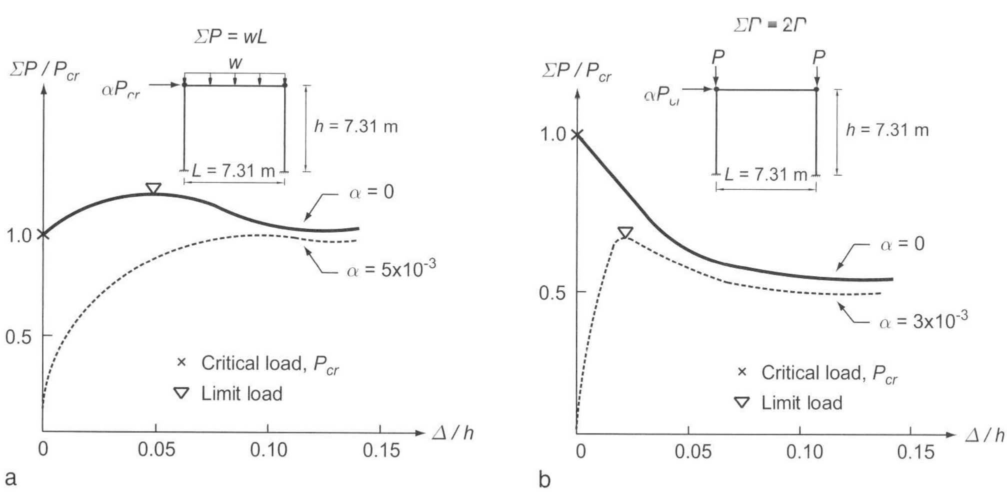

FIGURE 2.13 Behavior of a simple portal frame under different load types: (a) portal frame under a distribution load; (b) portal frame under concentrated loads.

the slope of the connection moment-rotation curve decreases at higher values of moment (Figure 2.1). At bifurcation, the connections undergo unloading as a result of joint rotation when the columns buckle. When a connection unloads, its stiffness increases because the unloading stiffness is larger than its instantaneous stiffness prior to unloading (Figure 2.2). This increase in connection stiffness enhances the frame stiffness. Hence, additional loads can be applied beyond bifurcation. On the other hand, if concentrated loads are applied on the columns, none of the connections undergo loading until bifurcation. Once bifurcation occurs, the connections begin to load and so their instantaneous stiffness and hence the stiffness of the frame will continue to degrade. Thus, no additional loads can be applied beyond bifurcation.

For the imperfect frame, except for the disappearance of the bifurcation point, the behavior when subjected to the two types of vertical loads is similar to its perfect counterpart. Note that when subjected to the uniformly distributed load, the limit load of the imperfect frame rises above the bifurcation point of its perfect counterpart, indicating that the bifurcation load of the perfect system is a conservative estimate of the stability limit state of the imperfect system. On the contrary, if the imperfect frame is subjected to the concentrated loads, the limit load always falls below the bifurcation point of its perfect counterpart, indicating that the bifurcation load is an unconservative estimate of the stability limit state of the frame. This observation is valid even for semi-rigid frames that are subject to a history of cyclic wind load before and after the application of the gravity load.

2.4.1.3 Allowance for the Formation of Plastic or Connection Hinges

When the bending moment at a particular location in the semi-rigid frame equals the plastic moment capacity of the cross-section, a plastic hinge will develop. Similarly, when the moment in the connection equals M_{cn}, a connection hinge is said to have formed. If strain hardening is neglected and if M_{cn} is taken as the maximum moment a connection can withstand, no additional moment beyond the hinge moment can be carried by the cross-section or the connection. Analysis of semi-rigid frames, taking into account the effect of concentrated plasticity in regions of plastic or connection hinges, can be carried out by casting k_{beam} and k_{column} of Equation 2.45 and Equation 2.48 in incremental form, and set the incremental moments at the hinge locations equal zero for any additional loading beyond M_p (for plastic hinge) or M_{cn} (for connection hinge) in subsequent cycles of analysis.

Example:

As an illustrative example, consider Figure 2.14 in which the load-deflection behavior of a four-story frame is shown. The beams are W16 × 40 sections, the columns for the bottom story are W12 × 79 and other stories are W10 × 60 sections. All sections have a yield stress of 34.2 ksi (246 MPa). Two analyses were performed: One assumed the connections were rigid (dashed line), and the other assumed the connections were semi-rigid (solid line) with a moment-rotation behavior depicted in Figure 2.15. The numbers on the curves and next to the × sign on the frames indicate the sequence of plastic hinge formation as the magnitude of the applied loads increases monotonically. Note that the location of hinge formation differs starting from the sixth hinge. This is because the connection flexibility influences the distribution of moments in the frame. Another interesting observation is that although the sequence of hinge formation differs, the maximum loads the two frames can carry do not differ significantly. It has been shown (Poggi and Zandonini, 1985; Chen and Zhou, 1987) that if the connections possess moment capacities in excess of their adjoining members (i.e., full strength connections), the ultimate load capacity of the semi-rigid frame will not differ significantly from that of a rigidly-jointed frame.

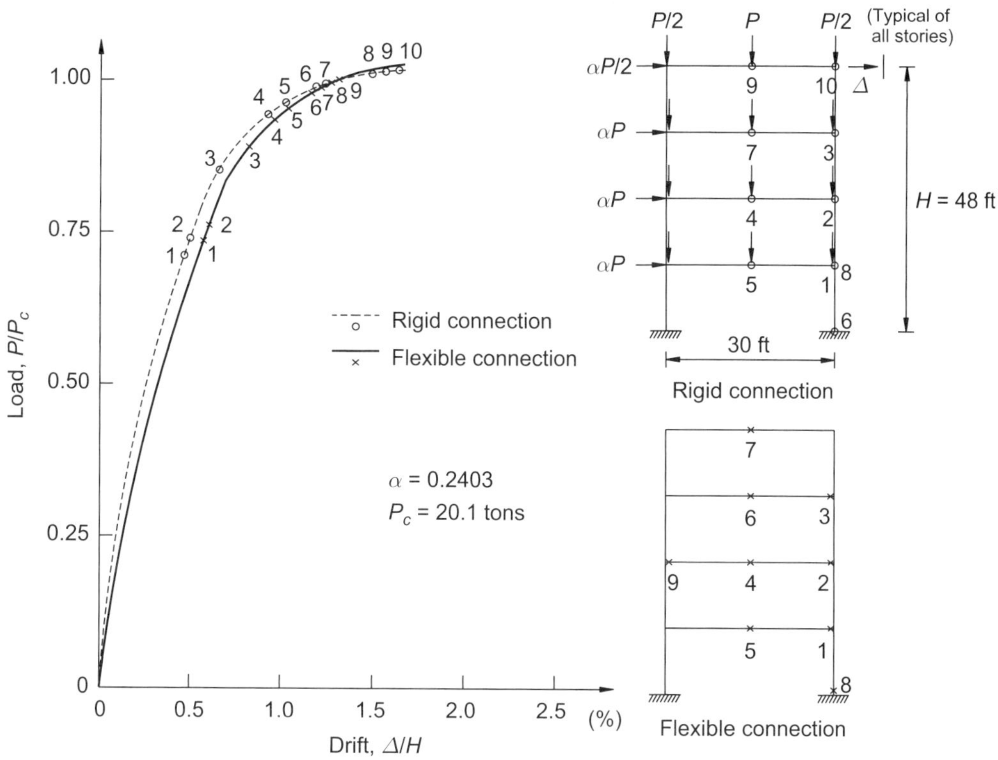

FIGURE 2.14 Comparative analysis of a four-story frame.

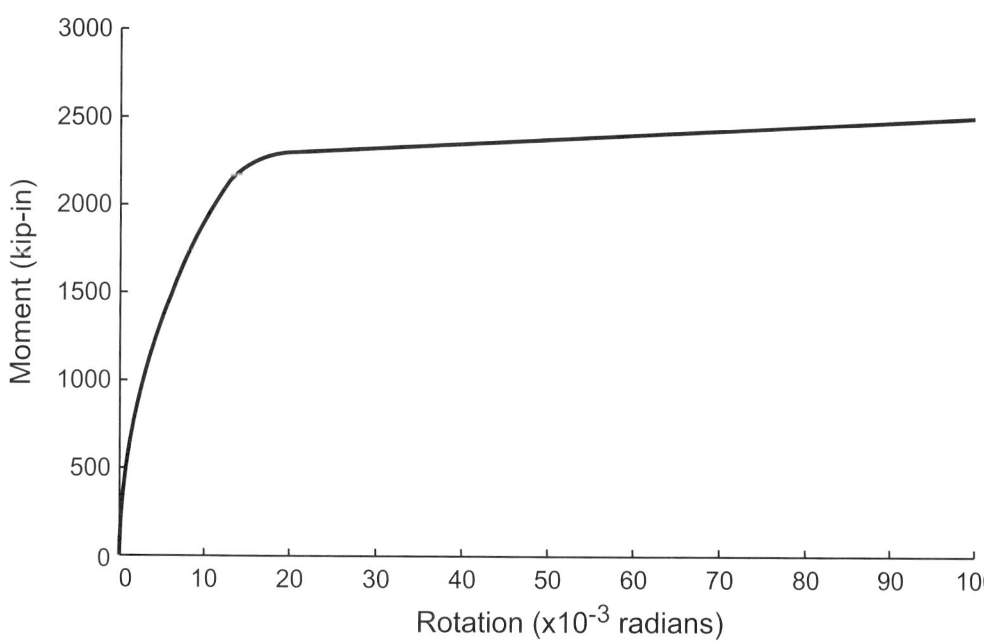

FIGURE 2.15 Connection moment rotation behavior (1 kip-in = 0.113 kN-m).

2.4.2 Design of Semi-rigid Frames

Over the years, a number of design approaches for semi-rigid frames have been proposed (Lui, 1988; King and Chen, 1993; Xu and Grierson, 1993; Xu et al., 1995; Cabrero and Bayo, 2005). Some of these design methods have been summarized by Chen and Lui (1991), Chen et al., (1996), Chen (2000), and Faella et al. (2000). Generally speaking, the design of semi-rigid frames is complicated by the fact that almost all connections exhibit some type of nonlinear M-θ_r behavior and allowance for this nonlinear behavior in the design will lead to a nonlinear problem. However, the design of semiframes can be greatly facilitated by the use of some simplifying assumptions. In this section, a simple semi-rigid frame design approach will be discussed.

2.4.2.1 Design of Sway-prevented Semi-rigid Frames

Sway-prevented or nonsway frames are defined here as frames for which sufficient bracings are provided to limit the lateral deflection to a negligible value. In terms of design, a sway-prevented frame differs from a sway-permitted frame in that the loading and unloading behavior of the connections in a nonsway frame is not an important design consideration, and the K factor for columns in a nonsway frame can conveniently and conservatively be taken as unity.

2.4.2.1.1 Design of Beams

Refer to Figure 2.16 in which the moment diagrams and maximum deflection values of a beam subjected to a uniformly distributed load of w under three different end support conditions (pinned, partially restrained, and fixed) are shown. From these diagrams, it can be seen that the design moment of the beam for the partially restrained case is always less than that of the pinned or fixed case. Since all types of connections offer a certain degree of partial restraint to the beam they connect, it is logical to design the beam as partially restrained members. The advantages involve not only a reduction in beam size but a reduction in beam deflection. Therefore, consideration of end restraint in design serves the dual purpose of reducing construction cost and improving the serviceability limit state of the structure.

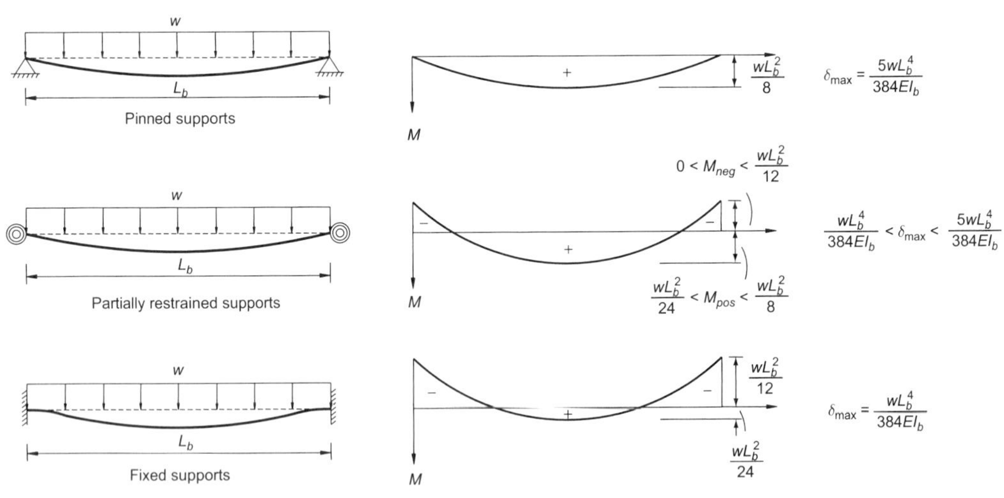

FIGURE 2.16 Effect of end restraint on beam moments and deflections: (a) beams with different support conditions; (b) moment diagrams; (c) deflection formulas.

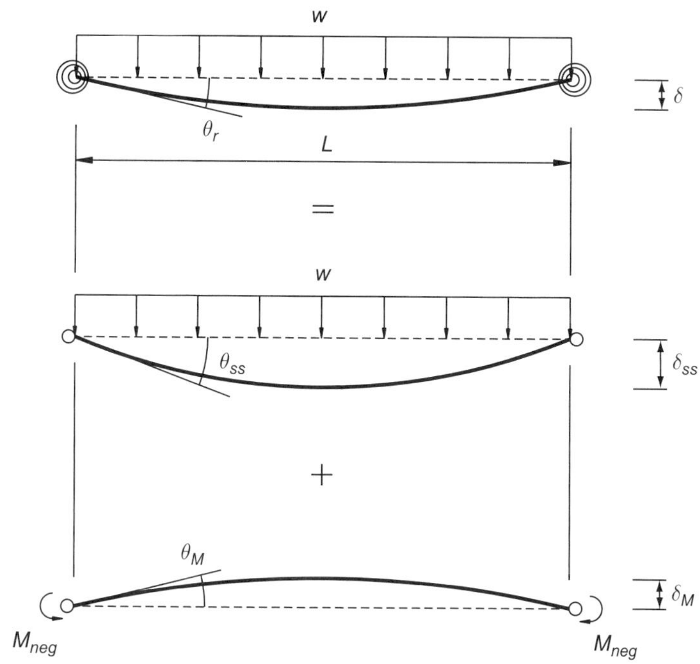

FIGURE 2.17 Load effects on a partially restrained beam.

If linearly elastic material behavior is assumed, the amount of decrease in design moment and deflection for the partially restrained beam shown in Figure 2.16 can be calculated readily using the principle of superposition. Consider Figure 2.17 in which the load effects on the partially restrained beam are decomposed into two load cases. The first load case corresponds to a simply-supported beam subjected to the uniformly distributed load. The second load case represents the restraining effect on the beam from the connections. Compatibility of rotation at the ends requires that:

$$\theta_r = \theta_{ss} + \theta_M \tag{2.58}$$

where θ_r is the rotational deformation of the connection, θ_{ss} is the end rotation of a simply-supported beam due to the uniformly distributed load and θ_M is the end rotation of the beam due to the restraining moment. Substituting:

$$\theta_r = \frac{M_{neg}}{R_k} \tag{2.59}$$

where M_{neg} is the negative beam end moment, R_k is the connection stiffness, and:

$$\theta_{ss} = \frac{wL_b^3}{24EI_b} \tag{2.60}$$

$$\theta_M = \frac{-M_{neg}L_b}{2EI_b} \tag{2.61}$$

into Equation 4.58 and solving for M_{neg}, we obtain:

$$M_{neg} = \frac{2\bar{R}_k}{3(\bar{R}_k + 2)} M_{ss} \tag{2.62}$$

In the above equation, $\bar{R}_k = R_k L_b / EI_b$ is the nondimensional connection stiffness, and $M_{ss} = wL_b^2/8$ is the maximum moment of a uniformly loaded simply-supported beam.

Once M_{neg} is calculated, the positive moment of the partially restrained beam can be evaluated from:

$$M_{pos} = M_{ss} - M_{neg} \tag{2.63}$$

Substituting Equation 2.62 for M_{neg} into Equation 2.63 gives:

$$M_{pos} = \left(\frac{\bar{R}_k + 6}{3\bar{R}_k + 6} \right) M_{ss} \tag{2.64}$$

Equations 2.62 and 2.64 are plotted in Figure 2.18. As can be seen, both the negative and the positive moments of the partially restrained beam fall below the simple beam moment M_{ss}. M_{neg}/M_{ss} will approach a value of 0.667, and M_{pos}/M_{ss} will approach a value of 0.333. These values correspond to the case when the ends of the beam are fixed.

From the standpoint of the beam, an optimal design can be achieved by enforcing the condition that $M_{neg} = M_{pos} = 0.5M_{ss}$. This condition is depicted in Figure 2.18 as the intersection of the two curves. The nondimensional connection stiffness that corresponds to this condition is 6 for the uniformly loaded beam case.

Also shown in Figure 2.18 is the range within which the semi-rigid nature of the connections should be taken into account. The range is $2 < \bar{R}_k < 20$. For $\bar{R}_k \leq 2$, the connection can practically be assumed as pinned, and for $\bar{R}_k \leq 20$, the connection can practically be assumed as rigid.

Although Equation 2.62 provides a convenient means to estimate M_{neg} for design, another approach to estimate M_{neg} was proposed by Nethercot et al. (1988). By performing a series of beam

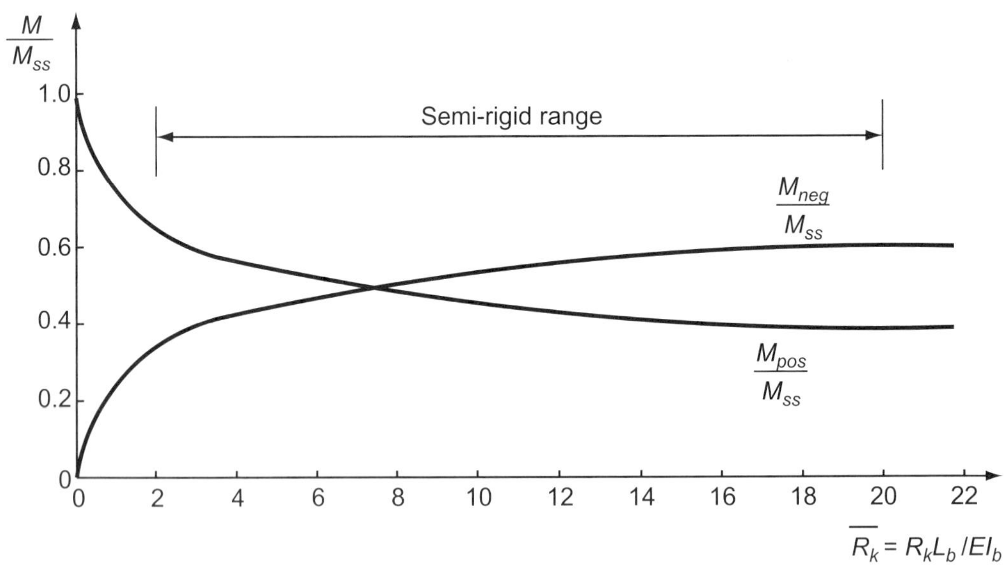

FIGURE 2.18 Relationships between moments and connection stiffness.

analyses using commonly encountered beam span-to-depth ratios with partial restraint provided by three types of connection: double web angles; flush end plate; and extended end plate, Nethercot et al. (1988) recommended that the following values for M_{neg} be used for the design of laterally supported beams:

Double web angles:	8% of M_{pb}
Flush end plate:	45% of M_{pb}
Extended end plate:	60% of M_{pb}

where M_{pb} is the plastic moment capacity of the beam.

In addition to reducing the design moment, the presence of connections also reduces the deflection of the beam. Referring once again to Figure 2.17 and using superposition, the midspan deflection of the partially restrained beam can be written as:

$$\delta = \delta_{ss} + \delta_M \tag{2.65}$$

where $\delta_{ss} = 5wL_b^4/384EI_b$ is the midspan deflection of a uniformly loaded simply-supported beam and $\delta_M = -M_{neg}L_b^2/8EI_b$ is the midspan deflection due to M_{neg}.

Substituting Equation 2.62 for M_{neg} into the expression for δ_M, Equation 2.65 can be written as:

$$\delta = \left(\frac{\overline{R}_k + 10}{5\overline{R}_k + 10}\right)\delta_{ss} \tag{2.66}$$

Equation 2.66 is plotted in Figure 2.19. Note that the decrease in deflection is rather rapid in the range $0 < \overline{R}_k < 10$. The curve levels off and becomes asymptotic to a value of 0.2 when the fixed-end condition is approached.

In designing the partially restrained beam shown in Figure 2.17, a proper value for R_k must be used. Generally, it is inadvisable to use the initial stiffness value R_{ki}, since this represents an upper bound value for the connection stiffness. For design purposes, a secant stiffness obtained by intersecting the connection moment-rotation curve by a beam line under factored gravity load (i.e., the moment created by the factored gravity load as indicated by M_g in Figure 2.3) can be used as the representative connection stiffness. Alternatively, if sufficient information on the connection moment-rotation behavior and a second-order analysis program are available, the actual connection stiffness and the amount of rotational restraint provided by the connections to the beam can be used for the design (Lindsey et al., 1985; Lindsey, 1987).

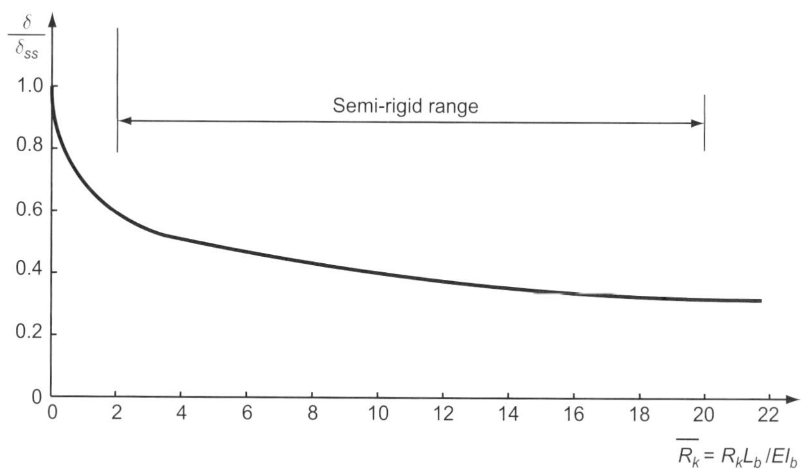

FIGURE 2.19 Relationship between deflection and connection stiffness.

2.4.2.1.2 Design of Columns

Once the beams have been designed, the columns of the nonsway frame can be proportioned using the interaction equations provided in the design specifications. The equations are given by AISC (2005) as follows:

For $P_u/\phi P_n \geq 0.2$:

$$\frac{P_u}{\phi_c P_n} + \frac{8}{9}\left(\frac{M_{ux}}{\phi_b M_{nx}} + \frac{M_{uy}}{\phi_b M_{ny}}\right) \leq 1.0 \tag{2.67a}$$

and for $P_u/\phi P_n < 0.2$:

$$\frac{P_u}{2\phi_c P_n} + \left(\frac{M_{ux}}{\phi_b M_{nx}} + \frac{M_{uy}}{\phi_b M_{ny}}\right) \leq 1.0 \tag{2.68b}$$

where the subscripts x and y refers to strong and weak axes of the member, respectively, and:

M_n = design flexural strength of the member (see Section 2.3.1.1)
M_u = maximum moment in the member accounting for second-order effect
P_n = design compressive strength of the member (see Section 2.2.2.5)
P_u = maximum axial force in the member accounting for second-order effect
ϕ_c = resistance factor for compression = 0.90
ϕ_b = resistance factor for flexure = 0.90

Herein, to compute P_n, a K factor of unity is recommended for the sway-prevented frame. The moments acting at the ends of the column can be obtained from the end restrained moments of the beam distributed according to the relative stiffness of the columns above and below the beam, i.e.:

$$M_c = DF \times \sum (\textit{Beam end moments on either side of the column}) \tag{2.68}$$

where DF is a stiffness distribution factor given by:

$$DF = \frac{(EI/L)_c}{\sum (EI/L)_c} \tag{2.69}$$

The summation in the denominator is to be carried out for all columns that are present above and below the beam-column joint.

2.4.2.2 Design of Sway-permitted Semi-rigid Frames

In the design of sway-permitted semi-rigid frames, the loading and unloading characteristics of the connections must be considered. To illustrate how the connections affect semi-rigid frame behavior, the next two paragraphs describe the response of a subassemblage (Figure 2.20a) representing a portion of a semi-rigid frame under gravity load subjected to one cycle of wind load.

Under the action of a factored gravity load, the moments at the ends of a typical girder can be obtained as the intersection of the connection moment-rotation curve and the beam line. This is shown as Point 1 in Figures 2.21a and 2.21b for connections A and B, respectively. When a wind or lateral force acts on the subassemblage (Figure 2.20b), connection A unloads while connection B loads. The states of moments for the two connections at the end of this stage are denoted by Point 2 in Figure 2.21. When the wind load is removed, the frame undergoes an elastic rebound (Figure 2.20c) and consequently, connection A will reload while connection B unloads. The states of moments for the connections that correspond to this stage are denoted by Point 3 in Figure 2.21. For connection A, Point 3 is below Point 1 because of the P-Δ effect. The vertical force acting through the lateral displacement caused by the wind load tends to prevent the frame from snap-

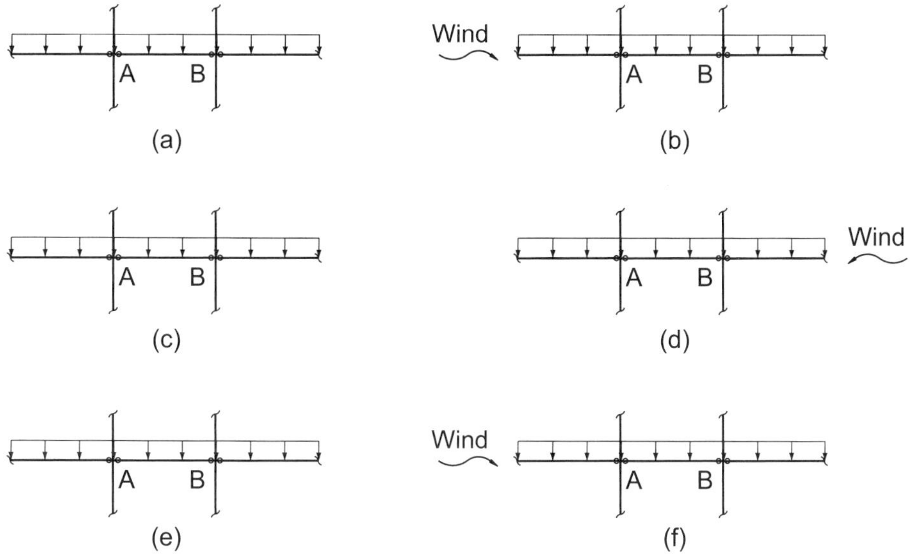

FIGURE 2.20 Semi-rigid frame subassemblage.

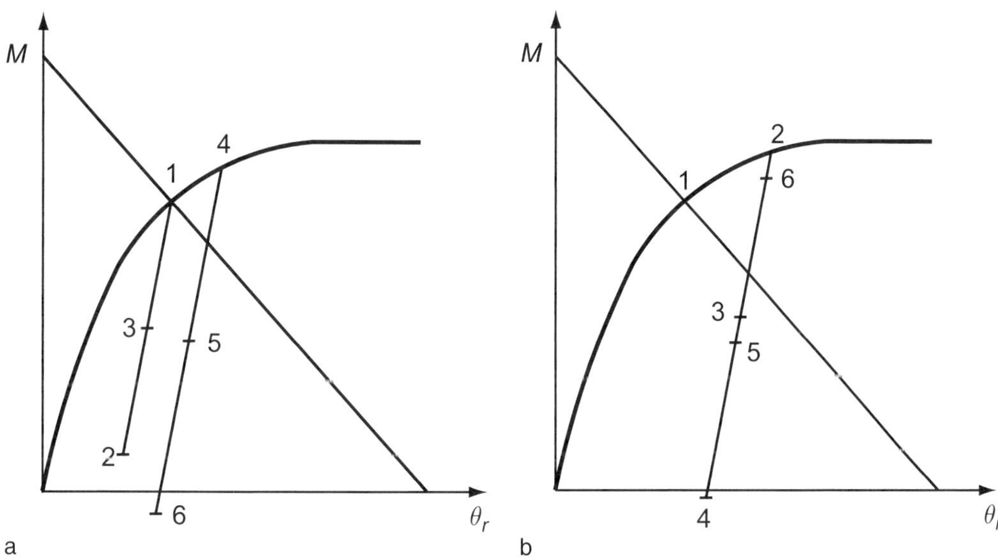

FIGURE 2.21 States of moments in connections A and B: (a) connection A; (b) connection B.

ping back to its original configuration. As a result, the equilibrium position of the frame is slightly swayed to the direction of the wind even after the wind load is removed. Because of this swaying, positive moment is induced in connection *A*, causing Point 3 to be below Point 1.

If the wind blows from the other direction as shown in Figure 2.20d, connection *A* will load and connection *B* will unload. The states of moments at the end of this stage are denoted as Point 4 in Figure 2.21. Upon removal of this wind load (Figure 2.20e), connection *A* will unload to Point 5 in Figure 2.21a and connection *B* will reload to Point 5 in Figure 2.21b. Point 5 is below Point

3 for connection *B* because of the *P*-Δ effect as explained earlier. Finally, to complete the cycle, when the wind blows again in the original direction (Figure 2.20f), connection *A* will unload and connection *B* will reload to Point 6 as shown in Figures 2.21a and 2.21b, respectively. Note that if the cycle is repeated, both connections will behave elastically with the initial connection stiffness governing the behavior of both connections.

Figure 2.22 shows the corresponding bending moment diagrams for the beam during the complete cycle of loading. Note that for all stages of loading, negative moment(s) is (are) likely to be present at the beam end(s). Therefore, to design the beam as if its ends were hinged is a rather conservative approach. It should be mentioned that Point 6 of Figure 2.21a for connection *A* and Point 4 of Figure 2.21b for connection *B* are usually small in real frames. Therefore, it is justified to reduce the positive moment in the design of the beam.

To simplify the design process, only two parameters R_{ki} and M_{cn} will be used. To begin the design, the connections are first proportioned to carry the shear from factored gravity loads on the beam. These connections must be designed to sustain large inelastic rotation without failure as stipulated in the AISC Specifications. R_{ki} and M_{cn} are then determined. As indicated earlier, R_{ki} can be conservatively taken as the secant stiffness of the connection under factored gravity load. As for M_{cn}, its value can be determined graphically as shown in Figure 2.23. Having determined these two parameters, the design of the beam can now proceed.

2.4.2.2.1 Design of Beams

If M_{cn} of the connection is smaller than M_{pb} of the adjoining beam, M_{neg} in Figure 2.22 can be approximated by M_{cn}. Otherwise, M_{neg} needs to be determined from structural analysis. In addition

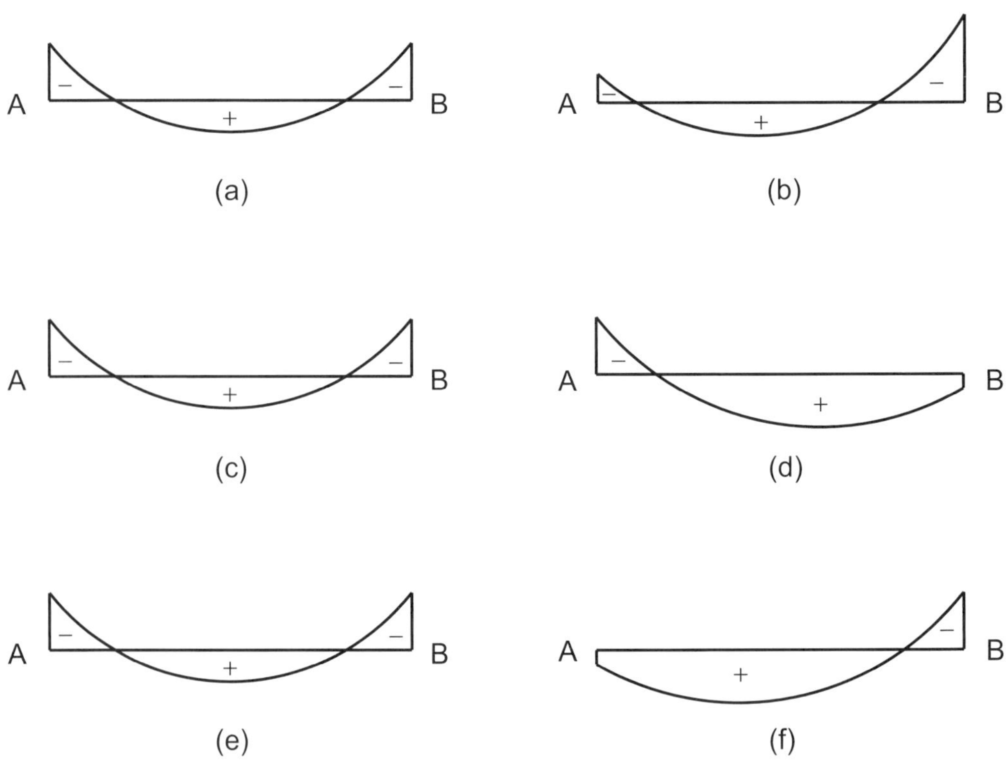

FIGURE 2.22 Bending moment diagrams.

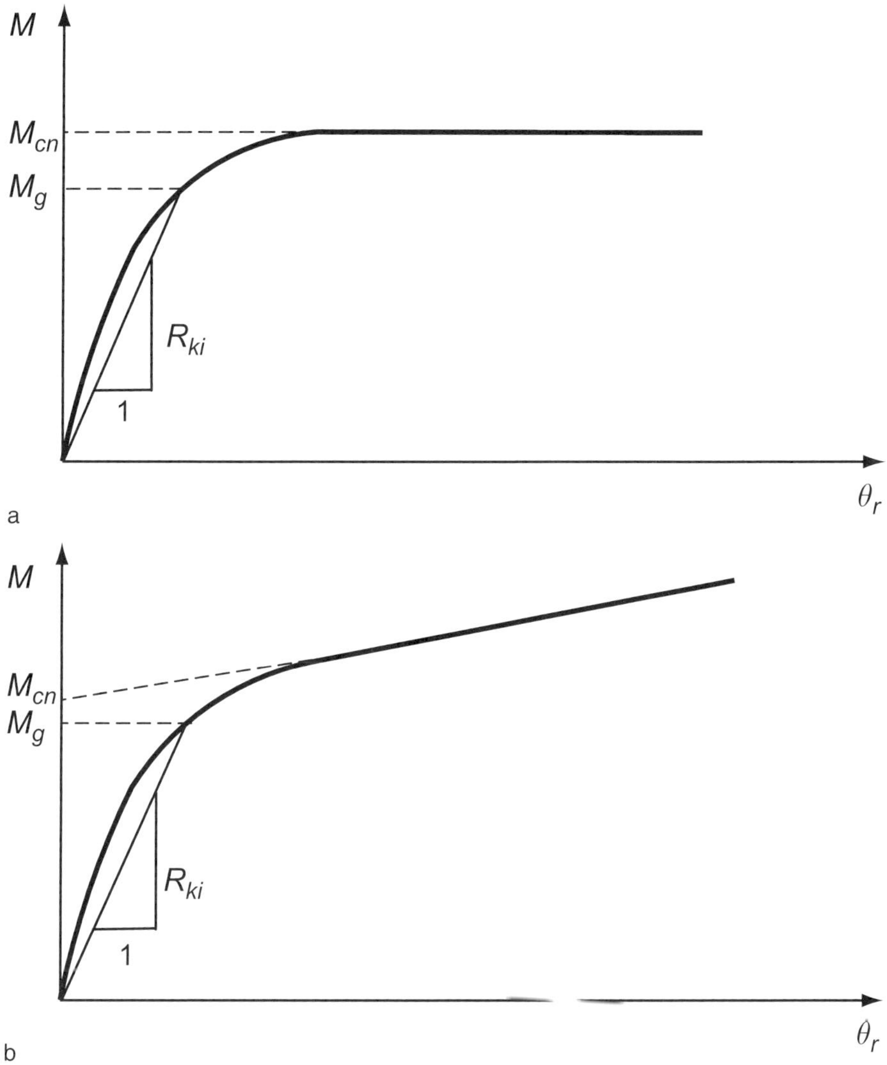

FIGURE 2.23 Determination of R_{ki} and M_u for design: (a) connection without strain-hardening stiffness; (b) connection with strain-hardening stiffness.

to knowing M_{neg} at one end, we need to know the moment at the other end of the beam to proceed with the beam design, so the amount of reduction in positive moment can be determined. From Figure 2.22, it can be seen that the most severe loading cases are those of Figure 2.2d and 2.2f. For design purposes, the positive end moment can be taken as zero. To approximate the positive end moment by zero does not seem conservative at first glance. However, it should be noted that this positive end moment (which corresponds to Point 6 or Point 4 in Figures. 2.21a and 2.21b, respectively) is in reality small and, additionally, it occurs as a result of frame sway. The amount of sway of a semi-rigid frame is controlled by the drift limit. In actual frameworks, the satisfaction of the drift requirement will limit this positive moment to a small value.

With the two beam end moments known, the maximum positive moment of the beam can be obtained easily using elementary structural analysis. The beam can then be proportioned according to this moment or M_{neg}, whichever is larger.

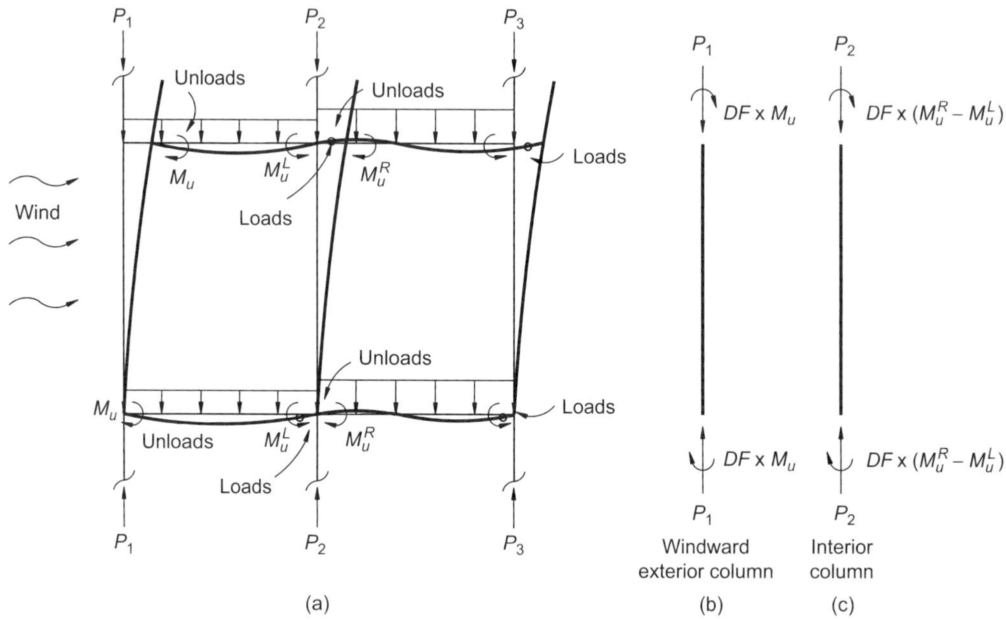

FIGURE 2.24 Determination of column end moments.

2.4.2.2.2 Design of Columns

For a sway-permitted exterior windward column, both the top and bottom connections undergo unloading when the frame sways and thus rotational restraints are offered to both ends of the column by the beams (Figure 2.24a). The free-body diagram for the design of this beam-column is shown in Figure 2.24b. In the diagram, *DF* is the stiffness distribution factor given in Equation 2.69, M_u is the moment from the beam, which can be taken as M_{cn} (if $M_{cn} \le M_p$) or determined from structural analysis (if $M_{cn} > M_p$).

The exterior leeward column, however, does not have any rotational restraint from the beams since both connections undergo loading as the frame sways (Figure 2.24a). As a result, the exterior leeward column cannot participate in frame stability and the stability of the frame needs to be provided by the remaining exterior windward and interior columns.

The behavior of the sway-permitted interior column is similar to that of the exterior windward column in that rotational restraints at both ends of the column are provided by the beams that frame into the leeward side of the column (Figure 2.24a). The free body diagram of the beam-column to be designed is shown in Figure 2.24c.

2.4.3 Drift of Semi-rigid Frames

Following the design of the beams and columns, the frame must be analyzed for drift to ensure that excessive lateral deflection does not occur. As a result of the connection flexibility, semi-rigid frames tend to deflect quite significantly under the action of lateral loads (Gerstle and Ackroyd, 1990). Studies have shown that most flexibly-connected frames designed on the basis of strength violate the drift requirement for serviceability. If the amount of drift exceeds the intended tolerance limit, the lateral stiffness of the frame must be increased. The lateral stiffness of semi-rigid frames can be enhanced by instituting one or a combination of the following measures: (1) provide bracings; (2) increase the member size; and/or (3) use stiffer connections.

Although a spreadsheet application for approximate drift calculations was proposed by Ackroyd (1990) for semi-rigid frames, the following prediction formulas (Cronembold and Ackroyd, 1986; Gerstle and Ackroyd, 1990) for drift can be used for a preliminary analysis.

For top and seat angles:

$$\frac{\Delta}{H}=\frac{W}{90+160\left(B/H\right)} \tag{2.70}$$

For flange plates

$$\frac{\Delta}{H}=\frac{W}{130+160\left(B/H\right)} \tag{2.71}$$

In the above formulas, Δ is the lateral deflection at the top story of the semi-rigid frame, H is the overall height of the frame, B is the overall width of the frame, and W is the lateral load intensity (in kips/ft of vertical height).

If an idealized connection model is used in which the unloading connection stiffness is taken as R_{ki} and the loading connection stiffness is taken as 0, frame drift can be obtained by analyzing the frame as a rigidly connected frame provided that I of all the beams with semi-rigid connections are replaced by I' computed using Equation 2.14.

Example:

The semi-rigid fixed based portal frame shown in Figure 2.25 is to be designed. The frame is braced against out-of-plane deformation at all beam-column joints. This means the lateral unbraced length of the beam is 20 ft. (6.1 m), and the effective length factor for weak axis (i.e., out-of-plane) bending of the columns can conveniently be taken as 1. The effective length factor for strong axis (i.e., in-plane) bending still needs to be computed using the procedure outlined in Section 2.2.2.3. The connections are assumed to possess a stiffness of $\overline{R}_{ki}$ =15 and M_{cn} = 250 k-ft (339 kN-m). The loads to be designed for are: uniformly distributed dead load, D = 2 k/ft (29.2 kN/m); uniformly distributed live load, L = 3 k/ft (43.8 kN/m); and wind load, W= 2.5 kips (11.1 kN). The lateral deflection of the frame is to be limited to H/300. Use A992 steel (F_y = 50 ksi or 345 MPa) and wide flange (W) shapes for all members.

From AISC (2005), the load combinations that most likely will control the design are:

$$1.2D + 1.6L$$
$$1.2D + 1.6W + 0.5L$$

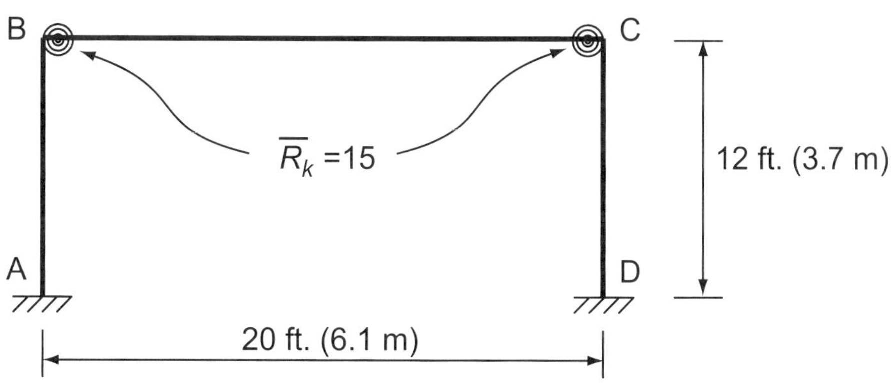

FIGURE 2.25 Design of a semi-rigid portal frame.

Thus, a factored uniform load of w_u=7.2 k/ft (105 kN/m) for the first load case, and a factored uniform load of w_u=3.9 k/ft (56.9 kN/m) and a factored wind load of W_u=4 kips (17.8 kN) for the second load case will be used for the design.

Under the first (gravity only) load case, the frame will not undergo any sidesway, as a result Equation 2.62 and Equation 2.64 can be used to estimate M_{neg} and M_{pos}, respectively. Substituting $\bar{R}_{ki}$ = 15 into these equations gives M_{neg} = 212 k-ft (287 kN-m) and M_{pos} = 182 k-ft (247 kN-m), thus M_{neg} controls. Since this value is less than M_{cn}, the beam will be selected based on M_{neg}, i.e., M_u is to be taken as M_{neg}. It is important to note that these moment values are only approximate because the bending stiffness of the columns has not been taken into consideration. If the effect of column flexibility is taken into consideration, the actual end restraint provided to the beam will be less than $\bar{R}_{ki}$ =15.

From simple equilibrium consideration and neglecting the self-weight of the members for the time being, P_u in the column can be determined to be 72 kips (320 kN).

The objective here is to select a beam section so that its design strength $\phi_b M_n$ (where ϕ_b = 0.9 and M_n is to be determined using the equations in Section 2.3.1.1) exceeds the required strength M_u, and to select a column section that satisfies the appropriate interaction equation in Section 2.4.2.1.2.

Using the equations from Section 2.3.1.1, a W12 × 53 section is selected for the beam; and using Equation 2.67b, a W10 × 60 section is selected for the columns.

Now, with the preliminary beam and column sections selected, a second order analysis taking into consideration the applied loads and the member weight is carried out for both load cases. The results for the required strengths are given in the following tables:

Load case 1 (1.2*D* + 1.6*L*) – Gravity only

Member	P_u kips (kN)	M_u k-ft (kN-m)
Beam	20 (89)	204 (277)
Left column	73.25 (326)	160 (217)
Right column	73.25 (326)	160 (217)

Load Case 2 (1.2*D* + 1.6*W* + 0.5*L*) – Gravity plus wind

Member	P_u kips (kN)	M_u k-ft (kN-m)
Beam	6 (26.7)	166 (225)
Left column	43.5 (194)	64.2 (87.1)
Right column	37.0 (165)	73.7 (100)

As can be seen, the gravity-only load case controls the design. It should be noted that in carrying out the analysis for load case 2, the connection at the leeward side of the frame, which is undergoing loading, is conservatively assumed to have zero stiffness. Also, because the required axial strength for the beam is relatively small, the beam can be designed as a flexural member, whereas the columns must be designed as beam-columns.

To complete the design the sections are to be checked for both load cases: For the W12 × 53 section, the design flexural strength $\phi_b M_n$ computed using Equation 2.23 with ϕ_b = 0.9 and C_b = 1 is 231 k-ft (313 kN/m). This exceeds the required flexural strength M_u = 204 k-ft (277 kN-m) for the first load case, and M_u = 166 k-ft (225 kN-m) for the second load case. Therefore, the section is adequate.

For the W10 × 60 section, the design axial strength $\phi_c P_n$ computed using Equation 2.19a with $\phi_c = 0.9$, $(KL)_y = 12$ ft. (3.66 m) for the gravity-only load case, and $(KL)_x = 25.2$ ft. (7.7 m) for the gravity-plus-wind load case, for the right column are 631 kips (2810 kN) and 560 kips (2500 kN), respectively. The design flexural strength $\phi_b M_n$ computed using Equation 2.23 with $\phi_b = 0.9$, $L_b = 12$ ft (3.7 m) and $C_b = 1$ is 269 k-ft (365 kN-m). Substituting these values with $P_u = 73.25$ kips (326 kN) and $M_u = 160$ k-ft (217 kN-m) for the first load case, and $P_u = 37.0$ kips (165 kN) and $M_u = 73.7$ k-ft (100 kN-m) for the second load case into Equation 2.67b gives 0.65 and 0.31 for the two load cases, respectively. Thus, the interaction equation is satisfied, and so the section is adequate.

Finally, under a service wind load of 2.5 kips (11.1 kN), the lateral deflection experienced by the semi-rigid frame is 0.4 in. (10 mm). This is less than the specified allowable lateral deflection value of $H/300=0.48$ in. (12.2 mm). Therefore, the serviceability limit state is also satisfied.

2.5 Summary and Conclusions

The analysis and design of semi-rigid frames are intrinsically more complex than their rigid counterparts because of the difficulty and uncertainty in predicting the moment-rotation behavior of the connections. Fortunately, studies by Goto and Chen (1987), Wu and Chen (1990), and others have demonstrated that minor errors in predicting the response of the connections will not noticeably affect the overall behavior of the semi-rigid frame.

Experiments carried out on steel beam-to-column connections over the past decades have demonstrated clearly that the response of a connection is nonlinear when the connection is undergoing loading but becomes more or less linear when it is undergoing unloading. Various mathematical models to represent the connection M-θ_r behavior have been proposed. Some of these models have been summarized in Section III of this handbook. There is no ironclad rule as to which model is the best. Each has its advantages and disadvantages. Nonetheless, all of these models can represent the M-θ_r behavior of the connection fairly well and they have all been successfully used in the analysis of semi-rigid framed structures.

Based on a number of studies on semi-rigid frame behavior under combined gravity and lateral loads, the following conclusions can be made:

1. If the nominal moment capacity of a connection is larger than the maximum moment it experiences during the entire load history of the frame, the maximum load-carrying capacity of the frame is independent of the types of connection used.
2. Although the maximum load-carrying capacity of the frame may not be affected by the connection, the frame with flexible connections does deform more.
3. Inelasticity in the column may cause the column to shed its moment to the beam provided that the beam remains elastic. The more flexible the connection is, the later the column will shed its moment to the beam because less moment is transmitted to the column from the beam.
4. Connection flexibility may cause early nonlinear behavior for the frame and the loading and unloading characteristics of the connection will alter the moment distribution in the frame.
5. The use of bracing can drastically reduce the drift of the frame and the presence of bracing seems to obscure the nonlinear behavior of the connections (Lui and Chen, 1988). Because of this, a linear elastic frame analysis may be sufficient for flexibly connected frames loaded in the service load range.

The simplest approach to design semi-rigid frames is to assume that the connections provide no restraint to the beams. The beams are therefore designed as simply-supported members insofar as gravity loads are concerned. On the other hand, the columns are designed assuming that the connections can provide full rotational restraint for lateral loads. This approach, proposed by Disque (1975), while historically safe, often results in a design in which the beams are oversized and the columns are undersized. A more rational approach, in light of the limit state philosophy should therefore be used.

If a second-order analysis program is available which can take into consideration connection flexibility, it should be used for design. In lieu of such a program, simplified design methods which are based on simplified behavioral models for connections and for semi-rigid frame action can be used. One such simplified method has been discussed in this chapter. The method requires the use of only two connection parameters R_{ki} (taken as the secant stiffness of the connection under factored gravity load) and M_{cn} (obtained graphically from the connection M-θ_r curve), and is considered a compromise between accuracy and simplicity.

It is important to note that the method proposed is strength based, which means that the design is based on strength, not serviceability. Since excessive frame drift is often a problem for semi-rigid frames, due consideration must be given to check the resulting design for serviceability requirements and proper measures must be taken to limit the drift of semi-rigid frames.

References

Ackroyd, M. H., Electronic spreadsheet tools for semi-rigid frames, *AISC Engineering Journal*, 27(2), 69–78, 1990.

AISC, *Steel Construction Manual*, 13th edition, American Institute of Steel Construction, Chicago, IL.

Aristizabal-Ochoa, J. D., Stability and second-order analysis of frames with semi-rigid connections under distributed loads, *Journal of Structural Engineering*, ASCE, 127(11) 1306–1315, 2001.

Aristizabal-Ochoa, J. D., Story stability of braced, partially-braced, and unbraced frames: Classical approach, *Journal of Structural Engineering*, ASCE, 123(6), 799–806, 1997.

Aristizabal-Ochoa, J. D., K-factor for columns in any type of construction: Nonparadoxical approach, *Journal of Structural Engineering*, ASCE, 120(4), 1272–1290, 1994.

ASCE, *Effective Length and Notional Load Approaches for Assessing Frame Stability: Implications for American Steel Design*, Task Committee on Effective Length, ASCE, 1997.

Cabrero, J. M. and Bayo, E., Development of practical design methods for steel structures with semi-rigid connections, *Engineering Structures*, 27(8), 1125–1137, 2005.

Cerfontaine, F. and Jaspart, J. P., Analytical study of the interaction between bending and axial force in bolted jointed, *In: Eurosteel, Coimbra*, 997–1006, 2002.

Chapius, J. and Galambos, T. V., Restrained crooked aluminum columns, *Journal of the Structural Division*, ASCE, 108(ST3), 511–524, 1982.

Chen, W. F. (editor), *Practical analysis for semi-rigid frame design*, *World Scientific*, Singapore, 2000.

Chen, W. F., Goto, Y., Liew, J. Y. R., *Stability design of semi-rigid frames*, John Wiley and Sons, New York, NY, 1996.

Chen, W. F. and Lui, E. M., *Stability design of steel frames*, CRC Press, 1991.

Chen, W. F. and Lui, E. M., *Structural stability—Theory and implementation*, Elsevier, 1987.

Chen, W. F. and Lui, E. M., Effects of joint flexibility on the behavior of steel frames, *Computers and Structures*, 26(5), 719–732, 1987.

Chen, W. F. and Zhou, S. P., Inelastic analysis of steel braced frames with flexible joints, *International Journal of Solids and Structures*, 23(5), 631–649, 1987.

Chikho, A. H. and Kirby, P. A., An approximate method for estimation of bending moments in continuous and semi-rigid frames, *Canadian Journal of Civil Engineering*, 22, 1120–1132, 1995.

Cronembold, J. R. and Ackroyd, M. H., Economy and safety of semi-rigid frame design: Case studies, *Stability of Tall Steel Buildings, Workshop Proceedings, Third International Conference on Tall Buildings*, Bethlehem, PA, Council on Tall Buildings and Urban Habitat, Chicago, IL, 113–171, 1986.

Cunningham, R., Some aspects of semi-rigid connections in structural steelwork, *The Structural Engineer*, 68(5), 85–92, 1990.

Disque, R. O., Directional moment connections—A proposed design method for unbraced steel frames, *AISC Engineering Journal*, 12(1), 14–18, 1975.

Faella, C., Piluso, V., and Rizzano, G., *Structural steel semi-rigid connections—Theory, design and software*, CRC Press, Boca Raton, FL, 2000.

Frye, M. J. and Morris, G. A., Analysis of flexibly connected steel frames, *Canadian Journal of Civil Engineers*, 2(3), 280–291, 1976.

Fukumoto, Y. and Itoh, Y., Statistical study of experiments on welded beams, *Journal of the Structural Division, ASCE*, 107(ST1), 89–103, 1981.

Fukumoto, Y., Itoh, Y., and Kubo, M., Strength variation of laterally unsupported beams, *Journal of the Structural Division*, ASCE, 106(ST1), 165–181, 1980.

Galambos, T. V., Inelastic lateral buckling of beams, *Journal of the Structural Division*, ASCE, 89(ST5), 217–242, 1963.

Gerstle, K. H. and Ackroyd, M. H., Behavior and design of flexibly-connected building frames, *AISC Engineering Journal*, 27(1), 22–30, 1990.

Geschwindner, L. F., A simplified look at partially restrained beams, *AISC Engineering Journal*, 28(2), 73–78, 1991.

Gizejowski, M. A., Branicki, C. J., Barszcz, A. M., and Krol, P., Advanced analysis of steel frames with effects of joint deformability and partial strength accounted for, *Journal of Civil Engineering and Management*, X(3), 199–208, 2004.

Goto, Y., Suzuki, S., and Chen, W. F., Analysis of critical behavior of semi-rigid frames with or without load history in connections, *International Journal of Solids and Structures*, 27(4), 467–483, 1991.

Goto, Y. and Chen, W. F., On the computer-based design analysis for flexibly-connected building frames, *Journal of Constructional Steel Research*, Special Issue on Joint Flexibility in Steel Frames, 8, 203–231, 1987.

Julian, O. G. and Lawrence, L. S., Notes on J and L nomographs for determination of effective lengths, *Jackson and Moreland Engineers*, USA, 1959 (unpublished report).

King, W. S. and Chen, W. F., LRFD Analysis for semi-rigid frame design, *AISC Engineering Journal*, 30(4), 130–140, 1993.

Kishi, N., Hasan, R., Chen, W. F., and Goto, Y., Power model for semi-rigid connections, *Steel Structures, Journal of Singapore Structural Steel Society*, 5(1), 37–48, 1994.

Kishi, N., Chen, W. F., Goto, Y., and Matsuoka, K. G., Analysis program for the design of flexibly jointed frames, *Computers and Structures*, 49(4), 705–714, 1993a.

Kishi, N., Chen, W. F., Goto, Y., and Matsuoka, K. G., Design aid of semi-rigid connections for frame analysis, *AISC Engineering Journal*, 30(3), 90–107, 1993b.

Kishi, N. and Chen, W. F., Moment-rotation relations of semi-rigid connections with angles, *Journal of Structural Engineering*, ASCE, 116(7), 1813 1834, 1990.

Johnston, B. G. (editor), *SSRC guide to stability design criteria for metal structures*, 3rd edition, John Wiley, New York, NY, 1976.

LeMessurier, W. J., A practical method of second order analysis, Part 2—Rigid frames, *AISC Engineering Journal*, 14(2), 49–67, 1972.

Leon, R. T., Composite semi-rigid construction, *AISC Engineering Journal*, 31(2), 57–67, 1994.

Lindsey, S. D., Design of frames with PR connections, *Journal of Constructional Steel Research*, 8, 251–260, 1987.

Lindsey, S. D., Loannides, S., and Goverdhan, A. V., LRFD analysis and design of beams with partially restrained connections, *AISC Engineering Journal*, 22(4), 157–162, 1985.

Lui, E. M., Stability design of partially restrained frames, *Fifth International Colloquium on Stability of Metal Structures*, Chicago, IL, 259–268, 1996.

Lui, E. M., Column effective length factor for semi-rigid frames, *Steel Structures, Journal of Singapore Structural Steel Society*, 6(1), 3–20, 1995.

Lui, E. M., A novel approach for K factor determination, *AISC Engineering Journal*, 29(4), 150–159, 1992.

Lui, E. M. and Chen, W. F., Behavior of braced and unbraced semi-rigid frames, *International Journal of Solids and Structures*, 24(9), 893–913, 1988.

Lui, E. M., A practical P-Delta analysis method for Type FR and PR frames, *AISC Engineering Journal*, 25(3) 85–98, 1988.

Lui, E. M. and Chen, W. F., Steel frame analysis with flexible joints, *Journal of Constructional Steel Research*, 8, 161–202, 1987.

Lui, E. M. and Chen, W. F., End restraint and column design using LRFD, *AISC Engineering Journal*, 20(1), 29–39, 1983.

Monforton, A. R. and Wu, T. S., Matrix analysis of semi-rigid connected frames, *Journal of the Structural Division*, 87(6), 13–42, 1963.

Nethercot, D. A., Davison, J. B., and Kirby, P. A., Connection flexibility and beam design in non-sway frames, *AISC Engineering Journal*, 25(3), 99–108, 1988.

Poggi, C. and Zandonini, R., Behavior of strength of steel frames with semi-rigid connections, *Proceedings in Connection Flexibility and Steel Frames*, ASCE Convention, Detroit, MI, October, ASCE, NY, 57–76, 1985.

Rodrigues, F. C., Saldanha, A. C., and Pfeil, M. S., Non-linear analysis of steel plane frames with semi-rigid connections, *Journal of Constructional Steel Research*, 46(1-3), 94–97, 1998.

Sekulovic, M. and Nefovska-Danilovic, M., Static inelastic analysis of steel frames with flexible connections, *Theoretical and Applied Mechanics*, 31(2), 101–134, 2004.

Timoshenko, S. P. and Gere, J. M., *Theory of elastic stability*, 2nd edition, McGraw-Hill, New York, 1961.

Trahair, N. S., *Flexural-torsional buckling of structures*, CRC Press, Boca Raton, FL, 1993.

Urbonas, K. and Daniunas, A., Behaviour of semi-rigid steel beam-to-beam joints under bending and axial forces, *Journal of Constructional Steel Research*, 62, 1244–1249, 2006.

Urbonas, K. and Daniunas, A., Component method extension to steel beam-to-beam and beam-to-column knee joints under bending and axial forces, *Journal of Civil Engineering and Management*, XI(3), 217–224, 2005.

van Keulen, D. C., Nethercot, D. A., Snijder, H. H., and Bakker, M. C. M., Frames analysis incorporating semi-rigid joint action: Applicability of the half initial secant stiffness approach, *Journal of Constructional Steel Research*, 59, 1083–1100, 2003.

Vinnakota, S., Inelastic stability of laterally unsupported I-beams, *Computers and Structures*, 7(3), 377–389, 1977.

Wu, F. H. and Chen, W. F., A design model for semi-rigid connections, *Engineering Structures*, 12(2), 88–97, 1990.

Xu, L. and Liu, Y., Story-based effective length factors for unbraced PR frames, *AISC Engineering Journal*, 39(1), 13–29, 2002.

Xu, L., Grierson, D. E., and Sherbourne, A. N., Optimal cost design of semi-rigid, low rise industrial frames, *AISC Engineering Journal*, 32(3), 87–97, 1995.

Xu, L. and Grierson, D. E., Computer-automated design of semi-rigid steel frameworks, *Journal of Structural Engineering*, ASCE, 119(6), 1740–1760, 1993.

Yau, C. Y. and Chan, S. L., Inelastic and stability analysis of flexibly connected steel frames by spring-in-series model, *Journal of Structural Engineering*, ASCE, 120(10), 2803–2820, 1994.

Yoshida, H. and Imoto, Y., Inelastic lateral buckling of restrained beams, *Journal of the Engineering Mechanics Division*, ASCE, 99(EM2), 343–366, 1973.

Yura, J. A., Galambos, T. V., and Ravindra, M. K., The bending resistance of steel beams, *Journal of the Structural Division*, ASCE, 104(ST9), 1355–1369, 1978.

3

Types of PR Connections

Norimitsu Kishi
Professor, Civil Engineering Unit, College of Environmental Engineering, Muroran Institute of Technology, Muroran 050-8585, Japan

Masato Komuro
Associate Professor, Civil Engineering Unit, College of Environmental Engineering, Muroran Institute of Technology, Muroran 050-8585, Japan

3.1 Single Web-angle Connections/Single Plate Connections

Single web-angle connections consist of an angle either bolted or welded to both the column and the beam web, as shown in Figure 3.1. On the other hand, single plate connections use the plate instead of the angle (Figure 3.2). This connection type requires less material than does a single web-angle connection. Generally, in designing these connections, the single web-angle connections have moment rigidity equal to about one-half of the double web-angle connections, and the single plate connections have rigidity equal to or greater than the single web-angle connections, since one side of the plate is fully welded with the column flange.

3.2 Double Web-angle Connections

Double web angle connections consist of two angles, either bolted or riveted to both the column and the beam web, as shown in Figure 3.3. Rivets were used as fasteners in the earliest tests on double web-angle connections conducted by Rathbun (1936). In the 1950s, most specifications for the design of steel structures allowed the use of high-strength bolts instead of rivets. To clarify the effect of high-strength bolts on connection behavior when used in conjunction with rivets, Bell et al. (1958) and Lewitt et al. (1966) conducted experiments on riveted and bolted beam-to-column connections. Today, high-strength bolts are used popularly as fasteners for this type

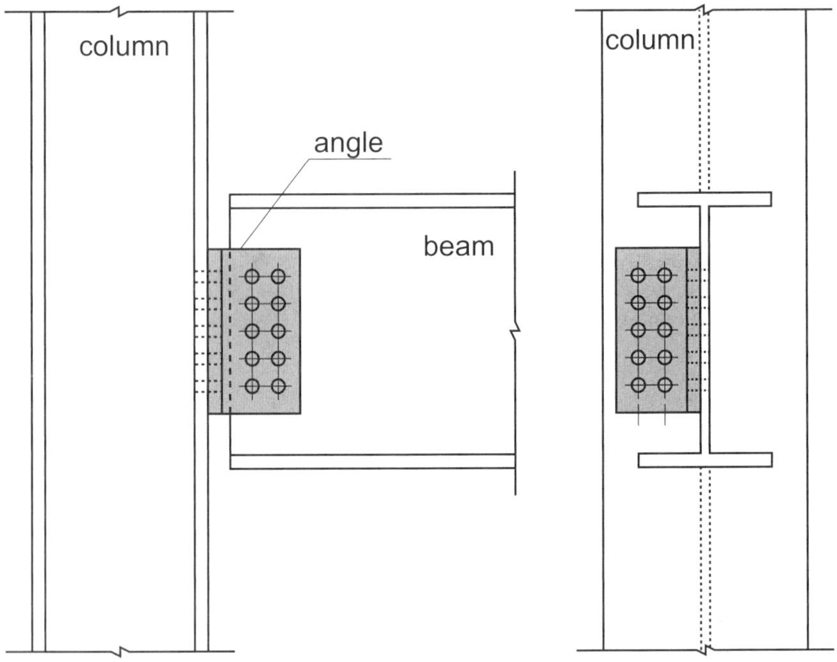

FIGURE 3.1 Single web-angle connection.

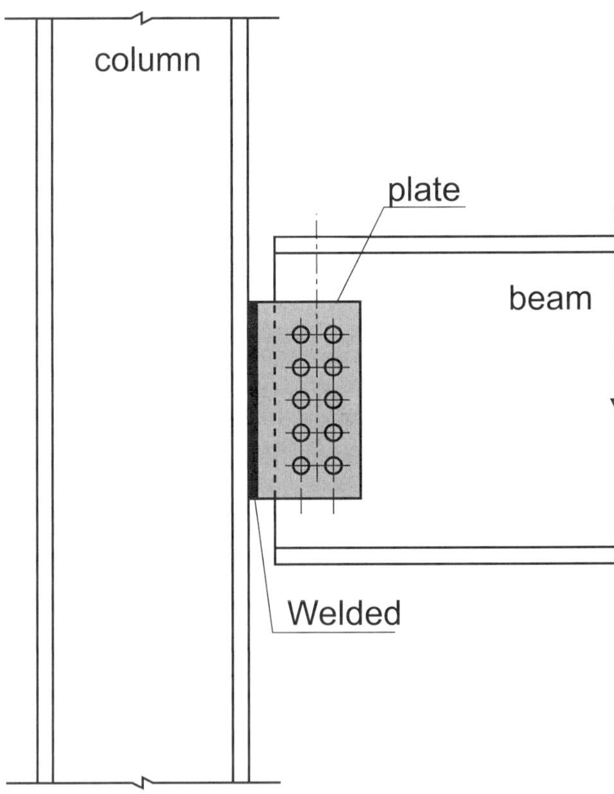

FIGURE 3.2 Single plate connection.

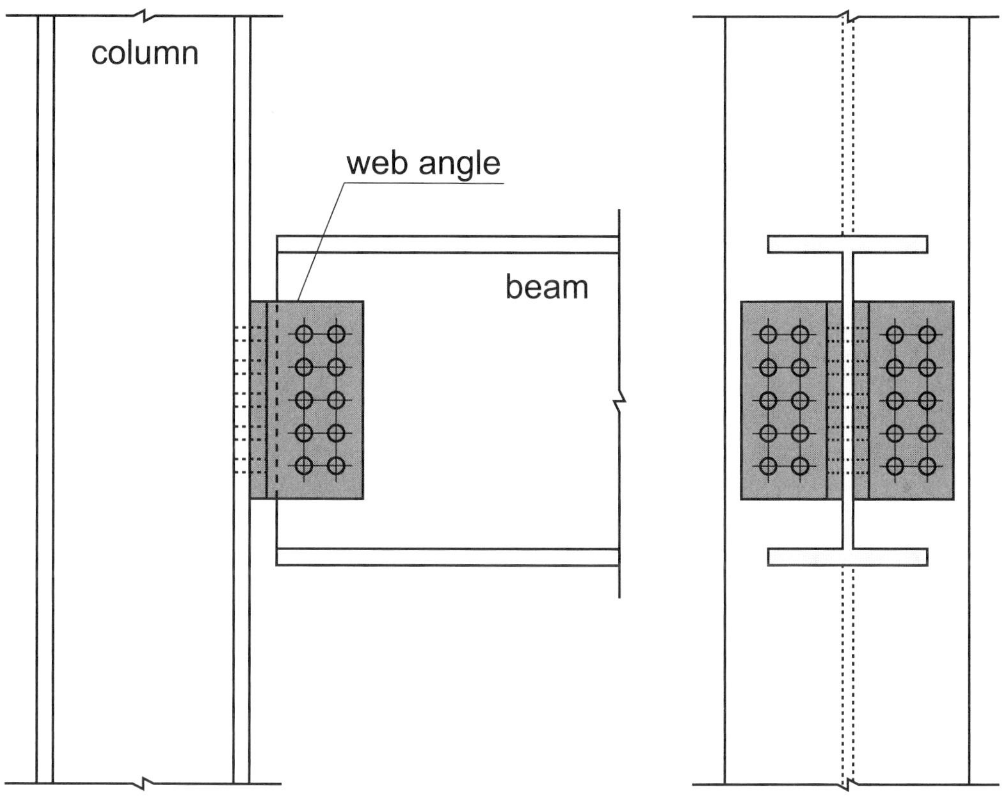

FIGURE 3.3 Double web-angle connection.

of connection. Though the connection rigidity of this type of connection is stiffer than those of single web-angle and single plate connections, the AISC-ASD specifications (1989) consider this connection type as a Type 2 construction (simple connection or shear connection).

3.3 Top- and Seat-angle with Double Web-angle Connections

This type of connection is a combination of top- and seat-angle connections and double web-angle connections. A typical top- and seat-angle connection with a double web-angle is shown in Figure 3.4. Double web angles are used to improve the connection restraint characteristics of top- and seat-angle connections, and for shear transfer. This type of connection is considered as Type 3 framing in the AISC-ASD specifications (1989) (i.e., a semi-rigid connection).

3.4 Top- and Seat-angle Connections

A typical top- and seat-angle connection is shown in Figure 3.5. The AISC-ASD specifications (1989) described this type of connection as follows: (1) the top angle is used to provide lateral support of the compression flange of a beam; and (2) the seat angle is to transfer only the vertical reaction of the beam to the column and should not apply a significant restraining moment to the end of the beam. However, according to the experimental results, these connections will be able to transfer not only the vertical reaction, but also some end moment of the beam to the column.

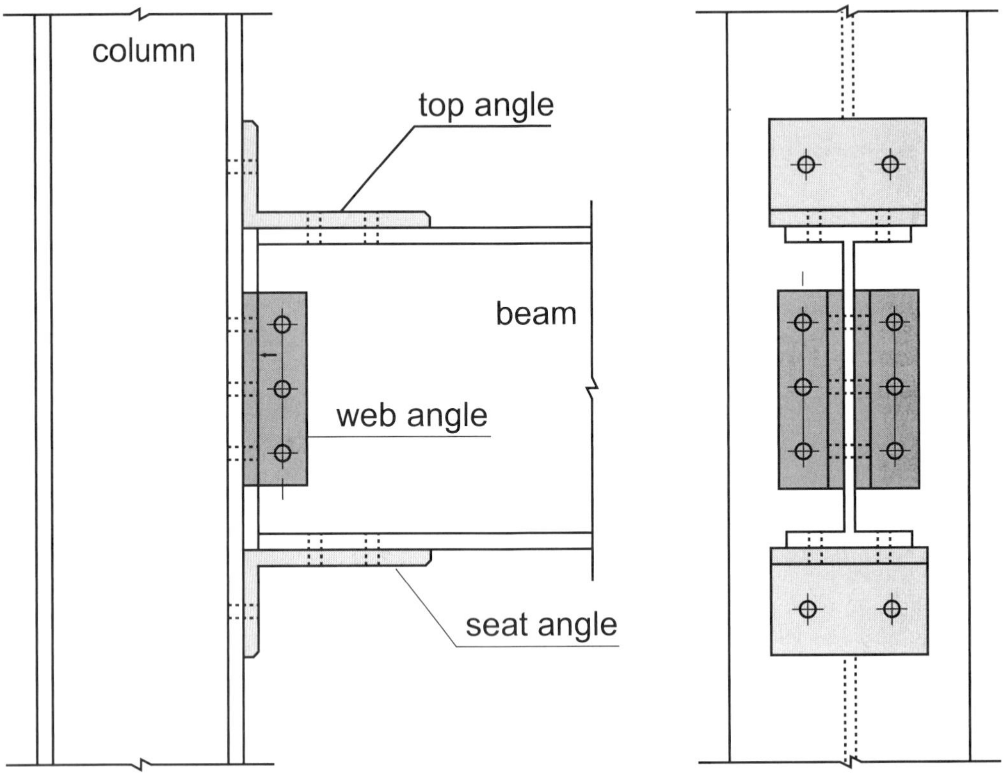

FIGURE 3.4 Top- and seat-angle connection with double web-angle.

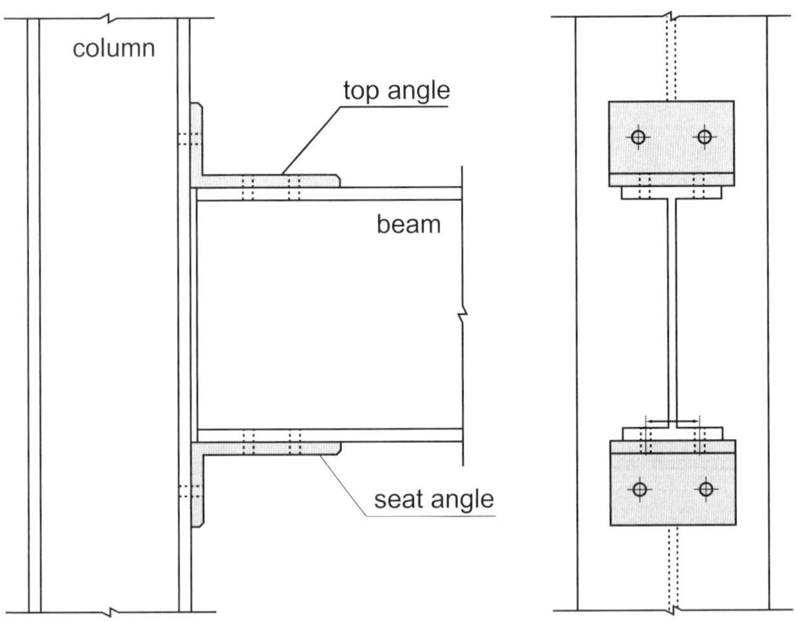

FIGURE 3.5 Top- and seat-angle connection.

3.5 Extended End-plate Connections/Flush End-plate Connections

In general, end-plate connections are always welded to the web and flanges of the beam in the fabricator's workshop and bolted to the column on site. The end-plate connection has been used extensively since the 1960s. The extended end-plate connections are classified into two types as end-plate either extended on the tension side only or on both the tension and compression sides as shown in Figures 3.6 and 3.7. A typical flush end-plate connection is shown in Figure 3.8. Since some end-plate connections are considered as a Type FR construction rather than a Type PR construction in the AISC-LRFD specifications (1994), they have often been used as a means of transferring beam end moment to the column. The extended end-plate connection on both sides is preferred when the connection is subjected to moment reversal such as during severe earthquake loading. While the flush end-plate connection is weaker than the extended end-plate connection, this connection type is often used in roof details. The behavior of the end-plate connection depends on the stiffness of the column flange near the connection. The stiffeners of the column flanges act to prevent flexural deformation, thereby influencing the behavior of the plate and the fasteners.

3.6 Header Plate Connections

A header plate connection is made with an end plate whose length is less than the depth of the beam. The end plate is always welded to the beam web (with fillet welds on each side) and is bolted

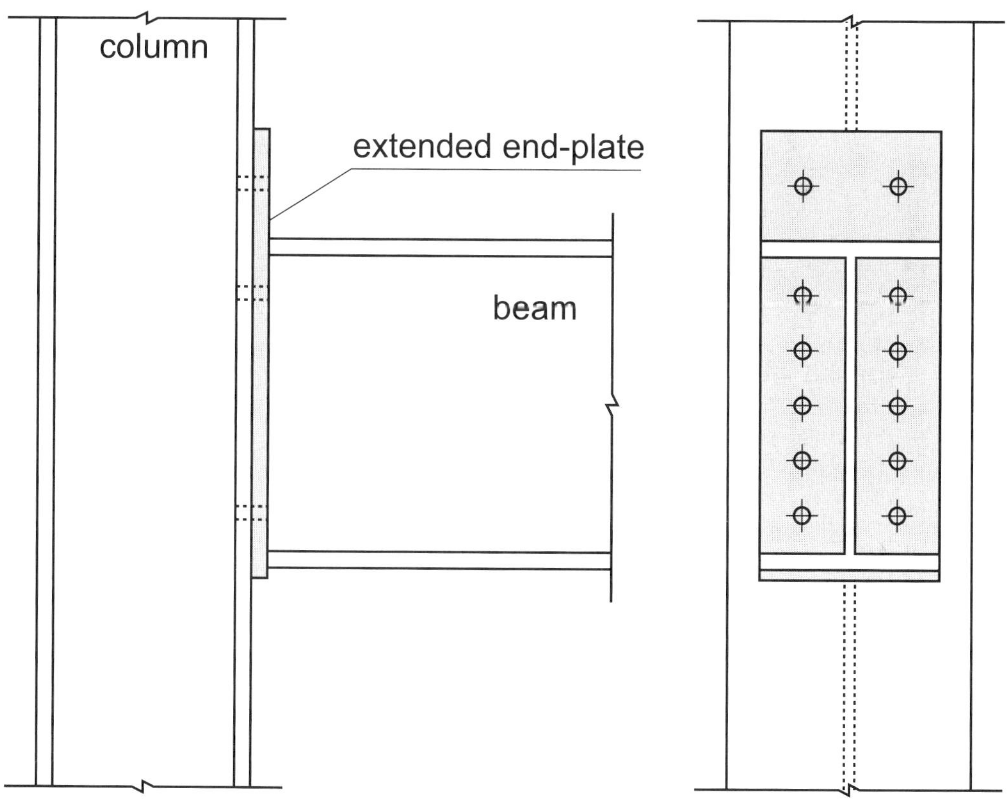

FIGURE 3.6 Extended end-plate connection only on tension side.

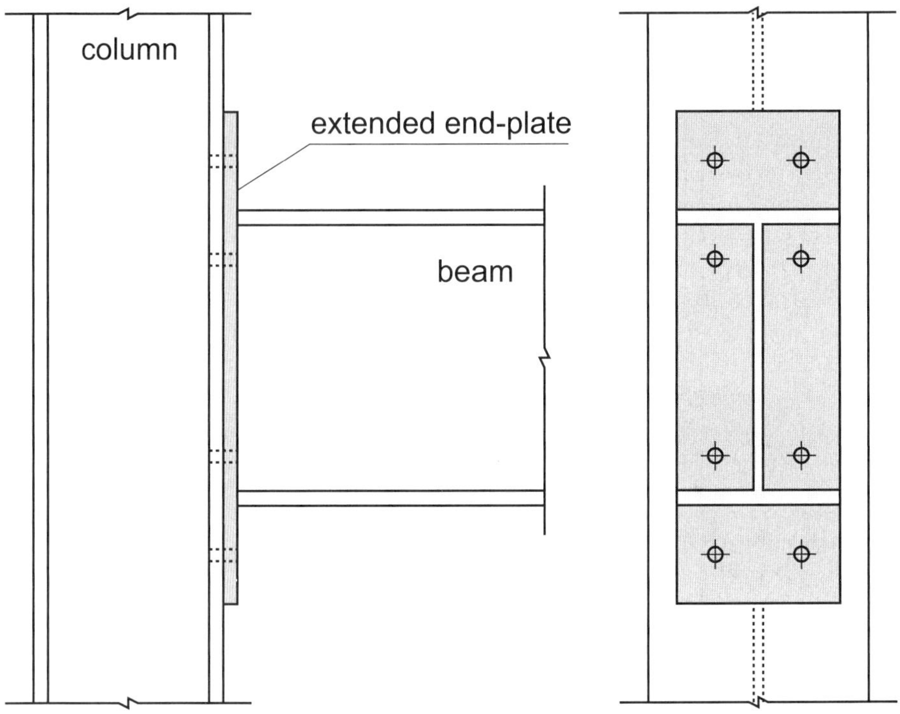

FIGURE 3.7 Extended end-plate connection on both tension and compression sides.

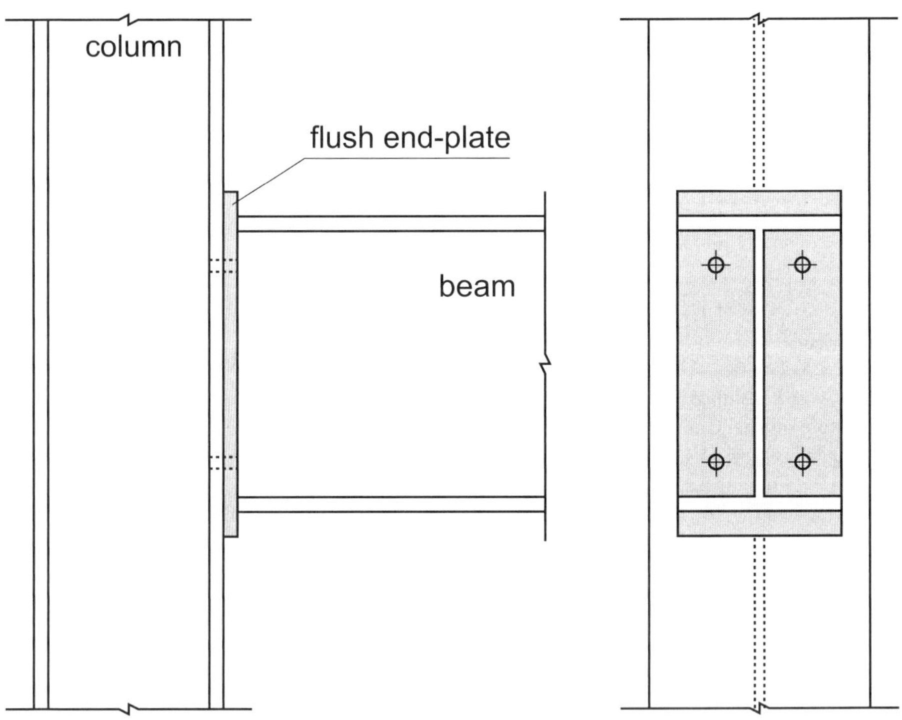

FIGURE 3.8 Flush end-plate connection.

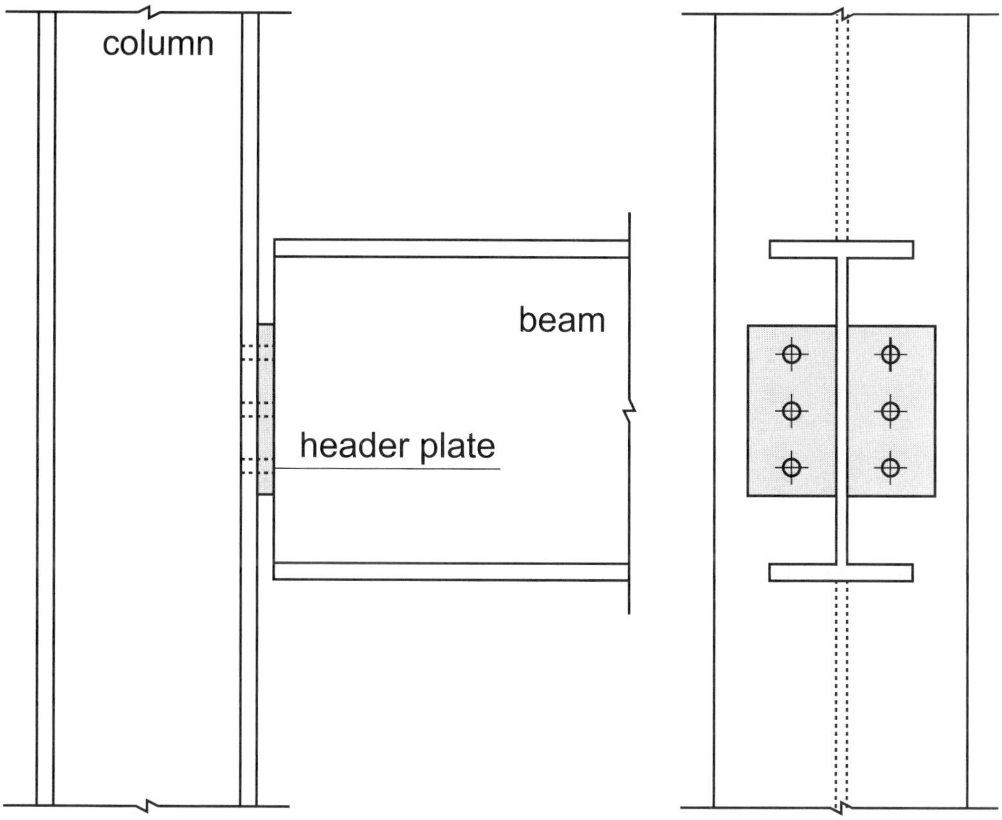

FIGURE 3.9 Header plate connection.

on site to the column as shown in Figure 3.9. This connection is also called a shear end-plate connection in the AISC-LRFD specifications (1994). The moment-rotation characteristics of these connections are similar to those of double web-angle connections. Accordingly, the header plate connections are used mainly to transfer the reaction of the beam to the column and are classified as Type 2 framing in the AISC-ASD specifications (1989).

References

AISC, *Load and Resistance Factor Design Specifications for Structural Steel Buildings*, 2nd edition, American Institute of Steel Construction, Chicago, 1994.

AISC, *Allowable Stress Design and Plastic Design Specifications for Structural Steel Buildings*, 9th edition, American Institute of Steel Construction, Chicago, 1989.

Bell, W. G., Chesson, E., Jr., and Munse, W. H., *Static tests of standard riveted and bolted beam-to-column connections*, University of Illinois Engineering Experiment Station, College of Engineering. University of Illinois at Urbana-Champaign, 1958.

Lewitt, C. W., Chesson, E., Jr., and Munse, W. H., *Restraint characteristics of flexible riveted and bolted beam-to-column connections*, Department of Civil Engineering, University of Illinois, 1966.

Rathbun, J. C., Elastic properties of riveted connections, *Transactions of* ASCE, 101, 524–563, 1936.

4

Modeling of Connections

Norimitsu Kishi
Professor, Civil Engineering Unit, College of Environmental Engineering, Muroran Institute of Technology, Muroran 050-8585, Japan

Masato Komuro
Associate Professor, Civil Engineering Unit, College of Environmental Engineering, Muroran Institute of Technology, Muroran 050-8585, Japan

4.1 General Remarks

In the conventional analysis and design of steel framework, the behavior of connections is ideal ized as perfectly rigid or ideally pinned. However, numerous experimental investigations have clearly shown that actual connections behave nonlinearly due to gradual yielding of connection components such as plates and angles, bolts, etc. A beam-to-column connection is generally subjected to axial force, shear force, and bending moment. For most connections, however, the axial and shear deformations are usually small comparing with the flexural deformation. For simplicity, only the rotational deformation of the connection due to flexural action is considered in this book. Typical moment-rotation (M–θ_r) curves for several commonly used connections are shown in Figure 4.1. In order to incorporate the M–θ_r curves more systematically and efficiently into a frame analysis computer program, the moment-rotation relationship is usually modeled by using mathematical functions.

The connection behavior can be simplified as a set of M–θ_r relationships. Mathematically, these relations can be expressed in the general form:

$$M = f(\theta_r) \tag{4.1}$$

or, inversely:

$$\theta_r = g(M) \tag{4.2}$$

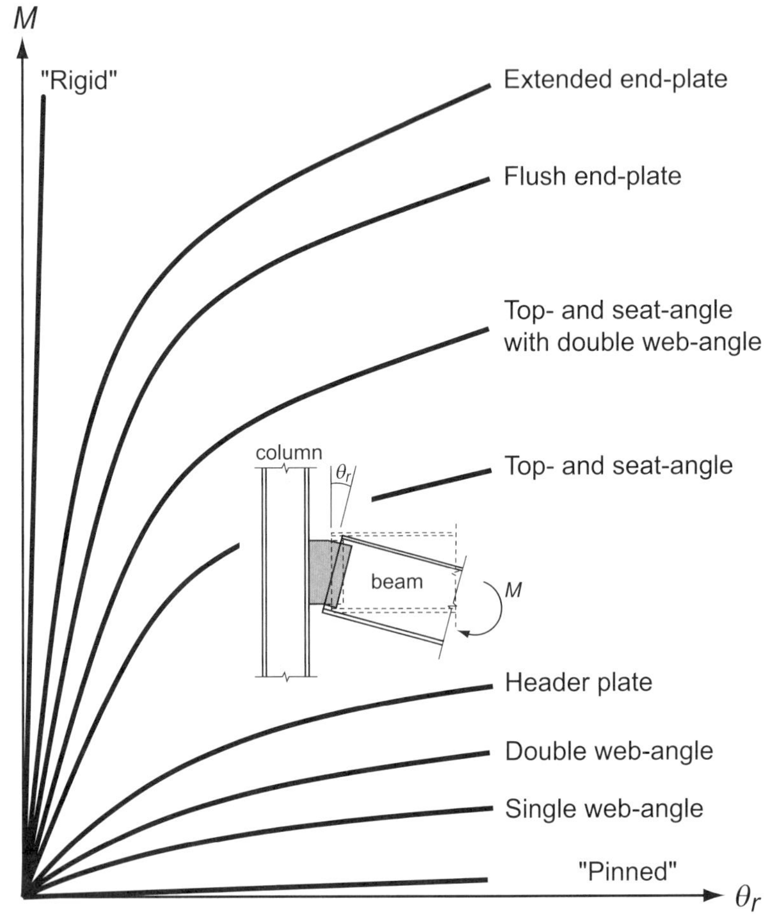

FIGURE 4.1 Typical moment-rotation (M–θ_r) curves of the beam-to-column connections.

where f and g are some mathematical functions. M is the moment at the connection, and θ_r is the relative rotation of the connection.

4.2 Behavior Under Monotonic Loading

4.2.1 Linear Model

The linear model (Rathbun, 1936; etc.) utilizes the initial connection stiffness R_{ki} to represent the connection behavior as shown in Figure 4.2a. Although this model is easy to apply the moment-rotation behavior, it overestimates the connection stiffness at finite rotation.

Tarpy and Cardinal (1981), Melchers and Kaur (1982), and Lui and Chen (1986) also proposed a bilinear model (see Figure 4.2b), in which the initial slope of moment-rotation curve is replaced by a shallower line at a certain transition moment.

Razzaq (1983) proposed a piecewise multilinear model (see Figure 4.2c), in which the nonlinear moment-rotation curve is approximated by a series of straight-line segments. Although these linear models are easy to use, the inaccuracies and the jumps in stiffness at the transition points make them undesirable.

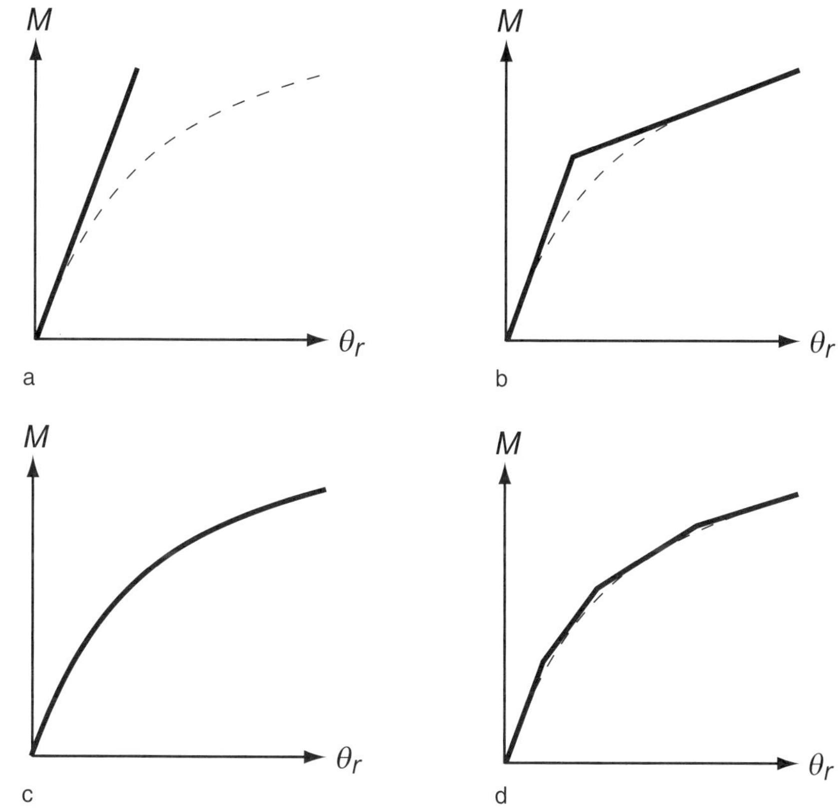

FIGURE 4.2 Different mathematical representations for the moment-rotation curve: (a) linear model; (b) bilinear model; (c) multilinear model; (d) nonlinear model.

4.2.2 Polynomial Model

The first mathematical model is proposed by Frye and Morris (1975), which is based on an odd power polynomial factor to evaluate the moment-rotation behavior of several types of connection. The Frye-Morris model was developed on the basis of a procedure formulated by Sommer (1969). They used the method of least squares to determine the constant of the polynomial. This model is represented by:

$$\theta_r = C_1(KM) + C_2(KM)^3 + C_3(KM)^5 \tag{4.3}$$

where K is a standardization parameter depending on the geometrical and mechanical properties of the connection, and C_1, C_2, and C_3 are curve-fitting constants.

The main drawback of this formulation is that the tangent connection stiffness may become negative at some value of connection moment *M*. This is physically unacceptable, and the negative stiffness may cause numerical difficulties in the analysis of frame structures if the tangent stiffness formulation is used.

Following the procedure of Frye-Morris (1975), Picard et al. (1976) and Altman et al. (1982) developed prediction equations to describe the M–θ_r curve for strap-angle connections and top- and seat-angle connection with double web angles, respectively. Goverdhan (1983) re-estimated the size parameters in the standardization constant *K* for flush end-plate connections to get a good agreement with moment-rotation curves obtained from experimental results. The curve-fitting constants

C_1, C_2, and C_3 and the standardization constant K for each connection type are summarized in Table 4.1. The size parameters for each type of connection are shown in Figure 4.3.

4.2.3 Exponential Model

Lui and Chen (1986) used an exponential function to curve-fit the experimental $M–\theta_r$ data. This model is a good representation of the monotonic nonlinear connection behavior. However, if there are some sharp changes in slope in the $M–\theta_r$ curve, this model cannot adequately represent it (Wu, 1989). Kishi and Chen (1986) refined the Lui-Chen exponential model to accommodate any sharp changes in slope in the M-θ_r experimental data (modified exponential model). The Kishi-Chen model can be used instead of the $M–\theta_r$ experimental data.

The modified exponential model is represented by a function of the following form:

$$M = M_0 + \sum_{j=1}^{m} C_j \left\{ 1 - \exp\left(-\frac{|\theta_r|}{2j\alpha} \right) \right\} + \sum_{k=1}^{n} D_k \left(|\theta_r| - |\theta_k| \right) H\left[|\theta_r| - |\theta_k| \right] \tag{4.4}$$

where M_0 is the initial connection moment, α is a scaling factor for the purpose of numerical stability, C_j and D_k are curve-fitting parameters, θ_k is the initial rotation of the k-th linear component given from the experimental $M–\theta_r$ curve, and $H[\theta]$ is Heaviside's step function (unity for $\theta \geq 0$, zero for $\theta < 0$).

Using the linear interpolation technique for the original $M–\theta_r$ experimental data, the weight function for each piece of $M–\theta_r$ data is nearly equal. The constants C_j and D_k for the exponential and linear terms of the function are determined by linear regression analysis.

The instantaneous connection stiffness R_k at an arbitrary relative rotation $|\theta_r|$ can be evaluated by differentiating Equation 4.4 with respect to $|\theta_r|$.

TABLE 4.1 (a) Curve-fitting constants and standardization constants for Frye-Morris polynomial model (all size parameters are in inches)

Connection types	Curve-fitting constants	Standardization constants
Single web-angle connection	$C_1 = 4.28 \times 10^{-3}$ $C_2 = 1.45 \times 10^{-9}$ $C_3 = 1.51 \times 10^{-16}$	$K = d_a^{-2.4}\ t_a^{-1.81}\ g^{0.15}$
Double web-angle connection	$C_1 = 3.66 \times 10^{-4}$ $C_2 = 1.15 \times 10^{-6}$ $C_3 = 4.57 \times 10^{-8}$	$K = d_a^{-2.4}\ t_a^{-1.81}\ g^{0.15}$
Top- and seat-angle with double web-angle connection	$C_1 = 2.23 \times 10^{-5}$ $C_2 = 1.85 \times 10^{-8}$ $C_3 = 3.19 \times 10^{-12}$	$K = d^{-1.287}\ t^{-1.128}\ t_c^{-0.415} l_a^{-0.694} (g - d_b/2)^{1.350}$
Top- and seat-angle connection	$C_1 = 8.46 \times 10^{-4}$ $C_2 = 1.01 \times 10^{-4}$ $C_3 = 1.24 \times 10^{-8}$	$K = d^{-1.5}\ t^{-0.5}\ l_a^{-0.7}\ d_b^{-1.1}$
Extended end-plate connection without column stiffeners	$C_1 = 1.83 \times 10^{-3}$ $C_2 = -1.04 \times 10^{-4}$ $C_3 = 6.38 \times 10^{-6}$	$K = d_g^{-2.4}\ t_p^{-0.4}\ d_b^{-1.5}$

Extended end-plate connection with column stiffeners	$C_1 = 1.79 \times 10^{-3}$ $C_2 = 1.76 \times 10^{-4}$ $C_3 = 2.04 \times 10^{-4}$	$K = d_g^{-2.4}\ t_p^{-0.6}$
Header plate connection	$C_1 = 5.10 \times 10^{-5}$ $C_2 = 6.20 \times 10^{-10}$ $C_3 = 2.40 \times 10^{-13}$	$K = t_p^{-1.6}\ g^{1.6}\ d_p^{-2.3}\ t_w^{-0.5}$
T-stub connection	$C_1 = 2.10 \times 10^{-4}$ $C_2 = 6.20 \times 10^{-6}$ $C_3 = -7.60 \times 10^{-9}$	$K = d^{-1.5}\ t^{-0.5}\ l_t^{-0.7}\ d_b^{-1.1}$

TABLE 4.1 (b) curve-fitting constants and standardization constants for Frye-Morris polynomial model (all size parameters are in centimeters)

Connection types	Curve-fitting constants	Standardization constants
Single web-angle connection	$C_1 = 1.67 \times 10^{-0}$ $C_2 = 8.56 \times 10^{-2}$ $C_3 = 1.35 \times 10^{-3}$	$K = d_a^{-2.4}\ t_a^{-1.81}\ g^{0.15}$
Double web-angle connection	$C_1 = 1.43 \times 10^{-1}$ $C_2 = 6.79 \times 10^{1}$ $C_3 = 4.09 \times 10^{5}$	$K = d_a^{-2.4}\ t_a^{-1.81}\ g^{0.15}$
Top- and seat-angle with double web-angle connection	$C_1 = 1.50 \times 10^{-3}$ $C_2 = 5.60 \times 10^{-3}$ $C_3 = 4.35 \times 10^{-3}$	$K = d^{-1.287}\ t^{-1.128}\ t_c^{-0.415} l_a^{-0.694} \left(g - d_b/2\right)^{1.350}$
Top- and seat-angle connection	$C_1 = 2.59 \times 10^{-1}$ $C_2 = 2.88 \times 10^{3}$ $C_3 = 3.31 \times 10^{4}$	$K = d^{-1.5}\ t^{-0.5}\ l_a^{-0.7}\ d_b^{-1.1}$
Extended end-plate connection without column stiffeners	$C_1 = 8.91 \times 10^{-1}$ $C_2 = -1.20 \times 10^{4}$ $C_3 = 1.75 \times 10^{8}$	$K = d_g^{-2.4}\ t_p^{-0.4}\ d_b^{-1.5}$
Extended end-plate connection with column stiffeners	$C_1 = 2.60 \times 10^{-1}$ $C_2 = 5.36 \times 10^{2}$ $C_3 = 1.31 \times 10^{7}$	$K = d_g^{-2.4}\ t_p^{-0.6}$
Header plate connection	$C_1 = 6.14 \times 10^{-3}$ $C_2 = 1.08 \times 10^{-3}$ $C_3 = 6.05 \times 10^{-3}$	$K = t_p^{-1.6}\ g^{1.6}\ d_p^{-2.3}\ t_w^{-0.5}$
T-stub connection	$C_1 = 6.42 \times 10^{-2}$ $C_2 = 1.77 \times 10^{2}$ $C_3 = -2.03 \times 10^{4}$	$K = d^{-1.5}\ t^{-0.5}\ l_t^{-0.7}\ d_b^{-1.1}$

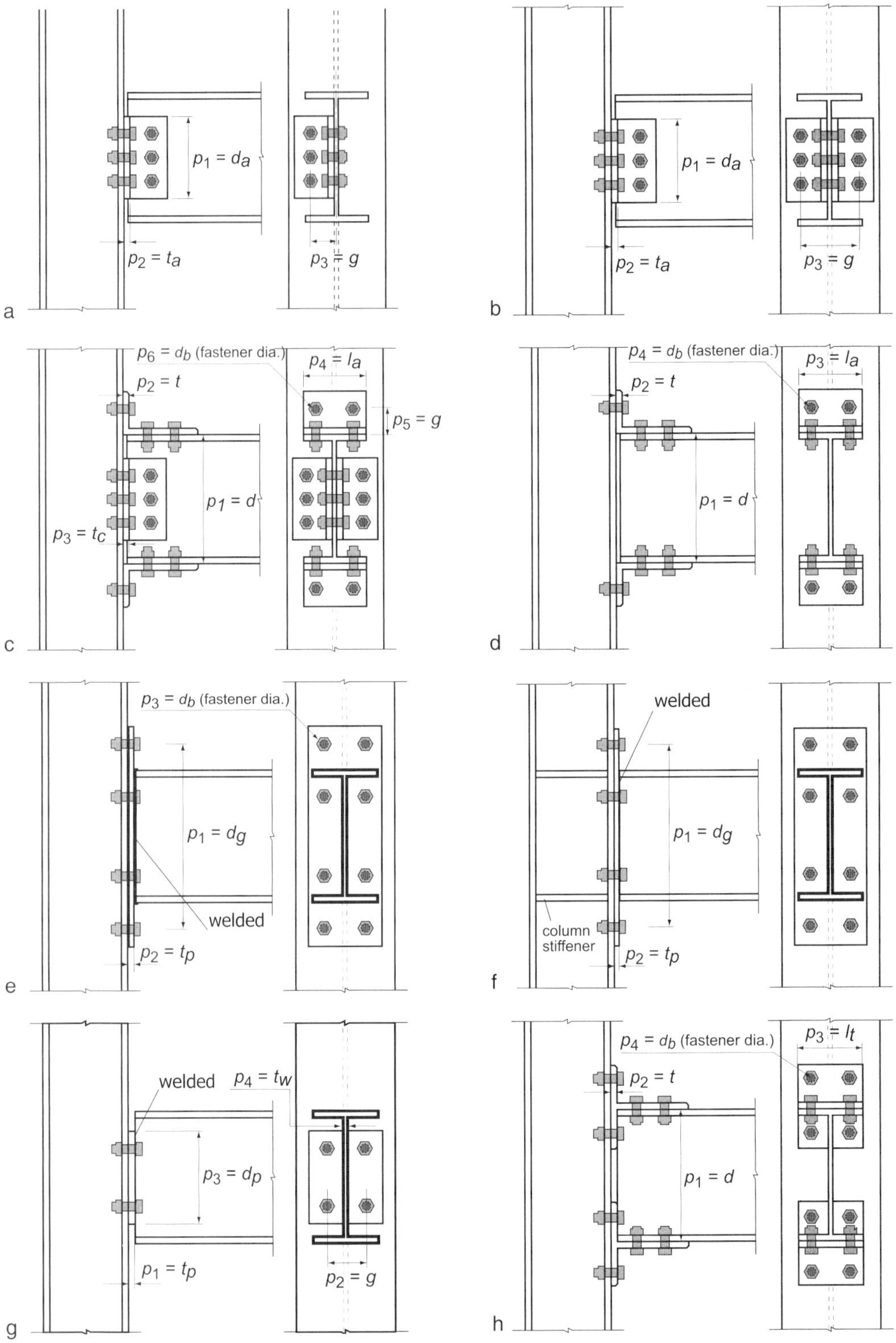

FIGURE 4.3 Size parameters for various connection types of the Frye-Morris polynomial model: (a) single web-angle connection; (b) double web-angle connection; (c) top- and seat-angle connection with double web-angle; (d) top- and seat-angle connection; (e) extended end-plate connection without column stiffener; (f) extended end-plate connection with column stiffener; (g) header plate connection; (h) T-stub connection.

When the connection is loaded, we obtain the equation as follows:

$$R_k = R_{kt} = \left.\frac{dM}{d|\theta_r|}\right|_{|\theta_r|=|\theta_r|} = \sum_{j=1}^{m} \frac{C_j}{2j\alpha} \exp\left(-\frac{|\theta_r|}{2j\alpha}\right) + \sum_{k=1}^{n} D_k H\left[|\theta_r| - |\theta_k|\right] \tag{4.5}$$

When connection is unloaded, we have:

$$R_k = R_{kt} = \left.\frac{dM}{d|\theta_r|}\right|_{|\theta_r|=0} = \sum_{j=1}^{m} \frac{C_j}{2j\alpha} + D_k H\left[|\theta_k|\right]\Big|_{k=1} \tag{4.6}$$

This model has the following merits:

1. The formulation is relatively simple and straightforward.
2. It can deal with connection loading and unloading for the full range of relative rotation in a second-order structural analysis with secant connection stiffness.
3. The abrupt changes in the connection stiffness among the sampling data is only generated from inherent experimental characteristics.

The curve-fitting and tangent connection stiffness values from the experimental data examined are calculated with $m = 6$ in Equations 4.4 through 4.6. The comparison between the Chen-Lui exponential model (1985) and the modified exponential model for the numerical example of the test data, including a linear component, is shown in Figure 4.4.

Yee and Melchers (1986) proposed a four-parameter exponential model:

$$M = M_p \left\{ 1 - \exp\left[-\frac{\left(R_{ki} - K_p + C\theta_r\right)\theta_r}{M_p} \right] \right\} + K_p \theta_r \tag{4.7}$$

where M_p is the plastic moment capacity, R_{ki} is the initial connection stiffness, K_p is the strain-hardening stiffness, and C is a constant controlling the slope of the curve.

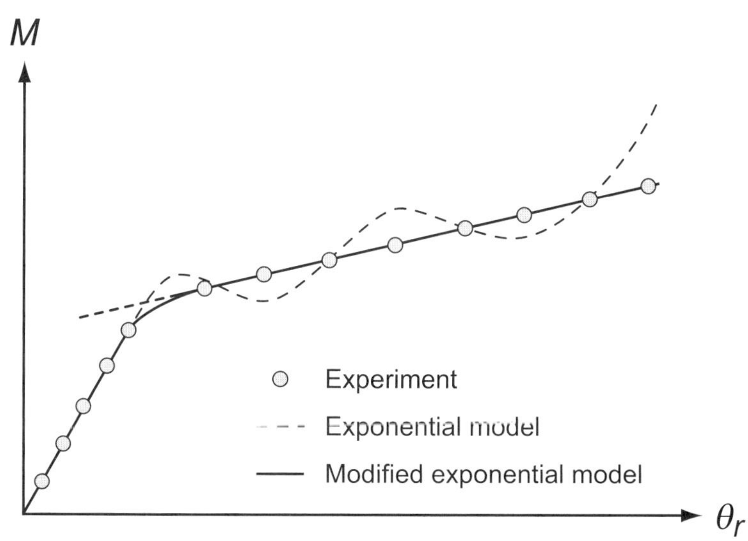

FIGURE 4.4 Comparison between results using exponential and modified exponential models for $M–\theta_r$ data including a linear term.

As with the other exponential model, the Wu-Chen model (1990) proposed a three-parameter exponential model in the form of:

$$\frac{M}{M_u} = n \ln\left(1 + \frac{\theta_r}{\theta_0}\right) \tag{4.8}$$

where M_u is an idealized elastic-plastic mechanism moment, θ_0 is a reference rotation ($= M_u/R_{ki}$), R_{ki} is the initial connection stiffness, and n is the shape parameter, which is determined empirically from test data.

4.2.4 Power Model

Several power models have been developed for the different types of connection. Two or three parameters are required in their functions.

A two-parameter model (Batho and Lash, 1936; Krishnamurthy et al., 1979) has the simple form:

$$\theta_r = aM^b \tag{4.9}$$

where a and b are two curve-fitting parameters with the condition $a > 0$ and $b > 1$.

Colson and Louveau (1983) introduced a three-parameter power model function as:

$$\theta_r = \frac{|M|}{R_{ki}} \frac{1}{\left(1 - |M/M_u|^n\right)} \tag{4.10}$$

where R_{ki} is the initial connection stiffness, M_u is the ultimate capacity of connection moment, and n is the shape parameter.

Kishi and Chen (1987a, 1987b) proposed a similar model removing the strain-hardening stiffness of the Richard-Abott model (Richard and Abott, 1975):

$$\theta_r = \frac{M}{R_{ki}\left\{1 - \left(\dfrac{M}{M_u}\right)^n\right\}^{1/n}} \quad \text{or} \quad M = \frac{R_{ki}\theta_r}{\left\{1 + \left(\dfrac{\theta_r}{\theta_0}\right)^n\right\}^{1/n}} \tag{4.11a, b}$$

where R_{ki}, M_u, and n are the same as those defined in the previous Equation 4.10, and θ_0 is a reference plastic rotation ($= M_u/R_{ki}$). Equation 4.11 has the shape shown in Figure 4.5. From this figure, it is seen that the larger the power index n, the steeper the curve. The shape parameter n can be determined using the method of least squares for the differences between the predicted moments and the experimental test data (Kishi et al., 1993, 1995).

This power model is an effective tool for designers to execute the second-order nonlinear structural analysis quickly and accurately. This is because the tangent connection stiffness R_k and relative rotation θ_r can be determined directly from Equation 4.11 without iteration, in which the tangent connection stiffness R_k in Equation 4.11 is:

$$R_k = \frac{dM}{d\theta_r} = \frac{R_{ki}}{\left\{1 + \left(\dfrac{\theta_r}{\theta_0}\right)^n\right\}^{(n+1)/n}} \tag{4.12}$$

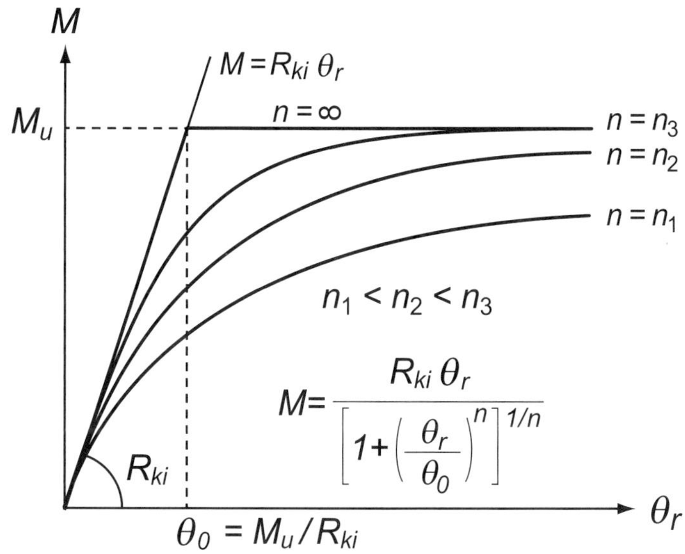

FIGURE 4.5 Three-parameter power model.

4.2.5 Bounding Line Model

Al-bermani et al. (1994) and Zhu et al. (1995) have proposed a bounding line model as shown in Figure 4.6. It requires four parameters and the concept of this model is to divide the curve into three segments, in which the first and third segments are linear elastic and plastic portions, respectively, and the second segment is a smooth transition portion. The form of this model is represented as:

$$M = \begin{cases} R_{ki}\theta_r & |M| < m_1 \\ R_{ki} + \dfrac{|M| - m_1}{M_c - M_y}\left(R_{kp} - R_{ki}\right) & m_1 \le |M| < m_2 \\ R_{kp}\theta_r & |M| \ge m_2 \end{cases} \tag{4.13}$$

where $m_1 = M_y + R_{kp}\theta_r$, $m_2 = M_c + R_{kp}\theta_r$, R_{ki} and R_{kp} are the initial connection stiffness and the bounding connection stiffness, respectively, M_y is the yielding connection moment, and M_c is the bounding connection moment.

4.2.6 Ramberg-Osgood Model

The Ramberg-Osgood model was originally proposed for nonlinear stress-strain relationships by Ramberg and Osgood (1943) and then standardized by Ang and Morris (1984). The M–θ_r curve of this model is represented as:

$$\frac{\theta_r}{\theta_0} = \frac{|KM|}{(KM)_0}\left\{1 + \left(\frac{|KM|}{(KM)_0}\right)^{n-1}\right\} \tag{4.14}$$

where $(KM)_0$ and θ_0 are constants defining the position of the intersection point A (see Figure 4.7), n is a parameter defining the sharpness of the curve, and K is a dimensionless factor depending on the connection type and geometry.

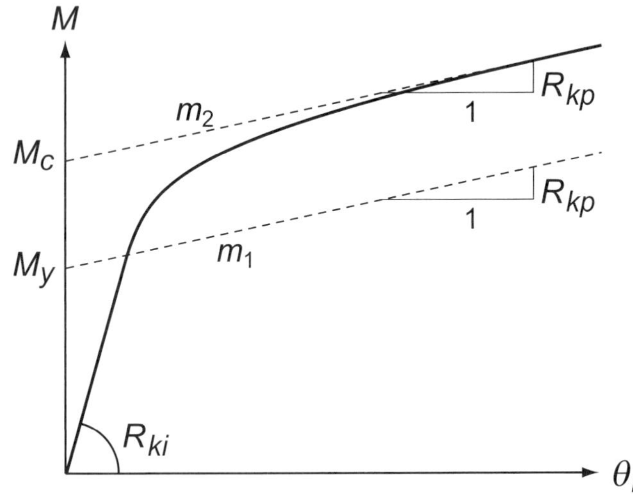

FIGURE 4.6 Bounding line model.

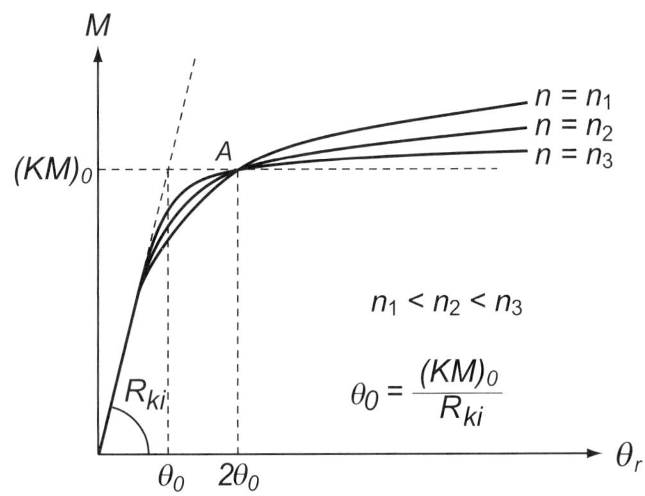

FIGURE 4.7 Ramberg-Osgood model.

4.2.7 Richard-Abbott Model

The four-parameter power model is proposed by Richard and Abbott (1975) for modeling elastic-plastic stress-strain relation as shown in Figure 4.8. In the virgin loading path, the M–θ_r curve is written by the following expression:

$$M = \frac{\left(R_{ki} - R_{kp}\right)\theta_r}{\left\{1 + \left(\theta_r / \theta_0\right)^n\right\}^{1/n}} + R_{kp}\theta_r \tag{4.15}$$

where R_{ki} is the initial connection stiffness, R_{kp} is the strain-hardening connection stiffness, n is the shape parameter, θ_0 is the reference relative rotation [$= M_0/(R_{ki} - R_{kp})$], and M_0 is the reference connection moment.

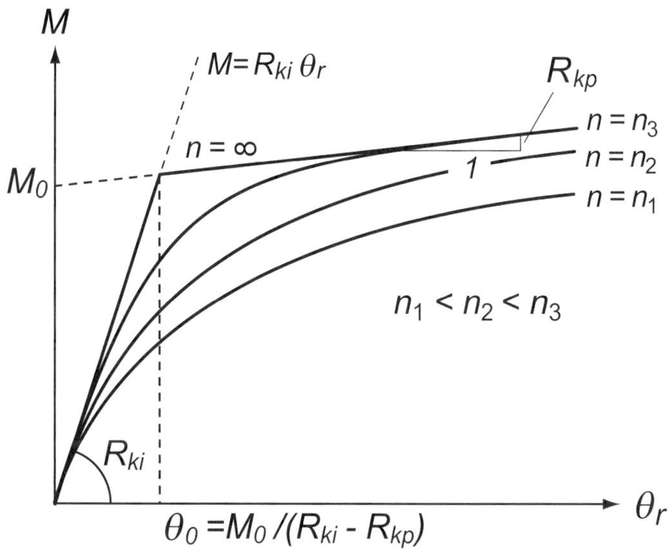

FIGURE 4.8 Richard-Abbott model.

4.3 Behavior Under Cyclic Loading

4.3.1 Independent Hardening Model

The independent hardening model (Chen and Saleeb, 1982) is a simple method for representing behavior under cyclic loading. The characteristics of connection behavior are assumed to be unchanged following the virgin M–θ_r curve under the loading and reloading conditions. Since the loops of the M–θ_r relation in each cycle are independent, no hardening effect is considered. The cyclic M–θ_r relation based on this model is schematically shown in Figure 4.9. In the case of

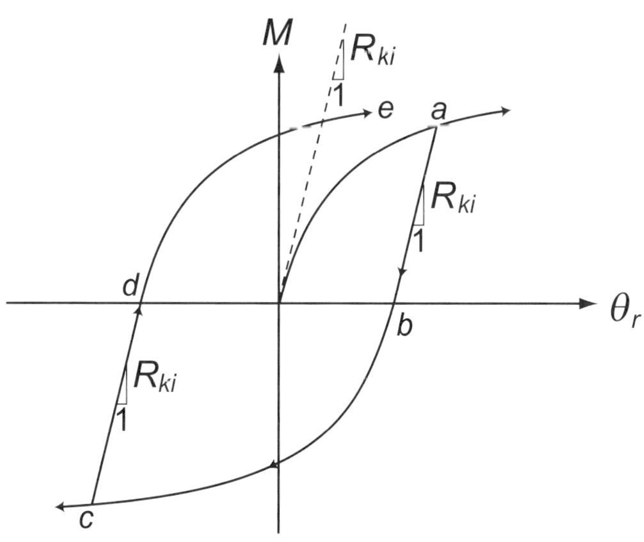

FIGURE 4.9 Independent hardening model.

unloading from the point a on the initial loading curve, the connection unloads linearly down to $M = 0$ with the initial connection stiffness R_{ki}. When the sign of the connection moment M changes at the point b in the unloading process, the connection enters into the reverse loading process. In this case, the connection behavior (excluding the residual plastic rotation) resulting from the previous loading process is assumed to coincide with that of the virgin connections under monotonic loading.

4.3.2 Kinematic Hardening Model

The kinematic hardening model is a modified independent hardening model taking into account the effect of material hardening. The behavior of this connection model is illustrated in Figure 4.10, which is represented by the hardening line with the slope R_b. In the case of reversal unloading, the path of the M–θ_r curve moves along the line with the slope of the initial connection stiffness R_{ki} (i.e., for line ab or cd) until it reaches the hardening line. For further reversal unloading, the path follows the virgin nonlinear M–θ_r curve of the connection under the monotonic loading (i.e., for line bc or de). If the hardening line has a zero slope (i.e., $R_b = 0$), the kinematic hardening model is exactly the same as the independent hardening model.

4.3.3 Bounding Surface Model with Internal Variables

The previously mentioned independent and kinematic hardening models are simple to use in frame analysis. Although these models can handle the M–θ_r relation under one cycle loading, unloading, and reversal loading, the connection behavior for a repetition of this loading cycle cannot be expressed with acceptable accuracy. As shown in Figure 4.11, one problem associated with the independent and kinematic hardening models is that the M–θ_r relation is completely different depending on whether the connection is reloaded from the region between a and b or from

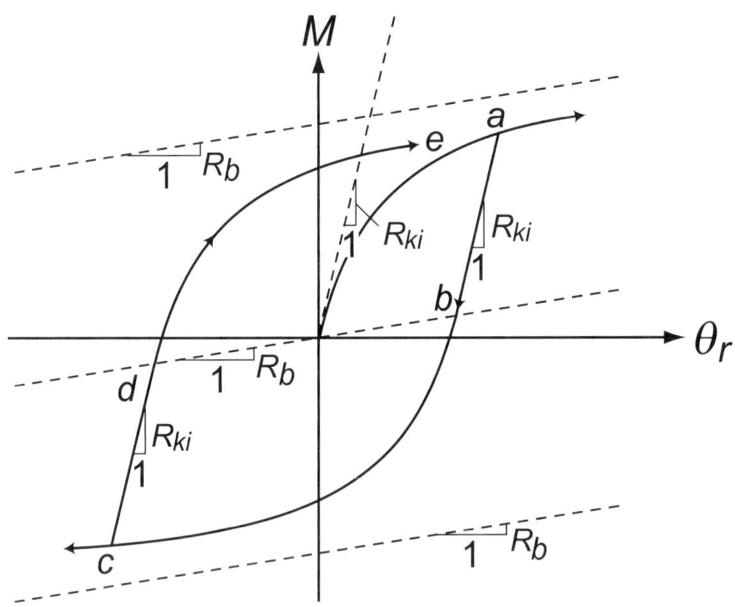

FIGURE 4.10 Kinematic hardening model.

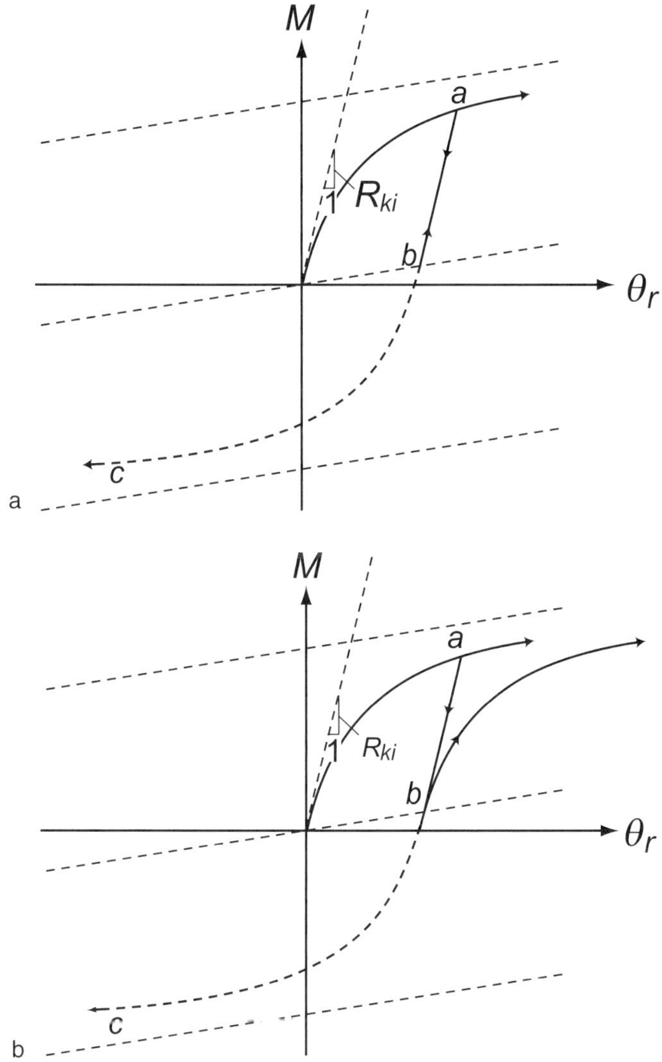

FIGURE 4.11 Problems associated with the kinematic hardening model: (a) reloading from the region between *a* and *b*; (b) reloading from the region between *b* and *c*.

the region between *b* and *c*. To overcome this deficiency, a bounding surface model with internal variables (Dafalias and Popov, 1976) may be used (Cook, 1983; Goto et al., 1991, 1993).

In this model, the M–θ_r relation is defined in the incremental form:

$$\Delta M = R_{kt} \Delta \theta_r \tag{4.16}$$

where R_{kt} is the tangent stiffness of the connections. The tangent connection stiffness R_{kt} is represented in terms of the initial connection stiffness R_{ki} and the plastic tangent connection stiffness R_{kp} as:

$$R_{kt} = \frac{R_{kt} R_{kp}}{R_{ki} + R_{kp}} \tag{4.17}$$

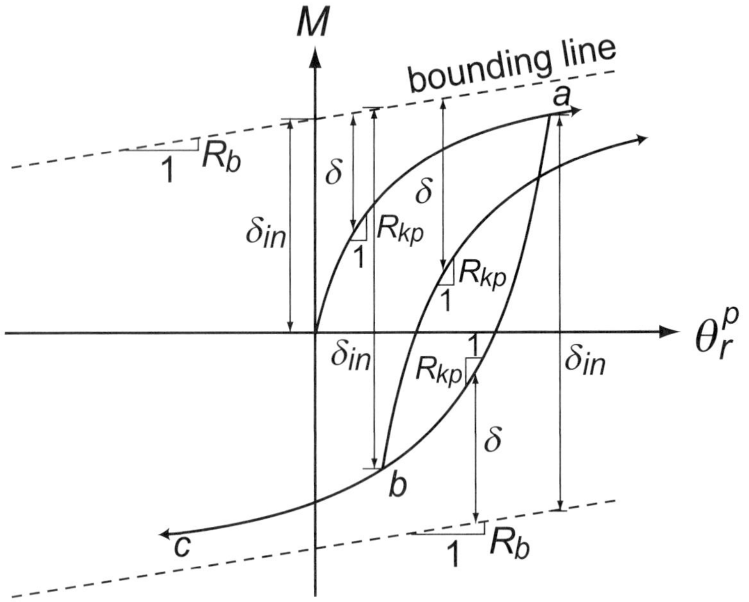

FIGURE 4.12 Bounding surface model.

The plastic tangent connection stiffness R_{kp} is represented by using the plastic internal variables δ and δ_{in}

$$R_{kp} = R_b + h\frac{\delta}{\delta_{in} - \delta} \tag{4.18}$$

where h is the hardening shape parameter, R_b is the slope of the bounding line, δ is the distance of the current moment state from the corresponding bound, and δ_{in} is the value of δ at the initiation of each loading process. These quantities are schematically shown in Figure 4.12 using the moment-plastic rotation (M–θ_r^p) curve.

References

Al-Bermani, F. G. A, Li B., Zhu K., and Kitipornchai S., Cyclic and seismic response of flexibly jointed frames, *Engineering Structures*, 16(4), 249–255, 1994.

Altman, W. G., Azizinamini, A., Bradburn, J. H., and Radziminski, J. B., *Moment-rotation characteristics of semi-rigid steel beam-column connections*, Final Report, South Carolina University, Columbia. Department of Civil Engineering, SC, 1982.

Ang, K. M. and Morris, G. A., Analysis of three-dimensional frames with flexible beam-column connections, *Canadian Journal of Civil Engineering*, 11, 245–254, 1984.

Batho, C. and Lash, S. D., *Further investigations on beam and stanchions connections, Including connections encased in concrete; Together with laboratory investigations on a full-scale steel frame*, Final report of the Steel Structures Research Committee, Department of Scientific and Industrial Research, His Majesty's Stationery Office, London, 1936.

Chen, W. F. and Lui, E. M., Column with end restraint and bending in load and resistance factor design, *AISC Engineering Journal*, 22(4), 105–132, 1985.

Chen, W. F. and Saleeb, A. F., *Uniaxial behavior and modeling in plasticity*, Structural Engineering Report No. CE-STR-82–35, School of Civil Engineering, Purdue University, W. Lafayette, IN, 1982.

Colson, A. and Louveau, J. M., Connections incidence on the inelastic behavior of steel structures, *Proceedings of the Euromech Colloquium*, 174, 1983.

Cook, N. E., *Strength of flexibly-connected steel frames under load histories*, Thesis presented to University of Colorado, in partial fulfillment of the requirements for the degree of Doctor of Philosophy, Boulder, CO, 1983.

Dafalias, Y. F. and Popov, E. P., Plastic internal variables formalism of cyclic plasticity, *Series E-Journal of Applied Mechanics*, Transactions, ASME, 43, 645–651, 1976.

Frye, M. J. and Morris, G. A., Analysis of flexibly connected steel frames, *Canadian Journal of Civil Engineering*, 2(3), 280–291, 1975.

Goto, Y., Suzuki, S., and Chen, W. F., Stability behaviour of semi-rigid sway frames, *Engineering Structures*, 15(3), 209–219, 1993.

Goto, Y., Suzuki, S., and Chen, W. F., Analysis of critical behavior of semi-rigid frames with or without load history in connections. *International Journal of Solids and Structures*, 27(4), 467–483, 1991.

Goverdhan, A. V., *A collection of experimental moment-rotation curves and evaluation of prediction equations for semi-rigid connections*, Thesis presented to Vanderbilt University, in partial fulfillment of the requirements for the degree of Master of Science, Nashville, TN, 1983.

Kishi, N., Hasan, R., Chen, W. F., and Goto, Y., Power model for semi-rigid connections, *Steel Structures, Journal of Singapore Structural Steel Society*, 5(1), 37–48, 1995.

Kishi, N., Chen, W. F., Goto, Y., and Matsuoka, K. G., Design aid of semi-rigid connections for frame analysis, *AISC Engineering Journal*, 30(3), 90–107, 1993.

Kishi, N., Chen, W. F., Matsuoka, K. G., and Nomachi, S. G., Moment-rotation relation of top- and seat-angle with double web-angle connections, *Proceedings of the State-of-the-Art Workshop on Connections and the Behavior, Strength and Design of Steel Structures*, R. Bjorhovde, J. Brozzetti, and A. Colson, eds., Ecole Normale Superieure, Cachan, France, May 25–27, 1987a.

Kishi, N., Chen, W. F., Matsuoka, K. G., and Nomachi, S. G., Moment-rotation relation of single/double web-angle connections, *Proceedings of the State-of-the-Art Workshop on Connections and the Behavior, Strength and Design of Steel Structures*, R. Bjorhovde, J. Brozzetti, and A. Colson, eds., Ecole Normale Superieure de Cachan, France, May 25–27, 1987b.

Kishi, N. and Chen, W. F., *Database of steel beam-to-column connections*, Structural Engineering Report No. CE-STR-86-26, School of Civil Engineering, Purdue University, West Lafayette, IN, 1986.

Krishnamurthy, N., Huang, H. T., Jefferey, P. K., and Avery, L. K., Analytical M-θ curves for end-plate connections, *Journal of the Structural Division*, ASCE, 105(1), 133–145, 1979.

Lui, E. M. and Chen, W. F., Analysis and behavior of flexibly-jointed frames, *Engineering Structures*, 8(2), 107–118, 1986.

Melchers, R. E. and Kaur, D., Behaviour of frames with flexible joints, *Proceedings of 8th Australian Conference Mechanical of Structural Materials*, Newcastle, Australia, 27.1–27.5, 1982.

Picard, A., Giroux, Y. M., and Brun, P., Analysis of flexibly connected steel frames: Discussion by Frye, M. J. and Morris, G. A., *Canadian Journal of Civil Engineering*, 3(2), 350–352, 1976.

Ramberg, W. and Osgood, W. R., *Description of stress-strain curves by three parameters*, National Advisory Committee for Aeronautics, Washington, DC, Technical notes No.902, 1943.

Rathbun, J. C., Elastic properties of riveted connections, *Transaction of the ASCE*, 101, 524–563, 1936.

Razzaq, Z., End restraint effect on steel column strength, *Journal of Structural Engineering*, ASCE, 109(2), 314–334, 1983.

Richard, R. M. and Abbott, B. J., Versatile elastic-plastic stress-strain formula, *Journal of the Engineering Mechanics Division*, ASCE, 101(4), 511–515, 1975.

Sommer W. H., *Behaviour of welded header-plate connections*, Thesis presented to University of Toronto, in partial fulfillment of the requirements for the degree of Master of Applied Science, Toronto, Canada, 1969.

Tarpy, T. S. and Cardinal, J. W., Behavior of semi-rigid beam-to-column end plate connections, *Proceedings of the International Conference: Joints in Structural Steelwork*, Teesside Polytechnic, Middlesbrough, Cleveland, UK, 2.3–2.25, 1981.

Wu, F. S. and Chen, W. F., A design model for semi-rigid connections, *Engineering Structures*, 12(2), 88–97, 1990.

Wu, F. S., *Semi-rigid connections in steel frames*, Thesis presented to Purdue University, in partial fulfillment of the requirements for the degree of Doctor of Philosophy, West Lafayette, IN, 1989.
Yee, Y. L. and Melchers, R. E., Moment-rotation curves for bolted connections, *Journal of Structural Engineering*, ASCE, 112(3), 615–635, 1986.
Zhu, K., Al-Bermani, F. G. A., Kitipornchai, S., and Li, B., Dynamic response of flexibly jointed frames, *Engineering Structures*, 17(8), 575–580, 1995.

5

Steel Connection Database

Norimitsu Kishi
Professor, Civil Engineering Unit, College of Environmental Engineering, Muroran Institute of Technology, Muroran 050-8585, Japan

Masato Komuro
Associate Professor, Civil Engineering Unit, College of Environmental Engineering, Muroran Institute of Technology, Muroran 050-8585, Japan

5.1 General Remarks

Conventional analysis and design of steel frameworks are performed using the assumption that the connections are either fully rigid or ideally pinned. The assumption of the fully rigid connection implies that no relative rotation of connection occurs and the end moment of the beam is completely transferred to the column. On the other hand, the pinned connection implies that no restraint on the rotation of connection exists and the connection moment is always zero. However, it is recognized that actual beam-to-column connections always provide some rigidity.

The primary distortion of a steel beam-to-column connection is its rotational deformation θ_r, caused by the in-plane bending moment M (Figure 5.1). The effect of this connection deformation has a destabilizing effect on the frame stability since additional drift will occur because of the decrease in the effective stiffness of the members to which the connections are attached. An increased frame drift will intensify the P-Δ effect, affecting the overall stability of the frame.

Thus, the nonlinear characteristics of the beam-to-column connection play a very important role in frame stability. New design method has shifted from the allowable stress design (ASD) method to a limit-state design method such as the AISC load and resistance factor design (LRFD) specifications (AISC, 1994). The limit-state design specifications refer to a more sophisticated method of analysis than the allowable stress design specifications. For example, AISC-LRFD

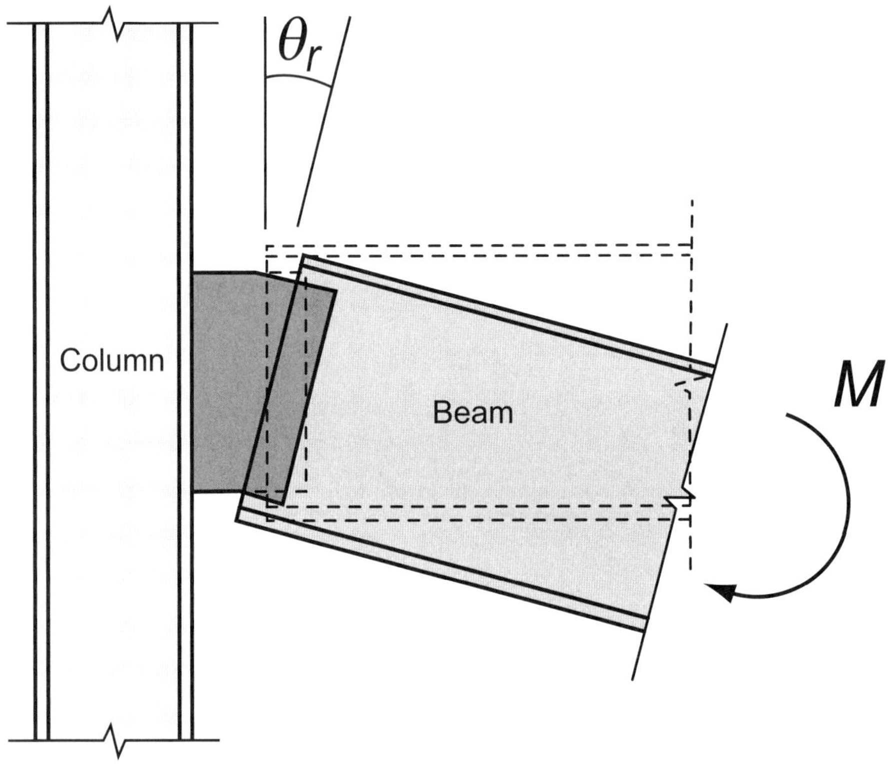

FIGURE 5.1 Rotational deformation of a connection.

specifications require that the limit state of the structural stability corresponds to the analysis method used for design as follows:

1. When structures are designed on the basis of plastic analysis, the required flexural strength M_u should be determined from a second-order plastic analysis.
2. When designed on the basis of elastic analysis, M_u for beam-columns, connections, and connected members should be determined from a second-order elastic analysis or approximate second-order analysis by means of the first-order elastic analysis method with B_1/B_2 amplification factors.

While two types of construction are defined in the specifications—fully restrained (FR) construction and partially restrained (PR) construction—the guidelines for the design of the PR construction are not given since it is difficult to evaluate the actual restraint of the semi-rigid connections used in engineering practice.

To establish the guideline for design of the PR frames, it is necessary to know the M–θ_r behavior of the actual beam-to-column connections and to formulate the appropriate M–θ_r model for use in the design and analysis of PR frames. In recent years, several researchers have published papers discussing the influences of connection rigidity on steel frame structures for all connection types. For example, Goverdhan (1983) collected much of the available test data on moment-rotation characteristics and tried to formulate a prediction equation for each type of connection. Nethercot (1985) conducted an extensive literature survey for steel beam-to-column connection test data and their corresponding $M–\theta_r$ curve representations for the period 1915–1985. Kishi and Chen (1986b) and Chen and Kishi (1989) also collected experimental test data published from 1936 to 1985 and compiled a database. After that, Kishi and his colleagues revised the database appending 93 experimental data for the period 1986–1998. Komuro and Kishi (2009) upgraded the connection database including the SCDB program. The connection data compiled in the database have now reached 486.

The aim of this chapter is to provide moment-rotation characteristics and corresponding parameters of the semi-rigid beam-to-column connections used frequently in steel construction. Based on the experimental test data collected by Kishi-Chen, a database on steel beam-to-column connections was developed at Purdue University Computer Center and then revised by appending newly performed experimental results. To control this database systematically, the Steel Connection Data Bank (SCDB) program has been developed (Kishi and Chen, 1986a). The SCDB program provides seven main functions including routines for tabulating the experimental test data and the tangent connection stiffness evaluated by using some prediction equations. Three prediction equations including the Frye-Morris model, modified exponential model, and three-parameter power model were chosen and installed in the program. Here, the program has been revised by appending a units conversion routine from the customary U.S. units to SI units.

The literature survey encompassed experimental data published from 1936 (Rathbun) to 2007 (Sato et al.), on riveted, bolted, and welded connections. All 486 experimental data collected so far are classified into seven types as shown in Table 5.1. The references and number of experimental data of each connection type are listed in Table 5.2. These experimental data with the SCDB program are installed on the J. Ross Web site (http://www.jrosspub.com/wav/). The items for each experimental data registered in the database are:

1. Connection type and fastening mode
2. Author, test I.D., and tested country
3. The material properties and size of fasteners
4. The material strength of the connection angles and plate
5. All parameters used in beam-to-column connection
6. Moment-rotation test data

Numerical values registered in the database were standardized in the customary U.S. units by executing a subprogram as the pre-processor. The connection type is classified as shown in Table 5.1, and the fastening mode is classified as shown in Table 5.3. The notations of the major parameters for each connection type are identified in the next section.

TABLE 5.1 Connection types

Type No.	Connection type
1	Single web-angle connection
2	Double web-angle connection
3	Top- and seat-angle connection with double web-angle
4	Top- and seat-angle connection
5	Extended end-plate connection
6	Flush end-plate connection
7	Header plate connection

TABLE 5.2 Authors and numbers of experimental M–θ_r curves for each connection type

Connection type	Reference (author, year)	Number of tests
Single web-angle and single plate connections	S. L. Lipson (1968)	30
	L. E. Thompson et al. (1970)	12
	S. L. Lipson et al. (1977)	8
	R. M. Richard et al. (1982)	4
Double web-angle connections	J. C. Rathbun (1936)	7
	W. G. Bell et al. (1958)	4
	C. W. Lewitt et al. (1966)	6
	W. H. Sommer (1969)	4
	L. E. Thompson et al. (1970)	48
	B. Bose (1981)	1
	J. B. Davison et al. (1987)	2
	J.G. Yang and G.Y. Lee (2007)	3
Top- and seat-angle with double web-angle connections	J. C. Rathbun (1936)	2
	A. Azizinamini et al. (1985)	20
	C. W. Roeder et al. (1996)	1
	A. S. Elnashai et al. (1998)	1
	Z. Fu et al. (1998)	4
	L. Calado et al. (2000)	3
	M. Komuro et al. (2002)	2
Top- and seat-angle connections	J. C. Rathbun (1936)	3
	R. A. Hechtman and B. G. Johnston (1947)	12
	S. M. Maxwell et al. (1981)	12
	M. J. Marley (1982)	26
	J. B. Davison et al. (1987)	1
	A. Azizinamini (1985)	2
	W. L. Harper, Jr. (1990)	1
	J. B. Mander et al. (1994)	4
	C. Bernuzzi et al. (1996)	1
	N. Kubo et al. (1999)	5
	M. Komuro et al. (2002)	1
	Y. Sato et al. (2007)	6
Extended end-plate connections	L. G. Johnson et al. (1960)	1
	A. N. Sherbourne (1961)	5
	J. R. Bailey (1970)	26
	J. O. Surtees and A. P. Mann (1970)	6
	J. A. Packer and L. J. Morris (1977)	3
	S. A. Ioannides (1978)	6
	R. J. Dews (1979)	3

(Continues)

TABLE 5.2 (Continued)

Extended end-plate connections (Continued)	P. Grundy et al. (1980)	2
	N. D. Johnstone and W. R. Walpole (1981)	8
	A. Mazroi (1983, 1984)	24
	Y. L. Yee (1984)	16
	D. B. Moore and P. A. C. Sims (1986)	2
	J. B. Davison et al. (1987)	1
	R. Zandonini and P. Zanon (1987)	9
	L. F. L. Ribeiro et al. (1998)	24
	B. Bose et al. (1996)	9
	L.R.O. Lima (2003)	7
	A. M. Girão Coelho (2004, 2007)	11
	J. M. Cabrero and E. Bayo (2007)	2
	G. Shi et al. (2007)	1
Flush end-plate connections	J. R. Ostrander (1970)	24
	J. Phillips and J. A. Packer (1981)	5
	P. Zoetemeijer (1981)	7
	J. B. Davison et al. (1987)	3
	C. Bernuzzi et al. (1996)	1
	B. Bose et al. (1996)	9
	N. D. Brown and D. Anderson (2001)	1
	A.W. Thomson and B.M. Broderick (2002)	3
	A. M. Girao Coelho and F. S. K. Bijlaard (2007)	4
	G. Shi et al. (2007)	1
Header plate connections	W. H. Sommer (1969)	20
	J. B. Davison et al. (1987)	1
	A. K. Aggarwal (1990)	5

TABLE 5.3 Fastening mode patterns

Pattern No.	Fastening mode pattern
1	All riveted
2	All bolted
3	Riveted-to-beam and bolted-to-column
4	Bolted-to-beam and riveted-to-column
5	Riveted-to-beam and welded-to-column
6	Bolted-to-beam and welded-to-column
7	Welded-to-beam and riveted-to-column
8	Welded-to-beam and bolted-to-column
9	All riveted without column stiffener
10	All riveted with column stiffener
11	All bolted without column stiffener
12	All bolted with column stiffener

5.2 Parameter Definition for Connection Type

5.2.1 Single Web-angle Connections/Single Plate Connections

Collected experimental moment-rotation data for a single web-angle with bolted beam-to-column connections are tabulated in Table 5.4. Single plate connections and single web-angle connections bolted to the beam and welded to the column are tabulated in Table 5.5. The notations for major parameters of these types of connections as shown in Figures 5.2 and 5.3 are identified as:

l_p Angle height
l_u Distance from the top edge of the angle to the tension flange of the beam
l_l Distance from the bottom edge of the angle to the compression flange of the beam
t_a Angle thickness
g_b Distance from the angle's heel to the center of the fastener holes in the leg adjacent to the beam web face
g_c Distance from the angle's heel to the center of the fastener holes in the leg adjacent to the column face
c_u Distance from the center of the upper fastener holes to the top edge of the angle in the beam web
c_l Distance from the center of the lower fastener holes to the bottom edge of the angle in the beam web
p_b Distance between two rows of fasteners in the beam web
p_c Distance between two rows of fasteners in the column
q_b Distance between two lines of fasteners in the beam web
q_c Distance between two lines of fasteners in the column
n_b Total number of fasteners in the beam
n_c Total number of fasteners in the column

TABLE 5.4 Collected test data for single web-angle bolted beam-to-column connections

No.	Author	Test id.	Angle	l_p (in)	g_c (in)	g_b (in)	p_c (in)	d_b (in)	n_c	n_b
1	S. L. Lipson (1968)	AA-2/1	4 × 3.5 × 1/4	5.5	2.56	2.25	3.0	3/4	1 × 2	1 × 2
2		AA-2/2	4 × 3.5 × 1/4	5.5	2.56	2.25	3.0	3/4	1 × 2	1 × 2
3		AA-3/1	4 × 3.5 × 1/4	8.5	2.56	2.25	3.0	3/4	1 × 3	1 × 3
4		AA-3/2	4 × 3.5 × 1/4	8.5	2.56	2.25	3.0	3/4	1 × 3	1 × 3
5		AA-4/1	4 × 3.5 × 1/4	11.5	2.56	2.25	3.0	3/4	1 × 4	1 × 4
6		AA-4/2	4 × 3.5 × 1/4	11.5	2.56	2.25	3.0	3/4	1 × 4	1 × 4
7		AA-5/1	4 × 3.5 × 1/4	14.5	2.56	2.25	3.0	3/4	1 × 5	1 × 5
8		AA-5/2	4 × 3.5 × 1/4	14.5	2.56	2.25	3.0	3/4	1 × 5	1 × 5
9		AA-6/1	4 × 3.5 × 1/4	17.5	2.56	2.25	3.0	3/4	1 × 6	1 × 6
10		AA-6/2	4 × 3.5 × 1/4	17.5	2.56	2.25	3.0	3/4	1 × 6	1 × 6
11		BB-4/1	4 × 3.5 × 5/16	11.5	2.56	2.25	3.0	3/4	1 × 4	1 × 4
12		BB-4/2	4 × 3.5 × 5/16	11.5	2.56	2.25	3.0	3/4	1 × 4	1 × 4
13		B2-4	4 × 3.5 × 5/16	11.5	2.56	2.25	3.0	3/4	1 × 4	1 × 4
14		B3-4	4 × 3.5 × 5/16	11.5	2.56	2.25	3.0	3/4	1 × 4	1 × 4
15		C-4	5 × 3.5 × 5/16	11.5	1.94	2.25	3.0	3/4	1 × 4	1 × 4
16		D-4	5 × 3.5 × 5/16	11.5	1.94	3.25	3.0	3/4	1 × 4	1 × 4

TABLE 5.5 Collected test data for single web-angle bolted beam and welded to column connections

No.	Author	Test id.	Angle	l_p (in)	g_b (in)	d_b (in)	n_b
1	S. L. Lipson (1968)	P1-2/1	—	5.5	2.50	3/4	1 × 2
2		P1-2/2	—	5.5	2.50	3/4	1 × 2
3		P1-3/1	—	8.5	2.50	3/4	1 × 3
4		P1-3/2	—	8.5	2.50	3/4	1 × 3
5		P1-4/1	—	11.5	2.50	3/4	1 × 4
6		P1-4/2	—	11.5	2.50	3/4	1 × 4
7		P1-5/1	—	14.5	2.50	3/4	1 × 5
8		P1-5/2	—	14.5	2.50	3/4	1 × 5
9		P1-6/1	—	17.5	2.50	3/4	1 × 6
10		P1-6/2	—	17.5	2.50	3/4	1 × 6
11		G-2/1	4 × 3.5 × 5/16	4.75	2.50	3/4	1 × 2
12		G-3/1	4 × 3.5 × 5/16	8.0	2.50	3/4	1 × 3
13		G-4/1	4 × 3.5 × 5/16	12.0	2.50	3/4	1 × 4
14		G-5/1	4 × 3.5 × 5/16	14.5	2.50	3/4	1 × 5
15	S. L. Lipson et al. (1977)	A1-5	4 × 3 × 3/8	14.5	2.50	3/4	1 × 5
16		A1-6	4 × 3 × 3/8	17.5	2.50	3/4	1 × 6
17		A1-7	4 × 3 × 3/8	20.5	2.50	3/4	1 × 7
18		A1-8	4 × 3 × 3/8	23.5	2.50	3/4	1 × 8
19		A1-9	4 × 3 × 3/8	26.5	2.50	3/4	1 × 9
20		A1-10	4 × 3 × 3/8	29.5	2.50	3/4	1 × 10
21		A1-11	4 × 3 × 3/8	32.5	2.50	3/4	1 × 11
22		A1-12	4 × 3 × 3/8	32.5	2.50	3/4	1 × 12
23	L. E. Thompson et al. (1970)	F-1 alt	—	11.5	2.25	3/4	2 × —
24		F-1 alt	—	11.5	2.25	3/4	2 ×
25		G-1 alt	—	11.5	2.25	3/4	2 × —
26		G-1 alt	—	11.5	2.25	3/4	2 ×
27		H-1 alt	—	11.5	2.25	3/4	2 × —
28		H-1 alt	—	11.5	2.25	3/4	2 × —
29		F-2-1 alt	—	11.5	2.25	3/4	2 × —
30		F-2-1 alt	—	11.5	2.25	3/4	2 × —
31		F-2-2 alt	—	11.5	2.25	3/4	2 × —
32		F-2-2 alt	—	11.5	2.25	3/4	2 × —
33		F-2-3 alt		11.5	2.25	3/4	2 × —
34		F-2-3 alt	—	11.5	2.25	3/4	2 × —
35	R. M. Richard et al. (1982)	RGKL2	—	6.0	—	3/4	1 × 2
36		RGKL2	—	9.0	—	3/4	1 × 3
37		RGKL2	—	15.0	—	3/4	1 × 5
38		RGKL2	—	21.5	—	7/8	1 × 7

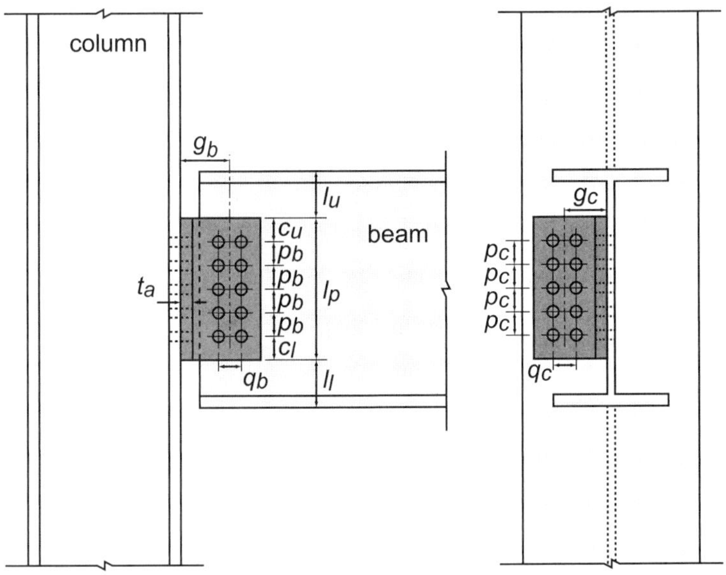

FIGURE 5.2 Size parameters for single web-angle connection.

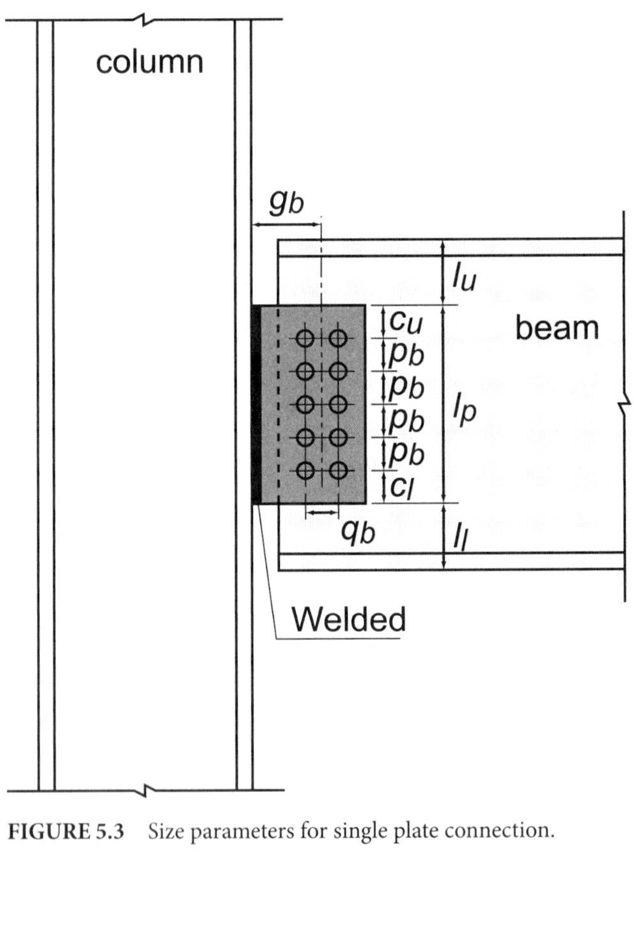

FIGURE 5.3 Size parameters for single plate connection.

5.2.2 Double Web-angle Connections

A total of 75 experimental data have been collected here. Collected experimental moment-rotation data for double web-angle connections are tabulated in Table 5.6. The notations for major parameters of this connection type as shown in Figure 5.4 are identified as:

- l_p Angle height
- l_u Distance from the top edge of the angle to the tension flange of the beam
- l_l Distance from the bottom edge of the angle to the compression flange of the beam
- t_a Angle thickness
- g_b Distance from the angle's heel to the center of the fastener holes in the leg adjacent to the beam web face
- g_c Distance from the angle's heel to the center of the fastener holes in the leg adjacent to the column face
- c_u Distance from the center of the upper fastener holes to the top edge of the angle in the beam web
- c_l Distance from the center of the lower fastener holes to the bottom edge of the angle in the beam web
- p_b Distance between two rows of fasteners in the beam web
- p_c Distance between two rows of fasteners in the column
- q_b Distance between two lines of fasteners in the beam web
- q_c Distance between two lines of fasteners in the column
- n_b Total number of fasteners in the beam
- n_c Total number of fasteners in the column

TABLE 5.6 Collected test data for double web-angle connections

No.	Author	Test id.	Angle	l_p (in)	g_c (in)	g_b (in)	p_c (in)	d_b	n_c	n_b
1	J. C. Rathbun (1936)	A-1	6 × 4 × 3/8	2.50	2.63	3.50	—	7/8″	1 × 1	2 × 1
2		A-2	6 × 4 × 3/8	6.00	2.63	3.50	3.00	7/8″	1 × 2	2 × 2
3		A-3	6 × 6 × 3/8	6.00	3.63	3.50	3.00	7/8″	2 × 2	2 × 2
4		A-4	4 × 3 1/2 × 3/8	9.00	2.56	2.25	3.00	7/8″	1 × 3	1 × 3
5		A-5	6 × 6 × 3/8	9.00	3.56	3.50	3.00	7/8″	2 × 3	2 × 3
6		A-6	4 × 3 1/2 × 3/8	15.00	2.50	2.25	3.00	7/8″	1 × 5	1 × 5
7		A-7	6 × 6 × 3/8	15.00	3.50	3.50	3.00	7/8″	2 × 5	2 × 5
8	W. G. Bell et al. (1958)	FK-4A	6 × 4 × 3/8	11.50	2.57	4.50	3.00	3/4″	1 × 4	2 × 4
9		FK-4B	6 × 4 × 3/8	11.50	2.57	4.50	3.00	3/4″	1 × 4	2 × 4
10		FK-4C	6 × 4 × 3/8	11.50	2.57	4.50	3.00	3/4″	1 × 4	2 × 4
11		FK-4R	6 × 4 × 3/8	11.50	2.57	4.50	3.00	3/4″	1 × 4	2 × 4
12	C. W. Lewitt et al. (1966)	FK-3	6 × 4 × 3/8	8.50	2.63	3.50	3.00	3/4″	1 × 3	2 × 3
13		FK-4AB-M	6 × 4 × 3/8	11.50	2.07	3.50	3.00	3/4″	2 × 4	2 × 4
14		FK-4P	6 × 4 × 3/8	11.50	2.07	3.50	3.00	3/4″	2 × 4	2 × 4
15		WK-4	6 × 4 × 3/8	11.50	2.57	3.50	3.00	3/4″	1 × 4	2 × 4
16		FB-4	6 × 3.5 × 3/8	11.50	2.57	2.25	3.00	3/4″	1 × 4	1 × 4
17		FK-5	6 × 4 × 7/16	14.50	2.55	3.50	3.00	3/4″	1 × 5	2 × 5

(Continues)

TABLE 5.6 (Continued)

No.	Author	Test id.	Angle	l_p (in)	g_c (in)	g_b (in)	p_c (in)	d_b	n_c	n_b
18	B. Bose (1981)	B-1	150 × 90 × 15	15.75	2.38	3.54	2.95	M20	1 × 5	2 × –
19	L. E. Thompson et al. (1970)	A1-1 ALT	4 × 3.5 × 5/16	11.50	2.55	2.25	—	3/4″	– × –	– × 2
20		A1-1 ALT	4 × 3.5 × 5/16	11.50	2.55	2.25	—	3/4″	– × –	– × 2
21		A1-2 ALT	4 × 3.5 × 5/16	11.50	2.55	2.25	—	3/4″	– × –	– × 2
22		A1-2 ALT	4 × 3.5 × 5/16	11.50	2.55	2.25	—	3/4″	– × –	– × 2
23		A1-3 ALT	4 × 3.5 × 5/16	11.50	2.55	2.25	—	3/4″	– × –	– × 2
24		A1-3 ALT	4 × 3.5 × 5/16	11.50	2.55	2.25	—	3/4″	– × –	– × 2
25		B1-1 ALT	4 × 3.5 × 1/2	11.50	2.55	2.25	—	3/4″	– × –	– × 2
26		B1-1 ALT	4 × 3.5 × 1/2	11.50	2.55	2.25	—	3/4″	– × –	– × 2
27		B1-2 ALT	4 × 3.5 × 1/2	11.50	2.55	2.25	—	3/4″	– × –	– × 2
28		B1-2 ALT	4 × 3.5 × 1/2	11.50	2.55	2.25	—	3/4″	– × –	– × 2
29		B1-3 ALT	4 × 3.5 × 1/2	11.50	2.55	2.25	—	3/4″	– × –	– × 2
30		B1-3 ALT	4 × 3.5 × 1/2	11.50	2.55	2.25	—	3/4″	– × –	– × 2
31		D1-1 ALT	4 × 3.5 × 5/16	11.50	1.80	2.25	—	3/4″	– × –	– × 2
32		D1-1 ALT	4 × 3.5 × 5/16	11.50	1.80	2.25	—	3/4″	– × –	– × 2
33		D1-2 ALT	4 × 3.5 × 5/16	11.50	1.80	2.25	—	3/4″	– × –	– × 2
34		D1-2 ALT	4 × 3.5 × 5/16	11.50	1.80	2.25	—	3/4″	– × –	– × 2
35		D1-3 ALT	4 × 3.5 × 5/16	11.50	1.80	2.25	—	3/4″	– × –	– × 2
36		D1-3 ALT	4 × 3.5 × 5/16	11.50	1.80	2.25	—	3/4″	– × –	– × 2
37		E1-1 ALT	4 × 3.5 × 1/2	11.50	1.80	2.25	—	3/4″	– × –	– × 2
38		E1-1 ALT	4 × 3.5 × 1/2	11.50	1.80	2.25	—	3/4″	– × –	– × 2
39		E1-2 ALT	4 × 3.5 × 1/2	11.50	1.80	2.25	—	3/4″	– × –	– × 2
40		E1-2 ALT	4 × 3.5 × 1/2	11.50	1.80	2.25	—	3/4″	– × –	– × 2
41		E1-3 ALT	4 × 3.5 × 1/2	11.50	1.80	2.25	—	3/4″	– × –	– × 2
42		E1-3 ALT	4 × 3.5 × 1/2	11.50	1.80	2.25	—	3/4″	– × –	– × 2
43		A2-1 ALT	4 × 3.5 × 5/16	11.50	2.55	2.25	—	3/4″	– × –	– × 2
44		A2-1 ALT	4 × 3.5 × 5/16	11.50	2.55	2.25	—	3/4″	– × –	– × 2
45		A2-2 ALT	4 × 3.5 × 5/16	11.50	2.55	2.25	—	3/4″	– × –	– × 2
46		A2-2 ALT	4 × 3.5 × 5/16	11.50	2.55	2.25	—	3/4″	– × –	– × 2
47		A2-3 ALT	4 × 3.5 × 5/16	11.50	2.55	2.25	—	3/4″	– × –	– × 2
48		A2-3 ALT	4 × 3.5 × 5/16	11.50	2.55	2.25	—	3/4″	– × –	– × 2
49		B2-1 ALT	4 × 3.5 × 5/16	11.50	1.80	2.25	—	3/4″	– × –	– × 2
50		B2-1 ALT	4 × 3.5 × 5/16	11.50	1.80	2.25	—	3/4″	– × –	– × 2
51		B2-2 ALT	4 × 3.5 × 5/16	11.50	1.80	2.25	—	3/4″	– × –	– × 2
52		B2-2 ALT	4 × 3.5 × 5/16	11.50	1.80	2.25	—	3/4″	– × –	– × 2
53		B2-3 ALT	4 × 3.5 × 5/16	11.50	1.80	2.25	—	3/4″	– × –	– × 2
54		B2-3 ALT	4 × 3.5 × 5/16	11.50	1.80	2.25	—	3/4″	– × –	– × 2
55		C2-1 ALT	4 × 3.5 × 1/2	11.50	2.55	2.25	—	3/4″	– × –	– × 2

(Continues)

TABLE 5.6 (Continued)

No.	Author	Test id.	Angle	l_p (in)	g_c (in)	g_b (in)	p_c (in)	d_b	n_c	n_b
56	L. E. Thompson et al. (1970) (Continued)	C2-1 ALT	4 × 3.5 × 1/2	11.50	2.55	2.25	—	3/4″	– × –	– × 2
57		C2-2 ALT	4 × 3.5 × 1/2	11.50	2.55	2.25	—	3/4″	– × –	– × 2
58		C2-2 ALT	4 × 3.5 × 1/2	11.50	2.55	2.25	—	3/4″	– × –	– × 2
59		C2-3 ALT	4 × 3.5 × 1/2	11.50	2.55	2.25	—	3/4″	– × –	– × 2
60		C2-3 ALT	4 × 3.5 × 1/2	11.50	2.55	2.25	—	3/4″	– × –	– × 2
61		D2-1 ALT	4 × 3.5 × 1/2	11.50	1.80	2.25	—	3/4″	– × –	– × 2
62		D2-1 ALT	4 × 3.5 × 1/2	11.50	1.80	2.25	—	3/4″	– × –	– × 2
63		D2-2 ALT	4 × 3.5 × 1/2	11.50	1.80	2.25	—	3/4″	– × –	– × 2
64		D2-2 ALT	4 × 3.5 × 1/2	11.50	1.80	2.25	—	3/4″	– × –	– × 2
65		D2-3 ALT	4 × 3.5 × 1/2	11.50	1.80	2.25	—	3/4″	– × –	– × 2
66		D2-3 ALT	4 × 3.5 × 1/2	11.50	1.80	2.25	—	3/4″	– × –	– × 2
67	W. H. Sommer (1969)	TEST21	3.5 × 3 × 3/8	9.00	1.83		3.00	3/4″	1 × 3	
68		TEST22	3.5 × 3 × 3/8	12.00	1.83		3.00	3/4″	1 × 4	
69		TEST23	4 × 3 × 3/8	15.00	2.53		3.00	3/4″	1 × 5	
70		TEST24	4 × 3 × 3/8	18.00	2.53		3.00	3/4″	1 × 6	
71	J. B. Davison et al. (1987)	JT/01B	RSA 80 × 60 × 8	6.30	1.38	2.17	3.94	M16	2 × 2	1 × 2
72		JT/06	RSA 80 × 60 × 8	6.30	1.38	2.17	3.94	M16	2 × 2	1 × 2
73	J. G. Yang and G. Y. Lee (2007)	3B-L-7-65	L125 × 75 × 7	8.27	2.56	—	2.76	M20	1 × 3	
74		4B-L-7-65	L125 × 75 × 7	11.02	2.56	—	2.76	M20	1 × 4	
75		5B-L-7-65	L125 × 75 × 7	13.78	2.56	—	2.76	M20	1 × 4	

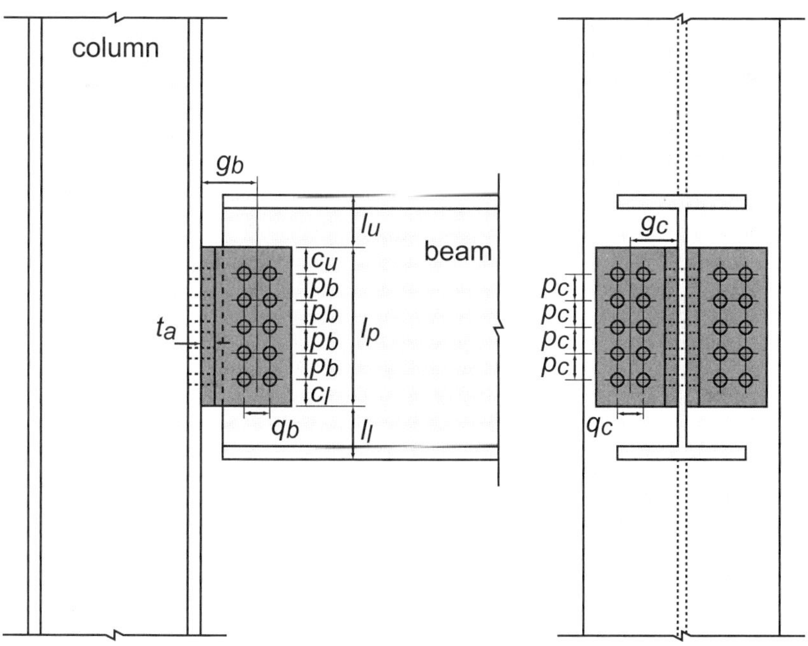

FIGURE 5.4 Size parameters for double web-angle connection.

5.2.3 Top- and Seat-angle with Double Web-angle Connections

The experimental studies on this connection type were mainly performed by Altman et al. (1982) and Azizinamini et al. (1985) at University of South Carolina. The test program included specimens that were subjected to both cyclic loading and static loading. A total of 33 experimental data has been collected. However, cyclic loading test data are excluded from the database, as listed in Table 5.7. The notations for major parameters of this connection type as shown in Figure 5.5 are as follows:

On the web angle:

- l_p Angle height
- l_u Distance from the top edge of the angle to the tension flange of the beam
- l_l Distance from the bottom edge of the angle to the compression flange of the beam
- t_a Web-angle thickness
- g_b Distance from the angle's heel to the center of the fastener holes in the leg adjacent to the beam web face
- g_c Distance from the angle's heel to the center of the fastener holes in the leg adjacent to the column face
- c_u Distance from the center of the upper fastener holes to the top edge of the angle in the beam web
- c_l Distance from the center of the lower fastener holes to the bottom edge of the angle in the beam web
- p_b Distance between two rows of fasteners in the beam web
- p_c Distance between two rows of fasteners in the column
- q_b Distance between two lines of fasteners in the beam web
- q_c Distance between two lines of fasteners in the column
- n_b Total number of fasteners in the beam
- n_c Total number of fasteners in the column

On the top and seat angles:

- l_t Width of the top angle along the column
- t_t Top-angle thickness
- l_s Width of the seat angle along the column
- t_s Seat-angle thickness
- g_t Gage from the top angle's heel to the center of the fastener holes in the leg seating on the tension-beam flange
- g_t' Gage from the top angle's heel to the center of the fastener holes in the leg adjacent to the column face
- g_s Gage from the seat angle's heel to the center of the fastener holes in the leg under the compression-beam flange
- g_s' Gage from the seat angle's heel to the center of the fastener holes in the leg adjacent to the column face
- q_t Distance between two inner lines of fasteners in the leg seating on the tension-beam flange
- q_{t2} Distance between the outer and inner lines of fasteners in the leg seating on the tension-beam flange where the number of fastener lines exceeds three
- r_t Distance between two inner lines of fasteners in the leg adjacent to the column face in the tension side
- r_{t2} Distance between the outer and inner lines of fasteners in the leg adjacent to the column face in the tension side where the number of fastener lines exceeds three

TABLE 5.7 Collected test data for top- and seat-angle with double web-angle connections

No	Author	Test id.	Beam	Column	F. Angle	W. Angle
1	A. Azizinamini et al. (1985)	8S1	W8 × 21	W12 × 58	6 × 3.5 × 5/16 × 6.0	4 × 3.5 × 1/4 × 5.5
2		8S2	W8 × 21	W12 × 58	6 × 3.5 × 3/8 × 6.0	4 × 3.5 × 1/4 × 5.5
3		8S3	W8 × 21	W12 × 58	6 × 3.5 × 5/16 × 8.0	4 × 3.5 × 1/4 × 5.5
4		8S4	W8 × 21	W12 × 58	6 × 6.0 × 3/8 × 6.0	4 × 3.5 × 1/4 × 5.5
5		8S5	W8 × 21	W12 × 58	6 × 4.0 × 3/8 × 8.0	4 × 3.5 × 1/4 × 5.5
6		8S6	W8 × 21	W12 × 58	6 × 4.0 × 5/16 × 6.0	4 × 3.5 × 1/4 × 5.5
7		8S7	W8 × 21	W12 × 58	6 × 4.0 × 3/8 × 6.0	4 × 3.5 × 1/4 × 5.5
8		8S8	W8 × 21	W12 × 58	6 × 3.5 × 5/16 × 6.0	4 × 3.5 × 1/4 × 5.5
9		8S9	W8 × 21	W12 × 58	6 × 3.5 × 3/8 × 6.0	4 × 3.5 × 1/4 × 5.5
10		8S10	W8 × 21	W12 × 58	6 × 3.5 × 1/2 × 6.0	4 × 3.5 × 1/4 × 5.5
11		14S1	W14 × 38	W12 × 96	6 × 4.0 × 3/8 × 8.0	4 × 3.5 × 1/4 × 8.5
12		14S2	W14 × 38	W12 × 96	6 × 4.0 × 1/2 × 8.0	4 × 3.5 × 1/4 × 8.5
13		14S3	W14 × 38	W12 × 96	6 × 4.0 × 3/8 × 8.0	4 × 3.5 × 1/4 × 5.5
14		14S4	W14 × 38	W12 × 96	6 × 4.0 × 3/8 × 8.0	4 × 3.5 × 3/8 × 8.5
15		14S5	W14 × 38	W12 × 96	6 × 4.0 × 3/8 × 8.0	4 × 3.5 × 1/4 × 8.5
16		14S6	W14 × 38	W12 × 96	6 × 4.0 × 1/2 × 8.0	4 × 3.5 × 1/4 × 8.5
17		14S8	W14 × 38	W12 × 96	6 × 4.0 × 5/8 × 8.0	4 × 3.5 × 1/4 × 8.5
18		14S9	W14 × 38	W12 × 96	6 × 4.0 × 1/2 × 8.0	4 × 3.5 × 1/4 × 8.5
19		14WS1	W14 × 38	W12 × 96	6 × 4.0 × 3/8 × 8.0	4 × 3.5 × 1/4 × 8.5
20		14WS2	W14 × 38	W12 × 96	6 × 4.0 × 1/2 × 8.0	4 × 3.5 × 1/4 × 8.5
21	J.C.Rathbun (1936)	B-11	W12 × 31.8	—	6 × 6.0 × 3/8 × 9 (top)	4 × 3.5 × 3/8 × 9
22		B-12	W12 × 31.8	—	6 × 6.0 × 3/8 × 14 (top)	4 × 3.5 × 3/8 × 9
23	C.W. Roeder et al. (1996)	L1	W14 × 22	W10 × 68	L5 × 3.5 × 5/8	L3.5 × 3 × 5/16

(Continues)

TABLE 5.7 (Continued)

No	Author	Test id.	Beam	Column	F. Angle	W. Angle
24	A.S. Elnashai et al. (1998)	SRB01	H250 × 130 × 9 × 9	H150 × 150 × 7 × 10	L75 × 75 × 9	L75 × 75 × 8
25	Z. Fu et al. (1998)	LM-P	H250 × 125 × 6 × 9	—	L75 × 75 × 9 × 8.5	L75 × 75 × 9 × 8.5
26		LM-T	H250 × 125 × 6 × 9	—	L75 × 75 × 9 × 8.5	L75 × 75 × 9 × 8.5
27		LM-P15	H250 × 125 × 6 × 9	—	L75 × 75 × 9 × 8.5	L75 × 75 × 9 × 8.5
28		LM-T15	H250 × 125 × 6 × 9	—	L75 × 75 × 9 × 8.5	L75 × 75 × 9 × 8.5
29	L.Calado et al. (2000)	BCC7-M	IPE300	HEB200	L120 × 120 × 10	L120 × 80 × 10
30		BCC9-M	IPE300	HEB160	L120 × 120 × 10	L120 × 80 × 10
31		BCC10-M	IPE300	HEB240	L120 × 120 × 10	L120 × 80 × 10
32	M. Komuro et al. (2002)	W18-m	H400 × 200 × 8 × 13	—	L150 × 100 × 12	L90 × 90 × 7
33		W29-m	H400 × 200 × 8 × 13	—	L150 × 100 × 12	L90 × 90 × 7

q_s Distance between two inner lines of fasteners in the leg under the compression-beam flange
q_{s2} Distance between the outer lines and the inner lines of fasteners in the leg under the compression-beam flange where the number of fastener lines exceeds three
r_s Distance between two inner lines of fasteners in the leg adjacent to the column face in the compression side
r_{s2} Distance between the outer and inner lines of fasteners in the leg adjacent to the column face in the compression side where the number of fastener lines exceeds three
p_t Distance between two rows of fasteners in the leg seating on the tension-beam flange
p_t' Distance between two rows of fasteners in the leg adjacent to the column face in the tension side
p_s Distance between two rows of fasteners in the leg under the compression-beam flange
p_s' Distance between two rows of fasteners in the leg adjacent to the column face in the compression side
n_t Total number of fasteners in the tension-beam flange
n_t' Total number of fasteners in the leg adjacent to the column face in the tension side
n_s Total number of fasteners in the compression-beam flange
n_s' Total number of fasteners in the leg adjacent to the column face in the compression side

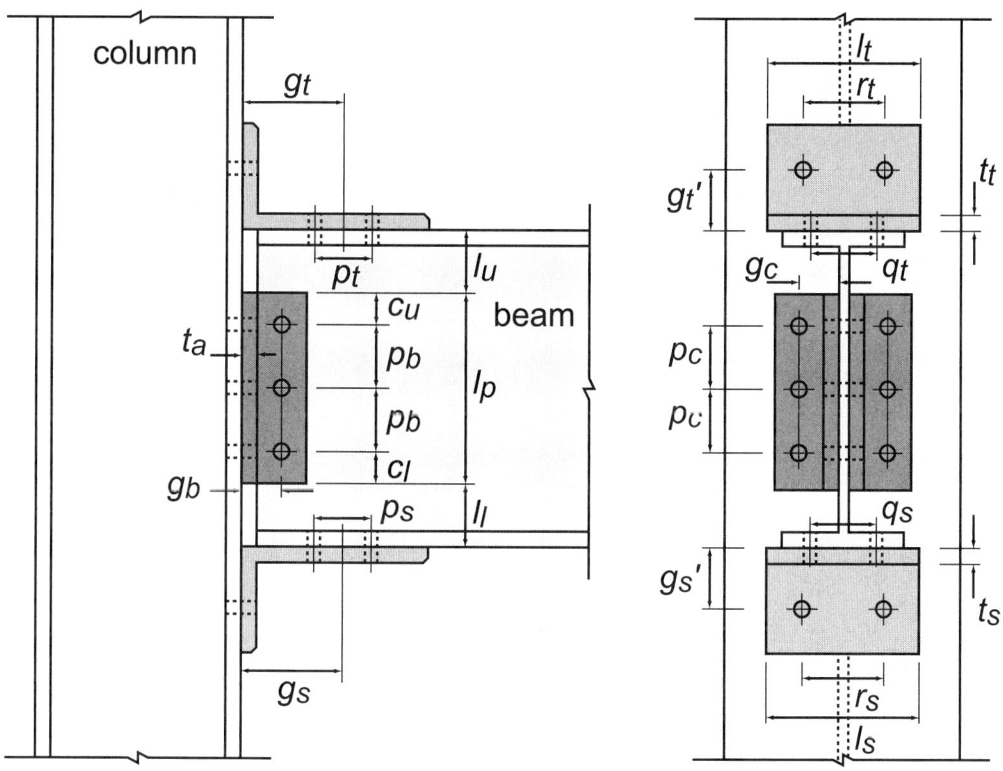

FIGURE 5.5 Size parameters for top- and seat-angle connection with double web-angle.

5.2.4 Top- and Seat-angle Connections

The collected experimental data for this connection type are listed in Table 5.8. A total of 74 experimental data have been collected. The notations for major parameters of this connection type as shown in Figure 5.6 are:

- l_t Width of the top angle along the column
- t_t Top-angle thickness
- l_s Width of the seat angle along the column
- t_s Seat-angle thickness
- g_t Gage from the top angle's heel to the center of the fastener holes in the leg seating on the tension-beam flange
- g_t' Gage from the top angle's heel to the center of the fastener holes in the leg adjacent to the column face
- g_s Gage from the seat angle's heel to the center of the fastener holes in the leg under the compression-beam flange
- g_s' Gage from the seat angle's heel to the center of the fastener holes in the leg adjacent to the column
- q_t Distance between two inner lines of fasteners in the leg seating on the tension-beam flange
- q_{t2} Distance between the outer and inner lines of fasteners in the leg seating on the tension-beam flange where the number of fastener lines exceeds three
- r_t Distance between two inner lines of fasteners in the leg adjacent to the column face in the tension side
- r_{t2} Distance between the outer and inner lines of fasteners in the leg adjacent to the column face in the tension side where the number of fastener lines exceeds three
- q_s Distance between two inner lines of fasteners in the leg under the compression-beam flange
- q_{s2} Distance between the outer lines and the inner lines of fasteners in the leg under the compression-beam flange where the number of fastener lines exceeds three
- r_s Distance between two inner lines of fasteners in the leg adjacent to the column face in the compression side
- r_{s2} Distance between the outer and inner lines of fasteners in the leg adjacent to the column face in the compression side where the number of fastener lines exceeds three
- p_t Distance between two rows of fasteners in the leg seating on the tension-beam flange
- p_t' Distance between two rows of fasteners in the leg adjacent to the column face in the tension side
- p_s Distance between two rows of fasteners in the leg under the compression-beam flange
- p_s' Distance between two rows of fasteners in the leg adjacent to the column face in the compression side
- n_t Total number of fasteners in the tension-beam flange
- n_t' Total number of fasteners in the leg adjacent to the column face in the tension side
- n_s Total number of fasteners in the compression-beam flange
- n_s' Total number of fasteners in leg adjacent to the column face in the compression side

5.2.5 Extended End-plate Connections

The type of extended end-plate connections includes both cases of extension on the tension and compression sides and extension on the tension side only. Collected experimental moment-rotation

TABLE 5.8 Collected test data for top- and seat-angle connections

No.	Author	Test id.	Beam	Column	Angle	l_t (in)	g_t (in)	t_t (in)
1	J. C. Rathbun (1936)	B-8	W12 × 1.8	—	6 × 6 × 3/8 × 6 (top)	6.00	3.50	0.375
2		B-9	W12 × 31.8	—	6 × 6 × 3/8 × 8 (top)	8.00	3.50	0.375
3		B-10	W12 × 31.8	—	6 × 6 × 3/8 × 14 (top)	14.00	3.50	0.375
4	R. A. Hechtman and B. G. Johnston (1947)	NO 2	W12 × 25	W10 × 49	6 × 4 × 5/8 × 6 3/4 (top)	6.75	3.50	0.625
5		NO 5	W18 × 47	W12 × 65	6 × 4 × 1/2 × 12 (top)	12.00	3.50	0.500
6		NO 9	W18 × 47	W12 × 65	6 × 4 × 5/8 × 12 (top)	12.00	3.50	0.625
7		NO 10	W18 × 47	W12 × 65	6 × 4 × 3/4 × 12 (top)	12.00	3.50	0.750
8		NO 11	W18 × 47	W14 × 58	6 × 4 × 1/2 × 10 (top)	10.00	3.50	0.500
9		NO 16	W12 × 25	W10 × 49	6 × 4 × 1/2 × 6 3/4 (top)	6.75	3.50	0.500
10		NO 17	W12 × 25	W10 × 49	6 × 4 × 1/2 × 6 3/4 (top)	6.75	3.50	0.500
11		NO 18	W12 × 50	W10 × 49	6 × 4 × 1/2 × 8 (top)	8.00	3.50	0.500
12		NO 20	W14 × 34	W12 × 65	6 × 4 × 5/8 × 12 (top)	12.00	3.50	0.625
13		NO 22	W16 × 40	W12 × 65	6 × 4 × 5/8 × 12 (top)	12.00	3.50	0.625
14		NO 23	W18 × 40	W14 × 58	6 × 4 × 5/8 × 10 (top)	10.00	3.50	0.625
15		NO 24	W18 × 47	W12 × 65	6 × 4 × 5/8 × 12 1/2 (top)	12.50	3.50	0.625
16	M. J. Marley (1982)	A-1/4-1	W5 × 16	W5 × 16	4 × 4 × 1/4 × 5.0	5.00	—	0.250
17		A-1/4-2	W5 × 16	W5 × 16	4 × 4 × 1/4 × 5.0	5.00	—	0.250
18		B-1/4-1	W5 × 16	W5 × 16	4 × 4 × 1/4 × 5.0	5.00	—	0.250
19		B-1/4-2	W5 × 16	W5 × 16	4 × 4 × 1/4 × 5.0	5.00	—	0.250
20		C1-1/4-2	W5 × 16	W5 × 16	4 × 4 × 1/4 × 5.0	5.00	—	0.250
21		C2-1/4-1	W5 × 16	W5 × 16	4 × 4 × 1/4 × 5.0	5.00	—	0.250
22		C2-1/4-2	W5 × 16	W5 × 16	4 × 4 × 1/4 × 5.0	5.00	—	0.250
23		D1-1/4-1	W5 × 16	W5 × 16	4 × 4 × 1/4 × 5.0	5.00	—	0.250
24		D1-1/4-2	W5 × 16	W5 × 16	4 × 4 × 1/4 × 5.0	5.00	—	0.250
25		D2-1/4-1	W5 × 16	W5 × 16	4 × 4 × 1/4 × 5.0	5.00	—	0.250

(Continues)

TABLE 5.8 (Continued)

No.	Author	Test id.	Beam	Column	Angle	l_t (in)	g_t (in)	t_t (in)
26	M. J. Marley (1982) (Continued)	D2-1/4-2	W5 × 16	W5 × 16	4 × 4 × 1/4 × 5.0	5.00	—	0.250
27		D3-1/4-1	W5 × 16	W5 × 16	4 × 4 × 1/4 × 5.0	5.00	—	0.250
28		D3-1/4-2	W5 × 16	W5 × 16	4 × 4 × 1/4 × 5.0	5.00	—	0.250
29		A-1/2-1	W5 × 16	W5 × 16	4 × 4 × 1/2 × 5.0	5.00	—	0.500
30		B-1/2-1	W5 × 16	W5 × 16	4 × 4 × 1/2 × 5.0	5.00	—	0.500
31		B-1/2-2	W5 × 16	W5 × 16	4 × 4 × 1/2 × 5.0	5.00	—	0.500
32		C1-1/2-1	W5 × 16	W5 × 16	4 × 4 × 1/2 × 5.0	5.00	—	0.500
33		C1-1/2-2	W5 × 16	W5 × 16	4 × 4 × 1/2 × 5.0	5.00	—	0.500
34		C2-1/2-1	W5 × 16	W5 × 16	4 × 4 × 1/2 × 5.0	5.00	—	0.500
35		C2-1/2-2	W5 × 16	W5 × 16	4 × 4 × 1/2 × 5.0	5.00	—	0.500
36		D1-1/2-1	W5 × 16	W5 × 16	4 × 4 × 1/2 × 5.0	5.00	—	0.500
37		D1-1/2-2	W5 × 16	W5 × 16	4 × 4 × 1/2 × 5.0	5.00	—	0.500
38		D2-1/2-1	W5 × 16	W5 × 16	4 × 4 × 1/2 × 5.0	5.00	—	0.500
39		D2-1/2-2	W5 × 16	W5 × 16	4 × 4 × 1/2 × 5.0	5.00	—	0.500
40		D3-1/2-1	W5 × 16	W5 × 16	4 × 4 × 1/2 × 5.0	5.00	—	0.500
41		D3-1/2-2	W5 × 16	W5 × 16	4 × 4 × 1/2 × 5.0	5.00	—	0.500
42	S. M. Maxwell et al. (1981)	A1	457 × 191 × UB	305 × 305 × UC	150 × 90 × 10 × 150	5.91	—	0.394
43		A2	457 × 191 × UB	305 × 305 × UC	150 × 90 × 10 × 200	7.87	—	0.394
44		A3	457 × 191 × UB	305 × 305 × UC	150 × 90 × 10 × 150	5.91	—	0.394
45		A4	457 × 191 × UB	305 × 305 × UC	150 × 90 × 10 × 200	7.87	—	0.394
46		B1	457 × 191 × UB	305 × 305 × UC	150 × 90 × 12 × 150	5.91	—	0.472
47		B2	457 × 191 × UB	305 × 305 × UC	150 × 90 × 12 × 200	7.87	—	0.472
48		B3	457 × 191 × UB	305 × 305 × UC	150 × 90 × 12 × 150	5.91	—	0.472
49		B4	457 × 191 × UB	305 × 305 × UC	150 × 90 × 12 × 200	7.87	—	0.472
50		C1	457 × 191 × UB	305 × 305 × UC	150 × 90 × 15 × 150	5.91	—	0.591

51		C2	457 × 191 × UB	305 × 305 × UC	150 × 90 × 15 × 200	7.87	—	0.591
52		C3	457 × 191 × UB	305 × 305 × UC	150 × 90 × 15 × 150	5.91	—	0.591
53		C4	457 × 191 × UB	305 × 305 × UC	150 × 90 × 15 × 200	7.87	—	0.591
54	J. B. Davison et al. (1987)	JT/08	UB 254 × 102 × 22	UC152 × 152 × 23	RSA 80 × 60 × 8 (top)	4.76	2.17	0.315
55	A. Azizinamini (1985)	—	W14 × 38	W12 × 96	6 × 4 × 3/8	8.00	3.00	0.375
56		—	W14 × 38	W12 × 96	6 × 4 × 1/2	8.00	3.00	0.500
57	W. L. Harper, Jr. (1990)	TEST3	W8 × 21	W8 × 24	6 × 3.5 × 3/8	6.00	3.00	0.375
58	J. B. Mander et al. (1994)	R1_01	W8 × 21	W8 × 31	6 × 4 × 3/8	6.50	3.63	0.375
59		R1_05	W8 × 21	W8 × 31	6 × 4 × 3/8	6.50	3.63	0.375
60		R1_06	W8 × 21	W8 × 31	6 × 4 × 3/8	6.50	3.63	0.375
61		R1_11	W8 × 21	W8 × 31	6 × 4 × 3/8	6.50	3.63	0.375
62	C. Bernuzzi et al. (1996)	TSC/M	IPE300	—	L120 × 120 × 12	7.09	2.36	0.465
63	N. Kubo et al. (1999)	No.1	H250 × 125 × 6 × 9	H250 × 250 × 9 × 14	L100 × 100 × 10	5.72	2.34	0.366
64		No.3	H250 × 125 × 6 × 9	H150 × 150 × 7 × 10	L100 × 100 × 10	5.71	2.38	0.362
65		No.5	H250 × 125 × 6 × 9	H250 × 250 × 9 × 14	L120 × 120 × 8	5.41	2.77	0.299
66		No.7	H250 × 125 × 6 × 9	H150 × 150 × 7 × 10	L120 × 120 × 8	5.41	2.77	0.307
67		No.9	H250 × 125 × 6 × 9	H175 × 175 × 7.5 × 11	L120 × 120 × 8	5.40	2.76	0.315
68	M. Komuro et al. (2002)	W00-m	H400 × 200 × 8 × 13	—	L150 × 100 × 12	7.87	3.25	0.472
69	Y. Sato et al. (2007)	G60	H400 × 200 × 8 × 13	—	L200 × 200 × 15	7.87	4.13	0.591
70		G105	H400 × 200 × 8 × 13	—	L200 × 200 × 15	7.87	4.13	0.591
71		G150	H400 × 200 × 8 × 13	—	L200 × 200 × 15	7.87	4.13	0.591
72		GW60	H400 × 200 × 8 × 13	—	L200 × 200 × 15	7.87	4.13	0.591
73		A40	H400 × 200 × 8 × 13	—	L200 × 100 × 15	7.87	4.13	0.591
74		A90	H400 × 200 × 8 × 13	—	L200 × 150 × 15	7.87	4.13	0.591

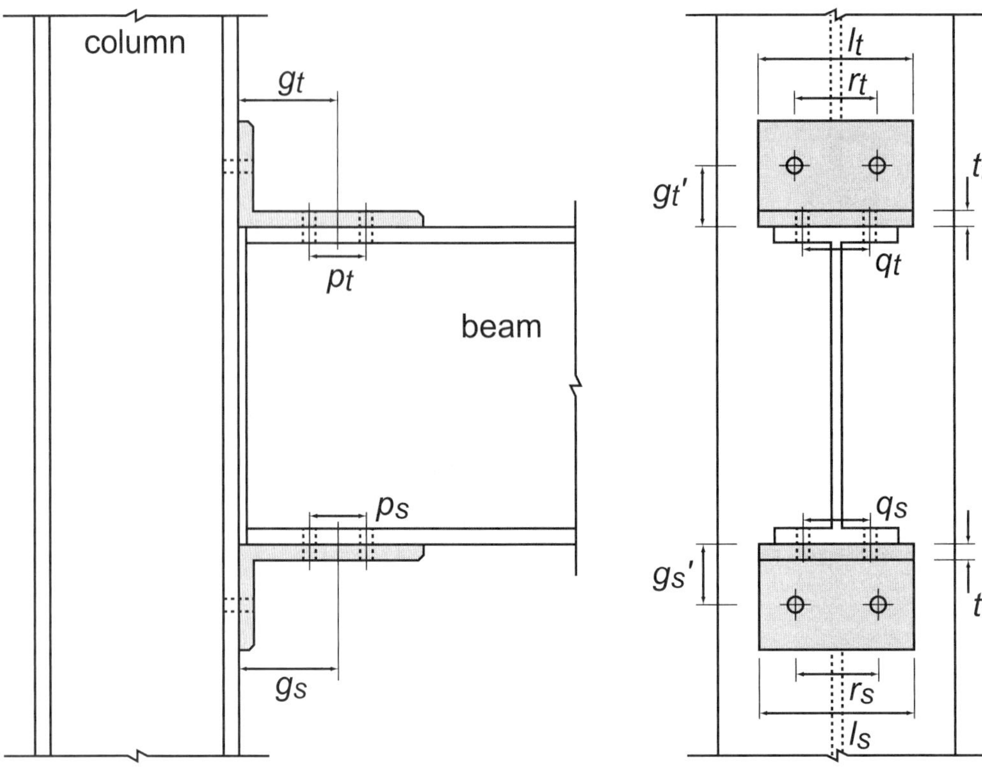

FIGURE 5.6 Size parameters for top- and seat-angle connection.

data for extended end-plate connections are tabulated in Table 5.9. This type of connection—apparently used quite often in construction—has been a subject of much experimental work. A total of 166 experimental moment-rotation data have been appended to the original connection database. The notations for major parameters of this connection type as shown in Figure 5.7 are identified as:

- c_t Distance from the tension-side edge of plate to the center of the outer tension-side fastener holes
- p_t Distance between two rows of tension-side fasteners
- l_i Distance from the center of the lower tension-side fastener holes to the center of the upper compression-side fasteners holes
- p_{it} Distance from the inner row of fasteners on the tension side to the upper row of central fasteners
- p_i Distance between two inner rows of central fasteners
- p_{ic} Distance from the inner row of fasteners on the compression side to the lower row of central fasteners
- g_t Distance between two inner lines of tension-side fasteners
- g_{t2} Distance from the outer line to the inner line of tension-side fasteners where the number of fastener lines exceeds three
- g_i Distance between two inner lines of central fasteners
- g_{i2} Distance from the outer line to the inner line of central fasteners where the number of fastener lines exceeds three
- g_c Distance between two inner lines of compression-side fasteners

TABLE 5.9 Collected test data for extended end-plate connections

No.	Author	Test id.	Beam	Column	l_p (in)	t_p (in)	g (in)	p_t (in)	d_b
1	S. A. Ioanrides (1978)	TEST 1	W14 × 22	W8 × 35	17.25	0.625	5.50	3.50	3/4″
2		TEST 2	W18 × 35	W10 × 49	21.50	0.750	5.50	3.75	7/8″
3		TEST 3	W24 × 55	W14 × 48	27.25	0.875	5.50	4.00	1″
4		TEST 4	W14 × 22	W8 × 35	17.25	0.875	5.50	3.50	3/4″
5		TEST 5	W18 × 35	W10 × 49	21.50	1.250	5.50	3.75	7/8″
6		TEST 6	W24 × 55	W14 × 48	27.25	1.250	5.50	4.00	1″
7	R. J. Dews (1979)	TEST 1	W14 × 22	W8 × 31	17.16	0.969	5.50	3.50	3/4″
8		TEST 2	W14 × 22	W8 × 31	17.27	0.938	5.53	3.50	3/4″
9		TEST 3	W16 × 26	W10 × 33	19.48	1.562	5.38	3.75	1″
10	A. N. Sherbourne (1961)	TEST A1	15 × 5 RSJ 42	8 × 8 UC 35	14.50	1.250	4.00	3.50	3/4″
11	J. R. Bailey (1970)	TEST A1-L	12 × 5 RSJ 32	8 × 8 UC 48	—	1.020	—	4.50	7/8″
12		TEST A1-R	12 × 5 RSJ 32	8 × 8 UC 48	—	1.020	—	4.50	7/8″
13		TEST A2-L	10 × 4 UB 17	8 × 8 UC 40	—	0.805	—	4.38	5/8″
14		TEST A2-R	10 × 4 UB 17	8 × 8 UC 40	—	0.805	—	4.38	5/8″
15		TEST A3-L	10 × 4 UB 19	8 × 8 UC 40	—	0.850	—	4.38	3/4″
16		TEST A3-R	10 × 4 UB 19	8 × 8 UC 40	—	0.850	—	4.38	3/4″
17	J. O. Surtees and A. P. Mann (1970)	TEST C1	12 × 5 UB 25	8 × 8 UC 48	11.50	0.750	3.50	4.50	3/4″
18		TEST C2	15 × 6 UB 40	10 × 10 UC 60	14.50	0.750	3.50	4.50	3/4″
19		TEST C3	15 × 6 UB 40	10 × 10 UC 60	14.50	0.750	3.50	4.50	1″
20		TEST C4	15 × 6 UB 40	10 × 10 UC 60	14.50	1.000	3.50	4.50	1″
21		TEST C5	15 × 6 UB 40	10 × 10 UC 60	14.50	1.000	3.50	4.50	1″
22	J. A. Packer and L. J. Morris (1977)	TEST J1	254 × 102 UB 22	152 × 152 UC 37	9.764	0.591	3.39	3.82	M16
23		TEST J2	254 × 102 UB 22	152 × 152 UC 30	9.764	0.591	3.39	3.82	M16
24		TEST J3	254 × 102 UB 22	152 × 152 UC 23	9.764	0.591	3.39	3.82	M16

(Continues)

TABLE 5.9 (Continued)

No.	Author	Test id.	Beam	Column	l_p (in)	t_p (in)	g (in)	p_t (in)	d_b
25	P. Grundy et al. (1980)	T1	610 UB 113	310 UC 240	25.78	1.000	3.50	3.75	7/8″
26		T2	610 UB 113	310 UC 240	25.78	1.250	3.50	3.75	7/8″
27	A. N. Sherbourne (1961)	TEST A2	15 × 5 RSJ 42	8 × 8 UC 35	14.50	1.250	4.00	3.50	3/4″
28		TEST A3	15 × 5 RSJ 42	8 × 8 UC 35	14.50	0.750	4.00	3.50	3/4″
29		TEST B1	15 × 5 RSJ 42	8 × 8 UC 35	14.00	1.000	4.00	5.00	7/8″
30		TEST B2	15 × 5 RSJ 42	8 × 8 UC 35	14.00	0.750	4.00	5.00	7/8″
31	L. G. Johnson et al. (1960)	TEST 5	10 × 4 1/2 RSJ	8 × 8 UC 45	9.750	0.500	3.50	3.75	3/4″
32	J. R. Bailey (1970)	B4-L	12 × 5 UB 32	8 × 8 UC 40	—	1.100	—	4.50	7/8″
33		B4-R	12 × 5 UB 32	8 × 8 UC 40	—	1.100	—	4.50	7/8″
34		B5-L	14 × 5 UB 26	8 × 8 UC 40	—	1.000	—	4.25	3/4″
35		B5-R	14 × 5 UB 26	8 × 8 UC 40	—	1.000	—	4.25	3/4″
36		B6-L	14 × 6 3/4 UB	8 × 8 UC 40	—	1.240	—	4.63	1″
37		B6-R	14 × 6 3/4 UB	8 × 8 UC 40	—	1.240	—	4.63	1″
38		B7-L	12 × 6 1/2 UB	8 × 8 UC 40	—	1.160	—	4.50	1″
39		B7-R	12 × 6 1/2 UB	8 × 8 UC 40	—	1.160	—	4.50	1″
40		B8-L	8 × 5 1/4 UB	8 × 8 UC 40	—	0.920	—	4.50	3/4″
41		B8-R	8 × 5 1/4 UB	8 × 8 UC 40	—	0.920	—	4.50	3/4″
42		C9-L	8 × 5 1/4 UB	8 × 8 UC 40	—	0.875	—	4.50	5/8″
43		C9-R	8 × 5 1/4 UB	8 × 8 UC 40	—	0.875	—	4.50	5/8″
44		C10-L	12 × 5 UB 32	8 × 8 UC 40	—	0.960	—	4.50	3/4″
45		C10-R	12 × 5 UB 32	8 × 8 UC 40	—	0.960	—	4.50	3/4″
46		C11-L	14 × 5 UB 26	8 × 8 UC 40	—	0.825	—	4.25	5/8″
47		C11-R	14 × 5 UB 26	8 × 8 UC 40	—	0.825	—	4.25	5/8″
48		C12-L	12 × 6 1/2 UB	8 × 8 UC 40	11.69	0.950	—	4.38	7/8″
49		C12-R	12 × 6 1/2 UB	8 × 8 UC 40	11.69	0.950	—	4.38	7/8″

50		C13-L	14 × 6 3/4 UB	8 × 8 UC 40	—	1.140	—	4.63	1″
51		C13-R	14 × 6 3/4 UB	8 × 8 UC 40	—	1.140	—	4.63	1″
52	J. O. Surtees and A. P. Mann (1970)	TEST C6	18 × 6 UB 55	10 × 10 UC 89	16.00	1.125	6.00	6.00	1 1/8″
53	N. D. Johnstone and W. R. Walpole (1981)	TEST 1-L	310 UB 46	250 UC 89	—	1.260	5.51	7.01	M30
54		TEST 1-R	310 UB 46	250 UC 89	—	1.260	5.51	7.01	M30
55		TEST 2-L	310 UB 46	250 UC 89	—	1.260	5.51	7.01	M30
56		TEST 2-R	310 UB 46	250 UC 89	—	1.260	5.51	7.01	M30
57		TEST 3-L	310 UB 46	250 UC 89	—	0.945	5.12	4.02	M24
58		TEST 3-R	310 UB 46	250 UC 89	—	0.945	5.12	4.02	M24
59		TEST 4-L	310 UB 46	250 UC 89	—	0.630	5.12	4.53	M24
60		TEST 4-R	310 UB 46	250 UC 89	—	0.630	5.12	4.53	M24
61	A. Mazroi (1983)	EP3	W24 × 100	—	26.75	0.750	5.50	3.50	7/8″
62		EP4	W24 × 100	—	26.75	1.000	5.50	3.50	7/8″
63		EP5	W27 × 114	—	30.25	1.000	5.50	3.93	1″
64		EP6	W27 × 114	—	31.25	1.250	6.50	4.93	1″
65		EP7	W24 × 162	—	28.25	1.500	5.50	4.47	1 1/8″
66		EP7 W	W24 × 162	—	28.25	1.500	5.50	4.47	1 1/8″
67		EP8	W24 × 162	—	29.25	1.750	6.50	5.47	1 1/8″
68		EP8 W	W24 × 162	—	29.25	1.750	6.50	5.47	1 1/8″
69	A. Mazroi (1984)	CF4-U12 × 87	W24 × 100	W12 × 87	26.75	1.000	5.50	3.50	7/8″
70		CF4-U12 × 106	W24 × 100	W12 × 106	26.75	1.000	5.50	3.50	7/8″
71		CF4-U12 × 120	W24 × 100	W12 × 120	26.75	1.000	5.50	3.50	7/8″
72		CF5-U10 × 68	W27 × 114	W10 × 68	30.25	1.000	5.50	3.93	1″
73		CF5-U14 × 61	W27 × 114	W14 × 61	30.25	1.000	5.50	3.93	1″
74		CF5-U10 × 49	W27 × 114	W10 × 49	30.25	1.000	5.50	3.93	1″
75		CF6-U12 × 96	W27 × 114	W12 × 96	31.25	1.250	6.50	4.93	1″
76		CF6-U14 × 158	W27 × 114	W14 × 158	31.25	1.250	6.50	4.93	1″

(Continues)

TABLE 5.9 (Continued)

No.	Author	Test id.	Beam	Column	l_p (in)	t_p (in)	g (in)	p_t (in)	d_b
77	A. Mazroi (1984) (Continued)	CF8-U14 × 159	W24 × 162	W14 × 159	29.25	1.750	6.50	5.47	1 1/8″
78		CF3-S12 × 65	W24 × 100	W12 × 65	26.75	0.750	5.50	3.50	7/8″
79		CF4-S12 × 87	W24 × 100	W12 × 87	26.75	1.000	5.50	3.50	7/8″
80		CF5-S10 × 68	W27 × 114	W10 × 68	30.25	1.000	5.50	3.93	1″
81		CF5-S12 × 96	W27 × 114	W12 × 96	30.25	1.000	5.50	3.93	1″
82		CF6-S12 × 79	W27 × 114	W12 × 79	31.25	1.250	6.50	4.93	1″
83		CF8-S14 × 145	W24 × 162	W14 × 145	29.25	1.750	6.50	5.47	1 1/8″
84		CF8-S14 × 158	W24 × 162	W14 × 158	29.25	1.750	6.50	5.47	1 1/8″
85	Y. L. Yee (1984)	K1B	460UB67.1	460UB67.1	22.60	1.260	5.12	5.22	M30
86		K9	460UB67.1	460UB67.1	22.60	0.787	5.12	5.22	M30
87		K4	460UB67.1	460UB67.1	22.60	0.787	5.12	5.22	M30
88		K7	460UB67.1	460UB67.1	22.60	0.787	5.12	5.22	M30
89		K5A	460UB67.1	460UB67.1	22.60	0.787	5.12	5.22	M24
90		K8	460UB67.1	460UB67.1	22.60	0.787	5.12	5.22	M30
91		K10	460UB67.1	460UB67.1	22.60	0.787	5.12	5.22	M24
92		K11	460UB67.1	460UB67.1	22.60	0.787	5.12	5.22	M24
93		K12	460UB67.1	460UB67.1	22.60	0.787	5.12	5.22	M24
94		B1	460UB67.1	310UC96.8	22.60	0.787	5.12	5.22	M24
95		B2	460UB67.1	310UC96.8	22.60	0.787	5.12	5.22	M24
96		B3	460UB67.1	310UC96.8	22.60	0.630	5.12	5.22	M24
97		C1	310UB40.4	310UB40.4	15.51	0.787	3.94	3.94	M20
98		C2	310UB40.4	310UB40.4	15.51	0.394	3.94	3.94	M16
99		D1	410UB59.7	460UB67.1	19.96	0.630	4.72	4.44	M20
100		D2	410UB59.7	460UB67.1	19.96	0.630	4.72	4.44	M20

101	J. B. Davison et al. (1987)	JT/13B	UB 254 × 102 × 22	UC 152 × 152 × 23	10.24	0.591	2.99	3.94	M16
102	D. B. Moore and P. A. C.Sims (1986)	J3	UB 254 × 102 × 22	UC 152 × 152 × 23	9.252	0.591	3.43	3.94	M16
103		J4	UB 254 × 102 × 22	UC 152 × 152 × 23	9.252	0.591	3.43	3.94	M16
104	R. Zandonini and P. Zanon (1988)	EP 1-1	IPE 300	—	11.81	0.472	4.13	4.72	M20
105		EP 1-2	IPE 300	—	11.81	0.591	4.13	4.72	M20
106		EP 1-3	IPE 300	—	11.81	0.709	4.13	4.72	M20
107		EP 1-4	IPE 300	—	11.81	0.866	4.13	4.72	M20
108		EP 1-5	IPE 300	—	11.81	0.984	4.13	4.72	M20
109		EPB 1-2	IPE 300	—	16.54	0.591	4.13	4.72	M20
110		EPB 1-3	IPE 300	—	16.54	0.709	4.13	4.72	M20
111		EPB 1-4	IPE 300	—	16.54	0.866	4.13	4.72	M20
112		EPB 1-5	IPE 300	—	16.54	0.984	4.13	4.72	M20
113	L. F. L. Ribeiro et al. (1998)	CT1A-1	VS250 × 37	CVS350 × 105	9.35	1.250	2.50	3.00	5/8″
114		CT1A-2	VS250 × 37	CVS350 × 105	9.35	1.000	2.50	3.00	5/8″
115		CT1A-3	VS250 × 37	CVS350 × 105	9.35	1.000	2.50	3.00	5/8″
116		CT1A-4	VS250 × 37	CVS350 × 105	9.35	0.875	2.50	3.00	5/8″
117		CT1A-5	VS250 × 37	CVS350 × 105	9.35	0.875	2.50	3.00	5/8″
118		CT1A-6	VS250 × 37	CVS350 × 105	9.35	0.750	3.00	3.49	3/4″
119		CT1B-1	VS250 × 37	CVS350 × 105	9.35	1.250	2.50	3.00	5/8″
120		CT1B-2	VS250 × 37	CVS350 × 105	9.35	1.000	3.00	3.49	3/4″
121		CT1B-3	VS250 × 37	CVS350 × 105	9.35	1.000	3.00	3.49	3/4″
122		CT1B-4	VS250 × 37	CVS350 × 105	9.35	0.875	3.00	3.49	3/4″
123		CT1B-5	VS250 × 37	CVS350 × 105	9.35	0.875	3.00	3.49	3/4″
124		CT1B-6	VS250 × 37	CVS350 × 105	9.35	0.750	3.00	3.49	3/4″
125		CT2A-1	VS350 × 58	CVS350 × 128	13.15	1.500	3.50	4.13	7/8″
126		CT2A-2	VS350 × 58	CVS350 × 128	13.15	1.250	3.50	4.13	7/8″

(Continues)

TABLE 5.9 (Continued)

No.	Author	Test id.	Beam	Column	l_p (in)	t_p (in)	g (in)	p_t (in)	d_b
127	L. F. L. Ribeiro et al. (1998) (Continued)	CT2A-3	VS350 × 58	CVS350 × 128	13.15	1.250	3.50	4.13	7/8″
128		CT2A-4	VS350 × 58	CVS350 × 128	13.15	1.000	3.50	4.13	7/8″
129		CT2A-5	VS350 × 58	CVS350 × 128	13.15	1.000	4.00	4.63	1″
130		CT2A-6	VS350 × 58	CVS350 × 128	13.15	0.875	4.00	4.63	1″
131		CT2B-1	VS350 × 58	CVS350 × 128	13.15	1.500	3.50	4.13	7/8″
132		CT2B-2	VS350 × 58	CVS350 × 128	13.15	1.250	3.50	4.13	7/8″
133		CT2B-3	VS350 × 58	CVS350 × 128	13.15	1.250	3.50	4.13	7/8″
134		CT2B-4	VS350 × 58	CVS350 × 128	13.15	1.000	3.50	4.13	7/8″
135		CT2B-5	VS350 × 58	CVS350 × 128	13.15	1.000	4.00	4.63	1″
136		CT2B-6	VS350 × 58	CVS350 × 128	13.15	0.875	4.00	4.63	1″
137	B. Bose et al. (1996)	7	457 × 191 UB 74	254 × 254 UC 89	21.14	0.787	3.54	3.94	M24
138		11	457 × 191 UB 74	254 × 254 UC 73	21.14	0.591	3.54	3.94	M24
139		12	457 × 191 UB 74	254 × 254 UC 89	21.14	0.591	3.54	3.94	M24
140		13	762 × 267 UB 147	254 × 254 UC 89	32.83	0.591	3.54	3.94	M24
141		14	457 × 191 UB 74	254 × 254 UC 132	21.14	0.591	3.54	3.94	M24
142		15	762 × 267 UB 147	254 × 254 UC 132	32.83	0.591	3.54	3.94	M24
143		16	457 × 191 UB 74	254 × 254 UC 89	23.22	0.591	3.54	3.94	M24
144		17	457 × 191 UB 74	254 × 254 UC 89	21.14	0.591	3.54	3.94	M20
145		18	686 × 254 UB 125	254 × 254 UC 89	29.84	0.787	3.54	3.94	M24
146	Ana M. Girao Coelho et al. (2004)	FS1a	IPE300	HE340M	11.61	0.409	3.54	3.54	M20
147		FS1b	IPE300	HE340M	11.61	0.409	3.54	3.54	M20
148		FS2a	IPE300	HE340M	11.61	0.591	3.54	3.54	M20
149		FS2b	IPE300	HE340M	11.61	0.591	3.54	3.54	M20
150		FS3a	IPE300	HE340M	11.61	0.788	3.54	3.54	M20

151		FS3b	IPE300	HE340M	11.61	0.788	3.54	3.54	M20
152		FS4a	IPE300	HE340M	11.61	0.396	3.54	3.54	M20
153		FS4b	IPE300	HE340M	11.61	0.396	3.54	3.54	M20
154		EEP_1E_2	HE320A	HE300M	11.61	0.576	5.91	5.32	M24
155		EEP_1C_2a	HE320A	HE300M	11.61	0.398	5.91	5.32	M24
156		EEP_1C_2b	HE320A	HE300M	11.61	0.398	5.91	5.32	M24
157	J. M. Cabrero and E. Bayo (2007)	A-M	IPE330	HEB160	12.60	0.652	3.94	3.54	M20
158		B-M	IPE330	HEB160	12.60	0.398	3.94	3.54	M20
159	G. Shi et al. (2007)	JD3	H300 × 200 × 8 × 12	H300 × 250 × 8 × 12	15.75	0.787	4.25	4.41	M20
160	L. R. O. Lima (2003)	EE1	IPE240	HEB240	9.055	0.591	3.78	2.91	M20
161		EE2	IPE240	HEB240	9.055	0.591	3.78	2.91	M20
162		EE3	IPE240	HEB240	9.055	0.591	3.78	2.91	M20
163		EE4	IPE240	HEB240	9.055	0.591	3.78	2.91	M20
164		EE5	IPE240	HEB240	9.055	0.591	3.78	2.91	M20
165		EE6	IPE240	HEB240	9.055	0.591	3.78	2.91	M20
166		EE7	IPE240	HEB240	9.055	0.591	3.78	2.91	M20

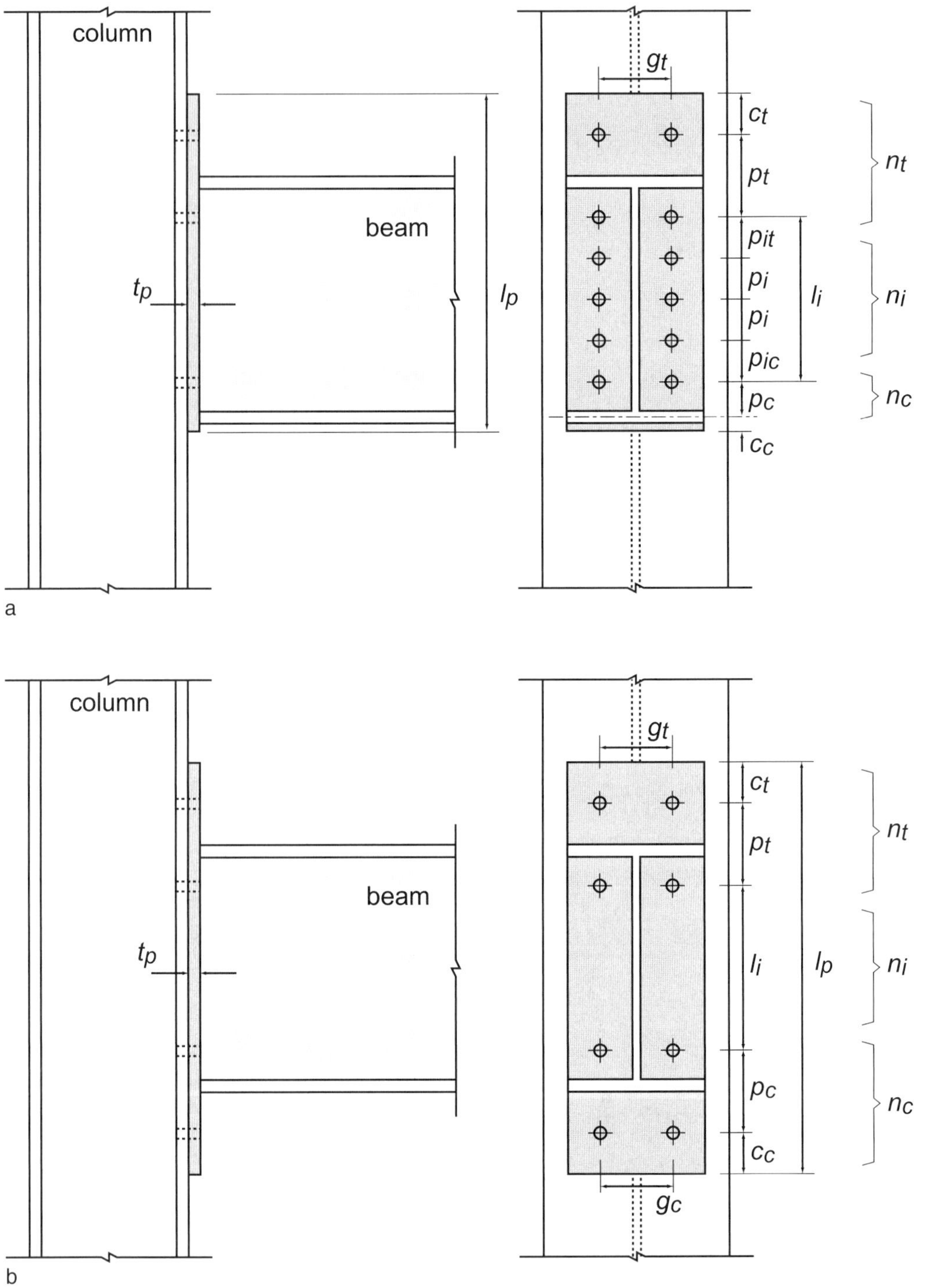

FIGURE 5.7 Size parameters for extended end-plate connection: (a) extended only on tension side; (b) extended on both tension and compression sides.

g_{c2} Distance from the outer line to the inner line of compression-side fasteners where the number of fastener lines exceeds three
l_p Plate length
t_p Plate thickness
n_t Total number of tension-side fasteners
n_i Total number of central fasteners
n_c Total number of compression-side fasteners

The case of end-plate connections extended only on the tension side:

p_c Distance from the inner row of fasteners on the compression side to the center of the compression-side flange of the beam
c_c Distance from the center of the compression side flange of the beam to the compression-side edge of the plate

The case of end-plate connections extended on both the tension and the compression sides:

p_c Distance between two rows of compression-side fasteners
c_c Distance from the outer row of fasteners on the compression side to the compression-side edge of the plate

5.2.6 Flush End-plate Connections

Originally, 24 experimental data conducted by Ostrander (1970) were installed in the database. Then, 15 experimental data conducted in the 1980s were appended. A total of 58 experimental data have been compiled in this database. Collected experimental moment-rotation data for flush end-plate connections are tabulated in Table 5.10. The notations for major parameters of this connection type as shown in Figure 5.8 are identified as follows:

c_t Distance from the tension-side edge of plate to the outer surface of the tension-side flange of the beam
p_t Distance from the outer surface of the tension-side flange to the row of tension-side fasteners
l_i Distance from the center of tension-side fastener holes to the center of the compression-side fastener holes
p_c Distance from the row of compression-side fasteners to the outer surface of the compression-side flange of the beam
c_c Distance from the outer surface of the compression-side flange of the beam to the compression-side edge of the plate
p_{it} Distance from the row of tension-side fasteners to the upper row of central fasteners
p_i Distance between two inner rows of central fasteners
p_{ic} Distance from the row of compression-side fasteners to the row of central fasteners
g_t Distance between two inner lines of tension-side fasteners
g_{t2} Distance from the outer line to the inner line of the tension-side fasteners where the number of fastener lines exceeds three
g_i Distance between two inner lines of central fasteners
g_{i2} Distance from the outer line to the inner line of the central fasteners where the number of fastener lines exceeds three
g_c Distance between two inner lines of the compression-side fasteners
g_{c2} Distance from the outer line to the inner line of the compression-side fasteners where the number of fastener lines exceeds three

TABLE 5.10 Collected test data for flush end-plate connections

No.	Author	Test id.	Beam	Column	l_i (in)	g (in)	t_p (in)	d_b
1	J. R. Ostrander (1970)	TEST 1	W10 × 21	W8 × 28	5.000	3.50	0.500	3/4″
2		TEST 3	W10 × 21	W8 × 28	5.000	3.50	0.375	3/4″
3		TEST 4	W10 × 21	W8 × 28	5.000	3.50	0.250	3/4″
4		TEST 9	W10 × 21	W8 × 28	5.000	3.50	0.750	3/4″
5		TEST 11	W12 × 27	W8 × 40	7.000	4.00	0.375	3/4″
6		TEST 12	W12 × 27	W8 × 40	7.000	4.00	0.500	3/4″
7		TEST 13	W12 × 27	W8 × 40	7.000	4.00	0.625	3/4″
8		TEST 17	W12 × 27	W8 × 24	7.000	4.00	0.375	3/4″
9		TEST 18	W12 × 27	W8 × 24	7.000	4.00	0.500	3/4″
10		TEST 19	W12 × 27	W8 × 24	7.000	4.00	0.625	3/4″
11		TEST 23	W12 × 27	W8 × 48	7.000	4.00	0.625	3/4″
12		TEST 2	W10 × 21	W8 × 28	5.000	3.50	0.500	3/4″
13		TEST 5	W10 × 21	W8 × 28	5.000	3.50	0.500	3/4″
14		TEST 6	W10 × 21	W8 × 28	5.000	3.50	0.375	3/4″
15		TEST 7	W10 × 21	W8 × 28	5.000	3.50	0.250	3/4″
16		TEST 8	W10 × 21	W8 × 28	5.000	3.50	0.250	3/4″
17		TEST 10	W10 × 21	W8 × 28	5.000	3.50	0.750	3/4″
18		TEST 14	W12 × 27	W8 × 40	7.000	4.00	0.375	3/4″
19		TEST 15	W12 × 27	W8 × 40	7.000	4.00	0.500	3/4″
20		TEST 16	W12 × 27	W8 × 40	7.000	4.00	0.625	3/4″
21		TEST 20	W12 × 27	W8 × 24	7.000	4.00	0.375	3/4″
22		TEST 21	W12 × 27	W8 × 24	7.000	4.00	0.500	3/4″
23		TEST 22	W12 × 27	W8 × 24	7.000	4.00	0.625	3/4″
24		TEST 24	W12 × 27	W8 × 48	7.000	4.00	0.625	3/4″

25	J. B. Davison et al. (1987)	JT/11	UB 254 × 102 × 22	UC 152 × 152 × 23	5.906	2.99	0.472	M16
26		JT/11B	UB 254 × 102 × 22	UC 152 × 152 × 23	5.906	2.99	0.472	M16
27		JT/12	UB 254 × 102 × 22	UC 152 × 152 × 23	5.906	2.99	0.472	M16
28	J. Phillips and J. A. Packer (1981)	BM1	W250 × 33	W200 × 100	6.693	3.15	0.375	7/8″
29		BM2	W250 × 33	W200 × 100	6.693	3.15	0.625	7/8″
30		BM3	W250 × 33	W200 × 100	6.693	3.15	0.750	7/8″
31		BM4	W250 × 33	W200 × 100	6.693	3.15	0.875	7/8″
32		BM5	W250 × 33	W200 × 100	6.693	3.15	1.000	7/8″
33	C. Bernuzzi et al. (1996)	FPC/M	IPE300	—	7.087	4.13	0.472	M20
34	B. Bose et al. (1996)	1	406 × 178 UB 60	254 × 254 UC 73	11.28	3.54	0.472	M20
35		2	406 × 178 UB 60	254 × 254 UC 89	11.28	3.54	0.472	M20
36		3	457 × 191 UB 74	254 × 254 UC 89	13.27	3.54	0.472	M20
37		4	406 × 178 UB 60	254 × 254 UC 89	11.28	3.54	0.591	M24
38		5	457 × 191 UB 74	254 × 254 UC 89	13.27	3.54	0.591	M24
39		6	686 × 254 UB 125	254 × 254 UC 89	21.97	3.54	0.591	M24
40		8	406 × 178 UB 60	254 × 254 UC 73	11.28	3.54	0.472	M20
41		9	457 × 191 UB 74	254 × 254 UC 89	13.27	3.54	0.472	M20
42		10	406 × 178 UB 60	254 × 254 UC 73	11.28	3.54	0.591	M24
43	N. D. Brown and D. Anderson (2001)	TEST1	457 × 152 UB 52	203 × 203 UC 52	13.78	3.54	0.591	M20
44	A. W. Thomson and B. M. Broderick (2002)	J1-M	254 × 146 UB 37	203 × 203 UC 86	5.354	3.54	0.394	M20
45		J5-M	254 × 146 UB 37	203 × 203 UC 52	5.354	3.54	0.472	M20
46		J7-M	254 × 146 UB 37	203 × 203 UC 86	5.354	3.54	0.472	M20
47	A. M. G. Coelho and F. S. K. Bijlaard (2007)	F1EP_15_2	HE320A	HE300M	6.299	5.91	0.581	M24
48		F2EP_15_2	HE320A	HE300M	6.299	5.91	0.576	M24
49		F1EP_10_2	HE320A	HE300M	6.299	5.91	0.400	M24
50		F2EP_10_2	HE320A	HE300M	6.299	5.91	0.404	M24

(Continues)

TABLE 5.10 (Continued)

No.	Author	Test id.	Beam	Column	l_i (in)	g (in)	t_p (in)	d_b
51	P. Zoetemeijer (1981)	16	HE300A	—	7.591	4.93	0.472	M24
52		17	HE300A	—	7.591	4.93	0.630	M24
53		18	HE300A	—	5.937	4.93	0.472	M24
54		19	HE300A	—	5.937	4.93	0.630	M24
55		31	IPE400	—	11.72	3.87	0.472	M24
56		34	IPE400	—	11.72	3.87	0.630	M24
57		37	IPE400	—	11.47	4.12	1.260	M24
58	G. Shi et al. (2007)	JD1	H300 × 200 × 8 × 12	H300 × 250 × 8 × 12	6.929	4.25	0.787	M20

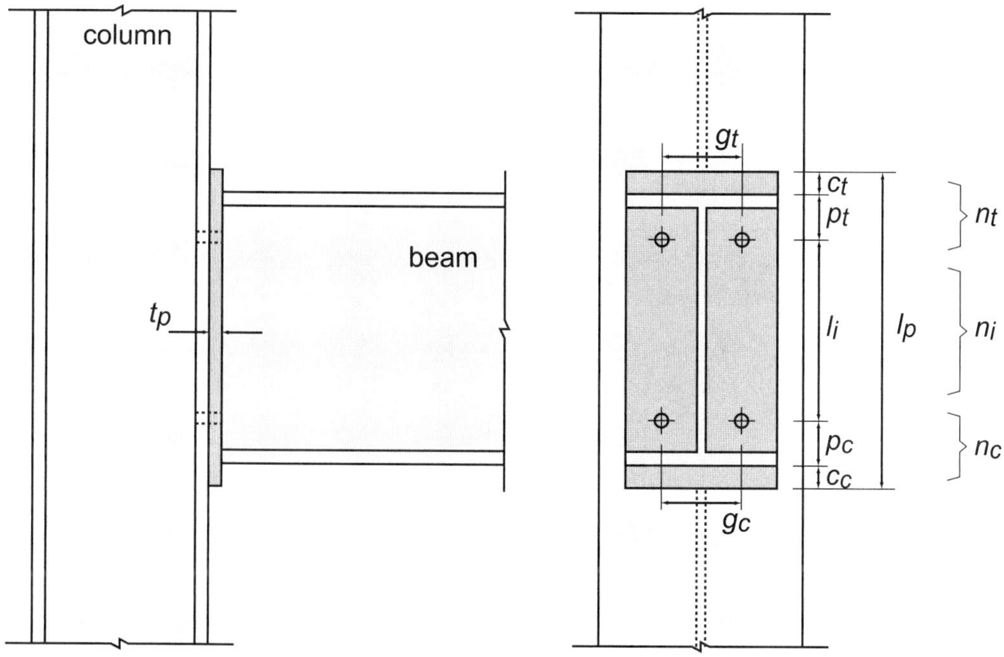

FIGURE 5.8 Size parameters for flush end-plate connection.

l_p Plate length
t_p Plate thickness
n_t Total number of tension-side fasteners
n_i Total number of central fasteners
n_c Total number of compression-side fasteners

5.2.7 Header Plate Connections

Moment-rotation tests on header plate connections were firstly performed by Sommer at the University of Toronto in 1969. He conducted experiments on 20 specimens. In the 1980s, two series of experimental investigations were conducted by Davison et al. (1987) and Aggarwal (1990). A total of 26 experimental moment-rotation data has been installed as listed in Table 5.11. The notations for the major parameters of this connection type as shown in Figure 5.9 are identified as follows:

l_t Distance from the tension-beam flange to the top edge of the plate
c_t Distance from the center of the upper fastener holes to the top edge of the plate
p Distance between two rows of fasteners
c_c Distance from the center of the lower fastener holes to the bottom edge of the plate
l_c Distance from the compression-beam flange to the bottom edge of the plate
g Distance between two lines of fasteners
l_p Plate depth
t_p Plate thickness
t_{bw} Beam-web thickness
n Total number of fasteners

TABLE 5.11 Collected test data for header-plate connections

No.	Author	Test Id.	Beam	Column	l_p (in)	g (in)	t_p (in)	d_b
1	W. H. Sommer (1969)	TEST 5	W18 × 45	—	15.000	4.00	0.250	3/4″
2		TEST 6	W24 × 76	—	9.000	4.00	0.250	3/4″
3		TEST 7	W24 × 76	—	12.000	4.00	0.250	3/4″
4		TEST 8	W24 × 76	—	15.000	4.00	0.250	3/4″
5		TEST 9	W24 × 76	—	18.000	4.00	0.250	3/4″
6		TEST 10	W18 × 45	—	9.000	4.00	0.375	3/4″
7		TEST 11	W18 × 45	—	12.000	4.00	0.375	3/4″
8		TEST 12	W18 × 45	—	15.000	4.00	0.375	3/4″
9		TEST 13	W24 × 76	—	9.000	4.00	0.375	3/4″
10		TEST 14	W24 × 76	—	12.000	4.00	0.375	3/4″
11		TEST 15	W24 × 76	—	15.000	5.50	0.375	3/4″
12		TEST 16	W24 × 76	—	18.000	5.50	0.375	3/4″
13		TEST 17	W24 × 76	—	18.000	5.50	0.250	3/4″
14		TEST 18	W24 × 76	—	15.000	5.50	0.250	3/4″
15		TEST 19	W24 × 76	—	12.000	5.50	0.500	3/4″
16		TEST 20	W24 × 76	—	15.000	5.50	0.500	3/4″
17		TEST 25	W18 × 45	—	12.000	4.00	0.250	3/4″
18		TEST 26	W18 × 45	—	9.000	4.00	0.250	3/4″
19		TEST 27	W12 × 27	—	9.000	4.00	0.250	3/4″
20		TEST 28	W12 × 27	—	6.000	4.00	0.250	3/4″
21	J. B. Davison et al. (1987)	JT/14	UB 254 × 102 × 22	UC 152 × 152 × 23	10.43	2.99	0.472	M16
22	A. K. Aggarwal (1990)	M 2	UB 203 × 133 × 25	UC 203 × 203 × 46	5.906	3.94	0.630	M20
23		M 3	UB 203 × 133 × 25	UC 203 × 203 × 46	5.906	3.94	0.472	M20
24		M 4	UB 203 × 133 × 25	UC 203 × 203 × 46	5.906	3.94	0.472	M20
25		M 5	UB 203 × 133 × 25	UC 203 × 203 × 46	5.906	3.94	0.630	M24
26		M 6	UB 203 × 133 × 25	UC 203 × 203 × 46	5.906	3.94	0.630	M16

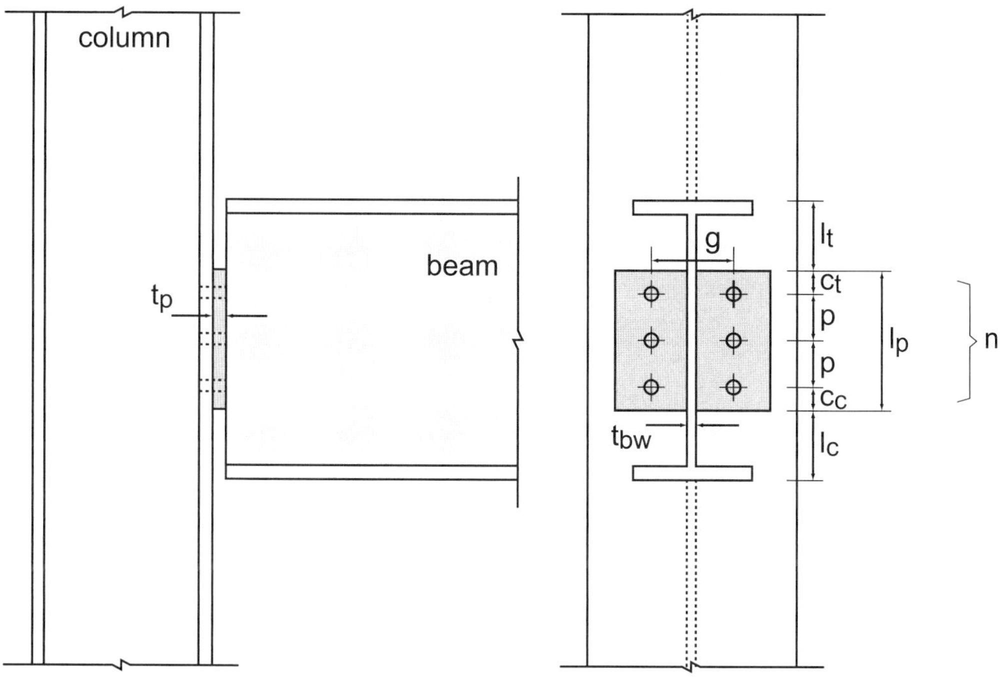

FIGURE 5.9 Size parameters for header-plate connection.

5.3 Steel Connection Data Bank Program

5.3.1 Outline of SCDB

The experimental connection test data collected in the previous section can be controlled by the Steel Connection Data Bank (SCDB) program. The first version of the SCDB program has seven main routines. For use with any workstation or PC with a FORTRAN compiler, the plotter routines in it have been removed. In this book, general views of M–θ_r curves of experimental test data for each connection type are shown in the Appendix. The user can then verify his own plotter routine by comparing it with these figures.

The latest version of SCDB program for this book consists of the following main routines:

1. Transform output unit system from original U.S. units to MKS units or SI units.
2. Set and print the selected test data.
3. Set and print general tables of test data concerning connection type and mode.
4. Determine and print the connection moment and tangent connection stiffness obtained from the three prediction equations for the selected test data.

In this program, three prediction equations are available: Frye-Morris polynomial model; modified exponential model; and three-parameter power model. The prediction curves obtained from the three-parameter power model are limited to the angle types of connection and header plate connections. In the three-parameter power model, the initial connection stiffness R_{ki} and the ultimate connection moment M_u are based on the values determined from the simple mechanics and mechanism formulated by Kishi and Chen (1990). The shape parameter n is determined numerically by applying the least mean squares technique for the connection moments between the experimental data and the model. The validity of either the three-parameter power model or the modified experimental model for practical use was numerically discussed by Kishi and Chen et al. (1991, 1993) and Kishi and Hasan et al. (1995). It was concluded that the three-parameter power model can replace the experimental data in adequately describing the connection M–θ_r behavior for practical use.

Depending on the user's requirements, it is possible to execute the SCDB program together with the above-mentioned routines. This program enables the user to make comparative studies of the role of different connection parameters on M–θ_r behavior.

5.3.2 User Manual for SCDB Program

The SCDB program is written in FORTRAN77, which is composed of four files (*scdb1.f-scdb4.f*). These four files are provided in the source list of the J. Ross Web site (http://www.jrosspub.com/wav/). The user can compile the files and then create the executable file, for which a hard disk is generally required. To execute the SCDB program, the files shown in Table 5.12 must be prepared. It should be noted that *scdb.d* and *BANK/no1-BANK/no7* are files for input, while *bank.d* is for output. The files of *BANK/no1-BANK/no7* are the files for the original database of each connection type and are provided on the J. Ross Web site. The file *scdb.d* should be defined in the same directory with an executable program, and input data should be created according to the user's purpose as follows:

Lines 1 to 7 (7a50)

File names with full pass of files *BANK/no1* to *BANK/no7*

Line 8 (4i5)

idsi: Index for unit name of output data
0: Customary U.S. units
1: MKS units except dimensions (cm) of connection
2: SI units except dimensions (cm) of connection
icnct: Identification number of the connection type referred to in Table 5.1
ilist: Flag for output
0: Print the data bank list
n: Print the table concerning the *n*th fastening mode referred to in Table 5.13. Numbers 9 through 12 or numbers 16 and 17 are for the types of end-plate connections.
nterm: Number of terms considered in the modified exponential curve-fitting equation. The case of nterm = 6 will give a good result.

0: By default, 6 is set

Line 9 (i5)

ngroup: Number of calculating group of experimental data for the type of connection given by icnct in Line 8

Line 10 (10i5)

istart (i), i = 1, ngroup:

Starting the sequential number of the experimental data in the *i*th group specified by the above variable ngroup

TABLE 5.12 List of files used in SCDB program

File name	Contents
scdb.d	input file
BANK/no1-BANK/no7	Original data base of each connection type
bank.d	Output file
temp1, temp2	Working file

TABLE 5.13 Fastening mode patterns for making tables

Pattern number	Fastening mode pattern
1	All riveted
2	All bolted
3	Riveted-to-beam and bolted-to-column
4	Bolted-to-beam and riveted-to-column
5	Riveted-to-beam and welded-to-column
6	Bolted-to-beam and welded-to-column
7	Welded-to-beam and riveted-to-column
8	Welded-to-beam and bolted-to-column
9	All riveted without column stiffener
10	All riveted with column stiffener
11	All bolted without column stiffener
12	All bolted with column stiffener
13	Riveted/bolted-to-beam and column (1, 2, 3, 4)
14	Riveted/bolted-to-beam and welded-to-column (5, 6)
15	Riveted/bolted-to-column and welded-to-beam (7, 8)
16	Riveted/bolted without column stiffener (9, 11)
17	Riveted/bolted with column stiffener (10, 12)
20	All modes included

Line 11 (10i5)

iend (i), i = 1, ngroup:

Finishing the sequential number of experimental data in the *i*th group specified by the above variable ngroup

5.3.3 Examples

Example 1: *Printing a database list*

The first example is for making a database list: connection parameters, experimental $M–\theta_r$ data, parameters for each prediction equation, and the comparison of the values obtained from experimental test data with the values from the prediction equations. The input file named *scdb.d* for this example is given in Table 5.14 and may be used for making a database list of the experimental

TABLE 5.14 Content of scdb.d for Example 1

```
BANK/no1
BANK/no2
BANK/no3
BANK/no4
BANK/no5
BANK/no6
BANK/no7
   2   3   0   0
   1
  18
  18
```

data of the sequential number 18 for the type of top- and seat-angle connection with double-web angle. The content of the output file *bank.d* is shown in Table 5.15 and can be used to verify the SCDB program upon installation on a computer system. Input and output files for this example are prepared in files of *scdb_ex1.d* and *bank_ex1.d*, respectively.

Example 2: Printing the general tables of test data
The second example is for making the general table of experimental data concerning connection type and mode. In this example, an attempt is made to produce a general table of experimental data concerning extended end-plate connection without column stiffener. The input file *scdb.d* for this example is given in Table 5.16. The output file *bank.d* obtained from the program SCDB is shown in Table 5.17. These files are provided in the files of *scdb_ex2.d* and *bank_ex2.d*, respectively.

5.4 Web Site of Connections

The latest information of the semi-rigid connection database is available on http://struct.ce.muroran-it.ac.jp/. This Web page also includes the list of journals, textbooks, monographs, and some useful programs that are related to the steel semi-rigid connection or frame structures. We invite readers who have other experimental results of the semi-rigid connections to contact Dr. Masato Komuro about including their results.

TABLE 5.15 Content of **bank.d** for Example 1

III - 18

Connection type : Top-and seat-angle connections with double web angle
Mode : All bolted

Tested by : A.Azizinamini et al. (1985) U.S.A.
Test id. : 14S9

Column : W12 × 96
Beam : W14 × 38
F.Angle : 6 × 4.0 × 1/2 × 8.0
W.Angle : 4 × 3.5 × 1/4 × 8.5

Fasteners : A325- -M22.2D
M23.8 Oversize holes
Material : A36
Fy = 272 MPa
Fu = 468 MPa

Major parameters (cm)

lp = 21.5900	lu = 7.1437	ll = 7.1437	ta = 0.6350
gb = 5.0800	gc = 6.5888	cu = 3.1750	cl = 3.1750
pb = 7.6200	pc = 7.6200	nb = 1 × 3	nc = 1 × 3
lt = 20.3200	tt = 1.2700	ts = 20.3200	ts = 1.2700
gt = 8.8900	gt′ = 6.3500	gs = 8.8900	gs′ = 6.3500
qt = 8.8900	rt = 13.9700	qs = 8.8900	rs = 13.9700
pt = 6.3500	ps = 6.3500		
nt = 2 × 2	nt′ = 2 × 1	ns = 2 × 2	ns′ = 2 × 1

Remark 1) Heavy hex high strength bolts used
2)

No	Moment (kN-m)	Rotation (radians) × 1/1000	No	Moment (kN-m)	Rotation (radians) × 1/1000
1	0.00	0.00	26	102.12	11.71
2	4.28	0.15	27	104.29	12.86
3	8.55	0.31	28	105.74	13.88
4	11.94	0.42	29	107.19	14.90
5	15.32	0.52	30	108.73	15.96
6	21.20	0.77	31	110.27	17.01
7	27.26	1.08	32	111.28	18.18
8	31.90	1.30	33	112.29	19.34
9	36.53	1.51	34	112.49	20.27
10	40.36	1.70	35	112.69	21.20
11	44.19	1.89	36	113.52	22.21
12	47.85	2.13	37	114.34	23.21
13	51.50	2.38	38	115.98	24.35
14	55.52	2.66	39	116.62	25.29
15	59.53	2.94	40	117.27	26.24
16	62.92	3.24	41	118.36	27.05
17	66.31	3.54	42	119.45	27.87
18	70.95	4.07	43	120.54	28.78
19	75.59	4.60	44	121.63	29.70
20	79.61	5.20			
21	83.63	5.79			
22	87.75	6.75			
23	91.87	7.70			
24	96.89	9.12			
25	99.95	10.56			

(Continues)

Table 5.15 (continued)

Moment-rotation prediction equations (R : × 1/1000 radians)

Frye and Morris polynomial model : R = Sum (Ai × (K*Bm)**Pi × 10**Qi)

t = 1.270000 d = 35.877500 tc = 0.635000 w = 20.320000 g = 5.238750
A1 = 1.498414 A2 = 5.596349 A3 = 4.344317 K = 0.010631
P1 = 1 P2 = 3 P3 = 5 Q1 = 0 Q2 = 0 Q3 = 0

Modified exponential model :

Bm = Sum (Ai × (1 − exp(− R/(2*i*C)))) + Sum (RKj × (R − Rj0)) + Bm0

C = 0.28596917E+00 Bmo= 0.00000000E+00 Nexp= 6 Nliner= 1

AI = −0.12052011E+02 0.26430784E+03 −0.17687217E+04 0.47893731E+04 −0.55406128E+04 0.23828771E+04

Rj0 = 23.2100

RKj = 0.10143474E+01

Power model : Bm = (Rki × R) / (1 + (R/R0)**rn)**(1/rn)
rn = 0.880 rki = 0.47207127E+02 rmu = 0.13448064E+03

No	Rotation (radians) × 1/1000	Moment (kN–m)				Connection stiffness (kN–m) × 1000			
		Expri.	Poly.	M.Expo.	P.Model.	Expri.	Poly.	M.Expo.	P.Model.
1	0.00	0.00	0.00	0.00	0.00	0.2777E+02	0.6277E+02	0.2954E+02	0.4721E+02
3	0.31	8.55	17.16	8.83	12.51	0.2996E+02	0.4519E+02	0.2768E+02	0.3560E+02
5	0.52	15.32	25.47	14.61	19.60	0.2915E+02	0.3305E+02	0.2606E+02	0.3060E+02
7	1.08	27.26	39.13	27.96	34.07	0.2063E+02	0.1861E+02	0.2191E+02	0.2212E+02
9	1.51	36.53	46.07	36.84	42.70	0.2097E+02	0.1388E+02	0.1916E+02	0.1793E+02
11	1.89	44.19	50.74	43.58	48.85	0.1808E+02	0.1144E+02	0.1715E+02	0.1528E+02
13	2.38	51.50	55.86	51.53	55.76	0.1451E+02	0.9292E+01	0.1484E+02	0.1261E+02
15	2.94	59.53	60.60	59.20	62.18	0.1292E+02	0.7705E+01	0.1263E+02	0.1041E+02
17	3.54	66.31	64.80	66.08	67.82	0.1047E+02	0.6550E+01	0.1062E+02	0.8678E+01
19	4.60	75.59	70.99	75.76	75.82	0.7799E+01	0.5195E+01	0.7731E+01	0.6529E+01
21	5.79	83.63	76.56	83.52	82.60	0.5815E+01	0.4245E+01	0.5458E+01	0.4976E+01

23	7.70	91.87	83.71	91.78	90.53	0.4000E+01	0.3309E+01	0.3443E+01	0.3455E+01
25	10.56	99.95	91.93	99.70	98.47	0.1989E+01	0.2518E+01	0.2293E+01	0.2236E+01
27	12.86	104.29	97.24	104.37	102.92	0.1639E+01	0.2126E+01	0.1792E+01	0.1679E+01
29	14.90	107.19	101.31	107.62	105.99	0.1440E+01	0.1873E+01	0.1388E+01	0.1344E+01
31	17.01	110.27	105.04	110.14	108.55	0.1179E+01	0.1673E+01	0.1011E+01	0.1096E+01
33	19.34	112.29	108.73	112.08	110.86	0.5056E+00	0.1500E+01	0.6709E+00	0.8961E+00
35	21.20	112.69	111.41	113.12	112.40	0.5073E+00	0.1387E+01	0.4657E+00	0.7741E+00
37	23.21	114.34	114.70	113.89	113.85	0.1113E+01	0.1285E+01	0.1318E+01	0.6686E+00
39	25.29	116.62	116.67	116.51	115.15	0.6857E+00	0.1195E+01	0.1205E+01	0.5813E+00
41	27.05	118.36	118.74	118.57	116.12	0.1332E+01	0.1128E+01	0.1140E+01	0.5203E+00
43	28.78	120.54	120.63	120.50	116.97	0.1196E+01	0.1070E+01	0.1097E+01	0.4695E+00

TABLE 5.16 Content of **scdb.d** for Example 2

```
BANK/no1
BANK/no2
BANK/no3
BANK/no4
BANK/no5
BANK/no6
BANK/no7
    2   5   11   0
    1
    1
  166
```

TABLE 5.17 Content of **bank.d** for Example 2

1 V. Extended end-plate connections

All bolted without column stiffener

No	Author	Test id.	Beam	Column	lp(cm)	tp(cm)	g(cm)	pt(cm)	db
1	S. A. Ioannides (1978)	TEST 1	W14 × 22	W8 × 35	43.8150	1.5875	13.9700	8.8900	M19.1
2		TEST 2	W18 × 35	W10 × 49	54.6100	1.9050	13.9700	9.5250	M22.2
3		TEST 3	W24 × 55	W14 × 48	69.2150	2.2225	13.9700	10.1600	M25.4
4		TEST 4	W14 × 22	W8 × 35	43.8150	2.2225	13.9700	8.8900	M19.1
5		TEST 5	W18 × 35	W10 × 49	54.6100	3.1750	13.9700	9.5250	M22.2
6		TEST 6	W24 × 55	W14 × 48	69.2150	3.1750	13.9700	10.1600	M25.4
7	R. J. Dews (1979)	TEST 1	W14 × 22	W8 × 31	43.5762	2.4613	13.9700	8.8900	M19.1
8		TEST 2	W14 × 22	W8 × 31	43.8556	2.3825	14.0487	8.8900	M19.1
9		TEST 3	W16 × 26	W10 × 33	49.4894	3.9675	13.6525	9.5250	M25.4
10	A. N. Sherbourne (1961)	TEST A1	15 × 5 RSJ 42	8 × 8 UC 35	36.8300	3.1750	10.1600	8.8900	M19.1
11	J. R. Bailey (1970)	TEST A1-L	12 × 5 RSJ 32	8 × 8 UC 48	—	2.5908	—	11.4300	M22.2
12		TEST A1-R	12 × 5 RSJ 32	8 × 8 UC 48	—	2.5908	—	11.4300	M22.2
13		TEST A2-L	10 × 4 UB 17	8 × 8 UC 40	—	2.0447	—	11.1125	M15.9
14		TEST A2-R	10 × 4 UB 17	8 × 8 UC 40	—	2.0447	—	11.1125	M15.9
15		TEST A3-L	10 × 4 UB 19	8 × 8 UC 40	—	2.1590	—	11.1125	M19.1
16		TEST A3-R	10 × 4 UB 19	8 × 8 UC 40	—	2.1590	—	11.1125	M19.1
17	J. O. Surtees and A. P. Mann (1970)	TEST C1	12 × 5 UB 25	8 × 8 UC 48	29.2100	1.9050	8.8900	11.4300	M19.1
18		TEST C2	15 × 6 UB 40	10 × 10 UC 60	36.8300	1.9050	8.8900	11.4300	M19.1
19		TEST C3	15 × 6 UB 40	10 × 10 UC 60	36.8300	1.9050	8.8900	11.4300	M25.4
20		TEST C4	15 × 6 UB 40	10 × 10 UC 60	36.8300	2.5400	8.8900	11.4300	M25.4
21		TEST C5	15 × 6 UB 40	10 × 10 UC 60	36.8300	2.5400	8.6900	11.4300	M25.4

22	J. A. Packer and L. J. Morris (1977)	TEST J1	254 × 102 UB 2	152 × 152 UC 3	24.8001	1.5001	8.5999	9.7000	M16
23		TEST J2	254 × 102 UB 2	152 × 152 UC 3	24.8001	1.5001	8.5999	9.7000	M16
24		TEST J3	254 × 102 UB 2	152 × 152 UC 2	24.8001	1.5001	8.5999	9.7000	M16
25	P. Grundy et al. (1980)	T1	610 UB 113	310 UC 240	65.4924	2.5400	8.8900	9.5250	M22.2
26		T2	610 UB 113	310 UC 240	65.4924	3.1750	8.8900	9.5250	M22.2
27	A. Mazroi (1983)	EP3	W24 × 100	—	67.9450	1.9050	13.9700	8.8900	M22.2
28		EP4	W24 × 100	—	67.9450	2.5400	13.9700	8.8900	M22.2
29		EP5	W27 × 114	—	76.8353	2.5400	13.9700	9.9822	M25.4
30		EP6	W27 × 114	—	79.3750	3.1750	16.5100	12.5222	M25.4
31		EP7	W24 × 162	—	71.7550	3.8100	13.9700	11.3538	M28.6
32		EP7 With shim	W24 × 162	—	71.7550	3.8100	13.9700	11.3538	M28.6
33		EP8	W24 × 162	—	74.2950	4.4450	16.5100	13.8938	M28.6
34		EP8 With shim	W24 × 162	—	74.2950	4.4450	16.5100	13.8938	M28.6
35	A. Mazroi (1984)	CF4-U12 × 8	W24 × 100	W12 × 87	67.9450	2.5400	13.9700	8.8900	M22.2
36		CF4-U12 × 1	W24 × 100	W12 × 106	67.9450	2.5400	13.9700	8.8900	M22.2
37		CF4-U12 × 1	W24 × 100	W12 × 120	67.9450	2.5400	13.9700	8.8900	M22.2
38		CF5-U10 × 6	W27 × 114	W10 × 68	76.8353	2.5400	13.9700	9.9822	M25.4
39		CF5-U14 × 6	W27 × 114	W14 × 61	76.8353	2.5400	13.9700	9.9822	M25.4
40		CF5-U10 × 4	W27 × 114	W10 × 49	76.8353	2.5400	13.9700	9.9822	M25.4
41		CF6-U12 × 9	W27 × 114	W12 × 96	79.3750	3.1750	16.5100	12.5222	M25.4
42		CF6-U14 × 1	W27 × 114	W14 × 158	79.3750	3.1750	16.5100	12.5222	M25.4
43		CF8-U14 × 1	W24 × 162	W14 × 159	74.2950	4.4450	16.5100	13.8938	M28.6
44	Y. L. Yee (1984)	B2	460UB67.1	310UC96.8	57.3999	2.0000	13.0000	13.2700	M24
45		B3	460UB67.1	310UC96.8	57.3999	1.5999	13.0000	13.2700	M24
46		D1	410UB59.7	460UB67.1	50.6997	1.5999	12.0000	11.2799	M20
47		D2	410UB59.7	460UB867.1	50.6997	1.5999	12.0000	11.2799	M20
48	J. B. Davison et al. (1987)	JT/13B	U3 254 × 102 × 22	UC 152 × 152 × 23	25.9999	1.5001	7.5999	10.0000	M16

(Continues)

TABLE 5.17 (Continued)

No	Author	Test id.	Beam	Column	lp(cm)	tp(cm)	g(cm)	pt(cm)	db
49	D. B. Moore and	J3	UB 254 × 102 × 22	UC 152 × 152 × 23	23.4998	1.5001	8.7000	10.0000	M16
50	P. A. C. Sims (1986)	J4	UB 254 × 102 × 22	UC 152 × 152 × 23	23.4998	1.5001	8.7000	10.0000	M16
51	B. Bose et al. (1996)	7	457 × 191 UB 74	254 × 254 UC 89	53.7000	2.0000	9.0000	10.0000	M24
52		11	457 × 191 UB 74	254 × 254 UC 73	53.7000	1.5001	9.0000	10.0000	M24
53		12	457 × 191 UB 74	254 × 254 UC 89	53.7000	1.5001	9.0000	10.0000	M24
54		13	762 × 267 UB 147	254 × 254 UC 89	83.3999	1.5001	9.0000	10.0000	M24
55		14	457 × 191 UB 74	254 × 254 UC 132	53.7000	1.5001	9.0000	10.0000	M24
56		15	762 × 267 UB 147	254 × 254 UC 132	83.3999	1.5001	9.0000	10.0000	M24
57		16	457 × 191 UB 74	254 × 254 UC 89	58.9751	1.5001	9.0000	10.0000	M24
58		17	457 × 191 UB 74	254 × 254 UC 89	53.7000	1.5001	9.0000	10.0000	M20
59		18	686 × 254 UB 125	254 × 254 UC 89	75.7900	2.0000	9.0000	10.0000	M24
60	Ana M. Girao	FS1a	IPE300	HE340M	29.5000	1.0399	9.0000	9.0000	M20
61	Coelho et al. (2004,	FS1b	IPE300	HE340M	29.5000	1.0399	9.0000	9.0000	M20
62	2007)	FS2a	IPE300	HE340M	29.5000	1.5010	9.0000	9.0000	M20
63		FS2b	IPE300	HE340M	29.5000	1.5010	9.0000	9.0000	M20
64		FS3a	IPE300	HE340M	29.5000	2.0020	9.0000	9.0000	M20
65		FS3b	IPE300	HE340M	29.5000	2.0020	9.0000	9.0000	M20
66		FS4a	IPE300	HE340M	29.5000	1.0060	9.0000	9.0000	M20
67		FS4b	IPE300	HE340M	29.5000	1.0060	9.0000	9.0000	M20
68		EEP_15_2	HE320A	HE300M	29.5001	1.4620	15.0000	13.5001	M24
69		EEP_10_2a	HE320A	HE300M	29.5001	1.0100	15.0000	13.5001	M24
70		EEP_10_2b	HE320A	HE300M	29.5001	1.0100	15.0000	13.5001	M24
71	J. M. Cabrero and	A–M	IPE330	HEB160	32.0000	1.6550	10.0000	9.0000	M20
72	E. Bayo (2007)	B–M	IPE330	HEB160	32.0000	1.0100	10.0000	9.0000	M20

73	L. R. O. Lima (2003)	EE1	IPE240	HEB240	23.0000	1.5000	9.6000	7.4000	M20
74		EE2	IPE240	HEB240	23.0000	1.5000	9.6000	7.4000	M20
75		EE3	IPE240	HEB240	23.0000	1.5000	9.6000	7.4000	M20
76		EE4	IPE240	HEB240	23.0000	1.5000	9.6000	7.4000	M20
77		EE5	IPE240	HEB240	23.0000	1.5000	9.6000	7.4000	M20
78		EE6	IPE240	HEB240	23.0000	1.5000	9.6000	7.4000	M20
79		EE7	IPE240	HEB240	23.0000	1.5000	9.6000	7.4000	M20

References

Aggarwal, A. K., Behaviour of flexible end-plate beam-to-column joints, *Journal of Constructional Steel Research*, 16(2), 111–134, 1990.

AISC, *Load and Resistance Factor Design Specifications for Structural Steel Buildings*, 2nd edition, American Institute of Steel Construction, Chicago, IL, 1994.

AISC, *Allowable Stress Design and Plastic Design Specifications for Structural Steel Buildings*, 9th edition, American Institute of Steel Construction, Chicago, IL, 1989.

Altman, W. G., Jr., Azizinamini, A., Bradburn, J. H., and Radziminski, J. B., *Moment-rotation characteristics of semi-rigid steel beam-to-column connection*, Department of Civil Engineering, University of South Carolina, Columbia, SC, 1982.

Azizinamini, A., Bradburn, J. H., and Radziminski, J. B., *Static and cyclic behavior of semi-rigid steel beam-column connections*, Department of Civil Engineering, University of South Carolina, Columbia, SC, 1985.

Bailey, J. R., Strength and rigidity of bolted beam-to-column connections, *Proceedings of the Conference on Joints in Structures*, University of Sheffield, Sheffield, England, 1(4), 1970.

Bell, W. G., Chesson, E. Jr., and Munse, W. H., *Static tests of standard riveted and bolted beam-to-column connections*, University of Illinois Engineering Experiment Station, Urbana, IL, 1958.

Bernuzzi, C., Zandonini, R., and Zanon, P., Experimental analysis and modeling of semi-rigid steel joints under cyclic reversal loading, *Journal of Constructional Steel Research*, 38(2), 95–123, 1996.

Bose, B., Youngson, G. K., and Wang, Z. M., An appraisal of the design rules in Eurocode 3 for bolted end-plate joints by comparison with experimental results, *Proceedings of the Institution of Civil Engineers, Structures and Buildings*, 116(2), 221–234, 1996.

Bose, B., Moment-rotation characteristics of semi-rigid joints in steel structures, *Journal of the Institution of Engineers* (India), Part CI, Civil Engineering Division, 62(2), 128–132, 1981.

Brown, N. D. and Anderson, D., Structural properties of composite major axis end plate connections, *Journal of Constructional Steel Research*, 57(3), 327–349, 2001.

Cabrero, J. M. and Bayo, E., The semi-rigid behaviour of three-dimensional steel beam-to-column joints subjected to proportional loading. Part I. Experimental evaluation, *Journal of Constructional Steel Research*, 63(9), 1241–1253, 2007.

Calado, L., De Matteis, G., and Landolfo, R., Experimental response of top and seat angle semi-rigid steel frame connections, *Materials and Structures*, 33(8), 499–510, 2000.

Chen, W. F. and Kishi, N., Semirigid steel beam-to-column connections: Data base and modeling, *Journal of Structural Engineering*, ASCE, 115(1), 105–119, 1989.

Davison, J. B., Kirby, P. A., and Nethercot, D. A., Rotational stiffness characteristics of steel beam-to-column connections, *Journal of Constructional Steel Research*, 8, 17–54, 1987.

Dews, R. J., *Experimental test results on experimental end-plate moment connections*, Thesis presented to Vanderbilt University, in partial fulfillment of the requirements for the degree of Master of Science, Nashville, TN, 1979.

Elnashai, A. S., Elghazouli, A. Y., and Denesh-Ashtiani, F. A., Response of semirigid steel frames to cyclic and earthquake loads, *Journal of Structural Engineering*, ASCE, 124(8), 857–867, 1998.

Fu, Z., Ohi, K., Takanashi, K., and Lin, X., Seismic behavior of steel frames with semi-rigid connections and braces, *Journal of Constructional Steel Research*, 46(1), 440–441(2), 1998.

Gang, S., Yongjiu, S., and Yuanqing, W., Behaviour of end-plate moment connections under earthquake loading, *Engineering Structures*, 29(5), 703–716, 2007.

Girão Coelho, A. M. and Bijlaard, F. S. K., Experimental behaviour of high strength steel end-plate connections, *Journal of Constructional Steel Research*, 63(9), 1228–1240, 2007.

Girão Coelho, A. M., Bijlaard, F. S. K., and Simões da Silva, L., Experimental assessment of the ductility of extended end plate connections, *Engineering Structures*, 26(9), 1185–1206, 2004.

Goverdhan, A. V., *A collection of experimental moment-rotation curves and evaluation of prediction equations for semi-rigid connections*, Thesis presented to Vanderbilt University, in partial fulfillment of the requirements for the degree of Master of Science, Nashville, TN, 1983.

Grundy, P., Thomas, I. R., and Bennetts, I. D., Beam-to-column moment connections, *Journal of the Structural Division*, ASCE, 106(1), 313–330, 1980.

Harper, W. L., Jr., Dynamic response of steel frames with semi-rigid connections, *Structural research studies*, Department of Civil Engineering, University of South Carolina, Columbia, SC, 1990.

Hechtman, R. A. and Johnston, B. G., *Riveted semi-rigid beam-to-column building connections*, Progress Report No. 1, AISC Committee of Steel Structures Research, Lehigh University, Bethlehem, PA, 1947.

Ioannides, S. A., *Flange behavior in bolted end-plate moment connections*, Thesis presented to Vanderbilt University, in partial fulfillment of the requirements for the degree of Doctor of Philosophy, Nashville, TN, 1978.

Johnson, L. G., Cannon, J. C., and Spooner, L. A., High tensile preloaded bolted joints for development of full plastic moments, *British Welding Journal*, 7, 560–569, 1960.

Johnstone, N. D. and Waldpole, W. R., *Bolted end-plate beam-to-column connections under earthquake type loading*, Research Report 81(7), Department of Civil Engineering, University of Canterbury, Christchurch, New Zealand, 1981.

Kishi, N., Hasan, R., Chen, W. F., and Goto, Y., Power model for semi-rigid connections, *Steel Structures, Journal of Singapore Structural Steel Society*, 5(1), 37–48, 1995.

Kishi, N., Chen, W. F., Goto, Y., and Matsuoka, K. G., Design aid of semi-rigid connections for frame analysis, *AISC Engineering Journal*, 30(3), 90–107, 1993.

Kishi, N., Chen, W. F., Goto, Y., and Matsuoka, K. G., Applicability of three-parameter power model to structural analysis of flexibly jointed frames, *Proceedings of Mechanics Computing in 1990's and Beyond*, H. Adeli and R. L. Sierakowski eds., Columbus, OH, May 20–22, 233–237, 1991.

Kishi, N. and Chen, W. F., Moment-rotation relations of semi-rigid connections with angles, *Journal of Structural Engineering*, ASCE, 116(7), 1813–1834, 1990.

Kishi, N. and Chen, W. F., *Steel connection data bank program*, Structural Engineering, CE-STR-86-18, School of Civil Engineering, Purdue University, West Lafayette, IN, 1986a.

Kishi, N. and Chen, W. F., *Data base of steel beam-to-column connections*, Structural Engineering, CE-STR-86-26, School of Civil Engineering, Purdue University, West Lafayette, IN, 1986b.

Komuro, M. and Kishi, N., Update of bolted semi-rigid connection database, *Summaries of Technical Papers of Annual Meeting*, Architectural Institute of Japan, CD-ROM, 2009. (Japanese)

Komuro, M., Kishi, N., and Matsuoka, K. G., Static loading tests for moment-rotation relation of top- and seat-angle connections, *Journal of Constructional Steel*, JSSC, 10, 57–64, 2002. (Japanese)

Kubo, N., Yoshida, T., Hashimoto, K., and Tanuma, Y., Column influence on the moment-rotation behaviour of semi-rigid angle connections, *Journal of Construction Steel*, JSSC, 7, 427–434, 1999. (Japanese)

Lewitt, C. W., Chesson, E., Jr., and Munse, W. H., *Restraint characteristics of flexible riveted and bolted beam-to-column connections*, Department of Civil Engineering, University of Illinois, Urbana, IL, 1966.

Lima, L. R. O. de, *Behaviour of structural steel endplate joints subjected to bending moment and axial force*, Ph. D. Thesis, Departamento de Engenharia Civil, Pontifícia Universidade Católica do Rio de Janeiro, 2003. (Portuguese)

Lipson, S. L., Single angle welded-bolted connections, *Journal of the Structural Division*, ASCE, 103(3), 559–571, 1977.

Lipson, S. L., Single-angle and single-plate beam framing connections, *Proceedings of Canadian Structural Engineering Conference*, Toronto, Canada, 141-162, 1968.

Mander, J. B., Chen, S. S., and Pekcan, G., Low-cycle fatigue behavior of semi-rigid top-and-seat angle connections, *AISC Engineering Journal*, 31(3), 111–122, 1994.

Marley, M. J. and Gerstle, K. H., *Analysis and tests of flexibly-connected steel frames*, Report to AISC under Project 199, AISC, Chicago, IL, 1982.

Maxwell, S. M., et al., *A realistic approach to the performance and application of semi-rigid jointed structures, Joints in Structural Steelwork*, J. H. Howlett, W. M. Jenkins, and R. Stainsby, eds., Pentech Press, United Kingdom, 2.71–2.98, 1981.

Mazroi, A., *Moment-rotation behavior of beam-to-column end-plate connections in multi-story frames*, Thesis presented to the University of Oklahoma, in partial fulfillment of the requirements for the degree of Doctor of Philosophy, Norman, OK, 1990.

Moore, D. B. and Sims P. A. C., Preliminary investigations into the behaviour of extended end-plate steel connections with backing plates, *Journal of Constructional Steel Research*, 6(2), 95–122, 1986.

Nethercot, D. A., *Steel beam-to-column connections-A review of test data*, CIRIA Project 338, London, 1985.

Ostrander, J. R., *An experimental investigation of end-plate connections*, Thesis presented to the University of Saskatchewan, in partial fulfillment of the requirements for the degree of Master of Science, Saskatchewan, Canada, 1970.

Packer, J. A. and Morris, L. J., A limit state design method for the tension region of bolted beam-column connections, *The Structural Engineering*, 55(10), 446–458, 1977.

Phillips, J. and Packer, J. A., *The effect of plate thickness on flush end-plate connections, Joints in Structural Steelwork*, J. H. Howlett, W. M. Jenkins, and R. Stainsby, eds., Pentech Press, UK, 6.77–6.92, 1981.

Rathbun, J. C., Elastic properties of riveted connections, *Transactions of ASCE*, 101, 524–563, 1936.

Ribeiro, L. F. L., Gonçalves R. M., and Castiglioni C. A., Beam-to-column end plate connections—An experimental analysis, *Journal of Constructional Steel Research*, 46(1), 264–266, 1998.

Richard, R. M., Kriegh, J. D., and Hormby, D. E., Design of single plate framing connections with A307 bolts, *AISC Engineering Journal*, 19(4), 209–213, 1982.

Roeder, C. W., Knechtel, B., Thomas, E., Vaneaton, A., Leon, R. T., and Preece, F. R., Seismic behavior of older steel structures, *Journal of Structural Engineering*, ASCE, 122(4), 365–373, 1996.

Sato, Y., Komuro, M., and Kishi, N., Experimental study on moment-rotation of top- and seat-angle connections, *Journal of Constructional Steel Research*, JSSC, 15, 121–128, 2007. (Japanese)

Sherbourne, A. N., Bolted beam-to-column connections, *The Structural Engineering*, 39(6), 203–210, 1961.

Sommer, W. H., *Behavior of welded-header-plate connections*, Thesis presented to University of Toronto, in partial fulfillment of the requirements for the degree of Master of Applied Science, Toronto, Canada, 1969.

Surtees, J. O. and Mann, A. P., End plate connection in plastically designed structures, *Proceedings of the International Conference on Joints in Structures*, 1(5), University of Sheffield, Sheffield, England, 1970.

Thompson, L. E., McKee, R. J., and Visintainer, D. A., *An investigation of rotation characteristic of web shear framed connections using A-36 and A-441 steels*, Technical report, Department of Civil Engineering, University of Missouri, Rolla, MO, 1970.

Thomson, A. W. and Broderick, B. M., Earthquake resistance of flush end-plate steel joints for moment frames, *Proceedings of Institution of Civil Engineers, Structures and Buildings*, 152(2), 157–165, 2002.

Yang, J. G. and Lee, G. Y., Analytical models for the initial stiffness and ultimate moment of a double angle connection, *Engineering Structures*, 29(4), 542–551, 2007.

Yee, Y. L., *Prediction of nonlinear behaviour of end-plate eave connections*, Thesis presented to Monash University, in partial fulfillment of the requirements for the degree of Doctor of Philosophy in the Department of Civil Engineering, Victoria, Australia, 1984.

Zandonini, R. and Zanon, P., Experimental analysis of end plate connections, *Connections in Steel Structures; Behavior, Strength and Design*, R. Bjorhovde, J. Brozzetti, and A. Colson, eds., Elsevier Applied Science Publishers, London, England, 41–51, 1988.

Zoetemeijer, P., *Semi-rigid bolted beam-to-column connections with stiffened column flanges and flush-end plates*, In *Joints in Structural Steelwork*, Pentech Press, London, 2.99–2.118, 1981.

6

Advanced Analysis of Steel and Composite Semi-rigid Frames

Hong Chen
Offshore Technology Development Pte Ltd, Keppel Offshore & Marine, Singapore

J. Y. Richard Liew
Department of Civil Engineering, National University of Singapore, Singapore

6.1 Structural Frames

A structural framing system generally refers to geometric arrangement of the frame elements such as beams, columns, braces (if any), and shear walls (or core). Connections are used as a means of joining together the framing elements. The structural framing system shall be capable of resisting the design loads including gravity and lateral loads. The building structure shall be a combination of braced frames, unbraced frames, and shear walls that are capable of resisting the design loads. The stability of a simple frame (an unbraced frame with simple connections) shall be provided by a lateral load resisting framing system. For more complicated three-dimensional (3-D) structures involving the interaction of different structural systems, simple models are useful for preliminary design and for checking computer results. These models should capture the behavior of individual subframes and their effects on the overall structures.

6.1.1 Rigid Frames

A rigid frame derives its lateral stiffness mainly from the bending rigidity of frame members interconnected by rigid joints. The joints shall be designed in such a manner that they have enough strength and stiffness and negligible deformation. The deformation must be small enough to have any significant influence on the distribution of internal forces and moments in the structure or on the overall frame deformation.

An unbraced rigid frame should be capable of resisting lateral loads without relying on an additional bracing system for stability. By itself, the frame has to resist the design forces including gravity and lateral forces. At the same time, it should have adequate lateral stiffness against side sway when subjected to horizontal wind or earthquake loads. Even though the detailing of the rigid joints results in a less economic structure, rigid unbraced frame systems perform better in a load reversal situation or in earthquakes. From architectural and functional points of view, it can be advantageous not to have any triangulated bracing systems or shear wall systems in the building.

6.1.2 Simple Frames

A simple frame referred to a structural system in which the beams and columns are pinned connected and the system is incapable of resisting any lateral loads. The stability of the entire structure must be provided by attaching the simple frame to some forms of bracing systems. The lateral loads are resisted by the bracing systems while the gravity loads are resisted by both the simple frame and the bracing system.

In most cases, the lateral load response of the bracing system is sufficiently small such that second-order effects may be neglected for the design of the frames. Thus, the simple frames that are attached to the bracing system may be classified as nonsway frames. Figure 6.1 shows the principal components—the simple frame and bracing system—of such a structure.

There are several reasons for adopting pinned (or flexible) connections in the design of multistory steel frames:

1. Pin-jointed frames are easier to fabricate and erect. For steel structures, it is more convenient to join the webs of the members without connecting the flanges.
2. Bolted connections are preferred over welded connections, which normally require weld inspection, weather protection, and surface preparation.
3. It is easier to design and analyze a building structure that can be separated into system-resisting vertical loads and system-resisting horizontal loads. For example, if all the girders are simply supported between the columns, the sizing of the simply supported girders and the columns is a straightforward task.
4. It is more cost effective to reduce the horizontal drift by means of bracing systems added to the simple framing than it is to use unbraced frame systems with rigid connections.

Actual connections in structures do not always fall within the categories of pinned or rigid connections. Practical connections are semi-rigid in nature and therefore the pinned and rigid conditions are only idealizations. Modern design codes allow the design of semi-rigid frames using the concept of wind-moment design (type-2 connections). In wind-moment design, the connection is assumed to be capable of transmitting only part of the bending moments (those due to the wind

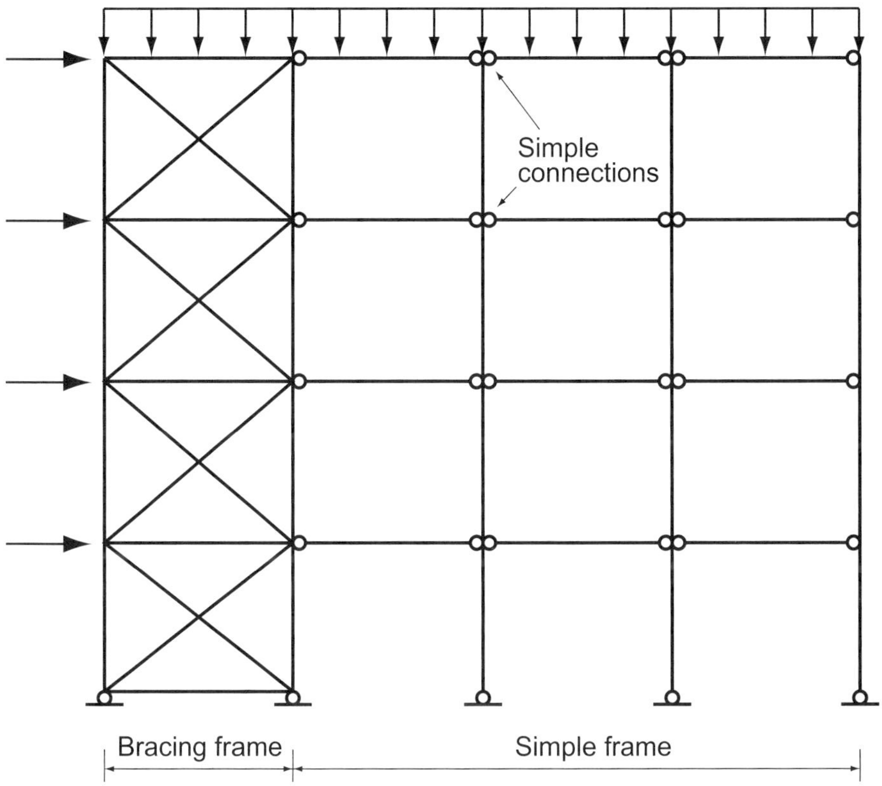

FIGURE 6.1 Simple braced frame.

only). Recent development in the analysis and design of semi-rigid steel and composite frames can be obtained from Chen et al (1996), Leon et al. (1996), and Leon (1998). Design guidance is also available in Section 6 and Annex J of Eurocode 3 (CEN, 1992).

6.1.3 Bracing Systems

Bracing systems provide lateral stability to the overall framework. It may be in the forms of triangulated frames, shear wall or cores, or rigid-jointed frames. It is common to find bracing systems represented as shown in Figure 6.2. In steel structures, it is common to have a triangulated vertical truss to provide bracing (see Figure 6.2a). Unlike concrete structures where all the joints are naturally continuous, the most direct way to make connections between steel members is to hinge one member to the other. For a stiff structure, shear wall or core wall is often used (Figure 6.2b). The efficiency of a building to resist lateral forces depends on the location and type of the bracing systems employed, and the presence or otherwise of shear walls and cores around lift shafts and stairwells.

6.1.4 Braced Versus Unbraced Frames

An unbraced frame comprises beams and columns without braces. Rigid connections shall be provided in an unbraced frame that is intended to carry lateral forces. A bracing frame resists lateral forces through the trussing action of diagonal braces or shear wall.

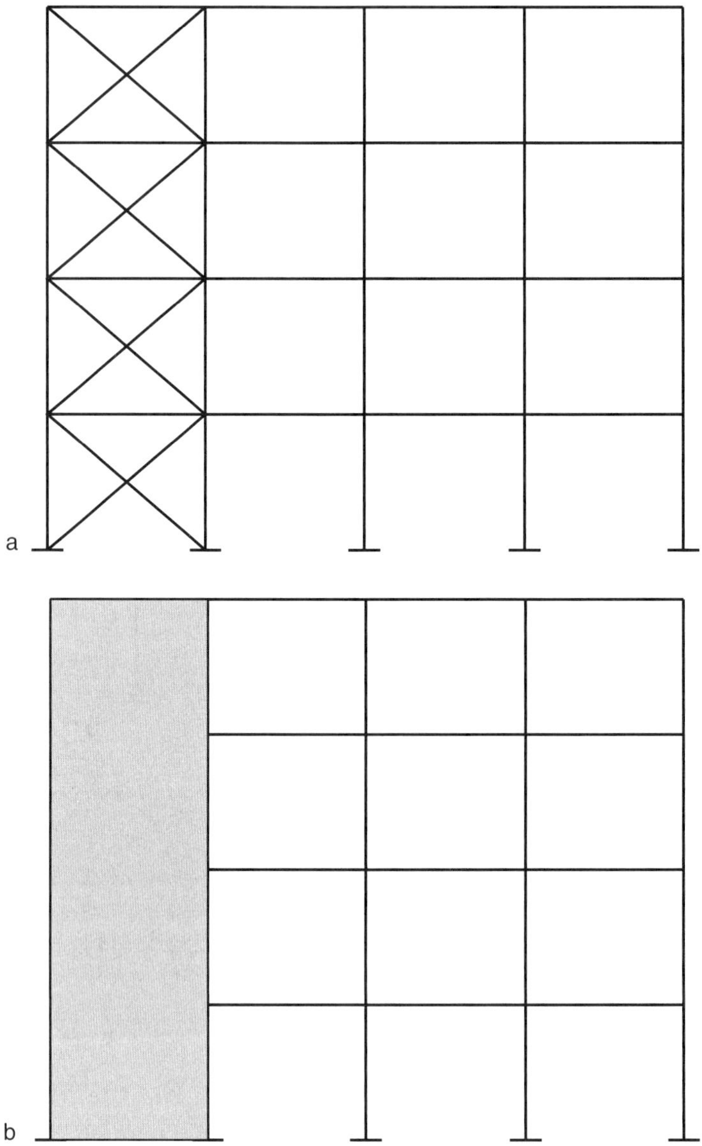

FIGURE 6.2 Common bracing systems: (a) vertical truss system; (b) shear wall system.

Figures 6.3 and 6.4 represent the structures that can be easily defined, within one system, two subassemblies identifying the bracing system and the system to be braced. For the structure shown in Figure 6.3, there is a clear separation of functions in which the gravity loads are resisted by thc hinged subassembly (Frame B) and the horizontal loads are resisted by the braced assembly (Frame A). In contrast, for the frame in Figure 6.4, since Frame B is able to resist horizontal actions as well as vertical actions, it is necessary to assume that practically all the horizontal actions are resisted by Frame A in order to define this system as braced.

According to Section 5.2 of Eurocode 3 (CEN, 1992) a frame may be classified as braced if its sway resistance is supplied by a bracing system in which its response to lateral loads is sufficiently

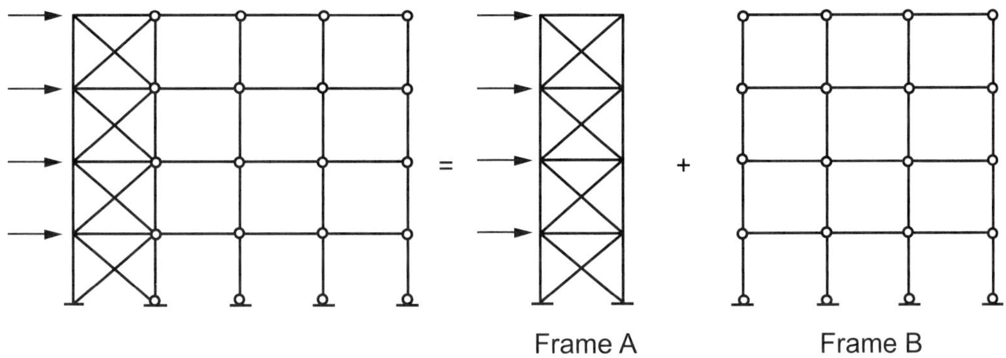

FIGURE 6.3 Frames split into two subassemblies: Frame A-bracing system. Frame B-simple frame.

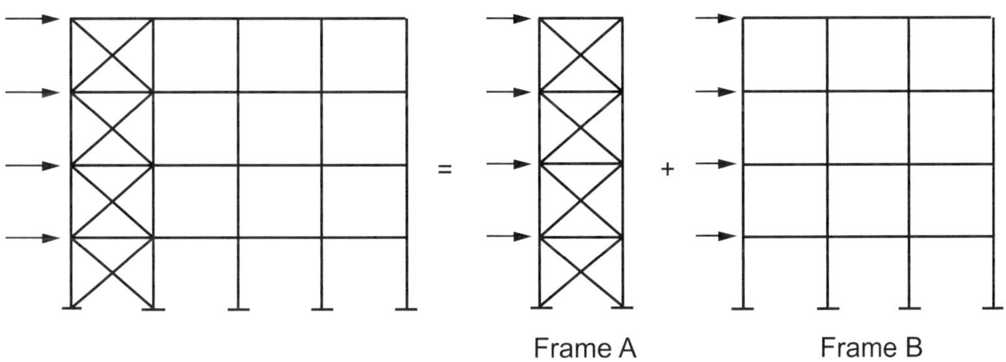

FIGURE 6.4 Mixed frames split into two subassemblies: Frame A-bracing frame, Frame B-unbraced frame.

stiff for it to be acceptably accurate to assume all horizontal loads are resisted by the bracing system. The frame can be classified as braced if the bracing system reduces its horizontal displacement by at least 80 percent.

For the frame shown in Figure 6.3, the simple frame (Frame B) has no lateral stiffness, and the Truss frame (Frame A) resists all lateral load. In this case, Frame B is considered to be braced by Frame A. For the frame shown in Figure 6.4, Frame B may be considered a braced frame if the following deflection criterion is satisfied:

$$\left(1-\frac{\Delta_A}{\Delta_B}\right) \geq 0.8 \tag{6.1}$$

where:

Δ_A = lateral deflection calculated from the truss frame (Frame A) alone
Δ_B = lateral deflection calculated from the simple frame (Frame B) alone

6.1.5 Sway Versus Nonsway Frames

A frame can be classified as nonsway if its response to in-plane horizontal forces is sufficiently stiff for it to be acceptable to neglect any additional internal forces or moments arising from horizontal displacements of the frame. In the design of a multistory building frame, it is convenient to isolate the columns from the frame and treat the stability of columns and the stability of frames as independent from each other. For a column in a braced frame, it is assumed the columns are prevented from horizontal displacements at their ends and therefore are only subjected to end moments and axial loads. It is then assumed that the frame, possibly by means of a bracing system, satisfies global stability checks and that the global stability of the frame does not affect column behavior. This gives the commonly assumed nonsway frame. The design of columns in a nonsway frame follows the conventional beam-column capacity check approach, and the column effective length may be evaluated based on the column end-restraint conditions.

Another reason for defining sway and nonsway frames is the need to adopt conventional analyses in which all the internal forces are computed based on the undeformed geometry of the structure. This assumption is valid if second-order effects are negligible. When there is an interaction between overall frame stability and column stability, it is impossible to isolate the column. The column and the frame have to act interactively in a sway mode. The design of sway frames has to consider the frame subassemblage or the structure as a whole. Moreover, the presence of inelasticity in the columns will render some doubts about the use of the familiar concept, elastic effective length (Liew et al., 1991, 1992).

Based on the above considerations, a definition can be established for sway and nonsway frames as:

> A frame can be classified as nonsway if its response to in-plane horizontal forces is sufficiently stiff for it to be acceptably accurate to neglect any additional internal forces or moments arising from horizontal displacements of its nodes.

Section 5.2 of Eurocode 3 (CEN, 1992) provides some guidelines to distinguish between sway and nonsway frames. It states that a frame may be classified as nonsway for a given load case if $P_{cr} / P \geq 10$ for that load case where P_{cr} is the elastic critical buckling value for sway buckling and P is the design value of the total vertical load. When the system buckling load factor is ten times more than the design load factor, the frame is said to be stiff enough to resist lateral load and is unlikely to be sensitive to side-sway deflections. AISC (1993) does not give specific guidance on frame classification. However, for frames to be classified as nonsway in AISC LRFD format, the moment amplification factor B_2 has to be small (a possible range is $B_2 \leq 1.10$) so that sway deflection would have negligible influence on the final value obtained from the beam-column capacity check.

6.1.6 Joint Representation

The Load and Resistance Factored Design Specification for Structural Steel Buildings of the American Institute of Steel Construction (AISC, 2001) classifies two types of constructions: Type FR (Fully Restrained) and Type PR (Partially Restrained). Type PR comprises two cases, depending on whether connection restraint is ignored. If it is ignored, the connection is called a simple connection. When the connection restraint is considered, it requires that the strength, stiffness, and ductility characteristic of the connection be incorporated into the analysis and design. The assumptions used in the analysis and design of the framing systems shall be consistent with the performance characteristics of the connections. The European standard on Design of Steel Structures or Eurocode 3 Section 5.2 (CEN, 1992) defines three framing types: simple, continuous, and semi-continuous. The standard precisely distinguishes between these three types of connections and also recognizes that the extent of semi-rigid action depends to a large extent on the type of structure, such as braced or unbraced and sway or nonsway frames. Typical moment-rotation (M-θ_r) curves for beam-to-column connections are available from several databases: Goverdhan (1983), Ang and Morris (1984), Nethercot (1985), and Kishi and Chen (1986). Care should be exercised when using tabulated M-θ_r curves not to extrapolate to sizes or conditions beyond those used to develop the database. When the connections to be modeled fall outside the range of the databases, it may be possible to derive them from tests, simple component modeling, or finite element studies. Examples of how to model connection behavior are given in numerous references such as Bjorhovde et al. (1988, 1990, 1992, 1996), Chen and Lui (1991), Liew et al. (1991, 1992), Lorenz et al. (1993), Chen and Toma (1994), Chen et al. (1996), and Leon et al. (1996).

6.1.6.1 Joint Classification

The purpose of joint classification is to select a suitable basis on which to conduct the frame analysis and design. Based on stiffness and strength of a joint, the criteria presented in the next three sections may be adopted.

6.1.6.1.1 Rotation Stiffness

Because the nonlinear behavior of the connection manifests even at low force levels, the initial stiffness of the connection R_{ki} does not adequately characterize connection response at service levels. The secant stiffness R_{ks} at service loads should be used. For a fully rigid connection, the connecting elements shall have sufficient strength and stiffness to effectively maintain the angles between intersecting members under the design loads. The secant rotational stiffness of the fully rigid connection at the service load level shall exceed $20EI_b/L_b$ of the beam. For simple connections, the secant rotational stiffness at the service load level shall not exceed $2EI_b/L_b$ of the beam. A semi-rigid beam-to-column connection shall have sufficient rotation capacity to avoid overloading the connecting elements, fasteners, or welds under the design loads.

6.1.6.1.2 Moment Capacity

The strength of a connection can be determined based on an ultimate limit-state model of the connection or from a physical test. If the moment–rotation response does not exhibit a peak load, then the strength can be taken as the moment at a rotation of 0.02 radians (Leon et al., 1996). The flexural strength of a full strength connection shall exceed the moment capacity of the beam. Connections that transmit less than 20 percent beam flexural capacity at a rotation of 0.02 may be generally classified as a simple connection. Connection with flexural capacity that falls within 20 percent and 80 percent of the moment capacity of the beam is called a partial strength connection.

6.1.6.1.3 Ductility Requirement

If the connection strength exceeds the beam section strength, then the ductility of the structural system is controlled by the beam and the connection can be considered elastic. If the beam strength exceeds the connection strength, then deformations can concentrate in the connection. The ductility required of a connection will depend on the particular application. For example, the ductility requirement for a braced frame in a nonseismic region will generally be less than the ductility required in a high seismic region.

The rotation capacity can be defined as the value of the connection rotation at the point where either (a) the resisting strength of the connection has dropped to 80 percent of the peak moment capacity or (b) the connection has deformed beyond 0.03 radians. Criterion (b) is intended to apply to ductile connections where there is no strength drop until the occurrence of large rotations.

The available rotation capacity should be compared with the rotation required under the full factored loads as determined by an analysis that takes into account the nonlinear behavior of the connection. In the absence of an accurate analysis, a rotation capacity of 0.03 radians is considered adequate. This rotation is equal to the minimum beam-to-column connection capacity as specified in the AISC seismic provisions for special moment frames.

6.1.7 Analysis and Design of Semi-rigid Frames

As discussed in the previous sections, the three types of framing are the simple, rigid (continuous), and semi-rigid (semicontinuous) framings corresponding to where pinned-joint, rigid-joint, and semi-rigid joint models are adopted, respectively.

Joints represented as rigid or pinned in the analysis must be designed to satisfy the rigid joint and nominally pinned joint classifications, respectively.

The semi-rigid model can be more or less sophisticated. Joints can be modeled as spiral springs having moment-rotational relationships that can range from the linear-elastic type to the nonlinear type, which allows for the degree of ductility of the joints. The use of a linear-elastic global analysis model requires that the joint behavior be modeled as being linear-elastic also. For an elastic perfectly plastic analysis, a bilinear joint model is needed. Therefore, the type of analysis used has a direct impact on the degree of complexity of the joint model to be adopted, particularly when plastic analysis is used and hinges are allowed to occur in the joints. A nonlinear joint modeling means that the results for different analysis cannot be superimposed.

When a joint is classified as semi-rigid or partial strength, the relevant response characteristics of the joint should be included in the analysis of the structure to determine the member and connection forces, displacements, and frame stability. An analysis that incorporates semi-rigid joint deformations must include consideration of:

1. Joint initial stiffness, rotation capacity, ductility, and energy absorption capability; these are critical factors for many structures, including rigidly joined frames. These factors are especially important for structures located in seismic regions.
2. Serviceability of the structure and its joints; the prime considerations here are the deformations and other stiffness-related characteristics of the joints, and their effects on the service load deformations of the frame.
3. Strength and stability of the frame and its component members; the effects of partial joint restraint on the ultimate strengths of the members and frame should be considered. The structure is acceptable with respect to strength if the nominal resistance of the structure is greater than the factored load effects.

The degree of sophistication of the analysis depends on the problem at hand. Usually, the design of semi-rigid frames requires separate analyses for the serviceability and ultimate-limit states. For serviceability, an analysis using linear springs with initial stiffness at a service load level is sufficient if the resistance demanded of the joint is well below the strength. Under factored loads, a more careful procedure is needed to ensure that the characteristics assumed in the analysis are consistent with those of the joint response. The response is especially nonlinear as the applied moment approaches the joint strength. In particular, the effect of the joint nonlinearity on second-order moments and other stability checks needs to be considered. A design procedure for semi-rigid frames based on advanced inelastic analysis is proposed below:

1. Perform preliminary analysis and design steps. The preliminary analysis should be carried out assuming rigid frame action. The magnitude of moment to be resisted at the beam ends may be estimated based on the gravity load combination. The engineer might use first-order elastic or plastic analysis with amplification factors, second-order elastic analysis, or any other technique judged to be appropriate in this preliminary stage of design.
2. Select the type of semi-rigid joint to be used in the construction. For example, the beam-to-column connection could be any type of standard connection, such as double web angles, single web angles, top-and-seat angles with or without double web angles, header

plates, flushed or extended end plates, etc. Based on the information in Step 1, make an initial guess of the various dimensions and thicknesses of the connecting material, and check the connections for the second-order moments by calculating the bolt tension, bolt shear, and stresses in the connecting angles, etc.

3. Determine the ultimate moment capacity and initial stiffness of the joint. The approximated shape of the M-θ_r curve of the joint should be known in advance. Since the currently available data do not include all conceivable types of connections, it is advisable to restrict connection designs to those for which M-θ_r curves have been well established.
4. Check the joints for ductility. In essence, this requires that all connection materials such as bolts, welds, and plates should not be loaded beyond the resisting strengths under the simultaneous actions of factored gravity and wind loads.
5. Perform second-order inelastic analysis incorporating the effects of semi-rigid joints in the factored load analysis.
6. Check strength-limit states. The structure is said to be satisfactory if its nominal resistance is greater than the factored load effects.
7. Check for serviceability limit states to ensure adequate stiffness for functionality at service loads.
8. Since the preliminary design is based on the assumption that joints are fully rigid, Steps 1 and 2 may yield a conservative design. The joint designs can be modified if the design resistance of the structure is greater than the factored load effects. Alternatively, the member sizes may be adjusted to provide a more efficient design. Iteration of Steps 2–8 gives results that will converge to the most cost effective design satisfying the various design constraints imposed on the analysis.

6.1.8 Frame Stability

There are many parameters and behavioral effects that influence stability of steel-framed structures. The extent to which these factors are modeled in analysis will affect the criteria that one applies in design of the frame, its members, and joints. Three basic aspects of behavioral effects due to geometric and material nonlinearities and initial imperfection will ultimately govern frame deformations under applied loads and the resulting internal load effects.

6.1.8.1 Imperfection Effects

Modern stability design provisions are based on the premise that the member forces are calculated by second-order elastic analyses, where equilibrium is satisfied on the deformed structure. When stability effects are significant, consideration must be given to member initial imperfections. Many code provisions for calibrating the stability requirements are based on initial geometric imperfections conservatively assumed as equal to the maximum fabrication and erection tolerances permitted by the code. For columns and frames, this would normally imply a member out-of-straightness equal to $0.001L$ where L is the member length between brace or framing points, and a frame out-of-plumb equal to $0.002H$ where H is the story height.

Many of the code's specified analysis and design approaches are calibrated against inelastic distributed-plasticity analyses that account for yielding through the member cross-section and along the member length. Thermal residual stresses in I- or H- (wide flange) shape members are assumed to have maximum values of $0.3f_y$, which are linearly varying across the flanges and uniform tension in the web. Detailed behavioral effects for frame analysis and design are shown in Table 6.1.

TABLE 6.1 Behavioral effects for frame analysis and design

1. Geometrical nonlinear effects
• Geometric interaction due to axial, flexural, torsional and warping deformations • *P*-Δ effect: effect of axial force acting through the relative horizontal displacements associated with member chord rotation • *P*-δ effect: effect of axial force acting through displacements associated with member curvature • Sway shortening: effect of large chord rotation on longitudinal displacements at member ends • Curvature shortening (bowing effect): effect of curvature on axial displacements at member ends • Initial out-of-straightness of member • Initial out-of-plumbness of frame • Behavior of moments under finite three-dimensional rotations • Diaphragm action • Finite joint size and panel zone deformation
2. Material nonlinear effects
• Zero-length plastic hinge ○ elastic-plastic ○ elastic unloading ○ elasto-plastic ○ strain hardening • Spread-of-plasticity in members • Inelastic interaction ○ axial force and biaxial bending ○ axial force, biaxial bending and warping • Connection flexibility • Composite action from floor slabs • Walls and infilling • Initial residual stresses • Local buckling and distortions • Shear yielding • Cyclic plasticity effects
3. Loading effects
• Proportional or non-proportional loading • Variable repeated loading • Dynamic loading

6.1.8.2 Limit States Design

Member strength may be controlled by any one of the following limit states: cross-section yielding, local buckling, flexural buckling, and torsional-flexural buckling. Most of the structural analyses envisioned for routine design use do not model limit states associated with local buckling of cross-section or torsional-flexural buckling of members. Therefore, these limits must be considered in separate member design checks. For inelastic analyses, the effect of cross-section yielding is incorporated directly in the analysis. For elastic analyses, the cross-section strength can be checked by an interaction equation that approximates the *P-M* yield surface. The ability of the analysis to capture in-plane flexural buckling depends on the extent to which the maximum moments are affected by distributed plasticity and member imperfections. Concerns regarding the ability of the analysis to capture these effects suggest the need to apply a member check for in-plane flexural buckling, even when an accurate second-order analysis is used. A major consideration for the in-plane flexural buckling check relates to the assumed buckling length used in calculating the design compression strength.

6.1.8.3 Second-order Analysis

In practice, there are alternative approaches one can employ for conducting second-order analyses, some of which are more rigorous than others. The difference between simplified and rigorous analyses depends on the extent to which P-δ effects due to curvature of the member relative to its chord are modeled and whether the problem is linearized to expedite the solution. Rigorous second-order analyses are those that accurately model all significant second-order effects such as to include the solution for the governing differential equation either through stability functions or computer frame analysis programs that model these effects.

Not all modern commercial computer programs are capable of rigorous analyses. Methods that modify first-order analysis results through second-order amplifiers are in some cases accurate enough to constitute a rigorous analysis, but this depends on the magnitude of second-order effects and other characteristics of the problem. A common type of approximate analyses are those which capture only P-Δ due to member end translations (e.g., interstory drift) and fail to capture P-δ effects. Where P-δ effects are significant, errors arise in approximate methods that do not accurately account for the effect of P-δ moments on amplification of both local member moments and the calculated global displacements.

6.1.8.4 Direct Analysis Approach

The direct analysis approach is developed with the goal to accurately model frame stability effects, and thereby eliminate the need for calculating effective buckling lengths for column designs. The new provisions in the AISC Specification (2005) involve reducing the nominal elastic stiffness and applying a notional load to the frame. The notional load approach can be found in steel standards such as Eurocode 3 Section 5.2 (CEN, 1992), but many aspects of the proposed provisions given in the new AISC Specification (2005) address known shortcomings of conventional notional load approaches in other standards.

In the reduced stiffness approach, the reduction on the flexural stiffness *EI* can be applied by modifying *E* in the analysis. For computer programs that perform semiautomated design checks, one should be sure that the elastic modulus is not reduced in design equations that involve *E* to evaluate the design strength. The reduced stiffness and notional load requirements only pertain to analysis of the strength-limit state, but do not apply to analyses of other serviceability conditions.

Notional load is applied to represent the destabilizing effect of a geometric imperfection and other effects such as yielding, nonidealized boundary, loading conditions, etc. Notional loads are applied as lateral loads at each floor level and are specified in terms of the gravity loads applied at that floor level. The notional load magnitude of 0.002 corresponds to a frame out-of-plumb equal to 0.002*H* (where *H* is the story height). Notional loads shall be applied in the direction that adds to the destabilizing effects under the specified strength load combination.

Sections 6.2 and 6.3 of this chapter deal with detailed modeling of steel and composite semi-rigid frames with an aim to apply direct analysis for frame design.

6.2 Advanced Analysis of Steel Frames

6.2.1 Development of Advanced Analysis

6.2.1.1 Assessment Approaches for Strength and Stability

For decades, structural engineers and researchers have been exploring various approaches for assessing the member and frame stability in the design of frame structures. Figure 6.5 shows the load-deflection curves of a rigid sway frame generated by various analysis methods (White and Chen, 1993; Chan, 2001). The basic distinctions between the analysis methods represented by each curve are (1) whether equilibrium is satisfied on the deformed or undeformed geometry of the structure, and (2) whether material remains elastic or inelastic.

First-order elastic analysis provides a simple estimate of the distribution of forces within the structural system. However, it does not provide any information about the influence of either plasticity or stability effects on the behavior of the frame.

Second-order elastic analysis, which is the preferred method in AISC (2001), is formulated based on the deformed configuration of the structure. It can account for elastic stability effects, but it does not provide any direct information on the actual inelastic strength of the frame. Therefore, in most contemporary design standards for frame structures, the combined stability and

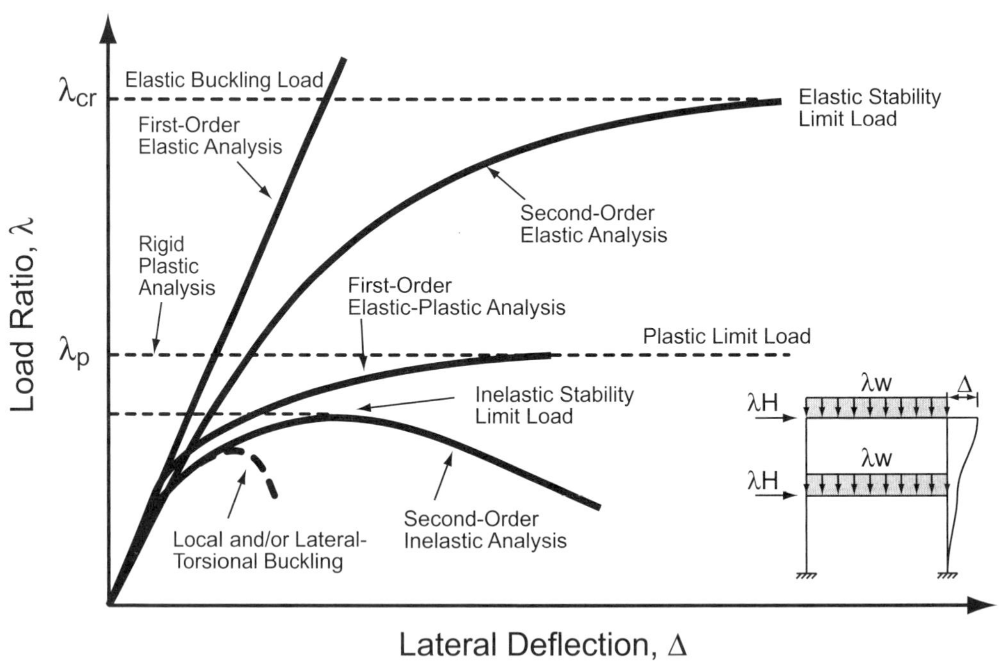

FIGURE 6.5 Types of analyses (White and Chen, 1993).

plasticity effects are considered indirectly through strength and stability criteria for beam-columns. Devices such as column effective length factors and notional loads are commonly used to approximate the member-frame interactions. However, such approaches usually introduce simplifying behavioral assumptions such as artificial distinctions between braced and unbraced frames, the likelihood of all columns in a story failing simultaneously, and the combination of in-plane and out-of-plane member stability effects.

Bifurcation analysis ignores the prebuckling deformation, initial imperfection, and material yielding and assumes a sudden interception of a secondary equilibrium path to the primary path. It performs the eigenvalue analysis to obtain the elastic buckling load, which represents an upper bound solution.

Rigid plastic analysis considers material yielding concentrating at zero length plastic hinges and ignores instability and large deflections. The plastic-limit load with assumed collapse mechanisms is obtained by equalizing the external work done to the internal strain energy for the most critical collapse mechanism. Merchant (1954) considered both the effects of yielding and buckling approximately by using the Merchant-Rankine formula obtained from the elastic buckling and plastic-limit load factors.

Second-order inelastic analysis considers the combined effects of stability and plasticity. The inelastic effects may be handled by techniques ranging from the simple elastic-plastic-hinge model to the more refined inelastic models, which can include the spread-of-plasticity or distributed plasticity effects. Subject to limitations of the behavioral effect considered in the analysis and accurate modeling of inelastic force redistribution, the limit load obtained by a second-order inelastic analysis is the most accurate representation to the true strength of the frame.

6.2.1.2 Direct Analysis of Frame's Stability and Strength

With the advancement of computer technology in recent decades, advanced analysis method has been fully developed for the design used in the steel construction (Chen and Toma, 1994; Chen et al., 1996; White, 1996; Chen and Kim, 1997; Liew et al., 2000a; Chen, 2008). Advanced analysis is defined as a method of second-order inelastic analysis that sufficiently represents the behavioral effects associated with member primary limit states such that separate member capacity checks are not required. When properly formulated, calibrated, and executed, the advanced analysis can rigorously assess the strength and stability interdependence of the member and system, the actual failure mode, and the maximum strength of the structural system.

Table 6.1 provides a summary of the behavioral phenomena and physical attributes that may be considered in the nonlinear analysis of frame members, either by explicit or implicit means (Ziemian, 1990; White and Chen, 1993; Chen et al., 1996; White, 1996). For an analysis method to be classified as advanced, not all of the behaviors need to be addressed or represented in the design and analysis situation. When certain attributions cannot be considered in the analysis, they may be viewed as the limitations of the analysis model, and these limitations must be considered in the final design. For example, at the current stage, most advanced analysis packages based on one or least elements per member are unable to model the more complex aspects of member behavior that involves inelastic lateral-torsional buckling and local buckling, which may significantly reduce the maximum load-carrying capacity of the structure (see Figure 6.5). Therefore, most practical approaches for stability design still require the separation of in-plane frame and member behavior from out-of-plane member stability checks. The members' slenderness ratios must be checked to ensure that inelastic lateral-torsional buckling does not occur. Additional checks are required to ensure that local buckling does not occur and that cross-sections have adequate rotation capacity to allow inelastic redistribution of forces between the frame members.

The theory and approaches for advanced analysis of plane frames composed of members with compact sections—fully braced out-of-plane—have been well developed (Chen and Toma, 1994;

Chen et al., 1996; Chen and Kim, 1997). Experimental calibration frames and numerical results from accurate distributed analyses are available to verify the advanced analysis (Chen and Toma, 1994; Toma et al. 1995; Liew et al., 1997b; Surovek, 2007). Thus, it is feasible to model inelastic member and frame stability directly in a single analysis of planar frames. Research on frames with semi-rigid connections has advanced to a stage where practical tools exist for application in design (Hsieh and Deierlein, 1991; Chen et al., 1996). Numerical tools for second-order inelastic analysis of 3-D frames have been developed and proposed by Orbison (1982), Ziemian et al. (1992), Chen et al. (2000), Liew et al. (2000b), Kim et al. (2001), and Liew and Chen (2004a). The Australian Standard (AS4100, 1990) and Eurocode 3 (CEN, 1992) explicitly permit the checking of in-plane member and frame stability based solely on the advanced analysis. In the 2005 AISC Specification, steps have been taken to allow designers to more explicitly assess the strength of steel structures. The most recent edition of the AISC Specification (AISC, 2005) in Appendix 7 includes a new method of analysis and design termed the *direct analysis method*, which permits the checking of steel frame strength accounting explicitly the key phenomena that affect the system and member strengths. By incorporating a nominal out-of-plumbness and a nominal stiffness reduction in the structural analysis, the direct analysis method eliminates the need to calculate and apply column effective length, and provides a direct path from elastic analysis and design to advanced analysis (Liew and Chen, 2004a; Lui and Ma, 2005; White et al., 2006; Surovek, 2007).

The primary benefits in directly assessing the capacity of a structure within the analysis is that it allows for a simplified design methodology that eliminates the need to check certain member interaction equations and determination of design approximations such as effective length factors. Furthermore, the member and system behavior can be directly assessed and hence can be tuned to better achieve design objectives. The advanced analysis provides better insight into the structural behavior up to failure by observing the load-displacement characteristic of the structure and the sequence of hinge formation in the frame. It is particularly useful for flexible and nonsymmetrical structures. The enhanced understanding of behavior provided by the advanced analysis improves the designer's ability to place materials where they are most effective. The evaluation of a system's limit state will provide a more uniform safety level compared to the conventional approach based on first member failure. Moreover, design economy can be realized if a higher proportion of the inelastic reserved strength built into the system can be utilized in design.

There is a worldwide trend for modern design specifications to depart from prescriptive methods and to move toward performance-based design approaches (Liew and Chen, 2004a). The gradual shift towards performance-based design increases the need to have efficient and robust nonlinear inelastic analysis methods to assess the global performance of structures. There is a logical relationship between this design approach and the use of advanced analysis method because the direction of performance-based design is to move away from prescriptive design equation checks of specification.

With the advanced analysis method, a structural engineer can now introduce the concept of structural fuse in his design process (Chen, 2000). These fuses can be designed to fail under extreme loads, while leaving the majority of the connections in satisfactory condition. One type of the fuses is the reduced beam section configuration, in which a portion of beam flanges near the beam-to-column joint is cut to allow a plastic hinge to form away from the joint. By using the advanced analysis, these structural fuses can be strategically located throughout a structure without the risk of the building, as a whole, falling down. This would not only limit the amount of post-quake repair, but would also indicate where the failed connections were and thus greatly reduce the expense of inspection procedures.

6.2.1.3 Application to Tubular Structures

The quest for oil and gas exploitation has been pushing the development of offshore platforms such as jackets and jackups that can operate in deeper and harsher water. During operation, offshore platforms are subjected to extreme loads caused by waves, currents, hurricanes, and earthquakes. Over the years, catastrophic events such as explosions, fires, ship impacts, and dropped objects have resulted in the increased focus on the structural resistance to accidental events. A rational evaluation of the structural strength considering these factors would require all the nonlinear effects to be taken directly into the analysis.

Amdahl et al. (1994), Hellan (1995), and Skallerud and Amdahl (2002) proposed methodologies for nonlinear analysis of 3-D tubular structures covering three main aspects: (1) basic nonlinear continuum mechanics and numerical procedures implemented in state-of-the-art computer codes, USFOS (Søreide et al., 1994; USFOS, 2009); (2) modeling of nonlinear tubular joint and member behavior on 3-D behavior and collapse mechanisms; (3) reserved and residual strength of 3-D frames and the redundancy and load-shedding characteristics system behavior. Their works, which includes many of the recent developments in assessing the capacity of offshore structures, have been adopted by DNV into the standard DNV-RP-C204 (2004).

DNV, SINTEF, and BOMEL carried out the ULTIGUIDE project (1999) to provide best practical guidance for engineers to select parameters and interpret results from a nonlinear analysis. The guide, which focuses on jacket structures, does not specify a recommended safety level, but is intended for use with existing codes and specifications. The topics covered in this guide address the question arising in connection with reassessment of existing structures, but the recommendations will be valid also in other design situations.

ICSS committee III.1 provides a summary report *Ultimate Strength* (2006) about the research progress on the ultimate strength of tubular members and joints, plates and stiffened plates, shells, ship structures, offshore structures, composite structures, aluminum structures, and relevant benchmarks in the International Ship and Offshore Structures Congress held every three years.

Experiments were carried out to benchmark the nonlinear analysis methods. To understand the frame cyclic inelastic behavior, which determines the survivability of the offshore structures in the event of severe seismic ground excitations, Zayas et al. (1980) carried out the first large scale frame test for two one-sixth scale two-dimensional (2-D) X-braced double-bay frames, which stand more than 8m high under cyclic loading. The Jointed Industry Project (JIP) organized by BOMEL and other organizations investigates experimentally two series of 2-D large scale frames which consist of six X-braced two bay frames and four K-braced single-bay frames, both under static loading (BOMEL, 1992). The subsequent phase of the JIP (Bolt and Billington, 2000) tests one large-scale 3-D frame under a series of static loading conditions. These are by far the largest frames tested worldwide. The findings indicated that within the confines of a frame, the structural component behaves differently than isolated tests on an individual component, on which engineering practice has been based. Furthermore, both analyses and tests have shown that structural systems exhibit significant reserve strength beyond design load levels. Besides, Qian (2005) and Choo et al. (2005) carried out research to extend the understanding to the thick-walled joints and developed a nonlinear joint model for the ultimate strength frame analysis of offshore structures.

6.2.2 Stiffness Formulation of Beam-column

6.2.2.1 Available Stiffness Formulations

Two popular methods have been used to derive the tangent stiffness matrix of a beam-column element, i.e., the beam-column approach and the finite element approach. The beam-column approach was originally proposed by Oran (1973), later improved by Kassimali and Abbasnia (1991) to include the effect of axial force on the torsional stiffness and some coupling stiffness terms, and recently by Chan and Gu (2000) to include the effect of the member initial out-of-straightness. In this approach, the tangent stiffness matrix is derived from the member basic force and deformation within the secant relationship, which satisfies the fourth-order beam-column equilibrium equation. The initial undeformed chord, which translates and rotates as a rigid body, is selected as the reference configuration and the member bowing effect can be captured. Chan and Zhou (1995) expressed their member force and deformation in the form similar to Chan and Gu (2000), and considered the bowing effect and member initial out-of-straightness. The stability and bowing functions in Chan and Zhou (1995) is based on the fifth-order polynomials, which can be used to approximate the accurate values of Chan and Gu (2000) with high accuracy. However, when an attempt is made to solve for the 3-D frame behavior, i.e., a direct extension is made from planar element to space frame element, some coupling terms between the flexural and torsional displacements are lost in the four aforementioned methods. Therefore, the lateral-torsional buckling cannot be predicted by using such an approach.

The other method is the finite element approach by using the cubic interpolation functions for lateral displacements (Orbison, 1982; Izzuddin and Elnashai, 1993a; Yang and Kuo, 1994). Since the stability criterion is based on the principles of energy, it eliminates the necessity of going through the seemingly ad hoc visualization required to account for changes in geometry when formulating the beam-column equilibrium equations for free-body diagrams. This approach can be extended to a wide range of problems, such as the 3-D inelastic analysis. Since the energy principles are able to define the nonlinear coupling between compression, bending, and torsion, the derived element can be used to predict the axial-torsional and lateral-torsional buckling, which are essential for an accurate estimate of the second-order effect in 3-D framework. However, the pitfall of using the cubic interpolation function approach is that one cubic element cannot be used to predict accurately the flexural buckling load of columns unless more elements per physical member are used. Since the straight, albeit stressed, element has been widely adopted in the element stiffness formulation, the frame members have to be subdivided into several elements to capture the member bowing effect in the nonlinear analysis. This will inevitably increase the complexity of structural modeling and the cost and time of computation.

To avoid the limitations of the beam-column approach and the cubic element approach, a 3-D beam-column element has been developed by using the stability interpolation functions that satisfy the fourth-order differential equation of beam-columns (Liew et al., 2000b). The flexural buckling load of columns and frames can be captured with one element per member. With some modifications, the proposed beam-column element considers the member bowing effect and initial out-of-straightness. Hence, the nonlinear behavior of frames composed of slender member subject to high axial force can be captured with less elements per member. The advantages of the proposed beam-column formulation over the conventional beam-column approach and the cubic element approach are highlighted by numerical examples given in Liew et al. (2000b).

Besides the beam-column approach and finite element approach, Yang et al. (2002) recently developed a methodologically new, physically interpretable approach for formulating the beam element using the equilibrium approach—called direct approach—by adhering strictly to the assumption of small strains, small displacement, and small rotations for each incremental step of nonlinear analysis.

6.2.2.2 Virtual Work Equation

In the incremental finite element analysis, three formulations are popularly used to derive the element stiffness, i.e., the Updated Lagrangian (UL) formulation, the Total Lagrangian (TL) formulation, and the Co-Rotational (CR) approach. In the UL formulation, the motion of element is described based on the previous (known) deformed configuration. Whereas, in the TL formulation, the motion of element is described based on the initial undeformed configuration. It has been concluded by Bathe and Bolourchi (1979) that the UL formulation is computationally more efficient than the TL formulation for frame structures. On the other hand, the reference configuration in a CR formulation continuously translates and rotates with the element, but does not deform with it (Teh and Clarke, 1996, 1998). The CR formulation was also called the Co-Rational Total Lagrangian formulation in Teh and Clarke (1996), the Eulerian formulation in Izzuddin and Elnashai (1993b) due to the co-rotational but undeformed reference configuration. The CR formulation has certain advantages such as reduction in degrees-of-freedom involved in the formulation by excluding the rigid body modes and volume integrations generally performed over the simple undeformed shape of the element. However, in the CR formulation, the approximate cubic interpolation functions for lateral displacement are normally adopted and special considerations should be taken to get the correct stability matrix.

Considering the above reasons, the motion of a beam-column element is described based on the UL formulation as shown in Figure 6.6, with C_0 representing the initial undeformed configuration, C_1 the previous (known) deformed equilibrium configuration, and C_2 the current (desired) deformed equilibrium configuration. The straight, albeit stressed, element at configuration C_1 has been widely adopted in the stiffness formulation. Hence, in the range of large deflection, the frame members have to be subdivided into several elements to capture the member bowing effect. Gattass and Abel (1987a) developed a 2-D curved beam-column element based on the curved and stressed element configuration. However, due to the complicated symbolic operations, an extension to a 3-D case is difficult.

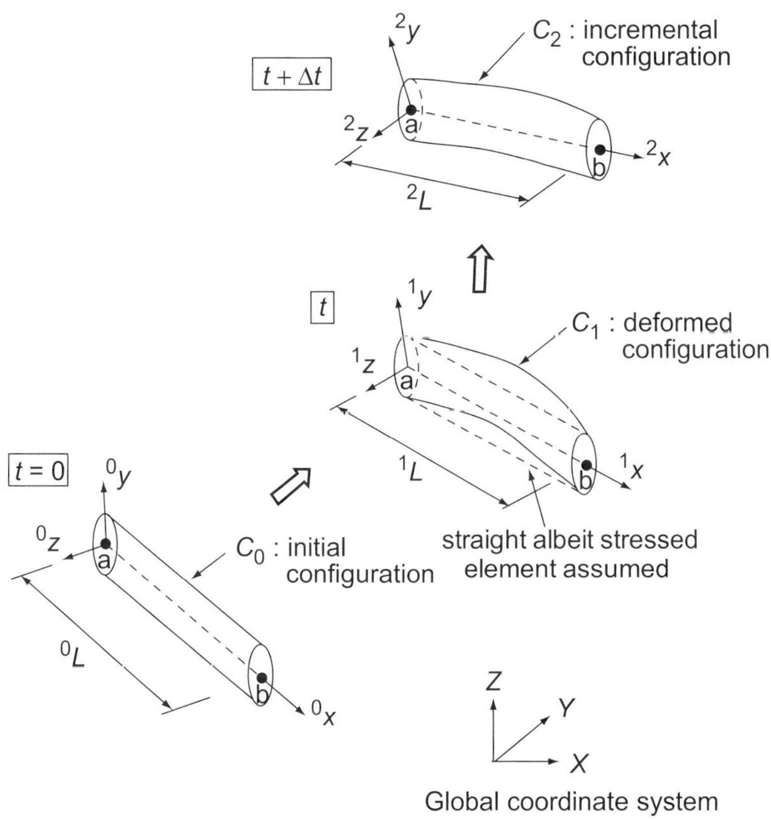

FIGURE 6.6 Motion of a beam-column element based on the UL formulation.

In the finite element approach, the element stiffness can be derived using either the principle of potential energy or the principle of virtual work. The principle of virtual work has advantages over the principle of potential energy in that it is not restricted to material laws or loading type and can consider the geometric nonlinearities easily by including the nonlinear strain components in the increment Lagarangian formulations (Yang and Kuo, 1994). Based on the principle of virtual work, the following equation can be written for an element in an incremental sense between the previous configuration C_1 and the current configuration C_2:

$$\begin{aligned}&\int_{^1V} {}_1C_{ijkl}\ {}_1e_{kl}\ \delta\ {}_1e_{ij}\ {}^1dV + \int_{^1V} {}_1C_{ijkl}\ {}_1\eta_{kl}\ \delta\ {}_1e_{ij}\ {}^1dV + \int_{^1V} {}_1C_{ijkl}\ {}_1e_{kl}\ \delta\ {}_1\eta_{ij}\ {}^1dV + \\ &\int_{^1V} {}_1C_{ijkl}\ {}_1\eta_{kl}\ \delta\ {}_1\eta_{ij}\ {}^1dV + \int_{^1V} {}_1^1\tau_{ij}\ \delta\ {}_1\eta_{ij}\ {}^1dV = {}_1\delta\mathbf{u}^T\left({}_1^2\mathbf{f} - {}_1^1\mathbf{f}\right)\end{aligned} \tag{6.2}$$

in which ${}_1C_{ijkl}$ is the incremental constitutive tensor; ${}_1e_{if}$ and ${}_1\eta_{if}$ are the linear and nonlinear components of Green incremental strain tensor, respectively; ${}_1^1\tau_{ij}$ is Cauchy stress tensor; **u** is element displacement vector; and **f** is the element force vector. In Equation 6.2, a left subscript in a symbol denotes the configuration in which the quantity is measured. A left superscript denotes the configuration in which the quantity occurs. The absence of a superscript means that the quantity is an increment between configurations C_1 and C_2.

If cubic interpolation functions are used for lateral displacements, the first term in the left side of Equation 6.2 yields the elastic stiffness matrix, and the fifth term yields the geometric stiffness matrix. The second to the fourth terms yield higher-order stiffness matrices, which have been considered by Yang and Leu (1994) for planar frame using the principle of virtual work, and by Chajes and Churchill (1987) for planar frame using the principle of potential energy. The third higher-order term has been considered by Al-Bermani and Kitipornchai (1990) for deriving the stiffness matrix of a 3-D beam element with thin-walled sections. However, inconsistent reference configurations were used for nodal displacements with translations referred to C_1 and rotations referred to C_2 as pointed out by Teh and Clarke (1996). Jiang et al. (2002) considered the first to the fifth terms in the left side of Equation 6.2 in their plastic zone analysis. However, the stiffness matrix is not given in explicit form, whereas it is integrated numerically.

Figure 6.7 shows the beam-column element with twelve degrees of freedom. The local coordinate is chosen with the x axis as the centroidal axis, and the y and z axes as the principal axes. There are six stress resultants at each end: three forces, F_x, F_y, and F_z; three moments, M_x, M_y, and

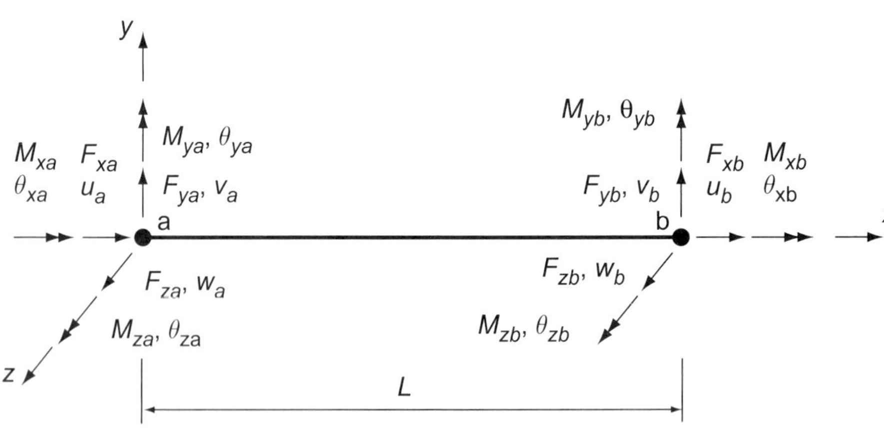

FIGURE 6.7 Force and displacement components of a beam-column element.

M_z. The corresponding displacements are translations u, v, and w; and rotations, θ_x, θ_y, and θ_z. The basic assumptions used in modeling the beam-column element are:

1. Elements are initially prismatic and cross-sections do not distort.
2. Plane sections remain plane after deformation.
3. Local buckling is not considered and all members are assumed to be compact.
4. Warping, shear, and inelastic torsional deformations are neglected.
5. Member deformations with respect to the chord length are assumed to be small, but global displacements and rotations can be moderately large.

Assumption 5 requires the rotational angles between the tangent at the member ends and the chord joining the two member ends to be small (see Figure 6.8). If this assumption is violated, the member should be divided into two or more elements to reduce the magnitude of the rotational angle. However, this condition is seldom encountered in practical structures.

By adopting the assumptions of beam-column and ignoring the second to the fourth terms on the left side of Equation 6.2 the linearized form of an incremental virtual work equation may be expressed as:

$$\begin{aligned}
&\frac{1}{2}\int_0^L\left[EA\delta\left(u'^2\right)+EI_z\delta\left(v''^2\right)+EI_y\delta\left(w''^2\right)+GJ\delta\left(\theta_x'^2\right)\right]\,dx \\
&+\int_0^L\frac{F_x}{2}\delta\left(v'^2+w'^2\right)\,dx-\int_0^L\left[F_y\delta\left(u'v'\right)+F_z\delta\left(u'w'\right)\right]\,dx+\int_0^L\frac{\bar{K}}{2}\delta\left(\theta_x'^2\right)\,dx \\
&-\int_0^L\left[F_z\delta\left(v'\theta_x\right)+M_y\delta\left(v'\theta_x'\right)\right]\,dx+\int_0^L\left[F_y\delta\left(w'\theta_x\right)-M_z\delta\left(w'\theta_x'\right)\right]\,dx \\
&-\int_0^L\frac{M_x}{2}\left[\delta\left(v'w''\right)-\delta\left(v''w'\right)\right]\,dx=\delta\mathbf{u}^T\left({}^2\mathbf{f}-{}^1\mathbf{f}\right)
\end{aligned} \tag{6.3}$$

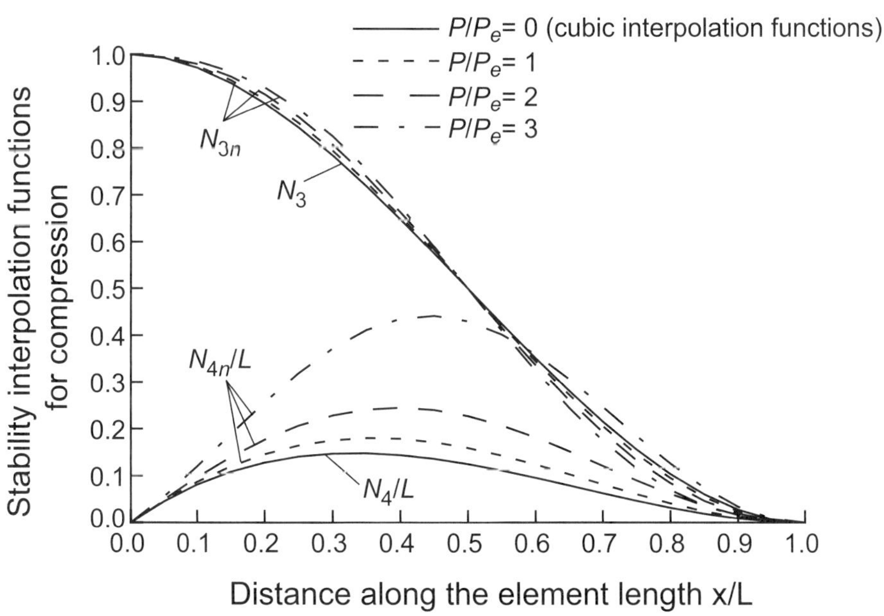

FIGURE 6.8 Stability interpolation functions.

in which E is the elastic modulus; G is the shear modulus; L is the element length at configuration C_1; A is the cross-sectional area; I_y and I_z, are moments of inertia, respectively; $\bar{K} = F_{xb}\,(I_y + I_z)$ is the Wagner coefficient; J is the torsional constant; and **f** and **u** are element force and displacement vectors given as:

$$\mathbf{f} = \left[F_{xa}\ F_{ya}\ F_{za}\ M_{xa}\ M_{ya}\ M_{za}\ F_{xb}\ F_{yb}\ F_{zb}\ M_{xb}\ M_{yb}\ M_{z} \right]_T \tag{6.4}$$

$$\mathbf{u} = \left[u_a\ v_a\ w_a\ \theta_{xa}\ \theta_{ya}\ \theta_{za}\ u_b\ v_b\ w_b\ \theta_{xb}\ \theta_{yb}\ \theta_{zb} \right]^T \tag{6.5}$$

The inclusion of the fourth integral in the left side of Equation 6.3 would lead to an accurate prediction of the axial-torsional instability of the column due to the Wagner effect. The inclusion of the fifth to the seventh integrals is important for predicting the lateral-torsional buckling.

6.2.2.3 Stability Interpolation Functions

The derivation of the element stiffness matrices is based on the displacement field assumed. Linear interpolation functions are adopted for the axial displacement u and the angle of twist θ_x. The stability interpolation functions are used for the lateral displacements v and w. The displacements at the centroid of an arbitrary cross-section can be written in terms of those at element ends:

$$u = \mathbf{N_u}\ \mathbf{u} \tag{6.6a}$$

$$v = \mathbf{N_v}\ \mathbf{u} \tag{6.6b}$$

$$w = \mathbf{N_w}\ \mathbf{u} \tag{6.6c}$$

$$\theta_x = \mathbf{N_\theta}\ \mathbf{u} \tag{6.6d}$$

where:

$$\mathbf{N_u} = \left[N_1\ 0\,0\,0\,0\,0\ N_2\ 0\,0\,0\,0\,0 \right] \tag{6.7a}$$

$$\mathbf{N_v} = \left[0\ \mathrm{N}_{3z}\ 0\,0\,0\ \mathrm{N}_{4z}\ 0\ \mathrm{N}_{5z}\ 0\,0\,0\ \mathrm{N}_{6z} \right] \tag{6.7b}$$

$$\mathbf{N_w} = [0\ 0\ \mathrm{N}_{3y}\ 0\ -\mathrm{N}_{4y}\ 0\,0\,0\ \mathrm{N}_{5y}\ 0\ -\mathrm{N}_{6y}\ 0] \tag{6.7c}$$

$$\mathbf{N_\theta} = \left[0\,0\,0\ \mathrm{N}_1\ 0\,0\,0\,0\,0\ \mathrm{N}_2\ 0\,0 \right] \tag{6.7d}$$

The linear interpolation functions are given by:

$$N_1 = 1 - \xi \tag{6.8a}$$

$$N_2 = \xi \tag{6.8b}$$

in which $\xi = x / L$.

The stability interpolation functions N_{nz} (where n = 3–6) can be derived from the fourth-order differential equation of a beam-column subject to the axial force P ($P < 0$ for compression, $P > 0$ for tension). Following the general procedures of the finite element formulation, the stability interpolation functions in both planes are assumed to be independent.

The stability interpolation functions for compression members are given by:

$$N_{3n} = \frac{1-c}{\Delta}\cos(kx) - \frac{s}{\Delta}\sin(kx) + \frac{s}{\Delta}kx + \frac{1-c-kLs}{\Delta} \tag{6.9a}$$

$$N_{4n} = \frac{s-kLc}{k\Delta}\cos(kx) + \frac{1-c-kLs}{k\Delta}\sin(kx) + \frac{1-c}{k\Delta}kx + \frac{kLc-s}{k\Delta} \tag{6.9b}$$

$$N_{5n} = \frac{c-1}{\Delta}\cos(kx) + \frac{s}{\Delta}\sin(kx) - \frac{s}{\Delta}kx + \frac{1-c}{\Delta} \tag{6.9c}$$

$$N_{6n} = \frac{kL-s}{k\Delta}\cos(kx) + \frac{c-1}{k\Delta}\sin(kx) + \frac{1-c}{k\Delta}kx + \frac{s-kL}{k\Delta} \tag{6.9d}$$

in which $c = \cos(kL)$, $s = \sin(kL)$, $\Delta = 2 - 2\cos(kL) - kL\sin(kL)$, and $k = \sqrt{|P/EI_n|}$ $(n = y, z)$.

The stability interpolation functions for tension members are given by:

$$N_{3n} = \frac{1-c}{\Delta}\cosh(kx) + \frac{s}{\Delta}\sinh(kx) - \frac{s}{\Delta}kx + \frac{1-c+kLs}{\Delta} \tag{6.10a}$$

$$N_{4n} = \frac{s-kLc}{k\Delta}\cosh(kx) + \frac{1-c+kLs}{k\Delta}\sinh(kx) + \frac{1-c}{k\Delta}kx + \frac{kLc-s}{k\Delta} \tag{6.10b}$$

$$N_{5n} = \frac{c-1}{\Delta}\cosh(kx) - \frac{s}{\Delta}\sinh(kx) + \frac{s}{\Delta}kx + \frac{1-c}{\Delta} \tag{6.10c}$$

$$N_{6n} = \frac{kL-s}{k\Delta}\cosh(kx) + \frac{c-1}{k\Delta}\sinh(kx) + \frac{1-c}{k\Delta}kx + \frac{s-kL}{k\Delta} \tag{6.10d}$$

in which $c = \cosh(kL)$, $s = \sinh(kL)$, and $\Delta = 2 - 2\cosh(kL) - kL\sinh(kL)$.

If the axial force P is small, they can be approximated by the cubic interpolation functions that are independent of the axial force:

$$N_3 = 1 - 3\xi^2 + 2\xi^3 \tag{6.11a}$$

$$N_4 = L\left(\xi - 2\xi^3 + \xi^3\right) \tag{6.11b}$$

$$N_5 = 3\xi^2 - 2\xi^3 \tag{6.11c}$$

$$N_6 = L\left(-\xi^2 + \xi^3\right) \tag{6.11d}$$

Figure 6.8 shows the comparisons of the cubic and stability interpolation functions. It can be seen that the errors for the cubic interpolation functions increase as the axial force increases. The main advantage of the stability interpolation functions is that they satisfy the governing differential equations of beam-column and thus can be used to predict accurately the lateral displacement under the effect of axial force (i.e., P-δ effect).

6.2.2.4 Elastic and Geometric Stiffness Matrices

The forces at the internal cross-section can be expressed in terms of those at the element ends:

$$F_x = F_{xb} \tag{6.12a}$$

$$F_y = -\left(M_{za} + M_{zb}\right)/L \tag{6.12b}$$

$$F_z = \left(M_{ya} + M_{yb}\right)/L \tag{6.12c}$$

$$M_x = M_{xb} \tag{6.12d}$$

$$M_y = -(1-\xi)M_{ya} + \xi M_{yb} \tag{6.12e}$$

$$M_z = -(1-\xi)M_{za} + \xi M_{zb} \tag{6.12f}$$

Based on the displacement fields and the expression for the forces at the internal cross-section, the element stiffness matrix **k** can be derived from the virtual work equation as:

$$\begin{aligned}\mathbf{k} = {} & EA\,K_{uu}^{110} + EI_z\,K_{vv}^{220} + EI_y\,K_{ww}^{220} + GJ\,K_{\theta\theta}^{110} + F_{xb}K_{vv}^{110} + F_{xb}K_{ww}^{110} \\ & + V_y\left(K_{uv}^{110} + K_{vu}^{110}\right) - V_z\left(K_{uw}^{110} + K_{wu}^{110}\right) + \bar{K}\,K_{\theta\theta}^{110} \\ & - \frac{M_{ya} + M_{yb}}{L}\left(K_{v\theta}^{100} + K_{\theta v}^{010} + K_{v\theta}^{111} + K_{\theta v}^{111}\right) + M_{ya}\left(K_{v\theta}^{110} + K_{\theta v}^{110}\right) \\ & - \frac{M_{za} + M_{zb}}{L}\left(K_{w\theta}^{100} + K_{\theta w}^{010} + K_{w\theta}^{111} + K_{\theta w}^{111}\right) + M_{za}\left(K_{w\theta}^{110} + K_{\theta w}^{110}\right) \\ & - \frac{M_{xb}}{2}\left(K_{vw}^{120} + K_{wv}^{210} - K_{vw}^{210} - K_{wv}^{120}\right)\end{aligned} \tag{6.13}$$

where $V_y = \left(M_{za} + M_{zb}\right)/L$, $V_z = \left(M_{ya} + M_{yb}\right)/L$, and matrix $\mathbf{K}_{\mathbf{gh}}^{\mathbf{stv}}$ is given by:

$$\mathbf{K}_{\mathbf{gh}}^{\mathbf{stv}} = \int_0^L \frac{d^s N_g^T}{dx^s}\frac{d^t N_h}{dx^t}\,x^v\,dx \tag{6.14}$$

in which the subscripts g and h represent u, v, w, or θ; the superscripts s and t represent the order of differentiation of the vector of interpolation functions **N**; and v is the order of exponent to x. The tenth to the fourteenth stiffness terms on the right side of Equation 6.13 includes the coupling effects between the lateral and torsional displacements, which can be used to predict the lateral-torsional instability.

If the cubic interpolation functions are used for the lateral displacements, the first to the fourth terms in the right side of Equation 6.13 will lead to the elastic stiffness matrix $\mathbf{k}_\mathbf{e}$, and the remaining terms will result in the cubic geometric stiffness matrix $\mathbf{k}_\mathbf{gc}$ (Orbison, 1982; Yang and Kuo, 1994).

If the stability interpolation functions are used for the lateral displacements, the first to the fourth terms on the right side of Equation 6.13 will contain terms that are functions of the axial force. The remaining terms must be evaluated together. Although the last term can be evaluated by using the stability interpolation function, the expressions are complex and not easy for use. For simplicity, the cubic interpolation functions are used instead. However, in evaluating the other terms, the stability interpolation functions for the lateral displacements are used consistently. The resulting element stiffness matrix **k** can be conceptually decomposed into two parts: the elastic stiffness matrix $\mathbf{k}_\mathbf{e}$ and the geometric stiffness matrix $\mathbf{k}_\mathbf{g}$, based on the stability interpolation functions. The elastic stiffness matrix can be modified to consider the shear deformation (Przemieniecki, 1968), and is given in Appendix 1. The geometric stiffness matrix is given in Appendix 2.

In the geometric stiffness matrix, the stability functions S_{1n} and S_{2n}, which account for the effect of axial force on the bending stiffness, are given by (Chen et al., 1996):

$$S_{1n} = \begin{cases} \dfrac{\phi_n \sin\phi_n - \phi_n^2 \cos\phi_n}{2 - 2\cos\phi_n - \phi_n \sin\phi_n} & P < 0 \\[2ex] \dfrac{\phi_n^2 - \cosh\phi_n - \phi_n \sinh\phi_n}{2 - 2\cosh\phi_n + \phi_n \sinh\phi_n} & P > 0 \end{cases} \tag{6.15}$$

$$S_{2n} = \begin{cases} \dfrac{\phi_n^2 - \phi_n \sin\phi_n}{2 - 2\cos\phi_n - \phi_n \sin\phi_n} & P < 0 \\[2ex] \dfrac{\phi_n \sinh\phi_n - \phi_n^2}{2 - 2\cosh\phi_n + \phi_n \sinh\phi_n} & P > 0 \end{cases} \tag{6.16}$$

in which $\phi_n = L\sqrt{\left|P/EI_n\right|}$ and $P = F_{xb}$.

If the axial force is small, they can be approximated as:

$$S_{1n} = 4 + \frac{2PL^2}{15EI_n} \tag{6.17}$$

$$S_{2n} = 2 - \frac{PL^2}{30EI_n} \tag{6.18}$$

Equations 6.17 and 6.18 can be called the third-order stability function because they can be derived if cubic interpolation functions are used for lateral displacement.

It should be noted that the cubic interpolation functions involve some approximation for the lateral displacements, and its range of application must be understood. Defining the percent error in the element stiffness matrix terms as:

$$\text{Error} = \left(\frac{\mathbf{k}_c - \mathbf{k}}{\mathbf{k}}\right) \times 100\ (\%) \tag{6.19}$$

in which $\mathbf{k}_c = \mathbf{k}_e + \mathbf{k}_{gc}$ and $\mathbf{k} = \mathbf{k}_e + \mathbf{k}_g$ are the element stiffness matrices based on the cubic and stability interpolation functions, respectively. The cubic geometric stiffness $\mathbf{k}_{gc}$ can be obtained if Equation 6.17 and 6.18 are substituted into the geometric stiffness $\mathbf{k}_g$ based on the stability interpolation functions. If $|P/P_e| < 0.82$ ($P_e = \pi^2 EI_n/L^2$ is the Euler's load), the largest error in any of the terms of $\mathbf{k}_c$ is less than 5 percent. If the cubic geometric stiffness matrix $\mathbf{k}_{gc}$ is used for cases of $|P/P_e| > 0.82$, the corresponding members should be divided into two or more elements to limit the error in stiffness terms to less than 5 percent (White et al., 1993; Liew et al., 2000b).

6.2.2.5 Beam-column Approach and Member Bowing Effect

In the updated Lagrangian approach, the assumption of straight, albeit stressed, element is widely adopted. It is difficult to account for the influence of displaced shape on the element stiffness, such as the curvature shortening effect (bowing effect) and member initial out-of-straightness. However, for slender members under high axial load and members that are loaded far into the post-collapse region, it is important to consider the bowing effect to capture the nonlinear behavior.

Considering a member with initial out-of-straightness as shown in Figure 6.9, the member basic force and deformation relationship can be written as (Oran, 1973; Kassimali and Abbasnia, 1991; Chan and Gu, 2000):

$$M_{na} = \frac{EI_n}{L_0}\left(S_{1n}\Theta_{na} + S_{2n}\Theta_{nb} + c_{0n}\frac{\delta_n}{L_0}\right) \tag{6.20a}$$

$$M_{nb} = \frac{EI_n}{L_0}\left(S_{2n}\Theta_{na} + S_{1n}\Theta_{nb} - c_{0n}\frac{\delta_n}{L_0}\right) \tag{6.20b}$$

$$M_x = \frac{GJ + \Pr_0^2}{L_0}\Theta_x \tag{6.20c}$$

$$P = EA\left\{\frac{e}{L_0} + \sum_{n=z,y}\left[b_{1n}\left(\Theta_{na} + \Theta_{nb}\right)^2 + b_{2n}\left(\Theta_{na} - \Theta_{nb}\right)^2 + b_{vsn}\frac{\delta_n}{L_0}\left(\Theta_{na} - \Theta_{nb}\right) + b_{vvn}\left(\frac{\delta_n}{L_0}\right)^2\right]\right\} \tag{6.20d}$$

where M_{na} and M_{nb} ($n = z, y$) are end moments; Θ_{na} and Θ_{nb} are the total end rotations; M_x is the torsional moment; Θ_x is the total twist; P is the axial force; e is the relative axial displacement; L_0 is the initial member length; $r_0 = \sqrt{(I_y + I_z)/A}$ is the polar radius of gyration; S_{1n} and S_{2n} are the stability functions; δ_y and δ_z, are the values of initial out-of-straightness at the midspan in the z and y directions, which can be worked out based on the direction and total amplitude of initial

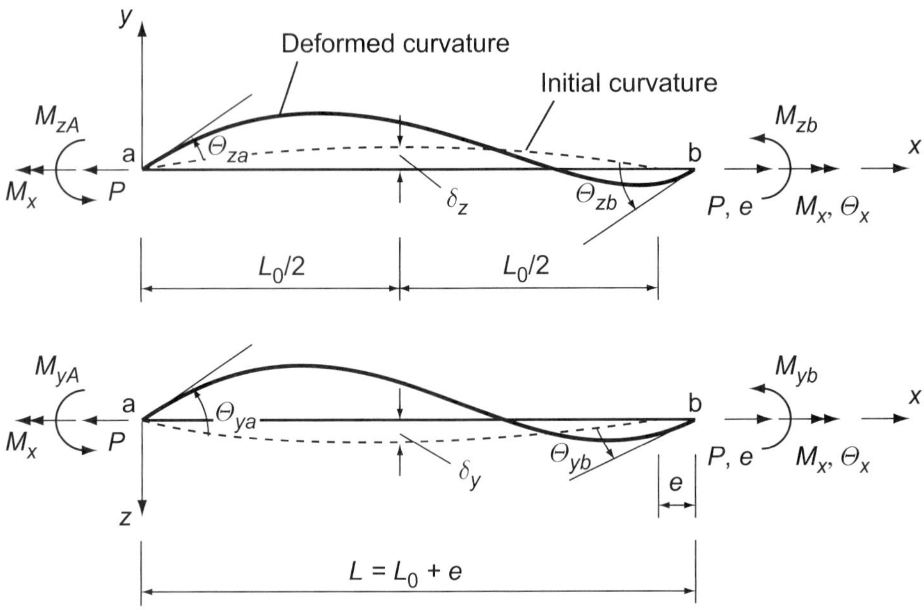

FIGURE 6.9 Member basic forces and deformations.

out-of-straightness at the midspan; b_{1n} and b_{2n} are the bowing functions that consider the change of member chord length due to curvature shortening, given by Oran (1973):

$$b_{1n} = -\frac{\left(S_{1n}+S_{2n}\right)\left(S_{2n}-2\right)}{8q_n} \tag{6.21}$$

$$b_{2n} = \frac{S_{2n}}{8\left(S_{1n}+S_{2n}\right)} \tag{6.22}$$

and stability function c_{0n} and bowing function b_{vsn} and b_{vvn}, considering the effects of member initial out-of-straightness are given by Chan and Gu (2000):

$$c_{0n} = \begin{cases} -\dfrac{\pi\phi_n^3 \sin\phi_n}{\left(\pi^2-\phi_n^2\right)\left(1-\cos\phi_n\right)} & P<0 \\ \dfrac{\pi\phi_n^3 \sinh\phi_n}{\left(\pi^2+\phi_n^2\right)\left(\quad \phi_n-1\right)} & P>0 \end{cases} \tag{6.23}$$

$$b_{vsn} = \frac{\pi^3\left(S_{1n}-S_{2n}\right)}{\left(\pi^2+q_n\right)^2} + \frac{4b_{2n}c_{0n}}{S_{1n}-S_{2n}} \tag{6.24}$$

$$b_{vvn} = -\frac{\pi^2 q_n\left(2\pi^3+q_n\right)}{4\left(\pi^2+q_n\right)^2} + \frac{2\pi^3 c_{0n}}{\left(\pi^2+q_n\right)^2} + \frac{4\pi^2 q_n^2 b_2}{\left(\pi^2+q\ \right)^2} \tag{6.25}$$

$$q_n = \frac{PL_0^2}{EI_n} \tag{6.26}$$

when axial force is zero, $b_{1n} = 1/40$, $b_{2n} = 1/24$, $c_{0n} = 0$, $b_{vsn} = 2/\pi$, $b_{vvn} = 0$.

The element tangent stiffness matrix can be determined as:

$$\mathbf{k}_t = \mathbf{T}^T \mathbf{t}\,\mathbf{T} + \mathbf{N} \tag{6.27}$$

in which **T** is a 12 × 6 transformation matrix relating the member basic forces to the element forces in the local coordinate **N**, a matrix accounting for the work done by the initial force and the translational displacements and **t**, the tangent stiffness matrix for member basic deformations given by:

$$t_{ij} = \frac{\partial S_i}{\partial d_j} + \frac{\partial S_i}{\partial q}\frac{\partial q}{\partial d_j} \quad \text{for } i, j = 1 \sim 6 \tag{6.28}$$

in which $\mathbf{S} = \left[M_{za}\ M_{zb}\ M_{ya}\ M_{yb}\ M_x\ P\right]^T$, $\mathbf{d} = \left[\Theta_{za}\ \Theta_{zb}\ \Theta_{ya}\ \Theta_{yb}\ \Theta_x\ e\right]^T$, $q = \dfrac{PL_0^2}{\pi^2 EI}$, and $I = I_y + I_z$.

The tangent stiffness matrix expressed by Equation 6.28 can be decomposed into three parts: the elastic stiffness matrix $\mathbf{k}_e$, the geometric stiffness matrix $\mathbf{k}'_g$, and the bowing matrix $\mathbf{k}_b$. The bowing matrix, which considers the influence of member bowing effect and initial out-of-straightness on the beam-column stiffness, is given in Appendix 3.

In the beam-column approach, the space element is a direct extension from the planar one, and some coupling terms between the flexural and torsional displacements are lost in the geometric stiffness $\mathbf{k}'_g$. However, these coupling terms have been included in the geometric stiffness matrix $\mathbf{k}_g$, which is derived from the virtual work equation by using the stability interpolation functions as given in Sec. 6.2.2.3. Hence, in the nonlinear analysis of frame structures, the geometric matrix $\mathbf{k}_g$ should be used instead of the matrix $\mathbf{k}'_g$.

Although Kondoh and Atluri (1986) considered bowing effect in their beam-column formulation, no allowance has been made for the member initial out-of-straightness. Al-Bermani and Kitipornchai (1990) proposed a deformation matrix to consider the effect of deformed geometry and allow the use of least elements in a nonlinear analysis. However, their formulation has not been tested for space frames in which the member is slender and subject to high axial force. Numerical examples in Liew et al. (2000b) show that the proposed bowing matrix is particularly useful for slender members subjected to high axial force. This method has widened the scope of advanced analysis for many types of space frame structures.

6.2.2.6 Stability and Bowing Functions

Figures 6.10–6.12 compare the different levels of approximation of the stability functions S_{1n}, S_{2n}, and c_{0n} and the bowing functions b_{1n}, b_{2n}, b_{vsn}, and b_{vvn}. It can be seen from the figures that these parameters vary considerably with the axial force, and thus influence the element stiffness. Chan and Zhou's expressions (1995) based on the polynomial functions of fifth-order provide good approximation to the accurate values except in the range of high compression load. The advantage of Chan and Zhou's expressions is that they have no singularity points. However, the exact expressions in Equations 6.15–6.16 and 6.21–6.25 encounter singularity at certain axial load values. To obtain the values at the singularity points of the

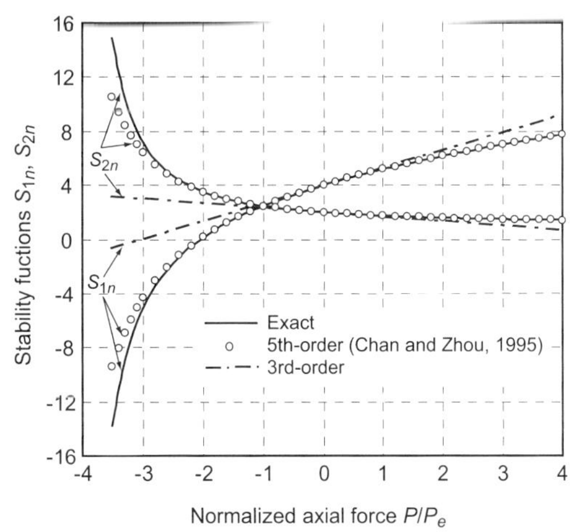

FIGURE 6.10 Exact and approximate stability functions S_{1n} and S_{2n}.

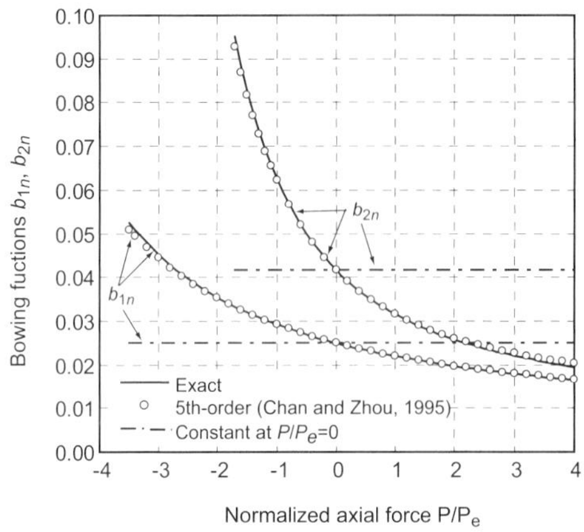

FIGURE 6.11 Exact and approximate bowing functions b_{1n} and b_{2n}.

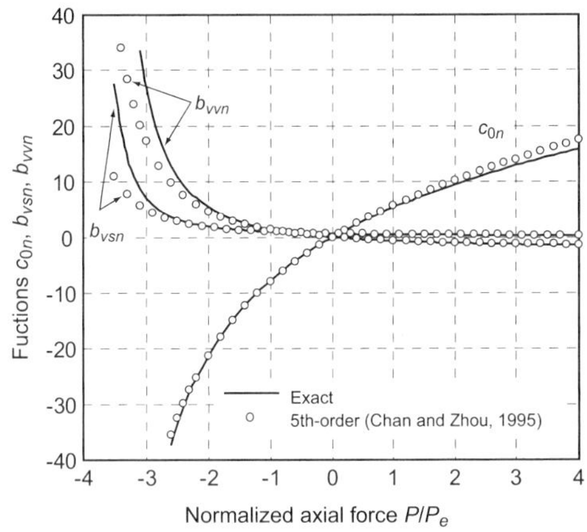

FIGURE 6.12 Exact and approximate stability function c_{0n} and bowing functions b_{vsn} and b_{vvn}.

accurate expression, L'Hospital's Rule will be used and linear approximations adopted for a region near the singularity points. In Chen et al. (1996), stability functions S_{1n} and S_{2n} are approximated by some approximated functions without singularity points. It should be noted that the secant stiffness relationship used by Meek and Loganathan (1989) is a degenerated form of Equation 6.20 by adopting the cubic stability functions S_{1n} and S_{2n} given in Equations 6.17 and 6.18, b_{1n} = 1/40, b_{2n} = 1/24, $c_{0n} = b_{vsn}$, b_{vvn} = 0.

6.2.2.7 Tangent Stiffness Matrix

The stiffness equation of a beam-column element can be expressed as:

$$\left(\mathbf{k}_e + \mathbf{k}_g + \mathbf{k}_b\right) d\mathbf{u} = \mathbf{k}_t \; d\mathbf{u} = d\mathbf{f} \tag{6.29}$$

in which $d\mathbf{u}$ is the incremental displacement vector, $d\mathbf{f}$ is the incremental force vector, $\mathbf{k}_e$ is the elastic stiffness matrix, $\mathbf{k}_g$ is the geometric stiffness matrix based on the stability interpolation functions, $\mathbf{k}_b$ is the bowing matrix, and $\mathbf{k}_t$ is the elastic tangent stiffness matrix. It is noted that in the cubic element approach (Orbison, 1982; Yang and Kuo, 1994), the bowing matrix $\mathbf{k}_b$ is not included, and the geometric stiffness matrix $\mathbf{k}_{gc}$, based on the cubic interpolation functions, are used instead of the matrix $\mathbf{k}_g$.

In the geometrically nonlinear analysis by using the beam-column element, an improper solution can be found if the joint rotations are represented by the displacement derivatives (Yang and Kuo, 1994). For spatial members, the induced moment matrix $\mathbf{k}_i$ should be added to the tangent stiffness matrix to make the usual degrees of freedom that are represented by the displacement derivatives in amounts equal to the commutative rotational degree of freedom, and hence to yield the true equilibrium condition and satisfy the rigid body tests. Due to the lack of conjugateness between bending moments and displacement derivatives, $\mathbf{k}_i$ is asymmetric for a single and discrete element. The symmetry will be restored when the element is connected to other elements. As a result, only the symmetric portion of the induced moment matrix, which is referred to as the joint moment matrix $\mathbf{k}_j$, need to be assembled to form the structural tangent stiffness matrix. More detailed research on this issue can be referred to Teh and Clarke (1997, 1999a) and Izzuddin (2002).

The induced moment matrix is given by:

$$\mathbf{k}_i = Diag[\mathbf{0}_{3x3}, \mathbf{k}_{ia}, \mathbf{0}_{3x3}, \mathbf{k}_{ib}] \tag{6.30}$$

in which:

$$\mathbf{k}_{in} = \begin{bmatrix} 0 & 0 & 0 \\ -M_{zn} & 0 & M_{xn}/2 \\ M_{yn} & -M_{xn}/2 & 0 \end{bmatrix} (n = a, b) \tag{6.31}$$

The joint moment matrix, $\mathbf{k}_j$, is given by:

$$\mathbf{k}_j = Diag[\mathbf{0}_{3\times3}, \mathbf{k}_{ja}, \mathbf{0}_{3\times3}, \mathbf{k}_{jb}] \tag{6.32}$$

in which:

$$\mathbf{k}_{jn} = \begin{bmatrix} 0 & -M_{zn}/2 & M_{yn}/2 \\ -M_{zn}/2 & 0 & 0 \\ M_{yn}/2 & 0 & 0 \end{bmatrix} (n = a, b) \tag{6.33}$$

Liew et al. (2000b) have demonstrated that the use of geometric stiffness $\mathbf{k}_g$ and induced moment matrix $\mathbf{k}_i$ enable the prediction of flexural buckling of columns with one element per member, and the prediction of elastic lateral-torsional buckling of beam with rectangular section. By updating the element forces with an additional bowing matrix $\mathbf{k}_b$ to consider the deformed shape effect together with the natural deformations based on the UL formation, the large-deflection geometrically nonlinear behavior of space frames can be captured with the least element per member. The approach can be naturally extended to the plastic hinge-based inelastic analysis.

Hellan (1995) investigated six different initial out-of-straightness patterns on the limit load of jackets using the pushover analyses. The out-of-straightness patterns are prescribed:

1. In the plane of the loading, acting in the direction of each member's distributed load
2. In the plane of loading, acting in the direction of global base shear
3. According to the (linear) solution vector of the external load
4. According to the elastic system buckling displacement vector
5. According to the displacement vector at system collapse
6. According to the lowest eigenvector of the system stiffness matrix

Hellan (1995) found in a series of numerical analyses that initial out-of-straightness assigned in the direction of each member's distributed line loads give a more consistent (small standard deviation) and slightly more conservative capacity prediction than do the other patterns investigated. Other patterns give more scatter in results and, in a number of cases, produce results where the presence of initial out-of-straightness leads to an increase in estimated collapse capacity compared with results for the initially straight structure.

In the past few decades, there has been a debate regarding the merits of the cubic element vis-à-vis the stability function-based beam-column for steel frame analysis (Teh, 2001; Teh et al., 2001; Chan and Gu, 2001). Teh (2001) demonstrated that in the practical analysis and design of steel frames, cubic element approach is sufficient to capture the elastic buckling and large-displacement behavior. However, his cubic beam element approach adopts the secant stiffness relationship to update the element forces, which is difficult to extend to the plastic-hinge analysis that is based on force and displacement increments. As pointed out by Chan and Gu (2001), the stability function-based element—while considering the bowing effect and initial out-of-straightness—enables the advanced analysis established on a more accurate and powerful foundation.

6.2.3 Modeling of Material Nonlinearity in Beam-column

6.2.3.1 Methods for Modeling Material Nonlinearity

Material nonlinearity in the form of cross-sectional plastification and spread of yielding along member length can be modeled by employing the finite element-based inelastic analysis methods or various numerical approaches that can be found in Chen and Atsuta (1976, 1977). Since the finite element method is the most versatile method for structural analysis, it is widely used in the inelastic analysis of frame structures. In general, these finite element-based inelastic models can be categorized into five main types, as covered in the next five sections.

6.2.3.1.1 Distributed Plasticity Analysis

This method is also called spread-of-plasticity analysis or elasto-plastic analysis. Distributed plasticity analysis such as the plastic zone analysis (White, 1986; Clarke, 1994; Pi and Trahair, 1994; Teh and Clarke, 1999b; Jiang et al., 2002) and 3-D shell element inelastic analysis (Avery and Mahendra, 2000a) involves explicit modeling of the gradual yielding in the volume of the structure through discretizing members along length and through cross-sections. It can model the residual stresses, geometric imperfections, material strain hardening, and elastic unloading. If correctly formulated to include all significant behavioral effects, it is considered to provide an analytically exact structural solution and can be classified as an advanced analysis technique in which design specification checks for the member and section capacities are not required.

However, distributed plasticity analysis is usually computationally intensive with the increase of structural size. The cost and effort of such methods are often so great that analyzing complete medium-to-large structures is often prohibitive. Therefore, this method is normally used to check and calibrate the accuracy of simplified inelastic analysis methods and to establish design charts and equations. One potential application of distributed plasticity analysis is to adopt the mixed element techniques in which the analysis might use coarser element discretization with plastic-hinge based procedures where these computational models suffice, and spread-of-plasticity approaches where more rigorous models are required to capture the performance (Jiang et al., 2002; Liew and Chen, 2004a and 2004b).

6.2.3.1.2 Plastic-hinge Analysis

The plastic-hinge analysis is also called the concentrated plastic-hinge analysis or elastic-plastic-hinge analysis. It can be traced to earlier works using a single plastic strength surface by Nigam (1970) and Porter and Powell (1971). Refinements to the method have been made in the process of implementing it for the inelastic analysis and design of transmission towers (Al-Bermani and Kitipornchai, 1990 and 1992; Kitipornchai and Al-Bermani, 1992), and frame and tubular structures (Orbison et al., 1982; Creus et al., 1984; Ueda et al., 1985; Bozzo and Gambarrotta, 1985; Wong and Tin-Loi, 1987; Ziemian et al., 1992; Chen and Toma, 1994; Papadrakakis and Papadopoulos, 1995; Chen et al., 1996; Krenk et al., 1999; Liew et al., 2000b). Plasticity is formulated based on the cross-sectional stress-resultant constitutive model, which represents the plastic interaction between axial force and biaxial moments. If the forces at any particular cross-section reach the plastic strength capacity, the cross-section changes abruptly from an elastic to fully plastic state, and a plastic hinge forms. The plastic-hinge analysis normally employs one beam-column element per member and is computationally more efficient and economical than the distributed plasticity analysis.

For slender structure in which elastic instability is the predominant mode of failure, both the plastic-hinge analysis and the distributed plasticity methods lead to almost identical results. For structures that exhibit significant yielding in the members, or when structural behavior is dominated by the instability of a few members, the plastic-hinge analysis often over-predicts the actual stiffness and strength of the structure. Due to the simplifying approximations inherent in the plastic-hinge analysis, it is generally not accurate enough to be classified as an advanced analysis. Therefore, some refinements to the basic plastic-hinge theory are necessary to generalize its application for the analysis of any structural system.

6.2.3.1.3 Modifications to Plastic-hinge Analysis

A significant quantity of research works have been done to propose modification to the plastic-hinge analysis to consider the degradation of stiffness due to the spread of plasticity. These alternative approaches are comparable to the plastic-hinge analysis in efficiency and simplicity, but provide results of accuracy comparable to the distributed plasticity solutions. These models include hardening plastic-hinge analysis (King et al., 1992), refined plastic-hinge analysis (Liew et al., 1993a), the notional load plastic-hinge analysis (Liew et al., 1994), quasi plastic-hinge analysis (Attalla et al., 1994) and improved quasi plastic-hinge analysis (Leu and Tsou, 2001; Leu et al., 2008), modified tangent modulus approach (Ziemian and McGuire, 2002), and two-surface plastic-hinge analysis (Himly and Abel, 1985; Powell and Chen, 1986; Deierlein et al., 1991; Hellan, 1995; Al-Bermani et al., 1995; Liew et al, 1998; El-Tawil and Deierlein, 1998; Krenk et al., 1999; Liew and Tang, 2000; Skallerud and Amdahl, 2002; Hajjar, 2003; Ma and Liew, 2004).

The notional load plastic-hinge analysis applies the notional lateral load of 0.5 percent of the story gravity loads to account for the effects of distributed plasticity and structural imperfection caused by residual stresses and initial member out-of-straightness and frame out-of-plumbness. The advantage of this method is that there is no need to modify the base plastic-hinge analysis. Refined plastic-hinge analysis considers the effect of gradual yielding, residual stresses, and geometric imperfections by using the tangent modulus and reduction functions for flexural stiffness. Both analysis methods have been verified by using many well-established calibration frames that range from simple beam-columns to braced and unbraced frames, from theoretical to experimental frames, and from plasticity-dominated to stability-dominated cases. They have been recommended for practical advanced analysis of steel plane frames composed of members with compact sections that are fully braced out-of-plane (Chen and Toma, 1994; Chen et al., 1996; Liew et al., 1997b).

The hardening plastic-hinge analysis degrades the member stiffness in a way similar to that of the refined plastic-hinge analysis. Quasi plastic-hinge analysis uses the inelastic stiffness matrix to consider the gradual plastification under combined action of bending moment and axial force. Attempts have been made by Attalla et al. (1996) for extending the quasi plastic-hinge analysis for the inelastic nonuniform torsion analysis. The modified tangent modulus approach adopts a relatively simple expression to reduce the modulus of elasticity of individual elements according to the amount of axial force and minor-axis bending moment. The two-surface plastic-hinge analysis extends the plastic-hinge method and allows for the gradual plastification effect by employing the yield surface and bounding surface in force space (Liew and Tang, 2000).

6.2.3.1.4 Flexibility-based Inelastic Analysis

Many of the nonlinear formulations presented in the literature are based on the stiffness or displacement method for its relative ease in implementation when compared with the flexibility or force method. The stress resultants obtained by a displacement-based model generally do not satisfy the governing differential equation of equilibrium for the inelastic member. Equilibrium is satisfied only in a variational sense. The assumed displacement interpolation may over-constraint the element in an inelastic state. If the element formulation is based on the interpolation of forces, then a model can be established in which the equilibrium, compatibility, and constitutive conditions are satisfied exactly. This type of model is referred to as an equilibrium-based or a flexibility-based element (Taucer et al., 1991). This method has been extended for the inelastic analysis of frame structures by Nukala and White (2004) and Alemdar and White (2005).

6.2.3.1.5 Mixed Elements Approach

Jiang et al. (2002) and Liew and Chen (2004) proposed a mixed-element approach that allows for adapting the refinement of the structural model to the expected response. In this approach, the structure model consists of different zones where the members may have different levels of discretization (Figure 6.13). A more refined modeling is adopted at critical zones where the members would exhibit significant inelastic and stability effects. Global analysis may be carried out on the structure primarily modeled by beam-column elements, whereas localized buckling effects may be modeled by the spread-of-plasticity fiber elements. The mixed element model is made possible by enforcing force and displacement compatibility at the connecting nodes of the members (Chen and Liew, 2005). The spread-of-plasticity elements and the plastic-hinge beam-column elements can coexist within the same model as illustrated in Figure 6.13. This approach has found great potential for analyzing large-scale 3-D framework subject to blast and fire attack (Liew, 2008).

6.2.3.2 Plastic Strength Surface

In theory, it is possible to develop exact expressions of cross-sectional plastic criteria (see Chen and Atsuta, 1976). However, most plastic-hinge methods rely on a simplified force interaction expression to approximate the plastic strength surface. For 2-D analysis, either piecewise linear or continuous functions are commonly used. For 3-D analysis, continuous functions are usually preferred over the multifaceted surfaces.

Single-equation plastic surfaces for solid rectangular and circular sections and circular hollow section have been theoretically derived in Duan and Chen (1990). Exact interaction formulas for box-type cross-sections are difficult to obtain. Instead, approximate expressions are derived numerically in Duan and Chen (1990). Single-equation plastic strength surfaces for channel, tee, and angle sections have been proposed by Al-Bermani and Kitipornchai (1990) and Kitipornchai et al. (1991), which were applied for the nonlinear analysis of transmission towers (Al-Bermani and Kitipornchai, 1992).

Single-equation plastic strength surfaces for compact wide-flange sections have been proposed by Orbison et al. (1982), Duan and Chen (1990), and Attalla et al. (1994). The three plastic strength surfaces are interaction functions of a member's axial force and biaxial bending moments. They are symmetric about the three coordinate planes. Duan and Chen's, as well as Attalla's, plastic strength surfaces are more accurate than Orbison's plastic strength surface. However,

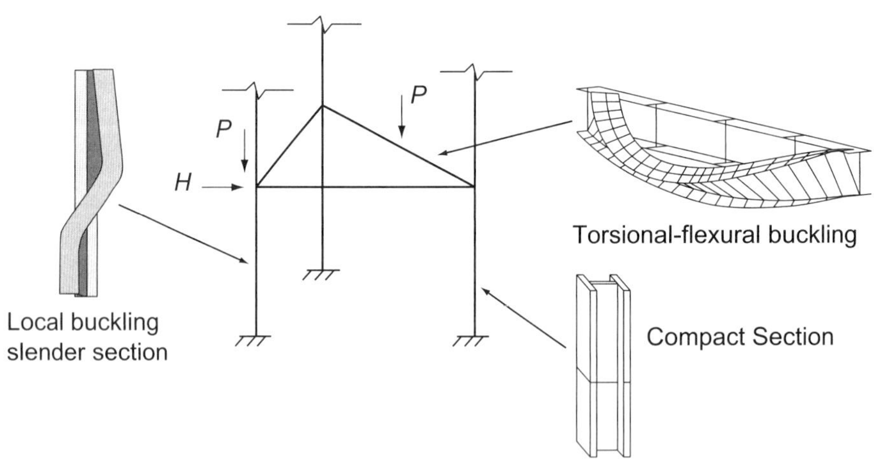

FIGURE 6.13 Use of mixed elements for inelastic analysis.

they have sharp vertices at the limits of pure axial force or at pure moment where the normal of the plastic strength surface cannot be uniquely defined. Numerical difficulties may arise when the force point of a plastic hinge crosses these vertices. An iteration scaling procedure proposed by Izzuddin and Elnashai (1993a) may be adopted to bring the force state back to the plastic strength surface. In the case of pure axial plasticity, a simple procedure to overcome such problems associated with shape vertice. Orbison's plastic strength surface has full slope continuity that eliminates the numerical difficulties associated with vertices. With the exception of the heavy wide-flange sections subject to the combined effect of the weak-axis bending moment and axial force, it gives a good prediction on the plastic strength of the light- to medium-weight wide-flange sections.

In the offshore industry, plastic strength surfaces for various shapes of chord section in the jackup leg structure (such as tubular chords with double central racks, split tubular chords with double central racks, triangular chords with single racks, etc.) have been proposed in SNAME (1994) and Marshall et al. (2001).

For frame structures, plastic-hinge analysis is usually based on a plastic criterion that considers the longitudinal normal stresses due to axial force and bending moments, often neglecting shear stresses due to shear force, torsion, and bimoments. By treating an I-section as a composition of three thin plates, the plastic strength surface diagrams have been established by Yang and Fan (1988) and Yang et al. (1989) for I-sections subjected to the axial force, two bending moments, torque, and bimoment. Based on an optimization formulation, Osterrieder and Kretzschmar (2006) developed a general approach to evaluate the plastic steel section capacity for any type of open thin-walled steel sections and any combination of internal forces.

6.2.3.3 Concentrated Plastic-hinge Formulation

Plastic hinges are introduced when the element end forces reach the plastic strength surface. Associated flow rule assumes that the state of forces at a plastic hinge keeps moving on the plastic strength surface and the incremental plastic displacement vector is normal to the plastic strength as shown in Figure 6.14. $\Phi = 0$ represents the full plastification of the cross-section, and $\Phi = -1$ is the value of a stress-free cross-section. In principle, a state of forces characterized by $\Phi > 0$ is illegal.

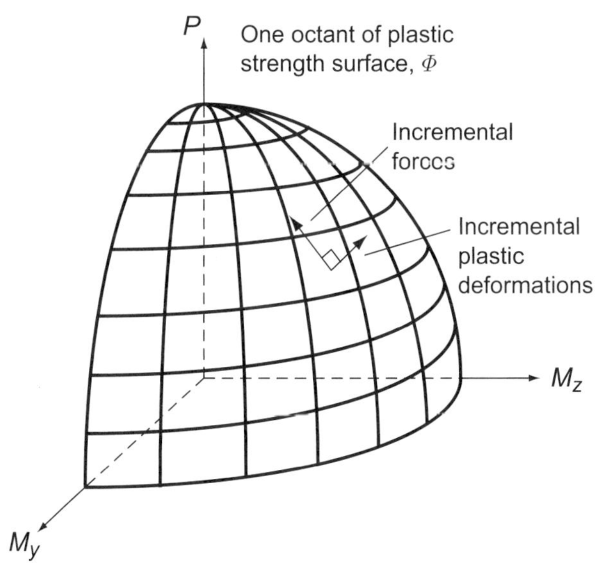

FIGURE 6.14 Cross-sectional force behavior at plastic hinge.

The flow rule for an element with plastic hinges formed at both ends is expressed as:

$$\mathbf{du}_p = \begin{bmatrix} g_a & 0 \\ 0 & g_b \end{bmatrix} \begin{Bmatrix} \Delta\lambda_a \\ \Delta\lambda_b \end{Bmatrix} = \mathbf{G}\,\boldsymbol{\Delta\lambda} \tag{6.34}$$

in which $\mathbf{du}_p$ is the incremental plastic displacement vector, $\boldsymbol{\Delta\lambda}_i$ $(i = a, b)$ is the plastic scalar, and $\mathbf{g}_i$ is the normal of plastic strength surface given by:

$$\mathbf{g}_i = \frac{\partial\Phi}{\partial\mathbf{f}_i} = \begin{bmatrix} \frac{\partial\Phi}{\partial F_x} & \frac{\partial\Phi}{\partial F_y} & \frac{\partial\Phi}{\partial F_z} & \frac{\partial\Phi}{\partial M_x} & \frac{\partial\Phi}{\partial M_y} & \frac{\partial\Phi}{\partial M_z} \end{bmatrix}^T \tag{6.35}$$

in which $\mathbf{f}_i$ is the force vector at element end.

The element incremental displacement vector is separated into an elastic component and a plastic component:

$$\mathbf{du} = \mathbf{du}_e + \mathbf{du}_p \tag{6.36}$$

The stiffness equation can be obtained by introducing the flow rule from Equation 6.34:

$$\mathbf{df} = \mathbf{k}_t\left(\mathbf{du} - \mathbf{du}_p\right) = \mathbf{k}_t\mathbf{du} - \mathbf{k}_t\mathbf{G}\,\Delta\lambda \tag{6.37}$$

For a perfectly plastic hinge, the plastic strength surface is fixed, and **df** must always be tangent to the surface, hence:

$$\mathbf{G}^T\mathbf{df} = \mathbf{G}^T\mathbf{k}_t\mathbf{du} - \mathbf{G}^T\mathbf{k}_t\mathbf{G}\Delta\lambda = 0 \tag{6.38}$$

and the vector of plastic scalars can be solved as:

$$\Delta\lambda = \left(\mathbf{G}^T\mathbf{k}_t\mathbf{G}\right)^{-1}\mathbf{G}^T\,\mathbf{k}_t\,\mathbf{du} \tag{6.39}$$

Substituting $\boldsymbol{\Delta\lambda}$ into Equation 6.37, the elastic-plastic stiffness equation is obtained:

$$\mathbf{df} = \left(\mathbf{k}_t + \mathbf{k}_p\right)\mathbf{du} = [\mathbf{k}_t - \mathbf{k}_t\mathbf{G}\left(\mathbf{G}^T\mathbf{k}_t\mathbf{G}\right)^{-1}\mathbf{G}^T\mathbf{k}_t]\,\mathbf{du} = \mathbf{k}_t^{EP}\mathbf{du} \tag{6.40}$$

in which $\mathbf{k}_p$ is the plastic reduction matrix and $\mathbf{k}_t^{EP}$ is the elastic-plastic tangent stiffness matrix.

6.2.3.4 Two-surface Plastic-hinge Formulation

A plasticity model that accounts for partial yielding and hardening employs two interaction surfaces—one yield surface and one bounding surface—as shown in Figure 6.15 (Liew and Tang, 2000; Ma and Liew, 2004). The bounding surface encompasses the cross-sectional force state and the yield surface at any stage during the yielding process. To keep the two surfaces from overlapping during translation, the initial yield surface is assumed to be a scaled-down version of the bounding surface.

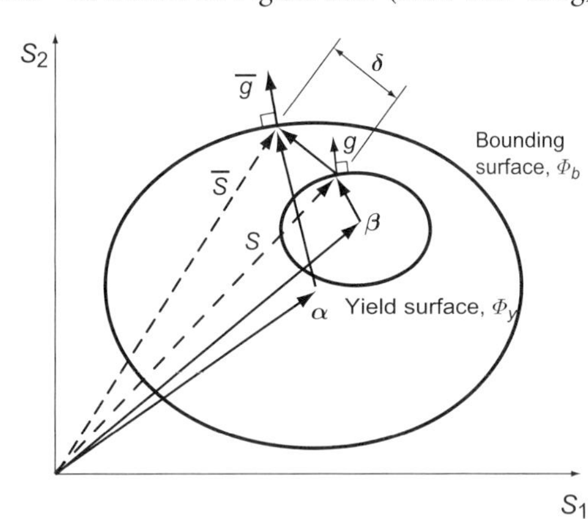

FIGURE 6.15 Schematic of initial yield surface and bounding surface.

In reality, the initial yield surface and the bounding surface are different in shape. Major deviation occurs in axial loading. Adoption of the same shape for the first yield and fully plastic surfaces would under-predict the axial stiffness. This may be remedied by tuning the hardening parameters that describe axial hardening by values other than those for the pure bending case, and then a steeper transition between the axial yield and full plasticity can be obtained (Skallerud and Amdahl, 2002). A study by El-Tawil and Deierlein (1998) shows that the simplified results are satisfactory for bisymmetrical sections under biaxial bending and under a combination of axial load and major-axis bending, but may not be as good for sections under a combination of axial load and minor-axis bending.

When the force state **S** remains within the initial yield surface, the member is elastic and there is no translation of the two surfaces. Once the force state **S** has reached the initial yield surface, the inner surface starts to translate so that the stress resultants remain on the yield surface during subsequent loading. This translation is uniquely defined by the history of the position vector $\boldsymbol{\beta}$, which marks the center of the yield surface. Translation of the bounding surface is at a much slower speed and is controlled by the kinematic hardening rule and defined by the history of center position vector $\boldsymbol{\alpha}$. To prevent the yield and bounding surface from intersecting during translation, Mroz's conjugate point approach is adopted to describe translation of the initial yield surface. For a given force state **S** that has reached the yield surface, a conjugate point $\bar{\mathbf{S}}$ on the bounding surface is defined so that **S** and $\bar{\mathbf{S}}$ have equal **g** and $\bar{\mathbf{g}}$ gradients. The yield surface translates parallel to the vector that connects the current force point and the conjugate point on the bounding surface.

In the two-surface model, the plastic hardening parameter and transition parameter are specific to each force component, and they may be determined by calibration with tests or more exact solutions. Numerical calibration work has been carried out by Punniyakotty (1998) to determine the appropriate parameters to predict the capacity of steel members. Liew and Tang (2000) investigated the effect of member imperfections and calibrated the inelastic beam-column model for advanced analysis of tubular structures.

6.2.3.5 Numerical Strategies for Plastic-hinge Analysis

6.2.3.5.1 Elastic Unloading at Plastic Hinge

The plastic scalar $\Delta\lambda$ defined in Equation 6.39, which characterizes the plastic displacement magnitude, is zero for an elastic state and positive for a plastic state. When a plastic hinge occurs at the cross-section, the value of the plastic scalar changes from zero to a positive value. When elastic unloading occurs at the plastic hinge, it changes from a positive value to a negative one. Then its actual value has no meaning and only its negativity indicates that elastic unloading occurs. If elastic unloading is detected, the structure is re-analyzed within the same load step after modifying the stiffness matrices of elements in which elastic unloading occurs.

6.2.3.5.2 Pure Axial Plasticity

Plastic hinges may form at both ends of the strut due to the action of pure axial force. These may be the cases when analyzing trusses and braced frames involving pin-ended members. Numerical difficulty may be encountered in calculating the plastic reduction matrix because the axial plastic deformations are of arbitrary magnitude at both ends of element, and the matrix $\mathbf{G}^T\mathbf{k}_t\mathbf{G}$, which appears in the plastic reduction matrix, is singular and cannot be inverted. In the present work, a simple technique has been developed to overcome this problem. Each element, with plastic hinges forming at both ends, is checked for the presence of bending moments prior to the calculation of the elastic-plastic tangent stiffness matrix. If the element is found to be in a plastic state due to pure compression or tension, the elastic-plastic tangent stiffness matrix is computed as the elastic tangent stiffness matrix, and the terms associated with axial stiffness are then removed. This method can capture the plastic behavior of element correctly with minimum computational effort.

6.2.3.5.3 Plastic Hinges at Common Node Adjoining Two Elements

Plastic hinges may occur simultaneously at a common node adjoining two elements. In this case, the rotational stiffness at the node will become very small. The severely ill conditioned stiffness matrix will result in large incorrect displacements of that node. The element formed with a common plastic hinge can be modeled using two parallel elements: a plastic-hinge element and an element that is always elastic and has a small elastic modulus to simulate strain hardening. The element tangent stiffness matrix is obtained by adding the stiffness matrices of the two parallel elements together. This model ensures a finite stiffness when all elements at a common node have formed plastic hinges. To reduce the unnecessary computational expense, this model is used only for the elements that have formed a plastic hinge.

6.2.3.5.4 Formation of Plastic Hinge Within the Member Length

A plastic hinge may sometimes form within the member length. A tedious and approximate procedure is to model each frame member with several elements. However, this method will increase the overall degrees of freedom of the structure, and becomes computationally expensive. Moreover, in most cases, only a few members in a structure will have plastic hinges forming between the member ends. Abdel-Ghaffar et al. (1991) developed a formulation that can model the plastic hinge forming at any location within the beam, which is under the uniformly distributed load as well as multiple concentrated loads on the member. Chan and Zhou (2004) proposed a method for plane steel frame in which plastic hinges are allowed to form in an element with two at the two ends and one at the location of maximum combined stress due to axial force and moment.

An efficient method for modeling the plastic hinge within the member length in space steel frames is to use the static condensation method with minimum computational effort as shown in Figure 6.16. Based on the member initial out-of-straightness, the deformed member shape, and the forces at member ends, the force state within the member length can be established by taking the equilibrium of axial force and moment at the internal cross-section. It is strictly not necessary to compute the exact location of the plastic hinge for an accurate estimate of member strength. As long as the exact location of the plastic hinge in a member is not more than a distance $L/5$ (L = member length) away from the assumed position, the difference in strength prediction is

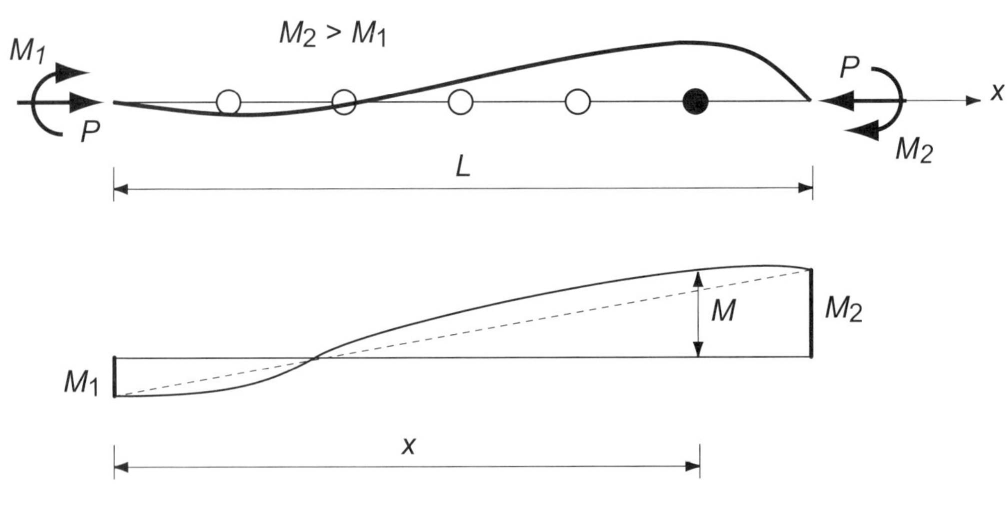

$$M(x) = \frac{(M_1 \cos kL + M_2)}{\sin kL} \sin kx - M_1 \cos kx \qquad kL = \sqrt{PL^2/EI}$$

FIGURE 6.16 Formation of plastic hinge within the member length.

no more than 5 percent (Chen et al., 1996). The element length is divided into six segments with equal length. The cross-sectional forces are then checked at five points between the member ends. A plastic hinge is said to have formed when the plastic strength is reached at any of these points. The analysis will automatically subdivide the original element into two subelements at the plastic hinge location as shown in Figure 6.16. The internal hinge is then modeled by an end hinge at one of the subelements. The stiffness matrices for the two subelements are determined. The inelastic stiffness properties for the original element are obtained by static condensation of the extra node at the location of the internal plastic hinge. Since the static condensation process is only performed at the element level, it does not involve much computational cost.

6.2.3.5.5 Load Increment Control

If a plastic hinge occurs, the load increment is scaled so that the element forces of that cross-section comply exactly with the plastic strength surface. This is indicated in Figure 6.17, which is a 2-D projection of the 3-D plastic strength surface. At the beginning of a load step, the state of a cross-section at Force Point A is elastic. However, at the end of the load step, the force state reaches Point B. A simple check of the plastic strength surface shows that a plastic hinge has formed during the load step. Consequently, the load increment is scaled down to find Force Point C, which lies on the plastic strength surface. The scaling factor γ is determined by the bisection method. Figure 6.18 shows that Force Point A lying on the plastic strength surface may have deviated excessively to Force Point B. The load increment is reduced to find Force Point C, which lies on the surface $\Phi = \varepsilon_2$. The tolerance for overshooting the plastic strength surface is selected as $\varepsilon_2 = 0.005$. The scaling factor γ is also determined by the bisection method. The final load increment is the minimum value determined by using the above constraints.

After the load increment is completed, the nodal coordinates, element forces, and geometry are updated. Then it is checked to see if new plastic hinges have formed. In some cases, plastic hinges may occur at approximately the same applied load level, resulting in a series of small load increments. In order to avoid small load increment in case of frequent occurrences of plastic hinges, an incipient plasticity tolerance $\varepsilon_1 = 0.005$ can been implemented. Once the force point at a cross-section locates outside the surface $\Phi = -\varepsilon_1$, a plastic hinge is inserted. In this way, exact scaling to the plastic strength surface is not always possible and several plastic hinges may be inserted during one load increment.

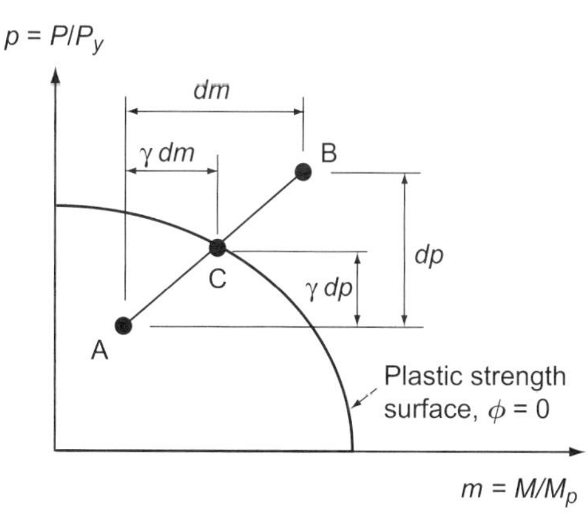

FIGURE 6.17 Load increment scaled back to the plastic strength surface.

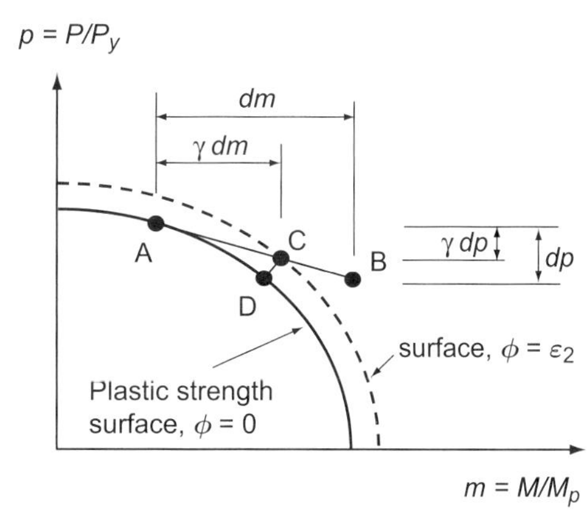

FIGURE 6.18 Load increment scaled down by the constraint of force increments at plastic hinge.

6.2.3.5.6 Control of Deviation of Yield Surface

When a plastic hinge forms at a cross-section, the force point is constrained by a plastic reduction matrix to move tangentially on the plastic strength surface. A drift of the force point from the plastic strength is inevitable. Various methods have been proposed to rectify the nonconformity of the plasticity condition. Hodge (1977) devised a technique to confine the force point movement to an area bounded between two yield surfaces, a procedure that Bozzo and Gambarotta (1985) later extended to include geometrical nonlinearity. Wong and Tin Loi (1987) proposed the piecewise linearization of the convex nonlinear plastic strength surface, thereby guaranteeing the satisfaction of both equilibrium and plasticity condition. Orbison et al. (1982) used a multistep iterative predictor-corrector-type procedure to force the force point back onto the plastic strength surface. Ma and Liew (2004) corrected the deviation from the plastic surface by applying an equivalent nodal load at element ends to regain plastic strength surface agreement in the succeeding increment. This method can be consistently extended for the inelastic analysis of steel frames under fire in which the plastic strength surface may contract due to the temperature effect.

6.2.3.6 Plasticity Formulation at Elevated Temperature

The thermal effects on the structural elements may be considered by conducting heat transfer analysis. The temperature history in each structural member is first calculated based on the fire model assumed. The equivalent nodal expansion forces for the line element are evaluated based on the incremental temperature change. The thermal effects on the structural element include the reduction of yield stress and elastic modulus, and thermal expansion at elevated temperatures. Consistent nodal forces are produced on an elastic element at elevated temperatures due to the axial expansion and temperature gradient increment over the cross-section.

Landesmann et al. (2005) extended the second-order refined plastic-hinge analysis proposed by Liew et al. (1993a) for plane steel-framed structures under fire conditions. Chan and Chan (2001) developed the plastic-hinge analysis of steel frames based on the spring-in-series model considering the combined effects of sectional plasticity and connection flexibility. Liew et al. (1998), Ma and Liew (2004), and Liew (2004b) extended the two-surface plastic-hinge model to model the behavior of 3-D steel frames under fire. At elevated temperatures, the reduction of yield strength causes degradation of the cross-section capacity. The yield and bounding surfaces thus have to be reduced in order to satisfy the yield condition. Degradation of the yield strength is based on the effective strength concept. At high temperatures, the stress-strain relationship of steel is highly nonlinear and does not exhibit a distinct yield plateau. The idea of effective yield strength is introduced to define a yield plateau at a relatively high level of strain.

For the two-surface plasticity model, the size of the bounding surface that corresponds to full cross-sectional plasticity follows the reduction curve for effective yield strength in Eurocode 3, Section 3.2 (CEN, 2001). The size of the initial yield surface is given by the temperature-invariant scaling factor z_y. At ambient temperature, z_y is usually taken as the reciprocal of the shape factor, and can be reduced further to account for the initial residual stress effect. At elevated temperatures, the initial yield surface is assumed to degrade proportionally to the bounding surface as shown in Figure 6.19. However, the initial yield strength decreases at a faster rate than the effective yield strength; therefore, the size of the initial yield surface is over-predicted at higher temperatures if a constant z_y value is adopted. In practice, this will not have any significant effect on the inelastic behavior of members on fire. For a more accurate prediction, the z_y parameter can be adjusted to a lower value by calibrating the size parameter close to the critical temperature.

After a yield hinge forms, the yield surface continues to contract with an increase in temperature (Figure 6.19). This will cause the resultant stress to leave the yield surface. This effect can be accounted for by including an additional term that represents the change of yield surface due to the temperature increment. The change of yield stress is considered to contribute to the increment

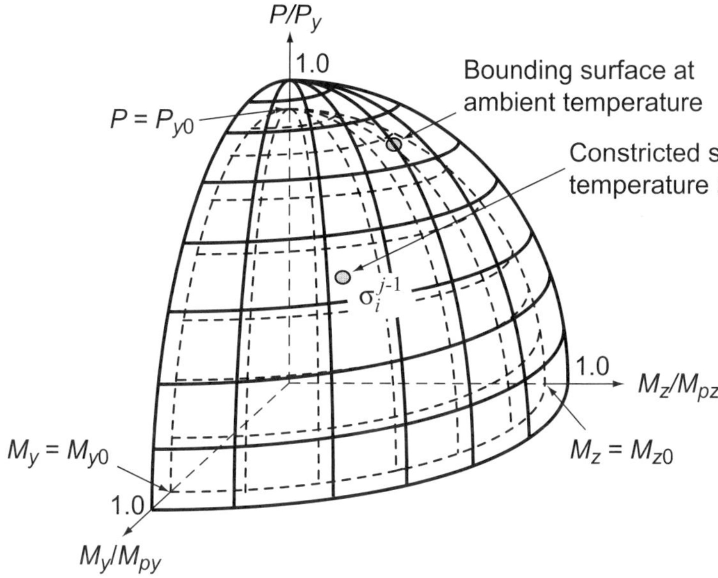

FIGURE 6.19 Constricting bounding surface at elevated temperatures.

of consistent nodal forces, which is required to make the force state remain on the contracting yield surface. Verification of the two-surface plasticity model at ambient and elevated temperatures for both components and frames over a wide range of parameters was reported by Liew et al. (1998) and Ma and Liew (2004).

6.2.3.7 Modeling of Lateral-torsion Buckling

As pointed out by White et al. (2006), although some progress has been made on 3-D distributed plasticity methods considering the nonuniform torsion (Pi and Trahair. 1994; Izzudin and Smith, 1996; Nukala and White, 2004), the complexity and cost of the analysis is significantly greater due to the appropriate handling of residual stresses, geometric imperfections, warping continuity at beam-to-column joints, local overall member buckling interactions, and restraint from and interaction with floor slabs. Other important attributes that can influence the 3-D response have not been studied thoroughly at this time.

A practical advanced analysis for planar steel frame design with consideration of out-of-plane buckling has been proposed by Wongkaew and Chen (2002) and Trahair and Chan (2003). The analysis can be simplified by separating it into two stages. First, an in-plane advanced plastic-hinge analysis is carried out to check whether the structure can resist the loads by in-plane behavior. If it cannot, the structure is inadequate and must be modified. If it can, the second stage out-of-plane buckling analysis would be carried out to check whether the structure has sufficient out-of-plane resistance. If it does not, the structure is inadequate and must be modified. If it does, the structure is adequate. The moment and axial force distributions found from in-plane analysis can be used directly in the out-of-plane analysis.

6.2.3.8 Modeling of Local Buckling

In the advanced analysis, a ductility check must be performed to ensure that beam-columns have adequate inelastic rotation to develop their full-plastic moment capacity. In most cases, local instability is forestalled if cross-sections are made compact because compact sections are capable of developing the full-plastic moment capacity and sustaining large hinge rotation before the onset of local buckling.

For steel frames composed of wide flange sections, Ziemian et al. (1992) proposed that two checks carried out to ensure that adequate ductility is provided at the hinged cross sections. The first check compares the required major axis inelastic rotation angles with the theoretical angular rotation capacity of the sections. The second check compares the required major axis inelastic rotation capacities with the available rotation capacities. In the analysis of a 22-story office building frame, Ziemian et al. (1992) found that a majority of the hinged cross sections had inelastic rotation demands in less than 50 percent of the theoretical rotation capacities. Recently, Van Long and Hung (2008) developed the local buckling check according to Section 5.3 of Eurocode 3 for plastic-hinge analysis of 3-D steel frames with I-shaped sections.

For fabricated circular hollow sections, simple critical strain criteria based on the outside diameter over wall thickness (D/t) ratio have been proposed based on the experimental results (Sohal and Chen 1988; Skallerud and Amdahl, 2002). To assess possible local failure modes such as local buckling and fracture, estimates of the plastic straining are required. Unfortunately, the plastic-hinge concept assumes zero hinge extension, and therefore implies infinite strains and is unable to provide the correct information of the strains within member length. Based on some engineering mechanics assumption, Moan et al. (1991) developed a procedure to predict the local strain of beam-column within the plastic-hinge framework.

Noncompact sections are often used in steel frames with cold-formed sections to achieve both construction economy and structural efficiency. For such frames, local buckling may occur before the structure reaches its maximum capacity. Avery and Mahendran (2000b) modified the refined plastic-hinge analysis (Liew et al., 1993a) to account for the effect of local buckling using simple equations derived from the codes.

In the offshore industry, in addition to normal functional and environmental loads, accidental loads due to collision, falling objects, blast, and fire, etc., might occur. The extent of damage caused by these loads ranges from the possibility of total structural collapse to small damages on structural components. Such damages may not have serious consequences at the time of accident, but may later affect the ability of the structure to resist loads. Taby and Moan (1985) and Søreide et al. (1994) investigated the ultimate strength and post-buckling behavior of circular and rectangular hollow steel tubes containing dent damage, which are useful for making optimal design decisions regarding safety and economy.

6.2.4 Updating of Element Forces and Geometry in Beam-column

An important phase in the nonlinear analysis is to update the element geometry and forces based on the displacement increments obtained in the solution phase. By summing the element forces at each node and comparing them with the external loads, the structural unbalanced forces can be computed and iteration is repeated until equilibrium is achieved. In their nonlinear geometric analyses of plane frames, Yang and Leu (1994) concluded that the accuracy of the numerical solutions depends primarily on the corrector or procedure for recovering the element forces.

6.2.4.1 Force Recovery Based on the Natural Deformation Approach

In the beam-column approach (Oran, 1973; Kassimali and Abbasnia, 1991; Chan and Gu, 2000), the member forces are updated using the secant stiffness that relates them to the member basic deformations. In contrast, the finite element approach based on the updated Lagrangian formulation leads to the tangent stiffness, which relates the incremental element forces to the incremental element displacements. Element forces can be rationally recovered by employing the direct approach, the external stiffness approach (Yang and Kuo, 1994), and the natural deformation approach (Gattass and Abel, 1987b; Yang and Kuo, 1994).

The direct approach calculates the incremental element forces referred to the previous configuration from the incremental element displacements by using the tangent stiffness matrix. Since it does not account for the effect of rigid-body rotation in the large displacement problems, it is often necessary to specify a sufficient number of load increments to ensure that the incremental displacements are reasonably small. In the external stiffness approach, the external stiffness terms associated with rigid-body rotation are isolated from the tangent elastic stiffness matrix so that the remaining stiffness matrix can be used to obtain the force increments referred to the current configuration. However, the external stiffness approach cannot be employed in the plastic-hinge analysis because it is difficult to isolate the terms associated with rigid-body rotation from the tangent elastic-plastic stiffness matrix.

In the natural deformation approach, the incremental element displacements can be conceptually decomposed into two parts: the rigid-body displacements and the natural deformations. The rigid-body displacements serve to rotate the initial forces acting on the element from the previous configuration to the current configuration. The natural deformations constitute the only source for generating the incremental forces. Based on the plastic-flow rule, the force point at the plastic hinge will move on the plastic strength surface. The natural deformation approach is adopted due to its consistence with the plastic-hinge analysis and its accuracy in updating the incremental element forces. An efficient formulation proposed by Conci (1992) may be used to calculate the natural deformations for the nonlinear analysis of frameworks.

The element incremental displacements between configurations C_1 and C_2, can be conceptually decomposed into two parts: the rigid-body displacements du_r and the natural deformations du_n. The basic idea in calculating the natural deformations is to define a configuration $C_{2'}$ if the element at C_2 is translated by $-u_a$, $-v_a$ and $-w_a$ along the ${}^1x$, ${}^1y$, and ${}^1z$ axes so that there are no rigid-body translations between C_1 and $C_{2'}$ (see Figure 6.20). The rigid-body rotations are given by (Conci, 1992):

$$\phi_x = \frac{1}{2}\left(\theta_{xa} + \theta_{xb}\right) \tag{6.41}$$

$$\phi_y = -\arctan\frac{w_b - w_a}{{}^1L + u_b - u_a} \tag{6.42}$$

$$\phi_z = \arctan\frac{v_b - v_a}{{}^1L + u_b - u_a} \tag{6.43}$$

in which the rigid-body axial rotation is taken as the average of the axial rotations at element ends A and B.

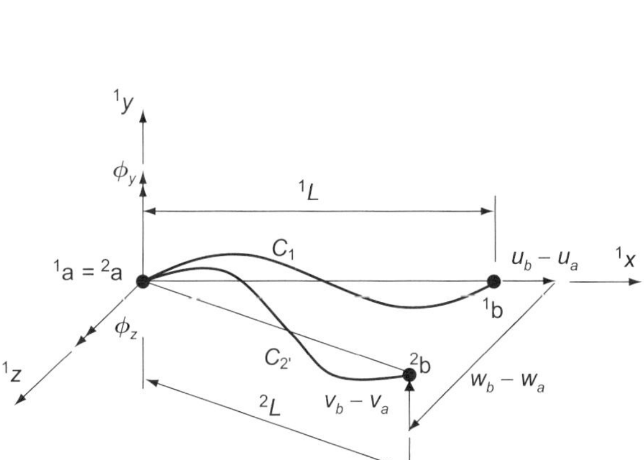

FIGURE 6.20 Rigid-body rotations.

The element natural deformations are calculated as:

$$\mathbf{du_n} = \begin{bmatrix} 0 & 0 & 0 & \alpha_{xa} & \alpha_{ya} & \alpha_{za} & \Delta_x & 0 & 0 & \alpha_{xb} & \alpha_{yb} & \alpha_{zb} \end{bmatrix}^T \tag{6.44}$$

in which $\alpha_{xa} = \theta_{xa} - \phi_x$, $\alpha_{ya} = \theta_{ya} - \phi_y$, $\alpha_{za} = \theta_{za} - \phi_z$, $\Delta_x = {}^2L - {}^1L$, $\alpha_{xb} = \theta_{xb} - \phi_x$, $\alpha_{yb} = \theta_{yb} - \phi_y$, $\alpha_{zb} = \theta_{zb} - \phi_z$, and ${}^1L$ and ${}^2L$ are the length of the element at configurations C_1 and C_2, respectively. These natural flexural deformations are illustrated in Figure 6.21.

The effect of rigid-body translations and rotations is to rotate the initial forces acting on the element at C_1 as the forces acting on the element at C_2 and referred to the axes of C_2. The result is the preservation of equilibrium of the element in the displaced configuration with no change on the magnitudes of the element forces at C_1. The incremental forces generated by the natural deformations $\mathbf{du_n}$, during the incremental step can be calculated as $\mathbf{k\,du_n}$. By adding it to the initial element forces ${}_1^1\mathbf{f}$ at configuration C_1, the total element forces ${}_2^2\mathbf{f}$ at configuration C_2 and referred to C_2, can be obtained as:

$${}_2^2\mathbf{f} = {}_1^1\mathbf{f} + \mathbf{k_t du_n} \tag{6.45}$$

6.2.4.2 Geometry Updating and Web-plane Vector

In nature, large spatial rotations are not commutative so that the geometry updating procedure for spatial frames is not a simple extension of that for planar frames. Oran (1973) and Kassimali and Abbasnia (1991) proposed the joint orientation approach, which employs the joint orientation matrices and the member orientation matrices to update the element configurations. Based on

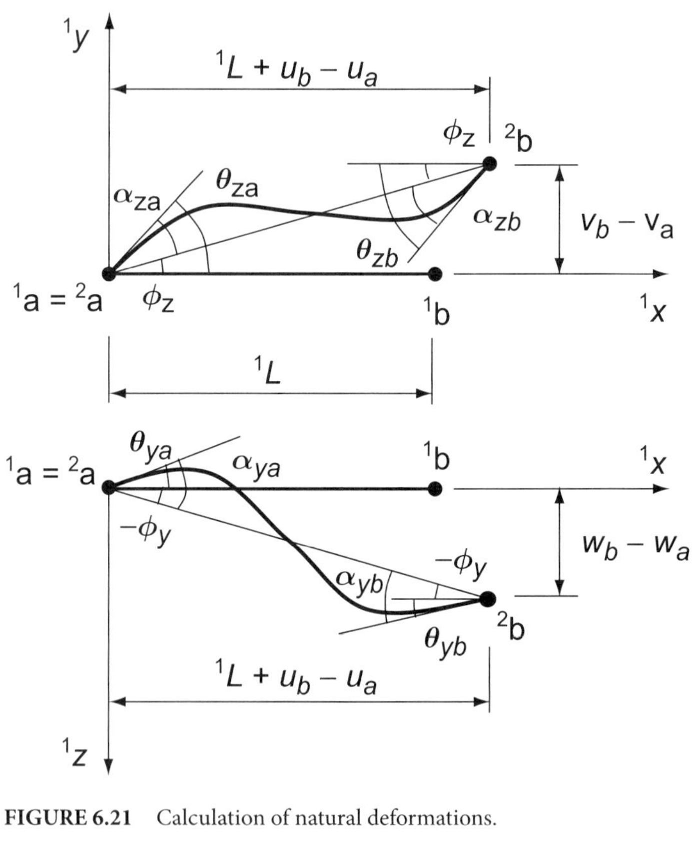

FIGURE 6.21 Calculation of natural deformations.

the joint orientation approach, Kuo et al. (1993) proposed a method to update the element configurations and calculate the natural deformation suitable for use in the updated Lagrangian formulation. Izzuddin and Elnashai (1993b) adopted a pair of end-section orientation matrices for each element and reduced the storage and computational effects of a pair of joint orientation matrices, the undeformed orientation matrix, and the current orientation matrix for each element in the method proposed by Kuo et al. (1993).

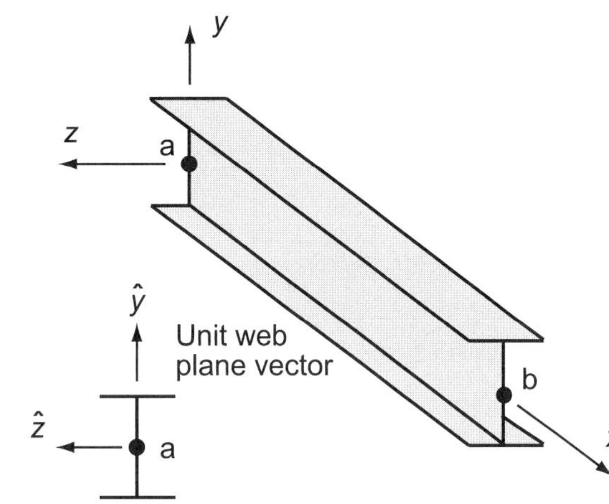

FIGURE 6.22 Web-plane vector of element.

The web-plane vector approach proposed by Orbison (1982) is adopted due to its highly computational efficiency and straightforward numerical implementation. In this approach, each element is assigned with a web-plane vector (located at the element end A) that lies along the positive local y-axis and is perpendicular to the longitudinal axis of element shown in Figure 6.22.

The updating of the web-plane vector is simplified by assuming that the element is straight at the start of each increment and does not develop plastic torsional deformations. However, if a long and torsionally flexible member is expected to undergo large elastic torsional rotations, it is recommended that the member be subdivided into a series of shorter elements. In updating the web-plane vector, the six element rigid-body motions need to be considered. It is clear that the three rigid-body translations have no effect on the components of web-plane vector, and only the three rigid-body rotations change the web-plane vector.

The rigid-body rotation about the element's longitudinal axis ϕ_x, as given by Equation 6.41, will alter the web-plane vector. In Figure 6.23, the element is observed in the negative local ${}^1x$ axis at configuration C_1. The web-plane vector at the start of the increment is identified as ${}^1\hat{y}$. After the axial rotation, the web-plane vector is partially updated as:

$$ {}^1\hat{y}' = (\cos\phi_x)\,{}^1\hat{y} + (\sin\phi_x)\,{}^1\hat{z} \tag{6.46} $$

The two remaining rigid-body rotations—one about the element ${}^1y$ axis and the other about the element ${}^1z$ axis—are considered next. A rotation about the ${}^1y$ axis has no effect on the components of the web-plane vector. Although a rotation about the ${}^1z$ axis will alter the partially updated web-plane vector, it will remain in the updated ${}^2x$-${}^2y$ plane.

The unit beam vector at configuration C_2 can be computed straightforward as:

$$ {}^2\hat{x} = \left[\frac{{}^2X_b - {}^2X_a}{{}^2L} \quad \frac{{}^2Y_b - {}^2Y_a}{{}^2L} \quad \frac{{}^2Z_b - {}^2Z_a}{{}^2L} \right]^T \tag{6.47} $$

in which ${}^2X_i$, ${}^2Y_i$, and ${}^2Z_i$ ($i = a, b$) are the updated global nodal coordinates obtained by adding the accumulated nodal translations to their respective coordinates.

A third vector perpendicular to the beam vector and the partially updated web-plane vector is created by the cross product:

$$ {}^2\bar{z} = {}^2\hat{x} \times {}^1\hat{y}' \tag{6.48} $$

As ${}^2\hat{x}$ will typically not be perpendicular to ${}^1\hat{y}'$, vector ${}^2\bar{z}$ will not be a unit vector and its components must be scaled back to generate the unit vector ${}^2\hat{z}$.

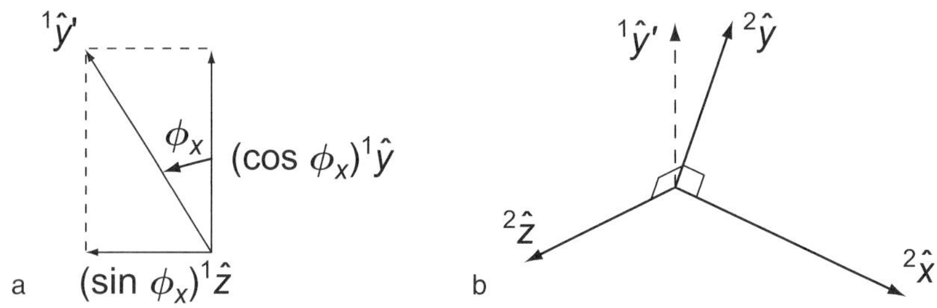

FIGURE 6.23 Updating of element local coordinate: (a) partially updated web-plane vector $^1\hat{y}'$; (b) construction of unit vectors $^2\hat{y}$ and $^2\hat{z}$.

In the final step, the fully updated unit web-plane vector is produced by the cross product:

$$^2\hat{y} = {}^2\hat{z} \times {}^2\hat{x} \tag{6.49}$$

The local coordinate at configuration C_2 can be updated based on the updated unit beam vector, the unit web-plane vector, and the unit third vector.

For the beam-column element with twelve degrees of freedom, the element local-to-global transformation matrix is given by:

$$\mathbf{\Gamma} = Diag\left[\gamma\ \gamma\ \gamma\ \gamma\right] \tag{6.50}$$

in which γ is the rotation matrix given by:

$$\gamma = \begin{bmatrix} \cos\alpha_x & \cos\beta_x & \cos\delta_x \\ \cos\alpha_y & \cos\beta_y & \cos\delta_y \\ \cos\alpha_z & \cos\beta_z & \cos\delta_z \end{bmatrix} \tag{6.51}$$

where α, β, and δ designate the angles between the subscriptal local axis and each of the global X, Y, and Z axes. The element stiffness matrix in the global coordinate **K** can be obtained by transforming the element stiffness matrix in the local coordinate **k** as:

$$\mathbf{K} = \mathbf{\Gamma}^T \mathbf{k}\, \mathbf{\Gamma} \tag{6.52}$$

6.2.5 Modeling of Truss Element

Murtha-Smith (1994) presented the chronological development of various analytical strut models and highlighted the limitation of their use in actual design practice. Blandford (1997) reviewed the extensive research on the progressive failure of space truss structures. The basic problem of various analytical models is that they do not satisfy the code requirements of member capacity checks, which therefore deters one from adopting them for use in actual design.

The advantage of the code-based approach for strut design is that the design formula includes implicitly the member imperfections such as member initial out-of-straightness of Span Length/1000 and initial residual stresses in the cross section. Liew et al. (1997a) developed an inelastic large-displacement advanced analysis capable of capturing individual member strength and overall buckling of structure using the strut and tie model. The strut model includes three distinct axial load-shortening curves as shown in Figure 6.24. The elastic loading stage is associated with the compression of an initially imperfect column until its maximum strength P_{max} is reached. The post-buckling stage is developed by assuming that a plastic hinge forms at the mid-span of the member. The post-buckling unloading stage is characterized by the fact that the strut

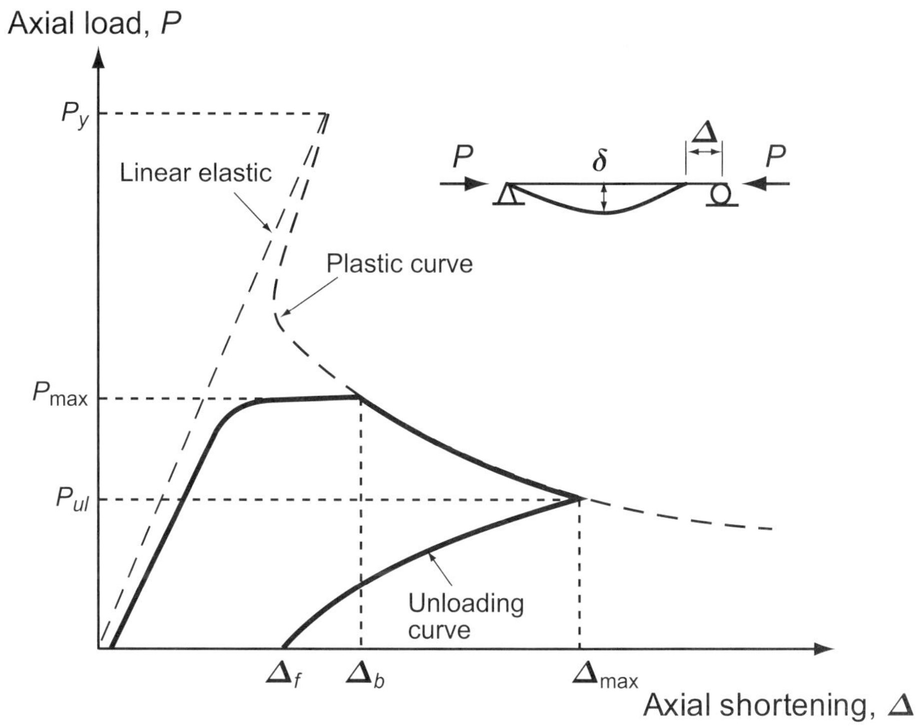

FIGURE 6.24 Axial load-shortening curve for a strut member (Liew et al., 1997a).

member is stretched due to the change in sign of the incremental force. It is assumed that the post-buckling unloading curve follows the elastic loading curve with the member's imperfection magnitude computed based on the initial shortening value Δ_f as shown in Figure 6.24. The maximum strength of an axially loaded strut P_{max} can be made to conform to the code formulae for the design of columns by assigning an equivalent initial imperfection in the member. For tie members, the axial load-elongation relationship as shown in Figure 6.25 is obtained based on the bilinear elastic-plastic stress-strain curve of the material.

The main advantage of the proposed model is that member axial load-deformation relationships are developed based on the strut equations from the design specification. A separate check for member stability is therefore not required. Liew et al. (1997a) also found that the conventional structural design of performing elastic large-displacement analysis followed by individual member capacity checks may sometimes lead to the over-prediction of the limit load. If the limit load of structure is governed by the instability of individual members and not due to the accelerated geometric instability of the system because of the inclusion of member imperfections, then the overestimation of limit load by the conventional design method is acceptable.

6.2.6 Modeling of Semi-rigid Connections

6.2.6.1 Connections in Building Frame

Traditional analysis and design of steel frameworks usually assumes that the connections are either fully rigid or ideally pinned. The design of member normally becomes a separate task from that of design of the connection, which is often performed at a later stage in the overall design

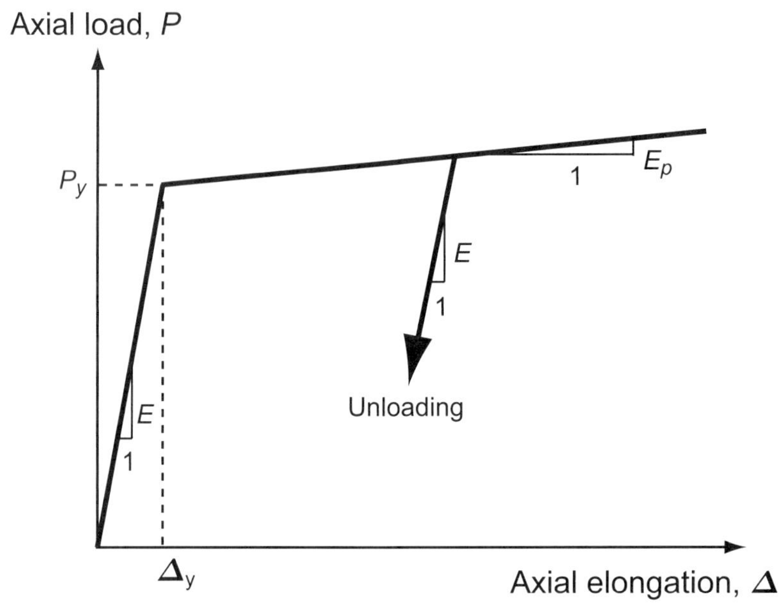

FIGURE 6.25 Axial load-elongation curve for a tie member (Liew et al., 1997a).

process by other personnel, such as the steel fabricator. However, as is evident from experimental observations, all beam-to-column connections used in the current practice possess some stiffness that falls between the two extreme cases of fully rigid and ideally pinned. The fact that a joint has sufficient strength does not mean it has sufficient stiffness for it to be reasonably modeled as rigid. The classification methods of connections have been proposed by Section 6 of Eurocode 3 (CEN, 1992), Bjorhvde et al. (1990), and Nethercot et al. (1998).

The use of semi-rigid connections in steel buildings can result in economical designs. Compared with the simple construction, semi-rigid design can reduce the floor depths and the overall volume of the building, increase the freedom for future modification and redevelopment, and result in aesthetically pleasing designs by eliminating the bracings for lateral stability. For medium and high multistory buildings, semi-rigid buildings can be designed with certain beam-to-column strength and stiffness for lateral stability. Furthermore, they are fast and simple to erect in contrast to the more traditional rigid-frame buildings, which typically require a substantial amount of field welding. Since semi-rigid design allows lower connection moment compared with rigid connection, it can readily lead to cost savings, possibly outweighing the extra cost incurred by increased members sizes and avoiding excessive deflections.

Beam-to-column connections in building frameworks can be subjected to various force combinations such as moment, shear, axial force, torque, bimoment, and their corresponding deformations. However, research on 3-D joint behavior is rare and there is limited study on joint behavior under moment-shear-axial force interaction.

Three main approaches to evaluate the mechanical properties of semi-rigid connections in terms of stiffness, strength, and ductility are by (1) experimental, (2) numerical, and (3) analytical means. The research on frame connections focuses primarily on their in-plane behavior. The most established data reported in the literature are the M-θ_r relationships of connections used in building construction.

Although numerical studies of the connection behavior using finite element techniques have been reported by Krishnamurthy and Graddy, (1976); Bursi and Jaspart, (1997); Sherbourne and

Bahaari, (1997); Swanson et al. (2000); van der Vegte and Makino, (2004); and Komuro et al. (2004), the time and cost involved as well as the uncertainty inherent in the analysis render these analytical techniques unacceptable for practical use.

The only practical option for the designers is the analytical approach. Analytical procedures have been developed that enable a prediction of the joint response based on the knowledge of the mechanical and geometrical properties of the joint component. Several analytical models have been developed to curve fit the experimental data with simple expressions, such as the B-spline model, polynomial model, and exponential models, etc., which are summarized in Nethercot and Zandonini (1989), Chen and Lui (1991), Chen et al. (1996), and Faella et al. (2000).

A three-parameter power model was proposed by Kishi and Chen (1986) and Kishi et al. (1993) for four types of connections—single web-angle connections, double web-angle connections, top-and-seat angles with double web-angle connections, and top-and-seat angles without double web-angle connections. The initial connection stiffness and ultimate moment capacity of the connections are determined by a simple analytical model, and the empirical equation for shape parameter is calibrated to experimental results. However, this three-parameter power model may not be suitable for estimating the M-θ_r behavior of end-plate type connections because clear-cut strain-hardening stiffness has been observed through testing. The four-parameter power model can be used instead to represent the M-θ_r curves of semi-rigid connections with strain-hardening stiffness. Kishi et al. (2004) extended the three-parameter power model to the four-parameter power model for end-plate connections. The four-parameter power model is given by (Richard and Abbott, 1975):

$$M = \frac{\left(R_{ki} - R_{kp}\right)\,\theta_r}{[1+\left|\left(R_{ki} - R_{kp}\right)\,\theta_r \,/\, M_0\right|^{\,n}]^{1/n}} + R_{kp}\theta_r \tag{6.53}$$

where R_{ki} is initial stiffness of connection, R_{kp} is strain hardening stiffness of connection, M_0 is a reference moment, and n is shape parameter (Figure 6.26).

The evaluation procedure for the four parameters in Equation 6.53 for the best representation of the M-θ_r curves has been provided by Attiogbe and Morris (1991). By differentiating Equation 6.53 with respect to θ_r, the tangent stiffness of the connection can be obtained. To allow for the

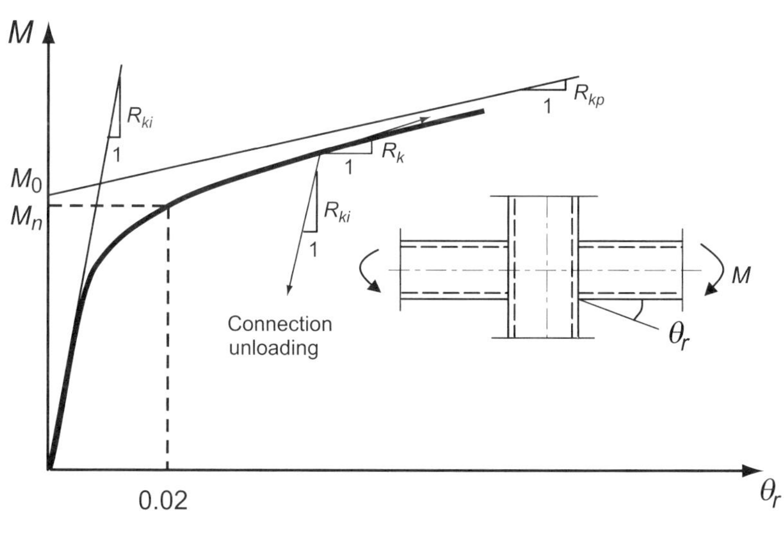

FIGURE 6.26 Four-parameter power model.

unloading of the connection associated with nonproportional loading and inelastic force redistribution, the unloading stiffness is assumed to be equal to the initial stiffness as shown in Figure 6.26.

6.2.6.1.1 Component Method

A more general approach that can be applied to any type of steel or composite joints is termed the component method (Annex J of Eurocode 3, 1992; Tschemmernegg and Huber, 1996; Faella et al. 2000; Liew et al., 2004). An attempt to extend the component method for predicting the 3-D behavior of steel joints under arbitrary loading is reported in Da Silva (2008). The component model considers any connection as a set of individual basic components, which is shown in Figure 6.27. The application of the component model includes three steps:

1. Identification of the active components in the connection being considered
2. Evaluation of the stiffness or resistance characteristics for each individual basic component (specific characteristics are initial stiffness, design resistance, or the whole deformability curve
3. Assembly of all the constituent components and evaluation of the stiffness or resistance characteristics of the whole joint

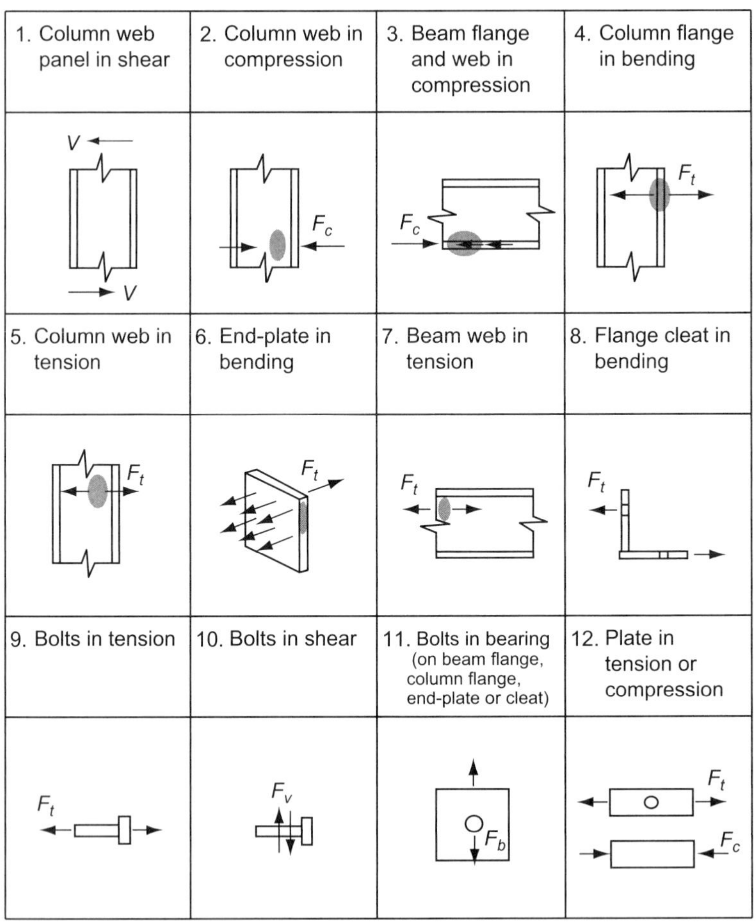

FIGURE 6.27 List of components covered by Eurocode 3 (1992).

The application of the component method requires sufficient knowledge of the behavior of the basic component. Those covered by Eurocode 3 (CEN, 1992) are listed in Figure 6.27. The application of the component method for a specific case of a welded beam-to-column connection is illustrated in Figure 6.28.

Based on different analysis method, the M-θ_r curves obtained by the component method can be idealized as:

1. A linear curve for elastic verification of the joint resistance and plastic verification for joint resistance
2. A rigid-plastic curve for a rigid-plastic analysis
3. A nonlinear curve for an elastic-plastic analysis

Similar developments for the design evaluation of semi-rigid steel frame in the framework of the AISC Direct Analysis was proposed by Surovek et al. (2005).

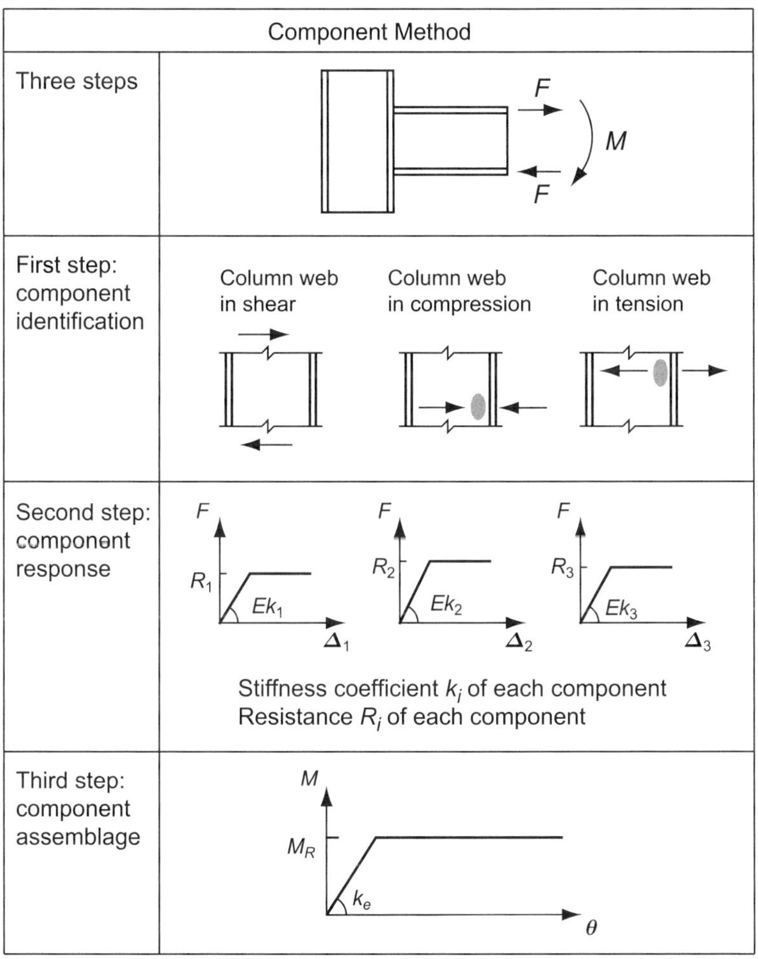

FIGURE 6.28 Application of the component method to a welded joint.

6.2.6.1.2 Computer Implementation

There are three different ways to incorporate the connection flexibility into the computer-based structural analysis of steel frameworks. The first is to introduce additional connection elements (or physical rotational spring elements) that model the beam-to-column connections directly. One drawback of this approach is that modeling effects are increased to incorporate an additional node at each element end with semi-rigid connection. In the second approach, the element stiffness matrix is modified to account for connection flexibility, i.e., in the refined plastic analysis proposed by Liew et al. (1993b). In the third approach (Hsieh and Deierlein, 1991; Chen et al., 2000), the corresponding local rotational degrees of freedom (DOF) between the member end and connection are treated as additional global unknowns of the structural system and are included in the global equilibrium equations. To introduce the local DOF into the global solution system, the conventional element transformation matrices are modified for the elements with semi-rigid connections. The modified matrices are used to transform the element stiffness matrices from the local to the global coordinate system with some of the unknowns retaining their local coordinate reference axes. A key advantage of this approach is that the existing nonlinear formulation for the beam-column elements is unaffected. This avoids the difficulty (particularly for 3-D problems) of directly formulating the connection flexibility into the nonlinear element stiffness matrices. The use of separate connection elements also facilitates further modifications to the connection model.

6.2.7 Modeling of Tubular Joints

The joints in tubular steel frames—particularly the simple and unstiffened joints—may exhibit considerable flexibility in both the elastic and elastic-plastic ranges. Some joints may reach their ultimate strength and cause excessive deflections and different internal force distribution in the structure. Depending on the joint geometry, the capacity of the connection between brace and chord may be less than the brace capacity. This means that the brace cannot be fully utilized. To take these into account, the conventional method (Qian, 2005) is to use shell or solid elements to model the joint. However, this treatment is normally for the detailed numerical study of the joint behavior and may not be suitable for the global nonlinear analysis of the frame system due to the time required to prepare the input as well as the extensive computer time required.

6.2.7.1 Joint Capacity

A joint model based on the simple elements with elastic-plastic behavior can be adopted to overcome this difficulty (Ueda and Rashed, 1991; Hellan, 1995; Dai, 2002). For thick-walled tubular sections used extensively in jackups, simplified joint stiffness and strength formulations have been developed by Choo et al. (2005) through the calibration to FEM analyses. For a K-joint as illustrated in Figure 6.29, two extra nodes are added at the brace to chord intersection face and two extra elements are generated at the end zone of the braces. A plastic hinge is formed at the extra node of the extra element is the forces satisfy the API joint capacity equation (API RP 2A-WSD, 2002):

$$\Phi = \cos\left(\frac{\pi}{2}\frac{P}{P_U}\right) - \sqrt{\left(\frac{M_i}{M_{iU}}\right)^2 + \left(\frac{M_o}{M_{oU}}\right)^2} = 0 \tag{6.54}$$

in which P_U is the ultimate axial capacity, M_{iU} is the in-plane ultimate moment capacity, M_{oU} is the out-of-plane ultimate moment capacity, P is axial force, M_i is the in-plane moment, and M_o is out-of-plane moment. For braces that carry part of their load as K-joints and part as T&Y or X-joints, the calculation of capacities P_U, M_{iU}, and M_{oU} should be based on the weighing factors according to the portions in the brace classification (API RP 2A-WSD, 2002).

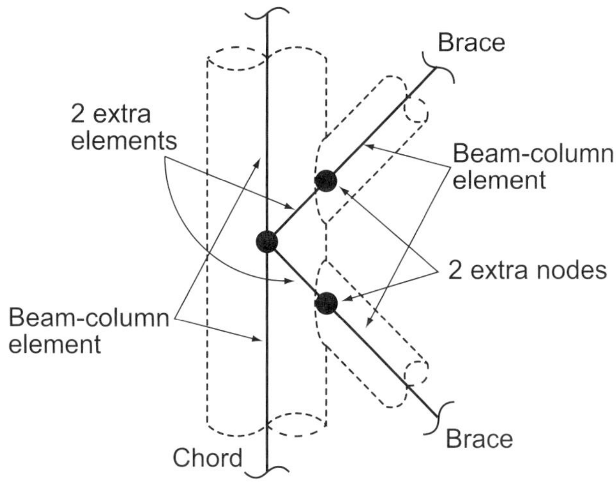

FIGURE 6.29 Modeling of a joint in tubular steel frame.

6.2.7.2 Parametric Equations for Local Joint Flexibility

When a tubular joint is loaded, the chord wall will deform and the local flexibility of the joint may be modeled by rotational and translation springs as shown in Figure 6.30. In principle, the local joint flexibility of a single-brace tubular joint should be given by a 6 × 6 matrix corresponding to three forces and three moments applied at the end of brace. However, the two shear forces and the torque are irrelevant to joint design, therefore reducing the flexibility matrix to a 3 × 3, which includes only the axial force and the in-plane and out-of-plane bending moments. The off-diagonal terms of this reduced matrix is either uncoupled, as is the case between in-plane and out-of-plane degrees of freedom, or small enough to be ignored for the analysis purpose. When two-brace joints are considered, the flexibility matrix is expanded to a 6 × 6 matrix and some carryover terms are included. The concept of the flexibility matrix also can be expanded to a joint with more than two braces (Dai, 2002).

The local joint flexibility is a function of the relative sizes of the through member (chord) and the attaching members (braces), as well as of the load pattern acting on the braces. Theoretical and experiment works have been carried out to find the method for determining the local flexibility of tubular joints. A number of parametric formulae for evaluating the local joint flexibility

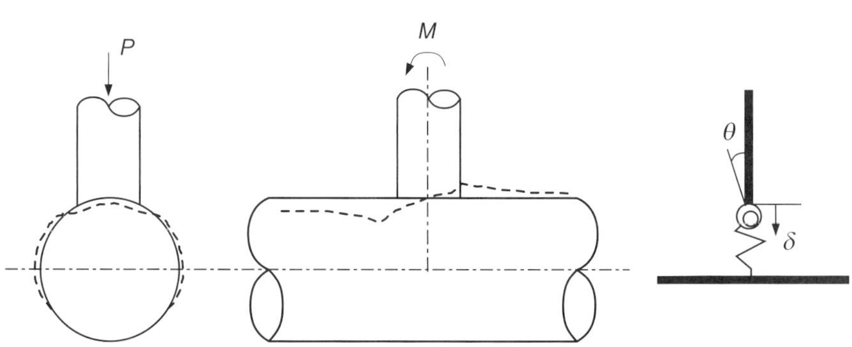

FIGURE 6.30 Local flexibility of a tubular joint modeled by rotational and translation springs.

have been proposed since the 1980s. The formulae of Fessler (Fessler et al., 1986a, b), Chen (Chen et al., 1990), and Buitrago (1993) cover a relatively wide range of parameters and can be utilized in the global structural analysis. The main limitation of Fessler's and Chen's equations is that they cannot deal with joints with equal brace and chord diameters, which is quite common for practical frames with X-joints. To solve this problem, Buitago's equations can be used.

6.2.7.3 Load-deformation Relationship

The load-deformation relationship of joint flexibility can be idealized as bilinear or trilinear models as shown in Figures 6.31 and 6.32 (Dai, 2002). P'_y and P'_u are the yielding axial strength and ultimate axial capacity of the joint when the joint is simultaneously subjected to an in-plane moment M and can be derived using the following equations:

$$\frac{M}{M_y} - \cos\left(\frac{\pi}{2}\frac{P'_y}{P_y}\right) = 0 \tag{6.55}$$

$$\frac{M}{M_U} - \cos\left(\frac{\pi}{2}\frac{P'_U}{P_U}\right) = 0 \tag{6.56}$$

The ratio $a = P_y/P_U = M_y/M_U = P'_y/P'_U = M'_y/M'_U$ is obtained from the test results and can be chosen as a constant. For K, Y/T-joints, $a = 0.8$. For X-joints, $a = 0.6$ (Dai, 2002). In the trilinear model shown in Figure 6.32, K'_{axial} and R'_{ki} can be chosen as a third of the elastic stiffness (Kamba, 1997).

For axially loaded joints, Yura et al. (1980) proposed a deformation limit beyond which joint failure is said to have occurred. The Yura deformation limit δ_{max} is taken to be equivalent to the change in length of an unconnected brace member of length $l_b = 30d$ under axial loading and with a strain of twice the uniaxial yield strain ε_y, i.e., $\delta_{max} = 2\varepsilon_y \times l_b$. For brace length of $30d$ and $\varepsilon_y = f_y/E$ (f_y = yield strength, E = Young's Modulus), $\delta_{max} = 60d\,f_y/E$. For in-plane bending loaded joints, the Yura rotation limit is taken to be equal to that occurring at the ends of a simple supported, unconnected brace member of length $l_b = 30d$ subjected to transverse, uniformly distributed load of sufficient magnitude to cause beam bending strain of $4\varepsilon_y$. Using the Yura recommendation that $l_b = 30d$, the resulting rotation limit $\theta_{max} = 80f_y/E$.

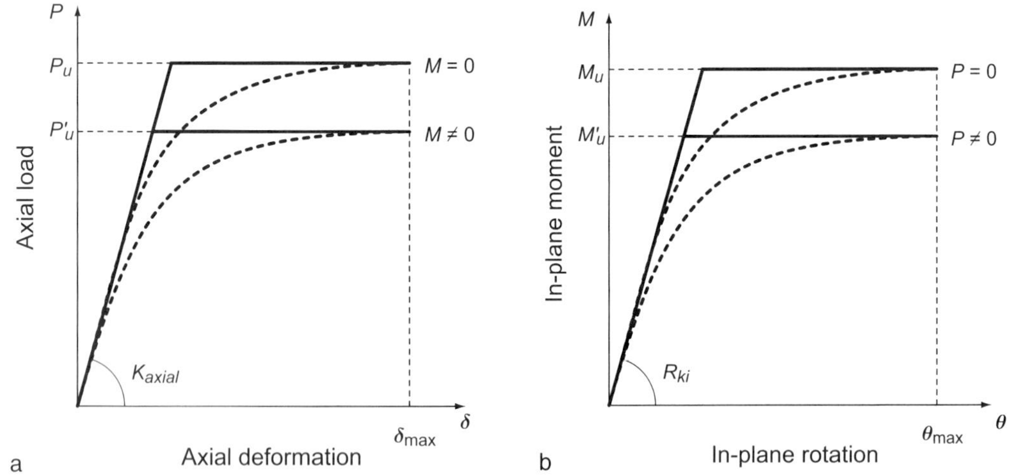

FIGURE 6.31 Bilinear load-deformation relationship: (a) axial load vs. axial deformation; (b) moment vs. rotation.

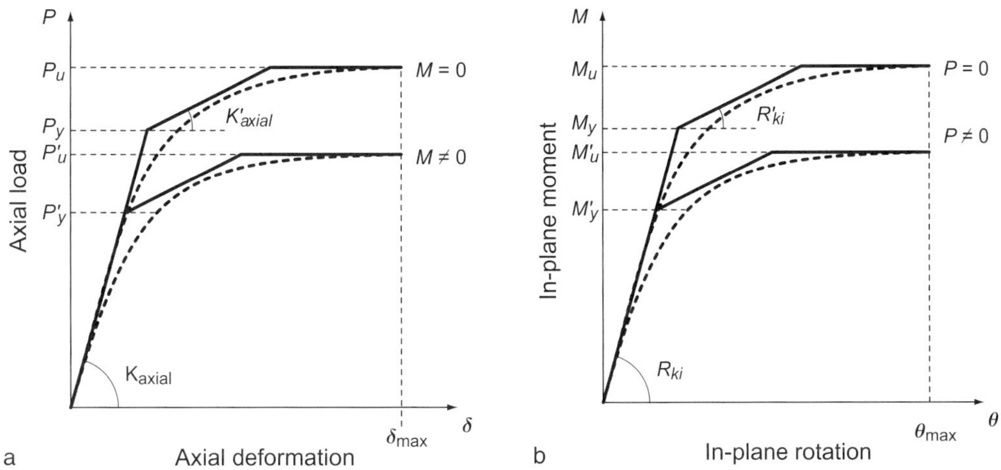

FIGURE 6.32 Trilinear load-deformation relationship: (a) axial load vs. axial deformation; (b) moment vs. rotation.

6.2.8 Tubular Frame with K-type Joints

Figure 6.33 shows a tubular frame with K-type joints tested by Bolt et al. (1994) and later analyzed by Dai (2002). The frame is subjected to a horizontal load *P* at the right top. The finite element models with and without considering initial imperfections are shown in Figure 6.34. The structure is analyzed by assuming both the rigid connection and semi-rigid connection with bilinear and trilinear models. The load displacement curves generated from the second-order plastic-hinge analysis are compared with the test results as shown in Figure 6.35. The initial stiffness, ultimate load and the load level of the first plastic hinge of the structure predicted by different analysis methods are compared to the test results in Table 6.2. It can be concluded that:

1. By neglecting the joint flexibility, the rigid frame analysis (rigid joint model) with and without considering the joint size predicts stiffer load deformation curves and higher ultimate strength than do the test results. The rigid frame analysis with zero joint size and without initial imperfection predicts the initial stiffness and ultimate load of 28.7 percent and 41.3 percent higher than those from the test.
2. For the semi-rigid models, there is little difference when bilinear or trilinear connection models are used. Both semi-rigid models can predict flexible load deformation curves that closely match the test result.
3. Modeling of member initial imperfection has shown some difference on the results predicted by the rigid joint model. However, for frames with semi-rigid joint models, the effects of member initial imperfection are negligible.
4. The semi-rigid models can predict the ultimate load with good accuracy in comparison with the test result with an error less than 2.5 percent.
5. The yielding of the K-joint is predicted as observed in the actual test when the joint performance is modeled in the analysis.
6. The effects of the joint performance are of most importance to this frame.

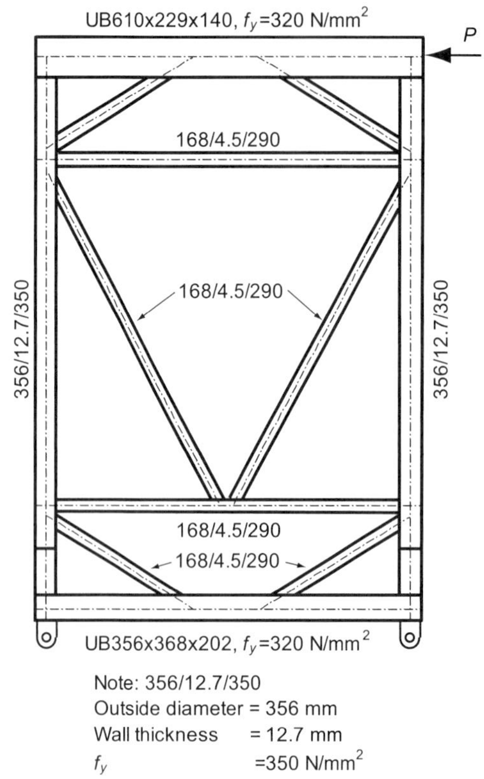

FIGURE 6.33 Frame with K-type joints

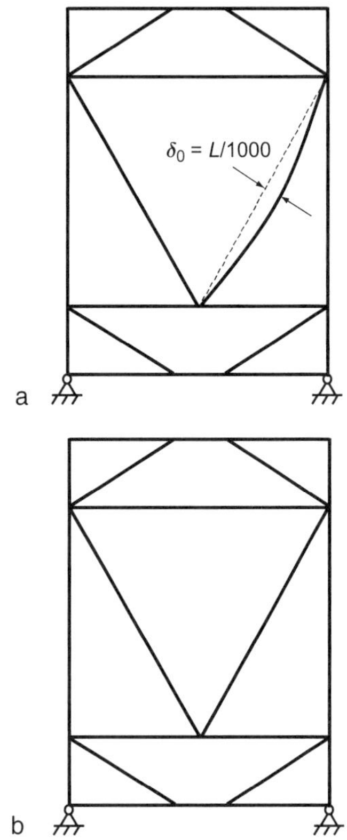

FIGURE 6.34 Finite element model: (a) with initial imperfection; (b) without initial imperfection.

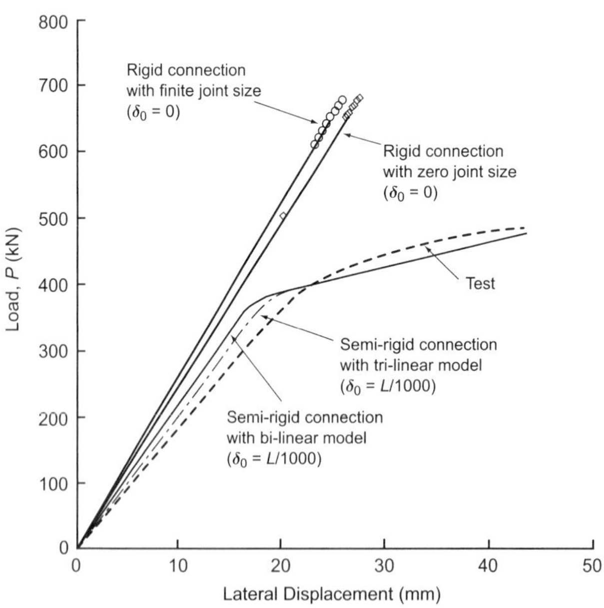

FIGURE 6.35 Global response of frame with K-type joints.

TABLE 6.2 Analysis results of frame with K-type joints

	Analysis without initial imperfection				Analysis with initial imperfection				Test
	Finite rigid	Zero rigid	Semi-rigid		Finite rigid	Zero rigid	Semi-rigid		
			Bi-linear	Tri-linear			Bi-linear	Tri-linear	
Max load (kN)	675 (40.5%)	679 (41.3%)	470.4 (−2.1%)	470.2 (−2.1%)	603.7 (25.6%)	603.7 (25.6%)	468.5 (−2.5%)	468.2 (−2.5%)	480.4
Initial stiffness (kN/m)	25903 (36.5%)	24417 (28.7%)	21500 (13.3%)	21500 (13.3%)	27300 (43.9%)	27300 (43.9%)	21500 (13.3%)	21500 (13.3%)	18970
Load at first plastic hinge (kN)	650	656	360	367	554	554	360	367	—

Note: values in bracket are percentage errors compared with the test results

6.2.9 Bowstring Column

Bowstring steel columns were first investigated by Liew et al. (2001a) where the effects of geometric imperfection and cable pretension forces were considered. Chan et al. (2002) presented a stability analysis and parameter study of prestressed cable-stay columns. In Liew and Li (2006), the nonlinear behavior of pretensioned bowstring columns is examined with advanced analysis. Different structural configurations, cable pretension forces, and length and spacing of horizontal struts are examined.

6.2.9.1 Pretensioned Steel Column

Pretensioned steel columns generally comprise main column and horizontal strut and cables, as shown in Figure 6.36. The column is the main component that resists axial load, while horizontal struts and pretensioned cables work together as the intermediate lateral support to the main column. The study in this paper is carried out using only circular hollow sections for the main column and horizontal struts, due to aesthetic reasons and the fact that circular hollow section offers better resistance to lateral torsional buckling compared to H or I sections. A summary of the material properties and section size used for analysis is shown in Table 6.3. Pretensioned steel columns with a length of 10, 15, and 20 m are selected in that their slenderness ratios L_V/r (83, 125, and 166, respectively) are within the general range of engineering application where L_v is the spacing of horizontal struts and r is the radius of gyration.

The bottom end of the pretensioned steel columns is assumed to be restrained in three translational directions and torsion while being free to rotate. However, the support at the top of the column is free to translate in the vertical direction and rotate, but it is restrained in the other two translational directions and torsion.

Initial geometric imperfection of $L_c/500$ magnitude is introduced to the main column (L_c is the column length). The imperfection pattern is assumed to be the same as the first buckling mode of the main column.

6.2.9.2 Structural Configurations

There are many types of cable shapes (or restraining configurations) for the pretensioned steel columns such as triangular, rectangular, and bowstring shapes as shown in Figure 6.36. The bowstring column configuration is defined by a second-order polynomial equation, which describes a parabolic curve. Configurations defined with other higher-order polynomial equations will not

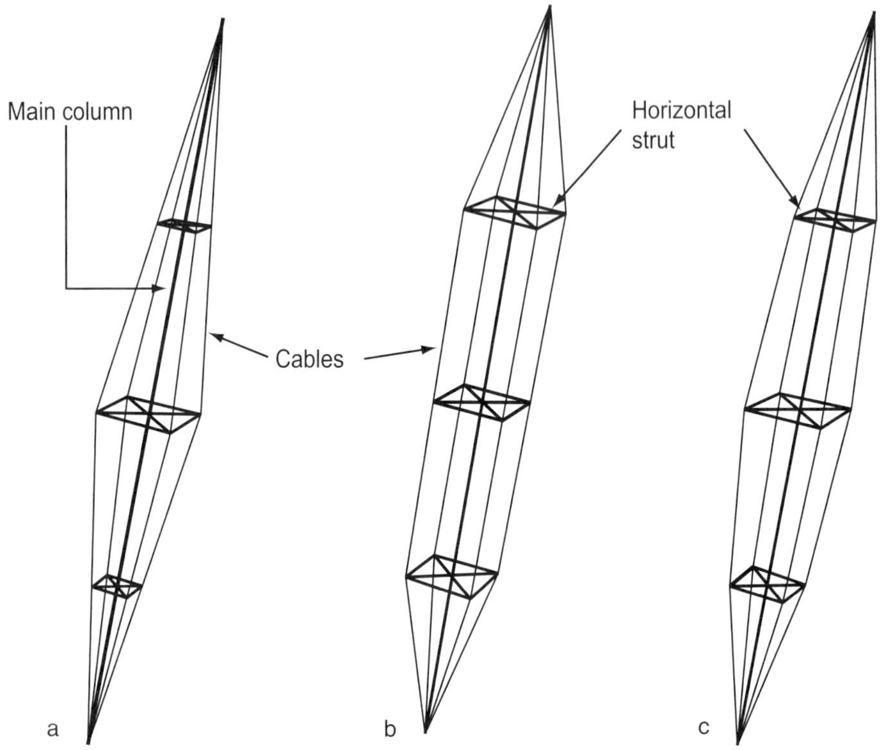

FIGURE 6.36 Three types of pretensioned steel column: (a) triangular; (b) rectangular; (c) bowstring.

TABLE 6.3 Summary of the material properties used for analysis of bowstring columns

Elements	Material properties	Section size
Vertical strut	f_y = 275 MPa E = 205 GPa	CHS 88.9 × 4.0 mm
Horizontal strut	f_y = 355 MPa E = 205 GPa	CHS 60.3 × 4.0 mm
Pre-tensioned cable	f_y = 460 MPa E = 150 GPa	Diameter 10 mm

f_y is the design strength and E is the Young's modulus

be included in the present analysis. The numerical examples are designed for triangular, rectangular, and bowstring columns with three lengths of 10 m, 15 m, and 20 m and with three horizontal struts uniformly spaced. The maximum length of horizontal struts at midspan is assumed as $L_c/16$ (L_c is the column length) for three configurations. The same cable forces of 11 kN are pretensioned for different column lengths and different column configurations. All pretensioned columns are subjected to initial geometric imperfection and axial compression force. The results of axial load capacities are listed in Table 6.4 and the corresponding failure modes are shown in Figure 6.37.

The bowstring column has the highest axial load capacity as compared to the triangular shape column with the lowest capacity. The rectangular columns can be the alternative if the column has a relatively large slenderness ratio (more than 100). For a triangular cable shape, the horizontal struts at quarter-span are not fully effective, and cannot provide the lateral restrain forces

TABLE 6.4 Axial load capacities of different column configurations with different lengths (Unit: kN)

Configuration \ Length	10 m	15 m	20 m
Triangular	122	94	59
Rectangular	165	115	69
Bowstring	183	121	74

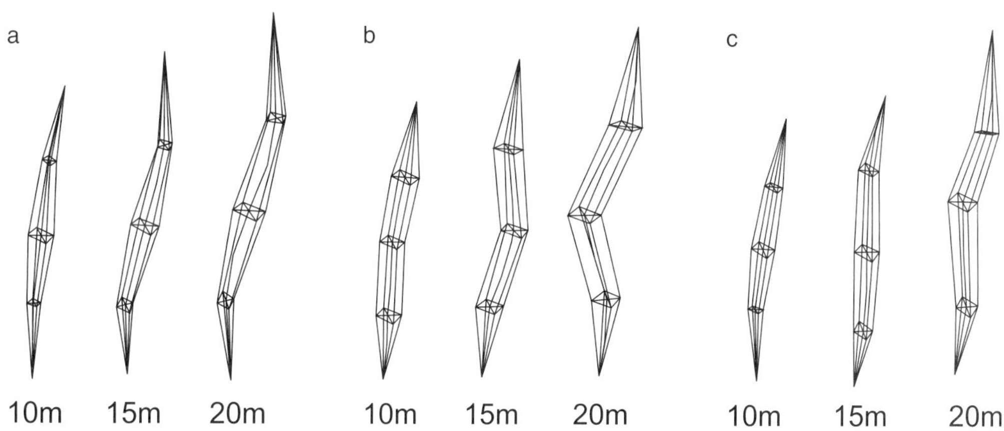

FIGURE 6.37 Failure modes of pretensioned steel columns with lengths of 10m, 15m, and 20m: (a) triangle; (b) rectangular; (c) bowstring.

to the main column that would enable earliest buckling. Although insufficient support also exists at midspan horizontal strut due to the equal length for all struts in rectangular cable shape, this disadvantageous effect will decrease with the development of lateral deflection at midspan. A bowstring cable shape can provide harmonious lateral restrain forces to both midspan and quarter-span horizontal struts, and can give the highest load carrying capacity. Since bowstring column is the most efficient configuration, it will be used in the subsequent sections to investigate the effects of cable pretension force, length, and spacing of horizontal struts.

6.2.9.3 Cable Pretension Force

Cable pretension force is an important parameter that affects the stiffness and strength of the pretensioned steel columns. The effects of cable pretensions on the load capacity of bowstring columns with lengths of 10 m, 15 m, and 20 m are investigated. The columns studied here have three horizontal struts uniformly spaced and the maximum length of horizontal struts at midspan is assumed as $L_c/16$ (L_c is the column length). It is assumed that the yield stress of cable material is 460 MPa. Since the diameter of cables is 10 mm, the yielding load of the cable is about 36 kN. The bowstring columns are analyzed using advanced analysis by assuming different pretensioning forces in the cables. The results are presented in Figure 6.38 showing the axial capacity of the column versus the pretension force, which is normalized by the cable yield load T_y. Cable pretensions equal to 5 percent of cable yield load can evidently improve the column capacity when compared to the cases without cable pretensions. The preferred range of cable pretension force is

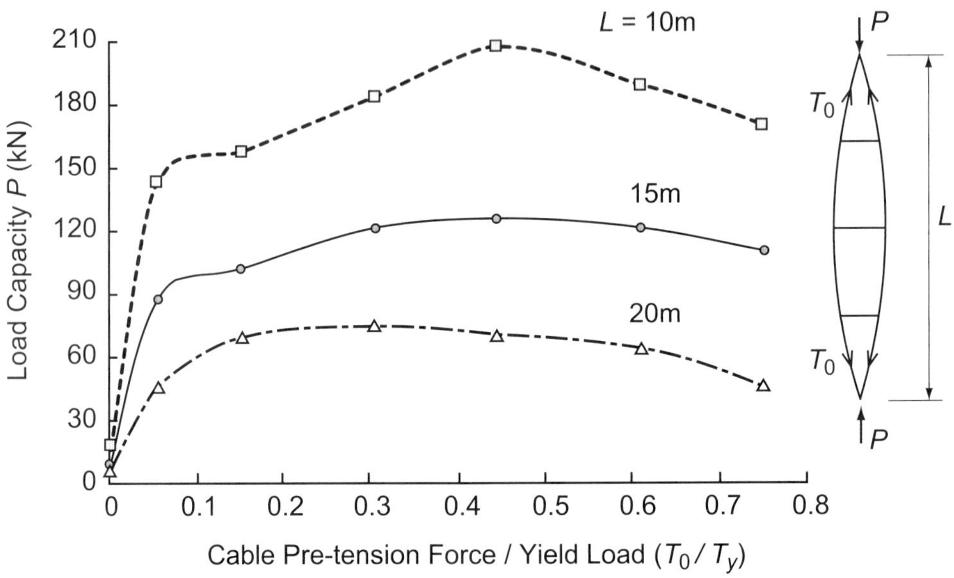

FIGURE 6.38 Axial load capacity vs. cable pretensions for bowstring columns.

related to the column slenderness and can be found generally between 30 percent and 50 percent of cable yield strength. When the pretension force is too high (more than 50 percent of the cable yield strength) the axial capacity of bowstring columns will reversely decrease, as shown in Figure 6.38. This is because high pretensioning force in the cables introduces a high compression load on the main column, and thus reduces its load-bearing capacity.

6.2.9.4 Length of Horizontal Strut

The length of horizontal struts in bowstring columns is limited for the sake of aesthetic reasons and restraint effectiveness. By changing the length of horizontal struts of bowstring columns, its effects on the column ultimate capacities can be examined. The numerical results are listed in Table 6.5 where the three lengths of horizontal struts studied are $L_c/8$, $L_c/16$, and $L_c/24$ (L_c is the column length). For the 10 m-long columns, yielding of horizontal struts is not observed in numerical analysis and the longer the horizontal struts, the higher the column capacity. However, in the case of $L_c/8$ for 15 and 20 m-long columns, yielding of horizontal struts occurs before column failure and causes the prematured collapse of columns when compared with the case of $L_c/16$, as

TABLE 6.5 Axial load capacities of a bowstring column of different lengths of horizontal strut and column (Unit: kN)

Column length (L_c) / Horizontal strut length	10 m	15 m	20 m
$L_c/8$	210	103	63
$L_c/16$	183	121	74
$L_c/24$	118	104	77

shown in Table 6.5. If the section of horizontal struts is strengthened and replaced by CHS 80.0 × 4.0, the capacities are 135 kN and 76 kN respectively for 15 and 20 m-long bowstring columns with $L_c/8$-long horizontal struts, which are higher than those of $L_c/16$ because yielding of horizontal struts is prevented. It can be drawn from investigation in this section that horizontal struts should be adequately designed to prevent premature yielding or buckling, and when longer struts are used, more attention should be paid to the possibility of yielding or buckling of the horizontal struts.

6.2.9.5 Spacing of Horizontal Strut

The spacing of horizontal struts L_V is varied as $L_c/4$, $L_c/5$, and $L_c/6$ for bowstring columns 10, 15, and 20 m in length. The length of horizontal struts is kept to be $L_c/16$ and the cable pretensioned to be 11 kN as above. The axial capacities obtained from advanced analysis are listed in Table 6.6. A general conclusion can be drawn that the more horizontal struts are provided, the higher compressive capacity—especially for the columns with relatively high slenderness ratio. This is because of the reduction in the effective buckling length of the main column between two lateral restrained points. Increasing the number of restrained points will not further increase the capacity, because the failure will be caused by the overall buckling of the bowstring column.

6.2.10 Bowstring Frame

The concept of bowstring columns can be extended to 3-D frames (Liew and Li, 2006) if the stability of support columns in the out-of-plane is ensured by bowstring cables, while that in the in-plane stability is provided by frame action. Based on the study of bowstring columns, structural behavior of a bowstring frame is analyzed in this section, where the structural layout, novel semi-rigid connection, structural nonlinearities under environmental loads, and effects of cable pretensions are addressed.

6.2.10.1 Structural Layout

A bowstring structure supporting a glass facade is shown in Figure 6.39. It consists of seven dependent frames (three on the left and four on the right) and one lateral bracing frame. The detailed model of a bracing frame is shown in Figure 6.40, which includes the elevation view in X-Y and Y-Z planes.

When the bowstring frames are assembled, pretension force is applied to the bowstring cables and bracing cables to provide lateral stiffness in the in-plane and out-of-plane direction of the frame. After the roof slab and glass facade are constructed, the bowstrings are under compression and a second-stage pretension force is applied to the cables so that they can have sufficient lateral

TABLE 6.6 Axial load capacities of a bowstring column of a different number of horizontal strut and column lengths (Unit: kN)

Column length (L_c) / Number of horizontal strut	10 m	15 m	20 m
3	183	121	74
4	194	154	96
5	196	163	113

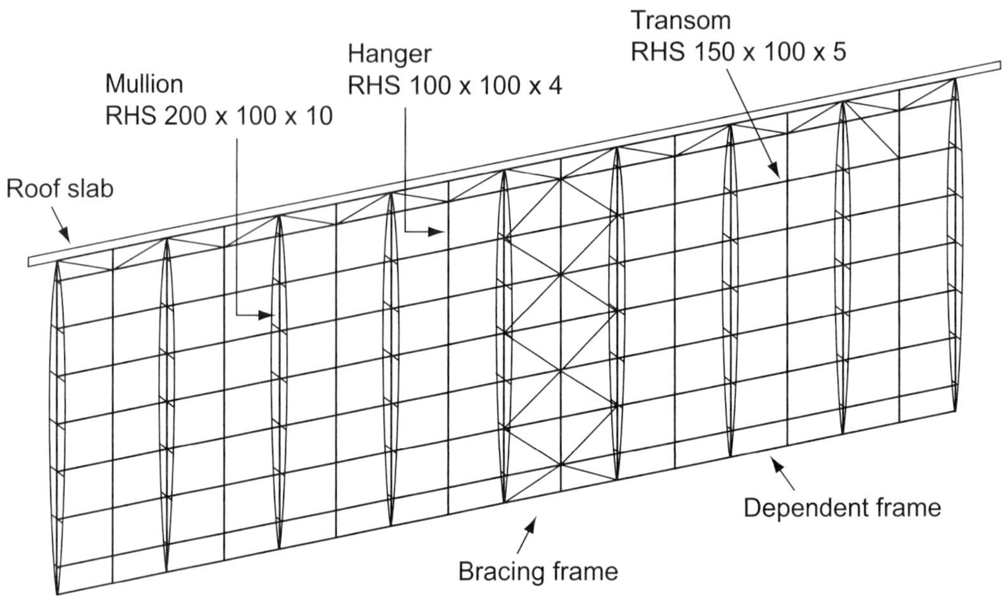

FIGURE 6.39 The structural model of bracing bowstring frame and a roof slab (Liew and Li, 2006).

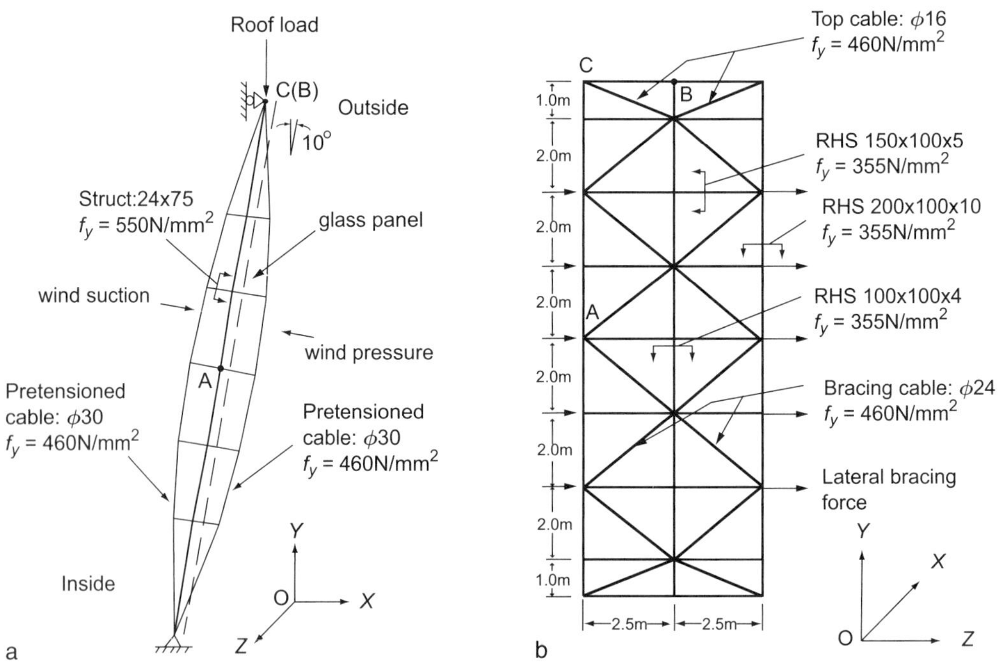

FIGURE 6.40 The structural model of bracing bowstring frame (Liew and Li, 2006): (a) X-Y plane; (b) Y-Z plane.

TABLE 6.7 Loads and loading sequence in analysis

Load sequence	Load type	Load description
1	1st stage pre-tension	a) Tension inside and outside cables to 52 kN and 50 kN respectively b) Tension bracing cables to 10 kN
2	Deal load	a) Apply dead loads = 301 kN, at upper column points b) Apply glass facade loads of 1.75 kN/m on vertical mullions and 1.4 kN/m on transoms c) Apply self-weight
3	2nd stage pre-tension	a) Tension the inside and outside cables to 78 kN and 82 kN respectively b) Tension bracing cables to 10 kN
4	Lateral bracing force (see Figure 6.40)	a) Apply bracing force at left side = 18.834 kN per node b) Apply bracing force at right side = 23.082 kN per node
5	Temperature effect	Temperature inside glass facade reduces 15°C and that outside increases 25°C simultaneously
6	Wind pressure	Apply wind pressure = 2.0 kN/m on mullions and 1.6 kN/m on transoms, up to failure
	Wind pressure	Apply wind suction = 1.4 kN/m on mullions and 1.12 kN/m on transoms, up to failure

stiffness to resist further loadings such as imposed roof load and wind load. The load sequences of the bowstring structures are tabulated in Table 6.7.

6.2.10.2 Testing of Mullion-to-transom and Transom-to-hanger Connections

Mullion-to-transom and transom-to-hanger connections of frames in-plane are designed as those of semi-rigidity. The rotational stiffness and moment capacity of these two connections are tested so that nonlinear analysis models of bowstring frames can be correctly established and then the structural behavior can be assessed.

The test setup and specimen instrumentations are illustrated respectively in Figure 6.41. Test Specimen 1 represents the hanger (100 × 100 × 4 mm RHS) to transom (150 × 100 × 5 mm RHS) connection, which consists of one M24 × 60 Grade 8.8 HEX. HD. bolt connecting the 20 mm plate welded on the inside of the hanger to the 30 mm plate welded on the surface of the transom (see Figure 6.41c). Test Specimen 2 represents the mullion (200 × 100 × 10 mm RHS) to transom (150 × 100 × 5 mm RHS) connection, which consists of two M24 × 60 Grade 8.8 HEX. HD. bolts connecting a 20 mm plate welded on the inside of the transom to a 30 mm plate welded on the surface of the mullion (see Figure 6.41c). The M-θ_r curves obtained from the tests are shown in Figure 6.42. The test results are fitted with the four-parameter connection model in Equation 6.53 and then used in the analysis as shown in Figure 6.42.

For Specimen 1, the left side of the connection is stiffer than the right. This could be due to nonsymmetry in the specimens and the fitting of connections. Failure of connection occurred

FIGURE 6.41 The test setup and specimen instrumentation of semi-rigid connections in bowstring frames (Liew and Li, 2006): (a) setup; (b) specimen instrumentation; (c) elevation view of connections; (d) A-A view of Specimen 1; (e) A-A view of Specimen 2.

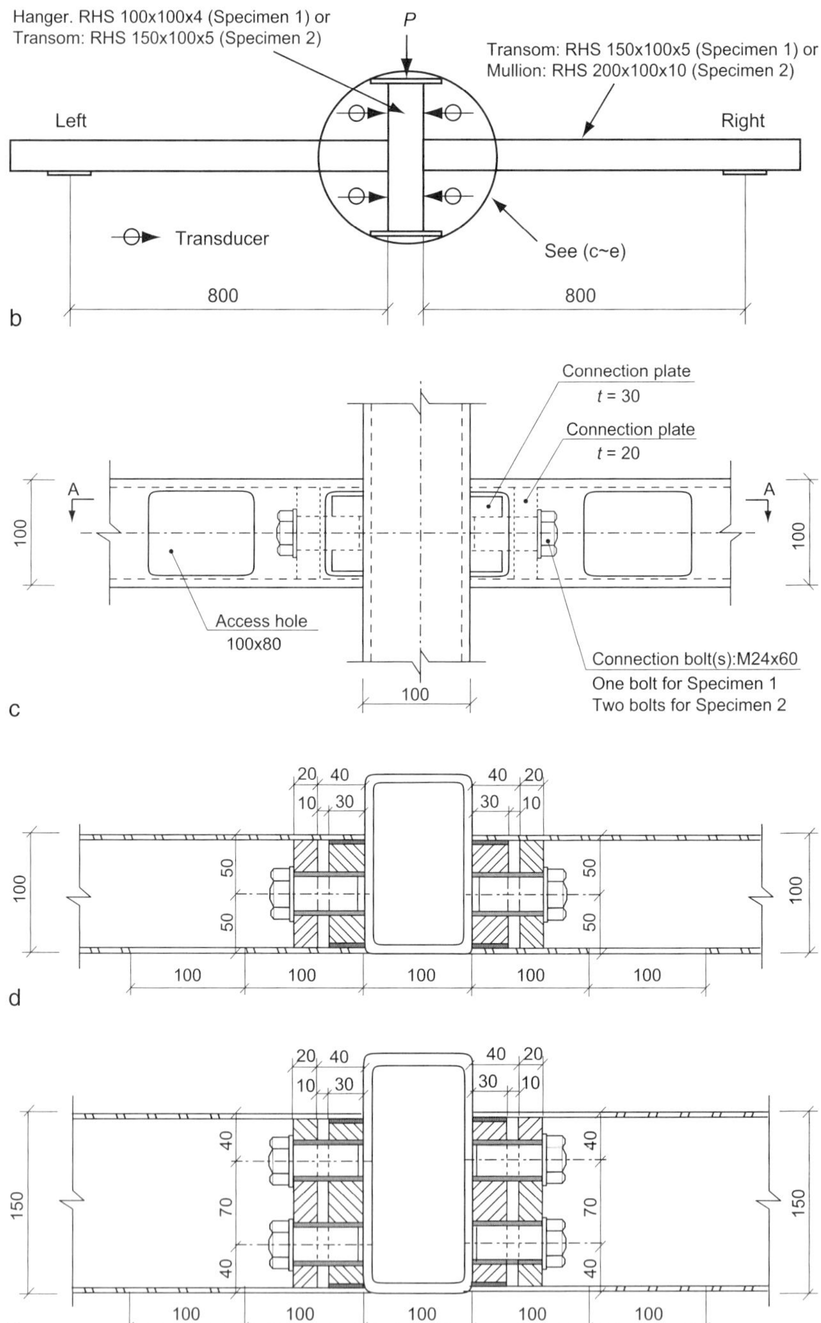

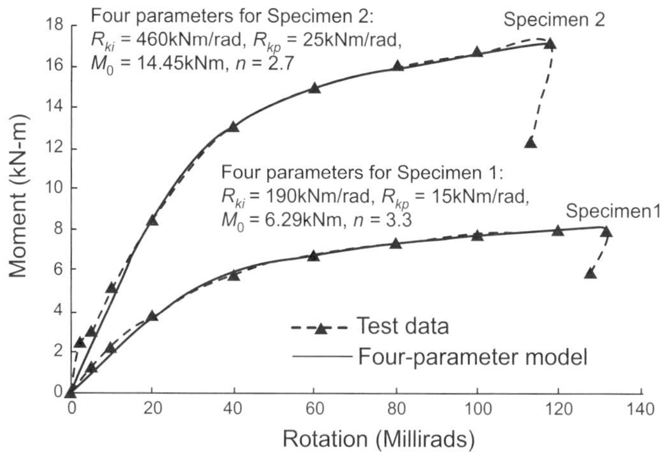

FIGURE 6.42 Moment-rotation curve of semi-rigid connections: test data vs. four-parameter model in Equation 6.53 for Specimen 1 (hanger-to-transom) and Specimen 2 (mullion-to-transom) (Liew and Li, 2006).

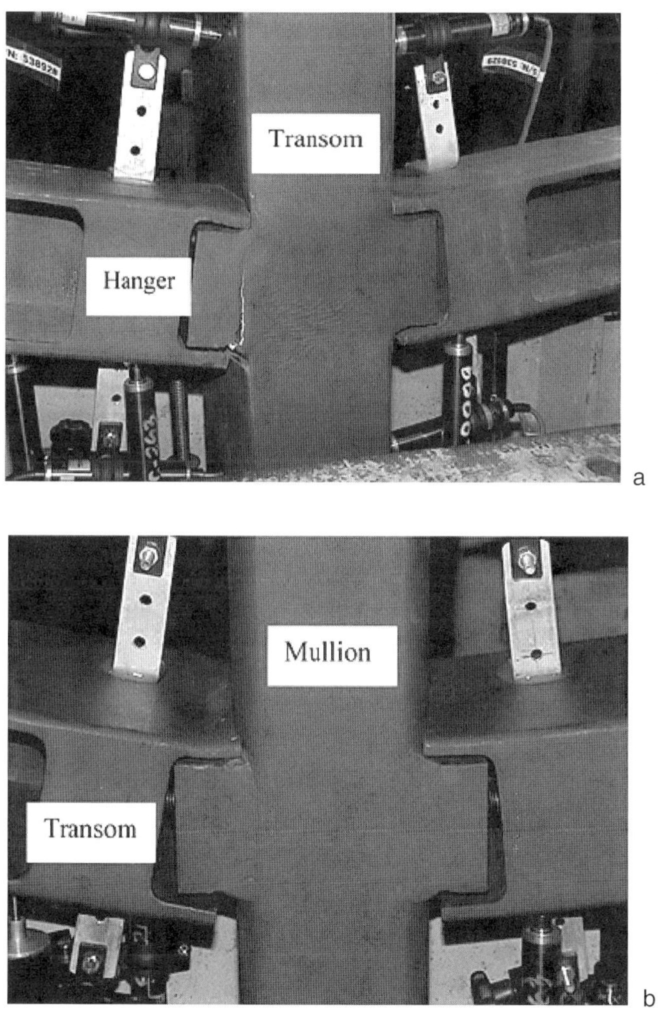

FIGURE 6.43 Connection detail and failure modes (Liew and Li, 2006): (a) Specimen 1 (hanger-to-transom); (b) Specimen 2 (mullion-to-transom).

due to yielding of the transom member caused by the compression force from the hanger beam. Weld failure, as shown in Figure 6.43, was observed upon subsequent loading after attaining the maximum load. The moment capacity is about 8 kN-m. For Specimen 2, the transom-to-mullion connection failed when the flange of the transom, which connects into the mullion, yielded under compression. The moment capacity of the connection is about 17 kN-m. Failure modes of two specimens are shown in Figure 6.43. Experimental results provided reliable parameters for modeling the nonlinear rotational spring elements used in the structural analysis.

6.2.10.3 Nonlinear Analysis

By preliminary calculation, two critical load combinations of the bracing frame are identified:

1. 1.0 (dead load) + lateral bracing load + ambient temperature difference + wind pressure
2. 1.4 (dead load) + lateral bracing load + ambient temperature difference + wind suction

Lateral bracing loads in the direction of *Z* axis (in-plane of frame) represent the required bracing forces provided by the bracing frame to the other seven dependent frames within one load-bearing unit (see Figure 6.40 and Table 6.7). According to BS5950-1 (2000), where three or more intermediate lateral restraints are provided, each intermediate lateral restraint should be capable of resisting a force of not less than 1 percent of the compression force.

For sake of simplification, only nonlinear analysis results of these two load cases are presented in this paper. The loads and loading sequences used in the analysis are listed in Table 6.7.

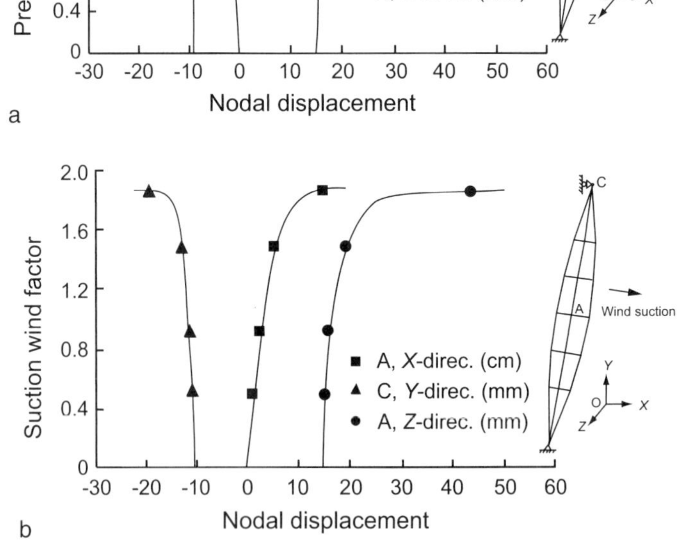

FIGURE 6.44 Nodal displacement versus load factor of wind at ambient temperature: (a) wind pressure; (b) wind suction.

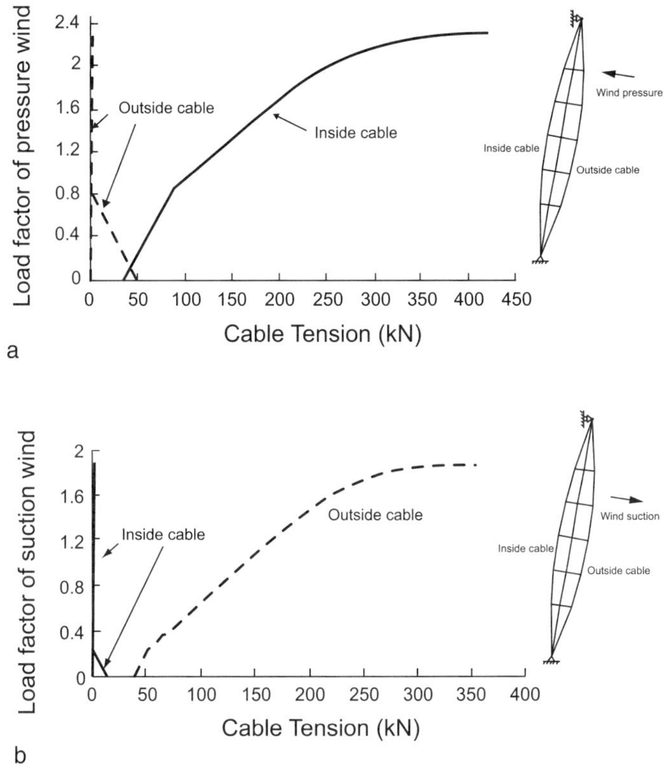

FIGURE 6.45 Cable tension vs. load factor of wind at ambient temperature: (a) wind pressure; (b) wind suction.

The typical load-displacement curves and cable tension variations are illustrated in Figures 6.44 and 6.45. Nonlinear structural response is observed when wind loads are applied. The nonlinearity is due to second-order effects associated with the axial force acting on the deflected geometry of the structure, the change in cable tension, and nonlinear behavior of semi-rigid connections. No plastic hinge is observed up to the structural failure, while buckling of vertical mullions occurs at the ultimate states. One of the reinforcing cables to the vertical mullions will slack soon after wind loads applied (as illustrated in Figure 6.45), but this does not trigger the collapse of the system because the other reinforcing cable is still in tension and the structure is capable of resisting more loads. Large deflections in both the X-direction and Z-direction are observed when instability occurs. But the deflection in X-direction is dominant rather than that in Z-direction, which implies that the bracing cables are providing effective lateral stiffness in the frame plane to other dependant frames.

TABLE 6.8 Five pretension cases for bowstring frames

Case	First stage pretension (kN)	Second stage pretension (kN)
a	0	0
b	50	0
c	50	45
d	50	80
e	50	115

6.2.10.4 Effects of Pretension

As mentioned above, the two-stage pretensioning technique is needed in the construction of the bowstring frames because the weight of the roof and glass facade will reduce the cable pretension forces, and the frame stiffness against wind loads will be insufficient if only first-stage pretensioning is applied. In this section, the ef-

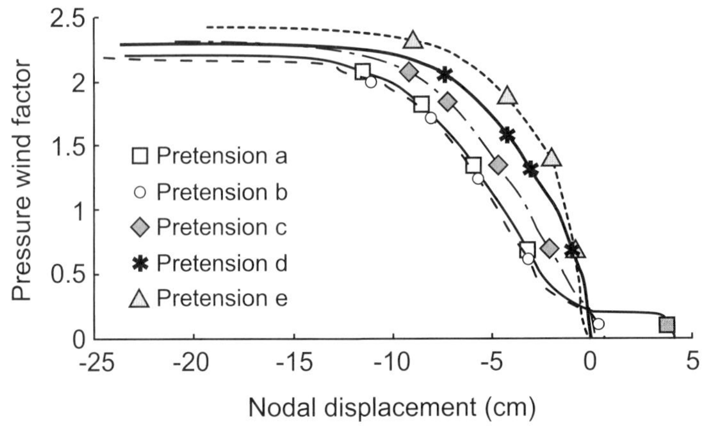

FIGURE 6.46 Effects of cable pretensions (see Table 6.8 for cable pretension cases).

fects of pretension forces on the capacity and stiffness of the bowstring frame, represented with five pretension cases (Case A to Case E as shown in Table 6.8) are studied.

Figure 6.46 illustrates the curves of displacement in *X*-direction of Node A with the pressure wind factor under different magnitudes of pretension forces. Local instability has been found from the numerical results for the states of Pretension A and Pretension B, which indicates that both the first- and second-stage pretensions are both essentially necessary for bowstring structures. Pretension E does not produce an evidently higher ultimate capacity against the wind pressure when compared to Pretension D and Pretension C. The rationality of Pretension D, namely the designed pretension force, is proved as well from Figure 6.46 where a satisfied structural stiffness at nominal pressure wind load (factor = 1) is shown.

6.2.11 Fire Analysis of Semi-rigid Frame

Advanced analyses on the semi-rigid portal frame and multistory frame subjected to localized fires show that overall structures, as opposed to single elements, have a higher resistance to failure due to fire (Liew et al., 1998). This reflects the conservative nature of the traditional fire-resistant design method in which loads transferred from the fire-weakened members to other parts of the structure remote from the fire is not considered.

Recent work by Bailey et al. (1996) and Liew et al. (1998) suggests that the progressive spread of fire can produce larger beam deflection than that predicted by assuming simultaneous heating of the same members in different compartments. Consequently, there exists the possibility that the structural integrity of a frame may remain intact up to its maximum temperature in a fire, but it may be lost during the cooling phase. The effect of progressive fire spread, where both cooling and heating are taking place simultaneously in different zones in a frame, is the subject for investigation. Such horizontal fire spread might happen due to nonclosure of fire doors, a breakdown of compartment walls, or even along expended surface materials such as carpets, wall linings, and ceiling linings.

The structural effects of horizontal fire spread are studied within the 2-D steel frame shown in Figure 6.47. All the internal beams are made of 610 × 229UB101 with design strength 275MPa, subject to a load ratio of 0.24. All columns are of 254 × 254UC132 with design strength 275MPa. The columns are assumed to be with initial out-of-straightness of *L*/750 (or 5mm) at midheight. The internal columns in the fire-affected story have a load ratio of 0.18. Load ratio is defined as the axial

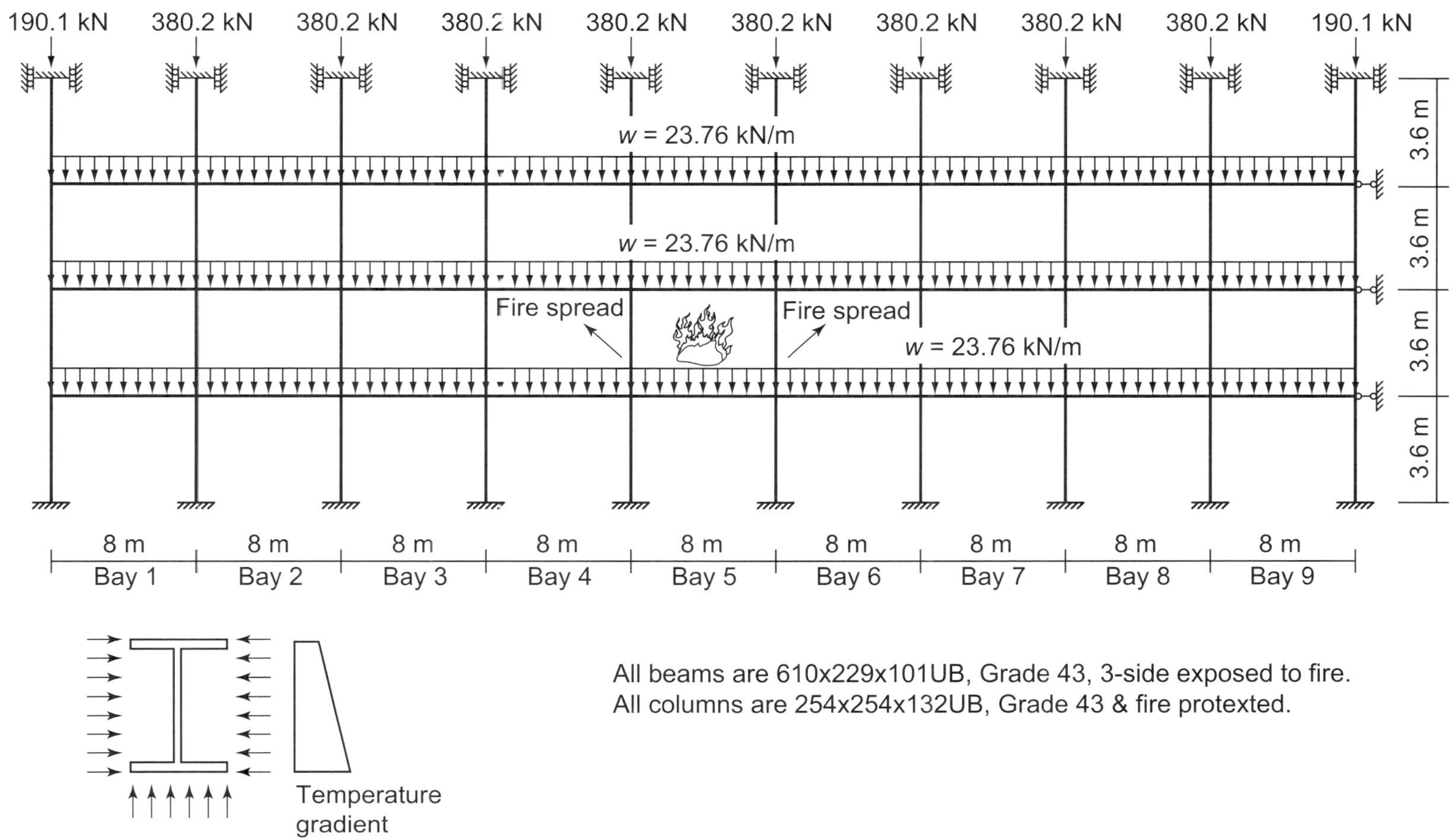

FIGURE 6.47 Modeling of horizontal fire-spread in a multistory semi-rigid frame.

force divided by the compression resistance of the member. It is assumed that all connections are fully protected so that the connections have been prevented from reaching their ultimate strength before fire spread. The M-θ_r curve for the beam-to-column connections is shown in Figure 6.48. The columns and connections are fully protected and the only heated members are the beams. The beams support the concrete slabs, and hence they are modeled as three sides exposed to fire. Two elements are used to subdivide the beam web plate in the heat transfer computation, and one element per member is adopted for modeling the structural frame.

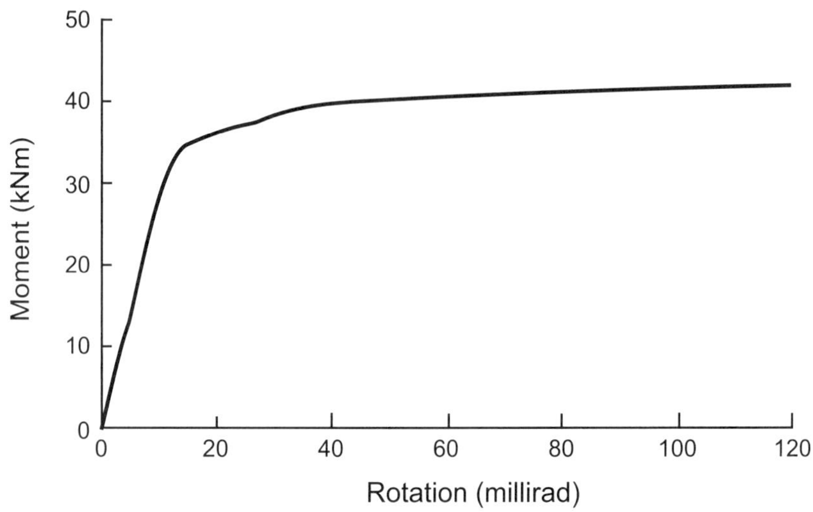

FIGURE 6.48 Moment-rotation curves for the beam-column connections in fire spread study (El-Rimawi et al., 1995).

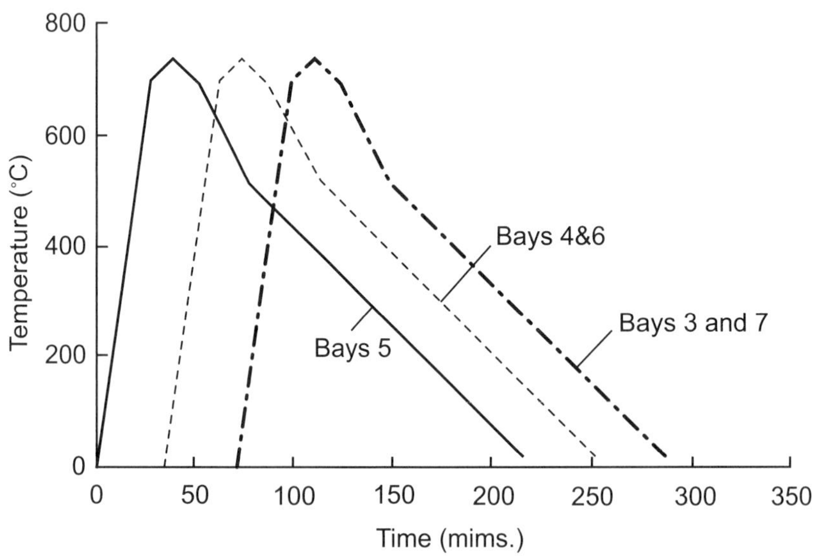

FIGURE 6.49 Temperature-time relationship of fire scenario 1: spread of fire in a story (Liew et al., 1998).

Two fire scenarios are considered. The first involves fire spreading from Bay 5 into Bays 4 and 6, continuing into Bays 3 and 7 in accordance with the time-temperature curves in Figure 6.49. The second fire scenario assumes a simultaneous heating of the beams in Bays 3 to 7 in the second story in accordance with the temperature-time curve in Figure 6.50; the beams in Bays 1, 2, 8, and 9 remain unheated.

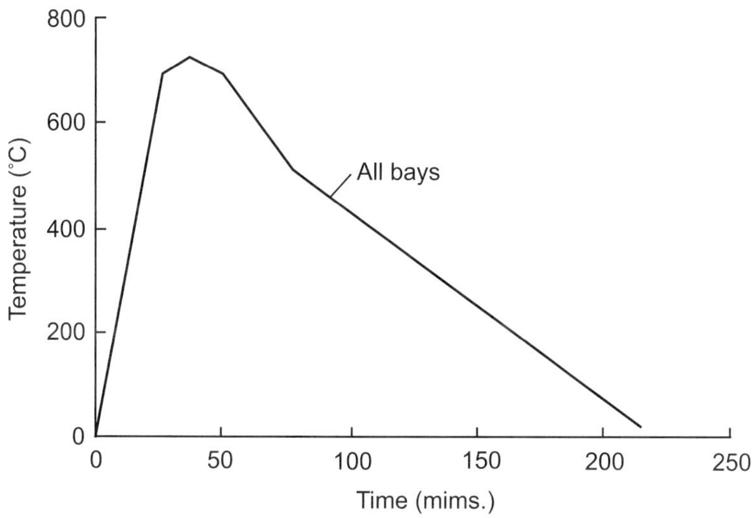

FIGURE 6.50 Temperature-time relationship of fire Scenario 2: simultaneous heating of all bays (Liew et al., 1998).

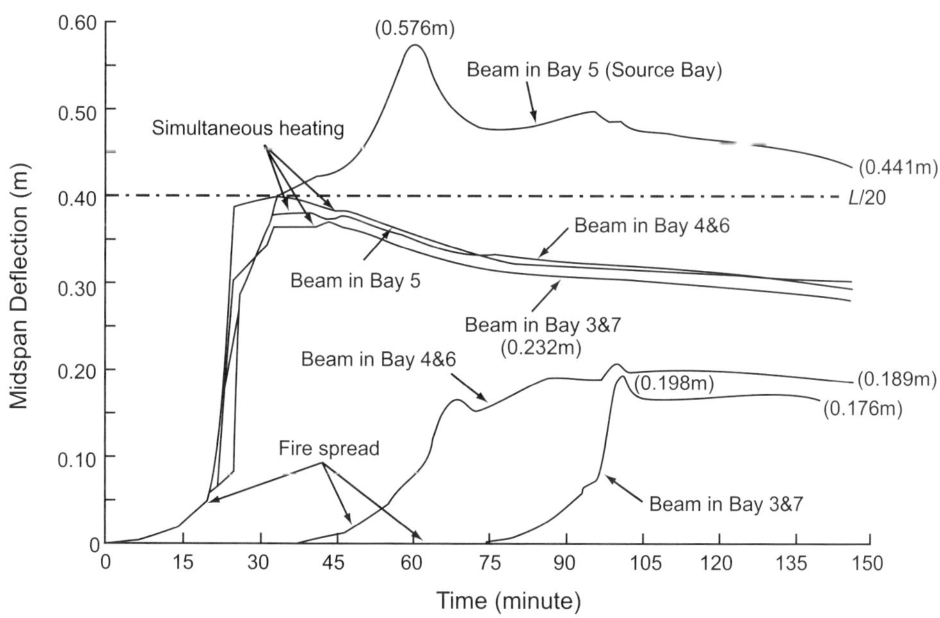

FIGURE 6.51 Comparison of midspan deflections of the heated beams.

The displacement-time relationships of the heated beams for the two fire scenarios are compared in Figure 6.51. When the spreading of fire is considered, the vertical displacement of the beam in Bay 5 does not reduce in magnitude after reaching the peak temperature in 36 minutes. In fact, the deflection continues to increase up to a calculated displacement of 576 mm at 62.6 minutes. An explanation of this behavior is given in Figure 6.52, which shows the axial force-time relationship of the heated members. In the case of fire spread, it can be observed that the axial compression in the beam at Bay 5 decreases when the temperature approaches its maximum value at 36 minutes. When Bay 5 enters into the cooling phase, the beam in Bay 4 begins to heat, adding compression to the beam in Bay 5, thus increasing its vertical displacement. The deflection of the beam in Bay 5 does not reduce until the axial forces in Bays 4 and 6 begin to reduce at 62.6 minutes. The final residual displacement of the beam in Bay 5 is about 441 mm. It is observed that the final residual displacements of the beams in Bays 4 and 6 and Bays 3 and 7 are not significantly affected by fire spread, with an average calculated deflection of approximately 180 mm.

In the case of uniform heating, the maximum beam deflections at Bay 5, Bays 4 and 6, and Bays 3 and 7 are 382, 397, and 366 mm, respectively. These occur at approximately 36, 34, and 39 min, respectively. The final residual beam deflections are 306 mm for Bay 5, 298 mm for Bays 4 and 6, and 286 mm for Bays 3 and 7, and they do not differ significantly. It can be observed from Figure 6.52 that all the heated beams in Bays 3 to 7 experience compression during the heating phase and the initial cooling phase for about 50 minutes. These beams are in tension in the subsequent cooling phase (i.e., when the time is greater than 50 minutes). It is noticed that the peak value of compression does not take place at the same time as the maximum temperature. The compression forces in the members increase with the temperature for the first 24 minutes, after which the compression forces start to decrease even as the temperature keeps increasing. This is because the plastic strength surface contracts at elevated temperatures. Since the force state (axial force and moment interaction) is not allowed to exceed the plastic strength surface after it has reached the cross-section plastic capacity, it has to remain on the plastic surface during subsequent heating.

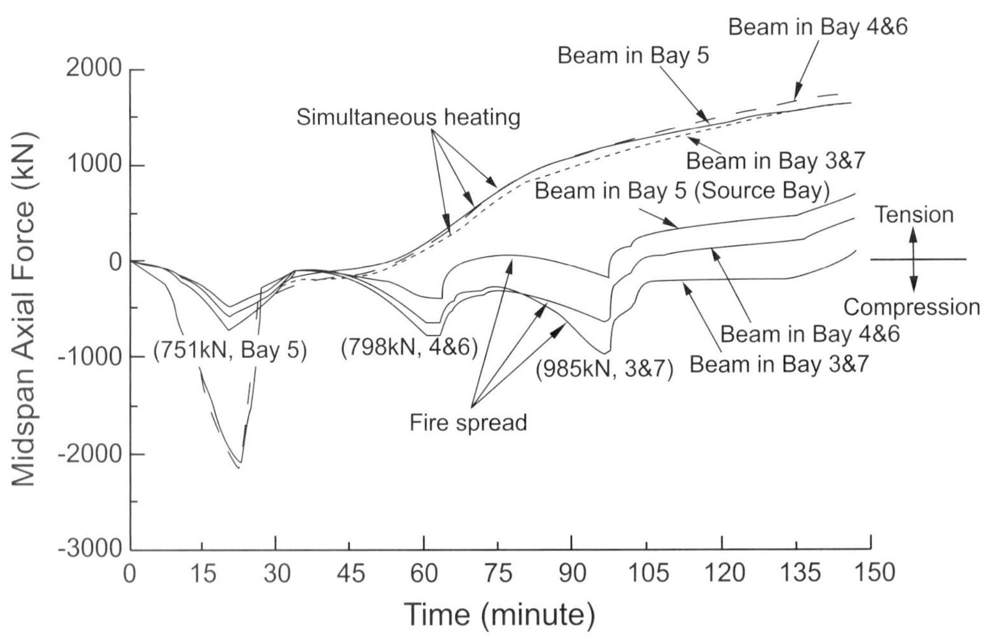

FIGURE 6.52 Comparison of midspan axial forces of the heated beams.

Hence, the compression forces decrease as a result of the contracting of the plastic surface during heating.

In comparison, the beam in Bay 5 has been significantly affected by fire spread. The maximum and residual deflections of Beam 5 are larger than the limiting deflection of span $L/20$ when the fire spread has been taken into account, but neither is so in the case where a constant time-temperature relationship is assumed. The analysis shows that the frame can survive in the fire scenario when all bays in the story are heated simultaneously, but not in the case when fire spreading is modeled. It is, therefore, important to consider the possibility of fire spread in the event of fire when the stability of the building is to be maintained for a reasonable period of time; otherwise the walls or partitions of the building must be so constructed that, in the event of fire, they are effective in preventing the spread of fire from one compartment to another.

6.3 Advanced Analysis of Composite Frames

Modern high-rise buildings constructed in recent years have adopted composite structural systems combining the use of concrete core wall and steel framing to provide a cost efficient design (Liew, 2001a, 2001b, 2004; Liew and Uy, 2001; Uy and Liew, 2002; Liew and Chen, 2004a). One of the distinctive features of composite floor construction is to design the steel beams to act compositely with the concrete floor slabs by means of the shear connectors. When building frames are subject to gravity and lateral loads, the distribution of the bending moment in the composite beams varies along the member length. In the hogging (negative) moment region, the concrete in tension is cracked and the steel reinforcement in the slab is subjected to tension offering resistance to hogging moment. In the sagging (positive) moment region, a large bending moment may cause partial yielding of the steel section and crushing of the part of the concrete slab. Consequently, the flexural stiffness of beams varies along the member length. To accurately predict the limit-state behavior of the overall system, the elasto-plastic nonlinear behavior of the composite members must be considered. The use of continuum finite elements for the analysis of composite beams is computationally too intensive for large-scale structures. The cost and effort of such a method is so great that they often prohibit analysis of a complete framework. To reduce the computational effort, Liew et al. (2001b) proposed a composite beam model to study the inelastic behavior of composite floor beams. The flexural stiffness of a composite beam segment can be evaluated, including the effects of partial shear stud interaction between the concrete slab and the floor beams.

For high-rise buildings, concrete core walls are often used to provide torsional and flexural rigidity and strength with or without the participation from the frame system. Conceptually, a central core with punched openings for lift access is used to provide a cantilever action for lateral load resistance. The floor framing is arranged in such a way that it distributes enough gravity loads to the core walls so that the design of the core is controlled by compressive stresses even under high lateral forces. Structural core walls provide an efficient type of lateral load resisting system up to a certain height premium because of their cantilever action. However, when it is used alone, the massiveness of the core increases with height, thereby inhabiting the free planning of interior spaces, especially in the core area. The space occupied by the core walls leads to loss of overall floor area efficiency as compared to tube system, which could otherwise be used. This section examines the nonlinear behavior of a combined system of concrete core wall and perimeter frames to form an efficient lateral system for high-rise construction.

Previous research on advanced analysis in steel framed structures is limited in its application unless the effects of concrete infill or encasement and concrete slabs can be considered. This section reviews the current trend in advanced analysis of composite members, composite connections, and frames. In undertaking this review, the types of advanced analysis for frames are summarized by tracing the progress that has been made in the advanced analysis of steel structures. This section also indicates the pertinent areas that need to be considered in order to bring the methods of advanced analysis of composite frames into wider circulation. Several illustrative examples are given to show the benefits of using advanced analysis for the design of real-size framework. Assessment of the building performance with respect to its load displacement behavior, serviceability, and ultimate-limit states are conducted.

6.3.1 Composite Members—Advanced Analysis

While there has been considerable research into the behavior of steel frames and advanced analysis, composite frames have received less attention until recently. Eurocode 4 (CEN, 1994) for example, makes no mention of this type of analysis procedure and it is largely a prescriptive-based code. Chen et al. (2000) developed a thin-walled beam-column element allowing for the advanced modeling of concrete core-walls and their interaction with the surrounding steel frames. Liew et al. (2001b) have developed an advanced inelastic analysis technique and applied it to considering multistory steel buildings with composite beams.

In trying to perceive future developments, it is worth highlighting some of the potential areas, which may need consideration in order to make the methods of advanced analysis robust and thus able to be used widely for the design of composite structures. Some of the potential areas that may need consideration include:

1. Local and overall buckling of composite members
2. Distortional buckling
3. Lateral torsional buckling
4. Limited slip capacity
5. Rotational capacities

FIGURE 6.53 Local buckling of concrete-filled steel columns (Uy, 2000).

6.3.1.1 Local and Overall Buckling of Composite Members

In order to model the inelastic response of composite structures accurately, the effects of local and overall buckling of composite members must be considered. Uy (2000, 2001) has considered the effects of local buckling on concrete-filled steel columns and for composite sections in a generic manner. Figure 6.53 shows a set of typical short column specimens that have failed by local

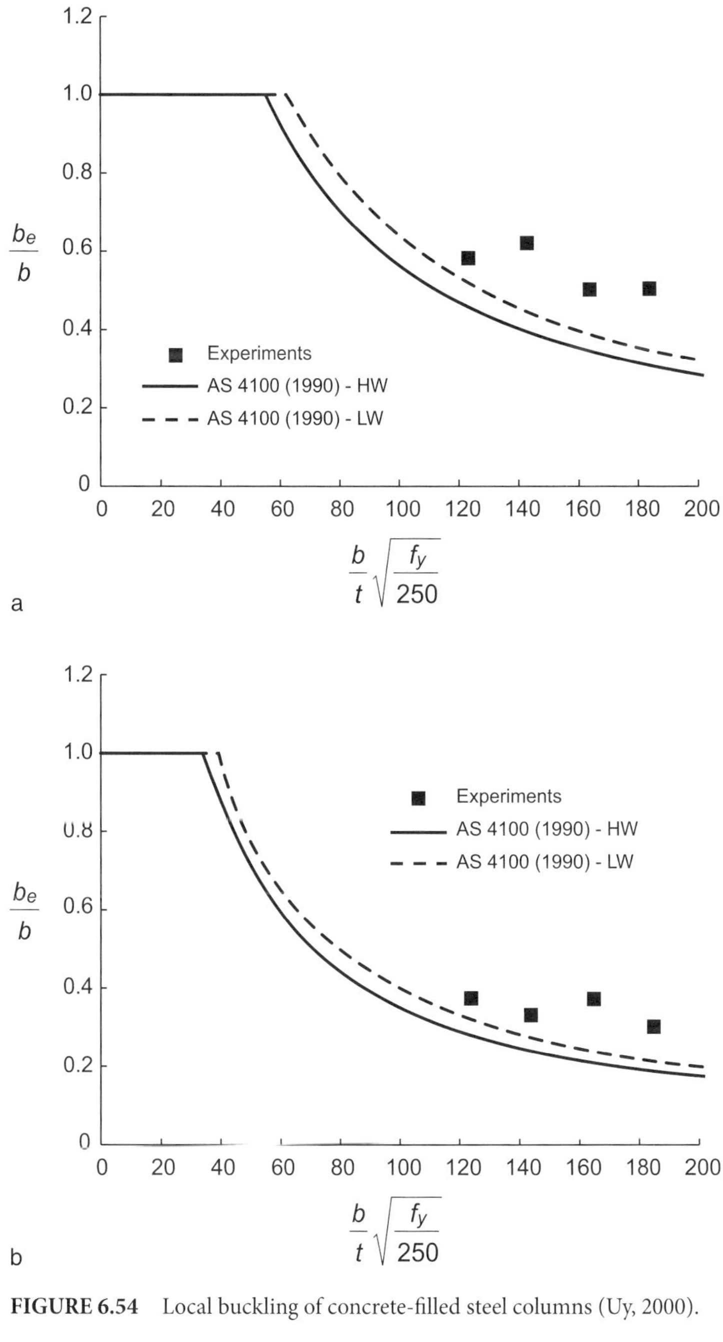

FIGURE 6.54 Local buckling of concrete-filled steel columns (Uy, 2000).

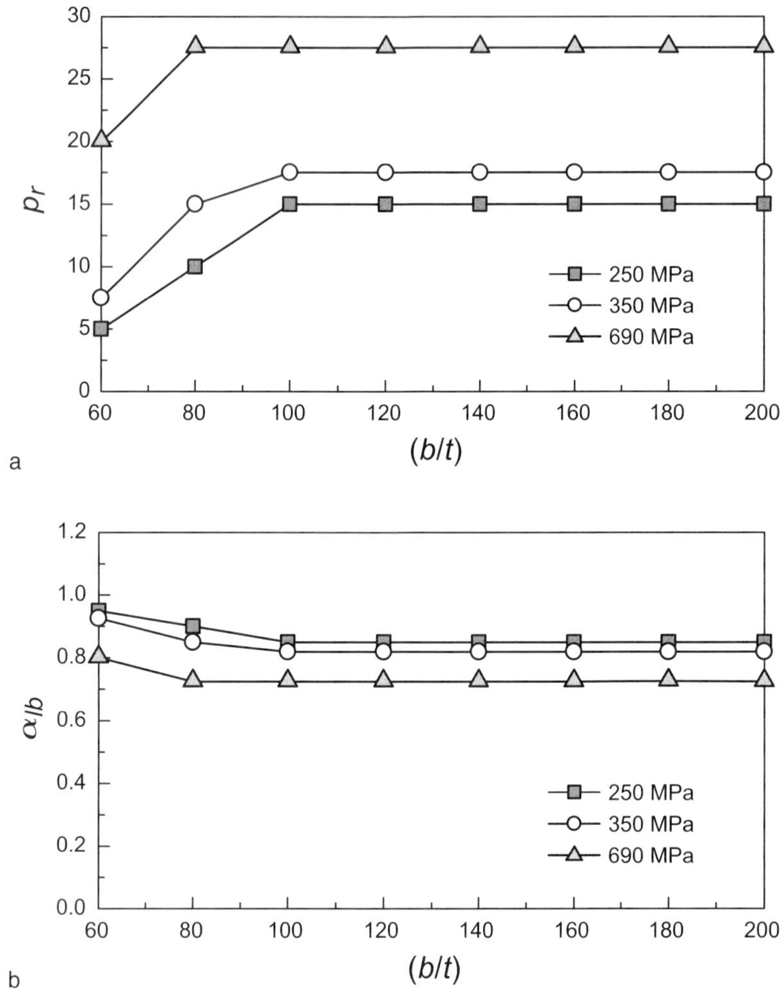

FIGURE 6.55 Interaction buckling of concrete filled steel columns (Vrcelj and Uy, 2002).

buckling. Figure 6.54 highlights the beneficial effects of including the concrete restraint when considering local buckling for these members. Vrcelj and Uy (2002) have considered the effects of interaction (coupled local and global) buckling effects for slender concrete-filled steel columns. Typical results are given in Figure 6.55 to highlight the effects of local buckling on the overall column buckling capacity.

Furthermore, local buckling effects may also need to be considered in composite beams in both sagging and hogging moment regions when noncompact sections are used (Bradford, 1986). In some cases, this can be more critical in continuous composite beams with semi-rigid connections as large compressive forces are entirely transferred through the lower compressive flanges.

6.3.1.2 Distortional Buckling

For intermediate buckling of half-wavelengths between those for localized buckling and lateral buckling, distortional modes can develop as illustrated in Figure 6.56. This is particularly the case when dealing with the hogging moment regions of continuous composite beams in frames with compact flanges. In these situations, the concrete slab provides sufficient restraint to the top flange and the web is thus forced to fail in a distortional manner. This has been receiving ongoing research, and significant developments are now currently available (Johnson and Bradford, 1983; Bradford and Gao, 1992; Vrcelj et al., 1999).

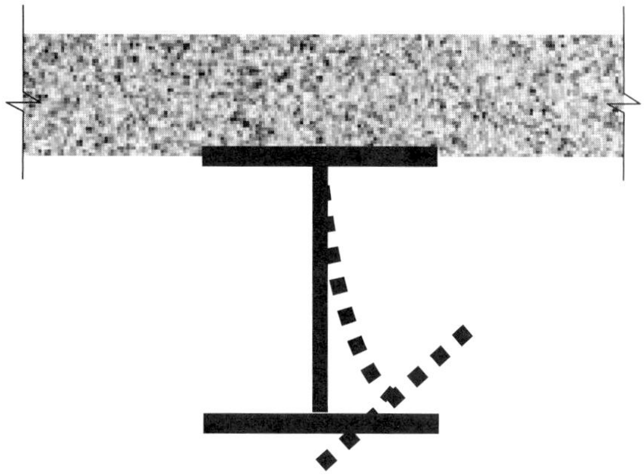

FIGURE 6.56 Interaction buckling of concrete-filled steel columns (Vrcelj et al., 1999).

6.3.1.3 Lateral Torsional Buckling

The effects of inelastic flexural-torsional buckling in steel frames have limited the development of advanced analysis and this will equally be the case for composite frames. Furthermore, the flexural-torsional behavior of composite beams in hogging moment regions needs attention in order for the proper development of advanced analysis procedures. Lindner (1998) has considered the effects of lateral torsional buckling of composite beams and has determined that this can only be critical for continuous composite beams near the internal supports. Methods of providing restraint to compression flange are shown in Figure 6.57, including a strut tied to the slab and a full-depth web stiffener to provide a moment resisting U-frame action.

6.3.1.4 Limited Slip Capacity

One other important factor that has only started to receive attention, is the effect of limited slip capacity. In many of the studies dealing with the strength of composite beams, it is assumed that the shear studs have unlimited ductility. Oehlers and Sved (1995) have considered the effects of limited slip capacity for simply supported beams. Figure 6.58 highlights the effects of limited slip capacity, which can lead to premature failure of the studs prior to reaching the plastic moment capacity of the beam. This has been recently augmented by Diedricks et al. (2000) for continuous members. In order to model the inelastic response of composite frames, it is important that slip displacements are checked and that the advanced analysis methods are amenable to this. A model developed by Fang et al. (1999, 2000) allows for these details to be determined using a single element.

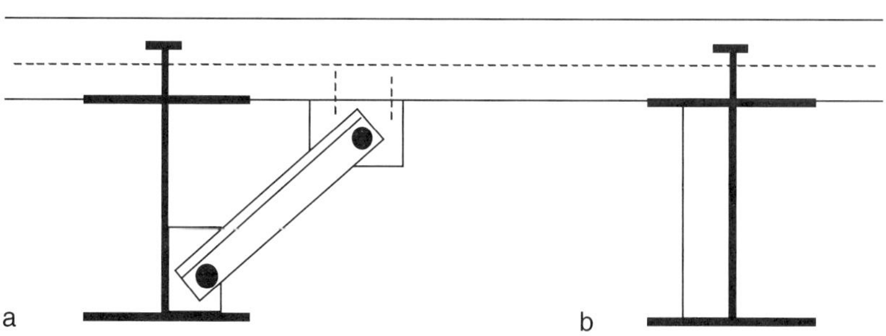

FIGURE 6.57 Methods of preventing lateral-torsional and distortional buckling: (a) strut tie to slab; (b) full-depth web stiffener.

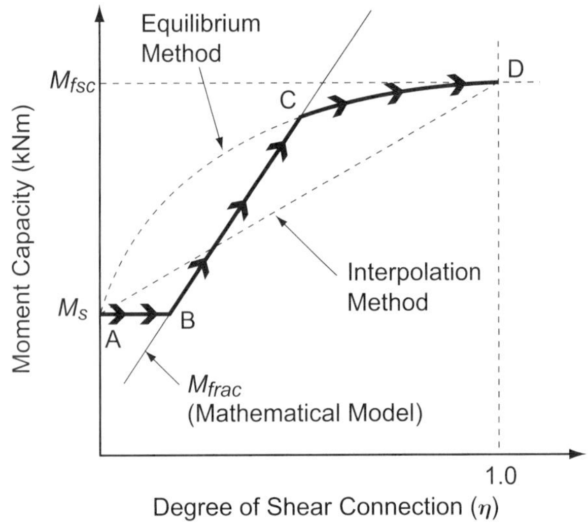

FIGURE 6.58 Limited slip capacity model (Oehlers and Sved, 1995).

6.3.1.5 Rotational Capacities

Another important issue that needs to be incorporated in the advanced analysis of composite frames is the issue of both available and required rotations. In order to ensure the plastic collapse behavior of beams in frames with semi-rigid or rigid connections, it is important to check that the rotations required to be developed in the hogging and sagging moment regions are able to be achieved by the connection and beam, respectively. Recent research by Kemp and Nethercot (2001) has provided an excellent platform for implementing this in advanced analysis methods.

6.3.2 Modeling of Composite Column

Shanmugam and Lakshmi (2001) carried out a state-of-the-art review on the research of steel-concrete composite columns. Emphasis is given on experimental and analytical work. Experimental data has been collected and compiled in a comprehensive format listing parameters involved in the study. The review also includes research work that has been carried out to date, accounting for the effects of local buckling, bond strength, seismic loading, confinement of concrete, and secondary stresses on the behavior of steel-concrete composite columns.

Saw and Liew (2000) presented the design assessment of concrete encased and concrete-filled composite columns based on the approaches given in the European (EC4), British (BS5400), and American codes (AISC-LRFD). This includes studies on the design parameters, comparison of the nominal strength predicted by the three codes, and comparison of the predicted strengths with the available test results. The studies concluded that the results obtained from the three codes vary considerably. This is because of the different design considerations adopted in these codes. However, the design methods are found to be mostly conservative when compared with the test results. The use of advanced analysis to capture the direct effect of plasticity and stability would alleviate the need to calculate individual member strength check, while providing more information on the structural performance than do the elastic analysis or design procedures.

Concrete-filled tubes are often preferred for the construction of high-rise buildings because of their high strength and stiffness compared to conventional reinforced concrete or steel columns. However, prior to infilling of concrete, the steel tubes are subjected to preloads from upper floors arising from construction loads and permanent loads of the building. These preloads cause initial stresses and deformations in the steel tubes, which would affect the load carrying capacity of the composite columns. Liew and Xiong (2009) studied the effects of preload on the axial capacity of concrete-filled composite columns and proposed a design method based on a modified Eurocode 4 approach, incorporating the effect of preload, to evaluate the axial capacity of concrete-filled composite columns. Eight full-scale composite column specimens were tested. Parameters studied included the strength of the concrete infill, slenderness of the columns, and the amount of preload applied on the steel tubes.

Hajjar and Gourley (1997) present a geometrically and materially nonlinear formulation to simulate the behavior of 3-D concrete-filled tubes (CFTs) subjected to monotonic static or cyclic seismic loading. The plasticity model adopts a two-surface plastic-hinge formulation with a polynomial expression for modeling the cross-section strength. Isotropic and kinetic hardening are proposed that are geared to capture the strength degradation, stiffness degradation, varnishing elastic zone, and Bauschinger effect exhibited by CFTs subjected to cyclic loading.

El-Tawil and Deierlein (2001) extended the two-surface plastic-hinge model for the advanced analysis of encased composite section. Key features of their proposed models are the flexibility-based element formulation and the bounding-surface representation of the cross-section response that permit efficient simulation of inelastic cyclic behavior. By relying on force-interpolation functions that are relatively invariant to inelastic action, the flexibility approach is more accurate than conventional stiffness formulations that rely on displacement shape functions. The bounding-surface representation of stress-resultant versus generalized strain response avoids the computationally intensive integrations through the member cross section. Summarized in this paper is the development of default modeling parameters for bisymmetric steel, reinforced concrete, and composite members representative of modern building design practices.

6.3.3 Modeling of Composite Beam

For a composite beam with sufficient skip and rotational capacities that is prevented from local, distortional, and lateral torsional buckling, Liew et al. (2001b) proposed a computational efficient model for studying the 3-D behavior of building frames. The composite beam is subdivided into several segments to capture the varying flexural stiffness along the member length. The instantaneous flexural stiffness of a composite segment is derived using the moment curvature (M-Φ) relationships, thus discretization of the cross section is not required. A closed form M-Φ relationship proposed by Li et al. (1993) for a composite beam section with full or partial shear connections has been adopted in the present analysis model. For the composite beam section under positive moment as shown in Figure 6.59a, the M-Φ relationship is given by:

$$\Phi(M) = \frac{M}{EI}, \quad 0 \le M \le M_y \tag{6.57a}$$

$$\Phi(M) = \frac{M}{EI} + \left(\Phi_u - \frac{M_u}{EI}\right)\left(\frac{M - M_y}{M_u - M_y}\right)^2, \quad M_y < M \le M_u \tag{6.57b}$$

where EI is elastic flexural stiffness of the composite section under the positive moment, E is the elastic modulus of steel, M_y and M_u are the yield moment and the ultimate moment, respectively, and Φ_u is the curvature at the ultimate moment given as:

$$\Phi_u = \beta\Phi_y = \beta\frac{M_y}{EI} \tag{6.58}$$

in which D and D_s are the depths of the steel beam and the slab, respectively.

Numerical analysis of composite frames with partial shear-stud interaction by one element per member has been carried out by Fang et al. (1999, 2000). In the present approach, the elastic

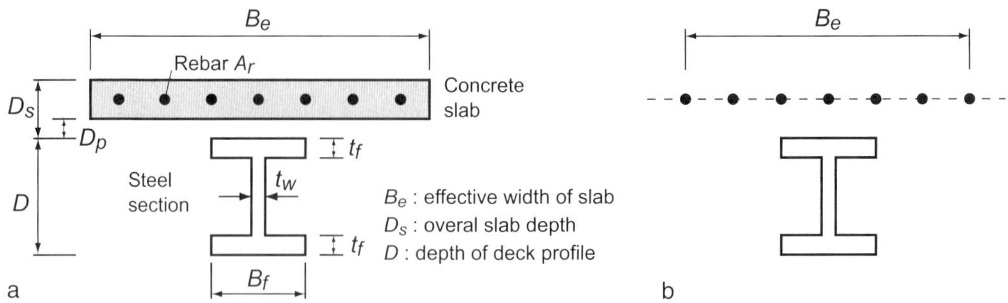

FIGURE 6.59 Composite beam sections: (a) positive moment region; (b) negative moment region.

flexural stiffness of composite beam with partial shear interaction under the positive moment is calculated based on the code specified equation (AISC, 1993):

$$EI = EI_s + \sqrt{\frac{N}{N_f}}\left(EI_f - EI_s\right) \tag{6.59}$$

in which EI_f is flexural stiffness of the composite beam with full shear connection, EI_s is flexural stiffness of the steel section, and N/N_f is degree of shear connection.

The instantaneous flexural stiffness of the composite beam section under the positive moment may be obtained by differentiating the moment with respect to the curvature as:

$$EI_t = EI, \quad 0 \le m \le \alpha_y \tag{6.60a}$$

$$EI_t = \frac{EI}{1 + 2\left(\beta\alpha_y - 1\right)\left(m - \alpha_y\right)/\left(1 - \alpha_y\right)^2}, \quad \alpha_y < m \le 1 \tag{6.60b}$$

in which m is the moment ratio (= M/M_u) and α_y is the yield moment ratio (= M_y/M_u).

For the negative moment region (see Figure 6.59b), it is assumed that once cracked, the concrete slab does not contribute to the beam's resistance. Applying the same rule for the positive moment region, the M-Φ relationship under the negative moment is (Liew et al. 2001b):

$$\Phi'\left(M\right) = \frac{M'}{EI'}, \quad 0 \le M' \le M'_y \tag{6.61a}$$

$$\Phi'\left(M\right) = \frac{M'}{EI'} + \left(\Phi'_u - \frac{M'_u}{EI'}\right)\left(\frac{M' - M'_y}{M'_u - M'_y}\right)^2, \quad M'_y < M' \le M' \tag{6.61b}$$

in which:

$$\Phi'_u = 5.3\Phi'_y + 2.4 = 5.3\frac{M'_y}{EI'} + 2.4 \tag{6.62}$$

The variables with prime (') have the same meaning as defined in Equation 6.57, except that here they refer to the negative moment region. In determining EI', M'_y, and M'_u, the contribution of the effectively anchored rebars located within the effective width of the concrete slab as shown in Figure 6.59 can be included and the concrete slab in tension is ignored. Equation 6.61b can be used to evaluate the flexural stiffness of the composite beam under the negative moment by differentiating M' with respect to Φ'.

Studies by Liew et al. (2001b) show that the ultimate-limit load of steel frames while considering the composite beam action is about 30 percent higher than that of the pure steel frame. The lateral stiffness can also be significantly enhanced by considering the composite beam action.

6.3.4 Modeling of Building Core Wall

A core-braced frame system represents an efficient form of a structural system up to a certain height where the predominant lateral load resistance is provided by concrete core walls (Iyengar, 1979; Iyengar et al., 1992; Liew, 2001a, 2004; Viest et al., 1997; Taranath, 1988, 1998). In prismatic office towers of normal floor size, core walls are usually planned at the central location. In office buildings with large floor areas, core walls may exit at different isolated locations. The basic concept is for the core walls to resist all the lateral forces. The steel frameworks around the core walls are generally designed only for gravity loads with simple connections between beams and columns. The steel beams are connected to core walls either by a typical corbel detail, or by bearing in a wall pocket or by anchor shear plate detail. In high seismic areas, the steel frameworks may be designed as moment frames to resist some part of the earthquake loads to provide a second line-of-defense.

The most rigorous way of modeling core wall is by the finite element method involving the use of plate and shell elements. However, this approach is computationally intensive for nonlinear analysis because it involves discretization of the wall into a large number of elements. Core wall is characterized by the fact that their cross-sectional and longitudinal dimensions are of different orders of magnitude. The wall thickness is usually small compared with the characteristic cross-sectional dimensions, which in turn are small compared with the height of the core wall. It is proposed that the core wall be modeled as a thin-walled beam-column element (referred as thin-walled element below) for its proportional similarity to Vlasov's thin-walled beam. The main advantage of such an approach is that the total degrees of freedom of the structure can be reduced tremendously, and the structural behavior can be predicted with sufficient accuracy (Taranath, 1988, 1998; Chen et al., 2000; Liew, 2001b; Liew and Uy, 2001).

The fundamental linear theory as well as stability problems of thin-walled members have been treated in the pioneer works by Vlasov (1961) and Timoshenko and Gere (1961). Taranath (1988, 1998) proposed a linear elastic formulation of thin-walled element for the analysis of core-braced frames. A considerable number of works have been presented in the fields of instability of thin-walled beams and frames using various finite element formulations and solution strategies (Barsoum and Gallagher, 1970; Yang and McGuire, 1986a, 1986b; Chan and Kitipornchai, 1987; Chen and Blandford, 1991; Conci, 1992; Izzuddin and Smith, 1996). Although the geometric nonlinear analysis using thin-walled element with asymmetric open cross-sections is well documented in the literature, the derivation of the geometric stiffness matrix appears to be different from one researcher to another based on different assumptions. In Conci (1992), the complete tangent stiffness matrix of a thin-walled element is derived from the virtual work equation based on the UL formulation. The approach includes both doubly symmetric and asymmetric core-wall sections. The centerline of a core wall is located on the shear center axis. Any significant twisting action should be analyzed to include both warping and torsional effects. Material nonlinearity of core walls is considered approximately by using the plastic-hinge formation. Plastic hinges are formed when the cross-sectional forces reach the cross-sectional capacity.

A single-story-high core wall is modeled as one thin-walled beam-column element as shown in Figure 6.60a. Figure 6.60c shows a generic thin-walled beam-column element that has an additional warping degree of freedom over the beam-column element at each end. The local coordinate is chosen with axis x lying on the shear center axis, and y and z axes paralleling to the principal y and z axes. The shear center of the core is selected as the reference point so that

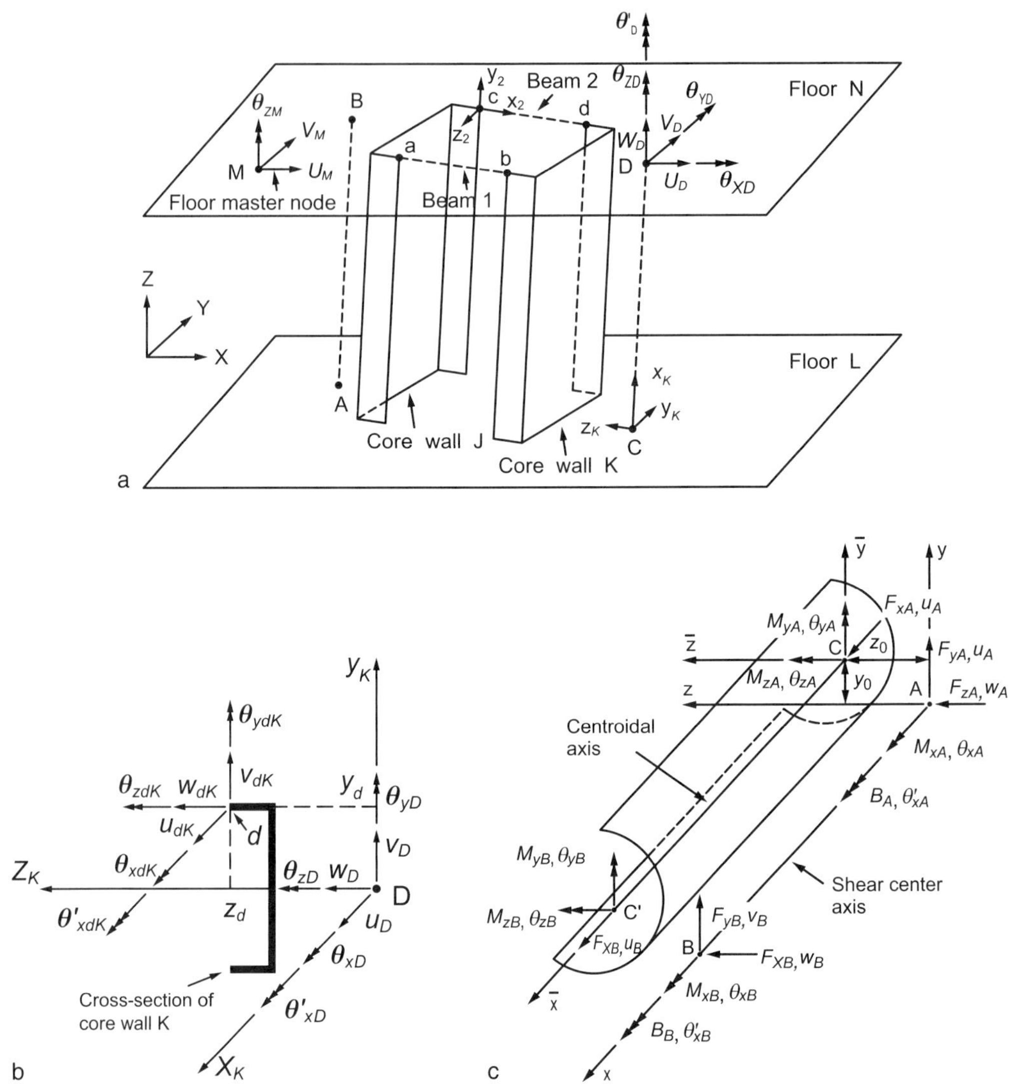

FIGURE 6.60 (a) A building core wall between two floors; (b) deformations of a thin-walled cross section; (c) a generic model for a thin-walled beam-column member.

appropriate transformation matrices can be established to relate the kinematics relationships between the point of contact between beams and columns with the core wall. For example in Figure 6.60b, the displacements at Node d of the core wall may be related to those at the shear center D in the local coordinate. Detailed derivation of the transformation matrices is given in Chen et al. (2000).

Material nonlinearity of the core wall is modeled approximately by using the concentrated plastic-hinge approach. The yielding of the core wall section would depend on the combined action of compression, biaxial bending, torsion, and warping. However, in the proposed formulation, the torsional shear and warping effects have been ignored in the plasticity formulation. It is necessary to model these factors if the core wall is subject to significant torsional and warping deformation. Such a situation is best avoided by selecting an appropriate shape and size of the core walls and an appropriate location for lateral load resistance.

6.3.5 Modeling of Composite Semi-rigid End-plate Connection

Composite joints consist of structural steelwork connections, a continuous reinforced slab and a column web panel. The influence of the slab on the steelwork connections and its interaction with the beam and the column increase the complexity of the analysis of composite joints over the bare steel joints. The M-θ_r response of a composite joint is influenced by the deformation of the connecting plates, bolt deformation and slippage, shear deformation of the column web panel, elongation of steel reinforcement in the slab, and interface slip of shear connectors. To simplify the modeling, a joint can be subdivided into three zones: tension, compression, and shear. Within each zone, the components that contribute to the joint's response are identified and their load-deformation characteristic determined. This methodology constitutes the basis of the component method (Anderson, 1999), which has been adopted as a practical approach for predicting the M-θ_r behavior of steel and composite joints.

Because of the nonlinear behavior of various components, evaluation of the M-θ_r response of steel or composite joints requires an incremental analysis. This approach has been successfully explored by Huber (1999) who developed a computer program to evaluate the M-θ_r response of composite joints. For design implementation, Eurocode 3 suggests a rotational spring model in which the non-linear M-θ_r response of joints may be approximated by a bilinear or trilinear rotational joint model. The Eurocode 3 model has been widely accepted for use in the analysis of braced frames in recent years. Tschemmernegg and Huber (1999) used a series of springs to model every component within a composite joint, and they were transformed into a typical rotational spring model by assembling all the component springs. The use of spring elements greatly simplified the complexity of composite joints behavior and allowed direct modeling of joint behavior in analysis of frames.

In the modeling of unbraced composite frames, it is necessary to establish a realistic model for beam-to-column composite joints, knowing that these joints behave differently when subjected to positive and negative moment. In Liew et al. (2004), the connection model developed for bare steel joints in Eurocode 3 is extended for composite joints. The methodology covers beam-to-column composite joints subjected to both positive and negative moments. It incorporates the effect of slab (concrete and steel reinforcement) and panel zone deformation and their interaction with the steelwork connections. It also explores the option of employing haunched joints in sway and nonsway composite frames by confirming its moment rotational characteristic through tests and to consider the composite action in haunch joint design previously designed as noncomposite. The procedure is validated through comparison with the test results.

6.3.5.1 Moment Resistance

Annex J of Eurocode 3 (CEN, 1992) provides an analytical procedure where the rotational behavior of a joint can be determined by taking into consideration all sources of deformability and resistance within the joint zone (Anderson, 1999). The advantage of this approach is that it can be applied to all types of joints with different configurations by decomposing the joint into relevant components. The M-θ_r characteristic of a joint can be represented by three properties, namely rotational stiffness, moment capacity, and rotation capacity. The formulae developed from component method for moment capacity and initial rotation stiffness are presented herein.

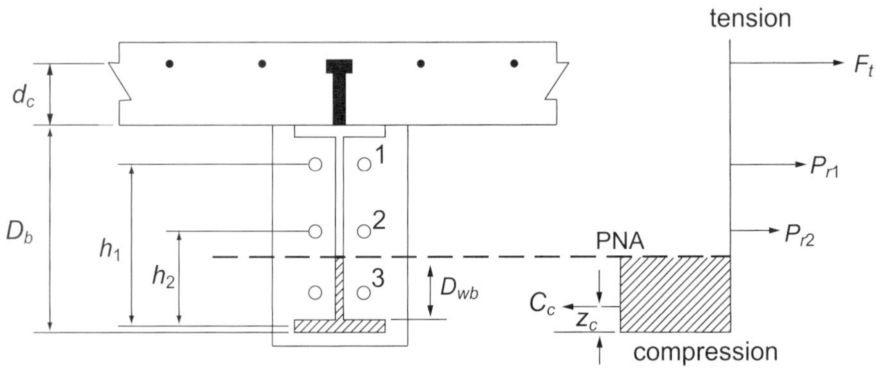

FIGURE 6.61 Force diagram in composite flush end-plate connection under negative moment.

6.3.5.1.1 Negative Moment Capacity

For composite end-plate connection subject to negative moment, the moment capacity can be determined using plastic analysis of compact steel section with force and stress distribution as shown in Figure 6.61. A similar approach may be used for joint under reversal of loading if panel joint resistance is accounted for when establishing the forces through equilibrium. As long as the compressive and tensile forces of individual components shown in the Figure 6.61 can be determined, the moment capacity of composite joints can be calculated by multiplying these forces with their respective lever arms. The location of Plastic Neutral Axis (PNA) is determined by equating the tensile force contributed by rebar and bolt rows to the compressive resistance. Tests carried out on a composite flush end-plate joint by Liew et al. (2000c) revealed that when the total tensile resistance F_s (inclusive of tensile force contributed by rebar and bolt rows) is larger than the compressive resistance of the bottom beam flange F_{fb}, PNA is always located above the bottom beam flange and a portion of the beam web D_{wb} is in compression. D_{wb} may be determined from force equilibrium as:

$$D_{wb} = \frac{F_s - F_{fb}}{1.2 f_{yb} t_{wb}} \tag{6.63}$$

It is recommended (SCI/BCSA, 1998) that the yield stress f_{yb} increases by a factor of 1.2 to account for strain hardening when the compression zone is extended into beam web. If only beam flange alone is adequate to resist total tensile resistance, a factor of 1.4 may be used to enhance the yield stress—20 percent each for the beneficial effects of strain hardening and dispersion into the web at the root of the section.

The center of compression from the bottom of beam flange Z_c may be determined as:

$$Z_c = \frac{t_{wb} D_{wb} \left[\left(\frac{D_{wb}}{2} + t_{fb} \right) \right] + \frac{b_{fb} t_{fb}^2}{2}}{t_{wb} D_{wb} + b_{fb} t_{fb}} \tag{6.64}$$

Taking moment about the center of the compression of all the forces shown in Figure 6.61, the negative moment capacity M_{neg} may be obtained as (see Figure 6.61):

$$M_{neg} = F_t \left(D_b + d_c - Z_c \right) + \sum_{i=1}^{n} \left\{ P_{ri} \left(h_i + \frac{t_{fb}}{2} - Z_c \right) \right\} \tag{6.65}$$

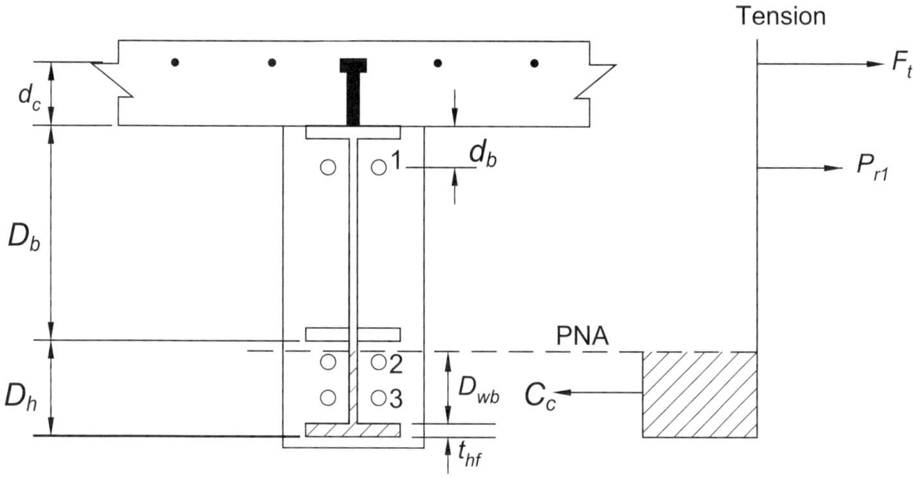

FIGURE 6.62 Force diagram in composite haunch connection under negative moment.

in which t_{wb} is the thickness of beam web, f_{yb} is the yield strength of the beam section, t_{fb} is the thickness of the beam flange, b_{fb} is the width of the beam flange, t_{fb} is the thickness of the beam flange, F_t is tensile force in the effective area of longitudinal reinforcement that stressed to yield strength, D_b is the height of the steel beam, d_c is the distance from the rebar to the upper beam flange, h_i is the distance from bolt rows i to the center of the bottom beam flange, and P_{ri} is the tensile force for the bolt row i.

For a composite haunch connection shown in Figure 6.62, the force equilibrium is established by considering the tensile forces contributed by rebar and bolts, and compressive forces contributed by the haunch flange and the web bearing against the column flange. The tests of haunch joints (Liew 2000c) recorded no tensile strain for bolt rows in the haunch indicating that the PNA is above these bolt rows. Examining the compressive resistances of the haunch flange F_{hf} and haunch web F_{hw}, it can be shown that:

$$F_{hf} < F_t + P_{r1} < F_{hf} + F_{hw} \tag{6.66}$$

Therefore, the PNA lies in haunch web as shown in Figure 6.62 and the depth of haunch web in compression, D_{hw} may be computed as:

$$D_{hw} = \frac{F_t + P_{r1} - F_{hf}}{1.2 f_{yb} t_{hw}} \tag{6.67}$$

The effective compressive resistance $F_{hw,eff}$ contributed by haunch web, is determined as:

$$F_{hw,eff} = 1.2 D_{hw} f_{yb} t_{hw} \tag{6.68}$$

Taking moment about the center of the bottom haunch flange, the moment resistance of the connection may be calculated as:

$$M_{neg} = F_t\left(D_b + d_c + D_h - \frac{t_{hf}}{2}\right) + P_{r1}\left(D_b - d_b + D_h - \frac{t_{hf}}{2}\right) - F_{hw,eff}\left(\frac{D_{hw}}{2} + \frac{t_{hf}}{2}\right) \tag{6.69}$$

in which t_{hf} is the thickness of the haunch flange, d_b is the distance between the first bolt row to the top of upper beam flange, and D_h is the depth of haunch. Other symbols have been defined earlier.

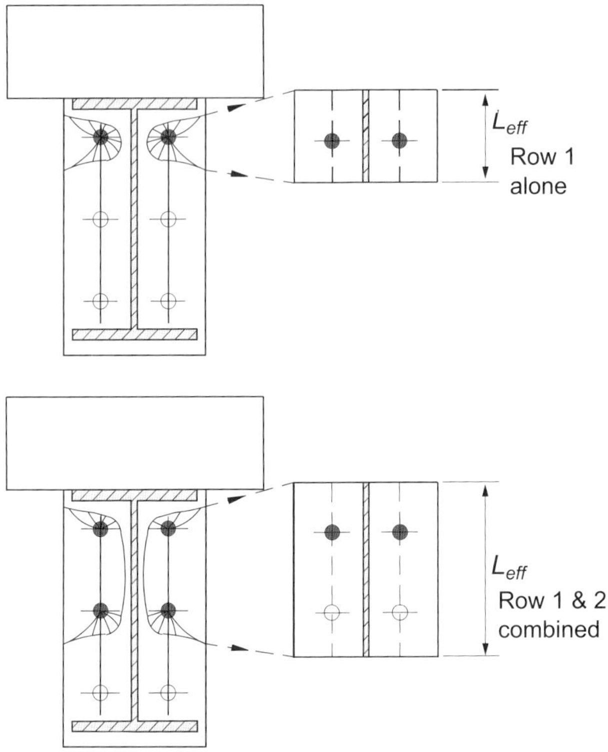

FIGURE 6.63 Equivalent T-stubs.

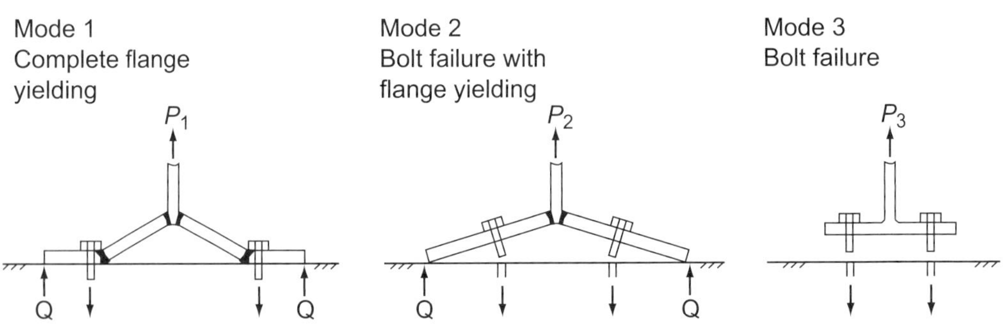

FIGURE 6.64 Three possible failure modes for column flange or end-plate bending.

6.3.5.1.2 Determination of Potential Bolt Forces P_{ri}

An important contribution of the EC model in predicting moment capacity of steel and composite end-plate joints is its ability to quantify the potential bolt force in each row P_{ri}. The bolt resistance may be governed either by the bending of the end plate or column flange or bolt failure, or by tension failure in the beam or column web. The moment capacity is then determined by the coupling tension in the bolt (and reinforcement if it is a composite joint) with compression in the lower part of the beam.

For column flange or end-plate bending, the component method in Eurocode converts the complex pattern of yield lines that occur round the bolt into a simple equivalent T-stub as shown in Figure 6.63. The capacity of the T-stub is then checked against three possible failure modes as shown in Figure 6.64.

Mode 1, complete flange or end-plate yielding:

$$P_1 = \frac{4M_p}{m} \tag{6.70}$$

Mode 2, bolt failure with flange or end-plate yielding:

$$P_2 = \frac{2M_p + n\left(\sum P_t'\right)}{m+n} \tag{6.71}$$

Mode 3, bolt failure:

$$P_3 = \sum P_t' \tag{6.72}$$

where:

P_i = potential resistance of bolt row or bolt group for mode i

P_t' = enhanced bolt tension capacity accounted for the prying effect

M_p = plastic moment capacity of the equivalent T-stub representing the column flange or end plate:

$$= \frac{L_{eff} t^2 p_y}{4} \tag{6.73}$$

L_{eff} = effective length of yield line in equivalent T-stub

t = column flange or end-plate thickness

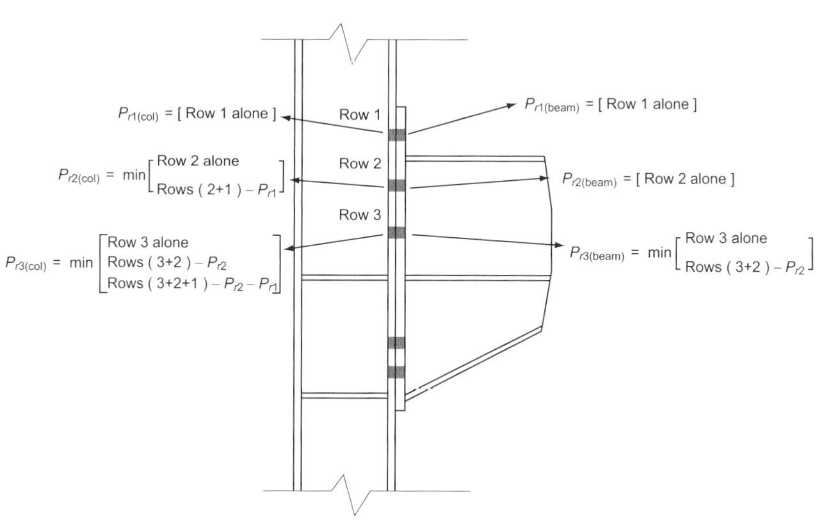

FIGURE 6.65 Steps in calculating the distribution of bolt forces.

m = distance from bolt center to 20 percent distance into column root or end-plate weld

n = effective edge distance

The potential tensile resistance of a row or a group of bolts due to tension failure in the beam or column web is taken as:

$$P_t = L_t t_w p_y \tag{6.74}$$

where L_t is the effective length of web assuming a maximum of spread at 60° from the bolts to the center of the web, t_w is the web thickness, and p_y is the design strength of the steel column or beam.

Therefore:

$$P_{ri} = \min(P_1, P_2, P_3, P_t) \tag{6.75}$$

The potential bolt force is calculated one row at one time, starting at the top and working down as illustrated in Figure 6.65.

One major improvement to the Eurocode 4 approach is not to restrict the PNA to the center of the bottom beam flange. In the proposed approach, the PNA is allowed to be located in the steel beam web if the steel beam section is compact and when the tensile resistance provided by the rebars and the bolts is higher than the compression resistance of the beam flange. This improvement is demonstrated in Table 6.9 for the test specimens reported by Liew et al. (2004). For the current specimens subjected to reversal moment, the beneficial effect was not obvious, since the majority of joint moment capacities were governed by shear resistance, and therefore the assumption of PNA located at the center of the bottom beam flange is still reasonably accurate.

Although the extended end plate that bears against the column flange could contribute marginally to the negative moment capacity, this beneficial effect has been ignored in the proposed calculation procedure.

6.3.5.1.3 Positive Moment Capacity

When a composite beam-to-column joint is subject to positive moment, its load transfer mechanisms are different from those subject to negative moment. In this case, the bottom bolt row(s) will be the primary contributor to tensile resistance. Compressive force, which is needed to equilibrate the tensile force, is now formed by the bearing between the concrete slab and the column flanges. The positive moment capacity of the composite joint can be calculated by multiplying the magnitude of the component forces with the respective lever arm. The following assumptions are made:

TABLE 6.9 Moment capacity: EC4 model versus the proposed model on negative moment capacity for Phase I specimens

Test	$M_{u,exp}$ (kNm)	EC4 Model $M_{u,pred}$ (kNm)	Proposed Model $M_{u,pred}$ (kNm)
SCCB1	271	256§	256§
SCCB2	441	298	395
SCCB3	449	298	445
SCCB4	186	130§	130§
SCCB5	410	298	395
SCCB6	423	298	395

§PNA located in mid of bottom beam flange in both models

1. The bolt-row(s) force that contributes to moment resistance is determined by the same procedures as those for negative moment as described in Section 6.3.5.2.
2. The compression resistance is taken by the bearing force developed between concrete slab and column flange, and the effective bearing width is taken as the width of the column flange. The depth of concrete slab in compression can be determined by comparing the available bolt-row tensile force. The bearing area is then taken as the product of the column flange width and the depth of the concrete slab in compression. This has been confirmed by the test measurement reported in Liew et al. (2000c). When the connection consists of a number of bottom bolt rows, there is a possibility that the upper beam flange is also in compression in order to achieve force equilibrium. Since the slab that bears against the column offers high compression resistance, it is unlikely that the beam web will be in compression.
3. The stress distribution in the concrete slab is assumed to be of a rectangular stress block with a mean stress of $0.85f_{ck}$ or $0.67f_{cu}$.
4. The crushing strain of concrete is taken as 0.0035.

Figure 6.66 shows the force diagram of a composite flush end-plate connection subjected to positive moment. If the moment capacity of the composite joint is governed by ductile failure modes such as column flange and end-plate in bending, a plastic stress block distribution may be assumed as shown in Figure 6.66b. On the other hand, the force distribution is assumed to be linear when brittle failure mode is detected, such as the fracture of bolts and column web in tension (Figure 6.66a).

The depth of the compression stress block from the top of the slab can be computed based on the equilibrium of forces:

$$c = \frac{P_{r2} + P_{r3}}{0.85 f_{ck} b_{fc}} \tag{6.76}$$

The moment resistance is then computed by taking moment about the center of the compression stress block as follows:

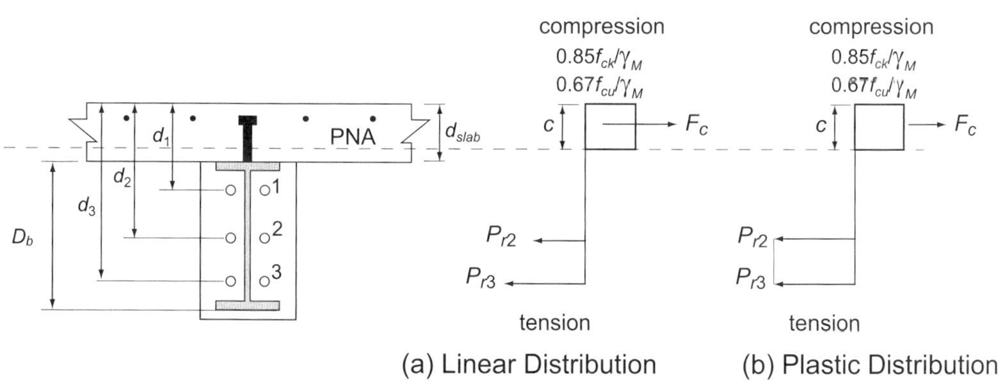

FIGURE 6.66 Stress and force distribution under positive moment.

$$M_{pos} = P_{r2}\left(d_2 - \frac{c}{2}\right) + P_{r3}\left(d_3 - \frac{c}{2}\right) \tag{6.77}$$

in which f_{ck} is the concrete cylinder compressive strength, b_{fc} is the width of the column flange, and d_i is the distance between bolt row i to the top of concrete slab. These formulae may be used to calculate positive moment capacity of the haunch joint since the force distribution mechanism is not different from the flush end-plate joint.

When establishing the force equilibrium as in Figures 6.61, 6.62, and 6.66, and calculating the moment resistance from Equations 6.65, 6.69, and 6.77, it is assumed that the effective tensile force is equal to or less than the panel zone shear resistance as the compression resistance is not critical for joints subject to reversal loads.

6.3.5.1.2 Panel Zone Shear Resistance

When the joints are subjected to reversal of loading, the joint panel zone may deform in shear and the panel zone shear resistance would affect the joint's moment capacity. Flush end-plate connections SJ1, CJ1, and CJ2 tested by Liew et al. (2000c) show that panel zone failure governed the ultimate capacity. In these test specimens, the column web was not strengthened by doubler plate or concrete encasement. The shear resistance of an unstiffened column web V_{wc} of a single-sided joint or for a double-sided joint in which the beam depths are similar is determined as:

$$V_{wc} = \frac{f_{y,wc} A_{vc}}{\beta\sqrt{3}} \tag{6.78}$$

in which $A_{vc} = A_c - 2b_{fc}t_{fc} + (t_{wc} + 2r_c)t_{fc}$ is the area of column web, A_c is the cross-section area of the column, b_{fc} is the overall breath of the column, t_{fc} is the thickness of the column flange, t_{wc} is the thickness of the column web, r_c is the root radius of the column, β is the transformation parameter accounting for the possible influence of the web panel in shear for adjacent connection.

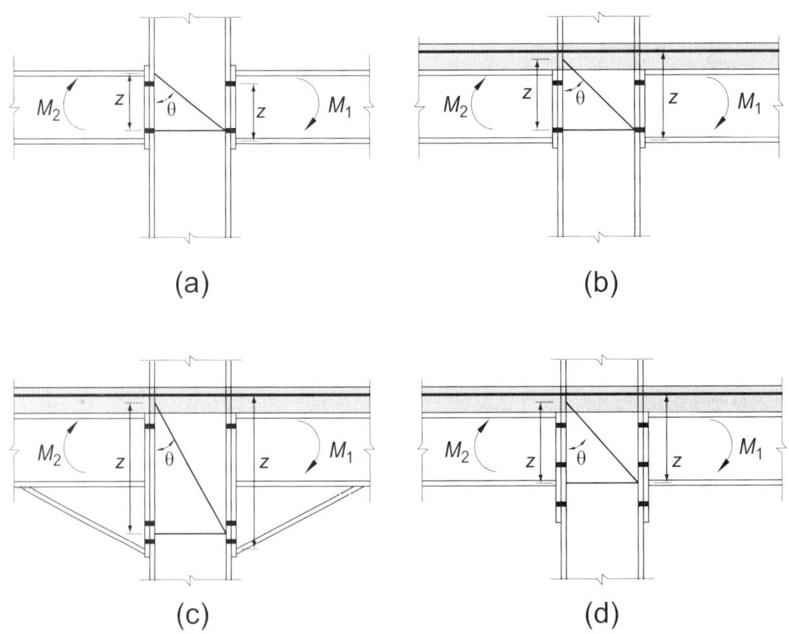

FIGURE 6.67 Lever arm z used for joints under reversal of moments: (a) bare steel joint; (b) flush end-plate composite joint; (c) haunched composite joint; (d) extended end-plate composite joint.

If a doubler plate is used, the shear area may be taken as:

$$A_{vc,dp} = A_c - 2b_{fc}t_{fc} + \left(t_{wc} + 2r_c\right)t_{fc} + b_s t_{wc} \tag{6.79}$$

where b_s is the width of the doubler plate.

When the steel column web is encased in concrete, the panel shear resistance of the column web may be increased to allow for the encasement:

$$V_{wc,con} = \frac{0.55\left(0.85 f_{ck}\right) A_{c,con} \sin\theta}{\beta} \tag{6.80}$$

where $A_{c,con} = 0.8\left(D_c - 2t_{fc}\right)\cos\theta(b_{fc} - t_{wc})$ and θ is taken as equal to $\tan^{-1}\left(\frac{D_c - 2t_{fc}}{z}\right)$, and z is the lever arm as shown Figure 6.67.

It was observed from the tests that the presence of concrete slab in a composite joint increased the panel zone shear resistance. This is due to the increase in effective depth of the panel zone compared to that of a steel joint. Larger effective depth would induce smaller panel shear force under the same applied moment. Therefore, less panel zone deformation is expected for composite joints than for steel joints.

6.3.5.2 Rotational Stiffness

6.3.5.2.1 Rotational Stiffness Under Negative Moment

The stiffness model by Anderson (1999) is taken as a basis to determine the initial rotational stiffness of a composite joint. Evaluation of joint elastic stiffness is derived from the elastic translational stiffnesses of the active joint components that form the spring model (Figure 6.68), as:

$$R_{ki} = \frac{Ez^2}{\sum \frac{1}{k_i}} \tag{6.81}$$

where:

- E = elastic modulus of steel
- z = level arm between compressive and tensile areas, or z_{eq} when there is more than one layer of tension, compression, or shear spring
- k_i = stiffness coefficient of component i, as determined by expressions in Table 6.10

As shown in Figure 6.68, when there is more than one component in a series at the same level r, these components are transformed into $k_{eff,r}$. Similarly, if there are components parallel to the same internal force type (tensile, compression, or shear), the lever arm must be redefined as equivalent lever arm $z_{eq,}$ and subsequently replaced these parallel springs as $k_{Equation}$. These transformed terms are determined by the expressions:

$$k_{eff,r} = \frac{1}{\sum_i \frac{1}{k_{i,r}}} \tag{6.82}$$

$$z_{eq} = \frac{\sum_r k_{eff,r} h_r^2}{\sum_r k_{eff,r} h_r} \tag{6.83}$$

$$k_{eq} = \frac{\sum_i k_{eff,r} h_r}{z_{eq}} \tag{6.84}$$

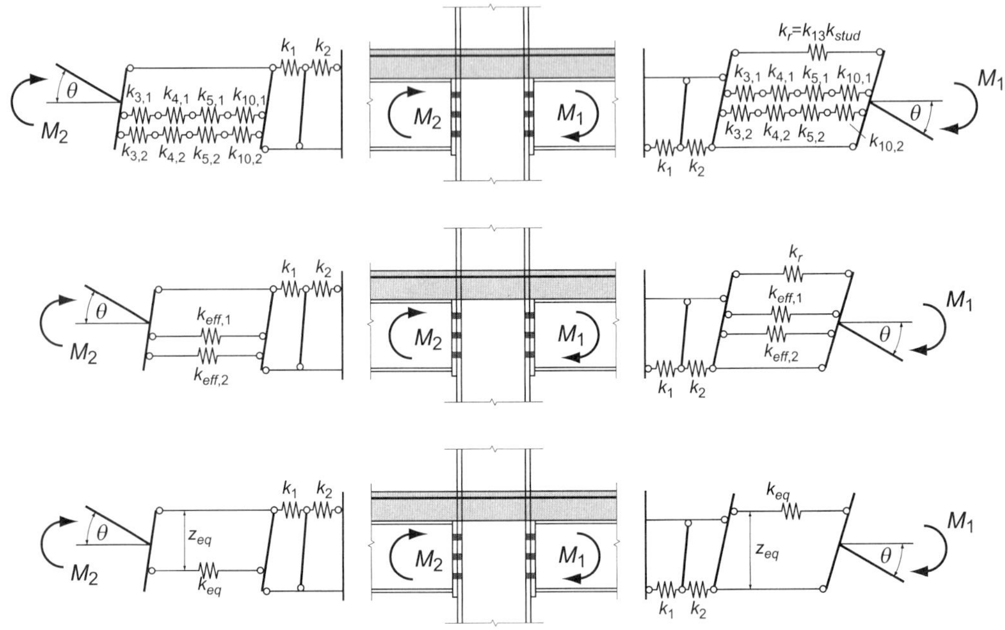

FIGURE 6.68 Spring model for negative and positive moments.

TABLE 6.10 Stiffness coefficient (Anderson, 1999; Eurocode 4, 1994)

Active components in end plate joint	Stiffness Coefficient, k_i
Column web panel in shear	Without concrete encasement: $k_1 = \frac{0.38A_{vc}}{\beta z}$ With concrete encasement: $0.06\frac{E_{cm}}{E}\frac{b_s h_c}{\beta z}$
Column web in compression	Without concrete encasement: $k_2 = \frac{0.7b_{eff,c,wc}t_{wc}}{d_c}$ With concrete encasement: $\frac{0.5b_{el}b_c}{h_c}\frac{E_{cm}}{E}$
Column web in tension	$k_3 = \frac{0.7b_{eff,t,wc}t_{wc}}{d_c}$
Column flange in bending	$k_4 = \frac{0.85l_{eff}t_{fc}^3}{m^3}$
End plate in bending	$k_5 = \frac{0.85l_{eff}t_p^3}{m^3}$

(Continues)

TABLE 6.10 (Continued)

Beam or column flange and web in compression	$k_7 = \infty$
Beam web in tension	$k_8 = \infty$
Bolts in tension	$k_{10} = \dfrac{1.6A_s}{L_b}$
Slab reinforcement in tension	For interior joint loaded with gravity load and M_1 equal to M_2: $k_{13} = \dfrac{A_r}{h_c\left(\dfrac{1+\beta}{2}+K_\beta\right)}$ For interior joint subjected to reversal load: $k_{13} = \dfrac{A_r}{h_c\left(\dfrac{1+\beta}{2}+K_\beta\right)}$ where $K_\beta = \beta\,(4.3\beta^2 - 8.9\beta + 7.2)$ for connection subjected to negative bending; $\beta = 1$ connection subjected to positive bending; $\beta = 0$ and $k_{13} = 0$.
Influence of shear stud slip	The influence of shear stud slip on joint stiffness to be imposed on k_{13} by the reduction factor, k_{stud} to be multiplied to k_{13}. $k_{stud} = \dfrac{1}{1+\dfrac{E_r k_{13}}{K_{sc}}}$ and $K_{sc} = \dfrac{Nk_{sc}}{v-\left(\dfrac{v-1}{v+\xi}\right)\dfrac{h_s}{d_s}}$ $v = \sqrt{\dfrac{(1+\xi)Nk_{sc}l_{eff,b}d_s^2}{EI}}$ and $\xi = \dfrac{EI}{d_c^2 E_c A_c}$

Note:
1. Symbols that are not defined in Table 10 may be referred to (Anderson, 1999; Eurocode 4, 1994)
2. Standard push out test on shear connector were carried out to determine the stiffness, k_{sc}

6.3.5.2.2 Rotational Stiffness Under Positive Moment

When a composite connection is subjected to positive moment, it behaves like an inverted steel joint with concrete slab in compression. The lever arm between tension and compression forces is smaller for composite connection under positive moment because the center of compression is located in the concrete slab as compared to the joint under negative moment, as shown in Figure 6.67. Therefore, by using the similar stiffness coefficients as those in the negative model and modified spring model (shown in Figure 6.68), and in Equation 6.81, the initial stiffness under positive moment can be determined.

TABLE 6.11 Moment capacity: experimental (Liew et al., 2004) versus analytical results

Test	$M_{u,exp}$(kNm)		$M_{u,pred}$ (kNm)		$M_{u,exp}/M_{u,pred}$	
	Negative	Positive	Negative	Positive	Negative	Positive
SJ1	125	110	94	94	1.33	1.17
CJ1	268	176	230	147	1.17	1.20
CJ2	273	167	230	154	1.19	1.08
CJ3	339	175	329	143	1.03	1.22
CJ4	318	226	329	195	0.97	1.16
CJ5	568	395	599	339	0.95	1.17
CJ6	632	461	751	429	0.84	1.07
CJ7	372	160	319	147	1.17	1.09

TABLE 6.12 Initial rotational stiffness: experimental (Liew et al., 2004) versus analytical results

Test	$R_{ki,exp}$ (kNm/rad)		$R_{ki,pred}$ (kNm/rad)		$R_{ki,exp}/R_{ki,pred}$	
	Negative	Positive	Negative	Positive	Negative	Positive
SJ1	13103	13091	13944	13944	0.94	0.94
CJ1	28784	17043	27001	15833	1.07	1.08
CJ2	27648	16987	27001	15833	1.02	1.07
CJ3	—	—	42430	24521	—	—
CJ4	42697	45118	42430	41501	1.01	1.09
CJ5	57000	47473	80293	57879	0.71	0.82
CJ6	107790	86737	109908	85550	0.98	1.01
CJ7	39784	23882	36925	20220	1.08	1.18

TABLE 6.13 Initial rotational stiffness: comparison of test results (Liew et al., 2000c) with predicted results for joints under negative moment

Test	$R_{ki,exp}$ (kNm/rad)	$R_{ki,pred}$ (kNm/rad)	$R_{ki,exp}/R_{ki,pred}$
SCCB1	36265	44164	0.82
SCCB2	56328	81271	0.69
SCCB3	83125	122951	0.68
SCCB4	37978	43930	0.86
SCCB5	49678	57670	0.86
SCCB6	50031	58980	0.85

6.3.5.3 Comparison with Test Results

The experimental results presented in Liew et al. (2004) are compared with the analytical predictions in Tables 6.11 and 6.12. The material properties obtained from the coupon tests were used to evaluate the strength and stiffness. Also, the partial safety factors for materials used in the Eurocode 3 and Eurocode 4 formulae are taken as unity. To further verify the accuracy of the analyti-

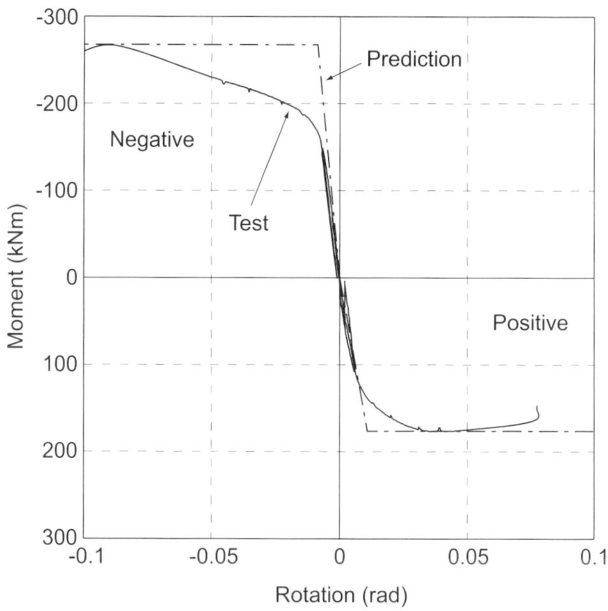

FIGURE 6.69 Comparison of experimental and analytical values for CJ1 (Liew et al., 2004).

cal model, the experimental values of initial rotational stiffness reported by Liew et al. (2004) for joints subject to symmetric loading (i.e., both sides of the beam-to-column connections subject to negative moments) are also compared with the results obtained by the analytical model as shown in Table 6.13.

It was observed from Table 6.13 that the EC model slightly over-predicts the initial rotational stiffness for joints loaded in symmetrical load only, but not for joints subject to reversal of loading as shown in Table 6.12. The probable reason is that the deformation behavior of composite joints under reversal of loading are dominated by the panel joint shear deformation, which is well captured by the panel zone model used in the Eurocode code.

Figure 6.69 shows the comparison of M-θ_r curves obtained experimentally with the corresponding analytical curves for CJ1. Bilinear idealization was adopted for the analytical M-θ_r curves as proposed by Eurocode 3. The comparison of moment capacities in Table 6.10 shows that the analytical method proposed for positive moment capacity is conservative, as the ratio $M_{u,exp}/M_{u,pred}$ is always greater than unity. The predicted negative moment capacity for CJ4, CJ5, and CJ6 are higher than the experimental values. This could be explained as the onset of premature failure due to bolt fracture on the connection side loaded with positive bending. Higher moment resistance could be developed if the fracture of the bolt can be avoided. Examination of the ratio of $R_{ki,exp}/R_{ki,pred}$ for positive bending shows good agreement between the predicted values and the experimental results. This indicates that the proposed modification to the EC3 joint model is adequate to provide a safe estimation of the joint's moment rotational curves. Tables 6.11 shows that the values predicted by the analytical model as proposed by Anderson (1999) are close to the experimental results, except for CJ5 in negative bending. For joints tested under symmetrical loading with negative applied moment, all the initial rotational stiffness obtained analytically exceeded the experimental results. The worst value obtained for the $R_{ki,exp}/R_{ki,pred}$ ratio is 0.68 for specimen SCCB3.

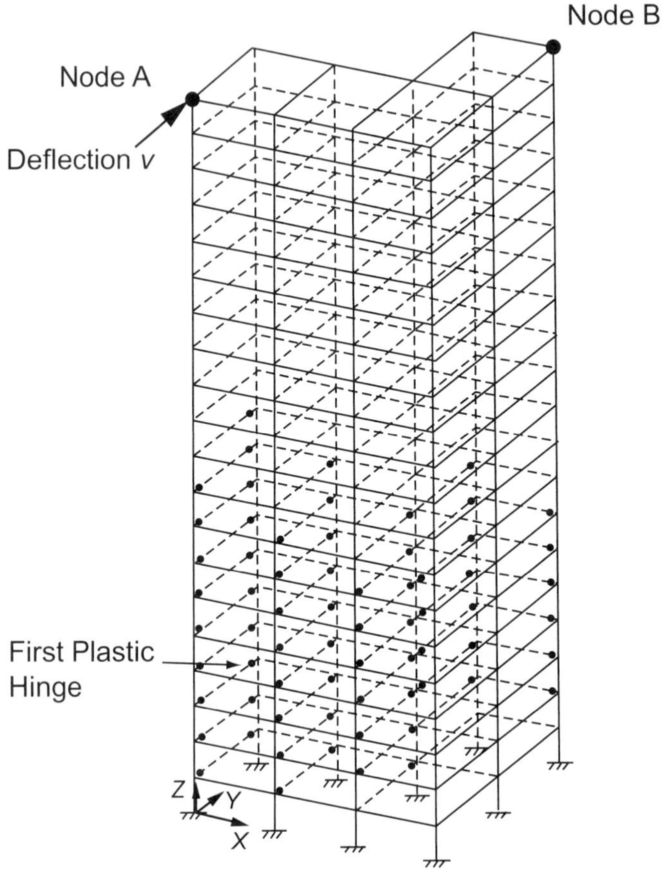

FIGURE 6.70 Formation of plastic hinges in a 20-story composite frame.

In the design of composite frames with semicontinuous joints, it is convenient to represent the overall joint behavior by rotational springs. In an unbraced frame, the joint model should take into account the behavior of the column web panel in shear as well as the M-ϕ behavior of the relevant connection. Liew et al. (2004) provides detailed studies on joints subject to reversal loading leading to a better understanding of the behavior of composite sway frames. The joint model developed can be used in the analysis of composite beams and composite frames.

6.3.6 Analysis of 20-Story Steel Frame with Composite Beams

Figure 6.70 shows a 20-story space frame with composite beams. Properties of steel column and composite beam sections are given in Liew et al. (2001b). A50 steel of yield strength 344.8 N/mm^2 is used. The overall slab depth is 127 mm and the cylinder strength of concrete is 27.6 N/mm^2. Full shear connection is assumed for the composite beams. The contribution of rebars is assumed to be negligible and therefore ignored in the calculation of the beam capacity. The frame is subject to a combination of gravity loads 4.8 kN/m^2 and wind loads 0.96 kN/m^2 acting in the Y-direction. Rigid-floor diaphragm action is assumed in the global analysis. Each steel column is modeled using one plastic-hinge beam-column element. Each beam is modeled using four elements. Anal-

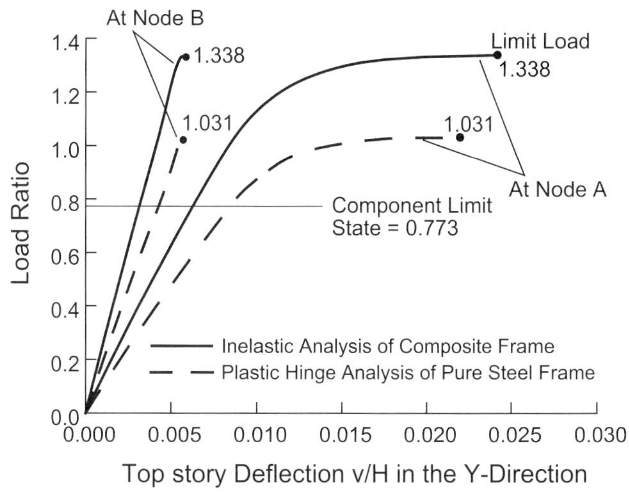

FIGURE 6.71 Load-deflection curves of the 20-story composite frame.

yses are carried out on the 3-D steel frame and composite frame so that their results can be compared.

The load-deflection curves of nodes A and B at the top of the frame are shown in Figure 6.71. The formation of plastic hinges is shown in Figure 6.70. Advanced analysis on the pure steel framework shows that the first plastic hinge occurs at a load ratio of 0.784 and the limit load of the frame is reached at 1.031. At the limit load, most plastic hinges occur in the beams on the right side of the frame.

In the analysis of a composite frame, the composite action between the steel beams and concrete slab is fully accounted, and the composite beams are analyzed using an inelastic approach based on the moment-curvature relationship given in Section 6.3.3. The frame collapses at a load ratio of 1.338, which represents a 30 percent increase of limiting strength compared with the pure steel frame. The component limit-state check approach predicts a limit load of 0.773. The analysis is based on a second order elastic analysis, and the limit load is similar to the load ratio of the first plastic hinge occurring in the steel frame. This is because the beam end is subject to hogging moment and behaves like a pure-steel section. The strength increase of the composite frame due to the inelastic force redistribution beyond the first member nominal strength is 73 percent. Due to higher redundancy in large frameworks, the strength increase of the 20-story frame considering the composite action is much higher than that of the small-scale frame.

In the analysis of 20-story frame with composite beams, Liew et al. (2001b) found that computing the elastic serviceability deflections for predominant lateral load using the averaged inertia of composite beam section:

$$I_c = 0.4I + 0.6I' \tag{6.85}$$

is more accurate than that using the following equation (Viest et al. 1997):

$$I_c = 0.6I + 0.4I' \tag{6.86}$$

in which I and I' are the moment of inertia of the composite beam section under the positive moment and negative moment respectively. This finding is consistent with the experimental evidence by Di Sarno and Elnashai (2002).

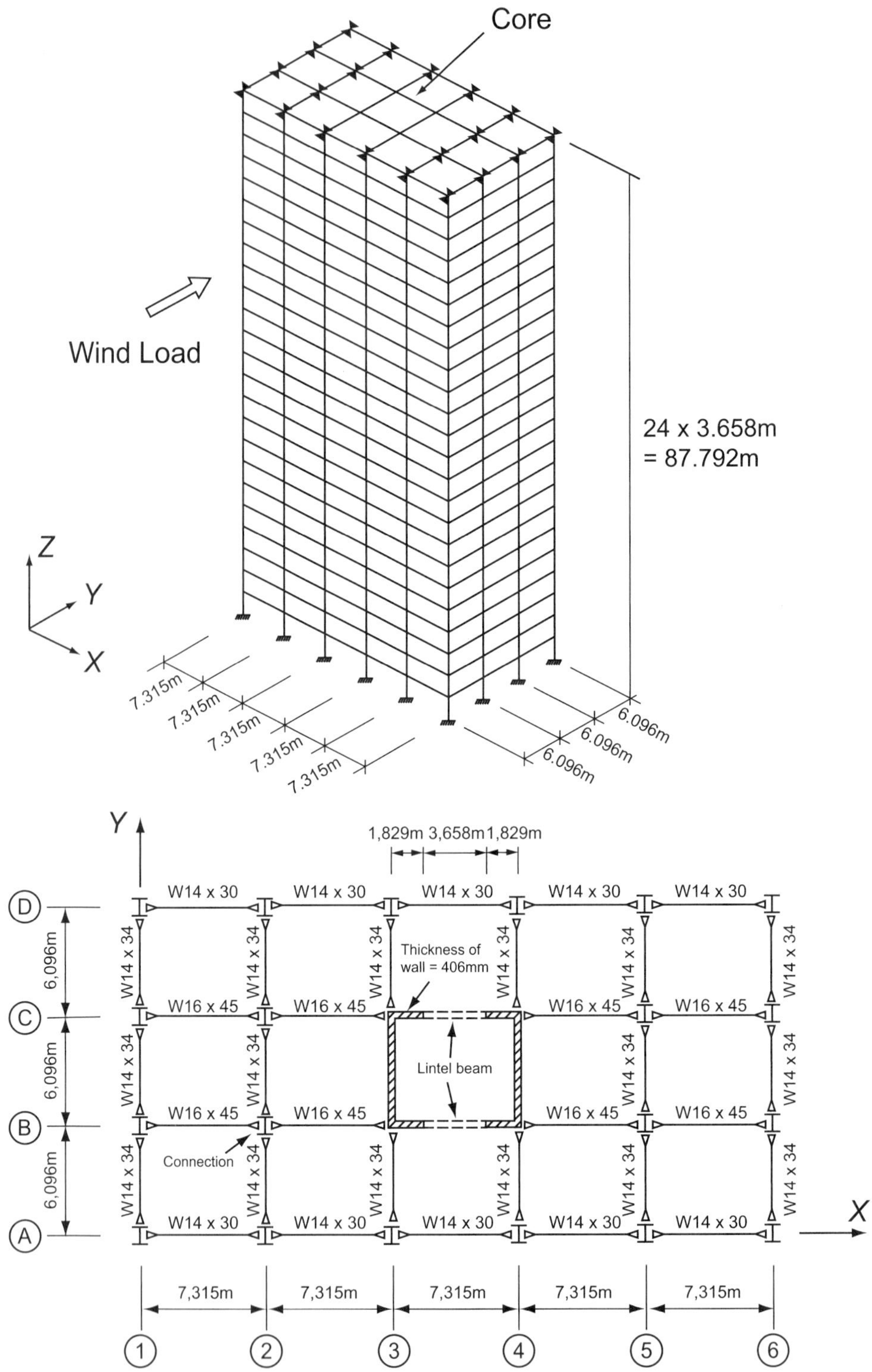

FIGURE 6.72 Multistory frame with strong core.

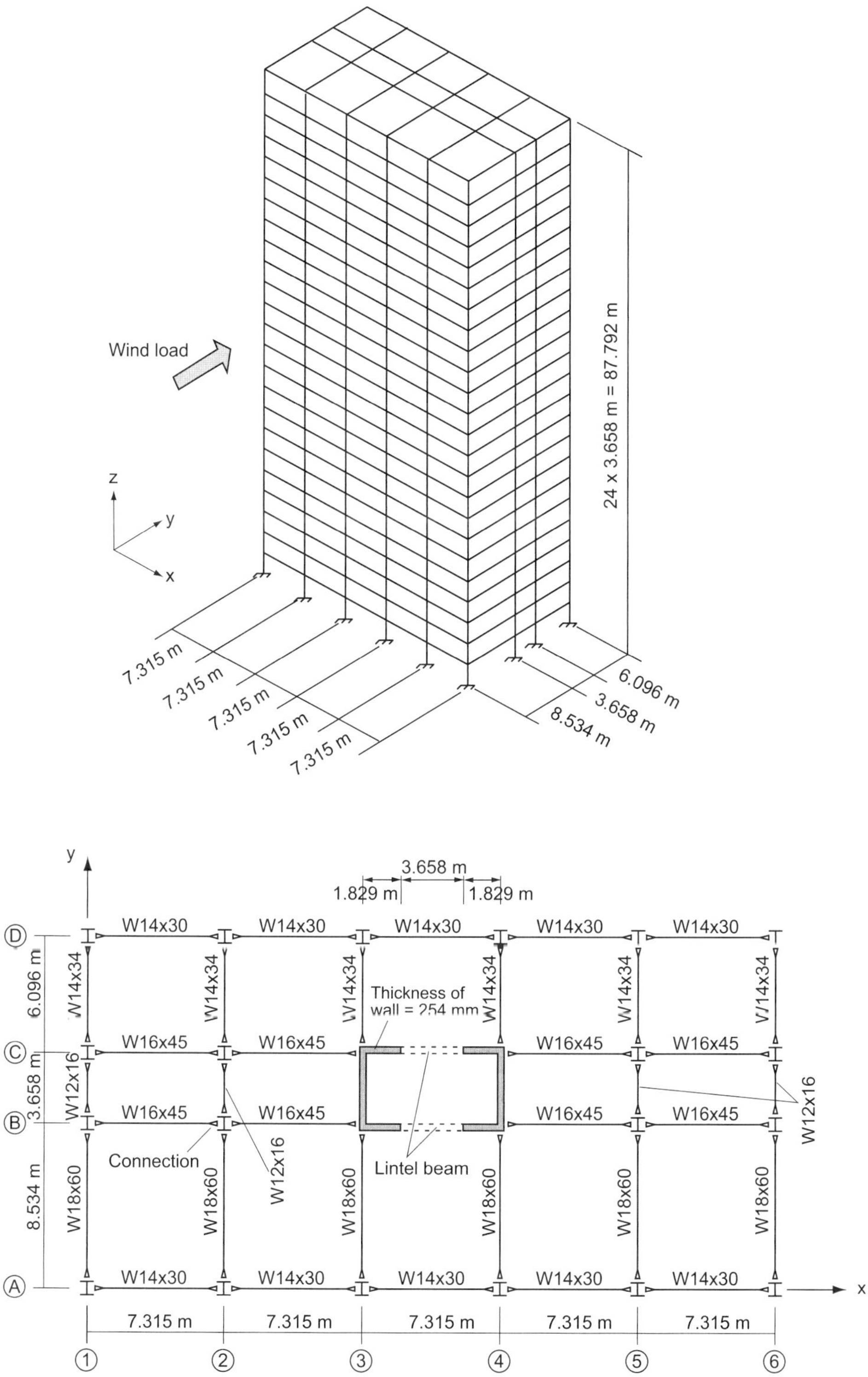

FIGURE 6.73 Multistory frame with weak core.

6.3.7 Analysis of Core-braced Multistory Frame

Figures 6.72 and 6.73 show the plan and isometric views of a 24-story core-braced frame with strong and weak core walls, respectively. The total height of the building frame is about 88 m. The wall thickness of the strong core is 406 mm, and the weak core-wall is 254 mm. Concrete lintel beams with a depth of 1.219 m are rigidly connected to the two *C*-shaped core-wall sections to enhance its resistance against torsion. All steel members are A36 steel. All floor plates are assumed to be rigid in plane to account for the diaphragm action of concrete slabs. The elastic modulus of concrete is 23,400 N/mm^2, and the compressive strength of concrete is 23.4 N/mm^2. The structure is analyzed for the most critical load combination of gravity loads, 4.8 kN/m^2, and wind loads, 0.96 kN/m^2, acting in the *Y* direction. Each steel column is modeled as one plastic-hinge beam-column element, and each beam as four plastic-hinge beam-column elements. The core-wall section is modeled using two *C*-shaped sections interconnected by lintel beams. Each *C*-shape core wall is modeled using one thin-walled element within a story height. The floor beams are assumed to be connected to the center dimensions of the concrete core wall. The finite width (thickness) of the core is not modeled in the analysis. This is because the half thickness of the core wall is only 128 mm, which is about 2 percent of the beam length. For this particular problem, the effect of finite wall thickness will not affect the computed results significantly. However, if the wall thickness is significant compared to the beam length, the finite thickness of the core wall may be modeled using a rigid link element to reflect the real behavior of a core-braced frame. The plastic moment resistance of the core-wall section has been reduced to approximately account for the tensile cracking; they are 1.57×10^5 kNm for the strong core and 4.8×10^4 kNm for the weak core wall.

6.3.7.1 Core-braced Frames with Strong Core Walls

Inelastic analyses are first carried out on the core-braced frame with strong core walls. If the steel frameworks are pin-connected to the concrete core, the whole building relies on the core to provide the lateral resistance. The system reaches its limit of resistance at a load ratio of 2.082 when a plastic hinge forms at the bottom of the core walls as shown in Figure 6.74a. If the presence of the core wall is ignored, the unbraced rigid frame can provide lateral load resistance up to a load ratio of 0.539 as shown in Figure 6.74a.

The limit load of the unbraced rigid framework is about 26 percent of the limit load of the central core walls. In the case of a strong core, the lateral load resistance of the core wall is higher than that of the rigid frameworks. To study the connection effects in core-braced frames, inelastic

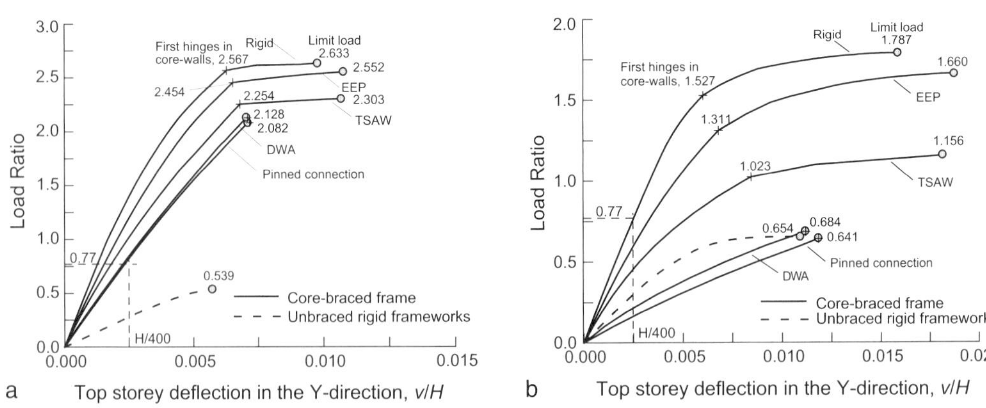

FIGURE 6.74 Nonlinear inelastic analysis of core-braced frames (a) frame with strong core walls; (b) frame with weak core walls.

TABLE 6.14 Parameters and values of M_n/M_{pb} for connections subjected to in-plane bending

Connection type	Connection name	$M'_0 = M_0/M_n$	$R'_{ki} = R_{kj}/M_n$	$R'_{kp} = R_{kp}/M_n$	n	M_n/M_{pb}	
						At the beam framing about the major axis of column	At the beam framing about the minor axis of column
SWA	Single web angle connection	0.98	110	10.6	1.44	0.023	0.011
DWA	Double web angle connection	1.03	301	5.0	1.06	0.053	0.026
TSAW	Top and seat angle connection with double web angles	0.94	363	6.9	1.11	0.467	0.233
TSA	Top and seat angle connection	1.05	448	7.3	0.80	0.302	0.151
EEP	Extended end plate connection without column stiffeners	0.97	309	5.5	1.20	1.040	0.520
EEPS	Extended end plate connection with column stiffeners	1.04	221	2.2	1.42	1.371	0.680
FEP	Flush end plate connection without column stiffeners	1.06	218	3.6	1.23	0.378	0.189
FEPS	Flush end plate connection with column stiffeners	1.06	314	4.5	1.03	0.429	0.214
HP	Header plate connection	0.91	142	13.2	1.20	0.078	0.039

Note: M_{pb} = the plastic capacity of the beam where the semi-rigid connection is located

TABLE 6.15 Comparison of limit loads and elastic lateral stiffness of core-braced frames with strong core walls

Connection types	Limit load	Initial lateral stiffness
DWA (Double web angle connection)	102%	104%
TSAW (Top and seat angle connection with double web angles)	111%	131%
EEP (Extended end plate without column stiffener)	123%	154%
Rigid connection	126%	185%

*All % values are compared with the core-braced frame with pinned connections

analyses are carried out for core-braced frames with DWA, TSAW, EEP, and rigid connections. The properties of DWA, TSAW, and EEP connections are given in Table 6.14. The four-parameters of the power model for semi-rigid connections are evaluated based on the statistic values obtained from the connection test database, so there is no need to know the connection details in order to perform the analysis (Hsieh and Deierlein, 1991).

As shown in Table 6.15, the limit load and elastic lateral stiffness of core-braced frame with DWA connections are respectively 1.02 times and 1.04 times those of the core-braced frame with pinned connections. The use of DWA connections does not increase the limit load and lateral stiffness significantly. If TSAW, EEP, and rigid connections are used, the surrounding frameworks will provide additional resistance to lateral loads, and the limit loads are increased by 11 percent, 23 percent, and 26 percent, respectively. Because the core-braced frame relies mainly on the strong core walls to provide lateral resistance, the frames collapse soon after the cross-section capacity of the core wall is reached. It is also found that the lateral stiffness of the building can be enhanced by 31 percent and 54 percent if TSAW and EEP connections are used (see Table 6.15).

Serviceability checks are also performed on the core-braced frames. All the floor beams satisfy the deflection requirement of span/360 under the service live load. It is observed from Figure 6.74a that the core-braced frames with strong core walls satisfies the lateral drift requirement, (story height)/400, for service wind load. Hence, several connections can be used if strong core walls are used for lateral load resistance. Since the pinned connection is the cheapest for construction, they are preferred for frames braced by the strong core walls.

6.3.7.2 Core-braced Frames with Weak Core Walls

Nonlinear analyses are also carried out on core-braced frames with weak core walls. Core-braced frames with pinned connections collapse at a load ratio of 0.641 as shown in Figure 6.74b. The

TABLE 6.16 Comparison of limit loads and elastic lateral stiffness of core-braced frames with weak core walls

Connection types	Limit load	Initial lateral stiffness
DWA (Double web angle connection)	107%	120%
TSAW (Top and seat angle connection with double web angles)	180%	306%
EEP (Extended end plate without column stiffener)	259%	440%
Rigid connection	279%	571%

*All % values are compared with the core-braced frame with pinned connections

limit load of the pure steel frameworks is 0.654, which is 2 percent higher than that of the core-braced frame with pinned connections. The elastic lateral stiffness of steel frameworks is 40 percent higher than that of the RC core. Inelastic analyses are carried out for core-braced frames with DWA, TSAW, EEP, and rigid connections. It is found that the use of DWA connections does not increase the limit load and lateral stiffness of core-braced frames significantly. If TSAW, EEP, and rigid connections are used, first hinges will form in the core walls at much higher load ratios. The frames have much strength reserve beyond the first plastic hinge, and the limit loads are respectively 1.8, 2.59, and 2.79 times those of the core-braced frame with pinned connections. It is found that the lateral stiffness can be respectively enhanced to 3.06, 4.4, and 5.71 times those of a core-braced frame with pinned connections if TSAW, EEP, and rigid connections are used (see Table 6.16). This is because the steel frameworks are much stiffer and stronger than the weak core. For core-braced frames with weak core walls, the strength and stiffness increase due to the use of semi-rigid connections is much more significant than core-braced frames with strong core walls. The combined use of core wall and rigid frame satisfies the deflection-limit requirement for lateral drift. But the core wall, when used together with other semi-rigid frames, does not satisfy the lateral drift requirement. It can be concluded from the study that if a weak core design is adopted, it is necessary to provide proper connection detailing in order to satisfy the requirements for strength and serviceability limit states.

6.3.7.3 Second-order Effects in the Core-braced Frame

If first-order elastic analyses are carried out on the strong core frame where the connections are designed as pin joints, the frame reaches a limit load ratio of 2.213 based on the plastic capacity check on the core wall section. From the second-order inelastic analysis, first plastic hinge forms at the base of core walls at a load ratio of 2.082. Since the surrounding steel framework relies on the building core to provide lateral stability, the limit load is reduced by 9 percent and the lateral deflection is increased by 10 percent due to the lean column effect and the second-order effect due to the cantilever action of the concrete core. For the core-braced frame with a weak core, the second-order effects also reduce the limit load by 9 percent and increase the lateral deflection by 10 percent. Hence, it is important to account for the geometrical nonlinear effects in the analysis of core-braced frames.

It is noted that a coarse inelastic model is adopted for core wall in the present analysis. Computers and Structures of Berkeley, California, has implemented a section designer in its integrated building design software, ETABS (Version 8, 2002). This section designer is able to predict the section capacity and moment-curvature curve of a concrete core-wall section under the combined action of axial force and biaxial mending moment. Similar developments can be found in El-Tawil et al. (1995). More research needs to be carried out to incorporate these features and to study the influence of axial load-moment interaction and core-wall ductility on the ultimate load capacity of core-braced frames.

References

Abdel-Ghaffar, M., White, D. W., and Chen, W. F., Simplified second-order inelastic analysis for steel frame design. Special Volume of Session on Approximate Methods and Verification Procedures of Structural Analysis and Design, *Proc. at Structures Congress 91*, New York, ASCE, 47–52, 1991.

AISC, *Load and Resistance Factor Design Specification for Structural Steel Buildings*, American Institute of Steel Construction, 2nd Ed., Chicago, IL, 1993.

AISC, *Load and Resistance Factor Specification for Structural Steel Buildings*, American Institute of Steel Construction, 3rd Ed, Chicago, IL, 2001.

AISC, *Specification for Structural Steel Buildings*, March 9, 2005, ANSI/AISC 360–05, 2005.

Alemdar, B. N. and White, D. W., Displacement, flexibility, and mixed beam-column finite element formulations for distributed plasticity analysis, *Journal of Structural Engineering*, ASCE, 131(12), 1811–1819, 2005.

Al-Bermani, F. G. A. and Kitipornchai, S., Nonlinear analysis of thin-walled structures using least element/member, *Journal of Structural Engineering*, ASCE, 116(1), 215–234, 1990.

Al-Bermani, F. G. A. and Kitipornchai, S., Nonlinear analysis of transmission towers, *Engineering Structures*, 14(3), 139–151, 1992.

Al-Bermani, F. G. A., Zhu, K., and Kitipornchai, S., Bounding-surface plasticity for nonlinear analysis of space structures, *International Journal for Numerical Methods in Engineering*, 38(5), 797–808, 1995.

Amdahl, J., Johansen, A., and Svanø, G., Ultimate capacity of jack-ups considering foundation behaviour, *Proc. 7th International Conference on the Behaviour of Offshore Structures*, BOSS'94, Boston, USA, 1994.

Anderson, D., *Composite steel-concrete joints in frames for buildings: design provision*, COST C1 Document; 1999.

Ang, K. M. and Morris, G. A., Analysis of three-dimensional frames with flexible beam-column connections, *Canadian Journal of Civil Engineering*, 11, 245–254, 1984.

API RP 2A-WSD, *Recommended practice for planning, designing and constructing fixed offshore platforms—Working stress design*, American Petroleum Institute, 2002.

AS4100, *Steel Structures*, Standards Australia, Sydney, 1990.

Attalla, M. R., Deierlein, G. G., and McGuire, W., Spread of plasticity: quasi-plastic-hinge approach, *Journal of Structural Engineering*, ASCE, 120(8), 2451–2473, 1994.

Attalla, M. R., Deierlein, G. G., and McGuire, W., Nonlinear frame analysis with torsional-flexural member behaviour. *In: Proceedings of the Fifth International Colloq. On Stability and Metal Structures*, SSRC, Lehigh University, 11–19, 1996.

Attiogbe, G. and Morris, G., Moment-rotation functions for steel connections, *Journal of Structural Engineering*, ASCE, 117(6), 1703–1718, 1991.

Avery, P. and Mahendran, M., Distributed plasticity analysis of steel frame structures comprising non-compact sections, *Engineering Structures*, 22, 901–919, 2000a.

Avery, P. and Mahendran, M., Refined plastic hinge analysis of steel frame structures comprising non-compact sections I: Formulation, *Advances in Structural Engineering*, 3(4), 291–307, 2000b.

Bailey, C. G., Burgess, I. W., and Plank, R. J., Analysis of the effects of cooling and fire spread on steel-concrete-framed buildings, *Fire Safety Journal*, 26, 273–293, 1996.

Bathe, K. J. and Bolourchi, S., Large displacement analysis of three-dimensional beam structures, *International Journal for Numerical Methods in Engineering*, 14, 961–986, 1979.

Barsoum, R. S. and Gallagher, R. H., Finite element analysis of torsional flexural stability problems, *International Journal for Numerical Methods in Engineering*, 2, 335–352, 1970.

Bjorhovde, R., Brozzetti, J., and Colson, A. (eds.), *Connections in steel structures: Behaviour, strength and design, Proceedings of the Workshop on Connections held at the Ecole Normale Superiere*, Cachan, France, May 1987, Elsevier Applied Science, London, 1988.

Bjorhovde, R., Brozzetti, R.J., and Colsen, A., A classification system for beam-to-column connections, *Journal of Structural Engineering*, ASCE, 116(ST11), 3059–3076, 1990.

Bjorhovde, R., Colson, A., Haaijer, G., and Stark, J. W. B. (eds.), *Connections in steel structures II: behaviour, strength and design*, Pittsburgh, April, 1991, AISC, Chicago, 1992.

Bjorhovde, R., Colson, A., and Zandonini, R. (eds.), *Connections in steel structures III: Behaviour, strength and design*, Proceedings of the Third International Workshop on Connections, Trento, Italy, May, 1995, Pergamon Press, London, 1996.

Blandford, G. E., Review of progressive failure analyses for truss structures, *Journal of Structural Engineering*, ASCE, 123(2), 122–129, 1997.

Bolt, H. M., Billington, C. J., and Ward, J. K., Results from large scale ultimate load tests on tubular jacket frame structures, *Offshore Technology Conference*, OTC 7451, 1994.

Bolt, H. M. and Billington, C. J., Results from ultimate load tests on 3-D jacket-type structures, *Proc. of Offshore Technology Conference*, OTC11941, Houston, Texas, USA, May 2000.

Bomel Limited, *Analytical and experimental investigation of the behavior of tubular frames*, Phase II. Part 4: Summary and Conclusions from Phase I and Phase II, Joint Industry Funded Research Programme, 1992.

Bozzo, E. and Gambarrotta, L., Inelastic analysis of steel frames for multistory buildings, *Computers and Structures*, 20(4), 707–713, 1985.

Bradford, M. A., Local buckling analysis of composite beams, *Civil Engineering Transactions, Institution of Engineers*, Australia, CE28(4), 312–317, 1986.

Bradford, M. A. and Gao, Z., Distortional buckling solutions for continuous composite beams, *Journal of Structural Engineering*, ASCE, 118(1), 73–89, 1992.

BS5400, *Steel, concrete and composite bridges*: Part 5: Code of practice for design of composite bridges, London: British Standards Institution, 1979.

BS5950-1, *Structural use for steelwork in building*, Code of practice for design-rolled and welded sections, British Standards Institution, 2000.

Bursi, O. S. and Jaspart, J. P., Benchmarks for finite element modelling of bolted steel connections, *Journal of Constructional Steel Research*, 43(1–3),17–42, 1997.

Buitrago, J., Healy, B. E., and Chang, T. Y., Local joint flexibility of tubular joints, *Proceedings of Offshore Mechanics and Arctic Engineering Conference*, ASME, 1, 405–411, 1993.

CEN, *Eurocode 3*, Design of Steel Structures, Part 1.1: General Rules and Rules for Buildings, ENV 1993-1-1: 1992, European Committee for Standardization, Brussels, Belgium, 1992.

CEN, *Eurocode 3*, Design of Steel Structures, Part 1.2: General Rules Structural Fire Design, ENV 1993-1-2: 2001, European Committee for Standardization, Brussels, 2001.

CEN, *Eurocode 4*, Design of Composite Steel and Concrete Structures, Part 1.1: General Rules and Rules for Buildings, ENV 1994-1-1, London, British Standards Institution, 1994.

Chajes, A. and Churchill, J. E., Nonlinear frame analysis by finite element method, *Journal of Structural Engineering*, ASCE, 113, 1221–1235, 1987.

Chan, S. L. and Kitipornchai, S., Geometric nonlinear analysis of asymmetric thin-walled beam-columns, *Engineering Structures*, 9(4), 243–254, 1987.

Chan, S. L. and Zhou, Z. H., Second-order elastic analysis of frames using single imperfect element per member, *Journal of Structural Engineering*, ASCE, 121(6):939–945, 1995.

Chan, S. L. and Gu, J. X., Exact tangent stiffness for imperfect beam-column members, *Journal of Structural Engineering*, ASCE, 126(9), 1094–1102, 2000.

Chan, S. L. and Gu, J. X., Exact tangent stiffness for imperfect beam-column members—Closure, *Journal of Structural Engineering*, ASCE, 127(12), 1492–1494, 2001.

Chan, S. L., Review: Non-linear behavior and design of steel structures, *Journal of Constructional Steel Research*, 57, 1217–1231, 2001.

Chan, S. L. and Chan, B. H. M., Refined plastic hinge analysis of steel frames under fire, *Steel and Composite Structures*, 1(1), 111–130, 2001.

Chan, S. L., Shu, G. P., and Lü Z. T., Stability analysis and parameter study of pre-stressed stayed columns, *Engineering Structures*, 24, 115–124, 2002.

Chan, S. L. and Zhou, Z. H., Elastoplastic and large deflection analysis of steel frames by one element per member. II: three hinges along member, *Journal of Structural Engineering*, ASCE, 130(4), 545–553, 2004.

Chen, B., Hu, Y., and Tan, M., Local joint flexibility of tubular joints of offshore structures, *Proceedings of Offshore Mechanics and Arctic Engineering Conference*, OMAE, Houston, Texas, 1990.

Chen, H. and Blandford, G. E., Thin-walled space frames, Parts I and II, *Journal of Structural Engineering*, ASCE, 117(8), 2499–2539, 1991.

Chen, H., *Nonlinear inelastic analysis of steel-concrete composite frames*, Thesis presented to National University of Singapore, in partial fulfillment of the requirements for the degree of Doctor of Philosophy, Singapore, 2000.

Chen, H., Liew, J. Y. R., and Shanmugam, N. E., Nonlinear inelastic analysis of building frames with thin-walled cores, *Thin-Walled Structure*, 37(3), 189–205, 2000.

Chen, H. and Liew, J. Y. R., Explosion and fire analysis of steel frames using mixed element method, *Journal of Engineering Mechanics*, ASCE, 131(6), 606–616, 2005.

Chen, W. F. and Atsuta, T., *Theory of beam-column. Vol.* 1, *Inplane behaviour and design*, McGraw-Hill, New York, 1976.

Chen, W. F. and Atsuta T., *Theory of beam-column. Vol.* 2, *Space behaviour and design*, McGraw-Hill, New York, 1977.

Chen, W. F. and Lui, E., *Stability design of steel frames*, CRC Press, Boca Raton, FL, 1991.

Chen, W. F. and Toma, S., *Advanced analysis of steel frames: Theory, software, and applications.* CRC Press, Boca Raton, FL, 1994.

Chen, W. F., Goto, Y., and Liew, J. Y. R., *Stability design of semi-rigid frames,* John Wiley & Sons Inc., New York, 1996.

Chen, W. F. and Kim, S.E., *LRFD steel design using advanced analysis*. CRC Press, Boca Raton, FL, 1997

Chen, W. F., Structural stability: from theory to practice, *Engineering Structures*, 22, 116–122, 2000.

Chen, W. F., Structural Engineering: seeing the big picture, *KSCE Journal of Civil Engineering*, 12(1), 25–29, 2008.

Choo, Y. S., Qian, X. D., and Foo, K. S., Nonlinear analysis of tubular space frame incorporating joint stiffness and strength, *10th International Jack-Up Platform Conference*, 13-14. Sept. London, 2005.

Clarke, M. J., *Plastic zone analysis of frames, in advanced analysis of steel frames: Theory, software, and applications*, Chen, W. F. and Toma, S. eds., CRC Press, Boca Raton, FL., 259–319, 1994.

Conci, A., Large displacement analysis of thin-walled beams with generic open section, *International Journal for Numerical Methods in Engineering*, 33, 2109–27, 1992.

Crues, G. J., Torres, P. L., and Groehs, A. G., Elastoplastic frame analysis with generalized yield function and finite displacements, *Computers and Structures*, 18(5), 925–929, 1984.

Da Silva, L. S., Towards a consistent design approach for steel joints under generalized loading, *Journal of Constructional Steel Research*, 64, 1059–1075, 2008.

Dai, Y. J., *The effect of joint performance on global behaviour of tubular steel structures*, Engineering thesis, Dept of Civil Engineering, National University of Singapore, Singapore, 2002.

Deierlein, G. G., Zhao, Y., and McGuire, W., A two-surface concentrated plasticity model for analysis of 3-D framed structures, *Annual Tech. Session Proc.*, SSRC, April 15–17, Chicago, IL, 423–432, 1991.

Di Sarno, L. and Elnashai, A.S., *Seismic retrofitting of steel and composite building structures*, Report No. 02-01, Mid-America Earthquake Center, Civil and Environmental Engineering Deptartment, University of Illinois at Urbana-Champaign, 2002.

Diedricks, A. A., Uy, B., Bradford, M. A., and Oehlers, D. J., Mixed analysis approach for semi-continuous steel-concrete composite beams under uniform loading, *Proceedings of Composite Construction in Steel and Concrete* IV *Conference*, May 29–June 2, 2000, Banff, Alberta, Canada, 2000.

DNV-RP-C204, Det Norske Veritas Recommended Practice—Design Against Accidental Loads, Nov., 2004.

Duan, L. and Chen, W. F., A yield surface equation for doubly symmetric sections, *Engineering Structures*, 12(2), 114–119, 1990.

El-Rimawi, J. A., Burgess, I. W., and Plank, R. J., The analysis of semi-rigid frames in fire—a secant approach, *Journal of Constructional Steel Research*, 22, 125–146, 1995.

El-Tawil, S., Sanz-Picon, C. F., and Deierlein, G. G., Evaluation of ACI 318 and AISC (LRFD) strength provisions for composite beam columns, *Journal of Constructional Steel Research*, 34, 103–123, 1995.

El-Tawil, S. and Deierlein, G. G., Stress-resultant plasticity for frame structures, *Journal of Engineering Mechanics*, ASCE, 24(12), 1360–1370, 1998.

El-Tawil, S. and Deierlein, G. G., Nonlinear analysis of mixed steel-concrete frames. Parts I: Element formulation, Part II: Implementation and verification, *Journal of Structural Engineering*, ASCE, 127(6), 647–665, 2001.

ETABS Version 8, *Integrated building design software—Section designer manual*, Computers and Structures, Berkeley, California, USA, 2002.

Faella, C., Piluso, V., and Rizzano, G., *Structural steel semirigid connections: Theory, design and software*. CRC Press, Boca Raton, 2000.

Fang, L. X., Chan, S. L., and Wong, Y. L., Strength analysis of semi-rigid steel concrete composite frames, *Journal of Constructional Steel Research*, 52, 269–291, 1999.

Fang, L. X., Chan, S. L., and Wong, Y. L., Numerical analysis of composite frames with partial shear-stud interaction by one element per member, *Engineering Structures*, 22, 1285–1300, 2000.

Fessler, H., Mockford, P. B., and Webster, J. J., Parametric equations for the flexibility matrices of single brace tubular joints in offshore structures, *Proceedings of the Institution of Civil Engineers*, London, ICE, 2, 659–673, 1986a.

Fessler, H., Mockford, P. B., and Webster, J. J., Parametric equations for the flexibility matrices of multi-brace tubular joints in offshore structures, *Proceedings of the Institution of Civil Engineers*, London, ICE, 2, 675–696, 1986b.

Gattass, M. and Abel, J. F., Convected systems for curved structural elements, *International Journal for Numerical Methods in Engineering*, 24, 2143–2166, 1987a.

Gattass, M. and Abel, J. F., Equilibrium considerations of the updated lagrangian formulation of beam-columns with natural concepts, *International Journal for Numerical Methods in Engineering*, 24, 2119–2141, 1987b.

Goverdhan, A. V., *A collection of experimental moment-rotation curves and evaluation of prediction equations for semi-rigid connections*, Thesis presented to Vanderbilt University, in partial fulfillment of the requirements for the degree of Master of Science, Nashville, TN, 1983.

Hajjar, J. F. and Gourley, B. C., A cyclic nonlinear model for concrete-filled tubes. I: Formulation, *Journal of Structural Engineering*, ASCE, 123(6), 736–744, 1997.

Hajjar, J. F., Evolution of stress-resultant loading and ultimate strength surfaces in cyclic plasticity of steel wide-flange cross-sections, *Journal of Constructional Steel Research*, 59, 713–750, 2003.

Hellan, Ø., *Nonlinear pushover and cyclic analysis in ultimate limit state design and re-assessment of tubular steel offshore structures*, Thesis presented to the Norwegian Institute of Technology, in partial fulfillment of the requirements for the degree of Doctor of Philosophy, Norway, 1995.

Himly, S. I. and Abel, J. F, Material and geometric nonlinear dynamics analysis of steel frames using computer graphics, *Computers and Structures*, 21(4), 825–840, 1985.

Hodge, P. G., Automatic piecewise linearization in ideal plasticity, *Computer Methods in Applied Mechanics and Engineering*, 10, 249–272, 1977.

Hsieh, S. H. and Deierlein, G. G., Nonlinear analysis of three-dimensional frames with semi-rigid connections, *Computers and Structures*, 41(5), 995–1009, 1991.

Huber, G., *Nonlinear calculations of composite sections and semi-continuous joints*, Thesis presented to University of Innsbruck, in partial fulfillment of the requirements for the degree of Doctor of Philosophy, Innsbruck, Austria, 1999.

Iyengar, S. H., Mixed steel-concrete high-rise systems. In: Sabnis GM, editor. *Handbook of composite construction engineering, Chapter 7*, Van Nostrand Reinhold Company, New York, 1979.

Iyengar, S. H., Baker, W. F., and Sinn, R., *Multi-storey buildings, in Chapter* 6.2, Constructional Steel Design, An International Guide, Dowling, P. J., Harding, J. E. and Bjorhovde (eds), R., Elsevier Applied Science, London and New York, 1992.

Izzuddin, B. A. and Elnashai, A. S., Adaptive space frame analysis, Part I: A plastic hinge approach, Part II. A distributed plasticity approach, *Proc Instn Civ Engrs Structs & Bldgs*, 99, 303–326, 1993a.

Izzuddin, B. A. and Elnashai, A. S., Eulerian formulation for large-displacement analysis of space frames, *Journal of Engineering Mechanics*, ASCE, 119(3), 549–569, 1993b.

Izzuddin, B. A. and Smith, D. L., Large displacement analysis of elastoplastic thin-walled frames. I: formulation and implementation, *Journal of Structural Engineering*, ASCE, 122(8), 905–914, 1996.

Izzuddin, B. A., Conceptual issues in geometrically nonlinear analysis of 3-D framed structures, *Computer Methods in Applied Mechanics and Engineering*, 191, 1029–1053, 2002.

Jiang, X. M., Chen, H., and Liew, J. Y. R., Spread-of-plasticity analysis of three-dimensional steel frames, *Journal of Constructional Steel Research*, 58, 193–212, 2002.

Johnson, R. P. and Bradford, M. A., Distortional lateral buckling of continuous composite bridge girders, In Morris, L. J. (ed.) *International Conference on Stability and Plastic Collapse of Steel Structures*, Granada, 569–580, 1983.

Kamba, T., Prediction equation for initial stiffness and ultimate capacity of unstiffened CHS column and H-beam connections, *Doc For Tubular Structure Committee of the AIJ*, 1997. (Japanese)

Kassimali, A. and Abbasnia, R., Large deformation analysis of elastic space frames. *Journal of Structural Engineering*, ASCE, 117(7), 2069–2087, 1991.

Kemp, A. R. and Nethercot, D. A., Required and available rotations in continuous composite beams with semi-rigid connections, *Journal of Constructional Steel Research*, 57(1–3), 375–400, 2001.

Kim, S. E., Park, M. H., and Choi, S. H., Direct design of three-dimensional frames using practical advanced analysis, *Engineering Structures*, 23, 1491–1502, 2001.

King, W. S., White, D. W., and Chen, W. F., Second-order inelastic analysis methods for steel-frame design, *Journal of Structural Engineering*, ASCE, 118(2), 408–428, 1992.

Kishi, N. and Chen, W. F., *Data base of steel beam–to–column connections, Vol.* 1 & 2, Structural Engineering Report No. CE-STR-86-26, School of Civil Engineering, Purdue University, West Lafayette, IN, 1986.

Kishi, N., Chen, W. F., Goto, Y., and Matsuoka, K., Design aid of semi-rigid connections for frame design, *AISC Engineering Journal*, 30(3), 90–107, 1993.

Kishi, N., Komuro, M., and Chen, W. F., Four-parameter power model for M-θ_r curves of end-plate connections. *Proceedings of Connections in Steel Structures V: Behaviour, Strength and Design*, Amsterdam, Netherlands June 3–4, 2004.

Kitipornchai, S., Zhu, K., Xiang, Y., and Al-Bermani, F. G. A., Single-equation yield surface for monosymmetric and asymmetric sections, *Engineering Structures*, 13, 366–370, 1991

Kitipornchai, S. and Al-Bermani, F. G. A., Nonlinear analysis of lattice structures, *Journal of Constructional Steel Research*, 23(1–3), 209–225, 1992.

Komuro, M., Kishi, N., and Chen, W. F., Elasto-plastic FE analysis on moment-rotation relations of top- and seat-angle connections, *Proceedings of Connections in Steel Structures V: Behaviour, Strength and Design*, Amsterdam, Netherlands June 3–4, 2004.

Kondoh, K. and Atluri, S. N., A simplified finite element method for large deformation, post-buckling analysis of large frame structures, using explicitly derived tangent stiffness matrices, *International Journal for Numerical Methods in Engineering*, 23, 69–90, 1986.

Krenk, S., Vissing-Jørgensen, C., and Thesbjerg, L., Efficient collapse analysis techniques for framed structures, *Computers and Structures*, 72, 481–496, 1999.

Krishnamurthy, N. and Graddy, D. E., Correlation between 2- and 3-dimensional finite element analysis of steel bolted end-plate connections, *Computers and Structures*, 6, 381–389, 1976.

Kuo, S. R., Yang, Y. B., and Chou, J. H., Nonlinear analysis of space frames with finite rotations, *Journal of Structural Engineering*, ASCE, 119(1), 1–15, 1993.

Landesmann, A., Batista, E. D. M, and Alves, J. L. D., Implementation of advanced analysis method for steel-framed structures under fire conditions, *Fire Safety Journal*, 40, 339–366, 2005.

Leon, R. T., Hoffman, J., and Staeger, T., *Design of partially-restrained composite connections*, AISC Design Guide 8, American Institute of Steel Construction, Chicago, 1996.

Leon, R. T., Design guide for partially retrained composite connections, *Journal of Structural Engineering*, ASCE, 124(10), 1099–1114, 1998.

Leu, L. J. and Tsou, C. H., Second-order analysis of planar steel frames considering the effect of spread-of-plasticity, *Structural Engineering & Mechanics*, 11(4), 423–442, 2001.

Leu, L. J., Yang, J. P., Tsai, M. H., and Yang, Y. B., Explicit inelastic stiffness for beam elements with uniform and nonuniform cross sections, *Journal of Structural Engineering*, ASCE, 134(4), 608–618, 2008.

Li, T. Q., Choo, B. S., and Nethercot, D. A., Moment curvature relations for steel and composite beams, *Steel Structures, Journal of Singapore Structural Steel Society*, 4(1), 35–51, 1993.

Liew, J. Y. R., White, D. W., and Chen, W. F., Beam-column design in steel frameworks-Insight on current methods and trends, *Journal of Constructional Steel Research*, 18, 259–308, 1991.

Liew, J. Y. R., White, D. W., and Chen, W. F., *Beam-columns, in constructional steel design, An international guide*, Dowling P. J. et al., Eds., Elsevier, England, 105–132, 1992.

Liew, J. Y. R., White, D. W., and Chen, W. F., Second-order refined plastic hinge analysis for framed design: Parts 1 and 2, *Journal of Structural Engineering*, ASCE, 119(11), 3196–3237, 1993a.

Liew, J. Y. R., White, D. W., and Chen, W. F., Limit states design of semi-rigid frames using advanced analysis, Part 1: Connection modelling and classification; Part 2: Analysis and design, *Journal of Constructional Steel Research*, 26(1), 1–57, 1993b.

Liew, J. Y. R., White, D. W., and Chen, W. F., Notional load plastic hinge method for frame design, *Journal of Structural Engineering*, ASCE , 120(5), 1434–1454, 1994.

Liew, J. Y. R., Punniyakotty, N. M., and Shanmugam, N. E., Advanced analysis and design of spatial structures, *Journal of Constructional Steel Research*, 42(1), 21–48, 1997a.

Liew, J. Y. R., Yu, C. H., and Ng, Y. H., and Shanmugam, N. E., Testing of semi-rigid unbraced frames for calibration of second-order inelastic analysis, *Journal of Constructional Steel Research*, 41(2/3), 159–195, 1997b.

Liew, J. Y. R, Tang, L. K., Holmaas, T., and Choo, Y. S., Advanced analysis for the assessment of steel frames in fire, *Journal of Constructional Steel Research*, 47(1–2), 19–45, 1998.

Liew, J. Y. R., Chen, W. F., and Chen, H., Advanced inelastic analysis of frame structures, *Journal of Constructional Steel Research*, 55(1–3), 245–265, 2000a.

Liew, J. Y. R., Chen, H., Shanmugam, N. E., and Chen, W. F., Improved nonlinear plastic hinge analysis of space frame structures, *Engineering Structures*, 22, 1324–1338, 2000b.

Liew, J. Y. R., Teo, T. H., Shanmugam, N. E., Testing of steel-concrete composite connections and appraisal of results, *Journal of Constructional Steel Research*, 56, 117–150, 2000c.

Liew, J. Y. R. and Tang, L. K., Advanced plastic hinge analysis for the design of tubular space frames, *Engineering Structures*, 22(7), 769–783, 2000.

Liew, J. Y. R., *A resource for structural steel design and construction*, SSSS/BCA Joint Publication, Singapore, 83pp., 2001a.

Liew, J. Y. R., State-of-the-art of advanced inelastic analysis of steel and composite structures, *Steel and Composite Structures*, 1(3), 341–354, 2001b.

Liew, J. Y. R., Punniyakotty, N. M., and Shanmugam, N. E., Limit-state analysis and design of cable-tensioned structures, *International Journal of Space Structures*, 16(2), 95–110, 2001a.

Liew, J. Y. R., Chen, H., and Shanmugam, N. E., Inelastic analysis of steel frames with composite beams, *Journal of Structural Engineering*, ASCE, 127(2), 361–370, 2001b.

Liew, J. Y. R. and Uy, B., Advanced analysis of composite frames, *Progress in Structural Engineering and Materials*, 3, 159–169, 2001.

Liew, J. Y. R., Buildable design of multi-storey and large-span steel structures, *Steel Structures*, 4, 53–70, 2004a.

Liew, J. Y. R., Performance based fire safety design of structures—A multi-dimensional integration, *Advances in Structural Engineering*, 7(4), 111–133, 2004b.

Liew, J. Y. R., Teo, T. H., and Shanmugam, N. E., Composite joints subject to reversal of loading, Part 1: Experimental study; Part 2: Analytical assessments, *Journal of Constructional Steel Research*, 60, 221–268, 2004.

Liew, J. Y. R. and Chen, H., Direct analysis for performance-based design of steel and composite structures, *Progress in Structural Engineering and Materials*, 6, 213–228, 2004a.

Liew, J. Y. R. and Chen, H., Explosion and fire analysis of steel frames using fiber element approach, *Journal of Structural Engineering*, ASCE, 130(7), 991–1000, 2004b.

Liew, J. Y. R. and Li, J. J., Advanced analysis of pre-tensioned bowstring structures, *Steel Structures*, 6, 153–162, 2006.

Liew, J. Y. R., Survivability of steel frame structures subject to blast and fire, *Journal of Constructional Steel Research*, 64(7–8), 854–866, 2008.

Liew, J. Y. R. and Xiong, D. X., Effect of preload on axial capacity of concrete-filled composite columns, *Journal of Constructional Steel Research*, 65, 709–722, 2009.

Lindner, J., Lateral torsional buckling of composite beams, *Journal of Constructional Steel Research*, 46(1–3), 222, 1998.

Lorenz, B., Kato, B., and Chen, W. F. (eds.), *Semi-rigid connections in steel frames*, Council on Tall Buildings and Urban Habitat, Bethlehem, PA, 1993.

Lui, E. M. and Ma, G., Analysis and design form stability in the U.S.—An overview, *Steel and Composite Structures*, 5(2–3), 103–126, 2005.

Ma, K. Y. and Liew, J. Y. R., Nonlinear plastic hinge analysis of 3-D steel frames in fire, *Journal of Structural Engineering*, ASCE, 130(7), 981–990, 2004.

Marshall, P., Lewis, D., and Stock, D., Yield interaction surface of an enhanced LeTourneau 116C chord, *Marine Structures*, 14, 379–396, 2001.

Meek, J. L., and Loganathan, S., Large displacement analysis of space-frame structures, *Computer Methods in Applied Mechanics and Engineering*, 72, 57–75, 1989.

Merchant, W., The failure load of rigid jointed frameworks as influenced by stability, *The Structural Engineer*, 32, 185–190, 1954.

Moan, T., Amdahl, J., Granli, T., and Hellan, Ø., Collapse behaviour of offshore structural systems, *2nd Int. Conf. on Advances in Marine Structures*, Dunfermline, Scotland, 21–24 May, 1991.

Murtha-Smith, E., Compression member models for space trusses: review, *Journal of Structural Engineering*, ASCE, 120(9), 2717–2147, 1994.

Nethercot, D. A., *Steel beam to column connections - A review of test data and their applicability to the evaluation of the joint behaviour of the performance of steel frames*, CIRIA, London, 1985.

Nethercot, D. A. and Zandonini, R., Methods of prediction of joint behaviour: beam-to-column connections, *in Structural Connections, Stability and Strength*, R. Narayanan, ed., Elsevier, London, 23–62, 1989.

Nethercot, D. A., Li, T. Q., and Ahmed, B., Unified classification system for beam to column connections, *Journal of Constructional Steel Research*, 45(1), 39–65, 1998.

Nigam, N. C., Yielding in framed structures under dynamic loads, *Journal of the Engineering Mechanics Division*, ASCE, EM5, 687–709, 1970.

Nukala, P. K. V. V. and White, D. W., A mixed finite element for three-dimensional nonlinear analysis of steel frames, *Computer Methods in Applied Mechanics and Engineering*, 193, 2507–2545, 2004.

Oehlers, D. J. and Sved, G., Flexural strength of composite beams with limited slip capacity shear connectors, *Journal of Structural Engineering*, ASCE, 121(6), 932–938, 1995.

Oran, C., Tangent stiffness in space frames, *Journal of the Structural Division*, ASCE, 99, 987–1001, 1973.

Orbison, J. G., *Nonlinear static analysis of three-dimensional steel frames*, Report No. 82–6, Department of Structural Engineering, Cornell University, Ithaca, NY., 1982.

Orbison, J. G., McGuire, W., and Abel, J. F., Yield surface applications in nonlinear steel frame analysis, *Computer Methods in Applied Mechanics and Engineering*, 33, 557–573, 1982.

Osterrieder, P. and Kretzschmar, J., First hinge analysis for lateral buckling design of open thin-walled steel members, *Journal of Constructional Steel Research*, 62, 35–43, 2006.

Papqdrakakis, M. and Papadopoulos, V., Computationally efficient method for the limit elasto plastic analysis of space frames, *Computational Mech*, 16,132–141, 1995.

Pi, Y. L. and Trahair, N. S., Nonlinear inelastic analysis of steel beam-column, I: Theory; II: Application, *Journal of Structural Engineering*, ASCE, 120(7), 2041–2085, 1994.

Porter, F. L. and Powell, G. H., *Static and dynamic analysis of inelastic frame structures*, Report No. EERC 71-3, Earthquake Engineering Research Centre, University of California, Berkeley, CA, 1971.

Powell, G. H. and Chen, P. F., 3D beam-column element with generalised plastic hinges, *Journal of Structural Engineering*, ASCE, 112(7), 627–641, 1986.

Przemieniecki, J. S., *Theory of matrix structural analysis*. McGraw-Hill, New York, 1968.

Punniyakotty, N. M., *Nonlinear analysis and design of spatial truss and flexible structures*, Thesis presented to National University of Singapore, in partial fulfillment of the requirements for the degree of Doctor of Philosophy, Singapore, 1998.

Qian, X. D., *Static strength of thick-walled CHS joints and global frame behaviour*, Thesis presented to National University of Singapore, in partial fulfillment of the requirements for the degree of Doctor of Philosophy, Singapore, 2005.

Richard, R. M. and Abbott, B. J., Versatile elastic-plastic stress-strain formula, *Journal of the Engineering Mechanics Division*, ASCE, 101(4), 511–515, 1975.

Saw, H. S. and Liew, J. Y. R., Assessment of current methods for the design of composite columns in buildings, *Journal of Constructional Steel Research*, 53(2), 121–147, 2000.

SCI/BCSA, *Joints in steel construction: Composite connection*, Connection Group Publication 213, Steel Construction Institute, UK, 1998.

Shanmugam, N. E. and Lakshmi, B., State of the art report on steel-concrete composite columns, *Journal of Constructional Steel Research*, 57, 1041–1080, 2001.

Sherbourne, A. N. and Bahaari, M. R., Finite element prediction of end-plate bolted connection behaviour. I: Parametric Study, *Journal of Structural Engineering*, ASCE, 123(2), 157–164, 1997.

Skallerud, B. and Amdahl, J., *Nonlinear analysis of offshore structures*. Research Studies Press, 2002.

SNAME, Guidelines for Site Specific Assessment of Mobile Jack-up Units, Technical & Research Bulletin 5–5A, 1994.

Sohal, I. S. and Chen, W. F., Local buckling and sectional behaviour of fabricated tubes, *Journal of Structural Engineering*, ASCE, 113(3), 1073–1090, 1988.

Surovek, A. E., White, D. W., and Leon, R. T., Direct analysis for design evaluation of partially restrained steel framing system, *Journal of Structural Engineering*, ASCE, 131(9), 1376–1389, 2005.

Surovek, A. E., *Guidelines for the use of direct second-order inelastic analysis in steel frame design*, Report of the Special Project Committee on Advanced Analysis Technical Committee on Compression and Flexural Members of the Structural Engineering Institute of ASCE, 2007.

Swanson, J. A., Kokan, D. S., and Leon, R. T., Advanced finite element modeling of bolted T-stub connection components, *Journal of Constructional Steel Research*, 58(5–8), 1015–1031, 2000.

Søreide, T. H., Amdahl, J., Eberg, E., Holmås, T., and Hellan, Ø., *USFOS—A computer program for progressive collapse analysis of steel offshore structures*, Theory Manual, Report STF71 F88038, 6th revision. SINTEF Structures and Concrete, Trondheim, 1994.

Taby, J. and Moan, T., Collapse and residual strength of damaged tubular members, *Behaviour of Offshore Structures*, BOSS'85, Amsterdam, 1985.

Taranath, B. S., *Structural analysis and design of tall buildings*. McGraw-Hill, Singapore, 1988.

Taranath, B. S., *Steel, concrete and composite design of tall buildings*, McGraw-Hill, New York, USA, 1998.

Taucer, F., Spacone, E., and Filippou, F. C., *A fiber beam-column element for seismic response analysis of concrete structures*, Report No. UCB/EERC-91/17, University of California, Berkeley, CA, 1991.

Teh, L. H. and Clarke, M. J., *The implications of linearisations, Approximations and computation algorithms in the second-order analysis of frames*, Research Report No. R733, Department of Civil Engineering, University of Sydney, Australia, 1996.

Teh, L. H. and Clarke, M. J., New definition of conservative internal moments in space frames, *Journal of Engineering Mechanics*, ASCE, 123(2), 97–106, 1997.

Teh, L. H. and Clarke, M. J., Corotational and Lagrangian formulations for elastic three dimensional beam finite elements, *Journal of Constructional Steel Research*, 48, 123–144, 1998.

Teh, L. H. and Clarke, M. J., Symmetry of tangent stiffness matrices of 3D elastic frame, *Journal of Structural Engineering*, ASCE, 125(2), 248–251, 1999a.

Teh, L. H. and Clarke, M. J., Plastic zone analysis of 3D steel frames using beam elements, *Journal of Structural Engineering*, ASCE, 125(11), 1328–1337, 1999b.

Teh, L. H., Cubic beam elements in practical analysis and design of steel frames, *Engineering Structures*, 23, 1243–1255, 2001.

Teh, L. H., Hsiao, K. M., White, D. W., and Ziemian, R. D., Exact tangent stiffness for imperfect beam-column members, *Journal of Structural Engineering*, ASCE, 127(12), 1490–1492, 2001.

Timoshenko, S. P. and Gere, J. M., *Theory of elastic stability*, 2nd ed., McGraw-Hill, New York, 1961.

Toma, S., Chen, W. F., and White, D. W., A selection of calibration frames in North America for second-order inelastic analysis, *Engineering Structures*, 17(2), 104–112, 1995.

Trahair, N. S. and Chan, S. L., Out-of-plane advanced analysis of steel structures, *Engineering Structures*, 25, 1627–1637, 2003.

Tschemmernegg, F. and Huber, G., *Joint transformation*, COST C1/WD2-96-09, 1996.

Ueda, Y., Rashed, S. M. H., and Nakacho, K., New efficient and accurate method of nonlinear analysis of offshore tubular frames (the idealized structural unit method), *Journal of Energy Resources Technology*, 107, 204–201, 1985.

Ueda, Y. and Rashed, S. M. H., Modern method of ultimate strength analysis of offshore structures, *International Journal of Offshore and Polar Engineering*, 1(1), 27–41, 1991.

ULTIGUIDE, *Best practice guidelines for use of non-linear analysis methods in documentation of ultimate limit states for jacket type offshore structures*, Prepared by DNV-SINTEF-BOMEL, 1999.

Uy, B. and Liew, J. Y. R., *Chapter 51: Composite steel-concrete structures, In civil engineering handbook: 2nd Edition*, Edited by Chen, W. F. and Liew, J. Y. R., CRC Press, Boca Raton, Florida, 2002.

Uy, B., Strength of short concrete filled steel box columns incorporating local buckling, *Journal of Structural Engineering*, ASCE, 126(3), 341–352, 2000.

Uy, B., Local and post-local buckling of fabricated steel and composite cross sections, *Journal of Structural Engineering*, ASCE, 127(6), 666–677, 2001.

van der Vegte, G. J. and Makino, Y., Numerical simulations of bolted connections: the implicit versus the explicit approach, *Proceedings of Connections in Steel Structures V: Behaviour, Strength and Design*, Amsterdam, Netherlands, June 3–4, 2004.

Ultimate Strength, *Proceedings of 16th International Ship and Offshore Structures Congress*, 20-25 Aug 2006, Southampton, UK, 2006.

USFOS, *Ultimate Strength for Framed Offshore Structures*, http://www.usfos.com, 2009.

Van Long, H., Hung, N. D., Local buckling check according to Eurocode-3 for plastic-hinge analysis of 3-D steel frames, *Engineering Structures*, 30, 3105–3113, 2008.

Viest, I. M., Colaco, J. P., Furlong, R. W., Griffis, L. G., Leon, R. T., and Wyllie, L. A., *Composite construction design for buildings*, McGraw Hill, New York, 1997.

Vlasov, V. Z., *Thin-walled elastic beams*, 2nd ed. Moscow, USSR; English translation, Israel Program for Scientific Translation, Jerusalem, Israel, 1961.

Vrcelj, Z., Bradford, M. A., and Uy, B., Elastic distortional buckling modes in continuous composite beams, *Proceedings of the 16th Australasian Conference on the Mechanics of Structures and Materials*, Sydney, Balkema, 327–333, 1999.

Vrcelj, Z. and Uy, B., Strength of slender concrete-filled steel box columns incorporating local buckling, *Journal of Constructional Steel Research*, 58(2), 275–300, 2002.

White, D. W., *Material and geometric nonlinear analysis of local planar behavior in steel frames using interactive computer graphics*, Report No. 86–4, Department of Structural Engineering, Cornell University, Ithaca, NY., 1986.

White, D. W. and Chen, W. F., *Introduction, in plastic hinge based methods for advanced analysis and design of steel frames—An assessment of the state-of-the-art*, Structural Stability Research Council, Lehigh University, Bethlehem, PA., 1–22, 1993.

White, D. W. Comprehensive performance assessment of building structural systems: research to practice, *Engineering Structures*, 18(10), 778–785, 1996.

White, D. W., Surovek, A. E., Alemdar, B. N., Chang, C-J., Kim, Y. D., and Kuchenbecker, G. H., Stability analysis and design of steel building frames using the 2005 AISC Specification, *Steel Structures*, 6, 71–91, 2006.

Wong, M. B. and Tin-Loi, F., Yield surface linearization in elastoplastic analysis, *Computers and Structures*, 26(6), 951–956, 1987.

Wongkaew, K. and Chen, W. F., Consideration of out-of-plane buckling in advanced analysis for planar steel frame design, *Journal of Constructional Steel Research*, 58, 943–965, 2002.

Yang, Y. B. and McGuire, W., Stiffness matrix for geometric nonlinear analysis, *Journal of Structural Engineering*, ASCE, 112(4), 853–877, 1986a.

Yang, Y. B. and McGuire, W., Joint rotation and geometric nonlinear analysis, *Journal of Structural Engineering*, ASCE, 112(4), 879–905, 1986b.

Yang, Y. B. and Fan, H. T., Yield surface modelling of I-sections with non-uniform torsion, *Journal of Structural Engineering*, ASCE, 114(6), 953–972, 1988.

Yang, Y. B., Chern, S. M., and Fan, H. T., Yield surface for I-sections with bimoments, *Journal of Structural Engineering*, ASCE, 115(12), 3044–3058, 1989.

Yang, Y. B., Kuo, S. R., *Theory and analysis of nonlinear framed structures*, Prentice Hall, Singapore, 1994.

Yang, Y. B., Leu, L. L., Nonlinear stiffnesses in analysis of planar frames, *Computer Methods in Applied Mechanics and Engineering*, 117, 233–247, 1994.

Yang, Y. B., Kuo, S. R., Wu, Y. S., Incrementally small-deformation theory for nonlinear analysis of structural frames, *Engineering Structures*, 24, 783–798, 2002.

Yura, J., Zettlemoyer, N., and Edwards, I. F., Ultimate capacity equation for circular tubular joints, *Proc of Offshore Technology Conference*, Paper OTC 3690, Texas, 113–126, 1980.

Zayas, V. A., Mahin, S. A., and Popov, E. P., *Cyclic inelastic behavior of steel offshore structures*, University of California. UCB/EERC-80/27. Aug., 1980.

Ziemian, R. D., *Advanced methods of inelastic analysis in the limit states design of steel structures*, Thesis presented to Cornell University, in partial fulfillment of the requirements for the degree of Doctor of Philosophy, Ithaca, NY., 1990.

Ziemian, R. D., McGuire, W., Deierlein, G. G., Inelastic limit states design. Part I: Planar frames studies. Part II. Three-dimensional frame design, *Journal of Structural Engineering*, ASCE, 118(9), 2532–2567, 1992.

Ziemian, R. D., McGuire, W., Modified tangent modulus approach, a contribution to plastic hinge analysis, *Journal of Structural Engineering*, ASCE, 128(10), 1301–1307, 2002.

Appendix 1 Elastic Stiffness Matrix, $\mathbf{k_e}$

$$\mathbf{k_e} = \begin{bmatrix} k_{e1} & k_{e2} & -k_{e1} & k_{e2} \\ & k_{e3} & -k_{e2}^T & k_{e4} \\ & & k_{e1} & -k_{e2} \\ Sym. & & & k_{e3} \end{bmatrix} \tag{A1.1}$$

in which:

$$\mathbf{k_{e1}} = \begin{bmatrix} EA/L & 0 & 0 \\ 0 & 12EI'_z/L^3 & 0 \\ 0 & 0 & 12EI'_y/L^3 \end{bmatrix} \tag{A1.2}$$

$$\mathbf{k_{e2}} = \begin{bmatrix} 0 & 0 & 0 \\ 0 & 0 & 6EI'_z/L^2 \\ 0 & -6EI'_y/L^2 & 0 \end{bmatrix} \tag{A1.3}$$

$$\mathbf{k_{e3}} = \begin{bmatrix} GJ/L & 0 & 0 \\ 0 & (4+\eta_y)\,EI'_y/L & 0 \\ 0 & 0 & (4+\eta_z)\,EI'_z/L \end{bmatrix} \tag{A1.4}$$

$$\mathbf{k_{e4}} = \begin{bmatrix} -GJ/L & 0 & 0 \\ 0 & (2-\eta_y)\,EI'_y/L & 0 \\ 0 & 0 & (2-\eta_z)\,EI'_z/L \end{bmatrix} \tag{A1.5}$$

where $EI'_n = \dfrac{EI_n}{1+\phi_n}$, $\eta_y = \dfrac{12EI_y}{GA_zL^2}$, $\eta_z = \dfrac{12EI_z}{GA_yL^2}$, and A_n is the shear area along axis n. If the effect of shear deformation is ignored, $\eta_y = \eta_z = 0$. L is the element length at C_1 based on the UL formulation.

Appendix 2 Geometric Stiffness Matrix, $\mathbf{k}_g$

$$\mathbf{k}_g = \begin{bmatrix} k_{g1} & k_{g2} & -k_{g1} & k_{g3} \\ & k_{g4} & k_{g5} & k_{g6} \\ & & k_{g1} & -k_{g3} \\ Sym. & & & k_{g4} \end{bmatrix} \tag{A2.1}$$

where:

$$\mathbf{k}_{g1} = \begin{bmatrix} 0 & \left(M_{za}+M_{zb}\right)/L^2 & -\left(M_{ya}+M_{yb}\right)/L^2 \\ & \phi_{1z}+F_{xb}/L & 0 \\ Sym. & & \phi_{1y}+F_{xb}/L \end{bmatrix} \tag{A2.2}$$

$$\mathbf{k}_{g2} = \begin{bmatrix} 0 & 0 & 0 \\ \underline{M_{ya}/L} & \underline{M_{xb}/L} & \phi_{2z} \\ \underline{M_{za}/L} & -\phi_{2y} & \underline{M_{xb}/L} \end{bmatrix} \tag{A2.3}$$

$$\mathbf{k}_{g3} = \begin{bmatrix} 0 & 0 & 0 \\ \underline{M_{yb}/L} & \underline{-M_{xb}/L} & \phi_{2z} \\ \underline{M_{zb}/L} & -\phi_{2y} & \underline{-M_{xb}/L} \end{bmatrix} \tag{A2.4}$$

$$\mathbf{k}_{g4} = \begin{bmatrix} \bar{K}/L & \underline{\left(M_{za}+M_{zb}\right)/6} & \underline{-\left(M_{ya}+M_{yb}\right)/6} \\ & \phi_{3y} & 0 \\ Sym. & & \phi_{3z} \end{bmatrix} \tag{A2.5}$$

$$\mathbf{k}_{g5} = \begin{bmatrix} 0 & \underline{-M_{ya}/L} & \underline{-M_{za}/L} \\ 0 & \underline{-M_{xb}/L} & \phi_{2y} \\ 0 & -\phi_{2z} & \underline{-M_{xb}/L} \end{bmatrix} \tag{A2.6}$$

$$\mathbf{k}_{g6} = \begin{bmatrix} -\bar{K}/L & \underline{-\left(M_{za}+M_{zb}\right)/6} & \underline{\left(M_{ya}+M_{yb}\right)/6} \\ \underline{-\left(M_{za}+M_{zb}\right)/6} & -\phi_{4y} & \underline{M_{xb}/2} \\ \underline{\left(M_{ya}+M_{yb}\right)/6} & \underline{-M_{xb}/2} & -\phi_{4z} \end{bmatrix} \tag{A2.7}$$

$$\phi_{1n} = 2EI_n\left(S_{1n}+S_{2n}-6\right)/L^3 \tag{A2.8}$$

$$\phi_{2n} = EI_n\left(S_{1n}+S_{2n}-6\right)/L^2 \tag{A2.9}$$

$$\phi_{3n} = EI_n\left(S_{1n}-4\right)/L \tag{A2.10}$$

$$\phi_{4n} = EI_n\left(2-S_{2n}\right)/L \tag{A2.11}$$

In the above matrices, the underlined coupling terms are lost in geometric stiffness matrix, $\mathbf{k}'_g$, based on the beam-column approach. L is the element length at C_1 based on the UL formulation.

Appendix 3 Bowing Matrix, $\mathbf{k}_\mathbf{b}$

$$\mathbf{k}_\mathbf{b} = \mathbf{k}_{\mathbf{b1}} + \mathbf{T}^T \mathbf{k}_{\mathbf{b2}} \mathbf{T} \tag{A3.1}$$

in which:

$$\mathbf{k}_{\mathbf{b1}} = Diag\left[\frac{EI}{HL_0^3} - \frac{EA}{L_0},\ 0,\ 0,\ 0,\ 0,\ 0,\ \frac{EI}{HL_0^3} - \frac{EA}{L_0},\ 0,\ 0,\ 0,\ 0,\ 0 \right] \tag{A3.2}$$

$$\mathbf{k}_{\mathbf{b2}} = \frac{EI}{HL_0} \begin{bmatrix} G_{1z}^2 & G_{1z}G_{2z} & G_{1z}G_{1y} & G_{1z}G_{2y} & 0 & G_{1z}/L_0 \\ & G_{2z}^2 & G_{2z}G_{1y} & G_{2z}G_{2y} & 0 & G_{2z}/L_0 \\ & & G_{1y}^2 & G_{1y}G_{2y} & 0 & G_{1y}/L_0 \\ & & & G_{2y}^2 & 0 & G_{2y}/L_0 \\ & & & & 0 & 0 \\ Sym. & & & & & 0 \end{bmatrix} \tag{A3.3}$$

$$\mathbf{T} = \begin{bmatrix} 0 & 1/L & 0 & 0 & 0 & 1 & 0 & -1/L & 0 & 0 & 0 & 0 \\ 0 & 1/L & 0 & 0 & 0 & 0 & 0 & -1/L & 0 & 0 & 0 & 1 \\ 0 & 0 & -1/L & 0 & 1 & 0 & 0 & 0 & 1/L & 0 & 0 & 0 \\ 0 & 0 & -1/L & 0 & 0 & 0 & 0 & 0 & 1/L & 0 & 1 & 0 \\ 0 & 0 & 0 & -1 & 0 & 0 & 0 & 0 & 0 & 1 & 0 & 0 \\ -1 & 0 & 0 & 0 & 0 & 0 & 1 & 0 & 0 & 0 & 0 & 0 \end{bmatrix} \tag{A3.4}$$

$$I = I_y + I_z \tag{A3.5}$$

$$G_{1n} = S'_{1n}\Theta_{na} + S'_{2n}\Theta_{nb} + c'_{0n}\left(\delta_n / L_0\right) \tag{A3.6}$$

$$G_{2n} = S'_{2n}\Theta_{na} + S'_{1n}\Theta_{nb} - c'_{0n}\left(\delta_n / L_0\right) \tag{A3.7}$$

$$S'_{1n} = 2\left(b_{1n} + b_{2n}\right) \tag{A3.8}$$

$$S'_{21n} = 2\left(b_{1n} - b_{2n}\right) \tag{A3.9}$$

$$c'_{0n} = b_{vsn} \tag{A3.10}$$

$$H = \frac{1}{AL_0^2} - \sum_{n=z,y} \frac{I}{I_n}\left[b'_{1n}\left(\Theta_{na} + \Theta_{nb}\right)^2 + b'_{2n}\left(\Theta_{na} - \Theta_{nb}\right)^2 + b'_{vsn}\left(\delta_n / L\right)\left(\Theta_{na} - \Theta_{nb}\right) + b'_{vvn}\left(\delta_n / L_0\right)^2 \right] \tag{A3.11}$$

In the above equations a prime superscript denotes the differentiation with respect to q_n, L_0 is the initial member length, L is the element length at C_1 based on the UL formulation.

7

Case Studies for Second-order (Direct) Analysis of Semi-rigid Frames in Hong Kong

Y. P. Liu
Department of Civil and Structural Engineering, The Hong Kong Polytechnic University, Hung Hom, Kowloon, Hong Kong.

Dennis Lam
Department of Civil Engineering, University of Leeds, U.K.

S. L. Chan
Department of Civil and Structural Engineering, The Hong Kong Polytechnic University, Hung Hom, Kowloon, Hong Kong.

7.1 Introduction

After decades of research by Liew et al. (1993), Zhou and Chan (1995), White and Hajjar (2000), Hetao and Li (2009), and others, the technology of the simulation-based design of steel structures has been matured and is practically feasible with recommendations given in various codes such as The Code of Practice for Structural Uses of Steel Hong Kong (COPHK, 2005) and the Eurocode 3 (CEN, 2005). While the theoretical background, previous works, and research in the topic of semi-rigid frames have been described in detail in previous chapters, the report of second-order analysis of semi-rigid frames is rather limited in literature because of several issues. First, the stiffness, ductility, and strength of the connections are not comprehensive and engineers are not fully aware of the tested data in the appendix of this book. Second, code requires engineers to consider either strength for rigid connections or ductility for rotational ductility for pinned connections. However, for elastic-plastic or advanced analysis of semi-rigid frames, both these effects are required in checking the adequacy of connections. Finally, the computational technique for semi-rigid connections is not yet well known to engineers.

It has been well recognized that semi-rigid connection not only provides a more accurate picture on the actual response of a structure, it also provides a more economical solution to a design because the moments at midspan and ends of a beam are more balanced. This chapter details some of the previous work by the authors on the design of bare steel frames in Hong Kong and Macau using the concept and the method.

One hemispherical dome constructed in Hong Kong and one shallow roof built in Macau are studied in Examples 1 and 2, respectively. Both are predominantly under strong wind and the use of the elastic critical load factor for frame classification is rather irrelevant. The buckling length of the member is difficult to estimate as the ends of the member are displaced. When using the proposed second-order or direct analysis in LRFD terminology, the member resistance can be computed directly by a simulation process, and the sectional strength is directly checked.

7.2 Second-order Integrated Design and Analysis

The conventional linear design method divides the whole design procedure into two stages: (1) determination of the internal forces and moments acting on each member of the structural system by elastic linear analysis, and (2) assessment of the strength and stability of each member treated as an isolated member by plastic analysis. Compatibility between the isolated member and the structural system is doubtable. There is an increase in awareness for the use of second-order analysis that simulates directly the behavior of structural members, connections, and other components in the determination of overall system response.

The new and advanced nonlinear integrated design and analysis method is significantly different from the conventional linear design since the nonlinear analysis model contains more factors, which may significantly affect the structure behaviors. Also, the interaction between the structural members and the structural system can be considered. Generally speaking, the nonlinear design method is a system-based approach in contrast to the traditional member-based design method. The special features of the second-order analysis are that the P-Δ and P-δ effects as well as the initial imperfections should be considered in the analysis stage.

7.2.1 P-Δ and P-δ Effects

When a structure deforms grossly, the original geometry can no longer be employed for the formulation of the transformation matrix simply because the coordinates have been changed. This effect, named the P-Δ effect, may become important when the deflection or the conjugate force is large, such as the case of a building under a heavy mass at the roof and a lateral wind load. An additional moment as the P-Δ moment will be induced due to this effect.

The P-δ effect is referred to as the second-order effect due to the deflection along a member and the axial force. It affects the state of stress as well as the stiffness of the member. Like the P-Δ effect, an additional moment named P-δ moment will be induced due to the P-δ effect. Its careful consideration is important for buckling analysis and design of slender skeletal structures.

In general, both the P-Δ (frame sidesway) and P-δ effects (member curvature) will occur in a structure under vertical and horizontal external forces. These effects are shown in Figure 7.1.

In this chapter, the P-Δ effect is automatically considered in the nonlinear incremental-iterative procedure while the P-δ effect is accounted for by the use of the Pointwise Equilibrium Polynomial (PEP) element (Chan and Zhou, 1995) at the element level.

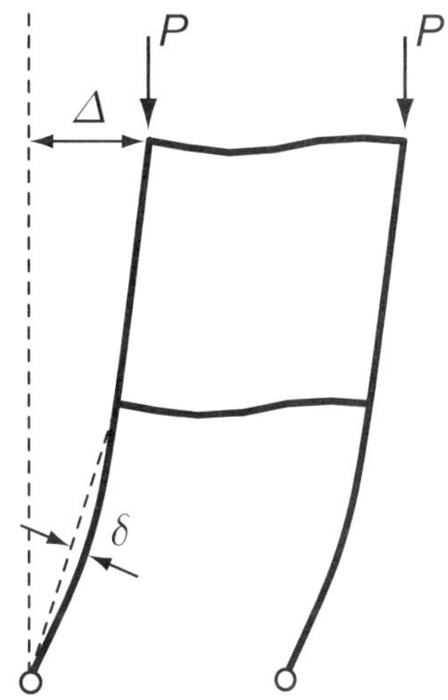

FIGURE 7.1 The P-Δ and P-δ effects.

7.2.2 Initial Imperfections

As there are no perfect structures or structures that are free from defects due to initial crookedness and residual stress, imperfections must be considered when using the second-order analysis and advanced analysis (COPHK, 2005) for practical design. Two types of imperfections should be included in the nonlinear analysis and design (i.e., the member and frame imperfections).

7.2.2.1 Member Imperfections

Member initial imperfections are due to either initial geometric imperfections or residual stresses or both. The initial geometric imperfections of members may be due to one or several aspects such as cambering, sweeping, twist, out-of-straightness and cross-section distortion. The residual stresses in members may be due to manufacturing and fabrication processes, erection out-of-fit, and construction sequencing. To account rigorously and exactly for all the imperfections seems impossible. Practically, they can be simulated in the analysis or design model by the equivalent initial bow imperfection, which may be slightly different in the national design codes due to the difference in steel products.

According to COPHK (2005), the equivalent initial bow imperfections for different sections are specified in Table 6.1. These values may be used in a second-order analysis for the steel members under compression. For composite column, the buckling curves and member imperfections are specified in Table 10.13 of COPHK (2005).

7.2.2.2 Frame Imperfections

The frame imperfections are mainly due to the out of plumbness of frame and column in the erection processes and construction sequence. This type of imperfection may increase the sway effect and induce P-Δ moments, especially when the structure is subjected to large vertical loads.

First-order linear analysis uses the moment amplification to enlarge the linear moment for sway effect, which can be due to wind load or notional force normally taken as 0.5 percent for permanent structures and 1 percent for temporary structures.

In second-order analysis, wind load or notional force are still used, but an alternative, more reliable, and convenient method is to use the elastic buckling mode as the imperfection mode with amplitude set equal to the out-of-plumbness normally taken as 1/200 of building height according to COPHK (2005) or other justified values.

7.3 Modeling of Semi-rigid Jointed Members

A semi-rigid connection must have the required strength, stiffness, and ductility. Some typical moment versus rotation (M-θ_r) curves and various analytical and mathematical models representing the M-θ_r relationships can be found in previous chapters. Conveniently, a joint can be considered in an analysis as dimensionless with the location at the intersection of the element center lines. Further, a rotational spring element satisfying the M-θ_r relationship is inserted into each end-of-beam element to model the connection behavior. The joint equilibrium condition can be expressed as:

$$M_e + M_i = 0 \tag{7.1}$$

in which M_e and M_i are the moments at the two ends of a connection (see Figure 7.2a). The corresponding external node is connected to the global node, and the internal node is joined to the beam element.

The stiffness of the connection S can be related to relative rotations at the two ends of the connection spring as:

$$S = \frac{M_e}{(\theta_e - \theta_i)} = \frac{M_i}{(\theta_i - \theta_e)} \tag{7.2}$$

where θ_e and θ_i are the conjugate rotations for the moments M_e and M_i (see Figure 7.2b). Rewriting Equation 7.2 in matrix form, the stiffness matrix of a connection spring can be written as:

$$\begin{bmatrix} S & -S \\ -S & S \end{bmatrix} \begin{bmatrix} \theta_e \\ \theta_i \end{bmatrix} = \begin{bmatrix} M_e \\ M_i \end{bmatrix} \tag{7.3}$$

A typical beam element bending stiffness matrix can be expressed as:

$$\begin{bmatrix} k_{11} & k_{12} \\ k_{21} & k_{22} \end{bmatrix} \begin{bmatrix} \theta_{1i} \\ \theta_{2i} \end{bmatrix} = \begin{bmatrix} M_{1i} \\ M_{2i} \end{bmatrix} \tag{7.4}$$

in which k_{ij} is the stiffness coefficient of a prismatic beam. Here, the imperfect PEP element proposed by Chan and Zhou (1995) is adopted and more details can be referred to the original reference. Therefore, a hybrid element can be obtained by directly adding the connection stiffness of the two ends to the PEP element bending stiffness matrix as:

$$\begin{bmatrix} S_1 & -S_1 & 0 & 0 \\ -S_1 & k_{11}+S_1 & k_{12} & 0 \\ 0 & k_{21} & k_{22}+S_2 & -S_2 \\ 0 & 0 & -S_2 & S_2 \end{bmatrix} \begin{bmatrix} \theta_{1e} \\ \theta_{1i} \\ \theta_{2i} \\ \theta_{2e} \end{bmatrix} = \begin{bmatrix} M_{1e} \\ M_{1i} \\ M_{2i} \\ M_{2e} \end{bmatrix} \tag{7.5}$$

in which the first subscript refers to node 1 or node 2. The internal degrees of freedom of the stiffness expression can be eliminated by a standard static condensation procedure. The stiff-

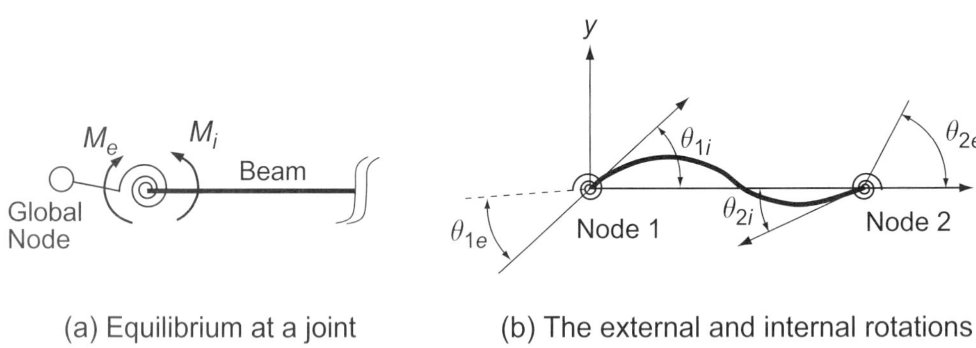

FIGURE 7.2 Modeling of semi-rigid jointed member.

ness expression of a beam element with both ends connected to a pair of springs can be finally written as:

$$[k_e][\theta_e]=[M_e] \tag{7.6}$$

in which:

$$[\theta_e]=\begin{bmatrix} \theta_{1e} & \theta_{2e} \end{bmatrix}^{\mathrm{T}},\ [M_e]=\begin{bmatrix} M_{1e} & M_{2e} \end{bmatrix}^{\mathrm{T}} \tag{7.7}$$

$$[k_e]=\frac{1}{\beta}\begin{bmatrix} S_1\beta - S_1^2(S_2+k_{22}) & k_{12}S_1S_2 \\ k_{21}S_1S_2 & S_2\beta - S_2^2(S_1+k_{11}) \end{bmatrix} \tag{7.8}$$

and β is given by:

$$\beta = \det\begin{vmatrix} k_{11}+S_1 & k_{12} \\ k_{21} & k_{22}+S_2 \end{vmatrix} \tag{7.9}$$

where S_1 and S_2 are the connection stiffness at nodes 1 and 2, respectively. When the spring stiffness S_i is zero, it means that the corresponding end is pinned end; when the spring stiffness S_i is infinite, it means that the corresponding end is the rigid end. For the semi-rigid case, the spring stiffness S_i can be determined by the given M-θ_r function.

7.4 Examples

7.4.1 Example 1: The Hemispherical Dome on the Roof of a Building

The design of a single-layered semispherical dome shaped steel roof of a building is described here. This type of dome structure is a common architectural feature, and is usually slender, because of aesthetic requirement, and under low loads. However, they can be under high wind load as they are widely constructed in open areas and at the roof of high-rise buildings. This structural form always poses a challenge to the structural engineer for special requirement as most of them cannot be classified as standard or typical structures such that the use of many formulae in codes are inapplicable. For example, the method of frame classification for rectilinear frames cannot be used here as buckling is due to horizontal wind loads causing tension in members on one side and compression in members on the other side, which is in contrast to typical building frames under gravitational dead and live loads. On the other hand, this type of dome structure is quite common to-date for construction as aesthetic and monumental building construction. The computer image of the dome is shown in Figure 7.3.

FIGURE 7.3 The single-layered semispherical dome at the roof of a building.

7.4.1.1 Structure Description

The dome frame is located at the roof top and a multistory building with a height of 178.25 m above mean sea level, where as the top roof of a services apartment and the members of the dome are circular hollow sections with a single bolt at the ends and connected to steel balls. The diameter of the dome is 18.95 m above concrete plinth level and the main member of the dome is 139.7 × 5 mm circular hollow sections of grade S235 steel.

Two inner frames are constructed with location at the top and the bottom levels of the dome in order to enhance the rigidity of the frame, and these inner frames are made of 48.3 × 3.2 mm-thick CHS sections of Grade S235. The dome is further supported by a ring beam of 254 × 254 × 132 UC of Grade S275 steel with 8 nos. of 457 × 16 mm-thick CHS sections of Grade S355 below the ring beam as the steel column supporting the dome frame. The 8 nos. of steel column are supported by the existing reinforced concrete (RC) beams. In addition, a 7 m-long steel pole is fixed at the top of the dome.

The main members of the dome are connected by a bolted ball joint, and the ball joint, with multiconnection holes, is adopted as semi-rigid connection of the main members. However, the ball joint at the bottom layer of the dome is connected to the ring beam by the bolt and assumed as a pinned connection. The circular steel column is connected to the ring beam by welding and the other end of the column is connected to the RC plinth by cast-in grade 8.8 steel bolts. In structural analysis, the steel columns are assumed to be pinned at the base and supported on the existing RC beam.

The pinned assumption for the dome leads to a mechanism, and therefore a larger size bolt of diameter, an M27 bolt, is used and assumed as a semi-rigid joint providing stability to the complete frame. The stiffness of this connection made of an M27 bolt is taken as $4EI/L$ in which E is the Young's modulus of elasticity for steel, I is the second moment of area of the bolt, and L is the extruded bolt length.

7.4.1.2 Load Path

Wind load and live load act along the members together with their self-weight. The applied loads from the main members will be transferred to the ring beam of the dome. The loading of the ring beam will then be passed to the steel columns and then to the existing RC elements. It is assumed the connections between main members of the dome are pinned, rigid, and semi-rigid connections. All base connections to concrete plinth are assumed pinned in order to transfer minimum moment to the supporting concrete structure.

7.4.1.3 Computer Method of Analysis

The three-dimensional analysis model is used in determining the forces in members and in connections. The structure is analyzed by second-order first-plastic hinge (P-Δ-δ) analysis. The first-order linear analysis requires assumption of effective length, which cannot be determined accurately and scientifically in the present structural form. Further, the snap-through type of buckling for modules under wind involves significant deformation, which violates the underlying assumption for the effective length method as the frame does not deform until elastic eigen-buckling occurs. Once the force is applied onto the frame, its members deflect and the second-order stress magnifies. The structure is imperfection sensitive and therefore the eigen-buckling load carries little physical significance here.

In the second-order analysis used by the authors, both the imperfections at local member and global frame levels are considered with values adopted directly from the code respectively as span/500 and height/200. These numbers are input into the software NIDA (2008), which automatically considers the effects by assuming appropriate member curvature and imperfect frame geometry before load is applied to the frame (Chan and Zhou, 2004). For checking of structural stability and adequacy, the design load under combined and factored load cases is used and applied to the frame in the analysis. The cross-sectional strength check in Equation 1 below is then applied to monitor the structural adequacy at every section along all members. For the first-plastic-hinge analysis, any sectional strength failure indicates the inadequacy of the frame,

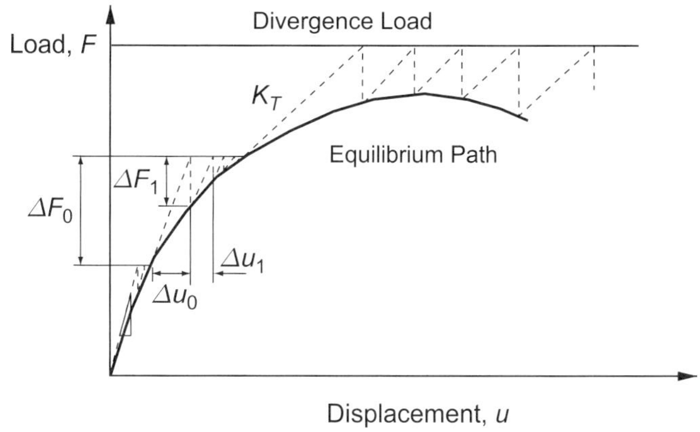

FIGURE 7.4 The constant load Newton-Raphson method.

and the member size or steel grade must be increased. Figure 7.4 shows the process of checking for stability and strength by the Newton-Raphson method. Note that this approach is adopted to deal with many combined load cases with each combined load case considered in a single and separate analysis. The method is used as a standard design procedure for any member checked under the ultimate-limit state.

$$\frac{F_c}{A_g P_y} + \frac{M_x + F_c\left(\Delta_x + \delta_x\right)}{M_{cx}} + \frac{M_y + F_c\left(\Delta_y + \delta_y\right)}{M_{cy}} = \phi \le 1 \tag{7.10}$$

in which F_c is the axial force, A_g is the cross-sectional area, p_y is the design strength, M_{cx} and M_{cy} are the moment capacities about the principal x- and y-axes, M_x and M_y are the bending moment about the principal axes, Δ and δ are the global frame and local member second-order deflection, and ϕ is the section capacity factor.

When the advanced analysis allowing for plastic hinges is used for determination of the ultimate resistance of the structure, the yielded section is assumed to possess a plastic hinge when zero tangential stiffness (taken as 0.1 percent of the elastic stiffness in computer to avoid numerical overflow), and plastic moment—reduced for presence of axial force is assumed in the section that can carry no further load with limiting capacity equal to the instance of force and moment in Equation 7.10. Owing to the change of geometric and material properties during a load increment, the applied force will not be absolutely balanced by the resistance. The iterations are needed to dissipate the unbalanced forces between the applied (action) and resistance forces. This unbalanced force is then used as a new set of external loads applied to the structure with the process repeated until the norm of the unbalanced forces is sufficiently small or less than 0.1 percent in our studies. The process is repeated and incremented until the collapse load is reached with the collapse load factor determined. The procedure is adopted for progressive collapse analysis of finding the true collapse load and the identification of key elements, which are defined as members with their failure causing more than a certain area of collapse. In some codes, the area is taken as 70 m^2.

7.4.1.4 Design Code

7.4.1.4.1 Material Properties

There are three types of structural steel, which should be conformed to the requirements of COPHK (2005) and are used for the steel members as shown in Table 7.1.

TABLE 7.1 Material properties of steel

Material	Elastic modulus (kN/m²)	Poisson's Ratio	Design strength p_y (kN/m²)	Unit weight (kN/m³)
S235	2.05×10^8	0.3	2.35×10^5	77
S275	2.05×10^8	0.3	2.75×10^5	77
S355	2.05×10^8	0.3	3.55×10^5	77

Note:
1. All welds must be compatible or over-matching to steel grade used
2. Design strength p_y for Hot Rolled section and other welded plates

Steel grade	Thickness less than or equal to (mm)	Design strength p_y (N/mm²)
S235	16	235
	40	225
S275	16	275
	40	265
S355	16	355
	40	345

3. For welded section under compression, p_y should be reduced by 20 N/mm² for reduction due to welding residual stress

7.4.1.4.2 Section Assignment and Connections

The size of members is CHS 139.7 × 5 for those in the dome, CHS219.1 × 8 for the pole, CHS48.3 × 3.2 for beams, and UC254 × 254 × 132 for columns.

For the steel bolts, the ISO grade 8.8 bolts are used and the weld grade of E42, in compliance with the code requirements, is adopted.

7.4.1.4.3 Load Combination

The basic load cases include dead loads, live loads, wind loads, and temperature loads. The average temperature in Hong Kong is coded from +0.1°C to +40.0°C. The basic design wind pressure is 3.2 kPa for the height 178.25 m above mean sea level according to the HKWC (2004). A shape pressure coefficient C_p of 2.0 for individual members of an open framework building is multiplied to the basic design wind pressure.

According to COPHK (2005), the following partial load factors and combinations for normal ultimate-limit state shall be applied to strength and stability for normal design situations. Table 7.2 is a summarization of the partial load factors and combination.

7.4.1.5 Boundary Conditions

The 8 nos. of steel columns made of 457 × 16 mm-thick CHS sections of Grade S355 are pinned to the RC beams and columns because the concrete cannot resist the large moments from the base of stanchions.

7.4.1.6 Connection Stiffness

A sensitivity study for elastic critical load factor λ_{cr} against the rigidity of the ball joint of the dome is carried out to investigate the effects of rotational stiffness of the joint. The computed results are summarized in Table 7.3.

TABLE 7.2 Partial load factors according to COPHK (2005)

Load combination	Partial load factors for load type					
	Dead G_k		Imposed Q_k		Wind W_k	Temperature T_k
	Adverse	Beneficial	Adverse	Beneficial		
1. Dead and imposed	1.4	1.0	1.6	0	—	1.2
2. Dead and lateral	1.4	1.0	—	—	1.4	—
3. Dead, lateral and imposed	1.2	1.0	1.2	0	1.2	1.2

TABLE 7.3 Numerical results

Connection type	λ_{cr}	Maximum end moment of the dome member (kNm)
All rigid	39.2	17.8
All pinned	0.08	0
Only bottom layer pinned and others are rigid	32	4.0
Semi-rigid with rotational stiffness = 135kNm/rad (M27 bolt)	7.8	1.3
Semi-rigid with rotational stiffness = 5kNm/rad (M12 bolt)	0.6	0.05
Semi-rigid with rotational stiffness = 1600kNm/rad (Hexagonal nut of diameter 50mm)	35.7	8.23

For semi-rigid connections with M27 bolts, the stiffness of semi-rigid connection is taken as 135 kNm/rad as the rotational stiffness of the joints. This value of joint stiffness is calculated from the length of cross-sectional diameter of the connecting bolt as $4EI/L$ of the bolt in which E is the Young's modulus of elasticity of steel, I is the second moment of area, and L is the length of the bolt. For rigid and pinned connections, respectively infinite and zero joint stiffness is assumed for the connection stiffness of every member.

The rotational stiffnesses are evaluated by $4EI/L$ where $L = 160$mm. The assumed length of bolt L is longer than the actual extruded length of the bolt in order to allow for fitting tolerance, which makes the connection more flexible.

All critical members are found in the bottom level of the dome under the combined load case of wind and dead loads.

7.4.1.7 Imperfections

There are no perfect structures or structures that are free from defects, due to initial crookedness and residual stress; imperfections must therefore be considered when using the second-order analysis and advanced analysis to COPHK (2005). Below is a description of codified consideration for member and for frame imperfections.

7.4.1.7.1 Member Imperfections

For hot-rolled hollow sections, Curve A in Table 6.1 in COPHK (2005) is used. Thus, member imperfection equal to span/500 is assumed. For other types of profiled steel section, a larger imperfection may be used and the values of these member types can be found by curve-fitting of buckling experimental curves or design curves from any codes.

7.4.1.7.2 Frame Imperfections
The out-of-plumbness equal to the height/200 to COPHK (2005) is assumed. For domes, the frame imperfection can also be taken as span/200 in accordance with the Chinese Technical Specification for latticed shells (JGJ61-2003).

7.4.1.8 Additional Vibration Frequency

According to clause 7.6 of the Code of Practice on Wind Effects in Hong Kong 2004 (HKWC, 2004), the natural frequency should be greater than 0.2 Hz for no resonance with wind action. The period of the whole dome frame is found to be 0.62 of a second (i.e., 1.61 Hz), and therefore it is acceptable.

7.4.1.9 Load Carrying Capacity

As mentioned before, three conditions for connection stiffness are assumed and their design adequacy is investigated. Only the case of semi-rigid connecting stiffness with a stiffness value equal to 135 kNm/radian is satisfactory for the design. Therefore, the M27 bolts are used. Further full-scale tests for connections are conducted to confirm the achievement of this value in the connection type.

It is of interest to note that the pinned assumption for connections is unable to provide a valid solution to the design because of low stiffness. On the other hand, the rigid assumption cannot be achievable in experimental tests. Further, because of the presence of large moment at member ends, the section capacity check cannot be passed for connecting members or the bolts. Therefore, the present approach represents a realistic assumption for the dome design. Also, no assumption for the effective length has been made for the complete design process.

7.4.2 Example 2: The Shallow Roof at the Top of a Building

7.4.2.1 Structure Description

The shallow roof steel frame, which is arch dome shaped with a 57.04 m span, 62.24 length, and 11.81 m height is located at the roof top of a three-story building. This steel roof is the supporting structure for the glass skylight and curtain walls at two ends (see Figures 7.5 and 7.6).

The arch main beams are of 660 mm height and designed for resisting the vertical loads such as dead loads, service loads, and wind loads. At the four edges of the roof, bracings are used to provide lateral stiffness and to resist the horizontal wind loads. The vertical loads are first transferred to the secondary purlins and then to the arch main beams. The lateral restraint due to glass panels is weak and is normally ignored. Therefore, the purlins are made of an RHS section so that the lateral-torsional buckling can be avoided.

It is noted that the roof structure cannot perfectly locate on the RC columns as the tensile strength of concrete is small. Thus, two continuous steel beams made of a box section of 650 mm height are used to support the arch main beams. The box section has a relatively large torsional constant and therefore the two continuous beams will not fail in torsional moments or by lateral-torsional buckling.

The curtain walls at two ends are of 11.81m height and subjected to large wind loads. It is easy to pass the ultimate-limit state checking, but the checking for serviceability limit state requires deep vertical beams to reduce the deflection. Therefore, the members in curtain walls are made of Chinese steel of Grade Q235 for economical design.

For architectural reasons, the arch main beams are continuous and the splice joints are made of full-strength butt weld. In such a case, the fully rigid connections for the arch main beams are

FIGURE 7.5 The single-layered shallow roof of a building (as built).

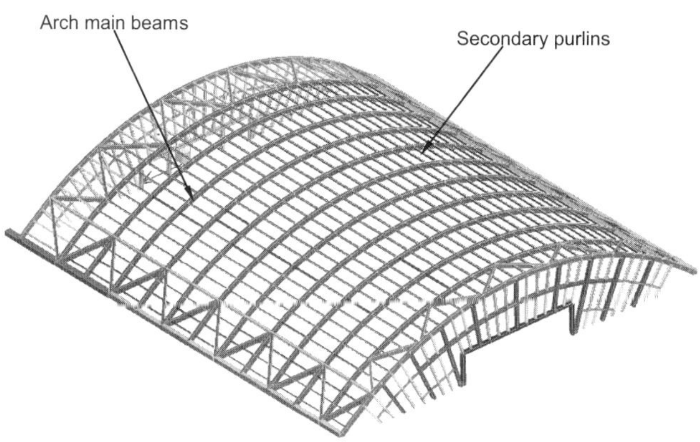

FIGURE 7.6 The single-layered shallow roof of a building (computer model).

reasonably in line with the current engineering practice. Also, welded connections rather than bolted connections are adopted for the connections between the secondary purlins to the arch main beams. For economical design, the welded connection is made of fillet weld, and therefore ultrasonic examination is avoided. However, the modeling of this type of welded connection in an analysis is debatable. Basically, two types of assumptions of this connection are widely made in practice (i.e., the fully rigid connection and the ideally pinned connection). This example aims to discuss the influence of the different modeling of this type of connection.

7.4.2.2 Load Path

For the skylight, the pressures due to the weight of glass panels and pipework, live loads, and wind loads are applied to the secondary purlins of rectangular hollow sections of galvanized mild steel, and then transferred to the arch main beams of the built-up galvanized H mild steel section. The service loads are added directly to the arch main beam as point loads.

For the curtain walls, the pressures due to the weight of glass panels and due to wind are applied to the secondary purlins and then transferred to the main arch beams, made of built-up galvanized mild steel H section.

The arch system is supported on the steel beams of box section and further pinned to the concrete structure. The two curtain walls are supported on concrete beams with the pinned assumption.

7.4.2.3 Computer Method of Analysis

The three-dimensional analysis model is used in determining the forces in members and in connections. The structure is analyzed by second-order elastic (P-Δ-δ) analysis. The same theory and procedure used in Example 1 are used in this example.

7.4.2.4 Design Code

7.4.2.4.1 Material Properties

Two types of structural steel conformed to the requirements of COPHK (2005) are used for the steel members as shown in Table 7.4.

7.4.2.4.2 Section Assignment and Connections

The size of the members are 660 × 350 × 20 × 12 I-section for the main beams, RHS250 × 150 × 5 for the secondary purlins, built-up box section 650 × 550 × 20 for two supporting beams, and 560 × 300 × 16 × 12 I-section for bracing members. The members in curtain walls are mainly made of RHS sections.

Fillet weld is adopted for the connections between arch main beams and the secondary purlins. All other member-to-member connections of the members with box sections are connected by fillet weld. The pinned supports for the two supporting beams are made of bolted joint.

TABLE 7.4 Material properties of steel

Material	Elastic modulus (kN/m^2)	Poisson's Ratio	Design strength p_y (kN/m^2)	Unit weight (kN/m^3)
Q235	2.05×10^8	0.3	2.15×10^5	77
Q345	2.05×10^8	0.3	3.10×10^5	77

Note:
1. All welds must be compatible or over-matching to steel grade used
2. Design strength p_y for Hot Rolled section and other welded plates

Steel grade	Thickness less than or equal to (mm)	Design strength p_y (N/mm2)
Q235	16	215
	35	205
Q345	16	310
	35	295

3. For welded section under compression, p_y should be reduced by 20 N/mm^2

7.4.2.4.3 Load Combination

The basic load cases include dead loads, live loads, wind loads, and temperature loads. The design temperature in Macau is set to be +30/−20°C for unprotected metal structures and +10/−10°C for structures with protection.

The wind pressure is obtained from a wind tunnel test due to the complex wind environment and the importance of structure. The wind pressure for wind tunnel test has a return period of 200 years.

According to loading code of Macau, the following partial load factors and combinations for the normal ultimate-limit state shall be applied to the strength and stability for normal design situations.

Table 7.5 summarizes the partial load factors and combination.

7.4.2.5 Boundary Conditions

All supports are pinned to the concrete beams and columns because the concrete cannot resist the large moments.

7.4.2.6 Connection Stiffness

As mentioned above, the modeling for the connections using fillet weld is controversial. For simplicity, only the connections between the secondary purlins and the arch main beams will be discussed here. The original design of this type of connection is assumed as pinned connections. To study the structural behavior more exactly, the assumption of semi-rigid connection is also used.

For semi-rigid connections, the stiffnesses of the semi-rigid connection are taken as 617 kNm/rad and 1360 kNm/rad for minor and major axes, respectively. The values of joint stiffnesses are directly calculated from *EI*/*L* of the member, in which *E* is the Young's modulus of elasticity of steel, *I* is the second moment of area, and *L* is the length of the member.

For the other member-to-member connections, fully rigid or ideally pinned are assumed.

7.4.2.7 Imperfections

There are no perfect structures or structures that are free from defects due to initial crookedness and residual stress. Imperfections must therefore be considered when using the second-order analysis and advanced analysis to COPHK (2005). Below is a description of codified consideration for member and for frame imperfections.

7.4.2.7.1 Member imperfections

For hot-rolled I-sections, Curves A and B of Table 6.1 in COPHK (2005) are used for major and minor axes, respectively. Thus, member imperfections equal to span/500 and span/400 are assumed,

TABLE 7.5 Partial load factors

<table>
<tr><th rowspan="2">Load combination</th><th colspan="6">Partial load factors for load type</th></tr>
<tr><th colspan="2">Dead G_k</th><th colspan="2">Imposed Q_k</th><th>Wind w_k</th><th>Temperature T_k</th></tr>
<tr><td></td><td>Adverse</td><td>Beneficial</td><td>Adverse</td><td>Beneficial</td><td></td><td></td></tr>
<tr><td>1. Dead and imposed</td><td>1.35</td><td>1.0</td><td>1.5</td><td>0</td><td>—</td><td>0.9</td></tr>
<tr><td rowspan="2">2. Dead and lateral</td><td rowspan="2">1.35</td><td rowspan="2">1.0</td><td rowspan="2">—</td><td rowspan="2">—</td><td>1.5</td><td>—</td></tr>
<tr><td>—</td><td>1.5</td></tr>
<tr><td rowspan="2">3. Dead, lateral and imposed</td><td rowspan="2">1.35</td><td rowspan="2">1.0</td><td>1.5</td><td rowspan="2">0</td><td>1.0</td><td>0.9</td></tr>
<tr><td>0.9</td><td>1.5</td><td>0.9</td></tr>
</table>

respectively. For other types of profiled steel section, a larger imperfection may be used and the values of these member types can be found by curve-fitting of buckling experimental curves or design curves from any other codes. Note that direct measurement of member imperfection is not usable here.

7.4.2.7.2 Frame Imperfections

The out-of-plumbness equal to the height/200 to COPHK (2005) is assumed. For domes, the frame imperfection can also be taken as span/200 in accordance with the Chinese Technical Specification for latticed shells (JGJ61-2003).

7.4.2.8 Additional Vibration Frequency

According to clause 7.6 of the Code of Practice on Wind Effects in Hong Kong 2004 (HKWC 2004), the natural frequency should be greater than 0.2 Hz for no resonance with wind action. The period of the whole dome frame is found to be 0.57 of a second (i.e., 1.74 Hz) and therefore it is acceptable.

7.4.2.9 Load Carrying Capacity

For the pinned connections case, the maximum section capacity factors are 0.874 and 0.818 for the arch main beams and the secondary purlins, respectively. The section capacity factor is determined from the Equation 7.1. When this factor is greater than 1.0, it means the member will fail under the load combination.

For the semi-rigid connections case, the maximum section capacity factors are 0.873 and 0.643 for the arch main beams and the secondary purlins, respectively.

As noted above, the modeling for the connections between the arch main beams and the secondary purlins will not affect the arch main beams. However, the maximum section capacity factor of the secondary purlins for the semi-rigid case is 0.175 smaller than that of a pinned connection case. Therefore, about a 20 percent reduction in strength for the secondary purlins can be made when considering the semi-rigid connection.

7.5 Conclusions

The direct or second-order analysis allowing for semi-rigid connections is adopted for the design of two dome structures in Hong Kong and Macau. Consideration to code recommendations for local member and global frame imperfections is made to account for geometric and material imperfections. Two types of analysis and design have been used, namely as design check by the Newton-Raphson method and the collapse analysis by the incremental-iterative elastic-plastic analysis. From the examples studied in this chapter, semi-rigid connections combines the benefit of using pinned and rigid connections in which structural stiffness is increased when compared to pinned structures, whereas the transferred moment through connections can be reduced when compared with rigid connections. Most importantly, the analysis and design represent a close reflection of the actual response of the structure, which is basically made of semi-rigid connections as no connections are ideally rigid or perfectly pinned.

References

BS 5950-1, *Structural use of steelwork in buildings*, Code of practice for design—Rolled and welded sections, British Standards Institution, 2000.

CEN, *Eurocode 3, Design of steel structures*, Part1-1: General rules and rules for buildings, EN 1993-1-1, European Committee for Standardization, Brussels, Belgium, 2005.

Chan, S. L. and Zhou, Z. H., 2nd-order elastic analysis of frames using single imperfect element per member, *Journal of Structural Engineering*, ACSE, 121(6), 939–945, 1995.

Chan, S. L. and Zhou, Z. H., Elastoplastic and large deflection analysis of steel frames by one element per member II: Three hinges along member, *Journal of Structural Engineering*, ASCE, 130(4), 545–553, 2004.

COPHK, *Code of Practice for the Structural Use of Steel 2005*, Buildings Department, Hong Kong SAR Government, 2005.

Hetao, H. and Li, G., Modified elastic approach for stability design of in-plane frame columns, *Advanced Steel Construction*, 5(1), 33–48, 2009.

HKWC, *Code of practice on wind effects in Hong Kong 2004*, Buildings Department, Hong Kong SAR Government, 2004.

JGJ61-2003, *Technical specification for latticed shells*, Chinese Planning Press, Beijing (Chinese).

Liew, J. Y. R., Chen, H., and Chen W. F., Second-order refined plastic hinge analysis of frame design Parts 1 and 2, *Journal of Structural Engineering*, ASCE, 119(11), 3196–3216, 1993.

NIDA, *User's Manual, Nonlinear Integrated Design and Analysis*, NIDA 8.0 HTML Online Documentation, (http://www.nida-naf.com), 2008.

White, D. W. and Hajjar, J. F., Stability of steel frames: the cases for simple elastic and rigorous inelastic analysis/design procedures, *Engineering Structures*, 22(2), 155–167, 2000.

Zhou, Z. H. and Chan, S. L., A self-equilibrium element for second-order analysis of semi-rigid jointed frames, *Journal of Engineering Mechanics*, ASCE, 121(8), 896–902, 1995

A.1 – 1

I - 1

Connection type : Single web-angle connections
Mode : All bolted

Tested by : S.L.Lipson (1968) Canada
Test Id. : AA-2/1

Column : --
Beam : W21X62
Angle : 4 X 3.5 X 1/4

Fasteners: A325- -3/4"D
13/16" Oversize holes
Material : A36
Fy = -- ksi
Fu = -- ksi

Major parameters

lp = 5.5000"	lu = 1.7500"	ll = 1.7500"	ta = 0.2500"
gb = 2.2500"	gc = 2.5625"	cu = 1.2500"	cl = 1.2500"
pb = 3.0000"	pc = 3.0000"	nb = 1 X 2	nc = 1 X 2

Remark 1) Connection subject to pure moment.
2) Hardened washers installed in both legs

No	Moment (k-in)	Rotation (radians) X 1/1000
1	0.00	0.00
2	0.74	1.38
3	1.49	2.76
4	1.93	3.60
5	2.38	4.44
6	2.82	5.28
7	3.27	6.12
8	3.69	7.23
9	4.11	8.34
10	4.70	9.58
11	5.28	10.82
12	5.84	12.40
13	6.39	13.98
14	7.06	15.66
15	7.72	17.34
16	8.38	19.02
17	8.94	20.59
18	9.49	22.16
19	10.05	23.72
20	10.73	25.16
21	11.42	26.59
22	12.10	28.02
23	12.61	30.17
24	13.12	32.32
25	13.61	34.67
26	14.10	37.02
27	14.44	39.24
28	14.77	41.46
29	15.12	43.50
30	15.46	45.55
31	15.49	47.36
32	15.51	49.18

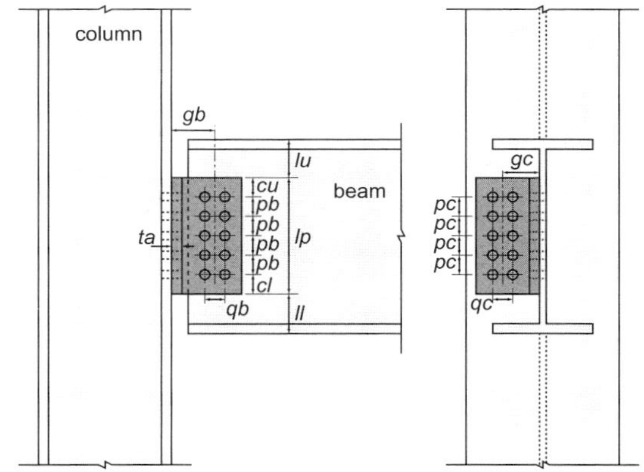

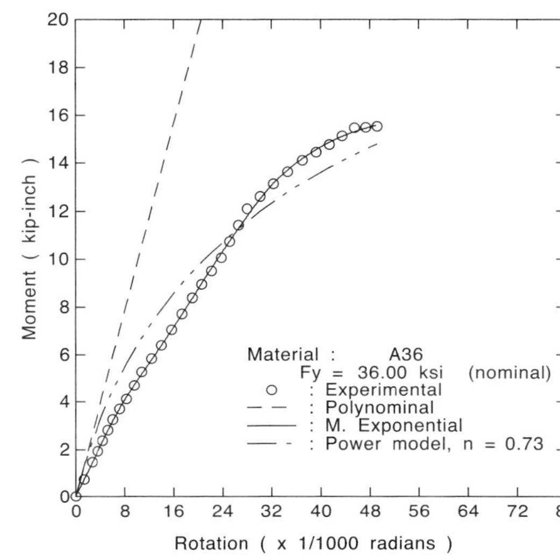

A.1 – 2

```
Moment-rotation prediction equations     ( R : X 1/1000 radians )

Frye and Morris polynominal model  :    R = Sum ( Ai X ( K*Bm )**Pi X 10**Qi )
     xd =  5.500000"   g =  2.762500"   t =  0.250000"
     A1 =  4.280000   A2 =  1.450000   A3 =  1.510000     K =  0.239361
     P1 =         1   P2 =         3   P3 =         5    Q1 =         0     Q2 =       -6     Q3 =      -13

Modified exponential model  :
     Bm = Sum ( Ai X ( 1 - exp( - R/( 2*i*C ) ) ) )  +      Sum ( RKj X ( R - Rj0 ) )  +  Bm0
       C =   0.75435250E+00               Bmo=   0.00000000E+00               Nexp=  6         Nliner=  2
      AI =   0.29149749E+01  -0.29963328E+02   0.73967184E+01   0.37852066E+03  -0.81221006E+03   0.47015516E+03
     Rj0 =          13.9800          32.3200
     RKj =   0.92425571E-02  -0.35940357E-01

Power model  :    Bm = ( Rki X R ) / ( 1 + ( R/R0 )**rn )**( 1/rn )
      rn = 0.731      rki =   0.11095362E+01          rmu =   0.28735590E+02
```

No	Rotation (radians) X 1/1000	Moment (kip-inch) Expri.	Poly.	M.Expo.	P.Model.	Connection stiffness (kip-inch) X 1000 Expri.	Poly.	M.Expo.	P.Model.
1	0.00	0.00	0.00	0.00	0.00	0.5397E+00	0.9761E+00	0.6272E+00	0.1110E+01
2	1.38	0.74	1.35	0.76	1.31	0.5397E+00	0.9761E+00	0.5196E+00	0.8535E+00
3	2.76	1.49	2.69	1.48	2.40	0.5335E+00	0.9761E+00	0.5311E+00	0.7284E+00
4	3.60	1.93	3.51	1.93	2.99	0.5298E+00	0.9761E+00	0.5326E+00	0.6715E+00
5	4.44	2.38	4.33	2.37	3.53	0.5298E+00	0.9761E+00	0.5230E+00	0.6237E+00
6	5.28	2.82	5.15	2.81	4.04	0.5298E+00	0.9761E+00	0.5045E+00	0.5826E+00
7	6.12	3.27	5.97	3.22	4.51	0.4659E+00	0.9761E+00	0.4810E+00	0.5468E+00
8	7.23	3.69	7.05	3.73	5.09	0.3818E+00	0.9761E+00	0.4484E+00	0.5058E+00
9	8.34	4.11	8.14	4.22	5.63	0.4230E+00	0.9761E+00	0.4195E+00	0.4705E+00
10	9.58	4.70	9.35	4.72	6.20	0.4693E+00	0.9761E+00	0.3952E+00	0.4361E+00
11	10.82	5.28	10.56	5.20	6.72	0.4177E+00	0.9761E+00	0.3802E+00	0.4063E+00
12	12.40	5.84	12.10	5.80	7.33	0.3523E+00	0.9761E+00	0.3728E+00	0.3735E+00
13	13.98	6.39	13.65	6.38	7.90	0.3725E+00	0.9761E+00	0.3745E+00	0.3452E+00
14	15.66	7.06	15.29	7.03	8.46	0.3941E+00	0.9761E+00	0.3900E+00	0.3192E+00
15	17.34	7.72	16.93	7.69	8.98	0.3941E+00	0.9761E+00	0.3966E+00	0.2965E+00
16	19.02	8.38	18.57	8.36	9.46	0.3740E+00	0.9761E+00	0.4004E+00	0.2766E+00
17	20.59	8.94	20.10	8.99	9.88	0.3553E+00	0.9761E+00	0.3999E+00	0.2600E+00
18	22.16	9.49	21.63	9.62	10.27	0.3554E+00	0.9761E+00	0.3950E+00	0.2451E+00
19	23.72	10.05	23.16	10.23	10.65	0.4182E+00	0.9761E+00	0.3857E+00	0.2316E+00
20	25.16	10.73	24.56	10.77	10.97	0.4757E+00	0.9761E+00	0.3738E+00	0.2204E+00
21	26.59	11.42	25.96	11.30	11.28	0.4757E+00	0.9761E+00	0.3592E+00	0.2100E+00
22	28.02	12.10	27.35	11.80	11.57	0.3802E+00	0.9761E+00	0.3423E+00	0.2005E+00
23	30.17	12.61	29.45	12.51	11.99	0.2372E+00	0.9761E+00	0.3143E+00	0.1876E+00
24	32.32	13.12	31.55	13.15	12.38	0.2241E+00	0.9761E+00	0.2485E+00	0.1760E+00
25	34.67	13.61	33.84	13.70	12.78	0.2098E+00	0.9761E+00	0.2155E+00	0.1647E+00
26	37.02	14.10	36.14	14.17	13.16	0.1792E+00	0.9760E+00	0.1836E+00	0.1546E+00
27	39.24	14.44	38.30	14.54	13.49	0.1503E+00	0.9760E+00	0.1553E+00	0.1460E+00
28	41.46	14.77	40.46	14.86	13.80	0.1601E+00	0.9760E+00	0.1295E+00	0.1381E+00
29	43.50	15.12	42.46	15.10	14.08	0.1693E+00	0.9760E+00	0.1079E+00	0.1315E+00
30	45.55	15.46	44.46	15.30	14.34	0.8635E-01	0.9760E+00	0.8853E-01	0.1254E+00
31	47.36	15.49	46.23	15.45	14.57	0.1294E-01	0.9760E+00	0.7328E-01	0.1204E+00
32	49.18	15.51	48.00	15.57	14.78	0.1291E-01	0.9760E+00	0.5972E-01	0.1157E+00

A.1 – 3

I - 2

Connection type : Single web-angle connections
Mode : All bolted

Tested by : S.L.Lipson (1968) Canada
Test Id. : AA-2/2

Column : --
Beam : W21X62
Angle : 4 X 3.5 X 1/4

Fasteners: A325- -3/4"D
13/16" Oversize holes
Material : A36
Fy = -- ksi
Fu = -- ksi

Major parameters

lp = 5.5000"	lu = 1.7500"	ll = 1.7500"	ta = 0.2500"
gb = 2.2500"	gc = 2.5625"	cu = 1.2500"	cl = 1.2500"
pb = 3.0000"	pc = 3.0000"	nb = 1 X 2	nc = 1 X 2

Remark 1) Connection subject to pure moment.
2) Hardened washers installed in both legs

No	Moment (k-in)	Rotation (radians) X 1/1000	No	Moment (k-in)	Rotation (radians) X 1/1000
1	0.00	0.00	26	15.04	25.71
2	0.46	0.64	27	15.19	26.77
3	0.92	1.28	28	15.33	27.82
4	1.54	2.19	29	15.47	28.87
5	2.15	3.10	30	16.22	30.19
6	2.76	3.97	31	16.97	31.50
7	3.38	4.85			
8	4.10	5.72			
9	4.83	6.60			
10	5.56	7.48			
11	6.18	8.33			
12	6.79	9.18			
13	7.41	10.03			
14	8.01	11.04			
15	8.62	12.05			
16	9.22	13.06			
17	9.62	14.48			
18	10.02	15.89			
19	10.60	17.10			
20	11.19	18.31			
21	11.78	19.52			
22	12.38	20.58			
23	12.98	21.63			
24	13.58	22.69			
25	14.31	24.20			

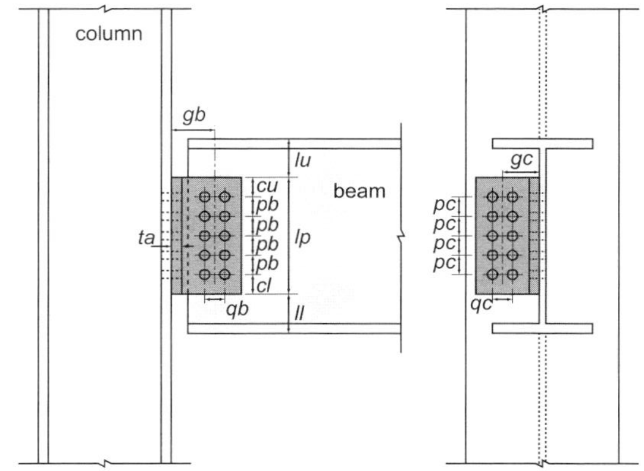

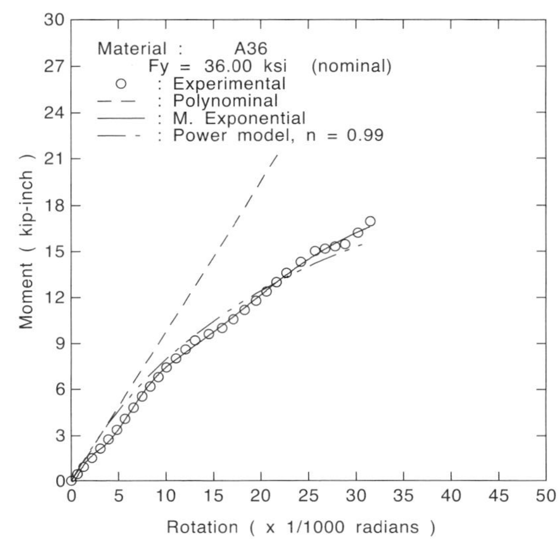

A.1 – 4

```
Moment-rotation prediction equations      ( R : X 1/1000 radians )

Frye and Morris polynominal model   :     R = Sum ( Ai X ( K*Bm )**Pi X 10**Qi )
     xd =  5.500000"    g =  2.762500"    t =  0.250000"
     A1 =  4.280000    A2 =  1.450000    A3 =  1.510000     K =  0.239361
     P1 =        1     P2 =        3     P3 =        5     Q1 =        0     Q2 =       -6     Q3 =      -13

Modified exponential model   :
     Bm = Sum ( Ai X ( 1 - exp( - R/( 2*i*C ) ) ) )  +      Sum ( RKj X ( R - Rj0 ) )  +  Bm0
       C =   0.42240750E+00                Bmo=   0.00000000E+00              Nexp=  6         Nliner=  1
      AI =  -0.25526455E+02   0.37619342E+03  -0.18708504E+04   0.41350051E+04  -0.42052409E+04   0.16063551E+04
     Rj0 =            11.5000
     RKj =   0.92198278E-01

Power model   :     Bm = ( Rki X R ) / ( 1 + ( R/R0 )**rn )**( 1/rn )
      rn = 0.989       rki =    0.11095362E+01           rmu =   0.28735590E+02
```

No	Rotation (radians) X 1/1000	Moment (kip-inch) Expri.	Poly.	M.Expo.	P.Model.	Connection stiffness (kip-inch) X 1000 Expri.	Poly.	M.Expo.	P.Model.
1	0.00	0.00	0.00	0.00	0.00	0.7224E+00	0.9761E+00	-0.7311E+00	0.1110E+01
2	0.64	0.46	0.62	0.31	0.69	0.7222E+00	0.9761E+00	0.1100E+01	0.1054E+01
3	1.28	0.92	1.25	1.01	1.35	0.7017E+00	0.9761E+00	0.9868E+00	0.1004E+01
4	2.19	1.54	2.14	1.68	2.23	0.6727E+00	0.9761E+00	0.5378E+00	0.9387E+00
5	3.10	2.15	3.02	2.12	3.06	0.6875E+00	0.9761E+00	0.4868E+00	0.8798E+00
6	3.97	2.76	3.88	2.62	3.81	0.7017E+00	0.9761E+00	0.6595E+00	0.8283E+00
7	4.85	3.38	4.73	3.28	4.51	0.7667E+00	0.9761E+00	0.8396E+00	0.7814E+00
8	5.72	4.10	5.59	4.06	5.18	0.8319E+00	0.9761E+00	0.9323E+00	0.7383E+00
9	6.60	4.83	6.44	4.88	5.80	0.8319E+00	0.9761E+00	0.9275E+00	0.6988E+00
10	7.48	5.56	7.30	5.66	6.40	0.7763E+00	0.9761E+00	0.8530E+00	0.6624E+00
11	8.33	6.18	8.13	6.35	6.95	0.7222E+00	0.9761E+00	0.7475E+00	0.6297E+00
12	9.18	6.79	8.96	6.94	7.47	0.7222E+00	0.9761E+00	0.6368E+00	0.5993E+00
13	10.03	7.41	9.79	7.44	7.97	0.6651E+00	0.9761E+00	0.5387E+00	0.5711E+00
14	11.04	8.01	10.78	7.94	8.53	0.5975E+00	0.9761E+00	0.4503E+00	0.5403E+00
15	12.05	8.62	11.77	8.41	9.07	0.5974E+00	0.9761E+00	0.4867E+00	0.5119E+00
16	13.06	9.22	12.75	8.89	9.57	0.4661E+00	0.9761E+00	0.4592E+00	0.4857E+00
17	14.48	9.62	14.13	9.53	10.23	0.2825E+00	0.9761E+00	0.4535E+00	0.4524E+00
18	15.89	10.02	15.51	10.18	10.85	0.3912E+00	0.9761E+00	0.4669E+00	0.4224E+00
19	17.10	10.60	16.69	10.75	11.35	0.4845E+00	0.9761E+00	0.4812E+00	0.3990E+00
20	18.31	11.19	17.87	11.34	11.82	0.4845E+00	0.9761E+00	0.4911E+00	0.3775E+00
21	19.52	11.78	19.06	11.94	12.26	0.5295E+00	0.9761E+00	0.4933E+00	0.3577E+00
22	20.58	12.38	20.08	12.45	12.63	0.5686E+00	0.9761E+00	0.4882E+00	0.3417E+00
23	21.63	12.98	21.11	12.96	12.98	0.5686E+00	0.9761E+00	0.4770E+00	0.3267E+00
24	22.69	13.58	22.14	13.46	13.32	0.5340E+00	0.9761E+00	0.4604E+00	0.3128E+00
25	24.20	14.31	23.62	14.13	13.78	0.4845E+00	0.9761E+00	0.4298E+00	0.2942E+00
26	25.71	15.04	25.10	14.76	14.21	0.2786E+00	0.9761E+00	0.3941E+00	0.2773E+00
27	26.77	15.19	26.13	15.16	14.50	0.1354E+00	0.9761E+00	0.3680E+00	0.2664E+00
28	27.82	15.33	27.16	15.53	14.77	0.1354E+00	0.9761E+00	0.3418E+00	0.2561E+00
29	28.87	15.47	28.18	15.88	15.04	0.3295E+00	0.9761E+00	0.3163E+00	0.2464E+00
30	30.19	16.22	29.46	16.27	15.35	0.5714E+00	0.9761E+00	0.2860E+00	0.2351E+00
31	31.50	16.97	30.75	16.63	15.65	0.5714E+00	0.9761E+00	0.2581E+00	0.2245E+00

A.1 – 5

I - 3

Connection type : Single web-angle connections
Mode : All bolted

Tested by : S.L.Lipson (1968) Canada
Test Id. : AA-3/1

Column : --
Beam : W21X62
Angle : 4 X 3.5 X 1/4

Fasteners: A325- -3/4"D
13/16" Oversize holes
Material : A36
Fy = -- ksi
Fu = -- ksi

Major parameters

lp = 8.5000"	lu = 1.7500"	ll = 1.7500"	ta = 0.2500"
gb = 2.2500"	gc = 2.5625"	cu = 1.2500"	cl = 1.2500"
pb = 3.0000"	pc = 3.0000"	nb = 1 X 3	nc = 1 X 3

Remark 1) Connection subject to pure moment.
2) Hardened washers installed in both legs

No	Moment (k-in)	Rotation (radians) X 1/1000	No	Moment (k-in)	Rotation (radians) X 1/1000
1	0.00	0.00	26	43.73	37.17
2	2.16	1.01	27	44.43	39.21
3	4.31	2.02	28	45.47	41.30
4	6.65	2.97	29	46.52	43.38
5	8.98	3.91	30	47.59	45.12
6	11.30	4.99	31	48.66	46.87
7	13.63	6.07	32	49.35	49.02
8	15.94	7.32	33	50.05	51.16
9	18.25	8.57			
10	20.72	9.95			
11	23.19	11.33			
12	25.15	12.71			
13	27.10	14.09			
14	28.88	15.64			
15	30.65	17.18			
16	32.24	18.83			
17	33.83	20.48			
18	35.42	22.13			
19	37.02	23.77			
20	37.73	25.69			
21	38.44	27.60			
22	39.66	29.68			
23	40.87	31.76			
24	41.95	33.44			
25	43.02	35.12			

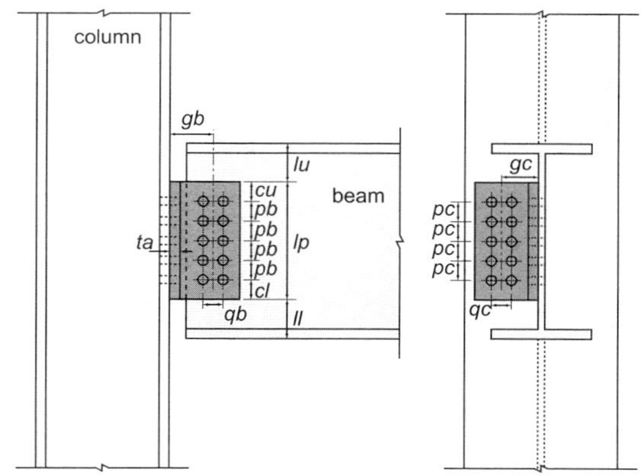

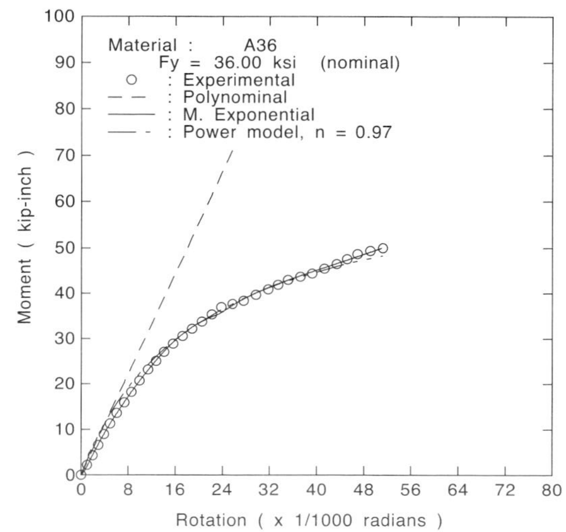

A.1 – 6

```
Moment-rotation prediction equations      ( R : X 1/1000 radians )

Frye and Morris polynominal model  :    R = Sum ( Ai X ( K*Bm )**Pi X 10**Qi )
     xd =  8.500000"    g =  2.762500"    t =  0.250000"
     A1 =  4.280000    A2 =  1.450000    A3 =  1.510000     K =  0.084201
     P1 =         1    P2 =         3    P3 =         5    Q1 =         0    Q2 =       -6    Q3 =      -13

Modified exponential model  :
     Bm = Sum ( Ai X ( 1 - exp( - R/( 2*i*C ) ) ) )  +      Sum ( RKj X ( R - Rj0 ) )  +  Bm0
       C =   0.67934167E+00                 Bmo=   0.00000000E+00           Nexp=  6        Nliner=  1
      AI =  -0.96571609E+01  -0.29073652E+01   0.47525019E+03  -0.17286658E+04   0.20828160E+04  -0.78091338E+03
     Rj0 =         23.5000
     RKj =   0.49381954E+00

Power model  :    Bm = ( Rki X R ) / ( 1 + ( R/R0 )**rn )**( 1/rn )
      rn = 0.966      rki =   0.34414089E+01          rmu =   0.68632939E+02
```

No	Rotation (radians) X 1/1000	Moment (kip-inch) Expri.	Moment Poly.	Moment M.Expo.	Moment P.Model.	Connection stiffness (kip-inch) X 1000 Expri.	Stiffness Poly.	Stiffness M.Expo.	Stiffness P.Model.
1	0.00	0.00	0.00	0.00	0.00	0.2131E+01	0.2775E+01	0.1141E+01	0.3441E+01
2	1.01	2.16	2.81	1.91	3.29	0.2131E+01	0.2775E+01	0.2363E+01	0.3079E+01
3	2.02	4.31	5.62	4.44	6.25	0.2305E+01	0.2775E+01	0.2543E+01	0.2784E+01
4	2.97	6.65	8.24	6.80	8.77	0.2468E+01	0.2775E+01	0.2416E+01	0.2549E+01
5	3.91	8.98	10.86	9.00	11.08	0.2321E+01	0.2775E+01	0.2240E+01	0.2344E+01
6	4.99	11.30	13.86	11.32	13.50	0.2152E+01	0.2775E+01	0.2076E+01	0.2141E+01
7	6.07	13.63	16.85	13.50	15.71	0.2013E+01	0.2775E+01	0.1965E+01	0.1964E+01
8	7.32	15.94	20.31	15.89	18.05	0.1853E+01	0.2775E+01	0.1883E+01	0.1786E+01
9	8.57	18.25	23.77	18.20	20.18	0.1823E+01	0.2775E+01	0.1819E+01	0.1633E+01
10	9.95	20.72	27.61	20.66	22.33	0.1789E+01	0.2775E+01	0.1743E+01	0.1485E+01
11	11.33	23.19	31.44	23.01	24.29	0.1603E+01	0.2775E+01	0.1647E+01	0.1357E+01
12	12.71	25.15	35.27	25.20	26.08	0.1418E+01	0.2775E+01	0.1528E+01	0.1246E+01
13	14.09	27.10	39.10	27.22	27.73	0.1289E+01	0.2775E+01	0.1390E+01	0.1148E+01
14	15.64	28.88	43.39	29.24	29.43	0.1145E+01	0.2775E+01	0.1221E+01	0.1051E+01
15	17.18	30.65	47.68	30.99	30.99	0.1059E+01	0.2775E+01	0.1047E+01	0.9665E+00
16	18.83	32.24	52.25	32.57	32.52	0.9668E+00	0.2775E+01	0.8653E+00	0.8873E+00
17	20.48	33.83	56.83	33.85	33.92	0.9668E+00	0.2775E+01	0.6961E+00	0.8176E+00
18	22.13	35.42	61.40	34.87	35.22	0.9668E+00	0.2775E+01	0.5442E+00	0.7559E+00
19	23.77	37.02	65.97	35.79	36.41	0.6920E+00	0.2775E+01	0.9060E+00	0.7011E+00
20	25.69	37.73	71.27	37.39	37.70	0.3728E+00	0.2775E+01	0.7782E+00	0.6447E+00
21	27.60	38.44	76.58	38.78	38.89	0.4739E+00	0.2775E+01	0.6767E+00	0.5949E+00
22	29.68	39.66	82.36	40.10	40.07	0.5840E+00	0.2775E+01	0.5927E+00	0.5471E+00
23	31.76	40.87	88.13	41.27	41.17	0.6149E+00	0.2775E+01	0.5322E+00	0.5049E+00
24	33.44	41.95	92.79	42.13	41.99	0.6399E+00	0.2775E+01	0.4972E+00	0.4743E+00
25	35.12	43.02	97.45	42.94	42.76	0.5063E+00	0.2775E+01	0.4723E+00	0.4465E+00
26	37.17	43.73	103.13	43.89	43.64	0.3434E+00	0.2775E+01	0.4526E+00	0.4158E+00
27	39.21	44.43	108.81	44.80	44.47	0.4218E+00	0.2775E+01	0.4417E+00	0.3882E+00
28	41.30	45.47	114.59	45.71	45.25	0.5015E+00	0.2775E+01	0.4370E+00	0.3629E+00
29	43.38	46.52	120.36	46.62	45.98	0.5619E+00	0.2775E+01	0.4367E+00	0.3400E+00
30	45.12	47.59	125.20	47.39	46.56	0.6125E+00	0.2775E+01	0.4387E+00	0.3224E+00
31	46.87	48.66	130.05	48.15	47.10	0.4830E+00	0.2775E+01	0.4420E+00	0.3062E+00
32	49.02	49.35	136.01	49.11	47.74	0.3237E+00	0.2774E+01	0.4470E+00	0.2879E+00
33	51.16	50.05	141.97	50.07	48.34	0.3237E+00	0.2774E+01	0.4525E+00	0.2712E+00

A.1 – 7

I - 4

```
Connection type : Single web-angle connections
           Mode : All bolted

Tested by   :  S.L.Lipson (1968)                           Canada
Test  Id.   :  AA-3/2

Column :  --                           Fasteners: A325- -3/4"D
Beam   : W21X62                                  13/16" Oversize holes
Angle  : 4 X 3.5 X 1/4                 Material : A36
                                            Fy =   --   ksi
                                            Fu =   --   ksi

                           Major parameters

     lp =  8.5000"    lu =  1.7500"    ll =  1.7500"    ta =  0.2500"
     gb =  2.2500"    gc =  2.5625"    cu =  1.2500"    cl =  1.2500"
     pb =  3.0000"    pc =  3.0000"    nb = 1 X  3      nc = 1 X  3

Remark  1)  Connection subject to pure moment.
        2)  Hardened washers installed in both legs
```

No	Moment (k-in)	Rotation (radians) X 1/1000
1	0.00	0.00
2	2.01	0.65
3	4.01	1.31
4	6.02	1.96
5	7.81	2.53
6	9.59	3.09
7	11.38	3.66
8	13.39	4.27
9	15.40	4.88
10	17.41	5.49
11	19.19	6.14
12	20.97	6.79
13	22.75	7.45
14	25.07	8.43
15	27.40	9.41
16	28.95	10.11
17	30.49	10.80
18	32.04	11.50
19	33.93	12.25
20	35.81	12.99
21	37.70	13.73
22	38.90	14.54
23	40.09	15.35
24	41.29	16.16
25	42.94	16.95
26	44.60	17.74
27	46.25	18.52
28	47.90	19.40
29	49.55	20.28

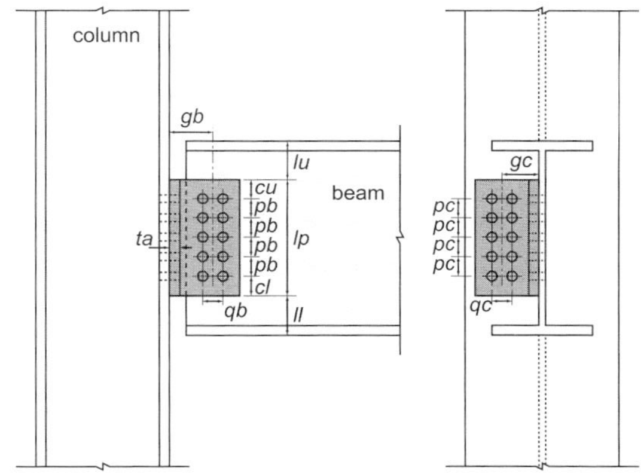

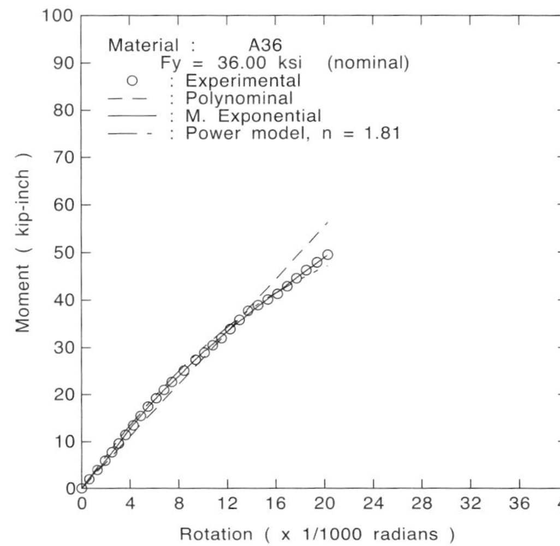

A.1 – 8

```
Moment-rotation prediction equations      ( R : X 1/1000 radians )

Frye and Morris polynominal model  :    R = Sum ( Ai X ( K*Bm )**Pi X 10**Qi )
     xd =  8.500000"   g =  2.762500"   t =  0.250000"
     A1 =  4.280000   A2 =  1.450000   A3 =  1.510000    K =  0.084201
     P1 =        1    P2 =        3    P3 =        5    Q1 =        0    Q2 =       -6    Q3 =      -13

Modified exponential model  :
     Bm = Sum ( Ai X ( 1 - exp( - R/( 2*i*C ) ) ) )  +      Sum ( RKj X ( R - Rj0 ) )  +  BmO
       C =   0.33263667E+00            Bmo=   0.00000000E+00            Nexp=  6         Nliner=  2
      AI =  -0.15457230E+01   0.60249888E+02  -0.31690963E+03   0.62739613E+03  -0.59688369E+03   0.24586727E+03
     Rj0 =          0.0000            8.0000
     RKj =   0.12980867E+01   0.44060924E+00

Power model  :    Bm = ( Rki X R ) / ( 1 + ( R/RO )**rn )**( 1/rn )
      rn = 1.814      rki =   0.34414089E+01         rmu =   0.68632939E+02
```

No	Rotation (radians) X 1/1000	Moment (kip-inch) Expri.	Poly.	M.Expo.	P.Model.	Connection stiffness (kip-inch) X 1000 Expri.	Poly.	M.Expo.	P.Model.
1	0.00	0.00	0.00	0.00	0.00	0.3067E+01	0.2775E+01	0.2094E+01	0.3441E+01
2	0.65	2.01	1.82	2.06	2.25	0.3067E+01	0.2775E+01	0.2977E+01	0.3431E+01
3	1.31	4.01	3.63	3.99	4.49	0.3067E+01	0.2775E+01	0.2967E+01	0.3404E+01
4	1.96	6.02	5.45	5.97	6.70	0.3119E+01	0.2775E+01	0.3113E+01	0.3363E+01
5	2.53	7.81	7.02	7.77	8.59	0.3163E+01	0.2775E+01	0.3236E+01	0.3319E+01
6	3.09	9.59	8.58	9.62	10.45	0.3163E+01	0.2775E+01	0.3305E+01	0.3267E+01
7	3.66	11.38	10.15	11.49	12.28	0.3228E+01	0.2775E+01	0.3306E+01	0.3209E+01
8	4.27	13.39	11.84	13.48	14.21	0.3297E+01	0.2775E+01	0.3239E+01	0.3139E+01
9	4.88	15.40	13.53	15.43	16.11	0.3297E+01	0.2775E+01	0.3119E+01	0.3064E+01
10	5.49	17.41	15.23	17.28	17.95	0.3019E+01	0.2775E+01	0.2966E+01	0.2984E+01
11	6.14	19.19	17.04	19.16	19.87	0.2720E+01	0.2775E+01	0.2784E+01	0.2894E+01
12	6.79	20.97	18.85	20.92	21.73	0.2720E+01	0.2775E+01	0.2602E+01	0.2801E+01
13	7.45	22.75	20.67	22.57	23.54	0.2581E+01	0.2775E+01	0.2428E+01	0.2706E+01
14	8.43	25.07	23.39	25.02	26.12	0.2373E+01	0.2775E+01	0.2639E+01	0.2561E+01
15	9.41	27.40	26.11	27.51	28.55	0.2282E+01	0.2775E+01	0.2451E+01	0.2416E+01
16	10.11	28.95	28.04	29.18	30.21	0.2216E+01	0.2775E+01	0.2342E+01	0.2314E+01
17	10.80	30.49	29.98	30.78	31.79	0.2216E+01	0.2775E+01	0.2252E+01	0.2214E+01
18	11.50	32.04	31.92	32.33	33.30	0.2371E+01	0.2775E+01	0.2177E+01	0.2115E+01
19	12.25	33.93	33.98	33.92	34.83	0.2536E+01	0.2775E+01	0.2111E+01	0.2013E+01
20	12.99	35.81	36.05	35.47	36.29	0.2536E+01	0.2775E+01	0.2058E+01	0.1914E+01
21	13.73	37.70	38.11	36.99	37.68	0.2029E+01	0.2775E+01	0.2014E+01	0.1819E+01
22	14.54	38.90	40.35	38.60	39.11	0.1478E+01	0.2775E+01	0.1975E+01	0.1720E+01
23	15.35	40.09	42.60	40.18	40.46	0.1478E+01	0.2775E+01	0.1942E+01	0.1625E+01
24	16.16	41.29	44.84	41.74	41.74	0.1793E+01	0.2775E+01	0.1915E+01	0.1535E+01
25	16.95	42.94	47.03	43.24	42.92	0.2100E+01	0.2775E+01	0.1893E+01	0.1452E+01
26	17.74	44.60	49.21	44.73	44.03	0.2100E+01	0.2775E+01	0.1874E+01	0.1373E+01
27	18.52	46.25	51.40	46.20	45.08	0.1995E+01	0.2775E+01	0.1857E+01	0.1298E+01
28	19.40	47.90	53.83	47.82	46.18	0.1878E+01	0.2775E+01	0.1841E+01	0.1220E+01
29	20.28	49.55	56.27	49.43	47.22	0.1878E+01	0.2775E+01	0.1827E+01	0.1147E+01

I - 5

```
Connection type : Single web-angle connections
           Mode : All bolted

Tested by   :  S.L.Lipson (1968)                         Canada
Test  Id.   :  AA-4/1

Column :  --                        Fasteners: A325- -3/4"D
Beam   : W21X62                           13/16" Oversize holes
Angle  : 4 X 3.5 X 1/4              Material : A36
                                         Fy =   --   ksi
                                         Fu =   --   ksi

                           Major parameters

   lp = 11.5000"    lu =  1.7500"    ll =  1.7500"    ta =  0.2500"
   gb =  2.2500"    gc =  2.5625"    cu =  1.2500"    cl =  1.2500"
   pb =  3.0000"    pc =  3.0000"    nb = 1 X  4      nc = 1 X  4

Remark  1)  Connection subject to pure moment.
        2)  Hardened washers installed in both legs
```

No	Moment (k-in)	Rotation (radians) X 1/1000	No	Moment (k-in)	Rotation (radians) X 1/1000
1	0.00	0.00	26	70.13	27.00
2	3.31	0.81	27	71.76	28.73
3	6.10	1.39	28	73.39	30.53
4	8.89	1.97	29	75.02	32.34
5	11.48	2.38	30	76.58	34.17
6	14.07	2.79	31	78.15	36.01
7	18.05	3.61	32	79.57	37.94
8	22.69	4.70	33	81.00	39.87
9	25.41	5.24	34	82.36	41.87
10	28.13	5.79	35	83.72	43.87
11	30.78	6.47	36	84.81	45.77
12	33.43	7.14	37	85.91	47.67
13	36.74	8.09	38	87.33	49.54
14	40.05	9.04	39	88.76	51.41
15	42.63	9.95	40	90.57	54.41
16	45.21	10.86	41	92.11	57.01
17	47.72	11.83	42	93.55	58.28
18	50.23	12.81	43	94.84	60.41
19	52.67	14.12			
20	55.11	15.42			
21	59.85	18.37			
22	62.15	20.11			
23	64.44	21.85			
24	66.47	23.55			
25	68.50	25.26			

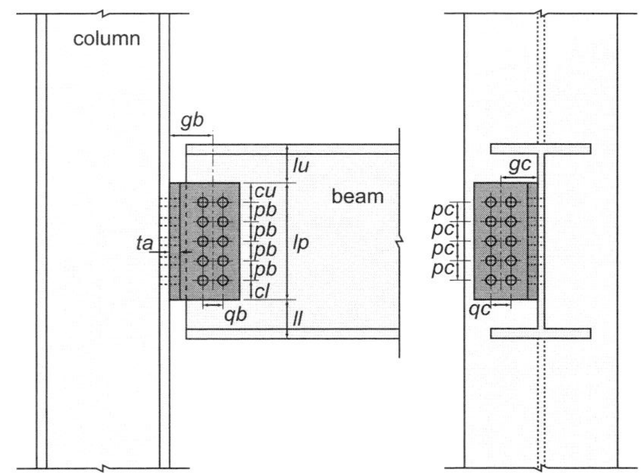

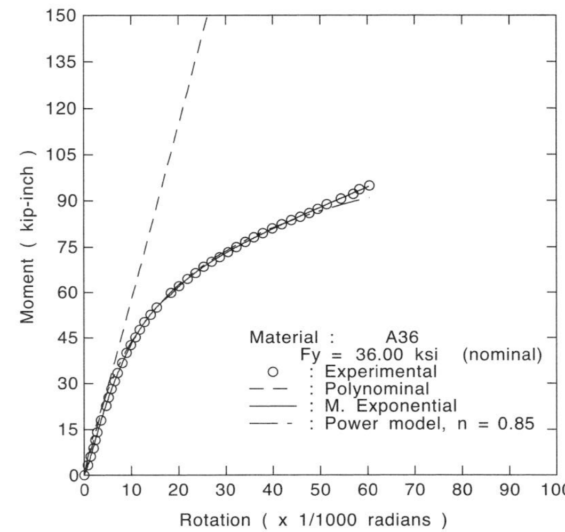

A.1 – 10

```
Moment-rotation prediction equations     ( R : X 1/1000 radians )

Frye and Morris polynominal model  :    R = Sum ( Ai X ( K*Bm )**Pi X 10**Qi )
     xd = 11.500000"    g =  2.762500"   t =  0.250000"
     A1 =  4.280000    A2 =  1.450000    A3 =  1.510000    K =  0.040761
     P1 =        1     P2 =        3     P3 =        5     Q1 =        0     Q2 =      -6     Q3 =      -13

Modified exponential model  :
     Bm = Sum ( Ai X ( 1 - exp( - R/( 2*i*C ) ) ) )  +      Sum ( RKj X ( R - Rj0 ) )  +  Bm0
      C =   0.70688417E+00           Bmo=   0.00000000E+00           Nexp=  6        Nliner=  1
     AI =  -0.25968016E+02   0.24808235E+03  -0.10304382E+04   0.18661063E+04  -0.15796917E+04   0.59070956E+03
    Rj0 =          18.3700
    RKj =   0.61845541E+00

Power model  :     Bm = ( Rki X R ) / ( 1 + ( R/R0 )**rn )**( 1/rn )
      rn = 0.853      rki =    0.79915454E+01         rmu =    0.12562915E+03
```

No	Rotation (radians) X 1/1000	Moment (kip-inch) Expri.	Moment (kip-inch) Poly.	Moment (kip-inch) M.Expo.	Moment (kip-inch) P.Model.	Connection stiffness (kip-inch) X 1000 Expri.	Connection stiffness (kip-inch) X 1000 Poly.	Connection stiffness (kip-inch) X 1000 M.Expo.	Connection stiffness (kip-inch) X 1000 P.Model.
1	0.00	0.00	0.00	0.00	0.00	0.4072E+01	0.5732E+01	0.2570E+01	0.7992E+01
3	1.39	6.10	7.98	6.10	9.68	0.4814E+01	0.5732E+01	0.5343E+01	0.6170E+01
5	2.38	11.48	13.66	11.49	15.38	0.6292E+01	0.5732E+01	0.5425E+01	0.5378E+01
7	3.61	18.05	20.70	17.93	21.51	0.4614E+01	0.5732E+01	0.5007E+01	0.4634E+01
9	5.24	25.41	30.05	25.57	28.44	0.4990E+01	0.5732E+01	0.4375E+01	0.3897E+01
11	6.47	30.78	37.06	30.66	32.93	0.3912E+01	0.5732E+01	0.3970E+01	0.3468E+01
13	8.09	36.74	46.37	36.72	38.18	0.3498E+01	0.5732E+01	0.3492E+01	0.3011E+01
15	9.95	42.63	57.02	42.73	43.37	0.2835E+01	0.5732E+01	0.2980E+01	0.2600E+01
17	11.83	47.72	67.83	47.87	47.95	0.2573E+01	0.5732E+01	0.2481E+01	0.2271E+01
19	14.12	52.67	80.92	52.89	52.76	0.1864E+01	0.5732E+01	0.1924E+01	0.1955E+01
21	18.37	59.85	105.31	59.25	60.11	0.1427E+01	0.5732E+01	0.1737E+01	0.1527E+01
23	21.85	64.44	125.25	64.50	64.97	0.1255E+01	0.5732E+01	0.1317E+01	0.1279E+01
25	25.26	68.50	144.79	68.52	69.00	0.1067E+01	0.5732E+01	0.1064E+01	0.1093E+01
27	28.73	71.76	164.68	71.93	72.53	0.9220E+00	0.5732E+01	0.9123E+00	0.9445E+00
29	32.34	75.02	185.34	75.04	75.71	0.8778E+00	0.5732E+01	0.8198E+00	0.8214E+00
31	36.01	78.15	206.38	77.94	78.53	0.7959E+00	0.5732E+01	0.7629E+00	0.7203E+00
33	39.87	81.00	228.55	80.81	81.14	0.7090E+00	0.5732E+01	0.7240E+00	0.6335E+00
35	43.87	83.72	251.48	83.64	83.52	0.6264E+00	0.5731E+01	0.6961E+00	0.5598E+00
37	47.67	85.91	273.25	86.25	85.54	0.6712E+00	0.5731E+01	0.6765E+00	0.5017E+00
39	51.41	88.76	294.66	88.75	87.32	0.7021E+00	0.5731E+01	0.6617E+00	0.4532E+00
41	57.01	92.11	326.74	92.40	89.68	0.9573E+00	0.5731E+01	0.6456E+00	0.3932E+00
43	60.41	94.84	346.24	94.59	90.96	0.6046E+00	0.5731E+01	0.6387E+00	0.3625E+00

A.1 – 11

I - 6

Connection type : Single web-angle connections
Mode : All bolted

Tested by : S.L.Lipson (1968) Canada
Test Id. : AA-4/2

Column : --
Beam : W21X62
Angle : 4 X 3.5 X 1/4

Fasteners: A325- -3/4"D
13/16" Oversize holes
Material : A36
Fy = -- ksi
Fu = -- ksi

Major parameters

lp = 11.5000"	lu = 1.7500"	ll = 1.7500"	ta = 0.2500"
gb = 2.2500"	gc = 2.5625"	cu = 1.2500"	cl = 1.2500"
pb = 3.0000"	pc = 3.0000"	nb = 1 X 4	nc = 1 X 4

Remark 1) Connection subject to pure moment.
2) Hardened washers installed in both legs

No	Moment (k-in)	Rotation (radians) X 1/1000	No	Moment (k-in)	Rotation (radians) X 1/1000
1	0.00	0.00	26	76.56	22.75
2	3.99	0.48	27	78.28	24.05
3	6.65	0.75	28	79.99	25.35
4	9.31	1.03	29	81.84	26.75
5	14.10	1.57	30	83.69	28.15
6	17.03	1.98	31	85.73	29.95
7	19.95	2.39	32	87.77	31.75
8	23.14	2.96	33	89.54	33.42
9	26.33	3.54	34	91.32	35.08
10	29.12	4.01	35	93.16	37.01
11	31.91	4.49	36	95.00	38.94
12	34.77	5.26	37	96.71	40.71
13	37.62	6.03	38	98.42	42.47
14	42.13	7.31	39	100.19	44.47
15	47.18	8.52	40	101.96	46.47
16	49.97	9.42	41	103.34	48.06
17	52.75	10.33	42	104.71	49.66
18	55.53	11.43			
19	58.32	12.54			
20	61.17	13.68			
21	64.02	14.81			
22	66.39	16.25			
23	68.77	17.68			
24	72.87	20.15			
25	74.72	21.45			

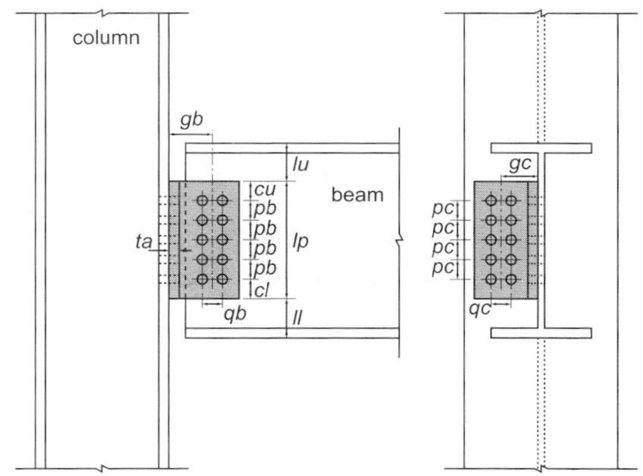

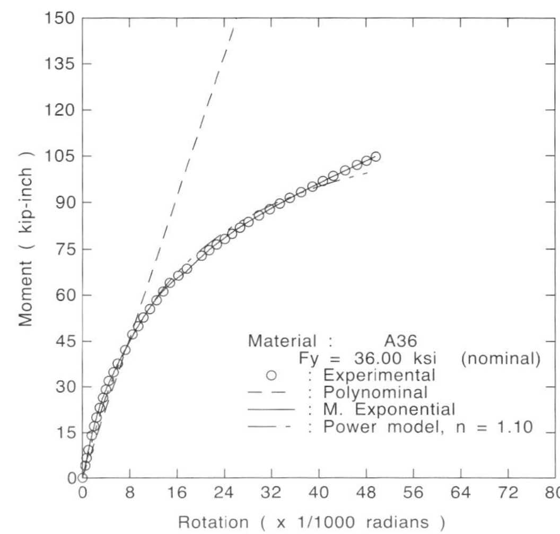

A.1 – 12

```
Moment-rotation prediction equations      ( R : X 1/1000 radians )

Frye and Morris polynominal model  :     R = Sum ( Ai X ( K*Bm )**Pi X 10**Qi )
     xd = 11.500000"    g =  2.762500"   t =  0.250000"
     A1 =  4.280000    A2 =  1.450000   A3 =  1.510000     K =  0.040761
     P1 =         1    P2 =         3   P3 =         5    Q1 =         0     Q2 =        -6     Q3 =      -13

Modified exponential model  :
     Bm = Sum ( Ai X ( 1 - exp( - R/( 2*i*C ) ) ) )  +      Sum ( RKj X ( R - Rj0 ) )  +  Bm0
       C =   0.53329417E+00           Bmo=   0.00000000E+00          Nexp=  6        Nliner=  1
      AI =  -0.38937503E+02   0.35913899E+03  -0.10980850E+04   0.16364935E+04  -0.13656148E+04   0.58384858E+03
     Rj0 =           17.6800
     RKj =   0.87892363E+00

Power model  :    Bm = ( Rki X R ) / ( 1 + ( R/R0 )**rn )**( 1/rn )
      rn = 1.096      rki =   0.79915454E+01          rmu =   0.12562915E+03
```

No	Rotation (radians) X 1/1000	Moment (kip-inch)				Connection stiffness (kip-inch) X 1000			
		Expri.	Poly.	M.Expo.	P.Model.	Expri.	Poly.	M.Expo.	P.Model.
1	0.00	0.00	0.00	0.00	0.00	0.8351E+01	0.5732E+01	0.7418E+01	0.7992E+01
3	0.75	6.65	4.31	6.60	5.82	0.9698E+01	0.5732E+01	0.9322E+01	0.7473E+01
5	1.57	14.10	9.02	13.92	11.72	0.7849E+01	0.5732E+01	0.8266E+01	0.6895E+01
7	2.39	19.95	13.69	20.02	17.12	0.6499E+01	0.5732E+01	0.6707E+01	0.6360E+01
9	3.54	26.33	20.28	26.73	24.04	0.5737E+01	0.5732E+01	0.5119E+01	0.5684E+01
11	4.49	31.91	25.72	31.21	29.18	0.5055E+01	0.5732E+01	0.4425E+01	0.5192E+01
13	6.03	37.62	34.57	37.61	36.66	0.3634E+01	0.5732E+01	0.3939E+01	0.4502E+01
15	8.52	47.18	48.83	46.87	46.72	0.3543E+01	0.5732E+01	0.3490E+01	0.3629E+01
17	10.33	52.75	59.21	52.80	52.83	0.2826E+01	0.5732E+01	0.3045E+01	0.3135E+01
19	12.54	58.32	71.87	58.84	59.20	0.2512E+01	0.5732E+01	0.2419E+01	0.2652E+01
21	14.81	64.02	84.91	63.64	64.76	0.2130E+01	0.5732E+01	0.1814E+01	0.2257E+01
23	17.68	68.77	101.36	67.94	70.66	0.1658E+01	0.5732E+01	0.2093E+01	0.1869E+01
25	21.45	74.72	122.97	74.79	76.95	0.1421E+01	0.5732E+01	0.1587E+01	0.1491E+01
27	24.05	78.28	137.88	78.61	80.56	0.1319E+01	0.5732E+01	0.1372E+01	0.1291E+01
29	26.75	81.84	153.35	82.10	83.81	0.1319E+01	0.5732E+01	0.1223E+01	0.1122E+01
31	29.95	85.73	171.68	85.81	87.13	0.1135E+01	0.5732E+01	0.1107E+01	0.9610E+00
33	33.42	89.54	191.54	89.50	90.21	0.1067E+01	0.5732E+01	0.1027E+01	0.8217E+00
35	37.01	93.16	212.15	93.09	92.95	0.9526E+00	0.5732E+01	0.9741E+00	0.7062E+00
37	40.71	96.71	233.33	96.62	95.38	0.9681E+00	0.5731E+01	0.9391E+00	0.6106E+00
39	44.47	100.19	254.89	100.11	97.52	0.8872E+00	0.5731E+01	0.9163E+00	0.5315E+00
41	48.06	103.34	275.49	103.38	99.32	0.8623E+00	0.5731E+01	0.9025E+00	0.4692E+00

A.1 – 13

I - 7

Connection type : Single web-angle connections
Mode : All bolted

Tested by : S.L.Lipson (1968) Canada
Test Id. : AA-5/1

Column : --
Beam : W21X62
Angle : 4 X 3.5 X 1/4

Fasteners: A325- -3/4"D
13/16" Oversize holes
Material : A36
Fy = -- ksi
Fu = -- ksi

Major parameters

lp = 14.5000"	lu = 1.7500"	ll = 1.7500"	ta = 0.2500"
gb = 2.2500"	gc = 2.5625"	cu = 1.2500"	cl = 1.2500"
pb = 3.0000"	pc = 3.0000"	nb = 1 X 5	nc = 1 X 5

Remark 1) Connection subject to pure moment.
2) Hardened washers installed in both legs

No	Moment (k-in)	Rotation (radians) X 1/1000	No	Moment (k-in)	Rotation (radians) X 1/1000
1	0.00	0.00	26	156.20	23.02
2	7.42	0.55	27	160.17	24.58
3	14.85	1.10	28	164.82	26.30
4	22.27	1.66	29	169.46	28.02
5	29.09	2.18	30	173.85	30.08
6	35.90	2.69	31	178.23	32.14
7	43.18	3.14	32	181.91	34.36
8	50.46	3.60	33	185.59	36.59
9	57.05	4.11	34	187.89	38.16
10	63.64	4.63	35	190.18	39.73
11	70.90	5.28	36	192.48	41.31
12	78.16	5.94	37	194.84	42.89
13	83.82	6.59	38	197.21	44.47
14	89.48	7.24	39	199.38	45.82
15	95.13	8.05	40	201.54	47.17
16	100.77	8.87			
17	105.95	9.75			
18	111.13	10.64			
19	120.09	12.74			
20	125.24	13.93			
21	130.39	15.11			
22	135.52	16.53			
23	140.66	17.95			
24	146.44	19.71			
25	152.23	21.47			

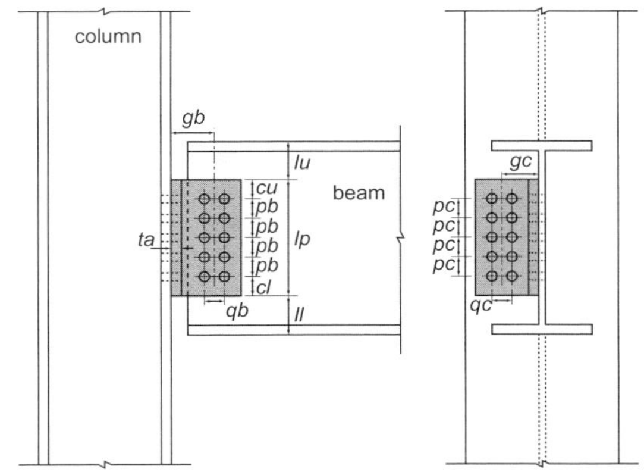

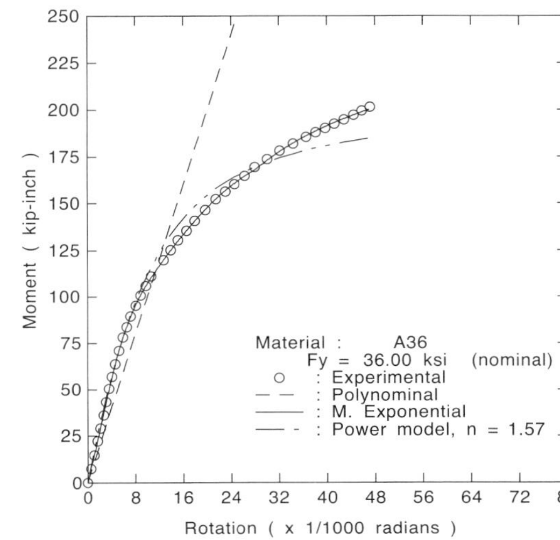

A.1 – 14

```
Moment-rotation prediction equations      ( R : X 1/1000 radians )

Frye and Morris polynominal model  :    R = Sum ( Ai X ( K*Bm )**Pi X 10**Qi )
     xd = 14.500000"    g =  2.762500"   t =  0.250000"
     A1 =  4.280000    A2 =  1.450000   A3 =  1.510000    K =  0.023369
     P1 =         1    P2 =         3   P3 =         5    Q1 =        0    Q2 =      -6    Q3 =     -13

Modified exponential model  :
     Bm = Sum ( Ai X ( 1 - exp( - R/( 2*i*C ) ) ) )  +     Sum ( RKj X ( R - Rj0 ) )  +  Bm0
      C =  0.53120333E+00              Bmo=  0.00000000E+00          Nexp=  6        Nliner=  1
     AI =  0.59232979E+02  -0.56866471E+03  0.73074258E+03  0.23528880E+04  -0.57731910E+04  0.33823696E+04
    Rj0 =         28.0200
    RKj =  0.95351928E+00

Power model  :    Bm = ( Rki X R ) / ( 1 + ( R/R0 )**rn )**( 1/rn )
     rn = 1.571      rki =  0.15543076E+02          rmu =  0.19972423E+03
```

No	Rotation (radians) X 1/1000	Moment (kip-inch) Expri.	Moment (kip-inch) Poly.	Moment (kip-inch) M.Expo.	Moment (kip-inch) P.Model.	Connection stiffness (kip-inch) X 1000 Expri.	Connection stiffness (kip-inch) X 1000 Poly.	Connection stiffness (kip-inch) X 1000 M.Expo.	Connection stiffness (kip-inch) X 1000 P.Model.
1	0.00	0.00	0.00	0.00	0.00	0.1344E+02	0.9998E+01	0.1487E+02	0.1554E+02
3	1.10	14.85	11.05	14.60	16.95	0.1344E+02	0.9998E+01	0.1306E+02	0.1502E+02
5	2.18	29.09	21.75	29.32	32.55	0.1317E+02	0.9998E+01	0.1435E+02	0.1410E+02
7	3.14	43.18	31.43	43.33	45.74	0.1613E+02	0.9998E+01	0.1435E+02	0.1311E+02
9	4.11	57.05	41.12	56.75	57.94	0.1274E+02	0.9998E+01	0.1323E+02	0.1207E+02
11	5.28	70.90	52.82	71.06	71.33	0.1112E+02	0.9998E+01	0.1118E+02	0.1082E+02
13	6.59	83.82	65.84	84.11	84.56	0.8712E+01	0.9998E+01	0.8898E+01	0.9513E+01
15	8.05	95.13	80.51	95.60	97.53	0.6904E+01	0.9998E+01	0.6891E+01	0.8184E+01
17	9.75	105.95	97.52	105.96	110.29	0.5861E+01	0.9998E+01	0.5426E+01	0.6859E+01
19	12.74	120.09	127.37	120.20	127.92	0.4315E+01	0.9998E+01	0.4321E+01	0.5055E+01
21	15.11	130.39	151.09	129.99	138.61	0.4010E+01	0.9998E+01	0.3967E+01	0.4005E+01
23	17.95	140.66	179.51	140.75	148.60	0.3467E+01	0.9998E+01	0.3591E+01	0.3075E+01
25	21.47	152.23	214.67	152.37	157.91	0.2901E+01	0.9998E+01	0.2992E+01	0.2270E+01
27	24.58	160.17	245.73	160.74	164.14	0.2623E+01	0.9998E+01	0.2396E+01	0.1771E+01
29	28.02	169.46	280.17	167.90	169.52	0.2438E+01	0.9998E+01	0.2725E+01	0.1374E+01
31	32.14	178.23	321.33	177.81	174.45	0.1902E+01	0.9997E+01	0.2117E+01	0.1039E+01
33	36.59	185.59	365.81	186.13	178.48	0.1540E+01	0.9997E+01	0.1654E+01	0.7899E+00
35	39.73	190.18	397.25	190.96	180.75	0.1459E+01	0.9997E+01	0.1431E+01	0.6601E+00
37	42.89	194.84	428.80	195.21	182.67	0.1496E+01	0.9997E+01	0.1273E+01	0.5573E+00
39	45.82	199.38	458.11	198.78	184.19	0.1604E+01	0.9997E+01	0.1171E+01	0.4802E+00

A.1 – 15

I - 8

Connection type : Single web-angle connections
Mode : All bolted

Tested by : S.L.Lipson (1968) Canada
Test Id. : AA-5/2

Column : --
Beam : W21X62
Angle : 4 X 3.5 X 1/4

Fasteners: A325- -3/4"D
13/16" Oversize holes
Material : A36
Fy = -- ksi
Fu = -- ksi

Major parameters

lp = 14.5000"	lu = 1.7500"	ll = 1.7500"	ta = 0.2500"
gb = 2.2500"	gc = 2.5625"	cu = 1.2500"	cl = 1.2500"
pb = 3.0000"	pc = 3.0000"	nb = 1 X 5	nc = 1 X 5

Remark 1) Connection subject to pure moment.
2) Hardened washers installed in both legs

No	Moment (k-in)	Rotation (radians) X 1/1000	No	Moment (k-in)	Rotation (radians) X 1/1000
1	0.00	0.00	26	170.52	17.35
2	7.60	0.33	27	175.42	18.64
3	15.21	0.66	28	178.84	19.58
4	22.81	0.99	29	182.26	20.53
5	30.79	1.38	30	185.67	21.48
6	38.77	1.76	31	190.12	22.86
7	46.05	2.21	32	194.56	24.25
8	53.34	2.67	33	197.87	25.43
9	61.31	3.12	34	201.19	26.61
10	69.28	3.57	35	203.18	27.32
11	76.11	4.06	36	205.17	28.03
12	82.93	4.54			
13	89.07	5.03			
14	95.20	5.51			
15	102.70	6.10			
16	110.20	6.68			
17	115.62	7.47			
18	121.04	8.25			
19	130.99	9.55			
20	136.63	10.40			
21	142.27	11.25			
22	147.88	12.40			
23	153.49	13.56			
24	159.55	14.81			
25	165.61	16.07			

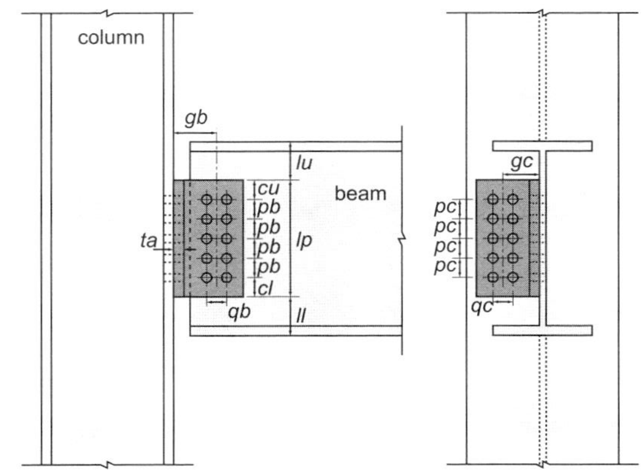

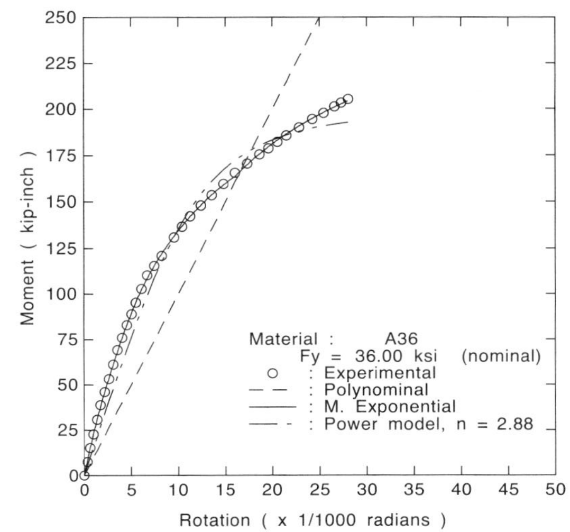

A.1 – 16

```
Moment-rotation prediction equations      ( R : X 1/1000 radians )

Frye and Morris polynominal model  :    R = Sum ( Ai X ( K*Bm )**Pi X 10**Qi )
     xd = 14.500000"    g =  2.762500"   t =  0.250000"
     A1 =  4.280000    A2 =  1.450000   A3 =  1.510000    K =  0.023369
     P1 =         1    P2 =         3   P3 =         5   Q1 =         0    Q2 =       -6     Q3 =      -13

Modified exponential model  :
     Bm = Sum ( Ai X ( 1 - exp( - R/( 2*i*C ) ) ) )  +      Sum ( RKj X ( R - Rj0 ) )  +  Bm0
      C =   0.31604667E+00              Bmo=   0.00000000E+00            Nexp=  6        Nliner=  1
     AI =  -0.39373981E+02   0.75585378E+03  -0.44985958E+04   0.11518106E+05  -0.13365073E+05   0.58129608E+04
    Rj0 =          16.0700
    RKj =   0.18634465E+01

Power model  :    Bm = ( Rki X R ) / ( 1 + ( R/R0 )**rn )**( 1/rn )
      rn = 2.877       rki =   0.15543076E+02          rmu =   0.19972423E+03
```

No	Rotation (radians) X 1/1000	Moment (kip-inch) Expri.	Moment (kip-inch) Poly.	Moment (kip-inch) M.Expo.	Moment (kip-inch) P.Model.	Connection stiffness (kip-inch) X 1000 Expri.	Connection stiffness (kip-inch) X 1000 Poly.	Connection stiffness (kip-inch) X 1000 M.Expo.	Connection stiffness (kip-inch) X 1000 P.Model.
1	0.00	0.00	0.00	0.00	0.00	0.2305E+02	0.9998E+01	0.2272E+02	0.1554E+02
3	0.66	15.21	6.59	15.45	10.25	0.2305E+02	0.9998E+01	0.2278E+02	0.1554E+02
5	1.38	30.79	13.76	30.65	21.37	0.2064E+02	0.9998E+01	0.1977E+02	0.1551E+02
7	2.21	46.05	22.14	46.24	34.35	0.1611E+02	0.9998E+01	0.1764E+02	0.1541E+02
9	3.12	61.31	31.20	61.56	48.22	0.1757E+02	0.9998E+01	0.1625E+02	0.1519E+02
11	4.06	76.11	40.58	76.12	62.31	0.1408E+02	0.9998E+01	0.1473E+02	0.1481E+02
13	5.03	89.07	50.26	89.51	76.38	0.1270E+02	0.9998E+01	0.1291E+02	0.1424E+02
15	6.10	102.70	60.95	102.24	91.18	0.1278E+02	0.9998E+01	0.1091E+02	0.1339E+02
17	7.47	115.62	74.66	115.67	108.63	0.6915E+01	0.9998E+01	0.8792E+01	0.1203E+02
19	9.55	130.99	95.46	131.59	131.20	0.7041E+01	0.9998E+01	0.6705E+01	0.9640E+01
21	11.25	142.27	112.48	142.04	145.95	0.5878E+01	0.9998E+01	0.5632E+01	0.7712E+01
23	13.56	153.49	135.54	153.69	161.06	0.4847E+01	0.9998E+01	0.4498E+01	0.5484E+01
25	16.07	165.61	160.63	163.56	172.45	0.4329E+01	0.9998E+01	0.3383E+01	0.3699E+01
27	18.64	175.42	186.34	175.71	180.27	0.3697E+01	0.9998E+01	0.4235E+01	0.2470E+01
29	20.53	182.26	205.28	183.12	184.33	0.3608E+01	0.9998E+01	0.3619E+01	0.1850E+01
31	22.86	190.12	228.60	190.84	187.98	0.3205E+01	0.9998E+01	0.3030E+01	0.1316E+01
33	25.43	197.87	254.26	198.00	190.82	0.2806E+01	0.9998E+01	0.2577E+01	0.9229E+00
35	27.32	203.18	273.17	202.65	192.36	0.2806E+01	0.9998E+01	0.2350E+01	0.7211E+00

A.1 – 17

I - 9

Connection type : Single web-angle connections
Mode : All bolted

Tested by : S.L.Lipson (1968) Canada
Test Id. : AA-6/1

Column : --
Beam : W21X62
Angle : 4 X 3.5 X 1/4

Fasteners: A325- -3/4"D
13/16" Oversize holes
Material : A36
Fy = -- ksi
Fu = -- ksi

Major parameters

lp = 17.5000"	lu = 1.7500"	ll = 1.7500"	ta = 0.2500"
gb = 2.2500"	gc = 2.5625"	cu = 1.2500"	cl = 1.2500"
pb = 3.0000"	pc = 3.0000"	nb = 1 X 6	nc = 1 X 6

Remark 1) Connection subject to pure moment.
2) Hardened washers installed in both legs

No	Moment (k-in)	Rotation (radians) X 1/1000	No	Moment (k-in)	Rotation (radians) X 1/1000
1	0.00	0.00	26	237.94	23.45
2	10.04	0.26	27	243.96	24.94
3	20.08	0.52	28	250.38	26.86
4	35.15	0.82	29	256.81	28.79
5	47.46	1.25	30	263.01	30.72
6	60.23	1.69	31	269.20	32.65
7	68.88	2.04	32	274.71	34.64
8	77.54	2.40	33	280.21	36.63
9	87.33	2.86	34	285.27	38.52
10	97.12	3.32	35	290.32	40.41
11	106.91	3.77	36	294.49	42.10
12	116.70	4.23	37	298.65	43.79
13	125.10	4.86	38	306.40	46.19
14	133.50	5.48	39	314.14	48.59
15	148.46	6.79			
16	160.24	7.89			
17	170.60	9.39			
18	181.84	11.23			
19	189.89	12.93			
20	196.85	14.15			
21	203.81	15.38			
22	210.73	17.01			
23	217.64	18.63			
24	224.79	20.29			
25	231.93	21.95			

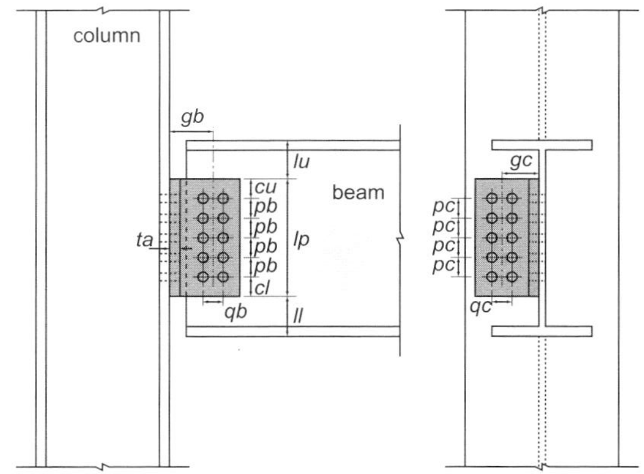

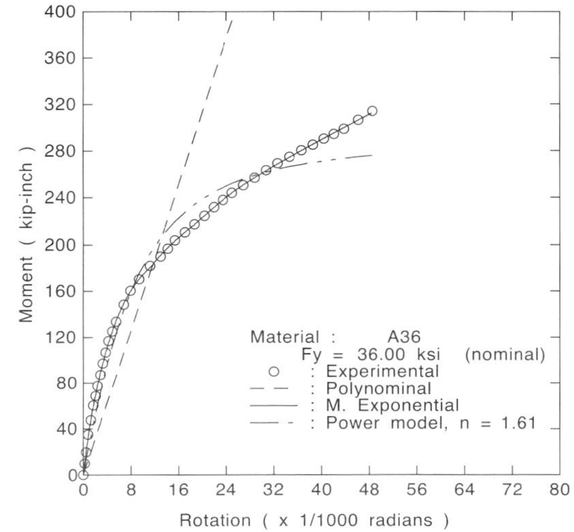

A.1 – 18

```
Moment-rotation prediction equations     ( R : X 1/1000 radians )

Frye and Morris polynominal model  :    R = Sum ( Ai X ( K*Bm )**Pi X 10**Qi )
     xd = 17.500000"   g =  2.762500"   t =  0.250000"
     A1 =  4.280000   A2 =  1.450000   A3 =  1.510000    K =  0.014881
     P1 =        1    P2 =        3    P3 =        5    Q1 =        0    Q2 =      -6    Q3 =     -13

Modified exponential model  :
     Bm = Sum ( Ai X ( 1 - exp( - R/( 2*i*C ) ) ) )  +      Sum ( RKj X ( R - Rj0 ) )  +  Bm0
      C =   0.46954583E+00              Bmo=   0.00000000E+00            Nexp=  6        Nliner=  3
     AI =   0.18867636E+02   0.46790588E+02  -0.88864194E+03   0.30074935E+04  -0.37645101E+04   0.17958368E+04
    Rj0 =          11.2300          18.6300          24.9400
    RKj =   0.21486458E+01   0.84925674E+00  -0.36341194E+00

Power model  :    Bm = ( Rki X R ) / ( 1 + ( R/R0 )**rn )**( 1/rn )
      rn = 1.610      rki =   0.26879042E+02         rmu =   0.29091817E+03
```

No	Rotation (radians) X 1/1000	Moment (kip-inch) Expri.	Poly.	M.Expo.	P.Model.	Connection stiffness (kip-inch) X 1000 Expri.	Poly.	M.Expo.	P.Model.
1	0.00	0.00	0.00	0.00	0.00	0.3884E+02	0.1570E+02	0.4720E+02	0.2688E+02
3	0.52	20.08	8.12	21.92	13.83	0.4379E+02	0.1570E+02	0.3828E+02	0.2656E+02
5	1.25	47.46	19.66	47.02	33.03	0.2901E+02	0.1570E+02	0.3066E+02	0.2558E+02
7	2.04	68.88	32.05	69.07	52.66	0.2432E+02	0.1570E+02	0.2562E+02	0.2416E+02
9	2.86	87.33	44.84	88.36	71.67	0.2132E+02	0.1570E+02	0.2187E+02	0.2246E+02
11	3.77	106.91	59.26	106.83	91.38	0.2132E+02	0.1570E+02	0.1846E+02	0.2046E+02
13	4.86	125.10	76.27	124.95	112.27	0.1346E+02	0.1570E+02	0.1510E+02	0.1812E+02
15	6.79	148.46	106.64	149.36	143.57	0.1102E+02	0.1570E+02	0.1041E+02	0.1436E+02
17	9.39	170.60	147.50	170.65	175.52	0.6549E+01	0.1570E+02	0.6328E+01	0.1040E+02
19	12.93	189.89	203.00	191.01	205.44	0.5291E+01	0.1570E+02	0.5647E+01	0.6815E+01
21	15.38	203.81	241.45	203.51	220.01	0.5069E+01	0.1570E+02	0.4648E+01	0.5183E+01
23	18.63	217.64	292.54	217.24	234.29	0.4275E+01	0.1570E+02	0.4706E+01	0.3701E+01
25	21.95	231.93	344.70	231.96	244.82	0.4160E+01	0.1570E+02	0.4189E+01	0.2705E+01
27	24.94	243.96	391.51	243.93	251.92	0.3729E+01	0.1570E+02	0.3856E+01	0.2090E+01
29	28.79	256.81	452.04	256.76	258.85	0.3274E+01	0.1570E+02	0.3181E+01	0.1541E+01
31	32.65	269.20	512.56	268.59	264.03	0.2992E+01	0.1570E+02	0.2970E+01	0.1169E+01
33	36.63	280.21	575.13	280.13	268.12	0.2717E+01	0.1570E+02	0.2831E+01	0.9012E+00
35	40.41	290.32	634.51	290.67	271.17	0.2565E+01	0.1570E+02	0.2749E+01	0.7182E+00
37	43.79	298.65	687.50	299.86	273.37	0.2781E+01	0.1570E+02	0.2704E+01	0.5950E+00
39	48.59	314.14	762.88	312.75	275.90	0.3227E+01	0.1570E+02	0.2668E+01	0.4645E+00

A.1 – 19

I - 10

Connection type : Single web-angle connections
Mode : All bolted

Tested by : S.L.Lipson (1968) Canada
Test Id. : AA-6/2

Column : --
Beam : W21X62
Angle : 4 X 3.5 X 1/4

Fasteners: A325- -3/4"D
13/16" Oversize holes
Material : A36
Fy = -- ksi
Fu = -- ksi

Major parameters

lp = 17.5000"	lu = 1.7500"	ll = 1.7500"	ta = 0.2500"
gb = 2.2500"	gc = 2.5625"	cu = 1.2500"	cl = 1.2500"
pb = 3.0000"	pc = 3.0000"	nb = 1 X 6	nc = 1 X 6

Remark 1) Connection subject to pure moment.
2) Hardened washers installed in both legs

No	Moment (k-in)	Rotation (radians) X 1/1000	No	Moment (k-in)	Rotation (radians) X 1/1000
1	0.00	0.00	26	247.76	15.81
2	14.82	0.34	27	255.82	17.24
3	29.64	0.68	28	263.88	18.67
4	45.15	0.92	29	270.81	20.03
5	54.26	1.18	30	277.73	21.39
6	63.38	1.44	31	283.74	22.79
7	73.62	1.76	32	289.75	24.18
8	83.87	2.09	33	295.30	25.64
9	93.65	2.49	34	300.84	27.10
10	103.44	2.88	35	307.05	28.69
11	111.85	3.27	36	313.27	30.29
12	120.26	3.66	37	319.69	32.02
13	129.80	4.12	38	326.12	33.75
14	139.35	4.58	39	329.18	34.96
15	147.98	5.08	40	332.24	36.18
16	156.61	5.57			
17	166.13	6.26			
18	175.65	6.96			
19	184.70	7.76			
20	193.75	8.55			
21	203.00	9.58			
22	212.25	10.61			
23	220.34	11.81			
24	228.43	13.01			
25	238.09	14.41			

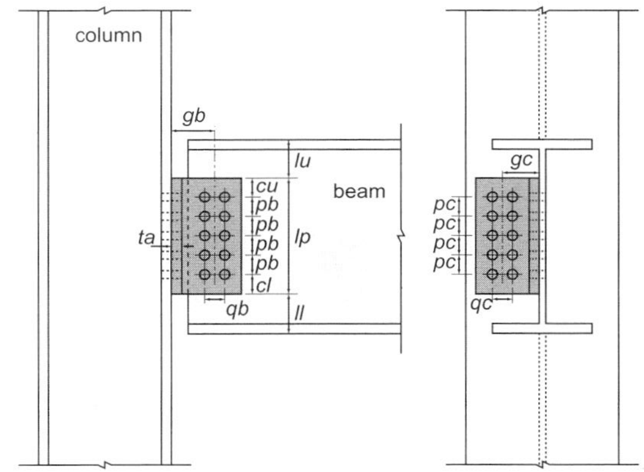

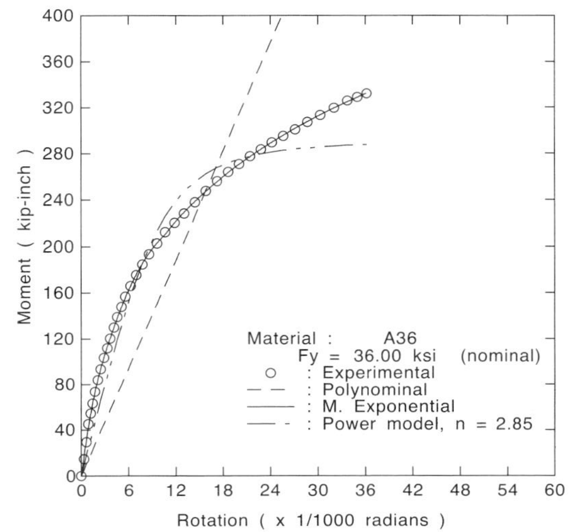

```
Moment-rotation prediction equations     ( R : X 1/1000 radians )

Frye and Morris polynominal model   :    R = Sum ( Ai X ( K*Bm )**Pi X 10**Qi )
     xd = 17.500000"    g =  2.762500"   t =  0.250000"
     A1 =  4.280000    A2 =  1.450000   A3 =  1.510000     K =  0.014881
     P1 =        1     P2 =        3    P3 =        5     Q1 =        0     Q2 =      -6     Q3 =     -13

Modified exponential model  :
     Bm = Sum ( Ai X ( 1 - exp( - R/( 2*i*C ) ) ) )  +      Sum ( RKj X ( R - Rj0 ) )  +  Bm0
      C =   0.38642583E+00            Bmo=   0.00000000E+00          Nexp=  6        Nliner=  2
     AI =  -0.14274756E+03   0.19325879E+04  -0.87476896E+04   0.18204321E+05  -0.17815485E+05   0.68285046E+04
     Rj0 =            10.6100           18.6700
     RKj =   0.30365395E+01  -0.16767093E+00

Power model  :    Bm = ( Rki X R ) / ( 1 + ( R/R0 )**rn )**( 1/rn )
      rn = 2.847      rki =   0.26879042E+02          rmu =   0.29091817E+03
```

No	Rotation (radians) X 1/1000	Moment (kip-inch) Expri.	Poly.	M.Expo.	P.Model.	Connection stiffness (kip-inch) X 1000 Expri.	Poly.	M.Expo.	P.Model.
1	0.00	0.00	0.00	0.00	0.00	0.4362E+02	0.1570E+02	0.4363E+02	0.2688E+02
3	0.68	29.64	10.67	31.91	18.26	0.5582E+02	0.1570E+02	0.4607E+02	0.2687E+02
5	1.18	54.26	18.50	53.30	31.65	0.3531E+02	0.1570E+02	0.3939E+02	0.2681E+02
7	1.76	73.62	27.70	74.03	47.33	0.3122E+02	0.1570E+02	0.3164E+02	0.2667E+02
9	2.49	93.65	39.05	94.35	66.49	0.2482E+02	0.1570E+02	0.2520E+02	0.2634E+02
11	3.27	111.85	51.37	112.50	86.94	0.2153E+02	0.1570E+02	0.2147E+02	0.2572E+02
13	4.12	129.80	64.74	129.78	108.44	0.2072E+02	0.1570E+02	0.1928E+02	0.2472E+02
15	5.08	147.98	79.70	147.25	131.29	0.1753E+02	0.1570E+02	0.1738E+02	0.2318E+02
17	6.26	166.13	98.35	166.40	157.44	0.1368E+02	0.1570E+02	0.1480E+02	0.2076E+02
19	7.76	184.70	121.77	185.87	185.83	0.1138E+02	0.1570E+02	0.1133E+02	0.1727E+02
21	9.58	203.00	150.45	203.08	213.46	0.8971E+01	0.1570E+02	0.7697E+01	0.1305E+02
23	11.81	220.34	185.42	220.43	237.58	0.6762E+01	0.1570E+02	0.7938E+01	0.8816E+01
25	14.41	238.09	226.22	238.77	255.77	0.6895E+01	0.1570E+02	0.6399E+01	0.5449E+01
27	17.24	255.82	270.70	255.72	267.82	0.5632E+01	0.1570E+02	0.5649E+01	0.3260E+01
29	20.03	270.81	314.54	270.55	275.04	0.5089E+01	0.1570E+02	0.4978E+01	0.2027E+01
31	22.79	283.74	357.78	283.61	279.56	0.4315E+01	0.1570E+02	0.4514E+01	0.1315E+01
33	25.64	295.30	402.55	295.84	282.62	0.3801E+01	0.1570E+02	0.4071E+01	0.8714E+00
35	28.69	307.05	450.48	307.65	284.81	0.3897E+01	0.1570E+02	0.3679E+01	0.5822E+00
37	32.02	319.69	502.67	319.33	286.40	0.3716E+01	0.1570E+02	0.3367E+01	0.3901E+00
39	34.96	329.18	548.94	328.95	287.38	0.2512E+01	0.1570E+02	0.3180E+01	0.2817E+00

I - 11

Connection type : Single web-angle connections
Mode : All bolted

Tested by : S.L.Lipson (1968) Canada
Test Id. : BB-4/1

Column : --
Beam : W21X62
Angle : 4 X 3.5 X 5/16

Fasteners: A325- -3/4"D
13/16" Oversize holes
Material : A36
Fy = -- ksi
Fu = -- ksi

Major parameters

lp = 11.5000"	lu = 1.7500"	ll = 1.7500"	ta = 0.3125"
gb = 2.2500"	gc = 2.5625"	cu = 1.2500"	cl = 1.2500"
pb = 3.0000"	pc = 3.0000"	nb = 1 X 4	nc = 1 X 4

Remark 1) Connection subject to pure moment.
2) Hardened washers installed in both legs

No	Moment (k-in)	Rotation (radians) X 1/1000	No	Moment (k-in)	Rotation (radians) X 1/1000
1	0.00	0.00	26	109.89	30.09
2	4.65	0.55	27	112.78	32.33
3	9.69	1.04	28	115.27	34.63
4	15.53	1.86	29	117.76	36.93
5	21.51	2.41	30	120.64	39.69
6	27.61	3.17	31	123.52	42.46
7	31.13	3.65	32	125.67	45.12
8	34.64	4.13	33	127.82	47.79
9	38.23	4.57	34	129.90	50.25
10	41.81	5.02	35	131.99	52.72
11	48.45	5.98	36	133.48	54.78
12	54.28	7.06	37	134.98	56.84
13	59.18	8.28	38	136.21	58.74
14	65.54	9.96	39	137.44	60.64
15	69.37	11.14			
16	73.21	12.32			
17	77.44	13.79			
18	81.67	15.27			
19	85.57	16.78			
20	89.47	18.29			
21	95.81	21.24			
22	98.77	22.91			
23	101.74	24.58			
24	104.37	26.22			
25	107.00	27.86			

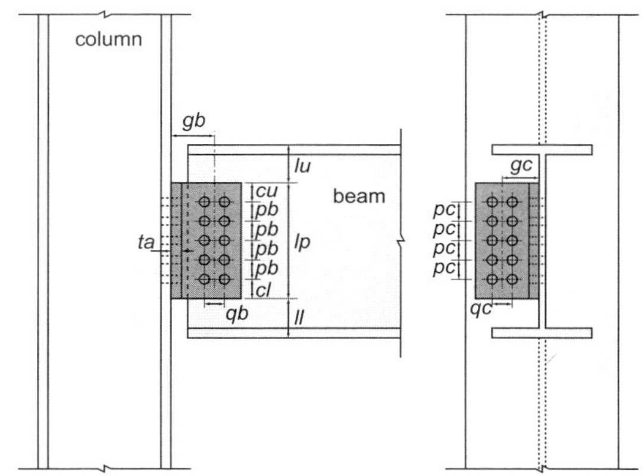

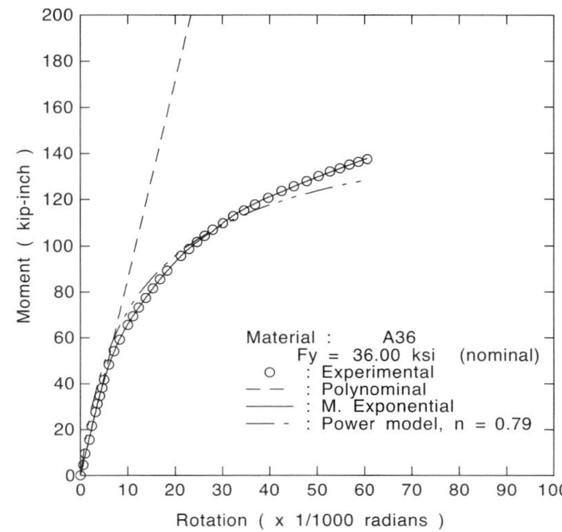

A.1 – 22

```
Moment-rotation prediction equations     ( R : X 1/1000 radians )

Frye and Morris polynominal model  :    R = Sum ( Ai X ( K*Bm )**Pi X 10**Qi )
     xd = 11.500000"   g =  2.762500"  t =  0.312500"
     A1 =  4.280000   A2 =  1.450000  A3 =  1.510000    K =  0.027217
     P1 =        1    P2 =        3   P3 =        5    Q1 =        0     Q2 =       -6     Q3 =      -13

Modified exponential model  :
     Bm = Sum ( Ai X ( 1 - exp( - R/( 2*i*C ) ) ) )  +     Sum ( RKj X ( R - Rj0 ) )  +  Bm0
       C =   0.64285167E+00          Bmo=   0.00000000E+00          Nexp=  6        Nliner=  2
      AI =   0.45751894E+02  -0.75951420E+03  0.33127018E+04  -0.57815929E+04   0.41378111E+04  -0.84062951E+03
     Rj0 =          27.8600          42.4600
     RKj =   0.64096720E+00   0.12267807E+00

Power model  :    Bm = ( Rki X R ) / ( 1 + ( R/R0 )**rn )**( 1/rn )
      rn = 0.786      rki =   0.18062801E+02         rmu =   0.16669996E+03
```

No	Rotation (radians) X 1/1000	Moment (kip-inch) Expri.	Poly.	M.Expo.	P.Model.	Connection stiffness (kip-inch) X 1000 Expri.	Poly.	M.Expo.	P.Model.
1	0.00	0.00	0.00	0.00	0.00	0.8444E+01	0.8584E+01	0.9557E+01	0.1806E+02
3	1.04	9.69	8.89	9.11	15.17	0.9173E+01	0.8584E+01	0.8522E+01	0.1242E+02
5	2.41	21.51	20.71	20.97	29.80	0.9620E+01	0.8584E+01	0.8649E+01	0.9164E+01
7	3.65	31.13	31.31	31.38	39.95	0.7338E+01	0.8584E+01	0.8098E+01	0.7392E+01
9	4.57	38.23	39.25	38.53	46.33	0.8030E+01	0.8584E+01	0.7315E+01	0.6430E+01
11	5.98	48.45	51.29	47.84	54.54	0.6202E+01	0.8584E+01	0.5965E+01	0.5336E+01
13	8.28	59.18	71.05	59.37	65.31	0.3923E+01	0.8584E+01	0.4202E+01	0.4114E+01
15	11.14	69.37	95.63	69.66	75.59	0.3259E+01	0.8584E+01	0.3156E+01	0.3142E+01
17	13.79	77.44	118.42	77.50	83.09	0.2863E+01	0.8584E+01	0.2807E+01	0.2540E+01
19	16.78	85.57	144.06	85.52	89.92	0.2583E+01	0.8584E+01	0.2555E+01	0.2061E+01
21	21.24	95.81	182.32	95.81	97.94	0.1910E+01	0.8584E+01	0.2032E+01	0.1576E+01
23	24.58	101.74	211.01	101.82	102.77	0.1690E+01	0.8584E+01	0.1562E+01	0.1323E+01
25	27.86	107.00	239.12	106.22	106.78	0.1475E+01	0.8584E+01	0.1131E+01	0.1133E+01
27	32.33	112.78	277.50	113.03	111.38	0.1189E+01	0.8584E+01	0.1308E+01	0.9363E+00
29	36.93	117.76	316.99	118.28	115.32	0.1063E+01	0.8584E+01	0.9963E+00	0.7856E+00
31	42.46	123.52	364.48	123.15	119.28	0.9219E+00	0.8584E+01	0.7911E+00	0.6502E+00
33	47.79	127.82	410.21	127.74	122.47	0.8278E+00	0.8583E+01	0.8217E+00	0.5518E+00
35	52.72	131.99	452.51	131.69	125.00	0.7803E+00	0.8583E+01	0.7843E+00	0.4804E+00
37	56.84	134.98	487.94	134.89	126.88	0.6854E+00	0.8583E+01	0.7704E+00	0.4312E+00
39	60.64	137.44	520.50	137.80	128.44	0.6495E+00	0.8583E+01	0.7648E+00	0.3927E+00

A.1 – 23

I - 12

Connection type : Single web-angle connections
Mode : All bolted

Tested by : S.L.Lipson (1968) Canada
Test Id. : BB-4/2

Column : --
Beam : W21X62
Angle : 4 X 3.5 X 5/16

Fasteners: A325- -3/4"D
13/16" Oversize holes
Material : A36
Fy = -- ksi
Fu = -- ksi

Major parameters

lp = 11.5000"	lu = 1.7500"	ll = 1.7500"	ta = 0.3125"
gb = 2.2500"	gc = 2.5625"	cu = 1.2500"	cl = 1.2500"
pb = 3.0000"	pc = 3.0000"	nb = 1 X 4	nc = 1 X 4

Remark 1) Connection subject to pure moment.
2) Hardened washers installed in both legs

No	Moment (k-in)	Rotation (radians) X 1/1000	No	Moment (k-in)	Rotation (radians) X 1/1000
1	0.00	0.00	26	113.81	27.59
2	4.25	0.41	27	117.58	29.66
3	9.16	0.96	28	120.74	31.60
4	13.81	1.64	29	123.91	33.53
5	20.85	2.26	30	127.34	35.73
6	26.96	3.01	31	130.77	37.93
7	32.93	3.83	32	132.92	39.51
8	36.58	4.27	33	135.07	41.09
9	40.24	4.71	34	137.22	42.67
10	46.74	5.66	35	140.71	44.97
11	53.38	6.62	36	144.21	47.27
12	59.35	7.70			
13	66.51	9.18			
14	70.36	10.12			
15	74.20	11.06			
16	77.84	12.17			
17	81.49	13.28			
18	85.46	14.59			
19	89.43	15.89			
20	92.87	17.33			
21	96.31	18.77			
22	99.61	20.18			
23	102.92	21.58			
24	106.49	23.55			
25	110.05	25.52			

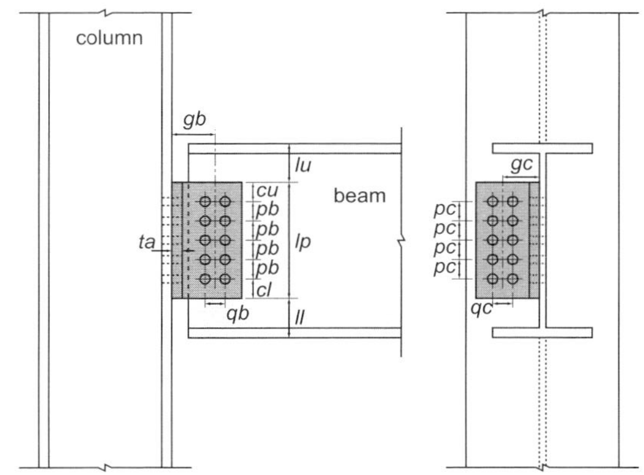

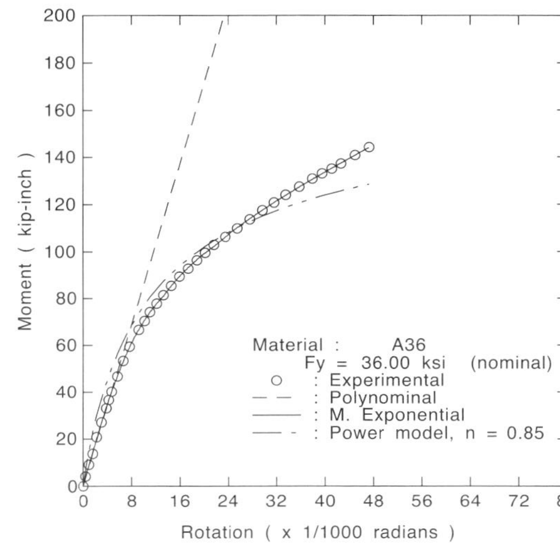

A.1 – 24

```
Moment-rotation prediction equations     ( R : X 1/1000 radians )

Frye and Morris polynominal model  :    R = Sum ( Ai X ( K*Bm )**Pi X 10**Qi )
     xd = 11.500000"    g =  2.762500"   t =  0.312500"
     A1 =  4.280000    A2 =  1.450000   A3 =  1.510000     K =  0.027217
     P1 =         1    P2 =         3   P3 =         5    Q1 =         0    Q2 =       -6    Q3 =      -13

Modified exponential model  :
     Bm = Sum ( Ai X ( 1 - exp( - R/( 2*i*C ) ) ) )  +      Sum ( RKj X ( R - Rj0 ) )  +  Bm0
       C =   0.49699333E+00            Bmo=   0.00000000E+00            Nexp=  6         Nliner=  2
      AI =   0.77717127E+01  -0.40026797E+02   0.33779186E+02   0.55997685E+02  -0.30075810E+03   0.34273301E+03
     Rj0 =          13.2800          21.5800
     RKj =   0.10031083E+01   0.41204916E+00

Power model  :    Bm = ( Rki X R ) / ( 1 + ( R/R0 )**rn )**( 1/rn )
      rn = 0.851      rki =   0.18062801E+02         rmu =   0.16669996E+03
```

No	Rotation (radians) X 1/1000	Moment (kip-inch) Expri.	Poly.	M.Expo.	P.Model.	Connection stiffness (kip-inch) X 1000 Expri.	Poly.	M.Expo.	P.Model.
1	0.00	0.00	0.00	0.00	0.00	0.1031E+02	0.8584E+01	0.1005E+02	0.1806E+02
3	0.96	9.16	8.24	8.92	14.77	0.8025E+01	0.8584E+01	0.8864E+01	0.1344E+02
5	2.26	20.85	19.40	20.22	29.93	0.9893E+01	0.8584E+01	0.8569E+01	0.1017E+02
7	3.83	32.93	32.85	33.25	43.85	0.7911E+01	0.8584E+01	0.7980E+01	0.7782E+01
9	4.71	40.24	40.46	40.10	50.31	0.7791E+01	0.8584E+01	0.7455E+01	0.6823E+01
11	6.62	53.38	56.79	53.06	61.77	0.6295E+01	0.8584E+01	0.6154E+01	0.5326E+01
13	9.18	66.51	78.81	66.67	73.63	0.4371E+01	0.8584E+01	0.4509E+01	0.4019E+01
15	11.06	74.20	94.98	74.21	80.55	0.3718E+01	0.8584E+01	0.3529E+01	0.3359E+01
17	13.28	81.49	113.99	80.98	87.32	0.3175E+01	0.8584E+01	0.2623E+01	0.2783E+01
19	15.89	89.43	136.44	89.37	93.90	0.2730E+01	0.8584E+01	0.2838E+01	0.2283E+01
21	18.77	96.31	161.14	96.61	99.86	0.2372E+01	0.8584E+01	0.2231E+01	0.1880E+01
23	21.58	102.92	185.25	102.27	104.71	0.2127E+01	0.8584E+01	0.2239E+01	0.1586E+01
25	25.52	110.05	219.07	110.32	110.32	0.1814E+01	0.8584E+01	0.1880E+01	0.1280E+01
27	29.66	117.58	254.61	117.62	115.12	0.1724E+01	0.8584E+01	0.1667E+01	0.1049E+01
29	33.53	123.91	287.84	123.84	118.85	0.1599E+01	0.8584E+01	0.1555E+01	0.8867E+00
31	37.93	130.77	325.63	130.51	122.43	0.1444E+01	0.8584E+01	0.1486E+01	0.7455E+00
33	41.09	135.07	352.72	135.16	124.65	0.1363E+01	0.8584E+01	0.1459E+01	0.6647E+00
35	44.97	140.71	386.02	140.77	127.07	0.1518E+01	0.8584E+01	0.1439E+01	0.5828E+00

A.1 – 25

I - 13

Connection type : Single web-angle connections
Mode : All bolted

Tested by : S.L.Lipson (1968) Canada
Test Id. : B2-4

Column : --
Beam : W21X62
Angle : 4 X 3.5 X 5/16

Fasteners: A325- -3/4"D
15/16" Oversize holes
Material : A36
Fy = -- ksi
Fu = -- ksi

Major parameters

lp = 11.5000"	lu = 1.7500"	ll = 1.7500"	ta = 0.3125"
gb = 2.2500"	gc = 2.5625"	cu = 1.2500"	cl = 1.2500"
pb = 3.0000"	pc = 3.0000"	nb = 1 X 4	nc = 1 X 4

Remark 1) Connection subject to moment and shear.
2)

No	Moment (k-in)	Rotation (radians) X 1/1000	No	Moment (k-in)	Rotation (radians) X 1/1000
1	0.00	0.00	26	93.22	12.32
2	3.06	0.24	27	95.81	13.16
3	6.12	0.48	28	98.40	14.00
4	11.31	0.97	29	100.12	14.70
5	14.31	1.14	30	101.84	15.40
6	17.30	1.32	31	103.56	16.24
7	20.63	1.59	32	105.28	17.07
8	23.95	1.87	33	106.60	17.98
9	27.61	2.22	34	107.92	18.88
10	31.27	2.56	35	109.33	19.77
11	34.13	2.83	36	110.74	20.66
12	36.99	3.11	37	112.15	21.54
13	40.39	3.52			
14	43.78	3.93			
15	48.03	4.44			
16	52.29	4.95			
17	56.28	5.50			
18	60.27	6.04			
19	64.79	6.72			
20	69.31	7.40			
21	73.90	8.18			
22	78.48	8.96			
23	82.86	9.97			
24	87.25	10.98			
25	90.24	11.65			

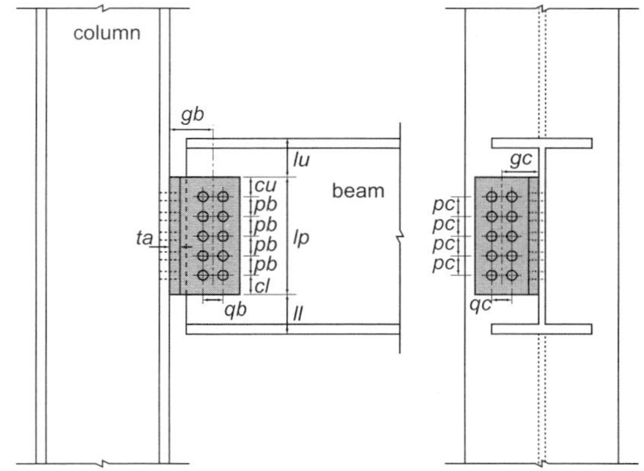

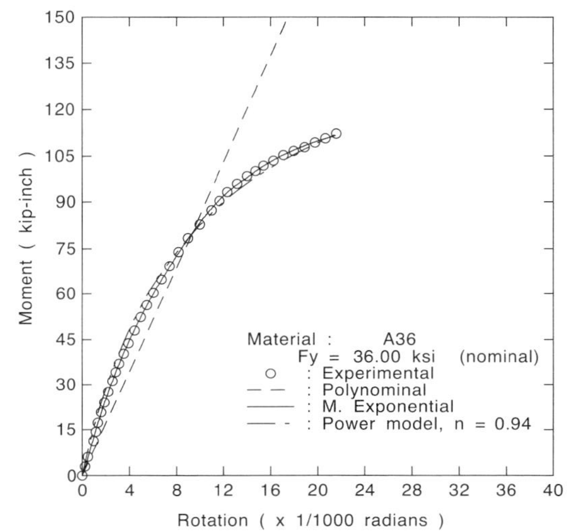

```
Moment-rotation prediction equations      ( R : X 1/1000 radians )

Frye and Morris polynominal model   :    R = Sum ( Ai X ( K*Bm )**Pi X 10**Qi )
     xd = 11.500000"    g =  2.762500"   t =  0.312500"
     A1 =  4.280000   A2 =  1.450000   A3 =  1.510000    K =  0.027217
     P1 =        1    P2 =        3    P3 =        5    Q1 =        0    Q2 =       -6    Q3 =      -13

Modified exponential model  :
     Bm = Sum ( Ai X ( 1 - exp( - R/( 2*i*C ) ) ) )  +      Sum ( RKj X ( R - Rj0 ) )  +  Bm0
      C =   0.24016333E+00             Bmo=   0.00000000E+00          Nexp=  6        Nliner=  1
     AI =  -0.92755206E+01   0.19015663E+03  -0.16869377E+04   0.56013411E+04  -0.76316737E+04   0.36438006E+04
    Rj0 =          14.0000
    RKj =   0.71459023E+00

Power model  :    Bm = ( Rki X R ) / ( 1 + ( R/R0 )**rn )**( 1/rn )
     rn = 0.937      rki =   0.18062801E+02         rmu =   0.16669996E+03
```

No	Rotation (radians) X 1/1000	Moment (kip-inch) Expri.	Moment (kip-inch) Poly.	Moment (kip-inch) M.Expo.	Moment (kip-inch) P.Model.	Connection stiffness (kip-inch) X 1000 Expri.	Connection stiffness (kip-inch) X 1000 Poly.	Connection stiffness (kip-inch) X 1000 M.Expo.	Connection stiffness (kip-inch) X 1000 P.Model.
1	0.00	0.00	0.00	0.00	0.00	0.1262E+02	0.8584E+01	0.9974E+01	0.1806E+02
3	0.48	6.12	4.16	5.84	8.20	0.1201E+02	0.8584E+01	0.1320E+02	0.1591E+02
5	1.14	14.31	9.81	14.60	17.92	0.1703E+02	0.8584E+01	0.1306E+02	0.1374E+02
7	1.59	20.63	13.69	20.34	23.86	0.1203E+02	0.8584E+01	0.1228E+02	0.1254E+02
9	2.22	27.61	19.02	27.59	31.19	0.1064E+02	0.8584E+01	0.1109E+02	0.1115E+02
11	2.83	34.13	24.33	34.11	37.73	0.1040E+02	0.8584E+01	0.9983E+01	0.1000E+02
13	3.52	40.39	30.21	40.57	44.20	0.8284E+01	0.8584E+01	0.8948E+01	0.8937E+01
15	4.44	48.03	38.12	48.31	51.87	0.8314E+01	0.8584E+01	0.7926E+01	0.7768E+01
17	5.50	56.28	47.18	56.26	59.49	0.7331E+01	0.8584E+01	0.7178E+01	0.6700E+01
19	6.72	64.79	57.68	64.66	67.07	0.6661E+01	0.8584E+01	0.6581E+01	0.5727E+01
21	8.18	73.90	70.19	73.75	74.73	0.5890E+01	0.8584E+01	0.5885E+01	0.4828E+01
23	9.97	82.86	85.55	83.38	82.58	0.4339E+01	0.8584E+01	0.4842E+01	0.3995E+01
25	11.65	90.24	100.00	90.62	88.78	0.4436E+01	0.8584E+01	0.3764E+01	0.3397E+01
27	13.16	95.81	112.98	95.61	93.59	0.3086E+01	0.8584E+01	0.2853E+01	0.2970E+01
29	14.70	100.12	126.21	99.88	97.88	0.2452E+01	0.8584E+01	0.2780E+01	0.2614E+01
31	16.24	103.56	139.41	103.66	101.67	0.2060E+01	0.8584E+01	0.2160E+01	0.2321E+01
33	17.98	106.60	154.31	106.94	105.45	0.1466E+01	0.8584E+01	0.1650E+01	0.2046E+01
35	19.77	109.33	169.67	109.55	108.90	0.1584E+01	0.8584E+01	0.1294E+01	0.1812E+01
37	21.54	112.15	184.95	111.63	111.94	0.1584E+01	0.8584E+01	0.1066E+01	0.1617E+01

I - 14

Connection type : Single web-angle connections
Mode : All bolted

Tested by : S.L.Lipson (1968) Canada
Test Id. : B3-4

Column : --
Beam : W21X62
Angle : 4 X 3.5 X 5/16

Fasteners: A325- -3/4"D
-5/16" Oversize holes
Material : A36
Fy = -- ksi
Fu = -- ksi

Major parameters

lp = 11.5000"	lu = 1.7500"	ll = 1.7500"	ta = 0.3125"
gb = 2.2500"	gc = 2.5625"	cu = 1.2500"	cl = 1.2500"
pb = 3.0000"	pc = 3.0000"	nb = 1 X 4	nc = 1 X 4

Remark 1) Connection subject to moment and shear.
2)

No	Moment (k-in)	Rotation (radians) X 1/1000	No	Moment (k-in)	Rotation (radians) X 1/1000
1	0.00	0.00	26	105.26	16.47
2	5.32	0.48	27	108.04	17.95
3	11.84	1.17	28	110.82	19.42
4	15.43	1.62	29	113.06	20.69
5	19.02	2.06	30	115.31	21.96
6	24.87	2.68	31	117.42	23.40
7	30.99	3.17	32	119.53	24.83
8	35.45	3.59	33	121.06	25.99
9	39.91	4.00	34	122.60	27.15
10	43.96	4.38	35	124.14	28.30
11	48.02	4.76	36	126.31	29.91
12	54.54	5.52	37	128.49	31.51
13	58.13	5.93	38	130.13	32.98
14	61.72	6.34	39	131.77	34.45
15	66.31	6.99			
16	70.89	7.64			
17	74.21	8.25			
18	77.54	8.86			
19	81.32	9.71			
20	85.10	10.55			
21	88.75	11.39			
22	92.41	12.24			
23	95.65	13.28			
24	98.90	14.32			
25	102.08	15.40			

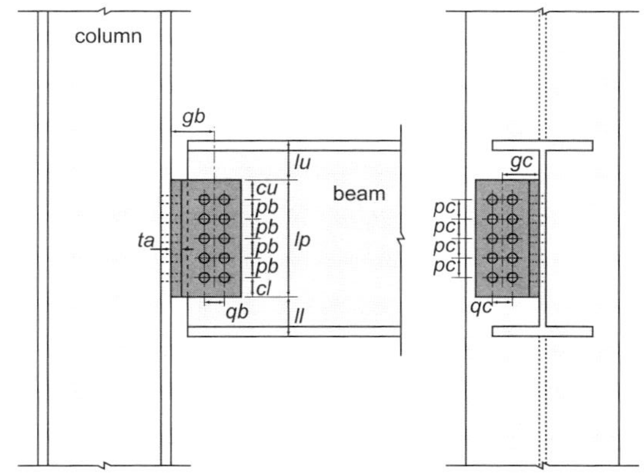

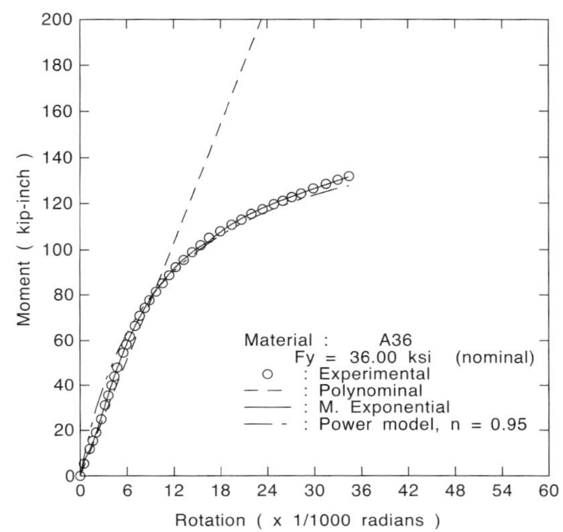

A.1 – 28

```
Moment-rotation prediction equations      ( R : X 1/1000 radians )

Frye and Morris polynominal model  :    R = Sum ( Ai X ( K*Bm )**Pi X 10**Qi )
     xd = 11.500000"    g =  2.762500"   t =  0.312500"
     A1 =  4.280000    A2 =  1.450000   A3 =  1.510000    K =  0.027217
     P1 =        1     P2 =        3    P3 =        5    Q1 =        0    Q2 =       -6    Q3 =      -13

Modified exponential model  :
     Bm = Sum ( Ai X ( 1 - exp( - R/( 2*i*C ) ) ) )  +      Sum ( RKj X ( R - Rj0 ) )  +  Bm0
      C =  0.40829917E+00             Bmo=  0.00000000E+00          Nexp=  6        Nliner=  1
     AI =  0.94259256E+02 -0.95165788E+03  0.37669182E+04 -0.71368433E+04  0.60801595E+04 -0.17436531E+04
    Rj0 =         16.4700
    RKj =  0.12200905E+01

Power model  :    Bm = ( Rki X R ) / ( 1 + ( R/R0 )**rn )**( 1/rn )
      rn = 0.946       rki =   0.18062801E+02          rmu =   0.16669996E+03
```

No	Rotation (radians) X 1/1000	Moment (kip-inch) Expri.	Poly.	M.Expo.	P.Model.	Connection stiffness (kip-inch) X 1000 Expri.	Poly.	M.Expo.	P.Model.
1	0.00	0.00	0.00	0.00	0.00	0.1097E+02	0.8584E+01	0.1871E+02	0.1806E+02
3	1.17	11.84	10.08	11.29	18.42	0.8612E+01	0.8584E+01	0.7370E+01	0.1374E+02
5	2.06	19.02	17.72	18.99	29.64	0.8639E+01	0.8584E+01	0.9939E+01	0.1156E+02
7	3.17	30.99	27.23	30.97	41.27	0.1156E+02	0.8584E+01	0.1120E+02	0.9537E+01
9	4.00	39.91	34.36	40.10	48.69	0.1069E+02	0.8584E+01	0.1063E+02	0.8367E+01
11	4.76	48.02	40.90	47.83	54.71	0.9980E+01	0.8584E+01	0.9641E+01	0.7481E+01
13	5.93	58.13	50.92	58.14	62.78	0.8713E+01	0.8584E+01	0.8056E+01	0.6382E+01
15	6.99	66.31	60.02	66.04	69.11	0.7073E+01	0.8584E+01	0.6883E+01	0.5587E+01
17	8.25	74.21	70.83	74.01	75.64	0.5441E+01	0.8584E+01	0.5839E+01	0.4827E+01
19	9.71	81.32	83.32	81.81	82.14	0.4480E+01	0.8584E+01	0.4920E+01	0.4131E+01
21	11.39	88.75	97.82	89.34	88.56	0.4325E+01	0.8584E+01	0.4013E+01	0.3500E+01
23	13.28	95.65	114.01	96.02	94.63	0.3117E+01	0.8584E+01	0.3080E+01	0.2955E+01
25	15.40	102.08	132.18	101.53	100.36	0.2959E+01	0.8584E+01	0.2146E+01	0.2485E+01
27	17.95	108.04	154.05	107.60	106.12	0.1885E+01	0.8584E+01	0.2477E+01	0.2056E+01
29	20.69	113.06	177.60	113.45	111.26	0.1767E+01	0.8584E+01	0.1835E+01	0.1709E+01
31	23.40	117.42	200.84	117.87	115.52	0.1469E+01	0.8584E+01	0.1468E+01	0.1448E+01
33	25.99	121.06	223.10	121.41	119.01	0.1328E+01	0.8584E+01	0.1286E+01	0.1250E+01
35	28.30	124.14	242.98	124.29	121.73	0.1340E+01	0.8584E+01	0.1208E+01	0.1106E+01
37	31.51	128.49	270.49	128.09	125.01	0.1233E+01	0.8584E+01	0.1172E+01	0.9456E+00
39	34.45	131.77	295.69	131.52	127.61	0.1119E+01	0.8584E+01	0.1172E+01	0.8276E+00

A.1 – 29

I - 15

```
Connection type : Single web-angle connections
           Mode : All bolted

Tested by  :  S.L.Lipson (1968)                          Canada
Test  Id.  :  C-4

Column :  --                         Fasteners: A325- -3/4"D
Beam   : W21X62                            15/16" Oversize holes
Angle  : 5 X 3.5 X 5/16              Material : A36
                                          Fy =   --   ksi
                                          Fu =   --   ksi

                          Major parameters

     lp = 11.5000"   lu =  1.7500'   ll =  1.7500"   ta =  0.3125"
     gb =  2.2500"   gc =  1.9375'   cu =  1.2500"   cl =  1.2500"
     pb =  3.0000"   pc =  3.0000'   nb = 1 X  4     nc = 1 X  4

Remark  1)  Connection subject to moment and shear.
        2)
```

No	Moment (k-in)	Rotation (radians) X 1/1000	No	Moment (k-in)	Rotation (radians) X 1/1000
1	0.00	0.00	26	130.88	8.25
2	6.60	0.11	27	133.22	8.79
3	13.21	0.23	28	135.57	9.33
4	19.81	0.34	29	137.91	9.87
5	25.66	0.69	30	140.47	10.50
6	31.51	1.05	31	143.03	11.13
7	37.62	1.16	32	145.59	11.76
8	43.74	1.27	33	147.54	12.41
9	49.85	1.38	34	149.50	13.07
10	57.43	1.67	35	151.45	13.72
11	61.82	1.79	36	153.40	14.38
12	66.21	1.91			
13	73.25	2.33			
14	78.10	2.65			
15	82.95	2.96			
16	90.52	3.32			
17	94.44	3.67			
18	98.36	4.02			
19	102.41	4.46			
20	106.46	4.91			
21	110.97	5.43			
22	115.49	5.94			
23	119.73	6.59			
24	123.98	7.23			
25	127.43	7.74			

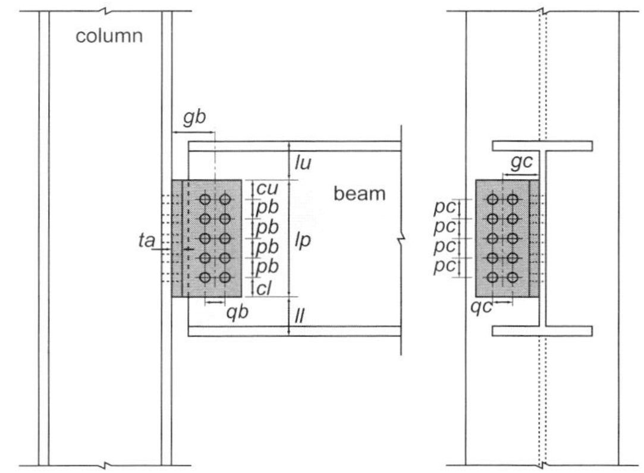

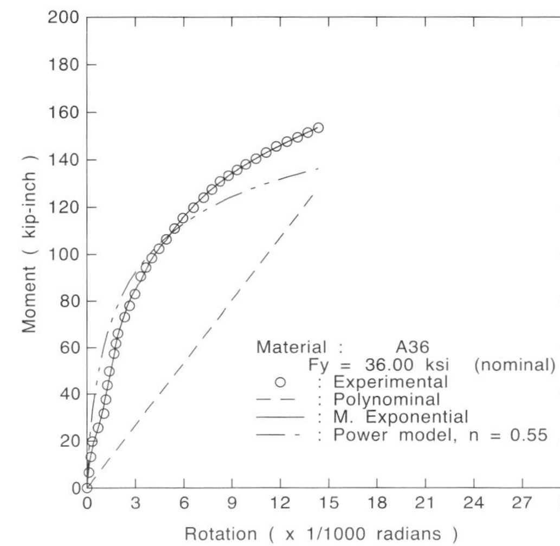

A.1 – 30

```
Moment-rotation prediction equations     ( R : X 1/1000 radians )

Frye and Morris polynominal model  :    R = Sum ( Ai X ( K*Bm )**Pi X 10**Qi )
     xd = 11.500000"   g =  2.137500"   t =  0.312500"
     A1 =  4.280000   A2 =  1.450000   A3 =  1.510000    K =  0.026190
     P1 =        1    P2 =        3    P3 =        5    Q1 =        0    Q2 =       -6    Q3 =      -13

Modified exponential model  :
     Bm = Sum ( Ai X ( 1 - exp( - R/( 2*i*C ) ) ) )  +     Sum ( RKj X ( R - Rj0 ) )  +  Bm0
      C =   0.14809083E+00            Bmo=   0.00000000E+00            Nexp=  6        Nliner=  4
     AI =   0.63363439E+02  0.13333942E+03 -0.20505799E+04  0.52958687E+04 -0.57682947E+04  0.24686708E+04
     Rj0 =          1.9100          3.3200          4.9100          9.8700
     RKj =  -0.91381997E+01  0.82933853E+01  0.35128252E+01  0.13386231E+00

Power model  :    Bm = ( Rki X R ) / ( 1 + ( R/R0 )**rn )**( 1/rn )
      rn = 0.551      rki =   0.22746147E+03          rmu =   0.19246395E+03
```

No	Rotation (radians) X 1/1000	Moment (kip-inch) Expri.	Poly.	M.Expo.	P.Model.	Connection stiffness (kip-inch) X 1000 Expri.	Poly.	M.Expo.	P.Model.
1	0.00	0.00	0.00	0.00	0.00	0.5839E+02	0.8921E+01	0.9541E+02	0.2275E+03
3	0.23	13.21	2.02	13.74	25.14	0.5837E+02	0.8921E+01	0.3714E+02	0.7487E+02
5	0.69	25.66	6.18	25.50	49.33	0.1653E+02	0.8921E+01	0.2411E+02	0.3754E+02
7	1.16	37.62	10.33	38.96	63.51	0.5493E+02	0.8921E+01	0.3321E+02	0.2506E+02
9	1.38	49.85	12.32	46.62	68.68	0.4694E+02	0.8921E+01	0.3524E+02	0.2154E+02
11	1.79	61.82	15.97	61.09	76.51	0.3785E+02	0.8921E+01	0.3460E+02	0.1702E+02
13	2.33	73.25	20.80	74.58	84.60	0.1586E+02	0.8921E+01	0.1991E+02	0.1321E+02
15	2.96	82.95	26.45	84.71	92.00	0.1799E+02	0.8921E+01	0.1219E+02	0.1036E+02
17	3.67	94.44	32.75	93.69	98.54	0.1131E+02	0.8921E+01	0.1370E+02	0.8275E+01
19	4.46	102.41	39.83	102.47	104.43	0.9069E+01	0.8921E+01	0.8862E+01	0.6687E+01
21	5.43	110.97	48.40	111.07	110.19	0.8773E+01	0.8921E+01	0.9132E+01	0.5370E+01
23	6.59	119.73	58.76	120.33	115.75	0.6569E+01	0.8921E+01	0.7021E+01	0.4291E+01
25	7.74	127.43	69.08	127.64	120.26	0.6757E+01	0.8921E+01	0.5694E+01	0.3543E+01
27	8.79	133.22	78.45	133.14	123.71	0.4339E+01	0.8921E+01	0.4807E+01	0.3039E+01
29	9.87	137.91	88.08	137.94	126.76	0.4216E+01	0.8921E+01	0.4255E+01	0.2637E+01
31	11.13	143.03	99.30	142.91	129.84	0.4073E+01	0.8921E+01	0.3691E+01	0.2273E+01
33	12.41	147.54	110.75	147.39	132.56	0.2983E+01	0.8921E+01	0.3320E+01	0.1982E+01
35	13.72	151.45	122.43	151.57	135.00	0.2983E+01	0.8921E+01	0.3090E+01	0.1745E+01

A.1 – 31

I - 16

Connection type : Single web-angle connections
Mode : All bolted

Tested by : S.L.Lipson (1968) Canada
Test Id. : D-4

Column : --
Beam : W21X62
Angle : 5 X 3.5 X 5/16

Fasteners: A325- -3/4"D
15/16" Oversize holes
Material : A36
Fy = -- ksi
Fu = -- ksi

Major parameters

lp = 11.5000"	lu = 1.7500"	ll = 1.7500"	ta = 0.3125"
gb = 3.2500"	gc = 1.9375"	cu = 1.2500"	cl = 1.2500"
pb = 3.0000"	pc = 3.0000"	nb = 1 X 4	nc = 1 X 4

Remark 1) Connection subject to moment and shear.
2)

No	Moment (k-in)	Rotation (radians) X 1/1000	No	Moment (k-in)	Rotation (radians) X 1/1000
1	0.00	0.00	26	125.71	8.73
2	6.62	0.20	27	128.72	9.22
3	13.24	0.40	28	131.74	9.72
4	19.86	0.60	29	133.76	10.22
5	26.65	0.88	30	135.79	10.73
6	33.45	1.17	31	137.82	11.23
7	37.65	1.28	32	139.84	11.73
8	41.85	1.40	33	141.57	12.29
9	47.38	1.65	34	143.29	12.85
10	52.91	1.90	35	145.02	13.41
11	57.24	2.05			
12	61.57	2.19			
13	66.63	2.54			
14	71.69	2.89			
15	76.56	3.17			
16	81.42	3.45			
17	86.88	3.83			
18	92.34	4.22			
19	97.40	4.80			
20	102.46	5.38			
21	106.92	5.96			
22	111.38	6.53			
23	115.15	7.10			
24	118.92	7.66			
25	122.69	8.23			

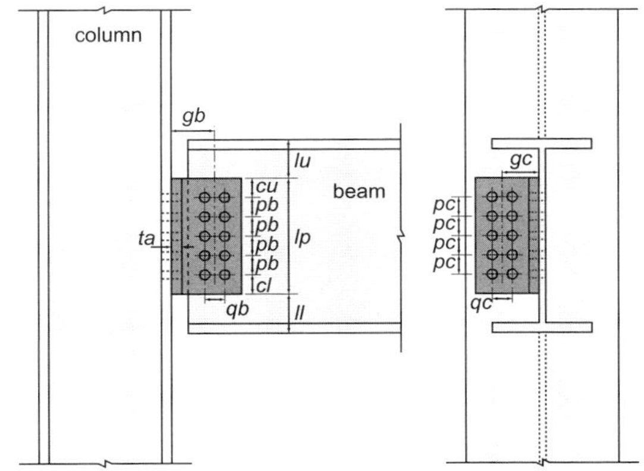

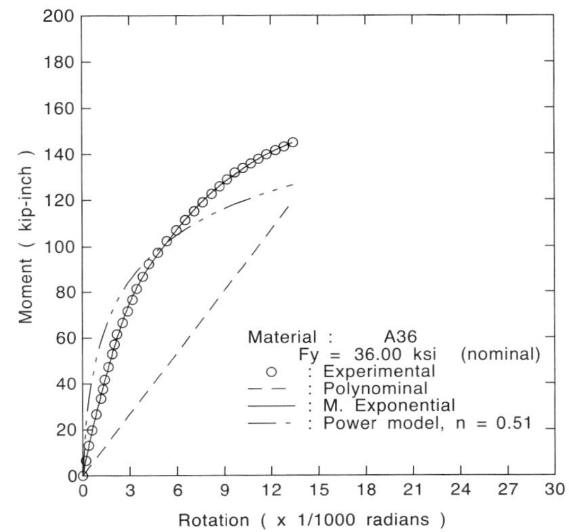

```
Moment-rotation prediction equations     ( R : X 1/1000 radians )

Frye and Morris polynominal model  :     R = Sum ( Ai X ( K*Bm )**Pi X 10**Qi )
     xd = 11.500000"   g =  2.137500"   t =  0.312500"
     A1 =  4.280000   A2 =  1.450000   A3 =  1.510000    K =  0.026190
     P1 =         1    P2 =         3   P3 =         5   Q1 =         0   Q2 =        -6   Q3 =       -13

Modified exponential model  :
     Bm = Sum ( Ai X ( 1 - exp( - R/( 2*i*C ) ) ) )  +      Sum ( RKj X ( R - Rj0 ) )  + Bm0
       C =   0.16141500E+00           Bmo=   0.00000000E+00          Nexp=  6         Nliner=  2
      AI =  -0.28483230E+02   0.70047265E+03  -0.44303349E+04   0.11346610E+05  -0.12869378E+05   0.54198283E+04
     Rj0 =          5.3800           9.7200
     RKj =   0.62835104E+00   0.10057174E+01

Power model  :    Bm = ( Rki X R ) / ( 1 + ( R/R0 )**rn )**( 1/rn )
     rn = 0.514      rki =   0.22746147E+03         rmu =   0.19246395E+03
```

No	Rotation (radians) X 1/1000	Moment (kip-inch) Expri.	Poly.	M.Expo.	P.Model.	Connection stiffness (kip-inch) X 1000 Expri.	Poly.	M.Expo.	P.Model.
1	0.00	0.00	0.00	0.00	0.00	0.3331E+02	0.8921E+01	0.3425E+02	0.2275E+03
3	0.40	13.24	3.55	13.18	33.02	0.3331E+02	0.8921E+01	0.3100E+02	0.4950E+02
5	0.88	26.65	7.89	27.19	51.06	0.2362E+02	0.8921E+01	0.2727E+02	0.2856E+02
7	1.28	37.65	11.46	37.86	60.89	0.3704E+02	0.8921E+01	0.2605E+02	0.2116E+02
9	1.65	47.38	14.71	47.12	67.79	0.2209E+02	0.8921E+01	0.2477E+02	0.1707E+02
11	2.05	57.24	18.25	56.57	73.93	0.2947E+02	0.8921E+01	0.2264E+02	0.1404E+02
13	2.54	66.63	22.67	66.97	80.19	0.1453E+02	0.8921E+01	0.1932E+02	0.1143E+02
15	3.17	76.56	28.29	77.80	86.64	0.1729E+02	0.8921E+01	0.1520E+02	0.9194E+01
17	3.83	86.88	34.21	86.74	92.18	0.1427E+02	0.8921E+01	0.1193E+02	0.7572E+01
19	4.80	97.40	42.80	96.76	98.65	0.8718E+01	0.8921E+01	0.9241E+01	0.5977E+01
21	5.96	106.92	53.14	106.86	104.79	0.7710E+01	0.8921E+01	0.8346E+01	0.4721E+01
23	7.10	115.15	63.34	115.75	109.68	0.6673E+01	0.8921E+01	0.7208E+01	0.3878E+01
25	8.23	122.69	73.42	123.21	113.70	0.6356E+01	0.8921E+01	0.5977E+01	0.3274E+01
27	9.22	128.72	82.27	128.59	116.74	0.6077E+01	0.8921E+01	0.4865E+01	0.2868E+01
29	10.22	133.76	91.20	133.43	119.44	0.4021E+01	0.8921E+01	0.4840E+01	0.2541E+01
31	11.23	137.82	100.19	137.86	121.86	0.4021E+01	0.8921E+01	0.3964E+01	0.2272E+01
33	12.29	141.57	109.66	141.66	124.15	0.3092E+01	0.8921E+01	0.3242E+01	0.2038E+01
35	13.41	145.02	119.62	144.96	126.31	0.3092E+01	0.8921E+01	0.2691E+01	0.1834E+01

A.1 – 33

I - 17

Connection type : Single web-angle connections
Mode : Bolted-to-beam and welded-to-column

Tested by : S.L.Lipson (1968) Canada
Test Id. : P1-2/1

Column : --
Beam : W21X62
Angle : --

Fasteners: A325- -3/4"D
15/16" Oversize holes
Material : A36
Fy = 48.75 ksi
Fu = 71.76 ksi

Major parameters

lp = 5.5000"	lu = 7.7500"	ll = 7.7500"	ta = 0.2500"
gb = 2.5000"	cu = 1.2500"	cl = 1.2500"	pb = 3.0000"
nb = 1 X 2			

Remark 1) Connection plate welded to 1" mounting plate.
2)

No	Moment (k-in)	Rotation (radians) X 1/1000	No	Moment (k-in)	Rotation (radians) X 1/1000
1	0.00	0.00	26	42.39	13.23
2	2.13	0.14	27	42.16	14.63
3	4.26	0.28	28	41.93	16.02
4	6.39	0.43	29	41.70	17.42
5	8.52	0.57	30	41.66	18.56
6	10.65	0.71	31	41.63	19.71
7	12.78	0.85	32	41.59	20.86
8	14.94	1.09	33	41.56	22.01
9	17.09	1.34	34	42.17	23.26
10	19.25	1.58	35	42.79	24.51
11	21.40	1.82	36	44.22	25.39
12	23.11	2.19	37	45.65	26.27
13	24.83	2.55	38	47.37	27.29
14	26.54	2.92	39	49.10	28.31
15	28.78	3.43			
16	31.02	3.95			
17	32.69	4.43			
18	34.37	4.90			
19	36.04	5.38			
20	37.83	6.23			
21	39.61	7.08			
22	40.58	8.30			
23	41.55	9.52			
24	41.83	10.76			
25	42.11	11.99			

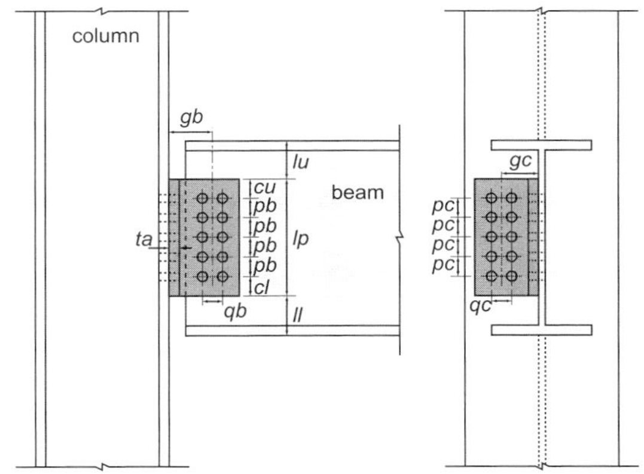

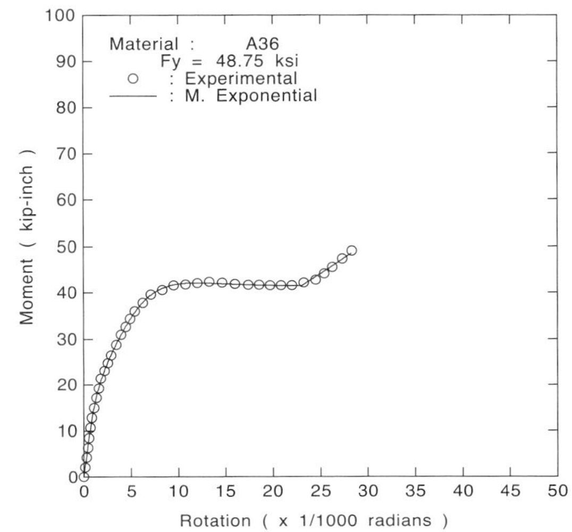

A.1 – 34

Moment-rotation prediction equations (R : X 1/1000 radians)

```
Modified exponential model  :
     Bm = Sum ( Ai X ( 1 - exp( - R/( 2*i*C ) ) ) )  +     Sum ( RKj X ( R - Rj0 ) )  +  Bm0
      C =   0.27137083E+00           Bmo=   0.00000000E+00          Nexp=  6        Nliner=  1
     AI =  -0.26682103E+02   0.25038758E+03  -0.64429600E+03   0.56908210E+03  -0.64390882E+01  -0.10066890E+03
    Rj0 =          23.0000
    RKj =   0.13656307E+01
```

No	Rotation (radians) X 1/1000	Moment (kip-inch) Expri.	Moment (kip-inch) M.Expo.	Connection stiffness (kip-inch) X 1000 Expri.	Connection stiffness (kip-inch) X 1000 M.Expo.
1	0.00	0.00	0.00	0.1503E+02	0.1465E+02
3	0.28	4.26	4.37	0.1503E+02	0.1559E+02
5	0.57	8.52	8.62	0.1502E+02	0.1416E+02
7	0.85	12.78	12.34	0.1275E+02	0.1203E+02
9	1.34	17.09	17.34	0.8859E+01	0.8683E+01
11	1.82	21.40	20.98	0.7194E+01	0.6469E+01
13	2.55	24.83	25.02	0.4686E+01	0.4848E+01
15	3.43	28.78	28.91	0.4357E+01	0.4102E+01
17	4.43	32.69	32.69	0.3506E+01	0.3508E+01
19	5.38	36.04	35.72	0.2999E+01	0.2828E+01
21	7.08	39.61	39.45	0.1564E+01	0.1584E+01
23	9.52	41.55	41.72	0.5141E+00	0.4274E+00
25	11.99	42.11	42.13	0.2250E+00	-0.7261E-02
27	14.63	42.16	41.95	-0.1658E+00	-0.9566E-01
29	17.42	41.70	41.70	-0.9162E-01	-0.7384E-01
31	19.71	41.63	41.56	-0.3054E-01	-0.4663E-01
33	22.01	41.56	41.48	0.2206E+00	-0.2677E-01
35	24.51	42.79	43.49	0.1153E+01	0.1352E+01
37	26.27	45.65	45.88	0.1654E+01	0.1357E+01
39	28.31	49.10	48.65	0.1696E+01	0.1361E+01

I - 18

Connection type : Single web-angle connections
Mode : Bolted-to-beam and welded-to-column

Tested by : S.L.Lipson (1968) Canada
Test Id. : P1-2/2

Column : --
Beam : W21X62
Angle : --

Fasteners: A325- -3/4"D
15/16" Oversize holes
Material : A36
Fy = 48.75 ksi
Fu = 71.76 ksi

Major parameters

lp = 5.5000" lu = 7.7500" ll = 7.7500" ta = 0.2500"
gb = 2.5000" cu = 1.2500" cl = 1.2500" pb = 3.0000"
nb = 1 X 2

Remark 1) Connection plate welded to 1" mounting plate.
2)

No	Moment (k-in)	Rotation (radians) X 1/1000	No	Moment (k-in)	Rotation (radians) X 1/1000
1	0.00	0.00	26	41.16	18.04
2	1.87	0.14	27	40.74	20.49
3	3.75	0.28	28	40.32	22.91
4	5.62	0.42	29	39.91	25.33
5	7.50	0.56	30	39.49	27.76
6	9.37	0.71	31	38.84	30.52
7	11.37	0.90	32	38.18	33.28
8	13.36	1.09	33	38.56	36.04
9	15.35	1.28	34	38.42	38.33
10	17.35	1.47	35	40.17	40.47
11	19.31	1.75	36	41.92	42.61
12	21.27	2.03	37	43.67	44.75
13	23.24	2.30	38	44.29	46.84
14	25.01	2.63	39	44.91	48.93
15	26.78	2.96	40	45.53	51.02
16	28.55	3.28	41	44.81	53.65
17	30.33	3.61	42	44.10	56.27
18	32.48	4.32	43	43.38	58.90
19	34.63	5.04			
20	36.73	5.89			
21	38.83	6.74			
22	40.82	8.83			
23	41.42	10.99			
24	42.01	13.15			
25	41.59	15.59			

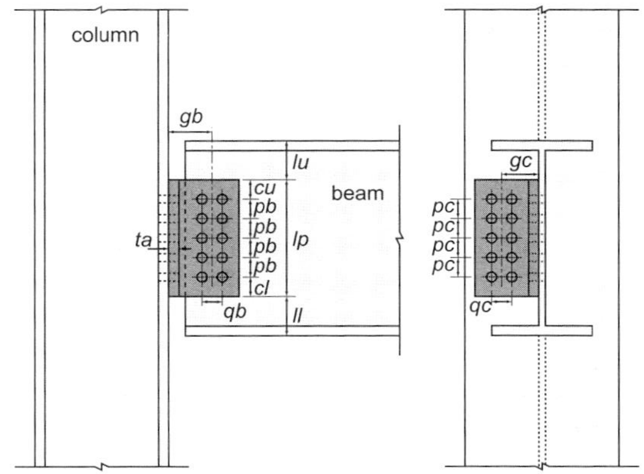

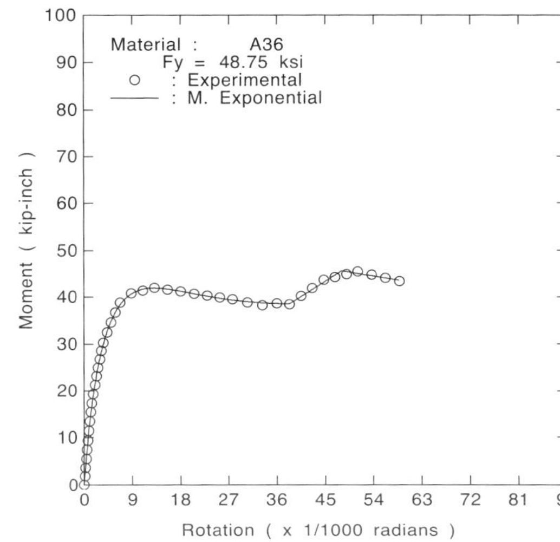

```
Moment-rotation prediction equations     ( R : X 1/1000 radians )

Modified exponential model  :
     Bm = Sum ( Ai X ( 1 - exp( - R/( 2*i*C ) ) ) )  +      Sum ( RKj X ( R - Rj0 ) )  +  Bm0
      C =   0.52609583E+00              Bmo=   0.00000000E+00            Nexp=  6        Nliner=  3
     AI =   0.67468569E+00   0.90064598E+01   0.11974384E+03  -0.50358102E+03   0.83982281E+03  -0.42836885E+03
    Rj0 =          16.8200          38.3000          48.0000
    RKj =   0.33234553E-01   0.74106593E+00  -0.95332683E+00
```

No	Rotation (radians) X 1/1000	Moment (kip-inch) Expri.	Moment (kip-inch) M.Expo.	Connection stiffness (kip-inch) X 1000 Expri.	Connection stiffness (kip-inch) X 1000 M.Expo.
1	0.00	0.00	0.00	0.1327E+02	0.1498E+02
3	0.28	3.75	4.01	0.1327E+02	0.1344E+02
5	0.56	7.50	7.61	0.1327E+02	0.1208E+02
7	0.90	11.37	11.40	0.1037E+02	0.1067E+02
9	1.28	15.35	15.22	0.1037E+02	0.9282E+01
11	1.75	19.31	19.23	0.7105E+01	0.7860E+01
13	2.30	23.24	23.18	0.6340E+01	0.6493E+01
15	2.96	26.78	26.98	0.5437E+01	0.5212E+01
17	3.61	30.33	30.04	0.4676E+01	0.4206E+01
19	5.04	34.63	34.86	0.2762E+01	0.2659E+01
21	6.74	38.83	38.34	0.2034E+01	0.1542E+01
23	10.99	41.42	41.70	0.2764E+00	0.2670E+00
25	15.59	41.59	41.71	-0.1736E+00	-0.1698E+00
27	20.49	40.74	40.70	-0.1727E+00	-0.2228E+00
29	25.33	39.91	39.71	-0.1718E+00	-0.1785E+00
31	30.52	38.84	38.97	-0.2364E+00	-0.1061E+00
33	36.04	38.56	38.56	0.2876E-01	-0.4521E-01
35	40.47	40.17	40.05	0.8175E+00	0.7277E+00
37	44.75	43.67	43.21	0.5542E+00	0.7471E+00
39	48.93	44.91	45.47	0.2969E+00	-0.1947E+00
41	53.65	44.81	44.57	-0.2736E+00	-0.1873E+00
43	58.90	43.38	43.60	-0.2737E+00	-0.1830E+00

I - 19

```
Connection type : Single web-angle connections
           Mode : Bolted-to-beam and welded-to-column

Tested by  :  S.L.Lipson (1968)                       Canada
Test  Id.  :  P1-3/1

Column :  --                          Fasteners: A325- -3/4"D
Beam   : W21X62                           15/16" Oversize holes
Angle  :  --                          Material : A36
                                          Fy =  48.75 ksi
                                          Fu =  71.76 ksi

                        Major parameters

     lp =  8.5000"    lu =  6.2500"    ll =  6.2500"    ta =  0.2500"
     gb =  2.5000"    cu =  1.2500"    cl =  1.2500"    pb =  3.0000"
     nb = 1 X  3

Remark  1)  Connection plate welded to 1" mounting plate.
        2)
```

No	Moment (k-in)	Rotation (radians) X 1/1000	No	Moment (k-in)	Rotation (radians) X 1/1000
1	0.00	0.00	26	131.53	13.61
2	6.69	0.11	27	133.45	14.78
3	13.38	0.23	28	135.36	15.95
4	20.07	0.34	29	137.27	17.13
5	26.76	0.46	30	140.09	18.33
6	33.44	0.57	31	142.91	19.53
7	40.22	0.83	32	145.73	20.72
8	47.01	1.09	33	148.32	21.92
9	54.14	1.31	34	150.91	23.12
10	61.27	1.54	35	153.50	24.32
11	66.35	1.75	36	155.87	25.49
12	71.43	1.97	37	158.24	26.67
13	77.18	2.22	38	160.61	27.84
14	82.93	2.48	39	162.71	29.15
15	87.27	2.86	40	164.81	30.46
16	91.61	3.25	41	166.90	31.77
17	99.58	4.08			
18	104.18	4.70			
19	108.79	5.32			
20	113.26	6.38			
21	117.72	7.44			
22	121.43	8.73			
23	125.13	10.02			
24	127.27	11.21			
25	129.40	12.41			

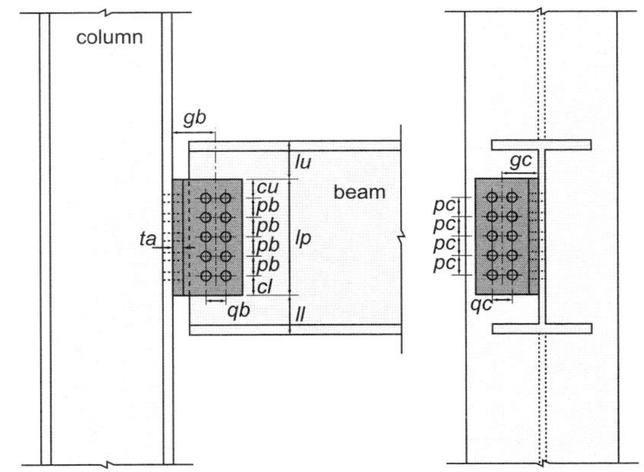

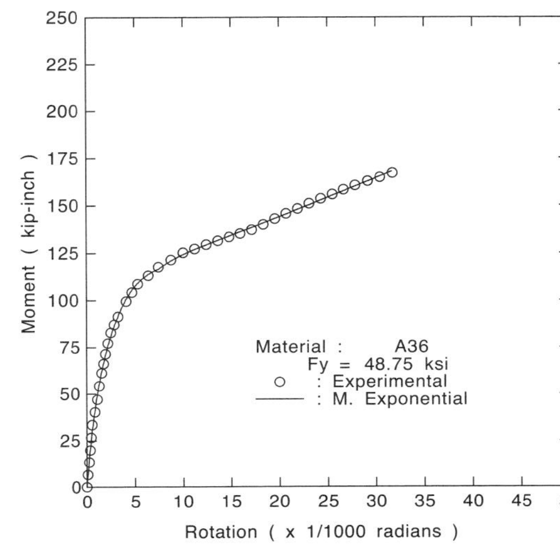

```
Moment-rotation prediction equations     ( R : X 1/1000 radians )

Modified exponential model  :
     Bm = Sum ( Ai X ( 1 - exp( - R/( 2*i*C ) ) ) )  +      Sum ( RKj X ( R - Rj0 ) )  +  Bm0
      C =   0.29347000E+00             Bmo=   0.00000000E+00          Nexp=  6        Nliner=  1
     AI =   0.14223767E+02   0.65462457E+02  -0.31896045E+03   0.47501272E+03  -0.26606636E+01  -0.11524991E+03
    Rj0 =           7.4400
    RKj =   0.20641279E+01
```

No	Rotation (radians) X 1/1000	Moment (kip-inch) Expri.	Moment (kip-inch) M.Expo.	Connection stiffness (kip-inch) X 1000 Expri.	Connection stiffness (kip-inch) X 1000 M.Expo.
1	0.00	0.00	0.00	0.5821E+02	0.6755E+02
3	0.23	13.38	14.03	0.5821E+02	0.5522E+02
5	0.46	26.76	25.62	0.5821E+02	0.4610E+02
7	0.83	40.22	40.71	0.2642E+02	0.3584E+02
9	1.31	54.14	55.76	0.3184E+02	0.2742E+02
11	1.75	66.35	66.67	0.2332E+02	0.2231E+02
13	2.22	77.18	76.20	0.2269E+02	0.1834E+02
15	2.86	87.27	86.55	0.1130E+02	0.1433E+02
17	4.08	99.58	100.50	0.8339E+01	0.8989E+01
19	5.32	108.79	109.27	0.6240E+01	0.5389E+01
21	7.44	117.72	116.54	0.3613E+01	0.1933E+01
23	10.02	125.13	124.34	0.2306E+01	0.2367E+01
25	12.41	129.40	129.40	0.1783E+01	0.1954E+01
27	14.78	133.45	133.93	0.1630E+01	0.1898E+01
29	17.13	137.27	138.42	0.1987E+01	0.1935E+01
31	19.53	142.91	143.12	0.2352E+01	0.1982E+01
33	21.92	148.32	147.91	0.2163E+01	0.2016E+01
35	24.32	153.50	152.77	0.2089E+01	0.2037E+01
37	26.67	158.24	157.57	0.2017E+01	0.2050E+01
39	29.15	162.71	162.67	0.1603E+01	0.2057E+01
41	31.77	166.90	168.06	0.1603E+01	0.2060E+01

I - 20

```
Connection type : Single web-angle connections
           Mode : Bolted-to-beam and welded-to-column

Tested by  :  S.L.Lipson (1968)                          Canada
Test  Id.  :  P1-3/2

Column :  --                        Fasteners: A325- -3/4"D
Beam   :  W21X62                          15/16" Oversize holes
Angle  :  --                        Material : A36
                                         Fy =  48.75 ksi
                                         Fu =  71.76 ksi

                          Major parameters

     lp =  8.5000"    lu =  6.2500"    ll =  6.2500"    ta =  0.2500"
     gb =  2.5000"    cu =  1.2500"    cl =  1.2500"    pb =  3.0000"
     nb = 1 X  3

Remark  1)  Connection plate welded to 1" mounting plate.
        2)
```

No	Moment (k-in)	Rotation (radians) X 1/1000	No	Moment (k-in)	Rotation (radians) X 1/1000
1	0.00	0.00	26	90.06	34.94
2	7.33	0.16	27	93.73	37.41
3	14.65	0.31	28	97.80	39.71
4	21.98	0.47	29	101.87	42.01
5	29.31	0.63	30	106.42	43.88
6	34.24	0.94	31	110.97	45.74
7	39.18	1.24	32	115.86	47.61
8	44.11	1.55	33	120.75	49.48
9	50.11	2.11	34	128.12	52.20
10	56.11	2.67	35	135.49	54.91
11	63.44	3.36	36	141.46	57.69
12	67.39	3.88	37	147.65	59.80
13	71.34	4.40			
14	77.17	5.49			
15	77.59	6.30			
16	78.01	8.85			
17	78.44	11.40			
18	78.86	13.94			
19	79.28	16.49			
20	79.70	19.04			
21	80.68	21.69			
22	81.66	24.34			
23	82.65	27.00			
24	84.52	29.73			
25	86.39	32.47			

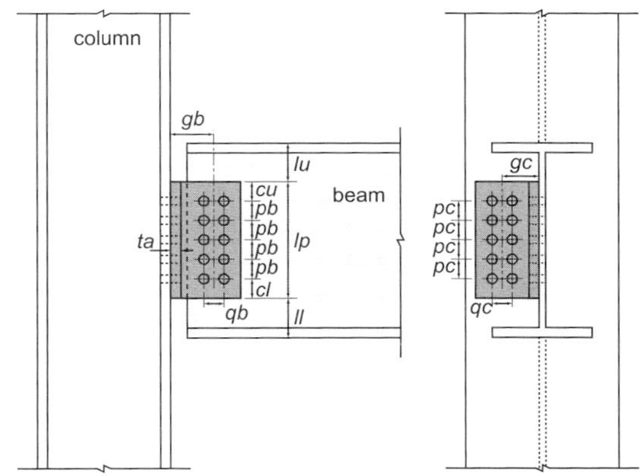

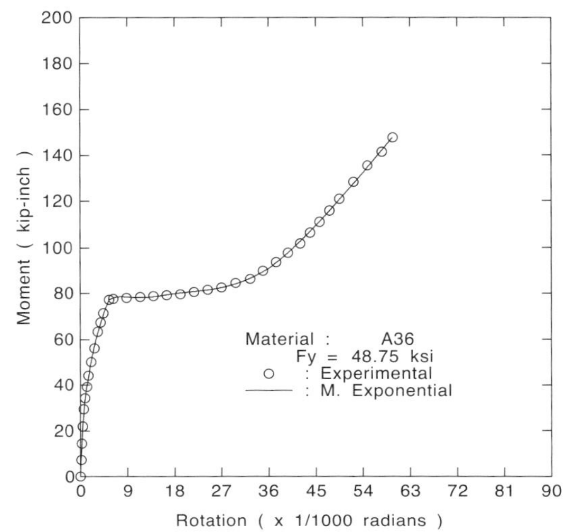

A.1 – 40

```
Moment-rotation prediction equations     ( R : X 1/1000 radians )

Modified exponential model  :
     Bm = Sum ( Ai X ( 1 - exp( - R/( 2*i*C ) ) ) )  +      Sum ( RKj X ( R - Rj0 ) )  +  Bm0
      C =   0.53766583E+00              Bmo=   0.00000000E+00           Nexp=  6          Nliner=  3
     AI =   0.18854298E+03  -0.17557310E+04   0.88496166E+04  -0.20536968E+05   0.22039214E+05  -0.88082437E+04
    Rj0 =           1.5500           5.4900          36.0000
    RKj =   0.64030424E+01  -0.38209022E+01   0.23020071E+00
```

No	Rotation (radians) X 1/1000	Moment (kip-inch) Expri.	Moment (kip-inch) M.Expo.	Connection stiffness (kip-inch) X 1000 Expri.	Connection stiffness (kip-inch) X 1000 M.Expo.
1	0.00	0.00	0.00	0.4658E+02	0.6148E+02
3	0.31	14.65	16.08	0.4658E+02	0.4197E+02
5	0.63	29.31	27.06	0.3625E+02	0.2870E+02
7	1.24	39.18	39.57	0.1607E+02	0.1395E+02
9	2.11	50.11	50.96	0.1076E+02	0.1200E+02
11	3.36	63.44	63.13	0.8869E+01	0.8123E+01
13	4.40	71.34	70.68	0.6894E+01	0.6554E+01
15	6.30	77.59	78.02	0.4405E+00	0.7680E+00
17	11.40	78.44	78.10	0.1653E+00	0.6244E-02
19	16.49	79.28	79.34	0.1653E+00	0.3426E+00
21	21.69	80.68	80.82	0.3703E+00	0.2403E+00
23	27.00	82.65	82.54	0.5250E+00	0.4844E+00
25	32.47	86.39	86.67	0.1106E+01	0.1050E+01
27	37.41	93.73	93.50	0.1631E+01	0.1798E+01
29	42.01	101.87	102.69	0.2137E+01	0.2172E+01
31	45.74	110.97	111.22	0.2529E+01	0.2389E+01
33	49.48	120.75	120.45	0.2657E+01	0.2540E+01
35	54.91	135.49	134.65	0.2433E+01	0.2674E+01
37	59.80	147.65	147.90	0.2937E+01	0.2740E+01

I - 21

Connection type : Single web-angle connections
Mode : Bolted-to-beam and welded-to-column

Tested by : S.L.Lipson (1968) Canada
Test Id. : P1-4/1

Column : --
Beam : W21X62
Angle : --

Fasteners: A325- -3/4"D
15/16" Oversize holes
Material : A36
Fy = 48.75 ksi
Fu = 71.76 ksi

Major parameters

lp = 11.5000" lu = 4.7500" ll = 4.7500" ta = 0.2500"
gb = 2.5000" cu = 1.2500" cl = 1.2500" pb = 3.0000"
nb = 1 X 4

Remark 1) Connection plate welded to 1" mounting plate.
2)

No	Moment (k-in)	Rotation (radians) X 1/1000	No	Moment (k-in)	Rotation (radians) X 1/1000
1	0.00	0.00	26	186.09	21.06
2	11.48	0.18	27	197.30	22.48
3	22.96	0.37	28	205.84	23.65
4	34.44	0.55	29	214.38	24.82
5	45.92	0.74	30	221.76	26.06
6	57.40	0.92	31	229.14	27.30
7	65.58	1.12	32	236.51	28.54
8	73.76	1.31	33	241.58	29.83
9	80.55	1.54	34	246.65	31.13
10	87.35	1.76	35	251.72	32.43
11	99.50	2.41	36	256.50	33.91
12	110.28	3.05	37	261.28	35.38
13	118.18	4.08	38	266.06	36.86
14	123.87	5.59			
15	126.21	6.81			
16	128.56	8.03			
17	131.31	9.41			
18	134.06	10.79			
19	136.81	12.17			
20	140.52	13.43			
21	144.22	14.69			
22	150.65	16.02			
23	157.08	17.36			
24	165.98	18.50			
25	174.88	19.64			

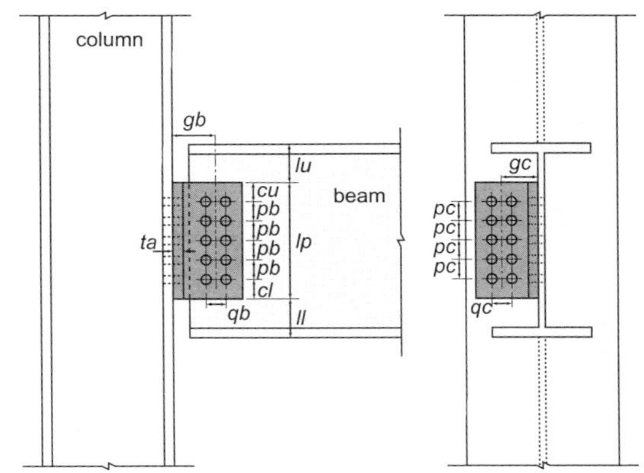

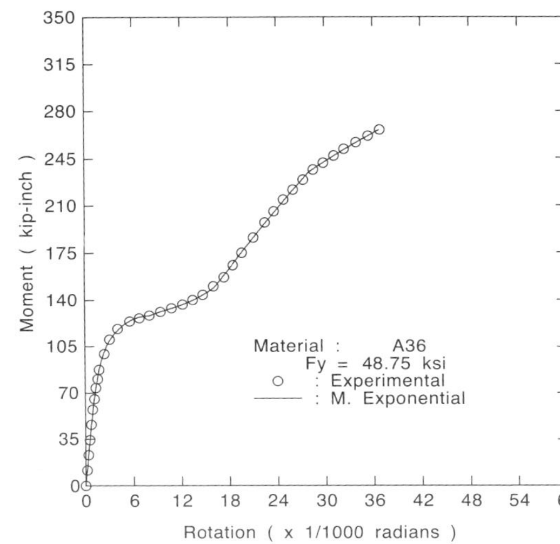

A.1 – 42

```
Moment-rotation prediction equations     ( R : X 1/1000 radians )

Modified exponential model  :
     Bm = Sum ( Ai X ( 1 - exp( - R/( 2*i*C ) ) ) )  +      Sum ( RKj X ( R - Rj0 ) )  +  Bm0
      C =   0.35337667E+00              Bmo=   0.00000000E+00           Nexp=  6        Nliner=  3
     AI =  -0.21038004E+03   0.27259985E+04  -0.12308029E+05   0.26201795E+05  -0.25438158E+05   0.91769926E+04
    Rj0 =          8.0300          16.0000          28.0000
    RKj =   0.27054835E+01   0.30864531E+01  -0.26459694E+01
```

No	Rotation (radians) X 1/1000	Moment (kip-inch) Expri.	Moment (kip-inch) M.Expo.	Connection stiffness (kip-inch) X 1000 Expri.	Connection stiffness (kip-inch) X 1000 M.Expo.
1	0.00	0.00	0.00	0.6209E+02	0.5979E+02
3	0.37	22.96	23.54	0.6210E+02	0.6428E+02
5	0.74	45.92	46.02	0.6210E+02	0.5622E+02
7	1.12	65.58	65.14	0.4203E+02	0.4444E+02
9	1.54	80.55	81.13	0.3034E+02	0.3232E+02
11	2.41	99.50	101.19	0.1781E+02	0.1583E+02
13	4.08	118.18	117.21	0.6056E+01	0.6013E+01
15	6.81	126.21	127.17	0.1920E+01	0.1567E+01
17	9.41	131.31	131.35	0.1993E+01	0.1970E+01
19	12.17	136.81	136.83	0.2490E+01	0.2330E+01
21	14.69	144.22	144.11	0.3851E+01	0.3444E+01
23	17.36	157.08	158.67	0.6423E+01	0.7298E+01
25	19.64	174.88	175.59	0.7845E+01	0.7464E+01
27	22.48	197.30	196.55	0.7563E+01	0.7269E+01
29	24.82	214.38	213.28	0.6638E+01	0.6972E+01
31	27.30	229.14	230.13	0.5962E+01	0.6655E+01
33	29.83	241.58	241.79	0.3906E+01	0.3737E+01
35	32.43	251.72	251.20	0.3593E+01	0.3530E+01
37	35.38	261.28	261.37	0.3237E+01	0.3372E+01

I - 22

Connection type : Single web-angle connections
Mode : Bolted-to-beam and welded-to-column

Tested by : S.L.Lipson (1968) Canada
Test Id. : P1-4/2

Column : --
Beam : W21X62
Angle : --

Fasteners: A325- -3/4"D
15/16" Oversize holes
Material : A36
Fy = 48.75 ksi
Fu = 71.76 ksi

Major parameters

lp = 11.5000" lu = 4.7500" ll = 4.7500" ta = 0.2500"
gb = 2.5000" cu = 1.2500" cl = 1.2500" pb = 3.0000"
nb = 1 X 4

Remark 1) Connection plate welded to 1" mounting plate.
2)

No	Moment (k-in)	Rotation (radians) X 1/1000	No	Moment (k-in)	Rotation (radians) X 1/1000
1	0.00	0.00	26	185.14	21.09
2	11.34	0.20	27	194.15	22.24
3	22.68	0.39	28	203.16	23.39
4	34.01	0.59	29	214.63	25.07
5	45.35	0.79	30	226.09	26.76
6	56.69	0.99	31	233.53	28.16
7	67.91	1.33	32	240.96	29.56
8	79.14	1.66	33	246.93	31.27
9	85.91	1.95	34	252.89	32.97
10	92.68	2.25	35	257.26	34.28
11	105.52	2.89	36	261.64	35.60
12	113.53	3.59	37	266.01	36.91
13	120.70	4.76			
14	125.82	5.92			
15	128.05	7.47			
16	129.03	8.65			
17	130.02	9.83			
18	132.37	11.02			
19	134.73	12.20			
20	139.74	13.67			
21	144.75	15.13			
22	149.89	16.22			
23	155.02	17.31			
24	165.58	18.63			
25	176.14	19.94			

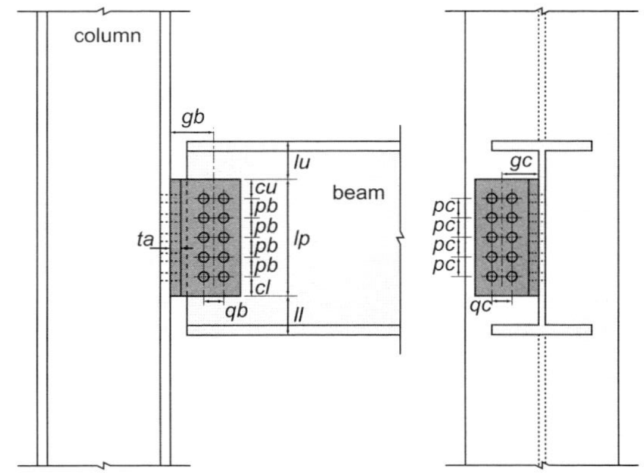

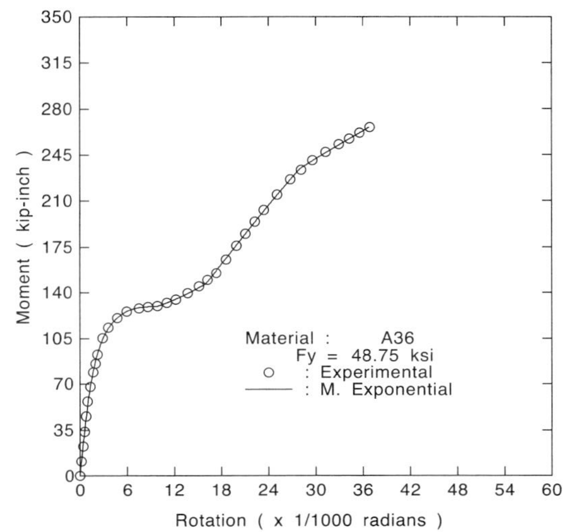

A.1 – 44

Moment-rotation prediction equations (R : X 1/1000 radians)

```
Modified exponential model  :
     Bm = Sum ( Ai X ( 1 - exp( - R/( 2*i*C ) ) ) )  +      Sum ( RKj X ( R - Rj0 ) )  +  Bm0
      C =    0.35696417E+00              Bmo=    0.00000000E+00          Nexp=  6        Nliner=  3
     AI =  -0.11022778E+03   0.14301288E+04  -0.67785006E+04   0.15504677E+05  -0.15886802E+05   0.59909372E+04
     Rj0 =           9.8300          16.0000          28.0000
     RKj =   0.21581363E+01   0.40211256E+01  -0.29021407E+01
```

No	Rotation (radians) X 1/1000	Moment (kip-inch) Expri.	Moment (kip-inch) M.Expo.	Connection stiffness (kip-inch) X 1000 Expri.	Connection stiffness (kip-inch) X 1000 M.Expo.
1	0.00	0.00	0.00	0.5741E+02	0.5972E+02
3	0.39	22.68	23.58	0.5742E+02	0.5794E+02
5	0.79	45.35	44.93	0.5742E+02	0.4963E+02
7	1.33	67.91	68.05	0.3316E+02	0.3683E+02
9	1.95	85.91	87.19	0.2330E+02	0.2472E+02
11	2.89	105.52	104.68	0.1583E+02	0.1383E+02
13	4.76	120.70	120.98	0.5274E+01	0.5299E+01
15	7.47	128.05	128.60	0.1094E+01	0.1044E+01
17	9.83	130.02	129.57	0.1410E+01	0.2252E+01
19	12.20	134.73	135.19	0.2632E+01	0.2607E+01
21	15.13	144.75	143.90	0.4154E+01	0.3311E+01
23	17.31	155.02	156.75	0.6220E+01	0.7581E+01
25	19.94	176.14	176.66	0.7933E+01	0.7549E+01
27	22.24	194.15	193.81	0.7832E+01	0.7349E+01
29	25.07	214.63	214.21	0.6804E+01	0.7040E+01
31	28.16	233.53	235.00	0.5300E+01	0.3838E+01
33	31.27	246.93	246.55	0.3506E+01	0.3618E+01
35	34.28	257.26	257.24	0.3327E+01	0.3477E+01
37	36.91	266.01	266.27	0.3327E+01	0.3400E+01

I - 23

Connection type : Single web-angle connections
Mode : Bolted-to-beam and welded-to-column

Tested by : S.L.Lipson (1968) Canada
Test Id. : P1-5/1

Column : --
Beam : W21X62
Angle : --

Fasteners: A325- -3/4"D
-5/16" Oversize holes
Material : A36
Fy = 48.75 ksi
Fu = 71.76 ksi

Major parameters

lp = 14.5000" lu = 3.2500" ll = 3.2500" ta = 0.2500"
gb = 2.5000" cu = 1.2500" cl = 1.2500" pb = 3.0000"
nb = 1 X 5

Remark 1) Connection plate welded to 1" mounting plate.
2)

No	Moment (k-in)	Rotation (radians) X 1/1000
1	0.00	0.00
2	20.83	0.16
3	41.66	0.32
4	62.49	0.48
5	83.32	0.64
6	104.15	0.80
7	120.35	0.99
8	135.54	1.17
9	152.74	1.36
10	170.99	1.58
11	189.24	1.79
12	207.48	2.01
13	229.29	2.55
14	251.09	3.09
15	273.48	3.93
16	295.87	4.78
17	314.71	5.98
18	333.54	7.19
19	344.47	8.44
20	355.39	9.68
21	362.35	10.78
22	369.32	11.88
23	376.28	12.98
24	382.55	14.10
25	388.82	15.22

No	Moment (k-in)	Rotation (radians) X 1/1000
26	395.09	16.34
27	400.66	17.51
28	406.23	18.67
29	411.80	19.83
30	417.29	21.24
31	422.78	22.65
32	428.27	24.06
33	431.81	25.13
34	435.35	26.19
35	438.89	27.26
36	441.95	28.39
37	445.01	29.53
38	448.07	30.66

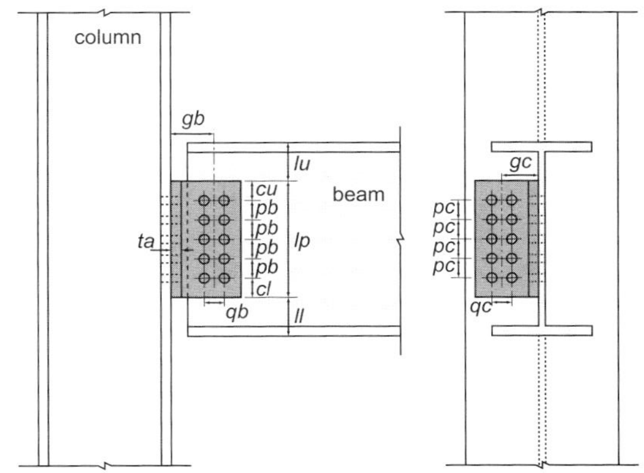

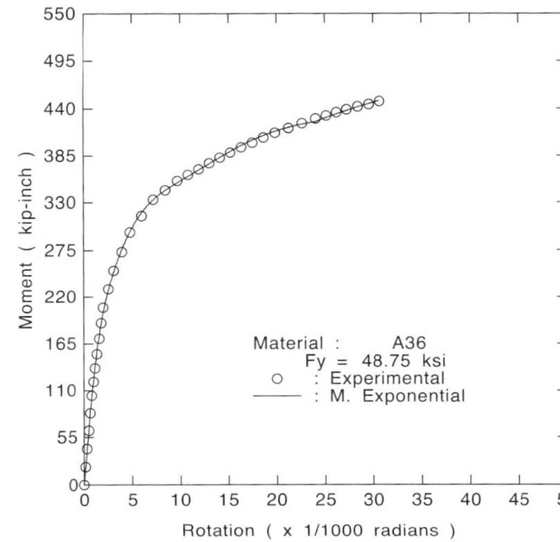

```
Moment-rotation prediction equations     ( R : X 1/1000 radians )

Modified exponential model  :
     Bm = Sum ( Ai X ( 1 - exp( - R/( 2*i*C ) ) ) )  +      Sum ( RKj X ( R - Rj0 ) )  +  Bm0
      C =   0.29560667E+00            Bmo=   0.00000000E+00         Nexp=  6        Nliner=  1
     AI =  -0.32058566E+03   0.39513903E+04  -0.18759767E+05   0.42359110E+05  -0.43757960E+05   0.16960721E+05
    Rj0 =          24.0600
    RKj =   0.27279078E+01
```

No	Rotation (radians) X 1/1000	Moment (kip-inch) Expri.	Moment (kip-inch) M.Expo.	Connection stiffness (kip-inch) X 1000 Expri.	Connection stiffness (kip-inch) X 1000 M.Expo.
1	0.00	0.00	0.00	0.1298E+03	0.1130E+03
3	0.32	41.66	40.60	0.1298E+03	0.1328E+03
5	0.64	83.32	82.24	0.1298E+03	0.1240E+03
7	0.99	120.35	121.94	0.8701E+02	0.1045E+03
9	1.36	152.74	156.81	0.8599E+02	0.8336E+02
11	1.79	189.24	188.30	0.8483E+02	0.6403E+02
13	2.55	229.29	228.03	0.4032E+02	0.4350E+02
15	3.93	273.48	275.61	0.2647E+02	0.2752E+02
17	5.98	314.71	318.10	0.1563E+02	0.1486E+02
19	8.44	344.47	344.04	0.8758E+01	0.7765E+01
21	10.78	362.35	360.52	0.6332E+01	0.6763E+01
23	12.98	376.28	375.52	0.5968E+01	0.6828E+01
25	15.22	388.82	390.33	0.5598E+01	0.6266E+01
27	17.51	400.66	403.36	0.4790E+01	0.5086E+01
29	19.83	411.80	413.60	0.4384E+01	0.3722E+01
31	22.65	422.78	422.01	0.3893E+01	0.2312E+01
33	25.13	431.81	429.48	0.3320E+01	0.4155E+01
35	27.26	438.89	437.75	0.3020E+01	0.3637E+01
37	29.53	445.01	445.55	0.2703E+01	0.3276E+01

A.1 – 47

I - 24

Connection type : Single web-angle connections
Mode : Bolted-to-beam and welded-to-column

Tested by : S.L.Lipson (1968) Canada
Test Id. : P1-5/2

Column : --
Beam : W21X62
Angle : --

Fasteners: A325- -3/4"D
15/16" Oversize holes
Material : A36
Fy = 48.75 ksi
Fu = 71.76 ksi

Major parameters

lp = 14.5000"	lu = 3.2500"	ll = 3.2500"	ta = 0.2500"
gb = 2.5000"	cu = 1.2500"	cl = 1.2500"	pb = 3.0000"
nb = 1 X 5			

Remark 1) Connection plate welded to 1" mounting plate.
2)

No	Moment (k-in)	Rotation (radians) X 1/1000	No	Moment (k-in)	Rotation (radians) X 1/1000
1	0.00	0.00	26	301.49	23.09
2	22.11	0.14	27	314.35	23.67
3	44.22	0.27	28	327.21	24.25
4	66.33	0.41	29	341.31	25.24
5	88.43	0.55	30	361.37	26.92
6	104.16	0.78	31	381.42	28.59
7	119.88	1.01	32	397.71	30.23
8	135.61	1.24	33	414.00	31.86
9	152.61	1.74	34	422.89	33.03
10	169.62	2.24	35	431.77	34.20
11	186.63	2.75	36	441.49	35.98
12	201.23	3.26	37	442.85	37.06
13	215.83	3.78	38	444.21	38.15
14	235.99	5.12			
15	239.82	6.01			
16	239.95	7.69			
17	240.89	8.99			
18	241.84	10.30			
19	242.78	11.60			
20	245.94	13.46			
21	249.10	15.32			
22	254.78	16.86			
23	260.46	18.39			
24	272.25	20.11			
25	290.87	21.98			

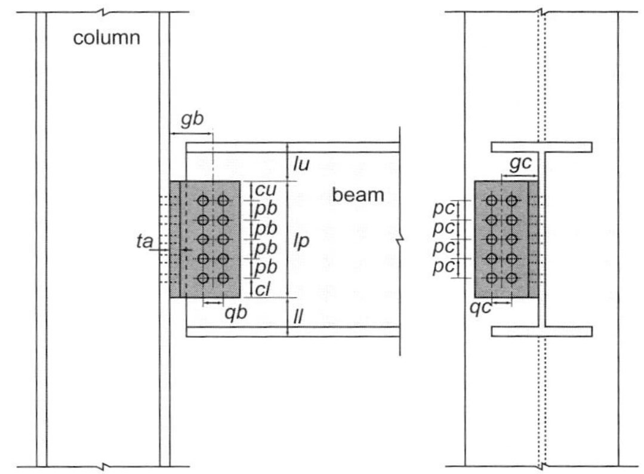

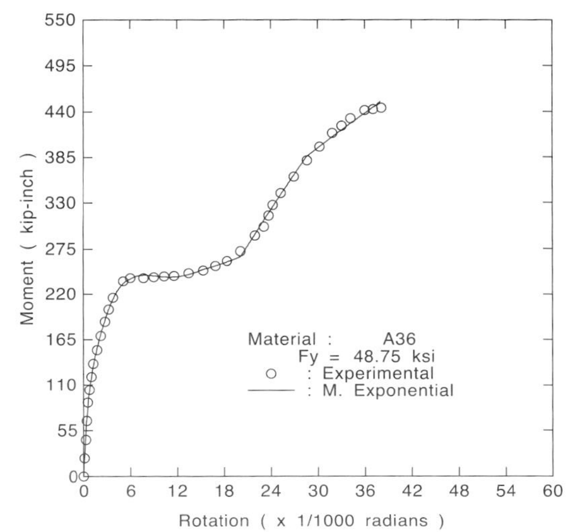

A.1 – 48

Moment-rotation prediction equations (R : X 1/1000 radians)

```
Modified exponential model  :
     Bm = Sum ( Ai X ( 1 - exp( - R/( 2*i*C ) ) ) )  +      Sum ( RKj X ( R - Rj0 ) )  +  Bm0
      C =   0.35222917E+00             Bmo=   0.00000000E+00           Nexp=  6        Nliner=  4
     AI =   0.24588172E+03   0.12527586E+03  -0.66192245E+04   0.21115721E+05  -0.23645436E+05   0.90181708E+04
    Rj0 =          6.0100          15.3200          20.1100          28.5900
    RKj =   0.34947968E+01  -0.18202207E+01   0.11214554E+02  -0.63126248E+01
```

No	Rotation (radians) X 1/1000	Moment (kip-inch) Expri.	M.Expo.	Connection stiffness (kip-inch) X 1000 Expri.	M.Expo.
1	0.00	0.00	0.00	0.1609E+03	0.2200E+03
3	0.27	44.22	49.58	0.1609E+03	0.1466E+03
5	0.55	88.43	83.14	0.1263E+03	0.1013E+03
7	1.01	119.88	119.20	0.6857E+02	0.6149E+02
9	1.74	152.61	154.30	0.3385E+02	0.3915E+02
11	2.75	186.63	188.09	0.3107E+02	0.2919E+02
13	3.78	215.83	213.84	0.2455E+02	0.2041E+02
15	6.01	239.82	238.56	0.2853E+01	0.3059E+01
17	8.99	240.89	242.62	0.7231E+00	-0.1081E+01
19	11.60	242.78	241.32	0.1126E+01	0.6145E+00
21	15.32	249.10	250.63	0.2800E+01	0.2266E+01
23	18.39	260.46	259.67	0.5192E+01	0.3379E+01
25	21.98	290.87	292.99	0.9701E+01	0.1454E+02
27	23.67	314.35	317.38	0.2221E+02	0.1432E+02
29	25.24	341.31	339.62	0.1342E+02	0.1409E+02
31	28.59	381.42	386.10	0.1095E+02	0.7326E+01
33	31.86	414.00	409.48	0.8583E+01	0.7012E+01
35	34.20	431.77	425.73	0.6739E+01	0.6862E+01
37	37.06	442.85	445.17	0.1258E+01	0.6742E+01

I - 25

Connection type : Single web-angle connections
Mode : Bolted-to-beam and welded-to-column

Tested by : S.L.Lipson (1968) Canada
Test Id. : P1-6/1

Column : --
Beam : W21X62
Angle : --

Fasteners: A325- -3/4"D
15/16" Oversize holes
Material : A36
Fy = 48.75 ksi
Fu = 71.76 ksi

Major parameters

lp = 17.5000" lu = 1.7500" ll = 1.7500" ta = 0.2500"
gb = 2.5000" cu = 1.2500" cl = 1.2500" pb = 3.0000"
nb = 1 X 6

Remark 1) Connection plate welded to 1" mounting plate.
2)

No	Moment (k-in)	Rotation (radians) X 1/1000	No	Moment (k-in)	Rotation (radians) X 1/1000
1	0.00	0.00	26	471.56	13.01
2	33.08	0.14	27	484.41	13.81
3	66.15	0.29	28	497.27	14.62
4	99.23	0.43	29	521.74	15.72
5	132.31	0.57	30	546.20	16.82
6	165.38	0.72	31	564.17	17.80
7	187.77	0.93	32	582.14	18.79
8	210.16	1.14	33	594.66	19.59
9	232.55	1.35	34	607.17	20.40
10	257.93	1.50	35	618.41	21.28
11	283.31	1.66	36	629.65	22.15
12	301.10	1.98	37	640.89	23.03
13	318.88	2.29	38	655.01	24.28
14	340.10	2.60	39	669.14	25.52
15	360.55	3.26			
16	372.78	3.82			
17	382.89	4.66			
18	394.99	5.76			
19	403.37	6.59			
20	411.76	7.42			
21	423.15	8.62			
22	434.54	9.83			
23	440.56	10.48			
24	446.59	11.13			
25	456.69	11.97			

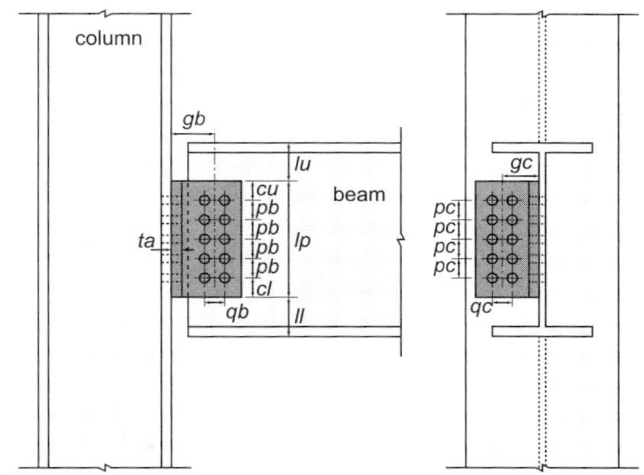

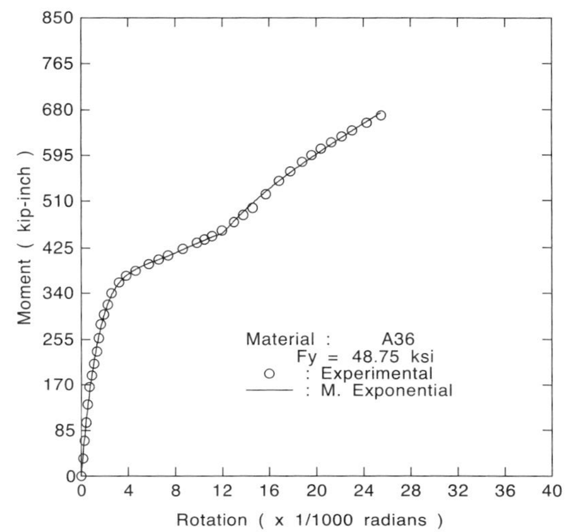

A.1 – 50

```
Moment-rotation prediction equations     ( R : X 1/1000 radians )

Modified exponential model  :
     Bm = Sum ( Ai X ( 1 - exp( - R/( 2*i*C ) ) ) )  +      Sum ( RKj X ( R - Rj0 ) )  +  Bm0
       C =   0.24862000E+00             Bmo=   0.00000000E+00          Nexp=  6       Nliner=  2
      AI =  -0.24043658E+01   0.10695429E+03  -0.28743541E+04   0.14516606E+05  -0.21549396E+05   0.10312831E+05
     Rj0 =          5.7600         12.0000
     RKj =  -0.20585343E+01   0.15255662E+02
```

No	Rotation (radians) X 1/1000	Moment (kip-inch) Expri.	Moment (kip-inch) M.Expo.	Connection stiffness (kip-inch) X 1000 Expri.	Connection stiffness (kip-inch) X 1000 M.Expo.
1	0.00	0.00	0.00	0.2302E+03	0.2635E+03
3	0.29	66.15	69.84	0.2301E+03	0.2231E+03
5	0.57	132.31	128.72	0.2301E+03	0.1873E+03
7	0.93	187.77	187.93	0.1072E+03	0.1496E+03
9	1.35	232.55	242.51	0.1366E+03	0.1131E+03
11	1.66	283.31	274.90	0.1244E+03	0.9052E+02
13	2.29	318.88	320.20	0.6190E+02	0.5760E+02
15	3.26	360.55	359.96	0.2609E+02	0.2785E+02
17	4.66	382.89	385.38	0.1163E+02	0.1212E+02
19	6.59	403.37	403.62	0.1014E+02	0.8127E+01
21	8.62	423.15	421.76	0.9451E+01	0.9493E+01
23	10.48	440.56	439.11	0.9216E+01	0.8863E+01
25	11.97	456.69	451.21	0.1307E+02	0.7328E+01
27	13.81	484.41	490.32	0.1599E+02	0.2027E+02
29	15.72	521.74	526.77	0.2223E+02	0.1808E+02
31	17.80	564.17	562.46	0.1825E+02	0.1625E+02
33	19.59	594.66	590.48	0.1558E+02	0.1516E+02
35	21.28	618.41	615.36	0.1278E+02	0.1446E+02
37	23.03	640.89	640.33	0.1219E+02	0.1398E+02
39	25.52	669.14	674.59	0.1135E+02	0.1358E+02

I - 26

```
Connection type : Single web-angle connections
           Mode : Bolted-to-beam and welded-to-column

Tested by   :  S.L.Lipson (1968)                            Canada
Test  Id.   :  P1-6/2

Column :  --                        Fasteners: A325- -3/4"D
Beam   : W21X62                           15/16" Oversize holes
Angle  :  --                        Material : A36
                                        Fy =  48.75 ksi
                                        Fu =  71.76 ksi

                          Major parameters

     lp = 17.5000"    lu =  1.7500"    ll =  1.7500"    ta =  0.2500"
     gb =  2.5000"    cu =  1.2500"    cl =  1.2500"    pb =  3.0000"
     nb = 1 X  6

Remark  1)  Connection plate welded to 1" mounting plate.
        2)
```

No	Moment (k-in)	Rotation (radians) X 1/1000	No	Moment (k-in)	Rotation (radians) X 1/1000
1	0.00	0.00	26	498.03	11.34
2	27.56	0.11	27	525.85	12.55
3	55.12	0.22	28	546.80	13.81
4	82.68	0.33	29	560.58	14.92
5	110.24	0.44	30	574.36	16.03
6	137.80	0.55	31	583.26	17.00
7	170.04	0.61	32	592.16	17.96
8	202.27	0.68	33	601.06	18.93
9	234.51	0.75	34	611.72	20.16
10	266.75	0.81	35	622.38	21.40
11	298.99	0.88	36	631.46	22.52
12	321.80	1.12	37	640.55	23.65
13	344.62	1.35	38	649.64	24.77
14	367.43	1.59	39	658.66	25.70
15	397.48	2.14	40	667.68	26.63
16	418.64	2.59			
17	437.55	3.78			
18	444.55	4.54			
19	451.56	5.30			
20	454.79	6.08			
21	458.02	6.86			
22	461.57	7.71			
23	465.13	8.56			
24	472.36	9.79			
25	485.19	10.56			

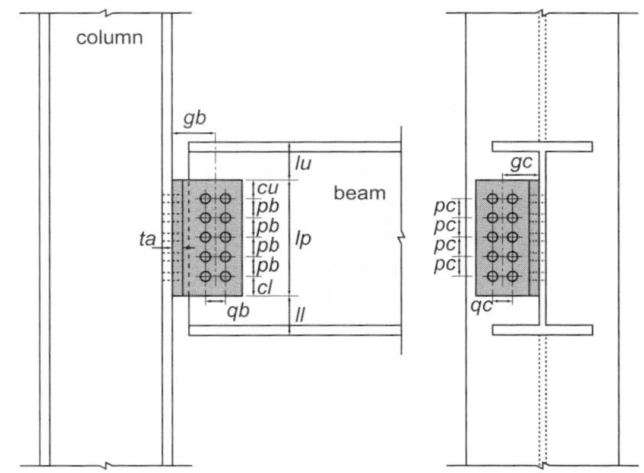

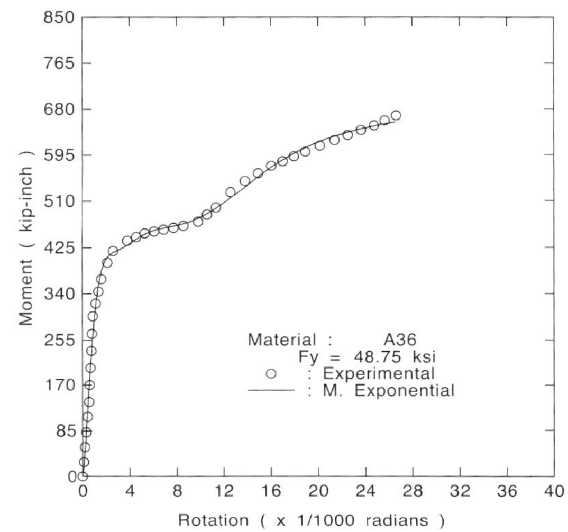

Moment-rotation prediction equations (R : X 1/1000 radians)

```
Modified exponential model  :
     Bm = Sum ( Ai X ( 1 - exp( - R/( 2*i*C ) ) ) )  +      Sum ( RKj X ( R - Rj0 ) )  +  Bm0
      C =   0.24920583E+00             Bmo=   0.00000000E+00          Nexp=  6       Nliner=  1
     AI =  -0.17315110E+04   0.17806165E+05  -0.72729613E+05   0.14799624E+06  -0.14271216E+06   0.51997757E+05
     Rj0 =          13.8100
     RKj =   0.26012966E+01
```

No	Rotation (radians) X 1/1000	Moment (kip-inch) Expri.	Moment (kip-inch) M.Expo.	Connection stiffness (kip-inch) X 1000 Expri.	Connection stiffness (kip-inch) X 1000 M.Expo.
1	0.00	0.00	0.00	0.2524E+03	0.1029E+03
3	0.22	55.12	49.05	0.2523E+03	0.3091E+03
5	0.44	110.24	123.38	0.2523E+03	0.3533E+03
7	0.61	170.04	184.23	0.4804E+03	0.3311E+03
9	0.75	234.51	226.54	0.4804E+03	0.2981E+03
11	0.88	298.99	263.98	0.3956E+03	0.2593E+03
13	1.35	344.62	354.77	0.9649E+02	0.1309E+03
15	2.14	397.48	408.05	0.5031E+02	0.2706E+02
17	3.78	437.55	429.53	0.1183E+02	0.1356E+02
19	5.30	451.56	449.81	0.6731E+01	0.1082E+02
21	6.86	458.02	460.97	0.4154E+01	0.4463E+01
23	8.56	465.13	468.77	0.4876E+01	0.6015E+01
25	10.56	485.19	487.13	0.1663E+02	0.1227E+02
27	12.55	525.85	515.39	0.1982E+02	0.1544E+02
29	14.92	560.58	554.20	0.1240E+02	0.1670E+02
31	17.00	583.26	585.57	0.9231E+01	0.1340E+02
33	18.93	601.06	608.31	0.8960E+01	0.1025E+02
35	21.40	622.38	629.50	0.8345E+01	0.7073E+01
37	23.65	640.55	643.09	0.8102E+01	0.5184E+01
39	25.70	658.66	652.53	0.9693E+01	0.4107E+01

A.1 – 53

I - 27

Connection type : Single web-angle connections
Mode : Bolted-to-beam and welded-to-column

Tested by : S.L.Lipson (1968) Canada
Test Id. : G-2/1

Column : --
Beam : W21X62
Angle : 4 X 3.5 X 5/16

Fasteners: A325- -3/4"D
-5/16" Oversize holes
Material : A36
Fy = 51.50 ksi
Fu = 78.00 ksi

Major parameters

lp = 4.7500" lu = 8.1250" ll = 8.1250" ta = 0.3125"
gb = 2.5000" cu = 1.2500" cl = 1.2500" pb = 2.2500"
nb = 1 X 2

Remark 1) 1/4" fillet weld at heel of angle.
2) Connection rotated under constant shear.

No	Moment (k-in)	Rotation (radians) X 1/1000	No	Moment (k-in)	Rotation (radians) X 1/1000
1	0.00	0.00	26	48.87	4.92
2	2.26	0.12	27	48.97	5.90
3	4.51	0.25	28	48.75	6.77
4	6.77	0.37	29	46.48	7.60
5	9.03	0.49	30	44.75	9.01
6	11.28	0.61	31	44.49	10.09
7	13.54	0.74	32	45.78	10.41
8	15.51	0.82	33	47.07	10.73
9	17.48	0.91	34	48.85	11.84
10	19.45	0.99	35	49.16	12.67
11	21.42	1.08	36	49.47	13.49
12	23.38	1.16	37	49.58	14.40
13	25.50	1.29	38	49.70	15.31
14	27.61	1.41	39	47.62	16.74
15	29.73	1.53	40	47.34	17.87
16	31.84	1.66	41	47.06	18.99
17	33.79	1.76	42	47.22	20.42
18	35.73	1.87	43	47.37	21.85
19	37.68	1.97	44	47.74	23.13
20	39.63	2.08	45	48.11	24.41
21	41.83	2.18	46	47.80	25.63
22	44.03	2.29	47	47.49	26.86
23	45.88	2.46	48	47.47	27.85
24	47.72	2.63	49	47.45	28.85
25	48.77	3.95	50	47.43	29.85

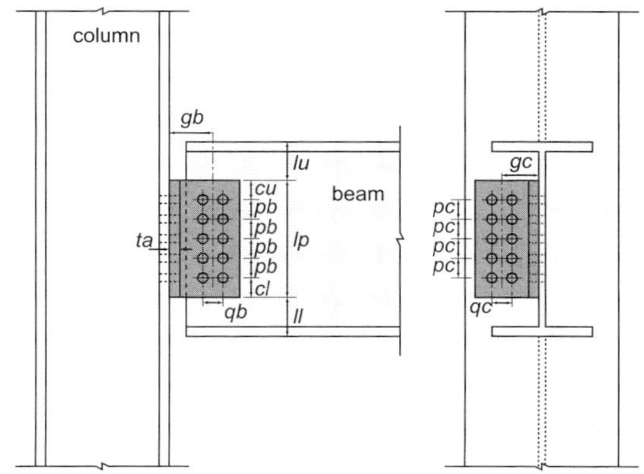

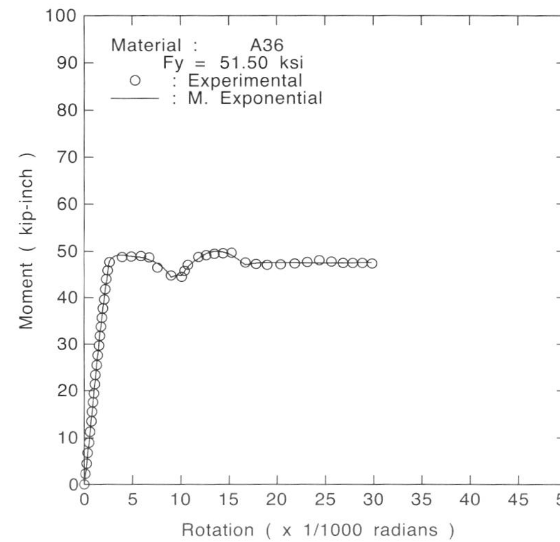

A.1 – 54

Moment-rotation prediction equations (R : X 1/1000 radians)

```
Modified exponential model  :
     Bm = Sum ( Ai X ( 1 - exp( - R/( 2*i*C ) ) ) )  +      Sum ( RKj X ( R - Rj0 ) )  +  Bm0
      C =   0.27947750E+00             Bmo=   0.00000000E+00         Nexp=  6        Nliner=  8
     AI =   0.45989844E+02  -0.11508069E+04   0.63338620E+04  -0.13397834E+05   0.11790868E+05  -0.34855290E+04
    Rj0 =           2.6300           5.9000           6.7700           9.0100          10.0900
                   11.8400          15.3100          16.7400
    RKj =  -0.10789485E+02  -0.37200075E+00   0.25173867E+00   0.42904666E+01   0.37731939E+01
            0.10007323E+01  -0.14713868E+00   0.21218320E+01
```

No	Rotation (radians) X 1/1000	Moment (kip-inch) Expri.	Moment (kip-inch) M.Expo.	Connection stiffness (kip-inch) X 1000 Expri.	Connection stiffness (kip-inch) X 1000 M.Expo.
1	0.00	0.00	0.00	0.1835E+02	0.1730E+02
3	0.25	4.51	4.36	0.1835E+02	0.1831E+02
5	0.49	9.03	9.03	0.1835E+02	0.1962E+02
7	0.74	13.54	13.97	0.2113E+02	0.2047E+02
9	0.91	17.48	17.49	0.2307E+02	0.2067E+02
11	1.08	21.42	21.02	0.2307E+02	0.2055E+02
13	1.29	25.50	25.24	0.1724E+02	0.2004E+02
15	1.53	29.73	30.05	0.1723E+02	0.1907E+02
17	1.76	33.79	34.27	0.1846E+02	0.1795E+02
19	1.97	37.68	37.94	0.1846E+02	0.1684E+02
21	2.18	41.83	41.39	0.2077E+02	0.1573E+02
23	2.46	45.88	45.57	0.1070E+02	0.1437E+02
25	3.95	48.77	49.22	0.3972E+00	-0.1327E+00
27	5.90	48.97	48.73	-0.8392E-01	-0.1860E+00
29	7.60	46.48	47.41	-0.2162E+01	-0.1269E+01
31	10.09	44.49	44.79	0.3026E+01	-0.4565E+00
33	10.73	47.07	46.60	0.3469E+01	0.2363E+01
35	12.67	49.16	49.54	0.3744E+00	0.9508E+00
37	14.40	49.58	49.84	0.1278E+00	-0.4868E+00
39	16.74	47.62	47.11	-0.7730E+00	0.4476E+00
41	18.99	47.06	47.60	-0.9157E-01	0.5973E-01
43	21.85	47.37	47.56	0.2033E+00	-0.4540E-01
45	24.41	48.11	47.48	0.1267E-01	-0.1576E-01
47	26.86	47.49	47.50	-0.1237E+00	0.3053E-01
49	28.85	47.45	47.59	-0.2082E-01	0.6236E-01

I - 28

```
Connection type : Single web-angle connections
           Mode : Bolted-to-beam and welded-to-column

Tested by   :  S.L.Lipson (1968)                       Canada
Test  Id.   :  G-3/1

Column  :  --                        Fasteners: A325- -3/4"D
Beam    : W21X62                             15/16" Oversize holes
Angle   : 4 X 3.5 X 5/16             Material : A36
                                           Fy =  51.50 ksi
                                           Fu =  78.00 ksi

                          Major parameters

     lp =  8.0000"     lu =  6.5000"     ll =  6.5000"     ta =  0.3125"
     gb =  2.5000"     cu =  1.7500"     cl =  1.7500"     pb =  2.2500"
     nb = 1 X  3

Remark  1)  1/4" fillet weld at heel of angle.
        2)  Connection rotated under constant shear.
```

No	Moment (k-in)	Rotation (radians) X 1/1000	No	Moment (k-in)	Rotation (radians) X 1/1000
1	0.00	0.00	26	92.36	10.07
2	5.10	0.14	27	92.85	11.53
3	10.20	0.28	28	93.34	13.00
4	15.31	0.42	29	94.73	14.30
5	20.41	0.57	30	96.12	15.60
6	25.51	0.71	31	97.56	16.72
7	30.61	0.85	32	98.99	17.84
8	36.05	1.03	33	101.08	19.07
9	41.48	1.21	34	103.17	20.31
10	46.91	1.39	35	103.59	21.37
11	51.54	1.61	36	104.01	22.43
12	56.17	1.82	37	105.13	23.46
13	60.80	2.03	38	106.24	24.49
14	65.53	2.28	39	108.19	25.61
15	70.26	2.53	40	110.14	26.73
16	75.00	2.78	41	113.99	27.77
17	79.03	3.07	42	117.84	28.81
18	83.07	3.37	43	121.69	29.86
19	87.77	3.72			
20	91.09	4.16			
21	92.67	4.67			
22	92.84	5.38			
23	93.60	6.48			
24	94.35	7.58			
25	93.36	8.82			

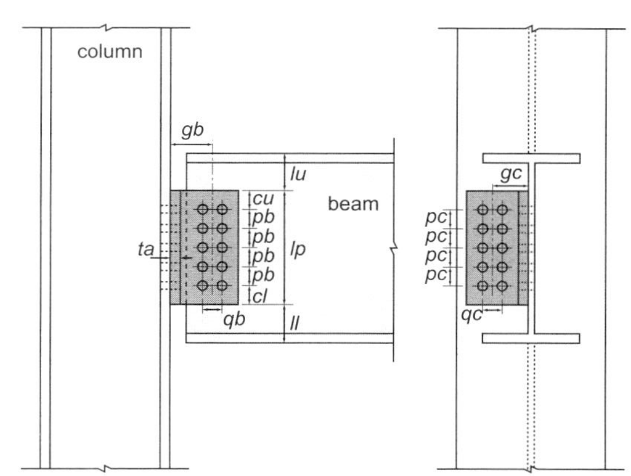

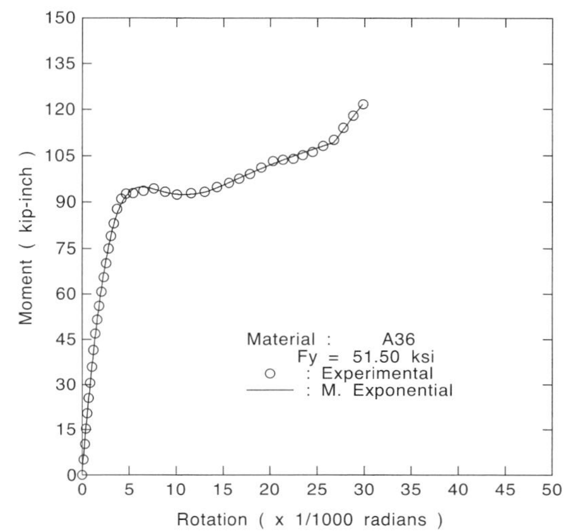

A.1 – 56

Moment-rotation prediction equations (R : X 1/1000 radians)

```
Modified exponential model  :
   Bm = Sum ( Ai X ( 1 - exp( - R/( 2*i*C ) ) ) )  +     Sum ( RKj X ( R - Rj0 ) )  +  Bm0
    C =   0.28418250E+00           Bmo=   0.00000000E+00          Nexp=  6          Nliner=  3
   AI =   0.12700454E+02   0.56826575E+02  -0.15822150E+04   0.51810900E+04  -0.57035544E+04   0.21284967E+04
  Rj0 =           4.7600          13.3000          26.7300
  RKj =   0.52764118E+00   0.35236740E+00   0.31026779E+01
```

No	Rotation (radians) X 1/1000	Moment (kip-inch) Expri.	Moment (kip-inch) M.Expo.	Connection stiffness (kip-inch) X 1000 Expri.	Connection stiffness (kip-inch) X 1000 M.Expo.
1	0.00	0.00	0.00	0.3606E+02	0.4050E+02
3	0.28	10.20	10.84	0.3606E+02	0.3642E+02
5	0.57	20.41	20.74	0.3606E+02	0.3365E+02
7	0.85	30.61	29.93	0.3336E+02	0.3137E+02
9	1.21	41.48	40.82	0.2992E+02	0.2855E+02
11	1.61	51.54	51.47	0.2168E+02	0.2533E+02
13	2.03	60.80	61.50	0.2051E+02	0.2167E+02
15	2.53	70.26	71.16	0.1913E+02	0.1742E+02
17	3.07	79.03	79.42	0.1362E+02	0.1304E+02
19	3.72	87.77	86.34	0.1084E+02	0.8560E+01
21	4.67	92.67	92.01	0.1909E+01	0.3660E+01
23	6.48	93.60	94.79	0.6877E+00	-0.1690E+00
25	8.82	93.36	93.14	-0.7987E+00	-0.7186E+00
27	11.53	92.85	92.50	0.3349E+00	0.2669E+00
29	14.30	94.73	94.57	0.1067E+01	0.1221E+01
31	16.72	97.56	97.71	0.1286E+01	0.1332E+01
33	19.07	101.08	100.79	0.1692E+01	0.1267E+01
35	21.37	103.59	103.58	0.3959E+00	0.1162E+01
37	23.46	105.13	105.92	0.1080E+01	0.1075E+01
39	25.61	108.19	108.15	0.1744E+01	0.1006E+01
41	27.77	113.99	113.50	0.3691E+01	0.4061E+01
43	29.86	121.69	121.93	0.3692E+01	0.4031E+01

I - 29

Connection type : Single web-angle connections
Mode : Bolted-to-beam and welded-to-column

Tested by : S.L.Lipson (1968) Canada
Test Id. : G-4/1

Column : --
Beam : W21X62
Angle : 4 X 3.5 X 5/16

Fasteners: A325- -3/4"D
15/16" Oversize holes
Material : A36
Fy = 51.50 ksi
Fu = 78.00 ksi

Major parameters

lp = 12.0000" lu = 4.8750" ll = 4.8750" ta = 0.3125"
gb = 2.5000" cu = 2.2500" cl = 2.2500" pb = 2.2500"
nb = 1 X 4

Remark 1) 1/4" fillet weld at heel of angle.
2) Connection rotated under constant shear.

No	Moment (k-in)	Rotation (radians) X 1/1000	No	Moment (k-in)	Rotation (radians) X 1/1000
1	0.00	0.00	26	136.98	11.58
2	8.82	0.12	27	138.15	12.35
3	17.63	0.23	28	139.13	13.26
4	26.45	0.35	29	140.10	14.18
5	35.26	0.46	30	140.31	15.41
6	44.08	0.58	31	140.52	16.64
7	52.28	0.68	32	142.31	17.68
8	60.48	0.79	33	143.99	18.50
9	67.41	0.95	34	145.68	19.31
10	74.34	1.11	35	149.35	20.38
11	81.27	1.27	36	153.03	21.45
12	87.92	1.41	37	157.75	22.45
13	94.57	1.54	38	162.47	23.46
14	101.68	1.89	39	169.53	24.68
15	108.79	2.25	40	176.60	25.91
16	118.62	2.71	41	181.98	27.00
17	127.72	3.34	42	187.36	28.10
18	132.70	3.99	43	192.78	29.07
19	135.63	4.60	44	198.19	30.04
20	134.78	5.27			
21	133.73	6.72			
22	133.66	7.70			
23	133.59	8.67			
24	134.69	9.74			
25	135.80	10.80			

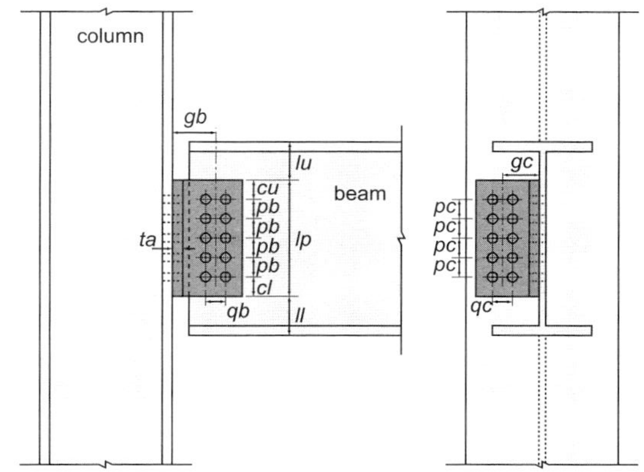

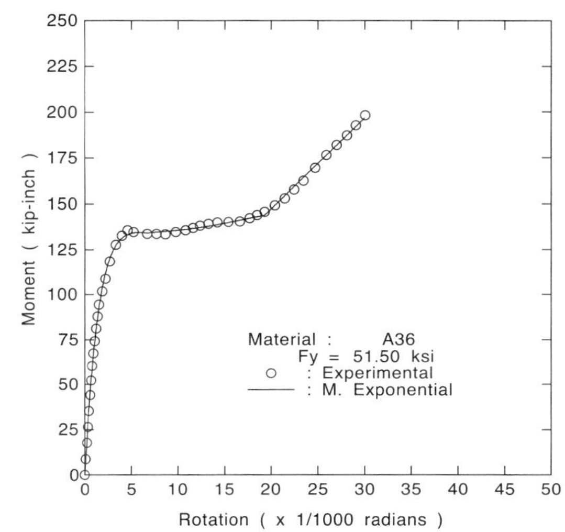

A.1 – 58

```
Moment-rotation prediction equations     ( R : X 1/1000 radians )

Modified exponential model  :
   Bm = Sum ( Ai X ( 1 - exp( - R/( 2*i*C ) ) ) )  +     Sum ( RKj X ( R - Rj0 ) )  +  Bm0
    C =   0.27921167E+00           Bmo=   0.00000000E+00          Nexp=  6          Nliner=  2
   AI =  -0.10082154E+02  -0.52516263E+02   0.58976630E+03  -0.84857185E+03   0.77994126E+03  -0.32977486E+03
  Rj0 =           6.7200          19.3100
  RKj =   0.11742254E+01   0.38042031E+01
```

No	Rotation (radians) X 1/1000	Moment (kip-inch) Expri.	Moment (kip-inch) M.Expo.	Connection stiffness (kip-inch) X 1000 Expri.	Connection stiffness (kip-inch) X 1000 M.Expo.
1	0.00	0.00	0.00	0.7632E+02	0.8798E+02
3	0.23	17.63	19.33	0.7636E+02	0.7924E+02
5	0.46	35.26	36.59	0.7632E+02	0.7029E+02
7	0.68	52.28	51.10	0.7894E+02	0.6204E+02
9	0.95	67.41	66.37	0.4262E+02	0.5273E+02
11	1.27	81.27	81.84	0.4651E+02	0.4266E+02
13	1.54	94.57	92.28	0.4155E+02	0.3550E+02
15	2.25	108.79	111.94	0.2062E+02	0.2109E+02
17	3.34	127.72	127.22	0.1113E+02	0.8517E+01
19	4.60	135.63	133.38	0.1894E+01	0.2216E+01
21	6.72	133.73	134.20	-0.3357E+00	-0.5689E+00
23	8.67	133.59	135.08	0.4595E+00	0.4291E+00
25	10.80	135.80	136.20	0.1314E+01	0.6292E+00
27	12.35	138.15	137.29	0.1309E+01	0.7700E+00
29	14.18	140.10	138.81	0.6847E+00	0.8940E+00
31	16.64	140.52	141.16	0.1002E+01	0.1005E+01
33	18.50	143.99	143.08	0.2072E+01	0.1060E+01
35	20.38	149.35	149.17	0.3437E+01	0.4903E+01
37	22.45	157.75	159.37	0.4698E+01	0.4931E+01
39	24.68	169.53	170.41	0.5747E+01	0.4951E+01
41	27.00	181.98	181.91	0.4938E+01	0.4963E+01
43	29.07	192.78	192.16	0.5566E+01	0.4969E+01

I - 30

```
Connection type : Single web-angle connections
           Mode : Bolted-to-beam and welded-to-column

Tested by  :  S.L.Lipson (1968)                          Canada
Test  Id.  :  G-5/1

Column :  --                          Fasteners: A325- -3/4"D
Beam   : W21X62                             15/16" Oversize holes
Angle  : 4 X 3.5 X 5/16               Material : A36
                                           Fy =  51.50 ksi
                                           Fu =  78.00 ksi

                         Major parameters

     lp = 14.5000"   lu =  3.2500"   ll =  3.2500"   ta =  0.3125"
     gb =  2.5000"   cu =  2.7500"   cl =  2.7500"   pb =  2.2500"
     nb = 1 X  5

Remark  1)  1/4" fillet weld at heel of angle.
        2)  Connection rotated under constant shear.
```

No	Moment (k-in)	Rotation (radians) X 1/1000	No	Moment (k-in)	Rotation (radians) X 1/1000
1	0.00	0.00	26	243.64	14.81
2	16.41	0.19	27	252.70	16.30
3	32.82	0.37	28	260.99	17.41
4	49.22	0.56	29	269.28	18.53
5	63.92	0.74	30	275.36	19.82
6	78.61	0.93	31	281.43	21.12
7	91.07	1.15	32	287.51	22.41
8	103.53	1.36	33	294.74	23.64
9	113.60	1.54	34	301.98	24.87
10	123.67	1.71	35	309.21	26.10
11	133.70	2.04	36	316.67	27.37
12	143.73	2.37	37	324.12	28.65
13	155.28	2.78	38	331.57	29.92
14	166.83	3.20			
15	177.14	3.81			
16	187.44	4.42			
17	191.81	4.77			
18	194.75	5.35			
19	199.33	6.24			
20	203.90	7.12			
21	209.91	8.47			
22	215.92	9.81			
23	222.14	10.98			
24	228.36	12.15			
25	234.58	13.33			

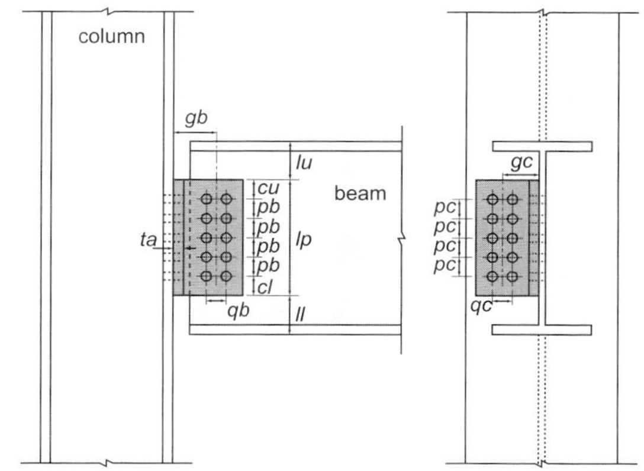

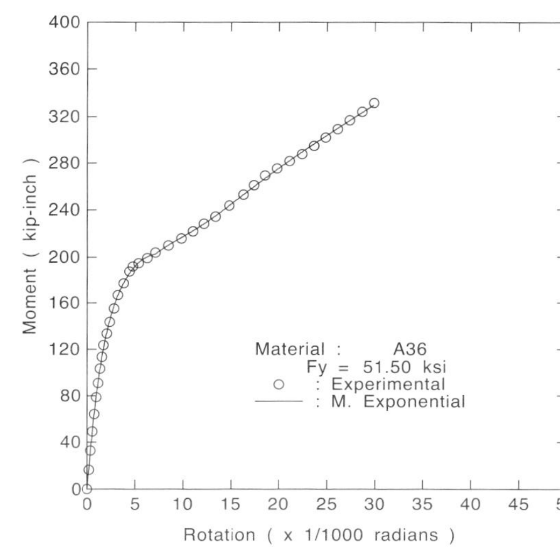

```
Moment-rotation prediction equations     ( R : X 1/1000 radians )

Modified exponential model  :
     Bm = Sum ( Ai X ( 1 - exp( - R/( 2*i*C ) ) ) )  +      Sum ( RKj X ( R - Rj0 ) )  +  Bm0
      C =   0.29571167E+00              Bmo=   0.00000000E+00          Nexp=  6        Nliner=  1
     AI =  -0.13294210E+03   0.17488474E+04  -0.82707462E+04   0.17932142E+05  -0.17151114E+05   0.60904488E+04
    Rj0 =          7.1200
    RKj =   0.50015897E+01
```

No	Rotation (radians) X 1/1000	Moment (kip-inch) Expri.	Moment (kip-inch) M.Expo.	Connection stiffness (kip-inch) X 1000 Expri.	Connection stiffness (kip-inch) X 1000 M.Expo.
1	0.00	0.00	0.00	0.8850E+02	0.8869E+02
3	0.37	32.82	33.40	0.8853E+02	0.8821E+02
5	0.74	63.92	64.14	0.7864E+02	0.7605E+02
7	1.15	91.07	91.66	0.5789E+02	0.6106E+02
9	1.54	113.60	113.00	0.5731E+02	0.4857E+02
11	2.04	133.70	134.23	0.3066E+02	0.3659E+02
13	2.78	155.28	156.84	0.2784E+02	0.2531E+02
15	3.81	177.14	177.74	0.1678E+02	0.1607E+02
17	4.77	191.81	190.21	0.9772E+01	0.1018E+02
19	6.24	199.33	200.23	0.5165E+01	0.4012E+01
21	8.47	209.91	210.43	0.4468E+01	0.5013E+01
23	10.98	222.14	222.40	0.5313E+01	0.4912E+01
25	13.33	234.58	234.86	0.5656E+01	0.5719E+01
27	16.30	252.70	252.76	0.6864E+01	0.6189E+01
29	18.53	269.28	266.52	0.6171E+01	0.6119E+01
31	21.12	281.43	282.03	0.4695E+01	0.5853E+01
33	23.64	294.74	296.45	0.5884E+01	0.5579E+01
35	26.10	309.21	309.90	0.5865E+01	0.5368E+01
37	28.65	324.12	323.38	0.5845E+01	0.5218E+01

I - 31

Connection type : Single web-angle connections
Mode : Bolted-to-beam and welded-to-column

Tested by : S.L.Lipson (1977) Canada
Test Id. : A1-5

Column : T=44",TW=5/8"
Beam : --
Angle : 4 X 3 X 3/8

Fasteners: A325- -3/4"D
13/16" Oversize holes
Material : G40.21
Fy = 55.40 ksi
Fu = 77.80 ksi

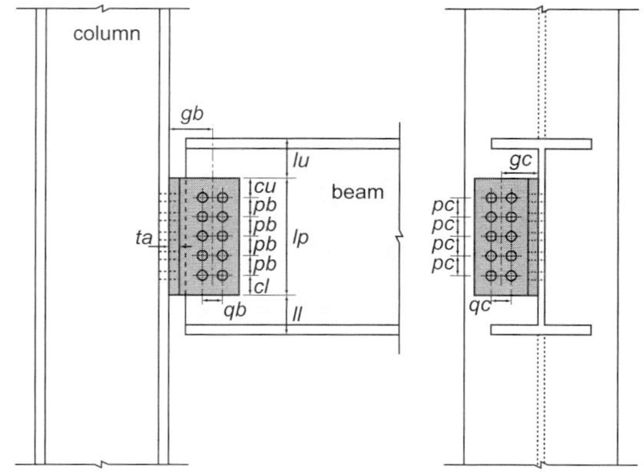

Major parameters

lp = 14.5000" lu = -- " ll = -- " ta = 0.3750"
gb = 2.5000" cu = 1.2500" cl = 1.2500" pb = 3.0000"
nb = 1 X 5

Remark 1) 1/4" fillet weld at heel, top and bottom of angle.
2) Connection subject to moment and shear.

No	Moment (k-in)	Rotation (radians) X 1/1000
1	0.00	0.00
2	12.89	0.11
3	25.79	0.23
4	38.68	0.34
5	51.57	0.46
6	64.46	0.57
7	74.51	0.70
8	84.56	0.83
9	94.61	0.96
10	104.66	1.09
11	114.10	1.30
12	123.54	1.51
13	132.99	1.73
14	142.49	1.97
15	151.98	2.22
16	158.74	2.62
17	165.49	3.01
18	163.89	3.81
19	159.22	4.58
20	147.99	5.06
21	136.76	5.54
22	127.53	6.01
23	118.30	6.49
24	109.07	6.96
25	102.73	7.64
26	96.38	8.32
27	88.77	9.13
28	81.16	9.95
29	88.57	10.72
30	95.98	11.48
31	107.75	12.16
32	119.53	12.83
33	130.38	13.27
34	141.23	13.71
35	152.08	14.15
36	164.95	14.59
37	177.82	15.03
38	190.68	15.46
39	201.89	15.82
40	213.11	16.18
41	224.32	16.54
42	235.54	16.90
43	245.83	17.59
44	256.12	18.28
45	263.83	19.22
46	271.54	20.15
47	269.78	20.86
48	268.02	21.57
49	264.69	22.25
50	261.37	22.92
51	259.82	23.76

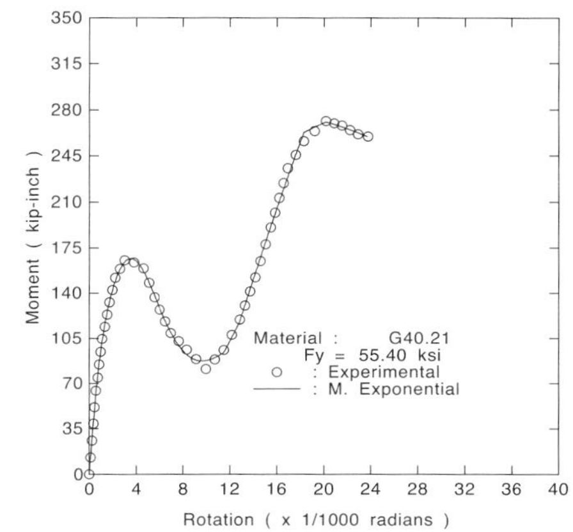

Moment-rotation prediction equations (R : X 1/1000 radians)

Modified exponential model :

```
Bm = Sum ( Ai X ( 1 - exp( - R/( 2*i*C ) ) ) )  +     Sum ( RKj X ( R - Rj0 ) )  +  Bm0
 C =   0.22658500E+00          Bmo=   0.00000000E+00          Nexp=  6        Nliner=  4
AI =  -0.28671862E+03   0.56504464E+04  -0.33184380E+05   0.78892620E+05  -0.79907960E+05   0.28984325E+05
Rj0 =         11.4800          12.8300          18.2800          20.1600
RKj =   0.11136377E+02   0.94927919E+01  -0.19178599E+02  -0.60417455E+01
```

No	Rotation (radians) X 1/1000	Moment (kip-inch) Expri.	Moment (kip-inch) M.Expo.	Connection stiffness (kip-inch) X 1000 Expri.	Connection stiffness (kip-inch) X 1000 M.Expo.
1	0.00	0.00	0.00	0.1126E+03	0.1089E+03
3	0.23	25.79	26.27	0.1126E+03	0.1156E+03
5	0.46	51.57	51.55	0.1126E+03	0.1037E+03
7	0.70	74.51	74.65	0.7798E+02	0.8600E+02
9	0.96	94.61	94.53	0.7798E+02	0.6889E+02
11	1.30	114.10	115.02	0.4422E+02	0.5191E+02
13	1.73	132.99	134.01	0.4174E+02	0.3803E+02
15	2.22	151.98	149.73	0.3054E+02	0.2672E+02
17	3.01	165.49	164.36	0.1064E+02	0.1002E+02
19	4.58	159.22	157.30	-0.1679E+02	-0.1616E+02
21	5.54	136.76	138.56	-0.2147E+02	-0.2156E+02
23	6.49	118.30	118.54	-0.1950E+02	-0.1982E+02
25	7.64	102.73	99.49	-0.9363E+01	-0.1279E+02
27	9.13	88.77	88.11	-0.9322E+01	-0.2761E+01
29	10.72	88.57	89.91	0.9653E+01	0.4282E+01
31	12.16	107.75	105.98	0.1746E+02	0.1816E+02
33	13.27	130.38	130.73	0.2473E+02	0.2807E+02
35	14.15	152.08	155.26	0.2704E+02	0.2776E+02
37	15.03	177.82	179.33	0.2934E+02	0.2712E+02
39	15.82	201.89	200.69	0.3121E+02	0.2641E+02
41	16.54	224.32	219.42	0.3121E+02	0.2573E+02
43	17.59	245.83	245.91	0.1492E+02	0.2476E+02
45	19.22	263.83	267.09	0.8232E+01	0.4288E+01
47	20.86	269.78	269.04	-0.2489E+01	-0.2737E+01
49	22.25	264.69	264.82	-0.4911E+01	-0.3329E+01
51	23.76	259.82	259.44	-0.1856E+01	-0.3778E+01

I - 32

Connection type : Single web-angle connections
Mode : Bolted-to-beam and welded-to-column

Tested by : S.L.Lipson (1977) Canada
Test Id. : A1-6

Column : T=44",TW=5/8"
Beam : --
Angle : 4 X 3 X 3/8

Fasteners: A325- -3/4"D
13/16" Oversize holes
Material : G40.21
Fy = 55.40 ksi
Fu = 77.80 ksi

Major parameters

lp = 17.5000" lu = -- " ll = -- " ta = 0.3750"
gb = 2.5000" cu = 1.2500" cl = 1.2500" pb = 3.0000"
nb = 1 X 6

Remark 1) 1/4" fillet weld at heel, top and bottom of angle.
2) Connection subject to moment and shear.

No	Moment (k-in)	Rotation (radians) X 1/1000	No	Moment (k-in)	Rotation (radians) X 1/1000
1	0.00	0.00	26	476.04	7.75
2	25.62	0.09	27	469.92	8.55
3	51.24	0.18	28	463.79	9.36
4	76.86	0.26	29	472.52	10.02
5	102.48	0.35	30	481.24	10.67
6	128.10	0.44	31	491.81	11.52
7	153.72	0.53	32	502.38	12.36
8	179.34	0.61	33	507.04	13.28
9	207.84	0.78	34	511.70	14.21
10	236.34	0.95	35	507.19	15.07
11	264.83	1.12	36	502.69	15.93
12	292.29	1.27	37	510.32	16.82
13	319.74	1.42	38	517.94	17.71
14	347.20	1.57	39	541.00	18.78
15	371.78	1.80	40	563.46	19.52
16	396.36	2.03	41	585.91	20.25
17	420.94	2.26	42	594.56	20.86
18	443.84	2.65	43	603.20	21.48
19	473.73	3.05	44	595.24	22.09
20	492.49	3.42	45	587.27	22.71
21	503.24	3.78			
22	523.01	4.70			
23	517.51	5.59			
24	489.06	6.58			
25	482.55	7.17			

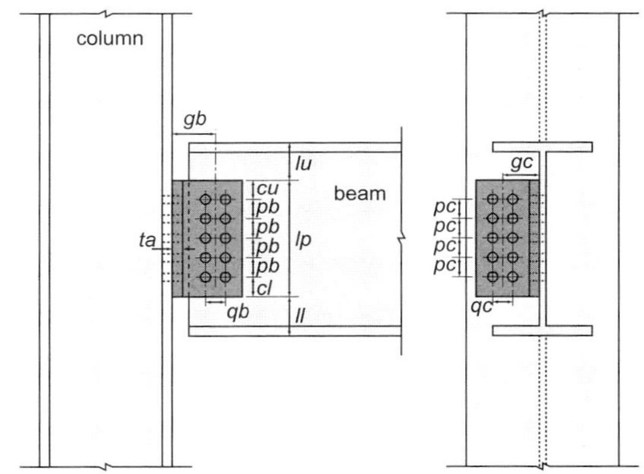

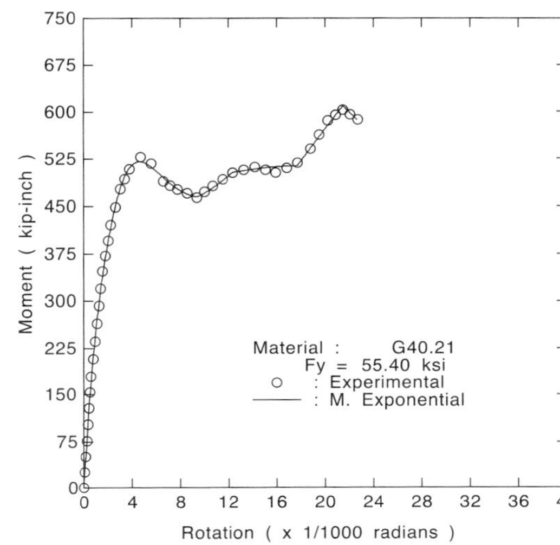

Moment-rotation prediction equations (R : X 1/1000 radians)

```
Modified expcnential model  :
     Bm = Sum ( Ai X ( 1 - exp( - R/( 2*i*C ) ) ) )  +      Sum ( RKj X ( R - Rj0 ) )  +  Bm0
      C =   0.21113083E+00            Bmo=   0.00000000E+00            Nexp=  6        Nliner=  4
     AI =  -0.37456435E+03   0.70517529E+04  -0.38287138E+05   0.84392080E+05  -0.79384219E+05   0.27105178E+05
     Rj0 =           9.3600          12.3600          17.7100          21.4800
     RKj =   0.10843076E+02  -0.13271110E+02   0.26121797E+02  -0.43187906E+02
```

No	Rotation (radians) X 1/1000	Moment (kip-inch) Expri.	Moment (kip-inch) M.Expo.	Connection stiffness (kip-inch) X 1000 Expri.	Connection stiffness (kip-inch) X 1000 M.Expo.
1	0.00	0.00	0.00	0.2928E+03	0.3023E+03
3	0.18	51.24	53.09	0.2926E+03	0.2994E+03
5	0.35	102.48	103.74	0.2926E+03	0.2775E+03
7	0.53	153.72	149.90	0.2928E+03	0.2495E+03
9	0.78	207.84	208.69	0.1686E+03	0.2099E+03
11	1.12	264.83	272.36	0.1771E+03	0.1691E+03
13	1.42	319.74	318.68	0.1846E+03	0.1437E+03
15	1.80	371.78	368.50	0.1063E+03	0.1197E+03
17	2.26	420.94	418.28	0.9315E+02	0.9598E+02
19	3.05	476.73	479.01	0.5634E+02	0.5822E+02
21	3.78	508.24	509.78	0.3690E+02	0.2680E+02
23	5.59	517.51	512.24	-0.1982E+02	-0.1477E+02
25	7.17	482.55	485.55	-0.1112E+02	-0.1577E+02
27	8.55	469.92	469.00	-0.7652E+01	-0.7762E+01
29	10.02	472.52	470.67	0.1323E+02	0.1065E+02
31	11.52	491.81	490.29	0.1252E+02	0.1486E+02
33	13.28	507.04	505.80	0.5059E+01	0.2830E+01
35	15.07	507.19	510.35	-0.5229E+01	0.2089E+01
37	16.82	510.32	512.93	0.8566E+01	0.8429E+00
39	18.78	541.00	541.34	0.2690E+02	0.2573E+02
41	20.25	585.91	578.57	0.2160E+02	0.2505E+02
43	21.48	603.20	609.04	0.6230E+00	0.2463E+02
45	22.71	587.27	586.07	-0.1292E+02	-0.1885E+02

I - 33

Connection type : Single web-angle connections
Mode : Bolted-to-beam and welded-to-column

Tested by : S.L.Lipson (1977) Canada
Test Id. : A1-7

Column : T=14",TW=5/8"
Beam : --
Angle : 4 X 3 X 3/8

Fasteners: A325- -3/4"D
13/16" Oversize holes
Material : G40.21
Fy = 55.40 ksi
Fu = 77.80 ksi

Major parameters

lp = 20.5000"	lu = -- "	ll = -- "	ta = 0.3750"
gb = 2.5000"	cu = 1.2500"	cl = 1.2500"	pb = 3.0000"
nb = 1 X 7			

Remark 1) 1/4" fillet weld at heel, top and bottom of angle.
2) Connection subject to moment and shear.

No	Moment (k-in)	Rotation (radians) X 1/1000	No	Moment (k-in)	Rotation (radians) X 1/1000
1	0.00	0.00	26	669.90	9.23
2	45.50	0.10	27	688.24	10.15
3	91.01	0.20	28	706.58	11.07
4	136.51	0.30	29	720.46	12.02
5	132.01	0.40	30	734.33	12.97
6	233.73	0.59	31	738.79	13.74
7	285.44	0.78	32	743.26	14.51
8	320.35	0.86	33	762.94	15.34
9	355.25	0.95	34	782.62	16.17
10	387.31	1.13	35	818.94	17.01
11	419.37	1.30	36	855.26	17.85
12	457.02	1.51	37	889.83	18.56
13	494.67	1.72	38	924.41	19.28
14	532.32	1.93	39	955.97	20.00
15	573.55	2.17	40	987.53	20.72
16	614.79	2.41	41	1016.70	21.77
17	654.97	2.89	42	1045.88	22.83
18	680.11	3.41			
19	690.22	3.96			
20	679.70	4.79			
21	676.32	5.41			
22	672.95	6.03			
23	666.87	6.81			
24	660.79	7.60			
25	665.34	8.42			

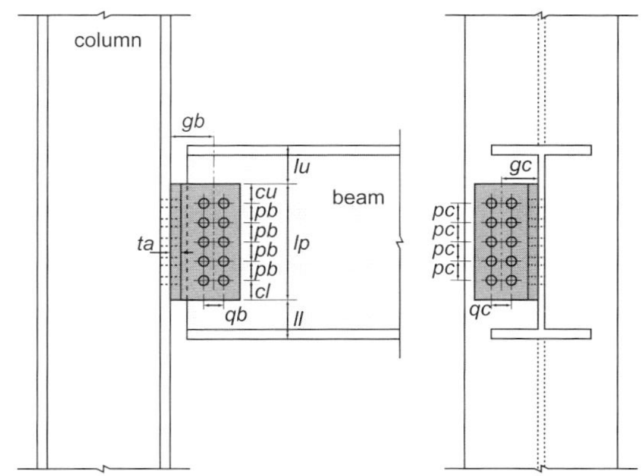

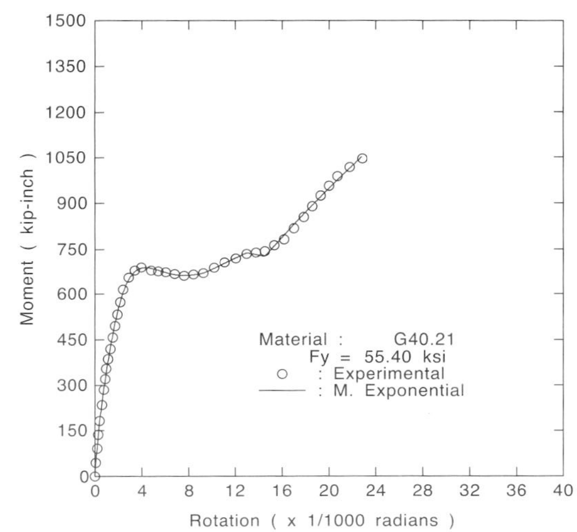

```
Moment-rotation prediction equations     ( R : X 1/1000 radians )

Modified exponential model  :
     Bm = Sum ( Ai X ( 1 - exp( - R/( 2*i*C ) ) ) )  +      Sum ( RKj X ( R - Rj0 ) )  +  Bm0
      C =   0.21496417E+00             Bmo=   0.00000000E+00            Nexp=  6        Nliner=  3
     AI =   0.33266444E+03   0.93120855E+02  -0.15974874E+05   0.56707555E+05  -0.67493208E+05   0.27092526E+05
    Rj0 =          9.2300          12.9700          14.5100
    RKj =   0.69192410E+01  -0.24058586E+02   0.52968014E+02
```

No	Rotation (radians) X 1/1000	Moment (kip-inch) Expri.	Moment (kip-inch) M.Expo.	Connection stiffness (kip-inch) X 1000 Expri.	Connection stiffness (kip-inch) X 1000 M.Expo.
1	0.00	0.00	0.00	0.4601E+03	0.5767E+03
3	0.20	91.01	98.40	0.4601E+03	0.4322E+03
5	0.40	182.01	175.10	0.3957E+03	0.3509E+03
7	0.78	285.44	292.08	0.3656E+03	0.2755E+03
9	0.95	355.25	337.46	0.3338E+03	0.2554E+03
11	1.30	419.37	422.25	0.1802E+03	0.2201E+03
13	1.72	494.67	505.47	0.1808E+03	0.1793E+03
15	2.17	573.55	575.50	0.1730E+03	0.1346E+03
17	2.89	654.97	648.56	0.6627E+02	0.7012E+02
19	3.96	690.22	687.61	0.5968E+01	0.1032E+02
21	5.41	676.32	680.03	-0.5473E+01	-0.1257E+02
23	6.81	666.87	665.33	-0.7727E+01	-0.6358E+01
25	8.42	665.34	664.68	0.5584E+01	0.5025E+01
27	10.15	688.24	686.26	0.1993E+02	0.1822E+02
29	12.02	720.46	721.44	0.1457E+02	0.1857E+02
31	13.74	738.79	732.97	0.5806E+01	-0.7805E+01
33	15.34	762.94	762.41	0.2378E+02	0.4262E+02
35	17.01	818.94	831.51	0.4325E+02	0.4036E+02
37	18.56	889.83	893.04	0.4823E+02	0.3879E+02
39	20.00	955.97	947.90	0.4399E+02	0.3777E+02
41	21.77	1016.70	1014.12	0.2761E+02	0.3695E+02

Connection type : Single web-angle connections
Mode : Bolted-to-beam and welded-to-column

Tested by : S.L.Lipson (1977) Canada
Test Id. : A1-8

Column : T=44",TW=5/8"
Beam : --
Angle : 4 X 3 X 3/8

Fasteners: A325- -3/4"D
13/16" Oversize holes
Material : G40.21
Fy = 55.40 ksi
Fu = 77.80 ksi

Major parameters

lp = 23.5000" lu = -- " ll = -- " ta = 0.3750"
gb = 2.5000" cu = 1.2500" cl = 1.2500" pb = 3.0000"
nb = 1 X 8

Remark 1) 1/4" fillet weld at heel, top and bottom of angle.
2) Connection subject to moment and shear.

No	Moment (k-in)	Rotation (radians) X 1/1000	No	Moment (k-in)	Rotation (radians) X 1/1000
1	0.00	0.00	26	1033.75	8.42
2	59.95	0.10	27	1007.69	8.98
3	119.91	0.20	28	970.35	9.94
4	179.86	0.29	29	976.49	10.75
5	239.81	0.39	30	982.62	11.56
6	299.76	0.49	31	1018.77	12.29
7	358.39	0.63	32	1054.92	13.02
8	417.02	0.77	33	1120.34	14.05
9	475.65	0.92	34	1185.33	14.87
10	528.29	1.08	35	1250.33	15.69
11	580.93	1.24	36	1297.07	16.43
12	633.56	1.41	37	1343.81	17.17
13	686.36	1.65	38	1389.46	18.13
14	739.16	1.90	39	1435.12	19.09
15	791.96	2.14	40	1440.92	19.73
16	854.57	2.51	41	1446.72	20.38
17	917.17	2.88	42	1433.40	21.27
18	960.26	3.31	43	1395.88	22.14
19	1003.34	3.73			
20	1071.41	4.57			
21	1103.59	5.58			
22	1098.68	6.15			
23	1093.76	6.72			
24	1076.79	7.29			
25	1059.81	7.86			

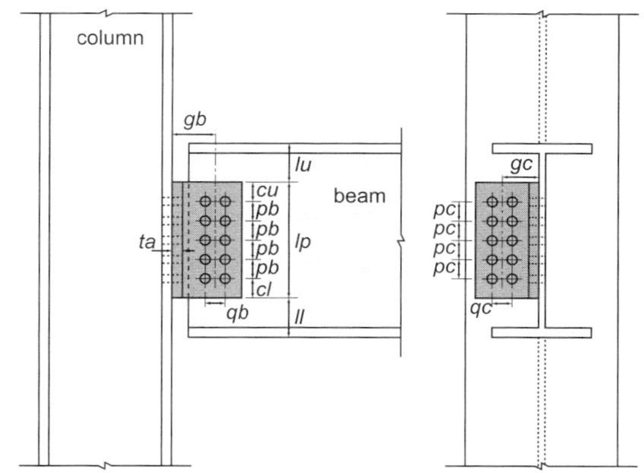

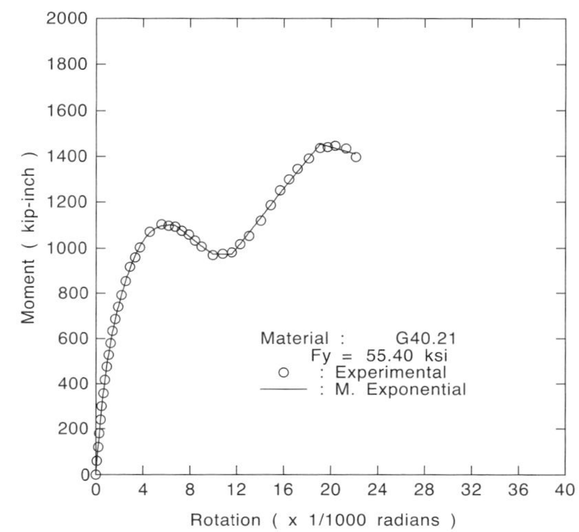

A.1 – 68

```
Moment-rotation prediction equations     ( R : X 1/1000 radians )

Modified exponential model  :
    Bm = Sum ( Ai X ( 1 - exp( - R/( 2*i*C ) ) ) )  +      Sum ( RKj X ( R - Rj0 ) )  +  Bm0
     C =   0.20905667E+00           Bmo=   0.00000000E+00         Nexp=  6        Nliner=  4
    AI =  -0.77242177E+03   0.11402540E+05  -0.50995818E+05   0.98826485E+05  -0.84588218E+05   0.27240109E+05
   Rj0 =          6.7200           9.9400          11.5600          19.0900
   RKj =  -0.33559306E+02   0.32743618E+02   0.61284992E+02  -0.76186241E+02
```

No	Rotation (radians) X 1/1000	Moment (kip-inch) Expri.	Moment (kip-inch) M.Expo.	Connection stiffness (kip-inch) X 1000 Expri.	Connection stiffness (kip-inch) X 1000 M.Expo.
1	0.00	0.00	0.00	0.6099E+03	0.6200E+03
3	0.20	119.91	122.67	0.6099E+03	0.6138E+03
5	0.39	239.81	237.86	0.6099E+03	0.5531E+03
7	0.63	358.39	359.61	0.4141E+03	0.4618E+03
9	0.92	475.65	476.42	0.3715E+03	0.3670E+03
11	1.24	580.93	582.59	0.3223E+03	0.2889E+03
13	1.65	686.36	687.34	0.2153E+03	0.2296E+03
15	2.14	791.96	788.96	0.1968E+03	0.1882E+03
17	2.88	917.17	911.25	0.1376E+03	0.1422E+03
19	3.73	1003.34	1009.71	0.9503E+02	0.9189E+02
21	5.58	1103.59	1097.09	0.6023E+01	0.1306E+02
23	6.72	1093.76	1100.13	-0.1914E+02	-0.4483E+01
25	7.86	1059.81	1053.99	-0.3829E+02	-0.4115E+02
27	8.98	1007.69	1009.34	-0.4382E+02	-0.3838E+02
29	10.75	976.49	973.15	0.7547E+01	0.4825E-01
31	12.29	1018.77	1020.52	0.4954E+02	0.6397E+02
33	14.05	1120.34	1133.94	0.7249E+02	0.6453E+02
35	15.69	1250.33	1238.83	0.7089E+02	0.6391E+02
37	17.17	1343.81	1333.19	0.5624E+02	0.6308E+02
39	19.09	1435.12	1453.12	0.2454E+02	0.6212E+02
41	20.38	1446.72	1434.77	-0.1011E+01	-0.1456E+02
43	22.14	1395.88	1408.67	-0.4324E+02	-0.1503E+02

I - 35

Connection type : Single web-angle connections
Mode : Bolted-to-beam and welded-to-column

Tested by : S.L.Lipson (1977) Canada
Test Id. : A1-9

Column : T=44",TW=5/8"
Beam : --
Angle : 4 X 3 X 3/8

Fasteners: A325- -3/4"D
13/16" Oversize holes
Material : G40.21
Fy = 55.40 ksi
Fu = 77.80 ksi

Major parameters

lp = 26.5000" lu = -- " ll = -- " ta = 0.3750"
gb = 2.5000" cu = 1.2500" cl = 1.2500" pb = 3.0000"
nb = 1 X 9

Remark 1) 1/4" fillet weld at heel, top and bottom of angle.
2) Connection subject to moment and shear.

No	Moment (k-in)	Rotation (radians) X 1/1000	No	Moment (k-in)	Rotation (radians) X 1/1000
1	0.00	0.00	26	2039.43	8.34
2	104.53	0.12	27	2004.13	8.96
3	209.05	0.24	28	1944.64	9.58
4	312.22	0.46	29	1854.91	10.21
5	415.38	0.67	30	1833.99	10.37
6	494.33	0.87	31	1757.43	10.72
7	573.28	1.07	32	1680.87	11.07
8	652.23	1.28	33	1621.23	11.60
9	750.39	1.52	34	1561.44	12.05
10	848.55	1.76	35	1520.19	12.73
11	946.72	2.00	36	1526.72	13.04
12	1053.02	2.29	37	1579.10	13.64
13	1159.33	2.58	38	1668.34	14.61
14	1265.64	2.87	39	1727.19	15.49
15	1357.83	3.16	40	1774.11	16.47
16	1450.03	3.45	41	1808.24	17.02
17	1542.23	3.75	42	1872.90	17.74
18	1626.40	4.06	43	1931.13	18.23
19	1710.57	4.37	44	1992.18	18.59
20	1794.73	4.69	45	2053.23	18.95
21	1855.69	4.99	46	2130.04	19.70
22	1916.64	5.28	47	2161.18	20.27
23	1996.28	5.92	48	2192.31	20.83
24	2033.83	6.71	49	2203.80	21.40
25	2050.41	7.63	50	2215.28	21.96

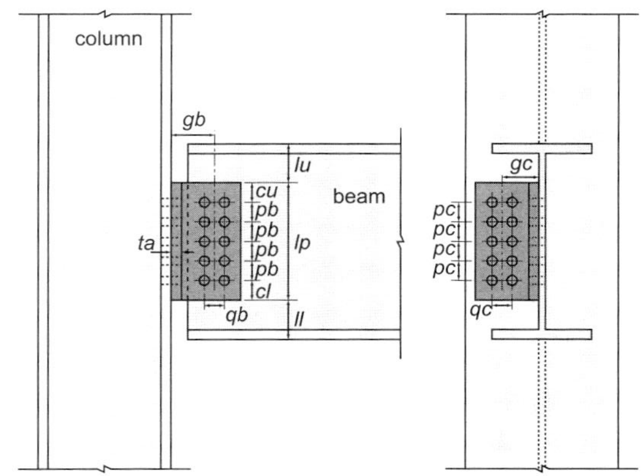

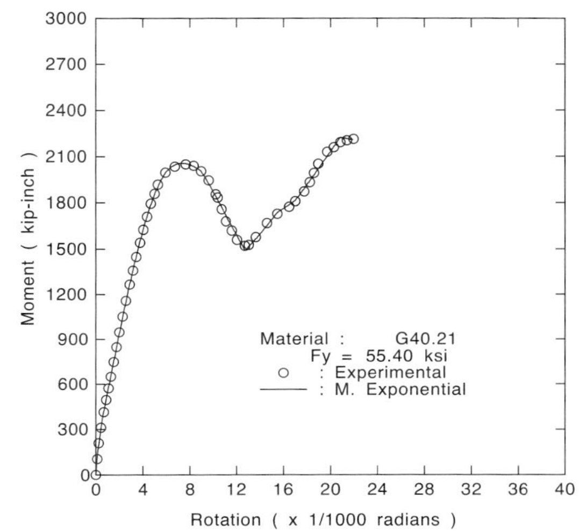

A.1 – 70

```
Moment-rotation prediction equations     ( R : X 1/1000 radians )

Modified exponential model  :
     Bm = Sum ( Ai X ( 1 - exp( - R/( 2*i*C ) ) ) )  +      Sum ( RKj X ( R - Rj0 ) )  +  Bm0
      C =   0.21332917E+00              Bmo=   0.00000000E+00          Nexp=  6         Nliner=  6
     AI =   0.92140114E+03  -0.11106162E+05   0.79466523E+05  -0.23294081E+06   0.27759167E+06  -0.11223354E+06
    Rj0 =           5.9800           8.9600          12.7300          15.4900          17.0200
                   20.8300
    RKj =  -0.24096718E+02  -0.64006495E+02   0.21384094E+03  -0.52055671E+02   0.44984522E+02
           -0.12792020E+03
```

No	Rotation (radians) X 1/1000	Moment (kip-inch) Expri.	Moment (kip-inch) M.Expo.	Connection stiffness (kip-inch) X 1000 Expri.	Connection stiffness (kip-inch) X 1000 M.Expo.
1	0.00	0.00	0.00	0.8624E+03	0.1019E+04
3	0.24	209.05	196.12	0.7245E+03	0.6470E+03
5	0.67	415.38	414.01	0.4351E+03	0.4283E+03
7	1.07	573.28	577.82	0.3903E+03	0.3928E+03
9	1.52	750.39	750.88	0.4087E+03	0.3907E+03
11	2.00	946.72	937.84	0.3896E+03	0.3862E+03
13	2.58	1159.33	1157.14	0.3664E+03	0.3673E+03
15	3.16	1357.83	1362.44	0.3153E+03	0.3357E+03
17	3.75	1542.23	1547.19	0.2924E+03	0.2947E+03
19	4.37	1710.57	1716.70	0.2679E+03	0.2440E+03
21	4.99	1855.69	1850.09	0.2049E+03	0.1920E+03
23	5.92	1996.28	1992.72	0.9100E+02	0.1160E+03
25	7.63	2050.41	2053.93	-0.8921E+00	-0.1314E+02
27	8.96	2004.13	2004.06	-0.7597E+02	-0.5670E+02
29	10.21	1854.91	1840.10	-0.1350E+03	-0.1376E+03
31	10.72	1757.43	1769.53	-0.2182E+03	-0.1398E+03
33	11.60	1621.23	1645.88	-0.1250E+03	-0.1394E+03
35	12.73	1520.19	1491.89	-0.3978E+01	-0.1341E+03
37	13.64	1579.10	1567.11	0.8855E+02	0.8585E+02
39	15.49	1727.19	1737.93	0.5786E+02	0.4655E+02
41	17.02	1808.24	1815.77	0.7372E+02	0.5520E+02
43	18.23	1931.13	1940.19	0.1490E+03	0.1054E+03
45	18.95	2053.23	2016.84	0.1480E+03	0.1079E+03
47	20.27	2161.18	2161.79	0.5522E+02	0.1115E+03
49	21.40	2203.80	2216.69	0.2025E+02	-0.1429E+02

I - 36

Connection type : Single web-angle connections
Mode : Bolted-to-beam and welded-to-column

Tested by : S.L.Lipson (1977) Canada
Test Id. : A1-10

Column : T=44",TW=5/8"
Beam : --
Angle : 4 X 3 X 3/8

Fasteners: A325- -3/4"D
13/16" Oversize holes
Material : G40.21
Fy = 55.40 ksi
Fu = 77.80 ksi

Major parameters

lp = 29.5000" lu = -- " ll = -- " ta = 0.3750"
gb = 2.5000" cu = 1.2500" cl = 1.2500" pb = 3.0000"
nb = 1 X 10

Remark 1) 1/4" fillet weld at heel, top and bottom of angle.
2) Connection subject to moment and shear.

No	Moment (k-in)	Rotation (radians) X 1/1000	No	Moment (k-in)	Rotation (radians) X 1/1000
1	0.00	0.00	26	774.26	7.66
2	76.51	0.04	27	730.74	7.88
3	153.01	0.08	28	671.15	8.44
4	229.52	0.12	29	657.16	9.12
5	306.03	0.16	30	703.91	9.91
6	375.84	0.27	31	786.76	10.58
7	445.65	0.38	32	849.45	10.95
8	515.47	0.50	33	912.14	11.33
9	562.63	0.64	34	974.96	11.78
10	609.80	0.79	35	1037.78	12.23
11	663.00	0.92	36	1112.83	12.75
12	716.19	1.05	37	1187.89	13.28
13	778.67	1.30	38	1246.16	13.73
14	841.16	1.55	39	1304.44	14.18
15	899.28	1.91	40	1385.86	14.89
16	957.40	2.27	41	1467.28	15.60
17	1040.46	3.05	42	1533.49	16.27
18	1062.58	3.60	43	1599.70	16.93
19	1084.70	4.15	44	1680.15	17.97
20	1071.18	5.11	45	1714.53	18.61
21	1015.13	5.98	46	1748.91	19.25
22	965.92	6.42	47	1756.10	19.92
23	916.72	6.86	48	1739.45	20.82
24	857.25	7.15			
25	817.78	7.43			

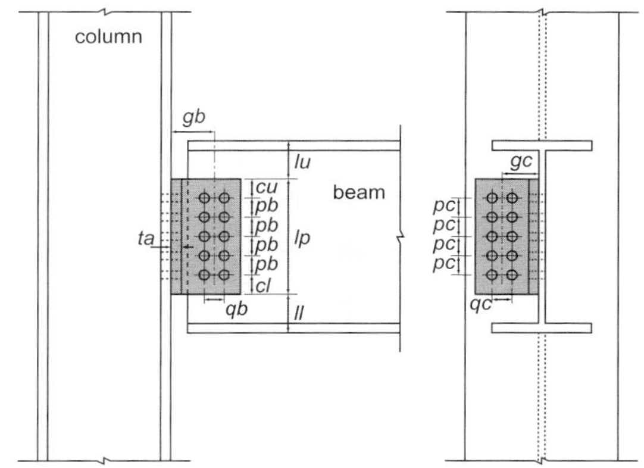

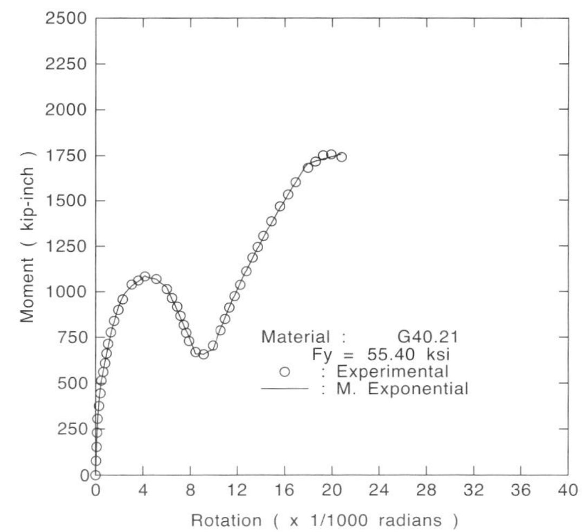

A.1 – 72

```
Moment-rotation prediction equations     ( R : X 1/1000 radians )

Modified exponential model  :
     Bm = Sum ( Ai X ( 1 - exp( - R/( 2*i*C ) ) ) )  +      Sum ( RKj X ( R - Rj0 ) )  +  Bm0
      C =   0.18361667E+00             Bmo=   0.00000000E+00           Nexp=  6        Nliner=  8
     AI =   0.29114720E+04  -0.19671758E+05   0.43580738E+05  -0.18655843E+04  -0.74300214E+05   0.51312782E+05
    Rj0 =           4.1500           5.1100           5.9800           7.4300           8.4400
                    9.1200           9.9200          17.9000
    RKj =  -0.10385205E+03  -0.67091556E+02  -0.10544885E+03  -0.34915436E+00   0.16797467E+03
            0.54517160E+02   0.15362829E+03  -0.83640760E+02
```

No	Rotation (radians) X 1/1000	Moment (kip-inch) Expri.	Moment (kip-inch) M.Expo.	Connection stiffness (kip-inch) X 1000 Expri.	Connection stiffness (kip-inch) X 1000 M.Expo.
1	0.00	0.00	0.00	0.1889E+04	0.2255E+04
3	0.08	153.01	157.12	0.1887E+04	0.1652E+04
5	0.16	306.03	272.37	0.1552E+04	0.1218E+04
7	0.38	445.65	460.42	0.6290E+03	0.5809E+03
9	0.64	562.63	573.12	0.3237E+03	0.3494E+03
11	0.92	663.00	660.17	0.4114E+03	0.2984E+03
13	1.30	778.67	769.43	0.2491E+03	0.2744E+03
15	1.91	899.28	912.18	0.1617E+03	0.1852E+03
17	3.05	1040.46	1032.40	0.6715E+02	0.5251E+02
19	4.15	1084.70	1088.69	0.2036E+02	-0.3909E+02
21	5.98	1015.13	1025.87	-0.9594E+02	-0.1471E+03
23	6.86	916.72	899.87	-0.1490E+03	-0.1423E+03
25	7.43	817.78	817.36	-0.1854E+03	-0.1476E+03
27	7.88	730.74	750.06	-0.1697E+03	-0.1544E+03
29	9.12	657.16	657.33	0.1617E+02	0.4240E+02
31	10.58	786.76	797.15	0.1522E+03	0.1652E+03
33	11.33	912.14	915.56	0.1548E+03	0.1518E+03
35	12.23	1037.78	1046.16	0.1409E+03	0.1384E+03
37	13.28	1187.89	1185.05	0.1354E+03	0.1264E+03
39	14.18	1304.44	1295.64	0.1235E+03	0.1188E+03
41	15.60	1467.28	1457.99	0.1069E+03	0.1107E+03
43	16.93	1599.70	1602.07	0.9099E+02	0.1061E+03
45	18.61	1714.53	1717.45	0.5367E+02	0.1914E+02
47	19.92	1756.10	1741.63	-0.1933E+01	0.1771E+02

A.1 – 73

I - 37

```
Connection type : Single web-angle connections
           Mode : Bolted-to-beam and welded-to-column

Tested by  :  S.L.Lipson (1977)                          Canada
Test  Id.  :  A1-11

Column : T=44",TW=5/8"              Fasteners: A325- -3/4"D
Beam   :  --                                  13/16" Oversize holes
Angle  : 4 X 3 X 3/8                Material : G40.21
                                         Fy =  55.40 ksi
                                         Fu =  77.80 ksi

                          Major parameters

     lp = 32.5000"    lu =   --   "    ll =   --   "    ta =  0.3750"
     gb =  2.5000"    cu =  1.2500"    cl =  1.2500"    pb =  3.0000"
     nb = 1 X 11

Remark  1)  1/4" fillet weld at heel, top and bottom of angle.
        2)  Connection subject to moment and shear.
```

No	Moment (k-in)	Rotation (radians) X 1/1000	No	Moment (k-in)	Rotation (radians) X 1/1000
1	0.00	0.00	26	1084.82	10.91
2	102.06	0.16	27	1131.24	11.58
3	204.13	0.32	28	1186.69	12.21
4	306.19	0.47	29	1287.63	12.94
5	379.00	0.62	30	1371.70	13.51
6	451.80	0.76	31	1455.77	14.08
7	524.61	0.90	32	1536.82	14.65
8	600.54	1.10	33	1617.86	15.22
9	676.46	1.31	34	1695.96	15.84
10	743.37	1.54	35	1774.06	16.46
11	810.27	1.78	36	1836.96	17.03
12	872.69	2.04	37	1899.87	17.61
13	935.11	2.31	38	1949.20	18.20
14	994.73	2.72	39	1998.52	18.79
15	1054.34	3.12	40	2069.67	19.80
16	1140.12	3.82	41	2125.46	20.65
17	1177.23	4.33	42	2172.32	21.60
18	1185.98	4.76			
19	1195.93	5.32			
20	1183.83	5.93			
21	1171.96	6.74			
22	1142.99	7.54			
23	1114.22	8.47			
24	1076.48	9.46			
25	1071.46	10.11			

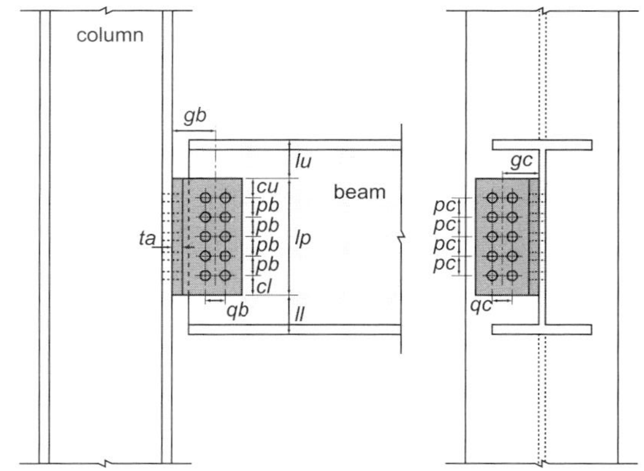

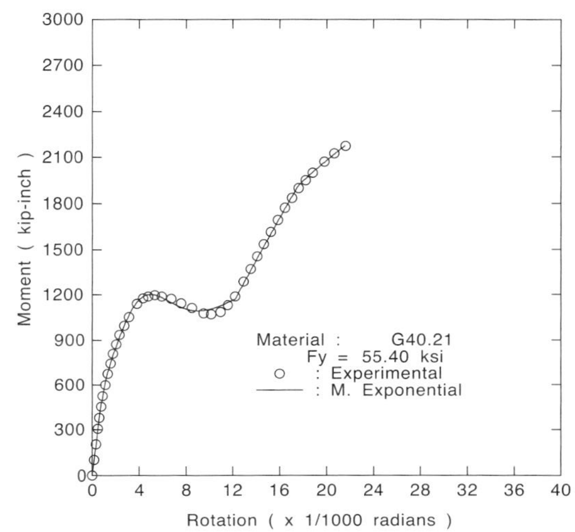

```
 Moment-rotation prediction equations     ( R : X 1/1000 radians )

Modified exponential model  :
     Bm = Sum ( Ai X ( 1 - exp( - R/( 2*i*C ) ) ) )  +      Sum ( RKj X ( R - Rj0 ) )  +  Bm0
      C =   0.21940250E+00           Bmo=   0.00000000E+00         Nexp=  6        Nliner=  2
     AI =  -0.26407291E+04   0.40404303E+05  -0.20238925E+06  0.43705241E+06  -0.41974398E+06   0.14871781E+06
    Rj0 =         12.2100          17.6100
    RKj =   0.10833524E+03  -0.55293194E+02
```

No	Rotation (radians) X 1/1000	Moment (kip-inch) Expri.	Moment (kip-inch) M.Expo.	Connection stiffness (kip-inch) X 1000 Expri.	Connection stiffness (kip-inch) X 1000 M.Expo.
1	0.00	0.00	0.00	0.6472E+03	0.4534E+03
3	0.32	204.13	194.92	0.6470E+03	0.6798E+03
5	0.62	379.00	386.13	0.5127E+03	0.5757E+03
7	0.90	524.61	529.87	0.4556E+03	0.4391E+03
9	1.31	676.46	677.64	0.3324E+03	0.3028E+03
11	1.78	810.27	801.16	0.2606E+03	0.2338E+03
13	2.31	935.11	917.13	0.1994E+03	0.2040E+03
15	3.12	1054.34	1064.84	0.1383E+03	0.1550E+03
17	4.33	1177.23	1187.58	0.4494E+02	0.4671E+02
19	5.32	1196.93	1198.47	0.3303E+01	-0.1795E+02
21	6.74	1171.96	1147.22	-0.2848E+02	-0.4215E+02
23	8.47	1114.22	1096.48	-0.3441E+02	-0.1161E+02
25	10.11	1071.46	1105.52	0.3213E+01	0.2041E+02
27	11.58	1131.24	1147.82	0.7836E+02	0.3486E+02
29	12.94	1287.63	1276.58	0.1439E+03	0.1458E+03
31	14.08	1455.77	1441.83	0.1445E+03	0.1430E+03
33	15.22	1617.86	1602.63	0.1343E+03	0.1381E+03
35	16.46	1774.06	1769.71	0.1176E+03	0.1321E+03
37	17.61	1899.87	1918.47	0.9657E+02	0.1269E+03
39	18.79	1998.52	2000.51	0.7866E+02	0.6693E+02
41	20.65	2125.46	2119.43	0.5787E+02	0.6148E+02

I - 38

```
Connection type : Single web-angle connections
           Mode : Bolted-to-beam and welded-to-column

Tested by  :  S.L.Lipson (1977)                          Canada
Test  Id.  :  A1-12

Column : T=44",TW=5/8"              Fasteners: A325- -3/4"D
Beam   :  --                               13/16" Oversize holes
Angle  : 4 X 3 X 3/8                Material : G40.21
                                         Fy =  55.40 ksi
                                         Fu =  77.80 ksi

                          Major parameters

     lp = 35.5000"   lu =    --   '    ll =    --   "    ta =  0.3750"
     gb =  2.5000"   cu =  1.2500'     cl =  1.2500"     pb =  3.0000"
     nb = 1 X 12

Remark  1)  1/4" Fillet weld at heel, top and bottom of angle.
        2)  Connection subject to moment and shear.
```

No	Moment (k-in)	Rotation (radians) X 1/1000	No	Moment (k-in)	Rotation (radians) X 1/1000
1	0.00	0.00	26	1635.90	10.06
2	130.22	0.10	27	1676.90	11.01
3	260.43	0.20	28	1721.78	11.58
4	390.65	0.30	29	1766.66	12.16
5	508.98	0.51	30	1843.17	12.67
6	627.30	0.72	31	1919.68	13.18
7	745.63	0.94	32	1997.38	13.50
8	855.95	1.19	33	2075.07	13.83
9	966.26	1.43	34	2162.11	14.30
10	1076.58	1.68	35	2249.15	14.78
11	1181.90	1.94	36	2351.69	15.47
12	1287.21	2.21	37	2454.23	16.15
13	1392.53	2.48	38	2541.71	16.88
14	1489.99	2.86	39	2629.18	17.60
15	1587.44	3.24	40	2692.64	18.42
16	1684.90	3.63	41	2756.10	19.24
17	1810.27	4.40	42	2800.07	20.15
18	1899.10	5.03	43	2814.80	20.78
19	1952.03	5.88	44	2829.52	21.41
20	1950.21	6.56			
21	1893.92	7.22			
22	1834.23	7.66			
23	1741.22	8.07			
24	1652.06	8.95			
25	1643.98	9.51			

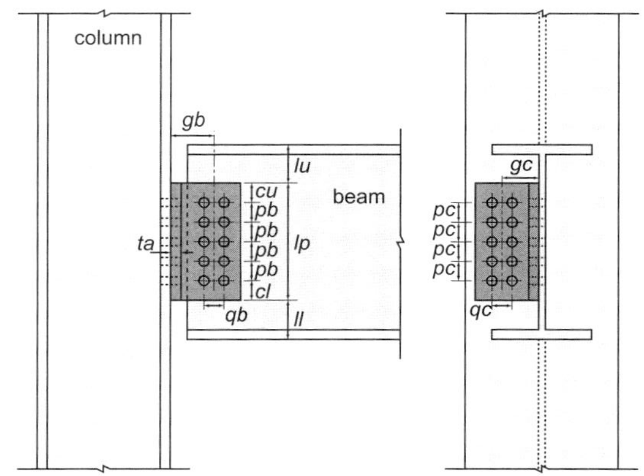

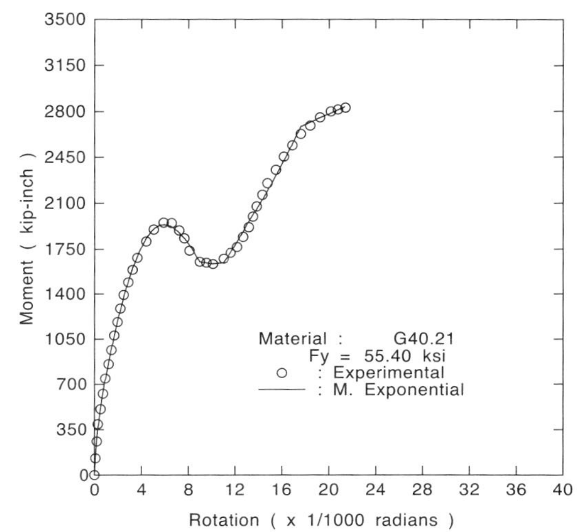

```
Moment-rotation prediction equations     ( R : X 1/1000 radians )

Modified exponential model  :
     Bm = Sum ( Ai X ( 1 - exp( - R/( 2*i*C ) ) ) )  +     Sum ( RKj X ( R - Rj0 ) )  +  Bm0
      C =   0.20312583E+00              Bmo=   0.00000000E+00          Nexp=  6         Nliner=  7
     AI =   0.18924539E+04  -0.23940891E+05   0.15445681E+06  -0.42886636E+06   0.50827611E+06  -0.21095107E+06
    Rj0 =           8.0700           8.9500          10.0600          11.0100          13.1800
                   17.6000          19.2400
    RKj =  -0.19912429E+02   0.14313016E+03   0.19103347E+02   0.10988871E+03  -0.39228704E+02
           -0.15187802E+03  -0.11465822E+02
```

No	Rotation (radians) X 1/1000	Moment (kip-inch) Expri.	Moment (kip-inch) M.Expo.	Connection stiffness (kip-inch) X 1000 Expri.	Connection stiffness (kip-inch) X 1000 M.Expo.
1	0.00	0.00	0.00	0.1315E+04	0.1693E+04
3	0.20	260.43	264.52	0.1315E+04	0.1052E+04
5	0.51	508.98	515.58	0.5532E+03	0.6320E+03
7	0.94	745.63	745.73	0.5042E+03	0.4818E+03
9	1.43	966.26	969.62	0.4482E+03	0.4326E+03
11	1.94	1181.90	1180.27	0.3936E+03	0.3862E+03
13	2.48	1392.53	1372.91	0.3366E+03	0.3334E+03
15	3.24	1587.44	1598.77	0.2553E+03	0.2585E+03
17	4.40	1810.27	1832.07	0.1502E+03	0.1436E+03
19	5.88	1952.03	1937.23	0.2604E+02	0.2397E+01
21	7.22	1893.92	1874.83	-0.1153E+03	-0.8826E+02
23	8.07	1741.22	1783.41	-0.1840E+03	-0.1412E+03
25	9.51	1643.98	1643.83	-0.1456E+02	-0.1510E+02
27	11.01	1676.90	1647.26	0.6483E+02	0.1827E+02
29	12.16	1766.66	1806.07	0.1164E+03	0.1486E+03
31	13.18	1919.68	1967.09	0.2050E+03	0.1679E+03
33	13.83	2075.07	2054.37	0.2168E+03	0.1403E+03
35	14.78	2249.15	2195.51	0.1690E+03	0.1558E+03
37	16.15	2454.23	2423.09	0.1354E+03	0.1739E+03
39	17.60	2629.18	2684.77	0.1007E+03	0.1877E+03
41	19.24	2756.10	2752.85	0.6361E+02	0.4616E+02
43	20.78	2814.80	2811.45	0.2346E+02	0.4075E+02

I - 39

```
Connection type : Single web-angle connections
           Mode : Bolted-to-beam and welded-to-column

Tested by  :  L.E.Thompson et al. (1970)                U.S.A.
Test  Id.  :  F-1 ALT

Column : W10X49                      Fasteners: A325- -3/4"D
Beam   : W21X62                            15/16" Oversize holes
Angle  : --                          Material : A36
                                          Fy =   --   ksi
                                          Fu =   --   ksi

                         Major parameters

     lp = 11.5000"    lu =   --   "    ll =   --   "    ta =  0.6250"
     gb =  2.2500"    cu =   --   "    cl =   --   "
     pb =  3.0000"    qb =   --   "    nb = 2 X  -

Remark  1)  Single plate connection.
        2)
```

No	Moment (k-in)	Rotation (radians) X 1/1000
1	0.00	0.00
2	43.00	0.35
3	85.99	0.70
4	123.11	1.05
5	160.22	1.40
6	192.57	1.75
7	224.92	2.10
8	264.97	2.65
9	305.03	3.20
10	337.73	3.75
11	370.44	4.30
12	407.18	5.10
13	441.00	6.60
14	459.38	8.95
15	477.76	11.30
16	490.00	13.43
17	502.25	15.57
18	514.50	17.70
19	523.69	20.75
20	532.87	23.80
21	545.12	26.13
22	557.38	28.47
23	569.63	30.80
24	588.00	33.85
25	606.37	36.90
26	612.25	39.74
27	618.13	42.58
28	624.01	45.42
29	629.89	48.26
30	635.77	51.10
31	657.82	53.00
32	679.87	54.90
33	734.26	55.70
34	788.65	56.50
35	846.72	57.30
36	904.79	58.10
37	958.44	58.90
38	1012.09	59.70
39	1060.97	60.45
40	1109.86	61.20
41	1144.40	61.75
42	1178.94	62.30
43	1231.13	63.10

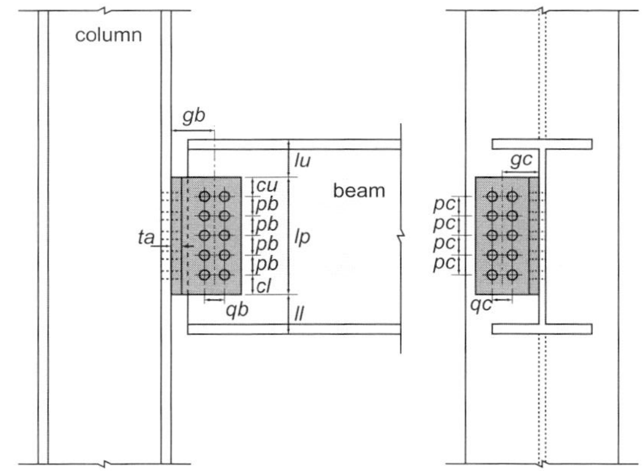

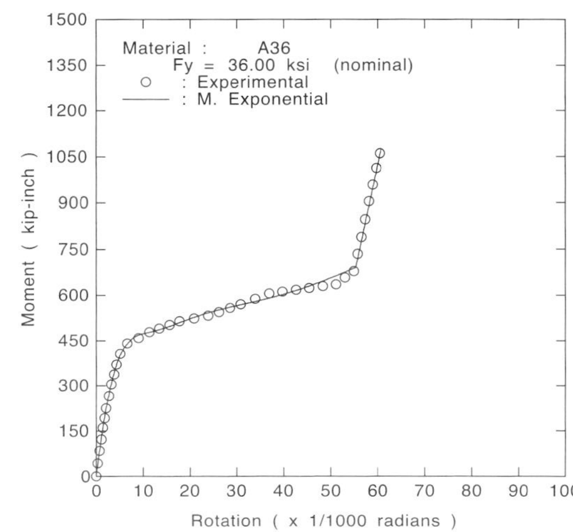

A.1 – 78

Moment-rotation prediction equations (R : X 1/1000 radians)

Modified exponential model :

```
Bm = Sum ( Ai X ( 1 - exp( - R/( 2*i*C ) ) ) )  +     Sum ( RKj X ( R - Rj0 ) )  +  Bm0
 C =  0.61333333E+00             Bmo=  0.00000000E+00          Nexp=  6         Nliner=  3
AI =  0.36346292E+03  -0.73841783E+04  0.38898994E+05  -0.83901630E+05  0.83369958E+05  -0.31024598E+05
Rj0 =         6.6000          11.3000         55.3000
RKj =  0.56149868E+01   0.20084087E+01  0.63868948E+02
```

No	Rotation (radians) X 1/1000	Moment (kip-inch) Expri.	Moment (kip-inch) M.Expo.	Connection stiffness (kip-inch) X 1000 Expri.	Connection stiffness (kip-inch) X 1000 M.Expo.
1	0.00	0.00	0.00	0.1228E+03	0.1349E+03
3	0.70	85.99	85.55	0.1144E+03	0.1120E+03
5	1.40	160.22	158.79	0.9923E+02	0.9796E+02
7	2.10	224.92	223.10	0.8480E+02	0.8584E+02
9	3.20	305.03	306.90	0.6615E+02	0.6638E+02
11	4.30	370.44	369.18	0.5395E+02	0.4708E+02
13	6.60	441.00	438.56	0.1681E+02	0.1597E+02
15	11.30	477.76	477.63	0.6731E+01	0.2402E+01
17	15.57	502.25	497.53	0.5741E+01	0.5383E+01
19	20.75	523.69	526.83	0.3012E+01	0.5404E+01
21	26.13	545.12	551.80	0.5251E+01	0.3907E+01
23	30.80	569.63	568.39	0.5586E+01	0.3377E+01
25	36.90	606.37	590.22	0.3977E+01	0.3952E+01
27	42.58	618.13	615.57	0.2070E+01	0.4991E+01
29	48.26	629.89	646.73	0.2070E+01	0.5945E+01
31	53.00	657.82	676.39	0.1161E+02	0.6539E+01
33	55.70	734.26	719.95	0.6799E+02	0.7066E+02
35	57.30	846.72	833.11	0.7258E+02	0.7079E+02
37	58.90	958.44	946.46	0.6706E+02	0.7090E+02
39	60.45	1060.97	1056.42	0.6518E+02	0.7099E+02
41	61.75	1144.40	1148.75	0.6280E+02	0.7105E+02
43	63.10	1231.13	1244.71	0.6523E+02	0.7111E+02

I - 40

Connection type : Single web-angle connections
Mode : Bolted-to-beam and welded-to-column

Tested by : L.E.Thompson et al. (1970) U.S.A.
Test Id. : F-1 ALT

Column : W10X49
Beam : W21X62
Angle : --

Fasteners: A325- -3/4"D
15/16" Oversize holes
Material : A36
Fy = -- ksi
Fu = -- ksi

Major parameters

lp = 11.5000"	lu = -- "	ll = -- "	ta = 0.6250"
gb = 2.2500"	cu = -- "	cl = -- "	
pb = 3.0000"	qb = -- "	nb = 2 X -	

Remark 1) Single plate connection
2)

No	Moment (k-in)	Rotation (radians) X 1/1000	No	Moment (k-in)	Rotation (radians) X 1/1000
1	0.00	0.00	26	633.94	41.10
2	43.00	0.30	27	661.50	44.20
3	85.99	0.60	28	666.09	47.20
4	123.11	0.90	29	670.69	50.20
5	160.22	1.20	30	675.28	53.20
6	192.57	1.50	31	679.87	56.20
7	224.92	1.80	32	707.44	58.90
8	264.97	2.25	33	735.00	61.60
9	305.03	2.70	34	787.00	62.50
10	337.73	3.15	35	839.00	63.40
11	370.44	3.60	36	899.46	64.15
12	405.72	4.30	37	959.92	64.90
13	441.00	5.00	38	1009.16	65.60
14	496.13	6.50	39	1058.40	66.30
15	532.87	9.60	40	1102.50	67.00
16	537.47	12.55	41	1146.60	67.70
17	542.06	15.50	42	1197.31	69.10
18	546.66	18.45	43	1231.13	71.20
19	551.26	21.40			
20	557.38	23.97			
21	563.50	26.53			
22	569.63	29.10			
23	581.88	32.07			
24	594.12	35.03			
25	606.37	38.00			

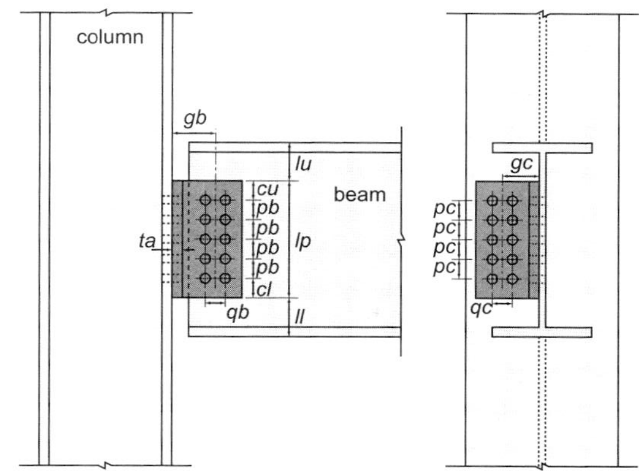

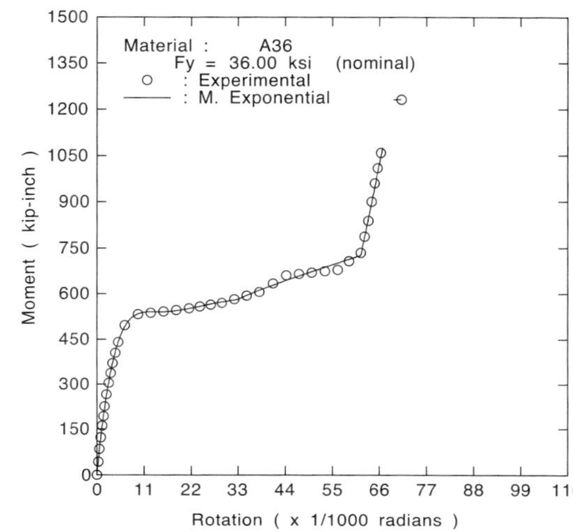

A.1 – 80

```
Moment-rotation prediction equations     ( R : X 1/1000 radians )

Modified exponential model  :
     Bm = Sum ( Ai X ( 1 - exp( - R/( 2*i*C ) ) ) )  +      Sum ( RKj X ( R - Rj0 ) )  +  Bm0
       C =   0.66833333E+00           Bmo=   0.00000000E+00          Nexp=  6        Nliner=  4
      AI =   0.21526746E+03  -0.35697020E+04  0.15275399E+05  -0.22781211E+05   0.14196025E+05  -0.27415056E+04
     Rj0 =          15.5000          32.7000         61.6000          69.1000
     RKj =   0.23576865E-01   0.45387460E+01  0.62844928E+02  -0.67902244E+02
```

No	Rotation (radians) X 1/1000	Moment (kip-inch) Expri.	Moment (kip-inch) M.Expo.	Connection stiffness (kip-inch) X 1000 Expri.	Connection stiffness (kip-inch) X 1000 M.Expo.
1	0.00	0.00	0.00	0.1433E+03	0.1565E+03
3	0.60	85.99	86.08	0.1335E+03	0.1319E+03
5	1.20	160.22	159.54	0.1158E+03	0.1138E+03
7	1.80	224.92	223.29	0.1003E+03	0.9912E+02
9	2.70	305.03	303.86	0.8085E+02	0.8039E+02
11	3.60	370.44	368.68	0.6396E+02	0.6399E+02
13	5.00	441.00	442.58	0.4606E+02	0.4239E+02
15	9.60	532.87	533.77	0.6578E+01	0.5279E+01
17	15.50	542.06	540.39	0.1558E+01	0.5248E+00
19	21.40	551.26	550.98	0.2001E+01	0.2761E+01
21	26.53	563.50	565.98	0.2386E+01	0.2834E+01
23	32.07	581.88	579.39	0.4129E+01	0.1954E+01
25	38.00	606.37	612.17	0.6458E+01	0.5579E+01
27	44.20	661.50	644.77	0.5151E+01	0.4999E+01
29	50.20	670.69	673.85	0.1531E+01	0.4728E+01
31	56.20	679.87	701.82	0.6098E+01	0.4613E+01
33	61.60	735.00	726.61	0.4589E+02	0.4574E+01
35	63.40	839.00	847.96	0.7023E+02	0.6741E+02
37	64.90	959.92	949.07	0.7530E+02	0.6741E+02
39	66.30	1058.40	1043.44	0.6667E+02	0.6741E+02
41	67.70	1146.60	1137.81	0.5407E+02	0.6740E+02
43	71.20	1231.13	1231.13	0.1610E+02	-0.4994E+00

I - 41

```
Connection type : Single web-angle connections
           Mode : Bolted-to-beam and welded-to-column

Tested by   :  L.E.Thompson et al. (1970)                 U.S.A.
Test  Id.   :  G-1 ALT

Column : W10X49                  Fasteners: A325- -3/4"D
Beam   : W21X62                        15/16" Oversize holes
Angle  : --                      Material : A36
                                      Fy =   --   ksi
                                      Fu =   --   ksi

                           Major parameters

     lp = 11.5000"    lu =    --    "    ll =    --    "    ta =  0.6250"
     gb =  2.2500"    cu =    --    "    cl =    --    "
     pb =  3.0000"    qb =    --    "    nb = 2 X  -

Remark  1)  Single plate connection.
        2)
```

No	Moment (k-in)	Rotation (radians) X 1/1000
1	0.00	0.00
2	42.26	0.35
3	84.53	0.70
4	121.64	1.05
5	158.76	1.40
6	194.04	1.80
7	229.32	2.20
8	265.33	2.65
9	301.34	3.10
10	338.83	3.90
11	376.32	4.70
12	404.26	6.20
13	420.05	9.35
14	435.85	12.50
15	449.82	15.87
16	463.79	19.23
17	477.76	22.60
18	493.19	25.65
19	508.62	28.70
20	535.08	31.83
21	561.54	34.97
22	588.00	38.10
23	615.56	41.50
24	643.13	44.90
25	647.72	47.65

No	Moment (k-in)	Rotation (radians) X 1/1000
26	652.31	50.40
27	656.91	53.15
28	661.50	55.90
29	702.66	59.30
30	743.82	62.70
31	781.31	63.30
32	818.80	63.90
33	859.22	64.43
34	899.64	64.97
35	940.07	65.50
36	988.21	66.30
37	1036.36	67.10
38	1049.21	69.70
39	1062.07	72.30

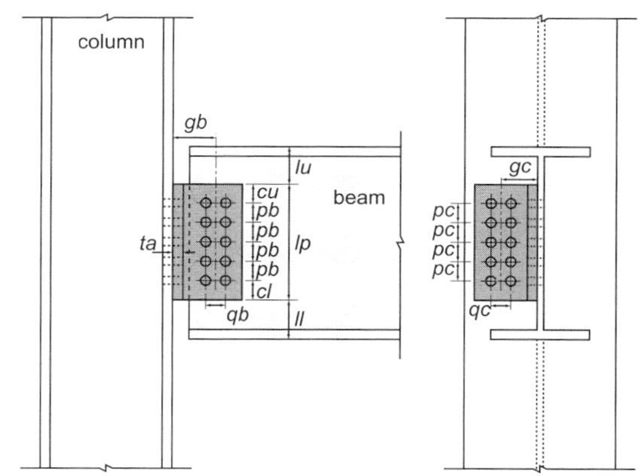

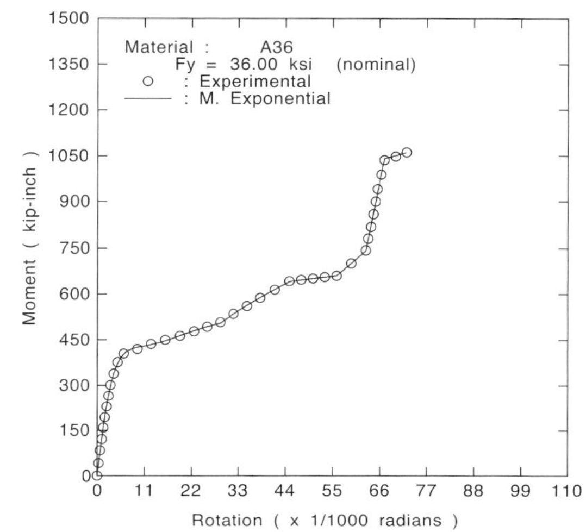

A.1 – 82

Moment-rotation prediction equations   (R : X 1/1000 radians)

```
Modified expcnential model  :
    Bm = Sum ( Ai X ( 1 - exp( - R/( 2*i*C ) ) ) )  +     Sum ( RKj X ( R - Rj0 ) )  +  Bm0
     C =   0.69000000E+00              Bmo=   0.00000000E+00          Nexp=  6        Nliner=  5
    AI =  -0.31620063E+03   0.10299317E+04   0.23005865E+04  -0.60054736E+04   0.30726978E+04   0.45478622E+03
   Rj0 =          28.7000          44.9000          55.9000          62.7000          67.1000
   RKj =   0.68061454E+01  -0.53201916E+01   0.10302492E+02   0.55990565E+02  -0.63941046E+02
```

No	Rotation (radians) X 1/1000	Moment (kip-inch) Expri.	Moment (kip-inch) M.Expo.	Connection stiffness (kip-inch) X 1000 Expri.	Connection stiffness (kip-inch) X 1000 M.Expo.
1	0.00	0.00	0.00	0.1208E+03	0.1120E+03
3	0.70	84.53	80.85	0.1134E+03	0.1153E+03
5	1.40	158.76	157.99	0.9772E+02	0.1036E+03
7	2.20	229.32	233.11	0.8435E+02	0.8378E+02
9	3.10	301.34	298.27	0.6809E+02	0.6141E+02
11	4.70	376.32	370.38	0.3704E+02	0.3109E+02
13	9.35	420.05	425.77	0.5015E+01	0.3062E+01
15	15.87	449.82	447.10	0.4149E+01	0.4557E+01
17	22.60	477.76	480.52	0.4627E+01	0.4753E+01
19	28.70	508.62	505.01	0.6729E+01	0.3218E+01
21	34.97	561.54	563.12	0.8445E+01	0.8606E+01
23	41.50	615.56	616.07	0.8107E+01	0.7694E+01
25	47.65	647.72	647.20	0.1670E+01	0.1918E+01
27	53.15	656.91	657.11	0.1670E+01	0.1706E+01
29	59.30	702.66	702.22	0.1211E+02	0.1189E+02
31	63.30	781.31	783.29	0.6248E+02	0.6784E+02
33	64.43	859.22	860.17	0.7579E+02	0.6783E+02
35	65.50	940.07	932.52	0.6955E+02	0.6783E+02
37	67.10	1036.36	1041.03	0.4718E+02	0.6782E+02
39	72.30	1062.07	1061.13	0.4945E+01	0.3857E+01

I - 42

```
Connection type : Single web-angle connections
           Mode : Bolted-to-beam and welded-to-column

Tested by  :  L.E.Thompson et al. (1970)              U.S.A.
Test  Id.  :  G-1 ALT

Column : W10X49                    Fasteners: A325- -3/4"D
Beam   : W21X62                          15/16" Oversize holes
Angle  : --                        Material : A36
                                        Fy =   --   ksi
                                        Fu =   --   ksi

                          Major parameters

     lp = 11.5000"    lu =   --   "    ll =   --   "    ta =  0.6250"
     gb =  2.2500"    cu =   --   "    cl =   --   "
     pb =  3.0000"    qb =   --   "    nb = 2 X  -

Remark  1)  Single plate connection.
        2)
```

No	Moment (k-in)	Rotation (radians) X 1/1000	No	Moment (k-in)	Rotation (radians) X 1/1000
1	0.00	0.00	26	680.62	45.70
2	42.26	0.30	27	707.81	49.20
3	84.53	0.60	28	735.00	52.70
4	121.64	0.90	29	737.21	56.02
5	158.76	1.20	30	739.41	59.35
6	194.04	1.50	31	741.62	62.67
7	229.32	1.80	32	743.82	66.00
8	265.33	2.25	33	781.31	66.45
9	301.34	2.70	34	818.80	66.90
10	338.83	3.30	35	864.37	67.40
11	376.32	3.90	36	909.94	67.90
12	404.26	4.60	37	955.50	68.45
13	420.05	8.10	38	1001.06	69.00
14	435.85	11.60	39	1031.57	69.55
15	446.33	14.35	40	1062.07	70.10
16	456.80	17.10			
17	467.28	19.85			
18	477.76	22.60			
19	488.04	25.00			
20	498.33	27.40			
21	508.62	29.80			
22	557.50	33.10			
23	606.37	36.40			
24	631.12	39.50			
25	655.87	42.60			

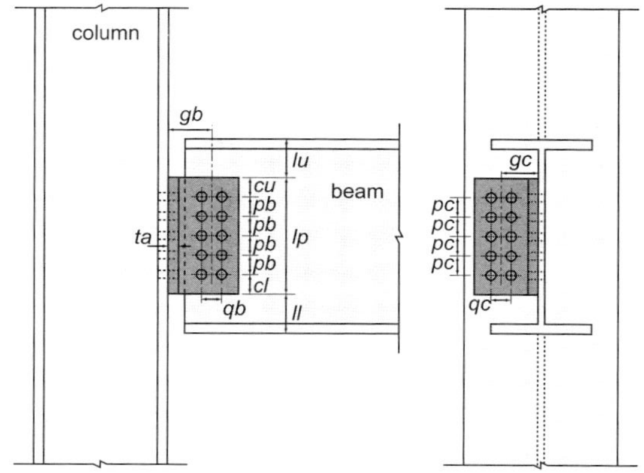

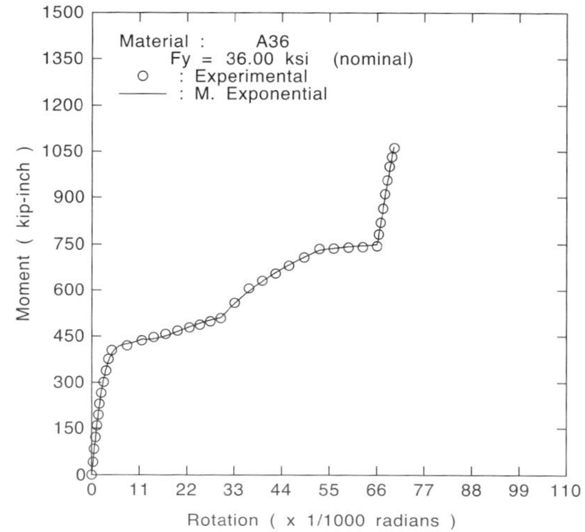

A.1 – 84

```
Moment-rotation prediction equations     ( R : X 1/1000 radians )

Modified exponential model  :
     Bm = Sum ( Ai X ( 1 - exp( - R/( 2*i*C ) ) ) )  +      Sum ( RKj X ( R - Rj0 ) )  +  Bm0
      C =   0.65916667E+00             Bmo=   0.00000000E+00          Nexp=  6        Nliner=  4
     AI =  -0.61339462E+03   0.52396380E+04  -0.19484630E+05   0.49397274E+05  -0.62163011E+05   0.28429141E+05
    Rj0 =          11.6000          29.8000          52.7000          66.0000
    RKj =  -0.83117480E+01   0.12462279E+02  -0.38694691E+01   0.79811769E+02
```

No	Rotation (radians) X 1/1000	Moment (kip-inch) Expri.	Moment (kip-inch) M.Expo.	Connection stiffness (kip-inch) X 1000 Expri.	Connection stiffness (kip-inch) X 1000 M.Expo.
1	0.00	0.00	0.00	0.1409E+03	0.1263E+03
3	0.60	84.53	80.18	0.1323E+03	0.1359E+03
5	1.20	158.76	158.91	0.1207E+03	0.1242E+03
7	1.80	229.32	227.62	0.1026E+03	0.1041E+03
9	2.70	301.34	306.62	0.7251E+02	0.7195E+02
11	3.90	376.32	371.38	0.5206E+02	0.3823E+02
13	8.10	420.05	425.80	0.4514E+01	0.2853E+01
15	14.35	446.33	441.96	0.3809E+01	0.2225E+01
17	19.85	467.28	465.46	0.3809E+01	0.5402E+01
19	25.00	488.04	492.38	0.4287E+01	0.4647E+01
21	29.80	508.62	509.76	0.8718E+01	0.2462E+01
23	36.40	606.37	596.67	0.1129E+02	0.1144E+02
25	42.60	655.87	658.70	0.7983E+01	0.8704E+01
27	49.20	707.81	709.05	0.7769E+01	0.6705E+01
29	56.02	737.21	737.33	0.6632E+00	0.1602E+01
31	62.67	741.62	745.58	0.6632E+00	0.9464E+00
33	66.45	781.31	784.62	0.8331E+02	0.8054E+02
35	67.40	864.37	861.11	0.9114E+02	0.8049E+02
37	68.45	955.50	945.60	0.8284E+02	0.8045E+02
39	69.55	1031.57	1034.08	0.5546E+02	0.8041E+02

I - 43

Connection type : Single web-angle connections
Mode : Bolted-to-beam and welded-to-column

Tested by : L.E.Thompson et al. (1970) U.S.A.
Test Id. : H-1 ALT

Column : W10X49
Beam : W21X62
Angle : --

Fasteners: A325- -3/4"D
15/16" Oversize holes
Material : A36
Fy = -- ksi
Fu = -- ksi

Major parameters

lp = 11.5000"	lu = -- "	ll = -- "	ta = 0.6250"
gb = 2.2500"	cu = -- "	cl = -- "	
pb = 3.0000"	qb = -- "	nb = 2 X -	

Remark 1) Single plate connection.
2)

No	Moment (k-in)	Rotation (radians) X 1/1000	No	Moment (k-in)	Rotation (radians) X 1/1000
1	0.00	0.00	26	588.37	30.40
2	26.09	0.20	27	606.37	32.80
3	52.19	0.40	28	615.25	35.35
4	89.30	0.70	29	624.12	37.90
5	119.81	1.00	30	642.81	39.60
6	159.49	1.40	31	661.50	41.30
7	196.25	1.80	32	664.44	43.88
8	232.26	2.30	33	667.38	46.45
9	267.17	2.70	34	670.32	49.02
10	301.34	3.20	35	673.26	51.60
11	338.83	3.90	36	698.26	52.10
12	372.65	4.60	37	735.73	52.60
13	393.96	5.20	38	765.13	52.90
14	415.27	5.80	39	794.53	53.20
15	424.34	8.07	40	817.50	53.50
16	433.40	10.33	41	840.47	53.80
17	442.46	12.60			
18	458.64	13.95			
19	474.82	15.30			
20	488.05	17.65			
21	501.28	20.00			
22	517.07	21.85			
23	532.87	23.70			
24	551.62	25.85			
25	570.36	28.00			

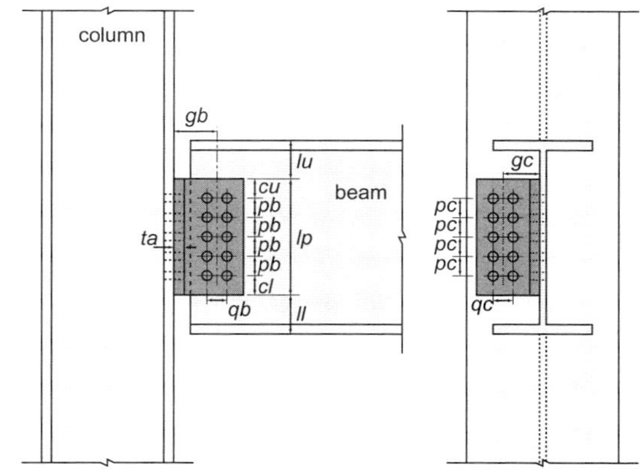

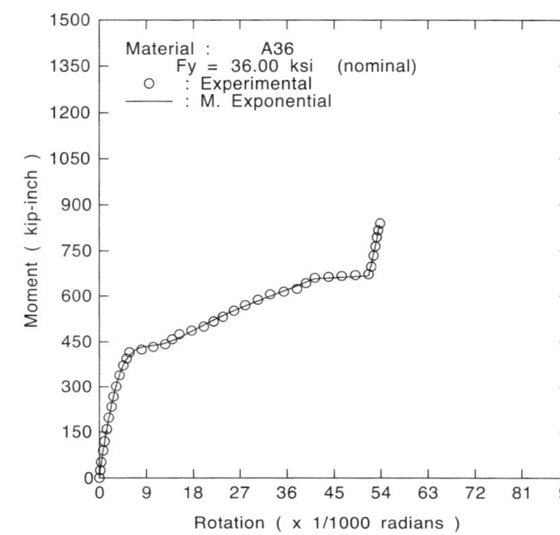

A.1 – 86

```
Moment-rotation prediction equations     ( R : X 1/1000 radians )

Modified exponential model  :
     Bm = Sum ( Ai X ( 1 - exp( - R/( 2*i*C ) ) ) )  +      Sum ( RKj X ( R - Rj0 ) )  +  Bm0
      C =   0.49833333E+00              Bmo=   0.00000000E+00         Nexp=  6        Nliner=  3
     AI =   0.59076959E+03  -0.85493789E+04   0.36239811E+05  -0.65243098E+05   0.55379735E+05  -0.18048075E+05
    Rj0 =           5.8000          41.3000          51.6000
    RKj =   0.79672902E+01  -0.65956494E+01   0.75680930E+02
```

No	Rotation (radians) X 1/1000	Moment (kip-inch) Expri.	Moment (kip-inch) M.Expo.	Connection stiffness (kip-inch) X 1000 Expri.	Connection stiffness (kip-inch) X 1000 M.Expo.
1	0.00	0.00	0.00	0.1305E+03	0.1537E+03
3	0.40	52.19	54.00	0.1278E+03	0.1203E+03
5	1.00	119.81	118.91	0.1006E+03	0.9967E+02
7	1.80	196.25	193.78	0.8306E+02	0.8835E+02
9	2.70	267.17	267.53	0.7886E+02	0.7472E+02
11	3.90	338.83	343.03	0.5093E+02	0.5050E+02
13	5.20	393.96	391.67	0.3552E+02	0.2523E+02
15	8.07	424.34	432.00	0.3999E+01	0.4718E+01
17	12.60	442.46	446.16	0.9002E+01	0.5181E+01
19	15.30	474.82	464.69	0.9664E+01	0.8197E+01
21	20.00	501.28	506.36	0.7258E+01	0.8787E+01
23	23.70	532.87	536.87	0.8622E+01	0.7683E+01
25	28.00	570.36	567.76	0.8144E+01	0.6815E+01
27	32.80	606.37	599.86	0.5552E+01	0.6674E+01
29	37.90	624.12	634.60	0.7988E+01	0.6984E+01
31	41.30	661.50	658.77	0.7076E+01	0.7235E+01
33	46.45	667.38	662.91	0.1142E+01	0.9487E+00
35	51.60	673.26	668.35	0.4205E+02	0.1147E+01
37	52.60	735.73	745.19	0.8936E+02	0.7686E+02
39	53.20	794.53	791.31	0.8728E+02	0.7687E+02
41	53.80	840.47	837.44	0.7656E+02	0.7688E+02

I - 44

```
Connection type : Single web-angle connections
           Mode : Bolted-to-beam and welded-to-column

Tested by  :  L.E.Thompson et al. (1970)                    U.S.A.
Test  Id.  :  H-1 ALT

Column : W10X49                        Fasteners: A325- -3/4"D
Beam   : W21X62                              15/16" Oversize holes
Angle  : --                            Material : A36
                                            Fy =    --   ksi
                                            Fu =    --   ksi

                              Major parameters

     lp = 11.5000"    lu =    --    "    ll =    --    "    ta =  0.6250"
     gb =  2.2500"    cu =    --    "    cl =    --    "
     pb =  3.0000"    qb =    --    "    nb = 2 X  -

Remark  1)  Single plate connection.
        2)
```

No	Moment (k-in)	Rotation (radians) X 1/1000	No	Moment (k-in)	Rotation (radians) X 1/1000
1	0.00	0.00	26	606.37	36.60
2	26.09	0.20	27	633.94	39.75
3	52.19	0.40	28	661.50	42.90
4	89.30	0.70	29	679.88	45.43
5	119.81	1.00	30	698.25	47.97
6	159.49	1.40	31	716.63	50.50
7	196.25	1.70	32	721.40	53.83
8	232.26	2.10	33	726.18	57.15
9	267.17	2.50	34	730.96	60.48
10	301.34	2.90	35	735.73	63.80
11	338.83	3.40	36	753.37	65.70
12	372.65	3.90	37	784.98	67.55
13	393.96	4.35	38	816.59	69.40
14	415.27	4.80	39	840.47	72.30
15	442.46	5.70			
16	474.82	8.90			
17	479.23	11.97			
18	483.64	15.03			
19	488.05	18.10			
20	492.46	21.17			
21	496.87	24.23			
22	501.28	27.30			
23	527.36	29.45			
24	553.45	31.60			
25	579.91	34.10			

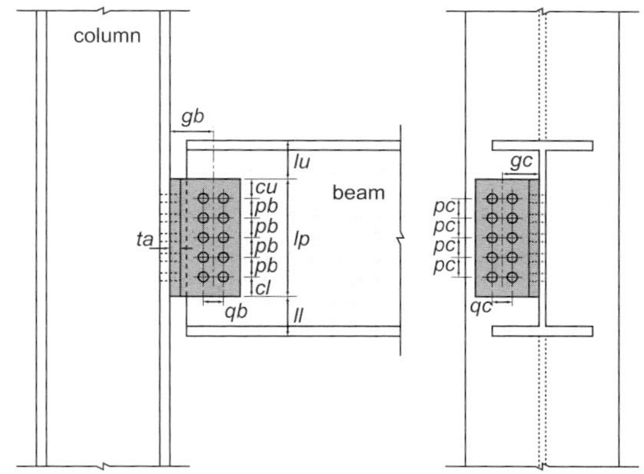

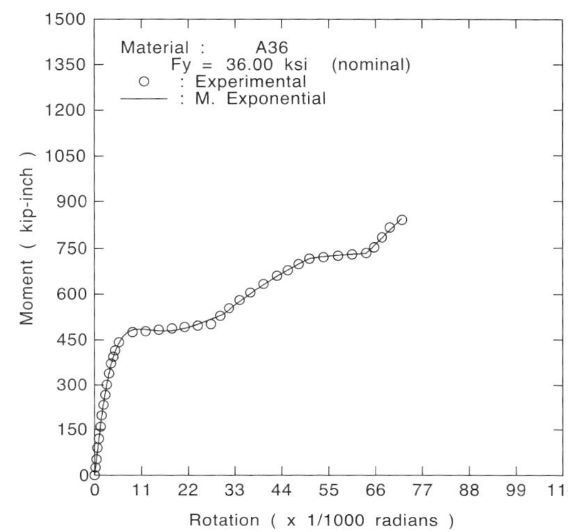

```
Moment-rotation prediction equations      ( R : X 1/1000 radians )

Modified exponential model  :
     Bm = Sum ( Ai X ( 1 - exp( - R/( 2*i*C ) ) ) )  +      Sum ( RKj X ( R - Rj0 ) )  +  Bm0
      C =   0.65250000E+00          Bmo=   0.00000000E+00          Nexp=  6        Nliner=  4
     AI =  -0.30030151E+03   0.24948491E+04  -0.14372117E+05   0.44707597E+05  -0.56148432E+05   0.24261064E+05
    Rj0 =          8.9000           30.1000           50.5000           63.8000
    RKj =  -0.70053065E+00   0.59135183E+01  -0.49308834E+01   0.12578773E+02
```

No	Rotation (radians) X 1/1000	Moment (kip-inch) Expri.	Moment (kip-inch) M.Expo.	Connection stiffness (kip-inch) X 1000 Expri.	Connection stiffness (kip-inch) X 1000 M.Expo.
1	0.00	0.00	0.00	0.1305E+03	0.1128E+03
3	0.40	52.19	46.82	0.1278E+03	0.1198E+03
5	1.00	119.81	118.53	0.1006E+03	0.1172E+03
7	1.70	196.25	196.43	0.1086E+03	0.1042E+03
9	2.50	267.17	272.25	0.8636E+02	0.8510E+02
11	3.40	338.83	339.27	0.7130E+02	0.6425E+02
13	4.35	393.96	391.21	0.4736E+02	0.4579E+02
15	5.70	442.46	439.18	0.2580E+02	0.2657E+02
17	11.97	479.23	484.00	0.1438E+01	-0.1342E+01
19	18.10	488.05	480.42	0.1438E+01	0.1338E+01
21	24.23	496.87	500.92	0.1438E+01	0.4864E+01
23	29.45	527.36	528.66	0.1213E+02	0.5440E+01
25	34.10	579.91	576.19	0.1058E+02	0.1061E+02
27	39.75	633.94	632.17	0.8750E+01	0.9174E+01
29	45.43	679.88	680.31	0.7254E+01	0.7816E+01
31	50.50	716.63	717.46	0.4738E+01	0.6896E+01
33	57.15	726.18	727.66	0.1436E+01	0.1175E+01
35	63.80	735.73	733.85	0.6430E+01	0.7323E+00
37	67.55	784.98	783.47	0.1709E+02	0.1316E+02
39	72.30	840.47	845.67	0.8234E+01	0.1304E+02

A.1 – 89

I - 45

Connection type : Single web-angle connections
Mode : Bolted-to-beam and welded-to-column

Tested by : L.E.Thompson et al. (1970) U.S.A.
Test Id. : F-2-1 ALT

Column : W10X49
Beam : W21X62
Angle : --

Fasteners: A325- -3/4"D
15/16" Oversize holes
Material : A441
Fy = -- ksi
Fu = -- ksi

Major parameters

lp = 11.5000"	lu = -- "	ll = -- "	ta = 0.6250"
gb = 2.2500"	cu = -- "	cl = -- "	
pb = 3.0000"	qb = -- "	nb = 2 X -	

Remark 1) Single plate connection.
2)

No	Moment (k-in)	Rotation (radians) X 1/1000	No	Moment (k-in)	Rotation (radians) X 1/1000
1	0.00	0.00	26	777.08	57.42
2	49.98	0.40	27	787.19	61.30
3	83.80	0.70	28	846.72	63.20
4	117.60	1.10	29	934.92	65.40
5	150.67	1.80	30	1019.45	65.40
6	188.16	2.90	31	1108.38	66.80
7	221.23	4.50	32	1179.67	67.90
8	257.26	6.30	33	1254.65	69.10
9	293.27	8.50	34	1325.21	70.30
10	332.95	10.60	35	1400.92	71.70
11	373.38	12.70	36	1470.00	73.10
12	405.72	14.90	37	1543.36	75.10
13	441.00	17.10	38	1617.00	77.30
14	481.43	20.10	39	1688.29	79.90
15	515.23	22.90	40	1767.67	83.10
16	551.99	26.20	41	1840.44	86.80
17	585.06	29.70			
18	607.54	32.70			
19	630.01	35.70			
20	652.49	38.70			
21	698.26	41.30			
22	722.51	43.55			
23	746.76	45.80			
24	756.87	49.67			
25	766.97	53.55			

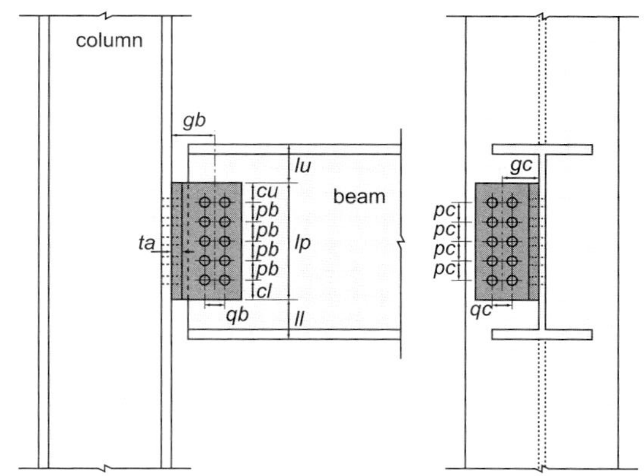

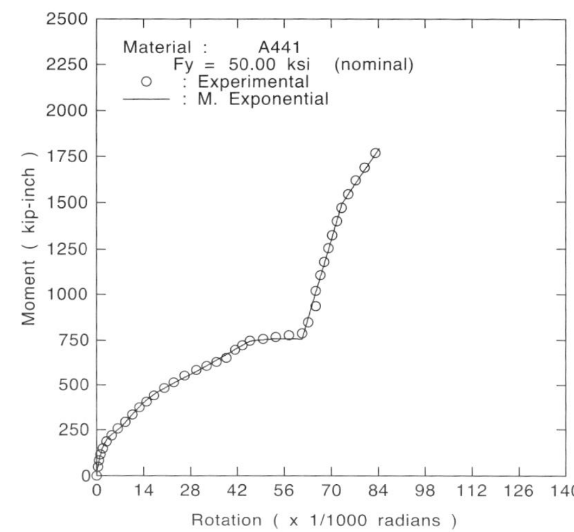

A.1 – 90

```
Moment-rotation prediction equations     ( R : X 1/1000 radians )

Modified exponential model  :
     Bm = Sum ( Ai X ( 1 - exp( - R/( 2*i*C ) ) ) )  +      Sum ( RKj X ( R - Rj0 ) )  +  Bm0
      C =   0.82333333E+00              Bmo=   0.00000000E+00         Nexp=  6        Nliner=  4
     AI =  -0.26245288E+03   0.10747418E+05  -0.61991417E+05   0.14646274E+06  -0.15453232E+06   0.60389800E+05
    Rj0 =          20.1000          45.8000          61.3000          73.1000
    RKj =   0.22357237E+01  -0.68483308E+01   0.64403281E+02  -0.33620085E+02
```

No	Rotation (radians) X 1/1000	Moment (kip-inch) Expri.	Moment (kip-inch) M.Expo.	Connection stiffness (kip-inch) X 1000 Expri.	Connection stiffness (kip-inch) X 1000 M.Expo.
1	0.00	0.00	0.00	0.1249E+03	0.1346E+03
3	0.70	83.80	78.82	0.1006E+03	0.9266E+02
5	1.80	150.67	154.56	0.4213E+02	0.4913E+02
7	4.50	221.23	224.62	0.2036E+02	0.1463E+02
9	8.50	293.27	288.82	0.1766E+02	0.1947E+02
11	12.70	373.38	374.68	0.1703E+02	0.1949E+02
13	17.10	441.00	446.48	0.1495E+02	0.1300E+02
15	22.90	515.23	510.17	0.1164E+02	0.1011E+02
17	29.70	585.06	576.74	0.8395E+01	0.1004E+02
19	35.70	630.01	639.50	0.7492E+01	0.1076E+02
21	41.30	698.26	699.33	0.1394E+02	0.1044E+02
23	45.80	746.76	744.46	0.7777E+01	0.9551E+01
25	53.55	766.97	757.59	0.2608E+01	0.6463E+00
27	61.30	787.19	754.93	0.2188E+02	-0.1259E+01
29	65.40	934.92	1012.13	0.8449E+05	0.6234E+02
31	66.80	1108.38	1099.24	0.6424E+02	0.6211E+02
33	69.10	1254.65	1241.67	0.6064E+02	0.6175E+02
35	71.70	1400.92	1401.78	0.5171E+02	0.6141E+02
37	75.10	1543.36	1542.68	0.3515E+02	0.2742E+02
39	79.90	1688.29	1673.29	0.2625E+02	0.2702E+02
41	86.80	1840.44	1858.36	0.1967E+02	0.2665E+02

A.1 – 91

I - 46

Connection type : Single web-angle connections
Mode : Bolted-to-beam and welded-to-column

Tested by : L.E.Thompson et al. (1970) U.S.A.
Test Id. : F-2-1 ALT

Column : W10X49
Beam : W21X62
Angle : --

Fasteners: A325- -3/4"D
15/16" Oversize holes
Material : A441
Fy = -- ksi
Fu = -- ksi

Major parameters

lp = 11.5000" lu = -- " ll = -- " ta = 0.6250"
gb = 2.2500" cu = -- " cl = -- "
pb = 3.0000" qb = -- " nb = 2 X -

Remark 1) Single plate connection.
2)

No	Moment (k-in)	Rotation (radians) X 1/1000	No	Moment (k-in)	Rotation (radians) X 1/1000
1	0.00	0.00	26	816.95	64.75
2	49.98	0.50	27	831.84	69.78
3	83.80	0.80	28	846.72	74.80
4	117.60	1.20	29	934.92	78.50
5	150.67	1.70	30	1019.45	78.50
6	188.16	3.00	31	1108.38	79.90
7	204.70	6.20	32	1179.67	81.00
8	221.23	9.40	33	1254.65	82.30
9	257.26	13.40	34	1325.21	83.70
10	293.27	16.20	35	1400.92	85.20
11	332.95	20.30	36	1470.00	86.80
12	373.38	23.00	37	1543.36	88.80
13	405.72	24.80	38	1617.00	91.00
14	441.00	26.50	39	1688.29	93.50
15	481.43	28.80	40	1767.67	96.90
16	515.23	31.10	41	1840.44	100.80
17	551.99	33.20			
18	585.06	35.70			
19	618.77	39.45			
20	652.49	43.20			
21	698.26	46.10			
22	722.51	48.70			
23	746.76	51.30			
24	787.19	54.70			
25	802.07	59.73			

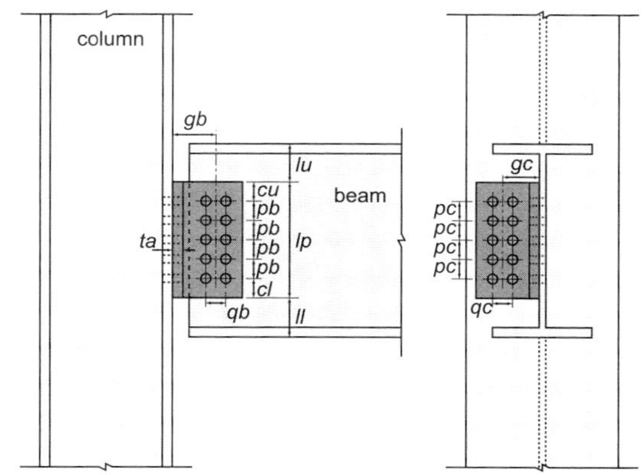

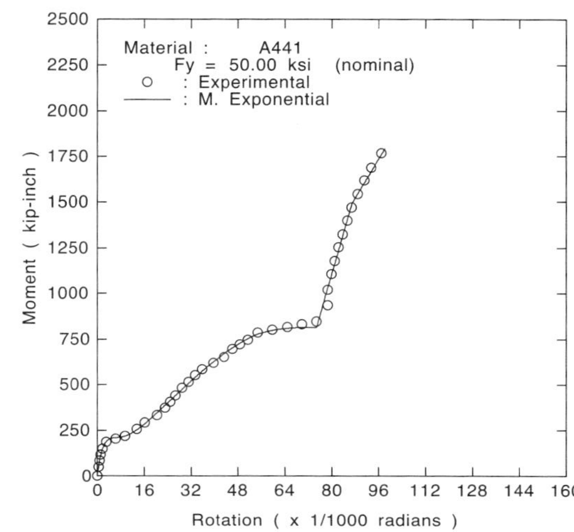

Moment-rotation prediction equations (R : X 1/1000 radians)

```
Modified exponential model  :
     Bm = Sum ( Ai X ( 1 - exp( - R/( 2*i*C ) ) ) )  +      Sum ( RKj X ( R - Rj0 ) )  +  Bm0
      C =   0.96500000E+00            Bmo=   0.00000000E+00            Nexp=  6         Nliner=  4
     AI =   0.49064675E+02   0.51202398E+04  -0.32048176E+05   0.88121944E+05  -0.10948755E+06   0.49171071E+05
    Rj0 =          13.4000          54.7000          74.8000          86.8000
    RKj =  -0.42630341E+00  -0.24887472E+01   0.57424653E+02  -0.28677574E+02
```

No	Rotation (radians) X 1/1000	Moment (kip-inch) Expri.	Moment (kip-inch) M.Expo.	Connection stiffness (kip-inch) X 1000 Expri.	Connection stiffness (kip-inch) X 1000 M.Expo.
1	0.00	0.00	0.00	0.9996E+02	0.1319E+03
3	0.80	83.80	85.54	0.1006E+03	0.8485E+02
5	1.70	150.67	144.85	0.5578E+02	0.4954E+02
7	6.20	204.70	210.56	0.5168E+01	0.1333E+01
9	13.40	257.26	252.43	0.1127E+02	0.1122E+02
11	20.30	332.95	344.35	0.1287E+02	0.1479E+02
13	24.80	405.72	412.27	0.1940E+02	0.1526E+02
15	28.80	481.43	473.22	0.1614E+02	0.1515E+02
17	33.20	551.99	538.72	0.1555E+02	0.1454E+02
19	39.45	618.77	624.93	0.8990E+01	0.1291E+02
21	46.10	698.26	703.00	0.1238E+02	0.1050E+02
23	51.30	746.76	752.37	0.1044E+02	0.8491E+01
25	59.73	802.07	798.54	0.2962E+01	0.3048E+01
27	69.78	831.84	815.22	0.2962E+01	0.4534E+00
29	78.50	934.92	1024.90	0.8451E+05	0.5644E+02
31	79.90	1108.38	1103.80	0.6424E+02	0.5627E+02
33	82.30	1254.65	1238.53	0.5417E+02	0.5601E+02
35	85.20	1400.92	1400.56	0.4694E+02	0.5574E+02
37	88.80	1543.36	1543.35	0.3515E+02	0.2679E+02
39	93.50	1688.29	1668.57	0.2633E+02	0.2651E+02
41	100.80	1840.44	1861.00	0.1866E+02	0.2623E+02

I - 47

Connection type : Single web-angle connections
Mode : Bolted-to-beam and welded-to-column

Tested by : L.E.Thompson et al. (1970) U.S.A.
Test Id. : F-2-2 ALT

Column : W10X49
Beam : W21X62
Angle : --

Fasteners: A325- -3/4"D
15/16" Oversize holes
Material : A441
Fy = -- ksi
Fu = -- ksi

Major parameters

lp = 11.5000"	lu = -- "	ll = -- "	ta = 0.6250"
gb = 2.2500"	cu = -- "	cl = -- "	
pb = 3.0000"	qb = -- "	nb = 2 X -	

Remark 1) Single plate connection.
2)

No	Moment (k-in)	Rotation (radians) X 1/1000	No	Moment (k-in)	Rotation (radians) X 1/1000
1	0.00	0.00	26	1202.83	52.90
2	51.44	0.10	27	1258.32	53.80
3	121.27	0.80	28	1310.50	54.90
4	191.10	1.40	29	1362.68	56.00
5	260.93	2.30	30	1418.55	57.35
6	329.28	4.40	31	1474.42	58.70
7	367.87	7.95	32	1528.07	60.25
8	406.45	11.50	33	1581.72	61.80
9	442.47	14.30	34	1655.22	64.30
10	478.49	17.10	35	1728.72	66.20
11	514.87	20.05	36	1800.76	69.40
12	551.26	23.00	37	1837.50	71.20
13	588.01	25.80	38	1877.20	73.30
14	624.76	28.60	39	1911.00	75.20
15	671.80	32.10	40	1949.22	78.00
16	692.86	36.37	41	1984.50	81.10
17	713.93	40.63	42	2024.20	85.30
18	735.00	44.90	43	2035.96	89.70
19	788.29	46.45			
20	841.57	48.00			
21	922.43	49.10			
22	979.39	49.80			
23	1036.36	50.50			
24	1091.84	51.25			
25	1147.33	52.00			

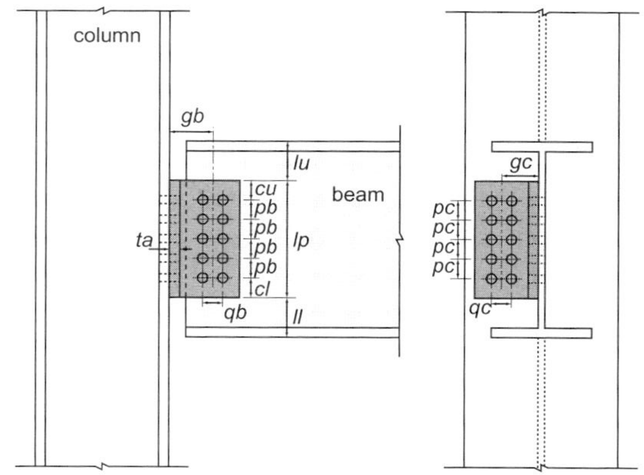

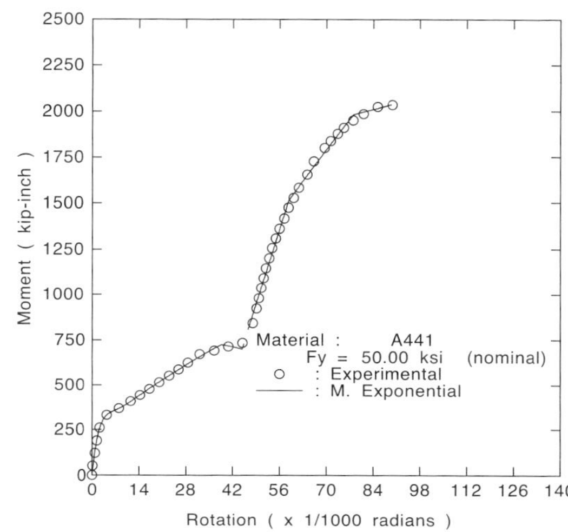

A.1 – 94

Moment-rotation prediction equations (R : X 1/1000 radians)

```
Modified exponential model  :
     Bm = Sum ( Ai X ( 1 - exp( - R/( 2*i*C ) ) ) )  +      Sum ( RKj X ( R - Rj0 ) )  +  Bm0
      C =   0.77250000E+00           Bmo=   0.00000000E+00           Nexp=  6        Nliner=  4
     AI =  -0.27697160E+03   0.82633999E+04  -0.42811640E+05   0.10292462E+06  -0.11464514E+06   0.47422691E+05
    Rj0 =         38.5000          44.9000          58.7000          78.0000
    RKj =  -0.11862347E+02   0.65817205E+02  -0.31107630E+02  -0.18313086E+02
```

No	Rotation (radians) X 1/1000	Moment (kip-inch) Expri.	Moment (kip-inch) M.Expo.	Connection stiffness (kip-inch) X 1000 Expri.	Connection stiffness (kip-inch) X 1000 M.Expo.
1	0.00	0.00	0.00	0.5144E+03	0.1878E+03
3	0.80	121.27	125.83	0.1087E+03	0.1293E+03
5	2.30	260.93	260.36	0.6408E+02	0.5775E+02
7	7.95	367.87	366.02	0.1087E+02	0.9207E+01
9	14.30	442.47	443.09	0.1286E+02	0.1348E+02
11	20.05	514.87	517.54	0.1233E+02	0.1234E+02
13	25.80	588.01	586.61	0.1312E+02	0.1179E+02
15	32.10	671.80	659.00	0.9609E+01	0.1105E+02
17	40.63	713.93	718.35	0.4938E+01	-0.3306E+01
19	46.45	788.29	794.97	0.3438E+02	0.6039E+02
21	49.10	922.43	953.79	0.7832E+02	0.5948E+02
23	50.50	1036.36	1036.74	0.7781E+02	0.5902E+02
25	52.00	1147.33	1124.92	0.6838E+02	0.5856E+02
27	53.80	1258.32	1229.85	0.5526E+02	0.5804E+02
29	56.00	1362.68	1356.89	0.4472E+02	0.5746E+02
31	58.70	1474.42	1511.16	0.3823E+02	0.5683E+02
33	61.80	1581.72	1589.94	0.3262E+02	0.2512E+02
35	66.20	1728.72	1698.92	0.3266E+02	0.2445E+02
37	71.20	1837.50	1819.69	0.1972E+02	0.2390E+02
39	75.20	1911.00	1914.63	0.1612E+02	0.2359E+02
41	81.10	1984.50	1996.05	0.1056E+02	0.4970E+01
43	89.70	2035.96	2037.63	0.2673E+01	0.4729E+01

I - 48

```
Connection type : Single web-angle connections
           Mode : Bolted-to-beam and welded-to-column

Tested by   :  L.E.Thompson et al. (1970)                U.S.A.
Test   Id.  :  F-2-2 ALT

Column : W10X49                    Fasteners: A325- -3/4"D
Beam   : W21X62                           15/16" Oversize holes
Angle  : --                        Material  : A441
                                         Fv =   --   ksi
                                         Fu =   --   ksi
```

Major parameters

lp = 11.5000"	lu = -- "	ll = -- "	ta = 0.6250"
gb = 2.2500"	cu = -- "	cl = -- "	
pb = 3.0000"	qb = -- "	nb = 2 X -	

Remark 1) Single plate connection.
2)

No	Moment (k-in)	Rotation (radians) X 1/1000	No	Moment (k-in)	Rotation (radians) X 1/1000
1	0.00	0.00	26	1202.83	55.80
2	51.44	0.30	27	1258.32	56.70
3	121.27	1.10	28	1310.50	57.65
4	191.10	1.90	29	1362.68	58.60
5	260.93	3.10	30	1418.55	59.80
6	329.28	5.10	31	1474.42	61.00
7	406.45	8.20	32	1528.07	62.35
8	478.49	11.70	33	1581.72	63.70
9	551.26	15.20	34	1655.22	65.80
10	588.01	17.55	35	1728.72	68.20
11	624.76	19.90	36	1800.76	71.20
12	671.80	23.30	37	1837.50	72.80
13	703.40	25.75	38	1877.20	74.90
14	735.00	28.20	39	1911.00	76.70
15	770.52	31.23	40	1949.22	79.10
16	806.05	34.27	41	1984.50	81.90
17	841.57	37.30	42	2024.20	85.50
18	861.79	40.83	43	2035.96	89.20
19	882.00	44.35			
20	902.21	47.88			
21	922.43	51.40			
22	979.39	52.40			
23	1035.36	53.40			
24	1091.84	54.15			
25	1147.33	54.90			

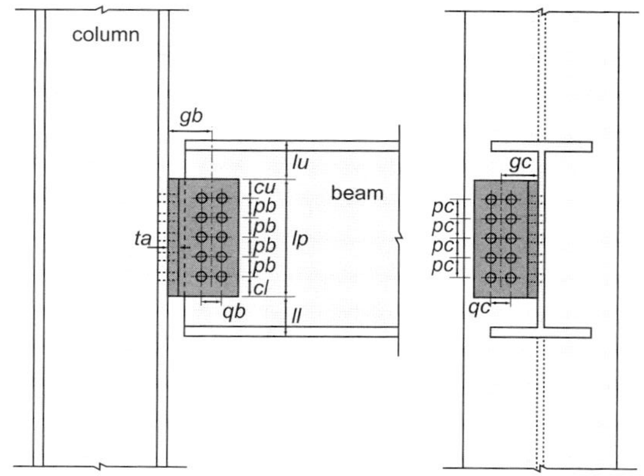

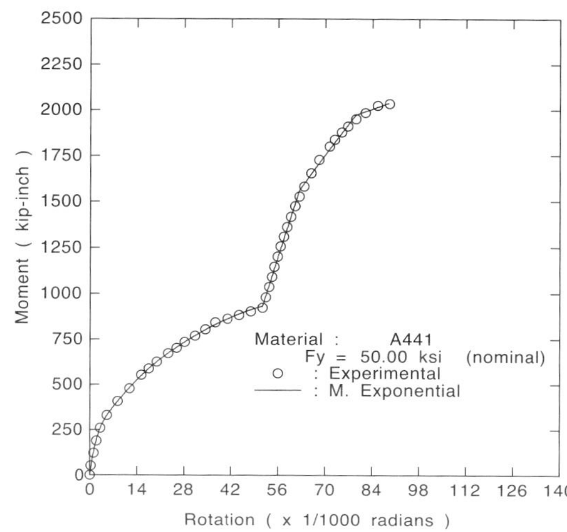

Moment-rotation prediction equations (R : X 1/1000 radians)

```
Modified exponential model  :
     Bm = Sum ( Ai X ( 1 - exp( - R/( 2*i*C ) ) ) )  +     Sum ( RKj X ( R - Rj0 ) )  +  Bm0
      C =   0.81833333E+00             Bmo=   0.00000000E+00          Nexp=  6        Nliner=  4
     AI =  -0.98575833E+02   0.47166556E+04  -0.26991299E+05   0.69507512E+05  -0.81465639E+05   0.35320167E+05
    Rj0 =         44.3500          51.4000          62.3500          79.1000
    RKj =   0.65432773E+00   0.53881867E+02  -0.31344512E+02  -0.16944877E+02
```

No	Rotation (radians) X 1/1000	Moment (kip-inch) Expri.	Moment (kip-inch) M.Expo.	Connection stiffness (kip-inch) X 1000 Expri.	Connection stiffness (kip-inch) X 1000 M.Expo.
1	0.00	0.00	0.00	0.1715E+03	0.1424E+03
3	1.10	121.27	127.41	0.8728E+02	0.9282E+02
5	3.10	260.93	258.69	0.4918E+02	0.4569E+02
7	8.20	406.45	405.91	0.2287E+02	0.2234E+02
9	15.20	551.26	547.94	0.1771E+02	0.1801E+02
11	19.90	624.76	625.17	0.1490E+02	0.1501E+02
13	25.75	703.40	705.53	0.1290E+02	0.1267E+02
15	31.23	770.52	770.48	0.1171E+02	0.1104E+02
17	37.30	841.57	831.72	0.8947E+01	0.9107E+01
19	44.35	882.00	887.41	0.5734E+01	0.6695E+01
21	51.40	922.43	931.31	0.4564E+02	0.5175E+01
23	53.40	1036.36	1048.89	0.6669E+02	0.5853E+02
25	54.90	1147.33	1136.40	0.6838E+02	0.5816E+02
27	56.70	1258.32	1240.71	0.5839E+02	0.5775E+02
29	58.60	1362.68	1350.06	0.5123E+02	0.5736E+02
31	61.00	1474.42	1487.18	0.4335E+02	0.5692E+02
33	63.70	1581.72	1597.95	0.3789E+02	0.2515E+02
35	68.20	1728.72	1709.76	0.2769E+02	0.2458E+02
37	72.80	1837.50	1821.79	0.2121E+02	0.2415E+02
39	76.70	1911.00	1915.45	0.1756E+02	0.2389E+02
41	81.90	1984.50	1991.54	0.1191E+02	0.6696E+01
43	89.20	2035.96	2039.57	0.3178E+01	0.6484E+01

I - 49

Connection type : Single web-angle connections
Mode : Bolted-to-beam and welded-to-column

Tested by : L.E.Thompson et al. (1970) U.S.A.
Test Id. : F-2-3 ALT

Column : W10X49
Beam : W21X62
Angle : --

Fasteners: A325- -3/4"D
15/16" Oversize holes
Material : A441
Fy = -- ksi
Fu = -- ksi

Major parameters

lp = 11.5000"	lu = -- "	ll = -- "	ta = 0.6250"
gb = 2.2500"	cu = -- "	cl = -- "	
pb = 3.0000"	qb = -- "	nb = 2 X -	

Remark 1) Single plate connection.
2)

No	Moment (k-in)	Rotation (radians) X 1/1000	No	Moment (k-in)	Rotation (radians) X 1/1000
1	0.00	0.00	26	1400.17	61.50
2	43.37	0.20	27	1454.38	62.80
3	129.36	1.00	28	1508.59	64.10
4	186.70	3.10	29	1560.22	65.65
5	221.98	5.90	30	1611.85	67.20
6	257.26	8.70	31	1665.51	69.30
7	330.76	13.50	32	1719.17	71.40
8	402.78	17.90	33	1802.95	74.90
9	437.69	20.45	34	1836.04	76.70
10	472.61	23.00	35	1874.99	78.80
11	511.93	25.60	36	1915.42	81.40
12	551.26	28.20	37	1950.70	84.10
13	624.76	32.40	38	1986.38	87.90
14	698.26	36.80	39	1997.20	91.55
15	738.68	40.90	40	2008.02	95.20
16	779.10	45.00	41	1984.50	97.80
17	819.52	49.10			
18	859.94	53.20			
19	954.77	54.20			
20	1022.39	55.00			
21	1090.01	55.80			
22	1180.42	57.40			
23	1234.80	58.35			
24	1289.18	59.30			
25	1344.68	60.40			

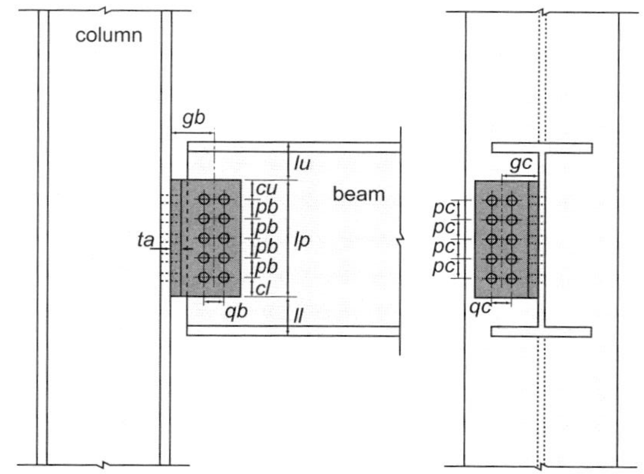

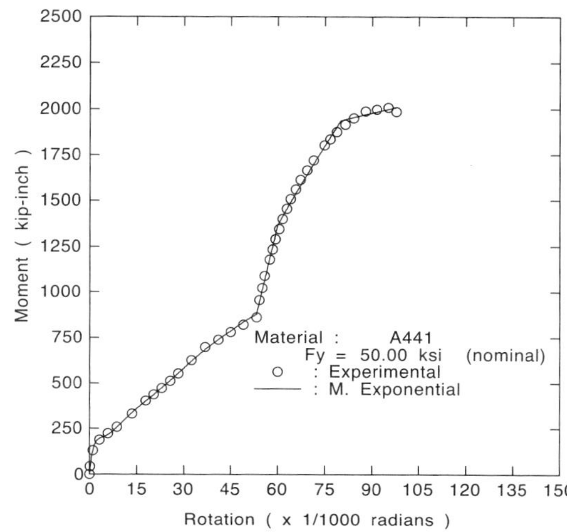

A.1 – 98

```
Moment-rotation prediction equations     ( R : X 1/1000 radians )

Modified exponential model  :
     Bm = Sum ( Ai X ( 1 - exp( - R/( 2*i*C ) ) ) )  +      Sum ( RKj X ( R - Rj0 ) )  +  Bm0
      C =   0.86500000E+00               Bmo=    0.00000000E+00           Nexp=  6        Nliner=  4
     AI =   0.80803403E+03  -0.44478488E+03  -0.18015560E+05   0.71838664E+05  -0.10020157E+06   0.46968989E+05
    Rj0 =          45.0000          53.2000          60.0000          80.0000
    RKj =   0.24044333E+01   0.64449177E+02  -0.40845677E+02  -0.22554838E+02
```

No	Rotation (radians) X 1/1000	Moment (kip-inch) Expri.	Moment (kip-inch) M.Expo.	Connection stiffness (kip-inch) X 1000 Expri.	Connection stiffness (kip-inch) X 1000 M.Expo.
1	0.00	0.00	0.00	0.2168E+03	0.1896E+03
3	1.00	129.36	126.23	0.8537E+02	0.7740E+02
5	5.90	221.98	216.56	0.1260E+02	0.1057E+02
7	13.50	330.76	335.26	0.1586E+02	0.1574E+02
9	20.45	437.69	437.24	0.1369E+02	0.1443E+02
11	25.60	511.93	513.84	0.1512E+02	0.1533E+02
13	32.40	624.76	619.25	0.1711E+02	0.1529E+02
15	40.90	738.68	738.52	0.9859E+01	0.1239E+02
17	49.10	819.52	834.33	0.9859E+01	0.1098E+02
19	54.20	954.77	949.18	0.8910E+02	0.7329E+02
21	55.80	1090.01	1065.96	0.7519E+02	0.7269E+02
23	58.35	1234.80	1250.20	0.5725E+02	0.7183E+02
25	60.40	1344.68	1380.45	0.5045E+02	0.3035E+02
27	62.80	1454.38	1452.50	0.4170E+02	0.2970E+02
29	65.65	1560.22	1536.17	0.3331E+02	0.2903E+02
31	69.30	1665.51	1640.81	0.2555E+02	0.2833E+02
33	74.90	1802.95	1797.07	0.2027E+02	0.2753E+02
35	78.80	1874.99	1903.61	0.1721E+02	0.2713E+02
37	84.10	1950.70	1953.81	0.1154E+02	0.4184E+01
39	91.55	1997.20	1983.60	0.2964E+01	0.3847E+01
41	97.80	1984.50	2007.10	-0.9046E+01	0.3684E+01

I - 50

Connection type : Single web-angle connections
Mode : Bolted-to-beam and welded-to-column

Tested by : L.E.Thompson et al. (1970) U.S.A.
Test Id. : F-2-3 ALT

Column : W10X49
Beam : W21X62
Angle : --

Fasteners: A325- -3/4"D
15/16" Oversize holes
Material : A441
Fy = -- ksi
Fu = -- ksi

Major parameters

lp = 11.5000" lu = -- " ll = -- " ta = 0.6250"
gb = 2.2500" cu = -- " cl = -- "
pb = 3.0000" qb = -- " nb = 2 X -

Remark 1) Single plate connection.
2)

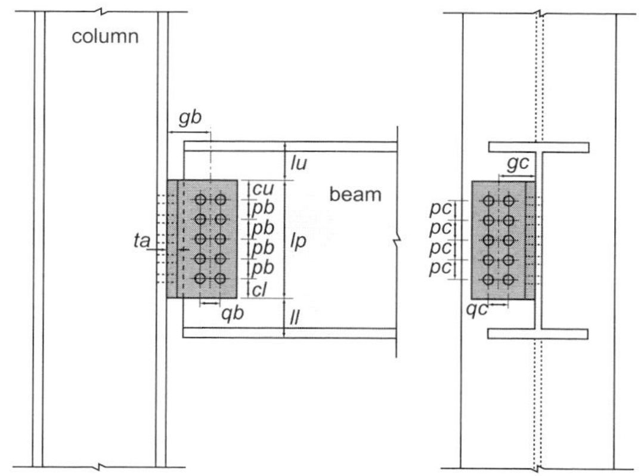

No	Moment (k-in)	Rotation (radians) X 1/1000	No	Moment (k-in)	Rotation (radians) X 1/1000
1	0.00	0.00	26	1719.17	83.70
2	43.37	0.20	27	1797.08	87.40
3	129.36	1.10	28	1874.99	91.10
4	186.70	1.90	29	1912.84	94.10
5	240.72	7.28	30	1950.70	97.10
6	294.74	12.65	31	1969.80	100.97
7	348.76	18.02	32	1988.91	104.83
8	402.78	23.40	33	2008.02	108.70
9	472.61	27.80	34	1984.50	111.70
10	551.26	33.00			
11	624.76	37.60			
12	698.26	42.60			
13	762.67	47.20			
14	827.08	51.80			
15	891.49	56.40			
16	923.13	61.65			
17	954.77	66.90			
18	1043.85	68.44			
19	1132.93	69.98			
20	1222.01	71.52			
21	1311.09	73.06			
22	1400.17	74.60			
23	1479.92	76.88			
24	1559.67	79.15			
25	1639.42	81.42			

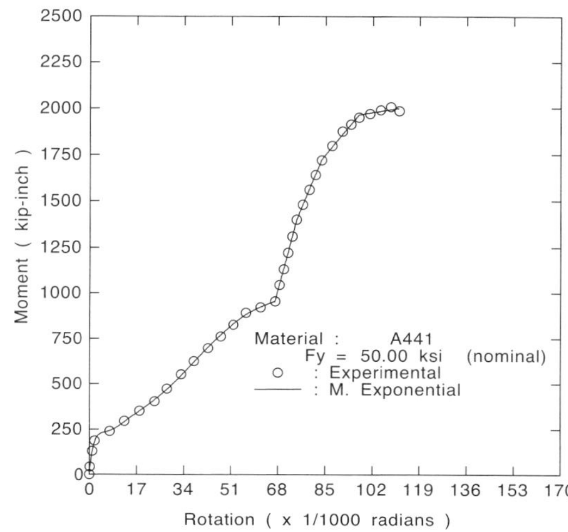

A.1 – 100

```
Moment-rotation prediction equations     ( R : X 1/1000 radians )

Modified exponential model   :
   Bm = Sum ( Ai X ( 1 - exp( - R/( 2*i*C ) ) ) )  +      Sum ( RKj X ( R - Rj0 ) )  +  Bm0
    C =   0.98083333E+00               Bmo=   0.00000000E+00           Nexp=  6        Nliner=  5
   AI =   0.50880834E+03   0.42427865E+04  -0.43606708E+05   0.13570385E+06  -0.17139460E+06   0.75610514E+05
  Rj0 =          56.4000          66.9000          74.6000          83.7000          97.1000
  RKj =  -0.13450369E+01   0.53964771E+02  -0.19664025E+02  -0.16519067E+02  -0.14603391E+02
```

No	Rotation (radians) X 1/1000	Moment (kip-inch) Expri.	Moment (kip-inch) M.Expo.	Connection stiffness (kip-inch) X 1000 Expri.	Connection stiffness (kip-inch) X 1000 M.Expo.
1	0.00	0.00	0.00	0.2168E+03	0.1751E+03
3	1.10	129.36	135.13	0.8291E+02	0.8109E+02
5	7.28	240.72	242.39	0.1005E+02	0.5675E+01
7	18.02	348.76	349.79	0.1005E+02	0.1086E+02
9	27.80	472.61	468.06	0.1553E+02	0.1403E+02
11	37.60	624.76	620.80	0.1537E+02	0.1636E+02
13	47.20	762.67	770.26	0.1400E+02	0.1419E+02
15	56.40	891.49	883.14	0.1028E+02	0.1026E+02
17	66.90	954.77	954.02	0.4609E+02	0.4777E+01
19	69.98	1132.93	1133.42	0.5784E+02	0.5777E+02
21	73.06	1311.09	1310.02	0.5784E+02	0.5692E+02
23	76.88	1479.92	1480.69	0.3505E+02	0.3637E+02
25	81.42	1639.42	1644.13	0.3505E+02	0.3551E+02
27	87.40	1797.08	1792.54	0.2106E+02	0.1815E+02
29	94.10	1912.84	1911.89	0.1262E+02	0.1752E+02
31	100.97	1969.80	1974.16	0.4942E+01	0.2495E+01
33	108.70	2008.02	1992.25	-0.2256E+01	0.2208E+01

I - 51

Connection type : Single web-angle connections
Mode : Bolted-to-beam and welded-to-column

Tested by : R.M.Richard et al. (1982) U.S.A.
Test Id. : RGKL2

Column : --
Beam : --
Angle : --

Fasteners: A325- -3/4"D
15/16" Oversize holes
Material : A36
Fy = -- ksi
Fu = -- ksi

Major parameters

lp = 6.0000" lu = -- " ll = -- " ta = 0.2500"
gb = -- " cu = -- " cl = -- " pb = 3.0000"
nb = 1 X 2

Remark 1) Single plate connection.
2)

No	Moment (k-in)	Rotation (radians) X 1/1000	No	Moment (k-in)	Rotation (radians) X 1/1000
1	0.00	0.00	26	44.78	35.68
2	1.95	0.27	27	45.86	38.32
3	3.90	0.53	28	46.94	40.96
4	5.84	0.80	29	48.02	43.60
5	8.46	1.12	30	49.16	45.87
6	11.07	1.43	31	50.30	48.13
7	13.68	1.75	32	51.44	50.40
8	16.29	2.07	33	52.30	52.67
9	18.90	2.38	34	53.16	54.93
10	21.52	2.70	35	54.01	57.20
11	23.32	3.20			
12	25.13	3.70			
13	26.94	4.20			
14	29.50	6.65			
15	32.06	9.10			
16	33.78	11.70			
17	35.49	14.30			
18	37.20	16.90			
19	38.15	18.90			
20	39.10	20.90			
21	40.04	22.90			
22	40.90	25.40			
23	41.76	27.90			
24	42.61	30.40			
25	43.69	33.04			

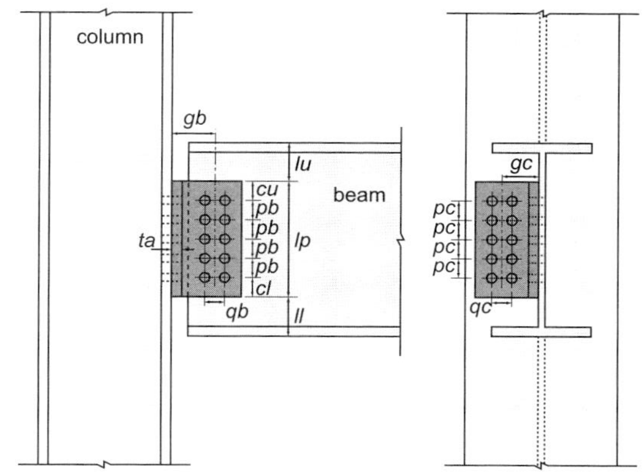

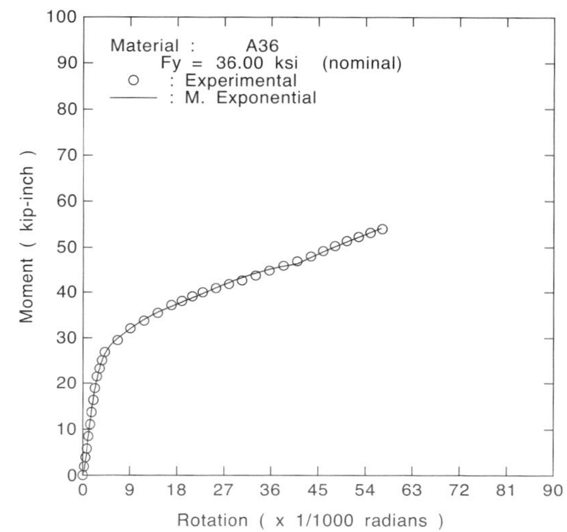

A.1 – 102

Moment-rotation prediction equations (R : X 1/1000 radians)

```
Modified exponential model  :
     Bm = Sum ( Ai X ( 1 - exp( - R/( 2*i*C ) ) ) )  +      Sum ( RKj X ( R - Rj0 ) )  +  Bm0
       C =    0.54334167E+00              Bmo=   0.00000000E+00          Nexp=  6        Nliner=  1
      AI =  -0.84633403E+02   0.77620941E+03  -0.30821072E+04   0.63921378E+04  -0.63745251E+04   0.24209010E+04
     Rj0 =          41.0000
     RKj =    0.39979525E+00
```

No	Rotation (radians) X 1/1000	Moment (kip-inch) Expri.	Moment (kip-inch) M.Expo.	Connection stiffness (kip-inch) X 1000 Expri.	Connection stiffness (kip-inch) X 1000 M.Expo.
1	0.00	0.00	0.00	0.7304E+01	0.2499E+01
3	0.53	3.90	3.20	0.7305E+01	0.8389E+01
5	1.12	8.46	8.61	0.8249E+01	0.9563E+01
7	1.75	13.68	14.34	0.8248E+01	0.8305E+01
9	2.38	18.90	18.99	0.8249E+01	0.6364E+01
11	3.20	23.32	23.23	0.3616E+01	0.4128E+01
13	4.20	26.94	26.38	0.3180E+01	0.2342E+01
15	9.10	32.06	32.06	0.8578E+00	0.8337E+00
17	14.30	35.49	35.63	0.6585E+00	0.5388E+00
19	18.90	38.15	37.86	0.4740E+00	0.4640E+00
21	22.90	40.04	39.75	0.4155E+00	0.4841E+00
23	27.90	41.76	42.14	0.3424E+00	0.4548E+00
25	33.04	43.69	44.23	0.4100E+00	0.3543E+00
27	38.32	45.86	45.79	0.4100E+00	0.2364E+00
29	43.60	48.02	47.81	0.4600E+00	0.5420E+00
31	48.13	50.30	50.14	0.5029E+00	0.4865E+00
33	52.67	52.30	52.25	0.3776E+00	0.4506E+00
35	57.20	54.01	54.24	0.3776E+00	0.4287E+00

A.1 – 103

I - 52

Connection type : Single web-angle connections
Mode : Bolted-to-beam and welded-to-column

Tested by : R.M.Richard et al. (1982) U.S.A.
Test Id. : RGKL3

Column : --
Beam : --
Angle : --

Fasteners: A325- -3/4"D
15/16" Oversize holes
Material : A36
Fy = -- ksi
Fu = -- ksi

Major parameters

lp = 9.0000" lu = -- ' ll = -- " ta = 0.2500"
gb = -- " cu = -- ' cl = -- " pb = 3.0000"
nb = 1 X 3

Remark 1) Single plate connection.
2)

No	Moment (k-in)	Rotation (radians) X 1/1000	No	Moment (k-in)	Rotation (radians) X 1/1000
1	0.00	0.00	26	119.86	27.95
2	6.54	0.20	27	122.47	30.40
3	12.64	0.30	28	124.22	33.07
4	18.74	0.40	29	125.96	35.73
5	24.41	0.55	30	127.70	38.40
6	30.07	0.70	31	128.86	41.00
7	35.74	1.00	32	130.02	43.60
8	41.40	1.30	33	131.18	46.20
9	47.50	1.60	34	132.06	48.73
10	53.60	1.90	35	132.93	51.25
11	59.71	2.65	36	133.80	53.77
12	65.81	3.40	37	134.68	56.30
13	71.48	4.70	38	135.11	58.25
14	77.15	6.00	39	135.54	60.20
15	82.37	7.60			
16	85.86	8.65			
17	89.35	9.70			
18	95.45	12.30			
19	101.54	14.70			
20	104.60	16.50			
21	107.65	18.30			
22	110.70	19.85			
23	113.75	21.40			
24	115.49	23.45			
25	117.24	25.50			

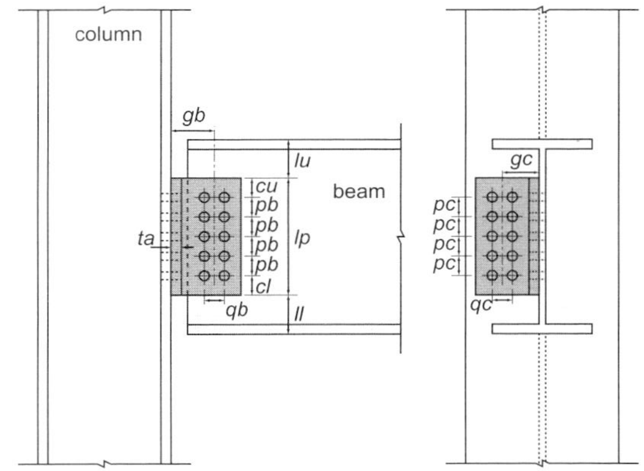

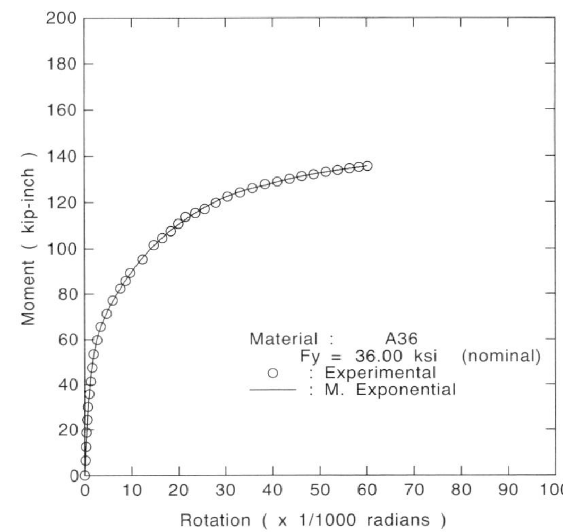

A.1 – 104

```
Moment-rotation prediction equations     ( R : X 1/1000 radians )

Modified exponential model  :
     Bm = Sum ( Ai X ( 1 - exp( - R/( 2*i*C ) ) ) )  +      Sum ( RKj X ( R - Rj0 ) )  +  Bm0
       C =   0.55166667E+00             Bmo=   0.00000000E+00          Nexp=  6        Nliner=  1
      AI =   0.23417194E+02   0.46235540E+03  -0.25057736E+04   0.61510335E+04  -0.70000263E+04   0.29996436E+04
     Rj0 =          38.4000
     RKj =   0.23802390E+00
```

No	Rotation (radians) X 1/1000	Moment (kip-inch) Expri.	Moment (kip-inch) M.Expo.	Connection stiffness (kip-inch) X 1000 Expri.	Connection stiffness (kip-inch) X 1000 M.Expo.
1	0.00	0.00	0.00	0.3270E+02	0.5169E+02
3	0.30	12.64	13.85	0.6102E+02	0.4109E+02
5	0.55	24.41	23.21	0.3776E+02	0.3400E+02
7	1.00	35.74	36.21	0.1888E+02	0.2434E+02
9	1.60	47.50	48.08	0.2034E+02	0.1591E+02
11	2.65	59.71	60.22	0.8136E+01	0.8337E+01
13	4.70	71.48	71.80	0.4362E+01	0.4190E+01
15	7.60	82.37	82.44	0.3301E+01	0.3352E+01
17	9.70	89.35	89.07	0.3043E+01	0.2945E+01
19	14.70	101.54	101.35	0.2058E+01	0.2035E+01
21	18.30	107.65	107.93	0.1842E+01	0.1647E+01
23	21.40	113.75	112.65	0.1486E+01	0.1408E+01
25	25.50	117.24	117.84	0.9501E+00	0.1125E+01
27	30.40	122.47	122.55	0.8696E+00	0.8032E+00
29	35.73	125.96	126.01	0.6540E+00	0.5086E+00
31	41.00	128.86	128.73	0.4462E+00	0.5381E+00
33	46.20	131.18	131.15	0.3952E+00	0.4063E+00
35	51.25	132.93	133.00	0.3457E+00	0.3301E+00
37	56.30	134.68	134.54	0.2757E+00	0.2869E+00
39	60.20	135.54	135.62	0.2215E+00	0.2675E+00

I - 53

Connection type : Single web-angle connections
Mode : Bolted-to-beam and welded-to-column

Tested by : R.M.Richard et al. (1982) U.S.A.
Test Id. : RGKL5

Column : --
Beam : --
Angle : --

Fasteners: A325- -3/4"D
15/16" Oversize holes
Material : A36
Fy = -- ksi
Fu = -- ksi

Major parameters

lp = 15.0000" lu = -- " ll = -- " ta = 0.2500"
gb = -- " cu = -- " cl = -- " pb = 3.0000"
nb = 1 X 5

Remark 1) Single plate connection.
2)

No	Moment (k-in)	Rotation (radians) X 1/1000	No	Moment (k-in)	Rotation (radians) X 1/1000
1	0.00	0.00	26	514.70	18.10
2	23.19	0.10	27	524.82	19.60
3	56.39	0.20	28	536.39	21.90
4	82.40	0.40	29	547.96	24.20
5	108.43	0.90	30	552.29	26.40
6	125.78	0.90	31	556.63	28.60
7	147.47	1.10	32	559.52	30.20
8	169.15	1.30	33	562.40	31.80
9	190.84	1.55	34	568.19	34.20
10	212.53	1.80	35	573.97	36.60
11	238.55	2.10	36	578.31	38.90
12	264.58	2.40	37	582.65	41.20
13	287.71	2.85	38	585.54	44.13
14	310.85	3.30	39	588.43	47.07
15	328.19	3.70	40	591.32	50.00
16	345.54	4.10	41	594.22	52.40
17	367.23	4.80	42	597.11	54.80
18	388.92	5.50	43	600.00	57.20
19	409.16	6.65	44	602.89	59.60
20	429.40	7.80			
21	455.42	10.50			
22	466.99	12.00			
23	478.56	13.50			
24	491.57	15.05			
25	504.58	16.60			

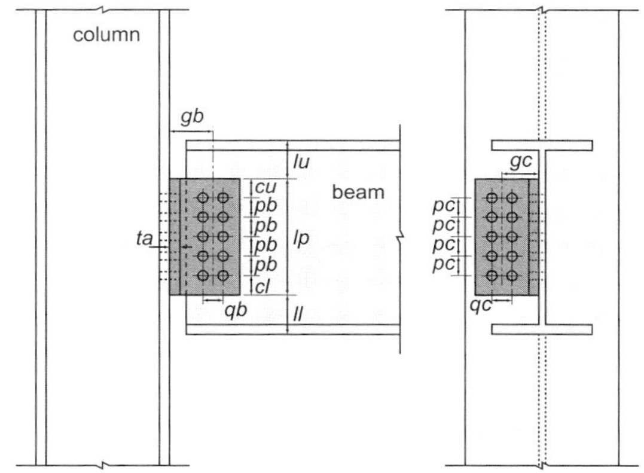

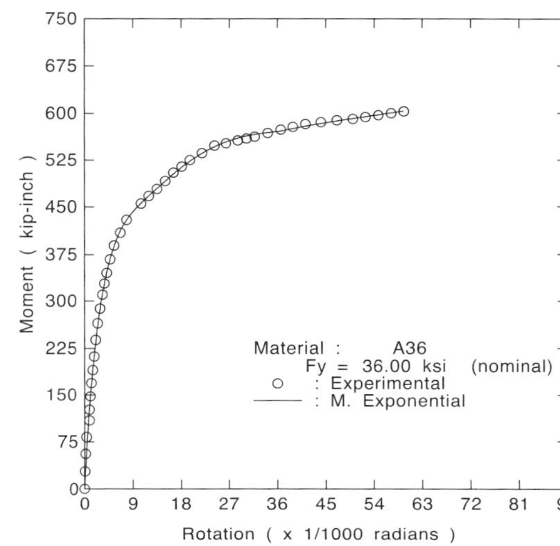

Moment-rotation prediction equations (R : X 1/1000 radians)

```
Modified exponential model  :
     Bm = Sum ( Ai X ( 1 - exp( - R/( 2*i*C ) ) ) )  +     Sum ( RKj X ( R - Rj0 ) )  +  Bm0
      C =   0.52166667E+00            Bmo=   0.00000000E+00         Nexp=  6        Nliner=  1
     AI =   0.23648353E+03  -0.23628038E+04   0.94110640E+04  -0.12939983E+05   0.58638075E+04   0.36704670E+03
    Rj0 =          36.4000
    RKj =   0.12056297E+01
```

No	Rotation (radians) X 1/1000	Moment (kip-inch) Expri.	Moment (kip-inch) M.Expo.	Connection stiffness (kip-inch) X 1000 Expri.	Connection stiffness (kip-inch) X 1000 M.Expo.
1	0.00	0.00	0.00	0.2819E+03	0.1831E+03
3	0.20	56.39	34.43	0.2313E+03	0.1619E+03
5	0.90	108.43	128.81	0.1732E+05	0.1130E+03
7	1.10	147.47	150.44	0.1084E+03	0.1037E+03
9	1.55	190.84	193.19	0.8676E+02	0.8719E+02
11	2.10	238.55	236.87	0.8674E+02	0.7236E+02
13	2.85	287.71	285.21	0.5141E+02	0.5727E+02
15	3.70	328.19	328.06	0.4337E+02	0.4408E+02
17	4.80	367.23	369.02	0.3099E+02	0.3106E+02
19	6.65	409.16	412.26	0.1760E+02	0.1718E+02
21	10.50	455.42	455.62	0.8401E+01	0.8289E+01
23	13.50	478.56	479.33	0.8047E+01	0.7741E+01
25	16.60	504.58	502.77	0.7557E+01	0.7265E+01
27	19.60	524.82	523.02	0.6070E+01	0.6147E+01
29	24.20	547.96	546.36	0.3467E+01	0.4004E+01
31	28.60	556.63	560.08	0.1875E+01	0.2330E+01
33	31.80	562.40	566.11	0.2046E+01	0.1491E+01
35	36.60	573.97	571.46	0.2142E+01	0.1927E+01
37	41.20	582.65	579.36	0.1490E+01	0.1550E+01
39	47.07	588.43	587.73	0.9859E+00	0.1335E+01
41	52.40	594.22	594.61	0.1205E+01	0.1257E+01
43	57.20	600.00	600.57	0.1205E+01	0.1228E+01

I - 54

Connection type : Single web-angle connections
Mode : Bolted-to-beam and welded-to-column

Tested by : R.M.Richard et al. (1982) U.S.A.
Test Id. : RGKL7

Column : --
Beam : --
Angle : --

Fasteners: A325- -7/8"D
1 1/16" Oversize holes
Material : A36
Fy = -- ksi
Fu = -- ksi

Major parameters

lp = 21.5000" lu = -- " ll = -- " ta = 0.3750"
gb = -- " cu = -- " cl = -- " pb = 3.0000"
nb = 1 X 7

Remark 1) Single plate connection.
2)

No	Moment (k-in)	Rotation (radians) X 1/1000	No	Moment (k-in)	Rotation (radians) X 1/1000
1	0.00	0.00	26	1131.11	22.00
2	53.58	0.23	27	1151.57	23.75
3	107.17	0.47	28	1172.03	25.50
4	160.75	0.70	29	1183.72	27.60
5	213.36	1.07	30	1195.40	29.70
6	265.97	1.43	31	1204.18	31.45
7	318.58	1.80	32	1212.95	33.20
8	375.08	2.13	33	1221.71	34.80
9	431.59	2.47	34	1230.48	36.40
10	488.10	2.80	35	1242.17	39.03
11	538.76	3.23	36	1253.86	41.67
12	589.42	3.67	37	1265.56	44.30
13	640.08	4.10	38	1271.40	47.07
14	683.92	4.55	39	1277.24	49.83
15	727.76	5.00	40	1283.09	52.60
16	768.68	5.55	41	1288.94	55.20
17	809.60	6.10	42	1294.78	57.80
18	853.45	7.30	43	1300.63	60.40
19	897.29	8.50			
20	932.36	9.80			
21	967.43	11.10			
22	1014.19	13.50			
23	1043.42	15.70			
24	1059.73	17.35			
25	1096.03	19.00			

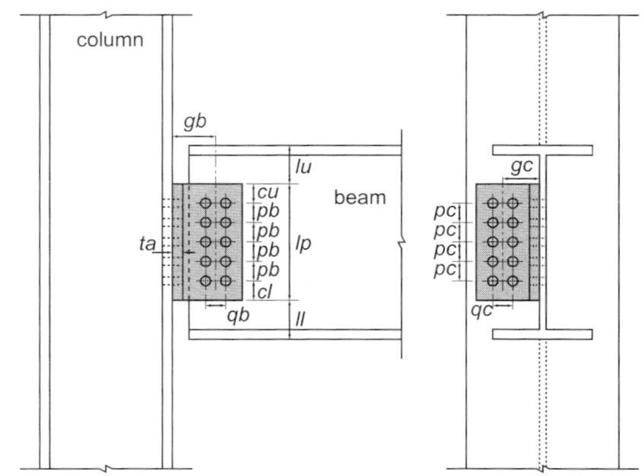

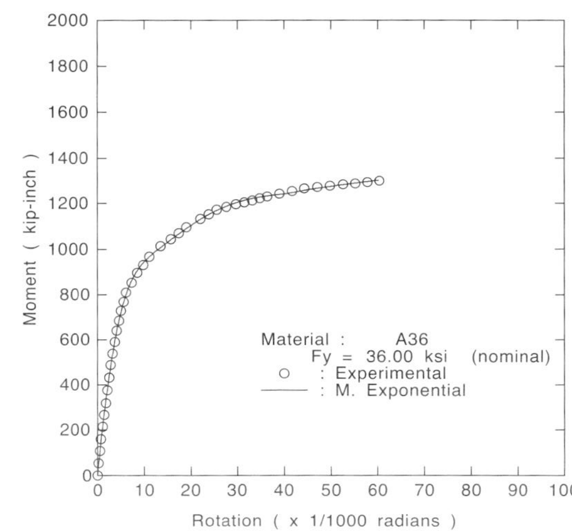

A.1 – 108

```
Moment-rotation prediction equations     ( R : X 1/1000 radians )

Modified exponential model  :
     Bm = Sum ( Ai X ( 1 - exp( - R/( 2*i*C ) ) ) )  +      Sum ( RKj X ( R - Rj0 ) )  +  Bm0
       C =   0.56165833E+00          Bmo=   0.00000000E+00          Nexp=  6        Nliner=  1
      AI =   0.43107636E+03  -0.58166408E+04   0.19301242E+05  -0.19492206E+05   0.10343154E+04   0.57968144E+04
     Rj0 =         40.3500
     RKj =   0.24794263E+01
```

No	Rotation (radians) X 1/1000	Moment (kip-inch) Expri.	Moment (kip-inch) M.Expo.	Connection stiffness (kip-inch) X 1000 Expri.	Connection stiffness (kip-inch) X 1000 M.Expo.
1	0.00	0.00	0.00	0.2297E+03	0.2283E+03
3	0.47	107.17	98.95	0.2296E+03	0.1985E+03
5	1.07	213.36	210.91	0.1435E+03	0.1766E+03
7	1.80	318.58	333.59	0.1571E+03	0.1586E+03
9	2.47	431.59	434.34	0.1695E+03	0.1435E+03
11	3.23	538.76	537.46	0.1169E+03	0.1253E+03
13	4.10	640.08	636.89	0.1074E+03	0.1042E+03
15	5.00	727.76	721.10	0.8706E+02	0.8332E+02
17	6.10	809.60	800.30	0.6250E+02	0.6148E+02
19	8.50	897.29	906.80	0.3195E+02	0.3108E+02
21	11.10	967.43	967.94	0.2434E+02	0.1847E+02
23	15.70	1043.42	1041.59	0.1480E+02	0.1510E+02
25	19.00	1096.03	1090.22	0.1443E+02	0.1418E+02
27	23.75	1151.57	1150.66	0.1169E+02	0.1099E+02
29	27.60	1183.72	1187.03	0.5566E+01	0.7929E+01
31	31.45	1204.18	1212.32	0.5013E+01	0.5319E+01
33	34.80	1221.71	1227.13	0.5479E+01	0.3606E+01
35	39.03	1242.17	1239.02	0.4440E+01	0.2124E+01
37	44.30	1265.56	1256.88	0.3305E+01	0.3535E+01
39	49.83	1277.24	1274.69	0.2112E+01	0.2970E+01
41	55.20	1288.94	1289.85	0.2249E+01	0.2708E+01
43	60.40	1300.63	1303.58	0.2249E+01	0.2587E+01

II - 1

Connection type : Double web-angle connections
Mode : All riveted

Tested by : J.C.Rathbun (1936) U.S.A.
Test Id. : A-1

Column : --
Beam : W6X12.5
Angle : 6 X 4 X 3/8

Fasteners: 10.9-X-7/8"D
15/16" Oversize holes
Material : --
Fy = 44.00 ksi
Fu = 64.80 ksi

Major parameters

lp = 2.5000"	lu = 1.7500"	ll = 1.7500"	ta = 0.3750"
gb = 3.5000"	gc = 2.6250"	cu = 1.2500"	cl = 1.2500"
qb = 2.5000"	nb = 2 X 1	nc = 1 X 1	

Remark 1) Connection angle riveted to 1/2" mounting plate.
2)

No	Moment (k-in)	Rotation (radians) X 1/1000	No	Moment (k-in)	Rotation (radians) X 1/1000
1	0.00	0.00	26	17.08	19.17
2	0.61	0.60	27	17.55	19.94
3	1.23	1.20	28	18.05	20.85
4	1.86	1.82	29	18.55	21.76
5	2.50	2.44	30	19.26	23.10
6	3.23	3.12	31	19.52	23.98
7	3.96	3.81	32	19.78	24.85
8	4.61	4.41	33	20.01	25.78
9	5.26	5.01	34	20.24	26.71
10	6.25	6.08	35	20.55	28.15
11	7.14	7.02	36	20.74	29.00
12	7.90	7.79	37	20.93	29.84
13	8.65	8.55			
14	9.35	9.31			
15	10.04	10.06			
16	10.80	10.86			
17	11.72	11.80			
18	12.45	12.67			
19	13.18	13.54			
20	13.71	14.21			
21	14.24	14.88			
22	14.90	15.85			
23	15.57	16.82			
24	16.09	17.61			
25	16.61	18.40			

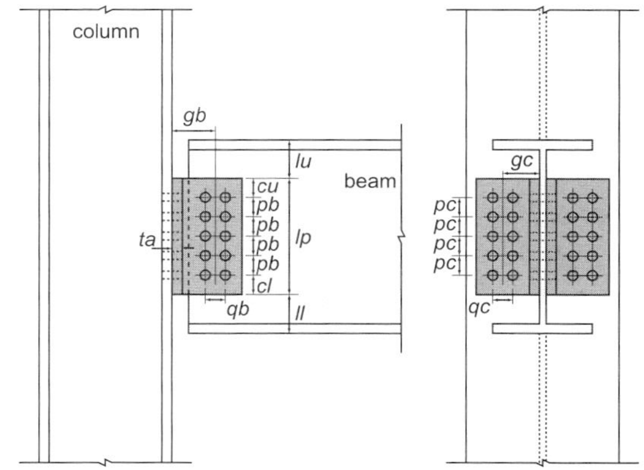

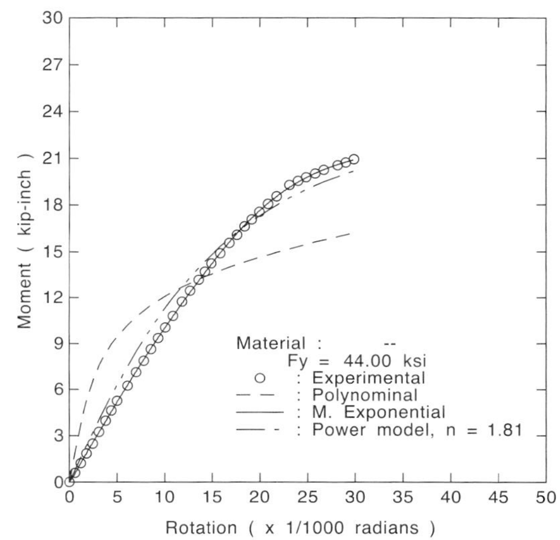

A.2 – 2

```
Moment-rotation prediction equations     ( R : X 1/1000 radians )

Frye and Morris polynominal model  :    R = Sum ( Ai X ( K*Bm )**Pi X 10**Qi )
     xd =  2.500000"   g =  5.500000"   t =  0.375000"
     A1 =  3.660000   A2 =  1.150000   A3 =  4.570000    K =  0.845282
     P1 =        1     P2 =        3    P3 =        5    Q1 =       -1    Q2 =       -3    Q3 =       -5

Modified exponential model  :
     Bm = Sum ( Ai X ( 1 - exp( - R/( 2*i*C ) ) ) )  +      Sum ( RKj X ( R - Rj0 ) )  +  Bm0
       C =   0.39904000E+00              Bmo=   0.00000000E+00          Nexp=  6        Nliner=  0
      AI =  -0.56289577E+01   0.10550336E+03  -0.69927895E+03   0.19892423E+04  -0.24849135E+04   0.11169381E+04

Power model  :    Bm = ( Rki X R ) / ( 1 + ( R/R0 )**rn )**( 1/rn )
      rn = 1.807      rki =   0.13188761E+01          rmu =   0.24538170E+02
```

No	Rotation (radians) X 1/1000	Moment (kip-inch) Expri.	Moment (kip-inch) Poly.	Moment (kip-inch) M.Expo.	Moment (kip-inch) P.Model.	Connection stiffness (kip-inch) X 1000 Expri.	Connection stiffness (kip-inch) X 1000 Poly.	Connection stiffness (kip-inch) X 1000 M.Expo.	Connection stiffness (kip-inch) X 1000 P.Model.
1	0.00	0.00	0.00	0.00	0.00	0.1019E+01	0.3232E+01	0.6440E+00	0.1319E+01
3	1.20	1.23	3.73	1.23	1.58	0.1024E+01	0.2799E+01	0.1109E+01	0.1304E+01
5	2.44	2.50	6.52	2.54	3.17	0.1047E+01	0.1736E+01	0.1017E+01	0.1268E+01
7	3.81	3.96	8.37	3.93	4.87	0.1073E+01	0.1064E+01	0.1027E+01	0.1210E+01
9	5.01	5.26	9.46	5.17	6.29	0.1023E+01	0.7775E+00	0.1034E+01	0.1148E+01
11	7.02	7.14	10.75	7.21	8.48	0.9703E+00	0.5358E+00	0.9863E+00	0.1031E+01
13	8.55	8.65	11.49	8.69	9.99	0.9527E+00	0.4344E+00	0.9390E+00	0.9378E+00
15	10.06	10.04	12.09	10.07	11.34	0.9321E+00	0.3676E+00	0.9037E+00	0.8478E+00
17	11.80	11.72	12.68	11.62	12.72	0.9088E+00	0.3133E+00	0.8700E+00	0.7494E+00
19	13.54	13.18	13.19	13.10	13.95	0.8136E+00	0.2737E+00	0.8293E+00	0.6591E+00
21	14.88	14.24	13.54	14.18	14.79	0.7470E+00	0.2498E+00	0.7872E+00	0.5959E+00
23	16.82	15.57	14.00	15.64	15.87	0.6690E+00	0.2222E+00	0.7081E+00	0.5143E+00
25	18.40	16.61	14.33	16.70	16.63	0.6361E+00	0.2040E+00	0.6325E+00	0.4564E+00
27	19.94	17.55	14.64	17.61	17.29	0.5832E+00	0.1893E+00	0.5538E+00	0.4064E+00
29	21.76	18.55	14.97	18.53	17.98	0.5405E+00	0.1747E+00	0.4619E+00	0.3552E+00
31	23.98	19.52	15.34	19.44	18.71	0.2999E+00	0.1598E+00	0.3580E+00	0.3023E+00
33	25.78	20.01	15.62	20.02	19.22	0.2456E+00	0.1496E+00	0.2845E+00	0.2660E+00
35	28.15	20.55	15.96	20.59	19.81	0.2234E+00	0.1383E+00	0.2048E+00	0.2259E+00
37	29.84	20.93	16.19	20.90	20.17	0.2279E+00	0.1312E+00	0.1597E+00	0.2018E+00

A.2 – 3

II - 2

Connection type : Double web-angle connections
Mode : All riveted

Tested by : J.C.Rathbun (1936) U.S.A.
Test Id. : A-2

Column : --
Beam : W8X18.4
Angle : 6 X 4 X 3/8

Fasteners: 10.9-X-7/8"D
15/16" Oversize holes
Material : --
Fy = 39.84 ksi
Fu = 62.32 ksi

Major parameters

lp = 6.0000"	lu = 1.0000"	ll = 1.0000"	ta = 0.3750"
gb = 3.5000"	gc = 2.6250"	cu = 1.5000"	cl = 1.5000"
pb = 3.0000"	pc = 3.0000"	qb = 2.5000"	
nb = 2 X 2	nc = 1 X 2		

Remark 1) Connection angle riveted to 1/2" mouning plate.
2)

No	Moment (k-in)	Rotation (radians) X 1/1000
1	0.00	0.00
2	4.84	0.16
3	9.45	0.39
4	12.29	0.50
5	15.14	0.62
6	19.50	0.81
7	22.16	1.01
8	24.83	1.20
9	29.67	1.60
10	32.34	1.88
11	35.00	2.16
12	39.49	2.55
13	44.45	3.07
14	49.66	3.60
15	52.27	3.94
16	54.88	4.29
17	59.36	4.91
18	62.15	5.41
19	64.94	5.90
20	69.67	6.86
21	74.64	7.98
22	77.01	8.69
23	79.38	9.40
24	81.44	10.02
25	83.51	10.65

No	Moment (k-in)	Rotation (radians) X 1/1000
26	86.91	11.84
27	89.46	12.89
28	91.41	13.75
29	93.35	14.61
30	95.18	15.50
31	97.00	16.39
32	98.34	17.45
33	99.69	18.51
34	100.66	19.33
35	101.64	20.16
36	102.38	21.25
37	103.11	22.34
38	103.67	23.28
39	104.22	24.22

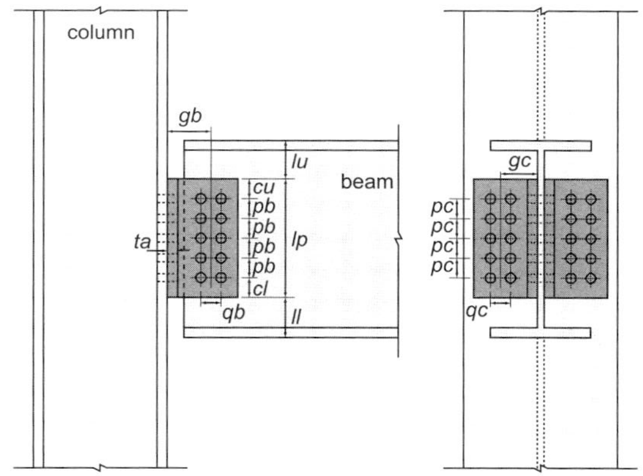

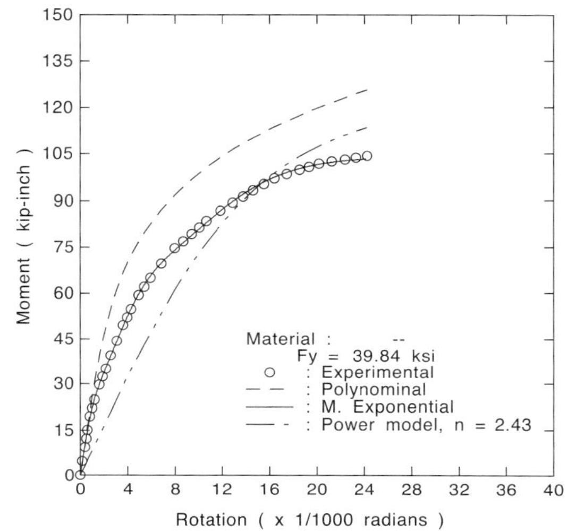

A.2 – 4

```
Moment-rotation prediction equations      ( R : X 1/1000 radians )

Frye and Morris polynominal model  :    R = Sum ( Ai X ( K*Bm )**Pi X 10**Qi )
     xd =  6.000000"   g =  5.500000"   t =  0.375000"
     A1 =  3.660000   A2 =  1.150000   A3 =  4.570000    K =  0.103394
     P1 =        1    P2 =        3    P3 =        5    Q1 =       -1     Q2 =       -3     Q3 =       -5

Modified exponential model  :
     Bm = Sum ( Ai X ( 1 - exp( - R/( 2*i*C ) ) ) )  +      Sum ( RKj X ( R - Rj0 ) )  +  Bm0
       C =   0.24226667E+00              Bmo=   0.00000000E+00          Nexp=  6        Nliner=  0
      AI =  -0.80265807E+02   0.12733183E+04  -0.64116498E+04   0.14480937E+05  -0.15055916E+05   0.58975965E+04

Power model  :     Bm = ( Rki X R ) / ( 1 + ( R/R0 )**rn )**( 1/rn )
      rn = 2.434       rki =    0.82298237E+01          rmu =   0.12797682E+03
```

No	Rotation (radians) X 1/1000	Moment (kip-inch) Expri.	Poly.	M.Expo.	P.Model.	Connection stiffness (kip-inch) X 1000 Expri.	Poly.	M.Expo.	P.Model.
1	0.00	0.00	0.00	0.00	0.00	0.2997E+02	0.2643E+02	0.2303E+02	0.8230E+01
3	0.39	9.45	10.26	9.87	3.21	0.2345E+02	0.2613E+02	0.2478E+02	0.8228E+01
5	0.62	15.14	16.14	15.12	5.07	0.2405E+02	0.2563E+02	0.2134E+02	0.8225E+01
7	1.01	22.16	25.89	22.25	8.29	0.1357E+02	0.2403E+02	0.1538E+02	0.8215E+01
9	1.60	29.67	38.96	29.68	13.13	0.1071E+02	0.2006E+02	0.1068E+02	0.8184E+01
11	2.16	35.00	48.99	35.28	17.68	0.1032E+02	0.1599E+02	0.9710E+01	0.8136E+01
13	3.07	44.45	61.25	44.18	25.10	0.9690E+01	0.1109E+02	0.9619E+01	0.8010E+01
15	3.94	52.27	69.60	52.19	32.00	0.7549E+01	0.8355E+01	0.8614E+01	0.7835E+01
17	4.91	59.36	76.74	59.66	39.48	0.6323E+01	0.6497E+01	0.6765E+01	0.7574E+01
19	5.90	64.94	82.52	65.48	46.80	0.5411E+01	0.5292E+01	0.5100E+01	0.7244E+01
21	7.98	74.64	91.80	74.06	60.98	0.3765E+01	0.3820E+01	0.3559E+01	0.6385E+01
23	9.40	79.38	96.77	78.93	69.57	0.3315E+01	0.3222E+01	0.3358E+01	0.5724E+01
25	10.65	83.51	100.56	83.07	76.37	0.3145E+01	0.2837E+01	0.3232E+01	0.5130E+01
27	12.89	89.46	106.34	89.83	86.73	0.2334E+01	0.2347E+01	0.2730E+01	0.4117E+01
29	14.61	93.35	110.12	94.05	93.19	0.2159E+01	0.2078E+01	0.2172E+01	0.3431E+01
31	16.39	97.00	113.63	97.39	98.75	0.1692E+01	0.1862E+01	0.1588E+01	0.2819E+01
33	18.51	99.69	117.34	100.11	104.05	0.1221E+01	0.1661E+01	0.1016E+01	0.2225E+01
35	20.16	101.64	119.98	101.51	107.41	0.9644E+00	0.1533E+01	0.6863E+00	0.1850E+01
37	22.34	103.11	123.16	102.66	110.99	0.6292E+00	0.1395E+01	0.3911E+00	0.1455E+01
39	24.22	104.22	125.70	103.23	113.48	0.5884E+00	0.1295E+01	0.2332E+00	0.1189E+01

A.2 – 5

II - 3

Connection type : Double web-angle connections
Mode : All riveted

Tested by : J.C.Rathbun (1936) U.S.A.
Test Id. : A-3

Column : --
Beam : W8X18.4
Angle : 6 X 6 X 3/8

Fasteners: 10.9-X-7/8"D
15/16" Oversize holes
Material : --
Fy = 39.84 ksi
Fu = 62.32 ksi

Major parameters

lp = 6.0000"	lu = 1.0000'	ll = 1.0000"	ta = 0.3750"
gb = 3.5000"	gc = 3.6250'	cu = 1.5000"	cl = 1.5000"
pb = 3.0000"	pc = 3.0000'	qb = 2.5000"	qc = 2.5000"
nb = 2 X 2	nc = 2 X 2		

Remark 1) Connection angle riveted to 1/2" mouning plate.
2)

No	Moment (k-in)	Rotation (radians) X 1/1000	No	Moment (k-in)	Rotation (radians) X 1/1000
1	0.00	0.00	26	111.43	17.49
2	4.86	0.18	27	113.02	18.69
3	9.72	0.35	28	114.61	19.88
4	14.76	0.63	29	116.33	21.25
5	19.80	0.90	30	118.05	22.62
6	24.78	1.25	31	119.89	23.77
7	29.76	1.59	32	121.72	24.92
8	34.68	1.99	33	123.38	26.29
9	39.60	2.40	34	125.04	27.66
10	44.65	2.91			
11	49.69	3.41			
12	54.68	4.00			
13	59.66	4.59			
14	64.65	5.36			
15	69.64	6.13			
16	74.68	6.92			
17	79.73	7.71			
18	84.78	8.63			
19	89.83	9.55			
20	94.95	10.83			
21	100.06	12.11			
22	102.81	13.20			
23	105.56	14.28			
24	107.69	15.29			
25	109.83	16.29			

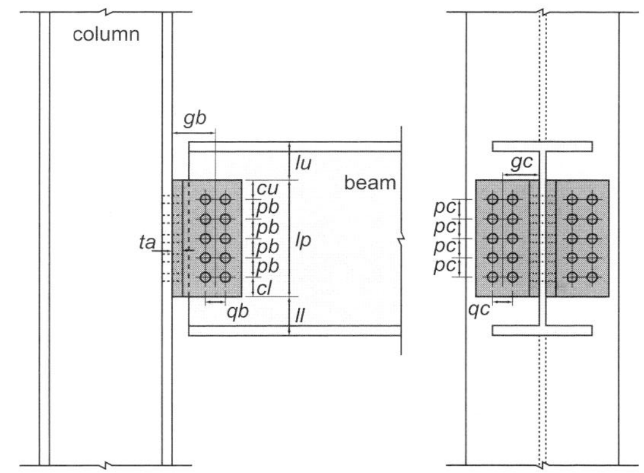

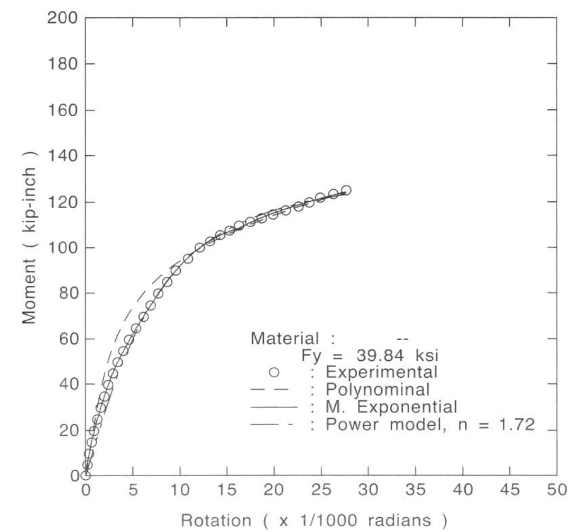

A.2 – 6

```
Moment-rotation prediction equations     ( R : X 1/1000 radians )

Frye and Morris polynominal model  :    R = Sum ( Ai X ( K*Bm )**Pi X 10**Qi )
     xd =  6.000000"   g =  7.500000"   t =  0.375000"
     A1 =  3.660000   A2 =  1.150000   A3 =  4.570000    K =  0.108318
     P1 =        1    P2 =        3    P3 =        5    Q1 =       -1    Q2 =       -3    Q3 =       -5

Modified exponential model  :
     Bm = Sum ( Ai X ( 1 - exp( - R/( 2*i*C ) ) ) )  +     Sum ( RKj X ( R - Rj0 ) )  +  Bm0
       C =   0.27465500E+00            Bmo=   0.00000000E+00          Nexp=  6        Nliner=  1
      AI =   0.44964010E+02  -0.45051047E+03   0.15995959E+04  -0.18351364E+04  -0.35014526E+02   0.78929535E+03
     Rj0 =          16.2900
     RKj =   0.10051603E+01

Power model  :    Bm = ( Rki X R ) / ( 1 + ( R/R0 )**rn )**( 1/rn )
      rn = 1.722      rki =   0.14124903E+02          rmu =   0.13428760E+03
```

No	Rotation (radians) X 1/1000	Moment (kip-inch) Expri.	Moment Poly.	Moment M.Expo.	Moment P.Model.	Connection stiffness (kip-inch) X 1000 Expri.	Stiffness Poly.	Stiffness M.Expo.	Stiffness P.Model.
1	0.00	0.00	0.00	0.00	0.00	0.2751E+02	0.2522E+02	0.3399E+02	0.1412E+02
3	0.35	9.72	8.88	9.54	4.98	0.2391E+02	0.2499E+02	0.2187E+02	0.1405E+02
5	0.90	19.80	22.30	19.56	12.64	0.1665E+02	0.2344E+02	0.1589E+02	0.1375E+02
7	1.59	29.76	36.98	29.54	21.84	0.1344E+02	0.1923E+02	0.1357E+02	0.1316E+02
9	2.40	39.60	50.34	39.66	32.21	0.1113E+02	0.1377E+02	0.1126E+02	0.1226E+02
11	3.41	49.69	61.86	49.75	44.00	0.9269E+01	0.9408E+01	0.8822E+01	0.1100E+02
13	4.59	59.66	71.17	59.12	56.07	0.7601E+01	0.6699E+01	0.7316E+01	0.9495E+01
15	6.13	69.64	79.91	69.75	69.27	0.6444E+01	0.4841E+01	0.6590E+01	0.7682E+01
17	7.71	79.73	86.62	79.67	80.09	0.5991E+01	0.3782E+01	0.5972E+01	0.6124E+01
19	9.55	89.83	92.83	89.70	89.97	0.4867E+01	0.3027E+01	0.4883E+01	0.4697E+01
21	12.11	100.06	99.69	99.99	100.08	0.3199E+01	0.2383E+01	0.3164E+01	0.3284E+01
23	14.28	105.56	104.46	105.51	106.28	0.2321E+01	0.2028E+01	0.1983E+01	0.2468E+01
25	16.29	109.83	108.28	108.68	110.66	0.1765E+01	0.1788E+01	0.2226E+01	0.1925E+01
27	18.69	113.02	112.29	113.27	114.68	0.1332E+01	0.1571E+01	0.1660E+01	0.1461E+01
29	21.25	116.33	116.08	117.06	117.95	0.1256E+01	0.1394E+01	0.1329E+01	0.1111E+01
31	23.77	119.89	119.42	120.17	120.43	0.1593E+01	0.1258E+01	0.1163E+01	0.8665E+00
33	26.29	123.38	122.45	122.99	122.38	0.1212E+01	0.1148E+01	0.1081E+01	0.6879E+00

A.2 – 7

II - 4

Connection type : Double web-angle connections
Mode : All riveted

Tested by : J.C.Rathbun (1936) U.S.A.
Test Id. : A-4

Column : --
Beam : W12X31.8
Angle : 4 X 3 1/2 X 3/8

Fasteners: 10.9-X-7/8"D
15/16" Oversize holes
Material : --
Fy = 47.12 ksi
Fu = 62.98 ksi

Major parameters

lp = 9.0000" lu = 1.5000" ll = 1.5000" ta = 0.3750"
gb = 2.2500" gc = 2.5625" cu = 1.5000" cl = 1.5000"
pb = 3.0000" pc = 3.0000" nb = 1 X 3 nc = 1 X 3

Remark 1) Connection angle riveted to 1/2" mouning plate.
2)

No	Moment (k-in)	Rotation (radians) X 1/1000	No	Moment (k-in)	Rotation (radians) X 1/1000
1	0.00	0.00	26	172.04	15.71
2	9.87	0.32	27	174.96	16.80
3	19.74	0.64	28	177.89	17.90
4	29.13	1.08	29	180.82	18.99
5	38.53	1.52	30	183.75	20.06
6	45.70	1.84	31	186.68	21.13
7	52.87	2.16	32	189.61	22.20
8	60.04	2.48	33	193.03	23.47
9	69.98	3.03	34	196.46	24.74
10	79.93	3.57	35	199.89	26.01
11	86.61	4.05			
12	93.30	4.54			
13	99.98	5.02			
14	106.81	5.53			
15	113.64	6.05			
16	120.47	6.57			
17	125.97	7.13			
18	131.46	7.69			
19	139.72	8.74			
20	146.03	9.85			
21	152.34	10.95			
22	155.96	11.74			
23	159.59	12.52			
24	163.74	13.58			
25	167.89	14.65			

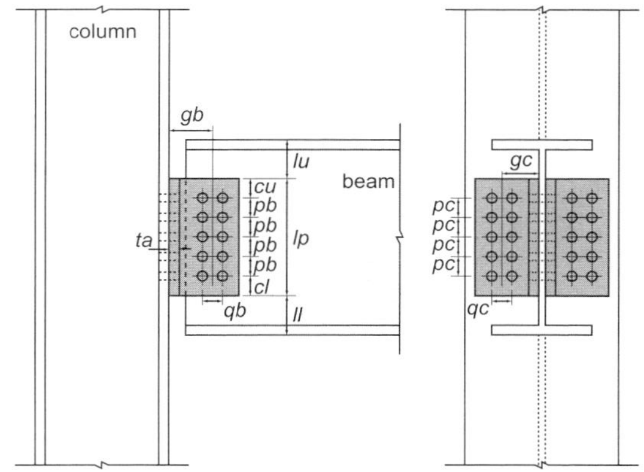

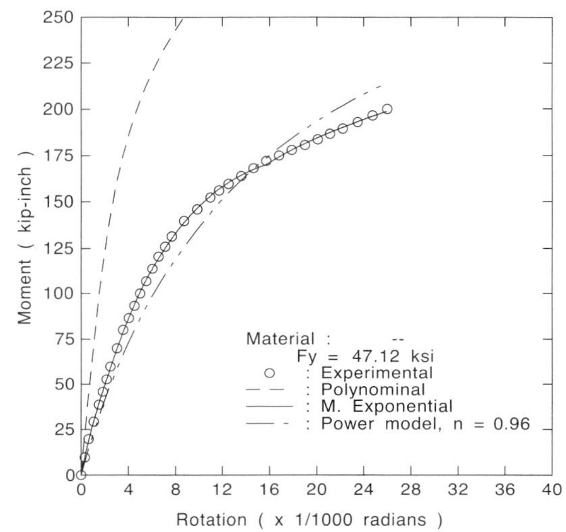

A.2 – 8

```
Moment-rotation prediction equations      ( R : X 1/1000 radians )

Frye and Morris polynominal model  :    R = Sum ( Ai X ( K*Bm )**Pi X 10**Qi )
     xd =  9.000000"    g =  5.500000"   t =  0.375000"
     A1 =  3.660000    A2 =  1.150000   A3 =  4.570000    K =  0.039073
     P1 =         1    P2 =         3   P3 =         5   Q1 =        -1    Q2 =        -3    Q3 =        -5

Modified exponential model  :
     Bm = Sum ( Ai X ( 1 - exp( - R/( 2*i*C ) ) ) )  +      Sum ( RKj X ( R - Rj0 ) )  +  Bm0
       C =   0.29668250E+00              Bmo=   0.00000000E+00          Nexp=  6        Nliner=  1
      AI =   0.84211801E+01   0.83415471E+02  -0.88747195E+03   0.29978135E+04  -0.43858447E+04   0.23641637E+04
     Rj0 =          15.7100
     RKj =   0.18670374E+01

Power model  :    Bm = ( Rki X R ) / ( 1 + ( R/R0 )**rn )**( 1/rn )
      rn = 0.964      rki =   0.23765806E+02          rmu =   0.34056481E+03
```

No	Rotation (radians) X 1/1000	Moment (kip-inch) Expri.	Moment (kip-inch) Poly.	Moment (kip-inch) M.Expo.	Moment (kip-inch) P.Model.	Connection stiffness (kip-inch) X 1000 Expri.	Connection stiffness (kip-inch) X 1000 Poly.	Connection stiffness (kip-inch) X 1000 M.Expo.	Connection stiffness (kip-inch) X 1000 P.Model.
1	0.00	0.00	0.00	0.00	0.00	0.3087E+02	0.6993E+02	0.3475E+02	0.2377E+02
3	0.64	19.74	44.25	18.80	14.45	0.2687E+02	0.6764E+02	0.2571E+02	0.2152E+02
5	1.52	38.53	98.85	39.36	32.25	0.2195E+02	0.5465E+02	0.2174E+02	0.1904E+02
7	2.16	52.87	129.80	52.77	43.95	0.2239E+02	0.4224E+02	0.2019E+02	0.1752E+02
9	3.03	69.98	160.65	69.42	58.35	0.1819E+02	0.2987E+02	0.1820E+02	0.1576E+02
11	4.05	86.61	186.57	86.86	73.62	0.1390E+02	0.2142E+02	0.1575E+02	0.1401E+02
13	5.02	99.98	204.78	100.96	86.43	0.1357E+02	0.1680E+02	0.1358E+02	0.1264E+02
15	6.05	113.64	220.38	113.89	98.81	0.1323E+02	0.1363E+02	0.1151E+02	0.1138E+02
17	7.13	125.97	233.82	125.30	110.48	0.9748E+01	0.1140E+02	0.9675E+01	0.1026E+02
19	8.74	139.72	250.25	139.03	125.84	0.6829E+01	0.9191E+01	0.7484E+01	0.8886E+01
21	10.95	152.34	268.34	152.98	143.82	0.5075E+01	0.7299E+01	0.5233E+01	0.7411E+01
23	12.52	159.59	279.02	160.18	154.75	0.4324E+01	0.6392E+01	0.4020E+01	0.6583E+01
25	14.65	167.89	291.60	167.31	167.74	0.3900E+01	0.5487E+01	0.2748E+01	0.5667E+01
27	16.80	174.96	302.67	174.22	179.13	0.2676E+01	0.4812E+01	0.3684E+01	0.4922E+01
29	18.99	180.82	312.60	181.51	189.20	0.2707E+01	0.4289E+01	0.3028E+01	0.4310E+01
31	21.13	186.68	321.34	187.50	197.88	0.2738E+01	0.3885E+01	0.2599E+01	0.3816E+01
33	23.47	193.03	329.99	193.20	206.28	0.2699E+01	0.3528E+01	0.2298E+01	0.3369E+01
35	26.01	199.89	338.54	198.77	214.31	0.2699E+01	0.3213E+01	0.2104E+01	0.2967E+01

II - 5

Connection type : Double web-angle connections
Mode : All riveted

Tested by : J.C.Rathbun (1936) U.S.A.
Test Id. : A-5

Column : --
Beam : W12X31.8
Angle : 6 X 6 X 3/8

Fasteners: 10.9-X-7/8"D
15/16" Oversize holes
Material : --
Fy = 47.12 ksi
Fu = 62.98 ksi

Major parameters

lp = 9.0000"	lu = 1.5000"	ll = 1.5000"	ta = 0.3750"
gb = 3.5000"	gc = 3.5625"	cu = 1.5000"	cl = 1.5000"
pb = 3.0000"	pc = 3.0000"	qb = 2.5000"	qc = 2.5000"
nb = 2 X 3	nc = 2 X 3		

Remark 1) Connection angle riveted to 1/2" mouning plate.
2)

No	Moment (k-in)	Rotation (radians) X 1/1000	No	Moment (k-in)	Rotation (radians) X 1/1000
1	0.00	0.00	26	235.03	13.67
2	13.50	0.09	27	239.78	14.57
3	27.01	0.17	28	243.58	15.42
4	40.51	0.26	29	247.39	16.27
5	50.40	0.40	30	251.20	17.12
6	60.28	0.55	31	255.47	18.15
7	70.25	0.88	32	259.75	19.18
8	80.22	1.21	33	263.75	20.25
9	90.53	1.55	34	267.75	21.32
10	100.84	1.89	35	271.75	22.39
11	110.57	2.32			
12	120.29	2.76			
13	130.57	3.27			
14	140.86	3.78			
15	150.67	4.36			
16	160.49	4.94			
17	170.72	5.72			
18	180.96	6.51			
19	190.59	7.49			
20	200.23	8.47			
21	206.84	9.36			
22	213.45	10.26			
23	220.06	11.16			
24	225.17	11.96			
25	230.28	12.76			

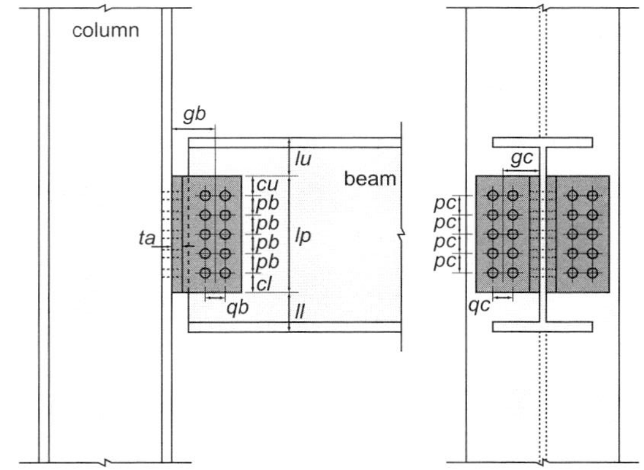

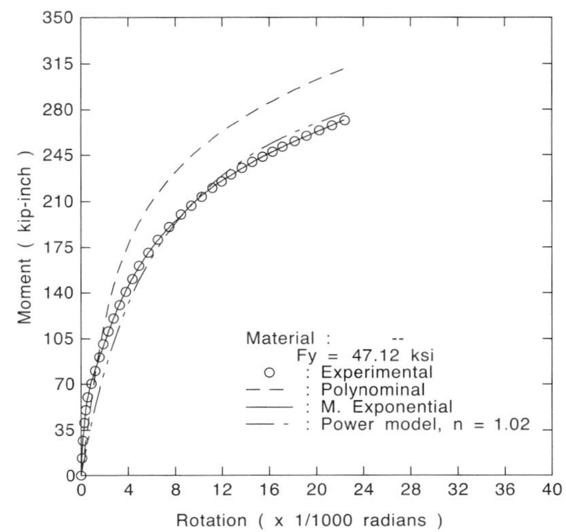

```
Moment-rotation prediction equations      ( R : X 1/1000 radians )

Frye and Morris polynominal model  :    R = Sum ( Ai X ( K*Bm )**Pi X 10**Qi )
     xd =  9.000000"    g =  7.500000"   t =  0.375000"
     A1 =  3.660000    A2 =  1.150000   A3 =  4.570000     K =  0.040934
     P1 =         1    P2 =         3   P3 =         5    Q1 =        -1    Q2 =        -3    Q3 =        -5

Modified exponential model  :
     Bm = Sum ( Ai X ( 1 - exp( - R/( 2*i*C ) ) ) )  +      Sum ( RKj X ( R - Rj0 ) )  +  Bm0
      C =   0.20832833E+00           Bmo=   0.00000000E+00          Nexp=  6          Nliner=  2
     AI =   0.18228595E+03  -0.74560665E+03   0.16399778E+04  -0.73894153E+03  -0.19767307E+04   0.18643013E+04
     Rj0 =           8.4700          12.7600
     RKj =  0.25688439E+01   0.11517704E+01

Power model  :    Bm = ( Rki X R ) / ( 1 + ( R/R0 )**rn )**( 1/rn )
     rn = 1.016       rki =   0.50682586E+02          rmu =   0.36243728E+03
```

No	Rotation (radians) X 1/1000	Moment (kip-inch) Expri.	Poly.	M.Expo.	P.Model.	Connection stiffness (kip-inch) X 1000 Expri.	Poly.	M.Expo.	P.Model.
1	0.00	0.00	0.00	0.00	0.00	0.1554E+03	0.6675E+02	0.2083E+03	0.5068E+02
3	0.17	27.01	11.60	28.51	8.61	0.1555E+03	0.6660E+02	0.1270E+03	0.4846E+02
5	0.40	50.40	26.81	50.24	19.41	0.6933E+02	0.6594E+02	0.6965E+02	0.4568E+02
7	0.88	70.25	57.53	71.84	39.94	0.2981E+02	0.6230E+02	0.3164E+02	0.4056E+02
9	1.55	90.53	95.98	90.39	65.06	0.3071E+02	0.5157E+02	0.2632E+02	0.3464E+02
11	2.32	110.57	130.32	110.19	89.70	0.2220E+02	0.3763E+02	0.2446E+02	0.2926E+02
13	3.27	130.57	160.03	131.37	114.92	0.2017E+02	0.2613E+02	0.2019E+02	0.2420E+02
15	4.36	150.67	184.07	151.04	138.74	0.1699E+02	0.1882E+02	0.1626E+02	0.1984E+02
17	5.72	170.72	206.01	170.88	162.94	0.1299E+02	0.1383E+02	0.1298E+02	0.1583E+02
19	7.49	190.59	227.01	190.62	187.50	0.9859E+01	0.1032E+02	0.9437E+01	0.1222E+02
21	9.36	206.84	244.16	207.39	207.78	0.7378E+01	0.8169E+01	0.8699E+01	0.9585E+01
23	11.16	220.06	257.53	220.72	223.26	0.6841E+01	0.6840E+01	0.6312E+01	0.7783E+01
25	12.76	230.28	267.80	229.61	234.74	0.5833E+01	0.5987E+01	0.6010E+01	0.6566E+01
27	14.57	239.78	277.97	239.49	245.64	0.4847E+01	0.5262E+01	0.4984E+01	0.5504E+01
29	16.27	247.39	286.45	247.45	254.31	0.4479E+01	0.4735E+01	0.4424E+01	0.4725E+01
31	18.15	255.47	294.89	255.39	262.51	0.4164E+01	0.4271E+01	0.4081E+01	0.4043E+01
33	20.25	263.75	303.40	263.73	270.34	0.3733E+01	0.3857E+01	0.3888E+01	0.3440E+01
35	22.39	271.75	311.28	271.95	277.17	0.3733E+01	0.3515E+01	0.3796E+01	0.2954E+01

A.2 – 11

II - 6

Connection type : Double web-angle connections
Mode : All riveted

Tested by : J.C.Rathbun (1936) U.S.A.
Test Id. : A-6

Column : --
Beam : W18X54.7
Angle : 4 X 3 1/2 X 3/8

Fasteners: 10.9-X-7/8"D
15/16" Oversize holes
Material : --
Fy = 44.81 ksi
Fu = 58.98 ksi

Major parameters

lp = 15.0000"	lu = 1.5000"	ll = 1.5000"	ta = 0.3750"
gb = 2.2500"	gc = 2.5000"	cu = 1.5000"	cl = 1.5000"
pb = 3.0000"	pc = 3.0000"	nb = 1 X 5	nc = 1 X 5

Remark 1) Connection angle riveted to 1/2" mouning plate.
2)

No	Moment (k-in)	Rotation (radians) X 1/1000	No	Moment (k-in)	Rotation (radians) X 1/1000
1	0.00	0.00	26	391.92	6.24
2	16.94	0.05	27	400.32	6.60
3	33.88	0.10	28	409.67	7.09
4	50.83	0.14	29	419.02	7.58
5	67.54	0.26	30	428.37	8.07
6	84.25	0.38	31	433.65	8.45
7	100.97	0.49	32	438.92	8.83
8	117.81	0.67	33	444.20	9.21
9	134.66	0.84	34	449.48	9.60
10	151.50	1.02	35	458.12	10.09
11	167.80	1.21			
12	184.10	1.41			
13	200.40	1.60			
14	217.18	1.86			
15	233.97	2.11			
16	250.75	2.36			
17	267.29	2.71			
18	283.83	3.05			
19	300.37	3.39			
20	316.86	3.81			
21	333.34	4.22			
22	349.82	4.64			
23	362.48	5.08			
24	375.14	5.52			
25	383.53	5.88			

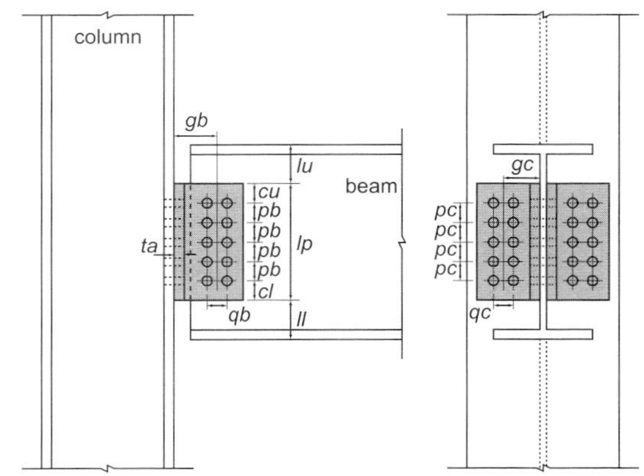

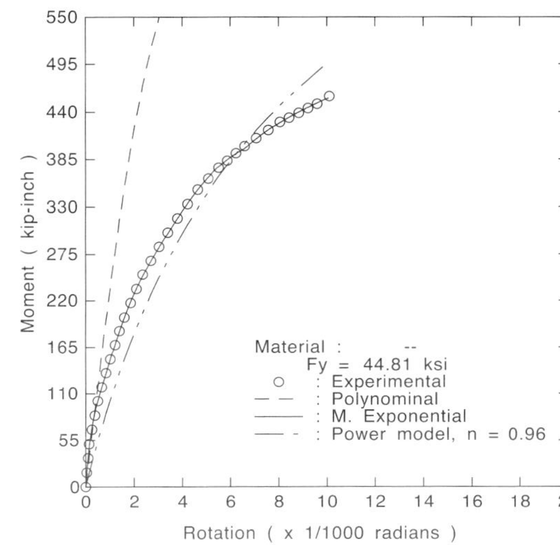

A.2 – 12

```
Moment-rotation prediction equations      ( R : X 1/1000 radians )

Frye and Morris polynominal model  :    R = Sum ( Ai X ( K*Bm )**Pi X 10**Qi )
     xd = 15.000000"    g =  5.500000"   t =  0.375000"
     A1 =  3.660000   A2 =  1.150000   A3 =  4.570000     K =  0.011467
     P1 =        1    P2 =        3    P3 =        5     Q1 =       -1    Q2 =       -3    Q3 =       -5

Modified exponential model  :
     Bm = Sum ( Ai X ( 1 - exp( - R/( 2*i*C ) ) ) )  +      Sum ( RKj X ( R - Rj0 ) )  +  Bm0
       C =   0.96163333E-01            Bmo=   0.00000000E+00          Nexp=  6        Nliner=  1
      AI =  -0.17530126E+02   0.22299255E+04  -0.14727152E+05   0.38949631E+05  -0.45275720E+05   0.19261428E+05
     Rj0 =           6.6000
     RKj =   0.10789658E+02

Power model  :    Bm = ( Rki X R ) / ( 1 + ( R/R0 )**rn )**( 1/rn )
      rn = 0.960      rki =   0.11575203E+03          rmu =   0.90952672E+03
```

No	Rotation (radians) X 1/1000	Moment (kip-inch) Expri.	Moment (kip-inch) Poly.	Moment (kip-inch) M.Expo.	Moment (kip-inch) P.Model.	Connection stiffness (kip-inch) X 1000 Expri.	Connection stiffness (kip-inch) X 1000 Poly.	Connection stiffness (kip-inch) X 1000 M.Expo.	Connection stiffness (kip-inch) X 1000 P.Model.
1	0.00	0.00	0.00	0.00	0.00	0.3508E+03	0.2383E+03	0.4206E+03	0.1158E+03
3	0.10	33.88	23.02	33.81	11.01	0.3508E+03	0.2381E+03	0.2888E+03	0.1124E+03
5	0.26	67.54	62.26	69.91	29.11	0.1434E+03	0.2371E+03	0.1658E+03	0.1072E+03
7	0.49	100.97	117.22	99.95	53.33	0.1247E+03	0.2338E+03	0.1060E+03	0.1008E+03
9	0.84	134.66	197.13	133.91	86.97	0.9656E+02	0.2238E+03	0.9422E+02	0.9230E+02
11	1.21	167.80	276.74	168.07	119.61	0.8354E+02	0.2057E+03	0.8926E+02	0.8456E+02
13	1.60	200.40	352.23	200.53	151.19	0.7603E+02	0.1805E+03	0.7635E+02	0.7747E+02
15	2.11	233.97	435.03	234.77	188.39	0.6628E+02	0.1470E+03	0.5984E+02	0.6961E+02
17	2.71	267.29	512.60	266.89	227.47	0.4829E+02	0.1151E+03	0.4956E+02	0.6186E+02
19	3.39	300.37	582.10	299.05	267.25	0.4438E+02	0.8961E+02	0.4486E+02	0.5451E+02
21	4.22	333.34	647.73	334.16	309.48	0.3962E+02	0.6967E+02	0.3903E+02	0.4726E+02
23	5.08	362.48	701.46	364.00	347.37	0.2862E+02	0.5642E+02	0.3017E+02	0.4124E+02
25	5.88	383.53	743.01	384.60	378.47	0.2344E+02	0.4791E+02	0.2145E+02	0.3662E+02
27	6.60	400.32	775.19	397.50	403.41	0.2160E+02	0.4223E+02	0.1482E+02	0.3314E+02
29	7.58	419.02	813.57	419.13	433.86	0.1908E+02	0.3638E+02	0.1908E+02	0.2913E+02
31	8.45	433.65	843.51	434.07	457.93	0.1381E+02	0.3244E+02	0.1547E+02	0.2615E+02
33	9.21	444.20	867.46	445.10	477.04	0.1381E+02	0.2962E+02	0.1354E+02	0.2391E+02
35	10.09	458.12	892.01	456.35	497.01	0.1746E+02	0.2702E+02	0.1225E+02	0.2168E+02

II - 7

Connection type : Double web-angle connections
Mode : All riveted

Tested by : J.C.Rathbun (1936) U.S.A.
Test Id. : A-7

Column : --
Beam : W18X54.7
Angle : 6 X 6 X 3/8

Fasteners: 10.9-X-7/8"D
15/16" Oversize holes
Material : --
Fy = 44.81 ksi
Fu = 58.98 ksi

Major parameters

lp = 15.0000"	lu = 1.5000"	ll = 1.5000"	ta = 0.3750"
gb = 3.5000"	gc = 3.5000"	cu = 1.5000"	cl = 1.5000"
pb = 3.0000"	pc = 3.0000"	qb = 2.5000"	qc = 2.5000"
nb = 2 X 5	nc = 2 X 5		

Remark 1) Connection angle riveted to 3/4" mouning plate.
2)

No	Moment (k-in)	Rotation (radians) X 1/1000	No	Moment (k-in)	Rotation (radians) X 1/1000
1	0.00	0.00	26	538.20	4.71
2	24.83	0.06	27	547.89	4.95
3	49.67	0.13	28	561.13	5.28
4	74.21	0.20	29	574.36	5.62
5	98.76	0.28	30	586.55	5.95
6	124.09	0.37	31	598.74	6.28
7	149.41	0.45	32	606.83	6.53
8	174.95	0.57	33	614.91	6.77
9	200.50	0.68	34	623.00	7.01
10	225.43	0.85			
11	250.37	1.01			
12	275.23	1.21			
13	300.10	1.42			
14	325.20	1.64			
15	350.31	1.86			
16	375.34	2.13			
17	400.37	2.39			
18	417.31	2.63			
19	434.25	2.86			
20	451.20	3.10			
21	467.21	3.36			
22	483.23	3.62			
23	499.24	3.88			
24	513.87	4.18			
25	528.51	4.48			

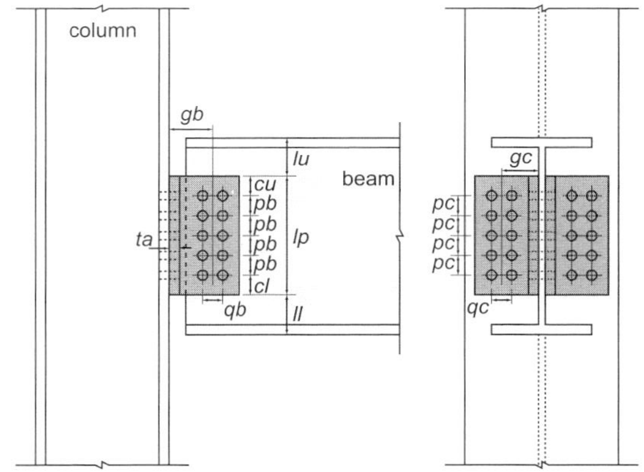

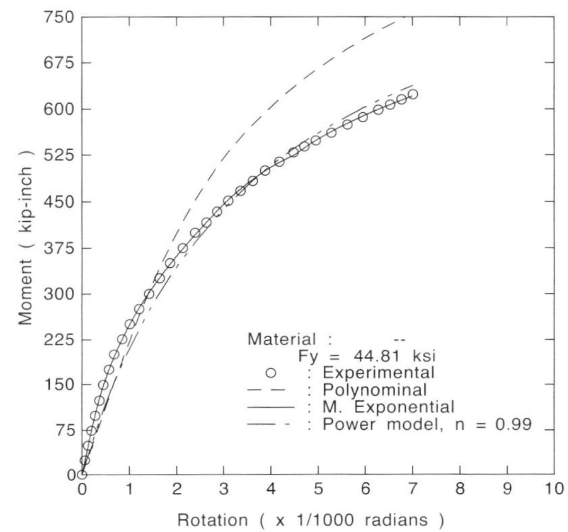

A.2 – 14

```
Moment-rotation prediction equations     ( R : X 1/1000 radians )

Frye and Morris polynominal model  :    R = Sum ( Ai X ( K*Bm )**Pi X 10**Qi )
     xd = 15.000000"   g =  7.500000"   t =  0.375000"
     A1 =  3.660000   A2 =  1.150000   A3 =  4.570000    K =  0.012013
     P1 =        1     P2 =        3    P3 =        5     Q1 =       -1     Q2 =       -3     Q3 =       -5

Modified exponential model  :
     Bm = Sum ( Ai X ( 1 - exp( - R/( 2*i*C ) ) ) )  +     Sum ( RKj X ( R - Rj0 ) )  + Bm0
      C =   0.74424167E-01             Bmo=   0.00000000E+00          Nexp=  6         Nliner=  1
     AI =  -0.13241697E+03   0.18746915E+04  -0.11695762E+05   0.33636330E+05  -0.42316337E+05   0.19199228E+05
     Rj0 =           4.4800
     RKj =   0.23428209E+02

Power model  :    Bm = ( Rki X R ) / ( 1 + ( R/R0 )**rn )**( 1/rn )
     rn = 0.992      rki =   0.26852496E+03         rmu =   0.97199298E+03
```

No	Rotation (radians) X 1/1000	Moment (kip-inch) Expri.	Poly.	M.Expo.	P.Model.	Connection stiffness (kip-inch) X 1000 Expri.	Poly.	M.Expo.	P.Model.
1	0.00	0.00	0.00	0.00	0.00	0.3886E+03	0.2274E+03	0.3495E+03	0.2685E+03
3	0.13	49.67	29.09	47.87	33.13	0.3590E+03	0.2272E+03	0.3747E+03	0.2500E+03
5	0.28	98.76	63.40	100.23	69.44	0.3082E+03	0.2262E+03	0.3133E+03	0.2306E+03
7	0.45	149.41	102.88	148.68	108.10	0.2599E+03	0.2239E+03	0.2432E+03	0.2110E+03
9	0.68	200.50	153.86	197.42	154.07	0.1935E+03	0.2188E+03	0.1858E+03	0.1888E+03
11	1.01	250.37	223.01	249.99	210.97	0.1394E+03	0.2068E+03	0.1431E+03	0.1632E+03
13	1.42	300.10	303.45	302.15	272.47	0.1175E+03	0.1841E+03	0.1141E+03	0.1376E+03
15	1.86	350.31	378.53	348.58	328.37	0.1050E+03	0.1556E+03	0.9779E+02	0.1163E+03
17	2.39	400.37	452.31	397.06	384.42	0.8258E+02	0.1248E+03	0.8628E+02	0.9678E+02
19	2.86	434.25	505.99	435.47	426.84	0.7168E+02	0.1034E+03	0.7594E+02	0.8321E+02
21	3.36	467.21	553.08	470.16	465.36	0.6109E+02	0.8639E+02	0.6296E+02	0.7179E+02
23	3.88	499.24	594.74	499.37	500.38	0.5564E+02	0.7319E+02	0.4850E+02	0.6214E+02
25	4.48	528.51	634.61	523.56	534.46	0.4489E+02	0.6220E+02	0.3373E+02	0.5343E+02
27	4.95	547.89	662.21	548.01	558.20	0.4038E+02	0.5552E+02	0.4764E+02	0.4776E+02
29	5.62	574.36	697.14	576.64	588.17	0.3800E+02	0.4807E+02	0.3769E+02	0.4105E+02
31	6.28	598.74	726.96	599.33	613.47	0.3478E+02	0.4251E+02	0.3155E+02	0.3579E+02
33	6.77	614.91	746.84	613.95	630.09	0.3320E+02	0.3919E+02	0.2867E+02	0.3253E+02

II - 8

Connection type : Double web-angle connections
Mode : Riveted-to-beam and bolted-to-column

Tested by : W.G.Bell et al. (1958) U.S.A.
Test Id. : FK-4A

Column : W12X65
Beam : W18X50
Angle : 6 X 4 X 3/8

Fasteners: A325- -3/4"D
13/16" Oversize holes
Material : --
Fy = 38.10 ksi
Fu = 64.50 ksi

Major parameters

lp = 11.5000" lu = 3.2500' ll = 3.2500" ta = 0.3750"
gb = 4.5000" gc = 2.5730' cu = 1.2500" cl = 1.2500"
pb = 3.0000" pc = 3.0000' qb = 3.0000"
nb = 2 X 4 nc = 1 X 4

Remark 1) High strength bolts used
2) Bolts : Fy=87.9 ksi, Fu=136.6 ksi, Rivets : Fy=36.5 ksi, Fu=63.2 ksi

No	Moment (k-in)	Rotation (radians) X 1/1000	No	Moment (k-in)	Rotation (radians) X 1/1000
1	0.00	0.00	26	411.69	44.27
2	19.30	0.16	27	421.44	47.19
3	38.60	0.32	28	431.20	50.11
4	55.90	0.83	29	438.74	52.94
5	73.20	1.33	30	446.27	55.76
6	93.00	1.49	31	453.81	58.58
7	112.79	1.65	32	461.34	61.41
8	136.51	2.50	33	469.87	64.13
9	160.23	3.34	34	478.40	66.85
10	183.95	4.18	35	486.93	69.58
11	203.21	5.19	36	495.46	72.30
12	222.47	6.20			
13	241.71	7.46			
14	260.96	8.71			
15	275.19	11.17			
16	289.43	13.62			
17	306.63	16.22			
18	323.83	18.82			
19	335.56	21.92			
20	347.29	25.02			
21	359.01	28.12			
22	370.07	31.55			
23	381.12	34.99			
24	392.17	38.42			
25	401.93	41.34			

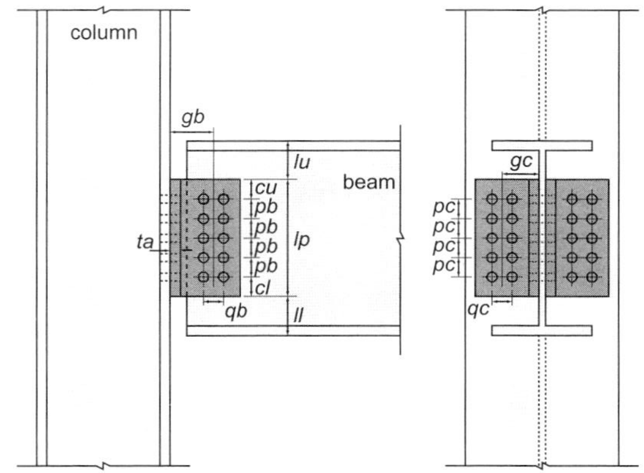

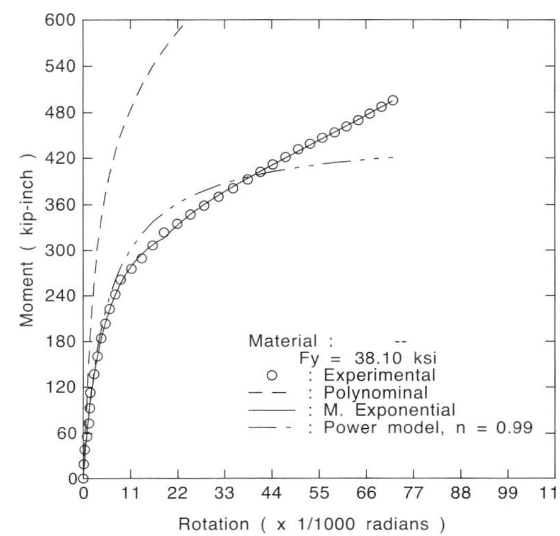

```
Moment-rotation prediction equations     ( R : X 1/1000 radians )

Frye and Morris polynominal model  :    R = Sum ( Ai X ( K*Bm )**Pi X 10**Qi )
     xd = 11.500000"   g =  5.504000"   t =  0.375000"
     A1 =  3.660000   A2 =  1.150000   A3 =  4.570000    K =  0.021699
     P1 =        1    P2 =        3    P3 =        5    Q1 =       -1     Q2 =       -3     Q3 =       -5

Modified exponential model  :
     Bm = Sum ( Ai X ( 1 - exp( - R/( 2*i*C ) ) ) )  +      Sum ( RKj X ( R - Rj0 ) )  +  Bm0
       C =   0.64234333E+00            Bmo=   0.00000000E+00            Nexp=  6        Nliner=  1
      AI =   0.89943955E+02  -0.55707191E+03   0.29428550E+04  -0.63897028E+04   0.62801071E+04  -0.20332224E+04
     Rj0 =          18.8200
     RKj =   0.30333621E+01

Power model  :    Bm = ( Rki X R ) / ( 1 + ( R/R0 )**rn )**( 1/rn )
      rn = 0.988      rki =    0.83485722E+02          rmu =    0.45369160E+03
```

No	Rotation (radians) X 1/1000	Moment (kip-inch) Expri.	Poly.	M.Expo.	P.Model.	Connection stiffness (kip-inch) X 1000 Expri.	Poly.	M.Expo.	P.Model.
1	0.00	0.00	0.00	0.00	0.00	0.1211E+03	0.1259E+03	0.8725E+02	0.8349E+02
3	0.32	38.60	40.03	25.73	25.07	0.1003E+03	0.1250E+03	0.7469E+02	0.7416E+02
5	1.33	73.20	159.18	87.18	88.88	0.1025E+03	0.1048E+03	0.4929E+02	0.5332E+02
7	1.65	112.79	190.86	102.08	105.13	0.1088E+03	0.9360E+02	0.4423E+02	0.4860E+02
9	3.34	160.23	305.21	161.31	171.32	0.2812E+02	0.4816E+02	0.2819E+02	0.3170E+02
11	5.19	203.21	374.01	204.65	219.81	0.1911E+02	0.2909E+02	0.1947E+02	0.2165E+02
13	7.46	241.71	427.61	241.06	260.27	0.1531E+02	0.1956E+02	0.1323E+02	0.1475E+02
15	11.17	275.19	486.04	278.64	302.84	0.5805E+01	0.1289E+02	0.7705E+01	0.8931E+01
17	16.22	306.63	539.92	307.44	337.53	0.6606E+01	0.8959E+01	0.4115E+01	0.5273E+01
19	21.92	335.56	584.10	333.56	361.38	0.3784E+01	0.6761E+01	0.4985E+01	0.3318E+01
21	28.12	359.01	621.41	360.10	378.18	0.3517E+01	0.5392E+01	0.3727E+01	0.2214E+01
23	34.99	381.12	654.92	383.33	390.82	0.3220E+01	0.4439E+01	0.3143E+01	0.1531E+01
25	41.34	401.93	681.11	402.67	399.24	0.3337E+01	0.3834E+01	0.2982E+01	0.1146E+01
27	47.19	421.44	702.26	420.02	405.20	0.3337E+01	0.3417E+01	0.2961E+01	0.9074E+00
29	52.94	438.74	720.93	437.07	409.91	0.2669E+01	0.3094E+01	0.2976E+01	0.7388E+00
31	58.58	453.81	737.65	453.93	413.71	0.2669E+01	0.2837E+01	0.2995E+01	0.6152E+00
33	64.13	469.87	752.79	470.58	416.86	0.3132E+01	0.2626E+01	0.3010E+01	0.5217E+00
35	69.58	486.93	766.60	487.01	419.49	0.3132E+01	0.2450E+01	0.3019E+01	0.4493E+00

A.2 – 17

II - 9

Connection type : Double web-angle connections
Mode : Riveted-to-beam and bolted-to-column

Tested by : W.G.Bell et al. (1958) U.S.A.
Test Id. : FK-4B

Column : W12X65
Beam : W18X50
Angle : 6 X 4 X 3/8

Fasteners: A325- -3/4"D
13/16" Oversize holes
Material : --
Fy = 38.10 ksi
Fu = 64.50 ksi

Major parameters

lp = 11.5000"	lu = 3.2500"	ll = 3.2500"	ta = 0.3750"
gb = 4.5000"	gc = 2.5730"	cu = 1.2500"	cl = 1.2500"
pb = 3.0000"	pc = 3.0000"	qb = 3.0000"	
nb = 2 X 4	nc = 1 X 4		

Remark 1) High strength bolts used
2) Bolts : Fy=87.9 ksi, Fu=136.6 ksi, Rivets : Fy=36.5 ksi, Fu=63.2 ksi

No	Moment (k-in)	Rotation (radians) X 1/1000	No	Moment (k-in)	Rotation (radians) X 1/1000
1	0.00	0.00	26	353.64	44.37
2	17.32	0.32	27	360.23	47.90
3	34.65	0.63	28	366.83	51.44
4	53.96	0.95	29	373.42	54.97
5	73.26	1.27	30	380.01	58.50
6	89.09	1.73	31	390.56	62.20
7	104.91	2.20	32	401.10	65.90
8	121.69	2.96	33	411.65	69.60
9	138.48	3.73	34	422.19	73.30
10	153.26	4.64	35	432.73	77.00
11	168.05	5.55			
12	185.34	6.17			
13	202.64	6.78			
14	218.84	8.44			
15	235.04	10.10			
16	251.25	11.67			
17	267.46	13.23			
18	283.50	16.48			
19	299.54	19.74			
20	308.78	23.24			
21	318.02	26.74			
22	327.27	30.24			
23	333.86	33.77			
24	340.45	37.30			
25	347.05	40.84			

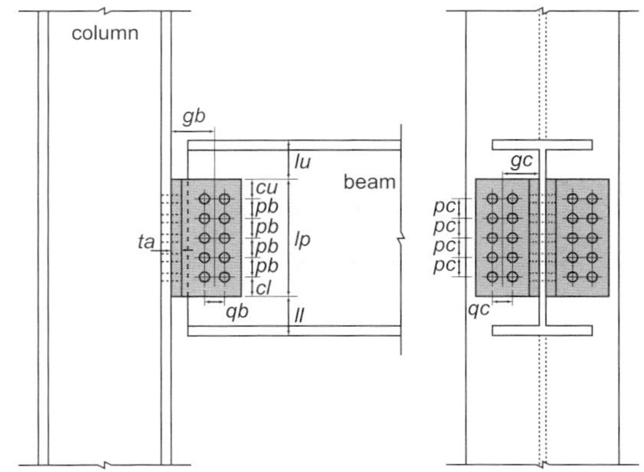

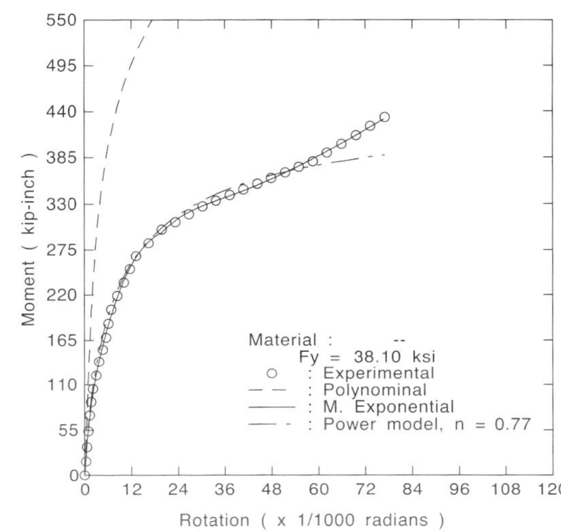

A.2 – 18

```
Moment-rotation prediction equations      ( R : X 1/1000 radians )

Frye and Morris polynominal model   :     R = Sum ( Ai X ( K*Bm )**Pi X 10**Qi )
     xd = 11.500000"    g =  5.504000"   t =  0.375000"
     A1 =  3.660000    A2 =  1.150000   A3 =  4.570000     K =  0.021699
     P1 =         1    P2 =         3   P3 =         5    Q1 =        -1    Q2 =       -3    Q3 =       -5

Modified exponential model  :
     Bm = Sum ( Ai X ( 1 - exp( - R/( 2*i*C ) ) ) )  +      Sum ( RKj X ( R - Rj0 ) )  + Bm0
      C =  0.72078083E+00          Bmo=  0.00000000E+00         Nexp=  6         Nliner=  1
     AI =  0.11789430E+03  -0.57827493E+03  0.37995230E+04  -0.11830491E+05  0.15260551E+05  -0.64964724E+04
    Rj0 =         19.7400
    RKj =  0.27577203E+01

Power model  :    Bm = ( Rki X R ) / ( 1 + ( R/R0 )**rn )**( 1/rn )
     rn = 0.771      rki =   0.83485722E+02          rmu =   0.45369160E+03
```

No	Rotation (radians)	Moment (kip-inch)				Connection stiffness (kip-inch) X 1000			
	X 1/1000	Expri.	Poly.	M.Expo.	P.Model.	Expri.	Poly.	M.Expo.	P.Model.
1	0.00	0.00	0.00	0.00	0.00	0.5475E+02	0.1259E+03	0.7423E+02	0.8349E+02
3	0.63	34.65	78.86	39.85	42.14	0.5770E+02	0.1219E+03	0.5339E+02	0.5596E+02
5	1.27	73.26	152.35	69.37	73.54	0.4983E+02	0.1069E+03	0.4034E+02	0.4370E+02
7	2.20	104.91	236.69	101.25	108.74	0.2949E+02	0.7482E+02	0.2951E+02	0.3303E+02
9	3.73	138.48	322.82	139.29	150.92	0.1934E+02	0.4246E+02	0.2148E+02	0.2317E+02
11	5.55	168.05	384.07	174.21	186.71	0.2331E+02	0.2699E+02	0.1725E+02	0.1668E+02
13	6.78	202.64	413.73	194.20	205.41	0.2314E+02	0.2166E+02	0.1529E+02	0.1385E+02
15	10.10	235.04	471.61	237.44	242.82	0.1008E+02	0.1425E+02	0.1088E+02	0.9194E+01
17	13.23	267.46	510.38	265.81	267.43	0.8613E+01	0.1090E+02	0.7403E+01	0.6770E+01
19	19.74	299.54	568.62	295.81	301.77	0.3825E+01	0.7449E+01	0.5041E+01	0.4126E+01
21	26.74	318.02	613.78	320.35	325.32	0.2641E+01	0.5643E+01	0.2394E+01	0.2754E+01
23	33.77	333.86	649.43	333.80	341.76	0.1866E+01	0.4580E+01	0.1649E+01	0.1987E+01
25	40.84	347.05	679.16	345.54	353.99	0.1866E+01	0.3875E+01	0.1749E+01	0.1510E+01
27	47.90	360.23	704.68	358.98	363.47	0.1866E+01	0.3373E+01	0.2062E+01	0.1193E+01
29	54.97	373.42	727.12	374.58	371.07	0.1866E+01	0.2996E+01	0.2339E+01	0.9697E+00
31	62.20	390.56	747.66	392.24	377.45	0.2850E+01	0.2695E+01	0.2529E+01	0.8029E+00
33	69.60	411.65	766.66	411.41	382.90	0.2850E+01	0.2449E+01	0.2641E+01	0.6749E+00
35	77.00	432.73	784.01	431.19	387.51	0.2850E+01	0.2248E+01	0.2701E+01	0.5764E+00

II - 10

Connection type : Double web-angle connections
Mode : Riveted-to-beam and bolted-to-column

Tested by : W.G.Bell et al. (1958) U.S.A.
Test Id. : FK-4C

Column : W12X65
Beam : W18X50
Angle : 6 X 4 X 3/8

Fasteners: A325- -3/4"D
13/16" Oversize holes
Material : --
Fy = 38.10 ksi
Fu = 64.50 ksi

Major parameters

lp = 11.5000"	lu = 3.2500"	ll = 3.2500"	ta = 0.3750"
gb = 4.5000"	gc = 2.5730"	cu = 1.2500"	cl = 1.2500"
pb = 3.0000"	pc = 3.0000"	qb = 3.0000"	
nb = 2 X 4	nc = 1 X 4		

Remark 1) High strength bolts used
2) Bolts : Fy=87.9 ksi, Fu=136.6 ksi, Rivets : Fy=36.5 ksi, Fu=63.2 ksi

No	Moment (k-in)	Rotation (radians) X 1/1000	No	Moment (k-in)	Rotation (radians) X 1/1000
1	0.00	0.00	26	410.32	44.97
2	26.80	0.57	27	420.37	48.46
3	42.16	1.01	28	430.43	51.94
4	57.52	1.45	29	453.28	56.09
5	71.88	1.89	30	476.13	60.25
6	86.24	2.33	31	481.74	63.56
7	112.85	3.59	32	487.35	66.86
8	127.74	3.93	33	510.84	68.62
9	142.62	4.27	34	524.98	72.70
10	158.51	4.61	35	539.12	76.78
11	174.39	4.95	36	553.26	80.86
12	188.79	5.24	37	567.40	84.94
13	203.19	5.53			
14	227.81	6.79			
15	255.16	9.06			
16	275.63	10.91			
17	290.45	13.98			
18	305.28	17.05			
19	320.10	20.11			
20	328.04	24.03			
21	335.99	27.94			
22	343.93	31.85			
23	351.87	35.77			
24	376.07	38.62			
25	400.27	41.48			

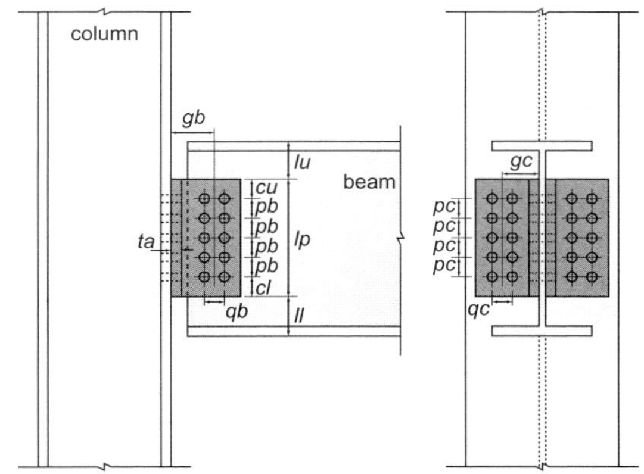

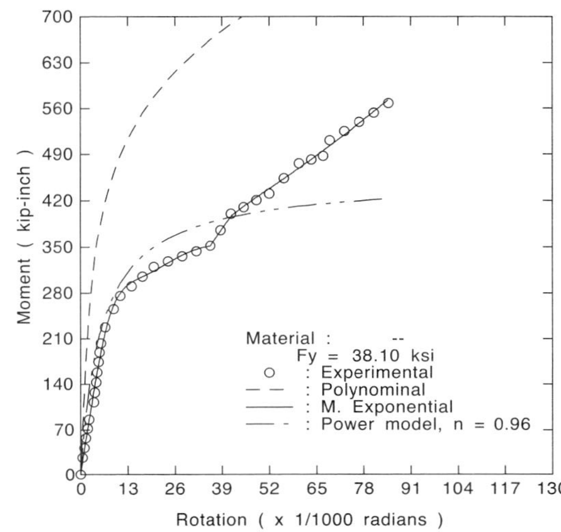

```
Moment-rotation prediction equations      ( R : X 1/1000 radians )

Frye and Morris polynominal model  :    R = Sum ( Ai X ( K*Bm )**Pi X 10**Qi )
     xd = 11.500000"    g =  5.146000"   t =  0.375000"
     A1 =  3.660000   A2 =  1.150000   A3 =  4.570000    K =  0.021481
     P1 =         1   P2 =         3   P3 =         5   Q1 =        -1    Q2 =        -3    Q3 =        -5

Modified exponential model  :
     Bm = Sum ( Ai X ( 1 - exp( - R/( 2*i*C ) ) ) )  +      Sum ( RKj X ( R - Rj0 ) )  +  Bm0
      C =   0.85040917E+00           Bmo=   0.00000000E+00          Nexp=  6         Nliner=  2
     AI =   0.43582679E+03  -0.71415566E+04   0.33242691E+05  -0.66136249E+05   0.60354956E+05  -0.20432854E+05
     Rj0 =           35.7700           41.4800
     RKj =   0.76301928E+01  -0.29302073E+01

Power model  :     Bm = ( Rki X R ) / ( 1 + ( R/R0 )**rn )**( 1/rn )
      rn = 0.962       rki =    0.83485722E+02          rmu =    0.45369160E+03
```

No	Rotation (radians) X 1/1000	Moment (kip-inch) Expri.	Poly.	M.Expo.	P.Model.	Connection stiffness (kip-inch) X 1000 Expri.	Poly.	M.Expo.	P.Model.
1	0.00	0.00	0.00	0.00	0.00	0.4700E+02	0.1272E+03	0.4551E+02	0.8349E+02
3	1.01	42.16	124.89	39.42	69.90	0.3490E+02	0.1156E+03	0.3518E+02	0.5774E+02
5	1.89	71.88	214.01	69.52	114.35	0.3283E+02	0.8605E+02	0.3397E+02	0.4449E+02
7	3.59	112.85	320.14	127.46	175.87	0.3907E+02	0.4476E+02	0.3350E+02	0.2928E+02
9	4.27	142.62	347.58	149.77	194.35	0.4519E+02	0.3670E+02	0.3218E+02	0.2537E+02
11	4.95	174.39	370.60	171.10	210.56	0.4840E+02	0.3099E+02	0.3016E+02	0.2219E+02
13	5.53	203.19	387.36	187.87	222.68	0.4436E+02	0.2738E+02	0.2804E+02	0.1996E+02
15	9.06	255.16	460.50	260.49	276.31	0.1151E+02	0.1610E+02	0.1339E+02	0.1157E+02
17	13.98	290.45	523.69	296.17	319.11	0.4834E+01	0.1043E+02	0.3505E+01	0.6556E+01
19	20.11	320.10	577.16	313.51	349.90	0.3602E+01	0.7396E+01	0.2958E+01	0.3847E+01
21	27.94	335.99	626.72	337.63	373.13	0.2030E+01	0.5479E+01	0.2738E+01	0.2290E+01
23	35.77	351.87	665.02	351.75	387.73	0.5748E+01	0.4398E+01	0.8224E+00	0.1521E+01
25	41.48	400.27	688.56	396.60	395.39	0.5950E+01	0.3861E+01	0.4411E+01	0.1181E+01
27	48.46	420.37	713.69	424.76	402.59	0.2884E+01	0.3374E+01	0.3788E+01	0.9024E+00
29	56.09	453.28	737.87	453.23	408.65	0.5504E+01	0.2974E+01	0.3740E+01	0.6972E+00
31	63.56	481.74	758.89	481.85	413.30	0.1695E+01	0.2673E+01	0.3952E+01	0.5580E+00
33	68.62	510.84	772.04	502.31	415.94	0.1038E+02	0.2503E+01	0.4121E+01	0.4860E+00
35	76.78	539.12	791.46	536.92	419.52	0.3466E+01	0.2276E+01	0.4350E+01	0.3965E+00
37	84.94	567.40	809.24	573.08	422.47	0.3466E+01	0.2090E+01	0.4504E+01	0.3297E+00

II - 11

Connection type : Double web-angle connections
Mode : All riveted

Tested by : W.G.Bell et al. (1958) U.S.A.
Test Id. : FK-4R

Column : W12X65
Beam : W18X50
Angle : 6 X 4 X 3/8

Fasteners: 10.9-X-3/4"D
13/16" Oversize holes
Material : --
Fy = 38.10 ksi
Fu = 64.50 ksi

Major parameters

lp = 11.5000"	lu = 3.2500"	ll = 3.2500"	ta = 0.3750"
gb = 4.5000"	gc = 2.5730"	cu = 1.2500"	cl = 1.2500"
pb = 3.0000"	pc = 3.0000"	qb = 3.0000"	
nb = 2 X 4	nc = 1 X 4		

Remark 1) High strength bolts used
2) Bolts : Fy=87.9 ksi, Fu=136.6 ksi, Rivets : Fy=36.5 ksi, Fu=63.2 ksi

No	Moment (k-in)	Rotation (radians) X 1/1000	No	Moment (k-in)	Rotation (radians) X 1/1000
1	0.00	0.00	26	373.40	53.64
2	18.91	0.75	27	382.68	56.91
3	37.82	1.50	28	391.97	60.17
4	55.13	2.05	29	399.90	63.69
5	72.43	2.61	30	407.83	67.21
6	92.39	3.45	31	415.75	70.72
7	112.36	4.30	32	423.68	74.24
8	128.30	5.11	33	433.30	78.41
9	144.25	5.93	34	442.91	82.57
10	163.96	7.22	35	452.53	86.73
11	183.67	8.52			
12	200.78	9.63			
13	217.89	10.74			
14	235.51	13.71			
15	253.13	16.67			
16	271.14	20.33			
17	289.15	24.00			
18	301.59	27.24			
19	314.04	30.49			
20	326.48	33.73			
21	333.57	37.08			
22	340.65	40.43			
23	347.74	43.77			
24	354.82	47.12			
25	364.11	50.38			

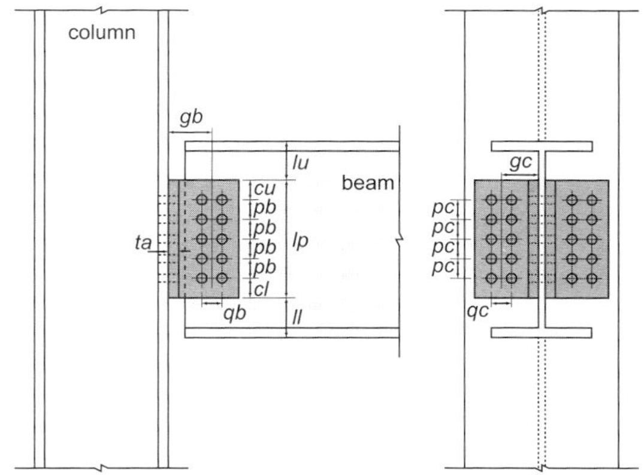

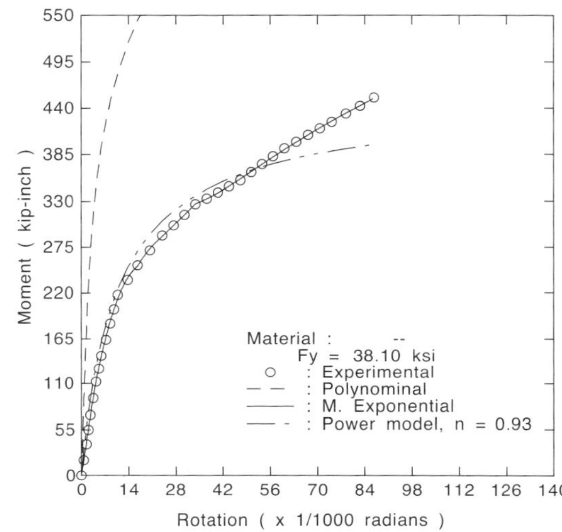

A.2 – 22

```
Moment-rotation prediction equations      ( R : X 1/1000 radians )

Frye and Morris polynominal model  :    R = Sum ( Ai X ( K*Bm )**Pi X 10**Qi )
     xd = 11.500000"    g =  5.504000"   t =  0.375000"
     A1 =  3.660000   A2 =  1.150000   A3 =  4.570000    K =  0.021699
     P1 =        1     P2 =        3     P3 =        5     Q1 =       -1     Q2 =       -3     Q3 =       -5

Modified exponential model  :
     Bm = Sum ( Ai X ( 1 - exp( - R/( 2*i*C ) ) ) )  +      Sum ( RKj X ( R - Rj0 ) )  +  Bm0
      C =   0.90996667E+00             Bmo=   0.00000000E+00           Nexp=  6        Nliner=  3
     AI =  -0.15575161E+03   0.24759510E+04  -0.14199667E+05   0.32981292E+05  -0.32432896E+05   0.11607501E+05
     Rj0 =          16.6700          24.0000          33.7300
     RKj =   0.39588087E+01   0.71014756E+00  -0.27299689E+01

Power model  :    Bm = ( Rki X R ) / ( 1 + ( R/R0 )**rn )**( 1/rn )
      rn = 0.928      rki =   0.45935212E+02         rmu =   0.45369160E+03
```

No	Rotation (radians) X 1/1000	Moment (kip-inch) Expri.	Moment (kip-inch) Poly.	Moment (kip-inch) M.Expo.	Moment (kip-inch) P.Model.	Connection stiffness (kip-inch) X 1000 Expri.	Connection stiffness (kip-inch) X 1000 Poly.	Connection stiffness (kip-inch) X 1000 M.Expo.	Connection stiffness (kip-inch) X 1000 P.Model.
1	0.00	0.00	0.00	0.00	0.00	0.2525E+02	0.1259E+03	0.2326E+02	0.4594E+02
3	1.50	37.82	175.95	39.94	57.90	0.2859E+02	0.9914E+02	0.2792E+02	0.3294E+02
5	2.61	72.43	265.13	70.32	91.11	0.2813E+02	0.6309E+02	0.2630E+02	0.2703E+02
7	4.30	112.36	345.12	111.27	131.15	0.2158E+02	0.3606E+02	0.2220E+02	0.2087E+02
9	5.93	144.25	393.86	144.39	161.68	0.1788E+02	0.2509E+02	0.1851E+02	0.1681E+02
11	8.52	183.67	447.00	186.01	199.23	0.1532E+02	0.1699E+02	0.1378E+02	0.1249E+02
13	10.74	217.89	480.39	212.75	224.07	0.1283E+02	0.1340E+02	0.1038E+02	0.1002E+02
15	16.67	253.13	543.90	251.85	270.70	0.5485E+01	0.8730E+01	0.3346E+01	0.6185E+01
17	24.00	289.15	597.56	288.27	306.65	0.4343E+01	0.6224E+01	0.3499E+01	0.3896E+01
19	30.49	314.04	633.70	313.88	328.06	0.3836E+01	0.5017E+01	0.3975E+01	0.2797E+01
21	37.08	333.57	663.97	332.99	344.04	0.2116E+01	0.4218E+01	0.1899E+01	0.2102E+01
23	43.77	347.74	690.20	347.71	356.45	0.2116E+01	0.3648E+01	0.2443E+01	0.1634E+01
25	50.38	364.11	712.86	364.72	366.12	0.2848E+01	0.3229E+01	0.2652E+01	0.1312E+01
27	56.91	382.68	732.83	382.02	373.88	0.2848E+01	0.2908E+01	0.2629E+01	0.1080E+01
29	63.69	399.90	751.62	399.43	380.57	0.2253E+01	0.2641E+01	0.2498E+01	0.8996E+00
31	70.72	415.75	769.39	416.46	386.38	0.2253E+01	0.2416E+01	0.2340E+01	0.7568E+00
33	78.41	433.30	787.14	433.86	391.71	0.2310E+01	0.2214E+01	0.2197E+01	0.6369E+00
35	86.73	452.53	804.80	451.66	396.58	0.2310E+01	0.2033E+01	0.2088E+01	0.5369E+00

A.2 – 23

II - 12

Connection type : Double web-angle connections
Mode : Riveted-to-beam and bolted-to-column

Tested by : C.W.Lewitt et al. (1966) U.S.A.
Test Id. : FK-3

Column : W10X49
Beam : W12X27
Angle : 6 X 4 X 3/8

Fasteners: A325- -3/4"D
13/16" Oversize holes
Material : --
Fy = 40.30 ksi
Fu = 64.00 ksi

Major parameters

lp = 8.5000"	lu = 1.7500"	ll = 1.7500"	ta = 0.3750"
gb = 3.5000"	gc = 2.6320"	cu = 1.2500"	cl = 1.2500"
pb = 3.0000"	pc = 3.0000"	qb = 2.5000"	
nb = 2 X 3	nc = 1 X 3		

Remark 1) High strength bolts used
2)

No	Moment (k-in)	Rotation (radians) X 1/1000
1	0.00	0.00
2	11.62	0.21
3	23.24	0.41
4	31.46	0.53
5	39.67	0.65
6	49.69	0.96
7	59.71	1.28
8	69.83	1.74
9	79.96	2.20
10	89.89	3.17
11	99.82	4.14
12	109.76	5.91
13	119.70	7.68
14	129.77	11.01
15	139.83	14.33
16	146.56	17.56
17	153.29	20.78
18	160.01	24.00
19	165.07	27.43
20	170.13	30.85
21	175.19	34.27
22	180.26	37.69
23	185.17	41.52
24	190.09	45.34
25	195.00	49.17

No	Moment (k-in)	Rotation (radians) X 1/1000
26	199.92	52.99
27	204.94	56.74
28	209.95	60.49
29	214.97	64.24
30	219.98	67.99
31	223.96	71.43
32	227.94	74.87
33	231.92	78.32
34	235.90	81.76
35	239.88	85.20

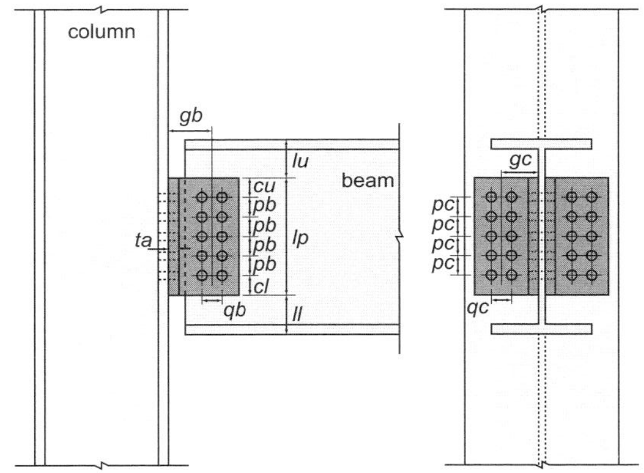

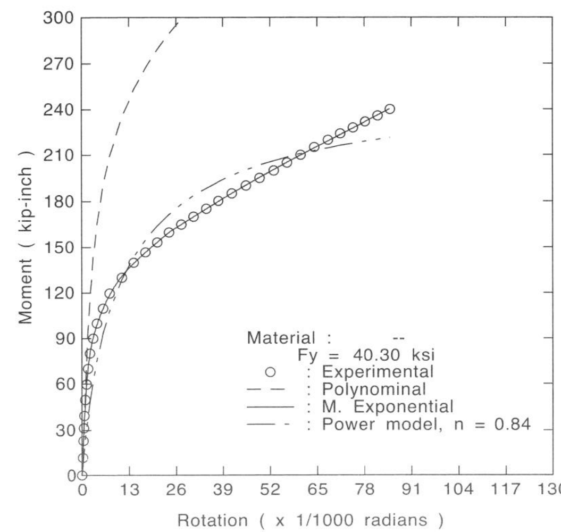

A.2 – 24

```
Moment-rotation prediction equations      ( R : X 1/1000 radians )

Frye and Morris polynominal model  :    R = Sum ( Ai X ( K*Bm )**Pi X 10**Qi )
     xd =  8.500000"    g =  5.501000"   t =  0.375000"
     A1 =  3.660000    A2 =  1.150000   A3 =  4.570000     K =  0.044819
     P1 =         1    P2 =         3   P3 =         5    Q1 =        -1     Q2 =        -3     Q3 =        -5

Modified exponential model  :
     Bm = Sum ( Ai X ( 1 - exp( - R/( 2*i*C ) ) ) )  +     Sum ( RKj X ( R - Rj0 ) )  +  Bm0
      C =  0.76157333E+00                Bmo=   0.00000000E+00          Nexp=  6         Nliner=  3
     AI =  0.16769802E+03  -0.48028817E+03   0.10266017E+04  -0.72048023E+03  -0.14632101E+03   0.28537063E+03
    Rj0 =          4.1400          14.3300          24.0000
    RKj =  0.15305401E+01   0.26467909E+00  -0.57159894E+00

Power model  :    Bm = ( Rki X R ) / ( 1 + ( R/R0 )**rn )**( 1/rn )
      rn = 0.843      rki =   0.30505012E+02          rmu =   0.25950039E+03
```

No	Rotation (radians) X 1/1000	Moment (kip-inch) Expri.	Poly.	M.Expo.	P.Model.	Connection stiffness (kip-inch) X 1000 Expri.	Poly.	M.Expo.	P.Model.
1	0.00	0.00	0.00	0.00	0.00	0.5633E+02	0.6096E+02	0.7086E+02	0.3051E+02
3	0.41	23.24	25.05	25.30	11.51	0.6441E+02	0.6019E+02	0.5268E+02	0.2588E+02
5	0.65	39.67	39.23	36.83	17.45	0.5883E+02	0.5889E+02	0.4450E+02	0.2406E+02
7	1.28	59.71	74.13	59.40	31.30	0.2792E+02	0.5166E+02	0.2884E+02	0.2039E+02
9	2.20	79.96	114.78	79.43	48.33	0.1809E+02	0.3614E+02	0.1592E+02	0.1661E+02
11	4.14	99.82	164.17	98.73	75.32	0.8628E+01	0.1825E+02	0.6319E+01	0.1179E+02
13	7.68	119.70	209.14	118.79	108.18	0.4716E+01	0.9170E+01	0.4355E+01	0.7349E+01
15	14.33	139.83	252.75	139.29	143.82	0.2548E+01	0.4881E+01	0.2466E+01	0.3932E+01
17	20.78	153.29	278.95	153.38	164.12	0.2087E+01	0.3439E+01	0.2058E+01	0.2529E+01
19	27.43	165.07	299.01	165.04	178.15	0.1479E+01	0.2670E+01	0.1477E+01	0.1764E+01
21	34.27	175.19	315.52	175.06	188.50	0.1479E+01	0.2189E+01	0.1440E+01	0.1299E+01
23	41.52	185.17	330.07	185.24	196.70	0.1285E+01	0.1849E+01	0.1369E+01	0.9864E+00
25	49.17	195.00	343.20	195.45	203.36	0.1285E+01	0.1597E+01	0.1305E+01	0.7680E+00
27	56.74	204.94	354.56	205.17	208.58	0.1338E+01	0.1411E+01	0.1266E+01	0.6180E+00
29	64.24	214.97	364.59	214.57	212.79	0.1338E+01	0.1269E+01	0.1244E+01	0.5101E+00
31	71.43	223.96	373.31	223.48	216.16	0.1156E+01	0.1160E+01	0.1234E+01	0.4318E+00
33	78.32	231.92	380.99	231.95	218.92	0.1156E+01	0.1073E+01	0.1229E+01	0.3731E+00
35	85.20	239.88	388.11	240.40	221.32	0.1156E+01	0.9992E+00	0.1226E+01	0.3260E+00

II - 13

Connection type : Double web-angle connections
Mode : All bolted

Tested by : C.W.Lewitt et al. (1966) U.S.A.
Test Id. : FK-4AB-M

Column : W12X65
Beam : W18X50
Angle : 6 X 4 X 3/8

Fasteners: A325- -3/4"D
13/16" Oversize holes
Material : --
Fy = 41.60 ksi
Fu = 61.60 ksi

Major parameters

lp = 11.5000"	lu = 3.2500"	ll = 3.2500"	ta = 0.3750"
gb = 3.5000"	gc = 2.0710"	cu = 1.2500"	cl = 1.2500"
pb = 3.0000"	pc = 3.0000"	qb = 2.5000"	qc = 2.0000"
nb = 2 X 4	nc = 2 X 4		

Remark 1) No washers used.
2) High strength bolts used

No	Moment (k-in)	Rotation (radians) X 1/1000	No	Moment (k-in)	Rotation (radians) X 1/1000
1	0.00	0.00	26	408.17	45.75
2	19.09	0.18	27	416.41	49.07
3	39.70	0.46	28	424.64	52.39
4	60.33	0.94	29	432.87	55.71
5	78.93	1.23	30	441.11	59.03
6	99.55	1.71	31	448.71	62.04
7	120.68	2.19	32	456.31	65.05
8	140.81	2.74	33	463.91	68.05
9	160.94	3.30	34	471.52	71.06
10	181.09	4.03	35	479.12	74.07
11	200.72	4.66			
12	220.90	5.74			
13	241.08	6.82			
14	261.82	8.58			
15	282.57	10.33			
16	301.93	13.49			
17	321.29	16.64			
18	334.63	19.84			
19	347.98	23.04			
20	361.32	26.24			
21	369.04	29.48			
22	376.77	32.72			
23	384.49	35.96			
24	392.22	39.20			
25	399.94	42.44			

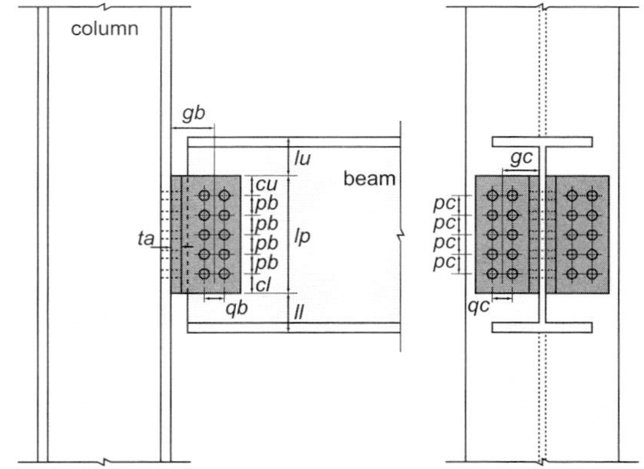

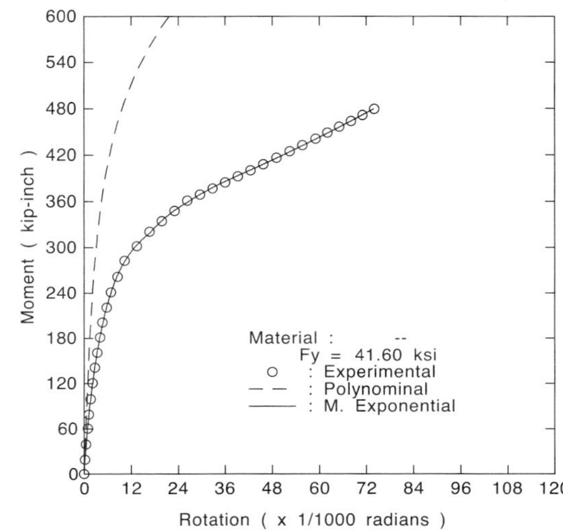

A.2 – 26

```
Moment-rotation prediction equations      ( R : X 1/1000 radians )

Frye and Morris polynominal model  :     R = Sum ( Ai X ( K*Bm )**Pi X 10**Qi )
    xd = 11.500000"   g =  4.500000"   t =  0.375000"
    A1 =  3.660000   A2 =  1.150000   A3 =  4.570000    K =  0.021053
    P1 =         1   P2 =         3   P3 =         5   Q1 =        -1    Q2 =        -3    Q3 =        -5

Modified exponential model  :
    Bm = Sum ( Ai X ( 1 - exp( - R/( 2*i*C ) ) ) )  +      Sum ( RKj X ( R - Rj0 ) )  +  Bm0
     C =   0.66292833E+00            Bmo=   0.00000000E+00           Nexp=  6        Nliner=  2
    AI =   0.15114560E+03  -0.17019443E+04   0.85397912E+04  -0.18338365E+05   0.18020863E+05  -0.63397921E+04
   Rj0 =          16.6400          26.2400
   RKj =   0.19102621E+01   0.79888153E+00
```

No	Rotation (radians) X 1/1000	Moment (kip-inch) Expri.	Poly.	M.Expo.	Connection stiffness (kip-inch) X 1000 Expri.	Poly.	M.Expo.
1	0.00	0.00	0.00	0.00	0.1046E+03	0.1298E+03	0.8275E+02
3	0.46	39.70	59.86	34.43	0.6209E+02	0.1277E+03	0.6680E+02
5	1.23	78.93	152.36	78.66	0.5728E+02	0.1116E+03	0.5059E+02
7	2.19	120.68	243.21	121.30	0.4037E+02	0.7742E+02	0.3922E+02
9	3.30	160.94	312.65	160.16	0.3246E+02	0.5030E+02	0.3118E+02
11	4.66	200.72	368.61	197.62	0.2651E+02	0.3387E+02	0.2413E+02
13	6.82	241.08	427.37	240.39	0.1607E+02	0.2217E+02	0.1599E+02
15	10.33	282.57	489.39	281.76	0.9793E+01	0.1436E+02	0.8608E+01
17	16.64	321.29	560.32	318.73	0.5162E+01	0.9012E+01	0.3999E+01
19	23.04	347.98	609.62	348.65	0.4169E+01	0.6658E+01	0.3578E+01
21	29.48	369.04	647.86	369.09	0.2385E+01	0.5328E+01	0.2925E+01
23	35.96	384.49	679.40	385.76	0.2385E+01	0.4466E+01	0.2342E+01
25	42.44	399.94	706.27	400.53	0.2432E+01	0.3863E+01	0.2268E+01
27	49.07	416.41	730.31	415.90	0.2481E+01	0.3405E+01	0.2376E+01
29	55.71	432.87	751.69	432.09	0.2481E+01	0.3052E+01	0.2501E+01
31	62.04	448.71	770.13	448.21	0.2526E+01	0.2784E+01	0.2588E+01
33	68.05	463.91	786.23	463.96	0.2526E+01	0.2573E+01	0.2641E+01
35	74.07	479.12	801.16	479.96	0.2526E+01	0.2394E+01	0.2672E+01

II - 14

Connection type : Double web-angle connections
Mode : Riveted-to-beam and bolted-to-column

Tested by : C.W.Lewitt et al. (1966) U.S.A.
Test Id. : FK-4P

Column : W12X65
Beam : W18X50
Angle : 6 X 4 X 3/8

Fasteners: A325- -3/4"D
13/16" Oversize holes
Material : --
Fy = 40.30 ksi
Fu = 64.00 ksi

Major parameters

lp = 11.5000"	lu = 3.2500"	ll = 3.2500"	ta = 0.3750"
gb = 3.5000"	gc = 2.0710"	cu = 1.2500"	cl = 1.2500"
pb = 3.0000"	pc = 3.0000"	qb = 2.5000"	qc = 2.0000"
nb = 2 X 4	nc = 2 X 4		

Remark 1) High strength bolts used
2)

No	Moment (k-in)	Rotation (radians) X 1/1000	No	Moment (k-in)	Rotation (radians) X 1/1000
1	0.00	0.00	26	376.91	45.11
2	18.52	0.24	27	386.76	48.40
3	39.02	0.32	28	396.62	51.69
4	53.55	0.76	29	404.63	54.63
5	77.58	1.14	30	412.63	57.57
6	98.09	1.33	31	420.64	60.52
7	118.63	1.87	32	428.64	63.46
8	138.68	2.60	33	436.65	66.40
9	159.25	3.54			
10	179.32	4.52			
11	198.92	6.01			
12	219.02	7.49			
13	239.22	10.28			
14	252.38	12.55			
15	265.54	14.82			
16	278.70	17.09			
17	288.87	19.61			
18	299.05	22.13			
19	309.23	24.64			
20	319.40	27.16			
21	328.85	30.00			
22	338.30	32.84			
23	347.75	35.68			
24	357.20	38.52			
25	367.05	41.82			

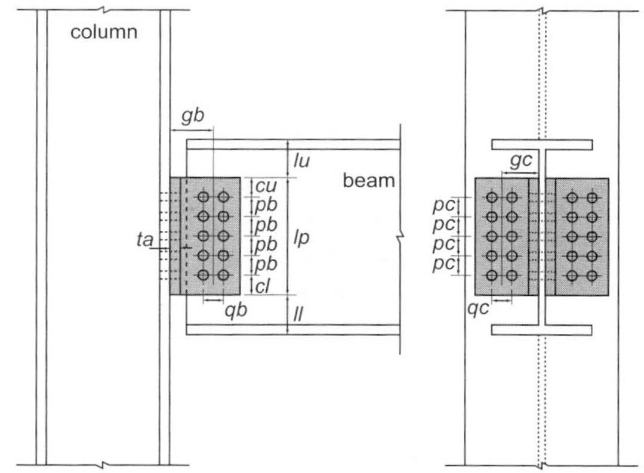

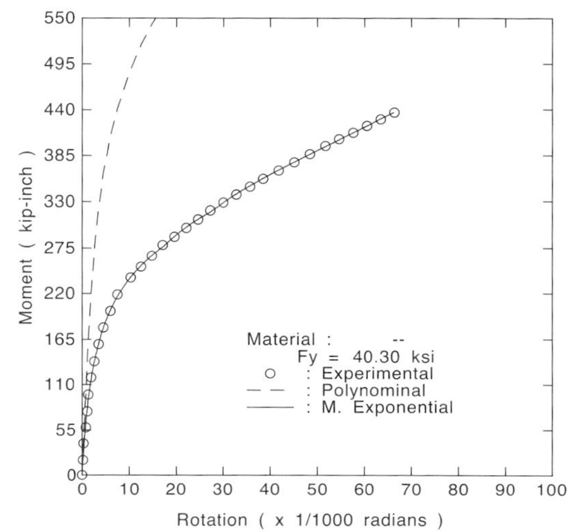

A.2 – 28

```
Moment-rotation prediction equations      ( R : X 1/1000 radians )

Frye and Morris polynominal model  :    R = Sum ( Ai X ( K*Bm )**Pi X 10**Qi )
     xd = 11.500000"    g =  4.500000"   t =  0.375000"
     A1 =  3.660000    A2 =  1.150000   A3 =  4.570000     K =  0.021053
     P1 =        1     P2 =        3    P3 =        5     Q1 =       -1     Q2 =       -3     Q3 =       -5

Modified exponential model  :
     Bm = Sum ( Ai X ( 1 - exp( - R/( 2*i*C ) ) ) )  +      Sum ( RKj X ( R - Rj0 ) )  +  Bm0
      C =   0.61256417E+00            Bmo=   0.00000000E+00          Nexp=  6        Nliner=  2
     AI =   0.14136882E+03  -0.71929538E+03   0.32986680E+04  -0.60678936E+04   0.45854200E+04  -0.92548673E+03
    Rj0 =          17.0900          27.1600
    RKj =   0.14154446E+01   0.13889879E+01
```

No	Rotation (radians) X 1/1000	Moment (kip-inch) Expri.	Moment (kip-inch) Poly.	Moment (kip-inch) M.Expo.	Connection stiffness (kip-inch) X 1000 Expri.	Connection stiffness (kip-inch) X 1000 Poly.	Connection stiffness (kip-inch) X 1000 M.Expo.
1	0.00	0.00	0.00	0.00	0.7816E+02	0.1298E+03	0.1038E+03
2	0.24	18.52	30.70	22.72	0.1967E+03	0.1293E+03	0.8855E+02
3	0.32	39.02	41.77	30.09	0.2076E+03	0.1288E+03	0.8375E+02
4	0.76	58.55	96.96	62.05	0.4715E+02	0.1236E+03	0.6403E+02
5	1.14	77.58	143.04	84.26	0.9064E+02	0.1142E+03	0.5159E+02
6	1.33	98.09	163.68	93.39	0.9194E+02	0.1082E+03	0.4686E+02
7	1.87	118.63	216.45	115.50	0.3369E+02	0.8859E+02	0.3648E+02
8	2.60	138.68	272.52	138.73	0.2494E+02	0.6532E+02	0.2746E+02
9	3.54	159.25	323.99	160.97	0.2120E+02	0.4652E+02	0.2072E+02
10	4.52	179.32	363.83	179.10	0.1751E+02	0.3505E+02	0.1640E+02
11	6.01	198.92	407.93	200.19	0.1337E+02	0.2549E+02	0.1235E+02
12	7.49	219.02	441.49	216.51	0.1135E+02	0.2005E+02	0.9797E+01
13	10.28	239.22	488.55	239.51	0.6448E+01	0.1445E+02	0.7049E+01
14	12.55	252.38	518.23	254.00	0.5790E+01	0.1183E+02	0.5796E+01
15	14.82	265.54	542.98	266.11	0.5790E+01	0.1007E+02	0.4890E+01
16	17.09	278.70	564.33	276.33	0.4962E+01	0.8787E+01	0.5527E+01
17	19.61	288.87	585.04	289.22	0.4044E+01	0.7726E+01	0.4734E+01
18	22.13	299.05	603.41	300.21	0.4044E+01	0.6910E+01	0.4014E+01
19	24.64	309.23	619.96	309.50	0.4044E+01	0.6263E+01	0.3388E+01
20	27.16	319.40	635.03	317.35	0.3707E+01	0.5735E+01	0.2871E+01
21	30.00	328.85	650.61	328.77	0.3325E+01	0.5245E+01	0.3804E+01
22	32.84	338.30	664.91	339.08	0.3325E+01	0.4839E+01	0.3470E+01
23	35.68	347.75	678.16	348.59	0.3325E+01	0.4496E+01	0.3235E+01
24	38.52	357.20	690.51	357.54	0.3172E+01	0.4203E+01	0.3075E+01
25	41.82	367.05	703.86	367.45	0.2995E+01	0.3912E+01	0.2957E+01
26	45.11	376.91	716.31	377.05	0.2995E+01	0.3663E+01	0.2886E+01
27	48.40	386.76	728.00	386.48	0.2995E+01	0.3446E+01	0.2846E+01
28	51.69	396.62	739.02	395.80	0.2850E+01	0.3256E+01	0.2823E+01
29	54.63	404.63	748.37	404.10	0.2721E+01	0.3104E+01	0.2813E+01
30	57.57	412.63	757.30	412.36	0.2721E+01	0.2967E+01	0.2807E+01
31	60.52	420.64	765.85	420.62	0.2721E+01	0.2843E+01	0.2804E+01
32	63.46	428.64	774.22	428.86	0.2721E+01	0.2728E+01	0.2803E+01
33	66.40	436.65	781.93	437.11	0.2721E+01	0.2627E+01	0.2802E+01

II - 15

Connection type : Double web-angle connections
Mode : Riveted-to-beam and bolted-to-column

Tested by : C.W.Lewitt et al. (1966) U.S.A.
Test Id. : WK-4

Column : W12X65
Beam : W18X50
Angle : 6 X 4 X 3/8

Fasteners: A325- -3/4"D
13/16" Oversize holes
Material : --
Fy = 40.30 ksi
Fu = 64.00 ksi

Major parameters

lp = 11.5000"	lu = 3.2500"	ll = 3.2500"	ta = 0.3750"
gb = 3.5000"	gc = 2.5710"	cu = 1.2500"	cl = 1.2500"
pb = 3.0000"	pc = 3.0000"	qb = 2.5000"	
nb = 2 X 4	nc = 1 X 4		

Remark 1) No washers used.
2) High strength bolts used

No	Moment (k-in)	Rotation (radians) X 1/1000	No	Moment (k-in)	Rotation (radians) X 1/1000
1	0.00	0.00	26	350.22	45.42
2	10.99	0.06	27	358.22	48.64
3	21.97	0.11	28	368.25	52.08
4	41.94	0.32	29	378.29	55.51
5	60.91	0.44	30	388.32	58.95
6	79.39	0.50	31	398.35	62.38
7	90.37	0.60	32	408.41	65.41
8	101.36	0.71	33	418.47	68.43
9	119.82	0.97	34	428.53	71.45
10	140.27	1.43	35	438.59	74.48
11	158.73	1.84			
12	178.66	2.75			
13	198.57	3.86			
14	220.42	5.92			
15	239.24	8.72			
16	252.86	11.92			
17	266.49	15.12			
18	280.12	18.32			
19	289.65	21.87			
20	299.17	25.42			
21	308.70	28.96			
22	318.23	32.51			
23	326.23	35.74			
24	334.22	38.96			
25	342.22	42.19			

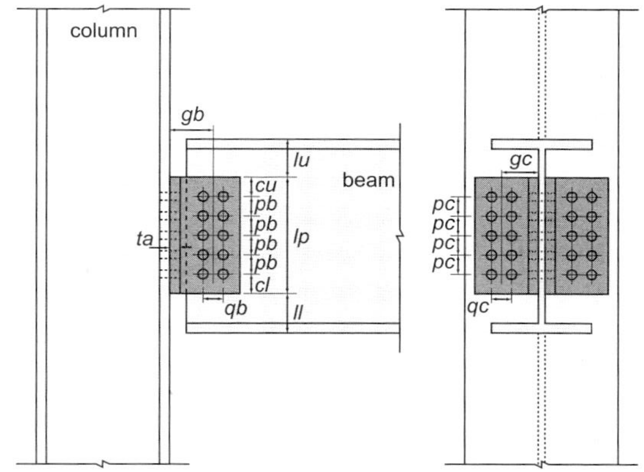

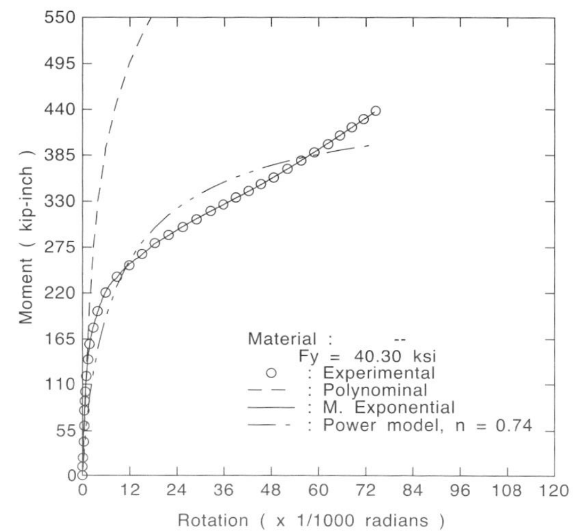

A.2 – 30

```
Moment-rotation prediction equations      ( R : X 1/1000 radians )

Frye and Morris polynominal model  :    R = Sum ( Ai X ( K*Bm )**Pi X 10**Qi )
     xd = 11.500000"   g =  5.500000"   t =  0.375000"
     A1 =  3.660000   A2 =  1.150000   A3 =  4.570000     K =  0.021696
     P1 =         1   P2 =         3   P3 =         5    Q1 =        -1     Q2 =        -3     Q3 =        -5

Modified exponential model  :
     Bm = Sum ( Ai X ( 1 - exp( - R/( 2*i*C ) ) ) )  +      Sum ( RKj X ( R - Rj0 ) )  +  Bm0
      C =   0.63477417E+00              Bmo=   0.00000000E+00          Nexp=  6        Nliner=  2
     AI =   0.72208756E+03  -0.50622271E+04   0.18792698E+05  -0.34122869E+05   0.30281047E+05  -0.10405301E+05
    Rj0 =           8.7200          18.3200
    RKj =   0.50367239E+01  -0.17654912E+01

Power model  :    Bm = ( Rki X R ) / ( 1 + ( R/R0 )**rn )**( 1/rn )
      rn = 0.741       rki =   0.83935443E+02          rmu =   0.48006048E+03
```

No	Rotation (radians) X 1/1000	Moment (kip-inch) Expri.	Moment Poly.	Moment M.Expo.	Moment P.Model.	Connection stiffness (kip-inch) X 1000 Expri.	Poly.	M.Expo.	P.Model.
1	0.00	0.00	0.00	0.00	0.00	0.1944E+03	0.1259E+03	0.1941E+03	0.8394E+02
3	0.11	21.97	14.22	20.65	8.82	0.1736E+03	0.1258E+03	0.1721E+03	0.7408E+02
5	0.44	60.91	54.63	67.60	30.33	0.2565E+03	0.1241E+03	0.1219E+03	0.6064E+02
7	0.60	90.37	75.20	86.28	40.06	0.1033E+03	0.1223E+03	0.1021E+03	0.5591E+02
9	0.97	119.82	118.94	117.37	59.03	0.6133E+02	0.1155E+03	0.6968E+02	0.4800E+02
11	1.84	158.73	207.83	158.44	95.18	0.3783E+02	0.8687E+02	0.3129E+02	0.3612E+02
13	3.86	198.57	328.26	197.33	152.54	0.1539E+02	0.4083E+02	0.1398E+02	0.2262E+02
15	8.72	239.24	450.40	237.61	228.82	0.5564E+01	0.1658E+02	0.7452E+01	0.1108E+02
17	15.12	266.49	529.84	266.21	281.13	0.4258E+01	0.9578E+01	0.3796E+01	0.6085E+01
19	21.87	289.65	583.80	289.49	313.82	0.2686E+01	0.6777E+01	0.2950E+01	0.3877E+01
21	28.96	308.70	625.96	309.39	336.68	0.2686E+01	0.5252E+01	0.2583E+01	0.2687E+01
23	35.74	326.23	658.29	325.92	352.45	0.2479E+01	0.4357E+01	0.2373E+01	0.2018E+01
25	42.19	342.22	684.40	341.58	364.03	0.2479E+01	0.3767E+01	0.2515E+01	0.1599E+01
27	48.64	358.22	707.24	358.60	373.35	0.2693E+01	0.3329E+01	0.2760E+01	0.1304E+01
29	55.51	378.29	728.82	378.35	381.49	0.2921E+01	0.2971E+01	0.2979E+01	0.1076E+01
31	62.38	398.35	748.23	399.34	388.27	0.3137E+01	0.2688E+01	0.3119E+01	0.9056E+00
33	68.43	418.47	763.85	418.44	393.38	0.3327E+01	0.2485E+01	0.3190E+01	0.7886E+00
35	74.48	438.59	778.34	437.86	397.85	0.3327E+01	0.2312E+01	0.3229E+01	0.6940E+00

II - 16

Connection type : Double web-angle connections
Mode : Riveted-to-beam and bolted-to-column

Tested by : C.W.Lewitt et al. (1966) U.S.A.
Test Id. : FB-4

Column : W12X65
Beam : W18X50
Angle : 6 X 3.5 X 3/8

Fasteners: A325- -3/4"D
13/16" Oversize holes
Material : --
Fy = 40.30 ksi
Fu = 64.00 ksi

Major parameters

lp = 11.5000"	lu = 3.2500"	ll = 3.2500"	ta = 0.3750"
gb = 2.2500"	gc = 2.5710"	cu = 1.2500"	cl = 1.2500"
pb = 3.0000"	pc = 3.0000"	nb = 1 X 4	nc = 1 X 4

Remark 1) No washers used.
2) High strength bolts used

No	Moment (k-in)	Rotation (radians) X 1/1000	No	Moment (k-in)	Rotation (radians) X 1/1000
1	0.00	0.00	26	369.40	42.25
2	20.45	0.16	27	379.45	45.79
3	15.95	0.36	28	389.51	49.33
4	30.92	0.38	29	399.56	52.87
5	45.89	0.41	30	406.23	55.94
6	60.86	0.43	31	412.89	59.02
7	80.30	0.79	32	419.56	62.10
8	101.25	1.00	33	426.23	65.17
9	120.18	1.51	34	432.89	68.25
10	140.12	2.02	35	439.56	71.33
11	160.55	2.58			
12	179.47	3.34			
13	199.89	4.15			
14	220.29	5.41			
15	240.17	7.02			
16	260.00	9.57			
17	230.80	12.88			
18	290.50	15.81			
19	300.21	18.74			
20	309.92	21.67			
21	319.63	24.59			
22	329.56	28.12			
23	339.49	31.65			
24	349.42	35.18			
25	359.35	38.71			

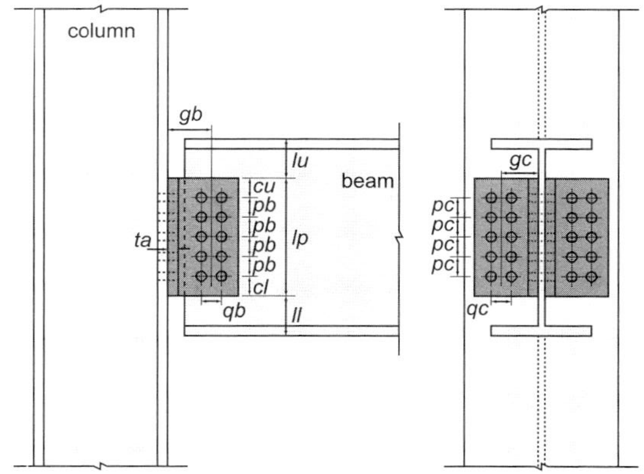

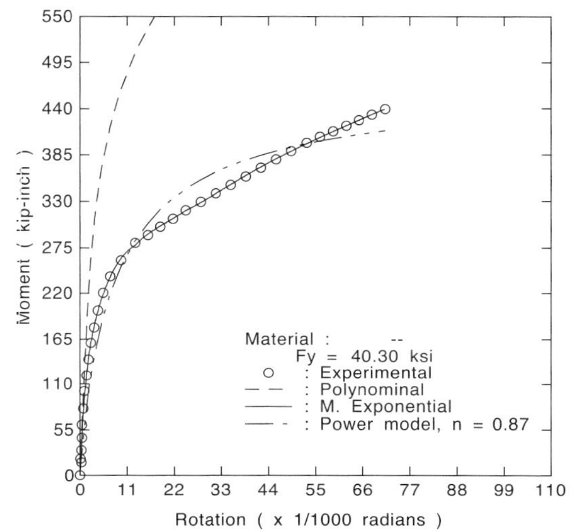

```
Moment-rotation prediction equations     ( R : X 1/1000 radians )

Frye and Morris polynominal model  :    R = Sum ( Ai X ( K*Bm )**Pi X 10**Qi )
     xd = 11.500000"   g =  5.500000"   t =  0.375000"
     A1 =  3.660000   A2 =  1.150000   A3 =  4.570000    K =  0.021696
     P1 =        1    P2 =        3    P3 =        5    Q1 =       -1    Q2 =       -3    Q3 =       -5

Modified exponential model  :
     Bm = Sum ( Ai X ( 1 - exp( - R/( 2*i*C ) ) ) )  +      Sum ( RKj X ( R - Rj0 ) )  +  Bm0
      C =   0.63440833E+00          Bmo=   0.00000000E+00           Nexp=  6        Nliner=  2
     AI =   0.76606601E+02   0.14280102E+04  -0.90745485E+04   0.21724215E+05  -0.21977329E+05   0.81312216E+04
     Rj0 =           12.8800          24.5900
     RKj =  0.28820759E+01  -0.77923603E+00

Power model  :    Bm = ( Rki X R ) / ( 1 + ( R/R0 )**rn )**( 1/rn )
      rn = 0.868       rki =   0.71313740E+02         rmu =   0.47488239E+03
```

No	Rotation (radians) X 1/1000	Moment (kip-inch) Expri.	Moment (kip-inch) Poly.	Moment (kip-inch) M.Expo.	Moment (kip-inch) P.Model.	Connection stiffness (kip-inch) X 1000 Expri.	Connection stiffness (kip-inch) X 1000 Poly.	Connection stiffness (kip-inch) X 1000 M.Expo.	Connection stiffness (kip-inch) X 1000 P.Model.
1	0.00	0.00	0.00	0.00	0.00	0.1277E+03	0.1259E+03	0.1234E+03	0.7131E+02
3	0.36	15.95	44.92	39.41	23.38	0.5512E+03	0.1247E+03	0.9799E+02	0.6056E+02
5	0.41	45.89	50.93	44.06	26.27	0.6198E+03	0.1244E+03	0.9505E+02	0.5947E+02
7	0.79	80.30	97.82	76.50	47.61	0.8281E+02	0.1194E+03	0.7501E+02	0.5211E+02
9	1.51	120.18	177.07	120.59	81.33	0.3815E+02	0.9876E+02	0.4976E+02	0.4225E+02
11	2.58	160.55	263.09	161.91	120.97	0.3158E+02	0.6393E+02	0.3008E+02	0.3261E+02
13	4.15	199.89	339.62	198.42	164.68	0.2169E+02	0.3757E+02	0.1847E+02	0.2388E+02
15	7.02	240.17	418.87	239.44	219.36	0.1058E+02	0.2086E+02	0.1119E+02	0.1528E+02
17	12.88	280.80	506.62	278.14	283.79	0.4710E+01	0.1119E+02	0.5704E+01	0.7942E+01
19	18.74	300.21	561.03	301.21	320.46	0.3316E+01	0.7821E+01	0.3019E+01	0.4948E+01
21	24.59	319.63	601.28	319.15	344.49	0.3089E+01	0.6088E+01	0.2509E+01	0.3408E+01
23	31.65	339.49	639.51	338.96	364.50	0.2815E+01	0.4852E+01	0.3011E+01	0.2364E+01
25	38.71	359.35	670.78	360.24	378.82	0.2827E+01	0.4062E+01	0.2952E+01	0.1744E+01
27	45.79	379.45	697.47	380.19	389.66	0.2840E+01	0.3508E+01	0.2673E+01	0.1343E+01
29	52.87	399.56	720.79	398.19	398.14	0.2480E+01	0.3098E+01	0.2426E+01	0.1069E+01
31	59.02	412.89	738.97	412.64	404.16	0.2167E+01	0.2819E+01	0.2284E+01	0.8949E+00
33	65.17	426.23	755.59	426.41	409.23	0.2167E+01	0.2590E+01	0.2199E+01	0.7611E+00
35	71.33	439.56	770.92	439.78	413.58	0.2167E+01	0.2399E+01	0.2152E+01	0.6559E+00

II - 17

Connection type : Double web-angle connections
Mode : Riveted-to-beam and bolted-to-column

Tested by : C.W.Lewitt et al. (1966) U.S.A.
Test Id. : FK-5

Column : W12X65
Beam : W18X50
Angle : 6 X 4 X 7/16

Fasteners: A325- -3/4"D
13/16" Oversize holes
Material : --
Fy = 36.80 ksi
Fu = 63.70 ksi

Major parameters

lp = 14.5000"	lu = 3.2500"	ll = 3.2500"	ta = 0.4380"
gb = 3.5000"	gc = 2.5500"	cu = 1.2500"	cl = 1.2500"
pb = 3.0000"	pc = 3.0000"	qb = 2.5000"	
nb = 2 X 5	nc = 1 X 5		

Remark 1) No washers used.
2)

No	Moment (k-in)	Rotation (radians) X 1/1000	No	Moment (k-in)	Rotation (radians) X 1/1000
1	0.00	0.00	26	611.63	10.75
2	37.06	0.18	27	638.41	12.19
3	57.84	0.30	28	654.50	13.64
4	78.62	0.41	29	670.59	15.09
5	98.65	0.55	30	686.68	16.54
6	118.68	0.69	31	702.77	17.99
7	153.24	0.88	32	718.87	19.44
8	178.28	1.07	33	730.59	20.82
9	198.31	1.26	34	742.32	22.21
10	218.34	1.40	35	754.05	23.60
11	238.37	1.54	36	765.77	24.98
12	258.41	1.73	37	777.50	26.37
13	278.44	1.92	38	789.23	27.76
14	298.48	2.09	39	800.95	29.14
15	318.51	2.25			
16	358.09	2.84			
17	378.13	3.10			
18	398.17	3.37			
19	418.47	3.86			
20	438.77	4.35			
21	458.81	4.77			
22	478.86	5.19			
23	518.47	6.52			
24	558.09	7.86			
25	584.86	9.30			

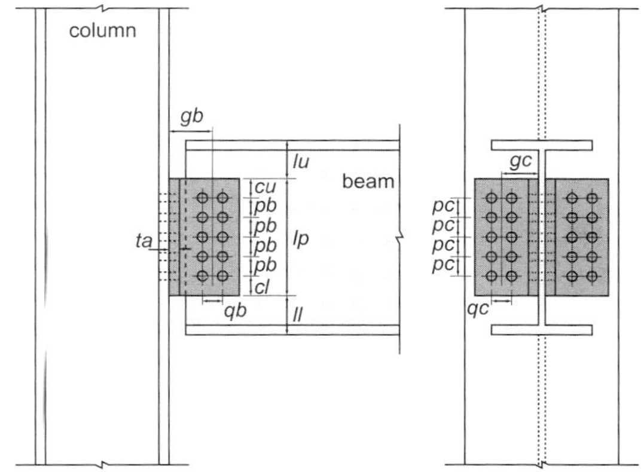

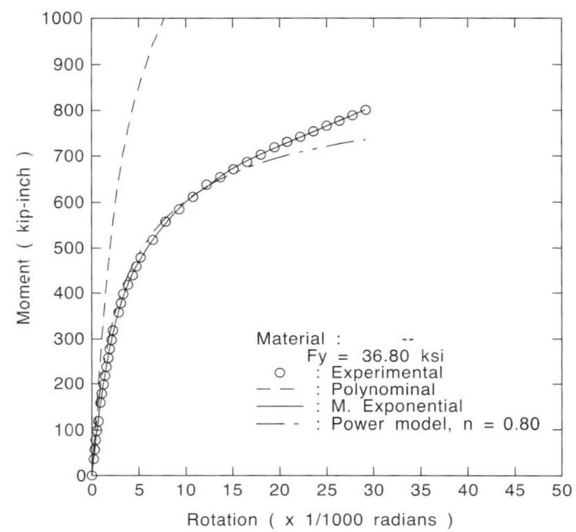

A.2 – 34

```
Moment-rotation prediction equations      ( R : X 1/1000 radians )

Frye and Morris polynominal model  :    R = Sum ( Ai X ( K*Bm )**Pi X 10**Qi )
     xd = 14.500000"   g =  5.458000"   t =  0.438000"
     A1 =  3.660000   A2 =  1.150000   A3 =  4.570000    K =  0.009380
     P1 =        1     P2 =        3     P3 =        5    Q1 =       -1     Q2 =       -3     Q3 =       -5

Modified exponential model  :
     Bm = Sum ( Ai X ( 1 - exp( - R/( 2*i*C ) ) ) )  +     Sum ( RKj X ( R - Rj0 ) )  +  Bm0
      C =   0.28847250E+00            Bmo=   0.00000000E+00         Nexp=  6        Nliner=  2
     AI =   0.15275796E+03  -0.28212114E+04  0.14360033E+05  -0.28862674E+05  0.25141353E+05  -0.73326268E+04
     Rj0 =          7.8600          12.1900
     RKj =   0.26231006E+01   0.63478854E+01

Power model  :     Bm = ( Rki X R ) / ( 1 + ( R/R0 )**rn )**( 1/rn )
      rn = 0.804       rki =   0.33244480E+03          rmu =   0.87001992E+03
```

No	Rotation (radians) X 1/1000	Moment (kip-inch) Expri.	Poly.	M.Expo.	P.Model.	Connection stiffness (kip-inch) X 1000 Expri.	Poly.	M.Expo.	P.Model.
1	0.00	0.00	0.00	0.00	0.00	0.2032E+03	0.2913E+03	0.2068E+03	0.3324E+03
3	0.30	57.84	86.50	57.00	81.05	0.1805E+03	0.2894E+03	0.1787E+03	0.2320E+03
5	0.55	98.65	159.96	100.57	134.49	0.1424E+03	0.2843E+03	0.1628E+03	0.1889E+03
7	0.88	158.24	249.69	150.32	189.08	0.1632E+03	0.2721E+03	0.1469E+03	0.1527E+03
9	1.26	198.31	349.36	203.20	241.49	0.1265E+03	0.2483E+03	0.1301E+03	0.1236E+03
11	1.54	238.37	416.15	238.14	273.96	0.1265E+03	0.2260E+03	0.1183E+03	0.1078E+03
13	1.92	278.44	496.40	280.31	311.80	0.1134E+03	0.1945E+03	0.1028E+03	0.9124E+02
15	2.25	318.51	556.64	312.33	340.15	0.1091E+03	0.1693E+03	0.9045E+02	0.8008E+02
17	3.10	378.13	678.31	377.63	398.92	0.7525E+02	0.1211E+03	0.6489E+02	0.5993E+02
19	3.86	418.47	759.35	420.71	439.63	0.4123E+02	0.9442E+02	0.4981E+02	0.4812E+02
21	4.77	458.81	835.24	460.62	478.65	0.4807E+02	0.7410E+02	0.3901E+02	0.3830E+02
23	6.52	518.47	944.00	518.99	534.52	0.2963E+02	0.5223E+02	0.2883E+02	0.2656E+02
25	9.30	584.86	1063.42	587.67	593.33	0.1854E+02	0.3588E+02	0.2059E+02	0.1690E+02
27	12.19	638.41	1153.46	632.50	633.88	0.1483E+02	0.2733E+02	0.1736E+02	0.1169E+02
29	15.09	670.59	1224.68	673.72	662.97	0.1110E+02	0.2221E+02	0.1176E+02	0.8627E+01
31	17.99	702.77	1283.81	703.82	684.94	0.1110E+02	0.1880E+02	0.9412E+01	0.6664E+01
33	20.82	730.59	1333.55	729.31	701.86	0.8456E+01	0.1640E+02	0.8729E+01	0.5347E+01
35	23.60	754.05	1376.47	753.34	715.34	0.8456E+01	0.1462E+02	0.8649E+01	0.4416E+01
37	26.37	777.50	1415.00	777.43	726.57	0.8456E+01	0.1322E+02	0.8729E+01	0.3717E+01
39	29.14	800.95	1450.04	801.77	736.10	0.8456E+01	0.1208E+02	0.8820E+01	0.3177E+01

II - 18

Connection type : Double web-angle connections
Mode : All bolted

Tested by : B.Bose (1981) U.K
Test Id. : B-1

Column : 305 X 305 X 97
Beam : 457 X 191 X 67
Angle : 150 X 90 X 15

Fasteners: G8.8- -M20
7/8" Oversize holes
Material : BS4360
Fy = -- ksi
Fu = -- ksi

Major parameters

lp = 15.7480" lu = 1.1220" ll = 1.1220" ta = 0.5906"
gb = 3.5433" gc = 2.3799" cu = 1.9685" cl = 1.9685"
pb = 2.9527" pc = 2.9527" qb = 2.3622"
nb = 2 X - nc = 1 X 5

Remark 1) Grade 8.8 bolts were used.
2) 4 and 5 bolts used either line of the cleat to column.

No	Moment (k-in)	Rotation (radians) X 1/1000	No	Moment (k-in)	Rotation (radians) X 1/1000
1	0.00	0.00	26	676.41	6.33
2	27.59	0.14	27	689.10	6.63
3	55.17	0.28	28	701.78	6.94
4	82.76	0.42	29	720.59	7.29
5	113.37	0.62	30	739.40	7.65
6	143.99	0.82	31	758.21	8.00
7	174.61	1.03	32	774.56	8.34
8	202.80	1.18	33	790.92	8.68
9	230.99	1.33	34	807.27	9.02
10	259.18	1.48			
11	289.19	1.68			
12	319.19	1.88			
13	349.20	2.07			
14	386.78	2.42			
15	424.36	2.76			
16	458.23	3.14			
17	492.10	3.51			
18	511.28	3.77			
19	530.47	4.04			
20	551.74	4.38			
21	573.02	4.72			
22	594.29	5.07			
23	617.44	5.39			
24	640.58	5.70			
25	663.72	6.02			

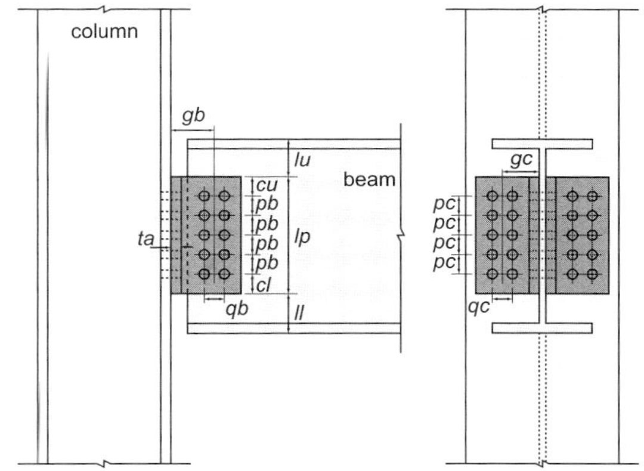

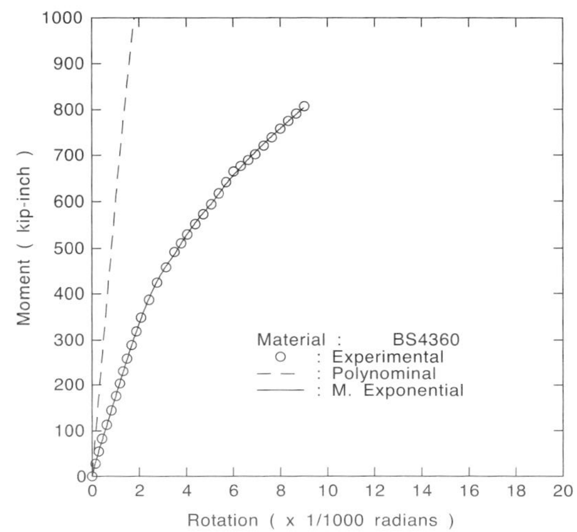

A.2 – 36

```
Moment-rotation prediction equations      ( R : X 1/1000 radians )

Frye and Morris polynominal model  :    R = Sum ( Ai X ( K*Bm )**Pi X 10**Qi )
     xd = 15.748000"   g =  5.118100"   t =  0.590600"
     A1 =  3.660000   A2 =  1.150000   A3 =  4.570000    K =  0.004436
     P1 =         1   P2 =         3   P3 =         5   Q1 =        -1    Q2 =        -3    Q3 =        -5

Modified exponential model  :
     Bm = Sum ( Ai X ( 1 - exp( - R/( 2*i*C ) ) ) )  +     Sum ( RKj X ( R - Rj0 ) )  +  Bm0
      C =   0.11030000E+00              Bmo=   0.00000000E+00          Nexp=  6        Nliner=  2
     AI =  -0.55340607E+02   0.15859125E+04  -0.80319600E+04   0.17543102E+05  -0.19700981E+05   0.92725157E+04
     Rj0 =          3.5100          5.0700
     RKj =   0.23717061E+02   0.16698006E+02
```

No	Rotation (radians) X 1/1000	Moment (kip-inch) Expri.	Moment (kip-inch) Poly.	Moment (kip-inch) M.Expo.	Connection stiffness (kip-inch) X 1000 Expri.	Connection stiffness (kip-inch) X 1000 Poly.	Connection stiffness (kip-inch) X 1000 M.Expo.
1	0.00	0.00	0.00	0.00	0.1963E+03	0.6159E+03	0.2325E+03
3	0.28	55.17	172.75	55.77	0.1963E+03	0.6124E+03	0.1737E+03
5	0.62	113.37	379.95	111.93	0.1519E+03	0.5969E+03	0.1619E+03
7	1.03	174.61	613.53	178.95	0.1716E+03	0.5579E+03	0.1703E+03
9	1.33	230.99	775.86	230.54	0.1863E+03	0.5136E+03	0.1693E+03
11	1.68	289.19	944.86	287.91	0.1516E+03	0.4535E+03	0.1577E+03
13	2.07	349.20	1110.74	346.36	0.1360E+03	0.3857E+03	0.1368E+03
15	2.76	424.36	1342.22	427.13	0.1003E+03	0.2907E+03	0.9812E+02
17	3.51	492.10	1531.31	487.49	0.8032E+02	0.2228E+03	0.6622E+02
19	4.04	530.47	1639.75	530.48	0.6793E+02	0.1897E+03	0.7398E+02
21	4.72	573.02	1759.11	576.03	0.6174E+02	0.1584E+03	0.5910E+02
23	5.39	617.44	1856.19	616.84	0.7323E+02	0.1366E+03	0.6577E+02
25	6.02	663.72	1937.32	656.09	0.5699E+02	0.1207E+03	0.5876E+02
27	6.63	689.10	2007.91	690.71	0.4111E+02	0.1085E+03	0.5367E+02
29	7.29	720.59	2075.94	724.75	0.5347E+02	0.9794E+02	0.4966E+02
31	8.00	758.21	2141.52	758.56	0.5068E+02	0.8882E+02	0.4662E+02
33	8.68	790.92	2199.52	789.60	0.4797E+02	0.8152E+02	0.4457E+02

A.2 – 37

II - 19

Connection type : Double web-angle connections
Mode : All bolted

Tested by : L.E.Thompson et al. (1970) U.S.A
Test Id. : A1-1 ALT

Column : W10X49
Beam : W21X62
Angle : 4 X 3.5 X 5/16

Fasteners: A325- -3/4"D
15/16" Oversize holes
Material : A36
Fy = -- ksi
Fu = -- ksi

Major parameters

lp = 11.5000"	lu = -- "	ll = -- "	ta = 0.3125"
gb = 2.2500"	gc = 2.5500"	cu = -- "	cl = -- "
pb = -- "	pc = -- "	qb = -- "	qc = -- "
nb = - X 2	nc = - X -		

Remark 1)
2)

No	Moment (k-in)	Rotation (radians) X 1/1000
1	0.00	0.00
2	24.50	1.72
3	49.00	3.43
4	73.50	5.15
5	98.00	6.87
6	122.50	8.58
7	147.00	10.30
8	163.16	15.30
9	175.48	20.65
10	187.79	26.00
11	200.10	31.35
12	212.41	36.70
13	223.62	42.05
14	234.83	47.40
15	246.04	52.75
16	257.26	58.10
17	250.93	63.42
18	254.60	68.74
19	258.28	74.06
20	271.95	79.38
21	275.63	84.70
22	234.81	89.75
23	294.00	94.80
24	312.38	96.35
25	330.76	97.90
26	349.13	99.05
27	367.50	100.20
28	385.88	101.40
29	404.26	102.60
30	422.63	103.00
31	441.00	103.40
32	459.38	103.75
33	477.76	104.10
34	496.13	105.05
35	514.50	106.00
36	532.88	107.30
37	551.26	108.60
38	569.63	111.75
39	588.00	114.90

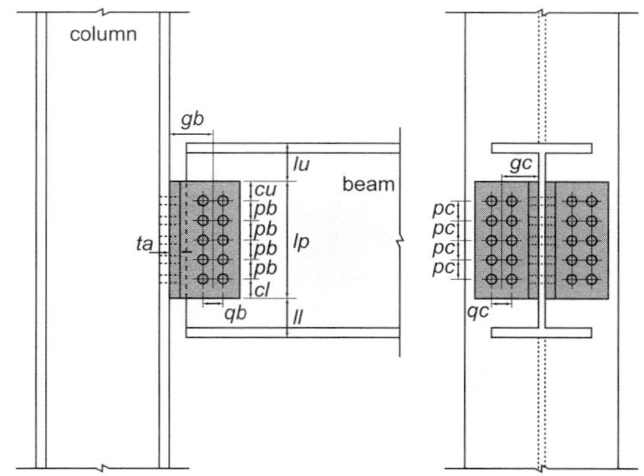

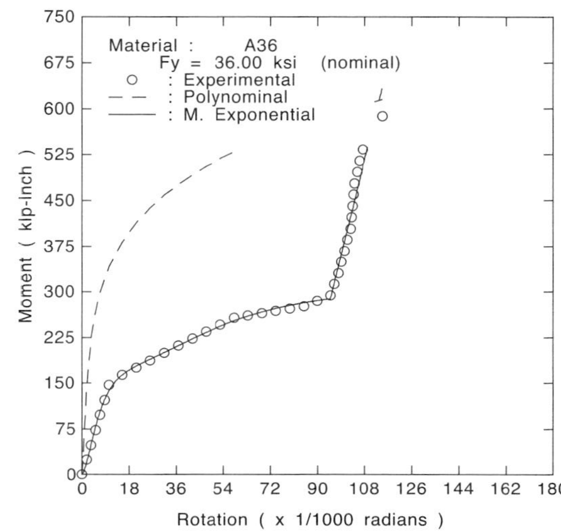

A.2 – 38

```
Moment-rotation prediction equations     ( R : X 1/1000 radians )

Frye and Morris polynominal model  :    R = Sum ( Ai X ( K*Bm )**Pi X 10**Qi )
     xd = 11.500000"    g =  5.500000"   t =  0.312500"
     A1 =   3.660000   A2 =  1.150000   A3 =  4.570000    K =  0.030179
     P1 =          1   P2 =         3   P3 =         5    Q1 =       -1     Q2 =       -3     Q3 =       -5

Modified exponential model  :
     Bm = Sum ( Ai X ( 1 - exp( - R/( 2*i*C ) ) ) )  +      Sum ( RKj X ( R - Rj0 ) )  +  Bm0
      C =    0.13866750E+01            Bmo=   0.00000000E+00          Nexp=  6         Nliner=  3
     AI =  -0.27629246E+03   0.15399204E+04  -0.50463775E+04   0.12162151E+05  -0.14796109E+05   0.66965150E+04
     Rj0 =           10.3000         58.1000          94.8000
     RKj =   0.26064200E+00  -0.12358625E+00   0.16928574E+02
```

No	Rotation (radians) X 1/1000	Moment (kip-inch) Expri.	Moment (kip-inch) Poly.	Moment (kip-inch) M.Expo.	Connection stiffness (kip-inch) X 1000 Expri.	Connection stiffness (kip-inch) X 1000 Poly.	Connection stiffness (kip-inch) X 1000 M.Expo.
1	0.00	0.00	0.00	0.00	0.1427E+02	0.9054E+02	0.3225E+01
3	3.43	49.00	222.73	48.94	0.1427E+02	0.3353E+02	0.1808E+02
5	6.87	98.00	298.78	103.34	0.1427E+02	0.1536E+02	0.1284E+02
7	10.30	147.00	341.07	137.27	0.1145E+02	0.1005E+02	0.7299E+01
9	20.65	175.48	413.70	178.12	0.2301E+01	0.5133E+01	0.2305E+01
11	31.35	200.10	458.72	200.86	0.2301E+01	0.3518E+01	0.2089E+01
13	42.05	223.62	491.66	222.97	0.2096E+01	0.2716E+01	0.2014E+01
15	52.75	246.04	517.94	243.37	0.2096E+01	0.2232E+01	0.1778E+01
17	63.42	260.93	539.91	260.03	0.6907E+00	0.1906E+01	0.1339E+01
19	74.06	268.28	558.87	272.55	0.6907E+00	0.1670E+01	0.1021E+01
21	84.70	275.63	575.65	281.91	0.1270E+01	0.1491E+01	0.7479E+00
23	94.80	294.00	590.01	288.41	0.9499E+01	0.1356E+01	0.5480E+00
25	97.90	330.76	594.18	342.50	0.1422E+02	0.1320E+01	0.1743E+02
27	100.20	367.50	597.20	382.55	0.1565E+02	0.1295E+01	0.1739E+02
29	102.60	404.26	600.30	424.25	0.3828E+02	0.1269E+01	0.1736E+02
31	103.40	441.00	601.31	438.14	0.4944E+02	0.1261E+01	0.1735E+02
33	104.10	477.76	602.19	450.28	0.4358E+02	0.1254E+01	0.1734E+02
35	106.00	514.50	604.56	483.21	0.1714E+02	0.1235E+01	0.1732E+02
37	108.60	551.26	607.76	528.20	0.1171E+02	0.1210E+01	0.1729E+02
39	114.90	588.00	615.13	636.95	0.5832E+01	0.1155E+01	0.1723E+02

II - 20

Connection type : Double web-angle connections
Mode : All bolted

Tested by : L.E.Thompson et al. (1970) U.S.A
Test Id. : A1-1 ALT

Column : W10X49
Beam : W21X62
Angle : 4 X 3.5 X 5/16

Fasteners: A325- -3/4"D
15/16" Oversize holes
Material : A36
Fy = -- ksi
Fu = -- ksi

Major parameters

lp = 11.5000"	lu = -- "	ll = -- "	ta = 0.3125"
gb = 2.2500"	gc = 2.5500"	cu = -- "	cl = -- "
pb = -- "	pc = -- "	qb = -- "	qc = -- "
nb = - X 2	nc = - X -		

Remark 1)
2)

No	Moment (k-in)	Rotation (radians) X 1/1000	No	Moment (k-in)	Rotation (radians) X 1/1000
1	0.00	0.00	26	330.76	106.60
2	24.50	1.93	27	349.13	107.55
3	49.00	3.87	28	367.50	108.50
4	73.50	5.80	29	385.88	109.60
5	98.00	7.73	30	404.26	110.70
6	122.50	9.67	31	422.63	111.65
7	147.00	11.60	32	441.00	112.60
8	163.16	17.70	33	459.38	113.55
9	173.01	22.88	34	477.76	114.50
10	182.86	28.06	35	496.13	115.75
11	192.71	33.24	36	514.50	117.00
12	202.56	38.42	37	532.88	118.75
13	212.41	43.60	38	551.26	120.50
14	218.02	49.56	39	569.63	123.45
15	223.62	55.52	40	588.00	126.40
16	229.23	61.49			
17	234.83	67.45			
18	240.44	73.41			
19	246.04	79.38			
20	251.65	85.34			
21	257.26	91.30			
22	266.44	96.90			
23	275.63	102.50			
24	294.00	104.80			
25	312.38	105.70			

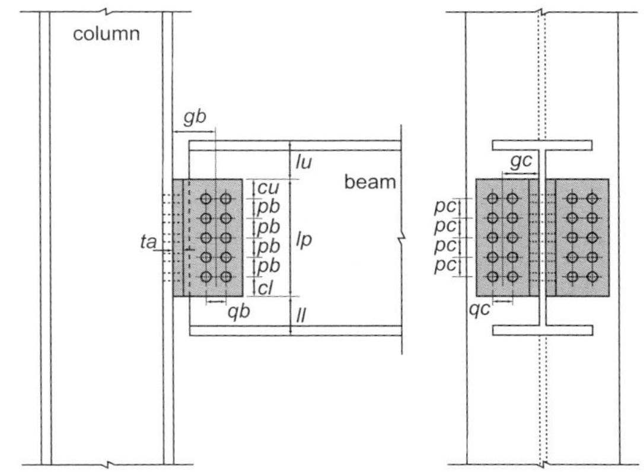

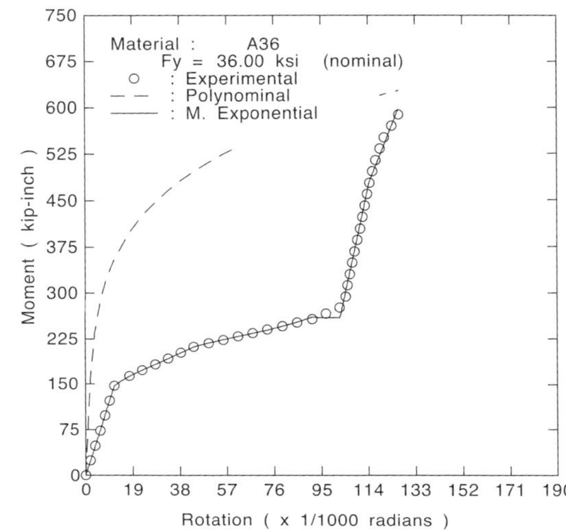

```
Moment-rotation prediction equations     ( R : X 1/1000 radians )

Frye and Morris polynominal model  :     R = Sum ( Ai X ( K*Bm )**Pi X 10**Qi )
     xd = 11.500000"   g =  5.500000"   t =  0.312500"
     A1 =  3.660000   A2 =  1.150000   A3 =  4.570000    K =  0.030179
     P1 =        1    P2 =        3    P3 =        5    Q1 =       -1    Q2 =       -3    Q3 =       -5

Modified exponential model  :
     Bm = Sum ( Ai X ( 1 - exp( - R/( 2*i*C ) ) ) )  +      Sum ( RKj X ( R - Rj0 ) )  +  Bm0
       C =   0.15366583E+01              Bmo=   0.00000000E+00           Nexp=  6          Nliner=  6
      AI =   0.10748012E+03  -0.21307681E+04   0.12642901E+05  -0.31975221E+05   0.36071279E+05  -0.14838656E+05
     Rj0 =           0.0000          11.6000          43.6000          91.3000         102.5000
                   114.5000
     RKj =   0.12284647E+02  -0.91816167E+01  -0.73467427E+00  -0.15979833E+01   0.18205512E+02
            -0.85128330E+01
```

No	Rotation (radians) X 1/1000	Moment (kip-inch) Expri.	Moment (kip-inch) Poly.	Moment (kip-inch) M.Expo.	Connection stiffness (kip-inch) X 1000 Expri.	Connection stiffness (kip-inch) X 1000 Poly.	Connection stiffness (kip-inch) X 1000 M.Expo.
1	0.00	0.00	0.00	0.00	0.1267E+02	0.9054E+02	0.1216E+01
3	3.87	49.00	236.21	48.81	0.1267E+02	0.2929E+02	0.1256E+02
5	7.73	98.00	311.26	98.12	0.1267E+02	0.1353E+02	0.1281E+02
7	11.60	147.00	353.38	146.88	0.1026E+02	0.8922E+01	0.1232E+02
9	22.88	173.01	424.57	173.30	0.1901E+01	0.4673E+01	0.1872E+01
11	33.24	192.71	465.19	192.58	0.1901E+01	0.3340E+01	0.1898E+01
13	43.60	212.41	495.80	212.15	0.1455E+01	0.2632E+01	0.1839E+01
15	55.52	223.62	523.99	224.06	0.9401E+00	0.2135E+01	0.9026E+00
17	67.45	234.83	547.39	234.43	0.9401E+00	0.1808E+01	0.8775E+00
19	79.38	246.04	567.49	245.88	0.9401E+00	0.1575E+01	0.1070E+01
21	91.30	257.26	585.18	260.36	0.1301E+01	0.1400E+01	0.1365E+01
23	102.50	275.63	600.11	259.33	0.6140E+01	0.1271E+01	0.4336E-01
25	105.70	312.38	604.16	317.84	0.2042E+02	0.1238E+01	0.1832E+02
27	107.55	349.13	606.44	351.77	0.1934E+02	0.1220E+01	0.1836E+02
29	109.60	385.88	608.93	389.45	0.1671E+02	0.1201E+01	0.1840E+02
31	111.65	422.63	611.38	427.21	0.1934E+02	0.1182E+01	0.1844E+02
33	113.55	459.38	613.62	462.28	0.1935E+02	0.1166E+01	0.1848E+02
35	115.75	496.13	616.18	492.33	0.1470E+02	0.1147E+01	0.1000E+02
37	118.75	532.88	619.60	522.41	0.1050E+02	0.1123E+01	0.1005E+02
39	123.45	569.63	624.84	569.80	0.6228E+01	0.1087E+01	0.1012E+02

II - 21

Connection type : Double web-angle connections
Mode : All bolted

Tested by : L.E.Thompson et al. (1970) U.S.A
Test Id. : A1-2 ALT

Column : W10X49
Beam : W21X62
Angle : 4 X 3.5 X 5/16

Fasteners: A325- -3/4"D
15/16" Oversize holes
Material : A36
Fy = -- ksi
Fu = -- ksi

Major parameters

lp = 11.5000"	lu = -- "	ll = -- "	ta = 0.3125"
gb = 2.2500"	gc = 2.5500"	cu = -- "	cl = -- "
pb = -- "	pc = -- "	qb = -- "	qc = -- "
nb = - X 2	nc = - X -		

Remark 1)
2)

No	Moment (k-in)	Rotation (radians) X 1/1000	No	Moment (k-in)	Rotation (radians) X 1/1000
1	0.00	0.00	26	308.70	70.24
2	20.58	0.55	27	312.37	74.50
3	41.16	1.10	28	321.56	79.00
4	60.27	1.65	29	330.76	83.50
5	79.38	2.20	30	349.13	83.80
6	110.99	3.30	31	367.50	84.80
7	128.63	4.20	32	388.08	85.80
8	147.00	6.00	33	408.11	86.50
9	165.37	8.30	34	428.14	87.20
10	183.76	11.80	35	447.80	89.05
11	202.13	14.50	36	467.46	90.90
12	211.31	17.45	37	499.07	91.10
13	220.50	20.40	38	515.97	91.35
14	229.69	23.70	39	532.87	91.60
15	238.87	27.00	40	551.25	91.80
16	248.06	31.00	41	569.63	92.00
17	257.26	35.00	42	588.00	92.20
18	266.44	38.70	43	606.37	92.40
19	275.63	42.40	44	624.75	92.70
20	281.75	46.00	45	643.13	93.00
21	287.88	49.60			
22	294.00	53.20			
23	297.67	57.46			
24	301.35	61.72			
25	305.02	65.98			

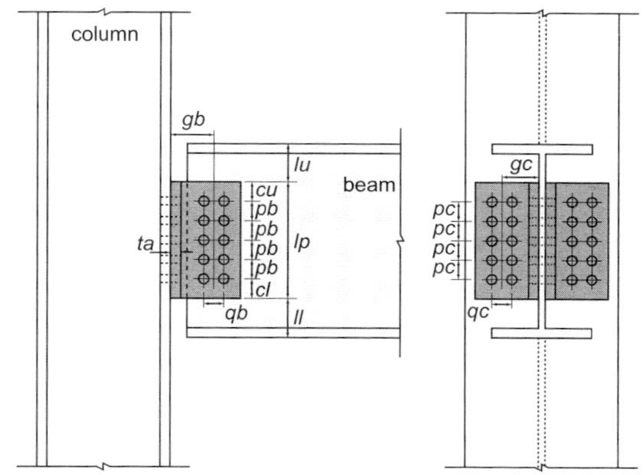

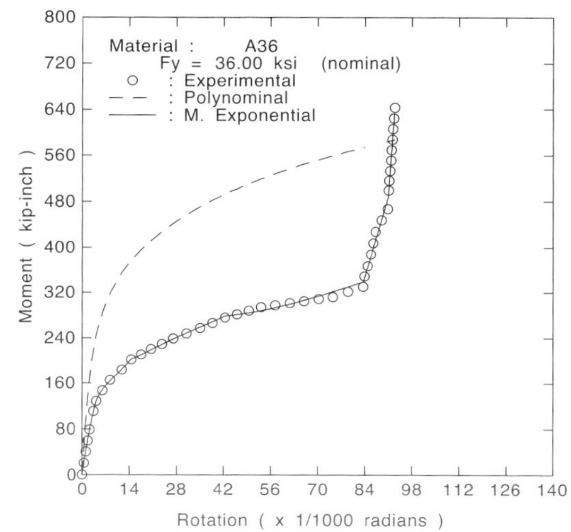

A.2 – 42

```
Moment-rotation prediction equations      ( R : X 1/1000 radians )

Frye and Morris polynominal model  :    R = Sum ( Ai X ( K*Bm )**Pi X 10**Qi )
    xd = 11.500000"   g =  5.500000"   t =  0.312500"
    A1 =  3.660000   A2 =  1.150000   A3 =  4.570000     K =  0.030179
    P1 =         1   P2 =         3   P3 =         5    Q1 =        -1    Q2 =       -3    Q3 =       -5

Modified exponential model  :
    Bm = Sum ( Ai X ( 1 - exp( - R/( 2*i*C ) ) ) )  +     Sum ( RKj X ( R - Rj0 ) )  +  Bm0
     C =   0.91250000E+00          Bmo=   0.00000000E+00          Nexp=  6        Nliner=  6
    AI =  -0.11619294E+03   0.86413062E+03   0.13989851E+03  -0.64135382E+04   0.10964574E+05  -0.54682413E+04
   Rj0 =           0.0000           4.2000          14.5000          42.4000          83.5000
                  91.0000
   RKj =   0.17052361E+02  -0.83067304E+01  -0.44987309E+01  -0.21426735E+01   0.17800315E+02
           0.60948662E+02
```

No	Rotation (radians) X 1/1000	Moment (kip-inch) Expri.	Moment (kip-inch) Poly.	Moment (kip-inch) M.Expo.	Connection stiffness (kip-inch) X 1000 Expri.	Connection stiffness (kip-inch) X 1000 Poly.	Connection stiffness (kip-inch) X 1000 M.Expo.
1	0.00	0.00	0.00	0.00	0.3742E+02	0.9054E+02	0.2228E+02
3	1.10	41.16	96.19	41.90	0.3608E+02	0.8056E+02	0.3607E+02
5	2.20	79.38	170.30	78.50	0.3274E+02	0.5374E+02	0.3037E+02
7	4.20	128.63	245.60	129.55	0.1647E+02	0.2661E+02	0.2130E+02
9	8.30	165.37	318.64	164.19	0.6903E+01	0.1256E+02	0.6071E+01
11	14.50	202.13	376.52	202.69	0.5041E+01	0.7171E+01	0.6875E+01
13	20.40	220.50	412.31	219.23	0.2958E+01	0.5196E+01	0.3003E+01
15	27.00	238.87	442.36	238.36	0.2564E+01	0.4022E+01	0.2739E+01
17	35.00	257.26	470.93	259.18	0.2394E+01	0.3191E+01	0.2545E+01
19	42.40	275.63	492.60	278.62	0.2087E+01	0.2696E+01	0.2757E+01
21	49.60	287.88	510.72	284.32	0.1701E+01	0.2353E+01	0.9754E+00
23	57.46	297.67	528.06	293.52	0.8625E+00	0.2074E+01	0.1356E+01
25	65.98	305.02	544.70	306.47	0.8625E+00	0.1843E+01	0.1663E+01
27	74.50	312.37	559.60	321.55	0.1436E+01	0.1662E+01	0.1860E+01
29	83.50	330.76	573.85	338.88	0.5754E+02	0.1509E+01	0.1979E+01
31	84.80	367.50	575.81	364.60	0.1948E+02	0.1490E+01	0.1979E+02
33	86.50	408.11	578.33	398.25	0.2861E+02	0.1465E+01	0.1980E+02
35	89.05	447.80	582.05	448.78	0.1063E+02	0.1429E+01	0.1982E+02
37	91.10	499.07	584.97	495.52	0.1178E+03	0.1402E+01	0.8078E+02
39	91.60	532.87	585.67	535.92	0.8110E+02	0.1396E+01	0.8079E+02
41	92.00	569.63	586.23	568.23	0.9187E+02	0.1390E+01	0.8079E+02
43	92.40	606.37	586.78	600.55	0.7962E+02	0.1385E+01	0.8079E+02
45	93.00	643.13	587.62	649.02	0.6126E+02	0.1378E+01	0.8079E+02

Connection type : Double web-angle connections
Mode : All bolted

Tested by : L.E.Thompson et al. (1970) U.S.A
Test Id. : A1-2 ALT

Column : W10X49
Beam : W21X62
Angle : 4 X 3.5 X 5/16

Fasteners: A325- -3/4"D
15/16" Oversize holes
Material : A36
Fy = -- ksi
Fu = -- ksi

Major parameters

lp = 11.5000"	lu = -- "	ll = -- "	ta = 0.3125"
gb = 2.2500"	gc = 2.5500"	cu = -- "	cl = -- "
pb = -- "	pc = -- "	qb = -- "	qc = -- "
nb = - X 2	nc = - X -		

Remark 1)
2)

No	Moment (k-in)	Rotation (radians) X 1/1000	No	Moment (k-in)	Rotation (radians) X 1/1000
1	0.00	0.00	26	319.73	72.90
2	20.58	0.60	27	323.40	77.60
3	41.16	1.20	28	327.08	82.30
4	60.27	1.70	29	330.76	87.00
5	79.38	2.20	30	339.94	89.90
6	110.99	3.50	31	349.13	92.80
7	128.63	4.70	32	367.50	94.90
8	147.00	6.60	33	388.08	96.30
9	165.37	8.80	34	408.11	97.10
10	183.76	11.80	35	428.14	97.90
11	202.13	14.20	36	447.80	98.85
12	211.31	17.45	37	467.46	99.80
13	220.50	20.70	38	499.07	99.90
14	229.69	24.35	39	515.97	100.05
15	233.87	28.00	40	532.87	100.20
16	257.26	32.80	41	551.25	100.30
17	263.38	36.27	42	569.63	100.40
18	269.50	39.73	43	588.00	100.60
19	275.63	43.20	44	606.37	100.80
20	281.75	46.93	45	624.75	101.15
21	287.88	50.67	46	643.13	101.50
22	294.00	54.40			
23	303.19	58.95			
24	312.37	63.50			
25	316.05	68.20			

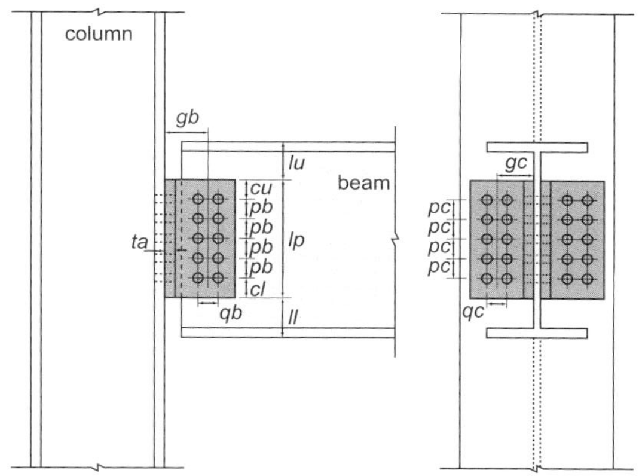

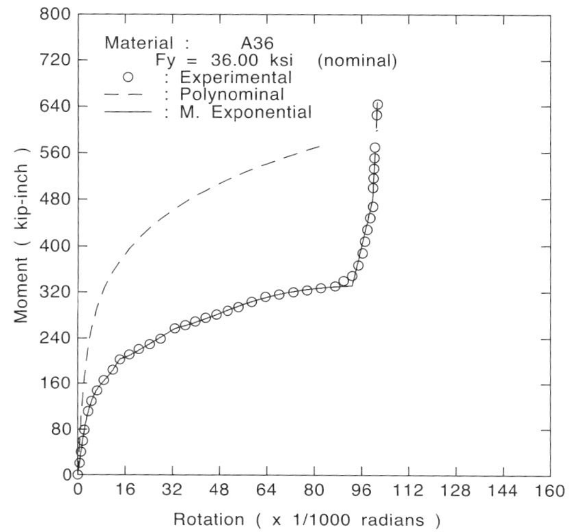

```
Moment-rotation prediction equations     ( R : X 1/1000 radians )

Frye and Morris polynominal model  :    R = Sum ( Ai X ( K*Bm )**Pi X 10**Qi )
     xd = 11.500000"   g =  5.500000"   t =  0.312500"
     A1 =  3.660000   A2 =  1.150000   A3 =  4.570000     K =  0.030179
     P1 =         1   P2 =         3   P3 =         5    Q1 =        -1    Q2 =        -3    Q3 =        -5

Modified exponential model  :
     Bm = Sum ( Ai X ( 1 - exp( - R/( 2*i*C ) ) ) )  +     Sum ( RKj X ( R - Rj0 ) )  +  Bm0
      C =   0.99583333E+00             Bmo=   0.00000000E+00          Nexp=  6          Nliner=  6
     AI =  -0.50780109E+03   0.63517255E+04  -0.26791101E+05   0.55681030E+05  -0.55424116E+05   0.20869391E+05
    Rj0 =           0.0000           4.7000          14.2000          32.8000          92.8000
                   99.8000
    RKj =   0.65524552E+01   0.33845655E+01  -0.80469602E+01  -0.19560629E+01   0.21284012E+02
            0.89466331E+02
```

No	Rotation (radians) X 1/1000	Moment (kip-inch) Expri.	Moment (kip-inch) Poly.	Moment (kip-inch) M.Expo.	Connection stiffness (kip-inch) X 1000 Expri.	Connection stiffness (kip-inch) X 1000 Poly.	Connection stiffness (kip-inch) X 1000 M.Expo.
1	0.00	0.00	0.00	0.00	0.3430E+02	0.9054E+02	0.2579E+02
3	1.20	41.16	104.15	43.32	0.3644E+02	0.7845E+02	0.3648E+02
5	2.20	79.38	170.28	77.05	0.3436E+02	0.5375E+02	0.3030E+02
7	4.70	128.63	257.99	128.45	0.1275E+02	0.2342E+02	0.1193E+02
9	8.80	165.37	324.73	164.18	0.7410E+01	0.1181E+02	0.5896E+01
11	14.20	202.13	374.34	203.45	0.5604E+01	0.7318E+01	0.9135E+01
13	20.70	220.50	413.86	219.03	0.2681E+01	0.5126E+01	0.3070E+01
15	28.00	238.87	446.32	241.87	0.3084E+01	0.3892E+01	0.3145E+01
17	36.27	263.38	474.91	262.36	0.1767E+01	0.3092E+01	0.1560E+01
19	43.20	275.63	494.74	274.44	0.1706E+01	0.2653E+01	0.1888E+01
21	50.67	287.88	513.20	288.86	0.1640E+01	0.2311E+01	0.1915E+01
23	58.95	303.19	531.12	303.62	0.2019E+01	0.2029E+01	0.1608E+01
25	68.20	316.05	548.74	316.25	0.7823E+00	0.1791E+01	0.1120E+01
27	77.60	323.40	564.67	324.62	0.7823E+00	0.1606E+01	0.6795E+00
29	87.00	330.76	579.04	329.44	0.2257E+01	0.1458E+01	0.3674E+00
31	92.80	349.13	587.38	331.16	0.6405E+01	0.1380E+01	0.2354E+00
33	96.30	388.08	592.04	406.37	0.2128E+02	0.1339E+01	0.2146E+02
35	97.90	428.14	594.17	440.68	0.2305E+02	0.1320E+01	0.2143E+02
37	99.80	467.46	596.67	481.38	0.2879E+03	0.1299E+01	0.2141E+02
39	100.05	515.97	597.00	509.10	0.1127E+03	0.1296E+01	0.1109E+03
41	100.30	551.25	597.32	536.82	0.1838E+03	0.1294E+01	0.1109E+03
43	100.60	588.00	597.71	570.08	0.9186E+02	0.1290E+01	0.1109E+03
45	101.15	624.75	598.42	631.05	0.5251E+02	0.1284E+01	0.1109E+03

II - 23

Connection type : Double web-angle connections
Mode : All bolted

Tested by : L.E.Thompson et al. (1970) U.S.A
Test Id. : A1-3 ALT

Column : W10X49
Beam : W21X62
Angle : 4 X 3.5 X 5/16

Fasteners: A325- -3/4"D
15/16" Oversize holes
Material : A36
Fy = -- ksi
Fu = -- ksi

Major parameters

lp = 11.5000"	lu = -- "	ll = -- "	ta = 0.3125"
gb = 2.2500"	gc = 2.5500"	cu = -- "	cl = -- "
pb = -- "	pc = -- "	qb = -- "	qc = -- "
nb = - X 2	nc = - X -		

Remark 1)
2)

No	Moment (k-in)	Rotation (radians) X 1/1000
1	0.00	0.00
2	22.19	0.45
3	44.39	0.90
4	77.17	1.60
5	110.26	2.80
6	128.63	4.00
7	147.00	5.90
8	165.37	8.40
9	183.76	12.70
10	202.13	17.40
11	211.31	20.70
12	220.50	24.00
13	229.69	28.30
14	238.87	32.60
15	248.06	36.65
16	257.26	40.70
17	266.81	45.00
18	276.36	49.30
19	280.58	54.00
20	284.81	58.70
21	296.75	62.85
22	308.70	67.00
23	313.85	70.15
24	319.00	73.30
25	321.94	78.33

No	Moment (k-in)	Rotation (radians) X 1/1000
26	324.88	83.35
27	327.82	88.38
28	330.76	93.40
29	343.25	96.85
30	355.74	100.30
31	383.66	102.00
32	406.45	103.10
33	429.24	104.20
34	462.31	104.60
35	480.32	104.80
36	498.34	105.00
37	517.81	105.20
38	537.29	105.40
39	554.19	105.55
40	571.09	105.70
41	588.73	105.95
42	606.37	106.20
43	635.77	106.60
44	668.84	107.10

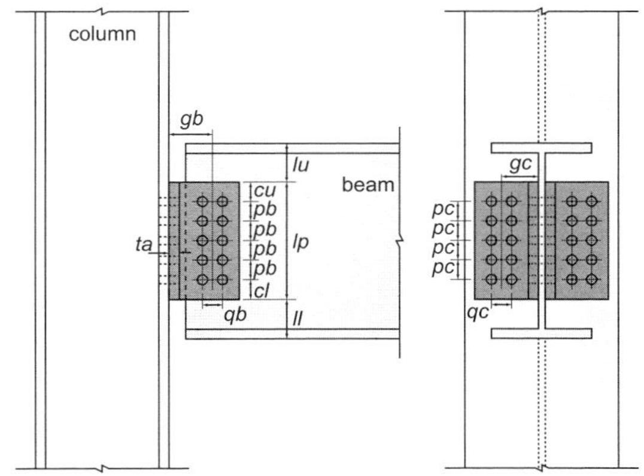

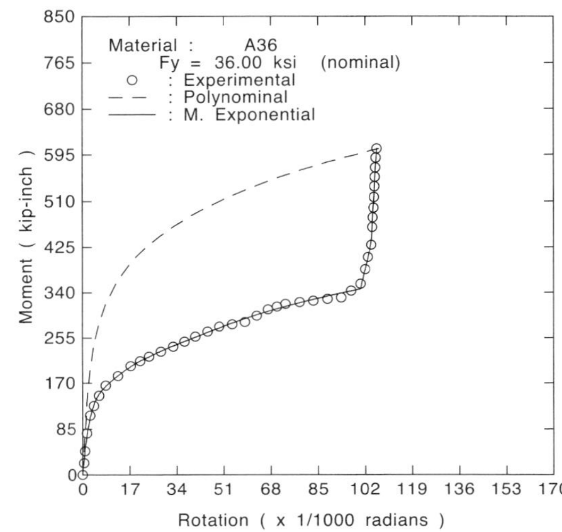

```
Moment-rotation prediction equations     ( R : X 1/1000 radians )

Frye and Morris polynominal model  :    R = Sum ( Ai X ( K*Bm )**Pi X 10**Qi )
     xd = 11.500000"   g =  5.500000"   t =  0.312500"
     A1 =  3.660000   A2 =  1.150000   A3 =  4.570000     K =  0.030179
     P1 =         1   P2 =         3   P3 =         5    Q1 =        -1    Q2 =        -3    Q3 =        -5

Modified exponential model  :
     Bm = Sum ( Ai X ( 1 - exp( - R/( 2*i*C ) ) ) )  +      Sum ( RKj X ( R - Rj0 ) )  +  Bm0
      C =   0.10050000E+01           Bmo=   0.00000000E+00            Nexp=  6         Nliner=  6
     AI =  -0.31272088E+02   0.19965538E+04  -0.96974738E+04   0.21391855E+05  -0.21792129E+05   0.83970871E+04
    Rj0 =           0.0000           4.0000           8.4000          24.0000         100.3000
                  104.2000
    RKj =   0.70485870E-02   0.11076194E+00   0.30486987E+00   0.57329150E+00   0.20926854E+02
            0.63557721E+02
```

No	Rotation (radians) X 1/1000	Moment (kip-inch) Expri.	Moment (kip-inch) Poly.	Moment (kip-inch) M.Expo.	Connection stiffness (kip-inch) X 1000 Expri.	Connection stiffness (kip-inch) X 1000 Poly.	Connection stiffness (kip-inch) X 1000 M.Expo.
1	0.00	0.00	0.00	0.00	0.4932E+02	0.9054E+02	0.6148E+02
3	0.90	44.39	79.70	47.24	0.4835E+02	0.8419E+02	0.4435E+02
5	2.80	110.26	198.83	107.75	0.2144E+02	0.4213E+02	0.2193E+02
7	5.90	147.00	282.66	149.33	0.8667E+01	0.1814E+02	0.7905E+01
9	12.70	183.76	362.74	183.83	0.4100E+01	0.8159E+01	0.4104E+01
11	20.70	211.31	413.86	211.58	0.2784E+01	0.5126E+01	0.2741E+01
13	28.30	229.69	447.49	230.55	0.2136E+01	0.3855E+01	0.2320E+01
15	36.65	248.06	476.09	248.38	0.2270E+01	0.3064E+01	0.2042E+01
17	45.00	266.81	499.43	265.25	0.2221E+01	0.2560E+01	0.1997E+01
19	54.00	280.58	520.70	282.70	0.8987E+00	0.2187E+01	0.1861E+01
21	62.85	296.75	538.82	298.24	0.2879E+01	0.1921E+01	0.1647E+01
23	70.15	313.85	552.30	309.61	0.1634E+01	0.1748E+01	0.1470E+01
25	78.33	321.94	565.83	320.93	0.5851E+00	0.1593E+01	0.1307E+01
27	88.38	327.82	581.03	333.31	0.5851E+00	0.1439E+01	0.1169E+01
29	96.85	343.25	592.83	342.88	0.3621E+01	0.1332E+01	0.1096E+01
31	102.00	383.66	599.49	384.03	0.1903E+02	0.1276E+01	0.2199E+02
33	104.20	429.24	602.29	432.40	0.6616E+02	0.1253E+01	0.2198E+02
35	104.80	480.32	603.04	483.73	0.9006E+02	0.1247E+01	0.8554E+02
37	105.20	517.81	603.54	517.94	0.9738E+02	0.1243E+01	0.8554E+02
39	105.55	554.19	603.97	547.88	0.1127E+03	0.1240E+01	0.8554E+02
41	105.95	588.73	604.47	582.09	0.7056E+02	0.1236E+01	0.8553E+02
43	106.60	635.77	605.27	637.69	0.7023E+02	0.1229E+01	0.8553E+02

II - 24

Connection type : Double web-angle connections
Mode : All bolted

Tested by : L.E.Thompson et al. (1970) U.S.A
Test Id. : A1-3 ALT

Column : W10X49
Beam : W21X62
Angle : 4 X 3.5 X 5/16

Fasteners: A325- -3/4"D
15/16" Oversize holes
Material : A36
Fy = -- ksi
Fu = -- ksi

Major parameters

lp = 11.5000"	lu = -- "	ll = -- "	ta = 0.3125"
gb = 2.2500"	gc = 2.5500"	cu = -- "	cl = -- "
pb = -- "	pc = -- "	qb = -- "	qc = -- "
nb = - X 2	nc = - X -		

Remark 1)
2)

No	Moment (k-in)	Rotation (radians) X 1/1000	No	Moment (k-in)	Rotation (radians) X 1/1000
1	0.00	0.00	26	319.00	92.60
2	22.19	0.40	27	330.76	95.60
3	44.39	0.80	28	355.74	100.90
4	77.17	1.70	29	383.66	102.60
5	110.26	3.10	30	406.45	103.35
6	128.63	4.40	31	429.24	104.10
7	147.00	6.40	32	462.31	104.30
8	165.37	9.40	33	480.32	104.45
9	183.76	14.60	34	498.34	104.60
10	202.13	19.70	35	517.81	104.75
11	211.31	23.20	36	537.29	104.90
12	220.50	26.70	37	554.19	105.05
13	229.69	31.10	38	571.09	105.20
14	238.87	35.50	39	588.73	105.45
15	248.06	39.65	40	606.37	105.70
16	257.26	43.80	41	635.77	106.60
17	266.81	48.35	42	668.84	108.60
18	276.36	52.90			
19	280.58	57.25			
20	284.81	61.60			
21	290.78	66.88			
22	296.75	72.15			
23	302.73	77.42			
24	308.70	82.70			
25	313.85	87.65			

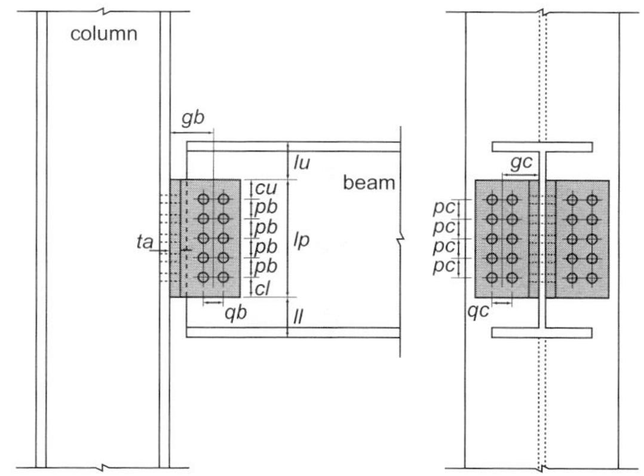

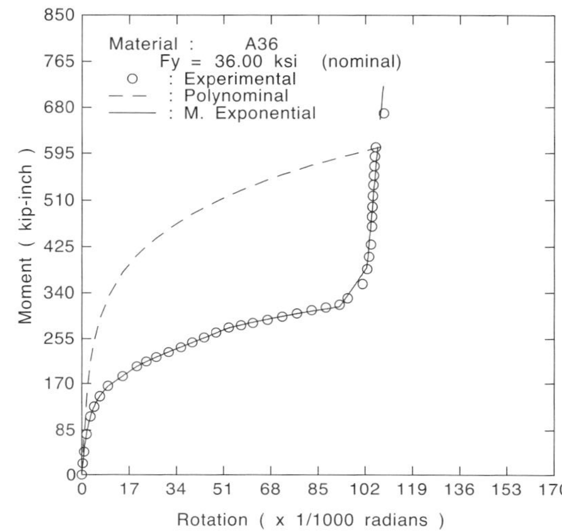

A.2 – 48

```
Moment-rotation prediction equations      ( R : X 1/1000 radians )

Frye and Morris polynominal model  :     R = Sum ( Ai X ( K*Bm )**Pi X 10**Qi )
     xd = 11.500000"    g =  5.500000"    t =  0.312500"
     A1 =  3.660000    A2 =  1.150000    A3 =  4.570000    K =  0.030179
     P1 =         1    P2 =         3    P3 =         5    Q1 =        -1    Q2 =        -3    Q3 =        -5

Modified exponential model  :
     Bm = Sum ( Ai X ( 1 - exp( - R/( 2*i*C ) ) ) )  +      Sum ( RKj X ( R - Rj0 ) )  +  Bm0
      C =  0.10050000E+01              Bmo=   0.00000000E+00            Nexp=  6        Nliner=  7
     AI =  0.19410764E+02   0.14923016E+04  -0.71067588E+04   0.15241550E+05  -0.15263434E+05   0.58015439E+04
    Rj0 =          0.0000           4.4000           9.4000          19.7000          52.9000
                  92.6000         102.6000
    RKj =  0.31096754E+01   0.11699326E+01  -0.22788616E+01  -0.53809066E+00  -0.77114484E+00
           0.63553186E+01   0.50586562E+02
```

No	Rotation (radians) X 1/1000	Moment (kip-inch) Expri.	Poly.	M.Expo.	Connection stiffness (kip-inch) X 1000 Expri.	Poly.	M.Expo.
1	0.00	0.00	0.00	0.00	0.5548E+02	0.9054E+02	0.6033E+02
3	0.80	44.39	71.21	43.19	0.4962E+02	0.8567E+02	0.4550E+02
5	3.10	110.26	210.80	110.56	0.1871E+02	0.3766E+02	0.1760E+02
7	6.40	147.00	291.33	146.87	0.7961E+01	0.1658E+02	0.7237E+01
9	14.60	183.76	377.22	183.76	0.3569E+01	0.7125E+01	0.3653E+01
11	23.20	211.31	426.02	211.58	0.2625E+01	0.4616E+01	0.2579E+01
13	31.10	229.69	457.83	229.96	0.2088E+01	0.3543E+01	0.2158E+01
15	39.65	248.06	484.97	248.01	0.2215E+01	0.2859E+01	0.2097E+01
17	48.35	266.81	507.74	266.22	0.2099E+01	0.2406E+01	0.2080E+01
19	57.25	280.58	527.63	281.00	0.9710E+00	0.2080E+01	0.1210E+01
21	66.88	290.78	546.34	291.92	0.1132E+01	0.1822E+01	0.1058E+01
23	77.42	302.73	564.39	302.27	0.1132E+01	0.1609E+01	0.9115E+00
25	87.65	313.85	579.99	311.06	0.1040E+01	0.1449E+01	0.8138E+00
27	95.60	330.76	591.14	336.39	0.4207E+01	0.1346E+01	0.7121E+01
29	102.60	383.66	600.26	386.13	0.2611E+02	0.1269E+01	0.7093E+01
31	104.10	429.24	602.15	472.65	0.1369E+03	0.1254E+01	0.5768E+02
33	104.45	480.32	602.59	492.83	0.1201E+03	0.1251E+01	0.5767E+02
35	104.75	517.81	602.97	510.14	0.1298E+03	0.1248E+01	0.5767E+02
37	105.05	554.19	603.34	527.44	0.1127E+03	0.1245E+01	0.5767E+02
39	105.45	588.73	603.84	550.51	0.7056E+02	0.1241E+01	0.5767E+02
41	106.60	635.77	605.26	616.83	0.2766E+02	0.1229E+01	0.5767E+02

A.2 – 49

II - 25

Connection type : Double web-angle connections
Mode : All bolted

Tested by : L.E.Thompson et al. (1970) U.S.A
Test Id. : B1-1 ALT

Column : W10X49
Beam : W21X62
Angle : 4 X 3.5 X 1/2

Fasteners: A325- -3/4"D
15/16" Oversize holes
Material : A36
Fy = -- ksi
Fu = -- ksi

Major parameters

lp = 11.5000"	lu = -- "	ll = -- "	ta = 0.5000"
gb = 2.2500"	gc = 2.5500"	cu = -- "	cl = -- "
pb = -- "	pc = -- "	qb = -- "	qc = -- "
nb = - X 2	nc = - X -		

Remark 1)
2)

No	Moment (k-in)	Rotation (radians) X 1/1000
1	0.00	0.00
2	41.85	0.40
3	83.70	0.80
4	119.39	1.45
5	155.09	2.10
6	190.00	2.95
7	224.92	3.80
8	275.63	5.50
9	312.37	7.80
10	349.86	10.40
11	388.08	14.60
12	405.35	17.20
13	422.63	19.80
14	429.98	24.22
15	437.33	28.64
16	444.67	33.06
17	452.02	37.48
18	459.37	41.90
19	468.56	45.90
20	477.76	49.90
21	481.43	54.08
22	485.10	58.26
23	488.78	62.44
24	492.45	66.62
25	496.13	70.80
26	505.31	73.90
27	514.50	77.00
28	538.02	80.00
29	588.00	81.90
30	626.59	83.20
31	664.44	84.40
32	699.72	85.45
33	735.00	86.50
34	771.75	87.75
35	808.50	89.00
36	845.25	90.75
37	882.00	92.50
38	918.75	94.60
39	955.50	96.70
40	992.25	99.10
41	1029.00	101.50

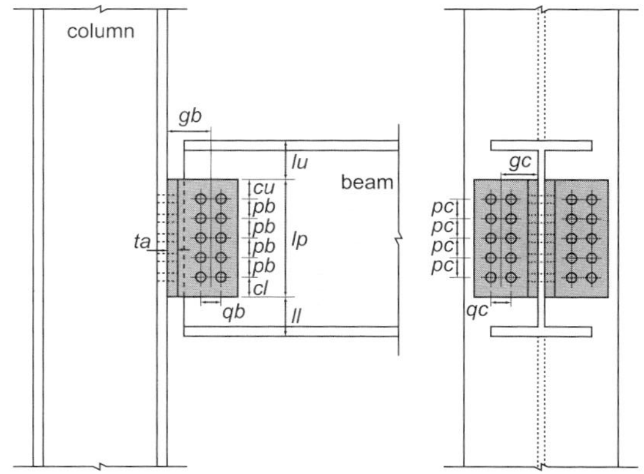

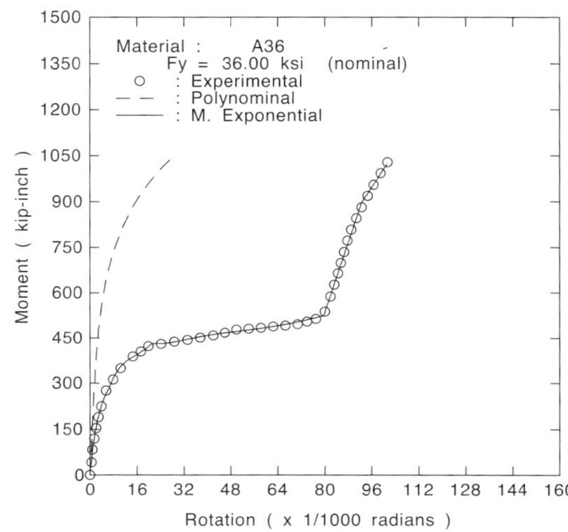

```
Moment-rotation prediction equations      ( R : X 1/1000 radians )

Frye and Morris polynominal model  :     R = Sum ( Ai X ( K*Bm )**Pi X 10**Qi )
     xd = 11.500000"   g =  5.500000"   t =  0.500000"
     A1 =  3.660000   A2 =  1.150000   A3 =  4.570000    K =  0.012890
     P1 =         1   P2 =         3   P3 =         5   Q1 =        -1    Q2 =        -3    Q3 =        -5

Modified exponential model  :
     Bm = Sum ( Ai X ( 1 - exp( - R/( 2*i*C ) ) ) )  +     Sum ( RKj X ( R - Rj0 ) )  +  Bm0
       C =   0.94583333E+00            Bmo=   0.00000000E+00          Nexp=  6        Nliner=  4
      AI =   0.57706361E+03  -0.78942103E+04   0.39226983E+05  -0.81152949E+05   0.77092146E+05  -0.27715797E+05
     Rj0 =           6.5000          21.2000          79.6000          92.5000
     RKj =   0.13247002E+02  -0.10043723E+02   0.26144943E+02  -0.14838231E+02
```

No	Rotation (radians) X 1/1000	Moment (kip-inch) Expri.	Poly.	M.Expo.	Connection stiffness (kip-inch) X 1000 Expri.	Poly.	M.Expo.
1	0.00	0.00	0.00	0.00	0.1046E+03	0.2120E+03	0.1145E+03
3	0.80	83.70	166.71	76.24	0.8569E+02	0.2006E+03	0.7928E+02
5	2.10	155.09	386.00	157.54	0.4892E+02	0.1312E+03	0.4963E+02
7	3.80	224.92	548.65	224.55	0.3733E+02	0.6989E+02	0.3120E+02
9	7.80	312.37	730.85	314.49	0.1524E+02	0.3139E+02	0.2052E+02
11	14.60	388.08	883.17	387.26	0.7583E+01	0.1668E+02	0.5921E+01
13	19.80	422.63	957.94	421.17	0.4799E+01	0.1251E+02	0.7842E+01
15	28.64	437.33	1050.76	436.87	0.1663E+01	0.8930E+01	0.1797E+01
17	37.48	452.02	1120.57	454.58	0.1663E+01	0.7033E+01	0.1842E+01
19	45.90	468.56	1174.66	467.35	0.2298E+01	0.5892E+01	0.1257E+01
21	54.08	481.43	1219.51	477.18	0.8790E+00	0.5114E+01	0.1239E+01
23	62.44	488.78	1259.68	488.99	0.8790E+00	0.4522E+01	0.1626E+01
25	70.80	496.13	1295.49	504.60	0.2076E+01	0.4063E+01	0.2106E+01
27	77.00	514.50	1319.79	518.65	0.5442E+01	0.3784E+01	0.2415E+01
29	81.90	588.00	1338.10	591.12	0.2831E+02	0.3589E+01	0.2876E+02
31	84.40	664.44	1346.72	663.13	0.3264E+02	0.3502E+01	0.2884E+02
33	86.50	735.00	1354.04	723.77	0.3168E+02	0.3430E+01	0.2891E+02
35	89.00	808.50	1362.56	796.12	0.2590E+02	0.3348E+01	0.2897E+02
37	92.50	882.00	1374.18	897.67	0.1941E+02	0.3241E+01	0.2905E+02
39	96.70	955.50	1387.66	957.53	0.1648E+02	0.3121E+01	0.1429E+02
41	101.50	1029.00	1402.13	1026.28	0.1531E+02	0.2999E+01	0.1435E+02

II - 26

Connection type : Double web-angle connections
Mode : All bolted

Tested by : L.E.Thompson et al. (1970) U.S.A
Test Id. : B1-1 ALT

Column : W10X49
Beam : W21X62
Angle : 4 X 3.5 X 1/2

Fasteners: A325- -3/4"D
15/16" Oversize holes
Material : A36
Fy = -- ksi
Fu = -- ksi

Major parameters

lp = 11.5000"	lu = -- "	ll = -- "	ta = 0.5000"
gb = 2.2500"	gc = 2.5500"	cu = -- "	cl = -- "
pb = -- "	pc = -- "	qb = -- "	qc = -- "
nb = - X 2	nc = - X -		

Remark 1)
2)

No	Moment (k-in)	Rotation (radians) X 1/1000	No	Moment (k-in)	Rotation (radians) X 1/1000
1	0.00	0.00	26	538.02	91.00
2	41.85	0.35	27	588.00	96.60
3	83.70	0.70	28	626.59	101.10
4	119.39	1.05	29	662.73	102.23
5	155.09	1.40	30	698.86	103.37
6	190.00	2.25	31	735.00	104.50
7	224.92	3.10	32	771.75	105.75
8	275.63	4.70	33	808.50	107.00
9	312.37	6.90	34	845.25	108.65
10	349.86	9.40	35	882.00	110.30
11	388.08	13.70	36	918.75	112.20
12	422.63	19.30	37	955.50	114.10
13	429.41	23.95	38	992.25	116.30
14	436.19	28.60	39	1029.00	118.50
15	443.92	33.90			
16	451.64	39.20			
17	459.37	44.50			
18	466.97	48.50			
19	474.56	52.50			
20	482.16	56.50			
21	498.33	62.30			
22	514.50	68.10			
23	520.38	73.83			
24	526.26	79.55			
25	532.14	85.28			

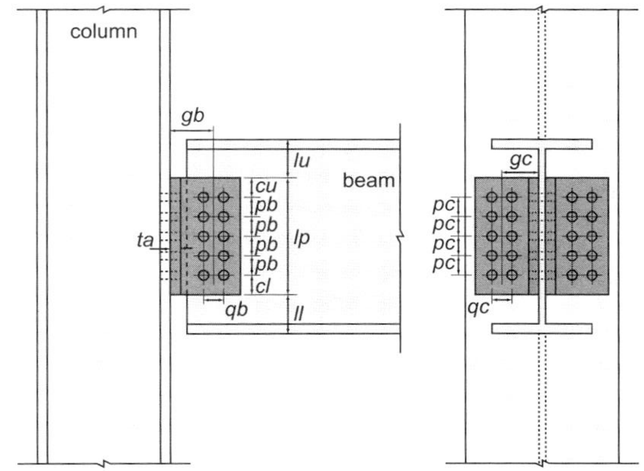

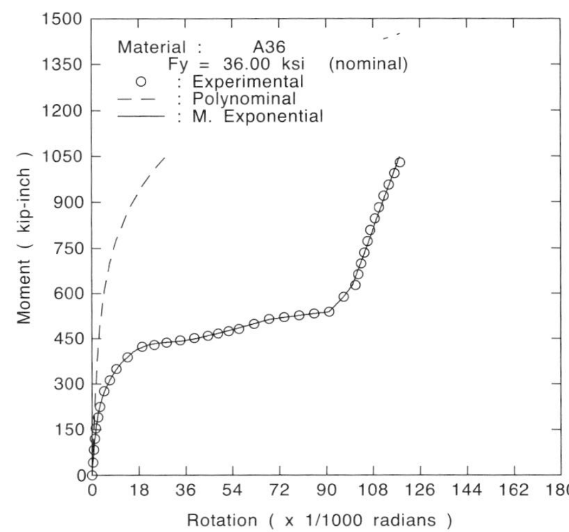

```
Moment-rotation prediction equations     ( R : X 1/1000 radians )

Frye and Morris polynominal model  :    R = Sum ( Ai X ( K*Bm )**Pi X 10**Qi )
     xd = 11.500000"    g =  5.500000"   t =  0.500000"
     A1 =  3.660000    A2 =  1.150000   A3 =  4.570000     K =  0.012890
     P1 =         1    P2 =         3   P3 =         5    Q1 =        -1    Q2 =        -3    Q3 =        -5

Modified exponential model  :
     Bm = Sum ( Ai X ( 1 - exp( - R/( 2*i*C ) ) ) ) +      Sum ( RKj X ( R - Rj0 ) ) + Bm0
       C =  0.10750000E+01             Bmo=  0.00000000E+00          Nexp=  6          Nliner=  8
      AI =  0.59013481E+03  -0.11809278E+04  -0.43164459E+04   0.19869415E+05  -0.23645803E+05   0.91388767E+04
     Rj0 =          0.0000          13.7000          19.3000          44.5000          68.1000
                   91.0000          96.6000         100.0000
     RKj = -0.30940337E+01   0.39469528E+01   0.16735861E+01  -0.69819829E+00  -0.10885376E+01
            0.80142414E+01   0.26254975E+01   0.11610873E+02
```

No	Rotation (radians) X 1/1000	Moment (kip-inch) Expri.	Moment (kip-inch) Poly.	Moment (kip-inch) M.Expo.	Connection stiffness (kip-inch) X 1000 Expri.	Connection stiffness (kip-inch) X 1000 Poly.	Connection stiffness (kip-inch) X 1000 M.Expo.
1	0.00	0.00	0.00	0.00	0.1196E+03	0.2120E+03	0.1499E+03
3	0.70	83.70	146.51	85.72	0.1108E+03	0.2035E+03	0.1012E+03
5	1.40	155.09	279.49	145.25	0.8422E+02	0.1727E+03	0.7094E+02
7	3.10	224.92	493.49	229.15	0.3782E+02	0.8819E+02	0.3412E+02
9	6.90	312.37	700.72	313.24	0.1590E+02	0.3578E+02	0.1620E+02
11	13.70	388.08	867.69	388.41	0.7707E+01	0.1774E+02	0.5698E+01
13	23.95	429.41	1005.46	432.92	0.1458E+01	0.1050E+02	0.1436E+01
15	33.90	443.92	1094.27	441.43	0.1458E+01	0.7684E+01	0.1022E+01
17	44.50	459.37	1166.31	459.70	0.1709E+01	0.6053E+01	0.2377E+01
19	52.50	474.56	1211.33	475.71	0.1899E+01	0.5246E+01	0.2242E+01
21	62.30	498.33	1259.05	498.83	0.2788E+01	0.4530E+01	0.2406E+01
23	73.83	520.38	1307.56	519.75	0.1027E+01	0.3921E+01	0.1191E+01
25	85.28	532.14	1349.77	532.37	0.1027E+01	0.3472E+01	0.1017E+01
27	96.60	588.00	1387.06	588.00	0.8731E+01	0.3126E+01	0.8906E+01
29	102.23	662.73	1404.53	678.76	0.3188E+02	0.2979E+01	0.2310E+02
31	104.50	735.00	1410.95	731.11	0.3070E+02	0.2927E+01	0.2309E+02
33	107.00	808.50	1418.23	788.80	0.2633E+02	0.2869E+01	0.2307E+02
35	110.30	882.00	1427.64	864.91	0.2091E+02	0.2797E+01	0.2306E+02
37	114.10	955.50	1438.19	952.50	0.1812E+02	0.2719E+01	0.2304E+02
39	118.50	1029.00	1450.06	1053.86	0.1670E+02	0.2634E+01	0.2303E+02

II - 27

Connection type : Double web-angle connections
Mode : All bolted

Tested by : L.E.Thompson et al. (1970) U.S.A.
Test Id. : B1-2 ALT

Column : W10X49
Beam : W21X62
Angle : 4 X 3.5 X 1/2

Fasteners: A325- -3/4"D
15/16" Oversize holes
Material : A36
Fy = -- ksi
Fu = -- ksi

Major parameters

lp = 11.5000"	lu = -- "	ll = -- "	ta = 0.5000"
gb = 2.2500"	gc = 2.5500"	cu = -- "	cl = -- "
pb = -- "	pc = -- "	qb = -- "	qc = -- "
nb = - X 2	nc = - X -		

Remark 1)
2)

No	Moment (k-in)	Rotation (radians) X 1/1000
1	0.00	0.00
2	42.26	0.35
3	84.53	0.70
4	120.17	1.25
5	155.82	1.80
6	138.16	2.60
7	222.71	3.50
8	260.18	5.00
9	277.09	9.30
10	296.57	13.75
11	316.04	18.20
12	332.22	23.80
13	351.34	27.80
14	360.16	32.60
15	368.98	37.40
16	388.08	44.30
17	405.35	49.45
18	422.63	54.60
19	429.98	59.90
20	437.32	65.20
21	444.67	70.50
22	478.49	77.40
23	515.79	79.45
24	553.09	81.50
25	590.39	82.75
26	627.70	84.00
27	664.07	85.10
28	700.45	86.20
29	737.20	87.50
30	773.95	88.80
31	809.97	90.35
32	845.99	91.90
33	882.37	94.00
34	918.76	96.10
35	955.51	99.10
36	992.26	102.10
37	1029.01	107.05
38	1065.76	112.00
39	1090.26	117.17
40	1114.76	122.33
41	1139.26	127.50
42	1151.50	134.93
43	1163.75	142.37
44	1176.00	149.80

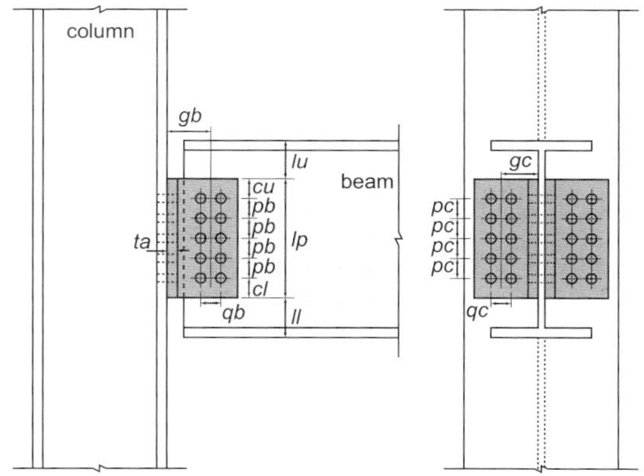

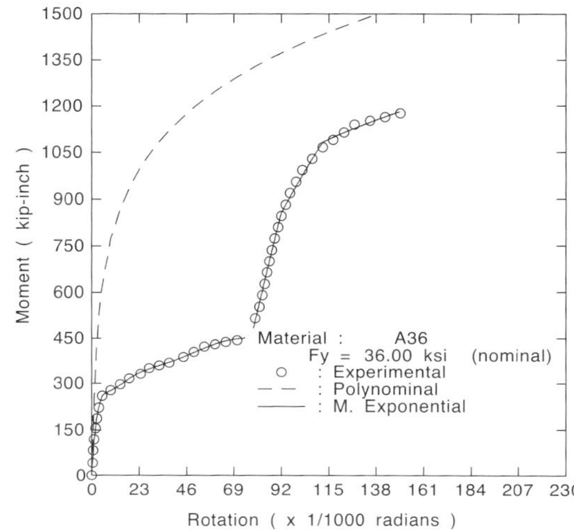

A.2 – 54

```
Moment-rotation prediction equations     ( R : X 1/1000 radians )

Frye and Morris polynominal model   :     R = Sum ( Ai X ( K*Bm )**Pi X 10**Qi )
     xd = 11.500000"   g =  5.500000"   t =  0.500000"
     A1 =  3.660000   A2 =  1.150000   A3 =  4.570000    K =  0.012890
     P1 =        1    P2 =        3    P3 =        5    Q1 =       -1    Q2 =       -3    Q3 =       -5

Modified exponential model  :
     Bm = Sum ( Ai X ( 1 - exp( - R/( 2*i*C ) ) ) )  +      Sum ( RKj X ( R - Rj0 ) )  +  Bm0
      C =   0.13358333E+01             Bmo=   0.00000000E+00           Nexp=  6        Nliner=  5
     AI =   0.41709333E+03   0.32570617E+04  -0.31328362E+05   0.93621778E+05  -0.11527189E+06   0.50211871E+05
    Rj0 =          0.0000           5.0000          77.4000          91.9000         112.0000
    RKj =   0.37647976E+01  -0.87339727E+01   0.29132991E+02  -0.14785694E+02  -0.72835694E+01
```

No	Rotation (radians) X 1/1000	Moment (kip-inch) Expri.	Moment (kip-inch) Poly.	Moment (kip-inch) M.Expo.	Connection stiffness (kip-inch) X 1000 Expri.	Connection stiffness (kip-inch) X 1000 Poly.	Connection stiffness (kip-inch) X 1000 M.Expo.
1	0.00	0.00	0.00	0.00	0.1208E+03	0.2120E+03	0.1207E+03
3	0.70	84.53	146.51	75.13	0.9900E+02	0.2035E+03	0.9181E+02
5	1.80	155.82	343.89	155.40	0.5488E+02	0.1486E+03	0.5698E+02
7	3.50	222.71	526.41	224.72	0.3336E+02	0.7687E+02	0.2844E+02
9	9.30	277.09	773.87	276.73	0.4151E+01	0.2611E+02	0.3269E+01
11	18.20	316.04	937.31	315.83	0.3718E+01	0.1352E+02	0.4507E+01
13	27.80	351.34	1043.15	348.47	0.3442E+01	0.9172E+01	0.2468E+01
15	37.40	368.98	1120.01	370.40	0.2220E+01	0.7046E+01	0.2387E+01
17	49.45	405.35	1194.92	403.18	0.3354E+01	0.5524E+01	0.2912E+01
19	59.90	429.98	1247.99	431.35	0.1386E+01	0.4685E+01	0.2300E+01
21	70.50	444.67	1294.27	448.61	0.2913E+01	0.4078E+01	0.8788E+00
23	79.45	515.79	1329.03	510.09	0.1820E+02	0.3684E+01	0.2865E+02
25	82.75	590.39	1341.09	603.84	0.2984E+02	0.3559E+01	0.2817E+02
27	85.10	664.07	1349.40	669.67	0.3307E+02	0.3475E+01	0.2785E+02
29	87.50	737.20	1357.69	736.12	0.2827E+02	0.3395E+01	0.2753E+02
31	90.35	809.97	1367.00	814.09	0.2324E+02	0.3306E+01	0.2718E+02
33	94.00	882.37	1378.97	881.49	0.1733E+02	0.3197E+01	0.1198E+02
35	99.10	955.51	1394.79	941.28	0.1225E+02	0.3060E+01	0.1148E+02
37	107.05	1029.01	1418.34	1029.90	0.7424E+01	0.2869E+01	0.1085E+02
39	117.17	1090.26	1446.27	1098.98	0.4742E+01	0.2660E+01	0.3002E+01
41	127.50	1139.26	1472.81	1127.95	0.3473E+01	0.2480E+01	0.2634E+01
43	142.37	1163.75	1508.00	1164.67	0.1648E+01	0.2263E+01	0.2341E+01

II - 28

Connection type : Double web-angle connections
Mode : All bolted

Tested by : L.E.Thompson et al. (1970) U.S.A.
Test Id. : B1-2 ALT

Column : W10X49
Beam : W21X62
Angle : 4 X 3.5 X 1/2

Fasteners: A325- -3/4"D
15/16" Oversize holes
Material : A36
Fy = -- ksi
Fu = -- ksi

Major parameters

lp = 11.5000"	lu = -- "	ll = -- "	ta = 0.5000"
gb = 2.2500"	gc = 2.5500"	cu = -- "	cl = -- "
pb = -- "	pc = -- "	qb = -- "	qc = -- "
nb = - X 2	nc = - X -		

Remark 1)
2)

No	Moment (k-in)	Rotation (radians) X 1/1000
1	0.00	0.00
2	42.26	0.35
3	84.53	0.70
4	120.17	1.15
5	155.82	1.60
6	188.16	2.40
7	222.71	3.20
8	228.65	6.20
9	234.60	9.20
10	246.48	15.10
11	260.18	27.00
12	268.64	38.00
13	277.09	49.00
14	295.47	66.60
15	313.85	84.20
16	332.22	101.80
17	350.60	119.40
18	368.98	137.00
19	379.02	152.72
20	389.06	168.43
21	399.11	184.15
22	409.15	199.87
23	419.20	215.58
24	429.24	231.30
25	446.73	246.00
26	464.23	260.70
27	481.72	275.40
28	499.21	290.10
29	516.71	304.80
30	553.09	306.10
31	590.39	307.50
32	627.70	308.90
33	664.07	310.10
34	700.45	311.30
35	737.20	312.70
36	773.95	314.10
37	809.97	315.60
38	845.99	317.10
39	882.37	319.85
40	918.76	322.60
41	955.51	324.90
42	992.26	327.20
43	1029.01	330.70
44	1065.76	334.20
45	1102.51	340.85
46	1139.26	347.50
47	1176.00	357.60

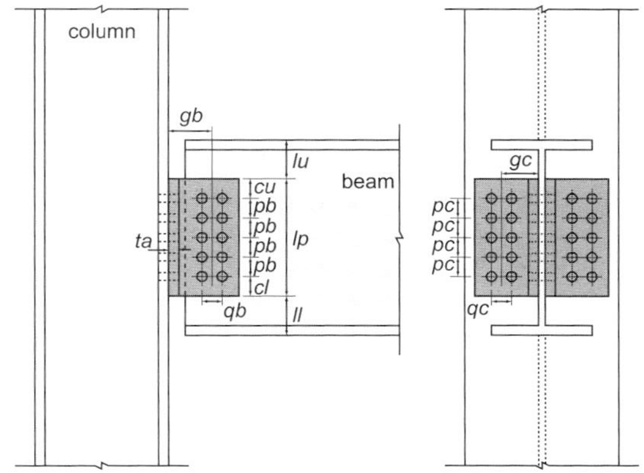

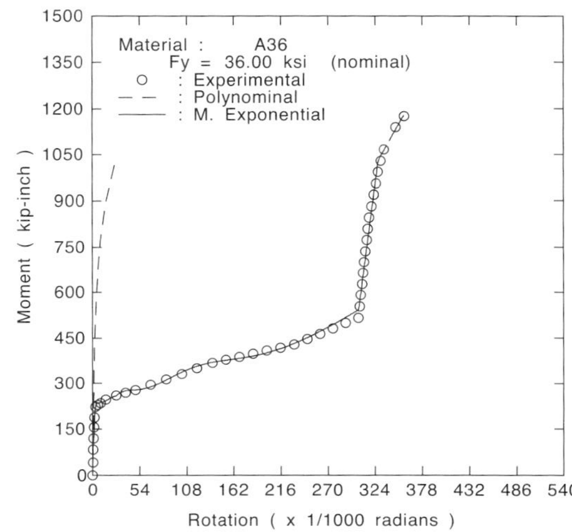

A.2 – 56

```
Moment-rotation prediction equations     ( R : X 1/1000 radians )

Frye and Morris polynominal model  :    R = Sum ( Ai X ( K*Bm )**Pi X 10**Qi )
     xd = 11.500000"    g =  5.500000"   t =  0.500000"
     A1 =  3.660000    A2 =  1.150000   A3 =  4.570000    K =  0.012890
     P1 =         1    P2 =         3   P3 =         5   Q1 =        -1    Q2 =        -3    Q3 =        -5

Modified exponential model  :
     Bm = Sum ( Ai X ( 1 - exp( - R/( 2*i*C ) ) ) ) +     Sum ( RKj X ( R - Rj0 ) ) +  Bm0
      C =   0.30675000E+01             Bmo=   0.00000000E+00            Nexp=  6        Nliner=  8
     AI =   0.10264832E+04  -0.13648205E+05   0.66734672E+05  -0.14599267E+06   0.14641175E+06  -0.54710561E+05
    Rj0 =           0.0000           0.7000           3.2000          27.0000          49.0000
                  137.0000         304.8000         327.2000
    RKj =   0.11605567E+03  -0.72069868E+02  -0.44166934E+02   0.18631459E+01   0.29660457E+00
           -0.97343272E-01   0.19920503E+02  -0.16639962E+02
```

No	Rotation (radians) X 1/1000	Moment (kip-inch) Expri.	Moment (kip-inch) Poly.	Moment (kip-inch) M.Expo.	Connection stiffness (kip-inch) X 1000 Expri.	Connection stiffness (kip-inch) X 1000 Poly.	Connection stiffness (kip-inch) X 1000 M.Expo.
1	0.00	0.00	0.00	0.00	0.1208E+03	0.2120E+03	0.1842E+02
3	0.70	84.53	146.51	92.66	0.1026E+03	0.2035E+03	0.1304E+03
5	1.60	155.82	312.88	143.24	0.6525E+02	0.1608E+03	0.5424E+02
7	3.20	222.71	502.15	225.76	0.3451E+02	0.8510E+02	0.4934E+02
9	9.20	234.60	771.11	236.21	0.1994E+01	0.2642E+02	0.5412E+00
11	27.00	260.18	1035.72	258.22	0.9525E+00	0.9416E+01	0.6687E+00
13	49.00	277.09	1192.52	277.66	0.8745E+00	0.5566E+01	0.4097E-01
15	84.20	313.85	1346.10	311.37	0.1044E+01	0.3508E+01	0.1518E+01
17	119.40	350.60	1452.18	357.75	0.1044E+01	0.2619E+01	0.9329E+00
19	152.72	379.02	1530.75	377.60	0.6391E+00	0.2135E+01	0.3821E+00
21	184.15	399.11	1592.77	391.50	0.6391E+00	0.1829E+01	0.5728E+00
23	215.58	419.20	1646.60	415.81	0.6391E+00	0.1607E+01	0.9809E+00
25	246.00	446.73	1692.86	451.25	0.1190E+01	0.1442E+01	0.1333E+01
27	275.40	481.72	1733.31	494.14	0.1190E+01	0.1314E+01	0.1568E+01
29	304.80	516.71	1770.37	542.54	0.2581E+02	0.1210E+01	0.1712E+01
31	307.50	590.39	1773.63	600.96	0.2664E+02	0.1201E+01	0.2164E+02
33	310.10	664.07	1776.75	657.24	0.3031E+02	0.1193E+01	0.2165E+02
35	312.70	737.20	1779.84	713.54	0.2625E+02	0.1185E+01	0.2166E+02
37	315.60	809.97	1783.27	776.37	0.2401E+02	0.1176E+01	0.2167E+02
39	319.85	882.37	1788.26	868.48	0.1323E+02	0.1163E+01	0.2168E+02
41	324.90	955.51	1794.11	978.00	0.1598E+02	0.1148E+01	0.2169E+02
43	330.70	1029.01	1800.75	1045.62	0.1050E+02	0.1132E+01	0.5066E+01
45	340.85	1102.51	1812.17	1097.15	0.5526E+01	0.1104E+01	0.5086E+01
47	357.60	1176.00	1830.49	1182.56	0.3638E+01	0.1061E+01	0.5111E+01

II - 29

Connection type : Double web-angle connections
Mode : All bolted

Tested by : L.E.Thompson et al. (1970) U.S.A
Test Id. : B1-3 ALT

Column : W10X49
Beam : W21X62
Angle : 4 X 3.5 X 1/2

Fasteners: A325- -3/4"D
15/16" Oversize holes
Material : A36
Fy = -- ksi
Fu = -- ksi

Major parameters

lp = 11.5000"	lu = -- "	ll = -- "	ta = 0.5000"
gb = 2.2500"	gc = 2.5500"	cu = -- "	cl = -- "
pb = -- "	pc = -- "	qb = -- "	qc = -- "
nb = - X 2	nc = - X -		

Remark 1)
2)

No	Moment (k-in)	Rotation (radians) X 1/1000	No	Moment (k-in)	Rotation (radians) X 1/1000
1	0.00	0.00	26	432.37	26.20
2	24.62	0.20	27	437.24	30.15
3	49.25	0.40	28	442.10	34.10
4	66.52	0.55	29	462.31	35.70
5	83.80	0.70	30	470.03	39.65
6	99.26	0.85	31	477.76	43.60
7	114.73	1.00	32	486.94	45.85
8	134.54	1.25	33	496.13	48.10
9	154.36	1.50	34	497.97	52.10
10	170.89	1.80	35	499.80	56.10
11	187.43	2.10	36	501.64	60.10
12	205.43	2.55	37	503.48	64.10
13	223.44	3.00	38	505.32	68.10
14	241.45	3.55	39	507.16	72.10
15	259.45	4.10	40	520.39	75.45
16	276.36	4.80	41	533.62	78.80
17	294.73	5.70	42	552.72	80.00
18	313.85	6.50	43	572.20	80.85
19	330.76	7.50	44	591.67	81.70
20	349.87	9.25			
21	368.98	11.00			
22	385.87	12.90			
23	404.26	15.10			
24	422.63	18.30			
25	427.50	22.25			

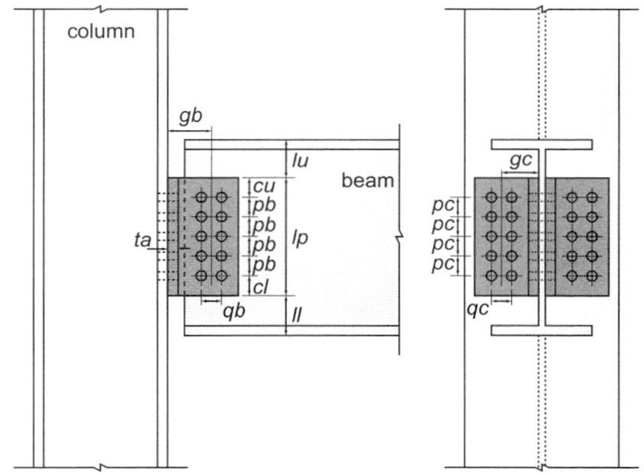

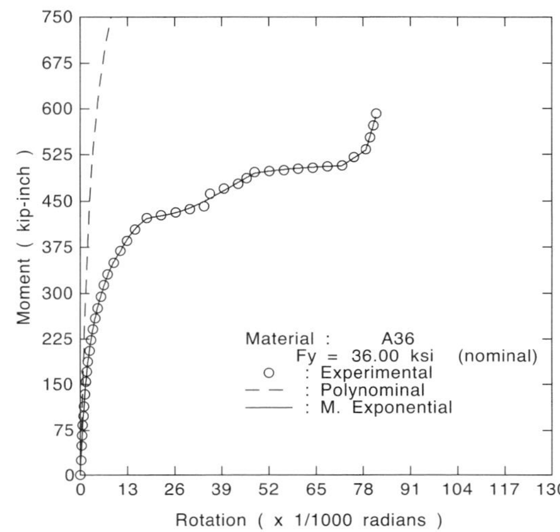

```
 Moment-rotation prediction equations     ( R : X 1/1000 radians )

Frye and Morris polynominal model  :    R = Sum ( Ai X ( K*Bm )**Pi X 10**Qi )
     xd = 11.500000"   g =  5.500000"  t =  0.500000"
     A1 =  3.660000   A2 =  1.150000   A3 =  4.570000    K =  0.012890
     P1 =        1    P2 =        3    P3 =        5    Q1 =       -1     Q2 =       -3     Q3 =       -5

Modified exponential model  :
     Bm = Sum ( Ai X ( 1 - exp( - R/( 2*i*C ) ) ) )  +      Sum ( RKj X ( R - Rj0 ) )  +  Bm0
      C =   0.73083333E+00              Bmo=   0.00000000E+00           Nexp=  6         Nliner=  5
     AI =   0.14051820E+03   0.16923399E+04  -0.12317928E+05   0.30693851E+05  -0.31092260E+05   0.11280378E+05
    Rj0 =          11.0000          18.3000          48.1000          72.1000          78.8000
    RKj =   0.52588977E+01  -0.28722734E+01  -0.23213804E+01   0.36489949E+01   0.16139291E+02
```

No	Rotation (radians) X 1/1000	Moment (kip-inch) Expri.	Moment (kip-inch) Poly.	Moment (kip-inch) M.Expo.	Connection stiffness (kip-inch) X 1000 Expri.	Connection stiffness (kip-inch) X 1000 Poly.	Connection stiffness (kip-inch) X 1000 M.Expo.
1	0.00	0.00	0.00	0.00	0.1231E+03	0.2120E+03	0.1476E+03
3	0.40	49.25	84.46	52.77	0.1186E+03	0.2094E+03	0.1176E+03
5	0.70	83.80	146.51	85.29	0.1091E+03	0.2035E+03	0.9978E+02
7	1.00	114.73	206.12	112.96	0.9417E+02	0.1931E+03	0.8522E+02
9	1.50	154.36	296.46	150.67	0.6828E+02	0.1668E+03	0.6663E+02
11	2.10	187.43	385.75	185.69	0.4908E+02	0.1313E+03	0.5114E+02
13	3.00	223.44	484.54	224.66	0.3674E+02	0.9146E+02	0.3683E+02
15	4.10	259.45	568.52	259.40	0.2896E+02	0.6411E+02	0.2734E+02
17	5.70	294.73	653.18	296.82	0.2226E+02	0.4411E+02	0.2018E+02
19	7.50	330.76	721.26	328.37	0.1473E+02	0.3272E+02	0.1512E+02
21	11.00	368.98	814.48	367.10	0.9948E+01	0.2203E+02	0.7334E+01
23	15.10	404.26	891.40	405.01	0.7291E+01	0.1615E+02	0.6549E+01
25	22.25	427.50	986.98	426.53	0.1233E+01	0.1123E+02	0.1027E+01
27	30.15	437.24	1063.92	439.30	0.1233E+01	0.8529E+01	0.2341E+01
29	35.70	462.31	1108.02	454.53	0.9553E+01	0.7335E+01	0.3061E+01
31	43.60	477.76	1160.81	480.09	0.3311E+01	0.6162E+01	0.3287E+01
33	48.10	496.13	1187.37	494.69	0.2778E+01	0.5658E+01	0.3185E+01
35	56.10	499.80	1229.68	500.49	0.4595E+00	0.4956E+01	0.5858E+00
37	64.10	503.48	1267.10	504.20	0.4595E+00	0.4422E+01	0.3558E+00
39	72.10	507.16	1300.73	506.42	0.2359E+01	0.4001E+01	0.2131E+00
41	78.80	533.62	1326.54	532.05	0.1276E+02	0.3711E+01	0.3794E+01
43	80.85	572.20	1334.10	572.90	0.2291E+02	0.3631E+01	0.1992E+02

II - 30

Connection type : Double web-angle connections
Mode : All bolted

Tested by : L.E.Thompson et al. (1970) U.S.A
Test Id. : B1-3 ALT

Column : W10X49
Beam : W21X62
Angle : 4 X 3.5 X 1/2

Fasteners: A325- -3/4"D
15/16" Oversize holes
Material : A36
Fy = -- ksi
Fu = -- ksi

Major parameters

lp = 11.5000"	lu = -- "	ll = -- "	ta = 0.5000"
gb = 2.2500"	gc = 2.5500"	cu = -- "	cl = -- "
pb = -- "	pc = -- "	qb = -- "	qc = -- "
nb = - X 2	nc = - X -		

Remark 1)
2)

No	Moment (k-in)	Rotation (radians) X 1/1000	No	Moment (k-in)	Rotation (radians) X 1/1000
1	0.00	0.00	26	435.61	33.87
2	24.62	0.20	27	442.10	39.50
3	49.25	0.40	28	462.31	41.60
4	66.52	0.60	29	477.76	46.40
5	83.80	0.80	30	480.82	51.17
6	99.26	1.10	31	483.88	55.93
7	114.73	1.40	32	486.94	60.70
8	134.54	1.80	33	490.00	65.47
9	154.36	2.20	34	493.07	70.23
10	170.89	2.65	35	496.13	75.00
11	187.43	3.10	36	497.05	79.35
12	205.43	3.85	37	497.96	83.70
13	223.44	4.60	38	498.88	88.05
14	241.45	5.65	39	499.80	92.40
15	259.45	6.70	40	503.48	97.25
16	276.36	7.90	41	507.16	102.10
17	294.73	9.10	42	520.39	105.80
18	313.85	10.30	43	533.62	109.50
19	330.76	11.50	44	552.72	111.50
20	349.87	13.45	45	572.20	112.45
21	368.98	15.40	43	591.67	113.40
22	385.87	17.40			
23	404.26	19.70			
24	422.63	22.60			
25	429.12	28.23			

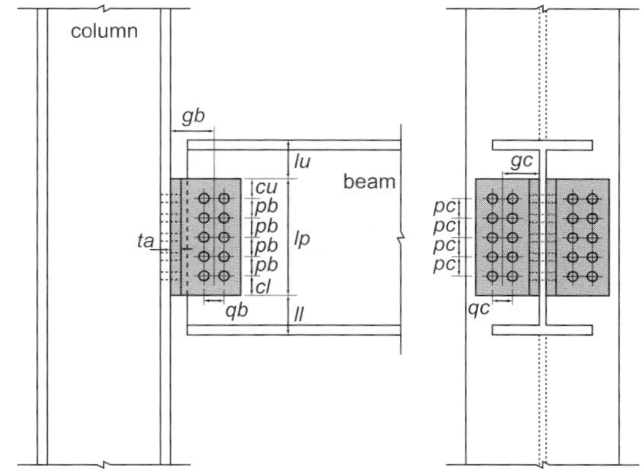

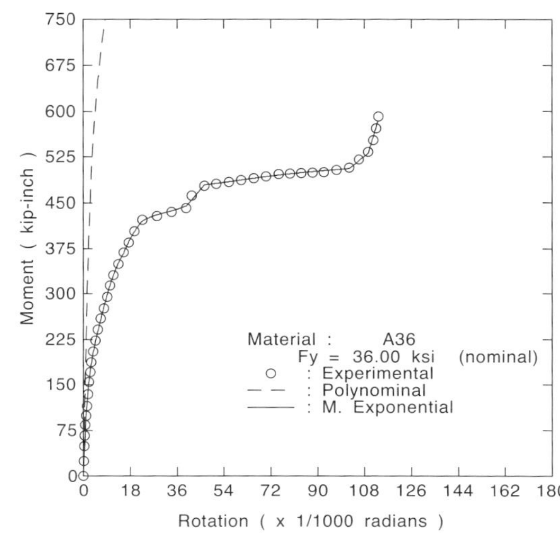

```
Moment-rotation prediction equations      ( R : X 1/1000 radians )

Frye and Morris polynominal model  :    R = Sum ( Ai X ( K*Bm )**Pi X 10**Qi )
     xd = 11.500000"   g =  5.500000"   t =  0.500000"
     A1 =  3.660000   A2 =  1.150000   A3 =  4.570000     K =  0.012890
     P1 =        1    P2 =        3    P3 =        5     Q1 =       -1     Q2 =       -3     Q3 =       -5

Modified exponential model  :
     Bm = Sum ( Ai X ( 1 - exp( - R/( 2*i*C ) ) ) )  +      Sum ( RKj X ( R - Rj0 ) )  +  Bm0
       C =  0.99500000E+00            Bmo=  0.00000000E+00          Nexp=  6        Nliner=  6
      AI =  0.45908760E+03  -0.74168500E+03  -0.37372130E+04   0.14186913E+05  -0.15463007E+05   0.56912986E+04
     Rj0 =         15.4000          22.6000          39.5000          46.4000         102.1000
              109.5000
     RKj =  0.41137346E+01  -0.22971675E+01   0.31992936E+01  -0.47344042E+01   0.31404400E+01
            0.10951706E+02
```

No	Rotation (radians) X 1/1000	Moment (kip-inch) Expri.	Moment (kip-inch) Poly.	Moment (kip-inch) M.Expo.	Connection stiffness (kip-inch) X 1000 Expri.	Connection stiffness (kip-inch) X 1000 Poly.	Connection stiffness (kip-inch) X 1000 M.Expo.
1	0.00	0.00	0.00	0.00	0.1231E+03	0.2120E+03	0.1232E+03
3	0.40	49.25	84.46	43.98	0.1047E+03	0.2094E+03	0.9777E+02
5	0.80	83.80	166.71	78.98	0.7245E+02	0.2006E+03	0.7804E+02
7	1.40	114.73	279.49	118.94	0.5069E+02	0.1727E+03	0.5658E+02
9	2.20	154.36	398.63	156.38	0.4351E+02	0.1259E+03	0.3866E+02
11	3.10	187.43	493.52	185.65	0.3197E+02	0.8818E+02	0.2760E+02
13	4.60	223.44	598.55	220.22	0.2115E+02	0.5618E+02	0.1997E+02
15	6.70	259.45	693.49	258.47	0.1572E+02	0.3693E+02	0.1708E+02
17	9.10	294.73	768.46	297.14	0.1562E+02	0.2671E+02	0.1503E+02
19	11.50	330.76	825.23	329.88	0.1246E+02	0.2107E+02	0.1213E+02
21	15.40	368.98	896.19	366.81	0.9133E+01	0.1585E+02	0.6887E+01
23	19.70	404.26	956.68	404.09	0.7260E+01	0.1257E+02	0.6692E+01
25	28.23	429.12	1047.15	430.87	0.1152E+01	0.9044E+01	0.1263E+01
27	39.50	442.10	1134.46	445.18	0.7323E+01	0.6716E+01	0.1542E+01
29	46.40	477.76	1177.60	479.27	0.1925E+01	0.5837E+01	0.5112E+01
31	55.93	483.88	1228.86	484.13	0.6424E+00	0.4968E+01	0.5882E+00
33	65.47	490.00	1273.09	489.76	0.6424E+00	0.4343E+01	0.5720E+00
35	75.00	496.13	1312.14	494.83	0.4168E+00	0.3869E+01	0.4875E+00
37	83.70	497.96	1344.26	498.74	0.2110E+00	0.3527E+01	0.4143E+00
39	92.40	499.80	1373.98	502.09	0.4698E+00	0.3242E+01	0.3612E+00
41	102.10	507.16	1403.85	505.40	0.2356E+01	0.2985E+01	0.3240E+00
43	109.50	533.62	1425.30	530.97	0.7455E+01	0.2815E+01	0.3448E+01
45	112.45	572.20	1433.56	573.44	0.2050E+02	0.2753E+01	0.1439E+02

II - 31

Connection type : Double web-angle connections
Mode : All bolted

Tested by : L.E.Thompson et al. (1970) U.S.A
Test Id. : D1-1 ALT

Column : W10X49
Beam : W21X62
Angle : 4 X 3.5 X 5/16

Fasteners: A325- -3/4"D
15/16" Oversize holes
Material : A36
Fy = -- ksi
Fu = -- ksi

Major parameters

lp = 11.5000"	lu = -- "	ll = -- "	ta = 0.3125"
gb = 2.2500"	gc = 1.8000"	cu = -- "	cl = -- "
pb = -- "	pc = -- "	qb = -- "	qc = -- "
nb = - X 2	nc = - X -		

Remark 1)
2)

No	Moment (k-in)	Rotation (radians) X 1/1000	No	Moment (k-in)	Rotation (radians) X 1/1000
1	0.00	0.00	26	337.55	60.67
2	23.52	0.30	27	339.56	64.70
3	47.04	0.60	28	354.27	67.75
4	83.05	1.10	29	368.98	70.80
5	115.39	1.70	30	410.14	72.00
6	152.88	2.60	31	446.88	73.27
7	169.42	3.10	32	483.63	74.53
8	183.76	3.60	33	520.38	75.80
9	202.13	4.40	34	556.03	77.00
10	220.50	5.70	35	591.67	78.20
11	240.35	7.10	36	627.32	79.25
12	258.72	8.70	37	662.96	80.30
13	267.17	12.70	38	699.71	82.25
14	275.63	16.70	39	736.46	84.20
15	284.81	19.15	40	772.48	85.40
16	294.00	21.60	41	808.50	86.60
17	297.82	25.70	42	845.25	88.70
18	301.64	29.80	43	882.00	90.80
19	305.46	33.90			
20	309.28	38.00			
21	313.10	42.10			
22	322.30	45.35			
23	331.49	48.60			
24	333.51	52.62			
25	335.53	56.65			

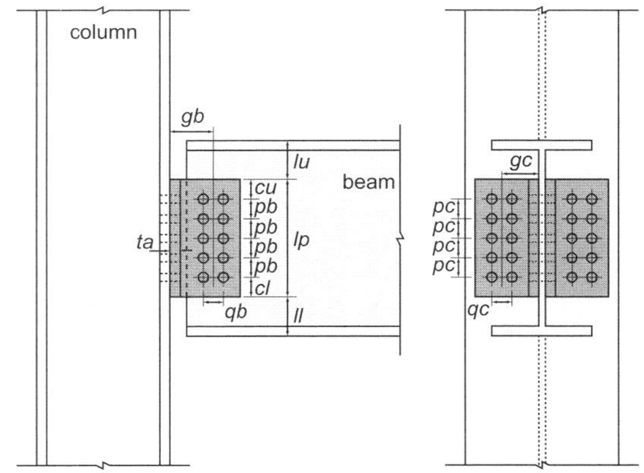

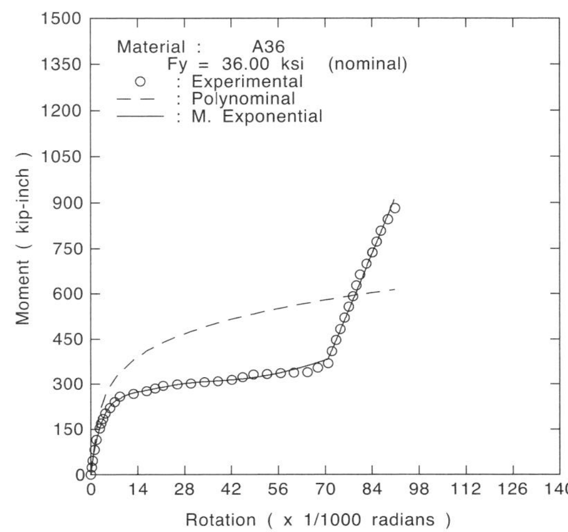

A.2 – 62

```
Moment-rotation prediction equations     ( R : X 1/1000 radians )

Frye and Morris polynominal model  :    R = Sum ( Ai X ( K*Bm )**Pi X 10**Qi )
     xd = 11.500000"   g =  4.000000"   t =  0.312500"
     A1 =  3.660000   A2 =  1.150000   A3 =  4.570000    K =  0.028771
     P1 =         1   P2 =         3   P3 =         5   Q1 =        -1   Q2 =        -3   Q3 =        -5

Modified exponential model  :
     Bm = Sum ( Ai X ( 1 - exp( - R/( 2*i*C ) ) ) )  +     Sum ( RKj X ( R - Rj0 ) )  +  Bm0
      C =   0.83166667E+00              Bmo=   0.00000000E+00          Nexp=  6       Nliner=  2
     AI =   0.27589841E+03  -0.41615162E+04   0.24636387E+05  -0.59507828E+05   0.64528811E+05  -0.25679924E+05
    Rj0 =           8.7000          70.8000
    RKj =   0.45298408E+01   0.22477930E+02
```

No	Rotation (radians) X 1/1000	Moment (kip-inch) Expri.	Moment (kip-inch) Poly.	Moment (kip-inch) M.Expo.	Connection stiffness (kip-inch) X 1000 Expri.	Connection stiffness (kip-inch) X 1000 Poly.	Connection stiffness (kip-inch) X 1000 M.Expo.
1	0.00	0.00	0.00	0.00	0.7840E+02	0.9496E+02	0.9384E+02
3	0.60	47.04	56.46	49.35	0.7601E+02	0.9227E+02	0.7208E+02
5	1.70	115.39	147.28	114.06	0.4900E+02	0.6929E+02	0.4807E+02
7	3.10	169.42	221.11	168.62	0.3088E+02	0.3950E+02	0.3145E+02
9	4.40	202.13	263.01	202.75	0.1960E+02	0.2648E+02	0.2162E+02
11	7.10	240.35	317.13	241.55	0.1292E+02	0.1554E+02	0.8349E+01
13	12.70	267.17	380.49	269.92	0.2113E+01	0.8559E+01	0.2775E+01
15	19.15	284.81	425.47	284.33	0.3749E+01	0.5780E+01	0.2194E+01
17	25.70	297.82	458.41	297.81	0.9319E+00	0.4411E+01	0.1741E+01
19	33.90	305.46	490.24	307.62	0.9319E+00	0.3443E+01	0.7651E+00
21	42.10	313.10	515.85	313.54	0.1990E+01	0.2846E+01	0.8696E+00
23	48.60	331.49	533.21	321.22	0.1789E+01	0.2512E+01	0.1543E+01
25	56.65	335.53	552.13	337.62	0.5016E+00	0.2202E+01	0.2520E+01
27	64.70	339.56	568.86	361.29	0.2959E+01	0.1965E+01	0.3316E+01
29	70.80	368.98	580.39	382.89	0.2598E+02	0.1820E+01	0.3742E+01
31	73.27	446.88	584.85	447.73	0.2901E+02	0.1768E+01	0.2635E+02
33	75.80	520.38	589.30	514.64	0.2937E+02	0.1717E+01	0.2647E+02
35	78.20	591.67	593.39	578.28	0.3197E+02	0.1672E+01	0.2656E+02
37	80.30	662.96	596.88	634.13	0.2866E+02	0.1635E+01	0.2663E+02
39	84.20	736.46	603.06	738.20	0.2576E+02	0.1572E+01	0.2673E+02
41	86.60	808.50	606.81	802.42	0.2546E+02	0.1535E+01	0.2678E+02
43	90.80	882.00	613.19	915.06	0.1750E+02	0.1474E+01	0.2685E+02

II - 32

```
Connection type : Double web-angle connections
           Mode : All bolted

Tested by  :  L.E.Thompson et al. (1970)                U.S.A
Test  Id.  :  D1-1 ALT

Column : W10X49                        Fasteners: A325- -3/4"D
Beam   : W21X62                                   15/16" Oversize holes
Angle  : 4 X 3.5 X 5/16                Material : A36
                                            Fy =   --   ksi
                                            Fu =   --   ksi

                          Major parameters

     lp = 11.5000"    lu =   --   "    ll =   --   "    ta =  0.3125"
     gb =  2.2500"    gc =  1.8000"    cu =   --   "    cl =   --   "
     pb =   --   "    pc =   --   "    qb =   --   "    qc =   --   "
     nb = - X  2      nc = - X  -

Remark  1)
        2)
```

No	Moment (k-in)	Rotation (radians) X 1/1000	No	Moment (k-in)	Rotation (radians) X 1/1000
1	0.00	0.00	26	368.98	55.90
2	23.52	0.30	27	370.81	59.48
3	47.04	0.60	28	372.65	63.05
4	83.05	1.30	29	374.48	66.62
5	115.39	2.10	30	376.32	70.20
6	152.88	3.20	31	393.23	73.55
7	169.42	3.70	32	410.14	76.90
8	183.76	4.30	33	448.34	78.80
9	202.13	5.20	34	484.36	79.75
10	220.50	6.70	35	520.38	80.70
11	240.35	8.00	36	556.03	81.70
12	258.72	9.50	37	591.67	82.70
13	275.63	11.90	38	627.32	83.75
14	284.81	14.95	39	662.96	84.80
15	294.00	18.00	40	699.71	86.05
16	298.78	22.27	41	736.46	87.30
17	303.55	26.55	42	772.48	88.95
18	308.33	30.82	43	808.50	90.60
19	313.10	35.10	44	845.25	92.50
20	322.30	38.05	45	882.00	94.40
21	331.49	41.00			
22	335.53	43.55			
23	339.56	46.10			
24	349.37	49.37			
25	359.17	52.63			

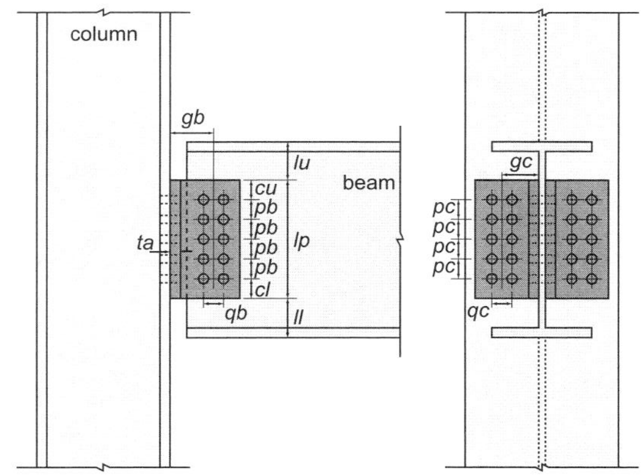

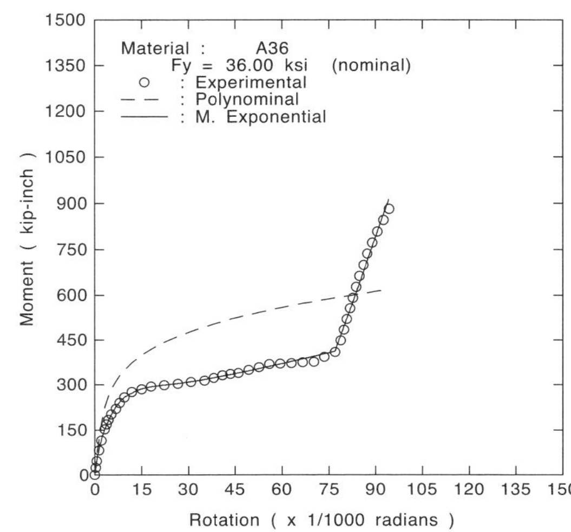

```
Moment-rotation prediction equations     ( R : X 1/1000 radians )

Frye and Morris polynominal model  :    R = Sum ( Ai X ( K*Bm )**Pi X 10**Qi )
     xd = 11.500000"   g =  4.000000"   t =  0.312500"
     A1 =  3.660000   A2 =  1.150000   A3 =  4.570000    K =  0.028771
     P1 =         1   P2 =         3   P3 =         5    Q1 =        -1    Q2 =        -3    Q3 =        -5

Modified exponential model   :
     Bm = Sum ( Ai X ( 1 - exp( - R/( 2*i*C ) ) ) )  +      Sum ( RKj X ( R - Rj0 ) )  +  Bm0
       C =   0.86166667E+00             Bmo=   0.00000000E+00          Nexp=  6        Nliner=  2
      AI =   0.26973158E+03  -0.18910005E+04   0.62402540E+04  -0.96001795E+04   0.82598252E+04  -0.30340123E+04
     Rj0 =          11.9000          76.9000
     RKj =   0.25603289E+01   0.26425703E+02
```

No	Rotation (radians) X 1/1000	Moment (kip-inch) Expri.	Moment (kip-inch) Poly.	Moment (kip-inch) M.Expo.	Connection stiffness (kip-inch) X 1000 Expri.	Connection stiffness (kip-inch) X 1000 Poly.	Connection stiffness (kip-inch) X 1000 M.Expo.
1	0.00	0.00	0.00	0.00	0.7840E+02	0.9496E+02	0.8737E+02
3	0.60	47.04	56.46	45.21	0.7031E+02	0.9227E+02	0.6491E+02
5	2.10	115.39	172.82	118.02	0.3775E+02	0.5882E+02	0.3687E+02
7	3.70	169.42	242.52	166.55	0.2890E+02	0.3231E+02	0.2538E+02
9	5.20	202.13	282.24	199.99	0.1735E+02	0.2190E+02	0.1957E+02
11	8.00	240.35	330.22	243.16	0.1387E+02	0.1369E+02	0.1161E+02
13	11.90	275.63	373.46	271.95	0.5269E+01	0.9122E+01	0.3855E+01
15	18.00	294.00	418.64	293.00	0.2223E+01	0.6124E+01	0.1671E+01
17	26.55	303.55	462.10	303.83	0.1117E+01	0.4283E+01	0.1318E+01
19	35.10	313.10	494.31	317.31	0.2300E+01	0.3338E+01	0.1806E+01
21	41.00	331.49	512.68	328.63	0.2294E+01	0.2913E+01	0.2014E+01
23	46.10	339.56	526.79	339.23	0.2205E+01	0.2630E+01	0.2139E+01
25	52.63	359.17	543.00	353.62	0.3001E+01	0.2345E+01	0.2259E+01
27	59.48	370.81	558.22	369.42	0.5136E+00	0.2112E+01	0.2355E+01
29	66.62	374.48	572.59	386.53	0.5136E+00	0.1917E+01	0.2429E+01
31	73.55	393.23	585.44	403.53	0.5047E+01	0.1761E+01	0.2478E+01
33	78.80	448.34	594.34	466.82	0.3198E+02	0.1662E+01	0.2893E+02
35	80.70	520.38	597.48	521.80	0.3681E+02	0.1629E+01	0.2894E+02
37	82.70	591.67	600.72	579.68	0.3482E+02	0.1595E+01	0.2894E+02
39	84.80	662.96	604.05	640.47	0.3187E+02	0.1562E+01	0.2895E+02
41	87.30	736.46	607.93	712.85	0.2614E+02	0.1524E+01	0.2896E+02
43	90.60	808.50	612.78	808.42	0.2067E+02	0.1478E+01	0.2896E+02
45	94.40	882.00	618.35	918.49	0.1934E+02	0.1428E+01	0.2897E+02

A.2 – 65

II - 33

Connection type : Double web-angle connections
Mode : All bolted

Tested by : L.E.Thompson et al. (1970) U.S.A
Test Id. : D1-2 ALT

Column : W10X49
Beam : W21X62
Angle : 4 X 3.5 X 5/16

Fasteners: A325- -3/4"D
15/16" Oversize holes
Material : A36
Fy = -- ksi
Fu = -- ksi

Major parameters

lp = 11.5000"	lu = -- "	ll = -- "	ta = 0.3125"
gb = 2.2500"	gc = 1.8000"	cu = -- "	cl = -- "
pb = -- "	pc = -- "	qb = -- "	qc = -- "
nb = - X 2	nc = - X -		

Remark 1)
2)

No	Moment (k-in)	Rotation (radians) X 1/1000
1	0.00	0.00
2	45.58	0.60
3	80.86	1.10
4	117.60	1.80
5	150.31	2.70
6	186.70	3.90
7	221.23	5.40
8	258.72	8.40
9	276.36	10.90
10	294.00	13.40
11	313.11	16.65
12	332.22	19.90
13	349.86	24.05
14	367.50	28.20
15	370.22	32.38
16	372.94	36.56
17	375.66	40.74
18	378.38	44.92
19	381.10	49.10
20	383.82	53.28
21	386.53	57.46
22	389.25	61.64
23	391.97	65.82
24	394.69	70.00
25	422.63	71.10
26	451.66	72.25
27	480.68	73.40
28	518.54	74.60
29	556.39	75.80
30	591.67	77.00
31	626.95	78.20
32	663.52	79.40
33	700.09	80.60
34	735.92	81.95
35	771.76	83.30
36	808.42	85.15
37	845.09	87.00
38	881.92	89.05
39	918.76	91.10

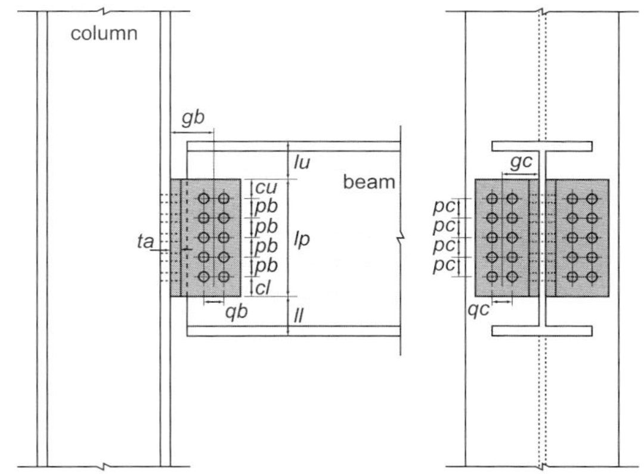

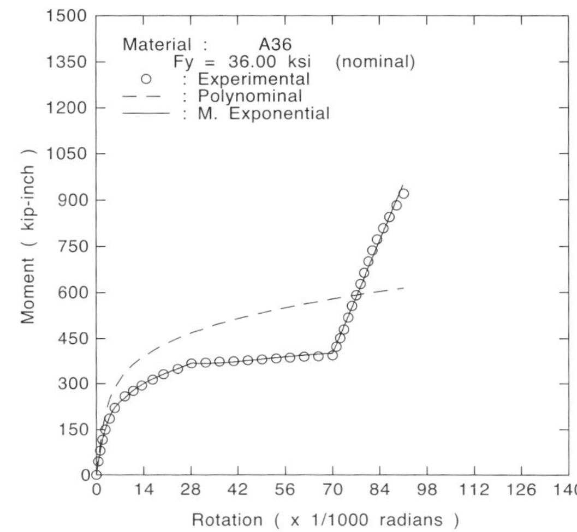

A.2 – 66

```
Moment-rotation prediction equations      ( R : X 1/1000 radians )

Frye and Morris polynominal model  :      R = Sum ( Ai X ( K*Bm )**Pi X 10**Qi )
     xd = 11.500000"    g =  4.000000"    t =  0.312500"
     A1 =  3.660000    A2 =  1.150000    A3 =  4.570000     K =  0.028771
     P1 =         1    P2 =         3    P3 =         5    Q1 =        -1    Q2 =        -3    Q3 =        -5

Modified exponential model  :
     Bm = Sum ( Ai X ( 1 - exp( - R/( 2*i*C ) ) ) )  +      Sum ( RKj X ( R - Rj0 ) )  +  Bm0
      C =   0.90916667E+00             Bmo=   0.00000000E+00          Nexp=  6        Nliner=  3
     AI =  -0.12614252E+03   0.28653679E+04  -0.14417115E+05   0.33700620E+05  -0.34763242E+05   0.13043299E+05
    Rj0 =          8.4000          28.2000          70.0000
    RKj =   0.52164825E+01  -0.51420096E+01   0.25794334E+02
```

No	Rotation (radians) X 1/1000	Moment (kip-inch) Expri.	Moment (kip-inch) Poly.	Moment (kip-inch) M.Expo.	Connection stiffness (kip-inch) X 1000 Expri.	Connection stiffness (kip-inch) X 1000 Poly.	Connection stiffness (kip-inch) X 1000 M.Expo.
1	0.00	0.00	0.00	0.00	0.7596E+02	0.9496E+02	0.8097E+02
3	1.10	80.86	100.90	77.35	0.6303E+02	0.8451E+02	0.6018E+02
5	2.70	150.31	204.05	153.88	0.3376E+02	0.4596E+02	0.3700E+02
7	5.40	221.23	286.50	220.89	0.1951E+02	0.2100E+02	0.1570E+02
9	10.90	276.36	363.97	277.09	0.7056E+01	0.9953E+01	0.8619E+01
11	16.65	313.11	410.08	315.80	0.5880E+01	0.6590E+01	0.5358E+01
13	24.05	349.86	450.91	350.01	0.4251E+01	0.4685E+01	0.4347E+01
15	32.38	370.22	484.90	367.85	0.6505E+00	0.3586E+01	0.9303E-01
17	40.74	375.66	511.92	372.68	0.6505E+00	0.2929E+01	0.9624E+00
19	49.10	381.10	534.47	382.06	0.6505E+00	0.2490E+01	0.1186E+01
21	57.46	386.53	553.90	391.40	0.6505E+00	0.2175E+01	0.1010E+01
23	65.82	391.97	571.04	398.66	0.6505E+00	0.1937E+01	0.7269E+00
25	71.10	422.63	580.95	430.43	0.2532E+02	0.1814E+01	0.2636E+02
27	73.40	480.68	585.09	490.99	0.2833E+02	0.1765E+01	0.2630E+02
29	75.80	556.39	589.30	554.03	0.3047E+02	0.1717E+01	0.2624E+02
31	78.20	626.95	593.39	616.93	0.2994E+02	0.1672E+01	0.2619E+02
33	80.60	700.09	597.38	679.72	0.2862E+02	0.1630E+01	0.2614E+02
35	83.30	771.76	601.61	750.23	0.2371E+02	0.1586E+01	0.2609E+02
37	87.00	845.09	607.42	846.68	0.1894E+02	0.1529E+01	0.2604E+02
39	91.10	918.76	613.63	953.36	0.1797E+02	0.1470E+01	0.2600E+02

II - 34

Connection type : Double web-angle connections
Mode : All bolted

Tested by : L.E.Thompson et al. (1970) U.S.A
Test Id. : D1-2 ALT

Column : W10X49
Beam : W21X62
Angle : 4 X 3.5 X 5/16

Fasteners: A325- -3/4"D
15/16" Oversize holes
Material : A36
Fy = -- ksi
Fu = -- ksi

Major parameters

lp = 11.5000"	lu = -- "	ll = -- "	ta = 0.3125"
gb = 2.2500"	gc = 1.8000"	cu = -- "	cl = -- "
pb = -- "	pc = -- "	cb = -- "	qc = -- "
nb = - X 2	nc = - X -		

Remark 1)
2)

No	Moment (k-in)	Rotation (radians) X 1/1000
1	0.00	0.00
2	45.58	0.60
3	80.86	1.10
4	117.60	1.90
5	150.31	2.70
6	186.70	3.90
7	221.23	5.30
8	258.72	7.90
9	294.00	11.70
10	313.11	14.20
11	332.22	16.70
12	349.86	19.80
13	367.50	22.90
14	370.90	27.36
15	374.30	31.82
16	377.70	36.29
17	381.10	40.75
18	384.50	45.21
19	387.89	49.67
20	391.29	54.14
21	394.69	58.60
22	401.68	63.42
23	408.66	68.25
24	415.64	73.08
25	422.63	77.90

No	Moment (k-in)	Rotation (radians) X 1/1000
26	451.66	79.65
27	480.68	81.40
28	518.54	82.50
29	556.39	83.60
30	591.67	84.60
31	626.95	85.60
32	663.52	86.70
33	700.09	87.80
34	735.92	89.00
35	771.76	90.20
36	808.42	92.05
37	845.09	93.90
38	881.92	95.70
39	918.76	97.50

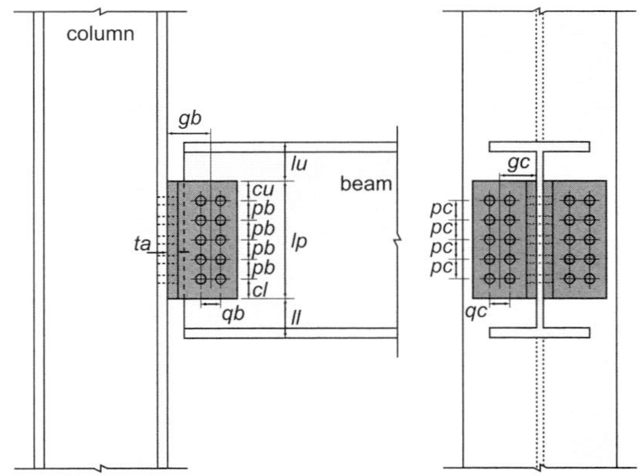

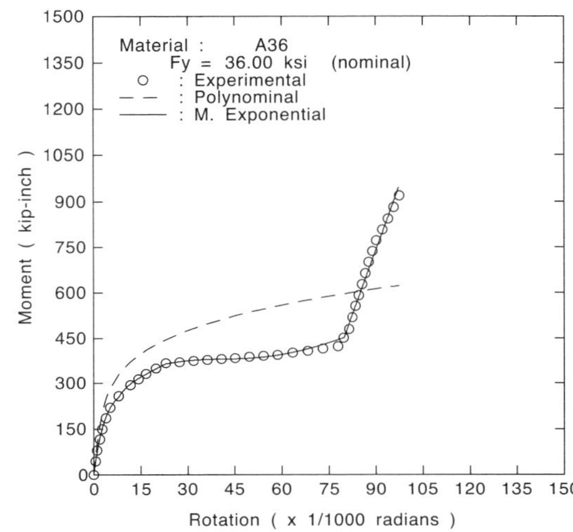

```
Moment-rotation prediction equations     ( R : X 1/1000 radians )

Frye and Morris polynominal model  :    R = Sum ( Ai X ( K*Bm )**Pi X 10**Qi )
     xd = 11.500000"    g =  4.000000"   t =  0.312500"
     A1 =  3.660000   A2 =  1.150000   A3 =  4.570000    K =  0.028771
     P1 =         1    P2 =         3   P3 =         5    Q1 =        -1    Q2 =        -3    Q3 =        -5

Modified exponential model  :
     Bm = Sum ( Ai X ( 1 - exp( - R/( 2*i*C ) ) ) )  +      Sum ( RKj X ( R - Rj0 ) )  +  Bm0
      C =   0.96250000E+00            Bmo=   0.00000000E+00          Nexp=  6        Nliner=  3
     AI =   0.24432865E+03  -0.40891858E+04   0.26031272E+05  -0.65655735E+05   0.73391191E+05  -0.29868773E+05
    Rj0 =           7.9000          22.9000          79.8000
    RKj =   0.83317469E+01  -0.37517650E+01   0.24255496E+02
```

No	Rotation (radians) X 1/1000	Moment (kip-inch) Expri.	Moment (kip-inch) Poly.	Moment (kip-inch) M.Expo.	Connection stiffness (kip-inch) X 1000 Expri.	Connection stiffness (kip-inch) X 1000 Poly.	Connection stiffness (kip-inch) X 1000 M.Expo.
1	0.00	0.00	0.00	0.00	0.7596E+02	0.9496E+02	0.8468E+02
3	1.10	80.86	100.90	77.23	0.6109E+02	0.8451E+02	0.5816E+02
5	2.70	150.31	204.05	151.45	0.3666E+02	0.4596E+02	0.3683E+02
7	5.30	221.23	284.38	220.77	0.2108E+02	0.2144E+02	0.1841E+02
9	11.70	294.00	371.58	297.69	0.8295E+01	0.9281E+01	0.8327E+01
11	16.70	332.22	410.39	330.66	0.6772E+01	0.6573E+01	0.5712E+01
13	22.90	367.50	445.39	365.20	0.3670E+01	0.4900E+01	0.5495E+01
15	31.82	374.30	482.90	377.17	0.7617E+00	0.3641E+01	0.8108E+00
17	40.75	381.10	511.95	380.71	0.7617E+00	0.2928E+01	0.1645E+00
19	49.67	387.89	535.89	383.54	0.7617E+00	0.2465E+01	0.6234E+00
21	58.60	394.69	556.36	393.51	0.1091E+01	0.2139E+01	0.1648E+01
23	68.25	408.66	575.68	414.85	0.1447E+01	0.1878E+01	0.2736E+01
25	77.90	422.63	592.80	445.33	0.1256E+02	0.1679E+01	0.3527E+01
27	81.40	480.68	598.62	496.86	0.2753E+02	0.1617E+01	0.2799E+02
29	83.60	556.39	602.16	558.57	0.3487E+02	0.1581E+01	0.2810E+02
31	85.60	626.95	605.30	614.87	0.3431E+02	0.1550E+01	0.2819E+02
33	87.80	700.09	608.69	676.99	0.3163E+02	0.1517E+01	0.2828E+02
35	90.20	771.76	612.31	744.97	0.2591E+02	0.1482E+01	0.2837E+02
37	93.90	845.09	617.61	850.13	0.2015E+02	0.1434E+01	0.2847E+02
39	97.50	918.76	622.73	952.79	0.2046E+02	0.1389E+01	0.2855E+02

Connection type : Double web-angle connections
Mode : All bolted

Tested by : L.E.Thompson et al. (1970) U.S.A
Test Id. : D1-3 ALT

Column : W10X49
Beam : W21X62
Angle : 4 X 3.5 X 5/16

Fasteners: A325- -3/4"D
15/16" Oversize holes
Material : A36
Fy = -- ksi
Fu = -- ksi

Major parameters

lp = 11.5000"	lu = -- "	ll = -- "	ta = 0.3125"
gb = 2.2500"	gc = 1.8000"	cu = -- "	cl = -- "
pb = -- "	pc = -- "	qb = -- "	qc = -- "
nb = - X 2	nc = - X -		

Remark 1)
2)

No	Moment (k-in)	Rotation (radians) X 1/1000	No	Moment (k-in)	Rotation (radians) X 1/1000
1	0.00	0.00	26	407.93	78.20
2	44.10	0.50	27	445.78	79.45
3	80.86	1.10	28	483.64	80.70
4	116.14	1.80	29	518.87	81.75
5	150.67	2.80	30	554.10	82.80
6	183.76	4.10	31	590.53	83.95
7	221.96	6.60	32	626.95	85.10
8	257.26	9.50	33	663.70	86.30
9	275.63	12.50	34	700.45	87.50
10	281.75	16.60	35	736.10	88.80
11	287.88	20.70	36	771.76	90.10
12	294.00	24.80	37	808.51	91.70
13	303.19	27.65	38	845.26	93.30
14	312.37	30.50	39	882.01	95.60
15	330.76	34.30	40	918.76	97.90
16	344.71	39.00			
17	356.11	41.80			
18	367.50	44.60			
19	373.62	48.43			
20	379.75	52.27			
21	385.87	56.10			
22	390.28	60.52			
23	394.69	64.94			
24	399.11	69.36			
25	403.52	73.78			

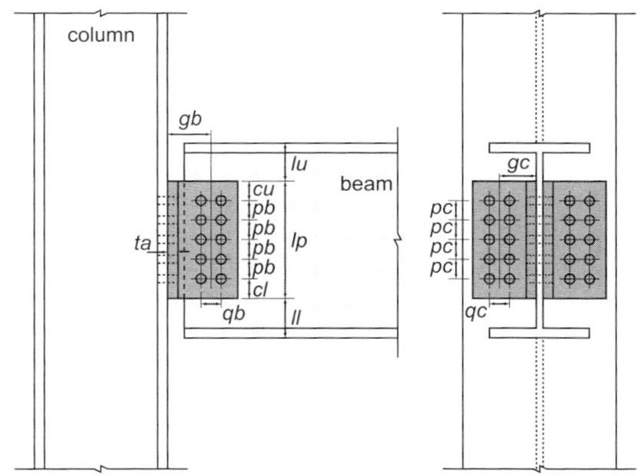

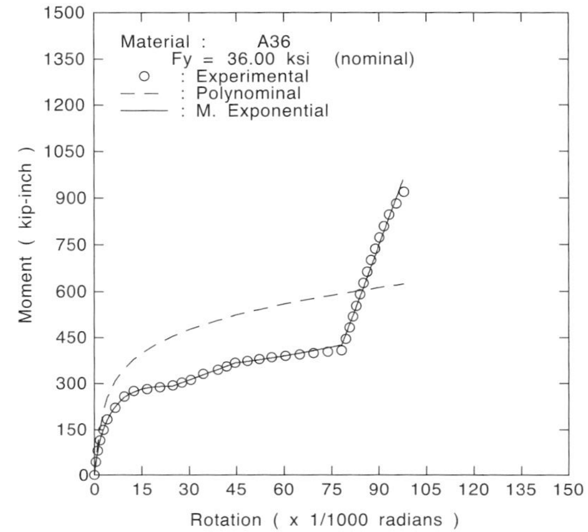

```
Moment-rotation prediction equations     ( R : X 1/1000 radians )

Frye and Morris polynominal model  :    R = Sum ( Ai X ( K*Bm )**Pi X 10**Qi )
     xd = 11.500000"   g =  4.000000"   t =  0.312500"
     A1 =  3.660000   A2 =  1.150000   A3 =  4.570000    K =  0.028771
     P1 =        1     P2 =        3     P3 =        5     Q1 =       -1     Q2 =       -3     Q3 =       -5

Modified exponential model  :
     Bm = Sum ( Ai X ( 1 - exp( - R/( 2*i*C ) ) ) )  +      Sum ( RKj X ( R - Rj0 ) )  +  Bm0
       C =   0.94083333E+00             Bmo=   0.00000000E+00           Nexp=  6        Nliner=  4
      AI =   0.33225987E+03  -0.23049668E+04   0.96618643E+04  -0.19865358E+05   0.20424157E+05  -0.80181347E+04
     Rj0 =          12.5000          24.8000          44.6000          78.2000
     RKj =   0.11063111E+01   0.38458198E+01  -0.25508858E+01   0.25057791E+02
```

No	Rotation (radians) X 1/1000	Moment (kip-inch) Expri.	Moment (kip-inch) Poly.	Moment (kip-inch) M.Expo.	Connection stiffness (kip-inch) X 1000 Expri.	Connection stiffness (kip-inch) X 1000 Poly.	Connection stiffness (kip-inch) X 1000 M.Expo.
1	0.00	0.00	0.00	0.00	0.8820E+02	0.9496E+02	0.9701E+02
3	1.10	80.86	100.90	81.73	0.5625E+02	0.8450E+02	0.5630E+02
5	2.80	150.67	208.56	150.78	0.3059E+02	0.4420E+02	0.2954E+02
7	6.60	221.96	309.01	224.96	0.1384E+02	0.1682E+02	0.1332E+02
9	12.50	275.63	378.86	272.27	0.4168E+01	0.8686E+01	0.3837E+01
11	20.70	287.88	434.13	290.16	0.1494E+01	0.5375E+01	0.6214E+00
13	27.65	303.19	466.72	302.84	0.3223E+01	0.4129E+01	0.3877E+01
15	34.30	330.76	491.61	328.11	0.4003E+01	0.3407E+01	0.3753E+01
17	41.80	356.11	514.99	356.25	0.4069E+01	0.2864E+01	0.3778E+01
19	48.43	373.62	532.79	371.95	0.1598E+01	0.2520E+01	0.1368E+01
21	56.10	385.87	550.91	383.32	0.1319E+01	0.2220E+01	0.1603E+01
23	64.94	394.69	569.33	398.72	0.9980E+00	0.1959E+01	0.1873E+01
25	73.78	403.52	585.72	416.24	0.9980E+00	0.1758E+01	0.2081E+01
27	79.45	445.78	595.39	459.65	0.3028E+02	0.1651E+01	0.2723E+02
29	81.75	518.87	599.17	522.33	0.3355E+02	0.1611E+01	0.2727E+02
31	83.95	590.53	602.69	582.34	0.3167E+02	0.1575E+01	0.2729E+02
33	86.30	663.70	606.37	646.51	0.3062E+02	0.1539E+01	0.2732E+02
35	88.80	736.10	610.20	714.83	0.2742E+02	0.1502E+01	0.2734E+02
37	91.70	808.51	614.52	794.15	0.2297E+02	0.1462E+01	0.2736E+02
39	95.60	882.01	620.04	900.91	0.1598E+02	0.1413E+01	0.2739E+02

Connection type : Double web-angle connections
Mode : All bolted

Tested by : L.E.Thompson et al. (1970) U.S.A
Test Id. : D1-3 ALT

Column : W10X49
Beam : W21X62
Angle : 4 X 3.5 X 5/16

Fasteners: A325- -3/4"D
15/16" Oversize holes
Material : A36
Fy = -- ksi
Fu = -- ksi

Major parameters

lp = 11.5000"	lu = -- "	ll = -- "	ta = 0.3125"
gb = 2.2500"	gc = 1.8000"	cu = -- "	cl = -- "
pb = -- "	pc = -- "	qb = -- "	qc = -- "
nb = - X 2	nc = - X -		

Remark 1)
2)

No	Moment (k-in)	Rotation (radians) X 1/1000	No	Moment (k-in)	Rotation (radians) X 1/1000
1	0.00	0.00	26	445.78	74.60
2	44.10	0.60	27	483.64	75.80
3	80.86	1.30	28	518.87	76.80
4	116.14	2.00	29	554.10	77.80
5	150.67	2.90	30	590.53	78.80
6	183.76	4.20	31	626.95	79.80
7	221.96	7.10	32	663.70	80.95
8	257.26	10.40	33	700.45	82.10
9	275.63	14.00	34	736.10	83.50
10	280.22	18.55	35	771.76	84.90
11	284.81	23.10	36	808.51	86.60
12	289.41	27.65	37	845.26	88.30
13	294.00	32.20	38	882.01	90.80
14	303.19	35.30	39	918.76	93.30
15	312.37	38.40			
16	330.76	42.50			
17	337.73	45.85			
18	344.71	49.20			
19	350.41	53.40			
20	356.11	57.60			
21	361.80	61.80			
22	367.50	66.00			
23	376.69	68.60			
24	385.87	71.20			
25	407.93	73.40			

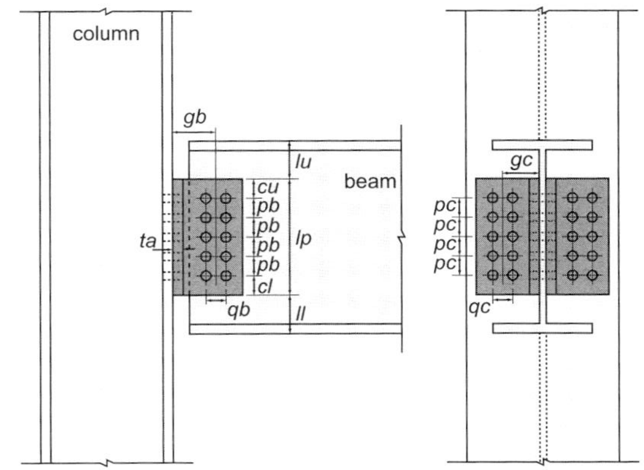

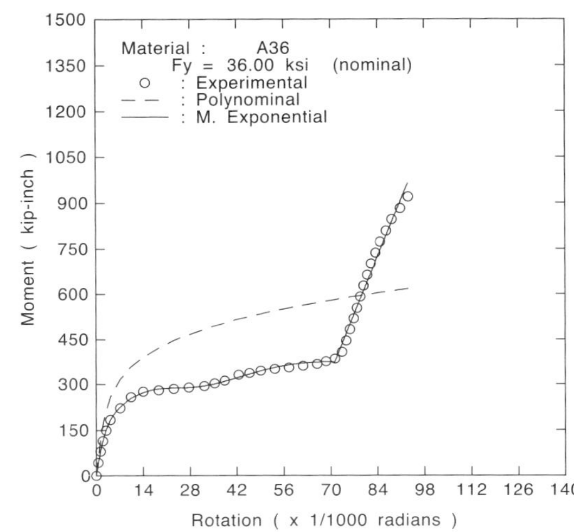

```
Moment-rotation prediction equations     ( R : X 1/1000 radians )

Frye and Morris polynominal model  :    R = Sum ( Ai X ( K*Bm )**Pi X 10**Qi )
     xd = 11.500000"   g =  4.000000"   t =  0.312500"
     A1 =  3.660000   A2 =  1.150000   A3 =  4.570000     K =  0.028771
     P1 =         1   P2 =         3   P3 =         5    Q1 =       -1     Q2 =       -3     Q3 =       -5

Modified exponential model  :
     Bm = Sum ( Ai X ( 1 - exp( - R/( 2*i*C ) ) ) )  +      Sum ( RKj X ( R - Rj0 ) )  +  Bm0
      C =   0.92750000E+00             Bmo=   0.00000000E+00           Nexp=  6       Nliner=  2
     AI =  -0.37072265E+03  0.80150287E+04  -0.45065995E+05  0.10954481E+06  -0.11809886E+06   0.46521406E+05
    Rj0 =        10.4000          71.2000
    RKj =  -0.23261339E+01  0.28151185E+02
```

No	Rotation (radians) X 1/1000	Moment (kip-inch) Expri.	Moment (kip-inch) Poly.	Moment (kip-inch) M.Expo.	Connection stiffness (kip-inch) X 1000 Expri.	Connection stiffness (kip-inch) X 1000 Poly.	Connection stiffness (kip-inch) X 1000 M.Expo.
1	0.00	0.00	0.00	0.00	0.7350E+02	0.9496E+02	0.7266E+02
3	1.30	80.86	117.38	82.54	0.5145E+02	0.7989E+02	0.5358E+02
5	2.90	150.67	212.89	150.00	0.3309E+02	0.4254E+02	0.3199E+02
7	7.10	221.96	317.11	224.84	0.1201E+02	0.1554E+02	0.1033E+02
9	14.00	275.63	391.09	272.37	0.3295E+01	0.7783E+01	0.4049E+01
11	23.10	284.81	446.39	287.62	0.1009E+01	0.4860E+01	0.2150E+00
13	32.20	294.00	484.26	294.43	0.2171E+01	0.3603E+01	0.1761E+01
15	38.40	312.37	504.89	309.71	0.3618E+01	0.3085E+01	0.3032E+01
17	45.85	337.73	526.13	334.08	0.2083E+01	0.2642E+01	0.3276E+01
19	53.40	350.41	544.79	356.07	0.1356E+01	0.2316E+01	0.2428E+01
21	61.80	361.80	563.05	370.78	0.1356E+01	0.2044E+01	0.1061E+01
23	68.60	376.69	576.42	374.44	0.3533E+01	0.1869E+01	0.4893E-01
25	73.40	407.93	585.10	435.16	0.2395E+02	0.1765E+01	0.2762E+02
27	75.80	483.64	589.31	501.14	0.3356E+02	0.1717E+01	0.2737E+02
29	77.80	554.10	592.73	555.68	0.3583E+02	0.1680E+01	0.2718E+02
31	79.80	626.95	596.07	609.88	0.3435E+02	0.1644E+01	0.2702E+02
33	82.10	700.45	599.82	671.82	0.2903E+02	0.1605E+01	0.2685E+02
35	84.90	771.76	604.14	746.73	0.2373E+02	0.1561E+01	0.2667E+02
37	88.30	845.26	609.41	837.09	0.1882E+02	0.1510E+01	0.2649E+02
39	93.30	918.76	616.73	969.00	0.1470E+02	0.1442E+01	0.2629E+02

II - 37

Connection type : Double web-angle connections
Mode : All bolted

Tested by : L.E.Thompson et al. (1970) U.S.A
Test Id. : E1-1 ALT

Column : W10X49
Beam : W21X62
Angle : 4 X 3.5 X 1/2

Fasteners: A325- -3/4"D
15/16" Oversize holes
Material : A36
Fy = -- ksi
Fu = -- ksi

Major parameters

lp = 11.5000"	lu = -- "	ll = -- "	ta = 0.5000"
gb = 2.2500"	gc = 1.8000"	cu = -- "	cl = -- "
pb = -- "	pc = -- "	qb = -- "	qc = -- "
nb = - X 2	nc = - X -		

Remark 1)
2)

No	Moment (k-in)	Rotation (radians) X 1/1000	No	Moment (k-in)	Rotation (radians) X 1/1000
1	0.00	0.00	26	485.83	63.10
2	35.65	0.20	27	514.50	67.10
3	71.29	0.40	28	514.50	67.50
4	106.94	0.70	29	540.96	68.00
5	142.60	1.00	30	602.89	69.15
6	176.95	1.35	31	664.81	70.30
7	211.31	1.70	32	713.32	71.20
8	249.90	2.20	33	761.83	72.10
9	298.42	3.00	34	810.34	73.00
10	332.22	3.80	35	873.73	74.55
11	334.24	5.30	36	937.13	76.10
12	336.24	6.80	37	992.25	78.05
13	337.58	10.80	38	1047.37	80.00
14	343.36	13.60	39	1109.11	83.20
15	349.13	16.40	40	1142.56	85.00
16	367.50	20.15	41	1176.00	86.80
17	385.87	23.90	42	1212.76	89.30
18	415.27	28.80	43	1249.50	92.90
19	437.32	33.45	44	1267.88	95.65
20	459.37	38.10	45	1286.26	98.40
21	468.56	41.45	46	1304.63	102.10
22	477.76	44.80	47	1323.00	105.80
23	479.77	49.38			
24	481.79	53.95			
25	483.81	58.52			

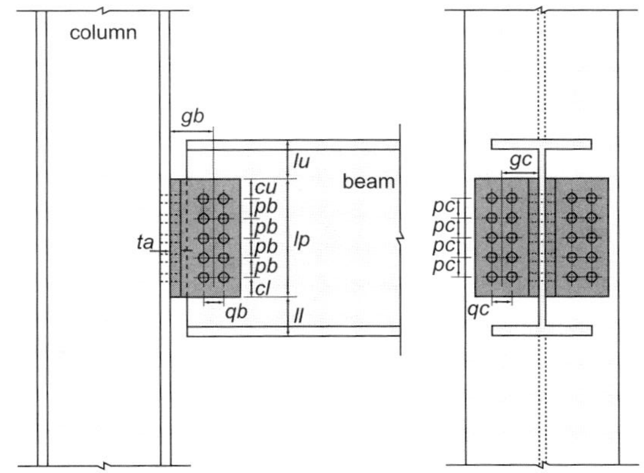

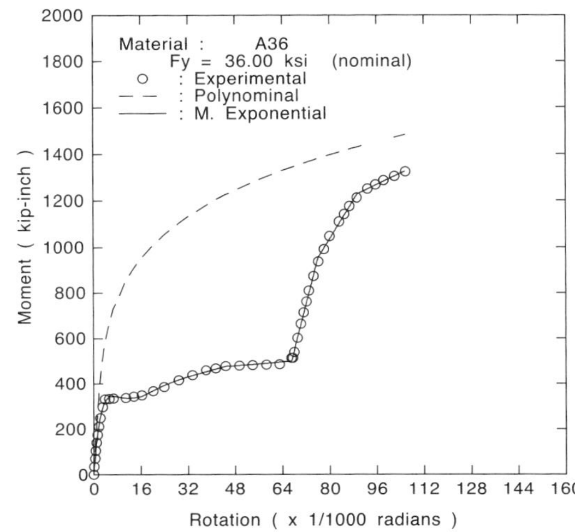

A.2 – 74

```
Moment-rotation prediction equations     ( R : X 1/1000 radians )

Frye and Morris polynominal model  :     R = Sum ( Ai X ( K*Bm )**Pi X 10**Qi )
     xd = 11.500000"   g =  4.000000"   t =  0.500000"
     A1 =  3.660000   A2 =  1.150000   A3 =  4.570000    K =  0.012288
     P1 =        1    P2 =        3    P3 =        5    Q1 =       -1    Q2 =       -3    Q3 =       -5

Modified exponential model  :
     Bm = Sum ( Ai X ( 1 - exp( - R/( 2*i*C ) ) ) )  +      Sum ( RKj X ( R - Rj0 ) )  +  Bm0
       C =   0.93166667E+00            Bmo=   0.00000000E+00            Nexp=  6          Nliner=  6
      AI =  -0.12130037E+03   0.46342497E+04  -0.14344499E+05   0.19390732E+05  -0.11753405E+05   0.25291306E+04
     Rj0 =           3.8000          16.4000          44.8000          67.1000          76.1000
                    89.3000
     RKj =   0.10705203E+01   0.32061988E+01  -0.29104256E+01   0.49332602E+02  -0.30222866E+02
            -0.14416609E+02
```

No	Rotation (radians) X 1/1000	Moment (kip-inch) Expri.	Moment (kip-inch) Poly.	Moment (kip-inch) M.Expo.	Connection stiffness (kip-inch) X 1000 Expri.	Connection stiffness (kip-inch) X 1000 Poly.	Connection stiffness (kip-inch) X 1000 M.Expo.
1	0.00	0.00	0.00	0.00	0.1782E+03	0.2223E+03	0.1786E+03
3	0.40	71.29	88.59	65.80	0.1545E+03	0.2197E+03	0.1510E+03
5	1.00	142.60	216.20	145.46	0.1093E+03	0.2026E+03	0.1157E+03
7	1.70	211.31	344.71	214.49	0.8952E+02	0.1623E+03	0.8289E+02
9	3.00	298.42	508.21	292.60	0.5145E+02	0.9595E+02	0.4088E+02
11	5.30	334.24	665.81	340.98	0.1340E+01	0.5020E+02	0.7176E+01
13	10.80	337.58	849.64	337.03	0.1351E+01	0.2354E+02	-0.1109E+01
15	16.40	349.13	956.20	345.03	0.3275E+01	0.1566E+02	0.3175E+01
17	23.90	385.87	1054.06	392.70	0.5376E+01	0.1103E+02	0.5649E+01
19	33.45	437.32	1144.16	436.16	0.4742E+01	0.8156E+01	0.3772E+01
21	41.45	468.56	1203.40	464.90	0.2744E+01	0.6754E+01	0.3554E+01
23	49.38	479.77	1252.94	480.59	0.4413E+00	0.5802E+01	0.8804E+00
25	58.52	483.81	1302.23	489.98	0.4413E+00	0.5013E+01	0.1151E+01
27	67.10	514.50	1342.74	500.52	0.6515E+00	0.4461E+01	0.1291E+01
29	68.00	540.96	1346.75	546.09	0.5320E+02	0.4410E+01	0.5063E+02
31	70.30	664.81	1356.82	662.57	0.5388E+02	0.4287E+01	0.5065E+02
33	72.10	761.83	1364.49	753.75	0.5390E+02	0.4195E+01	0.5066E+02
35	74.55	873.73	1374.70	877.90	0.4090E+02	0.4077E+01	0.5068E+02
37	78.05	992.25	1388.51	996.36	0.2827E+02	0.3924E+01	0.2047E+02
39	83.20	1109.11	1408.46	1101.80	0.1884E+02	0.3715E+01	0.2048E+02
41	86.80	1176.00	1421.38	1175.52	0.1696E+02	0.3587E+01	0.2048E+02
43	92.90	1249.50	1442.59	1248.56	0.8209E+01	0.3388E+01	0.6066E+01
45	98.40	1286.26	1461.00	1281.93	0.5951E+01	0.3227E+01	0.6066E+01
47	105.80	1323.00	1484.12	1326.81	0.4965E+01	0.3037E+01	0.6064E+01

II - 38

Connection type : Double web-angle connections
Mode : All bolted

Tested by : L.E.Thompson et al. (1970) U.S.A
Test Id. : E1-1 ALT

Column : W10X49
Beam : W21X62
Angle : 4 X 3.5 X 1/2

Fasteners: A325- -3/4"D
15/16' Oversize holes
Material : A36
Fy = -- ksi
Fu = -- ksi

Major parameters

lp = 11.5000"	lu = -- "	ll = -- "	ta = 0.5000"
gb = 2.2500'	gc = 1.8000"	cu = -- "	cl = -- "
pb = -- "	pc = -- "	cb = -- "	qc = -- "
nb = - X 2	nc = - X -		

Remark 1)
2)

No	Moment (k-in)	Rotation (radians) X 1/1000	No	Moment (k-in)	Rotation (radians) X 1/1000
1	0.00	0.00	26	530.38	86.78
2	35.65	0.20	27	535.67	93.34
3	71.29	0.40	28	540.96	99.90
4	106.94	0.70	29	602.89	101.35
5	142.60	1.00	30	664.81	102.80
6	176.95	1.35	31	713.32	103.80
7	211.31	1.70	32	761.83	104.80
8	249.90	2.30	33	810.34	105.80
9	298.42	3.10	34	873.73	107.40
10	332.22	3.80	35	937.13	109.00
11	349.13	4.30	36	992.25	111.10
12	385.87	5.70	37	1047.37	113.20
13	400.57	12.30	38	1109.11	116.60
14	415.27	18.90	39	1142.56	118.30
15	437.32	23.35	40	1176.00	120.00
16	459.37	27.80	41	1212.76	122.60
17	477.76	34.50	42	1249.50	126.30
18	485.83	40.30	43	1286.26	130.70
19	500.17	45.10	44	1323.00	137.10
20	514.50	49.90			
21	514.50	55.63			
22	514.50	61.37			
23	514.50	67.10			
24	519.79	73.66			
25	525.08	80.22			

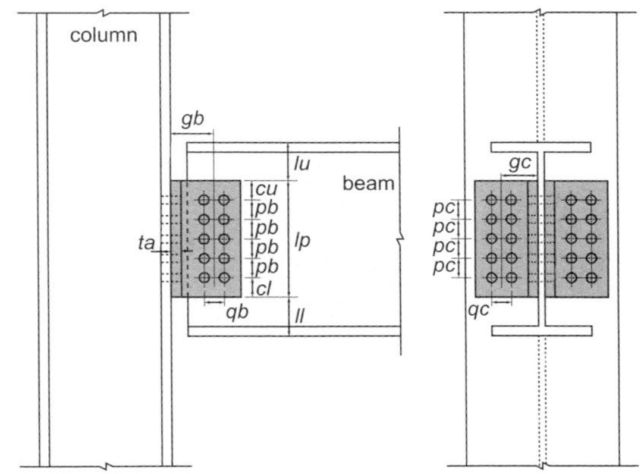

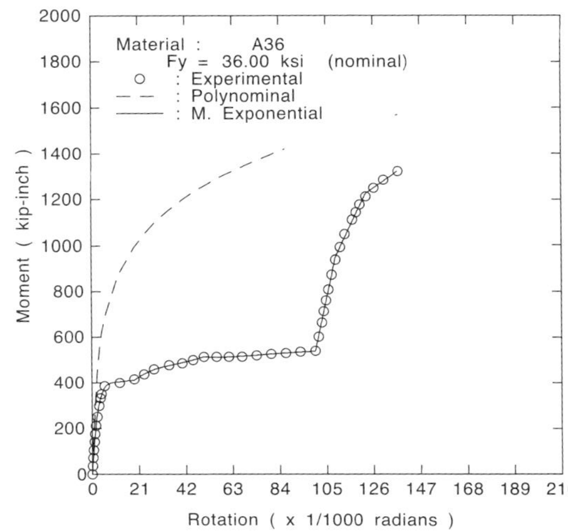

A.2 – 76

```
Moment-rotation prediction equations     ( R : X 1/1000 radians )

Frye and Morris polynominal model  :    R = Sum ( Ai X ( K*Bm )**Pi X 10**Qi )
     xd = 11.500000"    g =  4.000000"   t =  0.500000"
     A1 =  3.660000   A2 =  1.150000   A3 =  4.570000     K =  0.012288
     P1 =         1    P2 =         3   P3 =         5    Q1 =        -1    Q2 =        -3    Q3 =        -5

Modified exponential model  :
     Bm = Sum ( Ai X ( 1 - exp( - R/( 2*i*C ) ) ) )  +     Sum ( RKj X ( R - Rj0 ) )  +  Bm0
      C =   0.11925000E+01          Bmo=   0.00000000E+00          Nexp=  6        Nliner=  6
     AI =   0.50153364E+02   0.61228711E+04  -0.30064977E+05   0.61162931E+05  -0.55180581E+05   0.18352530E+05
    Rj0 =           6.5000          18.9000          49.9000          99.9000         109.0000
                  122.6000
    RKj =  -0.26679000E+01   0.66849581E+01  -0.38180954E+01   0.44736399E+02  -0.24775393E+02
           -0.13358764E+02
```

No	Rotation (radians) X 1/1000	Moment (kip-inch) Expri.	Moment (kip-inch) Poly.	Moment (kip-inch) M.Expo.	Connection stiffness (kip-inch) X 1000 Expri.	Connection stiffness (kip-inch) X 1000 Poly.	Connection stiffness (kip-inch) X 1000 M.Expo.
1	0.00	0.00	0.00	0.00	0.1782E+03	0.2223E+03	0.1691E+03
3	0.40	71.29	88.59	62.73	0.1545E+03	0.2197E+03	0.1451E+03
5	1.00	142.60	216.20	140.39	0.1093E+03	0.2026E+03	0.1148E+03
7	1.70	211.31	344.71	210.57	0.8569E+02	0.1623E+03	0.8686E+02
9	3.10	298.42	517.64	303.03	0.5406E+02	0.9251E+02	0.4859E+02
11	4.30	349.13	609.41	348.56	0.3182E+02	0.6367E+02	0.2889E+02
13	12.30	400.57	882.70	404.83	0.2227E+01	0.2068E+02	0.1134E+01
15	23.35	437.32	1047.96	439.42	0.4955E+01	0.1127E+02	0.5247E+01
17	34.50	477.76	1152.65	476.52	0.2019E+01	0.7935E+01	0.2154E+01
19	45.10	500.17	1227.16	500.15	0.2986E+01	0.6276E+01	0.2627E+01
21	55.63	514.50	1287.42	512.21	0.1492E-14	0.5235E+01	-0.4086E-01
23	67.10	514.50	1342.74	516.28	0.3762E+00	0.4461E+01	0.6333E+00
25	80.22	525.08	1396.93	525.82	0.8067E+00	0.3834E+01	0.7422E+00
27	93.34	535.67	1444.08	534.66	0.8067E+00	0.3375E+01	0.5910E+00
29	101.35	602.89	1470.21	603.84	0.4271E+02	0.3150E+01	0.4522E+02
31	103.80	713.32	1477.89	714.60	0.4851E+02	0.3087E+01	0.4519E+02
33	105.80	810.34	1484.04	804.97	0.4509E+02	0.3038E+01	0.4517E+02
35	109.00	937.13	1493.70	949.46	0.3384E+02	0.2963E+01	0.4514E+02
37	113.20	1047.37	1506.05	1034.91	0.2316E+02	0.2870E+01	0.2033E+02
39	118.30	1142.56	1520.21	1138.47	0.1967E+02	0.2768E+01	0.2029E+02
41	122.60	1212.76	1532.03	1225.65	0.1240E+02	0.2686E+01	0.2026E+02
43	130.70	1286.26	1553.10	1281.41	0.7289E+01	0.2548E+01	0.6866E+01

II - 39

Connection type : Double web-angle connections
Mode : All bolted

Tested by : L.E.Thompson et al. (1970) U.S.A
Test Id. : E1-2 ALT

Column : W10X49
Beam : W21X62
Angle : 4 X 3.5 X 1/2

Fasteners: A325- -3/4"D
15/16" Oversize holes
Material : A36
Fy = -- ksi
Fu = -- ksi

Major parameters

lp = 11.5000"	lu = -- "	ll = -- "	ta = 0.5000"
gb = 2.2500"	gc = 1.8000"	cu = -- "	cl = -- "
pb = -- "	pc = -- "	qb = -- "	qc = -- "
nb = - X 2	nc = - X -		

Remark 1)
2)

No	Moment (k-in)	Rotation (radians) X 1/1000
1	0.00	0.00
2	43.36	0.30
3	86.72	0.60
4	123.11	0.85
5	159.49	1.10
6	192.56	1.40
7	206.53	5.60
8	220.50	9.80
9	240.71	13.80
10	260.93	17.80
11	330.76	19.10
12	349.13	25.60
13	367.50	30.60
14	385.87	34.00
15	404.26	39.20
16	441.00	45.70
17	459.37	52.80
18	468.56	57.85
19	477.76	62.90
20	482.90	68.55
21	488.05	74.20
22	493.19	79.85
23	498.34	85.50
24	545.38	86.70
25	592.42	87.90

No	Moment (k-in)	Rotation (radians) X 1/1000
26	629.17	88.60
27	665.92	89.30
28	737.57	90.65
29	809.23	92.00
30	845.62	92.80
31	882.00	93.60
32	948.14	95.60
33	988.57	96.70
34	1029.00	97.80
35	1065.75	99.50
36	1102.50	101.20
37	1139.25	103.10
38	1176.00	105.00
39	1212.75	107.95
40	1249.50	110.90
41	1286.26	114.50
42	1318.51	118.65
43	1350.76	122.80
44	1392.01	128.00
45	1433.26	133.20
46	1470.00	139.80
47	1451.63	147.00

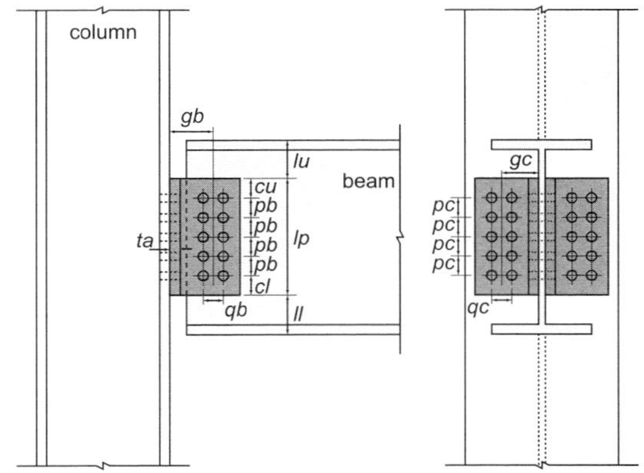

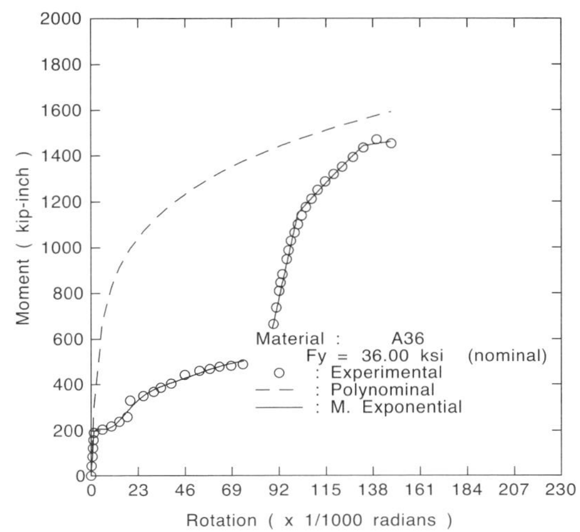

A.2 – 78

```
Moment-rotation prediction equations     ( R : X 1/1000 radians )

Frye and Morris polynominal model  :    R = Sum ( Ai X ( K*Bm )**Pi X 10**Qi )
     xd = 11.500000"   g =  4.000000"   t =  0.500000"
     A1 =  3.660000   A2 =  1.150000   A3 =  4.570000    K =  0.012288
     P1 =         1   P2 =         3   P3 =         5   Q1 =        -1    Q2 =        -3    Q3 =        -5

Modified exponential model  :
     Bm = Sum ( Ai X ( 1 - exp( - R/( 2*i*C ) ) ) )  +     Sum ( RKj X ( R - Rj0 ) )  +  Bm0
      C =   0.13000000E+01            Bmo=   0.00000000E+00          Nexp=  6         Nliner=  7
     AI =  -0.13097285E+03   0.73837308E+04  -0.46819402E+05   0.11513534E+06  -0.12484823E+06   0.49864171E+05
    Rj0 =          0.0000           1.4000          17.8000          57.8500          85.5000
                 101.2000         133.2000
    RKj =   0.11823233E+03  -0.12003844E+03  -0.92111515E-01  -0.56202712E+00   0.40912007E+02
           -0.30061901E+02  -0.73506487E+01
```

No	Rotation (radians) X 1/1000	Moment (kip-inch) Expri.	Moment (kip-inch) Poly.	Moment (kip-inch) M.Expo.	Connection stiffness (kip-inch) X 1000 Expri.	Connection stiffness (kip-inch) X 1000 Poly.	Connection stiffness (kip-inch) X 1000 M.Expo.
1	0.00	0.00	0.00	0.00	0.1445E+03	0.2223E+03	0.3050E+02
3	0.60	86.72	132.19	86.76	0.1451E+03	0.2160E+03	0.1407E+03
5	1.10	159.49	236.23	155.73	0.1295E+03	0.1979E+03	0.1353E+03
7	5.60	206.53	680.46	208.30	0.3326E+01	0.4719E+02	-0.4549E+00
9	13.80	240.71	912.00	245.59	0.5053E+01	0.1848E+02	0.9214E+01
11	19.10	330.76	995.46	297.62	0.4523E+02	0.1357E+02	0.9549E+01
13	30.60	367.50	1120.00	377.09	0.4704E+01	0.8828E+01	0.4402E+01
15	39.20	404.26	1187.84	407.29	0.4477E+01	0.7092E+01	0.3031E+01
17	52.80	459.37	1272.25	449.52	0.2139E+01	0.5476E+01	0.3263E+01
19	62.90	477.76	1323.49	479.00	0.1391E+01	0.4713E+01	0.2456E+01
21	74.20	488.05	1373.08	501.86	0.9106E+00	0.4096E+01	0.1511E+01
23	85.50	498.34	1416.63	512.41	0.3249E+02	0.3633E+01	0.3597E+00
25	87.90	592.42	1425.30	611.18	0.4760E+02	0.3549E+01	0.4104E+02
27	89.30	665.92	1430.25	668.54	0.5270E+02	0.3502E+01	0.4091E+02
29	92.00	809.23	1439.65	778.67	0.4831E+02	0.3415E+01	0.4067E+02
31	93.60	882.00	1445.09	843.64	0.4193E+02	0.3366E+01	0.4053E+02
33	96.70	988.57	1455.46	968.91	0.3675E+02	0.3275E+01	0.4029E+02
35	99.50	1065.75	1464.58	1081.42	0.2162E+02	0.3197E+01	0.4008E+02
37	103.10	1139.25	1475.66	1168.17	0.1934E+02	0.3105E+01	0.9784E+01
39	107.95	1212.75	1490.58	1214.93	0.1246E+02	0.2987E+01	0.9508E+01
41	114.50	1286.26	1509.52	1276.18	0.9077E+01	0.2845E+01	0.9209E+01
43	122.80	1350.76	1532.45	1351.40	0.7843E+01	0.2683E+01	0.8932E+01
45	133.20	1433.26	1559.42	1443.01	0.6890E+01	0.2508E+01	0.8705E+01
47	147.00	1451.63	1592.63	1460.42	-0.2552E+01	0.2311E+01	0.1188E+01

A.2 – 79

II - 40

Connection type : Double web-angle connections
Mode : All bolted

Tested by : L.E.Thompson et al. (1970) U.S.A
Test Id. : E1-2 ALT

Column : W10X49
Beam : W21X62
Angle : 4 X 3.5 X 1/2

Fasteners: A325- -3/4"D
15/16" Oversize holes
Material : A36
Fy = -- ksi
Fu = -- ksi

Major parameters

lp = 11.5000"	lu = -- "	ll = -- "	ta = 0.5000"
gb = 2.2500"	gc = 1.8000"	cu = -- "	cl = -- "
pb = -- "	pc = -- "	qb = -- "	qc = -- "
nb = - X 2	nc = - X -		

Remark 1)
2)

No	Moment (k-in)	Rotation (radians) X 1/1000	No	Moment (k-in)	Rotation (radians) X 1/1000
1	0.00	0.00	26	703.40	71.25
2	43.36	0.35	27	740.88	72.00
3	86.72	0.70	28	809.23	73.50
4	123.11	1.00	29	845.62	74.60
5	159.49	1.30	30	882.00	75.70
6	192.56	1.80	31	931.00	77.37
7	198.77	3.80	32	980.00	79.03
8	208.08	6.80	33	1029.00	80.70
9	220.50	10.80	34	1078.00	83.60
10	260.93	16.80	35	1127.00	86.50
11	312.37	23.30	36	1176.00	89.40
12	330.76	28.80	37	1212.75	93.55
13	367.50	35.10	38	1249.50	97.70
14	385.87	39.30	39	1286.26	101.40
15	422.63	45.70	40	1318.51	105.25
16	441.00	49.10	41	1350.76	109.10
17	446.10	55.05	42	1392.01	115.60
18	451.20	61.00	43	1433.26	122.10
19	459.37	63.00	44	1470.00	127.80
20	477.76	65.30	45	1460.81	134.15
21	498.34	66.50	46	1451.63	140.50
22	545.38	67.95			
23	592.42	69.40			
24	629.17	69.95			
25	665.92	70.50			

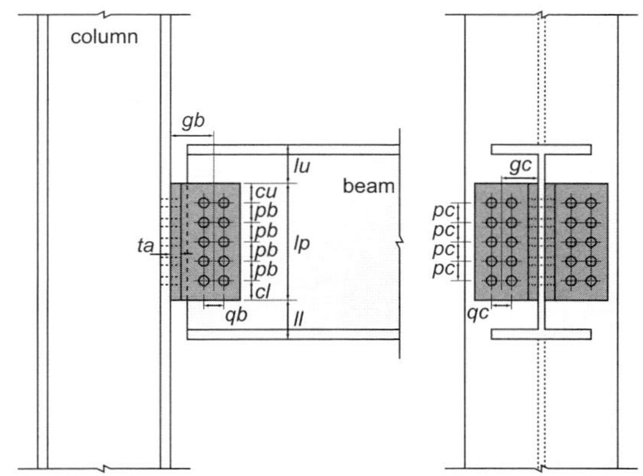

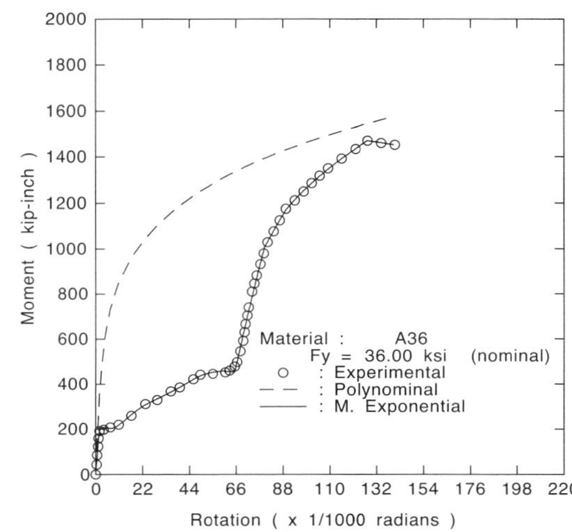

A.2 – 80

```
Moment-rotation prediction equations     ( R : X 1/1000 radians )

Frye and Morris polynominal model   :    R = Sum ( Ai X ( K*Bm )**Pi X 10**Qi )
     xd = 11.500000"    g =  4.000000"    t =  0.500000"
     A1 =  3.660000    A2 =  1.150000   A3 =  4.570000    K =  0.012288
     P1 =         1    P2 =         3   P3 =         5   Q1 =        -1    Q2 =        -3    Q3 =        -5

Modified exponential model  :
     Bm = Sum ( Ai X ( 1 - exp( - R/( 2*i*C ) ) ) )  +      Sum ( RKj X ( R - Rj0 ) )  +  Bm0
       C =   0.12583333E+01              Bmo=   0.00000000E+00           Nexp=  6         Nliner=  7
      AI =   0.33282284E+03   0.50544825E+04  -0.52687471E+05   0.16617462E+06  -0.21281962E+06   0.95505834E+05
     Rj0 =           1.8000          49.1000          65.3000          79.3000          89.4000
                   127.8000           0.0000
     RKj =  -0.73805566E+02  -0.16779921E+01   0.44259993E+02  -0.19255481E+02  -0.42915337E+01
            -0.73913835E+01   0.60141738E+02
```

No	Rotation (radians) X 1/1000	Moment (kip-inch) Expri.	Moment (kip-inch) Poly.	Moment (kip-inch) M.Expo.	Connection stiffness (kip-inch) X 1000 Expri.	Connection stiffness (kip-inch) X 1000 Poly.	Connection stiffness (kip-inch) X 1000 M.Expo.
1	0.00	0.00	0.00	0.00	0.1239E+03	0.2223E+03	0.7746E+02
3	0.70	86.72	153.67	87.10	0.1225E+03	0.2135E+03	0.1128E+03
5	1.30	159.49	274.75	150.09	0.1006E+03	0.1871E+03	0.9798E+02
7	3.80	198.77	575.24	207.26	0.3103E+01	0.7338E+02	-0.6592E+00
9	10.80	220.50	849.64	218.12	0.4558E+01	0.2354E+02	0.6642E+01
11	23.30	312.37	1047.34	308.47	0.5438E+01	0.1129E+02	0.5340E+01
13	35.10	367.50	1157.36	364.42	0.4958E+01	0.7815E+01	0.4964E+01
15	45.70	422.63	1230.91	420.68	0.5521E+01	0.6204E+01	0.5286E+01
17	55.05	446.10	1284.35	452.77	0.8571E+00	0.5283E+01	0.1703E+01
19	63.00	459.37	1324.09	455.68	0.5903E+01	0.4705E+01	-0.1078E+01
21	66.50	498.34	1340.06	502.65	0.2407E+02	0.4495E+01	0.4183E+02
23	69.40	592.42	1352.97	622.31	0.5736E+02	0.4333E+01	0.4070E+02
25	70.50	665.92	1357.72	666.85	0.5969E+02	0.4276E+01	0.4028E+02
27	72.00	740.88	1364.11	726.84	0.4851E+02	0.4200E+01	0.3971E+02
29	74.60	845.62	1374.94	828.82	0.3308E+02	0.4074E+01	0.3875E+02
31	77.37	931.00	1385.82	934.67	0.2940E+02	0.3953E+01	0.3777E+02
33	80.70	1029.00	1398.88	1031.77	0.2484E+02	0.3814E+01	0.1742E+02
35	86.50	1127.00	1420.25	1127.76	0.1690E+02	0.3598E+01	0.1573E+02
37	93.55	1212.75	1444.78	1214.71	0.8855E+01	0.3369E+01	0.9769E+01
39	101.40	1286.26	1470.34	1285.62	0.9171E+01	0.3149E+01	0.8373E+01
41	109.10	1350.76	1493.85	1346.11	0.7621E+01	0.2962E+01	0.7394E+01
43	122.10	1433.26	1530.57	1434.89	0.6400E+01	0.2696E+01	0.6373E+01
45	134.15	1460.81	1561.80	1461.45	-0.1447E+01	0.2493E+01	-0.1513E+01

Connection type : Double web-angle connections
Mode : All bolted

Tested by : L.E.Thompson et al. (1970) U.S.A
Test Id. : E1-3 ALT

Column : W10X49
Beam : W21X62
Angle : 4 X 3.5 X 1/2

Fasteners: A325- -3/4"D
1-5/16" Oversize holes
Material : A36
Fv = -- ksi
Fu = -- ksi

Major parameters

lp = 11.5000"	lu = -- "	ll = -- "	ta = 0.5000"
gb = 2.2500"	gc = 1.8000"	cu = -- "	cl = -- "
pb = -- "	pc = -- "	qb = -- "	qc = -- "
nb = - X 2	nc = - X -		

Remark 1)
2)

No	Moment (k-in)	Rotation (radians) X 1/1000	No	Moment (k-in)	Rotation (radians) X 1/1000
1	0.00	0.00	26	664.44	68.30
2	46.85	0.25	27	702.29	69.00
3	93.71	0.50	28	740.15	69.70
4	163.16	1.00	29	778.00	70.85
5	213.16	1.60	30	815.86	72.00
6	266.44	2.20	31	885.67	73.60
7	299.15	2.70	32	955.50	75.40
8	330.76	3.40	33	992.25	76.50
9	348.40	5.20	34	1029.00	77.60
10	371.92	5.50	35	1065.75	79.05
11	380.36	10.85	36	1102.50	80.50
12	388.81	16.20	37	1139.25	83.55
13	399.10	20.85	38	1176.00	86.60
14	409.39	25.50	39	1212.75	89.20
15	425.54	29.97	40	1249.50	91.80
16	443.70	34.43	41	1286.25	95.45
17	460.85	38.90	42	1323.00	99.10
18	469.30	42.45	43	1347.50	104.10
19	477.76	46.00	44	1372.00	109.10
20	496.13	50.70	45	1396.50	114.10
21	502.25	55.57	46	1450.12	120.00
22	508.38	60.43			
23	514.50	65.30			
24	555.29	66.15			
25	596.09	67.00			

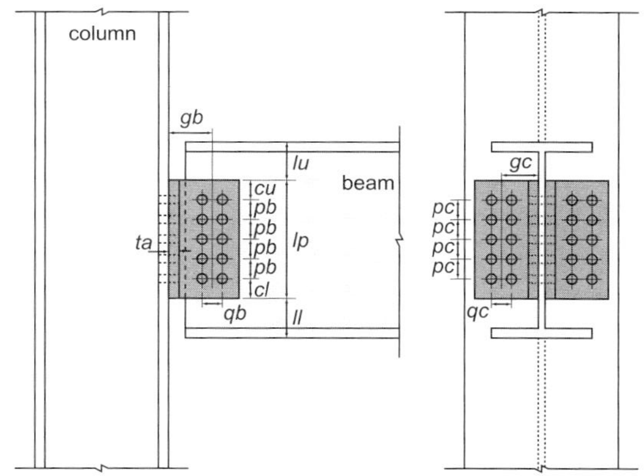

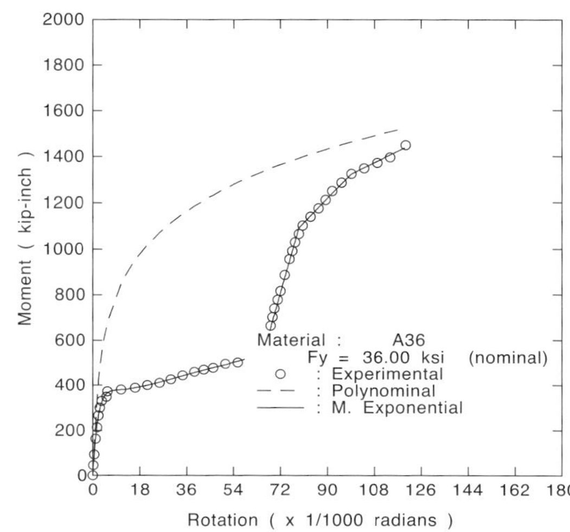

```
Moment-rotation prediction equations     ( R : X 1/1000 radians )

Frye and Morris polynominal model  :    R = Sum ( Ai X ( K*Bm )**Pi X 10**Qi )
     xd = 11.500000"   g =  4.000000"   t =  0.500000"
     A1 =  3.660000   A2 =  1.150000   A3 =  4.570000    K =  0.012288
     P1 =         1   P2 =         3   P3 =         5   Q1 =       -1    Q2 =       -3    Q3 =       -5

Modified exponential model   :
     Bm = Sum ( Ai X ( 1 - exp( - R/( 2*i*C ) ) ) )  +      Sum ( RKj X ( R - Rj0 ) )  +  Bm0
       C =   0.10625000E+01             Bmo=   0.00000000E+00          Nexp=  6         Nliner=  4
      AI =   0.47627138E+03   0.79562123E+03  -0.72281906E+04   0.21830806E+05  -0.28036302E+05   0.12707582E+05
     Rj0 =          50.7000          65.3000          79.0000          99.1000
     RKj =   0.10414642E+01   0.38485581E+02  -0.28421809E+02  -0.56531269E+01
```

No	Rotation (radians) X 1/1000	Moment (kip-inch) Expri.	Moment (kip-inch) Poly.	Moment (kip-inch) M.Expo.	Connection stiffness (kip-inch) X 1000 Expri.	Connection stiffness (kip-inch) X 1000 Poly.	Connection stiffness (kip-inch) X 1000 M.Expo.
1	0.00	0.00	0.00	0.00	0.1874E+03	0.2223E+03	0.2038E+03
3	0.50	93.71	110.48	89.93	0.1712E+03	0.2181E+03	0.1578E+03
5	1.60	213.16	328.25	222.16	0.8606E+02	0.1686E+03	0.8883E+02
7	2.70	299.15	477.81	295.94	0.5698E+02	0.1076E+03	0.4899E+02
9	5.20	348.40	660.75	361.20	0.6860E+02	0.5129E+02	0.1136E+02
11	10.85	380.36	850.81	379.38	0.1579E+01	0.2343E+02	0.1011E+01
13	20.85	399.10	1018.37	399.33	0.2213E+01	0.1250E+02	0.2746E+01
15	29.97	426.54	1114.34	427.12	0.3840E+01	0.8995E+01	0.3261E+01
17	38.90	460.85	1185.70	456.59	0.3027E+01	0.7140E+01	0.3247E+01
19	46.00	477.76	1232.75	478.52	0.3039E+01	0.6169E+01	0.2889E+01
21	55.57	502.25	1287.07	507.86	0.1258E+01	0.5241E+01	0.3206E+01
23	65.30	514.50	1334.62	535.43	0.4104E+02	0.4565E+01	0.2479E+01
25	67.00	596.09	1342.34	604.97	0.4981E+02	0.4466E+01	0.4085E+02
27	69.00	702.29	1351.22	686.55	0.5408E+02	0.4355E+01	0.4073E+02
29	70.85	778.00	1359.23	761.80	0.3292E+02	0.4258E+01	0.4062E+02
31	73.60	885.67	1370.85	873.30	0.4136E+02	0.4121E+01	0.4048E+02
33	76.50	992.25	1382.40	990.48	0.3341E+02	0.3991E+01	0.4034E+02
35	79.05	1065.75	1392.51	1091.78	0.2534E+02	0.3881E+01	0.1181E+02
37	83.55	1139.25	1409.48	1144.54	0.1205E+02	0.3705E+01	0.1165E+02
39	89.20	1212.75	1430.12	1209.89	0.1413E+02	0.3504E+01	0.1149E+02
41	95.45	1286.25	1451.33	1281.30	0.1007E+02	0.3311E+01	0.1137E+02
43	104.10	1347.50	1478.75	1350.83	0.4900E+01	0.3080E+01	0.5601E+01
45	114.10	1396.50	1508.38	1406.44	0.6821E+01	0.2853E+01	0.5528E+01

II - 42

Connection type : Double web-angle ccnnections
Mode : All bolted

Tested by : L.E.Thompson et al. (1970) U.S.A
Test Id. : E1-3 ALT

Column : W10X49
Beam : W21X62
Angle : 4 X 3.5 X 1/2

Fasteners: A325- -3/4"D
15/16" Oversize holes
Material : A36
Fy = -- ksi
Fu = -- ksi

Major parameters

lp = 11.5000"	lu = -- "	ll = -- "	ta = 0.5000"
gb = 2.2500"	gc = 1.8000"	cu = -- "	cl = -- "
pb = -- "	pc = -- "	qb = -- "	qc = -- "
nb = - X 2	nc = - X -		

Remark 1)
2)

No	Moment (k-in)	Rotation (radians) X 1/1000	No	Moment (k-in)	Rotation (radians) X 1/1000
1	0.00	0.00	26	702.29	69.35
2	46.85	0.25	27	740.15	69.80
3	93.71	0.50	28	778.00	70.80
4	163.16	1.10	29	815.86	71.80
5	213.16	1.70	30	885.67	73.00
6	265.44	2.60	31	955.50	74.70
7	299.15	3.30	32	992.25	75.95
8	330.76	4.20	33	1029.00	77.20
9	339.58	7.55	34	1065.75	78.80
10	348.40	10.90	35	1102.50	80.40
11	371.92	15.80	36	1139.25	81.45
12	388.81	20.50	37	1176.00	82.50
13	399.10	23.50	38	1212.75	86.20
14	409.39	26.50	39	1249.50	89.90
15	435.12	32.05	40	1286.25	93.35
16	460.85	37.60	41	1323.00	96.80
17	469.30	42.45	42	1359.75	101.30
18	477.76	47.30	43	1396.50	105.80
19	496.13	51.40	44	1423.31	111.35
20	502.25	56.20	45	1450.12	116.90
21	508.38	61.00			
22	514.50	65.80			
23	555.29	66.65			
24	596.09	67.50			
25	664.44	68.90			

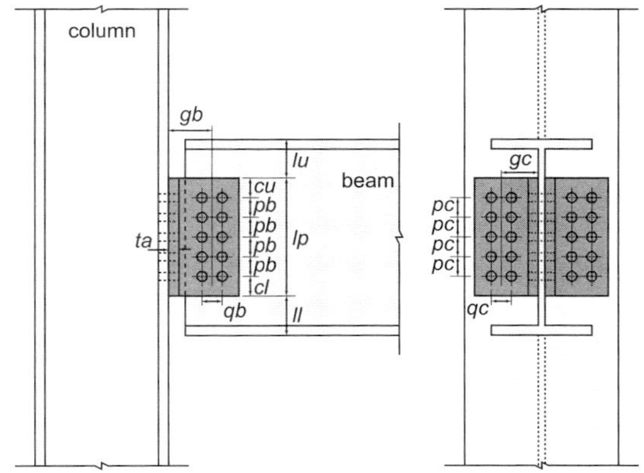

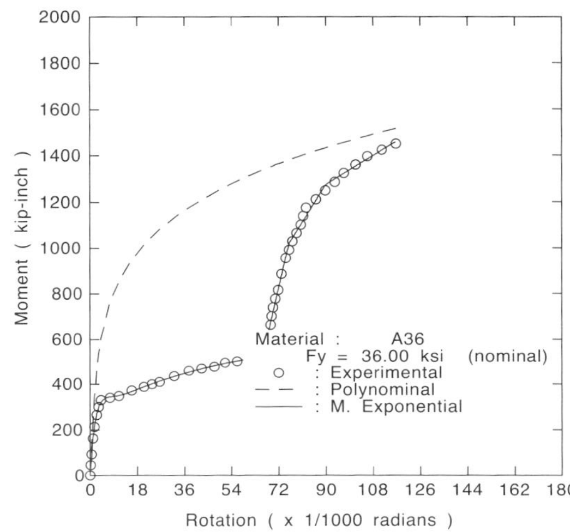

```
Moment-rotation prediction equations      ( R : X 1/1000 radians )

Frye and Morris polynominal model  :    R = Sum ( Ai X ( K*Bm )**Pi X 10**Qi )
     xd = 11.500000"   g =  4.000000"   t =  0.500000"
     A1 =  3.660000   A2 =  1.150000   A3 =  4.570000    K =  0.012288
     P1 =        1    P2 =        3    P3 =        5    Q1 =       -1    Q2 =       -3    Q3 =       -5

Modified exponential model  :
     Bm = Sum ( Ai X ( 1 - exp( - R/( 2*i*C ) ) ) )  +     Sum ( RKj X ( R - Rj0 ) )  +  Bm0
      C =   0.10366667E+01            Bmo=   0.00000000E+00          Nexp=  6        Nliner=  3
     AI =   0.37257876E+03   0.20564792E+04  -0.12530905E+05   0.30776580E+05  -0.34794350E+05   0.14662676E+05
    Rj0 =          65.8000          75.9000          89.9000
    RKj =   0.48428510E+02  -0.30773009E+02  -0.10948301E+02
```

No	Rotation (radians) X 1/1000	Moment (kip-inch) Expri.	Poly.	M.Expo.	Connection stiffness (kip-inch) X 1000 Expri.	Poly.	M.Expo.
1	0.00	0.00	0.00	0.00	0.1874E+03	0.2223E+03	0.1943E+03
3	0.50	93.71	110.48	85.52	0.1663E+03	0.2181E+03	0.1496E+03
5	1.70	213.16	344.82	217.86	0.7367E+02	0.1622E+03	0.7806E+02
7	3.30	299.15	535.68	299.65	0.4165E+02	0.8616E+02	0.3069E+02
9	7.55	339.58	758.36	343.71	0.2633E+01	0.3406E+02	0.1953E+01
11	15.80	371.92	946.60	367.17	0.4185E+01	0.1622E+02	0.4329E+01
13	23.50	399.10	1049.60	401.51	0.3430E+01	0.1120E+02	0.4255E+01
15	32.05	435.12	1132.54	434.69	0.4636E+01	0.8471E+01	0.3562E+01
17	42.45	469.30	1210.09	468.70	0.1743E+01	0.6615E+01	0.2979E+01
19	51.40	496.13	1264.49	492.74	0.3004E+01	0.5604E+01	0.2372E+01
21	61.00	508.38	1314.41	512.11	0.1276E+01	0.4838E+01	0.1667E+01
23	66.65	555.29	1340.75	561.61	0.4799E+02	0.4486E+01	0.4972E+02
25	68.90	664.44	1350.78	673.33	0.7553E+02	0.4360E+01	0.4959E+02
27	69.80	740.15	1354.69	717.93	0.6976E+02	0.4312E+01	0.4954E+02
29	71.80	815.86	1363.26	816.90	0.4709E+02	0.4210E+01	0.4943E+02
31	74.70	955.50	1375.37	960.03	0.3435E+02	0.4070E+01	0.4929E+02
33	77.20	1029.00	1385.16	1043.11	0.2658E+02	0.3961E+01	0.1840E+02
35	80.40	1102.50	1397.72	1101.80	0.3023E+02	0.3826E+01	0.1828E+02
37	82.50	1176.00	1405.71	1140.11	0.2946E+02	0.3743E+01	0.1821E+02
39	89.90	1249.50	1432.52	1274.07	0.1030E+02	0.3481E+01	0.1801E+02
41	96.80	1323.00	1455.75	1322.33	0.9574E+01	0.3272E+01	0.6935E+01
43	105.80	1396.50	1484.21	1384.24	0.6673E+01	0.3037E+01	0.6832E+01
45	116.90	1450.12	1516.62	1459.66	0.4830E+01	0.2793E+01	0.6765E+01

II - 43

```
Connection type : Double web-angle connections
           Mode : All bolted

Tested by  :  L.E.Thompson et al. (1970)                    U.S.A
Test  Id.  :  A2-1 ALT

Column : W10X49                      Fasteners: A325- -3/4"D
Beam   : W21X62                            15/16" Oversize holes
Angle  : 4 X 3.5 X 5/16              Material : A441
                                          Fy =   --   ksi
                                          Fu =   --   ksi

                        Major parameters

     lp = 11.5000"    lu =    --   "    ll =    --   "    ta =  0.3125"
     gb =  2.2500"    gc =  2.5500"    cu =    --   "    cl =    --   "
     pb =    --   "    pc =    --   "    qb =    --   "    qc =    --   "
     nb = - X  2      nc = - X  -

Remark  1)
        2)
```

No	Moment (k-in)	Rotation (radians) X 1/1000	No	Moment (k-in)	Rotation (radians) X 1/1000
1	0.00	0.00	26	661.51	133.55
2	39.68	1.10	27	698.26	139.70
3	73.50	3.70	28	733.53	146.50
4	91.88	8.80	29	768.80	153.30
5	110.26	13.90	30	807.03	159.60
6	128.63	20.55	31	845.26	165.90
7	147.00	27.20	32	882.01	171.30
8	155.38	33.40	33	918.76	176.70
9	183.76	39.60	34	955.51	182.70
10	199.55	47.20	35	992.26	188.70
11	215.35	54.80	36	1029.01	195.20
12	223.81	62.20	37	1065.76	201.70
13	232.26	69.60			
14	240.34	77.00			
15	248.43	84.40			
16	256.51	91.80			
17	291.98	93.90			
18	327.44	96.00			
19	366.95	97.60			
20	406.45	99.20			
21	457.17	102.80			
22	507.89	106.40			
23	553.01	113.30			
24	588.88	120.35			
25	624.76	127.40			

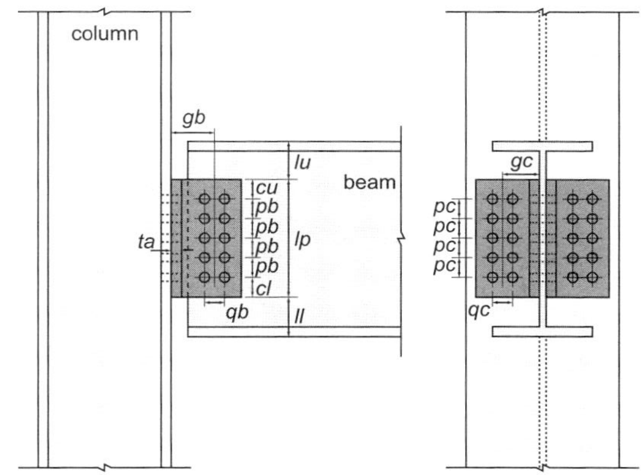

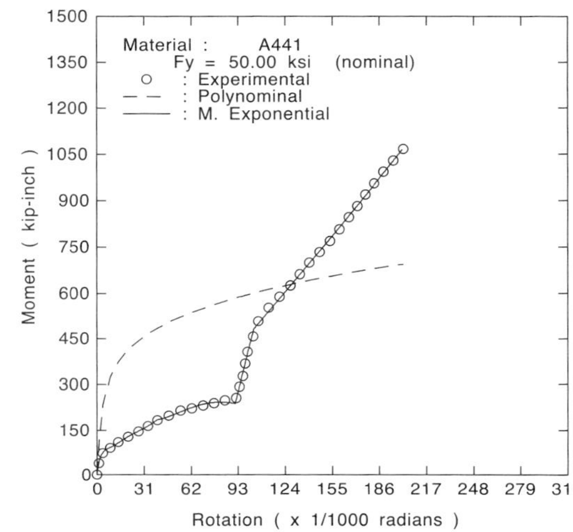

```
Moment-rotation prediction equations     ( R : X 1/1000 radians )

Frye and Morris polynominal model  :    R = Sum ( Ai X ( K*Bm )**Pi X 10**Qi )
     xd = 11.500000"   g =  5.500000"   t =  0.312500"
     A1 =  3.660000   A2 =  1.150000   A3 =  4.570000    K =  0.030179
     P1 =        1    P2 =        3    P3 =        5    Q1 =       -1    Q2 =       -3    Q3 =       -5

Modified exponential model  :
     Bm = Sum ( Ai X ( 1 - exp( - R/( 2*i*C ) ) ) )  +      Sum ( RKj X ( R - Rj0 ) )  +  Bm0
      C =   0.19558333E+01              Bmo=   0.00000000E+00          Nexp=  6          Nliner=  5
     AI =   0.10312842E+04  -0.10838972E+05   0.49060105E+05  -0.98558393E+05   0.88381059E+05  -0.28827470E+05
    Rj0 =           0.0000           3.7000          39.6000          91.2000         103.2000
    RKj =  -0.36397225E+01   0.51471853E+01  -0.29898281E+01   0.21360944E+02  -0.13635890E+02
```

No	Rotation (radians) X 1/1000	Moment (kip-inch) Expri.	Moment (kip-inch) Poly.	Moment (kip-inch) M.Expo.	Connection stiffness (kip-inch) X 1000 Expri.	Connection stiffness (kip-inch) X 1000 Poly.	Connection stiffness (kip-inch) X 1000 M.Expo.
1	0.00	0.00	0.00	0.00	0.3608E+02	0.9054E+02	0.5041E+02
3	3.70	73.50	231.23	73.33	0.9831E+01	0.3080E+02	0.3832E+01
5	13.90	110.26	372.11	108.08	0.3239E+01	0.7471E+01	0.3360E+01
7	27.20	147.00	443.24	147.64	0.2867E+01	0.3993E+01	0.2558E+01
9	39.60	183.76	484.83	186.12	0.2566E+01	0.2862E+01	0.3909E+01
11	54.80	215.35	522.44	211.61	0.1604E+01	0.2160E+01	0.2038E+01
13	69.60	232.26	551.22	236.66	0.1117E+01	0.1761E+01	0.1131E+01
15	84.40	248.43	575.20	242.70	0.1092E+01	0.1496E+01	-0.2921E+00
17	93.90	291.98	588.81	294.01	0.1689E+02	0.1367E+01	0.2035E+02
19	97.60	366.95	593.83	368.89	0.2469E+02	0.1323E+01	0.2013E+02
21	102.80	457.17	600.49	472.92	0.1409E+02	0.1267E+01	0.1989E+02
23	113.30	553.01	613.27	542.19	0.5822E+01	0.1169E+01	0.5940E+01
25	127.40	624.76	628.95	624.65	0.5562E+01	0.1060E+01	0.5797E+01
27	139.70	698.26	641.49	695.99	0.5601E+01	0.9817E+00	0.5815E+01
29	153.30	768.80	654.33	775.58	0.5644E+01	0.9089E+00	0.5895E+01
31	165.90	845.26	665.41	850.39	0.6465E+01	0.8515E+00	0.5979E+01
33	176.70	918.76	674.49	915.32	0.6483E+01	0.8077E+00	0.6044E+01
35	188.70	992.26	683.87	988.22	0.5899E+01	0.7654E+00	0.6103E+01
37	201.70	1065.76	693.49	1067.87	0.5654E+01	0.7248E+00	0.6150E+01

II - 44

Connection type : Double web-angle connections
Mode : All bolted

Tested by : L.E.Thompson et al. (1970) U.S.A
Test Id. : A2-1 ALT

Column : W10X49
Beam : W21X62
Angle : 4 X 3.5 X 5/16

Fasteners: A325- -3/4"D
15/16" Oversize holes
Material : A441
Fy = -- ksi
Fu = -- ksi

Major parameters

lp = 11.5000"	lu = -- "	ll = -- "	ta = 0.3125"
gb = 2.2500"	gc = 2.5500"	cu = -- "	cl = -- "
pb = -- "	pc = -- "	qb = -- "	qc = -- "
nb = - X 2	nc = - X -		

Remark 1)
2)

No	Moment (k-in)	Rotation (radians) X 1/1000	No	Moment (k-in)	Rotation (radians) X 1/1000
1	0.00	0.00	26	733.53	135.00
2	39.68	1.50	27	768.80	141.00
3	73.50	3.40	28	807.03	147.30
4	91.88	10.50	29	845.26	153.60
5	110.26	17.60	30	882.01	159.45
6	147.00	23.40	31	918.76	165.30
7	183.76	33.00	32	955.51	172.30
8	199.55	41.25	33	992.26	179.30
9	215.35	49.50	34	1029.01	187.05
10	223.81	54.70	35	1065.76	194.80
11	232.26	59.90			
12	240.34	67.90			
13	248.43	75.90			
14	256.51	83.90			
15	291.98	86.05			
16	327.44	88.20			
17	366.95	89.50			
18	406.45	90.80			
19	457.17	93.15			
20	507.89	95.50			
21	553.01	102.00			
22	583.88	108.95			
23	624.76	115.90			
24	661.51	122.45			
25	693.26	129.00			

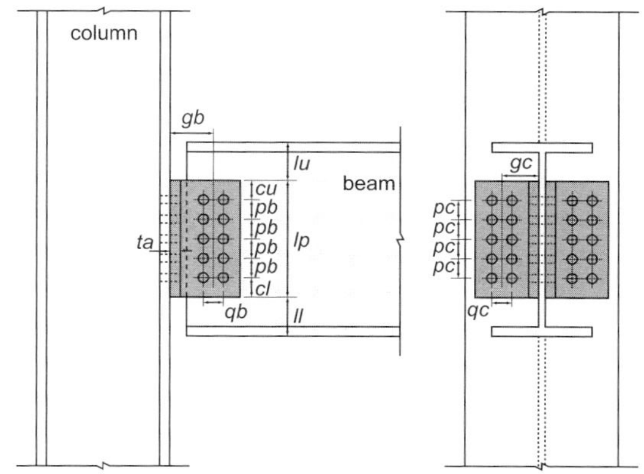

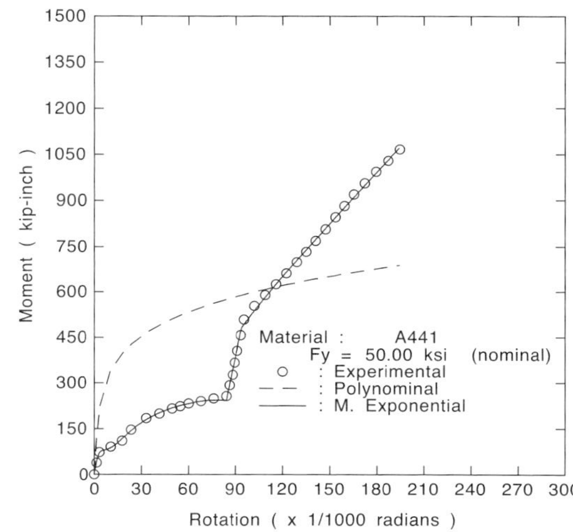

```
Moment-rotation prediction equations      ( R : X 1/1000 radians )

Frye and Morris polynominal model  :    R = Sum ( Ai X ( K*Bm )**Pi X 10**Qi )
     xd = 11.500000"   g =  5.500000"   t =  0.312500"
     A1 =  3.660000   A2 =  1.150000   A3 =  4.570000    K =  0.030179
     P1 =         1   P2 =         3   P3 =         5   Q1 =        -1    Q2 =        -3    Q3 =        -5

Modified exponential model  :
     Bm = Sum ( Ai X ( 1 - exp( - R/( 2*i*C ) ) ) )  +      Sum ( RKj X ( R - Rj0 ) )  +  Bm0
      C =   0.19983333E+01           Bmo=   0.00000000E+00           Nexp=  6         Nliner=  5
     AI =   0.54364181E+03  -0.42372141E+04   0.14806640E+05  -0.23829531E+05   0.15272050E+05  -0.20349877E+04
    Rj0 =           0.0000           3.4000          59.9000          83.9000          93.8000
    RKj =   0.55967636E+01  -0.87423765E+01   0.46560963E+00   0.25003926E+02  -0.16827610E+02
```

No	Rotation (radians) X 1/1000	Moment (kip-inch) Expri.	Moment (kip-inch) Poly.	Moment (kip-inch) M.Expo.	Connection stiffness (kip-inch) X 1000 Expri.	Connection stiffness (kip-inch) X 1000 Poly.	Connection stiffness (kip-inch) X 1000 M.Expo.
1	0.00	0.00	0.00	0.00	0.2646E+02	0.9054E+02	0.2964E+02
3	3.40	73.50	221.50	73.38	0.1459E+02	0.3394E+02	0.1327E+02
5	17.60	110.26	396.74	117.10	0.4651E+01	0.5963E+01	0.4736E+01
7	33.00	183.76	464.38	179.59	0.2799E+01	0.3361E+01	0.3073E+01
9	49.50	215.35	510.49	217.80	0.1738E+01	0.2357E+01	0.1684E+01
11	59.90	232.26	533.03	231.38	0.1383E+01	0.2001E+01	0.9140E+00
13	75.90	248.43	561.91	243.22	0.1011E+01	0.1636E+01	0.1006E+00
15	86.05	291.98	577.68	294.16	0.1650E+02	0.1471E+01	0.2437E+02
17	89.50	366.95	582.72	377.84	0.3039E+02	0.1423E+01	0.2414E+02
19	93.15	457.17	587.87	465.56	0.2158E+02	0.1376E+01	0.2393E+02
21	102.00	553.01	599.48	537.27	0.6081E+01	0.1276E+01	0.6652E+01
23	115.90	624.76	616.28	626.00	0.5393E+01	0.1147E+01	0.6154E+01
25	129.00	698.26	630.64	704.57	0.5751E+01	0.1049E+01	0.5866E+01
27	141.00	768.80	642.76	773.94	0.5971E+01	0.9742E+00	0.5708E+01
29	153.60	845.26	654.60	845.18	0.6179E+01	0.9075E+00	0.5610E+01
31	165.30	918.76	664.90	910.49	0.5812E+01	0.8540E+00	0.5559E+01
33	179.30	992.26	676.46	988.05	0.5009E+01	0.7986E+00	0.5526E+01
35	194.80	1065.76	688.42	1073.55	0.4742E+01	0.7459E+00	0.5508E+01

II - 45

Connection type : Double web-angle connections
Mode : All bolted

Tested by : L.E.Thompson et al. (1970) U.S.A
Test Id. : A2-2 ALT

Column : W10X49
Beam : W21X62
Angle : 4 X 3.5 X 5/16

Fasteners: A325- -3/4"D
15/16" Oversize holes
Material : A441
Fy = -- ksi
Fu = -- ksi

Major parameters

lp = 11.5000"	lu = -- "	ll = -- "	ta = 0.3125"
gb = 2.2500"	gc = 2.5500"	cu = -- "	cl = -- "
pb = -- "	pc = -- "	qb = -- "	qc = -- "
nb = - X 2	nc = - X -		

Remark 1)
2)

No	Moment (k-in)	Rotation (radians) X 1/1000	No	Moment (k-in)	Rotation (radians) X 1/1000
1	0.00	0.00	26	769.55	138.60
2	40.43	0.70	27	807.40	145.45
3	77.92	1.70	28	845.26	152.30
4	115.87	3.20	29	879.80	158.15
5	129.73	9.60	30	914.34	164.00
6	142.58	16.00	31	951.46	169.70
7	163.17	21.75	32	988.57	175.40
8	183.76	27.50	33	1027.53	181.70
9	198.64	36.85	34	1066.49	188.00
10	213.52	46.20	35	1102.87	195.00
11	238.69	55.30	36	1139.26	202.00
12	263.87	64.40			
13	272.93	74.10			
14	282.00	83.80			
15	291.06	93.50			
16	330.50	95.27			
17	369.95	97.03			
18	409.39	98.80			
19	464.89	100.85			
20	520.38	102.90			
21	573.30	106.00			
22	626.22	109.10			
23	662.24	117.40			
24	698.26	125.70			
25	733.90	132.15			

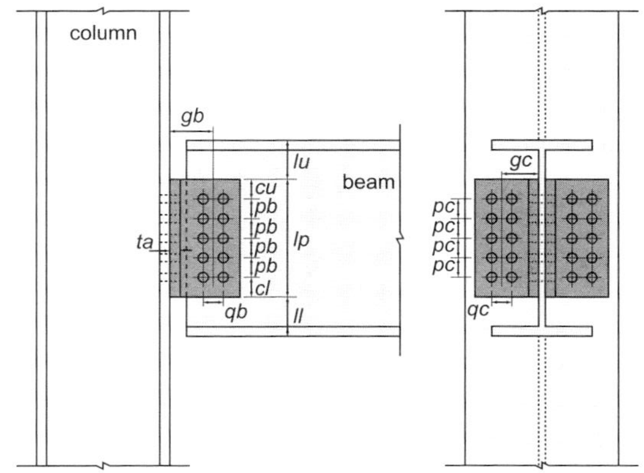

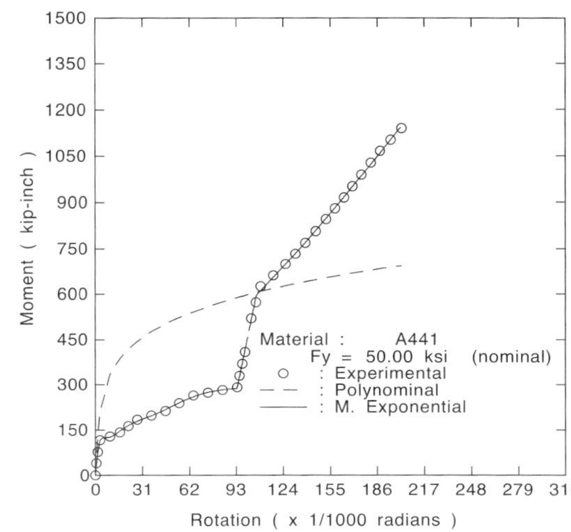

```
Moment-rotation prediction equations     ( R : X 1/1000 radians )

Frye and Morris polynominal model  :    R = Sum ( Ai X ( K*Bm )**Pi X 10**Qi )
     xd = 11.500000"   g =  5.500000"   t =  0.312500"
     A1 =  3.660000   A2 =  1.150000   A3 =  4.570000    K =  0.030179
     P1 =         1   P2 =         3   P3 =         5   Q1 =        -1    Q2 =        -3    Q3 =        -5

Modified exponential model  :
     Bm = Sum ( Ai X ( 1 - exp( - R/( 2*i*C ) ) ) )  +     Sum ( RKj X ( R - Rj0 ) )  +  Bm0
      C =   0.18583333E+01             Bmo=   0.00000000E+00          Nexp=  6       Nliner=  4
     AI =   0.11111110E+04  -0.10466538E+05   0.45096331E+05  -0.91784240E+05   0.86430439E+05  -0.30271856E+05
     Rj0 =          0.0000           3.2000          93.5000         106.2000
     RKj =  0.10862074E+02  -0.99348790E+01   0.24145778E+02  -0.18899266E+02
```

No	Rotation (radians) X 1/1000	Moment (kip-inch) Expri.	Poly.	M.Expo.	Connection stiffness (kip-inch) X 1000 Expri.	Poly.	M.Expo.
1	0.00	0.00	0.00	0.00	0.5775E+02	0.9054E+02	0.5507E+02
3	1.70	77.92	140.36	79.96	0.3288E+02	0.6608E+02	0.3226E+02
5	9.60	129.73	333.76	127.07	0.2009E+01	0.1080E+02	0.1422E+01
7	21.75	163.17	419.11	165.23	0.3580E+01	0.4897E+01	0.2921E+01
9	36.85	198.64	476.78	197.21	0.1591E+01	0.3047E+01	0.1988E+01
11	55.30	238.69	523.52	239.91	0.2767E+01	0.2143E+01	0.2351E+01
13	74.10	272.93	558.94	274.10	0.9344E+00	0.1670E+01	0.1186E+01
15	93.50	291.06	588.23	286.69	0.1903E+02	0.1372E+01	0.2520E+00
17	97.03	369.95	593.05	372.75	0.2233E+02	0.1330E+01	0.2432E+02
19	100.85	464.89	598.08	465.45	0.2707E+02	0.1287E+01	0.2426E+02
21	106.00	573.30	604.50	590.28	0.1707E+02	0.1235E+01	0.2422E+02
23	117.40	662.24	617.99	654.83	0.4340E+01	0.1134E+01	0.5358E+01
25	132.15	733.90	633.91	734.98	0.5527E+01	0.1028E+01	0.5523E+01
27	145.45	807.40	647.04	809.58	0.5526E+01	0.9494E+00	0.5694E+01
29	158.15	879.80	658.68	882.83	0.5905E+01	0.8858E+00	0.5836E+01
31	169.70	951.46	668.62	950.84	0.6512E+01	0.8357E+00	0.5937E+01
33	181.70	1027.53	678.50	1022.58	0.6184E+01	0.7893E+00	0.6015E+01
35	195.00	1102.87	688.65	1102.99	0.5198E+01	0.7449E+00	0.6074E+01

A.2 – 91

II - 46

```
Connection type : Double web-angle connections
           Mode : All bolted

Tested by   :  L.E.Thompson et al. (1970)                  U.S.A
Test  Id.   :  A2-2 ALT

Column : W10X49                      Fasteners: A325- -3/4"D
Beam   : W21X62                             15/16" Oversize holes
Angle  : 4 X 3.5 X 5/16              Material : A441
                                           Fy =   --   ksi
                                           Fu =   --   ksi

                          Major parameters

     lp = 11.5000"     lu =    --   "     cl =    --   "     ta =  0.3125"
     gb =  2.2500"     gc =  2.5500"     cu =    --   "     cl =    --   "
     pb =    --   "     pc =    --   "     qb =    --   "     qc =    --   "
     nb = - X  2        nc = - X  -

Remark  1)
        2)
```

No	Moment (k-in)	Rotation (radians) X 1/1000	No	Moment (k-in)	Rotation (radians) X 1/1000
1	0.00	0.00	26	807.40	132.40
2	40.43	1.00	27	845.26	139.80
3	77.92	2.20	28	879.80	145.95
4	116.87	3.60	29	914.34	152.10
5	142.58	8.20	30	951.46	158.60
6	163.17	13.35	31	988.57	165.10
7	183.76	18.50	32	1027.53	172.30
8	198.64	26.65	33	1066.49	179.50
9	213.52	34.80	34	1102.87	188.35
10	238.69	42.55	35	1139.26	197.20
11	263.87	50.30			
12	272.93	57.47			
13	282.00	64.63			
14	291.06	71.80			
15	330.50	74.17			
16	369.95	76.53			
17	409.39	78.90			
18	464.89	81.15			
19	520.38	83.40			
20	573.30	88.20			
21	626.22	93.00			
22	662.24	102.10			
23	698.26	111.20			
24	733.90	118.10			
25	769.55	125.00			

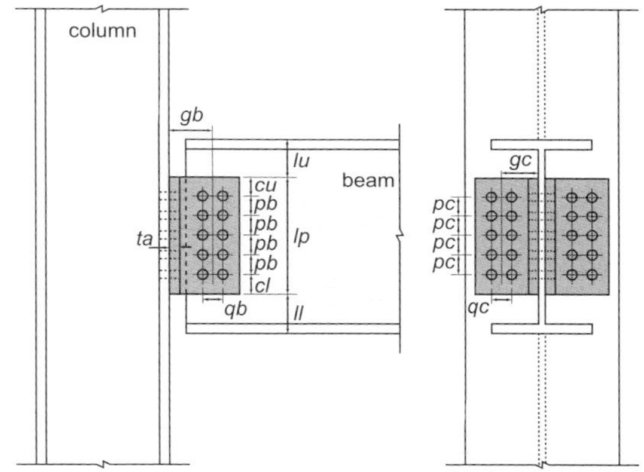

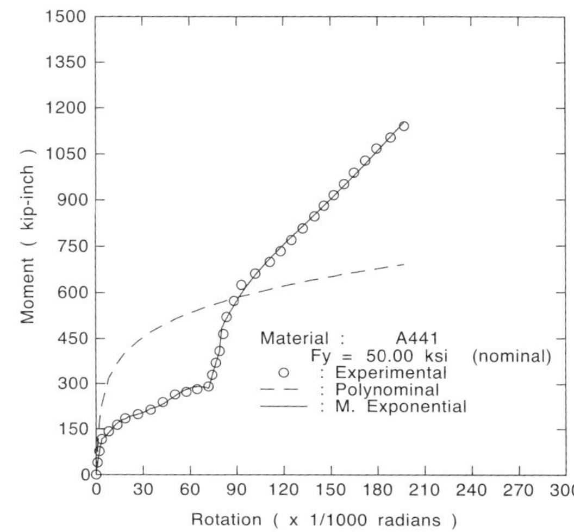

A.2 – 92

```
Moment-rotation prediction equations     ( R : X 1/1000 radians )

Frye and Morris polynominal model  :    R = Sum ( Ai X ( K*Bm )**Pi X 10**Qi )
     xd = 11.500000"   g =  5.500000"   t =  0.312500"
     A1 =  3.660000   A2 =  1.150000   A3 =  4.570000    K =  0.030179
     P1 =         1   P2 =         3   P3 =         5    Q1 =       -1    Q2 =       -3    Q3 =       -5

Modified exponential model  :
     Bm = Sum ( Ai X ( 1 - exp( - R/( 2*i*C ) ) ) )  +      Sum ( RKj X ( R - Rj0 ) )  +  Bm0
      C =   0.18933333E+01              Bmo=   0.00000000E+00         Nexp=  6         Nliner=  8
     AI =   0.13195329E+04  -0.21710605E+05   0.11048185E+06  -0.23128417E+06   0.20969660E+06  -0.67681611E+05
    Rj0 =           0.0000           3.6000          18.5000          34.8000          50.3000
                   71.8000          78.9000          80.2000
    RKj =   0.13415349E+02  -0.21797779E+02   0.27872844E+01  -0.32487291E+01  -0.40766691E+00
            0.20877948E+02   0.40549235E+02  -0.46359387E+02
```

No	Rotation (radians) X 1/1000	Moment (kip-inch) Expri.	Poly.	M.Expo.	Connection stiffness (kip-inch) X 1000 Expri.	Poly.	M.Expo.
1	0.00	0.00	0.00	0.00	0.4043E+02	0.9054E+02	0.3421E+02
3	2.20	77.92	170.25	79.35	0.2966E+02	0.5376E+02	0.2873E+02
5	8.20	142.58	317.38	138.30	0.4839E+01	0.1272E+02	0.5844E+01
7	18.50	183.76	401.98	182.23	0.3156E+01	0.5691E+01	0.8353E+00
9	34.80	213.52	470.35	217.25	0.2555E+01	0.3206E+01	0.3756E+01
11	50.30	263.87	512.36	259.42	0.2218E+01	0.2325E+01	0.3744E+01
13	64.63	282.00	542.20	290.46	0.1265E+01	0.1875E+01	0.3330E+00
15	74.17	330.50	559.10	328.13	0.1667E+02	0.1668E+01	0.1807E+02
17	78.90	409.39	566.74	410.14	0.2077E+02	0.1583E+01	0.1660E+02
19	83.40	520.38	573.78	516.06	0.2031E+02	0.1510E+01	0.9526E+01
21	93.00	626.22	587.55	596.66	0.8585E+01	0.1379E+01	0.7397E+01
23	111.20	698.26	610.79	709.00	0.4645E+01	0.1187E+01	0.5320E+01
25	125.00	769.55	626.38	778.64	0.5142E+01	0.1077E+01	0.4881E+01
27	139.80	845.26	641.59	850.69	0.5389E+01	0.9811E+00	0.4908E+01
29	152.10	914.34	653.24	912.01	0.5662E+01	0.9149E+00	0.5073E+01
31	165.10	988.57	664.73	979.23	0.5568E+01	0.8549E+00	0.5267E+01
33	179.50	1066.49	676.62	1056.45	0.4828E+01	0.7979E+00	0.5450E+01
35	197.20	1139.26	690.20	1154.43	0.4111E+01	0.7384E+00	0.5608E+01

Connection type : Double web-angle connections
Mode : All bolted

Tested by : L.E.Thompson et al. (1970) U.S.A
Test Id. : A2-3 ALT

Column : W10X49
Beam : W21X62
Angle : 4 X 3.5 X 5/16

Fasteners: A325- -3/4"D
15/16" Oversize holes
Material : A441
Fy = -- ksi
Fu = -- ksi

Major parameters

lp = 11.5000"	lu = -- "	ll = -- "	ta = 0.3125"
gb = 2.2500"	gc = 2.5500"	cu = -- "	cl = -- "
pb = -- "	pc = -- "	qb = -- "	qc = -- "
nb = - X 2	nc = - X -		

Remark 1)
2)

No	Moment (k-in)	Rotation (radians) X 1/1000	No	Moment (k-in)	Rotation (radians) X 1/1000
1	0.00	0.00	26	660.40	131.40
2	36.76	1.80	27	697.51	138.10
3	74.96	4.20	28	734.63	144.05
4	109.51	9.00	29	771.76	150.00
5	128.26	16.55	30	809.24	155.60
6	147.00	24.10	31	846.72	161.20
7	162.80	32.35	32	882.74	166.75
8	173.61	40.60	33	918.76	172.30
9	183.53	45.65	34	955.51	178.90
10	193.46	50.70	35	992.26	185.50
11	205.07	58.00	36	1025.33	190.65
12	211.68	65.30	37	1058.40	195.80
13	227.85	74.33			
14	244.02	83.37			
15	260.20	92.40			
16	296.94	93.93			
17	333.69	95.47			
18	370.44	97.00			
19	406.82	98.55			
20	443.21	100.10			
21	479.22	102.77			
22	515.24	105.43			
23	551.26	108.10			
24	587.27	116.40			
25	623.28	124.70			

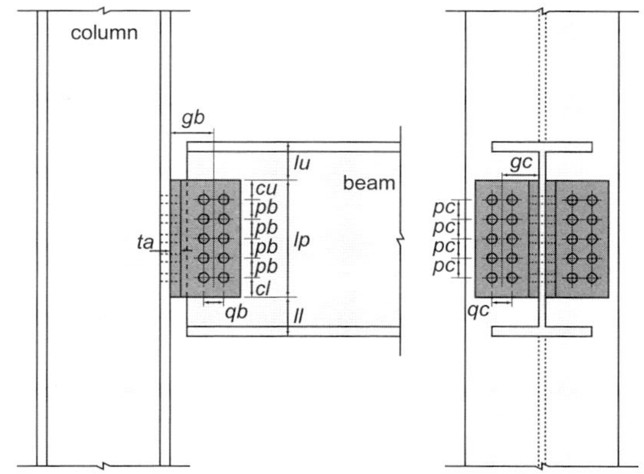

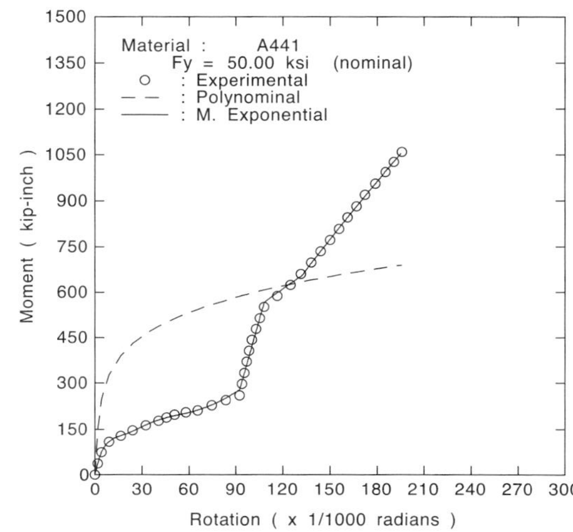

```
Moment-rotation prediction equations      ( R : X 1/1000 radians )

Frye and Morris polynominal model  :     R = Sum ( Ai X ( K*Bm )**Pi X 10**Qi )
     xd = 11.500000"   g =  5.500000"   t =  0.312500"
     A1 =  3.660000   A2 =  1.150000   A3 =  4.570000    K =  0.030179
     P1 =         1   P2 =         3   P3 =         5   Q1 =        -1    Q2 =        -3    Q3 =        -5

Modified exponential model  :
     Bm = Sum ( Ai X ( 1 - exp( - R/( 2*i*C ) ) ) )  +      Sum ( RKj X ( R - Rj0 ) )  + Bm0
      C =   0.20816667E+01            Bmo=   0.00000000E+00          Nexp=  6        Nliner=  4
     AI =  -0.43613438E+03   0.88133715E+04  -0.46666593E+05   0.10651642E+06  -0.10847300E+06   0.40464255E+05
    Rj0 =           9.0000          92.4000         108.1000         132.1000
    RKj =   0.19913544E+01   0.15072655E+02  -0.14831416E+02   0.33959388E+01
```

No	Rotation (radians) X 1/1000	Moment (kip-inch) Expri.	Moment (kip-inch) Poly.	Moment (kip-inch) M.Expo.	Connection stiffness (kip-inch) X 1000 Expri.	Connection stiffness (kip-inch) X 1000 Poly.	Connection stiffness (kip-inch) X 1000 M.Expo.
1	0.00	0.00	0.00	0.00	0.2042E+02	0.9054E+02	0.2247E+02
3	4.20	74.96	245.52	74.57	0.1301E+02	0.2664E+02	0.1234E+02
5	16.55	128.26	390.31	129.61	0.2483E+01	0.6319E+01	0.1702E+01
7	32.35	162.80	462.26	164.42	0.1916E+01	0.3419E+01	0.2343E+01
9	45.65	188.53	501.09	188.63	0.1965E+01	0.2529E+01	0.1309E+01
11	58.00	205.07	529.18	202.76	0.9058E+00	0.2057E+01	0.1166E+01
13	74.33	227.85	559.33	228.91	0.1790E+01	0.1665E+01	0.2140E+01
15	92.40	260.20	586.72	277.80	0.2075E+02	0.1386E+01	0.3169E+01
17	95.47	333.69	590.94	333.91	0.2397E+02	0.1348E+01	0.1835E+02
19	98.55	406.82	595.07	390.61	0.2347E+02	0.1313E+01	0.1843E+02
21	102.77	479.22	600.56	468.48	0.1351E+02	0.1267E+01	0.1850E+02
23	108.10	551.26	607.11	567.32	0.1128E+02	0.1215E+01	0.1855E+02
25	124.70	623.28	626.06	628.53	0.5003E+01	0.1079E+01	0.3602E+01
27	138.10	697.51	639.91	695.61	0.5910E+01	0.9911E+00	0.6754E+01
29	150.00	771.76	651.30	774.57	0.6473E+01	0.9255E+00	0.6517E+01
31	161.20	846.72	661.36	846.39	0.6591E+01	0.8719E+00	0.6313E+01
33	172.30	918.76	670.91	915.48	0.6069E+01	0.8247E+00	0.6142E+01
35	185.50	992.26	681.42	995.45	0.6048E+01	0.7762E+00	0.5982E+01
37	195.80	1058.40	689.16	1056.56	0.6422E+01	0.7427E+00	0.5888E+01

```
Connection type : Double web-angle connections
           Mode : All bolted

Tested by  :  L.E.Thompson et al. (1970)                U.S.A
Test  Id.  :  A2-3 ALT

Column : W10X49                        Fasteners: A325- -3/4"D
Beam   : W21X62                              15/16" Oversize holes
Angle  : 4 X 3.5 X 5/16                Material : A441
                                             Fy =   --   ksi
                                             Fu =   --   ksi

                           Major parameters

     lp = 11.5000"    lu =    --   "    ll =    --   "    ta =  0.3125"
     gb =  2.2500"    gc =  2.5500"    cu =    --   "    cl =    --   "
     pb =    --   "    pc =    --   "    qb =    --   "    qc =    --   "
     nb = - X  2       nc = - X  -

Remark  1)
        2)
```

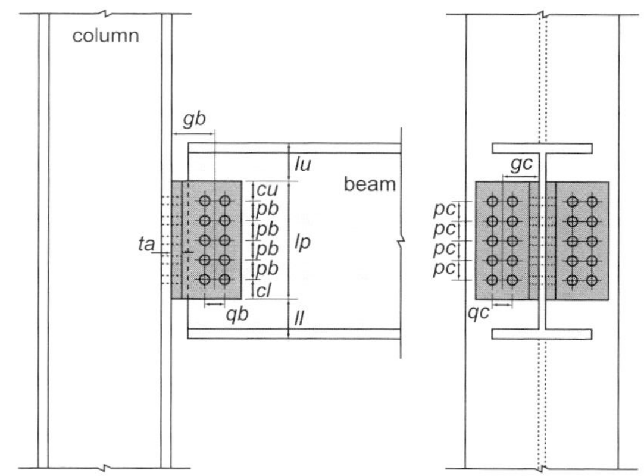

No	Moment (k-in)	Rotation (radians) X 1/1000	No	Moment (k-in)	Rotation (radians) X 1/1000
1	0.00	0.00	26	697.51	135.00
2	36.76	2.10	27	734.63	141.75
3	74.96	4.90	28	771.76	148.50
4	92.24	12.35	29	809.24	154.85
5	109.51	19.80	30	846.72	161.20
6	128.26	25.55	31	882.74	166.70
7	147.00	31.30	32	918.76	172.20
8	162.80	38.50	33	955.51	179.70
9	178.61	45.70	34	992.26	187.20
10	188.53	51.65	35	1014.30	194.30
11	198.46	57.60	36	1036.35	201.40
12	205.07	67.15	37	1058.40	208.50
13	211.68	76.70			
14	255.05	79.70			
15	298.42	82.70			
16	334.43	84.25			
17	370.44	85.80			
18	406.82	87.85			
19	443.21	89.90			
20	479.22	94.77			
21	515.24	99.63			
22	551.26	104.50			
23	587.27	112.55			
24	623.28	120.60			
25	660.40	127.80			

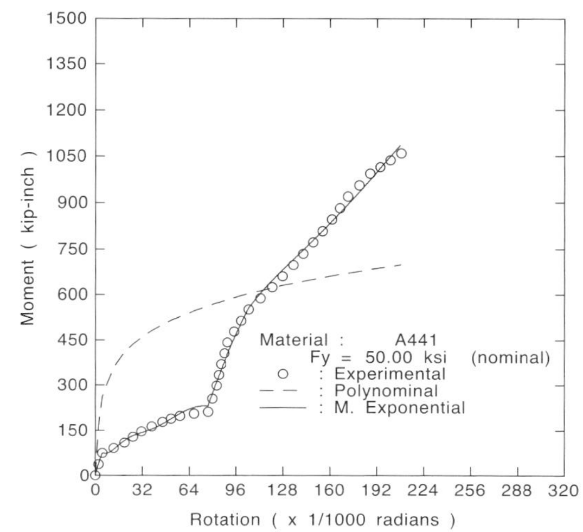

A.2 – 96

```
Moment-rotation prediction equations     ( R : X 1/1000 radians )

Frye and Morris polynominal model  :    R = Sum ( Ai X ( K*Bm )**Pi X 10**Qi )
     xd = 11.500000"   g =  5.500000"   t =  0.312500"
     A1 =  3.660000   A2 =  1.150000   A3 =  4.570000     K =  0.030179
     P1 =         1   P2 =         3   P3 =         5    Q1 =        -1    Q2 =        -3    Q3 =        -5

Modified exponential model  :
     Bm = Sum ( Ai X ( 1 - exp( - R/( 2*i*C ) ) ) )  +     Sum ( RKj X ( R - Rj0 ) )  +  Bm0
      C =   0.22625000E+01             Bmo=   0.00000000E+00         Nexp=  6        Nliner=  4
     AI =   0.10182338E+04  -0.16966204E+05   0.89834745E+05  -0.19996719E+06   0.19412829E+06  -0.67497619E+05
    Rj0 =          0.0000           4.9000          76.7000         104.5000
    RKj =   0.93787434E+01  -0.14965573E+02   0.14837732E+02  -0.32671910E+01
```

No	Rotation (radians) X 1/1000	Moment (kip-inch) Expri.	Moment (kip-inch) Poly.	Moment (kip-inch) M.Expo.	Connection stiffness (kip-inch) X 1000 Expri.	Connection stiffness (kip-inch) X 1000 Poly.	Connection stiffness (kip-inch) X 1000 M.Expo.
1	0.00	0.00	0.00	0.00	0.1750E+02	0.9054E+02	0.1421E+02
3	4.90	74.96	262.57	75.51	0.1055E+02	0.2234E+02	0.1340E+02
5	19.80	109.51	409.15	116.78	0.2850E+01	0.5342E+01	0.3455E+01
7	31.30	147.00	458.55	142.60	0.2787E+01	0.3523E+01	0.1582E+01
9	45.70	178.61	501.22	172.05	0.1906E+01	0.2526E+01	0.2684E+01
11	57.60	198.46	528.35	205.70	0.1293E+01	0.2069E+01	0.2660E+01
13	76.70	211.68	563.22	230.27	0.1117E+02	0.1622E+01	-0.4909E+00
15	82.70	298.42	572.64	312.61	0.2024E+02	0.1522E+01	0.1311E+02
17	85.80	370.44	577.32	352.28	0.2087E+02	0.1475E+01	0.1249E+02
19	89.90	443.21	583.30	401.88	0.1468E+02	0.1417E+01	0.1172E+02
21	99.63	515.24	596.43	508.08	0.7401E+01	0.1301E+01	0.1018E+02
23	112.55	587.27	612.39	603.63	0.4474E+01	0.1175E+01	0.5552E+01
25	127.80	660.40	629.37	681.68	0.5155E+01	0.1057E+01	0.4815E+01
27	141.75	734.63	643.63	747.44	0.5500E+01	0.9691E+00	0.4674E+01
29	154.85	809.24	655.84	809.23	0.5903E+01	0.9009E+00	0.4784E+01
31	166.70	882.74	666.16	866.96	0.6549E+01	0.8478E+00	0.4967E+01
33	179.70	955.51	676.78	932.96	0.4900E+01	0.7971E+00	0.5186E+01
35	194.30	1014.30	688.05	1010.35	0.3105E+01	0.7475E+00	0.5408E+01
37	208.50	1058.40	698.35	1088.41	0.3105E+01	0.7053E+00	0.5580E+01

II - 49

Connection type : Double web-angle connections
Mode : All bolted

Tested by : L.E.Thompson et al. (1970) U.S.A
Test Id. : B2-1 ALT

Column : W10X49
Beam : W21X62
Angle : 4 X 3.5 X 5/16

Fasteners: A325- -3/4"D
15/16" Oversize holes
Material : A441
Fy = -- ksi
Fu = -- ksi

Major parameters

lp = 11.5000"	lu = -- "	ll = -- "	ta = 0.3125"
gb = 2.2500"	gc = 1.8000"	cu = -- "	cl = -- "
pb = -- "	pc = -- "	qb = -- "	qc = -- "
nb = - X 2	nc = - X -		

Remark 1)
2)

No	Moment (k-in)	Rotation (radians) X 1/1000	No	Moment (k-in)	Rotation (radians) X 1/1000
1	0.00	0.00	26	848.20	96.70
2	44.83	0.70	27	901.85	104.25
3	77.92	1.80	28	955.50	111.80
4	82.14	3.60	29	990.04	114.90
5	86.36	5.50	30	1024.58	118.00
6	99.04	11.00	31	1063.54	123.30
7	111.72	16.60	32	1102.50	128.60
8	148.47	20.10	33	1138.88	132.95
9	185.22	23.60	34	1175.27	137.30
10	221.24	27.80	35	1211.65	142.50
11	257.26	32.00	36	1248.04	147.70
12	294.00	39.00	37	1284.79	155.15
13	314.21	45.50	38	1321.54	162.60
14	334.43	52.00	39	1359.75	172.70
15	354.09	61.50	40	1397.96	182.80
16	373.75	71.00	41	1384.37	192.50
17	426.85	73.00	42	1370.77	202.20
18	479.95	75.00			
19	528.71	76.70			
20	577.46	78.40			
21	626.22	80.10			
22	674.73	82.50			
23	723.24	84.90			
24	771.76	87.30			
25	809.98	92.00			

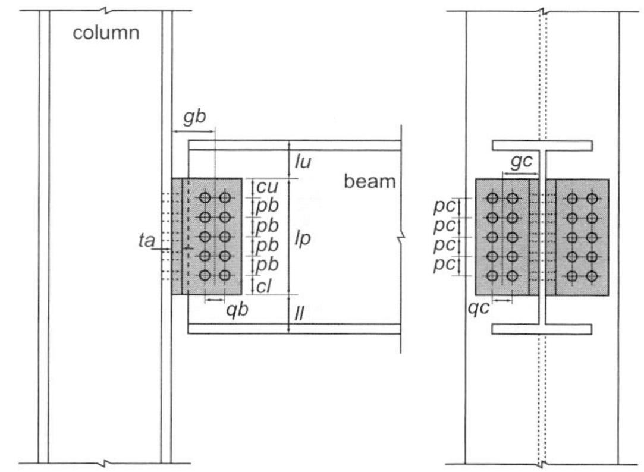

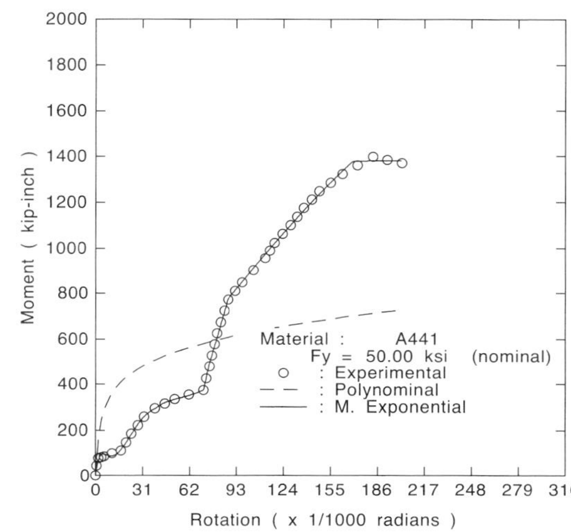

```
Moment-rotation prediction equations      ( R : X 1/1000 radians )

Frye and Morris polynominal model  :    R = Sum ( Ai X ( K*Bm )**Pi X 10**Qi )
     xd = 11.500000"   g =  4.000000"   t =  0.312500"
     A1 =  3.660000   A2 =  1.150000   A3 =  4.570000    K =  0.028771
     P1 =        1    P2 =        3    P3 =        5    Q1 =       -1    Q2 =       -3    Q3 =       -5

Modified exponential model  :
     Bm = Sum ( Ai X ( 1 - exp( - R/( 2*i*C ) ) ) )  +      Sum ( RKj X ( R - Rj0 ) )  +  Bm0
      C =   0.18600000E+01              Bmo=  0.00000000E+00             Nexp=  6          Nliner=  6
     AI =   0.84941189E+02  0.10680167E+05  -0.70559240E+05  0.16635651E+06  -0.16849309E+06  0.62417329E+05
    Rj0 =          16.6000         39.0000          71.0000         75.0000          87.3000
                  169.4000
    RKj =  -0.23230984E+00  0.23472608E+01  0.26956347E+02  -0.79854910E+01  -0.15400121E+02
           -0.59177646E+01
```

No	Rotation (radians) X 1/1000	Moment (kip-inch) Expri.	Moment (kip-inch) Poly.	Moment (kip-inch) M.Expo.	Connection stiffness (kip-inch) X 1000 Expri.	Connection stiffness (kip-inch) X 1000 Poly.	Connection stiffness (kip-inch) X 1000 M.Expo.
1	0.00	0.00	0.00	0.00	0.6405E+02	0.9496E+02	0.5341E+02
3	1.80	77.92	154.02	66.16	0.1956E+02	0.6660E+02	0.2304E+02
5	5.50	86.36	288.58	96.15	0.2244E+01	0.2057E+02	-0.6452E+00
7	16.60	111.72	409.72	115.00	0.7333E+01	0.6611E+01	0.8197E+01
9	23.60	185.22	448.78	183.86	0.9625E+01	0.4766E+01	0.1042E+02
11	32.00	257.26	483.53	258.35	0.7328E+01	0.3623E+01	0.6722E+01
13	45.50	314.21	525.21	317.50	0.3110E+01	0.2660E+01	0.2924E+01
15	61.50	354.09	562.44	350.01	0.2070E+01	0.2052E+01	0.2023E+01
17	73.00	426.85	584.39	433.46	0.2655E+02	0.1773E+01	0.3012E+02
19	76.70	528.71	590.88	532.02	0.2868E+02	0.1700E+01	0.2250E+02
21	80.10	626.22	596.47	609.05	0.2517E+02	0.1639E+01	0.2280E+02
23	84.90	723.24	604.25	719.33	0.2021E+02	0.1560E+01	0.2314E+02
25	92.00	809.98	614.84	812.45	0.8132E+01	0.1459E+01	0.8046E+01
27	104.25	901.85	631.79	911.98	0.7106E+01	0.1314E+01	0.8120E+01
29	114.90	990.04	645.27	997.33	0.1114E+02	0.1211E+01	0.7875E+01
31	123.30	1063.54	655.10	1062.30	0.7351E+01	0.1142E+01	0.7587E+01
33	132.95	1138.88	665.78	1133.79	0.8364E+01	0.1073E+01	0.7230E+01
35	142.50	1211.65	675.88	1201.22	0.6997E+01	0.1012E+01	0.6898E+01
37	155.15	1284.79	688.10	1286.05	0.4933E+01	0.9440E+00	0.6529E+01
39	172.70	1359.75	703.94	1377.68	0.3784E+01	0.8640E+00	0.2476E+00
41	192.50	1384.37	720.29	1379.95	-0.1402E+01	0.7901E+00	0.7025E-02

II - 50

Connection type : Double web-angle connections
Mode : All bolted

Tested by : L.E.Thompson et al. (1970) U.S.A
Test Id. : B2-1 ALT

Column : W10X49
Beam : W21X62
Angle : 4 X 3.5 X 5/16

Fasteners: A325- -3/4"D
15/16" Oversize holes
Material : A441
Fy = -- ksi
Fu = -- ksi

Major parameters

lp = 11.5000"	lu = -- "	ll = -- "	ta = 0.3125"
gb = 2.2500"	gc = 1.8000"	cu = -- "	cl = -- "
pb = -- "	pc = -- "	qb = -- "	qc = -- "
nb = - X 2	nc = - X -		

Remark 1)
2)

No	Moment (k-in)	Rotation (radians) X 1/1000	No	Moment (k-in)	Rotation (radians) X 1/1000
1	0.00	0.00	26	809.98	117.45
2	44.83	0.80	27	848.20	122.00
3	77.92	1.40	28	901.85	128.35
4	111.72	2.50	29	955.50	134.70
5	148.48	13.30	30	990.04	139.10
6	160.72	20.73	31	1024.58	143.50
7	172.97	28.17	32	1063.54	149.65
8	185.22	35.60	33	1102.50	155.80
9	202.13	41.45	34	1138.88	159.55
10	219.04	47.30	35	1175.27	163.30
11	256.52	53.45	36	1211.65	169.45
12	294.00	59.60	37	1248.04	175.60
13	317.71	66.25	38	1284.79	182.80
14	341.41	72.90	39	1321.54	190.00
15	352.19	81.17	40	1347.01	197.27
16	362.97	89.43	41	1372.49	204.53
17	373.75	97.70	42	1397.96	211.80
18	426.85	99.50	43	1370.77	217.60
19	479.95	101.30			
20	528.71	102.80			
21	577.46	104.30			
22	626.22	105.80			
23	674.73	108.17			
24	723.24	110.53			
25	771.76	112.90			

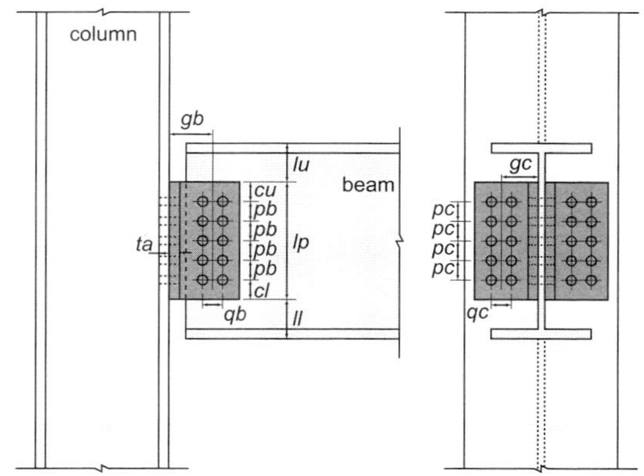

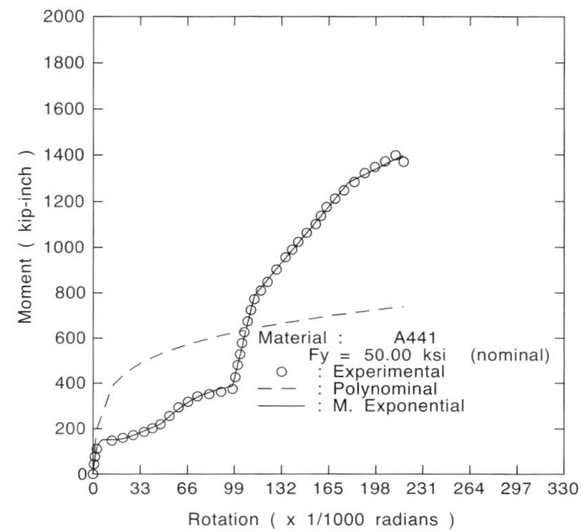

```
 Moment-rotation prediction equations      ( R : X 1/1000 radians )

Frye and Morris polynominal model  :    R = Sum ( Ai X ( K*Bm )**Pi X 10**Qi )
     xd = 11.500000"    g =  4.000000"   t =  0.312500"
     A1 =  3.660000    A2 =  1.150000   A3 =  4.570000    K =  0.028771
     P1 =         1    P2 =         3   P3 =         5   Q1 =        -1     Q2 =        -3     Q3 =        -5

Modified exponential model  :
     Bm = Sum  ( Ai X ( 1 - exp( - R/( 2*i*C ) ) ) )  +      Sum ( RKj X ( R - Rj0 ) )  +  Bm0
      C =   0.20133333E+01            Bmo=   0.00000000E+00          Nexp=  6         Nliner=  7
     AI =   0.63811479E+03  -0.44082834E+04   0.20476447E+05  -0.43316123E+05   0.38606337E+05  -0.11503899E+05
    Rj0 =           0.0000           2.5000          47.3000          71.5000          97.7000
                  112.9000         178.7000
    RKj =   0.67704419E+01  -0.96818873E+01   0.33081717E+01  -0.30203987E+00   0.25689238E+02
           -0.18241020E+02  -0.44880891E+01
```

No	Rotation (radians) X 1/1000	Moment (kip-inch) Expri.	Moment (kip-inch) Poly.	Moment (kip-inch) M.Expo.	Connection stiffness (kip-inch) X 1000 Expri.	Connection stiffness (kip-inch) X 1000 Poly.	Connection stiffness (kip-inch) X 1000 M.Expo.
1	0.00	0.00	0.00	0.00	0.5604E+02	0.9496E+02	0.5821E+02
3	1.40	77.92	125.24	73.79	0.4653E+02	0.7736E+02	0.4218E+02
5	13.30	148.48	385.56	151.24	0.2363E+01	0.8177E+01	-0.1454E+00
7	28.17	172.97	468.94	170.94	0.1648E+01	0.4058E+01	0.2323E+01
9	41.45	202.13	513.99	205.56	0.2890E+01	0.2885E+01	0.2682E+01
11	53.45	256.52	544.91	254.90	0.6095E+01	0.2314E+01	0.5325E+01
13	66.25	317.71	571.88	315.02	0.3565E+01	0.1926E+01	0.4018E+01
15	81.17	352.19	598.19	360.24	0.1304E+01	0.1621E+01	0.2183E+01
17	97.70	373.75	622.95	385.66	0.2446E+02	0.1387E+01	0.9957E+00
19	101.30	479.95	627.91	481.40	0.3114E+02	0.1346E+01	0.2651E+02
21	104.30	577.46	631.92	560.72	0.3250E+02	0.1313E+01	0.2638E+02
23	108.17	674.73	636.96	662.42	0.2050E+02	0.1273E+01	0.2623E+02
25	112.90	771.76	642.78	786.24	0.1636E+02	0.1229E+01	0.2609E+02
27	122.00	848.20	653.61	856.72	0.8421E+01	0.1152E+01	0.7657E+01
29	134.70	955.50	667.79	952.95	0.8095E+01	0.1060E+01	0.7518E+01
31	143.50	1024.58	676.80	1018.91	0.7218E+01	0.1007E+01	0.7477E+01
33	155.80	1102.50	688.81	1110.73	0.8427E+01	0.9402E+00	0.7460E+01
35	163.30	1175.27	695.63	1166.69	0.8268E+01	0.9049E+00	0.7462E+01
37	175.60	1248.04	706.43	1258.54	0.5542E+01	0.8522E+00	0.7475E+01
39	190.00	1321.54	718.31	1315.59	0.4309E+01	0.7986E+00	0.3005E+01
41	204.53	1372.49	729.57	1359.38	0.3506E+01	0.7516E+00	0.3021E+01
43	217.60	1370.77	739.26	1398.93	-0.4688E+01	0.7139E+00	0.3032E+01

Connection type : Double web-angle connections
Mode : All bolted

Tested by : L.E.Thompson et al. (1970) U.S.A
Test Id. : B2-2 ALT

Column : W1CX49
Beam : W21X62
Angle : 4 X 3.5 X 5/16

Fasteners: A325- -3/4"D
15/16" Oversize holes
Material : A441
Fy = -- ksi
Fu = -- ksi

Major parameters

lp = 11.5000"	lu = -- ▪	ll = -- "	ta = 0.3125"
gb = 2.2500"	gc = 1.8000▪	cu = -- "	cl = -- "
pb = -- "	pc = -- ▪	qb = -- "	qc = -- "
nb = - X 2	nc = - X -		

Remark 1)
2)

No	Moment (k-in)	Rotation (radians) X 1/1000	No	Moment (k-in)	Rotation (radians) X 1/1000
1	0.00	0.00	26	848.56	104.00
2	43.73	0.60	27	884.94	107.80
3	77.17	1.30	28	922.43	111.50
4	113.18	2.30	29	959.92	115.20
5	121.64	9.43	30	994.82	119.35
6	130.09	16.55	31	1029.73	123.50
7	138.55	23.68	32	1066.12	128.45
8	147.00	30.80	33	1102.50	133.40
9	183.75	38.15	34	1138.88	138.55
10	220.50	45.50	35	1175.27	143.70
11	266.07	52.95	36	1214.22	150.20
12	311.64	60.40	37	1253.17	156.70
13	325.61	69.00	38	1289.19	164.25
14	339.58	77.60	39	1325.21	171.80
15	392.50	79.40	40	1355.53	180.50
16	445.42	81.20	41	1385.84	189.20
17	482.53	82.30			
18	519.65	83.40			
19	556.40	84.60			
20	593.15	85.80			
21	629.16	87.25			
22	665.17	88.70			
23	729.12	93.60			
24	770.65	96.90			
25	812.17	100.20			

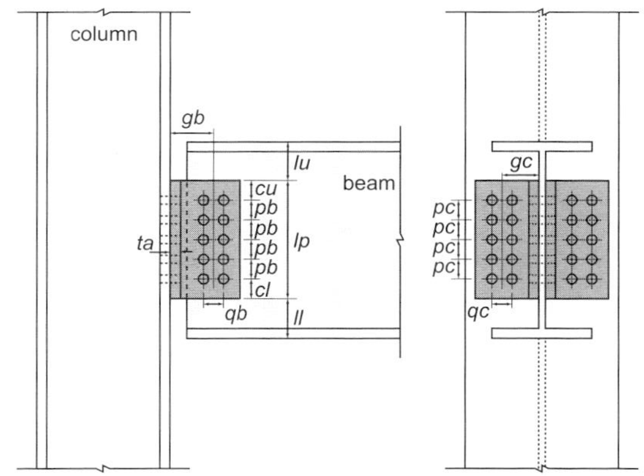

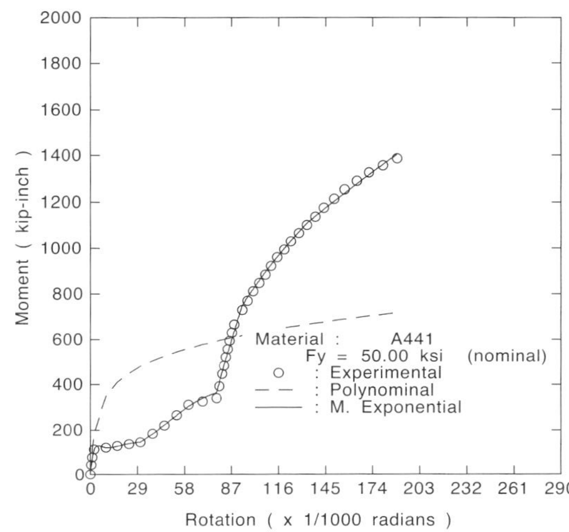

```
Moment-rotation prediction equations      ( R : X 1/1000 radians )

Frye and Morris polynominal model  :    R = Sum ( Ai X ( K*Bm )**Pi X 10**Qi )
     xd = 11.500000"    g =  4.000000"   t =  0.312500"
     A1 =  3.660000   A2 =  1.150000   A3 =  4.570000    K =  0.028771
     P1 =         1   P2 =         3   P3 =         5   Q1 =        -1    Q2 =        -3    Q3 =        -5

Modified exponential model  :
     Bm = Sum ( Ai X ( 1 - exp( - R/( 2*i*C ) ) ) )  +      Sum ( RKj X ( R - Rj0 ) )  +  Bm0
      C =   0.17266667E+01               Bmo=   0.00000000E+00         Nexp=  6        Nliner=  7
     AI =   0.10119327E+04  -0.61902968E+04   0.12792470E+05   0.50178128E+04  -0.37095509E+05   0.25507744E+05
    Rj0 =           0.0000           2.3000          30.8000          60.4000          77.6000
                   93.6000         133.4000
    RKj =   0.18539770E+01  -0.10348395E+02   0.37167784E+01  -0.34001318E+00   0.23517063E+02
           -0.11784420E+02  -0.17513957E+01
```

No	Rotation (radians) X 1/1000	Moment (kip-inch) Expri.	Moment (kip-inch) Poly.	Moment (kip-inch) M.Expo.	Connection stiffness (kip-inch) X 1000 Expri.	Connection stiffness (kip-inch) X 1000 Poly.	Connection stiffness (kip-inch) X 1000 M.Expo.
1	0.00	0.00	0.00	0.00	0.7288E+02	0.9496E+02	0.7749E+02
3	1.30	77.17	117.38	77.94	0.4293E+02	0.7989E+02	0.4389E+02
5	9.43	121.64	348.09	122.41	0.1187E+01	0.1154E+02	-0.1100E+01
7	23.68	138.55	449.13	140.01	0.1187E+01	0.4753E+01	0.1200E+01
9	38.15	183.75	504.14	182.47	0.5000E+01	0.3102E+01	0.5139E+01
11	52.95	266.07	543.75	264.80	0.6117E+01	0.2333E+01	0.5636E+01
13	69.00	325.61	577.08	339.98	0.1624E+01	0.1861E+01	0.3440E+01
15	79.40	392.50	595.33	408.88	0.2940E+02	0.1652E+01	0.2517E+02
17	82.30	482.53	600.08	481.16	0.3374E+02	0.1602E+01	0.2467E+02
19	84.60	556.40	603.75	537.46	0.3063E+02	0.1565E+01	0.2429E+02
21	87.25	629.16	607.87	601.25	0.2484E+02	0.1525E+01	0.2386E+02
23	93.60	729.12	617.16	749.64	0.1277E+02	0.1438E+01	0.2290E+02
25	100.20	812.17	626.51	820.03	0.1119E+02	0.1357E+01	0.1024E+02
27	107.80	884.94	636.47	894.54	0.9857E+01	0.1277E+01	0.9398E+01
29	115.20	959.92	645.65	961.55	0.9321E+01	0.1208E+01	0.8739E+01
31	123.50	1029.73	655.33	1031.58	0.7928E+01	0.1141E+01	0.8163E+01
33	133.40	1102.50	666.26	1109.77	0.7210E+01	0.1070E+01	0.7662E+01
35	143.70	1175.27	676.95	1168.68	0.6591E+01	0.1006E+01	0.5551E+01
37	156.70	1253.17	689.56	1238.81	0.5427E+01	0.9363E+00	0.5262E+01
39	171.80	1325.21	703.16	1316.67	0.4173E+01	0.8677E+00	0.5071E+01
41	189.20	1385.84	717.81	1403.82	0.3485E+01	0.8008E+00	0.4960E+01

Connection type : Double web-angle connections
Mode : All bolted

Tested by : L.E.Thompson et al. (1970) U.S.A
Test Id. : B2-2 ALT

Column : W10X49
Beam : W21X62
Angle : 4 X 3.5 X 5/16

Fasteners: A325- -3/4"D
15/16" Oversize holes
Material : A441
Fy = -- ksi
Fu = -- ksi

Major parameters

lp = 11.5000"	lu = -- "	ll = -- "	ta = 0.3125"
gb = 2.2500"	gc = 1.8000"	cu = -- "	cl = -- "
pb = -- "	pc = -- "	qb = -- "	qc = -- "
nb = - X 2	nc = - X -		

Remark 1)
2)

No	Moment (k-in)	Rotation (radians) X 1/1000	No	Moment (k-in)	Rotation (radians) X 1/1000
1	0.00	0.00	26	770.65	100.40
2	43.73	1.50	27	812.17	103.00
3	77.17	2.20	28	848.56	106.45
4	113.18	3.10	29	884.94	109.90
5	124.46	9.73	30	922.43	113.55
6	135.73	16.37	31	959.92	117.20
7	147.00	23.00	32	994.82	122.25
8	166.48	28.20	33	1029.73	127.30
9	185.95	33.40	34	1084.49	134.55
10	220.50	42.90	35	1139.26	141.80
11	255.41	49.40	36	1196.21	150.70
12	290.33	55.90	37	1253.17	159.60
13	311.64	61.60	38	1289.19	166.95
14	330.75	69.37	39	1325.21	174.30
15	349.86	77.13	40	1345.42	181.77
16	368.96	84.90	41	1365.63	189.23
17	407.19	86.10	42	1385.84	196.70
18	445.42	87.30			
19	482.53	88.40			
20	519.65	89.50			
21	556.40	90.60			
22	593.15	91.70			
23	629.16	93.00			
24	665.17	94.30			
25	729.12	97.80			

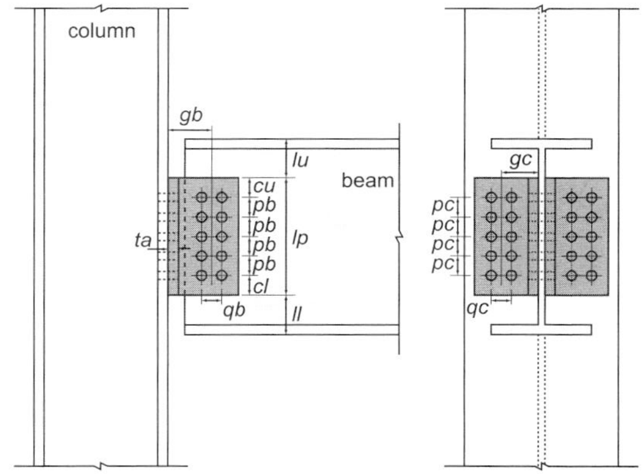

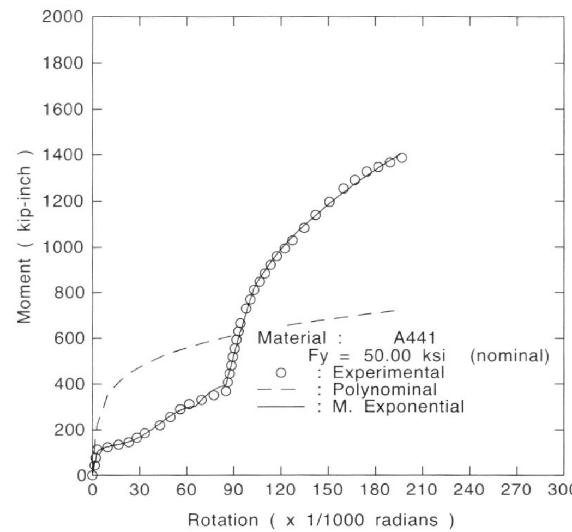

```
Moment-rotation prediction equations     ( R : X 1/1000 radians )

Frye and Morris polynominal model  :    R = Sum ( Ai X ( K*Bm )**Pi X 10**Qi )
     xd = 11.500000"   g =  4.000000"   t =  0.312500"
     A1 =  3.660000   A2 =  1.150000   A3 =  4.570000     K =  0.028771
     P1 =         1   P2 =         3   P3 =         5    Q1 =        -1    Q2 =        -3    Q3 =        -5

Modified exponential model  :
     Bm = Sum ( Ai X ( 1 - exp( - R/( 2*i*C ) ) ) )  +      Sum ( RKj X ( R - Rj0 ) )  +  Bm0
      C =   0.20141667E+01              Bmo=   0.00000000E+00           Nexp=  6        Nliner=  6
     AI =   0.29428361E+03  -0.37982615E+04   0.13258174E+05   0.28086046E+04  -0.50295948E+05   0.40229545E+05
    Rj0 =           3.1000          23.6000          65.6000          84.9000         103.0000
                  141.8000
    RKj =  -0.21951160E+02   0.27729796E+00   0.82782066E+01   0.25390050E+02  -0.91080026E+01
            0.88623991E+00
```

No	Rotation (radians) X 1/1000	Moment (kip-inch) Expri.	Moment (kip-inch) Poly.	Moment (kip-inch) M.Expo.	Connection stiffness (kip-inch) X 1000 Expri.	Connection stiffness (kip-inch) X 1000 Poly.	Connection stiffness (kip-inch) X 1000 M.Expo.
1	0.00	0.00	0.00	0.00	0.2915E+02	0.9496E+02	0.4032E+02
3	2.20	77.17	178.58	77.10	0.4438E+02	0.5640E+02	0.3097E+02
5	9.73	124.46	351.59	126.61	0.1699E+01	0.1117E+02	0.1814E+01
7	23.00	147.00	445.88	147.91	0.2846E+01	0.4880E+01	0.2123E+01
9	33.40	185.95	488.51	182.70	0.3707E+01	0.3488E+01	0.4313E+01
11	49.40	255.41	535.21	260.06	0.5371E+01	0.2477E+01	0.4507E+01
13	61.60	311.64	562.64	298.79	0.3198E+01	0.2049E+01	0.1483E+01
15	77.13	349.86	591.51	374.24	0.2460E+01	0.1693E+01	0.4084E+01
17	86.10	407.19	606.00	426.37	0.3186E+02	0.1543E+01	0.2618E+02
19	88.40	482.53	609.53	485.65	0.3374E+02	0.1509E+01	0.2538E+02
21	90.60	556.40	612.83	540.67	0.3341E+02	0.1478E+01	0.2464E+02
23	93.00	629.16	616.36	598.87	0.2770E+02	0.1445E+01	0.2386E+02
25	97.80	729.12	623.09	709.83	0.1695E+02	0.1386E+01	0.2240E+02
27	103.00	812.17	630.22	822.48	0.1364E+02	0.1327E+01	0.2096E+02
29	109.90	884.94	639.05	898.34	0.1041E+02	0.1257E+01	0.1018E+02
31	117.20	959.92	648.10	967.07	0.8862E+01	0.1191E+01	0.8698E+01
33	127.30	1029.73	659.71	1046.34	0.7176E+01	0.1112E+01	0.7075E+01
35	141.80	1139.26	675.02	1136.25	0.7036E+01	0.1017E+01	0.5448E+01
37	159.60	1253.17	692.25	1237.30	0.5579E+01	0.9222E+00	0.5136E+01
39	174.30	1325.21	705.32	1308.28	0.3812E+01	0.8575E+00	0.4570E+01
41	189.23	1365.63	717.69	1373.75	0.2707E+01	0.8013E+00	0.4229E+01

II - 53

Connection type : Double web-angle connections
Mode : All bolted

Tested by : L.E.Thompson et al. (1970) U.S.A
Test Id. : B2-3 ALT

Column : W10X49
Beam : W21X62
Angle : 4 X 3.5 X 5/16

Fasteners: A325- -3/4"D
15/16" Oversize holes
Material : A441
Fy = -- ksi
Fu = -- ksi

Major parameters

lp = 11.5000"	lu = -- '	ll = -- "	ta = 0.3125"
gb = 2.2500"	gc = 1.8000'	cu = -- "	cl = -- "
pb = -- "	pc = -- '	qb = -- "	qc = -- "
nb = - X 2	nc = - X -		

Remark 1)
2)

No	Moment (k-in)	Rotation (radians) X 1/1000
1	0.00	0.00
2	46.31	0.40
3	78.28	1.00
4	110.26	2.30
5	114.94	4.60
6	119.63	6.90
7	133.68	13.70
8	147.73	20.60
9	184.85	27.20
10	221.96	33.80
11	240.71	39.80
12	259.45	45.80
13	309.43	53.60
14	359.41	61.40
15	371.91	70.20
16	384.40	79.00
17	396.90	87.80
18	459.74	90.05
19	522.59	92.30
20	570.12	93.83
21	617.64	95.37
22	665.17	96.90
23	714.42	98.93
24	763.67	100.97
25	812.92	103.00

No	Moment (k-in)	Rotation (radians) X 1/1000
26	847.46	106.90
27	882.00	110.80
28	919.85	117.93
29	957.70	125.07
30	995.56	132.20
31	1049.39	137.65
32	1103.23	143.10
33	1139.25	148.15
34	1175.27	153.20
35	1230.76	162.25
36	1286.26	171.30
37	1323.37	179.60
38	1360.49	187.90
39	1356.32	196.13
40	1352.15	204.37
41	1347.98	212.60

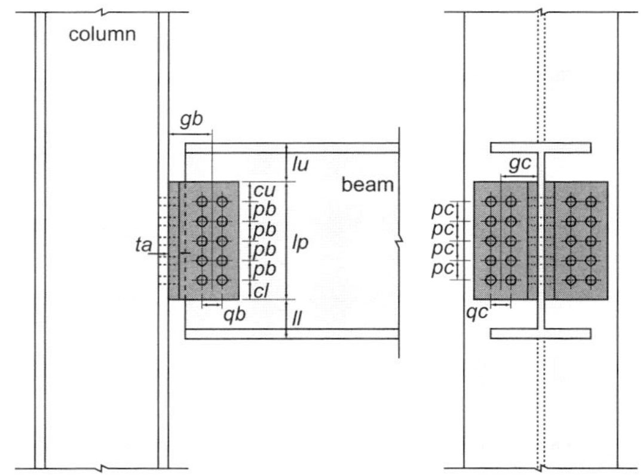

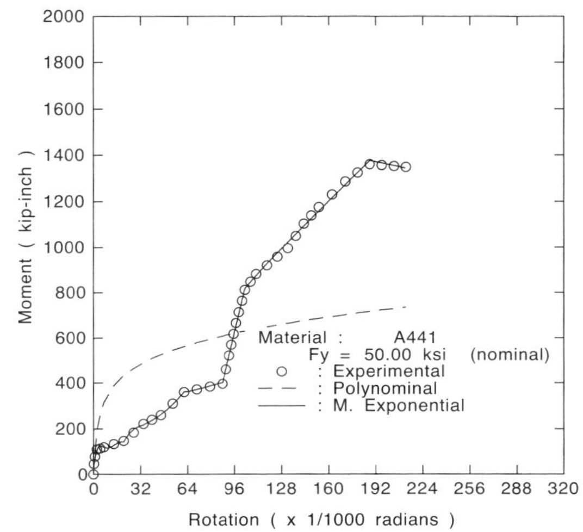

```
Moment-rotation prediction equations     ( R : X 1/1000 radians )

Frye and Morris polynominal model  :    R = Sum ( Ai X ( K*Bm )**Pi X 10**Qi )
     xd = 11.500000"   g =  4.000000"   t =  0.312500"
     A1 =  3.660000   A2 =  1.150000   A3 =  4.570000    K =  0.028771
     P1 =         1    P2 =         3    P3 =         5    Q1 =        -1    Q2 =        -3    Q3 =        -5

Modified exponential model  :
     Bm = Sum ( Ai X ( 1 - exp( - R/( 2*i*C ) ) ) )  +     Sum ( RKj X ( R - Rj0 ) )  +  Bm0
       C =   0.18716667E+01            Bmo=    0.00000000E+00         Nexp=  6          Nliner=  9
      AI =   0.22386121E+04  -0.17004689E+05   0.48662690E+05  -0.58343260E+05   0.26987414E+05  -0.21263764E+04
     Rj0 =           0.4000           2.3000          20.6000          33.8000          45.8000
                    61.4000          87.8000         103.0000         187.9000
     RKj =  -0.23107083E+02   0.15287431E+02   0.12466641E+02   0.20866644E+01  -0.15909861E+00
            -0.70202723E+01   0.26169329E+02  -0.19535683E+02  -0.75284642E+01
```

No	Rotation (radians) X 1/1000	Moment (kip-inch) Expri.	Moment (kip-inch) Poly.	Moment (kip-inch) M.Expo.	Connection stiffness (kip-inch) X 1000 Expri.	Connection stiffness (kip-inch) X 1000 Poly.	Connection stiffness (kip-inch) X 1000 M.Expo.
1	0.00	0.00	0.00	0.00	0.1158E+03	0.9496E+02	0.1107E+03
3	1.00	78.28	92.35	72.96	0.4422E+02	0.8652E+02	0.4305E+02
5	4.60	114.94	268.13	125.55	0.2037E+01	0.2518E+02	0.1115E+00
7	13.70	133.68	388.83	128.14	0.2052E+01	0.7942E+01	0.3958E+01
9	27.20	184.85	464.92	197.00	0.5624E+01	0.4188E+01	0.4968E+01
11	39.80	240.71	509.15	238.34	0.3124E+01	0.2989E+01	0.3589E+01
13	53.60	309.43	545.26	305.20	0.6408E+01	0.2309E+01	0.6359E+01
15	70.20	371.91	579.30	371.77	0.1420E+01	0.1833E+01	0.1593E+01
17	87.80	396.90	608.59	401.59	0.2253E+02	0.1518E+01	0.1545E+01
19	92.30	522.59	615.35	525.91	0.2975E+02	0.1455E+01	0.2754E+02
21	95.37	617.64	619.78	610.16	0.3100E+02	0.1415E+01	0.2740E+02
23	98.93	714.42	624.79	707.62	0.2422E+02	0.1372E+01	0.2725E+02
25	103.00	812.92	630.16	818.07	0.1896E+02	0.1327E+01	0.2707E+02
27	110.80	882.00	640.18	875.58	0.7602E+01	0.1249E+01	0.7220E+01
29	125.07	957.70	657.11	975.22	0.5306E+01	0.1129E+01	0.6778E+01
31	137.65	1049.39	670.85	1058.81	0.9879E+01	0.1042E+01	0.6528E+01
33	148.15	1139.25	681.44	1126.62	0.7132E+01	0.9804E+00	0.6396E+01
35	162.25	1230.76	694.68	1216.00	0.6132E+01	0.9097E+00	0.6292E+01
37	179.60	1323.37	709.95	1324.56	0.4472E+01	0.8359E+00	0.6231E+01
39	196.13	1356.32	723.26	1365.35	-0.5062E+00	0.7775E+00	-0.1323E+01
41	212.60	1347.98	735.63	1343.46	-0.5062E+00	0.7277E+00	-0.1334E+01

II - 54

Connection type : Double web-angle connections
Mode : All bolted

Tested by : L.E.Thompson et al. (1970) U.S.A
Test Id. : B2-3 ALT

Column : W10X49
Beam : W21X62
Angle : 4 X 3.5 X 5/16

Fasteners: A325- -3/4"D
15/16" Oversize holes
Material : A441
Fy = -- ksi
Fu = -- ksi

Major parameters

lp = 11.5000"	lu = -- "	ll = -- "	ta = 0.3125"
gb = 2.2500"	gc = 1.8000"	cu = -- "	cl = -- "
pb = -- "	pc = -- "	qb = -- "	qc = -- "
nb = - X 2	nc = - X -		

Remark 1)
2)

No	Moment (k-in)	Rotation (radians) X 1/1000	No	Moment (k-in)	Rotation (radians) X 1/1000
1	0.00	0.00	26	847.46	106.15
2	46.31	0.80	27	882.00	110.30
3	78.28	1.50	28	918.75	115.40
4	110.26	2.20	29	955.50	120.50
5	147.73	5.40	30	1004.74	127.00
6	172.48	12.53	31	1053.99	133.50
7	197.22	19.67	32	1103.23	140.00
8	221.96	26.80	33	1157.99	147.40
9	259.09	33.20	34	1212.76	154.80
10	296.21	39.60	35	1249.51	160.95
11	327.81	45.45	36	1286.26	167.10
12	359.41	51.30	37	1323.37	175.70
13	378.16	58.90	38	1360.49	184.30
14	396.90	66.50	39	1354.24	189.25
15	412.58	73.83	40	1347.98	194.20
16	428.26	81.17			
17	443.94	88.50			
18	493.92	90.13			
19	543.90	91.77			
20	593.88	93.40			
21	644.10	95.10			
22	694.33	96.80			
23	744.55	98.50			
24	778.73	100.25			
25	812.92	102.00			

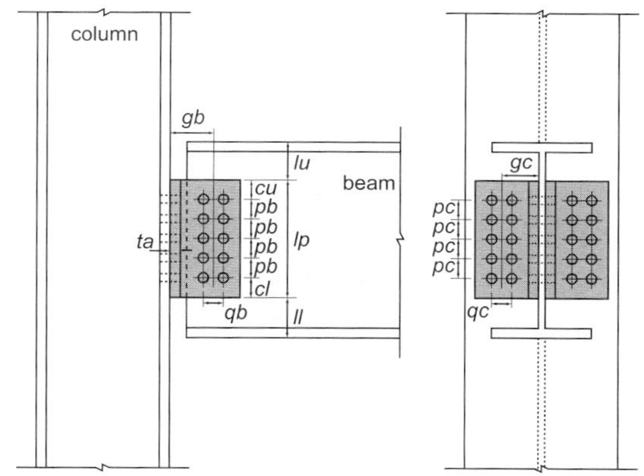

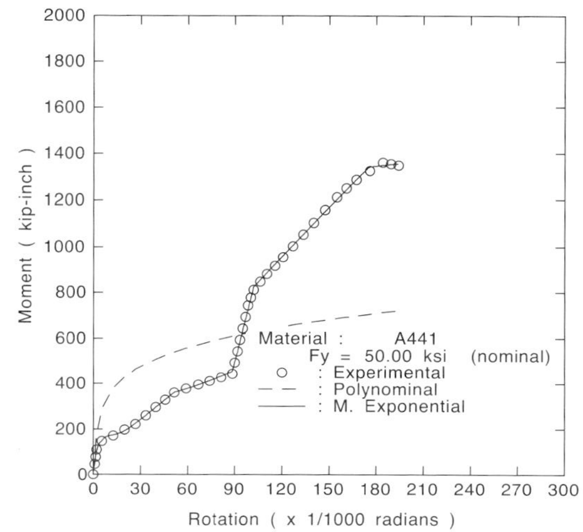

A.2 – 108

```
Moment-rotation prediction equations     ( R : X 1/1000 radians )

Frye and Morris polynominal model  :    R = Sum ( Ai X ( K*Bm )**Pi X 10**Qi )
     xd = 11.500000"   g =  4.000000"   t =  0.312500"
     A1 =  3.660000   A2 =  1.150000   A3 =  4.570000    K =  0.028771
     P1 =        1    P2 =        3    P3 =        5    Q1 =       -1    Q2 =       -3    Q3 =       -5

Modified exponential model  :
     Bm = Sum ( Ai X ( 1 - exp( - R/( 2*i*C ) ) ) )  +      Sum ( RKj X ( R - Rj0 ) )  +  Bm0
       C =   0.18183333E+01              Bmo=   0.00000000E+00          Nexp=  6         Nliner=  6
      AI =   0.29076022E+03   0.27129795E+04  -0.18068812E+05   0.44431940E+05  -0.47982576E+05   0.18825879E+05
     Rj0 =           0.0000          5.4000          51.3000          88.5000         102.0000
                   175.1000
     RKj =  -0.40704215E+01   0.96206349E+01  -0.38372854E+01   0.24638463E+02  -0.19766024E+02
            -0.58969789E+01
```

No	Rotation (radians) X 1/1000	Moment (kip-inch) Expri.	Moment (kip-inch) Poly.	Moment (kip-inch) M.Expo.	Connection stiffness (kip-inch) X 1000 Expri.	Connection stiffness (kip-inch) X 1000 Poly.	Connection stiffness (kip-inch) X 1000 M.Expo.
1	0.00	0.00	0.00	0.00	0.5788E+02	0.9496E+02	0.7519E+02
3	1.50	78.28	132.87	81.38	0.4568E+02	0.7471E+02	0.3969E+02
5	5.40	147.73	286.52	147.91	0.9159E+01	0.2099E+02	0.2037E+01
7	19.67	197.22	428.41	192.52	0.3469E+01	0.5638E+01	0.3708E+01
9	33.20	259.09	487.83	259.64	0.5800E+01	0.3506E+01	0.5511E+01
11	45.45	327.81	525.07	327.73	0.5402E+01	0.2662E+01	0.5617E+01
13	58.90	378.16	557.00	376.62	0.2466E+01	0.2129E+01	0.2198E+01
15	73.83	412.58	585.82	413.32	0.2138E+01	0.1757E+01	0.2675E+01
17	88.50	443.94	609.65	454.00	0.2542E+02	0.1508E+01	0.2816E+01
19	91.77	543.90	614.54	543.67	0.3060E+02	0.1462E+01	0.2744E+02
21	95.10	644.10	619.37	635.11	0.2954E+02	0.1418E+01	0.2742E+02
23	98.50	744.55	624.16	728.28	0.2461E+02	0.1377E+01	0.2739E+02
25	102.00	812.92	628.81	824.06	0.1621E+02	0.1338E+01	0.2734E+02
27	110.30	882.00	639.55	886.46	0.7822E+01	0.1253E+01	0.7453E+01
29	120.50	955.50	651.87	961.61	0.7369E+01	0.1164E+01	0.7284E+01
31	133.50	1053.99	666.37	1054.97	0.7576E+01	0.1069E+01	0.7084E+01
33	147.40	1157.99	680.63	1152.19	0.7400E+01	0.9849E+00	0.6914E+01
35	160.95	1249.51	693.59	1245.02	0.5976E+01	0.9153E+00	0.6795E+01
37	175.70	1323.37	706.52	1341.02	0.4316E+01	0.8518E+00	0.8124E+00
39	189.25	1354.24	717.71	1351.67	-0.1263E+01	0.8012E+00	0.7631E+00

```
Connection type : Double web-angle connections
           Mode : All bolted

Tested by  :  L.E.Thompson et al. (1970)                U.S.A
Test  Id.  :  C2-1 ALT

Column : W1CX49                      Fasteners: A325- -3/4"D
Beam   : W21X62                            15/16" Oversize holes
Angle  : 4 X 3.5 X 1/2               Material : A441
                                          Fy =  --   ksi
                                          Fu =  --   ksi

                          Major parameters

     lp = 11.5000"   lu =    --   "   ll =   --   "   ta =  0.5000"
     gb =  2.2500"   gc =  2.5500"   cu =   --   "   cl =   --   "
     pb =    --   "   pc =    --   "   qb =   --   "   qc =   --   "
     nb = - X  2      nc = - X  -

Remark  1)
        2)
```

No	Moment (k-in)	Rotation (radians) X 1/1000	No	Moment (k-in)	Rotation (radians) X 1/1000
1	0.00	0.00	26	844.51	107.30
2	38.95	0.30	27	900.01	114.35
3	77.17	0.90	28	955.50	121.40
4	93.71	3.70	29	992.25	126.35
5	110.26	6.40	30	1029.00	131.30
6	128.63	11.50	31	1084.13	140.75
7	147.00	16.60	32	1139.26	150.20
8	184.12	26.05	33	1177.10	157.50
9	221.23	35.50	34	1214.95	164.80
10	257.26	41.70	35	1250.60	171.40
11	294.00	50.50	36	1286.26	178.00
12	312.38	55.70	37	1304.87	185.80
13	330.76	60.90	38	1323.49	193.60
14	351.32	67.50	39	1342.10	201.40
15	362.23	74.57			
16	373.13	81.63			
17	384.04	88.70			
18	444.67	90.50			
19	487.67	91.70			
20	530.66	92.90			
21	579.17	94.50			
22	627.68	96.10			
23	682.45	98.20			
24	737.21	100.30			
25	790.86	103.80			

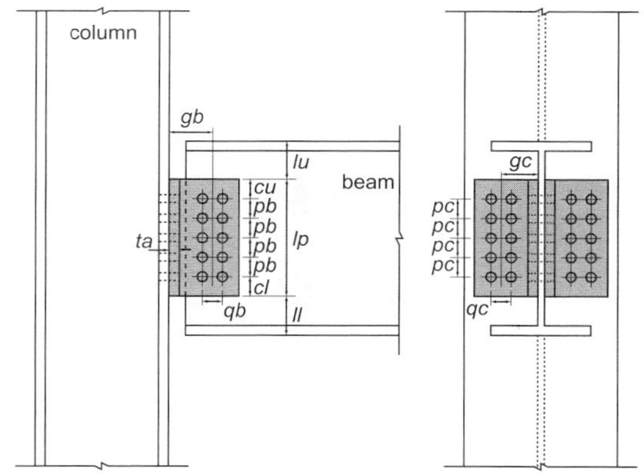

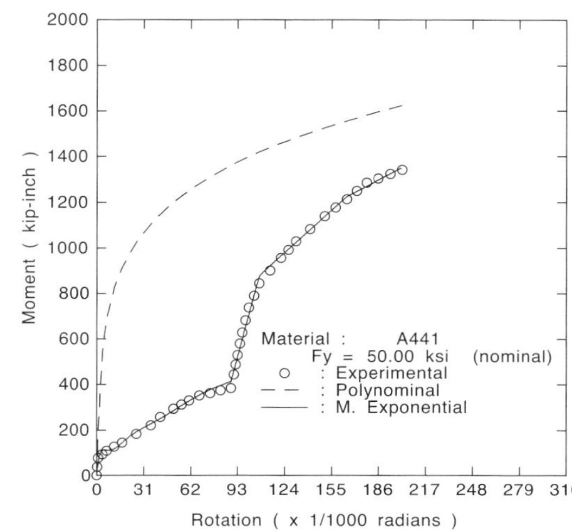

```
Moment-rotation prediction equations      ( R : X 1/1000 radians )

Frye and Morris polynominal model  :    R = Sum ( Ai X ( K*Bm )**Pi X 10**Qi )
     xd = 11.500000"    g =  5.500000"   t =  0.500000"
     A1 =  3.660000    A2 =  1.150000   A3 =  4.570000     K =  0.012890
     P1 =         1    P2 =         3   P3 =         5    Q1 =        -1     Q2 =        -3     Q3 =        -5

Modified exponential model  :
     Bm = Sum ( Ai X ( 1 - exp( - R/( 2*i*C ) ) ) )  +      Sum ( RKj X ( R - Rj0 ) )  +  Bm0
      C =   0.17533333E+01              Bmo=   0.00000000E+00         Nexp=  6        Nliner=  4
     AI =   0.12616871E+04  -0.92539293E+04   0.31198702E+05  -0.49539553E+05   0.34932742E+05  -0.82535234E+04
    Rj0 =           0.9000          88.7000         107.3000         164.8000
    RKj =   0.96593450E+00   0.23213840E+02  -0.18205631E+02  -0.24998156E+01
```

No	Rotation (radians) X 1/1000	Moment (kip-inch) Expri.	Poly.	M.Expo.	Connection stiffness (kip-inch) X 1000 Expri.	Poly.	M.Expo.
1	0.00	0.00	0.00	0.00	0.1298E+03	0.2120E+03	0.7425E+02
3	0.90	77.17	186.61	52.69	0.5350E+02	0.1971E+03	0.4478E+02
5	6.40	110.26	682.16	113.29	0.5254E+01	0.3882E+02	-0.1003E+01
7	16.60	147.00	914.53	144.45	0.3716E+01	0.1476E+02	0.5814E+01
9	35.50	221.23	1106.55	227.68	0.5064E+01	0.7371E+01	0.3534E+01
11	50.50	294.00	1200.69	285.61	0.3772E+01	0.5424E+01	0.4140E+01
13	60.90	330.76	1252.64	328.17	0.3350E+01	0.4619E+01	0.3941E+01
15	74.57	362.23	1310.46	376.42	0.1543E+01	0.3888E+01	0.3065E+01
17	88.70	384.04	1361.47	412.93	0.2716E+02	0.3358E+01	0.2140E+01
19	91.70	487.67	1371.47	488.75	0.3583E+02	0.3265E+01	0.2519E+02
21	94.50	579.17	1380.56	559.10	0.3032E+02	0.3183E+01	0.2506E+02
23	98.20	682.45	1392.24	651.51	0.2608E+02	0.3082E+01	0.2490E+02
25	103.80	790.86	1408.89	790.36	0.1533E+02	0.2944E+01	0.2470E+02
27	114.35	900.01	1438.70	921.09	0.7871E+01	0.2715E+01	0.6233E+01
29	126.35	992.25	1470.15	994.80	0.7424E+01	0.2497E+01	0.6071E+01
31	140.75	1084.13	1504.33	1081.50	0.5834E+01	0.2284E+01	0.5986E+01
33	157.50	1177.10	1541.13	1181.47	0.5185E+01	0.2079E+01	0.5958E+01
35	171.40	1250.60	1568.96	1247.77	0.5402E+01	0.1940E+01	0.3457E+01
37	185.80	1304.87	1596.03	1297.57	0.2387E+01	0.1815E+01	0.3461E+01
39	201.40	1342.10	1623.39	1351.60	0.2387E+01	0.1698E+01	0.3465E+01

II - 56

Connection type : Double web-angle connections
Mode : All bolted

Tested by : L.E.Thompson et al. (1970)   U.S.A
Test Id. : C2-1 ALT

Column : W10X49
Beam : W21X62
Angle : 4 X 3.5 X 1/2

Fasteners: A325- -3/4"D
15/16" Oversize holes
Material : A441
Fy = -- ksi
Fu = -- ksi

Major parameters

lp = 11.5000"	lu = -- "	ll = -- "	ta = 0.5000"
gb = 2.2500"	gc = 2.5500"	cu = -- "	cl = -- "
pb = -- "	pc = -- "	qb = -- "	qc = -- "
nb = - X 2	nc = - X -		

Remark 1)
2)

No	Moment (k-in)	Rotation (radians) X 1/1000	No	Moment (k-in)	Rotation (radians) X 1/1000
1	0.00	0.00	26	992.25	99.65
2	38.95	0.50	27	1029.00	104.60
3	77.17	1.20	28	1065.75	110.43
4	110.26	2.00	29	1102.50	116.27
5	147.00	6.20	30	1139.26	122.10
6	186.68	10.50	31	1177.10	129.75
7	221.23	14.80	32	1214.95	137.40
8	257.26	20.00	33	1250.60	143.90
9	294.00	26.20	34	1286.26	150.40
10	330.76	33.80	35	1314.18	157.45
11	351.32	40.00	36	1342.10	164.50
12	367.68	45.35			
13	384.04	50.70			
14	414.35	58.85			
15	444.67	67.00			
16	487.67	68.15			
17	530.66	69.30			
18	579.17	70.65			
19	627.68	72.00			
20	682.45	73.85			
21	737.21	75.70			
22	790.86	78.65			
23	844.51	81.60			
24	900.01	88.15			
25	955.50	94.70			

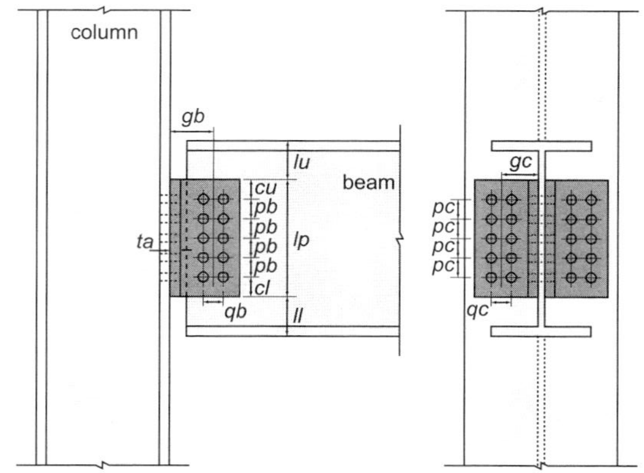

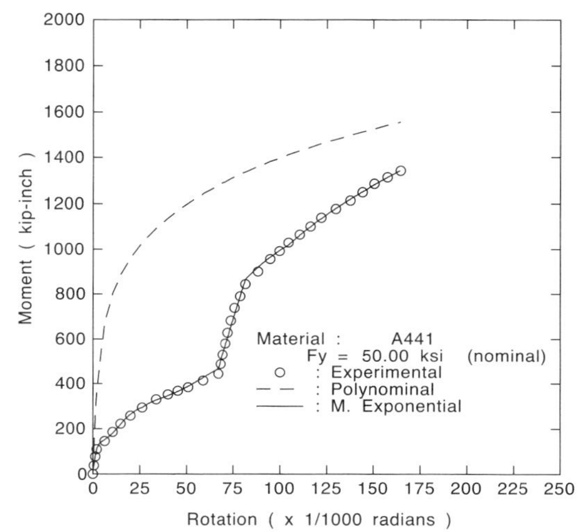

A.2 – 112

```
Moment-rotation prediction equations      ( R : X 1/1000 radians )

Frye and Morris polynominal model  :     R = Sum ( Ai X ( K*Bm )**Pi X 10**Qi )
     xd = 11.500000"    g =  5.500000"   t =  0.500000"
     A1 =  3.660000   A2 =  1.150000   A3 =  4.570000     K =  0.012890
     P1 =         1   P2 =         3   P3 =         5    Q1 =        -1     Q2 =        -3     Q3 =        -5

Modified exponential model  :
     Bm = Sum ( Ai X ( 1 - exp( - R/( 2*i*C ) ) ) )  +       Sum ( RKj X ( R - Rj0 ) )  +  Bm0
      C =   0.14958333E+01             Bmo=   0.00000000E+00          Nexp=  6         Nliner=  6
     AI =   0.32445663E+03   0.50684855E+04  -0.50920715E+05   0.15246565E+06  -0.18494286E+06   0.79142597E+05
    Rj0 =           0.0000           2.0000          26.2000          67.0000          81.6000
                  104.6000
    RKj =   0.97980893E+01  -0.17088897E+02   0.28391690E+01   0.23420307E+02  -0.17611771E+02
            0.26799252E+01
```

No	Rotation (radians) X 1/1000	Moment (kip-inch) Expri.	Moment (kip-inch) Poly.	Moment (kip-inch) M.Expo.	Connection stiffness (kip-inch) X 1000 Expri.	Connection stiffness (kip-inch) X 1000 Poly.	Connection stiffness (kip-inch) X 1000 M.Expo.
1	0.00	0.00	0.00	0.00	0.7790E+02	0.2120E+03	0.6799E+02
3	1.20	77.17	243.87	75.44	0.4842E+02	0.1837E+03	0.5041E+02
5	6.20	147.00	674.18	152.04	0.8986E+01	0.4021E+02	0.5278E+01
7	14.80	221.23	886.49	219.99	0.7533E+01	0.1647E+02	0.9468E+01
9	26.20	294.00	1028.21	294.30	0.5437E+01	0.9671E+01	0.2975E+01
11	40.00	351.32	1137.80	346.62	0.3178E+01	0.6643E+01	0.3240E+01
13	50.70	384.04	1201.75	387.86	0.3320E+01	0.5407E+01	0.4517E+01
15	67.00	444.67	1279.69	467.77	0.3322E+02	0.4258E+01	0.4740E+01
17	69.30	530.66	1289.41	532.30	0.3672E+02	0.4137E+01	0.2794E+02
19	72.00	627.68	1300.49	607.34	0.3326E+02	0.4004E+01	0.2763E+02
21	75.70	737.21	1314.84	708.65	0.2520E+02	0.3839E+01	0.2712E+02
23	81.60	844.51	1336.77	865.97	0.1517E+02	0.3603E+01	0.2619E+02
25	94.70	955.50	1381.07	963.97	0.7875E+01	0.3179E+01	0.6404E+01
27	104.60	1029.00	1411.54	1020.07	0.6908E+01	0.2922E+01	0.4978E+01
29	116.27	1102.50	1443.87	1101.47	0.6300E+01	0.2678E+01	0.6367E+01
31	129.75	1177.10	1478.34	1180.05	0.4947E+01	0.2444E+01	0.5369E+01
33	143.90	1250.60	1511.74	1251.20	0.5485E+01	0.2241E+01	0.4746E+01
35	157.45	1314.18	1541.01	1313.04	0.3961E+01	0.2080E+01	0.4413E+01

II - 57

Connection type : Double web-angle connections
Mode : All bolted

Tested by : L.E.Thompson et al. (1970) U.S.A
Test Id. : C2-2 ALT

Column : W10X49
Beam : W21X62
Angle : 4 X 3.5 X 1/2

Fasteners: A325- -3/4"D
15/16" Oversize holes
Material : A441
Fy = -- ksi
Fu = -- ksi

Major parameters

lp = 11.5000"	lu = -- "	ll = -- "	ta = 0.5000"
gb = 2.2500"	gc = 2.5500"	cu = -- "	cl = -- "
pb = -- "	pc = -- "	qb = -- "	qc = -- "
nb = - X 2	nc = - X -		

Remark 1)
2)

No	Mcment (k-in)	Rotation (radians) X 1/1000	No	Moment (k-in)	Rotation (radians) X 1/1000
1	0.00	0.00	26	742.34	96.70
2	40.43	0.10	27	792.32	99.30
3	76.63	0.60	28	842.30	101.90
4	112.82	1.10	29	898.90	107.20
5	122.09	3.00	30	955.50	112.50
6	131.36	4.90	31	992.25	117.40
7	140.65	6.80	32	1029.00	122.30
8	149.94	8.70	33	1065.75	129.65
9	166.85	13.60	34	1102.50	137.00
10	183.76	18.40	35	1138.15	142.90
11	220.50	24.90	36	1173.79	148.80
12	257.99	30.40	37	1215.69	156.85
13	294.37	37.95	38	1257.59	164.90
14	330.76	45.50	39	1280.00	170.30
15	349.13	52.45	40	1302.42	175.70
16	367.50	59.40			
17	390.29	65.80			
18	408.42	72.90			
19	426.54	80.00			
20	444.67	87.10			
21	433.26	88.25			
22	521.84	89.40			
23	578.44	91.10			
24	635.04	92.80			
25	688.69	94.75			

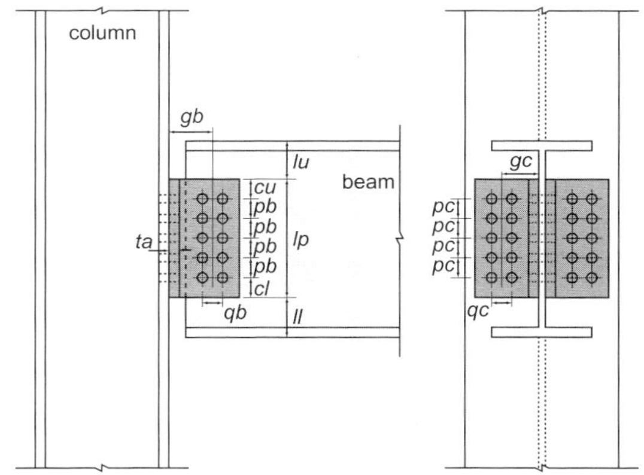

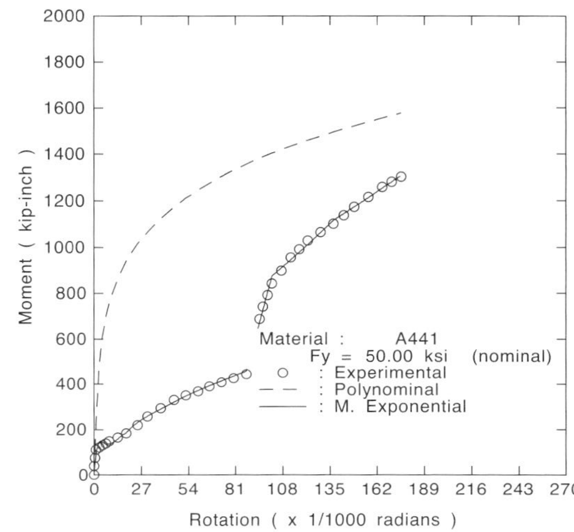

A.2 – 114

```
Moment-rotation prediction equations      ( R : X 1/1000 radians )

Frye and Morris polynominal model   :     R = Sum ( Ai X ( K*Bm )**Pi X 10**Qi )
     xd = 11.500000"   g =  5.500000"   t =  0.500000"
     A1 =  3.660000   A2 =  1.150000   A3 =  4.570000    K =  0.012890
     P1 =         1    P2 =         3    P3 =         5    Q1 =        -1    Q2 =        -3    Q3 =        -5

Modified exponential model  :
     Bm = Sum ( Ai X ( 1 - exp( - R/( 2*i*C ) ) ) )  +      Sum ( RKj X ( R - Rj0 ) )  +  Bm0
       C =   0.14891667E+01           Bmo=   0.00000000E+00           Nexp=  6        Nliner=  4
      AI =   0.16674541E+04  -0.12581421E+05   0.47139478E+05  -0.91151975E+05   0.86188246E+05  -0.31164474E+05
     Rj0 =          18.4000          87.1000         101.9000         137.0000
     RKj =   0.45405137E+01   0.24957477E+02  -0.21959573E+02  -0.24998683E+01
```

No	Rotation (radians) X 1/1000	Moment (kip-inch) Expri.	Moment (kip-inch) Poly.	Moment (kip-inch) M.Expo.	Connection stiffness (kip-inch) X 1000 Expri.	Connection stiffness (kip-inch) X 1000 Poly.	Connection stiffness (kip-inch) X 1000 M.Expo.
1	0.00	0.00	0.00	0.00	0.4043E+03	0.2120E+03	0.1160E+03
3	0.60	76.63	126.03	57.75	0.7240E+02	0.2060E+03	0.7842E+02
5	3.00	122.09	484.51	142.60	0.4879E+01	0.9147E+02	0.8418E+01
7	6.80	140.65	697.11	138.60	0.4888E+01	0.3635E+02	-0.2193E+01
9	13.60	166.85	865.92	160.07	0.3487E+01	0.1787E+02	0.6125E+01
11	24.90	220.50	1015.43	232.41	0.6283E+01	0.1012E+02	0.5533E+01
13	37.95	294.37	1123.91	289.74	0.4819E+01	0.6955E+01	0.4021E+01
15	52.45	349.13	1211.08	346.80	0.2643E+01	0.5251E+01	0.3726E+01
17	65.80	390.29	1274.54	392.52	0.3083E+01	0.4324E+01	0.3154E+01
19	80.00	426.54	1330.96	435.50	0.2553E+01	0.3664E+01	0.2994E+01
21	88.25	483.26	1359.97	489.34	0.3355E+02	0.3373E+01	0.2808E+02
23	91.10	578.44	1369.52	569.45	0.3329E+02	0.3283E+01	0.2814E+02
25	94.75	688.69	1381.41	672.33	0.2751E+02	0.3176E+01	0.2824E+02
27	99.30	792.32	1395.40	801.09	0.1922E+02	0.3055E+01	0.2836E+02
29	107.20	898.90	1418.77	909.65	0.1068E+02	0.2865E+01	0.6624E+01
31	117.40	992.25	1446.89	978.57	0.7500E+01	0.2656E+01	0.6883E+01
33	129.65	1065.75	1478.10	1064.44	0.5000E+01	0.2445E+01	0.7123E+01
35	142.90	1138.15	1509.45	1145.32	0.6042E+01	0.2254E+01	0.4798E+01
37	156.85	1215.69	1539.47	1213.11	0.5205E+01	0.2088E+01	0.4909E+01
39	170.30	1280.00	1566.75	1279.58	0.4151E+01	0.1950E+01	0.4970E+01

II - 58

Connection type : Double web-angle connections
Mode : All bolted

Tested by : L.E.Thompson et al. (1970) U.S.A
Test Id. : C2-2 ALT

Column : W10X49
Beam : W21X62
Angle : 4 X 3.5 X 1/2

Fasteners: A325- -3/4"D
15/16" Oversize holes
Material : A441
Fy = -- ksi
Fu = -- ksi

Major parameters

lp = 11.5000"	lu = -- "	ll = -- "	ta = 0.5000"
gb = 2.2500"	gc = 2.5500"	cu = -- "	cl = -- "
pb = -- "	pc = -- "	qb = -- "	qc = -- "
nb = - X 2	nc = - X -		

Remark 1)
2)

No	Moment (k-in)	Rotation (radians) X 1/1000	No	Moment (k-in)	Rotation (radians) X 1/1000
1	0.00	0.00	26	992.25	106.90
2	40.43	0.50	27	1029.00	113.20
3	76.63	1.10	28	1065.75	120.55
4	112.82	1.70	29	1102.50	127.90
5	149.94	3.00	30	1138.15	135.55
6	183.76	7.50	31	1173.79	143.20
7	220.50	12.50	32	1215.69	152.30
8	257.99	18.40	33	1257.59	161.40
9	294.37	26.25	34	1272.53	169.73
10	330.76	34.10	35	1287.48	178.07
11	349.13	41.25	36	1302.42	186.40
12	367.50	48.40			
13	390.29	56.50			
14	417.48	65.45			
15	444.67	74.40			
16	483.26	75.75			
17	521.84	77.10			
18	578.44	78.95			
19	635.04	80.80			
20	688.69	83.00			
21	742.34	85.20			
22	792.32	87.90			
23	842.30	90.60			
24	898.90	95.60			
25	955.50	100.60			

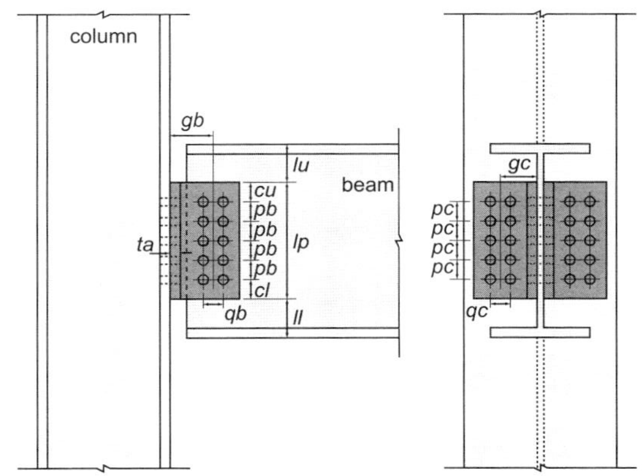

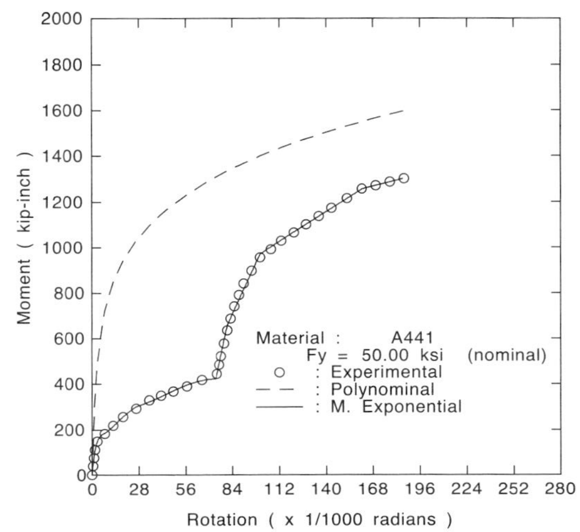

```
Moment-rotation prediction equations      ( R : X 1/1000 radians )

Frye and Morris polynominal model   :     R = Sum ( Ai X ( K*Bm )**Pi X 10**Qi )
     xd = 11.500000"    g =  5.500000"   t =  0.500000"
     A1 =  3.660000   A2 =  1.150000   A3 =  4.570000    K =  0.012890
     P1 =        1    P2 =         3   P3 =         5    Q1 =        -1    Q2 =        -3    Q3 =        -5

Modified exponential model  :
     Bm = Sum ( Ai X ( 1 - exp( - R/( 2*i*C ) ) ) )  +     Sum ( RKj X ( R - Rj0 ) )  +  Bm0
      C =   0.16783333E+01              Bmo=   0.00000000E+00          Nexp=  6        Nliner=  5
     AI =   0.13379513E+04  -0.11229766E+05   0.44252744E+05  -0.79501320E+05   0.63086998E+05  -0.17373702E+05
    Rj0 =          26.2500          74.0000          80.8000         100.6000         161.4000
    RKj =  -0.24466896E+01   0.34559634E+02  -0.17296991E+02  -0.10087633E+02  -0.28682584E+01
```

No	Rotation (radians) X 1/1000	Moment (kip-inch) Expri.	Poly.	M.Expo.	Connection stiffness (kip-inch) X 1000 Expri.	Poly.	M.Expo.
1	0.00	0.00	0.00	0.00	0.8086E+02	0.2120E+03	0.9546E+02
3	1.10	76.63	225.22	81.11	0.6033E+02	0.1886E+03	0.5530E+02
5	3.00	149.94	484.54	147.77	0.2384E+02	0.9146E+02	0.2030E+02
7	12.50	220.50	845.45	216.80	0.6892E+01	0.1940E+02	0.6963E+01
9	26.25	294.37	1028.55	302.21	0.4635E+01	0.9659E+01	0.5053E+01
11	41.25	349.13	1146.01	345.46	0.2570E+01	0.6466E+01	0.3407E+01
13	56.50	390.29	1231.67	395.66	0.2920E+01	0.4925E+01	0.2733E+01
15	74.40	444.67	1309.82	436.15	0.2523E+02	0.3896E+01	0.3478E+02
17	77.10	521.84	1320.25	529.60	0.2943E+02	0.3779E+01	0.3444E+02
19	80.80	635.04	1334.10	656.21	0.2776E+02	0.3631E+01	0.3401E+02
21	85.20	742.34	1349.60	728.73	0.2175E+02	0.3473E+01	0.1627E+02
23	90.60	842.30	1367.79	815.32	0.1599E+02	0.3299E+01	0.1582E+02
25	100.60	955.50	1399.35	970.41	0.8892E+01	0.3022E+01	0.1525E+02
27	113.20	1029.00	1435.57	1032.81	0.5449E+01	0.2738E+01	0.4795E+01
29	127.90	1102.50	1473.80	1101.95	0.4833E+01	0.2473E+01	0.4646E+01
31	143.20	1173.79	1509.88	1172.81	0.4634E+01	0.2252E+01	0.4629E+01
33	161.40	1257.59	1548.86	1257.30	0.3137E+01	0.2039E+01	0.4658E+01
35	178.07	1287.48	1581.81	1287.38	0.1793E+01	0.1879E+01	0.1818E+01

II - 59

Connection type : Double web-angle connections
Mode : All bolted

Tested by : L.E.Thompson et al. (1970) U.S.A
Test Id. : C2-3 ALT

Column : W10X49
Beam : W21X62
Angle : 4 X 3.5 X 1/2

Fasteners: A325- -3/4"D
15/16" Oversize holes
Material : A441
Fy = -- ksi
Fu = -- ksi

Major parameters

lp = 11.5000"	lu = -- "	ll = -- "	ta = 0.5000"
gb = 2.2500"	gc = 2.5500"	cu = -- "	cl = -- "
pb = -- "	pc = -- "	qb = -- "	qc = -- "
nb = - X 2	nc = - X -		

Remark 1)
2)

No	Moment (k-in)	Rotation (radians) X 1/1000	No	Moment (k-in)	Rotation (radians) X 1/1000
1	0.00	0.00	26	811.44	105.50
2	43.37	0.40	27	865.10	109.25
3	78.46	1.25	28	918.76	113.00
4	113.56	2.10	29	957.34	116.40
5	121.92	4.50	30	995.93	119.80
6	130.28	7.00	31	1031.57	125.30
7	147.00	11.80	32	1067.22	130.80
8	165.01	20.45	33	1103.24	137.85
9	183.01	29.10	34	1139.26	144.90
10	220.50	33.60	35	1176.01	153.50
11	257.25	38.65	36	1212.76	162.10
12	294.00	43.70	37	1250.24	169.25
13	330.75	51.40	38	1287.72	176.40
14	367.50	59.10	39	1312.71	181.95
15	394.33	65.55	40	1337.70	187.50
16	421.15	72.00			
17	446.39	79.77			
18	471.62	87.53			
19	496.86	95.30			
20	533.98	96.50			
21	571.09	97.70			
22	627.68	99.50			
23	668.48	100.65			
24	709.27	101.80			
25	760.36	103.65			

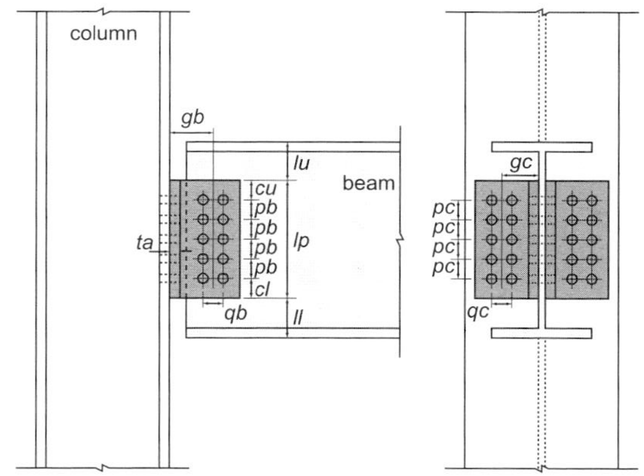

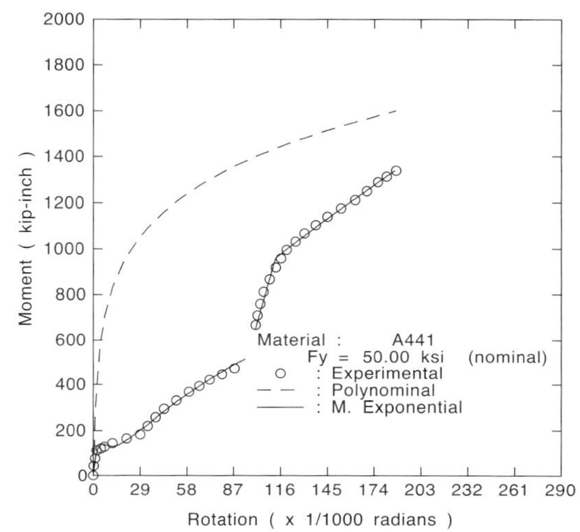

A.2 – 118

```
Moment-rotation prediction equations     ( R : X 1/1000 radians )

Frye and Morris polynominal model  :    R = Sum ( Ai X ( K*Bm )**Pi X 10**Qi )
     xd = 11.500000"   g =  5.500000"   t =  0.500000"
     A1 =  3.660000   A2 =  1.150000   A3 =  4.570000    K =  0.012890
     P1 =        1    P2 =        3    P3 =        5    Q1 =       -1    Q2 =       -3    Q3 =       -5

Modified exponential model  :
     Bm = Sum ( Ai X ( 1 - exp( - R/( 2*i*C ) ) ) )  +      Sum ( RKj X ( R - Rj0 ) )  +  Bm0
       C =  0.16625000E+01               Bmo=   0.00000000E+00           Nexp=  6         Nliner=  3
      AI =  0.48863739E+03   0.12591756E+04  -0.15248111E+05   0.43430506E+05  -0.51799902E+05   0.22329147E+05
     Rj0 =         33.6000          95.3000         113.0000
     RKj =  0.18144074E+01   0.21731977E+02  -0.18800917E+02
```

No	Rotation (radians) X 1/1000	Moment (kip-inch) Expri.	Moment (kip-inch) Poly.	Moment (kip-inch) M.Expo.	Connection stiffness (kip-inch) X 1000 Expri.	Connection stiffness (kip-inch) X 1000 Poly.	Connection stiffness (kip-inch) X 1000 M.Expo.
1	0.00	0.00	0.00	0.00	0.1084E+03	0.2120E+03	0.7660E+02
3	1.25	78.46	253.02	73.07	0.4129E+02	0.1810E+03	0.4292E+02
5	4.50	121.92	592.80	137.22	0.3417E+01	0.5762E+02	0.5037E+01
7	11.80	147.00	831.49	132.28	0.2983E+01	0.2054E+02	0.3346E-01
9	29.10	183.01	1054.82	203.11	0.6192E+01	0.8804E+01	0.4983E+01
11	38.65	257.25	1128.69	254.86	0.7277E+01	0.6846E+01	0.5785E+01
13	51.40	330.75	1205.52	323.56	0.4773E+01	0.5343E+01	0.5118E+01
15	65.55	394.33	1273.45	393.53	0.4159E+01	0.4338E+01	0.4768E+01
17	79.77	446.39	1330.11	457.92	0.3249E+01	0.3673E+01	0.4256E+01
19	95.30	496.86	1382.97	518.68	0.2723E+02	0.3162E+01	0.3562E+01
21	97.70	571.09	1390.52	579.25	0.3113E+02	0.3096E+01	0.2519E+02
23	100.65	668.48	1399.60	653.38	0.3547E+02	0.3020E+01	0.2506E+02
25	103.65	760.36	1408.60	728.38	0.2761E+02	0.2946E+01	0.2494E+02
27	109.25	865.10	1424.59	867.44	0.1431E+02	0.2820E+01	0.2473E+02
29	116.40	957.34	1444.49	979.43	0.1135E+02	0.2673E+01	0.5685E+01
31	125.30	1031.57	1467.58	1028.89	0.6481E+01	0.2514E+01	0.5439E+01
33	137.85	1103.24	1497.64	1095.44	0.5109E+01	0.2324E+01	0.5183E+01
35	153.50	1176.01	1532.41	1174.84	0.4273E+01	0.2126E+01	0.4982E+01
37	169.25	1250.24	1564.81	1252.32	0.5242E+01	0.1960E+01	0.4868E+01
39	181.95	1312.71	1588.86	1313.80	0.4503E+01	0.1847E+01	0.4817E+01

II - 60

Connection type : Double web-angle connections
Mode : All bolted

Tested by : L.E.Thompson et al. (1970) U.S.A
Test Id. : C2-3 ALT

Column : W10X49
Beam : W21X62
Angle : 4 X 3.5 X 1/2

Fasteners: A325- -3/4"D
15/16" Oversize holes
Material : A441
Fy = -- ksi
Fu = -- ksi

Major parameters

lp = 11.5000"	lu = -- "	ll = -- "	ta = 0.5000"
gb = 2.2500"	gc = 2.5500"	cu = -- "	cl = -- "
pb = -- "	pc = -- "	qb = -- "	qc = -- "
nb = - X 2	nc = - X -		

Remark 1)
2)

No	Moment (k-in)	Rotation (radians) X 1/1000	No	Moment (k-in)	Rotation (radians) X 1/1000
1	0.00	0.00	26	957.34	89.95
2	43.37	0.50	27	995.93	95.10
3	78.46	1.15	28	1031.57	100.20
4	113.56	1.80	29	1067.22	105.30
5	147.00	3.90	30	1103.24	110.60
6	183.01	8.10	31	1139.26	115.90
7	220.50	13.30	32	1176.01	123.95
8	257.25	18.65	33	1212.76	132.00
9	294.00	24.00	34	1237.74	137.73
10	318.50	29.53	35	1262.73	143.47
11	343.00	35.07	36	1287.72	149.20
12	367.50	40.60	37	1312.71	157.15
13	394.33	46.30	38	1337.70	165.10
14	421.15	52.00			
15	459.01	58.05			
16	496.86	64.10			
17	533.98	67.05			
18	571.09	70.00			
19	627.68	71.50			
20	668.48	72.65			
21	709.27	73.80			
22	760.36	75.65			
23	811.44	77.50			
24	865.10	81.15			
25	918.76	84.80			

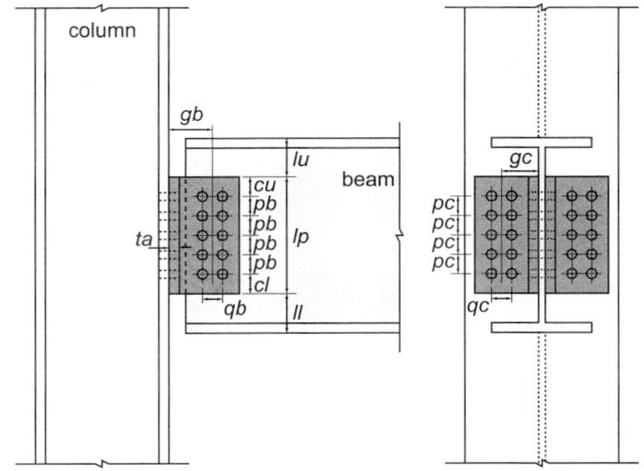

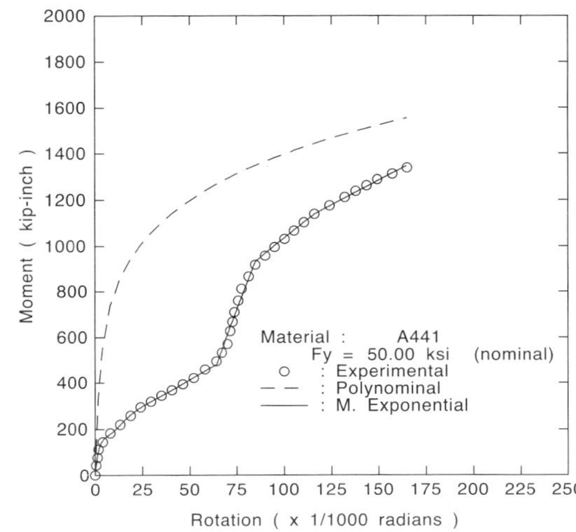

```
Moment-rotation prediction equations     ( R : X 1/1000 radians )

Frye and Morris polynominal model  :    R = Sum ( Ai X ( K*Bm )**Pi X 10**Qi )
     xd = 11.500000"   g =  5.500000"   t =  0.500000"
     A1 =  3.660000   A2 =  1.150000   A3 =  4.570000    K =  0.012890
     P1 =        1    P2 =        3    P3 =        5    Q1 =       -1    Q2 =       -3    Q3 =       -5

Modified exponential model  :
     Bm = Sum ( Ai X ( 1 - exp( - R/( 2*i*C ) ) ) )  +      Sum ( RKj X ( R - Rj0 ) )  +  Bm0
      C =   0.15008333E+01              Bmo=    0.00000000E+00             Nexp=  6          Nliner=  4
     AI =   0.98329307E+03  -0.57827983E+04   0.15678175E+05  -0.17367367E+05   0.47486922E+04   0.21461833E+04
    Rj0 =           3.9000          64.1000          84.8000         115.9000
    RKj =   0.21789294E+01   0.17970239E+02  -0.14070639E+02  -0.19224723E+01
```

No	Rotation (radians) X 1/1000	Moment (kip-inch) Expri.	Moment (kip-inch) Poly.	Moment (kip-inch) M.Expo.	Connection stiffness (kip-inch) X 1000 Expri.	Connection stiffness (kip-inch) X 1000 Poly.	Connection stiffness (kip-inch) X 1000 M.Expo.
1	0.00	0.00	0.00	0.00	0.8674E+02	0.2120E+03	0.9446E+02
3	1.15	78.46	234.60	81.23	0.5399E+02	0.1862E+03	0.5080E+02
5	3.90	147.00	555.30	151.02	0.1348E+02	0.6790E+02	0.1002E+02
7	13.30	220.50	860.50	219.64	0.7042E+01	0.1826E+02	0.8102E+01
9	24.00	294.00	1006.03	292.09	0.5669E+01	0.1047E+02	0.5157E+01
11	35.07	343.00	1103.11	342.50	0.4428E+01	0.7457E+01	0.4456E+01
13	46.30	394.33	1177.02	395.45	0.4706E+01	0.5848E+01	0.4923E+01
15	58.05	459.01	1239.21	452.96	0.6257E+01	0.4813E+01	0.4743E+01
17	67.05	533.98	1280.15	546.60	0.1258E+02	0.4252E+01	0.2223E+02
19	71.50	627.68	1298.38	644.92	0.3645E+02	0.4029E+01	0.2196E+02
21	73.80	709.27	1307.58	695.27	0.3246E+02	0.3921E+01	0.2182E+02
23	77.50	811.44	1321.94	775.62	0.2327E+02	0.3761E+01	0.2161E+02
25	84.80	918.76	1348.36	931.99	0.1171E+02	0.3486E+01	0.2124E+02
27	95.10	995.93	1382.34	1003.62	0.7240E+01	0.3168E+01	0.6765E+01
29	105.30	1067.22	1413.28	1071.15	0.6894E+01	0.2908E+01	0.6497E+01
31	115.90	1139.26	1443.18	1139.01	0.5910E+01	0.2682E+01	0.6321E+01
33	132.00	1212.76	1483.80	1208.55	0.4444E+01	0.2409E+01	0.4259E+01
35	143.47	1262.73	1510.48	1257.08	0.4358E+01	0.2248E+01	0.4211E+01
37	157.15	1312.71	1540.10	1314.47	0.3143E+01	0.2085E+01	0.4181E+01

II - 61

Connection type : Double web-angle connections
Mode : All bolted

Tested by : L.E.Thompson et al. (1970) U.S.A
Test Id. : D2-1 ALT

Column : W10X49
Beam : W21X62
Angle : 4 X 3.5 X 1/2

Fasteners: A325- -3/4"D
15/16" Oversize holes
Material : A441
Fy = -- ksi
Fu = -- ksi

Major parameters

lp = 11.5000"	lu = -- "	ll = -- "	ta = 0.5000"
gb = 2.2500"	gc = 1.8000"	cu = -- "	cl = -- "
pb = -- "	pc = -- "	qb = -- "	qc = -- "
nb = - X 2	nc = - X -		

Remark 1)
2)

No	Moment (k-in)	Rotation (radians) X 1/1000
1	0.00	0.00
2	38.22	0.30
3	76.44	0.60
4	144.79	3.00
5	164.27	6.15
6	183.76	9.30
7	220.13	13.75
8	256.51	18.20
9	294.00	23.10
10	330.76	28.90
11	349.13	32.20
12	367.50	35.50
13	404.26	40.60
14	422.63	44.25
15	441.00	47.90
16	456.43	52.00
17	471.86	56.10
18	493.18	60.58
19	514.50	65.05
20	535.82	69.53
21	557.14	74.00
22	596.82	74.60
23	636.50	75.20
24	692.00	76.45
25	747.49	77.70

No	Moment (k-in)	Rotation (radians) X 1/1000
26	800.05	78.85
27	852.60	80.00
28	905.89	81.40
29	959.17	82.80
30	1012.83	84.65
31	1066.49	86.50
32	1121.24	89.10
33	1176.00	91.70
34	1208.71	95.45
35	1241.41	99.20
36	1275.47	103.33
37	1309.52	107.47
38	1343.58	111.60
39	1396.50	115.80
40	1433.26	119.20

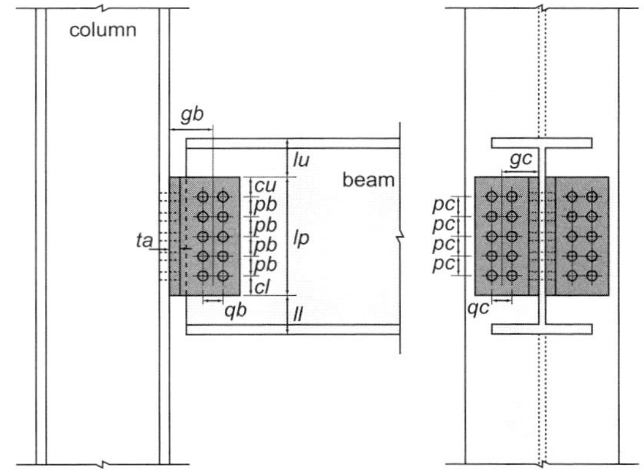

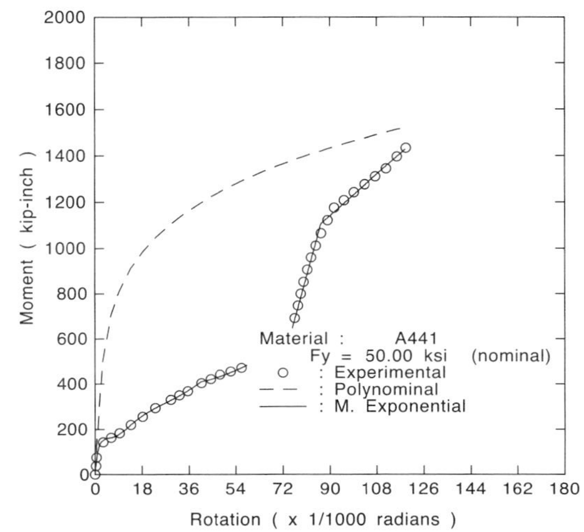

```
Moment-rotation prediction equations      ( R : X 1/1000 radians )

Frye and Morris polynominal model  :     R = Sum ( Ai X ( K*Bm )**Pi X 10**Qi )
     xd = 11.500000"   g =  4.000000"   t =  0.500000"
     A1 =  3.660000   A2 =  1.150000   A3 =  4.570000    K =  0.012288
     P1 =        1    P2 =        3    P3 =        5    Q1 =       -1    Q2 =       -3    Q3 =       -5

Modified exponential model  :
     Bm = Sum ( Ai X ( 1 - exp( - R/( 2*i*C ) ) ) )  +     Sum ( RKj X ( R - Rj0 ) )  +  Bm0
       C =  0.10683333E+01            Bmo=  0.00000000E+00          Nexp=  6        Nliner=  4
      AI =  0.71631411E+03  -0.16926351E+03  -0.89711221E+04   0.23544237E+05  -0.20435029E+05   0.51821644E+04
     Rj0 =          3.0000          40.6000          74.0000          86.5000
     RKj =  0.13091245E+02  -0.64977392E+01   0.35607281E+02  -0.32301638E+02
```

No	Rotation (radians) X 1/1000	Moment (kip-inch) Expri.	Moment (kip-inch) Poly.	Moment (kip-inch) M.Expo.	Connection stiffness (kip-inch) X 1000 Expri.	Connection stiffness (kip-inch) X 1000 Poly.	Connection stiffness (kip-inch) X 1000 M.Expo.
1	0.00	0.00	0.00	0.00	0.1274E+03	0.2223E+03	0.1423E+03
3	0.60	76.44	132.19	68.66	0.1164E+03	0.2160E+03	0.8969E+02
5	6.15	164.27	705.05	164.00	0.6185E+01	0.4255E+02	0.2406E+01
7	13.75	220.13	911.08	220.78	0.8175E+01	0.1854E+02	0.9724E+01
9	23.10	294.00	1045.17	293.90	0.7049E+01	0.1138E+02	0.6034E+01
11	32.20	349.13	1133.78	347.42	0.5567E+01	0.8437E+01	0.6408E+01
13	40.60	404.26	1197.61	410.52	0.5940E+01	0.6878E+01	0.8682E+01
15	47.90	441.00	1244.27	433.31	0.4436E+01	0.5956E+01	0.3975E+01
17	56.10	471.86	1289.85	471.88	0.4242E+01	0.5198E+01	0.5315E+01
19	65.05	514.50	1333.48	523.36	0.4764E+01	0.4580E+01	0.6093E+01
21	74.00	557.14	1372.26	579.67	0.5888E+02	0.4105E+01	0.6440E+01
23	75.20	636.50	1377.17	630.15	0.5909E+02	0.4049E+01	0.4207E+02
25	77.70	747.49	1387.22	735.39	0.4507E+02	0.3938E+01	0.4212E+02
27	80.00	852.60	1396.22	832.30	0.4225E+02	0.3842E+01	0.4215E+02
29	82.80	959.17	1406.90	950.36	0.3416E+02	0.3731E+01	0.4218E+02
31	86.50	1066.49	1420.25	1106.47	0.2570E+02	0.3598E+01	0.4220E+02
33	91.70	1176.00	1438.72	1158.00	0.1601E+02	0.3424E+01	0.9917E+01
35	99.20	1241.41	1463.55	1232.41	0.8492E+01	0.3205E+01	0.9923E+01
37	107.47	1309.52	1489.19	1314.42	0.8239E+01	0.2998E+01	0.9919E+01
39	115.80	1396.50	1513.39	1397.06	0.1161E+02	0.2817E+01	0.9913E+01

II - 62

Connection type : Double web-angle connections
Mode : All bolted

Tested by : L.E.Thompson et al. (1970) U.S.A
Test Id. : D2-1 ALT

Column : W10X49
Beam : W21X62
Angle : 4 X 3.5 X 1/2

Fasteners: A325- -3/4"D
15/16" Oversize holes
Material : A441
Fy = -- ksi
Fu = -- ksi

Major parameters

lp = 11.5000"	lu = -- "	ll = -- "	ta = 0.5000"
gb = 2.2500"	gc = 1.8000"	cu = -- "	cl = -- "
pb = -- "	pc = -- "	qb = -- "	qc = -- "
nb = - X 2	nc = - X -		

Remark 1)
2)

No	Moment (k-in)	Rotation (radians) X 1/1000	No	Moment (k-in)	Rotation (radians) X 1/1000
1	0.00	0.00	26	905.89	74.50
2	33.22	0.40	27	959.17	75.80
3	76.44	0.80	28	1012.83	77.65
4	144.79	5.70	29	1066.49	79.50
5	183.76	9.20	30	1121.24	82.05
6	220.13	14.40	31	1176.00	84.60
7	256.51	19.60	32	1208.71	88.95
8	294.00	23.00	33	1241.41	93.30
9	330.76	27.20	34	1275.47	97.43
10	367.50	32.60	35	1309.52	101.57
11	404.26	38.00	36	1343.58	105.70
12	422.63	41.30	37	1370.04	109.15
13	441.00	44.60	38	1396.50	112.60
14	448.72	50.05	39	1414.88	115.75
15	456.43	55.50	40	1433.26	118.90
16	464.15	60.95			
17	471.86	66.40			
18	514.50	67.05			
19	557.14	67.70			
20	596.82	68.40			
21	636.50	69.10			
22	692.00	70.05			
23	747.49	71.00			
24	800.05	72.10			
25	852.60	73.20			

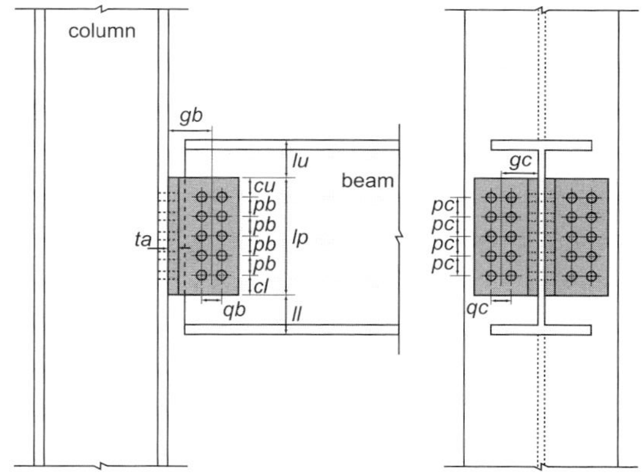

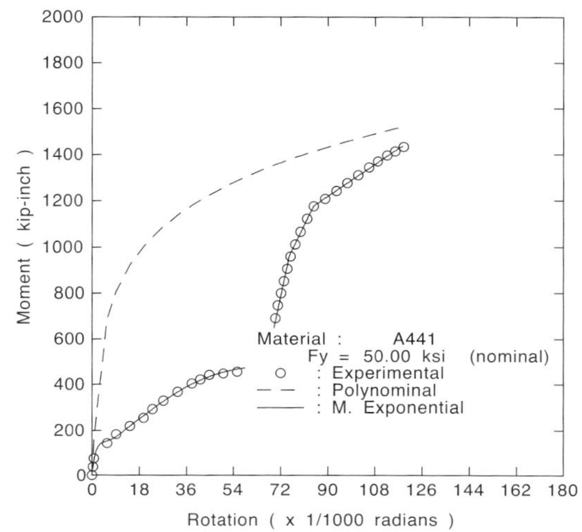

```
Moment-rotation prediction equations      ( R : X 1/1000 radians )

Frye and Morris polynominal model  :    R = Sum ( Ai X ( K*Bm )**Pi X 10**Qi )
     xd = 11.500000"   g =  4.000000"   t =  0.500000"
     A1 =  3.660000   A2 =  1.150000   A3 =  4.570000    K =  0.012288
     P1 =        1    P2 =        3    P3 =        5    Q1 =       -1     Q2 =       -3     Q3 =       -5

Modified exponential model  :
     Bm = Sum ( Ai X ( 1 - exp( - R/( 2*i*C ) ) ) )  +      Sum ( RKj X ( R - Rj0 ) )  +  Bm0
       C =   0.10908333E+01                Bmo=   0.00000000E+00         Nexp=  6        Nliner=  3
      AI =   0.80183409E+03  -0.48353336E+04   0.16714098E+05  -0.27293535E+05   0.18079608E+05  -0.29743531E+04
     Rj0 =          66.4000          75.3000          84.6000
     RKj =   0.53285117E+02  -0.30437815E+02  -0.15235249E+02
```

No	Rotation (radians) X 1/1000	Moment (kip-inch) Expri.	Poly.	M.Expo.	Connection stiffness (kip-inch) X 1000 Expri.	Poly.	M.Expo.
1	0.00	0.00	0.00	0.00	0.9555E+02	0.2223E+03	0.1157E+03
3	0.80	76.44	174.87	70.50	0.8939E+02	0.2104E+03	0.6490E+02
5	9.20	183.76	808.85	173.96	0.9468E+01	0.2771E+02	0.7798E+01
7	19.60	256.51	1002.16	265.34	0.9433E+01	0.1324E+02	0.8501E+01
9	27.20	330.76	1088.36	327.44	0.7900E+01	0.9811E+01	0.7802E+01
11	38.00	404.26	1179.21	401.69	0.6037E+01	0.7288E+01	0.5728E+01
13	44.60	441.00	1224.00	434.26	0.4002E+01	0.6337E+01	0.4153E+01
15	55.50	456.43	1286.72	467.36	0.1416E+01	0.5246E+01	0.2071E+01
17	66.40	471.86	1339.61	482.76	0.5876E+02	0.4501E+01	0.8904E+00
19	67.70	557.14	1345.43	553.13	0.6131E+02	0.4427E+01	0.5408E+02
21	69.10	636.50	1351.60	628.78	0.5742E+02	0.4350E+01	0.5399E+02
23	71.00	747.49	1359.82	731.27	0.5348E+02	0.4251E+01	0.5389E+02
25	73.20	852.60	1369.11	849.70	0.4467E+02	0.4141E+01	0.5378E+02
27	75.80	959.17	1379.80	974.17	0.3604E+02	0.4020E+01	0.2324E+02
29	79.50	1066.49	1394.15	1059.93	0.2584E+02	0.3863E+01	0.2312E+02
31	84.60	1176.00	1413.60	1177.57	0.1632E+02	0.3664E+01	0.2302E+02
33	93.30	1241.41	1444.25	1244.77	0.7888E+01	0.3374E+01	0.7680E+01
35	101.57	1309.52	1471.10	1308.06	0.8239E+01	0.3142E+01	0.7638E+01
37	109.15	1370.04	1494.14	1365.91	0.7670E+01	0.2959E+01	0.7621E+01
39	115.75	1414.88	1513.06	1416.18	0.5834E+01	0.2819E+01	0.7615E+01

II - 63

Connection type : Double web-angle connections
Mode : All bolted

Tested by : L.E.Thompson et al. (1970) U.S.A
Test Id. : D2-2 ALT

Column : W10X49
Beam : W21X62
Angle : 4 X 3.5 X 1/2

Fasteners: A325- -3/4"D
15/16" Oversize holes
Material : A441
Fy = -- ksi
Fu = -- ksi

Major parameters

lp = 11.5000"	lu = -- "	ll = -- "	ta = 0.5000"
gb = 2.2500"	gc = 1.8000"	cu = -- "	cl = -- "
pb = -- "	pc = -- "	qb = -- "	qc = -- "
nb = - X 2	nc = - X -		

Remark 1)
2)

No	Moment (k-in)	Rotation (radians) X 1/1000
1	0.00	0.00
2	36.68	0.30
3	74.20	0.90
4	111.72	1.50
5	118.58	7.87
6	125.44	14.23
7	132.30	20.60
8	153.03	26.65
9	183.76	32.70
10	221.42	36.15
11	259.09	39.60
12	313.30	45.50
13	367.50	51.40
14	385.51	57.35
15	403.51	63.30
16	421.52	69.25
17	439.52	75.20
18	502.49	76.37
19	565.46	77.53
20	628.43	78.70
21	683.18	79.75
22	737.94	80.80
23	795.64	82.10
24	853.33	83.40
25	906.99	84.85

No	Moment (k-in)	Rotation (radians) X 1/1000
26	960.65	86.30
27	1013.93	88.20
28	1067.22	90.10
29	1104.34	93.40
30	1141.45	96.70
31	1177.10	100.15
32	1212.76	103.60
33	1249.87	106.30
34	1286.99	109.00
35	1323.37	113.00
36	1359.76	117.00
37	1388.42	123.00
38	1417.08	129.00

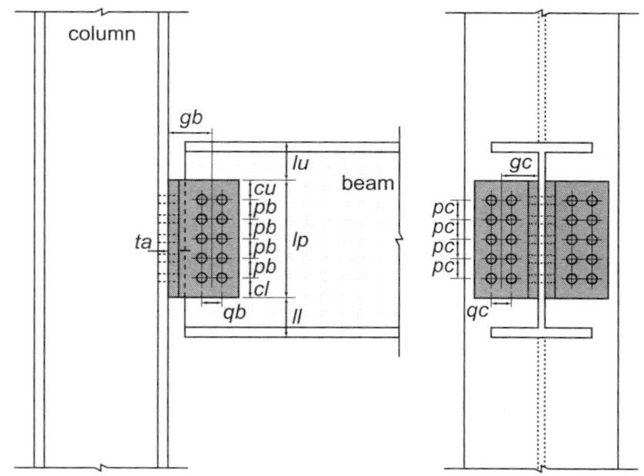

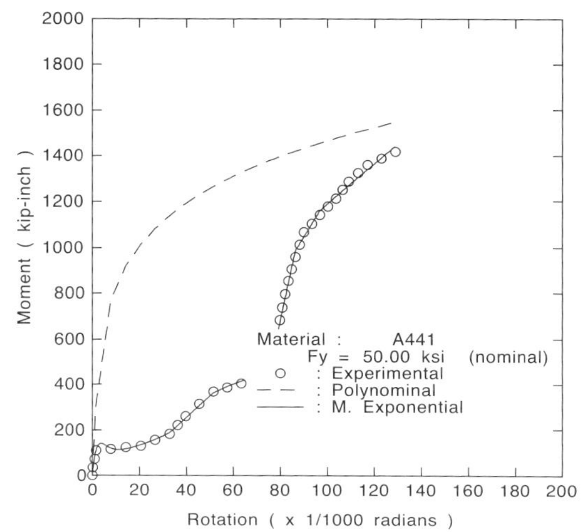

```
Moment-rotation prediction equations     ( R : X 1/1000 radians )

Frye and Morris polynominal model  :    R = Sum ( Ai X ( K*Bm )**Pi X 10**Qi )
     xd = 11.500000"   g =  4.000000"   t =  0.500000"
     A1 =  3.660000   A2 =  1.150000   A3 =  4.570000    K =  0.012288
     P1 =         1   P2 =         3   P3 =         5   Q1 =        -1    Q2 =        -3    Q3 =        -5

Modified exponential model  :
     Bm = Sum ( Ai X ( 1 - exp( - R/( 2*i*C ) ) ) )  +     Sum ( RKj X ( R - Rj0 ) )  +  Bm0
      C =  0.11500000E+01            Bmo=  0.00000000E+00         Nexp=  6         Nliner=  6
     AI =  0.62726472E+03  -0.49584357E+03  -0.87900708E+04   0.35812717E+05  -0.50159941E+05   0.23394682E+05
    Rj0 =          1.5000          32.7000          51.4000          75.2000          86.3000
                  96.7000
    RKj =  0.45692665E+00   0.37170790E+01  -0.39250297E+01   0.46599788E+02  -0.31192543E+02
          -0.71711048E+01
```

No	Rotation (radians) X 1/1000	Moment (kip-inch) Expri.	Moment (kip-inch) Poly.	Moment (kip-inch) M.Expo.	Connection stiffness (kip-inch) X 1000 Expri.	Connection stiffness (kip-inch) X 1000 Poly.	Connection stiffness (kip-inch) X 1000 M.Expo.
1	0.00	0.00	0.00	0.00	0.1223E+03	0.2223E+03	0.1172E+03
3	0.90	74.20	195.74	78.79	0.6253E+02	0.2067E+03	0.6238E+02
5	7.87	118.58	768.82	121.71	0.1077E+01	0.3262E+02	-0.4238E+01
7	20.60	132.30	1015.10	136.99	0.2706E+01	0.1265E+02	0.3417E+01
9	32.70	183.76	1138.01	188.66	0.8498E+01	0.8321E+01	0.5149E+01
11	39.60	259.09	1190.65	252.17	0.1028E+02	0.7030E+01	0.9434E+01
13	51.40	367.50	1264.49	362.19	0.6120E+01	0.5604E+01	0.8956E+01
15	63.30	403.51	1325.37	414.05	0.3026E+01	0.4688E+01	0.3633E+01
17	75.20	439.52	1377.15	449.05	0.4562E+02	0.4050E+01	0.2306E+01
19	77.53	565.46	1386.54	562.90	0.5397E+02	0.3945E+01	0.4869E+02
21	79.75	683.18	1395.23	670.62	0.5215E+02	0.3852E+01	0.4850E+02
23	82.10	795.64	1404.23	784.38	0.4438E+02	0.3758E+01	0.4832E+02
25	84.85	906.99	1414.49	916.99	0.3701E+02	0.3655E+01	0.4812E+02
27	88.20	1013.93	1426.31	1018.58	0.2805E+02	0.3540E+01	0.1672E+02
29	93.40	1104.34	1444.51	1104.84	0.1125E+02	0.3371E+01	0.1646E+02
31	100.15	1177.10	1466.55	1190.29	0.1033E+02	0.3180E+01	0.9032E+01
33	106.30	1249.87	1485.47	1245.30	0.1375E+02	0.3027E+01	0.8866E+01
35	113.00	1323.37	1505.49	1304.24	0.9096E+01	0.2874E+01	0.8738E+01
37	123.00	1388.42	1532.99	1390.98	0.4777E+01	0.2680E+01	0.8621E+01

II - 64

Connection type : Double web-angle connections
Mode : All bolted

Tested by : L.E.Thompson et al. (1970) U.S.A
Test Id. : D2-2 ALT

Column : W10X49
Beam : W21X62
Angle : 4 X 3.5 X 1/2

Fasteners: A325- -3/4"D
15/16" Oversize holes
Material : A441
Fy = -- ksi
Fu = -- ksi

Major parameters

lp = 11.5000"	lu = -- "	ll = -- "	ta = 0.5000"
gb = 2.2500"	gc = 1.8000"	cu = -- "	cl = -- "
pb = -- "	pc = -- "	qb = -- "	qc = -- "
nb = - X 2	nc = - X -		

Remark 1)
2)

No	Moment (k-in)	Rotation (radians) X 1/1000	No	Moment (k-in)	Rotation (radians) X 1/1000
1	0.00	0.00	26	849.29	75.50
2	36.68	0.20	27	904.97	76.90
3	74.20	0.60	28	960.65	78.30
4	111.72	1.00	29	1013.93	80.00
5	132.30	1.60	30	1067.22	81.70
6	158.03	5.60	31	1104.34	83.40
7	183.76	9.60	32	1141.45	85.10
8	203.23	12.70	33	1177.10	87.30
9	222.71	15.80	34	1212.76	89.50
10	240.90	19.90	35	1249.87	93.95
11	259.09	24.00	36	1286.99	98.40
12	294.92	28.40	37	1323.37	101.85
13	330.76	32.80	38	1359.76	105.30
14	359.05	37.30	39	1388.42	108.75
15	387.35	41.80	40	1417.08	112.20
16	413.44	46.55			
17	439.52	51.30			
18	465.98	56.90			
19	492.44	62.50			
20	518.90	68.10			
21	573.67	69.25			
22	628.43	70.40			
23	683.18	71.55			
24	737.94	72.70			
25	793.62	74.10			

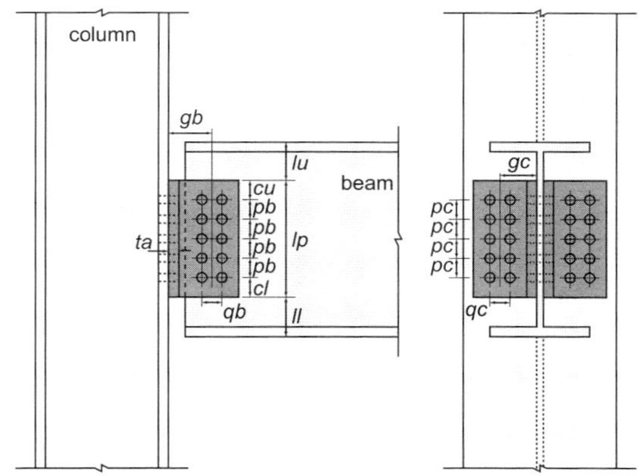

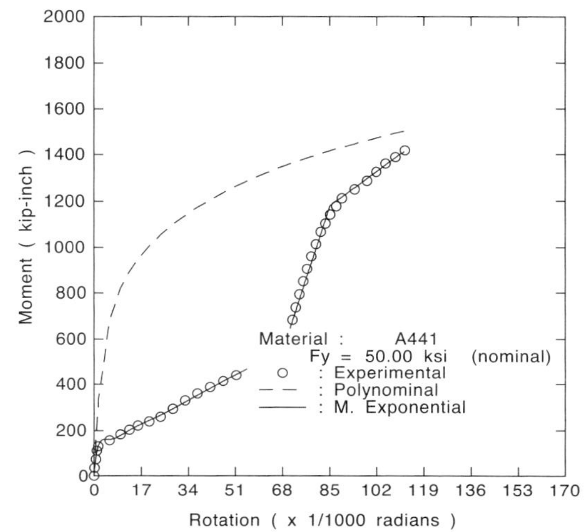

A.2 – 128

```
Moment-rotation prediction equations      ( R : X 1/1000 radians )

Frye and Morris polynominal model  :    R = Sum ( Ai X ( K*Bm )**Pi X 10**Qi )
     xd = 11.500000"    g =  4.000000"   t =  0.500000"
     A1 =  3.660000    A2 =  1.150000   A3 =  4.570000    K =  0.012288
     P1 =        1     P2 =        3    P3 =        5    Q1 =       -1     Q2 =       -3     Q3 =       -5

Modified exponential model  :
     Bm = Sum ( Ai X ( 1 - exp( - R/( 2*i*C ) ) ) )  +      Sum ( RKj X ( R - Rj0 ) )  +  Bm0
       C =   0.98500000E+00             Bmo=   0.00000000E+00          Nexp=  6       Nliner=  3
      AI =   0.11869952E+04  -0.73799885E+04   0.26073634E+05  -0.48377295E+05   0.44151185E+05  -0.15608367E+05
     Rj0 =           1.6000          68.1000           85.1000
     RKj =   0.73844694E+01   0.30186225E+02  -0.28783358E+02

     ------------------------------------------------------------------------------------------------
      No      Rotation             Moment                           Connection stiffness
             (radians)           (kip-inch)                          (kip-inch) X 1000
             X 1/1000      Expri.     Poly.   M.Expo.          Expri.          Poly.          M.Expo.
     ------------------------------------------------------------------------------------------------
       1         0.00        0.00      0.00      0.00       0.1834E+03     0.2223E+03     0.1638E+03
       3         0.60       74.20    132.19     76.13       0.9380E+02     0.2160E+03     0.9487E+02
       5         1.60      132.30    328.25    134.83       0.3067E+02     0.1686E+03     0.3081E+02
       7         9.60      183.76    819.72    179.81       0.6348E+01     0.2652E+02     0.7026E+01
       9        15.80      222.71    946.67    221.55       0.5488E+01     0.1622E+02     0.5849E+01
      11        24.00      259.09   1055.17    267.37       0.6225E+01     0.1099E+02     0.5861E+01
      13        32.80      330.76   1138.81    323.02       0.7226E+01     0.8299E+01     0.6606E+01
      15        41.80      387.35   1205.76    382.18       0.5901E+01     0.6705E+01     0.6467E+01
      17        51.30      439.52   1263.93    442.62       0.5140E+01     0.5614E+01     0.6307E+01
      19        62.50      492.44   1321.60    513.98       0.4725E+01     0.4739E+01     0.6485E+01
      21        69.25      573.67   1352.24    593.11       0.4762E+02     0.4342E+01     0.3686E+02
      23        71.55      683.18   1362.16    677.98       0.4761E+02     0.4223E+01     0.3693E+02
      25        74.10      793.62   1372.85    772.25       0.3977E+02     0.4098E+01     0.3700E+02
      27        76.90      904.97   1384.23    875.96       0.3977E+02     0.3971E+01     0.3708E+02
      29        80.00     1013.93   1396.08    991.01       0.3134E+02     0.3843E+01     0.3715E+02
      31        83.40     1104.34   1409.03   1117.44       0.2183E+02     0.3709E+01     0.3722E+02
      33        87.30     1177.10   1423.36   1199.43       0.1621E+02     0.3568E+01     0.8510E+01
      35        93.95     1249.87   1446.13   1256.35       0.8341E+01     0.3357E+01     0.8604E+01
      37       101.85     1323.37   1471.76   1324.64       0.1055E+02     0.3137E+01     0.8678E+01
      39       108.75     1388.42   1493.10   1384.67       0.8308E+01     0.2967E+01     0.8720E+01
     ------------------------------------------------------------------------------------------------
```

II - 65

Connection type : Double web-angle connections
Mode : All bolted

Tested by : L.E.Thompson et al. (1970) U.S.A
Test Id. : D2-3 ALT

Column : W10X49
Beam : W21X62
Angle : 4 X 3.5 X 1/2

Fasteners: A325- -3/4"D
15/16' Oversize holes
Material : A441
Fy = -- ksi
Fu = -- ksi

Major parameters

lp = 11.5000"	lu = -- "	ll = -- "	ta = 0.5000"
gb = 2.2500"	gc = 1.8000"	cu = -- "	cl = -- "
pb = -- "	pc = -- "	qb = -- "	qc = -- "
nb = - X 2	nc = - X -		

Remark 1)
2)

No	Moment (k-in)	Rotation (radians) X 1/1000	No	Moment (k-in)	Rotation (radians) X 1/1000
1	0.00	0.00	26	829.82	70.30
2	41.16	0.20	27	884.21	71.80
3	75.71	2.00	28	938.96	73.75
4	110.26	3.80	29	993.72	75.70
5	128.63	8.30	30	1048.11	79.45
6	147.00	12.80	31	1102.50	83.20
7	183.76	16.30	32	1139.25	87.63
8	211.31	19.65	33	1176.00	92.07
9	238.87	23.00	34	1212.76	96.50
10	266.44	26.95	35	1249.51	100.35
11	294.00	30.90	36	1286.26	104.20
12	317.28	35.90	37	1323.01	107.50
13	340.55	40.90	38	1359.76	110.80
14	363.83	45.90			
15	377.30	50.97			
16	390.78	56.03			
17	404.26	61.10			
18	443.57	61.80			
19	482.89	62.50			
20	518.54	63.10			
21	554.18	63.70			
22	610.04	64.95			
23	665.90	66.20			
24	720.67	67.50			
25	775.43	68.80			

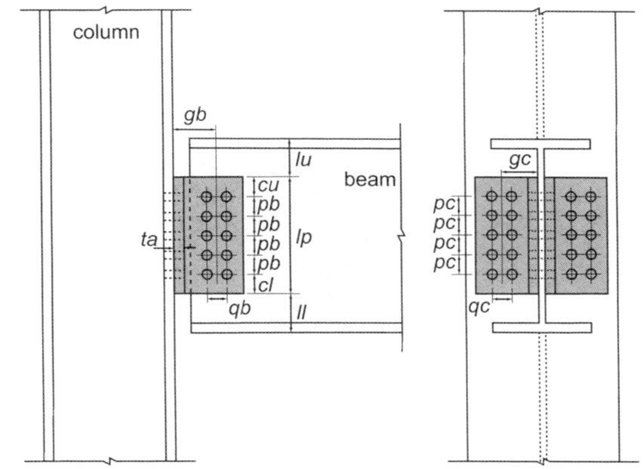

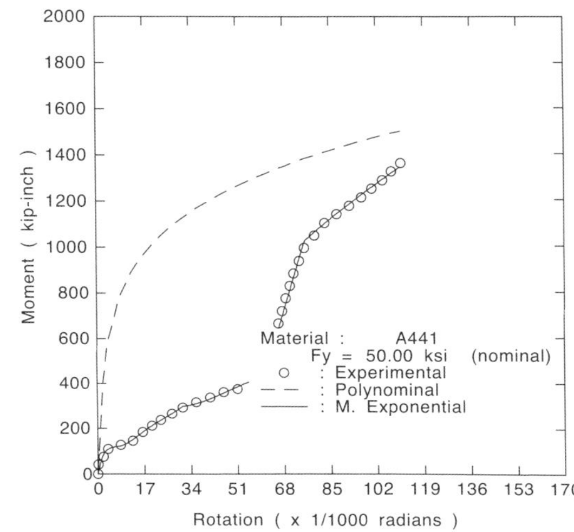

A.2 – 130

```
Moment-rotation prediction equations      ( R : X 1/1000 radians )

Frye and Morris polynominal model  :    R = Sum ( Ai X ( K*Bm )**Pi X 10**Qi )
     xd = 11.500000"    g =  4.000000"   t =  0.500000"
     A1 =  3.660000   A2 =  1.150000   A3 =  4.570000     K =  0.012288
     P1 =         1   P2 =         3   P3 =         5    Q1 =        -1     Q2 =        -3     Q3 =        -5

Modified exponential model  :
     Bm = Sum ( Ai X ( 1 - exp( - R/( 2*i*C ) ) ) )  +      Sum ( RKj X ( R - Rj0 ) )  +  Bm0
       C =   0.97333333E+00            Bmo=   0.00000000E+00          Nexp=  6          Nliner=  4
      AI =  -0.52673249E+03   0.12296715E+05  -0.69062063E+05   0.16535913E+06  -0.17655649E+06   0.68928548E+05
     Rj0 =           3.8000          30.9000          61.1000          75.7000
     RKj =   0.56342068E+01  -0.75866169E+01   0.38381161E+02  -0.28344471E+02
```

No	Rotation (radians) X 1/1000	Moment (kip-inch) Expri.	Moment (kip-inch) Poly.	Moment (kip-inch) M.Expo.	Connection stiffness (kip-inch) X 1000 Expri.	Connection stiffness (kip-inch) X 1000 Poly.	Connection stiffness (kip-inch) X 1000 M.Expo.
1	0.00	0.00	0.00	0.00	0.2058E+03	0.2223E+03	0.6037E+02
3	2.00	75.71	390.58	84.67	0.1919E+02	0.1436E+03	0.2475E+02
5	8.30	128.63	782.54	123.73	0.4083E+01	0.3084E+02	0.3281E+01
7	16.30	183.76	954.70	183.43	0.9339E+01	0.1574E+02	0.8900E+01
9	23.00	238.87	1043.95	236.77	0.7654E+01	0.1143E+02	0.7270E+01
11	30.90	294.00	1122.61	299.01	0.5953E+01	0.8752E+01	0.9103E+01
13	40.90	340.55	1199.67	333.43	0.4655E+01	0.6834E+01	0.4989E+01
15	50.97	377.30	1262.05	387.83	0.2660E+01	0.5645E+01	0.5289E+01
17	61.10	404.26	1314.90	433.19	0.4967E+02	0.4832E+01	0.3500E+01
19	62.50	482.89	1321.63	491.61	0.5791E+02	0.4739E+01	0.4159E+02
21	63.70	554.18	1327.29	541.36	0.5464E+02	0.4662E+01	0.4133E+02
23	66.20	665.90	1338.85	644.05	0.4343E+02	0.4510E+01	0.4082E+02
25	68.80	775.43	1350.48	749.50	0.3940E+02	0.4364E+01	0.4030E+02
27	71.80	884.21	1363.11	869.57	0.3270E+02	0.4211E+01	0.3975E+02
29	75.70	993.72	1379.35	1023.33	0.2344E+02	0.4025E+01	0.3911E+02
31	83.20	1102.50	1408.46	1100.26	0.1166E+02	0.3715E+01	0.9802E+01
33	92.07	1176.00	1439.75	1183.56	0.8290E+01	0.3414E+01	0.9050E+01
35	100.35	1249.51	1467.02	1256.63	0.9545E+01	0.3176E+01	0.8629E+01
37	107.50	1323.01	1489.40	1317.49	0.1114E+02	0.2996E+01	0.8410E+01

II - 66

Connection type : Double web-angle connections
Mode : All bolted

Tested by : L.E.Thompson et al. (1970) U.S.A
Test Id. : D2-3 ALT

Column : W10X49
Beam : W21X62
Angle : 4 X 3.5 X 1/2

Fasteners: A325- -3/4"D
15/16" Oversize holes
Material : A441
Fy = -- ksi
Fu = -- ksi

Major parameters

lp = 11.5000"	lu = -- "	ll = -- "	ta = 0.5000"
gb = 2.2500"	gc = 1.8000"	cu = -- "	cl = -- "
pb = -- "	pc = -- "	qb = -- "	qc = -- "
nb = - X 2	nc = - X -		

Remark 1)
2)

No	Moment (k-in)	Rotation (radians) X 1/1000
1	0.00	0.00
2	41.16	0.30
3	75.71	0.65
4	110.26	1.00
5	147.00	1.50
6	183.76	3.20
7	238.87	8.40
8	266.44	11.80
9	294.00	15.20
10	328.91	20.35
11	363.83	25.50
12	377.30	29.67
13	390.78	33.83
14	404.26	38.00
15	430.47	42.07
16	456.68	46.13
17	482.89	50.20
18	506.66	55.77
19	530.42	61.33
20	554.18	66.90
21	610.04	68.50
22	665.90	70.10
23	720.67	71.15
24	775.43	72.20
25	829.82	73.50
26	884.21	74.80
27	938.96	76.50
28	993.72	78.20
29	1048.11	80.65
30	1102.50	83.10
31	1157.63	87.45
32	1212.76	91.80
33	1237.26	96.50
34	1261.76	101.20
35	1286.26	105.90
36	1323.01	109.95
37	1359.76	114.00

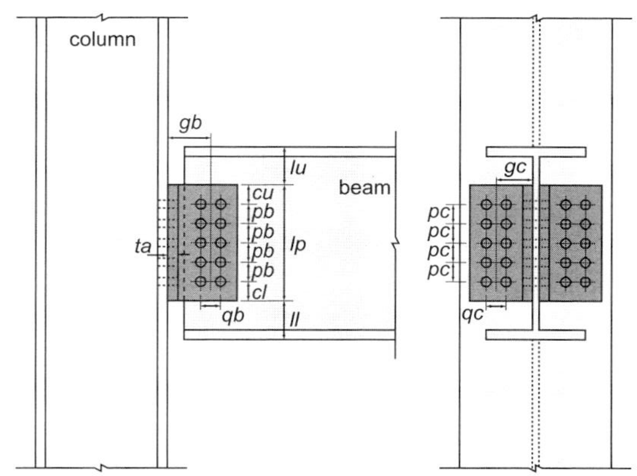

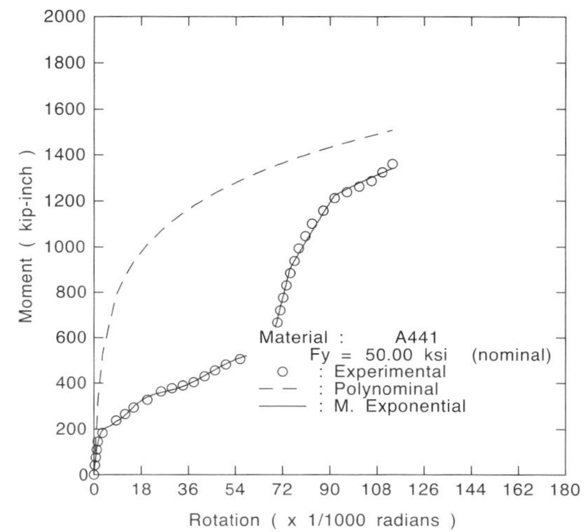

```
Moment-rotation prediction equations      ( R : X 1/1000 radians )

Frye and Morris polynominal model  :    R = Sum ( Ai X ( K*Bm )**Pi X 10**Qi )
     xd = 11.500000"   g =  4.000000"   t =  0.500000"
     A1 =  3.360000   A2 =  1.150000   A3 =  4.570000    K =  0.012288
     P1 =        1    P2 =        3    P3 =        5    Q1 =       -1    Q2 =       -3    Q3 =       -5

Modified exponential model  :
     Bm = Sum ( Ai X ( 1 - exp( - R/( 2*i*C ) ) ) )  +     Sum ( RKj X ( R - Rj0 ) )  +  Bm0
      C =   0.10250000E+01          Bmo=   0.00000000E+00           Nexp=  6        Nliner=  4
     AI =  -0.77817580E+02   0.14367129E+05  -0.96767098E+05   0.25052957E+06  -0.28160990E+06   0.11479016E+06
    Rj0 =           8.4000          66.9000          74.8000          91.8000
    RKj =  -0.93194222E+01   0.49334993E+02  -0.25101334E+02  -0.10808145E+02
```

No	Rotation (radians) X 1/1000	Moment (kip-inch) Expri.	Moment (kip-inch) Poly.	Moment (kip-inch) M.Expo.	Connection stiffness (kip-inch) X 1000 Expri.	Connection stiffness (kip-inch) X 1000 Poly.	Connection stiffness (kip-inch) X 1000 M.Expo.
1	0.00	0.00	0.00	0.00	0.1372E+03	0.2223E+03	0.1425E+03
3	0.65	75.71	142.97	76.96	0.9871E+02	0.2148E+03	0.9667E+02
5	1.50	147.00	311.01	140.29	0.6170E+02	0.1750E+03	0.5539E+02
7	8.40	238.87	785.71	234.68	0.9092E+01	0.3044E+02	0.1277E+02
9	15.20	294.00	936.82	298.28	0.7579E+01	0.1683E+02	0.1050E+02
11	25.50	363.83	1071.25	359.12	0.4820E+01	0.1040E+02	0.2718E+01
13	33.83	390.78	1147.26	385.79	0.3234E+01	0.8074E+01	0.4405E+01
15	42.07	430.47	1207.54	432.41	0.6446E+01	0.6668E+01	0.6518E+01
17	50.20	482.89	1257.70	484.17	0.5527E+01	0.5720E+01	0.5745E+01
19	61.33	530.42	1316.02	527.32	0.4269E+01	0.4816E+01	0.1715E+01
21	68.50	610.04	1349.00	608.09	0.3491E+02	0.4382E+01	0.4818E+02
23	71.15	720.67	1360.53	734.46	0.5215E+02	0.4242E+01	0.4721E+02
25	73.50	829.82	1370.43	844.44	0.4184E+02	0.4126E+01	0.4641E+02
27	76.50	938.96	1382.37	939.58	0.3221E+02	0.3991E+01	0.2038E+02
29	80.65	1048.11	1398.76	1021.75	0.2220E+02	0.3815E+01	0.1926E+02
31	87.45	1157.63	1423.65	1147.58	0.1267E+02	0.3565E+01	0.1783E+02
33	96.50	1237.26	1454.59	1251.91	0.5213E+01	0.3282E+01	0.5754E+01
35	105.90	1286.26	1484.25	1302.05	0.7287E+01	0.3036E+01	0.4987E+01
37	114.00	1359.76	1508.10	1340.78	0.9074E+01	0.2855E+01	0.4608E+01

II - 67

Connection type : Double web-angle connections
Mode : Welded-to-beam and bolted-to-column

Tested by : W.H.Sommer (1969) Canada
Test Id. : TEST21

Column : --
Beam : W18X45
Angle : 3.5 X 3 X 3/8

Fasteners: A325-X-3/4"D
15/16" Oversize holes
Material : G40.12
Fy = 50.00 ksi
Fu = 76.10 ksi

Major parameters

lp = 9.0000"	lu = 1.5000"	ll = 7.3750"	ta = 0.3750"
gc = 1.8325"	cu = 1.5000"	cl = 1.5000"	pc = 3.0000"
nc = 1 X 3			

Remark 1) Actual strength Fy = 50.0 ksi, Fu = 76.1 ksi for beam web.
2) No washers were used.

No	Moment (k-in)	Rotation (radians) X 1/1000	No	Moment (k-in)	Rotation (radians) X 1/1000
1	0.00	0.00	26	941.00	59.20
2	67.50	3.10	27	990.50	59.75
3	113.00	5.30	28	1040.00	60.30
4	165.00	7.60	29	1110.00	62.15
5	213.00	10.60	30	1180.00	64.00
6	262.00	14.00	31	1236.66	66.70
7	287.66	17.07	32	1293.33	69.40
8	313.33	20.13	33	1350.00	72.10
9	339.00	23.20	34	1410.00	75.50
10	356.00	26.95	35	1470.00	78.90
11	373.00	30.70			
12	390.00	34.45			
13	407.00	38.20			
14	415.00	41.20			
15	423.00	44.20			
16	431.00	47.20			
17	443.50	50.55			
18	456.00	53.90			
19	504.50	54.70			
20	553.00	55.50			
21	625.00	56.00			
22	686.00	56.75			
23	747.00	57.50			
24	811.67	58.07			
25	876.34	58.63			

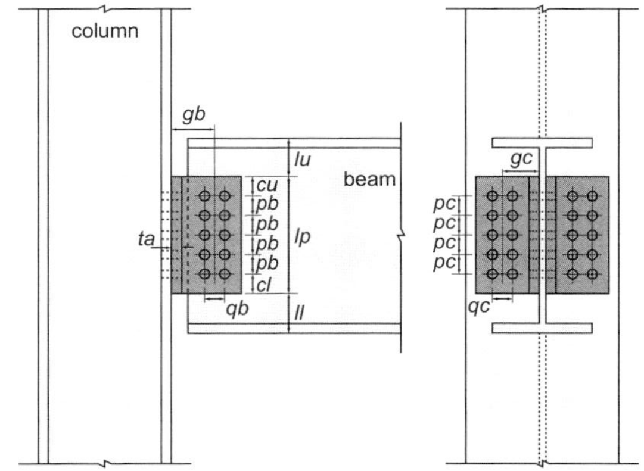

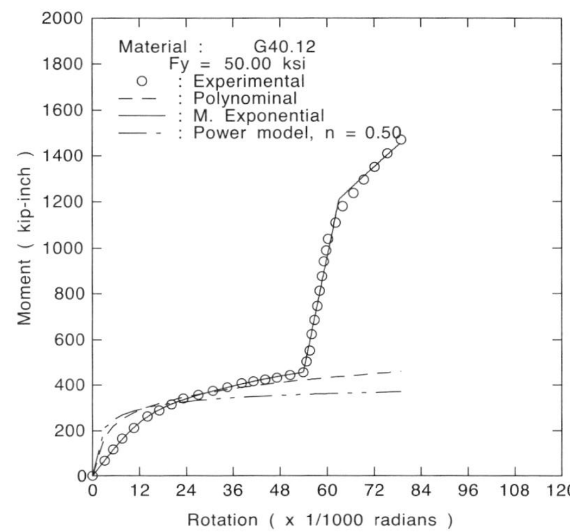

A.2 – 134

```
Moment-rotation prediction equations     ( R : X 1/1000 radians )

Frye and Morris polynominal model  :    R = Sum ( Ai X ( K*Bm )**Pi X 10**Qi )
     xd =  9.000000"    g =  4.000000"    t =  0.375000"
     A1 =  3.660000    A2 =  1.150000    A3 =  4.570000     K =  0.037250
     P1 =        1     P2 =        3     P3 =        5     Q1 =       -1     Q2 =      -3     Q3 =      -5

Modified exponential model  :
     Bm = Sum ( Ai X ( 1 - exp( - R/( 2*i*C ) ) ) )  +     Sum ( RKj X ( R - Rj0 ) )  +  Bm0
      C =   0.14325000E+01              Bmo=   0.00000000E+00          Nexp=  6        Nliner=  2
     AI =   0.91803008E+02  -0.22039393E+04   0.95878576E+04  -0.15216343E+05   0.91558206E+04  -0.90571919E+03
    Rj0 =          53.9000          63.0000
    RKj =   0.80486102E+02  -0.66166094E+02

Power model  :    Bm = ( Rki X R ) / ( 1 + ( R/R0 )**rn )**( 1/rn )
      rn = 0.498       rki =   0.76444761E+03         rmu =   0.43607210E+03
```

No	Rotation (radians) X 1/1000	Moment (kip-inch) Expri.	Poly.	M.Expo.	P.Model.	Connection stiffness (kip-inch) X 1000 Expri.	Poly.	M.Expo.	P.Model.
1	0.00	0.00	0.00	0.00	0.00	0.2177E+02	0.7335E+02	0.2161E+02	0.7644E+03
3	5.30	118.00	219.67	117.02	246.02	0.2172E+02	0.1656E+02	0.2127E+02	0.1151E+02
5	10.60	213.00	278.72	214.85	286.08	0.1525E+02	0.7912E+01	0.1529E+02	0.5109E+01
7	17.07	287.66	318.82	291.06	310.50	0.8370E+01	0.4973E+01	0.8844E+01	0.2830E+01
9	23.20	339.00	345.14	334.79	324.72	0.6644E+01	0.3739E+01	0.5849E+01	0.1911E+01
11	30.70	373.00	369.76	372.80	336.63	0.4534E+01	0.2904E+01	0.4534E+01	0.1326E+01
13	38.20	407.00	389.49	404.49	345.22	0.3496E+01	0.2393E+01	0.3944E+01	0.9922E+00
15	44.20	423.00	402.95	426.77	350.61	0.2667E+01	0.2107E+01	0.3471E+01	0.8162E+00
17	50.55	443.50	415.56	447.03	355.34	0.3731E+01	0.1876E+01	0.2903E+01	0.6811E+00
19	54.70	504.50	423.09	522.67	358.02	0.6062E+02	0.1752E+01	0.8300E+02	0.6120E+00
21	56.00	625.00	425.36	630.50	358.81	0.1189E+03	0.1717E+01	0.8289E+02	0.5928E+00
23	57.50	747.00	427.92	754.73	359.68	0.1000E+03	0.1678E+01	0.8275E+02	0.5719E+00
25	58.63	876.34	429.81	848.45	360.32	0.1141E+03	0.1650E+01	0.8265E+02	0.5569E+00
27	59.75	990.50	431.65	940.69	360.93	0.9000E+02	0.1624E+01	0.8255E+02	0.5427E+00
29	62.15	1110.00	435.52	1138.58	362.20	0.3784E+02	0.1570E+01	0.8235E+02	0.5143E+00
31	66.70	1236.66	442.44	1267.66	364.43	0.2099E+02	0.1478E+01	0.1584E+02	0.4670E+00
33	72.10	1350.00	450.15	1352.20	366.82	0.1951E+02	0.1384E+01	0.1548E+02	0.4197E+00
35	78.90	1470.00	459.23	1456.25	369.50	0.1765E+02	0.1282E+01	0.1513E+02	0.3707E+00

II - 68

Connection type : Double web-angle connections
Mode : Welded-to-beam and bolted-to-column

Tested by : W.H.Sommer (1969) Canada
Test Id. : TEST22

Column : --
Beam : W18X45
Angle : 3.5 X 3 X 3/8

Fasteners: A325-X-3/4"D
15/16" Oversize holes
Material : G40.12
Fy = 50.00 ksi
Fu = 76.10 ksi

Major parameters

lp = 12.0000" lu = 1.5000" ll = 4.3750" ta = 0.3750"
gc = 1.8325" cu = 1.5000" cl = 1.5000" pc = 3.0000"
nc = 1 X 4

Remark 1) Actual strength Fy = 50.0 ksi, Fu = 76.1 ksi for beam web.
2) No washers were used.

No	Moment (k-in)	Rotation (radians) X 1/1000	No	Moment (k-in)	Rotation (radians) X 1/1000
1	0.00	0.00	26	1039.50	76.80
2	55.00	0.93	27	1090.00	78.00
3	110.00	1.87	28	1135.00	78.95
4	165.00	2.80	29	1180.00	79.90
5	237.50	4.30	30	1240.00	81.30
6	310.00	5.80	31	1300.00	82.70
7	383.00	7.90	32	1365.00	84.40
8	456.00	10.00	33	1430.00	86.10
9	504.50	12.15	34	1475.00	87.85
10	553.00	14.30	35	1520.00	89.60
11	601.33	18.37	36	1530.00	93.10
12	649.67	22.43	37	1520.00	95.65
13	698.00	26.50	38	1510.00	98.20
14	722.50	30.05			
15	747.00	33.60			
16	771.00	38.00			
17	795.00	42.40			
18	811.17	47.27			
19	827.33	52.13			
20	843.50	57.00			
21	859.66	61.87			
22	875.83	66.73			
23	892.00	71.60			
24	921.00	74.00			
25	989.00	75.60			

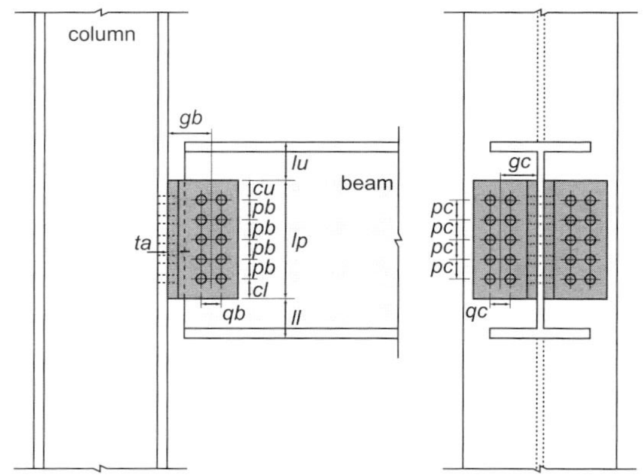

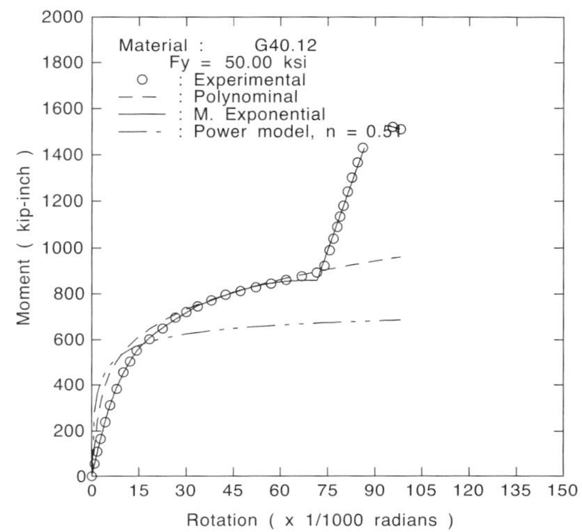

```
Moment-rotation prediction equations      ( R : X 1/1000 radians )

Frye and Morris polynominal model  :     R = Sum ( Ai X ( K*Bm )**Pi X 10**Qi )
     xd = 12.000000"   g =  4.000000"   t =  0.375000"
     A1 =  3.660000   A2 =  1.150000   A3 =  4.570000     K =  0.018676
     P1 =        1    P2 =        3    P3 =        5     Q1 =       -1     Q2 =       -3     Q3 =       -5

Modified exponential model  :
     Bm = Sum ( Ai X ( 1 - exp( - R/( 2*i*C ) ) ) )  +      Sum ( RKj X ( R - Rj0 ) )  +  Bm0
       C =   0.10516583E+01              Bmo=   0.00000000E+00          Nexp=  6          Nliner=  6
      AI =  -0.26461235E+03   0.38699400E+04  -0.25342389E+05   0.72286486E+05  -0.87772700E+05   0.38230398E+05
     Rj0 =           0.0000           5.8000          14.3000          33.6000          71.6000
                    89.6000
     RKj =   0.18294939E+01   0.65414756E+00  -0.28790736E+01  -0.28960281E+01   0.39473061E+02
            -0.40632824E+02

Power model  :     Bm = ( Rki X R ) / ( 1 + ( R/R0 )**rn )**( 1/rn )
      rn = 0.509       rki =   0.18009083E+04          rmu =   0.77523929E+03
```

No	Rotation (radians) X 1/1000	Moment (kip-inch) Expri.	Moment (kip-inch) Poly.	Moment (kip-inch) M.Expo.	Moment (kip-inch) P.Model.	Connection stiffness (kip-inch) X 1000 Expri.	Connection stiffness (kip-inch) X 1000 Poly.	Connection stiffness (kip-inch) X 1000 M.Expo.	Connection stiffness (kip-inch) X 1000 P.Model.
1	0.00	0.00	0.00	0.00	0.00	0.5893E+02	0.1463E+03	0.5311E+02	0.1801E+04
3	1.87	110.00	244.03	109.92	361.85	0.5893E+02	0.9986E+02	0.5899E+02	0.6232E+02
5	4.30	237.50	400.99	241.24	456.21	0.4833E+02	0.4189E+02	0.4811E+02	0.2510E+02
7	7.90	383.00	506.58	386.64	518.41	0.3476E+02	0.2137E+02	0.3363E+02	0.1216E+02
9	12.15	504.50	578.79	504.90	557.64	0.2256E+02	0.1377E+02	0.2304E+02	0.7087E+01
11	18.37	601.33	648.42	608.46	591.17	0.1189E+02	0.9255E+01	0.1222E+02	0.4149E+01
13	26.50	698.00	711.56	689.89	617.51	0.9224E+01	0.6609E+01	0.8692E+01	0.2548E+01
15	33.60	747.00	753.64	749.27	632.89	0.6255E+01	0.5346E+01	0.8177E+01	0.1848E+01
17	42.40	795.00	796.01	793.72	646.74	0.4442E+01	0.4357E+01	0.4700E+01	0.1344E+01
19	52.13	827.33	834.71	832.56	658.07	0.3322E+01	0.3643E+01	0.3151E+01	0.1010E+01
21	61.87	859.66	867.62	853.88	666.79	0.3322E+01	0.3146E+01	0.1235E+01	0.7959E+00
23	71.60	892.00	896.36	857.49	673.77	0.9191E+01	0.2778E+01	-0.4200E+00	0.6486E+00
25	75.60	989.00	907.35	1012.59	676.27	0.4226E+02	0.2651E+01	0.3851E+02	0.6008E+00
27	78.00	1090.00	913.67	1104.66	677.68	0.4503E+02	0.2582E+01	0.3822E+02	0.5750E+00
29	79.90	1180.00	918.55	1177.09	678.76	0.4554E+02	0.2529E+01	0.3802E+02	0.5558E+00
31	82.70	1300.00	925.37	1283.15	680.27	0.4077E+02	0.2459E+01	0.3775E+02	0.5294E+00
33	86.10	1430.00	933.66	1410.99	682.02	0.3207E+02	0.2376E+01	0.3746E+02	0.5001E+00
35	89.60	1520.00	941.90	1541.67	683.73	0.1809E+02	0.2297E+01	0.3722E+02	0.4727E+00
37	95.65	1520.00	955.33	1519.96	686.46	-0.3922E+01	0.2175E+01	-0.3742E+01	0.4309E+00

II - 69

Connection type : Double web-angle connections
Mode : Welded-to-beam and bolted-to-column

Tested by : W.H.Sommer (1969) Canada
Test Id. : TEST23

Column : --
Beam : W24X76
Angle : 4 X 3 X 3/8

Fasteners: A325-X-3/4"D
15/16" Oversize holes
Material : G40.12
Fy = 51.50 ksi
Fu = 77.90 ksi

Major parameters

lp = 15.0000" lu = 1.5000" ll = 7.3750" ta = 0.3750"
gc = 2.5300" cu = 1.5000" cl = 1.5000" pc = 3.0000"
nc = 1 X 5

Remark 1) Actual strength Fy = 51.5 ksi, Fu = 77.9 ksi for beam web.
2) No washers were used.

No	Moment (k-in)	Rotation (radians) X 1/1000	No	Moment (k-in)	Rotation (radians) X 1/1000
1	0.00	0.00	26	1240.00	69.40
2	75.80	1.10	27	1300.00	69.90
3	149.00	2.30	28	1360.00	70.40
4	221.00	3.40	29	1445.00	71.00
5	294.00	4.70	30	1530.00	71.60
6	367.00	6.00	31	1640.00	72.70
7	439.50	7.95	32	1750.00	73.80
8	512.00	9.90	33	1860.00	74.90
9	560.50	12.40	34	1970.00	76.00
10	609.00	14.90	35	2065.00	79.05
11	651.00	18.77	36	2160.00	82.10
12	693.00	22.63	37	2260.00	86.10
13	735.00	26.50	38	2285.00	88.50
14	764.25	30.43	39	2310.00	90.90
15	793.50	34.35			
16	822.75	38.27			
17	852.00	42.20			
18	876.25	46.20			
19	900.50	50.20			
20	924.75	54.20			
21	949.00	58.20			
22	974.50	62.60			
23	1000.00	67.00			
24	1090.00	68.10			
25	1165.00	68.75			

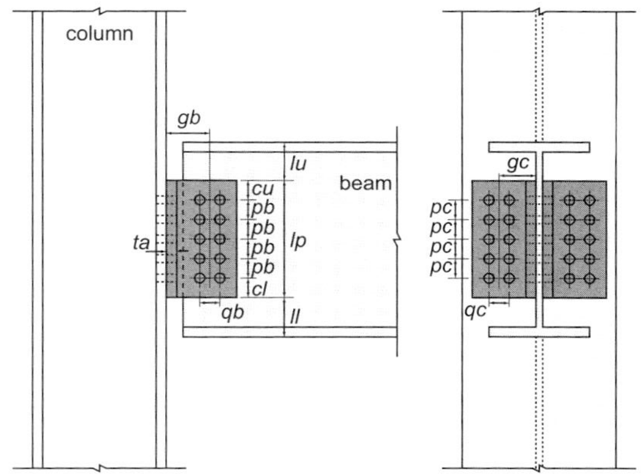

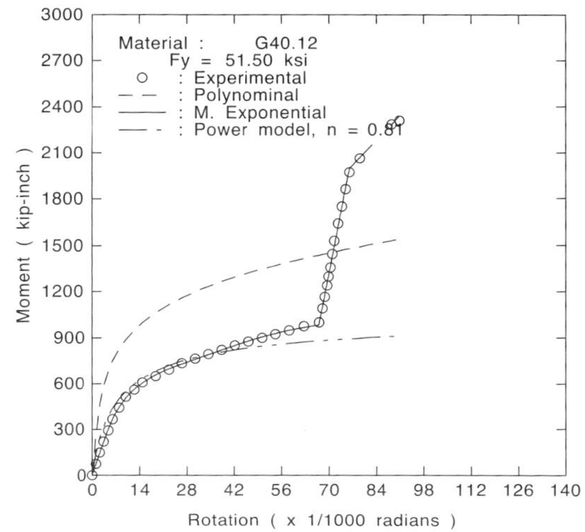

```
Moment-rotation prediction equations     ( R : X 1/1000 radians )

Frye and Morris polynominal model  :    R = Sum ( Ai X ( K*Bm )**Pi X 10**Qi )
     xd = 15.000000"   g =  5.500000"   t =  0.375000"
     A1 =  3.660000   A2 =  1.150000   A3 =  4.570000    K =  0.011467
     P1 =        1    P2 =        3    P3 =        5    Q1 =       -1    Q2 =       -3    Q3 =       -5

Modified exponential model  :
     Bm = Sum ( Ai X ( 1 - exp( - R/( 2*i*C ) ) ) )  +     Sum ( RKj X ( R - Rj0 ) )  +  Bm0
       C =  0.10325000E+01            Bmo=  0.00000000E+00        Nexp=  6        Nliner=  5
      AI = -0.12553982E+03  0.19219360E+04 -0.18196636E+05  0.60158318E+05 -0.77104029E+05  0.34419691E+05
     Rj0 =          0.0000          6.0000         14.9000         67.0000         76.0000
     RKj =  0.11846132E+01 -0.48141453E+01  0.33930542E+01  0.11036930E+03 -0.88439560E+02

Power model  :    Bm = ( Rki X R ) / ( 1 + ( R/R0 )**rn )**( 1/rn )
     rn = 0.808       rki =  0.17138319E+03         rmu =  0.10397592E+04
```

No	Rotation (radians) X 1/1000	Moment (kip-inch) Expri.	Poly.	M.Expo.	P.Model.	Connection stiffness (kip-inch) X 1000 Expri.	Poly.	M.Expo.	P.Model.
1	0.00	0.00	0.00	0.00	0.00	0.6891E+02	0.2383E+03	0.6066E+02	0.1714E+03
3	2.30	149.00	461.94	151.01	247.45	0.6332E+02	0.1357E+03	0.6543E+02	0.7385E+02
5	4.70	294.00	679.00	297.23	385.53	0.5615E+02	0.6164E+02	0.5558E+02	0.4523E+02
7	7.95	439.50	826.79	443.05	501.02	0.3718E+02	0.3458E+02	0.3543E+02	0.2808E+02
9	12.40	560.50	948.27	562.98	599.06	0.1940E+02	0.2198E+02	0.1966E+02	0.1737E+02
11	18.77	651.00	1062.01	658.92	684.65	0.1086E+02	0.1477E+02	0.1159E+02	0.1045E+02
13	26.50	735.00	1158.91	730.03	748.70	0.9170E+01	0.1076E+02	0.7942E+01	0.6584E+01
15	34.35	793.50	1233.92	791.12	791.66	0.7452E+01	0.8538E+01	0.7761E+01	0.4556E+01
17	42.20	852.00	1295.03	851.01	822.37	0.6764E+01	0.7126E+01	0.7356E+01	0.3364E+01
19	50.20	900.50	1347.83	905.42	845.96	0.6062E+01	0.6130E+01	0.6141E+01	0.2587E+01
21	58.20	949.00	1393.80	948.35	864.42	0.5935E+01	0.5398E+01	0.4580E+01	0.2059E+01
23	67.00	1000.00	1438.49	981.44	880.64	0.6661E+02	0.4787E+01	0.2992E+01	0.1651E+01
25	68.75	1165.00	1446.83	1179.58	883.47	0.1154E+03	0.4682E+01	0.1131E+03	0.1585E+01
27	69.90	1300.00	1452.19	1309.54	885.27	0.1200E+03	0.4616E+01	0.1129E+03	0.1543E+01
29	71.00	1445.00	1457.25	1433.66	886.94	0.1417E+03	0.4555E+01	0.1128E+03	0.1506E+01
31	72.70	1640.00	1464.96	1625.16	889.46	0.1000E+03	0.4464E+01	0.1125E+03	0.1450E+01
33	74.90	1860.00	1474.72	1872.42	892.57	0.1000E+03	0.4352E+01	0.1123E+03	0.1383E+01
35	79.05	2065.00	1492.20	2067.58	898.07	0.3115E+02	0.4160E+01	0.2337E+02	0.1268E+01
37	86.10	2260.00	1520.48	2230.16	906.41	0.1589E+02	0.3871E+01	0.2279E+02	0.1105E+01
39	90.90	2310.00	1538.85	2338.82	911.49	0.1042E+02	0.3696E+01	0.2250E+02	0.1012E+01

II - 70

Connection type : Double web-angle connections
Mode : Welded-to-beam and bolted-to-column

Tested by : W.H.Sommer (1969) Canada
Test Id. : TEST24

Column : --
Beam : W24X76
Angle : 4 X 3 X 3/8

Fasteners: A325-X-3/4"D
15/16" Oversize holes
Material : G40.12
Fy = 51.50 ksi
Fu = 77.90 ksi

Major parameters

lp = 18.0000" lu = 1.5000" ll = 4.3750" ta = 0.3750"
gc = 2.5300" cu = 1.5000" cl = 1.5000" pc = 3.0000"
nc = 1 X 6

Remark 1) Actual strength fy = 51.5 ksi fu = 77.9 ksi for beam web.
2) No washers were used.

No	Moment (k-in)	Rotation (radians) X 1/1000	No	Moment (k-in)	Rotation (radians) X 1/1000
1	0.00	0.00	26	1395.00	58.85
2	76.60	0.80	27	1427.50	62.08
3	138.80	1.45	28	1460.00	65.30
4	201.00	2.10	29	1480.00	67.50
5	259.00	2.70	30	1500.00	69.70
6	317.00	3.30	31	1580.00	73.30
7	390.00	4.05	32	1670.00	74.80
8	463.00	4.80	33	1755.00	76.00
9	550.00	6.05	34	1840.00	77.20
10	637.00	7.30	35	1900.00	78.65
11	719.00	9.40	36	1960.00	80.10
12	801.00	11.50	37	2020.00	83.70
13	874.00	14.65			
14	947.00	17.80			
15	993.50	20.70			
16	1040.00	23.60			
17	1090.00	27.15			
18	1140.00	30.70			
19	1165.00	34.20			
20	1190.00	37.70			
21	1225.00	41.38			
22	1260.00	45.05			
23	1295.00	48.73			
24	1330.00	52.40			
25	1362.50	55.62			

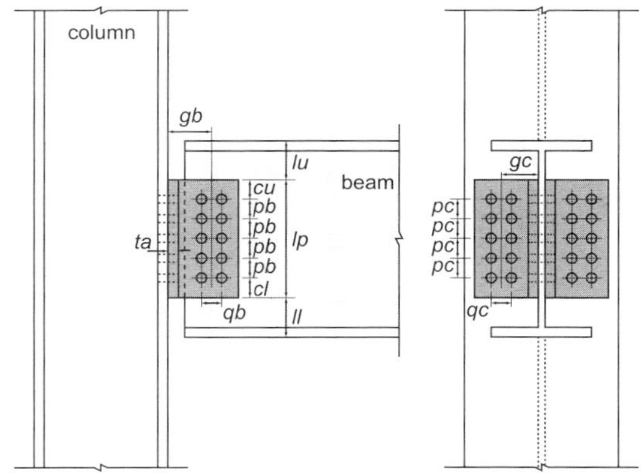

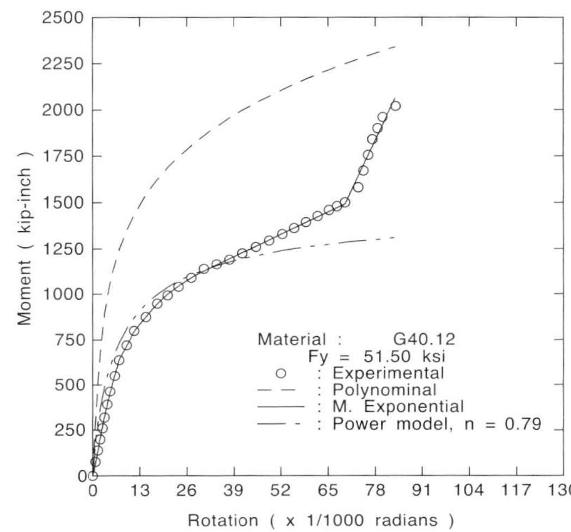

A.2 – 140

```
Moment-rotation prediction equations     ( R : X 1/1000 radians )

Frye and Morris polynominal model  :     R = Sum ( Ai X ( K*Bm )**Pi X 10**Qi )
     xd = 18.000000"    g =  5.500000"   t =  0.375000"
     A1 =  3.660000   A2 =  1.150000   A3 =  4.570000     K =  0.007403
     P1 =         1   P2 =         3   P3 =         5    Q1 =        -1    Q2 =        -3    Q3 =        -5

Modified exponential model   :
     Bm = Sum ( Ai X ( 1 - exp( - R/( 2*i*C ) ) ) )  +      Sum ( RKj X ( R - Rj0 ) )  +  Bm0
       C =   0.89750000E+00              Bmo=   0.00000000E+00          Nexp=  6        Nliner=  5
      AI =  -0.10611872E+04   0.12142415E+05  -0.47536429E+05   0.84924804E+05  -0.71717757E+05   0.23774134E+05
     Rj0 =           0.0000           7.3000          11.5000          23.6000          69.7000
     RKj =   0.58878830E+02  -0.36328754E+02  -0.16483134E+02   0.21707644E+01   0.32403594E+02

Power model  :     Bm = ( Rki X R ) / ( 1 + ( R/R0 )**rn )**( 1/rn )
      rn = 0.786       rki =   0.29274159E+03         rmu =   0.14972532E+04
```

No	Rotation (radians) X 1/1000	Moment (kip-inch) Expri.	Poly.	M.Expo.	P.Model.	Connection stiffness (kip-inch) X 1000 Expri.	Poly.	M.Expo.	P.Model.
1	0.00	0.00	0.00	0.00	0.00	0.9575E+02	0.3691E+03	0.8105E+01	0.2927E+03
3	1.45	138.80	501.69	135.83	284.09	0.9570E+02	0.2955E+03	0.1057E+03	0.1429E+03
5	2.70	259.00	793.30	265.27	432.93	0.9667E+02	0.1785E+03	0.9902E+02	0.9989E+02
7	4.05	390.00	984.41	390.01	548.75	0.9733E+02	0.1132E+03	0.8603E+02	0.7395E+02
9	6.05	550.00	1163.48	548.84	672.43	0.6960E+02	0.7187E+02	0.7476E+02	0.5191E+02
11	9.40	719.00	1352.01	718.37	810.73	0.3905E+02	0.4496E+02	0.3779E+02	0.3300E+02
13	14.65	874.00	1539.40	875.12	943.88	0.2318E+02	0.2894E+02	0.2359E+02	0.1960E+02
15	20.70	993.50	1687.13	998.97	1038.56	0.1603E+02	0.2090E+02	0.1659E+02	0.1254E+02
17	27.15	1090.00	1805.80	1089.48	1105.61	0.1408E+02	0.1631E+02	0.1200E+02	0.8637E+01
19	34.20	1165.00	1909.28	1162.57	1157.18	0.7143E+01	0.1328E+02	0.9401E+01	0.6204E+01
21	41.38	1225.00	1996.74	1228.75	1195.82	0.9523E+01	0.1123E+02	0.9228E+01	0.4682E+01
23	48.73	1295.00	2073.52	1297.42	1226.22	0.9523E+01	0.9742E+01	0.9437E+01	0.3656E+01
25	55.62	1362.50	2136.97	1362.62	1249.00	0.1008E+02	0.8692E+01	0.9420E+01	0.2983E+01
27	62.08	1427.50	2190.43	1422.85	1266.66	0.1008E+02	0.7913E+01	0.9240E+01	0.2514E+01
29	67.50	1480.00	2232.07	1472.45	1279.43	0.9090E+01	0.7364E+01	0.9043E+01	0.2204E+01
31	73.30	1580.00	2273.09	1640.94	1291.40	0.4889E+02	0.6869E+01	0.4124E+02	0.1934E+01
33	76.00	1755.00	2291.49	1752.18	1296.47	0.7083E+02	0.6660E+01	0.4116E+02	0.1826E+01
35	78.65	1900.00	2309.01	1861.15	1301.18	0.4138E+02	0.6469E+01	0.4108E+02	0.1728E+01
37	83.70	2020.00	2340.58	2068.29	1309.48	0.1667E+02	0.6141E+01	0.4096E+02	0.1564E+01

II - 71

Connection type : Double web-angle connections
Mode : All bolted

Tested by : J.B.Davison et al (1987) United Kingdom
Test Id. : JT/01B

Column : UC 152x152x23
Beam : UB 254x102x22
Angle : RSA 80x60x8

Fasteners: 10.9- -M16
23/32" Oversize holes
Material : --
Fy = -- ksi
Fu = -- ksi

Major parameters

lp = 6.2992"	lu = 1.1811"	ll = 1.1811"	ta = 0.3150"
gb = 2.1650"	gc = 1.3819"	cu = 1.1811"	cl = 1.1811"
pb = 3.9370"	pc = 3.9370"	qc = -- "	
nb = 1 X 2	nc = 2 X 2		

Remark 1) Bolts were tightened to a torque of 160 Nm
2) Random bolt tightness & minor axis connection

No	Moment (k-in)	Rotation (radians) X 1/1000
1	0.00	0.00
2	22.00	1.15
3	31.22	2.31
4	34.41	4.62
5	39.33	7.50
6	40.81	9.81
7	44.25	12.12
8	49.16	15.00
9	54.08	17.31
10	59.00	19.62
11	63.91	22.50
12	66.37	24.81
13	68.34	27.12
14	71.29	30.00
15	73.75	32.31
16	77.43	34.62
17	78.66	37.50
18	78.66	39.81
19	78.66	42.12
20	78.66	45.00
21	78.66	47.31
22	78.66	49.62
23	81.12	52.50
24	83.58	54.81
25	86.04	57.12
26	86.04	60.00
27	86.04	62.31
28	86.04	64.00

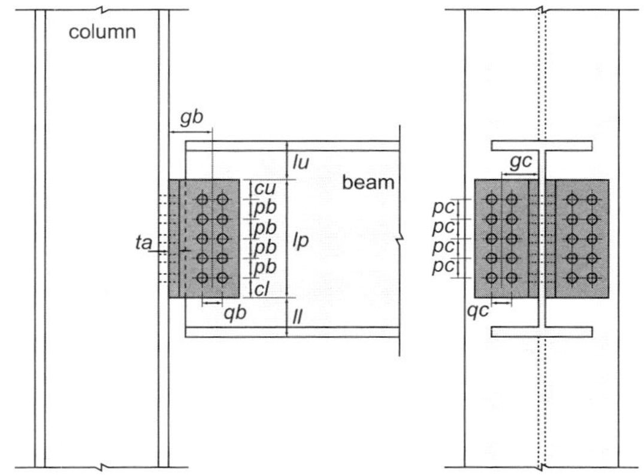

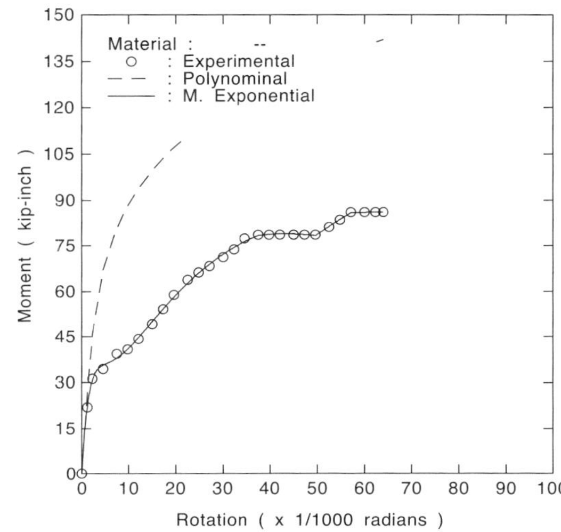

```
Moment-rotation prediction equations     ( R : X 1/1000 radians )

Frye and Morris polynominal model  :    R = Sum ( Ai X ( K*Bm )**Pi X 10**Qi )
     xd =  6.299200"    g =  2.968200"    t =  0.315000"
     A1 =  3.660000    A2 =  1.150000    A3 =  4.570000    K =  0.114987
     P1 =        1     P2 =        3     P3 =        5     Q1 =       -1     Q2 =       -3     Q3 =       -5

Modified exponential model  :
     Bm = Sum ( Ai X ( 1 - exp( - R/( 2*i*C ) ) ) )  +      Sum ( RKj X ( R - Rj0 ) )  + Bm0
       C =   0.82180833E+00            Bmo=   0.00000000E+00            Nexp=  6        Nliner=  3
      AI =   0.67676542E+02  -0.88134591E+01  -0.30327239E+03   0.10943033E+04  -0.18560520E+04   0.10930541E+04
     Rj0 =         37.5000          49.6200          57.1100
     RKj =  -0.43869604E+00   0.12486624E+01  -0.90457423E+00
```

No	Rotation (radians) X 1/1000	Moment (kip-inch) Expri.	Moment (kip-inch) Poly.	Moment (kip-inch) M.Expo.	Connection stiffness (kip-inch) X 1000 Expri.	Connection stiffness (kip-inch) X 1000 Poly.	Connection stiffness (kip-inch) X 1000 M.Expo.
1	0.00	0.00	0.00	0.00	0.1907E+02	0.2376E+02	0.2843E+02
2	1.15	22.00	26.38	21.96	0.1353E+02	0.2085E+02	0.1184E+02
3	2.31	31.22	46.17	30.85	0.5789E+01	0.1349E+02	0.4599E+01
4	4.62	34.41	67.19	35.56	0.1527E+01	0.6275E+01	0.8054E+00
5	7.50	39.33	80.85	37.87	0.1113E+01	0.3668E+01	0.1046E+01
6	9.81	40.81	88.20	40.88	0.1065E+01	0.2771E+01	0.1526E+01
7	12.12	44.25	93.93	44.74	0.1586E+01	0.2243E+01	0.1785E+01
8	15.00	49.16	99.75	50.04	0.1941E+01	0.1822E+01	0.1853E+01
9	17.31	54.08	103.67	54.26	0.2130E+01	0.1590E+01	0.1789E+01
10	19.62	59.00	107.12	58.26	0.1941E+01	0.1414E+01	0.1672E+01
11	22.50	63.91	110.95	62.83	0.1349E+01	0.1246E+01	0.1493E+01
12	24.81	66.37	113.70	66.10	0.9587E+00	0.1140E+01	0.1342E+01
13	27.12	68.34	116.22	69.03	0.9280E+00	0.1052E+01	0.1194E+01
14	30.00	71.29	119.12	72.21	0.1046E+01	0.9603E+00	0.1019E+01
15	32.31	73.75	121.26	74.42	0.1332E+01	0.8990E+00	0.8904E+00
16	34.62	77.43	123.27	76.33	0.1077E+01	0.8456E+00	0.7727E+00
17	37.50	78.66	125.63	78.37	0.1894E+00	0.7880E+00	0.6417E+00
18	39.81	78.66	127.40	78.73	0.1544E-15	0.7478E+00	0.1110E+00
19	42.12	78.66	129.08	78.89	0.0000E+00	0.7119E+00	0.2968E-01
20	45.00	78.66	131.08	78.84	0.0000E+00	0.6720E+00	-0.5778E-01
21	47.31	78.66	132.59	78.64	0.1544E-15	0.6435E+00	-0.1174E+00
22	49.62	78.66	134.08	78.31	0.3787E+00	0.6170E+00	-0.1688E+00
23	52.50	81.12	135.79	81.34	0.9706E+00	0.5881E+00	0.1026E+01
24	54.81	83.58	137.14	83.66	0.1065E+01	0.5664E+00	0.9898E+00
25	57.12	86.04	138.40	85.91	0.5918E+00	0.5470E+00	0.5477E-01
26	60.00	86.04	139.95	86.02	0.0000E+00	0.5244E+00	0.2333E-01
27	62.31	86.04	141.16	86.05	-0.8674E-18	0.5075E+00	0.2705E-02
28	64.00	86.04	142.02	86.04	0.0000E+00	0.4959E+00	-0.1022E-01

II - 72

Connection type : Double web-angle connections
Mode : All bolted

Tested by : J.B.Davison et al (1987) United Kingdom
Test Id. : JT/06

Column : UC 152x152x23
Beam : UB 254x102x22
Angle : RSA 80x60x8

Fasteners: 10.9- -M16
23/32" Oversize holes
Material : --
Fy = -- ksi
Fu = -- ksi

Major parameters

lp = 6.2992"	lu = 1.1811"	ll = 1.1811"	ta = 0.3150"
gb = 2.1650"	gc = 1.3819"	cu = 1.1811"	cl = 1.1811"
pb = 3.9370"	pc = 3.9370"	qc = -- "	
nb = 1 X 2	nc = 2 X 2		

Remark 1) Minor axis connection
2)

No	Moment (k-in)	Rotation (radians) X 1/1000	No	Moment (k-in)	Rotation (radians) X 1/1000
1	0.00	0.00	26	71.29	40.38
2	3.69	0.14	27	73.75	43.27
3	7.37	0.28	28	73.75	46.15
4	11.06	0.42	29	76.20	49.04
5	14.75	0.56	30	76.20	51.92
6	18.03	0.76	31	77.43	54.81
7	21.30	0.96	32	78.66	57.69
8	24.58	1.15	33	81.12	60.58
9	29.09	1.73	34	83.58	62.31
10	33.60	2.31	35	86.04	63.46
11	38.10	2.88	36	89.31	64.04
12	41.29	5.77	37	92.59	64.62
13	41.79	8.65	38	95.87	65.19
14	44.25	11.54			
15	46.71	14.42			
16	49.16	15.87			
17	51.13	17.31			
18	53.83	18.75			
19	56.54	20.19			
20	59.00	23.08			
21	61.46	25.96			
22	64.05	28.85			
23	66.99	31.73			
24	68.30	34.62			
25	69.44	37.50			

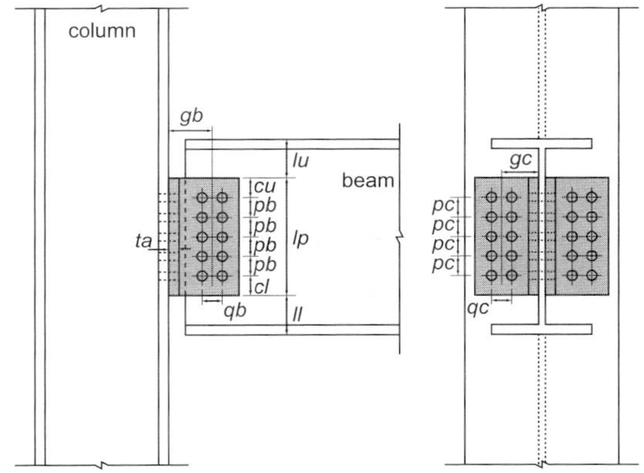

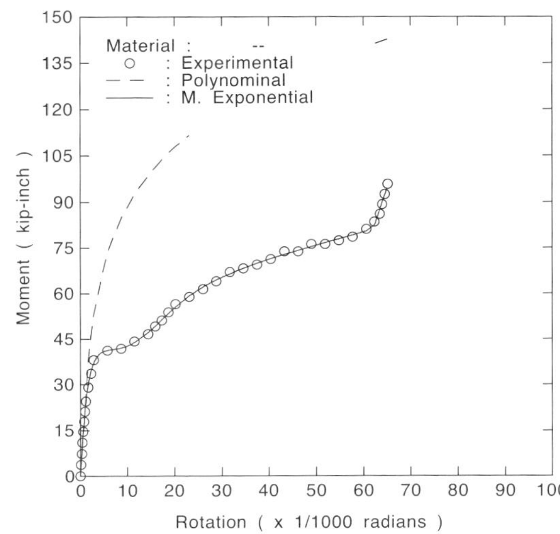

```
Moment-rotation prediction equations     ( R : X 1/1000 radians )

Frye and Morris polynominal model  :    R = Sum ( Ai X ( K*Bm )**Pi X 10**Qi )
     xd =  6.299200"    g =  2.968200"   t =  0.315000"
     A1 =  3.660000    A2 =  1.150000   A3 =  4.570000    K =  0.114987
     P1 =         1    P2 =         3   P3 =         5   Q1 =        -1    Q2 =        -3    Q3 =        -5

Modified exponential model  :
     Bm = Sum ( Ai X ( 1 - exp( - R/( 2*i*C ) ) ) )  +     Sum ( RKj X ( R - Rj0 ) )  +  Bm0
      C =  0.57811667E+00            Bmo=  0.00000000E+00          Nexp=  6         Nliner=  5
     AI =  0.36357639E+02  -0.25728135E+03  0.15758341E+04  -0.30939628E+04  0.22923441E+04  -0.49240619E+03
    Rj0 =          5.7700          15.8700          31.7300          60.5700          62.3100
    RKj =  0.14315325E+00   0.19786299E+00  0.99844468E-01   0.71093257E+00  0.31904206E+01
```

No	Rotation (radians) X 1/1000	Moment (kip-inch) Expri.	Moment (kip-inch) Poly.	Moment (kip-inch) M.Expo.	Connection stiffness (kip-inch) X 1000 Expri.	Connection stiffness (kip-inch) X 1000 Poly.	Connection stiffness (kip-inch) X 1000 M.Expo.
1	0.00	0.00	0.00	0.00	0.2645E+02	0.2376E+02	0.3105E+02
3	0.28	7.37	6.61	7.80	0.2644E+02	0.2363E+02	0.2513E+02
5	0.56	14.75	13.15	14.13	0.2235E+02	0.2319E+02	0.2040E+02
7	0.96	21.30	22.13	21.15	0.1650E+02	0.2186E+02	0.1524E+02
9	1.73	29.09	37.38	30.20	0.7811E+01	0.1713E+02	0.8723E+01
11	2.88	38.10	53.12	37.05	0.6694E+01	0.1070E+02	0.3793E+01
13	8.65	41.79	84.77	42.03	0.5122E+00	0.3155E+01	0.3813E+00
15	14.42	46.71	98.68	47.24	0.1420E+01	0.1892E+01	0.1288E+01
17	17.31	51.13	103.66	51.36	0.1619E+01	0.1590E+01	0.1524E+01
19	20.19	56.54	107.93	55.56	0.1534E+01	0.1377E+01	0.1369E+01
21	25.96	61.46	114.98	62.15	0.8760E+00	0.1094E+01	0.9177E+00
23	31.73	66.99	120.74	66.42	0.7363E+00	0.9135E+00	0.6946E+00
25	37.50	69.44	125.63	69.90	0.5181E+00	0.7880E+00	0.5340E+00
27	43.27	73.75	129.89	72.76	0.4261E+00	0.6953E+00	0.4687E+00
29	49.04	76.20	133.69	75.39	0.4261E+00	0.6238E+00	0.4463E+00
31	54.81	77.43	137.12	77.94	0.4261E+00	0.5668E+00	0.4401E+00
33	60.58	81.12	140.25	80.48	0.1207E+01	0.5202E+00	0.1150E+01
35	63.46	86.04	141.74	87.47	0.4498E+01	0.4997E+00	0.4341E+01
37	64.62	92.59	142.31	92.48	0.5681E+01	0.4921E+00	0.4341E+01

II - 73

```
Connection type : Double web-angle connections
           Mode : Welded-to-beam and bolted-to-column

Tested by  :  J.G. Yang and G.Y. Lee (2007)               Korea
Test   Id. :  3B-L-7-65

Column : H350x350x12x19             Fasteners: F10T- -M20
Beam   : H400x200x8x13                         7/8" Oversize holes
Angle  : L125x75x7                  Material :  --
                                         Fy =  46.20 ksi
                                         Fu =  68.08 ksi

                          Major parameters

     lp =  8.2677"    lu =  3.7402"    ll =  3.7402"    ta =  0.2756"
     gc =  2.5590"    cu =  1.3779"    cl =  1.3779"    pc =  2.7559"
     nc = 1 X  3
```

Remark 1) Actual strength Fy = 872.2 N/mm2, Fu = 925.1 N/mm2 for high strength bolt.
2) All bolts were tightened with 539 N-m torque (axial bolt tensile load: 147kN)

No	Moment (k-in)	Rotation (radians) X 1/1000	No	Moment (k-in)	Rotation (radians) X 1/1000
1	0.00	0.00	26	121.81	22.92
2	13.39	0.61	27	121.18	23.98
3	25.67	1.20	28	122.59	25.08
4	39.47	2.04	29	123.22	26.21
5	47.48	2.75	30	126.05	27.19
6	54.71	3.49			
7	61.94	4.22			
8	67.91	4.77			
9	71.21	5.47			
10	77.65	6.28			
11	82.52	6.90			
12	87.87	7.61			
13	92.27	8.36			
14	95.72	9.10			
15	100.12	10.22			
16	102.17	11.22			
17	102.80	12.14			
18	105.62	13.16			
19	109.71	14.62			
20	110.81	15.78			
21	112.22	16.93			
22	114.42	18.08			
23	115.68	19.30			
24	117.41	20.45			
25	120.39	21.69			

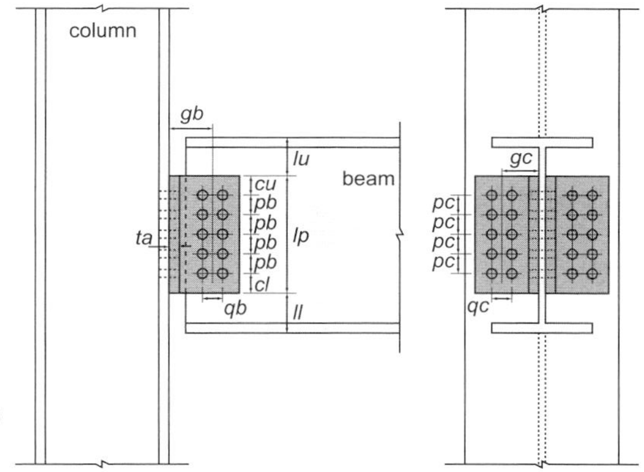

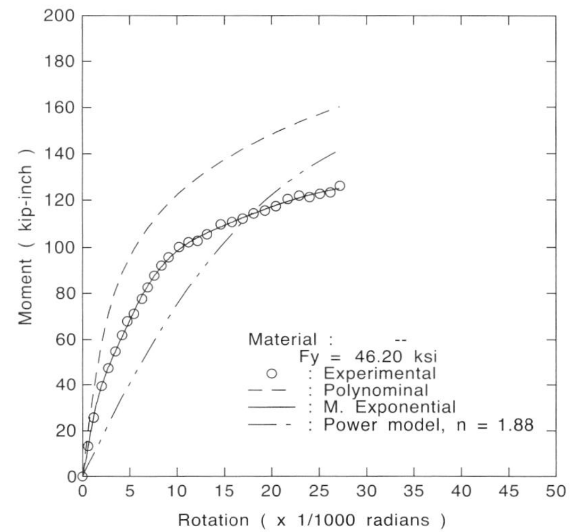

```
Moment-rotation prediction equations      ( R : X 1/1000 radians )

Frye and Morris polynominal model  :     R = Sum ( Ai X ( K*Bm )**Pi X 10**Qi )
     xd =  8.267720"    g =  5.432960"    t =  0.275600"
     A1 =  3.660000    A2 =  1.150000    A3 =  4.570000     K =  0.083488
     P1 =         1    P2 =         3    P3 =         5    Q1 =        -1     Q2 =        -3     Q3 =        -5

Modified exponential model  :
     Bm = Sum ( Ai X ( 1 - exp( - R/( 2*i*C ) ) ) )  +      Sum ( RKj X ( R - Rj0 ) )  +  Bm0
      C =   0.37978333E+00            Bmo=   0.00000000E+00            Nexp=  6          Nliner=  0
     AI =  -0.13258126E+03   0.17107253E+04  -0.76881296E+04   0.15874383E+05  -0.15343074E+05   0.57082826E+04

Power model  :     Bm = ( Rki X R ) / ( 1 + ( R/R0 )**rn )**( 1/rn )
     rn = 1.881      rki =   0.83726955E+01          rmu =   0.18733138E+03
```

No	Rotation (radians) X 1/1000	Moment (kip-inch) Expri.	Moment (kip-inch) Poly.	Moment (kip-inch) M.Expo.	Moment (kip-inch) P.Model.	Connection stiffness (kip-inch) X 1000 Expri.	Connection stiffness (kip-inch) X 1000 Poly.	Connection stiffness (kip-inch) X 1000 M.Expo.	Connection stiffness (kip-inch) X 1000 P.Model.
1	0.00	0.00	0.00	0.00	0.00	0.2185E+02	0.3273E+02	0.1505E+02	0.8373E+01
2	0.61	13.39	19.86	13.04	5.13	0.2131E+02	0.3175E+02	0.2352E+02	0.8358E+01
3	1.20	25.67	37.75	25.99	10.06	0.1900E+02	0.2833E+02	0.1961E+02	0.8320E+01
4	2.04	39.47	58.34	39.50	16.99	0.1370E+02	0.2078E+02	0.1315E+02	0.8233E+01
5	2.75	47.48	71.05	47.66	22.77	0.1059E+02	0.1555E+02	0.1034E+02	0.8131E+01
6	3.49	54.71	81.13	54.82	28.73	0.9784E+01	0.1191E+02	0.9258E+01	0.7999E+01
7	4.22	61.94	89.00	61.52	34.58	0.1050E+02	0.9560E+01	0.8922E+01	0.7845E+01
8	4.77	67.91	93.82	66.30	38.80	0.8265E+01	0.8331E+01	0.8722E+01	0.7719E+01
9	5.47	71.21	99.25	72.32	44.18	0.6233E+01	0.7129E+01	0.8296E+01	0.7541E+01
10	6.28	77.65	104.56	78.71	50.16	0.7895E+01	0.6124E+01	0.7535E+01	0.7322E+01
11	6.90	82.52	108.19	83.18	54.68	0.7696E+01	0.5524E+01	0.6801E+01	0.7142E+01
12	7.61	87.87	111.89	87.67	59.64	0.6727E+01	0.4976E+01	0.5908E+01	0.6931E+01
13	8.36	92.27	115.46	91.78	64.79	0.5246E+01	0.4504E+01	0.4975E+01	0.6697E+01
14	9.10	95.72	118.64	95.14	69.65	0.4385E+01	0.4124E+01	0.4153E+01	0.6464E+01
15	10.22	100.12	122.98	99.19	76.67	0.2939E+01	0.3664E+01	0.3157E+01	0.6107E+01
16	11.22	102.17	126.48	102.02	82.62	0.1333E+01	0.3335E+01	0.2528E+01	0.5786E+01
17	12.14	102.80	129.43	104.15	87.81	0.1678E+01	0.3084E+01	0.2142E+01	0.5495E+01
18	13.16	105.62	132.44	106.19	93.24	0.2786E+01	0.2849E+01	0.1878E+01	0.5180E+01
19	14.62	109.71	136.40	108.77	100.49	0.1766E+01	0.2573E+01	0.1689E+01	0.4745E+01
20	15.78	110.81	139.29	110.70	105.83	0.1088E+01	0.2390E+01	0.1621E+01	0.4415E+01
21	16.93	112.22	141.94	112.53	110.72	0.1572E+01	0.2236E+01	0.1575E+01	0.4108E+01
22	18.08	114.42	144.43	114.32	115.28	0.1487E+01	0.2102E+01	0.1526E+01	0.3818E+01
23	19.30	115.68	146.91	116.13	119.74	0.1276E+01	0.1978E+01	0.1458E+01	0.3532E+01
24	20.45	117.41	149.12	117.76	123.65	0.1930E+01	0.1874E+01	0.1375E+01	0.3280E+01
25	21.69	120.39	151.40	119.41	127.58	0.1767E+01	0.1775E+01	0.1268E+01	0.3026E+01
26	22.92	121.81	153.53	120.90	131.17	0.2061E+00	0.1687E+01	0.1149E+01	0.2795E+01
27	23.98	121.18	155.30	122.05	134.01	0.3221E+00	0.1618E+01	0.1042E+01	0.2613E+01
28	25.08	122.59	157.01	123.14	136.78	0.9256E+00	0.1554E+01	0.9308E+00	0.2436E+01
29	26.21	123.22	158.77	124.13	139.45	0.1794E+01	0.1492E+01	0.8189E+00	0.2267E+01
30	27.19	126.05	160.18	124.89	141.61	0.2871E+01	0.1445E+01	0.7263E+00	0.2131E+01

II - 74

Connection type : Double web-angle connections
Mode : Welded-to-beam and bolted-to-column

Tested by : J.G. Yang and G.Y. Lee (2007) Korea
Test Id. : 4B-L-7-65

Column : H350x350x12x19
Beam : H400x200x8x13
Angle : L125x75x7

Fasteners: F10T- -M20
7/8" Oversize holes
Material : --
Fy = 46.20 ksi
Fu = 68.08 ksi

Major parameters

lp = 11.0236"	lu = 2.3623"	ll = 2.3623"	ta = 0.2756"
gc = 2.5590"	cu = 1.3779"	cl = 1.3779"	pc = 2.7559"
nc = 1 X 4			

Remark 1) Actual strength Fy = 872.2 N/mm2, Fu = 925.1 N/mm2 for high strength bolt.
2) All bolts were tightened with 539 N-m torque (axial bolt tensile load: 147kN)

No	Moment (k-in)	Rotation (radians) X 1/1000
1	0.00	0.00
2	25.65	0.60
3	47.08	1.27
4	62.57	1.93
5	77.61	2.52
6	91.32	3.11
7	104.15	3.75
8	119.63	4.45
9	132.90	5.03
10	143.51	5.67
11	151.91	6.21
12	160.74	6.91
13	165.14	7.47
14	173.10	8.04
15	177.94	8.68
16	182.34	9.34
17	184.97	9.90
18	186.27	10.55
19	190.23	11.13
20	191.97	11.77
21	194.16	12.37
22	197.68	12.96
23	200.75	13.60
24	202.04	14.28
25	204.68	14.85

No	Moment (k-in)	Rotation (radians) X 1/1000
26	206.86	15.47
27	207.72	16.06
28	209.91	16.54

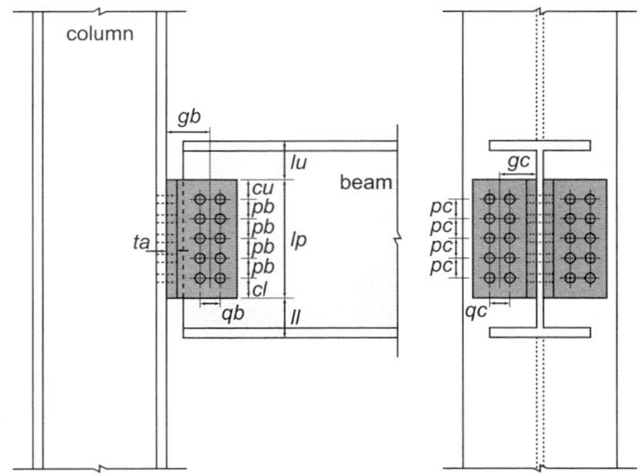

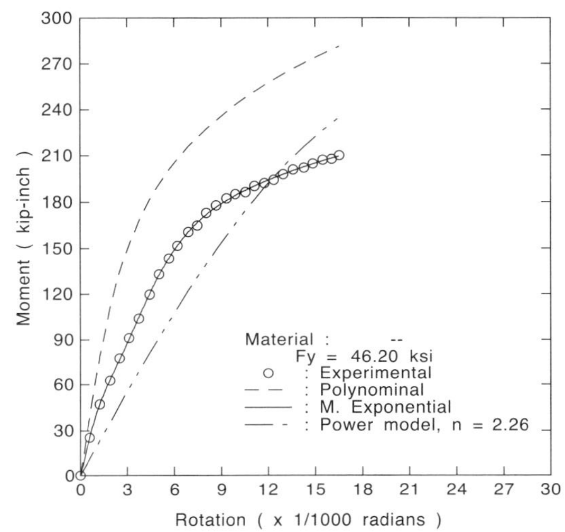

```
Moment-rotation prediction equations      ( R : X 1/1000 radians )

Frye and Morris polynominal model  :    R = Sum ( Ai X ( K*Bm )**Pi X 10**Qi )
     xd = 11.023620"   g =  5.432960"   t =  0.275600"
     A1 =  3.660000   A2 =  1.150000   A3 =  4.570000    K =  0.041857
     P1 =         1   P2 =         3   P3 =         5   Q1 =        -1    Q2 =        -3    Q3 =        -5

Modified exponential model  :
     Bm = Sum ( Ai X ( 1 - exp( - R/( 2*i*C ) ) ) )  +      Sum ( RKj X ( R - Rj0 ) )  +  Bm0
      C =  0.28748000E+00              Bmo=  0.00000000E+00             Nexp=  6          Nliner=  0
     AI =  -0.15192391E+03   0.24283127E+04  -0.11490715E+05   0.23893247E+05  -0.23211116E+05   0.87577342E+04

Power model  :    Bm = ( Rki X R ) / ( 1 + ( R/R0 )**rn )**( 1/rn )
     rn = 2.261       rki =   0.18579082E+02          rmu =   0.33303316E+03
```

No	Rotation (radians) X 1/1000	Moment (kip-inch) Expri.	Poly.	M.Expo.	P.Model.	Connection stiffness (kip-inch) X 1000 Expri.	Poly.	M.Expo.	P.Model.
1	0.00	0.00	0.00	0.00	0.00	0.4284E+02	0.6528E+02	0.3949E+02	0.1858E+02
2	0.60	25.65	38.72	25.29	11.12	0.3768E+02	0.6343E+02	0.3934E+02	0.1857E+02
3	1.27	47.08	79.08	47.41	23.59	0.2769E+02	0.5540E+02	0.2733E+02	0.1851E+02
4	1.93	62.57	111.61	63.45	35.70	0.2459E+02	0.4343E+02	0.2271E+02	0.1841E+02
5	2.52	77.61	134.32	76.71	46.53	0.2435E+02	0.3396E+02	0.2251E+02	0.1827E+02
6	3.11	91.32	152.21	90.17	57.26	0.2170E+02	0.2707E+02	0.2307E+02	0.1808E+02
7	3.75	104.15	167.76	104.92	68.75	0.2096E+02	0.2188E+02	0.2287E+02	0.1783E+02
8	4.45	119.63	181.72	120.57	81.20	0.2259E+02	0.1796E+02	0.2127E+02	0.1749E+02
9	5.03	132.90	191.32	132.20	91.15	0.2003E+02	0.1564E+02	0.1919E+02	0.1716E+02
10	5.67	143.51	200.67	143.63	102.00	0.1601E+02	0.1368E+02	0.1650E+02	0.1676E+02
11	6.21	151.91	207.70	151.94	110.97	0.1422E+02	0.1236E+02	0.1422E+02	0.1639E+02
12	6.91	160.74	215.92	160.99	122.35	0.9942E+01	0.1100E+02	0.1153E+02	0.1586E+02
13	7.47	165.14	221.80	166.90	131.07	0.1083E+02	0.1012E+02	0.9729E+01	0.1541E+02
14	8.04	173.10	227.39	172.04	139.78	0.1088E+02	0.9356E+01	0.8215E+01	0.1494E+02
15	8.68	177.94	233.13	176.85	149.15	0.7147E+01	0.8636E+01	0.6908E+01	0.1438E+02
16	9.34	182.34	238.59	181.04	158.40	0.5634E+01	0.8009E+01	0.5925E+01	0.1380E+02
17	9.90	184.97	242.92	184.17	165.95	0.3460E+01	0.7548E+01	0.5316E+01	0.1330E+02
18	10.55	186.27	247.71	187.48	174.48	0.4604E+01	0.7073E+01	0.4798E+01	0.1270E+02
19	11.13	190.23	251.67	190.13	181.62	0.4925E+01	0.6706E+01	0.4471E+01	0.1218E+02
20	11.77	191.97	255.84	192.90	189.22	0.3175E+01	0.6343E+01	0.4196E+01	0.1160E+02
21	12.37	194.16	259.59	195.38	196.10	0.4796E+01	0.6035E+01	0.3989E+01	0.1106E+02
22	12.96	197.68	263.07	197.69	202.48	0.5403E+01	0.5765E+01	0.3813E+01	0.1055E+02
23	13.60	200.75	266.67	200.07	209.05	0.3401E+01	0.5501E+01	0.3634E+01	0.1000E+02
24	14.28	202.04	270.29	202.45	215.59	0.3361E+01	0.5250E+01	0.3447E+01	0.9452E+01
25	14.85	204.68	273.30	204.38	220.88	0.4068E+01	0.5050E+01	0.3282E+01	0.8996E+01
26	15.47	206.86	276.33	206.37	226.34	0.2451E+01	0.4859E+01	0.3095E+01	0.8519E+01
27	16.06	207.72	279.20	208.14	231.24	0.3202E+01	0.4685E+01	0.2913E+01	0.8085E+01
28	16.54	209.91	281.35	209.49	235.01	0.4611E+01	0.4560E+01	0.2763E+01	0.7749E+01

II - 75

Connection type : Double web-angle connections
Mode : Welded-to-beam and bolted-to-column

Tested by : J.G. Yang and G.Y. Lee (2007) Korea
Test Id. : 5B-L-7-65

Column : H350x350x12x19
Beam : H400x200x8x13
Angle : L125x75x7

Fasteners: F10T- -M20
7/8" Oversize holes
Material : --
Fy = 46.20 ksi
Fu = 68.08 ksi

Major parameters

lp = 13.7795"	lu = 0.9843"	ll = 0.9843"	ta = 0.2756"
gc = 2.5590"	cu = 1.3779"	cl = 1.3779"	pc = 2.7559"
nc = 1 X 4			

Remark 1) Actual strength Fy = 872.2 N/mm2, Fu = 925.1 N/mm2 for high strength bolt.
2) All bolts were tightened with 539 N-m torque (axial bolt tensile load: 147kN)

No	Moment (k-in)	Rotation (radians) X 1/1000
1	0.00	0.00
2	30.97	0.35
3	52.36	0.63
4	65.63	0.93
5	87.02	1.42
6	111.36	2.11
7	132.74	2.65
8	151.18	3.17
9	167.40	3.80
10	181.42	4.34
11	195.43	5.00
12	207.23	5.64
13	218.29	6.26
14	228.61	6.88
15	234.51	7.48
16	242.63	8.20
17	247.05	8.80
18	251.47	9.26
19	255.16	9.81
20	258.85	10.36
21	261.80	10.96
22	267.70	11.56
23	269.91	12.17
24	274.34	12.76
25	274.34	13.32
26	279.50	13.94
27	283.19	14.61
28	288.35	15.16
29	290.56	15.99

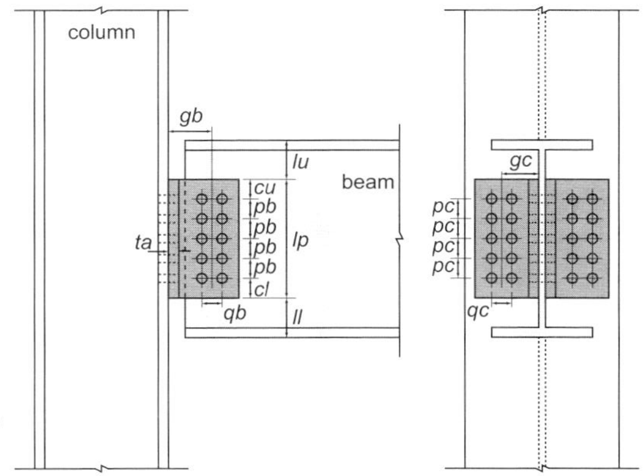

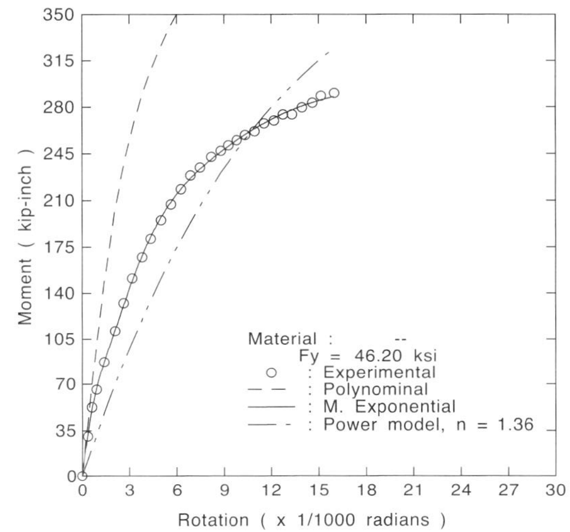

```
Moment-rotation prediction equations      ( R : X 1/1000 radians )

Frye and Morris polynominal model  :    R = Sum ( Ai X ( K*Bm )**Pi X 10**Qi )
     xd = 13.779530"    g =  5.432960"   t =  0.275600"
     A1 =  3.660000    A2 =  1.150000   A3 =  4.570000     K =  0.024501
     P1 =          1    P2 =         3   P3 =          5    Q1 =        -1     Q2 =       -3     Q3 =       -5

Modified exponential model  :
     Bm = Sum  ( Ai X ( 1 - exp( - R/( 2*i*C ) ) ) )  +      Sum  ( RKj X ( R - Rj0 ) )  +  Bm0
       C =   0.22094167E+00                 Bmo=   0.00000000E+00          Nexp=  6         Nliner=  0
      AI =  -0.11979496E+03   0.23451966E+04  -0.11937385E+05   0.26736805E+05  -0.27787270E+05   0.11059846E+05

Power model  :     Bm = ( Rki X R ) / ( 1 + ( R/R0 )**rn )**( 1/rn )
      rn = 1.362       rki =   0.35140220E+02         rmu =   0.52036469E+03
```

No	Rotation (radians) X 1/1000	Moment (kip-inch) Expri.	Poly.	M.Expo.	P.Model.	Connection stiffness (kip-inch) X 1000 Expri.	Poly.	M.Expo.	P.Model.
1	0.00	0.00	0.00	0.00	0.00	0.8827E+02	0.1115E+03	0.9896E+02	0.3514E+02
2	0.35	30.97	39.01	31.39	12.28	0.8156E+02	0.1105E+03	0.7822E+02	0.3477E+02
3	0.63	52.36	69.72	50.83	21.98	0.6100E+02	0.1080E+03	0.6091E+02	0.3433E+02
4	0.93	65.63	100.89	66.74	32.01	0.4448E+02	0.1032E+03	0.4807E+02	0.3379E+02
5	1.42	87.02	148.63	87.40	48.37	0.4014E+02	0.9031E+02	0.3798E+02	0.3278E+02
6	2.11	111.36	203.30	112.26	70.39	0.3744E+02	0.6891E+02	0.3532E+02	0.3125E+02
7	2.65	132.74	237.08	131.42	87.15	0.3724E+02	0.5496E+02	0.3452E+02	0.2997E+02
8	3.17	151.18	262.94	148.88	102.40	0.3107E+02	0.4521E+02	0.3251E+02	0.2876E+02
9	3.80	167.40	288.67	168.22	120.09	0.2601E+02	0.3676E+02	0.2855E+02	0.2729E+02
10	4.34	181.42	306.82	182.43	134.32	0.2402E+02	0.3162E+02	0.2470E+02	0.2608E+02
11	5.00	195.43	326.05	197.20	151.05	0.1974E+02	0.2691E+02	0.2019E+02	0.2463E+02
12	5.64	207.23	342.26	209.02	166.50	0.1809E+02	0.2348E+02	0.1658E+02	0.2326E+02
13	6.26	218.29	355.93	218.42	180.48	0.1732E+02	0.2093E+02	0.1398E+02	0.2202E+02
14	6.88	228.61	368.21	226.44	193.71	0.1321E+02	0.1889E+02	0.1212E+02	0.2084E+02
15	7.48	234.51	379.10	233.35	205.95	0.1049E+02	0.1726E+02	0.1084E+02	0.1974E+02
16	8.20	242.63	390.87	240.70	219.65	0.9162E+01	0.1567E+02	0.9768E+01	0.1852E+02
17	8.80	247.05	399.98	246.38	230.53	0.8588E+01	0.1455E+02	0.9084E+01	0.1756E+02
18	9.26	251.47	406.54	250.48	238.50	0.8262E+01	0.1380E+02	0.8627E+01	0.1685E+02
19	9.81	255.16	413.87	255.06	247.50	0.6737E+01	0.1301E+02	0.8118E+01	0.1606E+02
20	10.36	258.85	420.80	259.37	256.09	0.5857E+01	0.1231E+02	0.7618E+01	0.1531E+02
21	10.96	261.80	428.02	263.80	265.09	0.7332E+01	0.1163E+02	0.7059E+01	0.1454E+02
22	11.56	267.70	434.86	267.89	273.64	0.6721E+01	0.1102E+02	0.6490E+01	0.1380E+02
23	12.17	269.91	441.34	271.63	281.76	0.5609E+01	0.1048E+02	0.5914E+01	0.1312E+02
24	12.76	274.34	447.38	274.95	289.31	0.3661E+01	0.1001E+02	0.5355E+01	0.1249E+02
25	13.32	274.34	453.00	277.81	296.16	0.3981E+01	0.9584E+01	0.4833E+01	0.1192E+02
26	13.94	279.50	458.67	280.63	303.33	0.6979E+01	0.9179E+01	0.4282E+01	0.1133E+02
27	14.61	283.19	464.72	283.32	310.76	0.7657E+01	0.8770E+01	0.3718E+01	0.1073E+02
28	15.16	288.35	469.52	285.24	316.51	0.6741E+01	0.8460E+01	0.3292E+01	0.1027E+02
29	15.99	290.56	476.27	287.72	324.75	0.2672E+01	0.8046E+01	0.2710E+01	0.9628E+01

A.3 – 1

III - 1

Connection type : Top-and seat-angle connections with double web angle
Mode : All bolted

Tested by : A.Azizinamini et al. (1985) U.S.A.
Test Id. : 8S1

Column : W12X58
Beam : W8X21
F.Angle: 6 X 3.5 X 5/16 X 6.0
W.Angle: 4 X 3.5 X 1/4 X 5.5

Fasteners: A325- -3/4"D
13/16" Oversize holes
Material : A36
Fy = 40.65 ksi
Fu = 68.43 ksi

Major parameters

lp = 5.5000"	lu = 1.3750"	ll = 1.3750"	ta = 0.2500"
gb = 2.0000"	gc = 2.6250"	cu = 1.2500"	cl = 1.2500"
pb = 3.0000"	pc = 3.0000"	nb = 1 X 2	nc = 1 X 2
lt = 6.0000"	tt = 0.3125"	ls = 6.0000"	ts = 0.3125"
gt = 3.5000"	gt'= 2.0000"	gs = 3.5000"	gs'= 2.0000"
qt = 2.7500"	rt = 3.5000"	qs = 2.7500"	rs = 3.5000"
pt = 2.5000"	ps = 2.5000"		
nt = 2 X 2	nt'= 2 X 1	ns = 2 X 2	ns'= 2 X 1

Remark 1) Heavy hex high strength bolts used
2)

No	Moment (k-in)	Rotation (radians) X 1/1000	No	Moment (k-in)	Rotation (radians) X 1/1000
1	0.00	0.00	26	299.89	17.37
2	17.36	0.23	27	306.07	18.53
3	28.11	0.38	28	312.84	20.06
4	38.85	0.53	29	319.61	21.59
5	50.33	0.76	30	326.08	23.14
6	61.80	1.00	31	332.56	24.70
7	79.17	1.32	32	336.82	25.93
8	91.53	1.51	33	341.09	27.17
9	103.89	1.70	34	345.65	28.43
10	119.48	2.09	35	350.21	29.69
11	131.55	2.42	36	353.75	31.03
12	143.62	2.76	37	357.28	32.36
13	153.92	3.10	38	361.25	33.69
14	164.22	3.44	39	365.22	35.01
15	174.08	3.85	40	368.02	36.05
16	183.94	4.26	41	370.82	37.09
17	201.89	5.14	42	374.05	38.24
18	218.66	6.43	43	377.29	39.38
19	235.14	7.93	44	379.35	40.35
20	249.27	9.60			
21	264.28	11.31			
22	272.08	12.63			
23	279.88	13.95			
24	286.79	15.08			
25	293.71	16.21			

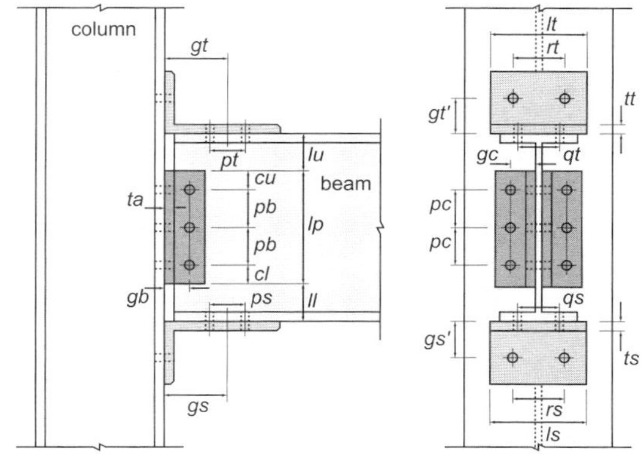

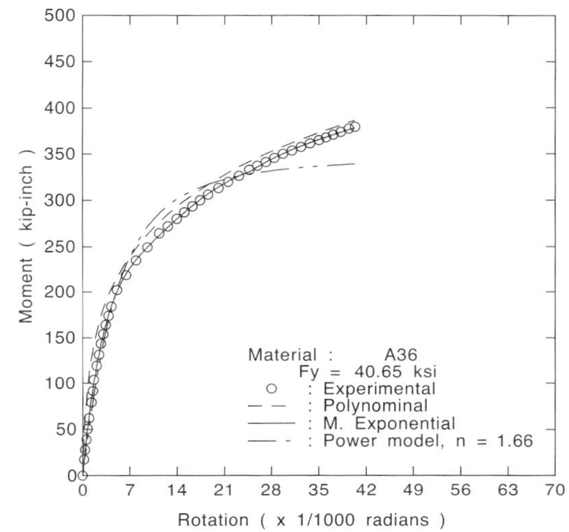

A.3 – 2

```
Moment-rotation prediction equations     ( R : X 1/1000 radians )

Frye and Morris polynominal model  :    R = Sum ( Ai X ( K*Bm )**Pi X 10**Qi )
      t =  0.312500"   d =  8.250000"  tc =  0.250000"   w =  6.000000"   g =  1.625000"
     A1 =  2.232429   A2 =  1.850728   A3 =  3.188976    K =  0.242339
     P1 =        1    P2 =        3    P3 =        5    Q1 =       -2    Q2 =       -5    Q3 =       -9

Modified exponential model  :
     Bm = Sum ( Ai X ( 1 - exp( - R/( 2*i*C ) ) ) )  +      Sum ( RKj X ( R - Rj0 ) )  +  Bm0
       C =   0.39499417E+00            Bmo=   0.00000000E+00           Nexp=  6         Nliner=  1
      AI =  -0.30968755E+02   0.62361146E+03  -0.50361983E+04   0.16127627E+05  -0.20637822E+05   0.93013975E+04
     Rj0 =          24.7000
     RKj =   0.20517930E+01

Power model  :    Bm = ( Rki X R ) / ( 1 + ( R/R0 )**rn )**( 1/rn )
      rn = 1.657      rki =   0.54559448E+02         rmu =   0.34855730E+03
```

No	Rotation (radians) X 1/1000	Moment (kip-inch) Expri.	Poly.	M.Expo.	P.Model.	Connection stiffness (kip-inch) X 1000 Expri.	Poly.	M.Expo.	P.Model.
1	0.00	0.00	0.00	0.00	0.00	0.7392E+02	0.1848E+03	0.7175E+02	0.5456E+02
3	0.38	28.11	59.82	26.52	20.71	0.7315E+02	0.1189E+03	0.6689E+02	0.5375E+02
5	0.76	50.33	95.23	50.94	40.93	0.4885E+02	0.7314E+02	0.6093E+02	0.5207E+02
7	1.32	79.17	127.35	82.44	69.09	0.6066E+02	0.4601E+02	0.5207E+02	0.4870E+02
9	1.70	103.89	143.01	101.24	87.18	0.5678E+02	0.3684E+02	0.4648E+02	0.4603E+02
11	2.42	131.55	165.57	131.29	118.39	0.3573E+02	0.2696E+02	0.3738E+02	0.4070E+02
13	3.10	153.92	181.86	154.14	144.20	0.3050E+02	0.2168E+02	0.3050E+02	0.3576E+02
15	3.85	174.08	196.59	174.63	169.05	0.2398E+02	0.1790E+02	0.2446E+02	0.3070E+02
17	5.14	201.89	216.89	201.08	203.78	0.1738E+02	0.1387E+02	0.1703E+02	0.2336E+02
19	7.93	235.14	248.68	235.77	253.15	0.9792E+01	0.9502E+01	0.9283E+01	0.1313E+02
21	11.31	264.28	276.05	262.17	285.96	0.7173E+01	0.6992E+01	0.7002E+01	0.7073E+01
23	13.95	279.88	292.91	279.97	301.16	0.6017E+01	0.5838E+01	0.6474E+01	0.4644E+01
25	16.21	293.71	305.27	293.99	310.12	0.5727E+01	0.5134E+01	0.5886E+01	0.3368E+01
27	18.53	306.07	316.51	306.71	316.86	0.4941E+01	0.4580E+01	0.5053E+01	0.2499E+01
29	21.59	319.61	329.61	320.26	323.27	0.4297E+01	0.4023E+01	0.3812E+01	0.1757E+01
31	24.70	332.56	341.43	330.26	327.94	0.3769E+01	0.3589E+01	0.2647E+01	0.1276E+01
33	27.17	341.09	350.02	340.89	330.74	0.3535E+01	0.3309E+01	0.3951E+01	0.1014E+01
35	29.69	350.21	358.06	350.09	333.04	0.3141E+01	0.3070E+01	0.3362E+01	0.8153E+00
37	32.36	357.28	365.97	358.43	335.00	0.2826E+01	0.2855E+01	0.2912E+01	0.6586E+00
39	35.01	365.22	373.26	365.70	336.58	0.2825E+01	0.2672E+01	0.2607E+01	0.5413E+00
41	37.09	370.82	378.64	370.95	337.63	0.2751E+01	0.2547E+01	0.2440E+01	0.4680E+00
43	39.38	377.29	384.40	376.39	338.62	0.2447E+01	0.2420E+01	0.2311E+01	0.4023E+00

III - 2

Connection type : Top-and seat-angle connections with double web angle
Mode : All bolted

Tested by : A.Azizinamini et al. (1985) U.S.A.
Test Id. : 8S2

Column : W12X58
Beam : W8X21
F.Angle: 6 X 3.5 X 3/8 X 6.0
W.Angle: 4 X 3.5 X 1/4 X 5.5

Fasteners: A325- -3/4"D
13/16" Oversize holes
Material : A36
Fy = 40.65 ksi
Fu = 68.43 ksi

Major parameters

lp = 5.5000"	lu = 1.3750"	ll = 1.3750"	ta = 0.2500"
gb = 2.0000"	gc = 2.6250"	cu = 1.2500"	cl = 1.2500"
pb = 3.0000"	pc = 3.0000"	nb = 1 X 2	nc = 1 X 2
lt = 6.0000"	tt = 0.3750"	ls = 6.0000"	ts = 0.3750"
gt = 3.5000"	gt'= 2.0000"	gs = 3.5000"	gs'= 2.0000"
qt = 2.7500"	rt = 3.5000"	qs = 2.7500"	rs = 3.5000"
pt = 2.5000"	ps = 2.5000"		
nt = 2 X 2	nt'= 2 X 1	ns = 2 X 2	ns'= 2 X 1

Remark 1) Heavy hex high strength bolts used
2)

No	Moment (k-in)	Rotation (radians) X 1/1000	No	Moment (k-in)	Rotation (radians) X 1/1000
1	0.00	0.00	26	329.31	7.17
2	13.14	0.05	27	339.41	8.20
3	26.29	0.11	28	349.52	9.23
4	39.43	0.16	29	356.69	10.20
5	50.17	0.21	30	363.86	11.18
6	60.91	0.26	31	368.82	12.12
7	74.74	0.37	32	373.79	13.06
8	88.56	0.48	33	378.58	14.40
9	104.15	0.66	34	383.38	15.74
10	119.74	0.84	35	385.10	16.76
11	133.18	1.01	36	386.83	17.77
12	146.61	1.17	37	383.55	18.77
13	160.04	1.33	38	386.01	19.56
14	173.57	1.49	39	388.48	20.36
15	187.10	1.64	40	389.32	21.39
16	200.63	1.79	41	386.04	22.30
17	213.13	2.02	42	385.71	23.19
18	225.63	2.24	43	385.38	24.09
19	241.80	2.66	44	386.99	25.50
20	252.82	3.09	45	388.59	26.91
21	263.83	3.52	46	390.20	28.32
22	282.34	4.24			
23	294.53	4.96			
24	306.71	5.69			
25	318.01	6.43			

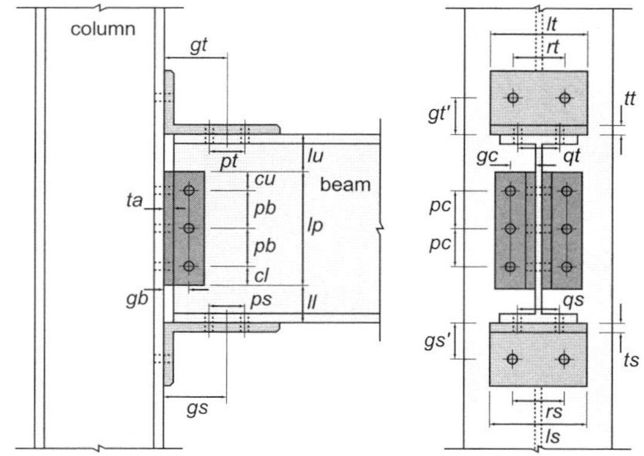

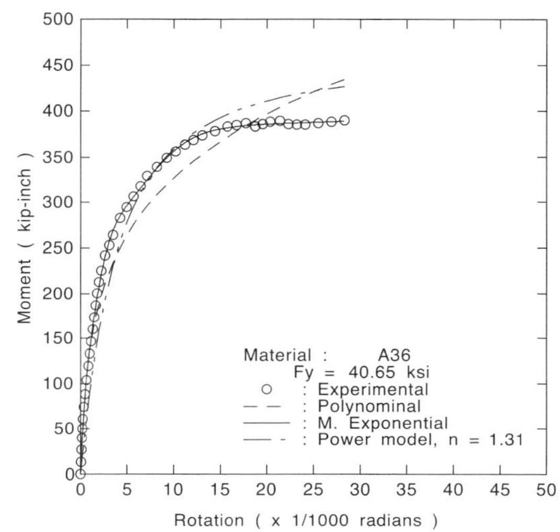

A.3 – 4

```
Moment-rotation prediction equations     ( R : X 1/1000 radians )

Frye and Morris polynominal model  :    R = Sum ( Ai X ( K*Bm )**Pi X 10**Qi )
      t =  0.375000"   d =  8.250000"  tc =  0.250000"   w =  6.000000"   g =  1.625000"
     A1 =  2.232429   A2 =  1.850728   A3 =  3.188976    K =  0.197288
     P1 =         1   P2 =         3   P3 =         5   Q1 =        -2   Q2 =        -5   Q3 =        -9

Modified exponential model  :
     Bm = Sum ( Ai X ( 1 - exp( - R/( 2*i*C ) ) ) )  +     Sum ( RKj X ( R - Rj0 ) )  +  Bm0
      C =   0.24967667E+00            Bmo=   0.00000000E+00        Nexp=  6        Nliner=  1
     AI =   0.42180824E+03  -0.38328612E+04   0.14038976E+05  -0.22251927E+05   0.15479094E+05  -0.34709931E+04
    Rj0 =          15.7400
    RKj =   0.38201717E+00

Power model  :    Bm = ( Rki X R ) / ( 1 + ( R/R0 )**rn )**( 1/rn )
      rn = 1.307      rki =   0.98019124E+02         rmu =   0.45685652E+03
```

No	Rotation (radians) X 1/1000	Moment (kip-inch) Expri.	Poly.	M.Expo.	P.Model.	Connection stiffness (kip-inch) X 1000 Expri.	Poly.	M.Expo.	P.Model.
1	0.00	0.00	0.00	0.00	0.00	0.2407E+03	0.2271E+03	0.2791E+03	0.9802E+02
3	0.11	26.29	24.33	27.12	10.64	0.2405E+03	0.2147E+03	0.2207E+03	0.9675E+02
5	0.21	50.17	45.22	47.73	20.55	0.2208E+03	0.1888E+03	0.1804E+03	0.9504E+02
7	0.37	74.74	71.71	72.60	35.29	0.1271E+03	0.1487E+03	0.1387E+03	0.9202E+02
9	0.66	104.15	107.35	106.53	61.25	0.8517E+02	0.1005E+03	0.9978E+02	0.8586E+02
11	1.01	133.18	136.48	137.51	89.63	0.8266E+02	0.7143E+02	0.8232E+02	0.7839E+02
13	1.33	160.04	157.00	162.88	113.99	0.8573E+02	0.5615E+02	0.7429E+02	0.7162E+02
15	1.64	187.10	172.65	184.57	134.95	0.8858E+02	0.4685E+02	0.6761E+02	0.6566E+02
17	2.02	213.13	188.79	208.46	158.50	0.5535E+02	0.3898E+02	0.5846E+02	0.5891E+02
19	2.66	241.80	210.93	240.82	193.16	0.3220E+02	0.3048E+02	0.4217E+02	0.4904E+02
21	3.52	263.83	234.05	269.34	230.75	0.2571E+02	0.2377E+02	0.2540E+02	0.3868E+02
23	4.96	294.53	263.32	295.88	277.18	0.1680E+02	0.1758E+02	0.1434E+02	0.2680E+02
25	6.43	318.01	286.25	315.50	310.43	0.1528E+02	0.1402E+02	0.1303E+02	0.1916E+02
27	8.20	339.41	308.55	337.88	338.84	0.9786E+01	0.1134E+02	0.1188E+02	0.1337E+02
29	10.20	356.69	329.19	358.54	361.30	0.7368E+01	0.9383E+01	0.8499E+01	0.9354E+01
31	12.12	368.82	345.86	371.46	376.73	0.5264E+01	0.8091E+01	0.5094E+01	0.6926E+01
33	14.40	378.58	362.99	379.60	390.23	0.3583E+01	0.6978E+01	0.2299E+01	0.5046E+01
35	16.76	385.10	378.37	383.44	400.51	0.1697E+01	0.6131E+01	0.1210E+01	0.3778E+01
37	18.77	383.55	390.12	385.24	407.32	0.2715E+00	0.5566E+01	0.6570E+00	0.3024E+01
39	20.36	388.48	398.66	386.11	411.75	0.2107E+01	0.5194E+01	0.4628E+00	0.2570E+01
41	22.30	386.04	408.42	386.90	416.29	-0.1974E+01	0.4805E+01	0.3684E+00	0.2137E+01
43	24.09	385.38	416.75	387.53	419.83	0.2150E+00	0.4500E+01	0.3453E+00	0.1823E+01
45	26.91	388.59	428.80	388.51	424.42	0.1139E+01	0.4100E+01	0.3501E+00	0.1448E+01

III - 3

Connection type : Top-and seat-angle connections with double web angle
Mode : All bolted

Tested by : A.Azizinamini et al. (1985) U.S.A.
Test Id. : 8S3

Column : W12X58
Beam : W8X21
F.Angle: 6 X 3.5 X 5/16 X 8.0
W.Angle: 4 X 3.5 X 1/4 X 5.5

Fasteners: A325- -3/4"D
13/16" Oversize holes
Material : A36
Fy = 40.65 ksi
Fu = 68.43 ksi

Major parameters

lp = 5.5000"	lu = 1.3750"	ll = 1.3750"	ta = 0.2500"
gb = 2.0000"	gc = 2.6250"	cu = 1.2500"	cl = 1.2500"
pb = 3.0000"	pc = 3.0000"	nb = 1 X 2	nc = 1 X 2
lt = 8.0000"	tt = 0.3125"	ls = 8.0000"	ts = 0.3125"
gt = 3.5000"	gt'= 2.0000"	gs = 3.5000"	gs'= 2.0000"
qt = 2.7500"	rt = 3.5000"	qs = 2.7500"	rs = 3.5000"
pt = 2.5000"	ps = 2.5000"		
nt = 2 X 2	nt'= 2 X 1	ns = 2 X 2	ns'= 2 X 1

Remark 1) Heavy hex high strength bolts used
2)

No	Moment (k-in)	Rotation (radians) X 1/1000	No	Moment (k-in)	Rotation (radians) X 1/1000
1	0.00	0.00	26	380.91	15.42
2	19.70	0.18	27	386.56	16.52
3	39.41	0.36	28	392.66	17.84
4	59.26	0.52	29	398.76	19.15
5	79.11	0.69	30	405.15	20.48
6	98.96	0.96	31	411.54	21.81
7	118.82	1.22	32	416.77	23.33
8	139.56	1.59	33	422.00	24.85
9	160.30	1.96	34	427.07	26.20
10	174.13	2.20	35	432.14	27.54
11	187.96	2.45	36	436.76	28.64
12	203.85	2.77	37	441.37	29.74
13	219.74	3.10	38	445.87	31.36
14	239.76	3.65	39	450.37	32.98
15	261.55	4.28	40	454.56	34.34
16	280.41	5.12	41	458.75	35.70
17	292.64	5.74	42	464.73	37.46
18	304.87	6.37	43	470.70	39.21
19	320.80	7.38			
20	334.10	8.61			
21	348.58	10.18			
22	356.14	11.25			
23	363.69	12.31			
24	369.48	13.32			
25	375.27	14.33			

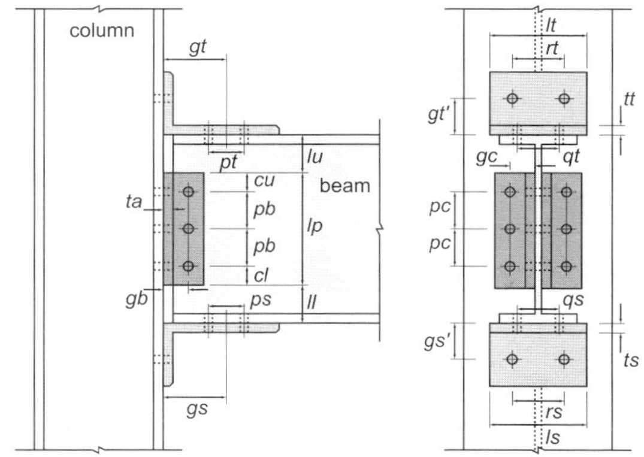

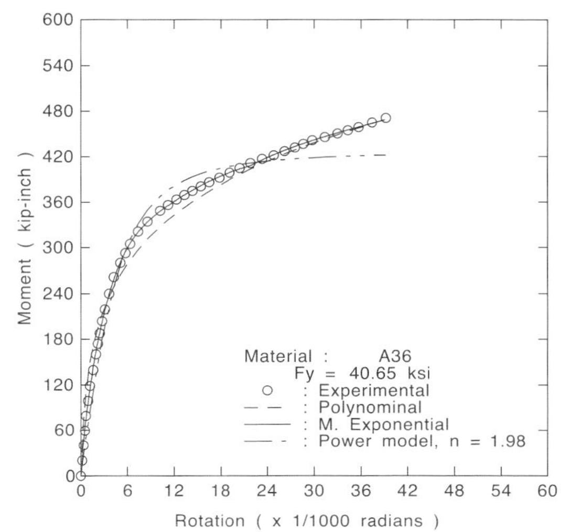

A.3 – 6

```
Moment-rotation prediction equations      ( R : X 1/1000 radians )

Frye and Morris polynominal model  :    R = Sum ( Ai X ( K*Bm )**Pi X 10**Qi )
      t =  0.312500"   d =  8.250000"  tc =  0.250000"   w =  8.000000"   g =  1.625000"
     A1 =  2.232429   A2 =  1.850728   A3 =  3.188976    K =  0.198473
     P1 =        1    P2 =        3    P3 =        5    Q1 =       -2    Q2 =       -5    Q3 =       -9

Modified expcnential model  :
     Bm = Sum ( Ai X ( 1 - exp( - R/( 2*i*C ) ) ) )  +      Sum ( RKj X ( R - Rj0 ) )  +  Bm0
       C =  0.37170833E+00              Bmo=   0.00000000E+00          Nexp=  6         Nliner=  1
      AI =  0.76539596E+02   0.22759533E+02  -0.29662357E+04   0.11705987E+05  -0.15662942E+05   0.72495441E+04
     Rj0 =          21.8100
     RKj =  0.25171042E+01

Power model  :    Bm = ( Rki X R ) / ( 1 + ( R/R0 )**rn )**( 1/rn )
     rn = 1.980      rki =   0.71606768E+02          rmu =   0.42715362E+03
```

No	Rotation (radians) X 1/1000	Moment (kip-inch) Expri.	Moment (kip-inch) Poly.	Moment (kip-inch) M.Expo.	Moment (kip-inch) P.Model.	Connection stiffness (kip-inch) X 1000 Expri.	Connection stiffness (kip-inch) X 1000 Poly.	Connection stiffness (kip-inch) X 1000 M.Expo.	Connection stiffness (kip-inch) X 1000 P.Model.
1	0.00	0.00	0.00	0.00	0.00	0.1096E+03	0.2257E+03	0.1363E+03	0.7161E+02
3	0.36	39.41	69.68	43.08	25.68	0.1153E+03	0.1503E+03	0.1059E+03	0.7120E+02
5	0.69	79.11	109.38	74.86	49.00	0.1028E+03	0.9684E+02	0.8809E+02	0.7013E+02
7	1.22	118.82	149.82	116.80	85.79	0.6663E+02	0.6007E+02	0.7030E+02	0.6716E+02
9	1.96	160.30	185.59	163.00	133.32	0.5682E+02	0.3967E+02	0.5590E+02	0.6114E+02
11	2.45	187.96	202.93	188.30	161.73	0.5351E+02	0.3264E+02	0.4905E+02	0.5645E+02
13	3.10	219.74	222.17	217.85	196.59	0.4400E+02	0.2644E+02	0.4117E+02	0.4973E+02
15	4.28	261.55	249.09	259.26	248.25	0.2929E+02	0.1989E+02	0.2962E+02	0.3819E+02
17	5.74	292.64	274.57	294.58	295.25	0.1963E+02	0.1537E+02	0.1941E+02	0.2666E+02
19	7.38	320.80	297.10	320.29	331.15	0.1349E+02	0.1234E+02	0.1264E+02	0.1776E+02
21	10.18	348.58	327.03	347.73	367.50	0.7943E+01	0.9344E+01	0.7931E+01	0.9296E+01
23	12.31	363.69	345.34	363.12	383.47	0.6398E+01	0.7934E+01	0.6660E+01	0.5990E+01
25	14.33	375.27	360.31	375.71	393.51	0.5460E+01	0.6966E+01	0.5847E+01	0.4118E+01
27	16.52	386.56	374.66	387.56	401.03	0.4918E+01	0.6168E+01	0.4961E+01	0.2851E+01
29	19.15	398.76	389.89	399.15	407.21	0.4722E+01	0.5438E+01	0.3848E+01	0.1921E+01
31	21.81	411.54	403.56	407.96	411.49	0.4168E+01	0.4869E+01	0.5316E+01	0.1345E+01
33	24.85	422.00	417.57	422.58	414.91	0.3614E+01	0.4359E+01	0.4344E+01	0.9344E+00
35	27.54	432.14	428.79	433.38	417.08	0.4013E+01	0.3996E+01	0.3717E+01	0.6987E+00
37	29.74	441.37	437.27	441.10	418.46	0.3630E+01	0.3746E+01	0.3349E+01	0.5615E+00
39	32.98	450.37	448.91	451.34	420.03	0.2939E+01	0.3433E+01	0.2985E+01	0.4169E+00
41	35.70	458.75	457.94	459.19	421.04	0.3222E+01	0.3211E+01	0.2800E+01	0.3316E+00
43	39.21	470.70	468.77	468.75	422.06	0.3405E+01	0.2968E+01	0.2661E+01	0.2526E+00

III - 4

Connection type : Top-and seat-angle connections with double web angle
Mode : All bolted

Tested by : A.Azizinamini et al. (1985) U.S.A.
Test Id. : 8S4

Column : W12X58
Beam : W8X21
F.Angle: 6 X 6.0 X 3/8 X 6.0
W.Angle: 4 X 3.5 X 1/4 X 5.5

Fasteners: A325- -3/4"D
13/16" Oversize holes
Material : A36
Fy = 40.65 ksi
Fu = 68.43 ksi

Major parameters

lp = 5.5000"	lu = 1.3750"	ll = 1.3750"	ta = 0.2500"
gb = 2.0000"	gc = 2.6250"	cu = 1.2500"	cl = 1.2500"
pb = 3.0000"	pc = 3.0000"	nb = 1 X 2	nc = 1 X 2
lt = 6.0000"	tt = 0.3750"	ls = 6.0000"	ts = 0.3750"
gt = 3.5000"	gt'= 4.5000"	gs = 3.5000"	gs'= 4.5000"
qt = 2.7500"	rt = 3.5000"	qs = 2.7500"	rs = 3.5000"
pt = 2.5000"	ps = 2.5000"		
nt = 2 X 2	nt'= 2 X 1	ns = 2 X 2	ns'= 2 X 1

Remark 1) Heavy hex high strength bolts used
2)

No	Moment (k-in)	Rotation (radians) X 1/1000	No	Moment (k-in)	Rotation (radians) X 1/1000
1	0.00	0.00	26	161.38	23.33
2	6.85	0.41	27	164.74	24.77
3	13.71	0.82	28	168.11	26.21
4	20.56	1.24	29	170.00	27.59
5	27.22	1.70	30	171.89	28.97
6	33.87	2.16	31	175.12	29.53
7	40.53	2.62	32	177.60	30.97
8	49.41	3.31	33	180.08	32.41
9	53.30	3.99	34	181.33	34.04
10	67.18	4.67	35	182.59	35.66
11	75.06	5.36	36	183.84	37.29
12	85.17	6.23	37	186.32	38.92
13	94.27	7.10	38	188.80	40.55
14	103.37	7.98			
15	110.26	8.90			
16	117.16	9.83			
17	124.78	11.43			
18	132.41	13.04			
19	137.10	14.15			
20	141.78	15.27			
21	144.86	16.45			
22	147.93	17.62			
23	151.29	19.09			
24	154.65	20.56			
25	153.02	21.94			

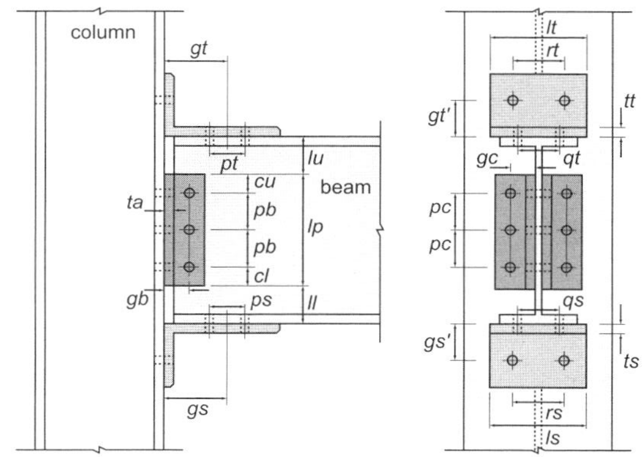

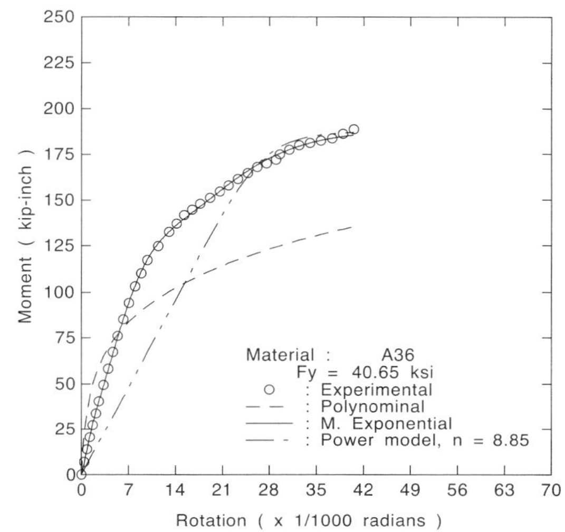

A.3 – 8

```
Moment-rotation prediction equations     ( R : X 1/1000 radians )

Frye and Morris polynominal model   :    R = Sum ( Ai X ( K*Bm )**Pi X 10**Qi )
     t =  0.375000"   d =  8.250000"  tc =  0.250000"   w =  6.000000"   g =  4.125000"
    A1 =  2.232429   A2 =  1.850728   A3 =  3.188976    K =  0.693822
    P1 =        1    P2 =        3    P3 =        5    Q1 =       -2    Q2 =       -5    Q3 =       -9

Modified exponential model   :
    Bm = Sum ( Ai X ( 1 - exp( - R/( 2*i*C ) ) ) )  +      Sum ( RKj X ( R - Rj0 ) )  +  Bm0
     C =  0.44100917E+00              Bmo=  0.00000000E+00           Nexp=  6       Nliner=  0
    AI = -0.16086198E+03   0.25470589E+04  -0.13077879E+05   0.29087866E+05  -0.29476580E+05   0.11268508E+05

Power model   :    Bm = ( Rki X R ) / ( 1 + ( R/R0 )**rn )**( 1/rn )
    rn = 8.851      rki =   0.68436672E+01         rmu =   0.18766589E+03
```

No	Rotation (radians) X 1/1000	Moment (kip-inch) Expri.	Poly.	M.Expo.	P.Model.	Connection stiffness (kip-inch) X 1000 Expri.	Poly.	M.Expo.	P.Model.
1	0.00	0.00	0.00	0.00	0.00	0.1662E+02	0.6456E+02	0.9195E+01	0.6844E+01
3	0.82	13.71	34.78	13.22	5.64	0.1662E+02	0.2400E+02	0.1843E+02	0.6844E+01
5	1.70	27.22	49.90	28.10	11.62	0.1443E+02	0.1289E+02	0.1526E+02	0.6844E+01
7	2.62	40.53	59.63	40.82	17.94	0.1385E+02	0.8783E+01	0.1272E+02	0.6844E+01
9	3.99	58.30	69.54	57.73	27.30	0.1298E+02	0.6056E+01	0.1237E+02	0.6844E+01
11	5.36	76.06	76.79	74.76	36.67	0.1186E+02	0.4673E+01	0.1234E+02	0.6844E+01
13	7.10	94.27	83.98	94.98	48.62	0.1042E+02	0.3652E+01	0.1052E+02	0.6844E+01
15	8.90	110.26	89.93	111.34	60.94	0.7442E+01	0.3002E+01	0.7646E+01	0.6843E+01
17	11.43	124.78	96.73	126.33	78.24	0.4756E+01	0.2419E+01	0.4511E+01	0.6840E+01
19	14.15	137.10	102.72	136.32	96.83	0.4197E+01	0.2014E+01	0.3121E+01	0.6822E+01
21	16.45	144.86	107.05	143.11	112.42	0.2612E+01	0.1771E+01	0.2879E+01	0.6762E+01
23	19.09	151.29	111.44	150.70	130.08	0.2287E+01	0.1560E+01	0.2862E+01	0.6547E+01
25	21.94	158.02	115.63	158.71	147.99	0.2434E+01	0.1386E+01	0.2720E+01	0.5920E+01
27	24.77	164.74	119.34	165.95	163.09	0.2334E+01	0.1251E+01	0.2379E+01	0.4684E+01
29	27.59	170.00	122.71	172.04	174.05	0.1371E+01	0.1142E+01	0.1930E+01	0.3070E+01
31	29.53	175.12	124.87	175.47	179.00	0.4641E+01	0.1078E+01	0.1613E+01	0.2072E+01
33	32.41	180.08	127.88	179.49	183.36	0.1275E+01	0.9958E+00	0.1183E+01	0.1050E+01
35	35.66	182.59	130.99	182.67	185.70	0.7709E+00	0.9187E+00	0.7939E+00	0.4635E+00
37	38.92	186.32	133.87	184.77	186.73	0.1519E+01	0.8538E+00	0.5113E+00	0.2069E+00

III - 5

Connection type : Top-and seat-angle connections with double web angle
Mode : All bolted

Tested by : A.Azizinamini et al. (1985) U.S.A.
Test Id. : 8S5

Column : W12X58
Beam : W8X21
F.Angle: 6 X 4.0 X 3/8 X 8.0
W.Angle: 4 X 3.5 X 1/4 X 5.5

Fasteners: A325- -3/4"D
13/16" Oversize holes
Material : A36
Fy = 40.65 ksi
Fu = 68.43 ksi

Major parameters

lp = 5.5000"	lu = 1.3750"	ll = 1.3750"	ta = 0.2500"
gb = 2.0000"	gc = 2.6250"	cu = 1.2500"	cl = 1.2500"
pb = 3.0000"	pc = 3.0000"	nb = 1 X 2	nc = 1 X 2
lt = 8.0000"	tt = 0.3750"	ls = 8.0000"	ts = 0.3750"
gt = 3.5000"	gt'= 2.5000"	gs = 3.5000"	gs'= 2.5000"
qt = 2.7500"	rt = 5.5000"	qs = 2.7500"	rs = 5.5000"
pt = 2.5000"	ps = 2.5000"		
nt = 2 X 2	nt'= 2 X 1	ns = 2 X 2	ns'= 2 X 1

Remark 1) Heavy hex high strength bolts used
2)

No	Moment (k-in)	Rotation (radians) X 1/1000
1	0.00	0.00
2	14.53	0.16
3	29.05	0.32
4	43.58	0.48
5	61.25	0.70
6	78.92	0.93
7	92.89	1.23
8	106.85	1.52
9	120.82	1.82
10	133.61	2.13
11	146.41	2.43
12	159.20	2.74
13	177.24	3.35
14	195.29	3.96
15	209.23	4.61
16	223.18	5.27
17	238.94	6.21
18	254.71	7.16
19	268.49	8.46
20	282.27	9.76
21	290.42	10.82
22	298.58	11.87
23	302.99	13.34
24	307.41	14.81
25	311.83	16.28

No	Moment (k-in)	Rotation (radians) X 1/1000
26	316.83	17.72
27	321.83	19.16
28	326.83	20.59
29	331.08	22.24
30	335.34	23.88
31	339.60	25.53
32	343.43	26.98
33	347.25	28.43
34	351.08	29.88
35	357.36	31.83
36	363.64	33.78
37	366.05	35.47
38	368.46	37.16
39	372.28	38.60
40	376.11	40.04

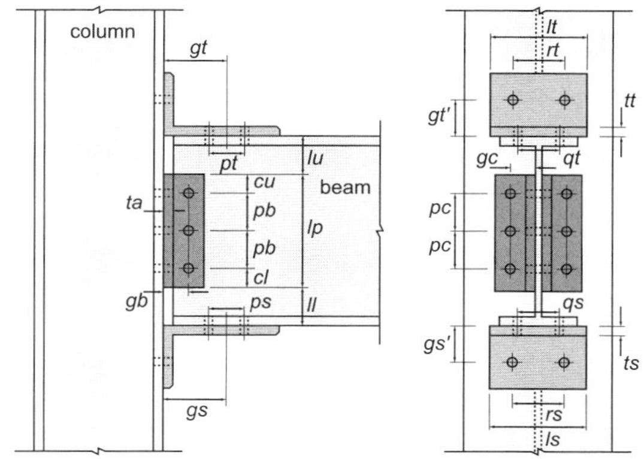

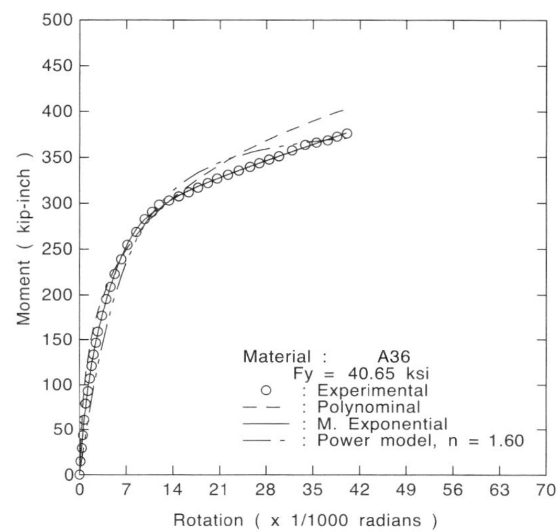

```
Moment-rotation prediction equations     ( R : X 1/1000 radians )

Frye and Morris polynominal model  :    R = Sum ( Ai X ( K*Bm )**Pi X 10**Qi )
      t =  0.375000"   d =  8.250000"  tc =  0.250000"   w =  8.000000"   g =  2.125000"
     A1 =  2.232429   A2 =  1.850728   A3 =  3.188976    K =  0.232088
     P1 =        1    P2 =        3    P3 =        5    Q1 =       -2    Q2 =       -5    Q3 =       -9

Modified exponential model  :
     Bm = Sum ( Ai X ( 1 - exp( - R/( 2*i*C ) ) ) )  +      Sum ( RKj X ( R - Rj0 ) )  +  Bm0
      C =   0.37343167E+00          Bmo=   0.00000000E+00          Nexp=  6        Nliner=  1
     AI =  -0.37271090E+02   0.84497851E+03  -0.30419542E+04   0.45055775E+04  -0.28173964E+04   0.85258295E+03
    Rj0 =           11.8700
    RKj =   0.24938829E+01

Power model  :    Bm = ( Rki X R ) / ( 1 + ( R/R0 )**rn )**( 1/rn )
      rn = 1.599      rki =   0.49060168E+02         rmu =   0.38809875E+03
```

No	Rotation (radians) X 1/1000	Moment (kip-inch) Expri.	Moment (kip-inch) Poly.	Moment (kip-inch) M.Expo.	Moment (kip-inch) P.Model.	Connection stiffness (kip-inch) X 1000 Expri.	Connection stiffness (kip-inch) X 1000 Poly.	Connection stiffness (kip-inch) X 1000 M.Expo.	Connection stiffness (kip-inch) X 1000 P.Model.
1	0.00	0.00	0.00	0.00	0.00	0.9130E+02	0.1930E+03	0.1021E+03	0.4906E+02
3	0.32	29.05	54.14	30.00	15.55	0.9130E+02	0.1368E+03	0.8681E+02	0.4860E+02
5	0.70	61.25	94.69	60.36	34.07	0.7816E+02	0.8151E+02	0.7138E+02	0.4744E+02
7	1.23	92.89	128.23	93.32	58.32	0.4708E+02	0.5129E+02	0.5568E+02	0.4527E+02
9	1.82	120.82	153.66	122.56	84.32	0.4442E+02	0.3629E+02	0.4377E+02	0.4231E+02
11	2.43	146.41	173.18	146.85	109.29	0.4167E+02	0.2804E+02	0.3594E+02	0.3899E+02
13	3.35	177.24	195.38	176.26	142.69	0.2967E+02	0.2113E+02	0.2896E+02	0.3401E+02
15	4.61	209.23	218.45	208.87	181.55	0.2130E+02	0.1594E+02	0.2304E+02	0.2769E+02
17	6.21	238.94	240.77	240.92	220.42	0.1664E+02	0.1227E+02	0.1711E+02	0.2112E+02
19	8.46	268.49	264.80	271.26	260.00	0.1059E+02	0.9377E+01	0.1018E+02	0.1453E+02
21	10.82	290.42	284.61	288.96	288.59	0.7729E+01	0.7584E+01	0.5264E+01	0.1007E+02
23	13.34	302.99	302.07	301.87	309.85	0.3009E+01	0.6333E+01	0.4887E+01	0.7023E+01
25	16.28	311.83	319.12	313.73	326.97	0.3244E+01	0.5342E+01	0.3424E+01	0.4815E+01
27	19.16	321.83	333.44	322.69	338.73	0.3475E+01	0.4650E+01	0.2891E+01	0.3458E+01
29	22.24	331.08	346.88	331.23	347.83	0.2589E+01	0.4096E+01	0.2685E+01	0.2513E+01
31	25.53	339.60	359.58	339.90	354.94	0.2617E+01	0.3644E+01	0.2600E+01	0.1851E+01
33	28.43	347.25	369.66	347.38	359.69	0.2641E+01	0.3327E+01	0.2562E+01	0.1449E+01
35	31.83	357.36	380.45	356.04	364.01	0.3217E+01	0.3025E+01	0.2535E+01	0.1114E+01
37	35.47	366.05	390.97	365.24	367.58	0.1425E+01	0.2761E+01	0.2517E+01	0.8621E+00
39	38.60	372.28	399.31	373.10	370.02	0.2657E+01	0.2572E+01	0.2508E+01	0.7041E+00

III - 6

```
Connection type : Top-and seat-angle connections with double web angle
           Mode : All bolted

Tested by  :  A.Azizinamini et al. (1985)                U.S.A.
Test  Id.  :  8S6

Column : W12X58                         Fasteners: A325- -3/4"D
Beam   : W8X21                                13/16" Oversize holes
F.Angle: 6 X 4.0 X 5/16 X 6.0           Material : A36
W.Angle: 4 X 3.5 X 1/4  X 5.5               Fy =  40.65 ksi
                                            Fu =  68.43 ksi

                       Major parameters

     lp =  5.5000"    lu =  1.3750"    ll =  1.3750"    ta =  0.2500"
     gb =  2.0000"    gc =  2.6250"    cu =  1.2500"    cl =  1.2500"
     pb =  3.0000"    pc =  3.0000"    nb = 1 X  2      nc = 1 X  2
     lt =  6.0000"    tt =  0.3125"    ls =  6.0000"    ts =  0.3125"
     gt =  2.5000"    gt'=  2.5000"    gs =  2.5000"    gs'=  2.5000"
     qt =  2.7500"    rt =  3.5000"    qs =  2.7500"    rs =  3.5000"
     pt =  2.5000"    ps =  2.5000"
     nt = 2 X  2      nt'= 2 X  1      ns = 2 X  2      ns'= 2 X  1

Remark  1)  Heavy hex high strength bolts used
        2)
```

No	Moment (k-in)	Rotation (radians) X 1/1000
1	0.00	0.00
2	11.48	0.19
3	22.97	0.38
4	31.65	0.59
5	40.34	0.79
6	52.41	1.18
7	64.48	1.56
8	73.02	1.82
9	81.56	2.09
10	94.32	2.61
11	107.07	3.12
12	119.83	3.64
13	131.02	4.23
14	142.20	4.82
15	150.30	5.41
16	158.39	5.99
17	170.46	7.25
18	182.52	8.52
19	191.35	9.98
20	200.17	11.45
21	208.11	13.36
22	216.05	15.26
23	223.70	17.23
24	231.34	19.19
25	238.10	20.92

No	Moment (k-in)	Rotation (radians) X 1/1000
26	244.87	22.65
27	250.45	24.41
28	256.04	26.17
29	260.88	28.10
30	265.73	30.02
31	269.69	31.66
32	273.66	33.30
33	276.89	34.90
34	280.12	36.50
35	283.93	38.48
36	287.75	40.46

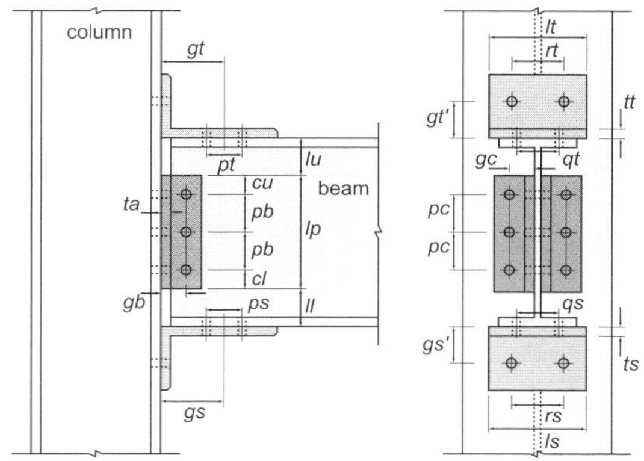

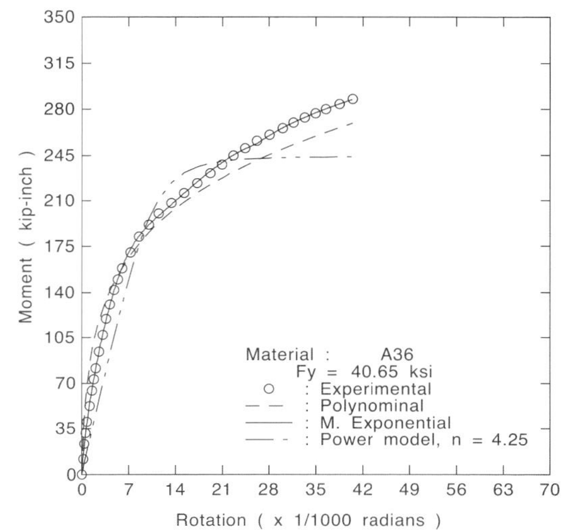

A.3 – 12

```
Moment-rotation prediction equations     ( R : X 1/1000 radians )

Frye and Morris polynominal model  :    R = Sum ( Ai X ( K*Bm )**Pi X 10**Qi )
     t =  0.312500"   d =  8.250000"  tc =  0.250000"   w =  6.000000"   g =  2.125000"
    A1 =  2.232429   A2 =  1.850728   A3 =  3.188976    K =  0.348096
    P1 =        1    P2 =        3    P3 =        5    Q1 =       -2    Q2 =       -5    Q3 =       -9

Modified exponential model  :
    Bm = Sum ( Ai X ( 1 - exp( - R/( 2*i*C ) ) ) )  +     Sum ( RKj X ( R - Rj0 ) )  +  Bm0
      C =  0.38497083E+00          Bmo=  0.00000000E+00          Nexp=  6        Nliner=  1
     AI =  0.69199691E+01  0.10688606E+04  -0.83632428E+04  0.22959468E+05  -0.26271474E+05  0.10866420E+05
    Rj0 =         26.1700
    RKj =  0.15020027E+01

Power model  :    Bm = ( Rki X R ) / ( 1 + ( R/R0 )**rn )**( 1/rn )
     rn = 4.255      rki =  0.22104764E+02         rmu =  0.24411998E+03
```

No	Rotation (radians) X 1/1000	Moment (kip-inch) Expri.	Poly.	M.Expo.	P.Model.	Connection stiffness (kip-inch) X 1000 Expri.	Poly.	M.Expo.	P.Model.
1	0.00	0.00	0.00	0.00	0.00	0.6003E+02	0.1287E+03	0.6526E+02	0.2210E+02
3	0.38	22.97	41.69	21.70	8.46	0.5144E+02	0.8273E+02	0.4947E+02	0.2210E+02
5	0.79	40.34	67.86	39.78	17.56	0.3848E+02	0.4930E+02	0.3926E+02	0.2210E+02
7	1.56	64.48	95.71	65.73	34.44	0.3201E+02	0.2773E+02	0.3013E+02	0.2210E+02
9	2.09	81.56	108.54	80.82	46.13	0.2968E+02	0.2143E+02	0.2717E+02	0.2208E+02
11	3.12	107.07	127.01	106.97	69.00	0.2459E+02	0.1498E+02	0.2338E+02	0.2198E+02
13	4.23	131.02	141.45	130.65	93.16	0.1905E+02	0.1146E+02	0.1938E+02	0.2165E+02
15	5.41	150.30	153.50	150.77	118.17	0.1379E+02	0.9243E+01	0.1492E+02	0.2087E+02
17	7.25	170.46	168.48	172.67	154.62	0.9563E+01	0.7148E+01	0.9166E+01	0.1826E+02
19	9.98	191.35	185.39	191.16	196.14	0.6017E+01	0.5420E+01	0.5206E+01	0.1190E+02
21	13.36	208.11	201.46	207.02	223.87	0.4164E+01	0.4219E+01	0.4560E+01	0.5165E+01
23	17.23	223.70	216.07	224.63	236.19	0.3890E+01	0.3393E+01	0.4415E+01	0.1796E+01
25	20.92	238.10	227.59	239.52	240.48	0.3907E+01	0.2876E+01	0.3552E+01	0.7110E+00
27	24.41	250.45	236.98	250.08	242.20	0.3173E+01	0.2523E+01	0.2499E+01	0.3281E+00
29	28.10	260.88	245.73	260.34	243.05	0.2523E+01	0.2241E+01	0.3051E+01	0.1598E+00
31	31.66	269.69	253.32	269.99	243.47	0.2414E+01	0.2026E+01	0.2408E+01	0.8610E-01
33	34.90	276.89	259.68	277.14	243.69	0.2020E+01	0.1865E+01	0.2033E+01	0.5183E-01
35	38.48	283.93	266.02	283.93	243.84	0.1928E+01	0.1719E+01	0.1786E+01	0.3113E-01

III - 7

```
Connection type : Top-and seat-angle connections with double web angle
           Mode : All bolted

Tested by  :  A.Azizinamini et al. (1985)                U.S.A.
Test  Id.  :  8S7

Column : W12X58                        Fasteners: A325- -3/4"D
Beam   : W8X21                                13/16" Oversize holes
F.Angle: 6 X 4.0 X 3/8  X 6.0          Material : A36
W.Angle: 4 X 3.5 X 1/4  X 5.5                Fy =  40.65 ksi
                                             Fu =  68.43 ksi
```

Major parameters

lp = 5.5000"	lu = 1.3750"	ll = 1.3750"	ta = 0.2500"
gb = 2.0000"	gc = 2.6250"	cu = 1.2500"	cl = 1.2500"
pb = 3.0000"	pc = 3.0000"	nb = 1 X 2	nc = 1 X 2
lt = 6.0000"	tt = 0.3750"	ls = 6.0000"	ts = 0.3750"
gt = 3.5000"	gt'= 2.5000"	gs = 3.5000"	gs'= 2.5000"
qt = 2.7500"	rt = 3.5000"	qs = 2.7500"	rs = 3.5000"
pt = 2.5000"	ps = 2.5000"		
nt = 2 X 2	nt'= 2 X 1	ns = 2 X 2	ns'= 2 X 1

Remark 1) Heavy hex high strength bolts used
2)

No	Moment (k-in)	Rotation (radians) X 1/1000
1	0.00	0.00
2	11.61	0.27
3	23.22	0.53
4	39.97	0.80
5	59.81	1.25
6	79.65	1.71
7	93.56	2.06
8	107.47	2.41
9	121.38	2.77
10	141.22	3.33
11	161.05	3.89
12	175.36	4.35
13	189.66	4.81
14	203.96	5.27
15	221.59	5.98
16	239.22	6.68
17	257.15	7.68
18	275.37	8.86
19	290.94	10.09
20	301.80	11.38
21	310.32	12.65
22	318.84	13.91
23	328.09	15.54
24	337.33	17.17
25	344.97	18.91

No	Moment (k-in)	Rotation (radians) X 1/1000
26	352.60	20.64
27	359.05	22.61
28	365.51	24.57
29	370.20	26.13
30	374.89	27.69
31	379.73	29.36
32	384.57	31.04
33	389.56	32.68
34	394.55	34.33
35	398.07	35.57
36	401.59	36.82
37	406.57	38.53
38	411.56	40.23

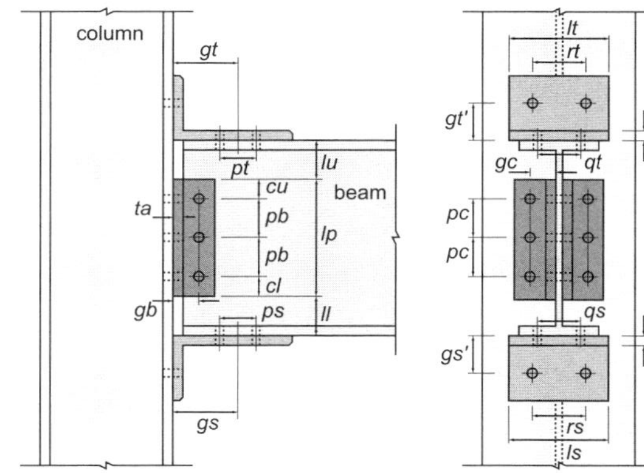

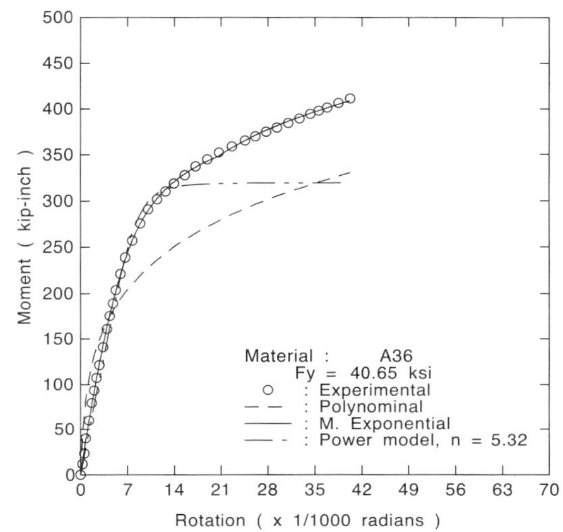

A.3 – 14

```
Moment-rotation prediction equations     ( R : X 1/1000 radians )

Frye and Morris polynominal model  :    R = Sum ( Ai X ( K*Bm )**Pi X 10**Qi )
     t =  0.375000"    d =  8.250000"  tc =  0.250000"    w =  6.000000"    g =  2.125000"
    A1 =  2.232429    A2 =  1.850728   A3 =  3.188976     K =  0.283384
    P1 =         1    P2 =         3   P3 =         5    Q1 =        -2    Q2 =        -5    Q3 =        -9

Modified exponential model  :
    Bm = Sum ( Ai X ( 1 - exp( - R/( 2*i*C ) ) ) )  +      Sum ( RKj X ( R - Rj0 ) )  +  Bm0
      C =   0.40150583E+00            Bmo=   0.00000000E+00          Nexp=  6        Nliner=  1
     AI =  -0.26558243E+02   0.78098961E+03  -0.51893405E+04   0.12956194E+05  -0.14245948E+05   0.60937604E+04
    Rj0 =         20.6400
    RKj =   0.20840430E+01

Power model  :    Bm = ( Rki X R ) / ( 1 + ( R/R0 )**rn )**( 1/rn )
     rn = 5.320      rki =   0.37662017E+02          rmu =   0.31951265E+03
```

No	Rotation (radians) X 1/1000	Moment (kip-inch) Expri.	Poly.	M.Expo.	P.Model.	Connection stiffness (kip-inch) X 1000 Expri.	Poly.	M.Expo.	P.Model.
1	0.00	0.00	0.00	0.00	0.00	0.4381E+02	0.1581E+03	0.4937E+02	0.3766E+02
3	0.53	23.22	64.71	25.73	19.96	0.5346E+02	0.8245E+02	0.4753E+02	0.3766E+02
5	1.25	59.81	106.07	58.88	47.14	0.4347E+02	0.4127E+02	0.4434E+02	0.3766E+02
7	2.06	93.56	132.65	93.42	77.63	0.3937E+02	0.2661E+02	0.4100E+02	0.3764E+02
9	2.77	121.38	149.09	121.39	104.20	0.3786E+02	0.2049E+02	0.3818E+02	0.3755E+02
11	3.89	161.05	168.71	161.49	145.94	0.3304E+02	0.1517E+02	0.3344E+02	0.3697E+02
13	4.81	189.66	181.43	190.41	179.48	0.3103E+02	0.1258E+02	0.2928E+02	0.3559E+02
15	5.98	221.59	194.75	221.49	219.03	0.2498E+02	0.1040E+02	0.2405E+02	0.3174E+02
17	7.68	257.15	210.60	256.60	265.15	0.1681E+02	0.8355E+01	0.1738E+02	0.2172E+02
19	10.09	290.94	228.44	289.99	300.04	0.1056E+02	0.6596E+01	0.1089E+02	0.8454E+01
21	12.65	310.32	243.70	312.46	312.80	0.6741E+01	0.5430E+01	0.7099E+01	0.2641E+01
23	15.54	328.09	258.06	329.45	317.17	0.5672E+01	0.4551E+01	0.4872E+01	0.7839E+00
25	18.91	344.97	272.12	343.13	318.67	0.4403E+01	0.3851E+01	0.3374E+01	0.2342E+00
27	22.61	359.05	285.31	357.49	319.19	0.3280E+01	0.3308E+01	0.4313E+01	0.7642E-01
29	26.13	370.20	296.27	371.21	319.36	0.3015E+01	0.2926E+01	0.3513E+01	0.3069E-01
31	29.36	379.73	305.27	381.66	319.43	0.2891E+01	0.2652E+01	0.2993E+01	0.1471E-01
33	32.68	389.56	313.67	390.96	319.47	0.3032E+01	0.2423E+01	0.2633E+01	0.7482E-02
35	35.57	398.07	320.48	398.26	319.48	0.2819E+01	0.2255E+01	0.2428E+01	0.4379E-02
37	38.53	406.57	326.95	405.21	319.49	0.2927E+01	0.2108E+01	0.2293E+01	0.2646E-02

III - 8

Connection type : Top-and seat-angle connections with double web angle
Mode : All bolted

Tested by : A.Azizinamini et al. (1985) U.S.A.
Test Id. : 8S8

Column : W12X58
Beam : W8X21
F.Angle: 6 X 3.5 X 5/16 X 6.0
W.Angle: 4 X 3.5 X 1/4 X 5.5

Fasteners: A325- -7/8"D
15/16" Oversize holes
Material : A36
Fy = 39.55 ksi
Fu = 67.95 ksi

Major parameters

lp = 5.5000"	lu = 1.3750"	ll = 1.3750"	ta = 0.2500"
gb = 2.0000"	gc = 2.6250"	cu = 1.2500"	cl = 1.2500"
pb = 3.0000"	pc = 3.0000"	nb = 1 X 2	nc = 1 X 2
lt = 6.0000"	tt = 0.3125"	ls = 6.0000"	ts = 0.3125"
gt = 3.5000"	gt'= 2.0000"	gs = 3.5000"	gs'= 2.0000"
qt = 2.7500"	rt = 3.5000"	qs = 2.7500"	rs = 3.5000"
pt = 2.5000"	ps = 2.5000"		
nt = 2 X 2	nt'= 2 X 1	ns = 2 X 2	ns'= 2 X 1

Remark 1) Heavy hex high strength bolts used
2)

No	Moment (k-in)	Rotation (radians) X 1/1000	No	Moment (k-in)	Rotation (radians) X 1/1000
1	0.00	0.00	26	367.49	21.18
2	14.66	0.24	27	375.68	23.00
3	29.31	0.47	28	380.41	24.35
4	43.49	0.72	29	386.08	25.77
5	57.68	0.97	30	390.49	27.40
6	70.28	1.18	31	399.95	29.35
7	82.89	1.38	32	404.36	30.57
8	95.65	1.65	33	409.09	31.67
9	108.42	1.92	34	413.81	32.77
10	120.87	2.17	35	417.12	33.76
11	133.32	2.42	36	420.43	34.75
12	152.23	2.89	37	424.53	36.17
13	169.24	3.52	38	431.15	37.80
14	188.47	4.24	39	438.08	39.47
15	207.70	5.06			
16	227.55	5.91			
17	246.78	7.10			
18	264.74	8.48			
19	286.49	10.09			
20	303.51	11.78			
21	321.16	13.70			
22	333.45	15.55			
23	344.79	17.22			
24	351.10	18.32			
25	357.40	19.42			

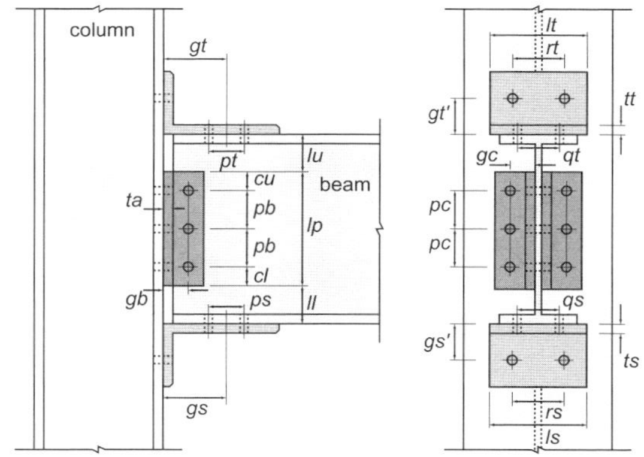

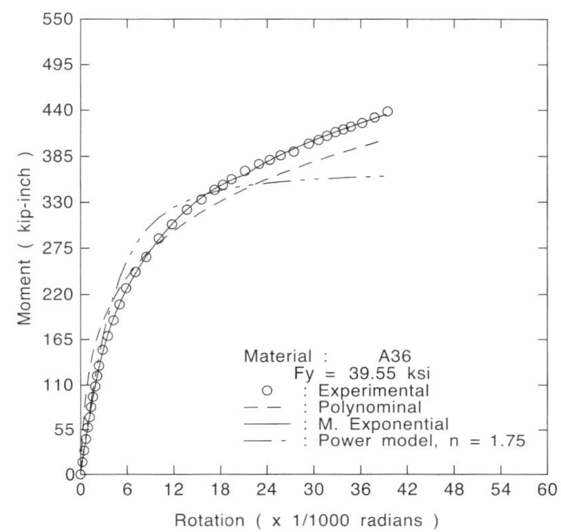

A.3 – 16

```
Moment-rotation prediction equations     ( R : X 1/1000 radians )

Frye and Morris polynominal model  :    R = Sum ( Ai X ( K*Bm )**Pi X 10**Qi )
      t =  0.312500"   d =  8.250000"  tc =  0.250000"   w =  6.000000"   g =  1.562500"
     A1 =  2.232429   A2 =  1.850728   A3 =  3.188976    K =  0.229842
     P1 =        1    P2 =        3    P3 =        5    Q1 =       -2    Q2 =       -5    Q3 =       -9

Modified exponential model  :
     Bm = Sum ( Ai X ( 1 - exp( - R/( 2*i*C ) ) ) )  +      Sum ( RKj X ( R - Rj0 ) )  +  Bm0
      C =   0.38780667E+00             Bmo=   0.00000000E+00          Nexp=  6       Nliner=  1
     AI =  -0.32502421E+02  -0.11554950E+03   0.33926834E+03   0.23135986E+04  -0.60854799E+04   0.39619303E+04
    Rj0 =          21.1800
    RKj =   0.29969626E+01

Power model  :    Bm = ( Rki X R ) / ( 1 + ( R/R0 )**rn )**( 1/rn )
     rn = 1.747      rki =   0.68456572E+02         rmu =   0.36734904E+03
```

No	Rotation (radians) X 1/1000	Moment (kip-inch) Expri.	Poly.	M.Expo.	P.Model.	Connection stiffness (kip-inch) X 1000 Expri.	Poly.	M.Expo.	P.Model.
1	0.00	0.00	0.00	0.00	0.00	0.6218E+02	0.1949E+03	0.5729E+02	0.6846E+02
3	0.47	29.31	73.56	28.40	32.00	0.5940E+02	0.1102E+03	0.6163E+02	0.6695E+02
5	0.97	57.68	115.11	58.91	64.82	0.5935E+02	0.6306E+02	0.5894E+02	0.6333E+02
7	1.38	82.89	137.15	82.10	89.92	0.5568E+02	0.4666E+02	0.5438E+02	0.5947E+02
9	1.92	108.42	158.64	109.32	120.19	0.4868E+02	0.3499E+02	0.4751E+02	0.5380E+02
11	2.42	133.32	174.48	131.61	145.85	0.4625E+02	0.2846E+02	0.4126E+02	0.4827E+02
13	3.52	169.24	200.85	170.56	192.59	0.2686E+02	0.2042E+02	0.3024E+02	0.3701E+02
15	5.06	207.70	227.49	209.23	239.67	0.2347E+02	0.1483E+02	0.2102E+02	0.2491E+02
17	7.10	246.78	253.49	245.58	279.39	0.1466E+02	0.1103E+02	0.1538E+02	0.1495E+02
19	10.09	286.49	281.62	285.44	311.73	0.1185E+02	0.8135E+01	0.1169E+02	0.7705E+01
21	13.70	321.16	307.28	321.60	331.81	0.7898E+01	0.6251E+01	0.8375E+01	0.3944E+01
23	17.22	344.79	327.19	345.90	342.44	0.6162E+01	0.5141E+01	0.5525E+01	0.2295E+01
25	19.42	357.40	337.92	356.41	346.80	0.5731E+01	0.4641E+01	0.4084E+01	0.1708E+01
27	23.00	375.68	353.38	373.20	351.76	0.3924E+01	0.4019E+01	0.5362E+01	0.1116E+01
29	25.77	386.08	363.96	386.75	354.43	0.3404E+01	0.3651E+01	0.4495E+01	0.8339E+00
31	29.35	399.95	376.33	401.48	356.96	0.4084E+01	0.3271E+01	0.3798E+01	0.5947E+00
33	31.67	409.09	383.69	409.97	358.21	0.4299E+01	0.3067E+01	0.3522E+01	0.4870E+00
35	33.76	417.12	390.02	417.14	359.14	0.3343E+01	0.2904E+01	0.3353E+01	0.4115E+00
37	36.17	424.53	396.78	425.03	360.05	0.3433E+01	0.2741E+01	0.3223E+01	0.3431E+00
39	39.47	438.08	405.52	435.47	361.06	0.4163E+01	0.2547E+01	0.3116E+01	0.2720E+00

III - 9

Connection type : Top-and seat-angle connections with double web angle
Mode : All bolted

Tested by : A.Azizinamini et al. (1985) U.S.A.
Test Id. : 8S9

Column : W12X58
Beam : W8X21
F.Angle: 6 X 3.5 X 3/8 X 6.0
W.Angle: 4 X 3.5 X 1/4 X 5.5

Fasteners: A325- -7/8"D
15/16" Oversize holes
Material : A36
Fy = 39.55 ksi
Fu = 67.95 ksi

Major parameters

lp = 5.5000"	lu = 1.3750"	ll = 1.3750"	ta = 0.2500"
gb = 2.0000"	gc = 2.6250"	cu = 1.2500"	cl = 1.2500"
pb = 3.0000"	pc = 3.0000"	nb = 1 X 2	nc = 1 X 2
lt = 6.0000"	tt = 0.3750"	ls = 6.0000"	ts = 0.3750"
gt = 3.5000"	gt'= 2.0000"	gs = 3.5000"	gs'= 2.0000"
qt = 2.7500"	rt = 3.5000"	qs = 2.7500"	rs = 3.5000"
pt = 2.5000"	ps = 2.5000"		
nt = 2 X 2	nt'= 2 X 1	ns = 2 X 2	ns'= 2 X 1

Remark 1) Heavy hex high strength bolts used
2)

No	Moment (k-in)	Rotation (radians) X 1/1000	No	Moment (k-in)	Rotation (radians) X 1/1000
1	0.00	0.00	26	381.37	16.71
2	16.66	0.14	27	387.58	17.71
3	33.32	0.29	28	392.85	18.79
4	47.79	0.45	29	398.13	19.86
5	62.25	0.61	30	405.15	21.13
6	74.67	0.77	31	412.17	22.40
7	87.09	0.93	32	418.24	23.59
8	100.14	1.08	33	424.31	24.78
9	113.19	1.22	34	431.36	26.48
10	129.87	1.49	35	438.42	28.19
11	146.54	1.76	36	442.76	29.41
12	160.39	2.09	37	447.11	30.63
13	174.24	2.43	38	451.44	31.65
14	197.23	3.10	39	455.76	32.67
15	214.09	3.71	40	458.84	33.76
16	230.94	4.33	41	461.92	34.86
17	254.60	5.41	42	464.99	35.86
18	278.26	6.58	43	468.06	36.87
19	299.74	8.07	44	474.16	38.50
20	322.51	9.78			
21	340.87	11.43			
22	350.06	12.46			
23	359.26	13.49			
24	367.21	14.60			
25	375.16	15.71			

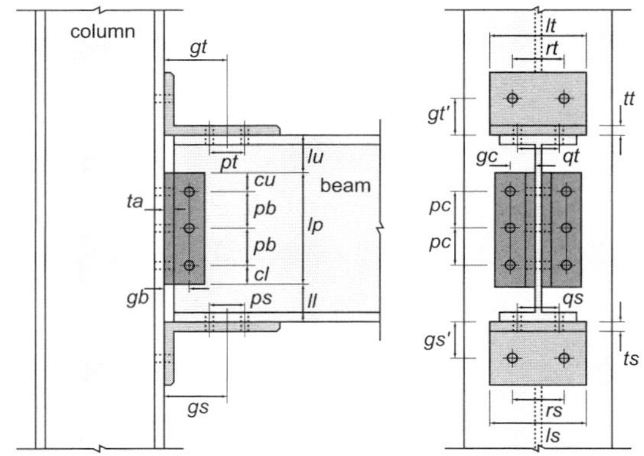

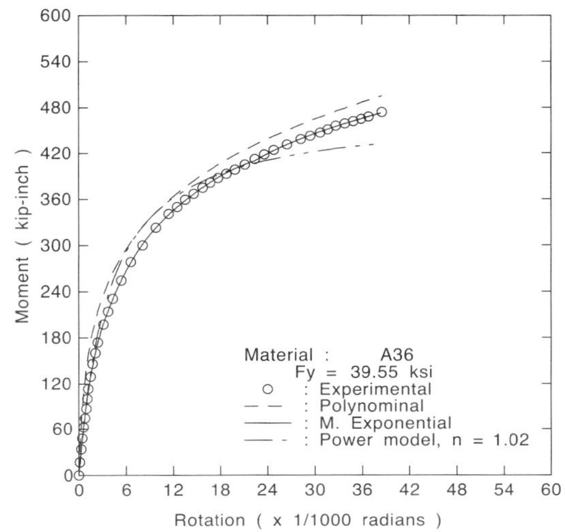

A.3 – 18

```
Moment-rotation prediction equations      ( R : X 1/1000 radians )

Frye and Morris polynominal model  :    R = Sum ( Ai X ( K*Bm )**Pi X 10**Qi )
     t =  0.375000"    d =  8.250000"  tc =  0.250000"   w =  6.000000"   g =  1.562500"
    A1 =  2.232429    A2 =  1.850728   A3 =  3.188976    K =  0.187114
    P1 =         1    P2 =         3   P3 =         5   Q1 =       -2    Q2 =       -5    Q3 =       -9

Modified exponential model  :
    Bm = Sum ( Ai X ( 1 - exp( - R/( 2*i*C ) ) ) )  +      Sum ( RKj X ( R - Rj0 ) )  +  Bm0
     C =   0.35652333E+00              Bmo=   0.00000000E+00          Nexp=  6        Nliner=  1
    AI =  -0.17624109E+03   0.24904635E+04  -0.11863388E+05   0.27776590E+05  -0.30481919E+05   0.12685490E+05
   Rj0 =          22.4000
   RKj =   0.26695801E+01

Power model  :    Bm = ( Rki X R ) / ( 1 + ( R/R0 )**rn )**( 1/rn )
    rn = 1.018      rki =   0.12350539E+03          rmu =   0.47287001E+03
```

No	Rotation (radians) X 1/1000	Moment (kip-inch) Expri.	Poly.	M.Expo.	P.Model.	Connection stiffness (kip-inch) X 1000 Expri.	Poly.	M.Expo.	P.Model.
1	0.00	0.00	0.00	0.00	0.00	0.1166E+03	0.2394E+03	0.1073E+03	0.1235E+03
3	0.29	33.32	61.56	30.40	33.00	0.1041E+03	0.1783E+03	0.1039E+03	0.1078E+03
5	0.61	62.25	107.16	62.13	65.16	0.8339E+02	0.1132E+03	0.9313E+02	0.9310E+02
7	0.93	87.09	137.98	90.26	93.34	0.8330E+02	0.8046E+02	0.8016E+02	0.8101E+02
9	1.22	113.19	158.84	112.00	115.65	0.7976E+02	0.6376E+02	0.6896E+02	0.7200E+02
11	1.76	146.54	187.91	144.29	150.62	0.5290E+02	0.4634E+02	0.5216E+02	0.5886E+02
13	2.43	174.24	214.63	174.19	185.73	0.3907E+02	0.3485E+02	0.3862E+02	0.4695E+02
15	3.71	214.09	251.48	214.97	235.75	0.2745E+02	0.2392E+02	0.2692E+02	0.3223E+02
17	5.41	254.60	285.63	254.88	280.24	0.2107E+02	0.1718E+02	0.2081E+02	0.2140E+02
19	8.07	299.74	323.81	301.00	324.36	0.1391E+02	0.1212E+02	0.1420E+02	0.1281E+02
21	11.43	340.87	358.64	339.97	357.80	0.9780E+01	0.8972E+01	0.9681E+01	0.7738E+01
23	13.49	359.26	375.81	358.36	371.81	0.8086E+01	0.7782E+01	0.8268E+01	0.5986E+01
25	15.71	375.16	391.96	375.29	383.56	0.6664E+01	0.6831E+01	0.7019E+01	0.4688E+01
27	17.71	387.58	404.94	388.22	392.05	0.5573E+01	0.6167E+01	0.5903E+01	0.3847E+01
29	19.86	398.13	417.61	399.67	399.59	0.5188E+01	0.5591E+01	0.4712E+01	0.3169E+01
31	22.40	412.17	431.06	409.95	406.84	0.5317E+01	0.5050E+01	0.3438E+01	0.2579E+01
33	24.78	424.31	442.56	423.25	412.45	0.4707E+01	0.4636E+01	0.5118E+01	0.2163E+01
35	28.19	438.42	457.53	438.83	419.04	0.3798E+01	0.4157E+01	0.4086E+01	0.1721E+01
37	30.63	447.11	467.34	448.18	422.94	0.3936E+01	0.3875E+01	0.3595E+01	0.1483E+01
39	32.67	455.76	475.11	455.18	425.79	0.3556E+01	0.3668E+01	0.3307E+01	0.1320E+01
41	34.86	461.92	482.90	462.19	428.52	0.2939E+01	0.3474E+01	0.3090E+01	0.1173E+01
43	36.87	468.06	489.65	468.24	430.76	0.3323E+01	0.3315E+01	0.2954E+01	0.1059E+01

III - 10

Connection type : Top-and seat-angle connections with double web angle
Mode : All bolted

Tested by : A.Azizinamini et al. (1985) U.S.A.
Test Id. : 8S10

Column : W12X58
Beam : W8X21
F.Angle: 6 X 3.5 X 1/2 X 6.0
W.Angle: 4 X 3.5 X 1/4 X 5.5

Fasteners: A325- -7/8"D
15/16" Oversize holes
Material : A36
Fy = 39.55 ksi
Fu = 67.95 ksi

Major parameters

lp = 5.5000"	lu = 1.3750"	ll = 1.3750"	ta = 0.2500"
gb = 2.0000"	gc = 2.6250"	cu = 1.2500"	cl = 1.2500"
pb = 3.0000"	pc = 3.0000"	nb = 1 X 2	nc = 1 X 2
lt = 6.0000"	tt = 0.5000"	ls = 6.0000"	ts = 0.5000"
gt = 3.5000"	gt'= 2.0000"	gs = 3.5000"	gs'= 2.0000"
qt = 2.7500"	rt = 3.5000"	qs = 2.7500"	rs = 3.5000"
pt = 2.5000"	ps = 2.5000"		
nt = 2 X 2	nt'= 2 X 1	ns = 2 X 2	ns'= 2 X 1

Remark 1) Heavy hex high strength bolts used
2)

No	Moment (k-in)	Rotation (radians) X 1/1000
1	0.00	0.00
2	27.53	0.05
3	55.05	0.10
4	82.58	0.15
5	113.47	0.24
6	144.36	0.34
7	171.78	0.48
8	199.20	0.62
9	222.11	0.75
10	245.01	0.89
11	267.92	1.02
12	293.81	1.30
13	329.71	1.58
14	355.88	2.02
15	382.05	2.46
16	400.97	3.06
17	419.90	3.65
18	436.62	4.36
19	453.34	5.06
20	440.12	5.69
21	455.91	6.75
22	471.70	7.80
23	485.59	8.74
24	499.48	9.68
25	511.80	10.80
26	524.12	11.92
27	535.91	12.82
28	547.70	13.72
29	559.49	14.62
30	569.29	15.69
31	579.09	16.75
32	585.11	17.90
33	591.13	19.05
34	597.14	20.20
35	603.16	21.34
36	625.25	22.51
37	630.95	23.44
38	636.65	24.37
39	642.35	25.30
40	645.84	26.12
41	649.32	26.94

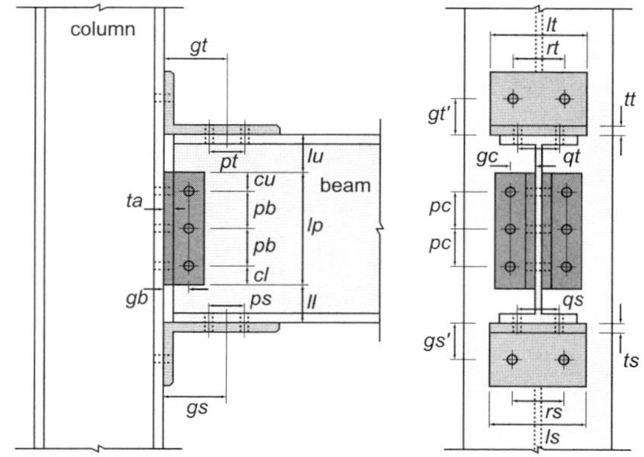

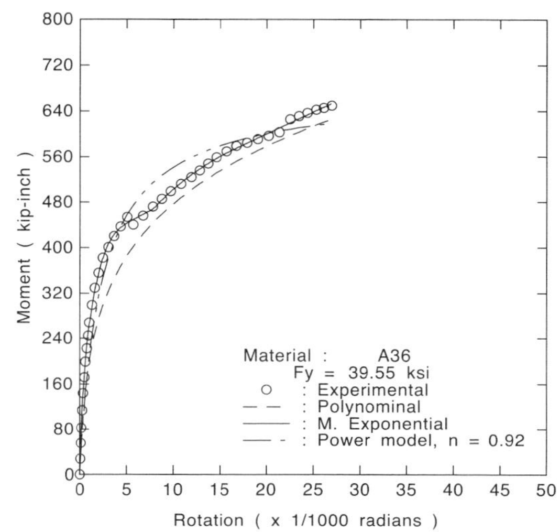

```
Moment-rotation prediction equations     ( R : X 1/1000 radians )

Frye and Morris polynominal model   :    R = Sum ( Ai X ( K*Bm )**Pi X 10**Qi )
      t =  0.500000"    d =  8.250000"  tc =  0.250000"   w =  6.000000"   g =  1.562500"
     A1 =  2.232429    A2 =  1.850728   A3 =  3.188976    K =  0.135258
     P1 =        1     P2 =        3    P3 =        5    Q1 =       -2    Q2 =       -5    Q3 =       -9

Modified exponential model   :
     Bm = Sum ( Ai X ( 1 - exp( - R/( 2*i*C ) ) ) )  +      Sum ( RKj X ( R - Rj0 ) )  +  Bm0
      C =   0.23713417E+00            Bmo=   0.00000000E+00          Nexp=  6        Nliner=  2
     AI =   0.57312009E+03  -0.38344392E+04   0.13317886E+05  -0.20484529E+05   0.14435851E+05  -0.35098248E+04
    Rj0 =           7.8000          14.6200
    RKj =   0.88619453E+01  -0.11806490E+01

Power model  :    Bm = ( Rki X R ) / ( 1 + ( R/R0 )**rn )**( 1/rn )
     rn = 0.921       rki =   0.32493559E+03          rmu =   0.68305103E+03
```

No	Rotation (radians) X 1/1000	Moment (kip-inch) Expri.	Poly.	M.Expo.	P.Model.	Connection stiffness (kip-inch) X 1000 Expri.	Poly.	M.Expo.	P.Model.
1	0.00	0.00	0.00	0.00	0.00	0.5440E+03	0.3312E+03	0.5825E+03	0.3249E+03
3	0.10	55.05	32.97	52.81	30.83	0.5440E+03	0.3155E+03	0.4664E+03	0.2871E+03
5	0.24	113.47	74.48	110.43	68.96	0.3347E+03	0.2629E+03	0.3475E+03	0.2484E+03
7	0.48	171.78	125.95	175.99	120.97	0.1962E+03	0.1858E+03	0.2295E+03	0.2025E+03
9	0.75	222.11	168.92	227.98	170.98	0.1700E+03	0.1328E+03	0.1579E+03	0.1642E+03
11	1.02	267.92	200.39	265.26	211.39	0.1505E+03	0.1031E+03	0.1222E+03	0.1369E+03
13	1.58	329.71	247.96	322.50	276.69	0.9034E+02	0.7044E+02	0.8591E+02	0.9883E+02
15	2.46	382.05	298.41	381.77	347.20	0.4784E+02	0.4767E+02	0.5101E+02	0.6550E+02
17	3.65	419.90	345.81	422.74	409.79	0.2805E+02	0.3359E+02	0.2143E+02	0.4213E+02
19	5.06	453.34	386.71	442.92	458.16	0.6040E-01	0.2518E+02	0.1016E+02	0.2785E+02
21	6.75	455.91	423.96	458.36	496.57	0.1500E+02	0.1960E+02	0.8827E+01	0.1874E+02
23	8.74	485.59	458.78	483.28	527.44	0.1474E+02	0.1565E+02	0.1630E+02	0.1279E+02
25	10.80	511.80	488.11	513.99	549.79	0.1105E+02	0.1303E+02	0.1349E+02	0.9227E+01
27	12.82	535.91	512.49	538.67	566.01	0.1309E+02	0.1125E+02	0.1115E+02	0.7023E+01
29	14.62	559.49	531.63	557.51	577.38	0.1130E+02	0.1005E+02	0.9877E+01	0.5667E+01
31	16.75	579.09	551.87	575.19	588.19	0.7278E+01	0.8943E+01	0.7956E+01	0.4518E+01
33	19.05	591.13	571.29	593.05	597.49	0.5245E+01	0.8016E+01	0.7674E+01	0.3639E+01
35	21.34	603.16	588.81	610.57	605.07	0.1208E+02	0.7277E+01	0.7613E+01	0.2996E+01
37	23.44	630.95	603.56	626.52	610.86	0.6110E+01	0.6718E+01	0.7619E+01	0.2549E+01
39	25.30	642.35	615.68	640.74	615.30	0.5128E+01	0.6297E+01	0.7636E+01	0.2231E+01
41	26.94	649.32	625.68	653.24	618.76	0.4268E+01	0.5973E+01	0.7649E+01	0.2000E+01

III - 11

Connection type : Top-and seat-angle connections with double web angle
Mode : All bolted

Tested by : A.Azizinamini et al. (1985) U.S.A.
Test Id. : 14S1

Column : W12X96
Beam : W14X38
F.Angle: 6 X 4.0 X 3/8 X 8.0
W.Angle: 4 X 3.5 X 1/4 X 8.5

Fasteners: A325- -3/4"D
13/16" Oversize holes
Material : A36
Fy = 40.65 ksi
Fu = 68.43 ksi

Major parameters

lp = 8.5000"	lu = 2.8125"	ll = 2.8125"	ta = 0.2500"
gb = 2.0000"	gc = 2.5940"	cu = 1.2500"	cl = 1.2500"
pb = 3.0000"	pc = 3.0000"	nb = 1 X 3	nc = 1 X 3
lt = 8.0000"	tt = 0.3750"	ls = 8.0000"	ts = 0.3750"
gt = 3.5000"	gt'= 2.5000"	gs = 3.5000"	gs'= 2.5000"
qt = 3.5000"	rt = 5.5000"	qs = 3.5000"	rs = 5.5000"
pt = 2.5000"	ps = 2.5000"		
nt = 2 X 2	nt'= 2 X 1	ns = 2 X 2	ns'= 2 X 1

Remark 1) Heavy hex high strength bolts used
2)

No	Moment (k-in)	Rotation (radians) X 1/1000	No	Moment (k-in)	Rotation (radians) X 1/1000
1	0.00	0.00	26	572.79	9.57
2	25.75	0.09	27	584.73	10.70
3	51.50	0.18	28	596.67	11.83
4	77.25	0.27	29	605.66	12.94
5	96.40	0.48	30	614.66	14.05
6	123.99	0.60	31	621.48	15.39
7	151.58	0.73	32	628.30	16.72
8	176.85	0.89	33	636.62	18.24
9	202.12	1.05	34	644.89	19.37
10	227.39	1.21	35	653.15	20.50
11	251.70	1.52	36	662.16	21.76
12	273.43	1.77	37	671.18	23.02
13	295.16	2.01	38	680.96	24.43
14	322.41	2.27	39	690.73	25.83
15	349.65	2.53	40	699.01	27.08
16	371.03	2.87	41	707.29	28.32
17	392.41	3.22	42	717.44	29.79
18	415.62	3.58	43	727.59	31.25
19	438.84	3.94			
20	469.85	4.76			
21	490.16	5.37			
22	510.48	5.98			
23	529.00	6.87			
24	547.51	7.76			
25	560.15	8.67			

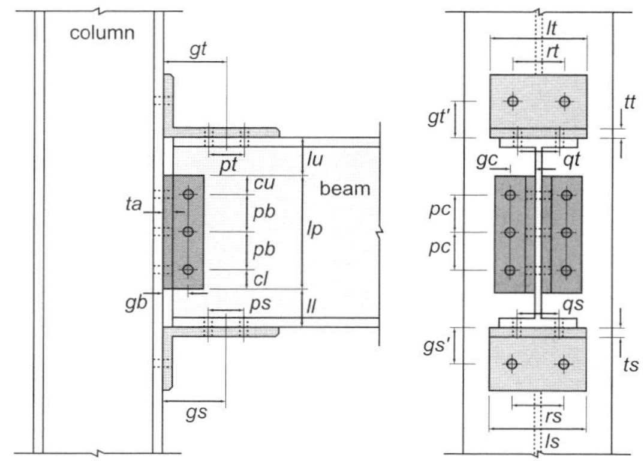

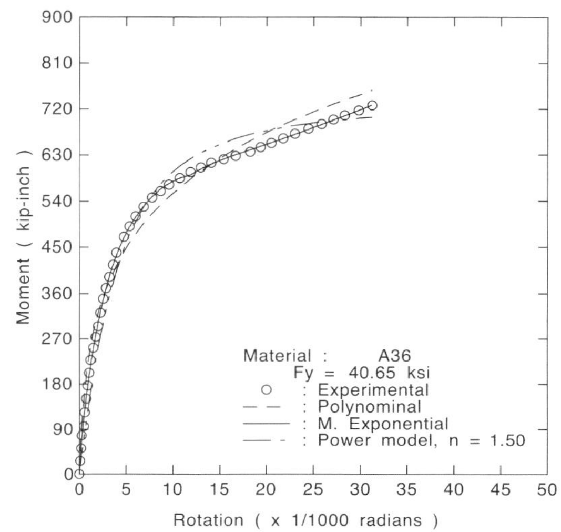

A.3 – 22

```
Moment-rotation prediction equations      ( R : X 1/1000 radians )

Frye and Morris polynominal model  :    R = Sum ( Ai X ( K*Bm )**Pi X 10**Qi )
     t =  0.375000"    d = 14.125000"  tc =  0.250000"    w =  8.000000"   g =  2.125000"
    A1 =  2.232429    A2 =  1.850728   A3 =  3.188976     K =  0.116167
    P1 =        1     P2 =        3    P3 =        5     Q1 =       -2     Q2 =       -5     Q3 =       -9

Modified exponential model  :
    Bm = Sum ( Ai X ( 1 - exp( - R/( 2*i*C ) ) ) )  +      Sum ( RKj X ( R - Rj0 ) )  +  Bm0
      C =   0.28331333E+00                Bmo=    0.00000000E+00          Nexp=  6        Nliner=  1
     AI =   0.29039464E+03  -0.29869790E+04   0.14275721E+05  -0.30433473E+05   0.29480450E+05  -0.10042707E+05
    Rj0 =          11.8300
    RKj =   0.74172040E+01

Power model  :    Bm = ( Rki X R ) / ( 1 + ( R/R0 )**rn )**( 1/rn )
     rn = 1.499      rki =   0.14477366E+03          rmu =   0.73429027E+03
```

No	Rotation (radians) X 1/1000	Moment (kip-inch) Expri.	Poly.	M.Expo.	P.Model.	Connection stiffness (kip-inch) X 1000 Expri.	Poly.	M.Expo.	P.Model.
1	0.00	0.00	0.00	0.00	0.00	0.2814E+03	0.3856E+03	0.2990E+03	0.1448E+03
3	0.18	51.50	67.18	48.87	26.39	0.2813E+03	0.3341E+03	0.2389E+03	0.1431E+03
5	0.48	96.40	146.41	109.34	67.52	0.1722E+03	0.2167E+03	0.1809E+03	0.1381E+03
7	0.73	151.58	192.81	150.83	101.46	0.1928E+03	0.1588E+03	0.1525E+03	0.1326E+03
9	1.05	202.12	236.41	195.83	142.67	0.1578E+03	0.1175E+03	0.1303E+03	0.1247E+03
11	1.52	251.70	283.82	252.66	199.20	0.8367E+02	0.8479E+02	0.1094E+03	0.1123E+03
13	2.01	295.16	320.30	302.02	250.95	0.9719E+02	0.6634E+02	0.9322E+02	0.9982E+02
15	2.53	349.65	351.11	346.12	299.16	0.8698E+02	0.5420E+02	0.7837E+02	0.8760E+02
17	3.22	392.41	384.80	394.45	354.79	0.6292E+02	0.4371E+02	0.6162E+02	0.7316E+02
19	3.94	438.84	413.63	433.94	402.97	0.5623E+02	0.3655E+02	0.4839E+02	0.6064E+02
21	5.37	490.16	458.93	490.18	475.51	0.3323E+02	0.2786E+02	0.3227E+02	0.4241E+02
23	6.87	529.00	496.41	530.86	529.12	0.2080E+02	0.2246E+02	0.2260E+02	0.2990E+02
25	8.67	560.15	532.81	563.65	573.70	0.1394E+02	0.1835E+02	0.1423E+02	0.2047E+02
27	10.70	584.73	566.82	584.86	608.19	0.1059E+02	0.1530E+02	0.7029E+01	0.1399E+02
29	12.94	605.66	598.34	602.64	634.27	0.8081E+01	0.1299E+02	0.9410E+01	0.9662E+01
31	15.39	621.48	627.81	622.03	654.09	0.5116E+01	0.1120E+02	0.6858E+01	0.6768E+01
33	18.24	636.62	657.53	640.22	670.23	0.6532E+01	0.9683E+01	0.6133E+01	0.4699E+01
35	20.50	653.15	678.30	654.15	679.56	0.7246E+01	0.8768E+01	0.6270E+01	0.3635E+01
37	23.02	671.18	699.32	670.36	687.60	0.7060E+01	0.7947E+01	0.6600E+01	0.2802E+01
39	25.83	690.73	720.60	689.40	694.52	0.6791E+01	0.7208E+01	0.6929E+01	0.2154E+01
41	28.32	707.29	737.86	706.93	699.34	0.6778E+01	0.6668E+01	0.7129E+01	0.1741E+01
43	31.25	727.59	756.78	728.04	703.89	0.6930E+01	0.6132E+01	0.7271E+01	0.1384E+01

III - 12

Connection type : Top-and seat-angle connections with double web angle
Mode : All bolted

Tested by : A.Azizinamini et al. (1985) U.S.A.
Test Id. : 14S2

Column : W12X96
Beam : W14X38
F.Angle: 6 X 4.0 X 1/2 X 8.0
W.Angle: 4 X 3.5 X 1/4 X 8.5

Fasteners: A325- -3/4"D
13/16" Oversize holes
Material : A36
Fy = 53.00 ksi
Fu = 80.00 ksi

Major parameters

lp = 8.5000"	lu = 2.8125"	ll = 2.8125"	ta = 0.2500"
gb = 2.0000"	gc = 2.5940"	cu = 1.2500"	cl = 1.2500"
pb = 3.0000"	pc = 3.0000"	nb = 1 X 3	nc = 1 X 3
lt = 8.0000"	tt = 0.5000"	ls = 8.0000"	ts = 0.5000"
gt = 3.5000"	gt'= 2.5000"	gs = 3.5000"	gs'= 2.5000"
qt = 3.5000"	rt = 5.5000"	qs = 3.5000"	rs = 5.5000"
pt = 2.5000"	ps = 2.5000"		
nt = 2 X 2	nt'= 2 X 1	ns = 2 X 2	ns'= 2 X 1

Remark 1) Heavy hex high strength bolts used
2)

No	Moment (k-in)	Rotation (radians) X 1/1000	No	Moment (k-in)	Rotation (radians) X 1/1000
1	0.00	0.00	26	859.30	12.68
2	49.28	0.11	27	849.79	13.04
3	98.56	0.21	28	862.47	14.21
4	147.50	0.46	29	875.15	15.38
5	196.43	0.72	30	889.33	16.73
6	245.73	0.95	31	903.51	18.08
7	295.04	1.19	32	919.88	19.33
8	329.87	1.41	33	920.68	19.75
9	364.71	1.62	34	910.48	20.39
10	399.55	1.84	35	921.43	21.46
11	449.62	2.28	36	932.38	22.52
12	499.68	2.73	37	943.32	23.59
13	527.30	3.01	38	954.27	24.62
14	554.92	3.30	39	965.21	25.66
15	595.44	3.82	40	976.15	26.69
16	621.98	4.25	41	989.07	27.83
17	643.51	4.69	42	1001.99	28.98
18	675.54	5.22	43	1014.91	30.12
19	704.56	5.76			
20	754.03	7.10			
21	776.22	7.96			
22	798.41	8.82			
23	816.96	9.92			
24	835.51	11.01			
25	847.41	11.85			

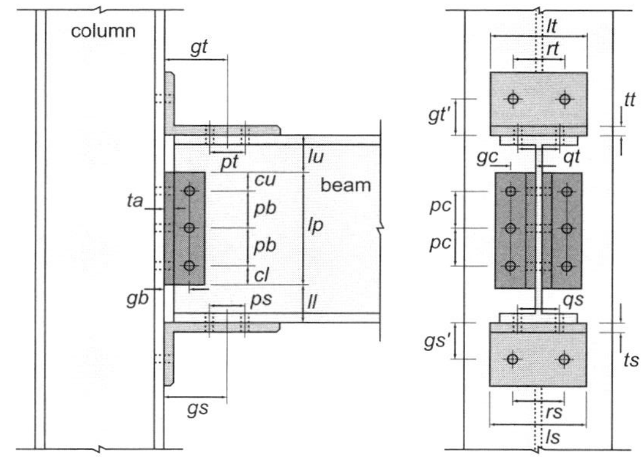

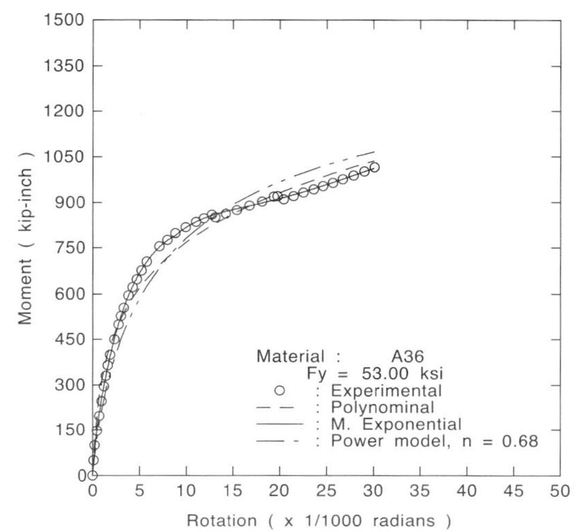

```
Moment-rotation prediction equations      ( R : X 1/1000 radians )

Frye and Morris polynominal model   :     R = Sum ( Ai X ( K*Bm )**Pi X 10**Qi )
      t =  0.500000"    d = 14.125000"  tc =  0.250000"    w =  8.000000"   g =  2.125000"
     A1 =  2.232429    A2 =  1.850728   A3 =  3.188976     K =  0.083973
     P1 =        1     P2 =        3    P3 =        5     Q1 =       -2     Q2 =       -5     Q3 =       -9

Modified exponential model  :
     Bm = Sum ( Ai X ( 1 - exp( - R/( 2*i*C ) ) ) )  +      Sum ( RKj X ( R - Rj0 ) )  +  Bm0
      C =   0.27734000E+00            Bmo=   0.00000000E+00          Nexp=  6         Nliner=  1
     AI =   0.48153487E+03  -0.48810830E+04   0.24147798E+05  -0.54773056E+05   0.56356541E+05  -0.20560226E+05
    Rj0 =           8.8200
    RKj =   0.11296354E+02

Power model  :    Bm = ( Rki X R ) / ( 1 + ( R/R0 )**rn )**( 1/rn )
     rn = 0.678      rki =   0.35272395E+03         rmu =   0.15101594E+04
```

No	Rotation (radians) X 1/1000	Moment (kip-inch) Expri.	Poly.	M.Expo.	P.Model.	Connection stiffness (kip-inch) X 1000 Expri.	Poly.	M.Expo.	P.Model.
1	0.00	0.00	0.00	0.00	0.00	0.4676E+03	0.5334E+03	0.4356E+03	0.3527E+03
3	0.21	98.56	105.51	79.24	62.11	0.3869E+03	0.4447E+03	0.3253E+03	0.2608E+03
5	0.72	196.43	264.42	207.48	171.95	0.2009E+03	0.2223E+03	0.2057E+03	0.1853E+03
7	1.19	295.04	349.26	294.42	250.45	0.1831E+03	0.1454E+03	0.1658E+03	0.1482E+03
9	1.62	364.71	404.07	361.61	309.47	0.1608E+03	0.1109E+03	0.1454E+03	0.1255E+03
11	2.28	449.62	466.72	449.08	383.88	0.1130E+03	0.8200E+02	0.1203E+03	0.1017E+03
13	3.01	527.30	519.45	527.62	451.16	0.9626E+02	0.6413E+02	0.9560E+02	0.8373E+02
15	3.82	595.44	565.87	595.44	512.66	0.6887E+02	0.5201E+02	0.7383E+02	0.6972E+02
17	4.69	648.51	607.04	651.76	568.18	0.5716E+02	0.4342E+02	0.5692E+02	0.5875E+02
19	5.76	704.56	649.53	704.57	625.62	0.4787E+02	0.3624E+02	0.4239E+02	0.4886E+02
21	7.96	776.22	718.46	774.04	717.35	0.2577E+02	0.2733E+02	0.2206E+02	0.3572E+02
23	9.92	816.96	767.01	816.29	779.66	0.1691E+02	0.2259E+02	0.2049E+02	0.2840E+02
25	11.85	847.41	807.36	847.31	829.38	0.1426E+02	0.1938E+02	0.1239E+02	0.2338E+02
27	13.04	849.79	829.54	860.18	855.71	-0.1805E+02	0.1785E+02	0.9504E+01	0.2098E+02
29	15.38	875.15	868.40	879.00	900.35	0.1068E+02	0.1550E+02	0.7157E+01	0.1732E+02
31	18.08	903.51	907.34	898.21	942.69	0.1180E+02	0.1350E+02	0.7400E+01	0.1427E+02
33	19.75	920.68	929.09	911.13	965.25	-0.4906E+01	0.1252E+02	0.8096E+01	0.1280E+02
35	21.46	921.43	949.75	925.65	986.06	0.1029E+02	0.1167E+02	0.8855E+01	0.1154E+02
37	23.59	943.32	973.83	945.40	1009.18	0.1043E+02	0.1076E+02	0.9663E+01	0.1023E+02
39	25.66	965.21	995.29	966.04	1029.25	0.1057E+02	0.1002E+02	0.1024E+02	0.9184E+01
41	27.83	989.07	1016.39	988.82	1048.21	0.1131E+02	0.9355E+01	0.1066E+02	0.8260E+01
43	30.12	1014.91	1037.04	1013.51	1066.12	0.1131E+02	0.8754E+01	0.1093E+02	0.7445E+01

III - 13

Connection type : Top-and seat-angle connections with double web angle
Mode : All bolted

Tested by : A.Azizinamini et al. (1985) U.S.A.
Test Id. : 14S3

Column : W12X96
Beam : W14X38
F.Angle: 6 X 4.0 X 3/8 X 8.0
W.Angle: 4 X 3.5 X 1/4 X 5.5

Fasteners: A325- -3/4"D
13/16" Oversize holes
Material : A36
Fy = 40.65 ksi
Fu = 68.43 ksi

Major parameters

lp = 5.5000"	lu = 4.3125'	ll = 4.3125"	ta = 0.2500"
gb = 2.0000"	gc = 2.5940'	cu = 1.2500"	cl = 1.2500"
pb = 3.0000"	pc = 3.0000'	nb = 1 X 2	nc = 1 X 2
lt = 8.0000"	tt = 0.3750"	ls = 8.0000"	ts = 0.3750"
gt = 3.5000"	gt'= 2.5000"	gs = 3.5000"	gs'= 2.5000"
qt = 3.5000"	rt = 5.5000"	qs = 3.5000"	rs = 5.5000"
pt = 2.5000"	ps = 2.5000"		
nt = 2 X 2	nt'= 2 X 1	ns = 2 X 2	ns'= 2 X 1

Remark 1) Heavy hex high strength bolts used
2)

No	Moment (k-in)	Rotation (radians) X 1/1000	No	Moment (k-in)	Rotation (radians) X 1/1000
1	0.00	0.00	26	567.75	13.83
2	25.90	0.20	27	580.31	15.22
3	51.80	0.41	28	592.88	16.60
4	77.70	0.61	29	604.88	18.12
5	104.59	0.89	30	616.88	19.64
6	131.48	1.16	31	625.49	20.81
7	158.37	1.43	32	634.11	21.97
8	186.44	1.75	33	642.72	23.14
9	214.51	2.07	34	653.54	24.65
10	242.59	2.39	35	664.36	26.15
11	269.29	2.74	36	674.28	27.52
12	296.00	3.09	37	684.20	28.89
13	322.71	3.44	38	693.81	30.15
14	343.91	3.70	39	703.41	31.40
15	356.90	4.05			
16	380.49	4.50			
17	404.08	4.96			
18	424.45	5.57			
19	444.83	6.18			
20	463.48	6.98			
21	482.12	7.77			
22	512.93	9.25			
23	527.21	10.32			
24	541.49	11.38			
25	554.62	12.61			

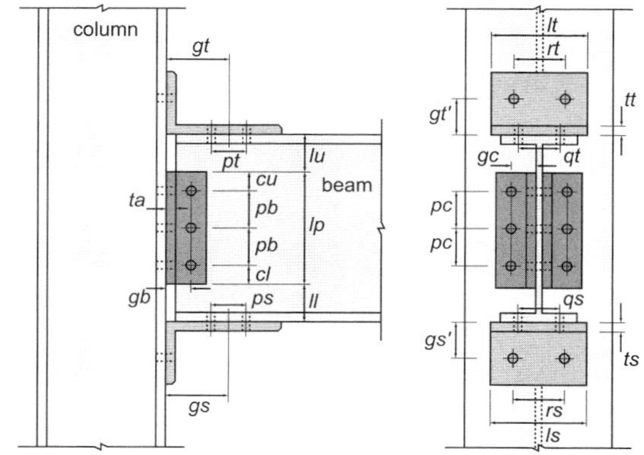

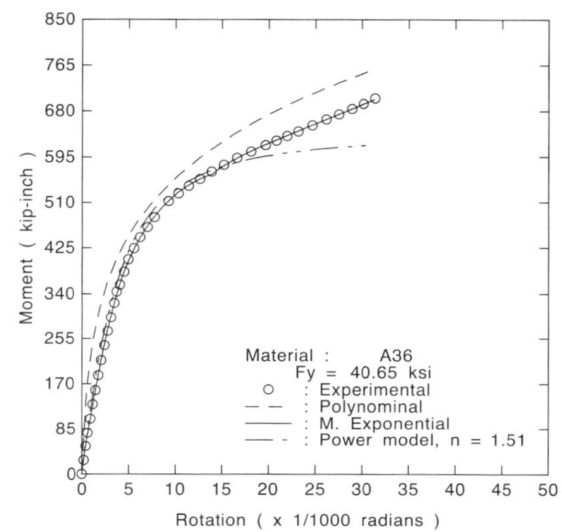

```
Moment-rotation prediction equations      ( R : X 1/1000 radians )

Frye and Morris polynominal model  :    R = Sum ( Ai X ( K*Bm )**Pi X 10**Qi )
       t =  0.375000"   d = 14.125000"  tc =  0.250000"   w =  8.000000"   g =  2.125000"
      A1 =  2.232429   A2 =  1.850728   A3 =  3.188976    K =  0.116167
      P1 =         1   P2 =         3   P3 =         5   Q1 =        -2    Q2 =       -5    Q3 =       -9

Modified exponential model  :
     Bm = Sum ( Ai X ( 1 - exp( - R/( 2*i*C ) ) ) )  +      Sum ( RKj X ( R - Rj0 ) )  +  Bm0
       C =   0.31282667E+00           Bmo=   0.00000000E+00          Nexp=  6        Nliner=  1
      AI =   0.34983723E+03  -0.39092877E+04   0.15600934E+05  -0.28670667E+05   0.24259144E+05  -0.70574845E+04
     Rj0 =          13.8300
     RKj =   0.73320691E+01

Power model  :    Bm = ( Rki X R ) / ( 1 + ( R/R0 )**rn )**( 1/rn )
      rn = 1.506      rki =   0.13915646E+03          rmu =   0.64019119E+03
```

No	Rotation (radians) X 1/1000	Moment (kip-inch) Expri.	Moment Poly.	Moment M.Expo.	Moment P.Model.	Connection stiffness (kip-inch) X 1000 Expri.	Stiffness Poly.	Stiffness M.Expo.	Stiffness P.Model.
1	0.00	0.00	0.00	0.00	0.00	0.1266E+03	0.3856E+03	0.1653E+03	0.1392E+03
3	0.41	51.80	131.32	54.38	55.94	0.1267E+03	0.2384E+03	0.1116E+03	0.1333E+03
5	0.89	104.59	216.21	103.25	116.86	0.9871E+02	0.1352E+03	0.9771E+02	0.1217E+03
7	1.43	158.37	275.74	155.87	179.14	0.9358E+02	0.8960E+02	0.9577E+02	0.1068E+03
9	2.07	214.51	324.27	215.85	242.13	0.8752E+02	0.6462E+02	0.9015E+02	0.8983E+02
11	2.74	269.29	362.45	272.67	297.05	0.7618E+02	0.5038E+02	0.7847E+02	0.7421E+02
13	3.44	322.71	394.31	322.74	344.17	0.8018E+02	0.4118E+02	0.6441E+02	0.6068E+02
15	4.05	356.90	417.36	358.10	377.72	0.4347E+02	0.3573E+02	0.5347E+02	0.5118E+02
17	4.96	404.08	447.29	400.94	419.24	0.4371E+02	0.2984E+02	0.4063E+02	0.3981E+02
19	6.18	444.83	480.13	443.21	460.67	0.2917E+02	0.2464E+02	0.3002E+02	0.2914E+02
21	7.77	482.12	515.63	484.26	499.33	0.2247E+02	0.2017E+02	0.2213E+02	0.2005E+02
23	10.32	527.21	560.87	529.50	538.85	0.1343E+02	0.1578E+02	0.1383E+02	0.1194E+02
25	12.61	554.62	593.94	554.18	561.26	0.1072E+02	0.1328E+02	0.8007E+01	0.8004E+01
27	15.22	580.31	625.89	578.84	578.43	0.9068E+01	0.1131E+02	0.1083E+02	0.5386E+01
29	18.12	604.88	656.32	606.01	591.37	0.7906E+01	0.9740E+01	0.8248E+01	0.3675E+01
31	20.81	625.49	680.96	626.69	599.83	0.7375E+01	0.8659E+01	0.7313E+01	0.2693E+01
33	23.14	642.72	700.28	643.42	605.40	0.7297E+01	0.7911E+01	0.7063E+01	0.2111E+01
35	26.15	664.36	722.86	664.59	610.92	0.7213E+01	0.7134E+01	0.7050E+01	0.1590E+01
37	28.89	684.20	741.63	684.04	614.80	0.7450E+01	0.6557E+01	0.7125E+01	0.1258E+01
39	31.40	703.41	757.52	702.02	617.66	0.7653E+01	0.6112E+01	0.7193E+01	0.1033E+01

III - 14

Connection type : Top-and seat-angle connections with double web angle
Mode : All bolted

Tested by : A.Azizinamini et al. (1985) U.S.A.
Test Id. : 14S4

Column : W12X96		Fasterers: A325- -3/4"D	
Beam : W14X38		13/16" Oversize holes	
F.Angle: 6 X 4.0 X 3/8 X 8.0		Material : A36	
W.Angle: 4 X 3.5 X 3/8 X 8.5		Fy = 40.65 ksi	
		Fu = 68.43 ksi	

Major parameters

lp = 8.5000"	lu = 2.8125"	ll = 2.8125"	ta = 0.3750"
gb = 2.0000"	gc = 2.5940"	cu = 1.2500"	cl = 1.2500"
pb = 3.0000"	pc = 3.0000"	nb = 1 X 3	nc = 1 X 3
lt = 8.0000"	tt = 0.3750"	ls = 8.0000"	ts = 0.3750"
gt = 3.5000"	gt'= 2.5000"	gs = 3.5000"	gs'= 2.5000"
qt = 3.5000"	rt = 5.5000"	qs = 3.5000"	rs = 5.5000"
pt = 2.5000"	ps = 2.5000"		
nt = 2 X 2	nt'= 2 X 1	ns = 2 X 2	ns'= 2 X 1

Remark 1) Heavy hex high strength bolts used
2)

No	Moment (k-in)	Rotation (radians) X 1/1000
1	0.00	0.00
2	39.44	0.18
3	78.87	0.35
4	118.90	0.55
5	158.92	0.74
6	198.66	0.94
7	238.39	1.15
8	278.43	1.47
9	318.48	1.80
10	358.23	2.16
11	397.99	2.53
12	439.24	3.07
13	480.49	3.62
14	515.88	4.31
15	551.26	5.00
16	577.52	5.57
17	603.77	6.15
18	626.82	6.92
19	649.86	7.69
20	681.23	9.08
21	694.30	10.03
22	707.37	10.98
23	719.89	12.25
24	732.42	13.52
25	742.57	14.64

No	Moment (k-in)	Rotation (radians) X 1/1000
26	752.72	15.76
27	762.87	16.87
28	772.44	18.01
29	782.00	19.15
30	791.57	20.29
31	802.33	21.54
32	813.08	22.78
33	823.84	24.03
34	833.80	25.20
35	843.77	26.36
36	853.73	27.53
37	862.50	28.59
38	871.27	29.65
39	880.04	30.71

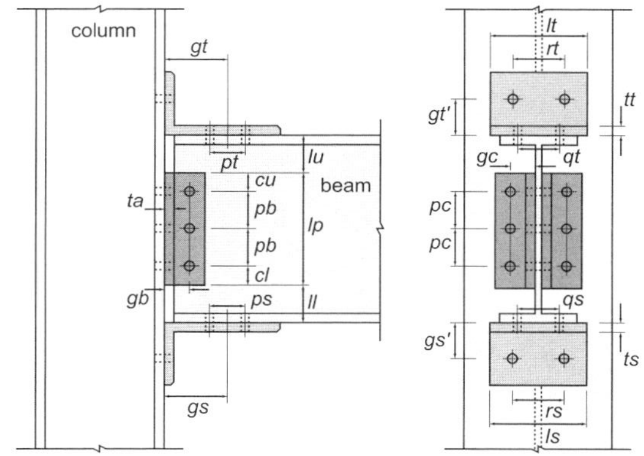

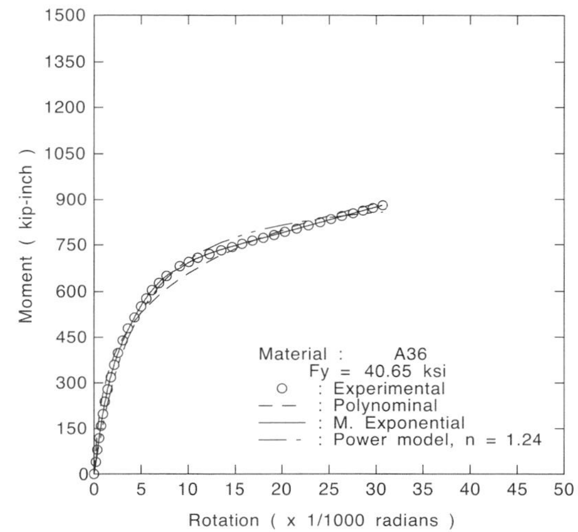

A.3 – 28

```
Moment-rotation prediction equations      ( R : X 1/1000 radians )

Frye and Morris polynominal model  :    R = Sum ( Ai X ( K*Bm )**Pi X 10**Qi )
      t =  0.375000"   d = 14.125000"  tc =  0.375000"   w =  8.000000"   g =  2.125000"
     A1 =  2.232429   A2 =  1.850728   A3 =  3.188976    K =  0.098194
     P1 =         1   P2 =         3   P3 =         5   Q1 =        -2   Q2 =        -5   Q3 =        -9

Modified exponential model  :
     Bm = Sum ( Ai X ( 1 - exp( - R/( 2*i*C ) ) ) )  +      Sum ( RKj X ( R - Rj0 ) )  +  Bm0
       C =   0.30022917E+00             Bmo=   0.00000000E+00          Nexp=  6        Nliner=  1
      AI =  -0.37246761E+03   0.34894158E+04  -0.98931248E+04   0.11366955E+05  -0.42321976E+04   0.33376875E+03
     Rj0 =          9.0800
     RKj =   0.87108777E+01

Power model  :    Bm = ( Rki X R ) / ( 1 + ( R/R0 )**rn )**( 1/rn )
      rn = 1.240      rki =   0.18726233E+03         rmu =   0.92915197E+03
```

No	Rotation (radians) X 1/1000	Moment (kip-inch) Expri.	Moment Poly.	Moment M.Expo.	Moment P.Model.	Connection stiffness (kip-inch) X 1000 Expri.	Stiffness Poly.	Stiffness M.Expo.	Stiffness P.Model.
1	0.00	0.00	0.00	0.00	0.00	0.2227E+03	0.4562E+03	0.2089E+03	0.1873E+03
3	0.35	78.87	139.28	78.18	64.37	0.2161E+03	0.3061E+03	0.2228E+03	0.1751E+03
5	0.74	158.92	230.29	159.44	128.42	0.2011E+03	0.1856E+03	0.1980E+03	0.1592E+03
7	1.15	238.39	293.53	233.61	190.63	0.1658E+03	0.1282E+03	0.1616E+03	0.1425E+03
9	1.80	318.48	361.22	322.08	275.06	0.1165E+03	0.8673E+02	0.1154E+03	0.1193E+03
11	2.53	397.99	415.69	394.65	354.43	0.9523E+02	0.6401E+02	0.8551E+02	0.9765E+02
13	3.62	480.49	474.69	474.93	446.65	0.6512E+02	0.4667E+02	0.6518E+02	0.7372E+02
15	5.00	551.26	530.36	553.95	533.25	0.4811E+02	0.3509E+02	0.4962E+02	0.5311E+02
17	6.15	603.77	567.25	604.05	587.39	0.3886E+02	0.2926E+02	0.3728E+02	0.4142E+02
19	7.69	649.86	607.96	649.68	642.23	0.2733E+02	0.2409E+02	0.2268E+02	0.3070E+02
21	10.03	694.30	658.10	692.22	701.21	0.1380E+02	0.1913E+02	0.1693E+02	0.2060E+02
23	12.25	719.89	696.95	721.94	740.03	0.9866E+01	0.1611E+02	0.1082E+02	0.1486E+02
25	14.64	742.57	732.55	744.57	770.46	0.9071E+01	0.1383E+02	0.8612E+01	0.1091E+02
27	16.87	762.87	761.65	763.23	791.94	0.8739E+01	0.1224E+02	0.8200E+01	0.8437E+01
29	19.15	782.00	788.06	781.95	809.03	0.8400E+01	0.1099E+02	0.8272E+01	0.6664E+01
31	21.54	802.33	812.97	801.87	823.24	0.8635E+01	0.9948E+01	0.8433E+01	0.5328E+01
33	24.03	823.84	836.62	823.05	835.17	0.8581E+01	0.9066E+01	0.8562E+01	0.4306E+01
35	26.36	843.77	856.98	843.14	844.35	0.8530E+01	0.8381E+01	0.8634E+01	0.3584E+01
37	28.59	862.50	875.03	862.44	851.72	0.8267E+01	0.7826E+01	0.8672E+01	0.3047E+01
39	30.71	880.04	891.14	880.86	857.74	0.8267E+01	0.7367E+01	0.8691E+01	0.2637E+01

III - 15

```
Connection type : Top-and seat-angle connections with double web angle
           Mode : All bolted

Tested by  :  A.Azizinamini et al. (1985)              U.S.A.
Test  Id.  :  14S5

Column : W12X96                   Fasteners: A325- -7/8"D
Beam   : W14X38                              15/16" Oversize holes
F.Angle: 6 X 4.0 X 3/8  X 8.0     Material : A36
W.Angle: 4 X 3.5 X 1/4  X 8.5                Fy =  39.55 ksi
                                             Fu =  67.95 ksi

                         Major parameters

      lp =  8.5000"    lu =  2.8125"    ll =  2.8125"    ta =  0.2500"
      gb =  2.0000"    gc =  2.5940"    cu =  1.2500"    cl =  1.2500"
      pb =  3.0000"    pc =  3.0000"    nb = 1 X  3      nc = 1 X  3
      lt =  8.0000"    tt =  0.3750"    ls =  8.0000"    ts =  0.3750"
      gt =  3.5000"    gt'=  2.5000"    gs =  3.5000"    gs'=  2.5000"
      qt =  3.5000"    rt =  5.5000"    qs =  3.5000"    rs =  5.5000"
      pt =  2.5000"    ps =  2.5000"
      nt = 2 X  2      nt'= 2 X  1      ns = 2 X  2      ns'= 2 X  1

Remark  1)  Heavy hex high strength bolts used
        2)
```

No	Moment (k-in)	Rotation (radians) X 1/1000	No	Moment (k-in)	Rotation (radians) X 1/1000
1	0.00	0.00	26	822.63	12.71
2	43.86	0.09	27	829.02	14.03
3	97.71	0.18	28	830.67	15.04
4	130.42	0.35	29	845.71	16.18
5	163.13	0.52	30	860.76	17.31
6	206.48	0.73	31	879.76	18.73
7	249.83	0.95	32	894.02	19.98
8	285.30	1.18	33	908.29	21.24
9	320.78	1.42	34	921.35	22.22
10	355.46	1.70	35	934.42	23.20
11	390.15	1.98	36	954.18	24.11
12	428.78	2.26	37	967.25	25.24
13	467.40	2.54	38	980.33	26.37
14	496.18	2.86	39	992.21	27.32
15	524.96	3.19	40	1004.09	28.26
16	573.06	3.69	41	1017.93	29.13
17	619.60	4.47	42	1031.78	29.99
18	657.46	5.10	43	1042.89	31.22
19	691.39	5.88			
20	724.53	6.67			
21	760.86	7.86			
22	787.74	9.22			
23	795.67	10.05			
24	803.60	10.88			
25	813.12	11.80			

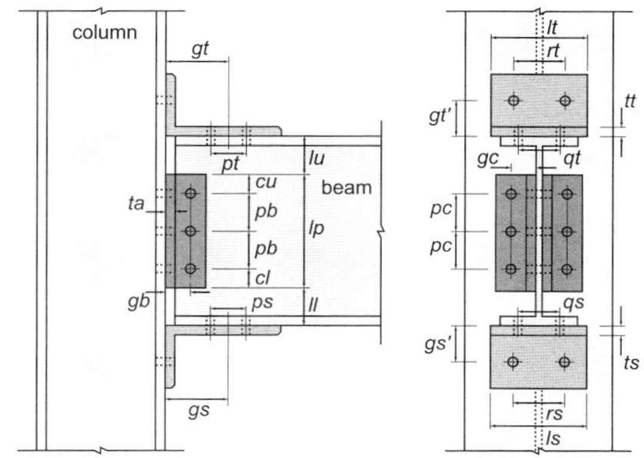

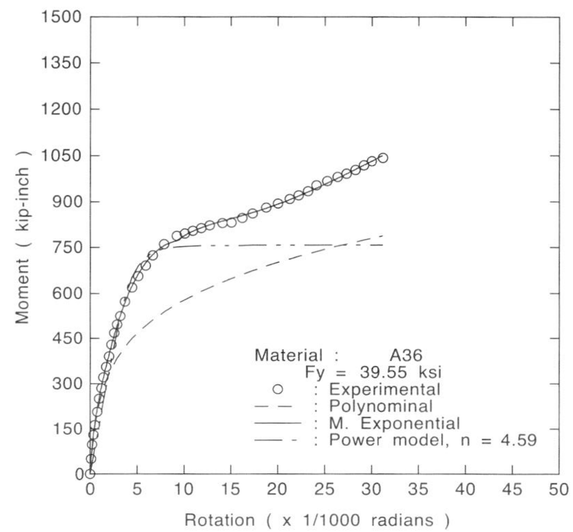

A.3 – 30

```
Moment-rotation prediction equations      ( R : X 1/1000 radians )

Frye and Morris polynominal model   :    R = Sum ( Ai X ( K*Bm )**Pi X 10**Qi )
      t =  0.375000"   d = 14.125000"  tc =  0.250000"   w =  8.000000"   g =  2.062500"
     A1 =  2.232429   A2 =  1.850728   A3 =  3.188976    K =  0.111579
     P1 =         1   P2 =         3   P3 =         5   Q1 =        -2   Q2 =        -5   Q3 =        -9

Modified exponential model   :
     Bm = Sum ( Ai X ( 1 - exp( - R/( 2*i*C ) ) ) )  +      Sum ( RKj X ( R - Rj0 ) )  +  Bm0
      C =   0.28253167E+00            Bmo=   0.00000000E+00          Nexp=  6        Nliner=  1
     AI =   0.60168489E+03  -0.47198808E+04   0.22671486E+05  -0.54998541E+05   0.61073874E+05  -0.23930773E+05
    Rj0 =            9.2200
    RKj =   0.15991268E+02

Power model   :    Bm = ( Rki X R ) / ( 1 + ( R/R0 )**rn )**( 1/rn )
      rn = 4.587      rki =   0.17086121E+03          rmu =   0.75715890E+03
```

No	Rotation (radians) X 1/1000	Moment (kip-inch) Expri.	Poly.	M.Expo.	P.Model.	Connection stiffness (kip-inch) X 1000 Expri.	Poly.	M.Expo.	P.Model.
1	0.00	0.00	0.00	0.00	0.00	0.5465E+03	0.4015E+03	0.4877E+03	0.1709E+03
3	0.18	97.71	68.48	74.92	30.57	0.4240E+03	0.3498E+03	0.3589E+03	0.1709E+03
5	0.52	163.13	162.03	171.05	88.62	0.1961E+03	0.2125E+03	0.2253E+03	0.1709E+03
7	0.95	249.83	234.03	251.73	162.44	0.1774E+03	0.1326E+03	0.1599E+03	0.1707E+03
9	1.42	320.78	285.71	319.78	241.69	0.1395E+03	0.9412E+02	0.1365E+03	0.1698E+03
11	1.98	390.15	331.03	392.35	335.89	0.1310E+03	0.7016E+02	0.1233E+03	0.1659E+03
13	2.54	467.40	366.07	457.88	426.27	0.1149E+03	0.5625E+02	0.1107E+03	0.1561E+03
15	3.19	524.96	399.31	525.19	521.99	0.9116E+02	0.4587E+02	0.9441E+02	0.1339E+03
17	4.47	619.60	449.77	626.53	654.05	0.5999E+02	0.3404E+02	0.6472E+02	0.7150E+02
19	5.88	691.39	492.20	699.77	718.44	0.4275E+02	0.2677E+02	0.4052E+02	0.2613E+02
21	7.86	760.86	538.71	755.65	745.76	0.2548E+02	0.2079E+02	0.1764E+02	0.6373E+01
23	10.05	795.67	579.49	788.94	753.35	0.9508E+01	0.1681E+02	0.1815E+02	0.1712E+01
25	11.80	813.12	606.87	814.25	755.32	0.1042E+02	0.1464E+02	0.1156E+02	0.7093E+00
27	14.03	829.02	637.23	835.78	756.33	0.3013E+01	0.1261E+02	0.8473E+01	0.2710E+00
29	16.18	845.71	662.63	853.81	756.72	0.1328E+02	0.1117E+02	0.8705E+01	0.1229E+00
31	18.73	879.76	689.36	878.13	756.94	0.1231E+02	0.9860E+01	0.1048E+02	0.5434E-01
33	21.24	908.29	712.88	906.93	757.03	0.1249E+02	0.8857E+01	0.1234E+02	0.2687E-01
35	23.20	934.42	729.53	932.19	757.08	0.1767E+02	0.8220E+01	0.1349E+02	0.1645E-01
37	25.24	967.25	745.75	960.75	757.10	0.1154E+02	0.7653E+01	0.1438E+02	0.1026E-01
39	27.32	992.21	761.12	991.30	757.12	0.1258E+02	0.7158E+01	0.1499E+02	0.6596E-02
41	29.13	1017.93	773.89	1018.76	757.13	0.1600E+02	0.6775E+01	0.1534E+02	0.4610E-02
43	31.22	1042.89	787.64	1051.17	757.14	0.9048E+01	0.6391E+01	0.1560E+02	0.3128E-02

III - 16

Connection type : Top-and seat-angle connections with double web angle
Mode : All bolted

Tested by : A.Azizinamini et al. (1985) U.S.A.
Test Id. : 14S6

Column : W12X96
Beam : W14X38
F.Angle: 6 X 4.0 X 1/2 X 8.0
W.Angle: 4 X 3.5 X 1/4 X 8.5

Fasteners: A325- -7/8"D
15/16" Oversize holes
Material : A36
Fy = 39.55 ksi
Fu = 67.95 ksi

Major parameters

lp = 8.5000"	lu = 2.8125"	ll = 2.8125"	ta = 0.2500"
gb = 2.0000"	gc = 2.5940"	cu = 1.2500"	cl = 1.2500"
pb = 3.0000"	pc = 3.0000"	nb = 1 X 3	nc = 1 X 3
lt = 8.0000"	tt = 0.5000"	ls = 8.0000"	ts = 0.5000"
gt = 3.5000"	gt'= 2.5000"	gs = 3.5000"	gs'= 2.5000"
qt = 3.5000"	rt = 5.5000"	qs = 3.5000"	rs = 5.5000"
pt = 2.5000"	ps = 2.5000"		
nt = 2 X 2	nt'= 2 X 1	ns = 2 X 2	ns'= 2 X 1

Remark 1) Heavy hex high strength bolts used
2)

No	Moment (k-in)	Rotation (radians) X 1/1000	No	Moment (k-in)	Rotation (radians) X 1/1000
1	0.00	0.00	26	886.80	11.99
2	39.43	0.15	27	904.37	13.03
3	78.86	0.31	28	921.95	14.07
4	121.45	0.50	29	934.73	14.85
5	164.05	0.69	30	947.52	15.64
6	206.64	0.89	31	960.32	16.46
7	249.23	1.08	32	973.11	17.28
8	291.84	1.33	33	985.17	18.33
9	334.44	1.58	34	997.23	19.39
10	375.49	1.90	35	1009.25	20.25
11	416.53	2.23	36	1021.27	21.12
12	455.22	2.59	37	1030.97	22.17
13	493.91	2.95	38	1040.66	23.23
14	532.62	3.36	39	1051.15	24.30
15	571.32	3.76	40	1061.64	25.37
16	606.09	4.23	41	1074.53	26.60
17	640.87	4.70	42	1087.42	27.83
18	672.51	5.23	43	1097.16	29.07
19	704.15	5.77	44	1106.90	30.32
20	753.30	7.09			
21	777.91	7.89			
22	802.52	8.69			
23	823.98	9.47			
24	845.43	10.26			
25	866.12	11.13			

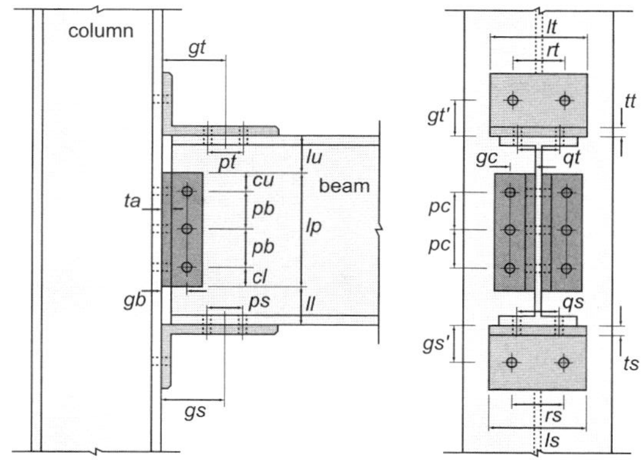

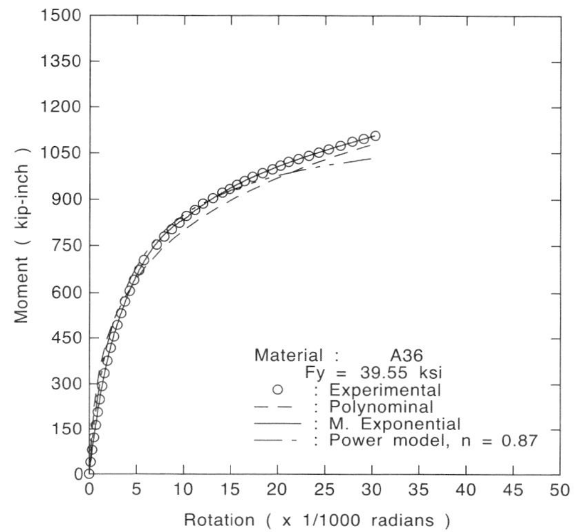

```
Moment-rotation prediction equations      ( R : X 1/1000 radians )

Frye and Morris polynominal model  :    R = Sum ( Ai X ( K*Bm )**Pi X 10**Qi )
     t =  0.500000"   d = 14.125000"  tc =  0.250000"   w =  8.000000"   g =  2.062500"
    A1 =  2.232429   A2 =  1.850728   A3 =  3.188976    K =  0.080657
    P1 =         1   P2 =         3   P3 =         5   Q1 =        -2   Q2 =        -5   Q3 =        -9

Modified exponential model  :
    Bm = Sum ( Ai X ( 1 - exp( - R/( 2*i*C ) ) ) )  +      Sum ( RKj X ( R - Rj0 ) )  +  Bm0
     C =   0.29113000E+00          Bmo=   0.00000000E+00          Nexp=  6        Nliner=  1
    AI =  -0.20614203E+03   0.24370657E+04  -0.10404093E+05   0.22543752E+05  -0.23819525E+05   0.10384737E+05
   Rj0 =          10.2600
   RKj =   0.86275181E+01

Power model  :    Bm = ( Rki X R ) / ( 1 + ( R/R0 )**rn )**( 1/rn )
    rn = 0.865      rki =   0.41765154E+03        rmu =   0.11897789E+04
```

No	Rotation (radians) X 1/1000	Moment (kip-inch) Expri.	Poly.	M.Expo.	P.Model.	Connection stiffness (kip-inch) X 1000 Expri.	Poly.	M.Expo.	P.Model.
1	0.00	0.00	0.00	0.00	0.00	0.2560E+03	0.5554E+03	0.2528E+03	0.4177E+03
3	0.31	78.86	151.84	77.46	109.91	0.2404E+03	0.3998E+03	0.2462E+03	0.3114E+03
5	0.69	164.05	270.38	167.44	215.10	0.2206E+03	0.2369E+03	0.2182E+03	0.2394E+03
7	1.08	249.23	346.20	245.57	297.94	0.1991E+03	0.1646E+03	0.1868E+03	0.1926E+03
9	1.58	334.44	415.19	329.46	382.75	0.1517E+03	0.1185E+03	0.1527E+03	0.1518E+03
11	2.23	416.53	481.37	418.29	469.63	0.1170E+03	0.8715E+02	0.1211E+03	0.1163E+03
13	2.95	493.91	536.48	496.58	543.34	0.1019E+03	0.6806E+02	0.9862E+02	0.9074E+02
15	3.76	571.32	586.00	569.19	608.62	0.8543E+02	0.5488E+02	0.8102E+02	0.7122E+02
17	4.70	640.87	632.62	637.67	667.84	0.6718E+02	0.4509E+02	0.6537E+02	0.5588E+02
19	5.77	704.15	676.49	699.40	720.67	0.5300E+02	0.3769E+02	0.5097E+02	0.4399E+02
21	7.89	777.91	745.93	784.20	797.13	0.3070E+02	0.2869E+02	0.3074E+02	0.2960E+02
23	9.47	823.98	787.93	825.15	838.65	0.2729E+02	0.2447E+02	0.2152E+02	0.2311E+02
25	11.13	866.12	825.58	862.70	872.73	0.2391E+02	0.2130E+02	0.2400E+02	0.1845E+02
27	13.03	904.37	863.43	903.72	904.05	0.1692E+02	0.1860E+02	0.1942E+02	0.1468E+02
29	14.85	934.73	895.50	936.29	928.35	0.1625E+02	0.1662E+02	0.1642E+02	0.1208E+02
31	16.46	960.32	921.02	961.00	946.28	0.1563E+02	0.1523E+02	0.1446E+02	0.1033E+02
33	18.33	985.17	948.25	986.38	964.09	0.1143E+02	0.1389E+02	0.1272E+02	0.8751E+01
35	20.25	1009.25	974.02	1009.50	979.64	0.1388E+02	0.1275E+02	0.1142E+02	0.7487E+01
37	22.17	1030.97	997.38	1030.50	993.02	0.9191E+01	0.1181E+02	0.1050E+02	0.6485E+01
39	24.30	1051.15	1021.53	1052.02	1005.83	0.9796E+01	0.1093E+02	0.9800E+01	0.5599E+01
41	26.60	1074.53	1045.71	1073.95	1017.79	0.1049E+02	0.1013E+02	0.9319E+01	0.4836E+01
43	29.07	1097.16	1069.82	1096.58	1028.91	0.7829E+01	0.9395E+01	0.9010E+01	0.4181E+01

III - 17

Connection type : Top-and seat-angle connections with double web angle
Mode : All bolted

Tested by : A.Azizinamini et al. (1985) U.S.A.
Test Id. : 14S8

Column : W12X96
Beam : W14X38
F.Angle: 6 X 4.0 X 5/8 X 8.0
W.Angle: 4 X 3.5 X 1/4 X 8.5

Fasteners: A325- -7/8"D
15/16" Oversize holes
Material : A36
Fy = 39.55 ksi
Fu = 67.95 ksi

Major parameters

lp = 8.5000"	lu = 2.8125"	ll = 2.8125"	ta = 0.2500"
gb = 2.0000"	gc = 2.5940"	cu = 1.2500"	cl = 1.2500"
pb = 3.0000"	pc = 3.0000"	nb = 1 X 3	nc = 1 X 3
lt = 8.0000"	tt = 0.6250"	ls = 8.0000"	ts = 0.6250"
gt = 3.5000"	gt'= 2.5000"	gs = 3.5000"	gs'= 2.5000"
qt = 3.5000"	rt = 5.5000"	qs = 3.5000"	rs = 5.5000"
pt = 2.5000"	ps = 2.5000"		
nt = 2 X 2	nt'= 2 X 1	ns = 2 X 2	ns'= 2 X 1

Remark 1) Heavy hex high strength bolts used
2)

No	Moment (k-in)	Rotation (radians) X 1/1000	No	Moment (k-in)	Rotation (radians) X 1/1000
1	0.00	0.00	26	1275.51	8.50
2	57.81	0.10	27	1310.48	9.64
3	115.61	0.20	28	1345.45	10.79
4	173.42	0.30	29	1372.55	11.97
5	233.33	0.42	30	1399.65	13.14
6	293.25	0.54	31	1407.84	14.34
7	353.16	0.67	32	1430.19	15.44
8	413.08	0.81	33	1452.54	16.54
9	462.76	0.94	34	1471.03	17.92
10	512.43	1.08	35	1489.51	19.31
11	568.42	1.26	36	1505.01	20.30
12	624.41	1.45	37	1520.50	21.30
13	667.79	1.61	38	1536.00	22.30
14	711.17	1.78	39	1552.04	23.37
15	764.83	2.08	40	1568.08	24.45
16	818.48	2.37	41	1584.12	25.53
17	869.78	2.68	42	1598.66	26.87
18	921.07	2.99	43	1613.19	28.21
19	966.89	3.44	44	1624.36	28.74
20	1012.71	3.89			
21	1060.15	4.52			
22	1107.59	5.14			
23	1152.69	5.86			
24	1197.79	6.58			
25	1236.65	7.54			

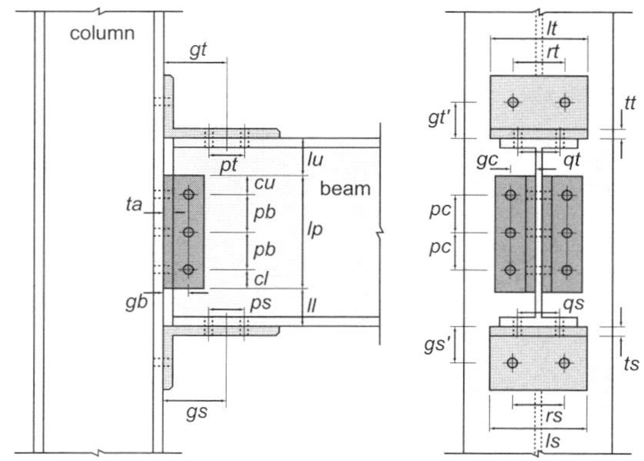

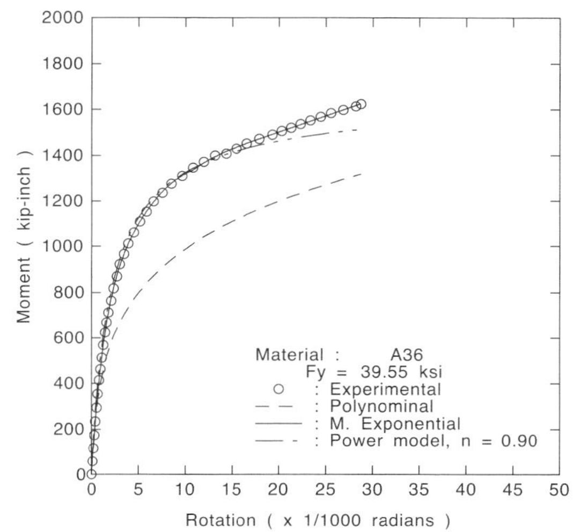

A.3 – 34

```
 Moment-rotation prediction equations      ( R : X 1/1000 radians )

Frye and Morris polynominal model  :     R = Sum ( Ai X ( K*Bm )**Pi X 10**Qi )
      t =  0.625000"   d = 14.125000"  tc =  0.250000"    w =  8.000000"   g =  2.125000"
     A1 =  2.232429   A2 =  1.850728   A3 =  3.188976    K =  0.065286
     P1 =         1   P2 =         3   P3 =         5   Q1 =        -2    Q2 =        -5    Q3 =        -9

Modified exponential model  :
     Bm = Sum ( Ai X ( 1 - exp( - R/( 2*i*C ) ) ) )  +      Sum ( RKj X ( R - Rj0 ) )  +  Bm0
      C =   0.26424583E+00              Bmo=   0.00000000E+00          Nexp=  6        Nliner=  1
     AI =   0.71983049E+02  -0.27760964E+04   0.19746232E+05  -0.44495995E+05   0.41698907E+05  -0.12884213E+05
     Rj0 =         10.7900
     RKj =   0.14714619E+02

Power model  :    Bm = ( Rki X R ) / ( 1 + ( R/R0 )**rn )**( 1/rn )
      rn = 0.904      rki =    0.87121352E+03          rmu =    0.16574884E+04
```

No	Rotation (radians) X 1/1000	Moment (kip-inch) Expri.	Poly.	M.Expo.	P.Model.	Connection stiffness (kip-inch) X 1000 Expri.	Poly.	M.Expo.	P.Model.
1	0.00	0.00	0.00	0.00	0.00	0.5845E+03	0.6861E+03	0.6328E+03	0.8712E+03
3	0.20	115.61	128.20	118.38	150.65	0.5842E+03	0.5826E+03	0.5655E+03	0.6745E+03
5	0.42	233.33	236.88	234.65	282.61	0.5007E+03	0.4195E+03	0.4997E+03	0.5415E+03
7	0.67	353.16	327.27	353.36	406.43	0.4424E+03	0.3004E+03	0.4322E+03	0.4353E+03
9	0.94	462.76	398.23	462.18	513.34	0.3639E+03	0.2281E+03	0.3698E+03	0.3555E+03
11	1.26	568.42	462.42	570.13	615.62	0.3058E+03	0.1778E+03	0.3076E+03	0.2883E+03
13	1.61	667.79	518.45	668.14	707.19	0.2571E+03	0.1433E+03	0.2516E+03	0.2351E+03
15	2.08	764.83	577.59	770.67	803.78	0.1828E+03	0.1147E+03	0.1949E+03	0.1858E+03
17	2.68	869.78	639.22	871.73	901.84	0.1657E+03	0.9151E+02	0.1440E+03	0.1424E+03
19	3.44	966.89	701.44	965.31	995.70	0.1014E+03	0.7333E+02	0.1056E+03	0.1068E+03
21	4.52	1060.15	771.25	1062.35	1092.60	0.7591E+02	0.5767E+02	0.7813E+02	0.7590E+02
23	5.86	1152.69	840.23	1154.53	1178.12	0.6267E+02	0.4589E+02	0.6038E+02	0.5335E+02
25	7.54	1236.65	908.98	1241.21	1252.72	0.4062E+02	0.3686E+02	0.4332E+02	0.3715E+02
27	9.64	1310.48	978.45	1311.85	1317.53	0.3052E+02	0.2978E+02	0.2459E+02	0.2561E+02
29	11.97	1372.55	1041.38	1368.28	1367.73	0.2300E+02	0.2472E+02	0.2506E+02	0.1823E+02
31	14.34	1407.84	1095.61	1417.73	1405.09	0.1387E+02	0.2116E+02	0.1759E+02	0.1359E+02
33	16.54	1452.54	1139.37	1452.84	1431.66	0.1724E+02	0.1873E+02	0.1479E+02	0.1073E+02
35	19.31	1489.51	1187.86	1492.10	1457.74	0.1464E+02	0.1642E+02	0.1388E+02	0.8275E+01
37	21.30	1520.50	1219.53	1519.73	1472.89	0.1556E+02	0.1509E+02	0.1391E+02	0.7001E+01
39	23.37	1552.04	1249.65	1548.79	1486.31	0.1486E+02	0.1395E+02	0.1410E+02	0.5967E+01
41	25.53	1584.12	1278.62	1579.47	1498.24	0.1308E+02	0.1294E+02	0.1431E+02	0.5120E+01
43	28.21	1613.19	1311.55	1617.99	1510.79	0.1805E+02	0.1191E+02	0.1449E+02	0.4304E+01

III - 18

Connection type : Top-and seat-angle connections with double web angle
Mode : All bolted

Tested by : A.Azizinamini et al. (1985) U.S.A.
Test Id. : 14S9

Column : W12X96
Beam : W14X38
F.Angle: 6 X 4.0 X 1/2 X 8.0
W.Angle: 4 X 3.5 X 1/4 X 8.5

Fasteners: A325- -7/8"D
15/16" Oversize holes
Material : A36
Fy = 39.55 ksi
Fu = 67.95 ksi

Major parameters

lp = 8.5000"	lu = 2.8125"	ll = 2.8125"	ta = 0.2500"
gb = 2.0000"	gc = 2.5940"	cu = 1.2500"	cl = 1.2500"
pb = 3.0000"	pc = 3.0000"	nb = 1 X 3	nc = 1 X 3
lt = 8.0000"	tt = 0.5000"	ls = 8.0000"	ts = 0.5000"
gt = 3.5000"	gt'= 2.5000"	gs = 3.5000"	gs'= 2.5000"
qt = 3.5000"	rt = 5.5000"	qs = 3.5000"	rs = 5.5000"
pt = 2.5000"	ps = 2.5000"		
nt = 2 X 2	nt'= 2 X 1	ns = 2 X 2	ns'= 2 X 1

Remark 1) Heavy hex high strength bolts used
2)

No	Moment (k-in)	Rotation (radians) X 1/1000	No	Moment (k-in)	Rotation (radians) X 1/1000
1	0.00	0.00	26	903.48	11.71
2	37.83	0.15	27	922.64	12.86
3	75.66	0.31	28	935.47	13.88
4	105.61	0.42	29	948.30	14.90
5	135.55	0.52	30	961.93	15.96
6	187.58	0.77	31	975.56	17.01
7	241.19	1.08	32	984.49	18.18
8	282.19	1.30	33	993.42	19.34
9	323.18	1.51	34	995.21	20.27
10	357.08	1.70	35	996.99	21.20
11	390.98	1.89	36	1004.31	22.21
12	423.32	2.13	37	1011.63	23.21
13	455.67	2.38	38	1026.06	24.35
14	491.16	2.66	39	1031.79	25.29
15	526.66	2.94	40	1037.52	26.24
16	556.65	3.24	41	1047.16	27.05
17	586.64	3.54	42	1056.79	27.87
18	627.71	4.07	43	1066.45	28.78
19	668.77	4.60	44	1076.11	29.70
20	704.34	5.20			
21	739.91	5.79			
22	776.35	6.75			
23	812.79	7.70			
24	857.21	9.12			
25	884.32	10.56			

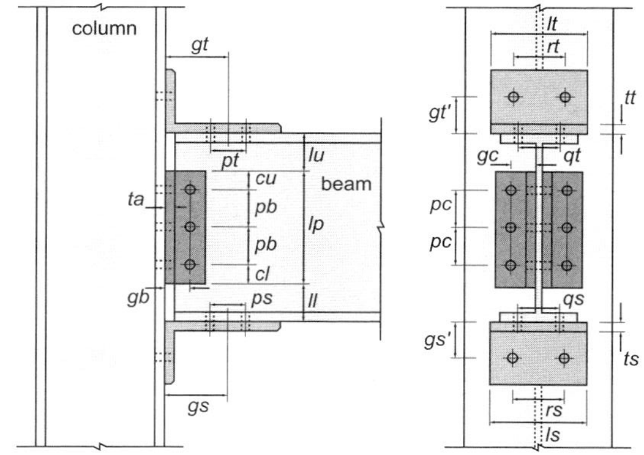

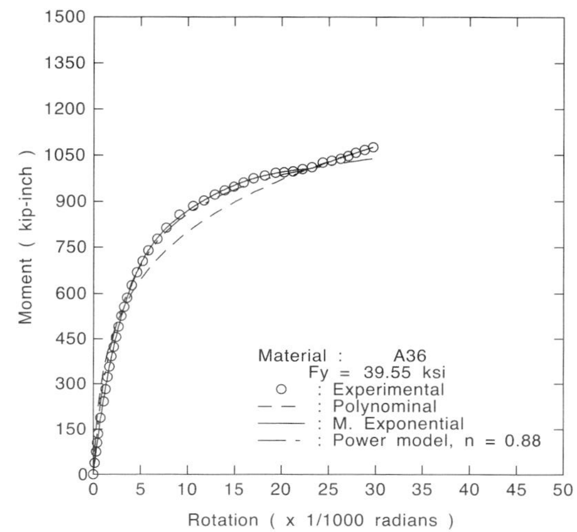

```
Moment-rotation prediction equations     ( R : X 1/1000 radians )

Frye and Morris polynominal model  :    R = Sum ( Ai X ( K*Bm )**Pi X 10**Qi )
     t =  0.500000"   d = 14.125000"  tc =  0.250000"   w =  8.000000"   g =  2.062500"
    A1 =  2.232429   A2 =  1.850728   A3 =  3.188976    K =  0.080657
    P1 =        1    P2 =        3    P3 =        5    Q1 =       -2    Q2 =      -5    Q3 =      -9

Modified exponential model  :
    Bm = Sum ( Ai X ( 1 - exp( - R/( 2*i*C ) ) ) )  +     Sum ( RKj X ( R - Rj0 ) )  +  Bm0
      C =   0.28596917E+00              Bmo=   0.00000000E+00          Nexp=  6        Nliner=  1
     AI =  -0.10662665E+03   0.23383866E+04  -0.15648250E+05   0.42372583E+05  -0.49018957E+05   0.21081811E+05
    Rj0 =           23.2100
    RKj =   0.89741431E+01

Power model  :    Bm = ( Rki X R ) / ( 1 + ( R/R0 )**rn )**( 1/rn )
     rn = 0.880       rki =   0.41765154E+03          rmu =   0.11897789E+04
```

No	Rotation (radians) X 1/1000	Moment (kip-inch) Expri.	Moment (kip-inch) Poly.	Moment (kip-inch) M.Expo.	Moment (kip-inch) P.Model.	Connection stiffness (kip-inch) X 1000 Expri.	Connection stiffness (kip-inch) X 1000 Poly.	Connection stiffness (kip-inch) X 1000 M.Expo.	Connection stiffness (kip-inch) X 1000 P.Model.
1	0.00	0.00	0.00	0.00	0.00	0.2457E+03	0.5554E+03	0.2613E+03	0.4177E+03
3	0.31	75.66	151.84	78.16	110.71	0.2650E+03	0.3998E+03	0.2449E+03	0.3150E+03
5	0.52	135.55	225.30	129.29	173.45	0.2579E+03	0.2924E+03	0.2306E+03	0.2707E+03
7	1.08	241.19	346.16	247.33	301.44	0.1825E+03	0.1647E+03	0.1939E+03	0.1957E+03
9	1.51	323.18	407.56	325.93	377.76	0.1855E+03	0.1228E+03	0.1695E+03	0.1586E+03
11	1.89	390.98	448.92	385.54	432.19	0.1600E+03	0.1012E+03	0.1517E+03	0.1352E+03
13	2.38	455.67	494.24	455.86	493.34	0.1284E+03	0.8221E+02	0.1313E+03	0.1116E+03
15	2.94	526.66	536.11	523.79	550.14	0.1143E+03	0.6817E+02	0.1117E+03	0.9210E+02
17	3.54	586.64	573.32	584.63	599.98	0.9261E+02	0.5795E+02	0.9396E+02	0.7678E+02
19	4.60	668.77	628.04	670.28	670.76	0.6900E+02	0.4596E+02	0.6840E+02	0.5776E+02
21	5.79	739.91	677.38	738.89	730.76	0.5144E+02	0.3756E+02	0.4829E+02	0.4402E+02
23	7.70	812.79	740.64	811.98	800.94	0.3539E+02	0.2927E+02	0.3046E+02	0.3057E+02
25	10.56	884.32	813.36	882.06	871.20	0.1760E+02	0.2227E+02	0.2029E+02	0.1978E+02
27	12.86	922.64	860.27	923.37	910.54	0.1450E+02	0.1881E+02	0.1585E+02	0.1485E+02
29	14.90	948.30	896.33	952.10	937.71	0.1274E+02	0.1657E+02	0.1228E+02	0.1189E+02
31	17.01	975.56	929.32	974.39	960.34	0.1043E+02	0.1480E+02	0.8946E+01	0.9700E+01
33	19.34	993.42	961.94	991.57	980.76	0.4473E+01	0.1327E+02	0.5936E+01	0.7928E+01
35	21.20	996.99	985.64	1000.83	994.45	0.4488E+01	0.1227E+02	0.4120E+01	0.6849E+01
37	23.21	1011.63	1009.43	1007.64	1007.27	0.9844E+01	0.1136E+02	0.1166E+02	0.5915E+01
39	25.29	1031.79	1032.18	1030.75	1018.72	0.6066E+01	0.1057E+02	0.1066E+02	0.5143E+01
41	27.05	1047.16	1050.54	1049.00	1027.30	0.1178E+02	0.9976E+01	0.1009E+02	0.4603E+01
43	28.78	1066.45	1067.23	1066.11	1034.86	0.1058E+02	0.9471E+01	0.9707E+01	0.4154E+01

III - 19

Connection type : Top-and seat-angle connections with double web angle
Mode : Welded-to-beam and bolted-to-column

Tested by : A.Azizinamini et al. (1985) U.S.A.
Test Id. : 14WS1

Column : W12X96
Beam : W14X38
F.Angle: 6 X 4.0 X 3/8 X 8.0
W.Angle: 4 X 3.5 X 1/4 X 8.5

Fasteners: A325- -3/4"D
15/16" Oversize holes
Material : A36
Fy = 39.55 ksi
Fu = 67.95 ksi

Major parameters

lp = 8.5000"	lu = 2.8125"	ll = 2.8125"	ta = 0.2500"
gc = 2.5940"	cu = 1.2500"	cl = 1.2500"	pc = 3.0000"
nc = 1 X 3			
lt = 8.0000"	tt = 0.3750"	ls = 8.0000"	ts = 0.3750"
gt'= 2.5000"	gs'= 2.5000"	rt = 5.5000"	rs = 5.5000"
nt'= 2 X 1	ns'= 2 X 1		

Remark 1) Heavy hex high strength bolts used
2)

No	Moment (k-in)	Rotation (radians) X 1/1000	No	Moment (k-in)	Rotation (radians) X 1/1000
1	0.00	0.00	26	788.15	13.84
2	38.01	0.01	27	803.64	14.82
3	76.03	0.02	28	819.54	15.97
4	114.04	0.04	29	835.44	17.11
5	142.00	0.28	30	848.97	18.14
6	169.95	0.52	31	862.49	19.18
7	208.53	0.83	32	880.80	20.62
8	251.84	1.16	33	890.38	21.58
9	295.96	1.66	34	899.96	22.53
10	338.51	2.21	35	911.88	23.35
11	365.68	2.47	36	923.80	24.16
12	392.85	2.73	37	940.54	25.63
13	417.68	3.12	38	944.60	26.48
14	442.52	3.51	39	948.66	27.33
15	489.02	4.22	40	958.32	28.77
16	530.82	4.99	41	965.93	29.71
17	577.36	5.92	42	973.55	30.65
18	616.82	6.83			
19	660.26	8.01			
20	682.80	8.84			
21	705.34	9.66			
22	723.15	10.46			
23	740.97	11.26			
24	756.82	12.06			
25	772.67	12.85			

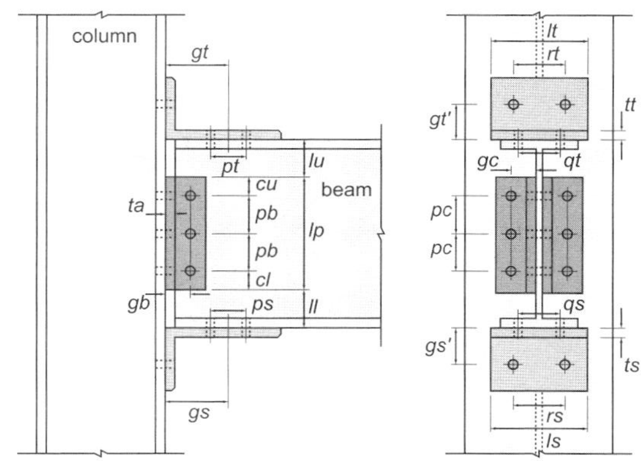

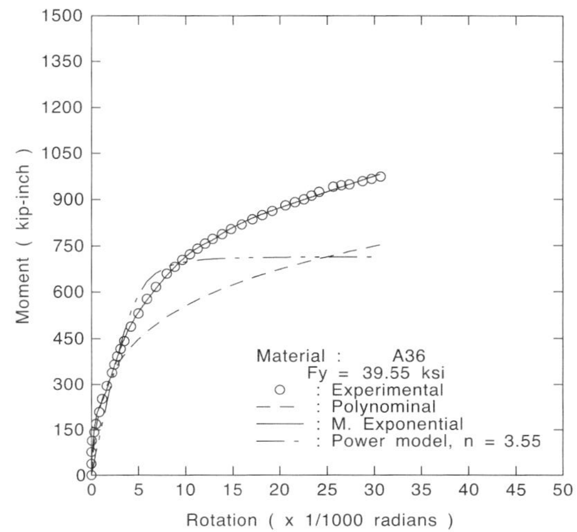

```
Moment-rotation prediction equations     ( R : X 1/1000 radians )

Frye and Morris polynominal model  :     R = Sum ( Ai X ( K*Bm )**Pi X 10**Qi )
     t =  0.375000"   d = 14.125000"  tc =  0.250000"   w =  8.000000"   g =  2.125000"
    A1 =  2.232429   A2 =  1.850728   A3 =  3.188976    K =  0.116167
    P1 =         1   P2 =         3   P3 =         5   Q1 =        -2   Q2 =        -5   Q3 =        -9

Modified exponential model  :
    Bm = Sum ( Ai X ( 1 - exp( - R/( 2*i*C ) ) ) )  +     Sum ( RKj X ( R - Rj0 ) )  +  Bm0
     C =   0.25838000E+00              Bmo=   0.00000000E+00          Nexp=  6        Nliner=  1
    AI =   0.15714337E+04  -0.11249199E+05   0.37235512E+05  -0.58430840E+05   0.40742084E+05  -0.90695243E+04
   Rj0 =           12.8500
   RKj =   0.10221404E+02

Power model  :    Bm = ( Rki X R ) / ( 1 + ( R/R0 )**rn )**( 1/rn )
     rn = 3.554      rki =   0.14477366E+03          rmu =   0.71442018E+03
```

No	Rotation (radians) X 1/1000	Moment (kip-inch) Expri.	Moment Poly.	Moment M.Expo.	Moment P.Model.	Connection stiffness (kip-inch) X 1000 Expri.	Stiffness Poly.	Stiffness M.Expo.	Stiffness P.Model.
1	0.00	0.00	0.00	0.00	0.00	0.3221E+04	0.3856E+03	0.7504E+03	0.1448E+03
3	0.02	76.03	9.06	17.01	3.40	0.3235E+04	0.3845E+03	0.6975E+03	0.1448E+03
5	0.28	142.00	97.37	139.33	40.52	0.1143E+03	0.2899E+03	0.3069E+03	0.1448E+03
7	0.83	208.53	208.38	222.06	120.02	0.1278E+03	0.1427E+03	0.7464E+02	0.1444E+03
9	1.66	295.96	294.69	284.39	238.46	0.8353E+02	0.7877E+02	0.8953E+02	0.1410E+03
11	2.47	365.68	348.16	362.92	349.75	0.1043E+03	0.5525E+02	0.9610E+02	0.1303E+03
13	3.12	417.68	380.35	420.82	429.48	0.6420E+02	0.4496E+02	0.8149E+02	0.1151E+03
15	4.22	489.02	423.41	495.24	537.63	0.5989E+02	0.3444E+02	0.5594E+02	0.8106E+02
17	5.92	577.36	473.79	574.62	634.78	0.4681E+02	0.2556E+02	0.4102E+02	0.3675E+02
19	8.01	660.26	520.34	654.18	682.12	0.3114E+02	0.1965E+02	0.3506E+02	0.1291E+02
21	9.66	705.34	550.26	706.31	697.00	0.2476E+02	0.1670E+02	0.2755E+02	0.6059E+01
23	11.26	740.97	575.16	743.64	704.06	0.2112E+02	0.1464E+02	0.1935E+02	0.3165E+01
25	12.85	772.67	597.20	768.68	707.87	0.1801E+02	0.1306E+02	0.2253E+02	0.1774E+01
27	14.82	803.64	621.41	806.66	710.44	0.1487E+02	0.1156E+02	0.1649E+02	0.9415E+00
29	17.11	835.44	646.25	839.45	712.02	0.1348E+02	0.1023E+02	0.1267E+02	0.4949E+00
31	19.18	862.49	666.38	863.78	712.82	0.1293E+02	0.9280E+01	0.1108E+02	0.2959E+00
33	21.58	890.38	687.68	889.32	713.36	0.1002E+02	0.8390E+01	0.1036E+02	0.1737E+00
35	23.35	911.88	701.99	907.48	713.62	0.1464E+02	0.7849E+01	0.1018E+02	0.1215E+00
37	25.63	940.54	719.15	930.68	713.85	0.7192E+01	0.7255E+01	0.1013E+02	0.7948E-01
39	27.33	948.66	731.27	947.86	713.96	0.5495E+01	0.6868E+01	0.1014E+02	0.5942E-01
41	29.71	965.93	747.01	972.05	714.08	0.8093E+01	0.6402E+01	0.1016E+02	0.4063E-01

III - 20

Connection type : Top-and seat-angle connections with double web angle
Mode : Welded-to-beam and bolted-to-column

Tested by : A.Azizinamini et al. (1985) U.S.A.
Test Id. : 14WS2

Column : W12X96
Beam : W14X38
F.Angle: 6 X 4.0 X 1/2 X 8.0
W.Angle: 4 X 3.5 X 1/4 X 8.5

Fasteners: A325- -3/4"D
15/16" Oversize holes
Material : A36
Fy = 39.55 ksi
Fu = 67.95 ksi

Major parameters

lp = 8.5000"	lu = 2.8125"	ll = 2.8125"	ta = 0.2500"
gc = 2.5940"	cu = 1.2500"	cl = 1.2500"	pc = 3.0000"
nc = 1 X 3			
lt = 8.0000"	tt = 0.5000"	ls = 8.0000"	ts = 0.5000"
gt'= 2.5000"	gs'= 2.5000"	rt = 5.5000"	rs = 5.5000"
nt'= 2 X 1	ns'= 2 X 1		

Remark 1) Heavy hex high strength bolts used
2)

No	Moment (k-in)	Rotation (radians) X 1/1000	No	Moment (k-in)	Rotation (radians) X 1/1000
1	0.00	0.00	26	1041.53	11.19
2	56.99	0.03	27	1067.24	12.25
3	113.98	0.07	28	1092.94	13.31
4	146.61	0.18	29	1108.01	14.20
5	179.25	0.30	30	1123.08	15.09
6	217.78	0.40	31	1135.77	15.86
7	256.30	0.50	32	1148.46	16.63
8	292.47	0.60	33	1156.08	17.62
9	328.64	0.71	34	1163.70	18.60
10	374.66	0.96	35	1176.80	19.48
11	420.68	1.22	36	1189.90	20.36
12	466.69	1.44	37	1198.67	21.22
13	512.71	1.67	38	1207.45	22.08
14	551.65	1.94	39	1218.97	22.89
15	590.60	2.21	40	1230.49	23.71
16	641.76	2.70	41	1240.04	24.51
17	682.31	3.15	42	1249.60	25.31
18	722.85	3.60	43	1256.03	26.28
19	785.84	4.30	44	1262.46	27.25
20	845.70	5.17	45	1269.73	28.54
21	890.65	6.16			
22	940.37	7.50			
23	968.81	8.42			
24	997.24	9.34			
25	1019.39	10.26			

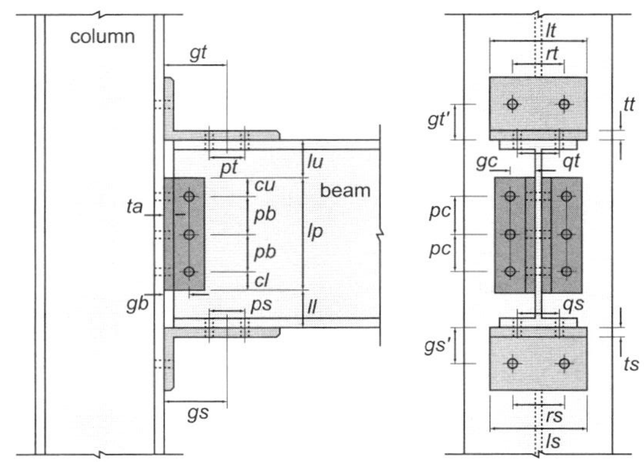

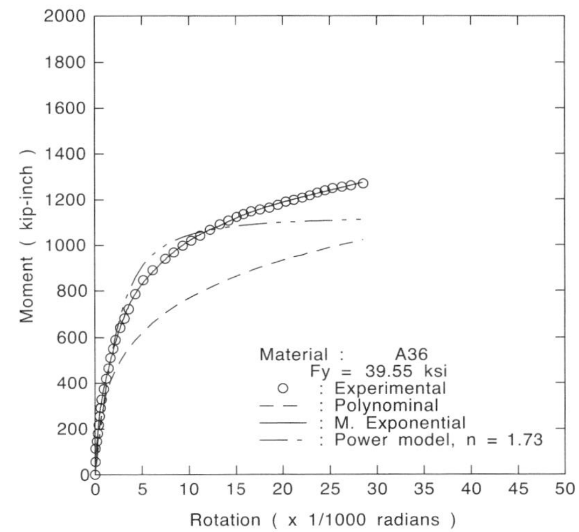

```
 Moment-rotation prediction equations      ( R : X 1/1000 radians )

Frye and Morris polynominal model  :    R = Sum ( Ai X ( K*Bm )**Pi X 10**Qi )
      t =  0.500000"   d = 14.125000"  tc =  0.250000"   w =  8.000000"   g =  2.125000"
     A1 =  2.232429   A2 =  1.850728   A3 =  3.188976    K =  0.083973
     P1 =        1    P2 =        3    P3 =        5    Q1 =       -2    Q2 =       -5    Q3 =       -9

Modified exponential model  :
     Bm = Sum ( Ai X ( 1 - exp( - R/( 2*i*C ) ) ) )  +      Sum ( RKj X ( R - Rj0 ) )  +  Bm0
      C =   0.24616667E+00           Bmo=   0.00000000E+00          Nexp=  6        Nliner=  2
     AI =   0.94859156E+03  -0.50312278E+04   0.12205454E+05  -0.93052156E+04  -0.39466054E+04   0.62342746E+04
    Rj0 =           6.1600          13.3100
    RKj =   0.31960308E+01   0.65089054E+01

Power model  :    Bm = ( Rki X R ) / ( 1 + ( R/R0 )**rn )**( 1/rn )
      rn = 1.725      rki =   0.35272395E+03          rmu =   0.11269208E+04
```

No	Rotation (radians) X 1/1000	Moment (kip-inch) Expri.	Moment (kip-inch) Poly.	Moment (kip-inch) M.Expo.	Moment (kip-inch) P.Model.	Connection stiffness (kip-inch) X 1000 Expri.	Connection stiffness (kip-inch) X 1000 Poly.	Connection stiffness (kip-inch) X 1000 M.Expo.	Connection stiffness (kip-inch) X 1000 P.Model.
1	0.00	0.00	0.00	0.00	0.00	0.1706E+04	0.5334E+03	0.8630E+03	0.3527E+03
3	0.07	113.98	35.32	53.40	23.51	0.1395E+04	0.5220E+03	0.7417E+03	0.3520E+03
5	0.30	179.25	143.58	190.45	105.58	0.3340E+03	0.3875E+03	0.4526E+03	0.3434E+03
7	0.50	256.30	210.88	266.85	173.51	0.3701E+03	0.2883E+03	0.3193E+03	0.3309E+03
9	0.71	328.64	262.32	323.24	238.97	0.3071E+03	0.2246E+03	0.2451E+03	0.3151E+03
11	1.22	420.68	353.22	426.73	388.51	0.1932E+03	0.1426E+03	0.1761E+03	0.2682E+03
13	1.67	512.71	408.81	501.21	499.39	0.1768E+03	0.1084E+03	0.1574E+03	0.2260E+03
15	2.21	590.60	460.90	582.06	610.03	0.1288E+03	0.8429E+02	0.1379E+03	0.1800E+03
17	3.15	682.31	528.05	693.49	748.69	0.9094E+02	0.6166E+02	0.1003E+03	0.1203E+03
19	4.30	785.84	589.77	787.26	858.72	0.8010E+02	0.4681E+02	0.6547E+02	0.7469E+02
21	6.16	890.65	663.73	882.68	958.60	0.4176E+02	0.3415E+02	0.4497E+02	0.3787E+02
23	8.42	968.81	730.79	970.68	1019.91	0.3091E+02	0.2603E+02	0.3374E+02	0.1915E+02
25	10.26	1019.39	774.72	1025.47	1047.95	0.2403E+02	0.2193E+02	0.2580E+02	0.1203E+02
27	12.25	1067.24	814.99	1068.81	1067.22	0.2420E+02	0.1884E+02	0.1817E+02	0.7810E+01
29	14.20	1108.01	849.50	1104.24	1079.92	0.1690E+02	0.1659E+02	0.1900E+02	0.5388E+01
31	15.86	1135.77	875.74	1132.81	1087.69	0.1655E+02	0.1510E+02	0.1566E+02	0.4066E+01
33	17.62	1156.08	901.05	1158.08	1093.93	0.7709E+01	0.1380E+02	0.1332E+02	0.3103E+01
35	19.48	1176.80	925.92	1181.36	1099.01	0.1494E+02	0.1266E+02	0.1178E+02	0.2389E+01
37	21.22	1198.67	947.12	1201.04	1102.73	0.1018E+02	0.1177E+02	0.1092E+02	0.1910E+01
39	22.89	1218.97	966.19	1218.90	1105.62	0.1415E+02	0.1104E+02	0.1043E+02	0.1564E+01
41	24.51	1240.04	983.47	1235.46	1107.93	0.1196E+02	0.1042E+02	0.1014E+02	0.1306E+01
43	26.28	1256.03	1001.28	1253.24	1110.04	0.6616E+01	0.9828E+01	0.9949E+01	0.1086E+01
45	28.54	1269.73	1022.71	1275.55	1112.24	0.5650E+01	0.9166E+01	0.9822E+01	0.8718E+00

III - 21

Connection type : Top-and seat-angle connections with double web angle
Mode : All riveted

Tested by : J.C.Rathbun (1936) U.S.A.
Test Id. : B-11

Column : --
Beam : W12X31.8
F.Angle: 6 X 6.0 X 3/8 X 9 (TOP)
W.Angle: 4 X 3.5 X 3/8 X 9

Fasteners: 10.9- -7/8"D
15/16" Oversize holes
Material : --
Fy = 47.12 ksi
Fu = 62.90 ksi

Major parameters

lp = 9.0000"	lu = 1.5000"	ll = 1.5900"	ta = 0.3750"
gb = 2.2500"	gc = 2.2500"	cu = 1.5000"	cl = 1.5000"
pb = 3.0000"	pc = 3.0000"	nb = 1 X 3	nc = 1 X 3
lt = 9.0000"	tt = 0.3750"	ls = 9.0000"	ts = 0.3750"
gt = 3.5000"	gt'= 3.7500"	gs = 3.5000"	gs'= 2.5000"
qt = 2.7500"	rt = 5.5000"	qs = 2.7500"	rs = 5.5000"
pt = 2.5000"	pt'= 2.5000"	ps = 2.5000"	
nt = 2 X 2	nt'= 2 X 2	ns = 2 X 2	ns'= 2 X 1

Remark 1) 6 X 4.0 X 3/8 X 9 (SEAT)
2) Connection cleat riveted to 1" mounting plate.

No	Moment (k-in)	Rotation (radians) X 1/1000	No	Moment (k-in)	Rotation (radians) X 1/1000
1	0.00	0.00	26	468.86	4.03
2	20.12	0.04	27	481.94	4.29
3	40.24	0.09	28	491.64	4.48
4	60.34	0.16	29	501.34	4.67
5	80.44	0.22	30	511.22	4.92
6	100.30	0.32	31	521.09	5.16
7	120.16	0.41	32	529.05	5.43
8	140.44	0.52	33	537.00	5.70
9	160.73	0.63	34	544.96	5.97
10	180.35	0.74	35	551.94	6.27
11	199.98	0.86	36	558.91	6.57
12	220.46	0.99			
13	240.94	1.13			
14	261.18	1.29			
15	281.42	1.45			
16	302.09	1.62			
17	322.76	1.80			
18	342.29	2.03			
19	361.81	2.26			
20	381.52	2.53			
21	401.23	2.81			
22	415.06	3.05			
23	428.88	3.29			
24	442.70	3.53			
25	455.78	3.78			

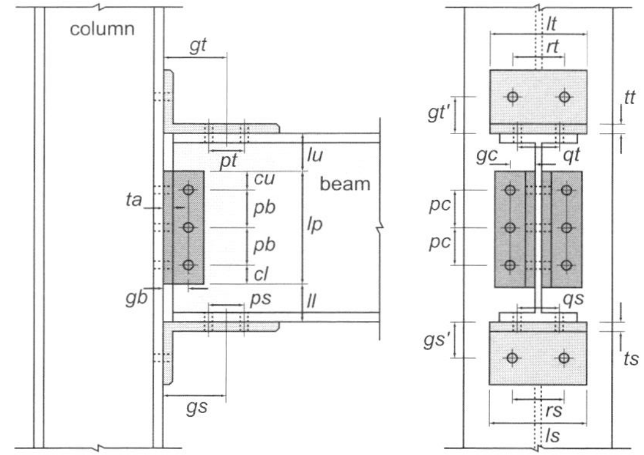

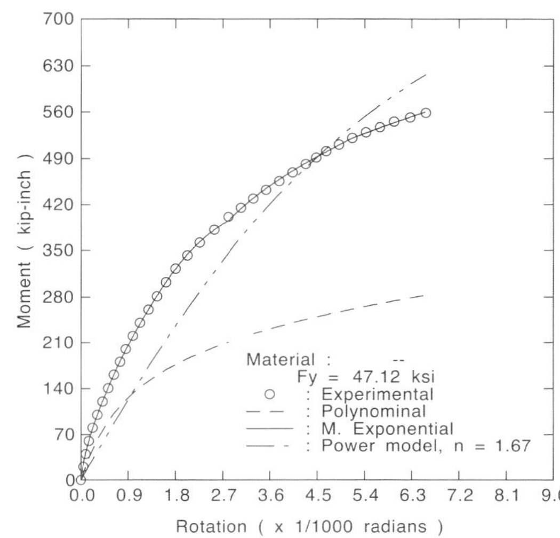

```
Moment-rotation prediction equations     ( R : X 1/1000 radians )

Frye and Morris polynominal model  :    R = Sum ( Ai X ( K*Bm )**Pi X 10**Qi )
     t =  0.375000"    d = 12.090000"  tc =  0.375000"    w =  9.000000"   g =  3.312500"
    A1 =  2.232429    A2 =  1.850728   A3 =  3.188976     K =  0.201280
    P1 =        1     P2 =        3    P3 =        5     Q1 =       -2     Q2 =       -5     Q3 =       -9

Modified exponential model  :
    Bm = Sum ( Ai X ( 1 - exp( - R/( 2*i*C ) ) ) )  +     Sum ( RKj X ( R - Rj0 ) )  +  Bm0
      C =   0.65654167E-01               Bmo=   0.00000000E+00            Nexp=  6        Nliner=  2
     AI =   0.13030326E+03  -0.12243151E+04   0.57311383E+04  -0.10463346E+05   0.65761576E+04  -0.31791970E+03
    Rj0 =           2.8100           5.1600
    RKj =   0.38285972E+02  -0.11044250E+02

Power model  :    Bm = ( Rki X R ) / ( 1 + ( R/R0 )**rn )**( 1/rn )
     rn = 1.666      rki =   0.13933198E+03          rmu =   0.95500412E+03
```

No	Rotation (radians) X 1/1000	Moment (kip-inch) Expri.	Poly.	M.Expo.	P.Model.	Connection stiffness (kip-inch) X 1000 Expri.	Poly.	M.Expo.	P.Model.
1	0.00	0.00	0.00	0.00	0.00	0.4625E+03	0.2225E+03	0.5707E+03	0.1393E+03
3	0.09	40.24	19.15	39.32	12.13	0.3964E+03	0.2146E+03	0.3614E+03	0.1392E+03
5	0.22	80.44	46.37	79.06	31.09	0.2602E+03	0.1821E+03	0.2451E+03	0.1386E+03
7	0.41	120.16	75.88	119.79	56.75	0.1991E+03	0.1375E+03	0.2013E+03	0.1373E+03
9	0.63	160.73	102.27	161.73	87.17	0.1790E+03	0.1020E+03	0.1760E+03	0.1352E+03
11	0.86	199.98	122.39	198.82	117.03	0.1644E+03	0.8027E+02	0.1577E+03	0.1326E+03
13	1.13	240.94	141.83	239.51	152.72	0.1390E+03	0.6356E+02	0.1415E+03	0.1290E+03
15	1.45	281.42	160.14	282.34	193.48	0.1226E+03	0.5110E+02	0.1243E+03	0.1241E+03
17	1.80	322.76	176.25	321.96	235.55	0.1042E+03	0.4230E+02	0.1036E+03	0.1183E+03
19	2.26	361.81	193.94	363.11	288.60	0.7873E+02	0.3451E+02	0.7395E+02	0.1102E+03
21	2.81	401.23	210.99	394.93	346.00	0.6437E+02	0.2850E+02	0.4435E+02	0.1006E+03
23	3.29	428.88	223.80	429.95	392.53	0.5725E+02	0.2476E+02	0.6436E+02	0.9222E+02
25	3.78	455.78	235.27	458.55	435.97	0.5200E+02	0.2189E+02	0.5264E+02	0.8407E+02
27	4.29	481.94	245.68	483.13	476.29	0.5097E+02	0.1960E+02	0.4577E+02	0.7628E+02
29	4.67	501.34	252.97	500.19	504.69	0.4576E+02	0.1817E+02	0.4272E+02	0.7071E+02
31	5.16	521.09	261.52	520.56	537.83	0.3518E+02	0.1664E+02	0.2947E+02	0.6415E+02
33	5.70	537.00	270.04	535.96	570.36	0.2977E+02	0.1526E+02	0.2827E+02	0.5769E+02
35	6.27	551.94	278.40	551.92	601.53	0.2294E+02	0.1404E+02	0.2769E+02	0.5153E+02

III - 22

Connection type : Top-and seat-angle connections with double web angle
Mode : All riveted

Tested by : J.C.Rathbun (1936) U.S.A.
Test Id. : B-12

Column : --
Beam : W12X31.8
F.Angle: 6 X 6.0 X 3/8 X 14 (TOP)
W.Angle: 4 X 3.5 X 3/8 X 9

Fasteners: 10.9- -7/8"D
15/16" Oversize holes
Material : --
Fy = 47.12 ksi
Fu = 62.90 ksi

Major parameters

lp = 9.0000"	lu = 1.5000"	ll = 1.5900"	ta = 0.3750"
gb = 2.2500"	gc = 2.2500"	cu = 1.5000"	cl = 1.5000"
pb = 3.0000"	pc = 3.0000"	nb = 1 X 3	nc = 1 X 3
lt = 14.0000"	tt = 0.3750"	ls = 14.0000"	ts = 0.3750"
gt = 3.5000"	gt'= 3.7500"	gs = 3.5000"	gs'= 2.5000"
qt = 2.7500"	rt = 5.5000"	rt2= 2.5000"	
qs = 2.7500"	rs = 5.5000"	rs2= 2.5000"	
pt = 2.5000"	pt'= 2.5000"	ps = 2.5000"	
nt = 2 X 2	nt'= 4 X 2	ns = 2 X 2	ns'= 4 X 1

Remark 1) 6 X 4.0 X 3/8 X 14 (SEAT)
2) Connection cleat riveted to 1" mounting plate.

No	Moment (k-in)	Rotation (radians) X 1/1000	No	Moment (k-in)	Rotation (radians) X 1/1000
1	0.00	0.00	26	618.86	6.13
2	25.29	0.05	27	635.24	6.57
3	50.58	0.09	28	651.63	7.00
4	74.67	0.17	29	668.19	7.53
5	98.76	0.24	30	684.76	8.05
6	124.04	0.35	31	701.32	8.58
7	149.32	0.46	32	716.10	9.12
8	174.29	0.60	33	730.87	9.67
9	199.26	0.73	34	740.91	10.11
10	224.23	0.91	35	750.94	10.54
11	249.20	1.09	36	760.98	10.98
12	275.05	1.27			
13	300.91	1.46			
14	326.46	1.70			
15	352.01	1.95			
16	376.96	2.18			
17	401.92	2.42			
18	428.34	2.73			
19	454.77	3.05			
20	478.21	3.40			
21	501.66	3.74			
22	528.95	4.21			
23	556.24	4.68			
24	579.36	5.19			
25	602.47	5.70			

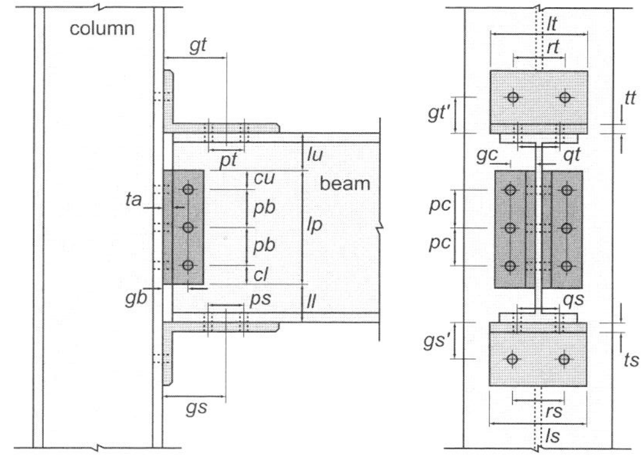

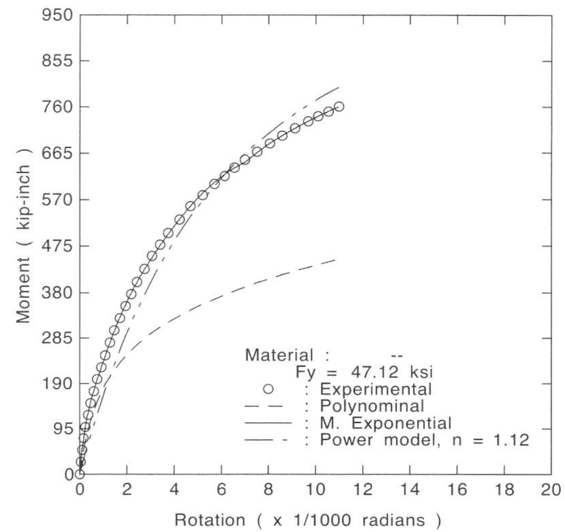

A.3 – 44

```
Moment-rotation prediction equations     ( R : X 1/1000 radians )

Frye and Morris polynominal model  :    R = Sum ( Ai X ( K*Bm )**Pi X 10**Qi )
      t =  0.375000"   d = 12.090000"  tc =  0.375000"   w = 14.000000"   g =  3.312500"
     A1 =  2.232429   A2 =  1.850728   A3 =  3.188976    K =  0.148119
     P1 =        1    P2 =        3    P3 =        5    Q1 =       -2     Q2 =       -5     Q3 =       -9

Modified exponential model  :
     Bm = Sum ( Ai X ( 1 - exp( - R/( 2*i*C ) ) ) )  +     Sum ( RKj X ( R - Rj0 ) )  +  Bm0
      C =   0.10326083E+00             Bmo=   0.00000000E+00         Nexp=  6        Nliner=  1
     AI =  -0.11199902E+03   0.38869945E+04  -0.23755029E+05   0.61402885E+05  -0.70809294E+05   0.30070293E+05
    Rj0 =           7.0200
    RKj =   0.20135712E+02

Power model  :    Bm = ( Rki X R ) / ( 1 + ( R/R0 )**rn )**( 1/rn )
      rn = 1.123      rki =   0.18302778E+03         rmu =   0.11852962E+04
```

No	Rotation (radians) X 1/1000	Moment (kip-inch) Expri.	Moment (kip-inch) Poly.	Moment (kip-inch) M.Expo.	Moment (kip-inch) P.Model.	Connection stiffness (kip-inch) X 1000 Expri.	Connection stiffness (kip-inch) X 1000 Poly.	Connection stiffness (kip-inch) X 1000 M.Expo.	Connection stiffness (kip-inch) X 1000 P.Model.
1	0.00	0.00	0.00	0.00	0.00	0.5369E+03	0.3024E+03	0.5507E+03	0.1830E+03
3	0.09	50.58	28.08	45.97	17.11	0.4541E+03	0.2899E+03	0.4300E+03	0.1801E+03
5	0.24	98.76	67.84	99.24	43.57	0.2874E+03	0.2403E+03	0.2950E+03	0.1746E+03
7	0.46	149.32	111.71	150.04	80.04	0.2119E+03	0.1744E+03	0.1952E+03	0.1666E+03
9	0.73	199.26	152.35	197.08	124.90	0.1631E+03	0.1234E+03	0.1533E+03	0.1563E+03
11	1.09	249.20	189.07	248.51	177.67	0.1410E+03	0.8921E+02	0.1414E+03	0.1441E+03
13	1.46	300.91	218.04	298.68	228.76	0.1245E+03	0.6919E+02	0.1287E+03	0.1323E+03
15	1.95	352.01	247.74	356.02	290.00	0.1052E+03	0.5359E+02	0.1054E+03	0.1184E+03
17	2.42	401.92	270.61	401.03	342.92	0.9625E+02	0.4422E+02	0.8698E+02	0.1066E+03
19	3.05	454.77	295.93	451.50	406.35	0.7574E+02	0.3594E+02	0.7345E+02	0.9308E+02
21	3.74	501.66	318.59	499.57	466.32	0.6370E+02	0.3000E+02	0.6612E+02	0.8084E+02
23	4.68	556.24	344.08	557.00	535.75	0.5210E+02	0.2463E+02	0.5542E+02	0.6747E+02
25	5.70	602.47	366.94	605.63	598.18	0.4132E+02	0.2074E+02	0.4033E+02	0.5627E+02
27	6.57	635.24	383.91	635.17	643.76	0.3761E+02	0.1832E+02	0.2779E+02	0.4862E+02
29	7.53	668.19	400.47	666.62	687.04	0.3158E+02	0.1627E+02	0.3714E+02	0.4178E+02
31	8.58	701.32	416.58	701.17	727.61	0.2934E+02	0.1452E+02	0.2939E+02	0.3578E+02
33	9.67	730.87	431.66	730.56	763.92	0.2480E+02	0.1309E+02	0.2476E+02	0.3075E+02
35	10.54	750.94	442.64	751.15	789.21	0.2305E+02	0.1215E+02	0.2272E+02	0.2744E+02

III - 23

Connection type : Top-and seat-angle connections with double web angle
Mode : All bolted

Tested by : C.W. Roeder et al. (1996) U.S.A.
Test Id. : L1

Column : W10x68
Beam : W14x22
F.Angle: L5x3.5x5/8
W.Angle: L3.5x3x5/16

Fasteners: A307- -3/4"D
15/16" Oversize holes
Material : --
Fy = 44.03 ksi
Fu = 65.05 ksi

Major parameters

lp = 10.0000"	lu = 1.8701"	ll = 1.8701"	ta = 0.3125"
gb = -- "	gc = -- "	cu = 2.0000"	cl = 2.0000"
pb = 2.0000"	pc = 2.0000"	nb = 1 X 4	nc = 1 X 4
lt = 5.5118"	tt = 0.6250"	ls = 5.5118"	ts = 0.6250"
gt = 3.0000"	gt'= 2.0000"	gs = 3.0000"	gs'= 2.0000"
qt = -- "	rt = -- "	qs = -- "	rs = -- "
pt = 2.0000"	ps = 2.0000"		
nt = 2 X 2	nt'= 2 X 1	ns = 2 X 2	ns'= 2 X 1

Remark 1) Fy = 289.8 MPa, Fu = 441.6 MPa for top- and seat angle
2)

No	Moment (k-in)	Rotation (radians) X 1/1000
1	0.00	0.00
2	53.05	0.28
3	107.13	0.86
4	151.40	1.52
5	188.83	2.41
6	223.33	3.37
7	262.76	4.47
8	312.95	5.30
9	358.22	6.14
10	403.53	7.16
11	437.05	8.15
12	470.60	9.32
13	496.33	10.72
14	525.03	12.28
15	551.77	13.93
16	569.62	15.15
17	590.45	16.58
18	609.32	18.06
19	625.25	19.59
20	640.22	21.25
21	650.27	22.88
22	659.45	25.20
23	662.66	26.98

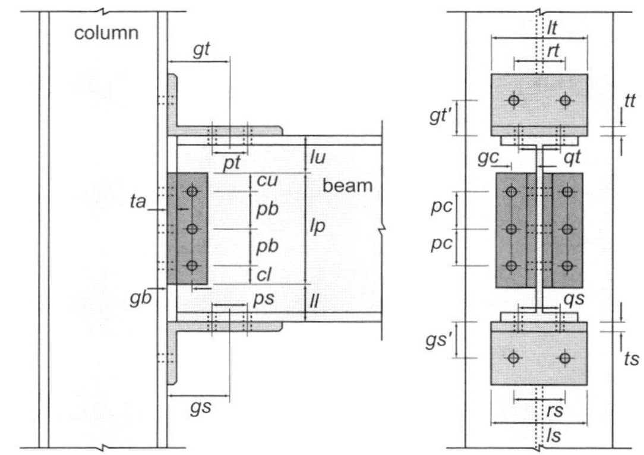

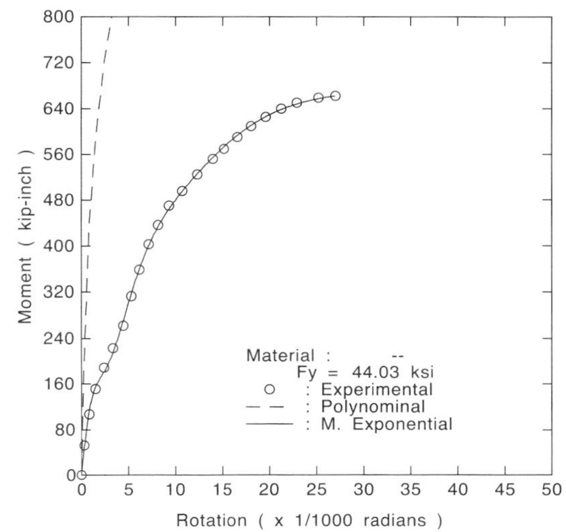

A.3 – 46

```
Moment-rotation prediction equations     ( R : X 1/1000 radians )

Frye and Morris polynominal model   :    R = Sum ( Ai X ( K*Bm )**Pi X 10**Qi )
      t =  0.625000"   d = 13.700000"  tc =  0.312500"   w =  5.511800"   g =  1.625000"
     A1 =  2.232429   A2 =  1.850728   A3 =  3.188976    K =  0.055815
     P1 =        1    P2 =        3    P3 =        5    Q1 =       -2    Q2 =       -5    Q3 =       -9

Modified exponential model   :
     Bm = Sum ( Ai X ( 1 - exp( - R/( 2*i*C ) ) ) )  +      Sum ( RKj X ( R - Rj0 ) )  +  Bm0
      C =   0.29386250E+00              Bmo=   0.00000000E+00         Nexp=  6        Nliner=  0
     AI =  -0.73314973E+03   0.11169676E+05  -0.50471544E+05   0.10462376E+06  -0.10337093E+06   0.39453457E+05
```

No	Rotation (radians) X 1/1000	Moment (kip-inch) Expri.	Moment (kip-inch) Poly.	Moment (kip-inch) M.Expo.	Connection stiffness (kip-inch) X 1000 Expri.	Connection stiffness (kip-inch) X 1000 Poly.	Connection stiffness (kip-inch) X 1000 M.Expo.
1	0.00	0.00	0.00	0.00	0.1922E+03	0.8025E+03	0.1449E+03
2	0.28	53.05	200.30	42.32	0.1603E+03	0.6071E+03	0.1529E+03
3	0.86	107.13	442.09	114.99	0.8082E+02	0.2888E+03	0.9158E+02
4	1.52	151.40	589.64	155.28	0.5647E+02	0.1771E+03	0.3839E+02
5	2.41	188.83	717.02	181.33	0.3902E+02	0.1178E+03	0.2853E+02
6	3.37	223.33	814.54	216.44	0.3587E+02	0.8731E+02	0.4470E+02
7	4.47	262.76	898.80	272.16	0.4957E+02	0.6811E+02	0.5457E+02
8	5.30	312.95	951.50	317.48	0.5698E+02	0.5860E+02	0.5257E+02
9	6.14	358.22	997.49	359.00	0.4976E+02	0.5154E+02	0.4612E+02
10	7.16	403.53	1046.45	401.23	0.3912E+02	0.4509E+02	0.3706E+02
11	8.15	437.05	1088.70	434.17	0.3144E+02	0.4029E+02	0.2970E+02
12	9.32	470.60	1133.15	465.26	0.2398E+02	0.3586E+02	0.2391E+02
13	10.72	496.33	1180.41	495.85	0.1842E+02	0.3178E+02	0.2021E+02
14	12.28	525.03	1226.95	525.49	0.1735E+02	0.2828E+02	0.1812E+02
15	13.93	551.77	1271.28	554.07	0.1526E+02	0.2537E+02	0.1639E+02
16	15.15	569.62	1301.20	573.24	0.1460E+02	0.2361E+02	0.1494E+02
17	16.58	590.45	1333.61	593.21	0.1370E+02	0.2186E+02	0.1301E+02
18	18.06	609.32	1364.76	610.88	0.1161E+02	0.2033E+02	0.1089E+02
19	19.59	625.25	1394.78	625.88	0.9754E+01	0.1897E+02	0.8758E+01
20	21.25	640.22	1425.14	638.63	0.7588E+01	0.1770E+02	0.6685E+01
21	22.88	650.27	1453.12	648.10	0.5251E+01	0.1663E+02	0.4978E+01
22	25.20	659.45	1490.15	657.40	0.2735E+01	0.1532E+02	0.3145E+01
23	26.98	662.66	1516.72	662.08	0.1797E+01	0.1446E+02	0.2153E+01

III - 24

Connection type : Top-and seat-angle connections with double web angle
Mode : All bolted

Tested by : A.S. Elnashai et al. (1998) U.K.
Test Id. : SRB01

Column : H150x150x7x10
Beam : H250x130x9x9
F.Angle: L75x75x9
W.Angle: L75x75x8

Fasteners: 10.9- -M16
23/32" Oversize holes
Material : --
Fy = 44.96 ksi
Fu = 65.27 ksi

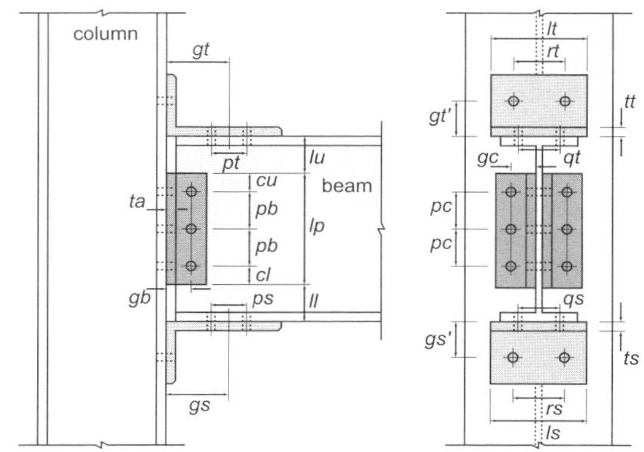

Major parameters

lp = 7.4803"	lu = 1.1811"	ll = 1.1811"	ta = 0.3150"
gb = -- "	gc = -- "	cu = 0.9842"	cl = 0.9842"
pb = 2.7559"	pc = -- "	nb = 1 X 3	nc = 1 X 3
lt = 5.9055"	tt = 0.3543"	ls = 5.9055"	ts = 0.3543"
gt = 1.5748"	gt'= 1.5748"	gs = 1.5748"	gs'= 1.5748"
qt = -- "	rt = -- "	qs = -- "	rs = -- "
pt = 0.0000"	ps = 0.0000"		
nt = 2 X 2	nt'= 2 X 1	ns = 2 X 2	ns'= 2 X 1

Remark 1) column stiffener (thickness 9 mm)
2)

No	Moment (k-in)	Rotation (radians) X 1/1000	No	Moment (k-in)	Rotation (radians) X 1/1000	No	Moment (k-in)	Rotation (radians) X 1/1000
1	0.00	0.00	26	413.91	41.96	51	654.77	120.17
2	32.03	0.25	27	428.24	44.42	52	662.18	125.07
3	61.78	0.49	28	440.32	46.39	53	669.62	129.32
4	87.10	0.69	29	455.43	48.53	54	674.79	133.24
5	126.21	1.00	30	466.74	50.99	55	678.42	137.64
6	174.65	1.38	31	481.81	54.11	56	682.82	141.72
7	199.70	1.74	32	492.35	56.57	57	686.46	145.80
8	230.83	2.11	33	504.37	60.17	58	689.36	149.55
9	251.34	2.54	34	518.65	64.26	59	693.01	153.30
10	277.89	3.16	35	527.66	66.88	60	696.67	157.05
11	301.42	3.68	36	535.90	69.99	61	699.61	159.66
12	318.08	4.84	37	547.22	71.80			
13	326.39	6.16	38	560.09	73.12			
14	329.31	9.26	39	573.67	75.58			
15	331.48	12.19	40	581.91	78.52			
16	338.22	14.81	41	586.34	81.95			
17	347.25	16.94	42	593.03	85.87			
18	348.66	19.71	43	595.15	89.95			
19	355.42	21.84	44	602.59	94.03			
20	362.14	24.78	45	611.60	96.81			
21	367.35	27.72	46	621.38	99.43			
22	377.11	30.67	47	626.53	103.84			
23	380.79	33.77	48	636.26	107.60			
24	389.81	36.39	49	642.19	111.36			
25	399.57	39.33	50	649.60	116.26			

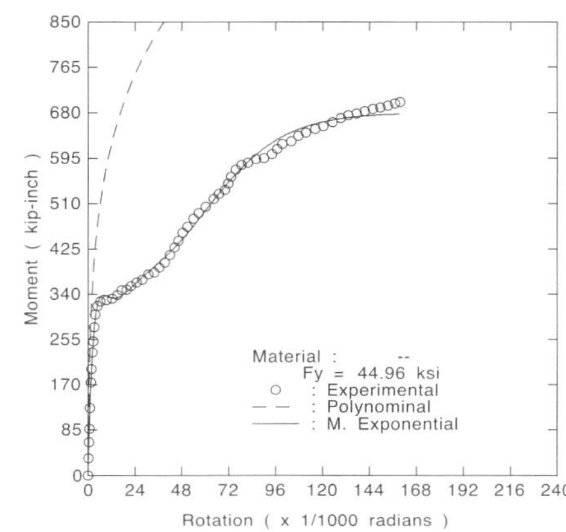

A.3 – 48

```
Moment-rotation prediction equations     ( R : X 1/1000 radians )

Frye and Morris polynominal model  :    R = Sum ( Ai X ( K*Bm )**Pi X 10**Qi )
      t =  0.354330"   d =  9.842520"  tc =  0.314960"   w =  5.905510"   g =  1.259850"
     A1 =  2.232429   A2 =  1.850728   A3 =  3.188976    K =  0.109196
     P1 =         1   P2 =         3   P3 =         5   Q1 =        -2   Q2 =        -5   Q3 =        -9

Modified exponential model  :
     Bm = Sum ( Ai X ( 1 - exp( - R/( 2*i*C ) ) ) )  +      Sum ( RKj X ( R - Rj0 ) )  +  Bm0
      C =   0.13939892E+01              Bmo=   0.00000000E+00          Nexp=  6         Nliner=  0
     AI =   0.27365324E+03   0.65456079E+04  -0.41423927E+05   0.10433520E+06  -0.11614065E+06   0.47089722E+05
```

No	Rotation (radians) X 1/1000	Moment (kip-inch) Expri.	Poly.	M.Expo.	Connection stiffness (kip-inch) X 1000 Expri.	Poly.	M.Expo.
1	0.00	0.00	0.00	0.00	0.1262E+03	0.4102E+03	0.1587E+03
3	0.49	61.78	159.00	70.53	0.1262E+03	0.2261E+03	0.1303E+03
5	1.00	126.21	245.61	130.51	0.1262E+03	0.1299E+03	0.1055E+03
7	1.74	199.70	320.72	197.75	0.7687E+02	0.8005E+02	0.7652E+02
9	2.54	251.34	374.41	248.99	0.4597E+02	0.5735E+02	0.5318E+02
11	3.68	301.42	429.39	295.33	0.3560E+02	0.4135E+02	0.3014E+02
13	6.16	326.39	510.25	334.24	0.4716E+01	0.2629E+02	0.5682E+01
15	12.19	331.48	626.01	332.79	0.1710E+01	0.1454E+02	-0.7401E+00
17	16.94	347.25	685.62	336.85	0.2619E+01	0.1097E+02	0.2267E+01
19	21.84	355.42	733.74	351.72	0.2802E+01	0.8841E+01	0.3479E+01
21	27.72	367.35	780.64	372.18	0.2541E+01	0.7224E+01	0.3357E+01
23	33.77	380.79	820.77	391.66	0.2409E+01	0.6117E+01	0.3154E+01
25	39.33	399.57	852.67	409.67	0.4447E+01	0.5381E+01	0.3379E+01
27	44.42	428.24	878.68	427.93	0.5990E+01	0.4859E+01	0.3811E+01
29	48.53	455.43	897.92	444.36	0.5916E+01	0.4512E+01	0.4177E+01
31	54.11	481.81	921.95	468.77	0.4530E+01	0.4120E+01	0.4548E+01
33	60.17	504.37	945.82	496.91	0.3408E+01	0.3771E+01	0.4692E+01
35	66.88	527.66	970.19	528.04	0.3080E+01	0.3451E+01	0.4531E+01
37	71.80	547.22	986.57	549.65	0.8270E+01	0.3254E+01	0.4246E+01
39	75.58	573.67	998.76	565.20	0.4278E+01	0.3117E+01	0.3963E+01
41	81.95	586.34	1017.95	588.74	0.1485E+01	0.2914E+01	0.3417E+01
43	89.95	595.15	1040.41	613.21	0.1171E+01	0.2697E+01	0.2707E+01
45	96.81	611.60	1058.25	629.83	0.3492E+01	0.2539E+01	0.2144E+01
47	103.84	626.53	1075.50	643.09	0.1934E+01	0.2397E+01	0.1647E+01
49	111.36	642.19	1093.00	653.78	0.1551E+01	0.2262E+01	0.1213E+01
51	120.17	654.77	1112.32	662.67	0.1405E+01	0.2124E+01	0.8259E+00
53	129.32	669.62	1131.15	668.84	0.1526E+01	0.1999E+01	0.5418E+00
55	137.64	678.42	1147.57	672.56	0.9567E+00	0.1898E+01	0.3630E+00
57	145.80	686.46	1162.58	675.00	0.8301E+00	0.1810E+01	0.2421E+00
59	153.30	693.01	1176.02	676.51	0.9742E+00	0.1736E+01	0.1652E+00
61	159.66	699.61	1186.70	677.41	0.1126E+01	0.1680E+01	0.1187E+00

III - 25

Connection type : Top-and seat-angle connections with double web angle
Mode : All bolted

Tested by : Z. Fu et al. (1998) Japan
Test Id. : LM-P

Column : --
Beam : H250x125x6x9
F.Angle: L75x75x9x8.5
W.Angle: L75x75x9x8.5

Fasteners: F10T- -M16
23/32" Oversize holes
Material : SS400
Fy = -- ksi
Fu = -- ksi

Major parameters

lp = 7.8740"	lu = 0.9843"	ll = 0.9843"	ta = 0.3347"
gb = 1.8701"	gc = 1.7716"	cu = 1.3780"	cl = 1.3780"
pb = 2.5591"	pc = 2.5591"	nb = 1 X 3	nc = 1 X 3
lt = 5.1181"	tt = 0.3347"	ls = 5.1181"	ts = 0.3347"
gt = 1.8701"	gt'= 1.7716"	gs = 1.8701"	gs'= 1.7716"
qt = 2.9528"	rt = 2.9528"	qs = 2.9528"	rs = 2.9528"
nt = 2 X 1	nt'= 2 X 1	ns = 2 X 1	ns'= 2 X 1

Remark 1) Beam was applied to 49 kN of compressive axial force
2)

No	Moment (k-in)	Rotation (radians) X 1/1000	No	Moment (k-in)	Rotation (radians) X 1/1000
1	0.00	0.00	26	499.07	42.12
2	38.62	0.32	27	519.62	44.63
3	80.86	0.84	28	497.94	46.33
4	117.20	1.43	29	513.62	47.44
5	161.41	2.17	30	523.38	49.29
6	210.52	2.84	31	526.28	50.84
7	253.72	4.02	32	502.67	51.36
8	290.05	5.06	33	475.13	51.95
9	332.25	6.91	34	497.71	52.77
10	361.67	8.68	35	528.13	54.10
11	375.33	11.42	36	549.67	56.32
12	383.10	14.16	37	569.26	58.24
13	382.99	17.19	38	581.96	60.46
14	396.67	19.33	39	602.54	62.16
15	406.43	21.48	40	618.17	64.68
16	348.43	21.70	41	634.80	66.97
17	374.94	22.51	42	650.46	68.97
18	396.52	23.62	43	662.17	71.11
19	410.22	25.40	44	672.90	73.63
20	423.90	27.69	45	683.62	76.14
21	436.60	29.91	46	697.28	78.88
22	445.36	32.35	47	699.20	80.28
23	460.02	34.65			
24	477.62	37.01			
25	491.30	39.53			

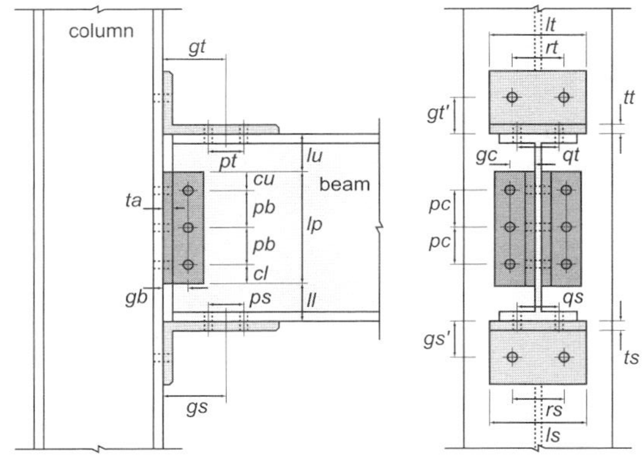

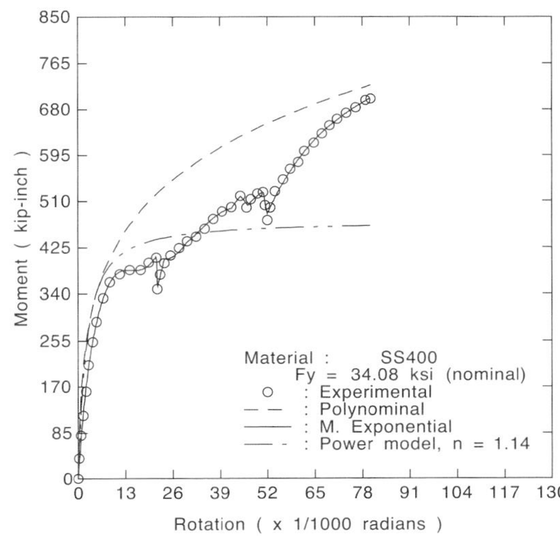

A.3 – 50

```
Moment-rotation prediction equations     ( R : X 1/1000 radians )

Frye and Morris polynominal model  :    R = Sum ( Ai X ( K*Bm )**Pi X 10**Qi )
     t =  0.334650"   d =  9.842520"  tc =  0.334650"   w =  5.118110"   g =  1.456700"
    A1 =  2.232429   A2 =  1.850728   A3 =  3.188976    K =  0.152595
    P1 =         1   P2 =         3   P3 =         5   Q1 =        -2   Q2 =        -5   Q3 =        -9

Modified exponential model  :
    Bm = Sum ( Ai X ( 1 - exp( - R/( 2*i*C ) ) ) )  +      Sum ( RKj X ( R - Rj0 ) )  +  Bm0
      C =   0.74900000E+00             Bmo=   0.00000000E+00             Nexp=  6          Nliner=  9
     AI =  -0.42443152E+03   0.13013118E+05  -0.97196347E+05   0.27814927E+06  -0.33051229E+06   0.13817269E+06
    Rj0 =          21.4800          21.7000          23.6200          29.9100          42.1200
                   44.6300          46.6300          50.8400          51.9500
    RKj =  -0.28059621E+03   0.29449691E+03  -0.30383615E+02  -0.38573329E+01   0.86825301E+01
           -0.14831602E+02   0.17156619E+02  -0.40076114E+02   0.52009854E+02

Power model  :     Bm = ( Rki X R ) / ( 1 + ( R/R0 )**rn )**( 1/rn )
    rn = 1.141       rki =   0.19182350E+03          rmu =   0.47255024E+03
```

No	Rotation (radians) X 1/1000	Moment (kip-inch) Expri.	Poly.	M.Expo.	P.Model.	Connection stiffness (kip-inch) X 1000 Expri.	Poly.	M.Expo.	P.Model.
1	0.00	0.00	0.00	0.00	0.00	0.1207E+03	0.2936E+03	0.9812E+02	0.1918E+03
3	0.84	80.86	159.85	76.38	128.64	0.7204E+02	0.1074E+03	0.8288E+02	0.1184E+03
5	2.17	161.41	251.58	169.41	241.03	0.6686E+02	0.4722E+02	0.5803E+02	0.5955E+02
7	4.02	253.72	317.02	254.38	317.90	0.3572E+02	0.2735E+02	0.3622E+02	0.2877E+02
9	6.91	332.25	378.70	332.19	373.40	0.1965E+02	0.1700E+02	0.1937E+02	0.1273E+02
11	11.42	375.33	439.72	379.77	410.64	0.3910E+01	0.1100E+02	0.3314E+01	0.5323E+01
13	17.19	382.99	492.61	383.55	431.60	0.3733E+01	0.7754E+01	0.1815E+01	0.2467E+01
15	21.48	406.43	522.79	409.59	440.12	-0.2387E+03	0.6416E+01	0.1094E+02	0.1597E+01
17	22.51	374.94	529.35	371.66	441.69	0.2712E+02	0.6162E+01	0.2730E+02	0.1455E+01
19	25.40	410.22	546.19	405.78	445.42	0.6942E+01	0.5566E+01	0.3125E+01	0.1144E+01
21	29.91	436.60	569.59	435.33	449.79	0.4705E+01	0.4848E+01	0.9128E+01	0.8234E+00
23	34.65	460.02	591.17	464.75	453.16	0.6911E+01	0.4283E+01	0.6352E+01	0.6106E+00
25	39.53	491.30	610.93	490.16	455.76	0.4229E+01	0.3834E+01	0.3595E+01	0.4662E+00
27	44.63	519.62	629.50	518.59	457.86	-0.4298E+01	0.3463E+01	0.7544E+01	0.3631E+00
29	47.44	513.62	639.11	508.12	458.82	0.1081E+02	0.3288E+01	0.7113E+01	0.3200E+00
31	50.84	526.28	649.92	526.87	459.83	-0.3352E+02	0.3105E+01	0.3968E+01	0.2773E+00
33	51.95	475.13	653.35	486.26	460.13	-0.1563E+02	0.3049E+01	-0.3706E+02	0.2651E+00
35	54.10	528.13	659.85	516.52	460.68	0.1794E+02	0.2946E+01	0.1323E+02	0.2437E+00
37	58.24	569.26	671.63	565.20	461.61	0.8124E+01	0.2771E+01	0.1040E+02	0.2090E+00
39	62.16	602.54	682.18	601.68	462.38	0.9728E+01	0.2625E+01	0.8302E+01	0.1824E+00
41	66.97	634.80	694.37	636.76	463.19	0.7563E+01	0.2468E+01	0.6398E+01	0.1561E+00
43	71.11	662.17	704.46	660.71	463.80	0.4915E+01	0.2346E+01	0.5234E+01	0.1377E+00
45	76.14	683.62	715.89	684.42	464.44	0.4613E+01	0.2217E+01	0.4262E+01	0.1193E+00
47	80.28	699.20	724.79	700.89	464.91	0.1369E+01	0.2123E+01	0.3725E+01	0.1067E+00

III - 26

Connection type : Top-and seat-angle connections with double web angle
Mode : All bolted

Tested by : Z. Fu et al. (1998) Japan
Test Id. : LM-T

Column : --
Beam : H250x125x6x9
F.Angle: L75x75x9x8.5
W.Angle: L75x75x9x8.5

Fasteners: F10T- -M16
23/32" Oversize holes
Material : SS400
Fy = -- ksi
Fu = -- ksi

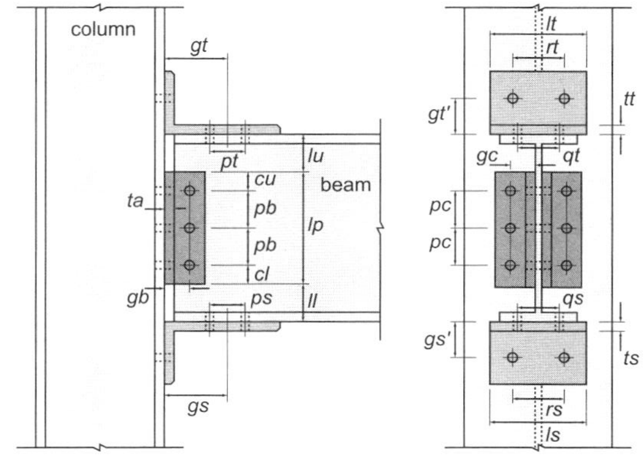

Major parameters

lp = 7.8740"	lu = 0.9843"	ll = 0.9843"	ta = 0.3347"
gb = 1.8701"	gc = 1.7716"	cu = 1.3780"	cl = 1.3780"
pb = 2.5591"	pc = 2.5591"	nb = 1 X 3	nc = 1 X 3
lt = 5.1181"	tt = 0.3347"	ls = 5.1181"	ts = 0.3347"
gt = 1.8701"	gt'= 1.7716"	gs = 1.8701"	gs'= 1.7716"
qt = 2.9528"	rt = 2.9528"	qs = 2.9528"	rs = 2.9528"
nt = 2 X 1	nt'= 2 X 1	ns = 2 X 1	ns'= 2 X 1

Remark 1) Beam was applied to 49 kN of tensile axial force
2)

No	Moment (k-in)	Rotation (radians) X 1/1000	No	Moment (k-in)	Rotation (radians) X 1/1000	No	Moment (k-in)	Rotation (radians) X 1/1000
1	0.00	0.00	26	412.76	36.94	51	633.64	71.93
2	44.51	0.47	27	441.24	37.68	52	642.43	73.70
3	80.86	0.92	28	461.84	38.79	53	659.06	75.92
4	119.17	1.51	29	472.60	40.05	54	668.82	77.84
5	148.64	1.88	30	477.48	41.08	55	680.54	79.92
6	184.98	2.54	31	471.56	41.89			
7	223.29	3.06	32	438.20	43.74			
8	257.65	4.24	33	492.09	44.85			
9	293.98	5.28	34	483.21	45.74			
10	323.42	6.32	35	499.89	46.70			
11	347.93	8.02	36	505.76	47.29			
12	365.57	9.64	37	476.24	48.48			
13	379.24	12.08	38	498.82	49.14			
14	384.08	14.30	39	511.56	50.18			
15	383.99	16.82	40	521.35	51.29			
16	394.71	19.33	41	509.53	52.10			
17	405.46	21.11	42	527.16	53.66			
18	412.27	23.11	43	541.84	55.36			
19	421.04	25.03	44	548.65	57.35			
20	425.88	27.32	45	564.30	59.43			
21	433.65	29.76	46	574.05	61.72			
22	454.22	31.83	47	586.74	64.23			
23	463.97	33.91	48	598.45	66.53			
24	471.76	35.98	49	615.11	67.93			
25	444.24	36.20	50	619.95	70.15			

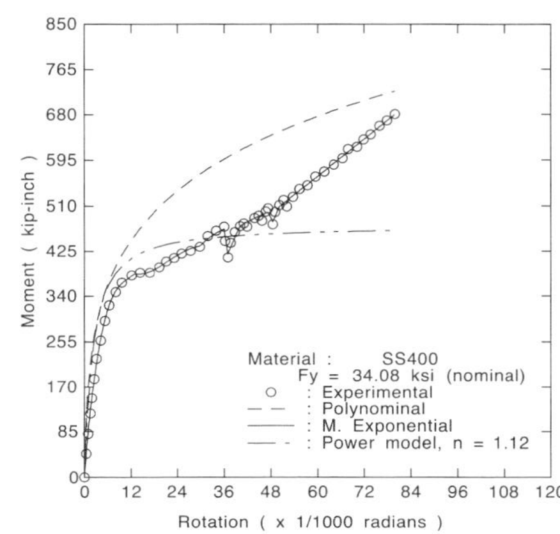

A.3 – 52

```
Moment-rotation prediction equations      ( R : X 1/1000 radians )

Frye and Morris polynominal model   :     R = Sum ( Ai X ( K*Bm )**Pi X 10**Qi )
      t =  0.334650"    d =  9.842520"  tc =  0.334650"    w =  5.118110"    g =  1.456700"
     A1 =  2.232429    A2 =  1.850728   A3 =  3.188976     K =  0.152595
     P1 =         1    P2 =         3   P3 =         5    Q1 =        -2    Q2 =        -5    Q3 =        -9

Modified exponential model  :
     Bm = Sum ( Ai X ( 1 - exp( - R/( 2*i*C ) ) ) )  +      Sum ( RKj X ( R - Rj0 ) )  +  Bm0
      C =   0.78350000E+00          Bmo=    0.00000000E+00          Nexp=  6         Nliner=  8
     AI =   0.39357637E+03  -0.74896626E+04   0.39521954E+05  -0.79971608E+05   0.69409910E+05  -0.21298244E+05
    Rj0 =          12.8000          29.7600          35.9800          36.9400          40.0500
                   47.2900          48.4800          50.1800
    RKj =  -0.57802723E+01   0.58556178E+01  -0.52378811E+02   0.66939980E+02  -0.11716399E+02
           -0.18072969E+02   0.32259309E+02  -0.10962812E+02

Power model  :     Bm = ( Rki X R ) / ( 1 + ( R/R0 )**rn )**( 1/rn )
      rn = 1.124       rki =    0.19182350E+03            rmu =    0.47255024E+03
```

No	Rotation (radians) X 1/1000	Moment (kip-inch) Expri.	Poly.	M.Expo.	P.Model.	Connection stiffness (kip-inch) X 1000 Expri.	Poly.	M.Expo.	P.Model.
1	0.00	0.00	0.00	0.00	0.00	0.9470E+02	0.2936E+03	0.1034E+03	0.1918E+03
3	0.92	80.86	168.07	81.60	136.90	0.7392E+02	0.9969E+02	0.7696E+02	0.1119E+03
5	1.88	148.64	237.02	147.60	220.58	0.7082E+02	0.5361E+02	0.6173E+02	0.6751E+02
7	3.06	223.29	287.50	212.38	282.44	0.6005E+02	0.3481E+02	0.4858E+02	0.4055E+02
9	5.28	293.98	347.52	296.99	344.99	0.3162E+02	0.2152E+02	0.2841E+02	0.1947E+02
11	8.02	347.93	396.31	349.53	383.33	0.1261E+02	0.1494E+02	0.1176E+02	0.1002E+02
13	12.08	379.24	446.77	378.01	411.79	0.3810E+01	0.1049E+02	0.5121E+01	0.4887E+01
15	16.82	383.99	489.70	386.13	428.83	0.2123E+01	0.7900E+01	0.2816E+01	0.2637E+01
17	21.11	405.46	520.40	403.71	437.92	0.4798E+01	0.6511E+01	0.4845E+01	0.1701E+01
19	25.03	421.04	544.11	421.83	443.56	0.3449E+01	0.5636E+01	0.4052E+01	0.1218E+01
21	29.76	433.65	568.86	434.89	448.41	0.6838E+01	0.4869E+01	0.1310E+01	0.8626E+00
23	33.91	463.97	587.97	459.14	451.55	0.4225E+01	0.4362E+01	0.4568E+01	0.6635E+00
25	36.20	444.24	597.69	456.62	452.97	-0.1062E+03	0.4128E+01	-0.4905E+02	0.5813E+00
27	37.68	441.24	603.74	433.05	453.80	0.3051E+02	0.3990E+01	0.1720E+02	0.5360E+00
29	40.05	472.60	612.91	472.67	454.99	0.6449E+01	0.3792E+01	0.1627E+02	0.4735E+00
31	41.89	471.56	619.82	480.50	455.83	-0.2349E+01	0.3651E+01	0.3977E+01	0.4320E+00
33	44.85	492.09	630.30	491.18	457.02	-0.3975E+01	0.3448E+01	0.3287E+01	0.3758E+00
35	46.70	499.89	636.63	496.97	457.69	0.1278E+02	0.3333E+01	0.2983E+01	0.3459E+00
37	48.48	476.24	642.52	480.57	458.28	0.1316E+02	0.3229E+01	-0.1531E+02	0.3204E+00
39	50.18	511.56	647.79	509.25	458.80	0.1059E+02	0.3140E+01	0.1680E+02	0.2985E+00
41	52.10	509.53	653.77	520.35	459.36	-0.5743E+01	0.3042E+01	0.5725E+01	0.2763E+00
43	55.36	541.84	663.40	538.83	460.20	0.6234E+01	0.2892E+01	0.5630E+01	0.2438E+00
45	59.43	564.30	674.82	561.70	461.12	0.5969E+01	0.2726E+01	0.5624E+01	0.2106E+00
47	64.23	586.74	687.49	588.85	462.06	0.5075E+01	0.2555E+01	0.5695E+01	0.1794E+00
49	67.93	615.11	696.84	610.05	462.68	0.8139E+01	0.2437E+01	0.5769E+01	0.1597E+00
51	71.93	633.64	706.29	633.29	463.28	0.6325E+01	0.2325E+01	0.5848E+01	0.1418E+00
53	75.92	659.06	715.46	656.76	463.82	0.6200E+01	0.2222E+01	0.5918E+01	0.1268E+00
55	79.92	680.54	724.07	680.56	464.30	0.5635E+01	0.2130E+01	0.5975E+01	0.1139E+00

III - 27

Connection type : Top-and seat-angle connections with double web angle
Mode : All bolted

Tested by : Z. Fu et al. (1998) Japan
Test Id. : LM-P15

Column : --
Beam : H250x125x6x9
F.Angle: L75x75x9x8.5
W.Angle: L75x75x9x8.5

Fasteners: F10T- -M16
23/32" Oversize holes
Material : SS400
Fy = -- ksi
Fu = -- ksi

Major parameters

lp = 7.8740"	lu = 0.9843"	ll = 0.9843"	ta = 0.3347"
gb = 1.8701"	gc = 1.7716"	cu = 1.3780"	cl = 1.3780"
pb = 2.5591"	pc = 2.5591"	nb = 1 X 3	nc = 1 X 3
lt = 5.1181"	tt = 0.3347"	ls = 5.1181"	ts = 0.3347"
gt = 1.8701"	gt'= 1.7716"	gs = 1.8701"	gs'= 1.7716"
qt = 2.9528"	rt = 2.9528"	qs = 2.9528"	rs = 2.9528"
nt = 2 X 1	nt'= 2 X 1	ns = 2 X 1	ns'= 2 X 1

Remark 1) Beam was applied to 147 kN of compressive axial force
2)

No	Moment (k-in)	Rotation (radians) X 1/1000	No	Moment (k-in)	Rotation (radians) X 1/1000
1	0.00	0.00	26	558.67	52.10
2	29.77	0.40	27	580.18	55.28
3	73.00	0.77	28	597.78	57.65
4	117.19	1.73	29	615.37	60.54
5	173.18	2.76	30	630.01	63.35
6	230.12	4.61	31	641.73	65.64
7	256.61	5.87	32	646.59	67.19
8	280.16	6.98	33	651.44	68.89
9	286.94	9.79	34	658.25	70.89
10	279.01	11.72	35	661.13	72.81
11	275.97	14.08	36	664.03	74.07
12	291.64	15.64	37	665.95	75.62
13	261.11	17.48	38	668.82	77.70
14	288.54	20.00			
15	319.91	22.22			
16	353.24	24.73			
17	381.64	27.62			
18	403.17	30.28			
19	427.65	32.87			
20	450.15	35.68			
21	463.85	37.38			
22	475.57	39.68			
23	493.14	42.93			
24	518.59	45.96			
25	537.14	49.29			

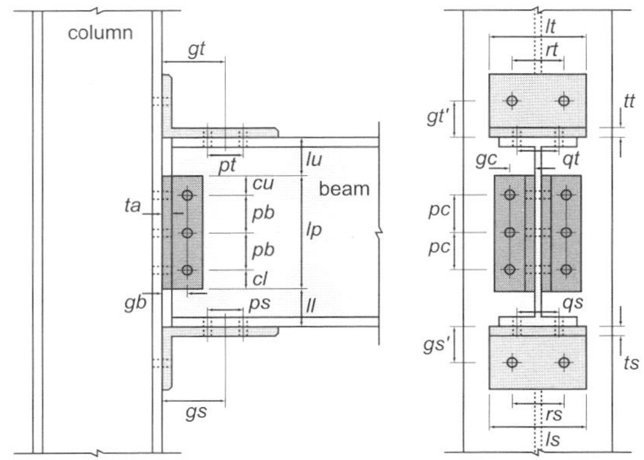

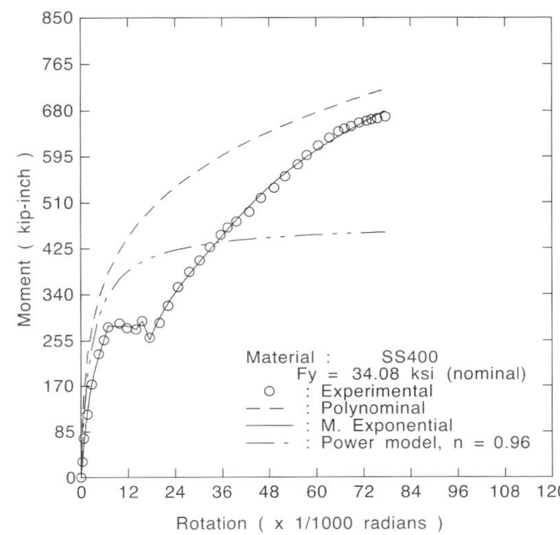

```
Moment-rotation prediction equations     ( R : X 1/1000 radians )

Frye and Morris polynominal model  :    R = Sum ( Ai X ( K*Bm )**Pi X 10**Qi )
     t =  0.334650"   d =  9.842520"  tc =  0.334650"   w =  5.118110"   g =  1.456700"
    A1 =  2.232429   A2 =  1.850728   A3 =  3.188976    K =  0.152595
    P1 =         1   P2 =         3   P3 =         5   Q1 =        -2   Q2 =        -5   Q3 =        -9

Modified exponential model  :
     Bm = Sum ( Ai X ( 1 - exp( - R/( 2*i*C ) ) ) )  +      Sum ( RKj X ( R - Rj0 ) )  +  Bm0
      C =   0.74750000E+00             Bmo=   0.00000000E+00            Nexp=  6        Nliner=  4
     AI =  -0.57148932E+03  0.10308593E+05  -0.48978076E+05  0.10169835E+06  -0.98096298E+05  0.36391486E+05
    Rj0 =           6.9800         14.0800          15.6400          17.4800
    RKj =  -0.26945572E+02  0.17257968E+02  -0.25947326E+02  0.39011916E+02

Power model  :     Bm = ( Rki X R ) / ( 1 + ( R/R0 )**rn )**( 1/rn )
    rn = 0.956      rki =   0.19182350E+03          rmu =   0.47255024E+03
```

No	Rotation (radians) X 1/1000	Moment (kip-inch) Expri.	Poly.	M.Expo.	P.Model.	Connection stiffness (kip-inch) X 1000 Expri.	Poly.	M.Expo.	P.Model.
1	0.00	0.00	0.00	0.00	0.00	0.7443E+02	0.2936E+03	0.8519E+02	0.1918E+03
3	0.77	73.00	151.99	62.53	109.66	0.9714E+02	0.1154E+03	0.7518E+02	0.1071E+03
5	2.76	173.18	276.56	172.51	241.77	0.4592E+02	0.3815E+02	0.3785E+02	0.4142E+02
7	5.87	256.61	359.65	256.57	323.53	0.2113E+02	0.1961E+02	0.2351E+02	0.1674E+02
9	9.79	286.94	420.49	282.37	368.71	-0.1453E+01	0.1257E+02	0.5482E+00	0.7950E+01
11	14.08	275.97	466.38	275.50	394.22	0.5536E+01	0.9197E+01	-0.5233E+01	0.4452E+01
13	17.48	261.11	494.82	256.34	406.85	-0.4997E+01	0.7645E+01	-0.2102E+02	0.3102E+01
15	22.22	319.91	527.47	324.26	418.81	0.1373E+02	0.6234E+01	0.1148E+02	0.2053E+01
17	27.62	381.64	558.11	377.69	427.97	0.8924E+01	0.5185E+01	0.8911E+01	0.1400E+01
19	32.87	427.65	583.37	422.79	434.25	0.8760E+01	0.4478E+01	0.8387E+01	0.1025E+01
21	37.38	463.85	602.63	459.97	438.37	0.6798E+01	0.4015E+01	0.8070E+01	0.8119E+00
23	42.93	493.14	623.52	503.01	442.34	0.6955E+01	0.3578E+01	0.7390E+01	0.6304E+00
25	49.29	537.14	644.98	546.83	445.87	0.6704E+01	0.3187E+01	0.6373E+01	0.4887E+00
27	55.28	580.18	663.16	582.19	448.50	0.7145E+01	0.2896E+01	0.5459E+01	0.3950E+00
29	60.54	615.37	677.83	609.13	450.41	0.5642E+01	0.2684E+01	0.4811E+01	0.3334E+00
31	65.64	641.73	691.06	632.40	451.98	0.3934E+01	0.2509E+01	0.4340E+01	0.2866E+00
33	68.89	651.44	699.13	646.12	452.87	0.3108E+01	0.2410E+01	0.4113E+01	0.2618E+00
35	72.81	661.13	708.28	661.81	453.85	0.1986E+01	0.2302E+01	0.3902E+01	0.2359E+00
37	75.62	665.95	714.70	672.61	454.49	0.1297E+01	0.2230E+01	0.3786E+01	0.2197E+00

III - 28

Connection type : Top-and seat-angle connections with double web angle
Mode : All bolted

Tested by : Z. Fu et al. (1998) Japan
Test Id. : LM-T15

Column : --
Beam : H250x125x6x9
F.Angle: L75x75x9x8.5
W.Angle: L75x75x9x8.5

Fasteners: F10T- -M16
23/32" Oversize holes
Material : SS400
Fy = -- ksi
Fu = -- ksi

Major parameters

lp = 7.8740"	lu = 0.9843"	ll = 0.9843"	ta = 0.3347"
gb = 1.8701"	gc = 1.7716"	cu = 1.3780"	cl = 1.3780"
pb = 2.5591"	pc = 2.5591"	nb = 1 X 3	nc = 1 X 3
lt = 5.1181"	tt = 0.3347"	ls = 5.1181"	ts = 0.3347"
gt = 1.8701"	gt'= 1.7716"	gs = 1.8701"	gs'= 1.7716"
qt = 2.9528"	rt = 2.9528"	qs = 2.9528"	rs = 2.9528"
nt = 2 X 1	nt'= 2 X 1	ns = 2 X 1	ns'= 2 X 1

Remark 1) Beam was applied to 147 kN of tensile axial force
2)

No	Moment (k-in)	Rotation (radians) X 1/1000	No	Moment (k-in)	Rotation (radians) X 1/1000
1	0.00	0.00	26	533.35	45.30
2	47.46	0.55	27	546.01	48.55
3	87.73	1.06	28	555.76	50.77
4	125.06	1.58	29	565.50	53.36
5	170.24	2.39	30	579.18	55.73
6	211.50	2.99	31	586.95	58.32
7	247.83	4.02	32	599.64	60.76
8	280.23	4.91	33	613.30	63.42
9	304.76	6.02	34	620.11	65.49
10	328.29	7.65	35	627.93	66.60
11	305.63	9.20	36	618.07	67.71
12	304.62	9.87	37	631.79	68.82
13	335.05	11.05	38	643.49	71.33
14	362.50	12.97	39	652.25	73.85
15	381.07	15.86	40	662.99	75.70
16	396.70	18.74	41	672.74	78.07
17	410.36	21.41			
18	425.04	23.33			
19	440.66	26.07			
20	455.32	28.58			
21	468.98	31.24			
22	479.70	33.91			
23	495.33	36.72			
24	504.07	39.53			
25	519.69	42.56			

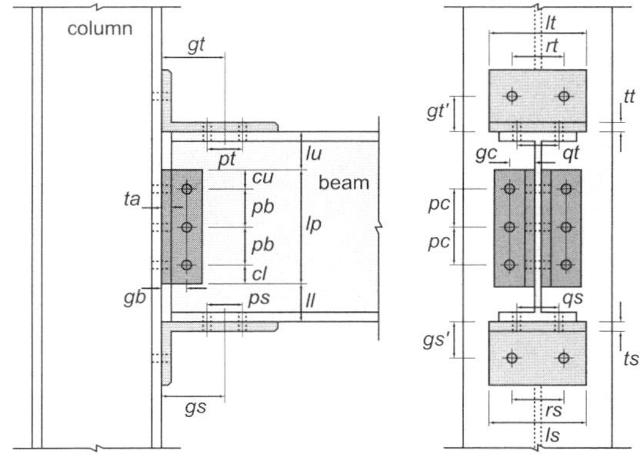

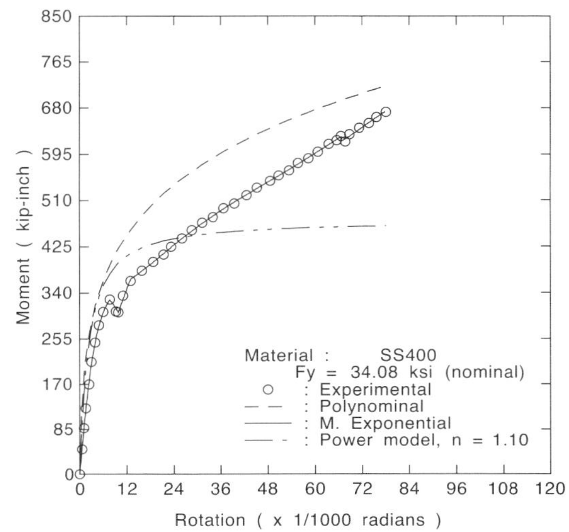

A.3 – 56

```
Moment-rotation prediction equations      ( R : X 1/1000 radians )

Frye and Morris polynominal model  :     R = Sum ( Ai X ( K*Bm )**Pi X 10**Qi )
      t =  0.334650"   d =  9.842520"  tc =  0.334650"   w =  5.118110"   g =  1.456700"
     A1 =  2.232429   A2 =  1.850728   A3 =  3.188976    K =  0.152595
     P1 =         1   P2 =         3   P3 =         5   Q1 =        -2   Q2 =        -5   Q3 =        -9

Modified exponential model  :
     Bm = Sum ( Ai X ( 1 - exp( - R/( 2*i*C ) ) ) )  +      Sum ( RKj X ( R - Rj0 ) )  +  Bm0
      C =   0.78808333E+00              Bmo=   0.00000000E+00          Nexp=  6        Nliner=  6
     AI =  -0.30849962E+02  -0.13314296E+04   0.10824861E+05  -0.24910686E+05   0.24691964E+05  -0.89014554E+04
    Rj0 =           7.6500           9.8700          12.9700          17.4800          66.6000
                   67.7100
    RKj =  -0.16829168E+02   0.34613873E+02  -0.13273467E+02   0.44251293E+00  -0.84243210E+01
            0.85826583E+01

Power model  :    Bm = ( Rki X R ) / ( 1 + ( R/R0 )**rn )**( 1/rn )
      rn = 1.098      rki =   0.19182350E+03         rmu =   0.47255024E+03
```

No	Rotation (radians) X 1/1000	Moment (kip-inch) Expri.	Moment (kip-inch) Poly.	Moment (kip-inch) M.Expo.	Moment (kip-inch) P.Model.	Connection stiffness (kip-inch) X 1000 Expri.	Connection stiffness (kip-inch) X 1000 Poly.	Connection stiffness (kip-inch) X 1000 M.Expo.	Connection stiffness (kip-inch) X 1000 P.Model.
1	0.00	0.00	0.00	0.00	0.00	0.8628E+02	0.2936E+03	0.8810E+02	0.1918E+03
3	1.06	87.73	181.21	87.14	150.04	0.7541E+02	0.8844E+02	0.7552E+02	0.1014E+03
5	2.39	170.24	261.53	175.12	247.55	0.6324E+02	0.4334E+02	0.5676E+02	0.5265E+02
7	4.02	247.83	317.11	250.36	310.81	0.3588E+02	0.2733E+02	0.3638E+02	0.2851E+02
9	6.02	304.76	362.54	304.15	353.60	0.1900E+02	0.1918E+02	0.1880E+02	0.1602E+02
11	9.20	305.63	412.89	312.37	389.81	-0.5461E+01	0.1327E+02	-0.1171E+02	0.8072E+01
13	11.05	335.05	435.52	328.40	402.56	0.2141E+02	0.1133E+02	0.1990E+02	0.5879E+01
15	15.86	381.07	481.94	380.46	422.97	0.5925E+01	0.8306E+01	0.5257E+01	0.3056E+01
17	21.41	410.36	522.34	412.61	435.75	0.6587E+01	0.6433E+01	0.6047E+01	0.1733E+01
19	26.07	440.66	549.88	440.21	442.43	0.5774E+01	0.5445E+01	0.5716E+01	0.1184E+01
21	31.24	468.98	575.92	468.09	447.56	0.4577E+01	0.4674E+01	0.5067E+01	0.8298E+00
23	36.72	495.33	599.82	494.28	451.43	0.4337E+01	0.4079E+01	0.4540E+01	0.6019E+00
25	42.56	519.69	622.19	519.96	454.46	0.5066E+01	0.3604E+01	0.4304E+01	0.4479E+00
27	48.55	546.01	642.60	545.67	456.81	0.4191E+01	0.3228E+01	0.4312E+01	0.3435E+00
29	53.36	565.50	657.52	566.62	458.32	0.4810E+01	0.2983E+01	0.4405E+01	0.2837E+00
31	58.32	586.95	671.77	588.76	459.60	0.4134E+01	0.2769E+01	0.4524E+01	0.2368E+00
33	63.42	613.30	685.41	612.14	460.71	0.4097E+01	0.2582E+01	0.4639E+01	0.1996E+00
35	66.60	627.93	693.54	626.99	461.31	-0.9206E+00	0.2478E+01	0.4700E+01	0.1806E+00
37	68.82	631.79	699.01	628.29	461.70	0.1000E+02	0.2411E+01	0.4896E+01	0.1689E+00
39	73.85	652.25	710.75	653.10	462.49	0.4821E+01	0.2274E+01	0.4965E+01	0.1462E+00
41	78.07	672.74	720.05	674.14	463.07	0.4112E+01	0.2172E+01	0.5007E+01	0.1305E+00

III - 29

Connection type : Top-and seat-angle connections with double web angle
Mode : All bolted

Tested by : L.Calado et al. (2000) Portugal
Test Id. : BCC7-M

Column : HEB200
Beam : IPE300
F.Angle: L120x120x10
W.Angle: L120x80x10

Fasteners: G8.8- -M16
23/32" Oversize holes
Material : S235
Fy = 36.58 ksi
Fu = 60.94 ksi

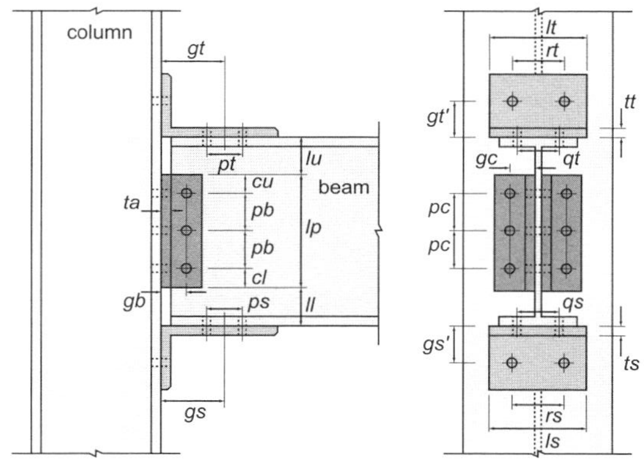

Major parameters

lp = 9.4488"	lu = 1.1811"	ll = 1.1811"	ta = 0.3937"
gb = 2.9528"	gc = 1.7716"	cu = 2.3622"	cl = 2.3622"
pb = 2.3622"	pc = 2.3622"	nb = 1 X 3	nc = 1 X 3
lt = 5.9055"	tt = 0.3937"	ls = 5.9055"	ts = 0.3937"
gt = 2.9528"	gt'= 2.9528"	gs = 2.9528"	gs'= 2.9528"
qt = 3.1496"	rt = 3.1496"	qs = 3.1496"	rs = 3.1496"
pt = 1.5748"	pt'= 1.5748"	ps = 1.5748"	ps'= 1.5748"
nt = 2 X 2	nt'= 2 X 2	ns = 2 X 2	ns'= 2 X 2

Remark 1) Continuity plate (12 mm in thickness) were used in column panel zone.
2) Fybf=304.7, Fubf=452.6, Fybw=315.7, Fubw=452.6 N/mm2 for beam, Hole size:17mm

No	Mcment (k-in)	Rotation (radians) X 1/1000	No	Moment (k-in)	Rotation (radians) X 1/1000
1	0.00	0.00	26	1100.88	71.87
2	61.74	0.25	27	1124.54	75.86
3	124.14	0.49	28	1148.55	79.85
4	185.71	0.75	29	1172.24	83.84
5	247.62	1.03	30	1195.97	87.84
6	3C9.52	1.38	31	1220.30	91.83
7	373.84	2.00	32	1242.16	95.82
8	424.49	2.51	33	1266.50	99.81
9	463.02	3.99	34	1282.30	104.00
10	500.51	7.99	35	1268.14	107.80
11	542.07	11.98			
12	559.65	15.97			
13	607.32	19.96			
14	654.47	23.96			
15	689.87	27.95			
16	726.79	31.94			
17	758.76	35.93			
18	800.45	39.93			
19	839.07	43.92			
20	893.81	47.91			
21	939.13	51.90			
22	979.17	55.90			
23	1015.96	59.89			
24	1045.14	63.88			
25	1074.39	67.87			

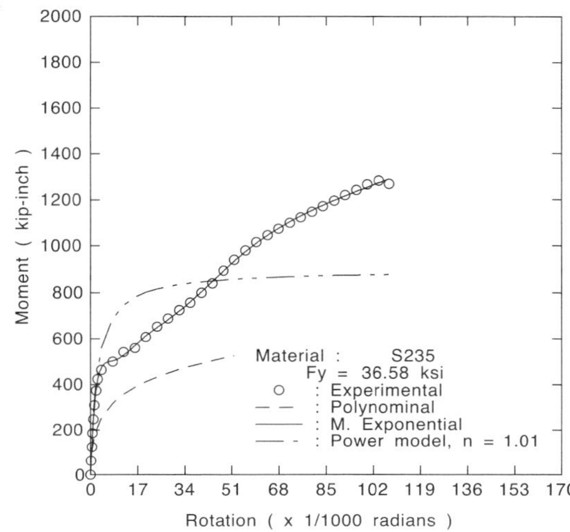

```
Moment-rotation prediction equations      ( R : X 1/1000 radians )

Frye and Morris polynominal model   :    R = Sum ( Ai X ( K*Bm )**Pi X 10**Qi )
     t =  0.393700"   d = 11.811000"  tc =  0.393700"   w =  5.905510"   g =  2.638295"
    A1 =  2.232429   A2 =  1.850728   A3 =  3.188976    K =  0.189603
    P1 =       1     P2 =       3     P3 =       5     Q1 =      -2     Q2 =      -5     Q3 =      -9

Modified exponential model  :
    Bm = Sum ( Ai X ( 1 - exp( - R/( 2*i*C ) ) ) )  +      Sum ( RKj X ( R - Rj0 ) )  +  Bm0
      C =   0.96083333E+00            Bmo=   0.00000000E+00          Nexp=  6        Nliner=  1
     AI =   0.25127828E+03   0.95551048E+04  -0.55594666E+05   0.13295256E+06  -0.14251120E+06   0.56214935E+05
    Rj0 =           3.9100
    RKj =   0.40815341E+01

Power model  :    Bm = ( Rki X R ) / ( 1 + ( R/R0 )**rn )**( 1/rn )
    rn = 1.006      rki =   0.37065233E+03          rmu =   0.89634495E+03
```

No	Rotation (radians) X 1/1000	Moment (kip-inch) Expri.	Moment Poly.	Moment M.Expo.	Moment P.Model.	Connection stiffness (kip-inch) X 1000 Expri.	Stiffness Poly.	Stiffness M.Expo.	Stiffness P.Model.
1	0.00	0.00	0.00	0.00	0.00	0.2470E+03	0.2363E+03	0.3134E+03	0.3707E+03
3	0.49	124.14	91.64	134.52	151.46	0.2489E+03	0.1301E+03	0.2386E+03	0.2575E+03
5	1.03	247.62	143.73	245.27	268.78	0.2014E+03	0.7290E+02	0.1745E+03	0.1833E+03
7	2.00	373.84	195.78	372.92	407.53	0.1013E+03	0.4085E+02	0.9531E+02	0.1116E+03
9	3.99	463.02	254.48	474.20	560.45	0.2153E+02	0.2216E+02	0.2439E+02	0.5290E+02
11	11.98	542.07	358.70	528.25	747.95	0.7411E+01	0.8503E+01	0.9288E+01	0.1040E+02
13	19.96	607.32	412.67	612.59	801.22	0.1187E+02	0.5495E+01	0.9915E+01	0.4286E+01
15	27.95	689.87	450.55	685.24	826.42	0.9062E+01	0.4131E+01	0.8912E+01	0.2321E+01
17	35.93	758.76	480.13	761.76	841.07	0.9216E+01	0.3343E+01	0.1035E+02	0.1452E+01
19	43.92	839.07	504.64	848.30	850.67	0.1170E+02	0.2825E+01	0.1105E+02	0.9931E+00
21	51.90	939.13	525.64	934.32	857.43	0.1069E+02	0.2457E+01	0.1032E+02	0.7218E+00
23	59.89	1015.96	544.11	1011.16	862.45	0.8268E+01	0.2180E+01	0.8864E+01	0.5480E+00
25	67.87	1074.39	560.61	1075.87	866.33	0.6977E+01	0.1964E+01	0.7387E+01	0.4302E+00
27	75.86	1124.54	575.59	1129.92	869.41	0.5973E+01	0.1791E+01	0.6204E+01	0.3466E+00
29	83.84	1172.24	589.43	1175.90	871.92	0.5935E+01	0.1647E+01	0.5375E+01	0.2852E+00
31	91.83	1220.30	602.08	1216.54	874.00	0.5788E+01	0.1527E+01	0.4839E+01	0.2387E+00
33	99.81	1266.50	613.84	1253.74	875.76	0.4964E+01	0.1425E+01	0.4512E+01	0.2028E+00
35	107.80	1268.14	624.86	1288.95	877.26	-0.3726E+01	0.1338E+01	0.4321E+01	0.1744E+00

III - 30

Connection type : Top-and seat-angle connections with double web angle
Mode : All bolted

Tested by : L.Calado et al. (2000) Portugal
Test Id. : BCC9-M

Column : HEB160
Beam : IPE300
F.Angle: L120x120x10
W.Angle: L120x80x10

Fasteners: G8.8- -M16
23/32" Oversize holes
Material : S235
Fy = 36.58 ksi
Fu = 60.94 ksi

Major parameters

lp = 9.4488"	lu = 1.1811"	ll = 1.1811"	ta = 0.3937"
gb = 2.9528"	gc = 1.7716"	cu = 2.3622"	cl = 2.3622"
pb = 2.3622"	pc = 2.3622"	nb = 1 X 3	nc = 1 X 3
lt = 5.9055"	tt = 0.3937"	ls = 5.9055"	ts = 0.3937"
gt = 2.9528"	gt'= 2.9528"	gs = 2.9528"	gs'= 2.9528"
qt = 3.1496"	rt = 3.1496"	qs = 3.1496"	rs = 3.1496"
pt = 1.5748"	pt'= 1.5748"	ps = 1.5748"	ps'= 1.5748"
nt = 2 X 2	nt'= 2 X 2	ns = 2 X 2	ns'= 2 X 2

Remark 1) Continuity plate (12 mm in thickness) were used in column panel zone.
2) Fybf=304.7, Fubf=452.6, Fybw=315.7, Fubw=452.6 N/mm2 for beam, Hole size:17mm

No	Moment (k-in)	Rotation (radians) X 1/1000	No	Moment (k-in)	Rotation (radians) X 1/1000
1	0.00	0.00	26	982.90	61.15
2	84.29	0.45	27	1002.23	64.21
3	168.59	0.93	28	1021.68	67.26
4	252.88	1.52	29	1038.16	70.32
5	337.17	2.21	30	1053.67	73.38
6	394.21	3.06	31	1067.91	76.44
7	421.47	4.25	32	1081.35	79.49
8	465.45	6.11	33	1094.38	82.55
9	509.80	9.17	34	1105.82	85.61
10	527.82	12.23	35	1117.25	88.67
11	540.42	15.29	36	1129.66	91.72
12	580.41	18.34	37	1142.35	94.78
13	622.32	21.40	38	1155.94	97.84
14	664.53	24.46	39	1170.85	100.90
15	705.88	27.52	40	1184.73	103.95
16	743.81	30.57	41	1193.81	108.00
17	780.06	33.63	42	1178.07	110.07
18	814.45	36.69	43	1059.41	110.56
19	844.67	39.75	44	1080.98	113.13
20	866.68	42.80	45	1106.69	116.18
21	888.68	45.86	46	1132.54	119.24
22	908.17	48.92	47	1150.94	122.30
23	927.46	51.98	48	1146.34	125.36
24	946.15	55.03			
25	964.53	58.09			

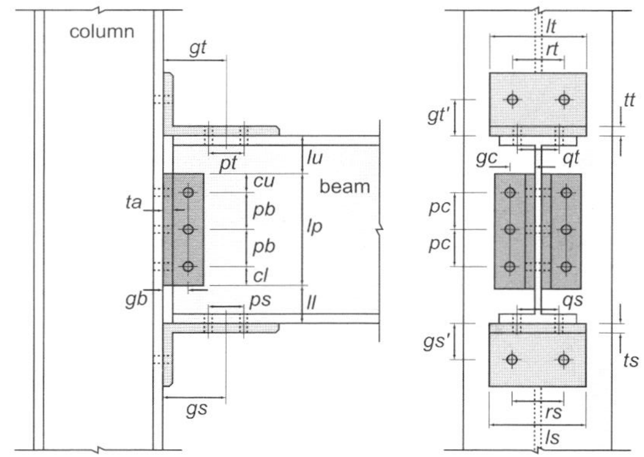

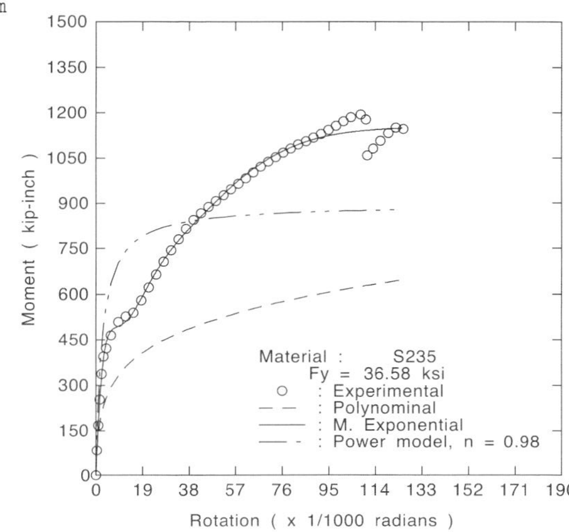

A.3 – 60

```
Moment-rotation prediction equations     ( R : X 1/1000 radians )

Frye and Morris polynominal model   :    R = Sum ( Ai X ( K*Bm )**Pi X 10**Qi )
      t =  0.393700"   d = 11.811000"  tc =  0.393700"   w =  5.905510"   g =  2.638295"
     A1 =  2.232429   A2 =  1.850728   A3 =  3.188976    K =  0.189603
     P1 =        1    P2 =        3    P3 =        5    Q1 =       -2    Q2 =       -5    Q3 =       -9

Modified exponential model  :
     Bm = Sum ( Ai X ( 1 - exp( - R/( 2*i*C ) ) ) )  +      Sum ( RKj X ( R - Rj0 ) )  +  Bm0
      C =   0.11571667E+01             Bmo=   0.00000000E+00          Nexp=  6        Nliner=  0
     AI =  -0.39785914E+03   0.14424905E+05  -0.70653935E+05   0.15599437E+06  -0.16187628E+06   0.63662023E+05

Power model  :     Bm = ( Rki X R ) / ( 1 + ( R/R0 )**rn )**( 1/rn )
      rn = 0.985       rki =   0.37065233E+03          rmu =   0.89634495E+03
```

No	Rotation (radians) X 1/1000	Moment (kip-inch) Expri.	Poly.	M.Expo.	P.Model.	Connection stiffness (kip-inch) X 1000 Expri.	Poly.	M.Expo.	P.Model.
1	0.00	0.00	0.00	0.00	0.00	0.1873E+03	0.2363E+03	0.2147E+03	0.3707E+03
3	0.93	168.59	136.06	169.72	246.66	0.1609E+03	0.7950E+02	0.1528E+03	0.1908E+03
5	2.21	337.17	203.98	322.79	423.38	0.9749E+02	0.3739E+02	0.9044E+02	0.1000E+03
7	4.25	421.47	260.08	441.70	565.44	0.2319E+02	0.2096E+02	0.3337E+02	0.4852E+02
9	9.17	509.80	331.96	495.60	703.59	0.1019E+02	0.1071E+02	0.3372E+01	0.1627E+02
11	15.29	540.42	384.01	543.52	769.07	0.8622E+01	0.6897E+01	0.1250E+02	0.7040E+01
13	21.40	622.32	420.34	630.36	801.16	0.1375E+02	0.5180E+01	0.1449E+02	0.3918E+01
15	27.52	705.88	448.76	711.44	820.28	0.1297E+02	0.4186E+01	0.1182E+02	0.2492E+01
17	33.63	780.06	472.22	776.09	832.95	0.1154E+02	0.3535E+01	0.9565E+01	0.1726E+01
19	39.75	844.67	492.36	831.01	841.99	0.8544E+01	0.3071E+01	0.8546E+01	0.1265E+01
21	45.86	888.68	510.02	881.75	848.75	0.6780E+01	0.2725E+01	0.8093E+01	0.9679E+00
23	51.98	927.46	525.83	929.90	854.02	0.6216E+01	0.2454E+01	0.7611E+01	0.7642E+00
25	58.09	964.53	540.14	974.32	858.22	0.6005E+01	0.2236E+01	0.6885E+01	0.6189E+00
27	64.21	1002.23	553.26	1013.67	861.66	0.6346E+01	0.2057E+01	0.5948E+01	0.5114E+00
29	70.32	1038.16	565.36	1046.90	864.53	0.5229E+01	0.1907E+01	0.4925E+01	0.4298E+00
31	76.44	1067.91	576.63	1073.95	866.95	0.4529E+01	0.1779E+01	0.3928E+01	0.3662E+00
33	82.55	1094.38	587.15	1095.17	869.03	0.3998E+01	0.1669E+01	0.3039E+01	0.3159E+00
35	88.67	1117.25	597.07	1111.40	870.84	0.3903E+01	0.1573E+01	0.2290E+01	0.2752E+00
37	94.78	1142.35	606.55	1123.49	872.41	0.4294E+01	0.1487E+01	0.1690E+01	0.2420E+00
39	100.90	1170.85	615.35	1132.34	873.81	0.4713E+01	0.1413E+01	0.1224E+01	0.2144E+00
41	108.00	1193.81	625.05	1139.54	875.23	-0.4274E+01	0.1336E+01	0.8265E+00	0.1880E+00
43	110.56	1059.41	628.46	1141.51	875.70	-0.2020E+03	0.1310E+01	0.7143E+00	0.1796E+00
45	116.18	1106.69	635.75	1144.94	876.67	0.8439E+01	0.1257E+01	0.5148E+00	0.1631E+00
47	122.30	1150.94	643.19	1147.58	877.62	0.2254E+01	0.1206E+01	0.3570E+00	0.1477E+00

III - 31

Connection type : Top-and seat-angle connections with double web angle
Mode : All bolted

Tested by : L.Calado et al. (2000) Portugal
Test Id. : BCC10-M

Column : HEB240
Beam : IPE300
F.Angle: L120x120x10
W.Angle: L120x80x10

Fasteners: G8.8- -M16
23/32" Oversize holes
Material : S235
Fy = 36.58 ksi
Fu = 60.94 ksi

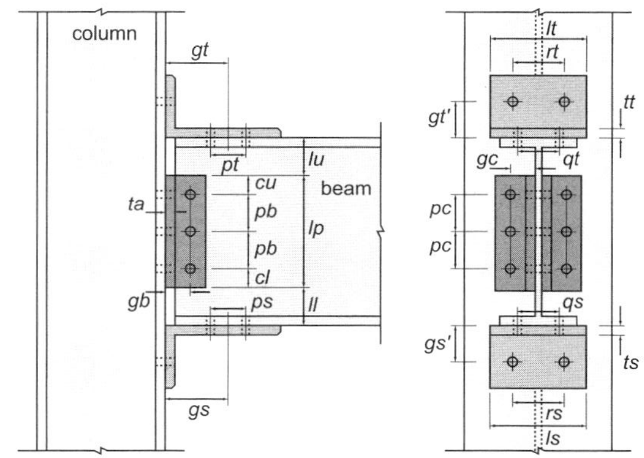

Major parameters

lp = 9.4488"	lu = 1.1811"	ll = 1.1811"	ta = 0.3937"
gb = 2.9528"	gc = 1.7716"	cu = 2.3622"	cl = 2.3622"
pb = 2.3622"	pc = 2.3622"	nb = 1 X 3	nc = 1 X 3
lt = 5.9055"	tt = 0.3937"	ls = 5.9055"	ts = 0.3937"
gt = 2.9528"	gt'= 2.9528"	gs = 2.9528"	gs'= 2.9528"
qt = 3.1496"	rt = 3.1496"	qs = 3.1496"	rs = 3.1496"
pt = 1.5748"	pt'= 1.5748"	ps = 1.5748"	ps'= 1.5748"
nt = 2 X 2	nt'= 2 X 2	ns = 2 X 2	ns'= 2 X 2

Remark 1) Continuity plate (12 mm in thickness) were used in column panel zone.
2) Fybf=304.7, Fubf=452.6, Fybw=315.7, Fubw=452.6 N/mm2 for beam, Hole size:17mm

No	Moment (k-in)	Rotation (radians) X 1/1000	No	Moment (k-in)	Rotation (radians) X 1/1000
1	0.00	0.00	26	1135.23	63.37
2	88.62	0.47	27	1161.07	66.39
3	177.24	0.98	28	1183.55	69.40
4	265.86	1.50	29	1203.09	72.42
5	354.48	2.30	30	1220.57	75.44
6	398.79	3.55	31	1238.63	78.46
7	464.90	6.04	32	1261.75	81.47
8	513.00	9.05	33	1279.03	84.49
9	542.36	12.07	34	1303.72	87.51
10	570.59	15.09	35	1322.50	90.53
11	604.89	18.11	36	1330.09	93.00
12	625.17	21.12	37	1312.93	95.97
13	654.15	24.14	38	1283.01	96.56
14	666.48	27.16			
15	711.41	30.18			
16	771.12	33.19			
17	830.76	36.21			
18	878.42	39.23			
19	918.67	42.25			
20	947.26	45.26			
21	972.75	48.28			
22	1001.46	51.30			
23	1036.59	54.32			
24	1074.33	57.33			
25	1105.42	60.35			

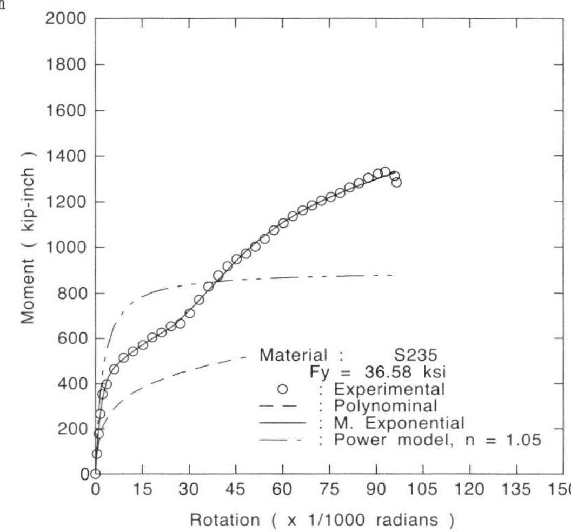

A.3 – 62

```
Moment-rotation prediction equations     ( R : X 1/1000 radians )

Frye and Morris polynominal model  :    R = Sum ( Ai X ( K*Bm )**Pi X 10**Qi )
      t =  0.393700"   d = 11.811000"  tc =  0.393700"   w =  5.905510"   g =  2.638295"
     A1 =  2.232429   A2 =  1.850728   A3 =  3.188976    K =  0.189603
     P1 =         1   P2 =         3   P3 =         5   Q1 =        -2   Q2 =        -5   Q3 =        -9

Modified exponential model  :
     Bm = Sum ( Ai X ( 1 - exp( - R/( 2*i*C ) ) ) )  +      Sum ( RKj X ( R - Rj0 ) )  +  Bm0
      C =   0.92216667E+00           Bmo=   0.00000000E+00           Nexp=  6         Nliner=  1
     AI =   0.20971707E+03   0.56475859E+04  -0.38190951E+05   0.10468557E+06  -0.12405813E+06   0.52749449E+05
    Rj0 =         25.0000
    RKj =   0.40504191E+01

Power model  :    Bm = ( Rki X R ) / ( 1 + ( R/R0 )**rn )**( 1/rn )
     rn = 1.047      rki =   0.37065233E+03         rmu =   0.89634495E+03
```

No	Rotation (radians) X 1/1000	Moment (kip-inch) Expri.	Poly.	M.Expo.	P.Model.	Connection stiffness (kip-inch) X 1000 Expri.	Poly.	M.Expo.	P.Model.
1	0.00	0.00	0.00	0.00	0.00	0.1886E+03	0.2363E+03	0.2464E+03	0.3707E+03
3	0.98	177.24	139.96	190.35	265.51	0.1721E+03	0.7608E+02	0.1499E+03	0.1951E+03
5	2.30	354.48	207.29	334.61	450.75	0.8138E+02	0.3609E+02	0.7693E+02	0.1006E+03
7	6.04	464.90	292.07	471.74	657.40	0.2177E+02	0.1540E+02	0.1651E+02	0.3017E+02
9	12.07	542.36	359.52	541.99	761.73	0.9535E+01	0.8444E+01	0.1005E+02	0.9887E+01
11	18.11	604.89	402.07	597.98	803.39	0.9044E+01	0.5970E+01	0.8608E+01	0.4805E+01
13	24.14	654.15	433.81	649.63	825.60	0.6839E+01	0.4677E+01	0.8843E+01	0.2821E+01
15	30.18	711.41	459.46	727.70	839.39	0.1736E+02	0.3872E+01	0.1411E+02	0.1848E+01
17	36.21	830.76	481.06	814.90	848.74	0.1777E+02	0.3322E+01	0.1462E+02	0.1302E+01
19	42.25	918.67	499.84	901.72	855.51	0.1141E+02	0.2918E+01	0.1395E+02	0.9648E+00
21	48.28	972.75	516.47	981.58	860.62	0.8973E+01	0.2610E+01	0.1245E+02	0.7432E+00
23	54.32	1036.59	531.47	1051.40	864.62	0.1209E+02	0.2365E+01	0.1066E+02	0.5894E+00
25	60.35	1105.42	545.11	1110.42	867.82	0.1008E+02	0.2166E+01	0.8956E+01	0.4788E+00
27	66.39	1161.07	557.80	1160.05	870.45	0.8012E+01	0.1999E+01	0.7532E+01	0.3963E+00
29	72.42	1203.09	569.41	1202.01	872.64	0.6129E+01	0.1860E+01	0.6439E+01	0.3334E+00
31	78.46	1238.63	580.27	1238.37	874.50	0.6833E+01	0.1740E+01	0.5644E+01	0.2842E+00
33	84.49	1279.03	590.44	1270.63	876.09	0.6947E+01	0.1637E+01	0.5091E+01	0.2452E+00
35	90.53	1322.50	600.04	1300.17	877.48	0.4488E+01	0.1546E+01	0.4717E+01	0.2135E+00
37	95.97	1312.93	608.18	1325.18	878.57	-0.4327E+02	0.1473E+01	0.4492E+01	0.1900E+00

III - 32

Connection type : Top-and seat-angle connections with double web angle
Mode : All bolted

Tested by : M. Komuro et al. (2002) Japan
Test Id. : W18-m

Column : --
Beam : H400x200x8x13
F.Angle: L150x100x12
W.Angle: L90x90x7

Fasteners: F10T- -M20
7/8" Oversize holes
Material : SS400
Fy = 40.90 ksi
Fu = 65.12 ksi

Major parameters

lp = 7.0866"	lu = 4.3307"	ll = 4.3307"	ta = 0.2756"
gb = 2.1654"	gc = 2.2047"	cu = 1.3780"	cl = 1.3780"
pb = 2.1654"	pc = 2.1654"	nb = 1 X 3	nc = 1 X 3
lt = 7.8740"	tt = 0.4724"	ls = 7.8740"	ts = 0.4724"
gt = 3.2480"	gt'= 2.1654"	gs = 3.2480"	gs'= 2.1654"
qt = 4.7244"	rt = 4.7244"	qs = 4.7244"	rs = 4.7244"
pt = 2.1654"	ps = 2.1654"		
nt = 2 X 2	nt'= 2 X 1	ns = 2 X 2	ns'= 2 X 1

Remark 1) Fy = 315 MPa, Fu = 469 MPa for web-angle
2)

No	Mcment (k-in)	Rotation (radians) X 1/1000	No	Moment (k-in)	Rotation (radians) X 1/1000
1	0.00	0.00	26	1315.09	43.99
2	135.89	0.25	27	1334.25	46.02
3	244.42	0.49	28	1353.07	48.03
4	364.97	0.74	29	1396.93	50.01
5	455.30	1.52	30	1406.36	52.01
6	544.66	3.00	31	1461.63	54.03
7	586.57	5.07	32	1466.81	56.01
8	625.89	7.01	33	1516.85	58.01
9	672.03	9.00	34	1520.75	60.03
10	707.12	12.00	35	1566.90	62.00
11	727.92	14.00	36	1591.61	64.03
12	764.63	16.03	37	1599.68	66.04
13	829.29	18.01	38	1651.35	68.04
14	867.96	19.99	39	1680.58	69.99
15	904.68	22.09	40	1691.99	72.10
16	954.07	24.01	41	1705.29	74.02
17	999.88	26.02	42	1742.67	76.00
18	1047.00	28.06	43	1779.07	77.99
19	1091.51	30.02	44	1792.05	80.03
20	1116.20	32.00	45	1798.21	82.01
21	1135.38	34.01	46	1823.59	84.00
22	1174.05	36.01	47	1861.90	86.03
23	1228.63	38.00	48	1880.78	88.01
24	1247.48	40.06	49	1909.69	90.11
25	1257.23	42.03			

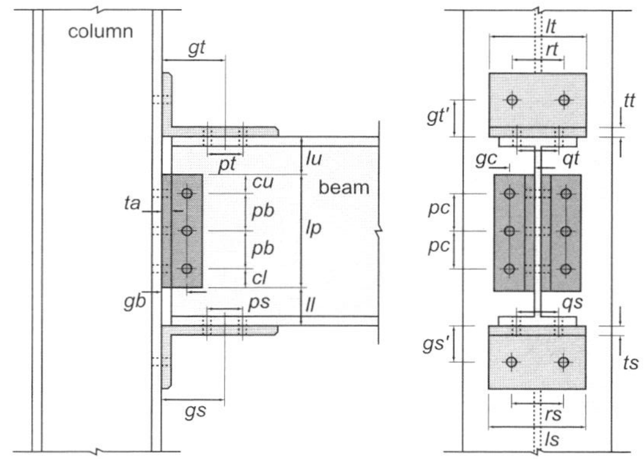

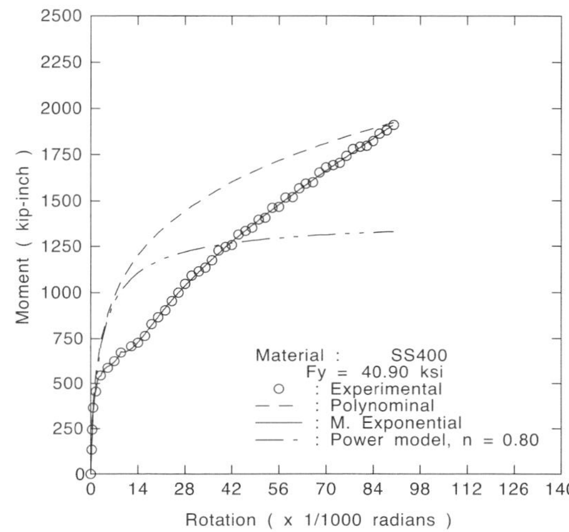

A.3 – 64

```
Moment-rotation prediction equations     ( R : X 1/1000 radians )

Frye and Morris polynominal model  :    R = Sum ( Ai X ( K*Bm )**Pi X 10**Qi )
      t =  0.472440"   d = 15.748000"  tc =  0.275590"   w =  7.874020"   g =  1.771650"
     A1 =  2.232429   A2 =  1.850728   A3 =  3.188976    K =  0.059121
     P1 =         1   P2 =         3   P3 =         5   Q1 =        -2   Q2 =        -5   Q3 =        -9

Modified exponential model  :
     Bm = Sum ( Ai X ( 1 - exp( - R/( 2*i*C ) ) ) )  +      Sum ( RKj X ( R - Rj0 ) )  +  Bm0
      C =  0.81341667E+00              Bmo=  0.00000000E+00           Nexp=  6       Nliner=  4
     AI =  0.10125710E+04   0.68878219E+04  -0.31898333E+05   0.59529603E+05  -0.56465008E+05   0.21652213E+05
    Rj0 =          0.7400           1.5200           3.0000           9.0000
    RKj = -0.11850784E+03   0.10460252E+03   0.68961362E+02  -0.43027321E+02

Power model  :     Bm = ( Rki X R ) / ( 1 + ( R/R0 )**rn )**( 1/rn )
     rn = 0.796      rki =    0.70099767E+03         rmu =    0.14122959E+04
```

No	Rotation (radians)	Moment (kip-inch)				Connection stiffness (kip-inch) X 1000			
	X 1/1000	Expri.	Poly.	M.Expo.	P.Model.	Expri.	Poly.	M.Expo.	P.Model.
1	0.00	0.00	0.00	0.00	0.00	0.5436E+03	0.7577E+03	0.6281E+03	0.7010E+03
3	0.49	244.42	293.88	256.29	241.21	0.4669E+03	0.4172E+03	0.4283E+03	0.3716E+03
5	1.52	455.30	557.16	455.30	509.23	0.9668E+02	0.1669E+03	0.4835E+02	0.1862E+03
7	5.07	586.57	885.04	594.96	862.85	0.2026E+02	0.5756E+02	0.1024E+02	0.5519E+02
9	9.00	672.03	1058.84	677.20	1011.62	0.1861E+02	0.3489E+02	0.3770E+02	0.2620E+02
11	14.00	727.92	1201.83	729.80	1106.88	0.1421E+02	0.2386E+02	0.2073E+02	0.1393E+02
13	18.01	829.29	1287.49	822.29	1153.06	0.2609E+02	0.1924E+02	0.2392E+02	0.9539E+01
15	22.09	904.68	1359.32	916.88	1186.28	0.2179E+02	0.1617E+02	0.2214E+02	0.6957E+01
17	26.02	999.88	1418.55	999.38	1210.30	0.2294E+02	0.1408E+02	0.1991E+02	0.5375E+01
19	30.02	1091.51	1471.51	1075.43	1229.47	0.1762E+02	0.1247E+02	0.1822E+02	0.4277E+01
21	34.01	1135.38	1518.70	1145.66	1244.89	0.1445E+02	0.1123E+02	0.1706E+02	0.3496E+01
23	38.00	1228.63	1561.44	1211.97	1257.63	0.1845E+02	0.1023E+02	0.1622E+02	0.2917E+01
25	42.03	1257.23	1601.33	1275.88	1268.45	0.1727E+02	0.9392E+01	0.1552E+02	0.2472E+01
27	46.02	1334.25	1637.39	1336.58	1277.60	0.9402E+01	0.8706E+01	0.1492E+02	0.2128E+01
29	50.01	1396.93	1670.89	1395.00	1285.52	0.1348E+02	0.8122E+01	0.1437E+02	0.1853E+01
31	54.03	1461.63	1702.48	1451.80	1292.51	0.1487E+02	0.7615E+01	0.1389E+02	0.1629E+01
33	58.01	1516.85	1731.87	1506.26	1298.61	0.1353E+02	0.7177E+01	0.1348E+02	0.1446E+01
35	62.00	1566.90	1759.52	1559.36	1304.07	0.1788E+02	0.6792E+01	0.1314E+02	0.1293E+01
37	66.04	1599.68	1786.59	1611.88	1309.02	0.1495E+02	0.6440E+01	0.1287E+02	0.1162E+01
39	69.99	1680.58	1811.23	1662.27	1313.39	0.1039E+02	0.6139E+01	0.1265E+02	0.1053E+01
41	74.02	1705.29	1835.68	1712.91	1317.44	0.1281E+02	0.5856E+01	0.1249E+02	0.9577E+00
43	77.99	1779.07	1858.15	1762.22	1321.07	0.1240E+02	0.5611E+01	0.1236E+02	0.8763E+00
45	82.01	1798.21	1880.46	1811.71	1324.45	0.7920E+01	0.5379E+01	0.1227E+02	0.8044E+00
47	86.03	1861.90	1901.43	1860.88	1327.55	0.1414E+02	0.5172E+01	0.1220E+02	0.7412E+00
49	90.11	1909.69	1922.33	1910.53	1330.46	0.1377E+02	0.4975E+01	0.1215E+02	0.6848E+00

III - 33

Connection type : Top-and seat-angle connections with double web angle
Mode : All bolted

Tested by : M. Komuro et al. (2002) Japan
Test Id. : W29-m

Column : --
Beam : H400x200x8x13
F.Angle: L150x100x12
W.Angle: L90x90x7

Fasteners: F10T- -M20
7/8" Oversize holes
Material : SS400
Fy = 40.90 ksi
Fu = 65.12 ksi

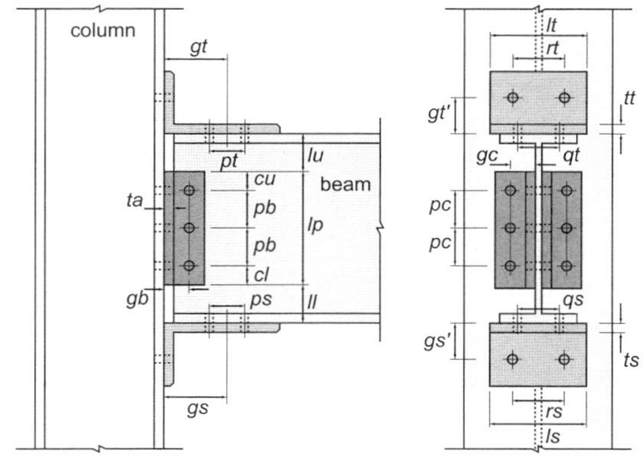

Major parameters

lp = 11.4173"	lu = 2.1654"	ll = 2.1654"	ta = 0.2756"
gb = 2.1654"	gc = 2.2047"	cu = 1.3780"	cl = 1.3780"
pb = 2.1654"	pc = 2.1654"	nb = 1 X 5	nc = 1 X 5
lt = 7.8740"	tt = 0.4724"	ls = 7.8740"	ts = 0.4724"
gt = 3.2480"	gt'= 2.1654"	gs = 3.2480"	gs'= 2.1654"
qt = 4.7244"	rt = 4.7244"	qs = 4.7244"	rs = 4.7244"
pt = 2.1654"	ps = 2.1654"		
nt = 2 X 2	nt'= 2 X 1	ns = 2 X 2	ns'= 2 X 1

Remark 1) Fy = 315 MPa, Fu = 469 MPa for web-angle
2)

No	Moment (k-in)	Rotation (radians) X 1/1000	No	Moment (k-in)	Rotation (radians) X 1/1000	No	Moment (k-in)	Rotation (radians) X 1/1000
1	0.00	0.00	26	1384.73	38.00	51	2139.22	88.04
2	134.10	0.26	27	1427.98	39.99	52	2147.02	90.02
3	258.23	0.50	28	1470.53	41.98	53	2185.04	92.04
4	322.56	0.73	29	1486.14	44.07	54	2234.74	94.51
5	408.67	1.02	30	1539.11	46.03			
6	529.22	1.52	31	1540.70	48.04			
7	579.58	2.01	32	1594.68	50.00			
8	663.42	3.05	33	1614.48	52.00			
9	698.18	4.02	34	1653.77	54.03			
10	748.55	6.01	35	1685.31	56.01			
11	782.01	7.96	36	1708.04	57.99			
12	808.01	10.05	37	1756.49	60.02			
13	835.30	12.08	38	1783.44	61.99			
14	858.70	14.02	39	1796.45	64.01			
15	860.32	15.99	40	1833.48	66.04			
16	903.53	17.99	41	1876.38	67.97			
17	953.58	20.00	42	1912.14	69.98			
18	910.36	22.04	43	1951.12	72.00			
19	1031.88	24.01	44	1978.70	73.96			
20	1094.60	26.03	45	2003.42	76.03			
21	1163.48	28.01	46	2022.27	78.03			
22	1214.50	30.01	47	2052.83	80.02			
23	1263.56	32.03	48	2069.71	82.08			
24	1291.18	34.02	49	2114.90	84.00			
25	1342.86	36.01	50	2127.53	86.02			

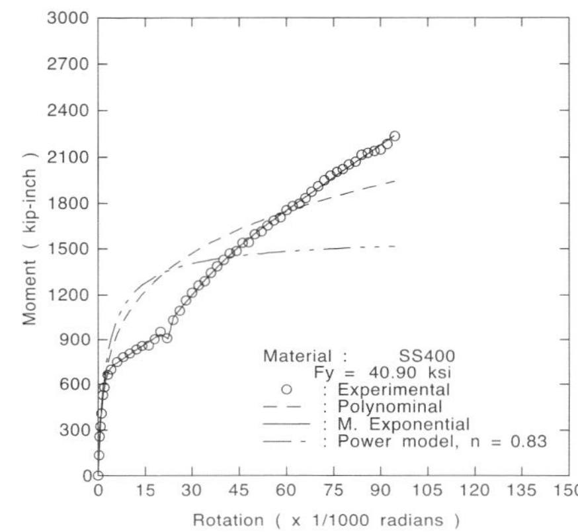

```
Moment-rotation prediction equations     ( R : X 1/1000 radians )

Frye and Morris polynominal model  :     R = Sum ( Ai X ( K*Bm )**Pi X 10**Qi )
      t =  0.472440"   d = 15.748000"  tc =  0.275590"   w =  7.874020"   g =  1.771650"
     A1 =  2.232429   A2 =  1.850728   A3 =  3.188976    K =  0.059121
     P1 =        1    P2 =        3    P3 =        5    Q1 =       -2    Q2 =       -5    Q3 =       -9

Modified exponential model  :
     Bm = Sum ( Ai X ( 1 - exp( - R/( 2*i*C ) ) ) )  +      Sum ( RKj X ( R - Rj0 ) )  +  Bm0
      C =   0.85258333E+00         Bmo=   0.00000000E+00          Nexp=  6         Nliner=  5
     AI =   0.21056726E+04  -0.88441289E+04   0.27582487E+05  -0.40415654E+05   0.22949281E+05  -0.21630372E+04
    Rj0 =           0.5000           1.5200          20.0000          22.4000          24.0100
    RKj =  -0.13368819E+02   0.13444719E+02  -0.25593484E+02   0.10115266E+03  -0.61765447E+02

Power model  :    Bm = ( Rki X R ) / ( 1 + ( R/R0 )**rn )**( 1/rn )
     rn = 0.834      rki =   0.72931872E+03          rmu =   0.15946059E+04
```

No	Rotation (radians) X 1/1000	Moment (kip-inch) Expri.	Moment (kip-inch) Poly.	Moment (kip-inch) M.Expo.	Moment (kip-inch) P.Model.	Connection stiffness (kip-inch) X 1000 Expri.	Connection stiffness (kip-inch) X 1000 Poly.	Connection stiffness (kip-inch) X 1000 M.Expo.	Connection stiffness (kip-inch) X 1000 P.Model.
1	0.00	0.00	0.00	0.00	0.00	0.5158E+03	0.7577E+03	0.5883E+03	0.7293E+03
3	0.50	258.23	298.03	244.71	268.19	0.3959E+03	0.4116E+03	0.4017E+03	0.4151E+03
5	1.02	408.67	458.39	409.57	446.97	0.2764E+03	0.2359E+03	0.2543E+03	0.2865E+03
7	2.01	579.58	629.32	586.21	665.50	0.9568E+02	0.1304E+03	0.1212E+03	0.1714E+03
9	4.02	698.18	818.41	708.32	906.50	0.3239E+02	0.7056E+02	0.2680E+02	0.8471E+02
11	7.96	782.01	1020.65	772.18	1122.34	0.1488E+02	0.3880E+02	0.1445E+02	0.3580E+02
13	12.08	835.30	1153.10	831.75	1231.69	0.1274E+02	0.2707E+02	0.1395E+02	0.1976E+02
15	15.99	860.32	1246.64	884.10	1294.13	0.1114E+02	0.2129E+02	0.1303E+02	0.1294E+02
17	20.00	953.58	1324.09	936.14	1337.65	0.2027E+01	0.1759E+02	0.1298E+02	0.9118E+01
19	24.01	1031.88	1389.28	1047.99	1369.23	0.4656E+02	0.1507E+02	0.8820E+02	0.6806E+01
21	28.01	1163.48	1445.69	1151.65	1393.24	0.3017E+02	0.1323E+02	0.2526E+02	0.5297E+01
23	32.03	1263.56	1495.90	1249.75	1412.30	0.1904E+02	0.1181E+02	0.2347E+02	0.4246E+01
25	36.01	1342.86	1540.62	1339.25	1427.63	0.2351E+02	0.1070E+02	0.2150E+02	0.3494E+01
27	39.99	1427.98	1581.37	1421.07	1440.36	0.2156E+02	0.9801E+01	0.1965E+02	0.2930E+01
29	44.07	1486.14	1619.74	1497.87	1451.37	0.1756E+02	0.9034E+01	0.1805E+02	0.2486E+01
31	48.04	1540.70	1654.33	1566.99	1460.54	0.1433E+02	0.8405E+01	0.1683E+02	0.2147E+01
33	52.00	1614.48	1686.52	1631.72	1468.49	0.1459E+02	0.7866E+01	0.1591E+02	0.1875E+01
35	56.01	1685.31	1717.09	1694.11	1475.55	0.1370E+02	0.7393E+01	0.1524E+02	0.1651E+01
37	60.02	1756.49	1746.30	1754.25	1481.78	0.1870E+02	0.6973E+01	0.1478E+02	0.1466E+01
39	64.01	1796.45	1773.17	1812.55	1487.32	0.1233E+02	0.6612E+01	0.1446E+02	0.1311E+01
41	67.97	1876.38	1798.53	1869.38	1492.25	0.2006E+02	0.6292E+01	0.1425E+02	0.1182E+01
43	72.00	1951.12	1823.58	1926.52	1496.78	0.1664E+02	0.5994E+01	0.1411E+02	0.1069E+01
45	76.03	2003.42	1846.91	1983.19	1500.89	0.1066E+02	0.5732E+01	0.1402E+02	0.9725E+00
47	80.02	2052.83	1869.53	2039.02	1504.60	0.1184E+02	0.5491E+01	0.1396E+02	0.8895E+00
49	84.00	2114.90	1890.72	2094.52	1507.99	0.1511E+02	0.5277E+01	0.1393E+02	0.8171E+00
51	88.04	2139.22	1911.83	2150.73	1511.16	0.4853E+01	0.5073E+01	0.1390E+02	0.7525E+00
53	92.04	2185.04	1931.93	2206.32	1514.05	0.1941E+02	0.4888E+01	0.1389E+02	0.6961E+00

A.4 – 1

IV - 1

Connection type : Top-and seat-angle connections
Mode : All riveted

Tested by : J.C.Rathbun (1936) U.S.A
Test Id. : B-8

Column : --
Beam : W12X31.8
Angle : 6 X 5 X 3/8 X 6 (TOP)

Fasteners: 10.9- -7/8"D
15/16" Oversize holes
Material : --
Fy = 47.12 ksi
Fu = 62.90 ksi

Major parameters

lt = 6.0000"	tt = 0.3750"	ls = 6.0000"	ts = 0.3750"
gt = 3.5000"	gt'= 3.5000"	gs = 3.5000"	gs'= 3.5000"
qt = 2.7500"	rt = 2.7500"	qs = 2.7500"	rs = 2.7500"
pt = 2.5000"	pt'= 2.5000"	ps = 2.5000"	
nt = 2 X 2	nt'= 2 X 2	ns = 2 X 2	ns'= 2 X 1

Remark 1) 6 X 4 X 3/8 X 6 (SEAT)
2) Connection cleat riveted to 1" mounting plate.

No	Moment (k-in)	Rotation (radians) X 1/1000
1	0.00	0.00
2	9.66	0.03
3	19.32	0.05
4	29.54	0.09
5	39.76	0.13
6	49.86	0.19
7	59.96	0.24
8	70.06	0.32
9	80.16	0.40
10	90.14	0.48
11	100.13	0.56
12	110.12	0.66
13	120.10	0.75
14	130.30	0.87
15	140.50	1.00
16	150.48	1.16
17	160.45	1.31
18	170.31	1.50
19	180.17	1.69
20	190.13	1.92
21	200.10	2.14
22	209.16	2.42
23	214.58	2.59
24	220.00	2.77
25	226.07	3.01

No	Moment (k-in)	Rotation (radians) X 1/1000
26	232.14	3.26
27	236.11	3.43
28	240.08	3.61
29	245.14	3.91
30	250.21	4.21
31	254.72	4.50
32	259.22	4.79
33	262.08	5.00
34	264.93	5.21
35	267.79	5.42
36	271.41	5.70
37	275.03	5.99

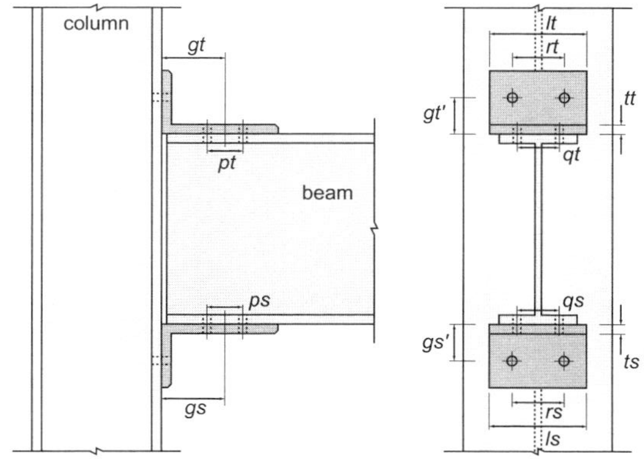

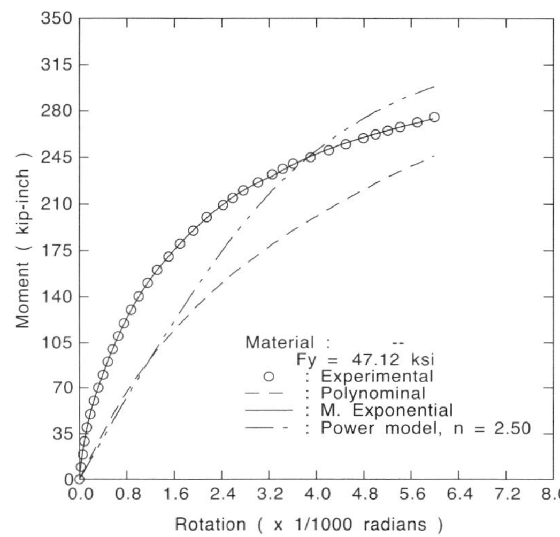

A.4 – 2

```
Moment-rotation prediction equations     ( R : X 1/1000 radians )

Frye and Morris polynominal model  :    R = Sum ( Ai X ( K*Bm )**Pi X 10**Qi )
     t =  0.375000"  xd = 12.000000"    f =  0.875000"  xl =  6.000000"
    A1 =  8.460000   A2 =  1.010000   A3 =  1.240000    K =  0.012981
    P1 =         1   P2 =         3   P3 =         5   Q1 =        -1     Q2 =        -1     Q3 =        -5

Modified exponential model  :
    Bm = Sum ( Ai X ( 1 - exp( - R/( 2*i*C ) ) ) )  +      Sum ( RKj X ( R - Rj0 ) )  +  Bm0
     C =  0.56531667E-01          Bmo=   0.00000000E+00          Nexp=  6        Nliner=  1
    AI =  0.54666872E+02   0.15328450E+03  -0.26722603E+04   0.94270326E+04  -0.12834518E+05   0.61192138E+04
   Rj0 =          3.2600
   RKj =  0.99491474E+01

Power model  :    Bm = ( Rki X R ) / ( 1 + ( R/R0 )**rn )**( 1/rn )
     rn = 2.504      rki =   0.78602809E+02         rmu =   0.34802630E+03
```

No	Rotation (radians) X 1/1000	Moment (kip-inch) Expri.	Moment (kip-inch) Poly.	Moment (kip-inch) M.Expo.	Moment (kip-inch) P.Model.	Connection stiffness (kip-inch) X 1000 Expri.	Connection stiffness (kip-inch) X 1000 Poly.	Connection stiffness (kip-inch) X 1000 M.Expo.	Connection stiffness (kip-inch) X 1000 P.Model.
1	0.00	0.00	0.00	0.00	0.00	0.3646E+03	0.9106E+02	0.4447E+03	0.7860E+02
3	0.05	19.32	4.83	19.70	4.17	0.3272E+03	0.9093E+02	0.3106E+03	0.7860E+02
5	0.13	39.76	11.57	38.66	10.01	0.2349E+03	0.9033E+02	0.2119E+03	0.7859E+02
7	0.24	59.96	22.05	59.30	19.21	0.1551E+03	0.8847E+02	0.1523E+03	0.7852E+02
9	0.40	80.16	35.29	80.40	31.19	0.1263E+03	0.8470E+02	0.1278E+03	0.7834E+02
11	0.56	100.13	48.99	100.31	44.23	0.1133E+03	0.7954E+02	0.1112E+03	0.7798E+02
13	0.75	120.10	63.38	119.53	58.85	0.9619E+02	0.7329E+02	0.9313E+02	0.7732E+02
15	1.00	140.50	80.37	139.82	77.65	0.7444E+02	0.6552E+02	0.7356E+02	0.7604E+02
17	1.31	160.45	99.68	160.34	101.33	0.5800E+02	0.5692E+02	0.5785E+02	0.7364E+02
19	1.69	180.17	119.70	180.21	128.60	0.4836E+02	0.4882E+02	0.4767E+02	0.6966E+02
21	2.14	200.10	140.03	199.69	158.72	0.3896E+02	0.4169E+02	0.3873E+02	0.6365E+02
23	2.59	214.58	157.56	215.11	185.84	0.3158E+02	0.3643E+02	0.2974E+02	0.5673E+02
25	3.01	226.07	171.97	225.84	208.15	0.2465E+02	0.3268E+02	0.2180E+02	0.5002E+02
27	3.43	236.11	185.07	235.28	227.80	0.2271E+02	0.2967E+02	0.2502E+02	0.4338E+02
29	3.91	245.14	198.47	245.73	246.70	0.1691E+02	0.2694E+02	0.1937E+02	0.3646E+02
31	4.50	254.72	213.58	255.74	265.99	0.1543E+02	0.2424E+02	0.1487E+02	0.2896E+02
33	5.00	262.08	225.33	262.59	279.14	0.1366E+02	0.2238E+02	0.1268E+02	0.2369E+02
35	5.42	267.79	234.39	267.64	288.25	0.1326E+02	0.2107E+02	0.1158E+02	0.2001E+02
37	5.99	275.03	246.00	273.98	298.44	0.1271E+02	0.1954E+02	0.1074E+02	0.1592E+02

A.4 – 3

IV - 2

Connection type : Top-and seat-angle connections
Mode : All riveted

Tested by : J.C.Rathbun (1936) U.S.A
Test Id. : B-9

Column : --
Beam : W12X31.8
Angle : 6 X 6 X 3/8 X 8 (TOP)

Fasteners: 10.9- -7/8"D
15/16" Oversize holes
Material : --
Fy = 47.12 ksi
Fu = 62.90 ksi

Major parameters

lt = 8.0000"	tt = 0.3750"	ls = 8.0000"	ts = 0.3750"
gt = 3.5000"	gt'= 3.5000"	gs = 3.5000"	gs'= 3.5000"
qt = 2.7500"	rt = 2.5000"	qs = 2.7500"	rs = 2.5000"
pt = 2.5000"	pt'= 2.5000"	ps = 2.5000"	
nt = 2 X 2	nt'= 3 X 2	ns = 2 X 2	ns'= 3 X 1

Remark 1) 6 X 4 X 3/8 X 8 (SEAT)
2) Connection cleat riveted to 1" mounting plate.

No	Moment (k-in)	Rotation (radians) X 1/1000
1	0.00	0.00
2	10.53	0.04
3	21.07	0.09
4	31.95	0.16
5	42.82	0.22
6	53.22	0.31
7	63.63	0.39
8	74.59	0.51
9	85.56	0.64
10	96.29	0.78
11	107.01	0.93
12	117.83	1.11
13	128.66	1.29
14	139.57	1.53
15	150.49	1.77
16	157.50	2.01
17	164.50	2.25
18	171.51	2.49
19	176.97	2.74
20	182.43	2.98
21	187.89	3.23
22	193.35	3.47
23	197.61	3.73
24	201.88	3.98
25	206.14	4.24
26	210.14	4.47
27	214.14	4.70
28	217.36	4.96
29	220.57	5.22
30	223.78	5.48
31	227.00	5.73
32	230.21	5.99

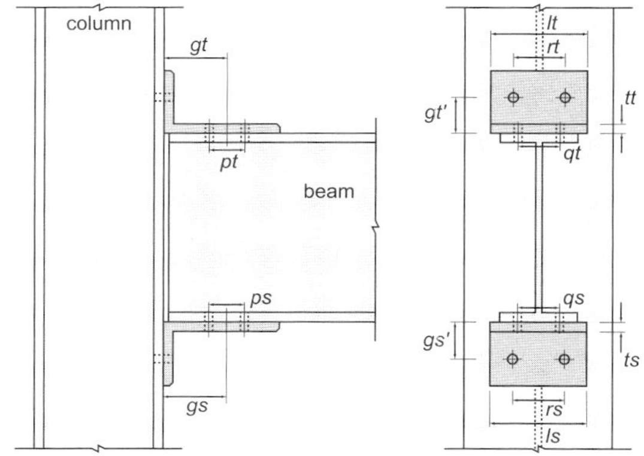

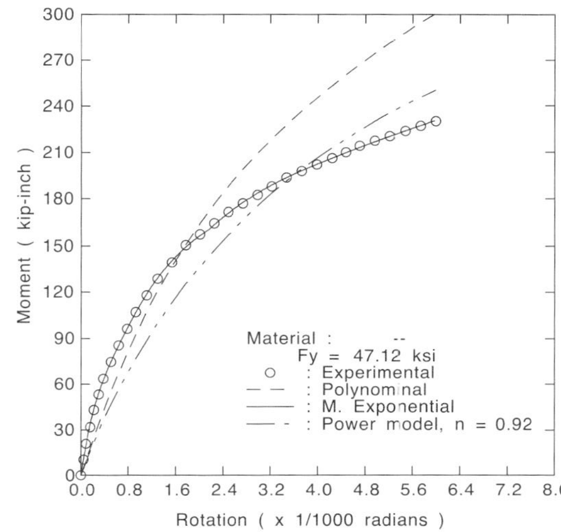

A.4 – 4

```
Moment-rotation prediction equations     ( R : X 1/1000 radians )

Frye and Morris polynominal model  :    R = Sum ( Ai X ( K*Bm )**Pi X 10**Qi )
      t =  0.375000"  xd = 12.000000"   f =  0.875000"  xl =  8.000000"
     A1 =  8.460000   A2 =  1.010000   A3 =  1.240000    K =  0.010613
     P1 =        1    P2 =        3    P3 =        5    Q1 =       -1     Q2 =       -1     Q3 =       -5

Modified exponential model  :
     Bm = Sum ( Ai X ( 1 - exp( - R/( 2*i*C ) ) ) )  +      Sum ( RKj X ( R - Rj0 ) )  +  Bm0
      C =   0.61155833E-01              Bmo=   0.00000000E+00             Nexp=  6        Nliner=  2
     AI =  -0.47672488E+02   0.69077972E+03  -0.28365484E+04   0.58675891E+04  -0.62594978E+04   0.27721739E+04
    Rj0 =           2.1600           4.1500
    RKj =   0.10860499E+02   0.14304094E+01

Power model  :    Bm = ( Rki X R ) / ( 1 + ( R/R0 )**rn )**( 1/rn )
     rn = 0.915      rki =   0.10480375E+03          rmu =   0.46403506E+03
```

No	Rotation (radians) X 1/1000	Moment (kip-inch) Expri.	Poly.	M.Expo.	P.Model.	Connection stiffness (kip-inch) X 1000 Expri.	Poly.	M.Expo.	P.Model.
1	0.00	0.00	0.00	0.00	0.00	0.2346E+03	0.1114E+03	0.2390E+03	0.1048E+03
2	0.04	10.53	5.00	10.43	4.63	0.2349E+03	0.1113E+03	0.2242E+03	0.1016E+03
3	0.09	21.07	9.99	20.03	9.12	0.2060E+03	0.1109E+03	0.2039E+03	0.9888E+02
4	0.16	31.95	17.38	32.61	15.62	0.1623E+03	0.1100E+03	0.1720E+03	0.9521E+02
5	0.22	42.82	24.71	43.18	21.89	0.1467E+03	0.1087E+03	0.1445E+03	0.9186E+02
6	0.31	53.22	33.48	53.86	29.21	0.1278E+03	0.1066E+03	0.1193E+03	0.8812E+02
7	0.39	63.63	42.05	62.83	36.24	0.1119E+03	0.1040E+03	0.1025E+03	0.8468E+02
8	0.51	74.59	54.81	74.66	46.56	0.8736E+02	0.9934E+02	0.8778E+02	0.7985E+02
9	0.64	85.56	66.97	85.17	56.31	0.8140E+02	0.9431E+02	0.8024E+02	0.7551E+02
10	0.78	96.29	80.11	96.27	66.83	0.7461E+02	0.8847E+02	0.7451E+02	0.7103E+02
11	0.93	107.01	92.42	106.60	76.75	0.6767E+02	0.8283E+02	0.6899E+02	0.6699E+02
12	1.11	117.83	107.04	118.58	88.64	0.5881E+02	0.7616E+02	0.6112E+02	0.6236E+02
13	1.29	128.66	120.50	129.07	99.74	0.5309E+02	0.7023E+02	0.5279E+02	0.5825E+02
14	1.53	139.57	136.46	140.44	113.08	0.4569E+02	0.6359E+02	0.4260E+02	0.5354E+02
15	1.77	150.49	150.97	149.55	125.37	0.3747E+02	0.5801E+02	0.3400E+02	0.4942E+02
16	2.01	157.50	164.29	156.84	136.77	0.2924E+02	0.5330E+02	0.2716E+02	0.4579E+02
17	2.25	164.50	176.58	163.67	147.35	0.2923E+02	0.4931E+02	0.3271E+02	0.4257E+02
18	2.49	171.51	187.98	171.00	157.20	0.2577E+02	0.4590E+02	0.2856E+02	0.3969E+02
19	2.74	176.97	198.87	177.59	166.62	0.2222E+02	0.4289E+02	0.2521E+02	0.3706E+02
20	2.98	182.43	209.08	183.44	175.43	0.2222E+02	0.4028E+02	0.2250E+02	0.3469E+02
21	3.23	187.89	218.69	188.68	183.69	0.2222E+02	0.3799E+02	0.2029E+02	0.3255E+02
22	3.47	193.35	227.77	193.43	191.44	0.1951E+02	0.3598E+02	0.1846E+02	0.3061E+02
23	3.73	197.61	236.73	197.95	199.03	0.1668E+02	0.3413E+02	0.1689E+02	0.2879E+02
24	3.98	201.88	245.23	202.09	206.17	0.1669E+02	0.3248E+02	0.1561E+02	0.2712E+02
25	4.24	206.14	253.34	206.07	212.90	0.1696E+02	0.3100E+02	0.1600E+02	0.2561E+02
26	4.47	210.14	260.40	209.70	218.70	0.1722E+02	0.2978E+02	0.1524E+02	0.2434E+02
27	4.70	214.14	267.31	213.16	224.22	0.1497E+02	0.2865E+02	0.1461E+02	0.2317E+02
28	4.96	217.36	274.42	216.85	230.03	0.1248E+02	0.2755E+02	0.1406E+02	0.2197E+02
29	5.22	220.57	281.51	220.41	235.55	0.1248E+02	0.2650E+02	0.1363E+02	0.2087E+02
30	5.48	223.78	288.08	223.88	240.79	0.1248E+02	0.2558E+02	0.1330E+02	0.1985E+02
31	5.73	227.00	294.67	227.27	245.78	0.1248E+02	0.2470E+02	0.1305E+02	0.1891E+02
32	5.99	230.21	300.82	230.60	250.53	0.1248E+02	0.2391E+02	0.1285E+02	0.1803E+02

Connection type : Top-and seat-angle connections
Mode : All riveted

Tested by : J.C.Rathbun (1936) U.S.A
Test Id. : B-10

Column : --
Beam : W12X31.8
Angle : 6 X 6 X 3/8 X 14 (TOP)

Fasteners: 10.9- -7/8"D
15/16" Oversize holes
Material : --
Fy = 47.12 ksi
Fu = 62.90 ksi

Major parameters

lt = 14.0000"	tt = 0.3750"	ls = 14.0000"	ts = 0.3750"
gt = 3.5000"	gt'= 3.5000"	gs = 3.5000"	gs'= 3.5000"
qt = 2.7500"	rt = 5.5000"	rt2= 2.5000"	
qs = 2.7500"	rs = 5.5000"	rs2= 2.5000"	
pt = 2.5000"	pt'= 2.5000"	ps = 2.5000"	
nt = 2 X 2	nt'= 4 X 2	ns = 2 X 2	ns'= 4 X 1

Remark 1) 6 X 4 X 3/8 X 14 (SEAT)
2) Connection cleat riveted to 1" mounting plate.

No	Moment (k-in)	Rotation (radians) X 1/1000	No	Moment (k-in)	Rotation (radians) X 1/1000
1	C.00	0.00	26	466.82	7.51
2	2C.07	0.03	27	479.81	8.12
3	4C.14	0.05	28	490.47	8.71
4	59.96	0.12	29	501.14	9.31
5	79.79	0.20	30	511.80	9.91
6	99.82	0.31	31	520.54	10.57
7	119.86	0.42	32	529.29	11.24
8	140.53	0.56	33	538.03	11.90
9	161.20	0.70	34	544.77	12.50
10	180.56	0.87	35	551.51	13.09
11	199.91	1.04	36	558.24	13.68
12	220.12	1.25			
13	240.33	1.45			
14	260.29	1.73			
15	280.26	2.01			
16	300.40	2.38			
17	320.54	2.75			
18	341.09	3.20			
19	361.63	3.66			
20	380.38	4.22			
21	399.13	4.79			
22	413.04	5.29			
23	426.94	5.79			
24	440.85	6.29			
25	453.84	6.90			

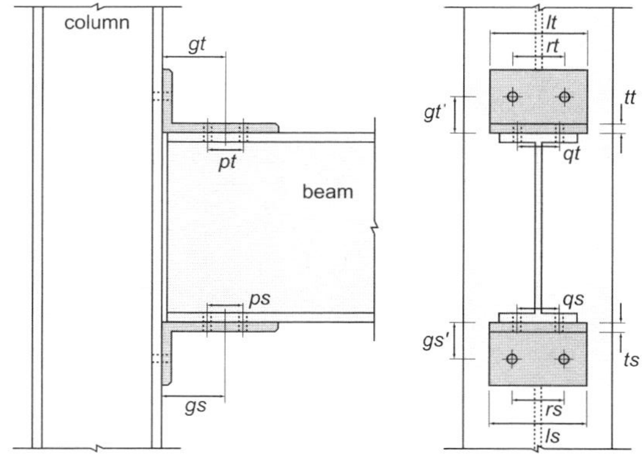

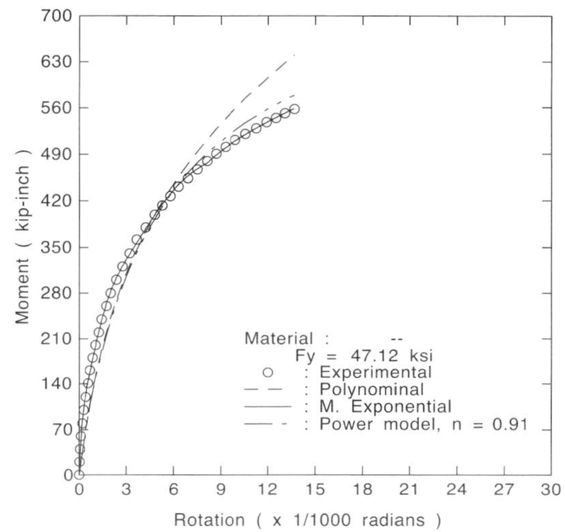

A.4 – 6

```
Moment-rotation prediction equations     ( R : X 1/1000 radians )

Frye and Morris polynominal model  :    R = Sum ( Ai X ( K*Bm )**Pi X 10**Qi )
     t =  0.375000"  xd = 12.000000"   f =  0.875000"  xl = 14.000000"
    A1 =  8.460000   A2 =  1.010000   A3 =  1.240000    K =  0.007173
    P1 =        1    P2 =        3    P3 =        5    Q1 =       -1     Q2 =       -1     Q3 =       -5

Modified exponential model  :
    Bm = Sum ( Ai X ( 1 - exp( - R/( 2*i*C ) ) ) )  +      Sum ( RKj X ( R - Rj0 ) )  +  Bm0
      C =   0.12025000E+00              Bmo=   0.00000000E+00          Nexp=  6        Nliner=  1
     AI =   0.28068559E+03  -0.69724209E+02  -0.72481404E+04   0.27265045E+05  -0.36266719E+05   0.16538218E+05
    Rj0 =           8.1200
    RKj =   0.10846612E+02

Power model  :    Bm = ( Rki X R ) / ( 1 + ( R/R0 )**rn )**( 1/rn )
     rn = 0.907      rki =   0.18340655E+03          rmu =   0.81206136E+03
```

No	Rotation (radians) X 1/1000	Moment (kip-inch) Expri.	Poly.	M.Expo.	P.Model.	Connection stiffness (kip-inch) X 1000 Expri.	Poly.	M.Expo.	P.Model.
1	0.00	0.00	0.00	0.00	0.00	0.8027E+03	0.1648E+03	0.6198E+03	0.1834E+03
3	0.05	40.14	8.24	27.61	9.00	0.6677E+03	0.1646E+03	0.4902E+03	0.1770E+03
5	0.20	79.79	32.45	80.99	34.10	0.2327E+03	0.1616E+03	0.2617E+03	0.1624E+03
7	0.42	119.86	67.35	123.01	68.14	0.1653E+03	0.1521E+03	0.1441E+03	0.1451E+03
9	0.70	161.20	108.21	158.88	106.73	0.1323E+03	0.1355E+03	0.1179E+03	0.1276E+03
11	1.04	199.91	149.98	196.95	146.29	0.1077E+03	0.1165E+03	0.1109E+03	0.1114E+03
13	1.45	240.33	194.48	239.73	189.45	0.8604E+02	0.9708E+02	0.9187E+02	0.9548E+02
15	2.01	280.26	242.86	282.45	237.67	0.6461E+02	0.7893E+02	0.6369E+02	0.7954E+02
17	2.75	320.54	294.88	320.71	290.35	0.5037E+02	0.6328E+02	0.4353E+02	0.6416E+02
19	3.66	361.63	346.56	356.94	342.32	0.3982E+02	0.5124E+02	0.3790E+02	0.5088E+02
21	4.79	399.13	398.75	398.30	393.02	0.3034E+02	0.4188E+02	0.3476E+02	0.3962E+02
23	5.79	426.94	437.68	430.15	428.94	0.2782E+02	0.3632E+02	0.2849E+02	0.3259E+02
25	6.90	453.84	475.38	457.06	461.77	0.2128E+02	0.3184E+02	0.1999E+02	0.2684E+02
27	8.12	479.81	511.91	476.36	491.53	0.1956E+02	0.2820E+02	0.1204E+02	0.2216E+02
29	9.31	501.14	543.89	500.23	515.83	0.1786E+02	0.2547E+02	0.1759E+02	0.1870E+02
31	10.57	520.54	574.57	520.15	537.59	0.1313E+02	0.2319E+02	0.1428E+02	0.1587E+02
33	11.90	538.03	604.13	537.80	557.13	0.1221E+02	0.2126E+02	0.1245E+02	0.1355E+02
35	13.09	551.51	628.42	552.00	572.15	0.1139E+02	0.1983E+02	0.1163E+02	0.1190E+02

IV - 4

Connection type : Top-and seat-angle connections
Mode : All riveted

Tested by : R.A.Hechtman & B.G.Johnston (1947) U.S.A
Test Id. : NO 2

Column : W10X49
Beam : W12X25
Angle : 6 X 4 X 5/8 X 6 3/4 (TOP)

Fasteners: 10.9- -3/4"D
13/16" Oversize holes
Material : --
Fy = 37.50 ksi
Fu = 62.50 ksi

Major parameters

lt = 6.7500"	tt = 0.6250"	ls = 6.7500"	ts = 0.5000"
gt = 3.5000"	gt'= 2.5000"	gs = 3.5000"	gs'= 3.5000"
qt = 4.2500"	rt = 4.2500"	qs = 4.2500"	rs = 4.2500"
pt = 2.5000"	ps = 2.5000"	ps'= 2.5000"	
nt = 2 X 2	nt'= 2 X 1	ns = 2 X 2	ns'= 2 X 2

Remark 1) 6 X 6 X 1/2 X 6 3/4 (SEAT)
2) 2 shims 5 X 1/8 X 7 are used on ten. flange.

No	Moment (k-in)	Rotation (radians) X 1/1000	No	Moment (k-in)	Rotation (radians) X 1/1000
1	0.00	0.00	26	497.25	14.09
2	24.87	0.29	27	505.02	14.86
3	49.73	0.59	28	514.03	15.73
4	73.79	0.83	29	523.03	16.60
5	97.85	1.08	30	531.63	17.50
6	122.72	1.38	31	540.24	18.41
7	147.59	1.69	32	548.09	19.47
8	173.67	2.05	33	555.93	20.53
9	199.75	2.40	34	564.19	21.65
10	226.66	2.85	35	572.45	22.78
11	253.56	3.30			
12	276.46	3.74			
13	299.36	4.18			
14	323.11	4.80			
15	346.85	5.43			
16	366.61	6.10			
17	386.36	6.78			
18	398.51	7.43			
19	410.66	8.07			
20	424.45	8.88			
21	438.23	9.68			
22	451.63	10.57			
23	465.04	11.45			
24	477.25	12.38			
25	489.47	13.31			

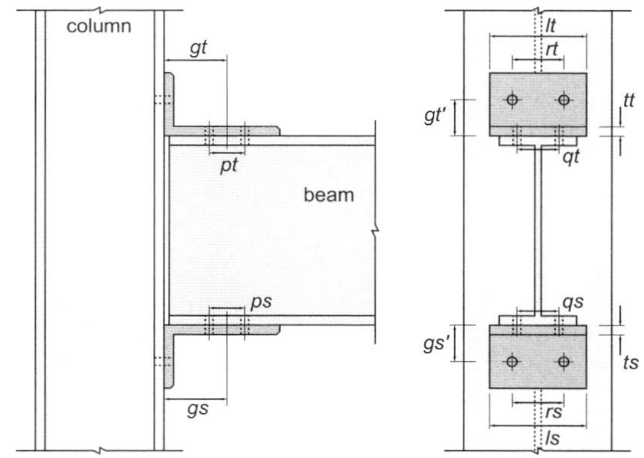

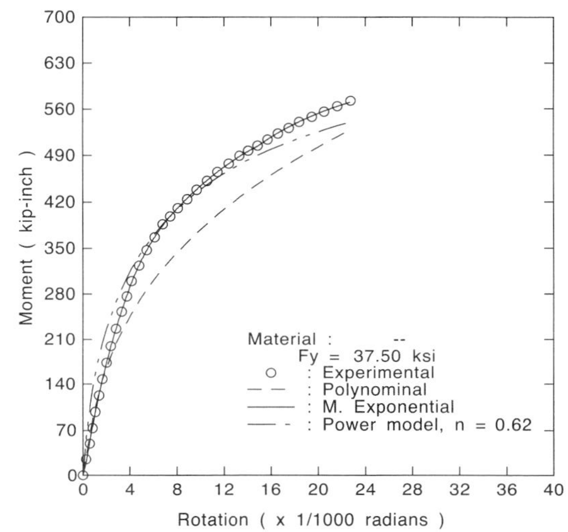

A.4 – 8

```
Moment-rotation prediction equations      ( R : X 1/1000 radians )

Frye and Morris polynominal model  :    R = Sum ( Ai X ( K*Bm )**Pi X 10**Qi )
      t =  0.625000"  xd = 12.250000"   f =  0.750000"  xl =  6.750000"
     A1 =  8.460000   A2 =  1.010000   A3 =  1.240000    K =  0.010636
     P1 =         1   P2 =         3   P3 =         5   Q1 =        -1     Q2 =        -1     Q3 =        -5

Modified exponential model  :
     Bm = Sum ( Ai X ( 1 - exp( - R/( 2*i*C ) ) ) )  +      Sum ( RKj X ( R - Rj0 ) )  +  Bm0
      C =   0.26294500E+00             Bmo=   0.00000000E+00            Nexp=  6        Nliner=  1
     AI =  -0.19523742E+03   0.29572405E+04  -0.16926191E+05   0.41722007E+05  -0.46089111E+05   0.19079206E+05
    Rj0 =          14.8600
    RKj =   0.37150218E+01

Power model  :    Bm = ( Rki X R ) / ( 1 + ( R/R0 )**rn )**( 1/rn )
      rn = 0.620      rki =   0.30143663E+03         rmu =   0.78474866E+03
```

No	Rotation (radians) X 1/1000	Moment (kip-inch) Expri.	Moment Poly.	Moment M.Expo.	Moment P.Model.	Connection stiffness (kip-inch) X 1000 Expri.	Stiffness Poly.	Stiffness M.Expo.	Stiffness P.Model.
1	0.00	0.00	0.00	0.00	0.00	0.8499E+02	0.1111E+03	0.6440E+02	0.3014E+03
3	0.59	49.73	61.84	51.14	102.97	0.9214E+02	0.9623E+02	0.9577E+02	0.1260E+03
5	1.08	97.85	104.24	96.67	155.39	0.9035E+02	0.7716E+02	0.8852E+02	0.9155E+02
7	1.69	147.59	146.03	147.55	203.98	0.7726E+02	0.5961E+02	0.7691E+02	0.6826E+02
9	2.40	199.75	183.56	198.53	246.48	0.6734E+02	0.4697E+02	0.6690E+02	0.5254E+02
11	3.30	253.56	221.14	253.86	287.89	0.5605E+02	0.3725E+02	0.5595E+02	0.4034E+02
13	4.18	299.36	250.85	298.08	319.52	0.4652E+02	0.3128E+02	0.4553E+02	0.3268E+02
15	5.43	346.85	286.26	346.66	355.56	0.3375E+02	0.2569E+02	0.3265E+02	0.2541E+02
17	6.78	386.36	318.12	384.05	386.17	0.2391E+02	0.2175E+02	0.2357E+02	0.2027E+02
19	8.07	410.66	344.43	411.17	410.03	0.1806E+02	0.1910E+02	0.1889E+02	0.1683E+02
21	9.68	438.23	373.12	438.91	434.51	0.1620E+02	0.1669E+02	0.1589E+02	0.1377E+02
23	11.45	465.04	400.83	464.95	456.63	0.1416E+02	0.1475E+02	0.1361E+02	0.1138E+02
25	13.31	489.47	426.83	488.05	476.03	0.1143E+02	0.1321E+02	0.1113E+02	0.9531E+01
27	14.86	505.02	446.49	503.66	489.85	0.1018E+02	0.1219E+02	0.1271E+02	0.8352E+01
29	16.60	523.03	466.85	523.76	503.43	0.9955E+01	0.1125E+02	0.1046E+02	0.7297E+01
31	18.41	540.24	486.57	540.78	515.77	0.8559E+01	0.1044E+02	0.8489E+01	0.6422E+01
33	20.53	555.93	507.70	556.83	528.48	0.7373E+01	0.9659E+01	0.6744E+01	0.5598E+01
35	22.78	572.45	528.80	570.49	540.26	0.7349E+01	0.8961E+01	0.5503E+01	0.4903E+01

A.4 – 9

IV - 5

Connection type : Top-and seat-angle connections
Mode : All riveted

Tested by : R.A.Hechtman & B.G.Johnston (1947) U.S.A
Test Id. : NO 5

Column : W12X65
Beam : W18X47
Angle : 6 X 4 X 1/2 X 12 (TOP)

Fasteners: 10.9- -3/4"D
13/16" Oversize holes
Material : --
Fy = 37.50 ksi
Fu = 62.50 ksi

Major parameters

lt = 12.0000"	tt = 0.5000"	ls = 7.5000"	ts = 0.8750"
gt = 3.5000"	gt'= 2.5000"	gs = 3.5000"	gs'= 3.5000"
qt = 4.5000"	rt = 4.5000"	rt2= 2.2500"	
qs = 4.5000"	rs = 4.5000"		
pt = 2.5000"	ps = 2.5000"	ps'= 2.5000"	
nt = 2 X 2	nt'= 4 X 1	ns = 2 X 2	ns'= 2 X 2

Remark 1) 6 X 6 X 7/8 X 7 1/2 (SEAT)
2) 1 shims 5 X 1/8 X 7 is used on ten. flange.

No	Moment (k-in)	Rotation (radians) X 1/1000	No	Moment (k-in)	Rotation (radians) X 1/1000
1	0.00	0.00	26	904.31	11.47
2	49.45	0.26	27	930.84	12.53
3	98.90	0.51	28	957.37	13.59
4	136.44	0.72	29	979.13	14.65
5	173.98	0.93	30	1000.90	15.70
6	218.08	1.21	31	1017.04	16.52
7	262.19	1.49	32	1033.18	17.34
8	303.33	1.80	33	1049.31	18.15
9	344.47	2.10	34	1063.88	19.01
10	386.21	2.43	35	1078.44	19.87
11	427.95	2.76	36	1093.00	20.73
12	465.54	3.12	37	1106.45	21.80
13	503.14	3.49	38	1119.90	22.87
14	540.76	3.93	39	1133.35	23.94
15	578.38	4.38			
16	607.05	4.76			
17	635.73	5.14			
18	666.21	5.60			
19	696.70	6.06			
20	728.43	6.66			
21	760.15	7.27			
22	805.68	8.24			
23	830.91	8.99			
24	856.15	9.74			
25	880.23	10.60			

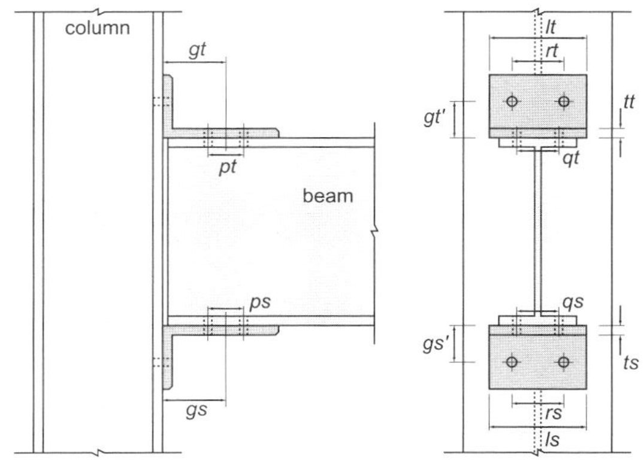

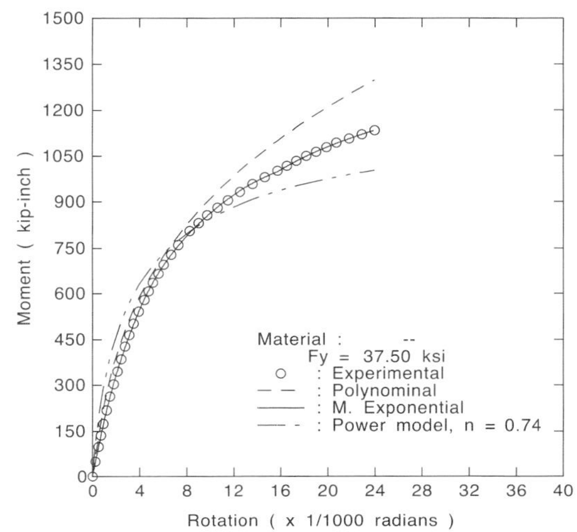

A.4 – 10

```
Moment-rotation prediction equations      ( R : X 1/1000 radians )

Frye and Morris polynominal model  :    R = Sum ( Ai X ( K*Bm )**Pi X 10**Qi )
     t =  0.500000"  xd = 18.125000"   f =  0.750000"  xl = 12.000000"
    A1 =  8.460000   A2 =  1.010000   A3 =  1.240000    K =  0.004417
    P1 =         1   P2 =         3   P3 =         5   Q1 =        -1    Q2 =        -1    Q3 =        -5

Modified exponential model  :
    Bm = Sum ( Ai X ( 1 - exp( - R/( 2*i*C ) ) ) )  +     Sum ( RKj X ( R - Rj0 ) )  +  Bm0
     C =   0.26342917E+00             Bmo=   0.00000000E+00         Nexp=  6        Nliner=  1
    AI =  -0.26617450E+03   0.43287546E+04  -0.25061086E+05   0.63864102E+05  -0.73495801E+05   0.31693988E+05
   Rj0 =          15.7000
   RKj =   0.95419433E+01

Power model  :    Bm = ( Rki X R ) / ( 1 + ( R/R0 )**rn )**( 1/rn )
     rn = 0.744      rki =   0.55318953E+03          rmu =   0.12401769E+04
```

No	Rotation (radians) X 1/1000	Moment (kip-inch) Expri.	Poly.	M.Expo.	P.Model.	Connection stiffness (kip-inch) X 1000 Expri.	Poly.	M.Expo.	P.Model.
1	0.00	0.00	0.00	0.00	0.00	0.1934E+03	0.2676E+03	0.1779E+03	0.5532E+03
3	0.51	98.90	131.56	97.77	192.18	0.1855E+03	0.2387E+03	0.1908E+03	0.2819E+03
5	0.93	173.98	223.15	173.86	293.15	0.1698E+03	0.1985E+03	0.1712E+03	0.2073E+03
7	1.49	262.19	321.59	262.42	392.16	0.1467E+03	0.1553E+03	0.1460E+03	0.1513E+03
9	2.10	344.47	406.06	345.19	472.44	0.1313E+03	0.1243E+03	0.1271E+03	0.1152E+03
11	2.76	427.95	480.03	423.67	539.35	0.1157E+03	0.1025E+03	0.1125E+03	0.9032E+02
13	3.49	503.14	548.45	500.41	598.04	0.9451E+02	0.8621E+02	0.9814E+02	0.7182E+02
15	4.38	578.38	619.26	580.87	655.09	0.7958E+02	0.7266E+02	0.8164E+02	0.5649E+02
17	5.14	635.73	670.92	637.79	694.15	0.7150E+02	0.6448E+02	0.6934E+02	0.4736E+02
19	6.06	696.70	726.75	695.89	733.83	0.6023E+02	0.5697E+02	0.5725E+02	0.3913E+02
21	7.27	760.15	790.95	757.88	776.16	0.5035E+02	0.4973E+02	0.4619E+02	0.3143E+02
23	8.99	830.91	869.84	828.45	823.42	0.3373E+02	0.4246E+02	0.3671E+02	0.2405E+02
25	10.60	880.23	934.31	882.68	858.28	0.2776E+02	0.3759E+02	0.3064E+02	0.1938E+02
27	12.53	930.84	1002.31	935.44	891.63	0.2507E+02	0.3326E+02	0.2422E+02	0.1549E+02
29	14.65	979.13	1068.74	979.64	921.04	0.2056E+02	0.2968E+02	0.1766E+02	0.1248E+02
31	16.52	1017.04	1122.42	1015.74	942.54	0.1975E+02	0.2718E+02	0.2218E+02	0.1053E+02
33	18.15	1049.31	1165.29	1048.99	958.62	0.1841E+02	0.2539E+02	0.1866E+02	0.9203E+01
35	19.87	1078.44	1207.40	1078.41	973.39	0.1700E+02	0.2380E+02	0.1582E+02	0.8076E+01
37	21.80	1106.45	1251.92	1106.58	987.95	0.1255E+02	0.2227E+02	0.1354E+02	0.7051E+01
39	23.94	1133.35	1298.02	1133.69	1002.04	0.1255E+02	0.2083E+02	0.1189E+02	0.6137E+01

A.4 – 11

IV - 6

Connection type : Top-and seat-angle connections
Mode : All riveted

Tested by : R.A.Hechtman & B.G.Johnston (1947) U.S.A
Test Id. : NO 9

Column : W12X65
Beam : W18X47
Angle : 6 X 4 X 5/8 X 12 (TOP)

Fasteners: 10.9- -3/4"D
13/16" Oversize holes
Material : --
Fy = 37.50 ksi
Fu = 62.50 ksi

Major parameters

lt = 12.0000"	tt = 0.6250"	ls = 7.5000"	ts = 0.8750"
gt = 3.5000"	gt'= 2.5000"	gs = 3.5000"	gs'= 3.5000"
qt = 4.5000"	rt = 4.5000"	rt2= 2.2500"	
qs = 4.5000"	rs = 4.5000"		
pt = 2.5000"	ps = 2.5000"	ps'= 2.5000"	
nt = 2 X 2	nt'= 4 X 1	ns = 2 X 2	ns'= 2 X 2

Remark 1) 6 X 6 X 7/8 X 7 1/2 (SEAT)
2) 1 shims 5 X 1/8 X 7 is used on ten. flange.

No	Moment (k-in)	Rotation (radians) X 1/1000	No	Moment (k-in)	Rotation (radians) X 1/1000
1	0.00	0.00	26	1083.05	12.49
2	45.21	0.12	27	1113.55	13.54
3	90.42	0.23	28	1141.52	14.69
4	141.90	0.41	29	1169.48	15.83
5	193.38	0.59	30	1191.21	16.79
6	242.33	0.82	31	1212.95	17.76
7	291.28	1.05	32	1230.29	18.68
8	342.09	1.34	33	1247.63	19.60
9	392.91	1.64	34	1265.57	20.64
10	444.96	2.04	35	1283.52	21.67
11	497.00	2.44	36	1293.39	22.32
12	545.90	2.87	37	1303.27	22.97
13	594.79	3.30			
14	641.79	3.78			
15	688.79	4.26			
16	734.51	4.82			
17	780.22	5.39			
18	821.51	6.06			
19	862.80	6.74			
20	902.78	7.61			
21	942.75	8.49			
22	973.34	9.25			
23	1003.92	10.02			
24	1028.23	10.73			
25	1052.55	11.44			

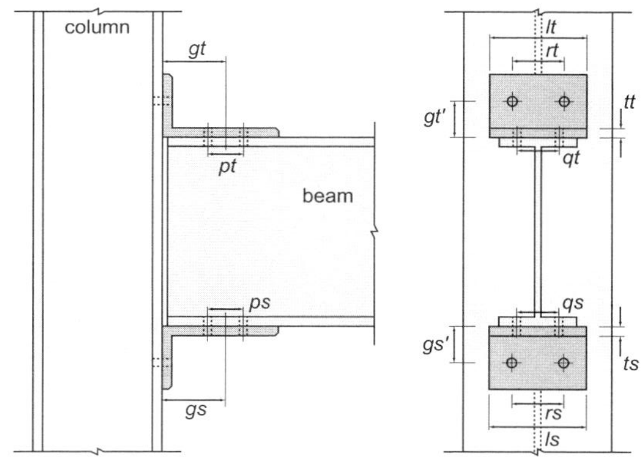

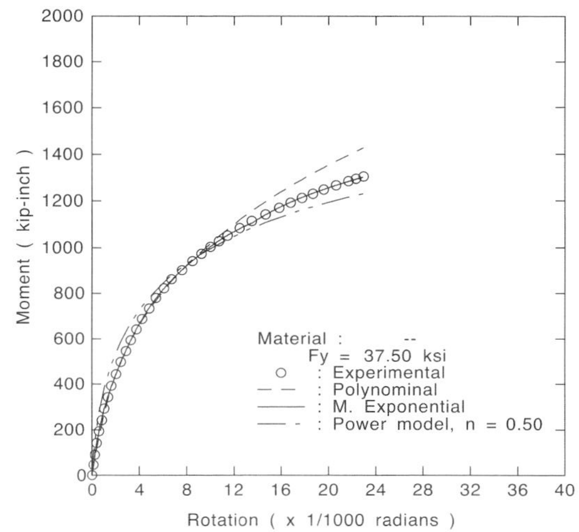

A.4 – 12

```
Moment-rotation prediction equations     ( R : X 1/1000 radians )

Frye and Morris polynominal model  :    R = Sum ( Ai X ( K*Bm )**Pi X 10**Qi )
      t =  0.625000"  xd = 18.125000"   f =  0.750000"  xl = 12.000000"
     A1 =  8.460000   A2 =  1.010000   A3 =  1.240000    K =  0.003950
     P1 =         1   P2 =         3   P3 =         5   Q1 =        -1    Q2 =        -1    Q3 =        -5

Modified exponential model  :
     Bm = Sum ( Ai X ( 1 - exp( - R/( 2*i*C ) ) ) )  +     Sum ( RKj X ( R - Rj0 ) )  +  Bm0
      C =   0.22062917E+00          Bmo=   0.00000000E+00          Nexp=  6        Nliner=  1
     AI =  -0.31778615E+03   0.64977494E+04  -0.35740764E+05   0.87923941E+05  -0.98860992E+05   0.41679716E+05
    Rj0 =          13.5400
    RKj =   0.13166499E+02

Power model  :    Bm = ( Rki X R ) / ( 1 + ( R/R0 )**rn )**( 1/rn )
      rn = 0.497      rki =   0.11629999E+04         rmu =   0.20140938E+04
```

No	Rotation (radians) X 1/1000	Moment (kip-inch) Expri.	Moment (kip-inch) Poly.	Moment (kip-inch) M.Expo.	Moment (kip-inch) P.Model.	Connection stiffness (kip-inch) X 1000 Expri.	Connection stiffness (kip-inch) X 1000 Poly.	Connection stiffness (kip-inch) X 1000 M.Expo.	Connection stiffness (kip-inch) X 1000 P.Model.
1	0.00	0.00	0.00	0.00	0.00	0.3867E+03	0.2992E+03	0.3918E+03	0.1163E+04
3	0.23	90.42	69.31	85.86	144.23	0.3474E+03	0.2914E+03	0.3403E+03	0.4505E+03
5	0.59	193.38	168.57	192.77	272.08	0.2544E+03	0.2582E+03	0.2571E+03	0.2889E+03
7	1.05	291.28	275.68	292.74	382.79	0.1962E+03	0.2100E+03	0.1861E+03	0.2044E+03
9	1.64	392.91	383.89	387.53	484.61	0.1552E+03	0.1640E+03	0.1455E+03	0.1503E+03
11	2.44	497.00	499.17	495.71	588.10	0.1217E+03	0.1250E+03	0.1251E+03	0.1101E+03
13	3.30	594.79	594.89	595.83	671.08	0.1067E+03	0.1004E+03	0.1082E+03	0.8554E+02
15	4.26	688.79	681.98	689.48	743.98	0.9033E+02	0.8303E+02	0.8779E+02	0.6821E+02
17	5.39	780.22	767.68	776.88	812.93	0.7186E+02	0.6958E+02	0.6815E+02	0.5475E+02
19	6.74	862.80	854.08	858.60	879.28	0.5433E+02	0.5882E+02	0.5410E+02	0.4402E+02
21	8.49	942.75	948.40	943.95	947.78	0.4269E+02	0.4952E+02	0.4412E+02	0.3486E+02
23	10.02	1003.92	1019.37	1005.71	996.52	0.3702E+02	0.4383E+02	0.3685E+02	0.2934E+02
25	11.44	1052.55	1078.76	1053.22	1035.43	0.3210E+02	0.3974E+02	0.2983E+02	0.2547E+02
27	13.54	1113.55	1157.07	1105.27	1084.14	0.2683E+02	0.3514E+02	0.2003E+02	0.2121E+02
29	15.83	1169.48	1233.22	1171.34	1128.75	0.2341E+02	0.3136E+02	0.2487E+02	0.1782E+02
31	17.76	1212.95	1291.00	1214.31	1160.82	0.2067E+02	0.2887E+02	0.2018E+02	0.1565E+02
33	19.60	1247.63	1342.40	1248.68	1188.12	0.1810E+02	0.2688E+02	0.1729E+02	0.1398E+02
35	21.67	1283.52	1396.07	1282.27	1215.41	0.1604E+02	0.2501E+02	0.1535E+02	0.1244E+02
37	22.97	1303.27	1428.16	1301.67	1231.00	0.1523E+02	0.2399E+02	0.1462E+02	0.1163E+02

IV - 7

```
Connection type : Top-and seat-angle connections
           Mode : All riveted

Tested by   :  R.A.Hechtman & B.G.Johnston (1947)          U.S.A
Test  Id.   :  NO 10

Column : W12X65                    Fasteners: 10.9- -3/4"D
Beam   : W18X47                          13/16" Oversize holes
Angle  : 6 X 4 X 3/4 X 12 ( TOP )  Material :  --
                                        Fy =  37.50 ksi
                                        Fu =  62.50 ksi

                      Major parameters

     lt = 12.0000"    tt =  0.7500"    ls =  7.5000"    ts =  0.8750"
     gt =  3.5000"    gt'=  2.5000"    gs =  3.5000"    gs'=  3.5000"
     qt =  4.5000"    rt =  4.5000"    rt2=  2.2500"
     qs =  4.5000"    rs =  4.5000"
     pt =  2.5000"    ps =  2.5000"    ps'=  2.5000"
     nt = 2 X  2      nt'= 4 X  1      ns = 2 X  2      ns'= 2 X  2

Remark  1)  6 X 6 X 7/8 X 7 1/2 ( SEAT )
        2)  1 shims 5 X 1/8 X 7 is used on ten. flange.
```

No	Moment (k-in)	Rotation (radians) X 1/1000	No	Moment (k-in)	Rotation (radians) X 1/1000
1	0.00	0.00	26	1155.71	11.82
2	53.83	0.09	27	1192.50	12.99
3	107.67	0.19	28	1217.04	13.89
4	152.64	0.36	29	1241.57	14.79
5	197.62	0.54	30	1268.14	15.80
6	254.17	0.81	31	1294.72	16.80
7	310.73	1.07	32	1317.21	17.88
8	354.34	1.38	33	1339.69	18.95
9	397.96	1.69	34	1357.41	19.93
10	447.70	2.00	35	1375.13	20.91
11	497.44	2.31	36	1390.12	21.74
12	549.23	2.62	37	1405.11	22.58
13	601.02	2.93	38	1420.10	23.41
14	652.13	3.41			
15	703.24	3.89			
16	749.57	4.39			
17	795.91	4.89			
18	842.93	5.41			
19	889.95	5.93			
20	960.82	7.09			
21	1019.42	8.05			
22	1053.49	8.76			
23	1087.56	9.48			
24	1103.24	10.07			
25	1118.91	10.66			

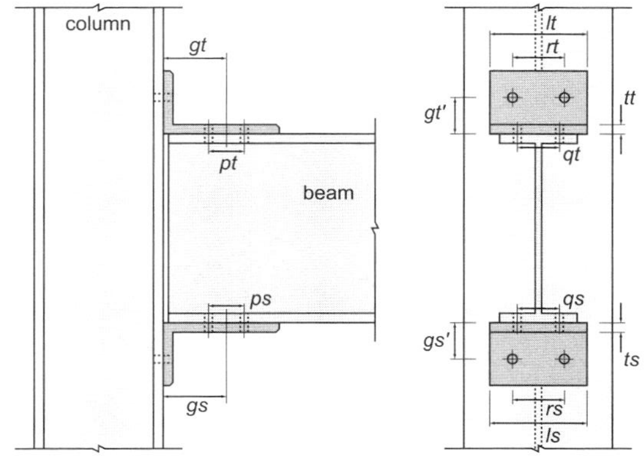

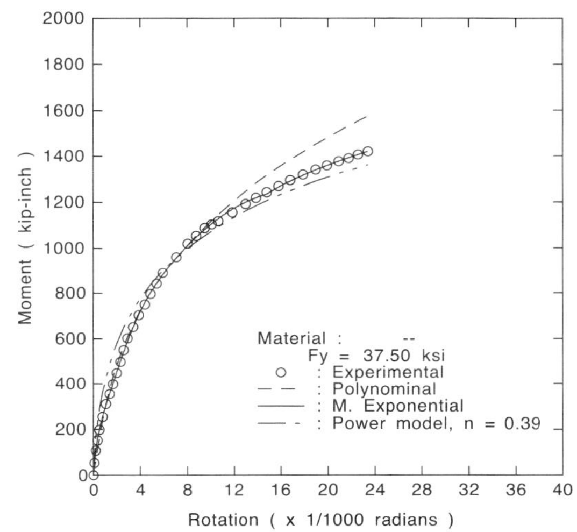

A.4 – 14

```
Moment-rotation prediction equations      ( R : X 1/1000 radians )

Frye and Morris polynominal model  :    R = Sum ( Ai X ( K*Bm )**Pi X 10**Qi )
      t =  0.750000"  xd = 18.125000"   f =  0.750000"  xl = 12.000000"
     A1 =  8.460000   A2 =  1.010000   A3 =  1.240000    K =  0.003606
     P1 =         1   P2 =         3   P3 =         5   Q1 =        -1    Q2 =        -1    Q3 =        -5

Modified exponential model  :
     Bm = Sum ( Ai X ( 1 - exp( - R/( 2*i*C ) ) ) )  +      Sum ( RKj X ( R - Rj0 ) )  +  Bm0
      C =   0.21856583E+00              Bmo=   0.00000000E+00           Nexp=  6        Nliner=  1
     AI =   0.33270893E+03  0.21681920E+04  -0.23007808E+05  0.68965650E+05  -0.85634839E+05  0.38462406E+05
    Rj0 =          14.7900
    RKj =   0.15954207E+02

Power model  :    Bm = ( Rki X R ) / ( 1 + ( R/R0 )**rn )**( 1/rn )
      rn = 0.387      rki =   0.21482616E+04          rmu =   0.28314087E+04
```

No	Rotation (radians) X 1/1000	Moment (kip-inch) Expri.	Poly.	M.Expo.	P.Model.	Connection stiffness (kip-inch) X 1000 Expri.	Poly.	M.Expo.	P.Model.
1	0.00	0.00	0.00	0.00	0.00	0.5727E+03	0.3278E+03	0.6231E+03	0.2148E+04
3	0.19	107.67	61.23	95.79	149.17	0.4629E+03	0.3221E+03	0.4146E+03	0.5399E+03
5	0.54	197.62	168.55	203.03	290.24	0.2392E+03	0.2895E+03	0.2318E+03	0.3168E+03
7	1.07	310.73	307.04	303.36	425.85	0.1784E+03	0.2277E+03	0.1641E+03	0.2062E+03
9	1.69	397.96	430.47	402.36	533.91	0.1512E+03	0.1759E+03	0.1593E+03	0.1502E+03
11	2.31	497.44	528.03	499.24	616.25	0.1645E+03	0.1425E+03	0.1528E+03	0.1190E+03
13	2.93	601.02	608.70	589.08	683.12	0.1436E+03	0.1202E+03	0.1372E+03	0.9885E+02
15	3.89	703.24	712.88	707.47	767.91	0.9962E+02	0.9725E+02	0.1082E+03	0.7826E+02
17	4.89	795.91	801.41	802.59	838.36	0.9161E+02	0.8201E+02	0.8495E+02	0.6449E+02
19	5.93	889.95	881.11	882.90	900.28	0.8096E+02	0.7089E+02	0.6981E+02	0.5438E+02
21	8.05	1019.42	1014.73	1011.84	1000.36	0.5330E+02	0.5642E+02	0.5337E+02	0.4119E+02
23	9.48	1087.56	1090.22	1081.46	1054.62	0.3623E+02	0.5001E+02	0.4448E+02	0.3537E+02
25	10.66	1118.91	1146.76	1129.58	1094.11	0.2821E+02	0.4586E+02	0.3700E+02	0.3162E+02
27	12.99	1192.50	1246.20	1199.52	1161.10	0.2912E+02	0.3966E+02	0.2334E+02	0.2609E+02
29	14.79	1241.57	1314.20	1233.82	1205.08	0.2686E+02	0.3609E+02	0.3114E+02	0.2295E+02
31	16.80	1294.72	1383.54	1289.66	1248.39	0.2376E+02	0.3290E+02	0.2480E+02	0.2019E+02
33	18.95	1339.69	1451.16	1338.07	1289.16	0.1945E+02	0.3017E+02	0.2067E+02	0.1787E+02
35	20.91	1375.13	1508.21	1376.27	1322.43	0.1805E+02	0.2811E+02	0.1852E+02	0.1615E+02
37	22.58	1405.11	1553.72	1406.13	1348.26	0.1802E+02	0.2661E+02	0.1745E+02	0.1491E+02

IV - 8

Connection type : Top-and seat-angle connections
Mode : All riveted

Tested by : R.A.Hechtman & B.G.Johnston (1947) U.S.A
Test Id. : NO 11

Column : W14X58
Beam : W18X47
Angle : 6 X 4 X 1/2 X 10 (TOP)

Fasteners: 10.9- -3/4"D
13/16" Oversize holes
Material : --
Fy = 37.50 ksi
Fu = 62.50 ksi

Major parameters

lt = 10.0000"	tt = 0.5000"	ls = 7.5000"	ts = 0.8750"
gt = 3.5000"	gt'= 2.5000"	gs = 3.5000"	gs'= 3.5000"
qt = 4.5000"	rt = 2.5000"	rt2= 2.5000"	
qs = 4.5000"	rs = 4.5000"		
pt = 2.5000"	ps = 2.5000"	ps'= 2.5000"	
nt = 2 X 2	nt'= 4 X 1	ns = 2 X 2	ns'= 2 X 2

Remark 1) 6 X 6 X 7/8 X 7 1/2 (SEAT). 1 shims 5 X 1/8 X 7 is used on ten. flange.
2) Connection cleat riveted to column web.

No	Moment (k-in)	Rotation (radians) X 1/1000	No	Moment (k-in)	Rotation (radians) X 1/1000
1	0.00	0.00	26	957.64	11.57
2	50.14	0.06	27	974.97	12.49
3	100.28	0.13	28	992.30	13.40
4	149.27	0.23	29	1009.63	14.32
5	198.25	0.32	30	1026.54	15.14
6	247.24	0.46	31	1043.46	15.96
7	296.24	0.59	32	1060.37	16.77
8	345.25	0.80	33	1078.52	17.87
9	394.26	1.02	34	1096.68	18.96
10	434.38	1.40	35	1114.83	20.05
11	474.50	1.78	36	1129.05	20.98
12	514.62	2.15	37	1143.28	21.92
13	555.52	2.55	38	1157.50	22.85
14	596.42	2.95			
15	647.84	3.46			
16	699.26	3.97			
17	732.01	4.43			
18	764.77	4.88			
19	797.57	5.50			
20	821.67	6.34			
21	845.78	7.17			
22	866.57	7.94			
23	887.36	8.72			
24	908.15	9.49			
25	932.89	10.53			

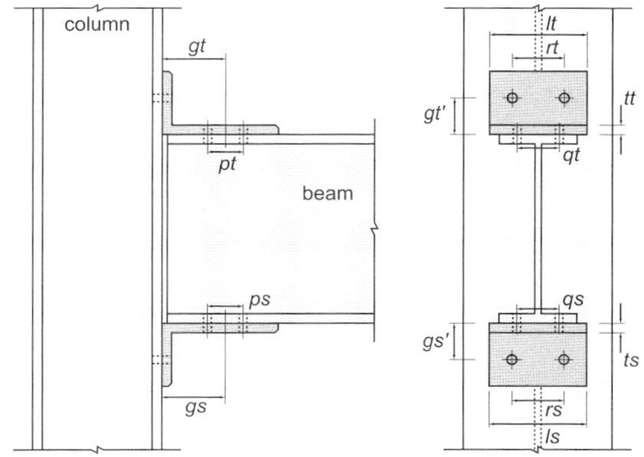

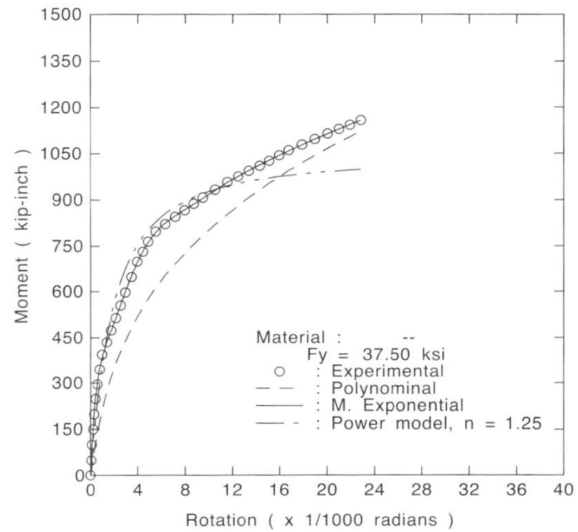

A.4 – 16

```
Moment-rotation prediction equations      ( R : X 1/1000 radians )

Frye and Morris polynominal model  :    R = Sum ( Ai X ( K*Bm )**Pi X 10**Qi )
     t =  0.500000"  xd = 18.125000"   f =  0.750000"  xl = 10.000000"
    A1 =  8.460000   A2 =  1.010000   A3 =  1.240000    K =  0.005018
    P1 =         1   P2 =         3   P3 =         5   Q1 =        -1    Q2 =        -1    Q3 =        -5

Modified exponential model  :
    Bm = Sum ( Ai X ( 1 - exp( - R/( 2*i*C ) ) ) )  +     Sum ( RKj X ( R - Rj0 ) )  +  Bm0
      C =   0.20641500E+00           Bmo=   0.00000000E+00          Nexp=  6        Nliner=  3
     AI =  -0.39319400E+03   0.95807303E+04  -0.41591381E+05   0.78478363E+05  -0.68456082E+05   0.23157730E+05
    Rj0 =           1.0200           5.5000          11.5700
    RKj =   0.21069231E+02  -0.15682501E+01  -0.43736478E+01

Power model  :    Bm = ( Rki X R ) / ( 1 + ( R/R0 )**rn )**( 1/rn )
     rn = 1.245      rki =   0.46099128E+03          rmu =   0.10424529E+04
```

No	Rotation (radians) X 1/1000	Moment (kip-inch) Expri.	Moment Poly.	Moment M.Expo.	Moment P.Model.	Connection stiffness (kip-inch) X 1000 Expri.	Stiffness Poly.	Stiffness M.Expo.	Stiffness P.Model.
1	0.00	0.00	0.00	0.00	0.00	0.7835E+03	0.2356E+03	0.7785E+03	0.4610E+03
3	0.13	100.28	30.05	90.69	57.67	0.6716E+03	0.2337E+03	0.6421E+03	0.4386E+03
5	0.32	198.25	75.25	198.97	139.84	0.4424E+03	0.2241E+03	0.4641E+03	0.3952E+03
7	0.59	296.24	132.24	297.63	237.22	0.3151E+03	0.2035E+03	0.2900E+03	0.3379E+03
9	1.02	394.26	211.32	384.78	364.33	0.1852E+03	0.1679E+03	0.1408E+03	0.2614E+03
11	1.78	474.50	319.82	475.28	524.61	0.1059E+03	0.1225E+03	0.1021E+03	0.1699E+03
13	2.55	555.52	403.50	556.00	633.29	0.1028E+03	0.9538E+02	0.1065E+03	0.1148E+03
15	3.46	647.84	480.86	650.32	718.92	0.1005E+03	0.7629E+02	0.9707E+02	0.7695E+02
17	4.43	732.01	548.03	732.26	780.95	0.7182E+02	0.6345E+02	0.7134E+02	0.5327E+02
19	5.50	797.57	610.66	794.13	828.90	0.4284E+02	0.5391E+02	0.4411E+02	0.3742E+02
21	7.17	845.78	692.02	849.28	878.55	0.2787E+02	0.4420E+02	0.2591E+02	0.2350E+02
23	8.72	887.36	755.32	886.53	908.83	0.2696E+02	0.3824E+02	0.2360E+02	0.1638E+02
25	10.53	932.89	819.89	930.40	933.60	0.2374E+02	0.3325E+02	0.2483E+02	0.1138E+02
27	12.49	974.97	881.02	975.52	952.44	0.1890E+02	0.2934E+02	0.2065E+02	0.8113E+01
29	14.32	1009.63	932.16	1012.52	965.40	0.1985E+02	0.2655E+02	0.1960E+02	0.6148E+01
31	15.96	1043.46	973.88	1043.57	974.40	0.2070E+02	0.2455E+02	0.1839E+02	0.4925E+01
33	17.87	1078.52	1018.87	1077.49	982.76	0.1661E+02	0.2262E+02	0.1717E+02	0.3895E+01
35	20.05	1114.83	1066.28	1113.90	990.31	0.1588E+02	0.2082E+02	0.1622E+02	0.3058E+01
37	21.92	1143.28	1103.88	1143.67	995.50	0.1526E+02	0.1953E+02	0.1574E+02	0.2534E+01

IV - 9

Connection type : Top-and seat-angle connections
Mode : All riveted

Tested by : R.A.Hechtman & B.G.Johnston (1947) U.S.A
Test Id. : NO 16

Column : W10X49 Fasteners: 10.9- -3/4"D
Beam : W12X25 13/16" Oversize holes
Angle : 6 X 4 X 1/2 X 6 3/4 (TOP) Material : --
Fy = 37.50 ksi
Fu = 62.50 ksi

Major parameters

lt = 6.7500"	tt = 0.5000"	ls = 6.7500"	ts = 0.5000"
gt = 3.5000"	gt'= 2.5000"	gs = 3.5000"	gs'= 3.5000"
qt = 4.2500"	rt = 4.2500"	qs = 4.2500"	rs = 4.2500"
pt = 2.5000"	ps = 2.5000"	ps'= 2.5000"	
nt = 2 X 2	nt'= 2 X 1	ns = 2 X 2	ns'= 2 X 2

Remark 1) 6 X 6 X 1/2 X 6 3/4 (SEAT)
2) 1 shims 5 X 1/8 X 6 3/4 is used on ten. flange.

No	Moment (k-in)	Rotation (radians) X 1/1000	No	Moment (k-in)	Rotation (radians) X 1/1000
1	0.00	0.00	26	384.86	13.94
2	16.74	0.11	27	392.33	14.98
3	33.48	0.21	28	399.80	16.02
4	50.22	0.32	29	405.20	16.84
5	66.56	0.46	30	410.60	17.66
6	82.91	0.60	31	416.00	18.48
7	99.25	0.75	32	420.43	19.34
8	116.35	0.96	33	424.86	20.20
9	133.45	1.17	34	429.29	21.06
10	150.55	1.39	35	434.26	22.03
11	166.09	1.70	36	439.24	23.00
12	181.62	2.01	37	444.22	23.98
13	197.16	2.32			
14	218.41	2.89			
15	239.67	3.46			
16	258.29	4.19			
17	276.91	4.92			
18	293.45	5.91			
19	309.98	6.90			
20	322.77	7.85			
21	335.56	8.79			
22	346.03	9.75			
23	356.51	10.71			
24	366.95	11.80			
25	377.40	12.89			

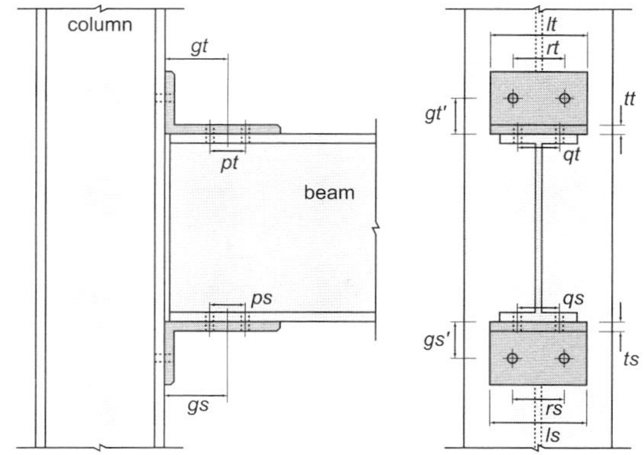

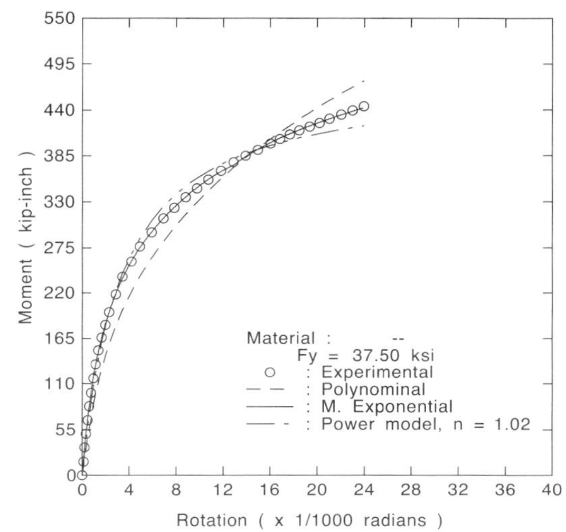

A.4 – 18

```
Moment-rotation prediction equations      ( R : X 1/1000 radians )

Frye and Morris polynominal model  :     R = Sum ( Ai X ( K*Bm )**Pi X 10**Qi )
      t =  0.500000"  xd = 12.125000"   f =  0.750000"  xl =  6.750000"
     A1 =  8.460000   A2 =  1.010000   A3 =  1.240000    K =  0.012076
     P1 =         1   P2 =         3   P3 =         5   Q1 =        -1    Q2 =       -1    Q3 =       -5

Modified exponential model  :
     Bm = Sum ( Ai X ( 1 - exp( - R/( 2*i*C ) ) ) )  +      Sum ( RKj X ( R - Rj0 ) )  +  Bm0
      C =   0.22618833E+00              Bmo=   0.00000000E+00           Nexp=  6          Nliner=  1
     AI =  -0.45198535E+02   0.12312500E+04  -0.73140193E+04   0.19663832E+05  -0.23396485E+05   0.10260252E+05
    Rj0 =           12.8900
    RKj =   0.39662141E+01

Power model  :     Bm = ( Rki X R ) / ( 1 + ( R/R0 )**rn )**( 1/rn )
     rn = 1.025       rki =   0.14014133E+03           rmu =   0.47668712E+03
```

No	Rotation (radians) X 1/1000	Moment (kip-inch) Expri.	Poly.	M.Expo.	P.Model.	Connection stiffness (kip-inch) X 1000 Expri.	Poly.	M.Expo.	P.Model.
1	0.00	0.00	0.00	0.00	0.00	0.1587E+03	0.9789E+02	0.1749E+03	0.1401E+03
3	0.21	33.48	20.49	33.95	27.98	0.1587E+03	0.9578E+02	0.1476E+03	0.1254E+03
5	0.46	66.56	43.59	67.21	57.28	0.1139E+03	0.8905E+02	0.1205E+03	0.1103E+03
7	0.75	99.25	67.72	98.19	86.82	0.1003E+03	0.7897E+02	0.9650E+02	0.9591E+02
9	1.17	133.45	98.25	133.70	123.86	0.8041E+02	0.6507E+02	0.7249E+02	0.7909E+02
11	1.70	166.09	128.84	166.73	161.09	0.4987E+02	0.5242E+02	0.5512E+02	0.6370E+02
13	2.32	197.16	158.16	197.05	196.44	0.4543E+02	0.4242E+02	0.4324E+02	0.5054E+02
15	3.46	239.67	199.76	238.42	244.40	0.3214E+02	0.3171E+02	0.3049E+02	0.3502E+02
17	4.92	276.91	240.14	275.28	286.42	0.2177E+02	0.2437E+02	0.2100E+02	0.2368E+02
19	6.90	309.98	282.54	310.20	324.24	0.1506E+02	0.1890E+02	0.1508E+02	0.1532E+02
21	8.79	335.56	315.12	335.85	348.65	0.1225E+02	0.1578E+02	0.1218E+02	0.1087E+02
23	10.71	356.51	343.14	356.55	366.57	0.1030E+02	0.1365E+02	0.9462E+01	0.8079E+01
25	12.89	377.40	370.99	373.89	381.83	0.8327E+01	0.1191E+02	0.1040E+02	0.6022E+01
27	14.98	392.33	394.49	393.01	392.94	0.7161E+01	0.1068E+02	0.8043E+01	0.4711E+01
29	16.84	405.20	413.70	406.50	400.90	0.6582E+01	0.9804E+01	0.6525E+01	0.3870E+01
31	18.48	416.00	429.24	416.40	406.76	0.5887E+01	0.9169E+01	0.5605E+01	0.3302E+01
33	20.20	424.86	444.48	425.43	412.01	0.5159E+01	0.8602E+01	0.4965E+01	0.2831E+01
35	22.03	434.26	459.77	434.10	416.81	0.5117E+01	0.8083E+01	0.4541E+01	0.2431E+01
37	23.98	444.22	475.02	442.66	421.20	0.5117E+01	0.7609E+01	0.4278E+01	0.2092E+01

IV - 10

Connection type : Top-and seat-angle connections
Mode : All riveted

Tested by : R.A.Hechtman & B.G.Johnston (1947) U.S.A
Test Id. : NO 17

Column : W10X49
Beam : W12X25
Angle : 6 X 4 X 1/2 X 6 3/4 (TOP)

Fasteners: 10.9- -3/4"D
13/16" Oversize holes
Material : --
Fy = 37.50 ksi
Fu = 62.50 ksi

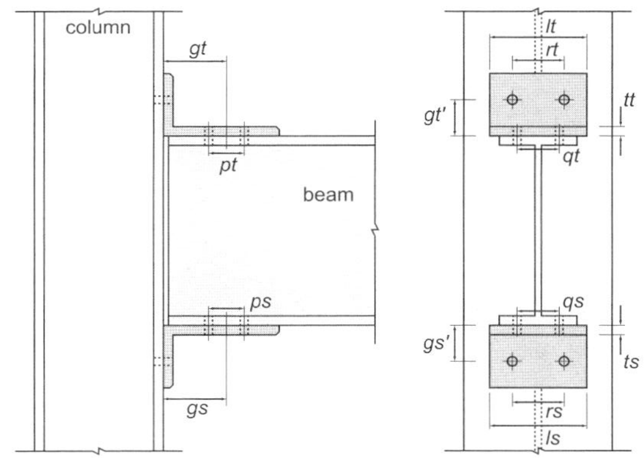

Major parameters

lt = 6.7500"	tt = 0.5000"	ls = 6.7500"	ts = 0.5000"
gt = 3.5000"	gt'= 2.5000"	gs = 3.5000"	gs'= 3.5000"
qt = 4.2500"	rt = 4.2500"	qs = 4.2500"	rs = 4.2500"
pt = 2.5000"	ps = 2.5000"	ps'= 2.5000"	
nt = 2 X 2	nt'= 2 X 1	ns = 2 X 2	ns'= 2 X 2

Remark 1) 6 X 6 X 1/2 X 6 3/4 (SEAT). 1 shims 5 X 1/8 X 6 3/4 is used on ten. flange.
2) Connection cleat riveted to column web.

No	Moment (k-in)	Rotation (radians) X 1/1000	No	Moment (k-in)	Rotation (radians) X 1/1000
1	0.00	0.00	26	366.09	10.00
2	19.79	0.07	27	375.24	11.01
3	39.58	0.13	28	384.39	12.03
4	59.38	0.20	29	391.28	12.85
5	79.17	0.30	30	398.17	13.67
6	98.97	0.40	31	405.06	14.49
7	115.67	0.55	32	410.41	15.41
8	132.37	0.70	33	415.77	16.32
9	149.07	0.85	34	421.12	17.23
10	166.16	1.06	35	426.47	18.14
11	183.26	1.27	36	432.23	19.21
12	200.35	1.48	37	437.99	20.28
13	215.22	1.72	38	443.75	21.35
14	230.09	1.96	39	448.53	22.39
15	244.68	2.31	40	453.32	23.42
16	259.27	2.65			
17	270.67	3.05			
18	282.06	3.46			
19	300.21	4.31			
20	307.84	4.84			
21	319.88	5.69			
22	331.91	6.54			
23	339.57	7.27			
24	347.22	7.99			
25	356.66	9.00			

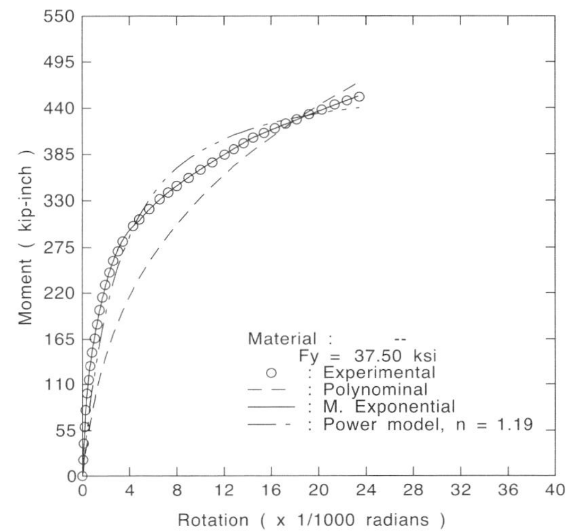

```
 Moment-rotation prediction equations     ( R : X 1/1000 radians )

Frye and Morris polynominal model  :    R = Sum ( Ai X ( K*Bm )**Pi X 10**Qi )
      t =  0.500000"  xd = 12.125000"   f =  0.750000"  xl =  6.750000"
     A1 =  8.460000   A2 =  1.010000   A3 =  1.240000    K =  0.012076
     P1 =        1    P2 =        3    P3 =        5    Q1 =       -1     Q2 =       -1     Q3 =       -5

Modified exponential model  :
     Bm = Sum ( Ai X ( 1 - exp( - R/( 2*i*C ) ) ) )  +      Sum ( RKj X ( R - Rj0 ) )  +  Bm0
       C =   0.21155083E+00              Bmo=   0.00000000E+00           Nexp=  6        Nliner=  2
      AI =   0.25323634E+03  -0.97727029E+03   0.17739097E+04  -0.80689464E+03  -0.49536732E+03   0.58524450E+03
     Rj0 =           4.8400          14.4900
     RKj =   0.76190372E+01  -0.22245866E+01

Power model  :    Bm = ( Rki X R ) / ( 1 + ( R/R0 )**rn )**( 1/rn )
      rn = 1.194      rki =   0.14014133E+03          rmu =   0.47668712E+03
```

No	Rotation (radians) X 1/1000	Moment (kip-inch) Expri.	Moment (kip-inch) Poly.	Moment (kip-inch) M.Expo.	Moment (kip-inch) P.Model.	Connection stiffness (kip-inch) X 1000 Expri.	Connection stiffness (kip-inch) X 1000 Poly.	Connection stiffness (kip-inch) X 1000 M.Expo.	Connection stiffness (kip-inch) X 1000 P.Model.
1	0.00	0.00	0.00	0.00	0.00	0.3026E+03	0.9789E+02	0.3608E+03	0.1401E+03
3	0.13	39.58	12.77	40.75	18.02	0.3029E+03	0.9706E+02	0.2681E+03	0.1350E+03
5	0.30	79.17	28.58	78.23	39.70	0.1979E+03	0.9388E+02	0.1915E+03	0.1272E+03
7	0.55	115.67	51.20	117.24	70.07	0.1108E+03	0.8610E+02	0.1274E+03	0.1151E+03
9	0.85	149.07	75.53	149.65	102.70	0.9825E+02	0.7541E+02	0.9256E+02	0.1017E+03
11	1.27	183.26	104.63	184.01	142.30	0.8056E+02	0.6227E+02	0.7224E+02	0.8540E+02
13	1.72	215.22	130.28	213.76	177.58	0.6205E+02	0.5187E+02	0.6002E+02	0.7129E+02
15	2.31	244.68	157.65	244.69	214.83	0.4247E+02	0.4258E+02	0.4619E+02	0.5714E+02
17	3.05	270.67	186.27	273.13	252.23	0.2835E+02	0.3478E+02	0.3070E+02	0.4397E+02
19	4.31	300.21	224.47	299.97	297.60	0.1706E+02	0.2692E+02	0.1415E+02	0.2972E+02
21	5.69	319.88	257.88	319.82	331.74	0.1412E+02	0.2185E+02	0.1412E+02	0.2048E+02
23	7.27	339.57	289.27	339.29	358.69	0.1054E+02	0.1819E+02	0.1108E+02	0.1421E+02
25	9.00	356.66	318.27	357.16	379.38	0.9409E+01	0.1552E+02	0.9748E+01	0.1006E+02
27	11.01	375.24	347.28	375.87	396.45	0.9016E+01	0.1337E+02	0.8870E+01	0.7108E+01
29	12.85	391.28	370.47	391.65	407.83	0.8379E+01	0.1194E+02	0.8356E+01	0.5392E+01
31	14.49	405.06	389.25	405.13	415.77	0.7191E+01	0.1094E+02	0.5835E+01	0.4319E+01
33	16.32	415.77	408.34	415.56	422.81	0.5872E+01	0.1004E+02	0.5635E+01	0.3456E+01
35	18.14	426.47	426.11	425.72	428.51	0.5649E+01	0.9292E+01	0.5522E+01	0.2821E+01
37	20.28	437.99	445.03	437.45	433.92	0.5387E+01	0.8583E+01	0.5454E+01	0.2271E+01
39	22.39	448.53	462.49	448.90	438.26	0.4608E+01	0.7996E+01	0.5422E+01	0.1869E+01

A.4 – 21

IV - 11

Connection type : Top-and seat-angle connections
Mode : All riveted

Tested by : R.A.Hechtman & B.G.Johnston (1947) U.S.A
Test Id. : NO 18

Column : W10X49
Beam : W12X50
Angle : 6 X 4 X 1/2 X 8 (TOP)

Fasteners: 10.9- -3/4"D
13/16" Oversize holes
Material : --
Fy = 37.50 ksi
Fu = 62.50 ksi

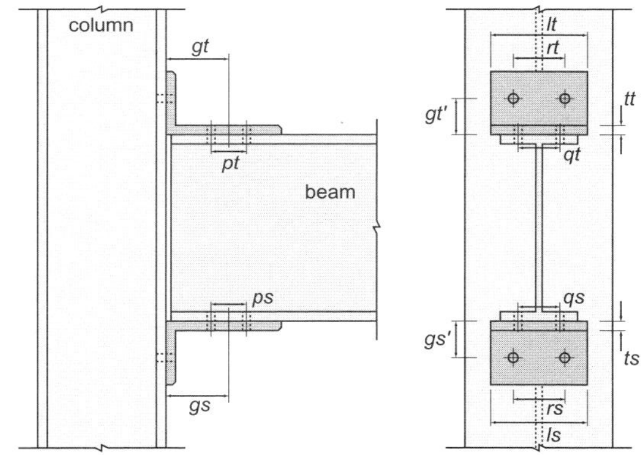

Major parameters

lt = 8.0000"	tt = 0.5000"	ls = 8.0000"	ts = 0.5000"
gt = 3.5000"	gt'= 2.5000"	gs = 3.5000"	gs'= 3.5000"
qt = 5.0000"	rt = 5.0000"	qs = 5.0000"	rs = 5.0000"
pt = 2.5000"	ps = 2.5000"	ps'= 2.5000"	
nt = 2 X 2	nt'= 2 X 1	ns = 2 X 2	ns'= 2 X 2

Remark 1) 6 X 6 X 1/2 X 8 (SEAT). 1 shims 5 X 1/8 X 8 is used on ten. flange.
2) Connection cleat riveted to column web.

No	Moment (k-in)	Rotation (radians) X 1/1000	No	Moment (k-in)	Rotation (radians) X 1/1000
1	0.00	0.00	26	396.75	10.51
2	19.69	0.06	27	404.85	11.31
3	39.37	0.12	28	411.81	12.20
4	59.06	0.19	29	418.78	13.08
5	78.75	0.27	30	425.74	13.96
6	98.45	0.36	31	432.75	15.04
7	114.87	0.49	32	439.76	16.11
8	131.30	0.62	33	446.77	17.19
9	147.72	0.75	34	451.82	18.14
10	164.93	0.92	35	456.87	19.08
11	182.13	1.08	36	461.92	20.03
12	199.34	1.24	37	466.43	21.11
13	218.79	1.56	38	470.93	22.20
14	238.25	1.87	39	475.44	23.28
15	253.69	2.32			
16	269.13	2.76			
17	284.00	3.24			
18	298.87	3.72			
19	314.67	4.45			
20	330.47	5.17			
21	343.69	6.02			
22	356.91	6.87			
23	368.73	7.88			
24	380.55	8.89			
25	388.65	9.70			

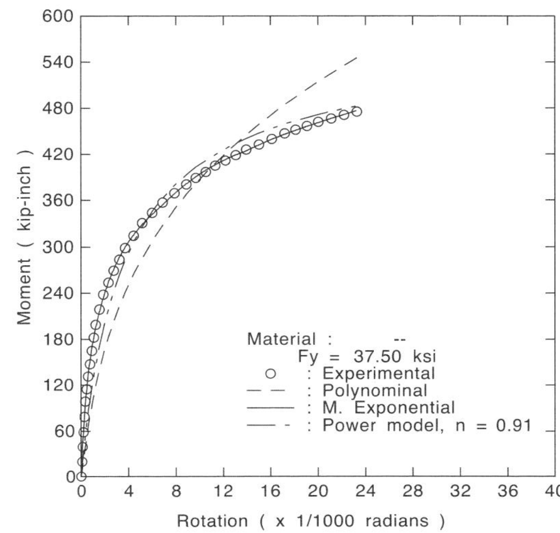

A.4 – 22

```
 Moment-rotation prediction equations     ( R : X 1/1000 radians )

Frye and Morris polynominal model  :    R = Sum ( Ai X ( K*Bm )**Pi X 10**Qi )
      t =  0.500000"  xd = 12.375000"  f =  0.750000"  xl =  8.000000"
     A1 =  8.460000   A2 =  1.010000   A3 =  1.240000   K =  0.010398
     P1 =        1    P2 =        3    P3 =        5    Q1 =       -1    Q2 =       -1    Q3 =       -5

Modified exponential model  :
     Bm = Sum ( Ai X ( 1 - exp( - R/( 2*i*C ) ) ) )  +     Sum ( RKj X ( R - Rj0 ) )  +  Bm0
      C =   0.20948333E+00              Bmo=   0.00000000E+00          Nexp=  6        Nliner=  2
     AI =   0.11835104E+03   0.31487609E+03  -0.35968147E+04   0.11940052E+05  -0.15571408E+05   0.72146069E+04
    Rj0 =           6.8700           13.9600
    RKj =   0.21889535E+01   0.23110283E+01

Power model  :    Bm = ( Rki X R ) / ( 1 + ( R/R0 )**rn )**( 1/rn )
      rn = 0.906      rki =   0.17273651E+03          rmu =   0.57484872E+03
```

No	Rotation (radians) X 1/1000	Moment (kip-inch) Expri.	Poly.	M.Expo.	P.Model.	Connection stiffness (kip-inch) X 1000 Expri.	Poly.	M.Expo.	P.Model.
1	0.00	0.00	0.00	0.00	0.00	0.3186E+03	0.1137E+03	0.3581E+03	0.1727E+03
3	0.12	39.37	14.00	39.43	20.20	0.3188E+03	0.1128E+03	0.2841E+03	0.1557E+03
5	0.27	78.75	30.35	76.13	41.89	0.2320E+03	0.1098E+03	0.2199E+03	0.1406E+03
7	0.49	114.87	53.45	116.53	70.44	0.1241E+03	0.1024E+03	0.1571E+03	0.1229E+03
9	0.75	147.72	79.12	151.63	100.69	0.1158E+03	0.9149E+02	0.1123E+03	0.1062E+03
11	1.08	182.13	106.82	182.78	132.61	0.1056E+03	0.7883E+02	0.8219E+02	0.9043E+02
13	1.56	218.79	140.90	216.19	171.60	0.6154E+02	0.6425E+02	0.5990E+02	0.7334E+02
15	2.32	253.69	183.61	254.29	219.97	0.3470E+02	0.4928E+02	0.4207E+02	0.5514E+02
17	3.24	284.00	223.82	286.81	264.01	0.3099E+02	0.3864E+02	0.2928E+02	0.4118E+02
19	4.45	314.67	264.96	315.71	306.20	0.2187E+02	0.3053E+02	0.1991E+02	0.2995E+02
21	6.02	343.69	307.66	342.16	345.81	0.1555E+02	0.2432E+02	0.1453E+02	0.2120E+02
23	7.88	368.73	348.52	368.33	379.14	0.1167E+02	0.1989E+02	0.1356E+02	0.1511E+02
25	9.70	388.65	381.94	390.50	403.04	0.1005E+02	0.1705E+02	0.1082E+02	0.1143E+02
27	11.31	404.85	407.89	406.03	419.57	0.9016E+01	0.1522E+02	0.8489E+01	0.9205E+01
29	13.08	418.78	433.38	419.06	434.21	0.7881E+01	0.1369E+02	0.6352E+01	0.7452E+01
31	15.04	432.75	458.86	432.22	447.36	0.6513E+01	0.1237E+02	0.6957E+01	0.6045E+01
33	17.19	446.77	484.20	445.83	459.10	0.5888E+01	0.1123E+02	0.5793E+01	0.4921E+01
35	19.08	456.87	504.83	456.18	467.68	0.5338E+01	0.1040E+02	0.5207E+01	0.4178E+01
37	21.11	466.43	525.03	466.37	475.51	0.4152E+01	0.9682E+01	0.4858E+01	0.3556E+01
39	23.28	475.44	545.31	476.68	482.64	0.4152E+01	0.9028E+01	0.4668E+01	0.3036E+01

IV - 12

```
Connection type : Top-and seat-angle connections
           Mode : All riveted

Tested by   :  R.A.Hechtman & B.G.Johnston (1947)          U.S.A
Test   Id.  :  NO 20

Column : W12X65                      Fasteners: 10.9- -3/4"D
Beam   : W14X34                            13/16" Oversize holes
Angle  : 6 X 4 X 5/8 X 12 ( TOP )    Material :  --
                                          Fy =  37.50 ksi
                                          Fu =  62.50 ksi

                          Major parameters

     lt = 12.0000"    tt =  0.6250"    ls =  7.2500"    ts =  0.6250"
     gt =  3.5000"    gt'=  2.5000"    gs =  3.5000"    gs'=  3.5000"
     qt =  4.2500"    rt =  4.2500"    rt2=  2.2500"
     qs =  4.2500"    rs =  4.2500"
     pt =  2.5000"    ps =  2.5000"    ps'=  2.5000"
     nt = 2 X  2      nt'= 4 X  1      ns = 2 X  2      ns'= 2 X  2

Remark  1)  6 X 6 X 5/8 X 7 1/4 ( SEAT )
        2)  2 shims 5 X 1/8 X 6 3/4 are used on ten. flange.
```

No	Moment (k-in)	Rotation (radians) X 1/1000	No	Moment (k-in)	Rotation (radians) X 1/1000
1	0.00	0.00	26	822.94	12.91
2	34.01	0.14	27	844.15	13.91
3	63.02	0.27	28	865.36	14.90
4	102.02	0.41	29	885.62	16.05
5	137.15	0.78	30	905.88	17.19
6	172.27	1.15	31	926.14	18.34
7	215.01	1.55	32	937.66	19.18
8	257.75	1.95	33	949.18	20.02
9	296.94	2.37	34	960.69	20.87
10	336.14	2.78	35	972.23	21.75
11	375.34	3.20	36	983.76	22.64
12	413.52	3.62	37	995.29	23.52
13	451.71	4.03			
14	489.89	4.45			
15	528.88	4.96			
16	567.86	5.46			
17	593.14	5.91			
18	618.41	6.36			
19	646.83	7.04			
20	675.25	7.71			
21	705.28	8.60			
22	735.31	9.48			
23	757.45	10.30			
24	779.59	11.11			
25	801.73	11.92			

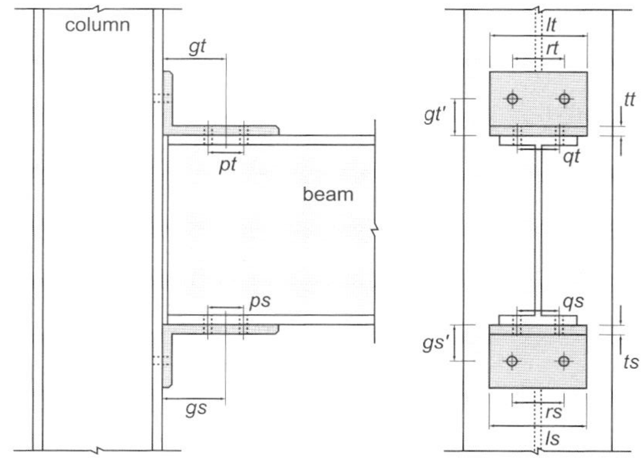

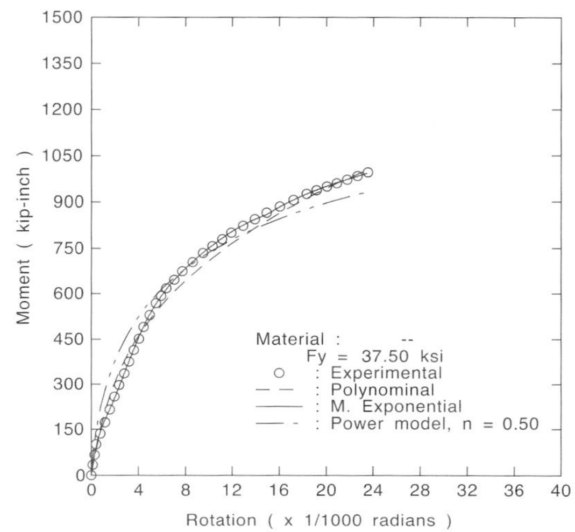

```
Moment-rotation prediction equations     ( R : X 1/1000 radians )

Frye and Morris polynominal model  :    R = Sum ( Ai X ( K*Bm )**Pi X 10**Qi )
     t =  0.625000"  xd = 14.250000"   f =  0.750000"  xl = 12.000000"
    A1 =  8.460000   A2 =  1.010000   A3 =  1.240000    K =  0.005667
    P1 =        1    P2 =        3    P3 =        5    Q1 =       -1    Q2 =       -1    Q3 =       -5

Modified exponential model  :
    Bm = Sum ( Ai X ( 1 - exp( - R/( 2*i*C ) ) ) )  +      Sum ( RKj X ( R - Rj0 ) )  +  Bm0
      C =   0.22981583E+00          Bmo=   0.00000000E+00          Nexp=  6        Nliner=  1
     AI =  -0.20675995E+03   0.67450312E+04  -0.37623270E+05   0.87814155E+05  -0.94136595E+05   0.38320823E+05
    Rj0 =          14.9000
    RKj =   0.98015388E+01

Power model  :    Bm = ( Rki X R ) / ( 1 + ( R/R0 )**rn )**( 1/rn )
     rn = 0.499      rki =   0.72230345E+03         rmu =   0.15953706E+04
```

No	Rotation (radians) X 1/1000	Moment (kip-inch) Expri.	Poly.	M.Expo.	P.Model.	Connection stiffness (kip-inch) X 1000 Expri.	Poly.	M.Expo.	P.Model.
1	0.00	0.00	0.00	0.00	0.00	0.2515E+03	0.2086E+03	0.2996E+03	0.7223E+03
3	0.27	68.02	55.72	67.44	106.90	0.2516E+03	0.2014E+03	0.2049E+03	0.2928E+03
5	0.78	137.15	149.38	143.25	220.57	0.9450E+02	0.1660E+03	0.1103E+03	0.1781E+03
7	1.55	215.01	257.67	214.39	330.52	0.1067E+03	0.1183E+03	0.8792E+02	0.1161E+03
9	2.37	296.94	341.24	291.68	411.55	0.9415E+02	0.8913E+02	0.1010E+03	0.8546E+02
11	3.20	375.34	407.57	377.19	474.57	0.9292E+02	0.7161E+02	0.1015E+03	0.6734E+02
13	4.03	451.71	462.19	457.05	525.39	0.9167E+02	0.6028E+02	0.8883E+02	0.5544E+02
15	4.96	528.88	513.64	530.67	572.11	0.7689E+02	0.5163E+02	0.7070E+02	0.4624E+02
17	5.91	593.14	559.78	590.47	612.85	0.5630E+02	0.4523E+02	0.5525E+02	0.3935E+02
19	7.04	646.83	607.36	645.42	653.56	0.4209E+02	0.3971E+02	0.4362E+02	0.3338E+02
21	8.60	705.28	664.81	706.15	700.77	0.3388E+02	0.3420E+02	0.3511E+02	0.2744E+02
23	10.30	757.45	719.09	760.60	743.29	0.2726E+02	0.2994E+02	0.2917E+02	0.2288E+02
25	11.92	801.73	765.13	803.51	777.71	0.2460E+02	0.2688E+02	0.2362E+02	0.1966E+02
27	13.91	844.15	815.58	843.68	813.69	0.2134E+02	0.2402E+02	0.1689E+02	0.1669E+02
29	16.05	885.62	864.31	884.21	846.72	0.1770E+02	0.2165E+02	0.2058E+02	0.1430E+02
31	18.34	926.14	911.52	925.64	877.13	0.1536E+02	0.1966E+02	0.1595E+02	0.1235E+02
33	20.02	949.18	943.67	950.55	896.96	0.1364E+02	0.1846E+02	0.1370E+02	0.1119E+02
35	21.75	972.23	974.63	972.83	915.41	0.1304E+02	0.1739E+02	0.1219E+02	0.1019E+02
37	23.52	995.29	1004.54	993.45	932.65	0.1304E+02	0.1645E+02	0.1121E+02	0.9317E+01

IV - 13

Connection type : Top-and seat-angle connections
Mode : All riveted

Tested by : R.A.Hechtman & B.G.Johnston (1947) U.S.A
Test Id. : NO 22

Column : W12X65
Beam : W16X40
Angle : 6 X 4 X 5/8 X 12 (TOP)

Fasteners: 10.9- -3/4"D
13/16" Oversize holes
Material : --
Fy = 37.50 ksi
Fu = 62.50 ksi

Major parameters

lt = 12.0000"	tt = 0.6250"	ls = 7.2500"	ts = 0.7500"
gt = 3.5000"	gt'= 2.5000"	gs = 3.5000"	gs'= 3.5000"
qt = 4.2500"	rt = 4.2500"	rt2= 2.2500"	
qs = 4.2500"	rs = 4.2500"		
pt = 2.5000"	ps = 2.5000"	ps'= 2.5000"	
nt = 2 X 2	nt'= 4 X 1	ns = 2 X 2	ns'= 2 X 2

Remark 1) 6 X 6 X 3/4 X 7 1/4 (SEAT)
2) 2 shims 5 X 1/8 X 6 3/4 are used on ten. flange.

No	Moment (k-in)	Rotation (radians) X 1/1000	No	Moment (k-in)	Rotation (radians) X 1/1000
1	0.00	0.00	26	1016.66	13.92
2	53.35	0.23	27	1037.67	14.85
3	106.70	0.45	28	1057.77	15.85
4	152.44	0.65	29	1077.86	16.84
5	198.18	0.85	30	1097.95	17.84
6	249.28	1.10	31	1113.34	18.81
7	300.38	1.34	32	1128.73	19.78
8	350.07	1.66	33	1144.12	20.75
9	399.77	1.98	34	1159.50	21.72
10	455.69	2.41	35	1172.12	22.60
11	511.60	2.85	36	1184.73	23.48
12	556.25	3.35			
13	600.90	3.85			
14	637.92	4.32			
15	674.94	4.78			
16	711.39	5.37			
17	747.84	5.97			
18	785.10	6.78			
19	824.37	7.59			
20	865.24	8.63			
21	906.12	9.68			
22	928.95	10.47			
23	951.79	11.26			
24	974.63	12.06			
25	995.64	12.99			

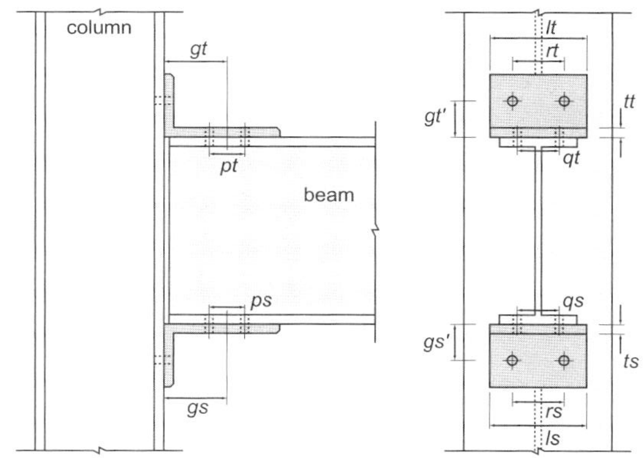

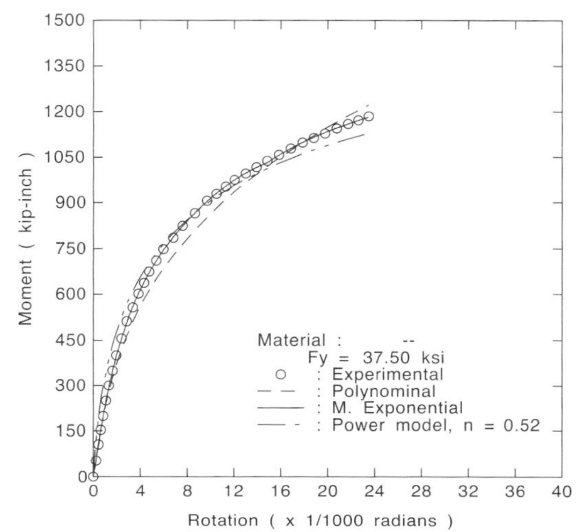

```
Moment-rotation prediction equations     ( R : X 1/1000 radians )

Frye and Morris polynominal model  :    R = Sum ( Ai X ( K*Bm )**Pi X 10**Qi )
      t =  0.625000"  xd = 16.250000"   f =  0.750000"  xl = 12.000000"
     A1 =  8.460000   A2 =  1.010000   A3 =  1.240000    K =  0.004654
     P1 =         1   P2 =         3   P3 =         5   Q1 =        -1    Q2 =        -1    Q3 =        -5

Modified exponential model  :
     Bm = Sum ( Ai X ( 1 - exp( - R/( 2*i*C ) ) ) )  +      Sum ( RKj X ( R - Rj0 ) )  +  Bm0
      C =   0.25230083E+00           Bmo=   0.00000000E+00            Nexp=  6         Nliner=  1
     AI =  -0.31639015E+03   0.38039807E+04  -0.21375714E+05   0.56906130E+05  -0.67939902E+05   0.30022960E+05
     Rj0 =           14.8500
     RKj =   0.10207116E+02

Power model  :    Bm = ( Rki X R ) / ( 1 + ( R/R0 )**rn )**( 1/rn )
      rn = 0.516      rki =   0.93649262E+03         rmu =   0.18088892E+04
```

No	Rotation (radians) X 1/1000	Moment (kip-inch) Expri.	Poly.	M.Expo.	P.Model.	Connection stiffness (kip-inch) X 1000 Expri.	Poly.	M.Expo.	P.Model.
1	0.00	0.00	0.00	0.00	0.00	0.2356E+03	0.2540E+03	0.2036E+03	0.9365E+03
3	0.45	106.70	111.44	106.07	200.23	0.2317E+03	0.2317E+03	0.2437E+03	0.3002E+03
5	0.85	198.18	197.06	199.70	300.83	0.2202E+03	0.1952E+03	0.2198E+03	0.2128E+03
7	1.34	300.38	282.11	296.96	389.83	0.1869E+03	0.1570E+03	0.1813E+03	0.1593E+03
9	1.98	399.77	370.56	399.24	477.73	0.1442E+03	0.1230E+03	0.1420E+03	0.1201E+03
11	2.85	511.60	464.46	507.21	567.96	0.1100E+03	0.9496E+02	0.1085E+03	0.8972E+02
13	3.85	600.90	549.54	603.17	646.94	0.8433E+02	0.7593E+02	0.8405E+02	0.6913E+02
15	4.78	674.94	614.10	673.09	704.74	0.7183E+02	0.6463E+02	0.6825E+02	0.5683E+02
17	5.97	747.84	685.00	745.69	765.73	0.5520E+02	0.5466E+02	0.5431E+02	0.4594E+02
19	7.59	824.37	765.55	823.58	831.51	0.4374E+02	0.4571E+02	0.4321E+02	0.3621E+02
21	9.68	906.12	852.80	904.27	898.26	0.3324E+02	0.3815E+02	0.3427E+02	0.2813E+02
23	11.26	951.79	909.83	953.77	939.28	0.2887E+02	0.3412E+02	0.2832E+02	0.2393E+02
25	12.99	995.64	965.62	997.11	977.43	0.2253E+02	0.3074E+02	0.2201E+02	0.2048E+02
27	14.85	1037.67	1020.20	1032.33	1012.87	0.2141E+02	0.2788E+02	0.2607E+02	0.1764E+02
29	16.84	1077.86	1073.11	1078.65	1045.50	0.2022E+02	0.2546E+02	0.2079E+02	0.1530E+02
31	18.81	1113.34	1121.16	1115.54	1073.70	0.1585E+02	0.2351E+02	0.1698E+02	0.1348E+02
33	20.75	1144.12	1165.22	1145.86	1098.41	0.1585E+02	0.2191E+02	0.1441E+02	0.1202E+02
35	22.60	1172.12	1204.59	1170.97	1119.58	0.1429E+02	0.2061E+02	0.1280E+02	0.1086E+02

IV - 14

Connection type : Top-and seat-angle connections
Mode : All riveted

Tested by : R.A.Hechtman & B.G.Johnston (1947) U.S.A
Test Id. : NO 23

Column : W14X58
Beam : W18X40
Angle : 6 X 4 X 5/8 X 10 (TOP)

Fasteners: 10.9- -3/4"D
13/16" Oversize holes
Material : --
Fy = 37.50 ksi
Fu = 62.50 ksi

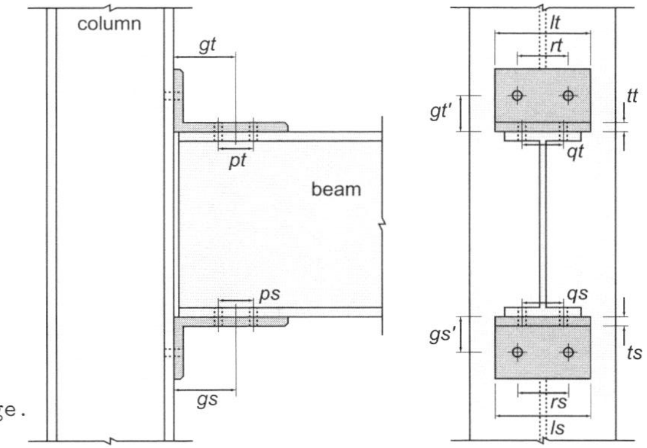

Major parameters

lt = 10.0000"	tt = 0.6250"	ls = 7.2500"	ts = 0.7500"
gt = 3.5000"	gt'= 2.5000"	gs = 3.5000"	gs'= 3.5000"
qt = 4.2500"	rt = 2.5000"	rt2= 2.5000"	
qs = 4.2500"	rs = 4.2500"		
pt = 2.5000"	ps = 2.5000"	ps'= 2.5000"	
nt = 2 X 2	nt'= 4 X 1	ns = 2 X 2	ns'= 2 X 2

Remark 1) 6 X 6 X 3/4 X 7 1/4 (SEAT). 2 shims 5 X 1/8 X 6 3/4 are used on ten. flange.
2) Connection cleat riveted to column web.

No	Moment (k-in)	Rotation (radians) X 1/1000	No	Moment (k-in)	Rotation (radians) X 1/1000
1	0.00	0.00	26	1070.34	11.71
2	46.59	0.04	27	1088.56	12.51
3	93.19	0.09	28	1106.79	13.31
4	139.78	0.13	29	1125.01	14.10
5	186.37	0.18	30	1143.24	14.90
6	244.72	0.39	31	1161.46	15.70
7	303.06	0.59	32	1177.42	16.65
8	350.80	0.82	33	1193.38	17.60
9	398.54	1.04	34	1209.34	18.54
10	453.86	1.36	35	1225.31	19.49
11	509.18	1.67			
12	554.65	1.99			
13	600.13	2.30			
14	635.75	2.59			
15	671.37	2.87			
16	713.06	3.32			
17	754.75	3.77			
18	798.73	4.55			
19	841.96	5.41			
20	885.19	6.27			
21	921.60	7.11			
22	958.01	7.95			
23	989.88	8.87			
24	1021.75	9.78			
25	1046.04	10.74			

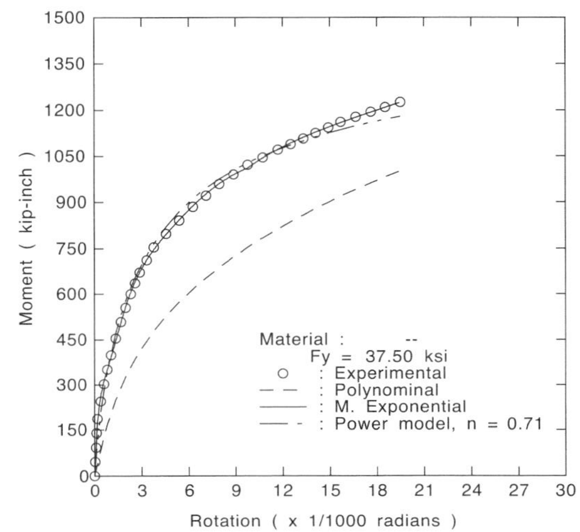

```
Moment-rotation prediction equations     ( R : X 1/1000 radians )

Frye and Morris polynominal model  :     R = Sum ( Ai X ( K*Bm )**Pi X 10**Qi )
     t =  0.625000"  xd = 16.250000"   f =  0.750000"  xl = 10.000000"
    A1 =  8.460000   A2 =  1.010000   A3 =  1.240000    K =  0.005287
    P1 =         1   P2 =         3   P3 =         5   Q1 =        -1    Q2 =        -1    Q3 =        -5

Modified exponential model  :
    Bm = Sum ( Ai X ( 1 - exp( - R/( 2*i*C ) ) ) )  +      Sum ( RKj X ( R - Rj0 ) )  +  Bm0
     C =  0.17345750E+00             Bmo=  0.00000000E+00            Nexp=  6       Nliner=  1
    AI =  0.13353919E+04  -0.65943872E+04  0.11882853E+05  0.60776308E+03  -0.20146339E+05  0.13982810E+05
   Rj0 =          9.7800
   RKj =  0.16198629E+02

Power model  :    Bm = ( Rki X R ) / ( 1 + ( R/R0 )**rn )**( 1/rn )
     rn = 0.711      rki =   0.78041052E+03          rmu =   0.15137797E+04
```

No	Rotation (radians) X 1/1000	Moment (kip-inch) Expri.	Poly.	M.Expo.	P.Model.	Connection stiffness (kip-inch) X 1000 Expri.	Poly.	M.Expo.	P.Model.
1	0.00	0.00	0.00	0.00	0.00	0.1057E+04	0.2236E+03	0.1304E+04	0.7804E+03
3	0.09	93.19	19.71	96.90	59.34	0.1057E+04	0.2227E+03	0.9165E+03	0.6055E+03
5	0.18	186.37	39.27	165.16	108.80	0.9212E+03	0.2202E+03	0.6480E+03	0.5218E+03
7	0.59	303.06	126.30	310.64	280.21	0.2469E+03	0.1928E+03	0.1977E+03	0.3289E+03
9	1.04	398.54	204.76	390.59	404.81	0.1973E+03	0.1575E+03	0.1785E+03	0.2359E+03
11	1.67	509.18	291.46	505.60	529.08	0.1601E+03	0.1208E+03	0.1775E+03	0.1664E+03
13	2.30	600.13	359.60	605.25	620.27	0.1342E+03	0.9739E+02	0.1364E+03	0.1265E+03
15	2.87	671.37	410.79	672.18	685.18	0.1125E+03	0.8308E+02	0.1003E+03	0.1027E+03
17	3.77	754.75	478.25	746.19	765.63	0.7932E+02	0.6790E+02	0.6900E+02	0.7791E+02
19	5.41	841.96	574.89	843.91	869.91	0.5049E+02	0.5181E+02	0.5405E+02	0.5232E+02
21	7.11	921.60	654.05	926.68	945.32	0.4328E+02	0.4224E+02	0.4260E+02	0.3782E+02
23	8.87	989.88	722.30	988.57	1003.17	0.3477E+02	0.3584E+02	0.2790E+02	0.2869E+02
25	10.74	1046.04	784.91	1044.04	1050.70	0.2526E+02	0.3110E+02	0.3156E+02	0.2235E+02
27	12.51	1088.56	836.67	1092.61	1086.26	0.2279E+02	0.2782E+02	0.2425E+02	0.1825E+02
29	14.10	1125.01	879.23	1128.09	1113.14	0.2279E+02	0.2548E+02	0.2048E+02	0.1549E+02
31	15.70	1161.46	918.77	1159.00	1136.12	0.2008E+02	0.2355E+02	0.1840E+02	0.1335E+02
33	17.60	1193.38	961.22	1192.54	1159.45	0.1685E+02	0.2171E+02	0.1718E+02	0.1137E+02
35	19.49	1225.31	1000.83	1224.48	1179.48	0.1685E+02	0.2017E+02	0.1662E+02	0.9833E+01

IV - 15

```
Connection type : Top-and seat-angle connections
           Mode : All riveted

Tested by  :  R.A.Hechtman & B.G.Johnston (1947)          U.S.A
Test  Id.  :  NO 24

Column : W12X65                         Fasteners: 10.9- -7/8"D
Beam   : W18X47                               15/16" Oversize holes
Angle  : 6 X 4 X 5/8 X 12 1/2 ( TOP     Material :  --
                                             Fy =  37.50 ksi
                                             Fu =  62.50 ksi

                         Major parameters

     lt = 12.5000"    tt =  0.6250"    ls =  7.5000"    ts =  0.8750"
     gt =  3.5000"    gt'=  2.5000"    gs =  3.5000"    gs'=  3.5000"
     qt =  4.2500"    rt =  4.2500"    rt2=  2.6250"
     qs =  4.2500"    rs =  4.2500"
     pt =  2.5000"    ps =  2.5000"    ps'=  2.5000"
     nt = 2 X  2      nt'= 4 X  1      ns = 2 X  2      ns'= 2 X  2

Remark  1)  6 X 6 X 7/8 X 7 1/2 ( SEAT )
        2)  2 shims 5 X 1/8 X 7 1/4 are used on ten. flange.
```

No	Moment (k-in)	Rotation (radians) X 1/1000	No	Moment (k-in)	Rotation (radians) X 1/1000
1	0.00	0.00	26	1185.73	8.80
2	50.63	0.13	27	1212.14	9.42
3	101.25	0.26	28	1249.71	10.49
4	150.26	0.44	29	1287.27	11.57
5	199.27	0.61	30	1315.49	12.49
6	254.71	0.79	31	1343.71	13.42
7	310.15	0.97	32	1371.93	14.34
8	355.95	1.15	33	1396.93	15.28
9	401.74	1.32	34	1421.93	16.22
10	457.16	1.58	35	1446.93	17.15
11	512.59	1.84	36	1468.67	18.33
12	559.18	2.06	37	1490.40	19.51
13	605.76	2.28	38	1512.14	20.69
14	662.78	2.63	39	1534.44	21.72
15	719.79	2.99	40	1556.74	22.75
16	763.13	3.35	41	1579.04	23.78
17	806.47	3.72			
18	862.66	4.18			
19	918.85	4.64			
20	962.96	5.18			
21	1007.07	5.72			
22	1051.96	6.35			
23	1096.86	6.98			
24	1128.09	7.58			
25	1159.32	8.19			

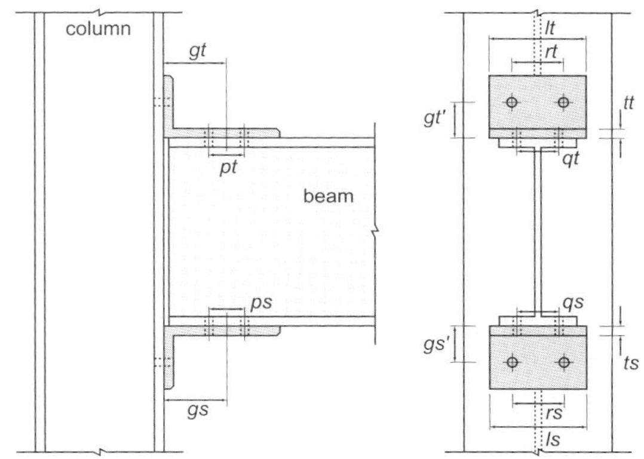

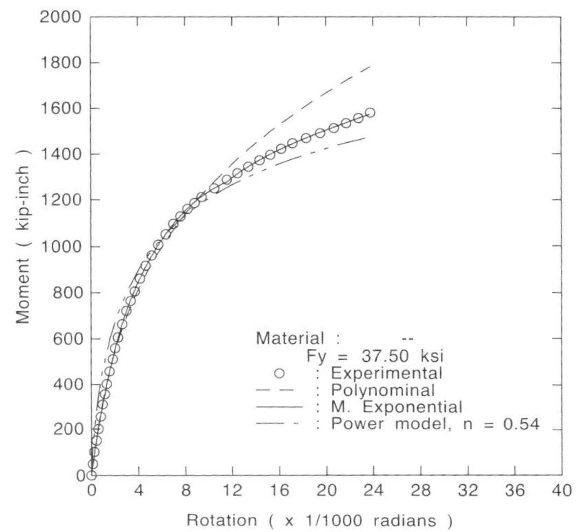

```
Moment-rotation prediction equations     ( R : X 1/1000 radians )

Frye and Morris polynominal model  :    R = Sum ( Ai X ( K*Bm )**Pi X 10**Qi )
      t =  0.625000"  xd = 18.250000"   f =  0.875000"  xl = 12.500000"
     A1 =  8.460000   A2 =  1.010000   A3 =  1.240000    K =  0.003207
     P1 =        1    P2 =        3    P3 =        5    Q1 =       -1    Q2 =       -1    Q3 =       -5

Modified exponential model  :
     Bm = Sum ( Ai X ( 1 - exp( - R/( 2*i*C ) ) ) )  +      Sum ( RKj X ( R - Rj0 ) )  + Bm0
      C =   0.23064667E+00                Bmo=   0.00000000E+00          Nexp=  6        Nliner=  1
     AI =   0.14458908E+03  -0.14982884E+04   0.32928151E+04   0.38555425E+04  -0.16127318E+05   0.11691611E+05
    Rj0 =          11.5700
    RKj =   0.17781039E+02

Power model  :    Bm = ( Rki X R ) / ( 1 + ( R/R0 )**rn )**( 1/rn )
      rn = 0.536      rki =   0.13554700E+04          rmu =   0.21896339E+04
```

No	Rotation (radians) X 1/1000	Moment (kip-inch) Expri.	Poly.	M.Expo.	P.Model.	Connection stiffness (kip-inch) X 1000 Expri.	Poly.	M.Expo.	P.Model.
1	0.00	0.00	0.00	0.00	0.00	0.3894E+03	0.3686E+03	0.3904E+03	0.1355E+04
3	0.26	101.25	94.74	94.08	194.26	0.3419E+03	0.3568E+03	0.3386E+03	0.5433E+03
5	0.61	199.27	214.26	206.35	348.05	0.2944E+03	0.3152E+03	0.2987E+03	0.3552E+03
7	0.97	310.15	317.82	307.24	457.19	0.2855E+03	0.2686E+03	0.2702E+03	0.2679E+03
9	1.32	401.74	405.63	398.26	542.07	0.2415E+03	0.2294E+03	0.2446E+03	0.2158E+03
11	1.84	512.59	511.83	514.52	639.51	0.2118E+03	0.1875E+03	0.2095E+03	0.1682E+03
13	2.28	605.76	589.76	602.07	708.41	0.1901E+03	0.1615E+03	0.1817E+03	0.1407E+03
15	2.99	719.79	692.47	716.57	796.22	0.1409E+03	0.1331E+03	0.1462E+03	0.1116E+03
17	3.72	806.47	782.08	813.13	869.95	0.1200E+03	0.1132E+03	0.1194E+03	0.9132E+02
19	4.64	918.85	877.97	911.97	945.59	0.1034E+03	0.9587E+02	0.9650E+02	0.7385E+02
21	5.72	1007.07	973.34	1005.60	1017.21	0.7680E+02	0.8195E+02	0.7829E+02	0.5994E+02
23	6.98	1096.86	1069.52	1094.18	1085.60	0.6118E+02	0.7055E+02	0.6215E+02	0.4871E+02
25	8.19	1159.32	1149.36	1160.98	1139.34	0.4747E+02	0.6268E+02	0.4933E+02	0.4111E+02
27	9.42	1212.14	1222.64	1214.62	1186.24	0.3998E+02	0.5649E+02	0.3802E+02	0.3526E+02
29	11.57	1287.27	1334.93	1278.89	1253.64	0.3256E+02	0.4856E+02	0.2278E+02	0.2800E+02
31	13.42	1343.71	1420.04	1345.19	1301.22	0.3047E+02	0.4356E+02	0.3171E+02	0.2360E+02
33	15.28	1396.93	1497.38	1398.50	1341.93	0.2670E+02	0.3964E+02	0.2598E+02	0.2027E+02
35	17.15	1446.93	1568.52	1443.57	1377.35	0.2305E+02	0.3645E+02	0.2246E+02	0.1766E+02
37	19.51	1490.40	1650.43	1493.26	1415.81	0.1846E+02	0.3323E+02	0.2002E+02	0.1512E+02
39	21.72	1534.44	1721.57	1536.07	1447.07	0.2164E+02	0.3075E+02	0.1888E+02	0.1326E+02
41	23.78	1579.04	1782.86	1574.37	1472.92	0.2164E+02	0.2882E+02	0.1834E+02	0.1186E+02

A.4 – 31

IV - 16

Connection type : Top-and seat-angle connections
Mode : All bolted

Tested by : M.J.Marley (1982) U.S.A
Test Id. : A-1/4-1

Column : W5X16
Beam : W5X16
Angle : 4 X 4 X 1/4 X 5.0

Fasteners: A325- -3/4"D
13/16" Oversize holes
Material : --
Fy = -- ksi
Fu = -- ksi

Major parameters

lt = 5.0000"	tt = 0.2500"	ls = 5.0000"	ts = 0.2500"
gt = -- "	gt'= -- "	gs = -- "	gs'= -- "
qt = -- "	rt = -- "	qs = -- "	rs = -- "
nt = 2 X 1	nt'= 2 X 1	ns = 2 X 1	ns'= 2 X 1

Remark 1)
2)

No	Moment (k-in)	Rotation (radians) X 1/1000	No	Moment (k-in)	Rotation (radians) X 1/1000
1	0.00	0.00	26	43.33	21.30
2	1.77	0.15	27	44.34	23.40
3	3.54	0.30	28	44.95	25.25
4	5.76	0.65	29	45.55	27.10
5	7.98	1.00	30	45.96	28.40
6	10.20	1.40	31	46.37	29.70
7	12.42	1.80	32	46.97	31.40
8	14.34	2.25	33	47.58	33.10
9	16.26	2.70	34	48.19	34.65
10	18.38	3.30	35	48.79	36.20
11	20.51	3.90	36	49.30	37.85
12	22.63	4.45	37	49.80	39.50
13	24.74	5.00	38	50.41	41.40
14	27.27	6.00	39	51.01	43.30
15	29.80	7.00			
16	31.51	8.10			
17	33.23	9.20			
18	35.05	10.60			
19	35.96	11.70			
20	36.86	12.80			
21	37.78	13.90			
22	38.69	15.00			
23	40.31	16.90			
24	42.32	18.70			
25	42.83	20.00			

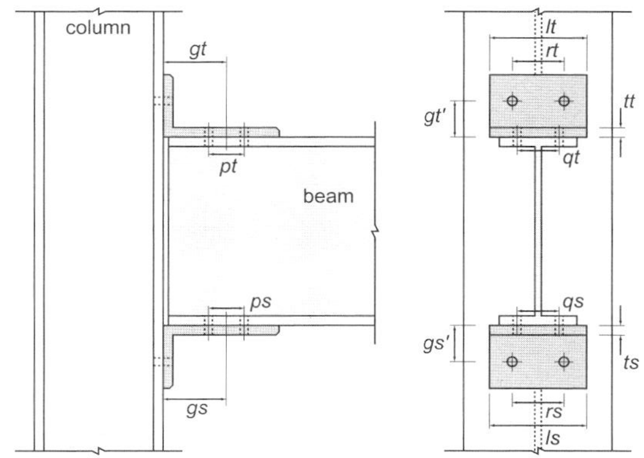

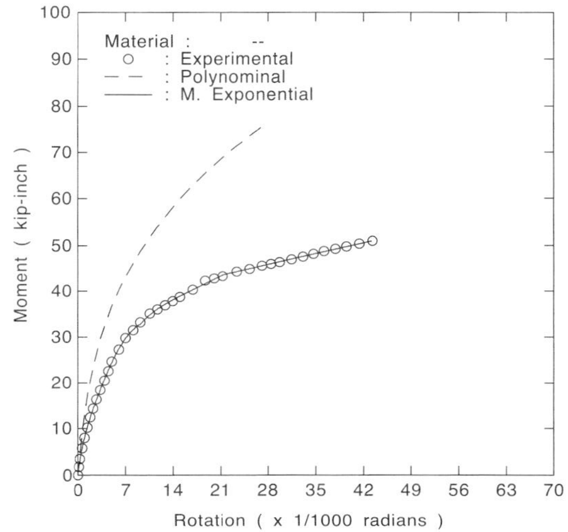

```
Moment-rotation prediction equations      ( R : X 1/1000 radians )

Frye and Morris polynominal model  :    R = Sum ( Ai X ( K*Bm )**Pi X 10**Qi )
     t =  0.250000"  xd =  5.000000"   f =  0.750000"  xl =  5.000000"
    A1 =  8.460000   A2 =  1.010000   A3 =  1.240000    K =  0.079566
    P1 =         1   P2 =         3   P3 =         5   Q1 =        -1     Q2 =        -1     Q3 =        -5

Modified exponential model  :
     Bm = Sum ( Ai X ( 1 - exp( - R/( 2*i*C ) ) ) )  +      Sum ( RKj X ( R - Rj0 ) )  +  Bm0
      C =   0.39833333E+00                Bmo=   0.00000000E+00           Nexp=  6         Nliner=  2
     AI =   0.39295103E+01   0.13730379E+03  -0.83853324E+03   0.16959029E+04  -0.14068127E+04   0.44311840E+03
    Rj0 =           9.2000          21.3000
    RKj =   0.75604296E+00  -0.43917392E+00
```

No	Rotation (radians) X 1/1000	Moment (kip-inch) Expri.	Moment (kip-inch) Poly.	Moment (kip-inch) M.Expo.	Connection stiffness (kip-inch) X 1000 Expri.	Connection stiffness (kip-inch) X 1000 Poly.	Connection stiffness (kip-inch) X 1000 M.Expo.
1	0.00	0.00	0.00	0.00	0.1180E+02	0.1486E+02	0.1197E+02
3	0.30	3.54	4.39	3.15	0.1016E+02	0.1423E+02	0.9189E+01
5	1.00	7.98	13.14	8.20	0.5973E+01	0.1067E+02	0.5804E+01
7	1.80	12.42	20.36	12.22	0.4946E+01	0.7655E+01	0.4511E+01
9	2.70	16.26	26.32	16.06	0.3955E+01	0.5775E+01	0.4092E+01
11	3.90	20.51	32.34	20.76	0.3702E+01	0.4401E+01	0.3710E+01
13	5.00	24.74	36.74	24.56	0.3381E+01	0.3653E+01	0.3176E+01
15	7.00	29.80	43.16	29.73	0.2066E+01	0.2839E+01	0.1985E+01
17	9.20	33.23	48.78	32.84	0.1447E+01	0.2317E+01	0.9196E+00
19	11.70	35.96	54.07	36.10	0.8236E+00	0.1941E+01	0.1020E+01
21	13.90	37.78	58.07	38.05	0.8291E+00	0.1711E+01	0.7953E+00
23	16.90	40.31	62.85	40.30	0.9899E+00	0.1485E+01	0.7346E+00
25	20.00	42.83	67.18	42.61	0.3877E+00	0.1316E+01	0.7578E+00
27	23.40	44.34	71.40	44.31	0.3990E+00	0.1176E+01	0.3426E+00
29	27.10	45.55	75.53	45.59	0.3195E+00	0.1059E+01	0.3483E+00
31	29.70	46.37	78.19	46.50	0.3323E+00	0.9922E+00	0.3447E+00
33	33.10	47.58	81.47	47.65	0.3745E+00	0.9183E+00	0.3372E+00
35	36.20	48.79	84.22	48.69	0.3495E+00	0.8622E+00	0.3307E+00
37	39.50	49.80	86.98	49.77	0.3117E+00	0.8109E+00	0.3255E+00
39	43.30	51.01	89.96	51.00	0.3189E+00	0.7602E+00	0.3216E+00

IV - 17

Connection type : Top-and seat-angle connections
Mode : All bolted

Tested by : M.J.Marley (1982) U.S.A
Test Id. : A-1/4-2

Column : W5X16
Beam : W5X16
Angle : 4 X 4 X 1/4 X 5.0

Fasteners: A325- -3/4"D
13/16" Oversize holes
Material : --
Fy = -- ksi
Fu = -- ksi

Major parameters

lt = 5.0000"	tt = 0.2500"	ls = 5.0000"	ts = 0.2500"
gt = -- "	gt'= -- "	gs = -- "	gs'= -- "
qt = -- "	rt = -- "	qs = -- "	rs = -- "
nt = 2 X 1	nt'= 2 X 1	ns = 2 X 1	ns'= 2 X 1

Remark 1)
2)

No	Moment (k-in)	Rotation (radians) X 1/1000
1	0.00	0.00
2	1.58	0.17
3	3.17	0.33
4	4.75	0.50
5	6.67	0.90
6	8.58	1.30
7	10.13	1.60
8	11.68	1.90
9	13.24	2.20
10	14.72	2.67
11	16.20	3.13
12	17.68	3.60
13	19.80	4.25
14	21.92	4.90
15	23.64	5.50
16	25.36	6.10
17	27.47	7.15
18	29.59	8.20
19	30.74	8.97
20	31.89	9.73
21	33.04	10.50
22	34.15	11.20
23	35.26	11.90
24	36.37	12.95
25	37.48	14.00
26	38.48	14.90
27	39.49	15.80
28	40.03	16.57
29	40.57	17.33
30	41.11	18.10
31	41.56	18.93
32	42.02	19.75
33	42.47	20.57
34	42.92	21.40

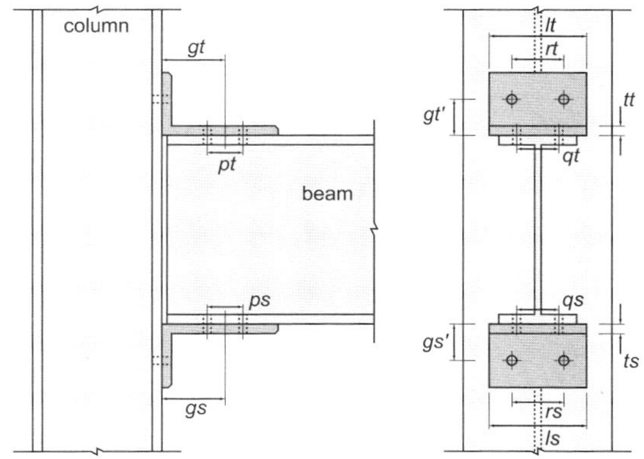

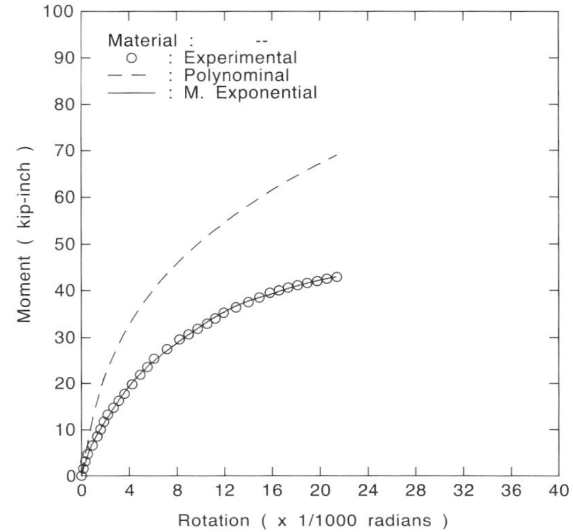

A.4 – 34

```
Moment-rotation prediction equations     ( R : X 1/1000 radians )

Frye and Morris polynominal model  :    R = Sum ( Ai X ( K*Bm )**Pi X 10**Qi )
      t =  0.250000"  xd =  5.000000"   f =  0.750000"  xl =  5.000000"
     A1 =  8.460000   A2 =  1.010000   A3 =  1.240000    K =  0.079566
     P1 =         1   P2 =         3   P3 =         5   Q1 =        -1    Q2 =        -1    Q3 =        -5

Modified exponential model  :
     Bm = Sum ( Ai X ( 1 - exp( - R/( 2*i*C ) ) ) )  +      Sum ( RKj X ( R - Rj0 ) )  +  Bm0
      C =   0.22000833E+00              Bmo=   0.00000000E+00           Nexp=  6       Nliner=  1
     AI =  -0.19313979E+02   0.36546377E+03  -0.20318315E+04   0.49569432E+04  -0.54864649E+04   0.22564048E+04
    Rj0 =          15.8000
    RKj =   0.38195435E+00
```

No	Rotation (radians) X 1/1000	Moment (kip-inch) Expri.	Poly.	M.Expo.	Connection stiffness (kip-inch) X 1000 Expri.	Poly.	M.Expo.
1	0.00	0.00	0.00	0.00	0.9502E+01	0.1486E+02	0.9434E+01
3	0.33	3.17	4.86	3.09	0.9505E+01	0.1410E+02	0.8605E+01
5	0.90	6.67	12.05	7.02	0.4785E+01	0.1117E+02	0.5475E+01
7	1.60	10.13	18.77	10.22	0.5173E+01	0.8257E+01	0.4086E+01
9	2.20	13.24	23.22	12.64	0.4390E+01	0.6682E+01	0.4004E+01
11	3.13	16.20	28.69	16.32	0.3172E+01	0.5179E+01	0.3816E+01
13	4.25	19.80	33.84	20.21	0.3268E+01	0.4127E+01	0.3100E+01
15	5.50	23.64	38.51	23.61	0.2860E+01	0.3401E+01	0.2398E+01
17	7.15	27.47	43.58	27.18	0.2017E+01	0.2794E+01	0.2012E+01
19	8.97	30.74	48.24	30.67	0.1497E+01	0.2361E+01	0.1833E+01
21	10.50	33.04	51.65	33.32	0.1544E+01	0.2102E+01	0.1609E+01
23	11.90	35.26	54.46	35.39	0.1374E+01	0.1917E+01	0.1335E+01
25	14.00	37.48	58.25	37.74	0.1091E+01	0.1702E+01	0.9070E+00
27	15.80	39.49	61.18	39.08	0.8955E+00	0.1559E+01	0.6008E+00
29	17.33	40.57	63.49	40.43	0.7044E+00	0.1458E+01	0.7867E+00
31	18.93	41.56	65.74	41.56	0.5491E+00	0.1369E+01	0.6417E+00
33	20.57	42.47	67.93	42.53	0.5491E+00	0.1289E+01	0.5414E+00

IV - 18

Connection type : Top-and seat-angle connections
Mode : All bolted

Tested by : M.J.Marley (1982) U.S.A
Test Id. : B-1/4-1

Column : W5X16
Beam : W5X16
Angle : 4 X 4 X 1/4 X 5.0

Fasteners: A325- -3/4"D
15/16" Oversize holes
Material : --
Fy = -- ksi
Fu = -- ksi

Major parameters

lt = 5.0000"	tt = 0.2500"	ls = 5.0000"	ts = 0.2500"
gt = -- "	gt'= -- "	gs = -- "	gs'= -- "
qt = -- "	rt = -- "	qs = -- "	rs = -- "
nt = 2 X 1	nt'= 2 X 1	ns = 2 X 1	ns'= 2 X 1

Remark 1)
2)

No	Moment (k-in)	Rotation (radians) X 1/1000
1	0.00	0.00
2	1.96	0.00
3	3.92	0.00
4	5.88	0.00
5	8.06	0.05
6	10.24	0.10
7	12.41	0.30
8	14.59	0.50
9	16.55	0.85
10	18.50	1.20
11	20.03	1.30
12	21.55	1.40
13	22.97	1.65
14	24.38	1.90
15	26.13	2.25
16	27.88	2.60
17	29.18	2.95
18	30.48	3.30
19	31.50	3.90
20	32.51	4.50
21	33.53	5.10
22	34.62	5.75
23	35.71	6.40
24	37.02	7.10
25	38.33	7.80

No	Moment (k-in)	Rotation (radians) X 1/1000
26	38.98	8.45
27	39.63	9.10
28	40.28	9.75
29	40.93	10.40
30	41.59	11.00
31	42.24	11.60
32	42.90	12.20
33	43.28	12.88
34	43.66	13.55
35	44.04	14.22
36	44.42	14.90
37	44.86	15.70
38	45.30	16.50
39	45.73	17.30

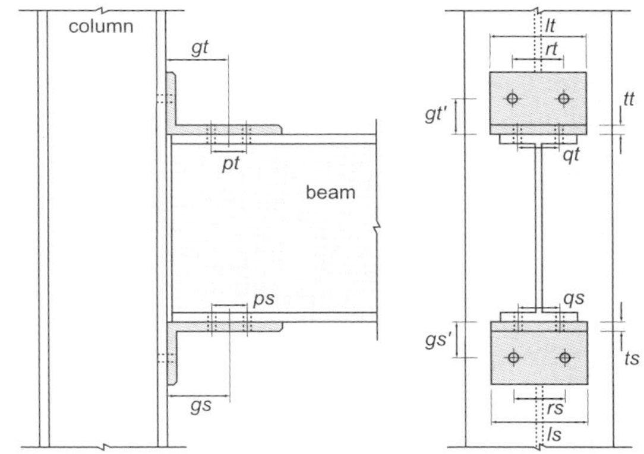

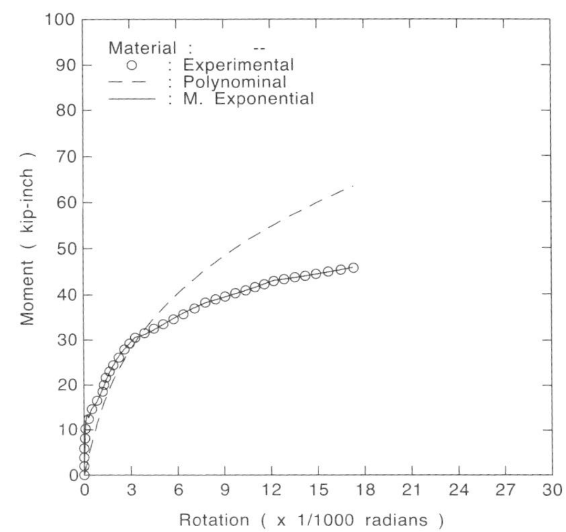

```
Moment-rotation prediction equations     ( R : X 1/1000 radians )

Frye and Morris polynominal model  :     R = Sum ( Ai X ( K*Bm )**Pi X 10**Qi )
      t =  0.250000"  xd =  5.000000"   f =  0.750000"  xl =  5.000000"
     A1 =  8.460000   A2 =  1.010000   A3 =  1.240000    K =  0.079566
     P1 =         1   P2 =         3   P3 =         5   Q1 =        -1    Q2 =        -1    Q3 =        -5

Modified exponential model  :
     Bm = Sum ( Ai X ( 1 - exp( - R/( 2*i*C ) ) ) )  +      Sum ( RKj X ( R - Rj0 ) )  +  Bm0
      C =  0.14419167E+00            Bmo=   0.00000000E+00          Nexp=  6        Nliner=  7
     AI =  0.16806293E+03  -0.11650263E+04  0.40541226E+04  -0.70652947E+04  0.60202375E+04  -0.19914183E+04
    Rj0 =          0.1000           0.3000          0.5000           1.9000          3.3000
                   7.8000          12.2000
    RKj = -0.37979664E+02   0.40563467E+02  0.14970847E+01   0.11615218E+01 -0.32664508E+01
          -0.41626388E+00  -0.90583219E+00
```

No	Rotation (radians) X 1/1000	Moment (kip-inch) Expri.	Moment (kip-inch) Poly.	Moment (kip-inch) M.Expo.	Connection stiffness (kip-inch) X 1000 Expri.	Connection stiffness (kip-inch) X 1000 Poly.	Connection stiffness (kip-inch) X 1000 M.Expo.
1	0.00	0.00	0.00	0.00	0.1960E+05	0.1486E+02	0.1482E+03
3	0.00	3.92	0.00	0.03	0.1960E+05	0.1486E+02	0.1481E+03
5	0.05	8.06	0.74	6.42	0.4369E+02	0.1484E+02	0.1103E+03
7	0.30	12.41	4.39	12.41	0.1089E+02	0.1423E+02	-0.1896E+02
9	0.85	16.55	11.49	16.44	0.5589E+01	0.1144E+02	0.6233E+01
11	1.30	20.03	16.14	20.01	0.1524E+02	0.9340E+01	0.8624E+01
13	1.65	22.97	19.18	22.81	0.5664E+01	0.8098E+01	0.7134E+01
15	2.25	26.13	23.55	26.42	0.4989E+01	0.6578E+01	0.4916E+01
17	2.95	29.18	27.72	29.27	0.3720E+01	0.5414E+01	0.3734E+01
19	3.90	31.50	32.34	31.23	0.1693E+01	0.4401E+01	0.1382E+01
21	5.10	33.53	37.11	33.44	0.1687E+01	0.3599E+01	0.2079E+01
23	6.40	35.71	41.40	36.01	0.1771E+01	0.3036E+01	0.1796E+01
25	7.80	38.33	45.34	38.25	0.1419E+01	0.2619E+01	0.1450E+01
27	9.10	39.63	48.55	39.54	0.1002E+01	0.2336E+01	0.9883E+00
29	10.40	40.93	51.44	40.88	0.1049E+01	0.2116E+01	0.1099E+01
31	11.60	42.24	53.88	42.28	0.1093E+01	0.1953E+01	0.1232E+01
33	12.88	43.28	56.27	43.32	0.5644E+00	0.1810E+01	0.4462E+00
35	14.22	44.04	58.63	43.99	0.5644E+00	0.1683E+01	0.5346E+00
37	15.70	44.86	61.02	44.82	0.5450E+00	0.1567E+01	0.5924E+00
39	17.30	45.73	63.44	45.80	0.5450E+00	0.1460E+01	0.6254E+00

IV - 19

Connection type : Top-and seat-angle connections
Mode : All bolted

Tested by : M.J.Marley (1982) U.S.A
Test Id. : B-1/4-2

Column : W5X16
Beam : W5X16
Angle : 4 X 4 X 1/4 X 5.0

Fasteners: A325- -3/4"D
13/16" Oversize holes
Material : --
Fy = -- ksi
Fu = -- ksi

Major parameters

lt = 5.0000"	tt = 0.2500"	ls = 5.0000"	ts = 0.2500"
gt = -- "	gt'= -- "	gs = -- "	gs'= -- "
qt = -- "	rt = -- "	qs = -- "	rs = -- "
nt = 2 X 1	nt'= 2 X 1	ns = 2 X 1	ns'= 2 X 1

Remark 1)
2)

No	Moment (k-in)	Rotation (radians) X 1/1000	No	Moment (k-in)	Rotation (radians) X 1/1000
1	0.00	0.00	26	37.48	21.50
2	1.74	0.07	27	38.48	23.00
3	3.49	0.15	28	39.49	24.50
4	5.23	0.23	29	40.17	25.60
5	6.97	0.30	30	40.84	26.70
6	8.92	0.50	31	41.52	27.80
7	10.88	0.70	32	42.32	29.05
8	12.83	0.90	33	43.13	30.30
9	14.85	1.30	34	44.04	31.30
10	16.87	1.70	35	44.95	32.30
11	18.89	2.10			
12	20.71	3.05			
13	22.52	4.00			
14	24.14	5.50			
15	25.75	7.00			
16	26.96	8.23			
17	28.18	9.47			
18	29.39	10.70			
19	30.70	12.30			
20	32.02	13.90			
21	33.10	15.33			
22	34.18	16.77			
23	35.26	18.20			
24	36.00	19.30			
25	36.74	20.40			

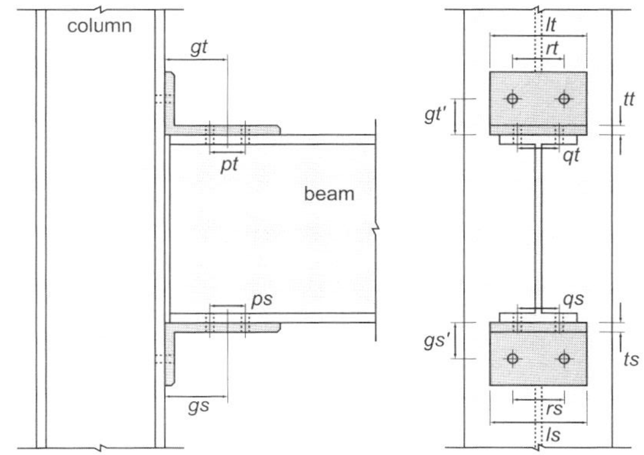

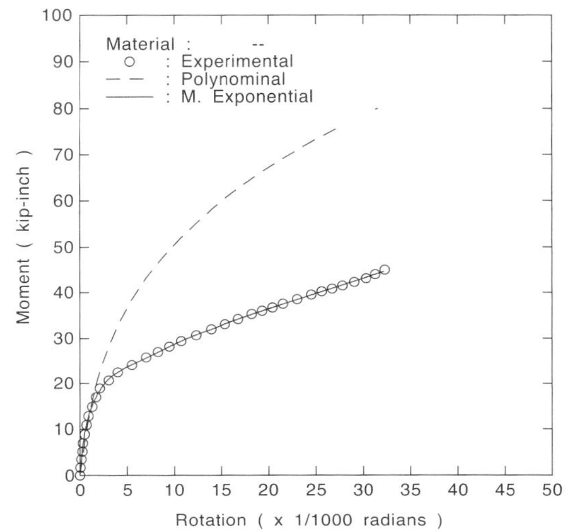

A.4 – 38

```
Moment-rotation prediction equations     ( R : X 1/1000 radians )

Frye and Morris polynominal model  :    R = Sum ( Ai X ( K*Bm )**Pi X 10**Qi )
      t =  0.250000"  xd =  5.000000"   f =  0.750000"  xl =  5.000000"
     A1 =  8.460000   A2 =  1.010000   A3 =  1.240000    K =  0.079566
     P1 =         1   P2 =         3   P3 =         5   Q1 =        -1    Q2 =        -1    Q3 =        -5

Modified exponential model  :
     Bm = Sum ( Ai X ( 1 - exp( - R/( 2*i*C ) ) ) )  +      Sum ( RKj X ( R - Rj0 ) )  +  Bm0
      C =  0.28791667E+00           Bmo=  0.00000000E+00          Nexp=  6        Nliner=  2
     AI =  0.41088807E+02  -0.33236183E+03  0.12978832E+04  -0.22616946E+04   0.18112808E+04  -0.52894032E+03
    Rj0 =          4.0000          13.9000
    RKj =  0.47769698E+00   0.21142155E+00
```

No	Rotation (radians) X 1/1000	Moment (kip-inch) Expri.	Moment (kip-inch) Poly.	Moment (kip-inch) M.Expo.	Connection stiffness (kip-inch) X 1000 Expri.	Connection stiffness (kip-inch) X 1000 Poly.	Connection stiffness (kip-inch) X 1000 M.Expo.
1	0.00	0.00	0.00	0.00	0.2324E+02	0.1486E+02	0.2815E+02
3	0.15	3.49	2.22	3.66	0.2324E+02	0.1469E+02	0.2102E+02
5	0.30	6.97	4.39	6.41	0.1956E+02	0.1423E+02	0.1598E+02
7	0.70	10.88	9.71	11.13	0.9760E+01	0.1224E+02	0.8724E+01
9	1.30	14.85	16.14	15.06	0.5050E+01	0.9338E+01	0.5136E+01
11	2.10	18.89	22.54	18.43	0.4121E+01	0.6901E+01	0.3466E+01
13	4.00	22.52	32.79	22.45	0.1589E+01	0.4318E+01	0.1056E+01
15	7.00	25.75	43.16	25.66	0.1025E+01	0.2838E+01	0.9928E+00
17	9.47	28.18	49.39	28.20	0.9827E+00	0.2268E+01	0.1019E+01
19	12.30	30.70	55.21	30.79	0.8212E+00	0.1871E+01	0.7940E+00
21	15.33	33.10	60.44	33.14	0.7535E+00	0.1594E+01	0.7832E+00
23	18.20	35.26	64.73	35.22	0.7078E+00	0.1408E+01	0.6872E+00
25	20.40	36.74	67.70	36.70	0.6727E+00	0.1297E+01	0.6641E+00
27	23.00	38.48	70.93	38.42	0.6720E+00	0.1190E+01	0.6619E+00
29	25.60	40.17	73.93	40.15	0.6145E+00	0.1102E+01	0.6686E+00
31	27.80	41.52	76.28	41.63	0.6280E+00	0.1039E+01	0.6748E+00
33	30.30	43.13	78.80	43.32	0.7925E+00	0.9778E+00	0.6804E+00
35	32.30	44.95	80.69	44.69	0.9120E+00	0.9350E+00	0.6834E+00

IV - 20

Connection type : Top-and seat-angle connections
Mode : All bolted

Tested by : M.J.Marley (1982) U.S.A
Test Id. : C1-1/4-2

Column : W5X16
Beam : W5X16
Angle : 4 X 4 X 1/4 X 5.0

Fasteners: A325- -3/4"D
13/16" Oversize holes
Material : --
Fy = -- ksi
Fu = -- ksi

Major parameters

lt = 5.0000"	tt = 0.2500"	ls = 5.0000"	ts = 0.2500"
gt = -- "	gt'= -- "	gs = -- "	gs'= -- "
qt = -- "	rt = -- "	qs = -- "	rs = -- "
nt = 2 X 1	nt'= 2 X 1	ns = 2 X 1	ns'= 2 X 1

Remark 1)
2)

No	Moment (k-in)	Rotation (radians) X 1/1000	No	Moment (k-in)	Rotation (radians) X 1/1000
1	0.00	0.00	26	41.92	21.70
2	2.26	0.10	27	43.23	23.55
3	4.51	0.20	28	44.54	25.40
4	6.77	0.30	29	45.35	27.35
5	8.79	0.55	30	46.16	29.30
6	10.81	0.80	31	46.77	31.50
7	12.93	1.10	32	47.38	33.70
8	15.05	1.40	33	48.08	35.55
9	17.17	1.65	34	48.79	37.40
10	19.30	1.90	35	49.09	39.30
11	21.31	2.40	36	49.39	41.20
12	23.33	2.90	37	49.80	43.25
13	24.64	3.25	38	50.21	45.30
14	25.96	3.60			
15	27.47	4.50			
16	28.99	5.40			
17	30.61	6.70			
18	32.22	8.00			
19	33.43	9.75			
20	34.64	11.50			
21	36.06	13.25			
22	37.48	15.00			
23	38.69	16.75			
24	39.90	18.50			
25	40.91	20.10			

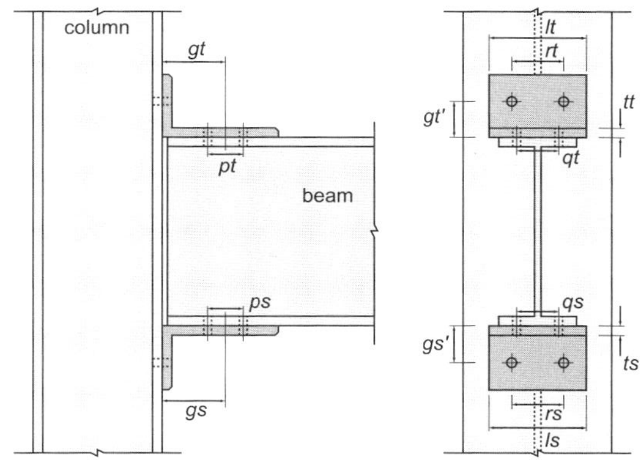

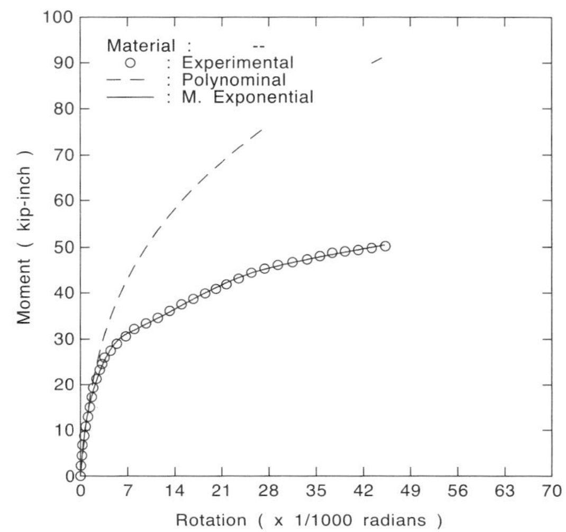

A.4 – 40

```
Moment-rotation prediction equations     ( R : X 1/1000 radians )

Frye and Morris polynominal model  :    R = Sum ( Ai X ( K*Bm )**Pi X 10**Qi )
      t =  0.250000"  xd =  5.000000"   f =  0.750000"  xl =  5.000000"
     A1 =  8.460000   A2 =  1.010000   A3 =  1.240000    K =  0.079566
     P1 =        1    P2 =        3    P3 =        5    Q1 =       -1    Q2 =       -1    Q3 =       -5

Modified exponential model  :
     Bm = Sum ( Ai X ( 1 - exp( - R/( 2*i*C ) ) ) )  +      Sum ( RKj X ( R - Rj0 ) )  +  Bm0
      C =   0.40250000E+00             Bmo=   0.00000000E+00          Nexp=  6        Nliner=  2
     AI =   0.30660901E+02  -0.14305337E+03   0.15959896E+03   0.52442488E+03  -0.11296924E+04   0.59780858E+03
    Rj0 =          8.0000         25.4000
    RKj =   0.34324344E+00  -0.10297035E+00
```

No	Rotation (radians) X 1/1000	Moment (kip-inch) Expri.	Moment (kip-inch) Poly.	Moment (kip-inch) M.Expo.	Connection stiffness (kip-inch) X 1000 Expri.	Connection stiffness (kip-inch) X 1000 Poly.	Connection stiffness (kip-inch) X 1000 M.Expo.
1	0.00	0.00	0.00	0.00	0.2256E+02	0.1486E+02	0.2129E+02
3	0.20	4.51	2.95	3.79	0.2256E+02	0.1457E+02	0.1681E+02
5	0.55	8.79	7.81	8.69	0.8088E+01	0.1305E+02	0.1164E+02
7	1.10	12.93	14.19	13.79	0.7060E+01	0.1020E+02	0.7496E+01
9	1.65	17.17	19.18	17.34	0.8496E+01	0.8098E+01	0.5631E+01
11	2.40	21.31	24.52	21.04	0.4032E+01	0.6285E+01	0.4369E+01
13	3.25	24.64	29.29	24.33	0.3754E+01	0.5041E+01	0.3409E+01
15	4.50	27.47	34.85	27.83	0.1687E+01	0.3953E+01	0.2221E+01
17	6.70	30.61	42.29	31.04	0.1242E+01	0.2933E+01	0.8736E+00
19	9.75	33.43	50.03	33.28	0.6926E+00	0.2219E+01	0.7289E+00
21	13.25	36.06	56.94	35.89	0.8091E+00	0.1772E+01	0.7823E+00
23	16.75	38.69	62.63	38.68	0.6926E+00	0.1495E+01	0.7870E+00
25	20.10	40.91	67.31	41.19	0.6300E+00	0.1311E+01	0.7018E+00
27	23.55	43.23	71.58	43.42	0.7103E+00	0.1170E+01	0.5923E+00
29	27.35	45.35	75.79	45.27	0.4154E+00	0.1052E+01	0.3902E+00
31	31.50	46.77	79.94	46.73	0.2755E+00	0.9517E+00	0.3196E+00
33	35.55	48.08	83.63	47.93	0.3827E+00	0.8738E+00	0.2807E+00
35	39.30	49.09	86.79	48.95	0.1579E+00	0.8142E+00	0.2612E+00
37	43.25	49.80	89.90	49.96	0.1990E+00	0.7612E+00	0.2505E+00

IV - 21

Connection type : Top-and seat-angle connections
Mode : All bolted

Tested by : M.J.Marley (1982) U.S.A
Test Id. : C2-1/4-1

Column : W5X16
Beam : W5X16
Angle : 4 X 4 X 1/4 X 5.0

Fasteners: A325- -3/4"D
13/16" Oversize holes
Material : --
Fy = -- ksi
Fu = -- ksi

Major parameters

lt = 5.0000"	tt = 0.2500"	ls = 5.0000"	ts = 0.2500"
gt = -- "	gt'= -- "	gs = -- "	gs'= -- "
qt = -- "	rt = -- "	qs = -- "	rs = -- "
nt = 2 X 1	nt'= 2 X 1	ns = 2 X 1	ns'= 2 X 1

Remark 1)
2)

No	Moment (k-in)	Rotation (radians) X 1/1000	No	Moment (k-in)	Rotation (radians) X 1/1000
1	0.00	0.00	26	40.91	22.40
2	1.99	0.07	27	41.82	24.10
3	3.98	0.13	28	42.73	25.80
4	5.96	0.20	29	43.94	27.40
5	8.12	0.37	30	45.16	29.00
6	10.27	0.53	31	45.86	30.40
7	12.42	0.70	32	46.56	31.80
8	14.17	1.03	33	46.97	33.05
9	15.92	1.37	34	47.38	34.30
10	17.68	1.70	35	47.88	36.35
11	20.10	2.45	36	48.38	38.40
12	22.52	3.20	37	48.89	39.95
13	24.14	3.70	38	49.39	41.50
14	25.75	4.20	39	49.80	43.25
15	27.37	5.25	40	50.21	45.00
16	28.99	6.30	41	50.91	47.25
17	30.61	7.70	42	51.61	49.50
18	32.22	9.10			
19	33.23	10.65			
20	34.25	12.20			
21	35.36	14.10			
22	36.47	16.00			
23	37.58	17.80			
24	38.69	19.60			
25	39.80	21.00			

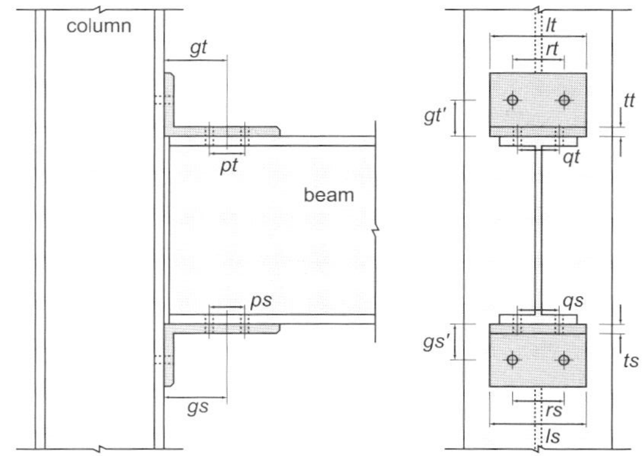

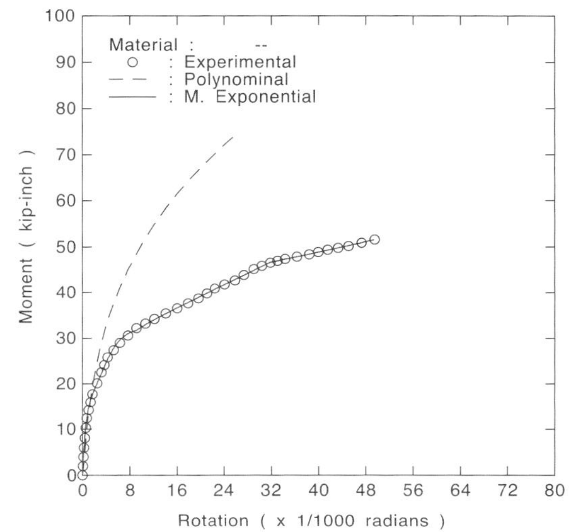

A.4 – 42

```
Moment-rotation prediction equations     ( R : X 1/1000 radians )

Frye and Morris polynominal model  :     R = Sum ( Ai X ( K*Bm )**Pi X 10**Qi )
     t =  0.250000"   xd =  5.000000"   f =  0.750000"  xl =  5.000000"
     A1 =  8.460000    A2 =  1.010000   A3 =  1.240000    K =  0.079566
     P1 =         1    P2 =         3   P3 =         5   Q1 =        -1    Q2 =        -1    Q3 =        -5

Modified exponential model  :
     Bm = Sum ( Ai X ( 1 - exp( - R/( 2*i*C ) ) ) )  +     Sum ( RKj X ( R - Rj0 ) )  +  Bm0
      C =   0.42917500E+00            Bmo=   0.00000000E+00          Nexp=  6         Nliner=  2
     AI =   0.72163807E+02  -0.38772069E+03  0.10526584E+04  -0.15122423E+04   0.11815342E+04  -0.37500991E+03
    Rj0 =           9.1000          31.8000
    RKj =   0.66494123E+00  -0.37846321E+00
```

No	Rotation (radians) X 1/1000	Moment (kip-inch) Expri.	Moment (kip-inch) Poly.	Moment (kip-inch) M.Expo.	Connection stiffness (kip-inch) X 1000 Expri.	Connection stiffness (kip-inch) X 1000 Poly.	Connection stiffness (kip-inch) X 1000 M.Expo.
1	0.00	0.00	0.00	0.00	0.2981E+02	0.1486E+02	0.2905E+02
3	0.13	3.98	1.97	3.48	0.2983E+02	0.1473E+02	0.2342E+02
5	0.37	8.12	5.33	8.04	0.1291E+02	0.1396E+02	0.1607E+02
7	0.70	12.42	9.71	12.20	0.1036E+02	0.1224E+02	0.9556E+01
9	1.37	15.92	16.75	16.41	0.5256E+01	0.9078E+01	0.4229E+01
11	2.45	20.10	24.84	20.02	0.3232E+01	0.6190E+01	0.3044E+01
13	3.70	24.14	31.45	23.81	0.3228E+01	0.4578E+01	0.2943E+01
15	5.25	27.37	37.64	27.77	0.1543E+01	0.3521E+01	0.2075E+01
17	7.70	30.61	45.07	31.01	0.1153E+01	0.2644E+01	0.7004E+00
19	10.65	33.23	51.96	32.99	0.6542E+00	0.2080E+01	0.7529E+00
21	14.10	35.36	58.42	35.35	0.5842E+00	0.1694E+01	0.6567E+00
23	17.80	37.58	64.16	37.76	0.6167E+00	0.1431E+01	0.6503E+00
25	21.00	39.80	68.47	39.83	0.7929E+00	0.1270E+01	0.6396E+00
27	24.10	41.82	72.22	41.80	0.5365E+00	0.1151E+01	0.6325E+00
29	27.40	43.94	75.84	43.88	0.7575E+00	0.1051E+01	0.6325E+00
31	30.40	45.86	78.90	45.79	0.5014E+00	0.9754E+00	0.6371E+00
33	33.05	46.97	81.41	47.01	0.3264E+00	0.9196E+00	0.2642E+00
35	36.35	47.88	84.32	47.89	0.2459E+00	0.8602E+00	0.2708E+00
37	39.95	48.89	87.34	48.88	0.3252E+00	0.8045E+00	0.2765E+00
39	43.25	49.80	89.92	49.80	0.2331E+00	0.7608E+00	0.2802E+00
41	47.25	50.91	92.88	50.92	0.3120E+00	0.7150E+00	0.2830E+00

A.4 – 43

IV - 22

Connection type : Top-and seat-angle connections
Mode : All bolted

Tested by : M.J.Marley (1982) U.S.A
Test Id. : C2-1/4-2

Column : W5X16
Beam : W5X16
Angle : 4 X 4 X 1/4 X 5.0

Fasteners: A325- -3/4"D
13/16" Oversize holes
Material : --
Fy = -- ksi
Fu = -- ksi

Major parameters

lt = 5.0000"	tt = 0.2500"	ls = 5.0000"	ts = 0.2500"
gt = -- "	gt'= -- "	gs = -- "	gs'= -- "
qt = -- "	rt = -- "	qs = -- "	rs = -- "
nt = 2 X 1	nt'= 2 X 1	ns = 2 X 1	ns'= 2 X 1

Remark 1)
2)

No	Moment (k-in)	Rotation (radians) X 1/1000	No	Moment (k-in)	Rotation (radians) X 1/1000
1	0.00	0.00	26	37.93	20.80
2	2.11	0.17	27	39.14	22.50
3	4.22	0.33	28	40.34	24.20
4	6.34	0.50	29	41.55	25.55
5	8.45	0.75	30	42.76	26.90
6	10.56	1.00	31	43.66	28.45
7	12.67	1.35	32	44.57	30.00
8	14.78	1.70	33	45.27	31.30
9	16.80	2.15	34	45.97	32.60
10	18.82	2.60	35	46.68	34.05
11	20.42	3.30	36	47.39	35.50
12	22.03	4.00	37	47.89	36.95
13	23.44	4.50	38	48.40	38.40
14	24.85	5.00	39	49.10	40.05
15	26.26	5.80	40	49.80	41.70
16	27.66	6.60	41	50.20	43.15
17	29.07	8.15	42	50.60	44.60
18	30.48	9.70	43	51.01	45.90
19	31.49	11.10	44	51.41	47.20
20	32.50	12.50	45	51.82	49.20
21	33.30	14.00			
22	34.10	15.50			
23	35.11	16.90			
24	36.12	18.30			
25	37.03	19.55			

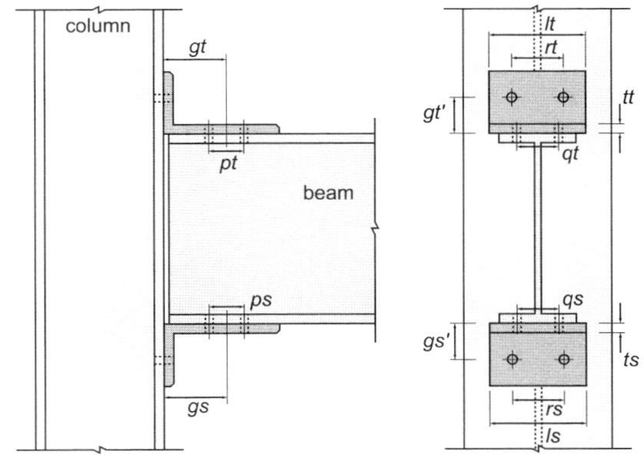

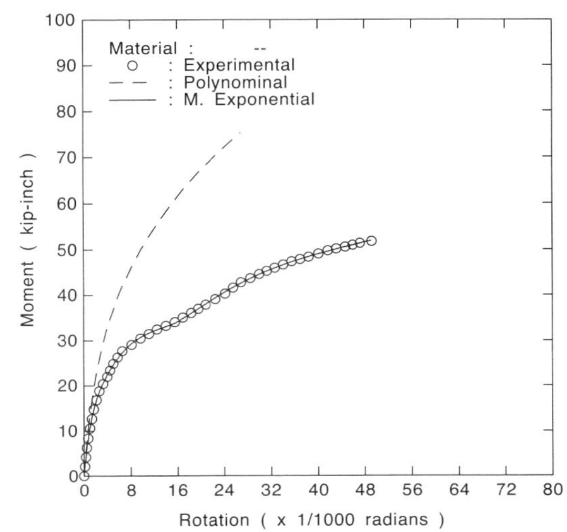

A.4 – 44

```
Moment-rotation prediction equations     ( R : X 1/1000 radians )

Frye and Morris polynominal model  :    R = Sum ( Ai X ( K*Bm )**Pi X 10**Qi )
      t =  0.250000"  xd =  5.000000"   f =  0.750000"  xl =  5.000000"
     A1 =  8.460000   A2 =  1.010000   A3 =  1.240000    K =  0.079566
     P1 =         1   P2 =         3   P3 =         5   Q1 =        -1     Q2 =        -1     Q3 =        -5

Modified exponential model  :
     Bm = Sum ( Ai X ( 1 - exp( - R/( 2*i*C ) ) ) )  +      Sum ( RKj X ( R - Rj0 ) )  +  Bm0
      C =   0.45167500E+00              Bmo=   0.00000000E+00          Nexp=  6        Nliner=  2
     AI =  -0.75034396E+01   0.31019671E+03  -0.18165213E+04   0.44649330E+04  -0.48213600E+04   0.19131051E+04
    Rj0 =          15.5000          32.6000
    RKj =   0.27837835E+00  -0.63863389E-02
```

No	Rotation (radians) X 1/1000	Moment (kip-inch) Expri.	Moment (kip-inch) Poly.	Moment (kip-inch) M.Expo.	Connection stiffness (kip-inch) X 1000 Expri.	Connection stiffness (kip-inch) X 1000 Poly.	Connection stiffness (kip-inch) X 1000 M.Expo.
1	0.00	0.00	0.00	0.00	0.1267E+02	0.1486E+02	0.1428E+02
3	0.33	4.22	4.86	4.28	0.1267E+02	0.1410E+02	0.1149E+02
5	0.75	8.45	10.31	8.47	0.8448E+01	0.1197E+02	0.8773E+01
7	1.35	12.67	16.60	12.87	0.6034E+01	0.9142E+01	0.6080E+01
9	2.15	16.80	22.88	16.82	0.4480E+01	0.6789E+01	0.4040E+01
11	3.30	20.42	29.54	20.61	0.2297E+01	0.4984E+01	0.2747E+01
13	4.50	23.44	34.85	23.52	0.2820E+01	0.3954E+01	0.2173E+01
15	5.80	26.26	39.51	26.06	0.1755E+01	0.3268E+01	0.1750E+01
17	8.15	29.07	46.24	29.36	0.9097E+00	0.2535E+01	0.1078E+01
19	11.10	31.49	52.88	31.68	0.7200E+00	0.2017E+01	0.5757E+00
21	14.00	33.30	58.25	33.13	0.5360E+00	0.1702E+01	0.4655E+00
23	16.90	35.11	62.85	34.91	0.7200E+00	0.1485E+01	0.7828E+00
25	19.55	37.03	66.58	37.03	0.7248E+00	0.1337E+01	0.8100E+00
27	22.50	39.14	70.33	39.40	0.7094E+00	0.1209E+01	0.7848E+00
29	25.55	41.55	73.85	41.69	0.8933E+00	0.1104E+01	0.7105E+00
31	28.45	43.66	76.93	43.62	0.5845E+00	0.1023E+01	0.6219E+00
33	31.30	45.27	79.77	45.27	0.5400E+00	0.9556E+00	0.5374E+00
35	34.05	46.68	82.32	46.64	0.4883E+00	0.9003E+00	0.4613E+00
37	36.95	47.89	84.86	47.89	0.3476E+00	0.8499E+00	0.4035E+00
39	40.05	49.10	87.42	49.07	0.4255E+00	0.8030E+00	0.3583E+00
41	43.15	50.20	89.84	50.13	0.2772E+00	0.7622E+00	0.3273E+00
43	45.90	51.01	91.88	51.00	0.3092E+00	0.7301E+00	0.3086E+00
45	49.20	51.82	94.26	51.99	0.2040E+00	0.6950E+00	0.2940E+00

A.4 – 45

IV - 23

Connection type : Top-and seat-angle connections
Mode : All bolted

Tested by : M.J.Marley (1982) U.S.A
Test Id. : D1-1/4-1

Column : W5X16
Beam : W5X16
Angle : 4 X 4 X 1/4 X 5.0

Fasteners: A325- -3/4"D
13/16" Oversize holes
Material : --
Fy = -- ksi
Fu = -- ksi

Major parameters

lt = 5.0000"	tt = 0.2500"	ls = 5.0000"	ts = 0.2500"
gt = -- "	gt'= -- "	gs = -- "	gs'= -- "
qt = -- "	rt = -- "	qs = -- "	rs = -- "
nt = 2 X 1	nt'= 2 X 1	ns = 2 X 1	ns'= 2 X 1

Remark 1)
2)

No	Moment (k-in)	Rotation (radians) X 1/1000	No	Moment (k-in)	Rotation (radians) X 1/1000
1	0.00	0.00	26	44.02	25.30
2	2.38	0.50	27	44.93	27.00
3	4.76	1.00	28	45.84	28.70
4	6.46	1.27	29	46.75	30.40
5	8.15	1.53	30	47.66	32.05
6	9.84	1.80	31	48.58	33.70
7	11.66	2.23	32	49.19	35.30
8	13.49	2.67	33	49.80	36.90
9	15.31	3.10			
10	17.75	3.65			
11	20.18	4.20			
12	22.62	4.85			
13	25.06	5.50			
14	26.88	6.25			
15	28.70	7.00			
16	30.63	8.35			
17	32.56	9.70			
18	34.18	11.50			
19	35.80	13.30			
20	37.22	15.00			
21	38.64	16.70			
22	39.86	18.40			
23	41.08	20.10			
24	42.09	21.85			
25	43.10	23.60			

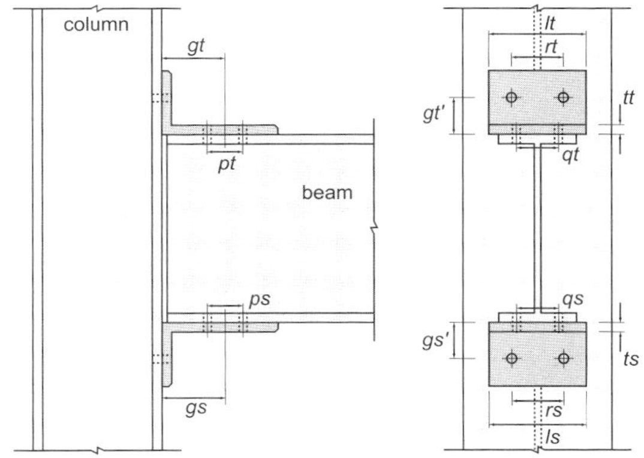

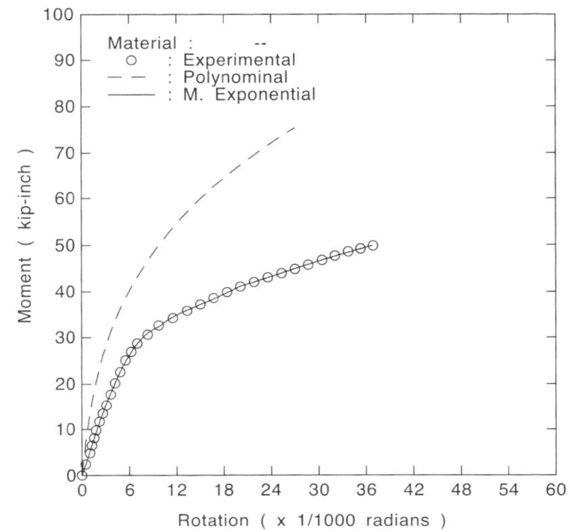

```
Moment-rotation prediction equations     ( R : X 1/1000 radians )

Frye and Morris polynominal model  :    R = Sum ( Ai X ( K*Bm )**Pi X 10**Qi )
      t =  0.250000"  xd =  5.000000"   f =  0.750000"  xl =  5.000000"
     A1 =  8.460000   A2 =  1.010000   A3 =  1.240000    K =  0.079566
     P1 =         1   P2 =         3   P3 =         5   Q1 =        -1     Q2 =        -1     Q3 =        -5

Modified exponential model  :
     Bm = Sum ( Ai X ( 1 - exp( - R/( 2*i*C ) ) ) )  +     Sum ( RKj X ( R - Rj0 ) )  +  Bm0
      C =   0.43250000E+00              Bmo=   0.00000000E+00              Nexp=  6         Nliner=  2
     AI =  -0.18939823E+02   0.33795070E+03  -0.19080798E+04   0.42458372E+04  -0.40609886E+04   0.14407735E+04
    Rj0 =           9.7000          20.1000
    RKj =   0.72937502E+00  -0.35925484E+00
```

No	Rotation (radians) X 1/1000	Moment (kip-inch) Expri.	Moment (kip-inch) Poly.	Moment (kip-inch) M.Expo.	Connection stiffness (kip-inch) X 1000 Expri.	Connection stiffness (kip-inch) X 1000 Poly.	Connection stiffness (kip-inch) X 1000 M.Expo.
1	0.00	0.00	0.00	0.00	0.4764E+01	0.1486E+02	0.3930E+01
2	0.50	2.38	7.15	2.37	0.4764E+01	0.1331E+02	0.5273E+01
3	1.00	4.76	13.15	5.09	0.5795E+01	0.1067E+02	0.5487E+01
4	1.27	6.46	15.82	6.55	0.6345E+01	0.9475E+01	0.5416E+01
5	1.53	8.15	18.21	7.97	0.6345E+01	0.8477E+01	0.5293E+01
6	1.80	9.84	20.36	9.37	0.5531E+01	0.7656E+01	0.5149E+01
7	2.23	11.66	23.44	11.55	0.4209E+01	0.6612E+01	0.4909E+01
8	2.67	13.49	26.13	13.62	0.4209E+01	0.5827E+01	0.4679E+01
9	3.10	15.31	28.52	15.60	0.4306E+01	0.5220E+01	0.4458E+01
10	3.65	17.75	31.22	17.98	0.4429E+01	0.4625E+01	0.4177E+01
11	4.20	20.18	33.63	20.19	0.4117E+01	0.4164E+01	0.3880E+01
12	4.85	22.62	36.19	22.59	0.3748E+01	0.3738E+01	0.3498E+01
13	5.50	25.06	38.51	24.73	0.3137E+01	0.3401E+01	0.3086E+01
14	6.25	26.88	40.94	26.86	0.2432E+01	0.3090E+01	0.2593E+01
15	7.00	28.70	43.16	28.62	0.2073E+01	0.2839E+01	0.2106E+01
16	8.35	30.63	46.74	30.92	0.1427E+01	0.2489E+01	0.1322E+01
17	9.70	32.56	49.92	32.28	0.1201E+01	0.2228E+01	0.7245E+00
18	11.50	34.18	53.68	34.41	0.9000E+00	0.1966E+01	0.9669E+00
19	13.30	35.80	57.03	35.93	0.8673E+00	0.1767E+01	0.7598E+00
20	15.00	37.22	59.91	37.17	0.8365E+00	0.1619E+01	0.7216E+00
21	16.70	38.64	62.55	38.42	0.7765E+00	0.1498E+01	0.7568E+00
22	18.40	39.86	65.01	39.76	0.7165E+00	0.1397E+01	0.8144E+00
23	20.10	41.08	67.31	41.19	0.6489E+00	0.1311E+01	0.8667E+00
24	21.85	42.09	69.54	42.11	0.5794E+00	0.1235E+01	0.5439E+00
25	23.60	43.10	71.64	43.08	0.5576E+00	0.1169E+01	0.5608E+00
26	25.30	44.02	73.57	44.04	0.5365E+00	0.1112E+01	0.5615E+00
27	27.00	44.93	75.42	44.98	0.5365E+00	0.1062E+01	0.5511E+00
28	28.70	45.84	77.19	45.91	0.5365E+00	0.1017E+01	0.5339E+00
29	30.40	46.75	78.88	46.80	0.5447E+00	0.9760E+00	0.5134E+00
30	32.05	47.66	80.49	47.63	0.5527E+00	0.9395E+00	0.4928E+00
31	33.70	48.58	81.98	48.42	0.4663E+00	0.9074E+00	0.4730E+00
32	35.30	49.19	83.44	49.17	0.3825E+00	0.8776E+00	0.4555E+00
33	36.90	49.80	84.80	49.88	0.3825E+00	0.8511E+00	0.4400E+00

A.4 – 47

IV - 24

Connection type : Top-and seat-angle connections
Mode : All bolted

Tested by : M.J.Marley (1982) U.S.A
Test Id. : D1-1/4-2

Column : W5X16
Beam : W5X16
Angle : 4 X 4 X 1/4 X 5.0

Fasteners: A325- -3/4"D
13/16" Oversize holes
Material : --
Fy = -- ksi
Fu = -- ksi

Major parameters

lt = 5.0000"	tt = 0.2500"	ls = 5.0000"	ts = 0.2500"
gt = -- "	gt'= -- "	gs = -- "	gs'= -- "
qt = -- "	rt = -- "	qs = -- "	rs = -- "
nt = 2 X 1	nt'= 2 X 1	ns = 2 X 1	ns'= 2 X 1

Remark 1)
2)

No	Moment (k-in)	Rotation (radians) X 1/1000
1	0.00	0.00
2	1.65	0.33
3	3.30	0.67
4	4.94	1.00
5	7.37	1.75
6	9.79	2.50
7	11.92	3.15
8	14.04	3.80
9	16.36	4.60
10	18.68	5.40
11	20.50	6.05
12	22.32	6.70
13	23.63	7.35
14	24.95	8.00
15	26.47	9.15
16	27.98	10.30
17	30.00	11.65
18	32.02	13.00
19	33.43	14.45
20	34.85	15.90
21	36.16	17.70
22	37.48	19.50
23	38.38	21.20
24	39.29	22.90
25	40.10	24.40
26	40.91	25.90
27	41.62	27.20
28	42.32	28.50
29	42.83	30.10
30	43.33	31.70
31	43.94	33.45
32	44.54	35.20
33	45.15	37.00
34	45.76	38.80
35	46.16	40.40
36	46.56	42.00
37	47.07	43.45
38	47.58	44.90
39	47.88	46.85
40	48.18	48.80
41	48.59	50.10

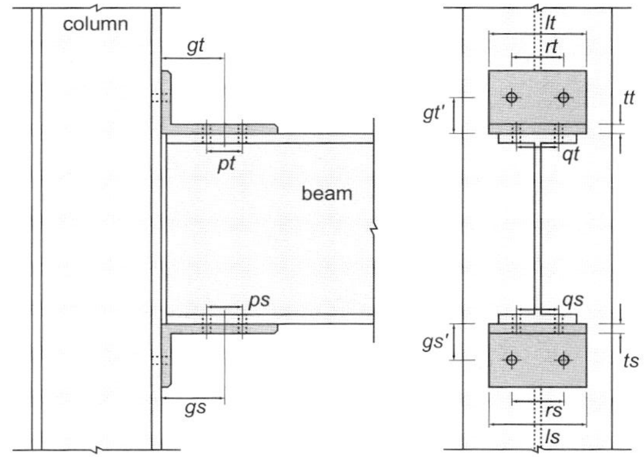

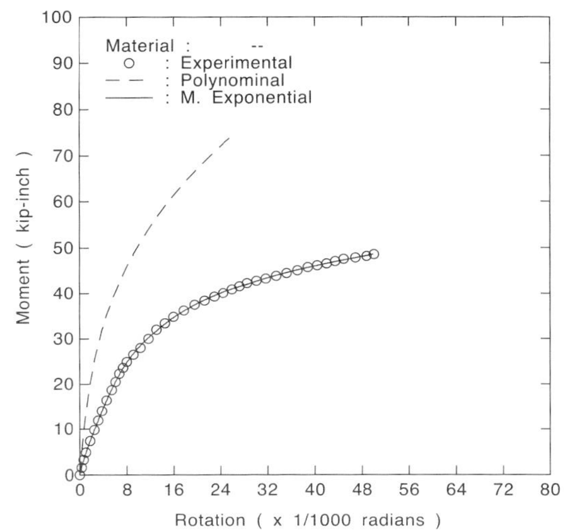

A.4 – 48

```
Moment-rotation prediction equations     ( R : X 1/1000 radians )

Frye and Morris polynominal model  :     R = Sum ( Ai X ( K*Bm )**Pi X 10**Qi )
      t =  0.250000"  xd =  5.000000"   f =  0.750000"  xl =  5.000000"
     A1 =  8.460000   A2 =  1.010000   A3 =  1.240000    K =  0.079566
     P1 =         1   P2 =         3   P3 =         5   Q1 =        -1     Q2 =       -1      Q3 =       -5

Modified exponential model  :
     Bm = Sum ( Ai X ( 1 - exp( - R/( 2*i*C ) ) ) )  +      Sum ( RKj X ( R - Rj0 ) )  +  Bm0
      C =  0.50082500E+00              Bmo=   0.00000000E+00           Nexp=  6        Nliner=  2
     AI =  0.10803469E+02  -0.28416414E+02  -0.17344039E+03   0.82037161E+03  -0.11917200E+04   0.60432313E+03
    Rj0 =         13.0000          28.5000
    RKj =  0.12951471E+00   0.92167996E-01
```

No	Rotation (radians) X 1/1000	Moment (kip-inch) Expri.	Moment (kip-inch) Poly.	Moment (kip-inch) M.Expo.	Connection stiffness (kip-inch) X 1000 Expri.	Connection stiffness (kip-inch) X 1000 Poly.	Connection stiffness (kip-inch) X 1000 M.Expo.
1	0.00	0.00	0.00	0.00	0.4944E+01	0.1486E+02	0.6241E+01
3	0.67	3.30	9.30	3.37	0.4944E+01	0.1242E+02	0.4211E+01
5	1.75	7.37	19.97	7.35	0.3232E+01	0.7798E+01	0.3387E+01
7	3.15	11.92	28.78	11.97	0.3268E+01	0.5159E+01	0.3241E+01
9	4.60	16.36	35.24	16.49	0.2903E+01	0.3889E+01	0.2950E+01
11	6.05	20.50	40.31	20.45	0.2797E+01	0.3166E+01	0.2506E+01
13	7.35	23.63	44.13	23.44	0.2022E+01	0.2737E+01	0.2098E+01
15	9.15	26.47	48.67	26.77	0.1320E+01	0.2326E+01	0.1624E+01
17	11.65	30.00	53.97	30.22	0.1493E+01	0.1947E+01	0.1174E+01
19	14.45	33.43	59.00	33.24	0.9766E+00	0.1664E+01	0.1001E+01
21	17.70	36.16	64.02	36.11	0.7300E+00	0.1436E+01	0.7820E+00
23	21.20	38.38	68.72	38.54	0.5329E+00	0.1262E+01	0.6121E+00
25	24.40	40.10	72.56	40.29	0.5400E+00	0.1141E+01	0.4884E+00
27	27.20	41.62	75.66	41.53	0.5446E+00	0.1055E+01	0.4000E+00
29	30.10	42.83	78.59	42.73	0.3150E+00	0.9829E+00	0.4189E+00
31	33.45	43.94	81.76	44.02	0.3463E+00	0.9121E+00	0.3550E+00
33	37.00	45.15	84.88	45.19	0.3367E+00	0.8495E+00	0.3073E+00
35	40.40	46.16	87.68	46.18	0.2513E+00	0.7985E+00	0.2765E+00
37	43.45	47.07	90.08	46.99	0.3517E+00	0.7582E+00	0.2579E+00
39	46.85	47.88	92.59	47.84	0.1538E+00	0.7193E+00	0.2442E+00
41	50.10	48.59	94.86	48.62	0.3138E+00	0.6865E+00	0.2358E+00

A.4 – 49

IV - 25

Connection type : Top-and seat-angle connections
Mode : All bolted

Tested by : M.J.Marley (1982) U.S.A
Test Id. : D2-1/4-1

Column : W5X16
Beam : W5X16
Angle : 4 X 4 X 1/4 X 5.0

Fasteners: A325- -3/4"D
13/16" Oversize holes
Material : --
Fy = -- ksi
Fu = -- ksi

Major parameters

lt = 5.0000"	tt = 0.2500"	ls = 5.0000"	ts = 0.2500"
gt = -- "	gt'= -- "	gs = -- "	gs'= -- "
qt = -- "	rt = -- "	qs = -- "	rs = -- "
nt = 2 X 1	nt'= 2 X 1	ns = 2 X 1	ns'= 2 X 1

Remark 1)
2)

No	Moment (k-in)	Rotation (radians) X 1/1000	No	Moment (k-in)	Rotation (radians) X 1/1000
1	0.00	0.00	26	40.79	22.70
2	1.82	0.27	27	42.11	24.40
3	3.64	0.53	28	43.52	25.65
4	5.46	0.80	29	44.94	26.90
5	7.15	1.13	30	45.85	28.35
6	8.84	1.47	31	46.76	29.80
7	10.52	1.80	32	47.47	30.90
8	12.95	2.25	33	48.18	32.00
9	15.38	2.70	34	48.68	33.30
10	17.81	3.30	35	49.19	34.60
11	20.24	3.90	36	49.70	35.80
12	21.66	4.20	37	50.21	37.00
13	23.08	4.50			
14	25.10	6.05			
15	27.12	7.60			
16	28.94	9.25			
17	30.77	10.90			
18	32.18	12.25			
19	33.60	13.60			
20	33.60	13.60			
21	34.68	14.97			
22	35.76	16.33			
23	36.84	17.70			
24	38.15	19.35			
25	39.47	21.00			

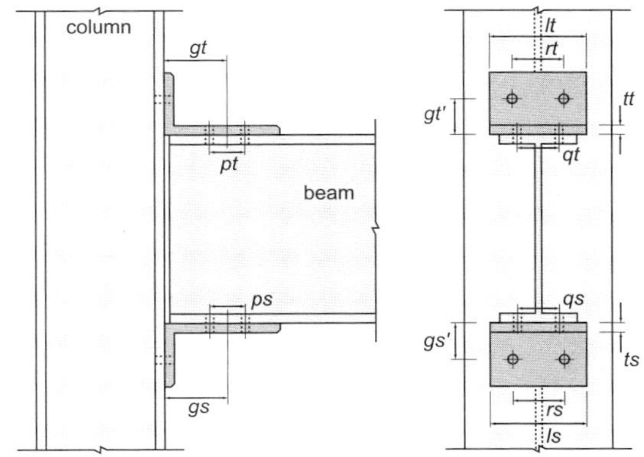

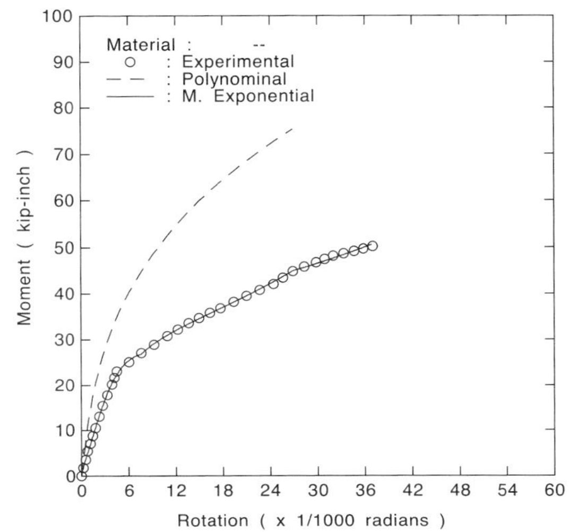

```
Moment-rotation prediction equations     ( R : X 1/1000 radians )

Frye and Morris polynominal model   :    R = Sum ( Ai X ( K*Bm )**Pi X 10**Qi )
     t =  0.250000"  xd =  5.000000"   f =  0.750000"  xl =  5.000000"
    A1 =  8.460000   A2 =  1.010000   A3 =  1.240000    K =  0.079566
    P1 =         1   P2 =         3   P3 =         5   Q1 =        -1     Q2 =        -1     Q3 =        -5

Modified exponential model  :
    Bm = Sum ( Ai X ( 1 - exp( - R/( 2*i*C ) ) ) )  +     Sum ( RKj X ( R - Rj0 ) )  +  Bm0
     C =   0.37500833E+00             Bmo=   0.00000000E+00          Nexp=  6        Nliner=  2
    AI =   0.29350233E+02  -0.32343998E+03   0.12893937E+04  -0.26247628E+04   0.27119744E+04  -0.10606729E+04
   Rj0 =           7.6000          26.9000
   RKj =   0.11527284E+01  -0.52034296E+00
```

No	Rotation (radians) X 1/1000	Moment (kip-inch) Expri.	Moment (kip-inch) Poly.	Moment (kip-inch) M.Expo.	Connection stiffness (kip-inch) X 1000 Expri.	Connection stiffness (kip-inch) X 1000 Poly.	Connection stiffness (kip-inch) X 1000 M.Expo.
1	0.00	0.00	0.00	0.00	0.6824E+01	0.1486E+02	0.9138E+01
3	0.53	3.64	7.59	3.75	0.6825E+01	0.1314E+02	0.5772E+01
5	1.13	7.15	14.52	7.01	0.5064E+01	0.1005E+02	0.5323E+01
7	1.80	10.52	20.36	10.60	0.5207E+01	0.7656E+01	0.5434E+01
9	2.70	15.38	26.32	15.35	0.4821E+01	0.5775E+01	0.4972E+01
11	3.90	20.24	32.34	20.50	0.4497E+01	0.4401E+01	0.3551E+01
13	4.50	23.08	34.85	22.40	0.4166E+01	0.3954E+01	0.2807E+01
15	7.60	27.12	44.81	26.84	0.1208E+01	0.2670E+01	0.4913E+00
17	10.90	30.77	52.48	30.97	0.1074E+01	0.2044E+01	0.1006E+01
19	13.60	33.60	57.56	33.44	0.7764E-03	0.1739E+01	0.8489E+00
21	14.97	34.68	59.85	34.57	0.7903E+00	0.1622E+01	0.8148E+00
23	17.70	36.84	64.02	36.77	0.7930E+00	0.1436E+01	0.8067E+00
25	21.00	39.47	68.47	39.51	0.7866E+00	0.1270E+01	0.8632E+00
27	24.40	42.11	72.56	42.59	0.9818E+00	0.1141E+01	0.9459E+00
29	26.90	44.94	75.31	45.03	0.8995E+00	0.1064E+01	0.1002E+01
31	29.80	46.76	78.29	46.50	0.6373E+00	0.9899E+00	0.5332E+00
33	32.00	48.18	80.44	47.71	0.5263E+00	0.9406E+00	0.5624E+00
35	34.60	49.19	82.81	49.20	0.4071E+00	0.8903E+00	0.5872E+00
37	37.00	50.21	84.88	50.63	0.4250E+00	0.8495E+00	0.6028E+00

A.4 – 51

IV - 26

Connection type : Top-and seat-angle connections
Mode : All bolted

Tested by : M.J.Marley (1982) U.S.A
Test Id. : D2-1/4-2

Column : W5X16
Beam : W5X16
Angle : 4 X 4 X 1/4 X 5.0

Fasteners: A325- -3/4"D
13/16" Oversize holes
Material : --
Fy = -- ksi
Fu = -- ksi

Major parameters

lt = 5.0000"	tt = 0.2500"	ls = 5.0000"	ts = 0.2500"
gt = -- "	gt'= -- "	gs = -- "	gs'= -- "
qt = -- "	rt = -- "	qs = -- "	rs = -- "
nt = 2 X 1	nt'= 2 X 1	ns = 2 X 1	ns'= 2 X 1

Remark 1)
2)

No	Moment (k-in)	Rotation (radians) X 1/1000	No	Moment (k-in)	Rotation (radians) X 1/1000
1	0.00	0.00	26	41.29	22.00
2	2.33	0.15	27	42.71	23.05
3	4.66	0.30	28	44.12	24.10
4	6.88	0.85	29	45.14	25.35
5	9.11	1.40	30	46.15	26.60
6	11.23	1.75	31	47.17	28.15
7	13.36	2.10	32	48.18	29.70
8	14.98	2.60	33	48.99	30.95
9	16.60	3.10	34	49.80	32.20
10	18.42	3.60			
11	20.24	4.10			
12	22.07	4.60			
13	23.89	5.30			
14	25.70	6.00			
15	27.22	7.45			
16	28.74	8.90			
17	29.96	10.35			
18	31.18	11.80			
19	32.09	13.15			
20	33.00	14.50			
21	34.01	15.85			
22	35.02	17.20			
23	36.53	18.45			
24	38.05	19.70			
25	39.67	20.85			

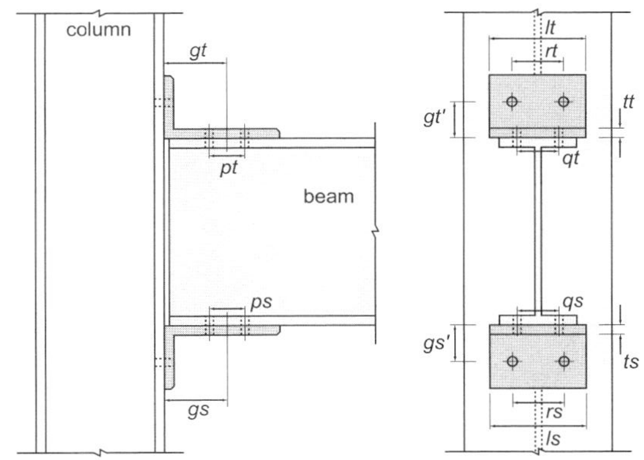

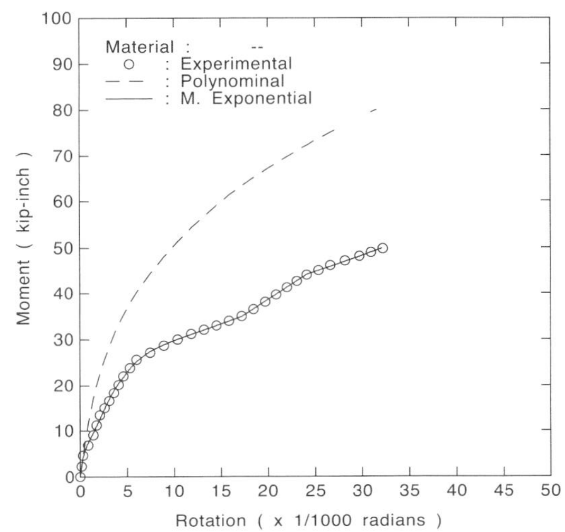

```
Moment-rotation prediction equations      ( R : X 1/1000 radians )

Frye and Morris polynominal model  :     R = Sum ( Ai X ( K*Bm )**Pi X 10**Qi )
      t =  0.250000"  xd =  5.000000"   f =  0.750000"  xl =  5.000000"
     A1 =  8.460000   A2 =  1.010000   A3 =  1.240000    K =  0.079566
     P1 =        1    P2 =        3    P3 =        5    Q1 =       -1    Q2 =       -1    Q3 =       -5

Modified exponential model  :
     Bm = Sum ( Ai X ( 1 - exp( - R/( 2*i*C ) ) ) )  +      Sum ( RKj X ( R - Rj0 ) )  +  Bm0
      C =   0.30583333E+00           Bmo=   0.00000000E+00          Nexp=  6         Nliner=  3
     AI =   0.40348787E+02  -0.22483971E+03  0.52472339E+03 -0.72918236E+03  0.66144220E+03 -0.24824849E+03
    Rj0 =           4.6000          17.2000          24.1000
    RKj =   0.80624586E+00   0.59373203E+00 -0.67863687E+00
```

No	Rotation (radians) X 1/1000	Moment (kip-inch) Expri.	Moment (kip-inch) Poly.	Moment (kip-inch) M.Expo.	Connection stiffness (kip-inch) X 1000 Expri.	Connection stiffness (kip-inch) X 1000 Poly.	Connection stiffness (kip-inch) X 1000 M.Expo.
1	0.00	0.00	0.00	0.00	0.1552E+02	0.1486E+02	0.1873E+02
3	0.30	4.66	4.39	4.06	0.1306E+02	0.1423E+02	0.9491E+01
5	1.40	9.11	17.06	9.56	0.5283E+01	0.8947E+01	0.3978E+01
7	2.10	13.36	22.54	12.55	0.4904E+01	0.6901E+01	0.4502E+01
9	3.10	16.60	28.52	16.96	0.3444E+01	0.5220E+01	0.4106E+01
11	4.10	20.24	33.21	20.51	0.3648E+01	0.4240E+01	0.2939E+01
13	5.30	23.89	37.82	23.77	0.2597E+01	0.3497E+01	0.2434E+01
15	7.45	27.22	44.41	27.43	0.1047E+01	0.2710E+01	0.1180E+01
17	10.35	29.96	51.33	30.04	0.8400E+00	0.2124E+01	0.7485E+00
19	13.15	32.09	56.77	32.04	0.6756E+00	0.1782E+01	0.6981E+00
21	15.85	34.01	61.25	33.93	0.7467E+00	0.1556E+01	0.7119E+00
23	18.45	36.53	65.08	36.56	0.1214E+01	0.1394E+01	0.1329E+01
25	20.85	39.67	68.28	39.77	0.1409E+01	0.1277E+01	0.1350E+01
27	23.05	42.71	71.01	42.76	0.1349E+01	0.1188E+01	0.1365E+01
29	25.35	45.14	73.66	45.07	0.8112E+00	0.1110E+01	0.6989E+00
31	28.15	47.17	76.62	47.04	0.6542E+00	0.1031E+01	0.7086E+00
33	30.95	48.99	79.43	49.03	0.6480E+00	0.9632E+00	0.7144E+00

A.4 – 53

IV - 27

Connection type : Top-and seat-angle connections
Mode : All bolted

Tested by : M.J.Marley (1982) U.S.A
Test Id. : D3-1/4-1

Column : W5X16
Beam : W5X16
Angle : 4 X 4 X 1/4 X 5.0

Fasteners: A325- -3/4"D
13/16" Oversize holes
Material : --
Fy = -- ksi
Fu = -- ksi

Major parameters

lt = 5.0000"	tt = 0.2500"	ls = 5.0000"	ts = 0.2500"
gt = -- "	gt'= -- "	gs = -- "	gs'= -- "
qt = -- "	rt = -- "	qs = -- "	rs = -- "
nt = 2 X 1	nt'= 2 X 1	ns = 2 X 1	ns'= 2 X 1

Remark 1)
2)

No	Moment (k-in)	Rotation (radians) X 1/1000	No	Moment (k-in)	Rotation (radians) X 1/1000
1	0.00	0.00	26	32.70	17.00
2	1.44	0.20	27	33.61	18.15
3	2.88	0.40	28	34.51	19.30
4	4.32	0.60	29	34.91	20.43
5	5.60	0.90	30	35.31	21.57
6	6.87	1.20	31	35.71	22.70
7	8.15	1.50	32	35.85	23.70
8	9.76	1.93	33	35.98	24.70
9	11.36	2.37	34	36.12	25.70
10	12.97	2.80	35	36.25	26.87
11	14.78	3.27	36	36.38	28.03
12	16.60	3.73	37	36.52	29.20
13	18.41	4.20			
14	19.75	4.73			
15	21.10	5.27			
16	22.44	5.80			
17	23.65	6.80			
18	24.85	7.80			
19	25.92	9.00			
20	27.00	10.20			
21	28.07	11.40			
22	28.94	12.47			
23	29.81	13.53			
24	30.68	14.60			
25	31.69	15.80			

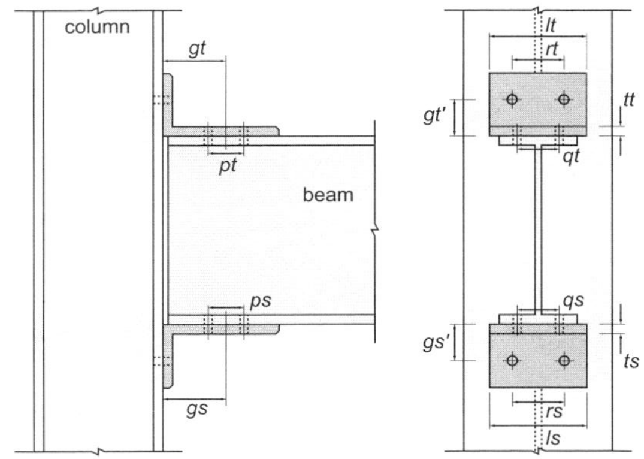

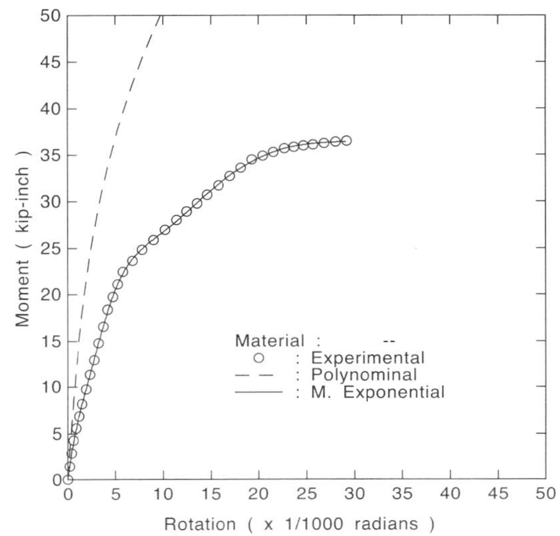

A.4 – 54

```
Moment-rotation prediction equations     ( R : X 1/1000 radians )

Frye and Morris polynominal model   :      R = Sum ( Ai X ( K*Bm )**Pi X 10**Qi )
      t =  0.250000"  xd =  5.000000"   f =  0.750000"  xl =  5.000000"
     A1 =  8.460000   A2 =  1.010000   A3 =  1.240000    K =  0.079566
     P1 =        1    P2 =        3    P3 =        5    Q1 =       -1     Q2 =       -1     Q3 =       -5

Modified exponential model  :
     Bm = Sum ( Ai X ( 1 - exp( - R/( 2*i*C ) ) ) )  +      Sum ( RKj X ( R - Rj0 ) )  +  Bm0
      C =   0.29333333E+00             Bmo=   0.00000000E+00            Nexp=  6         Nliner=  1
     AI =  -0.12409746E+02   0.38440288E+03  -0.23538034E+04   0.56873694E+04  -0.60174304E+04   0.23490185E+04
    Rj0 =          22.7000
    RKj =  -0.64953542E-01
```

No	Rotation (radians) X 1/1000	Moment (kip-inch) Expri.	Moment (kip-inch) Poly.	Moment (kip-inch) M.Expo.	Connection stiffness (kip-inch) X 1000 Expri.	Connection stiffness (kip-inch) X 1000 Poly.	Connection stiffness (kip-inch) X 1000 M.Expo.
1	0.00	0.00	0.00	0.00	0.7200E+01	0.1486E+02	0.8608E+01
3	0.40	2.88	5.80	2.96	0.7200E+01	0.1380E+02	0.6305E+01
5	0.90	5.60	12.05	5.62	0.4253E+01	0.1118E+02	0.4542E+01
7	1.50	8.15	17.93	8.08	0.4031E+01	0.8592E+01	0.3844E+01
9	2.37	11.36	24.31	11.40	0.3711E+01	0.6347E+01	0.3881E+01
11	3.27	14.78	29.37	14.88	0.3883E+01	0.5022E+01	0.3785E+01
13	4.20	18.41	33.63	18.18	0.3247E+01	0.4164E+01	0.3212E+01
15	5.27	21.10	37.70	21.13	0.2520E+01	0.3513E+01	0.2317E+01
17	6.80	23.65	42.58	23.83	0.1206E+01	0.2901E+01	0.1301E+01
19	9.00	25.92	48.31	25.97	0.8933E+00	0.2355E+01	0.8016E+00
21	11.40	28.07	53.48	27.93	0.8532E+00	0.1978E+01	0.8621E+00
23	13.53	29.81	57.44	29.84	0.8175E+00	0.1745E+01	0.9097E+00
25	15.80	31.69	61.18	31.83	0.8400E+00	0.1559E+01	0.8238E+00
27	18.15	33.61	64.66	33.56	0.7878E+00	0.1411E+01	0.6465E+00
29	20.43	34.91	67.74	34.83	0.3529E+00	0.1295E+01	0.4636E+00
31	22.70	35.71	70.60	35.70	0.2377E+00	0.1201E+01	0.3114E+00
33	24.70	35.98	72.92	36.09	0.1360E+00	0.1131E+01	0.1453E+00
35	26.87	36.25	75.30	36.31	0.1131E+00	0.1065E+01	0.6791E-01
37	29.20	36.52	77.71	36.40	0.1131E+00	0.1004E+01	0.1366E-01

```
Connection type : Top-and seat-angle connections
           Mode : All bolted

Tested by  :  M.J.Marley (1982)                          U.S.A
Test  Id.  :  D3-1/4-2

Column : W5X16                       Fasteners: A325- -3/4"D
Beam   : W5X16                              13/16" Oversize holes
Angle  : 4 X 4 X 1/4 X 5.0           Material :  --
                                          Fy =   --   ksi
                                          Fu =   --   ksi

                          Major parameters

     lt =  5.0000"    tt =  0.2500"    ls =  5.0000"    ts =  0.2500"
     gt =   --   "    gt'=   --   "    gs =   --   "    gs'=   --   "
     qt =   --   "    rt =   --   "    qs =   --   "    rs =   --   "
     nt = 2 X  1      nt'= 2 X  1      ns = 2 X  1      ns'= 2 X  1

Remark  1)
        2)
```

No	Moment (k-in)	Rotation (radians) X 1/1000	No	Moment (k-in)	Rotation (radians) X 1/1000
1	0.00	0.00	26	34.62	26.20
2	1.48	0.60	27	35.02	28.25
3	2.97	1.20	28	35.42	30.30
4	4.45	1.80	29	35.73	32.55
5	6.00	2.63	30	36.04	34.80
6	7.56	3.47	31	36.54	36.75
7	9.11	4.30	32	37.04	38.70
8	10.46	4.97	33	37.55	40.45
9	11.80	5.63	34	38.05	42.20
10	13.15	6.30	35	38.26	43.75
11	14.77	7.03	36	38.46	45.30
12	16.39	7.77	37	38.87	47.70
13	18.01	8.50	38	39.17	49.05
14	19.43	9.17	39	39.47	50.40
15	20.85	9.83			
16	22.27	10.50			
17	23.99	11.50			
18	25.70	12.50			
19	27.32	14.05			
20	28.94	15.60			
21	29.86	17.35			
22	30.77	19.10			
23	31.68	20.95			
24	32.59	22.80			
25	33.61	24.50			

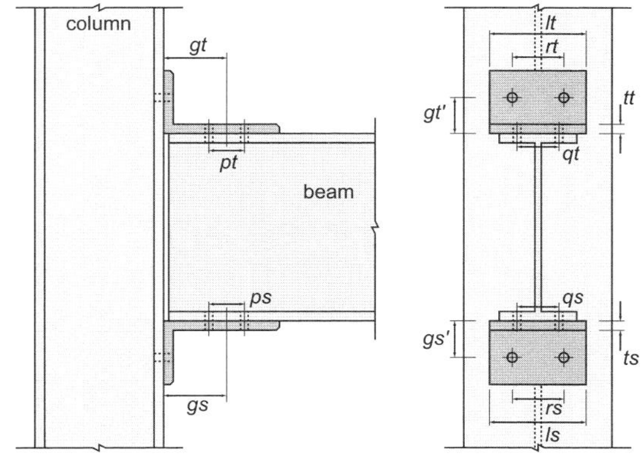

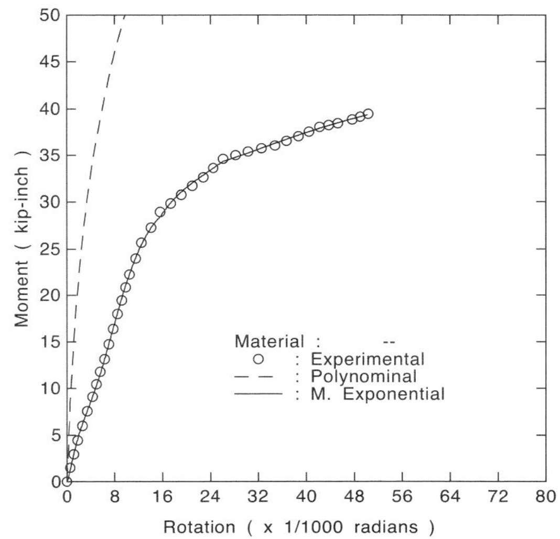

A.4 – 56

```
Moment-rotation prediction equations      ( R : X 1/1000 radians )

Frye and Morris polynominal model  :    R = Sum ( Ai X ( K*Bm )**Pi X 10**Qi )
     t =  0.250000"  xd =  5.000000"   f =  0.750000"  xl =  5.000000"
    A1 =  8.460000   A2 =  1.010000   A3 =  1.240000    K =  0.079566
    P1 =         1   P2 =         3   P3 =         5   Q1 =        -1    Q2 =       -1    Q3 =       -5

Modified exponential model  :
    Bm = Sum ( Ai X ( 1 - exp( - R/( 2*i*C ) ) ) )  +      Sum ( RKj X ( R - Rj0 ) )  +  Bm0
      C =   0.57000000E+00          Bmo=   0.00000000E+00          Nexp=  6        Nliner=  2
     AI =  -0.41341835E+02   0.60871564E+03  -0.26545084E+04   0.49126766E+04  -0.41827663E+04   0.13902802E+04
    Rj0 =          15.6000          26.2000
    RKj =   0.34979401E+00  -0.22765901E+00
```

No	Rotation (radians) X 1/1000	Moment (kip-inch) Expri.	Moment (kip-inch) Poly.	Moment (kip-inch) M.Expo.	Connection stiffness (kip-inch) X 1000 Expri.	Connection stiffness (kip-inch) X 1000 Poly.	Connection stiffness (kip-inch) X 1000 M.Expo.
1	0.00	0.00	0.00	0.00	0.2473E+01	0.1486E+02	0.1324E+01
3	1.20	2.97	15.18	2.98	0.2473E+01	0.9756E+01	0.2698E+01
5	2.63	6.00	25.95	6.19	0.1862E+01	0.5877E+01	0.1833E+01
7	4.30	9.11	34.04	9.06	0.1951E+01	0.4091E+01	0.1775E+01
9	5.63	11.80	38.96	11.63	0.2022E+01	0.3340E+01	0.2098E+01
11	7.03	14.77	43.25	14.75	0.2209E+01	0.2829E+01	0.2313E+01
13	8.50	18.01	47.11	18.14	0.2168E+01	0.2457E+01	0.2266E+01
15	9.83	20.85	50.21	21.02	0.2130E+01	0.2206E+01	0.2025E+01
17	11.50	23.99	53.68	24.04	0.1716E+01	0.1966E+01	0.1593E+01
19	14.05	27.32	58.33	27.24	0.1045E+01	0.1698E+01	0.9344E+00
21	17.35	29.86	63.51	29.91	0.5211E+00	0.1457E+01	0.7246E+00
23	20.95	31.68	68.41	31.99	0.4930E+00	0.1272E+01	0.4809E+00
25	24.50	33.61	72.67	33.58	0.5965E+00	0.1138E+01	0.4357E+00
27	28.25	35.02	76.73	34.77	0.1961E+00	0.1028E+01	0.2220E+00
29	32.55	35.73	80.93	35.76	0.1360E+00	0.9299E+00	0.2344E+00
31	36.75	36.54	84.67	36.73	0.2585E+00	0.8535E+00	0.2265E+00
33	40.45	37.55	87.72	37.54	0.2880E+00	0.7978E+00	0.2094E+00
35	43.75	38.26	90.28	38.20	0.1316E+00	0.7550E+00	0.1918E+00
37	47.70	38.87	93.17	38.92	0.2034E+00	0.7106E+00	0.1723E+00
39	50.40	39.47	95.07	39.37	0.2222E+00	0.6835E+00	0.1611E+00

A.4 – 57

IV - 29

Connection type : Top-and seat-angle connections
Mode : All bolted

Tested by : M.J.Marley (1982) U.S.A
Test Id. : A-1/2-1

Column : W5X16
Beam : W5X16
Angle : 4 X 4 X 1/2 X 5.0

Fasteners: A325- -3/4"D
13/16" Oversize holes
Material : --
Fy = -- ksi
Fu = -- ksi

Major parameters

lt = 5.0000"	tt = 0.5000"	ls = 5.0000"	ts = 0.5000"
gt = -- "	gt'= -- "	gs = -- "	gs'= -- "
qt = -- "	rt = -- "	qs = -- "	rs = -- "
nt = 2 X 1	nt'= 2 X 1	ns = 2 X 1	ns'= 2 X 1

Remark 1)
2)

No	Moment (k-in)	Rotation (radians) X 1/1000
1	0.00	0.00
2	8.03	0.47
3	16.06	0.93
4	24.10	1.40
5	32.40	1.87
6	40.70	2.33
7	49.00	2.80
8	56.22	3.33
9	63.45	3.87
10	70.68	4.40
11	77.11	5.10
12	83.54	5.80
13	89.96	6.50
14	98.80	7.65
15	107.63	8.80
16	115.66	10.20
17	123.70	11.60
18	129.32	13.05
19	134.94	14.50
20	139.76	15.85
21	144.58	17.20
22	144.58	17.20
23	149.80	18.95
24	155.02	20.70
25	157.43	22.70
26	159.84	24.70
27	162.25	26.55
28	164.66	28.40
29	166.54	29.80
30	168.41	31.20
31	170.28	32.60
32	172.29	34.25
33	174.30	35.90
34	176.30	37.45
35	178.31	39.00
36	179.51	40.05
37	180.72	41.10

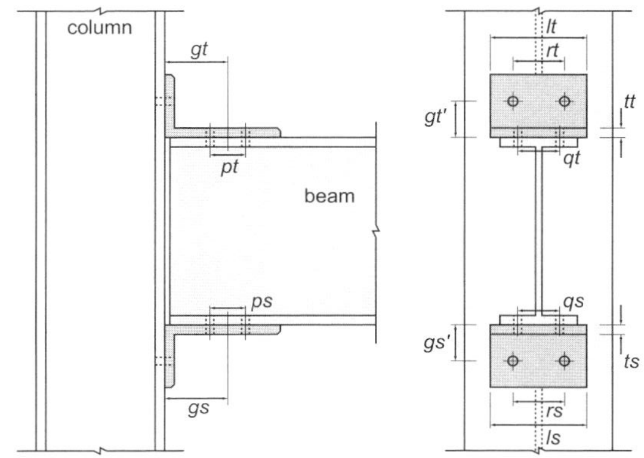

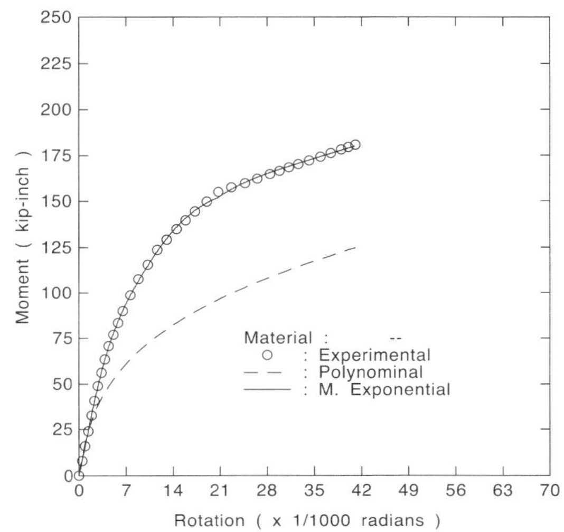

```
Moment-rotation prediction equations      ( R : X 1/1000 radians )

Frye and Morris polynominal model  :    R = Sum ( Ai X ( K*Bm )**Pi X 10**Qi )
     t =  0.500000"  xd =  5.000000"   f =  0.750000"  xl =  5.000000"
    A1 =  8.460000   A2 =  1.010000   A3 =  1.240000    K =  0.056262
    P1 =         1   P2 =         3   P3 =         5   Q1 =        -1    Q2 =        -1    Q3 =        -5

Modified exponential model  :
    Bm = Sum ( Ai X ( 1 - exp( - R/( 2*i*C ) ) ) )  +      Sum ( RKj X ( R - Rj0 ) )  +  Bm0
     C =  0.45917500E+00              Bmo=  0.00000000E+00         Nexp=  6         Nliner=  1
    AI =  0.38207135E+02  -0.88441293E+03  0.45379123E+04  -0.92751976E+04  0.80651471E+04  -0.23258602E+04
   Rj0 =        20.7000
   RKj =  0.11745676E+01
```

No	Rotation (radians) X 1/1000	Moment (kip-inch) Expri.	Moment (kip-inch) Poly.	Moment (kip-inch) M.Expo.	Connection stiffness (kip-inch) X 1000 Expri.	Connection stiffness (kip-inch) X 1000 Poly.	Connection stiffness (kip-inch) X 1000 M.Expo.
1	0.00	0.00	0.00	0.00	0.1721E+02	0.2101E+02	0.1658E+02
3	0.93	16.06	17.57	16.02	0.1721E+02	0.1556E+02	0.1778E+02
5	1.87	32.40	29.51	32.68	0.1779E+02	0.1057E+02	0.1757E+02
7	2.80	49.00	38.03	48.29	0.1581E+02	0.7954E+01	0.1569E+02
9	3.87	63.45	45.53	63.53	0.1355E+02	0.6265E+01	0.1289E+02
11	5.10	77.11	52.48	77.63	0.9183E+01	0.5090E+01	0.1011E+02
13	6.50	89.96	58.97	90.22	0.8614E+01	0.4243E+01	0.8058E+01
15	8.80	107.63	67.65	106.60	0.6804E+01	0.3386E+01	0.6413E+01
17	11.60	123.70	76.19	122.88	0.4824E+01	0.2762E+01	0.5236E+01
19	14.50	134.94	83.56	136.17	0.3718E+01	0.2347E+01	0.3904E+01
21	17.20	144.58	89.51	145.00	0.2642E-02	0.2074E+01	0.2654E+01
23	18.95	149.80	93.01	149.01	0.2983E+01	0.1934E+01	0.1952E+01
25	22.70	157.43	99.80	156.46	0.1206E+01	0.1699E+01	0.2039E+01
27	26.55	162.25	105.98	163.04	0.1304E+01	0.1519E+01	0.1451E+01
29	29.80	166.54	110.75	167.34	0.1337E+01	0.1399E+01	0.1227E+01
31	32.60	170.28	114.54	170.65	0.1283E+01	0.1313E+01	0.1145E+01
33	35.90	174.30	118.74	174.37	0.1257E+01	0.1226E+01	0.1116E+01
35	39.00	178.31	122.42	177.82	0.1207E+01	0.1157E+01	0.1118E+01
37	41.10	180.72	124.78	180.18	0.1149E+01	0.1116E+01	0.1125E+01

IV - 30

```
Connection type : Top-and seat-angle connections
           Mode : All bolted

Tested by   :  M.J.Marley (1982)                      U.S.A
Test  Id.   :  B-1/2-1

Column : W5X16                      Fasteners: A325- -3/4"D
Beam   : W5X16                          13/16" Oversize holes
Angle  : 4 X 4 X 1/2 X 5.0          Material :  --
                                         Fy =   --   ksi
                                         Fu =   --   ksi

                          Major parameters

     lt =  5.0000"    tt =  0.5000"    ls =  5.0000"    ts =  0.5000"
     gt =    --   "   gt'=    --   "   gs =    --   "   gs'=    --   "
     qt =    --   "   rt =    --   "   qs =    --   "   rs =    --   "
     nt = 2 X  1      nt'= 2 X  1      ns = 2 X  1      ns'= 2 X  1

Remark  1)
        2)
```

No	Moment (k-in)	Rotation (radians) X 1/1000	No	Moment (k-in)	Rotation (radians) X 1/1000
1	0.00	0.00	26	145.60	19.00
2	6.13	0.10	27	148.40	20.60
3	12.26	0.20	28	151.20	22.20
4	18.40	0.30	29	153.60	23.65
5	18.40	0.30	30	156.00	25.10
6	26.80	0.75	31	159.20	26.75
7	35.20	1.20	32	162.40	28.40
8	40.80	1.45	33	164.26	29.80
9	46.40	1.70	34	166.13	31.20
10	54.40	2.10	35	168.00	32.60
11	62.40	2.50	36	168.00	32.40
12	69.20	3.00	37	169.60	33.80
13	76.00	3.50	38	171.20	35.20
14	83.60	4.10	39	172.80	36.60
15	91.20	4.70	40	174.40	38.17
16	97.20	5.50	41	176.00	39.73
17	103.20	6.30	42	177.60	41.30
18	110.40	7.30			
19	117.60	8.30			
20	122.80	9.65			
21	128.00	11.00			
22	131.74	12.47			
23	135.47	13.93			
24	139.20	15.40			
25	142.40	17.20			

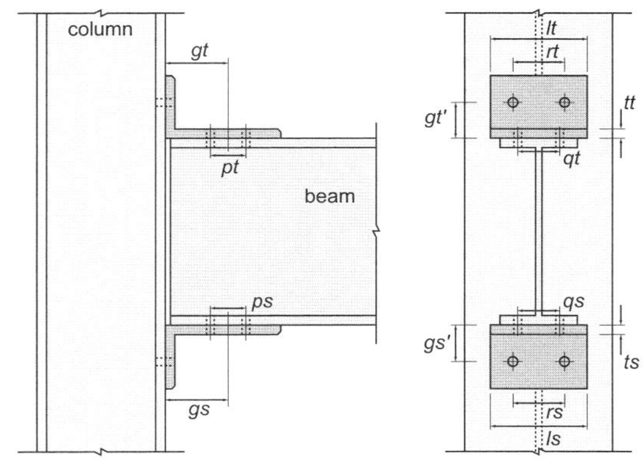

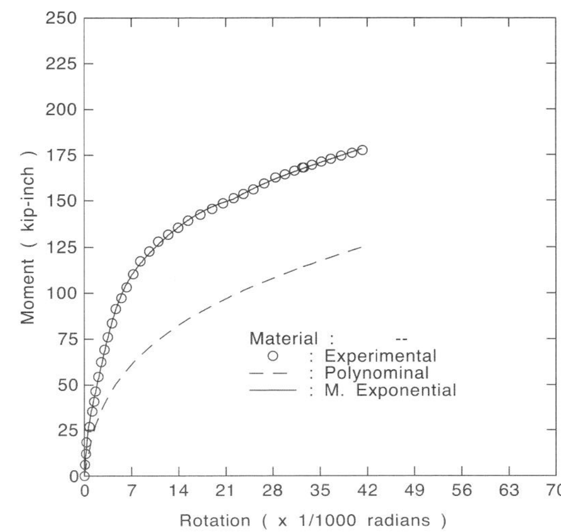

A.4 – 60

```
 Moment-rotation prediction equations      ( R : X 1/1000 radians )

Frye and Morris polynominal model  :    R = Sum ( Ai X ( K*Bm )**Pi X 10**Qi )
      t =  0.500000"  xd =  5.000000"   f =  0.750000"  xl =  5.000000"
     A1 =  8.460000   A2 =  1.010000   A3 =  1.240000    K =  0.056262
     P1 =        1    P2 =        3    P3 =        5    Q1 =       -1    Q2 =       -1    Q3 =       -5

Modified exponential model  :
     Bm = Sum ( Ai X ( 1 - exp( - R/( 2*i*C ) ) ) )  +      Sum ( RKj X ( R - Rj0 ) )  +  Bm0
      C =   0.36916667E+00             Bmo=   0.00000000E+00          Nexp=  6       Nliner=  1
     AI =   0.14299988E+03  -0.82495092E+03   0.15660586E+04  -0.75360051E+02  -0.23078156E+04   0.16552137E+04
    Rj0 =          22.2000
    RKj =   0.11726382E+01
```

No	Rotation (radians) X 1/1000	Moment (kip-inch) Expri.	Moment (kip-inch) Poly.	Moment (kip-inch) M.Expo.	Connection stiffness (kip-inch) X 1000 Expri.	Connection stiffness (kip-inch) X 1000 Poly.	Connection stiffness (kip-inch) X 1000 M.Expo.
1	0.00	0.00	0.00	0.00	0.6132E+02	0.2101E+02	0.6502E+02
3	0.20	12.26	4.17	11.08	0.6132E+02	0.2060E+02	0.4693E+02
5	0.30	18.40	6.21	15.43	0.4139E-01	0.2013E+02	0.4030E+02
7	1.20	35.20	21.48	38.12	0.2108E+02	0.1379E+02	0.1787E+02
9	1.70	46.40	27.69	46.64	0.2148E+02	0.1124E+02	0.1665E+02
11	2.50	62.40	35.55	59.90	0.1715E+02	0.8633E+01	0.1644E+02
13	3.50	76.00	43.15	75.59	0.1318E+02	0.6748E+01	0.1462E+02
15	4.70	91.20	50.38	91.00	0.1045E+02	0.5412E+01	0.1099E+02
17	6.30	103.20	58.11	105.12	0.7367E+01	0.4344E+01	0.6973E+01
19	8.30	117.60	65.92	116.19	0.5776E+01	0.3536E+01	0.4497E+01
21	11.00	128.00	74.50	126.60	0.3226E+01	0.2872E+01	0.3418E+01
23	13.93	135.47	82.21	135.69	0.2545E+01	0.2416E+01	0.2777E+01
25	17.20	142.40	89.51	143.48	0.1777E+01	0.2074E+01	0.1992E+01
27	20.60	148.40	96.11	148.95	0.1751E+01	0.1821E+01	0.1254E+01
29	23.65	153.60	101.39	153.70	0.1655E+01	0.1650E+01	0.1943E+01
31	26.75	159.20	106.29	159.18	0.1938E+01	0.1511E+01	0.1619E+01
33	29.80	164.26	110.75	163.80	0.1334E+01	0.1399E+01	0.1425E+01
35	32.60	168.00	114.54	167.62	-0.2224E+00	0.1313E+01	0.1319E+01
37	33.80	169.60	116.07	169.19	0.1143E+01	0.1281E+01	0.1288E+01
39	36.60	172.80	119.56	172.72	0.1085E+01	0.1211E+01	0.1238E+01
41	39.73	176.00	123.24	176.54	0.1021E+01	0.1143E+01	0.1207E+01

IV - 31

Connection type : Top-and seat-angle connections
Mode : All bolted

Tested by : M.J.Marley (1982) U.S.A
Test Id. : B-1/2-2

Column : W5X16
Beam : W5X16
Angle : 4 X 4 X 1/2 X 5.0

Fasteners: A325- -3/4"D
13/16" Oversize holes
Material : --
Fy = -- ksi
Fu = -- ksi

Major parameters

lt = 5.0000"	tt = 0.5000"	ls = 5.0000"	ts = 0.5000"
gt = -- "	gt'= -- "	gs = -- "	gs'= -- "
qt = -- "	rt = -- "	qs = -- "	rs = -- "
nt = 2 X 1	nt'= 2 X 1	ns = 2 X 1	ns'= 2 X 1

Remark 1)
2)

No	Moment (k-in)	Rotation (radians) X 1/1000	No	Moment (k-in)	Rotation (radians) X 1/1000
1	0.00	0.00	26	128.52	12.20
2	6.83	0.15	27	130.92	13.10
3	13.66	0.30	28	133.33	14.00
4	20.62	0.50	29	135.48	14.73
5	27.58	0.70	30	137.62	15.47
6	34.54	0.90	31	139.76	16.20
7	39.62	1.23	32	141.90	16.97
8	44.71	1.57	33	144.04	17.73
9	49.80	1.90	34	146.18	18.50
10	54.89	2.13	35	147.52	19.33
11	59.98	2.37	36	148.86	20.17
12	65.06	2.60	37	150.20	21.00
13	71.89	3.05			
14	78.72	3.50			
15	83.94	4.05			
16	89.16	4.60			
17	94.38	5.15			
18	99.60	5.70			
19	105.22	6.25			
20	110.84	6.80			
21	114.46	7.60			
22	118.07	8.40			
23	120.75	9.37			
24	123.43	10.33			
25	126.11	11.30			

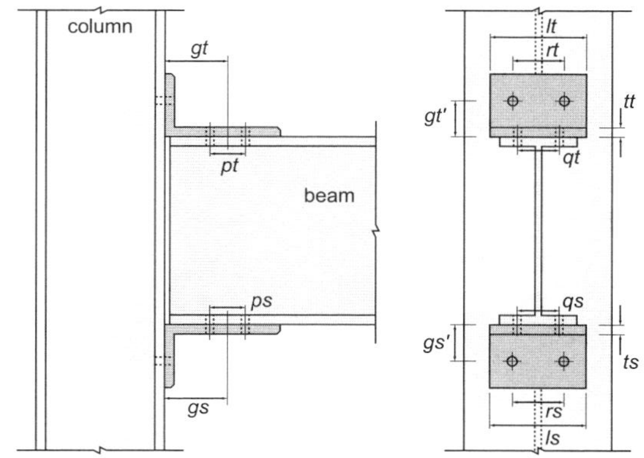

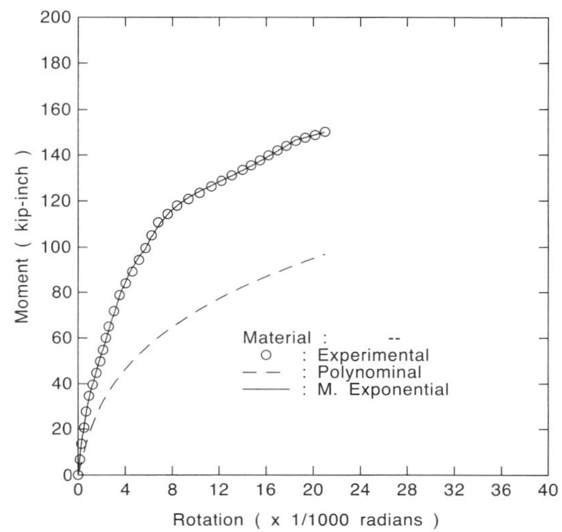

A.4 – 62

```
Moment-rotation prediction equations     ( R : X 1/1000 radians )

Frye and Morris polynominal model   :    R = Sum ( Ai X ( K*Bm )**Pi X 10**Qi )
     t =  0.500000"  xd =  5.000000"   f =  0.750000"  xl =  5.000000"
    A1 =  8.460000   A2 =  1.010000   A3 =  1.240000    K =  0.056262
    P1 =        1    P2 =        3    P3 =        5    Q1 =       -1     Q2 =       -1     Q3 =       -5

Modified exponential model  :
    Bm = Sum ( Ai X ( 1 - exp( - R/( 2*i*C ) ) ) )  +      Sum ( RKj X ( R - Rj0 ) )  +  Bm0
     C =   0.21250000E+00            Bmo=   0.00000000E+00          Nexp=  6        Nliner=  3
    AI =  -0.30346801E+02   0.30296947E+03   0.27381845E+03  -0.37044020E+04   0.57402228E+04  -0.24764451E+04
   Rj0 =           3.5000           5.7000          18.5000
   RKj =  -0.22977677E+01   0.58086230E+01  -0.18948032E+01
```

No	Rotation (radians) X 1/1000	Moment (kip-inch) Expri.	Moment (kip-inch) Poly.	Moment (kip-inch) M.Expo.	Connection stiffness (kip-inch) X 1000 Expri.	Connection stiffness (kip-inch) X 1000 Poly.	Connection stiffness (kip-inch) X 1000 M.Expo.
1	0.00	0.00	0.00	0.00	0.4552E+02	0.2101E+02	0.5086E+02
3	0.30	13.66	6.21	13.77	0.4093E+02	0.2013E+02	0.4063E+02
5	0.70	27.58	13.73	27.40	0.3480E+02	0.1731E+02	0.2826E+02
7	1.23	39.62	21.93	39.76	0.1526E+02	0.1360E+02	0.1949E+02
9	1.90	49.80	29.86	51.55	0.1911E+02	0.1045E+02	0.1678E+02
11	2.37	59.98	34.37	59.35	0.2180E+02	0.8977E+01	0.1674E+02
13	3.05	71.89	39.96	70.76	0.1517E+02	0.7472E+01	0.1649E+02
15	4.05	83.94	46.66	84.96	0.9491E+01	0.6051E+01	0.1176E+02
17	5.15	94.38	52.73	95.59	0.9491E+01	0.5053E+01	0.7449E+01
19	6.25	105.22	57.89	104.62	0.1022E+02	0.4370E+01	0.9094E+01
21	7.60	114.46	63.37	114.23	0.4515E+01	0.3776E+01	0.5439E+01
23	9.37	120.75	69.53	121.41	0.2772E+01	0.3233E+01	0.3080E+01
25	11.30	126.11	75.35	126.50	0.2722E+01	0.2816E+01	0.2407E+01
27	13.10	130.92	80.15	130.88	0.2676E+01	0.2527E+01	0.2512E+01
29	14.73	135.48	84.14	135.18	0.2924E+01	0.2318E+01	0.2764E+01
31	16.20	139.76	87.42	139.40	0.2859E+01	0.2164E+01	0.2982E+01
33	17.73	144.04	90.63	144.12	0.2791E+01	0.2028E+01	0.3161E+01
35	19.33	147.52	93.77	147.71	0.1608E+01	0.1905E+01	0.1398E+01
37	21.00	150.20	96.85	150.12	0.1608E+01	0.1796E+01	0.1487E+01

IV - 32

Connection type : Top-and seat-angle connections
Mode : All bolted

Tested by : M.J.Marley (1982) U.S.A
Test Id. : C1-1/2-1

Column : W5X16
Beam : W5X16
Angle : 4 X 4 X 1/2 X 5.0

Fasteners: A325- -3/4"D
13/16" Oversize holes
Material : --
Fy = -- ksi
Fu = -- ksi

Major parameters

lt = 5.0000"	tt = 0.5000"	ls = 5.0000"	ts = 0.5000"
gt = -- "	gt'= -- "	gs = -- "	gs'= -- "
qt = -- "	rt = -- "	qs = -- "	rs = -- "
nt = 2 X 1	nt'= 2 X 1	ns = 2 X 1	ns'= 2 X 1

Remark 1)
2)

No	Moment (k-in)	Rotation (radians) X 1/1000	No	Moment (k-in)	Rotation (radians) X 1/1000
1	0.00	0.00	26	163.32	19.23
2	6.96	0.20	27	166.26	20.60
3	13.92	0.40	28	168.40	21.93
4	20.88	0.60	29	170.55	23.27
5	28.11	0.77	30	172.69	24.60
6	35.34	0.93	31	174.30	25.83
7	42.58	1.10	32	175.91	27.07
8	51.41	1.55	33	177.52	28.30
9	60.24	2.00	34	179.39	29.57
10	69.08	2.65	35	181.26	30.83
11	77.92	3.30	36	183.13	32.10
12	84.74	4.20	37	184.47	33.53
13	91.57	5.10	38	185.81	34.97
14	98.80	6.10	39	187.15	36.40
15	106.02	7.10			
16	112.45	7.95			
17	118.87	8.80			
18	126.91	9.70			
19	134.94	10.60			
20	142.57	11.70			
21	150.20	12.80			
22	152.61	14.03			
23	155.02	15.27			
24	157.43	16.50			
25	160.37	17.87			

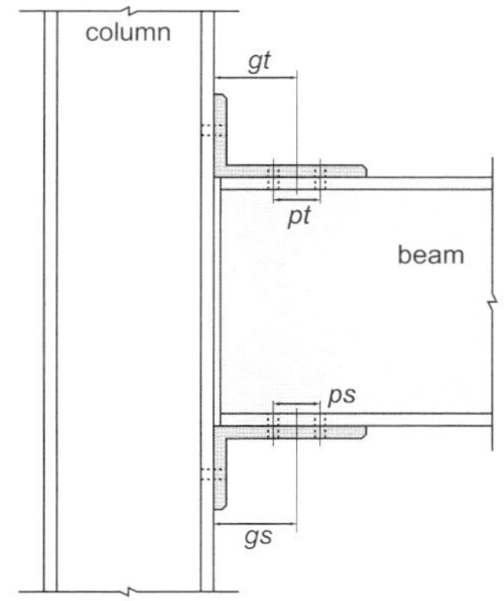

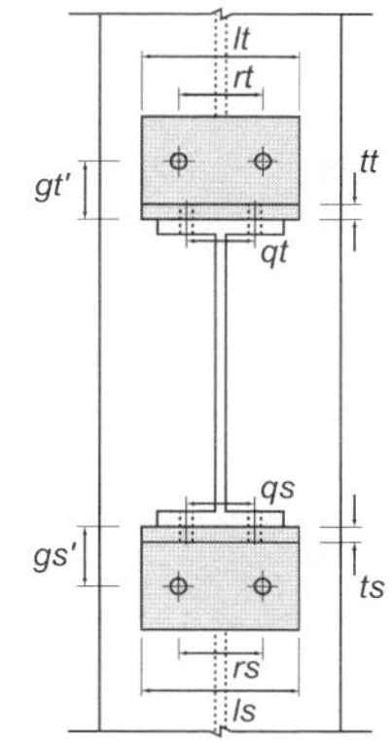

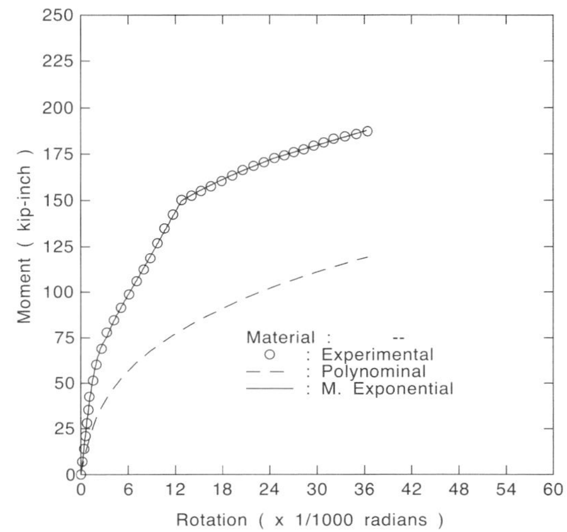

A.4 – 64

```
Moment-rotation prediction equations     ( R : X 1/1000 radians )

Frye and Morris polynominal model  :    R = Sum ( Ai X ( K*Bm )**Pi X 10**Qi )
     t =  0.500000"  xd =  5.000000"   f =  0.750000"  xl =  5.000000"
    A1 =  8.460000   A2 =  1.010000   A3 =  1.240000    K =  0.056262
    P1 =         1   P2 =         3   P3 =         5   Q1 =        -1     Q2 =        -1     Q3 =        -5

Modified exponential model  :
    Bm = Sum ( Ai X ( 1 - exp( - R/( 2*i*C ) ) ) )  +     Sum ( RKj X ( R - Rj0 ) )  +  Bm0
     C =   0.35333333E+00           Bmo=   0.00000000E+00           Nexp=  6        Nliner=  3
    AI =  -0.87549081E+02   0.40534324E+03  -0.38349960E+03   0.18071786E+03  -0.21108826E+03   0.18908018E+03
    Rj0 =          3.3000          12.8000          24.6000
    RKj =   0.64827822E+01  -0.49088134E+01  -0.32859336E+00
```

No	Rotation (radians) X 1/1000	Moment (kip-inch) Expri.	Moment (kip-inch) Poly.	Moment (kip-inch) M.Expo.	Connection stiffness (kip-inch) X 1000 Expri.	Connection stiffness (kip-inch) X 1000 Poly.	Connection stiffness (kip-inch) X 1000 M.Expo.
1	0.00	0.00	0.00	0.00	0.3480E+02	0.2101E+02	0.3080E+02
3	0.40	13.92	8.20	14.41	0.3480E+02	0.1952E+02	0.3870E+02
5	0.77	28.11	14.87	28.40	0.4340E+02	0.1680E+02	0.3673E+02
7	1.10	42.58	20.06	39.88	0.3696E+02	0.1443E+02	0.3187E+02
9	2.00	60.24	30.89	61.82	0.1716E+02	0.1009E+02	0.1735E+02
11	3.30	77.92	41.77	75.34	0.1108E+02	0.7049E+01	0.5357E+01
13	5.10	91.57	52.48	91.66	0.7415E+01	0.5090E+01	0.7655E+01
15	7.10	106.02	61.43	106.65	0.7406E+01	0.3972E+01	0.7546E+01
17	8.80	118.87	67.65	119.63	0.8224E+01	0.3386E+01	0.7689E+01
19	10.60	134.94	73.34	133.43	0.8032E+01	0.2951E+01	0.7612E+01
21	12.80	150.20	79.39	149.91	0.4588E+01	0.2570E+01	0.7358E+01
23	15.27	155.02	85.33	155.58	0.1952E+01	0.2261E+01	0.2157E+01
25	17.87	160.37	90.87	160.87	0.2154E+01	0.2018E+01	0.1929E+01
27	20.60	166.26	96.11	165.92	0.1878E+01	0.1821E+01	0.1778E+01
29	23.27	170.55	100.76	170.53	0.1608E+01	0.1669E+01	0.1690E+01
31	25.83	174.30	104.92	174.40	0.1304E+01	0.1548E+01	0.1312E+01
33	28.30	177.52	108.61	177.60	0.1390E+01	0.1451E+01	0.1284E+01
35	30.83	181.26	112.17	180.83	0.1478E+01	0.1366E+01	0.1268E+01
37	33.53	184.47	115.76	184.23	0.9349E+00	0.1287E+01	0.1258E+01
39	36.40	187.15	119.34	187.83	0.9349E+00	0.1215E+01	0.1252E+01

IV - 33

Connection type : Top-and seat-angle connections
Mode : All bolted

Tested by : M.J.Marley (1982) U.S.A
Test Id. : C1-1/2-2

Column : W5X16
Beam : W5X16
Angle : 4 X 4 X 1/2 X 5.0

Fasteners: A325- -3/4"D
13/16" Oversize holes
Material : --
Fy = -- ksi
Fu = -- ksi

Major parameters

lt = 5.0000"	tt = 0.5000"	ls = 5.0000"	ts = 0.5000"
gt = -- "	gt'= -- "	gs = -- "	gs'= -- "
qt = -- "	rt = -- "	qs = -- "	rs = -- "
nt = 2 X 1	nt'= 2 X 1	ns = 2 X 1	ns'= 2 X 1

Remark 1)
2)

No	Moment (k-in)	Rotation (radians) X 1/1000	No	Moment (k-in)	Rotation (radians) X 1/1000
1	0.00	0.00	26	162.67	22.10
2	8.80	0.00	27	165.60	23.40
3	17.60	0.00	28	168.80	25.00
4	26.80	0.00	29	172.00	26.60
5	36.00	0.00	30	174.00	28.25
6	43.20	0.27	31	176.00	29.90
7	50.40	0.53	32	178.00	31.45
8	57.60	0.80	33	180.00	33.00
9	66.80	1.20	34	181.60	34.60
10	76.00	1.60	35	183.20	36.20
11	83.20	2.65	36	184.80	37.30
12	90.40	3.70	37	186.40	38.40
13	96.00	4.65			
14	101.60	5.60			
15	107.20	6.50			
16	112.80	7.40			
17	121.20	8.60			
18	129.60	9.80			
19	134.40	11.25			
20	139.20	12.70			
21	144.00	14.20			
22	148.80	15.70			
23	152.80	17.60			
24	156.80	19.50			
25	159.74	20.80			

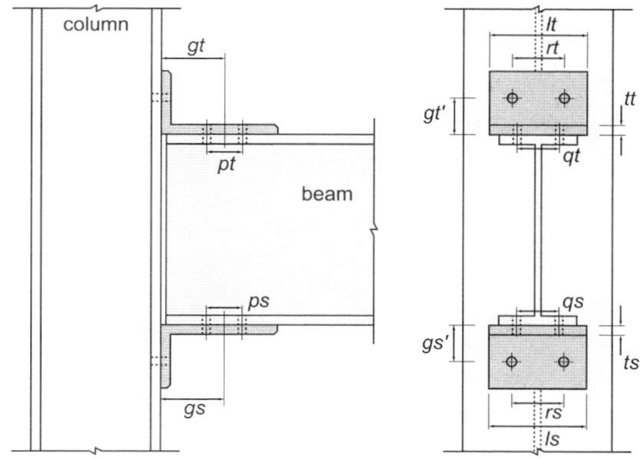

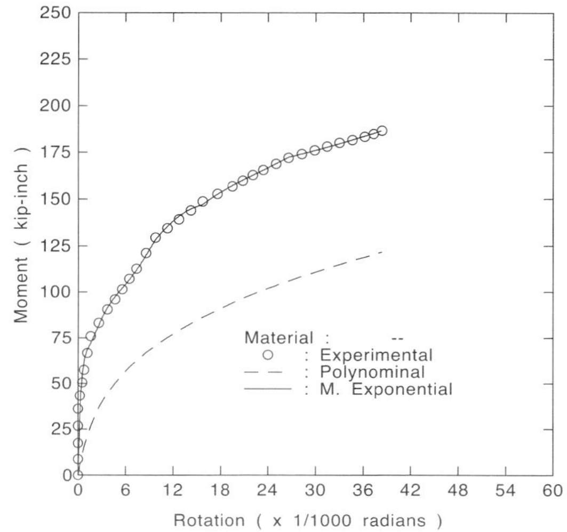

A.4 – 66

```
Moment-rotation prediction equations     ( R : X 1/1000 radians )

Frye and Morris polynominal model  :    R = Sum ( Ai X ( K*Bm )**Pi X 10**Qi )
      t =  0.500000"  xd =  5.000000"   f =  0.750000"  xl =  5.000000"
     A1 =  8.460000   A2 =  1.010000   A3 =  1.240000    K =  0.056262
     P1 =         1   P2 =         3   P3 =         5   Q1 =        -1     Q2 =       -1     Q3 =       -5

Modified exponential model  :
     Bm = Sum ( Ai X ( 1 - exp( - R/( 2*i*C ) ) ) )  +     Sum ( RKj X ( R - Rj0 ) )  +  Bm0
      C =  0.32002500E+00             Bmo=  0.00000000E+00           Nexp=  6        Nliner=  4
     AI =  0.40875745E+03  -0.32678685E+04  0.13152710E+05  -0.24698894E+05  0.21247533E+05  -0.67201040E+04
    Rj0 =          1.6000          9.8000         15.7000          26.6000
    RKj =  0.21477335E+01  -0.18203804E+01  0.22908483E+01  -0.12588609E+01
```

No	Rotation (radians) X 1/1000	Moment (kip-inch) Expri.	Moment (kip-inch) Poly.	Moment (kip-inch) M.Expo.	Connection stiffness (kip-inch) X 1000 Expri.	Connection stiffness (kip-inch) X 1000 Poly.	Connection stiffness (kip-inch) X 1000 M.Expo.
1	0.00	0.00	0.00	0.00	0.8802E+05	0.2101E+02	0.1778E+03
3	0.00	17.60	0.00	0.04	0.9000E+05	0.2101E+02	0.1778E+03
5	0.00	36.00	0.01	0.07	0.9195E+05	0.2101E+02	0.1777E+03
7	0.53	50.40	10.74	53.03	0.2700E+02	0.1858E+02	0.4729E+02
9	1.20	66.80	21.48	68.53	0.2299E+02	0.1379E+02	0.1049E+02
11	2.65	83.20	36.81	82.12	0.6857E+01	0.8279E+01	0.9670E+01
13	4.65	96.00	50.11	97.25	0.5899E+01	0.5455E+01	0.5528E+01
15	6.50	107.20	58.97	106.91	0.6220E+01	0.4243E+01	0.5542E+01
17	8.60	121.20	66.97	120.18	0.7000E+01	0.3444E+01	0.6896E+01
19	11.25	134.40	75.21	135.36	0.3310E+01	0.2825E+01	0.4304E+01
21	14.20	144.00	82.85	144.71	0.3200E+01	0.2383E+01	0.2033E+01
23	17.60	152.80	90.33	152.65	0.2106E+01	0.2040E+01	0.2611E+01
25	20.80	159.74	96.47	159.99	0.2255E+01	0.1809E+01	0.2106E+01
27	23.40	165.60	100.98	165.41	0.2140E+01	0.1663E+01	0.2099E+01
29	26.60	172.00	106.06	172.34	0.1613E+01	0.1517E+01	0.2241E+01
31	29.90	176.00	110.86	175.84	0.1253E+01	0.1397E+01	0.1133E+01
33	33.00	180.00	115.04	179.52	0.1147E+01	0.1302E+01	0.1232E+01
35	36.20	183.20	119.07	183.56	0.1268E+01	0.1220E+01	0.1293E+01
37	38.40	186.40	121.73	186.44	0.1451E+01	0.1170E+01	0.1318E+01

IV - 34

Connection type : Top-and seat-angle connections
Mode : All bolted

Tested by : M.J.Marley (1982) U.S.A
Test Id. : C2-1/2-1

Column : W5X16
Beam : W5X16
Angle : 4 X 4 X 1/2 X 5.0

Fasteners: A325- -3/4"D
13/16" Oversize holes
Material : --
Fy = -- ksi
Fu = -- ksi

Major parameters

lt = 5.0000"	tt = 0.5000"	ls = 5.0000"	ts = 0.5000"
gt = -- "	gt'= -- "	gs = -- "	gs'= -- "
qt = -- "	rt = -- "	qs = -- "	rs = -- "
nt = 2 X 1	nt'= 2 X 1	ns = 2 X 1	ns'= 2 X 1

Remark 1)
2)

No	Moment (k-in)	Rotation (radians) X 1/1000	No	Moment (k-in)	Rotation (radians) X 1/1000
1	0.00	0.00	26	157.43	22.40
2	7.23	0.10	27	160.10	23.77
3	14.46	0.20	28	162.78	25.13
4	21.42	0.33	29	165.46	26.50
5	28.38	0.47	30	167.60	27.80
6	35.34	0.60	31	169.74	29.10
7	43.78	1.10	32	171.89	30.40
8	52.21	1.60	33	174.70	32.15
9	58.90	2.07	34	177.52	33.90
10	65.60	2.53	35	178.85	35.23
11	72.29	3.00	36	180.19	36.57
12	79.78	3.67	37	181.52	37.90
13	87.28	4.33			
14	94.78	5.00			
15	103.61	5.95			
16	112.45	6.90			
17	118.07	7.65			
18	123.70	8.40			
19	128.51	10.05			
20	133.33	11.70			
21	138.15	13.35			
22	142.97	15.00			
23	146.18	16.85			
24	149.40	18.70			
25	153.41	20.55			

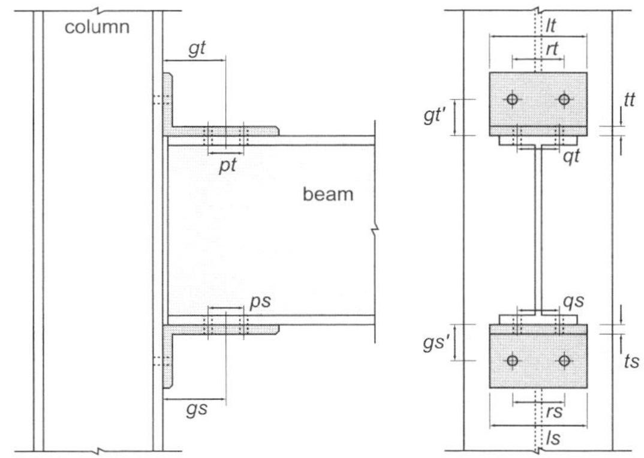

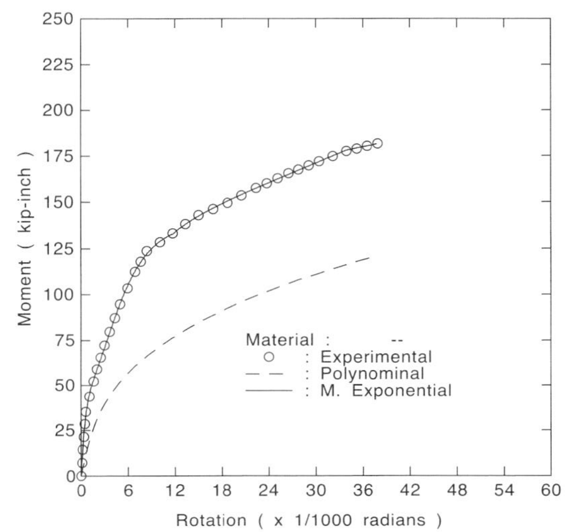

A.4 – 68

```
 Moment-rotation prediction equations      ( R : X 1/1000 radians )

Frye and Morris polynominal model  :    R = Sum ( Ai X ( K*Bm )**Pi X 10**Qi )
     t =  0.500000"  xd =  5.000000"   f =  0.750000"  xl =  5.000000"
    A1 =  8.460000   A2 =  1.010000   A3 =  1.240000    K =  0.056262
    P1 =         1   P2 =         3   P3 =         5   Q1 =        -1     Q2 =        -1     Q3 =        -5

Modified exponential model  :
    Bm = Sum ( Ai X ( 1 - exp( - R/( 2*i*C ) ) ) )  +      Sum ( RKj X ( R - Rj0 ) )  +  Bm0
      C =   0.34083333E+00              Bmo=   0.00000000E+00           Nexp=  6        Nliner=  2
     AI =   0.85980949E+02  -0.83572476E+01  -0.48594731E+03  0.60484836E+03  -0.94690385E+02   0.36777406E+02
    Rj0 =          11.7000          33.9000
    RKj =   0.17739571E+01  -0.92439601E+00
```

No	Rotation (radians) X 1/1000	Moment (kip-inch) Expri.	Moment (kip-inch) Poly.	Moment (kip-inch) M.Expo.	Connection stiffness (kip-inch) X 1000 Expri.	Connection stiffness (kip-inch) X 1000 Poly.	Connection stiffness (kip-inch) X 1000 M.Expo.
1	0.00	0.00	0.00	0.00	0.7230E+02	0.2101E+02	0.8541E+02
3	0.20	14.46	4.17	14.58	0.6369E+02	0.2060E+02	0.6178E+02
5	0.47	28.38	9.48	28.04	0.5219E+02	0.1907E+02	0.4084E+02
7	1.10	43.78	20.07	45.41	0.1687E+02	0.1442E+02	0.1856E+02
9	2.07	58.90	31.55	59.19	0.1434E+02	0.9866E+01	0.1247E+02
11	3.00	72.29	39.58	70.87	0.1306E+02	0.7563E+01	0.1269E+02
13	4.33	87.28	48.33	87.57	0.1124E+02	0.5752E+01	0.1201E+02
15	5.95	103.61	56.56	104.90	0.9303E+01	0.4534E+01	0.9264E+01
17	7.65	118.07	63.56	117.97	0.7496E+01	0.3757E+01	0.6194E+01
19	10.05	128.51	71.68	128.83	0.2920E+01	0.3069E+01	0.3129E+01
21	13.35	138.15	80.78	138.21	0.2920E+01	0.2493E+01	0.2875E+01
23	16.85	146.18	88.78	146.70	0.1738E+01	0.2105E+01	0.2119E+01
25	20.55	153.41	96.02	154.00	0.2170E+01	0.1825E+01	0.1874E+01
27	23.77	160.10	101.58	159.90	0.1958E+01	0.1644E+01	0.1809E+01
29	26.50	165.46	105.91	164.81	0.1800E+01	0.1522E+01	0.1789E+01
31	29.10	169.74	109.73	169.45	0.1649E+01	0.1424E+01	0.1781E+01
33	32.15	174.70	113.92	174.88	0.1608E+01	0.1327E+01	0.1777E+01
35	35.23	178.85	117.90	179.12	0.1002E+01	0.1243E+01	0.8508E+00
37	37.90	181.52	121.11	181.39	0.1002E+01	0.1181E+01	0.8502E+00

IV - 35

Connection type : Top-and seat-angle connections
Mode : All bolted

Tested by : M.J.Marley (1982) U.S.A
Test Id. : C2-1/2-2

Column : W5X16
Beam : W5X16
Angle : 4 X 4 X 1/2 X 5.0

Fasteners: A325- -3/4"D
13/16" Oversize holes
Material : --
Fy = -- ksi
Fu = -- ksi

Major parameters

lt = 5.0000"	tt = 0.5000"	ls = 5.0000"	ts = 0.5000"
gt = -- "	gt'= -- "	gs = -- "	gs'= -- "
qt = -- "	rt = -- "	qs = -- "	rs = -- "
nt = 2 X 1	nt'= 2 X 1	ns = 2 X 1	ns'= 2 X 1

Remark 1)
2)

No	Moment (k-in)	Rotation (radians) X 1/1000	No	Moment (k-in)	Rotation (radians) X 1/1000
1	0.00	0.00	26	154.03	23.05
2	8.87	0.40	27	157.26	25.00
3	17.74	0.80	28	160.48	26.55
4	26.21	1.35	29	163.70	28.10
5	34.68	1.90	30	165.59	29.53
6	43.15	2.50	31	167.47	30.97
7	51.61	3.10	32	169.36	32.40
8	58.87	3.60	33	171.77	34.10
9	66.13	4.10	34	174.19	35.80
10	72.58	4.75	35	175.54	37.30
11	79.03	5.40	36	176.88	38.80
12	85.08	5.85	37	178.22	40.30
13	91.13	6.30			
14	97.58	7.25			
15	104.03	8.20			
16	109.67	9.20			
17	115.32	10.20			
18	121.77	11.60			
19	128.22	13.00			
20	131.85	14.20			
21	135.48	15.40			
22	139.11	16.75			
23	142.74	18.10			
24	146.77	19.60			
25	150.80	21.10			

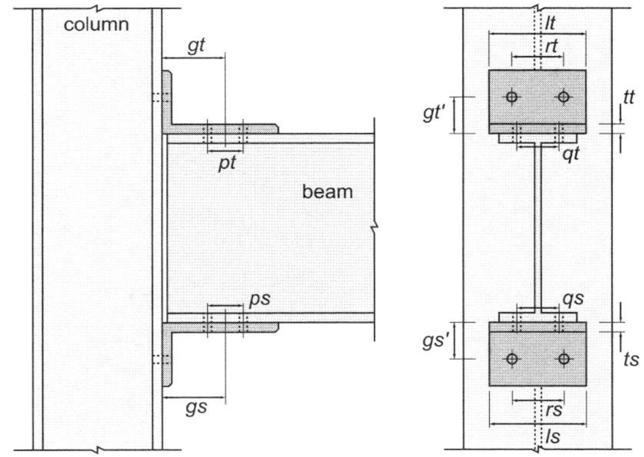

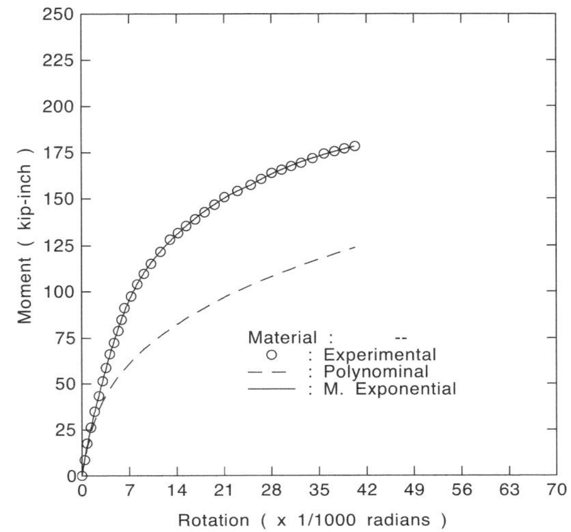

A.4 – 70

```
Moment-rotation prediction equations     ( R : X 1/1000 radians )

Frye and Morris polynominal model  :    R = Sum ( Ai X ( K*Bm )**Pi X 10**Qi )
      t =  0.500000"  xd =  5.000000"   f =  0.750000"  xl =  5.000000"
     A1 =  8.460000   A2 =  1.010000   A3 =  1.240000    K =  0.056262
     P1 =         1   P2 =         3   P3 =         5   Q1 =        -1     Q2 =        -1     Q3 =        -5

Modified exponential model  :
     Bm = Sum ( Ai X ( 1 - exp( - R/( 2*i*C ) ) ) )  +     Sum ( RKj X ( R - Rj0 ) )  +  Bm0
      C =   0.43583333E+00            Bmo=   0.00000000E+00          Nexp=  6        Nliner=  1
     AI =   0.23385693E+02   0.20643553E+03  -0.25479777E+04   0.79023950E+04  -0.97971758E+04   0.43812190E+04
    Rj0 =          25.0000
    RKj =   0.72460387E+00
```

No	Rotation (radians) X 1/1000	Moment (kip-inch) Expri.	Moment (kip-inch) Poly.	Moment (kip-inch) M.Expo.	Connection stiffness (kip-inch) X 1000 Expri.	Connection stiffness (kip-inch) X 1000 Poly.	Connection stiffness (kip-inch) X 1000 M.Expo.
1	0.00	0.00	0.00	0.00	0.2217E+02	0.2101E+02	0.2713E+02
3	0.80	17.74	15.43	17.16	0.1932E+02	0.1655E+02	0.1770E+02
5	1.90	34.68	29.87	34.61	0.1478E+02	0.1044E+02	0.1487E+02
7	3.10	51.61	40.33	51.89	0.1433E+02	0.7382E+01	0.1391E+02
9	4.10	66.13	46.96	65.23	0.1252E+02	0.5996E+01	0.1270E+02
11	5.40	79.03	53.97	80.44	0.1200E+02	0.4876E+01	0.1065E+02
13	6.30	91.13	58.11	89.36	0.1130E+02	0.4344E+01	0.9188E+01
15	8.20	104.03	65.57	104.24	0.6232E+01	0.3568E+01	0.6620E+01
17	10.20	115.32	72.14	115.60	0.5213E+01	0.3036E+01	0.4894E+01
19	13.00	128.22	79.90	127.36	0.3755E+01	0.2541E+01	0.3670E+01
21	15.40	135.48	85.63	135.46	0.2867E+01	0.2247E+01	0.3118E+01
23	18.10	142.74	91.34	143.21	0.2688E+01	0.1999E+01	0.2632E+01
25	21.10	150.80	97.01	150.31	0.2239E+01	0.1790E+01	0.2104E+01
27	25.00	157.26	103.58	157.22	0.1891E+01	0.1586E+01	0.1453E+01
29	28.10	163.70	108.29	163.28	0.1682E+01	0.1459E+01	0.1746E+01
31	30.97	167.47	112.33	167.81	0.1314E+01	0.1362E+01	0.1435E+01
33	34.10	171.77	116.45	171.90	0.1422E+01	0.1273E+01	0.1186E+01
35	37.30	175.54	120.40	175.40	0.8960E+00	0.1195E+01	0.1013E+01
37	40.30	178.22	123.89	178.27	0.8960E+00	0.1131E+01	0.9065E+00

A.4 – 71

IV - 36

Connection type : Top-and seat-angle connections
Mode : All bolted

Tested by : M.J.Marley (1982) U.S.A
Test Id. : D1-1/2-1

Column : W5X16
Beam : W5X16
Angle : 4 X 4 X 1/2 X 5.0

Fasteners: A325- -3/4"D
13/16" Oversize holes
Material : --
Fy = -- ksi
Fu = -- ksi

Major parameters

lt = 5.0000"	tt = 0.5000"	ls = 5.0000"	ts = 0.5000"
gt = -- "	gt'= -- "	gs = -- "	gs'= -- "
qt = -- "	rt = -- "	qs = -- "	rs = -- "
nt = 2 X 1	nt'= 2 X 1	ns = 2 X 1	ns'= 2 X 1

Remark 1)
2)

No	Moment (k-in)	Rotation (radians) X 1/1000	No	Moment (k-in)	Rotation (radians) X 1/1000
1	0.00	0.00	26	170.28	21.50
2	8.03	0.25	27	173.09	23.35
3	16.07	0.50	28	175.91	25.20
4	23.56	0.67	29	178.31	26.85
5	31.06	0.83	30	180.72	28.50
6	38.56	1.00	31	183.13	29.85
7	48.19	1.45	32	185.54	31.20
8	57.83	1.90	33	187.55	32.75
9	66.26	2.40	34	189.56	34.30
10	74.70	2.90	35	191.57	35.80
11	82.33	3.35	36	193.57	37.30
12	89.96	3.80	37	195.18	38.90
13	97.99	4.20	38	196.79	40.50
14	106.02	4.60			
15	112.45	5.40			
16	118.87	6.20			
17	125.70	7.10			
18	132.53	8.00			
19	137.75	9.50			
20	142.97	11.00			
21	147.39	12.60			
22	151.81	14.20			
23	156.63	16.15			
24	151.45	18.10			
25	155.86	19.80			

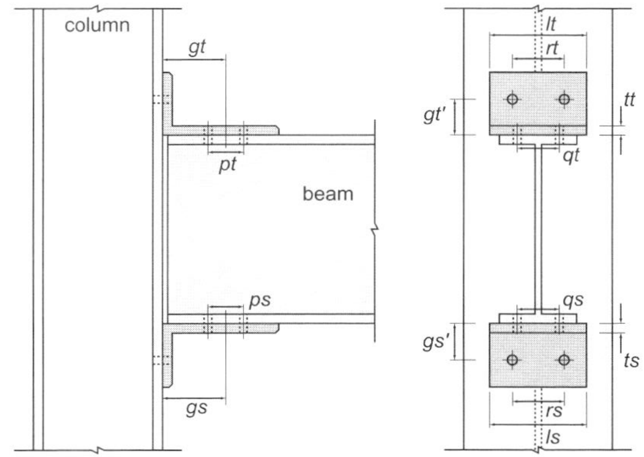

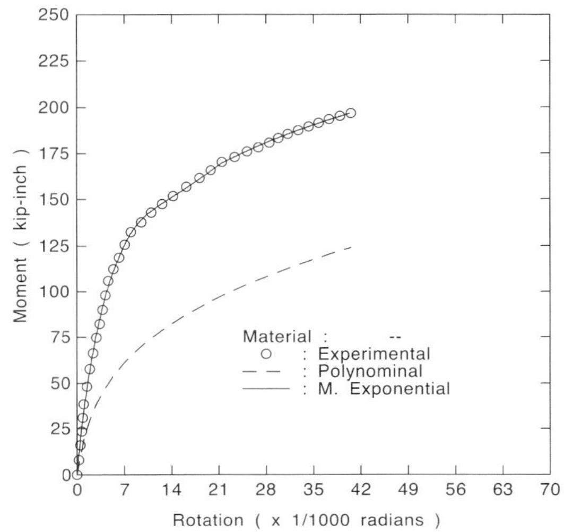

```
Moment-rotation prediction equations     ( R : X 1/1000 radians )

Frye and Morris polynominal model  :     R = Sum ( Ai X ( K*Bm )**Pi X 10**Qi )
      t =  0.500000"  xd =  5.000000"   f =  0.750000"  xl =  5.000000"
     A1 =  8.460000   A2 =  1.010000   A3 =  1.240000    K =  0.056262
     P1 =         1   P2 =         3   P3 =         5   Q1 =        -1    Q2 =        -1    Q3 =        -5

Modified exponential model  :
     Bm = Sum ( Ai X ( 1 - exp( - R/( 2*i*C ) ) ) )  +      Sum ( RKj X ( R - Rj0 ) )  +  Bm0
      C =   0.40000000E+00            Bmo=   0.00000000E+00          Nexp=  6        Nliner=  2
     AI =   0.30036805E+00   0.55046428E+03  -0.39607997E+04   0.99101686E+04  -0.10173159E+05   0.38229384E+04
    Rj0 =           6.2000          21.5000
    RKj =   0.17998646E+01  -0.75591617E+00
```

No	Rotation (radians) X 1/1000	Moment (kip-inch) Expri.	Moment (kip-inch) Poly.	Moment (kip-inch) M.Expo.	Connection stiffness (kip-inch) X 1000 Expri.	Connection stiffness (kip-inch) X 1000 Poly.	Connection stiffness (kip-inch) X 1000 M.Expo.
1	0.00	0.00	0.00	0.00	0.3214E+02	0.2101E+02	0.4417E+02
3	0.50	16.07	10.11	19.35	0.3983E+02	0.1883E+02	0.3410E+02
5	0.83	31.06	15.97	29.95	0.4498E+02	0.1630E+02	0.2971E+02
7	1.45	48.19	24.75	46.51	0.2141E+02	0.1240E+02	0.2447E+02
9	2.40	66.26	34.67	67.36	0.1687E+02	0.8888E+01	0.1983E+02
11	3.35	82.33	42.12	84.56	0.1696E+02	0.6971E+01	0.1647E+02
13	4.20	97.99	47.56	97.37	0.2007E+02	0.5889E+01	0.1368E+02
15	5.40	112.45	53.97	111.51	0.8032E+01	0.4876E+01	0.9948E+01
17	7.10	125.70	61.43	126.14	0.7587E+01	0.3972E+01	0.7389E+01
19	9.50	137.75	69.96	138.95	0.3480E+01	0.3200E+01	0.3770E+01
21	12.60	147.39	78.87	147.80	0.2764E+01	0.2600E+01	0.2360E+01
23	16.15	156.63	87.28	156.10	0.2471E+01	0.2171E+01	0.2423E+01
25	19.80	165.86	94.63	165.33	0.2598E+01	0.1874E+01	0.2592E+01
27	23.35	173.09	100.90	173.07	0.1521E+01	0.1665E+01	0.1773E+01
29	26.85	178.31	106.44	178.98	0.1458E+01	0.1507E+01	0.1596E+01
31	29.85	183.13	110.82	183.53	0.1787E+01	0.1397E+01	0.1439E+01
33	32.75	187.55	114.75	187.51	0.1297E+01	0.1309E+01	0.1313E+01
35	35.80	191.57	118.61	191.36	0.1336E+01	0.1229E+01	0.1216E+01
37	38.90	195.18	122.31	195.02	0.1005E+01	0.1159E+01	0.1149E+01

IV - 37

Connection type : Top-and seat-angle connections
Mode : All bolted

Tested by : M.J.Marley (1982) U.S.A
Test Id. : D1-1/2-2

Column : W5X16
Beam : W5X16
Angle : 4 X 4 X 1/2 X 5.0

Fasteners: A325- -3/4"D
13/16" Oversize holes
Material : --
Fy = -- ksi
Fu = -- ksi

Major parameters

lt = 5.0000"	tt = 0.5000"	ls = 5.0000"	ts = 0.5000"
gt = -- "	gt'= -- "	gs = -- "	gs'= -- "
qt = -- "	rt = -- "	qs = -- "	rs = -- "
nt = 2 X 1	nt'= 2 X 1	ns = 2 X 1	ns'= 2 X 1

Remark 1)
2)

No	Moment (k-in)	Rotation (radians) X 1/1000	No	Moment (k-in)	Rotation (radians) X 1/1000
1	0.00	0.00	26	167.87	19.33
2	7.76	0.13	27	171.89	21.20
3	15.53	0.27	28	175.90	23.25
4	23.29	0.40	29	179.92	25.30
5	28.51	0.45	30	183.13	27.30
6	33.73	0.50	31	186.35	29.30
7	42.57	0.95	32	189.16	30.85
8	51.41	1.40	33	191.96	32.40
9	58.64	1.80	34	193.97	33.95
10	65.87	2.20	35	195.98	35.50
11	73.90	2.45	36	197.19	36.55
12	81.92	2.70	37	198.40	37.60
13	89.15	3.15	38	201.20	39.65
14	96.38	3.60	39	204.01	41.70
15	104.42	4.20			
16	112.45	4.80			
17	121.69	5.70			
18	130.92	6.60			
19	136.94	8.05			
20	142.97	9.50			
21	147.39	11.15			
22	151.81	12.80			
23	155.83	14.20			
24	159.84	15.60			
25	163.86	17.47			

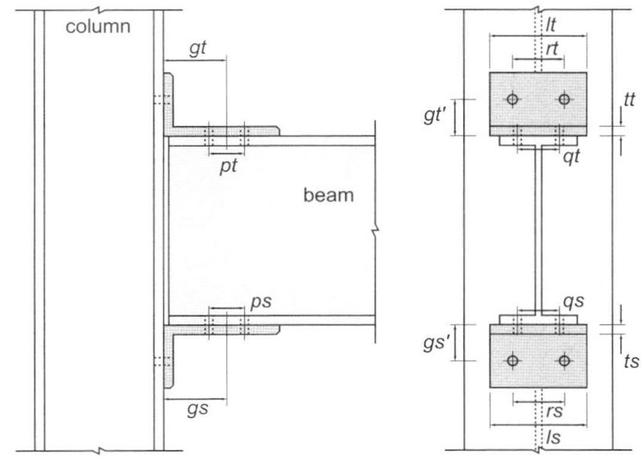

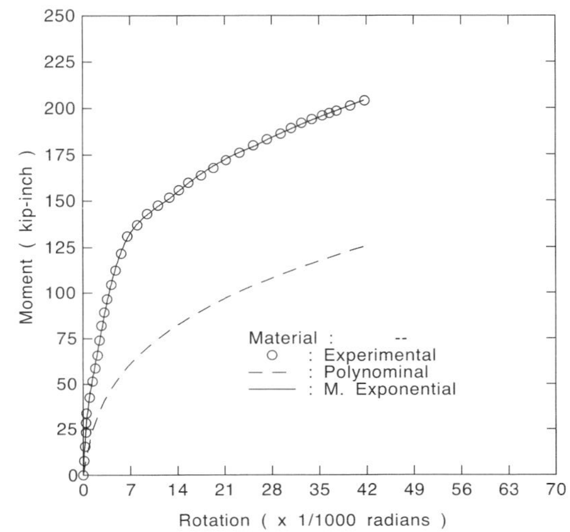

```
Moment-rotation prediction equations     ( R : X 1/1000 radians )

Frye and Morris polynominal model  :    R = Sum ( Ai X ( K*Bm )**Pi X 10**Qi )
     t =  0.500000"  xd =  5.000000"  f =  0.750000"  xl =  5.000000"
    A1 =  8.460000   A2 =  1.010000   A3 =  1.240000   K =  0.056262
    P1 =         1   P2 =         3   P3 =         5   Q1 =        -1     Q2 =        -1     Q3 =        -5

Modified exponential model  :
    Bm = Sum ( Ai X ( 1 - exp( - R/( 2*i*C ) ) ) )  +     Sum ( RKj X ( R - Rj0 ) )  +  Bm0
     C =   0.38082500E+00             Bmo=   0.00000000E+00         Nexp=  6        Nliner=  2
    AI =   0.19936402E+03  -0.11399414E+04   0.21183192E+04  -0.28086672E+03  -0.24220465E+04   0.16988301E+04
   Rj0 =          12.8000          25.3000
   RKj =   0.73057660E+00   0.58901091E+00
```

No	Rotation (radians) X 1/1000	Moment (kip-inch) Expri.	Moment (kip-inch) Poly.	Moment (kip-inch) M.Expo.	Connection stiffness (kip-inch) X 1000 Expri.	Connection stiffness (kip-inch) X 1000 Poly.	Connection stiffness (kip-inch) X 1000 M.Expo.
1	0.00	0.00	0.00	0.00	0.5824E+02	0.2101E+02	0.8404E+02
3	0.27	15.53	5.54	18.12	0.5822E+02	0.2030E+02	0.5444E+02
5	0.45	28.51	9.17	26.85	0.1044E+03	0.1918E+02	0.4163E+02
7	0.95	42.57	17.83	42.61	0.1964E+02	0.1544E+02	0.2470E+02
9	1.80	58.64	28.80	60.64	0.1807E+02	0.1083E+02	0.1996E+02
11	2.45	73.90	35.11	73.62	0.3211E+02	0.8759E+01	0.1993E+02
13	3.15	89.15	40.70	87.27	0.1607E+02	0.7296E+01	0.1881E+02
15	4.20	104.42	47.56	105.17	0.1339E+02	0.5888E+01	0.1503E+02
17	5.70	121.69	55.41	123.14	0.1026E+02	0.4682E+01	0.9144E+01
19	8.05	136.94	65.03	137.61	0.4154E+01	0.3617E+01	0.4053E+01
21	11.15	147.39	74.93	147.06	0.2680E+01	0.2844E+01	0.2548E+01
23	14.20	155.83	82.85	155.41	0.2867E+01	0.2383E+01	0.2992E+01
25	17.47	163.86	90.06	164.48	0.2151E+01	0.2051E+01	0.2523E+01
27	21.20	171.89	97.19	172.70	0.2059E+01	0.1784E+01	0.1891E+01
29	25.30	179.92	104.05	179.29	0.1781E+01	0.1573E+01	0.1359E+01
31	29.30	186.35	110.02	186.40	0.1723E+01	0.1417E+01	0.1637E+01
33	32.40	191.96	114.25	191.24	0.1554E+01	0.1319E+01	0.1499E+01
35	35.50	195.98	118.21	195.75	0.1208E+01	0.1237E+01	0.1419E+01
37	37.60	198.40	120.78	198.69	0.1223E+01	0.1187E+01	0.1386E+01
39	41.70	204.01	125.49	204.29	0.1370E+01	0.1104E+01	0.1349E+01

IV - 38

Connection type : Top-and seat-angle connections
Mode : All bolted

Tested by : M.J.Marley (1982) U.S.A
Test Id. : D2-1/2-1

Column : W5X16
Beam : W5X16
Angle : 4 X 4 X 1/2 X 5.0

Fasteners: A325- -3/4"D
13/16" Oversize holes
Material : --
Fy = -- ksi
Fu = -- ksi

Major parameters

lt = 5.0000"	tt = 0.5000"	ls = 5.0000"	ts = 0.5000"
gt = -- "	gt'= -- "	gs = -- "	gs'= -- "
qt = -- "	rt = -- "	qs = -- "	rs = -- "
nt = 2 X 1	nt'= 2 X 1	ns = 2 X 1	ns'= 2 X 1

Remark 1)
2)

No	Moment (k-in)	Rotation (radians) X 1/1000	No	Moment (k-in)	Rotation (radians) X 1/1000
1	0.00	0.00	26	161.94	20.90
2	9.71	0.10	27	165.59	22.75
3	19.43	0.20	28	169.24	24.60
4	28.74	0.25	29	172.07	26.20
5	38.05	0.30	30	174.90	27.80
6	44.80	0.37	31	177.33	29.20
7	51.55	0.43	32	179.76	30.60
8	58.30	0.50	33	182.19	32.55
9	66.40	0.60	34	184.62	34.50
10	74.50	0.70	35	186.24	36.35
11	83.00	1.05	36	187.86	38.20
12	91.50	1.40	37	189.47	39.85
13	97.57	1.75	38	191.09	41.50
14	103.64	2.10	39	191.90	43.30
15	109.72	2.45	40	192.71	45.10
16	115.79	2.80	41	193.93	46.75
17	122.27	4.50	42	195.14	48.40
18	126.32	5.80			
19	130.37	7.10			
20	136.03	9.40			
21	140.48	11.15			
22	144.94	12.90			
23	149.80	14.75			
24	154.66	16.60			
25	158.30	18.75			

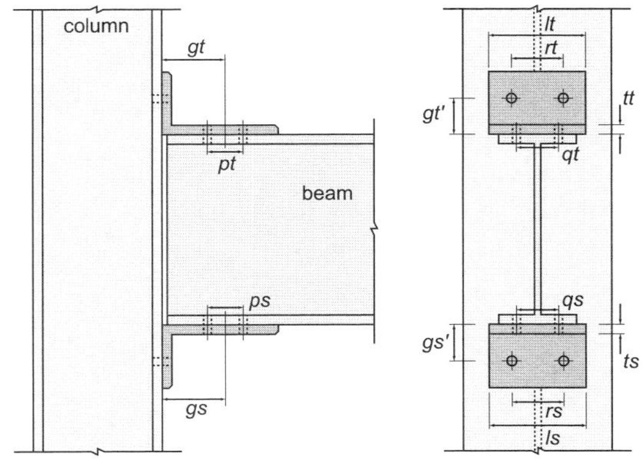

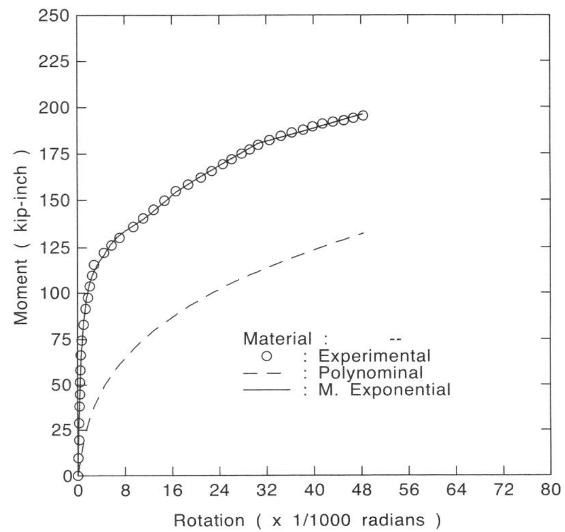

```
Moment-rotation prediction equations      ( R : X 1/1000 radians )

Frye and Morris polynominal model  :    R = Sum ( Ai X ( K*Bm )**Pi X 10**Qi )
     t =  0.500000"  xd =  5.000000"   f =  0.750000"  xl =  5.000000"
    A1 =  8.460000   A2 =  1.010000   A3 =  1.240000    K =  0.056262
    P1 =         1   P2 =         3   P3 =         5   Q1 =        -1    Q2 =        -1    Q3 =        -5

Modified exponential model  :
    Bm = Sum ( Ai X ( 1 - exp( - R/( 2*i*C ) ) ) )  +      Sum ( RKj X ( R - Rj0 ) )  +  Bm0
     C =   0.42833333E+00              Bmo=   0.00000000E+00            Nexp=  6         Nliner=  3
    AI =   0.28461400E+03  -0.14224620E+04   0.46685289E+04  -0.75511470E+04   0.60533836E+04  -0.19144837E+04
   Rj0 =           4.5000          16.6000          30.6000
   RKj =   0.27515709E+01  -0.77676420E+00  -0.10298021E+01
```

No	Rotation (radians) X 1/1000	Moment (kip-inch) Expri.	Moment (kip-inch) Poly.	Moment (kip-inch) M.Expo.	Connection stiffness (kip-inch) X 1000 Expri.	Connection stiffness (kip-inch) X 1000 Poly.	Connection stiffness (kip-inch) X 1000 M.Expo.
1	0.00	0.00	0.00	0.00	0.9714E+02	0.2101E+02	0.1557E+03
3	0.20	19.43	4.17	27.05	0.1565E+03	0.2060E+02	0.1167E+03
5	0.30	38.05	6.21	37.91	0.1498E+03	0.2013E+02	0.1010E+03
7	0.43	51.55	8.84	50.16	0.1012E+03	0.1930E+02	0.8336E+02
9	0.60	66.40	11.96	62.52	0.8100E+02	0.1808E+02	0.6570E+02
11	1.05	83.00	19.33	84.47	0.2429E+02	0.1476E+02	0.3529E+02
13	1.75	97.57	28.25	100.97	0.1735E+02	0.1103E+02	0.1547E+02
15	2.45	109.72	35.11	109.10	0.1735E+02	0.8759E+01	0.8918E+01
17	4.50	122.27	49.28	120.85	0.3417E+01	0.5592E+01	0.3442E+01
19	7.10	130.37	61.43	131.58	0.2880E+01	0.3972E+01	0.2600E+01
21	11.15	140.48	74.93	140.10	0.2544E+01	0.2844E+01	0.2199E+01
23	14.75	149.80	84.14	149.06	0.2627E+01	0.2318E+01	0.2705E+01
25	18.75	158.30	92.63	158.42	0.1694E+01	0.1949E+01	0.1974E+01
27	22.75	165.59	99.89	166.11	0.1972E+01	0.1696E+01	0.1870E+01
29	26.20	172.07	105.45	172.47	0.1770E+01	0.1534E+01	0.1829E+01
31	29.20	177.33	109.91	177.96	0.1736E+01	0.1419E+01	0.1834E+01
33	32.55	182.19	114.45	182.14	0.1246E+01	0.1315E+01	0.8315E+00
35	36.35	186.24	119.30	185.37	0.8757E+00	0.1216E+01	0.8673E+00
37	39.85	189.47	123.40	188.45	0.9782E+00	0.1140E+01	0.8945E+00
39	43.30	191.90	127.22	191.57	0.4500E+00	0.1075E+01	0.9137E+00
41	46.75	193.93	130.83	194.75	0.7382E+00	0.1019E+01	0.9263E+00

A.4 – 77

IV - 39

Connection type : Top-and seat-angle connections
Mode : All bolted

Tested by : M.J.Marley (1982) U.S.A
Test Id. : D2-1/2-2

Column : W5X16
Beam : W5X16
Angle : 4 X 4 X 1/2 X 5.0

Fasteners: A325- -3/4"D
13/16" Oversize holes
Material : --
Fy = -- ksi
Fu = -- ksi

Major parameters

lt = 5.0000"	tt = 0.5000"	ls = 5.0000"	ts = 0.5000"
gt = -- "	gt'= -- "	gs = -- "	gs'= -- "
qt = -- "	rt = -- "	qs = -- "	rs = -- "
nt = 2 X 1	nt'= 2 X 1	ns = 2 X 1	ns'= 2 X 1

Remark 1)
2)

No	Moment (k-in)	Rotation (radians) X 1/1000	No	Moment (k-in)	Rotation (radians) X 1/1000
1	0.00	0.00	26	162.97	16.00
2	7.11	0.23	27	166.60	17.05
3	14.22	0.47	28	170.22	18.10
4	21.32	0.70	29	173.04	19.60
5	29.78	1.10	30	175.86	21.10
6	38.23	1.50	31	177.07	22.95
7	44.67	1.83	32	178.27	24.80
8	51.10	2.17	33	180.68	26.65
9	57.54	2.50	34	183.10	28.50
10	66.40	2.90	35	184.30	29.80
11	75.25	3.30	36	185.51	31.10
12	82.49	3.65	37	186.72	32.95
13	89.74	4.00	38	187.93	34.80
14	96.58	4.55	39	188.74	36.10
15	103.42	5.10	40	189.54	37.40
16	109.05	5.75			
17	114.68	6.40			
18	121.12	7.45			
19	127.56	8.50			
20	136.42	10.20			
21	141.25	11.10			
22	146.08	12.00			
23	149.70	12.95			
24	153.32	13.90			
25	158.15	14.95			

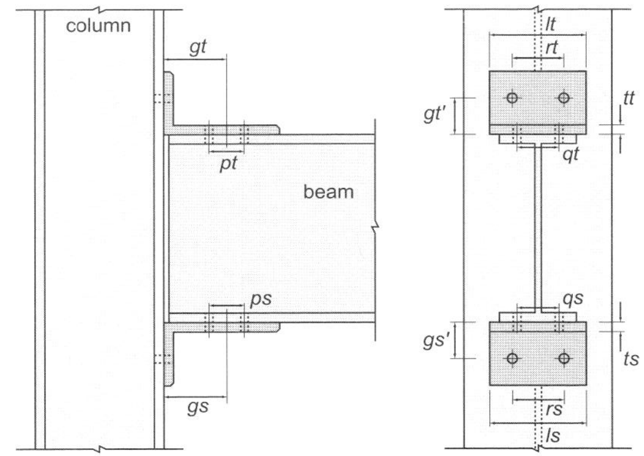

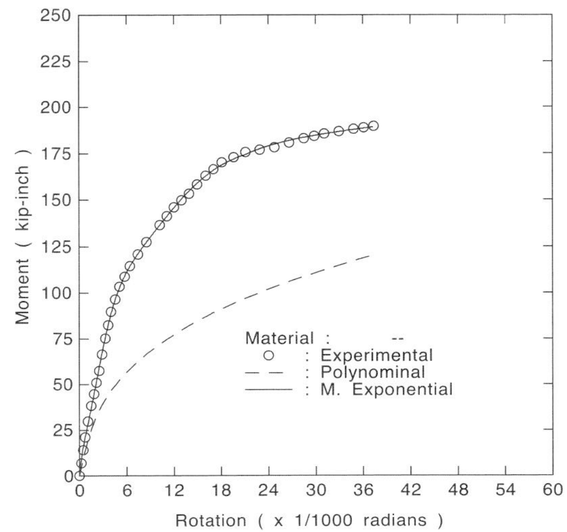

```
Moment-rotation prediction equations     ( R : X 1/1000 radians )

Frye and Morris polynominal model  :    R = Sum ( Ai X ( K*Bm )**Pi X 10**Qi )
     t =  0.500000"  xd =  5.000000"   f =  0.750000"  xl =  5.000000"
    A1 =  8.460000   A2 =  1.010000   A3 =  1.240000    K =  0.056262
    P1 =         1   P2 =         3   P3 =         5   Q1 =        -1    Q2 =        -1    Q3 =        -5

Modified exponential model  :
    Bm = Sum ( Ai X ( 1 - exp( - R/( 2*i*C ) ) ) )  +      Sum ( RKj X ( R - Rj0 ) )  +  Bm0
     C =  0.36999167E+00             Bmo=  0.00000000E+00          Nexp=  6          Nliner=  1
    AI =  0.18961942E+03  -0.18768433E+04  0.61029798E+04  -0.81537005E+04  0.40383298E+04  -0.11960750E+03
   Rj0 =         21.1000
   RKj =  0.51862842E+00
```

No	Rotation (radians) X 1/1000	Moment (kip-inch) Expri.	Moment (kip-inch) Poly.	Moment (kip-inch) M.Expo.	Connection stiffness (kip-inch) X 1000 Expri.	Connection stiffness (kip-inch) X 1000 Poly.	Connection stiffness (kip-inch) X 1000 M.Expo.
1	0.00	0.00	0.00	0.00	0.3047E+02	0.2101E+02	0.4707E+02
3	0.47	14.22	9.48	15.47	0.3046E+02	0.1907E+02	0.2412E+02
5	1.10	29.78	20.07	28.62	0.2113E+02	0.1442E+02	0.1989E+02
7	1.83	44.67	29.15	44.06	0.1931E+02	0.1070E+02	0.2221E+02
9	2.50	57.54	35.55	59.05	0.2060E+02	0.8633E+01	0.2229E+02
11	3.30	75.25	41.77	75.91	0.2137E+02	0.7049E+01	0.1946E+02
13	4.00	89.74	46.36	88.30	0.1748E+02	0.6108E+01	0.1592E+02
15	5.10	103.42	52.48	102.91	0.1071E+02	0.5090E+01	0.1090E+02
17	6.40	114.68	58.54	114.46	0.7698E+01	0.4293E+01	0.7309E+01
19	8.50	127.56	66.63	127.36	0.5779E+01	0.3474E+01	0.5511E+01
21	11.10	141.25	74.79	141.21	0.5367E+01	0.2853E+01	0.5168E+01
23	12.95	149.70	79.77	150.37	0.3815E+01	0.2548E+01	0.4674E+01
25	14.95	158.15	84.60	158.92	0.4594E+01	0.2295E+01	0.3842E+01
27	17.05	166.60	89.20	165.97	0.3451E+01	0.2087E+01	0.2882E+01
29	19.60	173.04	94.25	171.98	0.1880E+01	0.1888E+01	0.1870E+01
31	22.95	177.07	100.23	177.55	0.6519E+00	0.1686E+01	0.1481E+01
33	26.65	180.68	106.13	181.91	0.1304E+01	0.1515E+01	0.9426E+00
35	29.80	184.30	110.75	184.49	0.9277E+00	0.1399E+01	0.7199E+00
37	32.95	186.72	114.97	186.56	0.6551E+00	0.1304E+01	0.6112E+00
39	36.10	188.74	118.97	188.40	0.6185E+00	0.1222E+01	0.5602E+00

IV - 40

Connection type : Top-and seat-angle connections
Mode : All bolted

Tested by : M.J.Marley (1982) U.S.A
Test Id. : D3-1/2-1

Column : W5X16
Beam : W5X16
Angle : 4 X 4 X 1/2 X 5.0

Fasteners: A325- -3/4"D
13/16" Oversize holes
Material : --
Fy = -- ksi
Fu = -- ksi

Major parameters

lt = 5.0000"	tt = 0.5000"	ls = 5.0000"	ts = 0.5000"
gt = -- "	gt'= -- "	gs = -- "	gs'= -- "
qt = -- "	rt = -- "	qs = -- "	rs = -- "
nt = 2 X 1	nt'= 2 X 1	ns = 2 X 1	ns'= 2 X 1

Remark 1)
2)

No	Moment (k-in)	Rotation (radians) X 1/1000	No	Moment (k-in)	Rotation (radians) X 1/1000
1	0.00	0.00	26	147.39	15.30
2	7.63	0.15	27	150.20	16.70
3	15.26	0.30	28	155.02	18.90
4	23.29	0.55	29	158.23	20.70
5	31.32	0.80	30	160.64	21.85
6	38.55	1.20	31	163.06	23.00
7	45.78	1.60	32	165.46	24.65
8	51.00	1.95	33	167.87	26.30
9	56.22	2.30	34	170.28	27.65
10	62.24	2.60	35	172.69	29.00
11	68.27	2.90	36	174.70	30.40
12	75.90	3.35	37	176.71	31.80
13	83.53	3.80	38	178.72	33.25
14	90.36	4.20	39	180.72	34.70
15	97.19	4.60	40	181.93	36.10
16	103.21	5.30	41	183.13	37.50
17	109.24	6.00	42	185.14	38.80
18	115.26	6.60	43	187.15	40.10
19	121.28	7.20	44	188.75	42.05
20	126.10	8.10	45	190.36	44.00
21	130.92	9.00			
22	134.14	10.15			
23	137.35	11.30			
24	140.96	12.60			
25	144.58	13.90			

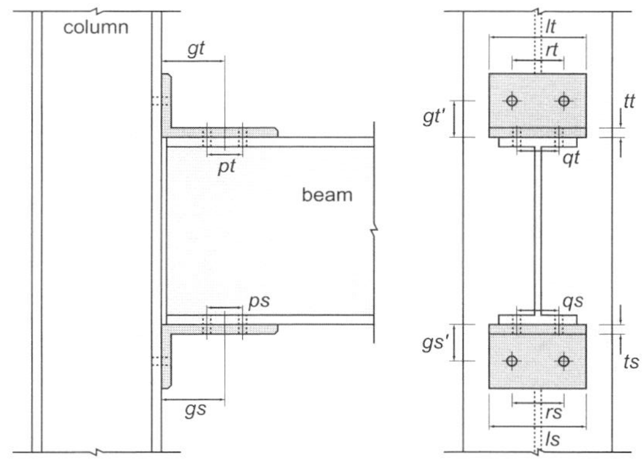

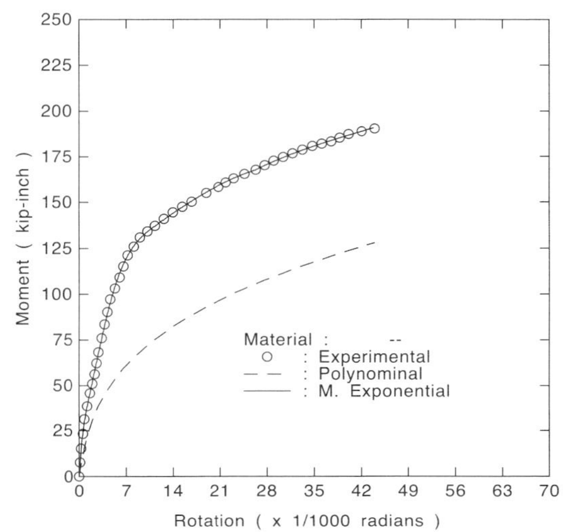

A.4 – 80

```
Moment-rotation prediction equations      ( R : X 1/1000 radians )

Frye and Morris polynominal model   :    R = Sum ( Ai X ( K*Bm )**Pi X 10**Qi )
      t =  0.500000"  xd =  5.000000"  f =  0.750000"  xl =  5.000000"
     A1 =  8.460000   A2 =  1.010000   A3 =  1.240000   K =  0.056262
     P1 =        1    P2 =        3    P3 =        5    Q1 =       -1     Q2 =       -1     Q3 =       -5

Modified exponential model  :
     Bm = Sum ( Ai X ( 1 - exp( - R/( 2*i*C ) ) ) )  +      Sum ( RKj X ( R - Rj0 ) )  +  Bm0
      C =  0.40416667E+00             Bmo=   0.00000000E+00          Nexp=  6        Nliner=  2
     AI =  0.12486170E+03  -0.32548556E+03  -0.14087398E+04   0.65744958E+04  -0.86131890E+04   0.38165957E+04
    Rj0 =         11.3000          26.3000
    RKj =  0.33721891E+00   0.64830197E+00
```

No	Rotation (radians) X 1/1000	Moment (kip-inch) Expri.	Moment (kip-inch) Poly.	Moment (kip-inch) M.Expo.	Connection stiffness (kip-inch) X 1000 Expri.	Connection stiffness (kip-inch) X 1000 Poly.	Connection stiffness (kip-inch) X 1000 M.Expo.
1	0.00	0.00	0.00	0.00	0.5088E+02	0.2101E+02	0.6139E+02
3	0.30	15.26	6.21	14.97	0.4384E+02	0.2013E+02	0.4028E+02
5	0.80	31.32	15.42	30.21	0.2671E+02	0.1655E+02	0.2352E+02
7	1.60	45.78	26.55	45.61	0.1639E+02	0.1168E+02	0.1723E+02
9	2.30	56.22	33.77	57.60	0.1770E+02	0.9160E+01	0.1722E+02
11	2.90	68.27	38.82	67.97	0.1883E+02	0.7753E+01	0.1723E+02
13	3.80	83.53	45.11	83.01	0.1702E+02	0.6347E+01	0.1594E+02
15	4.60	97.19	49.83	94.95	0.1399E+02	0.5500E+01	0.1381E+02
17	6.00	109.24	56.78	111.32	0.9378E+01	0.4506E+01	0.9611E+01
19	7.20	121.28	61.83	120.99	0.8165E+01	0.3931E+01	0.6647E+01
21	9.00	130.92	68.33	130.18	0.4231E+01	0.3330E+01	0.3884E+01
23	11.30	137.35	75.35	137.17	0.2788E+01	0.2816E+01	0.2478E+01
25	13.90	144.58	82.13	143.88	0.2408E+01	0.2420E+01	0.2444E+01
27	16.70	150.20	88.46	150.53	0.2079E+01	0.2119E+01	0.2297E+01
29	20.70	158.23	96.29	159.01	0.1976E+01	0.1815E+01	0.1904E+01
31	23.00	163.06	100.35	163.05	0.1835E+01	0.1682E+01	0.1612E+01
33	26.30	167.87	105.60	167.70	0.1639E+01	0.1530E+01	0.1216E+01
35	29.00	172.69	109.59	172.36	0.1614E+01	0.1427E+01	0.1601E+01
37	31.80	176.71	113.46	176.54	0.1409E+01	0.1337E+01	0.1395E+01
39	34.70	180.72	117.21	180.36	0.1117E+01	0.1257E+01	0.1246E+01
41	37.50	183.13	120.64	183.71	0.1216E+01	0.1190E+01	0.1150E+01
43	40.10	187.15	123.69	186.61	0.1256E+01	0.1135E+01	0.1091E+01
45	44.00	190.36	127.98	190.76	0.8215E+00	0.1063E+01	0.1039E+01

IV - 41

Connection type : Top-and seat-angle connections
Mode : All bolted

Tested by : M.J.Marley (1982) U.S.A
Test Id. : D3-1/2-2

Column : W5X16
Beam : W5X16
Angle : 4 X 4 X 1/2 X 5.0

Fasteners: A325- -3/4"D
13/16" Oversize holes
Material : --
Fy = -- ksi
Fu = -- ksi

Major parameters

lt = 5.0000"	tt = 0.5000"	ls = 5.0000"	ts = 0.5000"
gt = -- "	gt'= -- "	gs = -- "	gs'= -- "
qt = -- "	rt = -- "	qs = -- "	rs = -- "
nt = 2 X 1	nt'= 2 X 1	ns = 2 X 1	ns'= 2 X 1

Remark 1)
2)

No	Moment (k-in)	Rotation (radians) X 1/1000	No	Moment (k-in)	Rotation (radians) X 1/1000
1	0.00	0.00	26	155.82	19.30
2	8.03	0.30	27	159.04	20.65
3	16.07	0.60	28	162.25	22.00
4	23.30	0.85	29	165.06	23.40
5	30.53	1.10	30	167.87	24.80
6	37.35	1.45	31	171.08	26.15
7	44.17	1.80	32	174.30	27.50
8	52.61	2.25	33	177.11	29.10
9	61.04	2.70	34	179.92	30.70
10	61.04	2.70	35	181.52	32.25
11	67.87	3.15	36	183.13	33.80
12	74.70	3.60	37	184.74	35.10
13	81.12	4.17	38	186.35	36.40
14	87.55	4.73	39	187.96	38.05
15	93.97	5.30	40	189.56	39.70
16	100.80	5.95	41	190.36	41.20
17	107.63	6.60	42	191.16	42.70
18	115.67	8.20			
19	119.28	9.30			
20	122.89	10.40			
21	129.72	11.70			
22	136.55	13.00			
23	140.96	14.50			
24	145.38	16.00			
25	150.60	17.65			

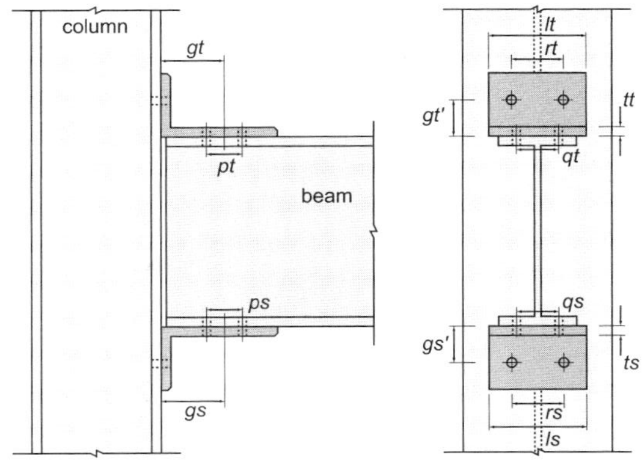

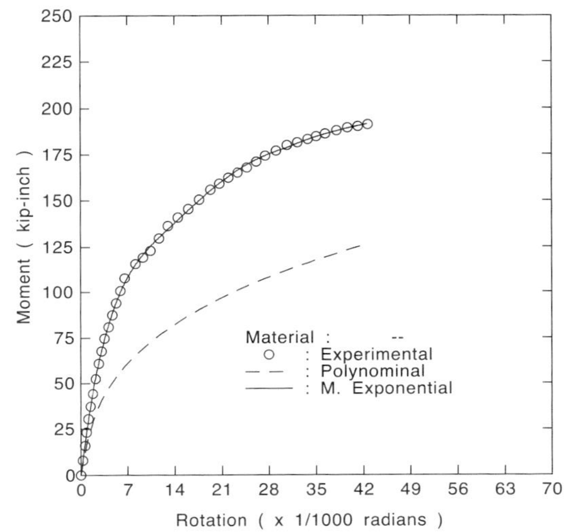

A.4 – 82

```
Moment-rotation prediction equations      ( R : X 1/1000 radians )

Frye and Morris polynominal model  :    R = Sum ( Ai X ( K*Bm )**Pi X 10**Qi )
     t =  0.500000"   xd =  5.000000"   f =  0.750000"  xl =  5.000000"
    A1 =  8.460000    A2 =  1.010000   A3 =  1.240000    K =  0.056262
    P1 =         1    P2 =         3   P3 =         5   Q1 =        -1    Q2 =        -1    Q3 =        -5

Modified exponential model  :
    Bm = Sum ( Ai X ( 1 - exp( - R/( 2*i*C ) ) ) )  +     Sum ( RKj X ( R - Rj0 ) )  +  Bm0
     C =   0.43083333E+00          Bmo=   0.00000000E+00          Nexp=  6        Nliner=  1
    AI =  -0.33963872E+02   0.79037673E+03  -0.54270361E+04   0.14951742E+05  -0.17483611E+05   0.73884783E+04
   Rj0 =          30.7000
   RKj =   0.56449810E+00
```

No	Rotation (radians) X 1/1000	Moment (kip-inch) Expri.	Moment (kip-inch) Poly.	Moment (kip-inch) M.Expo.	Connection stiffness (kip-inch) X 1000 Expri.	Connection stiffness (kip-inch) X 1000 Poly.	Connection stiffness (kip-inch) X 1000 M.Expo.
1	0.00	0.00	0.00	0.00	0.2678E+02	0.2101E+02	0.2883E+02
3	0.60	16.07	11.96	16.72	0.2795E+02	0.1808E+02	0.2651E+02
5	1.10	30.53	20.06	29.27	0.2499E+02	0.1443E+02	0.2370E+02
7	1.80	44.17	28.79	44.55	0.1917E+02	0.1083E+02	0.2007E+02
9	2.70	61.04	37.23	60.90	0.4157E-01	0.8167E+01	0.1644E+02
11	3.15	67.87	40.70	67.96	0.1517E+02	0.7296E+01	0.1499E+02
13	4.17	81.12	47.36	81.76	0.1134E+02	0.5924E+01	0.1225E+02
15	5.30	93.97	53.48	94.18	0.1095E+02	0.4945E+01	0.9756E+01
17	6.60	107.63	59.39	105.30	0.8922E+01	0.4195E+01	0.7436E+01
19	9.30	119.28	69.31	120.83	0.3284E+01	0.3250E+01	0.4480E+01
21	11.70	129.72	76.47	130.22	0.5252E+01	0.2745E+01	0.3542E+01
23	14.50	140.96	83.56	139.75	0.2944E+01	0.2347E+01	0.3336E+01
25	17.65	150.60	90.43	150.10	0.3164E+01	0.2036E+01	0.3203E+01
27	20.65	159.04	96.20	159.22	0.2382E+01	0.1818E+01	0.2834E+01
29	23.40	165.06	100.98	166.35	0.2006E+01	0.1663E+01	0.2342E+01
31	26.15	171.08	105.37	172.07	0.2382E+01	0.1536E+01	0.1819E+01
33	29.10	177.11	109.73	176.67	0.1755E+01	0.1424E+01	0.1312E+01
35	32.25	181.52	114.09	180.97	0.1037E+01	0.1323E+01	0.1446E+01
37	35.10	184.74	117.74	184.66	0.1237E+01	0.1246E+01	0.1160E+01
39	38.05	187.96	121.29	187.75	0.9745E+00	0.1178E+01	0.9501E+00
41	41.20	190.36	124.90	190.49	0.5320E+00	0.1114E+01	0.8014E+00

IV - 42

Connection type : Top-and seat-angle connections
Mode : All bolted

Tested by : S.M.Maxwell et al. (1981) U.S.A
Test Id. : A1

Column : 305 X 305 X UC 97
Beam : 457 X 191 X UB 67
Angle : 150 X 90 X 10 X 150 MM

Fasteners: G8.8- -M20
7/8" Oversize holes
Material : --
Fy = -- ksi
Fu = -- ksi

Major parameters

lt = 5.9055"	tt = 0.3937"	ls = 5.9055"	ts = 0.3937"
gt = -- "	gt'= -- "	gs = -- "	gs'= -- "
qt = -- "	rt = -- "	qs = -- "	rs = -- "
pt = -- "	ps = -- "		
nt = 2 X 2	nt'= 2 X 1	ns = 2 X 2	ns'= 2 X 1

Remark 1) Grade 8.8 bolts at column
2) Hsfg bolts in two rows at beam.

No	Moment (k-in)	Rotation (radians) X 1/1000	No	Moment (k-in)	Rotation (radians) X 1/1000
1	0.00	0.00	26	479.96	7.03
2	22.17	0.07	27	490.03	7.37
3	44.35	0.15	28	500.10	7.70
4	66.52	0.23	29	511.42	8.07
5	88.69	0.30	30	522.74	8.43
6	114.17	0.45	31	534.06	8.80
7	139.65	0.60	32	545.38	9.25
8	165.13	0.75	33	556.70	9.70
9	190.61	0.90			
10	210.74	1.13			
11	230.86	1.37			
12	250.99	1.60			
13	275.53	2.05			
14	300.06	2.50			
15	317.04	2.90			
16	334.02	3.30			
17	351.01	3.65			
18	367.99	4.00			
19	380.57	4.33			
20	393.15	4.67			
21	405.73	5.00			
22	420.83	5.40			
23	435.94	5.80			
24	452.92	6.25			
25	469.90	6.70			

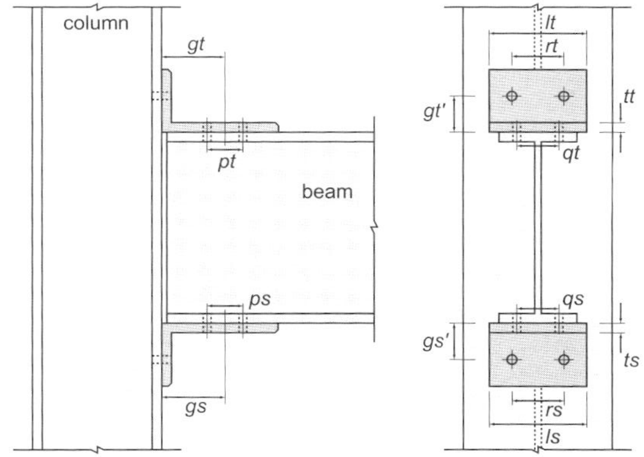

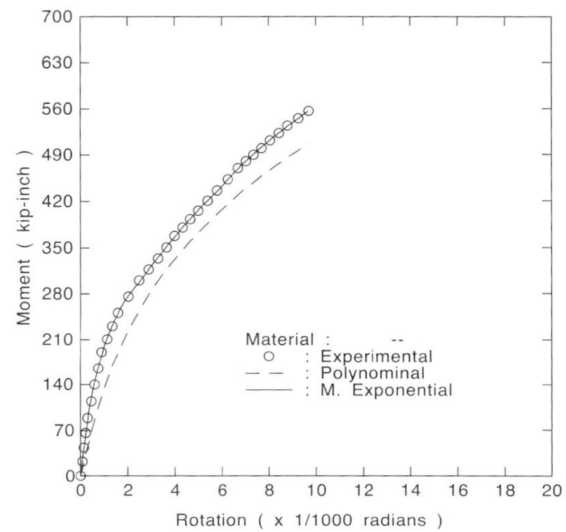

```
Moment-rotation prediction equations      ( R : X 1/1000 radians )

Frye and Morris polynominal model   :     R = Sum ( Ai X ( K*Bm )**Pi X 10**Qi )
      t =  0.393700"  xd = 17.992000"   f =  0.787000"  xl =  5.905500"
     A1 =  8.460000   A2 =  1.010000   A3 =  1.240000    K =  0.007841
     P1 =         1   P2 =         3   P3 =         5   Q1 =        -1   Q2 =        -1   Q3 =        -5

Modified exponential model  :
     Bm = Sum ( Ai X ( 1 - exp( - R/( 2*i*C ) ) ) )  +      Sum ( RKj X ( R - Rj0 ) )  +  Bm0
      C =   0.99583333E-01            Bmo=   0.00000000E+00          Nexp=  6        Nliner=  2
     AI =  -0.10573124E+03   0.19981288E+04  -0.11951490E+05   0.30972569E+05  -0.35206273E+05   0.14710534E+05
    Rj0 =          2.5000           5.8000
    RKj =   0.14561508E+02   0.93803764E+01
```

No	Rotation (radians) X 1/1000	Moment (kip-inch) Expri.	Moment (kip-inch) Poly.	Moment (kip-inch) M.Expo.	Connection stiffness (kip-inch) X 1000 Expri.	Connection stiffness (kip-inch) X 1000 Poly.	Connection stiffness (kip-inch) X 1000 M.Expo.
1	0.00	0.00	0.00	0.00	0.2956E+03	0.1508E+03	0.3171E+03
2	0.07	22.17	11.31	23.42	0.2956E+03	0.1503E+03	0.3053E+03
3	0.15	44.35	22.53	45.55	0.2956E+03	0.1491E+03	0.2839E+03
4	0.23	66.52	33.64	65.94	0.2956E+03	0.1471E+03	0.2594E+03
5	0.30	88.69	44.58	84.48	0.2537E+03	0.1444E+03	0.2353E+03
6	0.45	114.17	65.76	116.51	0.1699E+03	0.1377E+03	0.1936E+03
7	0.60	139.65	85.82	143.09	0.1699E+03	0.1297E+03	0.1624E+03
8	0.75	165.13	104.66	165.65	0.1699E+03	0.1215E+03	0.1395E+03
9	0.90	190.61	122.27	185.22	0.1372E+03	0.1134E+03	0.1222E+03
10	1.13	210.74	147.37	211.18	0.8626E+02	0.1020E+03	0.1013E+03
11	1.37	230.86	169.99	232.79	0.8626E+02	0.9213E+02	0.8436E+02
12	1.60	250.99	190.48	250.76	0.7543E+02	0.8379E+02	0.7014E+02
13	2.05	275.53	225.22	277.41	0.5452E+02	0.7120E+02	0.4987E+02
14	2.50	300.06	255.07	296.97	0.4813E+02	0.6195E+02	0.3833E+02
15	2.90	317.04	278.53	317.00	0.4245E+02	0.5564E+02	0.4788E+02
16	3.30	334.02	299.74	335.61	0.4569E+02	0.5058E+02	0.4547E+02
17	3.65	351.01	316.78	351.29	0.4853E+02	0.4693E+02	0.4415E+02
18	4.00	367.99	332.65	366.52	0.4301E+02	0.4383E+02	0.4285E+02
19	4.33	380.57	346.83	380.56	0.3774E+02	0.4128E+02	0.4133E+02
20	4.67	393.15	360.21	394.03	0.3774E+02	0.3904E+02	0.3947E+02
21	5.00	405.73	372.88	406.84	0.3775E+02	0.3707E+02	0.3733E+02
22	5.40	420.83	387.29	421.22	0.3775E+02	0.3499E+02	0.3454E+02
23	5.80	435.94	400.91	434.46	0.3774E+02	0.3316E+02	0.3169E+02
24	6.25	452.92	415.42	452.24	0.3773E+02	0.3136E+02	0.3800E+02
25	6.70	469.90	429.16	468.71	0.3341E+02	0.2977E+02	0.3523E+02
26	7.03	479.96	439.09	480.14	0.3020E+02	0.2869E+02	0.3342E+02
27	7.37	490.03	448.31	491.01	0.3020E+02	0.2773E+02	0.3182E+02
28	7.70	500.10	457.56	501.38	0.3052E+02	0.2682E+02	0.3044E+02
29	8.07	511.42	467.07	512.30	0.3087E+02	0.2592E+02	0.2915E+02
30	8.43	522.74	476.58	522.78	0.3087E+02	0.2506E+02	0.2809E+02
31	8.80	534.06	485.47	532.92	0.2831E+02	0.2430E+02	0.2722E+02
32	9.25	545.38	496.40	544.97	0.2516E+02	0.2340E+02	0.2637E+02
33	9.70	556.70	506.56	556.69	0.2516E+02	0.2261E+02	0.2573E+02

Connection type : Top-and seat-angle connections
Mode : All bolted

Tested by : S.M.Maxwell et al. (1981) U.S.A
Test Id. : A2

Column : 305 X 305 X UC 97
Beam : 457 X 191 X UB 67
Angle : 150 X 90 X 10 X 200 MM

Fasteners: G8.8- -M20
7/8" Oversize holes
Material : --
Fy = -- ksi
Fu = -- ksi

Major parameters

lt = 7.8740"	tt = 0.3937"	ls = 7.8740"	ts = 0.3937"
gt = -- "	gt'= -- "	gs = -- "	gs'= -- "
qt = -- "	rt = -- "	qs = -- "	rs = -- "
pt = -- "	ps = -- "		
nt = 2 X 2	nt'= 2 X 1	ns = 2 X 2	ns'= 2 X 1

Remark 1) Grade 8.8 bolts at column
2) Hsfg bolts in two rows at beam.

No	Moment (k-in)	Rotation (radians) X 1/1000	No	Moment (k-in)	Rotation (radians) X 1/1000
1	0.00	0.00	26	698.48	7.07
2	29.00	0.13	27	712.35	7.43
3	53.00	0.27	28	726.22	7.80
4	87.00	0.40	29	737.56	8.13
5	113.52	0.60	30	748.91	8.47
6	150.03	0.80	31	760.26	8.80
7	181.55	1.00	32	767.82	9.13
8	219.37	1.27	33	775.39	9.47
9	257.20	1.53	34	782.95	9.80
10	295.02	1.80			
11	329.06	2.10			
12	363.11	2.40			
13	397.15	2.70			
14	426.15	3.00			
15	455.14	3.30			
16	484.14	3.60			
17	509.36	3.93			
18	534.57	4.27			
19	559.79	4.60			
20	583.74	4.97			
21	607.70	5.33			
22	631.66	5.70			
23	649.31	6.03			
24	666.96	6.37			
25	684.61	6.70			

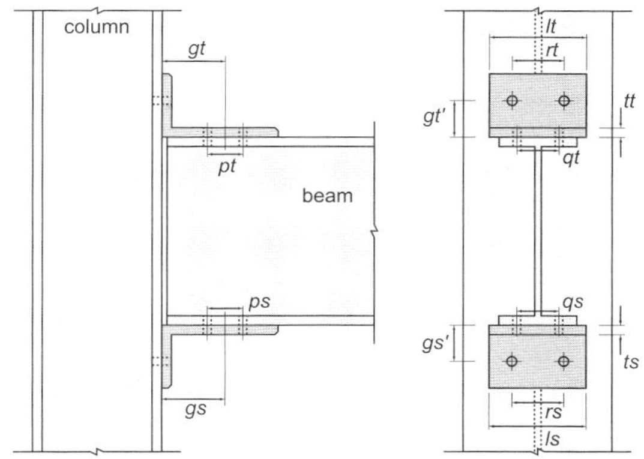

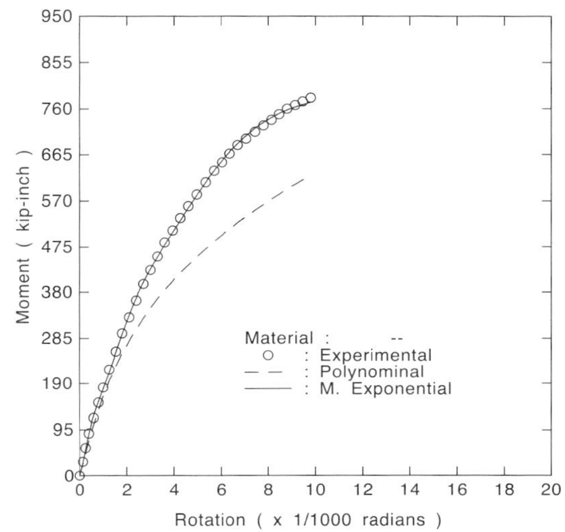

```
Moment-rotation prediction equations     ( R : X 1/1000 radians )

Frye and Morris polynominal model  :    R = Sum ( Ai X ( K*Bm )**Pi X 10**Qi )
     t =  0.393700"  xd = 17.992000"  f =  0.787000"  xl =  7.874000"
    A1 =  8.460000   A2 =  1.010000  A3 =  1.240000    K =  0.006410
    P1 =         1   P2 =         3  P3 =         5   Q1 =        -1    Q2 =        -1    Q3 =        -5

Modified exponential model  :
    Bm = Sum ( Ai X ( 1 - exp( - R/( 2*i*C ) ) ) )  +      Sum ( RKj X ( R - Rj0 ) )  + Bm0
     C =   0.11499167E+00             Bmo=  0.00000000E+00          Nexp=  6        Nliner=  0
    AI =  -0.62818444E+03   0.96451589E+04  -0.50899249E+05   0.12032673E+06  -0.13029428E+06   0.52643141E+05
```

No	Rotation (radians) X 1/1000	Moment (kip-inch) Expri.	Moment (kip-inch) Poly.	Moment (kip-inch) M.Expo.	Connection stiffness (kip-inch) X 1000 Expri.	Connection stiffness (kip-inch) X 1000 Poly.	Connection stiffness (kip-inch) X 1000 M.Expo.
1	0.00	0.00	0.00	0.00	0.2176E+03	0.1844E+03	0.1072E+03
3	0.27	58.00	48.61	58.05	0.2175E+03	0.1782E+03	0.2386E+03
5	0.60	118.52	104.96	122.88	0.1576E+03	0.1587E+03	0.1552E+03
7	1.00	181.55	163.10	178.58	0.1508E+03	0.1325E+03	0.1358E+03
9	1.53	257.20	226.05	254.56	0.1418E+03	0.1052E+03	0.1460E+03
11	2.10	329.06	279.76	333.11	0.1135E+03	0.8566E+02	0.1272E+03
13	2.70	397.15	326.70	400.65	0.1051E+03	0.7168E+02	0.9913E+02
15	3.30	455.14	366.60	454.62	0.9665E+02	0.6187E+02	0.8303E+02
17	3.93	509.36	403.27	504.95	0.7564E+02	0.5428E+02	0.7710E+02
19	4.60	559.79	437.36	555.36	0.7074E+02	0.4827E+02	0.7405E+02
21	5.33	607.70	470.82	607.70	0.6534E+02	0.4320E+02	0.6802E+02
23	6.03	649.31	499.67	652.25	0.5295E+02	0.3938E+02	0.5880E+02
25	6.70	684.61	525.14	688.01	0.4575E+02	0.3638E+02	0.4837E+02
27	7.43	712.35	550.82	719.25	0.3782E+02	0.3367E+02	0.3698E+02
29	8.13	737.56	573.54	741.69	0.3404E+02	0.3149E+02	0.2743E+02
31	8.80	760.26	593.92	757.41	0.2837E+02	0.2971E+02	0.2001E+02
33	9.47	775.39	613.18	768.74	0.2269E+02	0.2815E+02	0.1423E+02

IV - 44

```
Connection type : Top-and seat-angle connections
           Mode : All bolted

Tested by  :  S.M.Maxwell et al. (1981)                  U.S.A
Test  Id.  :  A3

Column : 305 X 305 X UC 97           Fasteners: G8.8- -M20
Beam   : 457 X 191 X UB 67                   7/8" Oversize holes
Angle  : 150 X 90  X 10 X 150 MM     Material :  --
                                          Fy =   --   ksi
                                          Fu =   --   ksi

                          Major parameters

     lt =  5.9055"    tt =  0.3937"    ls =  5.9055"    ts =  0.3937"
     gt =   --   "    gt'=   --   "    gs =   --   "    gs'=   --   "
     qt =   --   "    rt =   --   "    qs =   --   "    rs =   --   "
     pt =   --   "    ps =   --   "
     nt = 2 X  2      nt'= 2 X  1      ns = 2 X  2      ns'= 2 X  1

Remark  1)  Grade 8.8 bolts at column
        2)  Hsfg bolts in two rows at beam.
```

No	Moment (k-in)	Rotation (radians) X 1/1000
1	0.00	0.00
2	29.44	0.17
3	58.88	0.35
4	88.32	0.53
5	117.76	0.70
6	151.94	0.95
7	186.13	1.20
8	213.99	1.47
9	241.84	1.73
10	269.70	2.00
11	303.89	2.30
12	338.08	2.60
13	370.36	3.00
14	402.65	3.40
15	436.84	3.80
16	471.02	4.20
17	505.21	4.65
18	539.40	5.10
19	558.39	5.43
20	577.38	5.77
21	596.38	6.10
22	617.27	6.55
23	638.16	7.00
24	651.46	7.45
25	664.75	7.90
26	673.62	8.23
27	682.48	8.57
28	691.34	8.90
29	697.04	9.30
30	702.74	9.70

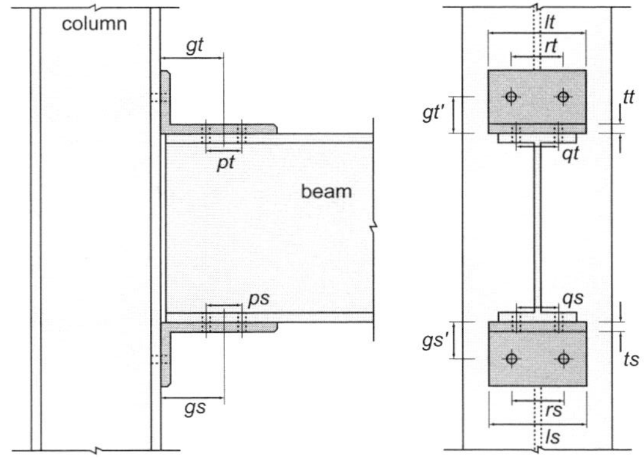

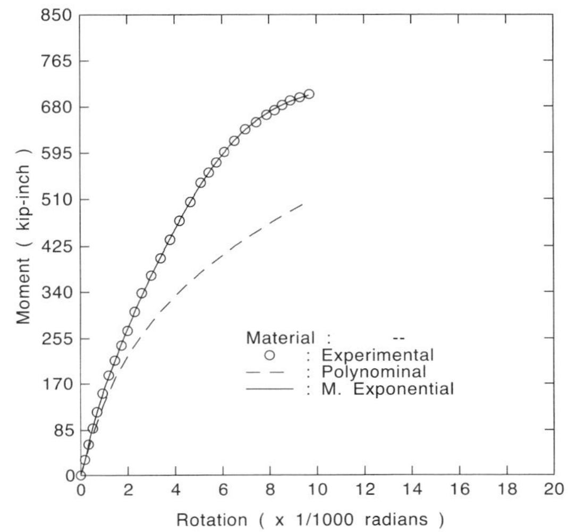

```
Moment-rotation prediction equations      ( R : X 1/1000 radians )

Frye and Morris polynominal model  :     R = Sum ( Ai X ( K*Bm )**Pi X 10**Qi )
      t =  0.393700"  xd = 17.992000"   f =  0.787000"  xl =  5.905500"
     A1 =  8.460000   A2 =  1.010000   A3 =  1.240000    K =  0.007841
     P1 =         1   P2 =         3   P3 =         5   Q1 =        -1     Q2 =        -1     Q3 =        -5

Modified exponential model  :
     Bm = Sum ( Ai X ( 1 - exp( - R/( 2*i*C ) ) ) )  +      Sum ( RKj X ( R - Rj0 ) )  +  Bm0
      C =   0.12458333E+00           Bmo=   0.00000000E+00            Nexp=  6        Nliner=  0
     AI =  -0.23872908E+03   0.34526670E+04  -0.20843442E+05   0.57496902E+05  -0.70876532E+05   0.31732406E+05
```

No	Rotation (radians) X 1/1000	Moment (kip-inch) Expri.	Moment (kip-inch) Poly.	Moment (kip-inch) M.Expo.	Connection stiffness (kip-inch) X 1000 Expri.	Connection stiffness (kip-inch) X 1000 Poly.	Connection stiffness (kip-inch) X 1000 M.Expo.
1	0.00	0.00	0.00	0.00	0.1682E+03	0.1508E+03	0.1101E+03
2	0.17	29.44	26.25	27.41	0.1682E+03	0.1485E+03	0.1829E+03
3	0.35	58.88	51.75	59.89	0.1682E+03	0.1424E+03	0.1824E+03
4	0.53	88.32	75.94	90.27	0.1682E+03	0.1338E+03	0.1640E+03
5	0.70	117.76	98.52	117.36	0.1553E+03	0.1242E+03	0.1463E+03
6	0.95	151.94	127.88	151.64	0.1368E+03	0.1108E+03	0.1296E+03
7	1.20	186.13	154.07	182.82	0.1211E+03	0.9900E+02	0.1207E+03
8	1.47	213.99	179.01	214.18	0.1045E+03	0.8838E+02	0.1149E+03
9	1.73	241.84	201.37	244.16	0.1045E+03	0.7963E+02	0.1101E+03
10	2.00	269.70	221.61	272.89	0.1089E+03	0.7241E+02	0.1053E+03
11	2.30	303.89	242.30	303.69	0.1140E+03	0.6573E+02	0.1001E+03
12	2.60	338.08	261.17	332.99	0.9971E+02	0.6023E+02	0.9535E+02
13	3.00	370.36	284.03	370.04	0.8072E+02	0.5427E+02	0.9007E+02
14	3.40	402.65	304.74	405.18	0.8309E+02	0.4948E+02	0.8579E+02
15	3.80	436.84	323.72	438.73	0.8547E+02	0.4554E+02	0.8196E+02
16	4.20	471.02	341.26	470.73	0.8100E+02	0.4225E+02	0.7800E+02
17	4.65	505.21	359.56	504.73	0.7597E+02	0.3915E+02	0.7297E+02
18	5.10	539.40	376.57	536.29	0.6506E+02	0.3652E+02	0.6718E+02
19	5.43	558.39	388.45	557.90	0.5697E+02	0.3483E+02	0.6246E+02
20	5.77	577.38	399.80	577.90	0.5697E+02	0.3331E+02	0.5750E+02
21	6.10	596.38	410.67	596.22	0.5249E+02	0.3193E+02	0.5242E+02
22	6.55	617.27	424.66	618.26	0.4643E+02	0.3028E+02	0.4559E+02
23	7.00	638.16	437.95	637.29	0.3799E+02	0.2881E+02	0.3904E+02
24	7.45	651.46	450.61	653.47	0.2955E+02	0.2750E+02	0.3296E+02
25	7.90	664.75	462.72	667.05	0.2785E+02	0.2633E+02	0.2748E+02
26	8.23	673.62	471.49	675.59	0.2659E+02	0.2552E+02	0.2385E+02
27	8.57	682.48	479.75	682.99	0.2659E+02	0.2479E+02	0.2057E+02
28	8.90	691.34	488.01	689.35	0.2098E+02	0.2409E+02	0.1766E+02
29	9.30	697.04	497.37	695.78	0.1425E+02	0.2332E+02	0.1460E+02
30	9.70	702.74	506.70	701.09	0.1425E+02	0.2260E+02	0.1201E+02

IV - 45

Connection type : Top-and seat-angle connections
Mode : All bolted

Tested by : S.M.Maxwell et al. (1981) U.S.A
Test Id. : A4

Column : 305 X 305 X UC 97
Beam : 457 X 191 X UB 67
Angle : 150 X 90 X 10 X 200 MM

Fasteners: G8.8- -M20
7/8" Oversize holes
Material : --
Fy = -- ksi
Fu = -- ksi

Major parameters

lt = 7.8740"	tt = 0.3937"	ls = 7.8740"	ts = 0.3937"
gt = -- "	gt'= -- "	gs = -- "	gs'= -- "
qt = -- "	rt = -- "	qs = -- "	rs = -- "
pt = -- "	ps = -- "		
nt = 2 X 2	nt'= 2 X 1	ns = 2 X 2	ns'= 2 X 1

Remark 1) Grade 8.8 bolts at column
2) Hsfg bolts in two rows at beam.

No	Moment (k-in)	Rotation (radians) X 1/1000	No	Moment (k-in)	Rotation (radians) X 1/1000
1	0.00	0.00	26	824.56	7.80
2	42.87	0.20	27	840.95	8.13
3	85.74	0.40	28	857.34	8.47
4	128.60	0.60	29	873.73	8.80
5	171.47	0.90	30	888.86	9.25
6	214.33	1.20	31	903.98	9.70
7	257.20	1.50			
8	295.02	1.77			
9	332.84	2.03			
10	370.67	2.30			
11	400.93	2.57			
12	431.19	2.83			
13	461.45	3.10			
14	501.16	3.50			
15	540.88	3.90			
16	584.38	4.35			
17	627.88	4.80			
18	654.35	5.13			
19	680.83	5.47			
20	707.30	5.80			
21	727.48	6.13			
22	747.65	6.47			
23	767.82	6.80			
24	786.73	7.13			
25	805.64	7.47			

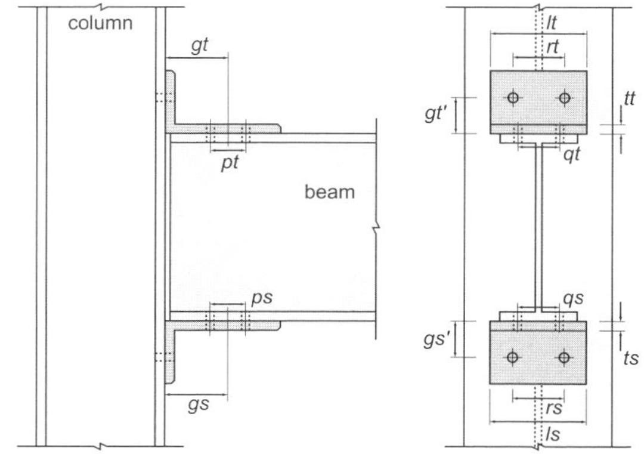

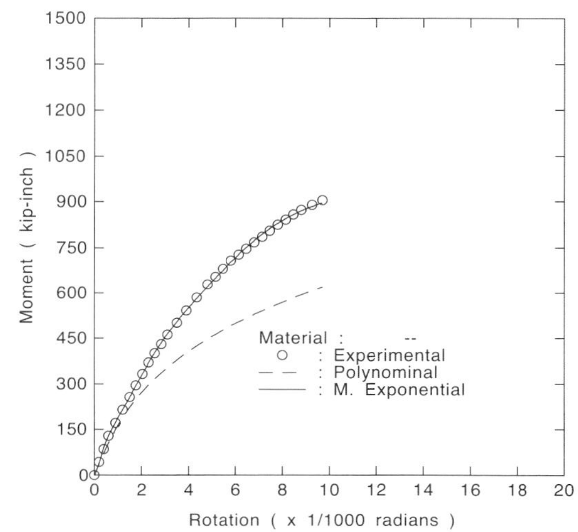

```
Moment-rotation prediction equations     ( R : X 1/1000 radians )

Frye and Morris polynominal model   :    R = Sum ( Ai X ( K*Bm )**Pi X 10**Qi )
      t =  0.393700"  xd = 17.992000"   f =  0.787000"  xl =  7.874000"
     A1 =  8.460000   A2 =  1.010000   A3 =  1.240000    K =  0.006410
     P1 =         1   P2 =         3   P3 =         5   Q1 =        -1     Q2 =        -1     Q3 =        -5

Modified exponential model  :
     Bm = Sum ( Ai X ( 1 - exp( - R/( 2*i*C ) ) ) )  +      Sum ( RKj X ( R - Rj0 ) )  +  Bm0
       C =   0.13083333E+00             Bmo=   0.00000000E+00           Nexp=  6         Nliner=  0
      AI =  -0.68782363E+03   0.10282632E+05  -0.54598678E+05   0.13144306E+06  -0.14516488E+06   0.59669778E+05
```

No	Rotation (radians) X 1/1000	Moment (kip-inch) Expri.	Moment (kip-inch) Poly.	Moment (kip-inch) M.Expo.	Connection stiffness (kip-inch) X 1000 Expri.	Connection stiffness (kip-inch) X 1000 Poly.	Connection stiffness (kip-inch) X 1000 M.Expo.
1	0.00	0.00	0.00	0.00	0.2143E+03	0.1844E+03	0.1020E+03
2	0.20	42.87	36.64	40.32	0.2143E+03	0.1808E+03	0.2482E+03
3	0.40	85.74	71.93	88.34	0.2143E+03	0.1713E+03	0.2212E+03
4	0.60	128.60	104.96	127.68	0.1858E+03	0.1587E+03	0.1739E+03
5	0.90	171.47	149.56	173.32	0.1429E+03	0.1387E+03	0.1381E+03
6	1.20	214.33	188.45	213.92	0.1429E+03	0.1211E+03	0.1358E+03
7	1.50	257.20	222.53	255.62	0.1423E+03	0.1066E+03	0.1420E+03
8	1.77	295.02	249.53	293.88	0.1418E+03	0.9619E+02	0.1440E+03
9	2.03	332.84	273.99	331.91	0.1418E+03	0.8757E+02	0.1405E+03
10	2.30	370.67	296.36	368.45	0.1276E+03	0.8039E+02	0.1330E+03
11	2.57	400.93	316.97	402.71	0.1135E+03	0.7435E+02	0.1237E+03
12	2.83	431.19	336.09	434.44	0.1135E+03	0.6921E+02	0.1145E+03
13	3.10	461.45	353.94	463.85	0.1078E+03	0.6480E+02	0.1063E+03
14	3.50	501.16	378.71	504.43	0.9929E+02	0.5922E+02	0.9722E+02
15	3.90	540.88	401.45	542.09	0.9805E+02	0.5463E+02	0.9157E+02
16	4.35	584.38	425.04	582.39	0.9667E+02	0.5034E+02	0.8785E+02
17	4.80	627.88	446.86	621.30	0.8677E+02	0.4675E+02	0.8512E+02
18	5.13	654.35	462.05	649.31	0.7942E+02	0.4445E+02	0.8293E+02
19	5.47	680.83	476.53	676.53	0.7942E+02	0.4240E+02	0.8024E+02
20	5.80	707.30	490.35	702.74	0.6998E+02	0.4056E+02	0.7692E+02
21	6.13	727.48	503.59	727.74	0.6051E+02	0.3890E+02	0.7297E+02
22	6.47	747.65	516.30	751.33	0.6051E+02	0.3739E+02	0.6849E+02
23	6.80	767.82	528.76	773.36	0.5863E+02	0.3598E+02	0.6362E+02
24	7.13	786.73	540.31	793.71	0.5673E+02	0.3474E+02	0.5850E+02
25	7.47	805.64	551.90	812.35	0.5673E+02	0.3356E+02	0.5329E+02
26	7.80	824.56	562.71	829.25	0.5296E+02	0.3250E+02	0.4813E+02
27	8.13	840.95	573.54	844.45	0.4917E+02	0.3149E+02	0.4311E+02
28	8.47	857.34	583.72	858.02	0.4917E+02	0.3058E+02	0.3833E+02
29	8.80	873.73	593.92	870.04	0.4256E+02	0.2971E+02	0.3386E+02
30	9.25	888.86	606.90	884.01	0.3361E+02	0.2864E+02	0.2835E+02
31	9.70	903.98	619.79	895.66	0.3361E+02	0.2764E+02	0.2350E+02

A.4 – 91

IV - 46

Connection type : Top-and seat-angle connections
Mode : All bolted

Tested by : S.M.Maxwell et al. (1981) U.S.A
Test Id. : B1

Column : 305 X 305 X UC 97
Beam : 457 X 191 X UB 67
Angle : 150 X 90 X 12 X 150 MM

Fasteners: G8.8- -M20
7/8" Oversize holes
Material : --
Fy = -- ksi
Fu = -- ksi

Major parameters

lt = 5.9055"	tt = 0.4724"	ls = 5.9055"	ts = 0.4724"
gt = -- "	gt'= -- "	gs = -- "	gs'= -- "
qt = -- "	rt = -- "	qs = -- "	rs = -- "
pt = -- "	ps = -- "		
nt = 2 X 2	nt'= 2 X 1	ns = 2 X 2	ns'= 2 X 1

Remark 1) Grade 8.8 bolts at column
2) Hsfg bolts in two rows at beam.

No	Moment (k-in)	Rotation (radians) X 1/1000	No	Moment (k-in)	Rotation (radians) X 1/1000
1	0.00	0.00	26	749.62	8.80
2	36.49	0.20	27	762.61	9.20
3	72.98	0.40	28	777.46	9.50
4	109.48	0.60	29	792.30	9.80
5	146.58	0.87			
6	183.69	1.13			
7	220.80	1.40			
8	254.20	1.70			
9	287.60	2.00			
10	321.00	2.30			
11	353.16	2.63			
12	385.33	2.97			
13	417.49	3.30			
14	454.60	3.75			
15	491.71	4.20			
16	523.25	4.65			
17	554.80	5.10			
18	580.77	5.55			
19	606.74	6.00			
20	630.87	6.45			
21	655.00	6.90			
22	675.40	7.30			
23	695.81	7.70			
24	716.22	8.05			
25	736.63	8.40			

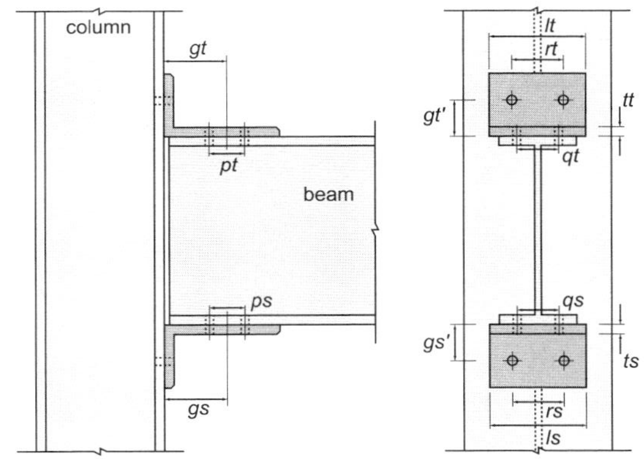

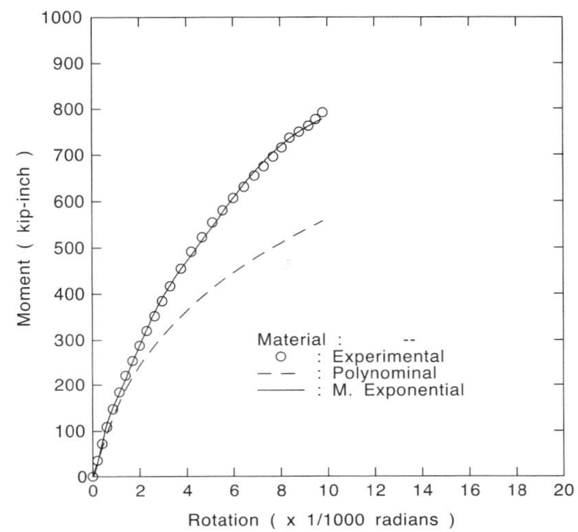

A.4 – 92

```
Moment-rotation prediction equations      ( R : X 1/1000 radians )

Frye and Morris polynominal model   :    R = Sum ( Ai X ( K*Bm )**Pi X 10**Qi )
      t =  0.472400"  xd = 17.992000"   f =  0.787000"  xl =  5.905500"
     A1 =  8.460000   A2 =  1.010000   A3 =  1.240000    K =  0.007158
     P1 =         1   P2 =         3   P3 =         5   Q1 =        -1     Q2 =        -1     Q3 =        -5

Modified exponential model   :
     Bm = Sum ( Ai X ( 1 - exp( - R/( 2*i*C ) ) ) )  +      Sum ( RKj X ( R - Rj0 ) )  +  Bm0
      C =   0.13166667E+00              Bmo=   0.00000000E+00          Nexp=  6        Nliner=  0
     AI =  -0.86111684E+03   0.12228876E+05  -0.63049499E+05   0.14708738E+06  -0.15732210E+06   0.62742492E+05
```

No	Rotation (radians) X 1/1000	Moment (kip-inch) Expri.	Moment (kip-inch) Poly.	Moment (kip-inch) M.Expo.	Connection stiffness (kip-inch) X 1000 Expri.	Connection stiffness (kip-inch) X 1000 Poly.	Connection stiffness (kip-inch) X 1000 M.Expo.
1	0.00	0.00	0.00	0.00	0.1825E+03	0.1651E+03	0.5048E+01
2	0.20	36.49	32.81	30.20	0.1825E+03	0.1619E+03	0.2268E+03
3	0.40	72.98	64.42	75.78	0.1825E+03	0.1535E+03	0.2131E+03
4	0.60	109.48	94.01	113.25	0.1639E+03	0.1421E+03	0.1624E+03
5	0.87	146.58	129.77	150.35	0.1392E+03	0.1262E+03	0.1225E+03
6	1.13	183.69	161.43	181.59	0.1392E+03	0.1117E+03	0.1157E+03
7	1.40	220.80	189.55	213.13	0.1261E+03	0.9951E+02	0.1215E+03
8	1.70	254.20	217.66	250.46	0.1113E+03	0.8832E+02	0.1263E+03
9	2.00	287.60	242.76	288.13	0.1113E+03	0.7932E+02	0.1236E+03
10	2.30	321.00	265.42	323.95	0.1043E+03	0.7200E+02	0.1145E+03
11	2.63	353.16	288.27	359.94	0.9649E+02	0.6537E+02	0.1013E+03
12	2.97	385.33	309.13	391.57	0.9649E+02	0.5994E+02	0.8880E+02
13	3.30	417.49	328.33	419.47	0.9053E+02	0.5541E+02	0.7920E+02
14	3.75	454.60	352.10	453.18	0.8247E+02	0.5038E+02	0.7161E+02
15	4.20	491.71	373.82	484.65	0.7628E+02	0.4629E+02	0.6884E+02
16	4.65	523.25	393.86	515.51	0.7009E+02	0.4288E+02	0.6854E+02
17	5.10	554.80	412.49	546.39	0.6391E+02	0.4001E+02	0.6868E+02
18	5.55	580.77	429.93	577.19	0.5772E+02	0.3755E+02	0.6797E+02
19	6.00	606.74	446.33	607.35	0.5567E+02	0.3542E+02	0.6587E+02
20	6.45	630.87	461.84	636.25	0.5361E+02	0.3355E+02	0.6234E+02
21	6.90	655.00	476.56	663.28	0.5224E+02	0.3190E+02	0.5764E+02
22	7.30	675.40	489.05	685.39	0.5102E+02	0.3058E+02	0.5281E+02
23	7.70	695.81	501.04	705.48	0.5491E+02	0.2939E+02	0.4764E+02
24	8.05	716.22	511.33	721.35	0.5832E+02	0.2842E+02	0.4304E+02
25	8.40	736.63	520.95	735.62	0.4626E+02	0.2755E+02	0.3851E+02
26	8.80	749.62	531.97	750.03	0.3248E+02	0.2660E+02	0.3356E+02
27	9.20	762.61	542.26	762.52	0.4219E+02	0.2575E+02	0.2895E+02
28	9.50	777.46	549.98	770.72	0.4948E+02	0.2514E+02	0.2576E+02
29	9.80	792.30	557.53	778.00	0.4948E+02	0.2457E+02	0.2282E+02

```
Connection type : Top-and seat-angle connections
           Mode : All bolted

Tested by  :  S.M.Maxwell et al. (1981)                 U.S.A
Test  Id.  :  B2

Column : 305 X 305 X UC 97            Fasteners: G8.8- -M20
Beam   : 457 X 191 X UB 67                     7/8" Oversize holes
Angle  : 150 X 90  X 12 X 200 MM      Material :  --
                                            Fy =    --   ksi
                                            Fu =    --   ksi

                          Major parameters

     lt =  7.8740"   tt =  0.4724'   ls =  7.8740"   ts =  0.4724"
     gt =   --   "   gt'=   --   '   gs =   --   "   gs'=   --   "
     qt =   --   "   rt =   --   '   qs =   --   "   rs =   --   "
     pt =   --   "   ps =   --   '
     nt = 2 X  2     nt'= 2 X  1     ns = 2 X  2     ns'= 2 X  1

Remark  1)  Grade 8.8 bolts at column
        2)  Hsfg bolts in two rows at beam.
```

No	Moment (k-in)	Rotation (radians) X 1/1000	No	Moment (k-in)	Rotation (radians) X 1/1000
1	0.00	0.00	26	792.30	8.00
2	34.33	0.25	27	807.14	8.33
3	68.65	0.50	28	821.99	8.67
4	102.05	0.70	29	836.83	9.00
5	135.45	0.90	30	851.68	9.33
6	168.85	1.10	31	866.52	9.67
7	211.53	1.40	32	881.36	10.00
8	254.21	1.70			
9	296.88	2.05			
10	339.55	2.40			
11	369.24	2.65			
12	398.93	2.90			
13	430.48	3.25			
14	462.02	3.60			
15	495.42	3.90			
16	528.82	4.20			
17	558.50	4.55			
18	588.19	4.90			
19	619.73	5.30			
20	651.28	5.70			
21	673.54	6.10			
22	695.81	6.50			
23	723.64	6.90			
24	751.48	7.30			
25	771.89	7.65			

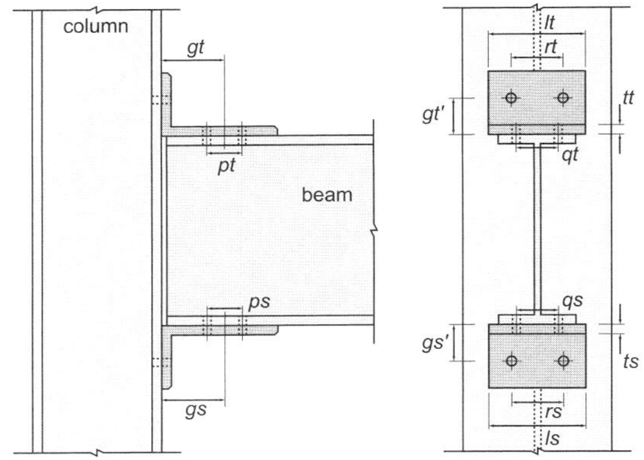

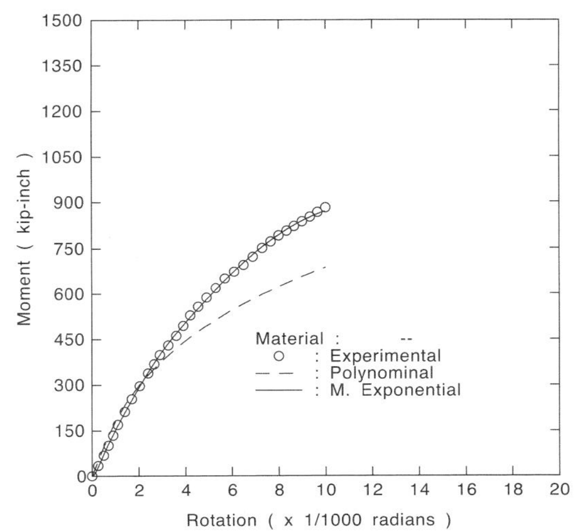

```
Moment-rotation prediction equations     ( R : X 1/1000 radians )

Frye and Morris polynominal model  :    R = Sum ( Ai X ( K*Bm )**Pi X 10**Qi )
      t =  0.472400"  xd = 17.992000"   f =  0.787000"  xl =  7.874000"
     A1 =  8.460000   A2 =  1.010000   A3 =  1.240000    K =  0.005852
     P1 =         1   P2 =         3   P3 =         5   Q1 =        -1    Q2 =       -1    Q3 =       -5

Modified exponential model  :
     Bm = Sum ( Ai X ( 1 - exp( - R/( 2*i*C ) ) ) )  +      Sum ( RKj X ( R - Rj0 ) )  +  Bm0
      C =  0.14583333E+00            Bmo=  0.00000000E+00          Nexp=  6         Nliner=  0
     AI = -0.52612105E+03   0.78286530E+04  -0.44652260E+05   0.11314424E+06  -0.12936884E+06   0.54510256E+05
```

No	Rotation (radians) X 1/1000	Moment (kip-inch) Expri.	Poly.	M.Expo.	Connection stiffness (kip-inch) X 1000 Expri.	Poly.	M.Expo.
1	0.00	0.00	0.00	0.00	0.1373E+03	0.2020E+03	0.4983E+01
2	0.25	34.33	49.99	28.14	0.1373E+03	0.1960E+03	0.1707E+03
3	0.50	68.65	97.24	72.74	0.1538E+03	0.1810E+03	0.1758E+03
4	0.70	102.05	131.99	106.36	0.1670E+03	0.1664E+03	0.1603E+03
5	0.90	135.45	163.81	137.11	0.1670E+03	0.1520E+03	0.1483E+03
6	1.10	168.85	192.85	166.07	0.1571E+03	0.1387E+03	0.1422E+03
7	1.40	211.53	231.84	208.18	0.1423E+03	0.1217E+03	0.1392E+03
8	1.70	254.21	266.22	249.69	0.1329E+03	0.1080E+03	0.1373E+03
9	2.05	296.88	301.73	296.89	0.1219E+03	0.9539E+02	0.1318E+03
10	2.40	339.55	333.31	341.53	0.1201E+03	0.8545E+02	0.1229E+03
11	2.65	369.24	353.91	371.36	0.1188E+03	0.7959E+02	0.1157E+03
12	2.90	398.93	373.16	399.39	0.1068E+03	0.7454E+02	0.1086E+03
13	3.25	430.48	398.17	435.83	0.9014E+02	0.6854E+02	0.9991E+02
14	3.60	462.02	421.26	469.56	0.1015E+03	0.6353E+02	0.9319E+02
15	3.90	495.42	439.75	496.85	0.1113E+03	0.5984E+02	0.8899E+02
16	4.20	528.82	457.21	523.07	0.9909E+02	0.5661E+02	0.8595E+02
17	4.55	558.50	476.43	552.68	0.8482E+02	0.5331E+02	0.8340E+02
18	4.90	588.19	494.57	581.51	0.8204E+02	0.5043E+02	0.8138E+02
19	5.30	619.73	514.16	613.62	0.7885E+02	0.4754E+02	0.7914E+02
20	5.70	651.28	532.66	644.77	0.6726E+02	0.4501E+02	0.7653E+02
21	6.10	673.54	550.21	674.77	0.5567E+02	0.4278E+02	0.7332E+02
22	6.50	695.81	566.91	703.34	0.6262E+02	0.4080E+02	0.6945E+02
23	6.90	723.64	582.87	730.24	0.6959E+02	0.3901E+02	0.6499E+02
24	7.30	751.48	598.15	755.27	0.6358E+02	0.3741E+02	0.6010E+02
25	7.65	771.89	611.24	775.52	0.5832E+02	0.3610E+02	0.5559E+02
26	8.00	792.30	623.45	794.18	0.5126E+02	0.3494E+02	0.5099E+02
27	8.33	807.14	635.09	810.44	0.4453E+02	0.3388E+02	0.4663E+02
28	8.67	821.99	646.05	825.28	0.4453E+02	0.3292E+02	0.4236E+02
29	9.00	836.83	657.03	838.70	0.4454E+02	0.3200E+02	0.3824E+02
30	9.33	851.68	667.40	850.79	0.4453E+02	0.3117E+02	0.3433E+02
31	9.67	866.52	677.79	861.62	0.4453E+02	0.3036E+02	0.3066E+02
32	10.00	881.36	687.91	871.26	0.4454E+02	0.2960E+02	0.2725E+02

Connection type : Top-and seat-angle connections
Mode : All bolted

Tested by : S.M.Maxwell et al. (1981) U.S.A
Test Id. : B3

Column : 305 X 305 X UC 97
Beam : 457 X 191 X UB 67
Angle : 150 X 90 X 12 X 150 MM

Fasteners: G8.8- -M20
7/8" Oversize holes
Material : --
Fy = -- ksi
Fu = -- ksi

Major parameters

lt = 5.9055"	tt = 0.4724"	ls = 5.9055"	ts = 0.4724"
gt = -- "	gt'= -- "	gs = -- "	gs'= -- "
qt = -- "	rt = -- "	qs = -- "	rs = -- "
pt = -- "	ps = -- "		
nt = 2 X 2	nt'= 2 X 1	ns = 2 X 2	ns'= 2 X 1

Remark 1) Grade 8.8 bolts at column
2) Hsfg bolts in two rows at beam.

No	Moment (k-in)	Rotation (radians) X 1/1000
1	0.00	0.00
2	40.91	0.30
3	79.34	0.53
4	117.76	0.77
5	156.19	1.00
6	195.86	1.27
7	235.53	1.53
8	275.20	1.80
9	312.38	2.07
10	349.57	2.33
11	386.76	2.60
12	418.99	2.83
13	451.22	3.07
14	483.44	3.30
15	518.77	3.65
16	554.10	4.00
17	587.57	4.35
18	621.04	4.70
19	647.07	5.05
20	673.10	5.40
21	697.28	5.80
22	721.45	6.20
23	740.05	6.55
24	758.64	6.90
25	777.23	7.35
26	795.83	7.80
27	814.42	8.25
28	833.02	8.70
29	844.17	9.07
30	855.33	9.43
31	866.48	9.80

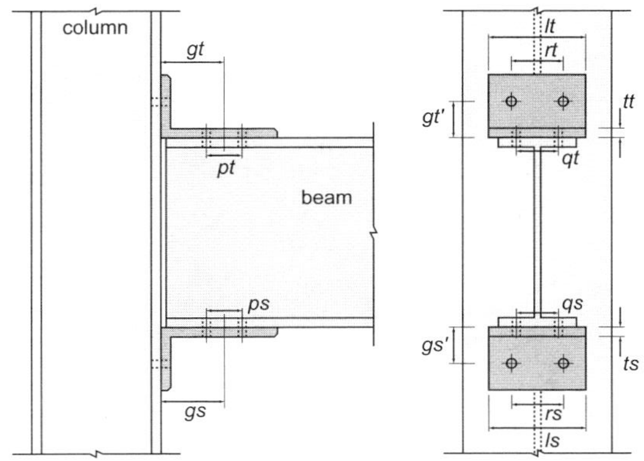

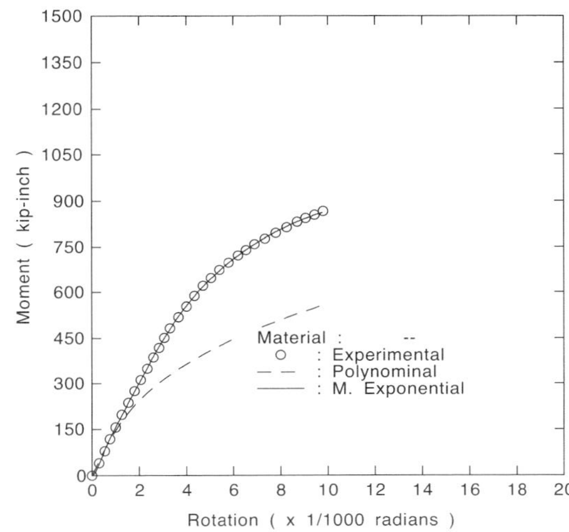

A.4 – 96

```
Moment-rotation prediction equations     ( R : X 1/1000 radians )

Frye and Morris polynominal model  :     R = Sum ( Ai X ( K*Bm )**Pi X 10**Qi )
      t =  0.472400"  xd = 17.992000"   f =  0.787000"  xl =  5.905500"
     A1 =  8.460000   A2 =  1.010000   A3 =  1.240000    K =  0.007158
     P1 =        1    P2 =        3    P3 =        5    Q1 =       -1     Q2 =       -1     Q3 =       -5

Modified exponential model  :
     Bm = Sum ( Ai X ( 1 - exp( - R/( 2*i*C ) ) ) )  +      Sum ( RKj X ( R - Rj0 ) )  +  Bm0
      C =   0.15666667E+00              Bmo=   0.00000000E+00             Nexp=  6          Nliner=  0
     AI =  -0.54660734E+03   0.69346711E+04  -0.33754509E+05   0.77364481E+05  -0.84807718E+05   0.35734590E+05
```

No	Rotation (radians) X 1/1000	Moment (kip-inch) Expri.	Moment (kip-inch) Poly.	Moment (kip-inch) M.Expo.	Connection stiffness (kip-inch) X 1000 Expri.	Connection stiffness (kip-inch) X 1000 Poly.	Connection stiffness (kip-inch) X 1000 M.Expo.
1	0.00	0.00	0.00	0.00	0.1364E+03	0.1651E+03	0.1458E+02
2	0.30	40.91	48.83	38.13	0.1523E+03	0.1582E+03	0.1816E+03
3	0.53	79.34	84.40	80.98	0.1647E+03	0.1461E+03	0.1790E+03
4	0.77	117.76	116.86	120.52	0.1647E+03	0.1320E+03	0.1601E+03
5	1.00	156.19	146.08	156.25	0.1573E+03	0.1187E+03	0.1478E+03
6	1.27	195.86	175.90	194.94	0.1488E+03	0.1053E+03	0.1438E+03
7	1.53	235.53	202.45	233.43	0.1488E+03	0.9424E+02	0.1455E+03
8	1.80	275.20	226.33	272.57	0.1441E+03	0.8511E+02	0.1478E+03
9	2.07	312.38	247.99	312.07	0.1395E+03	0.7756E+02	0.1480E+03
10	2.33	349.57	267.80	351.24	0.1395E+03	0.7127E+02	0.1453E+03
11	2.60	386.76	286.08	389.32	0.1387E+03	0.6598E+02	0.1399E+03
12	2.83	418.99	301.00	421.26	0.1381E+03	0.6199E+02	0.1336E+03
13	3.07	451.22	315.05	451.63	0.1381E+03	0.5849E+02	0.1265E+03
14	3.30	483.44	328.33	480.27	0.1233E+03	0.5541E+02	0.1190E+03
15	3.65	518.77	347.01	519.96	0.1009E+03	0.5141E+02	0.1079E+03
16	4.00	554.10	364.39	555.89	0.9828E+02	0.4801E+02	0.9766E+02
17	4.35	587.57	380.67	588.47	0.9562E+02	0.4508E+02	0.8869E+02
18	4.70	621.04	395.99	618.12	0.8500E+02	0.4254E+02	0.8100E+02
19	5.05	647.07	410.48	645.29	0.7438E+02	0.4031E+02	0.7443E+02
20	5.40	673.10	424.23	670.33	0.6787E+02	0.3833E+02	0.6878E+02
21	5.80	697.28	439.16	696.70	0.6043E+02	0.3633E+02	0.6315E+02
22	6.20	721.45	453.33	720.94	0.5654E+02	0.3456E+02	0.5812E+02
23	6.55	740.05	465.17	740.56	0.5313E+02	0.3316E+02	0.5405E+02
24	6.90	758.64	476.56	758.79	0.4796E+02	0.3190E+02	0.5018E+02
25	7.35	777.23	490.57	780.29	0.4132E+02	0.3043E+02	0.4542E+02
26	7.80	795.83	503.96	799.70	0.4132E+02	0.2911E+02	0.4085E+02
27	8.25	814.42	516.79	817.09	0.4132E+02	0.2792E+02	0.3649E+02
28	8.70	833.02	529.11	832.57	0.3532E+02	0.2684E+02	0.3234E+02
29	9.07	844.17	538.95	843.84	0.3043E+02	0.2602E+02	0.2915E+02
30	9.43	855.33	548.21	853.96	0.3043E+02	0.2528E+02	0.2614E+02
31	9.80	866.48	557.48	863.03	0.3042E+02	0.2457E+02	0.2333E+02

IV - 49

Connection type : Top-and seat-angle connections
Mode : All bolted

Tested by : S.M.Maxwell et al. (1981) U.S.A
Test Id. : B4

Column : 305 X 305 X UC 97
Beam : 457 X 191 X UB 67
Angle : 150 X 90 X 12 X 200 MM

Fasteners: G8.8- -M20
7/8" Oversize holes
Material : --
Fy = -- ksi
Fu = -- ksi

Major parameters

lt = 7.8740"	tt = 0.4724"	ls = 7.8740"	ts = 0.4724"
gt = -- "	gt'= -- "	gs = -- "	gs'= -- "
qt = -- "	rt = -- "	qs = -- "	rs = -- "
pt = -- "	ps = -- "		
nt = 2 X 2	nt'= 2 X 1	ns = 2 X 2	ns'= 2 X 1

Remark 1) Grade 8.8 bolts at column
2) Hsfg bolts in two rows at beam.

No	Moment (k-in)	Rotation (radians) X 1/1000	No	Moment (k-in)	Rotation (radians) X 1/1000
1	0.00	0.00	26	890.02	8.03
2	43.91	0.27	27	906.10	8.37
3	87.82	0.53	28	922.19	8.70
4	131.74	0.80	29	935.79	9.03
5	166.37	1.07	30	949.40	9.37
6	201.01	1.33	31	963.00	9.70
7	235.64	1.60			
8	269.05	1.80			
9	302.45	2.00			
10	344.50	2.30			
11	386.56	2.60			
12	428.62	2.90			
13	473.15	3.25			
14	517.68	3.60			
15	560.36	3.90			
16	603.04	4.20			
17	642.00	4.60			
18	680.96	5.00			
19	719.93	5.40			
20	758.90	5.80			
21	786.73	6.25			
22	814.56	6.70			
23	834.35	7.03			
24	854.14	7.37			
25	873.94	7.70			

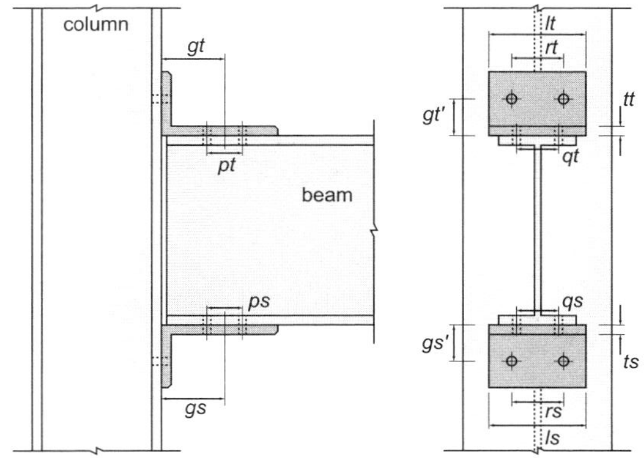

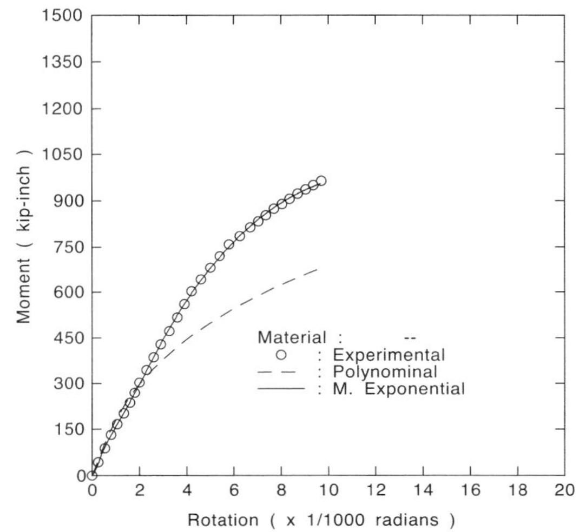

```
Moment-rotation prediction equations     ( R : X 1/1000 radians )

Frye and Morris polynominal model  :    R = Sum ( Ai X ( K*Bm )**Pi X 10**Qi )
     t =  0.472400"  xd = 17.992000"   f =  0.787000"  xl =  7.874000"
    A1 =  8.460000   A2 =  1.010000   A3 =  1.240000    K =  0.005852
    P1 =         1   P2 =         3   P3 =         5   Q1 =        -1     Q2 =        -1     Q3 =        -5

Modified exponential model  :
    Bm = Sum ( Ai X ( 1 - exp( - R/( 2*i*C ) ) ) )  +     Sum ( RKj X ( R - Rj0 ) )  +  Bm0
     C =   0.14750833E+00            Bmo=   0.00000000E+00           Nexp=  6          Nliner=  0
    AI =  -0.50283336E+03   0.68393197E+04  -0.34722567E+05   0.84628288E+05  -0.98205981E+05   0.42983539E+05
```

No	Rotation (radians) X 1/1000	Moment (kip-inch) Expri.	Moment (kip-inch) Poly.	Moment (kip-inch) M.Expo.	Connection stiffness (kip-inch) X 1000 Expri.	Connection stiffness (kip-inch) X 1000 Poly.	Connection stiffness (kip-inch) X 1000 M.Expo.
1	0.00	0.00	0.00	0.00	0.1646E+03	0.2020E+03	0.7605E+02
2	0.27	43.91	53.25	41.79	0.1647E+03	0.1952E+03	0.1908E+03
3	0.53	87.82	103.23	90.37	0.1647E+03	0.1786E+03	0.1674E+03
4	0.80	131.74	148.27	130.97	0.1473E+03	0.1591E+03	0.1398E+03
5	1.07	166.37	188.21	166.64	0.1299E+03	0.1408E+03	0.1305E+03
6	1.33	201.01	223.60	201.71	0.1299E+03	0.1252E+03	0.1339E+03
7	1.60	235.64	255.20	238.42	0.1511E+03	0.1123E+03	0.1415E+03
8	1.80	269.05	276.82	267.25	0.1670E+03	0.1041E+03	0.1465E+03
9	2.00	302.45	296.91	296.92	0.1563E+03	0.9701E+02	0.1498E+03
10	2.30	344.50	324.63	342.16	0.1402E+03	0.8806E+02	0.1510E+03
11	2.60	386.56	349.91	387.10	0.1402E+03	0.8069E+02	0.1481E+03
12	2.90	428.62	373.16	430.75	0.1342E+03	0.7454E+02	0.1426E+03
13	3.25	473.15	398.17	479.26	0.1272E+03	0.6854E+02	0.1344E+03
14	3.60	517.68	421.26	524.81	0.1353E+03	0.6353E+02	0.1258E+03
15	3.90	560.36	439.75	561.48	0.1423E+03	0.5984E+02	0.1187E+03
16	4.20	603.04	457.21	596.07	0.1230E+03	0.5661E+02	0.1120E+03
17	4.60	642.00	479.09	639.20	0.9741E+02	0.5288E+02	0.1038E+03
18	5.00	680.96	499.58	679.18	0.9742E+02	0.4967E+02	0.9624E+02
19	5.40	719.93	518.88	716.25	0.9742E+02	0.4688E+02	0.8917E+02
20	5.80	758.90	537.13	750.55	0.8068E+02	0.4443E+02	0.8237E+02
21	6.25	786.73	556.56	785.93	0.6184E+02	0.4201E+02	0.7490E+02
22	6.70	814.56	574.98	817.99	0.6043E+02	0.3988E+02	0.6758E+02
23	7.03	834.35	588.27	839.62	0.5937E+02	0.3844E+02	0.6227E+02
24	7.37	854.14	600.63	859.51	0.5937E+02	0.3715E+02	0.5708E+02
25	7.70	873.94	613.02	877.70	0.5382E+02	0.3593E+02	0.5206E+02
26	8.03	890.02	624.61	894.24	0.4825E+02	0.3483E+02	0.4724E+02
27	8.37	906.10	636.22	909.22	0.4825E+02	0.3378E+02	0.4266E+02
28	8.70	922.19	647.15	922.71	0.4454E+02	0.3283E+02	0.3834E+02
29	9.03	935.79	658.09	934.81	0.4081E+02	0.3192E+02	0.3430E+02
30	9.37	949.40	668.73	945.61	0.4081E+02	0.3106E+02	0.3055E+02
31	9.70	963.00	678.67	955.21	0.4082E+02	0.3029E+02	0.2710E+02

IV - 50

Connection type : Top-and seat-angle connections
Mode : All bolted

Tested by : S.M.Maxwell et al. (1981) U.S.A
Test Id. : C1

Column : 305 X 305 X UC 97
Beam : 457 X 191 X UB 67
Angle : 150 X 90 X 15 X 150 MM

Fasteners: G8.8- -M20
7/8" Oversize holes
Material : --
Fy = -- ksi
Fu = -- ksi

Major parameters

lt = 5.9055"	tt = 0.5906"	ls = 5.9055"	ts = 0.5906"
gt = -- "	gt'= -- "	gs = -- "	gs'= -- "
qt = -- "	rt = -- "	qs = -- "	rs = -- "
pt = -- "	ps = -- "		
nt = 2 X 2	nt'= 2 X 1	ns = 2 X 2	ns'= 2 X 1

Remark 1) Grade 8.8 bolts at column
2) Hsfg bolts in two rows at beam.

No	Moment (k-in)	Rotation (radians) X 1/1000	No	Moment (k-in)	Rotation (radians) X 1/1000
1	0.00	0.00	26	900.24	7.70
2	35.06	0.25	27	926.77	8.10
3	70.13	0.50	28	953.30	8.50
4	108.03	0.77	29	979.84	8.90
5	145.94	1.03	30	1000.69	9.35
6	183.84	1.30	31	1021.54	9.80
7	219.22	1.57			
8	254.59	1.83			
9	289.97	2.10			
10	332.93	2.43			
11	375.89	2.77			
12	418.85	3.10			
13	459.28	3.40			
14	499.71	3.70			
15	540.14	4.00			
16	574.26	4.30			
17	608.37	4.60			
18	642.48	4.90			
19	680.39	5.23			
20	718.30	5.57			
21	756.20	5.90			
22	784.00	6.23			
23	811.80	6.57			
24	839.59	6.90			
25	869.92	7.30			

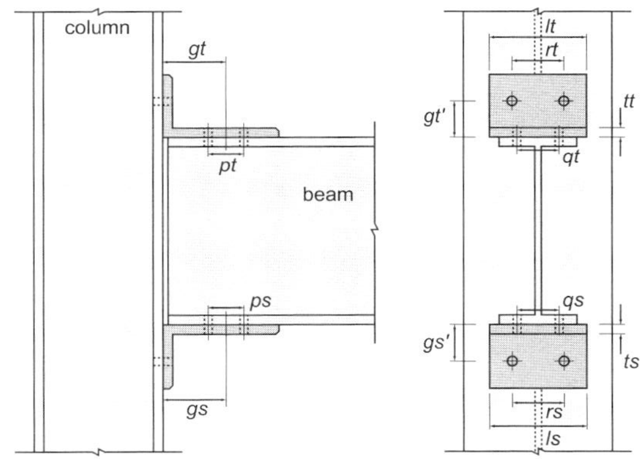

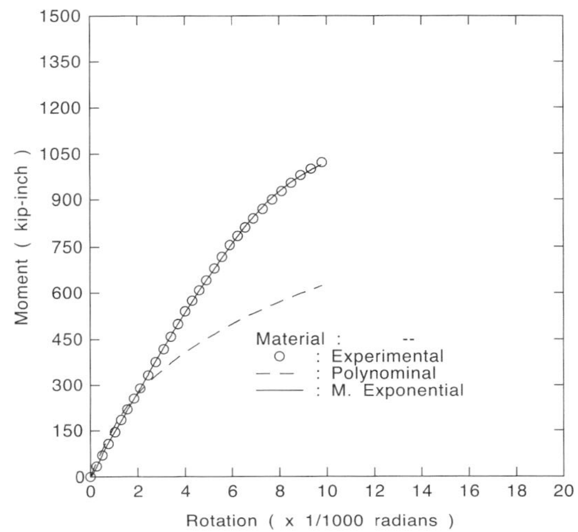

A.4 – 100

```
Moment-rotation prediction equations     ( R : X 1/1000 radians )

Frye and Morris polynominal model  :    R = Sum ( Ai X ( K*Bm )**Pi X 10**Qi )
     t =  0.590600"  xd = 17.992000"   f =  0.787000"  xl =  5.905500"
    A1 =  8.460000   A2 =  1.010000   A3 =  1.240000    K =  0.006402
    P1 =        1    P2 =        3    P3 =        5    Q1 =       -1    Q2 =       -1    Q3 =       -5

Modified exponential model  :
    Bm = Sum ( Ai X ( 1 - exp( - R/( 2*i*C ) ) ) )  +      Sum ( RKj X ( R - Rj0 ) )  +  Bm0
     C =   0.14416667E+00           Bmo=   0.00000000E+00          Nexp=  6        Nliner=  0
    AI =  -0.48428796E+03   0.75729089E+04  -0.44838508E+05   0.11949119E+06  -0.14328885E+06   0.62642506E+05
```

No	Rotation (radians) X 1/1000	Moment (kip-inch) Expri.	Moment (kip-inch) Poly.	Moment (kip-inch) M.Expo.	Connection stiffness (kip-inch) X 1000 Expri.	Connection stiffness (kip-inch) X 1000 Poly.	Connection stiffness (kip-inch) X 1000 M.Expo.
1	0.00	0.00	0.00	0.00	0.1403E+03	0.1846E+03	0.3968E+02
2	0.25	35.06	45.70	31.11	0.1403E+03	0.1792E+03	0.1661E+03
3	0.50	70.13	88.89	72.58	0.1412E+03	0.1655E+03	0.1582E+03
4	0.77	108.03	130.67	111.67	0.1421E+03	0.1476E+03	0.1365E+03
5	1.03	145.94	167.72	146.63	0.1421E+03	0.1307E+03	0.1281E+03
6	1.30	183.84	200.57	180.87	0.1374E+03	0.1161E+03	0.1296E+03
7	1.57	219.22	229.86	216.04	0.1327E+03	0.1040E+03	0.1342E+03
8	1.83	254.59	256.21	252.31	0.1327E+03	0.9402E+02	0.1375E+03
9	2.10	289.97	280.15	289.16	0.1310E+03	0.8578E+02	0.1385E+03
10	2.43	332.93	307.29	335.08	0.1289E+03	0.7736E+02	0.1366E+03
11	2.77	375.89	331.90	380.05	0.1289E+03	0.7052E+02	0.1330E+03
12	3.10	418.85	354.44	423.71	0.1320E+03	0.6489E+02	0.1290E+03
13	3.40	459.28	373.24	461.92	0.1348E+03	0.6060E+02	0.1258E+03
14	3.70	499.71	390.86	499.23	0.1348E+03	0.5690E+02	0.1230E+03
15	4.00	540.14	407.43	535.78	0.1242E+03	0.5368E+02	0.1207E+03
16	4.30	574.26	423.10	571.68	0.1137E+03	0.5085E+02	0.1186E+03
17	4.60	608.37	437.97	606.94	0.1137E+03	0.4834E+02	0.1164E+03
18	4.90	642.48	452.13	641.49	0.1137E+03	0.4610E+02	0.1139E+03
19	5.23	680.39	467.12	678.94	0.1137E+03	0.4388E+02	0.1107E+03
20	5.57	718.30	481.41	715.19	0.1137E+03	0.4189E+02	0.1067E+03
21	5.90	756.20	495.07	750.01	0.9857E+02	0.4010E+02	0.1021E+03
22	6.23	784.00	508.16	783.20	0.8338E+02	0.3848E+02	0.9691E+02
23	6.57	811.80	520.74	814.57	0.8338E+02	0.3701E+02	0.9119E+02
24	6.90	839.59	533.08	843.96	0.7995E+02	0.3564E+02	0.8511E+02
25	7.30	869.92	546.82	876.49	0.7581E+02	0.3420E+02	0.7754E+02
26	7.70	900.24	560.23	905.98	0.7107E+02	0.3286E+02	0.6989E+02
27	8.10	926.77	573.37	932.42	0.6633E+02	0.3163E+02	0.6237E+02
28	8.50	953.30	585.56	955.92	0.6633E+02	0.3054E+02	0.5516E+02
29	8.90	979.84	597.77	976.61	0.5692E+02	0.2950E+02	0.4838E+02
30	9.35	1000.69	610.61	996.78	0.4633E+02	0.2845E+02	0.4137E+02
31	9.80	1021.54	623.41	1013.95	0.4633E+02	0.2747E+02	0.3507E+02

A.4 – 101

IV - 51

```
Connection type : Top-and seat-angle connections
           Mode : All bolted

Tested by   :  S.M.Maxwell et al. (1981)                    U.S.A
Test  Id.   :  C2

Column : 305 X 305 X UC 97          Fasteners: G8.8- -M20
Beam   : 457 X 191 X UB 67                   7/8" Oversize holes
Angle  : 150 X 90  X 15 X 200 MM    Material :  --
                                         Fy =   --   ksi
                                         Fu =   --   ksi

                        Major parameters

     lt =  7.8740"    tt =  0.5906"    ls =  7.8740"    ts =  0.5906"
     gt =   --   "    gt'=   --   "    gs =   --   "    gs'=   --   "
     qt =   --   "    rt =   --   "    qs =   --   "    rs =   --   "
     pt =   --   "    ps =   --   "
     nt = 2 X  2      nt'= 2 X  1      ns = 2 X  2      ns'= 2 X  1

Remark  1)  Grade 8.8 bolts at column
        2)  Hsfg bolts in two rows at beam.
```

No	Moment (k-in)	Rotation (radians) X 1/1000	No	Moment (k-in)	Rotation (radians) X 1/1000
1	0.00	0.00	26	1038.26	7.95
2	26.47	0.20	27	1040.15	8.20
3	68.08	0.47	28	1040.15	8.50
4	109.69	0.73	29	1040.15	8.80
5	151.30	1.00	30	1042.04	9.10
6	202.36	1.30	31	1043.93	9.40
7	253.42	1.60	32	1043.93	9.65
8	298.81	1.90	33	1043.93	9.90
9	344.20	2.20			
10	389.59	2.45			
11	434.98	2.70			
12	486.04	3.10			
13	537.10	3.50			
14	586.27	3.80			
15	635.44	4.10			
16	673.94	4.45			
17	722.44	4.80			
18	760.26	5.10			
19	793.08	5.40			
20	837.80	5.75			
21	877.51	6.10			
22	917.23	6.45			
23	956.94	6.80			
24	996.65	7.25			
25	1036.37	7.70			

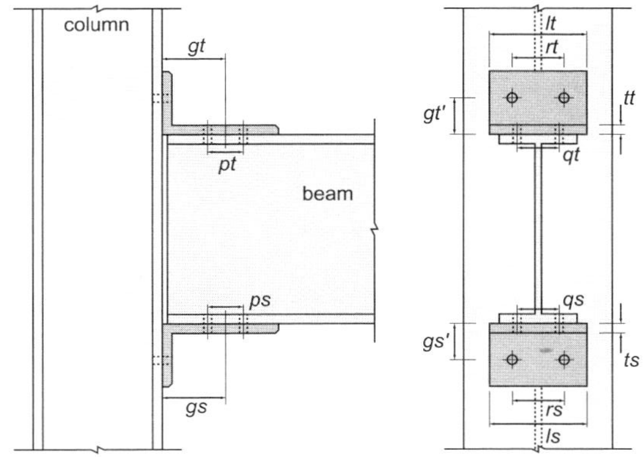

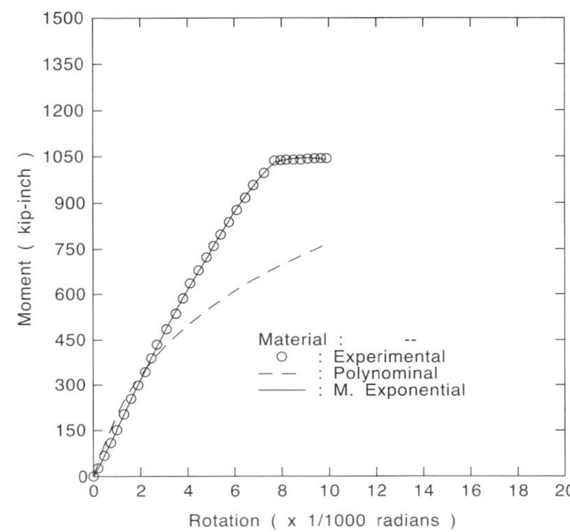

```
Moment-rotation prediction equations     ( R : X 1/1000 radians )

Frye and Morris polynominal model  :     R = Sum ( Ai X ( K*Bm )**Pi X 10**Qi )
     t =  0.590600"  xd = 17.992000"   f =  0.787000"  xl =  7.874000"
    A1 =  8.460000   A2 =  1.010000   A3 =  1.240000    K =  0.005234
    P1 =         1   P2 =         3   P3 =         5   Q1 =        -1    Q2 =        -1    Q3 =        -5

Modified exponential model  :
     Bm = Sum ( Ai X ( 1 - exp( - R/( 2*i*C ) ) ) )  +     Sum ( RKj X ( R - Rj0 ) )  +  Bm0
      C =   0.13250000E+00              Bmo=   0.00000000E+00          Nexp=  6          Nliner=  2
     AI =  -0.26098692E+03   0.49458822E+04  -0.32893155E+05   0.93926856E+05  -0.11818809E+06   0.53585655E+05
    Rj0 =           4.8000          7.7000
    RKj =   0.16202046E+02  -0.50918946E+02
```

No	Rotation (radians) X 1/1000	Moment (kip-inch) Expri.	Moment (kip-inch) Poly.	Moment (kip-inch) M.Expo.	Connection stiffness (kip-inch) X 1000 Expri.	Connection stiffness (kip-inch) X 1000 Poly.	Connection stiffness (kip-inch) X 1000 M.Expo.
1	0.00	0.00	0.00	0.00	0.1324E+03	0.2258E+03	0.8531E+02
2	0.20	26.47	44.87	25.76	0.1425E+03	0.2215E+03	0.1539E+03
3	0.47	68.08	101.94	68.35	0.1560E+03	0.2049E+03	0.1598E+03
4	0.73	109.69	153.74	110.39	0.1560E+03	0.1833E+03	0.1564E+03
5	1.00	151.30	199.78	152.34	0.1627E+03	0.1623E+03	0.1589E+03
6	1.30	202.36	245.32	200.79	0.1702E+03	0.1420E+03	0.1639E+03
7	1.60	253.42	285.35	250.42	0.1608E+03	0.1255E+03	0.1664E+03
8	1.90	298.81	320.95	300.24	0.1513E+03	0.1123E+03	0.1652E+03
9	2.20	344.20	352.98	349.27	0.1678E+03	0.1016E+03	0.1614E+03
10	2.45	389.59	377.42	389.13	0.1816E+03	0.9415E+02	0.1574E+03
11	2.70	434.98	400.14	427.97	0.1608E+03	0.8780E+02	0.1533E+03
12	3.10	486.04	433.51	488.04	0.1277E+03	0.7936E+02	0.1472E+03
13	3.50	537.10	463.84	545.83	0.1484E+03	0.7254E+02	0.1418E+03
14	3.80	586.27	484.94	587.81	0.1639E+03	0.6822E+02	0.1380E+03
15	4.10	635.44	504.83	628.63	0.1456E+03	0.6445E+02	0.1341E+03
16	4.45	678.94	526.70	674.68	0.1243E+03	0.6061E+02	0.1290E+03
17	4.80	722.44	547.32	718.81	0.1253E+03	0.5726E+02	0.1231E+03
18	5.10	760.26	564.11	759.77	0.1261E+03	0.5471E+02	0.1336E+03
19	5.40	798.08	580.17	798.95	0.1203E+03	0.5242E+02	0.1274E+03
20	5.75	837.80	598.09	842.20	0.1135E+03	0.5000E+02	0.1196E+03
21	6.10	877.51	615.20	882.65	0.1135E+03	0.4784E+02	0.1114E+03
22	6.45	917.23	631.59	920.19	0.1135E+03	0.4588E+02	0.1030E+03
23	6.80	956.94	647.34	954.78	0.1024E+03	0.4410E+02	0.9467E+02
24	7.25	996.65	666.71	995.01	0.8825E+02	0.4204E+02	0.8422E+02
25	7.70	1036.37	685.21	1030.68	0.3638E+02	0.4020E+02	0.7441E+02
26	7.95	1038.26	695.25	1035.91	0.7560E+01	0.3924E+02	0.1840E+02
27	8.20	1040.15	705.06	1039.90	0.4124E+01	0.3834E+02	0.1358E+02
28	8.50	1040.15	716.19	1043.15	0.0000E+00	0.3735E+02	0.8201E+01
29	8.80	1040.15	727.39	1044.86	0.3150E+01	0.3639E+02	0.3258E+01
30	9.10	1042.04	738.31	1045.15	0.6300E+01	0.3548E+02	-0.1249E+01
31	9.40	1043.93	748.56	1044.16	0.2864E+01	0.3467E+02	-0.5329E+01
32	9.65	1043.93	757.23	1042.43	0.0000E+00	0.3399E+02	-0.8416E+01
33	9.90	1043.93	765.72	1039.97	0.0000E+00	0.3335E+02	-0.1123E+02

```
Connection type : Top-and seat-angle connections
           Mode : All bolted

Tested by  :  S.M.Maxwell et al. (1981)                    U.S.A
Test  Id.  :  C3

Column : 305 X 305 X UC 97          Fasteners: G8.8- -M20
Beam   : 457 X 191 X UB 67                 7/8" Oversize holes
Angle  : 150 X 90  X 15 X 150 MM    Material :  --
                                         Fy =   --   ksi
                                         Fu =   --   ksi

                          Major parameters

     lt =  5.9055"   tt =  0.5906"   ls =  5.9055"   ts =  0.5906"
     gt =   --   "   gt'=   --   "   gs =   --   "   gs'=   --   "
     qt =   --   "   rt =   --   "   qs =   --   "   rs =   --   "
     pt =   --   "   ps =   --   "
     nt = 2 X  2     nt'= 2 X  1     ns = 2 X  2     ns'= 2 X  1

Remark  1)  Grade 8.8 bolts at column
        2)  Hsfg bolts in two rows at beam.
```

No	Moment (k-in)	Rotation (radians) X 1/1000
1	0.00	0.00
2	39.63	0.20
3	79.26	0.40
4	118.90	0.60
5	152.86	0.80
6	186.83	1.00
7	220.80	1.20
8	269.86	1.45
9	318.92	1.70
10	366.11	2.00
11	413.29	2.30
12	460.47	2.60
13	507.65	2.90
14	547.28	3.25
15	586.91	3.60
16	619.62	3.93
17	652.32	4.27
18	685.03	4.60
19	711.46	4.90
20	737.88	5.20
21	775.62	5.65
22	813.36	6.10
23	841.67	6.50
24	869.98	6.90
25	896.39	7.30
26	922.81	7.70
27	943.57	8.05
28	964.33	8.40
29	983.20	8.85
30	1002.07	9.30

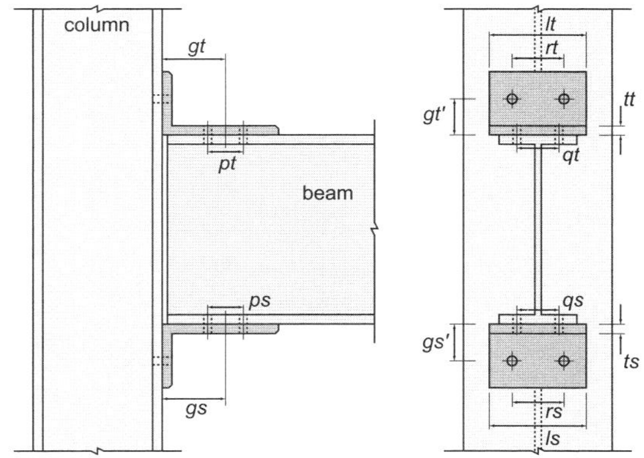

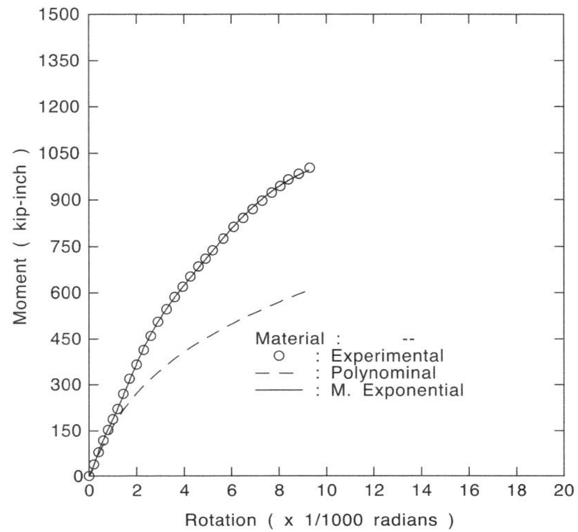

```
Moment-rotation prediction equations     ( R : X 1/1000 radians )

Frye and Morris polynominal model  :    R = Sum ( Ai X ( K*Bm )**Pi X 10**Qi )
     t =  0.590600"  xd = 17.992000"   f =  0.787000"  xl =  5.905500"
    A1 =  8.460000   A2 =  1.010000   A3 =  1.240000    K =  0.006402
    P1 =         1   P2 =         3   P3 =         5   Q1 =        -1   Q2 =       -1   Q3 =       -5

Modified exponential model  :
    Bm = Sum ( Ai X ( 1 - exp( - R/( 2*i*C ) ) ) )  +      Sum ( RKj X ( R - Rj0 ) )  +  Bm0
     C =   0.12750000E+00            Bmo=   0.00000000E+00             Nexp=  6        Nliner=  0
    AI =  -0.83679377E+03   0.13497945E+05  -0.72782299E+05   0.17244200E+06  -0.18612808E+06   0.74865824E+05
```

No	Rotation (radians) X 1/1000	Moment (kip-inch) Expri.	Poly.	M.Expo.	Connection stiffness (kip-inch) X 1000 Expri.	Poly.	M.Expo.
1	0.00	0.00	0.00	0.00	0.1982E+03	0.1846E+03	0.5464E+02
2	0.20	39.63	36.69	35.48	0.1982E+03	0.1811E+03	0.2362E+03
3	0.40	79.26	72.03	81.81	0.1982E+03	0.1716E+03	0.2156E+03
4	0.60	118.90	105.11	120.81	0.1840E+03	0.1589E+03	0.1771E+03
5	0.80	152.86	135.54	154.23	0.1698E+03	0.1454E+03	0.1612E+03
6	1.00	186.83	163.33	186.56	0.1698E+03	0.1327E+03	0.1643E+03
7	1.20	220.80	188.70	220.46	0.1816E+03	0.1213E+03	0.1750E+03
8	1.45	269.86	217.44	265.77	0.1962E+03	0.1090E+03	0.1863E+03
9	1.70	318.92	243.37	312.85	0.1785E+03	0.9876E+02	0.1887E+03
10	2.00	366.11	271.43	368.38	0.1573E+03	0.8869E+02	0.1797E+03
11	2.30	413.29	296.77	419.87	0.1573E+03	0.8051E+02	0.1628E+03
12	2.60	460.47	319.88	465.86	0.1573E+03	0.7377E+02	0.1438E+03
13	2.90	507.65	341.14	506.35	0.1369E+03	0.6814E+02	0.1267E+03
14	3.25	547.28	364.00	547.86	0.1132E+03	0.6266E+02	0.1115E+03
15	3.60	586.91	385.11	585.04	0.1055E+03	0.5808E+02	0.1018E+03
16	3.93	619.62	403.83	618.04	0.9812E+02	0.5436E+02	0.9673E+02
17	4.27	652.32	421.40	649.82	0.9812E+02	0.5115E+02	0.9419E+02
18	4.60	685.03	437.97	680.97	0.9284E+02	0.4834E+02	0.9288E+02
19	4.90	711.46	452.13	708.70	0.8808E+02	0.4610E+02	0.9190E+02
20	5.20	737.88	465.66	736.08	0.8639E+02	0.4409E+02	0.9057E+02
21	5.65	775.62	484.88	776.17	0.8387E+02	0.4142E+02	0.8733E+02
22	6.10	813.36	502.99	814.42	0.7693E+02	0.3911E+02	0.8238E+02
23	6.50	841.67	518.26	846.27	0.7077E+02	0.3729E+02	0.7673E+02
24	6.90	869.98	532.85	875.69	0.6841E+02	0.3567E+02	0.7023E+02
25	7.30	896.39	546.82	902.39	0.6605E+02	0.3420E+02	0.6326E+02
26	7.70	922.81	560.23	926.28	0.6246E+02	0.3286E+02	0.5615E+02
27	8.05	943.57	571.73	944.86	0.5931E+02	0.3178E+02	0.5005E+02
28	8.40	964.33	582.49	961.34	0.5171E+02	0.3081E+02	0.4421E+02
29	8.85	983.20	596.09	979.64	0.4193E+02	0.2964E+02	0.3724E+02
30	9.30	1002.07	609.42	994.96	0.4193E+02	0.2855E+02	0.3099E+02

Connection type : Top-and seat-angle connections
Mode : All bolted

Tested by : S.M.Maxwell et al. (1981) U.S.A
Test Id. : C4

Column : 305 X 305 X UC 97
Beam : 457 X 191 X UB 67
Angle : 150 X 90 X 15 X 200 MM

Fasteners: G8.8- -M20
7/8" Oversize holes
Material : --
Fy = -- ksi
Fu = -- ksi

Major parameters

lt = 7.8740"	tt = 0.5906"	ls = 7.8740"	ts = 0.5906"
gt = -- "	gt'= -- "	gs = -- "	gs'= -- "
qt = -- "	rt = -- "	qs = -- "	rs = -- "
pt = -- "	ps = -- "		
nt = 2 X 2	nt'= 2 X 1	ns = 2 X 2	ns'= 2 X 1

Remark 1) Grade 8.8 bolts at column
2) Hsfg bolts in two rows at beam

No	Moment (k-in)	Rotation (radians) X 1/1000	No	Moment (k-in)	Rotation (radians) X 1/1000
1	0.00	0.00	26	930.37	5.80
2	38.21	0.17	27	973.78	6.20
3	76.43	0.35	28	1017.18	6.60
4	114.64	0.53	29	1013.41	6.85
5	152.86	0.70	30	1009.63	7.10
6	186.82	0.87	31	1011.52	7.40
7	220.79	1.03	32	1013.40	7.70
8	254.76	1.20	33	1015.29	8.05
9	289.99	1.37	34	1017.18	8.40
10	325.22	1.53			
11	360.44	1.70			
12	394.41	1.90			
13	428.38	2.10			
14	462.35	2.30			
15	496.32	2.53			
16	530.29	2.77			
17	564.26	3.00			
18	598.23	3.23			
19	632.20	3.47			
20	666.17	3.70			
21	713.35	4.05			
22	760.52	4.40			
23	803.93	4.75			
24	847.33	5.10			
25	888.85	5.45			

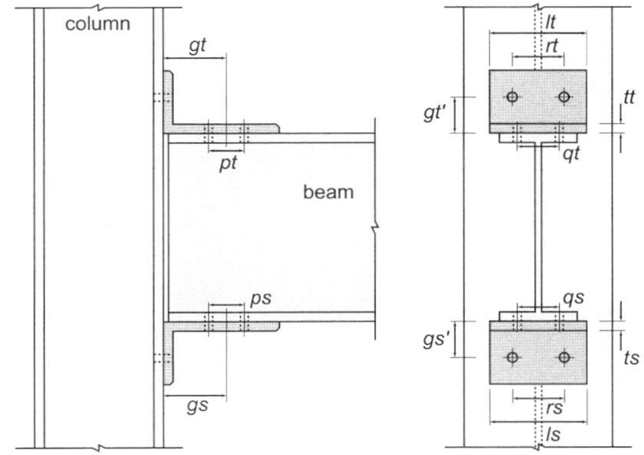

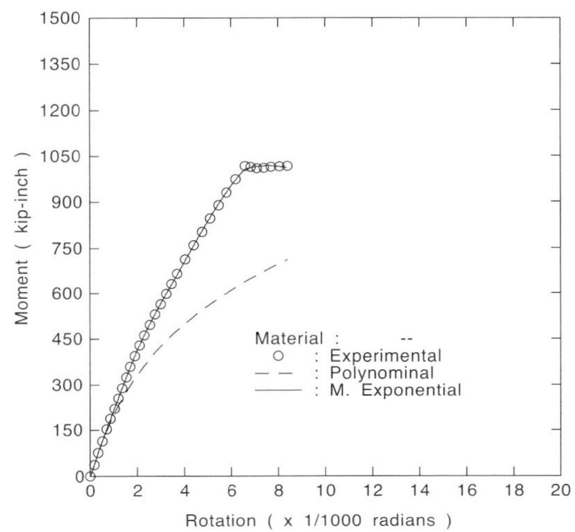

A.4 – 106

```
Moment-rotation prediction equations     ( R : X 1/1000 radians )

Frye and Morris polynominal model  :    R = Sum ( Ai X ( K*Bm )**Pi X 10**Qi )
      t =  0.590600"  xd = 17.992000"   f =  0.787000"  xl =  7.874000"
     A1 =  8.460000   A2 =  1.010000   A3 =  1.240000    K =  0.005234
     P1 =         1   P2 =         3   P3 =         5   Q1 =        -1     Q2 =        -1     Q3 =        -5

Modified exponential model  :
     Bm = Sum ( Ai X ( 1 - exp( - R/( 2*i*C ) ) ) )  +      Sum ( RKj X ( R - Rj0 ) )  +  Bm0
       C =  0.11375000E+00            Bmo=  0.00000000E+00         Nexp=  6        Nliner=  2
      AI = -0.36901634E+03  0.77909169E+04 -0.50577256E+05  0.13622470E+06 -0.16093239E+06  0.69004941E+05
     Rj0 =          3.7000          6.6000
     RKj =  0.68925982E+01 -0.57751556E+02
```

No	Rotation (radians) X 1/1000	Moment (kip-inch) Expri.	Moment (kip-inch) Poly.	Moment (kip-inch) M.Expo.	Connection stiffness (kip-inch) X 1000 Expri.	Connection stiffness (kip-inch) X 1000 Poly.	Connection stiffness (kip-inch) X 1000 M.Expo.
1	0.00	0.00	0.00	0.00	0.2184E+03	0.2258E+03	0.1665E+03
3	0.35	76.43	77.52	77.45	0.2184E+03	0.2133E+03	0.2241E+03
5	0.70	152.86	147.58	151.79	0.2109E+03	0.1861E+03	0.2065E+03
7	1.03	220.79	205.13	221.07	0.2038E+03	0.1598E+03	0.2097E+03
9	1.37	289.99	254.65	290.73	0.2114E+03	0.1380E+03	0.2061E+03
11	1.70	360.44	297.66	357.20	0.1925E+03	0.1208E+03	0.1915E+03
13	2.10	428.38	342.65	429.33	0.1698E+03	0.1049E+03	0.1693E+03
15	2.53	496.32	385.16	498.32	0.1456E+03	0.9193E+02	0.1506E+03
17	3.00	564.26	425.48	565.80	0.1456E+03	0.8130E+02	0.1401E+03
19	3.47	632.20	461.42	630.15	0.1456E+03	0.7306E+02	0.1362E+03
21	4.05	713.35	501.59	711.14	0.1348E+03	0.6505E+02	0.1398E+03
23	4.75	803.93	544.44	806.30	0.1240E+03	0.5772E+02	0.1308E+03
25	5.45	888.85	582.78	892.69	0.1186E+03	0.5205E+02	0.1150E+03
27	6.20	973.78	619.96	971.20	0.1085E+03	0.4726E+02	0.9398E+02
29	6.85	1013.41	649.69	1011.80	-0.1510E+02	0.4384E+02	0.1783E+02
31	7.40	1011.52	672.97	1017.64	0.6280E+01	0.4140E+02	0.3695E+01
33	8.05	1015.29	699.04	1015.28	0.5400E+01	0.3889E+02	-0.1049E+02

IV - 54

```
Connection type : Top-and seat-angle connections
           Mode : All bolted

Tested by  :  J.B.Davison et al. (1987)              United Kingdom
Test  Id.  :  JT/08

Column : UC 152x152x23              Fasteners: 10.9- -M16
Beam   : UB 254x102x22                   23/32"  Oversize holes
Angle  : RSA 80x60x8  TOP           Material :  --
                                         Fy =    --   ksi
                                         Fu =    --   ksi

                          Major parameters

     lt =  4.7638"    tt =  0.3150"    ls =  4.7638"    ts =  0.3150"
     gt =  2.1654"    gt'=  1.3780"    gs =  1.9685"    gs'=  1.7720"
     qt =  2.2441"    rt =  2.9921"    qs =  2.2441"    rs =  2.9921"
     ps'=  1.9685"
     nt = 2 X  1      nt'= 2 X  1      ns = 2 X  1      ns'= 2 X  2

Remark  1)  RSA 125x75x8 SEAT
        2)
```

No	Moment (k-in)	Rotation (radians) X 1/1000	No	Moment (k-in)	Rotation (radians) X 1/1000
1	0.00	0.00	26	136.53	27.98
2	8.43	0.06	27	139.06	29.85
3	16.85	0.12	28	141.59	31.71
4	25.28	0.19	29	144.11	33.58
5	31.60	0.47	30	146.64	35.45
6	37.92	0.75	31	149.17	37.31
7	44.25	0.93	32	151.70	39.18
8	50.57	1.12	33	154.23	41.04
9	55.62	1.49	34	156.76	42.91
10	60.68	1.87	35	159.28	44.77
11	68.27	2.24	36	161.81	46.64
12	75.85	3.17	37	164.34	48.50
13	80.91	4.48	38	166.87	50.37
14	88.49	5.60	39	169.40	51.86
15	96.08	7.46			
16	101.13	8.95			
17	107.45	11.19			
18	112.25	13.06			
19	113.77	14.92			
20	118.83	16.79			
21	121.36	18.66			
22	125.15	20.52			
23	127.68	22.39			
24	130.21	24.25			
25	134.00	26.12			

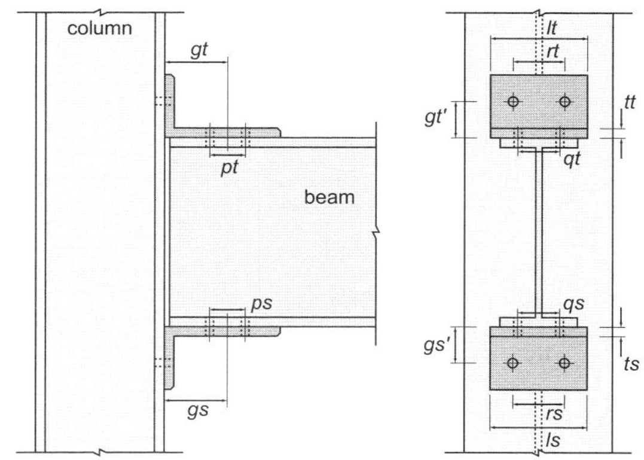

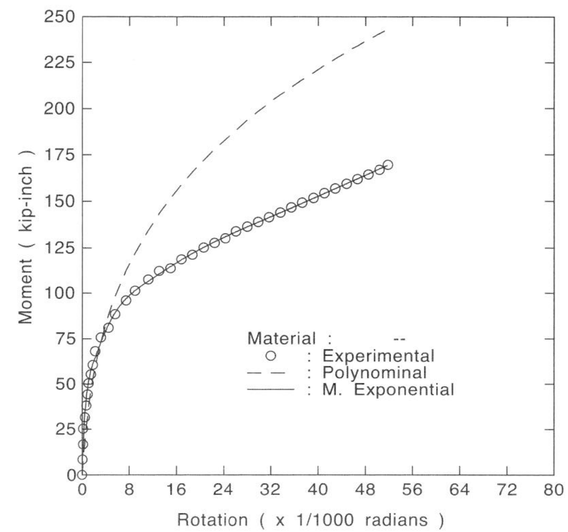

```
Moment-rotation prediction equations     ( R : X 1/1000 radians )

Frye and Morris polynominal model  :    R = Sum ( Ai X ( K*Bm )**Pi X 10**Qi )
     t =  0.315000"  xd = 10.000000"   f =  0.629900"  xl =  4.763800"
    A1 =  8.460000   A2 =  1.010000   A3 =  1.240000    K =  0.031411
    P1 =         1   P2 =         3   P3 =         5   Q1 =        -1    Q2 =        -1    Q3 =        -5

Modified exponential model  :
    Bm = Sum ( Ai X ( 1 - exp( - R/( 2*i*C ) ) ) )  +      Sum ( RKj X ( R - Rj0 ) )  +  Bm0
     C =   0.44771667E+00          Bmo=   0.00000000E+00          Nexp=  6        Nliner=  1
    AI =   0.21306380E+03  -0.12401823E+04   0.41420934E+04  -0.68852275E+04   0.55703721E+04  -0.17017404E+04
   Rj0 =           2.2400
   RKj =   0.14278643E+01
```

No	Rotation (radians) X 1/1000	Moment (kip-inch) Expri.	Moment (kip-inch) Poly.	Moment (kip-inch) M.Expo.	Connection stiffness (kip-inch) X 1000 Expri.	Connection stiffness (kip-inch) X 1000 Poly.	Connection stiffness (kip-inch) X 1000 M.Expo.
1	0.00	0.00	0.00	0.00	0.1355E+03	0.3763E+02	0.9249E+02
3	0.12	16.85	4.67	10.52	0.1355E+03	0.3734E+02	0.7709E+02
5	0.47	31.60	16.98	31.27	0.2260E+02	0.3415E+02	0.4685E+02
7	0.93	44.25	31.44	47.28	0.3388E+02	0.2789E+02	0.2454E+02
9	1.49	55.62	45.25	57.26	0.1355E+02	0.2183E+02	0.1297E+02
11	2.24	68.27	59.47	64.71	0.1685E+02	0.1672E+02	0.8123E+01
13	4.48	80.91	88.07	82.60	0.5438E+01	0.1005E+02	0.6717E+01
15	7.46	96.08	112.58	97.01	0.3689E+01	0.6854E+01	0.3277E+01
17	11.19	107.45	134.44	106.50	0.2684E+01	0.5079E+01	0.2226E+01
19	14.92	113.77	151.43	114.61	0.1765E+01	0.4117E+01	0.2110E+01
21	18.66	121.36	165.57	122.00	0.1693E+01	0.3503E+01	0.1832E+01
23	22.39	127.68	177.80	128.30	0.1356E+01	0.3074E+01	0.1557E+01
25	26.12	134.00	188.65	133.78	0.1694E+01	0.2754E+01	0.1402E+01
27	29.85	139.06	198.44	138.88	0.1355E+01	0.2505E+01	0.1348E+01
29	33.58	144.11	207.39	143.90	0.1353E+01	0.2304E+01	0.1347E+01
31	37.31	149.17	215.75	148.96	0.1358E+01	0.2138E+01	0.1365E+01
33	41.04	154.23	223.45	154.09	0.1355E+01	0.1999E+01	0.1385E+01
35	44.77	159.28	230.68	159.29	0.1355E+01	0.1881E+01	0.1401E+01
37	48.50	164.34	237.50	164.54	0.1356E+01	0.1779E+01	0.1412E+01
39	51.86	169.40	243.31	169.29	0.1697E+01	0.1697E+01	0.1418E+01

IV - 55

Connection type : Top-and seat-angle connections
Mode : All bolted

Tested by : A.Azizinamini (1985) U.S.A.
Test Id. : ---

Column : W12x96
Beam : W14x38
Angle : L6x4x3/8

Fasteners: A325- -7/8"D
7/8" Oversize holes
Material : --
Fy = 53.00 ksi
Fu = 80.00 ksi

Major parameters

lt = 8.0000"	tt = 0.3750"	ls = 8.0000"	ts = 0.3750"
gt = 3.0000"	gt'= 2.5000"	gs = 3.0000"	gs'= 2.5000"
qt = 3.5000"	rt = 5.5000"	qs = 3.5000"	rs = 5.5000"
pt = 2.5000"	ps = 2.5000"		
nt = 2 X 2	nt'= 2 X 1	ns = 2 X 2	ns'= 2 X 1

Remark 1)
2)

No	Moment (k-in)	Rotation (radians) X 1/1000	No	Moment (k-in)	Rotation (radians) X 1/1000
1	0.00	0.00	26	544.00	17.50
2	19.90	0.08	27	562.00	19.10
3	52.00	0.39	28	571.00	20.80
4	83.20	0.57	29	583.00	21.80
5	106.00	0.88	30	592.00	23.00
6	131.00	1.06	31	605.00	24.40
7	160.00	1.39	32	622.00	26.10
8	206.00	1.84			
9	232.00	2.19			
10	259.00	2.57			
11	283.00	2.96			
12	302.00	3.15			
13	318.00	3.55			
14	351.00	4.21			
15	373.00	4.85			
16	396.00	5.66			
17	412.00	6.46			
18	428.00	7.20			
19	445.00	7.99			
20	459.00	9.13			
21	479.00	10.50			
22	492.00	12.30			
23	504.00	13.00			
24	517.00	14.80			
25	531.00	16.10			

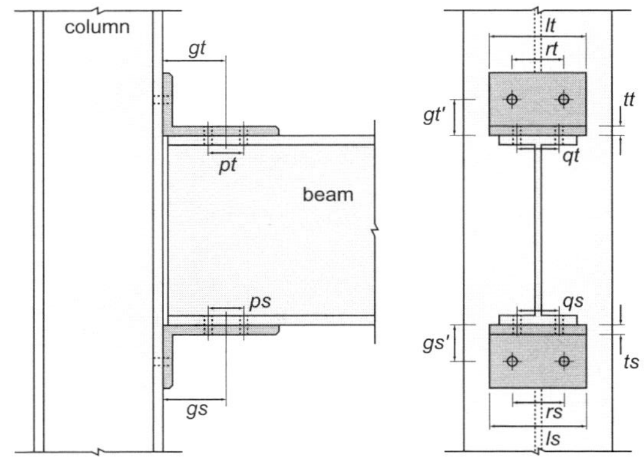

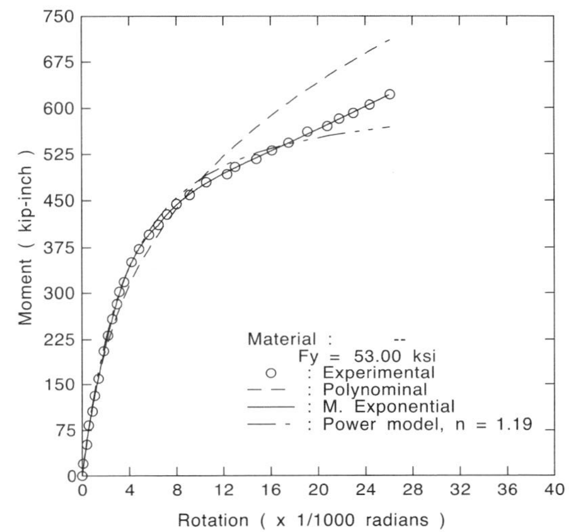

```
Moment-rotation prediction equations      ( R : X 1/1000 radians )

Frye and Morris polynominal model  :     R = Sum ( Ai X ( K*Bm )**Pi X 10**Qi )
     t =  0.375000"  xd = 14.100000"   f =  0.875000"  xl =  8.000000"
    A1 =  8.460000   A2 =  1.010000   A3 =  1.240000    K =  0.008333
    P1 =         1   P2 =         3   P3 =         5   Q1 =        -1     Q2 =       -1     Q3 =       -5

Modified exponential model  :
    Bm = Sum ( Ai X ( 1 - exp( - R/( 2*i*C ) ) ) )  +      Sum ( RKj X ( R - Rj0 ) )  +  Bm0
     C =  0.23775000E+00            Bmo=  0.00000000E+00          Nexp=  6        Nliner=  1
    AI =  0.10389105E+03  -0.53606330E+03  0.15605981E+04  -0.38889913E+04  0.50007319E+04  -0.17996149E+04
   Rj0 =        6.4600
   RKj =  0.92058973E+01

Power model  :    Bm = ( Rki X R ) / ( 1 + ( R/R0 )**rn )**( 1/rn )
    rn = 1.190       rki =   0.15175314E+03          rmu =   0.62120608E+03
```

No	Rotation (radians) X 1/1000	Moment (kip-inch) Expri.	Poly.	M.Expo.	P.Model.	Connection stiffness (kip-inch) X 1000 Expri.	Poly.	M.Expo.	P.Model.
1	0.00	0.00	0.00	0.00	0.00	0.2457E+03	0.1419E+03	0.1767E+03	0.1518E+03
2	0.08	19.90	11.48	13.63	12.20	0.2162E+03	0.1414E+03	0.1606E+03	0.1492E+03
3	0.39	52.00	53.49	56.26	55.77	0.1441E+03	0.1324E+03	0.1241E+03	0.1363E+03
4	0.57	83.20	77.31	78.24	80.36	0.1325E+03	0.1235E+03	0.1131E+03	0.1282E+03
5	0.88	106.00	112.59	111.06	117.50	0.1134E+03	0.1079E+03	0.1033E+03	0.1155E+03
6	1.06	131.00	131.51	129.61	137.99	0.1192E+03	0.9919E+02	0.9958E+02	0.1085E+03
7	1.39	160.00	161.97	161.55	171.83	0.9395E+02	0.8584E+02	0.9408E+02	0.9684E+02
8	1.84	206.00	197.34	202.17	212.25	0.8651E+02	0.7205E+02	0.8630E+02	0.8323E+02
9	2.19	232.00	221.07	231.20	239.77	0.7274E+02	0.6401E+02	0.7950E+02	0.7423E+02
10	2.57	259.00	244.04	259.93	266.33	0.6636E+02	0.5715E+02	0.7166E+02	0.6581E+02
11	2.96	283.00	265.20	286.30	290.52	0.8740E+02	0.5157E+02	0.6357E+02	0.5843E+02
12	3.15	302.00	274.77	298.01	301.32	0.8068E+02	0.4926E+02	0.5972E+02	0.5523E+02
13	3.55	318.00	293.61	320.34	322.17	0.4377E+02	0.4508E+02	0.5200E+02	0.4921E+02
14	4.21	351.00	321.51	350.83	351.86	0.4207E+02	0.3969E+02	0.4073E+02	0.4109E+02
15	4.85	373.00	345.57	373.91	376.10	0.3174E+02	0.3569E+02	0.3169E+02	0.3487E+02
16	5.66	396.00	372.83	395.78	401.75	0.2417E+02	0.3178E+02	0.2272E+02	0.2873E+02
17	6.46	412.00	397.01	411.18	422.78	0.2084E+02	0.2878E+02	0.1611E+02	0.2405E+02
18	7.20	428.00	417.45	428.15	439.26	0.2157E+02	0.2654E+02	0.2076E+02	0.2062E+02
19	7.99	445.00	437.61	443.05	454.35	0.1774E+02	0.2456E+02	0.1716E+02	0.1768E+02
20	9.13	459.00	464.24	460.45	472.53	0.1333E+02	0.2225E+02	0.1365E+02	0.1438E+02
21	10.50	479.00	493.17	477.31	490.14	0.1141E+02	0.2007E+02	0.1120E+02	0.1147E+02
22	12.30	492.00	527.22	495.90	508.21	0.1437E+02	0.1787E+02	0.9676E+01	0.8783E+01
23	13.00	504.00	539.47	502.56	514.07	0.1437E+02	0.1716E+02	0.9381E+01	0.7978E+01
24	14.80	517.00	568.91	519.06	526.88	0.9282E+01	0.1561E+02	0.9026E+01	0.6338E+01
25	16.10	531.00	588.59	530.74	534.51	0.1005E+02	0.1469E+02	0.8964E+01	0.5439E+01
26	17.50	544.00	608.54	543.29	541.56	0.1020E+02	0.1383E+02	0.8974E+01	0.4663E+01
27	19.10	562.00	629.97	557.69	548.43	0.8362E+01	0.1298E+02	0.9020E+01	0.3958E+01
28	20.80	571.00	651.38	573.07	554.64	0.9516E+01	0.1221E+02	0.9074E+01	0.3365E+01
29	21.80	583.00	663.59	582.15	557.86	0.9955E+01	0.1180E+02	0.9101E+01	0.3075E+01
30	23.00	592.00	677.28	593.09	561.36	0.8324E+01	0.1137E+02	0.9128E+01	0.2772E+01
31	24.40	605.00	692.86	605.89	565.03	0.9608E+01	0.1090E+02	0.9152E+01	0.2471E+01
32	26.10	622.00	710.94	621.47	568.96	0.1000E+02	0.1039E+02	0.9172E+01	0.2165E+01

A.4 – 111

IV - 56

Connection type : Top-and seat-angle connections
Mode : All bolted

Tested by : A.Azizinamini (1985) U.S.A.
Test Id. : ---

Column : W12x96
Beam : W14x38
Angle : L6x4x1/2

Fasteners: A325- -7/8"D
7/8" Oversize holes
Material : --
Fy = 39.55 ksi
Fu = 67.55 ksi

Major parameters

lt = 8.0000"	tt = 0.5000"	ls = 8.0000"	ts = 0.5000"
gt = 3.0000"	gt'= 2.5000"	gs = 3.0000"	gs'= 2.5000"
qt = 3.5000"	rt = 5.5000"	qs = 3.5000"	rs = 5.5000"
pt = 2.5000"	ps = 2.5000"		
nt = 2 X 2	nt'= 2 X 1	ns = 2 X 2	ns'= 2 X 1

Remark 1)
2)

No	Moment (k-in)	Rotation (radians) X 1/1000
1	0.00	0.00
2	118.00	0.02
3	142.00	0.10
4	175.00	0.10
5	206.00	0.20
6	248.00	0.25
7	283.00	0.41
8	310.00	0.47
9	341.00	0.66
10	387.00	0.97
11	437.00	1.53
12	473.00	2.06
13	516.00	3.25
14	555.00	4.68
15	581.00	6.31
16	608.00	7.96
17	629.00	9.44
18	661.00	11.30
19	676.00	12.60
20	694.00	13.80
21	716.00	15.50
22	750.00	17.80
23	785.00	20.00
24	805.00	21.50
25	825.00	23.20
26	847.00	24.90

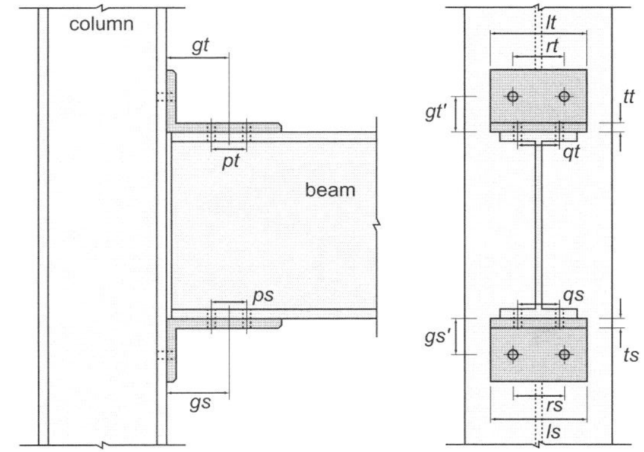

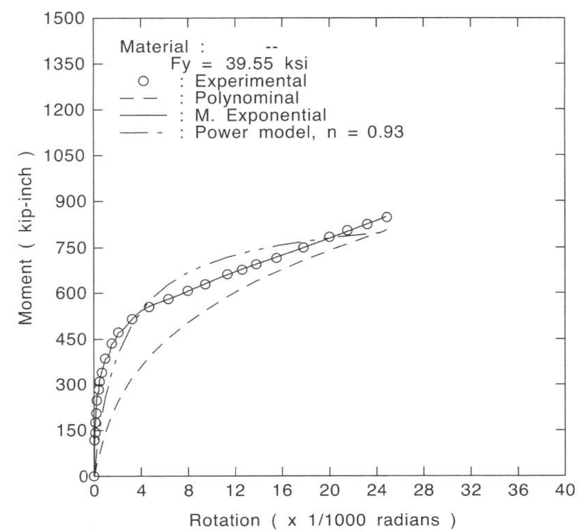

A.4 – 112

```
Moment-rotation prediction equations      ( R : X 1/1000 radians )

Frye and Morris polynominal model   :     R = Sum ( Ai X ( K*Bm )**Pi X 10**Qi )
      t =  0.500000"   xd = 14.100000"   f =  0.875000"   xl =  8.000000"
     A1 =  8.460000    A2 =  1.010000   A3 =  1.240000     K =  0.007216
     P1 =         1    P2 =         3   P3 =         5    Q1 =        -1     Q2 =        -1     Q3 =        -5

Modified exponential model  :
     Bm = Sum ( Ai X ( 1 - exp( - R/( 2*i*C ) ) ) )  +      Sum ( RKj X ( R - Rj0 ) )  +  Bm0
      C =   0.21275000E+00             Bmo=   0.00000000E+00             Nexp=  6            Nliner=  6
     AI =   0.35769220E+04  -0.16129222E+05   0.37458815E+05  -0.50995838E+05   0.36657703E+05  -0.10594418E+05
    Rj0 =           0.4100           0.4800           0.6600           0.9700           1.5300
                    2.0600
    RKj =   0.71486884E+03  -0.26968864E+03   0.66416616E+02  -0.18389540E+03  -0.19232220E+03
           -0.12083787E+03

Power model  :     Bm = ( Rki X R ) / ( 1 + ( R/R0 )**rn )**( 1/rn )
      rn = 0.927      rki =    0.39738425E+03          rmu =    0.89254120E+03
```

No	Rotation (radians) X 1/1000	Moment (kip-inch) Expri.	Poly.	M.Expo.	P.Model.	Connection stiffness (kip-inch) X 1000 Expri.	Poly.	M.Expo.	P.Model.
1	0.00	0.00	0.00	0.00	0.00	0.5619E+04	0.1638E+03	0.1916E+04	0.3974E+03
2	0.02	118.00	3.44	38.23	8.23	0.4523E+04	0.1638E+03	0.1727E+04	0.3867E+03
3	0.10	142.00	16.68	152.00	38.18	0.3260E+05	0.1630E+03	0.1111E+04	0.3542E+03
4	0.10	175.00	16.68	152.00	38.18	0.3265E+05	0.1630E+03	0.1111E+04	0.3542E+03
5	0.20	206.00	31.74	229.70	69.66	0.5740E+03	0.1608E+03	0.5897E+03	0.3236E+03
6	0.25	248.00	41.01	256.54	87.94	0.5877E+03	0.1588E+03	0.3451E+03	0.3070E+03
7	0.41	283.00	64.66	273.34	131.69	0.3367E+03	0.1520E+03	-0.7867E+02	0.2700E+03
8	0.47	310.00	75.17	310.00	150.07	0.3258E+03	0.1482E+03	0.5200E+03	0.2554E+03
9	0.66	341.00	101.87	341.00	194.62	0.1601E+03	0.1372E+03	0.1631E+03	0.2223E+03
10	0.97	387.00	140.99	387.00	255.87	0.1290E+03	0.1195E+03	0.1711E+03	0.1815E+03
11	1.53	437.00	200.62	437.00	342.87	0.7805E+02	0.9355E+02	0.1901E+03	0.1318E+03
12	2.06	473.00	245.50	472.74	404.32	0.5813E+02	0.7709E+02	0.1207E+03	0.1021E+03
13	3.25	516.00	322.90	515.86	500.35	0.3211E+02	0.5559E+02	0.4207E+02	0.6392E+02
14	4.68	555.00	391.93	556.54	573.69	0.2198E+02	0.4233E+02	0.1762E+02	0.4121E+02
15	6.31	581.00	453.49	580.21	628.67	0.1616E+02	0.3381E+02	0.1424E+02	0.2764E+02
16	7.96	608.00	504.50	605.84	667.26	0.1522E+02	0.2844E+02	0.1652E+02	0.1981E+02
17	9.44	629.00	543.94	630.41	693.06	0.1553E+02	0.2506E+02	0.1637E+02	0.1535E+02
18	11.30	661.00	587.51	659.53	717.89	0.1387E+02	0.2195E+02	0.1492E+02	0.1161E+02
19	12.60	676.00	614.91	678.36	731.73	0.1334E+02	0.2027E+02	0.1412E+02	0.9767E+01
20	13.80	694.00	638.43	695.04	742.62	0.1415E+02	0.1897E+02	0.1373E+02	0.8433E+01
21	15.50	716.00	669.33	718.21	755.66	0.1372E+02	0.1743E+02	0.1361E+02	0.6972E+01
22	17.80	750.00	707.45	749.72	769.94	0.1536E+02	0.1577E+02	0.1382E+02	0.5536E+01
23	20.00	785.00	740.70	780.43	780.98	0.1438E+02	0.1451E+02	0.1409E+02	0.4546E+01
24	21.50	805.00	761.90	801.68	787.39	0.1260E+02	0.1377E+02	0.1424E+02	0.4017E+01
25	23.20	825.00	784.68	825.98	793.79	0.1235E+02	0.1304E+02	0.1435E+02	0.3524E+01
26	24.90	847.00	806.29	850.45	799.42	0.1294E+02	0.1240E+02	0.1443E+02	0.3117E+01

IV - 57

Connection type : Top-and seat-angle connections
Mode : All bolted

Tested by : W.L.Harper,Jr. (1990) U.S.A.
Test Id. : TEST3

Column : W8x24
Beam : W8x21
Angle : L6x3.5x3/8

Fasteners: A325- -7/8"D
7/8" Oversize holes
Material : --
Fy = 43.00 ksi
Fu = 75.00 ksi

Major parameters

lt = 6.0000"	tt = 0.3750"	ls = 6.0000"	ts = 0.3750"
gt = 3.0000"	gt'= 2.0000"	gs = 3.0000"	gs'= 2.0000"
qt = 2.7500"	rt = 3.5000"	qs = 2.7500"	rs = 3.5000"
pt = 2.5000"	ps = 2.5000"		
nt = 2 X 2	nt'= 2 X 1	ns = 2 X 2	ns'= 2 X 1

Remark 1)
2)

No	Moment (k-in)	Rotation (radians) X 1/1000	No	Moment (k-in)	Rotation (radians) X 1/1000
1	0.00	0.00	26	364.00	39.30
2	19.00	0.16	27	367.00	41.30
3	35.70	0.43	28	374.00	43.50
4	49.60	0.75	29	383.00	46.80
5	76.70	1.32	30	388.00	51.00
6	110.00	2.11	31	392.00	53.30
7	136.00	2.85	32	399.00	56.40
8	155.00	3.58			
9	170.00	4.23			
10	195.00	5.34			
11	219.00	6.63			
12	234.00	7.51			
13	254.00	8.60			
14	269.00	9.88			
15	280.00	11.40			
16	287.00	13.80			
17	298.00	16.10			
18	306.00	18.40			
19	315.00	21.10			
20	320.00	23.90			
21	332.00	26.70			
22	341.00	29.30			
23	346.00	32.10			
24	352.00	34.60			
25	358.00	37.00			

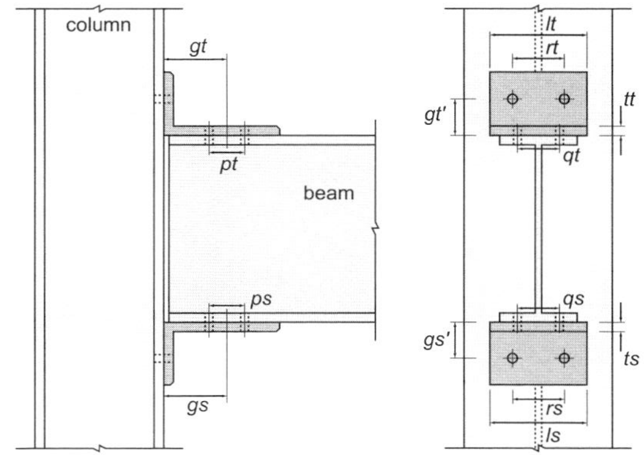

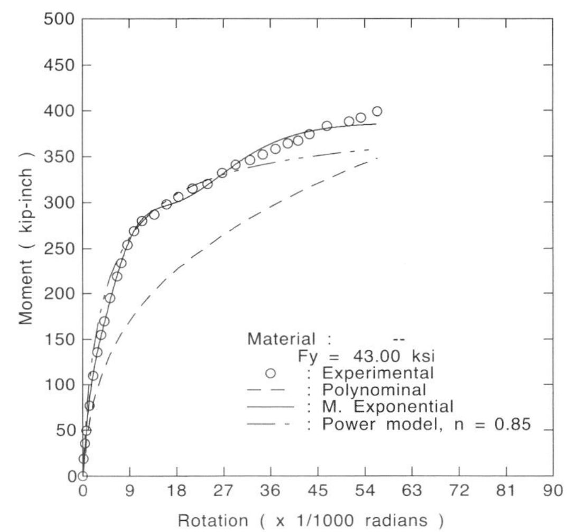

A.4 – 114

```
Moment-rotation prediction equations     ( R : X 1/1000 radians )

Frye and Morris polynominal model  :    R = Sum ( Ai X ( K*Bm )**Pi X 10**Qi )
     t =  0.375000"  xd =  8.280000"   f =  0.875000"  xl =  6.000000"
    A1 =  8.460000   A2 =  1.010000   A3 =  1.240000    K =  0.022648
    P1 =        1    P2 =        3    P3 =        5    Q1 =       -1      Q2 =      -1      Q3 =       -5

Modified exponential model  :
    Bm = Sum ( Ai X ( 1 - exp( - R/( 2*i*C ) ) ) )  +      Sum ( RKj X ( R - Rj0 ) )  +  Bm0
     C =   0.50975000E+00               Bmo=   0.00000000E+00          Nexp=  6         Nliner=  0
    AI =  -0.31788030E+03   0.62234952E+04  -0.32871795E+05   0.73378854E+05  -0.73121474E+05   0.27095389E+05

Power model  :    Bm = ( Rki X R ) / ( 1 + ( R/R0 )**rn )**( 1/rn )
    rn = 0.855      rki =   0.12029948E+03         rmu =   0.39501354E+03
```

No	Rotation (radians) X 1/1000	Moment (kip-inch) Expri.	Moment Poly.	Moment M.Expo.	Moment P.Model.	Connection stiffness (kip-inch) X 1000 Expri.	Stiffness Poly.	Stiffness M.Expo.	Stiffness P.Model.
1	0.00	0.00	0.00	0.00	0.00	0.1195E+03	0.5219E+02	0.7152E+02	0.1203E+03
2	0.16	19.00	8.26	11.34	17.57	0.9803E+02	0.5155E+02	0.7084E+02	0.1028E+03
3	0.43	35.70	21.90	30.22	42.96	0.5316E+02	0.4796E+02	0.6706E+02	0.8452E+02
4	0.75	49.60	36.19	50.47	67.33	0.4514E+02	0.4207E+02	0.6049E+02	0.7008E+02
5	1.32	76.70	57.34	81.30	102.07	0.4523E+02	0.3253E+02	0.4761E+02	0.5300E+02
6	2.11	110.00	79.43	113.07	137.83	0.3853E+02	0.2417E+02	0.3389E+02	0.3876E+02
7	2.85	136.00	95.48	135.20	163.20	0.3055E+02	0.1950E+02	0.2674E+02	0.3036E+02
8	3.58	155.00	108.53	153.38	183.15	0.2447E+02	0.1648E+02	0.2356E+02	0.2464E+02
9	4.23	170.00	118.58	168.28	197.89	0.2287E+02	0.1455E+02	0.2245E+02	0.2087E+02
10	5.34	195.00	133.36	192.70	218.32	0.2071E+02	0.1221E+02	0.2158E+02	0.1625E+02
11	6.63	219.00	147.86	219.50	236.81	0.1768E+02	0.1039E+02	0.1971E+02	0.1265E+02
12	7.51	234.00	156.57	235.95	247.13	0.1763E+02	0.9464E+01	0.1758E+02	0.1087E+02
13	8.60	254.00	166.37	253.38	258.00	0.1530E+02	0.8557E+01	0.1433E+02	0.9154E+01
14	9.88	269.00	176.77	269.17	268.70	0.9670E+01	0.7722E+01	0.1038E+02	0.7631E+01
15	11.40	280.00	187.90	281.77	279.22	0.5562E+01	0.6950E+01	0.6392E+01	0.6285E+01
16	13.80	287.00	203.43	292.02	292.40	0.3869E+01	0.6044E+01	0.2693E+01	0.4803E+01
17	16.10	298.00	216.56	296.76	302.27	0.4130E+01	0.5404E+01	0.1763E+01	0.3838E+01
18	18.40	306.00	228.40	301.20	310.26	0.3412E+01	0.4907E+01	0.2232E+01	0.3145E+01
19	21.10	315.00	241.00	308.60	317.91	0.2574E+01	0.4448E+01	0.3253E+01	0.2552E+01
20	23.90	320.00	252.90	318.89	324.40	0.3036E+01	0.4068E+01	0.4008E+01	0.2103E+01
21	26.70	332.00	263.85	330.51	329.79	0.3858E+01	0.3760E+01	0.4204E+01	0.1765E+01
22	29.30	341.00	273.30	341.21	334.05	0.2655E+01	0.3520E+01	0.3969E+01	0.1522E+01
23	32.10	346.00	282.84	351.64	338.01	0.2110E+01	0.3299E+01	0.3450E+01	0.1313E+01
24	34.60	352.00	290.87	359.58	341.10	0.2451E+01	0.3129E+01	0.2894E+01	0.1162E+01
25	37.00	358.00	298.20	365.88	343.74	0.2556E+01	0.2984E+01	0.2361E+01	0.1041E+01
26	39.30	364.00	304.92	370.76	346.01	0.2016E+01	0.2860E+01	0.1893E+01	0.9423E+00
27	41.30	367.00	310.64	374.18	347.82	0.2301E+01	0.2760E+01	0.1537E+01	0.8677E+00
28	43.50	374.00	316.51	377.18	349.65	0.3000E+01	0.2663E+01	0.1204E+01	0.7958E+00
29	46.80	383.00	325.07	380.48	352.12	0.2051E+01	0.2530E+01	0.8140E+00	0.7040E+00
30	51.00	388.00	335.38	383.14	354.87	0.1545E+01	0.2382E+01	0.4784E+00	0.6090E+00
31	53.30	392.00	340.86	384.09	356.22	0.1960E+01	0.2309E+01	0.3530E+00	0.5651E+00
32	56.40	399.00	347.80	384.98	357.89	0.2258E+01	0.2221E+01	0.2316E+00	0.5133E+00

A.4 – 115

IV - 58

Connection type : Top-and seat-angle connections
Mode : All bolted

Tested by : J.B. Mander et al. (1994) U.S.A.
Test Id. : R1_01

Column : W8x31
Beam : W8x21
Angle : L6x4x3/8

Fasteners: A325- -3/4"D
7/8" Oversize holes
Material : A36
Fy = 42.30 ksi
Fu = 66.60 ksi

Major parameters

lt = 6.5000"	tt = 0.3750"	ls = 6.5000"	ts = 0.3750"
gt = 3.6250"	gt'= 2.0000"	gs = 3.6250"	gs'= 2.0000"
qt = 2.5000"	rt = 3.5000"	qs = 2.5000"	rs = 3.5000"
pt = 2.3750"	ps = 2.3750"		
nt = 2 X 2	nt'= 2 X 1	ns = 2 X 2	ns'= 2 X 1

Remark 1) 1 in. thick column flange plate to minimize the column flange distortion
2) loading rate 0.4%/s

No	Moment (k-in)	Rotation (radians) X 1/1000
1	0.00	0.00
2	18.50	0.19
3	29.40	0.37
4	51.31	0.73
5	74.27	1.23
6	97.76	1.73
7	121.25	2.32
8	142.06	2.90
9	160.20	3.44
10	176.19	4.03
11	191.11	4.63
12	206.02	5.27
13	221.97	6.09
14	235.26	6.87
15	248.53	7.73
16	260.18	8.79
17	270.21	9.92
18	278.64	11.02
19	285.99	12.14
20	291.23	13.01
21	295.94	13.83
22	301.73	14.63
23	305.89	15.60
24	310.57	16.73

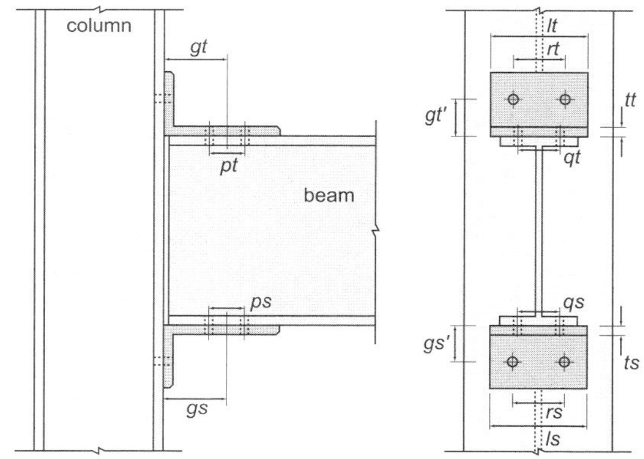

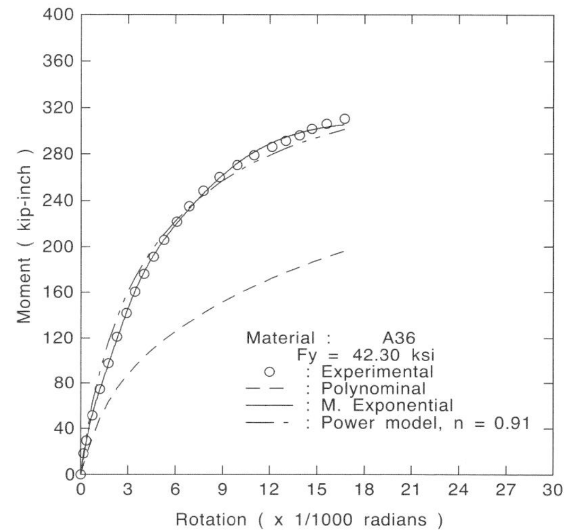

```
Moment-rotation prediction equations      ( R : X 1/1000 radians )

Frye and Morris polynominal model  :    R = Sum ( Ai X ( K*Bm )**Pi X 10**Qi )
      t =  0.375000"  xd =  8.280000"   f =  0.750000"  xl =  6.500000"
     A1 =  8.460000   A2 =  1.010000   A3 =  1.240000    K =  0.025371
     P1 =         1   P2 =         3   P3 =         5   Q1 =        -1     Q2 =        -1     Q3 =        -5

Modified exponential model  :
     Bm = Sum ( Ai X ( 1 - exp( - R/( 2*i*C ) ) ) )  +      Sum ( RKj X ( R - Rj0 ) )  + Bm0
       C =   0.18807333E+00               Bmo=   0.00000000E+00            Nexp=  6          Nliner=  0
      AI =  -0.10110810E+03   0.21959638E+04  -0.12327942E+05   0.29711134E+05  -0.32600490E+05   0.13431746E+05

Power model  :     Bm = ( Rki X R ) / ( 1 + ( R/R0 )**rn )**( 1/rn )
      rn = 0.907        rki =   0.10314466E+03           rmu =   0.38797606E+03
```

No	Rotation (radians) X 1/1000	Moment (kip-inch) Expri.	Poly.	M.Expo.	P.Model.	Connection stiffness (kip-inch) X 1000 Expri.	Poly.	M.Expo.	P.Model.
1	0.00	0.00	0.00	0.00	0.00	0.9531E+02	0.4659E+02	0.9004E+02	0.1031E+03
2	0.19	18.50	8.99	16.63	18.62	0.7779E+02	0.4574E+02	0.8025E+02	0.8983E+02
3	0.37	29.40	16.88	29.78	33.64	0.6000E+02	0.4372E+02	0.6902E+02	0.8096E+02
4	0.76	51.31	32.83	52.82	62.33	0.5268E+02	0.3732E+02	0.5050E+02	0.6616E+02
5	1.23	74.27	48.61	74.23	90.30	0.4619E+02	0.3016E+02	0.4268E+02	0.5376E+02
6	1.78	97.76	63.35	97.16	116.72	0.4301E+02	0.2419E+02	0.4176E+02	0.4356E+02
7	2.32	121.25	75.39	119.75	138.40	0.3981E+02	0.2016E+02	0.4058E+02	0.3617E+02
8	2.90	142.06	86.00	141.95	157.30	0.3487E+02	0.1721E+02	0.3685E+02	0.3038E+02
9	3.44	160.20	94.77	160.69	172.62	0.3028E+02	0.1516E+02	0.3198E+02	0.2612E+02
10	4.03	176.19	103.25	178.09	187.00	0.2603E+02	0.1346E+02	0.2676E+02	0.2245E+02
11	4.63	191.11	110.83	192.66	199.42	0.2418E+02	0.1214E+02	0.2252E+02	0.1954E+02
12	5.27	206.02	118.28	206.05	211.16	0.2149E+02	0.1101E+02	0.1927E+02	0.1699E+02
13	6.09	221.97	126.86	220.76	224.05	0.1827E+02	0.9876E+01	0.1669E+02	0.1442E+02
14	6.87	235.26	134.14	233.01	234.42	0.1609E+02	0.9033E+01	0.1517E+02	0.1253E+02
15	7.76	248.53	141.89	245.99	244.85	0.1317E+02	0.8241E+01	0.1374E+02	0.1076E+02
16	8.79	260.18	149.95	259.24	255.05	0.1019E+02	0.7516E+01	0.1207E+02	0.9184E+01
17	9.92	270.21	158.05	271.71	264.60	0.8270E+01	0.6874E+01	0.1004E+02	0.7826E+01
18	11.02	278.64	165.32	281.65	272.62	0.7099E+01	0.6362E+01	0.8032E+01	0.6778E+01
19	12.14	285.99	172.23	289.61	279.75	0.6247E+01	0.5924E+01	0.6134E+01	0.5914E+01
20	13.01	291.23	177.25	294.37	284.64	0.5722E+01	0.5632E+01	0.4860E+01	0.5358E+01
21	13.88	295.94	182.03	298.11	289.09	0.6398E+01	0.5372E+01	0.3782E+01	0.4878E+01
22	14.68	301.73	186.21	300.78	292.80	0.6014E+01	0.5159E+01	0.2966E+01	0.4496E+01
23	15.60	305.89	190.85	303.15	296.76	0.4320E+01	0.4937E+01	0.2207E+01	0.4107E+01
24	16.75	310.57	196.39	305.25	301.23	0.4062E+01	0.4689E+01	0.1497E+01	0.3690E+01

IV - 59

Connection type : Top-and seat-angle connections
Mode : All bolted

Tested by : J.B. Mander et al. (1994) U.S.A.
Test Id. : R1_05

Column : W8x31
Beam : W8x21
Angle : L6x4x3/8

Fasteners: A325- -3/4"D
7/8" Oversize holes
Material : A36
Fy = 42.30 ksi
Fu = 66.60 ksi

Major parameters

lt = 6.5000"	tt = 0.3750"	ls = 6.5000"	ts = 0.3750"
gt = 3.6250"	gt' = 2.0000"	gs = 3.6250"	gs' = 2.0000"
qt = 2.5000"	rt = 3.5000"	qs = 2.5000"	rs = 3.5000"
pt = 2.3750"	ps = 2.3750"		
nt = 2 X 2	nt' = 2 X 1	ns = 2 X 2	ns' = 2 X 1

Remark 1) 1 in. thick column flange plate to minimize the column flange distortion
2) loading rate 2.4%/s

No	Moment (k-in)	Rotation (radians) X 1/1000
1	0.00	0.00
2	62.72	0.46
3	117.60	0.86
4	161.27	1.22
5	209.41	1.59
6	259.77	2.10
7	307.85	2.78
8	355.96	3.52
9	400.66	4.74
10	434.14	6.11
11	465.33	7.87
12	494.26	9.96
13	516.45	12.20
14	538.67	14.12
15	564.19	16.68
16	590.82	19.40
17	614.05	22.43
18	640.65	25.39
19	661.68	28.03
20	684.92	30.91
21	703.71	33.62
22	718.01	36.34
23	733.45	38.81
24	750.02	41.21
25	764.34	43.68
26	774.19	46.08
27	781.78	48.71

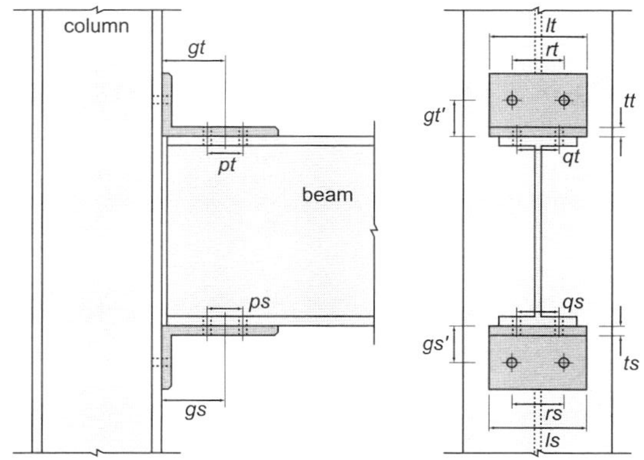

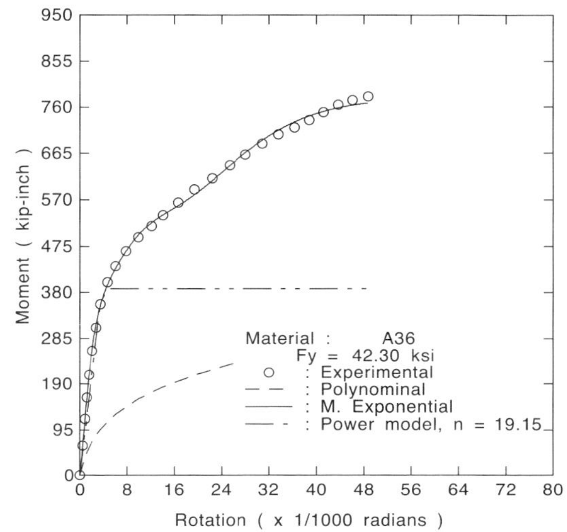

```
Moment-rotation prediction equations      ( R : X 1/1000 radians )

Frye and Morris polynominal model  :    R = Sum ( Ai X ( K*Bm )**Pi X 10**Qi )
      t =  0.375000"  xd =  8.280000"   f =  0.750000"  xl =  6.500000"
     A1 =  8.460000   A2 =  1.010000   A3 =  1.240000    K =  0.025371
     P1 =         1   P2 =         3   P3 =         5   Q1 =        -1    Q2 =        -1    Q3 =        -5

Modified exponential model  :
     Bm = Sum ( Ai X ( 1 - exp( - R/( 2*i*C ) ) ) )  +     Sum ( RKj X ( R - Rj0 ) )  +  Bm0
       C =   0.51984500E+00             Bmo=   0.00000000E+00          Nexp=  6        Nliner=  0
      AI =  -0.98306755E+03   0.12345926E+05  -0.58255081E+05   0.13045142E+06  -0.13440920E+06   0.51628781E+05

Power model  :    Bm = ( Rki X R ) / ( 1 + ( R/R0 )**rn )**( 1/rn )
      rn =19.155      rki =   0.10314466E+03        rmu =   0.38797606E+03
```

No	Rotation (radians) X 1/1000	Moment (kip-inch) Expri.	Moment (kip-inch) Poly.	Moment (kip-inch) M.Expo.	Moment (kip-inch) P.Model.	Connection stiffness (kip-inch) X 1000 Expri.	Connection stiffness (kip-inch) X 1000 Poly.	Connection stiffness (kip-inch) X 1000 M.Expo.	Connection stiffness (kip-inch) X 1000 P.Model.
1	0.00	0.00	0.00	0.00	0.00	0.1376E+03	0.4659E+02	0.1033E+03	0.1031E+03
2	0.46	62.72	20.57	57.56	47.01	0.1373E+03	0.4245E+02	0.1405E+03	0.1031E+03
3	0.86	117.60	36.26	114.75	88.34	0.1287E+03	0.3575E+02	0.1416E+03	0.1031E+03
4	1.22	161.27	48.13	163.98	125.50	0.1252E+03	0.3037E+02	0.1305E+03	0.1031E+03
5	1.59	209.41	58.59	209.59	163.90	0.1163E+03	0.2600E+02	0.1141E+03	0.1031E+03
6	2.10	259.77	70.72	261.88	216.70	0.8652E+02	0.2163E+02	0.9043E+02	0.1031E+03
7	2.78	307.85	84.00	313.89	286.70	0.6802E+02	0.1773E+02	0.6404E+02	0.1028E+03
8	3.52	355.96	96.00	353.48	358.44	0.5429E+02	0.1490E+02	0.4444E+02	0.7947E+02
9	4.74	400.66	112.18	396.20	387.73	0.3096E+02	0.1193E+02	0.2841E+02	0.9777E+00
10	6.11	434.14	126.99	430.08	387.97	0.2148E+02	0.9859E+01	0.2217E+02	0.5916E-02
11	7.87	465.33	142.80	465.72	387.98	0.1593E+02	0.8155E+01	0.1829E+02	0.3529E-04
12	9.96	494.26	158.34	498.90	387.98	0.1197E+02	0.6853E+01	0.1351E+02	0.3105E-06
13	12.20	516.45	172.56	523.98	387.98	0.1080E+02	0.5904E+01	0.9224E+01	0.5200E-08
14	14.12	538.67	183.30	539.71	387.98	0.1088E+02	0.5306E+01	0.7453E+01	0.2725E-09
15	16.68	564.19	196.07	558.39	387.98	0.9882E+01	0.4702E+01	0.7492E+01	0.9480E-11
16	19.40	590.82	208.17	580.52	387.98	0.8779E+01	0.4217E+01	0.8855E+01	0.4514E-12
17	22.43	614.05	220.32	609.54	387.98	0.8330E+01	0.3800E+01	0.1011E+02	0.2408E-13
18	25.39	640.65	231.07	639.98	387.98	0.8453E+01	0.3478E+01	0.1028E+02	0.1983E-14
19	28.03	661.68	239.92	666.35	387.98	0.8025E+01	0.3242E+01	0.9609E+01	0.2706E-15
20	30.91	684.92	248.93	692.24	387.98	0.7481E+01	0.3025E+01	0.8322E+01	0.3776E-16
21	33.62	703.71	256.90	712.92	387.98	0.6092E+01	0.2850E+01	0.6897E+01	0.6915E-17
22	36.34	718.01	264.42	729.71	387.98	0.5776E+01	0.2698E+01	0.5494E+01	0.1447E-17
23	38.81	733.45	270.95	741.85	387.98	0.6583E+01	0.2576E+01	0.4343E+01	0.3833E-18
24	41.21	750.02	276.99	751.08	387.98	0.6361E+01	0.2470E+01	0.3389E+01	0.1146E-18
25	43.68	764.34	282.98	758.43	387.98	0.4938E+01	0.2370E+01	0.2576E+01	0.3537E-19
26	46.08	774.19	288.54	763.81	387.98	0.3529E+01	0.2283E+01	0.1948E+01	0.1207E-19
27	48.71	781.78	294.43	768.20	387.98	0.2883E+01	0.2196E+01	0.1413E+01	0.3943E-20

IV - 60

Connection type : Top-and seat-angle connections
Mode : All bolted

Tested by : J.B. Mander et al. (1994) U.S.A.
Test Id. : R1_06

Column : W8x31
Beam : W8x21
Angle : L6x4x3/8

Fasteners: A325- -3/4"D
7/8" Oversize holes
Material : A36
Fy = 42.30 ksi
Fu = 66.60 ksi

Major parameters

lt = 6.5000"	tt = 0.3750"	ls = 6.5000"	ts = 0.3750"
gt = 3.6250"	gt'= 2.0000"	gs = 3.6250"	gs'= 2.0000"
qt = 2.5000"	rt = 3.5000"	qs = 2.5000"	rs = 3.5000"
pt = 2.3750"	ps = 2.3750"		
nt = 2 X 2	nt'= 2 X 1	ns = 2 X 2	ns'= 2 X 1

Remark 1) 1 in. thick column flange plate to minimize the column flange distortion
2) loading rate 0.32%/s

No	Moment (k-in)	Rotation (radians) X 1/1000
1	0.00	0.00
2	99.39	0.72
3	163.79	1.22
4	225.37	1.75
5	282.71	2.41
6	337.25	3.31
7	386.08	4.22
8	433.45	6.49
9	472.41	8.76
10	511.22	12.36
11	547.24	15.86
12	574.82	19.77
13	607.96	23.98
14	641.09	28.30
15	672.84	32.41
16	697.57	36.73
17	725.06	41.35
18	753.92	46.28
19	775.90	50.08
20	799.27	53.99
21	819.89	57.48
22	834.89	61.08
23	841.59	63.85
24	846.60	69.08

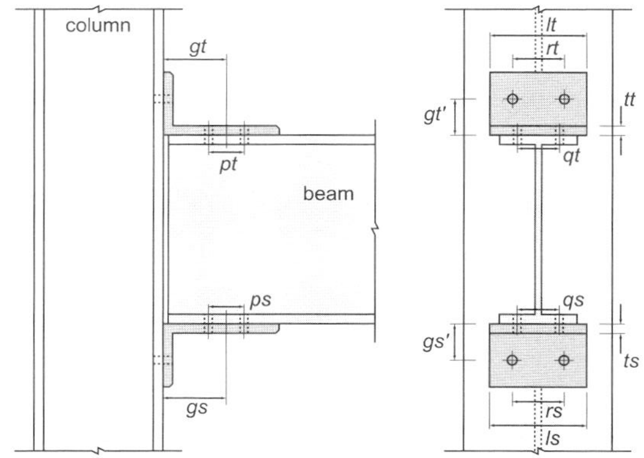

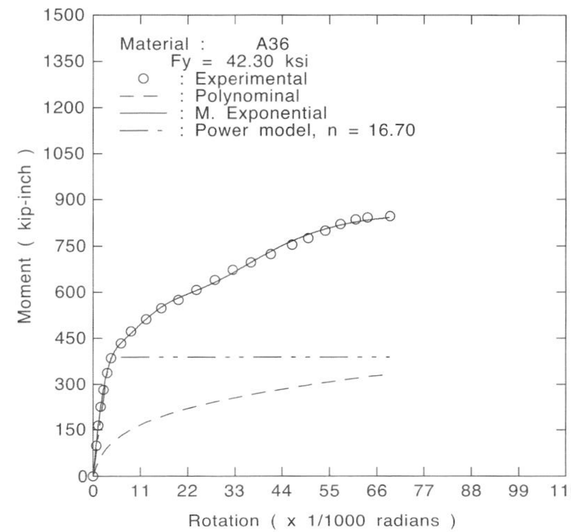

```
Moment-rotation prediction equations      ( R : X 1/1000 radians )

Frye and Morris polynominal model  :    R = Sum ( Ai X ( K*Bm )**Pi X 10**Qi )
      t =  0.375000"  xd =  8.280000"   f =  0.750000"  xl =  6.500000"
     A1 =  8.460000   A2 =  1.010000   A3 =  1.240000    K =  0.025371
     P1 =         1   P2 =         3   P3 =         5   Q1 =        -1    Q2 =        -1    Q3 =        -5

Modified exponential model  :
     Bm = Sum ( Ai X ( 1 - exp( - R/( 2*i*C ) ) ) )  +     Sum ( RKj X ( R - Rj0 ) )  +  Bm0
      C =   0.75507500E+00              Bmo=   0.00000000E+00          Nexp=  6          Nliner=  0
     AI =  -0.10295917E+04   0.14913611E+05  -0.69758410E+05   0.15336245E+06  -0.15619911E+06   0.59566842E+05

Power model  :    Bm = ( Rki X R ) / ( 1 + ( R/R0 )**rn )**( 1/rn )
     rn =16.701       rki =   0.10314466E+03          rmu =   0.38797606E+03
```

No	Rotation (radians) X 1/1000	Moment (kip-inch) Expri.	Moment Poly.	Moment M.Expo.	Moment P.Model.	Connection stiffness (kip-inch) X 1000 Expri.	Stiffness Poly.	Stiffness M.Expo.	Stiffness P.Model.
1	0.00	0.00	0.00	0.00	0.00	0.1385E+03	0.4659E+02	0.1344E+03	0.1031E+03
2	0.72	99.39	31.12	98.49	74.02	0.1325E+03	0.3809E+02	0.1347E+03	0.1031E+03
3	1.22	163.79	48.23	163.08	125.77	0.1228E+03	0.3032E+02	0.1216E+03	0.1031E+03
4	1.75	225.37	62.55	222.43	180.04	0.1034E+03	0.2449E+02	0.1037E+03	0.1031E+03
5	2.41	282.71	77.14	283.68	248.77	0.7520E+02	0.1964E+02	0.8042E+02	0.1031E+03
6	3.31	337.25	92.86	343.62	339.38	0.5716E+02	0.1558E+02	0.5390E+02	0.9149E+02
7	4.22	386.08	105.75	383.45	384.83	0.4434E+02	0.1301E+02	0.3508E+02	0.1159E+02
8	6.49	433.45	130.70	434.98	387.97	0.1902E+02	0.9419E+01	0.1531E+02	0.6570E-02
9	8.76	472.41	149.76	465.77	387.98	0.1470E+02	0.7532E+01	0.1302E+02	0.3266E-04
10	12.36	511.22	173.58	512.64	387.98	0.1053E+02	0.5843E+01	0.1248E+02	0.7356E-07
11	15.86	547.24	192.17	551.01	387.98	0.8767E+01	0.4876E+01	0.9260E+01	0.8934E-09
12	19.77	574.82	209.72	580.78	387.98	0.7445E+01	0.4160E+01	0.6362E+01	0.1814E-10
13	23.98	607.96	226.06	605.70	387.98	0.7769E+01	0.3623E+01	0.5859E+01	0.5925E-12
14	28.30	641.09	240.79	632.94	387.98	0.7697E+01	0.3220E+01	0.6853E+01	0.3162E-13
15	32.41	672.84	253.40	663.13	387.98	0.6749E+01	0.2925E+01	0.7745E+01	0.2864E-14
16	36.73	697.57	265.47	697.28	387.98	0.5834E+01	0.2678E+01	0.7940E+01	0.3134E-15
17	41.35	725.06	277.34	732.83	387.98	0.5900E+01	0.2464E+01	0.7318E+01	0.3842E-16
18	46.28	753.92	289.01	766.06	387.98	0.5813E+01	0.2276E+01	0.6097E+01	0.5228E-17
19	50.08	775.90	297.43	787.21	387.98	0.5883E+01	0.2154E+01	0.5032E+01	0.1293E-17
20	53.99	799.27	305.62	804.78	387.98	0.5941E+01	0.2044E+01	0.3985E+01	0.3423E-18
21	57.48	819.89	312.60	817.22	387.98	0.5050E+01	0.1956E+01	0.3154E+01	0.1128E-18
22	61.08	834.89	319.49	827.21	387.98	0.3181E+01	0.1875E+01	0.2430E+01	0.3856E-19
23	63.85	841.59	324.68	833.28	387.98	0.1911E+01	0.1817E+01	0.1963E+01	0.1758E-19
24	69.08	846.60	333.85	841.67	387.98	0.9587E+00	0.1721E+01	0.1282E+01	0.4361E-20

IV - 61

Connection type : Top-and seat-angle connections
Mode : All bolted

Tested by : J.B. Mander et al. (1994) U.S.A.
Test Id. : R1_11

Column : W8x31
Beam : W8x21
Angle : L6x4x3/8

Fasterers: A325- -3/4"D
7/8" Oversize holes
Material : A36
Fy = 42.30 ksi
Fu = 66.60 ksi

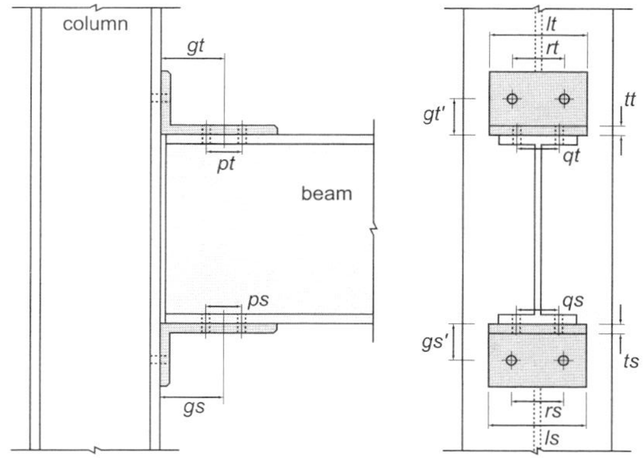

Major parameters

lt = 6.5000"	tt = 0.3750"	ls = 6.5000"	ts = 0.3750"
gt = 3.6250"	gt'= 2.0000"	gs = 3.6250"	gs'= 2.0000"
qt = 2.5000"	rt = 3.5000"	qs = 2.5000"	rs = 3.5000"
pt = 2.3750"	ps = 2.3750"		
nt = 2 X 2	nt'= 2 X 1	ns = 2 X 2	ns'= 2 X 1

Remark 1) 1 in. thick column flange plate to minimize the column flange distortion
2) loading rate 1.6%/s

No	Moment (k-in)	Rotation (radians) X 1/1000
1	0.00	0.00
2	15.35	0.14
3	30.72	0.29
4	44.70	0.42
5	55.17	0.60
6	69.10	0.80
7	86.54	1.06
8	107.49	1.36
9	130.52	1.70
10	145.89	2.00
11	164.68	2.33
12	184.89	2.75
13	208.61	3.21
14	229.54	3.59
15	256.03	4.12
16	278.95	5.17
17	296.90	6.76
18	307.08	8.84
19	315.91	10.56
20	325.38	12.74
21	334.92	14.41
22	343.76	16.13
23	351.13	18.26
24	359.92	20.29
25	366.63	22.19
26	376.17	23.91
27	367.70	24.45
28	373.18	24.50
29	392.03	25.37
30	399.49	26.95
31	410.44	28.58
32	417.88	30.26
33	419.78	31.61
34	427.95	33.11
35	431.90	34.73
36	437.24	36.45
37	442.56	38.30
38	447.89	40.07
39	453.22	41.87
40	453.52	43.82
41	461.07	45.49
42	463.58	46.71
43	470.43	48.38
44	475.83	49.74
45	477.03	51.04
46	481.72	52.40
47	484.96	54.16
48	483.86	56.15
49	494.22	57.73
50	493.83	59.63
51	500.02	61.07

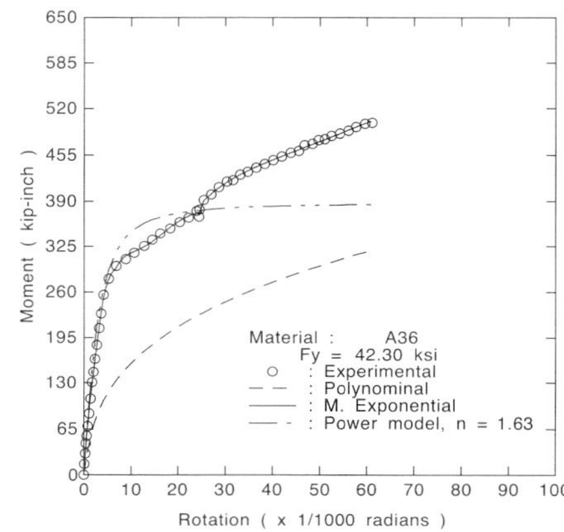

A.4 – 122

```
Moment-rotation prediction equations      ( R : X 1/1000 radians )

Frye and Morris polynominal model  :    R = Sum ( Ai X ( K*Bm )**Pi X 10**Qi )
      t =  0.375000"  xd =  8.280000"   f =  0.750000"  xl =  6.500000"
     A1 =  8.460000   A2 =  1.010000   A3 =  1.240000    K =  0.025371
     P1 =         1   P2 =         3   P3 =         5   Q1 =        -1   Q2 =        -1   Q3 =        -5

Modified exponential model  :
     Bm = Sum ( Ai X ( 1 - exp( - R/( 2*i*C ) ) ) )  +     Sum ( RKj X ( R - Rj0 ) )  +  Bm0
      C =   0.54489500E+00             Bmo=   0.00000000E+00          Nexp=  6        Nliner=  5
     AI =   0.53969977E+03  -0.69895766E+04   0.28455328E+05  -0.48457374E+05   0.36667042E+05  -0.97379111E+04
    Rj0 =           4.1200           6.6700          23.9100          24.4500          25.3700
    RKj =  -0.12965660E+02   0.10133997E+02  -0.37032176E+01   0.22463016E+02  -0.13242216E+02

Power model  :    Bm = ( Rki X R ) / ( 1 + ( R/R0 )**rn )**( 1/rn )
      rn = 1.635      rki =   0.10314466E+03          rmu =   0.38797606E+03
```

No	Rotation (radians) X 1/1000	Moment (kip-inch) Expri.	Poly.	M.Expo.	P.Model.	Connection stiffness (kip-inch) X 1000 Expri.	Poly.	M.Expo.	P.Model.
1	0.00	0.00	0.00	0.00	0.00	0.1067E+03	0.4659E+02	0.1157E+03	0.1031E+03
3	0.29	30.72	13.24	29.66	29.43	0.1067E+03	0.4478E+02	0.9217E+02	0.1007E+03
5	0.60	55.17	26.70	56.15	60.50	0.6296E+02	0.4001E+02	0.7664E+02	0.9530E+02
7	1.06	86.54	43.18	88.13	101.65	0.6855E+02	0.3258E+02	0.6544E+02	0.8518E+02
9	1.70	130.52	61.39	127.76	151.22	0.5937E+02	0.2493E+02	0.5954E+02	0.6993E+02
11	2.33	164.68	75.53	164.44	191.02	0.5306E+02	0.2012E+02	0.5638E+02	0.5619E+02
13	3.21	208.61	91.24	211.59	233.44	0.5324E+02	0.1595E+02	0.5033E+02	0.4101E+02
15	4.12	256.03	104.37	253.29	265.19	0.4088E+02	0.1326E+02	0.4150E+02	0.2984E+02
17	6.76	296.90	133.21	294.15	318.13	0.8498E+01	0.9134E+01	0.1363E+02	0.1304E+02
19	10.56	315.91	162.40	317.66	349.75	0.4792E+01	0.6561E+01	0.3004E+01	0.5164E+01
21	14.41	334.92	184.85	332.23	363.74	0.5417E+01	0.5228E+01	0.4865E+01	0.2526E+01
23	18.26	351.13	203.26	352.45	371.07	0.3903E+01	0.4405E+01	0.5132E+01	0.1428E+01
25	22.19	366.63	219.40	369.00	375.50	0.4586E+01	0.3829E+01	0.3067E+01	0.8807E+00
27	24.45	367.70	227.75	372.29	377.26	0.1730E+03	0.3573E+01	-0.2079E+01	0.6910E+00
29	25.37	392.03	231.03	390.73	377.86	0.1203E+02	0.3479E+01	0.1983E+02	0.6296E+00
31	28.58	410.44	241.70	409.15	379.60	0.5590E+01	0.3197E+01	0.4933E+01	0.4653E+00
33	31.61	419.78	251.10	422.34	380.84	0.3333E+01	0.2976E+01	0.3858E+01	0.3599E+00
35	34.73	431.90	260.10	433.22	381.84	0.2761E+01	0.2784E+01	0.3173E+01	0.2828E+00
37	38.30	442.56	269.71	443.72	382.73	0.2950E+01	0.2598E+01	0.2762E+01	0.2199E+00
39	41.87	453.22	278.69	453.22	383.43	0.2843E+01	0.2441E+01	0.2591E+01	0.1747E+00
41	45.49	461.07	287.24	462.47	384.00	0.4187E+01	0.2303E+01	0.2543E+01	0.1410E+00
43	48.38	470.43	293.71	469.83	384.38	0.2689E+01	0.2207E+01	0.2547E+01	0.1202E+00
45	51.04	477.03	299.53	476.64	384.67	0.2168E+01	0.2125E+01	0.2565E+01	0.1046E+00
47	54.16	484.96	305.97	484.67	384.98	0.1897E+01	0.2039E+01	0.2590E+01	0.8963E-01
49	57.73	494.22	313.17	493.96	385.27	0.2953E+01	0.1949E+01	0.2617E+01	0.7591E-01
51	61.07	500.02	319.50	502.74	385.51	0.8204E+00	0.1875E+01	0.2637E+01	0.6556E-01

IV - 62

Connection type : Top-and seat-angle connections
Mode : All bolted

Tested by : C. Bernuzzi et al. (1996) Italy
Test Id. : TSC/M

Column : --
Beam : IPE300
Angle : L120x120x12

Fasteners: G8.8- -M20
7/8" Oversize holes
Material : A36
Fy = 45.40 ksi
Fu = 66.57 ksi

Major parameters

lt = 7.0866"	tt = 0.4646"	ls = 7.0866"	ts = 0.4646"
gt = 2.3622"	gt'= 2.3622"	gs = 2.3622"	gs'= 2.3622"
qt = -- "	rt = 4.1339"	qs = -- "	rs = 4.1339"
nt = 2 X 1	nt'= 2 X 1	ns = 2 X 1	ns'= 2 X 1

Remark 1) Elastic and Plastic moments of beam are Me,b = 144 kNm, Mp,b = 172 kNm.
2)

No	Moment (k-in)	Rotation (radians) X 1/1000	No	Moment (k-in)	Rotation (radians) X 1/1000
1	0.00	0.00	26	455.07	34.13
2	123.40	1.00	27	477.45	36.14
3	187.93	1.52	28	497.74	38.14
4	216.87	2.01	29	494.68	40.15
5	235.48	3.01	30	511.72	42.16
6	242.66	4.02	31	527.41	44.17
7	241.67	5.02	32	548.07	46.17
8	247.71	6.02	33	550.44	48.18
9	261.92	7.03	34	552.04	50.19
10	298.71	8.03	35	566.49	52.20
11	317.86	9.03	36	568.47	54.21
12	330.12	10.04	37	586.98	56.21
13	350.55	11.04	38	595.53	58.22
14	368.06	12.05	39	604.42	60.23
15	375.10	13.05	40	605.13	62.24
16	376.44	14.05	41	613.93	64.24
17	381.03	16.06	42	629.74	66.25
18	385.43	18.07	43	637.30	68.26
19	392.19	20.08	44	634.23	70.27
20	400.70	22.08	45	644.67	72.27
21	405.31	24.09	46	656.13	74.28
22	406.86	26.10	47	666.01	76.29
23	408.26	28.11	48	673.40	78.30
24	426.35	30.11	49	674.34	79.30
25	451.37	32.12			

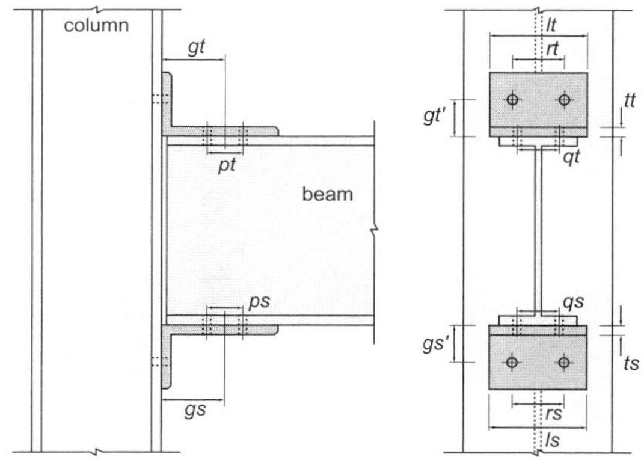

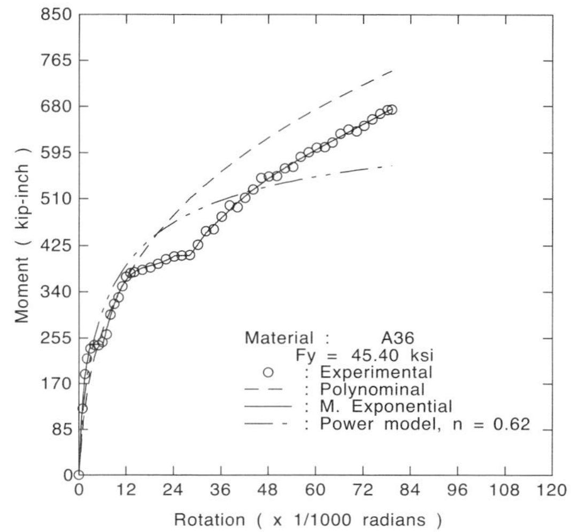

A.4 – 124

```
Moment-rotation prediction equations      ( R : X 1/1000 radians )

Frye and Morris polynominal model   :    R = Sum ( Ai X ( K*Bm )**Pi X 10**Qi )
      t =  0.464570"  xd = 11.811000"   f =  0.787400"  xl =  7.086600"
     A1 =  8.460000   A2 =  1.010000   A3 =  1.240000    K =  0.011938
     P1 =        1    P2 =         3   P3 =         5   Q1 =       -1     Q2 =       -1     Q3 =       -5

Modified exponential model  :
     Bm = Sum ( Ai X ( 1 - exp( - R/( 2*i*C ) ) ) )  +      Sum ( RKj X ( R - Rj0 ) )  +  Bm0
      C =   0.91083333E+00           Bmo=   0.00000000E+00          Nexp=  6        Nliner=  5
     AI =   0.15450686E+04  -0.12910879E+05   0.45054691E+05  -0.71754475E+05   0.53469773E+05  -0.14931887E+05
    Rj0 =           3.0100           6.0600          28.1100          24.4500          32.1200
    RKj =  -0.26454511E+02   0.30030566E+02   0.89656311E+01  -0.52458960E+01  -0.32302575E+01

Power model  :     Bm = ( Rki X R ) / ( 1 + ( R/R0 )**rn )**( 1/rn )
     rn = 0.625      rki =   0.21398471E+03        rmu =   0.70043587E+03
```

No	Rotation (radians) X 1/1000	Moment (kip-inch) Expri.	Poly.	M.Expo.	P.Model.	Connection stiffness (kip-inch) X 1000 Expri.	Poly.	M.Expo.	P.Model.
1	0.00	0.00	0.00	0.00	0.00	0.1234E+03	0.9902E+02	0.2056E+03	0.2140E+03
3	1.52	187.93	120.69	179.42	150.44	0.9061E+02	0.5678E+02	0.6188E+02	0.6113E+02
5	3.01	235.48	186.90	241.52	221.44	0.1289E+02	0.3556E+02	0.3021E+02	0.3775E+02
7	5.02	241.67	245.41	242.92	282.09	0.2522E+01	0.2427E+02	-0.2596E+00	0.2436E+02
9	7.03	261.92	288.21	269.61	323.64	0.2549E+02	0.1886E+02	0.2707E+02	0.1762E+02
11	9.03	317.86	322.49	317.51	354.57	0.1566E+02	0.1566E+02	0.2041E+02	0.1361E+02
13	11.04	350.55	351.66	350.87	379.07	0.1889E+02	0.1350E+02	0.1292E+02	0.1094E+02
15	13.05	375.10	377.15	370.59	399.08	0.4191E+01	0.1194E+02	0.7069E+01	0.9065E+01
17	16.06	381.03	410.43	383.94	423.23	0.2235E+01	0.1027E+02	0.2626E+01	0.7118E+01
19	20.08	392.19	448.45	392.56	448.20	0.3809E+01	0.8743E+01	0.2381E+01	0.5435E+01
21	24.09	405.31	481.26	405.89	467.64	0.1533E+01	0.7675E+01	0.4343E+01	0.4332E+01
23	28.11	408.26	510.43	407.68	483.42	0.4882E+01	0.6876E+01	0.7001E+00	0.3558E+01
25	32.12	451.37	536.71	447.97	496.49	0.7145E+01	0.6256E+01	0.1027E+02	0.2991E+01
27	36.14	477.45	560.83	476.11	507.61	0.1064E+02	0.5756E+01	0.6863E+01	0.2560E+01
29	40.15	494.68	583.26	502.56	517.18	0.3479E+01	0.5342E+01	0.6290E+01	0.2224E+01
31	44.17	527.41	604.00	526.49	525.56	0.9071E+01	0.4996E+01	0.5617E+01	0.1955E+01
33	48.18	550.44	623.42	547.77	532.95	0.9860E+00	0.4701E+01	0.5016E+01	0.1737E+01
35	52.20	566.49	641.78	566.95	539.56	0.4088E+01	0.4445E+01	0.4551E+01	0.1556E+01
37	56.21	586.98	659.13	584.51	545.48	0.6761E+01	0.4221E+01	0.4230E+01	0.1404E+01
39	60.23	604.42	675.59	601.07	550.86	0.2387E+01	0.4024E+01	0.4029E+01	0.1275E+01
41	64.24	613.93	691.54	616.98	555.75	0.6129E+01	0.3845E+01	0.3919E+01	0.1165E+01
43	68.26	637.30	706.59	632.62	560.23	0.1115E+01	0.3687E+01	0.3871E+01	0.1069E+01
45	72.27	644.67	721.22	648.12	564.35	0.5461E+01	0.3542E+01	0.3862E+01	0.9862E+00
47	76.29	666.01	735.05	663.66	568.16	0.4296E+01	0.3413E+01	0.3874E+01	0.9131E+00
49	79.30	674.34	745.26	675.35	570.84	0.9351E+00	0.3322E+01	0.3891E+01	0.8641E+00

IV - 63

Connection type : Top-and seat-angle connections
Mode : All bolted

Tested by : N. Kubo et al. (1999) Japan
Test Id. : No.1

Column : H250x250x9x14
Beam : H250x125x6x9
Angle : L100x100x10

Fasteners: F10T- -M20
7/8" Oversize holes
Material : SS400
Fy = 44.16 ksi
Fu = 65.57 ksi

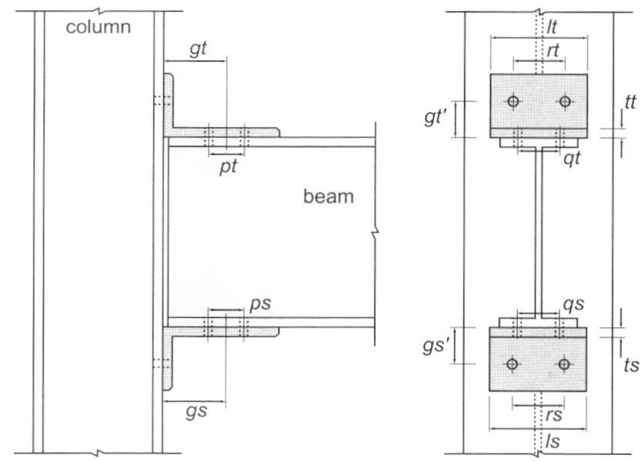

Major parameters

lt = 5.7165"	tt = 0.3661"	ls = 5.7165"	ts = 0.3661"
gt = 2.3386"	gt'= 2.3386"	gs = 2.3386"	gs'= 2.3386"
qt = 2.7953"	rt = 3.3622"	qs = 2.7953"	rs = 3.3622"
nt = 2 X 1	nt'= 2 X 1	ns = 2 X 1	ns'= 2 X 1

Remark 1) Plastic moment of beam: Mp,b = 90.2 kNm. Failure mode: local buckling of beam.
2) Max. axial force of beam at Nmax=118.36 kN

No	Moment (k-in)	Rotation (radians) X 1/1000	No	Moment (k-in)	Rotation (radians) X 1/1000
1	0.00	0.00	26	530.09	61.44
2	59.08	0.93	27	552.37	64.55
3	104.38	2.25	28	580.31	67.27
4	154.57	4.00	29	599.75	70.57
5	193.91	5.71	30	624.91	72.91
6	230.29	8.63	31	579.79	73.47
7	255.43	11.16	32	616.15	76.58
8	280.60	13.30	33	641.17	80.47
9	297.28	16.02	34	666.23	83.96
10	316.82	18.16	35	696.99	86.69
11	339.11	21.07	36	716.49	89.41
12	364.28	23.22	37	727.56	91.74
13	392.26	25.55	38	744.28	93.88
14	349.92	26.50	39	769.54	95.05
15	332.87	28.24	40	791.73	99.13
16	324.14	31.52	41	816.63	104.56
17	304.19	34.22	42	838.71	109.99
18	306.66	38.29	43	852.37	115.03
19	311.86	43.52			
20	348.32	45.47			
21	393.28	46.85			
22	426.92	48.80			
23	446.33	52.49			
24	471.36	56.38			
25	496.50	58.90			

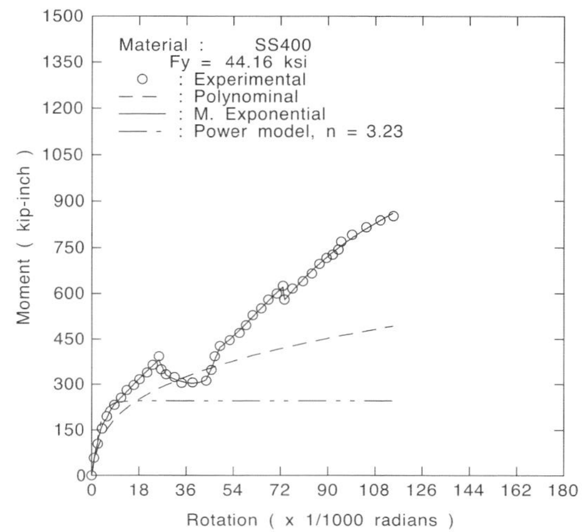

A.4 – 126

```
Moment-rotation prediction equations      ( R : X 1/1000 radians )

Frye and Morris polynominal model   :    R = Sum ( Ai X ( K*Bm )**Pi X 10**Qi )
     t =  0.366140"  xd =  9.842520"  f =  0.787400"  xl =  5.716500"
    A1 =  8.460000   A2 =  1.010000   A3 =  1.240000   K =  0.020545
    P1 =         1   P2 =         3   P3 =         5   Q1 =        -1   Q2 =        -1   Q3 =        -5

Modified exponential model  :
    Bm = Sum ( Ai X ( 1 - exp( - R/( 2*i*C ) ) ) )  +     Sum ( RKj X ( R - Rj0 ) )  +  Bm0
     C =   0.11910833E+01         Bmo=   0.00000000E+00        Nexp=  6        Nliner=  8
    AI =  -0.11924016E+04   0.20886304E+05  -0.11536080E+06   0.27726870E+06  -0.29366365E+06   0.11237926E+06
   Rj0 =           5.7100          25.5500          28.2400          34.2200          43.5200
                  48.8000          72.9100          73.4700
   RKj =   0.17502401E+02  -0.26677727E+02   0.12822062E+02  -0.15467006E+01   0.14889746E+02
          -0.16233923E+02  -0.84599901E+02   0.87162728E+02

Power model   :    Bm = ( Rki X R ) / ( 1 + ( R/R0 )**rn )**( 1/rn )
     rn = 3.230        rki =   0.57138773E+02          rmu =   0.24447802E+03
```

No	Rotation (radians) X 1/1000	Moment (kip-inch) Expri.	Poly.	M.Expo.	P.Model.	Connection stiffness (kip-inch) X 1000 Expri.	Poly.	M.Expo.	P.Model.
1	0.00	0.00	0.00	0.00	0.00	0.6353E+02	0.5753E+02	0.4679E+02	0.5714E+02
3	2.25	104.38	91.21	109.55	123.95	0.3189E+02	0.2547E+02	0.4125E+02	0.4895E+02
5	5.71	193.91	151.92	185.49	220.60	0.1911E+02	0.1280E+02	0.5581E+01	0.1091E+02
7	11.16	255.43	205.29	255.90	241.16	0.1093E+02	0.7781E+01	0.8880E+01	0.9342E+00
9	16.02	297.28	238.25	299.69	243.42	0.7810E+01	0.5979E+01	0.9179E+01	0.2106E+00
11	21.07	339.11	265.52	344.40	244.04	0.9989E+01	0.4908E+01	0.8377E+01	0.6679E-01
13	25.55	392.26	286.01	380.75	244.24	-0.2818E+02	0.4276E+01	0.8083E+01	0.2965E-01
15	28.24	332.87	297.10	331.43	244.31	-0.7325E+01	0.3983E+01	-0.1796E+02	0.1944E-01
17	34.22	304.19	319.32	309.93	244.39	-0.4200E+01	0.3476E+01	-0.1653E+01	0.8636E-02
19	43.52	311.86	348.96	314.61	244.44	0.1389E+02	0.2935E+01	0.4127E+01	0.3126E-02
21	46.85	393.28	358.61	381.65	244.44	0.2623E+02	0.2785E+01	0.2118E+02	0.2289E-02
23	52.49	446.33	373.58	449.08	244.46	0.5831E+01	0.2574E+01	0.7506E+01	0.1415E-02
25	58.90	496.50	389.51	502.16	244.46	0.1159E+02	0.2374E+01	0.8792E+01	0.8694E-03
27	64.55	552.37	402.53	552.29	244.47	0.8822E+01	0.2227E+01	0.8814E+01	0.5902E-03
29	70.57	599.75	415.53	603.53	244.47	0.8735E+01	0.2093E+01	0.8123E+01	0.4047E-03
31	73.47	579.79	421.40	579.03	244.47	-0.6650E+02	0.2036E+01	-0.7696E+02	0.3413E-03
33	80.47	641.17	435.20	645.91	244.47	0.6827E+01	0.1911E+01	0.8886E+01	0.2323E-03
35	86.69	696.99	446.93	697.52	244.47	0.9213E+01	0.1814E+01	0.7725E+01	0.1695E-03
37	91.74	727.56	455.81	734.35	244.47	0.6348E+01	0.1745E+01	0.6877E+01	0.1334E-03
39	95.05	769.54	461.55	756.28	244.47	0.1799E+02	0.1702E+01	0.6382E+01	0.1148E-03
41	104.56	816.63	477.30	811.23	244.48	0.4326E+01	0.1593E+01	0.5244E+01	0.7672E-04
43	115.03	852.37	493.40	861.48	244.48	0.2710E+01	0.1491E+01	0.4421E+01	0.5124E-04

IV - 64

Connection type : Top-and seat-angle connections
Mode : All bolted

Tested by : N. Kubo et al. (1999) Japan
Test Id. : No.3

Column : H150x150x7x10
Beam : H250x125x6x9
Angle : L100x100x10

Fasteners: F10T- -M20
7/8" Oversize holes
Material : SS400
Fy = 44.16 ksi
Fu = 65.57 ksi

Major parameters

lt = 5.7126"	tt = 0.3622"	ls = 5.7126"	ts = 0.3622"
gt = 2.3819"	gt'= 2.3819"	gs = 2.3819"	gs'= 2.3819"
qt = 2.8150"	rt = 3.3661"	qs = 2.8150"	rs = 3.3661"
nt = 2 X 1	nt'= 2 X 1	ns = 2 X 1	ns'= 2 X 1

Remark 1) Plastic moment of beam: Mp,b = 90.2 kNm. Failure mode: local buckling of beam.
2) Max. axial force of beam at Nmax=85.73 kN

No	Moment (k-in)	Rotation (radians) X 1/1000	No	Moment (k-in)	Rotation (radians) X 1/1000
1	C.00	0.00	26	490.65	77.56
2	45.70	1.00	27	504.34	83.18
3	84.45	1.93	28	523.15	90.41
4	109.11	2.97	29	540.27	97.51
5	135.00	4.36	30	555.71	104.85
6	158.78	7.57	31	577.89	111.73
7	194.42	11.71	32	595.03	119.41
8	221.61	16.07	33	618.98	128.93
9	252.18	20.09	34	639.50	136.61
10	292.92	25.26	35	661.75	145.67
11	320.16	31.34	36	685.72	155.76
12	357.56	37.77	37	719.87	166.77
13	383.08	42.93	38	750.34	176.51
14	400.14	47.86	39	788.25	186.60
15	390.07	50.61			
16	373.22	52.78			
17	361.45	55.53			
18	390.29	58.29			
19	415.72	60.36			
20	397.17	62.18			
21	424.28	63.80			
22	439.53	64.60			
23	446.39	67.81			
24	453.24	70.91			
25	471.91	72.98			

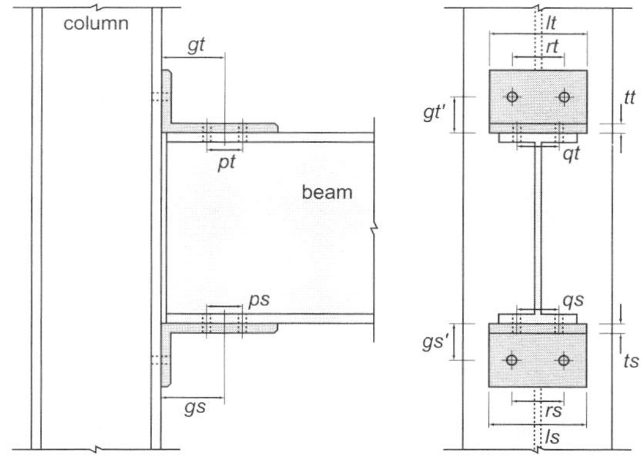

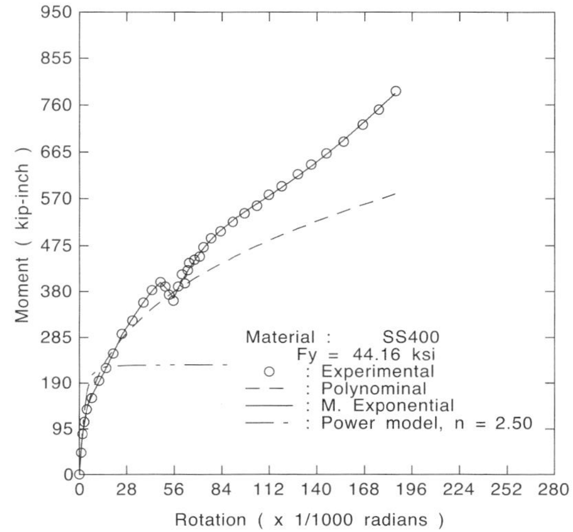

```
Moment-rotation prediction equations     ( R : X 1/1000 radians )

Frye and Morris polynominal model  :    R = Sum ( Ai X ( K*Bm )**Pi X 10**Qi )
     t =  0.362210"   xd =  9.842520"   f =  0.787400"  xl =  5.712600"
    A1 =  8.460000    A2 =  1.010000   A3 =  1.240000    K =  0.020666
    P1 =         1    P2 =         3   P3 =         5   Q1 =        -1    Q2 =        -1    Q3 =        -5

Modified exponential model  :
    Bm = Sum ( Ai X ( 1 - exp( - R/( 2*i*C ) ) ) )  +     Sum ( RKj X ( R - Rj0 ) )  +  Bm0
     C =  0.18050000E+01                Bmo=  0.00000000E+00          Nexp=  6         Nliner=  5
    AI =  0.64644953E+03  -0.67793173E+04  0.38550901E+05  -0.94110393E+05   0.98029810E+05  -0.35835709E+05
   Rj0 =          1.9300           4.3600         25.2600          47.8600          55.5300
   RKj =  0.32595528E+00  -0.11257062E+01 -0.23085068E+01  -0.64938775E+01   0.12955286E+02

Power model  :    Bm = ( Rki X R ) / ( 1 + ( R/R0 )**rn )**( 1/rn )
     rn = 2.502      rki =   0.50766416E+02         rmu =   0.22773212E+03
```

No	Rotation (radians) X 1/1000	Moment (kip-inch) Expri.	Poly.	M.Expo.	P.Model.	Connection stiffness (kip-inch) X 1000 Expri.	Poly.	M.Expo.	P.Model.
1	0.00	0.00	0.00	0.00	0.00	0.4570E+02	0.5720E+02	0.5896E+02	0.5077E+02
3	1.93	84.45	82.13	82.08	93.60	0.3319E+02	0.2814E+02	0.2968E+02	0.4325E+02
5	4.36	135.00	132.00	132.86	170.14	0.1524E+02	0.1559E+02	0.1440E+02	0.2021E+02
7	11.71	194.42	208.25	191.81	219.96	0.7454E+01	0.7468E+01	0.6719E+01	0.1562E+01
9	20.09	252.18	259.11	251.58	225.63	0.7725E+01	0.5048E+01	0.7484E+01	0.2579E+00
11	31.34	320.16	307.20	324.02	227.03	0.5130E+01	0.3677E+01	0.5410E+01	0.5555E-01
13	42.93	383.08	345.19	381.05	227.41	0.4186E+01	0.2946E+01	0.4138E+01	0.1857E-01
15	50.61	390.07	366.52	388.97	227.52	-0.5957E+01	0.2625E+01	-0.3984E+01	0.1045E-01
17	55.53	361.45	379.02	366.51	227.56	0.3072E+01	0.2460E+01	-0.5151E+01	0.7557E-02
19	60.36	415.72	390.55	401.46	227.60	0.3232E+00	0.2321E+01	0.6678E+01	0.5646E-02
21	63.80	424.28	398.46	423.13	227.61	0.1829E+02	0.2232E+01	0.5928E+01	0.4651E-02
23	67.81	446.39	407.15	445.27	227.63	0.2175E+01	0.2140E+01	0.5133E+01	0.3758E-02
25	72.98	471.91	418.08	469.47	227.65	0.7484E+01	0.2032E+01	0.4258E+01	0.2907E-02
27	83.18	504.34	437.74	506.16	227.67	0.2509E+01	0.1857E+01	0.3047E+01	0.1839E-02
29	97.51	540.27	462.90	543.43	227.69	0.2260E+01	0.1663E+01	0.2309E+01	0.1054E-02
31	111.73	577.89	485.44	575.30	227.70	0.2755E+01	0.1514E+01	0.2245E+01	0.6548E-03
33	128.93	618.98	510.20	615.93	227.71	0.2602E+01	0.1371E+01	0.2506E+01	0.3966E-03
35	145.67	661.75	532.19	660.39	227.72	0.2418E+01	0.1260E+01	0.2800E+01	0.2587E-03
37	166.77	719.87	557.54	722.54	227.72	0.3115E+01	0.1148E+01	0.3069E+01	0.1611E-03
39	186.60	788.25	579.42	784.97	227.72	0.3758E+01	0.1062E+01	0.3213E+01	0.1087E-03

IV - 65

Connection type : Top-and seat-angle connections
Mode : All bolted

Tested by : N. Kubo et al. (1999) Japan
Test Id. : No.5

Column : H250x250x9x14
Beam : H250x125x6x9
Angle : L120x120x8

Fasteners: F10T- -M16
23/32" Oversize holes
Material : SS400
Fy = 45.30 ksi
Fu = 62.37 ksi

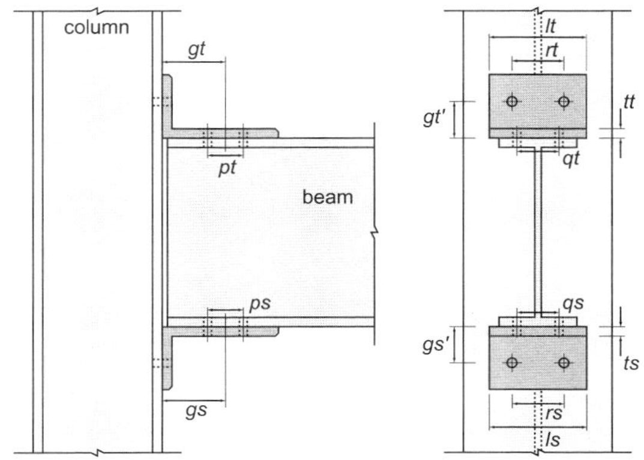

Major parameters

lt = 5.4094"	tt = 0.2992"	ls = 5.4094"	ts = 0.2992"
gt = 2.7717"	gt'= 2.7717"	gs = 2.7717"	gs'= 2.7717"
qt = 3.0551"	rt = 2.9921"	qs = 3.0551"	rs = 2.9921"
pt = 1.5827"	pt'= 1.5827"	ps = 1.5827"	ps'= 1.5827"
nt = 2 X 2	nt'= 2 X 2	ns = 2 X 2	ns'= 2 X 2

Remark 1) Plastic moment of beam: Mp,b = 90.2 kNm. Failure mode: local buckling of beam.
2) Max. axial force of beam at Nmax=225.25 kN

No	Moment (k-in)	Rotation (radians) X 1/1000
1	0.00	0.00
2	26.62	0.66
3	60.78	1.52
4	109.75	2.76
5	147.84	4.10
6	181.62	6.70
7	227.66	9.39
8	265.05	11.88
9	290.81	14.96
10	322.30	18.62
11	353.83	21.89
12	399.86	24.77
13	437.15	28.24
14	474.49	31.31
15	410.77	32.30
16	375.88	34.43
17	375.70	36.56
18	384.06	40.22
19	427.12	43.88
20	458.77	45.80
21	389.07	49.10
22	388.94	50.65
23	403.03	54.89
24	414.31	58.17
25	448.90	59.51
26	480.48	62.20
27	509.22	64.31
28	549.46	67.20
29	586.72	71.05
30	621.11	74.70
31	658.33	79.13
32	695.44	84.72
33	732.57	90.11
34	766.74	96.28
35	751.98	99.76
36	774.95	101.68
37	783.44	103.80
38	791.78	107.66
39	806.61	110.36
40	814.38	113.83
41	822.22	117.11
42	833.66	122.48
43	821.90	127.15

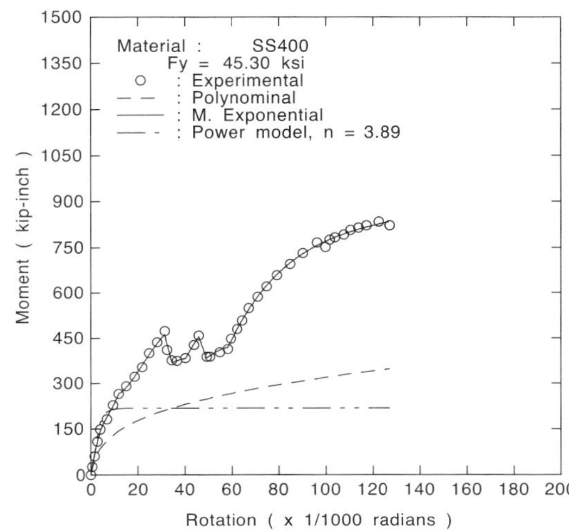

```
Moment-rotation prediction equations      ( R : X 1/1000 radians )

Frye and Morris polynominal model   :    R = Sum ( Ai X ( K*Bm )**Pi X 10**Qi )
     t =  0.299210"   xd =  9.842520"   f =  0.629900"   xl =  5.409450"
    A1 =  8.460000    A2 =  1.010000   A3 =  1.240000     K =  0.030196
    P1 =        1     P2 =        3    P3 =        5     Q1 =       -1     Q2 =       -1     Q3 =       -5

Modified exponential model  :
     Bm = Sum ( Ai X ( 1 - exp( - R/( 2*i*C ) ) ) ) +      Sum ( RKj X ( R - Rj0 ) )  + Bm0
      C =   0.12245833E+01          Bmo=   0.00000000E+00          Nexp=  6         Nliner=  8
     AI =  -0.54862691E+03   0.98489462E+04  -0.53183809E+05   0.13712546E+06  -0.16762922E+06   0.76201123E+05
    Rj0 =          4.1000          11.8800          31.3100          34.4300          40.2200
                  45.8000          49.1000          58.2000
    RKj =  -0.54513855E+01  -0.14192974E+02  -0.37801305E+02   0.34419161E+02   0.13650850E+02
           -0.32742595E+02   0.28956803E+02   0.14272473E+02

Power model  :    Bm = ( Rki X R ) / ( 1 + ( R/R0 )**rn )**( 1/rn )
     rn = 3.891       rki =   0.47315687E+02           rmu =   0.21746853E+03
```

No	Rotation (radians) X 1/1000	Moment (kip-inch) Expri.	Poly.	M.Expo.	P.Model.	Connection stiffness (kip-inch) X 1000 Expri.	Poly.	M.Expo.	P.Model.
1	0.00	0.00	0.00	0.00	0.00	0.4033E+02	0.3915E+02	0.4231E+02	0.4732E+02
3	1.52	60.78	47.69	63.05	71.67	0.3963E+02	0.2246E+02	0.3883E+02	0.4652E+02
5	4.10	147.84	87.50	145.23	170.80	0.2317E+02	0.1117E+02	0.2540E+02	0.2538E+02
7	9.39	227.66	129.73	224.95	214.13	0.1602E+02	0.6010E+01	0.1493E+02	0.1332E+01
9	14.96	290.81	157.69	286.32	216.91	0.8475E+01	0.4274E+01	0.8489E+01	0.1455E+00
11	21.89	353.83	183.36	361.77	217.34	0.1301E+02	0.3249E+01	0.1191E+02	0.2283E-01
13	28.24	437.15	202.15	433.02	217.42	0.1150E+02	0.2710E+01	0.1013E+02	0.6579E-02
15	32.30	410.77	212.66	432.93	217.44	-0.4913E+02	0.2463E+01	-0.2957E+02	0.3412E-02
17	36.56	375.70	222.67	375.75	217.45	0.7854E+00	0.2257E+01	0.2719E+01	0.1862E-02
19	43.88	427.12	238.15	432.10	217.46	0.1486E+02	0.1985E+01	0.1267E+02	0.7629E-03
21	49.10	389.07	248.11	383.30	217.46	-0.6809E+01	0.1835E+01	-0.2271E+02	0.4403E-03
23	54.89	403.03	258.32	411.00	217.46	0.3389E+01	0.1697E+01	0.3334E+01	0.2552E-03
25	59.51	448.90	265.96	439.88	217.47	0.2113E+02	0.1604E+01	0.1537E+02	0.1719E-03
27	64.31	509.22	273.44	508.31	217.47	0.1375E+02	0.1520E+01	0.1317E+02	0.1176E-03
29	71.05	586.72	283.33	587.57	217.47	0.9547E+01	0.1418E+01	0.1042E+02	0.7225E-04
31	79.13	658.33	294.36	660.33	217.47	0.7622E+01	0.1316E+01	0.7703E+01	0.4267E-04
33	90.11	732.57	308.16	729.25	217.47	0.6259E+01	0.1202E+01	0.5035E+01	0.2260E-04
35	99.76	751.98	319.40	769.93	217.47	0.6202E+01	0.1120E+01	0.3506E+01	0.1374E-04
37	103.80	783.44	323.90	783.13	217.47	0.3352E+01	0.1090E+01	0.3041E+01	0.1132E-04
39	110.36	806.61	330.84	801.06	217.47	0.4069E+01	0.1045E+01	0.2458E+01	0.8385E-05
41	117.11	822.22	337.82	816.13	217.47	0.2291E+01	0.1002E+01	0.2032E+01	0.6272E-05
43	127.15	821.90	347.57	834.31	217.47	-0.2517E+01	0.9470E+00	0.1625E+01	0.4195E-05

IV - 66

Connection type : Top-and seat-angle connections
Mode : All bolted

Tested by : N. Kubo et al. (1999) Japan
Test Id. : No.7

Column : H150x150x7x10
Beam : H250x125x6x9
Angle : L120x120x8

Fasteners: F10T- -M16
23/32" Oversize holes
Material : SS400
Fy = 45.30 ksi
Fu = 62.37 ksi

Major parameters

lt = 5.4055"	tt = 0.3071"	ls = 5.4055"	ts = 0.3071"
gt = 2.7697"	gt'= 2.7697"	gs = 2.7697"	gs'= 2.7697"
qt = 3.0276"	rt = 3.0039"	qs = 3.0276"	rs = 3.0039"
pt = 1.5472"	pt'= 1.5472"	ps = 1.5472"	ps'= 1.5472"
nt = 2 X 2	nt'= 2 X 2	ns = 2 X 2	ns'= 2 X 2

Remark 1) Plastic moment of beam: Mp,b = 90.2 kNm. Failure mode: bolt fracture
2) Max. axial force of beam at Nmax=79.32 kN

No	Moment (k-in)	Rotation (radians) X 1/1000	No	Moment (k-in)	Rotation (radians) X 1/1000
1	0.00	0.00	26	395.09	69.25
2	47.74	1.27	27	445.46	69.27
3	83.26	2.45	28	367.95	69.50
4	113.13	3.63	29	393.05	71.07
5	143.65	4.90	30	423.96	72.63
6	166.74	6.27	31	462.67	73.29
7	187.48	7.83	32	487.72	75.50
8	216.28	12.24	33	524.51	76.03
9	245.06	17.17	34	549.63	77.07
10	273.80	22.61	35	565.07	78.11
11	312.24	27.93	36	576.49	81.61
12	348.68	34.42	37	584.02	85.50
13	389.02	40.39	38	606.99	90.29
14	421.71	44.54	39	628.04	94.70
15	448.63	48.17	40	643.31	98.72
16	454.18	52.70	41	660.50	102.87
17	479.14	56.47	42	669.96	106.88
18	492.55	59.06	43	688.97	113.10
19	447.92	60.34	44	692.54	118.28
20	403.30	61.36	45	702.06	121.26
21	370.35	61.61	46	715.39	125.28
22	397.33	64.08	47	726.76	129.56
23	430.17	65.64	48	734.96	134.87
24	470.74	67.73	49	738.94	139.36
25	491.98	68.90	50	735.65	142.12

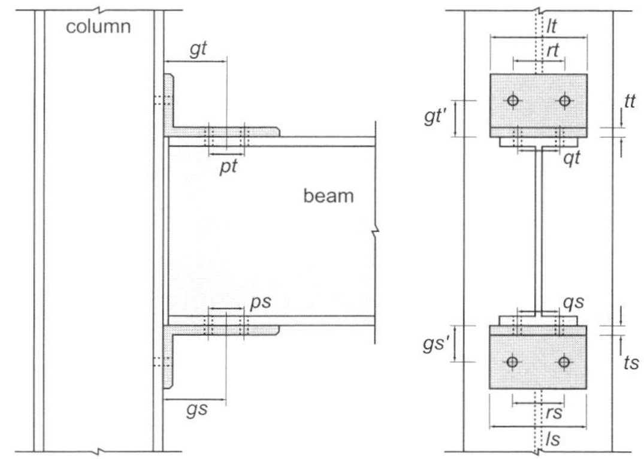

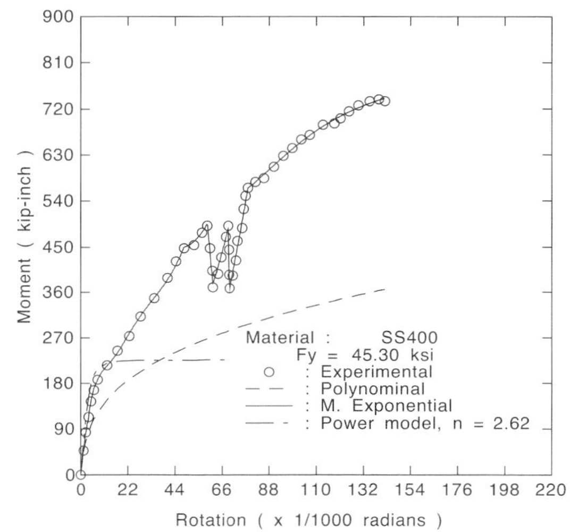

```
Moment-rotation prediction equations     ( R : X 1/1000 radians )

Frye and Morris polynominal model  :    R = Sum ( Ai X ( K*Bm )**Pi X 10**Qi )
      t =  0.307090"  xd =  9.842520"   f =  0.629900"  xl =  5.405510"
     A1 =  8.460000   A2 =  1.010000   A3 =  1.240000    K =  0.029821
     P1 =        1    P2 =        3    P3 =        5    Q1 =       -1     Q2 =       -1     Q3 =       -5

Modified exponential model  :
     Bm = Sum ( Ai X ( 1 - exp( - R/( 2*i*C ) ) ) )  +      Sum ( RKj X ( R - Rj0 ) )  +  Bm0
      C =   0.15018333E+01            Bmo=   0.00000000E+00             Nexp=  6         Nliner=  8
     AI =  -0.13091695E+04   0.23103518E+05  -0.12087614E+06   0.27199169E+06  -0.26932260E+06   0.96111921E+05
    Rj0 =           3.6300           7.8300          48.1700          59.0600          61.6100
                   68.9000          69.5000          78.1100
    RKj =   0.15129953E+02   0.45704082E+01  -0.94202616E+01  -0.54681558E+02   0.60043476E+02
           -0.22061243E+03   0.22601436E+03  -0.20580131E+02

Power model  :    Bm = ( Rki X R ) / ( 1 + ( R/R0 )**rn )**( 1/rn )
      rn = 2.622      rki =   0.49755155E+02         rmu =   0.22647891E+03
```

No	Rotation (radians) X 1/1000	Moment (kip-inch) Expri.	Poly.	M.Expo.	P.Model.	Connection stiffness (kip-inch) X 1000 Expri.	Poly.	M.Expo.	P.Model.
1	0.00	0.00	0.00	0.00	0.00	0.3759E+02	0.3964E+02	0.3416E+02	0.4976E+02
3	2.45	83.26	66.24	83.53	113.82	0.2771E+02	0.1652E+02	0.2795E+02	0.3881E+02
5	4.90	143.65	97.08	145.97	180.07	0.2058E+02	0.9892E+01	0.2270E+02	0.1661E+02
7	7.83	187.48	121.18	185.80	208.57	0.1153E+02	0.6967E+01	0.6493E+01	0.5176E+01
9	17.17	245.06	168.80	241.12	223.88	0.5574E+01	0.3916E+01	0.6284E+01	0.3893E+00
11	27.93	312.24	203.84	313.70	225.74	0.6500E+01	0.2765E+01	0.6043E+01	0.6886E-01
13	40.39	389.02	233.92	384.84	226.20	0.7419E+01	0.2131E+01	0.6671E+01	0.1823E-01
15	48.17	448.63	249.48	450.58	226.30	0.4661E+01	0.1883E+01	0.1050E+02	0.9649E-02
17	56.47	479.14	264.25	479.55	226.36	0.5766E+01	0.1685E+01	0.5860E+01	0.5431E-02
19	60.34	447.92	270.71	436.16	226.38	-0.3981E+02	0.1608E+01	-0.4684E+02	0.4273E-02
21	61.61	370.35	272.75	377.06	226.39	-0.1187E+03	0.1585E+01	-0.4625E+02	0.3963E-02
23	65.64	430.17	278.92	436.11	226.40	0.2035E+02	0.1517E+01	0.1545E+02	0.3151E-02
25	68.90	491.98	283.83	488.31	226.41	-0.2089E+03	0.1466E+01	0.1653E+02	0.2644E-02
27	69.27	445.46	284.37	412.82	226.41	0.2290E+04	0.1461E+01	-0.2040E+03	0.2593E-02
29	71.07	393.05	287.00	400.96	226.41	0.1791E+02	0.1435E+01	0.2253E+02	0.2363E-02
31	73.29	462.67	290.17	451.57	226.42	0.4777E+02	0.1404E+01	0.2304E+02	0.2114E-02
33	76.03	524.51	293.87	515.42	226.43	0.5413E+02	0.1370E+01	0.2354E+02	0.1851E-02
35	78.11	565.07	296.71	564.71	226.43	0.1219E+02	0.1344E+01	0.2384E+02	0.1679E-02
37	85.50	584.02	306.32	591.28	226.44	0.3217E+01	0.1263E+01	0.3814E+01	0.1210E-02
39	94.70	628.04	317.61	626.56	226.45	0.4263E+01	0.1176E+01	0.3752E+01	0.8360E-03
41	102.87	660.50	326.92	655.67	226.45	0.3235E+01	0.1110E+01	0.3341E+01	0.6195E-03
43	113.10	688.97	337.85	686.50	226.46	0.1765E+01	0.1040E+01	0.2681E+01	0.4395E-03
45	121.26	702.06	346.24	706.28	226.46	0.3245E+01	0.9905E+00	0.2176E+01	0.3415E-03
47	129.56	726.76	354.22	722.46	226.47	0.2161E+01	0.9465E+00	0.1737E+01	0.2687E-03
49	139.36	738.94	363.22	737.40	226.47	-0.4005E+00	0.9003E+00	0.1332E+01	0.2063E-03

A.4 – 133

IV - 67

Connection type : Top-and seat-angle connections
Mode : All bolted

Tested by : N. Kubo et al. (1999) Japan
Test Id. : No.9

Column : H175x175x7.5x11
Beam : H250x125x6x9
Angle : L120x120x8

Fasteners: F10T- -M16
23/32" Oversize holes
Material : SS400
Fy = 45.30 ksi
Fu = 62.37 ksi

Major parameters

lt = 5.4016"	tt = 0.3150"	ls = 5.4016"	ts = 0.3150"
gt = 2.7638"	gt'= 2.7638"	gs = 2.7638"	gs'= 2.7638"
qt = 3.0315"	rt = 3.0315"	qs = 3.0315"	rs = 3.0315"
pt = 1.5748"	pt'= 1.5748"	ps = 1.5748"	ps'= 1.5748"
nt = 2 X 2	nt'= 2 X 2	ns = 2 X 2	ns'= 2 X 2

Remark 1) Plastic moment of beam: Mp,b = 90.2 kNm. Failure mode: bolt fracture
2)

No	Moment (k-in)	Rotation (radians) X 1/1000	No	Moment (k-in)	Rotation (radians) X 1/1000
1	0.00	0.00	26	512.91	117.05
2	43.22	0.95	27	535.55	122.70
3	76.50	1.62	28	540.32	126.99
4	103.25	2.38	29	550.26	130.93
5	137.07	3.53	30	552.35	136.42
6	160.06	5.26	31	557.06	141.40
7	180.14	7.08	32	579.97	143.96
8	226.27	14.19	33	596.44	147.40
9	253.92	21.39	34	595.99	150.31
10	284.07	29.27	35	594.25	153.91
11	306.60	36.30	36	568.59	156.67
12	316.31	42.82			
13	327.86	49.16			
14	351.10	56.53			
15	366.46	61.74			
16	379.83	65.11			
17	387.36	68.86			
18	398.58	72.13			
19	423.72	78.64			
20	443.77	84.47			
21	458.73	89.78			
22	458.15	96.48			
23	457.71	101.62			
24	485.49	107.28			
25	500.49	112.08			

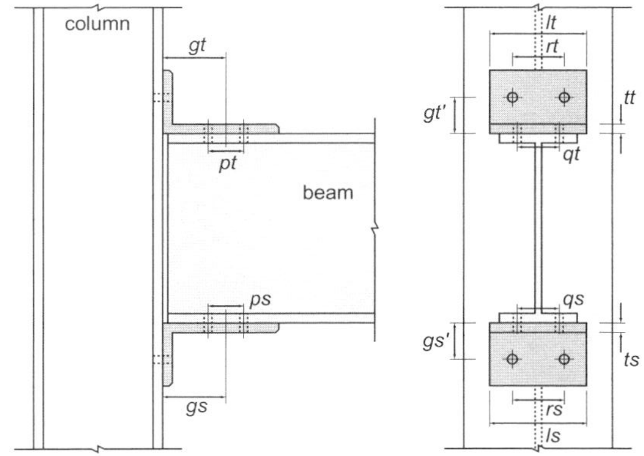

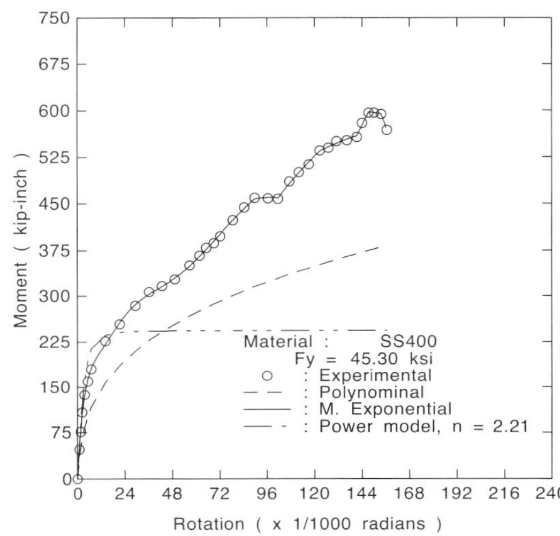

A.4 – 134

```
Moment-rotation prediction equations     ( R : X 1/1000 radians )

Frye and Morris polynominal model  :    R = Sum ( Ai X ( K*Bm )**Pi X 10**Qi )
      t =  0.314960"  xd =  9.842520"  f =  0.629900"  xl =  5.401570"
     A1 =  8.460000   A2 =  1.010000  A3 =  1.240000   K =  0.029461
     P1 =        1    P2 =        3   P3 =        5   Q1 =       -1    Q2 =      -1    Q3 =      -5

Modified exponential model  :
     Bm = Sum ( Ai X ( 1 - exp( - R/( 2*i*C ) ) ) )  +      Sum ( RKj X ( R - Rj0 ) )  +  Bm0
      C =   0.15430833E+01          Bmo=   0.00000000E+00          Nexp=  6        Nliner=  6
     AI =  -0.88223043E+02   0.62656791E+04  -0.39073844E+05   0.98471247E+05  -0.10837527E+06   0.43341618E+05
    Rj0 =          89.7800         101.6000         122.7000         141.4000         147.4000
                  153.9000
    RKj =  -0.22705036E+01   0.43785016E+01  -0.15082720E+01   0.53934246E+01  -0.68943041E+01
           -0.86030961E+01

Power model  :    Bm = ( Rki X R ) / ( 1 + ( R/R0 )**rn )**( 1/rn )
     rn = 2.209      rki =   0.56524873E+02         rmu =   0.24391942E+03
```

No	Rotation (radians) X 1/1000	Moment (kip-inch) Expri.	Poly.	M.Expo.	P.Model.	Connection stiffness (kip-inch) X 1000 Expri.	Poly.	M.Expo.	P.Model.
1	0.00	0.00	0.00	0.00	0.00	0.5076E+02	0.4012E+02	0.6039E+02	0.5652E+02
3	1.62	76.50	51.14	78.47	87.17	0.4201E+02	0.2213E+02	0.3803E+02	0.4827E+02
5	3.53	137.07	82.79	134.02	159.42	0.2036E+02	0.1281E+02	0.2163E+02	0.2751E+02
7	7.08	180.14	117.19	182.32	214.01	0.1011E+02	0.7600E+01	0.8258E+01	0.7585E+01
9	21.39	253.92	186.25	256.74	240.77	0.3834E+01	0.3386E+01	0.4535E+01	0.3187E+00
11	36.30	306.60	227.63	303.20	242.93	0.2315E+01	0.2325E+01	0.2033E+01	0.6008E-01
13	49.16	327.86	254.41	329.75	243.41	0.2438E+01	0.1879E+01	0.2399E+01	0.2285E-01
15	61.74	366.46	276.19	366.00	243.61	0.3567E+01	0.1603E+01	0.3287E+01	0.1103E-01
17	68.86	387.36	287.17	390.23	243.68	0.2768E+01	0.1485E+01	0.3470E+01	0.7776E-02
19	78.64	423.72	301.04	423.69	243.74	0.3639E+01	0.1354E+01	0.3304E+01	0.5082E-02
21	89.78	458.73	315.43	457.64	243.78	0.1533E+01	0.1235E+01	0.2750E+01	0.3324E-02
23	101.62	457.71	329.55	459.18	243.82	0.2290E+01	0.1133E+01	0.4146E+01	0.2235E-02
25	112.08	500.49	340.98	499.48	243.84	0.2818E+01	0.1059E+01	0.3576E+01	0.1632E-02
27	122.70	535.55	351.89	534.90	243.85	0.2361E+01	0.9946E+00	0.3114E+01	0.1221E-02
29	130.93	550.26	359.85	546.94	243.86	0.1628E+01	0.9512E+00	0.1332E+01	0.9916E-03
31	141.40	557.06	369.50	559.47	243.87	0.6231E+01	0.9022E+00	0.1077E+01	0.7748E-03
33	147.40	596.44	374.88	597.95	243.87	0.2110E+01	0.8765E+00	0.6363E+01	0.6781E-03
35	153.91	594.25	380.54	594.10	243.88	-0.5473E+01	0.8505E+00	-0.9226E+01	0.5903E-03

IV - 68

Connection type : Top-and seat-angle connections
Mode : All bolted

Tested by : M. Komuro et al. (2002) Japan
Test Id. : W00-m

Column : --
Beam : H400x200x8x13
Angle : L150x100x12

Fasteners: F10T- -M20
7/8" Oversize holes
Material : SS400
Fy = 45.69 ksi
Fu = 68.02 ksi

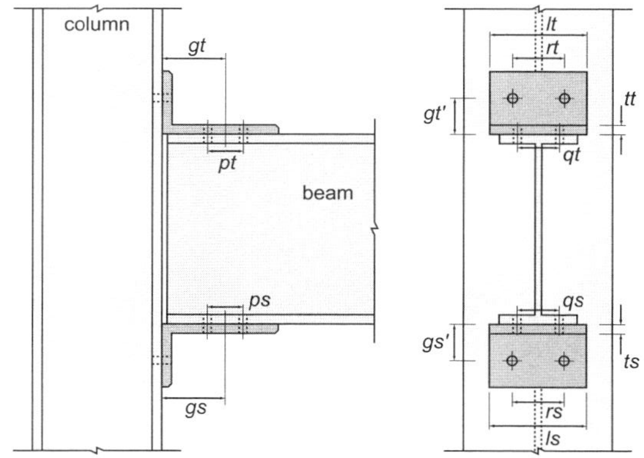

Major parameters

lt = 7.8740"	tt = 0.4724"	ls = 7.8740"	ts = 0.4724"
gt = 3.2480"	gt'= 2.1654"	gs = 3.2480"	gs'= 2.1654"
qt = 4.7244"	rt = 4.7244"	qs = 4.7244"	rs = 4.7244"
pt = 2.1654"	ps = 2.1654"		
nt = 2 X 2	nt'= 2 X 1	ns = 2 X 2	ns'= 2 X 1

Remark 1) Failure mode: Bolt fracture
2) Original T & S angles: L150x100x15

No	Moment (k-in)	Rotation (radians) X 1/1000	No	Moment (k-in)	Rotation (radians) X 1/1000	No	Moment (k-in)	Rotation (radians) X 1/1000
1	0.00	0.00	26	835.19	30.01	51	1392.76	80.08
2	171.36	0.26	27	859.56	31.97	52	1417.45	82.02
3	270.79	0.51	28	837.18	34.02	53	1427.19	83.97
4	329.60	0.75	29	907.97	36.03	54	1450.26	86.00
5	392.31	1.01	30	905.05	38.01	55	1459.39	88.00
6	413.43	1.25	31	944.36	40.04	56	1480.50	90.00
7	457.30	1.50	32	959.06	42.01	57	1488.65	92.02
8	497.59	2.01	33	991.81	43.98	58	1510.05	93.98
9	546.65	3.01	34	1018.12	46.01	59	1514.59	95.98
10	569.07	4.00	35	1044.77	48.00	60	1536.06	98.00
11	580.77	5.01	36	1074.01	50.01	61	1544.48	99.99
12	583.04	6.01	37	1094.48	52.04	62	1556.19	102.00
13	573.29	6.98	38	1114.63	54.01	63	1574.08	103.99
14	595.07	8.02	39	1131.52	55.98	64	1577.35	105.99
15	612.94	9.02	40	1159.47	57.97	65	1585.45	108.00
16	626.58	10.03	41	1189.36	60.03	66	1593.91	110.03
17	654.53	12.02	42	1201.71	62.01			
18	677.93	14.00	43	1223.48	64.02			
19	699.69	16.01	44	1260.52	66.01			
20	621.06	18.01	45	1272.54	67.99			
21	685.72	20.00	46	1290.09	70.01			
22	726.66	22.03	47	1321.31	71.99			
23	757.20	23.98	48	1333.63	74.01			
24	781.90	25.99	49	1357.66	76.03			
25	811.15	28.00	50	1379.80	77.98			

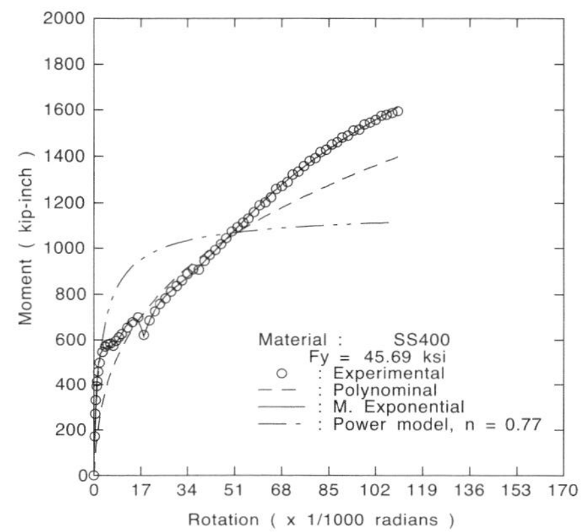

A.4 – 136

```
Moment-rotation prediction equations     ( R : X 1/1000 radians )

Frye and Morris polynominal model   :     R = Sum ( Ai X ( K*Bm )**Pi X 10**Qi )
      t =  0.472440"  xd = 15.748000"   f =  0.787400"  xl =  7.874020"
     A1 =  8.460000   A2 =  1.010000   A3 =  1.240000    K =  0.007142
     P1 =         1   P2 =         3   P3 =         5   Q1 =        -1     Q2 =       -1     Q3 =       -5

Modified exponential model  :
     Bm = Sum ( Ai X ( 1 - exp( - R/( 2*i*C ) ) ) )  +      Sum ( RKj X ( R - Rj0 ) )  +  Bm0
       C =   0.98191667E+00          Bmo=   0.00000000E+00           Nexp=  6       Nliner=  4
      AI =   0.38292850E+04  -0.81827185E+04  -0.30450376E+05   0.12478900E+06  -0.14800970E+06   0.59744923E+05
     Rj0 =           5.0100           6.9800          16.0100          18.0100
     RKj =  -0.37568114E+02  -0.16156171E+02  -0.19106488E+02   0.79344558E+02

Power model  :     Bm = ( Rki X R ) / ( 1 + ( R/R0 )**rn )**( 1/rn )
      rn = 0.767      rki =   0.65465433E+03          rmu =   0.11779658E+04
```

No	Rotation (radians) X 1/1000	Moment (kip-inch) Expri.	Poly.	M.Expo.	P.Model.	Connection stiffness (kip-inch) X 1000 Expri.	Poly.	M.Expo.	P.Model.
1	0.00	0.00	0.00	0.00	0.00	0.6591E+03	0.1655E+03	0.5808E+03	0.6547E+03
3	0.51	270.79	81.15	238.19	219.33	0.3198E+03	0.1477E+03	0.3667E+03	0.3116E+03
5	1.01	392.31	147.58	383.43	346.29	0.1615E+03	0.1184E+03	0.2237E+03	0.2088E+03
7	1.50	457.30	199.72	468.56	434.23	0.1437E+03	0.9571E+02	0.1303E+03	0.1548E+03
9	3.01	546.65	312.39	554.06	601.71	0.3579E+02	0.5943E+02	0.1479E+02	0.8048E+02
11	5.01	580.77	409.78	577.66	721.77	0.6904E+01	0.4063E+02	0.2181E+02	0.4512E+02
13	6.98	573.29	480.14	580.75	793.76	0.4902E+01	0.3169E+02	0.1771E+02	0.2971E+02
15	9.02	612.94	538.75	606.20	844.75	0.1570E+02	0.2619E+02	0.1998E+02	0.2108E+02
17	12.02	654.53	609.21	667.90	896.40	0.1293E+02	0.2120E+02	0.1665E+02	0.1410E+02
19	16.01	699.69	685.15	684.98	941.95	-0.1431E+02	0.1721E+02	-0.9216E+01	0.9271E+01
21	20.00	685.72	748.38	681.24	973.20	0.2639E+02	0.1466E+02	0.2822E+02	0.6629E+01
23	23.98	757.20	802.97	765.03	996.13	0.1400E+02	0.1287E+02	0.1547E+02	0.5012E+01
25	28.00	811.15	851.88	815.16	1013.97	0.1326E+02	0.1153E+02	0.1047E+02	0.3933E+01
27	31.97	859.56	895.51	854.38	1028.03	0.1294E+02	0.1049E+02	0.9745E+01	0.3188E+01
29	36.03	907.97	936.32	895.63	1039.79	0.4390E+01	0.9642E+01	0.1070E+02	0.2634E+01
31	40.04	944.36	973.90	940.98	1049.48	0.1590E+02	0.8946E+01	0.1189E+02	0.2222E+01
33	43.98	991.81	1007.95	989.56	1057.60	0.1224E+02	0.8376E+01	0.1269E+02	0.1908E+01
35	48.00	1044.77	1040.57	1041.32	1064.74	0.1397E+02	0.7878E+01	0.1298E+02	0.1654E+01
37	52.04	1094.48	1071.50	1093.57	1070.99	0.1016E+02	0.7446E+01	0.1282E+02	0.1449E+01
39	55.98	1131.52	1100.06	1143.20	1076.37	0.1130E+02	0.7076E+01	0.1234E+02	0.1285E+01
41	60.03	1189.36	1127.85	1191.86	1081.28	0.1029E+02	0.6741E+01	0.1167E+02	0.1145E+01
43	64.02	1223.48	1154.45	1236.97	1085.61	0.1474E+02	0.6443E+01	0.1094E+02	0.1029E+01
45	67.99	1272.54	1179.34	1278.96	1089.50	0.7365E+01	0.6180E+01	0.1022E+02	0.9313E+00
47	71.99	1321.31	1203.81	1318.47	1093.05	0.1098E+02	0.5937E+01	0.9548E+01	0.8467E+00
49	76.03	1357.66	1227.10	1355.81	1096.32	0.1162E+02	0.5718E+01	0.8951E+01	0.7729E+00
51	80.08	1392.76	1250.04	1391.00	1099.31	0.9578E+01	0.5514E+01	0.8441E+01	0.7086E+00
53	83.97	1427.19	1270.84	1423.02	1101.96	0.8117E+01	0.5338E+01	0.8034E+01	0.6544E+00
55	88.00	1459.39	1292.18	1454.68	1104.50	0.7560E+01	0.5166E+01	0.7691E+01	0.6048E+00
57	92.02	1488.65	1312.78	1485.03	1106.84	0.7528E+01	0.5007E+01	0.7418E+01	0.5610E+00
59	95.98	1514.59	1332.46	1513.96	1108.98	0.6430E+01	0.4862E+01	0.7205E+01	0.5225E+00
61	99.99	1544.48	1351.30	1542.50	1111.01	0.5023E+01	0.4729E+01	0.7037E+01	0.4877E+00
63	103.99	1574.08	1370.09	1570.38	1112.90	0.5322E+01	0.4601E+01	0.6908E+01	0.4564E+00
65	108.00	1585.45	1388.42	1597.88	1114.67	0.4100E+01	0.4482E+01	0.6809E+01	0.4281E+00

IV - 69

Connection type : Top-and seat-angle connections
Mode : All bolted

Tested by : Y. Sato et al. (2007) Japan
Test Id. : G60

Column : --
Beam : H400x200x8x13
Angle : L200x200x15

Fasteners: F10T- -M20
7/8" Oversize holes
Material : SS400
Fy = 45.69 ksi
Fu = 68.02 ksi

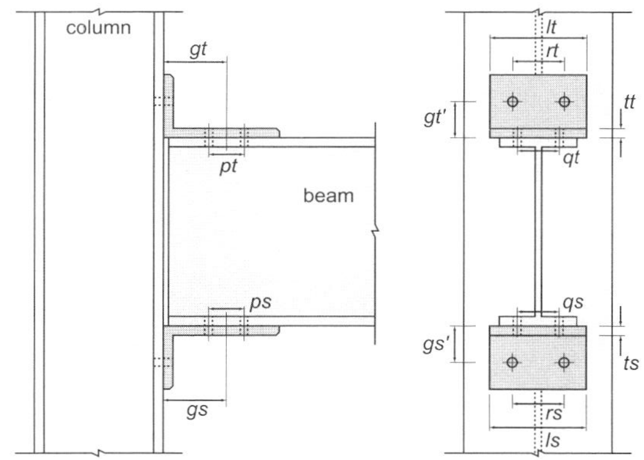

Major parameters

lt = 7.8740"	tt = 0.5906"	ls = 7.8740"	ts = 0.5906"
gt = 4.1339"	gt'= 2.3622"	gs = 4.1339"	gs'= 2.3622"
qt = 4.7244"	rt = 4.7244"	qs = 4.7244"	rs = 4.7244"
pt = 3.5433"	ps = 3.5433"		
nt = 2 X 2	nt'= 2 X 1	ns = 2 X 2	ns'= 2 X 1

Remark 1) Failure mode: Bolt fracture
2) Original T & S angles: L200x200x20

No	Moment (k-in)	Rotation (radians) X 1/1000	No	Moment (k-in)	Rotation (radians) X 1/1000	No	Moment (k-in)	Rotation (radians) X 1/1000
1	0.00	0.00	26	1232.72	28.01	51	1782.69	77.97
2	283.17	0.25	27	1262.73	30.00	52	1794.44	79.99
3	409.17	0.50	28	1296.89	32.01	53	1802.90	82.01
4	499.90	0.75	29	1328.23	34.04	54	1808.58	84.00
5	563.36	1.01	30	1358.42	36.01	55	1811.22	86.08
6	663.43	1.50	31	1387.83	38.02	56	1809.12	88.00
7	739.81	2.02	32	1417.06	40.01	57	1800.03	89.99
8	811.71	3.01	33	1441.27	42.01			
9	883.76	4.00	34	1468.66	44.00			
10	925.49	5.01	35	1498.31	46.02			
11	954.46	6.02	36	1518.26	47.98			
12	989.23	7.02	37	1541.76	50.01			
13	1008.97	8.01	38	1560.03	52.00			
14	1023.80	8.99	39	1572.37	54.00			
15	1034.49	10.00	40	1596.36	56.01			
16	1041.08	12.00	41	1616.40	58.01			
17	1055.77	14.02	42	1635.47	60.02			
18	1073.63	16.01	43	1655.77	61.98			
19	1084.40	17.64	44	1673.32	63.98			
20	972.90	18.03	45	1691.49	65.99			
21	898.66	18.38	46	1710.28	68.00			
22	922.97	20.02	47	1725.76	70.00			
23	1093.40	22.00	48	1742.87	71.99			
24	1143.85	24.02	49	1756.81	73.99			
25	1187.29	26.02	50	1772.68	76.00			

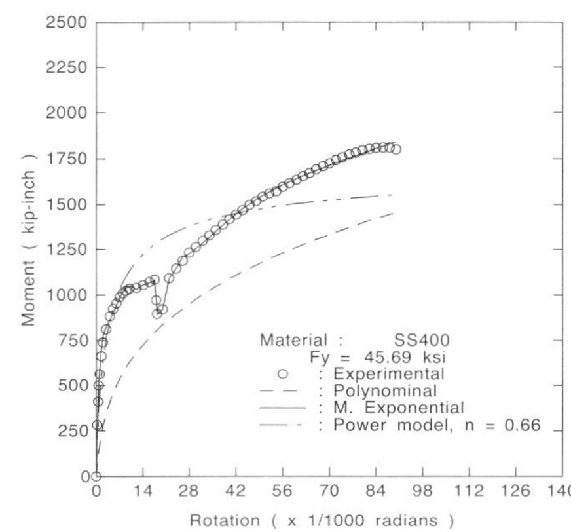

```
Moment-rotation prediction equations     ( R : X 1/1000 radians )

Frye and Morris polynominal model  :     R = Sum ( Ai X ( K*Bm )**Pi X 10**Qi )
     t =  0.590550"  xd = 15.748000"   f =  0.787400"  xl =  7.874020"
    A1 =  8.460000   A2 =  1.010000   A3 =  1.240000    K =  0.006388
    P1 =         1   P2 =         3   P3 =         5   Q1 =        -1     Q2 =        -1     Q3 =        -5

Modified exponential model  :
    Bm = Sum ( Ai X ( 1 - exp( - R/( 2*i*C ) ) ) )  +      Sum ( RKj X ( R - Rj0 ) )  +  Bm0
     C =  0.81241667E+00             Bmo=   0.00000000E+00          Nexp=  6        Nliner=  6
    AI =  0.61244397E+04  -0.41941031E+05   0.11853589E+06  -0.13825958E+06   0.59534420E+05  -0.30006035E+04
   Rj0 =         12.0000          17.6400          18.3800          20.0200          22.0000
                 26.0200
   RKj =  0.38906246E+02  -0.26688608E+03   0.26962832E+03   0.60784672E+02  -0.68154121E+02
         -0.28209955E+02

Power model  :    Bm = ( Rki X R ) / ( 1 + ( R/R0 )**rn )**( 1/rn )
    rn = 0.664      rki =   0.93453673E+03         rmu =   0.17350428E+04
```

No	Rotation (radians) X 1/1000	Moment (kip-inch) Expri.	Poly.	M.Expo.	P.Model.	Connection stiffness (kip-inch) X 1000 Expri.	Poly.	M.Expo.	P.Model.
1	0.00	0.00	0.00	0.00	0.00	0.1133E+04	0.1850E+03	0.9280E+03	0.9345E+03
3	0.50	409.17	89.08	362.97	275.98	0.4335E+03	0.1658E+03	0.5526E+03	0.3891E+03
5	1.01	563.36	165.00	579.25	436.97	0.2303E+03	0.1324E+03	0.3154E+03	0.2595E+03
7	2.02	739.81	273.82	767.24	636.76	0.1213E+03	0.8826E+02	0.9765E+02	0.1532E+03
9	4.00	883.76	408.36	863.14	854.23	0.5721E+02	0.5378E+02	0.3500E+02	0.8014E+02
11	6.02	954.46	500.90	944.41	983.32	0.3174E+02	0.3958E+02	0.4285E+02	0.5131E+02
13	8.01	1008.97	571.45	1017.98	1069.50	0.1752E+02	0.3198E+02	0.2766E+02	0.3668E+02
15	10.00	1034.49	629.95	1047.59	1133.12	0.8140E+01	0.2713E+02	0.1918E+01	0.2792E+02
17	14.02	1055.77	725.88	1055.51	1223.19	0.8130E+01	0.2118E+02	0.8567E+01	0.1807E+02
19	17.64	1084.40	796.28	1082.75	1279.21	-0.2294E+03	0.1794E+02	0.9971E+01	0.1329E+02
21	18.38	898.66	809.46	893.46	1288.77	-0.1722E+03	0.1741E+02	-0.2546E+03	0.1256E+02
23	22.00	1093.40	868.37	1093.05	1328.91	0.5583E+02	0.1530E+02	0.8985E+02	0.9802E+01
25	26.02	1187.29	926.25	1209.04	1363.91	0.2227E+02	0.1358E+02	0.3511E+02	0.7741E+01
27	30.00	1262.73	977.58	1253.16	1391.73	0.1603E+02	0.1227E+02	0.1431E+02	0.6317E+01
29	34.04	1328.23	1024.95	1317.42	1415.01	0.1538E+02	0.1122E+02	0.1684E+02	0.5263E+01
31	38.02	1387.83	1067.88	1384.27	1434.33	0.1466E+02	0.1038E+02	0.1643E+02	0.4478E+01
33	42.01	1441.27	1107.86	1446.69	1450.93	0.1294E+02	0.9676E+01	0.1475E+02	0.3867E+01
35	46.02	1498.31	1145.42	1501.84	1465.42	0.1239E+02	0.9076E+01	0.1276E+02	0.3378E+01
37	50.01	1541.76	1180.59	1549.05	1478.08	0.1037E+02	0.8563E+01	0.1096E+02	0.2984E+01
39	54.00	1572.37	1213.84	1589.75	1489.32	0.9046E+01	0.8115E+01	0.9505E+01	0.2659E+01
41	58.01	1616.40	1245.97	1625.55	1499.42	0.9754E+01	0.7715E+01	0.8411E+01	0.2387E+01
43	61.98	1655.77	1275.68	1657.31	1508.44	0.9574E+01	0.7369E+01	0.7635E+01	0.2160E+01
45	65.99	1691.49	1304.43	1686.77	1516.69	0.9195E+01	0.7056E+01	0.7094E+01	0.1963E+01
47	70.00	1725.76	1332.43	1714.44	1524.22	0.8169E+01	0.6769E+01	0.6729E+01	0.1795E+01
49	73.99	1756.81	1358.65	1740.78	1531.08	0.7431E+01	0.6516E+01	0.6490E+01	0.1649E+01
51	77.97	1782.69	1384.34	1766.28	1537.39	0.5445E+01	0.6281E+01	0.6335E+01	0.1521E+01
53	82.01	1802.90	1409.49	1791.65	1543.30	0.3516E+01	0.6063E+01	0.6234E+01	0.1408E+01
55	86.08	1811.22	1433.43	1816.88	1548.82	0.4147E-01	0.5865E+01	0.6170E+01	0.1306E+01
57	89.99	1800.03	1456.19	1840.93	1553.75	-0.4569E+01	0.5686E+01	0.6132E+01	0.1220E+01

Connection type : Top-and seat-angle connections
Mode : All bolted

Tested by : Y. Sato et al. (2007) Japan
Test Id. : G105

Column : --
Beam : H400x200x8x13
Angle : L200x200x15

Fasteners: F10T- -M20
7/8" Oversize holes
Material : SS400
Fy = 45.69 ksi
Fu = 68.02 ksi

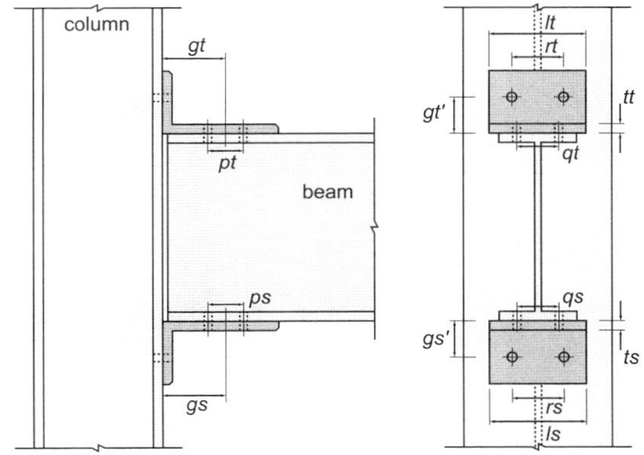

Major parameters

lt = 7.8740"	tt = 0.5906"	ls = 7.8740"	ts = 0.5906"
gt = 4.1339"	gt'= 4.1339"	gs = 4.1339"	gs'= 4.1339"
qt = 4.7244"	rt = 4.7244"	qs = 4.7244"	rs = 4.7244"
pt = 3.5433"	ps = 3.5433"		
nt = 2 X 2	nt'= 2 X 1	ns = 2 X 2	ns'= 2 X 1

Remark 1) Failure mode: Bolt fracture
2) Original T & S angles: L200x200x20

No	Moment (k-in)	Rotation (radians) X 1/1000	No	Moment (k-in)	Rotation (radians) X 1/1000	No	Moment (k-in)	Rotation (radians) X 1/1000
1	0.00	0.00	26	880.14	64.00	51	1516.98	153.61
2	57.51	0.25	27	918.32	68.01	52	1513.96	155.99
3	117.54	0.51	28	954.82	72.01	53	1497.60	160.00
4	159.46	0.75	29	994.89	76.02	54	1457.00	161.93
5	199.86	1.00	30	1025.00	80.01			
6	260.11	1.52	31	1052.99	83.99			
7	310.44	2.01	32	1083.13	88.00			
8	384.66	3.01	33	1116.65	92.01			
9	425.90	4.00	34	1155.52	96.02			
10	453.00	4.99	35	1177.36	100.00			
11	484.47	6.99	36	988.27	100.93			
12	508.37	9.00	37	954.09	102.01			
13	538.30	12.02	38	1033.51	104.00			
14	575.28	16.02	39	1094.67	106.01			
15	607.56	19.99	40	1143.29	110.00			
16	634.81	24.01	41	1182.13	114.01			
17	662.50	28.02	42	1222.25	117.99			
18	689.51	32.01	43	1267.15	122.00			
19	716.17	36.02	44	1303.68	126.00			
20	721.82	40.01	45	1346.76	130.00			
21	731.04	44.00	46	1385.91	133.98			
22	753.35	47.99	47	1425.23	138.02			
23	777.10	52.01	48	1459.00	142.00			
24	813.62	56.00	49	1488.45	146.02			
25	845.76	60.02	50	1510.31	149.99			

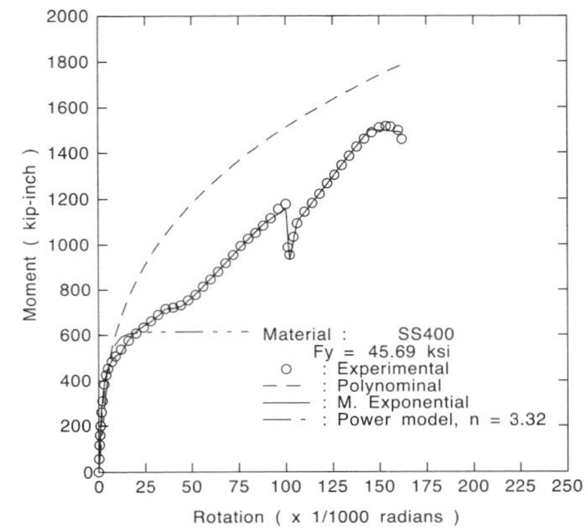

A.4 – 140

```
Moment-rotation prediction equations      ( R : X 1/1000 radians )

Frye and Morris polynominal model  :     R = Sum ( Ai X ( K*Bm )**Pi X 10**Qi )
      t =  0.590550"  xd = 15.748000"   f =  0.787400"  xl =  7.874020"
     A1 =  8.460000   A2 =  1.010000   A3 =  1.240000    K =  0.006388
     P1 =         1   P2 =         3   P3 =         5   Q1 =        -1    Q2 =        -1    Q3 =        -5

Modified exponential model  :
     Bm = Sum ( Ai X ( 1 - exp( - R/( 2*i*C ) ) ) )  +      Sum ( RKj X ( R - Rj0 ) )  +  Bm0
      C =  0.14119167E+01             Bmo=  0.00000000E+00          Nexp=  6         Nliner=  8
     AI =  0.18488477E+04  -0.10399070E+05  0.27634051E+05  -0.26642622E+05   0.47190750E+04   0.32665411E+04
    Rj0 =         12.0000          36.0200         44.0000          88.0000         100.0000
                 102.0000         106.0000        145.0000
    RKj =  0.13546827E+02  -0.93339015E+01  0.30832937E+01  -0.23017097E+01  -0.11805826E+03
           0.15449713E+03  -0.31313582E+02 -0.10757177E+02

Power model  :    Bm = ( Rki X R ) / ( 1 + ( R/R0 )**rn )**( 1/rn )
     rn = 3.317       rki =    0.92560280E+02          rmu =    0.61488186E+03
```

No	Rotation (radians) X 1/1000	Moment (kip-inch) Expri.	Poly.	M.Expo.	P.Model.	Connection stiffness (kip-inch) X 1000 Expri.	Poly.	M.Expo.	P.Model.
1	0.00	0.00	0.00	0.00	0.00	0.2300E+03	0.1850E+03	0.2437E+03	0.9256E+02
3	0.51	117.54	90.73	110.20	47.20	0.2017E+03	0.1652E+03	0.1905E+03	0.9254E+02
5	1.00	199.86	163.68	193.24	92.51	0.1467E+03	0.1330E+03	0.1500E+03	0.9234E+02
7	2.01	310.44	272.92	312.70	184.99	0.9335E+02	0.8856E+02	0.9130E+02	0.9032E+02
9	4.00	425.90	408.36	428.97	351.69	0.3452E+02	0.5378E+02	0.3501E+02	0.7414E+02
11	6.99	484.47	537.34	489.02	511.23	0.1382E+02	0.3538E+02	0.1136E+02	0.3349E+02
13	12.02	538.30	681.21	522.72	591.08	0.9624E+01	0.2369E+02	0.1670E+02	0.6036E+01
15	19.99	607.56	836.59	615.16	610.16	0.7458E+01	0.1639E+02	0.7262E+01	0.7704E+00
17	28.02	662.50	952.70	661.09	613.32	0.6837E+01	0.1288E+02	0.5520E+01	0.1834E+00
19	36.02	716.17	1046.73	716.74	614.20	0.4026E+01	0.1078E+02	0.8717E+01	0.6241E-01
21	44.00	731.04	1126.80	726.03	614.53	0.3951E+01	0.9367E+01	0.2769E+01	0.2637E-01
23	52.01	777.10	1197.48	782.05	614.68	0.7538E+01	0.8331E+01	0.7896E+01	0.1282E-01
25	60.02	845.76	1260.90	849.22	614.76	0.8317E+01	0.7538E+01	0.8713E+01	0.6914E-02
27	68.01	918.32	1318.53	919.55	614.80	0.9323E+01	0.6909E+01	0.8811E+01	0.4032E-02
29	76.02	994.89	1371.75	989.37	614.82	0.8766E+01	0.6395E+01	0.8594E+01	0.2494E-02
31	83.99	1052.99	1420.96	1056.68	614.84	0.7273E+01	0.5967E+01	0.8294E+01	0.1622E-02
33	92.01	1116.65	1467.31	1112.81	614.85	0.9027E+01	0.5601E+01	0.5712E+01	0.1094E-02
35	100.00	1177.36	1510.78	1157.52	614.86	-0.1638E+03	0.5287E+01	0.5489E+01	0.7639E-03
37	102.01	954.09	1521.37	932.75	614.86	-0.6473E+01	0.5215E+01	0.4188E+02	0.7010E-03
39	106.01	1094.67	1542.09	1099.80	614.86	0.2432E+02	0.5077E+01	0.1049E+02	0.5938E-03
41	114.01	1182.13	1581.52	1183.16	614.87	0.9883E+01	0.4828E+01	0.1036E+02	0.4338E-03
43	122.00	1267.15	1619.64	1265.58	614.87	0.1016E+02	0.4605E+01	0.1028E+02	0.3238E-03
45	130.00	1346.76	1655.45	1347.55	614.87	0.1030E+02	0.4408E+01	0.1022E+02	0.2462E-03
47	138.02	1425.23	1690.43	1429.36	614.87	0.9105E+01	0.4228E+01	0.1018E+02	0.1901E-03
49	146.02	1488.45	1723.24	1499.75	614.88	0.6410E+01	0.4068E+01	-0.5974E+00	0.1491E-03
51	153.61	1516.98	1753.85	1495.16	614.88	-0.3484E-01	0.3927E+01	-0.6117E+00	0.1198E-03
53	160.00	1497.60	1778.78	1491.23	614.88	-0.1552E+02	0.3818E+01	-0.6198E+00	0.1005E-03

IV - 71

Connection type : Top-and seat-angle connections
Mode : All bolted

Tested by : Y. Sato et al. (2007) Japan
Test Id. : G150

Column : --
Beam : H400x200x8x13
Angle : L200x200x15

Fasteners: F10T- -M20
7/8" Oversize holes
Material : SS400
Fy = 45.69 ksi
Fu = 68.02 ksi

Major parameters

lt = 7.8740"	tt = 0.5906"	ls = 7.8740"	ts = 0.5906"
gt = 4.1339"	gt'= 5.9055"	gs = 4.1339"	gs'= 5.9055"
qt = 4.7244"	rt = 4.7244"	qs = 4.7244"	rs = 4.7244"
pt = 3.5433"	ps = 3.5433"		
nt = 2 X 2	nt'= 2 X 1	ns = 2 X 2	ns'= 2 X 1

Remark 1) Original T & S angles: L200x200x20
2)

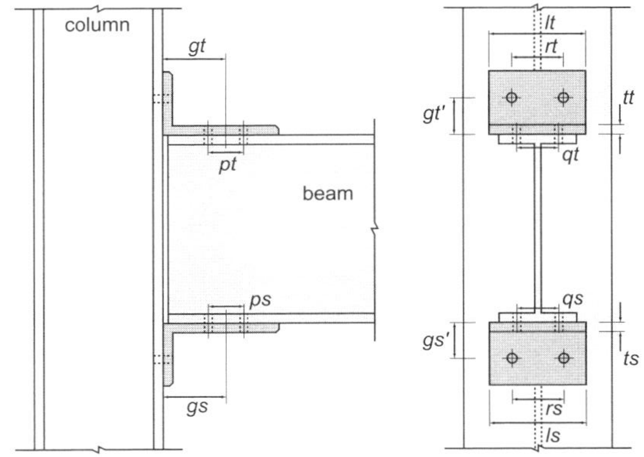

No	Moment (k-in)	Rotation (radians) X 1/1000
1	0.00	0.00
2	38.65	0.25
3	67.18	0.50
4	106.29	1.02
5	141.32	1.51
6	172.48	2.01
7	227.48	3.01
8	265.49	4.01
9	288.41	5.00
10	312.41	6.99
11	332.91	9.98
12	355.67	14.01
13	374.28	17.99
14	390.63	22.00
15	406.35	26.02
16	420.13	30.00
17	436.77	33.99
18	449.94	37.99
19	467.89	41.99
20	486.12	45.99
21	504.04	50.02
22	522.98	54.00
23	538.99	57.99
24	525.35	60.00
25	533.40	64.00
26	550.34	68.01
27	579.44	71.98
28	610.53	75.99
29	638.33	80.01
30	666.22	84.01
31	692.13	87.99
32	719.61	92.01
33	747.72	95.98
34	775.86	100.01
35	803.01	104.00
36	831.12	107.99
37	858.22	112.02
38	887.51	116.01
39	915.41	120.00
40	941.19	124.02
41	966.57	127.99
42	984.89	132.00
43	929.53	133.35
44	953.64	133.99
45	996.15	136.01
46	1017.34	140.01
47	986.32	141.99
48	940.61	143.99
49	974.04	146.02
50	1007.59	150.01
51	1038.13	153.99
52	1070.47	157.99
53	1100.03	162.00
54	1134.06	166.01
55	1169.53	169.76

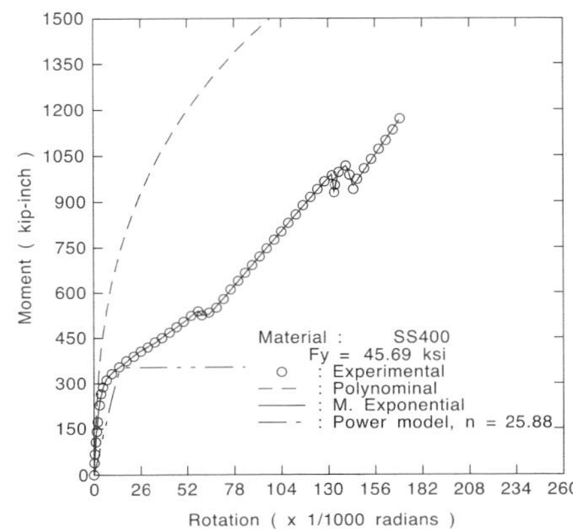

A.4 – 142

```
Moment-rotation prediction equations     ( R : X 1/1000 radians )

Frye and Morris polynominal model   :     R = Sum ( Ai X ( K*Bm )**Pi X 10**Qi )
     t =  0.590550"  xd = 15.748000"   f =  0.787400"  xl =  7.874020"
    A1 =  8.460000   A2 =  1.010000   A3 =  1.240000    K =  0.006388
    P1 =          1   P2 =          3   P3 =          5   Q1 =        -1     Q2 =        -1     Q3 =        -5

Modified exponential model  :
    Bm = Sum ( Ai X ( 1 - exp( - R/( 2*i*C ) ) ) )  +      Sum ( RKj X ( R - Rj0 ) )  +  Bm0
     C =  0.14771667E+01             Bmo=   0.00000000E+00          Nexp=  6        Nliner= 10
    AI =  0.28832471E+03  0.34139463E+04 -0.20282495E+05  0.45567403E+05 -0.44556145E+05  0.15937328E+05
   Rj0 =          4.0100         33.9900         57.9900         60.0000         64.0000
                132.0000        133.4000        136.0000        140.0000        144.0000
   RKj =  0.28305317E+01  0.22124109E+01 -0.12508825E+02  0.81095544E+01  0.57292255E+01
         -0.48390871E+02  0.66861062E+02 -0.19824867E+02 -0.21864440E+02  0.25205777E+02

Power model  :    Bm = ( Rki X R ) / ( 1 + ( R/R0 )**rn )**( 1/rn )
    rn =25.883       rki =   0.25131527E+02          rmu =   0.35426430E+03
```

No	Rotation (radians) X 1/1000	Moment (kip-inch) Expri.	Moment (kip-inch) Poly.	Moment (kip-inch) M.Expo.	Moment (kip-inch) P.Model.	Connection stiffness (kip-inch) X 1000 Expri.	Connection stiffness Poly.	Connection stiffness M.Expo.	Connection stiffness P.Model.
1	0.00	0.00	0.00	0.00	0.00	0.1546E+03	0.1850E+03	0.1257E+03	0.2513E+02
3	0.50	67.18	89.08	57.47	12.57	0.1015E+03	0.1658E+03	0.1048E+03	0.2513E+02
5	1.51	141.32	224.44	145.76	37.95	0.6696E+02	0.1065E+03	0.7196E+02	0.2513E+02
7	3.01	227.48	349.26	227.84	75.65	0.4650E+02	0.6645E+02	0.4025E+02	0.2513E+02
9	5.00	288.41	457.69	285.55	125.66	0.1947E+02	0.4549E+02	0.2054E+02	0.2513E+02
11	9.98	332.91	629.66	335.60	250.81	0.6342E+01	0.2715E+02	0.5037E+01	0.2513E+02
13	17.99	374.28	802.58	372.27	354.24	0.4378E+01	0.1768E+02	0.4782E+01	0.3564E-01
15	26.02	406.35	926.26	408.16	354.26	0.3686E+01	0.1358E+02	0.3871E+01	0.1754E-05
17	33.99	436.77	1024.40	433.29	354.26	0.3732E+01	0.1123E+02	0.2532E+01	0.1332E-08
19	41.99	467.89	1107.66	468.65	354.26	0.4523E+01	0.9679E+01	0.4247E+01	0.4536E-11
21	50.02	504.04	1180.67	503.27	354.26	0.4604E+01	0.8562E+01	0.4453E+01	0.4108E-13
23	57.99	538.99	1245.42	540.52	354.26	-0.3169E+01	0.7721E+01	0.4904E+01	0.7718E-15
25	64.00	533.40	1290.24	528.24	354.26	0.3117E+01	0.7208E+01	0.8213E+00	0.5448E-16
27	71.98	579.44	1345.42	581.78	354.26	0.7541E+01	0.6642E+01	0.6840E+01	0.2314E-17
29	80.01	638.33	1397.27	637.33	354.26	0.6944E+01	0.6167E+01	0.6973E+01	0.1347E-18
31	87.99	692.13	1444.83	693.09	354.26	0.6672E+01	0.5774E+01	0.6986E+01	0.1046E-19
33	95.98	747.72	1489.55	748.71	354.26	0.7033E+01	0.5437E+01	0.6929E+01	0.1011E-20
35	104.00	803.01	1531.93	803.93	354.26	0.6924E+01	0.5144E+01	0.6840E+01	0.1169E-21
37	112.02	858.22	1572.11	858.40	354.26	0.7034E+01	0.4886E+01	0.6744E+01	0.1587E-22
39	120.00	915.41	1609.93	911.87	354.26	0.6704E+01	0.4660E+01	0.6658E+01	0.2495E-23
41	127.99	966.57	1646.75	964.76	354.26	0.5486E+01	0.4455E+01	0.6585E+01	0.4411E-24
43	133.35	929.53	1669.92	934.62	354.26	0.1237E+02	0.4332E+01	-0.4185E+02	0.1464E-24
45	136.01	996.15	1681.41	997.59	354.26	0.1576E+02	0.4274E+01	0.5172E+01	0.8609E-25
47	141.99	986.32	1706.80	984.91	354.26	-0.1924E+02	0.4147E+01	-0.1673E+02	0.2708E-25
49	146.02	974.04	1723.43	968.38	354.26	0.1375E+02	0.4068E+01	0.8460E+01	0.1276E-25
51	153.99	1038.13	1754.92	1035.68	354.26	0.7879E+01	0.3923E+01	0.8430E+01	0.3058E-26
53	162.00	1100.03	1786.06	1103.11	354.26	0.7928E+01	0.3786E+01	0.8408E+01	0.7823E-27
55	169.76	1169.53	1815.20	1168.30	354.26	0.9459E+01	0.3665E+01	0.8393E+01	0.2224E-27

IV - 72

Connection type : Top-and seat-angle connections
Mode : All bolted

Tested by : Y. Sato et al. (2007) Japan
Test Id. : GW60

Column : --
Beam : H400x200x8x13
Angle : L200x200x15

Fasteners: F10T- -M20
7/8" Oversize holes
Material : SS400
Fy = 45.69 ksi
Fu = 68.02 ksi

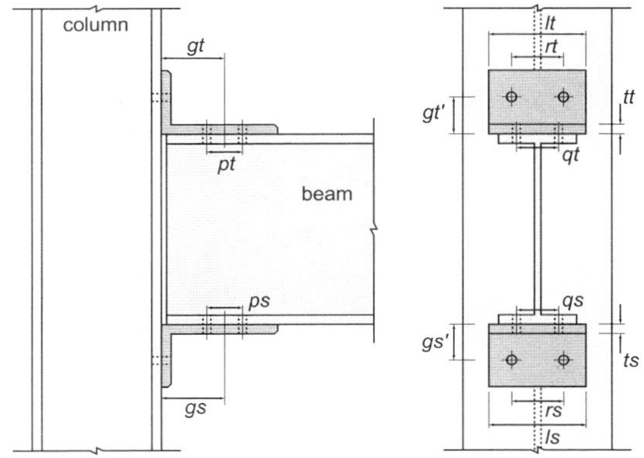

Major parameters

lt = 7.8740"	tt = 0.5906"	ls = 7.8740"	ts = 0.5906"
gt = 4.1339"	gt'= 4.1339"	gs = 4.1339"	gs'= 4.1339"
qt = 4.7244"	rt = 4.7244"	qs = 4.7244"	rs = 4.7244"
pt = 3.5433"	pt'= 3.5433"	ps = 3.5433"	ps'= 3.5433"
nt = 2 X 2	nt'= 2 X 2	ns = 2 X 2	ns'= 2 X 2

Remark 1) Failure mode: Bolt fracture
2) Original T & S angles: L200x200x20

No	Moment (k-in)	Rotation (radians) X 1/1000	No	Moment (k-in)	Rotation (radians) X 1/1000	No	Moment (k-in)	Rotation (radians) X 1/1000
1	0.00	0.00	26	1617.79	56.01	51	1029.11	157.99
2	174.65	0.25	27	1667.57	59.99	52	1075.93	162.01
3	268.45	0.50	28	1717.71	63.99	53	1131.44	166.00
4	366.74	0.75	29	1767.11	67.99	54	1162.76	170.01
5	439.11	1.00	30	1813.74	72.00	55	1209.44	173.99
6	572.33	1.51	31	1861.04	75.99	56	1255.52	178.01
7	669.61	2.01	32	1906.80	79.98	57	1305.66	182.00
8	784.26	3.02	33	1949.84	84.01	58	1355.14	186.00
9	846.39	4.01	34	1982.39	88.01	59	1452.48	190.00
10	933.23	6.00	35	1979.91	92.02	60	1515.75	193.88
11	994.80	8.00	36	1913.50	98.01			
12	1057.84	10.00	37	1829.82	101.71			
13	1128.56	14.00	38	1021.13	104.42			
14	1177.01	18.00	39	486.80	106.29			
15	1010.25	19.99	40	537.09	112.02			
16	1195.88	22.00	41	584.49	116.01			
17	1272.98	25.99	42	632.93	120.00			
18	1347.20	30.00	43	696.56	126.01			
19	1409.49	34.01	44	737.88	130.00			
20	1464.20	38.00	45	779.19	134.02			
21	1386.30	40.02	46	818.99	138.00			
22	1329.00	41.99	47	862.48	142.01			
23	1382.37	43.99	48	906.07	146.00			
24	1486.20	46.00	49	946.05	150.02			
25	1568.25	51.99	50	985.04	153.99			

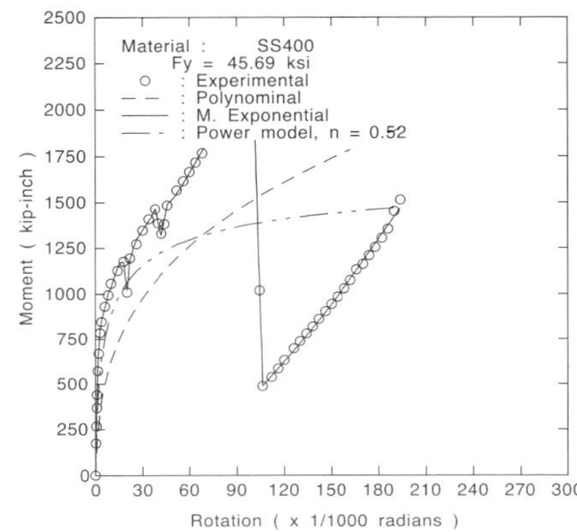

```
Moment-rotation prediction equations      ( R : X 1/1000 radians )

Frye and Morris polynominal model   :     R = Sum ( Ai X ( K*Bm )**Pi X 10**Qi )
      t =  0.590550"   xd = 15.748000"   f =  0.787400"   xl =  7.874020"
     A1 =  8.460000    A2 =  1.010000   A3 =  1.240000     K =  0.006388
     P1 =         1    P2 =         3   P3 =         5    Q1 =        -1     Q2 =        -1     Q3 =        -5

Modified exponential model  :
     Bm = Sum  ( Ai X ( 1 - exp( - R/( 2*i*C ) ) ) )  +      Sum ( RKj X ( R - Rj0 ) )  +  Bm0
       C =   0.16781667E+01            Bmo=  0.00000000E+00         Nexp=  6         Nliner= 10
      AI =   0.12266980E+05  -0.11668683E+06  0.43451023E+06  -0.74717843E+06  0.60669605E+06  -0.18819098E+06
     Rj0 =          8.0000          18.0000         20.0000          22.0000         38.0000
                   42.0000          46.0000         92.0000         101.7000        106.3000
     RKj =  -0.38186888E+02  -0.68939772E+02  0.18357869E+03  -0.44816276E+02  -0.59358014E+02
             0.74190477E+02  -0.40714248E+02  -0.20100143E+02  -0.27625754E+03  0.30342047E+03

Power model   :    Bm = ( Rki X R ) / ( 1 + ( R/R0 )**rn )**( 1/rn )
     rn = 0.520      rki =    0.93452773E+03          rmu =   0.17350388E+04
```

No	Rotation (radians) X 1/1000	Moment (kip-inch) Expri.	Poly.	M.Expo.	P.Model.	Connection stiffness (kip-inch) X 1000 Expri.	Poly.	M.Expo.	P.Model.
1	0.00	0.00	0.00	0.00	0.00	0.6986E+03	0.1850E+03	0.5779E+03	0.9345E+03
3	0.50	268.45	89.08	250.92	212.89	0.3842E+03	0.1658E+03	0.4318E+03	0.2828E+03
5	1.00	439.11	163.68	437.29	327.79	0.2802E+03	0.1330E+03	0.3186E+03	0.1900E+03
7	2.01	669.61	272.92	674.52	476.19	0.1677E+03	0.8856E+02	0.1658E+03	0.1160E+03
9	4.01	846.39	408.89	856.78	647.58	0.5641E+02	0.5368E+02	0.4525E+02	0.6478E+02
11	8.00	994.80	571.23	1003.85	829.92	0.3115E+02	0.3200E+02	0.4792E+02	0.3306E+02
13	14.00	1128.56	725.66	1140.14	975.18	0.1490E+02	0.2119E+02	0.2104E+02	0.1804E+02
15	19.99	1010.25	836.55	1010.25	1062.87	0.3837E+01	0.1639E+02	-0.8949E+02	0.1197E+02
17	25.99	1272.98	925.86	1290.22	1124.12	0.1892E+02	0.1359E+02	0.1886E+02	0.8743E+01
19	34.01	1409.49	1024.62	1400.08	1183.43	0.1462E+02	0.1123E+02	0.1379E+02	0.6281E+01
21	40.02	1386.30	1088.65	1387.54	1217.48	-0.3377E+02	0.1000E+02	-0.3676E+02	0.5121E+01
23	43.99	1382.37	1127.02	1402.24	1236.62	0.3914E+02	0.9363E+01	0.4387E+02	0.4541E+01
25	51.99	1568.25	1197.32	1549.77	1269.23	0.1288E+02	0.8333E+01	0.1218E+02	0.3665E+01
27	59.99	1667.57	1260.68	1661.63	1295.93	0.1252E+02	0.7541E+01	0.1486E+02	0.3043E+01
29	67.99	1767.11	1318.39	1776.47	1318.34	0.1199E+02	0.6911E+01	0.1339E+02	0.2582E+01
31	75.99	1861.04	1371.56	1871.71	1337.53	0.1166E+02	0.6396E+01	0.1032E+02	0.2229E+01
33	84.01	1949.84	1421.08	1942.01	1354.24	0.9404E+01	0.5966E+01	0.7297E+01	0.1950E+01
35	92.02	1979.91	1467.37	1990.38	1368.93	-0.4815E+01	0.5601E+01	-0.1507E+02	0.1726E+01
37	101.71	1829.82	1520.02	1833.02	1384.56	-0.1818E+03	0.5224E+01	-0.2929E+03	0.1508E+01
39	106.29	486.80	1543.77	490.65	1391.26	-0.2133E+03	0.5066E+01	-0.2933E+03	0.1421E+01
41	116.01	584.49	1591.36	584.85	1404.27	0.1201E+02	0.4769E+01	0.9952E+01	0.1261E+01
43	126.01	696.56	1637.47	685.43	1416.19	0.1045E+02	0.4506E+01	0.1022E+02	0.1127E+01
45	134.02	779.19	1673.17	768.70	1424.84	0.1014E+02	0.4316E+01	0.1058E+02	0.1036E+01
47	142.01	862.48	1706.60	854.84	1432.79	0.1089E+02	0.4148E+01	0.1098E+02	0.9563E+00
49	150.02	946.05	1739.51	944.31	1440.16	0.9882E+01	0.3993E+01	0.1135E+02	0.8867E+00
51	157.99	1029.11	1771.05	1036.14	1446.98	0.1133E+02	0.3851E+01	0.1168E+02	0.8255E+00
53	166.00	1131.44	1800.94	1130.83	1453.37	0.1087E+02	0.3724E+01	0.1195E+02	0.7709E+00
55	173.99	1209.44	1830.45	1227.24	1459.33	0.1160E+02	0.3604E+01	0.1217E+02	0.7222E+00
57	182.00	1305.66	1859.09	1325.43	1464.94	0.1247E+02	0.3493E+01	0.1234E+02	0.6783E+00
59	190.00	1452.48	1886.17	1424.69	1470.21	0.2026E+02	0.3392E+01	0.1247E+02	0.6389E+00

IV - 73

Connection type : Top-and seat-angle connections
Mode : All bolted

Tested by : Y. Sato et al. (2007) Japan
Test Id. : A40

Column : --
Beam : H400x200x8x13
Angle : L200x100x15

Fasteners: F10T- -M20
7/8" Oversize holes
Material : SS400
Fy = 45.69 ksi
Fu = 68.02 ksi

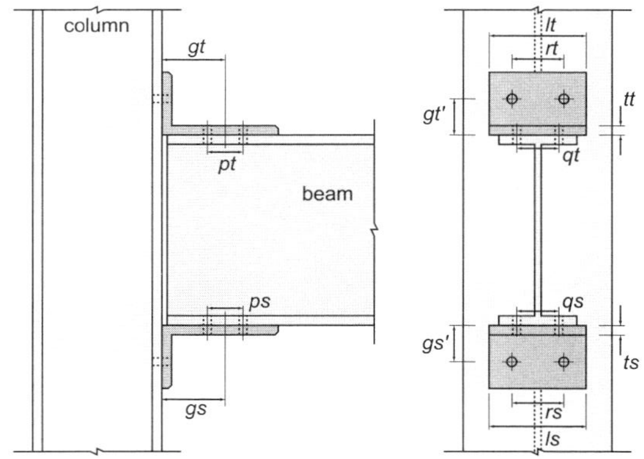

Major parameters

lt = 7.8740"	tt = 0.5906"	ls = 7.8740"	ts = 0.5906"
gt = 4.1339"	gt'= 2.3622"	gs = 4.1339"	gs'= 2.3622"
qt = 4.7244"	rt = 4.7244"	qs = 4.7244"	rs = 4.7244"
pt = 3.5433"	ps = 3.5433"		
nt = 2 X 2	nt'= 2 X 1	ns = 2 X 2	ns'= 2 X 1

Remark 1) Failure mode: Bolt fracture
2) Original T & S angles: L200x200x20

No	Moment (k-in)	Rotation (radians) X 1/1000	No	Moment (k-in)	Rotation (radians) X 1/1000	No	Moment (k-in)	Rotation (radians) X 1/1000
1	0.00	0.00	26	1242.31	31.99	51	1639.26	82.00
2	138.43	0.24	27	1268.56	34.01	52	1637.25	82.87
3	257.65	0.51	28	1295.82	36.00	53	1631.53	84.02
4	352.18	0.75	29	1321.01	38.01	54	1615.89	86.02
5	415.88	1.00	30	1345.97	40.00	55	1590.08	88.00
6	553.55	1.50	31	1369.03	42.02			
7	641.35	1.99	32	1393.02	44.00			
8	763.72	3.01	33	1415.08	46.00			
9	826.05	4.01	34	1439.33	47.99			
10	867.65	5.01	35	1463.87	49.99			
11	897.00	6.02	36	1485.54	52.02			
12	926.49	7.01	37	1507.46	54.01			
13	952.06	7.99	38	1526.43	56.00			
14	972.46	8.99	39	1543.53	58.02			
15	985.87	10.01	40	1571.29	60.02			
16	1009.07	12.00	41	1498.76	62.02			
17	1028.90	14.01	42	1524.36	64.01			
18	1051.69	16.02	43	1550.85	66.02			
19	1076.13	17.99	44	1573.57	68.00			
20	1098.61	19.99	45	1593.98	69.98			
21	1121.14	22.02	46	16[illegible]1.97	71.99			
22	1043.39	24.01	47	1625.39	74.02			
23	1146.11	26.01	48	1637.43	76.01			
24	1180.09	28.03	49	1640.70	77.98			
25	1213.83	30.01	50	1640.73	80.02			

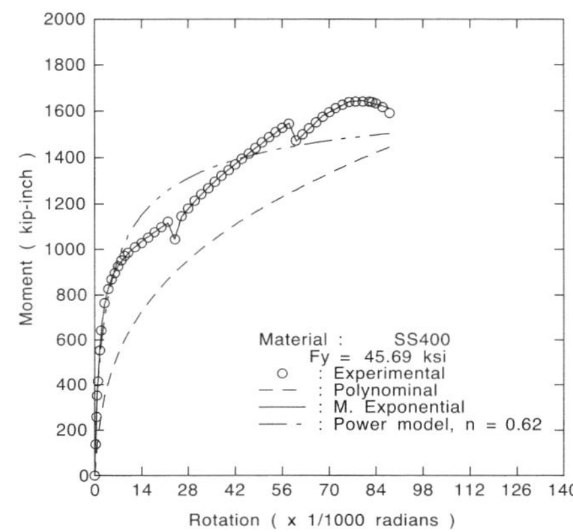

A.4 – 146

```
Moment-rotation prediction equations     ( R : X 1/1000 radians )

Frye and Morris polynominal model  :     R = Sum ( Ai X ( K*Bm )**Pi X 10**Qi )
      t =  0.590550"  xd = 15.748000"  f =  0.787400"  xl =  7.874020"
     A1 =  8.460000   A2 =  1.010000  A3 =  1.240000    K =  0.006388
     P1 =         1   P2 =         3  P3 =         5   Q1 =        -1    Q2 =       -1    Q3 =       -5

Modified exponential model  :
     Bm = Sum ( Ai X ( 1 - exp( - R/( 2*i*C ) ) ) )  +     Sum ( RKj X ( R - Rj0 ) )  +  Bm0
      C =  0.79333333E+00            Bmo=   0.00000000E+00           Nexp=  6         Nliner=  9
     AI =  0.15745252E+04  -0.75277842E+04   0.32435955E+05  -0.65473053E+05   0.59828878E+05  -0.19669285E+05
    Rj0 =         22.0200          24.0100          26.0100          58.0200          60.2000
                  68.0000          76.0100          80.0200          87.9900
    RKj = -0.44586288E+02   0.87236427E+02  -0.31090774E+02  -0.46519103E+02   0.49448742E+02
          -0.65112762E+01  -0.64425099E+01  -0.54302668E+01  -0.22580333E+04

Power model  :    Bm = ( Rki X R ) / ( 1 + ( R/R0 )**rn )**( 1/rn )
     rn = 0.616       rki =    0.93453673E+03          rmu =    0.17350428E+04
```

No	Rotation (radians) X 1/1000	Moment (kip-inch) Expri.	Poly.	M.Expo.	P.Model.	Connection stiffness (kip-inch) X 1000 Expri.	Poly.	M.Expo.	P.Model.
1	0.00	0.00	0.00	0.00	0.00	0.5768E+03	0.1850E+03	0.5936E+03	0.9345E+03
3	0.51	257.65	90.73	254.54	260.53	0.4163E+03	0.1652E+03	0.4155E+03	0.3521E+03
5	1.00	415.88	163.68	427.22	401.59	0.2616E+03	0.1330E+03	0.2964E+03	0.2386E+03
7	1.99	641.35	271.14	641.93	583.15	0.1600E+03	0.8916E+02	0.1545E+03	0.1434E+03
9	4.01	826.05	408.90	826.37	791.46	0.5196E+02	0.5368E+02	0.5255E+02	0.7570E+02
11	6.02	897.00	500.90	902.21	914.10	0.2943E+02	0.3958E+02	0.2799E+02	0.4955E+02
13	7.99	952.06	570.81	948.07	997.25	0.2328E+02	0.3204E+02	0.1965E+02	0.3609E+02
15	10.01	985.87	630.22	982.89	1061.05	0.1264E+02	0.2711E+02	0.1524E+02	0.2771E+02
17	14.01	1028.90	725.67	1034.68	1151.04	0.1060E+02	0.2119E+02	0.1143E+02	0.1836E+02
19	17.99	1076.13	802.52	1077.40	1213.23	0.1183E+02	0.1768E+02	0.1016E+02	0.1334E+02
21	22.02	1121.14	868.68	1115.69	1260.28	-0.1424E+02	0.1529E+02	0.8726E+01	0.1024E+02
23	26.01	1146.11	926.11	1143.11	1296.79	0.3418E+02	0.1358E+02	0.4930E+02	0.8189E+01
25	30.01	1213.83	977.70	1211.38	1326.49	0.1571E+02	0.1227E+02	0.1594E+02	0.6741E+01
27	34.01	1268.56	1024.62	1271.04	1351.22	0.1335E+02	0.1123E+02	0.1396E+02	0.5674E+01
29	38.01	1321.01	1067.78	1323.79	1372.22	0.1254E+02	0.1038E+02	0.1250E+02	0.4860E+01
31	42.02	1369.03	1107.95	1371.88	1390.38	0.1177E+02	0.9674E+01	0.1156E+02	0.4221E+01
33	46.00	1415.08	1145.24	1416.76	1406.13	0.1161E+02	0.9079E+01	0.1105E+02	0.3714E+01
35	49.99	1463.87	1180.42	1460.34	1420.10	0.1148E+02	0.8565E+01	0.1083E+02	0.3299E+01
37	54.01	1507.46	1213.92	1503.76	1432.64	0.1027E+02	0.8114E+01	0.1079E+02	0.2953E+01
39	58.02	1543.53	1246.05	1547.15	1443.89	-0.1394E+02	0.7714E+01	0.1086E+02	0.2664E+01
41	62.02	1498.76	1275.98	1494.70	1454.04	0.1330E+02	0.7366E+01	0.1389E+02	0.2419E+01
43	66.02	1550.85	1304.64	1550.49	1463.29	0.1232E+02	0.7054E+01	0.1401E+02	0.2209E+01
45	69.98	1593.98	1332.29	1593.27	1471.67	0.9635E+01	0.6771E+01	0.7599E+01	0.2029E+01
47	74.02	1625.39	1358.84	1624.17	1479.54	0.6329E+01	0.6514E+01	0.7691E+01	0.1869E+01
49	77.98	1640.70	1384.40	1642.08	1486.66	0.8514E+00	0.6281E+01	0.1322E+01	0.1732E+01
51	82.00	1639.26	1409.43	1636.76	1493.37	-0.1833E+01	0.6063E+01	-0.4051E+01	0.1608E+01
53	84.02	1631.53	1421.62	1628.61	1496.56	-0.6014E+01	0.5961E+01	-0.4027E+01	0.1551E+01
55	88.00	1590.08	1444.60	1590.08	1502.53	-0.1304E+02	0.5776E+01	-0.2262E+04	0.1449E+01

IV - 74

Connection type : Top-and seat-angle connections
Mode : All bolted

Tested by : Y. Sato et al. (2007) Japan
Test Id. : A90

Column : --
Beam : H400x200x8x13
Angle : L200x150x15

Fasteners: F10T- -M20
7/8" Oversize holes
Material : SS400
Fy = 45.69 ksi
Fu = 68.02 ksi

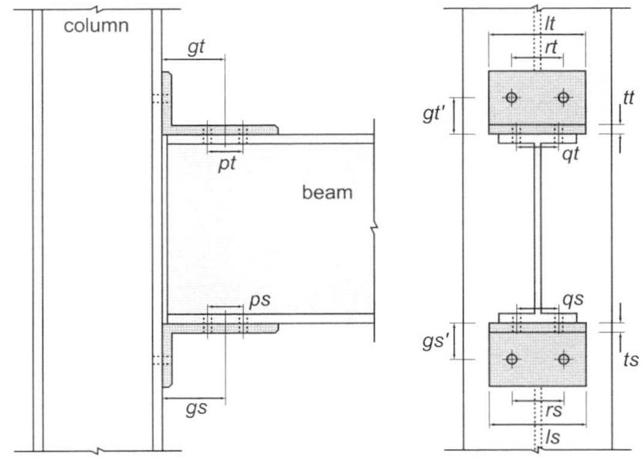

Major parameters

lt = 7.3740"	tt = 0.5906"	ls = 7.8740"	ts = 0.5906"
gt = 4.1339"	gt'= 2.3622"	gs = 4.1339"	gs'= 2.3622"
qt = 4.7244"	rt = 4.7244"	qs = 4.7244"	rs = 4.7244"
pt = 3.5433"	ps = 3.5433"		
nt = 2 X 2	nt'= 2 X 1	ns = 2 X 2	ns'= 2 X 1

Remark 1) Failure mode: Bolt fracture
2) Original T & S angles: L200x200x20

No	Moment (k-in)	Rotation (radians) X 1/1000	No	Moment (k-in)	Rotation (radians) X 1/1000	No	Moment (k-in)	Rotation (radians) X 1/1000
1	0.00	0.00	26	1184.69	28.02	51	1725.42	76.01
2	180.76	0.24	27	1220.45	30.02	52	1739.70	77.99
3	292.73	0.50	28	1252.58	32.01	53	1751.09	79.99
4	383.89	0.75	29	1280.78	34.01	54	1762.49	82.01
5	460.96	1.00	30	1312.38	36.00	55	1769.40	84.01
6	599.73	1.50	31	1345.16	38.00	56	1773.09	85.99
7	696.14	1.99	32	1375.04	40.00	57	1773.22	88.02
8	791.22	3.01	33	1405.20	41.98	58	1769.54	90.01
9	845.28	4.02	34	1436.09	44.01	59	1763.19	91.99
10	889.17	5.00	35	1462.73	46.00	60	1753.29	93.99
11	921.97	6.00	36	1489.97	48.01	61	1736.91	96.00
12	959.08	7.01	37	1515.90	50.02	62	1724.97	96.78
13	991.26	8.01	38	1540.96	52.02			
14	1007.71	8.98	39	1549.16	52.98			
15	1022.33	10.01	40	1108.65	53.96			
16	1043.73	12.01	41	1308.14	56.01			
17	1061.93	14.01	42	1492.22	57.99			
18	1080.82	16.00	43	1533.41	59.99			
19	1105.03	18.02	44	1567.40	62.01			
20	1127.17	20.00	45	1597.48	63.98			
21	1136.00	21.53	46	1624.05	65.99			
22	894.50	22.08	47	1648.47	68.01			
23	744.26	22.61	48	1673.40	70.01			
24	950.85	24.00	49	1693.19	72.02			
25	1142.68	26.01	50	1708.57	73.99			

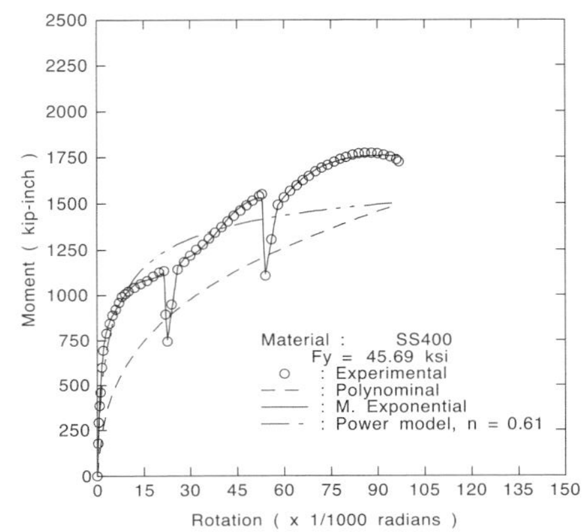

```
Moment-rotation prediction equations     ( R : X 1/1000 radians )

Frye and Morris polynominal model  :    R = Sum ( Ai X ( K*Bm )**Pi X 10**Qi )
      t =  0.590550"  xd = 15.748000"   f =  0.787400"  xl =  7.874020"
     A1 =  8.460000   A2 =  1.010000   A3 =  1.240000    K =  0.006388
     P1 =        1    P2 =        3    P3 =        5    Q1 =       -1     Q2 =       -1     Q3 =       -5

Modified exponential model  :
     Bm = Sum ( Ai X ( 1 - exp( - R/( 2*i*C ) ) ) )  +     Sum ( RKj X ( R - Rj0 ) )  +  Bm0
       C =   0.86650000E+00              Bmo=   0.00000000E+00              Nexp=  6          Nliner=  8
      AI =   0.18062051E+04   0.13262517E+05  -0.15860621E+06   0.50987380E+06  -0.63989072E+06   0.27656606E+06
     Rj0 =          21.5300          22.6100          26.0100          28.0200          38.0000
                    52.9800          53.9600          57.9900
     RKj =  -0.37130263E+03   0.46584234E+03  -0.10901728E+03  -0.22268161E+02   0.41668650E+01
            -0.42945318E+03   0.52714131E+03  -0.67286583E+02

Power model  :     Bm = ( Rki X R ) / ( 1 + ( R/R0 )**rn )**( 1/rn )
      rn = 0.606        rki =    0.93453673E+03           rmu =   0.17350428E+04
```

No	Rotation (radians) X 1/1000	Moment (kip-inch) Expri.	Poly.	M.Expo.	P.Model.	Connection stiffness (kip-inch) X 1000 Expri.	Poly.	M.Expo.	P.Model.
1	0.00	0.00	0.00	0.00	0.00	0.7532E+03	0.1850E+03	0.6655E+03	0.9345E+03
3	0.50	292.73	89.08	278.67	252.85	0.3970E+03	0.1658E+03	0.4609E+03	0.3484E+03
5	1.00	460.96	163.68	471.02	394.52	0.2980E+03	0.1330E+03	0.3170E+03	0.2338E+03
7	1.99	696.14	271.14	691.87	572.62	0.1632E+03	0.8916E+02	0.1499E+03	0.1408E+03
9	4.02	845.28	409.44	855.17	778.47	0.4909E+02	0.5358E+02	0.4209E+02	0.7455E+02
11	6.00	921.97	500.11	920.74	898.04	0.3476E+02	0.3968E+02	0.2913E+02	0.4929E+02
13	8.01	991.26	571.45	977.03	982.29	0.2446E+02	0.3198E+02	0.2655E+02	0.3578E+02
15	10.01	1022.33	630.22	1024.50	1045.10	0.1301E+02	0.2711E+02	0.2032E+02	0.2763E+02
17	14.01	1061.93	725.67	1075.36	1135.01	0.9296E+01	0.2119E+02	0.6327E+01	0.1838E+02
19	18.02	1105.03	803.05	1094.10	1197.80	0.1158E+02	0.1766E+02	0.5673E+01	0.1338E+02
21	21.53	1136.00	861.12	1129.60	1239.61	-0.3215E+03	0.1554E+02	0.1569E+02	0.1062E+02
23	22.61	744.26	877.76	747.73	1250.71	-0.1642E+03	0.1500E+02	-0.3515E+03	0.9960E+01
25	26.01	1142.68	926.11	1159.20	1281.56	0.5817E+02	0.1358E+02	0.1276E+03	0.8270E+01
27	30.02	1220.45	977.82	1216.29	1311.65	0.1701E+02	0.1226E+02	0.8992E+01	0.6818E+01
29	34.01	1280.78	1024.62	1267.99	1336.63	0.1499E+02	0.1123E+02	0.1592E+02	0.5751E+01
31	38.00	1345.16	1067.68	1335.84	1357.88	0.1566E+02	0.1038E+02	0.1725E+02	0.4936E+01
33	41.98	1405.20	1107.57	1416.44	1376.21	0.1523E+02	0.9681E+01	0.1852E+02	0.4298E+01
35	46.00	1462.73	1145.24	1480.39	1392.41	0.1347E+02	0.9079E+01	0.1299E+02	0.3781E+01
37	50.02	1515.90	1180.67	1519.30	1406.73	0.1272E+02	0.8562E+01	0.6275E+01	0.3360E+01
39	52.98	1549.16	1205.79	1530.34	1416.28	-0.2181E+03	0.8220E+01	0.1197E+01	0.3097E+01
41	56.01	1308.14	1229.94	1305.73	1425.30	0.9510E+02	0.7911E+01	0.9390E+02	0.2861E+01
43	59.99	1533.41	1261.05	1532.78	1436.14	0.1872E+02	0.7536E+01	0.2067E+02	0.2594E+01
45	63.98	1597.48	1290.25	1604.83	1446.02	0.1425E+02	0.7208E+01	0.1559E+02	0.2365E+01
47	68.01	1648.47	1318.99	1658.85	1455.13	0.1228E+02	0.6905E+01	0.1137E+02	0.2165E+01
49	72.02	1693.19	1345.82	1697.47	1463.46	0.8817E+01	0.6638E+01	0.8023E+01	0.1993E+01
51	76.01	1725.42	1372.06	1724.07	1471.11	0.7770E+01	0.6392E+01	0.5423E+01	0.1843E+01
53	79.99	1751.09	1396.69	1741.51	1478.18	0.5668E+01	0.6173E+01	0.3433E+01	0.1711E+01
55	84.01	1769.40	1421.29	1752.12	1484.82	0.2656E+01	0.5964E+01	0.1915E+01	0.1593E+01
57	88.02	1773.22	1445.01	1757.42	1490.99	-0.9028E+00	0.5773E+01	0.7880E+00	0.1488E+01
59	91.99	1763.19	1467.20	1758.83	1496.71	-0.4073E+01	0.5602E+01	-0.3584E-01	0.1395E+01
61	96.00	1736.91	1489.50	1757.40	1502.13	-0.1330E+02	0.5438E+01	-0.6447E+00	0.1310E+01

A.5 – 1

V - 1

Connection type : Extended end-plate connections
Mode : All bolted without column stiffener

Tested by : S.A.Ioannides (1978) U.S.A.
Test Id. : TEST 1

Column : W8x35
Beam : W14x22
Plate thickness : 5/8"

Fasteners: A325- -3/4"D
13/16" Oversize holes
Material : A36
Fy = -- ksi
Fu = -- ksi

Major parameters

End plate extended on both sides.
ct = 1.6250" pt = 3.5000" li = 10.2500" pc = 3.5000"
cc = 1.6250" gt = 5.5000" gc = 5.5000"
tp = 0.6250" nt = 2 X 2 nc = 2 X 2

Remark 1) End plate extended on tension and compression sides.
2)

No	Moment (k-in)	Rotation (radians) X 1/1000	No	Moment (k-in)	Rotation (radians) X 1/1000
1	0.00	0.00	26	820.37	14.44
2	41.66	0.19	27	839.93	15.69
3	83.32	0.38	28	859.50	16.93
4	124.97	0.57	29	873.09	17.94
5	165.18	0.79	30	886.69	18.94
6	205.39	1.00	31	900.29	19.95
7	245.60	1.21	32	913.88	20.95
8	279.02	1.36	33	925.31	22.02
9	312.44	1.51	34	936.74	23.10
10	345.87	1.65	35	948.17	24.17
11	385.83	1.84	36	959.60	25.24
12	425.79	2.02			
13	465.06	2.40			
14	497.78	2.72			
15	530.51	3.04			
16	571.26	3.63			
17	613.56	4.76			
18	645.68	5.85			
19	677.81	6.98			
20	699.01	7.86			
21	720.21	8.73			
22	738.52	9.71			
23	755.83	10.69			
24	773.82	11.94			
25	800.80	13.19			

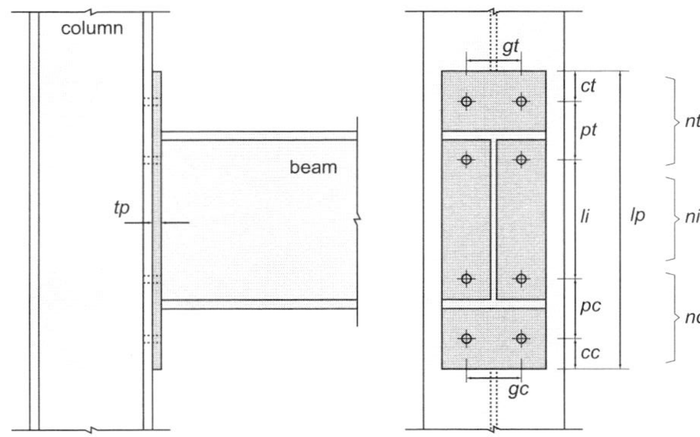

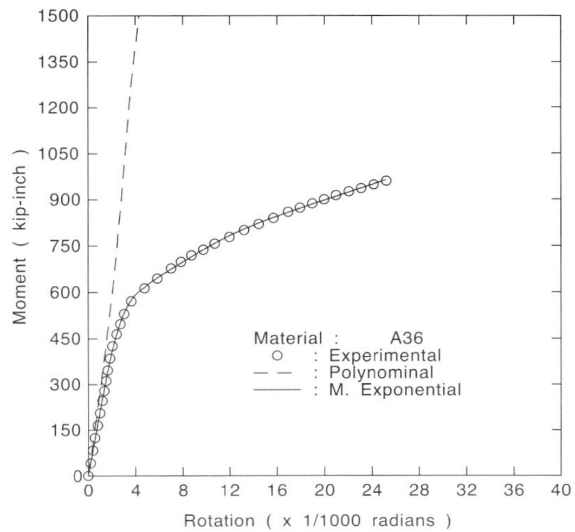

A.5 – 2

```
Moment-rotation prediction equations     ( R : X 1/1000 radians )

Frye and Morris polynominal model  :    R = Sum ( Ai X ( K*Bm )**Pi X 10**Qi )
     xd = 17.250000"   t =  0.625000"  tf =  0.750000"
     A1 =  1.830000   A2 = -1.040000   A3 =  6.380000    K =  0.001999
     P1 =         1   P2 =         3   P3 =         5   Q1 =         0   Q2 =        -1   Q3 =        -3

Modified exponential model  :
     Bm = Sum ( Ai X ( 1 - exp( - R/( 2*i*C ) ) ) )  +     Sum ( RKj X ( R - Rj0 ) )  +  Bm0
      C =   0.25807917E+00              Bmo=   0.00000000E+00         Nexp=  6         Nliner=  1
     AI =   0.47862372E+03  -0.68805673E+04   0.25126396E+05  -0.38886132E+05   0.28423489E+05  -0.75617989E+04
    Rj0 =           4.7600
    RKj =   0.12850051E+02
```

No	Rotation (radians) X 1/1000	Moment (kip-inch) Expri.	Moment (kip-inch) Poly.	Moment (kip-inch) M.Expo.	Connection stiffness (kip-inch) X 1000 Expri.	Connection stiffness (kip-inch) X 1000 Poly.	Connection stiffness (kip-inch) X 1000 M.Expo.
1	0.00	0.00	0.00	0.00	0.2182E+03	0.2734E+03	0.2260E+03
3	0.38	83.32	104.67	80.02	0.2182E+03	0.2754E+03	0.2046E+03
5	0.79	165.18	217.41	164.63	0.1879E+03	0.2823E+03	0.2133E+03
7	1.21	245.60	340.85	255.93	0.2124E+03	0.2957E+03	0.2101E+03
9	1.51	312.44	428.98	315.55	0.2292E+03	0.3092E+03	0.1974E+03
11	1.84	385.83	534.72	377.52	0.2155E+03	0.3302E+03	0.1757E+03
13	2.40	465.06	733.16	463.86	0.1029E+03	0.3828E+03	0.1323E+03
15	3.04	530.51	1006.36	533.99	0.9006E+02	0.4591E+03	0.8614E+02
17	4.76	613.56	1641.53	615.47	0.3329E+02	0.2308E+03	0.2262E+02
19	6.98	677.81	1980.07	675.08	0.2605E+02	0.1049E+03	0.2364E+02
21	8.73	720.21	2133.29	715.97	0.2164E+02	0.7465E+02	0.2295E+02
23	10.69	756.83	2260.49	758.54	0.1821E+02	0.5715E+02	0.2030E+02
25	13.19	800.80	2386.33	804.22	0.1662E+02	0.4450E+02	0.1632E+02
27	15.69	839.93	2487.02	841.51	0.1569E+02	0.3679E+02	0.1388E+02
29	17.94	873.09	2564.13	871.52	0.1354E+02	0.3198E+02	0.1293E+02
31	19.95	900.29	2624.96	897.13	0.1354E+02	0.2873E+02	0.1264E+02
33	22.02	925.31	2681.76	923.31	0.1065E+02	0.2606E+02	0.1260E+02
35	24.17	948.17	2735.20	950.39	0.1065E+02	0.2382E+02	0.1264E+02

A.5 – 3

V - 2

Connection type : Extended end-plate connections
Mode : All bolted without column stiffener

Tested by : S.A.Ioannides (1978) U.S.A.
Test Id. : TEST 2

Column : W10x49 Fasteners: A325- -7/8"D
Beam : W18x35 15/16" Oversize holes
Plate thickness : 3/4" Material : A36
Fy = -- ksi
Fu = -- ksi

Major parameters

End plate extended on both sides.
ct = 1.6250" pt = 3.7500" li = 14.0000" pc = 3.7500"
cc = 1.6250" gt = 5.5000" gc = 5.5000"
tp = 0.7500" nt = 2 X 2 nc = 2 X 2

Remark 1) End plate extended on tension and compression sides.
2)

No	Moment (k-in)	Rotation (radians) X 1/1000	No	Moment (k-in)	Rotation (radians) X 1/1000
1	0.00	0.00	26	1946.44	11.86
2	111.76	0.02	27	1981.78	12.68
3	223.52	0.04	28	2012.72	13.54
4	335.28	0.07	29	2043.66	14.41
5	447.03	0.12	30	2074.60	15.27
6	558.79	0.17	31	2106.60	16.04
7	657.09	0.24	32	2138.60	16.82
8	755.38	0.39	33	2170.61	17.59
9	842.11	0.52	34	2197.73	18.41
10	928.85	0.65	35	2224.86	19.23
11	1015.49	0.87	36	2251.99	20.04
12	1110.78	1.13	37	2287.06	21.12
13	1203.03	1.52	38	2322.12	22.19
14	1295.20	1.98			
15	1401.68	2.62			
16	1502.10	3.49			
17	1582.19	4.43			
18	1627.82	5.09			
19	1673.46	5.75			
20	1713.21	6.50			
21	1752.97	7.24			
22	1793.97	8.18			
23	1834.97	9.11			
24	1873.04	10.08			
25	1911.11	11.05			

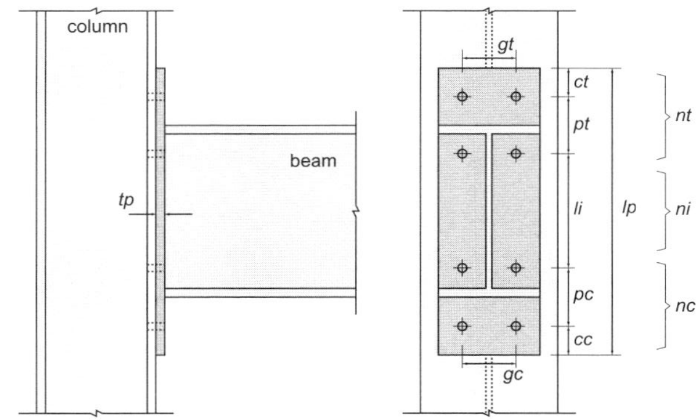

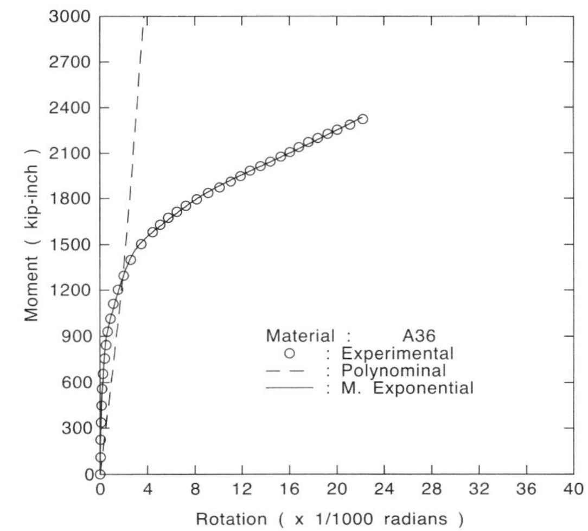

A.5 – 4

```
Moment-rotation prediction equations      ( R : X 1/1000 radians )

Frye and Morris polynominal model   :     R = Sum ( Ai X ( K*Bm )**Pi X 10**Qi )
     xd = 21.500000"    t =  0.750000"  tf =  0.875000"
     A1 =  1.830000   A2 = -1.040000   A3 =  6.380000    K =  0.000869
     P1 =        1    P2 =        3    P3 =        5    Q1 =        0    Q2 =       -1    Q3 =       -3

Modified exponential model  :
     Bm = Sum ( Ai X ( 1 - exp( - R/( 2*i*C ) ) ) )  +      Sum ( RKj X ( R - Rj0 ) )  +  Bm0
       C =   0.18989417E+00           Bmo=   0.00000000E+00          Nexp=  6        Nliner=  2
      AI =   0.37367515E+04  -0.22278426E+05   0.72104623E+05  -0.11782196E+06   0.94758148E+05  -0.29128075E+05
     Rj0 =           0.0100           0.0200
     RKj =   0.93224571E+04  -0.92831982E+04
```

No	Rotation (radians) X 1/1000	Moment (kip-inch) Expri.	Moment (kip-inch) Poly.	Moment (kip-inch) M.Expo.	Connection stiffness (kip-inch) X 1000 Expri.	Connection stiffness (kip-inch) X 1000 Poly.	Connection stiffness (kip-inch) X 1000 M.Expo.
1	0.00	0.00	0.00	0.00	0.5588E+04	0.6287E+03	0.3354E+04
3	0.04	223.52	28.29	233.18	0.4132E+04	0.6288E+03	0.2875E+04
5	0.12	447.03	75.45	421.51	0.2366E+04	0.6292E+03	0.2178E+04
7	0.24	657.09	148.15	626.19	0.1243E+04	0.6305E+03	0.1421E+04
9	0.52	842.11	328.88	881.67	0.6507E+03	0.6375E+03	0.5397E+03
11	0.87	1015.49	556.58	1013.52	0.3853E+03	0.6542E+03	0.2852E+03
13	1.52	1203.03	996.59	1185.13	0.2191E+03	0.7129E+03	0.2588E+03
15	2.62	1401.68	1885.02	1416.74	0.1454E+03	0.9412E+03	0.1511E+03
17	4.43	1582.19	3586.46	1581.05	0.7572E+02	0.6355E+03	0.6202E+02
19	5.75	1673.46	4199.17	1661.18	0.6170E+02	0.3450E+03	0.6060E+02
21	7.24	1752.97	4612.82	1749.06	0.4913E+02	0.2275E+03	0.5591E+02
23	9.11	1834.97	4969.13	1842.93	0.4162E+02	0.1618E+03	0.4462E+02
25	11.05	1911.11	5244.83	1921.36	0.4154E+02	0.1261E+03	0.3742E+02
27	12.68	1981.78	5433.61	1980.41	0.3973E+02	0.1071E+03	0.3564E+02
29	14.41	2043.66	5605.74	2042.08	0.3579E+02	0.9277E+02	0.3591E+02
31	16.04	2106.60	5748.97	2101.59	0.4140E+02	0.8260E+02	0.3679E+02
33	17.59	2170.61	5870.60	2159.11	0.3740E+02	0.7504E+02	0.3760E+02
35	19.23	2224.86	5989.17	2221.19	0.3316E+02	0.6848E+02	0.3825E+02
37	21.12	2287.06	6111.44	2293.94	0.3273E+02	0.6244E+02	0.3872E+02

A.5 – 5

V - 3

Connection type : Extended end-plate connections
Mcde : All bolted withou: column stiffener

Tested by : S.A.Ioannides (1978) U.S.A.
Test Id. : TEST 3

Column : W14x48
Beam : W24x55
Plate thickness : 7/8"

Fasteners: A325- -1"D
1 1/16" Oversize holes
Material : A36
Fy = -- ksi
Fu = -- ksi

Major parameters

End plate extended on both sides.
ct = 1.6250" pt = 4.0000" li = 19.2500" pc = 4.0000"
cc = 1.6250" gt = 5.5000" gc = 5.5000"
tp = 0.8750" nt = 2 X 2 nc = 2 X 2

Remark 1) End plate extended on tension and compression side.
2)

No	Moment (k-in)	Rotation (radians) X 1/1000	No	Moment (k-in)	Rotation (radians) X 1/1000
1	0.00	0.00	26	2232.71	4.63
2	97.85	0.02	27	2293.69	5.09
3	195.71	0.05	28	2354.67	5.55
4	293.56	0.07	29	2411.53	6.10
5	379.19	0.09	30	2468.40	6.66
6	464.81	0.12	31	2513.06	7.15
7	552.46	0.15	32	2557.71	7.64
8	640.12	0.19	33	2588.07	8.20
9	723.69	0.25	34	2618.43	8.76
10	807.26	0.31	35	2642.66	9.34
11	892.85	0.41	36	2666.90	9.92
12	978.44	0.51	37	2687.08	10.44
13	1072.17	0.64	38	2707.26	10.95
14	1165.90	0.78	39	2719.33	11.36
15	1273.87	1.00	40	2731.40	11.77
16	1381.84	1.22			
17	1495.92	1.46			
18	1610.00	1.70			
19	1701.60	2.06			
20	1793.21	2.41			
21	1862.42	2.69			
22	1931.63	2.97			
23	2006.91	3.35			
24	2082.19	3.74			
25	2157.45	4.18			

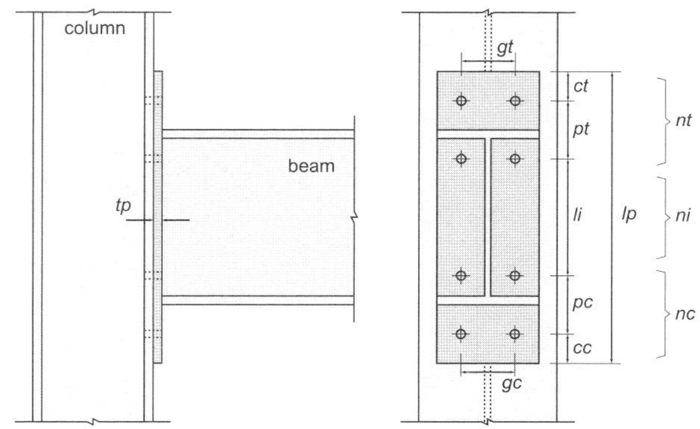

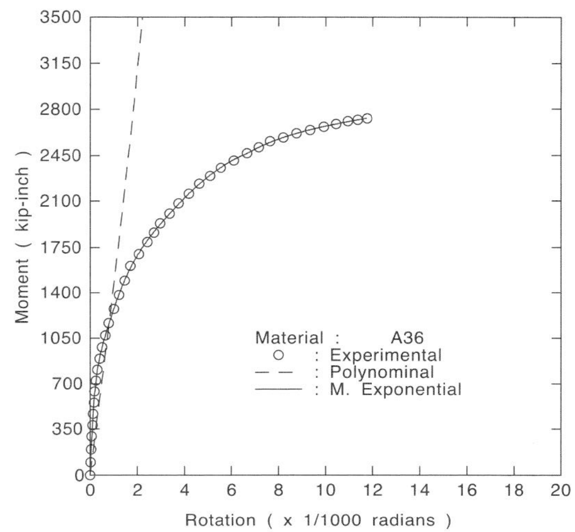

A.5 – 6

```
Moment-rotation prediction equations      ( R : X 1/1000 radians )

Frye and Morris polynominal model  :    R = Sum ( Ai X ( K*Bm )**Pi X 10**Qi )
     xd = 27.250000"    t =  0.875000"  tf =  1.000000"
     A1 =  1.830000   A2 = -1.040000   A3 =  6.380000     K =  0.000379
     P1 =         1   P2 =         3   P3 =         5    Q1 =         0    Q2 =        -1    Q3 =        -3

Modified exponential model  :
     Bm = Sum ( Ai X ( 1 - exp( - R/( 2*i*C ) ) ) )  +      Sum ( RKj X ( R - Rj0 ) )  +  Bm0
       C =   0.10390333E+00          Bmo=   0.00000000E+00          Nexp=  6        Nliner=  1
      AI =   0.15116299E+04   0.23445920E+04  -0.42161065E+05   0.13575855E+06  -0.17068572E+06   0.75823971E+05
     Rj0 =          6.6600
     RKj =   0.28246315E+02
```

No	Rotation (radians) X 1/1000	Moment (kip-inch) Expri.	Moment (kip-inch) Poly.	Moment (kip-inch) M.Expo.	Connection stiffness (kip-inch) X 1000 Expri.	Connection stiffness (kip-inch) X 1000 Poly.	Connection stiffness (kip-inch) X 1000 M.Expo.
1	0.00	0.00	0.00	0.00	0.4200E+04	0.1443E+04	0.5149E+04
3	0.05	195.71	67.24	212.76	0.4209E+04	0.1443E+04	0.4030E+04
5	0.09	379.19	136.65	384.34	0.3439E+04	0.1444E+04	0.3144E+04
7	0.15	552.46	222.85	546.54	0.2519E+04	0.1445E+04	0.2336E+04
9	0.25	723.69	358.99	724.49	0.1406E+04	0.1447E+04	0.1519E+04
11	0.41	892.85	588.50	905.69	0.8650E+03	0.1455E+04	0.8764E+03
13	0.64	1072.17	936.58	1073.65	0.6768E+03	0.1474E+04	0.6101E+03
15	1.00	1273.87	1468.95	1275.37	0.4963E+03	0.1521E+04	0.5349E+03
17	1.46	1495.92	2188.93	1498.09	0.4710E+03	0.1619E+04	0.4267E+03
19	2.06	1701.60	3216.39	1709.22	0.2575E+03	0.1837E+04	0.2895E+03
21	2.69	1862.42	4489.14	1866.60	0.2501E+03	0.2210E+04	0.2198E+03
23	3.35	2006.91	6069.74	2002.72	0.1954E+03	0.2463E+04	0.1954E+03
25	4.18	2157.45	7840.97	2156.71	0.1693E+03	0.1693E+04	0.1745E+03
27	5.09	2293.69	9034.99	2299.64	0.1327E+03	0.1036E+04	0.1398E+03
29	6.10	2411.53	9898.73	2418.97	0.1019E+03	0.7049E+03	0.9515E+02
31	7.15	2513.06	10537.84	2511.57	0.9133E+02	0.5331E+03	0.8542E+02
33	8.20	2588.07	11038.73	2586.61	0.5440E+02	0.4314E+03	0.5992E+02
35	9.34	2642.66	11486.92	2644.84	0.4158E+02	0.3592E+03	0.4383E+02
37	10.44	2687.08	11852.81	2687.98	0.3929E+02	0.3108E+03	0.3577E+02
39	11.36	2719.33	12127.71	2719.22	0.2944E+02	0.2796E+03	0.3222E+02

V - 4

Connection type : Extended end-plate connections
Mode : All bolted without column stiffener

Tested by : S.A.Ioannides (1978) U.S.A.
Test Id. : TEST 4

Column : W8x35
Beam : W14x22
Plate thickness : 7/8"

Fasteners: A325- -3/4"D
13/16" Oversize holes
Material : A36
Fy = -- ksi
Fu = -- ksi

Major parameters

End plate extended on both sides.

ct = 1.6250" pt = 3.5000" li = 10.2500" pc = 3.5000"
cc = 1.6250" gt = 5.5000" gc = 5.5000"
tp = 0.3750" nt = 2 X 2 nc = 2 X 2

Remark 1) End plate extended on tension and compression side.
2)

No	Moment (k-in)	Rotation (radians) X 1/1000	No	Moment (k-in)	Rotation (radians) X 1/1000
1	0.00	0.00	26	1113.14	22.57
2	53.94	0.08	27	1138.19	24.63
3	113.66	0.17	28	1163.25	26.69
4	173.39	0.27	29	1188.30	28.75
5	233.12	0.38	30	1213.36	30.81
6	269.53	0.50	31	1238.42	32.87
7	305.94	0.63	32	1263.47	34.92
8	363.62	0.77	33	1288.53	36.98
9	423.43	0.93	34	1313.58	39.04
10	492.72	1.20	35	1326.57	40.52
11	531.46	1.55	36	1339.55	42.00
12	570.20	1.90			
13	609.03	2.50			
14	647.87	3.09			
15	683.35	4.13			
16	723.82	5.18			
17	773.06	6.74			
18	823.30	8.29			
19	864.74	9.90			
20	903.18	11.52			
21	942.59	13.30			
22	979.00	15.09			
23	1014.72	16.96			
24	1050.44	18.83			
25	1081.79	20.70			

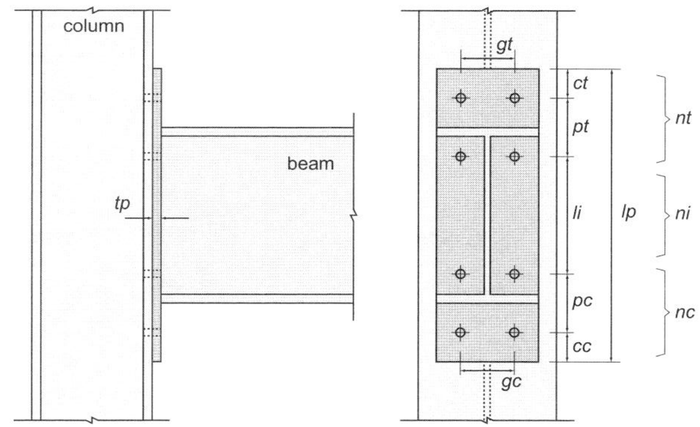

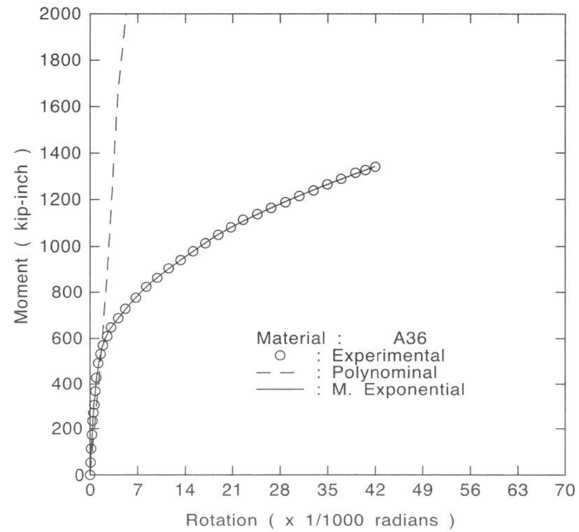

A.5 – 8

```
Moment-rotation prediction equations     ( R : X 1/1000 radians )

Frye and Morris polynominal model  :    R = Sum ( Ai X ( K*Bm )**Pi X 10**Qi )
     xd = 17.250000"   t =  0.875000"  tf =  0.750000"
     A1 =  1.830000   A2 = -1.040000   A3 =  6.380000    K =  0.001747
     P1 =         1   P2 =         3   P3 =         5    Q1 =         0    Q2 =        -1    Q3 =        -3

Modified exponential model  :
     Bm = Sum ( Ai X ( 1 - exp( - R/( 2*i*C ) ) ) )  +     Sum ( RKj X ( R - Rj0 ) )  +  Bm0
      C =   0.37001917E+00             Bmo=   0.00000000E+00          Nexp=  6        Nliner=  2
     AI =   0.15640815E+03   0.44408859E+04  -0.19834750E+05   0.41401260E+05  -0.41018889E+05   0.15678015E+05
    Rj0 =           0.0000          18.8300
    RKj =   0.14324318E+02  -0.35164673E+01
```

No	Rotation (radians) X 1/1000	Moment (kip-inch) Expri.	Moment (kip-inch) Poly.	Moment (kip-inch) M.Expo.	Connection stiffness (kip-inch) X 1000 Expri.	Connection stiffness (kip-inch) X 1000 Poly.	Connection stiffness (kip-inch) X 1000 M.Expo.
1	0.00	0.00	0.00	0.00	0.6742E+03	0.3128E+03	0.7092E+03
3	0.17	113.66	53.18	111.96	0.6322E+03	0.3132E+03	0.5977E+03
5	0.38	233.12	119.04	223.76	0.4284E+03	0.3151E+03	0.4721E+03
7	0.63	305.94	198.24	326.68	0.3608E+03	0.3192E+03	0.3567E+03
9	0.93	428.43	295.28	417.66	0.3239E+03	0.3272E+03	0.2557E+03
11	1.55	531.46	506.30	533.22	0.1097E+03	0.3565E+03	0.1322E+03
13	2.50	609.03	883.28	616.80	0.6533E+02	0.4514E+03	0.5901E+02
15	4.13	688.35	1681.37	687.51	0.3883E+02	0.3783E+03	0.3646E+02
17	6.74	776.06	2233.79	777.40	0.3032E+02	0.1278E+03	0.3224E+02
19	9.90	864.74	2532.64	866.66	0.2572E+02	0.7201E+02	0.2447E+02
21	13.30	942.59	2735.61	942.24	0.2040E+02	0.5044E+02	0.2082E+02
23	16.96	1014.72	2896.58	1016.03	0.1909E+02	0.3878E+02	0.1967E+02
25	20.70	1081.79	3027.44	1081.02	0.1675E+02	0.3168E+02	0.1499E+02
27	24.63	1138.19	3141.71	1137.18	0.1217E+02	0.2677E+02	0.1359E+02
29	28.75	1188.30	3244.08	1190.54	0.1217E+02	0.2315E+02	0.1240E+02
31	32.87	1238.42	3333.65	1239.87	0.1217E+02	0.2048E+02	0.1163E+02
33	36.98	1288.53	3413.55	1286.80	0.1217E+02	0.1841E+02	0.1121E+02
35	40.52	1326.57	3476.52	1326.07	0.8775E+01	0.1697E+02	0.1101E+02

A.5 – 9

V - 5

```
Connection type : Extended end-plate connections
           Mode : All bolted without column stiffener

Tested by   :  S.A.Ioannides (1978)                       U.S.A.
Test  Id.   :  TEST 5

Column : W10x49                        Fasteners: A325- -7/8"D
Beam   : W18x35                              15/16" Oversize holes
Plate thickness : 1 1/4"               Material : A36
                                             Fy =   --   ksi
                                             Fu =   --   ksi

                              Major parameters

     End plate extended on both sides.
     ct =  1.6250"    pt =  3.7500"    li = 14.0000"    pc =  3.7500"
     cc =  1.6250"    gt =  5.5000"    gc =  5.5000"
     tp =  1.2500"    nt = 2 X  2      nc = 2 X  2

Remark  1)  End plate extended on tension and compression side.
        2)
```

No	Moment (k-in)	Rotation (radians) X 1/1000
1	0.00	0.00
2	97.92	0.14
3	195.83	0.29
4	293.75	0.43
5	398.43	0.66
6	503.11	0.89
7	615.02	1.24
8	726.93	1.59
9	821.37	1.99
10	915.80	2.39
11	1016.03	2.88
12	1116.26	3.37
13	1210.62	4.02
14	1304.98	4.66
15	1396.36	5.50
16	1487.75	6.33
17	1566.03	7.22
18	1644.30	8.10
19	1709.38	9.30
20	1774.46	10.50
21	1845.29	11.90
22	1916.12	13.30
23	1978.19	14.81
24	2040.26	16.32
25	2086.41	17.58

No	Moment (k-in)	Rotation (radians) X 1/1000
26	2132.55	18.85
27	2190.22	20.46
28	2247.89	22.07
29	2290.98	23.77
30	2334.08	25.47
31	2378.57	27.32
32	2423.07	29.17
33	2452.69	30.55
34	2482.31	31.92
35	2511.93	33.29
36	2544.83	35.01
37	2577.74	36.72
38	2597.73	37.86
39	2617.73	39.01

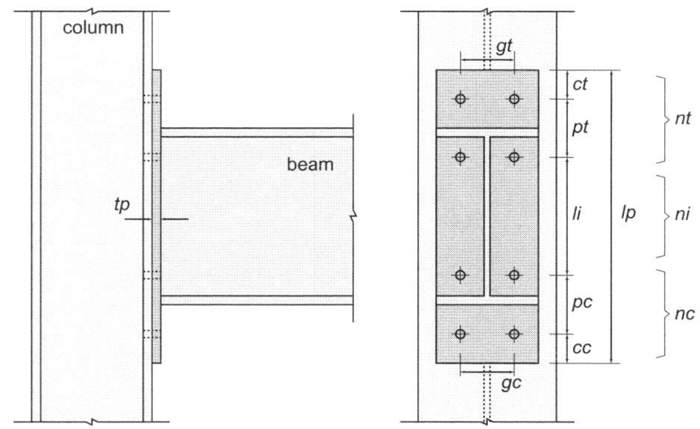

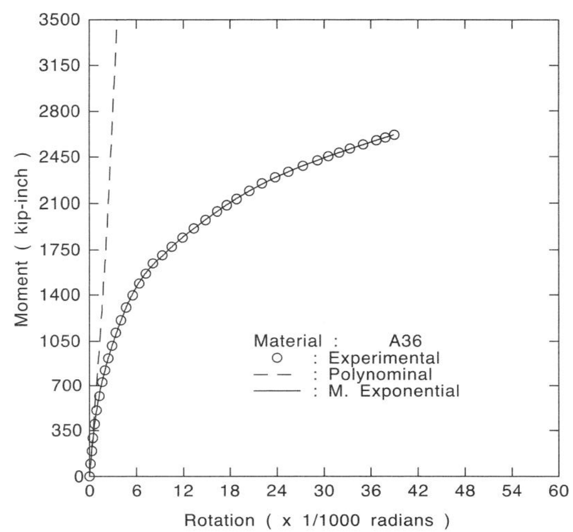

A.5 – 10

```
Moment-rotation prediction equations     ( R : X 1/1000 radians )

Frye and Morris polynominal model  :    R = Sum ( Ai X ( K*Bm )**Pi X 10**Qi )
     xd = 21.500000"   t =  1.250000"  tf =  0.875000"
     A1 =  1.830000   A2 = -1.040000   A3 =  6.380000    K =  0.000709
     P1 =         1   P2 =         3   P3 =         5   Q1 =         0   Q2 =        -1   Q3 =        -3

Modified exponential model  :
     Bm = Sum ( Ai X ( 1 - exp( - R/( 2*i*C ) ) ) )  +     Sum ( RKj X ( R - Rj0 ) )  +  Bm0
      C =   0.36078583E+00            Bmo=   0.00000000E+00           Nexp=  6        Nliner=  1
     AI =   0.32703876E+02   0.73065734E+04  -0.43543052E+05   0.10509319E+06  -0.11231733E+06   0.45519733E+05
    Rj0 =           9.3000
    RKj =   0.17874508E+02
```

No	Rotation (radians) X 1/1000	Moment (kip-inch) Expri.	Moment (kip-inch) Poly.	Moment (kip-inch) M.Expo.	Connection stiffness (kip-inch) X 1000 Expri.	Connection stiffness (kip-inch) X 1000 Poly.	Connection stiffness (kip-inch) X 1000 M.Expo.
1	0.00	0.00	0.00	0.00	0.6852E+03	0.7712E+03	0.7873E+03
3	0.29	195.83	220.53	199.44	0.6852E+03	0.7744E+03	0.6173E+03
5	0.66	398.43	511.56	398.82	0.4557E+03	0.7886E+03	0.4636E+03
7	1.24	615.02	981.98	622.72	0.3187E+03	0.8369E+03	0.3227E+03
9	1.99	821.37	1655.16	828.45	0.2348E+03	0.9659E+03	0.2354E+03
11	2.88	1016.03	2633.86	1014.93	0.2050E+03	0.1249E+04	0.1887E+03
13	4.02	1210.62	4034.15	1208.18	0.1464E+03	0.1001E+04	0.1541E+03
15	5.50	1396.36	5036.25	1407.24	0.1095E+03	0.4658E+03	0.1157E+03
17	7.22	1566.03	5651.23	1572.35	0.8836E+02	0.2807E+03	0.7802E+02
19	9.30	1709.38	6132.15	1700.97	0.5439E+02	0.1931E+03	0.4850E+02
21	11.90	1845.29	6559.50	1848.10	0.5043E+02	0.1415E+03	0.4947E+02
23	14.81	1978.19	6922.05	1979.93	0.4117E+02	0.1104E+03	0.4202E+02
25	17.58	2086.41	7201.00	2089.84	0.3648E+02	0.9208E+02	0.3734E+02
27	20.46	2190.22	7446.00	2190.46	0.3579E+02	0.7902E+02	0.3264E+02
29	23.77	2290.98	7688.70	2290.07	0.2539E+02	0.6828E+02	0.2773E+02
31	27.32	2378.57	7915.42	2380.97	0.2401E+02	0.5986E+02	0.2372E+02
33	30.55	2452.69	8100.34	2453.32	0.2156E+02	0.5393E+02	0.2130E+02
35	33.29	2511.93	8242.60	2509.89	0.2051E+02	0.4987E+02	0.1998E+02
37	36.72	2577.74	8404.86	2576.50	0.1818E+02	0.4569E+02	0.1898E+02
39	39.01	2617.73	8507.85	2619.38	0.1750E+02	0.4327E+02	0.1858E+02

V - 6

Connection type : Extended end-plate connections
Mode : All bolted without column stiffener

Tested by : S.A.Ioannides (1978) U.S.A.
Test Id. : TEST 6

Column : W14x48
Beam : W24x55
Plate thickness : 1 1/4"

Fasteners: A325- -1"D
1/16" Oversize holes
Material : A36
Fy = -- ksi
Fu = -- ksi

Major parameters

End plate extended on both sides.

ct = 1.6250" pt = 4.0000" li = 19.2500" pc = 4.0000"
cc = 1.6250" gt = 5.5000" gc = 5.5000"
tp = 1.2500" nt = 2 X 2 nc = 2 X 2

Remark 1) End plate extended on tension and compression side.
2)

No	Moment (k-in)	Rotation (radians) X 1/1000	No	Moment (k-in)	Rotation (radians) X 1/1000
1	0.00	0.00	26	2957.80	8.33
2	1511.59	0.12	27	3004.52	8.85
3	1605.15	0.15	28	3049.10	9.39
4	1653.91	0.24	29	3093.67	9.93
5	1710.49	0.34	30	3132.34	10.43
6	1762.99	0.45	31	3171.00	10.92
7	1823.48	0.63	32	3215.40	11.50
8	1891.74	0.90	33	3259.80	12.08
9	1973.05	1.23			
10	2055.96	1.64			
11	2132.09	1.98			
12	2185.82	2.30			
13	2239.54	2.62			
14	2334.92	3.22			
15	2383.41	3.60			
16	2441.90	3.97			
17	2503.29	4.41			
18	2564.69	4.84			
19	2614.01	5.24			
20	2663.32	5.64			
21	2718.42	6.11			
22	2773.51	6.58			
23	2819.36	6.99			
24	2865.22	7.40			
25	2911.07	7.81			

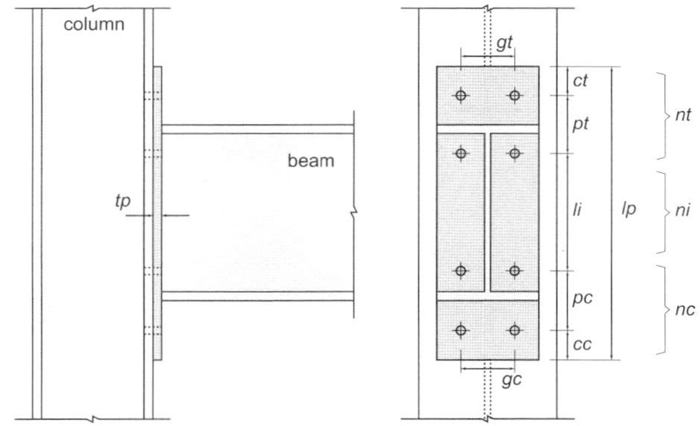

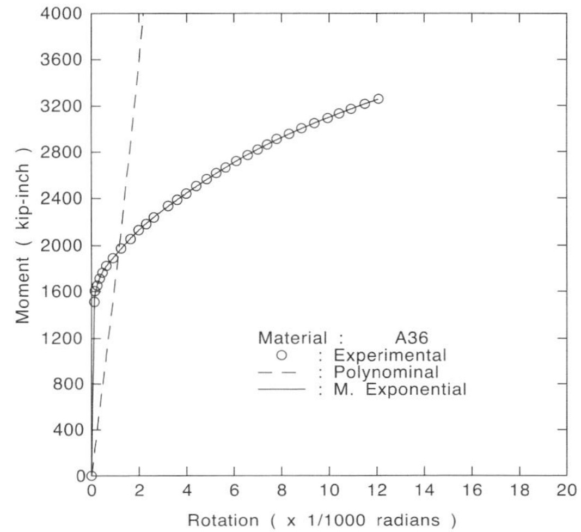

```
Moment-rotation prediction equations      ( R : X 1/1000 radians )

Frye and Morris polynominal model   :     R = Sum ( Ai X ( K*Bm )**Pi X 10**Qi )
      xd = 27.250000"    t =  1.250000"  tf =  1.000000"
      A1 =  1.830000    A2 = -1.040000   A3 =  6.380000    K =  0.000328
      P1 =         1    P2 =         3   P3 =         5   Q1 =         0    Q2 =        -1    Q3 =        -3

Modified exponential model   :
      Bm = Sum ( Ai X ( 1 - exp( - R/( 2*i*C ) ) ) )  +      Sum ( RKj X ( R - Rj0 ) )  +  Bm0
        C =   0.13070167E+00          Bmo=   0.00000000E+00          Nexp=  6         Nliner=  2
       AI =   0.10510584E+04  -0.23861587E+04  -0.37136495E+04   0.29473432E+05  -0.46638720E+05   0.23290045E+05
      Rj0 =          0.0000            0.1200
      RKj =   0.11221014E+05  -0.11150530E+05
```

No	Rotation (radians) X 1/1000	Moment (kip-inch) Expri.	Moment (kip-inch) Poly.	Moment (kip-inch) M.Expo.	Connection stiffness (kip-inch) X 1000 Expri.	Connection stiffness (kip-inch) X 1000 Poly.	Connection stiffness (kip-inch) X 1000 M.Expo.
1	0.00	0.00	0.00	0.00	0.1260E+05	0.1664E+04	0.2075E+04
2	0.12	1511.59	199.70	1536.93	0.5014E+04	0.1665E+04	0.1240E+05
3	0.15	1605.15	249.66	1572.18	0.2475E+04	0.1666E+04	0.1100E+04
4	0.24	1653.91	399.60	1654.56	0.5532E+03	0.1669E+04	0.7565E+03
5	0.34	1710.49	566.51	1717.47	0.5236E+03	0.1674E+04	0.5224E+03
6	0.45	1762.99	751.46	1766.13	0.4246E+03	0.1681E+04	0.3773E+03
7	0.63	1823.48	1052.68	1822.66	0.3056E+03	0.1698E+04	0.2758E+03
8	0.90	1891.74	1512.11	1890.46	0.2527E+03	0.1735E+04	0.2418E+03
9	1.23	1976.05	2108.11	1970.62	0.2259E+03	0.1804E+04	0.2335E+03
10	1.64	2055.96	2864.18	2061.60	0.2128E+03	0.1929E+04	0.2126E+03
11	1.98	2132.09	3537.47	2129.52	0.1958E+03	0.2076E+04	0.1908E+03
12	2.30	2185.82	4233.72	2187.89	0.1671E+03	0.2262E+04	0.1730E+03
13	2.62	2239.54	4997.48	2241.31	0.1639E+03	0.2494E+04	0.1601E+03
14	3.22	2334.92	6629.47	2333.40	0.1484E+03	0.2852E+04	0.1472E+03
15	3.60	2388.41	7682.31	2387.97	0.1423E+03	0.2680E+04	0.1435E+03
16	3.97	2441.90	8607.36	2441.41	0.1418E+03	0.2218E+04	0.1410E+03
17	4.41	2503.29	9457.17	2502.12	0.1411E+03	0.1703E+04	0.1380E+03
18	4.84	2564.69	10113.82	2561.39	0.1325E+03	0.1342E+04	0.1343E+03
19	5.24	2614.01	10596.47	2613.72	0.1247E+03	0.1116E+04	0.1302E+03
20	5.64	2663.32	11003.80	2664.29	0.1208E+03	0.9535E+03	0.1255E+03
21	6.11	2718.42	11420.65	2722.33	0.1162E+03	0.8118E+03	0.1193E+03
22	6.58	2773.51	11779.74	2777.38	0.1141E+03	0.7078E+03	0.1129E+03
23	6.99	2819.36	12054.05	2822.39	0.1123E+03	0.6382E+03	0.1075E+03
24	7.40	2865.22	12302.79	2865.26	0.1123E+03	0.5818E+03	0.1025E+03
25	7.81	2911.07	12530.57	2906.15	0.1025E+03	0.5351E+03	0.9782E+02
26	8.33	2957.80	12794.90	2955.46	0.9013E+02	0.4863E+03	0.9258E+02
27	8.85	3004.52	13036.34	3002.28	0.8620E+02	0.4463E+03	0.8811E+02
28	9.39	3049.10	13268.90	3049.03	0.8209E+02	0.4114E+03	0.8424E+02
29	9.93	3093.67	13484.06	3093.89	0.8012E+02	0.3819E+03	0.8110E+02
30	10.43	3132.34	13666.77	3133.34	0.7833E+02	0.3589E+03	0.7881E+02
31	10.92	3171.00	13838.88	3171.77	0.7736E+02	0.3388E+03	0.7697E+02
32	11.50	3215.40	14029.98	3216.08	0.7624E+02	0.3180E+03	0.7528E+02
33	12.08	3259.80	14209.76	3259.54	0.7624E+02	0.2998E+03	0.7400E+02

A.5 – 13

V - 7

Connection type : Extended end-plate connections
Mode : All bolted without column stiffener

Tested by : R.J.Dews (1979) U.S.A.
Test Id. : TEST 1

Column : W8x31
Beam : W14x22
Plate thickness : 1"

Fasteners: A325- -3/4"D
15/16" Oversize holes
Material : A36
Fy = 48.10 ksi
Fu = -- ksi

Major parameters

End plate extended on both sides.

ct = 1.5720" pt = 3.5000" li = 10.1560" pc = 3.5000"
cc = 1.5720" gt = 5.5000" gc = 5.5000"
tp = 0.9690" nt = 2 X 2 nc = 2 X 2

Remark 1) End plate extended on tension and compression side.
2)

No	Moment (k-in)	Rotation (radians) X 1/1000	No	Moment (k-in)	Rotation (radians) X 1/1000
1	0.00	0.00	26	932.02	38.83
2	38.18	0.68	27	951.80	41.63
3	76.36	1.37	28	971.58	44.44
4	110.54	2.00	29	988.19	47.07
5	144.72	2.61	30	1004.79	49.71
6	196.94	3.57	31	1021.40	52.35
7	249.16	4.53	32	1036.68	54.99
8	298.06	5.53	33	1051.96	57.63
9	346.97	6.33	34	1067.24	60.28
10	389.13	7.13	35	1081.29	63.17
11	431.30	7.93	36	1095.35	66.06
12	476.86	8.76	37	1109.40	68.95
13	522.42	9.65	38	1121.57	71.57
14	558.44	9.95	39	1133.74	74.19
15	605.46	11.16	40	1145.90	76.80
16	635.15	12.22			
17	662.33	13.65			
18	702.05	16.46			
19	745.91	19.63			
20	788.51	23.00			
21	812.04	25.23			
22	835.57	27.47			
23	860.72	30.40			
24	885.87	33.33			
25	908.94	36.08			

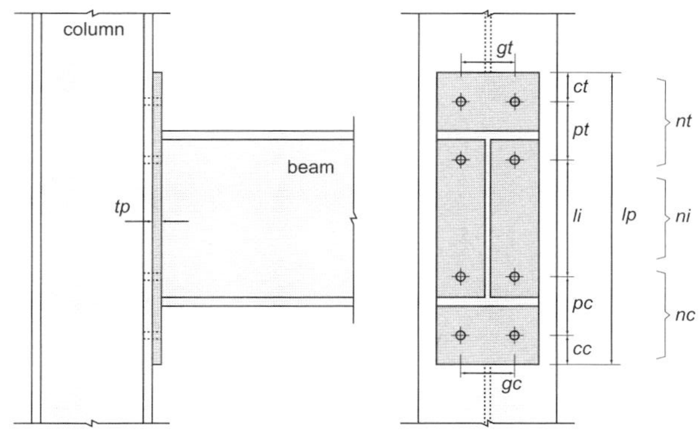

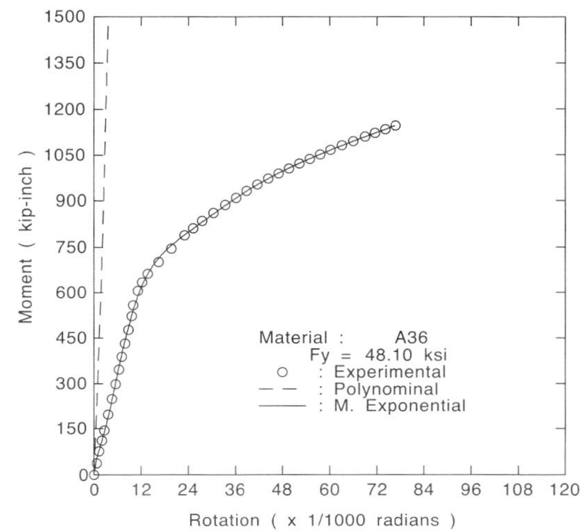

A.5 – 14

```
Moment-rotation prediction equations     ( R : X 1/1000 radians )

Frye and Morris polynominal model  :    R = Sum ( Ai X ( K*Bm )**Pi X 10**Qi )
     xd = 17.156000"   t =  0.969000"  tf =  0.750000"
     A1 =  1.830000   A2 = -1.040000   A3 =  6.380000    K =  0.001699
     P1 =         1   P2 =         3   P3 =         5   Q1 =         0   Q2 =        -1   Q3 =        -3

Modified exponential model  :
     Bm = Sum ( Ai X ( 1 - exp( - R/( 2*i*C ) ) ) )  +     Sum ( RKj X ( R - Rj0 ) )  +  Bm0
       C =   0.81043917E+00              Bmo=   0.00000000E+00          Nexp=  6        Nliner=  1
      AI =   0.93568907E+03  -0.75123633E+04  0.16377423E+05  -0.99766518E+04  -0.48641584E+04   0.58857043E+04
     Rj0 =           8.7600
     RKj =   0.44388797E+01
```

No	Rotation (radians) X 1/1000	Moment (kip-inch) Expri.	Moment (kip-inch) Poly.	Moment (kip-inch) M.Expo.	Connection stiffness (kip-inch) X 1000 Expri.	Connection stiffness (kip-inch) X 1000 Poly.	Connection stiffness (kip-inch) X 1000 M.Expo.
1	0.00	0.00	0.00	0.00	0.5601E+02	0.3216E+03	0.9415E+02
3	1.37	76.36	454.69	82.22	0.5504E+02	0.3555E+03	0.4350E+02
5	2.61	144.72	962.05	137.02	0.5510E+02	0.4808E+03	0.4751E+02
7	4.53	249.16	1865.45	242.35	0.5165E+02	0.3071E+03	0.6061E+02
9	6.33	346.97	2240.81	353.03	0.5696E+02	0.1466E+03	0.6024E+02
11	7.93	431.30	2433.51	443.45	0.5352E+02	0.1010E+03	0.5226E+02
13	9.65	522.42	2584.36	527.72	0.1030E+03	0.7663E+02	0.4533E+02
15	11.16	605.46	2689.55	588.70	0.3312E+02	0.6374E+02	0.3575E+02
17	13.65	662.33	2830.37	661.44	0.1731E+02	0.5036E+02	0.2343E+02
19	19.63	745.91	3076.93	755.22	0.1326E+02	0.3432E+02	0.1109E+02
21	25.23	812.04	3246.31	810.80	0.1052E+02	0.2690E+02	0.9373E+01
23	30.40	860.72	3373.50	858.43	0.8578E+01	0.2261E+02	0.9047E+01
25	36.08	908.94	3492.15	908.04	0.8394E+01	0.1935E+02	0.8348E+01
27	41.63	951.80	3592.89	951.85	0.7056E+01	0.1704E+02	0.7418E+01
29	47.07	988.19	3680.71	989.79	0.6294E+01	0.1531E+02	0.6548E+01
31	52.35	1021.40	3757.80	1022.48	0.6040E+01	0.1396E+02	0.5871E+01
33	57.63	1051.96	3829.24	1052.11	0.5785E+01	0.1284E+02	0.5376E+01
35	63.17	1081.29	3897.63	1080.81	0.4859E+01	0.1188E+02	0.5023E+01
37	68.95	1109.40	3963.77	1109.13	0.4750E+01	0.1103E+02	0.4787E+01
39	74.19	1133.74	4019.60	1133.81	0.4650E+01	0.1038E+02	0.4654E+01

```
Connection type : Extended end-plate connections
           Mode : All bolted without column stiffener

Tested by   :  R.J.Dews (1979)                            U.S.A.
Test   Id.  :  TEST 2

Column : W8x31                        Fasteners: A325- -3/4"D
Beam   : W14x22                             15/16" Oversize holes
Plate thickness : 1"                  Material : A36
                                           Fy =  36.90 ksi
                                           Fu =   --   ksi

                           Major parameters

     End plate extended on both sides.
     ct =  1.6170"    pt =  3.5000"    li = 10.2660"    pc =  3.5000"
     cc =  1.6170"    gt =  5.5310"    gc =  5.5310"
     tp =  0.9380"    nt = 2 X  2      nc = 2 X  2

Remark  1)  End plate extended on tension and compression side.
        2)
```

No	Moment (k-in)	Rotation (radians) X 1/1000
1	0.00	0.00
2	40.34	0.46
3	80.69	0.92
4	121.03	1.38
5	163.14	1.81
6	205.25	2.24
7	247.35	2.67
8	302.56	3.37
9	357.77	4.06
10	397.67	4.52
11	437.57	4.99
12	477.47	5.45
13	518.69	5.88
14	559.92	6.31
15	605.15	6.92
16	638.54	8.01
17	666.66	9.38
18	683.54	10.96
19	700.43	12.53
20	724.72	14.70
21	749.01	16.88
22	768.85	18.91
23	788.70	20.95
24	808.54	22.99
25	832.91	25.58
26	857.28	28.18
27	881.65	30.77
28	903.34	33.20
29	925.02	35.64
30	946.71	38.07
31	968.40	40.50
32	996.77	43.23
33	1025.15	45.96
34	1053.53	48.69
35	1081.91	51.42
36	1107.23	54.39
37	1132.55	57.36
38	1157.88	60.32

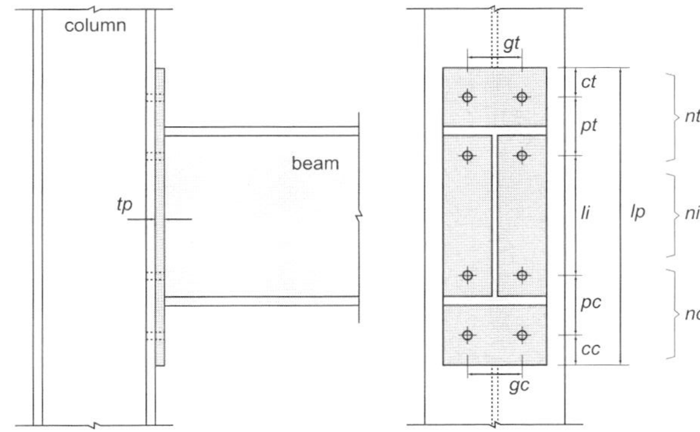

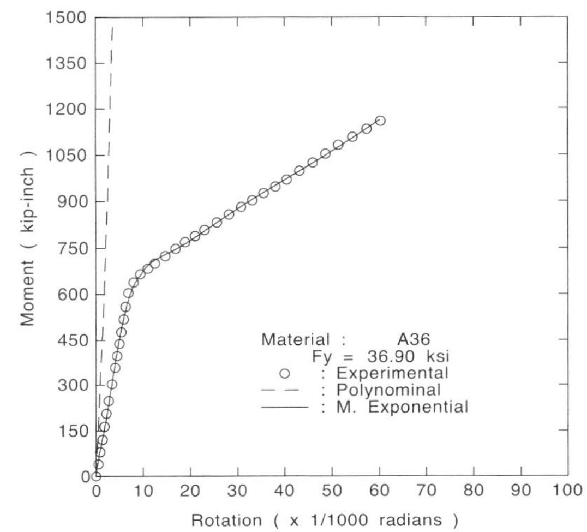

A.5 – 16

```
Moment-rotation prediction equations     ( R : X 1/1000 radians )

Frye and Morris polynominal model  :    R = Sum ( Ai X ( K*Bm )**Pi X 10**Qi )
     xd = 17.266000"   t =  0.938000"  tf =  0.750000"
     A1 =  1.830000   A2 = -1.040000   A3 =  6.380000    K =  0.001695
     P1 =         1   P2 =         3   P3 =         5   Q1 =         0   Q2 =        -1   Q3 =        -3

Modified exponential model  :
     Bm = Sum ( Ai X ( 1 - exp( - R/( 2*i*C ) ) ) )  +      Sum ( RKj X ( R - Rj0 ) )  +  Bm0
      C =   0.61798500E+00              Bmo=   0.00000000E+00         Nexp=  6        Nliner=  1
     AI =   0.13136093E+04  -0.15082644E+05   0.53593680E+05  -0.88152142E+05   0.72353074E+05  -0.23399691E+05
    Rj0 =           9.3800
    RKj =   0.10497956E+02
```

No	Rotation (radians) X 1/1000	Moment (kip-inch) Expri.	Moment (kip-inch) Poly.	Moment (kip-inch) M.Expo.	Connection stiffness (kip-inch) X 1000 Expri.	Connection stiffness (kip-inch) X 1000 Poly.	Connection stiffness (kip-inch) X 1000 M.Expo.
1	0.00	0.00	0.00	0.00	0.8748E+02	0.3223E+03	0.1371E+03
3	0.92	80.69	301.75	88.36	0.8747E+02	0.3369E+03	0.7568E+02
5	1.81	163.14	620.46	156.90	0.9828E+02	0.3871E+03	0.8229E+02
7	2.67	247.35	992.08	233.12	0.9102E+02	0.4904E+03	0.9469E+02
9	4.06	357.77	1705.18	368.75	0.8357E+02	0.4066E+03	0.9559E+02
11	4.99	437.57	1994.36	452.28	0.8643E+02	0.2425E+03	0.8425E+02
13	5.88	518.69	2173.97	520.73	0.9603E+02	0.1696E+03	0.6903E+02
15	6.92	605.15	2325.48	582.84	0.5830E+02	0.1258E+03	0.5034E+02
17	9.38	666.66	2569.80	661.17	0.1591E+02	0.7969E+02	0.2685E+02
19	12.53	700.43	2777.60	712.17	0.1092E+02	0.5566E+02	0.9563E+01
21	16.88	749.01	2981.75	746.60	0.1044E+02	0.4015E+02	0.8158E+01
23	20.95	788.70	3127.94	784.11	0.9740E+01	0.3226E+02	0.1005E+02
25	25.58	832.91	3263.30	831.83	0.9396E+01	0.2661E+02	0.1027E+02
27	30.77	881.65	3389.74	883.19	0.9147E+01	0.2241E+02	0.9513E+01
29	35.64	925.02	3491.63	928.24	0.8914E+01	0.1962E+02	0.9085E+01
31	40.50	968.40	3581.71	972.33	0.9611E+01	0.1750E+02	0.9102E+01
33	45.96	1025.15	3672.06	1022.77	0.1039E+02	0.1566E+02	0.9399E+01
35	51.42	1081.91	3753.48	1075.06	0.9504E+01	0.1421E+02	0.9746E+01
37	57.36	1132.55	3833.89	1133.83	0.8536E+01	0.1294E+02	0.1005E+02

V - 9

Connection type : Extended end-plate connections
Mode : All bolted without column stiffener

Tested by : R.J.Dews (1979) U.S.A.
Test Id. : TEST 3

Column : W10x33
Beam : W16x26
Plate thickness : 1 1/2"

Fasteners: A325- -1"D
1 1/4" Oversize holes
Material : A36
Fy = 42.30 ksi
Fu = -- ksi

Major parameters

End plate extended on both sides.

ct = 1.5705" pt = 3.7500" li = 11.9840" pc = 3.7500"
cc = 1.5705" gt = 5.3750" gc = 5.3750"
tp = 1.5620" nt = 2 X 2 nc = 2 X 2

Remark 1) End plate extended on tension and compression side.
2)

No	Moment (k-in)	Rotation (radians) X 1/1000
1	0.00	0.00
2	868.22	0.50
3	1031.43	1.36
4	1095.29	2.34
5	1133.49	3.37
6	1171.70	4.39
7	1209.90	5.42
8	1249.27	6.83
9	1288.64	8.25
10	1311.38	9.41
11	1334.12	10.57
12	1356.86	11.74
13	1380.62	13.09
14	1404.39	14.45
15	1428.15	15.80
16	1451.92	17.15
17	1471.77	18.50
18	1491.62	19.85
19	1511.46	21.19
20	1531.31	22.54
21	1551.16	23.88
22	1568.30	25.27
23	1585.43	26.65
24	1602.56	28.03
25	1619.70	29.42

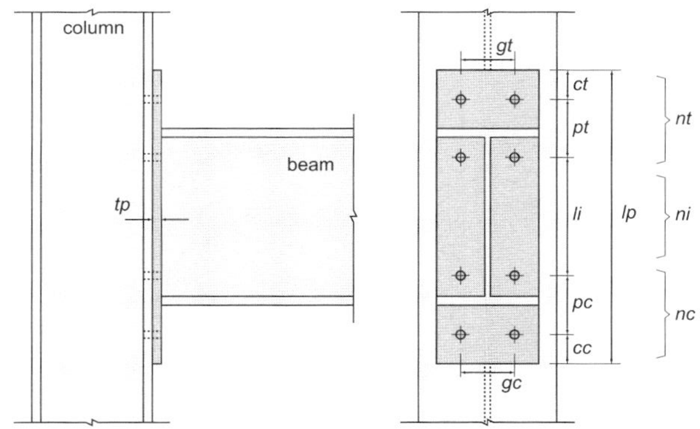

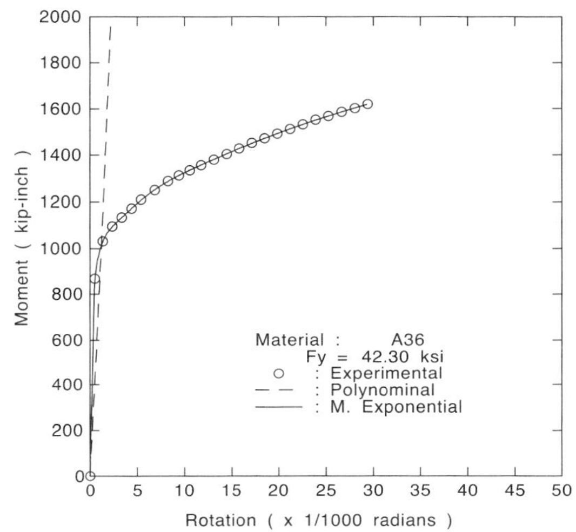

A.5 – 18

```
Moment-rotation prediction equations     ( R : X 1/1000 radians )

Frye and Morris polynominal model  :    R = Sum ( Ai X ( K*Bm )**Pi X 10**Qi )
     xd = 19.484000"   t =  1.562000"  tf =  1.000000"
     A1 =  1.830000   A2 = -1.040000   A3 =  6.380000    K =  0.000672
     P1 =        1    P2 =        3    P3 =        5    Q1 =        0    Q2 =       -1    Q3 =       -3

Modified exponential model  :
     Bm = Sum ( Ai X ( 1 - exp( - R/( 2*i*C ) ) ) )  +     Sum ( RKj X ( R - Rj0 ) )  +  Bm0
      C =   0.37015500E+00                Bmo=   0.00000000E+00          Nexp=  6        Nliner=  2
     AI =   0.69817012E+03   0.38399217E+03  -0.84105572E+04   0.24587247E+05  -0.28698087E+05   0.12137682E+05
     Rj0 =          0.0100           0.5000
     RKj =   0.12748553E+04  -0.12643643E+04
```

No	Rotation (radians) X 1/1000	Moment (kip-inch) Expri.	Moment (kip-inch) Poly.	Moment (kip-inch) M.Expo.	Connection stiffness (kip-inch) X 1000 Expri.	Connection stiffness (kip-inch) X 1000 Poly.	Connection stiffness (kip-inch) X 1000 M.Expo.
1	0.00	0.00	0.00	0.00	0.1736E+04	0.8133E+03	0.6981E+03
2	0.50	868.22	408.39	868.14	0.1169E+04	0.8238E+03	0.1598E+04
3	1.36	1031.43	1146.36	1031.94	0.1310E+03	0.8987E+03	0.9910E+02
4	2.34	1095.29	2115.06	1094.00	0.5166E+02	0.1119E+04	0.4425E+02
5	3.37	1133.49	3440.67	1134.60	0.3721E+02	0.1386E+04	0.3751E+02
6	4.39	1171.70	4607.41	1172.47	0.3721E+02	0.8410E+03	0.3609E+02
7	5.42	1209.90	5273.21	1208.09	0.3327E+02	0.5062E+03	0.3303E+02
8	6.83	1249.27	5840.64	1250.95	0.2785E+02	0.3242E+03	0.2760E+02
9	8.25	1288.64	6235.48	1286.60	0.2330E+02	0.2408E+03	0.2311E+02
10	9.41	1311.38	6489.21	1311.95	0.1956E+02	0.2004E+03	0.2066E+02
11	10.57	1334.12	6704.86	1335.02	0.1956E+02	0.1723E+03	0.1915E+02
12	11.74	1356.86	6892.59	1356.72	0.1862E+02	0.1517E+03	0.1827E+02
13	13.09	1380.62	7085.38	1381.05	0.1753E+02	0.1336E+03	0.1766E+02
14	14.45	1404.39	7256.61	1404.69	0.1753E+02	0.1196E+03	0.1722E+02
15	15.80	1428.15	7411.04	1427.75	0.1753E+02	0.1086E+03	0.1678E+02
16	17.16	1451.92	7551.98	1450.17	0.1614E+02	0.9960E+02	0.1629E+02
17	18.50	1471.77	7680.78	1471.72	0.1476E+02	0.9217E+02	0.1574E+02
18	19.85	1491.62	7800.38	1492.49	0.1476E+02	0.8589E+02	0.1515E+02
19	21.19	1511.46	7912.17	1512.45	0.1476E+02	0.8049E+02	0.1455E+02
20	22.54	1531.31	8017.18	1531.62	0.1476E+02	0.7580E+02	0.1397E+02
21	23.88	1551.16	8116.28	1550.03	0.1359E+02	0.7168E+02	0.1342E+02
22	25.27	1568.30	8212.86	1568.26	0.1238E+02	0.6794E+02	0.1292E+02
23	26.65	1585.43	8304.53	1585.82	0.1238E+02	0.6461E+02	0.1248E+02
24	28.03	1602.56	8391.85	1602.83	0.1238E+02	0.6162E+02	0.1210E+02
25	29.42	1619.70	8475.25	1619.34	0.1238E+02	0.5893E+02	0.1178E+02

V - 10

Connection type : Extended end-plate connections
Mode : All bolted without column stiffener

Tested by : A.N.Sherbourne (1961) England
Test Id. : TEST A1

Column : 8 x 8 UC 35
Beam : 15 x 5 RSJ 42
Plate thickness : 1 1/4"

Fasteners: A325- -3/4"D
15/16" Oversize holes
Material : --
Fy = 35.27 ksi
Fu = 64.62 ksi

Major parameters

End plate extended on tension side only.

ct = 1.5000"	pt = 3.5000"	li = 11.0000"	pc = 2.2500"
cc = 0.2500"	pit= 2.5000"	pi = 3.0000"	pic= 2.5000"
gt = 4.0000"	gi = 4.0000"	gc = 4.0000"	
tp = 1.2500"	nt = 2 X 2	ni = 2 X 3	nc = 2 X 1

Remark 1) End plate extended on tension side only.
2)

No	Moment (k-in)	Rotation (radians) X 1/1000	No	Moment (k-in)	Rotation (radians) X 1/1000
1	0.00	0.00	26	2695.00	7.10
2	140.00	0.15	27	2712.50	7.65
3	280.00	0.30	28	2730.00	8.20
4	420.00	0.40	29	2765.00	8.85
5	560.00	0.50	30	2800.00	9.50
6	700.00	0.65	31	2823.33	10.07
7	840.00	0.80	32	2846.67	10.63
8	962.50	1.00	33	2870.00	11.20
9	1085.00	1.20	34	2878.75	11.78
10	1225.00	1.30	35	2887.50	12.35
11	1365.00	1.40	36	2896.25	12.93
12	1505.00	1.65	37	2905.00	13.50
13	1645.00	1.90			
14	1785.00	2.10			
15	1925.00	2.30			
16	2047.50	2.75			
17	2170.00	3.20			
18	2257.50	3.65			
19	2345.00	4.10			
20	2397.50	4.50			
21	2450.00	4.90			
22	2485.00	5.30			
23	2520.00	5.70			
24	2555.00	6.15			
25	2590.00	6.60			

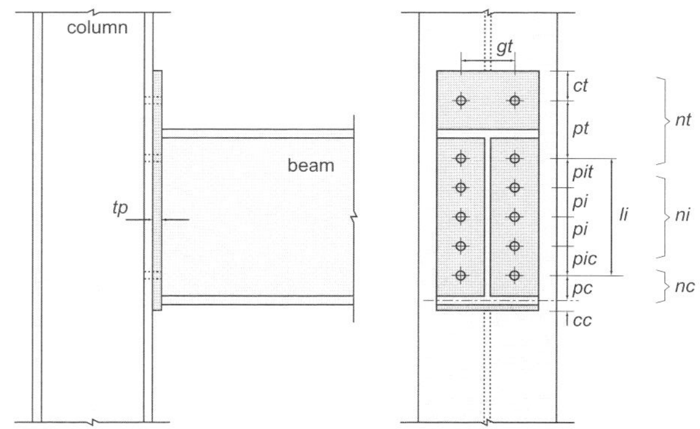

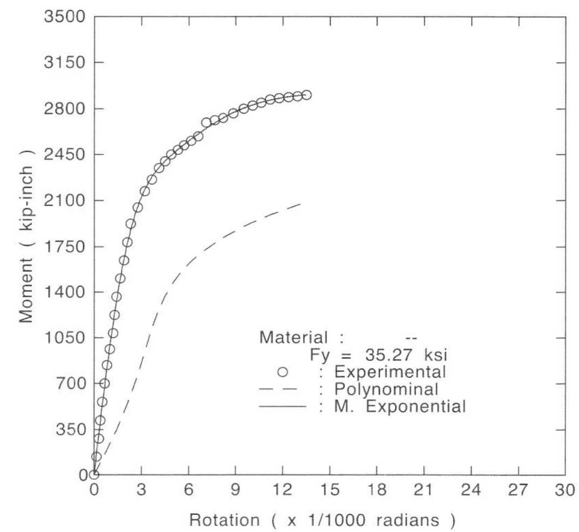

A.5 – 20

```
Moment-rotation prediction equations     ( R : X 1/1000 radians )

Frye and Morris polynominal model  :    R = Sum ( Ai X ( K*Bm )**Pi X 10**Qi )
     xd = 14.500000"   t =  1.250000"  tf =  0.750000"
     A1 =  1.830000   A2 = -1.040000   A3 =  6.380000    K =  0.002298
     P1 =        1    P2 =        3    P3 =        5    Q1 =        0    Q2 =       -1    Q3 =       -3

Modified exponential model  :
     Bm = Sum ( Ai X ( 1 - exp( - R/( 2*i*C ) ) ) )  +      Sum ( RKj X ( R - Rj0 ) )  +  Bm0
      C =   0.15000000E+00                 Bmo=   0.00000000E+00          Nexp=  6        Nliner=  0
     AI =  -0.20649841E+03   0.90072505E+04  -0.68913070E+05   0.18224414E+06  -0.20086352E+06   0.81660753E+05
```

No	Rotation (radians) X 1/1000	Moment (kip-inch) Expri.	Moment (kip-inch) Poly.	Moment (kip-inch) M.Expo.	Connection stiffness (kip-inch) X 1000 Expri.	Connection stiffness (kip-inch) X 1000 Poly.	Connection stiffness (kip-inch) X 1000 M.Expo.
1	0.00	0.00	0.00	0.00	0.9333E+03	0.2378E+03	0.1082E+04
3	0.30	280.00	71.38	317.15	0.1213E+04	0.2389E+03	0.1031E+04
5	0.50	560.00	119.35	519.72	0.1213E+04	0.2409E+03	0.9948E+03
7	0.80	840.00	192.34	809.42	0.7958E+03	0.2458E+03	0.9344E+03
9	1.20	1085.00	292.66	1163.02	0.1137E+04	0.2567E+03	0.8283E+03
11	1.40	1365.00	344.72	1322.42	0.1160E+04	0.2642E+03	0.7646E+03
13	1.90	1645.00	483.07	1661.89	0.6378E+03	0.2913E+03	0.5922E+03
15	2.30	1925.00	605.67	1871.96	0.5684E+03	0.3235E+03	0.4607E+03
17	3.20	2170.00	938.00	2180.05	0.2333E+03	0.4071E+03	0.2449E+03
19	4.10	2345.00	1268.44	2347.38	0.1610E+03	0.2936E+03	0.1424E+03
21	4.90	2450.00	1455.57	2445.34	0.1094E+03	0.1865E+03	0.1082E+03
23	5.70	2520.00	1581.02	2526.03	0.8292E+02	0.1331E+03	0.9536E+02
25	6.60	2590.00	1684.80	2607.84	0.1404E+03	0.1007E+03	0.8648E+02
27	7.65	2712.50	1778.11	2692.36	0.3182E+02	0.7892E+02	0.7391E+02
29	8.85	2765.00	1862.99	2770.74	0.5384E+02	0.6373E+02	0.5649E+02
31	10.07	2823.33	1933.99	2828.87	0.4118E+02	0.5363E+02	0.3947E+02
33	11.20	2870.00	1990.78	2865.99	0.2829E+02	0.4692E+02	0.2660E+02
35	12.35	2887.50	2041.64	2890.75	0.1521E+02	0.4176E+02	0.1703E+02
37	13.50	2905.00	2087.26	2906.33	0.1521E+02	0.3771E+02	0.1052E+02

V - 11

Connection type : Extended end-plate connections
Mode : All bolted without column stiffener

Tested by : J.R.Bailey (1970) England
Test Id. : TEST A1-L

Column : 8 x 8 UC 48
Beam : 12 x 5 RSJ 32
Plate thickness : 1"

Fasteners: 10.9- -7/8"D
1 1/16" Oversize holes
Material : --
Fy = -- ksi
Fu = -- ksi

Major parameters

End plate extended on tension side only.

ct =	-- "	pt =	4.5000"	li =	-- "	pc =	-- "
cc =	-- "	pit=	-- "	pi =	-- "	pic=	-- "
gt =	-- "	gi =	-- "	gc =	-- "		
tp =	1.0200"	nt =	2 X 2	ni =	2 X 1	nc =	2 X 1

Remark 1) Two 7/8" bolts used on either side of tension flange.
2) Four 1" bolts in compression. Material grade 50B (BS968).

No	Moment (k-in)	Rotation (radians) X 1/1000	No	Moment (k-in)	Rotation (radians) X 1/1000
1	0.00	0.00	26	1620.61	9.20
2	36.06	0.30	27	1705.45	10.00
3	72.12	0.60	28	1807.27	11.20
4	108.18	0.90	29	1883.64	12.50
5	144.24	1.20	30	1951.51	13.50
6	189.49	1.40	31	2010.91	14.90
7	234.74	1.60	32	2061.82	16.60
8	280.00	1.80	33	2112.73	18.10
9	347.87	2.15	34	2133.94	19.05
10	415.75	2.50	35	2155.15	20.00
11	483.64	2.85	36	2180.60	21.60
12	551.52	3.20	37	2193.33	22.50
13	640.61	3.65	38	2206.06	23.40
14	729.70	4.10	39	2214.55	24.30
15	797.57	4.45	40	2223.04	25.20
16	865.45	4.80	41	2231.52	26.50
17	937.57	5.20	42	2240.00	27.90
18	1009.69	5.60	43	2240.00	29.60
19	1094.54	6.05	44	2244.25	30.70
20	1179.40	6.50	45	2248.49	31.80
21	1238.79	6.85	46	2256.97	33.40
22	1298.18	7.20	47	2265.46	34.80
23	1361.82	7.65	48	2256.97	35.30
24	1425.46	8.10			
25	1527.28	8.80			

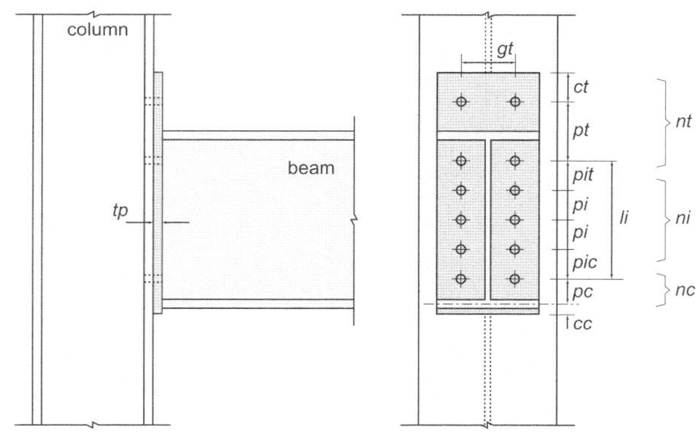

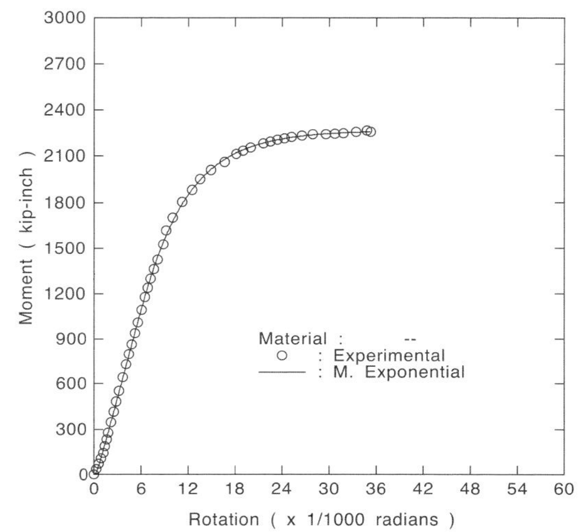

A.5 – 22

Moment-rotation prediction equations (R : X 1/1000 radians)

```
Modified exponential model  :
     Bm = Sum ( Ai X ( 1 - exp( - R/( 2*i*C ) ) ) )  +     Sum ( RKj X ( R - Rj0 ) )  +  Bm0
       C =   0.36916667E+00          Bmo=   0.00000000E+00          Nexp=  6       Nliner=  0
      AI =  -0.10367926E+04   0.11056913E+05  -0.42459275E+05   0.74907544E+05  -0.71231635E+05   0.31025695E+05
```

No	Rotation (radians) X 1/1000	Moment (kip-inch) Expri.	Moment (kip-inch) M.Expo.	Connection stiffness (kip-inch) X 1000 Expri.	Connection stiffness (kip-inch) X 1000 M.Expo.
1	0.00	0.00	0.00	0.1202E+03	-0.1344E+02
3	0.60	72.12	56.52	0.1202E+03	0.1603E+03
5	1.20	144.24	165.01	0.1838E+03	0.1905E+03
7	1.60	234.74	241.39	0.2263E+03	0.1905E+03
9	2.15	347.87	345.49	0.1939E+03	0.1884E+03
11	2.85	483.64	477.82	0.1940E+03	0.1907E+03
13	3.65	640.61	632.73	0.1980E+03	0.1966E+03
15	4.45	797.57	791.46	0.1939E+03	0.1993E+03
17	5.20	937.57	940.10	0.1803E+03	0.1960E+03
19	6.05	1094.54	1102.77	0.1886E+03	0.1856E+03
21	6.85	1238.79	1245.72	0.1697E+03	0.1711E+03
23	7.65	1361.82	1375.96	0.1414E+03	0.1542E+03
25	8.80	1527.28	1538.71	0.2014E+03	0.1289E+03
27	10.00	1705.45	1678.47	0.9757E+02	0.1045E+03
29	12.50	1883.64	1887.62	0.6390E+02	0.6561E+02
31	14.90	2010.91	2014.81	0.3679E+02	0.4226E+02
33	18.10	2112.73	2118.47	0.2683E+02	0.2437E+02
35	20.00	2155.15	2158.19	0.1993E+02	0.1781E+02
37	22.50	2193.33	2194.72	0.1414E+02	0.1182E+02
39	24.30	2214.55	2213.13	0.9433E+01	0.8775E+01
41	26.50	2231.52	2229.26	0.6302E+01	0.6049E+01
43	29.60	2240.00	2243.76	0.2341E+01	0.3519E+01
45	31.80	2248.49	2250.16	0.4446E+01	0.2364E+01
47	34.80	2265.46	2255.60	-0.1091E+02	0.1351E+01

A.5 – 23

V - 12

Connection type : Extended end-plate connections
Mode : All bolted without column stiffener

Tested by : J.R.Bailey (1970) England
Test Id. : TEST A1-R

Column : 8 x 8 UC 48
Beam : 12 x 5 RSJ 32
Plate thickness : 1"

Fasterers: 10.9- -7/8"D
1 1/16" Oversize holes
Material : --
Fy = -- ksi
Fu = -- ksi

Major parameters

End plate extended on tension side only.

ct = -- "	pt = 4.5000"	li = -- "	pc = -- "
cc = -- "	pit= -- "	pi = -- "	pic= -- "
gt = -- "	gi = -- "	gc = -- "	
tp = 1.0200"	nt = 2 X 2	ni = 2 X 1	nc = 2 X 1

Remark 1) Two 7/8" bolts used on either side of tension flange.
2) Four 1" bolts in compression. Material grade 50B (BS968).

No	Moment (k-in)	Rotation (radians) X 1/1000	No	Moment (k-in)	Rotation (radians) X 1/1000
1	0.00	0.00	26	1928.42	10.20
2	80.00	0.20	27	1970.53	11.40
3	160.00	0.40	28	2029.48	12.70
4	261.05	0.80	29	2063.16	13.70
5	324.21	1.00	30	2084.21	14.50
6	387.37	1.20	31	2105.27	15.30
7	458.95	1.45	32	2138.95	16.60
8	530.53	1.70	33	2172.64	18.00
9	640.00	2.20	34	2189.47	19.40
10	703.16	2.35	35	2206.32	20.90
11	766.32	2.50	36	2218.95	21.80
12	853.95	2.70	37	2231.58	22.70
13	963.42	3.20	38	2240.00	24.10
14	1027.37	3.40	39	2248.42	25.40
15	1085.31	3.60	40	2248.42	26.50
16	1149.47	3.80	41	2256.84	27.70
17	1212.64	4.00	42	2265.26	28.80
18	1280.00	4.20	43	2265.26	29.60
19	1347.37	4.40	44	2265.26	30.60
20	1423.15	4.80	45	2265.26	31.40
21	1524.22	5.30			
22	1616.84	5.90			
23	1726.32	6.80			
24	1802.10	7.80			
25	1877.89	9.00			

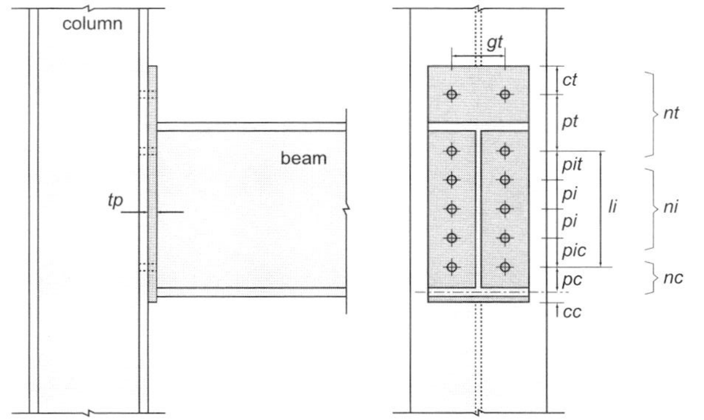

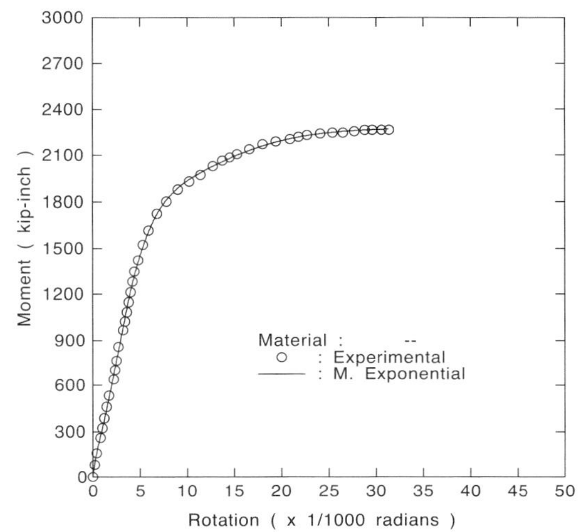

```
Moment-rotation prediction equations     ( R : X 1/1000 radians )

Modified exponential model  :
     Bm = Sum ( Ai X ( 1 - exp( - R/( 2*i*C ) ) ) )  +      Sum ( RKj X ( R - Rj0 ) )  +  Bm0
      C =   0.31166667E+00            Bmo=   0.00000000E+00         Nexp=  6        Nliner=  0
     AI =   0.75108790E+03   0.37033813E+04  -0.49664325E+05   0.13485678E+06  -0.14571315E+06   0.58344139E+05
```

No	Rotation (radians) X 1/1000	Moment (kip-inch) Expri.	Moment (kip-inch) M.Expo.	Connection stiffness (kip-inch) X 1000 Expri.	Connection stiffness (kip-inch) X 1000 M.Expo.
1	0.00	0.00	0.00	0.4000E+03	0.5512E+03
3	0.40	160.00	163.80	0.3509E+03	0.3118E+03
5	1.00	324.21	323.13	0.3158E+03	0.2522E+03
7	1.45	458.95	441.76	0.2863E+03	0.2780E+03
9	2.20	640.00	667.40	0.3744E+03	0.3181E+03
11	2.50	766.32	763.71	0.4391E+03	0.3228E+03
13	3.20	968.42	986.63	0.2731E+03	0.3091E+03
15	3.60	1086.31	1106.67	0.3053E+03	0.2901E+03
17	4.00	1212.64	1218.11	0.3263E+03	0.2665E+03
19	4.40	1347.37	1319.59	0.2877E+03	0.2406E+03
21	5.30	1524.22	1509.68	0.1804E+03	0.1826E+03
23	6.80	1726.32	1722.59	0.9992E+02	0.1069E+03
25	9.00	1877.89	1887.05	0.5264E+02	0.5191E+02
27	11.40	1970.53	1985.25	0.4001E+02	0.3394E+02
29	13.70	2063.16	2056.40	0.2959E+02	0.2852E+02
31	15.30	2105.27	2099.62	0.2616E+02	0.2546E+02
33	18.00	2172.64	2160.54	0.1804E+02	0.1953E+02
35	20.90	2206.32	2207.75	0.1298E+02	0.1317E+02
37	22.70	2231.58	2228.36	0.1090E+02	0.9832E+01
39	25.40	2248.42	2249.44	0.2966E+01	0.6024E+01
41	27.70	2256.84	2260.60	0.7353E+01	0.3819E+01
43	29.60	2265.26	2266.60	0.1110E-15	0.2566E+01
45	31.40	2265.26	2270.42	0.0000E+00	0.1736E+01

V - 13

Connection type : Extended end-plate connections
Mode : All bolted without column stiffener

Tested by : J.R.Bailey (1970) England
Test Id. : TEST A2-L

Column : 8 x 8 UC 40
Beam : 10 x 4 UB 17
Plate thickness : 3/4"

Fasteners: 10.9- -5/8"D
13/16" Oversize holes
Material : --
Fy = -- ksi
Fu = -- ksi

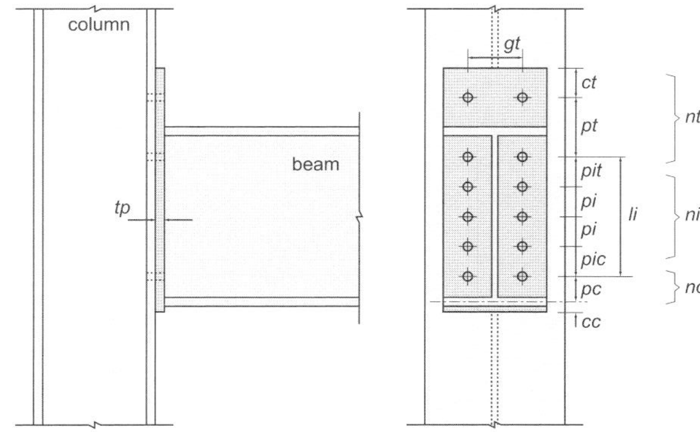

Major parameters

End plate extended on tension side only.

ct = -- "	pt = 4.3750"	li = -- "	pc = -- "
cc = -- "	pit= -- "	pi = -- "	pic= -- "
gt = -- "	gi = -- "	gc = -- "	
tp = 0.8050"	nt = 2 X 2	ni = 2 X 1	nc = 2 X 1

Remark 1) Two 5/8" bolts used on either side of tension flange.
2) Four 7/8" bolts in compression. Material grade 50B (BS968).

No	Moment (k-in)	Rotation (radians) X 1/1000	No	Moment (k-in)	Rotation (radians) X 1/1000	No	Moment (k-in)	Rotation (radians) X 1/1000
1	0.00	0.00	26	800.30	5.10	51	1240.68	24.10
2	8.94	0.10	27	834.18	5.40	52	1249.15	25.30
3	17.88	0.20	28	868.06	5.70	53	1240.68	26.10
4	26.82	0.30	29	901.93	6.00			
5	35.75	0.40	30	952.74	6.90			
6	44.69	0.50	31	986.62	7.30			
7	53.63	0.60	32	1020.49	7.70			
8	80.45	0.90	33	1062.84	8.90			
9	116.74	1.10	34	1096.72	9.80			
10	153.04	1.30	35	1122.12	11.00			
11	207.48	1.60	36	1139.05	12.30			
12	241.36	1.80	37	1151.75	13.10			
13	275.23	2.00	38	1164.46	13.90			
14	326.05	2.20	39	1177.16	14.65			
15	376.86	2.40	40	1189.87	15.40			
16	419.21	2.55	41	1202.57	16.10			
17	461.56	2.70	42	1215.28	16.80			
18	495.43	3.00	43	1215.28	17.55			
19	529.30	3.30	44	1215.28	18.30			
20	571.64	3.65	45	1219.51	19.15			
21	613.99	4.00	46	1223.75	20.00			
22	647.87	4.20	47	1223.75	20.75			
23	681.74	4.40	48	1223.75	21.50			
24	724.09	4.60	49	1240.68	22.50			
25	766.43	4.80	50	1240.68	23.30			

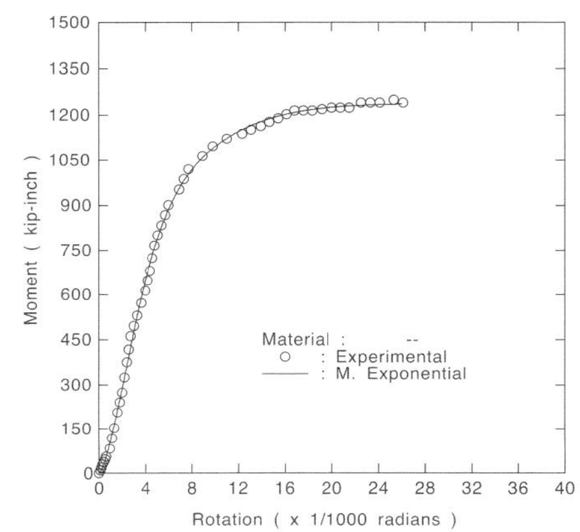

A.5 – 26

```
Moment-rotation prediction equations     ( R : X 1/1000 radians )

Modified exponential model  :
     Bm = Sum ( Ai X ( 1 - exp( - R/( 2*i*C ) ) ) )  +      Sum ( RKj X ( R - Rj0 ) )  + Bm0
      C =   0.24250000E+00              Bmo=   0.00000000E+00          Nexp=  6        Nliner=  0
     AI =  -0.26297662E+03   0.49813027E+04  -0.27461929E+05   0.59684768E+05  -0.60942890E+05   0.25239341E+05
```

No	Rotation (radians) X 1/1000	Moment (kip-inch) Expri.	Moment (kip-inch) M.Expo.	Connection stiffness (kip-inch) X 1000 Expri.	Connection stiffness (kip-inch) X 1000 M.Expo.
1	0.00	0.00	0.00	0.8940E+02	0.2653E+02
3	0.20	17.88	10.74	0.8940E+02	0.7667E+02
5	0.40	35.75	29.42	0.8934E+02	0.1078E+03
7	0.60	53.63	53.20	0.8940E+02	0.1287E+03
9	1.10	116.74	126.47	0.1815E+03	0.1613E+03
11	1.60	207.48	212.47	0.1742E+03	0.1810E+03
13	2.00	275.23	286.82	0.2117E+03	0.1896E+03
15	2.40	376.86	363.29	0.2702E+03	0.1917E+03
17	2.70	461.56	420.53	0.2258E+03	0.1894E+03
19	3.30	529.30	530.86	0.1166E+03	0.1769E+03
21	4.00	613.99	647.06	0.1518E+03	0.1542E+03
23	4.40	681.74	705.88	0.1905E+03	0.1398E+03
25	4.80	766.43	758.92	0.1722E+03	0.1254E+03
27	5.40	834.18	828.03	0.1129E+03	0.1053E+03
29	6.00	901.93	885.80	0.9880E+02	0.8775E+02
31	7.30	986.62	980.14	0.8469E+02	0.5952E+02
33	8.90	1062.84	1057.42	0.3663E+02	0.3913E+02
35	11.00	1122.12	1123.13	0.1726E+02	0.2494E+02
37	13.10	1151.75	1166.07	0.1588E+02	0.1653E+02
39	14.65	1177.16	1188.07	0.1694E+02	0.1205E+02
41	16.10	1202.57	1203.08	0.1815E+02	0.8792E+01
43	17.55	1215.28	1213.93	0.0000E+00	0.6281E+01
45	19.15	1219.51	1222.25	0.4984E+01	0.4235E+01
47	20.75	1223.75	1227.80	0.0000E+00	0.2794E+01
49	22.50	1240.68	1231.70	0.7525E+01	0.1735E+01
51	24.10	1240.68	1233.93	0.2824E+01	0.1104E+01
53	26.10	1240.68	1235.61	-0.1059E+02	0.6158E+00

V - 14

```
Connection type : Extended end-plate connections
           Mode : All bolted without column stiffener

Tested by  :  J.R.Bailey (1970)                         England
Test  Id.  :  TEST A2-R

Column : 8 x 8 UC 40                 Fasteners: 10.9- -5/8"D
Beam   : 10 x 4 UB 17                      13/16" Oversize holes
Plate thickness : 3/4"               Material :  --
                                          Fy =   --   ksi
                                          Fu =   --   ksi

                        Major parameters

     End plate extended on tension side only.
     ct =   --   "    pt =  4.3750"    li =   --   "    pc =   --   "
     cc =   --   "    pit=   --   "    pi =   --   "    pic=   --   "
     gt =   --   "    gi =   --   "    gc =   --   "
     tp =  0.3050"    nt = 2 X  2      ni = 2 X  1      nc = 2 X  1

Remark  1)  Two 5/8" bolts used on either side of tension flange.
        2)  Four 7/8" bolts in compression.   Material grade 50B ( BS968 ).
```

No	Moment (k-in)	Rotation (radians) X 1/1000	No	Moment (k-in)	Rotation (radians) X 1/1000
1	0.00	0.00	26	978.14	8.00
2	40.93	0.30	27	1020.49	9.20
3	81.86	0.60	28	1062.84	10.60
4	122.80	0.90	29	1088.24	11.90
5	173.61	1.25	30	1096.71	12.65
6	224.42	1.60	31	1105.18	13.40
7	271.00	1.75	32	1113.65	14.15
8	317.58	1.90	33	1122.12	14.90
9	359.92	2.10	34	1139.05	16.20
10	402.26	2.30	35	1143.29	17.20
11	453.08	2.35	36	1147.52	18.20
12	503.89	2.40	37	1147.52	18.95
13	533.53	2.65	38	1147.52	19.70
14	563.17	2.90	39	1151.76	20.65
15	597.05	3.15	40	1156.00	21.60
16	630.92	3.40	41	1156.00	22.40
17	660.56	3.60	42	1156.00	23.20
18	690.20	3.80	43	1160.23	24.00
19	728.32	4.10	44	1164.46	24.80
20	766.43	4.40	45	1160.23	25.65
21	817.24	5.20	46	1156.00	26.50
22	846.88	5.70	47	1164.46	27.55
23	876.52	6.20	48	1172.93	28.60
24	906.16	6.65			
25	935.81	7.10			

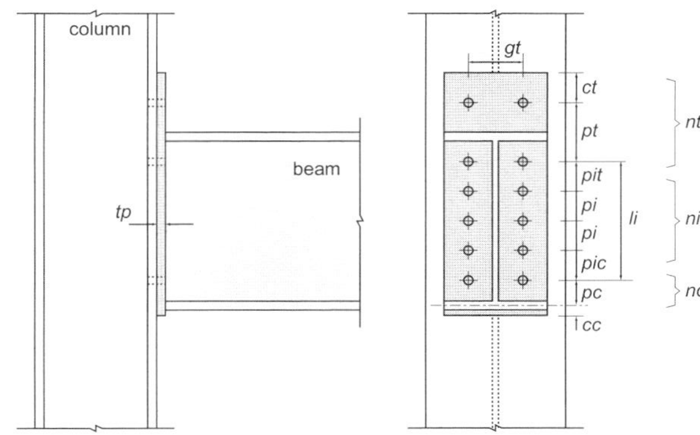

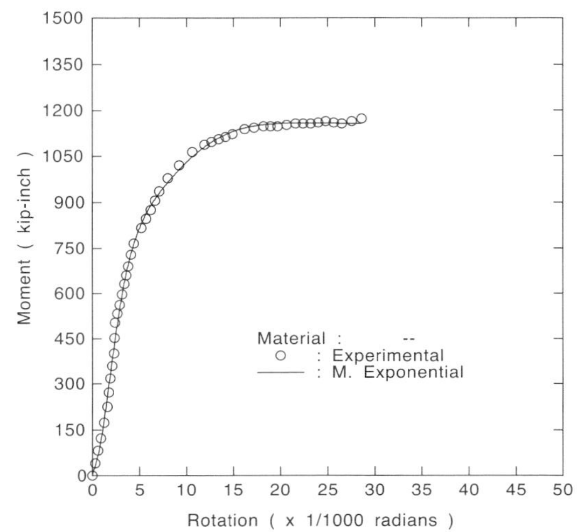

Moment-rotation prediction equations (R : X 1/1000 radians)

```
Modified exponential model  :
     Bm = Sum ( Ai X ( 1 - exp( - R/( 2*i*C ) ) ) )  +     Sum ( RKj X ( R - Rj0 ) )  +  Bm0
      C =   0.31333333E+00             Bmo=   0.00000000E+00         Nexp=  6       Nliner=  0
     AI =   0.11219745E+04  -0.14679275E+05   0.53382787E+05  -0.87033319E+05   0.66038723E+05  -0.17677445E+05
```

No	Rotation (radians) X 1/1000	Moment (kip-inch) Expri.	Moment (kip-inch) M.Expo.	Connection stiffness (kip-inch) X 1000 Expri.	Connection stiffness (kip-inch) X 1000 M.Expo.
1	0.00	0.00	0.00	0.1364E+03	0.1273E+03
3	0.60	81.86	67.42	0.1364E+03	0.1328E+03
5	1.25	173.61	181.20	0.1452E+03	0.2117E+03
7	1.75	271.00	294.58	0.3105E+03	0.2356E+03
9	2.10	359.92	376.93	0.2117E+03	0.2327E+03
11	2.35	453.08	434.08	0.1016E+04	0.2237E+03
13	2.65	533.53	498.90	0.1186E+03	0.2077E+03
15	3.15	597.05	594.82	0.1355E+03	0.1753E+03
17	3.60	660.56	666.91	0.1482E+03	0.1454E+03
19	4.10	728.32	732.00	0.1270E+03	0.1158E+03
21	5.20	817.24	831.77	0.6091E+02	0.7023E+02
23	6.20	876.52	890.26	0.6275E+02	0.4934E+02
25	7.10	935.81	930.21	0.5960E+02	0.4051E+02
27	9.20	1020.49	1005.48	0.3296E+02	0.3231E+02
29	11.90	1088.24	1080.22	0.1431E+02	0.2254E+02
31	13.40	1105.18	1109.48	0.1129E+02	0.1651E+02
33	14.90	1122.12	1130.12	0.1193E+02	0.1117E+02
35	17.20	1143.29	1148.43	0.4236E+01	0.5224E+01
37	18.95	1147.52	1154.96	0.0000E+00	0.2460E+01
39	20.65	1151.76	1157.67	0.4459E+01	0.8803E+00
41	22.40	1156.00	1158.37	0.0000E+00	0.2168E-01
43	24.00	1160.23	1158.08	0.5287E+01	-0.3396E+00
45	25.65	1160.23	1157.39	-0.4976E+01	-0.4695E+00
47	27.55	1164.46	1156.48	0.8063E+01	-0.4634E+00

A.5 – 29

V - 15

Connection type : Extended end-plate connections
Mcde : All bolted without column stiffener

Tested by : J.R.Bailey (1970) England
Test Id. : TEST A3-L

Column : 8 x 8 UC 40
Beam : 10 x 4 UB 19
Plate thickness : 3/4"

Fasteners: 10.9- -3/4"D
15/16" Oversize holes
Material : --
Fy = -- ksi
Fu = -- ksi

Major parameters

End plate extended on tension side only.

ct = -- "	pt = 4.3750"	li = -- "	pc = -- "
cc = -- "	pit= -- "	pi = -- "	pic= -- "
gt = -- "	gi = -- "	gc = -- "	
tp = 0.8500"	nt = 2 X 2	ni = 2 X 1	nc = 2 X 1

Remark 1) Two 3/4" bolts used on either side of tension flange.
2) Four 7/8" bolts in compression. Material grade 50B (BS968).

No	Moment (k-in)	Rotation (radians) X 1/1000	No	Moment (k-in)	Rotation (radians) X 1/1000
1	0.00	0.00	26	1113.62	7.55
2	19.55	0.20	27	1160.38	8.10
3	39.11	0.40	28	1194.38	8.70
4	58.66	0.60	29	1228.39	9.30
5	78.20	0.80	30	1253.89	10.15
6	97.76	1.00	31	1279.39	11.00
7	157.27	1.35	32	1300.64	11.60
8	216.78	1.70	33	1321.90	12.20
9	272.03	2.00	34	1334.65	12.95
10	327.29	2.30	35	1347.40	13.70
11	374.04	2.55	36	1355.90	14.55
12	420.79	2.80	37	1364.40	15.40
13	459.05	3.10	38	1372.90	16.20
14	497.30	3.40	39	1381.40	17.00
15	545.48	3.63	40	1384.24	17.80
16	593.65	3.87	41	1387.07	18.60
17	641.82	4.10	42	1389.90	19.40
18	697.07	4.45	43	1385.65	20.20
19	752.33	4.80	44	1381.40	21.00
20	799.09	5.10			
21	845.84	5.40			
22	901.10	5.85			
23	956.35	6.30			
24	1011.61	6.65			
25	1066.87	7.00			

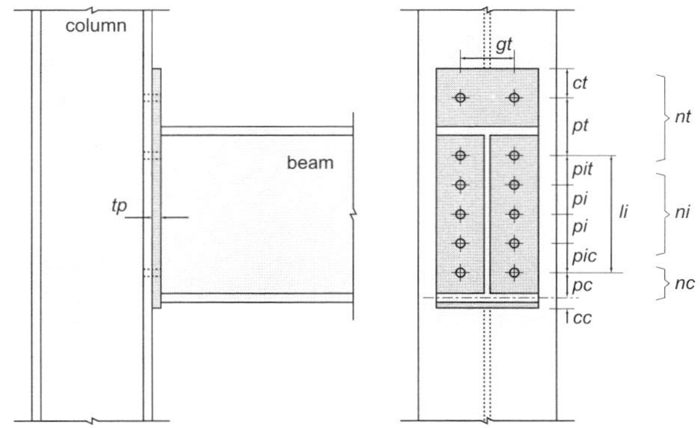

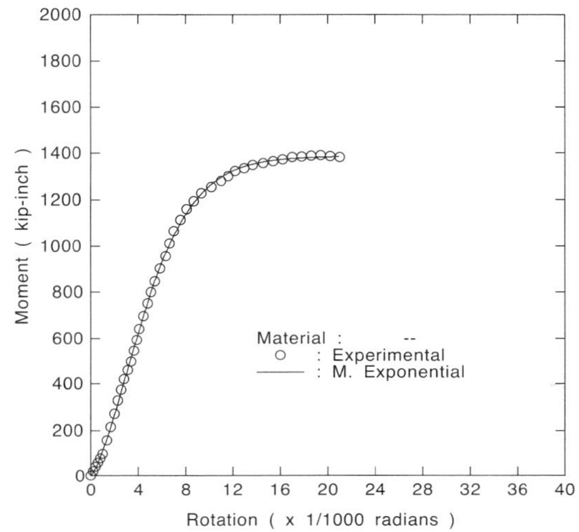

Moment-rotation prediction equations (R : X 1/1000 radians)

Modified exponential model :

```
Bm = Sum ( Ai X ( 1 - exp( - R/( 2*i*C ) ) ) )  +   Sum ( RKj X ( R - Rj0 ) )  + Bm0
 C =  0.22500000E+00            Bmo=  0.00000000E+00        Nexp= 6       Nliner= 0
 AI = -0.11957262E+03  0.73386630E+02  0.49286231E+04 -0.13983099E+05  0.61596652E+04  0.43268114E+04
```

No	Rotation (radians) X 1/1000	Moment (kip-inch) Expri.	Moment (kip-inch) M.Expo.	Connection stiffness (kip-inch) X 1000 Expri.	Connection stiffness (kip-inch) X 1000 M.Expo.
1	0.00	0.00	0.00	0.9774E+02	0.3842E+02
3	0.40	39.11	32.49	0.9777E+02	0.1108E+03
5	0.80	78.20	82.68	0.9777E+02	0.1363E+03
7	1.35	157.27	162.56	0.1700E+03	0.1529E+03
9	2.00	272.03	266.96	0.1842E+03	0.1677E+03
11	2.55	374.04	361.87	0.1870E+03	0.1766E+03
13	3.10	459.05	460.31	0.1275E+03	0.1804E+03
15	3.63	545.48	556.26	0.2064E+03	0.1786E+03
17	4.10	641.82	638.49	0.1870E+03	0.1732E+03
19	4.80	752.33	755.42	0.1568E+03	0.1600E+03
21	5.40	845.84	847.18	0.1426E+03	0.1455E+03
23	6.30	956.35	967.51	0.1425E+03	0.1217E+03
25	7.00	1066.87	1046.22	0.1295E+03	0.1033E+03
27	8.10	1160.38	1145.20	0.7146E+02	0.7735E+02
29	9.30	1228.39	1223.69	0.4564E+02	0.5445E+02
31	11.00	1279.39	1295.43	0.3318E+02	0.3169E+02
33	12.20	1321.90	1326.76	0.2723E+02	0.2116E+02
35	13.70	1347.40	1351.50	0.1372E+02	0.1253E+02
37	15.40	1364.40	1367.47	0.1032E+02	0.6804E+01
39	17.00	1381.40	1375.71	0.7084E+01	0.3784E+01
41	18.60	1387.07	1380.27	0.3540E+01	0.2088E+01
43	20.20	1385.65	1382.79	-0.5310E+01	0.1146E+01

A.5 – 31

V - 16

```
Connection type : Extended end-plate connections
           Mode : All bolted without column stiffener

Tested by  :  J.R.Bailey (1970)                          England
Test  Id.  :  TEST A3-R

Column : 8 x 8 UC 40                 Fasteners: 10.9- -3/4"D
Beam   : 10 x 4 UB 19                    15/16" Oversize holes
Plate thickness : 3/4"               Material : --
                                          Fy =   --   ksi
                                          Fu =   --   ksi

                         Major parameters

     End plate extended on tension side only.
     ct =   --   "    pt =  4.3750"    li =   --   "    pc =   --   "
     cc =   --   "    pit=   --   "    pi =   --   "    pic=   --   "
     gt =   --   "    gi =   --   "    gc =   --   "
     tp =  0.8500"    nt = 2 X  2      ni = 2 X  1      nc = 2 X  1

Remark  1)  Two 3/4" bolts used on either side of tension flange.
        2)  Four 7/8" bolts in compression.    Material grade 50B ( BS968 ).
```

No	Moment (k-in)	Rotation (radians) X 1/1000	No	Moment (k-in)	Rotation (radians) X 1/1000
1	0.00	0.00	26	1181.40	7.30
2	49.40	0.23	27	1181.40	7.30
3	98.80	0.47	28	1249.15	8.40
4	148.20	0.70	29	1270.32	9.05
5	199.01	1.00	30	1291.49	9.70
6	249.83	1.30	31	1333.84	10.80
7	300.64	1.60	32	1346.54	11.45
8	351.46	1.90	33	1359.24	12.10
9	402.27	2.15	34	1367.71	12.95
10	453.08	2.40	35	1376.18	13.80
11	503.90	2.55	36	1384.65	14.55
12	554.71	2.70	37	1393.12	15.30
13	597.05	2.95	38	1405.82	15.95
14	639.40	3.20	39	1418.52	16.60
15	639.40	3.20	40	1422.76	17.55
16	685.97	3.50	41	1426.99	18.50
17	732.55	3.80	42	1422.76	19.50
18	779.13	4.05	43	1418.52	20.50
19	825.71	4.30	44	1426.99	21.35
20	872.29	4.50	45	1435.46	22.20
21	918.86	4.70	46	1432.64	23.03
22	961.21	5.00	47	1429.82	23.87
23	1003.55	5.30	48	1426.99	24.70
24	1058.60	5.70			
25	1113.65	6.10			

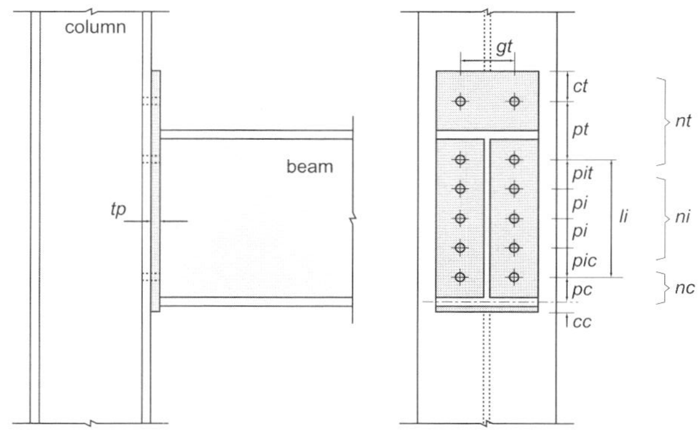

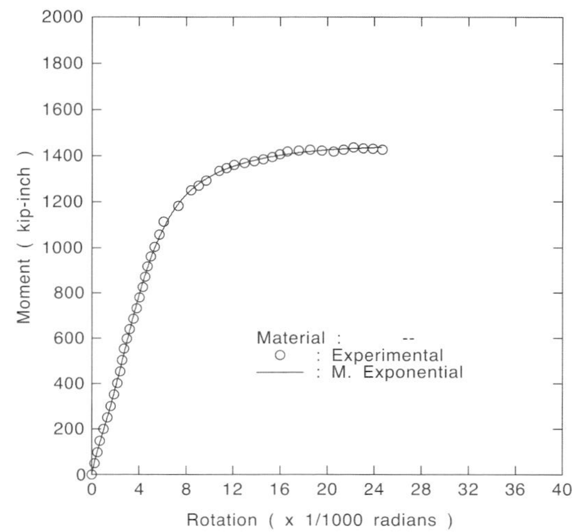

Moment-rotation prediction equations (R : X 1/1000 radians)

```
Modified exponential model  :
    Bm = Sum ( Ai X ( 1 - exp( - R/( 2*i*C ) ) ) )  +      Sum ( RKj X ( R - Rj0 ) )  +  Bm0
     C =   0.26415833E+00           Bmo=   0.00000000E+00         Nexp=  6        Nliner=  0
    AI =  -0.14262169E+03   0.62233522E+04  -0.33257096E+05   0.65788685E+05  -0.59990987E+05   0.22819350E+05
```

No	Rotation (radians) X 1/1000	Moment (kip-inch) Expri.	Moment (kip-inch) M.Expo.	Connection stiffness (kip-inch) X 1000 Expri.	Connection stiffness (kip-inch) X 1000 M.Expo.
1	0.00	0.00	0.00	0.2117E+03	0.2566E+03
3	0.47	98.80	100.86	0.2117E+03	0.1875E+03
5	1.00	199.01	194.99	0.1694E+03	0.1734E+03
7	1.60	300.64	303.37	0.1694E+03	0.1900E+03
9	2.15	402.27	412.62	0.2033E+03	0.2060E+03
11	2.55	503.90	496.31	0.3388E+03	0.2113E+03
13	2.95	597.05	580.84	0.1694E+03	0.2103E+03
15	3.20	639.40	633.01	0.5158E+00	0.2067E+03
17	3.80	732.55	752.70	0.1722E+03	0.1908E+03
19	4.30	825.71	843.67	0.2122E+03	0.1726E+03
21	4.70	918.86	909.50	0.1962E+03	0.1565E+03
23	5.30	1003.55	996.01	0.1396E+03	0.1320E+03
25	6.10	1113.65	1089.41	0.1173E+03	0.1022E+03
27	7.30	1181.40	1189.92	0.5594E-01	0.6745E+02
29	9.05	1270.32	1278.57	0.3257E+02	0.3734E+02
31	10.80	1333.84	1329.65	0.2658E+02	0.2279E+02
33	12.10	1359.24	1355.20	0.1539E+02	0.1698E+02
35	13.80	1376.18	1379.74	0.1067E+02	0.1225E+02
37	15.30	1393.12	1395.85	0.1571E+02	0.9363E+01
39	16.60	1418.52	1406.70	0.1341E+02	0.7387E+01
41	18.50	1426.99	1418.47	0.2229E+00	0.5115E+01
43	20.50	1418.52	1426.85	0.3441E+01	0.3361E+01
45	22.20	1435.46	1431.61	0.3223E+01	0.2292E+01
47	23.87	1429.82	1434.76	-0.3389E+01	0.1542E+01

A.5 – 33

V - 17

Connection type : Extended end-plate connections
Mode : All bolted without column stiffener

Tested by : J.O.Surtees & A.P.Mann (1970) England
Test Id. : TEST C1

Column : 8 x 8 UC 48
Beam : 12 x 5 UB 25
Plate thickness : 3/4"

Fasteners: 10.9- -3/4"D
15/16" Oversize holes
Material : --
Fy = -- ksi
Fu = -- ksi

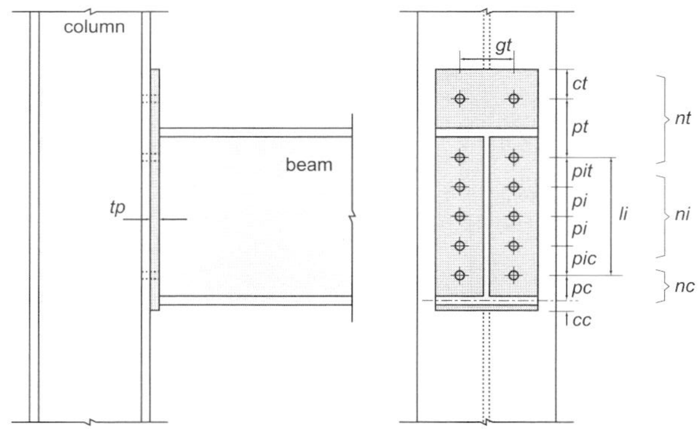

Major parameters

End plate extended on tension side only.

ct = 1.5000" pt = 4.5000" li = 7.0000" pc = 2.7500"
cc = 0.2500" gt = 3.5000" gc = 3.5000"
tp = 0.7500" nt = 2 X 2 nc = 2 X 1

Remark 1) 7" x 3" shear plate welded to column to support end plate.
2) Half length column web stiffner used in compression region.

No	Moment (k-in)	Rotation (radians) X 1/1000	No	Moment (k-in)	Rotation (radians) X 1/1000
1	0.00	0.00	26	1159.96	8.53
2	48.60	0.10	27	1185.88	9.07
3	97.20	0.20	28	1211.81	9.60
4	149.04	0.40	29	1229.09	10.00
5	200.89	0.60	30	1246.37	10.40
6	252.73	0.80	31	1263.65	10.80
7	308.89	1.00	32	1283.09	11.30
8	365.05	1.20	33	1302.53	11.80
9	421.21	1.40			
10	468.74	1.63			
11	516.26	1.87			
12	563.78	2.10			
13	628.58	2.50			
14	693.38	2.90			
15	758.18	3.45			
16	822.98	4.00			
17	866.19	4.50			
18	909.39	5.00			
19	952.60	5.50			
20	995.80	5.93			
21	1039.00	6.37			
22	1082.20	6.80			
23	1099.48	7.20			
24	1116.76	7.60			
25	1134.04	8.00			

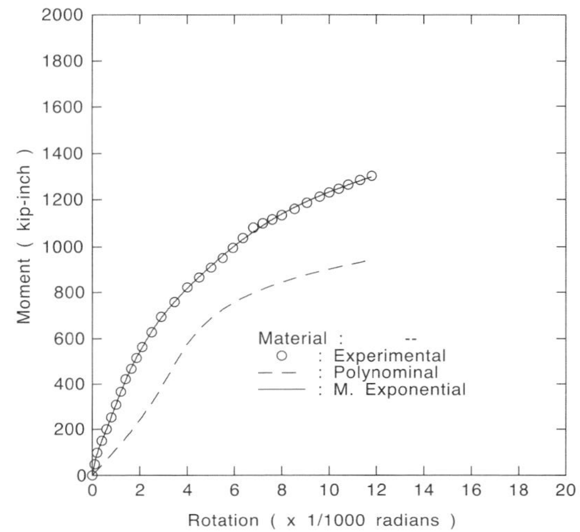

```
Moment-rotation prediction equations      ( R : X 1/1000 radians )

Frye and Morris polynominal model   :    R = Sum ( Ai X ( K*Bm )**Pi X 10**Qi )
     xd = 11.500000"    t =  0.750000"  tf =  0.750000"
     A1 =  1.830000   A2 = -1.040000   A3 =  6.380000    K =  0.004917
     P1 =        1    P2 =        3    P3 =        5    Q1 =        0    Q2 =       -1    Q3 =       -3

Modified exponential model  :
     Bm = Sum ( Ai X ( 1 - exp( - R/( 2*i*C ) ) ) )  +      Sum ( RKj X ( R - Rj0 ) )  +  Bm0
       C =   0.12333333E+00            Bmo=   0.00000000E+00         Nexp=  6       Nliner=  1
      AI =   0.35930801E+03   0.50596400E+03  -0.16256346E+05   0.54215906E+05  -0.68372854E+05   0.30554274E+05
     Rj0 =          2.1000
     RKj =   0.30727982E+02
```

No	Rotation (radians) X 1/1000	Moment (kip-inch) Expri.	Poly.	M.Expo.	Connection stiffness (kip-inch) X 1000 Expri.	Poly.	M.Expo.
1	0.00	0.00	0.00	0.00	0.4860E+03	0.1111E+03	0.6701E+03
2	0.10	48.60	11.11	54.65	0.4860E+03	0.1112E+03	0.4457E+03
3	0.20	97.20	22.23	92.71	0.4104E+03	0.1114E+03	0.3291E+03
4	0.40	149.04	44.57	148.95	0.2592E+03	0.1120E+03	0.2567E+03
5	0.60	200.89	67.09	200.33	0.2592E+03	0.1132E+03	0.2620E+03
6	0.80	252.73	89.89	254.24	0.2700E+03	0.1149E+03	0.2760E+03
7	1.00	308.89	113.08	309.94	0.2808E+03	0.1171E+03	0.2787E+03
8	1.20	365.05	136.78	364.84	0.2808E+03	0.1200E+03	0.2684E+03
9	1.40	421.21	161.11	416.69	0.2452E+03	0.1235E+03	0.2490E+03
10	1.63	468.74	190.50	471.54	0.2037E+03	0.1287E+03	0.2206E+03
11	1.87	516.26	221.25	519.59	0.2037E+03	0.1351E+03	0.1913E+03
12	2.10	563.78	253.66	561.02	0.1883E+03	0.1430E+03	0.1644E+03
13	2.50	628.58	314.20	631.34	0.1620E+03	0.1605E+03	0.1586E+03
14	2.90	693.38	382.47	689.52	0.1434E+03	0.1807E+03	0.1342E+03
15	3.45	758.18	485.80	757.53	0.1178E+03	0.1869E+03	0.1153E+03
16	4.00	822.98	578.72	817.96	0.1014E+03	0.1458E+03	0.1054E+03
17	4.50	866.19	641.70	869.07	0.8641E+02	0.1078E+03	0.9921E+02
18	5.00	909.39	688.87	917.23	0.8641E+02	0.8303E+02	0.9337E+02
19	5.50	952.60	726.04	962.36	0.9353E+02	0.6700E+02	0.8709E+02
20	5.93	995.80	752.84	998.85	0.9969E+02	0.5737E+02	0.8127E+02
21	6.37	1039.00	776.07	1032.78	0.9969E+02	0.5020E+02	0.7531E+02
22	6.80	1082.20	796.58	1064.13	0.7032E+02	0.4468E+02	0.6944E+02
23	7.20	1099.48	813.61	1090.86	0.4320E+02	0.4061E+02	0.6426E+02
24	7.60	1116.76	829.16	1115.59	0.4320E+02	0.3726E+02	0.5946E+02
25	8.00	1134.04	843.49	1138.49	0.4552E+02	0.3446E+02	0.5509E+02
26	8.53	1159.96	861.01	1166.47	0.4861E+02	0.3135E+02	0.5001E+02
27	9.07	1185.88	877.03	1191.98	0.4861E+02	0.2880E+02	0.4577E+02
28	9.60	1211.81	891.81	1215.44	0.4552E+02	0.2667E+02	0.4232E+02
29	10.00	1229.09	902.19	1231.93	0.4320E+02	0.2528E+02	0.4020E+02
30	10.40	1246.37	912.05	1247.64	0.4320E+02	0.2404E+02	0.3842E+02
31	10.80	1263.65	921.67	1262.70	0.4128E+02	0.2291E+02	0.3695E+02
32	11.30	1283.09	932.59	1280.79	0.3888E+02	0.2169E+02	0.3546E+02
33	11.80	1302.53	943.16	1298.22	0.3888E+02	0.2059E+02	0.3431E+02

V - 18

Connection type : Extended end-plate connections
Mode : All bolted without column stiffener

Tested by : J.O.Surtees & A.P.Mann (1970) England
Test Id. : TEST C2

Column : 10 x 10 UC 60
Beam : 15 x 6 UB 40
Plate thickness : 3/4"

Fasteners: 10.9- -3/4"D
15/16" Oversize holes
Material : --
Fy = -- ksi
Fu = -- ksi

Major parameters

End plate extended on tension side only.

ct = 1.7500"	pt = 4.5000"	li = 10.0000"	pc = 2.7500"
cc = 0.2500"	gt = 3.5000"	gc = 3.5000"	
tp = 0.7500"	nt = 2 X 2	nc = 2 X 1	

Remark 1) 7" x 3" shear plate welded to column to support end plate.
2) Half length column web stiffner used in compression region.

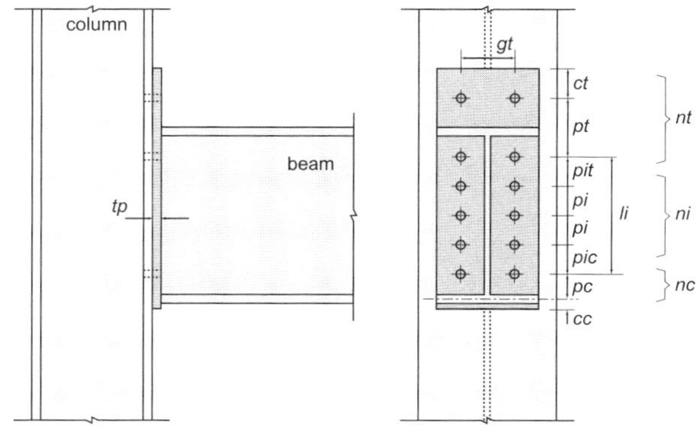

No	Moment (k-in)	Rotation (radians) X 1/1000	No	Moment (k-in)	Rotation (radians) X 1/1000
1	0.00	0.00	26	2993.16	10.10
2	120.11	0.20	27	3058.09	11.00
3	240.23	0.40	28	3097.04	12.10
4	337.62	0.70	29	3148.98	13.40
5	435.01	1.00	30	3174.96	14.80
6	545.39	1.25	31	3226.90	16.30
7	655.76	1.50	32	3265.86	17.40
8	779.13	1.80	33	3278.84	18.80
9	902.50	2.10	34	3291.83	20.00
10	1019.36	2.40	35	3304.81	21.40
11	1136.23	2.70	36	3330.78	22.50
12	1272.58	3.05	37	3343.76	23.80
13	1408.93	3.40	38	3356.75	24.90
14	1519.31	3.70	39	3369.74	26.20
15	1629.68	4.00	40	3408.70	27.50
16	1740.06	4.30	41	3395.71	28.50
17	1850.44	4.60	42	3408.70	29.90
18	1967.30	4.90	43	3434.66	31.50
19	2084.17	5.20	44	3447.65	33.00
20	2194.55	5.55	45	3473.63	34.40
21	2304.92	5.90	46	3499.60	35.50
22	2473.74	6.50	47	3525.56	37.00
23	2616.58	7.10	48	3551.53	38.80
24	2733.44	7.80	49	3564.52	39.90
25	2876.29	9.00			

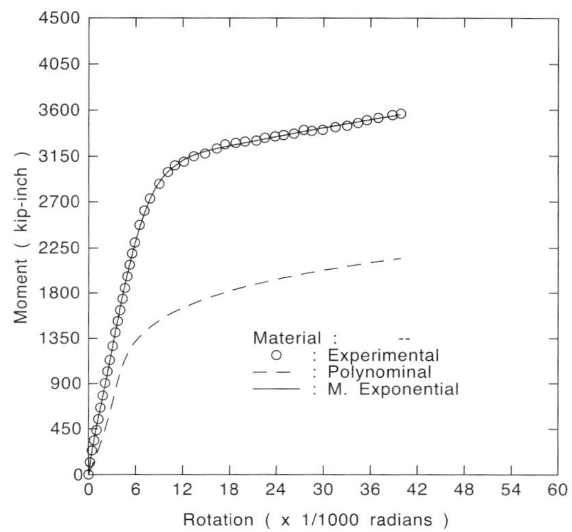

A.5 – 36

```
Moment-rotation prediction equations      ( R : X 1/1000 radians )

Frye and Morris polynominal model   :    R = Sum ( Ai X ( K*Bm )**Pi X 10**Qi )
     xd = 14.500000"    t =  0.750000"  tf =  0.750000"
     A1 =  1.830000    A2 = -1.040000   A3 =  6.380000     K =  0.002819
     P1 =         1    P2 =         3   P3 =         5    Q1 =         0    Q2 =        -1    Q3 =        -3

Modified exponential model   :
     Bm = Sum ( Ai X ( 1 - exp( - R/( 2*i*C ) ) ) )  +      Sum ( RKj X ( R - Rj0 ) )  +  Bm0
       C =   0.38250000E+00            Bmo=   0.00000000E+00          Nexp=  6        Nliner=  1
      AI =   0.30903923E+03   0.94249868E+04  -0.63131407E+05   0.12444647E+06  -0.99086146E+05   0.31252971E+05
     Rj0 =          11.0000
     RKj =   0.11797036E+02
```

No	Rotation (radians) X 1/1000	Moment (kip-inch) Expri.	Moment (kip-inch) Poly.	Moment (kip-inch) M.Expo.	Connection stiffness (kip-inch) X 1000 Expri.	Connection stiffness (kip-inch) X 1000 Poly.	Connection stiffness (kip-inch) X 1000 M.Expo.
1	0.00	0.00	0.00	0.00	0.6006E+03	0.1938E+03	0.6287E+03
3	0.40	240.23	77.75	215.20	0.4902E+03	0.1954E+03	0.4675E+03
5	1.00	435.01	197.24	464.13	0.3884E+03	0.2043E+03	0.3837E+03
7	1.50	655.76	302.73	653.32	0.4277E+03	0.2190E+03	0.3783E+03
9	2.10	902.50	442.45	884.46	0.4004E+03	0.2495E+03	0.3934E+03
11	2.70	1136.23	605.82	1124.68	0.3896E+03	0.2978E+03	0.4057E+03
13	3.40	1408.93	830.97	1409.17	0.3779E+03	0.3289E+03	0.4040E+03
15	4.00	1629.68	1009.10	1647.34	0.3679E+03	0.2544E+03	0.3879E+03
17	4.60	1850.44	1137.17	1872.52	0.3787E+03	0.1780E+03	0.3612E+03
19	5.20	2084.17	1229.02	2079.45	0.3553E+03	0.1323E+03	0.3277E+03
21	5.90	2304.92	1309.77	2293.75	0.3028E+03	0.1012E+03	0.2842E+03
23	7.10	2616.58	1411.99	2589.57	0.2052E+03	0.7247E+02	0.2098E+03
25	9.00	2876.29	1526.48	2892.27	0.1124E+03	0.5074E+02	0.1148E+03
27	11.00	3058.09	1615.14	3054.22	0.5562E+02	0.3910E+02	0.5332E+02
29	13.40	3148.98	1698.44	3162.21	0.2965E+02	0.3100E+02	0.3026E+02
31	16.30	3226.90	1779.02	3226.08	0.3508E+02	0.2505E+02	0.1697E+02
33	18.80	3278.84	1837.11	3264.97	0.1011E+02	0.2162E+02	0.1480E+02
35	21.40	3304.81	1889.75	3303.16	0.1730E+02	0.1900E+02	0.1469E+02
37	23.80	3343.76	1933.04	3338.45	0.1097E+02	0.1714E+02	0.1468E+02
39	26.20	3369.74	1972.32	3373.41	0.1998E+02	0.1565E+02	0.1442E+02
41	28.50	3395.71	2007.16	3406.11	-0.3710E+01	0.1446E+02	0.1400E+02
43	31.50	3434.66	2048.33	3447.19	0.1232E+02	0.1319E+02	0.1340E+02
45	34.40	3473.63	2085.07	3485.28	0.2139E+02	0.1218E+02	0.1289E+02
47	37.00	3525.56	2115.71	3518.32	0.1600E+02	0.1141E+02	0.1254E+02
49	39.90	3564.52	2147.83	3554.26	0.1180E+02	0.1066E+02	0.1226E+02

A.5 – 37

V - 19

Connection type : Extended end-plate connections
Mode : All bolted without column stiffener

Tested by : J.O.Surtees & A.P.Mann (1970) England
Test Id. : TEST C3

Column : 10 x 10 UC 60
Beam : 15 x 6 UB 40
Plate thickness : 3/4"

Fasteners: 10.9- -1"D
1 1/4" Oversize holes
Material : --
Fy = -- ksi
Fu = -- ksi

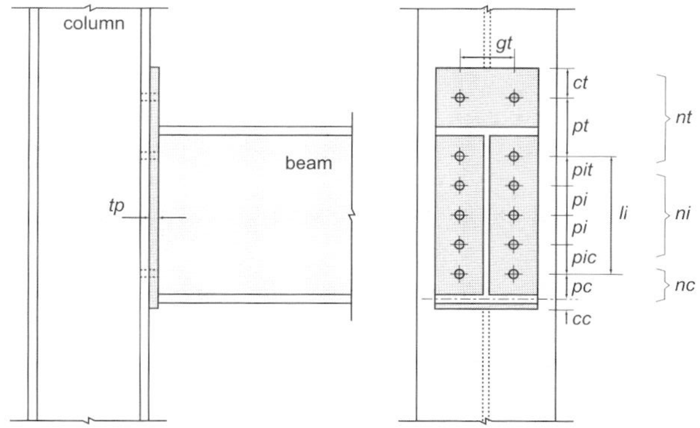

Major parameters

End plate extended on tension side only.
ct = 1.7500" pt = 4.5000" li = 10.0000" pc = 2.7500"
cc = 0.2500" gt = 3.5000" gc = 3.5000"
tp = 0.7500" nt = 2 X 2 nc = 2 X 1

Remark 1) 7" x 3" shear plate welded to column to support end plate.
2) Half length column web stiffner used in compression region.

No	Moment (k-in)	Rotation (radians) X 1/1000	No	Moment (k-in)	Rotation (radians) X 1/1000	No	Moment (k-in)	Rotation (radians) X 1/1000
1	0.00	0.00	26	3032.11	6.70	51	3473.63	38.40
2	100.64	0.15	27	3148.98	7.60	52	3473.63	40.00
3	201.28	0.30	28	3252.86	8.90			
4	363.59	0.45	29	3330.78	9.90			
5	525.91	0.60	30	3382.73	11.20			
6	649.27	0.75	31	3408.70	12.20			
7	772.63	0.90	32	3434.66	13.20			
8	896.00	1.10	33	3460.64	14.30			
9	1019.36	1.30	34	3473.63	15.40			
10	1129.74	1.45	35	3486.61	16.60			
11	1240.12	1.60	36	3525.56	17.90			
12	1395.94	1.80	37	3525.56	18.80			
13	1551.77	2.00	38	3512.58	20.20			
14	1681.62	2.20	39	3512.58	21.90			
15	1811.47	2.40	40	3512.58	23.20			
16	1921.85	2.65	41	3486.61	24.50			
17	2032.24	2.90	42	3486.61	25.90			
18	2149.10	3.20	43	3512.58	27.50			
19	2265.97	3.50	44	3512.58	28.80			
20	2389.33	3.90	45	3486.61	29.90			
21	2512.69	4.30	46	3493.10	31.25			
22	2603.59	4.60	47	3499.60	32.60			
23	2694.49	4.90	48	3493.10	33.65			
24	2850.32	5.80	49	3486.61	34.70			
25	2941.22	6.25	50	3486.61	36.50			

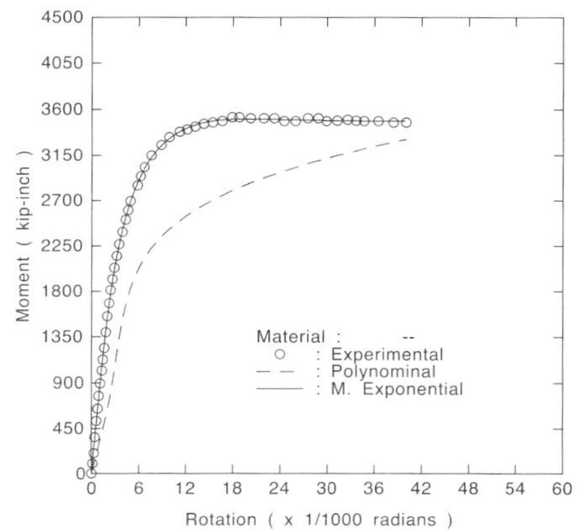

```
Moment-rotation prediction equations     ( R : X 1/1000 radians )

Frye and Morris polynominal model  :     R = Sum ( Ai X ( K*Bm )**Pi X 10**Qi )
     xd = 14.500000"   t =  0.750000"  tf =  1.000000"
     A1 =  1.830000   A2 = -1.040000   A3 =  6.380000    K =  0.001831
     P1 =         1   P2 =         3   P3 =         5   Q1 =         0   Q2 =        -1   Q3 =        -3

Modified exponential model  :
     Bm = Sum ( Ai X ( 1 - exp( - R/( 2*i*C ) ) ) )  +      Sum ( RKj X ( R - Rj0 ) )  +  Bm0
       C =   0.37083333E+00              Bmo=   0.00000000E+00         Nexp=  6        Nliner=  0
      AI =  -0.56948661E+03  -0.16720282E+04   0.19191749E+05  -0.47007293E+05   0.51737667E+05  -0.18194588E+05
```

No	Rotation (radians) X 1/1000	Moment (kip-inch) Expri.	Poly.	M.Expo.	Connection stiffness (kip-inch) X 1000 Expri.	Poly.	M.Expo.
1	0.00	0.00	0.00	0.00	0.6709E+03	0.2984E+03	0.7484E+03
3	0.30	201.28	89.58	238.59	0.8765E+03	0.2998E+03	0.8285E+03
5	0.60	525.91	180.00	490.62	0.9523E+03	0.3040E+03	0.8434E+03
7	0.90	772.63	272.39	740.80	0.7343E+03	0.3113E+03	0.8197E+03
9	1.30	1019.36	399.74	1056.84	0.6848E+03	0.3266E+03	0.7565E+03
11	1.60	1240.12	500.11	1275.26	0.7544E+03	0.3433E+03	0.6988E+03
13	2.00	1551.77	643.28	1538.83	0.7142E+03	0.3744E+03	0.6192E+03
15	2.40	1811.47	801.35	1771.09	0.5569E+03	0.4181E+03	0.5431E+03
17	2.90	2032.24	1027.14	2020.85	0.4179E+03	0.4852E+03	0.4580E+03
19	3.50	2265.97	1330.54	2269.20	0.3548E+03	0.4956E+03	0.3728E+03
21	4.30	2512.69	1661.29	2530.65	0.3053E+03	0.3262E+03	0.2854E+03
23	4.90	2694.49	1826.73	2686.38	0.2705E+03	0.2342E+03	0.2357E+03
25	6.25	2941.22	2068.11	2947.86	0.2020E+03	0.1395E+03	0.1582E+03
27	7.60	3148.98	2226.79	3126.49	0.1094E+03	0.1000E+03	0.1099E+03
29	9.90	3330.78	2415.98	3317.18	0.6141E+02	0.6878E+02	0.6076E+02
31	12.20	3408.70	2554.50	3421.51	0.2597E+02	0.5317E+02	0.3240E+02
33	14.30	3460.64	2656.32	3471.74	0.1771E+02	0.4440E+02	0.1672E+02
35	16.60	3486.61	2750.44	3497.53	0.2001E+02	0.3783E+02	0.6714E+01
37	18.80	3525.56	2828.42	3506.13	-0.3629E+01	0.3328E+02	0.1670E+01
39	21.90	3512.58	2923.93	3505.94	-0.9742E-13	0.2860E+02	-0.1201E+01
41	24.50	3486.61	2994.33	3501.91	-0.1036E+02	0.2567E+02	-0.1726E+01
43	27.50	3512.58	3067.20	3496.92	0.7276E+01	0.2302E+02	-0.1529E+01
45	29.90	3486.61	3120.75	3493.64	-0.1085E+02	0.2129E+02	-0.1195E+01
47	32.60	3499.60	3176.13	3490.91	-0.1374E+01	0.1967E+02	-0.8335E+00
49	34.70	3486.61	3216.05	3489.41	-0.3905E+01	0.1860E+02	-0.6050E+00
51	38.40	3473.63	3281.52	3487.74	-0.3124E+01	0.1699E+02	-0.3246E+00

A.5 – 39

V - 20

Connection type : Extended end-plate connections
Mode : All bolted without column stiffener

Tested by : J.O.Surtees & A.P.Mann (1970) England
Test Id. : TEST C4

Column : 10 x 10 UC 60
Beam : 15 x 6 UB 40
Plate thickness : 1"

Fasteners: 10.9- -1"D
1 1/4" Oversize holes
Material : --
Fy = -- ksi
Fu = -- ksi

Major parameters

End plate extended on tension side only.

ct = 1.7500"	pt = 4.5000'	li = 10.0000"	pc = 2.7500"
cc = 0.2500"	gt = 3.5000'	gc = 3.5000"	
tp = 1.0000"	nt = 2 X 2	nc = 2 X 1	

Remark 1) 7" x 3" shear plate welded to column to support end plate.
2) Half length column web stiffner used in compression region.

No	Moment (k-in)	Rotation (radians) X 1/1000	No	Moment (k-in)	Rotation (radians) X 1/1000
1	0.00	0.00	26	3298.44	11.30
2	126.37	0.40	27	3350.28	13.10
3	252.73	0.80	28	3382.68	14.35
4	408.26	1.25	29	3415.08	15.60
5	563.78	1.70	30	3433.96	16.75
6	680.42	2.05	31	3452.84	17.90
7	797.06	2.40	32	3472.84	19.10
8	939.63	2.80	33	3492.84	20.30
9	1082.20	3.20	34	3505.80	21.40
10	1231.24	3.65	35	3518.76	22.50
11	1380.29	4.10	36	3551.17	24.45
12	1496.93	4.45	37	3583.57	26.40
13	1613.58	4.80	38	3590.05	27.65
14	1756.15	5.20	39	3596.53	28.90
15	1898.71	5.60	40	3615.97	30.05
16	2034.79	5.95	41	3635.41	31.20
17	2170.87	6.30	42	3654.85	32.55
18	2313.44	6.70	43	3674.29	33.90
19	2456.00	7.10	44	3700.21	35.35
20	2579.13	7.40	45	3726.13	36.80
21	2702.26	7.70	46	3752.05	38.40
22	2818.90	8.15	47	3777.97	40.00
23	2935.55	8.60			
24	3078.11	9.20			
25	3194.75	10.00			

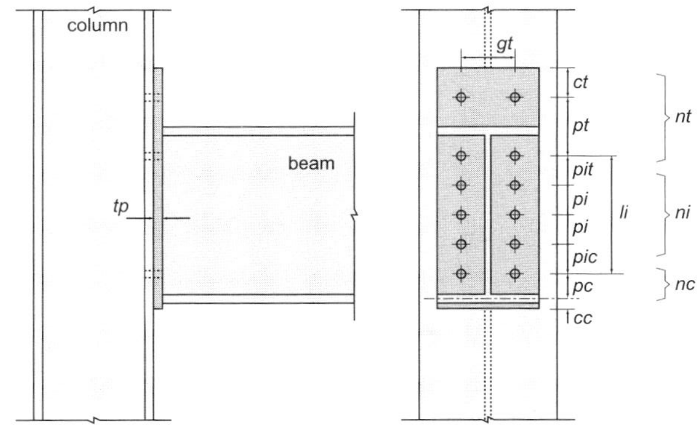

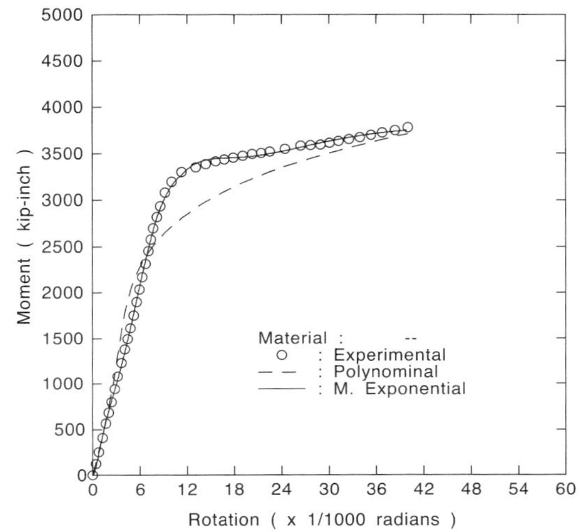

A.5 – 40

```
Moment-rotation prediction equations     ( R : X 1/1000 radians )

Frye and Morris polynominal model  :    R = Sum ( Ai X ( K*Bm )**Pi X 10**Qi )
     xd = 14.500000"   t =  1.000000"  tf =  1.000000"
     A1 =  1.830000   A2 = -1.040000   A3 =  6.380000     K =  0.001632
     P1 =         1   P2 =         3   P3 =         5    Q1 =         0    Q2 =        -1    Q3 =        -3

Modified exponential model  :
     Bm = Sum ( Ai X ( 1 - exp( - R/( 2*i*C ) ) ) )  +      Sum ( RKj X ( R - Rj0 ) )  +  Bm0
       C =   0.43333333E+00              Bmo=   0.00000000E+00             Nexp=  6          Nliner=  2
      AI =  -0.54195376E+04   0.79745538E+05  -0.35748357E+06   0.67325150E+06  -0.57559931E+06   0.18937370E+06
     Rj0 =           8.1500          11.3000
     RKj =  -0.40561455E+02   0.42214678E+02
```

No	Rotation (radians) X 1/1000	Moment (kip-inch) Expri.	Poly.	M.Expo.	Connection stiffness (kip-inch) X 1000 Expri.	Poly.	M.Expo.
1	0.00	0.00	0.00	0.00	0.3159E+03	0.3348E+03	0.5462E+02
3	0.80	252.73	270.84	245.52	0.3299E+03	0.3461E+03	0.4090E+03
5	1.70	563.78	599.98	587.88	0.3387E+03	0.3927E+03	0.3383E+03
7	2.40	797.06	899.07	808.51	0.3441E+03	0.4691E+03	0.3009E+03
9	3.20	1082.20	1320.84	1051.72	0.3446E+03	0.5732E+03	0.3147E+03
11	4.10	1380.29	1786.13	1354.04	0.3324E+03	0.4134E+03	0.3578E+03
13	4.80	1613.58	2022.63	1614.38	0.3441E+03	0.2763E+03	0.3835E+03
15	5.60	1898.71	2207.21	1925.86	0.3737E+03	0.1944E+03	0.3909E+03
17	6.30	2170.87	2328.07	2195.76	0.3737E+03	0.1542E+03	0.3773E+03
19	7.10	2456.00	2438.98	2485.12	0.3873E+03	0.1252E+03	0.3434E+03
21	7.70	2702.26	2509.33	2681.26	0.3499E+03	0.1100E+03	0.3095E+03
23	8.60	2935.55	2600.44	2916.15	0.2500E+03	0.9343E+02	0.2118E+03
25	10.00	3194.75	2718.27	3150.54	0.1206E+03	0.7617E+02	0.1248E+03
27	13.10	3350.28	2917.61	3391.56	0.2710E+02	0.5494E+02	0.4170E+02
29	15.60	3415.08	3042.02	3443.94	0.2097E+02	0.4533E+02	0.7685E+01
31	17.90	3452.84	3138.91	3454.42	0.1654E+02	0.3925E+02	0.4058E+01
33	20.30	3492.84	3227.24	3470.21	0.1412E+02	0.3458E+02	0.9685E+01
35	22.50	3518.76	3299.52	3497.85	0.1353E+02	0.3126E+02	0.1516E+02
37	26.40	3583.57	3412.31	3567.07	0.9651E+01	0.2684E+02	0.1901E+02
39	28.90	3596.53	3477.25	3613.76	0.1129E+02	0.2464E+02	0.1802E+02
41	31.20	3635.41	3531.81	3652.88	0.1575E+02	0.2297E+02	0.1588E+02
43	33.90	3674.29	3591.63	3691.73	0.1608E+02	0.2130E+02	0.1288E+02
45	36.80	3726.13	3651.17	3724.56	0.1708E+02	0.1979E+02	0.9835E+01
47	40.00	3777.97	3712.22	3751.49	0.1620E+02	0.1837E+02	0.7122E+01

A.5 – 41

V - 21

```
Connection type : Extended end-plate connections
           Mode : All bolted without column stiffener

Tested by   :  J.O.Surtees & A.P.Mann (1970)              England
Test  Id.   :  TEST C5

Column : 10 x 10 UC 60                Fasteners: 10.9- -1"D
Beam   : 15 x 6 UB 40                           1 1/4" Oversize holes
Plate thickness : 1"                  Material :  --
                                           Fy =   --   ksi
                                           Fu =   --   ksi
```

Major parameters

```
End plate extended on tension side orly.
ct =  1.7500"    pt =  4.5000"    li = 10.0000"     pc =  2.7500"
cc =  0.2500"    gt =  3.5000"    gc =  3.5000"
tp =  1.0000"    nt = 2 X  2      nc = 2 X  1
```

Remark 1) 7" x 3" shear plate welded to column to support end plate.
2) Half length column web stiffner used in compression region.

No	Moment (k-in)	Rotation (radians) X 1/1000	No	Moment (k-in)	Rotation (radians) X 1/1000
1	0.00	0.00	26	3279.16	12.90
2	164.93	0.25	27	3330.90	14.20
3	329.86	0.50	28	3408.50	15.70
4	446.27	0.75	29	3447.31	17.00
5	562.69	1.00	30	3486.12	18.30
6	769.67	1.50	31	3537.86	19.30
7	963.70	1.90	32	3576.67	21.20
8	1144.79	2.20	33	3628.42	22.60
9	1274.15	2.50	34	3654.29	23.90
10	1403.51	2.80	35	3693.08	25.50
11	1545.80	3.10	36	3770.70	26.80
12	1688.09	3.40	37	3796.57	28.30
13	1810.97	3.70	38	3848.32	29.70
14	1933.86	4.00	39	3887.12	30.75
15	2069.68	4.35	40	3925.93	31.80
16	2205.50	4.70	41	3951.80	33.10
17	2315.45	4.95	42	4029.41	35.00
18	2425.40	5.20	43	4055.28	36.30
19	2593.57	5.80	44	4094.09	37.70
20	2743.79	6.40	45	4132.90	38.90
21	2865.22	7.20	46	4158.77	40.00
22	2994.58	8.20			
23	3085.12	9.10			
24	3162.73	10.40			
25	3227.41	11.50			

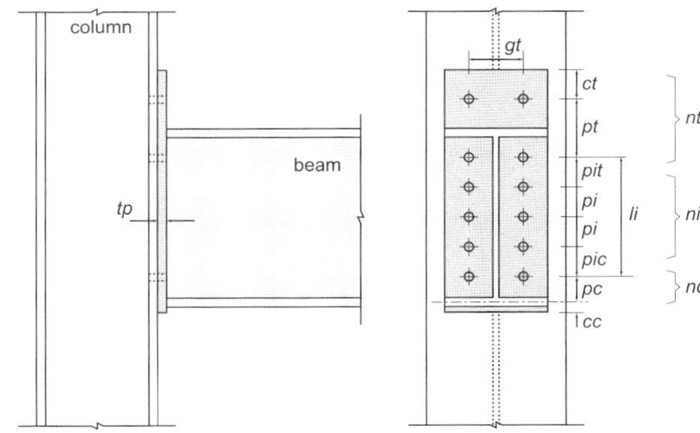

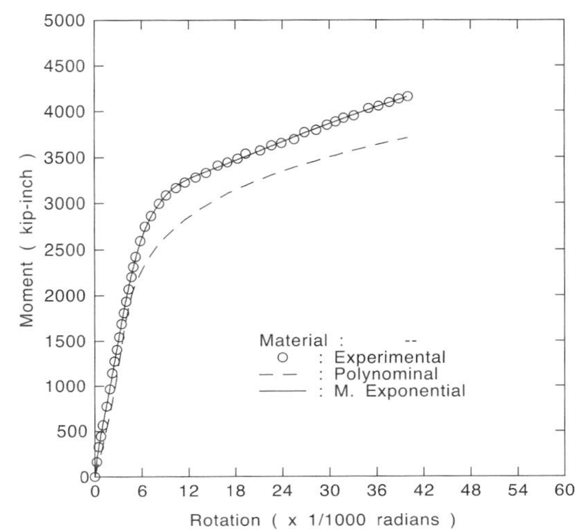

A.5 – 42

```
Moment-rotation prediction equations     ( R : X 1/1000 radians )

Frye and Morris polynominal model  :    R = Sum ( Ai X ( K*Bm )**Pi X 10**Qi )
     xd = 14.500000"   t =  1.000000"  tf =  1.000000"
     A1 =  1.830000   A2 = -1.040000   A3 =  6.380000    K =  0.001632
     P1 =        1     P2 =        3    P3 =        5     Q1 =        0     Q2 =       -1     Q3 =       -3

Modified exponential model  :
     Bm = Sum ( Ai X ( 1 - exp( - R/( 2*i*C ) ) ) )  +      Sum ( RKj X ( R - Rj0 ) )  +  Bm0
       C =   0.39583333E+00            Bmo=   0.00000000E+00           Nexp=  6         Nliner=  1
      AI =   0.20172765E+04  -0.68533653E+04  -0.20764685E+05   0.88421629E+05  -0.95281363E+05   0.35743883E+05
     Rj0 =          8.2000
     RKj =   0.27700585E+02
```

No	Rotation (radians) X 1/1000	Moment (kip-inch) Expri.	Moment (kip-inch) Poly.	Moment (kip-inch) M.Expo.	Connection stiffness (kip-inch) X 1000 Expri.	Connection stiffness (kip-inch) X 1000 Poly.	Connection stiffness (kip-inch) X 1000 M.Expo.
1	0.00	0.00	0.00	0.00	0.6597E+03	0.3348E+03	0.8532E+03
3	0.50	329.86	168.13	326.50	0.5627E+03	0.3392E+03	0.5181E+03
5	1.00	562.69	340.71	560.07	0.4484E+03	0.3528E+03	0.4410E+03
7	1.90	963.70	680.22	966.61	0.5528E+03	0.4102E+03	0.4726E+03
9	2.50	1274.15	946.70	1256.48	0.4312E+03	0.4836E+03	0.4894E+03
11	3.10	1545.80	1263.78	1548.27	0.4743E+03	0.5673E+03	0.4788E+03
13	3.70	1810.97	1599.32	1826.38	0.4096E+03	0.5171E+03	0.4451E+03
15	4.35	2069.68	1881.78	2099.21	0.3881E+03	0.3552E+03	0.3925E+03
17	4.95	2315.45	2062.46	2318.39	0.4398E+03	0.2564E+03	0.3377E+03
19	5.80	2593.57	2244.88	2572.32	0.2695E+03	0.1808E+03	0.2607E+03
21	7.20	2865.22	2451.48	2859.70	0.1383E+03	0.1223E+03	0.1553E+03
23	9.10	3085.12	2645.37	3085.71	0.8387E+02	0.8635E+02	0.9376E+02
25	11.50	3227.41	2822.81	3242.63	0.4919E+02	0.6399E+02	0.4583E+02
27	14.20	3330.90	2975.28	3343.23	0.4534E+02	0.5020E+02	0.3269E+02
29	17.00	3447.31	3102.63	3433.50	0.2985E+02	0.4140E+02	0.3261E+02
31	19.30	3537.86	3191.78	3509.97	0.4094E+02	0.3637E+02	0.3380E+02
33	22.60	3628.42	3302.64	3622.47	0.2811E+02	0.3112E+02	0.3407E+02
35	25.50	3693.08	3387.76	3720.07	0.4381E+02	0.2773E+02	0.3313E+02
37	28.30	3796.57	3461.65	3811.09	0.2744E+02	0.2515E+02	0.3187E+02
39	30.75	3887.12	3521.33	3887.83	0.3696E+02	0.2328E+02	0.3081E+02
41	33.10	3951.80	3574.38	3959.20	0.2841E+02	0.2177E+02	0.2996E+02
43	36.30	4055.28	3641.12	4053.60	0.2367E+02	0.2003E+02	0.2910E+02
45	38.90	4132.90	3691.52	4128.60	0.2774E+02	0.1884E+02	0.2862E+02

A.5 – 43

V - 22

Connection type : Extended end-plate connections
Mode : All bolted without column stiffener

Tested by : J.A.Packer & L.J.Morris (1977) England
Test Id. : TEST J1

Column : 152 x 152 UC 37
Beam : 254 x 102 UB 22
Plate thickness : 15 mm

Fasteners: 10.9- -M16
13/16" Oversize holes
Material : --
Fy = -- ksi
Fu = -- ksi

Major parameters

End plate extended on tension side only.
ct = 1.5749" pt = 3.8189" li = 5.9449" pc = 1.9685"
cc = 1.1811" gt = 3.3858" gc = 3.3858"
tp = 0.5906" nt = 2 X 2 nc = 2 X 1

Remark 1) End plate extended on tension side only. Tension side hole is 0.811" D.
2) Hsfg bolts used on the side of tension flange. Comp.side hole is 0.689" D.

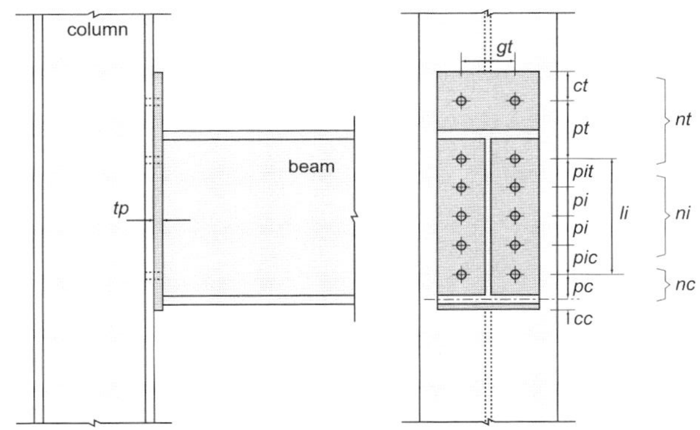

No	Moment (k-in)	Rotation (radians) X 1/1000
1	0.00	0.00
2	42.80	0.50
3	83.30	1.00
4	122.00	1.50
5	158.70	2.00
6	190.20	2.50
7	221.20	3.00
8	276.50	3.50
9	335.70	4.00
10	369.50	5.00
11	411.10	6.00
12	453.00	8.00
13	494.40	10.00
14	523.20	12.40
15	562.10	15.00
16	614.10	20.00
17	658.30	25.10
18	687.00	29.90
19	713.00	35.10
20	736.40	39.80
21	759.80	45.00
22	778.00	50.10
23	801.50	55.00
24	817.10	60.00
25	832.70	65.00

No	Moment (k-in)	Rotation (radians) X 1/1000
26	843.10	70.00
27	848.30	75.00
28	850.90	80.00
29	853.50	85.00
30	853.50	90.00
31	853.50	95.20
32	824.90	100.90
33	791.00	104.90
34	757.20	107.30

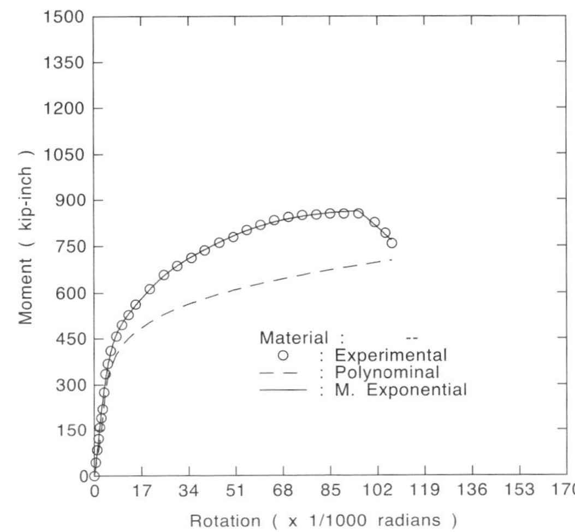

A.5 – 44

```
Moment-rotation prediction equations     ( R : X 1/1000 radians )

Frye and Morris polynominal model  :    R = Sum ( Ai X ( K*Bm )**Pi X 10**Qi )
     xd =  9.763800"    t =  0.590600"  tf =  0.630000"
     A1 =  1.830000   A2 = -1.040000   A3 =  6.380000    K =  0.010408
     P1 =        1    P2 =        3    P3 =        5    Q1 =        0    Q2 =       -1    Q3 =       -3

Modified exponential model  :
     Bm = Sum ( Ai X ( 1 - exp( - R/( 2*i*C ) ) ) )  +     Sum ( RKj X ( R - Rj0 ) )  +  Bm0
      C =   0.10191667E+01              Bmo=   0.00000000E+00          Nexp=  6        Nliner=  1
     AI =  -0.10564292E+04   0.10048125E+05  -0.38583394E+05   0.78494463E+05  -0.78286630E+05   0.30252601E+05
    Rj0 =           95.2000
    RKj =  -0.82945744E+01
```

No	Rotation (radians) X 1/1000	Moment (kip-inch) Expri.	Moment (kip-inch) Poly.	Moment (kip-inch) M.Expo.	Connection stiffness (kip-inch) X 1000 Expri.	Connection stiffness (kip-inch) X 1000 Poly.	Connection stiffness (kip-inch) X 1000 M.Expo.
1	0.00	0.00	0.00	0.00	0.8560E+02	0.5250E+02	0.5637E+02
3	1.00	83.30	53.42	74.32	0.7920E+02	0.5532E+02	0.8549E+02
5	2.00	158.70	113.16	161.64	0.6820E+02	0.6587E+02	0.8614E+02
7	3.00	221.20	189.32	242.78	0.8630E+02	0.8738E+02	0.7509E+02
9	4.00	335.70	273.37	310.82	0.9020E+02	0.6887E+02	0.6088E+02
11	6.00	411.10	357.43	406.54	0.3555E+02	0.2652E+02	0.3637E+02
13	10.00	494.40	426.21	498.96	0.1633E+02	0.1194E+02	0.1495E+02
15	15.00	562.10	472.65	560.06	0.1214E+02	0.7411E+01	0.1096E+02
17	25.10	658.30	529.47	655.17	0.7282E+01	0.4412E+01	0.7539E+01
19	35.10	713.00	567.02	716.12	0.4989E+01	0.3239E+01	0.5075E+01
21	45.00	759.80	595.65	761.49	0.4030E+01	0.2599E+01	0.4173E+01
23	55.00	801.50	619.43	798.97	0.3966E+01	0.2185E+01	0.3286E+01
25	65.00	832.70	639.75	826.81	0.2600E+01	0.1895E+01	0.2284E+01
27	75.00	848.30	657.57	845.15	0.7800E+00	0.1679E+01	0.1423E+01
29	85.00	853.50	673.49	856.13	0.2600E+00	0.1512E+01	0.8171E+00
31	95.20	853.50	688.20	862.35	-0.2394E+01	0.1376E+01	0.4363E+00
33	104.90	791.00	701.10	785.03	-0.1198E+02	0.1270E+01	-0.8064E+01

A.5 – 45

V - 23

Connection type : Extended end-plate connections
Mode : All bolted without column stiffener

Tested by : J.A.Packer & L.J.Morris (1977) England
Test Id. : TEST J2

Column : 152 x 152 UC 30
Beam : 254 x 102 UB 22
Plate thickness : 15 mm

Fasteners: 10.9- -M16
13/16" Oversize holes
Material : --
Fy = -- ksi
Fu = -- ksi

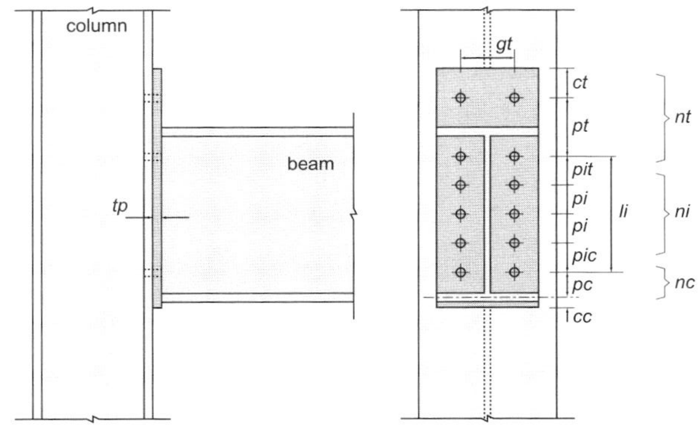

Major parameters

End plate extended on tension side only.
ct = 1.5749" pt = 3.8189" li = 5.9449" pc = 1.9685"
cc = 1.1811" gt = 3.3858" gc = 3.3858"
tp = 0.5906" nt = 2 X 2 nc = 2 X 1

Remark 1) End plate extended on tension side only. Tension side hole is 0.811" D.
2) Hsfg bolts used on the side of tension flange. Comp.side hole is 0.689" D.

No	Moment (k-in)	Rotation (radians) X 1/1000
1	0.00	0.00
2	34.00	0.50
3	65.10	1.00
4	99.00	1.50
5	137.90	2.00
6	180.00	2.50
7	252.40	3.00
8	296.60	4.00
9	338.30	6.00
10	372.10	8.00
11	395.50	10.00
12	418.90	12.00
13	442.40	14.00
14	460.60	15.90
15	481.40	18.10
16	499.60	20.10
17	541.20	25.00
18	572.50	30.00
19	595.90	35.00
20	616.70	40.10
21	632.30	45.10
22	645.30	50.10
23	658.30	55.10
24	673.90	60.00
25	687.00	65.00

No	Moment (k-in)	Rotation (radians) X 1/1000
26	700.00	70.00
27	713.00	75.00
28	726.00	80.00
29	739.00	85.20
30	746.80	90.00
31	757.20	95.20
32	754.60	100.20
33	757.20	105.10
34	757.20	109.00
35	733.80	112.90
36	694.80	117.10

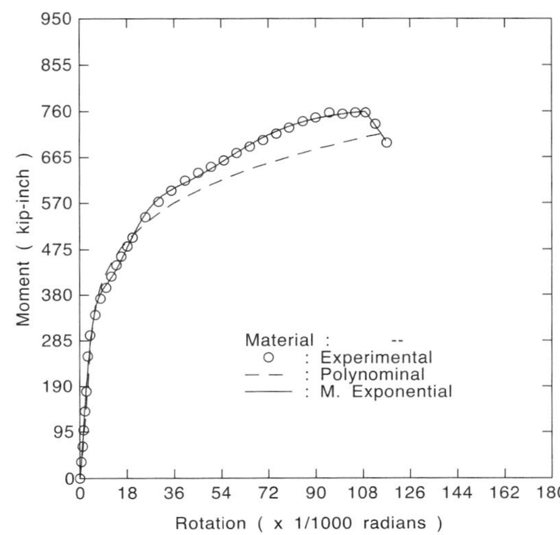

A.5 – 46

```
Moment-rotation prediction equations     ( R : X 1/1000 radians )

Frye and Morris polynominal model  :    R = Sum ( Ai X ( K*Bm )**Pi X 10**Qi )
     xd =  9.763800"   t =  0.590600"  tf =  0.630000"
     A1 =  1.830000   A2 = -1.040000   A3 =  6.380000    K =  0.010408
     P1 =         1   P2 =         3   P3 =         5   Q1 =         0    Q2 =        -1    Q3 =        -3

Modified exponential model  :
     Bm = Sum ( Ai X ( 1 - exp( - R/( 2*i*C ) ) ) )  +     Sum ( RKj X ( R - Rj0 ) )  +  Bm0
      C =   0.11008333E+01             Bmo=   0.00000000E+00          Nexp=  6        Nliner=  1
     AI =  -0.16078035E+04   0.19067276E+05  -0.78270076E+05   0.15355531E+06  -0.14337868E+06   0.51400241E+05
    Rj0 =          109.0000
    RKj =  -0.80619878E+01
```

No	Rotation (radians) X 1/1000	Moment (kip-inch) Expri.	Moment (kip-inch) Poly.	Moment (kip-inch) M.Expo.	Connection stiffness (kip-inch) X 1000 Expri.	Connection stiffness (kip-inch) X 1000 Poly.	Connection stiffness (kip-inch) X 1000 M.Expo.
1	0.00	0.00	0.00	0.00	0.6800E+02	0.5250E+02	0.5251E+02
3	1.00	65.10	53.42	69.73	0.6500E+02	0.5532E+02	0.7998E+02
5	2.00	137.90	113.16	150.41	0.8100E+02	0.6587E+02	0.7832E+02
7	3.00	252.40	189.32	222.59	0.1113E+03	0.8738E+02	0.6506E+02
9	6.00	338.30	357.44	350.65	0.1888E+02	0.2652E+02	0.2381E+02
11	10.00	395.50	426.21	401.11	0.1170E+02	0.1194E+02	0.7439E+01
13	14.00	442.40	464.93	434.03	0.1064E+02	0.7994E+01	0.9782E+01
15	18.10	481.40	493.39	478.88	0.9269E+01	0.6085E+01	0.1146E+02
17	25.00	541.20	529.02	547.89	0.7386E+01	0.4429E+01	0.7810E+01
19	35.00	595.90	566.69	597.56	0.4382E+01	0.3247E+01	0.3088E+01
21	45.10	632.30	595.91	625.86	0.2860E+01	0.2594E+01	0.2923E+01
23	55.10	658.30	619.65	657.95	0.2895E+01	0.2181E+01	0.3395E+01
25	65.00	687.00	639.75	690.73	0.2610E+01	0.1895E+01	0.3107E+01
27	75.00	713.00	657.57	718.17	0.2600E+01	0.1679E+01	0.2348E+01
29	85.20	739.00	673.79	737.95	0.2045E+01	0.1509E+01	0.1550E+01
31	95.20	757.20	688.20	750.25	0.7153E+00	0.1376E+01	0.9458E+00
33	105.10	757.20	701.40	757.49	0.2352E+00	0.1267E+01	0.5470E+00
35	112.90	733.80	710.94	729.47	-0.7582E+01	0.1195E+01	-0.7717E+01

A.5 – 47

V - 24

Connection type : Extended end-plate connections
Mode : All bolted without column stiffener

Tested by : J.A.Packer & L.J.Morris (1977) England
Test Id. : TEST J3

Column : 152 x 152 UC 23
Beam : 254 x 102 UB 22
Plate thickness : 15 mm

Fasteners: 10.9- -M16
13/16" Oversize holes
Material : --
Fy = -- ksi
Fu = -- ksi

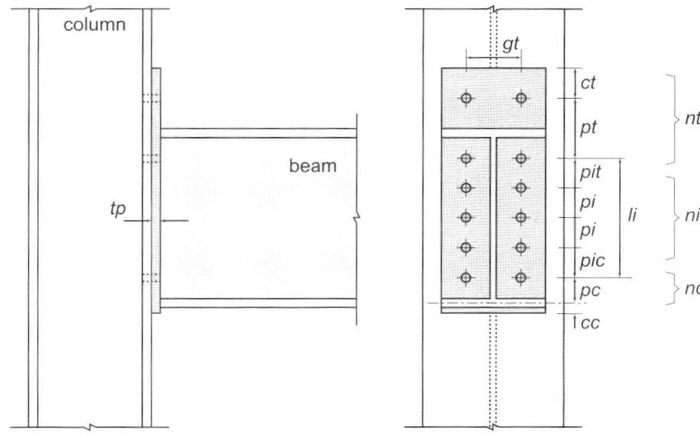

Major parameters

End plate extended on tension side only.
ct = 1.5749" pt = 3.8189" li = 5.9449" pc = 1.9685"
cc = 1.1811" gt = 3.3858" gc = 3.3858"
tp = 0.5906" nt = 2 X 2 nc = 2 X 1

Remark 1) End plate extended on tension side only. Tension side hole is 0.811" D.
2) Hsfg bolts used on the side of tension flange. Comp.side hloe is 0.689" D.

No	Moment (k-in)	Rotation (radians) X 1/1000	No	Moment (k-in)	Rotation (radians) X 1/1000
1	0.00	0.00	26	590.70	55.00
2	35.00	0.50	27	598.50	60.00
3	65.10	1.00	28	608.90	65.00
4	114.50	2.10	29	614.10	70.10
5	150.00	3.00	30	624.50	74.90
6	187.40	4.00	31	629.70	80.20
7	221.20	5.10	32	637.50	84.90
8	247.20	6.00	33	645.30	90.20
9	263.00	7.00	34	650.50	95.10
10	286.20	8.00	35	658.30	100.10
11	307.10	9.00	36	663.50	105.10
12	322.70	10.00	37	671.30	110.00
13	353.90	12.00	38	679.20	115.10
14	379.90	14.00	39	684.40	120.10
15	400.70	16.00	40	687.00	125.10
16	418.90	18.00	41	684.40	128.90
17	439.80	20.00	42	679.20	130.60
18	458.00	22.50			
19	478.80	25.10			
20	499.60	27.50			
21	517.80	29.90			
22	549.00	35.00			
23	564.70	40.00			
24	575.10	45.00			
25	585.50	50.20			

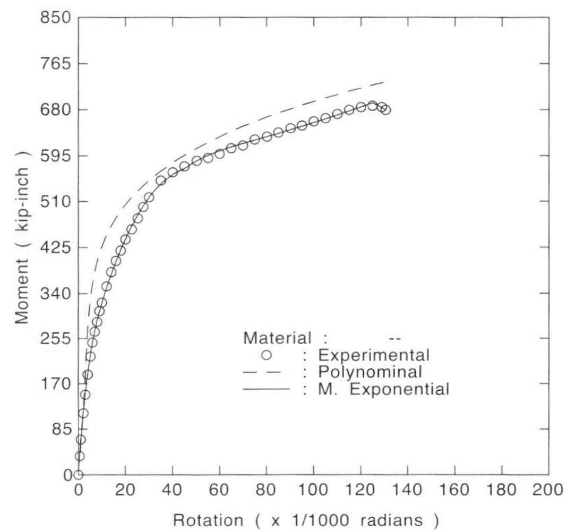

A.5 – 48

```
Moment-rotation prediction equations     ( R : X 1/1000 radians )

Frye and Morris polynominal model  :    R = Sum ( Ai X ( K*Bm )**Pi X 10**Qi )
     xd =  9.763800"    t =  0.590600"  tf =  0.630000"
     A1 =  1.830000   A2 = -1.040000   A3 =  6.380000    K =  0.010408
     P1 =        1    P2 =        3    P3 =        5    Q1 =        0    Q2 =       -1    Q3 =       -3

Modified exponential model  :
     Bm = Sum ( Ai X ( 1 - exp( - R/( 2*i*C ) ) ) )  +     Sum ( RKj X ( R - Rj0 ) )  +  Bm0
      C =   0.12133333E+01         Bmo=   0.00000000E+00         Nexp=  6        Nliner=  2
     AI =  -0.29381606E+02   0.13838610E+03   0.45781317E+04  -0.16771691E+05   0.20302012E+05  -0.76497973E+04
     Rj0 =          45.0000         125.1000
     RKj =   0.15443870E+01  -0.37571595E+01
```

No	Rotation (radians) X 1/1000	Moment (kip-inch) Expri.	Moment (kip-inch) Poly.	Moment (kip-inch) M.Expo.	Connection stiffness (kip-inch) X 1000 Expri.	Connection stiffness (kip-inch) X 1000 Poly.	Connection stiffness (kip-inch) X 1000 M.Expo.
1	0.00	0.00	0.00	0.00	0.7000E+02	0.5250E+02	0.6526E+02
3	1.00	65.10	53.42	59.99	0.5542E+02	0.5532E+02	0.5497E+02
5	3.00	150.00	189.22	152.98	0.3848E+02	0.8736E+02	0.3899E+02
7	5.10	221.20	329.23	222.24	0.2972E+02	0.3746E+02	0.2780E+02
9	7.00	268.00	380.50	268.43	0.1950E+02	0.2008E+02	0.2129E+02
11	9.00	307.10	413.42	306.32	0.1825E+02	0.1374E+02	0.1695E+02
13	12.00	353.90	447.52	351.10	0.1430E+02	0.9533E+01	0.1332E+02
15	16.00	400.70	479.78	399.16	0.9750E+01	0.6918E+01	0.1098E+02
17	20.00	439.80	504.38	440.02	0.9041E+01	0.5502E+01	0.9499E+01
19	25.10	478.80	529.43	483.94	0.8347E+01	0.4413E+01	0.7710E+01
21	29.90	517.80	548.92	516.78	0.7114E+01	0.3748E+01	0.5972E+01
23	40.00	564.70	581.97	560.12	0.2610E+01	0.2883E+01	0.2787E+01
25	50.20	585.50	608.53	585.49	0.1523E+01	0.2363E+01	0.2374E+01
27	60.00	598.50	629.95	603.94	0.1820E+01	0.2028E+01	0.1521E+01
29	70.10	614.10	649.10	617.55	0.1611E+01	0.1777E+01	0.1244E+01
31	80.20	629.70	666.05	629.94	0.1341E+01	0.1588E+01	0.1233E+01
33	90.20	645.30	681.16	642.60	0.1258E+01	0.1439E+01	0.1305E+01
35	100.10	658.30	694.80	655.91	0.1300E+01	0.1320E+01	0.1382E+01
37	110.00	671.30	707.37	669.90	0.1571E+01	0.1221E+01	0.1442E+01
39	120.10	684.40	719.27	684.70	0.7800E+00	0.1136E+01	0.1484E+01
41	128.90	684.40	728.98	683.59	-0.2325E+01	0.1072E+01	-0.2250E+01

V - 25

```
Connection type : Extended end-plate connections
           Mode : All bolted without column stiffener

Tested by  :  P.Grundy et al. (1980)                   Australia
Test  Id.  :  T1

Column : 310 UC 240                  Fasteners: 10.9- -7/8"D
Beam   : 610 UB 113                       1 1/16" Oversize holes
Plate thickness : 1"                 Material :  --
                                          Fy =    --   ksi
                                          Fu =    --   ksi

                          Major parameters

     End plate extended on tension side only.
     ct =  1.5000"    pt =  3.7500"    li = 20.1594"    pc =  1.8750"
     cc =  2.7156"    gt =  3.5000"    gt2=  2.5000"    gc =  6.0000"
     tp =  1.0000"    nt = 4 X  2      nc = 2 X  1

Remark  1)  End plate extended on tension side only.
        2)  Hs bolts used.
```

No	Moment (k-in)	Rotation (radians) X 1/1000	No	Moment (k-in)	Rotation (radians) X 1/1000
1	0.00	0.00	26	7341.30	3.20
2	322.52	0.10	27	7525.60	3.40
3	645.05	0.20	28	7709.89	3.60
4	952.22	0.30	29	7863.48	3.80
5	1259.39	0.40	30	8017.07	4.00
6	1566.55	0.50	31	8047.78	4.20
7	1873.72	0.60	32	8078.50	4.40
8	2180.89	0.70	33	8201.36	4.55
9	2488.06	0.80	34	8324.23	4.70
10	2795.22	0.90			
11	3102.38	1.00			
12	3440.27	1.10			
13	3778.15	1.20			
14	4023.89	1.30			
15	4269.62	1.40			
16	4576.79	1.55			
17	4883.96	1.70			
18	5191.13	1.85			
19	5498.29	2.00			
20	5805.46	2.15			
21	6112.63	2.30			
22	6419.80	2.50			
23	6726.96	2.70			
24	6931.74	2.87			
25	7136.52	3.03			

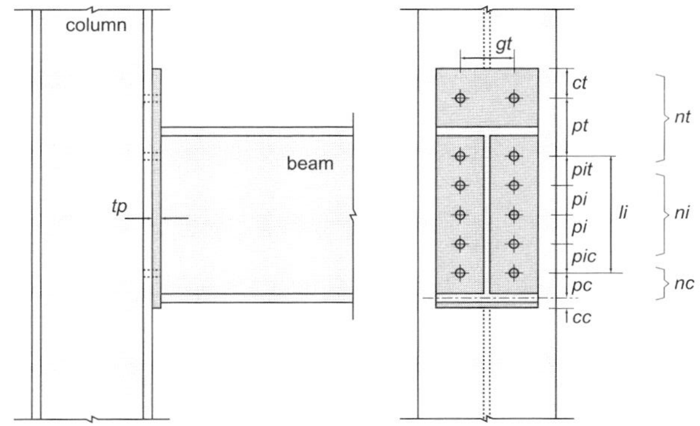

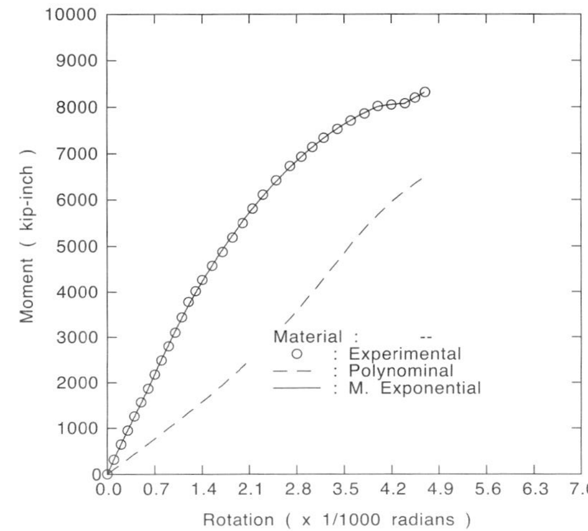

```
Moment-rotation prediction equations     ( R : X 1/1000 radians )

Frye and Morris polynominal model  :    R = Sum ( Ai X ( K*Bm )**Pi X 10**Qi )
     xd = 23.909400"   t =  1.000000"  tf =  0.875000"
     A1 =  1.830000   A2 = -1.040000   A3 =  6.380000    K =  0.000600
     P1 =         1   P2 =         3   P3 =         5   Q1 =         0   Q2 =        -1   Q3 =        -3

Modified exponential model  :
     Bm = Sum ( Ai X ( 1 - exp( - R/( 2*i*C ) ) ) )  +     Sum ( RKj X ( R - Rj0 ) )  +  Bm0
      C =   0.64166667E-01             Bmo=   0.00000000E+00          Nexp=  6        Nliner=  2
     AI =  -0.18874873E+04   0.40768165E+05  -0.25342062E+06   0.67233791E+06  -0.80586571E+06   0.35674603E+06
    Rj0 =           4.0000           4.4000
    RKj =  -0.37328426E+03   0.80642140E+03
```

No	Rotation (radians) X 1/1000	Moment (kip-inch) Expri.	Poly.	M.Expo.	Connection stiffness (kip-inch) X 1000 Expri.	Poly.	M.Expo.
1	0.00	0.00	0.00	0.00	0.3225E+04	0.9102E+03	0.3055E+04
3	0.20	645.05	182.08	647.91	0.3148E+04	0.9120E+03	0.3103E+04
5	0.40	1259.39	364.72	1250.08	0.3072E+04	0.9176E+03	0.3009E+04
7	0.60	1873.72	549.23	1868.90	0.3072E+04	0.9272E+03	0.3176E+04
9	0.80	2488.06	736.22	2509.72	0.3072E+04	0.9409E+03	0.3195E+04
11	1.00	3102.38	925.24	3133.84	0.3225E+04	0.9590E+03	0.3021E+04
13	1.20	3778.15	1119.60	3712.08	0.2918E+04	0.9824E+03	0.2755E+04
15	1.40	4269.62	1319.50	4235.51	0.2294E+04	0.1011E+04	0.2484E+04
17	1.70	4883.96	1630.94	4928.14	0.2048E+04	0.1068E+04	0.2150E+04
19	2.00	5498.29	1961.87	5534.33	0.2048E+04	0.1142E+04	0.1902E+04
21	2.30	6112.63	2318.31	6073.41	0.1828E+04	0.1238E+04	0.1696E+04
23	2.70	6726.96	2844.63	6698.94	0.1368E+04	0.1398E+04	0.1432E+04
25	3.03	7136.52	3332.98	7138.85	0.1229E+04	0.1525E+04	0.1208E+04
27	3.40	7525.60	3898.79	7537.36	0.9215E+03	0.1545E+04	0.9678E+03
29	3.80	7863.48	4484.65	7875.89	0.7679E+03	0.1338E+04	0.7304E+03
31	4.20	8047.78	4963.05	8052.21	0.1536E+03	0.1058E+04	0.1582E+03
33	4.55	8201.36	5296.93	8203.27	0.8191E+03	0.8597E+03	0.8255E+03

V - 26

Connection type : Extended end-plate connections
Mode : All bolted without column stiffener

Tested by : P.Grundy et al. (1980) Australia
Test Id. : T2

Column : 310 UC 240
Beam : 610 UB 113
Plate thickness : 1 1/4"

Fasteners: 10.9- -7/8"D
1 1/16" Oversize holes
Material : --
Fy = -- ksi
Fu = -- ksi

Major parameters

End plate extended on tension side only.

ct = 1.5000"	pt = 3.7500"	li = 20.1594"	pc = 1.8750"
cc = 2.7156"	gt = 3.5000"	gt2= 2.5000"	gc = 6.0000"
tp = 1.2500"	nt = 4 X 2	nc = 2 X 1	

Remark 1) End plate extended on tension side only.
2) Hs bolts used.

No	Moment (k-in)	Rotation (radians) X 1/1000	No	Moment (k-in)	Rotation (radians) X 1/1000
1	0.00	0.00	26	7767.12	3.20
2	308.22	0.10	27	7890.41	3.40
3	616.44	0.20	28	8013.70	3.60
4	924.66	0.30	29	8136.98	3.85
5	1232.88	0.40	30	8260.27	4.10
6	1664.38	0.53	31	8321.92	4.25
7	2095.89	0.67	32	8383.56	4.40
8	2527.39	0.80	33	8445.20	4.55
9	2835.61	0.90	34	8506.85	4.70
10	3143.83	1.00	35	8568.49	4.95
11	3452.05	1.05	36	8630.14	5.20
12	3760.27	1.10	37	8660.96	5.35
13	4099.31	1.20	38	8691.78	5.50
14	4438.36	1.30	39	8660.96	5.70
15	4777.40	1.40	40	8630.14	5.90
16	5116.44	1.50			
17	5393.84	1.60			
18	5671.24	1.70			
19	5948.63	1.80			
20	6226.03	1.90			
21	6503.43	2.10			
22	6780.83	2.30			
23	7058.22	2.55			
24	7335.61	2.80			
25	7551.37	3.00			

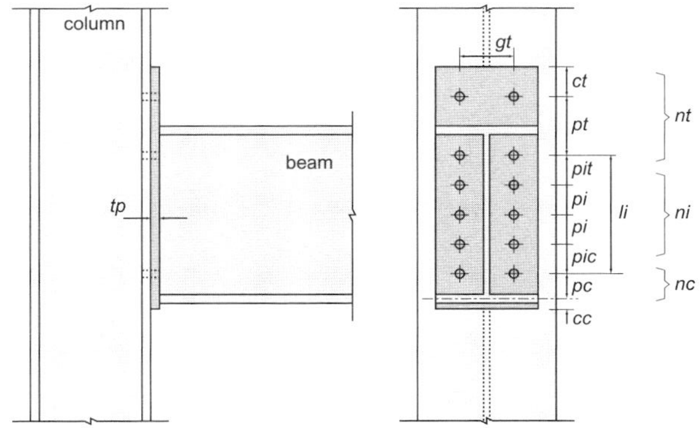

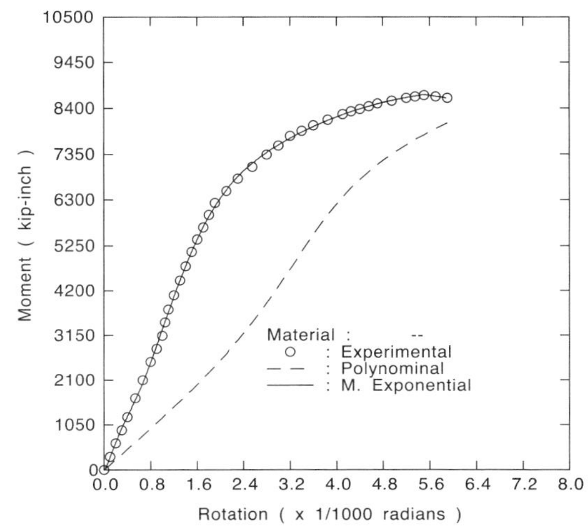

A.5 – 52

```
Moment-rotation prediction equations     ( R : X 1/1000 radians )

Frye and Morris polynominal model  :    R = Sum ( Ai X ( K*Bm )**Pi X 10**Qi )
     xd = 23.909400"    t =  1.250000"  tf =  0.875000"
     A1 =  1.830000    A2 = -1.040000   A3 =  6.380000    K =  0.000549
     P1 =        1     P2 =        3    P3 =        5    Q1 =        0     Q2 =       -1     Q3 =       -3

Modified exponential model  :
     Bm = Sum ( Ai X ( 1 - exp( - R/( 2*i*C ) ) ) )  +      Sum ( RKj X ( R - Rj0 ) )  +  Bm0
       C =   0.74166667E-01              Bmo=   0.00000000E+00          Nexp=  6        Nliner=  1
      AI =  -0.67783540E+04   0.99580227E+05  -0.44904916E+06   0.89263576E+06  -0.85158317E+06   0.32414391E+06
     Rj0 =           5.5000
     RKj =  -0.37967754E+03
```

No	Rotation (radians) X 1/1000	Moment (kip-inch) Expri.	Moment (kip-inch) Poly.	Moment (kip-inch) M.Expo.	Connection stiffness (kip-inch) X 1000 Expri.	Connection stiffness (kip-inch) X 1000 Poly.	Connection stiffness (kip-inch) X 1000 M.Expo.
1	0.00	0.00	0.00	0.00	0.3082E+04	0.9952E+03	0.1315E+04
3	0.20	616.44	199.08	635.00	0.3082E+04	0.9972E+03	0.3439E+04
5	0.40	1232.88	398.77	1242.02	0.3149E+04	0.1003E+04	0.2790E+04
7	0.67	2095.89	668.01	2039.31	0.3236E+04	0.1018E+04	0.3344E+04
9	0.90	2835.61	907.83	2891.26	0.3082E+04	0.1038E+04	0.3883E+04
11	1.05	3452.05	1065.19	3481.06	0.6164E+04	0.1055E+04	0.3941E+04
13	1.20	4099.31	1223.98	4062.88	0.3390E+04	0.1074E+04	0.3786E+04
15	1.40	4777.40	1441.87	4780.79	0.3390E+04	0.1106E+04	0.3364E+04
17	1.60	5393.84	1666.62	5401.88	0.2774E+04	0.1145E+04	0.2841E+04
19	1.80	5948.63	1899.93	5918.20	0.2774E+04	0.1192E+04	0.2330E+04
21	2.10	6503.43	2271.48	6517.95	0.1387E+04	0.1281E+04	0.1701E+04
23	2.55	7058.22	2886.10	7136.64	0.1110E+04	0.1460E+04	0.1113E+04
25	3.00	7551.37	3588.81	7562.84	0.1079E+04	0.1656E+04	0.8144E+03
27	3.40	7890.41	4266.35	7856.04	0.6164E+03	0.1689E+04	0.6619E+03
29	3.85	8136.98	4975.57	8124.59	0.4932E+03	0.1424E+04	0.5361E+03
31	4.25	8321.92	5483.33	8319.36	0.4110E+03	0.1122E+04	0.4393E+03
33	4.55	8445.20	5791.46	8441.02	0.4110E+03	0.9400E+03	0.3726E+03
35	4.95	8568.49	6129.77	8573.63	0.2466E+03	0.7621E+03	0.2923E+03
37	5.35	8660.96	6411.42	8676.37	0.2055E+03	0.6360E+03	0.2234E+03
39	5.70	8660.96	6616.54	8669.59	-0.1541E+03	0.5569E+03	-0.2064E+03

V - 27

Connection type : Extended end-plate connections
Mode : All bolted with column stiffener

Tested by : A.N.Sherbourne (1961) England
Test Id. : TEST A2

Column : 8 x 8 UC 35	Fasteners: A325- -3/4"D
Beam : 15 x 5 RSJ 42	15/16" Oversize holes
Plate thickness : 1 1/4"	Material : --
Stiffener thickness : 0.3125"	Fy = 37.22 ksi
	Fu = 65.75 ksi

Major parameters

End plate extended on tension side only.

ct = 1.5000"	pt = 3.5000"	li = 11.0000"	pc = 2.2500"
cc = 0.2500"	pit= 2.5000"	pi = 3.0000"	pic= 2.5000"
gt = 4.0000"	gi = 4.0000"	gc = 4.0000"	
tp = 1.2500"	nt = 2 X 2	ni = 2 X 3	nc = 2 X 1

Remark 1) End plate extended on tension side only.
2)

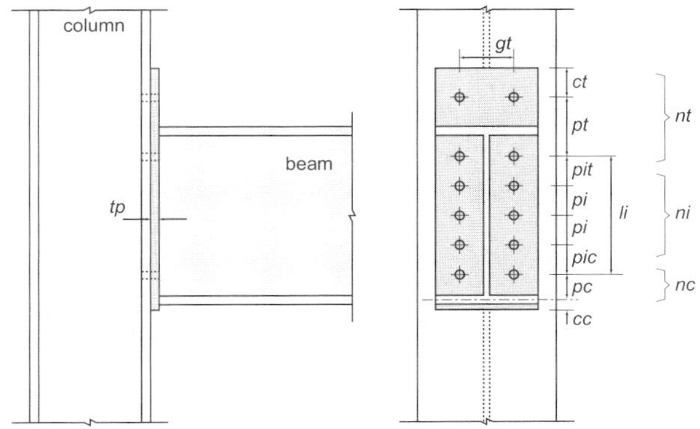

No	Moment (k-in)	Rotation (radians) X 1/1000	No	Moment (k-in)	Rotation (radians) X 1/1000	No	Moment (k-in)	Rotation (radians) X 1/1000
1	0.00	0.00	26	4186.81	6.30	51	5242.31	24.40
2	202.31	0.05	27	4327.54	7.00			
3	404.61	0.10	28	4397.90	8.10			
4	606.91	0.15	29	4538.64	9.20			
5	809.22	0.20	30	4609.01	9.95			
6	949.95	0.20	31	4679.38	10.70			
7	1090.68	0.20	32	4749.74	11.70			
8	1266.60	0.37	33	4820.10	12.70			
9	1442.51	0.53	34	4855.28	13.55			
10	1618.43	0.70	35	4890.47	14.40			
11	1776.76	0.80	36	4925.65	15.05			
12	1935.08	0.90	37	4960.84	15.70			
13	2075.81	1.10	38	4996.02	16.40			
14	2216.54	1.30	39	5031.20	17.10			
15	2462.83	1.30	40	5066.39	17.85			
16	2603.56	1.40	41	5101.57	18.60			
17	2744.29	1.50	42	5136.76	19.45			
18	2990.58	2.20	43	5171.94	20.30			
19	3131.31	2.40	44	5171.94	20.10			
20	3272.04	2.60	45	5207.12	20.95			
21	3412.78	2.95	46	5242.31	21.80			
22	3553.51	3.30	47	5242.31	21.60			
23	3729.42	3.80	48	5242.31	22.35			
24	3870.16	4.50	49	5242.31	23.10			
25	4046.08	5.30	50	5242.31	23.75			

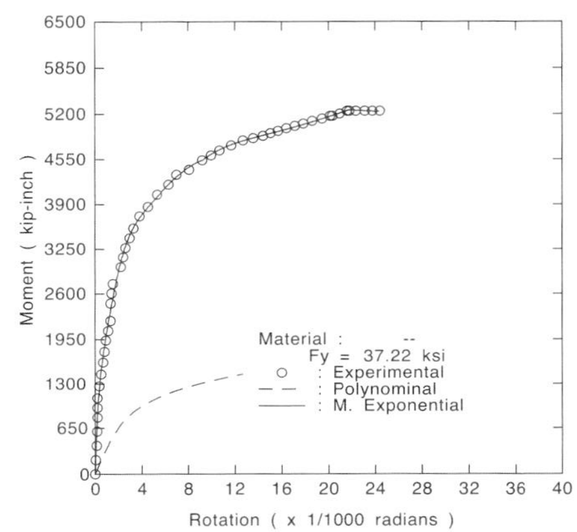

A.5 – 54

```
 Moment-rotation prediction equations      ( R : X 1/1000 radians )

Frye and Morris polynominal model  :     R = Sum ( Ai X ( K*Bm )**Pi X 10**Qi )
     xd = 14.500000"   t =  1.250000"
     A1 =  1.790000   A2 =  1.760000   A3 =  2.040000    K =  0.001427
     P1 =        1    P2 =        3    P3 =        5    Q1 =        0    Q2 =       -1    Q3 =       -1

Modified exponential model  :
     Bm = Sum ( Ai X ( 1 - exp( - R/( 2*i*C ) ) ) )  +     Sum ( RKj X ( R - Rj0 ) )  +  Bm0
       C =   0.21583333E+00          Bmo=   0.00000000E+00         Nexp=  6        Nliner=  2
      AI =   0.82728491E+04  -0.72304416E+05  0.28604551E+06  -0.52755294E+06  0.45408204E+06  -0.14406651E+06
     Rj0 =          9.2000          21.8000
     RKj =   0.60996590E+02  -0.66577822E+02
```

No	Rotation (radians) X 1/1000	Moment (kip-inch) Expri.	Moment (kip-inch) Poly.	Moment (kip-inch) M.Expo.	Connection stiffness (kip-inch) X 1000 Expri.	Connection stiffness (kip-inch) X 1000 Poly.	Connection stiffness (kip-inch) X 1000 M.Expo.
1	0.00	0.00	0.00	0.00	0.4046E+04	0.3914E+03	0.5528E+04
3	0.10	404.61	39.13	476.53	0.4046E+04	0.3910E+03	0.4084E+04
5	0.20	809.22	78.21	830.81	0.1380E+06	0.3899E+03	0.3062E+04
7	0.20	1090.68	78.21	830.81	0.1399E+06	0.3899E+03	0.3062E+04
9	0.53	1442.51	206.76	1525.59	0.1056E+04	0.3800E+03	0.1446E+04
11	0.80	1776.76	306.07	1853.22	0.1583E+04	0.3634E+03	0.1082E+04
13	1.10	2075.81	411.03	2156.88	0.7036E+03	0.3348E+03	0.9654E+03
15	1.30	2462.83	475.68	2345.94	0.2439E+06	0.3113E+03	0.9257E+03
17	1.50	2744.29	535.45	2526.81	0.1275E+04	0.2863E+03	0.8812E+03
19	2.40	3131.31	746.37	3187.38	0.7037E+03	0.1891E+03	0.5690E+03
21	2.95	3412.78	839.20	3449.07	0.4021E+03	0.1507E+03	0.3919E+03
23	3.80	3729.42	949.81	3707.38	0.2890E+03	0.1129E+03	0.2395E+03
25	5.30	4046.08	1089.39	4007.88	0.1847E+03	0.7756E+02	0.1832E+03
27	7.00	4327.54	1202.33	4297.27	0.1477E+03	0.5744E+02	0.1502E+03
29	9.20	4538.64	1311.62	4536.50	0.1077E+03	0.4331E+02	0.6597E+02
31	10.70	4679.38	1371.77	4690.22	0.8377E+02	0.3724E+02	0.8127E+02
33	12.70	4820.10	1440.16	4816.85	0.5470E+02	0.3150E+02	0.5058E+02
35	14.40	4890.47	1490.54	4895.14	0.4861E+02	0.2793E+02	0.4355E+02
37	15.70	4960.84	1525.37	4951.74	0.5227E+02	0.2573E+02	0.4406E+02
39	17.10	5031.20	1559.98	5015.25	0.4865E+02	0.2376E+02	0.4687E+02
41	18.60	5101.57	1594.23	5088.26	0.4432E+02	0.2197E+02	0.5046E+02
43	20.30	5171.94	1630.10	5177.18	-0.1274E+02	0.2027E+02	0.5400E+02
45	20.95	5207.12	1643.43	5212.64	0.4139E+02	0.1968E+02	0.5511E+02
47	21.60	5242.31	1655.73	5248.79	0.8704E-12	0.1916E+02	0.5607E+02
49	23.10	5242.31	1683.58	5247.73	0.1454E-12	0.1802E+02	-0.8764E+01
51	24.40	5242.31	1706.72	5237.06	0.0000E+00	0.1714E+02	-0.7719E+01

V - 28

Connection type : Extended end-plate connections
Mode : All bolted with cclumn stiffener

Tested by : A.N.Sherbourne (1961) England
Test Id. : TEST A3

Column : 8 x 8 UC 35
Beam : 15 x 5 RSJ 42
Plate thickness : 3/4"
Stiffener thickness : 0.6250"

Fasteners: A325- -3/4"D
15/16" Oversize holes
Material : --
Fy = 37.31 ksi
Fu = 64.76 ksi

Major parameters

End plate extended on tension side only.

ct = 1.5000"	pt = 3.5000"	li = 11.0000"	pc = 2.2500"
cc = 0.2500"	pit= 2.5000"	pi = 3.0000"	pic= 2.5000"
gt = 4.0000"	gi = 4.0000"	gc = 4.0000"	
tp = 0.7500"	nt = 2 X 2	ni = 2 X 3	nc = 2 X 1

Remark 1) End plate extended on tension side only.
2)

No	Moment (k-in)	Rotation (radians) X 1/1000	No	Moment (k-in)	Rotation (radians) X 1/1000
1	0.00	0.00	26	4977.78	13.00
2	284.45	0.20	27	5120.00	14.60
3	533.33	0.20	28	5191.12	15.65
4	782.22	0.50	29	5262.23	16.70
5	1066.67	0.70	30	5333.33	17.90
6	1351.12	0.70	31	5386.66	18.85
7	1564.44	0.70	32	5440.00	19.80
8	1724.44	0.80	33	5493.33	20.95
9	1884.44	0.90	34	5546.66	22.10
10	2168.89	0.90	35	5582.22	23.50
11	2417.77	0.90	36	5617.78	24.90
12	2666.66	1.10	37	5653.33	25.85
13	2951.11	1.80	38	5688.89	26.80
14	3200.00	2.00	39	5724.44	27.75
15	3360.00	2.20	40	5760.00	28.70
16	3520.00	2.40	41	5777.78	29.55
17	3804.44	3.30	42	5795.56	30.40
18	4088.89	4.80	43	5848.89	31.55
19	4195.56	5.30	44	5902.22	32.70
20	4302.23	6.60			
21	4480.00	7.30			
22	4622.22	8.50			
23	4675.55	9.45			
24	4728.89	10.40			
25	4871.11	11.50			

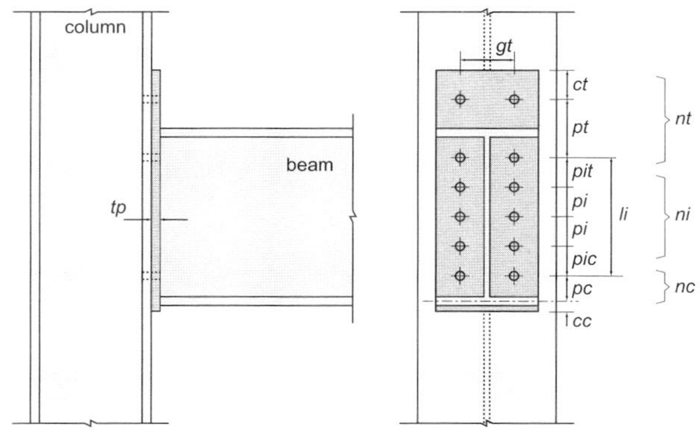

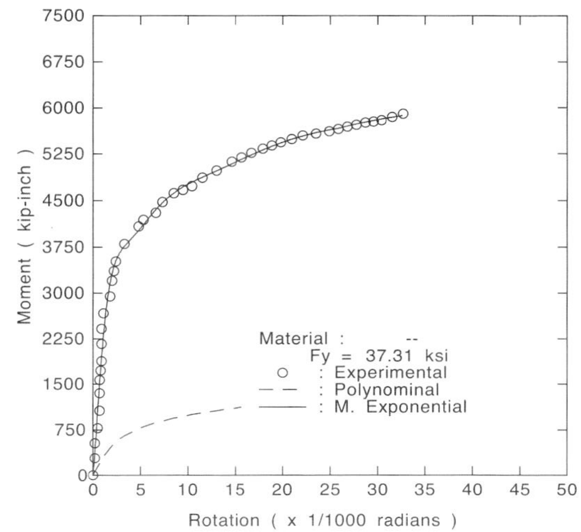

```
 Moment-rotation prediction equations      ( R : X 1/1000 radians )

Frye and Morris polynominal model  :     R = Sum ( Ai X ( K*Bm )**Pi X 10**Qi )
     xd = 14.500000"   t =  0.750000"
     A1 =  1.790000   A2 =  1.760000   A3 =  2.040000    K =  0.001939
     P1 =         1   P2 =         3   P3 =         5   Q1 =         0   Q2 =        -1   Q3 =        -1

Modified exponential model  :
     Bm = Sum ( Ai X ( 1 - exp( - R/( 2*i*C ) ) ) )  +       Sum ( RKj X ( R - Rj0 ) )  +  Bm0
       C =   0.32250000E+00             Bmo=   0.00000000E+00          Nexp=  6        Nliner=  1
      AI =  -0.10870078E+05   0.10861384E+06  -0.40928889E+06   0.78763426E+06  -0.73791188E+06   0.26761515E+06
     Rj0 =          25.8500
     RKj =   0.16366946E+02
```

No	Rotation (radians) X 1/1000	Moment (kip-inch) Expri.	Poly.	M.Expo.	Connection stiffness (kip-inch) X 1000 Expri.	Poly.	M.Expo.
1	0.00	0.00	0.00	0.00	0.1422E+04	0.2880E+03	0.1451E+04
3	0.20	533.33	57.57	367.53	0.2488E+07	0.2870E+03	0.2127E+04
5	0.70	1066.67	198.26	1514.85	0.2843E+07	0.2728E+03	0.2215E+04
7	0.70	1564.44	198.31	1515.30	0.2131E+07	0.2727E+03	0.2215E+04
9	0.90	1884.44	251.72	1935.23	0.2842E+07	0.2612E+03	0.1978E+04
11	0.90	2417.77	251.77	1935.63	0.2488E+07	0.2612E+03	0.1977E+04
13	1.80	2951.11	453.22	3169.46	0.1058E+04	0.1837E+03	0.8313E+03
15	2.20	3360.00	520.21	3434.33	0.8000E+03	0.1523E+03	0.5159E+03
17	3.30	3804.44	654.06	3765.36	0.2686E+03	0.9764E+02	0.1841E+03
19	5.30	4195.56	801.78	4116.35	0.1769E+03	0.5709E+02	0.1883E+03
21	7.30	4480.00	897.35	4467.05	0.2041E+03	0.4045E+02	0.1496E+03
23	9.45	4675.55	973.23	4718.31	0.5614E+02	0.3103E+02	0.8948E+02
25	11.50	4871.11	1030.81	4874.87	0.1047E+03	0.2553E+02	0.6871E+02
27	14.60	5120.00	1101.16	5086.54	0.7611E+02	0.2029E+02	0.6963E+02
29	16.70	5262.23	1141.11	5232.34	0.6377E+02	0.1787E+02	0.6817E+02
31	18.85	5386.66	1177.41	5371.59	0.5614E+02	0.1597E+02	0.6048E+02
33	20.95	5493.33	1209.35	5487.34	0.4638E+02	0.1450E+02	0.4944E+02
35	23.50	5582.22	1244.42	5595.41	0.2540E+02	0.1306E+02	0.3550E+02
37	25.85	5653.33	1274.01	5665.49	0.3743E+02	0.1198E+02	0.2456E+02
39	27.75	5724.44	1296.04	5736.39	0.3743E+02	0.1125E+02	0.3399E+02
41	29.55	5777.78	1315.69	5792.83	0.2092E+02	0.1064E+02	0.2896E+02
43	31.55	5848.89	1336.24	5846.40	0.4638E+02	0.1004E+02	0.2485E+02

V - 29

Connection type : Extended end-plate connections
Mode : All bolted with column stiffener

Tested by : A.N.Sherbourne (1961) England
Test Id. : TEST B1

Column : 8 x 8 UC 35
Beam : 15 x 5 RSJ 42
Plate thickness : 1"
Stiffener thickness : 0.3125"

Fasteners: A325- -7/8"D
1 1/16" Oversize holes
Material : --
Fy = -- ksi
Fu = -- ksi

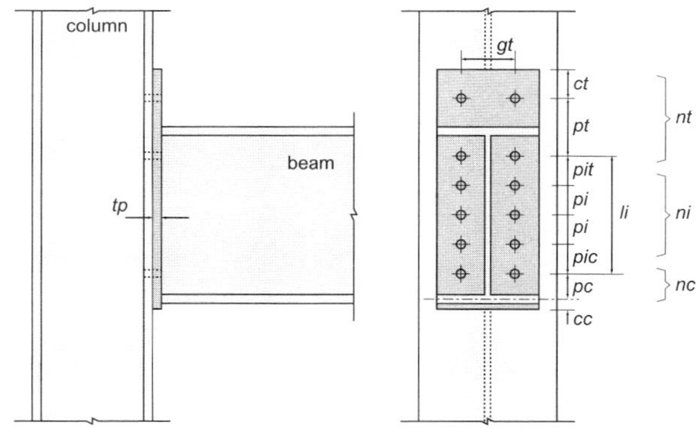

Major parameters

End plate extended on tension side only.

ct = 1.5000"	pt = 5.0000"	li = 9.0000"	pc = 3.5000"
cc = 0.0000"	pit= 3.0000"	pi = 3.0000"	pic= 3.0000"
gt = 4.0000"	gi = 4.0000"	gc = 4.0000"	
tp = 1.0000"	nt = 2 X 2	ni = 2 X 2	nc = 2 X 1

Remark 1) End plate extended on tension side only.
2)

No	Moment (k-in)	Rotation (radians) X 1/1000	No	Moment (k-in)	Rotation (radians) X 1/1000	No	Moment (k-in)	Rotation (radians) X 1/1000
1	0.00	0.00	26	4257.17	12.00	51	5453.40	61.60
2	263.87	0.10	27	4326.30	13.40			
3	527.75	0.20	28	4395.43	14.80			
4	756.44	0.35	29	4538.64	17.70			
5	985.13	0.50	30	4609.01	19.40			
6	1161.05	0.60	31	4679.38	21.10			
7	1336.97	0.70	32	4679.38	21.10			
8	1495.29	0.80	33	4734.92	23.60			
9	1653.61	0.90	34	4855.28	25.50			
10	1829.53	0.95	35	4837.76	27.40			
11	2005.44	1.00	36	4936.48	29.20			
12	2163.77	1.10	37	4936.25	30.80			
13	2322.10	1.20	38	4996.02	32.40			
14	2462.83	1.30	39	5101.57	35.30			
15	2603.56	1.40	40	5101.57	37.30			
16	2849.84	1.90	41	5136.76	38.90			
17	2990.58	2.20	42	5171.94	40.50			
18	3236.86	2.80	43	5189.53	42.10			
19	3377.59	3.60	44	5207.12	43.70			
20	3518.33	4.40	45	5277.49	46.50			
21	3660.26	5.45	46	5277.49	49.40			
22	3802.19	6.50	47	5312.68	51.70			
23	3909.24	7.70	48	5347.85	54.70			
24	4055.18	9.30	49	5418.22	57.70			
25	4156.18	10.70	50	5418.22	60.60			

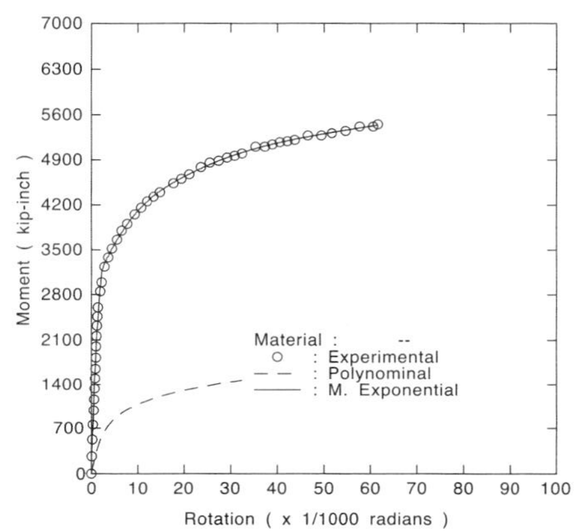

A.5 – 58

```
Moment-rotation prediction equations     ( R : X 1/1000 radians )

Frye and Morris polynominal model  :     R = Sum ( Ai X ( K*Bm )**Pi X 10**Qi )
     xd = 14.000000"     t =  1.000000"
     A1 =  1.790000   A2 =  1.760000   A3 =  2.040000    K =  0.001775
     P1 =         1   P2 =         3   P3 =         5   Q1 =         0   Q2 =       -1   Q3 =       -1

Modified exponential model  :
     Bm = Sum ( Ai X ( 1 - exp( - R/( 2*i*C ) ) ) )  +     Sum ( RKj X ( R - Rj0 ) )  +  Bm0
      C =  0.53833333E+00              Bmo=   0.00000000E+00          Nexp=  6       Nliner=  3
     AI =  0.30981049E+04   0.49161988E+04  -0.43789733E+05   0.11590296E+06  -0.13210784E+06   0.56208114E+05
    Rj0 =          0.5000           2.2000           3.6000
    RKj =  0.38931449E+03  -0.44515271E+03   0.66722531E+02
```

No	Rotation (radians) X 1/1000	Moment (kip-inch) Expri.	Moment (kip-inch) Poly.	Moment (kip-inch) M.Expo.	Connection stiffness (kip-inch) X 1000 Expri.	Connection stiffness (kip-inch) X 1000 Poly.	Connection stiffness (kip-inch) X 1000 M.Expo.
1	0.00	0.00	0.00	0.00	0.2639E+04	0.3147E+03	0.2677E+04
3	0.20	527.75	62.90	487.03	0.2193E+04	0.3135E+03	0.2209E+04
5	0.50	985.13	156.05	1062.94	0.1665E+04	0.3067E+03	0.1657E+04
7	0.70	1336.97	216.58	1442.65	0.1671E+04	0.2980E+03	0.1760E+04
9	0.90	1653.61	274.98	1770.23	0.2873E+04	0.2854E+03	0.1524E+04
11	1.00	2005.44	303.33	1917.47	0.2873E+04	0.2776E+03	0.1423E+04
13	1.20	2322.10	356.95	2183.90	0.1495E+04	0.2600E+03	0.1248E+04
15	1.40	2603.56	407.00	2418.66	0.1255E+04	0.2403E+03	0.1105E+04
17	2.20	2990.58	568.31	3142.41	0.4496E+03	0.1664E+03	0.7526E+03
19	3.60	3377.59	745.26	3395.05	0.1759E+03	0.9644E+02	0.1034E+03
21	5.45	3660.26	885.23	3650.38	0.1352E+03	0.6046E+02	0.1205E+03
23	7.70	3909.24	997.50	3899.11	0.9007E+02	0.4179E+02	0.1010E+03
25	10.70	4156.18	1102.97	4162.02	0.7501E+02	0.2994E+02	0.7465E+02
27	13.40	4326.30	1175.23	4337.79	0.4938E+02	0.2405E+02	0.5688E+02
29	17.70	4538.64	1265.60	4546.46	0.4435E+02	0.1849E+02	0.4233E+02
31	21.10	4679.38	1323.48	4679.96	0.2433E-01	0.1574E+02	0.3660E+02
33	23.60	4784.92	1360.85	4767.24	0.3927E+02	0.1421E+02	0.3328E+02
35	27.40	4887.76	1411.31	4884.65	0.2221E+02	0.1243E+02	0.2855E+02
37	30.80	4966.25	1451.41	4974.84	0.1861E+02	0.1121E+02	0.2457E+02
39	35.30	5101.57	1498.86	5074.87	0.1486E+02	0.9938E+01	0.2006E+02
41	38.90	5136.76	1533.41	5141.85	0.2199E+02	0.9124E+01	0.1727E+02
43	42.10	5189.53	1561.59	5193.98	0.1100E+02	0.8520E+01	0.1539E+02
45	46.50	5277.49	1597.28	5257.43	0.1279E+02	0.7823E+01	0.1358E+02
47	51.70	5312.68	1636.46	5324.41	0.1375E+02	0.7137E+01	0.1230E+02
49	57.70	5418.22	1676.99	5395.63	0.1153E+02	0.6503E+01	0.1153E+02
51	61.60	5453.40	1701.70	5440.04	0.3518E+02	0.6150E+01	0.1127E+02

A.5 – 59

V - 30

```
Connection type : Extended end-plate connections
           Mode : All bolted with column stiffener

Tested by  :  A.N.Sherbourne (1961)                    England
Test  Id.  :  TEST B2

Column : 8 x 8 UC 35                 Fasteners: A325- -7/8"D
Beam   : 15 x 5 RSJ 42                  1 1/16" Oversize holes
Plate thickness : 3/4"               Material :  --
Stiffener thickness : 0.5000"            Fy =   --   ksi
                                         Fu =   --   ksi

                          Major parameters

     End plate extended on tension side only.
     ct =  1.5000"    pt =  5.0000"    li =  9.0000"    pc =  3.5000"
     cc =  0.0000"    pit=  3.0000"    pi =  3.0000"    pic=  3.0000"
     gt =  4.0000"    gi =  4.0000"    gc =  4.0000"
     tp =  0.7500"    nt = 2 X  2      ni = 2 X  2      nc = 2 X  1

Remark  1)  End plate extended on tension side only.
        2)
```

No	Moment (k-in)	Rotation (radians) X 1/1000	No	Moment (k-in)	Rotation (radians) X 1/1000
1	0.00	0.00	26	4503.46	10.90
2	199.37	0.06	27	4609.01	12.60
3	398.74	0.13	28	4714.56	14.20
4	598.12	0.21	29	4749.73	15.70
5	785.76	0.30	30	4802.51	16.85
6	973.40	0.40	31	4855.28	18.00
7	1161.05	0.51	32	4890.47	19.40
8	1336.97	0.63	33	4925.65	20.80
9	1556.86	0.76	34	4960.84	21.75
10	1776.76	0.90	35	4996.02	22.70
11	1996.65	1.05	36	5031.20	24.40
12	2216.54	1.21	37	5031.20	25.40
13	2357.28	1.38	38	5031.20	26.40
14	2498.02	1.56	39	5066.39	28.10
15	2744.29	1.75	40	5101.57	29.80
16	2990.58	1.95	41	5136.76	31.30
17	3236.86	2.30	42	5171.94	33.00
18	3395.18	2.75	43	5171.94	34.25
19	3553.51	3.40	44	5171.94	35.50
20	3764.60	4.30			
21	3940.52	5.20			
22	4116.44	6.40			
23	4221.98	7.60			
24	4362.72	9.00			
25	4433.09	9.95			

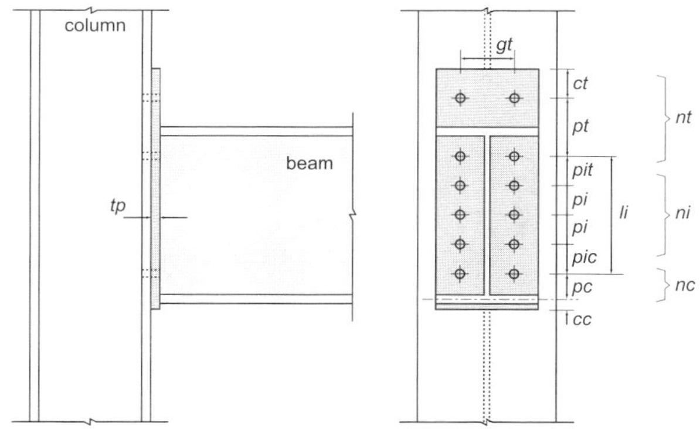

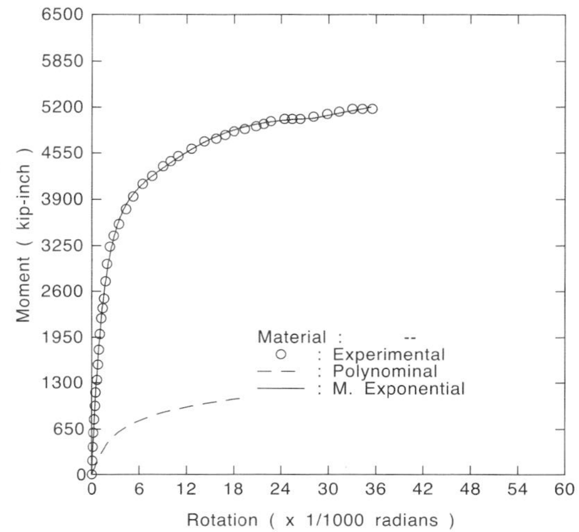

```
Moment-rotation prediction equations     ( R : X 1/1000 radians )

Frye and Morris polynominal model  :    R = Sum ( Ai X ( K*Bm )**Pi X 10**Qi )
     xd = 14.000000"   t =  0.750000"
     A1 =  1.790000   A2 =  1.760000   A3 =  2.040000    K =  0.002110
     P1 =        1    P2 =        3    P3 =        5    Q1 =        0    Q2 =      -1    Q3 =      -1

Modified exponential model  :
     Bm = Sum ( Ai X ( 1 - exp( - R/( 2*i*C ) ) ) )  +      Sum ( RKj X ( R - Rj0 ) )  +  Bm0
       C =   0.31083333E+00            Bmo=   0.00000000E+00          Nexp=  6        Nliner=  1
      AI =  -0.16696586E+03   0.69973746E+04  -0.36425458E+05   0.12020358E+06  -0.15902662E+06   0.73481307E+05
     Rj0 =            28.1000
     RKj =   0.18941563E+02
```

No	Rotation (radians) X 1/1000	Moment (kip-inch) Expri.	Moment (kip-inch) Poly.	Moment (kip-inch) M.Expo.	Connection stiffness (kip-inch) X 1000 Expri.	Connection stiffness (kip-inch) X 1000 Poly.	Connection stiffness (kip-inch) X 1000 M.Expo.
1	0.00	0.00	0.00	0.00	0.3323E+04	0.2648E+03	0.2706E+04
3	0.13	398.74	34.42	336.35	0.2682E+04	0.2644E+03	0.2472E+04
5	0.30	785.76	79.21	732.60	0.1986E+04	0.2625E+03	0.2195E+04
7	0.51	1161.05	133.89	1161.48	0.1591E+04	0.2578E+03	0.1897E+04
9	0.76	1556.86	197.21	1596.78	0.1633E+04	0.2479E+03	0.1595E+04
11	1.05	1996.65	266.68	2016.12	0.1422E+04	0.2301E+03	0.1307E+04
13	1.38	2357.28	338.42	2402.77	0.8055E+03	0.2039E+03	0.1047E+04
15	1.75	2744.29	408.08	2746.28	0.1265E+04	0.1729E+03	0.8205E+03
17	2.30	3236.86	491.91	3126.94	0.5497E+03	0.1338E+03	0.5797E+03
19	3.40	3553.51	610.19	3596.05	0.2398E+03	0.8669E+02	0.3082E+03
21	5.20	3940.52	731.79	3968.75	0.1745E+03	0.5358E+02	0.1400E+03
23	7.60	4221.98	835.83	4222.60	0.9375E+02	0.3564E+02	0.8658E+02
25	9.95	4433.09	908.52	4411.92	0.7407E+02	0.2708E+02	0.7633E+02
27	12.60	4609.01	972.25	4601.71	0.6409E+02	0.2147E+02	0.6602E+02
29	15.70	4749.73	1032.03	4779.91	0.3615E+02	0.1741E+02	0.4840E+02
31	18.00	4855.28	1069.56	4875.72	0.3653E+02	0.1532E+02	0.3516E+02
33	20.80	4925.65	1109.66	4954.91	0.3222E+02	0.1341E+02	0.2207E+02
35	22.70	4996.02	1134.15	4990.31	0.3118E+02	0.1239E+02	0.1550E+02
37	25.40	5031.20	1166.14	5022.74	0.0000E+00	0.1119E+02	0.9021E+01
39	28.10	5066.39	1194.81	5041.29	0.2070E+02	0.1023E+02	0.5064E+01
41	31.30	5136.76	1225.99	5113.49	0.2216E+02	0.9296E+01	0.2141E+02
43	34.25	5171.94	1252.52	5174.63	0.0000E+00	0.8583E+01	0.2018E+02

V - 31

Connection type : Extended end-plate connections
Mode : All bolted with column stiffener

Tested by : L.G.Johnson et al. (1960) England
Test Id. : TEST 5

Column : 8 x 8 UC 45
Beam : 10 x 4 1/2 RSJ 2
Plate thickness : 1/2"
Stiffener thickness : 0.5000"

Fasteners: A325- -3/4"D
15/16" Oversize holes
Material : --
Fy = -- ksi
Fu = -- ksi

Major parameters

End plate extended on tension side only.

ct = 1.2500"	pt = 3.7500"	li = 6.0000"	pc = 2.0000"
cc = 0.7500"	pit= 3.0000"	pi = 0.0000"	pic= 3.0000"
gt = 3.5000"	gi = 3.5000"	gc = 3.5000"	
tp = 0.5000"	nt = 2 X 2	ni = 2 X 1	nc = 2 X 1

Remark 1) End plate extended on tension side only.
2)

No	Moment (k-in)	Rotation (radians) X 1/1000	No	Moment (k-in)	Rotation (radians) X 1/1000
1	0.00	0.00	26	977.87	6.75
2	42.32	0.17	27	1002.50	7.30
3	84.65	0.33	28	1017.66	7.80
4	126.97	0.50	29	1030.93	8.40
5	168.66	0.67	30	1044.19	9.00
6	210.36	0.83	31	1047.98	9.45
7	252.05	1.00	32	1051.78	9.90
8	293.74	1.17	33	1066.94	10.45
9	335.43	1.33	34	1082.10	11.00
10	377.12	1.50	35	1085.89	11.65
11	421.34	1.70	36	1089.68	12.30
12	465.56	1.90	37	1097.26	12.77
13	509.78	2.10	38	1104.84	13.23
14	552.74	2.33	39	1112.42	13.70
15	595.70	2.57			
16	638.65	2.80			
17	677.82	3.13			
18	716.98	3.47			
19	756.14	3.80			
20	792.15	4.10			
21	828.16	4.40			
22	862.27	4.85			
23	896.38	5.30			
24	924.80	5.75			
25	953.23	6.20			

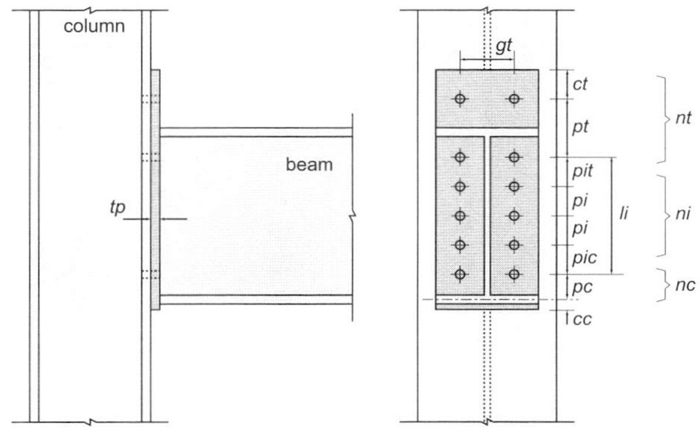

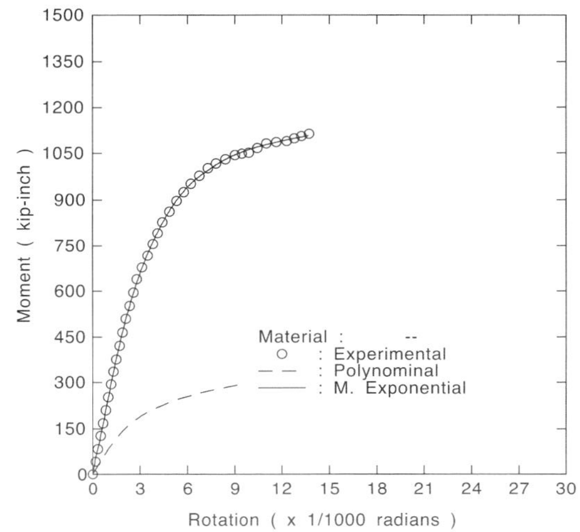

A.5 – 62

```
Moment-rotation prediction equations      ( R : X 1/1000 radians )

Frye and Morris polynominal model  :     R = Sum ( Ai X ( K*Bm )**Pi X 10**Qi )
     xd =  9.750000"   t =  0.500000"
     A1 =  1.790000   A2 =  1.760000   A3 =  2.040000    K =  0.006412
     P1 =         1   P2 =         3   P3 =         5   Q1 =         0   Q2 =        -1   Q3 =        -1

Modified exponential model  :
     Bm = Sum ( Ai X ( 1 - exp( - R/( 2*i*C ) ) ) )  +      Sum ( RKj X ( R - Rj0 ) )  +  Bm0
      C =   0.15584167E+00            Bmo=   0.00000000E+00          Nexp=  6        Nliner=  1
     AI =   0.38120846E+03  -0.41750673E+04   0.15921567E+05  -0.26882946E+05   0.17624885E+05  -0.18193381E+04
     Rj0 =          7.3000
     RKj =   0.90645329E+01
```

No	Rotation (radians) X 1/1000	Moment (kip-inch) Expri.	Moment (kip-inch) Poly.	Moment (kip-inch) M.Expo.	Connection stiffness (kip-inch) X 1000 Expri.	Connection stiffness (kip-inch) X 1000 Poly.	Connection stiffness (kip-inch) X 1000 M.Expo.
1	0.00	0.00	0.00	0.00	0.2539E+03	0.8712E+02	0.3269E+03
3	0.33	84.65	28.94	85.53	0.2540E+03	0.8619E+02	0.2313E+03
5	0.67	168.66	57.21	165.73	0.2502E+03	0.8297E+02	0.2522E+03
7	1.00	252.05	83.93	252.04	0.2501E+03	0.7688E+02	0.2616E+03
9	1.33	335.43	108.19	337.67	0.2502E+03	0.6839E+02	0.2494E+03
11	1.70	421.34	131.40	424.65	0.2211E+03	0.5822E+02	0.2240E+03
13	2.10	509.78	152.63	508.27	0.2040E+03	0.4826E+02	0.1945E+03
15	2.57	595.70	172.93	592.01	0.1841E+03	0.3917E+02	0.1655E+03
17	3.13	677.82	192.76	677.77	0.1175E+03	0.3133E+02	0.1385E+03
19	3.80	756.14	211.44	761.67	0.1188E+03	0.2513E+02	0.1140E+03
21	4.40	828.16	225.29	824.33	0.1023E+03	0.2126E+02	0.9518E+02
23	5.30	896.38	242.51	898.45	0.6949E+02	0.1727E+02	0.7015E+02
25	6.20	953.23	256.76	951.82	0.5490E+02	0.1455E+02	0.4921E+02
27	7.30	1002.50	271.41	994.72	0.3721E+02	0.1223E+02	0.2998E+02
29	8.40	1030.93	283.91	1030.14	0.2211E+02	0.1058E+02	0.2632E+02
31	9.45	1047.98	294.37	1053.52	0.8427E+01	0.9384E+01	0.1883E+02
33	10.45	1066.94	303.29	1070.03	0.2757E+02	0.8487E+01	0.1457E+02
35	11.65	1085.89	312.93	1085.62	0.5834E+01	0.7625E+01	0.1174E+02
37	12.77	1097.26	321.14	1097.90	0.1624E+02	0.6972E+01	0.1039E+02
39	13.70	1112.42	327.42	1107.29	0.1624E+02	0.6516E+01	0.9787E+01

V - 32

Connection type : Extended end-plate connections
Mode : All bolted with column stiffener

Tested by : J.R.Bailey (1970) England
Test Id. : B4-L

Column : 8 x 8 UC 40
Beam : 12 x 5 UB 32
Plate thickness : 1 1/4"
Stiffener thickness : 0.3125"

Fasteners: 10.9- -7/8"D
1 1/16" Oversize holes
Material : --
Fy = -- ksi
Fu = -- ksi

Major parameters

End plate extended on tension side only.

ct = -- "	pt = 4.5000"	li = -- "	pc = -- "
cc = -- "	pit= -- "	pi = -- "	pic= -- "
gt = -- "	gi = -- "	gc = -- "	
tp = 1.1000"	nt = 2 X 2	ni = 2 X 1	nc = 2 X 1

Remark 1) End plate extended on tension side only.
2) Material grade 43A (BS15)

No	Moment (k-in)	Rotation (radians) X 1/1000	No	Moment (k-in)	Rotation (radians) X 1/1000
1	0.00	0.00	26	1825.03	9.20
2	69.87	0.25	27	1918.19	10.80
3	139.74	0.50	28	1985.94	12.80
4	139.74	0.40	29	2045.22	14.80
5	21[illegible].72	0.60	30	2079.10	16.70
6	283.70	0.80	31	2129.90	18.80
7	372.63	1.00	32	2146.85	20.40
8	461.56	1.20	33	2180.72	22.30
9	525.07	1.50	34	2197.66	23.90
10	588.59	1.80	35	2214.59	25.80
11	652.10	2.10	36	2231.53	27.90
12	715.62	2.40	37	2256.94	29.70
13	779.14	2.50	38	2265.41	31.10
14	842.65	2.60	39	2282.34	32.70
15	923.10	2.90	40	2290.81	34.60
16	1003.55	3.20	41	2299.28	36.30
17	1071.30	3.30	42	2307.76	38.30
18	1139.05	3.40	43	2316.22	40.30
19	1219.51	3.70	44	2316.22	42.40
20	1299.96	4.00			
21	1410.06	4.20			
22	1477.81	4.90			
23	1587.90	5.50			
24	1681.06	6.60			
25	1757.28	7.70			

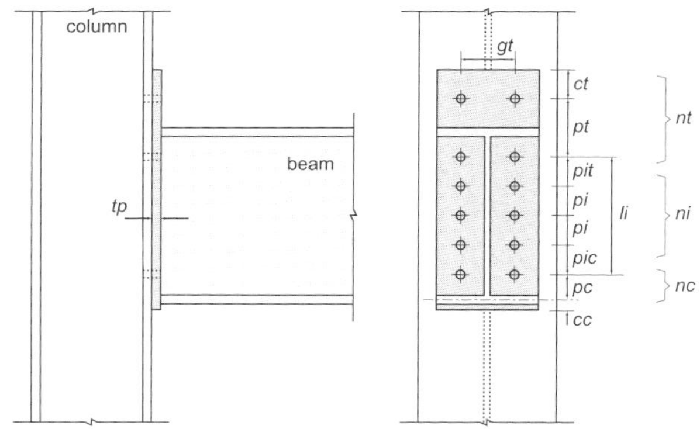

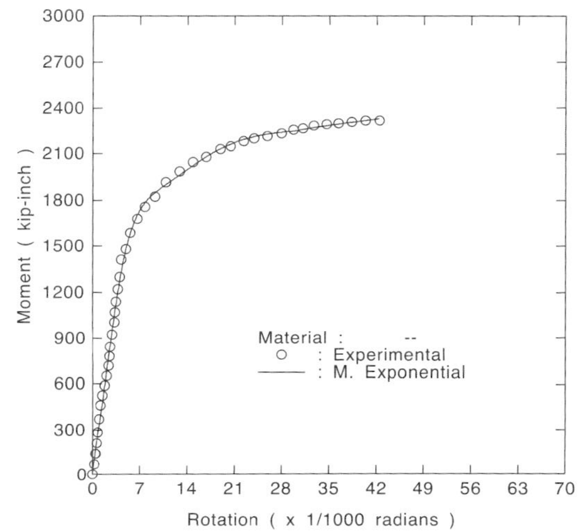

A.5 – 64

Moment-rotation prediction equations (R : X 1/1000 radians)

Modified exponential model :

```
    Bm = Sum ( Ai X ( 1 - exp( - R/( 2*i*C ) ) ) )  +     Sum ( RKj X ( R - Rj0 ) )  + Bm0
     C =  0.41583333E+00              Bmo=  0.00000000E+00         Nexp=  6        Nliner=  1
    AI =  0.25821585E+04  -0.26489368E+05  0.80767385E+05  -0.10238476E+06  0.53789999E+05  -0.60053143E+04
   Rj0 =        29.7000
   RKj =  0.54033264E+01
```

No	Rotation (radians) X 1/1000	Moment (kip-inch) Expri.	Moment (kip-inch) M.Expo.	Connection stiffness (kip-inch) X 1000 Expri.	Connection stiffness (kip-inch) X 1000 M.Expo.
1	0.00	0.00	0.00	0.2795E+03	0.5060E+03
3	0.50	139.74	190.11	-0.1863E+03	0.3034E+03
5	0.60	211.72	219.78	0.3599E+03	0.2910E+03
7	1.00	372.63	333.13	0.4446E+03	0.2845E+03
9	1.50	525.07	482.15	0.2117E+03	0.3135E+03
11	2.10	652.10	679.47	0.2117E+03	0.3399E+03
13	2.50	779.14	816.13	0.6352E+03	0.3411E+03
15	2.90	923.10	950.67	0.2682E+03	0.3297E+03
17	3.30	1071.30	1078.57	0.6775E+03	0.3085E+03
19	3.70	1219.51	1196.65	0.2682E+03	0.2811E+03
21	4.20	1410.06	1327.66	0.4497E+03	0.2425E+03
23	5.50	1587.90	1578.72	0.1486E+03	0.1472E+03
25	7.70	1757.28	1785.25	0.5909E+02	0.5587E+02
27	10.80	1918.19	1902.27	0.4740E+02	0.3045E+02
29	14.80	2045.22	2024.71	0.2358E+02	0.3010E+02
31	18.80	2129.90	2130.64	0.1647E+02	0.2187E+02
33	22.30	2180.72	2191.78	0.1390E+02	0.1332E+02
35	25.80	2214.59	2226.80	0.8511E+01	0.7156E+01
37	29.70	2256.94	2246.21	0.9579E+01	0.3238E+01
39	32.70	2282.34	2269.54	0.7783E+01	0.7069E+01
41	36.30	2299.28	2293.06	0.4640E+01	0.6116E+01
43	40.30	2316.22	2316.47	0.2167E+01	0.5663E+01

V - 33

```
Connection type : Extended end-plate connections
           Mode : All bolted with column stiffener

Tested by  :  J.R.Bailey (1970)                         England
Test  Id.  :  B4-R

Column : 8 x 8 UC 40                Fasteners: 10.9- -7/8"D
Beam   : 12 x 5 UB 32                  1 1/16" Oversize holes
Plate thickness : 1 1/4"            Material :  --
Stiffener thickness : 0.3125"            Fy =   --    ksi
                                         Fu =   --    ksi

                         Major parameters

     End plate extended on tension side only.
     ct =   --    "    pt =  4.5000"    li =   --    "    pc =   --    "
     cc =   --    "    pit=   --    "    pi =   --    "    pic=   --    "
     gt =   --    "    gi =   --    "    gc =   --    "
     tp =  1.1000"     nt = 2 X  2       ni = 2 X  1        nc = 2 X  1

Remark  1)  End plate extended on tension side only.
        2)  Material grade 43A ( BS15 )
```

No	Moment (k-in)	Rotation (radians) X 1/1000	No	Moment (k-in)	Rotation (radians) X 1/1000
1	0.00	0.00	26	1944.16	11.90
2	105.66	0.30	27	1969.51	13.20
3	211.32	0.60	28	2003.32	14.60
4	278.95	0.65	29	2028.68	16.20
5	346.57	0.70	30	2062.49	18.00
6	418.42	0.95	31	2096.30	19.80
7	490.26	1.20	32	2121.66	21.80
8	570.56	1.35	33	2147.02	23.60
9	650.87	1.50	34	2180.83	25.60
10	714.26	1.65	35	2197.74	27.50
11	777.66	1.80	36	2223.10	29.80
12	845.28	2.05	37	2240.00	31.50
13	912.90	2.30	38	2248.45	33.20
14	972.07	2.35	39	2265.36	35.30
15	1031.24	2.40	40	2273.81	37.00
16	1103.09	2.60	41	2282.27	39.00
17	1174.94	2.80	42	2282.27	40.90
18	1284.83	3.20	43	2290.72	42.70
19	1386.26	3.50	44	2299.16	44.20
20	1453.88	4.10	45	2307.62	45.70
21	1546.87	5.20	46	2307.62	47.50
22	1656.76	6.60	47	2307.62	48.80
23	1741.28	7.70			
24	1825.81	9.00			
25	1876.52	10.40			

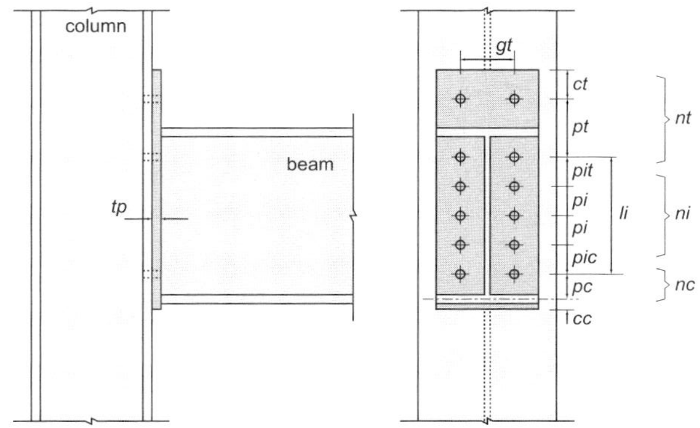

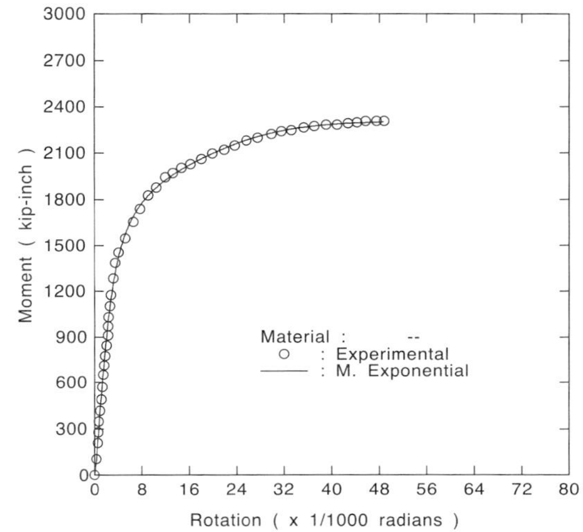

A.5 – 66

Moment-rotation prediction equations (R : X 1/1000 radians)

```
Modified exponential model  :
     Bm = Sum ( Ai X ( 1 - exp( - R/( 2*i*C ) ) ) )  +      Sum ( RKj X ( R - Rj0 ) )  +  Bm0
      C =   0.48166667E+00           Bmo=   0.00000000E+00          Nexp=  6        Nliner=  0
     AI =  -0.20917995E+04   0.14047644E+05  -0.50115114E+05   0.10908618E+06  -0.11497920E+06   0.46358011E+05
```

No	Rotation (radians) X 1/1000	Moment (kip-inch) Expri.	Moment (kip-inch) M.Expo.	Connection stiffness (kip-inch) X 1000 Expri.	Connection stiffness (kip-inch) X 1000 M.Expo.
1	0.00	0.00	0.00	0.3522E+03	0.2377E+03
3	0.60	211.32	227.60	0.1210E+04	0.4673E+03
5	0.70	346.57	274.99	0.1175E+04	0.4797E+03
7	1.20	490.26	518.76	0.4424E+03	0.4820E+03
9	1.50	650.87	659.64	0.4790E+03	0.4549E+03
11	1.80	777.66	790.79	0.3656E+03	0.4184E+03
13	2.30	912.90	983.14	0.1031E+04	0.3507E+03
15	2.40	1031.24	1017.54	0.1019E+04	0.3373E+03
17	2.80	1174.94	1142.14	0.3311E+03	0.2865E+03
19	3.50	1386.26	1315.47	0.2630E+03	0.2122E+03
21	5.20	1546.87	1573.98	0.8187E+02	0.1072E+03
23	7.70	1741.28	1761.79	0.7143E+02	0.5393E+02
25	10.40	1876.52	1877.38	0.4050E+02	0.3447E+02
27	13.20	1969.51	1959.38	0.2174E+02	0.2522E+02
29	16.20	2028.68	2027.20	0.1723E+02	0.2057E+02
31	19.80	2096.30	2095.43	0.1589E+02	0.1751E+02
33	23.60	2147.02	2156.25	0.1542E+02	0.1443E+02
35	27.50	2197.74	2205.81	0.9860E+01	0.1097E+02
37	31.50	2240.00	2242.83	0.7458E+01	0.7630E+01
39	35.30	2265.36	2266.76	0.6348E+01	0.5078E+01
41	39.00	2282.27	2281.99	0.2061E+01	0.3262E+01
43	42.70	2290.72	2291.61	0.5205E+01	0.2022E+01
45	45.70	2307.62	2296.59	0.3076E+01	0.1343E+01
47	48.80	2307.62	2299.97	0.0000E+00	0.8663E+00

V - 34

Connection type : Extended end-plate connections
Mode : All bolted with column stiffener

Tested by : J.R.Bailey (1970) England
Test Id. : B5-L

Column : 8 x 8 UC 40
Beam : 14 x 5 UB 26
Plate thickness : 1"
Stiffener thickness : 0.3125"

Fasteners: 10.9- -3/4"D
15/16" Oversize holes
Material : --
Fy = -- ksi
Fu = -- ksi

Major parameters

End plate extended on tension side only.

ct = -- "	pt = 4.2500"	li = -- "	pc = -- "
cc = -- "	pit= -- "	pi = -- "	pic= -- "
gt = -- "	gi = -- "	gc = -- "	
tp = 1.0000"	nt = 2 X 2	ni = 2 X 1	nc = 2 X 1

Remark 1) End plate extended on tension side only.
2) Material grade 43A (BS15)

No	Moment (k-in)	Rotation (radians) X 1/1000	No	Moment (k-in)	Rotation (radians) X 1/1000
1	0.00	0.00	26	1781.82	12.20
2	73.36	0.35	27	1824.24	13.80
3	152.72	0.70	28	1858.18	15.50
4	233.33	0.95	29	1892.12	17.60
5	313.94	1.20	30	1926.06	19.60
6	381.82	1.45	31	1943.03	20.85
7	449.70	1.70	32	1960.00	22.10
8	534.55	2.00	33	1976.96	24.10
9	619.39	2.30	34	1981.21	25.20
10	691.51	2.60	35	1985.46	26.30
11	763.63	2.90	36	1993.94	28.20
12	835.76	3.10	37	1993.94	30.10
13	907.88	3.30	38	1993.94	31.20
14	992.73	3.60	39	1993.94	32.30
15	1077.58	3.90	40	2010.91	34.20
16	1141.21	4.20	41	2010.91	35.55
17	1204.85	4.50	42	2010.91	36.90
18	1255.76	4.85	43	2010.91	38.60
19	1306.67	5.20	44	2010.91	40.50
20	1374.55	5.90	45	2019.40	42.20
21	1442.42	6.60	46	2027.88	43.50
22	1493.33	7.20			
23	1544.24	7.80			
24	1637.58	9.30			
25	1713.94	10.80			

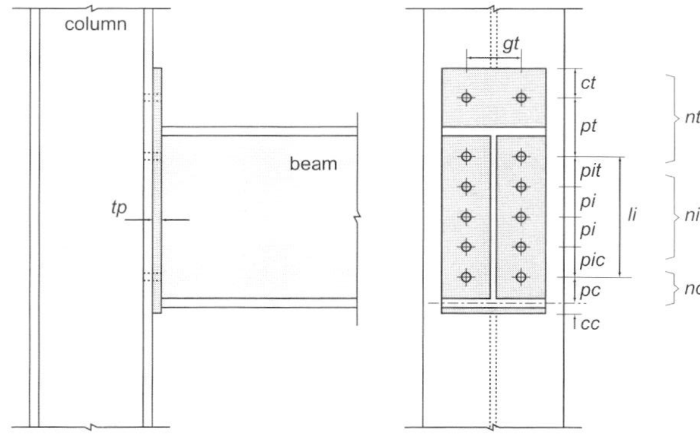

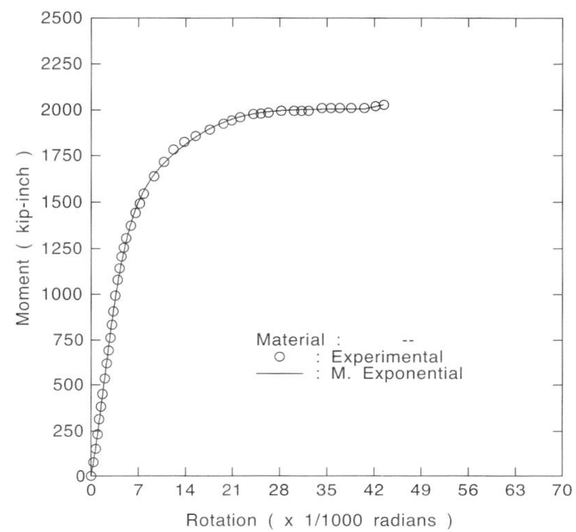

A.5 – 68

```
Moment-rotation prediction equations     ( R : X 1/1000 radians )

Modified exponential model  :
   Bm = Sum ( Ai X ( 1 - exp( - R/( 2*i*C ) ) ) )  +      Sum ( RKj X ( R - Rj0 ) )  +  Bm0
    C =   0.45000000E+00           Bmo=   0.00000000E+00         Nexp=  6        Nliner=  1
   AI =   0.82403414E+03  -0.14078649E+05   0.52150663E+05  -0.80907913E+05   0.57256775E+05  -0.13241895E+05
  Rj0 =         40.5000
  RKj =   0.81928014E+01
```

No	Rotation (radians) X 1/1000	Moment (kip-inch) Expri.	Moment (kip-inch) M.Expo.	Connection stiffness (kip-inch) X 1000 Expri.	Connection stiffness (kip-inch) X 1000 M.Expo.
1	0.00	0.00	0.00	0.2182E+03	0.2063E+03
3	0.70	152.72	160.88	0.2790E+03	0.2568E+03
5	1.20	313.94	297.79	0.2970E+03	0.2884E+03
7	1.70	449.70	446.36	0.2767E+03	0.3029E+03
9	2.30	619.39	627.83	0.2616E+03	0.2985E+03
11	2.90	763.63	801.06	0.3125E+03	0.2768E+03
13	3.30	907.88	907.84	0.3295E+03	0.2566E+03
15	3.90	1077.58	1051.73	0.2475E+03	0.2226E+03
17	4.50	1204.85	1174.94	0.1814E+03	0.1883E+03
19	5.20	1306.67	1293.68	0.1293E+03	0.1518E+03
21	6.60	1442.42	1464.53	0.9044E+02	0.9644E+02
23	7.80	1544.24	1561.30	0.7839E+02	0.6731E+02
25	10.80	1713.94	1708.78	0.4965E+02	0.3785E+02
27	13.80	1824.24	1807.22	0.2334E+02	0.2872E+02
29	17.60	1892.12	1898.60	0.1658E+02	0.1934E+02
31	20.85	1943.03	1949.05	0.1357E+02	0.1197E+02
33	24.10	1976.96	1978.55	0.5502E+01	0.6547E+01
35	26.30	1985.46	1990.07	0.4083E+01	0.4072E+01
37	30.10	1993.94	2000.17	-0.5537E-13	0.1554E+01
39	32.30	1993.94	2002.68	0.3275E+01	0.7921E+00
41	35.55	2010.91	2004.17	0.2603E-13	0.2060E+00
43	38.60	2010.91	2004.40	0.4186E-13	-0.1667E-01
45	42.20	2019.40	2018.09	0.5861E+01	0.8100E+01

V - 35

```
Connection type : Extended end-plate connections
           Mode : All bolted with column stiffener

Tested by  :  J.R.Bailey (1970)                    England
Test  Id.  :  B5-R

Column : 8 x 8 UC 40                  Fasteners: 10.9- -3/4"D
Beam   : 14 x 5 UB 26                        15/16" Oversize holes
Plate thickness : 1"                  Material :  --
Stiffener thickness : 0.3125"             Fy =    --   ksi
                                          Fu =    --   ksi

                         Major parameters

     End plate extended on tension side only.
     ct =    --    "   pt =  4.2500"   li =    --    "   pc =    --    "
     cc =    --    "   pit=    --   "   pi =    --    "   pic=    --    "
     gt =    --    "   gi =    --   "   gc =    --    "
     tp =  1.0000"     nt = 2 X  2      ni = 2 X  1      nc = 2 X  1

Remark  1)  End plate extended on tension side only.
        2)  Material grade 43A ( BS15 )
```

No	Moment (k-in)	Rotation (radians) X 1/1000	No	Moment (k-in)	Rotation (radians) X 1/1000
1	0.00	0.00	26	1704.44	7.30
2	72.26	0.13	27	1746.95	8.20
3	144.52	0.27	28	1789.45	9.30
4	216.78	0.40	29	1823.45	10.60
5	280.54	0.60	30	1848.96	12.10
6	344.29	0.80	31	1891.46	13.30
7	420.80	1.00	32	1916.96	14.90
8	497.30	1.20	33	1925.46	16.50
9	582.31	1.40	34	1950.97	17.80
10	667.32	1.60	35	1967.96	19.90
11	718.33	1.70	36	1976.47	21.80
12	769.33	1.80	37	1984.97	23.60
13	837.34	2.05	38	1993.48	25.40
14	905.35	2.30	39	1993.48	27.30
15	973.36	2.25	40	1993.48	29.20
16	1041.37	2.20	41	2001.97	31.10
17	1126.38	2.50	42	2001.97	33.00
18	1211.39	2.80	43	2001.97	34.40
19	1266.64	3.00	44	2001.97	36.00
20	1321.90	3.20	45	2010.48	38.00
21	1381.40	3.50	46	2010.48	39.20
22	1440.91	3.80	47	2018.98	40.70
23	1500.42	4.50	48	2010.48	42.70
24	1568.42	5.50	49	2010.48	44.30
25	1644.94	6.40	50	2018.98	46.10

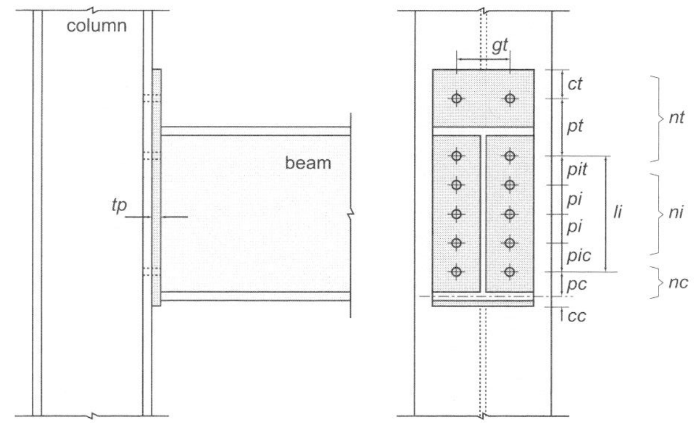

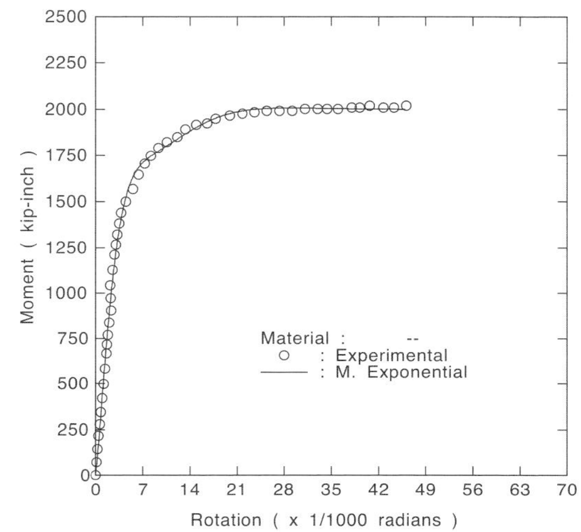

```
Moment-rotation prediction equations      ( R : X 1/1000 radians )

Modified exponential model  :
     Bm = Sum ( Ai X ( 1 - exp( - R/( 2*i*C ) ) ) )  +      Sum ( RKj X ( R - Rj0 ) )  +  Bm0
      C =   0.41749167E+00            Bmo=   0.00000000E+00         Nexp=  6        Nliner=  0
     AI =   0.64799572E+02  -0.12273793E+05   0.60234202E+05  -0.10561533E+06   0.82217741E+05  -0.22627650E+05
```

No	Rotation (radians) X 1/1000	Moment (kip-inch) Expri.	Moment (kip-inch) M.Expo.	Connection stiffness (kip-inch) X 1000 Expri.	Connection stiffness (kip-inch) X 1000 M.Expo.
1	0.00	0.00	0.00	0.5421E+03	0.3286E+03
3	0.27	144.52	98.78	0.5419E+03	0.4060E+03
5	0.60	280.54	244.05	0.3188E+03	0.4586E+03
7	1.00	420.80	432.27	0.3825E+03	0.4756E+03
9	1.40	582.31	620.15	0.4250E+03	0.4596E+03
11	1.70	718.33	754.45	0.5101E+03	0.4343E+03
13	2.05	837.34	900.01	0.2720E+03	0.3964E+03
15	2.25	973.36	976.92	-0.1360E+04	0.3726E+03
17	2.50	1126.38	1066.24	0.2834E+03	0.3419E+03
19	3.00	1266.64	1221.82	0.2763E+03	0.2809E+03
21	3.50	1381.40	1347.96	0.1984E+03	0.2248E+03
23	4.50	1500.42	1525.38	0.7801E+02	0.1356E+03
25	6.40	1644.94	1685.45	0.7557E+02	0.4825E+02
27	8.20	1746.95	1746.41	0.4336E+02	0.2535E+02
29	10.60	1823.45	1802.66	0.2191E+02	0.2350E+02
31	13.30	1891.46	1867.17	0.2707E+02	0.2343E+02
33	16.50	1925.46	1933.45	0.1321E+02	0.1726E+02
35	19.90	1967.96	1978.17	0.6194E+01	0.9271E+01
37	23.60	1984.97	2000.49	0.4723E+01	0.3382E+01
39	27.30	1993.48	2007.05	-0.8060E-13	0.5822E+00
41	31.10	2001.97	2007.00	0.2236E+01	-0.3970E+00
43	34.40	2001.97	2005.34	0.8338E-13	-0.5467E+00
45	38.00	2010.48	2003.50	0.1595E+01	-0.4557E+00
47	40.70	2018.98	2002.41	0.1416E+01	-0.3483E+00
49	44.30	2010.48	2001.40	0.2221E+01	-0.2216E+00

V - 36

Connection type : Extended end-plate connections
Mode : All bolted with column stiffener

Tested by : J.R.Bailey (1970) England
Test Id. : B6-L

Column : 8 x 8 UC 40
Beam : 14 x 6 3/4 UB 45
Plate thickness : 1 3/8"
Stiffener thickness : 0.3125"

Fasteners: 10.9- -1"D
1 1/4" Oversize holes
Material : --
Fy = -- ksi
Fu = -- ksi

Major parameters

End plate extended on tension side only.

ct = -- "	pt = 4.6250"	li = -- "	pc = -- "
cc = -- "	pit= -- "	pi = -- "	pic= -- "
gt = -- "	gi = -- "	gc = -- "	
tp = 1.2400"	nt = 2 X 2	ni = 2 X 1	nc = 2 X 1

Remark 1) End plate extended on tension side only.
2) Material grade 43A (BS15)

No	Moment (k-in)	Rotation (radians) X 1/1000	No	Moment (k-in)	Rotation (radians) X 1/1000
1	C.00	0.00	26	3134.21	13.80
2	153.31	0.65	27	3176.88	14.95
3	306.61	1.30	28	3219.55	16.10
4	485.48	1.75	29	3304.67	17.90
5	664.34	2.20	30	3338.72	19.60
6	792.10	2.50	31	3372.77	21.10
7	919.86	2.80	32	3440.87	22.70
8	1039.09	3.15	33	3474.91	24.50
9	1158.32	3.50	34	3491.94	26.30
10	1286.08	3.80	35	3517.48	27.50
11	1413.84	4.10	36	3543.01	28.70
12	1533.08	4.45	37	3560.26	30.60
13	1652.32	4.80	38	3568.77	31.80
14	1754.53	5.05	39	3577.28	33.00
15	1856.74	5.30	40	3628.36	35.10
16	1950.43	5.65	41	3636.86	36.25
17	2044.12	6.00	42	3645.37	37.40
18	2188.87	6.70	43	3696.44	39.40
19	2333.63	7.40	44	3747.52	41.40
20	2435.89	7.80	45	3747.52	43.30
21	2538.14	8.20	46	3747.52	45.10
22	2691.36	9.50			
23	2785.10	10.25			
24	2878.85	11.00			
25	2998.02	12.30			

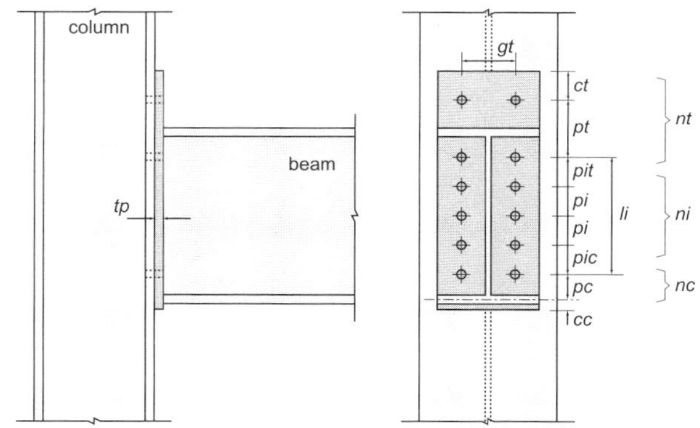

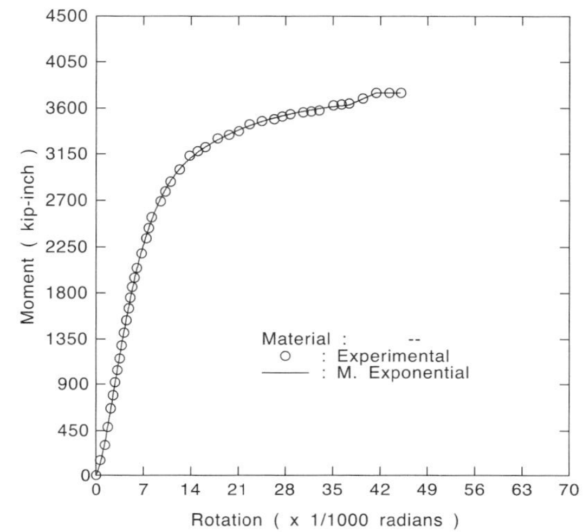

A.5 – 72

Moment-rotation prediction equations (R : X 1/1000 radians)

```
Modified exponential model  :
     Bm = Sum ( Ai X ( 1 - exp( - R/( 2*i*C ) ) ) ) +      Sum ( RKj X ( R - Rj0 ) ) + Bm0
       C =   0.53833333E+00            Bmo=   0.00000000E+00          Nexp=  6        Nliner=  2
      AI =   0.28220155E+03  -0.48764105E+04  -0.68611992E+04   0.56606448E+05  -0.79944920E+05   0.38474808E+05
     Rj0 =          37.4000          41.4000
     RKj =   0.24798874E+02  -0.28666815E+02
```

No	Rotation (radians) X 1/1000	Moment (kip-inch) Expri.	Moment (kip-inch) M.Expo.	Connection stiffness (kip-inch) X 1000 Expri.	Connection stiffness (kip-inch) X 1000 M.Expo.
1	0.00	0.00	0.00	0.2359E+03	0.1226E+03
3	1.30	306.61	325.42	0.3314E+03	0.3447E+03
5	2.20	664.34	664.61	0.4145E+03	0.3982E+03
7	2.80	919.86	906.14	0.3865E+03	0.4036E+03
9	3.50	1158.32	1184.97	0.3865E+03	0.3902E+03
11	4.10	1413.84	1412.78	0.3865E+03	0.3680E+03
13	4.80	1652.32	1659.21	0.3804E+03	0.3353E+03
15	5.30	1856.74	1820.56	0.3500E+03	0.3100E+03
17	6.00	2044.12	2024.97	0.2474E+03	0.2742E+03
19	7.40	2333.63	2361.47	0.2379E+03	0.2083E+03
21	8.20	2538.14	2515.01	0.2232E+03	0.1763E+03
23	10.25	2785.10	2808.29	0.1250E+03	0.1145E+03
25	12.30	2998.02	3000.85	0.9126E+02	0.7668E+02
27	14.95	3176.88	3164.87	0.3711E+02	0.5020E+02
29	17.90	3304.67	3289.15	0.3327E+02	0.3572E+02
31	21.10	3372.77	3388.60	0.3231E+02	0.2714E+02
33	24.50	3474.91	3469.69	0.1419E+02	0.2084E+02
35	27.50	3517.48	3525.23	0.2128E+02	0.1631E+02
37	30.60	3560.26	3569.47	0.7862E+01	0.1235E+02
39	33.00	3577.28	3595.95	0.1336E+02	0.9788E+01
41	36.25	3636.86	3622.99	0.7398E+01	0.6977E+01
43	39.40	3696.44	3691.16	0.2554E+02	0.2971E+02
45	43.30	3747.52	3748.80	-0.5201E-13	-0.7656E+00

V - 37

```
Connection type : Extended end-plate connections
           Mode : All bolted with column stiffener

Tested by  :  J.R.Bailey (1970)                        England
Test  Id.  :  B6-R

Column : 8 x 8 UC 40                 Fasteners: 10.9- -1"D
Beam   : 14 x 6 3/4 UB 45                  1 1/4" Oversize holes
Plate thickness : 1 3/8"             Material :  --
Stiffener thickness : 0.3125"             Fy =   --   ksi
                                          Fu =   --   ksi

                         Major parameters

     End plate extended on tension side only.
     ct =   --   "   pt =  4.6250"   li =   --   "   pc =   --   "
     cc =   --   "   pit=   --   "   pi =   --   "   pic=   --   "
     gt =   --   "   gi =   --   "   gc =   --   "
     tp =  1.2400"   nt = 2 X  2     ni = 2 X  1      nc = 2 X  1

Remark  1)  End plate extended on tension side only.
        2)  Material grade 43A ( BS15 )
```

No	Moment (k-in)	Rotation (radians) X 1/1000
1	0.00	0.00
2	130.10	0.33
3	260.20	0.67
4	390.30	1.00
5	568.48	1.20
6	746.66	1.40
7	865.45	1.80
8	984.23	2.20
9	1145.44	2.60
10	1306.66	3.00
11	1476.36	3.35
12	1646.06	3.70
13	1756.36	4.05
14	1866.66	4.40
15	1976.97	4.65
16	2087.28	4.90
17	2231.51	5.40
18	2375.75	5.90
19	2494.58	6.55
20	2613.41	7.20
21	2715.22	7.90
22	2817.02	8.60
23	2969.80	9.70
24	3139.36	11.10
25	3241.28	12.60
26	3342.97	14.40
27	3377.03	16.20
28	3444.90	17.80
29	3529.79	20.00
30	3546.59	22.10
31	3580.64	24.10
32	3631.49	25.70
33	3631.49	27.90
34	3665.53	30.20
35	3682.33	32.00
36	3716.39	34.10
37	3724.90	35.30
38	3733.40	36.50
39	3733.40	38.20
40	3767.23	39.80
41	3750.20	41.00
42	3733.40	42.80
43	3750.20	44.20
44	3750.20	45.70
45	3767.23	46.50

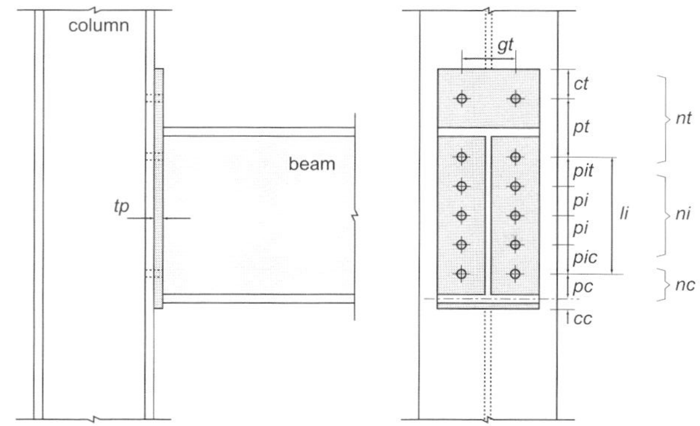

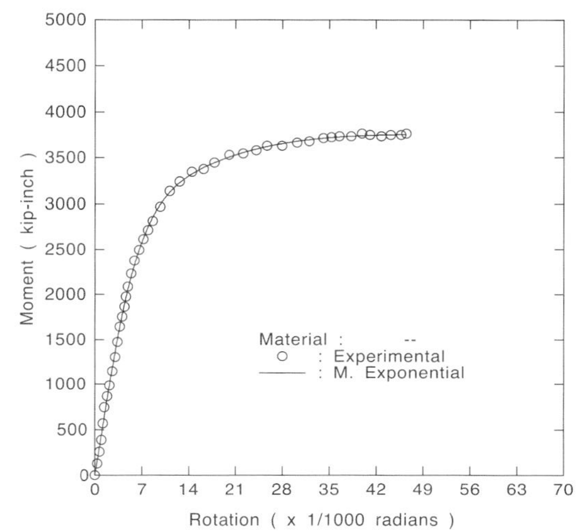

```
Moment-rotation prediction equations     ( R : X 1/1000 radians )

Modified exponential model  :
     Bm = Sum ( Ai X ( 1 - exp( - R/( 2*i*C ) ) ) )  +      Sum ( RKj X ( R - Rj0 ) )  +  Bm0
      C =   0.47082500E+00          Bmo=   0.00000000E+00         Nexp=  6         Nliner=  0
     AI =  -0.69750624E+02    0.32794216E+04  -0.38094187E+05   0.11097001E+06  -0.12498024E+06   0.52657486E+05
```

No	Rotation (radians) X 1/1000	Moment (kip-inch) Expri.	Moment (kip-inch) M.Expo.	Connection stiffness (kip-inch) X 1000 Expri.	Connection stiffness (kip-inch) X 1000 M.Expo.
1	0.00	0.00	0.00	0.3903E+03	0.4191E+03
3	0.67	260.20	295.38	0.3903E+03	0.4604E+03
5	1.20	568.48	544.17	0.8909E+03	0.4697E+03
7	1.80	865.45	824.77	0.2970E+03	0.4631E+03
9	2.60	1145.44	1185.38	0.4030E+03	0.4357E+03
11	3.35	1476.36	1498.76	0.4849E+03	0.3987E+03
13	4.05	1756.36	1764.30	0.3151E+03	0.3595E+03
15	4.65	1976.97	1969.63	0.4412E+03	0.3249E+03
17	5.40	2231.51	2197.23	0.2885E+03	0.2824E+03
19	6.55	2494.58	2486.96	0.1828E+03	0.2229E+03
21	7.90	2715.22	2747.36	0.1454E+03	0.1652E+03
23	9.70	2969.80	2991.26	0.1311E+03	0.1096E+03
25	12.60	3241.28	3227.54	0.6274E+02	0.5993E+02
27	16.20	3377.03	3391.70	0.3136E+02	0.3546E+02
29	20.00	3529.79	3504.85	0.2294E+02	0.2529E+02
31	24.10	3580.64	3593.91	0.2522E+02	0.1843E+02
33	27.90	3631.49	3653.57	0.7236E+01	0.1312E+02
35	32.00	3682.33	3697.51	0.1251E+02	0.8525E+01
37	35.30	3724.90	3720.87	0.7090E+01	0.5767E+01
39	38.20	3733.40	3734.88	0.1089E+02	0.3984E+01
41	41.00	3750.20	3744.18	-0.1225E+02	0.2733E+01
43	44.20	3750.20	3751.24	0.6207E+01	0.1742E+01
45	46.50	3767.23	3754.65	0.2128E+02	0.1247E+01

V - 38

```
Connection type : Extended end-plate connections
           Mode : All bolted with column stiffener

Tested by  :  J.R.Bailey (1970)                        England
Test  Id.  :  B7-L

Column : 8 x 8 UC 40                Fasteners: 10.9- -1"D
Beam   : 12 x 6 1/2 UB 36                 1 1/4" Oversize holes
Plate thickness : 1 1/4"            Material :  --
Stiffener thickness : 0.3125"           Fy =   --   ksi
                                        Fu =   --   ksi

                         Major parameters

     End plate extended on tension side only.
     ct =   --    "   pt =  4.5000"   li =   --    "   pc =   --    "
     cc =   --    "   pit=   --   "   pi =   --    "   pic=   --    "
     gt =   --    "   gi =   --   "   gc =   --    "
     tp =  1.1600"    nt = 2 X  2     ni = 2 X  1      nc = 2 X  1

Remark  1)  End plate extended on tension side only.
        2)  Material grade 43A ( BS15 )
```

No	Moment (k-in)	Rotation (radians) X 1/1000	No	Moment (k-in)	Rotation (radians) X 1/1000
1	0.00	0.00	26	2304.00	15.20
2	112.00	0.50	27	2320.00	16.80
3	224.00	1.00	28	2368.00	18.90
4	325.34	1.37	29	2384.00	21.10
5	426.67	1.73	30	2416.00	23.50
6	528.00	2.10	31	2448.00	25.70
7	640.00	2.55	32	2456.00	26.95
8	752.00	3.00	33	2464.00	28.20
9	848.00	3.35	34	2480.00	30.50
10	944.00	3.70	35	2512.00	32.90
11	1048.00	3.90	36	2512.00	34.80
12	1152.00	4.10	37	2504.00	36.25
13	1248.00	4.35	38	2496.00	37.70
14	1344.00	4.60	39	2528.00	39.70
15	1432.00	5.10	40	2528.00	41.50
16	1520.00	5.60	41	2560.00	43.40
17	1608.00	6.05	42	2544.00	45.40
18	1696.00	6.50	43	2544.00	46.90
19	1784.00	7.20	44	2560.00	48.30
20	1872.00	7.90			
21	1936.00	8.65			
22	2000.00	9.40			
23	2080.00	10.40			
24	2160.00	11.40			
25	2256.00	13.50			

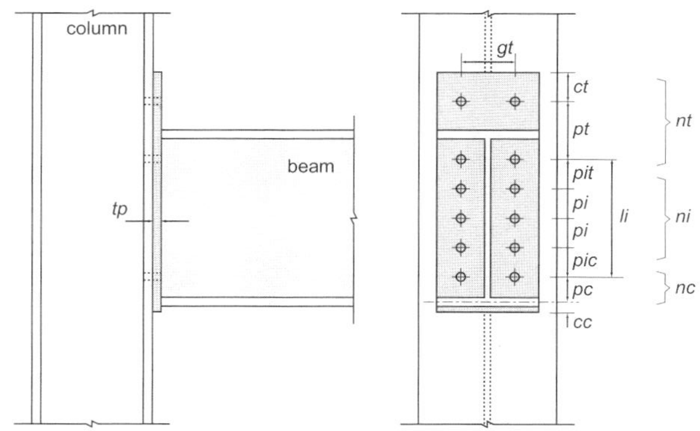

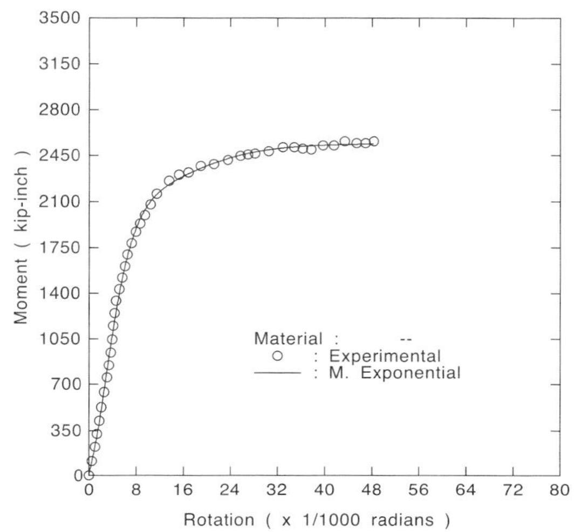

Moment-rotation prediction equations (R : X 1/1000 radians)

```
Modified exponential model  :
   Bm = Sum ( Ai X ( 1 - exp( - R/( 2*i*C ) ) ) )  +    Sum ( RKj X ( R - Rj0 ) )  + Bm0
    C =  0.52750000E+00           Bmo=  0.00000000E+00        Nexp= 6       Nliner= 0
   AI =  0.15839181E+04 -0.16906598E+05  0.42449467E+05 -0.38081779E+05  0.73381420E+04  0.61577559E+04
```

No	Rotation (radians) X 1/1000	Moment (kip-inch) Expri.	Moment (kip-inch) M.Expo.	Connection stiffness (kip-inch) X 1000 Expri.	Connection stiffness (kip-inch) X 1000 M.Expo.
1	0.00	0.00	0.00	0.2240E+03	0.2407E+03
3	1.00	224.00	221.00	0.2542E+03	0.2337E+03
5	1.73	426.67	409.53	0.2764E+03	0.2798E+03
7	2.55	640.00	652.52	0.2489E+03	0.3100E+03
9	3.35	848.00	902.02	0.2743E+03	0.3092E+03
11	3.90	1048.00	1068.59	0.5200E+03	0.2950E+03
13	4.35	1248.00	1197.64	0.3840E+03	0.2779E+03
15	5.10	1432.00	1393.42	0.1760E+03	0.2433E+03
17	6.05	1608.00	1602.31	0.1955E+03	0.1966E+03
19	7.20	1784.00	1797.93	0.1257E+03	0.1452E+03
21	8.65	1936.00	1970.30	0.8534E+02	0.9562E+02
23	10.40	2080.00	2101.65	0.8000E+02	0.5809E+02
25	13.50	2256.00	2229.06	0.3605E+02	0.2978E+02
27	16.80	2320.00	2310.45	0.1556E+02	0.2120E+02
29	21.10	2384.00	2390.17	0.1017E+02	0.1608E+02
31	25.70	2448.00	2452.22	0.9351E+01	0.1094E+02
33	28.20	2464.00	2476.36	0.6596E+01	0.8430E+01
35	32.90	2512.00	2506.87	0.5890E+01	0.4798E+01
37	36.25	2504.00	2519.87	-0.5516E+01	0.3075E+01
39	39.70	2528.00	2528.29	0.7580E+01	0.1894E+01
41	43.40	2560.00	2533.72	0.4738E+01	0.1102E+01
43	46.90	2544.00	2536.72	0.5910E+01	0.6511E+00

Connection type : Extended end-plate connections
Mode : All bolted with column stiffener

Tested by : J.R.Bailey (1970) England
Test Id. : B7-R

Column : 8 x 8 UC 40
Beam : 12 x 6 1/2 UB 36
Plate thickness : 1 1/4"
Stiffener thickness : 0.3125"

Fasteners: 10.9- -1"D
1 1/4" Oversize holes
Material : --
Fy = -- ksi
Fu = -- ksi

Major parameters

End plate extended on tension side only.

ct = -- "	pt = 4.5000"	li = -- "	pc = -- "
cc = -- "	pit= -- "	pi = -- "	pic= -- "
gt = -- "	gi = -- "	gc = -- "	
tp = 1.1600"	nt = 2 X 2	ni = 2 X 1	nc = 2 X 1

Remark 1) End plate extended on tension side only.
2) Material grade 43A (BS15)

No	Moment (k-in)	Rotation (radians) X 1/1000	No	Moment (k-in)	Rotation (radians) X 1/1000
1	0.00	0.00	26	2392.55	14.30
2	90.99	0.33	27	2432.69	15.30
3	181.98	0.67	28	2472.83	16.30
4	272.98	1.00	29	2504.95	17.45
5	393.41	1.40	30	2537.06	18.60
6	513.84	1.80	31	2545.09	19.85
7	626.24	2.10	32	2553.12	21.10
8	738.64	2.40	33	2577.20	22.25
9	834.98	2.60	34	2601.29	23.40
10	931.33	2.80	35	2617.34	25.00
11	1027.68	3.13	36	2617.34	26.10
12	1124.02	3.47	37	2617.34	27.20
13	1220.36	3.80	38	2617.34	28.15
14	1316.71	4.17	39	2617.34	29.10
15	1413.05	4.53	40	2625.37	30.40
16	1509.40	4.90	41	2633.40	31.70
17	1605.74	5.45	42	2649.47	33.40
18	1702.08	6.00	43	2649.47	34.60
19	1814.48	6.70	44	2649.47	35.80
20	1926.88	7.40	45	2649.47	37.10
21	2055.34	8.70			
22	2183.80	10.00			
23	2256.05	11.10			
24	2328.31	12.20			
25	2360.43	13.25			

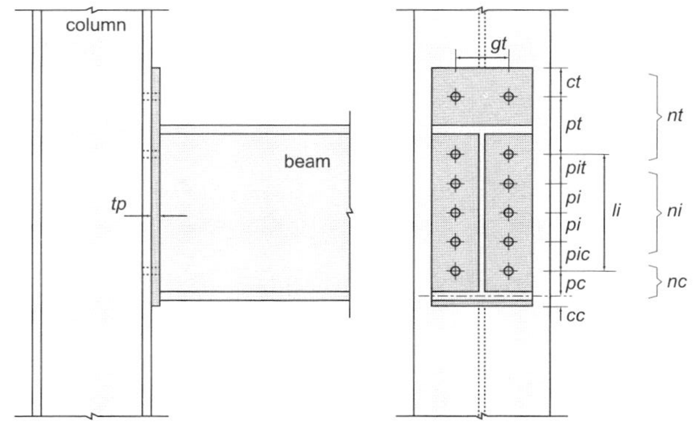

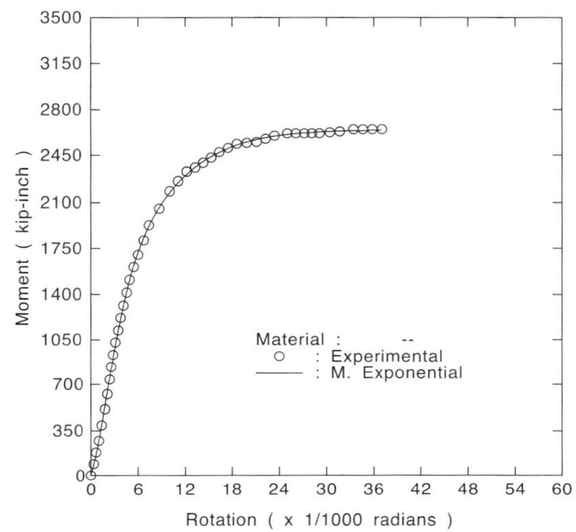

Moment-rotation prediction equations (R : X 1/1000 radians)

Modified exponential model :

Bm = Sum (Ai X (1 - exp(- R/(2*i*C)))) + Sum (RKj X (R - Rj0)) + Bm0

C = 0.39249167E+00 Bmo= 0.00000000E+00 Nexp= 6 Nliner= 0

AI = 0.68318500E+03 -0.83032592E+04 0.20316865E+05 -0.15759793E+05 -0.40177899E+04 0.97263586E+04

No	Rotation (radians) X 1/1000	Moment (kip-inch) Expri.	Moment (kip-inch) M.Expo.	Connection stiffness (kip-inch) X 1000 Expri.	Connection stiffness (kip-inch) X 1000 M.Expo.
1	0.00	0.00	0.00	0.2730E+03	0.2311E+03
3	0.67	181.98	168.58	0.2730E+03	0.2833E+03
5	1.40	393.41	400.45	0.3011E+03	0.3430E+03
7	2.10	626.24	648.73	0.3747E+03	0.3598E+03
9	2.60	834.98	826.94	0.4817E+03	0.3507E+03
11	3.13	1027.68	1008.47	0.2890E+03	0.3285E+03
13	3.80	1220.36	1215.48	0.2765E+03	0.2915E+03
15	4.53	1413.05	1413.27	0.2628E+03	0.2480E+03
17	5.45	1605.74	1617.26	0.1752E+03	0.1983E+03
19	6.70	1814.48	1830.31	0.1606E+03	0.1456E+03
21	8.70	2055.34	2063.92	0.9882E+02	0.9358E+02
23	11.10	2256.05	2245.99	0.6569E+02	0.6168E+02
25	13.25	2360.43	2359.55	0.3059E+02	0.4507E+02
27	15.30	2432.69	2439.79	0.4014E+02	0.3373E+02
29	17.45	2504.95	2502.04	0.2793E+02	0.2458E+02
31	19.85	2545.09	2551.28	0.6422E+01	0.1686E+02
33	22.25	2577.20	2584.67	0.2094E+02	0.1128E+02
35	25.00	2617.34	2609.28	0.4088E+01	0.6935E+01
37	27.20	2617.34	2621.84	-0.1453E-12	0.4624E+01
39	29.10	2617.34	2629.22	0.2607E+01	0.3228E+01
41	31.70	2633.40	2635.82	0.7595E+01	0.1953E+01
43	34.60	2649.47	2640.14	0.1643E-13	0.1101E+01
45	37.10	2649.47	2642.31	0.0000E+00	0.6672E+00

A.5 – 79

V - 40

Connection type : Extended end-plate connections
Mode : All bolted with column stiffener

Tested by : J.R.Bailey (1970) England
Test Id. : B8-L

Column : 8 x 8 UC 40
Beam : 8 x 5 1/4 UB 20
Plate thickness : 3/4"
Stiffener thickness : 0.3125"

Fasteners: 10.9- -3/4"D
15/16" Oversize holes
Material : --
Fy = -- ksi
Fu = -- ksi

Major parameters

End plate extended on tension side only.

ct = -- "	pt = 4.5000"	li = -- "	pc = -- "
cc = -- "	pit= -- "	pi = -- "	pic= -- "
gt = -- "	gi = -- "	gc = -- "	
tp = 0.9200"	nt = 2 X 2	ni = 2 X 1	nc = 2 X 1

Remark 1) End plate extended on tension side only.
2) Material grade 43A (BS15)

No	Moment (k-in)	Rotation (radians) X 1/1000	No	Moment (k-in)	Rotation (radians) X 1/1000
1	0.00	0.00	26	933.34	7.90
2	39.60	0.17	27	946.06	8.60
3	79.19	0.33	28	958.79	9.30
4	118.79	0.50	29	967.27	10.10
5	155.56	0.60	30	975.76	10.90
6	192.32	0.70	31	980.00	11.55
7	229.09	0.80	32	984.24	12.20
8	280.00	0.95	33	989.90	12.90
9	330.91	1.10	34	995.55	13.60
10	373.33	1.30	35	1001.21	14.30
11	415.75	1.50	36	1005.45	15.00
12	458.18	1.57	37	1009.69	15.70
13	500.60	1.65	38	1015.35	16.37
14	543.03	1.73	39	1021.01	17.03
15	585.46	1.80	40	1026.67	17.70
16	632.12	2.05			
17	678.79	2.30			
18	729.70	3.10			
19	780.61	3.90			
20	806.06	4.40			
21	831.52	4.90			
22	856.97	5.40			
23	882.42	5.90			
24	890.90	6.50			
25	899.39	7.10			

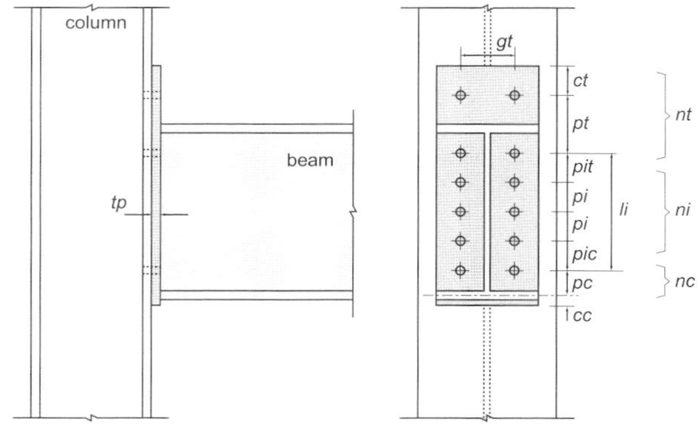

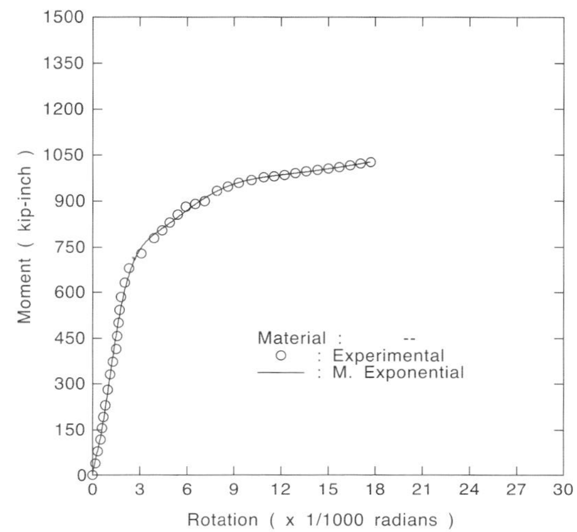

Moment-rotation prediction equations (R : X 1/1000 radians)

```
Modified exponential model  :
     Bm = Sum ( Ai X ( 1 - exp( - R/( 2*i*C ) ) ) )  +      Sum ( RKj X ( R - Rj0 ) )  +  Bm0
      C =   0.18917500E+00             Bmo=   0.00000000E+00          Nexp=  6         Nliner=  1
     AI =   0.15249145E+04  -0.19584925E+05   0.76041993E+05  -0.13411390E+06   0.11277635E+06  -0.35800988E+05
    Rj0 =           2.3000
    RKj =   0.11520570E+02
```

No	Rotation (radians) X 1/1000	Moment (kip-inch) Expri.	Moment (kip-inch) M.Expo.	Connection stiffness (kip-inch) X 1000 Expri.	Connection stiffness (kip-inch) X 1000 M.Expo.
1	0.00	0.00	0.00	0.2375E+03	0.3694E+03
3	0.33	79.19	84.05	0.2376E+03	0.2230E+03
5	0.60	155.56	152.11	0.3677E+03	0.2926E+03
7	0.80	229.09	215.72	0.3564E+03	0.3400E+03
9	1.10	330.91	322.82	0.2848E+03	0.3642E+03
11	1.50	415.75	462.20	0.4692E+03	0.3221E+03
13	1.65	500.60	508.45	0.5657E+03	0.2940E+03
15	1.80	585.46	550.27	0.4782E+03	0.2633E+03
17	2.30	678.79	656.26	0.1574E+03	0.1634E+03
19	3.90	780.61	792.56	0.5580E+02	0.4019E+02
21	4.90	831.52	829.59	0.5090E+02	0.3685E+02
23	5.90	882.42	867.30	0.3419E+02	0.3779E+02
25	7.10	899.39	909.62	0.2627E+02	0.3156E+02
27	8.60	946.06	947.75	0.1818E+02	0.1934E+02
29	10.10	967.27	969.47	0.1060E+02	0.1051E+02
31	11.55	980.00	981.50	0.6526E+01	0.6764E+01
33	12.90	989.90	989.99	0.8080E+01	0.6137E+01
35	14.30	1001.21	999.00	0.7070E+01	0.6864E+01
37	15.70	1009.69	1009.39	0.7304E+01	0.8005E+01
39	17.03	1021.01	1020.77	0.8490E+01	0.9028E+01

A.5 – 81

V - 41

```
Connection type : Extended end-plate connections
           Mode : All bolted with column stiffener

Tested by   :  J.R.Bailey (1970)                          England
Test  Id.   :  B8-R

Column : 8 x 8 UC 40                 Fasteners: 10.9- -3/4"D
Beam   : 8 x 5 1/4 UB 20                   15/16" Oversize holes
Plate thickness : 3/4"               Material :  --
Stiffener thickness : 0.3125"             Fy =   --   ksi
                                          Fu =   --   ksi

                         Major parameters

     End plate extended on tension side only.
     ct =   --   "    pt =  4.5000"   li =   --   "    pc =   --   "
     cc =   --   "    pit=   --   "   pi =   --   "    pic=   --   "
     gt =   --   "    gi =   --   "   gc =   --   "
     tp =  0.9200"    nt = 2 X  2     ni = 2 X  1      nc = 2 X  1

Remark  1)  End plate extended on tension side only.
        2)  Material grade 43A ( BS15 )
```

No	Moment (k-in)	Rotation (radians) X 1/1000	No	Moment (k-in)	Rotation (radians) X 1/1000
1	0.00	0.00	26	834.18	12.10
2	31.76	0.45	27	851.12	12.90
3	63.52	0.90	28	868.06	13.70
4	97.39	1.23	29	884.99	14.75
5	131.27	1.57	30	901.93	15.80
6	165.14	1.90	31	910.40	16.65
7	199.02	2.13	32	918.86	17.50
8	232.90	2.37	33	931.57	18.40
9	266.77	2.60	34	944.27	19.30
10	306.29	2.87	35	956.98	20.15
11	345.81	3.13	36	969.68	21.00
12	385.33	3.40	37	969.68	22.00
13	419.21	3.85	38	969.68	23.00
14	453.08	4.30	39	978.15	23.75
15	499.66	4.80	40	986.62	24.50
16	546.24	5.30	41	990.85	25.40
17	592.82	5.75	42	995.09	26.30
18	639.40	6.20	43	999.32	27.25
19	664.80	6.80	44	1003.55	28.20
20	690.20	7.40			
21	719.84	8.15			
22	749.48	8.90			
23	774.89	9.45			
24	800.30	10.00			
25	817.24	11.05			

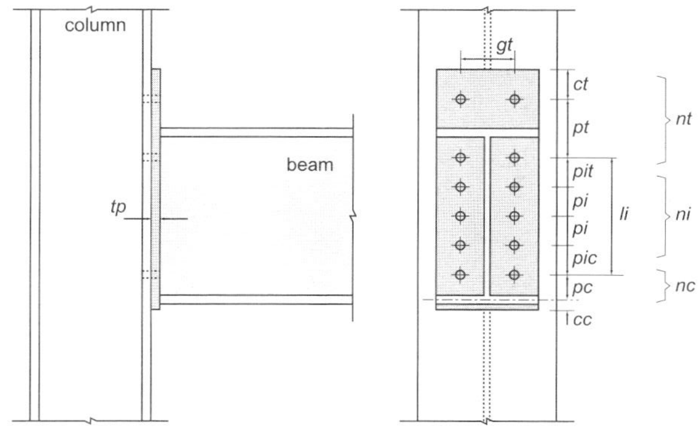

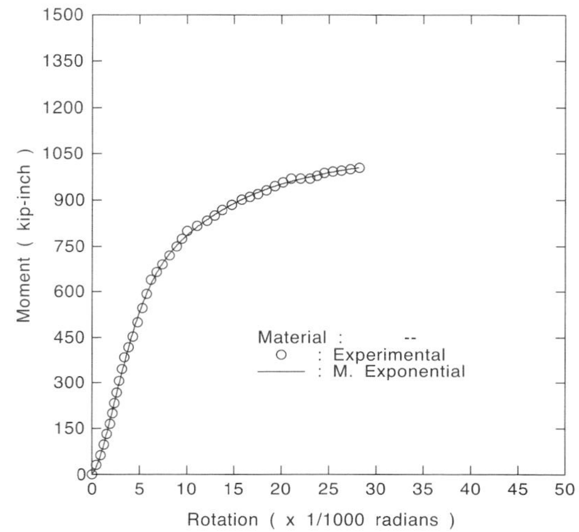

```
Moment-rotation prediction equations     ( R : X 1/1000 radians )

Modified exponential model  :
    Bm = Sum ( Ai X ( 1 - exp( - R/( 2*i*C ) ) ) )  +      Sum ( RKj X ( R - Rj0 ) )  +  Bm0
     C =   0.34750000E+00              Bmo=   0.00000000E+00          Nexp=  6        Nliner=  1
    AI =   0.30493559E+02   0.15157035E+04  -0.17283016E+05   0.49291274E+05  -0.56898018E+05   0.24352052E+05
   Rj0 =          23.0000
   RKj =   0.16281411E+01
```

No	Rotation (radians) X 1/1000	Moment (kip-inch) Expri.	Moment (kip-inch) M.Expo.	Connection stiffness (kip-inch) X 1000 Expri.	Connection stiffness (kip-inch) X 1000 M.Expo.
1	0.00	0.00	0.00	0.7057E+02	0.4206E+02
3	0.90	63.52	62.71	0.8842E+02	0.9505E+02
5	1.57	131.27	135.02	0.1016E+03	0.1192E+03
7	2.13	199.02	205.48	0.1452E+03	0.1278E+03
9	2.60	266.77	265.46	0.1466E+03	0.1283E+03
11	3.13	345.81	332.86	0.1482E+03	0.1236E+03
13	3.85	419.21	417.52	0.7528E+02	0.1119E+03
15	4.80	499.66	514.72	0.9316E+02	0.9250E+02
17	5.75	592.82	593.46	0.1035E+03	0.7363E+02
19	6.80	664.80	661.28	0.4234E+02	0.5627E+02
21	8.15	719.84	725.70	0.3952E+02	0.4032E+02
23	9.45	774.89	771.31	0.4620E+02	0.3063E+02
25	11.05	817.24	814.20	0.1613E+02	0.2367E+02
27	12.90	851.12	853.43	0.2117E+02	0.1915E+02
29	14.75	884.99	885.91	0.1613E+02	0.1610E+02
31	16.65	910.40	913.93	0.9960E+01	0.1343E+02
33	18.40	931.57	935.39	0.1411E+02	0.1112E+02
35	20.15	956.98	952.95	0.1495E+02	0.8986E+01
37	22.00	969.68	967.68	0.0000E+00	0.6981E+01
39	23.75	978.15	979.67	0.1129E+02	0.7001E+01
41	25.40	990.85	990.15	0.4707E+01	0.5751E+01
43	27.25	999.32	999.72	0.4453E+01	0.4637E+01

V - 42

Connection type : Extended end-plate connections
Mode : All bolted with column stiffener

Tested by : J.R.Bailey (1970) England
Test Id. : C9-L

Column : 8 x 8 UC 40
Beam : 8 x 5 1/4 UB 20
Plate thickness : 3/4"
Stiffener thickness : 0.3125"

Fasteners: 10.9- -5/8"D
13/16" Oversize holes
Material : --
Fy = -- ksi
Fu = -- ksi

Major parameters

End plate extended on tension side only.

ct = -- "	pt = 4.5000"	li = -- "	pc = -- "
cc = -- "	pit= -- "	pi = -- "	pic= -- "
gt = -- "	gi = -- "	gc = -- "	
tp = 0.8750"	nt = 2 X 2	ni = 2 X 1	nc = 2 X 1

Remark 1) End plate extended on tension side only.
2) Material grade 43A (BS15)

No	Moment (k-in)	Rotation (radians) X 1/1000	No	Moment (k-in)	Rotation (radians) X 1/1000
1	0.00	0.00	26	608.79	7.10
2	33.94	0.35	27	623.64	7.60
3	67.88	0.70	28	634.24	7.90
4	101.82	1.00	29	644.84	8.20
5	135.76	1.30	30	655.45	8.70
6	165.46	1.60	31	666.06	9.20
7	195.16	1.90	32	676.67	9.75
8	216.37	2.05	33	687.28	10.30
9	237.58	2.20	34	695.76	10.77
10	260.91	2.30	35	704.24	11.23
11	284.24	2.40	36	712.73	11.70
12	309.70	2.70			
13	335.15	3.00			
14	358.48	3.15			
15	381.82	3.30			
16	407.27	3.55			
17	432.73	3.80			
18	453.94	4.00			
19	475.15	4.20			
20	492.12	4.50			
21	509.09	4.80			
22	534.55	5.20			
23	560.00	5.60			
24	576.97	6.10			
25	593.94	6.60			

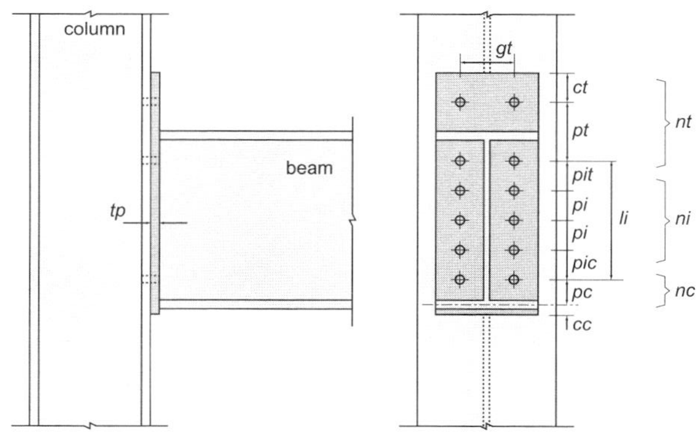

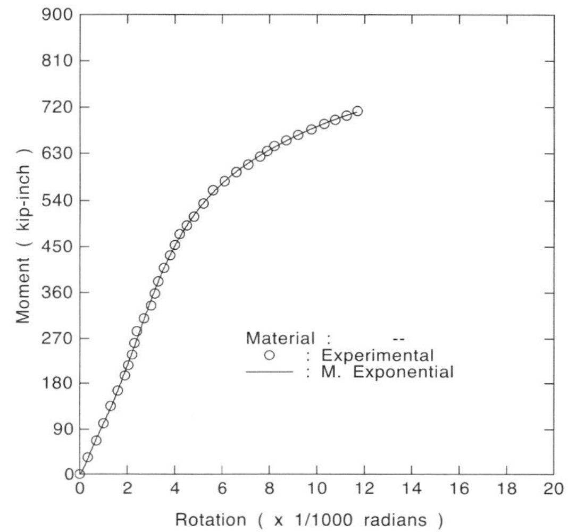

Moment-rotation prediction equations (R : X 1/1000 radians)

```
Modified exponential model  :
    Bm = Sum ( Ai X ( 1 - exp( - R/( 2*i*C ) ) ) )  +      Sum ( RKj X ( R - Rj0 ) )  +  Bm0
     C =   0.18500000E+00           Bmo=   0.00000000E+00          Nexp=  6       Nliner=  0
    AI =  -0.52664490E+03   0.83875028E+04  -0.40782372E+05   0.85816353E+05  -0.84701843E+05   0.32564163E+05
```

No	Rotation (radians) X 1/1000	Moment (kip-inch) Expri.	Moment (kip-inch) M.Expo.	Connection stiffness (kip-inch) X 1000 Expri.	Connection stiffness (kip-inch) X 1000 M.Expo.
1	0.00	0.00	0.00	0.9698E+02	0.3801E+02
3	0.70	67.88	70.93	0.1057E+03	0.1040E+03
5	1.30	135.76	131.99	0.1061E+03	0.1064E+03
7	1.90	195.16	202.28	0.1273E+03	0.1271E+03
9	2.20	237.58	241.32	0.1966E+03	0.1324E+03
11	2.40	284.24	267.91	0.1962E+03	0.1331E+03
13	3.00	335.15	345.57	0.1320E+03	0.1232E+03
15	3.30	381.82	381.18	0.1354E+03	0.1138E+03
17	3.80	432.73	433.64	0.1042E+03	0.9579E+02
19	4.20	475.15	469.07	0.8625E+02	0.8154E+02
21	4.80	509.09	512.27	0.5960E+02	0.6322E+02
23	5.60	560.00	555.46	0.5044E+02	0.4610E+02
25	6.60	593.94	594.94	0.3182E+02	0.3424E+02
27	7.60	623.64	625.86	0.3322E+02	0.2821E+02
29	8.20	644.84	642.03	0.3004E+02	0.2576E+02
31	9.20	666.06	666.00	0.2030E+02	0.2224E+02
33	10.30	687.28	688.40	0.1869E+02	0.1847E+02
35	11.23	704.24	704.15	0.1818E+02	0.1531E+02

V - 43

```
Connection type : Extended end-plate connections
           Mode : All bolted with column stiffener

Tested by   :  J.R.Bailey (1970)                         England
Test  Id.   :  C9-R

Column : 8 x 8 UC 40                  Fasteners: 10.9- -5/8"D
Beam   : 8 x 5 1/4 UB 20                    13/16" Oversize holes
Plate thickness : 3/4"                Material :  --
Stiffener thickness : 0.3125"              Fy =   --   ksi
                                           Fu =   --   ksi

                         Major parameters

     End plate extended on tension side only.
     ct =   --   "   pt =  4.5000"   li =   --   "   pc =   --   "
     cc =   --   "   pit=   --   "   pi =   --   "   pic=   --   "
     gt =   --   "   gi =   --   "   gi =   --   "
     tp =  0.8750"   nt = 2 X  2     ni = 2 X  1     nc = 2 X  1

Remark  1)  End plate extended on tension side only.
        2)  Material grade 43A ( BS15 )
```

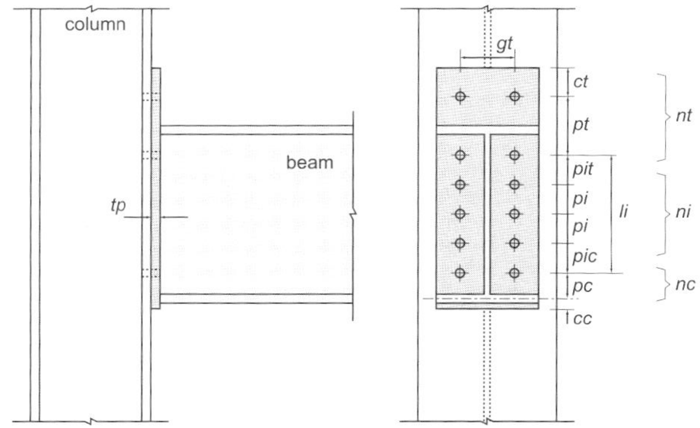

No	Moment (k-in)	Rotation (radians) X 1/1000	No	Moment (k-in)	Rotation (radians) X 1/1000
1	0.00	0.00	26	627.88	9.80
2	25.45	0.30	27	638.48	10.40
3	50.90	0.60	28	649.09	11.00
4	78.48	0.85	29	661.82	11.60
5	106.06	1.10	30	674.54	12.20
6	135.76	1.40	31	683.03	13.05
7	165.46	1.70	32	691.51	13.90
8	190.91	1.95	33	700.00	14.70
9	216.36	2.20	34	708.48	15.50
10	239.69	2.45	35	710.60	16.30
11	263.03	2.70	36	712.73	17.10
12	284.24	2.90	37	716.97	17.95
13	305.46	3.10	38	721.21	18.80
14	335.15	3.40	39	723.34	19.50
15	364.85	3.70	40	725.46	20.20
16	392.42	3.95	41	725.46	21.20
17	420.00	4.20			
18	447.58	4.50			
19	475.15	4.80			
20	500.60	5.30			
21	526.06	5.80			
22	549.39	6.55			
23	572.72	7.30			
24	606.67	8.30			
25	617.27	9.05			

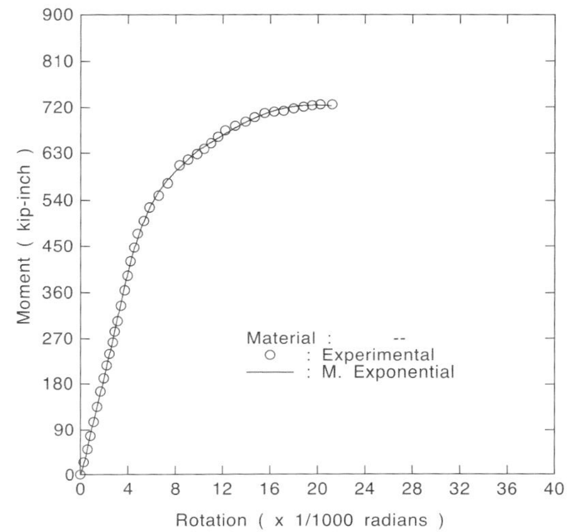

```
Moment-rotation prediction equations     ( R : X 1/1000 radians )

Modified exponential model  :
     Bm = Sum ( Ai X ( 1 - exp( - R/( 2*i*C ) ) ) )  +      Sum ( RKj X ( R - Rj0 ) )  +  Bm0
      C =   0.25166667E+00             Bmo=   0.00000000E+00          Nexp=  6        Nliner=  1
     AI =  -0.76422444E+03   0.12654987E+05  -0.65806308E+05   0.14518474E+06  -0.14520100E+06   0.54760414E+05
     Rj0 =           9.8000
     RKj =  -0.74660940E+01
```

No	Rotation (radians) X 1/1000	Moment (kip-inch) Expri.	Moment (kip-inch) M.Expo.	Connection stiffness (kip-inch) X 1000 Expri.	Connection stiffness (kip-inch) X 1000 M.Expo.
1	0.00	0.00	0.00	0.8484E+02	0.2097E+02
3	0.60	50.90	54.04	0.9873E+02	0.1114E+03
5	1.10	106.06	106.44	0.1052E+03	0.9788E+02
7	1.70	165.46	163.42	0.1005E+03	0.9532E+02
9	2.20	216.36	212.72	0.9757E+02	0.1022E+03
11	2.70	263.03	265.31	0.1004E+03	0.1073E+03
13	3.10	305.46	308.32	0.1032E+03	0.1070E+03
15	3.70	364.85	370.55	0.1052E+03	0.9909E+02
17	4.20	420.00	417.38	0.1019E+03	0.8774E+02
19	4.80	475.15	465.31	0.7654E+02	0.7190E+02
21	5.80	526.06	524.63	0.4299E+02	0.4778E+02
23	7.30	572.72	578.22	0.3233E+02	0.2688E+02
25	9.05	617.27	618.17	0.1414E+02	0.2092E+02
27	10.40	638.48	642.05	0.1768E+02	0.1383E+02
29	11.60	661.82	658.93	0.2121E+02	0.1416E+02
31	13.05	683.03	678.97	0.9981E+01	0.1320E+02
33	14.70	700.00	698.66	0.1061E+02	0.1044E+02
35	16.30	710.60	712.62	0.2655E+01	0.6956E+01
37	17.95	716.97	721.10	0.4991E+01	0.3382E+01
39	19.50	723.34	724.02	0.3034E+01	0.4779E+00
41	21.20	725.46	722.60	0.0000E+00	-0.2032E+01

A.5 – 87

V - 44

Connection type : Extended end-plate connections
Mode : All bolted with column stiffener

Tested by : J.R.Bailey (1970) England
Test Id. : C10-L

Column : 8 x 8 UC 40
Beam : 12 x 5 UB 32
Plate thickness : 13/16"
Stiffener thickness : 0.3125"

Fasteners: 10.9- -3/4"D
15/16" Oversize holes
Material : --
Fy = -- ksi
Fu = -- ksi

Major parameters

End plate extended on tension side only.

ct = -- "	pt = 4.5000"	li = -- "	pc = -- "
cc = -- "	pit= -- "	pi = -- "	pic= -- "
gt = -- "	gi = -- "	gc = -- "	
tp = 0.9600"	nt = 2 X 2	ni = 2 X 1	nc = 2 X 1

Remark 1) End plate extended on tension side only.
2) Material grade 43A (BS15)

No	Moment (k-in)	Rotation (radians) X 1/1000	No	Moment (k-in)	Rotation (radians) X 1/1000
1	0.00	0.00	26	1564.58	7.10
2	94.04	0.20	27	1607.33	7.65
3	138.09	0.40	28	1650.07	8.20
4	255.04	0.55	29	1672.87	8.73
5	341.99	0.70	30	1695.67	9.27
6	401.83	0.80	31	1718.47	9.80
7	451.68	0.90	32	1752.67	10.50
8	525.80	1.10	33	1786.87	11.20
9	539.92	1.30	34	1809.67	11.77
10	658.32	1.35	35	1832.47	12.33
11	726.72	1.40	36	1855.27	12.90
12	799.39	1.60	37	1876.64	13.60
13	872.06	1.80	38	1898.02	14.30
14	931.91	2.05	39	1919.39	15.05
15	991.75	2.30	40	1940.76	15.80
16	1047.32	2.40			
17	1102.90	2.50			
18	1171.30	3.00			
19	1239.70	3.50			
20	1291.00	4.00			
21	1342.30	4.50			
22	1419.24	5.20			
23	1457.71	5.60			
24	1496.18	6.00			
25	1530.38	6.55			

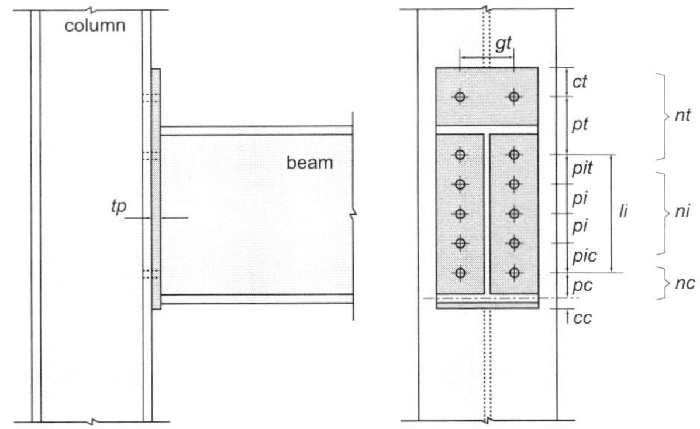

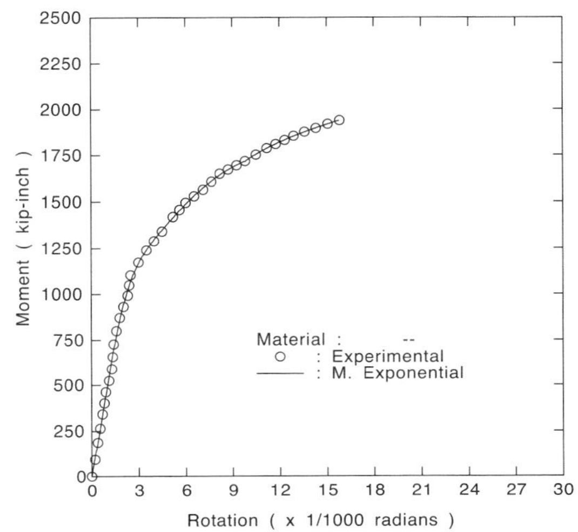

```
Moment-rotation prediction equations     ( R : X 1/1000 radians )

Modified exponential model  :
    Bm = Sum ( Ai X ( 1 - exp( - R/( 2*i*C ) ) ) )  +      Sum ( RKj X ( R - Rj0 ) )  +  Bm0
     C =   0.18166667E+00              Bmo=   0.00000000E+00          Nexp=  6        Nliner=  1
    AI =   0.10249853E+04  -0.11379480E+05   0.31632922E+05  -0.26429941E+05  -0.66498071E+04   0.13612022E+05
   Rj0 =          9.8000
   RKj =   0.22851654E+02
```

No	Rotation (radians) X 1/1000	Moment (kip-inch) Expri.	Moment (kip-inch) M.Expo.	Connection stiffness (kip-inch) X 1000 Expri.	Connection stiffness (kip-inch) X 1000 M.Expo.
1	0.00	0.00	0.00	0.4702E+03	0.5801E+03
3	0.40	188.09	192.32	0.4947E+03	0.4641E+03
5	0.70	341.99	338.42	0.5643E+03	0.5082E+03
7	0.90	461.68	441.54	0.5058E+03	0.5195E+03
9	1.30	589.92	644.19	0.1159E+04	0.4822E+03
11	1.40	726.72	691.51	0.1167E+04	0.4638E+03
13	1.80	872.06	859.84	0.3083E+03	0.3755E+03
15	2.30	991.75	1019.84	0.4653E+03	0.2680E+03
17	2.50	1102.90	1069.80	0.4859E+03	0.2325E+03
19	3.50	1239.70	1240.75	0.1197E+03	0.1278E+03
21	4.50	1342.30	1350.36	0.1057E+03	0.9824E+02
23	5.60	1457.71	1452.49	0.9618E+02	0.8827E+02
25	6.55	1530.38	1532.09	0.6218E+02	0.7871E+02
27	7.65	1607.33	1610.81	0.7772E+02	0.6398E+02
29	8.73	1672.87	1671.73	0.4275E+02	0.4862E+02
31	9.80	1718.47	1716.23	0.4539E+02	0.3521E+02
33	11.20	1786.87	1787.55	0.4409E+02	0.4464E+02
35	12.33	1832.47	1833.62	0.4024E+02	0.3713E+02
37	13.60	1876.64	1876.85	0.3053E+02	0.3153E+02
39	15.05	1919.39	1919.49	0.2850E+02	0.2765E+02

V - 45

Connection type : Extended end-plate connections
Mode : All bolted with column stiffener

Tested by : J.R.Bailey (1970) England
Test Id. : C10-R

Column : 8 x 8 UC 40
Beam : 12 x 5 UB 32
Plate thickness : 13/16"
Stiffener thickness : 0.3125"

Fasteners: 10.9- -3/4"D
15/16" Oversize holes
Material : --
Fy = -- ksi
Fu = -- ksi

Major parameters

End plate extended on tension side only.

ct = -- "	pt = 4.5000"	li = -- "	pc = -- "
cc = -- "	pit= -- "	pi = -- "	pic= -- "
gt = -- "	gi = -- "	gc = -- "	
tp = 0.9600"	nt = 2 X 2	ni = 2 X 1	nc = 2 X 1

Remark 1) End plate extended on tension side only.
2) Material grade 43A (BS15)

No	Moment (k-in)	Rotation (radians) X 1/1000
1	0.00	0.00
2	74.10	0.20
3	148.19	0.40
4	222.29	0.60
5	303.51	0.80
6	384.73	1.00
7	453.13	1.30
8	521.53	1.60
9	615.58	1.70
10	675.42	1.95
11	735.26	2.20
12	790.84	2.35
13	846.41	2.50
14	910.53	2.75
15	974.65	3.00
16	1021.68	3.15
17	1068.71	3.30
18	1120.00	3.55
19	1171.30	3.80
20	1235.42	4.35
21	1299.54	4.90
22	1350.84	5.50
23	1402.14	6.10
24	1453.43	6.80
25	1504.73	7.50

No	Moment (k-in)	Rotation (radians) X 1/1000
26	1551.76	8.35
27	1598.78	9.20
28	1620.16	9.85
29	1641.53	10.50
30	1675.73	11.25
31	1709.93	12.00
32	1739.85	12.95
33	1769.77	13.90
34	1791.14	14.90
35	1812.52	15.90
36	1829.62	16.80
37	1846.72	17.70
38	1863.82	18.70
39	1880.92	19.70
40	1880.92	20.45
41	1880.92	21.20

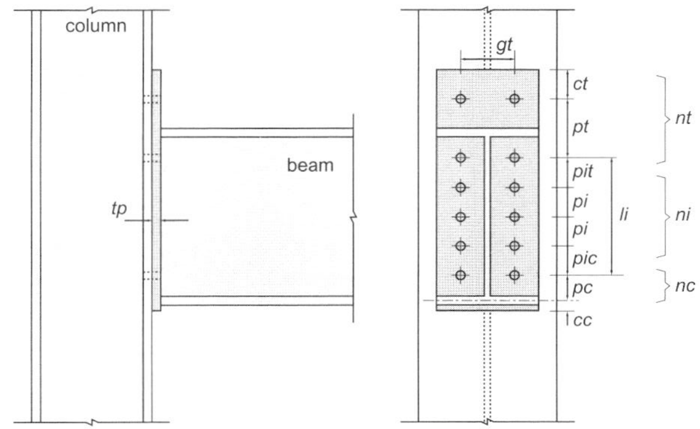

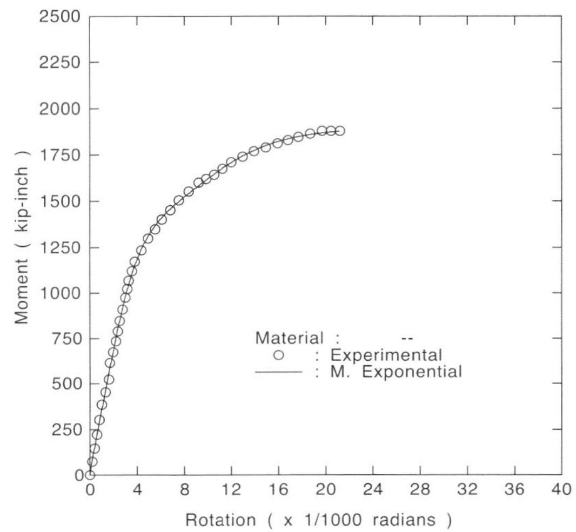

A.5 – 90

```
Moment-rotation prediction equations     ( R : X 1/1000 radians )

Modified exponential model  :
     Bm = Sum ( Ai X ( 1 - exp( - R/( 2*i*C ) ) ) )  +      Sum ( RKj X ( R - Rj0 ) )  +  Bm0
      C =   0.22666667E+00           Bmo=   0.00000000E+00         Nexp=  6       Nliner=  0
     AI =  -0.21216329E+02   0.79056963E+04  -0.61206632E+05   0.16027726E+06  -0.17641579E+06   0.71354482E+05
```

No	Rotation (radians) X 1/1000	Moment (kip-inch) Expri.	Moment (kip-inch) M.Expo.	Connection stiffness (kip-inch) X 1000 Expri.	Connection stiffness (kip-inch) X 1000 M.Expo.
1	0.00	0.00	0.00	0.3705E+03	0.4588E+03
3	0.40	148.19	157.21	0.3705E+03	0.3504E+03
5	0.80	303.51	291.84	0.4061E+03	0.3306E+03
7	1.30	453.13	458.14	0.2280E+03	0.3352E+03
9	1.70	615.58	592.06	0.7401E+03	0.3324E+03
11	2.20	735.26	753.82	0.3213E+03	0.3115E+03
13	2.50	846.41	844.35	0.3277E+03	0.2912E+03
15	3.00	974.65	979.94	0.2921E+03	0.2501E+03
17	3.30	1068.71	1051.05	0.2729E+03	0.2239E+03
19	3.80	1171.30	1152.32	0.1775E+03	0.1818E+03
21	4.90	1299.54	1309.77	0.1017E+03	0.1102E+03
23	6.10	1402.14	1414.13	0.7986E+02	0.6947E+02
25	7.50	1504.73	1497.54	0.6517E+02	0.5331E+02
27	9.20	1598.78	1583.58	0.4261E+02	0.4875E+02
29	10.50	1641.53	1644.97	0.3879E+02	0.4537E+02
31	12.00	1709.93	1708.57	0.3938E+02	0.3904E+02
33	13.90	1769.77	1773.41	0.2656E+02	0.2909E+02
35	15.90	1812.52	1821.50	0.2012E+02	0.1932E+02
37	17.70	1846.72	1849.86	0.1810E+02	0.1254E+02
39	19.70	1880.92	1869.37	0.7329E+01	0.7356E+01
41	21.20	1880.92	1878.36	0.0000E+00	0.4795E+01

V - 46

```
Connection type : Extended end-plate connections
           Mode : All bolted with column stiffener

Tested by   :  J.R.Bailey (1970)                         England
Test  Id.   :  C11-L

Column : 8 x 3 UC 40                 Fasteners: 10.9- -5/8"D
Beam   : 14 x 5 UB 26                      13/16" Oversize holes
Plate thickness : 7/8"               Material :  --
Stiffener thickness : 0.3125"            Fy =   --   ksi
                                         Fu =   --   ksi

                          Major parameters

     End plate extended on tension side only.
     ct =   --   "    pt =  4.2500"   li =   --   "    pc =   --   "
     cc =   --   "    pit=   --   "   pi =   --   "    pic=   --   "
     gt =   --   "    gi =   --   "   gc =   --   "
     tp =  0.8250"    nt = 2 X  2     ni = 2 X  1      nc = 2 X  1

Remark  1)  End plate extended on tension side only.
        2)  Material grade 43A ( BS15 )
```

No	Moment (k-in)	Rotation (radians) X 1/1000	No	Moment (k-in)	Rotation (radians) X 1/1000
1	0.00	0.00	26	1228.39	4.40
2	61.63	0.15	27	1245.39	4.70
3	123.26	0.30	28	1262.39	5.00
4	182.77	0.55	29	1270.89	5.27
5	242.28	0.80	30	1279.39	5.53
6	297.53	0.95	31	1287.89	5.80
7	352.79	1.10	32	1313.39	6.15
8	408.04	1.25	33	1338.90	6.50
9	463.30	1.40	34	1355.90	6.80
10	518.56	1.65	35	1372.91	7.10
11	573.82	1.90	36	1381.40	7.45
12	616.32	2.00	37	1389.90	7.80
13	658.82	2.10			
14	709.83	2.25			
15	760.84	2.40			
16	816.09	2.60			
17	871.34	2.80			
18	909.60	3.00			
19	947.86	3.20			
20	986.11	3.25			
21	1024.37	3.30			
22	1066.87	3.50			
23	1109.38	3.70			
24	1168.88	4.00			
25	1198.64	4.20			

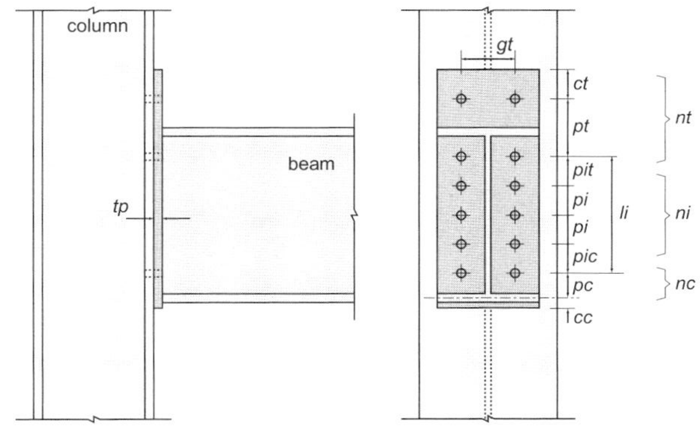

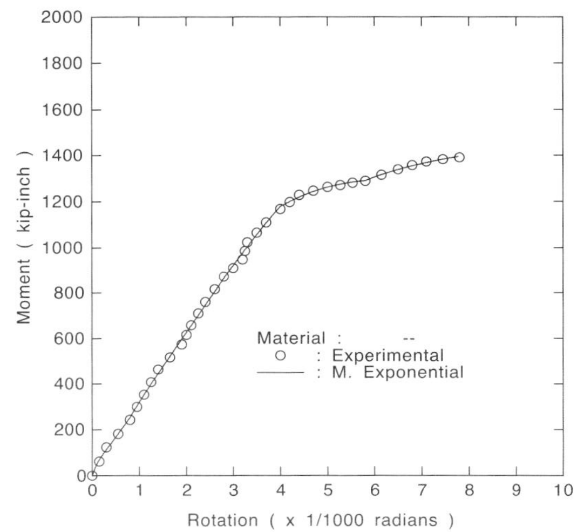

A.5 – 92

Moment-rotation prediction equations (R : X 1/1000 radians)

```
Modified exponential model  :
     Bm = Sum ( Ai X ( 1 - exp( - R/( 2*i*C ) ) ) )  +      Sum ( RKj X ( R - Rj0 ) )  +  Bm0
       C =   0.10250000E+00            Bmo=   0.00000000E+00          Nexp=  6       Nliner=  3
      AI =   0.46143422E+03  -0.43921227E+04   0.13090131E+05  -0.57413758E+04  -0.20897500E+05   0.18698841E+05
     Rj0 =           0.8000           4.0000           5.8000
     RKj =   0.74035942E+02  -0.11633230E+03   0.60741891E+02
```

No	Rotation (radians) X 1/1000	Moment (kip-inch) Expri.	Moment (kip-inch) M.Expo.	Connection stiffness (kip-inch) X 1000 Expri.	Connection stiffness (kip-inch) X 1000 M.Expo.
1	0.00	0.00	0.00	0.4109E+03	0.6360E+03
3	0.30	123.26	114.07	0.3461E+03	0.2766E+03
5	0.80	242.28	251.10	0.3195E+03	0.2716E+03
7	1.10	352.79	351.88	0.3684E+03	0.3252E+03
9	1.40	463.30	446.59	0.3131E+03	0.3078E+03
11	1.90	573.82	597.43	0.3668E+03	0.2989E+03
13	2.10	658.82	657.18	0.3910E+03	0.2987E+03
15	2.40	760.84	746.63	0.3127E+03	0.2973E+03
17	2.80	871.34	864.27	0.2338E+03	0.2896E+03
19	3.20	947.86	977.24	0.6504E+03	0.2741E+03
21	3.30	1024.37	1004.41	0.6546E+03	0.2692E+03
23	3.70	1109.38	1107.76	0.2069E+03	0.2469E+03
25	4.20	1198.64	1200.30	0.1488E+03	0.9986E+02
27	4.70	1245.39	1242.69	0.5666E+02	0.7012E+02
29	5.27	1270.89	1273.85	0.3188E+02	0.4078E+02
31	5.80	1287.89	1289.43	0.4960E+02	0.1851E+02
33	6.50	1338.90	1336.86	0.6415E+02	0.5761E+02
35	7.10	1372.91	1367.34	0.4172E+02	0.4475E+02
37	7.80	1389.90	1394.88	0.2427E+02	0.3467E+02

A.5 – 93

V - 47

Connection type : Extended end-plate connections
Mode : All bolted with column stiffener

Tested by : J.R.Bailey (1970) England
Test Id. : C11-R

Column : 8 x 8 UC 40
Beam : 14 x 5 UB 26
Plate thickness : 7/8"
Stiffener thickness : 0.3125"

Fasteners: 10.9- -5/8"D
13/16" Oversize holes
Material : --
Fy = -- ksi
Fu = -- ksi

Major parameters

End plate extended on tension side only.

ct = -- "	pt = 4.2500"	li = -- "	pc = -- "
cc = -- "	pit= -- "	pi = -- "	pic= -- "
gt = -- "	gi = -- "	gc = -- "	
tp = 0.8250"	nt = 2 X 2	ni = 2 X 1	nc = 2 X 1

Remark 1) End plate extended on tension side only.
2) Material grade 43A (BS15)

No	Moment (k-in)	Rotation (radians) X 1/1000	No	Moment (k-in)	Rotation (radians) X 1/1000
1	0.00	0.00	26	1214.22	5.17
2	63.76	0.27	27	1236.89	5.40
3	127.51	0.53	28	1258.14	5.70
4	191.27	0.80	29	1279.39	6.00
5	245.11	1.00			
6	298.95	1.20			
7	352.79	1.40			
8	412.30	1.65			
9	471.80	1.90			
10	518.56	2.05			
11	565.31	2.20			
12	612.07	2.40			
13	658.82	2.60			
14	701.33	2.80			
15	743.83	3.00			
16	807.59	3.15			
17	871.34	3.30			
18	926.60	3.55			
19	981.86	3.80			
20	1015.87	4.00			
21	1049.87	4.20			
22	1083.88	4.40			
23	1126.38	4.55			
24	1168.88	4.70			
25	1191.55	4.93			

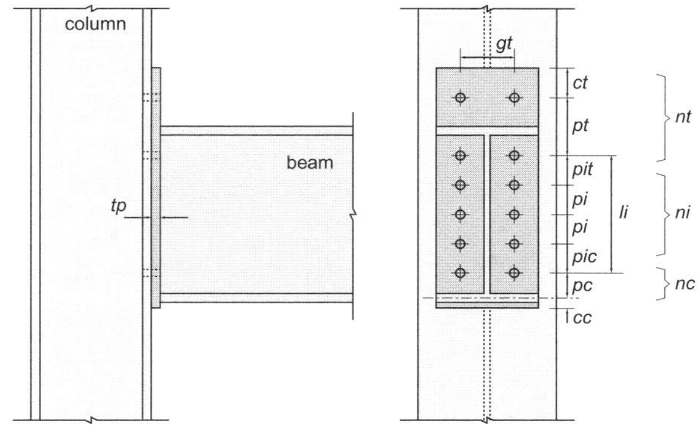

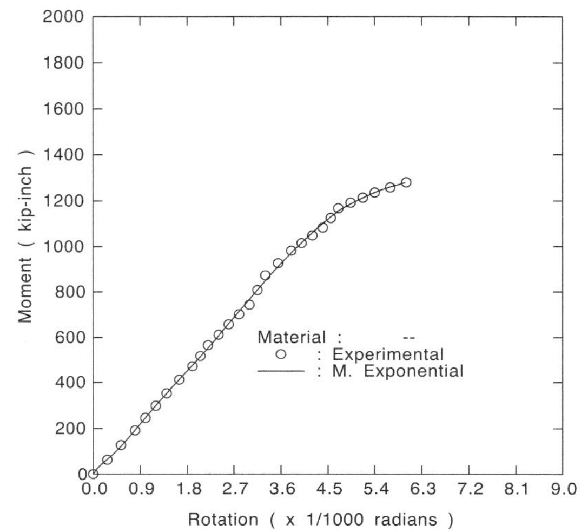

A.5 – 94

```
Moment-rotation prediction equations     ( R : X 1/1000 radians )

Modified exponential model  :
     Bm = Sum ( Ai X ( 1 - exp( - R/( 2*i*C ) ) ) )  +      Sum ( RKj X ( R - Rj0 ) )  +  Bm0
      C =   0.11667500E+00             Bmo=   0.00000000E+00           Nexp=  6        Nliner=  1
     AI =   0.60812984E+03  -0.77200477E+04   0.29620777E+05  -0.37101265E+05   0.19943345E+04   0.14075023E+05
    Rj0 =          4.7000
    RKj =  -0.35622686E+02
```

No	Rotation (radians) X 1/1000	Moment (kip-inch) Expri.	Moment (kip-inch) M.Expo.	Connection stiffness (kip-inch) X 1000 Expri.	Connection stiffness (kip-inch) X 1000 M.Expo.
1	0.00	0.00	0.00	0.2391E+03	0.3903E+03
2	0.27	63.76	65.44	0.2391E+03	0.1972E+03
3	0.53	127.51	123.74	0.2391E+03	0.2438E+03
4	0.80	191.27	193.02	0.2563E+03	0.2690E+03
5	1.00	245.11	246.85	0.2692E+03	0.2675E+03
6	1.20	298.95	299.61	0.2692E+03	0.2598E+03
7	1.40	352.79	350.86	0.2553E+03	0.2532E+03
8	1.65	412.30	413.72	0.2380E+03	0.2508E+03
9	1.90	471.80	476.75	0.2841E+03	0.2543E+03
10	2.05	518.56	515.16	0.3117E+03	0.2580E+03
11	2.20	565.31	554.16	0.2783E+03	0.2621E+03
12	2.40	612.07	607.08	0.2338E+03	0.2669E+03
13	2.60	658.82	660.80	0.2232E+03	0.2700E+03
14	2.80	701.33	714.90	0.2125E+03	0.2706E+03
15	3.00	743.83	768.85	0.3340E+03	0.2685E+03
16	3.15	807.59	808.88	0.4250E+03	0.2650E+03
17	3.30	871.34	848.28	0.3485E+03	0.2601E+03
18	3.55	926.60	911.99	0.2210E+03	0.2490E+03
19	3.80	981.86	972.53	0.1927E+03	0.2349E+03
20	4.00	1015.87	1018.25	0.1700E+03	0.2222E+03
21	4.20	1049.87	1061.33	0.1700E+03	0.2085E+03
22	4.40	1083.88	1101.62	0.2348E+03	0.1943E+03
23	4.55	1126.38	1129.96	0.2834E+03	0.1836E+03
24	4.70	1168.88	1156.70	0.2105E+03	0.1729E+03
25	4.93	1191.55	1186.79	0.9714E+02	0.1208E+03
26	5.17	1214.22	1213.14	0.9714E+02	0.1051E+03
27	5.40	1236.89	1235.90	0.8565E+02	0.9016E+02
28	5.70	1258.14	1260.25	0.7084E+02	0.7245E+02
29	6.00	1279.39	1279.55	0.7084E+02	0.5654E+02

A.5 – 95

V - 48

Connection type : Extended end-plate connections
Mode : All bolted with column stiffener

Tested by : J.R.Bailey (1970) England
Test Id. : C12-L

Column : 8 x 8 UC 40
Beam : 12 x 6 1/2 UB 36
Plate thickness : 1 1/16"
Stiffener thickness : 0.3125"

Fasterers: 10.9- -7/8"D
1 1/16" Oversize holes
Material : --
Fy = -- ksi
Fu = -- ksi

Major parameters

End plate extended on tension side only.

ct = 1.6875"	pt = 4.3750"	li = 7.3125"	pc = 2.5000"
cc = 0.7500"	pit= 3.8125"	pi = 0.0000"	pic= 3.5000"
gt = 4.0000"	gi = 4.0000"	gc = 4.0000"	
tp = 0.9500"	nt = 2 X 2	ni = 2 X 1	nc = 2 X 1

Remark 1) End plate extended on tension side only.
2) Material grade 43A (BS15)

No	Moment (k-in)	Rotation (radians) X 1/1000	No	Moment (k-in)	Rotation (radians) X 1/1000
1	0.00	0.00	26	2549.11	19.50
2	93.16	0.37	27	2599.92	21.20
3	186.31	0.73	28	2599.92	23.30
4	279.47	1.10	29	2650.74	25.10
5	398.03	1.40	30	2667.67	27.40
6	516.60	1.70	31	2684.62	29.60
7	618.22	2.10	32	2701.55	31.10
8	719.84	2.50	33	2710.02	32.35
9	838.41	2.95	34	2718.49	33.60
10	956.98	3.40	35	2718.49	35.50
11	1067.07	3.70	36	2735.42	37.30
12	1177.16	4.00	37	2735.42	39.20
13	1278.79	4.45	38	2735.42	41.20
14	1380.42	4.90	39	2769.30	43.40
15	1507.45	5.40	40	2752.37	45.20
16	1634.48	5.90	41	2786.24	47.40
17	1727.64	6.45	42	2769.30	48.50
18	1820.80	7.00			
19	1905.49	7.70			
20	1990.18	8.40			
21	2125.67	9.90			
22	2261.17	11.50			
23	2345.86	13.30			
24	2430.55	15.60			
25	2481.36	17.70			

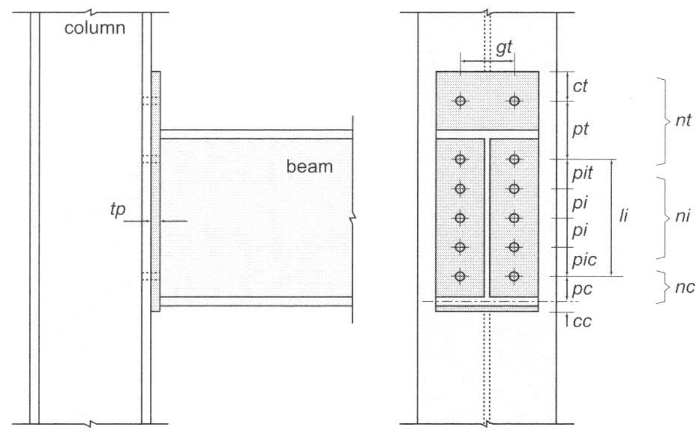

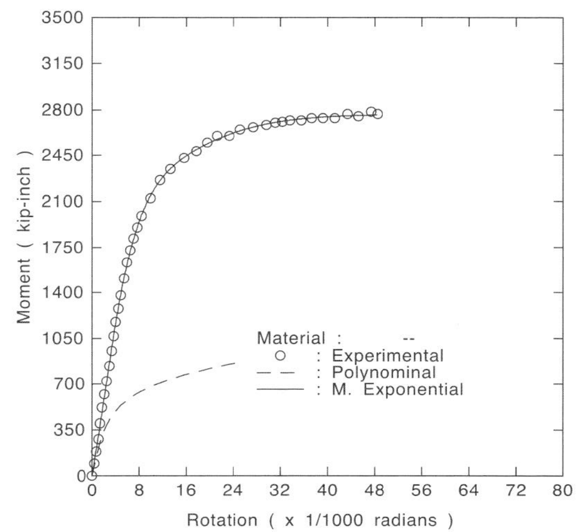

```
Moment-rotation prediction equations      ( R : X 1/1000 radians )

Frye and Morris polynominal model  :    R = Sum ( Ai X ( K*Bm )**Pi X 10**Qi )
     xd = 11.687500"   t =  0.950000"
     A1 =  1.790000   A2 =  1.760000   A3 =  2.040000    K =  0.002824
     P1 =        1    P2 =        3    P3 =        5    Q1 =        0    Q2 =       -1    Q3 =       -1

Modified exponential model  :
     Bm = Sum ( Ai X ( 1 - exp( - R/( 2*i*C ) ) ) )  +      Sum ( RKj X ( R - Rj0 ) )  +  Bm0
       C =   0.49584167E+00            Bmo=   0.00000000E+00         Nexp=  6        Nliner=  0
      AI =   0.60311678E+03  -0.41210082E+04  -0.59860340E+04   0.46259627E+05  -0.65095319E+05   0.31106334E+05
```

No	Rotation (radians) X 1/1000	Moment (kip-inch) Expri.	Moment (kip-inch) Poly.	Moment (kip-inch) M.Expo.	Connection stiffness (kip-inch) X 1000 Expri.	Connection stiffness (kip-inch) X 1000 Poly.	Connection stiffness (kip-inch) X 1000 M.Expo.
1	0.00	0.00	0.00	0.00	0.2540E+03	0.1978E+03	0.2798E+03
3	0.73	186.31	142.39	199.45	0.2541E+03	0.1862E+03	0.2740E+03
5	1.40	398.03	255.90	388.20	0.3952E+03	0.1511E+03	0.2925E+03
7	2.10	618.22	346.63	598.14	0.2541E+03	0.1096E+03	0.3052E+03
9	2.95	838.41	424.25	857.42	0.2635E+03	0.7618E+02	0.3015E+03
11	3.70	1067.07	474.34	1077.51	0.3670E+03	0.5882E+02	0.2836E+03
13	4.45	1278.79	513.99	1280.69	0.2258E+03	0.4767E+02	0.2573E+03
15	5.40	1507.45	554.56	1507.14	0.2541E+03	0.3841E+02	0.2191E+03
17	6.45	1727.64	591.12	1715.14	0.1694E+03	0.3168E+02	0.1776E+03
19	7.70	1905.49	627.13	1909.63	0.1210E+03	0.2628E+02	0.1350E+03
21	9.90	2125.67	677.83	2143.86	0.8760E+02	0.2034E+02	0.8252E+02
23	13.30	2345.86	737.38	2347.51	0.4256E+02	0.1523E+02	0.4357E+02
25	17.70	2481.36	795.69	2494.20	0.3143E+02	0.1163E+02	0.2622E+02
27	21.20	2599.92	833.09	2573.85	0.1652E+02	0.9850E+01	0.1971E+02
29	25.10	2650.74	868.63	2639.65	0.1907E+02	0.8454E+01	0.1422E+02
31	29.60	2684.62	903.93	2691.79	0.9834E+01	0.7297E+01	0.9186E+01
33	32.35	2710.02	923.37	2713.66	0.6778E+01	0.6742E+01	0.6804E+01
35	35.50	2718.49	943.61	2731.61	0.4830E+01	0.6218E+01	0.4697E+01
37	39.20	2735.42	965.63	2745.54	0.1044E-12	0.5702E+01	0.2949E+01
39	43.40	2769.30	988.54	2755.05	0.1756E+01	0.5220E+01	0.1684E+01
41	47.40	2786.24	1008.63	2760.22	-0.5136E+01	0.4838E+01	0.9632E+00

V - 49

Connection type : Extended end-plate connections
Mode : All bolted with column stiffener

Tested by : J.R.Bailey (1970) England
Test Id. : C12-R

Column : 8 x 8 UC 40
Beam : 12 x 6 1/2 UB 36
Plate thickness : 1 1/16"
Stiffener thickness : 0.3125"

Fasteners: 10.9- -7/8"D
1 1/16" Oversize holes
Material : --
Fy = -- ksi
Fu = -- ksi

Major parameters

End plate extended on tension side only.

ct = 1.6875"	pt = 4.3750"	li = 7.3125"	pc = 2.5000"
cc = 0.7500"	pit= 3.8125"	pi = 0.0000"	pic= 3.5000"
gt = 4.0000"	gi = 4.0000"	gc = 4.0000"	
tp = 0.9500"	nt = 2 X 2	ni = 2 X 1	nc = 2 X 1

Remark 1) End plate extended on tension side only.
2) Material grade 43A (BS15)

No	Moment (k-in)	Rotation (radians) X 1/1000	No	Moment (k-in)	Rotation (radians) X 1/1000
1	0.00	0.00	26	2725.48	15.90
2	144.79	0.35	27	2793.61	17.90
3	289.58	0.70	28	2844.72	20.10
4	425.86	1.05	29	2853.23	21.30
5	562.13	1.40	30	2861.75	22.50
6	689.89	1.65	31	2912.86	24.70
7	817.64	1.90	32	2929.88	26.80
8	945.40	2.20	33	2946.92	28.60
9	1073.16	2.50	34	2946.92	30.40
10	1175.36	2.75	35	2946.92	32.40
11	1277.57	3.00	36	2980.99	34.20
12	1371.26	3.40	37	2980.99	36.50
13	1464.95	3.80	38	2998.02	38.30
14	1592.70	4.05	39	2998.02	40.20
15	1720.45	4.30	40	2980.99	42.10
16	1822.66	4.80	41	3015.06	44.20
17	1924.87	5.30	42	3015.06	45.90
18	2001.52	5.90	43	3015.06	47.20
19	2078.17	6.50			
20	2180.38	7.35			
21	2282.58	8.20			
22	2359.24	9.00			
23	2435.89	9.80			
24	2572.16	11.90			
25	2657.34	13.70			

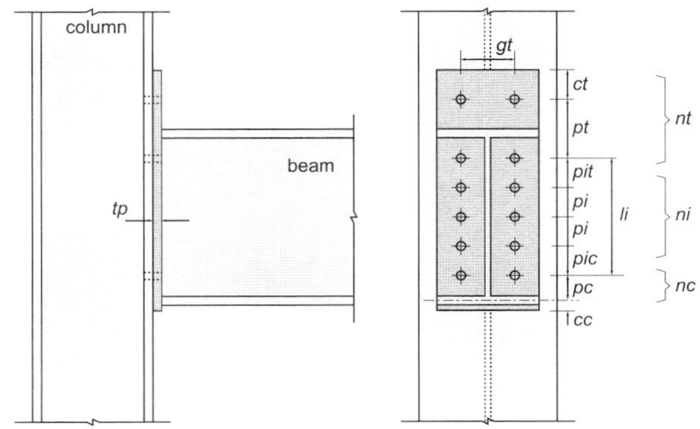

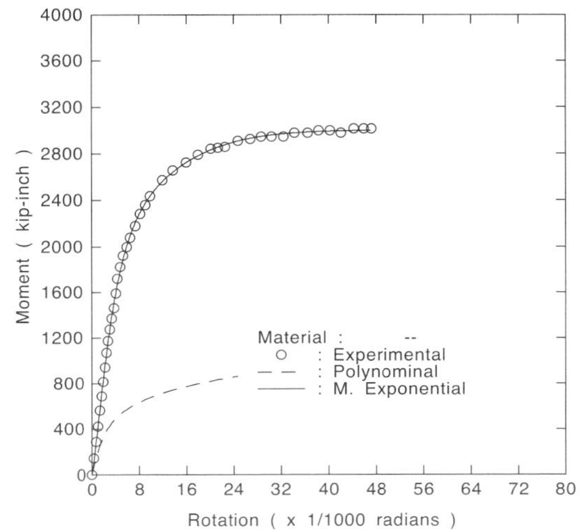

```
Moment-rotation prediction equations     ( R : X 1/1000 radians )

Frye and Morris polynominal model  :    R = Sum ( Ai X ( K*Bm )**Pi X 10**Qi )
     xd = 11.687500"   t =  0.950000"
     A1 =  1.790000   A2 =  1.760000    A3 =  2.040000    K =  0.002824
     P1 =        1    P2 =        3     P3 =        5     Q1 =        0     Q2 =      -1     Q3 =      -1

Modified exponential model  :
     Bm = Sum ( Ai X ( 1 - exp( - R/( 2*i*C ) ) ) )  +     Sum ( RKj X ( R - Rj0 ) )  +  Bm0
      C =   0.48083333E+00            Bmo=   0.00000000E+00           Nexp=  6        Nliner=  0
     AI =  -0.40678379E+03  -0.27634475E+04   0.93328384E+04   0.15024946E+04  -0.19163554E+05   0.14501571E+05
```

No	Rotation (radians) X 1/1000	Moment (kip-inch) Expri.	Moment (kip-inch) Poly.	Moment (kip-inch) M.Expo.	Connection stiffness (kip-inch) X 1000 Expri.	Connection stiffness (kip-inch) X 1000 Poly.	Connection stiffness (kip-inch) X 1000 M.Expo.
1	0.00	0.00	0.00	0.00	0.4137E+03	0.1978E+03	0.2935E+03
3	0.70	289.58	136.17	267.75	0.4015E+03	0.1873E+03	0.4415E+03
5	1.40	562.13	255.91	589.15	0.4603E+03	0.1511E+03	0.4632E+03
7	1.90	817.64	323.62	816.41	0.4723E+03	0.1204E+03	0.4426E+03
9	2.50	1073.16	386.66	1070.01	0.4166E+03	0.9153E+02	0.4008E+03
11	3.00	1277.57	427.99	1260.44	0.3417E+03	0.7475E+02	0.3607E+03
13	3.80	1464.95	480.14	1523.31	0.4046E+03	0.5705E+02	0.2973E+03
15	4.30	1720.45	506.70	1662.78	0.4088E+03	0.4956E+02	0.2612E+03
17	5.30	1924.87	550.68	1892.14	0.1696E+03	0.3921E+02	0.2001E+03
19	6.50	2078.17	592.70	2097.82	0.1246E+03	0.3142E+02	0.1460E+03
21	8.20	2282.58	639.84	2300.70	0.1077E+03	0.2462E+02	0.9722E+02
23	9.80	2435.89	675.76	2433.18	0.8729E+02	0.2055E+02	0.7070E+02
25	13.70	2657.34	743.39	2641.47	0.3996E+02	0.1480E+02	0.4116E+02
27	17.90	2793.61	798.00	2781.54	0.2891E+02	0.1151E+02	0.2677E+02
29	21.30	2853.23	834.07	2858.22	0.7095E+01	0.9808E+01	0.1870E+02
31	24.70	2912.86	865.22	2910.91	0.1549E+02	0.8577E+01	0.1260E+02
33	28.60	2946.92	896.52	2949.77	0.4733E+01	0.7523E+01	0.7668E+01
35	32.40	2946.92	923.55	2972.55	0.9961E+01	0.6737E+01	0.4560E+01
37	36.50	2980.99	949.75	2986.70	0.5307E+01	0.6068E+01	0.2525E+01
39	40.20	2998.02	971.27	2993.88	-0.4481E+01	0.5579E+01	0.1450E+01
41	44.20	3015.06	992.68	2998.22	0.7258E+01	0.5139E+01	0.7831E+00
43	47.20	3015.06	1007.88	3000.09	0.0000E+00	0.4852E+01	0.4887E+00

Connection type : Extended end-plate connections
Mode : All bolted with column stiffener

Tested by : J.R.Bailey (1970) England
Test Id. : C13-L

Column : 8 x 3 UC 40
Beam : 14 x 6 3/4 UB 45
Plate thickness : 1 1/4"
Stiffener thickness : 0.3125"

Fasteners: 10.9- -1"D
1 1/4" Oversize holes
Material : --
Fy = -- ksi
Fu = -- ksi

Major parameters

End plate extended on tension side only.

ct = -- "	pt = 4.6250"	li = -- "	pc = -- "
cc = -- "	pit= -- "	pi = -- "	pic= -- "
gt = -- "	gi = -- "	gc = -- "	
tp = 1.1400"	nt = 2 X 2	ni = 2 X 1	nc = 2 X 1

Remark 1) End plate extended on tension side only.
2) Material grade 43A (BS15)

No	Moment (k-in)	Rotation (radians) X 1/1000	No	Moment (k-in)	Rotation (radians) X 1/1000
1	0.00	0.00	26	2282.27	7.40
2	42.26	0.10	27	2417.51	8.20
3	84.53	0.20	28	2518.94	9.00
4	126.79	0.30	29	2578.12	9.60
5	169.06	0.40	30	2637.29	10.20
6	211.32	0.50	31	2679.55	10.70
7	253.58	0.60	32	2721.82	11.20
8	292.22	0.70	33	2755.63	11.70
9	330.86	0.80	34	2789.44	12.20
10	403.16	1.00	35	2823.25	12.80
11	524.08	1.30	36	2857.06	13.40
12	642.41	1.70	37	2882.41	13.95
13	760.75	2.10	38	2907.77	14.50
14	896.00	2.50	39	2916.22	15.10
15	1031.24	2.90	40	2924.68	15.70
16	1166.49	3.30	41	2933.13	16.25
17	1301.74	3.70	42	2941.58	16.80
18	1420.07	4.00			
19	1538.41	4.30			
20	1648.30	4.55			
21	1758.19	4.80			
22	1893.43	5.30			
23	2011.78	5.80			
24	2130.11	6.60			
25	2206.19	7.00			

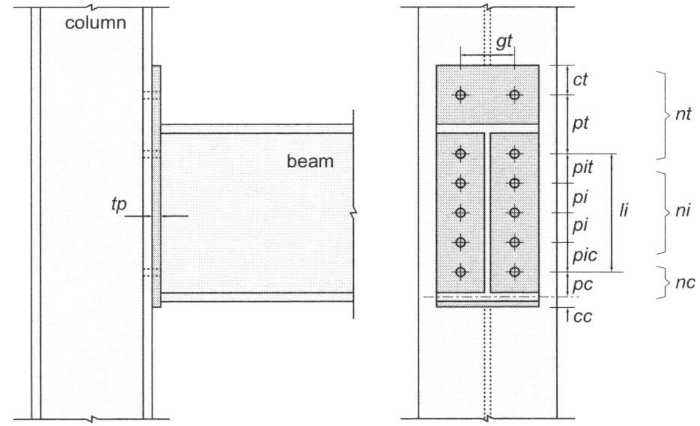

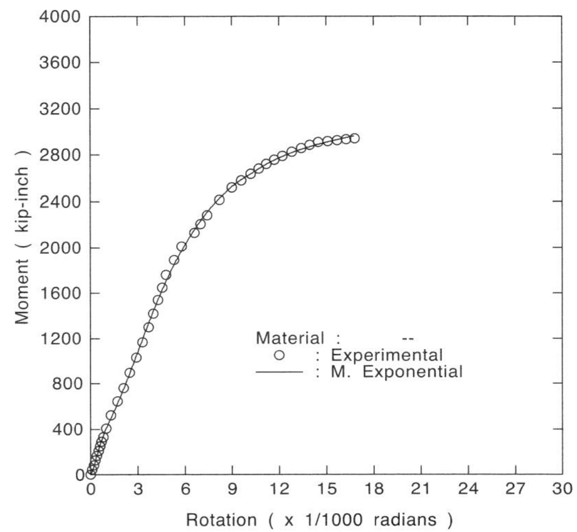

A.5 – 100

```
Moment-rotation prediction equations     ( R : X 1/1000 radians )

Modified exponential model  :
    Bm = Sum ( Ai X ( 1 - exp( - R/( 2*i*C ) ) ) )  +      Sum ( RKj X ( R - Rj0 ) )  +  Bm0
      C =   0.16500000E+00             Bmo=   0.00000000E+00          Nexp=  6        Nliner=  1
     AI =  -0.14105246E+04   0.15072138E+05  -0.66181746E+05   0.15887820E+06  -0.19435922E+06   0.90827187E+05
    Rj0 =          10.2000
    RKj =   0.22370666E+02
```

No	Rotation (radians) X 1/1000	Moment (kip-inch) Expri.	Moment (kip-inch) M.Expo.	Connection stiffness (kip-inch) X 1000 Expri.	Connection stiffness (kip-inch) X 1000 M.Expo.
1	0.00	0.00	0.00	0.4226E+03	0.1531E+03
3	0.20	84.53	70.00	0.4226E+03	0.4712E+03
5	0.40	169.06	169.56	0.4226E+03	0.4978E+03
7	0.60	253.58	263.43	0.4045E+03	0.4354E+03
9	0.80	330.86	343.44	0.3864E+03	0.3670E+03
11	1.30	524.08	502.50	0.3476E+03	0.2934E+03
13	2.10	760.75	752.22	0.3170E+03	0.3419E+03
15	2.90	1031.24	1042.94	0.3381E+03	0.3752E+03
17	3.70	1301.74	1338.80	0.3703E+03	0.3584E+03
19	4.30	1538.41	1545.55	0.4191E+03	0.3296E+03
21	4.80	1758.19	1703.51	0.3832E+03	0.3021E+03
23	5.80	2011.78	1977.79	0.2025E+03	0.2469E+03
25	7.00	2206.19	2237.23	0.1902E+03	0.1868E+03
27	8.20	2417.51	2429.85	0.1479E+03	0.1358E+03
29	9.60	2578.12	2585.52	0.9862E+02	0.8891E+02
31	10.70	2679.55	2678.72	0.8453E+02	0.8389E+02
33	11.70	2755.63	2752.87	0.6762E+02	0.6535E+02
35	12.80	2823.25	2816.21	0.5635E+02	0.5070E+02
37	13.95	2882.41	2868.13	0.4610E+02	0.4031E+02
39	15.10	2916.22	2910.29	0.1409E+02	0.3352E+02
41	16.25	2933.13	2946.16	0.1537E+02	0.2920E+02

A.5 – 101

V - 51

Connection type : Extended end-plate connections
Mode : All bolted with column stiffener

Tested by : J.R.Bailey (1970) England
Test Id. : C13-R

Column : 8 x 8 UC 40
Beam : 14 x 6 3/4 UB 45
Plate thickness : 1 1/4"
Stiffener thickness : 0.3125"

Fasteners: 10.9- -1"D
1 1/4" Oversize holes
Material : --
Fy = -- ksi
Fu = -- ksi

Major parameters

End plate extended on tension side only.

ct = -- "	pt = 4.6250"	li = -- "	pc = -- "
cc = -- "	pit= -- "	pi = -- "	pic= -- "
gt = -- "	gi = -- "	gc = -- "	
tp = 1.1400"	nt = 2 X 2	ni = 2 X 1	nc = 2 X 1

Remark 1) End plate extended on tension side only.
2) Material grade 43A (BS15)

No	Moment (k-in)	Rotation (radians) X 1/1000	No	Moment (k-in)	Rotation (radians) X 1/1000
1	0.00	0.00	26	2546.80	10.70
2	130.28	0.20	27	2597.23	11.45
3	260.57	0.40	28	2647.66	12.20
4	386.65	0.80	29	2681.27	13.05
5	512.72	1.20	30	2714.89	13.90
6	638.80	1.50	31	2748.52	14.65
7	764.88	1.80	32	2782.14	15.40
8	890.96	2.05	33	2798.95	16.20
9	1017.04	2.30	34	2815.76	17.00
10	1117.90	2.60	35	2824.17	17.85
11	1218.76	2.90	36	2832.58	18.70
12	1319.62	3.15	37	2832.58	19.50
13	1420.49	3.40	38	2832.58	20.30
14	1554.97	3.70			
15	1630.62	3.95			
16	1706.27	4.20			
17	1790.32	4.80			
18	1907.99	5.40			
19	2008.86	5.90			
20	2143.34	6.90			
21	2227.39	7.60			
22	2311.44	8.30			
23	2370.28	8.95			
24	2429.12	9.60			
25	2487.96	10.15			

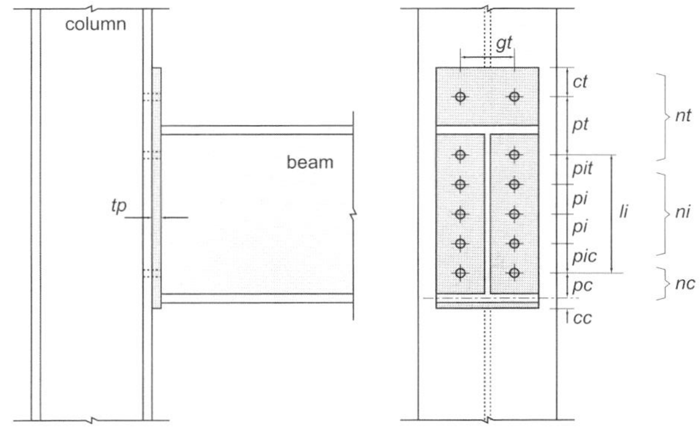

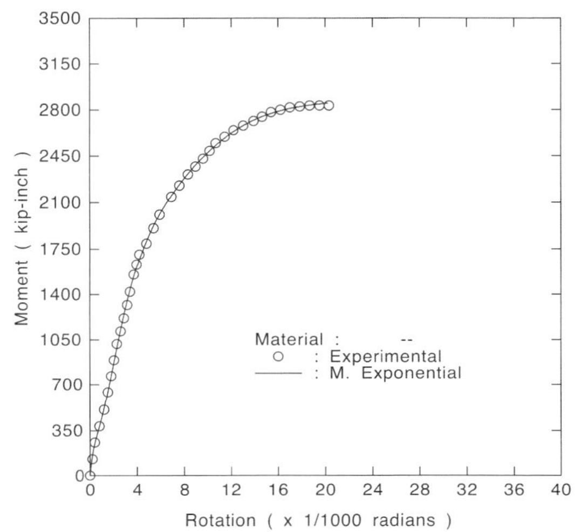

```
Moment-rotation prediction equations     ( R : X 1/1000 radians )

Modified exponential model  :
     Bm = Sum ( Ai X ( 1 - exp( - R/( 2*i*C ) ) ) )  +      Sum ( RKj X ( R - Rj0 ) )  +  Bm0
       C =   0.21916667E+00            Bmo=   0.00000000E+00          Nexp=  6        Nliner=  0
      AI =   0.11331078E+04   0.25406302E+04  -0.53055826E+05   0.16374997E+06  -0.19888217E+06   0.87384835E+05
```

No	Rotation (radians) X 1/1000	Moment (kip-inch) Expri.	Moment (kip-inch) M.Expo.	Connection stiffness (kip-inch) X 1000 Expri.	Connection stiffness (kip-inch) X 1000 M.Expo.
1	0.00	0.00	0.00	0.6514E+03	0.1011E+04
3	0.40	260.57	247.25	0.5393E+03	0.3840E+03
5	1.20	512.72	517.77	0.3752E+03	0.3807E+03
7	1.80	764.88	770.61	0.4661E+03	0.4488E+03
9	2.30	1017.04	995.98	0.4279E+03	0.4446E+03
11	2.90	1218.76	1249.21	0.3729E+03	0.3943E+03
13	3.40	1420.49	1432.68	0.4238E+03	0.3391E+03
15	3.95	1630.62	1602.77	0.3026E+03	0.2806E+03
17	4.80	1790.32	1809.32	0.1681E+03	0.2099E+03
19	5.90	2008.86	2006.73	0.1793E+03	0.1550E+03
21	7.60	2227.39	2232.97	0.1201E+03	0.1167E+03
23	8.95	2370.28	2378.27	0.9053E+02	0.9907E+02
25	10.15	2487.96	2488.32	0.1070E+03	0.8432E+02
27	11.45	2597.23	2587.48	0.6724E+02	0.6827E+02
29	13.05	2681.27	2681.64	0.3955E+02	0.4983E+02
31	14.65	2748.52	2748.61	0.4483E+02	0.3448E+02
33	16.20	2798.95	2792.83	0.2102E+02	0.2316E+02
35	17.85	2824.17	2823.56	0.9889E+01	0.1463E+02
37	19.50	2832.58	2842.69	0.0000E+00	0.8975E+01

V - 52

```
Connection type : Extended end-plate connections
           Mode : All bolted with column stiffener

Tested by   :  J.O.Surtees & A.P.Mann (1970)                England
Test   Id.  :  TEST C6

Column : 10 x 10 UC 89                  Fasteners: 10.9- -1 1/8"D
Beam    : 18 x 6 UB 55                      1 7/16" Oversize holes
Plate thickness : 1 1/8"                Material :  --
Stiffener thickness : 0.5000"              Fy =    --   ksi
                                           Fu =    --   ksi

                            Major parameters

     End plate extended on tension side only.
     ct =  3.0000"    pt =  6.0000"    li = 10.0000"    pc =  5.0000"
     cc =  0.0000"    gt =  6.0000"    gc =  6.0000"
     tp =  1.1250"    nt = 2 X  2      nc = 2 X  1

Remark  1)  End plate extended on tension side only.
        2)  Hsfg bolts used either side of beam tension flange.
```

No	Moment (k-in)	Rotation (radians) X 1/1000	No	Moment (k-in)	Rotation (radians) X 1/1000
1	0.00	0.00	26	4696.21	6.10
2	207.57	0.15	27	4851.89	7.10
3	415.14	0.30	28	4942.70	8.40
4	570.82	0.50	29	4994.59	9.90
5	726.49	0.70	30	5059.46	11.50
6	960.01	0.95	31	5085.41	13.20
7	1193.52	1.20	32	5137.30	14.80
8	1381.63	1.40	33	5176.21	16.10
9	1569.73	1.60	34	5215.14	17.60
10	1738.38	1.80	35	5228.11	19.10
11	1907.03	2.00	36	5254.06	20.90
12	2121.08	2.25	37	5267.03	22.20
13	2335.14	2.50	38	5280.00	23.70
14	2523.25	2.70	39	5292.97	25.70
15	2711.35	2.90	40	5331.89	27.20
16	2899.46	3.10	41	5344.86	28.90
17	3087.56	3.30	42	5357.84	30.20
18	3230.27	3.50	43	5383.79	31.30
19	3372.97	3.70	44	5396.76	32.60
20	3541.62	3.85	45	5409.73	33.90
21	3710.27	4.00	46	5422.70	35.10
22	3898.37	4.20	47	5448.65	36.30
23	4086.48	4.40	48	5461.62	37.50
24	4332.97	4.80	49	5474.59	38.60
25	4514.59	5.20	50	5500.54	40.00

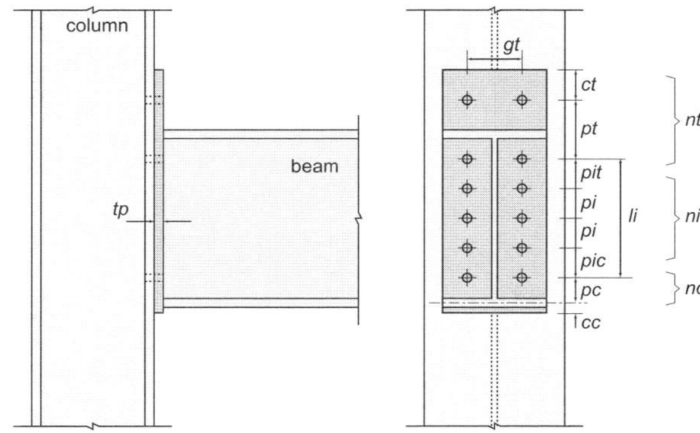

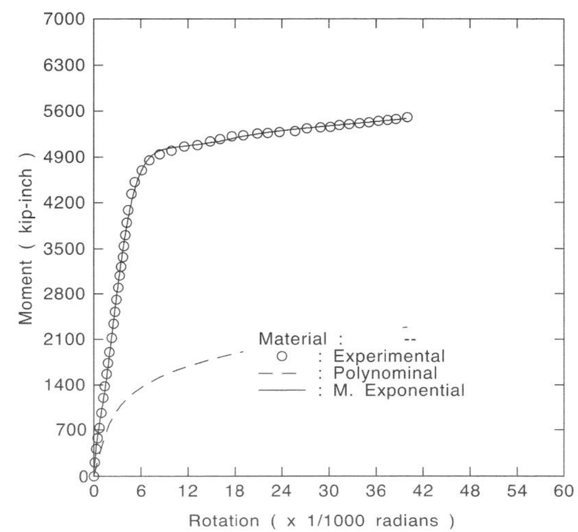

```
Moment-rotation prediction equations     ( R : X 1/1000 radians )

Frye and Morris polynominal model  :    R = Sum ( Ai X ( K*Bm )**Pi X 10**Qi )
     xd = 16.000000"   t =  1.125000"
     A1 =  1.790000   A2 =  1.760000   A3 =  2.040000    K =  0.001201
     P1 =         1   P2 =         3   P3 =         5   Q1 =         0   Q2 =        -1   Q3 =        -1

Modified exponential model  :
     Bm = Sum ( Ai X ( 1 - exp( - R/( 2*i*C ) ) ) )  +     Sum ( RKj X ( R - Rj0 ) )  +  Bm0
      C =   0.37083333E+00              Bmo=   0.00000000E+00          Nexp=  6        Nliner=  1
     AI =   0.85477149E+04  -0.72655018E+05   0.19976764E+06  -0.25355569E+06   0.16432032E+06  -0.41224464E+05
     Rj0 =           16.1000
     RKj =   0.11665979E+02
```

No	Rotation (radians) X 1/1000	Moment (kip-inch) Expri.	Moment (kip-inch) Poly.	Moment (kip-inch) M.Expo.	Connection stiffness (kip-inch) X 1000 Expri.	Connection stiffness (kip-inch) X 1000 Poly.	Connection stiffness (kip-inch) X 1000 M.Expo.
1	0.00	0.00	0.00	0.00	0.1384E+04	0.4653E+03	0.1906E+04
3	0.30	415.14	139.19	433.73	0.1124E+04	0.4613E+03	0.1096E+04
5	0.70	726.49	320.25	783.78	0.8476E+03	0.4406E+03	0.7477E+03
7	1.20	1193.52	527.82	1155.64	0.9377E+03	0.3844E+03	0.7827E+03
9	1.60	1569.73	669.91	1491.09	0.8919E+03	0.3255E+03	0.8946E+03
11	2.00	1907.03	788.74	1867.25	0.8490E+03	0.2701E+03	0.9781E+03
13	2.50	2335.14	909.37	2366.83	0.9031E+03	0.2153E+03	0.1006E+04
15	2.90	2711.35	988.66	2763.56	0.9405E+03	0.1827E+03	0.9706E+03
17	3.30	3087.56	1056.52	3138.46	0.8270E+03	0.1577E+03	0.8990E+03
19	3.70	3372.97	1115.58	3479.79	0.9483E+03	0.1383E+03	0.8050E+03
21	4.00	3710.27	1155.26	3709.73	0.1046E+04	0.1265E+03	0.7273E+03
23	4.40	4086.48	1203.19	3979.46	0.8324E+03	0.1136E+03	0.6215E+03
25	5.20	4514.59	1285.81	4396.31	0.3764E+03	0.9419E+02	0.4260E+03
27	7.10	4851.89	1436.28	4886.90	0.1184E+03	0.6727E+02	0.1318E+03
29	9.90	4994.59	1594.16	5042.37	0.3747E+02	0.4783E+02	0.1795E+02
31	13.20	5085.41	1730.62	5091.17	0.2411E+02	0.3608E+02	0.1636E+02
33	16.10	5176.21	1825.63	5140.30	0.2809E+02	0.2988E+02	0.1610E+02
35	19.10	5228.11	1908.31	5215.18	0.1127E+02	0.2549E+02	0.2179E+02
37	22.20	5267.03	1982.00	5273.10	0.9361E+01	0.2222E+02	0.1595E+02
39	25.70	5292.97	2054.71	5321.69	0.1760E+02	0.1947E+02	0.1233E+02
41	28.90	5344.86	2113.78	5358.82	0.8966E+01	0.1753E+02	0.1112E+02
43	31.30	5383.79	2154.38	5385.17	0.1735E+02	0.1634E+02	0.1090E+02
45	33.90	5409.73	2195.38	5413.53	0.1041E+02	0.1523E+02	0.1095E+02
47	36.30	5448.65	2230.83	5439.97	0.1622E+02	0.1434E+02	0.1109E+02
49	38.60	5474.59	2263.37	5465.64	0.1476E+02	0.1358E+02	0.1123E+02

A.5 – 105

V - 53

Connection type : Extended end-plate connections
Mode : All bolted with column stiffener

Tested by : N.D.Johnstone & W.R.Walpole (1981) New-Zealand
Test Id. : TEST 1-L

Column : 250 UC 89
Beam : 310 UB 46
Plate thickness : 32 mm
Stiffener thickness : 0.6300"

Fasteners: G8.8- -M30
7/8" Oversize holes
Material : --
Fy = -- ksi
Fu = -- ksi

Major parameters

End plate extended on both sides.
ct = -- " pt = 7.0078" li = -- " pc = 7.0078"
cc = -- " gt = 5.5118" gc = 5.5118"
tp = 1.2600" nt = 2 X 2 nc = 2 X 2

Remark 1) End plate extended on tension and compression side.
2) 16 mm web doubler plates on both sides of column used.

No	Moment (k-in)	Rotation (radians) X 1/1000	No	Moment (k-in)	Rotation (radians) X 1/1000
1	0.00	0.00	26	2277.85	21.70
2	100.68	0.05	27	2303.02	24.00
3	201.36	0.10	28	2328.19	26.30
4	302.03	0.35	29	2347.07	28.60
5	402.71	0.60	30	2365.94	30.90
6	503.39	0.87	31	2372.24	33.00
7	604.07	1.13	32	2378.53	35.10
8	704.75	1.40	33	2391.11	38.70
9	824.30	1.35	34	2416.28	42.00
10	943.86	1.30	35	2422.57	43.90
11	1038.25	1.50	36	2428.86	45.80
12	1132.63	1.70	37	2441.45	49.40
13	1239.60	2.30	38	2454.04	53.10
14	1346.57	2.90	39	2466.62	55.45
15	1440.95	3.50	40	2479.20	57.80
16	1535.34	4.10	41	2504.38	61.30
17	1636.02	4.65	42	2510.67	63.70
18	1736.70	5.20	43	2516.96	66.10
19	1805.92	6.05	44	2523.25	68.35
20	1875.13	6.90	45	2529.54	70.60
21	1988.40	9.70	46	2535.83	72.60
22	2063.90	12.00	47	2542.13	74.60
23	2126.83	14.65			
24	2189.76	17.30			
25	2233.81	19.50			

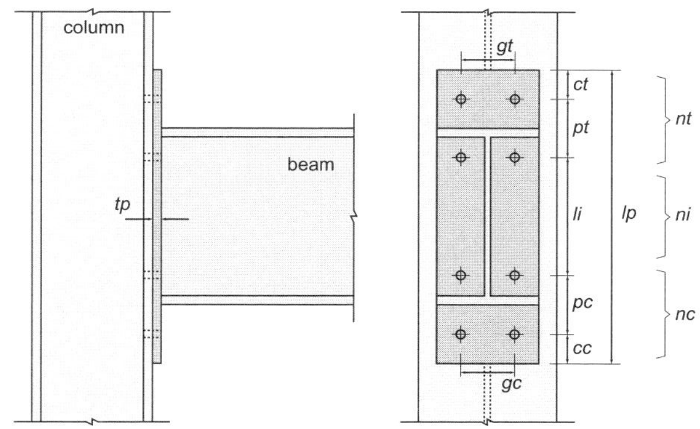

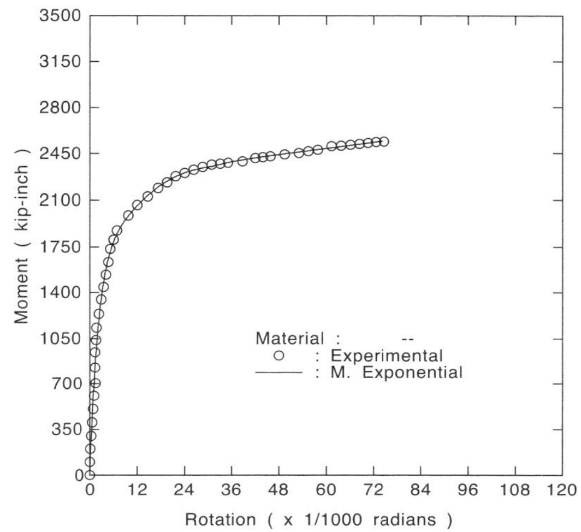

Moment-rotation prediction equations (R : X 1/1000 radians)

```
Modified exponential model  :
     Bm = Sum ( Ai X ( 1 - exp( - R/( 2*i*C ) ) ) )  +      Sum ( RKj X ( R - Rj0 ) )  +  Bm0
      C =   0.63416667E+00               Bmo=   0.00000000E+00          Nexp=  6         Nliner=  1
     AI =   0.18662481E+03   0.73369929E+03   0.10452855E+05  -0.28148011E+05   0.26010953E+05  -0.68637164E+04
    Rj0 =          30.9000
    RKj =   0.39636409E+01
```

No	Rotation (radians) X 1/1000	Moment (kip-inch) Expri.	Moment (kip-inch) M.Expo.	Connection stiffness (kip-inch) X 1000 Expri.	Connection stiffness (kip-inch) X 1000 M.Expo.
1	0.00	0.00	0.00	0.2014E+04	0.8349E+03
3	0.10	201.36	81.59	0.1745E+04	0.7973E+03
5	0.60	402.71	437.91	0.3905E+03	0.6348E+03
7	1.13	604.07	739.06	0.3776E+03	0.5005E+03
9	1.35	824.30	842.51	-0.2391E+04	0.4551E+03
11	1.50	1038.25	908.59	0.4719E+03	0.4263E+03
13	2.30	1239.60	1197.28	0.1783E+03	0.3031E+03
15	3.50	1440.95	1484.53	0.1573E+03	0.1864E+03
17	4.65	1636.02	1658.15	0.1831E+03	0.1213E+03
19	6.05	1805.92	1793.68	0.8143E+02	0.7705E+02
21	9.70	1988.40	1983.43	0.3627E+02	0.3747E+02
23	14.65	2126.83	2135.36	0.2375E+02	0.2579E+02
25	19.50	2233.81	2240.48	0.2002E+02	0.1757E+02
27	24.00	2303.02	2303.64	0.1094E+02	0.1077E+02
29	28.60	2347.07	2340.96	0.8207E+01	0.5813E+01
31	33.00	2372.24	2367.90	0.2997E+01	0.6875E+01
33	38.70	2391.11	2400.88	0.5651E+01	0.4966E+01
35	43.90	2422.57	2424.55	0.3309E+01	0.4246E+01
37	49.40	2441.45	2447.01	0.3450E+01	0.3973E+01
39	55.45	2466.62	2470.76	0.5354E+01	0.3902E+01
41	61.30	2504.38	2493.59	0.4482E+01	0.3906E+01
43	66.10	2516.96	2512.37	0.2711E+01	0.3921E+01
45	70.60	2529.54	2530.05	0.2981E+01	0.3934E+01
47	74.60	2542.13	2545.80	0.3147E+01	0.3943E+01

V - 54

Connection type : Extended end-plate connections
Mode : All bolted with column stiffener

Tested by : N.D.Johnstone & W.R.Walpole (1981) New-Zealand
Test Id. : TEST 1-R

Column : 250 UC 89
Beam : 310 UB 46
Plate thickness : 32 mm
Stiffener thickness : 0.6300"

Fasteners: G8.8- -M30
7/8" Oversize holes
Material : --
Fy = -- ksi
Fu = -- ksi

Major parameters

End plate extended on both sides.

ct = -- " pt = 7.0078" li = -- " pc = 7.0078"
cc = -- " gt = 5.5118" gc = 5.5118"
tp = 1.2600" nt = 2 X 2 nc = 2 X 2

Remark 1) End plate extended on tension and compression side.
2) 16 mm web doubler plates on both sides of column used.

No	Moment (k-in)	Rotation (radians) X 1/1000	No	Moment (k-in)	Rotation (radians) X 1/1000
1	0.00	0.00	26	2378.15	23.15
2	114.81	0.12	27	2403.38	25.60
3	229.61	0.24	28	2416.00	27.85
4	344.42	0.36	29	2428.62	30.10
5	459.23	0.48	30	2447.54	32.25
6	574.03	0.60	31	2466.47	34.40
7	693.89	0.75	32	2479.08	36.25
8	813.74	0.90	33	2491.69	38.10
9	920.98	1.10	34	2516.93	40.35
10	1028.22	1.30	35	2542.16	42.60
11	1116.53	1.55	36	2567.39	44.50
12	1204.85	1.80	37	2592.62	46.40
13	1337.32	2.25	38	2617.85	48.70
14	1469.78	2.70	39	2643.08	51.00
15	1602.25	3.40	40	2668.32	53.45
16	1734.72	4.10	41	2693.56	55.90
17	1829.35	4.75	42	2699.86	58.10
18	1923.97	5.40	43	2706.17	60.30
19	2012.28	7.25	44	2756.64	63.90
20	2100.59	9.10	45	2756.64	67.50
21	2226.76	12.30	46	2737.72	70.40
22	2264.60	14.30	47	2718.79	73.30
23	2302.45	16.30			
24	2327.68	18.50			
25	2352.91	20.70			

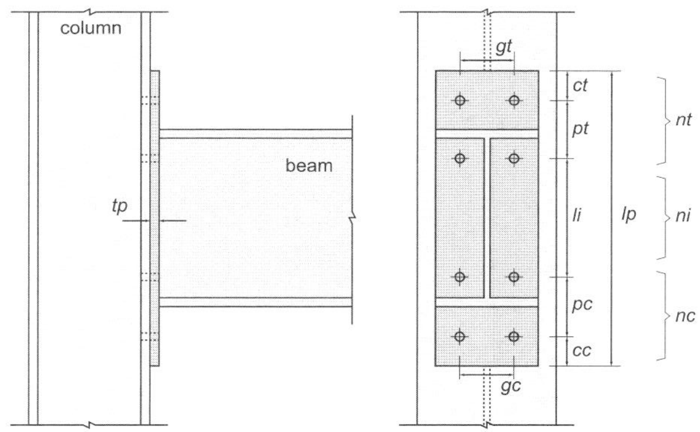

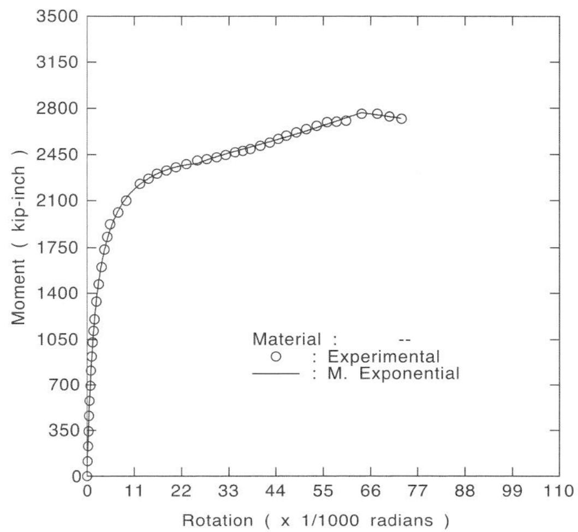

A.5 – 108

```
Moment-rotation prediction equations     ( R : X 1/1000 radians )

Modified exponential model  :
    Bm = Sum ( Ai X ( 1 - exp( - R/( 2*i*C ) ) ) )  +      Sum ( RKj X ( R - Rj0 ) )  +  Bm0
     C =  0.64083333E+00             Bmo=  0.00000000E+00          Nexp=  6        Nliner=  2
    AI =  0.18514098E+04  -0.87905922E+04  0.43088043E+05  -0.93103367E+05  0.92394948E+05  -0.33122459E+05
   Rj0 =         25.6000          63.9000
   RKj =  0.11605834E+02  -0.15993373E+02
```

No	Rotation (radians) X 1/1000	Moment (kip-inch) Expri.	Moment (kip-inch) M.Expo.	Connection stiffness (kip-inch) X 1000 Expri.	Connection stiffness (kip-inch) X 1000 M.Expo.
1	0.00	0.00	0.00	0.9567E+03	0.1172E+04
3	0.24	229.61	258.20	0.9567E+03	0.9862E+03
5	0.48	459.23	476.04	0.9567E+03	0.8341E+03
7	0.75	693.89	681.79	0.7990E+03	0.6951E+03
9	1.10	920.98	899.36	0.5362E+03	0.5547E+03
11	1.55	1116.53	1117.65	0.3533E+03	0.4232E+03
13	2.25	1337.32	1363.63	0.2944E+03	0.2911E+03
15	3.40	1602.25	1624.54	0.1892E+03	0.1772E+03
17	4.75	1829.35	1815.94	0.1456E+03	0.1138E+03
19	7.25	2012.28	2024.60	0.4773E+02	0.6089E+02
21	12.30	2226.76	2221.45	0.2681E+02	0.2535E+02
23	16.30	2302.45	2302.08	0.1537E+02	0.1586E+02
25	20.70	2352.91	2355.06	0.1092E+02	0.8462E+01
27	25.60	2403.38	2380.14	0.7854E+01	0.2192E+01
29	30.10	2428.62	2433.76	0.7241E+01	0.1041E+02
31	34.40	2466.47	2475.00	0.7735E+01	0.9023E+01
33	38.10	2491.69	2507.74	0.8802E+01	0.8779E+01
35	42.60	2542.16	2547.79	0.1233E+02	0.9092E+01
37	46.40	2592.62	2583.19	0.1223E+02	0.9554E+01
39	51.00	2643.08	2628.48	0.1065E+02	0.1012E+02
41	55.90	2693.56	2679.36	0.6384E+01	0.1062E+02
43	60.30	2706.17	2726.88	0.7097E+01	0.1095E+02
45	67.50	2756.64	2749.51	-0.3614E+01	-0.4702E+01
47	73.30	2718.79	2722.71	-0.6526E+01	-0.4556E+01

Connection type : Extended end-plate connections
Mode : All bolted with column stiffener

Tested by : N.D.Johnstone & W.R.Walpole (1981) New-Zealand
Test Id. : TEST 2-L

Column : 250 UC 89
Beam : 310 UB 46
Plate thickness : 32 mm
Stiffener thickness : 0.6300"

Fasteners: G8.8- -M30
7/8" Oversize holes
Material : --
Fy = -- ksi
Fu = -- ksi

Major parameters

End plate extended on both sides.
ct = -- " pt = 7.0078" li = -- " pc = 7.0078"
cc = -- " gt = 5.5118" gc = 5.5118"
tp = 1.2600" nt = 2 X 2 nc = 2 X 2

Remark 1) End plate extended on tension and compression side.
2) 16 mm web doubler plates on both sides of column used.

No	Moment (k-in)	Rotation (radians) X 1/1000	No	Moment (k-in)	Rotation (radians) X 1/1000
1	0.00	0.00	26	2331.89	27.97
2	123.01	0.40	27	2352.91	31.00
3	246.01	0.80	28	2378.15	34.25
4	351.15	1.00	29	2403.38	37.50
5	456.28	1.20	30	2409.69	40.45
6	561.42	1.40	31	2416.00	43.40
7	674.96	1.80	32	2428.61	46.55
8	788.51	2.20	33	2441.23	49.70
9	902.05	2.60	34	2460.16	52.65
10	1002.98	2.90	35	2479.08	55.60
11	1103.92	3.20	36	2487.49	58.47
12	1204.85	3.50	37	2495.90	61.33
13	1293.16	3.90	38	2504.32	64.20
14	1381.47	4.30	39	2521.14	66.87
15	1469.78	4.70	40	2537.96	69.53
16	1583.33	5.25	41	2554.78	72.20
17	1696.87	5.80	42	2567.39	75.15
18	1810.42	6.75	43	2580.01	78.10
19	1923.97	7.70	44	2592.62	80.00
20	2031.21	10.45			
21	2138.45	13.20			
22	2188.91	16.10			
23	2239.38	19.00			
24	2289.84	21.90			
25	2310.86	24.93			

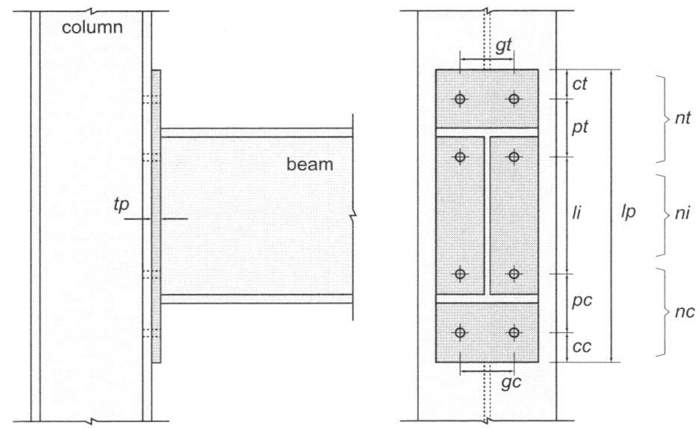

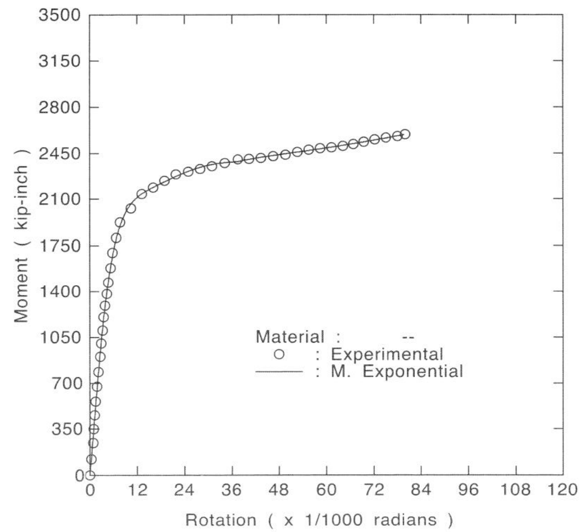

A.5 – 110

Moment-rotation prediction equations (R : X 1/1000 radians)

```
Modified exponential model  :
     Bm = Sum ( Ai X ( 1 - exp( - R/( 2*i*C ) ) ) ) +      Sum ( RKj X ( R - Rj0 ) ) + Bm0
       C =   0.76666667E+00             Bmo=   0.00000000E+00          Nexp=  6        Nliner=  1
      AI =  -0.34278274E+03  -0.95675851E+04   0.63257115E+05  -0.13129218E+06   0.11977575E+06  -0.39486424E+05
     Rj0 =          37.5000
     RKj =   0.57611366E+01
```

No	Rotation (radians) X 1/1000	Moment (kip-inch) Expri.	Moment (kip-inch) M.Expo.	Connection stiffness (kip-inch) X 1000 Expri.	Connection stiffness (kip-inch) X 1000 M.Expo.
1	0.00	0.00	0.00	0.3075E+03	0.3327E+03
3	0.80	246.01	286.07	0.4530E+03	0.3719E+03
5	1.20	456.28	435.18	0.5257E+03	0.3720E+03
7	1.80	674.96	654.62	0.2839E+03	0.3571E+03
9	2.60	902.05	926.67	0.3139E+03	0.3209E+03
11	3.20	1103.92	1109.55	0.3364E+03	0.2882E+03
13	3.90	1293.16	1297.50	0.2208E+03	0.2488E+03
15	4.70	1469.78	1479.04	0.2147E+03	0.2056E+03
17	5.80	1696.87	1675.71	0.1746E+03	0.1536E+03
19	7.70	1923.97	1900.31	0.9885E+02	0.8787E+02
21	13.20	2138.45	2139.14	0.2849E+02	0.2023E+02
23	19.00	2239.38	2232.61	0.1740E+02	0.1468E+02
25	24.93	2310.86	2312.41	0.6931E+01	0.1154E+02
27	31.00	2352.91	2365.33	0.7334E+01	0.5850E+01
29	37.50	2403.38	2386.58	0.4815E+01	0.1111E+01
31	43.40	2416.00	2419.98	0.3041E+01	0.4764E+01
33	49.70	2441.23	2447.34	0.5250E+01	0.4121E+01
35	55.60	2479.08	2471.83	0.4650E+01	0.4248E+01
37	61.33	2495.90	2497.13	0.2934E+01	0.4590E+01
39	66.87	2521.14	2523.48	0.6308E+01	0.4927E+01
41	72.20	2554.78	2550.49	0.5344E+01	0.5192E+01
43	78.10	2580.01	2581.79	0.5713E+01	0.5404E+01

A.5 – 111

V - 56

Connection type : Extended end-plate connections
Mode : All bolted with column stiffener

Tested by : N.D.Johnstone & W.R.Walpole (1981) New-Zealand
Test Id. : TEST 2-R

Column : 250 UC 89
Beam : 310 UB 46
Plate thickness : 32 mm
Stiffener thickness : 0.6300"

Fasteners: G8.8- -M30
7/8" Oversize holes
Material : --
Fy = -- ksi
Fu = -- ksi

Major parameters

End plate extended on both sides.

ct = -- "	pt = 7.0078"	li = -- "	pc = 7.0078"
cc = -- "	gt = 5.5118"	gc = 5.5118"	
tp = 1.2600"	nt = 2 X 2	nc = 2 X 2	

Remark 1) End plate extended on tension and compression side.
2) 16 mm web doubler plates on both sides of column used.

No	Moment (k-in)	Rotation (radians) X 1/1000	No	Moment (k-in)	Rotation (radians) X 1/1000
1	0.00	0.00	26	2174.52	24.40
2	112.85	0.20	27	2174.52	24.50
3	225.70	0.40	28	2199.88	27.40
4	338.54	0.60	29	2225.23	30.30
5	451.39	0.80	30	2244.25	33.30
6	564.24	1.00	31	2263.27	36.30
7	648.77	1.43	32	2275.96	40.00
8	733.30	1.87	33	2288.64	43.70
9	817.82	2.30	34	2301.32	46.50
10	910.80	2.80	35	2314.00	49.30
11	1003.78	3.30	36	2301.32	52.30
12	1096.76	3.80	37	2288.64	55.30
13	1181.30	4.33	38	2284.41	58.40
14	1265.83	4.87	39	2280.18	61.50
15	1350.36	5.40	40	2275.96	64.60
16	1451.80	5.95	41	2282.30	68.35
17	1553.23	6.50	42	2288.64	72.10
18	1633.53	7.23	43	2301.32	74.35
19	1713.83	7.97	44	2314.00	76.60
20	1794.13	8.70	45	2314.00	79.90
21	1901.91	10.95			
22	2009.69	13.20			
23	2060.41	15.85			
24	2111.12	18.50			
25	2142.82	21.45			

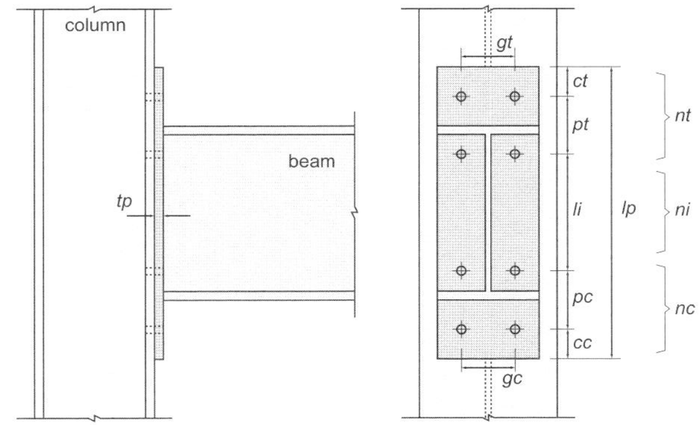

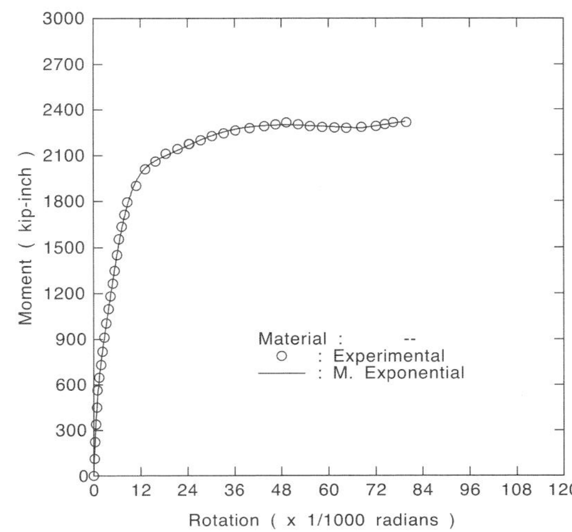

Moment-rotation prediction equations (R : X 1/1000 radians)

```
Modified exponential model  :
     Bm = Sum ( Ai X ( 1 - exp( - R/( 2*i*C ) ) ) )  +     Sum ( RKj X ( R - Rj0 ) )  +  Bm0
       C =   0.71583333E+00             Bmo=   0.00000000E+00         Nexp=  6        Nliner=  2
      AI =   0.39545646E+04  -0.30542232E+05   0.96307069E+05  -0.14075392E+06   0.98516720E+05  -0.25189553E+05
     Rj0 =          49.3000          68.3500
     RKj =  -0.10165896E+01   0.49572881E+01
```

No	Rotation (radians) X 1/1000	Moment (kip-inch) Expri.	Moment (kip-inch) M.Expo.	Connection stiffness (kip-inch) X 1000 Expri.	Connection stiffness (kip-inch) X 1000 M.Expo.
1	0.00	0.00	0.00	0.5642E+03	0.7700E+03
3	0.40	225.70	258.12	0.5642E+03	0.5371E+03
5	0.80	451.39	440.67	0.5642E+03	0.3870E+03
7	1.43	648.77	639.13	0.1951E+03	0.2568E+03
9	2.30	817.82	827.10	0.1908E+03	0.1909E+03
11	3.30	1003.78	1007.66	0.1860E+03	0.1751E+03
13	4.33	1181.30	1186.36	0.1585E+03	0.1705E+03
15	5.40	1350.36	1363.02	0.1713E+03	0.1592E+03
17	6.50	1553.23	1527.77	0.1523E+03	0.1392E+03
19	7.97	1713.83	1708.21	0.1095E+03	0.1064E+03
21	10.95	1901.91	1934.43	0.4790E+02	0.4958E+02
23	15.85	2060.41	2066.54	0.1914E+02	0.1422E+02
25	21.45	2142.82	2133.38	0.1075E+02	0.1164E+02
27	24.50	2174.52	2169.06	0.2914E+00	0.1160E+02
29	30.30	2225.23	2229.72	0.7562E+01	0.8846E+01
31	36.30	2263.27	2270.62	0.5036E+01	0.4860E+01
33	43.70	2288.64	2292.91	0.4054E+01	0.1541E+01
35	49.30	2314.00	2297.80	0.3019E+00	0.3623E+00
37	55.30	2288.64	2292.12	-0.2818E+01	-0.1148E+01
39	61.50	2280.18	2284.53	-0.1364E+01	-0.1260E+01
41	68.35	2282.30	2275.99	0.1691E+01	-0.1220E+01
43	74.35	2301.32	2298.60	0.5635E+01	0.3799E+01
45	79.90	2314.00	2319.82	0.0000E+00	0.3847E+01

V - 57

Connection type : Extended end-plate connections
Mode : All bolted with column stiffener

Tested by : N.D.Johnstone & W.R.Walpole (1981) New-Zealand
Test Id. : TEST 3-L

Column : 250 UC 89
Beam : 310 UB 46
Plate thickness : 24 mm
Stiffener thickness : 0.4720"

Fasteners: G8.8- -M24
33/32" Oversize holes
Material : --
Fy = -- ksi
Fu = -- ksi

Major parameters

End plate extended on both sides.
ct = -- " pt = 4.0160" li = -- " pc = 4.0160"
cc = -- " gt = 5.1180" gc = 5.1180"
tp = 0.9449" nt = 2 X 2 nc = 2 X 2

Remark 1) End plate extended on tension and compression side.
2) 12 mm web doubler plates on both sides of column used.

No	Moment (k-in)	Rotation (radians) X 1/1000	No	Moment (k-in)	Rotation (radians) X 1/1000
1	0.00	0.00	26	2125.82	36.30
2	86.21	0.23	27	2151.06	41.00
3	172.42	0.47	28	2176.29	46.75
4	258.64	0.70	29	2201.52	52.50
5	351.15	0.87	30	2220.44	58.05
6	443.67	1.03	31	2239.37	63.60
7	536.18	1.20	32	2245.68	67.30
8	624.50	1.70	33	2251.99	71.00
9	712.82	2.20	34	2277.22	77.25
10	801.13	2.70	35	2302.45	83.50
11	885.24	3.23	36	2315.07	88.33
12	969.35	3.77	37	2327.68	93.17
13	1053.46	4.30	38	2340.30	98.00
14	1137.56	5.00	39	2371.84	103.65
15	1221.67	5.70	40	2403.38	109.30
16	1305.78	6.40	41	2409.69	114.05
17	1419.32	7.85	42	2416.00	118.80
18	1532.87	9.30	43	2411.79	123.30
19	1627.49	10.85	44	2407.59	127.80
20	1722.11	12.40	45	2403.38	132.30
21	1823.04	15.60			
22	1923.97	18.80			
23	2037.52	24.70			
24	2069.05	28.15			
25	2100.59	31.60			

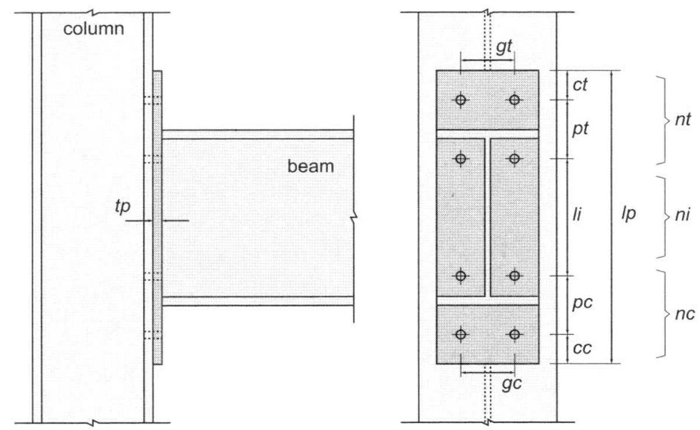

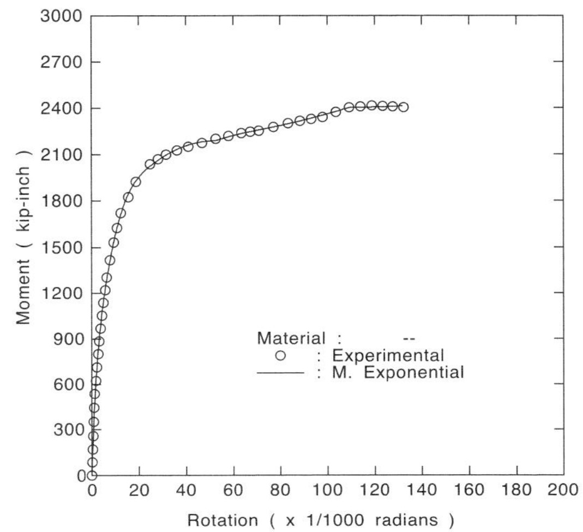

A.5 – 114

Moment-rotation prediction equations (R : X 1/1000 radians)

```
Modified exponential model  :
     Bm = Sum ( Ai X ( 1 - exp( - R/( 2*i*C ) ) ) )  +     Sum ( RKj X ( R - Rj0 ) )  + Bm0
      C =   0.11608250E+01          Bmo=   0.00000000E+00         Nexp=  6        Nliner=  2
     AI =   0.26422427E+04  -0.21514379E+05   0.89985699E+05  -0.17496769E+06   0.16094642E+06  -0.54989907E+05
    Rj0 =          52.5000         109.3000
    RKj =   0.50953998E+01  -0.43413576E+01
```

No	Rotation (radians) X 1/1000	Moment (kip-inch) Expri.	Moment (kip-inch) M.Expo.	Connection stiffness (kip-inch) X 1000 Expri.	Connection stiffness (kip-inch) X 1000 M.Expo.
1	0.00	0.00	0.00	0.3695E+03	0.5008E+03
3	0.47	172.42	210.75	0.3695E+03	0.4064E+03
5	0.87	351.15	360.14	0.5552E+03	0.3429E+03
7	1.20	536.18	467.05	0.4604E+03	0.2999E+03
9	2.20	712.82	717.96	0.1766E+03	0.2103E+03
11	3.23	885.24	905.70	0.1577E+03	0.1580E+03
13	4.30	1053.46	1056.44	0.1415E+03	0.1274E+03
15	5.70	1221.67	1217.15	0.1202E+03	0.1043E+03
17	7.85	1419.32	1416.92	0.7831E+02	0.8289E+02
19	10.85	1627.49	1629.90	0.6105E+02	0.5976E+02
21	15.60	1823.04	1843.35	0.3154E+02	0.3233E+02
23	24.70	2037.52	2023.50	0.1287E+02	0.1261E+02
25	31.60	2100.59	2096.73	0.7544E+01	0.9013E+01
27	41.00	2151.06	2162.18	0.4928E+01	0.4815E+01
29	52.50	2201.52	2189.29	0.3890E+01	0.2419E+00
31	63.60	2239.37	2235.09	0.2387E+01	0.3307E+01
33	71.00	2251.99	2258.12	0.2573E+01	0.3019E+01
35	83.50	2302.45	2297.66	0.3232E+01	0.3412E+01
37	93.17	2327.68	2333.03	0.2610E+01	0.3903E+01
39	103.65	2371.84	2376.42	0.5583E+01	0.4353E+01
41	114.05	2409.69	2402.78	0.1328E+01	0.3196E+00
43	123.30	2411.79	2406.61	-0.9342E+00	0.4949E+00
45	132.30	2403.38	2411.58	-0.9342E+00	0.6013E+00

V - 58

Connection type : Extended end-plate connections
Mode : All bolted with column stiffener

Tested by : N.D.Johnstone & W.R.Walpole (1981) New-Zealand
Test Id. : TEST 3-R

Column : 250 UC 89
Beam : 310 UB 46
Plate thickness : 24 mm
Stiffener thickness : 0.4720"

Fasteners: G8.8- -M24
33/32" Oversize holes
Material : --
Fy = -- ksi
Fu = -- ksi

Major parameters

End plate extended on both sides.

ct = -- "	pt = 4.0160"	li = -- "	pc = 4.0160"
cc = -- "	gt = 5.1180"	gc = 5.1180"	
tp = 0.9449"	nt = 2 X 2	nc = 2 X 2	

Remark 1) End plate extended on tension and compression side.
2) 12 mm web doubler plates on both sides of column used.

No	Moment (k-in)	Rotation (radians) X 1/1000	No	Moment (k-in)	Rotation (radians) X 1/1000
1	0.00	0.00	26	1933.86	23.13
2	79.70	0.20	27	1963.22	25.00
3	159.41	0.40	28	1992.59	27.33
4	239.11	0.60	29	2021.95	29.67
5	310.42	0.70	30	2051.32	32.00
6	381.74	0.80	31	2072.29	34.13
7	453.05	0.90	32	2093.27	36.27
8	553.73	1.35	33	2114.24	38.40
9	654.41	1.80	34	2114.24	40.20
10	748.79	2.65	35	2114.24	42.00
11	843.18	3.50	36	2110.05	43.80
12	943.86	4.20	37	2105.85	45.60
13	1044.54	4.90	38	2101.66	47.40
14	1115.85	5.40	39	2114.24	48.50
15	1187.16	5.90			
16	1258.48	6.40			
17	1346.57	7.65			
18	1434.67	8.90			
19	1522.76	10.25			
20	1610.86	11.60			
21	1686.37	13.40			
22	1761.88	15.20			
23	1813.50	17.30			
24	1875.13	19.40			
25	1904.50	21.27			

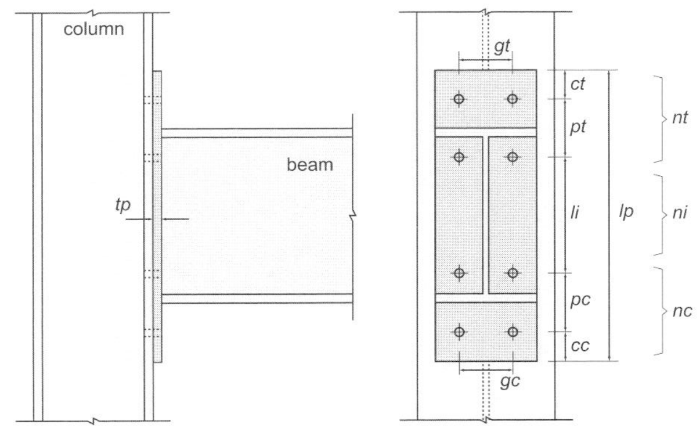

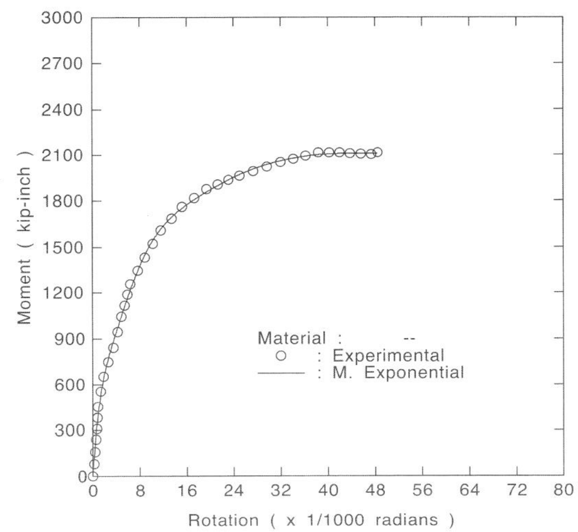

```
Moment-rotation prediction equations     ( R : X 1/1000 radians )

Modified exponential model  :
     Bm = Sum ( Ai X ( 1 - exp( - R/( 2*i*C ) ) ) )  +      Sum ( RKj X ( R - Rj0 ) )  +  Bm0
       C =   0.45416667E+00              Bmo=   0.00000000E+00          Nexp=  6        Nliner=  1
      AI =  -0.79365517E+03   0.18449274E+05  -0.90990931E+05   0.19657341E+06  -0.19657279E+06   0.75464810E+05
     Rj0 =          38.4000
     RKj =  -0.14972027E+01
```

No	Rotation (radians) X 1/1000	Moment (kip-inch) Expri.	Moment (kip-inch) M.Expo.	Connection stiffness (kip-inch) X 1000 Expri.	Connection stiffness (kip-inch) X 1000 M.Expo.
1	0.00	0.00	0.00	0.3985E+03	0.5581E+03
3	0.40	159.41	201.36	0.3985E+03	0.4500E+03
5	0.70	310.42	325.17	0.7131E+03	0.3770E+03
7	0.90	453.05	396.19	0.6241E+03	0.3340E+03
9	1.80	654.41	629.39	0.1847E+03	0.1996E+03
11	3.50	843.18	882.13	0.1290E+03	0.1225E+03
13	4.90	1044.54	1047.73	0.1431E+03	0.1159E+03
15	5.90	1187.16	1161.82	0.1426E+03	0.1116E+03
17	7.65	1346.57	1343.68	0.7048E+02	0.9447E+02
19	10.25	1522.76	1546.00	0.6525E+02	0.6167E+02
21	13.40	1686.37	1694.37	0.4195E+02	0.3585E+02
23	17.30	1818.50	1807.80	0.2697E+02	0.2478E+02
25	21.27	1904.50	1897.84	0.1573E+02	0.2087E+02
27	25.00	1963.22	1968.89	0.1433E+02	0.1704E+02
29	29.67	2021.95	2035.76	0.1258E+02	0.1163E+02
31	34.13	2072.29	2077.34	0.9832E+01	0.7196E+01
33	38.40	2114.24	2101.25	0.4500E+01	0.4214E+01
35	42.00	2114.24	2107.86	-0.1166E+01	0.1073E+01
37	45.60	2105.85	2109.69	-0.2331E+01	0.2367E-01
39	48.50	2114.24	2108.92	0.1144E+02	-0.5179E+00

```
Connection type : Extended end-plate connections
           Mode : All bolted with column stiffener

Tested by   :  N.D.Johnstone & W.R.Walpole (1981)          New-Zealand
Test  Id.   :  TEST 4-L

Column : 250 UC 89                  Fasteners: G8.8- -M24
Beam   : 310 UB 46                          33/32"  Oversize holes
Plate thickness : 16 mm             Material :  --
Stiffener thickness : 0.4720"            Fy =   --   ksi
                                         Fu =   --   ksi

                        Major parameters

     End plate extended on both sides.
     ct =   --   "    pt =  4.5270"    li =   --   "    pc =  4.5270"
     cc =   --   "    gt =  5.1180"    gc =  5.1180"
     tp =  0.6300"    nt = 2 X  2      nc = 2 X  2

Remark  1)  End plate extended on tension and compression side.
        2)  12 mm web doubler plates on both sides of column used.
```

No	Moment (k-in)	Rotation (radians) X 1/1000	No	Moment (k-in)	Rotation (radians) X 1/1000
1	0.00	0.00	26	2029.30	40.90
2	97.53	0.13	27	2067.05	44.75
3	195.06	0.27	28	2104.81	48.60
4	292.60	0.40	29	2156.72	52.55
5	410.58	1.05	30	2208.64	56.50
6	528.56	1.70	31	2227.51	60.15
7	610.36	2.23	32	2246.39	63.80
8	692.16	2.77	33	2279.42	68.40
9	773.96	3.30	34	2312.46	73.00
10	882.51	3.85	35	2345.50	77.00
11	991.06	4.40	36	2378.53	81.00
12	1061.84	5.45	37	2392.69	84.40
13	1132.63	6.50	38	2406.84	87.80
14	1212.86	8.15	39	2397.40	92.60
15	1293.08	9.80	40	2387.96	97.40
16	1373.32	11.35	41	2387.96	100.10
17	1453.55	12.90	42	2387.96	102.80
18	1538.50	15.25			
19	1623.44	17.60			
20	1727.27	21.80			
21	1783.90	24.45			
22	1840.52	27.10			
23	1897.16	30.55			
24	1953.79	34.00			
25	1991.54	37.45			

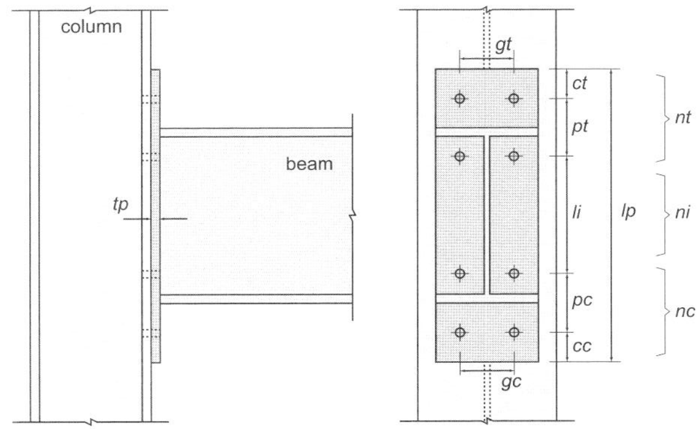

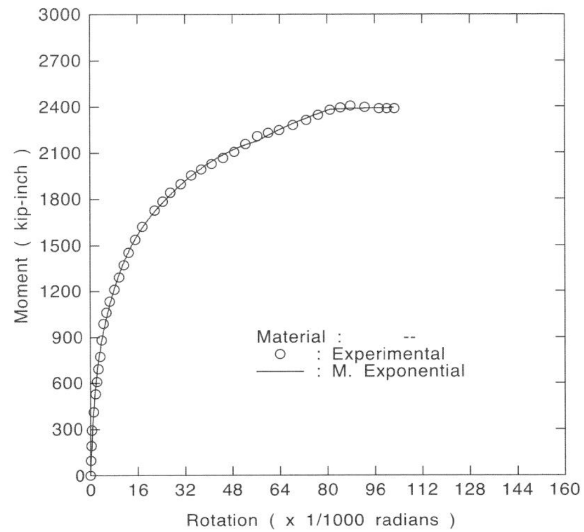

```
Moment-rotation prediction equations      ( R : X 1/1000 radians )

Modified exponential model  :
     Bm = Sum ( Ai X ( 1 - exp( - R/( 2*i*C ) ) ) )  +      Sum ( RKj X ( R - Rj0 ) )  +  Bm0
      C =   0.88999167E+00               Bmo=   0.00000000E+00          Nexp=  6        Nliner=  2
     AI =   0.11767348E+04  -0.17454906E+04  -0.97758674E+04   0.51372133E+05  -0.76488962E+05   0.37711768E+05
    Rj0 =          56.5000          81.0000
    RKj =   0.58228459E+01  -0.57401723E+01
```

No	Rotation (radians) X 1/1000	Moment (kip-inch) Expri.	Moment (kip-inch) M.Expo.	Connection stiffness (kip-inch) X 1000 Expri.	Connection stiffness (kip-inch) X 1000 M.Expo.
1	0.00	0.00	0.00	0.7317E+03	0.4921E+03
3	0.27	195.06	122.18	0.7314E+03	0.4260E+03
5	1.05	410.58	396.48	0.1815E+03	0.2860E+03
7	2.23	610.36	659.01	0.1534E+03	0.1721E+03
9	3.30	773.96	813.15	0.1750E+03	0.1225E+03
11	4.40	991.06	931.86	0.1527E+03	0.9617E+02
13	6.50	1132.63	1106.05	0.6011E+02	0.7303E+02
15	9.80	1293.08	1313.62	0.5024E+02	0.5402E+02
17	12.90	1453.55	1459.45	0.4556E+02	0.4069E+02
19	17.60	1623.44	1617.18	0.3205E+02	0.2787E+02
21	24.45	1783.90	1777.38	0.2137E+02	0.2030E+02
23	30.55	1897.16	1891.34	0.1642E+02	0.1722E+02
25	37.45	1991.54	1998.88	0.1094E+02	0.1390E+02
27	44.75	2067.05	2086.77	0.9807E+01	0.1020E+02
29	52.55	2156.72	2152.27	0.1314E+02	0.6724E+01
31	60.15	2227.51	2214.44	0.5172E+01	0.1001E+02
33	68.40	2279.42	2288.95	0.7182E+01	0.8195E+01
35	77.00	2345.50	2354.15	0.8259E+01	0.7075E+01
37	84.40	2392.69	2384.77	0.4163E+01	0.7852E+00
39	92.60	2397.40	2389.67	-0.1966E+01	0.4443E+00
41	100.10	2387.96	2392.31	0.1543E-13	0.2764E+00

V - 60

Connection type : Extended end-plate connections
Mode : All bolted with column stiffener

Tested by : N.D.Johnstone & W.R.Walpole (1981) New-Zealand
Test Id. : TEST 4-R

Column : 250 UC 89
Beam : 310 UB 46
Plate thickness : 16 mm
Stiffener thickness : 0.4720"

Fasteners: G8.8- -M24
3/32" Oversize holes
Material : --
Fy = -- ksi
Fu = -- ksi

Major parameters

End plate extended on both sides.
ct = -- " pt = 4.5270" li = -- " pc = 4.5270"
cc = -- " gt = 5.1180" gc = 5.1180"
tp = 0.6300" nt = 2 X 2 nc = 2 X 2

Remark 1) End plate extended on tension and compression side.
2) 12 mm web doubler plates on both sides of column used.

No	Moment (k-in)	Rotation (radians) X 1/1000	No	Moment (k-in)	Rotation (radians) X 1/1000
1	0.00	0.00	26	1624.61	26.20
2	80.75	0.60	27	1640.44	28.00
3	161.51	1.20	28	1656.28	29.80
4	242.26	1.80	29	1672.12	31.60
5	323.02	2.40	30	1694.28	33.93
6	402.19	2.93	31	1716.45	36.27
7	481.36	3.47	32	1738.62	38.60
8	560.53	4.00	33	1744.95	40.27
9	630.20	4.70	34	1751.28	41.93
10	699.88	5.40	35	1757.62	43.60
11	769.55	6.10	36	1729.12	45.75
12	836.05	6.97	37	1700.62	47.90
13	902.56	7.83			
14	969.06	8.70			
15	1032.40	9.80			
16	1095.74	10.90			
17	1159.08	12.00			
18	1216.08	12.63			
19	1273.09	13.27			
20	1330.09	13.90			
21	1391.84	15.55			
22	1453.60	17.20			
23	1505.85	19.50			
24	1558.10	21.80			
25	1591.36	24.00			

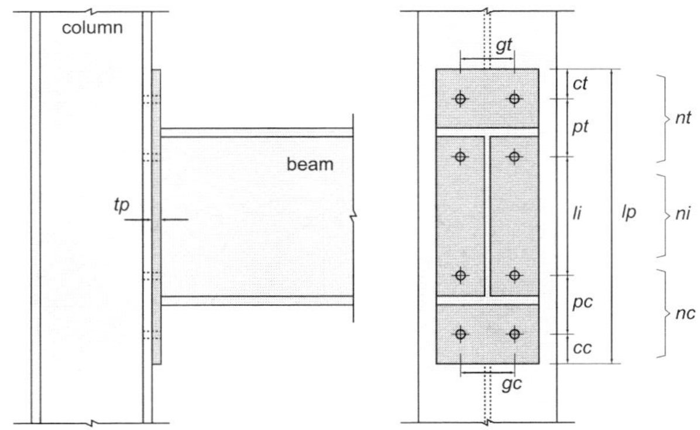

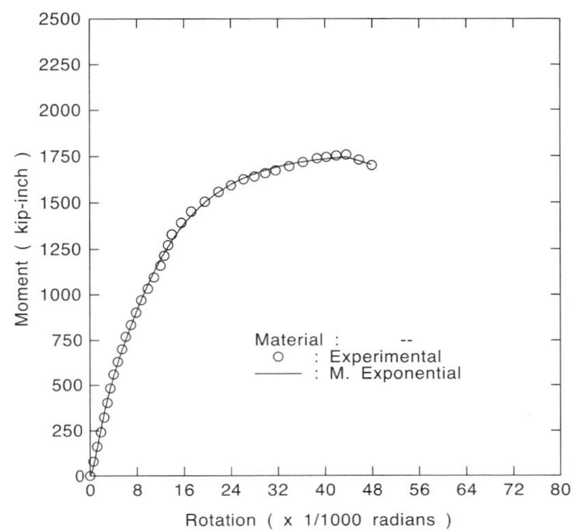

```
Moment-rotation prediction equations     ( R : X 1/1000 radians )

Modified exponential model  :
     Bm = Sum ( Ai X ( 1 - exp( - R/( 2*i*C ) ) ) )  +      Sum ( RKj X ( R - Rj0 ) )  +  Bm0
      C =   0.54916667E+00          Bmo=   0.00000000E+00          Nexp=  6        Nliner=  1
     AI =  -0.99492214E+03   0.83067583E+04  -0.30868642E+05   0.60853031E+05  -0.63369115E+05   0.27835092E+05
    Rj0 =          43.6000
    RKj =  -0.11392705E+02
```

No	Rotation (radians) X 1/1000	Moment (kip-inch) Expri.	Moment (kip-inch) M.Expo.	Connection stiffness (kip-inch) X 1000 Expri.	Connection stiffness (kip-inch) X 1000 M.Expo.
1	0.00	0.00	0.00	0.1346E+03	0.4327E+02
3	1.20	161.51	152.87	0.1346E+03	0.1644E+03
5	2.40	323.02	343.37	0.1419E+03	0.1463E+03
7	3.47	481.36	485.13	0.1484E+03	0.1205E+03
9	4.70	630.20	620.72	0.9953E+02	0.1013E+03
11	6.10	769.55	753.67	0.8934E+02	0.8987E+02
13	7.83	902.56	901.47	0.7674E+02	0.8098E+02
15	9.80	1032.40	1051.03	0.5758E+02	0.7090E+02
17	12.00	1159.08	1193.58	0.7816E+02	0.5864E+02
19	13.27	1273.09	1263.46	0.9000E+02	0.5176E+02
21	15.55	1391.84	1368.69	0.3743E+02	0.4076E+02
23	19.50	1505.85	1500.18	0.2272E+02	0.2686E+02
25	24.00	1591.36	1597.26	0.1511E+02	0.1712E+02
27	28.00	1640.44	1654.00	0.8798E+01	0.1161E+02
29	31.60	1672.12	1689.20	0.9104E+01	0.8135E+01
31	36.27	1716.45	1719.38	0.9501E+01	0.5015E+01
33	40.27	1744.95	1735.64	0.3799E+01	0.3232E+01
35	43.60	1757.62	1744.60	-0.3649E+01	0.2203E+01
37	47.90	1700.62	1703.03	-0.1326E+02	-0.1008E+02

V - 61

Connection type : Extended end-plate connections
Mode : All bolted without column stiffener

Tested by : A.Mazroi (1983) U.S.A
Test Id. : EP3

Column : --	Fasteners: A325- -7/8"D
Beam : W24x100	15/16" Oversize holes
Plate thickness : 3/4"	Material : A36
Stiffener thickness : 0.6250"	Fy = 38.70 ksi
	Fu = -- ksi

Major parameters

End plate extended on both sides.

ct = 4.1250"	pt = 3.5000"	li = 19.7500"	pc = 3.5000"
cc = 4.1250"	pit= 2.6250"	pi = 14.5000"	pic= 2.6250"
gt = 5.5000"	gi = 5.5000"	gc = 5.5000"	
tp = 0.7500"	nt = 2 X 3	ni = 2 X 2	nc = 2 X 3

Remark 1) End plate was stiffened in the plane of beam web.
2) Test was conducted for beam-to-beam connection.

No	Moment (k-in)	Rotation (radians) X 1/1000	No	Moment (k-in)	Rotation (radians) X 1/1000
1	0.00	0.00	26	7283.66	1.89
2	306.93	0.02	27	7432.78	2.04
3	613.85	0.03	28	7515.32	2.20
4	920.78	0.05	29	7597.87	2.35
5	1227.71	0.06	30	7680.41	2.51
6	1619.42	0.10	31	7762.96	2.67
7	2011.14	0.14	32	7845.50	2.83
8	2402.86	0.18	33	7888.04	2.99
9	2790.25	0.22	34	7930.58	3.15
10	3177.64	0.27	35	7973.12	3.31
11	3565.03	0.31	36	8015.66	3.47
12	3875.95	0.37			
13	4186.87	0.42			
14	4497.78	0.48			
15	4808.70	0.53			
16	5116.72	0.62			
17	5424.74	0.70			
18	5732.77	0.79			
19	6040.79	0.87			
20	6256.26	1.01			
21	6471.74	1.14			
22	6687.22	1.27			
23	6836.33	1.43			
24	6985.44	1.58			
25	7134.55	1.73			

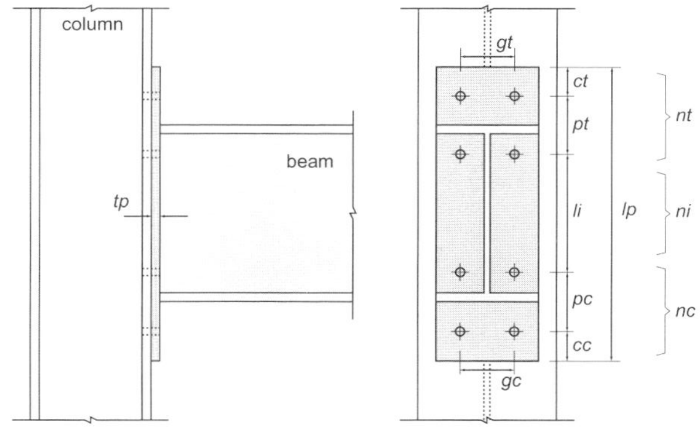

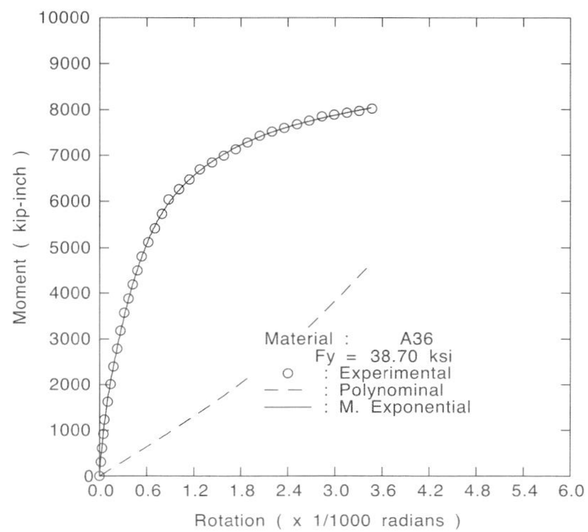

A.5 – 122

```
Moment-rotation prediction equations     ( R : X 1/1000 radians )

Frye and Morris polynominal model  :    R = Sum ( Ai X ( K*Bm )**Pi X 10**Qi )
     xd = 26.750000"   t =  0.750000"  tf =  0.875000"
     A1 =  1.830000   A2 = -1.040000   A3 =  6.380000    K =  0.000515
     P1 =         1   P2 =         3   P3 =         5   Q1 =         0   Q2 =        -1   Q3 =        -3

Modified exponential model  :
     Bm = Sum ( Ai X ( 1 - exp( - R/( 2*i*C ) ) ) )  +     Sum ( RKj X ( R - Rj0 ) )  +  Bm0
       C =   0.32683333E-01               Bmo=   0.00000000E+00          Nexp=  6       Nliner=  1
      AI =   0.31636033E+03   0.16551092E+05  -0.10326562E+06   0.25123998E+06  -0.27371995E+06   0.11587001E+06
     Rj0 =           0.0000
     RKj =   0.30541742E+03
```

No	Rotation (radians) X 1/1000	Moment (kip-inch) Expri.	Moment (kip-inch) Poly.	Moment (kip-inch) M.Expo.	Connection stiffness (kip-inch) X 1000 Expri.	Connection stiffness (kip-inch) X 1000 Poly.	Connection stiffness (kip-inch) X 1000 M.Expo.
1	0.00	0.00	0.00	0.00	0.2019E+05	0.1062E+04	0.2368E+05
3	0.03	613.85	32.39	640.59	0.2013E+05	0.1062E+04	0.1840E+05
5	0.06	1227.71	64.79	1141.66	0.1725E+05	0.1062E+04	0.1471E+05
7	0.14	2011.14	146.28	2059.96	0.1021E+05	0.1063E+04	0.1002E+05
9	0.22	2790.25	235.19	2806.36	0.8542E+04	0.1065E+04	0.8112E+04
11	0.31	3565.03	331.81	3489.88	0.7226E+04	0.1067E+04	0.7023E+04
13	0.42	4186.87	450.41	4205.66	0.5602E+04	0.1072E+04	0.5877E+04
15	0.53	4808.70	569.54	4795.33	0.4821E+04	0.1078E+04	0.4760E+04
17	0.70	5424.74	754.11	5476.70	0.3624E+04	0.1090E+04	0.3331E+04
19	0.87	6040.79	939.96	5955.20	0.2842E+04	0.1105E+04	0.2374E+04
21	1.14	6471.74	1238.32	6468.75	0.1616E+04	0.1138E+04	0.1583E+04
23	1.43	6836.33	1571.43	6857.12	0.9759E+03	0.1185E+04	0.1172E+04
25	1.73	7134.55	1943.65	7170.01	0.9759E+03	0.1254E+04	0.8904E+03
27	2.04	7432.78	2340.20	7408.48	0.7528E+03	0.1345E+04	0.6808E+03
29	2.35	7597.87	2784.48	7597.99	0.5218E+03	0.1468E+04	0.5274E+03
31	2.67	7762.96	3272.00	7748.04	0.5218E+03	0.1617E+04	0.4292E+03
33	2.99	7888.04	3809.56	7874.24	0.2671E+03	0.1763E+04	0.3712E+03
35	3.31	7973.12	4378.61	7986.82	0.2671E+03	0.1820E+04	0.3391E+03

A.5 – 123

V - 62

```
Connection type : Extended end-plate connections
           Mode : All bolted without column stiffener

Tested by  :  A.Mazroi (1983)                          U.S.A
Test  Id.  :  EP4

Column :   --                         Fasteners: A325- -7/8"D
Beam   : W24x100                                15/16" Oversize holes
Plate thickness : 1"                  Material : A36
Stiffener thickness : 0.6250"              Fy =  40.30 ksi
                                           Fu =   --   ksi

                          Major parameters

     End plate extended on both sides.
     ct =  4.1250"   pt =  3.5000"    li = 19.7500"    pc =  3.5000"
     cc =  4.1250"   pit=  2.6250"    pi = 14.5000"    pic=  2.6250"
     gt =  5.5000"   gi =  5.5000"    gc =  5.5000"
     tp =  1.0000"   nt = 2 X  3      ni = 2 X  2      nc = 2 X  3

Remark  1)  End plate was stiffened in the plane of beam web.
        2)  Test was conducted for beam-to-beam connection.
```

No	Moment (k-in)	Rotation (radians) X 1/1000	No	Moment (k-in)	Rotation (radians) X 1/1000
1	0.00	0.00	26	7460.38	0.80
2	335.44	0.01	27	7668.94	0.86
3	670.88	0.03	28	7990.69	1.00
4	1006.32	0.04	29	8217.04	1.15
5	1416.95	0.07	30	8443.39	1.29
6	1642.10	0.07	31	8669.75	1.44
7	1867.26	0.09	32	8896.10	1.58
8	2125.63	0.10			
9	2383.99	0.12			
10	2714.99	0.13			
11	3046.00	0.15			
12	3357.26	0.16			
13	3668.52	0.18			
14	3953.45	0.20			
15	4238.39	0.23			
16	4536.78	0.27			
17	4835.17	0.30			
18	5219.97	0.36			
19	5604.77	0.41			
20	5890.15	0.46			
21	6175.52	0.50			
22	6427.94	0.56			
23	6680.36	0.61			
24	6966.09	0.67			
25	7251.82	0.73			

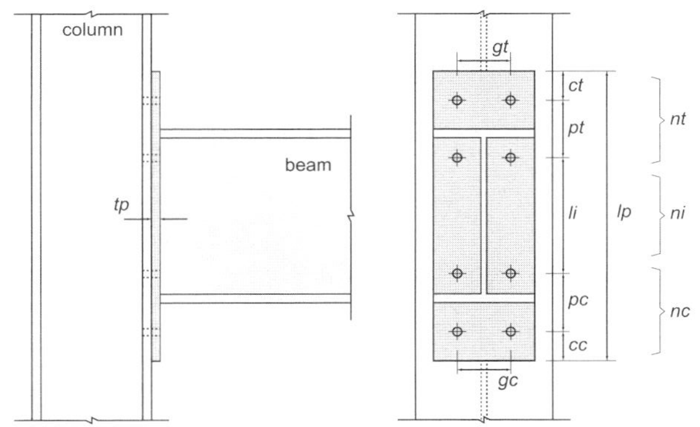

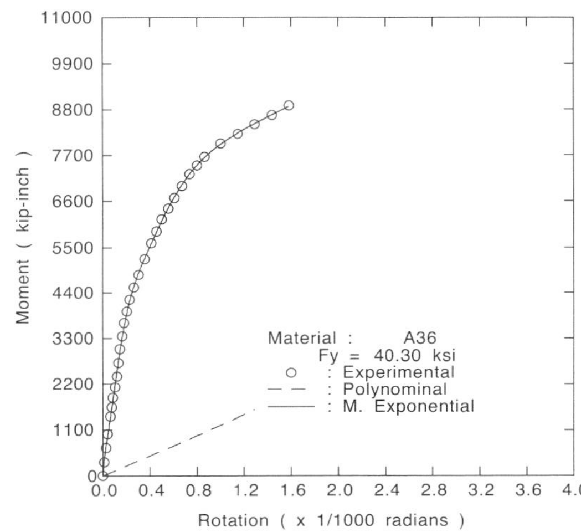

A.5 – 124

```
Moment-rotation prediction equations     ( R : X 1/1000 radians )

Frye and Morris polynominal model   :    R = Sum ( Ai X ( K*Bm )**Pi X 10**Qi )
     xd = 26.750000"    t =  1.000000"  tf =  0.875000"
     A1 =  1.830000   A2 = -1.040000   A3 =  6.380000    K =  0.000459
     P1 =         1   P2 =         3   P3 =         5   Q1 =         0    Q2 =       -1    Q3 =       -3

Modified exponential model  :
     Bm = Sum ( Ai X ( 1 - exp( - R/( 2*i*C ) ) ) )  +      Sum ( RKj X ( R - Rj0 ) )  +  Bm0
       C =   0.16533333E-01            Bmo=   0.00000000E+00          Nexp=  6        Nliner=  1
      AI =   0.69938572E+04  -0.72635585E+05   0.25505486E+06  -0.39084089E+06   0.25726413E+06  -0.49163623E+05
     Rj0 =           0.0000
     RKj =   0.13920519E+04
```

No	Rotation (radians) X 1/1000	Moment (kip-inch) Expri.	Poly.	M.Expo.	Connection stiffness (kip-inch) X 1000 Expri.	Poly.	M.Expo.
1	0.00	0.00	0.00	0.00	0.2522E+05	0.1192E+04	0.3759E+05
2	0.01	335.44	15.85	412.30	0.2513E+05	0.1192E+04	0.2490E+05
3	0.03	670.88	31.82	701.24	0.2513E+05	0.1192E+04	0.1923E+05
4	0.04	1006.32	47.67	944.75	0.2217E+05	0.1192E+04	0.1785E+05
5	0.07	1416.95	77.46	1403.10	0.2078E+05	0.1192E+04	0.1916E+05
6	0.07	1642.10	89.38	1598.21	0.2252E+05	0.1192E+04	0.1984E+05
7	0.09	1867.26	101.30	1799.24	0.1960E+05	0.1192E+04	0.2033E+05
8	0.10	2125.63	122.75	2168.36	0.1435E+05	0.1192E+04	0.2054E+05
9	0.12	2383.99	144.21	2533.80	0.2151E+05	0.1193E+04	0.1994E+05
10	0.13	2714.99	159.12	2778.43	0.2648E+05	0.1193E+04	0.1916E+05
11	0.15	3046.00	174.03	3011.88	0.2271E+05	0.1193E+04	0.1817E+05
12	0.16	3357.26	195.50	3324.48	0.1729E+05	0.1193E+04	0.1654E+05
13	0.18	3668.52	216.98	3606.96	0.1514E+05	0.1194E+04	0.1485E+05
14	0.20	3953.45	244.43	3924.54	0.1239E+05	0.1194E+04	0.1280E+05
15	0.23	4238.39	271.90	4198.09	0.1073E+05	0.1195E+04	0.1104E+05
16	0.27	4536.78	316.11	4564.30	0.8065E+04	0.1196E+04	0.8905E+04
17	0.30	4835.17	360.36	4866.95	0.7635E+04	0.1197E+04	0.7572E+04
18	0.36	5219.97	426.20	5252.30	0.6996E+04	0.1199E+04	0.6589E+04
19	0.41	5604.77	492.17	5602.16	0.6636E+04	0.1202E+04	0.6183E+04
20	0.46	5890.15	546.26	5875.32	0.6342E+04	0.1204E+04	0.5960E+04
21	0.50	6175.52	600.46	6138.05	0.5617E+04	0.1207E+04	0.5707E+04
22	0.56	6427.94	664.43	6430.86	0.4763E+04	0.1211E+04	0.5326E+04
23	0.61	6680.36	728.60	6701.22	0.4643E+04	0.1214E+04	0.4866E+04
24	0.67	6966.09	805.71	6991.31	0.4500E+04	0.1220E+04	0.4267E+04
25	0.73	7251.82	883.16	7243.41	0.3882E+04	0.1225E+04	0.3680E+04
26	0.80	7460.38	962.64	7461.52	0.3259E+04	0.1232E+04	0.3149E+04
27	0.86	7668.94	1041.47	7648.23	0.2969E+04	0.1239E+04	0.2701E+04
28	1.00	7990.69	1212.65	7968.42	0.1960E+04	0.1256E+04	0.2032E+04
29	1.15	8217.04	1397.77	8234.75	0.1548E+04	0.1277E+04	0.1657E+04
30	1.29	8443.39	1586.46	8463.12	0.1548E+04	0.1303E+04	0.1488E+04
31	1.44	8669.75	1779.05	8674.97	0.1548E+04	0.1333E+04	0.1421E+04
32	1.58	8896.10	1976.32	8880.58	0.1548E+04	0.1367E+04	0.1397E+04

V - 63

```
Connection type : Extended end-plate connections
           Mode : All bolted without column stiffener

Tested by  :  A.Mazroi (1983)                          U.S.A
Test  Id.  :  EP5

Column :   --                     Fasteners: A325- -1"D
Beam   : W27x114                           - 1/16" Oversize holes
Plate thickness : 1"              Material : A36
Stiffener thickness : 0.6250"         Fy =  38.20 ksi
                                      Fu =    --   ksi
```

Major parameters

End plate extended on both sides.

ct = 4.3750"	pt = 3.9300"	li = 22.3901"	pc = 3.9300"
cc = 4.3750"	pit= 2.6667"	pi = 17.0567"	pic= 2.6667"
gt = 5.5000"	gi = 5.5000"	gc = 5.5000"	
tp = 1.0000"	nt = 2 X 3	ni = 2 X 2	nc = 2 X 3

Remark 1) End plate was stiffened in the plane of beam web by an end-plate.
2) Test was conducted for beam-to-beam connection.

No	Moment (k-in)	Rotation (radians) X 1/1000
1	0.00	0.00
2	388.48	0.01
3	776.95	0.02
4	1165.43	0.03
5	1522.83	0.05
6	1880.24	0.06
7	2237.64	0.08
8	2628.07	0.10
9	3018.49	0.12
10	3428.71	0.14
11	3838.92	0.16
12	4275.41	0.18
13	4711.90	0.21
14	5128.01	0.25
15	5544.12	0.29
16	6006.40	0.34
17	6468.67	0.38
18	6970.91	0.42
19	7473.16	0.47
20	7798.76	0.50
21	8124.36	0.54
22	8449.96	0.58
23	8812.91	0.62
24	9175.87	0.66
25	9587.79	0.76
26	9999.70	0.85
27	10411.62	0.94

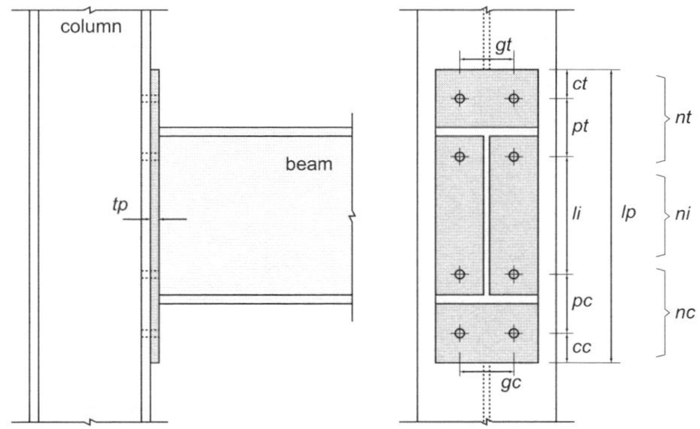

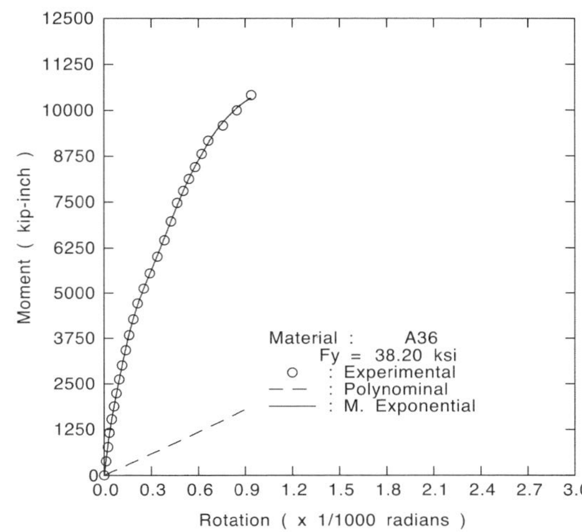

```
Moment-rotation prediction equations      ( R : X 1/1000 radians )

Frye and Morris polynominal model   :    R = Sum ( Ai X ( K*Bm )**Pi X 10**Qi )
     xd = 30.250100"    t =  1.000000"  tf =  1.000000"
     A1 =  1.830000   A2 = -1.040000   A3 =  6.380000    K =  0.000279
     P1 =        1    P2 =        3    P3 =        5    Q1 =        0    Q2 =       -1    Q3 =       -3

Modified exponential model  :
     Bm = Sum ( Ai X ( 1 - exp( - R/( 2*i*C ) ) ) )  +      Sum ( RKj X ( R - Rj0 ) )  +  Bm0
       C =   0.10650000E-01           Bmo=   0.00000000E+00          Nexp=  6         Nliner=  1
      AI =  -0.35556629E+04   0.81082554E+05  -0.52248609E+06   0.13792423E+07  -0.15864402E+07   0.66234351E+06
     Rj0 =             0.6200
     RKj =   0.11484273E+04
```

No	Rotation (radians) X 1/1000	Moment (kip-inch) Expri.	Moment (kip-inch) Poly.	Moment (kip-inch) M.Expo.	Connection stiffness (kip-inch) X 1000 Expri.	Connection stiffness (kip-inch) X 1000 Poly.	Connection stiffness (kip-inch) X 1000 M.Expo.
1	0.00	0.00	0.00	0.00	0.3438E+05	0.1956E+04	0.3459E+05
2	0.01	388.48	22.10	401.80	0.3423E+05	0.1956E+04	0.3540E+05
3	0.02	776.95	44.39	789.33	0.3423E+05	0.1956E+04	0.3232E+05
4	0.03	1165.43	66.49	1134.34	0.3000E+05	0.1956E+04	0.2882E+05
5	0.05	1522.83	95.24	1531.00	0.2440E+05	0.1956E+04	0.2542E+05
6	0.06	1880.24	123.80	1886.56	0.2440E+05	0.1956E+04	0.2349E+05
7	0.08	2237.64	152.55	2223.67	0.2289E+05	0.1956E+04	0.2247E+05
8	0.10	2628.07	188.74	2631.48	0.2110E+05	0.1957E+04	0.2164E+05
9	0.12	3018.49	224.94	3023.33	0.2017E+05	0.1957E+04	0.2068E+05
10	0.14	3428.71	267.02	3452.20	0.1908E+05	0.1958E+04	0.1914E+05
11	0.16	3838.92	309.10	3844.09	0.1779E+05	0.1958E+04	0.1728E+05
12	0.18	4275.41	361.97	4278.10	0.1617E+05	0.1959E+04	0.1490E+05
13	0.21	4711.90	414.87	4652.26	0.1384E+05	0.1960E+04	0.1291E+05
14	0.25	5128.01	493.27	5127.19	0.1040E+05	0.1962E+04	0.1107E+05
15	0.29	5544.12	571.75	5554.30	0.1024E+05	0.1964E+04	0.1045E+05
16	0.34	6006.40	662.11	6035.39	0.1005E+05	0.1967E+04	0.1055E+05
17	0.38	6468.67	752.59	6527.35	0.1130E+05	0.1970E+04	0.1082E+05
18	0.42	6970.91	832.40	6966.44	0.1240E+05	0.1974E+04	0.1082E+05
19	0.47	7473.16	912.33	7398.91	0.1045E+05	0.1977E+04	0.1048E+05
20	0.50	7798.76	986.88	7783.78	0.8648E+04	0.1981E+04	0.9892E+04
21	0.54	8124.36	1061.37	8141.28	0.8648E+04	0.1985E+04	0.9096E+04
22	0.58	8449.96	1136.21	8466.99	0.8591E+04	0.1989E+04	0.8168E+04
23	0.62	8812.91	1220.77	8791.26	0.8540E+04	0.1995E+04	0.8209E+04
24	0.66	9175.87	1305.54	9116.77	0.7256E+04	0.2000E+04	0.7117E+04
25	0.76	9587.79	1489.58	9672.44	0.4477E+04	0.2014E+04	0.5047E+04
26	0.85	9999.70	1677.09	10062.86	0.4477E+04	0.2030E+04	0.3535E+04
27	0.94	10411.62	1863.85	10338.85	0.4477E+04	0.2048E+04	0.2541E+04

A.5 – 127

V - 64

Connection type : Extended end-plate connections
Mode : All bolted without column stiffener

Tested by : A.Mazroi (1983) U.S.A
Test Id. : EP6

Column : --
Beam : W27x114
Plate thickness : 1 1/4"
Stiffener thickness : 0.6250"

Fasteners: A325- -1"D
1 1/16" Oversize holes
Material : A36
Fy = 39.20 ksi
Fu = -- ksi

Major parameters

End plate extended on both sides.

ct = 4.6250"	pt = 4.9300"	li = 21.3900"	pc = 4.9300"
cc = 4.6250"	pit= 3.0000"	pi = 15.3900"	pic= 3.0000"
gt = 6.5000"	gi = 6.5000"	gc = 6.5000"	
tp = 1.2500"	nt = 2 X 3	ni = 2 X 2	nc = 2 X 3

Remark 1) End plate was stiffened in the plane of beam web by an end-plate.
2) Test was conducted for beam-to-beam connection.

No	Moment (k-in)	Rotation (radians) X 1/1000
1	0.00	0.00
2	345.52	0.00
3	1036.56	0.00
4	1568.21	0.01
5	2206.22	0.04
6	2764.52	0.05
7	3336.04	0.07
8	3934.28	0.10
9	4452.82	0.13
10	4984.57	0.16
11	5649.32	0.21
12	6181.27	0.25
13	6646.81	0.32
14	7258.60	0.40
15	7737.46	0.47
16	8181.10	0.56
17	8575.94	0.65
18	8926.51	0.74
19	9281.60	0.84
20	9494.81	0.93
21	9734.66	1.03
22	10001.15	1.13

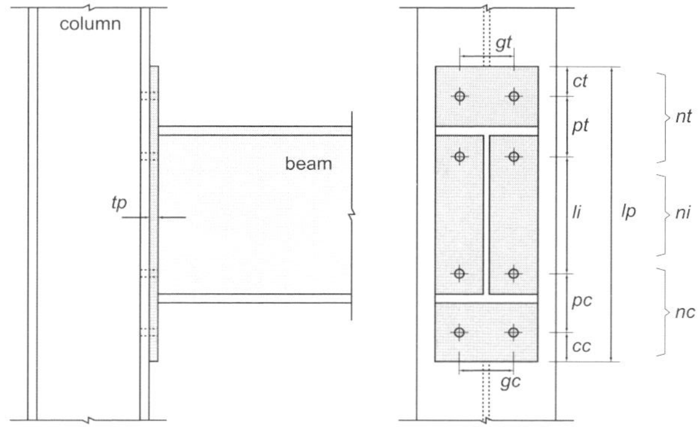

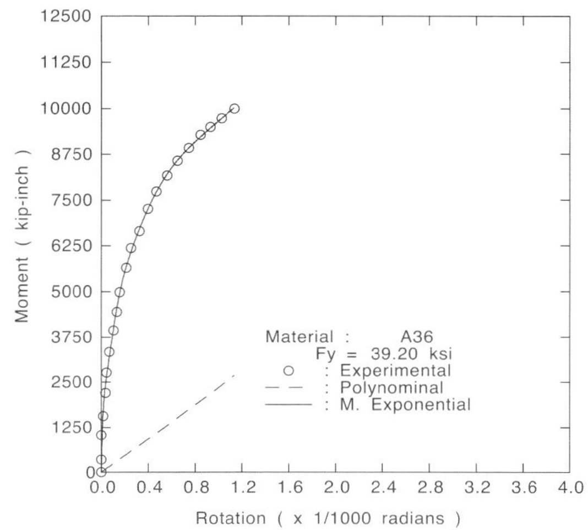

A.5 – 128

```
Moment-rotation prediction equations      ( R : X 1/1000 radians )

Frye and Morris polynominal model  :    R = Sum ( Ai X ( K*Bm )**Pi X 10**Qi )
     xd = 31.250000"   t =  1.250000"  tf =  1.000000"
     A1 =  1.830000   A2 = -1.040000   A3 =  6.380000    K =  0.000236
     P1 =        1    P2 =        3    P3 =        5    Q1 =        0    Q2 =       -1    Q3 =       -3

Modified exponential model  :
     Bm = Sum ( Ai X ( 1 - exp( - R/( 2*i*C ) ) ) )  +      Sum ( RKj X ( R - Rj0 ) )  +  Bm0
       C =   0.95250000E-02           Bmo=   0.00000000E+00          Nexp=  6       Nliner=  1
      AI =   0.74280345E+04  -0.23899132E+05   0.50597183E+04   0.12323549E+06  -0.22025181E+06   0.11551149E+06
     Rj0 =           0.0000
     RKj =   0.25995267E+04
```

No	Rotation (radians) X 1/1000	Moment (kip-inch) Expri.	Moment (kip-inch) Poly.	Moment (kip-inch) M.Expo.	Connection stiffness (kip-inch) X 1000 Expri.	Connection stiffness (kip-inch) X 1000 Poly.	Connection stiffness (kip-inch) X 1000 M.Expo.
1	0.00	0.00	0.00	0.00	0.1152E+07	0.2312E+04	0.1667E+06
2	0.00	345.52	0.69	50.28	0.1102E+07	0.2312E+04	0.1659E+06
3	0.00	1036.56	2.31	163.75	0.9420E+06	0.2312E+04	0.1583E+06
4	0.01	1568.21	34.68	1612.13	0.3547E+05	0.2312E+04	0.6415E+05
5	0.04	2206.22	80.91	2417.11	0.4406E+05	0.2312E+04	0.2675E+05
6	0.05	2764.52	106.35	2685.44	0.4266E+05	0.2312E+04	0.2293E+05
7	0.07	3336.04	154.90	3159.79	0.2322E+05	0.2312E+04	0.2280E+05
8	0.10	3934.28	240.46	3997.37	0.1839E+05	0.2313E+04	0.2160E+05
9	0.13	4452.82	300.60	4520.95	0.1982E+05	0.2314E+04	0.1855E+05
10	0.16	4984.57	363.07	4976.58	0.1728E+05	0.2315E+04	0.1528E+05
11	0.21	5649.32	485.75	5660.17	0.1261E+05	0.2317E+04	0.1103E+05
12	0.25	6181.27	583.06	6086.04	0.1041E+05	0.2319E+04	0.9434E+04
13	0.32	6646.81	745.42	6696.65	0.7378E+04	0.2324E+04	0.8143E+04
14	0.40	7258.60	919.72	7266.35	0.7476E+04	0.2331E+04	0.7039E+04
15	0.47	7737.46	1082.86	7721.41	0.5957E+04	0.2338E+04	0.5966E+04
16	0.56	8181.10	1299.57	8213.89	0.4600E+04	0.2349E+04	0.4711E+04
17	0.65	8575.94	1511.24	8593.08	0.4097E+04	0.2363E+04	0.3840E+04
18	0.74	8926.51	1731.67	8922.37	0.3646E+04	0.2379E+04	0.3269E+04
19	0.84	9281.60	1971.19	9232.45	0.2961E+04	0.2399E+04	0.2927E+04
20	0.93	9494.81	2180.04	9476.82	0.2476E+04	0.2418E+04	0.2772E+04
21	1.03	9734.66	2414.61	9740.75	0.2470E+04	0.2443E+04	0.2681E+04
22	1.13	10001.15	2681.30	10027.42	0.2467E+04	0.2474E+04	0.2634E+04

V - 65

```
Connection type : Extended end-plate connections
           Mode : All bolted without column stiffener

Tested by  :  A.Mazroi (1983)                              U.S.A
Test  Id.  :  EP7

Column :   --                          Fasteners: A325- -1 1/8"D
Beam   : W24x162                             1 1/4" Oversize holes
Plate thickness : 1 1/2"               Material : A36
Stiffener thickness : 0.7500"                Fy =  33.00 ksi
                                             Fu =   --   ksi
```

Major parameters

End plate extended on both sides.

ct = 4.6250"	pt = 4.4700"	li = 19.3100"	pc = 4.4700"
cc = 4.6250"	pit= 3.0000"	pi = 13.3100"	pic= 3.0000"
gt = 5.5000"	gi = 5.5000"	gc = 5.5000"	
tp = 1.5000"	nt = 2 X 3	ni = 2 X 2	nc = 2 X 3

Remark 1) End plate was stiffened in the plane of beam web by an end-plate.
2) Test was conducted for beam-to-beam connection.

No	Moment (k-in)	Rotation (radians) X 1/1000	No	Moment (k-in)	Rotation (radians) X 1/1000
1	0.00	0.00	26	11188.21	1.15
2	396.75	0.01	27	11524.91	1.27
3	793.50	0.02	28	11798.72	1.39
4	1190.24	0.03			
5	1758.68	0.06			
6	2327.11	0.09			
7	2800.82	0.11			
8	3274.54	0.13			
9	3633.01	0.14			
10	3991.48	0.16			
11	4363.74	0.18			
12	4736.00	0.20			
13	5155.89	0.23			
14	5575.78	0.26			
15	6036.50	0.30			
16	6497.22	0.34			
17	6958.24	0.39			
18	7419.67	0.46			
19	7908.14	0.52			
20	8396.62	0.59			
21	8735.87	0.64			
22	9374.06	0.74			
23	9938.09	0.85			
24	10398.34	0.95			
25	10788.62	1.05			

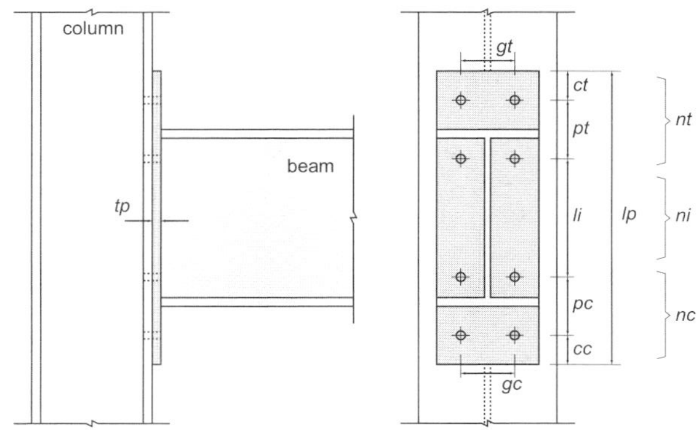

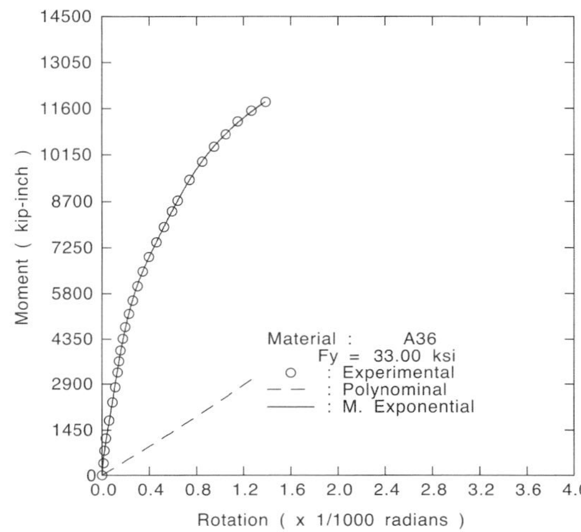

```
Moment-rotation prediction equations     ( R : X 1/1000 radians )

Frye and Morris polynominal model  :     R = Sum ( Ai X ( K*Bm )**Pi X 10**Qi )
     xd = 28.250000"   t =  1.500000"  tf =  1.125000"
     A1 =  1.830000   A2 = -1.040000   A3 =  6.380000    K =  0.000235
     P1 =        1    P2 =        3    P3 =        5    Q1 =        0    Q2 =       -1    Q3 =       -3

Modified exponential model  :
     Bm = Sum ( Ai X ( 1 - exp( - R/( 2*i*C ) ) ) )  +     Sum ( RKj X ( R - Rj0 ) )  +  Bm0
       C =   0.14316667E-01          Bmo=   0.00000000E+00          Nexp=  6         Nliner=  1
      AI =   0.30366525E+04   0.78667161E+04  -0.17518144E+06   0.57332455E+06  -0.71262014E+06   0.31224055E+06
     Rj0 =          0.0000
     RKj =   0.23144874E+04
```

No	Rotation (radians) X 1/1000	Moment (kip-inch) Expri.	Moment (kip-inch) Poly.	Moment (kip-inch) M.Expo.	Connection stiffness (kip-inch) X 1000 Expri.	Connection stiffness (kip-inch) X 1000 Poly.	Connection stiffness (kip-inch) X 1000 M.Expo.
1	0.00	0.00	0.00	0.00	0.3607E+05	0.2329E+04	0.4971E+05
2	0.01	396.75	25.62	477.49	0.3607E+05	0.2329E+04	0.3632E+05
3	0.02	793.50	51.24	825.39	0.3607E+05	0.2329E+04	0.2787E+05
4	0.03	1190.24	76.86	1106.56	0.3167E+05	0.2329E+04	0.2380E+05
5	0.06	1758.68	140.91	1721.43	0.2067E+05	0.2329E+04	0.2210E+05
6	0.09	2327.11	204.96	2344.99	0.2115E+05	0.2330E+04	0.2323E+05
7	0.11	2800.82	256.22	2859.92	0.2153E+05	0.2330E+04	0.2343E+05
8	0.13	3274.54	307.49	3368.70	0.2693E+05	0.2331E+04	0.2268E+05
9	0.14	3633.01	335.46	3636.75	0.2987E+05	0.2331E+04	0.2196E+05
10	0.16	3991.48	363.44	3895.08	0.2555E+05	0.2332E+04	0.2107E+05
11	0.18	4363.74	411.24	4309.58	0.1816E+05	0.2333E+04	0.1933E+05
12	0.20	4736.00	459.06	4687.02	0.1632E+05	0.2334E+04	0.1749E+05
13	0.23	5155.89	531.40	5187.16	0.1354E+05	0.2335E+04	0.1483E+05
14	0.26	5575.78	603.79	5610.82	0.1250E+05	0.2337E+04	0.1258E+05
15	0.30	6036.50	700.77	6083.99	0.1110E+05	0.2340E+04	0.1037E+05
16	0.34	6497.22	797.87	6482.98	0.1008E+05	0.2343E+04	0.8978E+04
17	0.39	6958.24	920.87	6926.81	0.8096E+04	0.2348E+04	0.8050E+04
18	0.46	7419.67	1069.94	7420.96	0.7305E+04	0.2354E+04	0.7586E+04
19	0.52	7908.14	1226.48	7916.75	0.7345E+04	0.2362E+04	0.7332E+04
20	0.59	8396.62	1383.56	8394.82	0.7184E+04	0.2371E+04	0.7026E+04
21	0.64	8735.87	1497.37	8725.14	0.6817E+04	0.2378E+04	0.6726E+04
22	0.74	9374.06	1740.92	9368.20	0.5793E+04	0.2396E+04	0.5917E+04
23	0.85	9938.09	1997.29	9951.56	0.4887E+04	0.2417E+04	0.4993E+04
24	0.95	10398.34	2243.38	10418.97	0.4272E+04	0.2441E+04	0.4214E+04
25	1.05	10788.62	2482.97	10796.79	0.3877E+04	0.2467E+04	0.3627E+04
26	1.15	11188.21	2749.37	11161.13	0.3337E+04	0.2499E+04	0.3150E+04
27	1.27	11524.91	3038.10	11499.71	0.2626E+04	0.2537E+04	0.2815E+04
28	1.39	11798.72	3342.58	11823.40	0.2282E+04	0.2583E+04	0.2598E+04

A.5 – 131

V - 66

```
Connection type : Extended end-plate connections
           Mode : All bolted without column stiffener

Tested by  :  A.Mazroi (1983)                         U.S.A
Test  Id.  :  EP7 With Shim

Column :   --                     Fasteners: A325- -1 1/8"D
Beam   : W24x162                        1 1/4" Oversize holes
Plate thickness : 1 1/2"          Material : A36
Stiffener thickness : 0.7500"          Fy =  33.00 ksi
                                       Fu =   --   ksi
```

Major parameters

End plate extended on both sides.

```
ct =  4.6250"    pt =  4.4700"    li = 19.3100"    pc =  4.4700"
cc =  4.6250"    pit=  3.0000"    pi = 13.3100"    pic=  3.0000"
gt =  5.5000"    gi =  5.5000"    gc =  5.5000"
tp =  1.5000"    nt = 2 X  3      ni = 2 X  2      nc = 2 X  3
```

Remark 1) End plate was stiffened in the plane of beam web by an end-plate.
2) Test was conducted for beam-to-beam connection.

No	Moment (k-in)	Rotation (radians) X 1/1000	No	Moment (k-in)	Rotation (radians) X 1/1000
1	0.00	0.00	26	12011.33	0.35
2	392.84	0.01	27	12284.28	0.44
3	764.36	0.03	28	12781.66	0.59
4	1135.88	0.04	29	13103.89	0.75
5	1853.04	0.05	30	13392.87	0.93
6	2238.26	0.05	31	13605.23	1.12
7	2623.48	0.05	32	13781.47	1.29
8	3049.43	0.06	33	13955.81	1.49
9	3475.39	0.07	34	14130.14	1.68
10	3962.26	0.08	35	14319.84	1.89
11	4449.12	0.10	36	14509.53	2.09
12	4854.71	0.10	37	14674.43	2.29
13	5260.30	0.10	38	14814.54	2.46
14	5868.68	0.11			
15	6456.89	0.12			
16	7045.10	0.13			
17	7639.99	0.14			
18	8234.87	0.15			
19	8836.83	0.17			
20	9438.79	0.19			
21	9810.82	0.20			
22	10182.84	0.21			
23	10791.62	0.23			
24	11224.92	0.26			
25	11685.80	0.31			

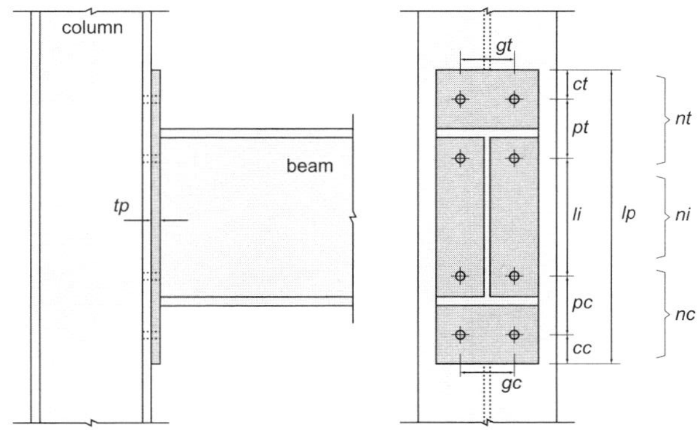

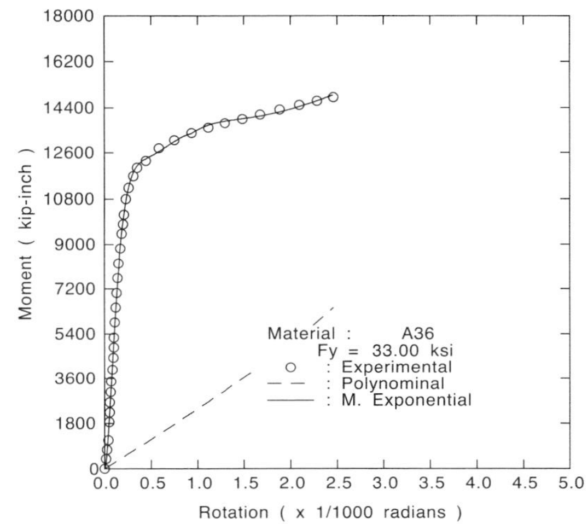

```
Moment-rotation prediction equations     ( R : X 1/1000 radians )

Frye and Morris polynominal model  :    R = Sum ( Ai X ( K*Bm )**Pi X 10**Qi )
     xd = 28.250000"   t =  1.500000"  tf =  1.125000"
     A1 =  1.830000   A2 = -1.040000   A3 =  6.380000    K =  0.000235
     P1 =        1    P2 =        3    P3 =        5    Q1 =        0    Q2 =       -1    Q3 =       -3

Modified exponential model  :
     Bm = Sum ( Ai X ( 1 - exp( - R/( 2*i*C ) ) ) )  +     Sum ( RKj X ( R - Rj0 ) )  +  Bm0
       C =  0.23758333E-01          Bmo=  0.00000000E+00          Nexp=  6        Nliner=  1
      AI = -0.26969202E+04  -0.13697045E+06   0.79073889E+06  -0.16426884E+07   0.15307921E+07  -0.52660514E+06
     Rj0 =          0.9300
     RKj =  0.14929365E+04
```

No	Rotation (radians) X 1/1000	Moment (kip-inch) Expri.	Moment (kip-inch) Poly.	Moment (kip-inch) M.Expo.	Connection stiffness (kip-inch) X 1000 Expri.	Connection stiffness (kip-inch) X 1000 Poly.	Connection stiffness (kip-inch) X 1000 M.Expo.
1	0.00	0.00	0.00	0.00	0.3022E+05	0.2329E+04	0.2447E+04
3	0.03	764.36	60.55	679.26	0.2858E+05	0.2329E+04	0.4457E+05
5	0.05	1853.04	121.11	2106.75	0.3616E+06	0.2329E+04	0.6216E+05
7	0.05	2623.48	125.77	2231.75	0.3514E+06	0.2329E+04	0.6283E+05
9	0.07	3475.39	167.70	3396.31	0.4514E+05	0.2330E+04	0.6571E+05
11	0.10	4449.12	221.28	4892.02	0.9872E+05	0.2330E+04	0.6352E+05
13	0.10	5260.30	237.59	5331.50	0.1038E+06	0.2330E+04	0.6200E+05
15	0.12	6456.89	281.86	6461.31	0.5347E+05	0.2331E+04	0.5669E+05
17	0.14	7639.99	326.15	7479.40	0.7436E+05	0.2331E+04	0.5038E+05
19	0.17	8836.83	390.26	8733.44	0.3087E+05	0.2332E+04	0.4084E+05
21	0.20	9810.82	463.74	9856.22	0.3100E+05	0.2334E+04	0.3067E+05
23	0.23	10791.62	536.09	10671.45	0.2518E+05	0.2335E+04	0.2222E+05
25	0.31	11685.80	718.33	11804.32	0.8639E+04	0.2340E+04	0.8544E+04
27	0.44	12284.28	1034.57	12356.21	0.3162E+04	0.2352E+04	0.1971E+04
29	0.75	13103.89	1763.69	13053.36	0.1786E+04	0.2398E+04	0.2349E+04
31	1.12	13605.23	2654.60	13709.13	0.1075E+04	0.2487E+04	0.1243E+04
33	1.49	13955.81	3599.12	13965.74	0.9104E+03	0.2625E+04	0.4949E+03
35	1.89	14319.84	4689.71	14235.32	0.9102E+03	0.2843E+04	0.9027E+03
37	2.29	14674.43	5886.82	14675.95	0.8279E+03	0.3155E+04	0.1260E+04

V - 67

```
Connection type : Extended end-plate connections
           Mode : All bolted without column stiffener

Tested by  :  A.Mazroi (1983)                          U.S.A
Test  Id.  :  EP8

Column :   --                         Fasteners: A325- -1 1/8"D
Beam   : W24x162                            1 1/4" Oversize holes
Plate thickness : 1 3/4"              Material : A36
Stiffener thickness : 0.7500"              Fy =  44.60 ksi
                                           Fu =   --   ksi
```

Major parameters

```
End plate extended on both sides.
ct =  4.8750"    pt =  5.4700"    li = 18.3100"    pc =  5.4700"
cc =  4.8750"    pit=  3.3750"    pi = 11.5600"    pic=  3.3750"
gt =  6.5000"    gi =  6.5000"    gc =  6.5000"
tp =  1.7500"    nt = 2 X  3      ni = 2 X  2      nc = 2 X  3
```

Remark 1) End plate was stiffened in the plane of beam web by an end-plate.
2) Test was conducted for beam-to-beam connection.

No	Moment (k-in)	Rotation (radians) X 1/1000
1	0.00	0.00
2	580.35	0.02
3	1160.70	0.04
4	1741.15	0.06
5	2321.60	0.08
6	2935.64	0.10
7	3549.68	0.12
8	4144.01	0.16
9	4738.33	0.19
10	5366.75	0.25
11	5995.16	0.30
12	6610.06	0.35
13	7224.96	0.40
14	7820.03	0.47
15	8415.11	0.54
16	8816.60	0.59
17	9218.09	0.65
18	9619.58	0.70
19	10039.47	0.77
20	10459.36	0.84
21	10879.25	0.92
22	11440.01	1.05
23	11904.17	1.18
24	12271.74	1.31
25	12619.13	1.45
26	12946.36	1.58

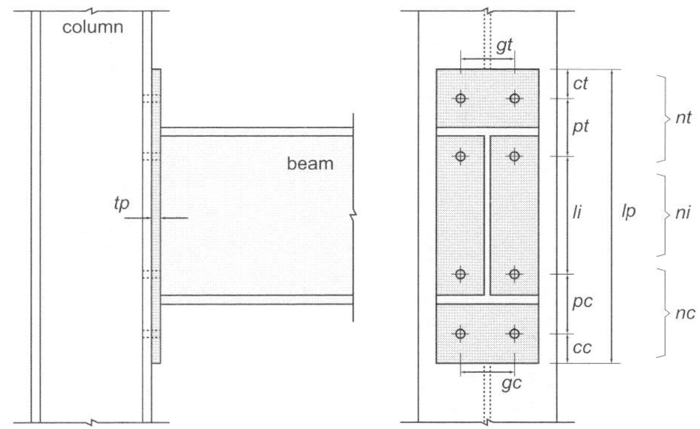

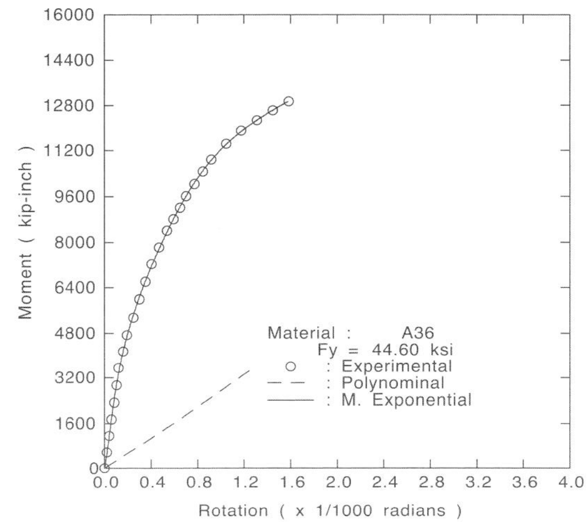

```
Moment-rotation prediction equations     ( R : X 1/1000 radians )

Frye and Morris polynominal model  :     R = Sum ( Ai X ( K*Bm )**Pi X 10**Qi )
     xd = 29.250000"    t =  1.750000"  tf =  1.125000"
     A1 =  1.830000   A2 = -1.040000    A3 =  6.380000     K =  0.000203
     P1 =         1   P2 =         3    P3 =         5    Q1 =         0    Q2 =        -1    Q3 =        -3

Modified exponential model  :
     Bm = Sum ( Ai X ( 1 - exp( - R/( 2*i*C ) ) ) )  +      Sum ( RKj X ( R - Rj0 ) )  +  Bm0
      C =    0.18083333E-01            Bmo=   0.00000000E+00           Nexp=  6         Nliner=  1
     AI =   -0.19054302E+04   0.47831520E+04  -0.28651897E+05   0.15716662E+06  -0.27758001E+06   0.15600466E+06
     Rj0 =            0.0000
     RKj =    0.20196656E+04
```

No	Rotation (radians) X 1/1000	Moment (kip-inch) Expri.	Moment (kip-inch) Poly.	Moment (kip-inch) M.Expo.	Connection stiffness (kip-inch) X 1000 Expri.	Connection stiffness (kip-inch) X 1000 Poly.	Connection stiffness (kip-inch) X 1000 M.Expo.
1	0.00	0.00	0.00	0.00	0.2976E+05	0.2693E+04	0.1969E+05
2	0.02	580.35	52.51	513.73	0.2976E+05	0.2693E+04	0.2961E+05
3	0.04	1160.70	105.02	1115.74	0.2792E+05	0.2693E+04	0.3145E+05
4	0.06	1741.15	165.61	1811.95	0.2580E+05	0.2693E+04	0.3004E+05
5	0.08	2321.60	226.21	2457.12	0.2933E+05	0.2694E+04	0.2719E+05
6	0.10	2935.64	277.39	2948.59	0.3232E+05	0.2694E+04	0.2455E+05
7	0.12	3549.68	328.58	3390.90	0.2686E+05	0.2695E+04	0.2205E+05
8	0.16	4144.01	425.59	4110.57	0.1651E+05	0.2696E+04	0.1812E+05
9	0.19	4738.33	522.65	4709.22	0.1466E+05	0.2698E+04	0.1531E+05
10	0.25	5366.75	664.29	5438.64	0.1197E+05	0.2701E+04	0.1273E+05
11	0.30	5995.16	806.10	6065.17	0.1184E+05	0.2705E+04	0.1127E+05
12	0.35	6610.06	948.12	6632.37	0.1171E+05	0.2710E+04	0.1040E+05
13	0.40	7224.96	1090.38	7161.41	0.1052E+05	0.2715E+04	0.9782E+04
14	0.47	7820.03	1269.59	7785.12	0.9016E+04	0.2723E+04	0.9127E+04
15	0.54	8415.11	1449.33	8366.31	0.8061E+04	0.2733E+04	0.8481E+04
16	0.59	8816.60	1600.45	8819.80	0.7254E+04	0.2742E+04	0.7917E+04
17	0.65	9218.09	1752.33	9242.41	0.7254E+04	0.2751E+04	0.7338E+04
18	0.70	9619.58	1904.47	9632.16	0.6661E+04	0.2762E+04	0.6759E+04
19	0.77	10039.47	2103.85	10088.07	0.5885E+04	0.2777E+04	0.6036E+04
20	0.84	10459.36	2302.16	10494.68	0.5885E+04	0.2794E+04	0.5364E+04
21	0.92	10879.25	2501.40	10855.23	0.5331E+04	0.2813E+04	0.4762E+04
22	1.05	11440.01	2870.83	11413.25	0.3920E+04	0.2852E+04	0.3866E+04
23	1.18	11904.17	3246.73	11877.11	0.3141E+04	0.2898E+04	0.3212E+04
24	1.31	12271.74	3639.80	12273.55	0.2658E+04	0.2952E+04	0.2763E+04
25	1.45	12619.13	4040.69	12627.19	0.2465E+04	0.3015E+04	0.2468E+04
26	1.58	12946.36	4465.96	12953.92	0.2371E+04	0.3089E+04	0.2283E+04

V - 68

Connection type : Extended end-plate connections
Mode : All bolted without column stiffener

Tested by : A.Mazroi (1983) U.S.A
Test Id. : EP8 With Shim

Column : --
Beam : W24x162
Plate thickness : 1 3/4"
Stiffener thickness : 0.7500"

Fasteners: A325- -1 1/8"D
1 1/4" Oversize holes
Material : A36
Fy = 44.60 ksi
Fu = -- ksi

Major parameters

End plate extended on both sides.

ct = 4.8750"	pt = 5.4700"	li = 18.3100"	pc = 5.4700"
cc = 4.8750"	pit= 3.3750"	pi = 11.5600"	pic= 3.3750"
gt = 6.5000"	gi = 6.5000"	gc = 6.5000"	
tp = 1.7500"	nt = 2 X 3	ni = 2 X 2	nc = 2 X 3

Remark 1) End plate was stiffened in the plane of beam web by an end-plate.
2) Test was conducted for beam-to-beam connection.

No	Moment (k-in)	Rotation (radians) X 1/1000
1	0.00	0.00
2	565.02	0.00
3	1130.03	0.00
4	1695.05	0.00
5	2438.45	0.01
6	3231.86	0.01
7	3892.55	0.03
8	4503.24	0.05
9	5113.93	0.07
10	5728.85	0.07
11	6343.77	0.08
12	6958.69	0.09
13	7718.57	0.10
14	8478.46	0.12
15	9211.21	0.14
16	9943.96	0.16
17	10629.74	0.20
18	11315.53	0.24
19	11975.60	0.33
20	12635.68	0.42
21	13157.49	0.53
22	13699.31	0.65
23	13974.04	0.78
24	14248.76	0.92
25	14523.49	1.05
26	14798.22	1.19
27	14981.71	1.33
28	15165.20	1.48
29	15348.70	1.62
30	15532.19	1.77
31	15646.95	1.91
32	15761.71	2.04
33	15876.47	2.18
34	15991.23	2.31
35	16105.99	2.45
36	16215.03	2.60
37	16324.07	2.75
38	16433.12	2.90
39	16542.16	3.06

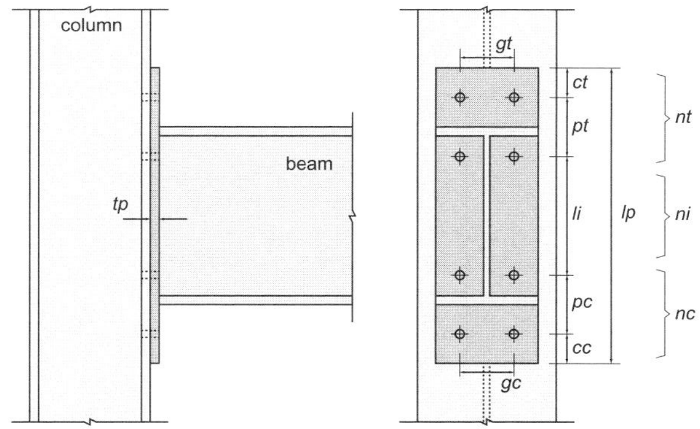

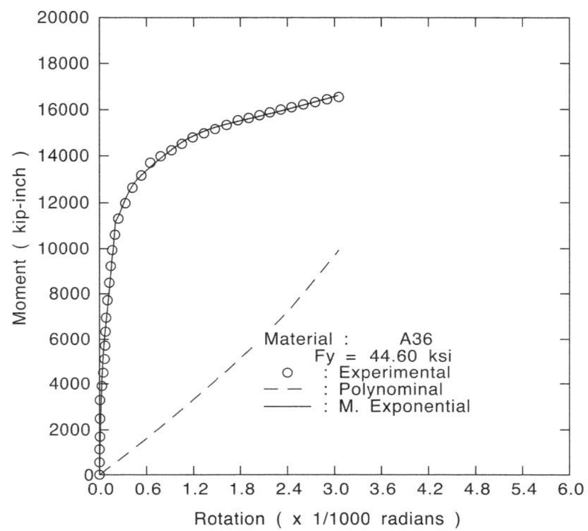

```
Moment-rotation prediction equations      ( R : X 1/1000 radians )

Frye and Morris polynominal model  :      R = Sum ( Ai X ( K*Bm )**Pi X 10**Qi )
     xd = 29.250000"    t =  1.750000"  tf =  1.125000"
     A1 =  1.830000    A2 = -1.040000   A3 =  6.380000     K =  0.000203
     P1 =        1     P2 =        3    P3 =        5     Q1 =        0     Q2 =       -1     Q3 =       -3

Modified exponential model  :
     Bm = Sum ( Ai X ( 1 - exp( - R/( 2*i*C ) ) ) )  +      Sum ( RKj X ( R - Rj0 ) )  +  Bm0
       C =   0.25783333E-01            Bmo=   0.00000000E+00           Nexp=  6           Nliner=  2
      AI =   0.35662055E+05  -0.21193927E+06  0.63736748E+06  -0.10139707E+07   0.79208594E+06  -0.23414558E+06
     Rj0 =           0.0000           0.2000
     RKj =   0.44030936E+05  -0.43071183E+05
```

No	Rotation (radians) X 1/1000	Moment (kip-inch) Expri.	Poly.	M.Expo.	Connection stiffness (kip-inch) X 1000 Expri.	Poly.	M.Expo.
1	0.00	0.00	0.00	0.00	0.4346E+06	0.2693E+04	0.1561E+06
3	0.00	1130.03	7.27	519.85	0.4197E+06	0.2693E+04	0.1852E+06
5	0.01	2488.45	16.16	1102.70	0.3967E+06	0.2693E+04	0.1684E+06
7	0.03	3892.55	73.51	3787.31	0.3156E+05	0.2693E+04	0.9191E+05
9	0.07	5113.93	177.73	6076.90	0.7628E+05	0.2693E+04	0.3901E+05
11	0.08	6343.77	213.55	6557.46	0.9248E+05	0.2694E+04	0.3393E+05
13	0.10	7718.57	281.43	7377.54	0.4107E+05	0.2694E+04	0.3244E+05
15	0.14	9211.21	382.48	8688.37	0.3857E+05	0.2696E+04	0.3815E+05
17	0.20	10629.74	536.15	11143.53	0.1805E+05	0.2698E+04	0.4724E+05
19	0.33	11975.60	884.43	12134.28	0.7254E+04	0.2708E+04	0.7593E+04
21	0.53	13167.49	1439.48	13171.80	0.4706E+04	0.2732E+04	0.3274E+04
23	0.78	13974.04	2123.02	13888.52	0.2031E+04	0.2779E+04	0.2802E+04
25	1.05	14523.49	2884.24	14583.01	0.2031E+04	0.2854E+04	0.2191E+04
27	1.33	14981.71	3701.89	15068.29	0.1255E+04	0.2961E+04	0.1310E+04
29	1.62	15348.70	4589.05	15375.56	0.1255E+04	0.3112E+04	0.8741E+03
31	1.91	15646.95	5493.48	15606.37	0.8438E+03	0.3304E+04	0.7953E+03
33	2.18	15876.47	6423.01	15826.57	0.8438E+03	0.3540E+04	0.8310E+03
35	2.45	16105.99	7424.30	16059.48	0.7862E+03	0.3831E+04	0.8804E+03
37	2.75	16324.07	8636.78	16331.74	0.7221E+03	0.4202E+04	0.9194E+03
39	3.06	16542.16	9946.96	16613.01	0.7221E+03	0.4528E+04	0.9409E+03

A.5 – 137

V - 69

```
Connection type : Extended end-plate connections
           Mode : All bolted without column stiffener

Tested by  :  A.Mazroi (1984)                              U.S.A
Test  Id.  :  CF4-U12x87

Column : W12x87                      Fasteners: A325- -7/8"D
Beam   : W24x100                           15/16" Oversize holes
Plate thickness : 1"                 Material : A36
                                          Fy =  46.80 ksi
                                          Fu =   --   ksi

                          Major parameters

     End plate extended on both sides.
     ct =  4.1250"    pt =  3.5000"    li = 19.7500"    pc =  3.5000"
     cc =  4.1250"    pit=  2.6250"    pi = 14.5000"    pic=  2.6250"
     gt =  5.5000"    gi =  5.5000"    gc =  5.5000"
     tp =  1.0000"    nt = 2 X  3      ni = 2 X  2      nc = 2 X  3

Remark  1)  End plate was stiffened in the plane of beam web by an end-plate.
        2)  pic=poc and pit=pot
```

No	Moment (k-in)	Rotation (radians) X 1/1000
1	0.00	0.00
2	759.14	0.12
3	1438.19	0.20
4	2064.18	0.31
5	2690.36	0.43
6	3249.78	0.54
7	3716.72	0.74
8	4158.41	1.13
9	4467.32	1.57
10	4736.23	1.99
11	4979.16	2.52
12	5248.46	2.99
13	5492.20	3.63
14	5723.18	4.36
15	5913.37	4.95
16	6063.65	5.55
17	6253.93	6.15
18	6405.60	6.95
19	6596.88	7.71
20	6774.53	8.41
21	6898.63	9.06
22	6996.26	9.74

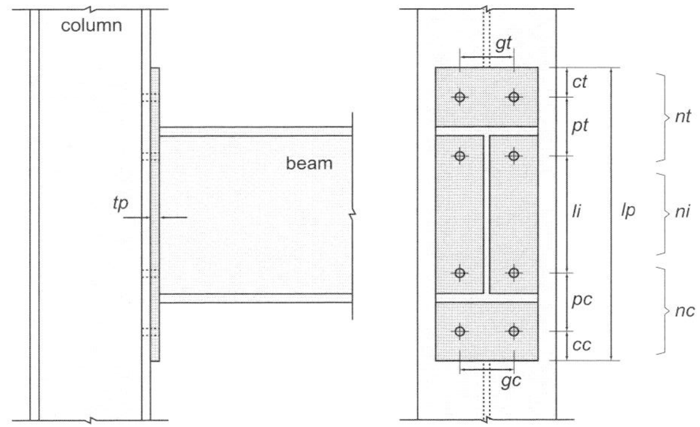

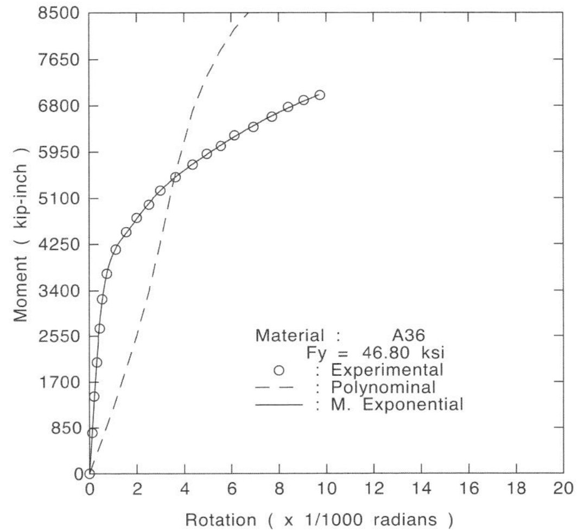

A.5 – 138

```
Moment-rotation prediction equations     ( R : X 1/1000 radians )

Frye and Morris polynominal model  :     R = Sum ( Ai X ( K*Bm )**Pi X 10**Qi )
     xd = 26.750000"   t =  1.000000"  tf =  0.875000"
     A1 =  1.830000   A2 = -1.040000   A3 =  6.380000    K =  0.000459
     P1 =         1   P2 =         3   P3 =         5   Q1 =         0   Q2 =        -1   Q3 =        -3

Modified exponential model  :
     Bm = Sum ( Ai X ( 1 - exp( - R/( 2*i*C ) ) ) )  +      Sum ( RKj X ( R - Rj0 ) )  +  Bm0
       C =   0.11070833E+00           Bmo=   0.00000000E+00          Nexp=  6         Nliner=  1
      AI =  -0.83567297E+04   0.73858696E+05  -0.24437766E+06   0.43919761E+06  -0.39852334E+06   0.14374615E+06
     Rj0 =          0.0000
     RKj =   0.15453575E+03
```

No	Rotation (radians) X 1/1000	Moment (kip-inch) Expri.	Moment (kip-inch) Poly.	Moment (kip-inch) M.Expo.	Connection stiffness (kip-inch) X 1000 Expri.	Connection stiffness (kip-inch) X 1000 Poly.	Connection stiffness (kip-inch) X 1000 M.Expo.
1	0.00	0.00	0.00	0.00	0.6433E+04	0.1192E+04	0.5265E+04
2	0.12	759.14	140.61	778.46	0.7338E+04	0.1192E+04	0.7218E+04
3	0.20	1438.19	241.97	1389.15	0.7042E+04	0.1194E+04	0.7020E+04
4	0.31	2064.18	369.74	2091.95	0.5497E+04	0.1197E+04	0.6036E+04
5	0.43	2690.36	517.62	2748.42	0.5164E+04	0.1203E+04	0.4634E+04
6	0.54	3249.78	646.36	3182.58	0.4221E+04	0.1210E+04	0.3510E+04
7	0.74	3716.72	888.92	3715.43	0.1937E+04	0.1226E+04	0.1973E+04
8	1.13	4158.41	1371.87	4185.35	0.9348E+03	0.1274E+04	0.7520E+03
9	1.57	4467.32	1953.10	4460.32	0.6652E+03	0.1363E+04	0.5800E+03
10	1.99	4736.23	2558.01	4710.48	0.5576E+03	0.1493E+04	0.5925E+03
11	2.52	4979.16	3394.76	5007.45	0.5159E+03	0.1729E+04	0.5321E+03
12	2.99	5248.46	4285.22	5240.84	0.4846E+03	0.1981E+04	0.4425E+03
13	3.63	5492.20	5568.43	5492.37	0.3522E+03	0.1893E+04	0.3520E+03
14	4.36	5723.18	6703.81	5726.68	0.3192E+03	0.1260E+04	0.3047E+03
15	4.95	5913.37	7345.97	5903.10	0.2850E+03	0.9096E+03	0.2880E+03
16	5.55	6063.65	7821.95	6071.66	0.2850E+03	0.7042E+03	0.2757E+03
17	6.15	6253.93	8200.29	6232.30	0.2637E+03	0.5743E+03	0.2620E+03
18	6.95	6405.60	8610.14	6432.70	0.2208E+03	0.4621E+03	0.2405E+03
19	7.71	6596.88	8934.38	6608.61	0.2522E+03	0.3907E+03	0.2196E+03
20	8.41	6774.53	9190.04	6756.01	0.2211E+03	0.3434E+03	0.2026E+03
21	9.06	6898.63	9402.28	6883.73	0.1671E+03	0.3091E+03	0.1896E+03
22	9.74	6996.26	9603.52	7009.65	0.1427E+03	0.2804E+03	0.1791E+03

V - 70

```
Connection type : Extended end-plate connections
           Mode : All bolted without column stiffener

Tested by   :  A.Mazroi (1984)                           U.S.A
Test  Id.   :  CF4-U12x106

Column : W12x106                    Fasteners: A325- -7/8"D
Beam   : W24x100                          15/16" Oversize holes
Plate thickness : 1"                Material : A36
                                         Fy =  38.50 ksi
                                         Fu =   --   ksi
```

Major parameters

```
End plate extended on both sides.
ct =  4.1250"    pt =  3.5000"    li = 19.7500"    pc =  3.5000"
cc =  4.1250"    pit=  2.6250"    pi = 14.5000"    pic=  2.6250"
gt =  5.5000"    gi =  5.5000"    gc =  5.5000"
tp =  1.0000"    nt = 2 X  3      ni = 2 X  2      nc = 2 X  3
```

Remark 1) End plate was stiffened in the plane of beam web by an end-plate.
2) pic=poc and pit=pot

No	Moment (k-in)	Rotation (radians) X 1/1000	No	Moment (k-in)	Rotation (radians) X 1/1000
1	0.00	0.00	26	9310.57	6.15
2	343.55	0.03	27	9547.39	6.77
3	687.10	0.06	28	9787.58	7.52
4	1030.64	0.09	29	9909.37	7.95
5	1437.64	0.14	30	10112.26	8.80
6	1844.63	0.19			
7	2251.62	0.24			
8	2691.62	0.36			
9	3131.62	0.47			
10	3608.23	0.63			
11	4084.84	0.80			
12	4432.96	0.96			
13	4781.08	1.13			
14	5129.21	1.30			
15	5572.69	1.56			
16	6016.16	1.83			
17	6437.01	2.15			
18	6857.86	2.48			
19	7236.06	2.86			
20	7614.26	3.25			
21	7883.57	3.60			
22	8152.87	3.94			
23	8418.74	4.34			
24	8684.60	4.74			
25	8973.26	5.28			

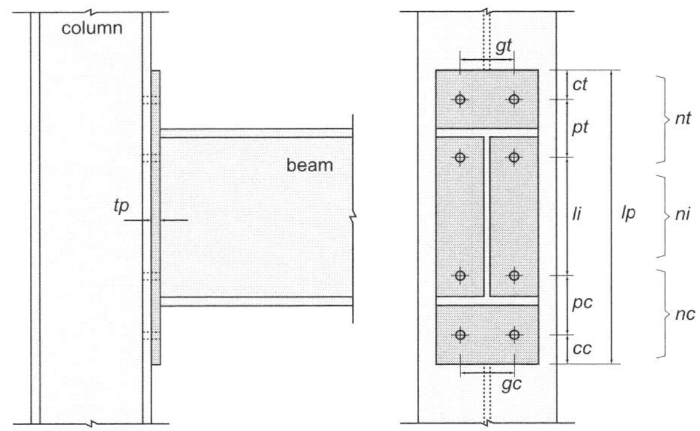

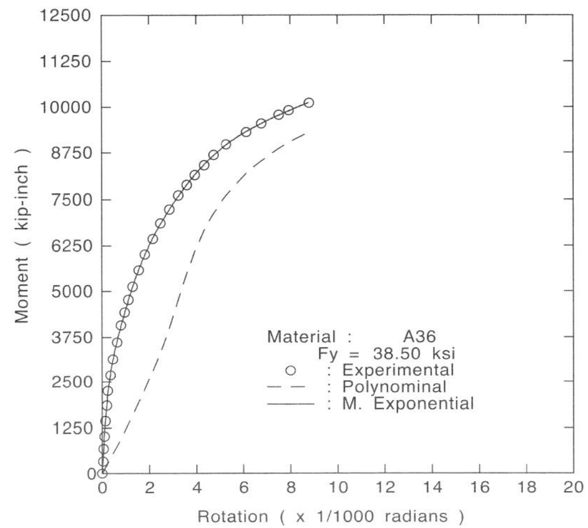

```
Moment-rotation prediction equations     ( R : X 1/1000 radians )

Frye and Morris polynominal model  :    R = Sum ( Ai X ( K*Bm )**Pi X 10**Qi )
     xd = 26.750000"   t =  1.000000"  tf =  0.875000"
     A1 =  1.830000   A2 = -1.040000   A3 =  6.380000    K =  0.000459
     P1 =         1   P2 =         3   P3 =         5   Q1 =        0    Q2 =       -1    Q3 =       -3

Modified exponential model  :
     Bm = Sum ( Ai X ( 1 - exp( - R/( 2*i*C ) ) ) )  +     Sum ( RKj X ( R - Rj0 ) )  +  Bm0
      C =   0.80883333E-01           Bmo=   0.00000000E+00         Nexp=  6        Nliner=  1
     AI =  -0.11448162E+04   0.37884699E+05  -0.19388405E+06   0.46152260E+06  -0.51276215E+06   0.21636978E+06
    Rj0 =           0.0000
    RKj =   0.24489256E+03
```

No	Rotation (radians) X 1/1000	Moment (kip-inch) Expri.	Moment (kip-inch) Poly.	Moment (kip-inch) M.Expo.	Connection stiffness (kip-inch) X 1000 Expri.	Connection stiffness (kip-inch) X 1000 Poly.	Connection stiffness (kip-inch) X 1000 M.Expo.
1	0.00	0.00	0.00	0.00	0.1134E+05	0.1192E+04	0.1273E+05
2	0.03	343.55	36.11	373.93	0.1132E+05	0.1192E+04	0.1173E+05
3	0.06	687.10	72.33	712.73	0.1132E+05	0.1192E+04	0.1058E+05
4	0.09	1030.64	108.45	1017.54	0.1009E+05	0.1192E+04	0.9554E+04
5	0.14	1437.64	169.24	1465.38	0.7980E+04	0.1193E+04	0.8055E+04
6	0.19	1844.63	230.08	1843.64	0.7980E+04	0.1194E+04	0.6820E+04
7	0.24	2251.62	290.97	2165.01	0.6709E+04	0.1195E+04	0.5818E+04
8	0.36	2691.62	426.03	2726.71	0.3894E+04	0.1199E+04	0.4257E+04
9	0.47	3131.62	562.18	3151.28	0.3491E+04	0.1205E+04	0.3343E+04
10	0.63	3608.23	760.71	3637.16	0.2906E+04	0.1217E+04	0.2676E+04
11	0.80	4084.84	961.41	4048.66	0.2504E+04	0.1232E+04	0.2379E+04
12	0.96	4432.96	1167.40	4428.61	0.2097E+04	0.1251E+04	0.2209E+04
13	1.13	4781.08	1376.97	4783.57	0.2097E+04	0.1275E+04	0.2068E+04
14	1.30	5129.21	1590.89	5114.96	0.1932E+04	0.1303E+04	0.1924E+04
15	1.56	5572.69	1944.95	5594.90	0.1667E+04	0.1361E+04	0.1686E+04
16	1.83	6016.16	2316.56	6013.56	0.1500E+04	0.1436E+04	0.1468E+04
17	2.15	6437.01	2801.76	6454.64	0.1295E+04	0.1556E+04	0.1257E+04
18	2.48	6857.86	3331.40	6837.52	0.1152E+04	0.1710E+04	0.1108E+04
19	2.86	7236.06	4028.81	7238.04	0.9836E+03	0.1918E+04	0.9828E+03
20	3.25	7614.26	4796.62	7596.83	0.8730E+03	0.2043E+04	0.8862E+03
21	3.60	7883.57	5494.99	7891.87	0.7728E+03	0.1921E+04	0.8077E+03
22	3.94	8152.87	6114.41	8160.10	0.7249E+03	0.1617E+04	0.7321E+03
23	4.34	8418.74	6685.35	8433.86	0.6705E+03	0.1271E+04	0.6495E+03
24	4.74	8684.60	7135.57	8675.95	0.6133E+03	0.1016E+04	0.5728E+03
25	5.28	8973.26	7617.16	8959.56	0.4790E+03	0.7866E+03	0.4825E+03
26	6.15	9310.57	8201.44	9329.41	0.3820E+03	0.5740E+03	0.3758E+03
27	6.77	9547.39	8528.46	9548.37	0.3528E+03	0.4824E+03	0.3263E+03
28	7.52	9787.58	8857.08	9776.00	0.2979E+03	0.4065E+03	0.2895E+03
29	7.95	9909.37	9024.29	9897.39	0.2682E+03	0.3732E+03	0.2759E+03
30	8.80	10112.26	9319.08	10124.45	0.2384E+03	0.3220E+03	0.2596E+03

A.5 – 141

V - 71

Connection type : Extended end-plate connections
Mode : All bolted without column stiffener

Tested by : A.Mazroi (1984) U.S.A
Test Id. : CF4-U12x120

Column : W12x120
Beam : W24x100
Plate thickness : 1"

Fasteners: A325- -7/8"D
15/16" Oversize holes
Material : A36
Fy = 52.50 ksi
Fu = -- ksi

Major parameters

End plate extended on both sides.

ct = 4.1250"	pt = 3.5000"	li = 19.7500"	pc = 3.5000"
cc = 4.1250"	pit= 2.6250"	pi = 14.5000"	pic= 2.6250"
gt = 5.5000"	gi = 5.5000"	gc = 5.5000"	
tp = 1.0000"	nt = 2 X 3	ni = 2 X 2	nc = 2 X 3

Remark 1) End plate was stiffened in the plane of beam web by an end-plate.
2) pic=poc and pit=pot

No	Moment (k-in)	Rotation (radians) X 1/1000	No	Moment (k-in)	Rotation (radians) X 1/1000
1	0.00	0.00	26	8715.68	2.80
2	384.92	0.02	27	8864.62	3.00
3	769.85	0.05	28	9013.56	3.19
4	1121.78	0.08	29	9127.40	3.41
5	1473.72	0.12	30	9241.25	3.62
6	1839.25	0.17	31	9355.09	3.84
7	2204.78	0.23	32	9442.08	4.03
8	2683.25	0.30	33	9529.08	4.23
9	3161.71	0.37	34	9616.07	4.42
10	3640.27	0.45	35	9671.68	4.59
11	4118.83	0.52	36	9727.30	4.75
12	4551.37	0.62	37	9782.92	4.92
13	4983.90	0.72			
14	5263.53	0.80			
15	5543.16	0.88			
16	6023.06	1.03			
17	6276.40	1.12			
18	6529.74	1.21			
19	7009.94	1.40			
20	7437.77	1.62			
21	7618.70	1.75			
22	7799.63	1.87			
23	8062.00	2.11			
24	8324.36	2.35			
25	8520.02	2.58			

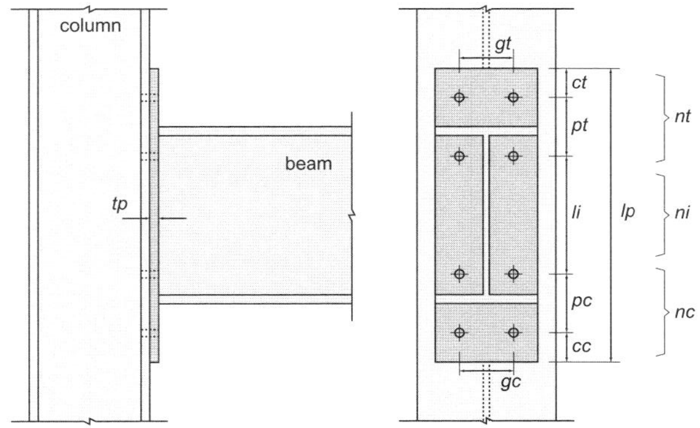

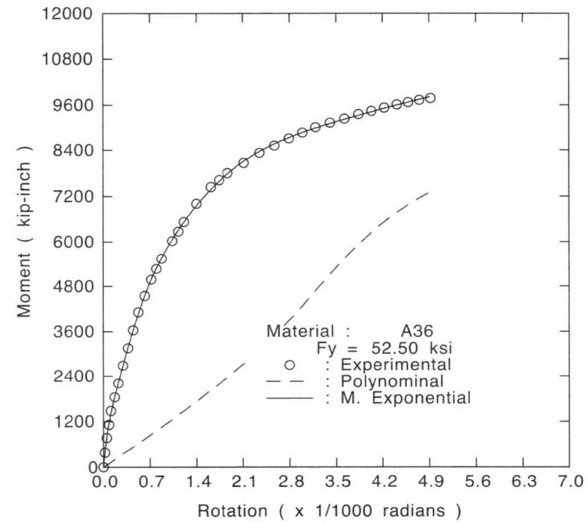

```
Moment-rotation prediction equations     ( R : X 1/1000 radians )

Frye and Morris polynominal model  :    R = Sum ( Ai X ( K*Bm )**Pi X 10**Qi )
     xd = 26.750000"   t =  1.000000"  tf =  0.875000"
     A1 =  1.830000   A2 = -1.040000   A3 =  6.380000    K =  0.000459
     P1 =        1    P2 =        3    P3 =        5    Q1 =        0    Q2 =       -1    Q3 =       -3

Modified exponential model  :
     Bm = Sum ( Ai X ( 1 - exp( - R/( 2*i*C ) ) ) )  +      Sum ( RKj X ( R - Rj0 ) )  +  Bm0
      C =   0.46883333E-01          Bmo=   0.00000000E+00          Nexp=  6        Nliner=  1
     AI =   0.44436107E+04  -0.27650188E+05   0.75336715E+05  -0.80203819E+05   0.61139139E+04   0.29674462E+05
    Rj0 =           0.0000
    RKj =   0.42889352E+03
```

No	Rotation (radians) X 1/1000	Moment (kip-inch) Expri.	Moment (kip-inch) Poly.	Moment (kip-inch) M.Expo.	Connection stiffness (kip-inch) X 1000 Expri.	Connection stiffness (kip-inch) X 1000 Poly.	Connection stiffness (kip-inch) X 1000 M.Expo.
1	0.00	0.00	0.00	0.00	0.1638E+05	0.1192E+04	0.1971E+05
3	0.05	769.85	56.01	749.63	0.1392E+05	0.1192E+04	0.1265E+05
5	0.12	1473.72	137.06	1431.84	0.8921E+04	0.1192E+04	0.8249E+04
7	0.23	2204.78	269.47	2233.87	0.6632E+04	0.1195E+04	0.6722E+04
9	0.37	3161.71	440.48	3162.71	0.6462E+04	0.1200E+04	0.6240E+04
11	0.52	4118.83	625.56	4057.76	0.5415E+04	0.1208E+04	0.5331E+04
13	0.72	4983.90	865.24	4983.26	0.3953E+04	0.1224E+04	0.4108E+04
15	0.88	5543.16	1055.54	5561.40	0.3409E+04	0.1240E+04	0.3392E+04
17	1.12	6276.40	1368.11	6303.93	0.2863E+04	0.1274E+04	0.2648E+04
19	1.40	7009.94	1726.18	6955.79	0.2282E+04	0.1324E+04	0.2114E+04
21	1.75	7618.70	2196.28	7595.76	0.1436E+04	0.1410E+04	0.1605E+04
23	2.11	8062.00	2739.90	8105.57	0.1084E+04	0.1539E+04	0.1186E+04
25	2.58	8520.02	3500.64	8564.83	0.8873E+03	0.1762E+04	0.8299E+03
27	3.00	8864.62	4287.42	8869.79	0.7484E+03	0.1981E+04	0.6415E+03
29	3.41	9127.40	5126.83	9111.87	0.5295E+03	0.2019E+04	0.5385E+03
31	3.84	9355.09	5939.01	9330.03	0.4872E+03	0.1716E+04	0.4825E+03
33	4.23	9529.08	6533.85	9511.37	0.4492E+03	0.1363E+04	0.4567E+03
35	4.59	9671.68	6976.06	9673.53	0.3330E+03	0.1102E+04	0.4438E+03
37	4.92	9782.92	7313.23	9820.55	0.3330E+03	0.9255E+03	0.4372E+03

A.5 – 143

V - 72

Connection type : Extended end-plate connections
Mode : All bolted without column stiffener

Tested by : A.Mazroi (1984) U.S.A
Test Id. : CF5-U10x68

Column : W10x68
Beam : W27x114
Plate thickness : 1"

Fasteners: A325- -1"D
1 1/16" Oversize holes
Material : A36
Fy = 40.60 ksi
Fu = -- ksi

Major parameters

End plate extended on both sides.

ct = 4.3750"	pt = 3.9300"	li = 22.3901"	pc = 3.9300"
cc = 4.3750"	pit= 2.6667"	pi = 17.0567"	pic= 2.6667"
gt = 5.5000"	gi = 5.5000"	gc = 5.5000"	
tp = 1.0000"	nt = 2 X 3	ni = 2 X 2	nc = 2 X 3

Remark 1) End plate was stiffened in the plane of beam web by an end-plate.
2) pic=poc and pit=pot

No	Moment (k-in)	Rotation (radians) X 1/1000	No	Moment (k-in)	Rotation (radians) X 1/1000
1	0.00	0.00	26	6185.08	3.04
2	343.97	0.02	27	6290.53	3.22
3	687.94	0.04	28	6402.52	3.43
4	1018.50	0.09	29	6514.50	3.63
5	1349.26	0.12	30	6619.85	3.86
6	1706.34	0.17	31	6725.20	4.09
7	2023.74	0.24	32	6810.76	4.30
8	2367.55	0.31	33	6896.32	4.52
9	2592.37	0.35	34	6997.20	4.76
10	2817.18	0.40	35	7098.08	5.00
11	3055.20	0.46	36	7198.97	5.25
12	3293.22	0.53	37	7317.38	5.54
13	3484.93	0.57	38	7435.80	5.83
14	3676.63	0.62			
15	4046.78	0.77			
16	4390.31	0.96			
17	4667.71	1.16			
18	4918.70	1.37			
19	5182.86	1.57			
20	5380.80	1.80			
21	5578.74	2.04			
22	5717.15	2.26			
23	5855.56	2.49			
24	5967.59	2.68			
25	6079.62	2.87			

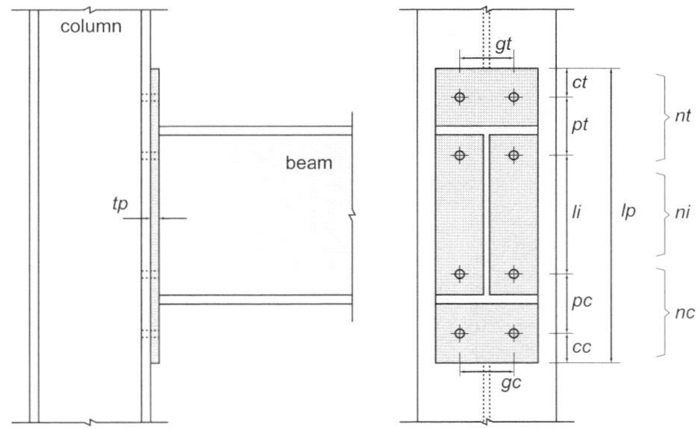

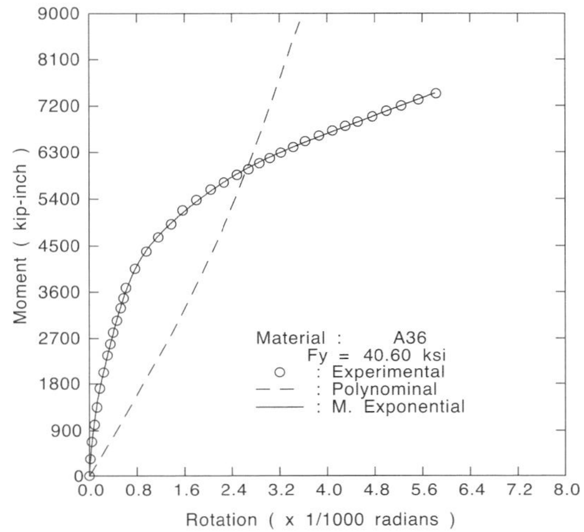

A.5 – 144

```
Moment-rotation prediction equations     ( R : X 1/1000 radians )

Frye and Morris polynominal model  :    R = Sum ( Ai X ( K*Bm )**Pi X 10**Qi )
     xd = 30.250100"   t =  1.000000"  tf =  1.000000"
     A1 =  1.830000   A2 = -1.040000   A3 =  6.380000    K =  0.000279
     P1 =        1    P2 =        3    P3 =        5    Q1 =        0    Q2 =      -1    Q3 =      -3

Modified exponential model  :
     Bm = Sum ( Ai X ( 1 - exp( - R/( 2*i*C ) ) ) )  +      Sum ( RKj X ( R - Rj0 ) )  +  Bm0
      C =  0.53316667E-01          Bmo=  0.00000000E+00         Nexp=  6       Nliner=  1
     AI =  0.51753792E+04  -0.34557292E+05   0.10037433E+06  -0.13079205E+06   0.74449638E+05  -0.96633456E+04
     Rj0 =          0.0000
     RKj =  0.42238404E+03
```

No	Rotation (radians) X 1/1000	Moment (kip-inch) Expri.	Moment (kip-inch) Poly.	Moment (kip-inch) M.Expo.	Connection stiffness (kip-inch) X 1000 Expri.	Connection stiffness (kip-inch) X 1000 Poly.	Connection stiffness (kip-inch) X 1000 M.Expo.
1	0.00	0.00	0.00	0.00	0.1810E+05	0.1956E+04	0.1816E+05
3	0.04	687.94	74.32	585.68	0.1492E+05	0.1956E+04	0.1274E+05
5	0.12	1349.26	242.54	1378.11	0.8085E+04	0.1957E+04	0.6879E+04
7	0.24	2023.74	463.79	2026.66	0.4937E+04	0.1961E+04	0.5100E+04
9	0.35	2592.37	691.48	2588.07	0.4887E+04	0.1968E+04	0.4610E+04
11	0.46	3055.20	909.18	3069.42	0.3690E+04	0.1977E+04	0.4073E+04
13	0.57	3484.93	1126.98	3481.90	0.4213E+04	0.1989E+04	0.3417E+04
15	0.77	4046.78	1521.31	4041.95	0.2157E+04	0.2017E+04	0.2333E+04
17	1.16	4667.71	2316.43	4694.61	0.1305E+04	0.2099E+04	0.1257E+04
19	1.57	5182.86	3200.94	5135.62	0.1125E+04	0.2236E+04	0.9609E+03
21	2.04	5578.74	4311.54	5550.44	0.7227E+03	0.2478E+04	0.7976E+03
23	2.49	5855.56	5501.20	5877.79	0.6040E+03	0.2816E+04	0.6579E+03
25	2.87	6079.62	6623.19	6107.23	0.5950E+03	0.3151E+04	0.5674E+03
27	3.22	6290.53	7777.65	6297.45	0.5696E+03	0.3351E+04	0.5084E+03
29	3.63	6514.50	9141.86	6498.37	0.5057E+03	0.3106E+04	0.4663E+03
31	4.09	6725.20	10395.60	6702.60	0.4317E+03	0.2437E+04	0.4422E+03
33	4.52	6896.32	11317.96	6890.08	0.4055E+03	0.1883E+04	0.4312E+03
35	5.00	7098.08	12124.91	7098.64	0.4141E+03	0.1460E+04	0.4256E+03
37	5.54	7317.38	12819.53	7325.30	0.4076E+03	0.1163E+04	0.4234E+03

V - 73

Connection type : Extended end-plate connections
Mode : All bolted without column stiffener

Tested by : A.Mazroi (1984) U.S.A
Test Id. : CF5-U14x61

Column : W14x61
Beam : W27x114
Plate thickness : 1"

Fasteners: A325- -1"D
1 1/16" Oversize holes
Material : A36
Fy = 44.60 ksi
Fu = -- ksi

Major parameters

End plate extended on both sides.

ct = 4.3750"	pt = 3.9300"	li = 22.3901"	pc = 3.9300"
cc = 4.3750"	pit= 2.6667"	pi = 17.0567"	pic= 2.6667"
gt = 5.5000"	gi = 5.5000"	gc = 5.5000"	
tp = 1.0000"	nt = 2 X 3	ni = 2 X 2	nc = 2 X 3

Remark 1) End plate was stiffened in the plane of beam web by an end-plate.
2) pic=poc and pit=pot

No	Moment (k-in)	Rotation (radians) X 1/1000	No	Moment (k-in)	Rotation (radians) X 1/1000
1	0.00	0.00	26	4093.55	2.74
2	167.58	0.03	27	4206.25	3.01
3	335.16	0.05	28	4297.71	3.23
4	529.63	0.09	29	4389.17	3.45
5	724.10	0.12	30	4474.48	3.69
6	885.04	0.15	31	4559.80	3.94
7	1045.97	0.17	32	4692.89	4.23
8	1273.92	0.21	33	4825.99	4.51
9	1501.87	0.25	34	4898.00	4.76
10	1683.25	0.29	35	4970.00	5.00
11	1864.62	0.34	36	5042.00	5.25
12	2006.01	0.39	37	5141.63	5.54
13	2147.40	0.44	38	5241.25	5.83
14	2309.14	0.51	39	5340.88	6.12
15	2470.87	0.57			
16	2606.02	0.64			
17	2741.16	0.72			
18	2958.27	0.87			
19	3175.38	1.03			
20	3394.58	1.28			
21	3614.69	1.58			
22	3692.71	1.79			
23	3770.72	2.01			
24	3875.78	2.24			
25	3980.84	2.47			

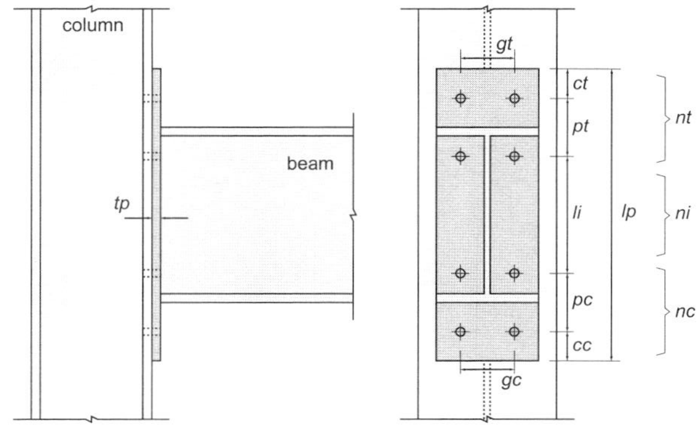

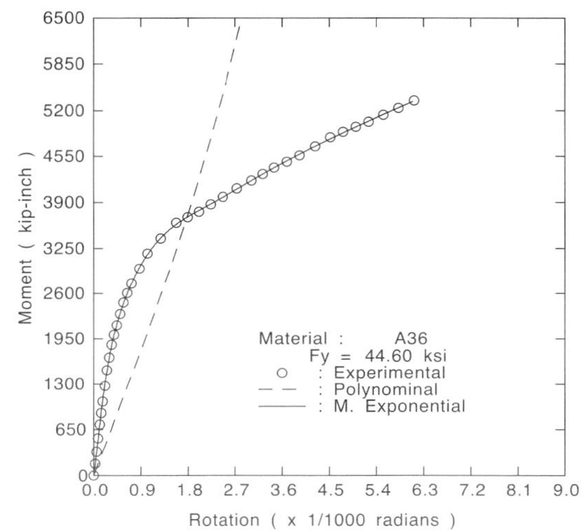

A.5 – 146

```
Moment-rotation prediction equations     ( R : X 1/1000 radians )

Frye and Morris polynominal model  :    R = Sum ( Ai X ( K*Bm )**Pi X 10**Qi )
     xd = 30.250100"   t =  1.000000"  tf =  1.000000"
     A1 =  1.830000   A2 = -1.040000   A3 =  6.380000    K =  0.000279
     P1 =        1    P2 =        3    P3 =        5    Q1 =        0    Q2 =       -1    Q3 =       -3

Modified exponential model  :
     Bm = Sum ( Ai X ( 1 - exp( - R/( 2*i*C ) ) ) )  +     Sum ( RKj X ( R - Rj0 ) )  +  Bm0
      C =   0.57591667E-01              Bmo=   0.00000000E+00        Nexp=  6        Nliner=  1
     AI =  -0.31913830E+04   0.33391990E+05  -0.13916109E+06   0.28140415E+06  -0.26285576E+06   0.93860535E+05
     Rj0 =          0.0000
     RKj =   0.30982765E+03
```

No	Rotation (radians) X 1/1000	Moment (kip-inch) Expri.	Poly.	M.Expo.	Connection stiffness (kip-inch) X 1000 Expri.	Poly.	M.Expo.
1	0.00	0.00	0.00	0.00	0.6324E+04	0.1956E+04	0.4696E+04
3	0.05	335.16	103.65	311.51	0.6212E+04	0.1956E+04	0.6431E+04
5	0.12	724.10	228.85	728.94	0.5800E+04	0.1957E+04	0.6412E+04
7	0.17	1045.97	342.38	1082.69	0.5818E+04	0.1959E+04	0.5734E+04
9	0.25	1501.87	487.37	1468.10	0.5186E+04	0.1962E+04	0.4684E+04
11	0.34	1864.62	666.01	1841.54	0.3414E+04	0.1967E+04	0.3576E+04
13	0.44	2147.40	866.86	2159.43	0.2655E+04	0.1975E+04	0.2721E+04
15	0.57	2470.87	1122.08	2465.30	0.2205E+04	0.1989E+04	0.2084E+04
17	0.72	2741.16	1412.64	2735.29	0.1714E+04	0.2008E+04	0.1674E+04
19	1.03	3175.38	2047.75	3164.78	0.1191E+04	0.2067E+04	0.1114E+04
21	1.58	3614.69	3230.03	3593.42	0.5225E+03	0.2241E+04	0.5323E+03
23	2.01	3770.72	4230.12	3787.25	0.4094E+03	0.2457E+04	0.4097E+03
25	2.47	3980.84	5433.86	3973.81	0.4367E+03	0.2795E+04	0.4105E+03
27	3.01	4206.25	7091.76	4202.68	0.4155E+03	0.3263E+04	0.4236E+03
29	3.45	4389.17	8552.63	4386.57	0.3861E+03	0.3288E+04	0.4134E+03
31	3.94	4559.80	10012.98	4581.98	0.4007E+03	0.2666E+04	0.3897E+03
33	4.51	4825.99	11316.40	4798.98	0.3699E+03	0.1884E+04	0.3607E+03
35	5.00	4970.00	12127.40	4970.83	0.2939E+03	0.1459E+04	0.3418E+03
37	5.54	5141.63	12820.92	5149.26	0.3451E+03	0.1162E+04	0.3280E+03
39	6.12	5340.88	13427.05	5335.82	0.3451E+03	0.9522E+03	0.3192E+03

V - 74

Connection type : Extended end-plate connections
Mode : All bolted without column stiffener

Tested by : A.Mazroi (1984) U.S.A
Test Id. : CF5-U10x49

Column : W10x49
Beam : W27x114
Plate thickness : 1"

Fasteners: A325- -1"D
1 1/16" Oversize holes
Material : A36
Fy = 40.90 ksi
Fu = -- ksi

Major parameters

End plate extended on both sides.

ct = 4.3750"	pt = 3.9300"	li = 22.3901"	pc = 3.9300"
cc = 4.3750"	pit= 2.6667"	pi = 17.0567"	pic= 2.6667"
gt = 5.5000"	gi = 5.5000"	gc = 5.5000"	
tp = 1.0000"	nt = 2 X 3	ni = 2 X 2	nc = 2 X 3

Remark 1) End plate was stiffened in the plane of beam web by an end-plate.
2) pic=poc and pit=pot

No	Moment (k-in)	Rotation (radians) X 1/1000	No	Moment (k-in)	Rotation (radians) X 1/1000
1	0.00	0.00	26	2946.92	2.66
2	146.07	0.02	27	3004.98	2.83
3	292.14	0.04	28	3063.04	2.99
4	444.86	0.06	29	3143.47	3.23
5	597.59	0.09	30	3223.90	3.47
6	717.20	0.12	31	3286.56	3.68
7	836.81	0.15	32	3349.22	3.89
8	989.63	0.20	33	3411.88	4.09
9	1142.46	0.25	34	3465.62	4.30
10	1268.72	0.29	35	3519.37	4.50
11	1394.98	0.34	36	3573.11	4.70
12	1501.38	0.41	37	3626.86	4.90
13	1607.78	0.48			
14	1767.46	0.60			
15	1887.41	0.73			
16	2007.36	0.87			
17	2133.92	1.01			
18	2260.49	1.14			
19	2393.84	1.33			
20	2527.20	1.51			
21	2614.14	1.72			
22	2701.08	1.92			
23	2763.67	2.11			
24	2826.26	2.31			
25	2888.86	2.50			

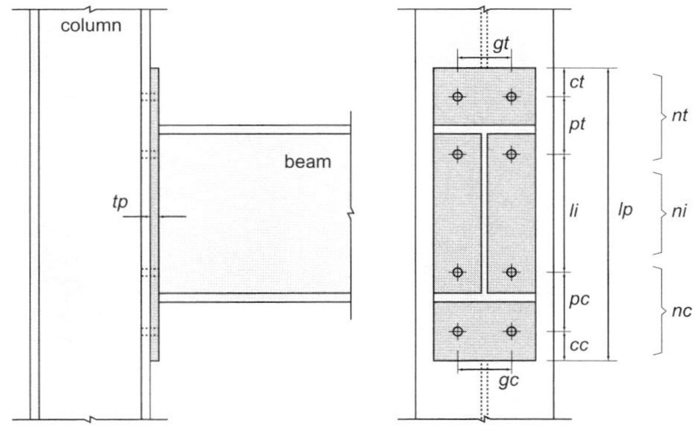

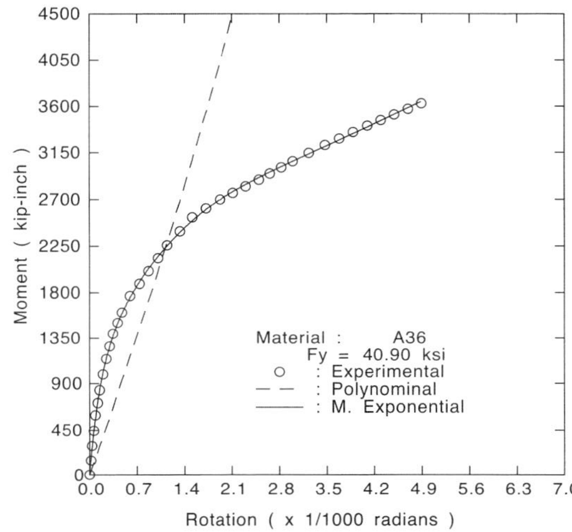

A.5 – 148

```
Moment-rotation prediction equations     ( R : X 1/1000 radians )

Frye and Morris polynominal model  :     R = Sum ( Ai X ( K*Bm )**Pi X 10**Qi )
     xd = 30.250100"   t =  1.000000"  tf =  1.000000"
     A1 =  1.830000   A2 = -1.040000   A3 =  6.380000    K =  0.000279
     P1 =        1    P2 =        3    P3 =        5    Q1 =        0    Q2 =       -1    Q3 =       -3

Modified exponential model  :
     Bm = Sum ( Ai X ( 1 - exp( - R/( 2*i*C ) ) ) )  +      Sum ( RKj X ( R - Rj0 ) )  +  Bm0
       C =   0.45425000E-01            Bmo=   0.00000000E+00           Nexp=  6        Nliner=  1
      AI =   0.80091581E+03  -0.72977487E+04   0.40203354E+05  -0.86622142E+05   0.81128919E+05  -0.26111967E+05
     Rj0 =           0.0000
     RKj =   0.31526591E+03
```

No	Rotation (radians) X 1/1000	Moment (kip-inch) Expri.	Moment (kip-inch) Poly.	Moment (kip-inch) M.Expo.	Connection stiffness (kip-inch) X 1000 Expri.	Connection stiffness (kip-inch) X 1000 Poly.	Connection stiffness (kip-inch) X 1000 M.Expo.
1	0.00	0.00	0.00	0.00	0.7896E+04	0.1956E+04	0.8491E+04
3	0.04	292.14	72.36	285.84	0.7181E+04	0.1956E+04	0.6779E+04
5	0.09	597.59	168.20	572.42	0.5151E+04	0.1956E+04	0.5058E+04
7	0.15	836.81	293.43	849.63	0.3505E+04	0.1958E+04	0.3728E+04
9	0.25	1142.46	483.43	1151.94	0.2889E+04	0.1962E+04	0.2623E+04
11	0.34	1394.98	671.88	1371.07	0.2192E+04	0.1967E+04	0.1991E+04
13	0.48	1607.78	937.83	1600.41	0.1495E+04	0.1979E+04	0.1457E+04
15	0.73	1887.41	1445.55	1903.02	0.8756E+03	0.2011E+04	0.9979E+03
17	1.01	2133.92	2002.52	2147.76	0.9238E+03	0.2062E+04	0.8103E+03
19	1.33	2393.84	2676.32	2381.96	0.7307E+03	0.2149E+04	0.6577E+03
21	1.72	2614.14	3541.09	2604.66	0.4200E+03	0.2301E+04	0.4923E+03
23	2.11	2763.67	4500.95	2776.14	0.3266E+03	0.2527E+04	0.3787E+03
25	2.50	2888.86	5523.77	2909.50	0.3401E+03	0.2823E+04	0.3247E+03
27	2.83	3004.98	6504.05	3013.09	0.3519E+03	0.3119E+04	0.3063E+03
29	3.23	3143.47	7821.81	3135.41	0.3365E+03	0.3352E+04	0.3013E+03
31	3.68	3286.56	9277.31	3270.56	0.3012E+03	0.3049E+04	0.3040E+03
33	4.09	3411.88	10417.42	3397.84	0.2845E+03	0.2424E+04	0.3078E+03
35	4.50	3519.37	11279.10	3521.73	0.2683E+03	0.1906E+04	0.3107E+03
37	4.90	3626.86	11964.11	3646.58	0.2685E+03	0.1538E+04	0.3126E+03

V - 75

Connection type : Extended end-plate connections
Mode : All bolted without column stiffener

Tested by : A.Mazroi (1984) U.S.A
Test Id. : CF6-U12x96

Column : W12x96
Beam : W27x114
Plate thickness : 1 1/4"

Fasteners: A325- -1"D
1 1/16" Oversize holes
Material : A36
Fy = 36.70 ksi
Fu = -- ksi

Major parameters

End plate extended on both sides.

ct = 4.5250"	pt = 4.9300"	li = 21.3900"	pc = 4.9300"
cc = 4.5250"	pit= 3.0000"	pi = 15.3900"	pic= 3.0000"
gt = 6.5000"	gi = 6.5000"	gc = 6.5000"	
tp = 1.2500"	nt = 2 X 3	ni = 2 X 2	nc = 2 X 3

Remark 1) End plate was stiffened in the plane of beam web by an end-plate.
2) pic=poc and pit=pot

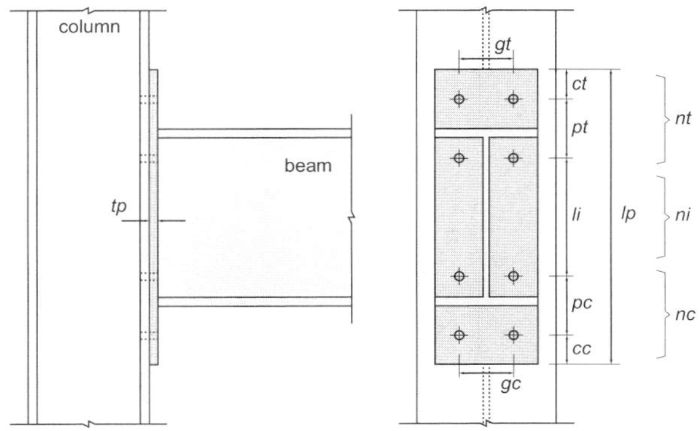

No	Moment (k-in)	Rotation (radians) X 1/1000
1	0.00	0.00
2	275.00	0.03
3	550.01	0.05
4	825.01	0.07
5	1190.95	0.12
6	1556.88	0.17
7	1809.76	0.21
8	2062.63	0.26
9	2315.51	0.31
10	2601.68	0.37
11	2887.85	0.44
12	3233.97	0.53
13	3580.09	0.62
14	3833.04	0.71
15	4085.98	0.79
16	4338.92	0.88
17	4592.00	1.00
18	4845.08	1.13
19	5098.16	1.25
20	5338.00	1.42
21	5577.84	1.59
22	5817.68	1.76
23	5986.63	1.96
24	6155.58	2.16
25	6324.53	2.35
26	6449.13	2.55
27	6573.73	2.74
28	6698.33	2.94
29	6804.16	3.11
30	6909.99	3.29
31	7015.82	3.46
32	7121.65	3.64
33	7209.48	3.81
34	7297.31	3.98
35	7385.15	4.16
36	7472.98	4.33

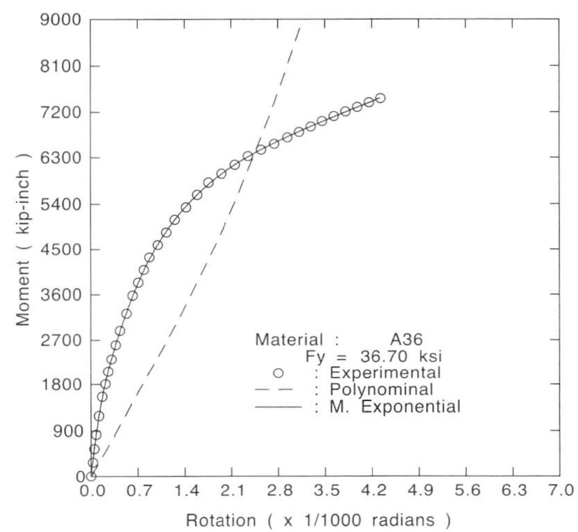

A.5 – 150

```
Moment-rotation prediction equations     ( R : X 1/1000 radians )

Frye and Morris polynominal model  :    R = Sum ( Ai X ( K*Bm )**Pi X 10**Qi )
     xd = 31.250000"    t =  1.250000"  tf =  1.000000"
     A1 =  1.830000   A2 = -1.040000   A3 =  6.380000    K =  0.000236
     P1 =        1    P2 =        3    P3 =        5    Q1 =        0    Q2 =       -1    Q3 =       -3

Modified exponential model  :
     Bm = Sum ( Ai X ( 1 - exp( - R/( 2*i*C ) ) ) )  +      Sum ( RKj X ( R - Rj0 ) )  +  Bm0
       C =   0.42366667E-01             Bmo=   0.00000000E+00            Nexp=  6         Nliner=  1
      AI =  -0.18328697E+03   0.66748673E+04  -0.25723877E+05   0.55658463E+05  -0.68218982E+05   0.37044841E+05
     Rj0 =             0.0000
     RKj =   0.51509839E+03
```

No	Rotation (radians) X 1/1000	Moment (kip-inch) Expri.	Moment (kip-inch) Poly.	Moment (kip-inch) M.Expo.	Connection stiffness (kip-inch) X 1000 Expri.	Connection stiffness (kip-inch) X 1000 Poly.	Connection stiffness (kip-inch) X 1000 M.Expo.
1	0.00	0.00	0.00	0.00	0.1100E+05	0.2312E+04	0.1209E+05
3	0.05	550.01	115.59	565.92	0.1100E+05	0.2312E+04	0.1015E+05
5	0.12	1190.95	279.77	1194.82	0.7955E+04	0.2314E+04	0.7745E+04
7	0.21	1809.76	494.31	1817.98	0.5421E+04	0.2317E+04	0.5886E+04
9	0.31	2315.51	710.62	2314.27	0.5004E+04	0.2323E+04	0.4849E+04
11	0.44	2887.85	1010.62	2882.71	0.4130E+04	0.2334E+04	0.4040E+04
13	0.62	3580.09	1452.23	3568.96	0.3316E+04	0.2359E+04	0.3303E+04
15	0.79	4085.98	1852.72	4082.31	0.2988E+04	0.2388E+04	0.2774E+04
17	1.00	4592.00	2361.38	4605.65	0.2014E+04	0.2437E+04	0.2220E+04
19	1.25	5098.16	2983.04	5095.86	0.1765E+04	0.2514E+04	0.1709E+04
21	1.59	5577.84	3844.79	5583.92	0.1436E+04	0.2654E+04	0.1250E+04
23	1.96	5986.63	4855.27	5981.35	0.8462E+03	0.2869E+04	0.9445E+03
25	2.35	6324.53	6062.76	6316.41	0.7427E+03	0.3196E+04	0.7519E+03
27	2.74	6573.73	7378.94	6585.58	0.6423E+03	0.3597E+04	0.6455E+03
29	3.11	6804.16	8771.41	6811.92	0.6056E+03	0.3923E+04	0.5876E+03
31	3.46	7015.82	10149.99	7011.31	0.6056E+03	0.3878E+04	0.5559E+03
33	3.81	7209.48	11426.35	7201.96	0.5033E+03	0.3380E+04	0.5377E+03
35	4.16	7385.15	12495.72	7387.67	0.5033E+03	0.2754E+04	0.5275E+03

A.5 – 151

V - 76

```
Connection type : Extended end-plate connections
           Mode : All bolted without column stiffener

Tested by   :  A.Mazroi (1984)                          U.S.A
Test  Id.   :  CF6-U14x158

Column : W14x158                        Fasteners: A325- -1"D
Beam   : W27x114                               1/16" Oversize holes
Plate thickness : 1 1/4"                Material : A36
                                             Fy =  37.70 ksi
                                             Fu =   --   ksi
```

Major parameters

```
End plate extended on both sides.
ct =  4.6250"   pt =  4.9300"   li = 21.3900"   pc =  4.9300"
cc =  4.6250"   pit=  3.0000"   pi = 15.3900"   pic=  3.0000"
gt =  6.5000"   gi =  6.5000"   gc =  6.5000"
tp =  1.2500"   nt = 2 X  3     ni = 2 X  2     nc = 2 X  3
```

Remark 1) End plate was stiffened in the plane of beam web by an end-plate.
2) pic=poc and pit=pot

No	Moment (k-in)	Rotation (radians) X 1/1000
1	0.00	0.00
2	472.27	0.03
3	944.53	0.05
4	1380.12	0.09
5	1815.72	0.13
6	2251.31	0.17
7	2839.07	0.23
8	3426.84	0.30
9	3856.10	0.37
10	4285.35	0.44
11	4714.61	0.51
12	5249.03	0.66
13	5783.46	0.81
14	6205.94	0.97
15	6628.43	1.13
16	7021.00	1.33
17	7413.58	1.54
18	7799.26	1.83
19	8184.95	2.12
20	8524.37	2.41
21	9116.74	2.98
22	9591.13	3.57
23	10022.24	4.21
24	10408.62	4.86
25	10761.22	5.60
26	11078.13	6.36
27	11360.18	7.07
28	11595.14	7.92
29	11881.62	8.75

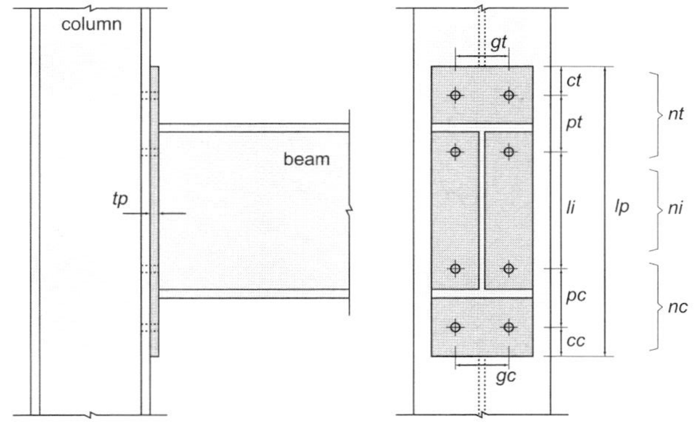

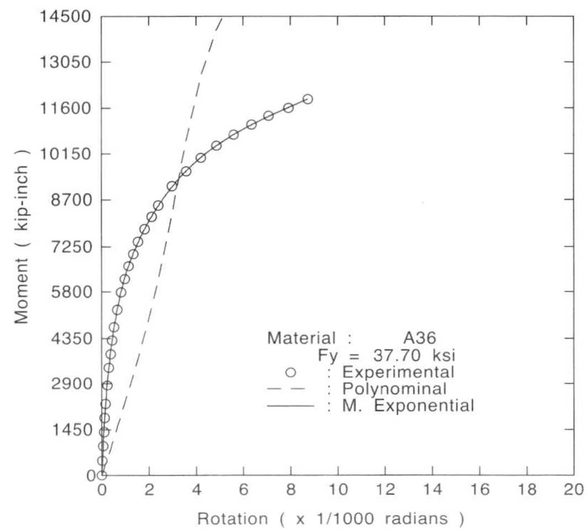

```
Moment-rotation prediction equations     ( R : X 1/1000 radians )

Frye and Morris polynominal model  :    R = Sum ( Ai X ( K*Bm )**Pi X 10**Qi )
     xd = 31.250000"   t =  1.250000"  tf =  1.000000"
     A1 =  1.830000   A2 = -1.040000   A3 =  6.380000    K =  0.000236
     P1 =         1   P2 =         3   P3 =         5   Q1 =         0    Q2 =        -1    Q3 =        -3

Modified exponential model  :
     Bm = Sum ( Ai X ( 1 - exp( - R/( 2*i*C ) ) ) )  +      Sum ( RKj X ( R - Rj0 ) )  +  Bm0
      C =   0.79525000E-01               Bmo=   0.00000000E+00          Nexp=  6         Nliner=  1
     AI =   0.80942701E+03   0.10384200E+05  -0.67130525E+05   0.20602027E+06  -0.26577205E+06   0.12486623E+06
    Rj0 =           0.0000
    RKj =   0.31025796E+03
```

No	Rotation (radians) X 1/1000	Moment (kip-inch) Expri.	Moment (kip-inch) Poly.	Moment (kip-inch) M.Expo.	Connection stiffness (kip-inch) X 1000 Expri.	Connection stiffness (kip-inch) X 1000 Poly.	Connection stiffness (kip-inch) X 1000 M.Expo.
1	0.00	0.00	0.00	0.00	0.1782E+05	0.2312E+04	0.1752E+05
2	0.03	472.27	61.26	450.39	0.1782E+05	0.2312E+04	0.1620E+05
3	0.05	944.53	122.53	860.19	0.1521E+05	0.2312E+04	0.1476E+05
4	0.09	1380.12	210.39	1386.54	0.1146E+05	0.2313E+04	0.1299E+05
5	0.13	1815.72	298.28	1851.36	0.1146E+05	0.2314E+04	0.1151E+05
6	0.17	2251.31	386.20	2264.53	0.1055E+05	0.2315E+04	0.1027E+05
7	0.23	2839.07	537.84	2878.32	0.8974E+04	0.2318E+04	0.8555E+04
8	0.30	3426.84	689.68	3394.24	0.7539E+04	0.2322E+04	0.7256E+04
9	0.37	3856.10	856.89	3875.82	0.5962E+04	0.2328E+04	0.6171E+04
10	0.44	4285.35	1024.50	4288.98	0.5962E+04	0.2335E+04	0.5341E+04
11	0.51	4714.61	1192.63	4649.33	0.5184E+04	0.2343E+04	0.4694E+04
12	0.66	5249.03	1546.70	5275.91	0.3563E+04	0.2365E+04	0.3730E+04
13	0.81	5783.46	1903.44	5783.81	0.3123E+04	0.2393E+04	0.3081E+04
14	0.97	6205.94	2286.68	6232.61	0.2657E+04	0.2429E+04	0.2590E+04
15	1.13	6628.43	2676.35	6614.20	0.2339E+04	0.2474E+04	0.2226E+04
16	1.33	7021.00	3185.32	7029.31	0.1934E+04	0.2543E+04	0.1882E+04
17	1.54	7413.58	3710.07	7384.60	0.1687E+04	0.2629E+04	0.1631E+04
18	1.83	7799.26	4491.48	7817.39	0.1335E+04	0.2785E+04	0.1381E+04
19	2.12	8184.95	5324.50	8190.68	0.1253E+04	0.2988E+04	0.1212E+04
20	2.41	8524.37	6226.95	8523.72	0.1123E+04	0.3245E+04	0.1090E+04
21	2.98	9116.74	8265.13	9097.63	0.9164E+03	0.3832E+04	0.9133E+03
22	3.57	9591.13	10582.10	9594.55	0.7420E+03	0.3755E+04	0.7663E+03
23	4.21	10022.24	12628.62	10036.61	0.6353E+03	0.2673E+04	0.6324E+03
24	4.86	10408.62	14085.21	10413.72	0.5369E+03	0.1847E+04	0.5226E+03
25	5.60	10761.22	15246.71	10764.51	0.4477E+03	0.1339E+04	0.4355E+03
26	6.36	11078.13	16142.63	11072.91	0.4067E+03	0.1045E+04	0.3791E+03
27	7.07	11360.18	16815.79	11329.66	0.3428E+03	0.8699E+03	0.3482E+03
28	7.92	11595.14	17490.03	11615.84	0.3120E+03	0.7274E+03	0.3282E+03
29	8.75	11881.62	18048.82	11882.87	0.3464E+03	0.6297E+03	0.3187E+03

A.5 – 153

V - 77

Connection type : Extended end-plate connections
Mode : All bolted without column stiffener

Tested by : A.Mazroi (1984) U.S.A
Test Id. : CF8-U14x159

Column : W14x159
Beam : W24x162
Plate thickness : 1 3/4"

Fasteners: A325- -1 1/8"D
1 1/4" Oversize holes
Material : A36
Fy = 35.00 ksi
Fu = -- ksi

Major parameters

End plate extended on both sides.

ct = 4.8750"	pt = 5.4700"	li = 18.3100"	pc = 5.4700"
cc = 4.8750"	pit= 3.3750"	pi = 11.5600"	pic= 3.3750"
gt = 6.5000"	gi = 6.5000"	gc = 6.5000"	
tp = 1.7500"	nt = 2 X 3	ni = 2 X 2	nc = 2 X 3

Remark 1) End plate was stiffened in the plane of beam web by an end-plate.
2) pic=poc and pit=pot

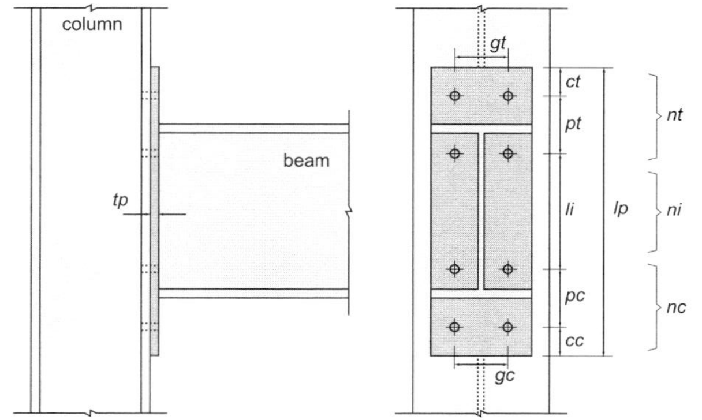

No	Moment (k-in)	Rotation (radians) X 1/1000	No	Moment (k-in)	Rotation (radians) X 1/1000
1	0.00	0.00	26	12577.40	5.23
2	485.36	0.04	27	12942.43	5.86
3	970.72	0.07	28	13281.74	6.48
4	1456.08	0.11	29	13595.34	7.09
5	2032.05	0.17	30	13917.26	7.75
6	2608.02	0.24	31	14164.23	8.42
7	3184.24	0.34	32	14336.24	9.08
8	3760.45	0.44			
9	4316.87	0.55			
10	4873.30	0.66			
11	5357.08	0.79			
12	5840.86	0.92			
13	6377.73	1.07			
14	6914.60	1.23			
15	7544.66	1.47			
16	8174.72	1.71			
17	8805.13	2.01			
18	9435.53	2.30			
19	9907.33	2.62			
20	10379.14	2.94			
21	10753.20	3.23			
22	11137.26	3.53			
23	11543.54	3.95			
24	11949.82	4.36			
25	12263.61	4.80			

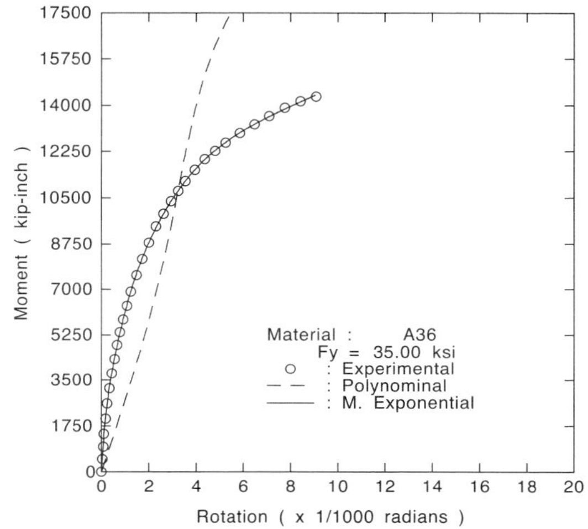

```
Moment-rotation prediction equations     ( R : X 1/1000 radians )

Frye and Morris polynominal model  :    R = Sum ( Ai X ( K*Bm )**Pi X 10**Qi )
     xd = 29.250000"    t =  1.750000"  tf =  1.125000"
     A1 =  1.830000   A2 = -1.040000   A3 =  6.380000    K =  0.000203
     P1 =        1    P2 =        3    P3 =        5    Q1 =        0    Q2 =       -1    Q3 =       -3

Modified exponential model  :
     Bm = Sum ( Ai X ( 1 - exp( - R/( 2*i*C ) ) ) )  +     Sum ( RKj X ( R - Rj0 ) )  +  Bm0
      C =   0.84625000E-01           Bmo=   0.00000000E+00           Nexp=  6        Nliner=  1
     AI =   0.17661531E+04   0.87633598E+04  -0.80422583E+05   0.25101755E+06  -0.33275375E+06   0.16258949E+06
    Rj0 =           0.0000
    RKj =   0.38013188E+03
```

No	Rotation (radians) X 1/1000	Moment (kip-inch) Expri.	Moment (kip-inch) Poly.	Moment (kip-inch) M.Expo.	Connection stiffness (kip-inch) X 1000 Expri.	Connection stiffness (kip-inch) X 1000 Poly.	Connection stiffness (kip-inch) X 1000 M.Expo.
1	0.00	0.00	0.00	0.00	0.1360E+05	0.2693E+04	0.1561E+05
2	0.04	485.36	96.13	531.08	0.1361E+05	0.2693E+04	0.1384E+05
3	0.07	970.72	192.00	991.50	0.1361E+05	0.2693E+04	0.1209E+05
4	0.11	1456.08	288.15	1396.84	0.1186E+05	0.2694E+04	0.1067E+05
5	0.17	2032.05	468.67	2040.38	0.8597E+04	0.2697E+04	0.8671E+04
6	0.24	2608.02	649.37	2572.74	0.7468E+04	0.2701E+04	0.7306E+04
7	0.34	3184.24	918.09	3229.59	0.5791E+04	0.2709E+04	0.6011E+04
8	0.44	3760.45	1187.61	3785.74	0.5433E+04	0.2720E+04	0.5229E+04
9	0.55	4316.87	1490.05	4330.91	0.5035E+04	0.2735E+04	0.4677E+04
10	0.66	4873.30	1792.27	4825.54	0.4465E+04	0.2754E+04	0.4295E+04
11	0.79	5357.08	2144.70	5348.61	0.3809E+04	0.2781E+04	0.3955E+04
12	0.92	5840.86	2497.86	5832.21	0.3617E+04	0.2813E+04	0.3667E+04
13	1.07	6377.73	2950.82	6389.14	0.3377E+04	0.2861E+04	0.3344E+04
14	1.23	6914.60	3410.25	6897.25	0.3075E+04	0.2919E+04	0.3052E+04
15	1.47	7544.66	4124.96	7583.17	0.2620E+04	0.3029E+04	0.2663E+04
16	1.71	8174.72	4869.39	8183.40	0.2403E+04	0.3167E+04	0.2339E+04
17	2.01	8805.13	5834.57	8824.51	0.2137E+04	0.3386E+04	0.2021E+04
18	2.30	9435.53	6873.28	9382.71	0.1825E+04	0.3667E+04	0.1773E+04
19	2.62	9907.33	8090.81	9909.41	0.1491E+04	0.4037E+04	0.1563E+04
20	2.94	10379.14	9430.31	10375.93	0.1379E+04	0.4419E+04	0.1390E+04
21	3.23	10758.20	10781.38	10768.45	0.1274E+04	0.4616E+04	0.1252E+04
22	3.53	11137.26	12138.48	11122.21	0.1151E+04	0.4434E+04	0.1129E+04
23	3.95	11543.54	13826.87	11558.56	0.9802E+03	0.3648E+04	0.9802E+03
24	4.36	11949.82	15166.20	11937.82	0.8532E+03	0.2837E+04	0.8534E+03
25	4.80	12263.61	16260.80	12284.88	0.7197E+03	0.2225E+04	0.7423E+03
26	5.23	12577.40	17133.96	12588.18	0.6630E+03	0.1811E+04	0.6524E+03
27	5.86	12942.43	18141.33	12965.65	0.5645E+03	0.1424E+04	0.5556E+03
28	6.48	13281.74	18940.93	13288.31	0.5306E+03	0.1178E+04	0.4908E+03
29	7.09	13595.34	19605.46	13574.27	0.5009E+03	0.1009E+04	0.4487E+03
30	7.75	13917.26	20225.72	13860.53	0.4297E+03	0.8757E+03	0.4202E+03
31	8.42	14164.23	20772.41	14133.47	0.3148E+03	0.7750E+03	0.4030E+03
32	9.08	14336.24	21262.67	14399.03	0.2575E+03	0.6961E+03	0.3930E+03

A.5 – 155

V - 78

Connection type : Extended end-plate connections
Mode : All bolted with column stiffener

Tested by : A.Mazroi (1983) U.S.A
Test Id. : CF3-S12x65

Column : W12x65
Beam : W24x100
Plate thickness : 3/4"
Stiffener thickness : 0.3750"

Fasteners: A325- -7/8"D
15/16" Oversize holes
Material : A36
Fy = 59.30 ksi
Fu = -- ksi

Major parameters

End plate extended on both sides.

ct = 4.1250"	pt = 3.5000"	li = 19.7500"	pc = 3.5000"
cc = 4.1250"	pit= 2.6250"	pi = 14.5000"	pic= 2.6250"
gt = 5.5000"	gi = 5.5000"	gc = 5.5000"	
tp = 0.7500"	nt = 2 X 3	ni = 2 X 2	nc = 2 X 3

Remark 1) End plate was stiffened in the plane of beam web.
2) pic=poc and pit=pot

No	Moment (k-in)	Rotation (radians) X 1/1000
1	0.00	0.00
2	279.09	0.02
3	558.18	0.04
4	764.26	0.07
5	970.33	0.10
6	1222.93	0.15
7	1475.53	0.20
8	1721.52	0.26
9	1967.51	0.33
10	2180.28	0.40
11	2393.04	0.47
12	2605.81	0.53
13	2825.21	0.60
14	3044.60	0.67
15	3230.80	0.74
16	3417.00	0.81
17	3649.76	0.90
18	3882.53	0.99
19	4075.39	1.07
20	4268.24	1.16
21	4434.55	1.25
22	4600.86	1.34
23	4753.90	1.44
24	4906.94	1.53
25	5066.60	1.62

No	Moment (k-in)	Rotation (radians) X 1/1000
26	5226.25	1.71
27	5379.29	1.81
28	5532.32	1.92
29	5647.67	2.00
30	5763.02	2.08
31	5878.37	2.16
32	5993.72	2.24
33	6109.06	2.32
34	6224.41	2.40

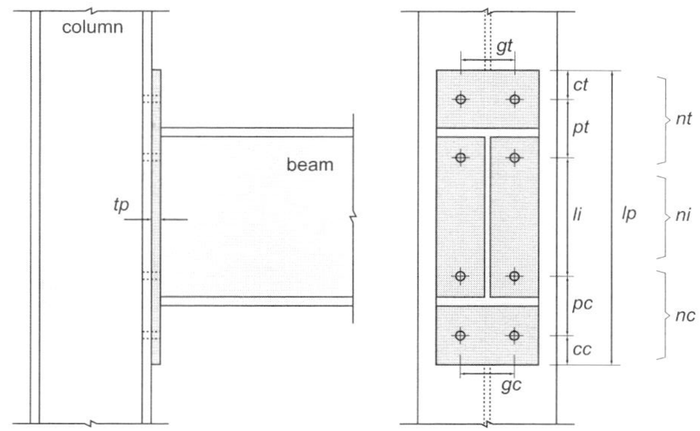

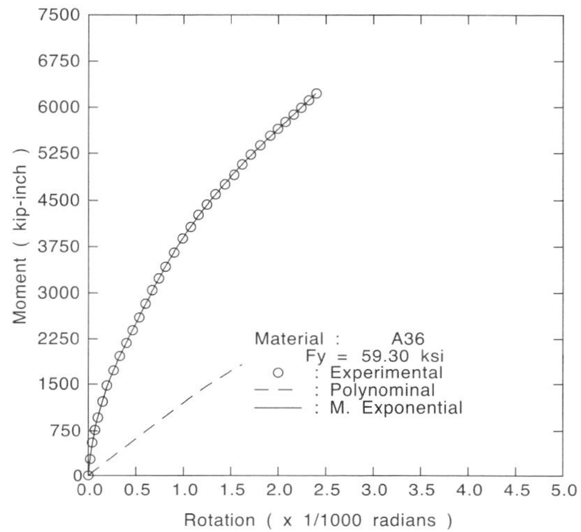

```
Moment-rotation prediction equations      ( R : X 1/1000 radians )

Frye and Morris polynominal model  :    R = Sum ( Ai X ( K*Bm )**Pi X 10**Qi )
     xd = 26.750000"   t =  0.750000"
     A1 =  1.790000   A2 =  1.760000   A3 =  2.040000    K =  0.000446
     P1 =        1    P2 =        3    P3 =        5    Q1 =        0     Q2 =      -1     Q3 =      -1

Modified exponential model  :
     Bm = Sum ( Ai X ( 1 - exp( - R/( 2*i*C ) ) ) )  +      Sum ( RKj X ( R - Rj0 ) )  +  Bm0
       C =   0.24875000E-01            Bmo=   0.00000000E+00          Nexp=  6        Nliner=  1
      AI =   0.16554739E+04  -0.13655847E+05   0.56274000E+05  -0.90923748E+05   0.52124086E+05  -0.26872007E+04
     Rj0 =            0.0000
     RKj =   0.14352674E+04
```

No	Rotation (radians) X 1/1000	Moment (kip-inch) Expri.	Moment (kip-inch) Poly.	Moment (kip-inch) M.Expo.	Connection stiffness (kip-inch) X 1000 Expri.	Connection stiffness (kip-inch) X 1000 Poly.	Connection stiffness (kip-inch) X 1000 M.Expo.
1	0.00	0.00	0.00	0.00	0.1431E+05	0.1252E+04	0.1671E+05
3	0.04	558.18	48.84	520.73	0.1135E+05	0.1252E+04	0.9880E+04
5	0.10	970.33	126.47	979.72	0.6123E+04	0.1251E+04	0.5871E+04
7	0.20	1475.53	245.28	1451.32	0.4627E+04	0.1248E+04	0.4336E+04
9	0.33	1967.51	413.50	1965.39	0.3403E+04	0.1239E+04	0.3418E+04
11	0.47	2393.04	579.42	2402.55	0.3159E+04	0.1225E+04	0.3141E+04
13	0.60	2825.21	743.05	2819.32	0.3250E+04	0.1205E+04	0.3041E+04
15	0.74	3230.80	907.54	3227.88	0.2641E+04	0.1178E+04	0.2861E+04
17	0.90	3649.76	1093.34	3662.81	0.2586E+04	0.1136E+04	0.2547E+04
19	1.07	4075.39	1286.35	4074.70	0.2296E+04	0.1081E+04	0.2194E+04
21	1.25	4434.55	1468.94	4430.60	0.1848E+04	0.1017E+04	0.1911E+04
23	1.44	4753.90	1653.65	4769.76	0.1554E+04	0.9423E+03	0.1703E+04
25	1.62	5066.60	1821.52	5073.31	0.1835E+04	0.8679E+03	0.1580E+04
27	1.81	5379.29	1980.19	5367.71	0.1471E+04	0.7942E+03	0.1510E+04
29	2.00	5647.67	2121.83	5642.79	0.1430E+04	0.7276E+03	0.1473E+04
31	2.16	5878.37	2234.86	5878.86	0.1429E+04	0.6749E+03	0.1456E+04
33	2.32	6109.06	2339.76	6112.84	0.1430E+04	0.6271E+03	0.1446E+04

A.5 – 157

V - 79

Connection type : Extended end-plate connections
Mode : All bolted with column stiffener

Tested by : A.Mazroi (1984) U.S.A
Test Id. : CF4-S12x87

Column : W12x87
Beam : W24x100
Plate thickness : 1"
Stiffener thickness : 0.3750"

Fasteners: A325- -7/8"D
15/16" Oversize holes
Material : A36
Fy = 56.80 ksi
Fu = -- ksi

Major parameters

End plate extended on both sides.

ct = 4.1250"	pt = 3.5000"	li = 19.7500"	pc = 3.5000"
cc = 4.1250"	pit= 2.6250"	pi = 14.5000"	pic= 2.6250"
gt = 5.5000"	gi = 5.5000"	gc = 5.5000"	
tp = 1.0000"	nt = 2 X 3	ni = 2 X 2	nc = 2 X 3

Remark 1) End plate was stiffened in the plane of beam web.
2) pic=poc and pit=pot

No	Moment (k-in)	Rotation (radians) X 1/1000	No	Moment (k-in)	Rotation (radians) X 1/1000
1	0.00	0.00	26	8079.12	1.28
2	403.95	0.02	27	8254.88	1.41
3	807.90	0.03	28	8430.64	1.53
4	1211.86	0.05	29	8606.40	1.66
5	1504.94	0.07	30	8781.44	1.80
6	1798.02	0.08	31	8956.49	1.95
7	2097.88	0.10	32	9131.53	2.10
8	2397.73	0.12	33	9218.61	2.24
9	2711.17	0.16	34	9305.69	2.39
10	3024.61	0.19	35	9392.77	2.53
11	3311.33	0.22	36	9479.86	2.68
12	3598.04	0.25	37	9566.94	2.83
13	3911.93	0.30	38	9654.02	2.97
14	4225.81	0.35	39	9741.10	3.12
15	4519.39	0.39	40	9828.18	3.26
16	4812.97	0.43			
17	5113.67	0.49			
18	5414.38	0.55			
19	5723.31	0.60			
20	6042.25	0.66			
21	6335.53	0.73			
22	6630.80	0.80			
23	7032.47	0.90			
24	7474.54	1.03			
25	7776.83	1.16			

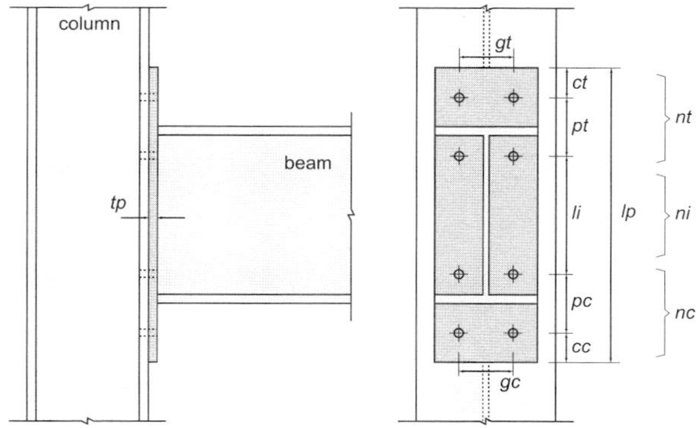

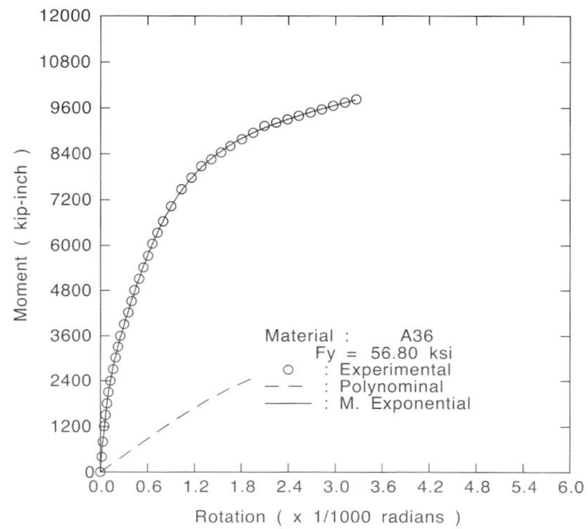

```
Moment-rotation prediction equations      ( R : X 1/1000 radians )

Frye and Morris polynominal model  :    R = Sum ( Ai X ( K*Bm )**Pi X 10**Qi )
     xd = 26.750000"   t =  1.000000"
     A1 =  1.790000   A2 =  1.760000   A3 =  2.040000    K =  0.000375
     P1 =         1   P2 =         3   P3 =         5   Q1 =         0   Q2 =        -1   Q3 =        -1

Modified exponential model  :
     Bm = Sum ( Ai X ( 1 - exp( - R/( 2*i*C ) ) ) )  +     Sum ( RKj X ( R - Rj0 ) )  +  Bm0
      C =   0.31275000E-01               Bmo=   0.00000000E+00          Nexp=  6         Nliner=  1
     AI =  -0.11138741E+04   0.22689708E+05  -0.84184518E+05   0.16885147E+06  -0.18917284E+06   0.91074773E+05
    Rj0 =            0.0000
    RKj =   0.51688531E+03
```

No	Rotation (radians) X 1/1000	Moment (kip-inch) Expri.	Moment (kip-inch) Poly.	Moment (kip-inch) M.Expo.	Connection stiffness (kip-inch) X 1000 Expri.	Connection stiffness (kip-inch) X 1000 Poly.	Connection stiffness (kip-inch) X 1000 M.Expo.
1	0.00	0.00	0.00	0.00	0.2478E+05	0.1488E+04	0.2761E+05
3	0.03	807.90	48.67	825.55	0.2471E+05	0.1488E+04	0.2252E+05
5	0.07	1504.94	98.97	1503.80	0.1675E+05	0.1488E+04	0.1781E+05
7	0.10	2097.88	155.50	2101.25	0.1463E+05	0.1487E+04	0.1386E+05
9	0.16	2711.17	236.51	2749.05	0.9219E+04	0.1485E+04	0.1026E+05
11	0.22	3311.33	328.53	3310.63	0.1024E+05	0.1482E+04	0.8121E+04
13	0.30	3911.93	447.70	3909.34	0.5979E+04	0.1475E+04	0.6942E+04
15	0.39	4519.39	583.24	4519.09	0.7432E+04	0.1466E+04	0.6371E+04
17	0.49	5113.67	724.35	5114.23	0.5230E+04	0.1452E+04	0.5893E+04
19	0.60	5728.31	887.91	5744.36	0.5657E+04	0.1431E+04	0.5242E+04
21	0.73	6336.53	1063.11	6346.06	0.4296E+04	0.1402E+04	0.4462E+04
23	0.90	7032.47	1303.37	7033.99	0.3593E+04	0.1349E+04	0.3454E+04
25	1.16	7776.83	1637.60	7771.47	0.2418E+04	0.1248E+04	0.2364E+04
27	1.41	8254.88	1935.01	8271.34	0.1410E+04	0.1133E+04	0.1690E+04
29	1.66	8606.40	2202.55	8634.83	0.1312E+04	0.1014E+04	0.1257E+04
31	1.95	8956.49	2479.37	8951.79	0.1199E+04	0.8845E+03	0.9404E+03
33	2.24	9218.61	2720.80	9196.45	0.5964E+03	0.7722E+03	0.7521E+03
35	2.53	9392.77	2932.19	9398.72	0.5964E+03	0.6786E+03	0.6436E+03
37	2.83	9566.94	3118.73	9576.99	0.5964E+03	0.6016E+03	0.5834E+03
39	3.12	9741.10	3284.85	9742.11	0.5964E+03	0.5383E+03	0.5510E+03

V - 80

Connection type : Extended end-plate connections
Mode : All bolted with column stiffener

Tested by : A.Mazroi (1984) U.S.A
Test Id. : CF5-S10x68

Column : W10x68
Beam : W27x114
Plate thickness : 1"
Stiffener thickness : 0.3750"

Fasteners: A325- -1"D
1 1/16" Oversize holes
Material : A36
Fy = 39.00 ksi
Fu = -- ksi

Major parameters

End plate extended on both sides.

ct = 4.3750"	pt = 3.9300"	li = 22.3901"	pc = 3.9300"
cc = 4.3750"	pit= 2.6667"	pi = 17.0567"	pic= 2.6667"
gt = 5.5000"	gi = 5.5000"	gc = 5.5000"	
tp = 1.0000"	nt = 2 X 3	ni = 2 X 2	nc = 2 X 3

Remark 1) End plate was stiffened in the plane of beam web by an end-plate.
2) pic=poc and pit=pot

No	Moment (k-in)	Rotation (radians) X 1/1000
1	0.00	0.00
2	282.46	0.01
3	564.92	0.02
4	969.02	0.04
5	1171.42	0.06
6	1373.82	0.08
7	1656.98	0.12
8	1994.54	0.16
9	2568.11	0.22
10	2750.46	0.25
11	3101.77	0.30
12	3385.75	0.35
13	3723.80	0.41
14	4007.98	0.46
15	4333.06	0.53
16	4549.98	0.58
17	4820.78	0.64
18	5091.59	0.70
19	5342.14	0.75
20	5592.68	0.80
21	5890.58	0.87
22	6188.47	0.93
23	6345.14	0.99
24	6501.80	1.05
25	6680.38	1.15
26	6858.95	1.26

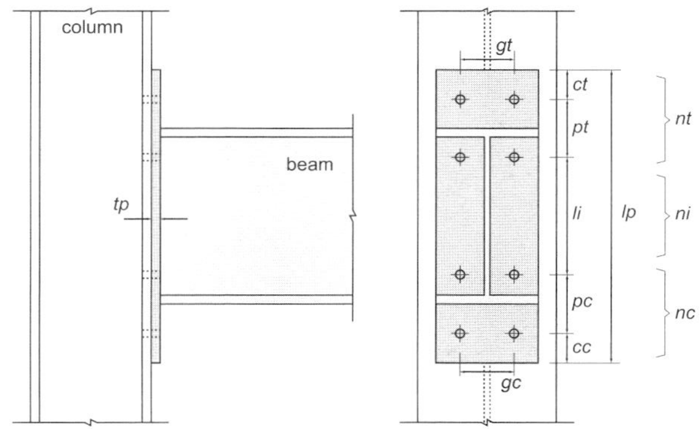

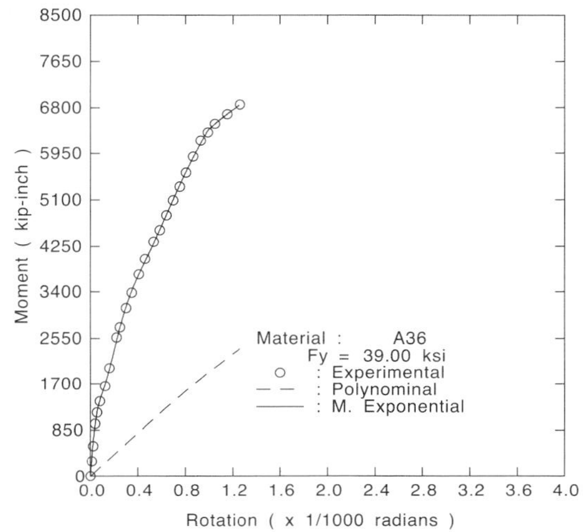

```
Moment-rotation prediction equations     ( R : X 1/1000 radians )

Frye and Morris polynominal model  :    R = Sum ( Ai X ( K*Bm )**Pi X 10**Qi )
     xd = 30.250100"   t =  1.000000"
     A1 =  1.790000   A2 =  1.760000   A3 =  2.040000    K =  0.000279
     P1 =         1   P2 =         3   P3 =         5   Q1 =         0   Q2 =      -1   Q3 =      -1

Modified exponential model  :
     Bm = Sum ( Ai X ( 1 - exp( - R/( 2*i*C ) ) ) )  +     Sum ( RKj X ( R - Rj0 ) )  +  Bm0
       C =   0.13000000E-01             Bmo=   0.00000000E+00         Nexp=  6       Nliner=  2
      AI =  -0.47667464E+04   0.76513777E+05  -0.35453545E+06   0.74008136E+06  -0.71934246E+06   0.26564812E+06
     Rj0 =           0.0400           0.9900
     RKj =   0.31078137E+04  -0.18242942E+04
```

No	Rotation (radians) X 1/1000	Moment (kip-inch) Expri.	Moment (kip-inch) Poly.	Moment (kip-inch) M.Expo.	Connection stiffness (kip-inch) X 1000 Expri.	Connection stiffness (kip-inch) X 1000 Poly.	Connection stiffness (kip-inch) X 1000 M.Expo.
1	0.00	0.00	0.00	0.00	0.2825E+05	0.1999E+04	0.2839E+05
2	0.01	282.46	19.99	286.25	0.2541E+05	0.1999E+04	0.2801E+05
3	0.02	564.92	45.99	619.43	0.2415E+05	0.1999E+04	0.2278E+05
4	0.04	969.02	75.97	906.03	0.1989E+05	0.1999E+04	0.1553E+05
5	0.06	1171.42	109.96	1158.14	0.1028E+05	0.1999E+04	0.1219E+05
6	0.08	1373.82	161.93	1400.40	0.7386E+04	0.1998E+04	0.7360E+04
7	0.12	1656.98	245.85	1686.28	0.8008E+04	0.1997E+04	0.7124E+04
8	0.16	1994.54	319.72	1977.45	0.9200E+04	0.1995E+04	0.8544E+04
9	0.22	2568.11	442.39	2524.20	0.7617E+04	0.1990E+04	0.8736E+04
10	0.25	2750.46	495.13	2747.86	0.6797E+04	0.1988E+04	0.8104E+04
11	0.30	3101.77	600.48	3136.46	0.6255E+04	0.1982E+04	0.6566E+04
12	0.35	3385.75	695.61	3423.64	0.5877E+04	0.1976E+04	0.5479E+04
13	0.41	3723.80	810.21	3718.86	0.5535E+04	0.1967E+04	0.4811E+04
14	0.46	4007.98	915.37	3973.18	0.4993E+04	0.1957E+04	0.4663E+04
15	0.53	4333.06	1052.35	4302.18	0.4330E+04	0.1942E+04	0.4761E+04
16	0.58	4549.98	1155.25	4557.18	0.4432E+04	0.1928E+04	0.4854E+04
17	0.64	4820.78	1262.90	4832.59	0.4793E+04	0.1912E+04	0.4880E+04
18	0.70	5091.59	1370.94	5106.91	0.4737E+04	0.1894E+04	0.4816E+04
19	0.75	5342.14	1472.27	5361.34	0.4683E+04	0.1875E+04	0.4687E+04
20	0.80	5592.68	1571.07	5607.62	0.4758E+04	0.1855E+04	0.4515E+04
21	0.87	5890.58	1685.13	5878.61	0.4844E+04	0.1829E+04	0.4296E+04
22	0.93	6188.47	1797.59	6136.05	0.3714E+04	0.1800E+04	0.4078E+04
23	0.99	6345.14	1903.11	6374.83	0.2611E+04	0.1771E+04	0.3884E+04
24	1.05	6501.80	2009.37	6498.72	0.2274E+04	0.1739E+04	0.1892E+04
25	1.15	6680.38	2190.54	6687.20	0.1677E+04	0.1679E+04	0.1662E+04
26	1.26	6858.95	2366.01	6855.47	0.1677E+04	0.1615E+04	0.1509E+04

A.5 – 161

V - 81

Connection type : Extended end-plate connections
Mode : All bolted with column stiffener

Tested by : A.Mazroi (1984) U.S.A
Test Id. : CF5-S12x96

Column : W12x96
Beam : W27x114
Plate thickness : 1"
Stiffener thickness : 0.3750"

Fasteners: A325- -1"D
1 1/16" Oversize holes
Material : A36
Fy = 40.00 ksi
Fu = -- ksi

Major parameters

End plate extended on both sides.

ct = 4.3750"	pt = 3.9300"	li = 22.3901"	pc = 3.9300"
cc = 4.3750"	pit= 2.6667"	pi = 17.0567"	pic= 2.6667"
gt = 5.5000"	gi = 5.5000"	gc = 5.5000"	
tp = 1.0000"	nt = 2 X 3	ni = 2 X 2	nc = 2 X 3

Remark 1) End plate was stiffened in the plane of beam web by an end-plate.
2) pic=poc and pit=pot

No	Moment (k-in)	Rotation (radians) X 1/1000	No	Moment (k-in)	Rotation (radians) X 1/1000
1	0.00	0.00	26	10095.24	1.77
2	361.09	0.02	27	10302.37	1.91
3	722.18	0.05	28	10509.49	2.06
4	1232.30	0.09	29	10667.68	2.21
5	1754.80	0.13	30	10825.88	2.35
6	2237.15	0.17	31	10984.07	2.50
7	2673.03	0.21	32	11087.38	2.64
8	3108.91	0.25	33	11190.70	2.78
9	3584.93	0.30	34	11294.01	2.92
10	4060.94	0.34	35	11397.32	3.06
11	4517.19	0.39	36	11480.42	3.19
12	4973.44	0.44	37	11563.52	3.33
13	5436.77	0.50	38	11646.61	3.46
14	5900.10	0.56	39	11729.71	3.60
15	6343.77	0.62			
16	6787.44	0.69			
17	7231.95	0.78			
18	7676.46	0.86			
19	8021.28	0.95			
20	8366.10	1.05			
21	8670.78	1.14			
22	8975.46	1.23			
23	9261.17	1.35			
24	9546.88	1.46			
25	9821.06	1.61			

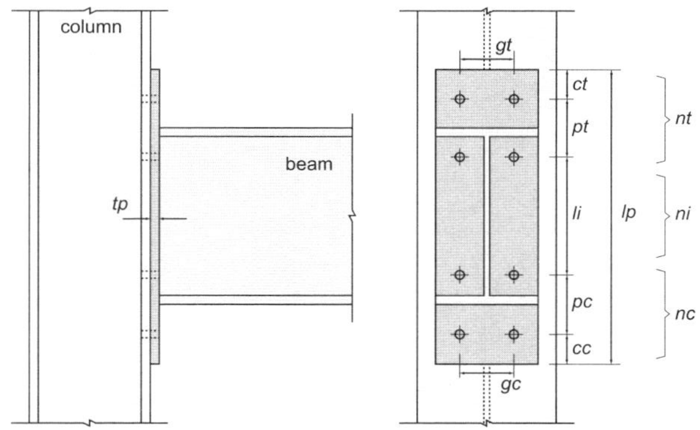

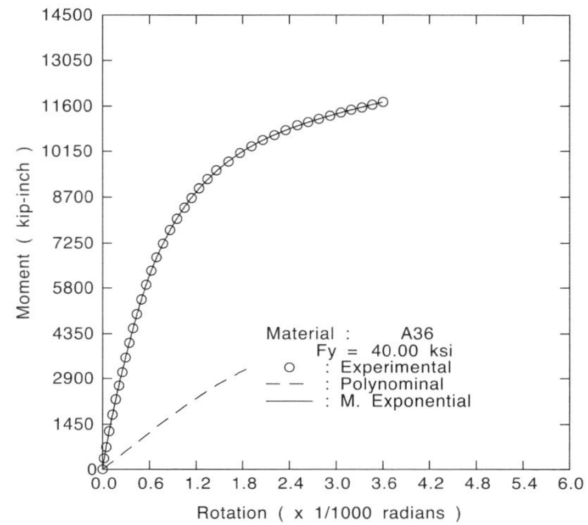

A.5 – 162

```
Moment-rotation prediction equations      ( R : X 1/1000 radians )

Frye and Morris polynominal model  :     R = Sum ( Ai X ( K*Bm )**Pi X 10**Qi )
     xd = 30.250100"    t =  1.000000"
     A1 =  1.790000    A2 =  1.760000    A3 =  2.040000    K =  0.000279
     P1 =         1    P2 =         3    P3 =         5    Q1 =         0    Q2 =        -1    Q3 =        -1

Modified exponential model  :
     Bm = Sum ( Ai X ( 1 - exp( - R/( 2*i*C ) ) ) )  +     Sum ( RKj X ( R - Rj0 ) )  +  Bm0
      C =   0.34991667E-01               Bmo=   0.00000000E+00          Nexp=  6        Nliner=  1
     AI =   0.26558002E+03   0.12695529E+05  -0.97971704E+05   0.26572263E+06  -0.32130148E+06   0.15033877E+06
     Rj0 =             0.0000
     RKj =   0.55535261E+03
```

No	Rotation (radians) X 1/1000	Moment (kip-inch) Expri.	Moment (kip-inch) Poly.	Moment (kip-inch) M.Expo.	Connection stiffness (kip-inch) X 1000 Expri.	Connection stiffness (kip-inch) X 1000 Poly.	Connection stiffness (kip-inch) X 1000 M.Expo.
1	0.00	0.00	0.00	0.00	0.1805E+05	0.1999E+04	0.1690E+05
3	0.05	722.18	93.97	732.70	0.1312E+05	0.1999E+04	0.1411E+05
5	0.13	1754.80	251.85	1740.64	0.1274E+05	0.1996E+04	0.1184E+05
7	0.21	2673.03	416.46	2676.30	0.1026E+05	0.1991E+04	0.1094E+05
9	0.30	3584.93	589.52	3596.30	0.1070E+05	0.1983E+04	0.1019E+05
11	0.39	4517.19	777.59	4519.66	0.9035E+04	0.1969E+04	0.9215E+04
13	0.50	5436.77	989.58	5451.61	0.7988E+04	0.1949E+04	0.7954E+04
15	0.62	6343.77	1231.31	6353.08	0.6672E+04	0.1917E+04	0.6553E+04
17	0.78	7231.95	1519.33	7236.60	0.5199E+04	0.1866E+04	0.5135E+04
19	0.95	8021.28	1846.19	8043.09	0.3668E+04	0.1787E+04	0.3929E+04
21	1.14	8670.78	2169.50	8690.44	0.3312E+04	0.1687E+04	0.3087E+04
23	1.35	9261.17	2503.40	9255.34	0.2517E+04	0.1561E+04	0.2449E+04
25	1.61	9821.06	2897.40	9827.24	0.1786E+04	0.1390E+04	0.1869E+04
27	1.91	10302.37	3286.76	10314.81	0.1414E+04	0.1209E+04	0.1408E+04
29	2.21	10667.68	3618.51	10679.47	0.1074E+04	0.1054E+04	0.1094E+04
31	2.50	10984.07	3909.38	10968.27	0.9038E+03	0.9242E+03	0.8814E+03
33	2.78	11190.70	4151.82	11194.02	0.7433E+03	0.8231E+03	0.7523E+03
35	3.06	11397.32	4368.62	11390.99	0.6781E+03	0.7392E+03	0.6712E+03
37	3.33	11563.52	4559.09	11565.51	0.6142E+03	0.6711E+03	0.6231E+03
39	3.60	11729.71	4732.63	11729.83	0.6146E+03	0.6136E+03	0.5943E+03

A.5 – 163

V - 82

```
Connection type : Extended end-plate connections
           Mode : All bolted with column stiffener

Tested by  :  A.Mazroi (1984)                    U.S.A
Test  Id.  :  CF6-S12x79

Column : W12x79                        Fasteners: A325- -1"D
Beam   : W27x114                                1 1/16" Oversize holes
Plate thickness : 1 1/4"               Material : A36
Stiffener thickness : 0.3750"              Fy =  39.80 ksi
                                           Fu =   --   ksi
```

Major parameters

```
End plate extended on both sides.
ct =  4.6250"    pt =  4.9300"    li = 21.3900"    pc =  4.9300"
cc =  4.6250"    pit=  3.0000"    pi = 15.3900"    pic=  3.0000"
gt =  6.5000"    gi =  6.5000"    gc =  6.5000"
tp =  1.2500"    nt = 2 X  3      ni = 2 X  2      nc = 2 X  3
```

Remark 1) End plate was stiffened in the plane of beam web by an end-plate.
2) pic=poc and pit=pot

No	Moment (k-in)	Rotation (radians) X 1/1000	No	Moment (k-in)	Rotation (radians) X 1/1000
1	0.00	0.00	26	7217.70	2.18
2	434.63	0.01	27	7369.42	2.36
3	869.26	0.03	28	7508.98	2.57
4	1204.49	0.06	29	7648.55	2.78
5	1539.73	0.09	30	7788.12	2.98
6	1841.73	0.12	31	7924.33	3.21
7	2143.73	0.15	32	8060.54	3.44
8	2452.66	0.19	33	8196.74	3.66
9	2761.60	0.23	34	8296.02	3.88
10	3064.29	0.28	35	8395.30	4.10
11	3366.98	0.32	36	8494.58	4.32
12	3636.89	0.38	37	8577.86	4.54
13	3906.79	0.44	38	8661.14	4.75
14	4170.15	0.50	39	8744.42	4.97
15	4433.51	0.56			
16	4867.39	0.69			
17	5131.85	0.77			
18	5396.30	0.85			
19	5724.81	1.01			
20	6053.32	1.17			
21	6246.89	1.30			
22	6440.46	1.43			
23	6677.37	1.62			
24	6914.28	1.82			
25	7065.99	2.00			

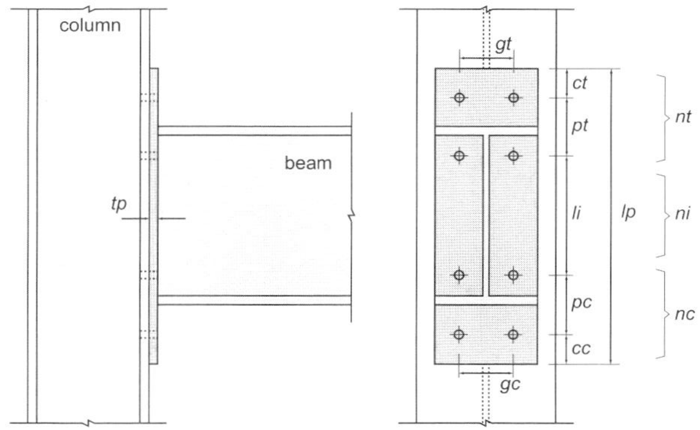

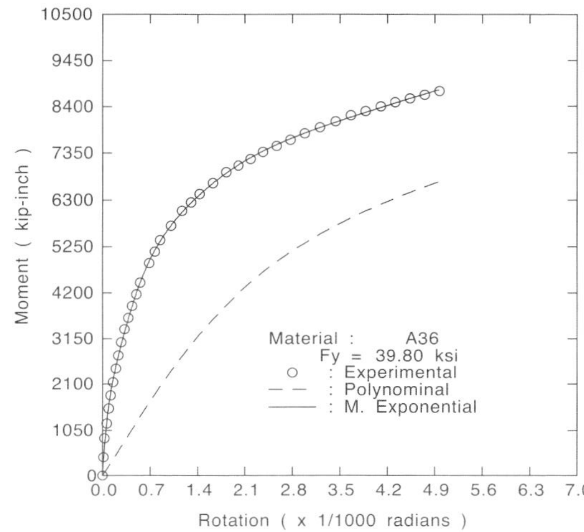

A.5 – 164

```
Moment-rotation prediction equations     ( R : X 1/1000 radians )

Frye and Morris polynominal model  :    R = Sum ( Ai X ( K*Bm )**Pi X 10**Qi )
     xd = 31.250000"    t =  1.250000"
     A1 =  1.790000   A2 =  1.760000   A3 =  2.040000    K =  0.000226
     P1 =         1   P2 =         3   P3 =         5   Q1 =         0    Q2 =       -1     Q3 =       -1

Modified exponential model  :
     Bm = Sum ( Ai X ( 1 - exp( - R/( 2*i*C ) ) ) )  +      Sum ( RKj X ( R - Rj0 ) )  +  Bm0
      C =   0.45191667E-01            Bmo=   0.00000000E+00          Nexp=  6         Nliner=  1
     AI =   0.45222545E+04  -0.21031166E+05   0.47298017E+05  -0.32944240E+05  -0.17497345E+05   0.26158313E+05
    Rj0 =           0.0000
    RKj =   0.45813126E+03
```

No	Rotation (radians) X 1/1000	Moment (kip-inch) Expri.	Moment (kip-inch) Poly.	Moment (kip-inch) M.Expo.	Connection stiffness (kip-inch) X 1000 Expri.	Connection stiffness (kip-inch) X 1000 Poly.	Connection stiffness (kip-inch) X 1000 M.Expo.
1	0.00	0.00	0.00	0.00	0.2898E+05	0.2471E+04	0.2652E+05
3	0.03	869.26	74.14	689.53	0.2304E+05	0.2471E+04	0.1951E+05
5	0.09	1539.73	222.39	1584.82	0.1027E+05	0.2469E+04	0.1148E+05
7	0.15	2143.73	382.85	2200.21	0.8803E+04	0.2466E+04	0.8007E+04
9	0.23	2761.60	567.68	2736.37	0.7494E+04	0.2459E+04	0.6532E+04
11	0.32	3366.98	793.65	3297.55	0.5756E+04	0.2447E+04	0.5735E+04
13	0.44	3906.79	1070.92	3905.15	0.4567E+04	0.2425E+04	0.4920E+04
15	0.56	4433.51	1360.98	4443.07	0.4051E+04	0.2392E+04	0.4052E+04
17	0.77	5131.85	1861.68	5162.81	0.3245E+04	0.2309E+04	0.2809E+04
19	1.01	5724.81	2400.38	5730.69	0.2066E+04	0.2175E+04	0.2004E+04
21	1.30	6246.89	3000.85	6233.29	0.1483E+04	0.1967E+04	0.1523E+04
23	1.62	6677.37	3598.55	6674.14	0.1218E+04	0.1711E+04	0.1210E+04
25	2.00	7065.99	4189.30	7078.06	0.8336E+03	0.1434E+04	0.9485E+03
27	2.36	7369.42	4669.77	7387.86	0.7608E+03	0.1214E+04	0.7644E+03
29	2.78	7648.55	5127.15	7671.84	0.6787E+03	0.1019E+04	0.6277E+03
31	3.21	7924.33	5533.27	7923.69	0.6000E+03	0.8652E+03	0.5447E+03
33	3.66	8196.74	5897.16	8159.51	0.5256E+03	0.7435E+03	0.4993E+03
35	4.10	8395.30	6201.83	8372.60	0.4541E+03	0.6536E+03	0.4778E+03
37	4.54	8577.86	6471.39	8578.84	0.3814E+03	0.5827E+03	0.4674E+03
39	4.97	8744.42	6712.98	8781.72	0.3815E+03	0.5257E+03	0.4624E+03

V - 83

Connection type : Extended end-plate connections
Mode : All bolted with column stiffener

Tested by : A.Mazroi (1984) U.S.A
Test Id. : CF8-S14x145

Column : W14x145
Beam : W24x162
Plate thickness : 1 3/4"
Stiffener thickness : 0.5000"

Fasteners: A325- -1 1/8"D
1 1/4" Oversize holes
Material : A36
Fy = 36.70 ksi
Fu = -- ksi

Major parameters

End plate extended on both sides.

ct = 4.8750"	pt = 5.4700"	li = 18.3100"	pc = 5.4700"
cc = 4.8750"	pit= 3.3750"	pi = 11.5600"	pic= 3.3750"
gt = 6.5000"	gi = 6.5000"	gc = 6.5000"	
tp = 1.7500"	nt = 2 X 3	ni = 2 X 2	nc = 2 X 3

Remark 1) End plate was stiffened in the plane of beam web by an end-plate.
2) pic=poc and pit=pot

No	Moment (k-in)	Rotation (radians) X 1/1000
1	0.00	0.00
2	499.28	0.02
3	998.56	0.04
4	1497.84	0.07
5	2107.55	0.09
6	2717.27	0.12
7	3321.44	0.17
8	3925.61	0.22
9	4556.55	0.28
10	5187.49	0.34
11	5780.08	0.44
12	6372.67	0.53
13	6870.63	0.63
14	7368.59	0.73
15	7866.55	0.83
16	8508.48	1.00
17	9150.41	1.17
18	9688.12	1.38
19	10225.84	1.60
20	10624.99	1.82
21	11024.14	2.04
22	11450.90	2.29
23	11877.67	2.54
24	12102.20	2.75
25	12326.74	2.96

No	Moment (k-in)	Rotation (radians) X 1/1000
26	12551.27	3.17
27	12758.84	3.40
28	12966.42	3.63
29	13174.00	3.86
30	13364.42	4.10
31	13554.84	4.34
32	13745.27	4.58
33	13872.38	4.78
34	13999.49	4.99
35	14126.60	5.19
36	14231.38	5.38
37	14336.16	5.58
38	14440.93	5.77

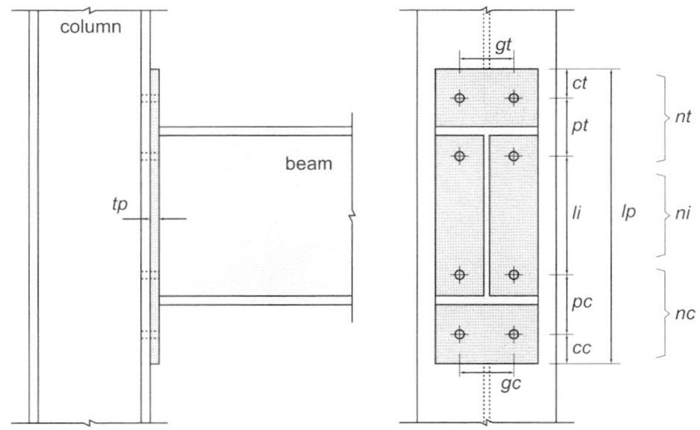

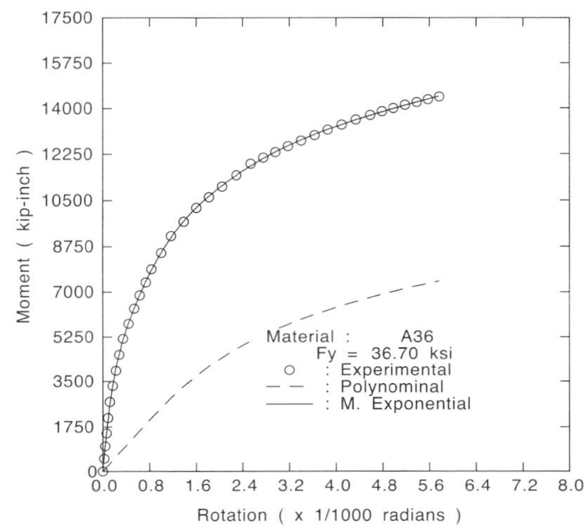

A.5 – 166

```
Moment-rotation prediction equations     ( R : X 1/1000 radians )

Frye and Morris polynominal model  :    R = Sum ( Ai X ( K*Bm )**Pi X 10**Qi )
     xd = 29.250000"   t =  1.750000"
     A1 =  1.790000   A2 =  1.760000   A3 =  2.040000    K =  0.000217
     P1 =         1   P2 =         3   P3 =         5    Q1 =         0    Q2 =        -1    Q3 =        -1

Modified exponential model  :
     Bm = Sum ( Ai X ( 1 - exp( - R/( 2*i*C ) ) ) )  +     Sum ( RKj X ( R - Rj0 ) )  +  Bm0
      C =   0.53516667E-01              Bmo=   0.00000000E+00         Nexp=  6        Nliner=  1
     AI =  -0.25951853E+04   0.53472939E+05  -0.24440294E+06   0.53782721E+06  -0.56386008E+06   0.23079997E+06
     Rj0 =          0.0000
     RKj =   0.56066978E+03
```

No	Rotation (radians) X 1/1000	Moment (kip-inch) Expri.	Poly.	M.Expo.	Connection stiffness (kip-inch) X 1000 Expri.	Poly.	M.Expo.
1	0.00	0.00	0.00	0.00	0.2301E+05	0.2580E+04	0.2639E+05
3	0.04	998.56	111.73	1078.97	0.2306E+05	0.2580E+04	0.2292E+05
5	0.09	2107.55	234.77	2074.09	0.2345E+05	0.2578E+04	0.1890E+05
7	0.17	3321.44	437.10	3342.88	0.1151E+05	0.2573E+04	0.1373E+05
9	0.28	4556.55	727.59	4609.56	0.1043E+05	0.2561E+04	0.9183E+04
11	0.44	5780.08	1124.46	5790.07	0.6205E+04	0.2531E+04	0.6409E+04
13	0.63	6870.63	1611.33	6894.78	0.5030E+04	0.2471E+04	0.5136E+04
15	0.83	7866.55	2092.00	7832.70	0.4565E+04	0.2379E+04	0.4358E+04
17	1.17	9150.41	2867.01	9109.01	0.3217E+04	0.2153E+04	0.3164E+04
19	1.60	10225.84	3707.80	10217.94	0.2180E+04	0.1809E+04	0.2173E+04
21	2.04	11024.14	4428.98	11051.89	0.1762E+04	0.1472E+04	0.1669E+04
23	2.54	11877.67	5086.80	11801.22	0.1356E+04	0.1175E+04	0.1349E+04
25	2.96	12326.74	5544.46	12328.24	0.1057E+04	0.9893E+03	0.1138E+04
27	3.40	12758.84	5946.69	12788.13	0.9104E+03	0.8447E+03	0.9572E+03
29	3.86	13174.00	6304.85	13190.20	0.8519E+03	0.7313E+03	0.8140E+03
31	4.34	13554.84	6634.28	13555.69	0.7901E+03	0.6396E+03	0.7101E+03
33	4.78	13872.38	6902.40	13855.71	0.6293E+03	0.5733E+03	0.6491E+03
35	5.19	14126.60	7123.71	14110.32	0.5829E+03	0.5238E+03	0.6140E+03
37	5.58	14336.16	7319.59	14344.92	0.5383E+03	0.4837E+03	0.5929E+03

A.5 – 167

V - 84

Connection type : Extended end-plate connections
Mode : All bolted with column stiffener

Tested by : A.Mazroi (1984) U.S.A
Test Id. : CF8-S14x158

Column : W14x158
Beam : W24x162
Plate thickness : 1 3/4"
Stiffener thickness : 0.5000"

Fasteners: A325- -1 1/8"D
1 1/4" Oversize holes
Material : A36
Fy = 37.30 ksi
Fu = -- ksi

Major parameters

End plate extended on both sides.

ct = 4.8750"	pt = 5.4700"	li = 18.3100"	pc = 5.4700"
cc = 4.8750"	pit= 3.3750"	pi = 11.5600"	pic= 3.3750"
gt = 6.5000"	gi = 6.5000"	gc = 6.5000"	
tp = 1.7500"	nt = 2 X 3	ni = 2 X 2	nc = 2 X 3

Remark 1) End plate was stiffened in the plane of beam web by an end-plate.
2) pic=poc and pit=pot

No	Moment (k-in)	Rotation (radians) X 1/1000	No	Moment (k-in)	Rotation (radians) X 1/1000
1	0.00	0.00	26	13161.16	2.18
2	585.98	0.02	27	13420.85	2.31
3	1171.96	0.04	28	13654.39	2.44
4	1919.50	0.10	29	13887.94	2.57
5	2667.05	0.16	30	14121.48	2.71
6	3288.41	0.22	31	14346.66	2.87
7	3909.78	0.28	32	14571.83	3.03
8	4531.30	0.34	33	14797.01	3.20
9	5152.81	0.40	34	15022.19	3.36
10	5721.30	0.47			
11	6289.79	0.53			
12	6884.91	0.59			
13	7480.03	0.66			
14	8056.87	0.75			
15	8633.71	0.85			
16	9117.41	0.94			
17	9601.12	1.03			
18	10120.00	1.16			
19	10638.89	1.29			
20	11104.95	1.42			
21	11571.01	1.55			
22	11927.94	1.68			
23	12284.87	1.80			
24	12641.80	1.92			
25	12901.48	2.05			

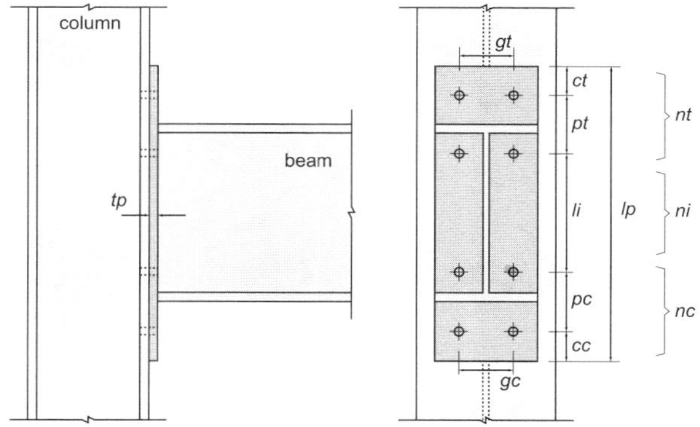

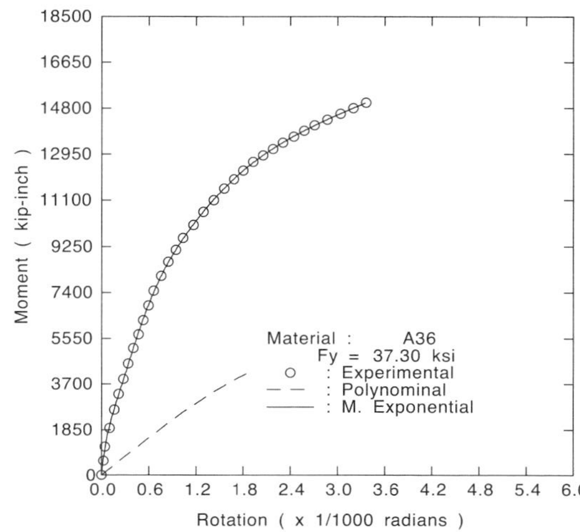

A.5 – 168

```
Moment-rotation prediction equations      ( R : X 1/1000 radians )

Frye and Morris polynominal model  :     R = Sum ( Ai X ( K*Bm )**Pi X 10**Qi )
     xd = 29.250000"   t =  1.750000"
     A1 =  1.790000   A2 =  1.760000   A3 =  2.040000    K =  0.000217
     P1 =         1   P2 =         3   P3 =         5   Q1 =         0    Q2 =        -1    Q3 =        -1

Modified exponential model  :
     Bm = Sum ( Ai X ( 1 - exp( - R/( 2*i*C ) ) ) )  +       Sum ( RKj X ( R - Rj0 ) )  +  Bm0
      C =   0.33366667E-01            Bmo=   0.00000000E+00          Nexp=  6         Nliner=  1
     AI =   0.75887301E+03   0.36594318E+05  -0.24368346E+06   0.61787971E+06  -0.70466334E+06   0.30411878E+06
    Rj0 =           0.0000
    RKj =   0.12061902E+04
```

No	Rotation (radians) X 1/1000	Moment (kip-inch) Expri.	Moment (kip-inch) Poly.	Moment (kip-inch) M.Expo.	Connection stiffness (kip-inch) X 1000 Expri.	Connection stiffness (kip-inch) X 1000 Poly.	Connection stiffness (kip-inch) X 1000 M.Expo.
1	0.00	0.00	0.00	0.00	0.2725E+05	0.2580E+04	0.3075E+05
3	0.04	1171.96	110.95	1086.20	0.2337E+05	0.2580E+04	0.1992E+05
5	0.16	2667.05	410.10	2666.45	0.1172E+05	0.2574E+04	0.1061E+05
7	0.28	3909.78	713.54	3878.54	0.1032E+05	0.2561E+04	0.1026E+05
9	0.40	5152.81	1026.85	5132.89	0.9445E+04	0.2540E+04	0.9977E+04
11	0.53	6289.79	1356.01	6359.67	0.8987E+04	0.2506E+04	0.8795E+04
13	0.66	7480.03	1676.21	7401.85	0.7960E+04	0.2460E+04	0.7371E+04
15	0.85	8633.71	2132.37	8625.14	0.5649E+04	0.2370E+04	0.5686E+04
17	1.03	9601.12	2562.78	9578.14	0.4730E+04	0.2254E+04	0.4648E+04
19	1.29	10638.89	3111.55	10642.58	0.3796E+04	0.2062E+04	0.3803E+04
21	1.55	11571.01	3632.92	11570.42	0.3188E+04	0.1842E+04	0.3169E+04
23	1.80	12284.87	4060.40	12284.96	0.2910E+04	0.1645E+04	0.2669E+04
25	2.05	12901.48	4448.45	12896.45	0.2039E+04	0.1463E+04	0.2238E+04
27	2.31	13420.85	4799.90	13420.74	0.1893E+04	0.1301E+04	0.1896E+04
29	2.57	13887.94	5129.54	13893.15	0.1738E+04	0.1157E+04	0.1637E+04
31	2.87	14346.66	5452.93	14349.80	0.1384E+04	0.1025E+04	0.1452E+04
33	3.20	14797.01	5766.65	14801.28	0.1384E+04	0.9071E+03	0.1334E+04

V - 85

Connection type : Extended end-plate connections
Mode : All bolted with column stiffener

Tested by : Y.L.Yee (1984) Australia
Test Id. : K1B

Column : 460UB67.1
Beam : 460UB67.1
Plate thickness : 32 mm
Stiffener thickness : 0.6299"

Fasteners: G8.8- -M30
1 1/4" Oversize holes
Material : G250
Fy = 37.70 ksi
Fu = -- ksi

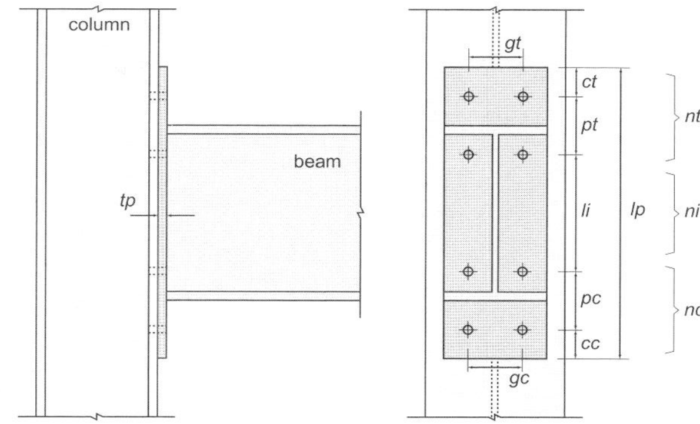

Major parameters

End plate extended on both sides.

ct = 2.3622" pt = 5.2244" li = 9.7874" pc = 7.5866"
cc = 2.3622" gt = 5.1181" gc = 5.1181"
tp = 1.2598" nt = 2 X 2 nc = 2 X 2

Remark 1) 32 mm Doubler plates were welded to flange on both sides of column web.
2) End-plate yield stress (Fy) is mentioned.

No	Moment (k-in)	Rotation (radians) X 1/1000	No	Moment (k-in)	Rotation (radians) X 1/1000
1	0.00	0.00	26	3456.97	13.47
2	143.14	0.24	27	3519.74	14.20
3	285.28	0.49	28	3559.45	15.08
4	433.09	0.68	29	3599.16	15.97
5	589.06	0.88	30	3638.87	16.85
6	740.04	1.08	31	3663.01	17.92
7	891.01	1.28	32	3687.15	18.99
8	1043.16	1.87	33	3711.28	20.06
9	1195.30	2.46	34	3718.91	20.98
10	1347.45	3.05	35	3726.54	21.89
11	1495.94	3.44	36	3734.17	22.80
12	1645.42	3.84	37	3741.79	23.71
13	1795.91	4.23			
14	1945.51	4.76			
15	2085.52	5.40			
16	2227.53	6.03			
17	2379.36	6.29			
18	2529.15	7.00			
19	2683.80	7.67			
20	2803.54	8.16			
21	2970.85	9.00			
22	3107.09	9.88			
23	3243.33	10.76			
24	3313.76	11.75			
25	3394.20	12.74			

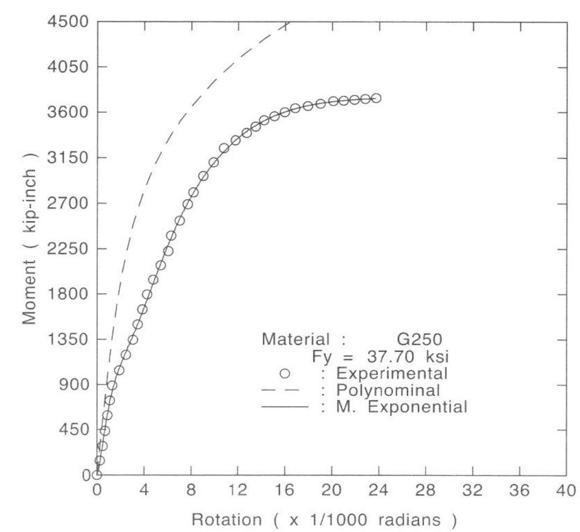

A.5 – 170

```
Moment-rotation prediction equations     ( R : X 1/1000 radians )

Frye and Morris polynominal model  :     R = Sum ( Ai X ( K*Bm )**Pi X 10**Qi )
     xd = 22.598400"   t =  1.259800"
     A1 =  1.790000   A2 =  1.760000   A3 =  2.040000    K =  0.000490
     P1 =        1    P2 =        3    P3 =        5    Q1 =        0    Q2 =       -1    Q3 =       -1

Modified exponential model  :
     Bm = Sum ( Ai X ( 1 - exp( - R/( 2*i*C ) ) ) )  +     Sum ( RKj X ( R - Rj0 ) )  +  Bm0
       C =   0.25873333E+00             Bmo=   0.00000000E+00          Nexp=  6        Nliner=  1
      AI =  -0.33054652E+04   0.29137180E+05  -0.99899131E+05   0.19123202E+06  -0.20269263E+06   0.89353719E+05
     Rj0 =          0.0000
     RKj =  -0.26688460E+01
```

No	Rotation (radians) X 1/1000	Moment (kip-inch) Expri.	Moment (kip-inch) Poly.	Moment (kip-inch) M.Expo.	Connection stiffness (kip-inch) X 1000 Expri.	Connection stiffness (kip-inch) X 1000 Poly.	Connection stiffness (kip-inch) X 1000 M.Expo.
1	0.00	0.00	0.00	0.00	0.5854E+03	0.1141E+04	0.2419E+03
3	0.49	286.28	553.39	308.91	0.6953E+03	0.1113E+04	0.7886E+03
5	0.88	589.06	978.76	601.20	0.7562E+03	0.1039E+04	0.6674E+03
7	1.28	891.01	1369.90	830.55	0.6300E+03	0.9137E+03	0.4855E+03
9	2.46	1195.30	2208.88	1228.26	0.2583E+03	0.5364E+03	0.2691E+03
11	3.44	1496.94	2643.93	1500.75	0.3791E+03	0.3682E+03	0.2933E+03
13	4.23	1795.91	2901.54	1741.42	0.3384E+03	0.2909E+03	0.3136E+03
15	5.40	2086.52	3196.11	2104.48	0.2215E+03	0.2217E+03	0.3031E+03
17	6.29	2379.36	3378.79	2363.81	0.4821E+03	0.1875E+03	0.2730E+03
19	7.67	2688.80	3610.08	2700.56	0.2417E+03	0.1522E+03	0.2179E+03
21	9.00	2970.85	3797.44	2958.99	0.1741E+03	0.1290E+03	0.1697E+03
23	10.76	3243.33	4004.39	3212.07	0.1180E+03	0.1079E+03	0.1210E+03
25	12.74	3394.20	4200.67	3410.45	0.8176E+02	0.9152E+02	0.8187E+02
27	14.20	3519.74	4327.69	3513.91	0.6732E+02	0.8246E+02	0.6045E+02
29	15.97	3599.16	4465.23	3602.42	0.4502E+02	0.7382E+02	0.4090E+02
31	17.92	3663.01	4601.80	3666.33	0.2252E+02	0.6628E+02	0.2548E+02
33	20.06	3711.28	4736.56	3707.98	0.1487E+02	0.5972E+02	0.1422E+02
35	21.89	3726.54	4841.27	3727.88	0.8359E+01	0.5516E+02	0.8018E+01
37	23.71	3741.79	4938.32	3738.55	0.8359E+01	0.5130E+02	0.3969E+01

V - 86

Connection type : Extended end-plate connections
Mode : All bolted with column stiffener

Tested by : Y.L.Yee (1984) Australia
Test Id. : K9

Column : 460UB67.1
Beam : 460UB67.1
Plate thickness : 20 mm
Stiffener thickness : 0.6299"

Fasteners: G8.8- -M30
1 1/4" Oversize holes
Material : G250
Fy = 37.70 ksi
Fu = -- ksi

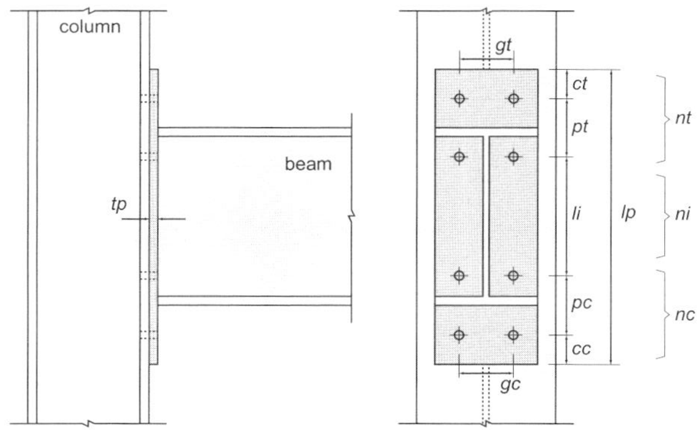

Major parameters

End plate extended on both sides.
ct = 2.3622" pt = 5.2244" li = 9.7874" pc = 7.5866"
cc = 2.3622" gt = 5.1181" gc = 5.1181"
tp = 0.7874" nt = 2 X 2 nc = 2 X 2

Remark 1) 32 mm Doubler plates were welded to flange on both sides of column web.
2) End-plate yield stress (Fy) is mentioned.

No	Moment (k-in)	Rotation (radians) X 1/1000	No	Moment (k-in)	Rotation (radians) X 1/1000
1	0.00	0.00	26	3251.08	15.17
2	153.16	0.30	27	3307.27	16.13
3	306.33	0.61	28	3363.47	17.08
4	463.25	0.95	29	3440.88	18.74
5	620.17	1.30	30	3483.30	20.15
6	767.82	1.72			
7	915.47	2.14			
8	1063.12	2.56			
9	1210.77	2.98			
10	1358.50	3.52			
11	1506.23	4.06			
12	1653.96	4.59			
13	1801.70	5.13			
14	1952.72	5.95			
15	2103.75	6.76			
16	2254.78	7.58			
17	2405.80	8.39			
18	2556.83	9.21			
19	2707.86	10.02			
20	2797.02	10.70			
21	2886.19	11.37			
22	2975.36	12.05			
23	3044.29	12.83			
24	3113.22	13.61			
25	3182.15	14.39			

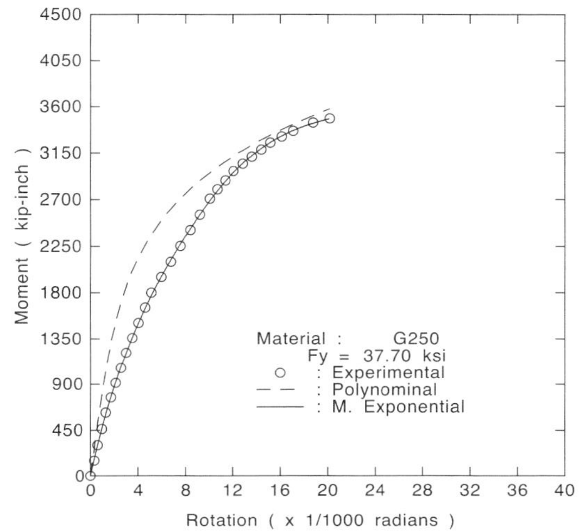

A.5 – 172

```
Moment-rotation prediction equations     ( R : X 1/1000 radians )

Frye and Morris polynominal model  :     R = Sum ( Ai X ( K*Bm )**Pi X 10**Qi )
     xd = 22.598400"   t =  0.787400"
     A1 =  1.790000   A2 =  1.760000   A3 =  2.040000    K =  0.000649
     P1 =         1   P2 =         3   P3 =         5   Q1 =         0   Q2 =        -1   Q3 =        -1

Modified exponential model  :
     Bm = Sum ( Ai X ( 1 - exp( - R/( 2*i*C ) ) ) )  +      Sum ( RKj X ( R - Rj0 ) )  +  Bm0
       C =   0.24392500E+00             Bmo=   0.00000000E+00            Nexp=  6       Nliner=  1
      AI =  -0.36393147E+03   0.97331858E+04  -0.72155120E+05   0.21421460E+06  -0.27210251E+06   0.12401441E+06
     Rj0 =           0.0000
     RKj =   0.10099667E+02
```

No	Rotation (radians) X 1/1000	Moment (kip-inch) Expri.	Moment (kip-inch) Poly.	Moment (kip-inch) M.Expo.	Connection stiffness (kip-inch) X 1000 Expri.	Connection stiffness (kip-inch) X 1000 Poly.	Connection stiffness (kip-inch) X 1000 M.Expo.
1	0.00	0.00	0.00	0.00	0.5038E+03	0.8603E+03	0.5190E+03
2	0.30	153.16	260.77	158.33	0.5038E+03	0.8527E+03	0.5074E+03
3	0.61	306.33	516.60	307.09	0.4809E+03	0.8269E+03	0.4709E+03
4	0.95	463.25	792.87	462.81	0.4548E+03	0.7693E+03	0.4332E+03
5	1.30	620.17	1044.34	606.95	0.4080E+03	0.6848E+03	0.4038E+03
6	1.72	767.82	1308.45	770.88	0.3509E+03	0.5698E+03	0.3765E+03
7	2.14	915.47	1526.02	924.37	0.3510E+03	0.4678E+03	0.3537E+03
8	2.56	1063.12	1705.19	1068.65	0.3509E+03	0.3878E+03	0.3324E+03
9	2.98	1210.77	1855.02	1204.12	0.3175E+03	0.3274E+03	0.3116E+03
10	3.52	1358.50	2014.96	1364.64	0.2750E+03	0.2709E+03	0.2862E+03
11	4.06	1506.23	2148.98	1512.07	0.2750E+03	0.2303E+03	0.2632E+03
12	4.59	1653.96	2264.17	1648.04	0.2750E+03	0.2000E+03	0.2436E+03
13	5.13	1801.70	2365.13	1774.50	0.2393E+03	0.1768E+03	0.2277E+03
14	5.95	1952.72	2497.91	1952.25	0.1853E+03	0.1504E+03	0.2096E+03
15	6.76	2103.75	2612.17	2117.49	0.1853E+03	0.1310E+03	0.1965E+03
16	7.58	2254.78	2712.71	2273.25	0.1852E+03	0.1162E+03	0.1857E+03
17	8.39	2405.80	2802.55	2420.44	0.1853E+03	0.1046E+03	0.1752E+03
18	9.21	2556.83	2883.79	2558.73	0.1853E+03	0.9514E+02	0.1640E+03
19	10.02	2707.86	2958.07	2687.42	0.1562E+03	0.8737E+02	0.1516E+03
20	10.70	2797.02	3015.17	2786.16	0.1320E+03	0.8189E+02	0.1407E+03
21	11.37	2886.19	3068.82	2877.38	0.1320E+03	0.7710E+02	0.1294E+03
22	12.05	2975.36	3119.43	2960.89	0.1117E+03	0.7287E+02	0.1179E+03
23	12.83	3044.29	3174.67	3047.95	0.8820E+02	0.6857E+02	0.1049E+03
24	13.61	3113.22	3226.74	3125.06	0.8820E+02	0.6478E+02	0.9252E+02
25	14.39	3182.15	3276.03	3192.78	0.8820E+02	0.6142E+02	0.8095E+02
26	15.17	3251.08	3322.84	3251.85	0.7501E+02	0.5842E+02	0.7040E+02
27	16.13	3307.27	3377.00	3313.48	0.5889E+02	0.5516E+02	0.5904E+02
28	17.08	3363.47	3428.24	3365.08	0.5442E+02	0.5227E+02	0.4935E+02
29	18.74	3440.88	3511.29	3435.45	0.3770E+02	0.4796E+02	0.3621E+02
30	20.15	3483.30	3576.68	3480.56	0.3009E+02	0.4487E+02	0.2820E+02

V - 87

Connection type : Extended end-plate connections
Mode : All bolted with column stiffener

Tested by : Y.L.Yee (1984) Australia
Test Id. : K4

Column : 460UB67.1
Beam : 460UB67.1
Plate thickness : 20 mm
Stiffener thickness : 0.6299"

Fasteners: G8.8- -M30
1 1/4" Oversize holes
Material : G250
Fy = 37.70 ksi
Fu = -- ksi

Major parameters

End plate extended on both sides.
ct = 2.3622" pt = 5.2244" li = 9.7874" pc = 7.5866"
cc = 2.3622" gt = 5.1181" gc = 5.1181"
tp = 0.7874" nt = 2 X 2 nc = 2 X 2

Remark 1) 32 mm Doubler plates were welded to flange on both sides of column web.
2) End-plate yield stress (Fy) is mentioned.

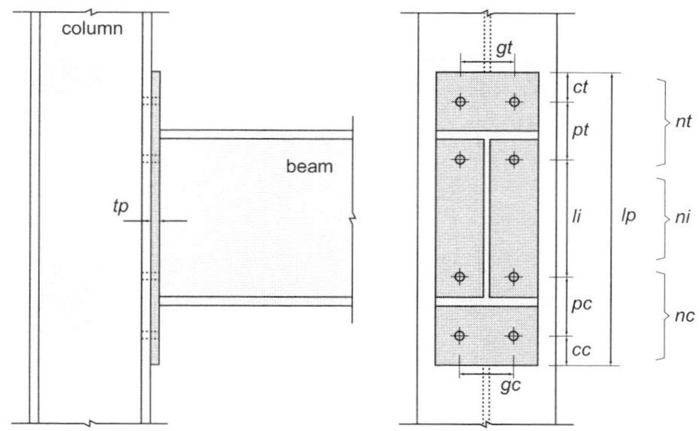

No	Moment (k-in)	Rotation (radians) X 1/1000
1	0.00	0.00
2	143.70	0.20
3	290.28	0.55
4	453.94	0.83
5	617.60	1.12
6	748.01	1.43
7	878.42	1.74
8	1029.08	2.08
9	1179.74	2.43
10	1330.40	2.77
11	1485.37	3.20
12	1640.34	3.63
13	1795.32	4.06
14	1937.09	4.47
15	2078.86	4.88
16	2230.91	5.46
17	2390.39	6.05
18	2520.04	6.55
19	2674.87	7.25
20	2829.82	8.00
21	2907.60	8.51
22	2985.39	9.01
23	3056.92	9.47
24	3128.45	9.94
25	3195.37	10.55
26	3262.30	11.15
27	3306.73	11.81
28	3351.16	12.46
29	3395.59	13.12
30	3414.53	13.80
31	3433.48	14.47
32	3515.11	15.53
33	3552.11	16.51
34	3488.67	16.82
35	3521.88	17.27

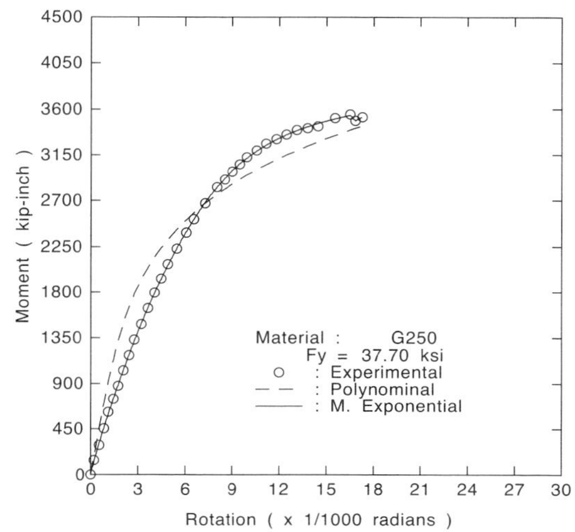

A.5 – 174

```
Moment-rotation prediction equations      ( R : X 1/1000 radians )

Frye and Morris polynominal model  :    R = Sum ( Ai X ( K*Bm )**Pi X 10**Qi )
     xd = 22.598400"    t =  0.787400"
     A1 =  1.790000   A2 =  1.760000   A3 =  2.040000    K =  0.000649
     P1 =        1    P2 =        3    P3 =        5    Q1 =        0    Q2 =       -1    Q3 =       -1

Modified exponential model  :
     Bm = Sum ( Ai X ( 1 - exp( - R/( 2*i*C ) ) ) )  +      Sum ( RKj X ( R - Rj0 ) )  +  Bm0
      C =  0.19416667E+00           Bmo=   0.00000000E+00            Nexp=  6        Nliner=  3
     AI =  0.15502046E+03   0.87394276E+02  -0.12251356E+05   0.62402584E+05  -0.10809195E+06   0.60910928E+05
    Rj0 =          0.0000          16.5100          16.8200
    RKj =  0.21746774E+02  -0.20185285E+03   0.24438839E+03
```

No	Rotation (radians) X 1/1000	Moment (kip-inch) Expri.	Moment (kip-inch) Poly.	Moment (kip-inch) M.Expo.	Connection stiffness (kip-inch) X 1000 Expri.	Connection stiffness (kip-inch) X 1000 Poly.	Connection stiffness (kip-inch) X 1000 M.Expo.
1	0.00	0.00	0.00	0.00	0.7149E+03	0.8603E+03	0.6412E+03
3	0.55	290.28	470.11	314.88	0.5072E+03	0.8333E+03	0.5176E+03
5	1.12	617.60	916.03	595.62	0.5032E+03	0.7317E+03	0.4807E+03
7	1.74	878.42	1319.74	884.99	0.4274E+03	0.5645E+03	0.4494E+03
9	2.43	1179.74	1653.03	1182.99	0.4363E+03	0.4104E+03	0.4137E+03
11	3.20	1485.37	1924.76	1488.52	0.3621E+03	0.3017E+03	0.3773E+03
13	4.06	1795.32	2149.73	1795.88	0.3535E+03	0.2301E+03	0.3414E+03
15	4.88	2078.86	2319.67	2062.60	0.3103E+03	0.1869E+03	0.3084E+03
17	6.05	2390.39	2513.19	2395.12	0.2649E+03	0.1476E+03	0.2608E+03
19	7.25	2674.87	2674.36	2679.77	0.2137E+03	0.1216E+03	0.2119E+03
21	8.51	2907.60	2814.37	2915.19	0.1553E+03	0.1031E+03	0.1651E+03
23	9.47	3056.92	2908.96	3059.71	0.1527E+03	0.9242E+02	0.1338E+03
25	10.55	3195.37	3002.73	3187.03	0.1111E+03	0.8305E+02	0.1050E+03
27	11.81	3306.73	3101.70	3302.05	0.6752E+02	0.7432E+02	0.7882E+02
29	13.12	3395.59	3194.57	3392.11	0.4805E+02	0.6709E+02	0.5922E+02
31	14.47	3433.48	3280.90	3462.13	0.4708E+02	0.6110E+02	0.4553E+02
33	16.51	3552.11	3398.09	3540.44	-0.1478E+03	0.5395E+02	-0.1686E+03
35	17.27	3521.88	3438.46	3521.87	0.7413E+02	0.5172E+02	0.7303E+02

V - 88

Connection type : Extended end-plate connections
Mode : All bolted with column stiffener

Tested by : Y.L.Yee (1984) Australia
Test Id. : K7

Column : 460UB67.1
Beam : 460UB67.1
Plate thickness : 20 mm
Stiffener thickness : 0.6299"

Fasteners: G8.8- -M30
1 1/4" Oversize holes
Material : G250
Fy = 37.70 ksi
Fu = -- ksi

Major parameters

End plate extended on both sides.
ct = 2.3622" pt = 5.2244" li = 9.7874" pc = 7.5866"
cc = 2.3622" gt = 5.1181" gc = 5.1181"
tp = 0.7874" nt = 2 X 2 nc = 2 X 2

Remark 1) 20 mm Doubler plates were welded to flange on both sides of column web.
2) End-plate yield stress (Fy) is mentioned.

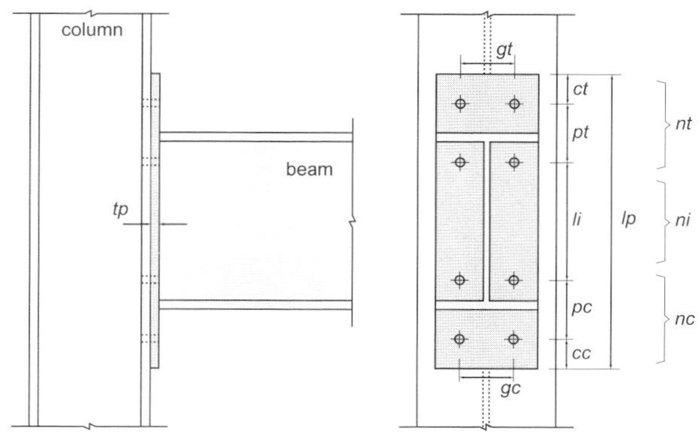

No	Moment (k-in)	Rotation (radians) X 1/1000
1	0.00	0.00
2	144.05	0.30
3	288.09	0.60
4	434.63	0.92
5	581.17	1.24
6	723.95	1.52
7	866.72	1.80
8	1021.91	2.09
9	1177.09	2.38
10	1326.13	2.71
11	1477.78	3.17
12	1629.44	3.63
13	1774.73	3.94
14	1920.03	4.24
15	2076.69	4.74
16	2233.35	5.24
17	2387.59	5.78
18	2541.83	6.33
19	2691.26	6.99
20	2840.69	7.66
21	2972.97	8.50
22	3105.26	9.34
23	3165.37	9.90
24	3263.41	11.10
25	3329.55	12.56
26	3392.89	13.76
27	3434.03	15.03
28	3463.60	16.30
29	3481.60	17.57

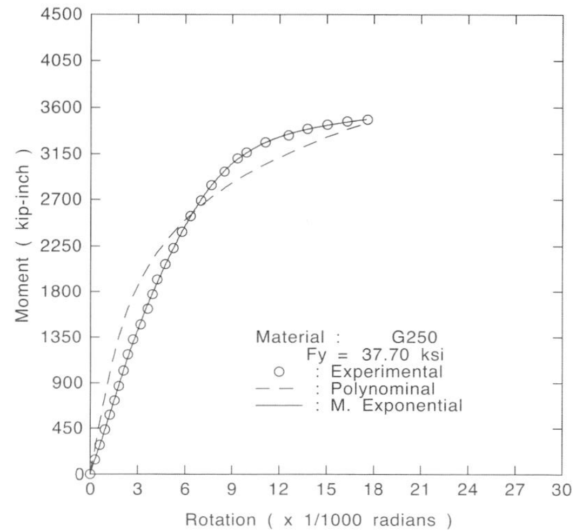

```
Moment-rotation prediction equations     ( R : X 1/1000 radians )

Frye and Morris polynominal model  :    R = Sum ( Ai X ( K*Bm )**Pi X 10**Qi )
     xd = 22.598400"   t =  0.787400"
     A1 =  1.790000   A2 =  1.760000   A3 =  2.040000    K =  0.000649
     P1 =         1   P2 =         3   P3 =         5   Q1 =         0    Q2 =        -1    Q3 =        -1

Modified exponential model  :
     Bm = Sum ( Ai X ( 1 - exp( - R/( 2*i*C ) ) ) )  +      Sum ( RKj X ( R - Rj0 ) )  +  Bm0
       C =   0.22151667E+00              Bmo=   0.00000000E+00          Nexp=  6        Nliner=  1
      AI =   0.13095853E+04  -0.18720455E+05   0.89536583E+05  -0.18163515E+06   0.15318572E+06  -0.40591747E+05
     Rj0 =          0.0000
     RKj =   0.22449164E+02
```

No	Rotation (radians) X 1/1000	Moment (kip-inch) Expri.	Moment (kip-inch) Poly.	Moment (kip-inch) M.Expo.	Connection stiffness (kip-inch) X 1000 Expri.	Connection stiffness (kip-inch) X 1000 Poly.	Connection stiffness (kip-inch) X 1000 M.Expo.
1	0.00	0.00	0.00	0.00	0.4794E+03	0.8603E+03	0.5822E+03
2	0.30	144.05	257.79	149.44	0.4794E+03	0.8529E+03	0.4396E+03
3	0.60	288.09	510.81	280.81	0.4710E+03	0.8277E+03	0.4473E+03
4	0.92	434.63	765.81	429.04	0.4623E+03	0.7766E+03	0.4873E+03
5	1.24	581.17	1000.64	587.83	0.4841E+03	0.7016E+03	0.5106E+03
6	1.52	723.95	1188.69	733.16	0.5036E+03	0.6243E+03	0.5119E+03
7	1.80	866.72	1354.76	876.78	0.5178E+03	0.5481E+03	0.4997E+03
8	2.09	1021.91	1504.07	1019.62	0.5324E+03	0.4780E+03	0.4795E+03
9	2.38	1177.09	1634.47	1156.11	0.4975E+03	0.4185E+03	0.4568E+03
10	2.71	1326.13	1761.38	1300.52	0.4055E+03	0.3643E+03	0.4323E+03
11	3.17	1477.78	1914.33	1491.66	0.3308E+03	0.3054E+03	0.4025E+03
12	3.63	1629.44	2043.82	1670.57	0.4146E+03	0.2617E+03	0.3787E+03
13	3.94	1774.73	2120.79	1785.25	0.4710E+03	0.2384E+03	0.3650E+03
14	4.24	1920.03	2191.21	1895.87	0.4109E+03	0.2187E+03	0.3523E+03
15	4.74	2076.69	2293.77	2066.82	0.3136E+03	0.1929E+03	0.3321E+03
16	5.24	2233.35	2384.85	2227.52	0.2998E+03	0.1726E+03	0.3111E+03
17	5.78	2387.59	2473.35	2389.36	0.2848E+03	0.1549E+03	0.2863E+03
18	6.33	2541.83	2553.24	2537.29	0.2577E+03	0.1406E+03	0.2598E+03
19	6.99	2691.26	2642.06	2699.22	0.2242E+03	0.1264E+03	0.2260E+03
20	7.66	2840.69	2722.34	2838.66	0.1945E+03	0.1149E+03	0.1926E+03
21	8.50	2972.97	2814.02	2984.16	0.1570E+03	0.1032E+03	0.1535E+03
22	9.34	3105.26	2896.82	3098.91	0.1277E+03	0.9372E+02	0.1198E+03
23	9.90	3165.37	2947.29	3160.02	0.9994E+02	0.8845E+02	0.1010E+03
24	11.10	3263.41	3047.68	3261.25	0.6518E+02	0.7895E+02	0.6937E+02
25	12.56	3329.55	3156.23	3343.22	0.4949E+02	0.6997E+02	0.4505E+02
26	13.76	3392.89	3236.29	3389.62	0.4295E+02	0.6412E+02	0.3356E+02
27	15.03	3434.03	3314.61	3427.57	0.2778E+02	0.5893E+02	0.2671E+02
28	16.30	3463.60	3386.61	3459.03	0.1876E+02	0.5460E+02	0.2328E+02
29	17.57	3481.60	3453.28	3487.38	0.1424E+02	0.5093E+02	0.2177E+02

V - 89

Connection type : Extended end-plate connections
Mode : All bolted with column stiffener

Tested by : Y.L.Yee (1984) Australia
Test Id. : K5A

Column : 460UB67.1
Beam : 460UB67.1
Plate thickness : 20 mm
Stiffener thickness : 0.6299"

Fasteners: G8.8- -M24
1 1/16" Oversize holes
Material : G250
Fy = 37.70 ksi
Fu = -- ksi

Major parameters

End plate extended on both sides.

ct = 2.3622" pt = 5.2244" li = 9.7874" pc = 7.5866"
cc = 2.3622" gt = 5.1181" gc = 5.1181"
tp = 0.7874" nt = 2 X 2 nc = 2 X 2

Remark 1) 20 mm Doubler plates were welded to flange on both sides of column web.
2) End-plate yield stress (Fy) is mentioned.

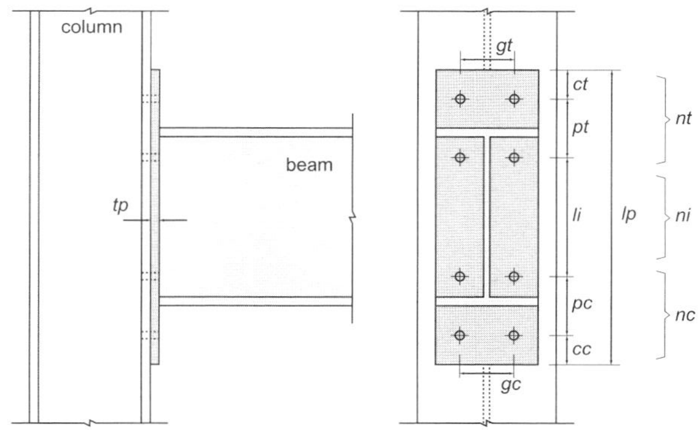

No	Moment (k-in)	Rotation (radians) X 1/1000	No	Moment (k-in)	Rotation (radians) X 1/1000
1	0.00	0.00	26	3256.88	11.52
2	143.30	0.17	27	3289.92	12.35
3	281.77	0.41	28	3322.96	13.18
4	431.45	0.72	29	3356.00	14.01
5	581.14	1.02	30	3389.04	14.84
6	724.67	1.33	31	3397.08	15.66
7	885.38	1.58	32	3405.13	16.49
8	1018.95	1.84	33	3413.18	17.31
9	1177.32	2.18	34	3446.78	18.18
10	1323.13	2.38	35	3480.52	19.13
11	1471.71	2.75	36	3483.82	19.60
12	1610.45	3.15			
13	1771.11	3.38			
14	1922.10	3.72			
15	2055.66	3.96			
16	2211.92	4.50			
17	2356.89	4.92			
18	2501.86	5.35			
19	2635.97	5.90			
20	2802.41	6.61			
21	2944.46	7.47			
22	3028.33	8.19			
23	3112.20	8.91			
24	3168.02	9.80			
25	3223.84	10.69			

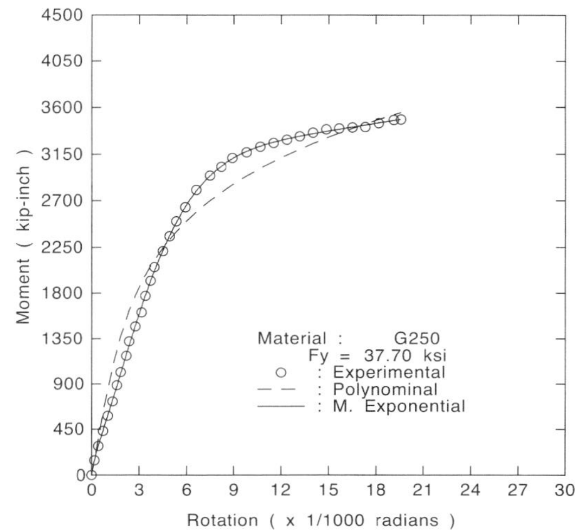

```
Moment-rotation prediction equations     ( R : X 1/1000 radians )

Frye and Morris polynominal model  :     R = Sum ( Ai X ( K*Bm )**Pi X 10**Qi )
     xd = 22.598400"   t =  0.787400"
     A1 =  1.790000   A2 =  1.760000   A3 =  2.040000    K =  0.000649
     P1 =         1   P2 =         3   P3 =         5   Q1 =         0    Q2 =        -1    Q3 =        -1

Modified exponential model  :
     Bm = Sum ( Ai X ( 1 - exp( - R/( 2*i*C ) ) ) )  +      Sum ( RKj X ( R - Rj0 ) )  +  Bm0
       C =  0.20630000E+00              Bmo=   0.00000000E+00           Nexp=  6        Nliner=  1
      AI =  0.68256450E+03  -0.54263598E+04   0.31011608E+05  -0.80803842E+05   0.80190657E+05  -0.22667049E+05
     Rj0 =          0.0000
     RKj =  0.25257832E+02
```

No	Rotation (radians) X 1/1000	Moment (kip-inch) Expri.	Moment (kip-inch) Poly.	Moment (kip-inch) M.Expo.	Connection stiffness (kip-inch) X 1000 Expri.	Connection stiffness (kip-inch) X 1000 Poly.	Connection stiffness (kip-inch) X 1000 M.Expo.
1	0.00	0.00	0.00	0.00	0.8331E+03	0.8603E+03	0.8869E+03
3	0.41	281.77	352.52	279.82	0.5390E+03	0.8459E+03	0.5415E+03
5	1.02	581.14	845.44	586.08	0.4799E+03	0.7542E+03	0.4987E+03
7	1.58	885.38	1228.37	873.09	0.5761E+03	0.6065E+03	0.5219E+03
9	2.18	1177.32	1543.26	1183.77	0.6172E+03	0.4598E+03	0.5191E+03
11	2.75	1471.71	1777.66	1476.91	0.3781E+03	0.3577E+03	0.4923E+03
13	3.38	1771.11	1977.32	1772.78	0.5807E+03	0.2834E+03	0.4482E+03
15	3.96	2055.66	2126.84	2018.40	0.4744E+03	0.2366E+03	0.4011E+03
17	4.92	2356.89	2327.67	2365.01	0.3427E+03	0.1851E+03	0.3197E+03
19	5.90	2635.97	2491.66	2640.12	0.2385E+03	0.1515E+03	0.2429E+03
21	7.47	2944.46	2700.01	2939.98	0.1388E+03	0.1180E+03	0.1469E+03
23	8.91	3112.20	2854.96	3107.04	0.9239E+02	0.9837E+02	0.8962E+02
25	10.69	3223.84	3014.49	3227.79	0.5086E+02	0.8195E+02	0.5079E+02
27	12.35	3289.92	3141.11	3296.44	0.3981E+02	0.7115E+02	0.3416E+02
29	14.01	3356.00	3252.17	3346.52	0.3981E+02	0.6302E+02	0.2726E+02
31	15.66	3397.08	3351.00	3389.26	0.9761E+01	0.5670E+02	0.2487E+02
33	17.31	3413.18	3440.15	3429.67	0.2389E+02	0.5163E+02	0.2432E+02
35	19.13	3480.52	3529.68	3473.91	0.1645E+02	0.4707E+02	0.2443E+02

V - 90

Connection type : Extended end-plate connections
Mode : All bolted with column stiffener

Tested by : Y.L.Yee (1984) Australia
Test Id. : K8

Column : 460UB67.1
Beam : 460UB67.1
Plate thickness : 20 mm
Stiffener thickness : 0.6299"

Fasteners: G8.8- -M30
1 1/4" Oversize holes
Material : G250
Fy = 37.70 ksi
Fu = -- ksi

Major parameters

End plate extended on both sides.

ct = 2.3622" pt = 5.2244" li = 9.7874" pc = 7.5866"
cc = 2.3622" gt = 5.1181" gc = 5.1181"
tp = 0.7874" nt = 2 X 2 nc = 2 X 2

Remark 1) 20 mm Doubler plates were welded to flange on both sides of column web.
2) End-plate yield stress (Fy) is mentioned.

No	Moment (k-in)	Rotation (radians) X 1/1000	No	Moment (k-in)	Rotation (radians) X 1/1000
1	0.00	0.00	26	3198.35	16.83
2	144.24	0.35	27	3243.93	17.67
3	288.49	0.70	28	3289.51	18.51
4	428.99	1.03	29	3335.09	19.35
5	569.48	1.37	30	3360.30	20.46
6	664.01	1.62	31	3385.51	21.56
7	758.53	1.87	32	3410.72	22.67
8	894.07	2.21	33	3456.88	24.03
9	1029.62	2.54			
10	1182.65	2.99			
11	1335.67	3.44			
12	1486.35	4.00			
13	1637.02	4.55			
14	1785.32	5.20			
15	1933.62	5.85			
16	2074.42	6.46			
17	2215.23	7.07			
18	2366.32	8.00			
19	2517.40	8.93			
20	2665.05	10.11			
21	2812.71	11.29			
22	2903.18	12.35			
23	2993.65	13.42			
24	3084.13	14.48			
25	3141.24	15.66			

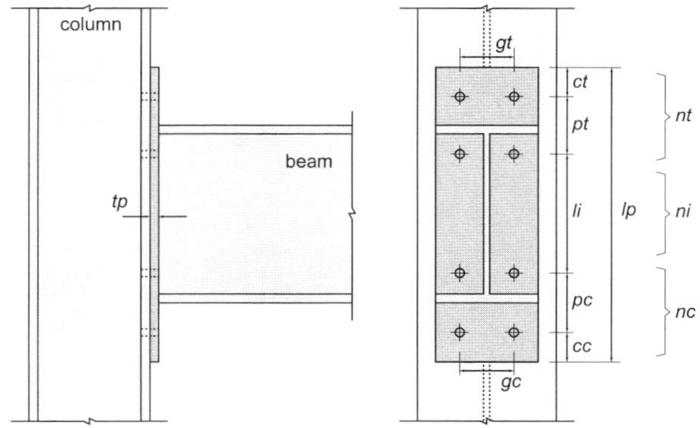

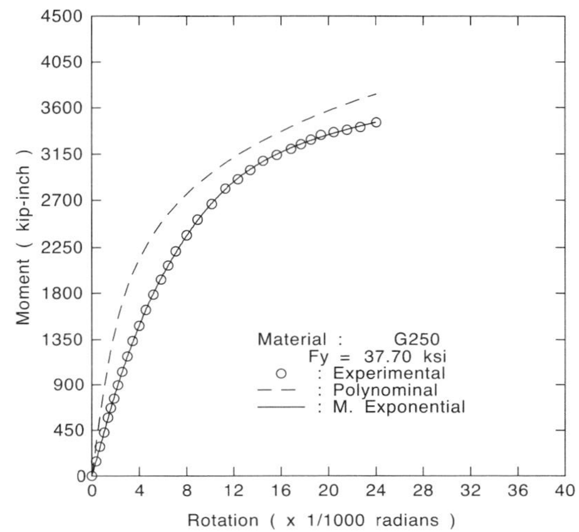

```
Moment-rotation prediction equations     ( R : X 1/1000 radians )

Frye and Morris polynominal model  :    R = Sum ( Ai X ( K*Bm )**Pi X 10**Qi )
     xd = 22.598400"    t =  0.787400"
     A1 =  1.790000   A2 =  1.760000   A3 =  2.040000    K =  0.000649
     P1 =        1    P2 =        3    P3 =        5    Q1 =        0    Q2 =      -1    Q3 =      -1

Modified exponential model  :
     Bm = Sum ( Ai X ( 1 - exp( - R/( 2*i*C ) ) ) )  +      Sum ( RKj X ( R - Rj0 ) )  +  Bm0
       C =    0.28775000E+00           Bmo=   0.00000000E+00          Nexp=  6        Nliner=  1
      AI =    0.29736676E+03  -0.43939313E+04   0.14359963E+05  -0.10303598E+05  -0.17386863E+05   0.20350648E+05
     Rj0 =            0.0000
     RKj =    0.22880921E+02
```

No	Rotation (radians) X 1/1000	Moment (kip-inch) Expri.	Moment (kip-inch) Poly.	Moment (kip-inch) M.Expo.	Connection stiffness (kip-inch) X 1000 Expri.	Connection stiffness (kip-inch) X 1000 Poly.	Connection stiffness (kip-inch) X 1000 M.Expo.
1	0.00	0.00	0.00	0.00	0.4121E+03	0.8603E+03	0.3919E+03
2	0.35	144.24	299.94	143.50	0.4121E+03	0.8501E+03	0.4094E+03
3	0.70	288.49	592.13	287.78	0.4168E+03	0.8146E+03	0.4152E+03
4	1.03	428.99	854.10	426.79	0.4213E+03	0.7516E+03	0.4174E+03
5	1.37	569.48	1090.96	565.42	0.3946E+03	0.6661E+03	0.4128E+03
6	1.62	664.01	1250.33	668.74	0.3743E+03	0.5966E+03	0.4051E+03
7	1.87	758.53	1392.49	769.73	0.3879E+03	0.5303E+03	0.3945E+03
8	2.21	894.07	1556.50	898.75	0.4058E+03	0.4537E+03	0.3776E+03
9	2.54	1029.62	1697.21	1021.81	0.3781E+03	0.3912E+03	0.3591E+03
10	2.99	1182.65	1857.67	1177.37	0.3408E+03	0.3264E+03	0.3339E+03
11	3.44	1335.67	1992.84	1321.84	0.3093E+03	0.2782E+03	0.3100E+03
12	4.00	1486.35	2135.06	1487.12	0.2703E+03	0.2342E+03	0.2835E+03
13	4.55	1637.02	2256.19	1638.69	0.2510E+03	0.2020E+03	0.2608E+03
14	5.20	1785.32	2377.70	1800.36	0.2287E+03	0.1741E+03	0.2384E+03
15	5.85	1933.62	2483.45	1948.62	0.2303E+03	0.1530E+03	0.2193E+03
16	6.46	2074.42	2571.53	2076.90	0.2318E+03	0.1376E+03	0.2033E+03
17	7.07	2215.23	2651.17	2195.90	0.2043E+03	0.1250E+03	0.1886E+03
18	8.00	2366.32	2760.23	2361.71	0.1623E+03	0.1099E+03	0.1678E+03
19	8.93	2517.40	2856.84	2508.87	0.1459E+03	0.9815E+02	0.1485E+03
20	10.11	2665.05	2965.61	2670.74	0.1251E+03	0.8662E+02	0.1261E+03
21	11.29	2812.71	3062.36	2807.57	0.1040E+03	0.7766E+02	0.1062E+03
22	12.35	2903.18	3141.40	2912.02	0.8503E+02	0.7113E+02	0.9054E+02
23	13.42	2993.65	3214.09	3001.04	0.8503E+02	0.6568E+02	0.7718E+02
24	14.48	3084.13	3281.45	3077.02	0.6770E+02	0.6106E+02	0.6600E+02
25	15.66	3141.24	3350.67	3148.53	0.4854E+02	0.5672E+02	0.5593E+02
26	16.83	3198.35	3415.15	3209.46	0.5191E+02	0.5299E+02	0.4798E+02
27	17.67	3243.93	3458.62	3247.75	0.5431E+02	0.5064E+02	0.4340E+02
28	18.51	3289.51	3500.23	3282.53	0.5431E+02	0.4851E+02	0.3959E+02
29	19.35	3335.09	3540.94	3314.39	0.4071E+02	0.4653E+02	0.3644E+02
30	20.46	3360.30	3590.33	3352.80	0.2278E+02	0.4425E+02	0.3314E+02
31	21.56	3385.51	3638.12	3388.00	0.2278E+02	0.4218E+02	0.3060E+02
32	22.67	3410.72	3683.73	3420.74	0.2779E+02	0.4030E+02	0.2867E+02
33	24.03	3456.88	3737.10	3458.48	0.3394E+02	0.3823E+02	0.2693E+02

V - 91

Connection type : Extended end-plate connections
Mode : All bolted with column stiffener

Tested by : Y.L.Yee (1984) Australia
Test Id. : K10

Column : 460UB67.1
Beam : 460UB67.1
Plate thickness : 20 mm
Stiffener thickness : 0.6299"

Fasteners: G8.8- -M24
1 1/16" Oversize holes
Material : G250
Fy = 37.70 ksi
Fu = -- ksi

Major parameters

End plate extended on both sides.

ct = 2.3622" pt = 5.2244" li = 9.7874" pc = 7.5866"
cc = 2.3622" gt = 5.1181" gc = 5.1181"
tp = 0.7874" nt = 2 X 2 nc = 2 X 2

Remark 1) 20 mm Doubler plates were welded to flange on both sides of column web.
2) End-plate yield stress (Fy) is mentioned.

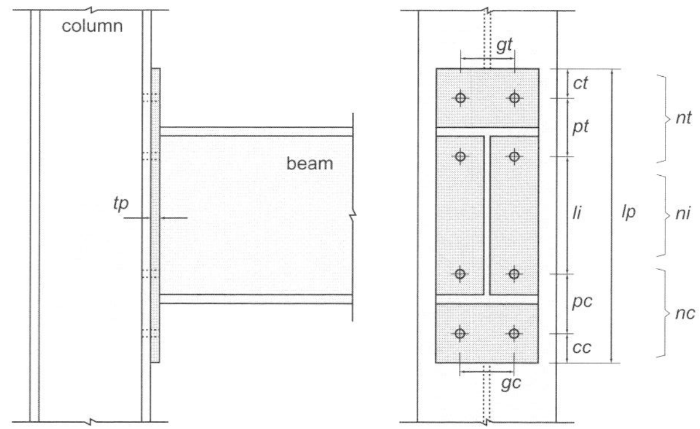

No	Moment (k-in)	Rotation (radians) X 1/1000
1	0.00	0.00
2	147.18	0.18
3	294.36	0.36
4	444.02	0.54
5	593.67	0.71
6	741.52	1.33
7	889.36	1.95
8	1053.56	1.86
9	1195.83	2.08
10	1348.99	2.94
11	1498.76	3.17
12	1646.97	4.04
13	1782.54	4.78
14	1955.66	5.67
15	2100.74	6.11
16	2251.04	6.70
17	2373.50	7.00
18	2524.74	8.24
19	2680.34	9.04
20	2829.49	9.88
21	2946.65	10.82
22	3038.51	12.12
23	3148.38	13.85
24	3224.77	15.12
25	3294.41	16.74
26	3364.06	18.36
27	3407.84	19.88
28	3451.62	21.41
29	3495.40	22.93
30	3533.66	24.66
31	3571.93	26.38
32	3610.20	28.11
33	3632.97	29.60
34	3655.74	31.09
35	3678.51	32.58
36	3701.28	34.07
37	3715.09	35.54
38	3728.90	37.00
39	3742.71	38.47

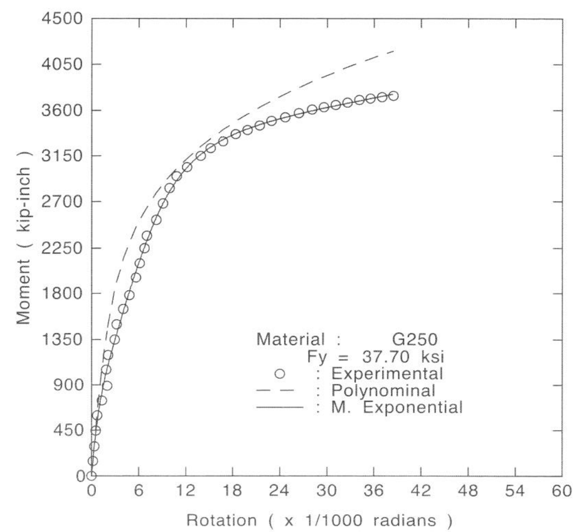

```
Moment-rotation prediction equations     ( R : X 1/1000 radians )

Frye and Morris polynominal model  :    R = Sum ( Ai X ( K*Bm )**Pi X 10**Qi )
     xd = 22.598400"   t =  0.787400"
     A1 =  1.790000   A2 =  1.760000   A3 =  2.040000    K =  0.000649
     P1 =        1    P2 =        3    P3 =        5    Q1 =        0    Q2 =       -1    Q3 =       -1

Modified exponential model  :
     Bm = Sum ( Ai X ( 1 - exp( - R/( 2*i*C ) ) ) )  +     Sum ( RKj X ( R - Rj0 ) )  +  Bm0
       C =   0.36596667E+00            Bmo=  0.00000000E+00         Nexp=  6       Nliner=  1
      AI =  -0.40902779E+03   0.10144863E+05  -0.34179593E+05   0.49506533E+05  -0.40002179E+05   0.18141309E+05
     Rj0 =          0.0000
     RKj =   0.14567243E+02
```

No	Rotation (radians) X 1/1000	Moment (kip-inch) Expri.	Poly.	M.Expo.	Connection stiffness (kip-inch) X 1000 Expri.	Poly.	M.Expo.
1	0.00	0.00	0.00	0.00	0.8109E+03	0.8603E+03	0.9153E+03
3	0.36	294.36	310.98	300.03	0.8400E+03	0.8493E+03	0.7309E+03
5	0.71	593.67	598.64	525.74	0.7308E+03	0.8135E+03	0.5843E+03
7	1.95	889.36	1434.09	1054.08	-0.2100E+04	0.5107E+03	0.3211E+03
9	2.08	1195.83	1498.04	1094.74	0.5430E+03	0.4808E+03	0.3096E+03
11	3.17	1498.76	1915.89	1402.04	0.5283E+03	0.3049E+03	0.2644E+03
13	4.78	1782.54	2301.68	1813.84	0.1889E+03	0.1911E+03	0.2478E+03
15	6.11	2100.74	2522.14	2128.47	0.2985E+03	0.1460E+03	0.2248E+03
17	7.00	2373.50	2643.63	2320.46	0.3427E+03	0.1262E+03	0.2031E+03
19	9.04	2680.34	2868.34	2679.66	0.1866E+03	0.9686E+02	0.1493E+03
21	10.82	2946.65	3025.09	2907.38	0.1018E+03	0.8098E+02	0.1092E+03
23	13.85	3148.38	3242.39	3163.04	0.6150E+02	0.6369E+02	0.6387E+02
25	16.74	3294.41	3410.19	3311.15	0.4311E+02	0.5327E+02	0.4120E+02
27	19.88	3407.84	3564.49	3418.59	0.2869E+02	0.4543E+02	0.2864E+02
29	22.93	3495.40	3694.27	3495.41	0.2563E+02	0.3988E+02	0.2235E+02
31	26.38	3571.93	3823.38	3565.39	0.2217E+02	0.3516E+02	0.1861E+02
33	29.60	3632.97	3930.72	3621.97	0.1528E+02	0.3173E+02	0.1676E+02
35	32.58	3678.51	4022.25	3670.35	0.1528E+02	0.2913E+02	0.1580E+02
37	35.54	3715.09	4105.03	3716.20	0.9416E+01	0.2699E+02	0.1526E+02
39	38.47	3742.71	4181.45	3760.47	0.9416E+01	0.2519E+02	0.1495E+02

V - 92

Connection type : Extended end-plate connections
Mode : All bolted with column stiffener

Tested by : Y.L.Yee (1984) Australia
Test Id. : K11

Column : 460UB67.1
Beam : 460UB67.1
Plate thickness : 20 mm
Stiffener thickness : 0.6299"

Fasteners: G8.8- -M24
1 1/16" Oversize holes
Material : G250
Fy = 37.70 ksi
Fu = -- ksi

Major parameters

End plate extended on both sides.

ct = 2.3622"	pt = 5.2244"	li = 9.7874"	pc = 7.5866"
cc = 2.3622"	gt = 5.1181"	gc = 5.1181"	
tp = 0.7874"	nt = 2 X 2	nc = 2 X 2	

Remark 1) The beam, 3 deg. to horizontal was subjected to cantilever loading.
2) End-plate yield stress (Fy) is mentioned.

No	Moment (k-in)	Rotation (radians) X 1/1000	No	Moment (k-in)	Rotation (radians) X 1/1000
1	0.00	0.00	26	2666.31	24.08
2	144.75	0.30	27	2710.98	25.74
3	289.50	0.60	28	2755.65	27.40
4	434.25	0.90	29	2794.04	29.04
5	578.99	1.20	30	2832.43	30.68
6	732.19	1.74	31	2870.83	32.31
7	902.43	1.93	32	2909.22	33.95
8	1045.25	2.18			
9	1198.16	2.51			
10	1343.88	3.06			
11	1486.88	3.47			
12	1628.25	4.51			
13	1772.65	5.98			
14	1908.27	7.42			
15	2043.90	8.86			
16	2122.02	10.29			
17	2200.14	11.73			
18	2269.46	13.12			
19	2338.77	14.51			
20	2387.07	15.76			
21	2435.37	17.00			
22	2483.66	18.24			
23	2529.66	19.64			
24	2575.65	21.03			
25	2621.64	22.43			

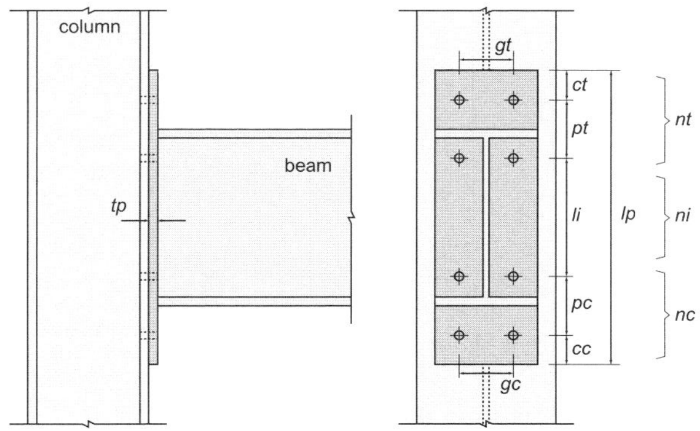

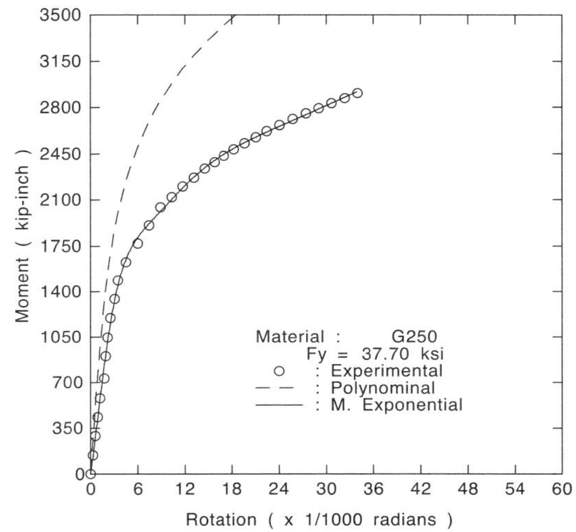

A.5 – 184

```
Moment-rotation prediction equations     ( R : X 1/1000 radians )

Frye and Morris polynominal model  :     R = Sum ( Ai X ( K*Bm )**Pi X 10**Qi )
     xd = 22.598400"   t =  0.787400"
     A1 =  1.790000   A2 =  1.760000   A3 =  2.040000    K =  0.000649
     P1 =         1   P2 =         3   P3 =         5   Q1 =         0    Q2 =       -1    Q3 =       -1

Modified exponential model  :
     Bm = Sum ( Ai X ( 1 - exp( - R/( 2*i*C ) ) ) )  +      Sum ( RKj X ( R - Rj0 ) )  +  Bm0
      C =   0.35770833E+00            Bmo=   0.00000000E+00          Nexp=  6        Nliner=  1
     AI =   0.10903435E+04  -0.22081175E+05   0.98835091E+05  -0.18231918E+06   0.15454497E+06  -0.48142728E+05
    Rj0 =           0.0000
    RKj =   0.29061468E+02
```

No	Rotation (radians) X 1/1000	Moment (kip-inch) Expri.	Moment (kip-inch) Poly.	Moment (kip-inch) M.Expo.	Connection stiffness (kip-inch) X 1000 Expri.	Connection stiffness (kip-inch) X 1000 Poly.	Connection stiffness (kip-inch) X 1000 M.Expo.
1	0.00	0.00	0.00	0.00	0.4836E+03	0.8603E+03	0.4196E+03
2	0.30	144.75	256.76	135.59	0.4837E+03	0.8529E+03	0.4607E+03
3	0.60	289.50	508.74	276.46	0.4837E+03	0.8280E+03	0.4805E+03
4	0.90	434.25	750.08	422.40	0.4837E+03	0.7806E+03	0.4928E+03
5	1.20	578.99	973.78	570.13	0.4126E+03	0.7115E+03	0.4923E+03
6	1.74	732.19	1318.62	828.38	0.7365E+03	0.5650E+03	0.4599E+03
7	1.93	902.43	1420.74	914.01	0.7507E+03	0.5170E+03	0.4410E+03
8	2.18	1045.25	1546.41	1024.05	0.5180E+03	0.4584E+03	0.4114E+03
9	2.51	1198.16	1684.20	1150.38	0.3957E+03	0.3968E+03	0.3704E+03
10	3.06	1343.88	1881.99	1336.36	0.3135E+03	0.3173E+03	0.2978E+03
11	3.47	1486.88	2001.94	1447.35	0.2908E+03	0.2752E+03	0.2485E+03
12	4.51	1628.25	2247.61	1651.23	0.1203E+03	0.2041E+03	0.1511E+03
13	5.98	1772.65	2503.38	1815.23	0.9620E+02	0.1494E+03	0.8358E+02
14	7.42	1908.27	2694.81	1919.02	0.9428E+02	0.1187E+03	0.6554E+02
15	8.86	2043.90	2850.26	2011.35	0.7435E+02	0.9891E+02	0.6387E+02
16	10.29	2122.02	2981.53	2102.91	0.5450E+02	0.8507E+02	0.6352E+02
17	11.73	2200.14	3095.76	2191.99	0.5210E+02	0.7482E+02	0.6022E+02
18	13.12	2269.46	3194.35	2272.11	0.4978E+02	0.6711E+02	0.5456E+02
19	14.51	2338.77	3283.35	2343.49	0.4399E+02	0.6094E+02	0.4792E+02
20	15.76	2387.07	3356.24	2399.49	0.3882E+02	0.5638E+02	0.4220E+02
21	17.00	2435.37	3423.91	2448.80	0.3882E+02	0.5251E+02	0.3724E+02
22	18.24	2483.66	3487.10	2492.55	0.3607E+02	0.4917E+02	0.3327E+02
23	19.64	2529.66	3553.35	2536.53	0.3299E+02	0.4594E+02	0.3003E+02
24	21.03	2575.65	3615.40	2576.78	0.3299E+02	0.4315E+02	0.2790E+02
25	22.43	2621.64	3673.80	2614.72	0.3022E+02	0.4070E+02	0.2666E+02
26	24.08	2666.31	3739.19	2658.33	0.2691E+02	0.3816E+02	0.2601E+02
27	25.74	2710.98	3800.64	2701.37	0.2691E+02	0.3594E+02	0.2594E+02
28	27.40	2755.65	3858.63	2744.59	0.2518E+02	0.3398E+02	0.2618E+02
29	29.04	2794.04	3912.78	2787.72	0.2347E+02	0.3228E+02	0.2657E+02
30	30.68	2832.43	3964.30	2831.53	0.2347E+02	0.3075E+02	0.2700E+02
31	32.31	2870.83	4013.43	2876.02	0.2347E+02	0.2937E+02	0.2740E+02
32	33.95	2909.22	4060.43	2921.14	0.2347E+02	0.2812E+02	0.2776E+02

V - 93

Connection type : Extended end-plate connections
Mode : All bolted with column stiffener

Tested by : Y.L.Yee (1984) Australia
Test Id. : K12

Column : 460UB67.1
Beam : 460UB67.1
Plate thickness : 20 mm

Fasteners: G8.8- -M24
1 1/16" Oversize holes
Material : G250
Fy = 37.70 ksi
Fu = -- ksi

Major parameters

End plate extended on both sides.
ct = 2.3622" pt = 5.2244" li = 9.7874" pc = 7.5866"
cc = 2.3622" gt = 5.1181" gc = 5.1181"
tp = 0.7874" nt = 2 X 2 nc = 2 X 2

Remark 1) The beam, 3 deg. to horizontal was subjected to cantilever loading.
2) End-plate yield stress (Fy) is mentioned.

No	Moment (k-in)	Rotation (radians) X 1/1000
1	0.00	0.00
2	89.43	0.37
3	178.87	0.73
4	268.30	1.10
5	351.61	1.38
6	434.91	1.66
7	518.21	1.95
8	601.51	2.23
9	672.93	2.48
10	744.35	2.73
11	813.62	3.25
12	882.88	3.77
13	956.14	4.49
14	1029.41	5.20
15	1078.14	6.86
16	1126.88	8.53
17	1175.62	10.19
18	1254.45	11.35
19	1333.28	12.51
20	1406.58	14.20
21	1479.87	15.88
22	1543.09	17.50
23	1606.31	19.12
24	1661.52	20.64
25	1716.73	22.17
26	1771.95	23.70
27	1815.56	25.26
28	1859.17	26.82
29	1902.78	28.38
30	1935.45	30.00
31	1968.12	31.61
32	2000.79	33.23
33	2033.46	34.84

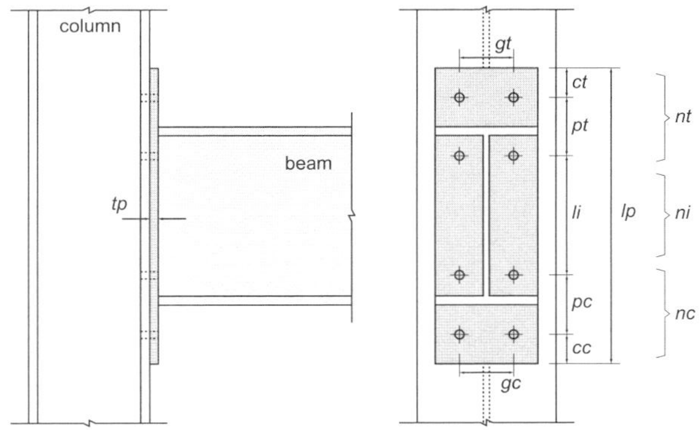

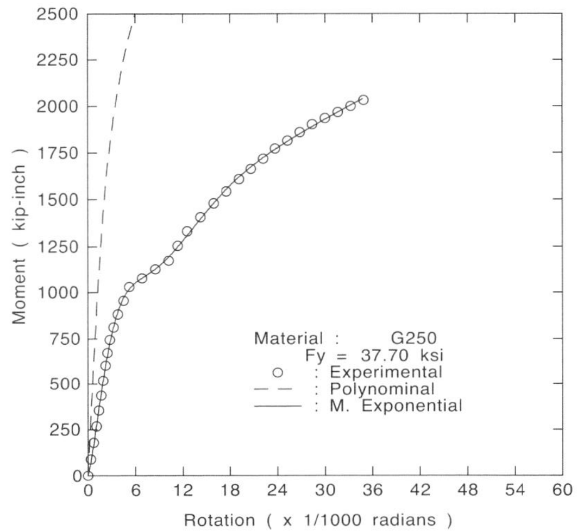

A.5 – 186

```
Moment-rotation prediction equations     ( R : X 1/1000 radians )

Frye and Morris polynominal model   :     R = Sum ( Ai X ( K*Bm )**Pi X 10**Qi )
     xd = 22.598400"    t =  0.787400"
     A1 =  1.790000   A2 =  1.760000   A3 =  2.040000    K =  0.000649
     P1 =        1     P2 =        3    P3 =        5    Q1 =        0    Q2 =       -1    Q3 =       -1

Modified exponential model  :
     Bm = Sum ( Ai X ( 1 - exp( - R/( 2*i*C ) ) ) )  +      Sum ( RKj X ( R - Rj0 ) )  +  Bm0
       C =   0.38192500E+00              Bmo=   0.00000000E+00          Nexp=  6        Nliner=  1
      AI =   0.66360683E+03  -0.15649297E+05  0.64402048E+05  -0.98504023E+05   0.61481416E+05  -0.11144305E+05
     Rj0 =          0.0000
     RKj =   0.22759215E+02
```

No	Rotation (radians) X 1/1000	Moment (kip-inch) Expri.	Moment (kip-inch) Poly.	Moment (kip-inch) M.Expo.	Connection stiffness (kip-inch) X 1000 Expri.	Connection stiffness (kip-inch) X 1000 Poly.	Connection stiffness (kip-inch) X 1000 M.Expo.
1	0.00	0.00	0.00	0.00	0.2442E+03	0.8603E+03	0.1561E+03
2	0.37	89.43	313.79	76.51	0.2441E+03	0.8491E+03	0.2370E+03
3	0.73	178.87	618.69	171.44	0.2441E+03	0.8097E+03	0.2777E+03
4	1.10	268.30	902.84	277.33	0.2731E+03	0.7361E+03	0.2969E+03
5	1.38	351.61	1100.23	361.51	0.2954E+03	0.6623E+03	0.2983E+03
6	1.66	434.91	1276.04	444.71	0.2954E+03	0.5848E+03	0.2904E+03
7	1.95	518.21	1430.56	524.62	0.2954E+03	0.5124E+03	0.2753E+03
8	2.23	601.51	1565.94	599.53	0.2882E+03	0.4494E+03	0.2553E+03
9	2.48	672.93	1673.59	661.70	0.2817E+03	0.4014E+03	0.2348E+03
10	2.73	744.35	1770.05	718.47	0.2331E+03	0.3608E+03	0.2130E+03
11	3.25	813.62	1940.12	817.52	0.1332E+03	0.2963E+03	0.1683E+03
12	3.77	882.88	2081.45	894.21	0.1205E+03	0.2500E+03	0.1277E+03
13	4.49	956.14	2242.46	968.53	0.1030E+03	0.2054E+03	0.8376E+02
14	5.20	1029.41	2376.77	1016.75	0.8097E+02	0.1743E+03	0.5419E+02
15	6.86	1078.14	2625.39	1078.99	0.2925E+02	0.1290E+03	0.2870E+02
16	8.53	1126.88	2817.06	1129.19	0.2925E+02	0.1028E+03	0.3402E+02
17	10.19	1175.62	2973.18	1195.04	0.5209E+02	0.8588E+02	0.4474E+02
18	11.35	1254.45	3067.43	1250.12	0.6799E+02	0.7722E+02	0.4981E+02
19	12.51	1333.28	3152.78	1309.44	0.5801E+02	0.7024E+02	0.5206E+02
20	14.20	1406.58	3264.03	1396.95	0.4352E+02	0.6222E+02	0.5119E+02
21	15.88	1479.87	3363.33	1480.18	0.4125E+02	0.5596E+02	0.4734E+02
22	17.50	1543.09	3449.82	1552.99	0.3907E+02	0.5111E+02	0.4261E+02
23	19.12	1606.31	3529.16	1618.11	0.3758E+02	0.4709E+02	0.3795E+02
24	20.64	1661.52	3598.55	1673.03	0.3616E+02	0.4389E+02	0.3411E+02
25	22.17	1716.73	3663.39	1722.63	0.3616E+02	0.4112E+02	0.3099E+02
26	23.70	1771.95	3724.29	1768.00	0.3210E+02	0.3872E+02	0.2856E+02
27	25.26	1815.56	3783.01	1811.07	0.2794E+02	0.3656E+02	0.2672E+02
28	26.82	1859.17	3838.55	1851.68	0.2794E+02	0.3465E+02	0.2539E+02
29	28.38	1902.78	3891.28	1890.54	0.2415E+02	0.3294E+02	0.2446E+02
30	30.00	1935.45	3943.20	1929.50	0.2022E+02	0.3136E+02	0.2382E+02
31	31.61	1968.12	3992.69	1967.61	0.2022E+02	0.2994E+02	0.2339E+02
32	33.23	2000.79	4040.01	2005.16	0.2022E+02	0.2865E+02	0.2312E+02
33	34.84	2033.46	4085.34	2042.36	0.2022E+02	0.2748E+02	0.2295E+02

V - 94

Connection type : Extended end-plate connections
Mode : All bolted with column stiffener

Tested by : Y.L.Yee (1984) Australia
Test Id. : B1

Column : 310UC96.8
Beam : 460UB67.1
Plate thickness : 20 mm
Stiffener thickness : 0.4724"

Fasteners: G8.8- -M24
1 1/16" Oversize holes
Material : G250
Fy = 37.70 ksi
Fu = -- ksi

Major parameters

End plate extended on both sides.
ct = 2.3622" pt = 5.2244" li = 9.7874" pc = 7.5866"
cc = 2.3622" gt = 5.1181" gc = 5.1181"
tp = 0.7874" nt = 2 X 2 nc = 2 X 2

Remark 1) The beam stub, 3 deg. to horizontal was subjected to cantilever loading.
2) End-plate yield stress (Fy) is mentioned.

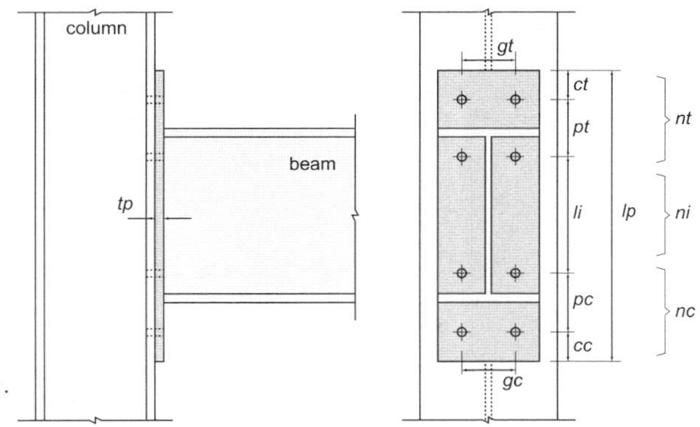

No	Moment (k-in)	Rotation (radians) X 1/1000	No	Moment (k-in)	Rotation (radians) X 1/1000
1	0.00	0.00	26	3169.69	23.26
2	151.71	0.32	27	3220.67	25.11
3	303.42	0.65	28	3257.98	26.41
4	445.18	0.96	29	3295.29	27.70
5	586.94	1.28	30	3341.30	29.18
6	733.67	1.66	31	3387.31	30.67
7	880.40	2.04	32	3408.45	32.23
8	1037.09	2.37	33	3429.58	33.78
9	1193.77	2.70	34	3464.40	35.76
10	1342.99	3.43	35	3499.21	37.74
11	1492.20	4.16	36	3521.62	39.82
12	1636.45	4.87			
13	1780.70	5.57			
14	1918.73	6.47			
15	2056.76	7.38			
16	2223.38	8.65			
17	2358.93	10.23			
18	2494.48	11.80			
19	2606.39	13.23			
20	2718.30	14.66			
21	2830.21	16.10			
22	2901.09	17.31			
23	2971.98	18.52			
24	3045.34	19.96			
25	3118.71	21.40			

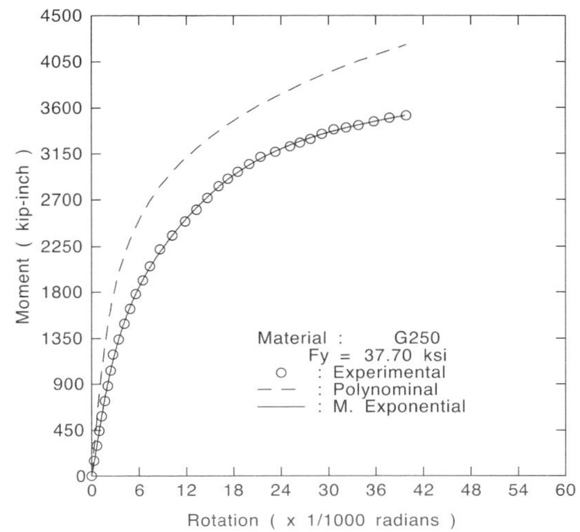

A.5 – 188

```
Moment-rotation prediction equations      ( R : X 1/1000 radians )

Frye and Morris polynominal model  :     R = Sum ( Ai X ( K*Bm )**Pi X 10**Qi )
     xd = 22.598400"    t =  0.787400"
     A1 =  1.790000    A2 =  1.760000    A3 =  2.040000    K =  0.000649
     P1 =         1    P2 =         3    P3 =         5    Q1 =         0    Q2 =        -1    Q3 =        -1

Modified exponential model  :
     Bm = Sum ( Ai X ( 1 - exp( - R/( 2*i*C ) ) ) )  +     Sum ( RKj X ( R - Rj0 ) )  +  Bm0
      C =   0.41269167E+00             Bmo=   0.00000000E+00           Nexp=  6        Nliner=  1
     AI =  -0.75465901E+03   0.55337071E+04  -0.28446140E+05   0.84545757E+05  -0.11132805E+06   0.53502543E+05
     Rj0 =            0.0000
     RKj =   0.12109669E+02
```

No	Rotation (radians) X 1/1000	Moment (kip-inch) Expri.	Moment (kip-inch) Poly.	Moment (kip-inch) M.Expo.	Connection stiffness (kip-inch) X 1000 Expri.	Connection stiffness (kip-inch) X 1000 Poly.	Connection stiffness (kip-inch) X 1000 M.Expo.
1	0.00	0.00	0.00	0.00	0.4690E+03	0.8603E+03	0.3854E+03
3	0.65	303.42	548.76	295.47	0.4597E+03	0.8220E+03	0.4829E+03
5	1.28	586.94	1029.19	591.82	0.4209E+03	0.6908E+03	0.4500E+03
7	2.04	880.40	1477.80	906.91	0.4317E+03	0.4902E+03	0.3746E+03
9	2.70	1193.77	1758.46	1134.20	0.3889E+03	0.3655E+03	0.3128E+03
11	4.16	1492.20	2173.14	1514.64	0.2046E+03	0.2236E+03	0.2177E+03
13	5.57	1780.70	2439.47	1781.72	0.1820E+03	0.1614E+03	0.1661E+03
15	7.38	2056.76	2689.34	2044.70	0.1436E+03	0.1195E+03	0.1286E+03
17	10.23	2358.93	2976.06	2361.64	0.8606E+02	0.8560E+02	0.9718E+02
19	13.23	2606.39	3202.07	2622.65	0.7820E+02	0.6655E+02	0.7771E+02
21	16.10	2830.21	3375.28	2824.03	0.6757E+02	0.5526E+02	0.6340E+02
23	18.52	2971.98	3500.44	2964.37	0.5506E+02	0.4850E+02	0.5278E+02
25	21.40	3118.71	3631.24	3100.34	0.4063E+02	0.4247E+02	0.4181E+02
27	25.11	3220.67	3777.73	3234.01	0.2828E+02	0.3675E+02	0.3079E+02
29	27.70	3295.29	3868.64	3306.02	0.2986E+02	0.3366E+02	0.2517E+02
31	30.67	3387.31	3964.05	3373.40	0.2250E+02	0.3075E+02	0.2053E+02
33	33.78	3429.58	4056.81	3431.96	0.1533E+02	0.2821E+02	0.1728E+02
35	37.74	3499.21	4162.36	3495.00	0.1427E+02	0.2563E+02	0.1480E+02

A.5 – 189

V - 95

Connection type : Extended end-plate connections
Mode : All bolted without column stiffener

Tested by : Y.L.Yee (1984) Australia
Test Id. : B2

Column : 310UC96.8
Beam : 460UB67.1
Plate thickness : 20 mm

Fasteners: G8.8- -M24
1 1/16" Oversize holes
Material : G250
Fy = 37.70 ksi
Fu = -- ksi

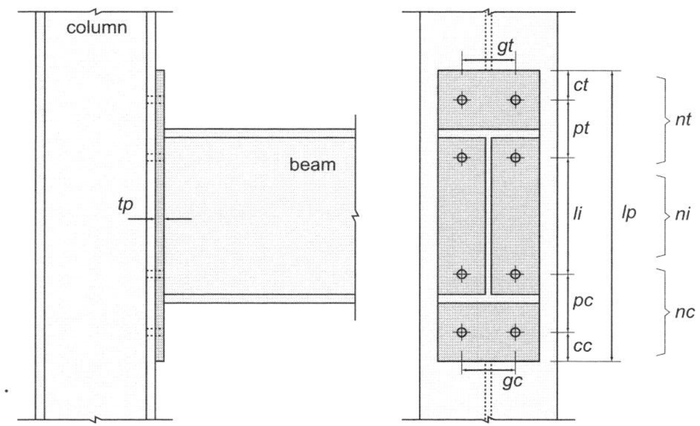

Major parameters

End plate extended on both sides.

ct = 2.3622" pt = 5.2244" li = 9.7874" pc = 7.5866"
cc = 2.3622" gt = 5.1181" gc = 5.1181"
tp = 0.7874" nt = 2 X 2 nc = 2 X 2

Remark 1) The beam stub, 3 deg. to horizontal was subjected to cantilever loading.
2) End-plate yield stress (Fy) is mentioned.

No	Moment (k-in)	Rotation (radians) X 1/1000	No	Moment (k-in)	Rotation (radians) X 1/1000
1	0.00	0.00	26	1917.15	20.49
2	98.72	0.28	27	1944.36	21.88
3	197.45	0.55	28	1971.56	23.28
4	296.17	0.83	29	1998.77	24.67
5	402.38	1.21	30	2025.98	26.07
6	508.59	1.60	31	2053.16	27.38
7	614.81	1.99	32	2080.35	28.69
8	701.11	2.32	33	2107.53	30.00
9	787.40	2.65	34	2134.72	31.30
10	873.70	2.98	35	2161.90	32.61
11	978.32	3.65			
12	1082.94	4.31			
13	1187.56	4.97			
14	1261.11	5.68			
15	1334.66	6.38			
16	1409.49	7.28			
17	1484.32	8.18			
18	1550.53	9.39			
19	1616.74	10.60			
20	1668.01	11.74			
21	1719.29	12.88			
22	1761.95	14.43			
23	1804.62	15.98			
24	1847.28	17.54			
25	1889.94	19.09			

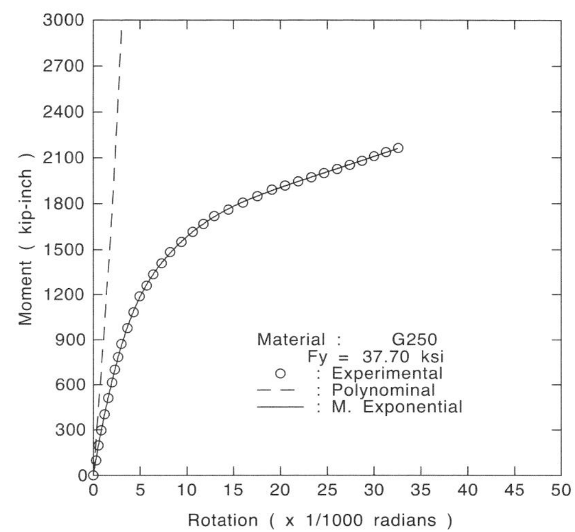

A.5 – 190

```
Moment-rotation prediction equations     ( R : X 1/1000 radians )

Frye and Morris polynominal model  :    R = Sum ( Ai X ( K*Bm )**Pi X 10**Qi )
     xd = 22.598400"   t =  0.787400"  tf =  0.944900"
     A1 =  1.830000   A2 = -1.040000   A3 =  6.380000    K =  0.000674
     P1 =         1   P2 =         3   P3 =         5   Q1 =         0   Q2 =        -1   Q3 =        -3

Modified exponential model  :
     Bm = Sum ( Ai X ( 1 - exp( - R/( 2*i*C ) ) ) )  +     Sum ( RKj X ( R - Rj0 ) )  +  Bm0
      C =   0.34078333E+00              Bmo=   0.00000000E+00          Nexp=  6       Nliner=  1
     AI =   0.43598843E+03  -0.59383754E+04   0.25693345E+05  -0.50219847E+05   0.45958650E+05  -0.14455978E+05
    Rj0 =           0.0000
    RKj =   0.21014309E+02
```

No	Rotation (radians) X 1/1000	Moment (kip-inch) Expri.	Moment (kip-inch) Poly.	Moment (kip-inch) M.Expo.	Connection stiffness (kip-inch) X 1000 Expri.	Connection stiffness (kip-inch) X 1000 Poly.	Connection stiffness (kip-inch) X 1000 M.Expo.
1	0.00	0.00	0.00	0.00	0.3577E+03	0.8108E+03	0.3795E+03
3	0.55	197.45	449.88	197.55	0.3577E+03	0.8235E+03	0.3291E+03
5	1.21	402.38	1010.37	403.92	0.2749E+03	0.8768E+03	0.2970E+03
7	1.99	614.81	1734.35	620.49	0.2670E+03	0.1014E+04	0.2624E+03
9	2.65	787.40	2472.45	783.17	0.2602E+03	0.1227E+04	0.2277E+03
11	3.65	978.32	3803.20	983.98	0.1577E+03	0.1282E+04	0.1771E+03
13	4.97	1187.56	5007.52	1182.67	0.1318E+03	0.6143E+03	0.1260E+03
15	6.38	1334.66	5668.22	1334.81	0.9510E+02	0.3644E+03	0.9259E+02
17	8.18	1484.32	6197.35	1477.02	0.7120E+02	0.2433E+03	0.6832E+02
19	10.60	1616.74	6688.79	1617.22	0.4968E+02	0.1712E+03	0.4883E+02
21	12.88	1719.29	7034.47	1713.65	0.3762E+02	0.1357E+03	0.3667E+02
23	15.98	1804.62	7407.40	1810.18	0.2747E+02	0.1070E+03	0.2658E+02
25	19.09	1889.94	7709.89	1884.47	0.2327E+02	0.8903E+02	0.2195E+02
27	21.88	1944.36	7941.87	1943.24	0.1949E+02	0.7777E+02	0.2045E+02
29	24.67	1998.77	8146.56	1999.70	0.1949E+02	0.6927E+02	0.2012E+02
31	27.38	2053.16	8324.85	2054.22	0.2076E+02	0.6280E+02	0.2023E+02
33	30.00	2107.53	8484.02	2107.46	0.2076E+02	0.5766E+02	0.2044E+02
35	32.61	2161.90	8629.13	2161.23	0.2076E+02	0.5342E+02	0.2062E+02

V - 96

Connection type : Extended end-plate connections
Mode : All bolted without column stiffener

Tested by : Y.L.Yee (1984) Australia
Test Id. : B3

Column : 310UC96.8
Beam : 460UB67.1
Plate thickness : 16 mm

Fasteners: G8.8- -M24
1 1/16" Oversize holes
Material : G250
Fy = 37.70 ksi
Fu = -- ksi

Major parameters

End plate extended on both sides.
ct = 2.3622" pt = 5.2244" li = 9.7874" pc = 7.5866"
cc = 2.3622" gt = 5.1181" gc = 5.1181"
tp = 0.6299" nt = 2 X 2 nc = 2 X 2

Remark 1) The beam stub, 3 deg. to horizontal was subjected to cantilever loading.
2) End-plate yield stress (Fy) is mentioned.

No	Moment (k-in)	Rotation (radians) X 1/1000
1	0.00	0.00
2	80.95	0.23
3	161.90	0.46
4	242.85	0.70
5	323.80	0.93
6	415.96	1.24
7	508.13	1.56
8	600.29	1.87
9	671.27	2.10
10	742.25	2.32
11	813.25	2.62
12	884.25	2.92
13	966.47	3.33
14	1048.69	3.75
15	1148.38	4.52
16	1248.07	5.30
17	1347.76	6.08
18	1436.28	7.18
19	1524.81	8.27
20	1589.67	9.30
21	1654.53	10.33
22	1694.52	11.54
23	1734.51	12.74
24	1774.49	13.95
25	1799.54	15.12

No	Moment (k-in)	Rotation (radians) X 1/1000
26	1824.58	16.29
27	1840.50	17.53
28	1856.43	18.78
29	1872.35	20.02
30	1902.45	21.78
31	1932.54	23.53
32	1962.64	25.28
33	1992.73	27.03
34	2022.83	28.79
35	2046.41	30.23
36	2069.99	31.67
37	2093.58	33.11
38	2117.16	34.55
39	2140.75	35.99

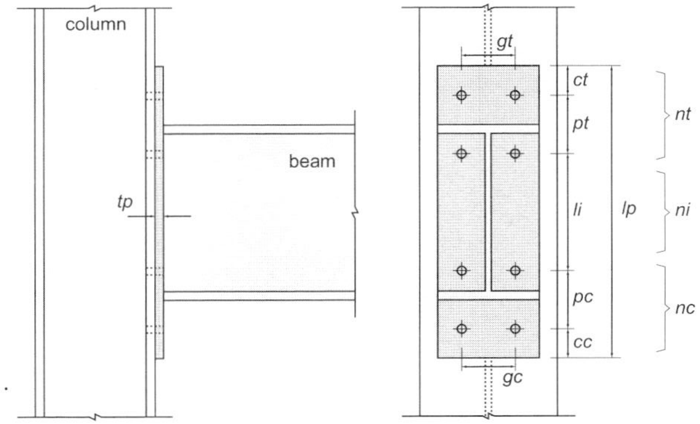

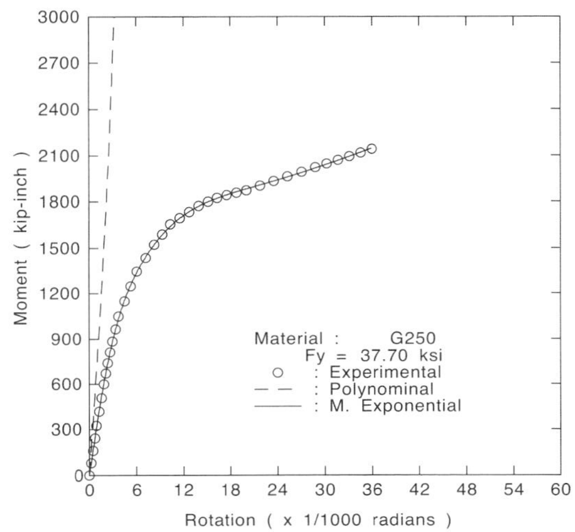

A.5 – 192

```
Moment-rotation prediction equations     ( R : X 1/1000 radians )

Frye and Morris polynominal model   :    R = Sum ( Ai X ( K*Bm )**Pi X 10**Qi )
     xd = 22.598400"   t =  0.629900"  tf =  0.944900"
     A1 =  1.830000   A2 = -1.040000   A3 =  6.380000    K =  0.000737
     P1 =        1    P2 =        3    P3 =        5    Q1 =        0    Q2 =       -1    Q3 =       -3

Modified exponential model  :
     Bm = Sum ( Ai X ( 1 - exp( - R/( 2*i*C ) ) ) )  +      Sum ( RKj X ( R - Rj0 ) )  +  Bm0
       C =   0.35790000E+00           Bmo=   0.00000000E+00          Nexp=  6        Nliner=  1
      AI =   0.18035825E+03  -0.61336166E+04   0.35501817E+05  -0.83586010E+05   0.87947344E+05  -0.32478596E+05
     Rj0 =           0.0000
     RKj =   0.19799227E+02
```

No	Rotation (radians) X 1/1000	Moment (kip-inch) Expri.	Moment (kip-inch) Poly.	Moment (kip-inch) M.Expo.	Connection stiffness (kip-inch) X 1000 Expri.	Connection stiffness (kip-inch) X 1000 Poly.	Connection stiffness (kip-inch) X 1000 M.Expo.
1	0.00	0.00	0.00	0.00	0.3489E+03	0.7415E+03	0.3176E+03
3	0.46	161.90	345.33	158.88	0.3489E+03	0.7497E+03	0.3446E+03
5	0.93	323.80	698.48	317.29	0.3251E+03	0.7756E+03	0.3357E+03
7	1.56	508.13	1206.32	519.79	0.2929E+03	0.8463E+03	0.3052E+03
9	2.10	671.27	1688.74	675.17	0.3169E+03	0.9534E+03	0.2711E+03
11	2.62	813.25	2225.25	807.73	0.2382E+03	0.1111E+04	0.2370E+03
13	3.33	966.47	3090.95	961.14	0.1984E+03	0.1268E+04	0.1948E+03
15	4.52	1148.38	4297.59	1159.62	0.1279E+03	0.7105E+03	0.1414E+03
17	6.08	1347.76	5080.56	1344.23	0.1083E+03	0.3643E+03	0.9958E+02
19	8.27	1524.81	5689.95	1524.10	0.7168E+02	0.2187E+03	0.6773E+02
21	10.33	1654.53	6074.53	1642.29	0.4930E+02	0.1618E+03	0.4817E+02
23	12.74	1734.51	6417.00	1737.26	0.3316E+02	0.1256E+03	0.3164E+02
25	15.12	1799.54	6687.68	1799.16	0.2140E+02	0.1038E+03	0.2141E+02
27	17.53	1840.50	6919.17	1843.65	0.1279E+02	0.8880E+02	0.1617E+02
29	20.02	1872.35	7125.61	1881.13	0.1461E+02	0.7763E+02	0.1440E+02
31	23.53	1932.54	7376.69	1931.93	0.1717E+02	0.6630E+02	0.1494E+02
33	27.03	1992.73	7594.04	1986.91	0.1717E+02	0.5812E+02	0.1646E+02
35	30.23	2046.41	7770.14	2041.52	0.1637E+02	0.5241E+02	0.1769E+02
37	33.11	2093.58	7916.36	2093.71	0.1637E+02	0.4819E+02	0.1849E+02
39	35.99	2140.75	8049.88	2147.80	0.1637E+02	0.4470E+02	0.1902E+02

A.5 – 193

V - 97

Connection type : Extended end-plate connections
Mode : All bolted with column stiffener

Tested by : Y.L.Yee (1984) Australia
Test Id. : C1

Column : 310UB40.4
Beam : 310UB40.4
Plate thickness : 20 mm
Stiffener thickness : 0.3937"

Fasteners: G8.8- -M20
13/16" Oversize holes
Material : G250
Fy = 36.69 ksi
Fu = -- ksi

Major parameters

End plate extended on both sides.
ct = 1.7716" pt = 3.9449" li = 5.8504" pc = 5.7165"
cc = 1.7717" gt = 3.9370" gc = 3.9370"
tp = 0.7874" nt = 2 X 2 nc = 2 X 2

Remark 1) 12 mm Doubler plates were welded to flange on both sides of column web.
2) End-plate yield stress (Fy) is mentioned.

No	Moment (k-in)	Rotation (radians) X 1/1000
1	0.00	0.00
2	80.45	0.38
3	160.90	0.76
4	233.91	1.10
5	306.93	1.45
6	370.99	1.98
7	435.06	2.52
8	515.46	2.94
9	595.86	3.36
10	648.88	3.77
11	701.90	4.18
12	761.00	4.70
13	820.10	5.23
14	886.45	5.93
15	952.79	6.63
16	1010.37	7.38
17	1067.94	8.14
18	1123.67	9.38
19	1179.39	10.62
20	1237.78	11.72
21	1296.16	12.82
22	1333.38	13.93
23	1370.60	15.03
24	1407.81	16.14
25	1445.03	17.25
26	1472.45	19.21
27	1488.41	21.39
28	1534.59	23.30
29	1546.82	25.48
30	1560.05	26.86
31	1628.45	27.91

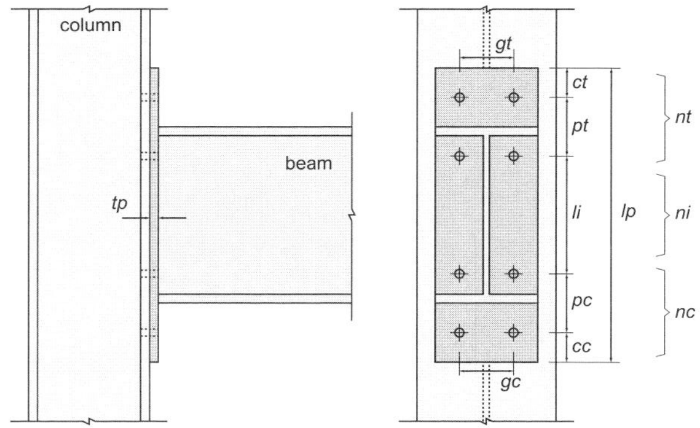

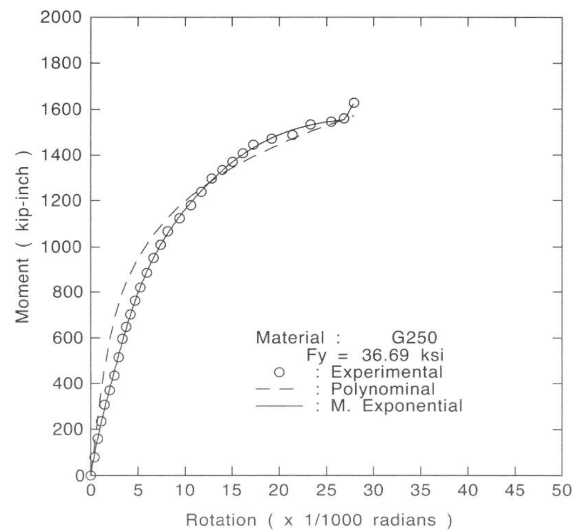

```
Moment-rotation prediction equations      ( R : X 1/1000 radians )

Frye and Morris polynominal model  :     R = Sum ( Ai X ( K*Bm )**Pi X 10**Qi )
     xd = 15.511800"    t =  0.787400"
     A1 =  1.790000    A2 =  1.760000    A3 =  2.040000     K =  0.001602
     P1 =        1     P2 =        3     P3 =        5     Q1 =        0     Q2 =       -1     Q3 =       -1

Modified exponential model  :
     Bm = Sum ( Ai X ( 1 - exp( - R/( 2*i*C ) ) ) )  +      Sum ( RKj X ( R - Rj0 ) )  +  Bm0
      C =   0.32720000E+00            Bmo=   0.00000000E+00            Nexp=  6         Nliner=  2
     AI =  -0.57825469E+02   0.50430818E+04  -0.37413447E+05   0.10263112E+06  -0.12140708E+06   0.52831025E+05
    Rj0 =           0.0000          26.8600
    RKj =  -0.17145241E+01   0.67750780E+02
```

No	Rotation (radians) X 1/1000	Moment (kip-inch) Expri.	Moment (kip-inch) Poly.	Moment (kip-inch) M.Expo.	Connection stiffness (kip-inch) X 1000 Expri.	Connection stiffness (kip-inch) X 1000 Poly.	Connection stiffness (kip-inch) X 1000 M.Expo.
1	0.00	0.00	0.00	0.00	0.2125E+03	0.3487E+03	0.2660E+03
2	0.38	80.45	131.38	89.94	0.2125E+03	0.3438E+03	0.2154E+03
3	0.76	160.90	258.73	165.74	0.2122E+03	0.3267E+03	0.1879E+03
4	1.10	233.91	366.69	227.91	0.2119E+03	0.2981E+03	0.1744E+03
5	1.45	306.93	463.18	286.58	0.1758E+03	0.2612E+03	0.1670E+03
6	1.98	370.99	587.62	374.18	0.1196E+03	0.2040E+03	0.1608E+03
7	2.52	435.06	684.44	458.91	0.1599E+03	0.1601E+03	0.1556E+03
8	2.94	515.46	746.00	523.21	0.1914E+03	0.1349E+03	0.1504E+03
9	3.36	595.86	798.48	585.07	0.1600E+03	0.1159E+03	0.1440E+03
10	3.77	648.88	842.93	642.61	0.1293E+03	0.1016E+03	0.1366E+03
11	4.18	701.90	882.16	696.94	0.1218E+03	0.9027E+02	0.1284E+03
12	4.70	761.00	926.58	761.65	0.1122E+03	0.7894E+02	0.1174E+03
13	5.23	820.10	965.73	820.56	0.1046E+03	0.7014E+02	0.1064E+03
14	5.93	886.45	1011.65	890.44	0.9451E+02	0.6110E+02	0.9290E+02
15	6.63	952.79	1052.01	951.43	0.8589E+02	0.5418E+02	0.8119E+02
16	7.38	1010.37	1090.43	1008.41	0.7666E+02	0.4838E+02	0.7097E+02
17	8.14	1067.94	1124.96	1058.58	0.6469E+02	0.4375E+02	0.6298E+02
18	9.38	1123.67	1175.42	1130.54	0.4490E+02	0.3785E+02	0.5368E+02
19	10.62	1179.39	1219.52	1193.10	0.4924E+02	0.3343E+02	0.4751E+02
20	11.72	1237.78	1254.54	1243.01	0.5308E+02	0.3035E+02	0.4336E+02
21	12.82	1296.16	1286.48	1288.66	0.4338E+02	0.2782E+02	0.3968E+02
22	13.93	1333.38	1316.07	1330.59	0.3361E+02	0.2569E+02	0.3607E+02
23	15.03	1370.60	1343.49	1368.51	0.3361E+02	0.2389E+02	0.3242E+02
24	16.14	1407.81	1369.06	1402.37	0.3361E+02	0.2234E+02	0.2874E+02
25	17.25	1445.03	1393.04	1432.17	0.2653E+02	0.2100E+02	0.2509E+02
26	19.21	1472.45	1432.21	1475.33	0.1082E+02	0.1900E+02	0.1901E+02
27	21.39	1488.41	1471.73	1510.31	0.1634E+02	0.1722E+02	0.1320E+02
28	23.30	1534.59	1503.29	1531.44	0.1554E+02	0.1593E+02	0.9134E+01
29	25.48	1546.82	1536.59	1547.29	0.8048E+01	0.1470E+02	0.5614E+01
30	26.86	1560.05	1556.41	1553.83	0.4109E+02	0.1402E+02	0.3923E+01
31	27.91	1628.45	1571.15	1628.45	0.6508E+02	0.1354E+02	0.7063E+02

V - 98

Connection type : Extended end-plate connections
Mode : All bolted with column stiffener

Tested by : Y.L.Yee (1984) Australia
Test Id. : C2

Column : 310UB40.4
Beam : 310UB40.4
Plate thickness : 10 mm

Fasteners: G8.8- -M16
13/16" Oversize holes
Material : G250
Fy = 36.69 ksi
Fu = -- ksi

Major parameters

End plate extended on both sides.

ct = 1.7716"	pt = 3.9449"	li = 5.8504"	pc = 5.7165"
cc = 1.7717"	gt = 3.9370"	gc = 3.9370"	
tp = 0.3937"	nt = 2 X 2	nc = 2 X 2	

Remark 1) The beam, 3 deg. to horizontal was subjected to cantilever loading.
2) End-plate yield stress (Fy) is mentioned.

No	Moment (k-in)	Rotation (radians) X 1/1000	No	Moment (k-in)	Rotation (radians) X 1/1000
1	0.00	0.00	26	666.40	26.13
2	24.54	0.33	27	677.16	27.99
3	49.09	0.67	28	687.93	29.85
4	73.63	1.00	29	691.76	31.53
5	108.82	1.30	30	695.59	33.21
6	144.00	1.60	31	699.42	34.89
7	179.19	1.90	32	703.25	36.58
8	214.37	2.20	33	707.08	38.26
9	245.28	2.43	34	710.90	39.94
10	276.19	2.65			
11	307.10	2.88			
12	338.01	3.10			
13	371.40	4.70			
14	404.78	6.29			
15	438.17	7.88			
16	471.56	9.47			
17	504.95	11.06			
18	538.34	12.66			
19	571.73	14.25			
20	587.32	15.82			
21	602.91	17.40			
22	618.50	18.98			
23	634.10	20.56			
24	644.86	22.41			
25	655.63	24.27			

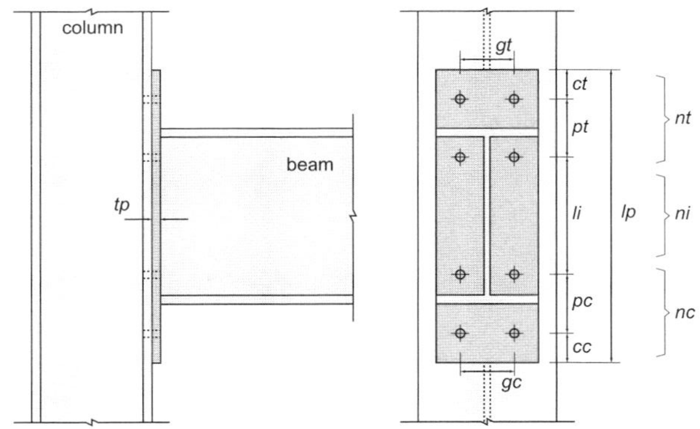

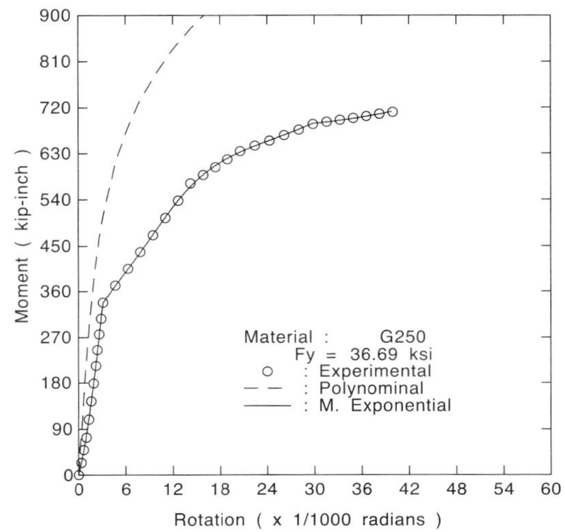

```
Moment-rotation prediction equations     ( R : X 1/1000 radians )

Frye and Morris polynominal model  :    R = Sum ( Ai X ( K*Bm )**Pi X 10**Qi )
     xd = 15.511800"   t =  0.393700"
     A1 =  1.790000   A2 =  1.760000   A3 =  2.040000    K =  0.002428
     P1 =         1   P2 =         3   P3 =         5   Q1 =         0   Q2 =        -1   Q3 =        -1

Modified exponential model  :
     Bm = Sum ( Ai X ( 1 - exp( - R/( 2*i*C ) ) ) )  +      Sum ( RKj X ( R - Rj0 ) )  +  Bm0
      C =   0.41606667E+00              Bmo=   0.00000000E+00          Nexp=  6        Nliner=  3
     AI =   0.35593354E+03  -0.61248543E+04   0.31199497E+05  -0.69520313E+05   0.68481041E+05  -0.24352252E+05
    Rj0 =           0.0000           3.1000          29.8500
    RKj =   0.12759736E+03  -0.11874868E+03  -0.52038667E+01
```

No	Rotation (radians) X 1/1000	Moment (kip-inch) Expri.	Moment (kip-inch) Poly.	Moment (kip-inch) M.Expo.	Connection stiffness (kip-inch) X 1000 Expri.	Connection stiffness (kip-inch) X 1000 Poly.	Connection stiffness (kip-inch) X 1000 M.Expo.
1	0.00	0.00	0.00	0.00	0.7370E+02	0.2301E+03	-0.5917E+02
3	0.67	49.09	150.92	48.29	0.7370E+02	0.2191E+03	0.8208E+02
5	1.30	108.82	279.54	108.16	0.1171E+03	0.1830E+03	0.1066E+03
7	1.90	179.19	376.39	177.86	0.1171E+03	0.1399E+03	0.1241E+03
9	2.43	245.28	441.71	245.81	0.1369E+03	0.1099E+03	0.1333E+03
11	2.88	307.10	486.85	307.04	0.1369E+03	0.9109E+02	0.1375E+03
13	4.70	371.40	610.92	371.48	0.2097E+02	0.5217E+02	0.2105E+02
15	7.88	438.17	734.79	437.73	0.2097E+02	0.2982E+02	0.2149E+02
17	11.06	504.95	814.24	506.87	0.2097E+02	0.2118E+02	0.2095E+02
19	14.25	571.73	873.70	565.80	0.1540E+02	0.1658E+02	0.1563E+02
21	17.40	602.91	921.20	606.05	0.9887E+01	0.1374E+02	0.1017E+02
23	20.56	634.10	961.27	632.42	0.8008E+01	0.1178E+02	0.6953E+01
25	24.27	655.63	1001.83	655.44	0.5793E+01	0.1013E+02	0.5805E+01
27	27.99	677.16	1037.13	677.40	0.5793E+01	0.8918E+01	0.6151E+01
29	31.53	691.76	1067.06	691.69	0.2277E+01	0.8024E+01	0.1676E+01
31	34.89	699.42	1092.86	698.46	0.2277E+01	0.7340E+01	0.2324E+01
33	38.26	707.08	1116.56	707.14	0.2277E+01	0.6773E+01	0.2813E+01

A.5 – 197

V - 99

Connection type : Extended end-plate connections
Mode : All bolted without column stiffener

Tested by : Y.L.Yee (1984) Australia
Test Id. : D1

Column : 460UB67.1
Beam : 410UB59.7
Plate thickness : 16 mm

Fasteners: G8.8- -M20
13/16" Oversize holes
Material : G250
Fy = 46.26 ksi
Fu = -- ksi

Major parameters

End plate extended on both sides.
ct = 1.9685" pt = 4.4409" li = 9.1102" pc = 6.4094"
cc = 1.9685" gt = 4.7244" gc = 4.7244"
tp = 0.6299" nt = 2 X 2 nc = 2 X 2

Remark 1) The beam, 3 deg. to horizontal was subjected to cantilever loading.
2) End-plate yield stress (Fy) is mentioned.

No	Moment (k-in)	Rotation (radians) X 1/1000	No	Moment (k-in)	Rotation (radians) X 1/1000
1	0.00	0.00	26	1469.87	26.50
2	70.85	0.35	27	1490.73	28.15
3	141.70	0.70	28	1511.60	29.80
4	212.56	1.06	29	1532.47	31.44
5	283.41	1.41	30	1553.34	33.09
6	357.37	1.78	31	1574.21	34.73
7	431.33	2.14	32	1595.08	36.38
8	505.29	2.51	33	1615.95	38.02
9	579.25	2.88			
10	654.14	3.67			
11	729.03	4.47			
12	803.92	5.26			
13	878.81	6.05			
14	952.62	7.46			
15	1026.43	8.87			
16	1079.35	10.40			
17	1132.27	11.92			
18	1185.20	13.45			
19	1223.35	14.99			
20	1261.50	16.52			
21	1299.65	18.06			
22	1337.80	19.59			
23	1370.81	21.32			
24	1403.83	23.05			
25	1436.85	24.78			

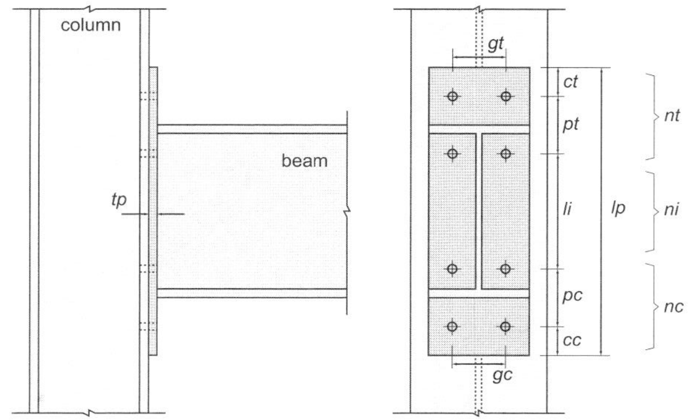

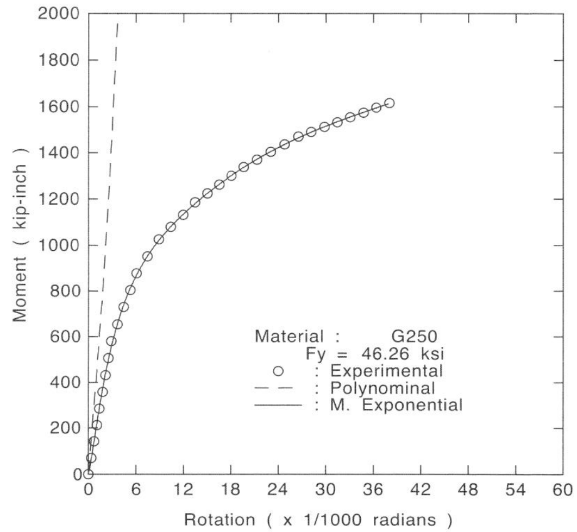

```
Moment-rotation prediction equations     ( R : X 1/1000 radians )

Frye and Morris polynominal model   :    R = Sum ( Ai X ( K*Bm )**Pi X 10**Qi )
     xd = 19.960500"   t =  0.629900"  tf =  0.787400"
     A1 =  1.330000   A2 = -1.040000   A3 =  6.380000    K =  0.001305
     P1 =        1    P2 =        3    P3 =        5    Q1 =        0    Q2 =       -1    Q3 =       -3

Modified exponential model  :
     Bm = Sum ( Ai X ( 1 - exp( - R/( 2*i*C ) ) ) )  +    Sum ( RKj X ( R - Rj0 ) )  +  Bm0
      C =   0.40481667E+00            Bmo=   0.00000000E+00          Nexp=  6        Nliner=  1
     AI =  -0.44913538E+03  0.32335489E+04  -0.16270536E+05  0.43270304E+05  -0.50812867E+05  0.22248429E+05
     Rj0 =          0.0000
     RKj =   0.10501049E+02
```

No	Rotation (radians) X 1/1000	Moment (kip-inch) Expri.	Moment (kip-inch) Poly.	Moment (kip-inch) M.Expo.	Connection stiffness (kip-inch) X 1000 Expri.	Connection stiffness (kip-inch) X 1000 Poly.	Connection stiffness (kip-inch) X 1000 M.Expo.
1	0.00	0.00	0.00	0.00	0.2014E+03	0.4188E+03	0.1324E+03
2	0.35	70.85	147.63	61.96	0.2014E+03	0.4214E+03	0.2012E+03
3	0.70	141.70	297.11	137.15	0.2014E+03	0.4296E+03	0.2218E+03
4	1.06	212.56	450.58	215.63	0.2014E+03	0.4440E+03	0.2219E+03
5	1.41	283.41	610.33	292.02	0.2011E+03	0.4659E+03	0.2114E+03
6	1.78	357.37	787.73	367.05	0.2007E+03	0.4989E+03	0.1954E+03
7	2.14	431.33	979.67	435.82	0.2007E+03	0.5453E+03	0.1779E+03
8	2.51	505.29	1191.54	498.20	0.2007E+03	0.6072E+03	0.1608E+03
9	2.88	579.25	1428.31	554.53	0.1670E+03	0.6775E+03	0.1452E+03
10	3.67	654.14	1983.33	657.97	0.9444E+02	0.6542E+03	0.1170E+03
11	4.47	729.03	2403.43	741.96	0.9444E+02	0.4148E+03	0.9586E+02
12	5.26	803.92	2671.95	811.40	0.9444E+02	0.2785E+03	0.8001E+02
13	6.05	878.81	2862.80	869.86	0.7930E+02	0.2077E+03	0.6796E+02
14	7.46	952.62	3104.55	954.17	0.5242E+02	0.1445E+03	0.5291E+02
15	8.87	1026.43	3283.38	1021.34	0.4390E+02	0.1118E+03	0.4320E+02
16	10.40	1079.35	3436.55	1081.78	0.3466E+02	0.9062E+02	0.3648E+02
17	11.92	1132.27	3563.49	1133.98	0.3466E+02	0.7664E+02	0.3219E+02
18	13.45	1185.20	3672.49	1180.79	0.2976E+02	0.6669E+02	0.2929E+02
19	14.99	1223.35	3768.91	1224.02	0.2483E+02	0.5919E+02	0.2707E+02
20	16.52	1261.50	3855.17	1264.14	0.2483E+02	0.5334E+02	0.2519E+02
21	18.06	1299.65	3933.39	1301.49	0.2483E+02	0.4865E+02	0.2345E+02
22	19.59	1337.80	4005.07	1336.24	0.2214E+02	0.4479E+02	0.2180E+02
23	21.32	1370.81	4079.25	1372.35	0.1911E+02	0.4120E+02	0.2003E+02
24	23.05	1403.83	4147.74	1405.52	0.1911E+02	0.3819E+02	0.1840E+02
25	24.78	1436.85	4211.45	1436.02	0.1911E+02	0.3563E+02	0.1694E+02
26	26.50	1469.87	4271.06	1464.16	0.1582E+02	0.3344E+02	0.1567E+02
27	28.15	1490.73	4324.54	1489.06	0.1268E+02	0.3161E+02	0.1463E+02
28	29.80	1511.60	4375.20	1512.41	0.1268E+02	0.2999E+02	0.1377E+02
29	31.44	1532.47	4423.33	1534.46	0.1268E+02	0.2855E+02	0.1306E+02
30	33.09	1553.34	4469.22	1555.46	0.1268E+02	0.2725E+02	0.1249E+02
31	34.73	1574.21	4513.09	1575.61	0.1268E+02	0.2608E+02	0.1203E+02
32	36.38	1595.08	4556.01	1595.10	0.1268E+02	0.2500E+02	0.1167E+02
33	38.02	1615.95	4595.48	1614.06	0.1268E+02	0.2405E+02	0.1139E+02

A.5 – 199

V -100

Connection type : Extended end-plate connections
Mode : All bolted without column stiffener

Tested by : Y.L.Yee (1984) Australia
Test Id. : D2

Column : 460UB67.1
Beam : 410UB59.7
Plate thickness : 16 mm

Fasteners: G8.8- -M20
13/16" Oversize holes
Material : G250
Fy = 46.26 ksi
Fu = -- ksi

Major parameters

End plate extended on both sides.

ct = 1.9685"	pt = 4.4409"	li = 9.1102"	pc = 6.4094"
cc = 1.9685"	gt = 4.7244"	gc = 4.7244"	
tp = 0.6299"	nt = 2 X 2	nc = 2 X 2	

Remark 1) 10 mm Doubler plates were welded to flange on both sides of column web.
2) End-plate yield stress (Fy) is mentioned.

No	Moment (k-in)	Rotation (radians) X 1/1000	No	Moment (k-in)	Rotation (radians) X 1/1000
1	0.00	0.00	26	1879.36	19.20
2	96.59	0.44	27	1909.87	20.30
3	193.19	0.88	28	1934.34	21.57
4	289.78	1.32	29	1958.81	22.84
5	392.23	1.75	30	1983.28	24.11
6	494.69	2.18	31	2007.75	25.38
7	597.14	2.60	32	2032.22	26.65
8	681.99	3.10	33	2056.69	27.92
9	766.85	3.59	34	2081.16	29.20
10	833.10	3.82			
11	899.36	4.06			
12	976.72	4.51			
13	1054.09	4.96			
14	1124.81	5.92			
15	1195.53	6.88			
16	1273.86	7.71			
17	1352.19	8.54			
18	1425.49	9.41			
19	1498.79	10.28			
20	1577.97	11.66			
21	1657.14	13.04			
22	1722.49	14.48			
23	1787.84	15.91			
24	1818.35	17.01			
25	1848.85	18.11			

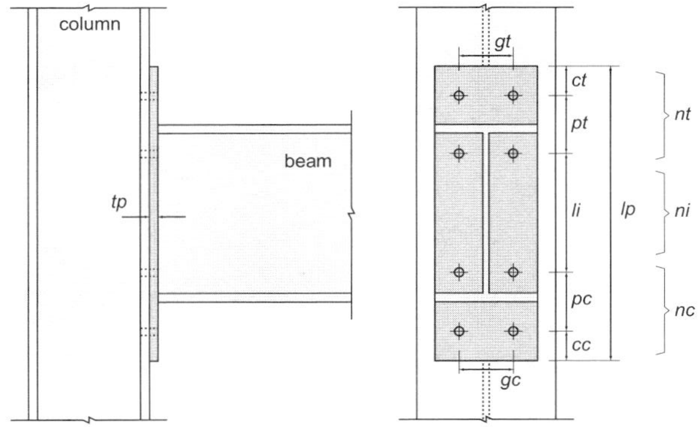

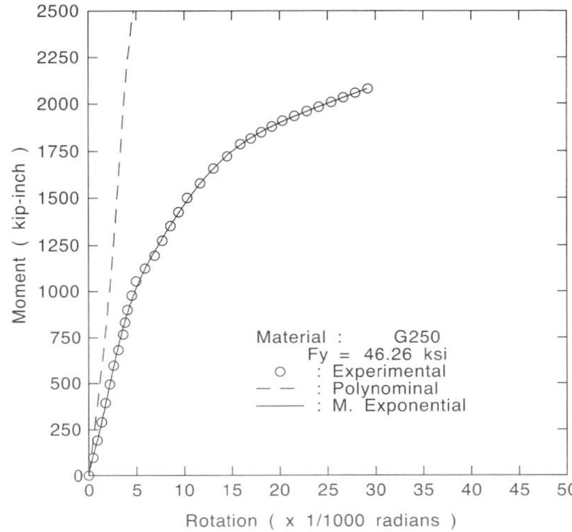

```
Moment-rotation prediction equations      ( R : X 1/1000 radians )

Frye and Morris polynominal model  :    R = Sum ( Ai X ( K*Bm )**Pi X 10**Qi )
     xd = 19.960500"   t =  0.629900"  tf =  0.787400"
     A1 =  1.830000   A2 = -1.040000   A3 =  6.380000    K =  0.001305
     P1 =        1    P2 =        3    P3 =        5    Q1 =        0    Q2 =       -1    Q3 =       -3

Modified exponential model  :
     Bm = Sum ( Ai X ( 1 - exp( - R/( 2*i*C ) ) ) )  +      Sum ( RKj X ( R - Rj0 ) )  +  Bm0
      C =   0.35371667E+00            Bmo=   0.00000000E+00          Nexp=  6        Nliner=  1
     AI =   0.15989988E+04  -0.20542298E+05   0.85693635E+05  -0.16096374E+06   0.13850643E+06  -0.42855109E+05
     Rj0 =           0.0000
     RKj =   0.21575371E+02
```

No	Rotation (radians) X 1/1000	Moment (kip-inch) Expri.	Moment (kip-inch) Poly.	Moment (kip-inch) M.Expo.	Connection stiffness (kip-inch) X 1000 Expri.	Connection stiffness (kip-inch) X 1000 Poly.	Connection stiffness (kip-inch) X 1000 M.Expo.
1	0.00	0.00	0.00	0.00	0.2187E+03	0.4188E+03	0.2972E+03
3	0.88	193.19	374.91	191.75	0.2187E+03	0.4361E+03	0.1981E+03
5	1.75	392.23	775.19	384.68	0.2409E+03	0.4963E+03	0.2420E+03
7	2.60	597.14	1245.99	591.21	0.2089E+03	0.6240E+03	0.2352E+03
9	3.59	766.85	1927.17	802.07	0.2460E+03	0.6765E+03	0.1888E+03
11	4.06	899.36	2212.29	885.30	0.2436E+03	0.5302E+03	0.1650E+03
13	4.96	1054.09	2584.17	1016.54	0.1400E+03	0.3184E+03	0.1276E+03
15	6.88	1195.53	3016.11	1217.24	0.8498E+02	0.1647E+03	0.8921E+02
17	8.54	1352.19	3244.96	1355.01	0.8969E+02	0.1180E+03	0.7877E+02
19	10.28	1498.79	3425.30	1484.86	0.7391E+02	0.9200E+02	0.7048E+02
21	13.04	1657.14	3644.93	1656.86	0.5147E+02	0.6905E+02	0.5316E+02
23	15.91	1787.84	3822.01	1783.78	0.3551E+02	0.5550E+02	0.3614E+02
25	18.11	1848.85	3935.63	1852.63	0.2784E+02	0.4852E+02	0.2735E+02
27	20.30	1909.87	4035.96	1906.41	0.2386E+02	0.4325E+02	0.2226E+02
29	22.84	1958.81	4139.65	1959.08	0.1925E+02	0.3853E+02	0.1963E+02
31	25.38	2007.75	4232.73	2007.92	0.1925E+02	0.3483E+02	0.1901E+02
33	27.92	2056.69	4317.35	2056.48	0.1925E+02	0.3184E+02	0.1927E+02

V -101

Connection type : Extended end-plate connections
Mode : All bolted without column stiffener

Tested by : J.B.Davison et al. (1987) United Kingdom
Test Id. : JT/13B

Column : UC 152x152x23
Beam : UB 254x102x22
Plate thickness : 15 mm

Fasteners: G8.8- -M16
23/32" Oversize holes
Material : 43A
Fy = -- ksi
Fu = -- ksi

Major parameters

End plate extended on tension side only.

ct = 1.5748"	pt = 3.9370"	li = 6.2992"	pc = 1.6142"
cc = 0.3543"	gt = 2.9921"	gc = 2.9921"	
tp = 0.5906"	nt = 2 X 2	nc = 2 X 1	

Remark 1) Slip occurred at a relatively high moment
2)

No	Moment (k-in)	Rotation (radians) X 1/1000
1	0.00	0.00
2	22.12	0.38
3	36.87	0.58
4	51.62	0.58
5	73.75	0.87
6	95.87	1.35
7	110.62	1.54
8	147.49	1.73
9	169.62	1.92
10	184.37	2.12
11	221.24	2.69
12	235.99	2.88
13	258.11	3.08
14	276.55	3.17
15	311.58	3.61
16	346.61	4.04
17	376.10	4.33
18	405.60	4.62
19	442.48	5.19
20	457.23	5.58
21	475.66	6.15
22	494.10	6.73
23	499.63	7.40
24	505.16	8.08
25	507.00	8.85
26	508.85	9.62
27	494.10	10.19

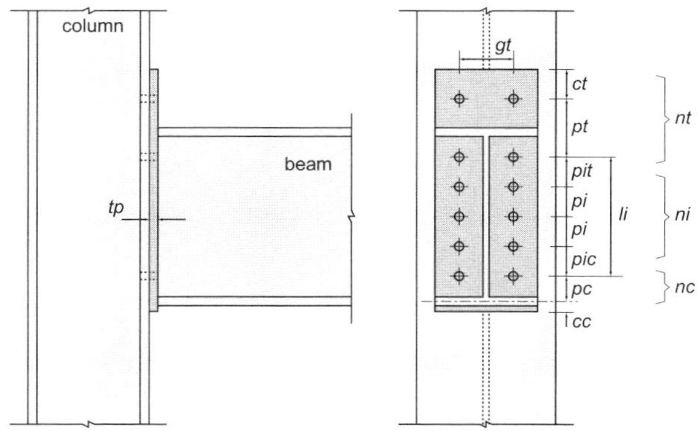

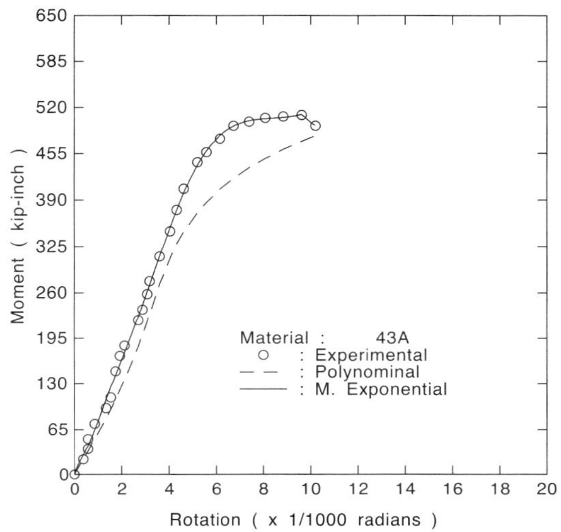

A.5 – 202

```
Moment-rotation prediction equations     ( R : X 1/1000 radians )

Frye and Morris polynominal model  :    R = Sum ( Ai X ( K*Bm )**Pi X 10**Qi )
     xd = 10.236200"    t =  0.590600"  tf =  0.629900"
     A1 =  1.830000   A2 = -1.040000   A3 =  6.380000    K =  0.009295
     P1 =        1    P2 =        3    P3 =        5    Q1 =        0    Q2 =       -1    Q3 =       -3

Modified exponential model  :
     Bm = Sum ( Ai X ( 1 - exp( - R/( 2*i*C ) ) ) )  +     Sum ( RKj X ( R - Rj0 ) )  +  Bm0
       C =  0.18108250E+00              Bmo=   0.00000000E+00         Nexp=  6        Nliner=  2
      AI =  0.52417211E+03  -0.12026646E+05   0.81543008E+05  -0.22066330E+06   0.25394053E+06  -0.10380330E+06
     Rj0 =          0.0000          9.6200
     RKj =  0.75899391E+02  -0.38153377E+02
```

No	Rotation (radians) X 1/1000	Moment (kip-inch) Expri.	Poly.	M.Expo.	Connection stiffness (kip-inch) X 1000 Expri.	Poly.	M.Expo.
1	0.00	0.00	0.00	0.00	0.5752E+02	0.5879E+02	0.3751E+02
2	0.38	22.12	22.67	28.40	0.7031E+02	0.5924E+02	0.6453E+02
3	0.58	36.87	34.11	41.70	0.1467E+05	0.5980E+02	0.7437E+02
4	0.58	51.62	34.11	41.70	0.1470E+05	0.5980E+02	0.7437E+02
5	0.87	73.75	51.54	65.12	0.6519E+02	0.6113E+02	0.8649E+02
6	1.35	95.87	81.72	107.92	0.6794E+02	0.6479E+02	0.8891E+02
7	1.54	110.62	94.38	124.87	0.1340E+03	0.6687E+02	0.8736E+02
8	1.73	147.49	107.50	141.57	0.1532E+03	0.6940E+02	0.8611E+02
9	1.92	169.62	121.10	158.05	0.9599E+02	0.7241E+02	0.8571E+02
10	2.12	184.37	135.35	174.56	0.7353E+02	0.7597E+02	0.8625E+02
11	2.69	221.24	183.01	225.69	0.7363E+02	0.9008E+02	0.9150E+02
12	2.88	235.99	200.82	243.43	0.9579E+02	0.9519E+02	0.9333E+02
13	3.08	258.11	219.59	261.55	0.1661E+03	0.9926E+02	0.9468E+02
14	3.17	276.55	229.14	270.68	0.1715E+03	0.1004E+03	0.9511E+02
15	3.61	311.58	272.06	311.83	0.8096E+02	0.9443E+02	0.9437E+02
16	4.04	346.61	309.02	351.65	0.9375E+02	0.7538E+02	0.8895E+02
17	4.33	376.10	328.92	376.49	0.1023E+03	0.6326E+02	0.8300E+02
18	4.62	405.60	345.71	399.40	0.8947E+02	0.5354E+02	0.7561E+02
19	5.19	442.48	372.52	438.14	0.4858E+02	0.4022E+02	0.5833E+02
20	5.58	457.23	386.74	458.26	0.3579E+02	0.3443E+02	0.4638E+02
21	6.15	475.66	404.73	480.14	0.3196E+02	0.2829E+02	0.2991E+02
22	6.73	494.10	419.75	493.43	0.2100E+02	0.2406E+02	0.1681E+02
23	7.40	499.63	434.69	501.05	0.8218E+01	0.2054E+02	0.6818E+01
24	8.08	505.16	447.61	503.87	0.5501E+01	0.1797E+02	0.2432E+01
25	8.85	507.00	460.54	505.67	0.2396E+01	0.1577E+02	0.3105E+01
26	9.62	508.85	471.99	509.75	-0.1358E+02	0.1408E+02	0.8065E+01
27	10.19	494.10	479.81	494.10	-0.2557E+02	0.1305E+02	-0.2465E+02

V -102

```
Connection type : Extended end-plate connections
           Mode : All bolted without column stiffener

Tested by   :  D.B.Moore & P.A.C.Sims (1986)              United Kingdom
Test  Id.   :  J3

Column : UC 152x152x23                Fasteners: 10.9- -M16
Beam   : UB 254x102x22                     23/32"  Oversize holes
Plate thickness : 15 mm               Material :  --
                                          Fy =    --   ksi
                                          Fu =    --   ksi

                              Major parameters

     End plate extended on tension side only.
     ct =  1.5748"    pt =  3.9370"    li =  5.3149"    pc =  2.7166"
     cc =  0.4330"    gt =  3.4252"    gc =  3.4252"
     tp =  0.5906"    nt = 2 X  2      nc = 2 X  1

Remark  1)  Faiulre of column flange at 221.24 lb in
        2)  Web flange fracture followed by bolt failure at 677 lb in
```

No	Moment (k-in)	Rotation (radians) X 1/1000
1	0.00	0.00
2	44.25	0.38
3	88.50	0.75
4	132.74	1.50
5	176.99	1.88
6	212.39	2.25
7	238.19	3.75
8	295.00	4.50
9	324.51	7.50
10	339.27	9.38
11	354.03	12.75
12	371.73	15.00
13	389.42	18.75
14	413.01	22.50
15	436.61	26.25
16	460.20	30.00
17	483.80	33.75
18	507.39	37.50
19	525.09	41.25
20	548.68	45.00
21	572.27	48.75
22	595.87	52.50
23	625.36	56.25
24	643.06	60.00
25	660.75	63.75
26	678.45	67.50
27	696.14	72.00

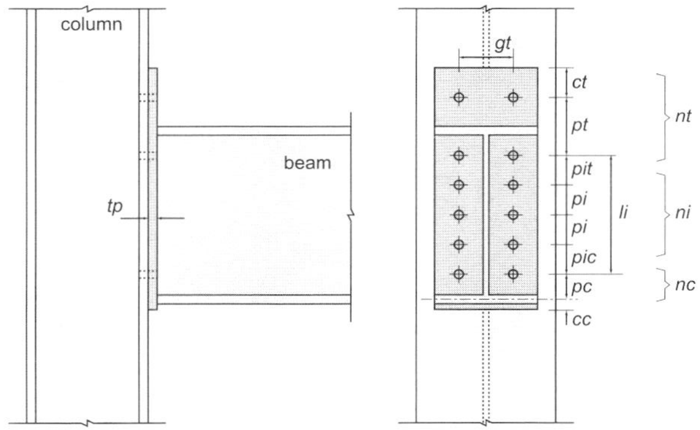

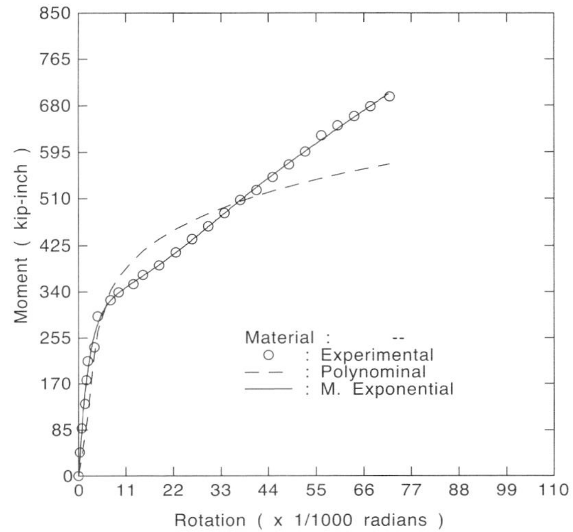

A.5 – 204

```
Moment-rotation prediction equations     ( R : X 1/1000 radians )

Frye and Morris polynominal model  :    R = Sum ( Ai X ( K*Bm )**Pi X 10**Qi )
     xd =  9.251900"   t =  0.590600"  tf =  0.629900"
     A1 =  1.830000   A2 = -1.040000   A3 =  6.380000    K =  0.011847
     P1 =         1   P2 =         3   P3 =         5   Q1 =         0    Q2 =        -1    Q3 =        -3

Modified exponential model  :
     Bm = Sum ( Ai X ( 1 - exp( - R/( 2*i*C ) ) ) )  +     Sum ( RKj X ( R - Rj0 ) )  +  Bm0
       C =   0.69375000E+00           Bmo=   0.00000000E+00          Nexp=  6          Nliner=  1
      AI =  -0.11133370E+03   0.18936344E+04  -0.73484438E+04   0.16750003E+05  -0.17950995E+05   0.70949723E+04
     Rj0 =           0.0000
     RKj =   0.52114503E+01
```

No	Rotation (radians) X 1/1000	Moment (kip-inch) Expri.	Moment (kip-inch) Poly.	Moment (kip-inch) M.Expo.	Connection stiffness (kip-inch) X 1000 Expri.	Connection stiffness (kip-inch) X 1000 Poly.	Connection stiffness (kip-inch) X 1000 M.Expo.
1	0.00	0.00	0.00	0.00	0.1180E+03	0.4612E+02	0.1195E+03
2	0.38	44.25	17.34	44.18	0.1180E+03	0.4646E+02	0.1110E+03
3	0.75	88.50	34.93	83.32	0.9833E+02	0.4749E+02	0.9792E+02
4	1.50	132.74	72.03	147.81	0.9833E+02	0.5211E+02	0.7480E+02
5	1.88	176.99	92.29	173.97	0.1062E+03	0.5617E+02	0.6492E+02
6	2.25	212.39	114.37	196.65	0.7896E+02	0.6185E+02	0.5618E+02
7	3.75	238.19	223.83	260.29	0.5623E+02	0.6956E+02	0.3110E+02
8	4.50	295.00	266.22	280.53	0.6256E+02	0.4483E+02	0.2328E+02
9	7.50	324.51	342.59	324.08	0.8628E+01	0.1579E+02	0.9053E+01
10	9.38	339.27	367.60	338.34	0.6621E+01	0.1142E+02	0.6559E+01
11	12.75	354.03	399.20	357.91	0.6468E+01	0.7807E+01	0.5405E+01
12	15.00	371.73	415.23	369.96	0.6685E+01	0.6512E+01	0.5351E+01
13	18.75	389.42	436.92	390.42	0.5505E+01	0.5156E+01	0.5601E+01
14	22.50	413.01	454.53	412.11	0.6291E+01	0.4305E+01	0.5966E+01
15	26.25	436.61	469.50	435.08	0.6292E+01	0.3716E+01	0.6263E+01
16	30.00	460.20	482.58	458.90	0.6292E+01	0.3282E+01	0.6412E+01
17	33.75	483.80	494.24	482.98	0.6292E+01	0.2948E+01	0.6413E+01
18	37.50	507.39	504.77	506.86	0.5506E+01	0.2682E+01	0.6308E+01
19	41.25	525.09	514.41	530.22	0.5505E+01	0.2465E+01	0.6147E+01
20	45.00	548.68	523.30	552.94	0.6291E+01	0.2283E+01	0.5970E+01
21	48.75	572.27	531.57	575.01	0.6292E+01	0.2130E+01	0.5803E+01
22	52.50	595.87	539.30	596.49	0.7078E+01	0.1998E+01	0.5658E+01
23	56.25	625.36	546.57	617.48	0.6292E+01	0.1883E+01	0.5541E+01
24	60.00	643.06	553.44	638.08	0.4719E+01	0.1782E+01	0.5449E+01
25	63.75	660.75	559.95	658.38	0.4719E+01	0.1692E+01	0.5380E+01
26	67.50	678.45	566.14	678.45	0.4361E+01	0.1612E+01	0.5330E+01
27	72.00	696.14	573.20	702.34	0.3932E+01	0.1527E+01	0.5287E+01

V -103

Connection type : Extended end-plate connections
Mode : All bolted without column stiffener

Tested by : D.B.Moore & P.A.C.Sims (1986) United Kingdom
Test Id. : J4

Column : UC 152x152x23
Beam : UB 254x102x22
Plate thickness : 15 mm

Fasteners: 10.9- -M16
23/32" Oversize holes
Material : --
Fy = -- ksi
Fu = -- ksi

Major parameters

End plate extended on tension side only.

ct = 1.5748" pt = 3.9370" li = 5.3149" pc = 2.7166"
cc = 0.4330" gt = 3.4252" gc = 3.4252"
tp = 0.5906" nt = 2 X 2 nc = 2 X 1

Remark 1) Tested with backing plates
2)

No	Moment (k-in)	Rotation (radians) X 1/1000
1	0.00	0.00
2	44.25	0.30
3	88.50	0.60
4	132.74	0.90
5	176.99	1.35
6	212.39	1.50
7	265.49	1.88
8	283.19	3.00
9	303.86	3.75
10	324.51	4.80
11	339.27	5.62
12	354.03	6.90
13	365.83	7.50
14	401.22	9.00
15	413.01	9.75
16	472.00	12.75
17	501.49	15.00
18	530.99	16.50
19	560.48	18.75
20	581.12	20.62
21	601.77	22.50
22	619.46	24.38
23	637.16	26.25
24	648.96	28.12
25	660.75	30.00
26	672.05	31.88
27	683.35	33.75

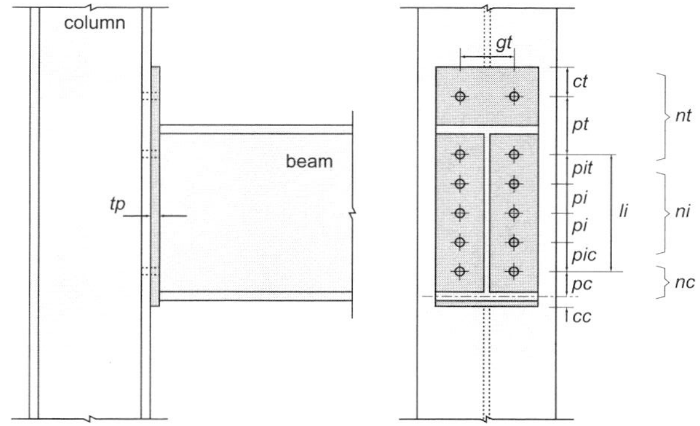

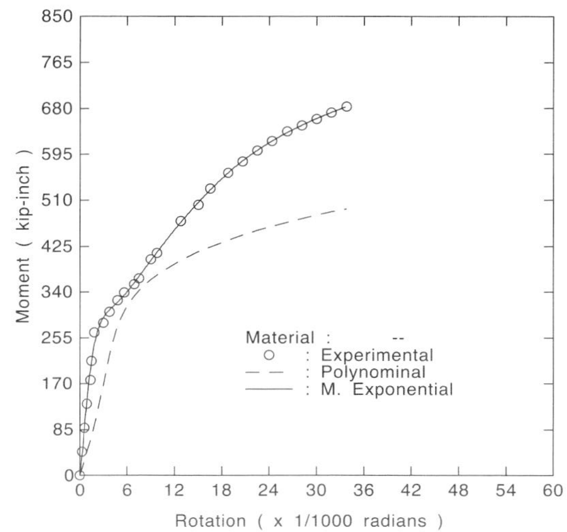

```
Moment-rotation prediction equations     ( R : X 1/1000 radians )

Frye and Morris polynominal model  :    R = Sum ( Ai X ( K*Bm )**Pi X 10**Qi )
     xd =  9.251900"   t =  0.590600"  tf =  0.629900"
     A1 =  1.830000   A2 = -1.040000   A3 =  6.380000    K =  0.011847
     P1 =         1   P2 =         3   P3 =         5   Q1 =         0    Q2 =        -1    Q3 =        -3

Modified exponential model  :
     Bm = Sum ( Ai X ( 1 - exp( - R/( 2*i*C ) ) ) )  +      Sum ( RKj X ( R - Rj0 ) )  + Bm0
      C =  0.35625000E+00          Bmo=  0.00000000E+00         Nexp=  6        Nliner=  1
     AI = -0.77092845E+03  0.67536770E+04 -0.24399180E+05  0.50085842E+05 -0.52244852E+05  0.21106132E+05
    Rj0 =          0.0000
    RKj =  0.46426962E+01
```

No	Rotation (radians) X 1/1000	Moment (kip-inch) Expri.	Moment (kip-inch) Poly.	Moment (kip-inch) M.Expo.	Connection stiffness (kip-inch) X 1000 Expri.	Connection stiffness (kip-inch) X 1000 Poly.	Connection stiffness (kip-inch) X 1000 M.Expo.
1	0.00	0.00	0.00	0.00	0.1475E+03	0.4612E+02	0.8846E+02
2	0.30	44.25	13.86	38.75	0.1475E+03	0.4634E+02	0.1539E+03
3	0.60	88.50	27.85	87.36	0.1475E+03	0.4699E+02	0.1644E+03
4	0.90	132.74	42.10	134.97	0.1278E+03	0.4811E+02	0.1504E+03
5	1.35	176.99	64.31	194.72	0.2016E+03	0.5086E+02	0.1139E+03
6	1.50	212.39	72.03	210.86	0.2090E+03	0.5211E+02	0.1013E+03
7	1.88	265.49	92.29	243.31	0.1101E+03	0.5617E+02	0.7274E+02
8	3.00	283.19	166.14	292.79	0.2283E+02	0.7673E+02	0.2413E+02
9	3.75	303.86	223.69	306.63	0.2427E+02	0.6963E+02	0.1479E+02
10	4.80	324.51	278.63	321.21	0.1867E+02	0.3805E+02	0.1430E+02
11	5.62	339.27	304.71	333.95	0.1541E+02	0.2653E+02	0.1666E+02
12	6.90	354.03	332.44	357.16	0.1707E+02	0.1809E+02	0.1940E+02
13	7.50	365.83	342.57	368.99	0.2079E+02	0.1579E+02	0.1994E+02
14	9.00	401.22	363.26	399.03	0.1835E+02	0.1207E+02	0.1986E+02
15	9.75	413.01	371.78	413.78	0.1652E+02	0.1084E+02	0.1943E+02
16	12.75	472.00	399.20	468.89	0.1592E+02	0.7807E+01	0.1731E+02
17	15.00	501.49	415.23	506.08	0.1704E+02	0.6512E+01	0.1573E+02
18	16.50	530.99	424.50	528.84	0.1704E+02	0.5884E+01	0.1461E+02
19	18.75	560.48	436.88	559.71	0.1196E+02	0.5158E+01	0.1282E+02
20	20.62	581.12	446.09	582.34	0.1101E+02	0.4689E+01	0.1133E+02
21	22.50	601.77	454.51	602.27	0.1022E+02	0.4306E+01	0.9948E+01
22	24.38	619.46	462.28	619.75	0.9438E+01	0.3987E+01	0.8738E+01
23	26.25	637.16	469.49	635.16	0.7865E+01	0.3716E+01	0.7730E+01
24	28.12	648.96	476.24	648.87	0.6292E+01	0.3484E+01	0.6924E+01
25	30.00	660.75	482.58	661.24	0.6158E+01	0.3282E+01	0.6300E+01
26	31.88	672.05	488.56	672.59	0.6025E+01	0.3105E+01	0.5828E+01
27	33.75	683.35	494.24	683.17	0.6025E+01	0.2948E+01	0.5481E+01

V -104

Connection type : Extended end-plate connections
Mode : All bolted with column stiffener

Tested by : R.Zandonini & P.Zanon (1987) Italy
Test Id. : EP 1-1

Column : ----
Beam : IPE 300
Plate thickness : 12 mm

Fasteners: 10.9- -M20
7/8" Oversize holes
Material : --
Fy = -- ksi
Fu = -- ksi

Major parameters

End plate extended on tension side only.
ct = 1.9685" pt = 4.7244" li = 7.0866" pc = 2.3622"
cc = 0.3937" gt = 4.1339" gc = 4.1339"
tp = 0.4724" nt = 2 X 2 nc = 2 X 1

Remark 1) The specimen was connected to a counter beam with negligible deformation
2) End-plate extended on both tension & compression sides

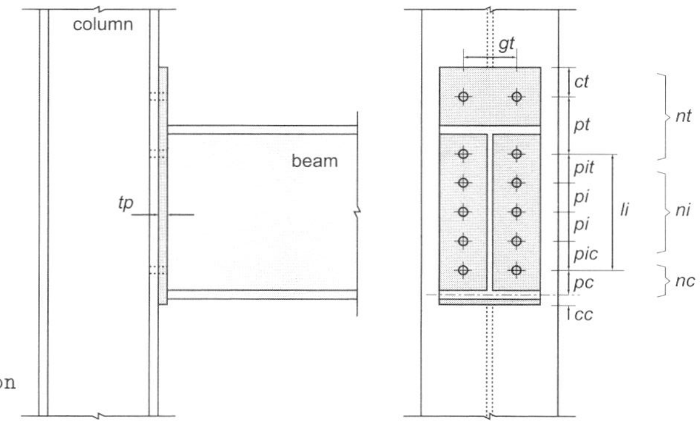

No	Moment (k-in)	Rotation (radians) X 1/1000	No	Moment (k-in)	Rotation (radians) X 1/1000
1	0.00	0.00	26	1465.01	43.09
2	74.37	0.28	27	1487.32	45.53
3	148.73	0.48	28	1502.19	48.78
4	223.10	0.72	29	1524.50	52.84
5	297.46	0.98	30	1546.81	56.91
6	371.83	1.36	31	1561.69	60.98
7	446.20	1.76	32	1576.56	65.04
8	520.56	2.20	33	1579.54	69.11
9	594.93	2.68	34	1582.51	73.17
10	669.29	3.15	35	1594.41	77.23
11	743.66	3.70	36	1606.31	81.30
12	788.28	4.55	37	1606.31	85.36
13	840.34	5.74	38	1606.31	89.43
14	892.39	6.95			
15	966.76	8.46			
16	1041.12	10.13			
17	1115.49	12.20			
18	1189.86	16.26			
19	1219.60	18.70			
20	1249.35	21.14			
21	1293.97	24.39			
22	1338.59	28.45			
23	1368.33	32.52			
24	1412.95	36.59			
25	1442.70	40.65			

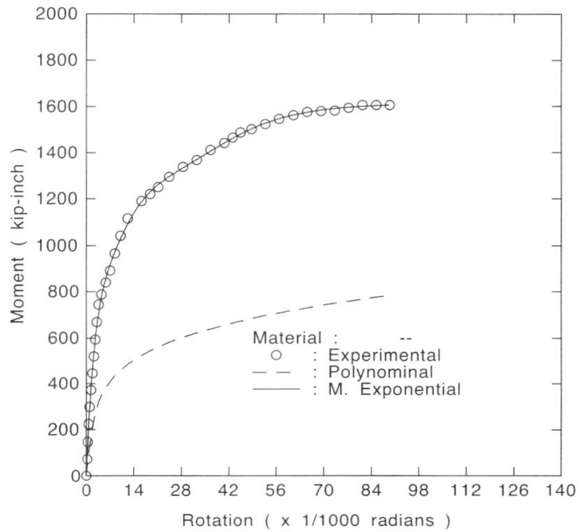

```
Moment-rotation prediction equations      ( R : X 1/1000 radians )

Frye and Morris polynominal model  :     R = Sum ( Ai X ( K*Bm )**Pi X 10**Qi )
     xd = 11.811000"   t =  0.472400"
     A1 =  1.790000   A2 =  1.760000   A3 =  2.040000    K =  0.004187
     P1 =         1   P2 =         3   P3 =         5    Q1 =         0    Q2 =       -1    Q3 =       -1

Modified exponential model  :
     Bm = Sum ( Ai X ( 1 - exp( - R/( 2*i*C ) ) ) )  +     Sum ( RKj X ( R - Rj0 ) )  +  Bm0
       C =   0.81525000E+00              Bmo=   0.00000000E+00           Nexp=  6        Nliner=  1
      AI =  -0.47084596E+03   0.12513638E+05  -0.62322283E+05   0.14185367E+06  -0.14752577E+06   0.57545170E+05
     Rj0 =          0.0000
     RKj =   0.17760766E+00
```

No	Rotation (radians) X 1/1000	Moment (kip-inch) Expri.	Moment (kip-inch) Poly.	Moment (kip-inch) M.Expo.	Connection stiffness (kip-inch) X 1000 Expri.	Connection stiffness (kip-inch) X 1000 Poly.	Connection stiffness (kip-inch) X 1000 M.Expo.
1	0.00	0.00	0.00	0.00	0.2656E+03	0.1334E+03	0.3441E+03
3	0.48	148.73	63.56	152.80	0.3437E+03	0.1303E+03	0.2933E+03
5	0.98	297.46	126.17	287.43	0.2493E+03	0.1184E+03	0.2462E+03
7	1.76	446.20	206.52	454.87	0.1779E+03	0.8668E+02	0.1857E+03
9	2.68	594.93	271.49	599.94	0.1566E+03	0.5719E+02	0.1328E+03
11	3.70	743.66	319.91	713.80	0.1027E+03	0.3966E+02	0.9336E+02
13	5.74	840.34	382.56	856.85	0.4328E+02	0.2422E+02	0.5347E+02
15	8.46	966.76	435.86	974.66	0.4701E+02	0.1607E+02	0.3650E+02
17	12.20	1115.49	485.46	1093.03	0.3005E+02	0.1117E+02	0.2753E+02
19	18.70	1219.60	544.24	1228.05	0.1220E+02	0.7454E+01	0.1482E+02
21	24.39	1293.97	581.70	1295.46	0.1250E+02	0.5850E+01	0.9891E+01
23	32.52	1368.33	623.38	1370.66	0.9147E+01	0.4529E+01	0.9093E+01
25	40.65	1442.70	656.70	1442.82	0.8461E+01	0.3726E+01	0.8405E+01
27	45.53	1487.32	673.99	1481.32	0.7187E+01	0.3378E+01	0.7318E+01
29	52.84	1524.50	697.15	1527.64	0.5488E+01	0.2972E+01	0.5327E+01
31	60.98	1561.69	719.87	1562.58	0.3659E+01	0.2631E+01	0.3357E+01
33	69.11	1579.54	740.13	1583.85	0.7318E+00	0.2366E+01	0.1976E+01
35	77.23	1594.41	758.47	1596.18	0.2927E+01	0.2154E+01	0.1134E+01
37	85.36	1606.31	775.26	1603.29	0.5043E-03	0.1980E+01	0.6642E+00

A.5 – 209

V -105

Connection type : Extended end-plate connections
Mode : All bolted with column stiffener

Tested by : R.Zandonini & P.Zanon (1987) Italy
Test Id. : EP 1-2

Column : ---
Beam : IPE 300
Plate thickness : 15 mm

Fasteners: 10.9- -M20
7/8" Oversize holes
Material : --
Fy = -- ksi
Fu = -- ksi

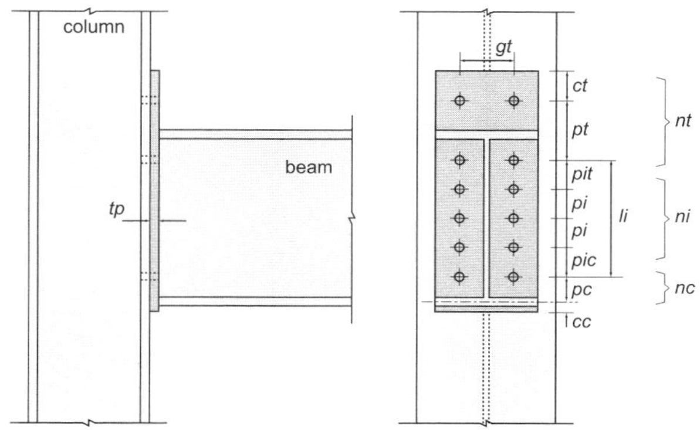

Major parameters

End plate extended on tension side only.
ct = 1.9685" pt = 4.7244" li = 7.0866" pc = 2.3622"
cc = 0.3937" gt = 4.1339" gc = 4.1339"
tp = 0.5906" nt = 2 X 2 nc = 2 X 1

Remark 1) The specimen was connected to a counter beam with negligible deformability
2) End-plate extended on tension side only

No	Moment (k-in)	Rotation (radians) X 1/1000	No	Moment (k-in)	Rotation (radians) X 1/1000
1	0.00	0.00	26	1576.56	22.36
2	81.13	0.14	27	1591.43	24.39
3	162.25	0.29	28	1613.74	26.02
4	243.38	0.45	29	1636.05	27.64
5	324.51	0.62	30	1662.83	30.08
6	405.63	0.80	31	1689.60	32.52
7	486.76	0.99	32	1707.44	34.96
8	567.89	1.19	33	1725.29	37.40
9	649.01	1.40	34	1728.27	39.02
10	730.14	1.63	35	1731.24	40.65
11	811.27	1.86	36	1743.14	42.68
12	892.39	2.09	37	1755.04	44.72
13	966.76	2.35	38	1758.02	46.75
14	1041.12	2.60	39	1760.99	48.78
15	1085.74	3.27	40	1754.05	51.49
16	1130.36	3.95	41	1747.10	54.20
17	1189.86	4.88	42	1740.16	56.91
18	1249.35	6.50	43	1725.29	58.95
19	1308.84	8.13	44	1710.42	60.98
20	1353.47	9.76			
21	1398.10	11.38			
22	1427.84	13.82			
23	1457.57	16.26			
24	1487.32	17.89			
25	1561.69	20.32			

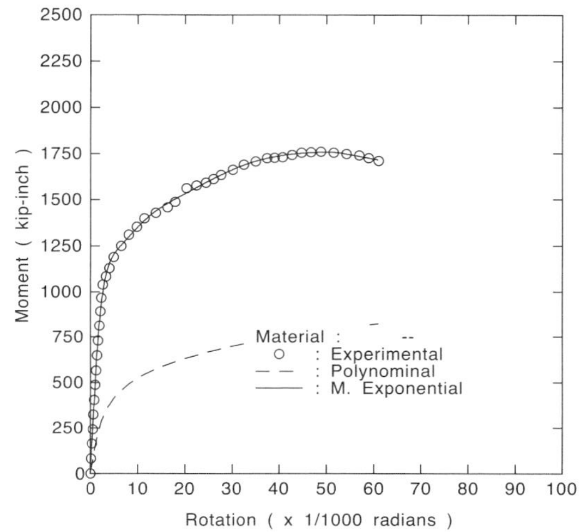

```
Moment-rotation prediction equations     ( R : X 1/1000 radians )

Frye and Morris polynominal model  :    R = Sum ( Ai X ( K*Bm )**Pi X 10**Qi )
     xd = 11.811000"    t =  0.590600"
     A1 =  1.790000   A2 =  1.760000   A3 =  2.040000    K =  0.003662
     P1 =        1    P2 =        3    P3 =        5    Q1 =        0    Q2 =       -1    Q3 =       -1

Modified exponential model  :
     Bm = Sum ( Ai X ( 1 - exp( - R/( 2*i*C ) ) ) )  +     Sum ( RKj X ( R - Rj0 ) )  +  Bm0
      C =   0.54316667E+00             Bmo=  0.00000000E+00          Nexp=  6        Nliner=  3
     AI =  -0.13155022E+04   0.21109539E+05  -0.87033819E+05   0.18152881E+06  -0.18190643E+06   0.69610626E+05
    Rj0 =           0.0000          0.8100          51.4900
    RKj =  -0.16763051E+03   0.16592072E+03  -0.38258506E+01
```

No	Rotation (radians) X 1/1000	Moment (kip-inch) Expri.	Moment (kip-inch) Poly.	Moment (kip-inch) M.Expo.	Connection stiffness (kip-inch) X 1000 Expri.	Connection stiffness (kip-inch) X 1000 Poly.	Connection stiffness (kip-inch) X 1000 M.Expo.
1	0.00	0.00	0.00	0.00	0.5795E+03	0.1525E+03	0.7647E+03
3	0.29	162.25	44.12	162.14	0.5245E+03	0.1513E+03	0.5198E+03
5	0.62	324.51	93.36	318.36	0.4643E+03	0.1464E+03	0.4268E+03
7	0.99	486.76	145.61	487.17	0.4166E+03	0.1350E+03	0.4913E+03
9	1.40	649.01	197.31	667.27	0.3737E+03	0.1165E+03	0.3896E+03
11	1.86	811.27	245.70	823.31	0.3468E+03	0.9461E+02	0.2935E+03
13	2.35	966.76	286.90	945.70	0.2922E+03	0.7552E+02	0.2128E+03
15	3.27	1085.74	344.77	1090.67	0.6613E+02	0.5227E+02	0.1121E+03
17	4.88	1189.86	411.33	1201.71	0.5421E+02	0.3314E+02	0.4154E+02
19	8.13	1308.84	492.09	1299.54	0.3202E+02	0.1915E+02	0.2765E+02
21	11.38	1398.10	544.32	1385.39	0.2135E+02	0.1366E+02	0.2386E+02
23	16.26	1457.57	600.12	1478.62	0.1585E+02	0.9705E+01	0.1550E+02
25	20.32	1561.69	635.60	1537.34	0.1785E+02	0.7894E+01	0.1393E+02
27	24.39	1591.43	665.09	1593.20	0.1087E+02	0.6689E+01	0.1345E+02
29	27.64	1636.05	685.64	1635.21	0.1262E+02	0.5980E+01	0.1224E+02
31	32.52	1689.60	712.75	1687.87	0.9146E+01	0.5178E+01	0.9182E+01
33	37.40	1725.29	736.50	1724.34	0.4023E+01	0.4581E+01	0.5813E+01
35	40.65	1731.24	750.85	1739.99	0.3619E+01	0.4260E+01	0.3875E+01
37	44.72	1755.04	767.63	1751.67	0.3659E+01	0.3920E+01	0.1973E+01
39	48.78	1760.99	782.96	1756.79	-0.2616E+00	0.3637E+01	0.6275E+00
41	54.20	1747.10	801.67	1746.49	-0.2562E+01	0.3325E+01	-0.4317E+01
43	58.95	1725.29	816.98	1724.58	-0.7307E+01	0.3094E+01	-0.4869E+01

V -106

Connection type : Extended end-plate connections
Mode : All bolted with column stiffener

Tested by : R.Zandonini & P.Zanon (1987) Italy
Test Id. : EP 1-3

Column : ----
Beam : IPE 300
Plate thickness : 18 mm

Fasteners: 10.9- -M20
7/8" Oversize holes
Material : --
Fy = -- ksi
Fu = -- ksi

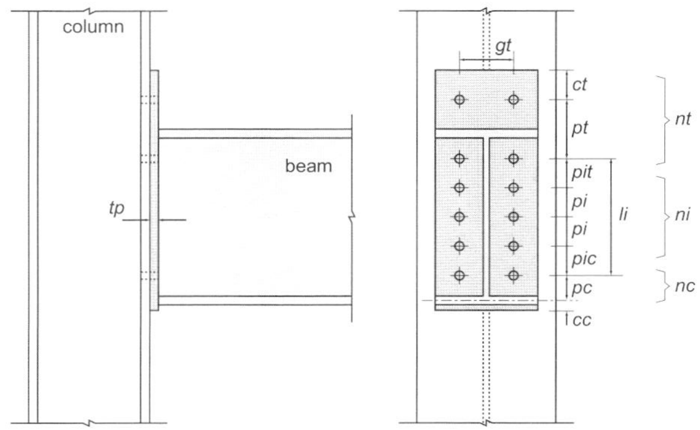

Major parameters

End plate extended on tension side only.
ct = 1.9685" pt = 4.7244" li = 7.0866" pc = 2.3622"
cc = 0.3937" gt = 4.1339" gc = 4.1339"
tp = 0.7087" nt = 2 X 2 nc = 2 X 1

Remark 1) The specimen was connected to a counter beam with negligible deformability
2) End-plate extended on tension side only

No	Moment (k-in)	Rotation (radians) X 1/1000	No	Moment (k-in)	Rotation (radians) X 1/1000
1	0.00	0.00	26	1685.63	14.63
2	74.37	0.06	27	1710.42	16.26
3	148.73	0.13	28	1732.73	17.43
4	223.10	0.21	29	1755.04	18.60
5	297.46	0.30	30	1769.91	20.28
6	371.83	0.40	31	1784.78	21.95
7	446.20	0.51	32	1792.22	23.17
8	520.56	0.63	33	1799.66	24.39
9	594.93	0.77	34	1804.61	25.75
10	669.29	0.92	35	1809.55	27.10
11	743.66	1.08	36	1814.50	28.45
12	818.03	1.26	37	1824.43	29.81
13	892.39	1.46	38	1834.35	31.16
14	966.76	1.68	39	1844.28	32.52
15	1041.12	1.93	40	1844.28	34.15
16	1115.49	2.21	41	1844.28	35.77
17	1189.86	2.52			
18	1264.22	2.96			
19	1338.59	3.25			
20	1412.95	4.88			
21	1487.32	6.50			
22	1561.69	8.13			
23	1598.87	9.76			
24	1636.05	11.38			
25	1660.84	13.01			

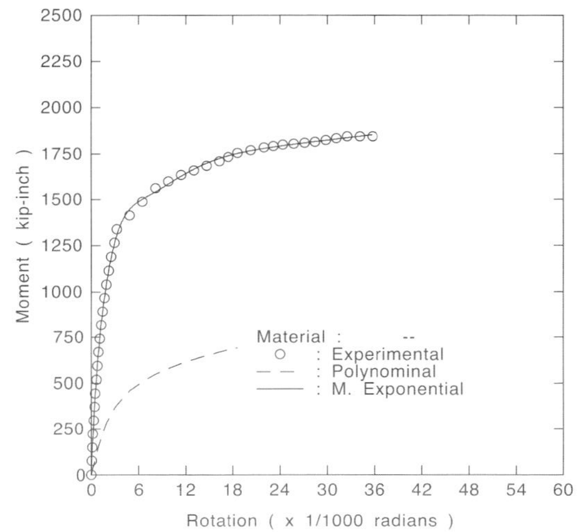

```
Moment-rotation prediction equations      ( R : X 1/1000 radians )

Frye and Morris polynominal model  :    R = Sum ( Ai X ( K*Bm )**Pi X 10**Qi )
     xd = 11.811000"   t =  0.708700"
     A1 =  1.790000   A2 =  1.760000   A3 =  2.040000    K =  0.003283
     P1 =         1   P2 =         3   P3 =         5    Q1 =         0    Q2 =        -1    Q3 =        -1

Modified exponential model  :
     Bm = Sum ( Ai X ( 1 - exp( - R/( 2*i*C ) ) ) )  +     Sum ( RKj X ( R - Rj0 ) )  +  Bm0
       C =   0.31310000E+00              Bmo=   0.00000000E+00          Nexp=  6        Nliner=  1
      AI =   0.12195463E+04  -0.10701683E+05   0.40177713E+05  -0.59637181E+05   0.38155160E+05  -0.75278263E+04
     Rj0 =          2.9600
     RKj =   0.51273088E+01
```

No	Rotation (radians) X 1/1000	Moment (kip-inch) Expri.	Moment (kip-inch) Poly.	Moment (kip-inch) M.Expo.	Connection stiffness (kip-inch) X 1000 Expri.	Connection stiffness (kip-inch) X 1000 Poly.	Connection stiffness (kip-inch) X 1000 M.Expo.
1	0.00	0.00	0.00	0.00	0.1239E+04	0.1702E+03	0.1163E+04
3	0.13	148.73	22.12	139.32	0.1000E+04	0.1699E+03	0.9873E+03
5	0.30	297.46	50.91	291.58	0.7871E+03	0.1687E+03	0.8132E+03
7	0.51	446.20	86.05	445.31	0.6491E+03	0.1657E+03	0.6604E+03
9	0.77	594.93	128.34	599.32	0.5141E+03	0.1590E+03	0.5331E+03
11	1.08	743.66	175.81	748.10	0.4405E+03	0.1465E+03	0.4337E+03
13	1.46	892.39	227.82	896.32	0.3557E+03	0.1267E+03	0.3514E+03
15	1.93	1041.12	281.46	1043.22	0.2824E+03	0.1021E+03	0.2770E+03
17	2.52	1189.86	334.18	1184.07	0.2106E+03	0.7806E+02	0.2032E+03
19	3.25	1338.59	383.64	1306.85	0.2229E+03	0.5866E+02	0.1369E+03
21	6.50	1487.32	509.95	1501.20	0.4574E+02	0.2701E+02	0.2582E+02
23	9.76	1598.87	580.54	1584.50	0.2287E+02	0.1775E+02	0.2696E+02
25	13.01	1660.84	630.43	1666.63	0.1525E+02	0.1338E+02	0.2206E+02
27	16.26	1710.42	669.47	1724.81	0.1747E+02	0.1083E+02	0.1397E+02
29	18.60	1755.04	693.25	1752.37	0.1488E+02	0.9554E+01	0.9883E+01
31	21.95	1784.78	722.88	1779.51	0.7269E+01	0.8209E+01	0.6763E+01
33	24.39	1799.66	741.96	1794.66	0.4940E+01	0.7463E+01	0.5777E+01
35	27.10	1809.55	761.24	1809.58	0.3651E+01	0.6790E+01	0.5308E+01
37	29.81	1824.43	779.05	1823.70	0.7325E+01	0.6233E+01	0.5144E+01
39	32.52	1844.28	795.29	1837.56	0.3996E+01	0.5772E+01	0.5100E+01
41	35.77	1844.28	813.33	1854.14	0.0000E+00	0.5308E+01	0.5099E+01

V -107

Connection type : Extended end-plate connections
Mode : All bolted with column stiffener

Tested by : R.Zandonini & P.Zanon (1987) Italy
Test Id. : EP 1-4

Column : ----
Beam : IPE 300
Plate thickness : 22 mm

Fasteners: 10.9- -M20
7/8" Oversize holes
Material : --
Fy = -- ksi
Fu = -- ksi

Major parameters

End plate extended on tension side only.

ct = 1.9685" pt = 4.7244" li = 7.0866" pc = 2.3622"
cc = 0.3937" gt = 4.1339" gc = 4.1339"
tp = 0.3661" nt = 2 X 2 nc = 2 X 1

Remark 1) The specimen was connected to a counter beam with negligible deformability
2) End-plate extended on tension side only

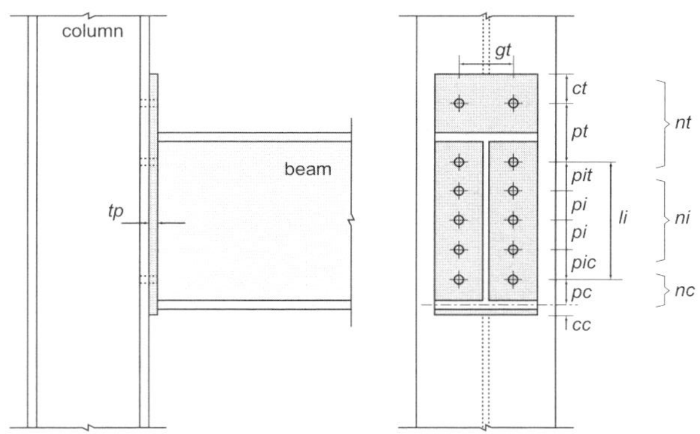

No	Moment (k-in)	Rotation (radians) X 1/1000	No	Moment (k-in)	Rotation (radians) X 1/1000
1	0.00	0.00	26	1814.53	8.94
2	94.65	0.04	27	1844.28	9.76
3	189.30	0.10	28	1866.59	10.98
4	283.94	0.18	29	1888.90	12.20
5	378.59	0.28	30	1911.21	13.41
6	473.24	0.40	31	1933.52	14.63
7	567.89	0.54	32	1936.49	15.45
8	662.53	0.60	33	1939.46	16.26
9	757.18	0.78	34	1949.13	17.48
10	851.83	0.98	35	1958.80	18.70
11	946.48	1.20	36	1968.47	19.92
12	1041.12	1.44	37	1978.14	21.14
13	1115.49	1.70	38	1973.89	22.53
14	1189.86	1.98	39	1969.64	23.93
15	1264.22	2.28	40	1965.39	25.32
16	1338.59	2.60	41	1961.14	26.71
17	1412.95	2.94	42	1956.89	28.11
18	1487.32	3.30	43	1952.64	29.50
19	1561.69	3.68	44	1948.39	30.89
20	1598.87	4.07			
21	1636.05	4.88			
22	1680.67	5.69			
23	1725.29	6.50			
24	1755.04	7.32			
25	1784.78	8.13			

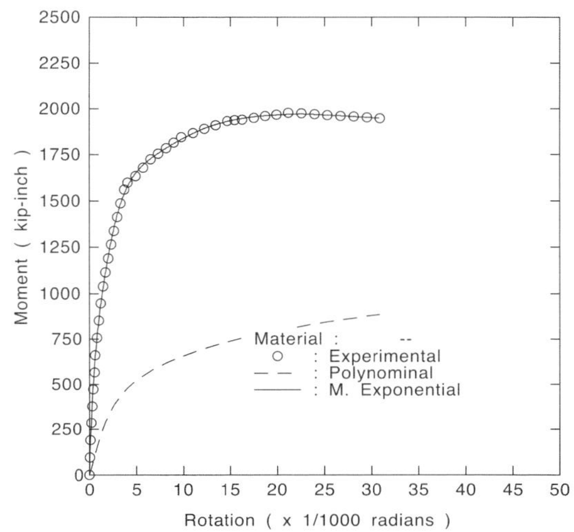

```
Moment-rotation prediction equations     ( R : X 1/1000 radians )

Frye and Morris polynominal model  :    R = Sum ( Ai X ( K*Bm )**Pi X 10**Qi )
     xd = 11.811000"   t =  0.866100"
     A1 =  1.790000   A2 =  1.760000   A3 =  2.040000    K =  0.002911
     P1 =        1    P2 =        3    P3 =        5    Q1 =        0    Q2 =       -1    Q3 =       -1

Modified exponential model  :
     Bm = Sum ( Ai X ( 1 - exp( - R/( 2*i*C ) ) ) )  +      Sum ( RKj X ( R - Rj0 ) )  +  Bm0
      C =   0.26745000E+00             Bmo=   0.00000000E+00          Nexp=  6       Nliner=  1
     AI =   0.19526412E+04  -0.11099948E+05   0.29878519E+05  -0.30693535E+05   0.79804635E+04   0.39580254E+04
    Rj0 =           22.5300
    RKj =  -0.32842818E+01
```

No	Rotation (radians) X 1/1000	Moment (kip-inch) Expri.	Moment (kip-inch) Poly.	Moment (kip-inch) M.Expo.	Connection stiffness (kip-inch) X 1000 Expri.	Connection stiffness (kip-inch) X 1000 Poly.	Connection stiffness (kip-inch) X 1000 M.Expo.
1	0.00	0.00	0.00	0.00	0.2366E+04	0.1919E+03	0.1766E+04
3	0.10	189.30	19.19	160.28	0.1408E+04	0.1918E+03	0.1452E+04
5	0.28	378.59	53.61	382.50	0.8748E+03	0.1905E+03	0.1046E+04
7	0.54	567.89	102.66	604.88	0.1307E+04	0.1862E+03	0.7009E+03
9	0.78	757.18	146.54	750.52	0.5009E+03	0.1790E+03	0.5295E+03
11	1.20	946.48	217.74	938.78	0.4131E+03	0.1586E+03	0.3886E+03
13	1.70	1115.49	289.49	1112.64	0.2762E+03	0.1283E+03	0.3143E+03
15	2.28	1264.22	354.63	1276.78	0.2404E+03	0.9788E+02	0.2528E+03
17	2.94	1412.95	410.85	1421.91	0.2128E+03	0.7420E+02	0.1879E+03
19	3.68	1561.69	459.06	1537.41	0.1465E+03	0.5742E+02	0.1270E+03
21	4.88	1636.05	517.51	1648.74	0.5031E+02	0.4171E+02	0.6647E+02
23	6.50	1725.29	575.17	1728.20	0.4574E+02	0.3046E+02	0.3806E+02
25	8.13	1784.78	619.10	1784.02	0.3659E+02	0.2410E+02	0.3190E+02
27	9.76	1844.28	654.75	1832.95	0.2927E+02	0.2002E+02	0.2809E+02
29	12.20	1888.90	698.38	1891.99	0.1830E+02	0.1607E+02	0.2001E+02
31	14.63	1933.52	734.22	1930.79	0.9512E+01	0.1349E+02	0.1212E+02
33	16.26	1939.46	755.08	1947.12	0.5365E+01	0.1221E+02	0.8161E+01
35	18.70	1958.80	782.96	1961.81	0.7928E+01	0.1072E+02	0.4245E+01
37	21.14	1978.14	807.67	1969.27	0.2805E+01	0.9582E+01	0.2104E+01
39	23.93	1969.64	832.90	1968.67	-0.3049E+01	0.8564E+01	-0.2375E+01
41	26.71	1961.14	855.61	1961.21	-0.3049E+01	0.7757E+01	-0.2901E+01
43	29.50	1952.64	876.29	1952.77	-0.3049E+01	0.7101E+01	-0.3125E+01

A.5 – 215

V -108

```
Connection type : Extended end-plate connections
           Mode : All bolted with column stiffener

Tested by  :  R.Zandonini & P.Zanon (1987)                 Italy
Test  Id.  :  EP 1-5

Column : ----                        Fasteners: 10.9- -M20
Beam   : IPE 300                             7/8" Oversize holes
Plate thickness : 25 mm              Material :  --
                                         Fy =    --   ksi
                                         Fu =    --   ksi

                         Major parameters

     End plate extended on tension side only.
     ct =  1.9685"    pt =  4.7244"    li =  7.0866"    pc =  2.3622"
     cc =  0.3937"    gt =  4.1339"    gc =  4.1339"
     tp =  0.9843"    nt = 2 X  2      nc = 2 X  1

Remark  1)  The specimen was connected to a counter beam with negligible deformability
        2)  End-plate extended on tension side only
```

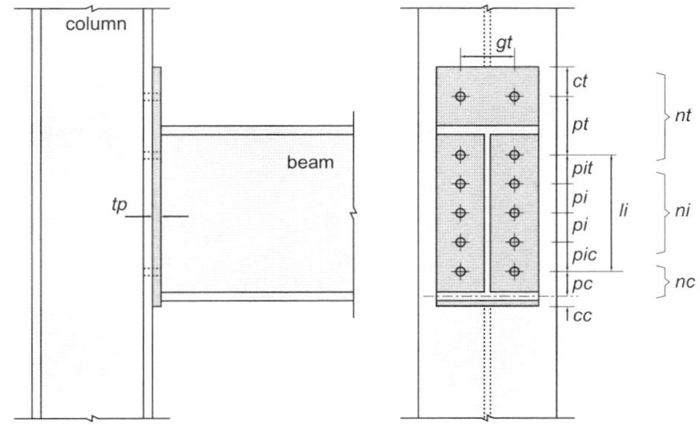

No	Moment (k-in)	Rotation (radians) X 1/1000	No	Moment (k-in)	Rotation (radians) X 1/1000
1	0.00	0.00	26	1888.90	10.57
2	94.65	0.03	27	1911.21	11.79
3	189.30	0.07	28	1933.52	13.01
4	283.94	0.12	29	1948.39	14.09
5	378.59	0.18	30	1963.27	15.18
6	473.24	0.25	31	1978.14	16.26
7	567.89	0.33	32	1987.06	17.56
8	662.53	0.42	33	1995.99	18.86
9	757.18	0.52	34	2004.91	20.16
10	851.83	0.63	35	2013.84	21.46
11	946.48	0.85	36	2022.76	22.76
12	1041.12	0.93	37	2019.79	23.58
13	1115.49	1.06	38	2016.81	24.39
14	1189.86	1.21	39	2009.67	25.69
15	1264.22	1.36	40	2002.53	26.99
16	1338.59	1.63	41	1995.39	28.29
17	1412.95	2.03	42	1988.25	29.59
18	1487.32	2.44	43	1981.11	30.89
19	1561.69	3.25			
20	1636.05	3.58			
21	1710.42	5.04			
22	1784.78	6.50			
23	1814.53	7.32			
24	1844.28	8.13			
25	1866.59	9.35			

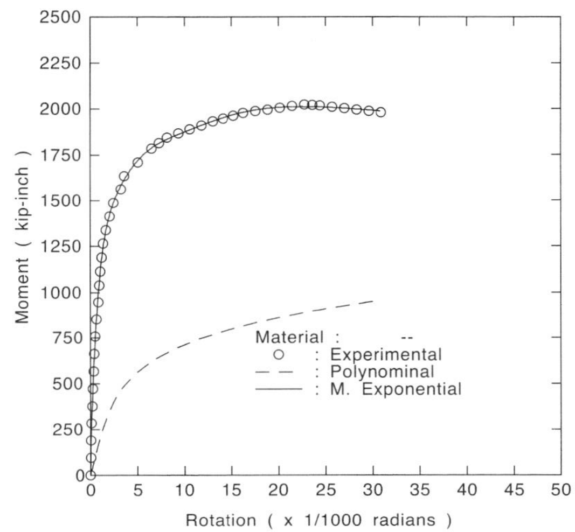

```
Moment-rotation prediction equations     ( R : X 1/1000 radians )

Frye and Morris polynominal model  :    R = Sum ( Ai X ( K*Bm )**Pi X 10**Qi )
     xd = 11.811000"   t =  0.984300"
     A1 =  1.790000   A2 =  1.760000   A3 =  2.040000    K =  0.002695
     P1 =        1    P2 =        3    P3 =        5    Q1 =        0    Q2 =       -1    Q3 =       -1

Modified exponential model  :
     Bm = Sum ( Ai X ( 1 - exp( - R/( 2*i*C ) ) ) )  +     Sum ( RKj X ( R - Rj0 ) )  +  Bm0
       C =  0.26497500E+00           Bmo=  0.00000000E+00          Nexp=  6        Nliner=  3
      AI = -0.59847312E+03   0.13627606E+05  -0.60136345E+05   0.13072404E+06  -0.13287468E+06   0.51277147E+05
     Rj0 =          0.0000           0.0010           0.1000
     RKj =  0.15675475E+05  -0.14617133E+05  -0.10632191E+04
```

No	Rotation (radians) X 1/1000	Moment (kip-inch) Expri.	Moment (kip-inch) Poly.	Moment (kip-inch) M.Expo.	Connection stiffness (kip-inch) X 1000 Expri.	Connection stiffness (kip-inch) X 1000 Poly.	Connection stiffness (kip-inch) X 1000 M.Expo.
1	0.00	0.00	0.00	0.00	0.3144E+04	0.2073E+03	0.1551E+04
3	0.07	189.30	14.55	194.29	0.2151E+04	0.2072E+03	0.2510E+04
5	0.18	378.59	37.37	377.08	0.1471E+04	0.2066E+03	0.1297E+04
7	0.33	567.89	68.26	557.61	0.1121E+04	0.2051E+03	0.1111E+04
9	0.52	757.18	106.96	748.92	0.9046E+03	0.2015E+03	0.9060E+03
11	0.85	946.48	171.83	999.54	0.9811E+03	0.1902E+03	0.6277E+03
13	1.06	1115.49	211.09	1118.10	0.5439E+03	0.1794E+03	0.4951E+03
15	1.36	1264.22	260.98	1241.97	0.4160E+03	0.1612E+03	0.3600E+03
17	2.03	1412.95	355.27	1419.46	0.1829E+03	0.1185E+03	0.1902E+03
19	3.25	1561.69	467.21	1582.40	0.1896E+03	0.7144E+02	0.9974E+02
21	5.04	1710.42	565.94	1720.53	0.5081E+02	0.4343E+02	0.5839E+02
23	7.32	1814.53	646.16	1814.38	0.3659E+02	0.2903E+02	0.2851E+02
25	9.35	1866.59	698.03	1862.62	0.1829E+02	0.2256E+02	0.2096E+02
27	11.79	1911.21	746.96	1911.32	0.1829E+02	0.1793E+02	0.1909E+02
29	14.09	1948.39	784.78	1951.93	0.1373E+02	0.1510E+02	0.1576E+02
31	16.26	1978.14	815.32	1981.11	0.1061E+02	0.1319E+02	0.1105E+02
33	18.86	1995.99	847.31	2002.37	0.6860E+01	0.1149E+02	0.5449E+01
35	21.46	2013.84	875.45	2010.68	0.6860E+01	0.1020E+02	0.1202E+01
37	23.58	2019.79	896.23	2010.60	-0.3659E+01	0.9366E+01	-0.1122E+01
39	25.69	2009.67	915.16	2006.51	-0.5489E+01	0.8674E+01	-0.2636E+01
41	28.29	1995.39	936.76	1998.08	-0.5489E+01	0.7958E+01	-0.3732E+01
43	30.89	1981.11	956.67	1987.53	-0.5489E+01	0.7361E+01	-0.4310E+01

V -109

```
Connection type : Extended end-plate connections
           Mode : All bolted with column stiffener

Tested by   :  R.Zandonini & P.Zanon (1987)                  Italy
Test  Id.   :  EPB 1-2

Column : ----                        Fasteners: 10.9- -M20
Beam   : IPE 300                             7/8" Oversize holes
Plate thickness : 15 mm              Material :  --
                                         Fy =    --   ksi
                                         Fu =    --   ksi
```

Major parameters

```
End plate extended on both sides.
ct =  1.9685"    pt =  4.7244"    li =  7.0866"    pc =  4.7244"
cc =  1.9685"    gt =  4.1339"    gc =  4.1339"
tp =  0.5906"    nt = 2 X  2      nc = 2 X  2
```

Remark 1) End-plate extended on both tension & compression sides
2)

No	Moment (k-in)	Rotation (radians) X 1/1000	No	Moment (k-in)	Rotation (radians) X 1/1000
1	0.00	0.00	26	1665.80	21.82
2	84.99	0.02	27	1695.55	24.24
3	169.98	0.06	28	1710.42	26.26
4	254.97	0.14	29	1725.29	28.28
5	339.96	0.24	30	1732.72	30.30
6	424.95	0.36	31	1740.16	32.32
7	509.94	0.50	32	1755.03	35.01
8	594.93	0.66	33	1769.91	37.71
9	669.29	0.84	34	1784.78	40.40
10	743.66	1.05	35	1784.78	43.09
11	818.03	1.26	36	1784.78	45.79
12	892.39	1.62	37	1784.78	48.48
13	966.76	2.02	38	1774.87	51.17
14	1041.12	2.42	39	1764.95	53.87
15	1115.49	2.83	40	1755.04	56.56
16	1189.86	3.23			
17	1264.22	4.44			
18	1338.59	5.66			
19	1418.90	8.08			
20	1453.11	9.70			
21	1487.32	11.31			
22	1524.50	13.74			
23	1561.69	16.16			
24	1598.87	17.78			
25	1636.05	19.39			

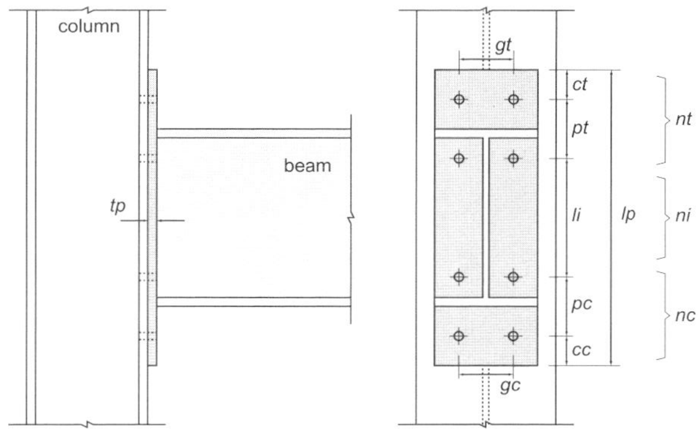

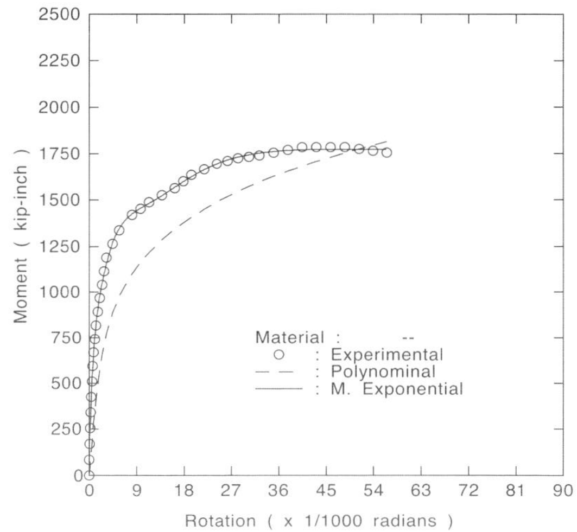

```
Moment-rotation prediction equations     ( R : X 1/1000 radians )

Frye and Morris polynominal model   :    R = Sum ( Ai X ( K*Bm )**Pi X 10**Qi )
     xd = 16.535400"   t =  0.590600"
     A1 =  1.790000   A2 =  1.760000   A3 =  2.040000    K =  0.001633
     P1 =         1   P2 =         3   P3 =         5   Q1 =         0   Q2 =        -1   Q3 =        -1

Modified exponential model  :
     Bm = Sum ( Ai X ( 1 - exp( - R/( 2*i*C ) ) ) ) +     Sum ( RKj X ( R - Rj0 ) ) + Bm0
       C =  0.47635833E+00             Bmo=  0.00000000E+00          Nexp=  6        Nliner=  3
      AI =  0.19627880E+04  -0.92608889E+04  0.22682472E+05  -0.14286554E+05  -0.12503519E+05   0.13004920E+05
     Rj0 =          0.0000           0.0010          4.4400
     RKj =  0.10192511E+06  -0.10190452E+06  -0.20980010E+02
```

No	Rotation (radians) X 1/1000	Moment (kip-inch) Expri.	Moment (kip-inch) Poly.	Moment (kip-inch) M.Expo.	Connection stiffness (kip-inch) X 1000 Expri.	Connection stiffness (kip-inch) X 1000 Poly.	Connection stiffness (kip-inch) X 1000 M.Expo.
1	0.00	0.00	0.00	0.00	0.4228E+04	0.3421E+03	0.1037E+04
3	0.06	169.98	20.63	163.48	0.1762E+04	0.3420E+03	0.9853E+03
5	0.24	339.96	82.23	323.68	0.7843E+03	0.3402E+03	0.7989E+03
7	0.50	509.94	169.94	503.75	0.5710E+03	0.3334E+03	0.5962E+03
9	0.84	669.29	281.59	675.26	0.3864E+03	0.3144E+03	0.4158E+03
11	1.26	818.03	405.36	818.73	0.2957E+03	0.2761E+03	0.2838E+03
13	2.02	966.76	583.90	984.53	0.1841E+03	0.1967E+03	0.1723E+03
15	2.83	1115.49	717.07	1103.08	0.1841E+03	0.1381E+03	0.1278E+03
17	4.44	1264.22	888.21	1275.35	0.6136E+02	0.8255E+02	0.6842E+02
19	8.08	1418.90	1101.16	1424.85	0.2595E+02	0.4322E+02	0.2208E+02
21	11.31	1487.32	1218.42	1480.97	0.1884E+02	0.3081E+02	0.1600E+02
23	16.16	1561.69	1343.51	1566.91	0.1994E+02	0.2189E+02	0.1882E+02
25	19.39	1636.05	1408.38	1625.59	0.1871E+02	0.1848E+02	0.1700E+02
27	24.24	1695.55	1489.13	1694.74	0.9594E+01	0.1508E+02	0.1134E+02
29	28.28	1725.29	1545.94	1731.40	0.5521E+01	0.1314E+02	0.6996E+01
31	32.32	1740.16	1595.93	1753.04	0.4470E+01	0.1167E+02	0.3935E+01
33	37.71	1769.91	1654.64	1767.18	0.5522E+01	0.1020E+02	0.1587E+01
35	43.09	1784.78	1706.44	1772.33	0.1366E-13	0.9082E+01	0.4709E+00
37	48.48	1784.78	1752.91	1773.35	-0.1840E+01	0.8205E+01	-0.2554E-01
39	53.87	1764.95	1795.14	1772.56	-0.3681E+01	0.7496E+01	-0.2377E+00

V -110

```
Connection type : Extended end-plate connections
           Mode : All bolted with column stiffener

Tested by   :  R.Zandonini & P.Zanon (1987)                  Italy
Test  Id.   :  EPB 1-3

Column : ----                           Fasteners: 10.9- -M20
Beam   : IPE 300                                7/8" Oversize holes
Plate thickness : 18 mm                 Material :  --
                                            Fy =    --   ksi
                                            Fu =    --   ksi

                          Major parameters

     End plate extended on both sides.
     ct =  1.9685"    pt =  4.7244"    li =  7.0866"     pc =  4.7244"
     cc =  1.9685"    gt =  4.1339"    gc =  4.1339"
     tp =  0.7087"    nt = 2 X  2      nc = 2 X  2

Remark  1)  End-plate extended on both tension & compression sides
        2)
```

No	Moment (k-in)	Rotation (radians) X 1/1000	No	Moment (k-in)	Rotation (radians) X 1/1000
1	0.00	0.00	26	1636.05	9.70
2	84.99	0.13	27	1660.84	10.77
3	169.98	0.15	28	1685.63	11.85
4	254.97	0.17	29	1710.42	12.93
5	339.96	0.19	30	1731.24	13.90
6	424.95	0.23	31	1752.06	14.87
7	509.94	0.28	32	1772.89	15.84
8	594.93	0.33	33	1793.71	16.81
9	669.29	0.50	34	1814.53	17.78
10	743.66	0.65	35	1808.58	18.75
11	818.03	0.75	36	1802.63	19.72
12	892.39	0.89	37	1796.68	20.68
13	966.76	1.04	38	1790.73	21.65
14	1041.12	1.22	39	1784.78	22.62
15	1115.49	1.42			
16	1189.86	1.62			
17	1264.22	2.42			
18	1338.59	3.23			
19	1412.95	4.04			
20	1487.32	4.85			
21	1512.11	5.66			
22	1536.90	6.46			
23	1561.69	7.27			
24	1586.47	8.08			
25	1611.26	8.89			

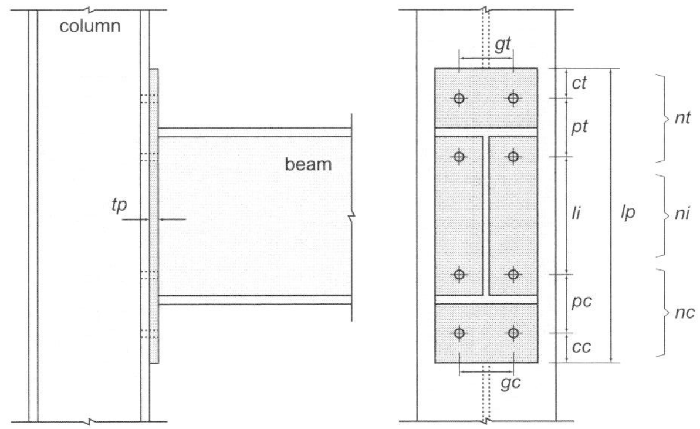

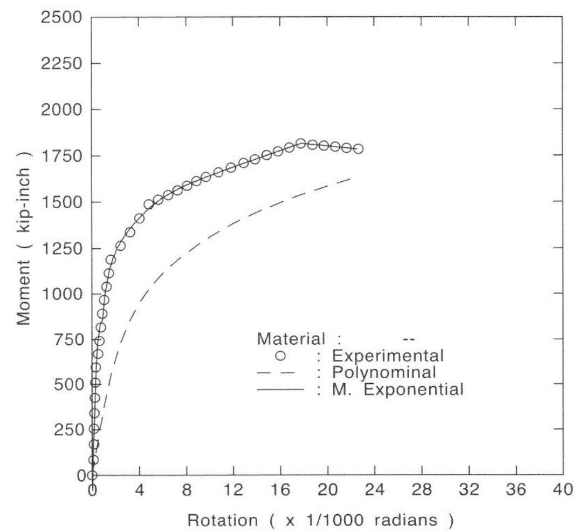

```
Moment-rotation prediction equations      ( R : X 1/1000 radians )

Frye and Morris polynominal model  :    R = Sum ( Ai X ( K*Bm )**Pi X 10**Qi )
     xd = 16.535400"    t =  0.708700"
     A1 =  1.790000   A2 =  1.760000   A3 =  2.040000    K =  0.001464
     P1 =         1   P2 =         3   P3 =         5   Q1 =         0    Q2 =        -1    Q3 =        -1

Modified exponential model  :
     Bm = Sum ( Ai X ( 1 - exp( - R/( 2*i*C ) ) ) )  +     Sum ( RKj X ( R - Rj0 ) )  +  Bm0
       C =   0.22105833E+00            Bmo=   0.00000000E+00          Nexp=  6        Nliner=  4
      AI =   0.13994857E+04   0.40835793E+04  -0.12968799E+05   0.19675116E+05  -0.14254777E+05   0.44636387E+04
     Rj0 =           0.0000           0.0010           0.8800          17.8000
     RKj =  -0.23731739E+06   0.23651831E+06   0.82027693E+03  -0.27948323E+02
```

No	Rotation (radians) X 1/1000	Moment (kip-inch) Expri.	Moment (kip-inch) Poly.	Moment (kip-inch) M.Expo.	Connection stiffness (kip-inch) X 1000 Expri.	Connection stiffness (kip-inch) X 1000 Poly.	Connection stiffness (kip-inch) X 1000 M.Expo.
1	0.00	0.00	0.00	0.00	0.6533E+03	0.3816E+03	0.4366E+04
3	0.15	169.98	55.44	195.51	0.5004E+04	0.3809E+03	0.2438E+04
5	0.19	339.96	72.69	299.33	0.2889E+04	0.3803E+03	0.2151E+04
7	0.28	509.94	106.85	469.72	0.1703E+04	0.3787E+03	0.1656E+04
9	0.50	669.29	190.92	732.62	0.4772E+03	0.3718E+03	0.7611E+03
11	0.75	818.03	280.21	839.13	0.6503E+03	0.3581E+03	0.1615E+03
13	1.04	966.76	381.15	951.77	0.4568E+03	0.3327E+03	0.5702E+03
15	1.42	1115.49	500.43	1111.55	0.3764E+03	0.2886E+03	0.2994E+03
17	2.42	1264.22	732.17	1282.87	0.9204E+02	0.1825E+03	0.1058E+03
19	4.04	1412.95	951.84	1415.09	0.9204E+02	0.1026E+03	0.6738E+02
21	5.66	1512.11	1088.22	1506.13	0.3068E+02	0.7042E+02	0.4606E+02
23	7.27	1561.69	1187.31	1568.12	0.3068E+02	0.5380E+02	0.3214E+02
25	8.89	1611.26	1265.52	1614.11	0.3068E+02	0.4373E+02	0.2574E+02
27	10.77	1660.84	1340.25	1659.71	0.2301E+02	0.3607E+02	0.2316E+02
29	12.93	1710.42	1411.23	1708.54	0.2220E+02	0.3019E+02	0.2233E+02
31	14.87	1752.06	1465.93	1751.51	0.2147E+02	0.2642E+02	0.2201E+02
33	16.81	1793.71	1514.26	1793.94	0.2147E+02	0.2354E+02	0.2176E+02
35	18.75	1808.58	1557.62	1809.51	-0.6136E+01	0.2127E+02	-0.6382E+01
37	20.68	1796.68	1597.03	1796.99	-0.6136E+01	0.1943E+02	-0.6524E+01
39	22.62	1784.78	1633.19	1784.24	-0.6136E+01	0.1791E+02	-0.6618E+01

V -111

Connection type : Extended end-plate connections
Mode : All bolted with column stiffener

Tested by : R.Zandonini & P.Zanon (1987) Italy
Test Id. : EPB 1-4

Column : ----
Beam : IPE 300
Plate thickness : 22 mm

Fasteners: 10.9- -M20
7/8" Oversize holes
Material : --
Fy = -- ksi
Fu = -- ksi

Major parameters

End plate extended on both sides.

ct = 1.9685"	pt = 4.7244"	li = 7.0866"	pc = 4.7244"
cc = 1.9685"	gt = 4.1339"	gc = 4.1339"	
tp = 0.8661"	nt = 2 X 2	nc = 2 X 2	

Remark 1) End-plate extended on both tension & compression sides
2)

No	Moment (k-in)	Rotation (radians) X 1/1000	No	Moment (k-in)	Rotation (radians) X 1/1000
1	0.00	0.00	26	1710.42	5.25
2	89.24	0.02	27	1747.60	5.86
3	178.48	0.05	28	1784.78	6.46
4	267.72	0.06	29	1808.58	7.11
5	356.96	0.09	30	1832.38	7.76
6	446.20	0.12	31	1856.18	8.40
7	535.44	0.15	32	1879.97	9.05
8	624.67	0.18	33	1903.77	9.70
9	713.91	0.22	34	1911.21	10.30
10	803.15	0.26	35	1918.64	10.91
11	892.39	0.30	36	1926.08	11.51
12	965.76	0.35	37	1933.52	12.12
13	1041.12	0.40	38	1928.56	12.66
14	1115.49	0.49	39	1923.60	13.20
15	1159.86	0.58	40	1918.64	13.74
16	1214.22	0.75			
17	1280.59	0.97			
18	1363.17	1.45			
19	1437.74	1.94			
20	1487.32	2.42			
21	1524.50	2.83			
22	1561.69	3.23			
23	1593.87	3.64			
24	1635.05	4.04			
25	1673.23	4.65			

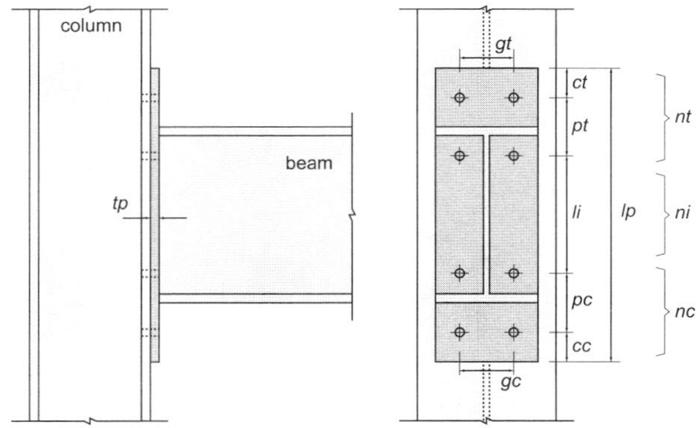

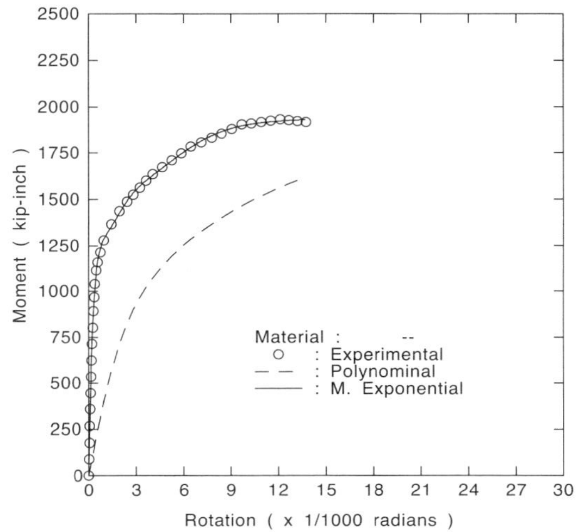

```
Moment-rotation prediction equations     ( R : X 1/1000 radians )

Frye and Morris polynominal model  :    R = Sum ( Ai X ( K*Bm )**Pi X 10**Qi )
     xd = 16.535400"   t =  0.866100"
     A1 =  1.790000   A2 =  1.760000   A3 =  2.040000    K =  0.001298
     P1 =        1    P2 =        3    P3 =        5    Q1 =        0    Q2 =      -1    Q3 =      -1

Modified exponential model  :
     Bm = Sum ( Ai X ( 1 - exp( - R/( 2*i*C ) ) ) )  +     Sum ( RKj X ( R - Rj0 ) )  +  Bm0
      C =   0.12021667E+00            Bmo=   0.00000000E+00         Nexp=  6       Nliner=  2
     AI =   0.10877717E+02   0.15772581E+05  -0.74966989E+05   0.16354001E+06  -0.16662334E+06   0.64208168E+05
    Rj0 =           0.0000           0.0010
    RKj =  -0.22203113E+05   0.22204129E+05
```

No	Rotation (radians) X 1/1000	Moment (kip-inch) Expri.	Moment (kip-inch) Poly.	Moment (kip-inch) M.Expo.	Connection stiffness (kip-inch) X 1000 Expri.	Connection stiffness (kip-inch) X 1000 Poly.	Connection stiffness (kip-inch) X 1000 M.Expo.
1	0.00	0.00	0.00	0.00	0.3880E+04	0.4304E+03	0.4866E+04
3	0.05	178.48	21.09	197.23	0.8248E+04	0.4303E+03	0.4109E+04
5	0.09	356.96	38.73	354.20	0.2709E+04	0.4301E+03	0.3560E+04
7	0.15	535.44	64.79	548.35	0.3268E+04	0.4295E+03	0.2870E+04
9	0.22	713.91	94.06	721.88	0.2174E+04	0.4285E+03	0.2242E+04
11	0.30	892.39	130.53	885.83	0.1847E+04	0.4266E+03	0.1636E+04
13	0.40	1041.12	171.68	1018.32	0.1249E+04	0.4236E+03	0.1136E+04
15	0.58	1159.86	248.40	1167.52	0.4171E+03	0.4151E+03	0.5648E+03
17	0.97	1280.59	403.04	1287.85	0.2605E+03	0.3831E+03	0.1721E+03
19	1.94	1437.74	714.44	1426.22	0.1229E+03	0.2570E+03	0.1422E+03
21	2.83	1524.50	902.26	1534.42	0.9204E+02	0.1738E+03	0.9760E+02
23	3.64	1598.87	1023.69	1600.32	0.9203E+02	0.1305E+03	0.7036E+02
25	4.65	1673.23	1138.08	1667.39	0.6136E+02	0.9879E+02	0.6520E+02
27	5.86	1747.60	1243.08	1745.33	0.6136E+02	0.7645E+02	0.6179E+02
29	7.11	1808.58	1329.22	1814.90	0.3681E+02	0.6214E+02	0.4806E+02
31	8.40	1856.18	1402.75	1865.92	0.3681E+02	0.5223E+02	0.3116E+02
33	9.70	1903.77	1465.48	1897.19	0.2415E+02	0.4518E+02	0.1806E+02
35	10.91	1918.64	1517.09	1913.96	0.1227E+02	0.4018E+02	0.1027E+02
37	12.12	1933.52	1563.31	1923.42	0.9014E+00	0.3625E+02	0.5778E+01
39	13.20	1923.60	1600.77	1928.34	-0.9208E+01	0.3338E+02	0.3570E+01

V -112

```
Connection type : Extended end-plate connections
           Mode : All bolted with column stiffener

Tested by   :  R.Zandonini & P.Zanon (1987)               Italy
Test  Id.   :  EPB 1-5

Column : ----                        Fasteners: 10.9- -M20
Beam   : IPE 300                             7/8" Oversize holes
Plate thickness : 25 mm              Material :  --
                                          Fy =    --   ksi
                                          Fu =    --   ksi
```

Major parameters

```
End plate extended on both sides.
ct =  1.9685"    pt =  4.7244"    li =  7.0866"     pc =  4.7244"
cc =  1.9685"    gt =  4.1339"    gc =  4.1339"
tp =  0.9843"    nt = 2 X  2      nc = 2 X  2
```

Remark 1) End-plate extended on both tension & compression sides
2)

No	Moment (k-in)	Rotation (radians) X 1/1000
1	0.00	0.00
2	99.15	0.09
3	198.29	0.12
4	297.44	0.16
5	396.58	0.18
6	495.73	0.21
7	594.87	0.24
8	694.02	0.28
9	793.17	0.33
10	892.31	0.36
11	966.72	0.40
12	1041.12	0.44
13	1140.28	0.53
14	1239.43	0.72
15	1338.59	0.91
16	1412.95	1.21
17	1487.32	1.82
18	1561.68	2.42
19	1636.05	3.23
20	1673.23	4.04
21	1710.42	4.85
22	1784.78	5.66
23	1824.44	6.46
24	1864.11	7.27
25	1903.77	8.08

No	Moment (k-in)	Rotation (radians) X 1/1000
26	1916.52	9.00
27	1929.27	9.93
28	1942.02	10.85
29	1954.76	11.77
30	1967.51	12.70
31	1980.26	13.62
32	1993.01	14.54
33	1998.96	15.51
34	2004.91	16.48
35	2010.86	17.45
36	2016.81	18.42
37	2022.76	19.39

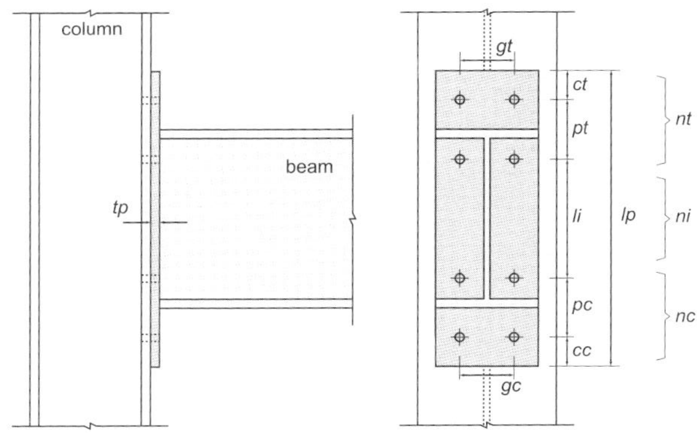

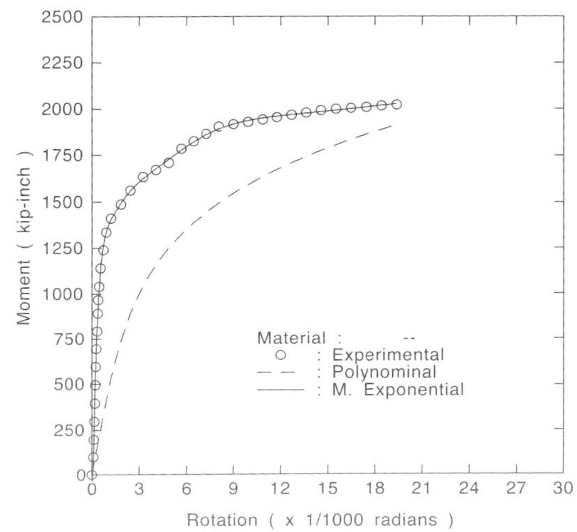

```
Moment-rotation prediction equations      ( R : X 1/1000 radians )

Frye and Morris polynominal model   :     R = Sum ( Ai X ( K*Bm )**Pi X 10**Qi )
     xd = 16.535400"    t =  0.984300"
     A1 =  1.790000    A2 =  1.760000    A3 =  2.040000     K =  0.001202
     P1 =         1    P2 =         3    P3 =         5    Q1 =         0     Q2 =        -1     Q3 =        -1

Modified exponential model   :
     Bm = Sum ( Ai X ( 1 - exp( - R/( 2*i*C ) ) ) )  +      Sum ( RKj X ( R - Rj0 ) )  +  Bm0
      C =   0.18412500E+00              Bmo=   0.00000000E+00              Nexp=  6          Nliner=  3
     AI =   0.49978100E+04  -0.20854607E+05   0.69186809E+05  -0.11739429E+06   0.93852283E+05  -0.27461415E+05
    Rj0 =           0.0000           0.0010           0.1200
    RKj =  -0.18284392E+06   0.18015901E+06   0.26954054E+04
```

No	Rotation (radians) X 1/1000	Moment (kip-inch) Expri.	Moment (kip-inch) Poly.	Moment (kip-inch) M.Expo.	Connection stiffness (kip-inch) X 1000 Expri.	Connection stiffness (kip-inch) X 1000 Poly.	Connection stiffness (kip-inch) X 1000 M.Expo.
1	0.00	0.00	0.00	0.00	0.1100E+04	0.4647E+03	0.6728E+04
3	0.12	198.29	56.78	168.60	0.3001E+04	0.4641E+03	0.4418E+04
5	0.18	396.58	84.71	408.41	0.3663E+04	0.4633E+03	0.3577E+04
7	0.24	594.87	111.71	596.92	0.2902E+04	0.4622E+03	0.2911E+04
9	0.33	793.17	151.50	813.02	0.2569E+04	0.4600E+03	0.2141E+04
11	0.40	966.72	185.40	951.85	0.1856E+04	0.4574E+03	0.1643E+04
13	0.53	1140.28	244.45	1122.74	0.9305E+03	0.4513E+03	0.1030E+04
15	0.91	1338.59	409.50	1341.15	0.4183E+03	0.4206E+03	0.3029E+03
17	1.82	1487.32	735.97	1492.06	0.1227E+03	0.2942E+03	0.1236E+03
19	3.23	1636.05	1044.45	1625.63	0.6903E+02	0.1613E+03	0.6954E+02
21	4.85	1710.42	1250.00	1727.41	0.6903E+02	0.1017E+03	0.6168E+02
23	6.46	1824.44	1389.69	1822.09	0.4909E+02	0.7423E+02	0.5261E+02
25	8.08	1903.77	1496.07	1891.91	0.3262E+02	0.5872E+02	0.3359E+02
27	9.93	1929.27	1593.54	1937.20	0.1381E+02	0.4765E+02	0.1712E+02
29	11.77	1954.76	1674.29	1960.99	0.1381E+02	0.4026E+02	0.9926E+01
31	13.62	1980.26	1743.51	1977.20	0.1381E+02	0.3497E+02	0.8192E+01
33	15.51	1998.96	1805.69	1992.85	0.6137E+01	0.3090E+02	0.8499E+01
35	17.45	2010.86	1862.37	2010.03	0.6137E+01	0.2768E+02	0.9221E+01
37	19.39	2022.76	1913.46	2028.50	0.6137E+01	0.2511E+02	0.9788E+01

V -113

```
Connection type : Extended end-plate connections
           Mode : All bolted with column stiffener

Tested by  :  L.F.L.Ribeiro & R.M.Goncalves (1996)         Brazil
Test  Id.  :  CT1A-1

Column : CVS350x105                  Fasteners: A325- -5/8"D
Beam   : VS250x37                            7/8" Oversize holes
Plate thickness :      1.25"         Material : A36
                                          Fy =  36.24 ksi
                                          Fu =  57.99 ksi
```

Major parameters

```
End plate extended on tension side only.
ct =  1.2503"    pt =  2.9929"    li =  6.3575"     pc =  1.4965"
cc =  0.6980"    gt =  2.5008"    gc =  2.5008"
tp =  1.2500"    nt = 2 X  2      nc = 2 X  1
```

Remark 1) End plate extended on tension side only.
2)

No	Moment (k-in)	Rotation (radians) X 1/1000
1	0.00	0.00
2	66.37	0.04
3	136.05	0.09
4	199.10	0.14
5	268.78	0.29
6	338.47	0.42
7	398.20	0.54
8	466.22	0.70
9	532.59	0.86
10	598.95	1.04
11	666.98	1.25
12	728.37	1.44
13	794.73	1.67
14	871.05	2.02
15	939.08	2.39
16	1010.42	2.93
17	1071.81	3.43
18	1136.52	4.56
19	1159.75	5.66
20	1182.97	6.76

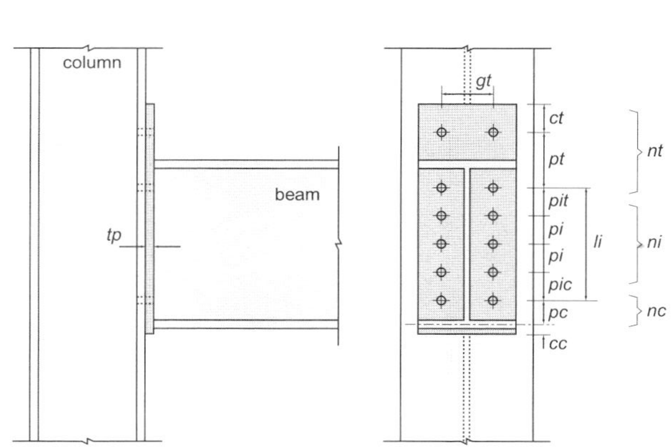

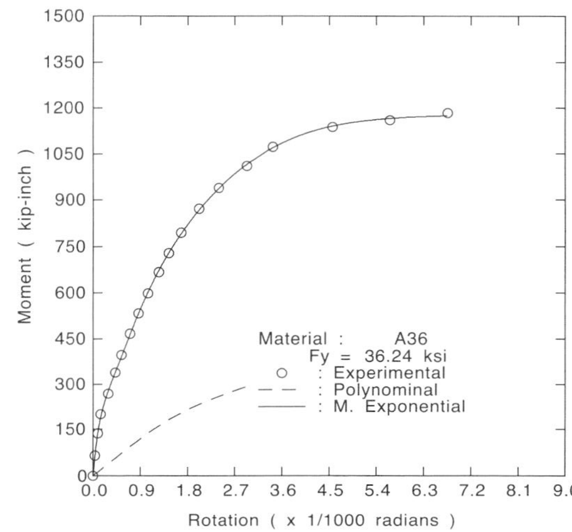

```
Moment-rotation prediction equations     ( R : X 1/1000 radians )

Frye and Morris polynominal model   :     R = Sum ( Ai X ( K*Bm )**Pi X 10**Qi )
     xd =  9.350400"   t =  1.250000"
     A1 =  1.790000   A2 =  1.760000   A3 =  2.040000    K =  0.004091
     P1 =         1   P2 =         3   P3 =         5   Q1 =         0   Q2 =        -1   Q3 =        -1

Modified exponential model   :
     Bm = Sum ( Ai X ( 1 - exp( - R/( 2*i*C ) ) ) )  +      Sum ( RKj X ( R - Rj0 ) )  +  Bm0
      C =   0.66333333E-01           Bmo=   0.00000000E+00          Nexp=  6       Nliner=  0
     AI =   0.33229944E+03   0.14225995E+04  -0.15512702E+05   0.47356483E+05  -0.60907710E+05   0.28487093E+05
```

No	Rotation (radians) X 1/1000	Moment (kip-inch) Expri.	Moment (kip-inch) Poly.	Moment (kip-inch) M.Expo.	Connection stiffness (kip-inch) X 1000 Expri.	Connection stiffness (kip-inch) X 1000 Poly.	Connection stiffness (kip-inch) X 1000 M.Expo.
1	0.00	0.00	0.00	0.00	0.1659E+04	0.1365E+03	0.2096E+04
2	0.04	66.37	5.46	72.52	0.1541E+04	0.1365E+03	0.1561E+04
3	0.09	136.05	12.29	138.49	0.1327E+04	0.1364E+03	0.1112E+04
4	0.14	199.10	19.11	186.45	0.1062E+04	0.1363E+03	0.8292E+03
5	0.29	268.78	39.49	279.37	0.5029E+03	0.1355E+03	0.4964E+03
6	0.42	338.47	57.03	339.72	0.5161E+03	0.1342E+03	0.4474E+03
7	0.54	398.20	73.03	393.05	0.4666E+03	0.1325E+03	0.4430E+03
8	0.70	466.22	93.98	463.38	0.4200E+03	0.1293E+03	0.4336E+03
9	0.86	532.59	114.35	530.92	0.3931E+03	0.1251E+03	0.4082E+03
10	1.04	598.95	136.34	600.90	0.3480E+03	0.1191E+03	0.3684E+03
11	1.25	666.98	160.48	673.14	0.3235E+03	0.1107E+03	0.3201E+03
12	1.44	728.37	180.75	730.22	0.3075E+03	0.1025E+03	0.2818E+03
13	1.67	794.73	203.18	790.54	0.2606E+03	0.9252E+02	0.2443E+03
14	2.02	871.05	233.07	868.25	0.2014E+03	0.7851E+02	0.2021E+03
15	2.39	939.08	259.77	936.53	0.1628E+03	0.6628E+02	0.1681E+03
16	2.93	1010.42	291.79	1015.68	0.1273E+03	0.5297E+02	0.1260E+03
17	3.43	1071.81	315.96	1069.92	0.1027E+03	0.4427E+02	0.9181E+02
18	4.56	1136.52	358.32	1139.80	0.3895E+02	0.3201E+02	0.3756E+02
19	5.66	1159.75	389.53	1165.70	0.2112E+02	0.2517E+02	0.1319E+02
20	6.76	1182.97	414.63	1174.37	0.2112E+02	0.2079E+02	0.4153E+01

A.5 – 227

V -114

Connection type : Extended end-plate connections
Mode : All bolted with column stiffener

Tested by : L.F.L.Ribeiro & R.M.Goncalves (1996) Brazil
Test Id. : CT1A-2

Column : CVS350x105
Beam : VS250x37
Plate thickness : 1.00"

Fasteners: A325- -5/8"D
7/8" Oversize holes
Material : A36
Fy = 36.24 ksi
Fu = 57.99 ksi

Major parameters

End plate extended on tension side only.

ct = 1.2503" pt = 2.9929" li = 6.3575" pc = 1.4965"
cc = 0.6980" gt = 2.5008" gc = 2.5008"
tp = 1.0000" nt = 2 X 2 nc = 2 X 1

Remark 1) End plate extended on tension side only.
2)

No	Moment (k-in)	Rotation (radians) X 1/1000
1	0.00	0.00
2	68.58	0.03
3	142.30	0.11
4	207.45	0.18
5	277.74	0.29
6	346.32	0.39
7	409.75	0.49
8	476.62	0.61
9	545.20	0.74
10	617.20	0.86
11	687.50	1.03
12	756.07	1.20
13	836.65	1.46
14	894.95	1.68
15	961.81	1.97
16	996.96	2.37
17	1032.10	2.76
18	1066.39	3.10
19	1100.68	3.45

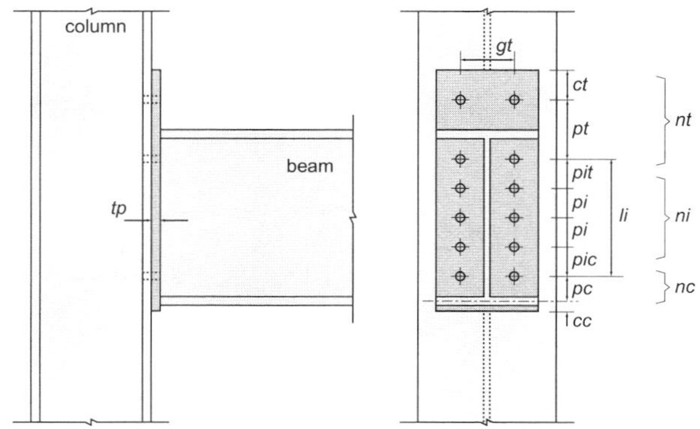

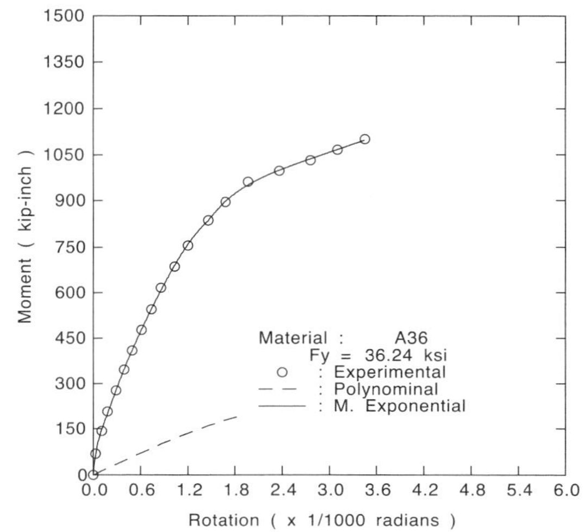

A.5 – 228

```
Moment-rotation prediction equations     ( R : X 1/1000 radians )

Frye and Morris polynominal model  :     R = Sum ( Ai X ( K*Bm )**Pi X 10**Qi )
     xd =  9.350400"    t =  1.000000"
     A1 =  1.790000   A2 =  1.760000   A3 =  2.040000    K =  0.004677
     P1 =        1    P2 =        3    P3 =        5    Q1 =        0    Q2 =       -1    Q3 =       -1

Modified exponential model  :
     Bm = Sum ( Ai X ( 1 - exp( - R/( 2*i*C ) ) ) )  +      Sum ( RKj X ( R - Rj0 ) )  +  Bm0
       C =   0.36250000E-01            Bmo=   0.00000000E+00          Nexp=  6        Nliner=  1
      AI =   0.83731356E+03  -0.90021720E+04   0.40614775E+05  -0.79442460E+05   0.66171289E+05  -0.18267614E+05
     Rj0 =          1.4600
     RKj =   0.92774988E+02
```

No	Rotation (radians) X 1/1000	Moment (kip-inch) Expri.	Moment (kip-inch) Poly.	Moment (kip-inch) M.Expo.	Connection stiffness (kip-inch) X 1000 Expri.	Connection stiffness (kip-inch) X 1000 Poly.	Connection stiffness (kip-inch) X 1000 M.Expo.
1	0.00	0.00	0.00	0.00	0.2286E+04	0.1194E+03	0.2807E+04
2	0.03	68.58	3.58	65.12	0.1914E+04	0.1194E+03	0.1660E+04
3	0.11	142.30	13.14	149.35	0.9264E+03	0.1193E+03	0.7566E+03
4	0.18	207.45	21.49	199.88	0.8173E+03	0.1191E+03	0.7181E+03
5	0.29	277.74	34.55	279.64	0.6635E+03	0.1185E+03	0.7158E+03
6	0.39	346.32	46.39	348.06	0.6601E+03	0.1177E+03	0.6483E+03
7	0.49	409.75	58.07	409.53	0.5993E+03	0.1165E+03	0.5853E+03
8	0.61	476.62	71.95	477.07	0.5430E+03	0.1148E+03	0.5464E+03
9	0.74	545.20	86.71	546.82	0.5652E+03	0.1123E+03	0.5281E+03
10	0.86	617.20	100.02	609.04	0.5229E+03	0.1094E+03	0.5066E+03
11	1.03	687.50	118.21	690.72	0.4084E+03	0.1045E+03	0.4488E+03
12	1.20	756.07	135.48	760.14	0.3664E+03	0.9868E+02	0.3652E+03
13	1.46	836.65	159.89	837.16	0.2856E+03	0.8892E+02	0.2294E+03
14	1.68	894.95	178.53	897.32	0.2501E+03	0.8055E+02	0.2294E+03
15	1.97	961.81	200.38	951.27	0.1706E+03	0.7031E+02	0.1505E+03
16	2.37	996.96	225.76	999.56	0.8898E+02	0.5862E+02	0.1027E+03
17	2.76	1032.10	247.05	1036.87	0.9454E+02	0.4956E+02	0.8956E+02
18	3.10	1066.39	263.03	1067.40	0.9939E+02	0.4340E+02	0.8819E+02
19	3.45	1100.68	277.12	1097.99	0.9939E+02	0.3847E+02	0.8927E+02

V -115

```
Connection type : Extended end-plate connections
           Mode : All bolted with column stiffener

Tested by  :  L.F.L.Ribeiro & R.M.Goncalves (1996)        Brazil
Test  Id.  :  CT1A-3

Column : CVS350x105                  Fasteners: A325- -5/8"D
Beam   : VS250x37                            7/8" Oversize holes
Plate thickness :       1.00"        Material : A36
                                          Fy =  36.24 ksi
                                          Fu =  57.99 ksi

                         Major parameters

     End plate extended on tension side only.
     ct =  1.2503"    pt =  2.9929"    li =  6.3575"    pc =  1.4965"
     cc =  0.6980"    gt =  2.5008"    gc =  2.5008"
     tp =  1.0000"    nt = 2 X  2      nc = 2 X  1

Remark  1)  End plate extended on tension side only.
        2)
```

No	Moment (k-in)	Rotation (radians) X 1/1000
1	0.00	0.00
2	68.58	0.03
3	137.16	0.09
4	205.73	0.16
5	276.03	0.23
6	341.18	0.31
7	414.90	0.40
8	480.05	0.51
9	546.91	0.64
10	617.20	0.81
11	684.07	0.98
12	756.07	1.22
13	826.37	1.52
14	896.66	1.87
15	968.67	2.29
16	1028.67	2.67
17	1100.68	3.24
18	1167.54	4.00
19	1211.26	4.71
20	1254.98	5.43
21	1279.84	6.12
22	1304.70	6.81

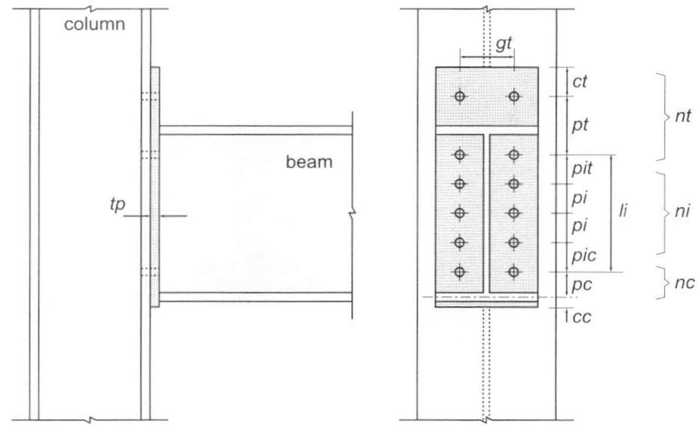

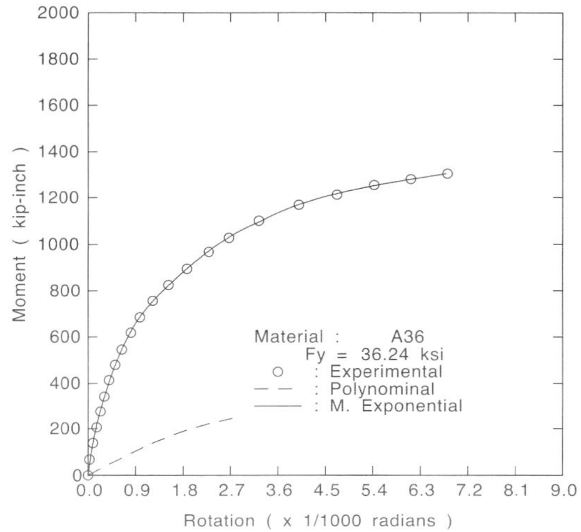

```
Moment-rotation prediction equations     ( R : X 1/1000 radians )

Frye and Morris polynominal model  :    R = Sum ( Ai X ( K*Bm )**Pi X 10**Qi )
     xd =  9.350400"   t =  1.000000"
     A1 =  1.790000   A2 =  1.760000   A3 =  2.040000    K =  0.004677
     P1 =        1    P2 =        3    P3 =        5    Q1 =        0    Q2 =       -1    Q3 =       -1

Modified exponential model  :
     Bm = Sum ( Ai X ( 1 - exp( - R/( 2*i*C ) ) ) )  +      Sum ( RKj X ( R - Rj0 ) )  +  Bm0
      C =   0.64250000E-01             Bmo=   0.00000000E+00          Nexp=  6       Nliner=  1
     AI =   0.42776412E+03  -0.28551951E+04   0.49582235E+04   0.83036211E+04  -0.26188668E+05   0.16543149E+05
    Rj0 =           3.2600
    RKj =   0.33077578E+02
```

No	Rotation (radians) X 1/1000	Moment (kip-inch) Expri.	Moment (kip-inch) Poly.	Moment (kip-inch) M.Expo.	Connection stiffness (kip-inch) X 1000 Expri.	Connection stiffness (kip-inch) X 1000 Poly.	Connection stiffness (kip-inch) X 1000 M.Expo.
1	0.00	0.00	0.00	0.00	0.2286E+04	0.1194E+03	0.1932E+04
2	0.03	68.58	3.58	53.12	0.1905E+04	0.1194E+03	0.1625E+04
3	0.09	137.16	10.75	137.55	0.1068E+04	0.1193E+03	0.1229E+04
4	0.16	205.73	19.10	214.05	0.9919E+03	0.1192E+03	0.9840E+03
5	0.23	276.03	27.44	277.91	0.9156E+03	0.1188E+03	0.8526E+03
6	0.31	341.18	36.91	342.19	0.8166E+03	0.1183E+03	0.7604E+03
7	0.40	414.90	47.56	406.99	0.7170E+03	0.1176E+03	0.6818E+03
8	0.51	480.05	60.40	477.18	0.5565E+03	0.1163E+03	0.5952E+03
9	0.64	546.91	75.39	548.27	0.4706E+03	0.1142E+03	0.5000E+03
10	0.81	617.20	94.52	624.01	0.4034E+03	0.1106E+03	0.3956E+03
11	0.98	684.07	112.95	684.45	0.3546E+03	0.1060E+03	0.3204E+03
12	1.22	756.07	137.45	752.92	0.2708E+03	0.9796E+02	0.2573E+03
13	1.52	826.37	165.16	823.83	0.2189E+03	0.8661E+02	0.2203E+03
14	1.87	896.66	193.19	896.53	0.1875E+03	0.7369E+02	0.1961E+03
15	2.29	968.67	221.30	972.84	0.1643E+03	0.6062E+02	0.1663E+03
16	2.67	1028.67	242.50	1030.21	0.1453E+03	0.5142E+02	0.1353E+03
17	3.24	1100.68	268.79	1094.17	0.1099E+03	0.4133E+02	0.9027E+02
18	4.00	1167.54	296.60	1169.20	0.7415E+02	0.3247E+02	0.7938E+02
19	4.71	1211.26	317.70	1216.62	0.6115E+02	0.2696E+02	0.5571E+02
20	5.43	1254.98	335.49	1251.60	0.4836E+02	0.2305E+02	0.4354E+02
21	6.12	1279.84	350.38	1279.45	0.3603E+02	0.2023E+02	0.3787E+02
22	6.81	1304.70	363.55	1304.54	0.3603E+02	0.1805E+02	0.3522E+02

V -116

```
Connection type : Extended end-plate connections
           Mode : All bolted with column stiffener

Tested by   :  L.F.L.Ribeiro & R.M.Goncalves (1996)        Brazil
Test  Id.   :  CT1A-4

Column : CVS350x105                  Fasteners: A325- -5/8"D
Beam   : VS250x37                            7/8" Oversize holes
Plate thickness :       0.875"       Material : A36
                                          Fy =  36.24 ksi
                                          Fu =  57.99 ksi

                       Major parameters

     End plate extended on tension side only.
     ct =  1.2503"    pt =  2.9929"    li =  6.3575"    pc =  1.4965"
     cc =  0.6980"    gt =  2.5008"    gc =  2.5008"
     tp =  0.8752"    nt = 2 X  2      nc = 2 X  1

Remark  1)  End plate extended on tension side only.
        2)
```

No	Moment (k-in)	Rotation (radians) X 1/1000
1	0.00	0.00
2	66.81	0.05
3	136.96	0.14
4	208.78	0.25
5	277.25	0.37
6	334.04	0.45
7	405.86	0.57
8	477.68	0.72
9	539.48	0.85
10	604.62	1.01
11	678.11	1.21
12	733.22	1.37
13	811.72	1.67
14	866.84	1.97
15	945.34	2.36
16	1008.81	2.85
17	1080.63	3.50
18	1104.84	4.03
19	1129.06	4.56
20	1162.47	5.13

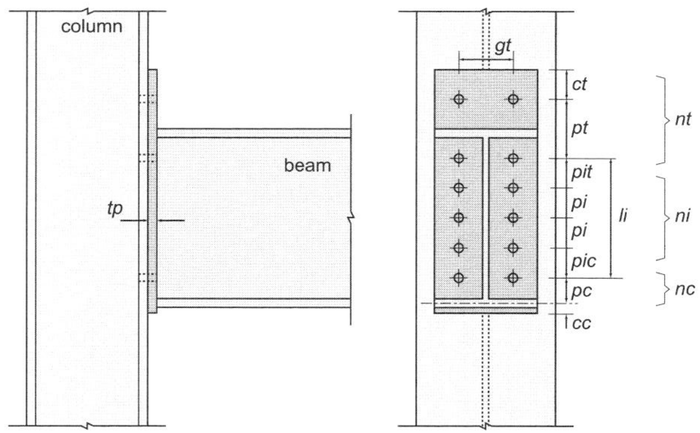

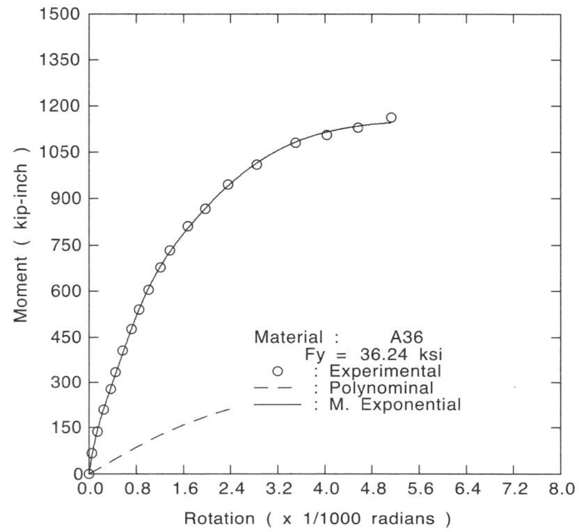

```
Moment-rotation prediction equations     ( R : X 1/1000 radians )

Frye and Morris polynominal model  :     R = Sum ( Ai X ( K*Bm )**Pi X 10**Qi )
     xd =  9.350400"   t =  0.875200"
     A1 =  1.790000   A2 =  1.760000   A3 =  2.040000    K =  0.005067
     P1 =         1   P2 =         3   P3 =         5    Q1 =         0    Q2 =       -1    Q3 =       -1

Modified exponential model  :
     Bm = Sum ( Ai X ( 1 - exp( - R/( 2*i*C ) ) ) )  +     Sum ( RKj X ( R - Rj0 ) )  +  Bm0
      C =   0.55250000E-01           Bmo=   0.00000000E+00          Nexp=  6        Nliner=  0
     AI =  -0.66026774E+02   0.50804474E+04  -0.34404372E+05   0.90383888E+05  -0.10432098E+06   0.44483011E+05
```

No	Rotation (radians) X 1/1000	Moment (kip-inch) Expri.	Moment (kip-inch) Poly.	Moment (kip-inch) M.Expo.	Connection stiffness (kip-inch) X 1000 Expri.	Connection stiffness (kip-inch) X 1000 Poly.	Connection stiffness (kip-inch) X 1000 M.Expo.
1	0.00	0.00	0.00	0.00	0.1336E+04	0.1103E+03	0.1373E+04
2	0.05	66.81	5.51	60.23	0.1137E+04	0.1102E+03	0.1057E+04
3	0.14	136.96	15.43	139.30	0.7225E+03	0.1101E+03	0.7433E+03
4	0.25	208.78	27.51	211.97	0.6136E+03	0.1096E+03	0.6076E+03
5	0.37	277.25	40.62	282.62	0.6542E+03	0.1088E+03	0.5796E+03
6	0.45	334.04	49.32	328.84	0.6653E+03	0.1080E+03	0.5761E+03
7	0.57	405.86	62.17	397.27	0.5453E+03	0.1065E+03	0.5614E+03
8	0.72	477.68	77.97	478.49	0.4770E+03	0.1040E+03	0.5170E+03
9	0.85	539.48	91.32	542.29	0.4448E+03	0.1012E+03	0.4633E+03
10	1.01	604.62	107.19	610.84	0.3895E+03	0.9701E+02	0.3941E+03
11	1.21	678.11	125.98	681.99	0.3547E+03	0.9076E+02	0.3207E+03
12	1.37	733.22	140.07	729.67	0.3157E+03	0.8526E+02	0.2776E+03
13	1.67	811.72	164.06	804.65	0.2227E+03	0.7470E+02	0.2278E+03
14	1.97	866.84	184.98	868.52	0.1914E+03	0.6490E+02	0.1999E+03
15	2.36	945.34	208.14	940.79	0.1695E+03	0.5423E+02	0.1707E+03
16	2.85	1008.81	232.12	1014.73	0.1213E+03	0.4413E+02	0.1305E+03
17	3.50	1080.63	257.63	1082.23	0.7480E+02	0.3492E+02	0.7887E+02
18	4.03	1104.84	274.66	1115.18	0.4569E+02	0.2973E+02	0.4735E+02
19	4.56	1129.06	289.33	1134.34	0.5191E+02	0.2584E+02	0.2662E+02
20	5.13	1162.47	303.12	1145.43	0.5860E+02	0.2266E+02	0.1358E+02

V -117

```
Connection type : Extended end-plate connections
           Mode : All bolted with column stiffener

Tested by  :  L.F.L.Ribeiro & R.M.Goncalves (1996)        Brazil
Test  Id.  :  CT1A-5

Column : CVS350x105                  Fasteners: A325- -5/8"D
Beam   : VS250x37                              7/8" Oversize holes
Plate thickness :       0.875"       Material : A36
                                          Fy =  36.24 ksi
                                          Fu =  57.99 ksi

                         Major parameters

     End plate extended on tension side only.
     ct =  1.2503"    pt =  2.9929"    li =  6.3575"     pc =  1.4965"
     cc =  0.6980"    gt =  2.5008"    gc =  2.5008"
     tp =  0.8752"    nt = 2 X  2      nc = 2 X  1

Remark  1)  End plate extended on tension side only.
        2)
```

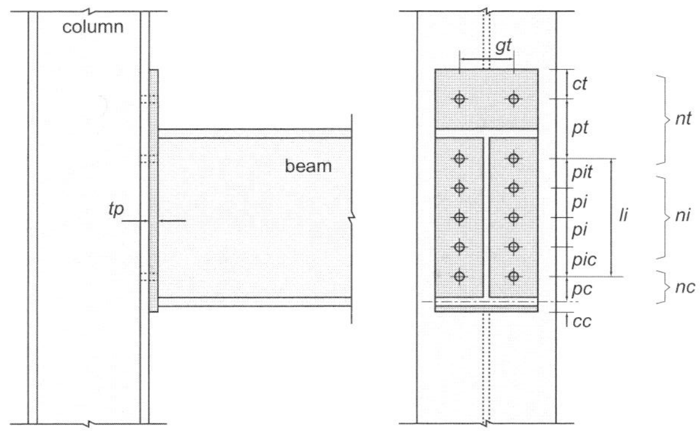

No	Moment (k-in)	Rotation (radians) X 1/1000
1	0.00	0 00
2	68.48	0 04
3	135.29	0 11
4	203.77	0 20
5	270.57	0 31
6	339.05	0 44
7	404.19	0 60
8	467.66	0 77
9	554.51	1 05
10	602.95	1 22
11	668.08	1 45
12	743.24	1 81
13	808.38	2 10
14	870.18	2 45
15	935.32	2 85
16	1012.15	3 56
17	1068.93	3 99
18	1129.06	4 84
19	1170.82	5 59
20	1202.55	6 06
21	1232.62	7 17
22	1259.34	8 59

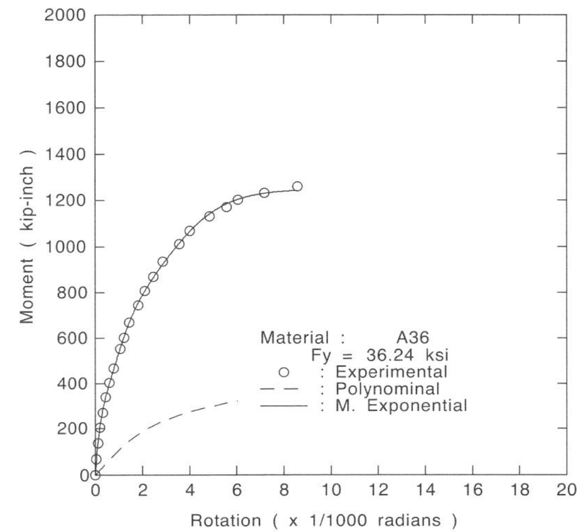

A.5 – 234

```
Moment-rotation prediction equations     ( R : X 1/1000 radians )

Frye and Morris polynominal model  :     R = Sum ( Ai X ( K*Bm )**Pi X 10**Qi )
     xd =  9.350400"   t =  0.875200"
     A1 =  1.790000   A2 =  1.760000    A3 =  2.040000    K =  0.005067
     P1 =        1     P2 =        3     P3 =        5     Q1 =        0     Q2 =       -1     Q3 =       -1

Modified exponential model  :
     Bm = Sum ( Ai X ( 1 - exp( - R/( 2*i*C ) ) ) )  +      Sum ( RKj X ( R - Rj0 ) )  +  Bm0
       C =   0.81583333E-01            Bmo=  0.00000000E+00         Nexp=  6        Nliner=  0
      AI =  -0.24307076E+03   0.79491783E+04  -0.46347806E+05   0.11527509E+06  -0.12905493E+06   0.53669518E+05
```

No	Rotation (radians) X 1/1000	Moment (kip-inch) Expri.	Moment (kip-inch) Poly.	Moment (kip-inch) M.Expo.	Connection stiffness (kip-inch) X 1000 Expri.	Connection stiffness (kip-inch) X 1000 Poly.	Connection stiffness (kip-inch) X 1000 M.Expo.
1	0.00	0.00	0.00	0.00	0.1712E+04	0.1103E+03	0.1440E+04
2	0.04	68.48	4.41	53.89	0.1436E+04	0.1102E+03	0.1258E+04
3	0.11	135.29	12.13	132.15	0.8697E+03	0.1101E+03	0.9891E+03
4	0.20	203.77	22.04	208.99	0.6918E+03	0.1098E+03	0.7345E+03
5	0.31	270.57	34.08	278.01	0.5704E+03	0.1092E+03	0.5385E+03
6	0.44	339.05	48.21	339.23	0.4731E+03	0.1081E+03	0.4191E+03
7	0.60	404.19	65.36	400.86	0.3907E+03	0.1061E+03	0.3621E+03
8	0.77	467.66	83.15	460.54	0.3495E+03	0.1030E+03	0.3431E+03
9	1.05	554.51	111.05	553.58	0.2945E+03	0.9583E+02	0.3192E+03
10	1.22	602.95	126.89	605.98	0.2842E+03	0.9043E+02	0.2962E+03
11	1.45	668.08	146.78	669.90	0.2542E+03	0.8244E+02	0.2591E+03
12	1.81	743.24	174.20	753.08	0.2175E+03	0.6997E+02	0.2054E+03
13	2.10	808.38	193.16	807.95	0.2028E+03	0.6107E+02	0.1751E+03
14	2.45	870.18	212.92	865.05	0.1702E+03	0.5213E+02	0.1534E+03
15	2.85	935.32	232.10	923.44	0.1432E+03	0.4414E+02	0.1398E+03
16	3.56	1012.15	259.72	1016.04	0.1231E+03	0.3424E+02	0.1205E+03
17	3.99	1068.93	273.46	1064.76	0.1115E+03	0.3007E+02	0.1057E+03
18	4.84	1129.06	296.36	1140.65	0.6273E+02	0.2417E+02	0.7275E+02
19	5.59	1170.82	313.06	1185.17	0.6296E+02	0.2061E+02	0.4698E+02
20	6.06	1202.55	322.32	1204.16	0.5549E+02	0.1888E+02	0.3434E+02
21	7.17	1232.62	341.46	1230.32	0.2346E+02	0.1578E+02	0.1500E+02
22	8.59	1259.34	361.83	1242.91	0.1882E+02	0.1308E+02	0.4585E+01

A.5 – 235

V -118

```
Connection type : Extended end-plate connections
           Mode : All bolted with column stiffener

Tested by  :  L.F.L.Ribeiro & R.M.Goncalves (1996)        Brazil
Test  Id.  :  CT1A-6

Column : CVS350x105                   Fasteners: A325- -3/4"D
Beam   : VS250x37                              13/16" Oversize holes
Plate thickness :       0.75"         Material : A36
                                           Fy =  36.24 ksi
                                           Fu =  57.99 ksi

                           Major parameters

     End plate extended on tension side only.
     ct =  1.5001"    pt =  3.4921"    li =  5.8583"     pc =  1.7461"
     cc =  0.7894"    gt =  3.0000"    gc =  3.0000"
     tp =  0.7500"    nt = 2 X  2      nc = 2 X  1

Remark  1)  End plate extended on tension side only.
        2)
```

No	Moment (k-in)	Rotation (radians) X 1/1000
1	0.00	0.00
2	78.50	0.04
3	138.63	0.13
4	205.44	0.23
5	270.57	0.37
6	337.38	0.54
7	399.18	0.78
8	471.00	1.15
9	541.15	1.52
10	611.30	1.87
11	673.09	2.09
12	741.57	2.38
13	801.70	2.62
14	878.53	3.02
15	942.00	3.42
16	1003.80	3.74
17	1068.93	4.33
18	1139.08	4.94
19	1209.23	5.70
20	1267.69	6.57
21	1324.48	7.37
22	1353.94	8.45
23	1372.91	9.50
24	1397.97	10.80

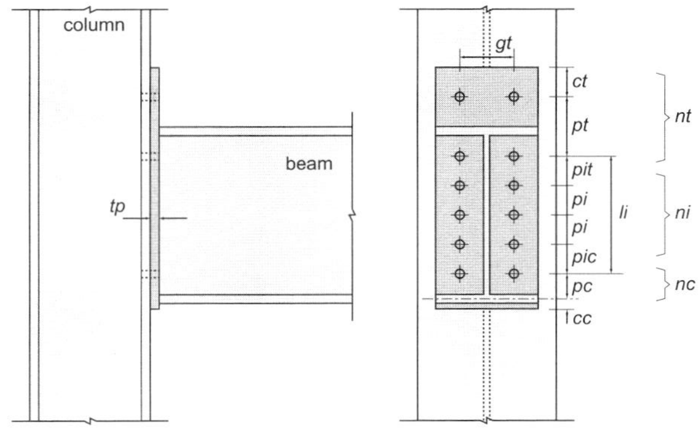

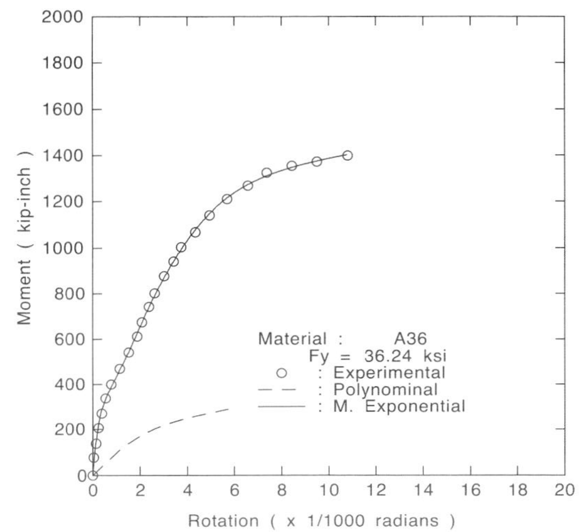

A.5 – 236

```
Moment-rotation prediction equations      ( R : X 1/1000 radians )

Frye and Morris polynominal model  :     R = Sum ( Ai X ( K*Bm )**Pi X 10**Qi )
     xd =  9.350400"    t =  0.750000"
     A1 =  1.790000   A2 =  1.760000   A3 =  2.040000    K =  0.005559
     P1 =        1    P2 =        3    P3 =        5    Q1 =        0    Q2 =       -1    Q3 =       -1

Modified exponential model  :
     Bm = Sum ( Ai X ( 1 - exp( - R/( 2*i*C ) ) ) )  +      Sum ( RKj X ( R - Rj0 ) )  +  Bm0
      C =   0.10000000E+00           Bmo=   0.00000000E+00          Nexp=  6        Nliner=  2
     AI =  -0.31214923E+03   0.64424949E+04  -0.28987755E+05   0.68538201E+05  -0.81445945E+05   0.36968709E+05
    Rj0 =          0.5400          1.1500
    RKj =   0.51562638E+02  -0.33871971E+02
```

No	Rotation (radians) X 1/1000	Moment (kip-inch) Expri.	Moment (kip-inch) Poly.	Moment (kip-inch) M.Expo.	Connection stiffness (kip-inch) X 1000 Expri.	Connection stiffness (kip-inch) X 1000 Poly.	Connection stiffness (kip-inch) X 1000 M.Expo.
1	0.00	0.00	0.00	0.00	0.1962E+04	0.1005E+03	0.1267E+04
2	0.04	78.50	4.02	48.08	0.1564E+04	0.1005E+03	0.1138E+04
3	0.13	138.63	13.06	138.19	0.6681E+03	0.1003E+03	0.8711E+03
4	0.23	205.44	23.10	212.64	0.5836E+03	0.1000E+03	0.6281E+03
5	0.37	270.57	37.02	282.53	0.4326E+03	0.9916E+02	0.3892E+03
6	0.54	337.38	53.75	333.45	0.3368E+03	0.9749E+02	0.2298E+03
7	0.78	399.18	76.73	389.60	0.2326E+03	0.9371E+02	0.2073E+03
8	1.15	471.00	109.82	467.40	0.1918E+03	0.8452E+02	0.2244E+03
9	1.52	541.15	138.98	544.42	0.1952E+03	0.7288E+02	0.2228E+03
10	1.87	611.30	162.56	624.67	0.2498E+03	0.6201E+02	0.2323E+03
11	2.09	673.09	175.51	675.57	0.2616E+03	0.5593E+02	0.2295E+03
12	2.38	741.57	190.70	740.82	0.2440E+03	0.4900E+02	0.2197E+03
13	2.62	801.70	201.86	792.26	0.2286E+03	0.4418E+02	0.2087E+03
14	3.02	878.53	218.19	871.79	0.1754E+03	0.3769E+02	0.1887E+03
15	3.42	942.00	232.22	943.27	0.1778E+03	0.3270E+02	0.1688E+03
16	3.74	1003.80	242.15	994.80	0.1640E+03	0.2952E+02	0.1534E+03
17	4.33	1068.93	258.16	1077.41	0.1127E+03	0.2498E+02	0.1270E+03
18	4.94	1139.08	272.30	1147.28	0.1049E+03	0.2154E+02	0.1026E+03
19	5.70	1209.23	287.41	1215.19	0.8059E+02	0.1839E+02	0.7705E+02
20	6.57	1267.69	302.21	1272.07	0.6917E+02	0.1578E+02	0.5500E+02
21	7.37	1324.48	314.07	1310.11	0.5239E+02	0.1397E+02	0.4099E+02
22	8.45	1353.94	328.12	1347.49	0.2261E+02	0.1213E+02	0.2939E+02
23	9.50	1372.91	340.11	1374.87	0.1860E+02	0.1077E+02	0.2342E+02
24	10.80	1397.97	353.23	1402.73	0.1927E+02	0.9475E+01	0.1995E+02

A.5 – 237

V -119

```
Connection type : Extended end-plate connections
           Mode : All bolted with column stiffener

Tested by  :  L.F.L.Ribeiro & R.M.Goncalves (1996)        Brazil
Test  Id.  :  CT1B-1

Column : CVS350x105                  Fasteners: A325- -5/8"D
Beam   : VS250x37                            7/8" Oversize holes
Plate thickness :      1.25"         Material : A36
                                          Fy =  36.24 ksi
                                          Fu =  57.99 ksi

                          Major parameters

     End plate extended on tension side only.
     ct =  1.2503"    pt =  2.9929"    li =  6.3575"    pc =  1.4965"
     cc =  0.6980"    gt =  2.5008"    gc =  2.5008"
     tp =  1.2500"    nt = 2 X  2      nc = 2 X  1

Remark   1)  End plate extended on tension side only.
         2)
```

No	Moment (k-in)	Rotation (radians) X 1/1000
1	0.00	0.00
2	199.10	0.03
3	298.65	0.08
4	404.83	0.15
5	502.72	0.24
6	600.61	0.36
7	696.84	0.55
8	796.39	0.84
9	894.28	1.19
10	997.15	1.66
11	1056.88	2.09
12	1143.15	2.97
13	1184.63	4.10
14	1221.13	5.18
15	1236.07	6.46
16	1239.38	8.04

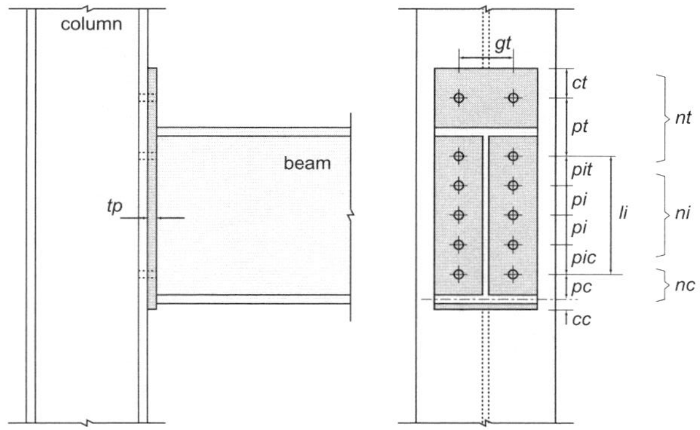

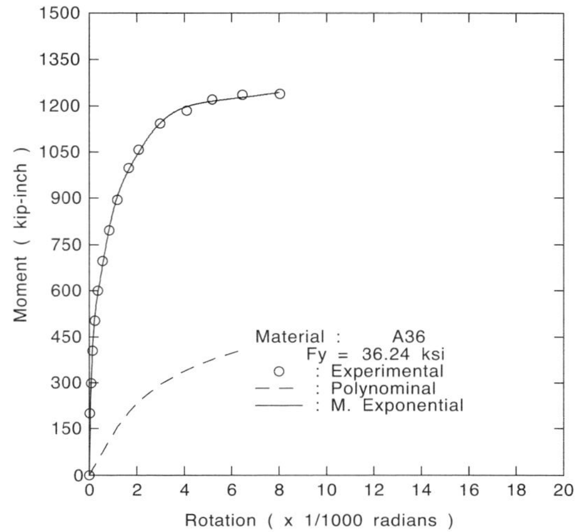

A.5 – 238

```
Moment-rotation prediction equations     ( R : X 1/1000 radians )

Frye and Morris polynominal model  :     R = Sum ( Ai X ( K*Bm )**Pi X 10**Qi )
     xd =  9.350400"   t =  1.250000"
     A1 =  1.790000   A2 =  1.760000   A3 =  2.040000    K =  0.004091
     P1 =        1    P2 =        3    P3 =        5    Q1 =        0    Q2 =       -1    Q3 =       -1

Modified exponential model  :
     Bm = Sum ( Ai X ( 1 - exp( - R/( 2*i*C ) ) ) )  +      Sum ( RKj X ( R - Rj0 ) )  +  Bm0
      C =   0.74500000E-01           Bmo=   0.00000000E+00         Nexp=  6       Nliner=  1
     AI =   0.21350244E+04  -0.12880191E+05   0.43643310E+05  -0.73354681E+05   0.57974852E+05  -0.16345175E+05
    Rj0 =           2.0900
    RKj =   0.11717671E+02
```

No	Rotation (radians) X 1/1000	Moment (kip-inch) Expri.	Moment (kip-inch) Poly.	Moment (kip-inch) M.Expo.	Connection stiffness (kip-inch) X 1000 Expri.	Connection stiffness (kip-inch) X 1000 Poly.	Connection stiffness (kip-inch) X 1000 M.Expo.
1	0.00	0.00	0.00	0.00	0.6637E+04	0.1365E+03	0.5200E+04
2	0.03	199.10	4.10	136.60	0.4894E+04	0.1365E+03	0.3962E+04
3	0.08	298.65	10.92	296.01	0.1793E+04	0.1365E+03	0.2525E+04
4	0.15	404.83	20.48	428.31	0.1329E+04	0.1363E+03	0.1386E+04
5	0.24	502.72	32.74	519.51	0.9711E+03	0.1358E+03	0.7457E+03
6	0.36	600.61	48.96	590.42	0.6960E+03	0.1348E+03	0.5010E+03
7	0.55	696.84	74.35	679.82	0.4419E+03	0.1323E+03	0.4573E+03
8	0.84	796.39	111.84	802.25	0.3145E+03	0.1256E+03	0.3707E+03
9	1.19	894.28	153.77	906.48	0.2537E+03	0.1132E+03	0.2329E+03
10	1.66	997.15	202.30	992.17	0.1771E+03	0.9292E+02	0.1487E+03
11	2.09	1056.88	238.50	1049.65	0.1255E+03	0.7597E+02	0.1211E+03
12	2.97	1143.15	293.84	1143.34	0.7119E+02	0.5219E+02	0.7841E+02
13	4.10	1184.63	342.69	1198.11	0.3522E+02	0.3611E+02	0.2579E+02
14	5.18	1221.13	376.88	1215.63	0.2367E+02	0.2774E+02	0.1064E+02
15	6.46	1236.07	408.27	1227.71	0.7385E+01	0.2182E+02	0.9397E+01
16	8.04	1239.38	438.92	1243.80	0.2100E+01	0.1734E+02	0.1088E+02

A.5 – 239

V -120

Connection type : Extended end-plate connections
Mode : All bolted with column stiffener

Tested by : L.F.L.Ribeiro & R.M.Goncalves (1996) Brazil
Test Id. : CT1B-2

Column : CVS350x105
Beam : VS250x37
Plate thickness : 1.00"

Fasteners: A325- -3/4"D
13/16" Oversize holes
Material : A36
Fy = 36.24 ksi
Fu = 57.99 ksi

Major parameters

End plate extended on tension side only.

ct = 1.5001" pt = 3.4921" li = 5.8583" pc = 1.7461"
cc = 0.7894" gt = 3.0000" gc = 3.0000"
tp = 1.0000" nt = 2 X 2 nc = 2 X 1

Remark 1) End plate extended on tension side only.
2)

No	Moment (k-in)	Rotation (radians) X 1/1000
1	0.00	0.00
2	104.58	0.03
3	212.59	0.11
4	310.32	0.20
5	416.61	0.33
6	517.77	0.49
7	618.92	0.69
8	721.79	0.90
9	893.23	1.31
10	1032.10	1.67
11	1131.54	1.92
12	1268.70	2.33
13	1376.71	2.73
14	1448.71	3.14
15	1549.87	3.67
16	1623.59	4.10
17	1697.31	4.76
18	1724.74	5.42
19	1783.03	6.04
20	1803.61	6.56
21	1819.04	7.01

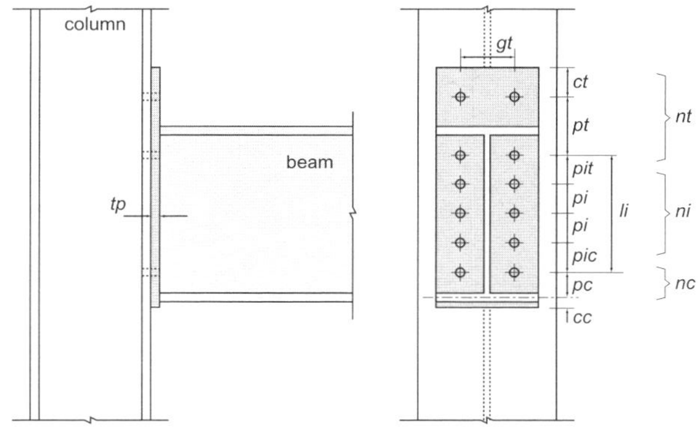

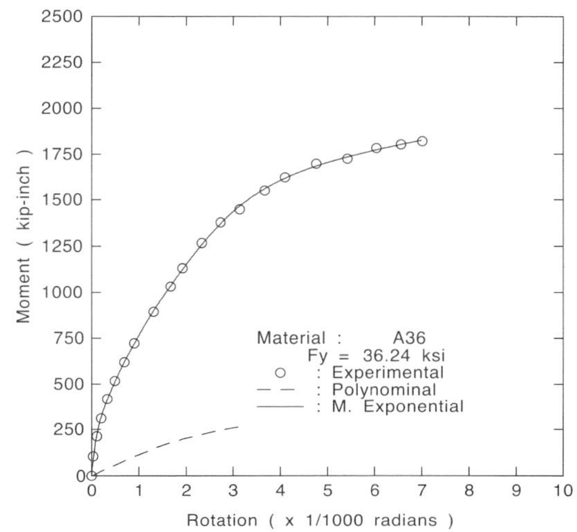

```
Moment-rotation prediction equations      ( R : X 1/1000 radians )

Frye and Morris polynominal model  :    R = Sum ( Ai X ( K*Bm )**Pi X 10**Qi )
     xd =  9.350400"    t =  1.000000"
     A1 =  1.790000    A2 =  1.760000    A3 =  2.040000    K =  0.004677
     P1 =        1     P2 =        3     P3 =        5     Q1 =        0     Q2 =       -1     Q3 =       -1

Modified exponential model  :
     Bm = Sum ( Ai X ( 1 - exp( - R/( 2*i*C ) ) ) )  +     Sum ( RKj X ( R - Rj0 ) )  +  Bm0
      C =   0.65916667E-01               Bmo=   0.00000000E+00         Nexp=  6        Nliner=  1
     AI =   0.76098344E+03  -0.33147918E+04   0.30474727E+04   0.19389850E+05  -0.45559318E+05   0.27210156E+05
    Rj0 =           0.9200
    RKj =   0.48321838E+02
```

No	Rotation (radians) X 1/1000	Moment (kip-inch) Expri.	Moment (kip-inch) Poly.	Moment (kip-inch) M.Expo.	Connection stiffness (kip-inch) X 1000 Expri.	Connection stiffness (kip-inch) X 1000 Poly.	Connection stiffness (kip-inch) X 1000 M.Expo.
1	0.00	0.00	0.00	0.00	0.3486E+04	0.1194E+03	0.2959E+04
2	0.03	104.58	3.58	78.94	0.2904E+04	0.1194E+03	0.2334E+04
3	0.11	212.59	13.14	220.84	0.1226E+04	0.1193E+03	0.1344E+04
4	0.20	310.32	23.87	317.39	0.9761E+03	0.1190E+03	0.8750E+03
5	0.33	416.61	39.28	413.88	0.7345E+03	0.1182E+03	0.6581E+03
6	0.49	517.77	58.07	512.12	0.5760E+03	0.1165E+03	0.5806E+03
7	0.69	618.92	81.07	621.17	0.4980E+03	0.1133E+03	0.5092E+03
8	0.90	721.79	104.37	720.33	0.4656E+03	0.1083E+03	0.4375E+03
9	1.31	893.23	146.13	898.68	0.4009E+03	0.9463E+02	0.4014E+03
10	1.67	1032.10	177.73	1036.55	0.3928E+03	0.8092E+02	0.3671E+03
11	1.92	1131.54	196.82	1125.73	0.3738E+03	0.7198E+02	0.3459E+03
12	2.33	1268.70	223.70	1259.21	0.3019E+03	0.5954E+02	0.3035E+03
13	2.73	1376.71	245.55	1371.02	0.2234E+03	0.5017E+02	0.2550E+03
14	3.14	1448.71	264.55	1465.38	0.1823E+03	0.4285E+02	0.2059E+03
15	3.67	1549.87	285.31	1559.69	0.1801E+03	0.3584E+02	0.1523E+03
16	4.10	1623.59	299.75	1617.77	0.1479E+03	0.3158E+02	0.1195E+03
17	4.76	1697.31	318.90	1684.59	0.7663E+02	0.2668E+02	0.8604E+02
18	5.42	1724.74	335.26	1734.60	0.6861E+02	0.2310E+02	0.6740E+02
19	6.04	1783.03	348.75	1773.19	0.6440E+02	0.2052E+02	0.5808E+02
20	6.56	1803.61	358.95	1802.17	0.3674E+02	0.1878E+02	0.5378E+02
21	7.01	1819.04	367.11	1825.84	0.3429E+02	0.1750E+02	0.5159E+02

A.5 – 241

V -121

Connection type : Extended end-plate connections
Mode : All bolted with column stiffener

Tested by : L.F.L.Ribeiro & R.M.Goncalves (1996) Brazil
Test Id. : CT1B-3

Column : CVS350x105
Beam : VS250x37
Plate thickness : 1.00"

Fasteners: A325- -3/4"D
13/16" Oversize holes
Material : A36
Fy = 36.24 ksi
Fu = 57.99 ksi

Major parameters

End plate extended on tension side only.

ct = 1.5001"	pt = 3.4921"	li = 5.8583"	pc = 1.7461"
cc = 0.7894"	gt = 3.0000"	gc = 3.0000"	
tp = 1.0000"	nt = 2 X 2	nc = 2 X 1	

Remark 1) End plate extended on tension side only.
2)

No	Moment (k-in)	Rotation (radians) X 1/1000
1	0.00	0.00
2	140.59	0.07
3	281.17	0.20
4	425.18	0.37
5	557.20	0.61
6	694.35	0.89
7	826.37	1.20
8	963.52	1.57
9	1095.54	2.03
10	1239.55	2.59
11	1328.70	3.05
12	1419.57	3.67
13	1481.29	4.30
14	1532.72	5.12
15	1536.15	6.19
16	1553.30	7.45
17	1572.16	9.24
18	1591.01	11.03
19	1615.02	12.56
20	1621.87	13.96

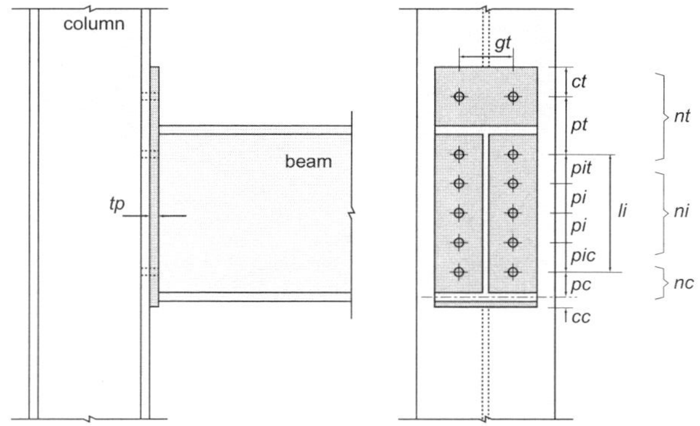

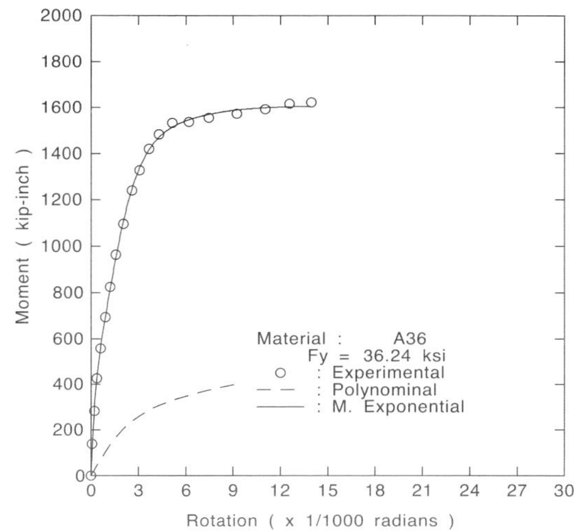

A.5 – 242

```
Moment-rotation prediction equations     ( R : X 1/1000 radians )

Frye and Morris polynominal model  :    R = Sum ( Ai X ( K*Bm )**Pi X 10**Qi )
     xd =  9.350400"    t =  1.000000"
     A1 =  1.790000   A2 =  1.760000   A3 =  2.040000    K =  0.004677
     P1 =         1   P2 =         3   P3 =         5   Q1 =         0   Q2 =        -1   Q3 =        -1

Modified exponential model  :
     Bm = Sum ( Ai X ( 1 - exp( - R/( 2*i*C ) ) ) )  +      Sum ( RKj X ( R - Rj0 ) )  +  Bm0
       C =   0.13383333E+00              Bmo=   0.00000000E+00          Nexp=  6        Nliner=  0
      AI =   0.29453522E+03   0.43419750E+04  -0.24163748E+05   0.48435405E+05  -0.43933606E+05   0.16631414E+05
```

No	Rotation (radians) X 1/1000	Moment (kip-inch) Expri.	Poly.	M.Expo.	Connection stiffness (kip-inch) X 1000 Expri.	Poly.	M.Expo.
1	0.00	0.00	0.00	0.00	0.2008E+04	0.1194E+03	0.1887E+04
2	0.07	140.59	8.36	118.65	0.1684E+04	0.1194E+03	0.1519E+04
3	0.20	281.17	23.88	283.09	0.9799E+03	0.1190E+03	0.1049E+04
4	0.37	425.18	44.00	428.89	0.7240E+03	0.1178E+03	0.7049E+03
5	0.61	557.20	71.95	568.33	0.5223E+03	0.1148E+03	0.4929E+03
6	0.89	694.35	103.29	692.99	0.4595E+03	0.1086E+03	0.4137E+03
7	1.20	826.37	135.49	816.27	0.4007E+03	0.9868E+02	0.3851E+03
8	1.57	963.52	169.45	953.39	0.3334E+03	0.8470E+02	0.3542E+03
9	2.03	1095.54	204.58	1104.04	0.2735E+03	0.6833E+02	0.2980E+03
10	2.59	1239.55	238.39	1248.88	0.2224E+03	0.5314E+02	0.2195E+03
11	3.05	1328.70	260.64	1336.24	0.1737E+03	0.4428E+02	0.1621E+03
12	3.67	1419.57	285.32	1417.61	0.1225E+03	0.3583E+02	0.1042E+03
13	4.30	1481.29	305.91	1470.44	0.8266E+02	0.2991E+02	0.6667E+02
14	5.12	1532.72	328.13	1512.82	0.3690E+02	0.2459E+02	0.3983E+02
15	6.19	1536.15	351.81	1545.64	0.7982E+01	0.1998E+02	0.2374E+02
16	7.45	1553.30	374.59	1569.35	0.1234E+02	0.1642E+02	0.1482E+02
17	9.24	1572.16	400.85	1588.91	0.1054E+02	0.1316E+02	0.7679E+01
18	11.03	1591.01	422.35	1598.65	0.1331E+02	0.1103E+02	0.3613E+01
19	12.56	1615.02	438.17	1602.61	0.1005E+02	0.9716E+01	0.1758E+01
20	13.96	1621.87	451.09	1604.38	0.4898E+01	0.8778E+01	0.8672E+00

A.5 – 243

V -122

```
Connection type : Extended end-plate connections
           Mode : All bolted with column stiffener

Tested by  :  L.F.L.Ribeiro & R.M.Goncalves (1996)        Brazil
Test  Id.  :  CT1B-4

Column : CVS350x105                Fasteners: A325- -3/4"D
Beam   : VS250x37                        13/16" Oversize holes
Plate thickness :       0.875"     Material : A36
                                        Fy =  36.24 ksi
                                        Fu =  57.99 ksi

                         Major parameters

     End plate extended on tension side only.
     ct =  1.5001"    pt =  3.4921"    li =  5.8583"    pc =  1.7461"
     cc =  0.7894"    gt =  3.0000"    gc =  3.0000"
     tp =  0.8752"    nt = 2 X  2      nc = 2 X  1

Remark  1)  End plate extended on tension side only.
        2)
```

No	Moment (k-in)	Rotation (radians) X 1/1000
1	0.00	0.00
2	146.98	0.18
3	275.58	0.42
4	410.87	0.70
5	541.15	1.01
6	671.42	1.34
7	800.03	1.73
8	942.00	2.25
9	1073.95	2.84
10	1202.55	3.68
11	1331.16	4.88
12	1416.34	6.08
13	1491.50	7.73
14	1536.59	9.27
15	1600.06	10.55
16	1628.46	12.08

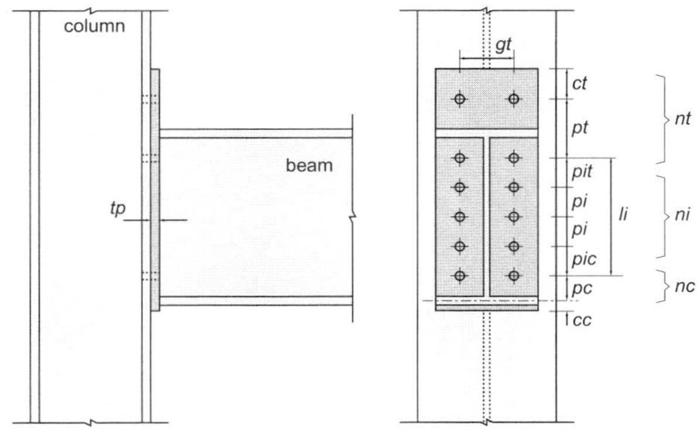

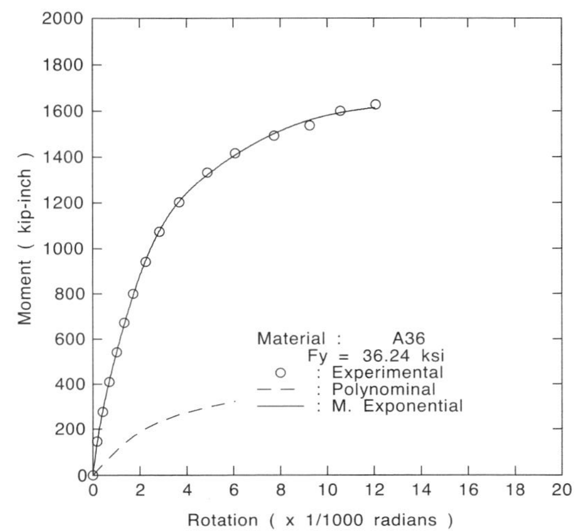

```
Moment-rotation prediction equations     ( R : X 1/1000 radians )

Frye and Morris polynominal model  :     R = Sum ( Ai X ( K*Bm )**Pi X 10**Qi )
     xd =  9.350400"    t =  0.875200"
     A1 =  1.790000    A2 =  1.760000    A3 =  2.040000    K =  0.005067
     P1 =         1    P2 =         3    P3 =         5    Q1 =         0    Q2 =        -1    Q3 =        -1

Modified exponential model  :
     Bm = Sum ( Ai X ( 1 - exp( - R/( 2*i*C ) ) ) )  +      Sum ( RKj X ( R - Rj0 ) )  +  Bm0
      C =   0.14566667E+00              Bmo=   0.00000000E+00           Nexp=  6        Nliner=  0
     AI =  -0.22238086E+03   0.78584508E+04  -0.47770928E+05   0.11649863E+06  -0.12605162E+06   0.51325988E+05
```

No	Rotation (radians) X 1/1000	Moment (kip-inch) Expri.	Poly.	M.Expo.	Connection stiffness (kip-inch) X 1000 Expri.	Poly.	M.Expo.
1	0.00	0.00	0.00	0.00	0.8165E+03	0.1103E+03	0.8645E+03
2	0.18	146.98	19.83	138.37	0.6963E+03	0.1099E+03	0.6827E+03
3	0.42	275.58	46.05	281.69	0.5115E+03	0.1083E+03	0.5267E+03
4	0.70	410.87	75.89	414.61	0.4533E+03	0.1044E+03	0.4347E+03
5	1.01	541.15	107.19	541.35	0.4079E+03	0.9701E+02	0.3886E+03
6	1.34	671.42	137.50	664.40	0.3650E+03	0.8631E+02	0.3579E+03
7	1.73	800.03	168.50	796.69	0.3054E+03	0.7264E+02	0.3191E+03
8	2.25	942.00	202.08	946.84	0.2499E+03	0.5696E+02	0.2573E+03
9	2.84	1073.95	231.72	1077.99	0.1945E+03	0.4429E+02	0.1892E+03
10	3.68	1202.55	263.69	1205.83	0.1342E+03	0.3299E+02	0.1217E+03
11	4.88	1331.16	297.29	1322.20	0.8908E+02	0.2395E+02	0.8005E+02
12	6.08	1416.34	322.74	1408.01	0.6028E+02	0.1880E+02	0.6456E+02
13	7.73	1491.50	350.00	1500.44	0.3714E+02	0.1458E+02	0.4718E+02
14	9.27	1536.59	370.38	1560.20	0.4037E+02	0.1211E+02	0.3076E+02
15	10.55	1600.06	384.88	1592.17	0.3545E+02	0.1064E+02	0.1972E+02
16	12.08	1628.46	400.10	1614.89	0.1856E+02	0.9314E+01	0.1071E+02

V -123

Connection type : Extended end-plate connections
Mode : All bolted with column stiffener

Tested by : L.F.L.Ribeiro & R.M.Goncalves (1996) Brazil
Test Id. : CT1B-5

Column : CVS350x105
Beam : VS250x37
Plate thickness : 0.875"

Fasteners: A325- -3/4"D
13/16" Oversize holes
Material : A36
Fy = 36.24 ksi
Fu = 57.99 ksi

Major parameters

End plate extended on tension side only.

ct = 1.5001"	pt = 3.4921"	li = 5.8583"	pc = 1.7461"
cc = 0.7894"	gt = 3.0000"	gc = 3.0000"	
tp = 0.8752"	nt = 2 X 2	nc = 2 X 1	

Remark 1) End plate extended on tension side only.
2)

No	Moment (k-in)	Rotation (radians) X 1/1000
1	0.00	0.00
2	1.67	0.07
3	143.64	0.10
4	277.25	0.27
5	405.86	0.47
6	537.81	0.74
7	676.43	1.06
8	810.05	1.45
9	945.34	1.90
10	1070.60	2.41
11	1205.89	3.26
12	1329.49	4.66
13	1386.28	5.87
14	1458.09	7.91
15	1503.19	9.56
16	1534.92	11.64
17	1551.63	13.52
18	1563.32	15.58
19	1575.01	17.64

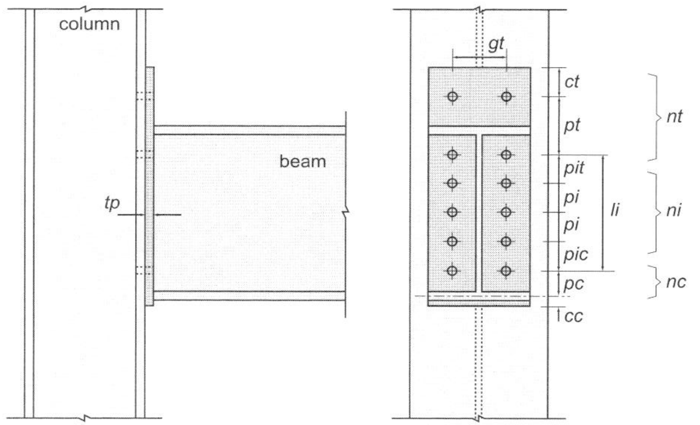

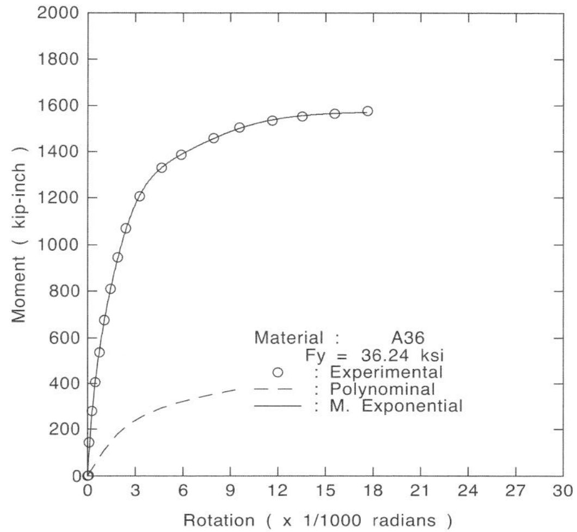

A.5 – 246

```
Moment-rotation prediction equations     ( R : X 1/1000 radians )

Frye and Morris polynominal model  :    R = Sum ( Ai X ( K*Bm )**Pi X 10**Qi )
     xd =  9.350400"   t =  0.875200"
     A1 =  1.790000   A2 =  1.760000   A3 =  2.040000    K =  0.005067
     P1 =         1   P2 =         3   P3 =         5   Q1 =         0   Q2 =      -1   Q3 =      -1

Modified exponential model  :
     Bm = Sum ( Ai X ( 1 - exp( - R/( 2*i*C ) ) ) )  +     Sum ( RKj X ( R - Rj0 ) )  +  Bm0
      C =   0.16450000E+00            Bmo=   0.00000000E+00          Nexp=  6       Nliner=  0
     AI =  -0.13638466E+03   0.72699564E+04  -0.41730807E+05   0.97431947E+05  -0.10036832E+06   0.39106961E+05
```

No	Rotation (radians) X 1/1000	Moment (kip-inch) Expri.	Moment (kip-inch) Poly.	Moment (kip-inch) M.Expo.	Connection stiffness (kip-inch) X 1000 Expri.	Connection stiffness (kip-inch) X 1000 Poly.	Connection stiffness (kip-inch) X 1000 M.Expo.
1	0.00	0.00	0.00	0.00	0.2385E+02	0.1103E+03	0.1187E+04
2	0.07	1.67	7.72	78.63	0.3320E+04	0.1102E+03	0.1063E+04
3	0.10	143.64	11.02	109.77	0.4140E+04	0.1102E+03	0.1014E+04
4	0.27	277.25	29.70	261.98	0.7203E+03	0.1095E+03	0.7896E+03
5	0.47	405.86	51.45	400.76	0.5774E+03	0.1078E+03	0.6112E+03
6	0.74	537.81	80.05	544.46	0.4633E+03	0.1036E+03	0.4677E+03
7	1.06	676.43	112.00	678.18	0.3924E+03	0.9553E+02	0.3778E+03
8	1.45	810.05	146.79	812.77	0.3231E+03	0.8243E+02	0.3173E+03
9	1.90	945.34	180.39	943.72	0.2748E+03	0.6706E+02	0.2658E+03
10	2.41	1070.60	210.86	1065.18	0.2132E+03	0.5303E+02	0.2108E+03
11	3.26	1205.89	248.85	1208.53	0.1324E+03	0.3790E+02	0.1302E+03
12	4.66	1329.49	291.88	1331.22	0.6610E+02	0.2522E+02	0.5750E+02
13	5.87	1386.28	318.73	1386.94	0.4257E+02	0.1953E+02	0.3861E+02
14	7.91	1458.09	352.53	1455.95	0.3085E+02	0.1424E+02	0.3007E+02
15	9.56	1503.19	373.84	1499.81	0.2199E+02	0.1174E+02	0.2278E+02
16	11.64	1534.92	395.95	1536.96	0.1191E+02	0.9656E+01	0.1324E+02
17	13.52	1551.63	412.80	1555.61	0.7353E+01	0.8355E+01	0.7073E+01
18	15.58	1563.32	428.87	1565.78	0.5675E+01	0.7302E+01	0.3227E+01
19	17.64	1575.01	443.05	1570.27	0.5675E+01	0.6501E+01	0.1373E+01

V -124

Connection type : Extended end-plate connections
Mode : All bolted with column stiffener

Tested by : L.F.L.Ribeiro & R.M.Goncalves (1996) Brazil
Test Id. : CT1B-6

Column : CVS350x105
Beam : VS250x37
Plate thickness : 0.75"

Fasteners: A325- -3/4"D
13/16" Oversize holes
Material : A36
Fy = 36.24 ksi
Fu = 57.99 ksi

Major parameters

End plate extended on tension side only.

ct = 1.5001"	pt = 3.4921"	li = 5.8583"	pc = 1.7461"
cc = 0.7894"	gt = 3.0000"	gc = 3.0000"	
tp = 0.7500"	nt = 2 X 2	nc = 2 X 1	

Remark 1) End plate extended on tension side only.
2)

No	Moment (k-in)	Rotation (radians) X 1/1000
1	0.00	0.00
2	108.56	0.08
3	208.78	0.21
4	302.31	0.35
5	410.87	0.58
6	514.42	0.83
7	607.96	1.09
8	708.17	1.45
9	806.71	1.93
10	906.92	2.58
11	1003.80	3.38
12	1117.37	4.75
13	1210.90	6.51
14	1271.03	7.73
15	1344.52	9.75
16	1384.60	11.03
17	1428.03	12.61
18	1484.82	14.38
19	1546.62	17.17
20	1559.14	19.37
21	1571.67	21.57

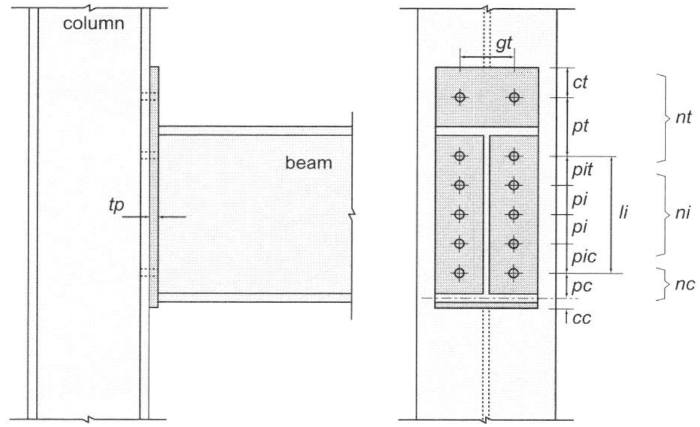

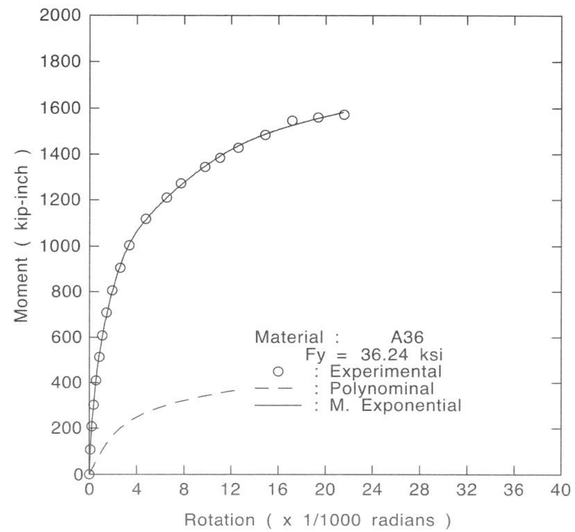

```
Moment-rotation prediction equations     ( R : X 1/1000 radians )

Frye and Morris polynominal model  :    R = Sum ( Ai X ( K*Bm )**Pi X 10**Qi )
     xd =  9.350400"    t =  0.750000"
     A1 =  1.790000   A2 =  1.760000   A3 =  2.040000    K =  0.005559
     P1 =        1    P2 =        3    P3 =        5    Q1 =        0    Q2 =       -1    Q3 =       -1

Modified expcnential model  :
     Bm = Sum ( Ai X ( 1 - exp( - R/( 2*i*C ) ) ) )  +     Sum ( RKj X ( R - Rj0 ) )  +  Bm0
      C =   0.19975000E+00          Bmo=   0.00000000E+00         Nexp=  6        Nliner=  1
     AI =   0.41590348E+03   0.18575484E+04  -0.19805563E+05   0.58904132E+05  -0.69942263E+05   0.29971058E+05
     Rj0 =          4.7500
     RKj =   0.10961984E+02
```

No	Rotation (radians) X 1/1000	Moment (kip-inch) Expri.	Moment (kip-inch) Poly.	Moment (kip-inch) M.Expo.	Connection stiffness (kip-inch) X 1000 Expri.	Connection stiffness (kip-inch) X 1000 Poly.	Connection stiffness (kip-inch) X 1000 M.Expo.
1	0.00	0.00	0.00	0.00	0.1357E+04	0.1005E+03	0.1190E+04
2	0.08	108.56	8.04	87.82	0.1134E+04	0.1004E+03	0.1011E+04
3	0.21	208.78	21.08	204.07	0.7214E+03	0.1001E+03	0.7897E+03
4	0.35	302.31	35.04	302.18	0.5939E+03	0.9931E+02	0.6220E+03
5	0.58	410.87	57.64	423.53	0.4443E+03	0.9698E+02	0.4501E+03
6	0.83	514.42	81.39	521.95	0.3875E+03	0.9269E+02	0.3476E+03
7	1.09	607.96	104.69	603.86	0.3256E+03	0.8625E+02	0.2878E+03
8	1.45	708.17	133.80	698.01	0.2471E+03	0.7514E+02	0.2393E+03
9	1.93	806.71	166.27	801.94	0.1836E+03	0.6026E+02	0.1957E+03
10	2.58	906.92	200.08	913.02	0.1393E+03	0.4493E+02	0.1475E+03
11	3.38	1003.80	230.89	1011.15	0.1070E+03	0.3315E+02	0.1004E+03
12	4.75	1117.37	268.11	1112.78	0.6988E+02	0.2250E+02	0.5485E+02
13	6.51	1210.90	301.25	1209.16	0.5086E+02	0.1593E+02	0.4806E+02
14	7.73	1271.03	318.99	1265.09	0.4443E+02	0.1329E+02	0.4392E+02
15	9.75	1344.52	342.81	1347.19	0.3328E+02	0.1049E+02	0.3702E+02
16	11.03	1384.60	355.39	1391.29	0.2960E+02	0.9281E+01	0.3185E+02
17	12.61	1428.03	369.11	1436.68	0.2647E+02	0.8144E+01	0.2574E+02
18	14.88	1484.82	386.17	1486.96	0.2600E+02	0.6951E+01	0.1903E+02
19	17.17	1546.62	401.04	1525.42	0.1613E+02	0.6077E+01	0.1496E+02
20	19.37	1559.14	413.67	1555.79	0.5694E+01	0.5435E+01	0.1288E+02
21	21.57	1571.67	425.05	1582.84	0.5694E+01	0.4926E+01	0.1184E+02

V -125

```
Connection type : Extended end-plate connections
           Mode : All bolted with column stiffener

Tested by  :  L.F.L.Ribeiro & R.M.Goncalves (1996)        Brazil
Test  Id.  :  CT2A-1

Column : CVS350x128                  Fasteners: A325- -7/8"D
Beam   : VS350x58                          15/16" Oversize holes
Plate thickness :       1.5"         Material : A36
                                           Fy =  36.24 ksi
                                           Fu =  57.99 ksi

                         Major parameters

     End plate extended on tension side only.
     ct =  1.7504"    pt =  4.1307"    li =  9.0189"    pc =  2.0654"
     cc =  0.7512"    gt =  3.5008"    gc =  3.5008"
     tp =  1.5000"    nt = 2 X  2      nc = 2 X  1

Remark  1)  End plate extended on tension side only.
        2)
```

No	Moment (k-in)	Rotation (radians) X 1/1000
1	0.00	0.00
2	201.74	0.13
3	411.75	0.34
4	598.61	0.55
5	800.35	0.82
6	1002.09	1.09
7	1165.80	1.30
8	1326.20	1.51
9	1501.49	1.60
10	1656.93	1.97
11	1828.90	2.22
12	1987.65	2.46
13	2154.67	2.78
14	2326.64	3.14
15	2473.81	3.47
16	2655.71	3.97
17	2776.43	4.30
18	2920.29	4.99
19	3057.54	5.87
20	3153.45	6.75
21	3251.02	7.80
22	3315.51	9.03

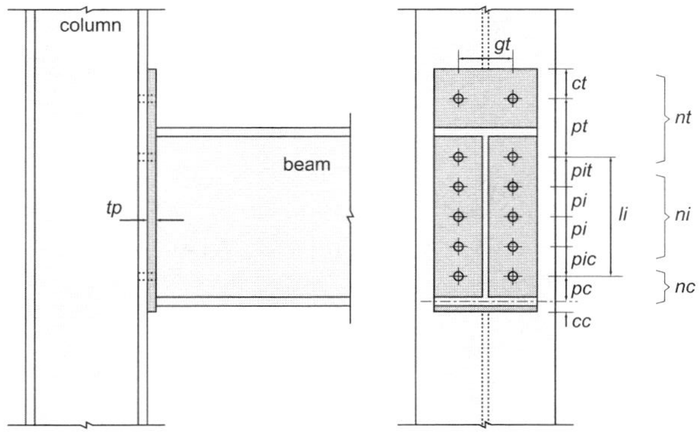

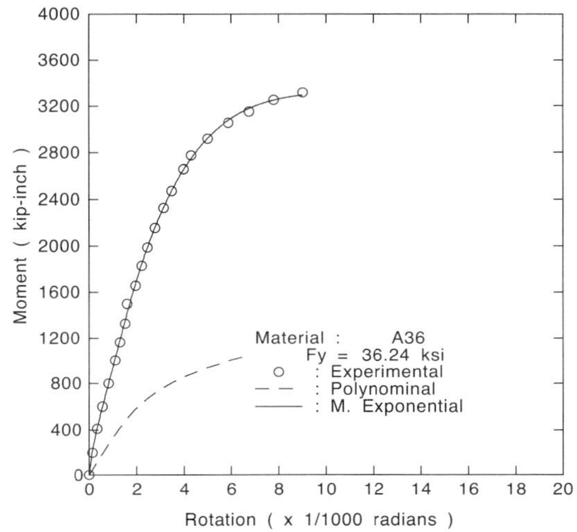

A.5 – 250

```
Moment-rotation prediction equations     ( R : X 1/1000 radians )

Frye and Morris polynominal model  :     R = Sum ( Ai X ( K*Bm )**Pi X 10**Qi )
     xd = 13.149600"    t =  1.500000"
     A1 =  1.790000   A2 =  1.760000   A3 =  2.040000    K =  0.001618
     P1 =         1   P2 =         3   P3 =         5   Q1 =         0   Q2 =        -1   Q3 =        -1

Modified exponential model  :
     Bm = Sum ( Ai X ( 1 - exp( - R/( 2*i*C ) ) ) )  +      Sum ( RKj X ( R - Rj0 ) )  +  Bm0
      C =   0.10775000E+00             Bmo=   0.00000000E+00           Nexp=  6        Nliner=  0
     AI =   0.18048972E+03   0.54908272E+04  -0.39450077E+05   0.11184871E+06  -0.14627830E+06   0.71537676E+05
```

No	Rotation (radians) X 1/1000	Moment (kip-inch) Expri.	Poly.	M.Expo.	Connection stiffness (kip-inch) X 1000 Expri.	Poly.	M.Expo.
1	0.00	0.00	0.00	0.00	0.1552E+04	0.3453E+03	0.1881E+04
2	0.13	201.74	44.89	200.62	0.1341E+04	0.3448E+03	0.1273E+04
3	0.34	411.75	116.97	417.43	0.9449E+03	0.3414E+03	0.8738E+03
4	0.55	598.61	188.02	588.55	0.8274E+03	0.3345E+03	0.7835E+03
5	0.82	800.35	276.45	800.10	0.7472E+03	0.3192E+03	0.7915E+03
6	1.09	1002.09	359.70	1016.05	0.7654E+03	0.2963E+03	0.8041E+03
7	1.30	1165.80	419.69	1184.17	0.7717E+03	0.2747E+03	0.7940E+03
8	1.51	1326.20	474.95	1348.39	0.1592E+04	0.2515E+03	0.7676E+03
9	1.60	1501.49	497.14	1416.81	0.1649E+04	0.2416E+03	0.7525E+03
10	1.97	1656.93	579.33	1681.84	0.5799E+03	0.2032E+03	0.6776E+03
11	2.22	1828.90	627.24	1844.41	0.6744E+03	0.1810E+03	0.6229E+03
12	2.46	1987.65	668.40	1987.72	0.6017E+03	0.1625E+03	0.5718E+03
13	2.78	2154.67	717.05	2160.41	0.5011E+03	0.1421E+03	0.5085E+03
14	3.14	2326.64	764.80	2331.80	0.4612E+03	0.1239E+03	0.4450E+03
15	3.47	2473.81	803.39	2470.00	0.4133E+03	0.1105E+03	0.3935E+03
16	3.97	2655.71	854.52	2649.25	0.3650E+03	0.9471E+02	0.3253E+03
17	4.30	2776.43	884.36	2749.95	0.3149E+03	0.8649E+02	0.2856E+03
18	4.99	2920.29	939.22	2921.45	0.1854E+03	0.7316E+02	0.2139E+03
19	5.87	3057.54	998.03	3076.61	0.1325E+03	0.6119E+02	0.1423E+03
20	6.75	3153.45	1047.92	3177.68	0.1017E+03	0.5267E+02	0.9053E+02
21	7.80	3251.02	1099.12	3249.78	0.7427E+02	0.4525E+02	0.5018E+02
22	9.03	3315.51	1150.69	3293.44	0.5243E+02	0.3894E+02	0.2376E+02

A.5 – 251

V -126

Connection type : Extended end-plate connections
Mode : All bolted with column stiffener

Tested by : L.F.L.Ribeiro & R.M.Goncalves (1996) Brazil
Test Id. : CT2A-2

Column : CVS350x128 Fasteners: A325- -7/8"D
Beam : VS350x58 15/16" Oversize holes
Plate thickness : 1.25" Material : A36
Fy = 36.24 ksi
Fu = 57.99 ksi

Major parameters

End plate extended on tension side only.
ct = 1.7504" pt = 4.1307" li = 9.0189" pc = 2.0654"
cc = 0.7512" gt = 3.5008" gc = 3.5008"
tp = 1.2500" nt = 2 X 2 nc = 2 X 1

Remark 1) End plate extended on tension side only.
2)

No	Moment (k-in)	Rotation (radians) X 1/1000
1	0.00	0.00
2	174.21	0.08
3	341.78	0.21
4	506.04	0.35
5	666.98	0.48
6	834.55	0.64
7	1008.76	0.81
8	1185.74	1.00
9	1338.93	1.16
10	1493.23	1.34
11	1669.10	1.56
12	1841.66	1.81
13	1989.32	2.04
14	2158.55	2.44
15	2322.81	2.82
16	2487.07	3.24
17	2652.98	3.79
18	2822.21	4.43
19	2979.83	5.34
20	3046.20	5.76
21	3144.09	6.95
22	3218.75	7.55
23	3280.14	8.73
24	3334.89	9.94

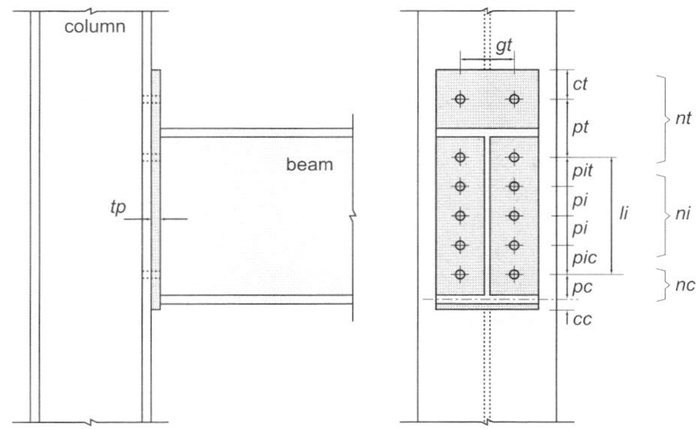

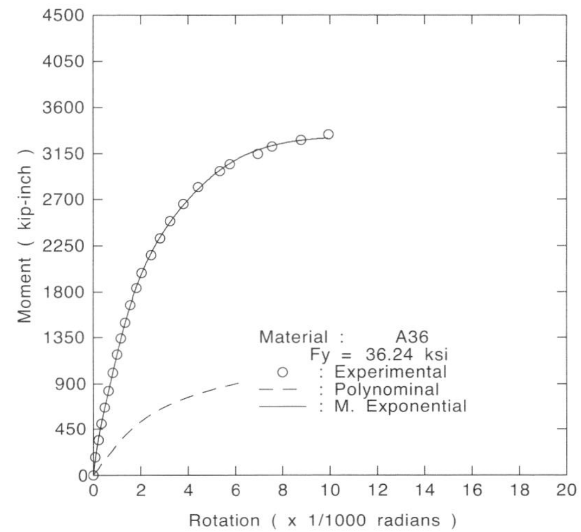

```
Moment-rotation prediction equations     ( R : X 1/1000 radians )

Frye and Morris polynominal model  :    R = Sum ( Ai X ( K*Bm )**Pi X 10**Qi )
     xd = 13.149600"   t =  1.250000"
     A1 =  1.790000   A2 =  1.760000   A3 =  2.040000    K =  0.001805
     P1 =        1    P2 =        3    P3 =        5    Q1 =        0    Q2 =       -1    Q3 =       -1

Modified exponential model  :
     Bm = Sum ( Ai X ( 1 - exp( - R/( 2*i*C ) ) ) )  +      Sum ( RKj X ( R - Rj0 ) )  +  Bm0
       C =   0.10283333E+00            Bmo=   0.00000000E+00         Nexp=  6        Nliner=  0
      AI =   0.76141510E+02   0.93652519E+04  -0.71312877E+05   0.19548763E+06  -0.23111977E+06   0.10082128E+06
```

No	Rotation (radians) X 1/1000	Moment (kip-inch) Expri.	Moment (kip-inch) Poly.	Moment (kip-inch) M.Expo.	Connection stiffness (kip-inch) X 1000 Expri.	Connection stiffness (kip-inch) X 1000 Poly.	Connection stiffness (kip-inch) X 1000 M.Expo.
1	0.00	0.00	0.00	0.00	0.2178E+04	0.3095E+03	0.2136E+04
2	0.08	174.21	24.76	153.14	0.1839E+04	0.3093E+03	0.1721E+04
3	0.21	341.78	64.91	348.61	0.1233E+04	0.3082E+03	0.1333E+04
4	0.35	506.04	107.91	520.00	0.1207E+04	0.3058E+03	0.1143E+04
5	0.48	666.98	147.46	663.09	0.1153E+04	0.3023E+03	0.1069E+04
6	0.64	834.55	195.36	830.70	0.1036E+04	0.2961E+03	0.1031E+04
7	0.81	1008.76	244.93	1003.56	0.9898E+03	0.2867E+03	0.1002E+04
8	1.00	1185.74	297.08	1186.15	0.9423E+03	0.2734E+03	0.9582E+03
9	1.16	1338.93	340.80	1339.09	0.8975E+03	0.2594E+03	0.9048E+03
10	1.34	1493.23	385.97	1495.75	0.8312E+03	0.2423E+03	0.8344E+03
11	1.56	1669.10	436.89	1669.13	0.7483E+03	0.2205E+03	0.7414E+03
12	1.81	1841.66	488.98	1841.51	0.6651E+03	0.1965E+03	0.6391E+03
13	2.04	1989.32	531.80	1978.72	0.5621E+03	0.1763E+03	0.5561E+03
14	2.44	2158.55	596.27	2177.50	0.4278E+03	0.1470E+03	0.4448E+03
15	2.82	2322.81	647.82	2332.14	0.4127E+03	0.1254E+03	0.3740E+03
16	3.24	2487.07	696.47	2477.66	0.3524E+03	0.1071E+03	0.3225E+03
17	3.79	2652.98	750.30	2641.65	0.2845E+03	0.8952E+02	0.2762E+03
18	4.43	2822.21	802.67	2803.96	0.2268E+03	0.7495E+02	0.2315E+03
19	5.34	2979.83	864.05	2986.60	0.1628E+03	0.6083E+02	0.1703E+03
20	5.76	3046.20	888.48	3052.45	0.1382E+03	0.5600E+02	0.1436E+03
21	6.96	3144.09	949.14	3184.76	0.1101E+03	0.4569E+02	0.8067E+02
22	7.56	3218.75	975.31	3226.07	0.1000E+03	0.4190E+02	0.5800E+02
23	8.78	3280.14	1022.61	3276.50	0.4872E+02	0.3591E+02	0.2781E+02
24	9.94	3334.89	1061.69	3299.20	0.4720E+02	0.3169E+02	0.1302E+02

V -127

```
Connection type : Extended end-plate connections
           Mode : All bolted with column stiffener

Tested by  :  L.F.L.Ribeiro & R.M.Goncalves (1996)        Brazil
Test  Id.  :  CT2A-3

Column : CVS350x128                  Fasteners: A325- -7/8"D
Beam   : VS350x58                          15/16" Oversize holes
Plate thickness :      1.25"         Material : A36
                                          Fy =  36.24 ksi
                                          Fu =  57.99 ksi

                          Major parameters

     End plate extended on tension side only.
     ct =  1.7504"    pt =  4.1307"    li =  9.0189"    pc =  2.0654"
     cc =  1.1449"    gt =  3.5008"    gc =  3.5008"
     tp =  1.2500"    nt = 2 X  2      nc = 2 X  1

Remark  1)  End plate extended on tension side only.
        2)
```

No	Moment (k-in)	Rotation (radians) X 1/1000
1	0.00	0.00
2	165.91	0.17
3	335.15	0.42
4	502.72	0.63
5	675.27	0.84
6	831.23	1.08
7	993.83	1.31
8	1166.38	1.57
9	1370.46	1.85
10	1506.51	2.06
11	1659.15	2.30
12	1828.38	2.66
13	1989.32	3.22
14	2151.92	3.71
15	2322.81	4.29
16	2477.11	4.85
17	2664.59	5.48
18	2794.01	5.97
19	2981.49	6.69
20	3112.57	7.35
21	3227.05	7.93
22	3389.64	8.90
23	3480.90	9.67
24	3501.64	11.01
25	3522.38	12.35

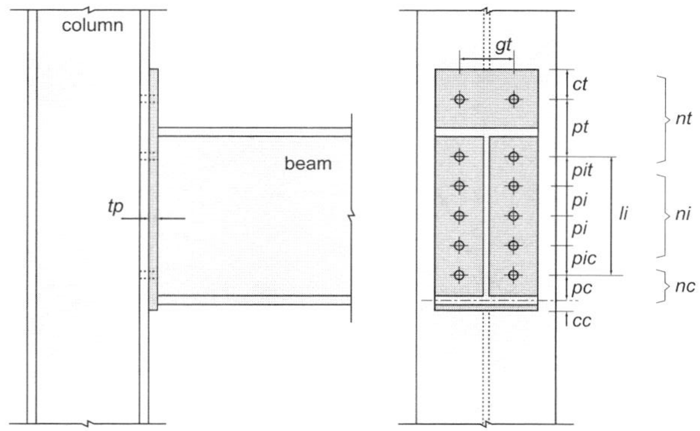

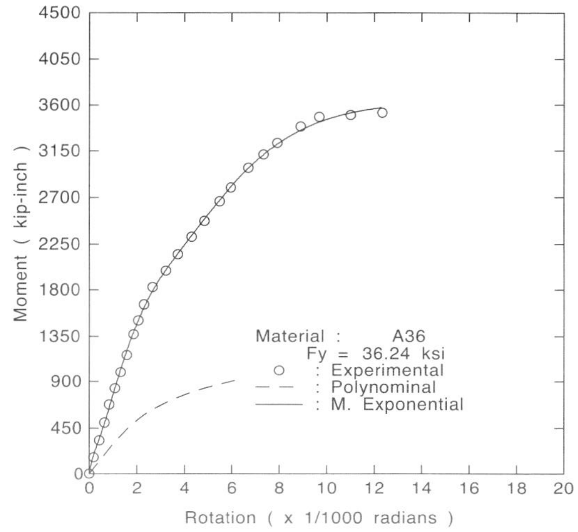

A.5 – 254

```
Moment-rotation prediction equations      ( R : X 1/1000 radians )

Frye and Morris polynominal model  :     R = Sum ( Ai X ( K*Bm )**Pi X 10**Qi )
     xd = 13.149600"   t =  1.250000"
     A1 =  1.790000   A2 =  1.760000   A3 =  2.040000    K =  0.001805
     P1 =         1   P2 =         3   P3 =         5   Q1 =         0    Q2 =        -1    Q3 =        -1

Modified exponential model  :
     Bm = Sum ( Ai X ( 1 - exp( - R/( 2*i*C ) ) ) )  +      Sum ( RKj X ( R - Rj0 ) )  + Bm0
      C =   0.14541667E+00              Bmo=  0.00000000E+00          Nexp=  6        Nliner=  0
     AI =   0.13864588E+04  -0.29667709E+04  -0.44134747E+05   0.19392017E+06  -0.27511782E+06   0.13054566E+06
```

No	Rotation (radians) X 1/1000	Moment (kip-inch) Expri.	Poly.	M.Expo.	Connection stiffness (kip-inch) X 1000 Expri.	Poly.	M.Expo.
1	0.00	0.00	0.00	0.00	0.9760E+03	0.3095E+03	0.1395E+04
2	0.17	165.91	52.62	177.17	0.8549E+03	0.3087E+03	0.7986E+03
3	0.42	335.15	129.26	347.57	0.7427E+03	0.3041E+03	0.6436E+03
4	0.63	502.72	192.40	488.33	0.8098E+03	0.2965E+03	0.7050E+03
5	0.84	675.27	253.51	643.59	0.7415E+03	0.2848E+03	0.7685E+03
6	1.08	831.23	319.76	831.90	0.6790E+03	0.2665E+03	0.7910E+03
7	1.31	993.83	378.66	1011.48	0.6866E+03	0.2452E+03	0.7640E+03
8	1.57	1166.38	439.09	1202.11	0.6950E+03	0.2195E+03	0.6980E+03
9	1.85	1370.46	496.77	1385.50	0.6826E+03	0.1928E+03	0.6111E+03
10	2.06	1506.51	535.31	1507.02	0.6423E+03	0.1747E+03	0.5471E+03
11	2.30	1659.15	574.99	1630.31	0.5696E+03	0.1564E+03	0.4821E+03
12	2.66	1828.38	627.09	1789.48	0.3986E+03	0.1338E+03	0.4067E+03
13	3.22	1989.32	694.39	1995.69	0.3111E+03	0.1079E+03	0.3386E+03
14	3.71	2151.92	743.04	2154.29	0.3148E+03	0.9173E+02	0.3124E+03
15	4.29	2322.81	791.97	2331.21	0.2849E+03	0.7773E+02	0.2993E+03
16	4.85	2477.11	832.55	2496.25	0.2859E+03	0.6770E+02	0.2898E+03
17	5.48	2664.59	872.38	2674.20	0.2788E+03	0.5913E+02	0.2739E+03
18	5.97	2794.01	900.01	2804.25	0.2626E+03	0.5386E+02	0.2562E+03
19	6.69	2981.49	936.47	2977.40	0.2282E+03	0.4766E+02	0.2238E+03
20	7.35	3112.57	966.39	3114.24	0.1979E+03	0.4315E+02	0.1907E+03
21	7.93	3227.05	990.43	3216.38	0.1862E+03	0.3987E+02	0.1617E+03
22	8.90	3389.64	1026.87	3351.26	0.1402E+03	0.3542E+02	0.1176E+03
23	9.67	3480.90	1053.01	3430.28	0.8091E+02	0.3258E+02	0.8858E+02
24	11.01	3501.64	1093.91	3522.27	0.1548E+02	0.2863E+02	0.5141E+02
25	12.35	3522.38	1130.15	3574.42	0.1548E+02	0.2559E+02	0.2842E+02

V -128

```
Connection type : Extended end-plate connections
           Mode : All bolted with column stiffener

Tested by   :  L.F.L.Ribeiro & R.M.Goncalves (1996)        Brazil
Test  Id.   :  CT2A-4

Column : CVS350x128                 Fasteners: A325- -7/8"D
Beam   : VS35Cx58                         15/16" Oversize holes
Plate thickness :      1.00"        Material : A36
                                         Fy =  36.24 ksi
                                         Fu =  57.99 ksi

                          Major parameters

     End plate extended on tension side only.
     ct =  1.7504"    pt =  4.1307"    li =  9.0189"    pc =  2.0654"
     cc =  0.7512"    gt =  3.5008"    gc =  3.5008"
     tp =  1.0000"    nt = 2 X  2      nc = 2 X  1

Remark  1)  End plate extended on tension side only.
        2)
```

No	Moment (k-in)	Rotation (radians) X 1/1000
1	0.00	0.00
2	169.80	0.07
3	334.60	0.21
4	512.72	0.33
5	677.53	0.53
6	843.99	0.71
7	998.81	0.91
8	1175.26	1.27
9	1336.74	1.71
10	1496.55	2.09
11	1664.68	2.59
12	1831.15	3.10
13	2005.94	3.64
14	2209.03	4.43
15	2358.85	5.95
16	2460.40	7.45
17	2526.98	8.56
18	2585.25	9.70
19	2635.19	10.72
20	2675.14	11.73
21	2730.08	12.83
22	2778.35	13.94
23	2859.92	15.32

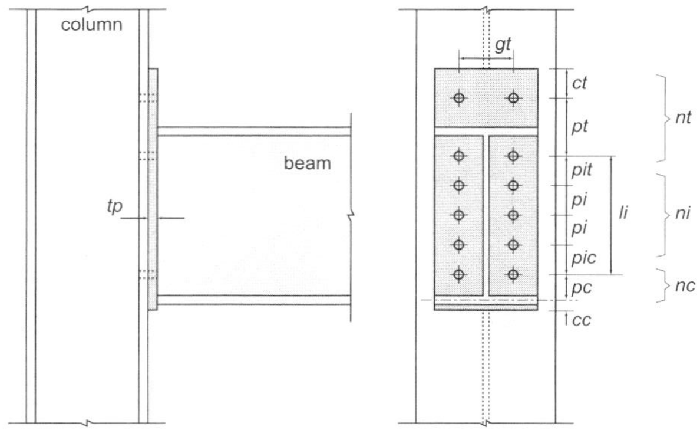

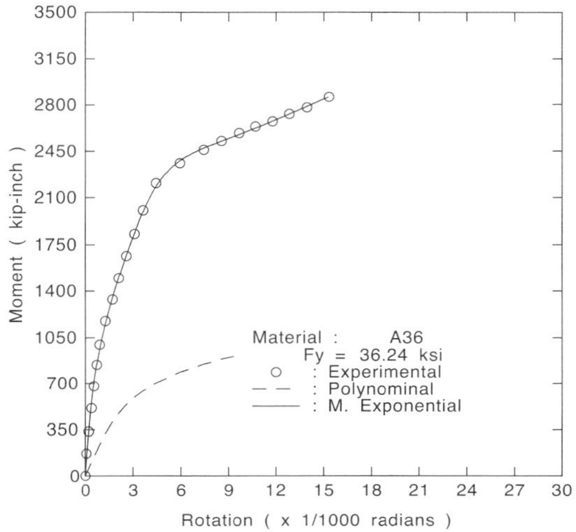

```
Moment-rotation prediction equations      ( R : X 1/1000 radians )

Frye and Morris polynominal model   :    R = Sum ( Ai X ( K*Bm )**Pi X 10**Qi )
     xd = 13.149600"   t =  1.000000"
     A1 =  1.790000   A2 =  1.760000   A3 =  2.040000    K =  0.002064
     P1 =        1    P2 =        3    P3 =        5    Q1 =        0    Q2 =      -1    Q3 =      -1

Modified exponential model   :
     Bm = Sum ( Ai X ( 1 - exp( - R/( 2*i*C ) ) ) )  +      Sum ( RKj X ( R - Rj0 ) )  +  Bm0
       C =  0.14516667E+00            Bmo=  0.00000000E+00         Nexp=  6         Nliner=  1
      AI = -0.57748394E+03  0.49251119E+04  0.29221423E+04 -0.44462986E+05  0.65714142E+05 -0.26336300E+05
     Rj0 =          3.1000
     RKj =  0.55079880E+02
```

No	Rotation (radians) X 1/1000	Moment (kip-inch) Expri.	Poly.	M.Expo.	Connection stiffness (kip-inch) X 1000 Expri.	Poly.	M.Expo.
1	0.00	0.00	0.00	0.00	0.2426E+04	0.2707E+03	0.1711E+04
2	0.07	169.80	18.95	116.68	0.2010E+04	0.2706E+03	0.1619E+04
3	0.21	334.60	56.78	328.03	0.1119E+04	0.2696E+03	0.1396E+04
4	0.38	512.72	102.40	541.89	0.1075E+04	0.2669E+03	0.1125E+04
5	0.53	677.53	142.16	694.89	0.1020E+04	0.2629E+03	0.9220E+03
6	0.71	843.99	188.90	843.11	0.8534E+03	0.2559E+03	0.7351E+03
7	0.91	998.81	239.03	974.95	0.6727E+03	0.2449E+03	0.5937E+03
8	1.27	1175.26	322.58	1161.24	0.4347E+03	0.2179E+03	0.4616E+03
9	1.71	1336.74	410.19	1349.40	0.3957E+03	0.1800E+03	0.4038E+03
10	2.09	1496.55	472.83	1497.76	0.3842E+03	0.1506E+03	0.3777E+03
11	2.59	1664.68	540.28	1677.08	0.3314E+03	0.1205E+03	0.3373E+03
12	3.10	1831.15	595.79	1835.59	0.3251E+03	0.9853E+02	0.2825E+03
13	3.64	2005.94	644.27	2000.60	0.2966E+03	0.8200E+02	0.2734E+03
14	4.43	2209.03	702.14	2181.44	0.2029E+03	0.6554E+02	0.1874E+03
15	5.95	2358.85	786.29	2377.95	0.8303E+02	0.4728E+02	0.8541E+02
16	7.45	2460.40	849.16	2474.18	0.6327E+02	0.3720E+02	0.5040E+02
17	8.56	2526.98	887.46	2525.97	0.5561E+02	0.3224E+02	0.4456E+02
18	9.70	2585.25	921.92	2576.76	0.4998E+02	0.2840E+02	0.4513E+02
19	10.72	2635.19	949.46	2623.86	0.4343E+02	0.2571E+02	0.4731E+02
20	11.78	2675.14	975.47	2675.27	0.4504E+02	0.2344E+02	0.4963E+02
21	12.83	2730.08	999.08	2728.40	0.4803E+02	0.2158E+02	0.5147E+02
22	13.94	2778.35	1022.08	2786.34	0.5045E+02	0.1992E+02	0.5284E+02
23	15.32	2859.92	1048.36	2860.06	0.5911E+02	0.1822E+02	0.5390E+02

A.5 – 257

V -129

```
Connection type : Extended end-plate connections
           Mode : All bolted with column stiffener

Tested by   :  L.F.L.Ribeiro & R.M.Goncalves (1996)         Brazil
Test   Id.  :  CT2A-5

Column : CVS350x128                     Fasteners: A325- -1"D
Beam   : VS350x58                             1 1/16" Oversize holes
Plate thickness :  ,   1.00"            Material : A36
                                             Fy =  36.24 ksi
                                             Fu =  57.99 ksi

                            Major parameters

     End plate extended on tension side only.
     ct =  2.0001"     pt =  4.6299"    li =  8.5197"     pc =  2.3150"
     cc =  0.6457"     gt =  4.0000"    gc =  4.0000"
     tp =  1.0000"     nt = 2 X  2      nc = 2 X  1

Remark  1)  End plate extended on tension side only.
        2)
```

No	Moment (k-in)	Rotation (radians) X 1/1000
1	0.00	0.00
2	179.79	0.11
3	352.91	0.25
4	506.06	0.38
5	679.19	0.52
6	834.00	0.68
7	1005.47	0.87
8	1171.93	1.07
9	1333.41	1.28
10	1513.19	1.54
11	1671.34	1.78
12	1832.81	2.05
13	1995.95	2.44
14	2174.07	2.96
15	2352.19	3.61
16	2530.31	4.20
17	2678.47	4.81
18	2853.26	5.68
19	3016.40	6.52
20	3094.64	7.33
21	3292.74	8.56
22	3477.52	9.84
23	3562.42	10.63
24	3629.00	11.85
25	3737.21	13.33
26	3752.19	14.54

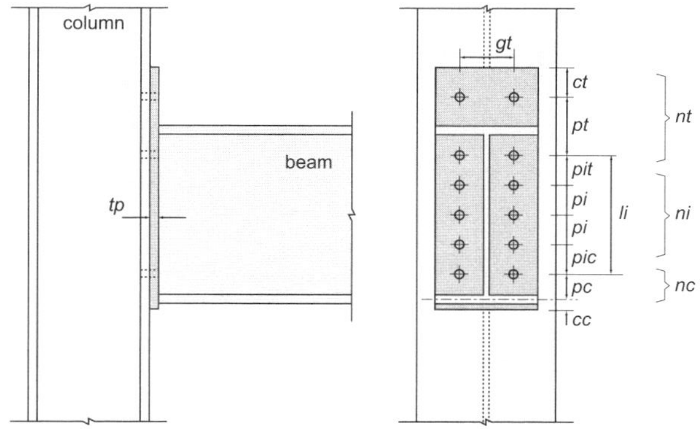

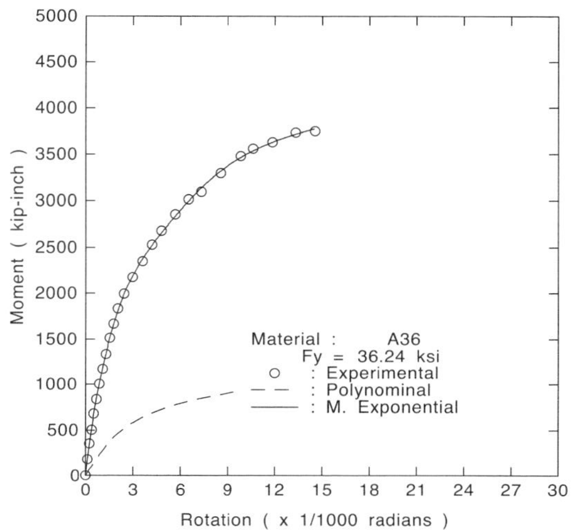

```
Moment-rotation prediction equations      ( R : X 1/1000 radians )

Frye and Morris polynominal model  :    R = Sum ( Ai X ( K*Bm )**Pi X 10**Qi )
     xd = 13.149600"   t =  1.000000"
     A1 =  1.790000   A2 =  1.760000   A3 =  2.040000    K =  0.002064
     P1 =        1    P2 =        3    P3 =        5    Q1 =        0    Q2 =       -1    Q3 =       -1

Modified exponential model  :
     Bm = Sum ( Ai X ( 1 - exp( - R/( 2*i*C ) ) ) )  +     Sum ( RKj X ( R - Rj0 ) )  +  Bm0
       C =   0.14866667E+00             Bmo=   0.00000000E+00         Nexp=  6       Nliner=  1
      AI =  -0.25698638E+03   0.11543695E+05  -0.83461222E+05   0.22869260E+06  -0.26587938E+06   0.11272780E+06
     Rj0 =           2.9600
     RKj =   0.36988841E+02
```

No	Rotation (radians) X 1/1000	Moment (kip-inch) Expri.	Moment (kip-inch) Poly.	Moment (kip-inch) M.Expo.	Connection stiffness (kip-inch) X 1000 Expri.	Connection stiffness (kip-inch) X 1000 Poly.	Connection stiffness (kip-inch) X 1000 M.Expo.
1	0.00	0.00	0.00	0.00	0.1634E+04	0.2707E+03	0.1613E+04
2	0.11	179.79	29.78	168.08	0.1459E+04	0.2704E+03	0.1448E+04
3	0.25	352.91	67.55	358.34	0.1206E+04	0.2691E+03	0.1277E+04
4	0.38	506.06	102.40	516.11	0.1206E+04	0.2669E+03	0.1155E+04
5	0.52	679.19	139.53	670.33	0.1111E+04	0.2633E+03	0.1053E+04
6	0.68	834.00	181.20	831.27	0.9378E+03	0.2573E+03	0.9629E+03
7	0.87	1005.47	229.19	1006.05	0.8683E+03	0.2474E+03	0.8801E+03
8	1.07	1171.93	277.36	1174.67	0.8014E+03	0.2338E+03	0.8078E+03
9	1.28	1333.41	324.74	1337.05	0.7343E+03	0.2170E+03	0.7394E+03
10	1.54	1513.19	378.27	1518.76	0.6746E+03	0.1946E+03	0.6588E+03
11	1.78	1671.34	422.51	1668.19	0.6303E+03	0.1743E+03	0.5868E+03
12	2.05	1832.81	466.71	1816.14	0.5245E+03	0.1535E+03	0.5100E+03
13	2.44	1995.95	521.54	1995.21	0.3858E+03	0.1285E+03	0.4113E+03
14	2.96	2174.07	581.64	2181.11	0.3121E+03	0.1038E+03	0.3098E+03
15	3.61	2352.19	641.85	2378.43	0.2886E+03	0.8277E+02	0.2684E+03
16	4.20	2530.31	686.54	2524.77	0.2729E+03	0.6964E+02	0.2318E+03
17	4.81	2678.47	725.85	2659.77	0.2256E+03	0.5977E+02	0.2129E+03
18	5.68	2853.26	773.23	2837.98	0.1975E+03	0.4973E+02	0.1975E+03
19	6.52	3016.40	811.95	2997.62	0.1445E+03	0.4285E+02	0.1819E+03
20	7.33	3094.64	844.54	3137.56	0.1222E+03	0.3785E+02	0.1630E+03
21	8.56	3292.74	887.46	3318.34	0.1529E+03	0.3224E+02	0.1308E+03
22	9.84	3477.52	925.87	3465.41	0.1215E+03	0.2800E+02	0.1000E+03
23	10.63	3562.42	947.14	3538.11	0.8668E+02	0.2593E+02	0.8454E+02
24	11.85	3629.00	977.11	3629.55	0.6295E+02	0.2330E+02	0.6651E+02
25	13.33	3737.21	1009.67	3716.82	0.3970E+02	0.2080E+02	0.5268E+02
26	14.54	3752.19	1033.80	3776.22	0.1238E+02	0.1914E+02	0.4605E+02

A.5 – 259

V -130

```
Connection type : Extended end-plate connections
           Mode : All bolted with column stiffener

Tested by  :  L.F.L.Ribeiro & R.M.Goncalves (1996)        Brazil
Test  Id.  :  CT2A-6

Column : CVS350x128                  Fasteners: A325- -1"D
Beam   : VS350x58                          - 1/16" Oversize holes
Plate thickness :       0.875"       Material : A36
                                          Fy =  36.24 ksi
                                          Fu =  57.99 ksi

                           Major parameters

     End plate extended on tension side only.
     ct =  2.0001"    pt =  4.6299"    li =  8.5197"    pc =  2.3150"
     cc =  0.6457"    gt =  4.0000"    gc =  4.0000"
     tp =  0.8752"    nt = 2 X  2      nc = 2 X  1

Remark  1)  End plate extended on tension side only.
        2)
```

No	Moment (k-in)	Rotation (radians) X 1/1000
1	0.00	0.00
2	186.44	0.10
3	342.92	0.23
4	511.06	0.39
5	702.49	0.57
6	842.33	0.72
7	1010.46	0.93
8	1178.59	1.14
9	1345.06	1.37
10	1504.87	1.59
11	1669.67	1.80
12	1841.14	1.97
13	2030.91	2.17
14	2167.41	2.33
15	2332.22	2.55
16	2495.36	2.80
17	2683.46	3.12
18	2834.95	3.46
19	2998.09	3.90
20	3122.94	4.39
21	3212.83	5.30
22	3241.13	6.31
23	3346.01	8.10
24	3450.88	9.70
25	3500.82	11.47
26	3538.28	13.93
27	3575.73	16.39
28	3627.34	17.95

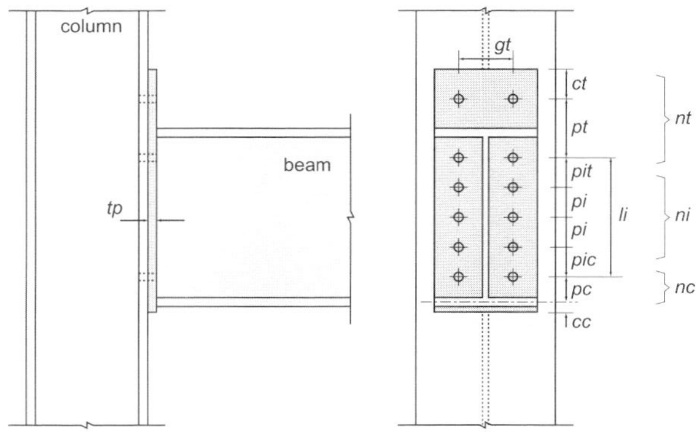

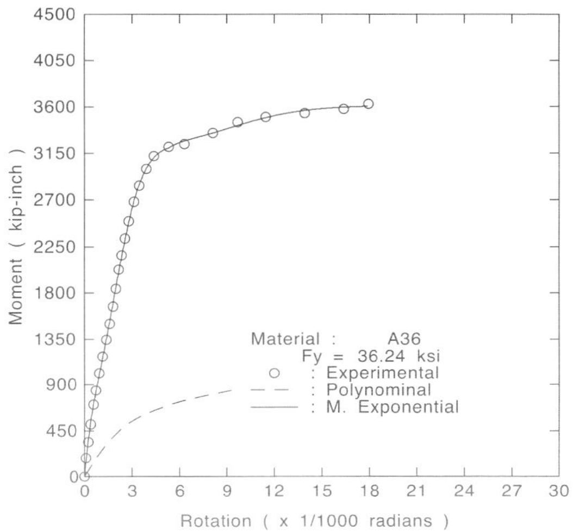

```
Moment-rotation prediction equations     ( R : X 1/1000 radians )

Frye and Morris polynominal model  :    R = Sum ( Ai X ( K*Bm )**Pi X 10**Qi )
     xd = 13.149600"   t =  0.875200"
     A1 =  1.790000   A2 =  1.760000   A3 =  2.040000    K =  0.002235
     P1 =         1   P2 =         3   P3 =         5   Q1 =         0    Q2 =        -1    Q3 =        -1

Modified exponential model  :
     Bm = Sum ( Ai X ( 1 - exp( - R/( 2*i*C ) ) ) )  +     Sum ( RKj X ( R - Rj0 ) )  +  Bm0
       C =   0.17458333E+00           Bmo=   0.00000000E+00         Nexp=  6        Nliner=  2
      AI =  -0.14185441E+04   0.38807430E+05  -0.20240242E+06   0.41548989E+06  -0.37985836E+06   0.13315405E+06
     Rj0 =           3.1200           5.3000
     RKj =  -0.60192112E+02   0.58323646E+02
```

No	Rotation (radians) X 1/1000	Moment (kip-inch) Expri.	Moment (kip-inch) Poly.	Moment (kip-inch) M.Expo.	Connection stiffness (kip-inch) X 1000 Expri.	Connection stiffness (kip-inch) X 1000 Poly.	Connection stiffness (kip-inch) X 1000 M.Expo.
1	0.00	0.00	0.00	0.00	0.1864E+04	0.2499E+03	0.1749E+04
2	0.10	186.44	24.99	164.12	0.1577E+04	0.2497E+03	0.1535E+04
3	0.23	342.92	57.39	347.06	0.1135E+04	0.2487E+03	0.1287E+04
4	0.39	511.06	97.00	533.08	0.1057E+04	0.2462E+03	0.1052E+04
5	0.57	702.49	140.92	705.94	0.9919E+03	0.2415E+03	0.8843E+03
6	0.72	842.33	176.74	832.32	0.8774E+03	0.2358E+03	0.8093E+03
7	0.93	1010.46	225.17	997.43	0.8006E+03	0.2249E+03	0.7739E+03
8	1.14	1178.59	270.98	1160.58	0.7639E+03	0.2109E+03	0.7849E+03
9	1.37	1345.06	317.49	1344.52	0.7251E+03	0.1933E+03	0.8155E+03
10	1.59	1504.87	358.07	1526.91	0.7563E+03	0.1756E+03	0.8406E+03
11	1.80	1669.67	393.23	1704.73	0.9085E+03	0.1594E+03	0.8500E+03
12	1.97	1841.14	419.27	1848.93	0.9812E+03	0.1471E+03	0.8445E+03
13	2.17	2030.91	447.36	2015.91	0.8957E+03	0.1340E+03	0.8228E+03
14	2.33	2167.41	468.04	2145.45	0.8093E+03	0.1246E+03	0.7949E+03
15	2.55	2332.22	494.16	2314.97	0.7039E+03	0.1132E+03	0.7442E+03
16	2.80	2495.36	521.03	2492.47	0.6242E+03	0.1021E+03	0.6741E+03
17	3.12	2683.46	551.75	2692.45	0.5188E+03	0.9031E+02	0.5748E+03
18	3.46	2834.95	580.68	2849.19	0.4130E+03	0.8024E+02	0.4081E+03
19	3.90	2998.09	613.63	3000.49	0.3159E+03	0.6996E+02	0.2829E+03
20	4.39	3122.94	645.66	3110.36	0.2002E+03	0.6115E+02	0.1708E+03
21	5.30	3212.83	695.71	3200.50	0.6524E+02	0.4952E+02	0.4352E+02
22	6.31	3241.13	741.11	3271.27	0.3905E+02	0.4096E+02	0.4950E+02
23	8.10	3346.01	805.32	3347.85	0.6226E+02	0.3148E+02	0.4349E+02
24	9.70	3450.88	851.09	3420.50	0.4782E+02	0.2622E+02	0.4563E+02
25	11.47	3500.82	893.72	3494.68	0.2278E+02	0.2221E+02	0.3675E+02
26	13.93	3538.28	943.38	3563.36	0.1523E+02	0.1840E+02	0.1945E+02
27	16.39	3575.73	985.23	3595.31	0.2615E+02	0.1578E+02	0.7708E+01
28	17.95	3627.34	1008.80	3603.81	0.3308E+02	0.1450E+02	0.3532E+01

A.5 – 261

V -131

Connection type : Extended end-plate connections
Mode : All bolted with column stiffener

Tested by : L.F.L.Ribeiro & R.M.Goncalves (1996) Brazil
Test Id. : CT2B-1

Column : CVS350x128
Beam : VS350x58
Plate thickness : 1.5"

Fasteners: A325- -7/8"D
15/16" Oversize holes
Material : A36
Fy = 36.24 ksi
Fu = 57.99 ksi

Major parameters

End plate extended on tension side only.

ct = 1.7504" pt = 4.1307" li = 9.0189" pc = 2.0654"
cc = 0.7512" gt = 3.5008" gc = 3.5008"
tp = 1.5000" nt = 2 X 2 nc = 2 X 1

Remark 1) End plate extended on tension side only.
2)

No	Moment (k-in)	Rotation (radians) X 1/1000
1	0.00	0.00
2	138.90	0.09
3	269.54	0.19
4	398.52	0.30
5	535.77	0.43
6	668.06	0.55
7	793.74	0.71
8	929.33	0.88
9	1059.97	1.04
10	1202.18	1.21
11	1324.55	1.36
12	1466.76	1.54
13	1600.70	1.72
14	1724.72	1.89
15	1867.19	2.08
16	1999.23	2.29
17	2116.63	2.48
18	2258.84	2.75
19	2381.21	3.00
20	2513.50	3.32
21	2650.75	3.72
22	2776.43	4.14
23	2865.72	4.54
24	2953.36	5.01
25	3044.31	5.64
26	3136.92	6.45
27	3183.22	6.98
28	3221.25	7.61
29	3251.02	8.37
30	3257.63	9.19

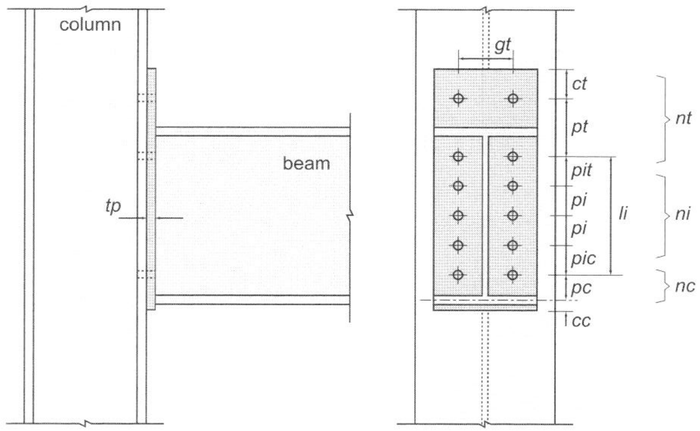

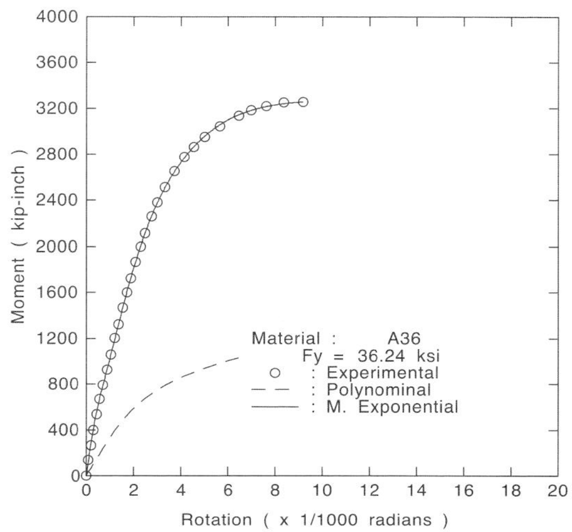

```
Moment-rotation prediction equations     ( R : X 1/1000 radians )

Frye and Morris polynominal model  :    R = Sum ( Ai X ( K*Bm )**Pi X 10**Qi )
     xd = 13.149600"    t =  1.500000"
     A1 =  1.790000   A2 =  1.760000   A3 =  2.040000    K =  0.001618
     P1 =        1    P2 =        3    P3 =        5    Q1 =        0    Q2 =       -1    Q3 =       -1

Modified exponential model  :
     Bm = Sum ( Ai X ( 1 - exp( - R/( 2*i*C ) ) ) )  +      Sum ( RKj X ( R - Rj0 ) )  +  Bm0
       C =   0.99083333E-01             Bmo=   0.00000000E+00          Nexp=  6       Nliner=  0
      AI =  -0.11773224E+04   0.16822417E+05  -0.76791291E+05   0.17051588E+06  -0.18950441E+06   0.83412618E+05
```

No	Rotation (radians) X 1/1000	Moment (kip-inch) Expri.	Moment (kip-inch) Poly.	Moment (kip-inch) M.Expo.	Connection stiffness (kip-inch) X 1000 Expri.	Connection stiffness (kip-inch) X 1000 Poly.	Connection stiffness (kip-inch) X 1000 M.Expo.
1	0.00	0.00	0.00	0.00	0.1543E+04	0.3453E+03	0.1347E+04
2	0.09	138.90	31.08	127.62	0.1431E+04	0.3450E+03	0.1443E+04
3	0.19	269.54	65.58	268.05	0.1243E+04	0.3441E+03	0.1345E+04
4	0.30	398.52	103.30	406.51	0.1119E+04	0.3423E+03	0.1170E+04
5	0.43	535.77	147.60	546.32	0.1080E+04	0.3390E+03	0.9904E+03
6	0.55	668.06	188.02	658.12	0.9666E+03	0.3345E+03	0.8823E+03
7	0.71	793.74	240.93	792.95	0.7914E+03	0.3264E+03	0.8154E+03
8	0.88	929.33	295.46	929.93	0.8073E+03	0.3147E+03	0.8026E+03
9	1.04	1059.97	344.76	1058.94	0.8262E+03	0.3011E+03	0.8112E+03
10	1.21	1202.18	394.54	1197.65	0.8255E+03	0.2842E+03	0.8192E+03
11	1.36	1324.55	435.98	1320.45	0.8041E+03	0.2681E+03	0.8164E+03
12	1.54	1466.76	482.45	1466.00	0.7671E+03	0.2482E+03	0.7982E+03
13	1.72	1600.70	525.35	1606.94	0.7366E+03	0.2286E+03	0.7657E+03
14	1.89	1724.72	562.70	1733.82	0.7391E+03	0.2111E+03	0.7258E+03
15	2.08	1867.19	601.07	1867.00	0.6923E+03	0.1931E+03	0.6753E+03
16	2.29	1999.23	639.71	2002.72	0.6231E+03	0.1753E+03	0.6172E+03
17	2.48	2116.63	671.64	2115.08	0.5803E+03	0.1611E+03	0.5658E+03
18	2.75	2258.84	712.75	2258.56	0.5074E+03	0.1439E+03	0.4983E+03
19	3.00	2381.21	746.99	2376.05	0.4561E+03	0.1305E+03	0.4429E+03
20	3.32	2513.50	786.39	2507.74	0.3822E+03	0.1162E+03	0.3820E+03
21	3.72	2650.75	829.94	2647.53	0.3217E+03	0.1020E+03	0.3192E+03
22	4.14	2776.43	870.23	2769.88	0.2603E+03	0.9030E+02	0.2652E+03
23	4.54	2865.72	904.49	2867.05	0.2063E+03	0.8134E+02	0.2219E+03
24	5.01	2953.36	940.65	2960.87	0.1685E+03	0.7285E+02	0.1786E+03
25	5.64	3044.31	983.62	3057.81	0.1312E+03	0.6392E+02	0.1310E+03
26	6.45	3136.92	1031.73	3144.13	0.9803E+02	0.5528E+02	0.8475E+02
27	6.98	3183.22	1059.80	3182.86	0.7503E+02	0.5083E+02	0.6235E+02
28	7.61	3221.25	1090.39	3215.52	0.5076E+02	0.4643E+02	0.4241E+02
29	8.37	3251.02	1123.96	3241.05	0.2421E+02	0.4207E+02	0.2596E+02
30	9.19	3257.63	1156.85	3257.41	0.8066E+01	0.3825E+02	0.1488E+02

V -132

Connection type : Extended end-plate connections
Mode : All bolted with column stiffener

Tested by : L.F.L.Ribeiro & R.M.Goncalves (1996) Brazil
Test Id. : CT2B-2

Column : CVS350x128
Beam : VS350x58
Plate thickness : 1.25"

Fasteners: A325- -7/8"D
15/16" Oversize holes
Material : A36
Fy = 36.24 ksi
Fu = 57.99 ksi

Major parameters

End plate extended on tension side only.

ct = 1.7504"	pt = 4.1307"	li = 9.0189"	pc = 2.0654"
cc = 0.7512"	gt = 3.5008"	gc = 3.5008"	
tp = 1.2500"	nt = 2 X 2	nc = 2 X 1	

Remark 1) End plate extended on tension side only.
2)

No	Moment (k-in)	Rotation (radians) X 1/1000
1	0.00	0.00
2	175.87	0.14
3	341.78	0.33
4	502.72	0.52
5	671.96	0.71
6	829.58	0.96
7	995.49	1.19
8	1161.40	1.42
9	1335.62	1.67
10	1491.58	1.89
11	1659.15	2.18
12	1825.06	2.45
13	2000.93	2.82
14	2166.85	3.24
15	2322.81	3.70
16	2488.72	4.14
17	2654.64	4.67
18	2837.15	5.37
19	2959.92	6.00
20	3071.09	6.70
21	3178.93	7.65
22	3213.77	8.47
23	3263.55	9.55
24	3288.44	10.54
25	3298.39	11.29

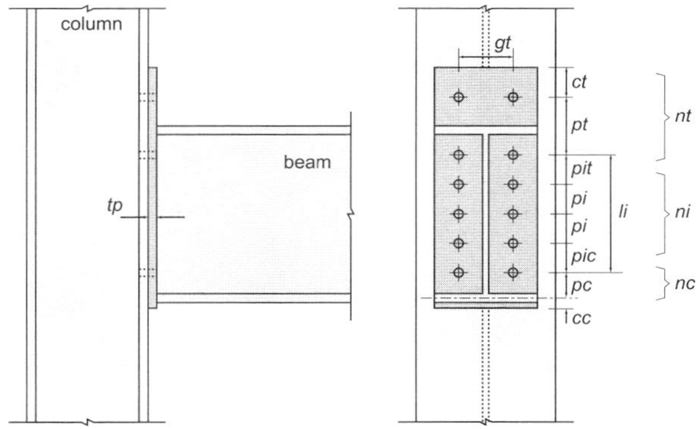

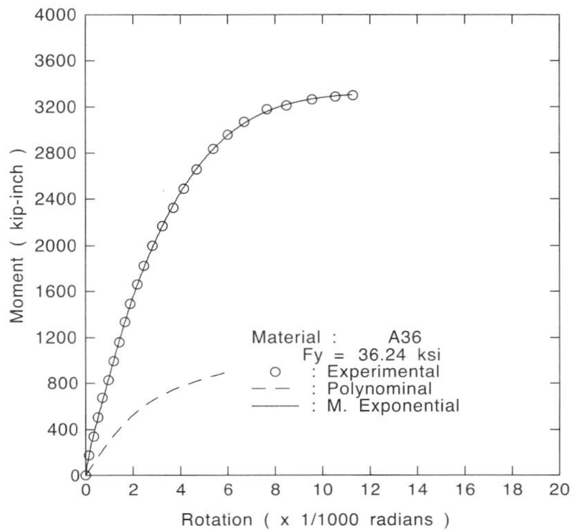

```
Moment-rotation prediction equations      ( R : X 1/1000 radians )

Frye and Morris polynominal model  :     R = Sum ( Ai X ( K*Bm )**Pi X 10**Qi )
     xd = 13.149600"    t =  1.250000"
     A1 =  1.790000    A2 =  1.760000    A3 =  2.040000    K =  0.001805
     P1 =        1     P2 =        3     P3 =        5     Q1 =        0     Q2 =       -1     Q3 =       -1

Modified exponential model  :
     Bm = Sum ( Ai X ( 1 - exp( - R/( 2*i*C ) ) ) )  +      Sum ( RKj X ( R - Rj0 ) )  +  Bm0
      C =   0.12908333E+00           Bmo=   0.00000000E+00           Nexp=  6        Nliner=  0
     AI =   0.73788969E+03  -0.23725189E+04  -0.81783936E+04   0.55884807E+05  -0.98116310E+05   0.55373703E+05
```

No	Rotation (radians) X 1/1000	Moment (kip-inch) Expri.	Poly.	M.Expo.	Connection stiffness (kip-inch) X 1000 Expri.	Poly.	M.Expo.
1	0.00	0.00	0.00	0.00	0.1256E+04	0.3095E+03	0.1559E+04
2	0.14	175.87	43.33	178.81	0.1094E+04	0.3090E+03	0.1062E+04
3	0.33	341.78	101.79	352.11	0.8601E+03	0.3063E+03	0.8148E+03
4	0.52	502.72	159.52	500.33	0.8689E+03	0.3010E+03	0.7618E+03
5	0.71	671.96	215.96	644.75	0.7783E+03	0.2926E+03	0.7612E+03
6	0.96	829.58	287.19	835.29	0.6778E+03	0.2763E+03	0.7600E+03
7	1.19	995.49	348.54	1008.13	0.7214E+03	0.2566E+03	0.7399E+03
8	1.42	1161.40	405.05	1174.48	0.7096E+03	0.2344E+03	0.7049E+03
9	1.67	1335.62	460.55	1345.01	0.7033E+03	0.2097E+03	0.6586E+03
10	1.89	1491.58	504.40	1485.25	0.6524E+03	0.1892E+03	0.6164E+03
11	2.18	1659.15	555.71	1656.24	0.5968E+03	0.1652E+03	0.5637E+03
12	2.45	1825.06	597.69	1802.35	0.5558E+03	0.1463E+03	0.5195E+03
13	2.82	2000.93	647.81	1984.55	0.4377E+03	0.1254E+03	0.4668E+03
14	3.24	2166.85	696.47	2169.57	0.3683E+03	0.1071E+03	0.4156E+03
15	3.70	2322.81	742.11	2349.19	0.3585E+03	0.9202E+02	0.3663E+03
16	4.14	2488.72	780.05	2500.80	0.3480E+03	0.8094E+02	0.3233E+03
17	4.67	2654.64	820.10	2659.28	0.2905E+03	0.7063E+02	0.2754E+03
18	5.37	2837.15	865.81	2831.50	0.2261E+03	0.6046E+02	0.2179E+03
19	6.00	2959.92	901.63	2954.12	0.1778E+03	0.5357E+02	0.1725E+03
20	6.70	3071.09	936.95	3059.53	0.1396E+03	0.4759E+02	0.1300E+03
21	7.65	3178.93	979.07	3160.79	0.7540E+02	0.4139E+02	0.8558E+02
22	8.47	3213.77	1011.25	3219.05	0.4404E+02	0.3726E+02	0.5804E+02
23	9.55	3263.55	1049.08	3267.56	0.3516E+02	0.3299E+02	0.3374E+02
24	10.54	3288.44	1080.15	3293.62	0.1839E+02	0.2989E+02	0.2001E+02
25	11.29	3298.39	1101.82	3305.95	0.1327E+02	0.2793E+02	0.1330E+02

V -133

Connection type : Extended end-plate connections
Mode : All bolted with column stiffener

Tested by : L.F.L.Ribeiro & R.M.Goncalves (1996) Brazil
Test Id. : CT2B-3

Column : CVS350x128
Beam : VS350x58
Plate thickness : 1.25"

Fasteners: A325- -7/8"D
15/16" Oversize holes
Material : A36
Fy = 36.24 ksi
Fu = 57.99 ksi

Major parameters

End plate extended on tension side only.

ct = 1.7504"	pt = 4.1307"	li = 9.0189"	pc = 2.0654"
cc = 1.1449"	gt = 3.5008"	gc = 3.5008"	
tp = 1.2500"	nt = 2 X 2	nc = 2 X 1	

Remark 1) End plate extended on tension side only.
2)

No	Moment (k-in)	Rotation (radians) X 1/1000
1	0.00	0.00
2	184.17	0.12
3	345.10	0.27
4	516.00	0.42
5	665.32	0.56
6	829.58	0.75
7	1010.42	0.99
8	1161.40	1.18
9	1327.32	1.41
10	1494.89	1.68
11	1670.76	1.95
12	1825.06	2.20
13	2014.21	2.56
14	2170.17	2.88
15	2329.45	3.25
16	2511.47	3.75
17	2654.64	4.26
18	2833.83	4.74
19	2953.29	5.22
20	3120.86	5.94
21	3222.07	6.50
22	3356.46	7.16
23	3429.46	7.96
24	3475.92	8.77
25	3522.38	9.80
26	3553.90	10.49
27	3583.76	11.07
28	3640.17	11.75

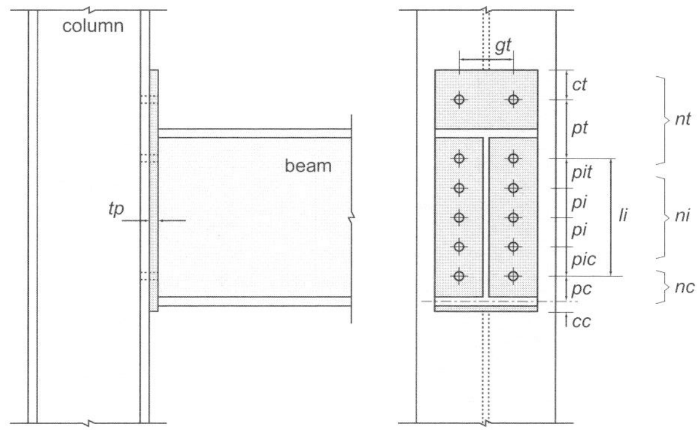

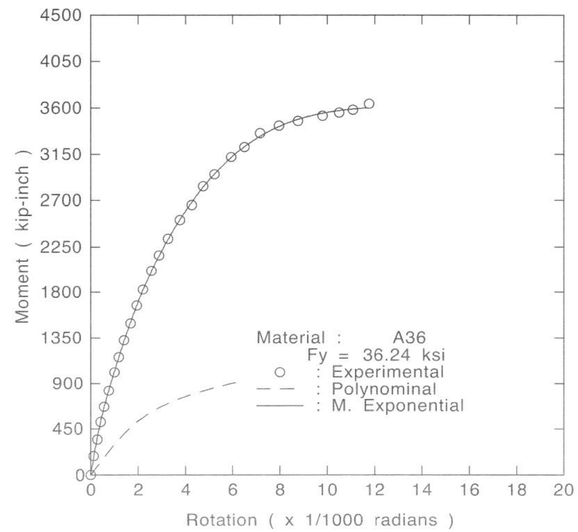

```
Moment-rotation prediction equations     ( R : X 1/1000 radians )

Frye and Morris polynominal model  :    R = Sum ( Ai X ( K*Bm )**Pi X 10**Qi )
     xd = 13.149600"   t =  1.250000"
     A1 =  1.790000   A2 =  1.760000    A3 =  2.040000    K =  0.001805
     P1 =         1   P2 =         3    P3 =         5    Q1 =         0    Q2 =       -1    Q3 =       -1

Modified exponential model  :
     Bm = Sum ( Ai X ( 1 - exp( - R/( 2*i*C ) ) ) )  +     Sum ( RKj X ( R - Rj0 ) )  +  Bm0
       C =   0.12791667E+00            Bmo=   0.00000000E+00          Nexp=  6        Nliner=  0
      AI =  -0.44665491E+03   0.11405697E+05  -0.71779494E+05   0.19523498E+06  -0.23871403E+06   0.10793069E+06
```

No	Rotation (radians) X 1/1000	Moment (kip-inch) Expri.	Moment (kip-inch) Poly.	Moment (kip-inch) M.Expo.	Connection stiffness (kip-inch) X 1000 Expri.	Connection stiffness (kip-inch) X 1000 Poly.	Connection stiffness (kip-inch) X 1000 M.Expo.
1	0.00	0.00	0.00	0.00	0.1535E+04	0.3095E+03	0.1501E+04
2	0.12	184.17	37.14	170.85	0.1329E+04	0.3091E+03	0.1345E+04
3	0.27	345.10	83.38	358.43	0.1106E+04	0.3074E+03	0.1162E+04
4	0.42	516.00	129.26	521.41	0.1102E+04	0.3041E+03	0.1019E+04
5	0.56	665.32	171.53	656.90	0.9809E+03	0.2995E+03	0.9226E+03
6	0.75	829.58	227.62	823.37	0.8155E+03	0.2903E+03	0.8367E+03
7	0.99	1010.42	295.44	1015.76	0.7765E+03	0.2739E+03	0.7720E+03
8	1.18	1161.40	345.97	1158.92	0.7615E+03	0.2576E+03	0.7362E+03
9	1.41	1327.32	402.70	1323.81	0.6750E+03	0.2354E+03	0.6978E+03
10	1.68	1494.89	462.65	1506.06	0.6360E+03	0.2087E+03	0.6518E+03
11	1.95	1670.76	515.60	1675.55	0.6336E+03	0.1839E+03	0.6035E+03
12	2.20	1825.06	558.99	1820.82	0.5796E+03	0.1637E+03	0.5588E+03
13	2.56	2014.21	613.43	2010.94	0.5053E+03	0.1395E+03	0.4985E+03
14	2.88	2170.17	655.24	2162.79	0.4610E+03	0.1225E+03	0.4517E+03
15	3.25	2329.45	697.53	2321.22	0.4022E+03	0.1068E+03	0.4062E+03
16	3.75	2511.47	746.69	2511.51	0.3228E+03	0.9061E+02	0.3570E+03
17	4.26	2654.64	789.62	2682.94	0.3284E+03	0.7835E+02	0.3164E+03
18	4.74	2833.83	825.00	2826.64	0.3111E+03	0.6946E+02	0.2828E+03
19	5.22	2953.29	856.58	2954.71	0.2424E+03	0.6239E+02	0.2511E+03
20	5.94	3120.86	898.41	3119.05	0.2035E+03	0.5415E+02	0.2058E+03
21	6.50	3222.07	927.27	3224.90	0.1912E+03	0.4915E+02	0.1726E+03
22	7.16	3356.46	958.07	3326.84	0.1528E+03	0.4436E+02	0.1370E+03
23	7.96	3429.46	991.63	3421.23	0.7441E+02	0.3972E+02	0.1002E+03
24	8.77	3475.92	1022.23	3489.98	0.5196E+02	0.3596E+02	0.7079E+02
25	9.80	3522.38	1057.22	3548.07	0.4545E+02	0.3214E+02	0.4379E+02
26	10.49	3553.90	1078.65	3573.69	0.4884E+02	0.3004E+02	0.3111E+02
27	11.07	3583.76	1095.61	3589.31	0.6598E+02	0.2848E+02	0.2311E+02
28	11.75	3640.17	1114.42	3602.52	0.8296E+02	0.2686E+02	0.1613E+02

V -134

Connection type : Extended end-plate connections
Mode : All bolted with column stiffener

Tested by : L.F.L.Ribeiro & R.M.Goncalves (1996) Brazil
Test Id. : CT2B-4

Column : CVS350x128
Beam : VS350x58
Plate thickness : 1.00"

Fasteners: A325- -7/8"D
15/16" Oversize holes
Material : A36
Fy = 36.24 ksi
Fu = 57.99 ksi

Major parameters

End plate extended on tension side only.

ct = 1.7504"	pt = 4.1307"	li = 9.0189"	pc = 2.0654"
cc = 0.7512"	gt = 3.5008"	gc = 3.5008"	
tp = 1.0000"	nt = 2 X 2	nc = 2 X 1	

Remark 1) End plate extended on tension side only.
2)

No	Moment (k-in)	Rotation (radians) X 1/1000
1	0.00	0.00
2	174.79	0.11
3	351.25	0.26
4	507.73	0.41
5	677.53	0.57
6	842.33	0.78
7	1008.80	1.00
8	1170.27	1.22
9	1336.74	1.46
10	1499.88	1.59
11	1664.68	1.97
12	1837.81	2.36
13	2022.59	2.79
14	2165.75	3.11
15	2345.53	3.51
16	2503.68	3.96
17	2671.81	4.54
18	2846.60	5.33
19	2946.48	5.39
20	3023.06	6.53
21	3106.29	7.33
22	3206.17	8.41
23	3269.43	9.52
24	3287.74	10.53

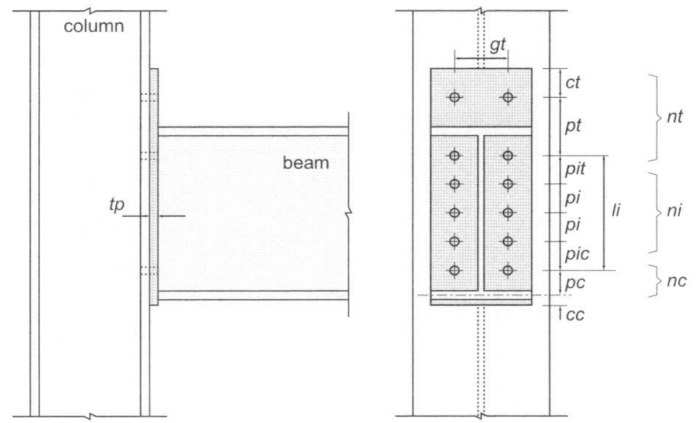

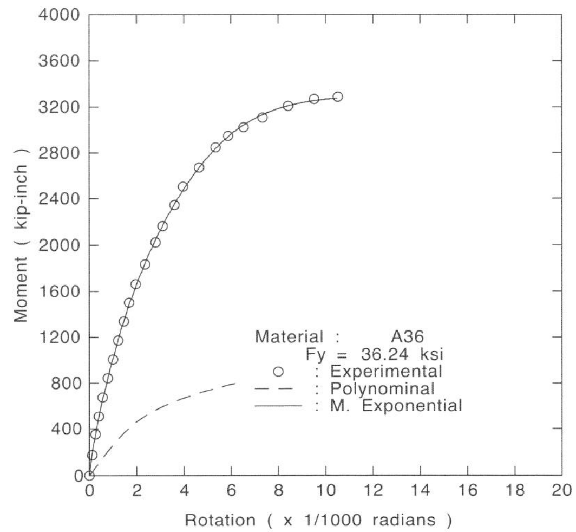

```
Moment-rotation prediction equations     ( R : X 1/1000 radians )

Frye and Morris polynominal model  :    R = Sum ( Ai X ( K*Bm )**Pi X 10**Qi )
     xd = 13.149600"   t =  1.000000"
     A1 =  1.790000   A2 =  1.760000   A3 =  2.040000    K =  0.002064
     P1 =         1   P2 =         3   P3 =         5    Q1 =         0    Q2 =       -1    Q3 =       -1

Modified exponential model  :
     Bm = Sum ( Ai X ( 1 - exp( - R/( 2*i*C ) ) ) )  +      Sum ( RKj X ( R - Rj0 ) )  +  Bm0
      C =   0.11525000E+00             Bmo=   0.00000000E+00          Nexp=  6        Nliner=  0
     AI =   0.14146039E+02   0.58447904E+04  -0.47422101E+05   0.14609341E+06  -0.19165874E+06   0.90430718E+05
```

No	Rotation (radians) X 1/1000	Moment (kip-inch) Expri.	Poly.	M.Expo.	Connection stiffness (kip-inch) X 1000 Expri.	Poly.	M.Expo.
1	0.00	0.00	0.00	0.00	0.1589E+04	0.2707E+03	0.1703E+04
2	0.11	174.79	29.78	169.12	0.1414E+04	0.2704E+03	0.1393E+04
3	0.26	351.25	70.24	356.67	0.1110E+04	0.2690E+03	0.1132E+04
4	0.41	507.73	110.40	514.33	0.1052E+04	0.2662E+03	0.9834E+03
5	0.57	677.53	152.66	663.63	0.9417E+03	0.2616E+03	0.8908E+03
6	0.78	842.33	206.69	842.36	0.7711E+03	0.2524E+03	0.8170E+03
7	1.00	1008.80	260.81	1015.75	0.7453E+03	0.2389E+03	0.7609E+03
8	1.22	1170.27	311.56	1177.48	0.7147E+03	0.2220E+03	0.7097E+03
9	1.46	1336.74	362.42	1341.23	0.7016E+03	0.2016E+03	0.6550E+03
10	1.69	1499.88	406.49	1486.00	0.6549E+03	0.1817E+03	0.6043E+03
11	1.97	1664.68	454.19	1647.11	0.5281E+03	0.1594E+03	0.5476E+03
12	2.36	1837.81	511.08	1847.22	0.4372E+03	0.1332E+03	0.4812E+03
13	2.79	2022.59	563.35	2041.38	0.4398E+03	0.1110E+03	0.4245E+03
14	3.11	2165.75	596.71	2171.66	0.4131E+03	0.9819E+02	0.3907E+03
15	3.61	2345.53	641.78	2355.64	0.4138E+03	0.8279E+02	0.3464E+03
16	3.96	2503.68	669.23	2471.91	0.3823E+03	0.7448E+02	0.3182E+03
17	4.64	2671.81	715.50	2670.26	0.2503E+03	0.6222E+02	0.2654E+03
18	5.33	2846.60	755.19	2835.33	0.2119E+03	0.5333E+02	0.2135E+03
19	5.89	2946.48	783.44	2943.67	0.1510E+03	0.4781E+02	0.1740E+03
20	6.53	3023.06	812.36	3041.89	0.1127E+03	0.4278E+02	0.1339E+03
21	7.33	3106.29	844.54	3131.94	0.9912E+02	0.3785E+02	0.9297E+02
22	8.41	3206.17	882.58	3209.62	0.7498E+02	0.3283E+02	0.5377E+02
23	9.52	3269.43	916.76	3254.39	0.3664E+02	0.2894E+02	0.2911E+02
24	10.53	3287.74	944.54	3276.63	0.1813E+02	0.2617E+02	0.1609E+02

A.5 – 269

V -135

```
Connection type : Extended end-plate connections
           Mode : All bolted with column stiffener

Tested by  :  L.F.L.Ribeiro & R.M.Goncalves (1996)        Brazil
Test  Id.  :  CT2B-5

Column : CVS350x128                  Fasteners: A325- -1"D
Beam   : VS350x58                      1 1/16" Oversize holes
Plate thickness :       1.00"        Material : A36
                                          Fy =  36.24 ksi
                                          Fu =  57.99 ksi

                         Major parameters

     End plate extended on tension side only.
     ct =  2.0001"    pt =  4.6299"    li =  8.5197"    pc =  2.3150"
     cc =  0.6457"    gt =  4.0000"    gc =  4.0000"
     tp =  1.0000"    nt = 2 X  2      nc = 2 X  1

Remark  1)  End plate extended on tension side only.
        2)
```

No	Moment (k-in)	Rotation (radians) X 1/1000
1	0.00	0.00
2	173.13	0.07
3	344.59	0.18
4	506.06	0.29
5	679.19	0.42
6	839.00	0.58
7	1003.80	0.77
8	1178.59	0.96
9	1333.41	1.14
10	1504.87	1.38
11	1673.00	1.63
12	1834.48	1.91
13	2002.61	2.24
14	2179.07	2.65
15	2342.21	3.14
16	2502.01	3.88
17	2666.82	4.83
18	2829.96	6.68
19	2998.09	8.80
20	3101.30	9.97
21	3244.46	12.52
22	3332.69	13.99
23	3435.90	16.31
24	3514.14	18.44
25	3580.22	20.74
26	3622.34	23.05

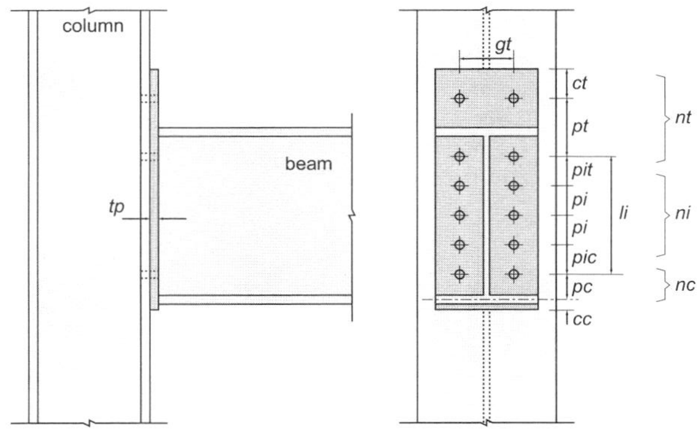

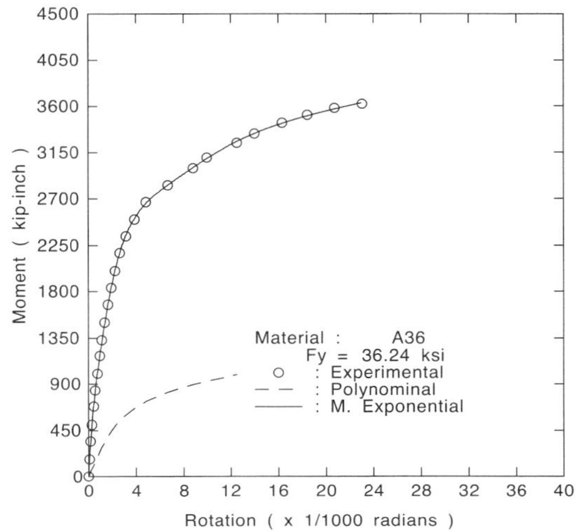

A.5 – 270

```
Moment-rotation prediction equations     ( R : X 1/1000 radians )

Frye and Morris polynominal model  :     R = Sum ( Ai X ( K*Bm )**Pi X 10**Qi )
     xd = 13.149600"    t =  1.000000"
     A1 =  1.790000   A2 =  1.760000   A3 =  2.040000    K =  0.002064
     P1 =        1    P2 =        3    P3 =        5    Q1 =        0    Q2 =      -1    Q3 =      -1

Modified exponential model  :
     Bm = Sum ( Ai X ( 1 - exp( - R/( 2*i*C ) ) ) )  +      Sum ( RKj X ( R - Rj0 ) )  +  Bm0
       C =   0.20958333E+00          Bmo=   0.00000000E+00          Nexp=  6          Nliner=  1
      AI =   0.14845109E+04  -0.35843302E+04  -0.21002013E+05   0.97957371E+05  -0.12849419E+06   0.56802926E+05
     Rj0 =         3.8800
     RKj =   0.24701840E+02
```

No	Rotation (radians) X 1/1000	Moment (kip-inch) Expri.	Poly.	M.Expo.	Connection stiffness (kip-inch) X 1000 Expri.	Poly.	M.Expo.
1	0.00	0.00	0.00	0.00	0.2473E+04	0.2707E+03	0.2265E+04
2	0.07	173.13	18.95	147.93	0.2118E+04	0.2706E+03	0.1972E+04
3	0.18	344.59	48.72	344.43	0.1513E+04	0.2699E+03	0.1621E+04
4	0.29	506.06	78.30	508.12	0.1406E+04	0.2686E+03	0.1370E+04
5	0.42	679.19	113.06	671.96	0.1182E+04	0.2660E+03	0.1165E+04
6	0.58	839.00	155.27	844.21	0.9387E+03	0.2613E+03	0.1002E+04
7	0.77	1003.80	204.17	1022.41	0.8937E+03	0.2529E+03	0.8841E+03
8	0.96	1178.59	251.20	1182.87	0.8892E+03	0.2416E+03	0.8097E+03
9	1.14	1333.41	293.54	1323.74	0.7977E+03	0.2285E+03	0.7574E+03
10	1.38	1504.87	346.02	1498.20	0.6939E+03	0.2085E+03	0.6972E+03
11	1.63	1673.00	395.44	1664.89	0.6273E+03	0.1868E+03	0.6361E+03
12	1.91	1834.48	444.50	1833.19	0.5458E+03	0.1639E+03	0.5657E+03
13	2.24	2002.61	494.63	2006.02	0.4742E+03	0.1406E+03	0.4819E+03
14	2.65	2179.07	547.36	2183.02	0.3860E+03	0.1175E+03	0.3831E+03
15	3.14	2342.21	599.68	2345.01	0.2863E+03	0.9711E+02	0.2817E+03
16	3.88	2502.01	663.33	2509.54	0.1974E+03	0.7620E+02	0.1716E+03
17	4.83	2666.82	727.17	2655.28	0.1445E+03	0.5946E+02	0.1208E+03
18	6.68	2829.96	818.70	2831.18	0.8405E+02	0.4176E+02	0.8304E+02
19	8.80	2998.09	895.34	3002.44	0.8505E+02	0.3131E+02	0.7860E+02
20	9.97	3101.30	929.48	3091.69	0.7813E+02	0.2763E+02	0.7350E+02
21	12.52	3244.46	992.41	3258.49	0.5860E+02	0.2208E+02	0.5683E+02
22	13.99	3332.69	1023.08	3335.12	0.5399E+02	0.1986E+02	0.4768E+02
23	16.31	3435.90	1065.89	3432.45	0.4044E+02	0.1717E+02	0.3704E+02
24	18.44	3514.14	1100.41	3504.54	0.3289E+02	0.1532E+02	0.3119E+02
25	20.74	3580.22	1133.77	3571.92	0.2349E+02	0.1375E+02	0.2778E+02
26	23.05	3622.34	1164.03	3633.91	0.1823E+02	0.1249E+02	0.2610E+02

V -136

```
Connection type : Extended end-plate connections
           Mode : All bolted with column stiffener

Tested by  :  L.F.L.Ribeiro & R.M.Goncalves (1996)        Brazil
Test  Id.  :  CT2B-6

Column : CVS350x128                  Fasteners: A325- -1"D
Beam   : VS350x58                         1 1/16" Oversize holes
Plate thickness :        0.875"      Material : A36
                                          Fy =  36.24 ksi
                                          Fu =  57.99 ksi

                         Major parameters

     End plate extended on tension side only.
     ct =  2.0001"    pt =  4.6299"    li =  8.5197"    pc =  2.3150"
     cc =  0.6457"    gt =  4.0000"    gc =  4.0000"
     tp =  0.8752"    nt = 2 X  2      nc = 2 X  1

Remark  1)  End plate extended on tension side only.
        2)
```

No	Moment (k-in)	Rotation (radians) X 1/1000
1	0.00	0.00
2	181.45	0.19
3	344.59	0.42
4	511.06	0.68
5	669.20	0.90
6	837.33	1.23
7	1008.80	1.74
8	1171.93	2.23
9	1336.74	2.81
10	1496.55	3.32
11	1676.33	3.88
12	1854.45	4.40
13	2012.60	4.90
14	2162.42	5.35
15	2343.87	6.03
16	2497.02	6.56
17	2675.14	7.42
18	2828.29	8.28
19	2994.76	9.60
20	3089.65	10.70
21	3162.89	11.69
22	3236.14	12.85
23	3309.38	14.12
24	3410.93	15.85
25	3475.85	17.00
26	3529.12	18.59
27	3547.43	20.39

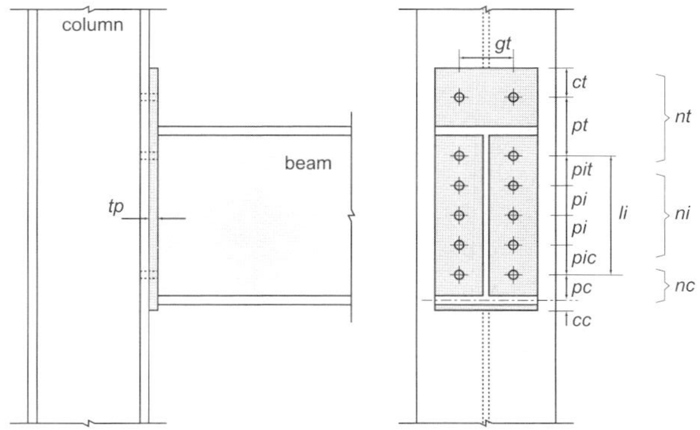

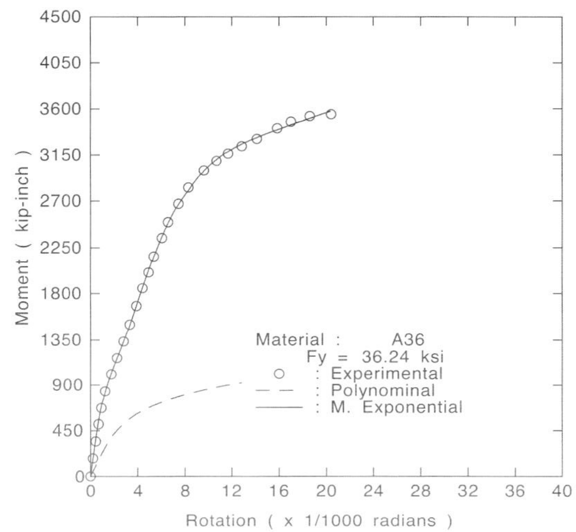

A.5 – 272

```
Moment-rotation prediction equations     ( R : X 1/1000 radians )

Frye and Morris polynominal model   :     R = Sum ( Ai X ( K*Bm )**Pi X 10**Qi )
     xd = 13.149600"   t =  0.875200"
     A1 =  1.790000   A2 =  1.760000    A3 =  2.040000    K =  0.002235
     P1 =         1    P2 =         3    P3 =         5    Q1 =         0    Q2 =        -1    Q3 =        -1

Modified exponential model  :
     Bm = Sum ( Ai X ( 1 - exp( - R/( 2*i*C ) ) ) )  +      Sum ( RKj X ( R - Rj0 ) )  +  Bm0
       C =   0.21741667E+00             Bmo=   0.00000000E+00         Nexp=  6       Nliner=  1
      AI =  -0.80018863E+03   0.75759677E+02   0.39119664E+05  -0.11812755E+06   0.11389702E+06  -0.31081473E+05
     Rj0 =           8.2800
     RKj =   0.40489105E+02
```

No	Rotation (radians) X 1/1000	Moment (kip-inch) Expri.	Moment (kip-inch) Poly.	Moment (kip-inch) M.Expo.	Connection stiffness (kip-inch) X 1000 Expri.	Connection stiffness (kip-inch) X 1000 Poly.	Connection stiffness (kip-inch) X 1000 M.Expo.
1	0.00	0.00	0.00	0.00	0.9550E+03	0.2499E+03	0.7931E+03
2	0.19	181.45	47.43	157.63	0.8439E+03	0.2491E+03	0.8439E+03
3	0.42	344.59	104.37	346.35	0.6769E+03	0.2456E+03	0.7832E+03
4	0.68	511.06	167.27	534.28	0.6828E+03	0.2375E+03	0.6587E+03
5	0.90	669.20	218.40	667.29	0.6351E+03	0.2267E+03	0.5524E+03
6	1.23	837.33	289.67	827.34	0.4414E+03	0.2042E+03	0.4255E+03
7	1.74	1008.80	383.68	1013.25	0.3345E+03	0.1638E+03	0.3205E+03
8	2.23	1171.93	455.42	1161.81	0.3106E+03	0.1303E+03	0.2946E+03
9	2.81	1336.74	522.20	1335.09	0.2997E+03	0.1016E+03	0.3069E+03
10	3.32	1496.55	569.22	1496.39	0.3170E+03	0.8411E+02	0.3252E+03
11	3.88	1676.33	612.25	1682.28	0.3322E+03	0.7037E+02	0.3365E+03
12	4.40	1854.45	646.28	1857.25	0.3292E+03	0.6099E+02	0.3343E+03
13	4.90	2012.60	674.96	2021.53	0.3251E+03	0.5404E+02	0.3212E+03
14	5.35	2162.42	698.11	2161.98	0.3066E+03	0.4903E+02	0.3021E+03
15	6.03	2343.87	729.32	2355.21	0.2793E+03	0.4302E+02	0.2650E+03
16	6.56	2497.02	751.10	2487.20	0.2578E+03	0.3930E+02	0.2329E+03
17	7.42	2675.14	782.75	2664.94	0.1926E+03	0.3450E+02	0.1810E+03
18	8.28	2828.29	810.76	2800.14	0.1576E+03	0.3079E+02	0.1346E+03
19	9.60	2994.76	848.43	2992.72	0.1044E+03	0.2649E+02	0.1200E+03
20	10.70	3089.65	876.01	3106.40	0.7980E+02	0.2378E+02	0.8870E+02
21	11.69	3162.89	898.54	3184.19	0.6899E+02	0.2180E+02	0.6971E+02
22	12.85	3236.14	922.68	3256.16	0.6053E+02	0.1989E+02	0.5559E+02
23	14.12	3309.38	946.81	3320.64	0.5811E+02	0.1817E+02	0.4688E+02
24	15.85	3410.93	976.55	3396.32	0.5735E+02	0.1629E+02	0.4151E+02
25	17.00	3475.85	994.67	3443.15	0.4682E+02	0.1525E+02	0.4012E+02
26	18.59	3529.12	1017.93	3506.32	0.2256E+02	0.1404E+02	0.3951E+02
27	20.39	3547.43	1042.14	3577.42	0.1017E+02	0.1289E+02	0.3956E+02

A.5 – 273

V -137

Connection type : Extended end-plate connections
Mode : All bolted without column stiffener

Tested by : B.Bose et al. (1996) U.K.
Test Id. : 7

Column : 254x254 UC 89
Beam : 457x191 UB 74
Plate thickness : 20 mm

Fasteners: G8.8- -M24
33/32" Oversize holes
Material : S275
Fy = 43.66 ksi
Fu = 70.49 ksi

Major parameters

End plate extended on both sides.
ct = 2.0276" pt = 3.9370" li = 15.3445" pc = 1.8602"
cc = 2.0276" gt = 3.5433" gc = 3.5433"
tp = 0.7874" nt = 2 X 2 nc = 2 X 1

Remark 1) End-plate: 640x200x20 mm
2) Failure mode: End-plate fracture

No	Moment (k-in)	Rotation (radians) X 1/1000	No	Moment (k-in)	Rotation (radians) X 1/1000
1	0.00	0.00	26	2084.59	19.90
2	185.62	0.16	27	2110.10	20.89
3	380.07	0.59	28	2128.24	21.89
4	584.02	0.98	29	2146.63	22.88
5	724.79	1.42	30	2159.93	23.88
6	868.73	1.97	31	2167.78	24.88
7	972.28	2.43	32	2175.71	25.87
8	1080.29	2.97	33	2183.93	26.87
9	1149.06	3.32	34	2189.46	27.86
10	1247.93	3.96	35	2193.63	28.86
11	1349.01	4.96	36	2202.83	29.85
12	1450.67	5.95	37	2214.16	30.85
13	1538.15	6.95			
14	1605.19	7.95			
15	1657.83	8.94			
16	1709.23	9.94			
17	1760.82	10.93			
18	1817.41	11.93			
19	1853.17	12.93			
20	1890.13	13.92			
21	1928.86	14.92			
22	1963.01	15.91			
23	1997.27	16.91			
24	2026.61	17.90			
25	2056.77	18.90			

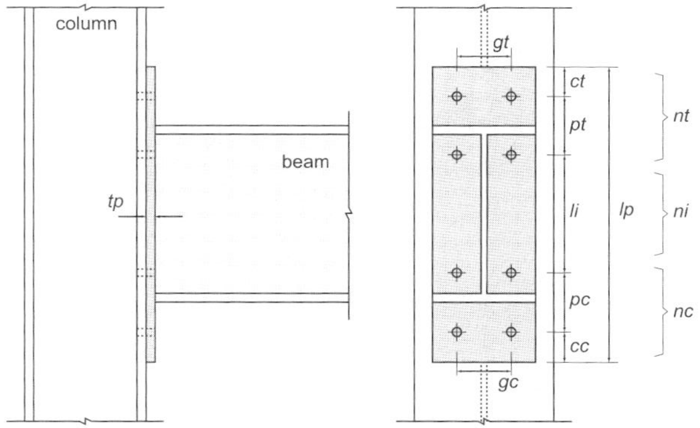

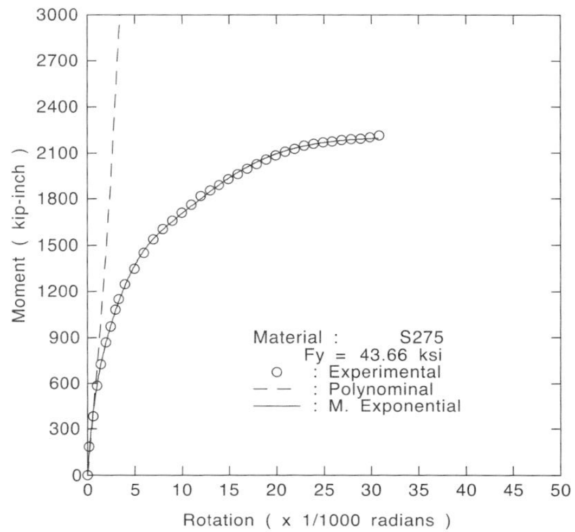

A.5 – 274

```
Moment-rotation prediction equations     ( R : X 1/1000 radians )

Frye and Morris polynominal model   :    R = Sum ( Ai X ( K*Bm )**Pi X 10**Qi )
     xd = 21.141740"   t =  0.787400"  tf =  0.944880"
     A1 =  1.830000   A2 = -1.040000   A3 =  6.380000     K =  0.000791
     P1 =         1    P2 =         3    P3 =         5    Q1 =         0    Q2 =        -1    Q3 =        -3

Modified exponential model  :
     Bm = Sum ( Ai X ( 1 - exp( - R/( 2*i*C ) ) ) )  +      Sum ( RKj X ( R - Rj0 ) )  +  Bm0
      C =   0.29708333E+00              Bmo=   0.00000000E+00           Nexp=  6        Nliner=  0
      AI =  -0.29232745E+03   0.13187717E+05  -0.80542385E+05   0.20287942E+06  -0.22407204E+06   0.91045408E+05
```

No	Rotation (radians) X 1/1000	Moment (kip-inch) Expri.	Poly.	M.Expo.	Connection stiffness (kip-inch) X 1000 Expri.	Poly.	M.Expo.
1	0.00	0.00	0.00	0.00	0.1160E+04	0.6909E+03	0.8983E+03
3	0.59	380.07	410.08	407.29	0.4893E+03	0.7034E+03	0.5223E+03
5	1.42	724.79	1017.01	728.43	0.2941E+03	0.7703E+03	0.2921E+03
7	2.43	972.28	1884.30	970.73	0.2136E+03	0.9768E+03	0.2047E+03
9	3.32	1149.06	2867.62	1136.59	0.1816E+03	0.1183E+04	0.1697E+03
11	4.96	1349.01	4261.30	1368.17	0.1019E+03	0.5264E+03	0.1135E+03
13	6.95	1538.15	4994.32	1541.50	0.7726E+02	0.2674E+03	0.6641E+02
15	8.94	1657.83	5429.84	1653.74	0.5229E+02	0.1826E+03	0.5035E+02
17	10.93	1760.82	5747.16	1750.97	0.5434E+02	0.1405E+03	0.4823E+02
19	12.93	1853.17	6000.82	1846.25	0.3655E+02	0.1151E+03	0.4657E+02
21	14.92	1928.86	6211.96	1934.43	0.3660E+02	0.9813E+02	0.4150E+02
23	16.91	1997.27	6394.47	2009.79	0.3194E+02	0.8591E+02	0.3400E+02
25	18.90	2056.77	6555.81	2069.44	0.2899E+02	0.7664E+02	0.2600E+02
27	20.89	2110.10	6700.80	2113.84	0.2197E+02	0.6935E+02	0.1884E+02
29	22.88	2146.63	6834.16	2145.36	0.1595E+02	0.6340E+02	0.1309E+02
31	24.88	2167.78	6955.83	2166.99	0.7932E+01	0.5852E+02	0.8776E+01
33	26.87	2183.93	7067.99	2181.25	0.6899E+01	0.5443E+02	0.5745E+01
35	28.86	2193.63	7172.66	2190.49	0.6741E+01	0.5093E+02	0.3682E+01
37	30.85	2214.16	7270.86	2196.37	0.1133E+02	0.4790E+02	0.2320E+01

V -138

```
Connection type : Extended end-plate connections
           Mode : All bolted without column stiffener

Tested by  :  B.Bose et al. (1996)                        U.K.
Test  Id.  :  11

Column : 254x254 UC 73                Fasteners: G8.8- -M24
Beam   : 457x191 UB 74                     33/32"  Oversize holes
Plate thickness : 15 mm               Material :  S275
                                           Fy =  44.96 ksi
                                           Fu =  74.11 ksi

                         Major parameters

     End plate extended on both sides.
     ct =  2.0276"    pt =  3.9370"    li = 13.2677"    pc =  3.9370"
     cc =  2.0276"    pit=  3.5433"    pi =  0.0000"    pic=  9.7244"
     gt =  3.5433"    gi =  3.5433"    gc =  3.5433"
     tp =  0.5906"    nt = 2 X  2      ni = 2 X  1      nc = 2 X  2

Remark  1)  End-plate: 640x200x15 mm
        2)  Failure mode: Column web buckling
```

No	Moment (k-in)	Rotation (radians) X 1/1000	No	Moment (k-in)	Rotation (radians) X 1/1000
1	0.00	0.00	26	2482.59	15.00
2	175.21	0.25	27	2485.90	16.00
3	350.42	0.50	28	2487.80	17.00
4	527.14	1.00	29	2489.73	18.00
5	650.78	1.25	30	2490.72	19.00
6	774.42	1.50	31	2491.15	20.00
7	939.65	1.61	32	2491.20	21.00
8	1104.88	1.71	33	2491.61	22.00
9	1211.78	2.02	34	2492.05	23.00
10	1318.68	2.33	35	2492.93	24.00
11	1468.85	2.53	36	2494.70	25.00
12	1619.02	2.74	37	2495.58	26.00
13	1819.78	3.14			
14	1986.30	3.45			
15	2048.09	4.00			
16	2120.56	5.00			
17	2188.49	6.00			
18	2261.53	7.00			
19	2326.80	8.00			
20	2370.01	9.00			
21	2424.75	10.00			
22	2453.52	11.00			
23	2464.29	12.00			
24	2470.72	13.00			
25	2477.48	14.00			

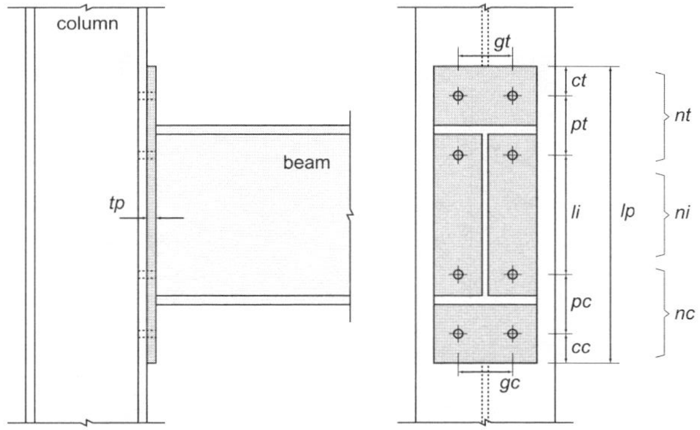

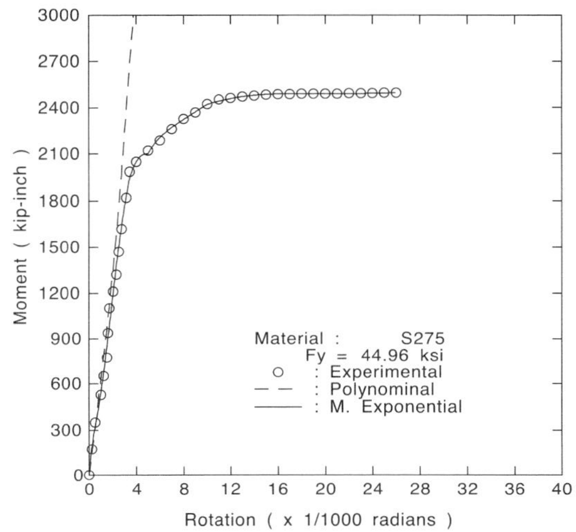

A.5 – 276

```
Moment-rotation prediction equations     ( R : X 1/1000 radians )

Frye and Morris polynominal model  :    R = Sum ( Ai X ( K*Bm )**Pi X 10**Qi )
     xd = 21.141740"   t =  0.590600"  tf =  0.944880"
     A1 =  1.830000   A2 = -1.040000   A3 =  6.380000    K =  0.000887
     P1 =        1    P2 =        3    P3 =        5    Q1 =        0    Q2 =       -1    Q3 =       -3

Modified exponential model  :
     Bm = Sum ( Ai X ( 1 - exp( - R/( 2*i*C ) ) ) )  +      Sum ( RKj X ( R - Rj0 ) )  +  Bm0
       C =  0.27916667E+00           Bmo=  0.00000000E+00          Nexp=  6       Nliner=  3
      AI =  0.46357451E+04  -0.44798500E+05  0.14376184E+06  -0.22131829E+06  0.17262680E+06  -0.52440757E+05
     Rj0 =          3.4500           5.0000          10.0000
     RKj = -0.12710724E+03   0.15737161E+03  -0.26136579E+02
```

No	Rotation (radians) X 1/1000	Moment (kip-inch) Expri.	Moment (kip-inch) Poly.	Moment (kip-inch) M.Expo.	Connection stiffness (kip-inch) X 1000 Expri.	Connection stiffness (kip-inch) X 1000 Poly.	Connection stiffness (kip-inch) X 1000 M.Expo.
1	0.00	0.00	0.00	0.00	0.7008E+03	0.6158E+03	0.1097E+04
3	0.50	350.42	309.24	316.82	0.5850E+03	0.6238E+03	0.4241E+03
5	1.25	650.78	791.30	683.78	0.4946E+03	0.6693E+03	0.5970E+03
7	1.61	939.65	1038.33	912.35	0.1581E+04	0.7096E+03	0.6625E+03
9	2.02	1211.78	1342.92	1186.63	0.3448E+03	0.7765E+03	0.6633E+03
11	2.53	1468.85	1768.03	1507.70	0.7334E+03	0.8978E+03	0.5851E+03
13	3.14	1819.78	2366.21	1822.36	0.5218E+03	0.1049E+04	0.4432E+03
15	4.00	2048.09	3206.94	2049.12	0.9819E+02	0.8078E+03	0.1279E+03
17	6.00	2188.49	4192.78	2206.76	0.7048E+02	0.3111E+03	0.7934E+02
19	8.00	2326.80	4674.50	2324.33	0.5424E+02	0.1909E+03	0.5010E+02
21	10.00	2424.75	4999.57	2422.15	0.4175E+02	0.1401E+03	0.4753E+02
23	12.00	2464.29	5249.15	2458.91	0.8598E+01	0.1119E+03	0.1474E+02
25	14.00	2477.48	5453.62	2480.52	0.5937E+01	0.9379E+02	0.7121E+01
27	16.00	2485.90	5627.87	2489.20	0.2606E+01	0.8115E+02	0.2095E+01
29	18.00	2489.73	5780.38	2490.86	0.1460E+01	0.7179E+02	-0.2033E-01
31	20.00	2491.15	5916.43	2490.39	0.2359E+00	0.6454E+02	-0.2309E+00
33	22.00	2491.61	6039.52	2490.52	0.4270E+00	0.5876E+02	0.4427E+00
35	24.00	2492.93	6152.16	2492.31	0.1328E+01	0.5402E+02	0.1351E+01
37	26.00	2495.58	6256.14	2495.87	0.8706E+00	0.5007E+02	0.2182E+01

V -139

```
Connection type : Extended end-plate connections
           Mode : All bolted without column stiffener

Tested by  :  B.Bose et al. (1996)                     U.K.
Test  Id.  :  12

Column : 254x254 UC 89              Fasteners: G8.8- -M24
Beam   : 457x191 UB 74                   33/32"  Oversize holes
Plate thickness : 15 mm             Material :  S275
                                         Fy =  44.96 ksi
                                         Fu =  74.11 ksi

                         Major parameters

     End plate extended on both sides.
     ct =  2.0276"    pt =  3.9370"    li = 13.2677"    pc =  3.9370"
     cc =  2.0276"    pit=  3.5433"    pi =  0.0000"    pic=  9.7244"
     gt =  3.5433"    gi =  3.5433"    gc =  3.5433"
     tp =  0.5906"    nt = 2 X  2      ni = 2 X  1      nc = 2 X  2

Remark  1)  End-plate: 640x200x15 mm
        2)  Failure mode: Column web buckling and end-plate fracture
```

No	Moment (k-in)	Rotation (radians) X 1/1000	No	Moment (k-in)	Rotation (radians) X 1/1000
1	0.00	0.00	26	3517.87	34.49
2	294.43	0.15	27	3564.27	36.52
3	538.85	0.29	28	3602.07	38.55
4	732.63	0.64	29	3622.97	40.58
5	976.40	1.00	30	3632.30	42.61
6	1191.80	1.51	31	3636.20	44.64
7	1407.19	2.03	32	3640.70	46.67
8	1576.57	2.51	33	3640.37	48.70
9	1745.94	2.99	34	3641.14	50.72
10	1862.53	3.99	35	3641.06	52.75
11	1979.12	4.99	36	3649.43	54.78
12	2193.53	6.09	37	3649.80	56.81
13	2425.65	8.12	38	3651.66	58.84
14	2612.31	10.15	39	3653.10	60.87
15	2759.80	12.17			
16	2893.08	14.20			
17	2977.49	16.23			
18	3074.18	18.26			
19	3153.79	20.29			
20	3216.89	22.32			
21	3271.15	24.35			
22	3322.24	26.38			
23	3375.66	28.41			
24	3422.42	30.44			
25	3471.48	32.46			

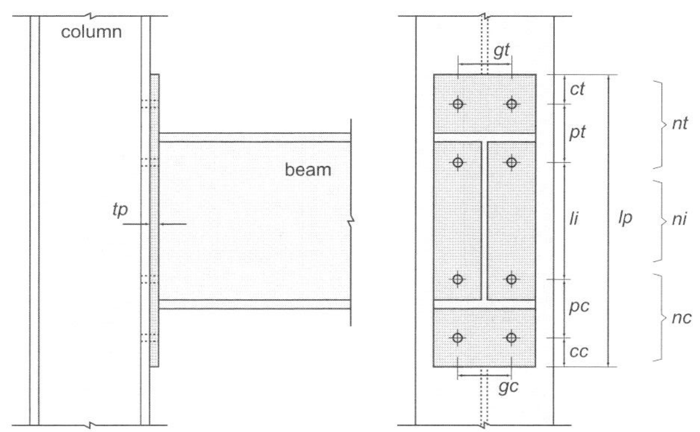

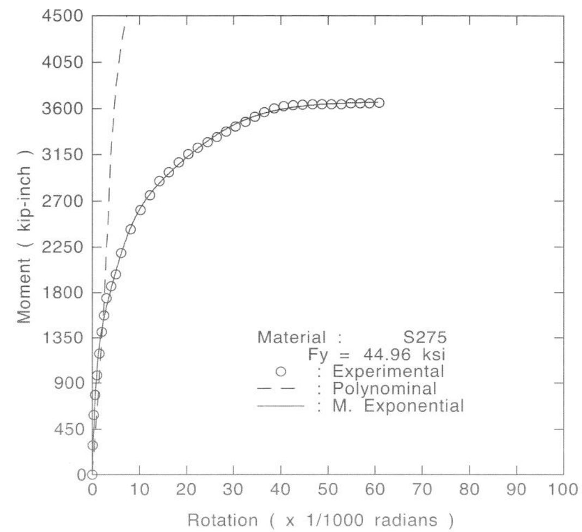

A.5 – 278

```
Moment-rotation prediction equations     ( R : X 1/1000 radians )

Frye and Morris polynominal model  :     R = Sum ( Ai X ( K*Bm )**Pi X 10**Qi )
     xd = 21.141740"   t =  0.590600"  tf =  0.944880"
     A1 =  1.830000   A2 = -1.040000   A3 =  6.380000    K =  0.000887
     P1 =        1    P2 =        3    P3 =        5    Q1 =        0    Q2 =       -1    Q3 =       -3

Modified exponential model  :
     Bm = Sum ( Ai X ( 1 - exp( - R/( 2*i*C ) ) ) )  +      Sum ( RKj X ( R - Rj0 ) )  +  Bm0
      C =   0.54475000E+00            Bmo=   0.00000000E+00           Nexp=  6        Nliner=  0
     AI =   0.32616599E+04  -0.50016507E+04  -0.26243993E+05   0.12028649E+06  -0.16464439E+06   0.76006112E+05
```

No	Rotation (radians) X 1/1000	Moment (kip-inch) Expri.	Poly.	M.Expo.	Connection stiffness (kip-inch) X 1000 Expri.	Poly.	M.Expo.
1	0.00	0.00	0.00	0.00	0.1963E+04	0.6158E+03	0.1673E+04
3	0.29	588.85	178.69	417.08	0.1660E+04	0.6185E+03	0.1227E+04
5	1.00	976.40	626.64	1033.84	0.4903E+03	0.6490E+03	0.5966E+03
7	2.03	1407.19	1350.64	1444.35	0.3823E+03	0.7785E+03	0.2708E+03
9	2.99	1745.94	2210.55	1659.05	0.2762E+03	0.1023E+04	0.1941E+03
11	4.99	1979.12	3812.19	2010.68	0.1539E+03	0.4625E+03	0.1636E+03
13	8.12	2425.65	4696.90	2443.18	0.1031E+03	0.1868E+03	0.1097E+03
15	12.17	2759.80	5268.41	2768.07	0.6935E+02	0.1100E+03	0.5861E+02
17	16.23	2977.49	5646.47	2968.27	0.4461E+02	0.7993E+02	0.4358E+02
19	20.29	3153.79	5935.03	3134.77	0.3515E+02	0.6362E+02	0.3873E+02
21	24.35	3271.15	6170.95	3281.14	0.2595E+02	0.5328E+02	0.3303E+02
23	28.41	3375.66	6371.87	3400.85	0.2468E+02	0.4609E+02	0.2582E+02
25	32.46	3471.48	6547.27	3490.72	0.2357E+02	0.4078E+02	0.1869E+02
27	36.52	3564.27	6704.15	3554.02	0.2074E+02	0.3667E+02	0.1274E+02
29	40.58	3622.97	6846.12	3596.21	0.7447E+01	0.3339E+02	0.8290E+01
31	44.64	3636.20	6977.36	3623.20	0.2069E+01	0.3068E+02	0.5209E+01
33	48.70	3640.37	7097.12	3639.93	0.1076E+00	0.2844E+02	0.3184E+01
35	52.75	3641.06	7208.27	3650.04	0.2043E+01	0.2655E+02	0.1907E+01
37	56.81	3649.80	7312.58	3656.06	0.5483E+00	0.2491E+02	0.1121E+01
39	60.87	3653.10	7410.71	3659.57	0.7101E+00	0.2349E+02	0.6502E+00

V -140

```
Connection type : Extended end-plate connections
           Mode : All bolted without column stiffener

Tested by  :  B.Bose et al. (1996)                        U.K.
Test  Id.  :  13

Column : 254x254 UC 89               Fasteners: G8.8- -M24
Beam   : 762x267 UB 147                   33/32"  Oversize holes
Plate thickness : 15 mm              Material :  S275
                                          Fy =  47.43 ksi
                                          Fu =  70.92 ksi

                          Major parameters

     End plate extended on both sides.
     ct =  1.8898"    pt =  3.9370"    li = 24.9606"    pc =  3.9370"
     cc =  1.8898"    pit=  3.5433"    pi =  0.0000"    pic= 21.4173"
     gt =  3.5433"    gi =  3.5433"    gc =  3.5433"
     tp =  0.5906"    nt = 2 X  2      ni = 2 X  1      nc = 2 X  2

Remark  1)  End-plate: 930x250x15 mm
        2)  Failure mode: Column web buckling
```

No	Moment (k-in)	Rotation (radians) X 1/1000	No	Moment (k-in)	Rotation (radians) X 1/1000
1	0.00	0.00	26	6095.58	19.00
2	498.69	0.25			
3	997.38	0.49			
4	1396.19	0.74			
5	1794.99	0.98			
6	2318.95	1.23			
7	2842.91	1.48			
8	3204.01	1.73			
9	3565.12	1.97			
10	4025.22	2.95			
11	4318.64	3.94			
12	4588.56	4.92			
13	4793.87	5.90			
14	4957.08	6.89			
15	5110.08	7.87			
16	5243.73	8.86			
17	5363.50	9.84			
18	5481.54	10.82			
19	5575.85	11.81			
20	5671.53	12.79			
21	5770.17	13.77			
22	5844.80	14.76			
23	5905.30	15.74			
24	5970.08	16.73			
25	6039.82	17.71			

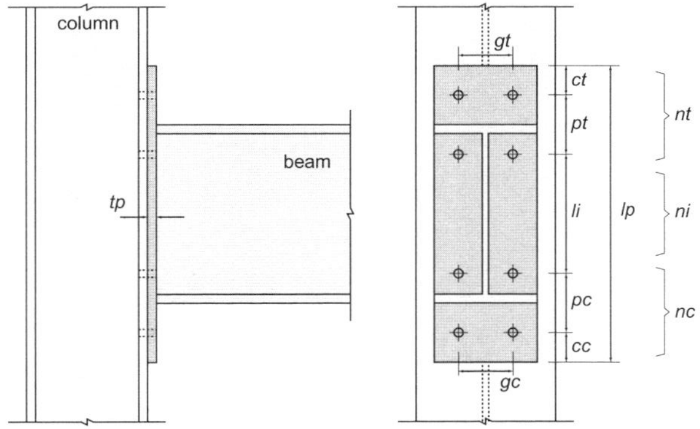

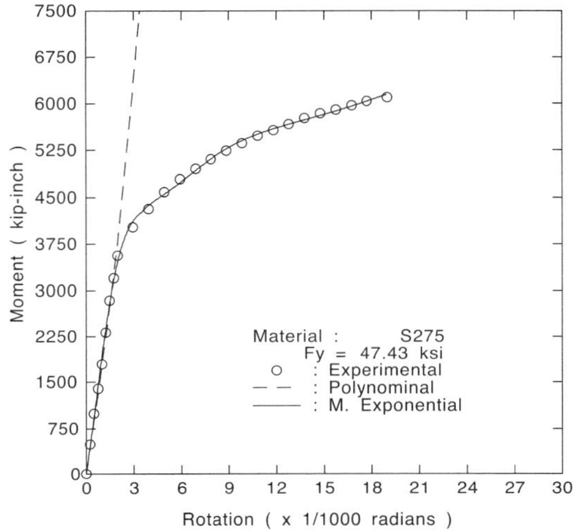

A.5 – 280

```
Moment-rotation prediction equations     ( R : X 1/1000 radians )

Frye and Morris polynominal model  :     R = Sum ( Ai X ( K*Bm )**Pi X 10**Qi )
     xd = 32.834590"    t =  0.590600"  tf =  0.944880"
     A1 =  1.830000   A2 = -1.040000   A3 =  6.380000    K =  0.000308
     P1 =        1     P2 =        3    P3 =        5     Q1 =        0     Q2 =       -1     Q3 =       -3

Modified exponential model  :
     Bm = Sum ( Ai X ( 1 - exp( - R/( 2*i*C ) ) ) )  +      Sum ( RKj X ( R - Rj0 ) )  +  Bm0
      C =   0.22083333E+00               Bmo=   0.00000000E+00          Nexp=  6          Nliner=  1
     AI =   0.42722375E+04  -0.78063282E+05   0.35087244E+06  -0.66821934E+06   0.59339288E+06  -0.19779365E+06
    Rj0 =           3.5000
    RKj =   0.10510306E+03
```

No	Rotation (radians) X 1/1000	Moment (kip-inch) Expri.	Moment (kip-inch) Poly.	Moment (kip-inch) M.Expo.	Connection stiffness (kip-inch) X 1000 Expri.	Connection stiffness (kip-inch) X 1000 Poly.	Connection stiffness (kip-inch) X 1000 M.Expo.
1	0.00	0.00	0.00	0.00	0.1995E+04	0.1771E+04	0.1939E+04
2	0.25	498.69	443.33	469.64	0.2037E+04	0.1777E+04	0.1876E+04
3	0.49	997.38	871.57	930.30	0.1841E+04	0.1793E+04	0.1965E+04
4	0.74	1396.19	1323.27	1428.62	0.1629E+04	0.1822E+04	0.2006E+04
5	0.98	1794.99	1765.22	1904.68	0.1874E+04	0.1863E+04	0.1945E+04
6	1.23	2318.95	2237.70	2373.16	0.2096E+04	0.1920E+04	0.1789E+04
7	1.48	2842.91	2726.59	2794.28	0.1770E+04	0.1995E+04	0.1572E+04
8	1.73	3204.01	3236.76	3156.97	0.1475E+04	0.2091E+04	0.1327E+04
9	1.97	3565.12	3751.87	3447.07	0.1301E+04	0.2206E+04	0.1092E+04
10	2.95	4025.22	6236.80	4125.35	0.3834E+03	0.2915E+04	0.3736E+03
11	3.94	4318.64	9079.48	4378.07	0.2859E+03	0.2409E+04	0.2091E+03
12	4.92	4588.56	10870.86	4556.10	0.2425E+03	0.1376E+04	0.1739E+03
13	5.90	4793.87	11969.39	4734.88	0.1873E+03	0.9248E+03	0.1916E+03
14	6.89	4957.08	12763.41	4928.28	0.1605E+03	0.6958E+03	0.1951E+03
15	7.87	5110.08	13373.75	5111.94	0.1456E+03	0.5628E+03	0.1769E+03
16	8.86	5243.73	13883.78	5272.93	0.1286E+03	0.4740E+03	0.1476E+03
17	9.84	5363.50	14315.90	5403.04	0.1213E+03	0.4115E+03	0.1186E+03
18	10.82	5481.54	14695.15	5507.42	0.1079E+03	0.3647E+03	0.9569E+02
19	11.81	5575.85	15037.25	5593.98	0.9645E+02	0.3279E+03	0.8052E+02
20	12.79	5671.53	15343.81	5668.54	0.9915E+02	0.2988E+03	0.7271E+02
21	13.77	5770.17	15624.48	5738.28	0.8808E+02	0.2748E+03	0.7036E+02
22	14.76	5844.80	15886.22	5808.34	0.6852E+02	0.2546E+03	0.7165E+02
23	15.74	5905.30	16127.12	5880.08	0.6358E+02	0.2375E+03	0.7499E+02
24	16.73	5970.08	16354.77	5956.41	0.6831E+02	0.2227E+03	0.7929E+02
25	17.71	6039.82	16566.65	6036.29	0.5910E+02	0.2100E+03	0.8370E+02
26	19.00	6095.58	16827.90	6147.79	0.4322E+02	0.1955E+03	0.8904E+02

V -141

```
Connection type : Extended end-plate connections
           Mode : All bolted without column stiffener

Tested by   :  B.Bose et al. (1996)                    U.K.
Test  Id.   :  14

Column : 254x254 UC 132               Fasteners: G8.8- -M24
Beam   : 457x191 UB 74                     33/32"  Oversize holes
Plate thickness : 15 mm               Material :  S275
                                           Fy =  47.43 ksi
                                           Fu =  70.92 ksi

                           Major parameters

     End plate extended on both sides.
     ct =  2.0276"    pt =  3.9370"    li = 13.2677"    pc =  3.9370"
     cc =  2.0276"    pit=  3.5433"    pi =  0.0000"    pic=  9.7244"
     gt =  3.5433"    gi =  3.5433"    gc =  3.5433"
     tp =  0.5906"    nt = 2 X  2      ni = 2 X  1      nc = 2 X  2

Remark  1)  End-plate: 640x200x15 mm
        2)  Failure mode: Bolt fracture
```

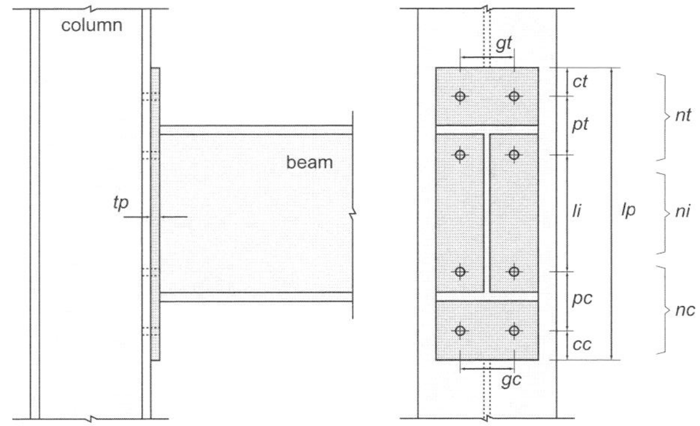

No	Moment (k-in)	Rotation (radians) X 1/1000	No	Moment (k-in)	Rotation (radians) X 1/1000
1	0.00	0.00	26	3314.33	20.86
2	173.59	0.50	27	3360.22	21.85
3	365.30	0.99	28	3407.29	22.84
4	646.33	1.49	29	3437.45	23.84
5	947.24	1.99	30	3462.75	24.83
6	1114.24	2.24	31	3482.16	25.82
7	1281.24	2.48	32	3504.64	26.82
8	1474.04	2.98	33	3527.57	27.81
9	1692.72	3.97	34	3551.77	28.80
10	1951.89	4.97	35	3576.60	29.79
11	2155.74	5.96	36	3603.03	30.79
12	2295.96	6.95	37	3621.28	31.78
13	2416.89	7.95	38	3636.94	32.77
14	2533.44	8.94	39	3651.58	33.77
15	2647.23	9.93	40	3656.34	34.76
16	2747.92	10.92	41	3657.10	35.75
17	2829.36	11.92	42	3657.35	36.75
18	2907.28	12.91	43	3657.52	37.74
19	2978.07	13.90	44	3658.41	38.74
20	3036.45	14.90			
21	3089.56	15.89			
22	3142.61	16.88			
23	3196.93	17.88			
24	3240.29	18.87			
25	3279.62	19.86			

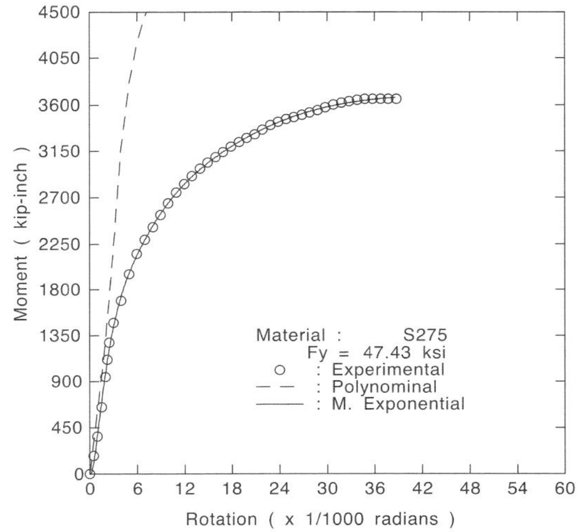

A.5 – 282

```
Moment-rotation prediction equations     ( R : X 1/1000 radians )

Frye and Morris polynominal model  :    R = Sum ( Ai X ( K*Bm )**Pi X 10**Qi )
     xd = 21.141740"    t =  0.590600"  tf =  0.944880"
     A1 =  1.830000   A2 = -1.040000   A3 =  6.380000    K =  0.000887
     P1 =         1   P2 =         3   P3 =         5   Q1 =         0   Q2 =        -1   Q3 =        -3

Modified exponential model  :
     Bm = Sum ( Ai X ( 1 - exp( - R/( 2*i*C ) ) ) )  +     Sum ( RKj X ( R - Rj0 ) )  +  Bm0
      C =   0.44783333E+00            Bmo=   0.00000000E+00         Nexp=  6        Nliner=  0
     AI =  -0.39974463E+04   0.29134139E+05  -0.11270906E+06   0.24149486E+06  -0.25283648E+06   0.10262462E+06
```

No	Rotation (radians) X 1/1000	Moment (kip-inch) Expri.	Moment (kip-inch) Poly.	Moment (kip-inch) M.Expo.	Connection stiffness (kip-inch) X 1000 Expri.	Connection stiffness (kip-inch) X 1000 Poly.	Connection stiffness (kip-inch) X 1000 M.Expo.
1	0.00	0.00	0.00	0.00	0.3472E+03	0.6158E+03	-0.9997E+02
3	0.99	365.30	620.15	386.46	0.4758E+03	0.6483E+03	0.6174E+03
5	1.99	947.24	1319.67	983.31	0.6459E+03	0.7708E+03	0.5330E+03
7	2.48	1281.24	1723.48	1223.61	0.5952E+03	0.8841E+03	0.4481E+03
9	3.97	1692.72	3181.62	1734.82	0.2399E+03	0.8228E+03	0.2603E+03
11	5.96	2155.74	4180.25	2139.80	0.1738E+03	0.3152E+03	0.1650E+03
13	7.95	2416.89	4664.92	2426.54	0.1193E+03	0.1927E+03	0.1266E+03
15	9.93	2647.23	4989.72	2649.68	0.1083E+03	0.1414E+03	0.9986E+02
17	11.92	2829.36	5240.17	2827.05	0.8007E+02	0.1128E+03	0.7951E+02
19	13.90	2978.07	5444.20	2969.67	0.6498E+02	0.9454E+02	0.6547E+02
21	15.89	3089.56	5618.91	3089.85	0.5362E+02	0.8175E+02	0.5589E+02
23	17.88	3196.93	5771.74	3193.78	0.4903E+02	0.7228E+02	0.4883E+02
25	19.86	3279.62	5907.36	3284.47	0.3723E+02	0.6500E+02	0.4288E+02
27	21.85	3360.22	6030.68	3364.17	0.4695E+02	0.5915E+02	0.3726E+02
29	23.84	3437.45	6143.49	3432.88	0.2785E+02	0.5437E+02	0.3183E+02
31	25.82	3482.16	6247.10	3490.76	0.2104E+02	0.5040E+02	0.2669E+02
33	27.81	3527.57	6343.93	3539.04	0.2380E+02	0.4701E+02	0.2191E+02
35	29.79	3576.60	6434.08	3578.15	0.2575E+02	0.4411E+02	0.1768E+02
37	31.78	3621.28	6520.53	3609.58	0.1712E+02	0.4154E+02	0.1402E+02
39	33.77	3651.58	6600.29	3634.32	0.9700E+01	0.3933E+02	0.1094E+02
41	35.75	3657.10	6675.65	3653.42	0.5091E+00	0.3737E+02	0.8442E+01
43	37.74	3657.52	6749.10	3668.14	0.5300E+00	0.3559E+02	0.6429E+01

A.5 – 283

V -142

Connection type : Extended end-plate connections
Mode : All bolted without column stiffener

Tested by : B.Bose et al. (1996) U.K.
Test Id. : 15

Column : 254x254 UC 132
Beam : 762x267 UB 147
Plate thickness : 15 mm

Fasteners: G8.8- -M24
33/32" Oversize holes
Material : S275
Fy = 47.43 ksi
Fu = 70.92 ksi

Major parameters

End plate extended on both sides.

ct = 1.8898"	pt = 3.9370"	li = 24.9606"	pc = 3.9370"
cc = 1.8898"	pit= 3.5433"	pi = 0.0000"	pic= 21.4173"
gt = 3.5433"	gi = 3.5433"	gc = 3.5433"	
tp = 0.5906"	nt = 2 X 2	ni = 2 X 1	nc = 2 X 2

Remark 1) End-plate: 930x250x15 mm
2) Failure mode: Bolt fracture, thread stripping and end-plate fracture

No	Moment (k-in)	Rotation (radians) X 1/1000
1	0.00	0.00
2	648.44	0.24
3	1041.89	0.48
4	1582.86	0.73
5	2151.77	0.97
6	2833.70	1.21
7	3304.39	1.45
8	3871.69	1.94
9	4068.13	2.42
10	4265.19	2.91
11	4555.95	3.39
12	4744.68	3.88
13	4930.06	4.36
14	5107.36	4.85
15	5203.48	5.33
16	5310.51	5.82
17	5430.42	6.30
18	5543.20	6.79
19	5676.53	7.50
20	5794.69	8.16
21	5883.19	9.00

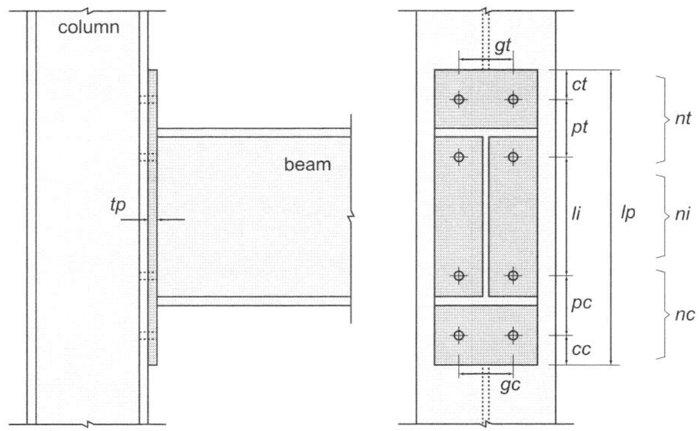

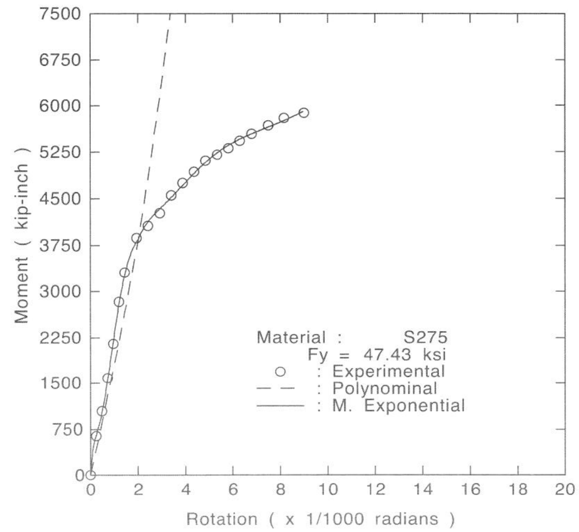

```
Moment-rotation prediction equations     ( R : X 1/1000 radians )

Frye and Morris polynominal model  :     R = Sum ( Ai X ( K*Bm )**Pi X 10**Qi )
     xd = 32.834590"    t =  0.590600"  tf =  0.944880"
     A1 =  1.830000   A2 = -1.040000   A3 =  6.380000     K =  0.000308
     P1 =        1    P2 =        3    P3 =        5     Q1 =        0     Q2 =       -1     Q3 =       -3

Modified exponential model  :
     Bm = Sum ( Ai X ( 1 - exp( - R/( 2*i*C ) ) ) )  +      Sum ( RKj X ( R - Rj0 ) )  +  Bm0
      C =   0.13500000E+00               Bmo=   0.00000000E+00             Nexp=  6        Nliner=  1
     AI =   0.15972516E+05  -0.18840363E+06   0.78411636E+06  -0.15368518E+07   0.14406880E+07  -0.51289172E+06
     Rj0 =          2.0000
     RKj =   0.39606589E+03
```

No	Rotation (radians) X 1/1000	Moment (kip-inch) Expri.	Poly.	M.Expo.	Connection stiffness (kip-inch) X 1000 Expri.	Poly.	M.Expo.
1	0.00	0.00	0.00	0.00	0.2702E+04	0.1771E+04	0.5872E+04
2	0.24	648.44	425.15	662.92	0.2171E+04	0.1777E+04	0.1322E+04
3	0.48	1041.89	853.64	1011.07	0.1896E+04	0.1792E+04	0.1828E+04
4	0.73	1582.86	1305.05	1576.52	0.2269E+04	0.1821E+04	0.2594E+04
5	0.97	2151.77	1746.60	2219.72	0.2606E+04	0.1861E+04	0.2660E+04
6	1.21	2833.70	2199.35	2814.30	0.2401E+04	0.1915E+04	0.2242E+04
7	1.45	3304.39	2666.91	3281.61	0.1697E+04	0.1985E+04	0.1643E+04
8	1.94	3871.69	3685.82	3810.82	0.7796E+03	0.2190E+04	0.5999E+03
9	2.42	4068.13	4806.08	4127.65	0.4057E+03	0.2497E+04	0.5049E+03
10	2.91	4265.19	6125.23	4333.74	0.5050E+03	0.2887E+04	0.3768E+03
11	3.39	4555.95	7563.86	4515.08	0.4966E+03	0.3010E+04	0.3872E+03
12	3.88	4744.68	8933.47	4710.16	0.3857E+03	0.2493E+04	0.4048E+03
13	4.36	4930.06	9977.14	4902.03	0.3741E+03	0.1866E+04	0.3887E+03
14	4.85	5107.36	10774.77	5081.99	0.2802E+03	0.1424E+04	0.3421E+03
15	5.33	5203.48	11385.21	5232.43	0.2092E+03	0.1143E+04	0.2841E+03
16	5.82	5310.51	11894.84	5357.61	0.2343E+03	0.9501E+03	0.2284E+03
17	6.30	5430.42	12316.77	5456.61	0.2401E+03	0.8158E+03	0.1867E+03
18	6.79	5543.20	12690.34	5541.09	0.2129E+03	0.7140E+03	0.1610E+03
19	7.50	5676.53	13157.03	5650.50	0.1833E+03	0.6063E+03	0.1523E+03
20	8.16	5794.69	13531.84	5754.89	0.1466E+03	0.5333E+03	0.1669E+03
21	9.00	5883.19	13949.20	5909.23	0.1054E+03	0.4638E+03	0.2026E+03

V -143

Connection type : Extended end-plate connections
Mode : All bolted without column stiffener

Tested by : B.Bose et al. (1996) U.K.
Test Id. : 16

Column : 254x254 UC 89
Beam : 457x191 UB 74
Plate thickness : 15 mm

Fasteners: G8.8- -M24
33/32" Oversize holes
Material : S275
Fy = 47.43 ksi
Fu = 70.92 ksi

Major parameters

End plate extended on both sides.
ct = 2.0276" pt = 3.9370" li = 15.3445" pc = 3.9370"
cc = 2.0276" gt = 3.5433" gc = 3.5433"
tp = 0.5906" nt = 2 X 2 nc = 2 X 1

Remark 1) End-plate: 640x200x15 mm
2) Failure mode: Bolt fracture and End-plate fracture

No	Moment (k-in)	Rotation (radians) X 1/1000	No	Moment (k-in)	Rotation (radians) X 1/1000
1	0.00	0.00	26	1550.29	20.79
2	186.98	0.50	27	1569.20	21.78
3	322.33	0.99	28	1584.08	22.77
4	456.47	1.49	29	1601.24	23.76
5	581.73	1.98	30	1617.57	24.75
6	672.28	2.23	31	1626.73	25.74
7	762.83	2.47	32	1629.27	26.73
8	861.95	2.97	33	1633.42	27.72
9	973.86	3.96	34	1634.07	28.71
10	1055.68	4.95	35	1634.69	29.70
11	1128.69	5.94	36	1634.78	30.69
12	1175.70	6.93	37	1634.96	32.00
13	1242.15	7.92	38	1635.04	33.00
14	1268.90	8.91	39	1635.40	34.00
15	1316.18	9.90			
16	1349.26	10.89			
17	1373.11	11.88			
18	1393.12	12.87			
19	1415.68	13.86			
20	1435.90	14.85			
21	1453.12	15.84			
22	1473.11	16.83			
23	1491.54	17.82			
24	1509.36	18.81			
25	1529.98	19.80			

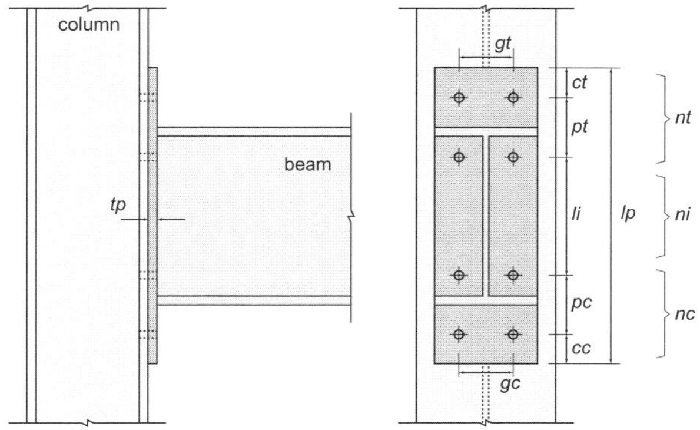

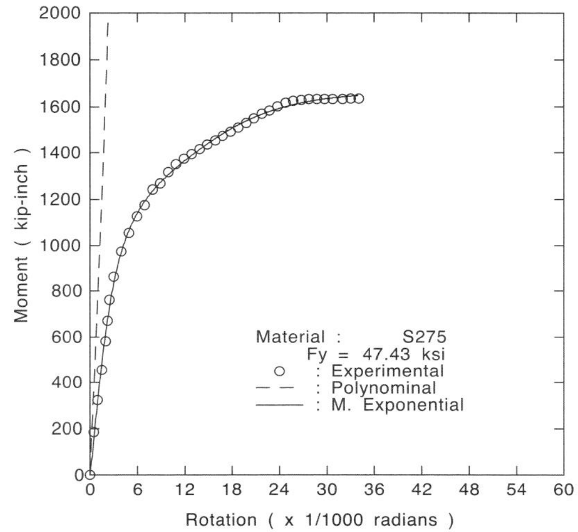

A.5 – 286

```
Moment-rotation prediction equations     ( R : X 1/1000 radians )

Frye and Morris polynominal model  :     R = Sum ( Ai X ( K*Bm )**Pi X 10**Qi )
     xd = 23.218540"    t =  0.590600"  tf =  0.944880"
     A1 =  1.830000   A2 = -1.040000   A3 =  6.380000    K =  0.000709
     P1 =         1   P2 =         3   P3 =         5   Q1 =         0   Q2 =        -1   Q3 =        -3

Modified exponential model  :
     Bm = Sum ( Ai X ( 1 - exp( - R/( 2*i*C ) ) ) )  +      Sum ( RKj X ( R - Rj0 ) )  +  Bm0
      C =   0.40833333E+00             Bmo=   0.00000000E+00          Nexp=  6       Nliner=  0
     AI =  -0.53614725E+03   0.26400236E+04  -0.12761059E+05   0.42931223E+05  -0.58142945E+05   0.27532183E+05
```

No	Rotation (radians) X 1/1000	Moment (kip-inch) Expri.	Moment (kip-inch) Poly.	Moment (kip-inch) M.Expo.	Connection stiffness (kip-inch) X 1000 Expri.	Connection stiffness (kip-inch) X 1000 Poly.	Connection stiffness (kip-inch) X 1000 M.Expo.
1	0.00	0.00	0.00	0.00	0.3740E+03	0.7711E+03	0.2732E+03
3	0.99	322.33	776.54	324.39	0.2723E+03	0.8118E+03	0.3328E+03
5	1.98	581.73	1642.83	617.11	0.3262E+03	0.9628E+03	0.2532E+03
7	2.47	762.83	2147.06	730.82	0.3192E+03	0.1104E+04	0.2116E+03
9	3.96	973.86	3973.52	970.27	0.9784E+02	0.1036E+04	0.1188E+03
11	5.94	1128.69	5226.50	1139.57	0.6061E+02	0.3973E+03	0.6122E+02
13	7.92	1242.15	5834.04	1235.82	0.4707E+02	0.2427E+03	0.3952E+02
15	9.90	1316.18	6242.69	1304.72	0.4059E+02	0.1777E+03	0.3134E+02
17	11.88	1373.11	6555.96	1362.80	0.2215E+02	0.1418E+03	0.2768E+02
19	13.86	1415.68	6812.36	1415.00	0.2161E+02	0.1188E+03	0.2508E+02
21	15.84	1453.12	7030.75	1462.02	0.1879E+02	0.1027E+03	0.2237E+02
23	17.82	1491.54	7221.81	1503.37	0.1831E+02	0.9082E+02	0.1937E+02
25	19.80	1529.98	7392.17	1538.64	0.2067E+02	0.8163E+02	0.1625E+02
27	21.78	1569.20	7546.29	1567.81	0.1707E+02	0.7430E+02	0.1325E+02
29	23.76	1601.24	7687.29	1591.28	0.1691E+02	0.6830E+02	0.1052E+02
31	25.74	1626.73	7817.42	1609.73	0.5906E+01	0.6329E+02	0.8174E+01
33	27.72	1633.42	7938.43	1623.92	0.2426E+01	0.5904E+02	0.6231E+01
35	29.70	1634.69	8051.63	1634.66	0.3576E+00	0.5539E+02	0.4674E+01
37	32.00	1634.96	8174.71	1643.73	0.1086E+00	0.5173E+02	0.3290E+01
39	34.00	1635.40	8276.72	1649.37	0.3540E+00	0.4892E+02	0.2393E+01

V -144

Connection type : Extended end-plate connections
Mode : All bolted without column stiffener

Tested by : B.Bose et al. (1996) U.K.
Test Id. : 17

Column : 254x254 UC 89
Beam : 457x191 UB 74
Plate thickness : 15 mm

Fasteners: G8.8- -M20
7/8" Oversize holes
Material : S275
Fy = 43.51 ksi
Fu = 670.92 ksi

Major parameters

End plate extended on both sides.

ct = 2.0276"	pt = 3.9370"	li = 13.2677"	pc = 3.9370"
cc = 2.0276"	pit= 3.5433"	pi = 0.0000"	pic= 9.7244"
gt = 3.5433"	gi = 3.5433"	gc = 3.5433"	
tp = 0.5906"	nt = 2 X 2	ni = 2 X 1	nc = 2 X 2

Remark 1) End-plate: 640x200x15 mm
2) Failure mode: Thread stripping

No	Moment (k-in)	Rotation (radians) X 1/1000
1	0.00	0.00
2	146.29	0.24
3	292.57	0.49
4	414.14	0.73
5	535.71	0.98
6	647.37	1.22
7	759.04	1.46
8	876.97	1.71
9	994.90	1.95
10	1079.46	2.20
11	1164.01	2.44
12	1219.43	2.68
13	1274.85	2.92
14	1368.33	3.41
15	1454.87	3.90
16	1529.05	4.39
17	1599.90	4.87
18	1672.82	5.36
19	1727.70	5.85
20	1776.77	6.34
21	1819.91	6.82
22	1863.05	7.31
23	1906.19	7.80
24	1938.53	8.29
25	1969.34	8.77
26	2004.24	9.26
27	2039.99	9.75
28	2077.27	10.24
29	2110.62	10.72
30	2139.38	11.21
31	2168.14	11.70
32	2177.88	12.40
33	2187.61	13.00

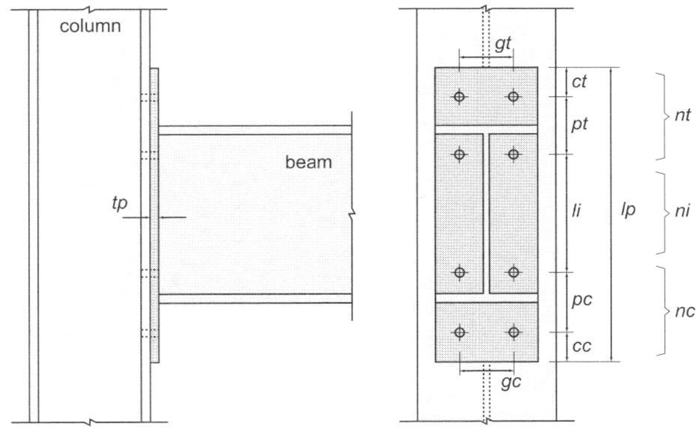

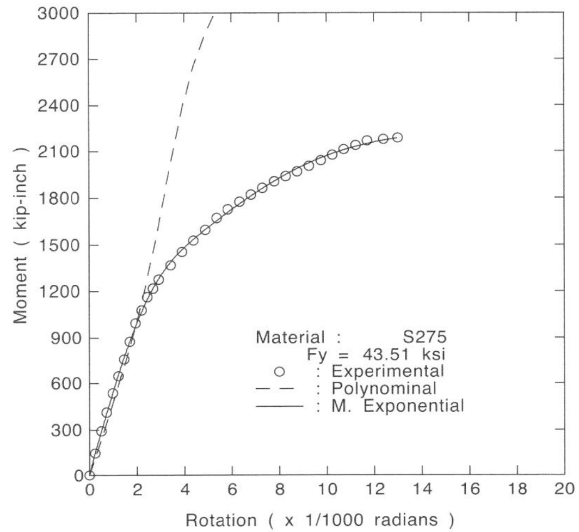

A.5 – 288

```
Moment-rotation prediction equations     ( R : X 1/1000 radians )

Frye and Morris polynominal model  :    R = Sum ( Ai X ( K*Bm )**Pi X 10**Qi )
     xd = 21.141740"    t =  0.590600"  tf =  0.787410"
     A1 =  1.830000   A2 = -1.040000   A3 =  6.380000    K =  0.001166
     P1 =        1     P2 =        3     P3 =        5     Q1 =        0     Q2 =       -1     Q3 =       -3

Modified exponential model  :
     Bm = Sum ( Ai X ( 1 - exp( - R/( 2*i*C ) ) ) )  +      Sum ( RKj X ( R - Rj0 ) )  +  Bm0
       C =   0.16833333E+00              Bmo=   0.00000000E+00          Nexp=  6        Nliner=  0
      AI =   0.40411450E+02   0.65160606E+04  -0.57994506E+05   0.16680956E+06  -0.19734643E+06   0.84219023E+05
```

No	Rotation (radians) X 1/1000	Moment (kip-inch) Expri.	Moment (kip-inch) Poly.	Moment (kip-inch) M.Expo.	Connection stiffness (kip-inch) X 1000 Expri.	Connection stiffness (kip-inch) X 1000 Poly.	Connection stiffness (kip-inch) X 1000 M.Expo.
1	0.00	0.00	0.00	0.00	0.6095E+03	0.4685E+03	0.7026E+03
2	0.24	146.29	112.44	150.30	0.5976E+03	0.4699E+03	0.5713E+03
3	0.49	292.57	230.51	286.27	0.5450E+03	0.4743E+03	0.5257E+03
4	0.73	414.14	345.15	410.58	0.4966E+03	0.4816E+03	0.5123E+03
5	0.98	535.71	466.85	537.46	0.4756E+03	0.4926E+03	0.5021E+03
6	1.22	647.37	586.73	656.15	0.4653E+03	0.5070E+03	0.4855E+03
7	1.46	759.04	710.57	769.78	0.4684E+03	0.5257E+03	0.4601E+03
8	1.71	876.97	844.99	880.65	0.4818E+03	0.5506E+03	0.4258E+03
9	1.95	994.90	980.62	978.42	0.4164E+03	0.5807E+03	0.3884E+03
10	2.20	1079.46	1130.47	1070.45	0.3454E+03	0.6195E+03	0.3478E+03
11	2.44	1164.01	1284.39	1149.29	0.2916E+03	0.6644E+03	0.3095E+03
12	2.68	1219.43	1449.86	1219.22	0.2309E+03	0.7153E+03	0.2738E+03
13	2.92	1274.85	1627.70	1281.00	0.2177E+03	0.7656E+03	0.2417E+03
14	3.41	1368.33	2016.32	1385.84	0.1837E+03	0.7937E+03	0.1892E+03
15	3.90	1454.87	2375.81	1469.26	0.1640E+03	0.6519E+03	0.1539E+03
16	4.39	1529.05	2653.44	1538.87	0.1495E+03	0.4848E+03	0.1322E+03
17	4.87	1599.90	2857.04	1599.14	0.1482E+03	0.3729E+03	0.1201E+03
18	5.36	1672.82	3020.09	1656.13	0.1304E+03	0.2987E+03	0.1132E+03
19	5.85	1727.70	3153.34	1710.49	0.1061E+03	0.2487E+03	0.1089E+03
20	6.34	1776.77	3266.01	1762.97	0.9496E+02	0.2133E+03	0.1053E+03
21	6.82	1819.91	3361.87	1812.58	0.8897E+02	0.1874E+03	0.1013E+03
22	7.31	1863.05	3448.50	1861.05	0.8804E+02	0.1671E+03	0.9635E+02
23	7.80	1906.19	3526.25	1906.84	0.7701E+02	0.1509E+03	0.9042E+02
24	8.29	1938.53	3596.88	1949.52	0.6508E+02	0.1378E+03	0.8367E+02
25	8.77	1969.34	3660.40	1987.99	0.6768E+02	0.1272E+03	0.7653E+02
26	9.26	2004.24	3720.40	2023.64	0.7209E+02	0.1180E+03	0.6899E+02
27	9.75	2039.99	3776.25	2055.60	0.7452E+02	0.1102E+03	0.6147E+02
28	10.24	2077.27	3828.52	2083.92	0.7275E+02	0.1034E+03	0.5418E+02
29	10.72	2110.62	3876.72	2108.29	0.6414E+02	0.9757E+02	0.4743E+02
30	11.21	2139.38	3923.22	2129.95	0.5870E+02	0.9235E+02	0.4105E+02
31	11.70	2168.14	3967.31	2148.62	0.4025E+02	0.8770E+02	0.3525E+02
32	12.40	2177.88	4026.62	2170.69	0.1515E+02	0.8189E+02	0.2800E+02
33	13.00	2187.61	4074.43	2185.87	0.1622E+02	0.7755E+02	0.2276E+02

V -145

Connection type : Extended end-plate connections
Mode : All bolted without column stiffener

Tested by : B.Bose et al. (1996) U.K.
Test Id. : 18

Column : 254x254 UC 89
Beam : 686x254 UB 125
Plate thickness : 20 mm

Fasteners: G8.8- -M24
33/32" Oversize holes
Material : S275
Fy = 43.51 ksi
Fu = 67.30 ksi

Major parameters

End plate extended on both sides.

ct = 2.0098"	pt = 3.9370"	li = 21.9646"	pc = 3.9370"
cc = 2.0098"	pit= 3.5433"	pi = 14.8780"	pic= 3.5433"
gt = 3.5433"	gi = 3.5433"	gc = 3.5433"	
tp = 0.7874"	nt = 2 X 2	ni = 2 X 2	nc = 2 X 2

Remark 1) End-plate: 860x250x20 mm
2) Failure mode: Column web buckling

No	Moment (k-in)	Rotation (radians) X 1/1000
1	0.00	0.00
2	325.23	0.25
3	650.46	0.50
4	924.81	0.75
5	1199.17	1.00
6	1557.66	1.24
7	1916.16	1.49
8	2211.46	1.74
9	2506.77	1.99
10	2832.21	2.24
11	3157.65	2.49
12	3394.29	2.74
13	3630.93	2.99
14	3855.20	3.24
15	4079.46	3.49
16	4363.43	3.98
17	4579.96	4.48
18	4716.58	4.98
19	4798.81	5.48
20	4850.41	5.98
21	4895.74	6.47
22	4932.74	6.97

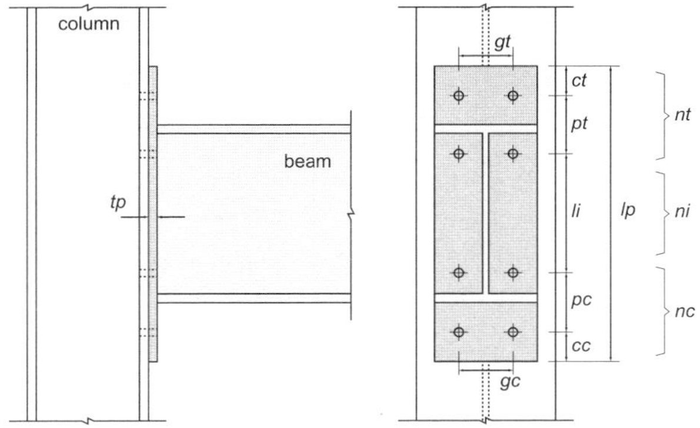

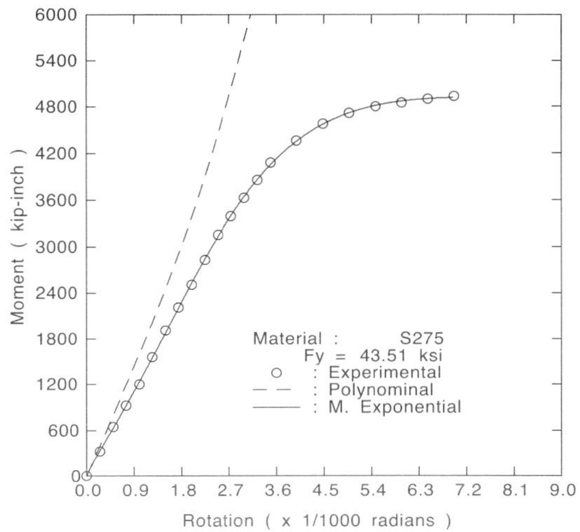

A.5 – 290

```
Moment-rotation prediction equations      ( R : X 1/1000 radians )

Frye and Morris polynominal model  :    R = Sum ( Ai X ( K*Bm )**Pi X 10**Qi )
     xd = 29.838600"    t =  0.787400"  tf =  0.944880"
     A1 =  1.830000   A2 = -1.040000   A3 =  6.380000    K =  0.000346
     P1 =        1    P2 =        3    P3 =        5    Q1 =        0    Q2 =       -1    Q3 =       -3

Modified exponential model  :
     Bm = Sum ( Ai X ( 1 - exp( - R/( 2*i*C ) ) ) )  +      Sum ( RKj X ( R - Rj0 ) )  +  Bm0
      C =   0.12058333E+00              Bmo=   0.00000000E+00          Nexp=  6        Nliner=  0
     AI =   0.25450039E+03  -0.69359149E+04   0.69376587E+05  -0.20674464E+06   0.21961892E+06  -0.70695673E+05
```

No	Rotation (radians) X 1/1000	Moment (kip-inch) Expri.	Moment (kip-inch) Poly.	Moment (kip-inch) M.Expo.	Connection stiffness (kip-inch) X 1000 Expri.	Connection stiffness (kip-inch) X 1000 Poly.	Connection stiffness (kip-inch) X 1000 M.Expo.
1	0.00	0.00	0.00	0.00	0.1301E+04	0.1580E+04	0.1522E+04
2	0.25	325.23	395.33	334.01	0.1301E+04	0.1585E+04	0.1219E+04
3	0.50	650.46	793.20	630.18	0.1199E+04	0.1600E+04	0.1173E+04
4	0.75	924.81	1196.26	927.46	0.1097E+04	0.1626E+04	0.1212E+04
5	1.00	1199.17	1607.35	1237.28	0.1300E+04	0.1665E+04	0.1267E+04
6	1.24	1557.66	2012.55	1546.75	0.1464E+04	0.1714E+04	0.1309E+04
7	1.49	1916.16	2449.19	1876.94	0.1308E+04	0.1782E+04	0.1328E+04
8	1.74	2211.46	2904.97	2208.11	0.1181E+04	0.1868E+04	0.1316E+04
9	1.99	2506.77	3385.10	2532.71	0.1241E+04	0.1977E+04	0.1276E+04
10	2.24	2832.21	3895.68	2843.83	0.1302E+04	0.2112E+04	0.1210E+04
11	2.49	3157.65	4443.48	3135.87	0.1124E+04	0.2275E+04	0.1124E+04
12	2.74	3394.29	5034.59	3404.77	0.9466E+03	0.2456E+04	0.1025E+04
13	2.99	3630.93	5670.39	3648.02	0.9218E+03	0.2624E+04	0.9198E+03
14	3.24	3855.20	6339.59	3864.54	0.8971E+03	0.2708E+04	0.8123E+03
15	3.49	4079.46	7016.70	4054.37	0.7898E+03	0.2630E+04	0.7070E+03
16	3.98	4363.43	8183.42	4353.35	0.5070E+03	0.2097E+04	0.5180E+03
17	4.48	4579.96	9090.38	4570.94	0.3532E+03	0.1551E+04	0.3585E+03
18	4.98	4716.58	9767.95	4717.95	0.2189E+03	0.1191E+04	0.2356E+03
19	5.48	4798.81	10300.89	4812.06	0.1338E+03	0.9599E+03	0.1460E+03
20	5.98	4850.41	10738.91	4868.51	0.9780E+02	0.8030E+03	0.8386E+02
21	6.47	4895.74	11103.91	4898.99	0.8335E+02	0.6930E+03	0.4341E+02
22	6.97	4932.74	11428.42	4913.64	0.7401E+02	0.6091E+03	0.1732E+02

V -146

Connection type : Extended end-plate connections
Mode : All bolted without column stiffener

Tested by : Ana M. Girao Coelho et al. (2004) Portugal
Test Id. : FS1a

Column : HE340M
Beam : IPE300
Plate thickness : 10 mm

Fasteners: G8.8- -M20
7/8" Oversize holes
Material : S355
Fy = 49.72 ksi
Fu = 73.66 ksi

Major parameters

End plate extended on tension side only.

ct = 1.1811"	pt = 3.5433"	li = 8.0709"	pc = 1.5472"
cc = 1.4055"	gt = 3.5433"	gc = 3.5433"	
tp = 0.4094"	nt = 2 X 2	nc = 2 X 1	

Remark 1) End-plate: 400x150x10 mm
2) Failure mode: Weld failure of assembly beam-end plate

No	Moment (k-in)	Rotation (radians) X 1/1000	No	Moment (k-in)	Rotation (radians) X 1/1000
1	0.00	0.00	26	1177.19	37.49
2	61.50	0.11	27	1194.44	39.46
3	122.99	0.22	28	1206.55	41.43
4	217.52	0.61	29	1218.30	43.41
5	312.05	1.00	30	1221.24	45.38
6	394.29	1.49	31	1221.24	47.35
7	476.53	1.97	32	1226.03	49.33
8	572.13	3.01	33	1231.18	51.30
9	648.61	3.95	34	1232.33	53.27
10	743.41	5.92	35	1233.20	55.24
11	801.83	7.89	36	1235.07	57.22
12	841.57	9.87	37	1248.57	59.19
13	878.52	11.84	38	1263.36	61.55
14	912.41	13.81	39	1255.73	63.14
15	942.03	15.78	40	1223.44	64.12
16	970.86	17.76	41	1191.15	65.11
17	997.84	19.73			
18	1023.84	21.70			
19	1048.84	23.68			
20	1072.14	25.65			
21	1092.55	27.62			
22	1112.70	29.60			
23	1130.90	31.57			
24	1147.66	33.54			
25	1161.29	35.51			

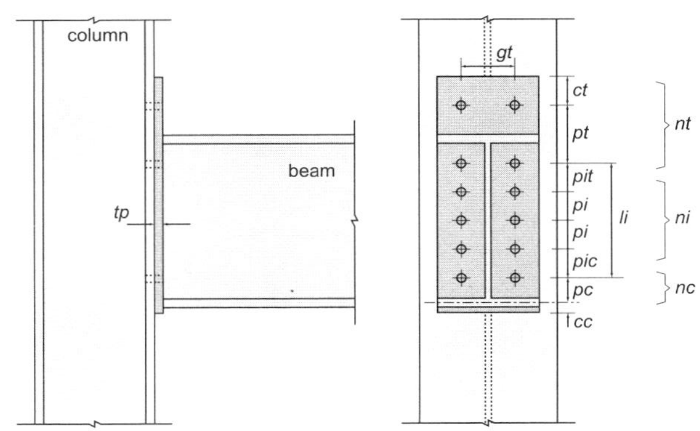

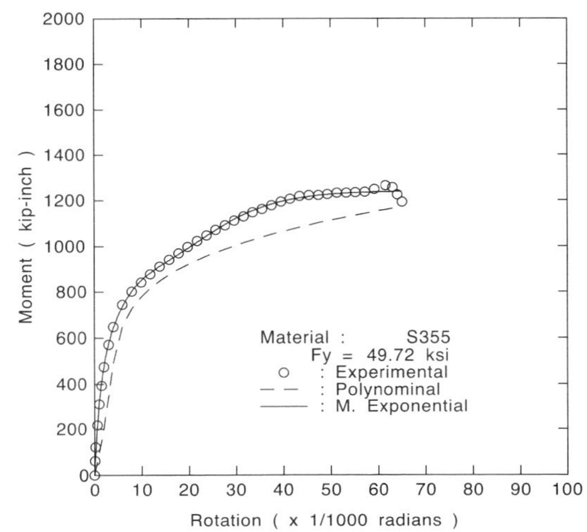

A.5 – 292

```
Moment-rotation prediction equations     ( R : X 1/1000 radians )

Frye and Morris polynominal model  :    R = Sum ( Ai X ( K*Bm )**Pi X 10**Qi )
     xd = 11.614170"    t =  0.409400"  tf =  0.787410"
     A1 =  1.830000   A2 = -1.040000   A3 =  6.380000    K =  0.005687
     P1 =        1    P2 =        3    P3 =        5    Q1 =        0    Q2 =      -1    Q3 =      -3

Modified exponential model  :
     Bm = Sum ( Ai X ( 1 - exp( - R/( 2*i*C ) ) ) )  +      Sum ( RKj X ( R - Rj0 ) )  +  Bm0
      C =   0.57008333E+00            Bmo=   0.00000000E+00          Nexp=  6        Nliner=  0
     AI =   0.40982251E+03   0.26167374E+04  -0.24100504E+05   0.73184093E+05  -0.89740221E+05   0.38869523E+05
```

No	Rotation (radians) X 1/1000	Moment (kip-inch) Expri.	Moment (kip-inch) Poly.	Moment (kip-inch) M.Expo.	Connection stiffness (kip-inch) X 1000 Expri.	Connection stiffness (kip-inch) X 1000 Poly.	Connection stiffness (kip-inch) X 1000 M.Expo.
1	0.00	0.00	0.00	0.00	0.5591E+03	0.9609E+02	0.4481E+03
3	0.22	122.99	21.15	90.72	0.4894E+03	0.9633E+02	0.3788E+03
5	1.00	312.05	97.78	316.12	0.2093E+03	0.1013E+03	0.2174E+03
7	1.97	476.53	203.51	474.71	0.1463E+03	0.1197E+03	0.1226E+03
9	3.95	648.61	493.85	640.37	0.7062E+02	0.1299E+03	0.5967E+02
11	7.89	801.83	726.02	802.66	0.2487E+02	0.3042E+02	0.2820E+02
13	11.84	878.52	816.28	884.76	0.1798E+02	0.1773E+02	0.1559E+02
15	15.78	942.03	875.33	939.19	0.1480E+02	0.1285E+02	0.1309E+02
17	19.73	997.84	920.41	991.86	0.1345E+02	0.1021E+02	0.1364E+02
19	23.68	1048.84	957.22	1045.75	0.1223E+02	0.8539E+01	0.1340E+02
21	27.62	1092.55	988.46	1095.77	0.1027E+02	0.7382E+01	0.1180E+02
23	31.57	1130.90	1015.85	1137.87	0.8873E+01	0.6526E+01	0.9452E+01
25	35.51	1161.29	1040.21	1170.31	0.7474E+01	0.5867E+01	0.7048E+01
27	39.46	1194.44	1062.30	1193.90	0.7452E+01	0.5340E+01	0.4972E+01
29	43.41	1218.30	1082.72	1210.21	0.3707E+01	0.4905E+01	0.3363E+01
31	47.35	1221.24	1101.30	1221.04	0.1207E+01	0.4547E+01	0.2205E+01
33	51.30	1231.18	1118.63	1228.07	0.1600E+01	0.4243E+01	0.1408E+01
35	55.24	1233.20	1134.81	1232.51	0.6929E+00	0.3981E+01	0.8815E+00
37	59.19	1248.57	1150.07	1235.27	0.6586E+01	0.3753E+01	0.5422E+00
39	63.14	1255.73	1164.45	1236.96	-0.2222E+02	0.3553E+01	0.3291E+00
41	65.11	1191.15	1171.40	1237.53	-0.3262E+02	0.3461E+01	0.2554E+00

V -147

Connection type : Extended end-plate connections
Mode : All bolted without column stiffener

Tested by : Ana M. Girao Coelho et al. (2004) Portugal
Test Id. : FS1b

Column : HE340M
Beam : IPE300
Plate thickness : 10 mm

Fasteners: G8.8- -M20
7/8" Oversize holes
Material : S355
Fy = 49.33 ksi
Fu = 69.69 ksi

Major parameters

End plate extended on tension side only.

ct = 1.1811"	pt = 3.5433"	li = 8.0709"	pc = 1.5472"
cc = 1.4055"	gt = 3.5433"	gc = 3.5433"	
tp = 0.4094"	nt = 2 X 2	nc = 2 X 1	

Remark 1) End-plate: 400x150x10 mm
2) Failure mode: Weld failure of the assembly beam-end plate

No	Moment (k-in)	Rotation (radians) X 1/1000	No	Moment (k-in)	Rotation (radians) X 1/1000
1	0.00	0.00	26	1363.25	64.95
2	87.50	0.50	27	1384.24	68.89
3	175.01	1.00	28	1404.52	72.83
4	251.44	1.48	29	1426.28	77.05
5	327.87	1.97	30	1336.28	80.70
6	401.88	2.48			
7	475.88	2.99			
8	564.42	3.94			
9	645.68	4.98			
10	711.79	5.90			
11	787.21	7.87			
12	837.42	9.84			
13	913.81	13.78			
14	972.91	17.71			
15	1024.68	21.65			
16	1071.80	25.59			
17	1111.13	29.52			
18	1145.58	33.46			
19	1180.12	37.40			
20	1210.03	41.33			
21	1237.60	45.27			
22	1265.64	49.21			
23	1292.25	53.14			
24	1315.30	57.08			
25	1339.97	61.02			

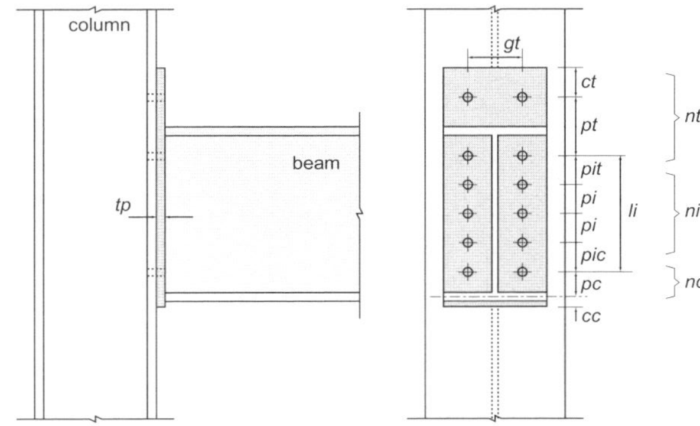

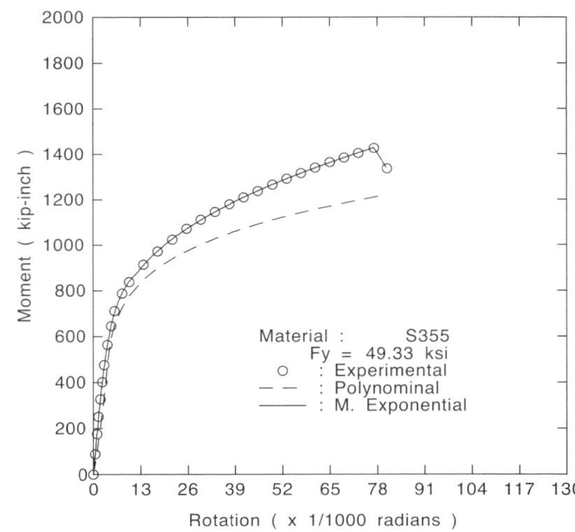

A.5 – 294

```
Moment-rotation prediction equations     ( R : X 1/1000 radians )

Frye and Morris polynominal model  :    R = Sum ( Ai X ( K*Bm )**Pi X 10**Qi )
     xd = 11.614170"    t =  0.409400"  tf =  0.787410"
     A1 =  1.830000   A2 = -1.040000   A3 =  6.380000    K =  0.005687
     P1 =         1   P2 =         3   P3 =         5   Q1 =         0    Q2 =        -1    Q3 =        -3

Modified exponential model  :
     Bm = Sum ( Ai X ( 1 - exp( - R/( 2*i*C ) ) ) )  +     Sum ( RKj X ( R - Rj0 ) )  +  Bm0
       C =   0.79750000E+00            Bmo=   0.00000000E+00          Nexp=  6        Nliner=  2
      AI =  -0.65170532E+03   0.48988229E+04  -0.14494565E+05   0.28580425E+05  -0.29194176E+05   0.11958089E+05
     Rj0 =          12.0000          77.0000
     RKj =   0.51213903E+01  -0.29981501E+02
```

No	Rotation (radians) X 1/1000	Moment (kip-inch) Expri.	Moment (kip-inch) Poly.	Moment (kip-inch) M.Expo.	Connection stiffness (kip-inch) X 1000 Expri.	Connection stiffness (kip-inch) X 1000 Poly.	Connection stiffness (kip-inch) X 1000 M.Expo.
1	0.00	0.00	0.00	0.00	0.1750E+03	0.9609E+02	0.1664E+03
2	0.50	87.50	48.25	86.07	0.1750E+03	0.9733E+02	0.1753E+03
3	1.00	175.01	97.78	173.37	0.1670E+03	0.1013E+03	0.1723E+03
4	1.48	251.44	147.90	254.01	0.1576E+03	0.1082E+03	0.1629E+03
5	1.97	327.87	203.51	330.67	0.1507E+03	0.1197E+03	0.1497E+03
6	2.48	401.88	268.90	403.12	0.1451E+03	0.1379E+03	0.1343E+03
7	2.99	475.88	344.91	467.68	0.1270E+03	0.1596E+03	0.1189E+03
8	3.94	564.42	492.53	567.71	0.8647E+02	0.1307E+03	0.9240E+02
9	4.98	646.68	594.11	651.05	0.7468E+02	0.7251E+02	0.6891E+02
10	5.90	711.79	649.27	706.85	0.6043E+02	0.5017E+02	0.5312E+02
11	7.87	787.21	725.41	788.08	0.3189E+02	0.3054E+02	0.3174E+02
12	9.84	837.42	776.55	838.89	0.2346E+02	0.2232E+02	0.2108E+02
13	13.78	913.81	847.69	911.40	0.1721E+02	0.1489E+02	0.1782E+02
14	17.71	972.91	898.79	974.03	0.1409E+02	0.1138E+02	0.1438E+02
15	21.65	1024.68	939.17	1026.10	0.1255E+02	0.9311E+01	0.1218E+02
16	25.59	1071.80	972.95	1070.87	0.1098E+02	0.7930E+01	0.1064E+02
17	29.52	1111.13	1002.06	1110.45	0.9504E+01	0.6940E+01	0.9559E+01
18	33.46	1146.58	1027.86	1146.46	0.8754E+01	0.6190E+01	0.8753E+01
19	37.40	1180.12	1051.04	1179.63	0.8061E+01	0.5601E+01	0.8108E+01
20	41.33	1210.03	1072.09	1210.39	0.7305E+01	0.5126E+01	0.7560E+01
21	45.27	1237.60	1091.48	1239.22	0.7184E+01	0.4732E+01	0.7084E+01
22	49.21	1266.64	1109.46	1266.30	0.6943E+01	0.4401E+01	0.6673E+01
23	53.14	1292.25	1126.18	1291.82	0.6184E+01	0.4118E+01	0.6327E+01
24	57.08	1315.30	1141.91	1316.17	0.6056E+01	0.3873E+01	0.6042E+01
25	61.02	1339.97	1156.74	1339.51	0.6093E+01	0.3659E+01	0.5813E+01
26	64.95	1363.25	1170.74	1361.99	0.5625E+01	0.3470E+01	0.5634E+01
27	68.89	1384.24	1184.08	1383.90	0.5237E+01	0.3301E+01	0.5497E+01
28	72.83	1404.52	1196.78	1405.35	0.5152E+01	0.3150E+01	0.5393E+01
29	77.05	1426.28	1209.76	1426.42	-0.1083E+02	0.3005E+01	-0.2467E+02
30	80.70	1336.28	1220.73	1336.28	-0.2466E+02	0.2888E+01	-0.2472E+02

A.5 – 295

V -148

Connection type : Extended end-plate connections
Mode : All bolted without column stiffener

Tested by : Ana M. Girao Coelho et al. (2004) Portugal
Test Id. : FS2a

Column : HE340M
Beam : IPE300
Plate thickness : 15 mm

Fasteners: G8.8- -M20
7/8" Oversize holes
Material : S355
Fy = 49.33 ksi
Fu = 69.69 ksi

Major parameters

End plate extended on tension side only.

ct = 1.1811"	pt = 3.5433"	li = 8.0709"	pc = 1.5472"
cc = 1.4055"	gt = 3.5433"	gc = 3.5433"	
tp = 0.5909"	nt = 2 X 2	nc = 2 X 1	

Remark 1) End-plate: 400x150x15 mm
2) Failure mode: Nut stripping of bolt #4 and weld failure

No	Moment (k-in)	Rotation (radians) X 1/1000
1	0.00	0.00
2	110.64	0.50
3	235.24	1.00
4	341.76	1.50
5	453.35	2.01
6	568.09	2.51
7	647.47	3.01
8	794.99	4.01
9	901.76	5.01
10	996.32	6.02
11	1084.47	7.02
12	1163.90	8.02
13	1226.58	9.02
14	1289.26	10.03
15	1369.52	12.03
16	1426.73	14.04
17	1471.10	16.04
18	1509.78	18.05
19	1544.48	20.05
20	1569.98	22.06
21	1593.17	24.07
22	1613.18	26.07
23	1631.31	28.08
24	1644.80	30.08
25	1662.00	32.09
26	1680.80	34.09
27	1688.34	36.10
28	1695.87	38.10
29	1701.16	40.11
30	1708.50	41.72
31	1668.85	42.92
32	1629.20	44.12

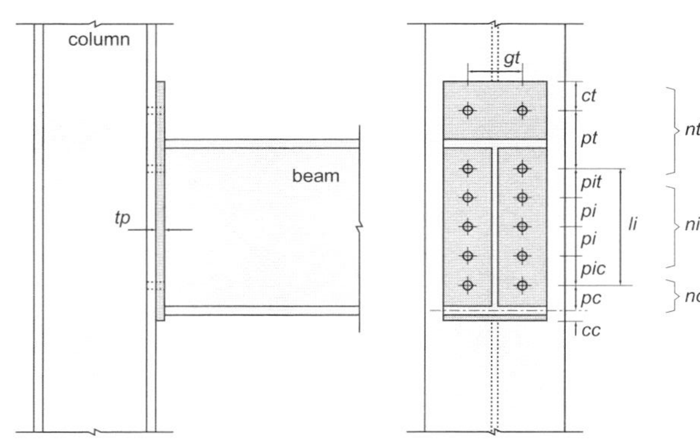

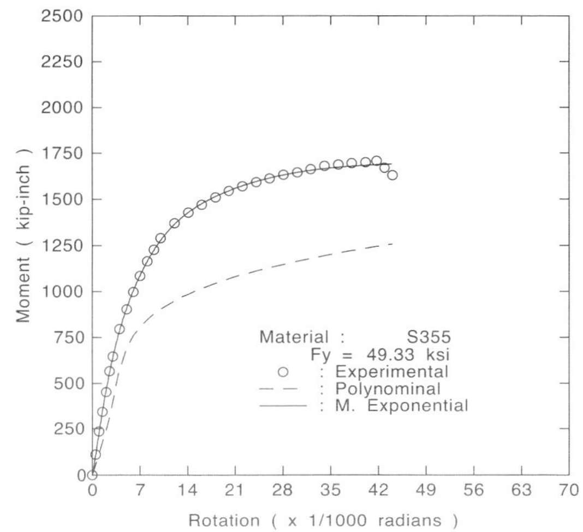

A.5 – 296

```
Moment-rotation prediction equations     ( R : X 1/1000 radians )

Frye and Morris polynominal model  :    R = Sum ( Ai X ( K*Bm )**Pi X 10**Qi )
     xd = 11.614170"    t =  0.590940"  tf =  0.787410"
     A1 =  1.830000   A2 = -1.040000   A3 =  6.380000    K =  0.004910
     P1 =        1    P2 =        3    P3 =        5    Q1 =        0    Q2 =       -1    Q3 =       -3

Modified exponential model  :
     Bm = Sum ( Ai X ( 1 - exp( - R/( 2*i*C ) ) ) )  +     Sum ( RKj X ( R - Rj0 ) )  +  Bm0
       C =   0.49266667E+00              Bmo=   0.00000000E+00            Nexp=  6        Nliner=  0
      AI =  -0.76227445E+03   0.73438484E+04  -0.32058211E+05   0.70087774E+05  -0.72363521E+05   0.29450682E+05
```

No	Rotation (radians) X 1/1000	Moment (kip-inch) Expri.	Poly.	M.Expo.	Connection stiffness (kip-inch) X 1000 Expri.	Poly.	M.Expo.
1	0.00	0.00	0.00	0.00	0.2213E+03	0.1113E+03	0.1840E+03
2	0.50	110.64	55.88	108.73	0.2352E+03	0.1127E+03	0.2393E+03
3	1.00	235.24	113.24	231.12	0.2311E+03	0.1173E+03	0.2449E+03
4	1.50	341.76	173.80	350.16	0.2159E+03	0.1257E+03	0.2291E+03
5	2.01	453.35	241.26	461.17	0.2242E+03	0.1400E+03	0.2057E+03
6	2.51	568.09	316.23	558.15	0.1941E+03	0.1612E+03	0.1824E+03
7	3.01	647.47	403.16	644.04	0.1550E+03	0.1856E+03	0.1616E+03
8	4.01	794.99	580.89	788.46	0.1271E+03	0.1451E+03	0.1293E+03
9	5.01	901.76	690.54	905.97	0.1002E+03	0.8280E+02	0.1071E+03
10	6.02	996.32	758.78	1005.54	0.9088E+02	0.5585E+02	0.9084E+02
11	7.02	1084.47	807.40	1089.80	0.8379E+02	0.4236E+02	0.7810E+02
12	8.02	1163.90	845.39	1162.38	0.7105E+02	0.3437E+02	0.6731E+02
13	9.02	1226.58	876.91	1224.87	0.6237E+02	0.2905E+02	0.5788E+02
14	10.03	1289.26	904.21	1279.04	0.5470E+02	0.2522E+02	0.4958E+02
15	12.03	1369.52	949.22	1364.39	0.3431E+02	0.2015E+02	0.3649E+02
16	14.04	1426.73	986.20	1427.96	0.2531E+02	0.1689E+02	0.2736E+02
17	16.04	1471.10	1017.58	1476.22	0.2072E+02	0.1463E+02	0.2132E+02
18	18.05	1509.78	1045.20	1514.74	0.1830E+02	0.1294E+02	0.1727E+02
19	20.05	1544.48	1069.71	1546.31	0.1502E+02	0.1163E+02	0.1445E+02
20	22.06	1569.98	1092.01	1573.09	0.1211E+02	0.1059E+02	0.1227E+02
21	24.07	1593.17	1112.41	1595.89	0.1077E+02	0.9735E+01	0.1047E+02
22	26.07	1613.18	1131.15	1615.21	0.9512E+01	0.9025E+01	0.8882E+01
23	28.08	1631.31	1148.66	1631.60	0.7879E+01	0.8419E+01	0.7455E+01
24	30.08	1644.80	1164.97	1645.22	0.7648E+01	0.7900E+01	0.6185E+01
25	32.09	1662.00	1180.38	1656.50	0.8982E+01	0.7446E+01	0.5061E+01
26	34.09	1680.80	1194.87	1665.62	0.6583E+01	0.7049E+01	0.4093E+01
27	36.10	1688.34	1208.68	1673.00	0.3758E+01	0.6695E+01	0.3267E+01
28	38.10	1695.87	1221.75	1678.83	0.3200E+01	0.6380E+01	0.2583E+01
29	40.11	1701.16	1234.28	1683.43	0.3700E+01	0.6095E+01	0.2020E+01
30	41.72	1708.50	1244.09	1686.38	-0.1698E+02	0.5884E+01	0.1648E+01
31	42.92	1668.85	1250.90	1688.21	-0.3304E+02	0.5742E+01	0.1411E+01
32	44.12	1629.20	1257.79	1689.78	-0.3304E+02	0.5603E+01	0.1205E+01

V -149

Connection type : Extended end-plate connections
Mode : All bolted without column stiffener

Tested by : Ana M. Girao Coelho et al. (2004) Portugal
Test Id. : FS2b

Column : HE340M
Beam : IPE300
Plate thickness : 15 mm

Fasteners: G8.8- -M20
7/8" Oversize holes
Material : S355
Fy = 49.33 ksi
Fu = 69.69 ksi

Major parameters

End plate extended on tension side only.

ct = 1.1811" pt = 3.5433" li = 8.0709" pc = 1.5472"
cc = 1.4055" gt = 3.5433" gc = 3.5433"
tp = 0.5909" nt = 2 X 2 nc = 2 X 1

Remark 1) End-plate: 400x150x15 mm
2) Failure mode: Nut stripping of bolt #1 and #4

No	Moment (k-in)	Rotation (radians) X 1/1000	No	Moment (k-in)	Rotation (radians) X 1/1000
1	0.00	0.00	26	1721.01	34.71
2	111.32	0.50	27	1731.40	36.75
3	222.63	1.00	28	1739.42	38.80
4	329.03	1.52	29	1746.11	40.30
5	435.44	2.04	30	1687.55	41.59
6	517.77	2.52	31	1628.99	42.88
7	600.11	2.99			
8	691.55	3.54			
9	783.00	4.08			
10	893.70	4.99			
11	1017.98	6.13			
12	1096.82	6.98			
13	1189.02	8.17			
14	1322.02	10.21			
15	1409.63	12.25			
16	1467.43	14.29			
17	1514.87	16.34			
18	1555.54	18.38			
19	1587.13	20.42			
20	1614.64	22.46			
21	1638.31	24.50			
22	1660.61	26.54			
23	1682.00	28.59			
24	1696.78	30.63			
25	1710.16	32.67			

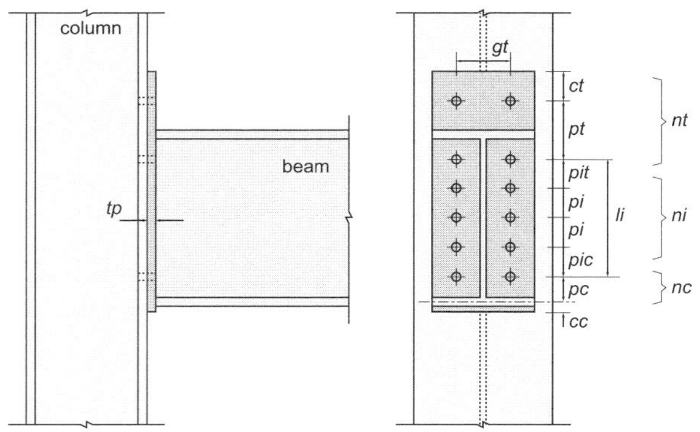

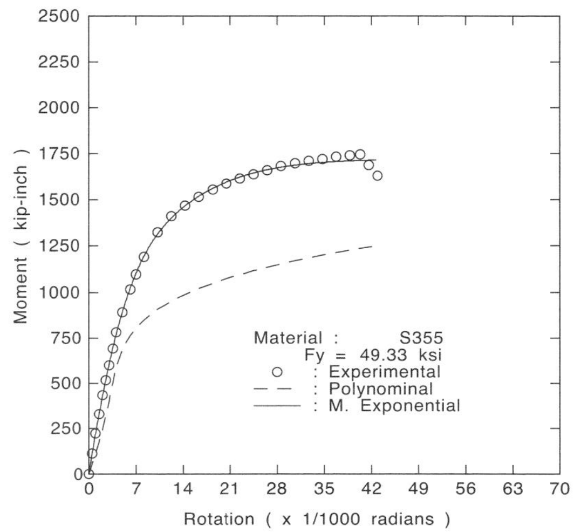

```
Moment-rotation prediction equations      ( R : X 1/1000 radians )

Frye and Morris polynominal model  :    R = Sum ( Ai X ( K*Bm )**Pi X 10**Qi )
     xd = 11.614170"   t =  0.590940"  tf =  0.787410"
     A1 =  1.830000   A2 = -1.040000   A3 =  6.380000    K =  0.004910
     P1 =         1   P2 =         3   P3 =         5   Q1 =         0   Q2 =        -1   Q3 =        -3

Modified exponential model  :
     Bm = Sum ( Ai X ( 1 - exp( - R/( 2*i*C ) ) ) )  +     Sum ( RKj X ( R - Rj0 ) )  +  Bm0
       C =   0.48233333E+00              Bmo=   0.00000000E+00            Nexp=  6          Nliner=  0
      AI =   0.12211565E+03  -0.89325822E+03  -0.26163096E+04   0.18186416E+05  -0.28270715E+05   0.15192439E+05
```

No	Rotation (radians) X 1/1000	Moment (kip-inch) Expri.	Moment (kip-inch) Poly.	Moment (kip-inch) M.Expo.	Connection stiffness (kip-inch) X 1000 Expri.	Connection stiffness (kip-inch) X 1000 Poly.	Connection stiffness (kip-inch) X 1000 M.Expo.
1	0.00	0.00	0.00	0.00	0.2226E+03	0.1113E+03	0.2363E+03
2	0.50	111.32	55.88	114.09	0.2226E+03	0.1127E+03	0.2213E+03
3	1.00	222.63	113.24	222.07	0.2138E+03	0.1173E+03	0.2110E+03
4	1.52	329.03	176.32	329.25	0.2046E+03	0.1262E+03	0.2012E+03
5	2.04	435.44	245.48	431.23	0.1874E+03	0.1410E+03	0.1909E+03
6	2.52	517.77	317.85	520.42	0.1734E+03	0.1617E+03	0.1806E+03
7	2.99	600.11	399.46	602.78	0.1711E+03	0.1848E+03	0.1698E+03
8	3.54	691.55	503.13	692.62	0.1678E+03	0.1828E+03	0.1568E+03
9	4.08	783.00	590.99	773.82	0.1516E+03	0.1390E+03	0.1440E+03
10	4.99	893.70	688.86	895.31	0.1160E+03	0.8359E+02	0.1233E+03
11	6.13	1017.98	764.82	1022.40	0.9970E+02	0.5395E+02	0.1003E+03
12	6.98	1096.82	805.62	1101.36	0.8639E+02	0.4279E+02	0.8585E+02
13	8.17	1189.02	850.53	1193.29	0.7295E+02	0.3343E+02	0.6933E+02
14	10.21	1322.02	908.93	1312.87	0.5407E+02	0.2462E+02	0.4945E+02
15	12.25	1409.63	953.61	1400.01	0.3564E+02	0.1972E+02	0.3692E+02
16	14.29	1467.43	990.38	1466.43	0.2574E+02	0.1657E+02	0.2873E+02
17	16.34	1514.87	1021.93	1519.07	0.2153E+02	0.1434E+02	0.2292E+02
18	18.38	1555.54	1049.43	1561.18	0.1771E+02	0.1270E+02	0.1854E+02
19	20.42	1587.13	1073.98	1595.28	0.1449E+02	0.1142E+02	0.1501E+02
20	22.46	1614.64	1096.21	1622.83	0.1255E+02	0.1041E+02	0.1209E+02
21	24.50	1638.31	1116.56	1644.94	0.1127E+02	0.9572E+01	0.9658E+01
22	26.54	1660.61	1135.36	1662.52	0.1068E+02	0.8875E+01	0.7639E+01
23	28.59	1682.00	1152.92	1676.41	0.8834E+01	0.8280E+01	0.5974E+01
24	30.63	1696.78	1169.28	1687.18	0.6903E+01	0.7770E+01	0.4633E+01
25	32.67	1710.16	1184.67	1695.50	0.5940E+01	0.7326E+01	0.3560E+01
26	34.71	1721.01	1199.21	1701.86	0.5205E+01	0.6935E+01	0.2713E+01
27	36.75	1731.40	1212.99	1706.69	0.4504E+01	0.6589E+01	0.2052E+01
28	38.80	1739.42	1226.18	1710.35	0.4226E+01	0.6278E+01	0.1539E+01
29	40.30	1746.11	1235.59	1712.43	-0.2235E+02	0.6067E+01	0.1243E+01
30	41.59	1687.55	1243.16	1713.89	-0.4539E+02	0.5904E+01	0.1031E+01
31	42.88	1628.99	1250.77	1715.10	-0.4539E+02	0.5745E+01	0.8536E+00

V -150

Connection type : Extended end-plate connections
Mode : All bolted without column stiffener

Tested by : Ana M. Girao Coelho et al. (2004) Portugal
Test Id. : FS3a

Column : HE340M
Beam : IPE300
Plate thickness : 20 mm

Fasteners: G8.8- -M20
7/8" Oversize holes
Material : S355
Fy = 49.69 ksi
Fu = 72.89 ksi

Major parameters

End plate extended on tension side only.

ct = 1.1811"	pt = 3.5433"	li = 8.0709"	pc = 1.5472"
cc = 1.4055"	gt = 3.5433"	gc = 3.5433"	
tp = 0.7882"	nt = 2 X 2	nc = 2 X 1	

Remark 1) End-plate: 400x150x20 mm
2) Failure mode: Nut stripping of bolt #3 and #4 and some weld failure

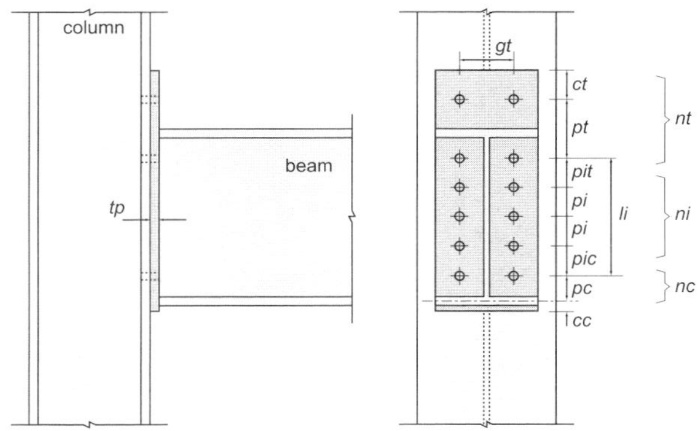

No	Moment (k-in)	Rotation (radians) X 1/1000
1	C.00	0.00
2	132.60	0.50
3	257.01	1.00
4	369.04	1.50
5	473.68	2.00
6	567.29	2.50
7	659.11	3.00
8	752.66	3.50
9	841.20	4.00
10	1001.27	5.00
11	1123.16	6.00
12	1223.36	7.00
13	1307.56	8.00
14	1378.02	9.00
15	1445.61	10.00
16	1506.38	11.00
17	1551.24	12.00
18	1583.93	13.00
19	1613.78	14.00
20	1642.31	15.00
21	1666.89	16.00
22	1688.36	17.00
23	1689.64	18.00
24	1718.44	19.00
25	1752.96	20.00

No	Moment (k-in)	Rotation (radians) X 1/1000
26	1790.82	21.00
27	1814.99	21.64
28	1685.38	21.67
29	1745.87	22.00
30	1771.40	23.00
31	1785.71	24.00
32	1795.66	25.00
33	1777.47	26.00
34	1760.43	27.00
35	1738.94	28.00

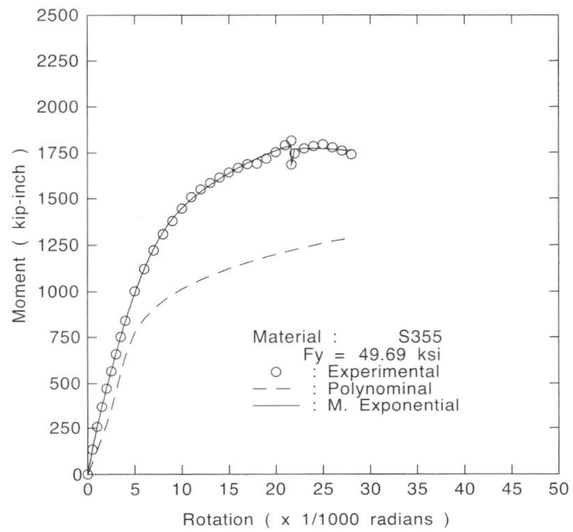

A.5 – 300

```
Moment-rotation prediction equations     ( R : X 1/1000 radians )

Frye and Morris polynominal model  :    R = Sum ( Ai X ( K*Bm )**Pi X 10**Qi )
     xd = 11.614170"   t =  0.788190"  tf =  0.787410"
     A1 =  1.830000   A2 = -1.040000   A3 =  6.380000    K =  0.004376
     P1 =        1    P2 =        3    P3 =        5    Q1 =        0    Q2 =       -1    Q3 =       -3

Modified exponential model  :
     Bm = Sum ( Ai X ( 1 - exp( - R/( 2*i*C ) ) ) )  +    Sum ( RKj X ( R - Rj0 ) )  +  Bm0
      C =  0.35833333E+00             Bmo=  0.00000000E+00          Nexp=  6        Nliner=  3
     AI = -0.53240204E+03  0.10285404E+05 -0.57414723E+05  0.13409549E+06 -0.14167113E+06  0.57129709E+05
     Rj0 =          21.6400         21.6700         22.0000
     RKj = -0.34789284E+04  0.37184686E+04 -0.25169043E+03
```

No	Rotation (radians) X 1/1000	Moment (kip-inch) Expri.	Moment (kip-inch) Poly.	Moment (kip-inch) M.Expo.	Connection stiffness (kip-inch) X 1000 Expri.	Connection stiffness (kip-inch) X 1000 Poly.	Connection stiffness (kip-inch) X 1000 M.Expo.
1	0.00	0.00	0.00	0.00	0.2652E+03	0.1249E+03	0.2558E+03
3	1.00	257.01	127.06	256.23	0.2364E+03	0.1316E+03	0.2381E+03
5	2.00	473.68	269.16	473.42	0.1982E+03	0.1567E+03	0.2004E+03
7	3.00	659.11	450.31	663.38	0.1854E+03	0.2078E+03	0.1812E+03
9	4.00	841.20	650.23	836.35	0.1714E+03	0.1638E+03	0.1643E+03
11	6.00	1123.16	850.17	1122.18	0.1110E+03	0.6308E+02	0.1200E+03
13	8.00	1307.56	947.85	1318.50	0.7733E+02	0.3871E+02	0.7850E+02
15	10.00	1445.61	1013.77	1446.85	0.6418E+02	0.2841E+02	0.5246E+02
17	12.00	1551.24	1064.37	1536.87	0.3877E+02	0.2268E+02	0.3913E+02
19	14.00	1613.78	1105.83	1607.58	0.2919E+02	0.1902E+02	0.3222E+02
21	16.00	1666.89	1141.17	1667.19	0.2302E+02	0.1646E+02	0.2755E+02
23	18.00	1689.64	1172.09	1718.04	0.1504E+02	0.1456E+02	0.2333E+02
25	20.00	1752.96	1199.68	1760.53	0.3619E+02	0.1309E+02	0.1917E+02
27	21.64	1814.99	1220.43	1789.27	-0.4125E+04	0.1210E+02	0.1591E+02
29	22.00	1745.87	1224.78	1769.55	0.1442E+03	0.1191E+02	0.2548E+03
31	24.00	1785.71	1247.71	1772.11	0.1213E+02	0.1095E+02	-0.4389E+00
33	26.00	1777.47	1268.76	1768.17	-0.1762E+02	0.1015E+02	-0.3395E+01
35	28.00	1738.94	1288.34	1758.92	-0.2149E+02	0.9467E+01	-0.5761E+01

A.5 – 301

V -151

Connection type : Extended end-plate connections
Mode : All bolted without column stiffener

Tested by : Ana M. Girao Coelho et al. (2004) Portugal
Test Id. : FS3b

Column : HE340M
Beam : IPE300
Plate thickness : 20 mm

Fasteners: G8.8- -M20
7/8" Oversize holes
Material : S355
Fy = 49.69 ksi
Fu = 72.89 ksi

Major parameters

End plate extended on tension side only.

ct = 1.1811" pt = 3.5433" li = 8.0709" pc = 1.5472"
cc = 1.4055" gt = 3.5433" gc = 3.5433"
tp = 0.7882" nt = 2 X 2 nc = 2 X 1

Remark 1) End-plate: 400x150x20 mm
2) Failure mode: Nut stripping of bolt #3

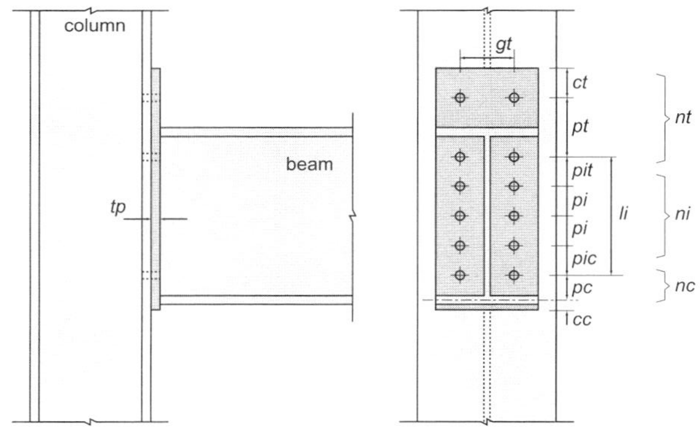

No	Moment (k-in)	Rotation (radians) X 1/1000	No	Moment (k-in)	Rotation (radians) X 1/1000
1	0.00	0.00	26	1817.76	20.54
2	113.58	0.49	27	1831.98	21.52
3	235.29	0.98	28	1841.02	22.50
4	328.76	1.47	29	1850.07	23.48
5	454.79	1.96	30	1859.11	24.45
6	565.99	2.45	31	1868.16	25.43
7	645.51	2.93	32	1877.22	26.41
8	738.62	3.42	33	1883.92	27.39
9	838.93	3.91	34	1887.60	28.37
10	997.53	4.89	35	1891.28	29.35
11	1112.79	5.87	36	1896.90	29.99
12	1223.55	6.85	37	1864.43	31.30
13	1316.26	7.83	38	1836.28	32.28
14	1393.38	8.80			
15	1460.52	9.78			
16	1521.57	10.76			
17	1574.20	11.74			
18	1618.69	12.72			
19	1660.86	13.69			
20	1694.41	14.67			
21	1727.97	15.65			
22	1748.32	16.63			
23	1766.48	17.61			
24	1784.64	18.59			
25	1801.26	19.56			

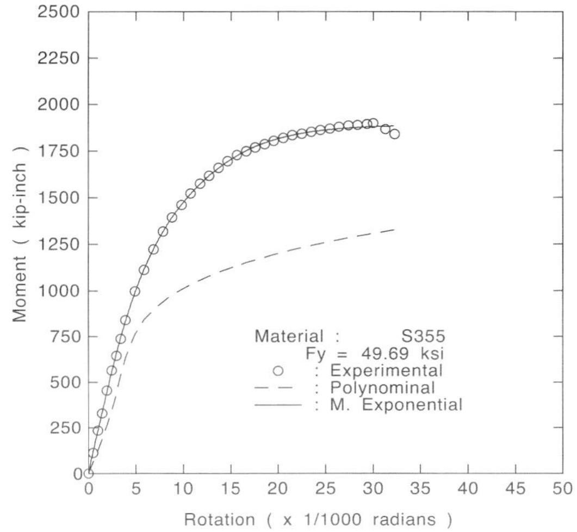

```
Moment-rotation prediction equations     ( R : X 1/1000 radians )

Frye and Morris polynominal model  :    R = Sum ( Ai X ( K*Bm )**Pi X 10**Qi )
     xd = 11.614170"   t =  0.788190"  tf =  0.787410"
     A1 =  1.830000   A2 = -1.040000   A3 =  6.380000    K =  0.004376
     P1 =        1    P2 =        3    P3 =        5    Q1 =        0    Q2 =       -1    Q3 =       -3

Modified exponential model  :
     Bm = Sum ( Ai X ( 1 - exp( - R/( 2*i*C ) ) ) )  +     Sum ( RKj X ( R - Rj0 ) )  +  Bm0
      C =   0.39150000E+00          Bmo=   0.00000000E+00          Nexp=  6       Nliner=  0
     AI =   0.36981670E+03  -0.43130610E+04   0.13599049E+05  -0.15753994E+05   0.18414000E+04   0.61449226E+04
```

No	Rotation (radians) X 1/1000	Moment (kip-inch) Expri.	Moment (kip-inch) Poly.	Moment (kip-inch) M.Expo.	Connection stiffness (kip-inch) X 1000 Expri.	Connection stiffness (kip-inch) X 1000 Poly.	Connection stiffness (kip-inch) X 1000 M.Expo.
1	0.00	0.00	0.00	0.00	0.2318E+03	0.1249E+03	0.2557E+03
3	0.98	235.29	124.44	228.53	0.2196E+03	0.1313E+03	0.2259E+03
5	1.96	454.79	262.92	448.58	0.2421E+03	0.1552E+03	0.2208E+03
7	2.93	645.51	435.87	653.82	0.1777E+03	0.2046E+03	0.2001E+03
9	3.91	838.93	635.00	835.90	0.1904E+03	0.1728E+03	0.1711E+03
11	5.87	1112.79	841.79	1117.32	0.1153E+03	0.6585E+02	0.1191E+03
13	7.83	1316.26	941.17	1315.38	0.8702E+02	0.3998E+02	0.8580E+02
15	9.78	1460.52	1007.43	1461.03	0.6540E+02	0.2924E+02	0.6498E+02
17	11.74	1574.20	1058.40	1573.05	0.4955E+02	0.2328E+02	0.5001E+02
19	13.69	1660.86	1099.86	1658.70	0.3888E+02	0.1950E+02	0.3825E+02
21	15.65	1727.97	1135.34	1723.95	0.2750E+02	0.1685E+02	0.2867E+02
23	17.61	1766.48	1166.35	1772.36	0.1853E+02	0.1489E+02	0.2104E+02
25	19.56	1801.26	1193.86	1807.39	0.1699E+02	0.1338E+02	0.1516E+02
27	21.52	1831.98	1218.86	1832.53	0.1187E+02	0.1217E+02	0.1072E+02
29	23.48	1850.07	1241.72	1850.19	0.9276E+01	0.1119E+02	0.7473E+01
31	25.43	1868.16	1262.71	1862.38	0.9239E+01	0.1037E+02	0.5159E+01
33	27.39	1883.92	1282.33	1870.79	0.5297E+01	0.9669E+01	0.3522E+01
35	29.35	1891.28	1300.68	1876.51	0.6796E+01	0.9068E+01	0.2386E+01
37	31.30	1864.43	1317.84	1880.36	-0.2704E+02	0.8548E+01	0.1610E+01

A.5 – 303

V -152

Connection type : Extended end-plate connections
Mode : All bolted without column stiffener

Tested by : Ana M. Girao Coelho et al. (2004) Portugal
Test Id. : FS4a

Column : HE340M
Beam : IPE300
Plate thickness : 10 mm

Fasteners: G8.8- -M20
7/8" Oversize holes
Material : S690
Fy = 101.32 ksi
Fu = 107.51 ksi

Major parameters

End plate extended on tension side only.

ct = 1.1811"	pt = 3.5433"	li = 8.0709"	pc = 1.5472"
cc = 1.4055"	gt = 3.5433"	gc = 3.5433"	
tp = 0.3961"	nt = 2 X 2	nc = 2 X 1	

Remark 1) End-plate: 400x150x10 mm
2) Failure mode: Fracture of bolt #4 and some weld failure

No	Moment (k-in)	Rotation (radians) X 1/1000	No	Moment (k-in)	Rotation (radians) X 1/1000
1	0.00	0.00	26	1613.35	43.31
2	208.51	1.01	27	1596.90	45.32
3	360.23	2.01	28	1575.93	47.34
4	510.52	3.02	29	1554.35	49.35
5	652.60	4.03	30	1525.85	51.37
6	779.89	5.04	31	1497.35	53.38
7	882.85	6.04			
8	980.04	7.05			
9	1133.72	9.06			
10	1243.38	11.08			
11	1320.31	13.09			
12	1383.37	15.11			
13	1437.70	17.12			
14	1473.09	19.14			
15	1507.41	21.15			
16	1533.51	23.16			
17	1559.60	25.18			
18	1579.04	27.19			
19	1595.72	29.21			
20	1611.07	31.22			
21	1619.39	33.24			
22	1627.71	35.25			
23	1640.00	37.70			
24	1635.81	39.28			
25	1628.74	41.29			

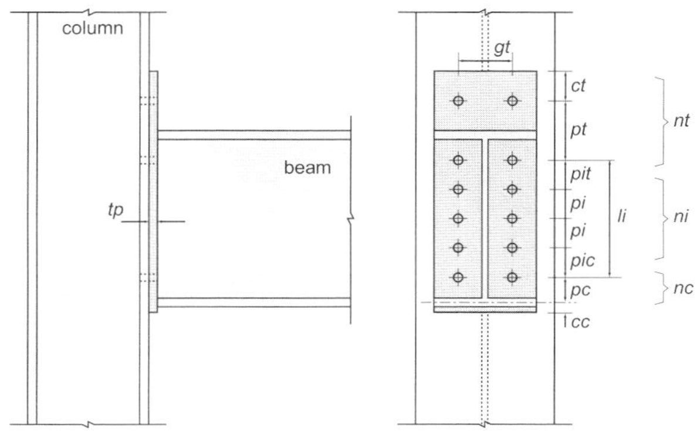

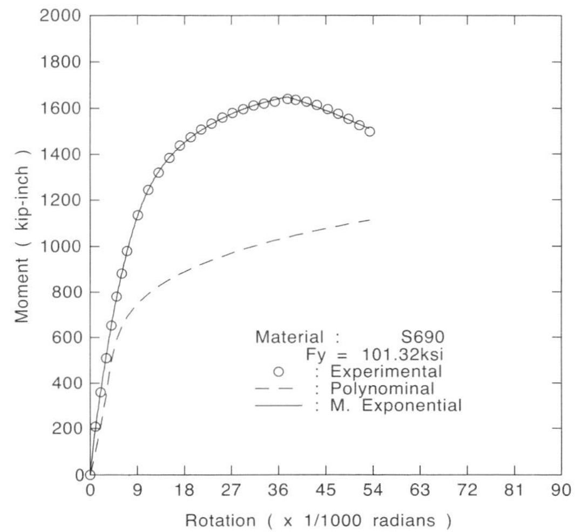

A.5 – 304

```
Moment-rotation prediction equations      ( R : X 1/1000 radians )

Frye and Morris polynominal model  :     R = Sum ( Ai X ( K*Bm )**Pi X 10**Qi )
     xd = 11.614170"   t =  0.396060"  tf =  0.787410"
     A1 =  1.830000   A2 = -1.040000   A3 =  6.380000    K =  0.005763
     P1 =         1   P2 =         3   P3 =         5   Q1 =         0   Q2 =        -1   Q3 =        -3

Modified exponential model  :
     Bm = Sum ( Ai X ( 1 - exp( - R/( 2*i*C ) ) ) )  +     Sum ( RKj X ( R - Rj0 ) )  +  Bm0
       C =   0.69733333E+00             Bmo=   0.00000000E+00         Nexp=  6        Nliner=  1
      AI =   0.49227858E+03  -0.41252348E+04   0.90436386E+04  -0.21140945E+04  -0.97447298E+04   0.81414850E+04
     Rj0 =          37.7000
     RKj =  -0.11115117E+02
```

No	Rotation (radians) X 1/1000	Moment (kip-inch) Expri.	Moment (kip-inch) Poly.	Moment (kip-inch) M.Expo.	Connection stiffness (kip-inch) X 1000 Expri.	Connection stiffness (kip-inch) X 1000 Poly.	Connection stiffness (kip-inch) X 1000 M.Expo.
1	0.00	0.00	0.00	0.00	0.2064E+03	0.9483E+02	0.2321E+03
2	1.01	208.51	97.49	201.69	0.1789E+03	0.1000E+03	0.1764E+03
3	2.01	360.23	205.52	365.74	0.1503E+03	0.1193E+03	0.1544E+03
4	3.02	510.52	344.85	514.77	0.1447E+03	0.1584E+03	0.1412E+03
5	4.03	652.60	497.46	651.22	0.1333E+03	0.1222E+03	0.1289E+03
6	5.04	779.89	590.51	774.77	0.1144E+03	0.6956E+02	0.1156E+03
7	6.04	882.85	647.50	883.57	0.9961E+02	0.4729E+02	0.1020E+03
8	7.05	980.04	689.06	979.65	0.8962E+02	0.3584E+02	0.8840E+02
9	9.06	1133.72	748.18	1132.61	0.6540E+02	0.2460E+02	0.6470E+02
10	11.08	1243.38	791.74	1244.22	0.4626E+02	0.1895E+02	0.4682E+02
11	13.09	1320.31	826.13	1325.20	0.3476E+02	0.1557E+02	0.3457E+02
12	15.11	1383.37	855.12	1386.23	0.2912E+02	0.1328E+02	0.2643E+02
13	17.12	1437.70	880.08	1433.64	0.2229E+02	0.1164E+02	0.2110E+02
14	19.14	1473.09	902.26	1472.36	0.1730E+02	0.1039E+02	0.1746E+02
15	21.15	1507.41	922.11	1504.72	0.1503E+02	0.9403E+01	0.1486E+02
16	23.16	1533.51	940.19	1532.49	0.1295E+02	0.8608E+01	0.1284E+02
17	25.18	1559.60	956.89	1556.69	0.1129E+02	0.7947E+01	0.1116E+02
18	27.19	1579.04	972.29	1577.64	0.8966E+01	0.7393E+01	0.9712E+01
19	29.21	1595.72	986.73	1595.93	0.7944E+01	0.6916E+01	0.8414E+01
20	31.22	1611.07	1000.21	1611.65	0.5881E+01	0.6506E+01	0.7254E+01
21	33.24	1619.39	1012.98	1625.23	0.4131E+01	0.6145E+01	0.6209E+01
22	35.25	1627.71	1025.00	1636.76	0.4535E+01	0.5828E+01	0.5283E+01
23	37.70	1640.00	1038.86	1648.47	0.3537E+00	0.5488E+01	0.4301E+01
24	39.28	1635.81	1047.53	1637.26	-0.3034E+01	0.5287E+01	-0.7366E+01
25	41.29	1628.74	1057.77	1621.82	-0.5563E+01	0.5063E+01	-0.7982E+01
26	43.31	1613.35	1067.78	1605.14	-0.7901E+01	0.4854E+01	-0.8513E+01
27	45.32	1596.90	1077.54	1587.57	-0.9280E+01	0.4662E+01	-0.8963E+01
28	47.34	1575.93	1086.59	1569.07	-0.1056E+02	0.4491E+01	-0.9344E+01
29	49.35	1554.35	1095.62	1549.96	-0.1242E+02	0.4329E+01	-0.9661E+01
30	51.37	1525.85	1104.06	1530.16	-0.1415E+02	0.4184E+01	-0.9927E+01
31	53.38	1497.35	1112.47	1509.98	-0.1418E+02	0.4046E+01	-0.1015E+02

V -153

```
Connection type : Extended end-plate connections
           Mode : All bolted without column stiffener

Tested by  :  Ana M. Girao Coelho et al. (2004)          Portugal
Test   Id. :  FS4b

Column : HE340M                          Fasteners: G8.8- -M20
Beam   : IPE300                                   7/8" Oversize holes
Plate thickness : 10 mm                  Material :  S690
                                              Fy = 101.32 ksi
                                              Fu = 107.51 ksi
```

Major parameters

```
     End plate extended on tension side only.
     ct =  1.1811"    pt =  3.5433"    li =  8.0709"    pc =  1.5472"
     cc =  1.4055"    gt =  3.5433"    gc =  3.5433"
     tp =  0.3961"    nt = 2 X  2      nc = 2 X  1

Remark  1)  End-plate: 400x150x10 mm
        2)  Failure mode:  Fracture of bolt #3
```

No	Moment (k-in)	Rotation (radians) X 1/1000	No	Moment (k-in)	Rotation (radians) X 1/1000
1	0.00	0.00	26	1650.10	39.86
2	170.64	1.00	27	1653.94	41.86
3	344.53	1.99	28	1660.80	43.85
4	508.08	2.99	29	1656.36	45.84
5	655.12	3.99	30	1648.65	47.84
6	776.53	4.98	31	1632.57	49.83
7	885.57	5.98	32	1613.97	51.82
8	979.04	6.98	33	1592.41	53.81
9	1063.25	7.97	34	1569.72	55.81
10	1127.41	8.97	35	1546.57	57.80
11	1189.31	9.97	36	1523.19	59.79
12	1290.55	11.96	37	1490.26	61.79
13	1349.61	13.95	38	1447.79	63.78
14	1395.65	15.95			
15	1440.02	17.94			
16	1480.86	19.93			
17	1507.75	21.92			
18	1528.79	23.92			
19	1551.73	25.91			
20	1573.37	27.90			
21	1589.52	29.90			
22	1603.36	31.89			
23	1618.56	33.88			
24	1632.65	35.88			
25	1641.75	37.87			

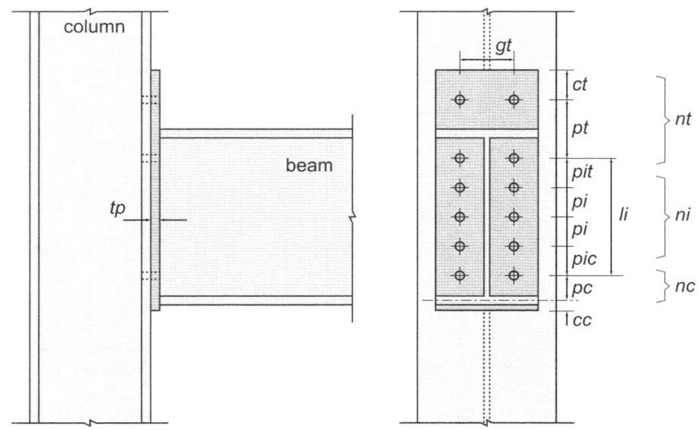

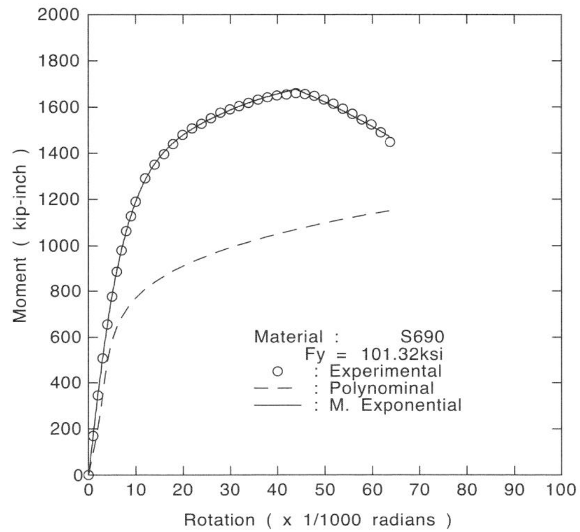

A.5 – 306

```
Moment-rotation prediction equations     ( R : X 1/1000 radians )

Frye and Morris polynominal model  :    R = Sum ( Ai X ( K*Bm )**Pi X 10**Qi )
     xd = 11.614170"   t =  0.396060"  tf =  0.787410"
     A1 =  1.830000   A2 = -1.040000   A3 =  6.380000     K =  0.005763
     P1 =        1    P2 =        3    P3 =        5     Q1 =        0     Q2 =       -1     Q3 =       -3

Modified exponential model  :
     Bm = Sum ( Ai X ( 1 - exp( - R/( 2*i*C ) ) ) )  +      Sum ( RKj X ( R - Rj0 ) )  +  Bm0
      C =   0.78150000E+00             Bmo=   0.00000000E+00          Nexp=  6        Nliner=  1
     AI =  -0.34982775E+03   0.27241538E+04  -0.17211005E+05   0.50121412E+05  -0.59058729E+05   0.25516057E+05
     Rj0 =          43.8500
     RKj =  -0.12909449E+02
```

No	Rotation (radians) X 1/1000	Moment (kip-inch) Expri.	Moment (kip-inch) Poly.	Moment (kip-inch) M.Expo.	Connection stiffness (kip-inch) X 1000 Expri.	Connection stiffness (kip-inch) X 1000 Poly.	Connection stiffness (kip-inch) X 1000 M.Expo.
1	0.00	0.00	0.00	0.00	0.1706E+03	0.9483E+02	0.1577E+03
3	1.99	344.53	203.15	344.52	0.1696E+03	0.1187E+03	0.1710E+03
5	3.99	655.12	492.43	653.08	0.1348E+03	0.1252E+03	0.1357E+03
7	5.98	885.57	644.63	887.44	0.1013E+03	0.4821E+02	0.1010E+03
9	7.97	1063.25	718.88	1060.42	0.7466E+02	0.2956E+02	0.7419E+02
11	9.97	1189.31	769.17	1188.04	0.5821E+02	0.2165E+02	0.5449E+02
13	13.95	1349.61	839.03	1351.39	0.2636E+02	0.1450E+02	0.3029E+02
15	17.94	1440.02	889.39	1445.35	0.2141E+02	0.1109E+02	0.1826E+02
17	21.92	1507.75	929.23	1505.39	0.1202E+02	0.9080E+01	0.1263E+02
19	25.91	1551.73	962.61	1549.84	0.1171E+02	0.7735E+01	0.9958E+01
21	29.90	1589.52	991.45	1586.36	0.7013E+01	0.6769E+01	0.8450E+01
23	33.88	1618.56	1016.88	1617.63	0.7342E+01	0.6040E+01	0.7288E+01
25	37.87	1641.75	1039.79	1644.51	0.4386E+01	0.5466E+01	0.6189E+01
27	41.86	1653.94	1060.64	1667.05	0.2684E+01	0.5002E+01	0.5118E+01
29	45.84	1656.36	1079.76	1659.71	-0.3040E+01	0.4619E+01	-0.8794E+01
31	49.83	1632.57	1097.53	1622.79	-0.8713E+01	0.4296E+01	-0.9688E+01
33	53.81	1592.41	1114.07	1582.68	-0.1109E+02	0.4021E+01	-0.1044E+02
35	57.80	1546.57	1129.62	1539.74	-0.1169E+02	0.3781E+01	-0.1106E+02
37	61.79	1490.26	1144.49	1494.59	-0.1891E+02	0.3569E+01	-0.1155E+02

V -154

Connection type : Extended end-plate connections
Mode : All bolted without column stiffener

Tested by : Ana M. Girao Coelho et al. (2007) Portugal
Test Id. : EEP_15_2

Column : HE300M
Beam : HE320A
Plate thickness : 15 mm

Fasteners: 12.9- -M24
33/32" Oversize holes
Material : S690
Fy = 112.26 ksi
Fu = 118.06 ksi

Major parameters

End plate extended on tension side only.

ct = 1.5748"	pt = 5.3150"	li = 6.2992"	pc = 2.6476"
cc = 1.2894"	gt = 5.9055"	gc = 5.9055"	
tp = 0.5756"	nt = 2 X 2	nc = 2 X 1	

Remark 1) End-plate: 435x300x15 mm
2)

No	Moment (k-in)	Rotation (radians) X 1/1000
1	0.00	0.00
2	258.93	0.97
3	527.10	1.94
4	823.75	2.91
5	1099.10	3.89
6	1393.04	4.86
7	1676.63	5.83
8	1955.75	6.80
9	2110.96	7.77
10	2222.70	8.74
11	2338.47	9.71
12	2447.37	10.69
13	2552.28	11.66
14	2651.26	12.63
15	2743.36	13.60
16	2829.33	14.57
17	2911.87	15.54
18	2991.07	16.51
19	3065.57	17.49
20	3131.48	18.46
21	3190.48	19.43
22	3238.94	20.00
23	3088.50	20.40

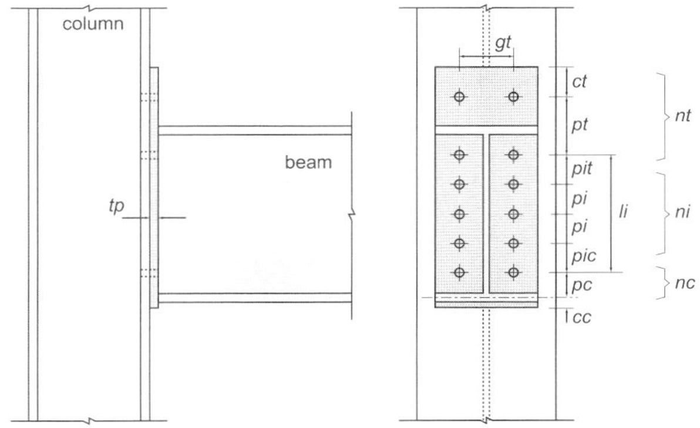

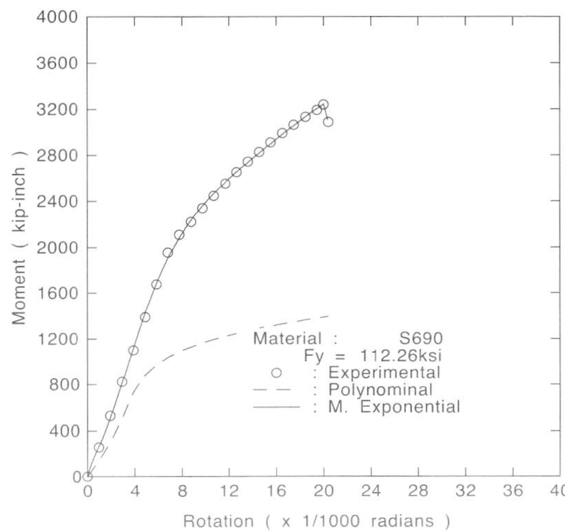

```
Moment-rotation prediction equations      ( R : X 1/1000 radians )

Frye and Morris polynominal model  :    R = Sum ( Ai X ( K*Bm )**Pi X 10**Qi )
     xd = 11.614210"   t =  0.575590"  tf =  0.944880"
     A1 =  1.830000   A2 = -1.040000   A3 =  6.380000    K =  0.003775
     P1 =        1    P2 =        3    P3 =        5    Q1 =        0    Q2 =       -1    Q3 =       -3

Modified exponential model  :
     Bm = Sum ( Ai X ( 1 - exp( - R/( 2*i*C ) ) ) )  +      Sum ( RKj X ( R - Rj0 ) )  +  Bm0
      C =   0.41250000E+00            Bmo=   0.00000000E+00          Nexp=  6         Nliner=  1
     AI =   0.35029862E+03   0.13828143E+05  -0.12668677E+06   0.35690645E+06  -0.41570783E+06   0.17516869E+06
    Rj0 =          20.0000
    RKj =  -0.45533706E+03
```

No	Rotation (radians) X 1/1000	Moment (kip-inch) Expri.	Moment (kip-inch) Poly.	Moment (kip-inch) M.Expo.	Connection stiffness (kip-inch) X 1000 Expri.	Connection stiffness (kip-inch) X 1000 Poly.	Connection stiffness (kip-inch) X 1000 M.Expo.
1	0.00	0.00	0.00	0.00	0.2669E+03	0.1448E+03	0.3822E+03
2	0.97	258.93	142.72	268.95	0.2717E+03	0.1521E+03	0.2363E+03
3	1.94	527.10	301.12	514.71	0.2911E+03	0.1790E+03	0.2783E+03
4	2.91	823.75	500.17	806.62	0.2935E+03	0.2359E+03	0.3179E+03
5	3.89	1099.10	731.31	1121.99	0.2921E+03	0.2030E+03	0.3194E+03
6	4.86	1393.04	881.47	1420.01	0.2977E+03	0.1159E+03	0.2915E+03
7	5.83	1676.63	972.74	1683.12	0.2901E+03	0.7739E+02	0.2499E+03
8	6.80	1955.75	1037.80	1904.41	0.2239E+03	0.5814E+02	0.2068E+03
9	7.77	2110.96	1088.21	2086.11	0.1376E+03	0.4689E+02	0.1691E+03
10	8.74	2222.70	1129.83	2235.03	0.1173E+03	0.3948E+02	0.1394E+03
11	9.71	2338.47	1165.43	2359.28	0.1153E+03	0.3422E+02	0.1181E+03
12	10.69	2447.37	1196.93	2467.40	0.1096E+03	0.3025E+02	0.1037E+03
13	11.66	2552.28	1224.73	2563.28	0.1051E+03	0.2721E+02	0.9477E+02
14	12.63	2651.26	1249.90	2652.43	0.9849E+02	0.2477E+02	0.8952E+02
15	13.60	2743.36	1272.93	2737.64	0.9179E+02	0.2277E+02	0.8643E+02
16	14.57	2829.33	1294.18	2820.43	0.8686E+02	0.2110E+02	0.8437E+02
17	15.54	2911.87	1313.94	2901.41	0.8337E+02	0.1968E+02	0.8258E+02
18	16.51	2991.07	1332.42	2980.56	0.7885E+02	0.1845E+02	0.8056E+02
19	17.49	3065.57	1349.96	3058.31	0.7197E+02	0.1738E+02	0.7802E+02
20	18.46	3131.48	1366.36	3132.53	0.6438E+02	0.1644E+02	0.7492E+02
21	19.43	3190.48	1381.90	3203.47	0.7606E+02	0.1562E+02	0.7126E+02
22	20.00	3238.94	1390.80	3243.42	-0.1860E+03	0.1516E+02	0.6889E+02
23	20.40	3088.50	1396.87	3088.50	-0.3761E+03	0.1487E+02	-0.3882E+03

A.5 – 309

V -155

Connection type : Extended end-plate connections
Mode : All bolted without column stiffener

Tested by : Ana M. Girao Coelho et al. (2007) Portugal
Test Id. : EEP_10_2a

Column : HE300M
Beam : HE320A
Plate thickness : 10 mm

Fasteners: 12.9- -M24
33/32" Oversize holes
Material : S690
Fy = 101.24 ksi
Fu = 108.63 ksi

Major parameters

End plate extended on tension side only.

ct = 1.5748"	pt = 5.3150"	li = 6.2992"	pc = 2.6476"
cc = 1.2894"	gt = 5.9055"	gc = 5.9055"	
tp = 0.3976"	nt = 2 X 2	nc = 2 X 1	

Remark 1) End-plate: 435x300x10 mm
2)

No	Moment (k-in)	Rotation (radians) X 1/1000
1	0.00	0.00
2	143.88	1.00
3	278.82	2.00
4	407.01	3.00
5	537.04	4.00
6	684.04	5.00
7	846.88	6.00
8	1031.47	7.00
9	1147.58	8.00
10	1319.43	10.00
11	1456.82	12.00
12	1558.52	14.00
13	1650.35	16.00
14	1729.52	18.00
15	1797.32	20.00
16	1866.39	22.00
17	1925.61	24.00
18	1981.74	26.00
19	2037.46	28.00
20	2094.02	30.00
21	2132.42	32.00
22	2153.90	34.00
23	2159.29	36.00
24	2112.66	38.00
25	2091.38	40.00
26	2072.52	42.00
27	2044.25	45.00

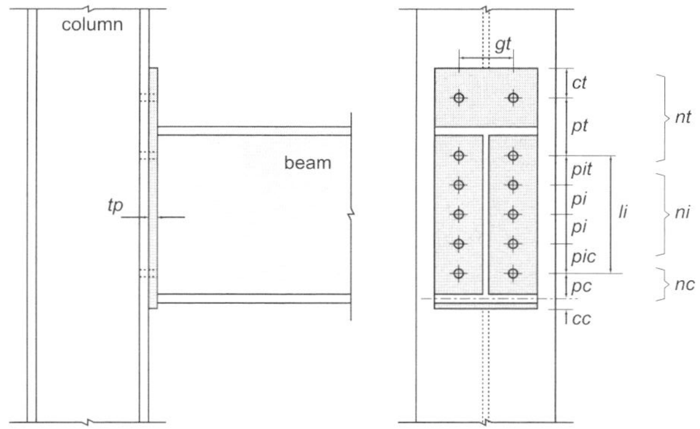

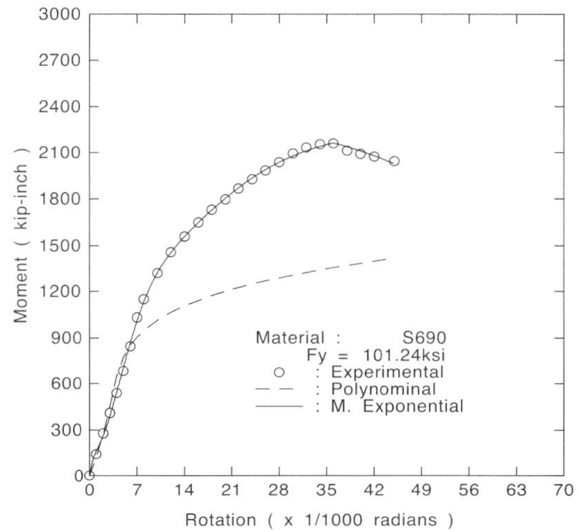

A.5 – 310

```
Moment-rotation prediction equations     ( R : X 1/1000 radians )

Frye and Morris polynominal model  :    R = Sum ( Ai X ( K*Bm )**Pi X 10**Qi )
     xd = 11.614210"   t =  0.397640"  tf =  0.944880"
     A1 =  1.830000   A2 = -1.040000   A3 =  6.380000    K =  0.004377
     P1 =        1    P2 =        3    P3 =        5    Q1 =        0    Q2 =       -1    Q3 =       -3

Modified exponential model  :
     Bm = Sum ( Ai X ( 1 - exp( - R/( 2*i*C ) ) ) )  +      Sum ( RKj X ( R - Rj0 ) )  +  Bm0
       C =   0.62500000E+00          Bmo=   0.00000000E+00         Nexp=  6        Nliner=  1
      AI =   0.24807763E+04  -0.21182463E+05   0.57823082E+05  -0.61981405E+05   0.15126107E+05   0.99837663E+04
     Rj0 =          36.0000
     RKj =  -0.21598053E+02
```

No	Rotation (radians) X 1/1000	Moment (kip-inch) Expri.	Moment (kip-inch) Poly.	Moment (kip-inch) M.Expo.	Connection stiffness (kip-inch) X 1000 Expri.	Connection stiffness (kip-inch) X 1000 Poly.	Connection stiffness (kip-inch) X 1000 M.Expo.
1	0.00	0.00	0.00	0.00	0.1439E+03	0.1248E+03	0.2862E+03
2	1.00	143.88	127.04	164.81	0.1394E+03	0.1316E+03	0.1005E+03
3	2.00	278.82	269.02	261.93	0.1316E+03	0.1566E+03	0.1068E+03
4	3.00	407.01	449.88	387.28	0.1291E+03	0.2077E+03	0.1431E+03
5	4.00	537.04	650.02	542.60	0.1385E+03	0.1638E+03	0.1640E+03
6	5.00	684.04	773.78	708.41	0.1549E+03	0.9333E+02	0.1648E+03
7	6.00	846.88	850.00	867.57	0.1737E+03	0.6307E+02	0.1519E+03
8	7.00	1031.47	904.86	1010.20	0.1504E+03	0.4775E+02	0.1328E+03
9	8.00	1147.58	947.66	1132.77	0.1060E+03	0.3870E+02	0.1125E+03
10	10.00	1319.43	1013.55	1321.93	0.7731E+02	0.2840E+02	0.7874E+02
11	12.00	1456.82	1064.23	1456.57	0.5977E+02	0.2267E+02	0.5794E+02
12	14.00	1558.52	1105.64	1560.36	0.4838E+02	0.1901E+02	0.4710E+02
13	16.00	1650.35	1140.95	1648.57	0.4275E+02	0.1645E+02	0.4170E+02
14	18.00	1729.52	1171.86	1728.56	0.3674E+02	0.1455E+02	0.3848E+02
15	20.00	1797.32	1199.44	1802.75	0.3422E+02	0.1308E+02	0.3570E+02
16	22.00	1866.39	1224.39	1871.20	0.3207E+02	0.1191E+02	0.3268E+02
17	24.00	1925.61	1247.22	1933.24	0.2884E+02	0.1095E+02	0.2931E+02
18	26.00	1981.74	1268.30	1988.29	0.2796E+02	0.1015E+02	0.2572E+02
19	28.00	2037.46	1287.91	2036.12	0.2807E+02	0.9470E+01	0.2212E+02
20	30.00	2094.02	1306.25	2076.89	0.2374E+02	0.8885E+01	0.1869E+02
21	32.00	2132.42	1323.50	2111.08	0.1497E+02	0.8375E+01	0.1555E+02
22	34.00	2153.90	1339.79	2139.34	0.6718E+01	0.7927E+01	0.1277E+02
23	36.00	2159.29	1355.24	2162.41	-0.1031E+02	0.7529E+01	0.1037E+02
24	38.00	2112.66	1369.93	2137.85	-0.1698E+02	0.7174E+01	-0.1326E+02
25	40.00	2091.38	1383.96	2109.59	-0.1003E+02	0.6855E+01	-0.1495E+02
26	42.00	2072.52	1397.37	2078.26	-0.9426E+01	0.6566E+01	-0.1633E+02
27	45.00	2044.25	1416.48	2026.74	-0.9426E+01	0.6181E+01	-0.1793E+02

V -156

Connection type : Extended end-plate connections
Mode : All bolted without column stiffener

Tested by : Ana M. Girao Coelho et al. (2007) Portugal
Test Id. : EEP_10_2b

Column : HE300M
Beam : HE320A
Plate thickness : 10 mm

Fasteners: G8.8- -M24
33/32" Oversize holes
Material : S690
Fy = 101.24 ksi
Fu = 108.63 ksi

Major parameters

End plate extended on tension side only.

ct = 1.5748"	pt = 5.3150"	li = 6.2992"	pc = 2.6476"
cc = 1.2894"	gt = 5.9055"	gc = 5.9055"	
tp = 0.3976"	nt = 2 X 2	nc = 2 X 1	

Remark 1) End-plate: 435x300x10 mm
2)

No	Moment (k-in)	Rotation (radians) X 1/1000	No	Moment (k-in)	Rotation (radians) X 1/1000
1	0.00	0.00	26	2229.47	43.00
2	159.57	1.00	27	2229.20	45.62
3	319.13	2.00	28	2115.04	46.00
4	481.07	3.00			
5	665.61	4.00			
6	836.32	5.00			
7	1009.21	6.00			
8	1186.59	7.50			
9	1324.32	9.00			
10	1478.74	11.00			
11	1610.10	13.00			
12	1713.11	15.00			
13	1792.39	17.00			
14	1863.67	19.00			
15	1925.63	21.00			
16	1980.76	23.00			
17	2033.91	25.00			
18	2084.04	27.00			
19	2128.81	29.00			
20	2167.68	31.00			
21	2200.52	33.00			
22	2221.68	35.00			
23	2230.09	37.00			
24	2229.88	39.00			
25	2229.68	41.00			

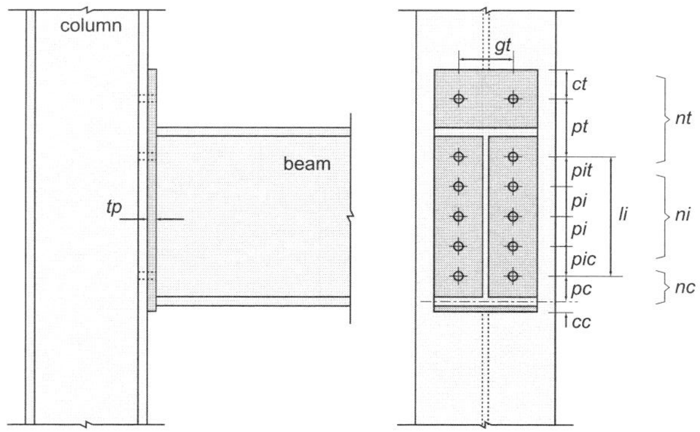

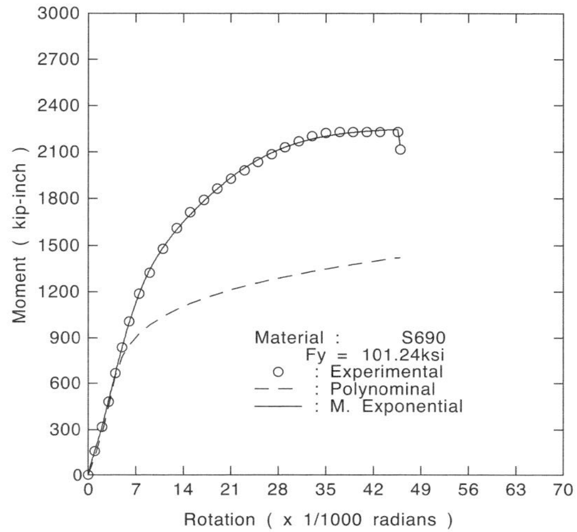

A.5 – 312

```
Moment-rotation prediction equations     ( R : X 1/1000 radians )

Frye and Morris polynominal model  :    R = Sum ( Ai X ( K*Bm )**Pi X 10**Qi )
     xd = 11.614210"    t =  0.397640"  tf =  0.944880"
     A1 =  1.830000   A2 = -1.040000   A3 =  6.380000    K =  0.004377
     P1 =        1     P2 =        3    P3 =        5    Q1 =        0    Q2 =       -1    Q3 =       -3

Modified exponential model  :
     Bm = Sum ( Ai X ( 1 - exp( - R/( 2*i*C ) ) ) )  +     Sum ( RKj X ( R - Rj0 ) )  +  Bm0
       C =   0.63333333E+00            Bmo=   0.00000000E+00          Nexp=  6        Nliner=  1
      AI =   0.15806948E+04  -0.18979121E+05   0.67944585E+05  -0.10480812E+06   0.69067237E+05  -0.12551899E+05
     Rj0 =          45.6200
     RKj =  -0.34478933E+03
```

No	Rotation (radians) X 1/1000	Moment (kip-inch) Expri.	Moment (kip-inch) Poly.	Moment (kip-inch) M.Expo.	Connection stiffness (kip-inch) X 1000 Expri.	Connection stiffness (kip-inch) X 1000 Poly.	Connection stiffness (kip-inch) X 1000 M.Expo.
1	0.00	0.00	0.00	0.00	0.1596E+03	0.1248E+03	0.2043E+03
2	1.00	159.57	127.04	162.09	0.1596E+03	0.1316E+03	0.1452E+03
3	2.00	319.13	269.02	314.07	0.1608E+03	0.1566E+03	0.1622E+03
4	3.00	481.07	449.88	485.80	0.1732E+03	0.2077E+03	0.1788E+03
5	4.00	665.61	650.02	666.06	0.1776E+03	0.1638E+03	0.1790E+03
6	5.00	836.32	773.78	839.18	0.1718E+03	0.9333E+02	0.1656E+03
7	6.00	1009.21	850.00	994.96	0.1510E+03	0.6307E+02	0.1454E+03
8	7.50	1186.59	927.27	1188.63	0.1050E+03	0.4274E+02	0.1133E+03
9	9.00	1324.32	983.27	1337.93	0.8556E+02	0.3267E+02	0.8703E+02
10	11.00	1478.74	1040.42	1486.91	0.7145E+02	0.2517E+02	0.6413E+02
11	13.00	1610.10	1085.84	1601.55	0.5859E+02	0.2066E+02	0.5185E+02
12	15.00	1713.11	1123.93	1698.11	0.4557E+02	0.1763E+02	0.4532E+02
13	17.00	1792.39	1156.88	1784.30	0.3764E+02	0.1544E+02	0.4105E+02
14	19.00	1863.67	1186.01	1862.59	0.3331E+02	0.1377E+02	0.3723E+02
15	21.00	1925.63	1212.21	1933.06	0.2927E+02	0.1247E+02	0.3319E+02
16	23.00	1980.76	1236.05	1995.19	0.2707E+02	0.1141E+02	0.2891E+02
17	25.00	2033.91	1257.96	2048.66	0.2582E+02	0.1053E+02	0.2457E+02
18	27.00	2084.04	1278.27	2093.58	0.2372E+02	0.9797E+01	0.2040E+02
19	29.00	2128.81	1297.22	2130.48	0.2091E+02	0.9167E+01	0.1658E+02
20	31.00	2167.68	1315.00	2160.20	0.1793E+02	0.8621E+01	0.1322E+02
21	33.00	2200.52	1331.75	2183.70	0.1350E+02	0.8144E+01	0.1036E+02
22	35.00	2221.68	1347.61	2201.97	0.7392E+01	0.7722E+01	0.7993E+01
23	37.00	2230.09	1362.67	2215.97	0.2052E+01	0.7347E+01	0.6078E+01
24	39.00	2229.88	1377.03	2226.55	-0.1026E+00	0.7011E+01	0.4558E+01
25	41.00	2229.68	1390.74	2234.43	-0.1027E+00	0.6707E+01	0.3373E+01
26	43.00	2229.47	1403.87	2240.22	-0.1027E+00	0.6432E+01	0.2464E+01
27	45.62	2229.20	1420.29	2245.48	-0.2624E+03	0.6107E+01	0.1599E+01
28	46.00	2115.04	1422.61	2115.04	-0.3004E+03	0.6063E+01	-0.3433E+03

A.5 – 313

V -157

Connection type : Extended end-plate connections
Mode : All bolted without column stiffener

Tested by : J.M.Cabrero & E.Bayo (2007) Spain
Test Id. : A-M

Column : HEB160
Beam : IPE330
Plate thickness : 16 mm

Fasteners: G8.8- -M20
7/8" Oversize holes
Material : S275
Fy = 55.57 ksi
Fu = 75.56 ksi

Major parameters

End plate extended on tension side only.

ct = 1.1811" pt = 3.5433" li = 9.0551" pc = 1.7421"
cc = 0.6201" gt = 3.9370" gc = 3.9370"
tp = 0.6516" nt = 2 X 2 nc = 2 X 1

Remark 1) End-plate: 410x160x16 mm
2) Failure mode: Nut stripping of bolt on tension side

No	Moment (k-in)	Rotation (radians) X 1/1000	No	Moment (k-in)	Rotation (radians) X 1/1000
1	0.00	0.00	26	1836.60	17.96
2	157.33	0.50	27	1856.62	18.96
3	268.15	0.75	28	1876.64	19.95
4	378.97	1.00	29	1896.66	20.95
5	469.55	1.25	30	1912.67	21.95
6	560.13	1.50	31	1926.97	22.95
7	643.72	1.75	32	1940.20	23.94
8	727.32	2.00	33	1949.78	24.94
9	860.37	2.49	34	1959.36	25.94
10	985.75	2.99	35	1974.16	27.50
11	1082.36	3.49	36	1885.65	28.93
12	1178.96	3.99	37	1793.81	29.93
13	1318.59	4.99			
14	1409.65	5.99			
15	1472.80	6.98			
16	1528.37	7.98			
17	1568.61	8.98			
18	1608.85	9.98			
19	1641.99	10.97			
20	1672.60	11.97			
21	1703.40	12.97			
22	1734.93	13.97			
23	1766.47	14.96			
24	1790.93	15.96			
25	1813.99	16.96			

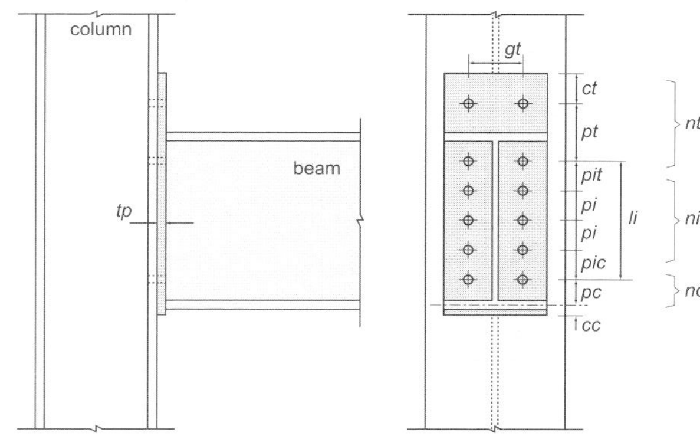

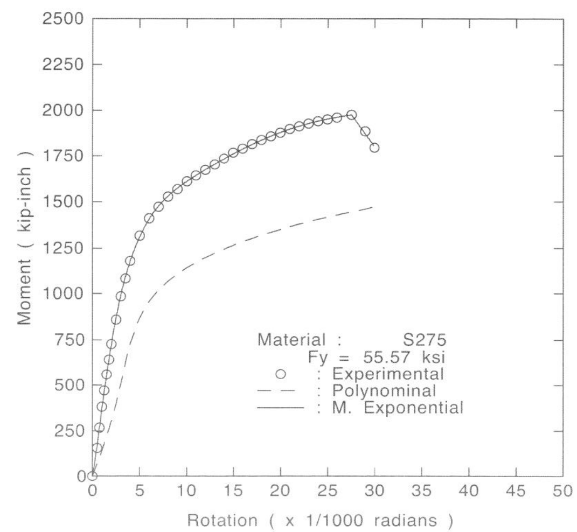

```
Moment-rotation prediction equations     ( R : X 1/1000 radians )

Frye and Morris polynominal model  :    R = Sum ( Ai X ( K*Bm )**Pi X 10**Qi )
     xd = 12.598420"   t =  0.651570"  tf =  0.787400"
     A1 =  1.830000   A2 = -1.040000   A3 =  6.380000    K =  0.003885
     P1 =        1    P2 =        3    P3 =        5    Q1 =        0    Q2 =       -1    Q3 =       -3

Modified exponential model  :
     Bm = Sum ( Ai X ( 1 - exp( - R/( 2*i*C ) ) ) )  +     Sum ( RKj X ( R - Rj0 ) )  +  Bm0
      C =   0.37441667E+00             Bmo=   0.00000000E+00         Nexp=  6        Nliner=  1
     AI =  -0.65595384E+03   0.50387780E+04  -0.31036684E+05   0.90218736E+05  -0.10842357E+06   0.46875192E+05
     Rj0 =           27.5000
     RKj =  -0.77319216E+02
```

No	Rotation (radians) X 1/1000	Moment (kip-inch) Expri.	Moment (kip-inch) Poly.	Moment (kip-inch) M.Expo.	Connection stiffness (kip-inch) X 1000 Expri.	Connection stiffness (kip-inch) X 1000 Poly.	Connection stiffness (kip-inch) X 1000 M.Expo.
1	0.00	0.00	0.00	0.00	0.3147E+03	0.1407E+03	0.2676E+03
3	0.75	268.15	106.52	269.15	0.4433E+03	0.1448E+03	0.3989E+03
5	1.25	469.55	180.74	465.39	0.3623E+03	0.1529E+03	0.3793E+03
7	1.75	643.72	260.35	644.54	0.3344E+03	0.1667E+03	0.3354E+03
9	2.49	860.37	395.66	866.45	0.2612E+03	0.2025E+03	0.2652E+03
11	3.49	1082.36	624.31	1090.65	0.1932E+03	0.2343E+03	0.1873E+03
13	4.99	1318.59	870.74	1309.75	0.1153E+03	0.1056E+03	0.1119E+03
15	6.98	1472.80	1018.40	1475.05	0.5970E+02	0.5406E+02	0.6107E+02
17	8.98	1568.61	1106.94	1572.60	0.4024E+02	0.3694E+02	0.3984E+02
19	10.97	1641.99	1171.20	1643.03	0.3205E+02	0.2847E+02	0.3229E+02
21	12.97	1703.40	1222.64	1704.45	0.3117E+02	0.2335E+02	0.2948E+02
23	14.96	1766.47	1265.49	1761.04	0.2817E+02	0.1992E+02	0.2734E+02
25	16.96	1813.99	1302.72	1813.03	0.2283E+02	0.1744E+02	0.2452E+02
27	18.96	1856.62	1335.63	1858.66	0.2012E+02	0.1555E+02	0.2103E+02
29	20.95	1896.66	1365.06	1896.81	0.1802E+02	0.1408E+02	0.1730E+02
31	22.95	1926.97	1392.27	1927.77	0.1383E+02	0.1287E+02	0.1371E+02
33	24.94	1949.78	1416.85	1951.82	0.9576E+01	0.1189E+02	0.1054E+02
35	27.50	1974.16	1445.68	1974.41	-0.2775E+02	0.1085E+02	0.7251E+01
37	29.93	1793.81	1471.18	1801.17	-0.9184E+02	0.1003E+02	-0.7239E+02

A.5 – 315

V -158

```
Connection type : Extended end-plate connections
           Mode : All bolted without column stiffener

Tested by  :  J.M.Cabrero & E.Bayo (2007)              Spain
Test  Id.  :  B-M

Column : HEB130                       Fasteners: G8.8- -M20
Beam   : IPE330                                 7/8" Oversize holes
Plate thickness : 10 mm               Material :  S275
                                           Fy =  55.57 ksi
                                           Fu =  75.56 ksi

                         Major parameters

     End plate extended on tension side only.
     ct =  1.1811"    pt =  3.5433"    li =  9.0551"    pc =  1.7421"
     cc =  0.5201"    gt =  3.9370"    gc =  3.9370"
     tp =  0.3976"    nt = 2 X  2      nc = 2 X  1

Remark  1)  End-plate: 410x160x10 mm
        2)  Failure mode: Nut stripping of bolt on tension side
```

No	Moment (k-in)	Rotation (radians) X 1/1000	No	Moment (k-in)	Rotation (radians) X 1/1000
1	0.00	0.00	26	1295.30	38.91
2	73.39	0.50	27	1314.01	40.91
3	159.85	1.00	28	1332.63	42.90
4	253.19	1.50	29	1349.50	44.90
5	350.55	2.00	30	1366.38	46.89
6	449.55	2.49	31	1381.45	48.89
7	555.29	2.99	32	1396.30	50.88
8	659.24	3.49	33	1409.81	52.88
9	742.01	4.99	34	1423.00	54.87
10	811.19	6.98	35	1436.08	56.87
11	867.07	8.98	36	1447.56	58.86
12	913.16	10.97	37	1459.05	60.86
13	950.19	12.97	38	1470.15	62.85
14	986.52	14.97	39	1479.79	64.85
15	1020.49	16.96	40	1489.43	66.85
16	1043.08	18.96	41	1498.33	68.84
17	1087.58	20.95	42	1505.21	70.84
18	1111.47	22.95	43	1512.09	72.83
19	1135.40	24.94	44	1518.89	74.83
20	1159.40	26.94	45	1525.38	76.82
21	1183.79	28.93	46	1531.86	78.82
22	1209.22	30.93	47	1536.09	80.81
23	1234.65	32.92	48	1539.97	82.81
24	1255.78	34.92	49	1545.30	85.55
25	1275.81	36.91	50	1249.09	85.80

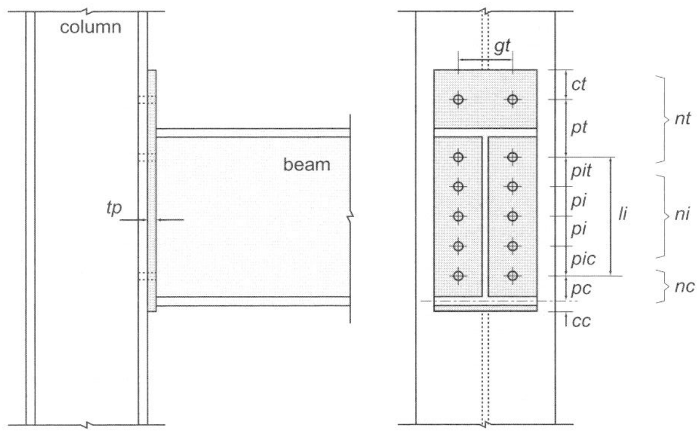

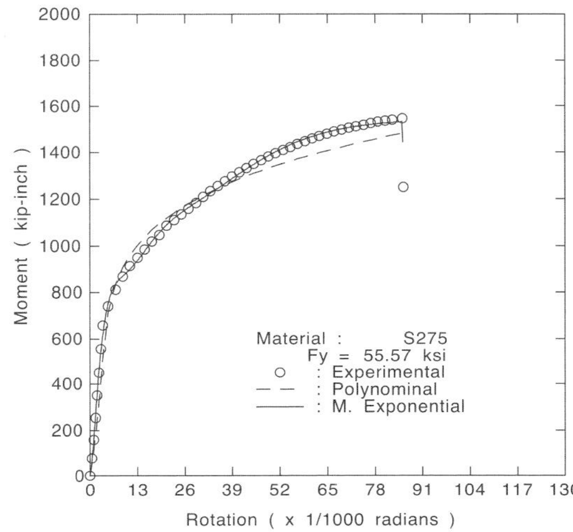

A.5 – 316

```
Moment-rotation prediction equations     ( R : X 1/1000 radians )

Frye and Morris polynominal model  :    R = Sum ( Ai X ( K*Bm )**Pi X 10**Qi )
     xd = 12.598420"    t =  0.397640"  tf =  0.787400"
     A1 =  1.830000   A2 = -1.040000   A3 =  6.380000     K =  0.004733
     P1 =        1    P2 =        3    P3 =        5     Q1 =        0    Q2 =       -1    Q3 =       -3

Modified exponential model  :
     Bm = Sum ( Ai X ( 1 - exp( - R/( 2*i*C ) ) ) )  +     Sum ( RKj X ( R - Rj0 ) )  +  Bm0
      C =   0.84000000E+00          Bmo=   0.00000000E+00          Nexp=  6        Nliner=  1
     AI =  -0.33587238E+04   0.34084266E+05  -0.13407891E+06   0.26617397E+06  -0.25571020E+06   0.94431502E+05
    Rj0 =         85.5500
    RKj =  -0.11297671E+04
```

No	Rotation (radians) X 1/1000	Moment (kip-inch) Expri.	Poly.	M.Expo.	Connection stiffness (kip-inch) X 1000 Expri.	Poly.	M.Expo.
1	0.00	0.00	0.00	0.00	0.1568E+03	0.1155E+03	0.7766E+02
3	1.00	159.85	117.48	159.23	0.1748E+03	0.1217E+03	0.2061E+03
5	2.00	350.55	248.84	368.54	0.1984E+03	0.1449E+03	0.2001E+03
7	2.99	555.29	414.40	544.71	0.2097E+03	0.1918E+03	0.1532E+03
9	4.99	742.01	714.63	755.05	0.4640E+02	0.8673E+02	0.6426E+02
11	8.98	867.07	908.50	870.22	0.2554E+02	0.3032E+02	0.1457E+02
13	12.97	950.19	1003.53	937.86	0.1834E+02	0.1916E+02	0.2007E+02
15	16.96	1020.49	1069.22	1020.65	0.1544E+02	0.1431E+02	0.1999E+02
17	20.95	1087.58	1120.38	1091.11	0.1591E+02	0.1156E+02	0.1517E+02
19	24.94	1135.40	1162.68	1143.36	0.1201E+02	0.9764E+01	0.1142E+02
21	28.93	1183.79	1198.98	1185.51	0.1248E+02	0.8496E+01	0.1005E+02
23	32.92	1234.65	1230.90	1225.40	0.1168E+02	0.7548E+01	0.1007E+02
25	36.91	1275.81	1259.48	1266.25	0.9906E+01	0.6809E+01	0.1038E+02
27	40.91	1314.01	1285.49	1307.86	0.9356E+01	0.6214E+01	0.1034E+02
29	44.90	1349.50	1309.52	1348.16	0.8459E+01	0.5722E+01	0.9782E+01
31	48.89	1381.45	1331.49	1385.36	0.7500E+01	0.5314E+01	0.8808E+01
33	52.88	1409.81	1351.96	1418.14	0.6689E+01	0.4966E+01	0.7598E+01
35	56.87	1436.08	1371.15	1445.91	0.6156E+01	0.4666E+01	0.6323E+01
37	60.86	1459.05	1389.22	1468.68	0.5659E+01	0.4404E+01	0.5108E+01
39	64.85	1479.79	1406.20	1486.85	0.4821E+01	0.4174E+01	0.4026E+01
41	68.84	1498.33	1422.65	1501.03	0.3956E+01	0.3966E+01	0.3109E+01
43	72.83	1512.09	1437.99	1511.88	0.3430E+01	0.3784E+01	0.2360E+01
45	76.82	1525.38	1452.92	1520.06	0.3250E+01	0.3616E+01	0.1765E+01
47	80.81	1536.09	1466.88	1526.14	0.2033E+01	0.3468E+01	0.1304E+01
49	85.55	1545.30	1483.13	1531.31	-0.1086E+04	0.3305E+01	0.8978E+00

A.5 – 317

V -159

Connection type : Extended end-plate connections
Mode : All bolted with column stiffener

Tested by : G. Shi et al (2007) China
Test Id. : JD3

Column : H300x250x8x12
Beam : H300x200x8x12
Plate thickness : 20 mm
Stiffener thickness : 0.3937"

Fasteners: 10.9- -M20
7/8" Oversize holes
Material : Q345
Fy = 54.04 ksi
Fu = 77.89 ksi

Major parameters

End plate extended on both sides.

ct = 1.9685"	pt = 4.4094"	li = 6.9291"	pc = 4.4094"
cc = 1.9685"	gt = 4.2520"	gc = 4.2520"	
tp = 0.7874"	nt = 2 X 2	nc = 2 X 2	

Remark 1) Plate size: 500x200x20 mm
2) Column flange thickness is taken the same as the end-plate thickness

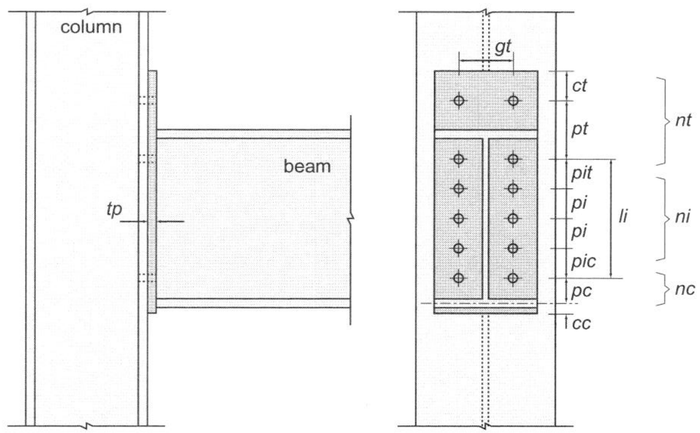

No	Moment (k-in)	Rotation (radians) X 1/1000	No	Moment (k-in)	Rotation (radians) X 1/1000
1	0.00	0.00	26	2501.19	39.07
2	169.81	0.36	27	2521.76	41.08
3	272.67	0.51	28	2556.05	43.55
4	415.66	0.82	29	2583.47	46.48
5	560.66	1.02	30	2624.62	49.87
6	725.23	1.50	31	2652.04	52.34
7	889.80	1.90	32	2679.47	55.27
8	1047.51	2.67	33	2693.19	58.20
9	1170.93	3.14	34	2706.90	60.82
10	1314.93	3.91	35	2727.47	63.44
11	1469.21	4.76	36	2734.33	66.68
12	1625.92	6.07			
13	1722.92	7.61			
14	1812.06	8.77			
15	1894.35	10.77			
16	1969.77	12.93			
17	2069.20	15.94			
18	2130.91	18.41			
19	2192.62	20.41			
20	2233.77	22.88			
21	2254.34	25.66			
22	2315.05	28.28			
23	2370.91	30.59			
24	2405.19	33.37			
25	2465.91	36.45			

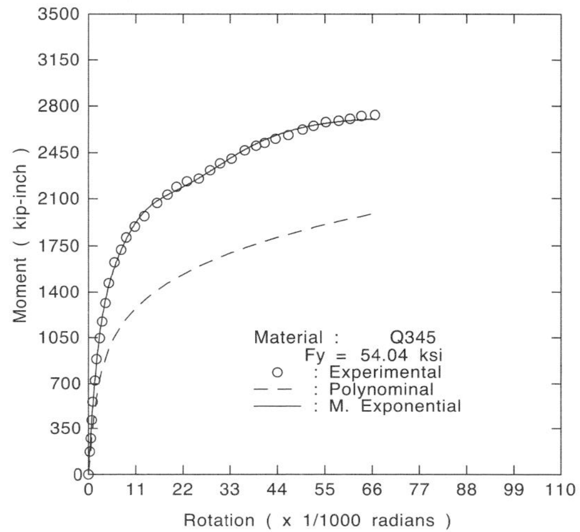

A.5 – 318

```
Moment-rotation prediction equations     ( R : X 1/1000 radians )

Frye and Morris polynominal model  :    R = Sum ( Ai X ( K*Bm )**Pi X 10**Qi )
     xd = 15.748020"   t =  0.787400"
     A1 =  1.790000   A2 =  1.760000   A3 =  2.040000    K =  0.001545
     P1 =        1    P2 =        3    P3 =        5    Q1 =        0    Q2 =       -1    Q3 =       -1

Modified expcnential model  :
     Bm = Sum ( Ai X ( 1 - exp( - R/( 2*i*C ) ) ) )  +      Sum ( RKj X ( R - Rj0 ) )  +  Bm0
       C =   0.64566667E+00              Bmo=   0.00000000E+00           Nexp=  6         Nliner=  0
      AI =  -0.16818908E+04   0.27602883E+05  -0.13635068E+06   0.31147295E+06  -0.32287919E+06   0.12455287E+06
```

No	Rotation (radians) X 1/1000	Moment (kip-inch) Expri.	Poly.	M.Expo.	Connection stiffness (kip-inch) X 1000 Expri.	Poly.	M.Expo.
1	0.00	0.00	0.00	0.00	0.4717E+03	0.3616E+03	0.5580E+03
3	0.51	272.67	182.84	274.74	0.6136E+03	0.3520E+03	0.5138E+03
5	1.02	560.66	354.70	520.57	0.6091E+03	0.3172E+03	0.4485E+03
7	1.90	889.80	591.53	863.10	0.3408E+03	0.2199E+03	0.3322E+03
9	3.14	1170.93	800.91	1192.60	0.2340E+03	0.1297E+03	0.2089E+03
11	4.76	1469.21	965.51	1452.88	0.1575E+03	0.8074E+02	0.1241E+03
13	7.61	1722.92	1141.99	1719.71	0.7061E+02	0.4859E+02	0.7377E+02
15	10.77	1894.35	1269.89	1911.22	0.3815E+02	0.3418E+02	0.4858E+02
17	15.94	2069.20	1415.06	2083.04	0.2861E+02	0.2343E+02	0.2172E+02
19	20.41	2192.62	1508.13	2163.87	0.2450E+02	0.1864E+02	0.1665E+02
21	25.66	2254.34	1596.12	2257.72	0.1572E+02	0.1515E+02	0.1947E+02
23	30.59	2370.91	1665.08	2358.10	0.1857E+02	0.1295E+02	0.2068E+02
25	36.45	2466.91	1735.26	2472.95	0.1628E+02	0.1110E+02	0.1792E+02
27	41.08	2521.76	1784.04	2547.38	0.1187E+02	0.1001E+02	0.1414E+02
29	46.48	2583.47	1835.23	2611.59	0.1065E+02	0.8993E+01	0.9744E+01
31	52.34	2652.04	1885.27	2657.07	0.1031E+02	0.8121E+01	0.5989E+01
33	58.20	2693.19	1930.73	2684.21	0.4973E+01	0.7417E+01	0.3463E+01
35	63.44	2727.47	1968.19	2698.34	0.5287E+01	0.6893E+01	0.2042E+01

A.5 – 319

V -160

Connection type : Extended end-plate connections
Mode : All bolted without column stiffener

Tested by : L.R.O. Lima (2003) Brazil
Test Id. : EE1

Column : HEB240
Beam : IPE240
Plate thickness : 15 mm

Fasteners: 10.9- -M20
7/8" Oversize holes
Material : S275
Fy = 53.58 ksi
Fu = 73.02 ksi

Major parameters

End plate extended on tension side orly.

ct = 1.1811" pt = 2.9134" li = 6.1417" pc = 1.4606"
cc = 0.6653" gt = 3.7795" gc = 3.7795"
tp = 0.5906" nt = 2 X 2 nc = 2 X 1

Remark 1) Plate size: 314x160x15 mm
2) Axial force of beam: 0 kN

No	Moment (k-in)	Rotation (radians) X 1/1000	No	Moment (k-in)	Rotation (radians) X 1/1000
1	0.00	0.00	26	1002.81	13.66
2	80.51	0.45	27	1018.80	15.14
3	128.23	0.71	28	1029.96	16.61
4	172.59	0.97	29	1039.66	18.09
5	220.56	1.23	30	1047.38	19.57
6	271.81	1.49	31	1053.67	21.05
7	308.06	1.75	32	1061.00	22.53
8	353.29	2.01	33	1064.88	24.01
9	386.84	2.27	34	1070.69	25.48
10	423.45	2.53	35	1075.54	26.96
11	470.93	2.79	36	1078.45	28.44
12	528.77	3.04	37	1087.18	29.92
13	578.64	3.30	38	1088.15	31.40
14	631.98	3.56	39	1092.03	32.88
15	686.98	3.82	40	1095.91	34.35
16	734.58	4.08	41	1097.85	35.83
17	782.32	4.34	42	1100.76	37.31
18	824.66	4.86	43	1095.91	38.79
19	860.72	5.38	44	1095.91	40.27
20	891.32	6.15	45	1098.82	41.75
21	922.29	7.45	46	1101.73	43.22
22	941.75	8.49	47	1103.67	44.70
23	963.05	9.78	48	1107.55	46.18
24	982.44	11.34	49	1109.49	47.66
25	1000.69	12.63	50	1104.64	49.88

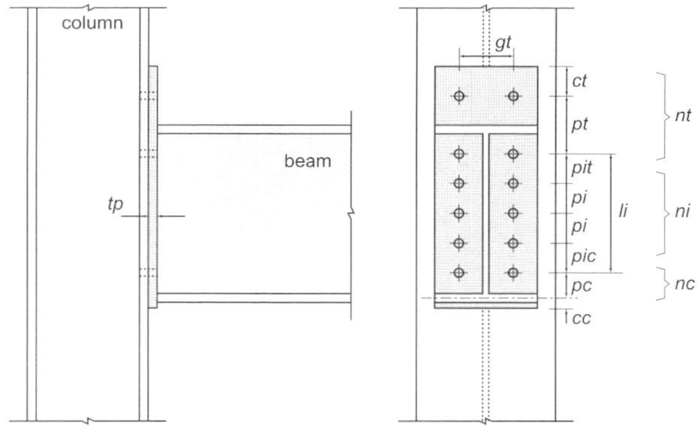

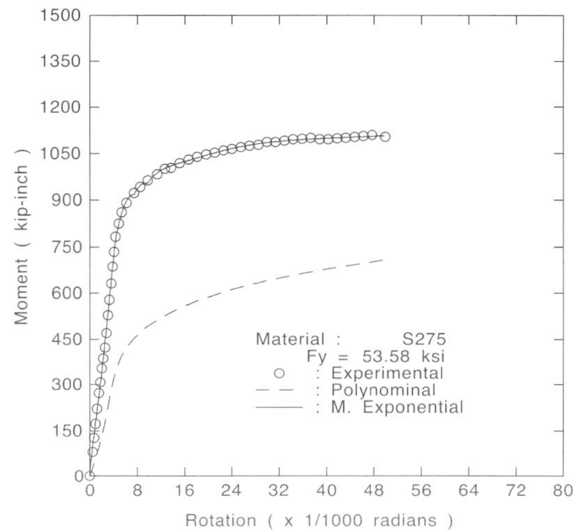

```
Moment-rotation prediction equations     ( R : X 1/1000 radians )

Frye and Morris polynominal model  :    R = Sum ( Ai X ( K*Bm )**Pi X 10**Qi )
     xd =  9.055120"   t =  0.590550"  tf =  0.787400"
     A1 =  1.830000   A2 = -1.040000   A3 =  6.380000    K =  0.008926
     P1 =         1   P2 =         3   P3 =         5   Q1 =         0   Q2 =        -1   Q3 =        -3

Modified exponential model  :
     Bm = Sum ( Ai X ( 1 - exp( - R/( 2*i*C ) ) ) )  +      Sum ( RKj X ( R - Rj0 ) )  +  Bm0
      C =  0.52816667E+00            Bmo=  0.00000000E+00          Nexp=  6        Nliner=  3
     AI =  0.17232495E+04  -0.16867090E+05  0.49449304E+05  -0.65614717E+05  0.43750804E+05  -0.11060821E+05
    Rj0 =          4.3400           7.4500          9.7800
    RKj = -0.76487548E+02   0.44698640E+02  0.32725634E+02
```

No	Rotation (radians) X 1/1000	Moment (kip-inch) Expri.	Poly.	M.Expo.	Connection stiffness (kip-inch) X 1000 Expri.	Poly.	M.Expo.
1	0.00	0.00	0.00	0.00	0.1789E+03	0.6122E+02	0.2611E+03
3	0.71	128.23	43.84	136.55	0.1771E+03	0.6283E+02	0.1544E+03
5	1.23	220.56	77.33	214.78	0.1908E+03	0.6635E+02	0.1518E+03
7	1.75	308.06	113.31	297.44	0.1567E+03	0.7255E+02	0.1672E+03
9	2.27	386.84	153.44	388.59	0.1349E+03	0.8256E+02	0.1827E+03
11	2.79	470.93	199.91	486.20	0.2075E+03	0.9658E+02	0.1914E+03
13	3.30	578.64	251.99	584.26	0.1985E+03	0.1049E+03	0.1919E+03
15	3.82	686.98	303.44	682.56	0.1973E+03	0.8906E+02	0.1851E+03
17	4.34	782.32	343.38	775.79	0.1495E+03	0.6536E+02	0.1728E+03
19	5.38	860.72	395.43	859.12	0.5741E+02	0.3871E+02	0.6277E+02
21	7.45	922.29	453.63	917.21	0.2098E+02	0.2118E+02	-0.3705E+01
23	9.78	963.05	493.90	953.91	0.1466E+02	0.1433E+02	-0.3845E+01
25	12.63	1000.69	528.61	1001.88	0.7424E+01	0.1048E+02	0.9234E+01
27	15.14	1018.80	552.31	1019.41	0.9194E+01	0.8558E+01	0.5789E+01
29	18.09	1039.66	575.26	1035.95	0.5884E+01	0.7100E+01	0.5590E+01
31	21.05	1053.67	594.71	1052.21	0.4600E+01	0.6099E+01	0.5279E+01
33	24.01	1064.88	611.63	1066.56	0.3292E+01	0.5368E+01	0.4347E+01
35	26.96	1075.54	626.61	1077.74	0.2621E+01	0.4811E+01	0.3234E+01
37	29.92	1087.18	640.17	1085.83	0.3276E+01	0.4367E+01	0.2272E+01
39	32.88	1092.03	652.67	1091.47	0.2630E+01	0.4003E+01	0.1586E+01
41	35.83	1097.85	664.02	1095.46	0.1638E+01	0.3705E+01	0.1161E+01
43	38.79	1095.91	674.59	1098.52	-0.1638E+01	0.3452E+01	0.9296E+00
45	41.75	1098.82	684.47	1101.09	0.1973E+01	0.3235E+01	0.8243E+00
47	44.70	1103.67	693.73	1103.46	0.1966E+01	0.3047E+01	0.7921E+00
49	47.66	1109.49	702.43	1105.81	-0.8737E-01	0.2883E+01	0.7972E+00

V -161

```
Connection type : Extended end-plate connections
           Mcde : All bolted without column stiffener

Tested by  :  L.R.O. Lima (2003)                          Brazil
Test  Id.  :  EE2

Column : HEB240                    Fasteners: 10.9- -M20
Beam   : IPE240                             7/8" Oversize holes
Plate thickness : 15 mm            Material :  S275
                                        Fy =  53.58 ksi
                                        Fu =  73.02 ksi

                        Major parameters

     End plate extended on tension side only.
     ct =  1.1772"    pt =  2.9134"    li =  6.1417"    pc =  1.4606"
     cc =  0.6653"    gt =  3.7795"    gc =  3.7795"
     tp =  0.5906"    nt = 2 X  2      nc = 2 X  1

Remark  1) Plate size: 314x160x15 mm
        2) Axial force of beam: -135.94 kN (N / Ny = -10 %)
```

No	Moment (k-in)	Rotation (radians) X 1/1000	No	Moment (k-in)	Rotation (radians) X 1/1000
1	0.00	0.00	26	1190.57	31.53
2	36.87	0.32	27	1199.68	33.71
3	118.87	1.05	28	1206.47	35.89
4	204.27	1.77	29	1213.40	38.07
5	290.61	2.50	30	1214.23	40.24
6	370.46	3.22	31	1214.23	42.42
7	437.14	3.95	32	1215.87	44.60
8	516.58	4.68	33	1218.11	46.78
9	587.29	5.40	34	1226.78	48.95
10	656.91	6.13	35	1228.77	51.13
11	745.28	6.85	36	1234.59	53.31
12	809.84	7.58	37	1238.47	55.49
13	848.61	8.30	38	1239.44	57.66
14	908.53	9.76	39	1245.23	59.84
15	948.32	11.21	40	1243.47	62.02
16	990.23	12.66	41	1252.05	64.20
17	1026.23	14.11	42	1253.02	66.37
18	1044.77	15.56	43	1256.90	68.55
19	1063.70	17.01	44	1261.75	70.73
20	1083.80	18.47	45	1244.29	72.18
21	1111.13	20.64			
22	1131.68	22.82			
23	1149.31	25.00			
24	1165.97	27.18			
25	1178.76	29.35			

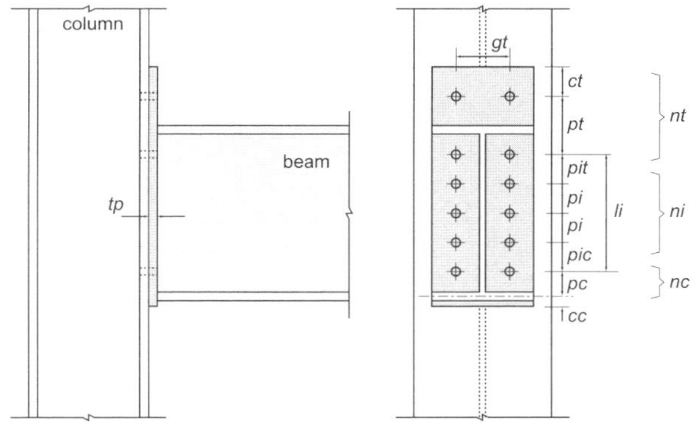

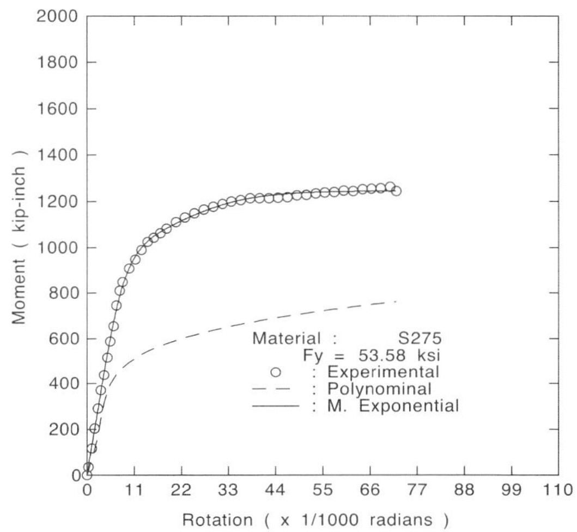

```
Moment-rotation prediction equations     ( R : X 1/1000 radians )

Frye and Morris polynominal model  :    R = Sum ( Ai X ( K*Bm )**Pi X 10**Qi )
     xd =  9.055120"    t =  0.590550"  tf =  0.787400"
     A1 =  1.830000    A2 = -1.040000   A3 =  6.380000    K =  0.008926
     P1 =         1    P2 =         3   P3 =         5    Q1 =        0    Q2 =       -1    Q3 =       -3

Modified exponential model  :
     Bm = Sum ( Ai X ( 1 - exp( - R/( 2*i*C ) ) ) )  +      Sum ( RKj X ( R - Rj0 ) )  +  Bm0
       C =   0.68150000E+00            Bmo=   0.00000000E+00            Nexp=  6       Nliner=  0
      AI =   0.77505002E+03  -0.80459956E+04   0.21462241E+05  -0.20190140E+05   0.36859126E+04   0.35578789E+04
```

No	Rotation (radians) X 1/1000	Moment (kip-inch) Expri.	Moment (kip-inch) Poly.	Moment (kip-inch) M.Expo.	Connection stiffness (kip-inch) X 1000 Expri.	Connection stiffness (kip-inch) X 1000 Poly.	Connection stiffness (kip-inch) X 1000 M.Expo.
1	0.00	0.00	0.00	0.00	0.1152E+03	0.6122E+02	0.1385E+03
3	1.05	118.87	65.53	122.83	0.1155E+03	0.6487E+02	0.1070E+03
5	2.50	290.61	173.05	281.11	0.1146E+03	0.8841E+02	0.1131E+03
7	3.95	437.14	314.71	446.83	0.1001E+03	0.8272E+02	0.1129E+03
9	5.40	587.29	396.35	601.81	0.9680E+02	0.3834E+02	0.9921E+02
11	6.85	745.28	440.06	731.63	0.1057E+03	0.2430E+02	0.7947E+02
13	8.30	848.61	470.20	832.40	0.4961E+02	0.1798E+02	0.5990E+02
15	11.21	948.32	512.67	961.11	0.2817E+02	0.1207E+02	0.3128E+02
17	14.11	1026.23	543.16	1028.58	0.1881E+02	0.9243E+01	0.1719E+02
19	17.01	1063.70	567.35	1068.94	0.1341E+02	0.7565E+01	0.1157E+02
21	20.64	1111.13	592.19	1105.85	0.1101E+02	0.6219E+01	0.9205E+01
23	25.00	1149.31	616.84	1142.85	0.7866E+01	0.5165E+01	0.7795E+01
25	29.35	1178.76	637.66	1173.48	0.5656E+01	0.4445E+01	0.6247E+01
27	33.71	1199.68	655.83	1197.13	0.3647E+01	0.3917E+01	0.4621E+01
29	38.07	1213.40	671.99	1214.08	0.1777E+01	0.3512E+01	0.3203E+01
31	42.42	1214.23	686.55	1225.53	0.3775E+00	0.3192E+01	0.2119E+01
33	46.78	1218.11	699.87	1232.99	0.2513E+01	0.2930E+01	0.1352E+01
35	51.13	1228.77	712.13	1237.69	0.1793E+01	0.2712E+01	0.8428E+00
37	55.49	1238.47	723.55	1240.59	0.1112E+01	0.2528E+01	0.5151E+00
39	59.84	1245.23	734.19	1242.35	0.9234E+00	0.2370E+01	0.3110E+00
41	64.20	1252.05	744.22	1243.41	0.2188E+01	0.2233E+01	0.1857E+00
43	68.55	1256.90	753.79	1244.04	0.2002E+01	0.2110E+01	0.1102E+00
45	72.18	1244.29	761.19	1244.37	-0.1204E+02	0.2022E+01	0.7100E-01

V -162

Connection type : Extended end-plate connections
Mode : All bolted without column stiffener

Tested by : L.R.O. Lima (2003) Brazil
Test Id. : EE3

Column : HEB240
Beam : IPE240
Plate thickness : 15 mm

Fasteners: 10.9- -M20
7/8" Oversize holes
Material : S275
Fy = 53.58 ksi
Fu = 73.02 ksi

Major parameters

End plate extended on tension side only.

ct = 1.1772" pt = 2.9134" li = 6.1417" pc = 1.4606"
cc = 0.6653" gt = 3.7795" gc = 3.7795"
tp = 0.5906" nt = 2 X 2 nc = 2 X 1

Remark 1) Plate size: 314x160x15 mm
2) Axial force of beam: -259.2 kN (N / Ny = -20 %)

No	Mcment (k-in)	Rotation (radians) X 1/1000	No	Moment (k-in)	Rotation (radians) X 1/1000
1	0.00	0.00	26	1015.99	21.68
2	50.53	0.40	27	1032.83	23.50
3	108.81	0.83	28	1050.97	25.33
4	168.05	1.26	29	1064.92	27.16
5	223.42	1.70	30	1074.67	28.98
6	282.67	2.35	31	1083.45	30.81
7	330.26	2.56	32	1093.19	32.64
8	372.02	2.78	33	1101.97	34.46
9	417.68	3.28	34	1110.74	36.29
10	459.45	3.93	35	1113.69	38.12
11	497.33	4.37	36	1118.58	39.95
12	533.27	4.80	37	1125.42	41.77
13	569.21	5.23	38	1128.37	43.60
14	600.29	5.31	39	1130.34	45.43
15	657.47	5.84	40	1134.26	47.25
16	689.02	6.45	41	1138.18	49.08
17	720.95	7.06	42	1138.22	50.91
18	763.56	8.28	43	1137.28	52.73
19	796.74	8.89	44	1133.44	54.56
20	843.98	10.71	45	1132.03	56.39
21	891.22	12.54	46	1128.86	58.22
22	925.83	14.37	47	1125.77	60.04
23	957.71	16.20	48	1123.87	61.87
24	979.40	18.02	49	1121.96	63.70
25	999.71	19.85	50	1119.08	65.52

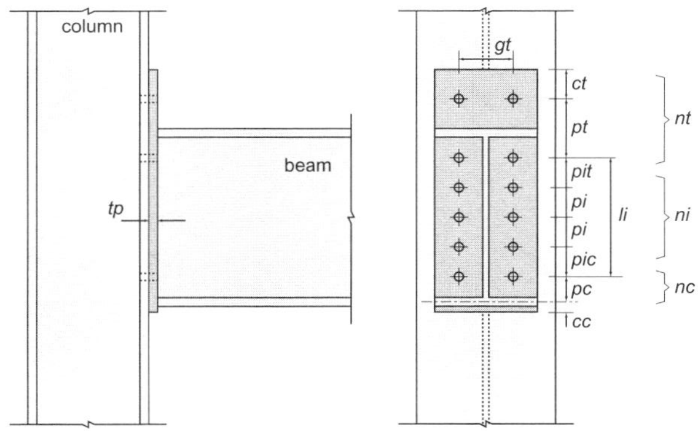

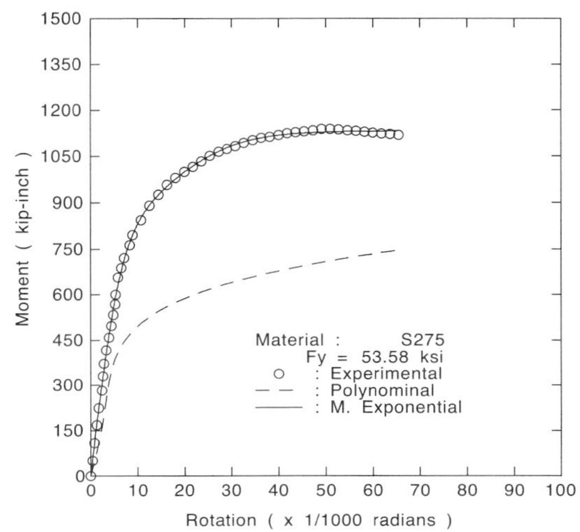

A.5 – 324

```
Moment-rotation prediction equations     ( R : X 1/1000 radians )

Frye and Morris polynominal model   :     R = Sum ( Ai X ( K*Bm )**Pi X 10**Qi )
     xd =  9.055120"   t =  0.590550"  tf =  0.787400"
     A1 =  1.830000   A2 = -1.040000   A3 =  6.380000     K =  0.008926
     P1 =         1   P2 =         3   P3 =         5    Q1 =         0    Q2 =        -1    Q3 =        -3

Modified exponential model  :
     Bm = Sum ( Ai X ( 1 - exp( - R/( 2*i*C ) ) ) )  +     Sum ( RKj X ( R - Rj0 ) )  +  Bm0
      C =   0.64600000E+00           Bmo=   0.00000000E+00          Nexp=  6        Nliner=  0
     AI =   0.39559175E+03  -0.56611984E+04   0.20011549E+05  -0.26889792E+05   0.13900111E+05  -0.62403035E+03
```

No	Rotation (radians) X 1/1000	Moment (kip-inch) Expri.	Moment (kip-inch) Poly.	Moment (kip-inch) M.Expo.	Connection stiffness (kip-inch) X 1000 Expri.	Connection stiffness (kip-inch) X 1000 Poly.	Connection stiffness (kip-inch) X 1000 M.Expo.
1	0.00	0.00	0.00	0.00	0.1263E+03	0.6122E+02	0.1463E+03
3	0.83	108.81	51.42	112.65	0.1367E+03	0.6345E+02	0.1286E+03
5	1.70	223.42	109.70	221.31	0.1118E+03	0.7181E+02	0.1218E+03
7	2.56	330.26	178.45	323.39	0.2087E+03	0.9007E+02	0.1152E+03
9	3.28	417.68	249.89	403.74	0.7955E+02	0.1050E+03	0.1076E+03
11	4.37	497.33	345.41	513.38	0.8483E+02	0.6414E+02	0.9309E+02
13	5.23	569.21	389.42	588.04	0.3406E+03	0.4123E+02	0.8051E+02
15	5.84	657.47	411.74	634.46	0.8177E+02	0.3261E+02	0.7171E+02
17	7.06	720.95	445.02	711.91	0.4654E+02	0.2310E+02	0.5569E+02
19	8.89	796.74	480.29	795.95	0.4725E+02	0.1630E+02	0.3737E+02
21	12.54	891.22	527.68	893.26	0.2236E+02	0.1056E+02	0.1926E+02
23	16.20	957.71	561.06	951.76	0.1466E+02	0.7962E+01	0.1377E+02
25	19.85	999.71	587.18	997.54	0.9998E+01	0.6464E+01	0.1147E+02
27	23.50	1032.83	608.86	1035.61	0.9580E+01	0.5480E+01	0.9365E+01
29	27.16	1064.92	627.57	1065.85	0.6488E+01	0.4778E+01	0.7163E+01
31	30.81	1083.45	644.00	1088.23	0.5061E+01	0.4251E+01	0.5154E+01
33	34.46	1101.97	658.74	1103.94	0.4809E+01	0.3840E+01	0.3527E+01
35	38.12	1113.69	672.17	1114.52	0.2142E+01	0.3508E+01	0.2318E+01
37	41.77	1125.42	684.46	1121.35	0.2686E+01	0.3235E+01	0.1479E+01
39	45.43	1130.34	695.87	1125.67	0.1619E+01	0.3006E+01	0.9202E+00
41	49.08	1138.18	706.47	1128.32	0.1081E+01	0.2810E+01	0.5619E+00
43	52.73	1137.28	716.41	1129.93	-0.1307E+01	0.2641E+01	0.3377E+00
45	56.39	1132.03	725.94	1130.89	-0.1251E+01	0.2491E+01	0.1999E+00
47	60.04	1125.77	734.72	1131.46	-0.1369E+01	0.2363E+01	0.1171E+00
49	63.70	1121.96	743.10	1131.79	-0.1312E+01	0.2247E+01	0.6786E-01

V -163

Connection type : Extended end-plate connections
Mode : All bolted without column stiffener

Tested by : L.R.O. Lima (2003) Brazil
Test Id. : EE4

Column : HEB240
Beam : IPE240
Plate thickness : 15 mm

Fasteners: 10.9- -M20
7/8" Oversize holes
Material : S275
Fy = 53.58 ksi
Fu = 73.02 ksi

Major parameters

End plate extended on tension side only.

ct = 1.1772" pt = 2.9134" li = 6.1417" pc = 1.4606"
cc = 0.6653" gt = 3.7795" gc = 3.7795"
tp = 0.5906" nt = 2 X 2 nc = 2 X 1

Remark 1) Plate size: 314x160x15 mm
2) Axial force of beam: -363.5 kN (N / Ny = -27 %)

No	Moment (k-in)	Rotation (radians) X 1/1000	No	Moment (k-in)	Rotation (radians) X 1/1000
1	0.00	0.00	26	1110.69	33.58
2	90.80	0.84	27	1123.35	35.67
3	162.68	1.54	28	1133.99	37.76
4	236.43	2.24	29	1138.97	39.85
5	312.72	2.93	30	1142.69	41.94
6	392.68	3.63	31	1151.42	44.03
7	465.66	4.33	32	1154.63	46.12
8	514.46	5.02	33	1160.50	48.21
9	569.25	5.72	34	1163.99	50.30
10	623.18	6.42	35	1168.35	52.38
11	664.80	7.11	36	1170.89	54.47
12	703.60	7.81	37	1168.43	56.56
13	754.98	9.20	38	1168.47	58.65
14	802.18	10.59	39	1166.57	60.74
15	845.13	11.99	40	1166.61	62.83
16	888.73	13.38	41	1160.83	64.92
17	918.18	14.77	42	1162.81	67.01
18	947.24	16.86	43	1148.28	69.10
19	974.45	18.95			
20	1002.44	21.04			
21	1031.18	23.13			
22	1052.52	25.22			
23	1071.93	27.31			
24	1087.30	29.40			
25	1101.91	31.49			

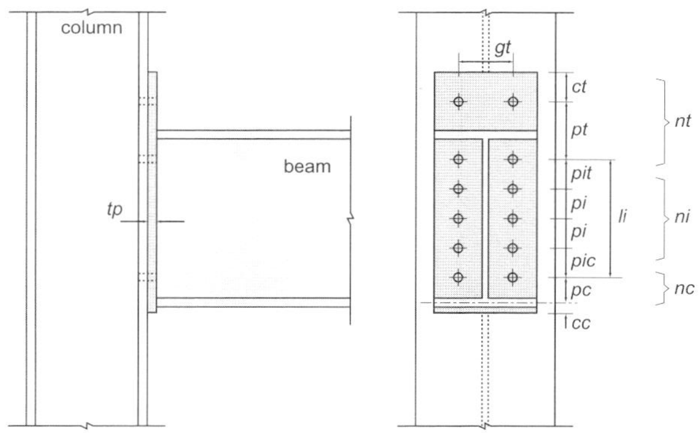

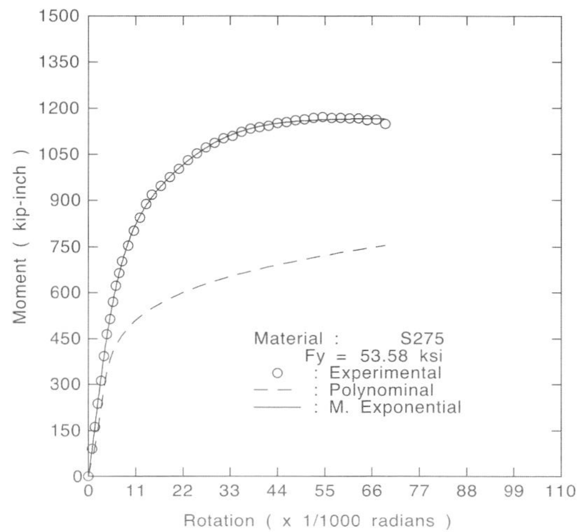

```
Moment-rotation prediction equations     ( R : X 1/1000 radians )

Frye and Morris polynominal model   :    R = Sum ( Ai X ( K*Bm )**Pi X 10**Qi )
     xd =  9.055120"   t =  0.590550"  tf =  0.787400"
     A1 =  1.830000   A2 = -1.040000   A3 =  6.380000    K =  0.008926
     P1 =        1    P2 =        3    P3 =        5    Q1 =        0    Q2 =       -1    Q3 =       -3

Modified exponential model  :
     Bm = Sum ( Ai X ( 1 - exp( - R/( 2*i*C ) ) ) )  +     Sum ( RKj X ( R - Rj0 ) )  +  Bm0
      C =   0.78583333E+00             Bmo=   0.00000000E+00          Nexp=  6        Nliner=  0
     AI =  -0.70923615E+01  -0.35977323E+04   0.17559869E+05  -0.29533615E+05   0.20578509E+05  -0.38341548E+04
```

No	Rotation (radians) X 1/1000	Moment (kip-inch) Expri.	Moment (kip-inch) Poly.	Moment (kip-inch) M.Expo.	Connection stiffness (kip-inch) X 1000 Expri.	Connection stiffness (kip-inch) X 1000 Poly.	Connection stiffness (kip-inch) X 1000 M.Expo.
1	0.00	0.00	0.00	0.00	0.1081E+03	0.6122E+02	0.8946E+02
3	1.54	162.68	98.38	162.81	0.1040E+03	0.6966E+02	0.1138E+03
5	2.93	312.72	213.65	318.24	0.1124E+03	0.1003E+03	0.1072E+03
7	4.33	465.66	342.71	457.44	0.8737E+02	0.6575E+02	0.9090E+02
9	5.72	569.25	407.82	571.37	0.7765E+02	0.3398E+02	0.7316E+02
11	7.11	664.80	446.17	661.77	0.5790E+02	0.2283E+02	0.5738E+02
13	9.20	754.98	485.27	761.76	0.3546E+02	0.1555E+02	0.3950E+02
15	11.99	845.13	521.70	850.17	0.3102E+02	0.1113E+02	0.2544E+02
17	14.77	918.18	549.10	910.17	0.1828E+02	0.8791E+01	0.1856E+02
19	18.95	974.45	581.22	976.79	0.1321E+02	0.6773E+01	0.1395E+02
21	23.13	1031.18	606.82	1029.26	0.1198E+02	0.5564E+01	0.1125E+02
23	27.31	1071.93	628.28	1071.08	0.8320E+01	0.4753E+01	0.8775E+01
25	31.49	1101.91	646.87	1102.83	0.5595E+01	0.4167E+01	0.6463E+01
27	35.67	1123.35	663.32	1125.60	0.5574E+01	0.3723E+01	0.4501E+01
29	39.85	1138.97	678.12	1141.09	0.2082E+01	0.3373E+01	0.2987E+01
31	44.03	1151.42	691.60	1151.17	0.2857E+01	0.3089E+01	0.1900E+01
33	48.21	1160.50	704.01	1157.46	0.2237E+01	0.2854E+01	0.1163E+01
35	52.38	1168.35	715.49	1161.25	0.1659E+01	0.2657E+01	0.6880E+00
37	56.56	1168.43	726.23	1163.45	-0.5790E+00	0.2487E+01	0.3914E+00
39	60.74	1166.57	736.31	1164.68	-0.4452E+00	0.2340E+01	0.2135E+00
41	64.92	1160.83	745.95	1165.34	-0.9099E+00	0.2210E+01	0.1103E+00
43	69.10	1148.28	754.88	1165.67	-0.6951E+01	0.2097E+01	0.5272E-01

A.5 – 327

V -164

Connection type : Extended end-plate connections
Mode : All bolted without column stiffener

Tested by : L.R.O. Lima (2003) Brazil
Test Id. : EE5

Column : HEB240
Beam : IPE240
Plate thickness : 15 mm

Fasteners: 10.9- -M20
7/8" Oversize holes
Material : S275
Fy = 53.58 ksi
Fu = 73.02 ksi

Major parameters

End plate extended on tension side only.

ct = 1.1772" pt = 2.9134" li = 6.1417" pc = 1.4606"
cc = 0.6653" gt = 3.7795" gc = 3.7795"
tp = 0.5906" nt = 2 X 2 nc = 2 X 1

Remark 1) Plate size: 314x160x15 mm
2) Axial force of beam: -193.3 kN (N / Ny = -15 %)

No	Moment (k-in)	Rotation (radians) X 1/1000
1	0.00	0.00
2	102.19	0.74
3	198.72	1.45
4	296.79	2.16
5	367.88	2.87
6	443.82	3.57
7	513.99	4.28
8	576.10	4.99
9	627.27	5.69
10	675.80	6.40
11	758.10	7.82
12	813.41	9.23
13	865.46	10.65
14	922.59	12.77
15	965.63	14.89
16	1001.75	17.01
17	1029.81	19.14
18	1054.13	21.26
19	1074.88	23.38
20	1093.50	25.50
21	1104.87	27.63
22	1115.00	29.75
23	1125.23	31.87
24	1135.27	33.99
25	1141.52	36.11
26	1145.74	38.24
27	1152.58	40.36
28	1158.45	42.48
29	1157.26	44.60
30	1157.24	46.73
31	1166.34	48.85
32	1166.38	50.97
33	1161.72	53.09
34	1167.91	55.22
35	1168.45	57.34
36	1168.49	59.46
37	1169.09	61.58
38	1170.52	63.70
39	1169.38	65.83
40	1170.60	67.95
41	1171.61	70.07

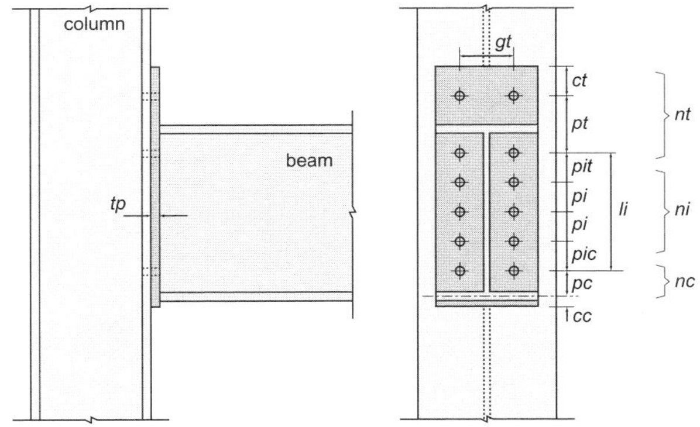

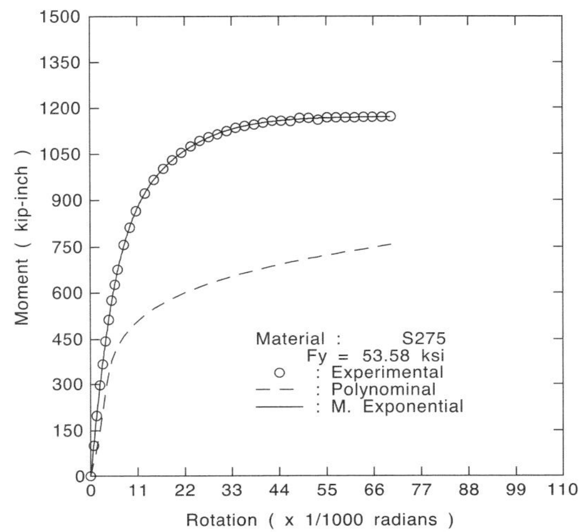

A.5 – 328

```
Moment-rotation prediction equations      ( R : X 1/1000 radians )

Frye and Morris polynominal model  :     R = Sum ( Ai X ( K*Bm )**Pi X 10**Qi )
     xd =  9.055120"    t =  0.590550"  tf =  0.787400"
     A1 =  1.830000   A2 = -1.040000   A3 =  6.380000    K =  0.008926
     P1 =        1    P2 =        3    P3 =        5    Q1 =        0    Q2 =       -1    Q3 =       -3

Modified exponential model  :
     Bm = Sum ( Ai X ( 1 - exp( - R/( 2*i*C ) ) ) )  +      Sum ( RKj X ( R - Rj0 ) )  +  Bm0
       C =   0.76891667E+00            Bmo=   0.00000000E+00         Nexp=  6        Nliner=  0
      AI =  -0.80650755E+02  -0.86382663E+03   0.51920865E+04  -0.71700426E+04   0.31515255E+04   0.94197911E+03
```

No	Rotation (radians) X 1/1000	Moment (kip-inch) Expri.	Moment (kip-inch) Poly.	Moment (kip-inch) M.Expo.	Connection stiffness (kip-inch) X 1000 Expri.	Connection stiffness (kip-inch) X 1000 Poly.	Connection stiffness (kip-inch) X 1000 M.Expo.
1	0.00	0.00	0.00	0.00	0.1381E+03	0.6122E+02	0.1385E+03
3	1.45	198.72	92.16	199.25	0.1370E+03	0.6859E+02	0.1320E+03
5	2.87	367.88	207.67	372.82	0.1043E+03	0.9874E+02	0.1115E+03
7	4.28	513.99	339.37	514.58	0.9315E+02	0.6777E+02	0.8980E+02
9	5.69	627.27	406.79	627.42	0.7074E+02	0.3435E+02	0.7085E+02
11	7.82	758.10	461.18	753.87	0.4856E+02	0.1964E+02	0.4926E+02
13	10.65	865.46	505.70	865.64	0.3276E+02	0.1285E+02	0.3144E+02
15	14.89	965.63	550.17	967.11	0.1867E+02	0.8713E+01	0.1826E+02
17	19.14	1029.81	582.50	1030.29	0.1232E+02	0.6705E+01	0.1213E+02
19	23.38	1074.88	608.20	1073.68	0.9285E+01	0.5507E+01	0.8586E+01
21	27.63	1104.87	629.80	1104.64	0.5059E+01	0.4702E+01	0.6113E+01
23	31.87	1125.23	648.44	1126.49	0.4781E+01	0.4122E+01	0.4286E+01
25	36.11	1141.52	664.94	1141.66	0.2466E+01	0.3682E+01	0.2938E+01
27	40.36	1152.58	679.83	1151.97	0.2997E+01	0.3335E+01	0.1969E+01
29	44.60	1157.26	693.35	1158.80	-0.2866E+00	0.3054E+01	0.1297E+01
31	48.85	1166.34	705.83	1163.27	0.2157E+01	0.2822E+01	0.8400E+00
33	53.09	1161.72	717.36	1166.15	0.3467E+00	0.2626E+01	0.5381E+00
35	57.34	1168.45	728.16	1167.99	0.1374E+00	0.2458E+01	0.3410E+00
37	61.58	1169.09	738.26	1169.15	0.4775E+00	0.2313E+01	0.2147E+00
39	65.83	1169.38	747.95	1169.88	0.2200E-01	0.2184E+01	0.1343E+00
41	70.07	1171.61	756.90	1170.33	0.4775E+00	0.2073E+01	0.8370E-01

A.5 – 329

V -165

```
Connection type : Extended end-plate connections
           Mode : All bolted without column stiffener

Tested by  :  L.R.O. Lima (2003)                           Brazil
Test  Id.  :  EE6

Column : HEB240                       Fasteners: 10.9- -M20
Beam   : IPE240                                7/8" Oversize holes
Plate thickness : 15 mm               Material :  S275
                                           Fy =  53.58 ksi
                                           Fu =  73.02 ksi
```

Major parameters

```
     End plate extended on tension side only.
     ct =  1.1772"     pt =  2.9134"    li =  6.1417"     pc =  1.4606"
     cc =  0.6653"     gt =  3.7795"    gc =  3.7795"
     tp =  0.5906"     nt = 2 X  2      nc = 2 X  1

Remark  1) Plate size: 314x160x15 mm
        2) Axial force of beam: 127.2 kN (N / Ny = +10 %)
```

No	Moment (k-in)	Rotation (radians) X 1/1000	No	Moment (k-in)	Rotation (radians) X 1/1000
1	0.00	0.00	26	1054.21	25.66
2	50.68	0.75	27	1069.79	27.80
3	112.45	1.46	28	1082.46	29.93
4	182.40	2.17	29	1092.83	32.07
5	250.93	2.88	30	1101.96	34.20
6	301.31	3.59	31	1109.57	36.34
7	341.86	4.31	32	1115.64	38.47
8	386.68	5.02	33	1120.54	40.61
9	431.65	5.73	34	1125.94	42.74
10	476.89	6.44	35	1130.42	44.88
11	523.94	7.15	36	1134.26	47.01
12	566.67	7.87	37	1137.21	49.15
13	608.62	8.58	38	1138.05	51.29
14	645.67	9.29	39	1137.89	53.42
15	678.13	10.00	40	1140.25	55.56
16	712.70	10.71	41	1144.18	57.69
17	741.69	11.42	42	1144.34	59.83
18	773.08	12.14	43	1145.53	61.96
19	803.32	12.85	44	1153.04	64.10
20	855.24	14.27	45	1158.91	66.23
21	898.42	15.70	46	1159.92	68.37
22	931.68	17.12	47	1163.85	70.50
23	975.60	19.25			
24	1008.25	21.39			
25	1034.10	23.53			

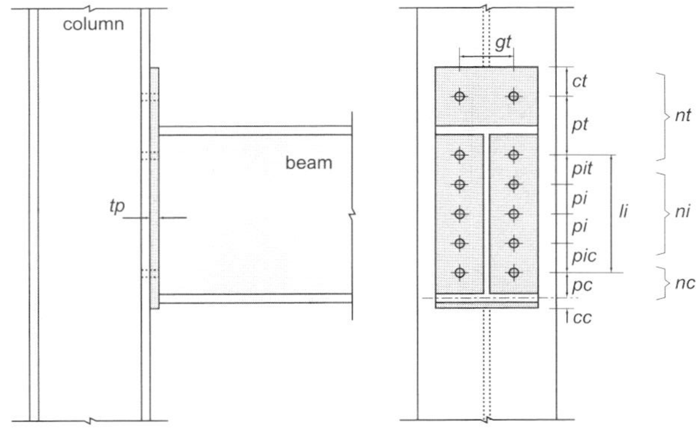

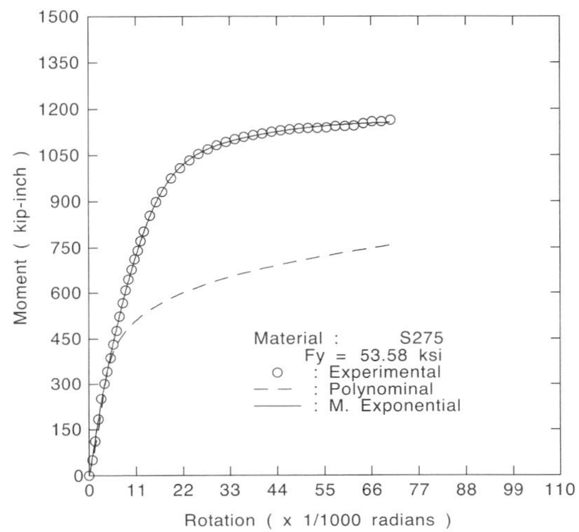

A.5 – 330

```
Moment-rotation prediction equations      ( R : X 1/1000 radians )

Frye and Morris polynominal model  :     R = Sum ( Ai X ( K*Bm )**Pi X 10**Qi )
     xd =  9.055120"    t =  0.590550"  tf =  0.787400"
     A1 =  1.830000    A2 = -1.040000   A3 =  6.380000     K =  0.008926
     P1 =        1     P2 =        3    P3 =        5     Q1 =        0     Q2 =       -1     Q3 =       -3

Modified exponential model  :
     Bm = Sum ( Ai X ( 1 - exp( - R/( 2*i*C ) ) ) )  +      Sum ( RKj X ( R - Rj0 ) )  +  Bm0
       C =   0.77500000E+00              Bmo=   0.00000000E+00            Nexp=  6        Nliner=  0
      AI =  -0.48684161E+03   0.64367873E+04  -0.29290020E+05   0.57082510E+05  -0.51170427E+05   0.18588579E+05
```

No	Rotation (radians) X 1/1000	Moment (kip-inch) Expri.	Moment (kip-inch) Poly.	Moment (kip-inch) M.Expo.	Connection stiffness (kip-inch) X 1000 Expri.	Connection stiffness (kip-inch) X 1000 Poly.	Connection stiffness (kip-inch) X 1000 M.Expo.
1	0.00	0.00	0.00	0.00	0.6758E+02	0.6122E+02	0.6636E+02
3	1.46	112.45	92.85	118.97	0.9276E+02	0.6870E+02	0.8705E+02
5	2.88	250.93	208.66	238.17	0.8374E+02	0.9901E+02	0.7948E+02
7	4.31	341.86	341.39	344.81	0.5975E+02	0.6655E+02	0.7002E+02
9	5.73	431.65	408.16	439.08	0.6353E+02	0.3386E+02	0.6315E+02
11	7.15	523.94	447.08	525.00	0.6284E+02	0.2262E+02	0.5805E+02
13	8.58	608.62	475.11	604.71	0.5563E+02	0.1714E+02	0.5345E+02
15	10.00	678.13	496.99	677.29	0.4720E+02	0.1393E+02	0.4872E+02
17	11.42	741.69	515.16	742.93	0.4221E+02	0.1180E+02	0.4369E+02
19	12.85	803.32	530.89	801.65	0.4058E+02	0.1027E+02	0.3843E+02
21	15.70	898.42	557.01	896.47	0.2680E+02	0.8231E+01	0.2829E+02
23	19.25	975.60	583.24	977.78	0.1794E+02	0.6666E+01	0.1807E+02
25	23.53	1034.10	609.03	1036.62	0.1076E+02	0.5473E+01	0.1019E+02
27	27.80	1069.79	630.59	1070.36	0.6612E+01	0.4675E+01	0.6091E+01
29	32.07	1092.83	649.26	1091.64	0.4567E+01	0.4099E+01	0.4119E+01
31	36.34	1109.57	665.79	1106.91	0.3202E+01	0.3661E+01	0.3132E+01
33	40.61	1120.54	680.66	1118.92	0.2412E+01	0.3317E+01	0.2532E+01
35	44.88	1130.42	694.21	1128.72	0.1948E+01	0.3038E+01	0.2068E+01
37	49.15	1137.21	706.67	1136.66	0.8857E+00	0.2807E+01	0.1661E+01
39	53.42	1137.89	718.23	1142.97	0.5129E+00	0.2612E+01	0.1301E+01
41	57.69	1144.18	729.02	1147.85	0.9609E+00	0.2445E+01	0.9926E+00
43	61.96	1145.53	739.14	1151.52	0.2032E+01	0.2301E+01	0.7387E+00
45	66.23	1158.91	748.82	1154.23	0.1617E+01	0.2173E+01	0.5380E+00
47	70.50	1163.85	757.79	1156.19	0.1843E+01	0.2062E+01	0.3846E+00

A.5 – 331

V -166

Connection type : Extended end-plate connections
Mode : All bolted without column stiffener

Tested by : L.R.O. Lima (2003) Brazil
Test Id. : EE7

Column : HEB240
Beam : IPE240
Plate thickness : 15 mm

Fasteners: 10.9- -M20
7/8" Oversize holes
Material : S275
Fy = 53.58 ksi
Fu = 73.02 ksi

Major parameters

End plate extended on tension side only.

ct = 1.1772"	pt = 2.9134"	li = 6.1417"	pc = 1.4606"
cc = 0.6653"	gt = 3.7795"	gc = 3.7795"	
tp = 0.5906"	nt = 2 X 2	nc = 2 X 1	

Remark 1) Plate size: 314x160x15 mm
2) Axial force of beam: 257.9 kN (N / Ny = +20 %)

No	Moment (k-in)	Rotation (radians) X 1/1000	No	Moment (k-in)	Rotation (radians) X 1/1000
1	0.00	0.00	26	997.17	27.90
2	47.40	0.72	27	1018.32	30.16
3	93.75	1.47	28	1037.72	32.43
4	129.56	2.23	29	1047.17	34.69
5	170.85	2.98	30	1061.00	36.96
6	217.51	3.74	31	1064.88	39.22
7	264.31	4.49	32	1077.05	41.49
8	309.17	5.25	33	1070.65	43.75
9	360.54	6.00	34	1088.15	46.02
10	424.89	6.76	35	1095.66	48.28
11	467.41	7.51	36	1096.04	50.55
12	504.54	8.27	37	1103.67	52.81
13	536.06	9.02	38	1109.49	55.08
14	564.57	9.78	39	1107.24	57.35
15	599.86	10.53	40	1115.02	59.61
16	637.13	11.29	41	1110.26	61.88
17	669.06	12.04	42	1123.06	64.14
18	721.65	13.55	43	1128.88	66.41
19	770.90	15.06	44	1128.46	68.67
20	817.25	16.57	45	1140.52	70.94
21	844.20	18.08	46	1136.64	73.20
22	887.48	19.59	47	1097.85	74.71
23	914.56	21.10			
24	941.52	23.37			
25	962.08	25.63			

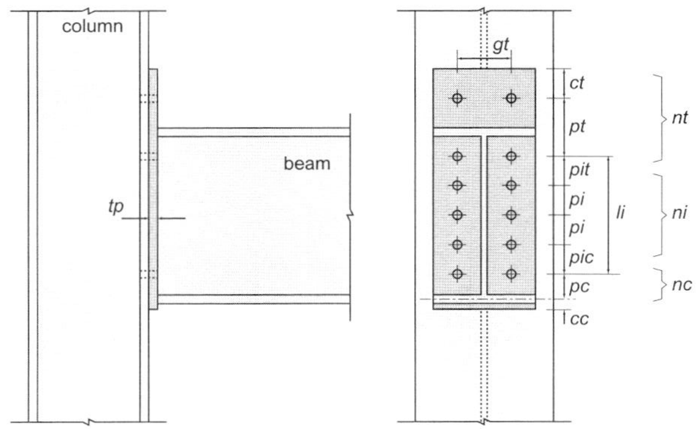

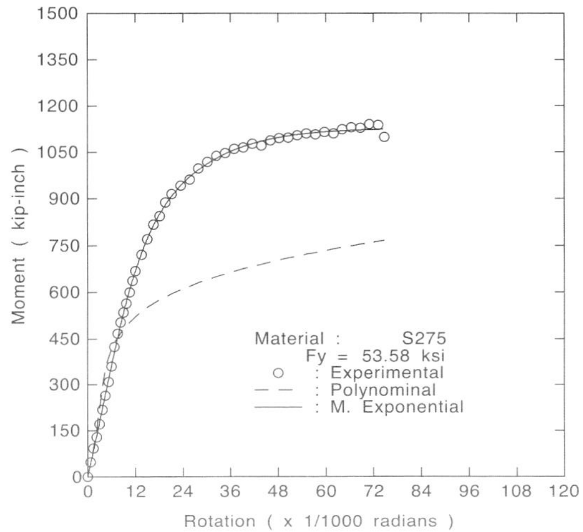

A.5 – 332

```
Moment-rotation prediction equations     ( R : X 1/1000 radians )

Frye and Morris polynominal model  :    R = Sum ( Ai X ( K*Bm )**Pi X 10**Qi )
     xd =  9.055120"    t =  0.590550"  tf =  0.787400"
     A1 =  1.830000   A2 = -1.040000   A3 =  6.380000    K =  0.008926
     P1 =        1    P2 =        3    P3 =        5    Q1 =        0    Q2 =       -1    Q3 =       -3

Modified exponential model  :
     Bm = Sum ( Ai X ( 1 - exp( - R/( 2*i*C ) ) ) )  +     Sum ( RKj X ( R - Rj0 ) )  +  Bm0
      C =  0.80258333E+00             Bmo=   0.00000000E+00          Nexp=  6        Nliner=  0
     AI =  0.14352488E+03  -0.27703054E+03  -0.50847483E+04   0.17985519E+05  -0.22975845E+05   0.11334783E+05
```

No	Rotation (radians) X 1/1000	Moment (kip-inch) Expri.	Moment (kip-inch) Poly.	Moment (kip-inch) M.Expo.	Connection stiffness (kip-inch) X 1000 Expri.	Connection stiffness (kip-inch) X 1000 Poly.	Connection stiffness (kip-inch) X 1000 M.Expo.
1	0.00	0.00	0.00	0.00	0.6584E+02	0.6122E+02	0.6257E+02
3	1.47	93.75	93.54	86.20	0.5451E+02	0.6882E+02	0.5774E+02
5	2.98	170.85	218.68	175.83	0.5820E+02	0.1014E+03	0.6112E+02
7	4.49	264.31	352.75	269.79	0.6073E+02	0.5984E+02	0.6276E+02
9	6.00	360.54	416.88	363.65	0.7652E+02	0.3090E+02	0.6104E+02
11	7.51	467.41	454.91	452.86	0.5280E+02	0.2091E+02	0.5679E+02
13	9.02	536.06	482.39	534.49	0.3979E+02	0.1598E+02	0.5120E+02
15	10.53	599.86	504.13	607.26	0.4804E+02	0.1304E+02	0.4518E+02
17	12.04	669.06	522.25	671.04	0.4001E+02	0.1108E+02	0.3934E+02
19	15.06	770.90	551.63	774.03	0.3166E+02	0.8607E+01	0.2929E+02
21	18.08	844.20	575.19	850.54	0.2325E+02	0.7104E+01	0.2179E+02
23	21.10	914.56	595.01	907.85	0.1552E+02	0.6085E+01	0.1648E+02
25	25.63	962.08	620.06	969.80	0.1227E+02	0.5045E+01	0.1131E+02
27	30.16	1018.32	641.21	1013.24	0.8952E+01	0.4335E+01	0.8101E+01
29	34.69	1047.17	659.62	1044.75	0.5134E+01	0.3817E+01	0.5938E+01
31	39.22	1064.88	675.98	1067.93	0.3536E+01	0.3421E+01	0.4374E+01
33	43.75	1070.65	690.74	1084.96	0.2427E+01	0.3106E+01	0.3201E+01
35	48.28	1095.66	704.21	1097.36	0.1748E+01	0.2851E+01	0.2314E+01
37	52.81	1103.67	716.63	1106.26	0.2971E+01	0.2638E+01	0.1650E+01
39	57.35	1107.24	728.18	1112.58	0.1232E+01	0.2458E+01	0.1159E+01
41	61.88	1110.26	738.96	1116.98	0.1793E+01	0.2303E+01	0.8037E+00
43	66.41	1128.88	749.08	1120.01	0.1185E+01	0.2169E+01	0.5508E+00
45	70.94	1140.52	758.77	1122.08	0.1790E+01	0.2050E+01	0.3735E+00
47	74.71	1097.85	766.23	1123.28	-0.2569E+02	0.1964E+01	0.2684E+00

A.6 – 1

VI - 1

Connection type : Flush end-plate connections
Mode : All bolted without column stiffener

Tested by : J.R.Ostrander (1970) Canada
Test Id. : TEST 1

Column : W8x28
Beam : W10x21
Plate thickness : 1/2"

Fasteners: A325- -3/4"D
13/16" Oversize holes
Material : A36
Fy = 45.40 ksi
Fu = 70.30 ksi

Major parameters

ct = 0.5000"	pt = 2.5000"	li = 5.0000"	pc = 2.5000"
cc = 0.5000"	gt = 3.5000"	gc = 3.5000"	
tp = 0.5000"	nt = 2 X 1	nc = 2 X 1	

Remark 1) Actual strength fy = 41.3 ksi, fu = 66.9 ksi for beam.
2) Actual strength fy = 51.1 ksi, fu = 77.3 ksi for column.

No	Mcment (k-in)	Rotation (radians) X 1/1000	No	Moment (k-in)	Rotation (radians) X 1/1000
1	0.00	0.00	26	497.00	29.10
2	24.00	0.30	27	502.50	31.55
3	47.00	0.60	28	508.00	34.00
4	70.50	0.85	29	513.50	36.45
5	94.00	1.10	30	519.00	38.90
6	117.50	1.40	31	523.83	41.58
7	141.00	1.70	32	528.67	44.27
8	164.50	2.05	33	533.50	46.95
9	188.00	2.40	34	538.34	49.63
10	211.50	2.95	35	543.17	52.32
11	235.00	3.50	36	548.00	55.00
12	258.50	4.30			
13	282.00	5.10			
14	305.50	6.05			
15	329.00	7.00			
16	352.50	8.35			
17	376.00	9.70			
18	399.50	11.65			
19	423.00	13.60			
20	437.33	15.77			
21	451.66	17.93			
22	466.00	20.10			
23	473.75	22.35			
24	481.50	24.60			
25	489.25	26.85			

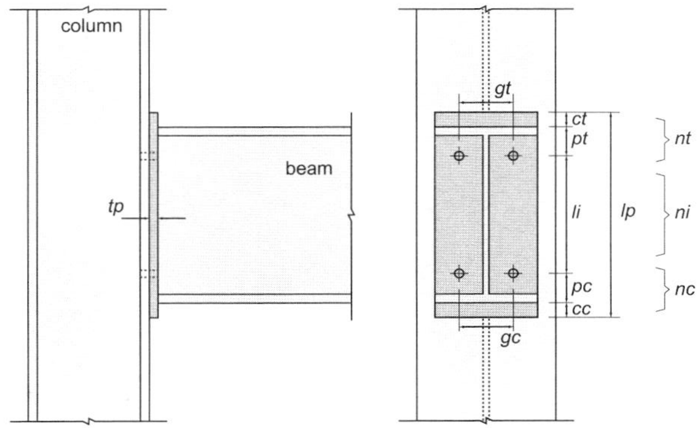

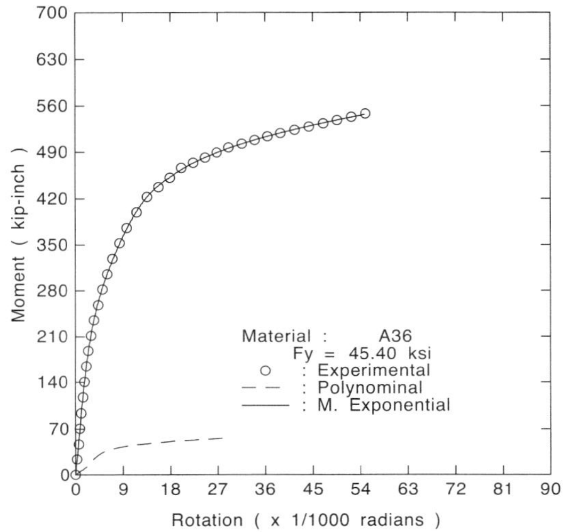

A.6 – 2

```
Moment-rotation prediction equations      ( R : X 1/1000 radians )

Frye and Morris polynominal model   :    R = Sum ( Ai X ( K*Bm )**Pi X 10**Qi )
     xd =  3.500000"   t =  0.500000"  tf =  0.750000"
     A1 =  1.830000   A2 = -1.040000   A3 =  6.380000    K =  0.100475
     P1 =        1    P2 =        3    P3 =        5    Q1 =        0    Q2 =       -1    Q3 =       -3

Modified exponential model  :
     Bm = Sum ( Ai X ( 1 - exp( - R/( 2*i*C ) ) ) )  +      Sum ( RKj X ( R - Rj0 ) )  +  Bm0
      C =  0.53333333E+00             Bmo=   0.00000000E+00         Nexp=  6        Nliner=  1
     AI = -0.38435629E+03   0.35983655E+04  -0.12102493E+05   0.20663917E+05  -0.17676278E+05   0.63934589E+04
    Rj0 =          17.9300
    RKj =  0.14898700E+01
```

No	Rotation (radians) X 1/1000	Moment (kip-inch) Expri.	Moment (kip-inch) Poly.	Moment (kip-inch) M.Expo.	Connection stiffness (kip-inch) X 1000 Expri.	Connection stiffness (kip-inch) X 1000 Poly.	Connection stiffness (kip-inch) X 1000 M.Expo.
1	0.00	0.00	0.00	0.00	0.8000E+02	0.5439E+01	0.7215E+02
3	0.60	47.00	3.28	49.79	0.8612E+02	0.5540E+01	0.8818E+02
5	1.10	94.00	6.11	93.21	0.8687E+02	0.5797E+01	0.8392E+02
7	1.70	141.00	9.75	140.01	0.7317E+02	0.6379E+01	0.7149E+02
9	2.40	188.00	14.60	184.51	0.5765E+02	0.7620E+01	0.5593E+02
11	3.50	235.00	24.23	235.19	0.3728E+02	0.9037E+01	0.3774E+02
13	5.10	282.00	34.10	283.79	0.2726E+02	0.3881E+01	0.2512E+02
15	7.00	329.00	39.42	325.87	0.2171E+02	0.2081E+01	0.2001E+02
17	9.70	376.00	43.78	374.15	0.1521E+02	0.1287E+01	0.1574E+02
19	13.60	423.00	47.83	422.82	0.9477E+01	0.8554E+00	0.9348E+01
21	17.93	451.66	51.01	451.96	0.6615E+01	0.6364E+00	0.6124E+01
23	22.35	473.75	53.52	473.58	0.3445E+01	0.5110E+00	0.3983E+01
25	26.85	489.25	55.62	489.18	0.3445E+01	0.4290E+00	0.3067E+01
27	31.55	502.50	57.49	502.29	0.2244E+01	0.3696E+00	0.2552E+01
29	36.45	513.50	59.18	513.85	0.2244E+01	0.3244E+00	0.2183E+01
31	41.58	523.83	60.75	524.30	0.1801E+01	0.2886E+00	0.1908E+01
33	46.95	533.50	62.22	534.00	0.1802E+01	0.2595E+00	0.1722E+01
35	52.32	543.17	63.55	542.92	0.1801E+01	0.2362E+00	0.1612E+01

A.6 – 3

VI - 2

Connection type : Flush end-plate connections
Mode : All bolted without column stiffener

Tested by : J.R.Ostrander (1970) Canada
Test Id. : TEST 3

Column : W8x23
Beam : W10x21
Plate thickness : 3/8"

Fasteners: A325- -3/4"D
13/16" Oversize holes
Material : A36
Fy = 45.50 ksi
Fu = 68.50 ksi

Major parameters

ct = 0.5000"	pt = 2.5000"	li = 5.0000"	pc = 2.5000"
cc = 0.5000"	gt = 3.5000"	gc = 3.5000"	
tp = 0.3750"	nt = 2 X 1	nc = 2 X 1	

Remark 1) Actual strength fy = 41.3 ksi, fu = 66.9 ksi for beam.
2) Actual strength fy = 51.1 ksi, fu = 77.3 ksi for column.

No	Moment (k-in)	Rotation (radians) X 1/1000	No	Moment (k-in)	Rotation (radians) X 1/1000
1	0.00	0.00	26	454.67	35.13
2	23.50	0.25	27	461.00	37.50
3	47.00	0.50	28	465.17	40.00
4	70.50	0.85	29	469.34	42.50
5	94.00	1.20	30	473.50	45.00
6	117.50	1.60	31	477.67	47.50
7	141.00	2.00	32	481.83	50.00
8	164.50	2.50	33	486.00	52.50
9	188.00	3.00			
10	211.50	3.70			
11	235.00	4.40			
12	258.50	5.35			
13	282.00	6.30			
14	305.50	7.80			
15	329.00	9.30			
16	352.50	11.75			
17	376.00	14.20			
18	387.75	16.48			
19	399.50	18.75			
20	411.25	21.02			
21	423.00	23.30			
22	429.33	25.67			
23	435.67	28.03			
24	442.00	30.40			
25	448.34	32.77			

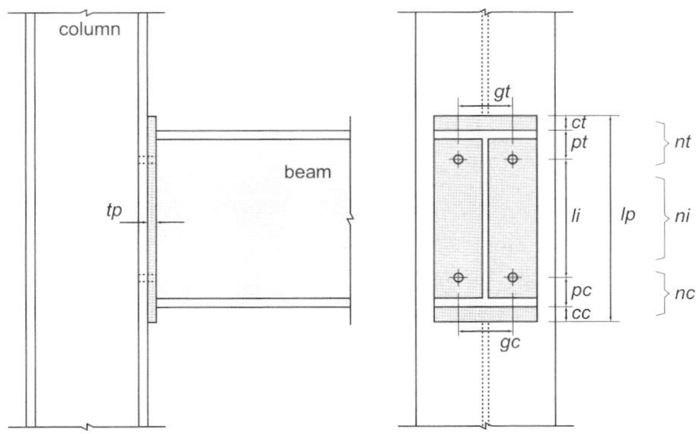

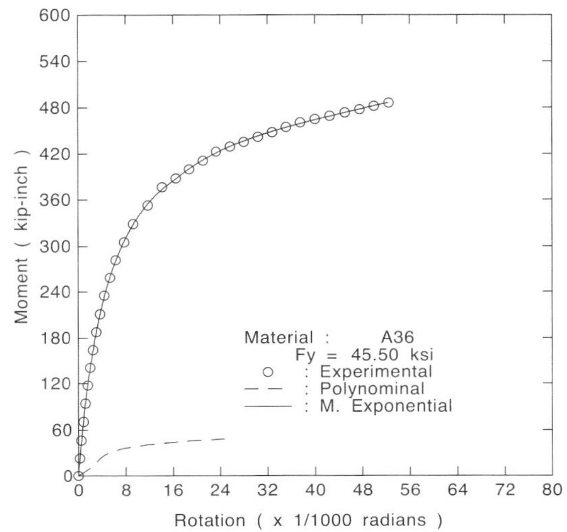

A.6 – 4

```
Moment-rotation prediction equations     ( R : X 1/1000 radians )

Frye and Morris polynominal model  :    R = Sum ( Ai X ( K*Bm )**Pi X 10**Qi )
     xd =  3.500000"    t =  0.375000"  tf =  0.750000"
     A1 =  1.830000   A2 = -1.040000   A3 =  6.380000    K =  0.112728
     P1 =        1    P2 =        3    P3 =        5    Q1 =        0     Q2 =       -1     Q3 =       -3

Modified exponential model  :
     Bm = Sum ( Ai X ( 1 - exp( - R/( 2*i*C ) ) ) )  +      Sum ( RKj X ( R - Rj0 ) )  +  Bm0
       C =   0.50000000E+00                 Bmo=   0.00000000E+00          Nexp=  6         Nliner=  1
      AI =   0.15702656E+02   0.27165523E+01   0.28144999E+02   0.10355672E+04  -0.23692699E+04   0.17129093E+04
     Rj0 =          16.4800
     RKj =   0.16923591E+01
```

No	Rotation (radians) X 1/1000	Moment (kip-inch) Expri.	Moment (kip-inch) Poly.	Moment (kip-inch) M.Expo.	Connection stiffness (kip-inch) X 1000 Expri.	Connection stiffness (kip-inch) X 1000 Poly.	Connection stiffness (kip-inch) X 1000 M.Expo.
1	0.00	0.00	0.00	0.00	0.9401E+02	0.4847E+01	0.9697E+02
2	0.25	23.50	1.21	23.14	0.9401E+02	0.4863E+01	0.8836E+02
3	0.50	47.00	2.43	44.27	0.8281E+02	0.4910E+01	0.8089E+02
4	0.85	70.50	4.17	70.98	0.6713E+02	0.5033E+01	0.7200E+02
5	1.20	94.00	5.97	94.84	0.6322E+02	0.5232E+01	0.6454E+02
6	1.60	117.50	8.12	119.18	0.5876E+02	0.5576E+01	0.5740E+02
7	2.00	141.00	10.45	140.91	0.5353E+02	0.6082E+01	0.5139E+02
8	2.50	164.50	13.70	164.99	0.4700E+02	0.7001E+01	0.4513E+02
9	3.00	188.00	17.48	186.21	0.4140E+02	0.8068E+01	0.3993E+02
10	3.70	211.50	23.13	212.01	0.3357E+02	0.7498E+01	0.3400E+02
11	4.40	235.00	27.49	234.09	0.2982E+02	0.4994E+01	0.2926E+02
12	5.35	258.50	31.22	259.38	0.2474E+02	0.3104E+01	0.2421E+02
13	6.30	282.00	33.72	280.45	0.2122E+02	0.2229E+01	0.2032E+02
14	7.80	305.50	36.49	307.33	0.1567E+02	0.1562E+01	0.1579E+02
15	9.30	329.00	38.55	328.48	0.1336E+02	0.1213E+01	0.1258E+02
16	11.75	352.50	41.10	354.59	0.9590E+01	0.9018E+00	0.9000E+01
17	14.20	376.00	43.08	373.54	0.7296E+01	0.7265E+00	0.6611E+01
18	16.48	387.75	44.60	386.68	0.5165E+01	0.6193E+00	0.5012E+01
19	18.75	399.50	45.91	400.48	0.5165E+01	0.5420E+00	0.5494E+01
20	21.02	411.25	47.08	411.88	0.5165E+01	0.4835E+00	0.4566E+01
21	23.30	423.00	48.12	421.42	0.3945E+01	0.4375E+00	0.3851E+01
22	25.67	429.33	49.11	429.83	0.2676E+01	0.3989E+00	0.3284E+01
23	28.03	435.67	50.02	437.07	0.2676E+01	0.3673E+00	0.2856E+01
24	30.40	442.00	50.85	443.43	0.2676E+01	0.3408E+00	0.2538E+01
25	32.77	448.34	51.63	449.15	0.2676E+01	0.3182E+00	0.2302E+01
26	35.13	454.67	52.36	454.38	0.2676E+01	0.2988E+00	0.2129E+01
27	37.50	461.00	53.05	459.26	0.2185E+01	0.2819E+00	0.2003E+01
28	40.00	465.17	53.73	464.14	0.1666E+01	0.2662E+00	0.1908E+01
29	42.50	469.34	54.38	468.83	0.1666E+01	0.2523E+00	0.1842E+01
30	45.00	473.50	55.00	473.37	0.1666E+01	0.2400E+00	0.1795E+01
31	47.50	477.67	55.58	477.81	0.1666E+01	0.2289E+00	0.1763E+01
32	50.00	481.83	56.14	482.19	0.1666E+01	0.2190E+00	0.1740E+01
33	52.50	486.00	56.68	486.52	0.1666E+01	0.2099E+00	0.1725E+01

VI - 3

Connection type : Flush end-plate connections
Mode : All bolted without column stiffener

Tested by : J.R.Ostrander (1970) Canada
Test Id. : TEST 4

Column : W8x28
Beam : W10x21
Plate thickness : 1/4"

Fasteners: A325- -3/4"D
13/16" Oversize holes
Material : A36
Fy = 46.50 ksi
Fu = 71.20 ksi

Major parameters

ct = 0.5000"	pt = 2.5000"	li = 5.0000"	pc = 2.5000"
cc = 0.5000"	gt = 3.5000"	gc = 3.5000"	
tp = 0.2500"	nt = 2 X 1	nc = 2 X 1	

Remark 1) Actual strength fy = 41.3 ksi, fu = 66.9 ksi for beam.
2) Actual strength fy = 51.1 ksi, fu = 77.3 ksi for column.

No	Moment (k-in)	Rotation (radians) X 1/1000	No	Moment (k-in)	Rotation (radians) X 1/1000
1	0.00	0.00	26	235.00	22.60
2	11.75	0.25	27	238.62	24.30
3	23.50	0.50	28	242.25	26.00
4	35.25	0.75	29	245.87	27.70
5	47.00	1.00	30	249.50	29.40
6	58.75	1.27	31	253.12	31.10
7	70.50	1.55	32	256.75	32.80
8	82.25	1.82	33	260.37	34.50
9	94.00	2.10	34	264.00	36.20
10	105.75	2.55			
11	117.50	3.00			
12	129.25	3.45			
13	141.00	3.90			
14	152.75	5.25			
15	164.50	6.60			
16	176.25	7.95			
17	188.00	9.30			
18	194.00	10.78			
19	200.00	12.25			
20	206.00	13.72			
21	212.00	15.20			
22	216.60	16.68			
23	221.20	18.16			
24	225.80	19.64			
25	230.40	21.12			

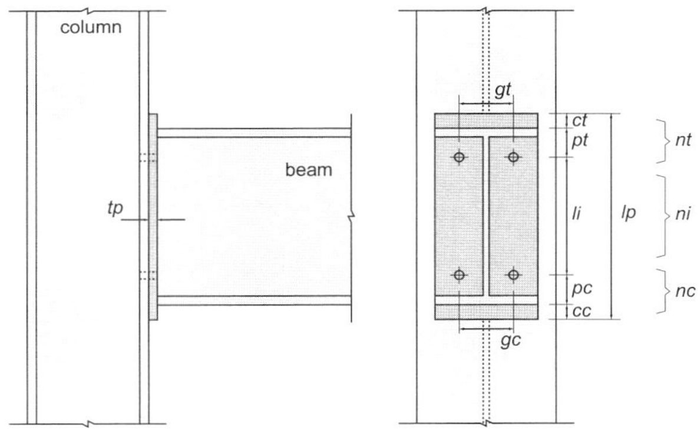

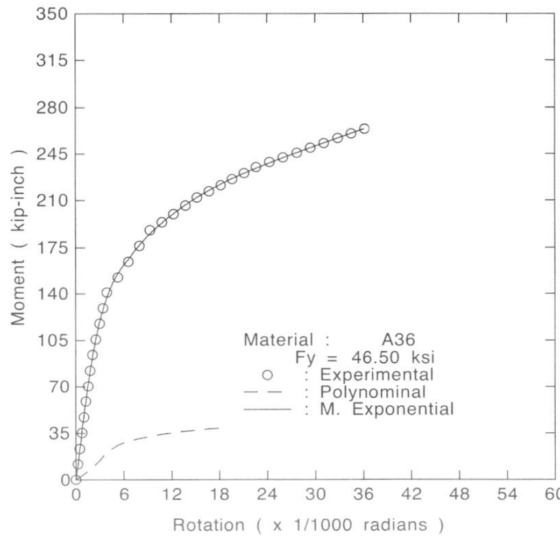

A.6 – 6

```
Moment-rotation prediction equations     ( R : X 1/1000 radians )

Frye and Morris polynominal model  :    R = Sum ( Ai X ( K*Bm )**Pi X 10**Qi )
     xd =  3.500000"   t =  0.250000"  tf =  0.750000"
     A1 =  1.830000   A2 = -1.040000   A3 =  6.380000    K =  0.132577
     P1 =        1    P2 =        3    P3 =        5    Q1 =        0     Q2 =       -1     Q3 =       -3

Modified exponential model  :
     Bm = Sum ( Ai X ( 1 - exp( - R/( 2*i*C ) ) ) )  +     Sum ( RKj X ( R - Rj0 ) )  +  Bm0
       C =   0.36416667E+00            Bmo=   0.00000000E+00          Nexp=  6        Nliner=  1
      AI =  -0.28791407E+02  -0.20816541E+03   0.11815713E+04  -0.15307941E+04   0.60733340E+03   0.18066502E+03
     Rj0 =           6.6000
     RKj =   0.21088583E+01
```

No	Rotation (radians) X 1/1000	Moment (kip-inch) Expri.	Poly.	M.Expo.	Connection stiffness (kip-inch) X 1000 Expri.	Poly.	M.Expo.
1	0.00	0.00	0.00	0.00	0.4700E+02	0.4122E+01	0.4100E+02
3	0.50	23.50	2.07	22.93	0.4700E+02	0.4175E+01	0.4855E+02
5	1.00	47.00	4.19	47.07	0.4496E+02	0.4343E+01	0.4706E+02
7	1.55	70.50	6.67	71.48	0.4272E+02	0.4698E+01	0.4132E+02
9	2.10	94.00	9.41	92.34	0.3642E+02	0.5305E+01	0.3452E+02
11	3.00	117.50	14.86	118.72	0.2611E+02	0.6860E+01	0.2448E+02
13	3.90	141.00	20.90	137.21	0.2176E+02	0.5735E+01	0.1706E+02
15	6.60	164.50	29.20	166.25	0.8704E+01	0.1745E+01	0.6657E+01
17	9.30	188.00	32.78	185.25	0.6489E+01	0.1032E+01	0.5842E+01
19	12.25	200.00	35.32	200.32	0.4068E+01	0.7307E+00	0.4499E+01
21	15.20	212.00	37.22	212.21	0.3588E+01	0.5737E+00	0.3614E+01
23	18.16	221.20	38.76	221.93	0.3107E+01	0.4761E+00	0.2997E+01
25	21.12	230.40	40.07	230.17	0.3107E+01	0.4093E+00	0.2604E+01
27	24.30	238.62	41.28	238.03	0.2133E+01	0.3573E+00	0.2361E+01
29	27.70	245.87	42.42	245.80	0.2133E+01	0.3158E+00	0.2227E+01
31	31.10	253.12	43.44	253.26	0.2133E+01	0.2838E+00	0.2163E+01
33	34.50	260.37	44.36	260.55	0.2133E+01	0.2583E+00	0.2133E+01

A.6 – 7

VI - 4

Connection type : Flush end-plate connections
Mode : All bolted without column stiffener

Tested by : J.R.Ostrander (1970) Canada
Test Id. : TEST 9

Column : W8x28
Beam : W10x21
Plate thickness : 3/4"

Fasteners: A325- -3/4"D
13/16" Oversize holes
Material : A36
Fy = 32.00 ksi
Fu = 58.50 ksi

Major parameters

ct = 0.5000" pt = 2.5000" li = 5.0000" pc = 2.5000"
cc = 0.5000" gt = 3.5000" gc = 3.5000"
tp = 0.7500" nt = 2 X 1 nc = 2 X 1

Remark 1) Actual strength fy = 42.8 ksi, fu = 48.2 ksi for beam.
2) Actual strength fy = 45.9 ksi, fu = 74.9 ksi for column.

No	Moment (k-in)	Rotation (radians) X 1/1000	No	Moment (k-in)	Rotation (radians) X 1/1000
1	0.00	0.00	26	532.33	26.23
2	23.50	0.15	27	538.50	28.80
3	47.00	0.30	28	544.67	31.37
4	70.50	0.40	29	550.84	33.93
5	94.00	0.50	30	557.00	36.50
6	117.50	0.70	31	562.38	39.33
7	141.00	0.90	32	567.75	42.15
8	164.50	1.10	33	573.13	44.98
9	188.00	1.30	34	578.50	47.80
10	211.50	1.55	35	583.88	50.62
11	235.00	1.80	36	589.25	53.45
12	258.50	2.10	37	594.63	56.27
13	282.00	2.40	38	600.00	59.10
14	305.50	2.95			
15	329.00	3.50			
16	352.50	4.35			
17	376.00	5.20			
18	396.50	6.40			
19	417.00	7.60			
20	441.50	10.00			
21	466.00	12.40			
22	484.00	15.30			
23	502.00	18.20			
24	520.00	21.10			
25	526.16	23.67			

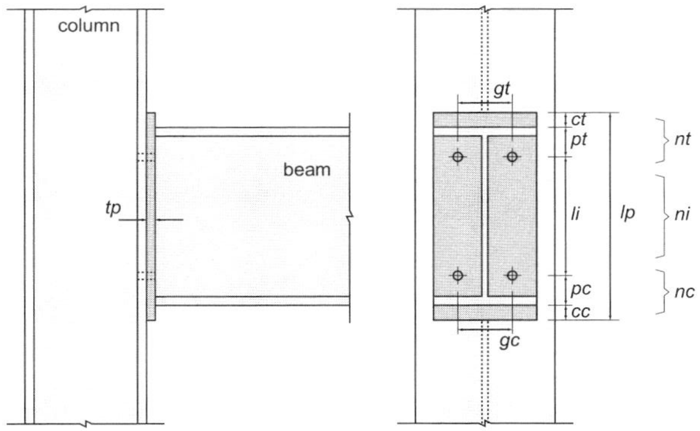

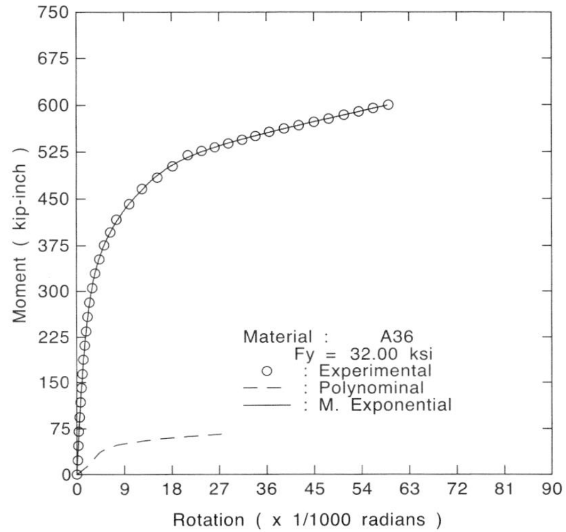

A.6 – 8

```
Moment-rotation prediction equations     ( R : X 1/1000 radians )

Frye and Morris polynominal model  :    R = Sum ( Ai X ( K*Bm )**Pi X 10**Qi )
     xd =  3.500000"   t =  0.750000"  tf =  0.750000"
     A1 =  1.830000   A2 = -1.040000   A3 =  6.380000    K =  0.085432
     P1 =         1   P2 =         3   P3 =         5   Q1 =         0   Q2 =        -1   Q3 =        -3

Modified exponential model  :
     Bm = Sum ( Ai X ( 1 - exp( - R/( 2*i*C ) ) ) )  +     Sum ( RKj X ( R - Rj0 ) )  +  Bm0
       C =   0.53000000E+00             Bmo=   0.00000000E+00          Nexp=  6        Nliner=  1
      AI =  -0.10740994E+03   0.18555252E+04  -0.49551065E+04   0.71443319E+04  -0.54274899E+04   0.20049151E+04
     Rj0 =          12.4000
     RKj =   0.18211741E+01
```

No	Rotation (radians) X 1/1000	Moment (kip-inch) Expri.	Poly.	M.Expo.	Connection stiffness (kip-inch) X 1000 Expri.	Poly.	M.Expo.
1	0.00	0.00	0.00	0.00	0.1567E+03	0.6396E+01	0.1919E+03
3	0.30	47.00	1.92	53.99	0.2036E+03	0.6426E+01	0.1684E+03
5	0.50	94.00	3.21	86.19	0.1958E+03	0.6479E+01	0.1537E+03
7	0.90	141.00	5.84	142.26	0.1175E+03	0.6672E+01	0.1273E+03
9	1.30	188.00	8.57	188.52	0.1071E+03	0.7001E+01	0.1047E+03
11	1.80	235.00	12.22	234.88	0.8687E+02	0.7661E+01	0.8167E+02
13	2.40	282.00	17.17	277.27	0.6577E+02	0.8962E+01	0.6075E+02
15	3.50	329.00	28.50	329.43	0.3680E+02	0.1063E+02	0.3665E+02
17	5.20	376.00	40.55	375.51	0.2326E+02	0.4366E+01	0.2040E+02
19	7.60	417.00	47.73	414.34	0.1479E+02	0.2142E+01	0.1338E+02
21	12.40	466.00	54.98	463.88	0.8396E+01	0.1117E+01	0.7610E+01
23	18.20	502.00	60.19	503.96	0.6207E+01	0.7371E+00	0.4916E+01
25	23.67	526.16	63.71	525.14	0.2403E+01	0.5687E+00	0.3120E+01
27	28.80	538.50	66.37	539.24	0.2403E+01	0.4727E+00	0.2465E+01
29	33.93	550.84	68.61	551.04	0.2403E+01	0.4068E+00	0.2170E+01
31	39.33	562.38	70.66	562.27	0.1902E+01	0.3565E+00	0.2009E+01
33	44.98	573.13	72.56	573.34	0.1902E+01	0.3168E+00	0.1918E+01
35	50.62	583.88	74.26	584.02	0.1902E+01	0.2858E+00	0.1869E+01
37	56.27	594.63	75.80	594.50	0.1902E+01	0.2610E+00	0.1844E+01

A.6 – 9

VI - 5

Connection type : Flush end-plate connections
Mode : All bolted without column stiffener

Tested by : J.R.Ostrander (1970) Canada
Test Id. : TEST 11

Column : W8x40
Beam : W12x27
Plate thickness : 3/8"

Fasteners: A325- -3/4"D
13/16" Oversize holes
Material : A36
Fy = 62.90 ksi
Fu = 109.10 ksi

Major parameters

ct = 0.5000"	pt = 2.5000"	li = 7.0000"	pc = 2.5000"
cc = 0.5000"	gt = 4.0000"	gc = 4.0000"	
tp = 0.3750"	nt = 2 X 1	nc = 2 X 1	

Remark 1) Actual strength fy = 58.0 ksi, fu = 78.8 ksi for beam.
2) Actual strength fy = 42.0 ksi, fu = 71.6 ksi for column.

No	Moment (k-in)	Rotation (radians) X 1/1000
1	0.00	0.00
2	23.50	0.20
3	47.00	0.40
4	70.50	0.60
5	94.00	0.80
6	117.50	1.05
7	141.00	1.30
8	164.50	1.65
9	188.00	2.00
10	211.50	2.35
11	235.00	2.70
12	258.50	3.10
13	282.00	3.50
14	305.50	4.00
15	329.00	4.50
16	352.50	5.30
17	376.00	6.10
18	397.00	7.35
19	418.00	8.60
20	443.00	10.70
21	468.00	12.80
22	482.67	15.33
23	497.34	17.87
24	512.00	20.40
25	524.67	22.90

No	Moment (k-in)	Rotation (radians) X 1/1000
26	537.33	25.40
27	550.00	27.90
28	558.43	30.29
29	566.85	32.67
30	575.28	35.06
31	583.71	37.44
32	592.14	39.83
33	600.57	42.21
34	609.00	44.60
35	616.50	46.80
36	624.00	49.00
37	631.50	51.20
38	639.00	53.40

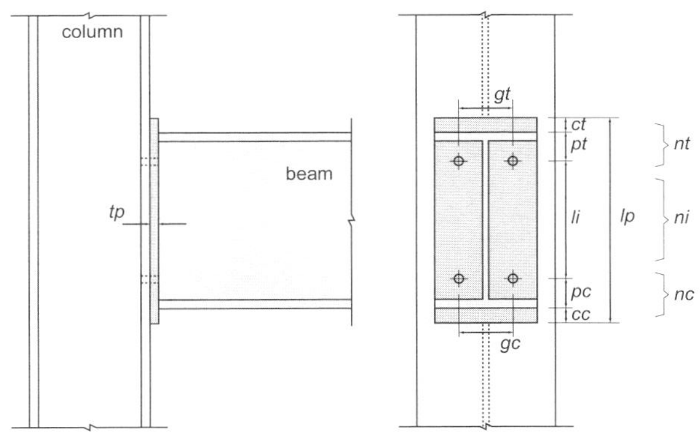

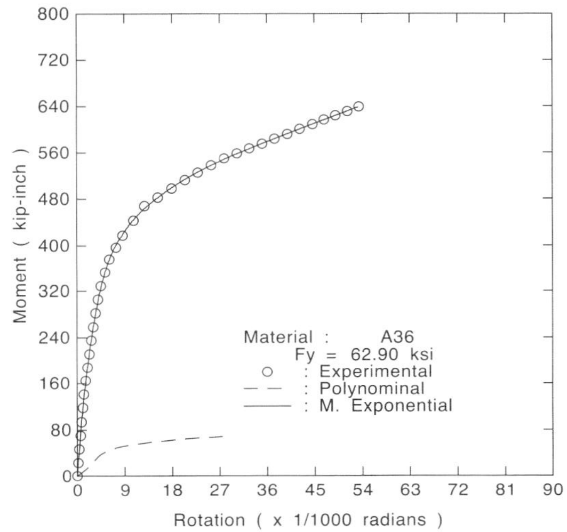

```
Moment-rotation prediction equations     ( R : X 1/1000 radians )

Frye and Morris polynominal model  :    R = Sum ( Ai X ( K*Bm )**Pi X 10**Qi )
     xd =  4.000000"   t =  0.375000"  tf =  0.750000"
     A1 =  1.830000   A2 = -1.040000   A3 =  6.380000     K =  0.081819
     P1 =        1    P2 =        3    P3 =        5     Q1 =        0     Q2 =       -1     Q3 =       -3

Modified exponential model  :
     Bm = Sum ( Ai X ( 1 - exp( - R/( 2*i*C ) ) ) )  +      Sum ( RKj X ( R - Rj0 ) )  +  Bm0
      C =   0.49500000E+00              Bmo=   0.00000000E+00            Nexp=  6         Nliner=  1
     AI =   0.16149455E+03  -0.15299202E+04   0.50575888E+04  -0.63643608E+04   0.35437970E+04  -0.38221181E+03
     Rj0 =           8.6000
     RKj =   0.34078830E+01
```

No	Rotation (radians) X 1/1000	Moment (kip-inch) Expri.	Moment (kip-inch) Poly.	Moment (kip-inch) M.Expo.	Connection stiffness (kip-inch) X 1000 Expri.	Connection stiffness (kip-inch) X 1000 Poly.	Connection stiffness (kip-inch) X 1000 M.Expo.
1	0.00	0.00	0.00	0.00	0.1175E+03	0.6679E+01	0.1377E+03
3	0.40	47.00	2.68	49.85	0.1175E+03	0.6734E+01	0.1133E+03
5	0.80	94.00	5.40	91.76	0.1070E+03	0.6904E+01	0.9730E+02
7	1.30	141.00	8.95	136.83	0.8282E+02	0.7310E+01	0.8383E+02
9	2.00	188.00	14.40	190.69	0.6714E+02	0.8380E+01	0.7079E+02
11	2.70	235.00	20.87	236.55	0.6322E+02	0.1026E+02	0.6050E+02
13	3.50	282.00	29.73	280.73	0.5353E+02	0.1110E+02	0.5015E+02
15	4.50	329.00	38.56	325.05	0.4022E+02	0.6482E+01	0.3877E+02
17	6.10	376.00	45.81	375.03	0.2446E+02	0.3265E+01	0.2456E+02
19	8.60	418.00	51.87	418.36	0.1497E+02	0.1863E+01	0.1177E+02
21	12.80	468.00	57.87	463.65	0.9134E+01	0.1125E+01	0.8131E+01
23	17.87	497.34	62.59	498.64	0.5790E+01	0.7844E+00	0.6025E+01
25	22.90	524.67	66.06	525.86	0.5066E+01	0.6129E+00	0.4861E+01
27	27.90	550.00	68.85	548.14	0.4281E+01	0.5083E+00	0.4114E+01
29	32.67	566.85	71.10	566.74	0.3533E+01	0.4396E+00	0.3727E+01
31	37.44	583.71	73.07	584.02	0.3533E+01	0.3889E+00	0.3541E+01
33	42.21	600.57	74.83	600.70	0.3533E+01	0.3497E+00	0.3460E+01
35	46.80	616.50	76.36	616.48	0.3409E+01	0.3195E+00	0.3428E+01
37	51.20	631.50	77.71	631.53	0.3409E+01	0.2956E+00	0.3415E+01

A.6 – 11

VI - 6

Connection type : Flush end-plate connections
Mode : All bolted without column stiffener

Tested by : J.R.Ostrander (1970) Canada
Test Id. : TEST 12

Column : W8x40
Beam : W12x27
Plate thickness : 1/2"

Fasteners: A325- -3/4"D
13/16" Oversize holes
Material : A36
Fy = 42.40 ksi
Fu = 71.10 ksi

Major parameters

ct = 0.5000" pt = 2.5000" li = 7.0000" pc = 2.5000"
cc = 0.5000" gt = 4.0000" gc = 4.0000"
tp = 0.5000" nt = 2 X 1 nc = 2 X 1

Remark 1) Actual strength fy = 58.0 ksi, fu = 78.8 ksi for beam.
2) Actual strength fy = 42.0 ksi, fu = 71.6 ksi for column.

No	Moment (k-in)	Rotation (radians) X 1/1000	No	Moment (k-in)	Rotation (radians) X 1/1000
1	0.00	0.00	26	685.33	20.10
2	31.33	0.17	27	700.00	22.30
3	62.66	0.33	28	715.00	24.90
4	94.00	0.50	29	730.00	27.50
5	125.33	0.70	30	745.00	30.10
6	156.67	0.90	31	750.83	32.52
7	188.00	1.10	32	756.66	34.93
8	219.34	1.30	33	762.50	37.35
9	250.67	1.50	34	768.33	39.77
10	282.00	1.70	35	774.17	42.18
11	313.33	1.97	36	780.00	44.60
12	344.66	2.23	37	789.00	47.13
13	375.00	2.50	38	798.00	49.67
14	407.33	3.00	39	807.00	52.20
15	433.67	3.50			
16	470.00	4.00			
17	493.50	4.80			
18	517.00	5.60			
19	541.00	6.85			
20	565.00	8.10			
21	588.00	9.60			
22	611.00	11.10			
23	633.50	13.40			
24	656.00	15.70			
25	670.67	17.90			

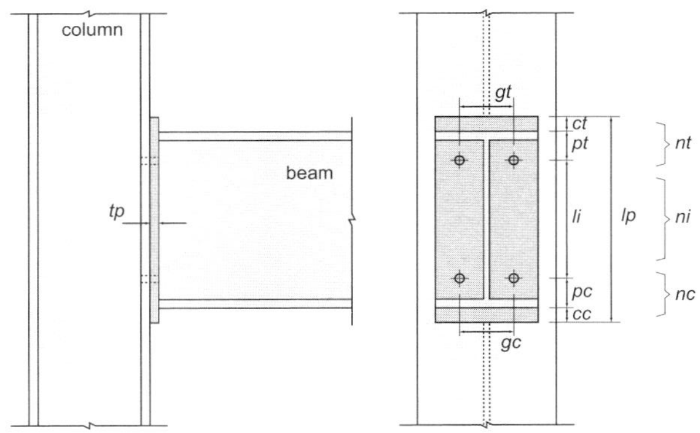

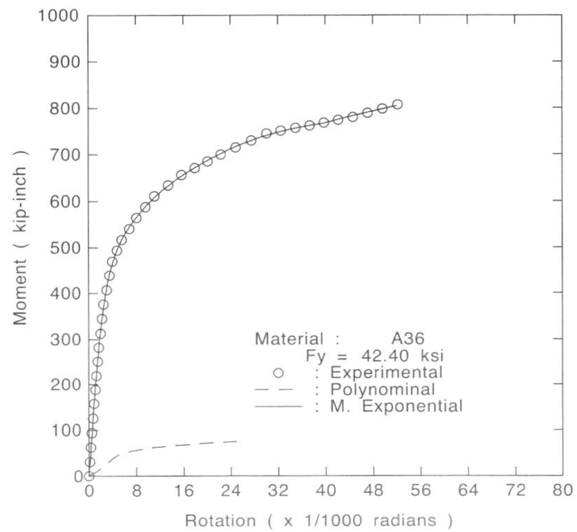

```
Moment-rotation prediction equations      ( R : X 1/1000 radians )

Frye and Morris polynominal model  :    R = Sum ( Ai X ( K*Bm )**Pi X 10**Qi )
     xd =  4.000000"   t =  0.500000"  tf =  0.750000"
     A1 =  1.830000   A2 = -1.040000   A3 =  6.380000    K =  0.072925
     P1 =         1   P2 =         3   P3 =         5   Q1 =         0    Q2 =        -1    Q3 =        -3

Modified exponential model  :
     Bm = Sum ( Ai X ( 1 - exp( - R/( 2*i*C ) ) ) )  +     Sum ( RKj X ( R - Rj0 ) )  +  Bm0
      C =   0.47667500E+00              Bmo=   0.00000000E+00             Nexp=  6        Nliner=  2
     AI =  -0.56965542E+03   0.50262466E+04  -0.18334234E+05   0.38645076E+05  -0.40087878E+05   0.16059878E+05
    Rj0 =          39.7700          0.0000
    RKj =   0.16933040E+01   0.87051240E+00
```

No	Rotation (radians) X 1/1000	Moment (kip-inch) Expri.	Moment (kip-inch) Poly.	Moment (kip-inch) M.Expo.	Connection stiffness (kip-inch) X 1000 Expri.	Connection stiffness (kip-inch) X 1000 Poly.	Connection stiffness (kip-inch) X 1000 M.Expo.
1	0.00	0.00	0.00	0.00	0.1880E+03	0.7493E+01	0.1599E+03
3	0.33	62.66	2.50	57.56	0.1880E+03	0.7536E+01	0.1803E+03
5	0.70	125.33	5.29	123.96	0.1567E+03	0.7685E+01	0.1789E+03
7	1.10	188.00	8.42	192.75	0.1567E+03	0.7987E+01	0.1634E+03
9	1.50	250.67	11.70	253.95	0.1567E+03	0.8466E+01	0.1421E+03
11	1.97	313.33	15.84	314.20	0.1175E+03	0.9325E+01	0.1163E+03
13	2.50	376.00	21.19	369.06	0.9841E+02	0.1082E+02	0.9020E+02
15	3.50	438.67	33.38	440.23	0.6267E+02	0.1245E+02	0.5518E+02
17	4.80	493.50	45.26	494.71	0.2937E+02	0.6184E+01	0.3187E+02
19	6.85	541.00	53.86	544.27	0.1920E+02	0.2975E+01	0.1920E+02
21	9.60	588.00	60.14	588.67	0.1534E+02	0.1797E+01	0.1381E+02
23	13.40	633.50	65.66	632.81	0.9783E+01	0.1198E+01	0.9829E+01
25	17.90	670.67	70.25	671.40	0.6665E+01	0.8784E+00	0.7618E+01
27	22.30	700.00	73.70	701.87	0.6255E+01	0.7056E+00	0.6249E+01
29	27.50	730.00	77.01	730.08	0.5769E+01	0.5780E+00	0.4596E+01
31	32.52	750.83	79.69	749.43	0.2414E+01	0.4954E+00	0.3176E+01
33	37.35	762.50	81.94	762.24	0.2414E+01	0.4373E+00	0.2194E+01
35	42.18	774.17	83.94	775.32	0.2414E+01	0.3926E+00	0.3274E+01
37	47.13	789.00	85.79	790.57	0.3553E+01	0.3563E+00	0.2921E+01
39	52.20	807.00	87.52	804.83	0.3553E+01	0.3261E+00	0.2733E+01

A.6 – 13

VI - 7

Connection type : Flush end-plate connections
Mode : All bolted without column stiffener

Tested by : J.R.Ostrander (1970) Canada
Test Id. : TEST 13

Column : W8x40
Beam : W12x27
Plate thickness : 5/8"

Fasteners: A325- -3/4"D
13/16" Oversize holes
Material : A36
Fy = 40.90 ksi
Fu = 73.10 ksi

Major parameters

ct = 0.5000" pt = 2.5000" li = 7.0000" pc = 2.5000"
cc = 0.5000" gt = 4.0000" gc = 4.0000"
tp = 0.6250" nt = 2 X 1 nc = 2 X 1

Remark 1) Actual strength fy = 58.0 ksi, fu = 78.8 ksi for beam.
2) Actual strength fy = 42.0 ksi, fu = 71.6 ksi for column.

No	Moment (k-in)	Rotation (radians) X 1/1000	No	Moment (k-in)	Rotation (radians) X 1/1000
1	0.00	0.00	26	749.00	17.80
2	31.33	0.27	27	762.67	19.13
3	62.66	0.53	28	776.33	20.47
4	94.00	0.80	29	790.00	21.80
5	125.33	1.13	30	801.50	23.60
6	156.67	1.47	31	813.00	25.40
7	188.00	1.80	32	824.50	27.20
8	219.34	2.13	33	836.00	29.00
9	250.67	2.47	34	846.00	30.88
10	282.00	2.80	35	856.00	32.76
11	313.33	3.23	36	866.00	34.64
12	344.66	3.67	37	876.00	36.52
13	376.00	4.10	38	886.00	38.40
14	407.33	4.63			
15	438.67	5.17			
16	470.00	5.70			
17	501.34	6.47			
18	532.67	7.23			
19	564.00	8.00			
20	587.50	8.80			
21	611.00	9.60			
22	655.00	11.40			
23	680.00	12.70			
24	705.00	14.00			
25	727.00	15.90			

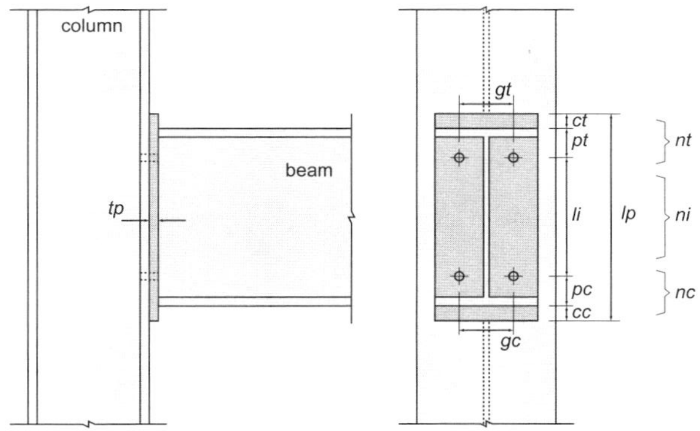

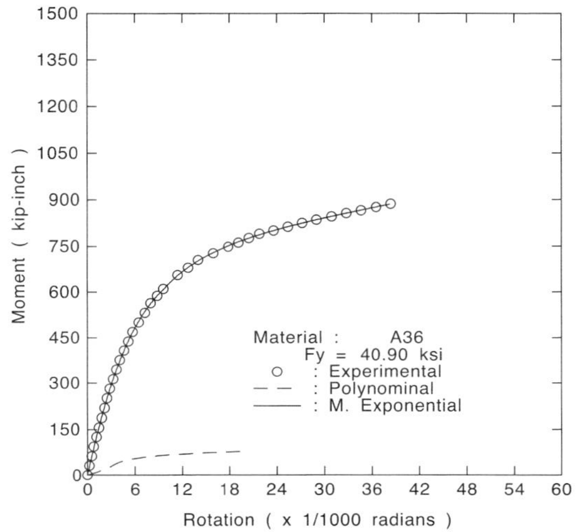

A.6 – 14

```
Moment-rotation prediction equations     ( R : X 1/1000 radians )

Frye and Morris polynominal model  :    R = Sum ( Ai X ( K*Bm )**Pi X 10**Qi )
     xd =  4.000000"    t =  0.625000"  tf =  0.750000"
     A1 =  1.830000   A2 = -1.040000   A3 =  6.380000    K =  0.066698
     P1 =        1    P2 =        3    P3 =        5    Q1 =        0    Q2 =       -1    Q3 =       -3

Modified exponential model  :
     Bm = Sum ( Ai X ( 1 - exp( - R/( 2*i*C ) ) ) )  +     Sum ( RKj X ( R - Rj0 ) )  +  Bm0
      C =   0.38667500E+00            Bmo=   0.00000000E+00         Nexp=  6        Nliner=  1
     AI =  -0.66543769E+00   0.39647840E+03  -0.30104260E+04   0.76360307E+04  -0.88822514E+04   0.46045542E+04
     Rj0 =           9.6000
     RKj =   0.49582692E+01
```

No	Rotation (radians) X 1/1000	Moment (kip-inch) Expri.	Moment (kip-inch) Poly.	Moment (kip-inch) M.Expo.	Connection stiffness (kip-inch) X 1000 Expri.	Connection stiffness (kip-inch) X 1000 Poly.	Connection stiffness (kip-inch) X 1000 M.Expo.
1	0.00	0.00	0.00	0.00	0.1175E+03	0.8193E+01	0.1217E+03
3	0.53	62.66	4.39	61.92	0.1175E+03	0.8313E+01	0.1111E+03
5	1.13	125.33	9.50	125.72	0.9400E+02	0.8768E+01	0.1019E+03
7	1.80	188.00	15.65	190.75	0.9401E+02	0.9813E+01	0.9330E+02
9	2.47	250.67	22.77	250.28	0.9399E+02	0.1171E+02	0.8533E+02
11	3.23	313.33	32.79	312.28	0.7230E+02	0.1405E+02	0.7649E+02
13	4.10	376.00	43.70	374.38	0.6624E+02	0.1012E+02	0.6687E+02
15	5.17	438.67	51.77	439.72	0.5875E+02	0.5670E+01	0.5582E+02
17	6.47	501.34	57.58	504.42	0.4087E+02	0.3596E+01	0.4405E+02
19	8.00	564.00	62.18	563.03	0.3524E+02	0.2540E+01	0.3288E+02
21	9.60	611.00	65.75	608.29	0.2786E+02	0.1966E+01	0.2415E+02
23	12.70	680.00	70.84	680.30	0.1923E+02	0.1393E+01	0.1848E+02
25	15.90	727.00	74.76	729.21	0.1158E+02	0.1087E+01	0.1269E+02
27	19.13	762.67	77.95	764.54	0.1025E+02	0.8976E+00	0.9467E+01
29	21.80	790.00	80.19	787.49	0.8606E+01	0.7887E+00	0.7861E+01
31	25.40	813.00	82.82	813.19	0.6390E+01	0.6810E+00	0.6544E+01
33	29.00	836.00	85.13	835.30	0.5866E+01	0.6016E+00	0.5808E+01
35	32.76	856.00	87.26	856.27	0.5318E+01	0.5380E+00	0.5391E+01
37	36.52	876.00	89.19	876.09	0.5318E+01	0.4878E+00	0.5174E+01

VI - 8

```
Connection type : Flush end-plate connections
           Mode : All bolted without column stiffener

Tested by  :  J.R.Ostrander (1970)                         Canada
Test   Id.  :  TEST 17

Column : W8x24                        Fasteners: A325- -3/4"D
Beam   : W12x27                                 13/16" Oversize holes
Plate thickness : 3/8"                Material : A36
                                           Fy =  62.90 ksi
                                           Fu = 109.10 ksi

                          Major parameters

     ct =  0.5000"    pt =  2.5000"    li =  7.0000"    pc =  2.5000"
     cc =  0.5000"    gt =  4.0000"    gc =  4.0000"
     tp =  0.3750"    nt = 2 X  1      nc = 2 X  1

Remark  1)  Actual strength  fy = 58.0 ksi, fu = 78.8 ksi for beam.
        2)
```

No	Moment (k-in)	Rotation (radians) X 1/1000	No	Moment (k-in)	Rotation (radians) X 1/1000
1	0.00	0.00	26	513.75	34.08
2	23.50	0.20	27	522.50	37.15
3	47.00	0.40	28	531.25	40.23
4	70.50	0.60	29	540.00	43.30
5	94.00	0.80	30	545.80	46.90
6	117.50	1.10	31	551.60	50.50
7	141.00	1.40	32	557.40	54.10
8	164.50	1.70	33	563.20	57.70
9	188.00	2.00	34	569.00	61.30
10	211.50	2.48	35	571.25	64.62
11	235.00	2.95	36	573.50	67.95
12	258.50	3.42	37	575.75	71.28
13	282.00	3.90	38	578.00	74.60
14	305.50	5.03			
15	329.00	6.15			
16	352.50	7.28			
17	376.00	8.40			
18	399.50	10.60			
19	423.00	12.80			
20	444.00	16.20			
21	465.00	19.60			
22	475.00	22.45			
23	485.00	25.30			
24	495.00	28.15			
25	505.00	31.00			

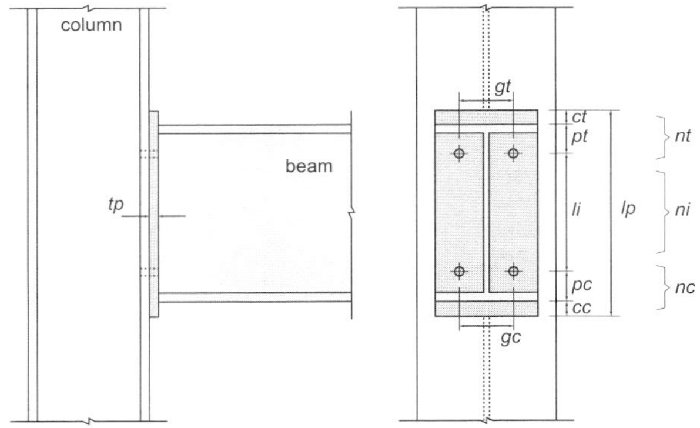

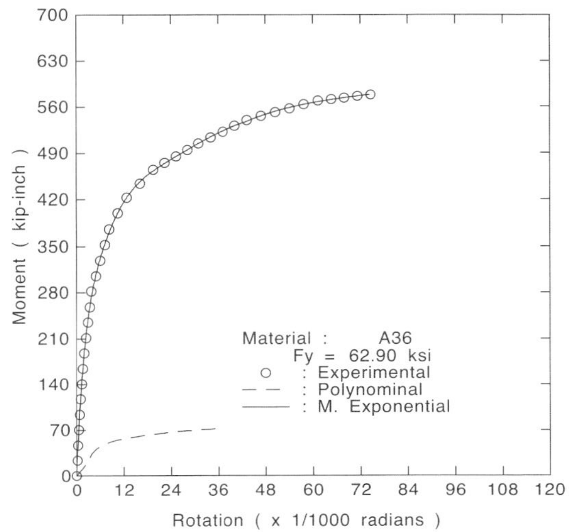

```
Moment-rotation prediction equations     ( R : X 1/1000 radians )

Frye and Morris polynominal model  :    R = Sum ( Ai X ( K*Bm )**Pi X 10**Qi )
     xd =  4.000000"   t =  0.375000"  tf =  0.750000"
     A1 =  1.830000   A2 = -1.040000   A3 =  6.380000     K =  0.081819
     P1 =        1    P2 =        3    P3 =        5     Q1 =        0     Q2 =       -1     Q3 =       -3

Modified exponential model  :
     Bm = Sum ( Ai X ( 1 - exp( - R/( 2*i*C ) ) ) )  +      Sum ( RKj X ( R - Rj0 ) )  +  Bm0
       C =   0.67166667E+00               Bmo=   0.00000000E+00           Nexp=  6         Nliner=  1
      AI =  -0.17179684E+03   0.36254161E+04  -0.17824441E+05   0.40979856E+05  -0.42967384E+05   0.16919100E+05
     Rj0 =           43.3000
     RKj =   0.59864861E+00
```

No	Rotation (radians) X 1/1000	Moment (kip-inch) Expri.	Poly.	M.Expo.	Connection stiffness (kip-inch) X 1000 Expri.	Poly.	M.Expo.
1	0.00	0.00	0.00	0.00	0.1175E+03	0.6679E+01	0.1271E+03
3	0.40	47.00	2.68	47.90	0.1175E+03	0.6734E+01	0.1124E+03
5	0.80	94.00	5.40	90.02	0.1018E+03	0.6904E+01	0.9834E+02
7	1.40	141.00	9.68	143.21	0.7834E+02	0.7422E+01	0.7950E+02
9	2.00	188.00	14.40	186.07	0.6717E+02	0.8380E+01	0.6391E+02
11	2.95	235.00	23.53	237.54	0.4947E+02	0.1099E+02	0.4566E+02
13	3.90	282.00	33.87	274.84	0.4099E+02	0.9293E+01	0.3376E+02
15	6.15	329.00	45.96	332.86	0.2089E+02	0.3218E+01	0.2020E+02
17	8.40	376.00	51.49	371.60	0.1744E+02	0.1927E+01	0.1483E+02
19	12.80	423.00	57.87	422.46	0.8912E+01	0.1125E+01	0.8723E+01
21	19.60	465.00	63.88	463.21	0.4725E+01	0.7141E+00	0.4251E+01
23	25.30	485.00	67.46	484.97	0.3508E+01	0.5572E+00	0.3594E+01
25	31.00	505.00	70.35	504.79	0.3190E+01	0.4612E+00	0.3326E+01
27	37.15	522.50	72.95	523.50	0.2846E+01	0.3916E+00	0.2706E+01
29	43.30	540.00	75.20	537.73	0.2277E+01	0.3420E+00	0.1922E+01
31	50.50	551.60	77.50	552.90	0.1611E+01	0.2991E+00	0.1732E+01
33	57.70	563.20	79.53	563.31	0.1611E+01	0.2667E+00	0.1206E+01
35	64.62	571.25	81.29	570.55	0.6767E+00	0.2422E+00	0.9114E+00
37	71.28	575.75	82.85	576.04	0.6767E+00	0.2228E+00	0.7573E+00

A.6 – 17

VI - 9

Connection type : Flush end-plate connections
Mode : All bolted without column stiffener

Tested by : J.R.Ostrander (1970) Canada
Test Id. : TEST 18

Column : W8x24
Beam : W12x27
Plate thickness : 1/2"

Fasteners: A325- -3/4"D
13/16" Oversize holes
Material : A36
Fy = 42.40 ksi
Fu = 71.10 ksi

Major parameters

ct = 0.5000" pt = 2.5000" li = 7.0000" pc = 2.5000"
cc = 0.5000" gt = 4.0000" gc = 4.0000"
tp = 0.5000" nt = 2 X 1 nc = 2 X 1

Remark 1) Actual strength fy = 58.0 ksi, fu = 78.8 ksi for beam.
2)

No	Moment (k-in)	Rotation (radians) X 1/1000	No	Moment (k-in)	Rotation (radians) X 1/1000
1	0.00	0.00	26	651.00	47.60
2	31.33	0.43	27	655.80	51.52
3	62.66	0.87	28	660.60	55.44
4	94.00	1.30	29	665.40	59.36
5	125.33	1.80	30	670.20	63.28
6	156.67	2.30	31	675.00	67.20
7	188.00	2.80	32	671.75	70.20
8	219.34	3.43	33	668.50	73.20
9	250.67	4.07	34	665.25	76.20
10	282.00	4.70	35	662.00	79.20
11	313.33	5.80			
12	344.66	6.90			
13	376.00	8.00			
14	405.66	9.77			
15	435.33	11.53			
16	465.00	13.30			
17	490.50	15.70			
18	516.00	18.10			
19	540.00	21.95			
20	564.00	25.80			
21	578.00	29.27			
22	592.00	32.73			
23	606.00	36.20			
24	621.00	40.00			
25	636.00	43.80			

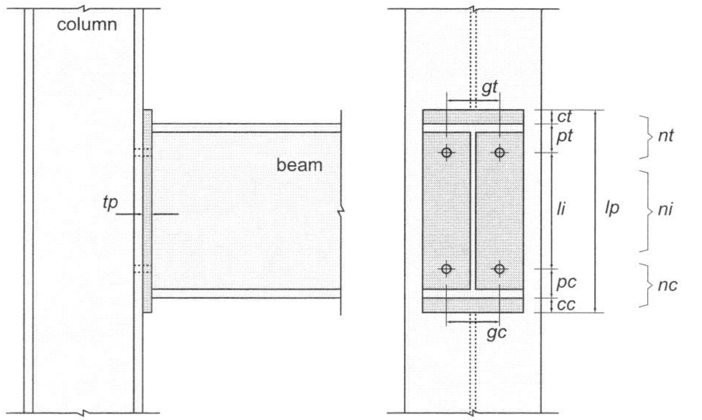

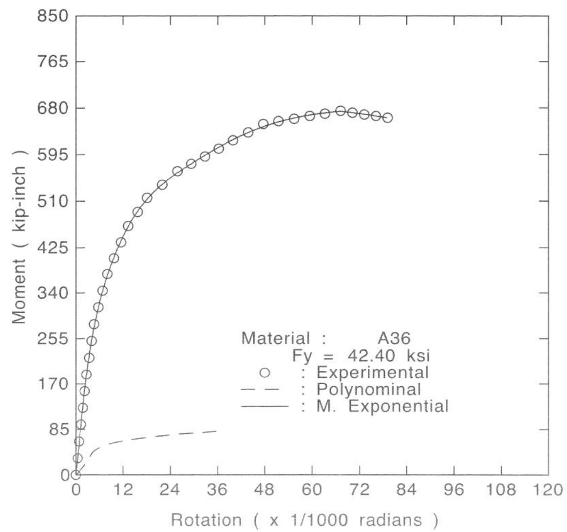

A.6 – 18

```
Moment-rotation prediction equations      ( R : X 1/1000 radians )

Frye and Morris polynominal model  :    R = Sum ( Ai X ( K*Bm )**Pi X 10**Qi )
     xd =  4.000000"   t =  0.500000"  tf =  0.750000"
     A1 =  1.830000   A2 = -1.040000   A3 =  6.380000     K =  0.072925
     P1 =         1   P2 =         3   P3 =         5    Q1 =         0    Q2 =        -1    Q3 =        -3

Modified exponential model  :
     Bm = Sum ( Ai X ( 1 - exp( - R/( 2*i*C ) ) ) )  +     Sum ( RKj X ( R - Rj0 ) )  + Bm0
      C =   0.76832500E+00            Bmo=   0.00000000E+00           Nexp=  6        Nliner=  1
     AI =  -0.39633538E+03   0.47978544E+04  -0.22886482E+05   0.52468668E+05  -0.55288247E+05   0.21985719E+05
    Rj0 =          67.2000
    RKj =  -0.14056746E+01
```

No	Rotation (radians) X 1/1000	Moment (kip-inch) Expri.	Moment (kip-inch) Poly.	Moment (kip-inch) M.Expo.	Connection stiffness (kip-inch) X 1000 Expri.	Connection stiffness (kip-inch) X 1000 Poly.	Connection stiffness (kip-inch) X 1000 M.Expo.
1	0.00	0.00	0.00	0.00	0.7231E+02	0.7493E+01	0.6349E+02
3	0.87	62.66	6.58	60.76	0.7230E+02	0.7792E+01	0.7297E+02
5	1.80	125.33	14.31	126.90	0.6267E+02	0.8975E+01	0.6726E+02
7	2.80	188.00	24.59	188.69	0.5685E+02	0.1186E+02	0.5611E+02
9	4.07	250.67	39.68	251.12	0.4947E+02	0.9432E+01	0.4298E+02
11	5.80	313.33	50.23	314.16	0.2848E+02	0.4048E+01	0.3085E+02
13	8.00	376.00	56.88	371.94	0.2400E+02	0.2322E+01	0.2260E+02
15	11.53	435.33	63.22	438.46	0.1679E+02	0.1427E+01	0.1568E+02
17	15.70	490.50	68.18	491.80	0.1062E+02	0.1007E+01	0.1026E+02
19	21.95	540.00	73.45	540.34	0.6234E+01	0.7164E+00	0.5996E+01
21	29.27	578.00	78.00	578.13	0.4038E+01	0.5455E+00	0.4663E+01
23	36.20	606.00	81.43	608.38	0.3995E+01	0.4497E+00	0.4033E+01
25	43.80	636.00	84.57	635.46	0.3947E+01	0.3798E+00	0.3051E+01
27	51.52	655.80	87.29	654.88	0.1224E+01	0.3298E+00	0.2001E+01
29	59.36	665.40	89.73	667.12	0.1224E+01	0.2921E+00	0.1171E+01
31	67.20	675.00	91.90	674.02	-0.8264E-01	0.2629E+00	0.6350E+00
33	73.20	668.50	93.42	668.58	-0.1083E+01	0.2446E+00	-0.1023E+01
35	79.20	662.00	94.84	661.93	-0.1083E+01	0.2290E+00	-0.1181E+01

A.6 – 19

VI - 10

Connection type : Flush end-plate connections
Mode : All bolted without column stiffener

Tested by : J.R.Ostrander (1970) Canada
Test Id. : TEST 19

Column : W8x24
Beam : W12x27
Plate thickness : 5/8"

Fasteners: A325- -3/4"D
13/16" Oversize holes
Material : A36
Fy = 40.90 ksi
Fu = 73.10 ksi

Major parameters

ct = 0.5000"	pt = 2.5000"	li = 7.0000"	pc = 2.5000"
cc = 0.5000"	gt = 4.0000"	gc = 4.0000"	
tp = 0.6250"	nt = 2 X 1	nc = 2 X 1	

Remark 1) Actual strength fy = 58.0 ksi, fu = 78.8 ksi for beam.
2)

No	Moment (k-in)	Rotation (radians) X 1/1000	No	Moment (k-in)	Rotation (radians) X 1/1000
1	0.00	0.00	26	562.00	21.20
2	23.50	0.20	27	568.00	23.14
3	47.00	0.40	28	574.00	25.08
4	70.50	0.65	29	580.00	27.02
5	94.00	0.90	30	586.00	28.96
6	117.50	1.15	31	592.00	30.90
7	141.00	1.40	32	597.83	33.10
8	164.50	1.70	33	603.66	35.30
9	188.00	2.00	34	609.50	37.50
10	211.50	2.35	35	615.33	39.70
11	235.00	2.70	36	621.17	41.90
12	258.50	3.15	37	627.00	44.10
13	282.00	3.60			
14	305.50	4.15			
15	329.00	4.70			
16	352.50	5.40			
17	376.00	6.10			
18	393.00	7.10			
19	420.00	8.10			
20	445.00	9.50			
21	470.00	10.90			
22	493.50	12.80			
23	517.00	14.70			
24	532.00	16.87			
25	547.00	19.03			

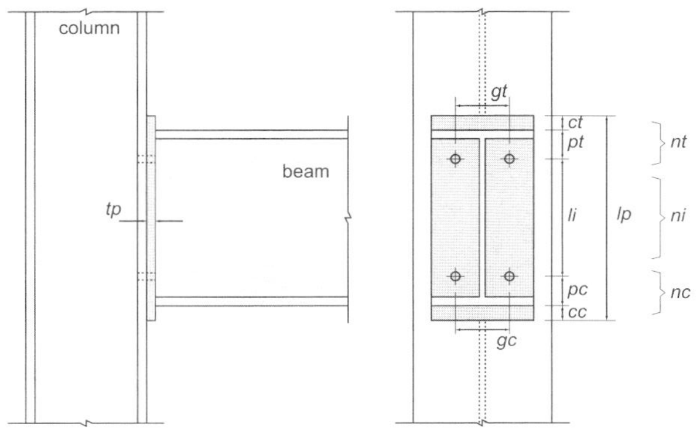

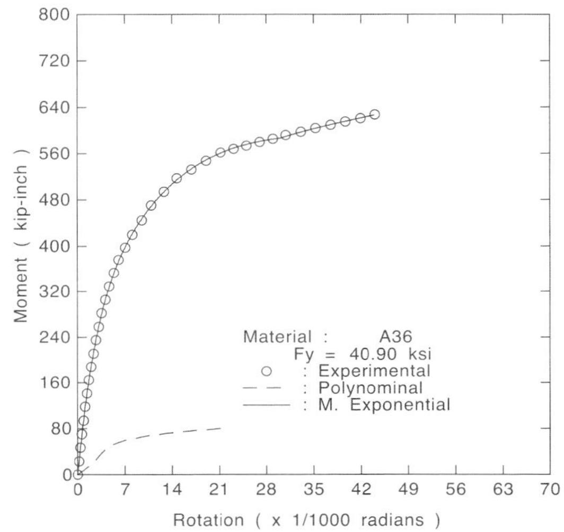

```
Moment-rotation prediction equations     ( R : X 1/1000 radians )

Frye and Morris polynominal model  :    R = Sum ( Ai X ( K*Bm )**Pi X 10**Qi )
     xd =  4.000000"   t =  0.625000"  tf =  0.750000"
     A1 =  1.830000   A2 = -1.040000   A3 =  6.380000    K =  0.066698
     P1 =        1    P2 =        3    P3 =        5    Q1 =        0    Q2 =       -1    Q3 =       -3

Modified exponential model  :
     Bm = Sum ( Ai X ( 1 - exp( - R/( 2*i*C ) ) ) )  +      Sum ( RKj X ( R - Rj0 ) )  +  Bm0
      C =   0.41750000E+00          Bmo=   0.00000000E+00         Nexp=  6        Nliner=  1
     AI =  -0.36083238E+02   0.41903858E+03  -0.30513975E+04   0.10892415E+05  -0.15359762E+05   0.77315636E+04
     Rj0 =          30.0900
     RKj =   0.22580243E+01
```

No	Rotation (radians) X 1/1000	Moment (kip-inch) Expri.	Moment (kip-inch) Poly.	Moment (kip-inch) M.Expo.	Connection stiffness (kip-inch) X 1000 Expri.	Connection stiffness (kip-inch) X 1000 Poly.	Connection stiffness (kip-inch) X 1000 M.Expo.
1	0.00	0.00	0.00	0.00	0.1175E+03	0.8193E+01	0.1150E+03
3	0.40	47.00	3.29	44.57	0.1071E+03	0.8260E+01	0.1074E+03
5	0.90	94.00	7.48	95.46	0.9400E+02	0.8546E+01	0.9599E+02
7	1.40	141.00	11.88	140.57	0.8689E+02	0.9105E+01	0.8455E+02
9	2.00	188.00	17.66	187.49	0.7317E+02	0.1028E+02	0.7214E+02
11	2.70	235.00	25.61	233.60	0.6061E+02	0.1258E+02	0.6002E+02
13	3.60	282.00	37.84	281.90	0.4795E+02	0.1319E+02	0.4791E+02
15	4.70	329.00	48.82	328.36	0.3870E+02	0.7113E+01	0.3722E+02
17	6.10	376.00	56.19	373.61	0.2880E+02	0.4007E+01	0.2808E+02
19	8.10	420.00	62.44	421.15	0.2028E+02	0.2493E+01	0.2022E+02
21	10.90	470.00	68.10	468.60	0.1553E+02	0.1671E+01	0.1431E+02
23	14.70	517.00	73.40	513.83	0.9823E+01	0.1182E+01	0.9856E+01
25	19.03	547.00	77.86	548.55	0.6923E+01	0.9023E+00	0.6342E+01
27	23.14	568.00	81.22	569.29	0.3093E+01	0.7443E+00	0.3892E+01
29	27.02	580.00	83.90	581.09	0.3093E+01	0.6425E+00	0.2303E+01
31	30.90	592.00	86.24	589.74	0.2886E+01	0.5674E+00	0.3554E+01
33	35.30	603.66	88.58	603.80	0.2652E+01	0.5029E+00	0.2902E+01
35	39.70	615.33	90.68	615.74	0.2652E+01	0.4529E+00	0.2566E+01
37	44.10	627.00	92.58	626.62	0.2652E+01	0.4128E+00	0.2401E+01

VI - 11

Connection type : Flush end-plate connections
Mode : All bolted without column stiffener

Tested by : J.R.Ostrander (1970) Canada
Test Id. : TEST 23

Column : W8x48	Fasteners: A325- -3/4"D
Beam : W12x27	13/16" Oversize holes
Plate thickness : 5/8"	Material : A36
	Fy = 40.90 ksi
	Fu = 73.10 ksi

Major parameters

ct = 0.5000"	pt = 2.5000"	li = 7.0000"	pc = 2.5000"
cc = 0.5000"	gt = 4.0000"	gc = 4.0000"	
tp = 0.6250"	nt = 2 X 1	nc = 2 X 1	

Remark 1) Actual strength fy = 58.0 ksi, fu = 78.8 ksi for beam.
2) Actual strength fy = 38.4 ksi, fu = 65.7 ksi for column.

No	Moment (k-in)	Rotation (radians) X 1/1000	No	Moment (k-in)	Rotation (radians) X 1/1000
1	0.00	0.00	26	690.50	8.15
2	31.33	0.17	27	721.00	9.30
3	62.66	0.33	28	739.50	10.28
4	94.00	0.50	29	758.00	11.25
5	125.33	0.67	30	776.50	12.22
6	156.67	0.83	31	795.00	13.20
7	188.00	1.00	32	806.25	14.18
8	219.34	1.27	33	817.50	15.15
9	250.67	1.53	34	828.75	16.12
10	282.00	1.80	35	840.00	17.10
11	313.33	2.07	36	848.83	18.22
12	344.66	2.33	37	857.67	19.33
13	376.00	2.60	38	866.50	20.45
14	399.50	2.80	39	875.34	21.57
15	423.00	3.00	40	884.17	22.68
16	446.50	3.25	41	893.00	23.80
17	470.00	3.50			
18	493.50	3.80			
19	517.00	4.10			
20	540.50	4.45			
21	564.00	4.80			
22	586.50	5.25			
23	609.00	5.70			
24	634.50	6.35			
25	660.00	7.00			

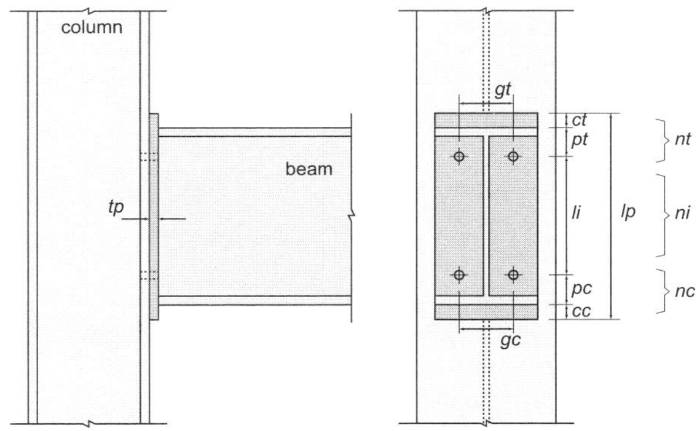

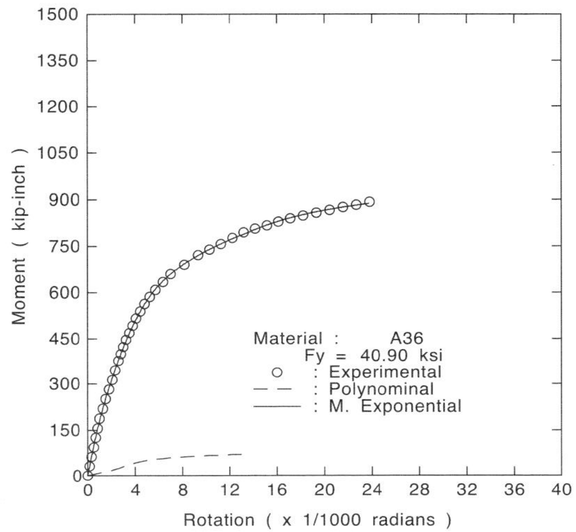

```
Moment-rotation prediction equations     ( R : X 1/1000 radians )

Frye and Morris polynominal model  :    R = Sum ( Ai X ( K*Bm )**Pi X 10**Qi )
     xd =  4.000000"   t =  0.625000"  tf =  0.750000"
     A1 =  1.830000   A2 = -1.040000   A3 =  6.380000    K =  0.066698
     P1 =        1    P2 =        3    P3 =        5    Q1 =        0    Q2 =       -1    Q3 =       -3

Modified exponential model  :
     Bm = Sum ( Ai X ( 1 - exp( - R/( 2*i*C ) ) ) )  +     Sum ( RKj X ( R - Rj0 ) )  +  Bm0
      C =   0.24000833E+00              Bmo=   0.00000000E+00          Nexp=  6         Nliner=  1
     AI =  -0.24545037E+03   0.47089901E+04  -0.26054167E+05   0.61136680E+05  -0.64934311E+05   0.26240233E+05
    Rj0 =           13.2000
    RKj =   0.37463731E+01
```

No	Rotation (radians) X 1/1000	Moment (kip-inch) Expri.	Moment (kip-inch) Poly.	Moment (kip-inch) M.Expo.	Connection stiffness (kip-inch) X 1000 Expri.	Connection stiffness (kip-inch) X 1000 Poly.	Connection stiffness (kip-inch) X 1000 M.Expo.
1	0.00	0.00	0.00	0.00	0.1880E+03	0.8193E+01	0.1979E+03
3	0.33	62.66	2.74	65.85	0.1880E+03	0.8239E+01	0.1916E+03
5	0.67	125.33	5.50	126.11	0.1880E+03	0.8383E+01	0.1694E+03
7	1.00	188.00	8.34	179.07	0.1609E+03	0.8634E+01	0.1493E+03
9	1.53	250.67	13.10	252.69	0.1175E+03	0.9311E+01	0.1289E+03
11	2.07	313.33	18.35	318.12	0.1175E+03	0.1045E+02	0.1172E+03
13	2.60	376.00	24.37	378.02	0.1175E+03	0.1220E+02	0.1074E+03
15	3.00	423.00	29.55	419.42	0.1071E+03	0.1364E+02	0.9949E+02
17	3.50	470.00	36.53	466.49	0.8688E+02	0.1360E+02	0.8867E+02
19	4.10	517.00	43.69	515.65	0.7316E+02	0.1012E+02	0.7521E+02
21	4.80	564.00	49.49	563.06	0.5965E+02	0.6761E+01	0.6058E+02
23	5.70	609.00	54.47	610.45	0.4559E+02	0.4584E+01	0.4552E+02
25	7.00	660.00	59.37	659.84	0.3464E+02	0.3136E+01	0.3191E+02
27	9.30	721.00	65.15	719.72	0.2244E+02	0.2051E+01	0.2197E+02
29	11.25	758.00	68.67	758.23	0.1898E+02	0.1608E+01	0.1767E+02
31	13.20	795.00	71.52	788.69	0.1526E+02	0.1333E+01	0.1357E+02
33	15.15	817.50	73.93	818.59	0.1154E+02	0.1145E+01	0.1343E+02
35	17.10	840.00	76.02	841.49	0.9848E+01	0.1007E+01	0.1021E+02
37	19.33	857.67	78.13	861.12	0.7911E+01	0.8883E+00	0.7554E+01
39	21.57	875.34	80.01	875.96	0.7911E+01	0.7970E+00	0.5875E+01
41	23.80	893.00	81.70	887.88	0.7911E+01	0.7245E+00	0.4890E+01

VI - 12

```
Connection type : Flush end-plate connections
           Mode : All bolted with column stiffener

Tested by  :  J.R.Ostrander (1970)                    Canada
Test  Id.  :  TEST 2

Column : W8x28                        Fasteners: A325- -3/4"D
Beam   : W10x21                                  13/16" Oversize holes
Plate thickness : 1/2"                Material  : A36
Stiffener thickness : 0.3750"              Fy =  45.40 ksi
                                           Fu =  70.30 ksi

                              Major parameters

     ct =  0.5000"    pt =  2.5000"    li =  5.0000"    pc =  2.5000"
     cc =  0.5000"    gt =  3.5000"    gc =  3.5000"
     tp =  0.5000"    nt = 2 X  1      nc = 2 X  1

Remark  1) 3 x 7 1/8 x 3/8 column web stiffeners used, fy = 50.5 ksi, fu = 72.5 ksi.
        2) fy = 41.3, fu = 66.9 ksi for beam, fy = 51.1, fu = 77.3 ksi for column.
```

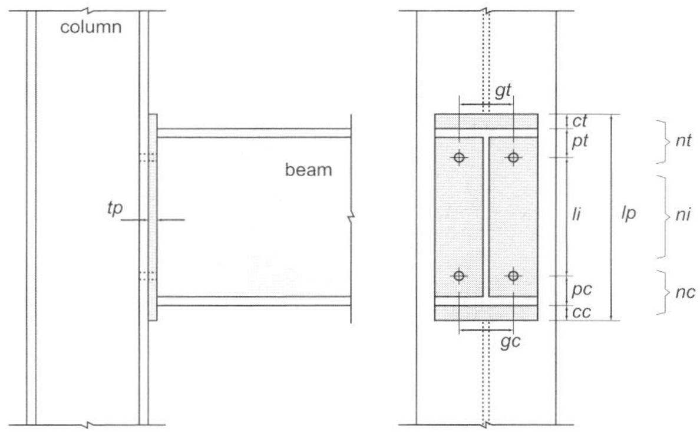

No	Moment (k-in)	Rotation (radians) X 1/1000	No	Moment (k-in)	Rotation (radians) X 1/1000
1	0.00	0.00	26	544.40	21.44
2	24.00	0.20	27	553.20	23.52
3	47.00	0.50	28	562.00	25.60
4	70.50	0.75	29	568.20	27.82
5	94.00	1.00	30	574.40	30.04
6	117.50	1.30	31	580.60	32.26
7	141.00	1.60	32	586.80	34.48
8	164.50	1.90	33	593.00	36.70
9	188.00	2.20	34	598.00	38.98
10	211.50	2.50	35	603.00	41.27
11	235.00	2.80	36	608.00	43.55
12	258.50	3.25	37	613.00	45.83
13	282.00	3.70	38	618.00	48.12
14	305.50	4.25	39	623.00	50.40
15	329.00	4.80			
16	352.50	5.50			
17	376.00	6.20			
18	399.50	7.25			
19	423.00	8.30			
20	444.50	9.70			
21	466.00	11.10			
22	492.00	13.15			
23	518.00	15.20			
24	526.80	17.28			
25	535.60	19.36			

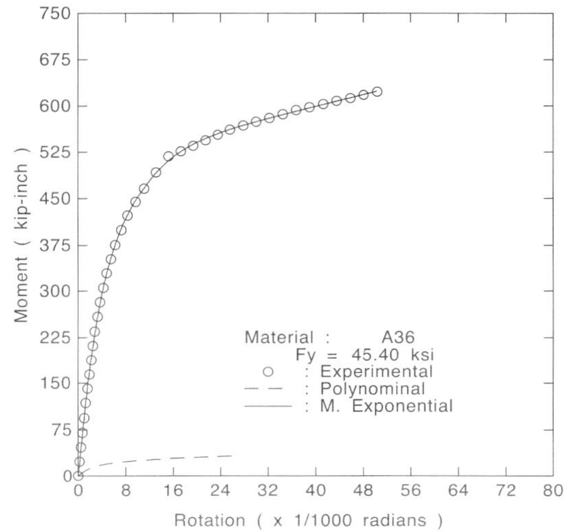

A.6 – 24

```
Moment-rotation prediction equations      ( R : X 1/1000 radians )

Frye and Morris polynominal model  :    R = Sum ( Ai X ( K*Bm )**Pi X 10**Qi )
     xd =  3.500000"    t =  0.500000"
     A1 =  1.790000   A2 =  1.760000   A3 =  2.040000    K =  0.074964
     P1 =        1    P2 =        3    P3 =        5    Q1 =        0    Q2 =       -1    Q3 =       -1

Modified exponential model  :
     Bm = Sum ( Ai X ( 1 - exp( - R/( 2*i*C ) ) ) )  +      Sum ( RKj X ( R - Rj0 ) )  +  Bm0
       C =    0.47000000E+00              Bmo=    0.00000000E+00          Nexp=  6        Nliner=  1
      AI =  -0.70992732E+02  -0.67417244E+02   0.18829482E+04  -0.48316907E+04   0.49852414E+04  -0.13692119E+04
     Rj0 =          11.2000
     RKj =    0.24321406E+01
```

No	Rotation (radians) X 1/1000	Moment (kip-inch) Expri.	Poly.	M.Expo.	Connection stiffness (kip-inch) X 1000 Expri.	Poly.	M.Expo.
1	0.00	0.00	0.00	0.00	0.1200E+03	0.7452E+01	0.8923E+02
3	0.50	47.00	3.70	46.23	0.8612E+02	0.7263E+01	0.9383E+02
5	1.00	94.00	7.18	92.50	0.8687E+02	0.6576E+01	0.9038E+02
7	1.60	141.00	10.73	144.27	0.7834E+02	0.5214E+01	0.8170E+02
9	2.20	188.00	13.46	190.29	0.7833E+02	0.3941E+01	0.7167E+02
11	2.80	235.00	15.54	230.36	0.6788E+02	0.3043E+01	0.6205E+02
13	3.70	282.00	17.87	280.44	0.4795E+02	0.2216E+01	0.4972E+02
15	4.80	329.00	19.96	328.51	0.3870E+02	0.1649E+01	0.3831E+02
17	6.20	376.00	21.96	374.67	0.2909E+02	0.1245E+01	0.2833E+02
19	8.30	423.00	24.19	423.59	0.1937E+02	0.9159E+00	0.1909E+02
21	11.10	466.00	26.40	466.06	0.1427E+02	0.6839E+00	0.1188E+02
23	15.20	518.00	28.80	510.91	0.8488E+01	0.5052E+00	0.8304E+01
25	19.36	535.60	30.67	537.99	0.4230E+01	0.4032E+00	0.5070E+01
27	23.52	553.20	32.20	555.38	0.4230E+01	0.3377E+00	0.3490E+01
29	27.82	568.20	33.55	568.63	0.2794E+01	0.2905E+00	0.2781E+01
31	32.26	580.60	34.75	580.27	0.2794E+01	0.2547E+00	0.2510E+01
33	36.70	593.00	35.82	591.19	0.2496E+01	0.2275E+00	0.2428E+01
35	41.27	603.00	36.81	602.23	0.2190E+01	0.2054E+00	0.2412E+01
37	45.83	613.00	37.70	613.25	0.2190E+01	0.1876E+00	0.2416E+01
39	50.40	623.00	38.53	624.29	0.2190E+01	0.1727E+00	0.2422E+01

VI - 13

Connection type : Flush end-plate connections
Mode : All bolted with column stiffener

Tested by : J.R.Ostrander (1970) Canada
Test Id. : TEST 5

Column : W8x28
Beam : W10x21
Plate thickness : 1/2"
Stiffener thickness : 0.2500"

Fasteners: A325- -3/4"D
13/16" Oversize holes
Material : A36
Fy = 45.40 ksi
Fu = 70.30 ksi

Major parameters

ct = 0.5000"	pt = 2.5000"	li = 5.0000"	pc = 2.5000"
cc = 0.5000"	gt = 3.5000"	gc = 3.5000"	
tp = 0.5000"	nt = 2 X 1	nc = 2 X 1	

Remark 1) 3 x 7 1/8 x 1/4 column web stiffeners used, fy = 49.6 ksi, fu = 71.9 ksi.
2) fy = 41.3, fu = 66.9 ksi for beam, fy = 51.1, fu = 77.3 ksi for column.

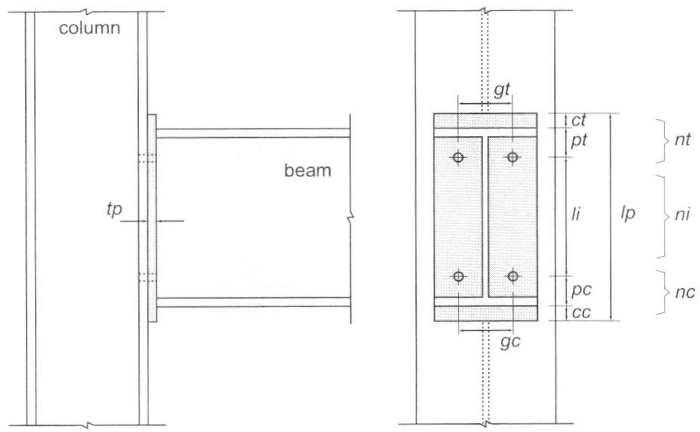

No	Moment (k-in)	Rotation (radians) X 1/1000	No	Moment (k-in)	Rotation (radians) X 1/1000
1	0.00	0.00	26	561.00	31.80
2	23.50	0.25	27	570.25	36.60
3	47.00	0.50	28	579.50	41.40
4	70.50	0.75	29	588.75	46.20
5	94.00	1.00	30	598.00	51.00
6	117.50	1.25	31	604.54	55.64
7	141.00	1.50	32	611.09	60.27
8	164.50	1.75	33	617.63	64.91
9	188.00	2.00	34	624.18	69.55
10	211.50	2.35	35	630.72	74.18
11	235.00	2.70	36	637.27	78.82
12	258.50	3.15	37	643.81	83.45
13	282.00	3.60	38	650.36	88.09
14	305.50	4.20	39	656.91	92.73
15	329.00	4.80	40	663.45	97.36
16	352.50	5.60	41	670.00	102.00
17	376.00	6.40			
18	399.50	7.70			
19	423.00	9.00			
20	446.50	10.95			
21	470.00	12.90			
22	493.50	16.10			
23	517.00	19.30			
24	531.66	23.47			
25	546.33	27.63			

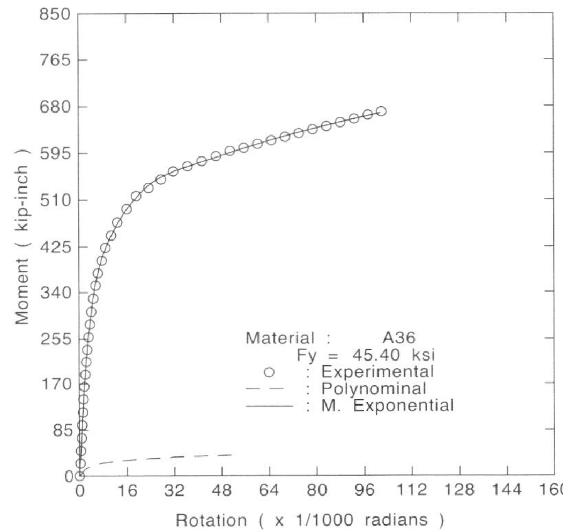

A.6 – 26

```
Moment-rotation prediction equations     ( R : X 1/1000 radians )

Frye and Morris polynominal model  :    R = Sum ( Ai X ( K*Bm )**Pi X 10**Qi )
     xd =  3.500000"   t =  0.500000"
     A1 =  1.790000   A2 =  1.760000   A3 =  2.040000    K =  0.074964
     P1 =         1   P2 =         3   P3 =         5   Q1 =         0    Q2 =      -1     Q3 =      -1

Modified exponential model  :
     Bm = Sum ( Ai X ( 1 - exp( - R/( 2*i*C ) ) ) )  +      Sum ( RKj X ( R - Rj0 ) )  +  Bm0
       C =   0.91250000E+00          Bmo=   0.00000000E+00          Nexp=  6        Nliner=  1
      AI =  -0.38855111E+03   0.47657155E+04  -0.17303404E+05   0.31420830E+05  -0.27195312E+05   0.92730532E+04
     Rj0 =          19.3000
     RKj =   0.11628567E+01
```

No	Rotation (radians) X 1/1000	Moment (kip-inch) Expri.	Moment (kip-inch) Poly.	Moment (kip-inch) M.Expo.	Connection stiffness (kip-inch) X 1000 Expri.	Connection stiffness (kip-inch) X 1000 Poly.	Connection stiffness (kip-inch) X 1000 M.Expo.
1	0.00	0.00	0.00	0.00	0.9401E+02	0.7452E+01	0.1031E+03
3	0.50	47.00	3.70	50.68	0.9400E+02	0.7263E+01	0.9898E+02
5	1.00	94.00	7.18	98.51	0.9400E+02	0.6576E+01	0.9204E+02
7	1.50	141.00	10.20	142.48	0.9401E+02	0.5452E+01	0.8368E+02
9	2.00	188.00	12.63	182.12	0.8281E+02	0.4325E+01	0.7487E+02
11	2.70	235.00	15.23	230.29	0.6061E+02	0.3169E+01	0.6287E+02
13	3.60	282.00	17.64	280.56	0.4663E+02	0.2286E+01	0.4926E+02
15	4.80	329.00	19.96	330.75	0.3497E+02	0.1649E+01	0.3519E+02
17	6.40	376.00	22.21	376.48	0.2507E+02	0.1203E+01	0.2309E+02
19	9.00	423.00	24.81	422.55	0.1567E+02	0.8432E+00	0.1389E+02
21	12.90	470.00	27.54	466.40	0.1027E+02	0.5908E+00	0.9410E+01
23	19.30	517.00	30.65	513.30	0.5682E+01	0.4043E+00	0.5327E+01
25	27.63	546.33	33.49	550.28	0.3520E+01	0.2922E+00	0.2883E+01
27	36.60	570.25	35.80	569.90	0.1927E+01	0.2280E+00	0.1823E+01
29	46.20	588.75	37.77	586.92	0.1927E+01	0.1863E+00	0.1766E+01
31	55.64	604.54	39.39	603.41	0.1412E+01	0.1589E+00	0.1708E+01
33	64.91	617.63	40.77	618.61	0.1412E+01	0.1393E+00	0.1562E+01
35	74.18	630.72	41.99	632.38	0.1412E+01	0.1245E+00	0.1414E+01
37	83.45	643.81	43.09	644.96	0.1412E+01	0.1127E+00	0.1306E+01
39	92.73	656.91	44.09	656.73	0.1412E+01	0.1032E+00	0.1239E+01
41	102.00	670.00	45.02	668.02	0.1412E+01	0.9522E-01	0.1201E+01

VI - 14

Connection type : Flush end-plate connections
Mode : All bolted with column stiffener

Tested by : J.R.Ostrander (1970) Canada
Test Id. : TEST 6

Column : W8x28
Beam : W10x21
Plate thickness : 3/8"
Stiffener thickness : 0.2500"

Fasteners: A325- -3/4"D
13/16" Oversize holes
Material : A36
Fy = 45.50 ksi
Fu = 68.50 ksi

Major parameters

ct = 0.5000"	pt = 2.5000"	li = 5.0000"	pc = 2.5000"
cc = 0.5000"	gt = 3.5000"	gc = 3.5000"	
tp = 0.3750"	nt = 2 X 1	nc = 2 X 1	

Remark 1) 3 x 7 1/8 x 1/4 column web stiffeners used, fy = 49.6 ksi, fu = 71.9 ksi.
2) fy = 41.3, fu = 66.9 ksi for bean, fy = 51.1, fu = 77.3 ksi for column.

No	Moment (k-in)	Rotation (radians) X 1/1000	No	Moment (k-in)	Rotation (radians) X 1/1000
1	0.00	0.00	26	492.00	40.50
2	23.50	0.15	27	497.60	43.72
3	47.00	0.30	28	503.20	46.94
4	70.50	0.50	29	508.80	50.16
5	94.00	0.70	30	514.40	53.38
6	117.50	0.90	31	520.00	56.60
7	141.00	1.10	32	523.40	59.96
8	164.50	1.55	33	526.80	63.32
9	188.00	2.00	34	530.20	66.68
10	211.50	2.55	35	533.60	70.04
11	235.00	3.10	36	537.00	73.40
12	258.50	3.90			
13	282.00	4.70			
14	305.50	6.05			
15	329.00	7.40			
16	352.50	9.65			
17	376.00	11.90			
18	399.50	15.35			
19	423.00	18.80			
20	434.67	22.07			
21	446.34	25.33			
22	458.00	28.60			
23	466.50	31.57			
24	475.00	34.55			
25	483.50	37.52			

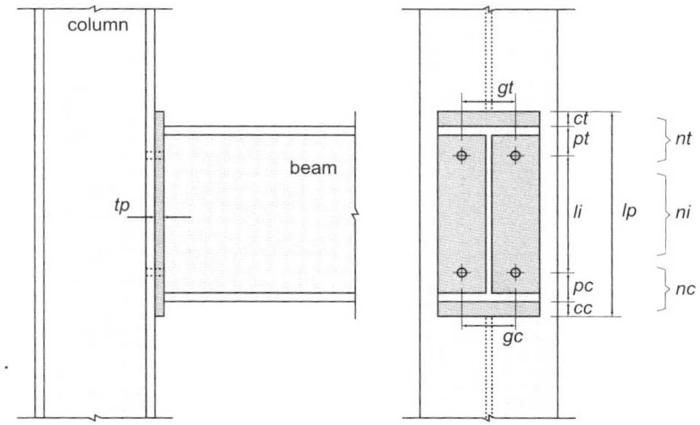

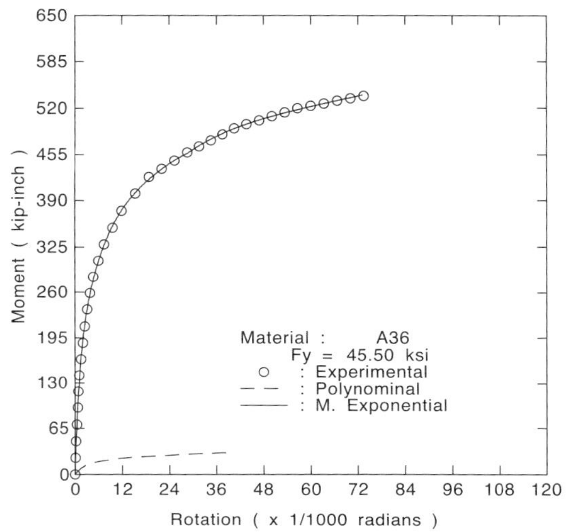

```
Moment-rotation prediction equations     ( R : X 1/1000 radians )

Frye and Morris polynominal model  :    R = Sum ( Ai X ( K*Bm )**Pi X 10**Qi )
     xd =  3.500000"    t =  0.375000"
     A1 =  1.790000   A2 =  1.760000   A3 =  2.040000    K =  0.089088
     P1 =        1    P2 =        3    P3 =        5    Q1 =        0    Q2 =      -1    Q3 =      -1

Modified exponential model  :
     Bm = Sum ( Ai X ( 1 - exp( - R/( 2*i*C ) ) ) )  +      Sum ( RKj X ( R - Rj0 ) )  +  Bm0
      C =   0.64916667E+00            Bmo=   0.00000000E+00           Nexp=  6        Nliner=  1
     AI =   0.30344010E+03  -0.94380150E+03   0.18320424E+04   0.66458173E+03  -0.46171460E+04   0.32479328E+04
     Rj0 =           28.6000
     RKj =   0.11522170E+01
```

No	Rotation (radians) X 1/1000	Moment (kip-inch) Expri.	Moment (kip-inch) Poly.	Moment (kip-inch) M.Expo.	Connection stiffness (kip-inch) X 1000 Expri.	Connection stiffness (kip-inch) X 1000 Poly.	Connection stiffness (kip-inch) X 1000 M.Expo.
1	0.00	0.00	0.00	0.00	0.1567E+03	0.6271E+01	0.1743E+03
3	0.30	47.00	1.88	46.92	0.1399E+03	0.6217E+01	0.1400E+03
5	0.70	94.00	4.32	95.77	0.1175E+03	0.5938E+01	0.1061E+03
7	1.10	141.00	6.59	133.12	0.9742E+02	0.5364E+01	0.8204E+02
9	2.00	188.00	10.63	190.86	0.4795E+02	0.3639E+01	0.5023E+02
11	3.10	235.00	13.80	235.20	0.3728E+02	0.2282E+01	0.3284E+02
13	4.70	282.00	16.66	278.26	0.2492E+02	0.1420E+01	0.2251E+02
15	7.40	329.00	19.62	327.31	0.1480E+02	0.8679E+00	0.1466E+02
17	11.90	376.00	22.66	376.51	0.9010E+01	0.5374E+00	0.8169E+01
19	18.80	423.00	25.62	419.57	0.5148E+01	0.3485E+00	0.4967E+01
21	25.33	446.34	27.60	446.85	0.3572E+01	0.2658E+00	0.3429E+01
23	31.57	466.50	29.10	467.65	0.2857E+01	0.2185E+00	0.3334E+01
25	37.52	483.50	30.30	484.71	0.2857E+01	0.1878E+00	0.2453E+01
27	43.72	497.60	31.39	497.93	0.1739E+01	0.1644E+00	0.1864E+01
29	50.16	508.80	32.38	508.69	0.1739E+01	0.1461E+00	0.1514E+01
31	56.60	520.00	33.28	517.78	0.1383E+01	0.1317E+00	0.1330E+01
33	63.32	526.80	34.12	526.36	0.1012E+01	0.1197E+00	0.1235E+01
35	70.04	533.60	34.89	534.49	0.1012E+01	0.1099E+00	0.1190E+01

VI - 15

Connection type : Flush end-plate connections
Mode : All bolted with column stiffener

Tested by : J.R.Ostrander (1970) Canada
Test Id. : TEST 7

Column : W8x28
Beam : W10x21
Plate thickness : 1/4"
Stiffener thickness : 0.2500"

Fasteners: A325- -3/4"D
13/16" Oversize holes
Material : A36
Fy = 46.50 ksi
Fu = 71.20 ksi

Major parameters

ct = 0.5000"	pt = 2.5000"	li = 5.0000"	pc = 2.5000"
cc = 0.5000"	gt = 3.5000"	gc = 3.5000"	
tp = 0.2500"	nt = 2 X 1	nc = 2 X 1	

Remark 1) 3 x 7 1/8 x 1/4 column web stiffeners used, fy = 49.6 ksi, fu = 71.9 ksi.
2) fy = 41.3, fu = 66.9 ksi for beam, fy = 51.1, fu = 77.3 ksi for column.

No	Moment (k-in)	Rotation (radians) X 1/1000	No	Moment (k-in)	Rotation (radians) X 1/1000
1	0.00	0.00	26	258.00	25.80
2	11.75	0.23	27	264.00	28.00
3	23.50	0.45	28	270.00	30.20
4	35.25	0.68	29	276.00	32.40
5	47.00	0.90	30	282.00	34.60
6	58.75	1.18	31	286.80	36.74
7	70.50	1.45	32	291.60	38.88
8	82.25	1.73	33	296.40	41.02
9	94.00	2.00	34	301.20	43.16
10	105.75	2.40	35	306.00	45.30
11	117.50	2.80	36	308.00	47.03
12	129.25	3.20	37	310.00	48.77
13	141.00	3.60	38	312.00	50.50
14	152.75	4.60			
15	164.50	5.60			
16	176.25	6.60			
17	188.00	7.60			
18	200.00	9.65			
19	212.00	11.70			
20	219.67	13.60			
21	227.33	15.50			
22	235.00	17.40			
23	240.75	19.50			
24	246.50	21.60			
25	252.25	23.70			

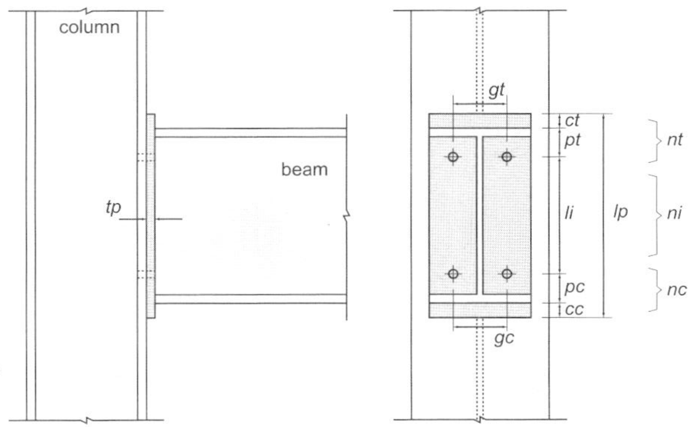

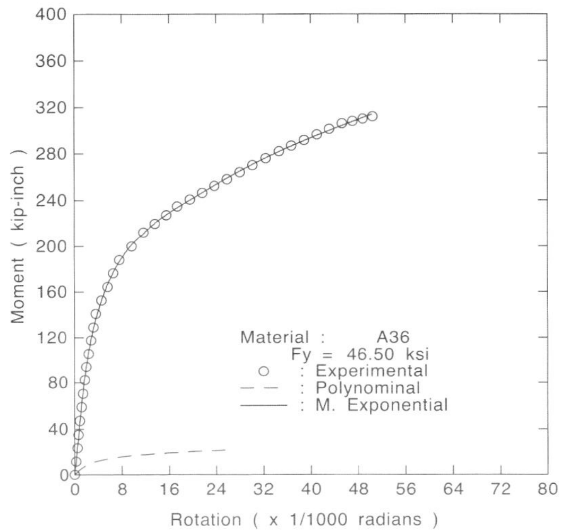

```
Moment-rotation prediction equations     ( R : X 1/1000 radians )

Frye and Morris polynominal model  :    R = Sum ( Ai X ( K*Bm )**Pi X 10**Qi )
     xd =  3.500000"   t =  0.250000"
     A1 =  1.790000   A2 =  1.760000   A3 =  2.040000    K =  0.113625
     P1 =        1    P2 =        3    P3 =        5    Q1 =        0    Q2 =       -1    Q3 =       -1

Modified exponential model  :
     Bm = Sum ( Ai X ( 1 - exp( - R/( 2*i*C ) ) ) )  +      Sum ( RKj X ( R - Rj0 ) )  +  Bm0
       C =   0.47708333E+00            Bmo=   0.00000000E+00           Nexp=  6        Nliner=  1
      AI =  -0.19287047E+03   0.20552146E+04  -0.83747242E+04   0.16847515E+05  -0.15917024E+05   0.58269461E+04
     Rj0 =          11.7000
     RKj =   0.17921142E+01
```

No	Rotation (radians) X 1/1000	Moment (kip-inch) Expri.	Moment (kip-inch) Poly.	Moment (kip-inch) M.Expo.	Connection stiffness (kip-inch) X 1000 Expri.	Connection stiffness (kip-inch) X 1000 Poly.	Connection stiffness (kip-inch) X 1000 M.Expo.
1	0.00	0.00	0.00	0.00	0.5223E+02	0.4917E+01	0.4485E+02
3	0.45	23.50	2.20	22.45	0.5223E+02	0.4817E+01	0.5265E+02
5	0.90	47.00	4.30	45.97	0.4795E+02	0.4459E+01	0.5086E+02
7	1.45	70.50	6.55	72.10	0.4272E+02	0.3676E+01	0.4373E+02
9	2.00	94.00	8.33	93.92	0.3728E+02	0.2854E+01	0.3569E+02
11	2.80	117.50	10.25	118.39	0.2938E+02	0.2008E+01	0.2600E+02
13	3.60	141.00	11.64	136.39	0.2434E+02	0.1508E+01	0.1950E+02
15	5.60	164.50	13.97	166.45	0.1175E+02	0.9170E+00	0.1193E+02
17	7.60	188.00	15.52	186.66	0.9817E+01	0.6619E+00	0.8517E+01
19	11.70	212.00	17.68	210.89	0.4909E+01	0.4284E+00	0.3720E+01
21	15.50	227.33	19.10	227.39	0.4034E+01	0.3272E+00	0.3539E+01
23	19.50	240.75	20.27	240.27	0.2739E+01	0.2643E+00	0.3043E+01
25	23.70	252.25	21.29	252.78	0.2739E+01	0.2213E+00	0.2919E+01
27	28.00	264.00	22.17	264.95	0.2727E+01	0.1905E+00	0.2725E+01
29	32.40	276.00	22.95	276.37	0.2727E+01	0.1674E+00	0.2463E+01
31	36.74	286.80	23.64	286.53	0.2243E+01	0.1500E+00	0.2228E+01
33	41.02	296.40	24.25	295.68	0.2243E+01	0.1362E+00	0.2057E+01
35	45.30	306.00	24.81	304.22	0.1641E+01	0.1250E+00	0.1944E+01
37	48.77	310.00	25.23	310.85	0.1154E+01	0.1173E+00	0.1886E+01

A.6 - 31

VI - 16

Connection type : Flush end-plate connections
Mode : All bolted with column stiffener

Tested by : J.R.Ostrander (1970) Canada
Test Id. : TEST 8

Column : W8x28
Beam : W10x21
Plate thickness : 1/4"
Stiffener thickness : 0.3750"

Fasteners: A325- -3/4"D
13/16" Oversize holes
Material : A36
Fy = 46.50 ksi
Fu = 71.20 ksi

Major parameters

ct = 0.5000" pt = 2.5000" li = 5.0000" pc = 2.5000"
cc = 0.5000" gt = 3.5000" gc = 3.5000"
tp = 0.2500" nt = 2 X 1 nc = 2 X 1

Remark 1) 3 x 7 1/8 x 3/8 column web stiffeners used, fy = 50.5 ksi, fu = 72.5 ksi.
2) fy = 41.3, fu = 66.9 ksi for beam, fy = 51.1, fu = 77.3 ksi for column.

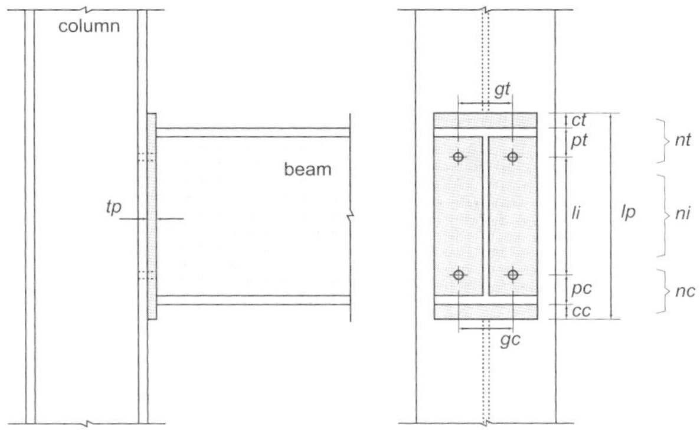

No	Moment (k-in)	Rotation (radians) X 1/1000	No	Moment (k-in)	Rotation (radians) X 1/1000
1	0.00	0.00	26	258.00	26.10
2	11.75	0.12	27	264.00	28.30
3	23.50	0.25	28	270.00	30.50
4	35.25	0.38	29	276.00	32.70
5	47.00	0.50	30	282.00	34.90
6	58.75	0.78	31	285.80	36.80
7	70.50	1.05	32	289.60	38.70
8	82.25	1.32	33	293.40	40.60
9	94.00	1.60	34	297.20	42.50
10	105.75	1.98	35	301.00	44.40
11	117.50	2.35			
12	129.25	2.73			
13	141.00	3.10			
14	152.75	4.10			
15	164.50	5.10			
16	176.25	6.10			
17	188.00	7.10			
18	200.00	9.15			
19	212.00	11.20			
20	219.67	13.27			
21	227.33	15.33			
22	235.00	17.40			
23	240.75	19.57			
24	246.50	21.75			
25	252.25	23.93			

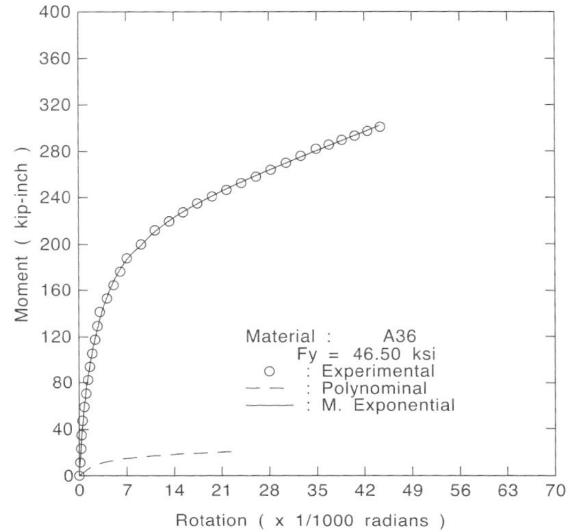

```
Moment-rotation prediction equations     ( R : X 1/1000 radians )

Frye and Morris polynominal model  :     R = Sum ( Ai X ( K*Bm )**Pi X 10**Qi )
     xd =  3.500000"   t =  0.250000"
     A1 =  1.790000   A2 =  1.760000   A3 =  2.040000    K =  0.113625
     P1 =         1   P2 =         3   P3 =         5   Q1 =         0   Q2 =        -1   Q3 =        -1

Modified exponential model  :
     Bm = Sum ( Ai X ( 1 - exp( - R/( 2*i*C ) ) ) )  +      Sum ( RKj X ( R - Rj0 ) )  +  Bm0
      C =   0.40125000E+00            Bmo=   0.00000000E+00            Nexp=  6        Nliner=  1
     AI =   0.59903542E+02   0.77518132E+02  -0.91889153E+03   0.26097716E+04  -0.28088008E+04   0.12027615E+04
    Rj0 =           9.1500
    RKj =   0.22701273E+01
```

No	Rotation (radians) X 1/1000	Moment (kip-inch) Expri.	Poly.	M.Expo.	Connection stiffness (kip-inch) X 1000 Expri.	Poly.	M.Expo.
1	0.00	0.00	0.00	0.00	0.9401E+02	0.4917E+01	0.1041E+03
3	0.25	23.50	1.23	23.32	0.9401E+02	0.4888E+01	0.8349E+02
5	0.50	47.00	2.44	42.18	0.7798E+02	0.4792E+01	0.6821E+02
7	1.05	70.50	4.95	73.18	0.4272E+02	0.4274E+01	0.4672E+02
9	1.60	94.00	7.08	95.31	0.3790E+02	0.3440E+01	0.3486E+02
11	2.35	117.50	9.26	117.77	0.3134E+02	0.2429E+01	0.2595E+02
13	3.10	141.00	10.82	135.12	0.2599E+02	0.1790E+01	0.2071E+02
15	5.10	164.50	13.49	167.50	0.1175E+02	0.1016E+01	0.1248E+02
17	7.10	188.00	15.18	187.09	0.9817E+01	0.7108E+00	0.7494E+01
19	11.20	212.00	17.46	210.75	0.4785E+01	0.4471E+00	0.4953E+01
21	15.33	227.33	19.04	227.60	0.3708E+01	0.3306E+00	0.3481E+01
23	19.57	240.75	20.29	241.07	0.2644E+01	0.2633E+00	0.2944E+01
25	23.93	252.25	21.33	253.20	0.2644E+01	0.2194E+00	0.2658E+01
27	28.30	264.00	22.22	264.41	0.2727E+01	0.1887E+00	0.2482E+01
29	32.70	276.00	23.00	275.09	0.2727E+01	0.1661E+00	0.2379E+01
31	36.80	285.80	23.65	284.72	0.2000E+01	0.1497E+00	0.2326E+01
33	40.60	293.40	24.19	293.50	0.2000E+01	0.1375E+00	0.2299E+01
35	44.40	301.00	24.70	302.21	0.2000E+01	0.1272E+00	0.2285E+01

A.6 – 33

VI - 17

Connection type : Flush end-plate connections
Mode : All bolted with column stiffener

Tested by : J.R.Ostrander (1970) Canada
Test Id. : TEST 10

Column : W8x23
Beam : W10x21
Plate thickness : 3/4"
Stiffener thickness : 0.3750"

Fasteners: A325- -3/4"D
13/16" Oversize holes
Material : A36
Fy = 32.00 ksi
Fu = 58.50 ksi

Major parameters

ct = 0.5000" pt = 2.5000" li = 5.0000" pc = 2.5000"
cc = 0.5000" gt = 3.5000" gc = 3.5000"
tp = 0.7500" nt = 2 X 1 nc = 2 X 1

Remark 1) 3 x 7 1/8 x 3/8 column web stiffeners used, fy = 40.5 ksi, fu = 60.3 ksi.
2) fy = 42.8, fu = 48.2 ksi for beam, fy = 45.9, fu = 74.9 ksi for column.

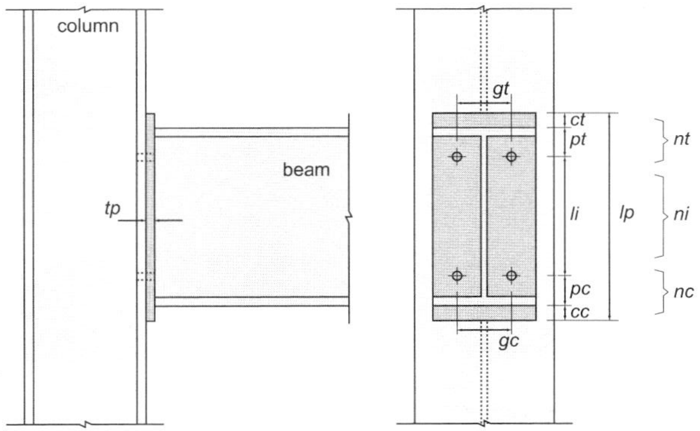

No	Moment (k-in)	Rotation (radians) X 1/1000	No	Moment (k-in)	Rotation (radians) X 1/1000
1	0.00	0.00	26	583.50	12.25
2	23.50	0.05	27	606.00	14.60
3	47.00	0.10	28	621.00	16.80
4	70.50	0.25	29	636.00	19.00
5	94.00	0.40	30	651.00	21.20
6	117.50	0.55	31	658.50	23.28
7	141.00	0.70	32	666.00	25.37
8	164.50	0.85	33	673.50	27.45
9	188.00	1.00	34	681.00	29.53
10	211.50	1.15	35	688.50	31.62
11	235.00	1.30	36	696.00	33.70
12	258.50	1.50	37	702.71	35.79
13	282.00	1.70	38	709.43	37.87
14	305.50	2.00	39	716.14	39.96
15	329.00	2.30	40	722.86	42.04
16	352.50	2.60	41	729.57	44.13
17	376.00	2.90	42	736.29	46.21
18	399.50	3.35	43	743.00	48.30
19	423.00	3.80			
20	446.50	4.50			
21	470.00	5.20			
22	493.50	6.20			
23	517.00	7.20			
24	539.00	8.55			
25	561.00	9.90			

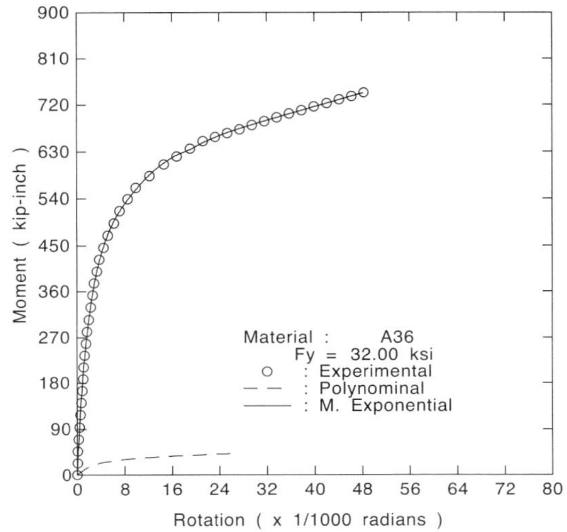

A.6 – 34

```
Moment-rotation prediction equations     ( R : X 1/1000 radians )

Frye and Morris polynominal model  :    R = Sum ( Ai X ( K*Bm )**Pi X 10**Qi )
     xd =  3.500000"   t =  0.750000"
     A1 =  1.790000   A2 =  1.760000   A3 =  2.040000    K =  0.058776
     P1 =        1    P2 =        3    P3 =        5    Q1 =        0    Q2 =       -1    Q3 =       -1

Modified exponential model  :
     Bm = Sum ( Ai X ( 1 - exp( - R/( 2*i*C ) ) ) )  +      Sum ( RKj X ( R - Rj0 ) )  +  Bm0
       C =   0.41500000E+00             Bmo=   0.00000000E+00         Nexp=  6        Nliner=  1
      AI =   0.23623043E+03  -0.24354922E+04   0.11996080E+05  -0.22550110E+05   0.18289595E+05  -0.48902596E+04
     Rj0 =          19.0000
     RKj =   0.33075336E+01
```

No	Rotation (radians) X 1/1000	Moment (kip-inch) Expri.	Moment (kip-inch) Poly.	Moment (kip-inch) M.Expo.	Connection stiffness (kip-inch) X 1000 Expri.	Connection stiffness (kip-inch) X 1000 Poly.	Connection stiffness (kip-inch) X 1000 M.Expo.
1	0.00	0.00	0.00	0.00	0.4700E+03	0.9505E+01	0.2681E+03
3	0.10	47.00	0.95	25.79	0.3917E+03	0.9496E+01	0.2481E+03
5	0.40	94.00	3.78	92.62	0.1567E+03	0.9355E+01	0.2003E+03
7	0.70	141.00	6.54	147.29	0.1567E+03	0.9000E+01	0.1660E+03
9	1.00	188.00	9.16	193.10	0.1567E+03	0.8387E+01	0.1406E+03
11	1.30	235.00	11.55	232.27	0.1399E+03	0.7561E+01	0.1213E+03
13	1.70	282.00	14.33	276.61	0.1018E+03	0.6351E+01	0.1014E+03
15	2.30	329.00	17.66	330.52	0.7833E+02	0.4803E+01	0.7947E+02
17	2.90	376.00	20.20	373.08	0.6788E+02	0.3732E+01	0.6313E+02
19	3.80	423.00	23.07	421.36	0.4493E+02	0.2741E+01	0.4527E+02
21	5.20	470.00	26.27	471.57	0.2942E+02	0.1924E+01	0.2829E+02
23	7.20	517.00	29.48	515.79	0.2043E+02	0.1354E+01	0.1780E+02
25	9.90	561.00	32.56	556.89	0.1384E+02	0.9770E+00	0.1341E+02
27	14.60	606.00	36.34	608.12	0.8151E+01	0.6693E+00	0.8231E+01
29	19.00	636.00	38.93	633.86	0.6818E+01	0.5232E+00	0.3747E+01
31	23.28	658.50	40.97	658.12	0.3600E+01	0.4346E+00	0.4585E+01
33	27.45	673.50	42.65	674.81	0.3600E+01	0.3749E+00	0.3587E+01
35	31.62	688.50	44.11	688.95	0.3600E+01	0.3307E+00	0.3273E+01
37	35.79	702.71	45.42	702.44	0.3219E+01	0.2966E+00	0.3218E+01
39	39.96	716.14	46.60	715.89	0.3219E+01	0.2694E+00	0.3236E+01
41	44.13	729.57	47.67	729.45	0.3219E+01	0.2472E+00	0.3263E+01
43	48.30	743.00	48.67	743.10	0.3219E+01	0.2287E+00	0.3283E+01

A.6 - 35

VI - 18

Connection type : Flush end-plate connections
Mode : All bolted with column stiffener

Tested by : J.R.Ostrander (1970) Canada
Test Id. : TEST 14

Column : W8x40
Beam : W12x27
Plate thickness : 3/8"
Stiffener thickness : 0.2500"

Fasteners: A325- -3/4"D
13/16" Oversize holes
Material : A36
Fy = 62.90 ksi
Fu = 109.10 ksi

Major parameters

ct = 0.5000"	pt = 2.5000"	li = 7.0000"	pc = 2.5000"
cc = 0.5000"	gt = 4.0000"	gc = 4.0000"	
tp = 0.3750"	nt = 2 X 1	nc = 2 X 1	

Remark 1) 4 x 7 x 1/4 column web stiffeners used, fy = 52.5 ksi, fu = 75.3 ksi.
2) fy = 58.0, fu = 78.8 ksi for beam, fy = 42.0, fu = 71.6 ksi for column.

No	Moment (k-in)	Rotation (radians) X 1/1000	No	Moment (k-in)	Rotation (radians) X 1/1000
1	0.00	0.00	26	619.25	29.98
2	23.50	0.10	27	630.50	32.25
3	47.00	0.20	28	641.75	34.52
4	78.34	0.87	29	653.00	36.80
5	109.67	1.53	30	664.75	39.27
6	141.00	2.20	31	676.50	41.75
7	172.33	2.80	32	688.25	44.23
8	203.66	3.40	33	700.00	46.70
9	235.00	4.00	34	704.66	48.80
10	266.33	4.67	35	709.33	50.90
11	297.67	5.33	36	714.00	53.00
12	329.00	6.00			
13	360.34	6.90			
14	391.67	7.80			
15	423.00	8.70			
16	446.50	9.90			
17	470.00	11.10			
18	493.50	12.85			
19	517.00	14.60			
20	538.00	16.85			
21	559.00	19.10			
22	571.00	21.50			
23	583.00	23.90			
24	595.50	25.80			
25	608.00	27.70			

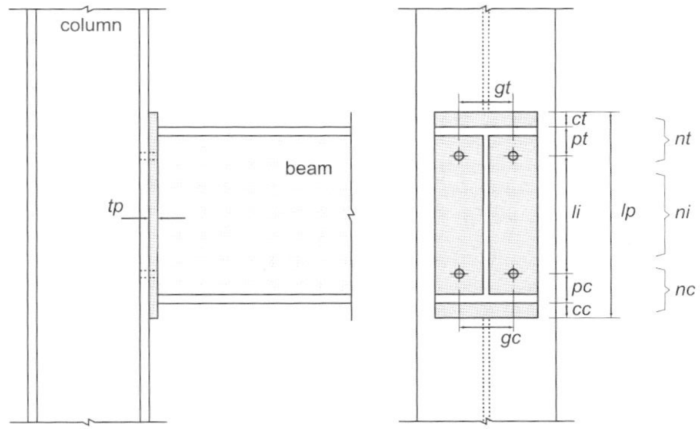

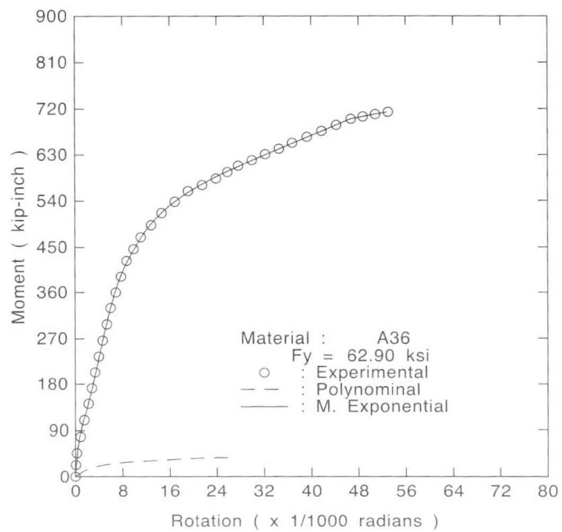

```
Moment-rotation prediction equations      ( R : X 1/1000 radians )

Frye and Morris polynominal model  :     R = Sum ( Ai X ( K*Bm )**Pi X 10**Qi )
     xd =  4.000000"    t =  0.375000"
     A1 =  1.790000    A2 =  1.760000    A3 =  2.040000     K =  0.064660
     P1 =         1    P2 =         3    P3 =         5    Q1 =         0    Q2 =        -1    Q3 =        -1

Modified exponential model  :
     Bm = Sum ( Ai X ( 1 - exp( - R/( 2*i*C ) ) ) )  +      Sum ( RKj X ( R - Rj0 ) )  +  Bm0
       C =   0.46666667E+00              Bmo=   0.00000000E+00          Nexp=  6         Nliner=  2
      AI =   0.58847722E+03  -0.30202634E+04   0.62077957E+04  -0.76748848E+04   0.59405316E+04  -0.15209577E+04
     Rj0 =           9.9000          46.7000
     RKj =   0.48812517E+01  -0.27399079E+01
```

No	Rotation (radians) X 1/1000	Moment (kip-inch) Expri.	Poly.	M.Expo.	Connection stiffness (kip-inch) X 1000 Expri.	Poly.	M.Expo.
1	0.00	0.00	0.00	0.00	0.2350E+03	0.8640E+01	0.1752E+03
3	0.20	47.00	1.73	30.11	0.2105E+03	0.8607E+01	0.1285E+03
5	1.53	109.67	12.03	114.63	0.4700E+02	0.6229E+01	0.3914E+02
7	2.80	172.33	18.02	169.23	0.5222E+02	0.3526E+01	0.4879E+02
9	4.00	235.00	21.45	231.59	0.4975E+02	0.2349E+01	0.5329E+02
11	5.33	297.67	24.11	299.77	0.4701E+02	0.1701E+01	0.4768E+02
13	6.90	360.34	26.42	365.51	0.3481E+02	0.1287E+01	0.3600E+02
15	8.70	423.00	28.46	418.86	0.2829E+02	0.1012E+01	0.2385E+02
17	11.10	470.00	30.61	468.06	0.1708E+02	0.7929E+00	0.1817E+02
19	14.60	517.00	33.03	516.88	0.1164E+02	0.6085E+00	0.1082E+02
21	19.10	559.00	35.44	556.06	0.7237E+01	0.4733E+00	0.7186E+01
23	23.90	583.00	37.48	586.42	0.5882E+01	0.3859E+00	0.5693E+01
25	27.70	608.00	38.85	606.98	0.5837E+01	0.3380E+00	0.5195E+01
27	32.25	630.50	40.29	629.97	0.4945E+01	0.2954E+00	0.4951E+01
29	36.80	653.00	41.56	652.29	0.4850E+01	0.2631E+00	0.4876E+01
31	41.75	676.50	42.79	676.38	0.4747E+01	0.2358E+00	0.4862E+01
33	46.70	700.00	43.90	700.45	0.3381E+01	0.2140E+00	0.4866E+01
35	50.90	709.33	44.77	709.39	0.2223E+01	0.1987E+00	0.2132E+01

VI - 19

Connection type : Flush end-plate connections
Mcde : All bolted with column stiffener

Tested by : J.R.Ostrander (1970) Canada
Test Id. : TEST 15

Column : W8x40
Beam : W12x27
Plate thickness : 1/2"
Stiffener thickness : 0.2500"

Fasteners: A325- -3/4"D
13/16" Oversize holes
Material : A36
Fy = 42.40 ksi
Fu = 71.10 ksi

Major parameters

ct = 0.5000"	pt = 2.5000"	li = 7.0000"	pc = 2.5000"
cc = 0.5000"	gt = 4.0000"	gc = 4.0000"	
tp = 0.5000"	nt = 2 X 1	nc = 2 X 1	

Remark 1) 4 x 7 x 1/4 column web stiffeners used, fy = 52.5 ksi, fu = 75.3 ksi.
2) fy = 58.0, fu = 78.8 ksi for beam, fy = 42.0, fu = 71.6 ksi for column.

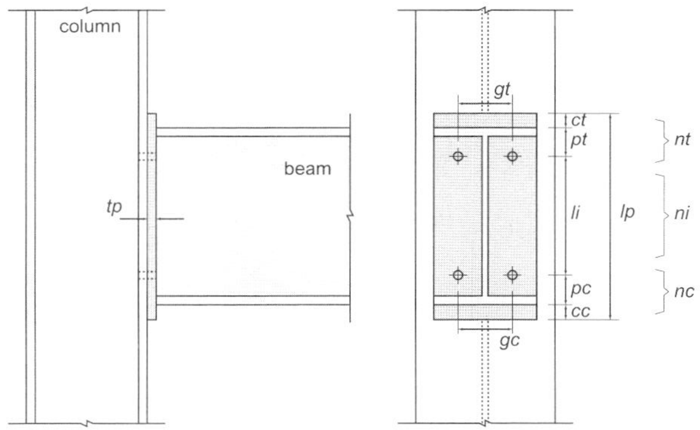

No	Moment (k-in)	Rotation (radians) X 1/1000	No	Moment (k-in)	Rotation (radians) X 1/1000
1	0.00	0.00	26	703.00	10.80
2	31.33	0.13	27	718.66	11.73
3	62.66	0.27	28	734.33	12.67
4	94.00	0.40	29	750.00	13.60
5	125.33	0.60	30	761.25	14.57
6	156.67	0.80	31	772.50	15.55
7	188.00	1.00	32	783.75	16.52
8	219.34	1.13	33	795.00	17.50
9	250.67	1.27	34	805.00	18.45
10	282.00	1.40	35	815.00	19.40
11	313.33	1.77	36	822.75	20.60
12	344.66	2.13	37	830.50	21.80
13	376.00	2.50	38	838.25	23.00
14	407.33	2.93	39	846.00	24.20
15	438.67	3.37			
16	470.00	3.80			
17	493.50	4.20			
18	517.00	4.60			
19	540.50	5.05			
20	564.00	5.50			
21	587.50	6.15			
22	611.00	6.80			
23	634.50	7.70			
24	658.00	8.60			
25	680.50	9.70			

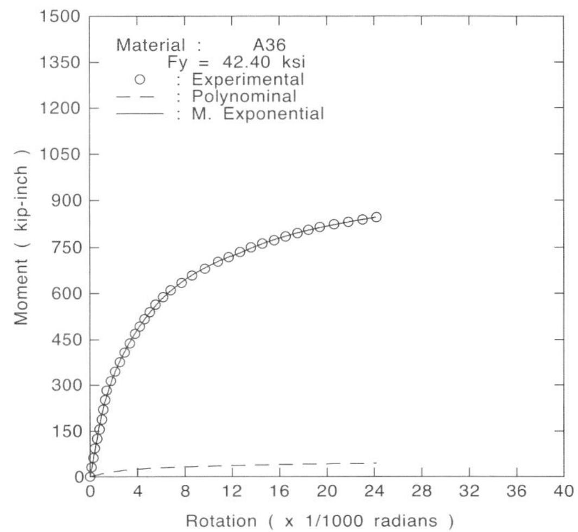

A.6 – 38

```
Moment-rotation prediction equations     ( R : X 1/1000 radians )

Frye and Morris polynominal model  :    R = Sum ( Ai X ( K*Bm )**Pi X 10**Qi )
     xd =  4.000000"    t =  0.500000"
     A1 =  1.790000   A2 =  1.760000   A3 =  2.040000    K =  0.054409
     P1 =        1    P2 =        3    P3 =        5    Q1 =        0    Q2 =       -1    Q3 =       -1

Modified exponential model  :
     Bm = Sum ( Ai X ( 1 - exp( - R/( 2*i*C ) ) ) )  +     Sum ( RKj X ( R - Rj0 ) )  +  Bm0
      C =   0.23499167E+00             Bmo=   0.00000000E+00           Nexp=  6       Nliner=  1
     AI =  -0.31428279E+03   0.32995252E+04  -0.14796717E+05   0.34626259E+05  -0.38947497E+05   0.16915189E+05
     Rj0 =          12.6700
     RKj =   0.57230902E+01
```

No	Rotation (radians) X 1/1000	Moment (kip-inch) Expri.	Poly.	M.Expo.	Connection stiffness (kip-inch) X 1000 Expri.	Poly.	M.Expo.
1	0.00	0.00	0.00	0.00	0.2350E+03	0.1027E+02	0.1904E+03
3	0.27	62.66	2.73	56.48	0.2350E+03	0.1020E+02	0.2223E+03
5	0.60	125.33	6.09	128.77	0.1567E+03	0.9880E+01	0.2059E+03
7	1.00	188.00	9.89	203.40	0.2037E+03	0.9060E+01	0.1668E+03
9	1.27	250.67	12.21	244.61	0.2350E+03	0.8274E+01	0.1429E+03
11	1.77	313.33	15.94	307.28	0.8545E+02	0.6650E+01	0.1104E+03
13	2.50	376.00	20.07	377.52	0.7943E+02	0.4750E+01	0.8435E+02
15	3.37	438.67	23.54	443.02	0.7231E+02	0.3401E+01	0.6813E+02
17	4.20	493.50	26.04	494.96	0.5874E+02	0.2642E+01	0.5691E+02
19	5.05	540.50	28.06	539.15	0.5223E+02	0.2147E+01	0.4735E+02
21	6.15	587.50	30.17	585.60	0.3616E+02	0.1730E+01	0.3759E+02
23	7.70	634.50	32.55	636.02	0.2611E+02	0.1364E+01	0.2818E+02
25	9.70	680.50	34.96	684.02	0.2045E+02	0.1077E+01	0.2033E+02
27	11.73	718.66	36.95	719.12	0.1679E+02	0.8924E+00	0.1441E+02
29	13.60	750.00	38.50	747.12	0.1422E+02	0.7737E+00	0.1577E+02
31	15.55	772.50	39.92	774.28	0.1154E+02	0.6813E+00	0.1227E+02
33	17.50	795.00	41.17	795.63	0.1103E+02	0.6100E+00	0.9786E+01
35	19.40	815.00	42.28	812.59	0.8730E+01	0.5545E+00	0.8183E+01
37	21.80	830.50	43.54	830.63	0.6457E+01	0.4984E+00	0.6974E+01
39	24.20	846.00	44.68	846.52	0.6458E+01	0.4534E+00	0.6336E+01

VI - 20

Connection type : Flush end-plate connections
Mode : All bolted with column stiffener

Tested by : J.R.Ostrander (1970) Canada
Test Id. : TEST 16

Column : W8x40
Beam : W12x27
Plate thickness : 5/8"
Stiffener thickness : 0.2500"

Fasteners: A325- -3/4"D
13/16" Oversize holes
Material : A36
Fy = 40.90 ksi
Fu = 73.10 ksi

Major parameters

ct = 0.5000"	pt = 2.5000"	li = 7.0000"	pc = 2.5000"
cc = 0.5000"	gt = 4.0000"	gc = 4.0000"	
tp = 0.6250"	nt = 2 X 1	nc = 2 X 1	

Remark 1) 4 x 7 x 1/4 column web stiffeners used, fy = 52.5 ksi, fu = 75.3 ksi.
2) fy = 58.0, fu = 78.8 ksi for beam, fy = 42.0, fu = 71.6 ksi for column.

No	Moment (k-in)	Rotation (radians) X 1/1000	No	Moment (k-in)	Rotation (radians) X 1/1000
1	0.00	0.00	26	905.00	15.70
2	47.00	0.15	27	919.00	17.10
3	94.00	0.30	28	933.00	18.50
4	141.00	0.45	29	941.40	19.72
5	188.00	0.60	30	949.80	20.94
6	235.00	0.75	31	958.20	22.16
7	282.00	0.90	32	966.60	23.38
8	329.00	1.05	33	975.00	24.60
9	376.00	1.20	34	981.25	25.85
10	423.00	1.55	35	987.50	27.10
11	470.00	1.90	36	993.75	28.35
12	517.00	2.30	37	1000.00	29.60
13	564.00	2.80			
14	611.00	3.40			
15	658.00	4.10			
16	702.00	5.10			
17	727.00	5.80			
18	752.00	6.50			
19	773.00	7.35			
20	794.00	8.20			
21	817.00	9.45			
22	840.00	10.70			
23	857.00	11.90			
24	874.00	13.10			
25	891.00	14.30			

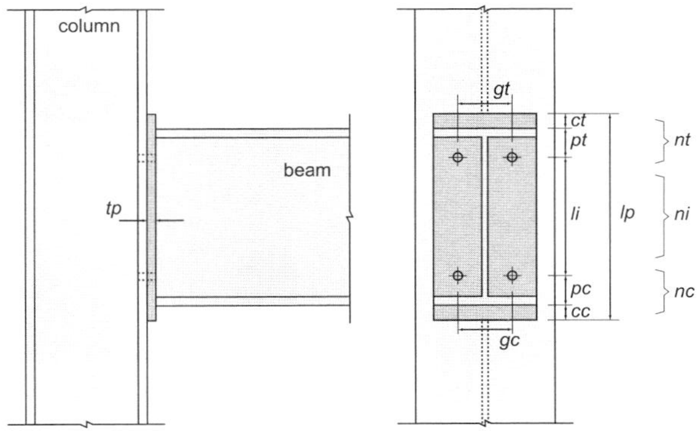

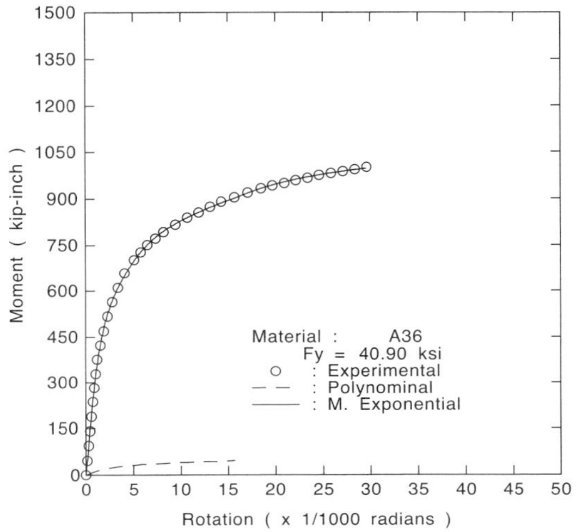

```
Moment-rotation prediction equations      ( R : X 1/1000 radians )

Frye and Morris polynominal model  :      R = Sum ( Ai X ( K*Bm )**Pi X 10**Qi )
     xd =  4.000000"    t =  0.625000"
     A1 =  1.790000    A2 =  1.760000    A3 =  2.040000    K =  0.047591
     P1 =         1    P2 =         3    P3 =         5    Q1 =         0    Q2 =        -1    Q3 =        -1

Modified exponential model  :
     Bm = Sum ( Ai X ( 1 - exp( - R/( 2*i*C ) ) ) )  +      Sum ( RKj X ( R - Rj0 ) )  +  Bm0
       C =   0.28416667E+00              Bmo=   0.00000000E+00          Nexp=  6         Nliner=  1
      AI =  -0.83615726E+03   0.84068037E+04  -0.33451774E+05   0.68940155E+05  -0.68440840E+05   0.26339772E+05
     Rj0 =           15.7000
     RKj =   0.29715859E+01
```

No	Rotation (radians) X 1/1000	Moment (kip-inch) Expri.	Moment (kip-inch) Poly.	Moment (kip-inch) M.Expo.	Connection stiffness (kip-inch) X 1000 Expri.	Connection stiffness (kip-inch) X 1000 Poly.	Connection stiffness (kip-inch) X 1000 M.Expo.
1	0.00	0.00	0.00	0.00	0.3133E+03	0.1174E+02	0.2701E+03
3	0.30	94.00	3.51	94.13	0.3133E+03	0.1164E+02	0.3368E+03
5	0.60	188.00	6.96	193.85	0.3133E+03	0.1130E+02	0.3197E+03
7	0.90	282.00	10.26	283.08	0.3133E+03	0.1065E+02	0.2729E+03
9	1.20	376.00	13.32	357.19	0.2596E+03	0.9698E+01	0.2217E+03
11	1.90	470.00	19.20	477.66	0.1264E+03	0.7141E+01	0.1310E+03
13	2.80	564.00	24.48	567.60	0.8689E+02	0.4792E+01	0.7793E+02
15	4.10	658.00	29.46	650.40	0.5761E+02	0.3103E+01	0.5398E+02
17	5.80	727.00	33.78	727.71	0.3572E+02	0.2108E+01	0.3749E+02
19	7.35	773.00	36.65	776.14	0.2471E+02	0.1637E+01	0.2574E+02
21	9.45	817.00	39.66	819.88	0.1840E+02	0.1264E+01	0.1721E+02
23	11.90	857.00	42.42	856.91	0.1417E+02	0.1006E+01	0.1357E+02
25	14.30	891.00	44.62	886.70	0.1224E+02	0.8432E+00	0.1123E+02
27	17.10	919.00	46.79	918.09	0.1000E+02	0.7125E+00	0.1115E+02
29	19.72	941.40	48.54	943.62	0.6885E+01	0.6245E+00	0.8425E+01
31	22.16	958.20	49.98	961.64	0.6885E+01	0.5614E+00	0.6443E+01
33	24.60	975.00	51.29	975.56	0.5954E+01	0.5108E+00	0.5064E+01
35	27.10	987.50	52.51	987.00	0.4999E+01	0.4683E+00	0.4165E+01
37	29.60	1000.00	53.63	996.69	0.4999E+01	0.4330E+00	0.3629E+01

A.6 – 41

VI - 21

Connection type : Flush end-plate connections
Mode : All bolted with column stiffener

Tested by : J.R.Ostrander (1970) Canada
Test Id. : TEST 20

Column : W8x24
Beam : W12x27
Plate thickness : 3/8"
Stiffener thickness : 0.2500"

Fasteners: A325- -3/4"D
13/16" Oversize holes
Material : A36
Fy = 62.90 ksi
Fu = 71.10 ksi

Major parameters

ct = 0.5000"	pt = 2.5000"	li = 7.0000"	pc = 2.5000"
cc = 0.5000"	gt = 4.0000"	gc = 4.0000"	
tp = 0.3750"	nt = 2 X 1	nc = 2 X 1	

Remark 1) 3 x 7 1/8 x 1/4 column web stiffeners used, fy = 41.4 ksi, fu = 57.0 ksi.
2) fy = 58.0, fu = 78.8 ksi for beam.

No	Moment (k-in)	Rotation (radians) X 1/1000	No	Moment (k-in)	Rotation (radians) X 1/1000
1	0.00	0.00	26	568.50	41.85
2	31.33	0.37	27	575.25	44.58
3	62.66	0.73	28	582.00	47.30
4	94.00	1.10	29	588.50	50.25
5	125.33	1.43	30	595.00	53.20
6	156.67	1.77	31	601.50	56.15
7	188.00	2.10	32	608.00	59.10
8	219.34	2.63	33	615.34	62.57
9	250.67	3.17	34	622.67	66.03
10	282.00	3.70	35	630.00	69.50
11	313.33	4.80			
12	344.66	5.90			
13	376.00	7.00			
14	399.50	8.60			
15	423.00	10.20			
16	446.50	12.85			
17	470.00	15.50			
18	483.34	18.40			
19	496.67	21.30			
20	510.00	24.20			
21	521.25	27.25			
22	532.50	30.30			
23	543.75	33.35			
24	555.00	36.40			
25	561.75	39.12			

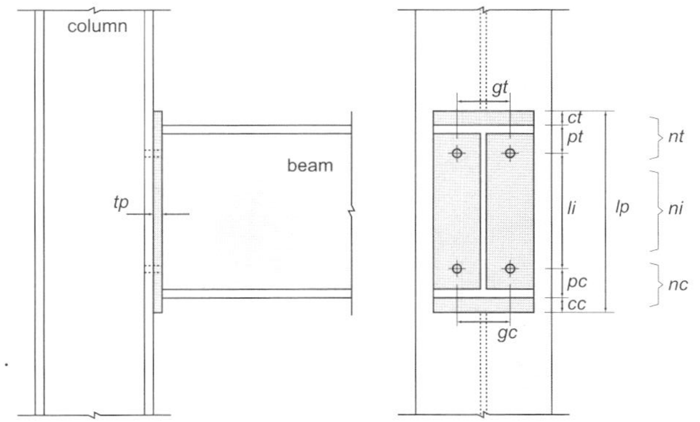

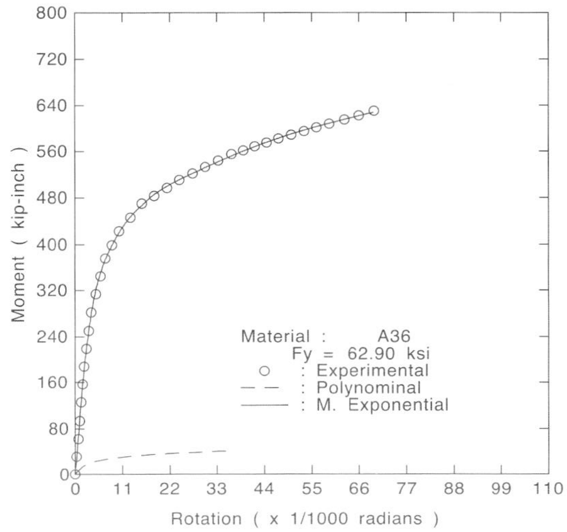

```
Moment-rotation prediction equations     ( R : X 1/1000 radians )

Frye and Morris polynominal model  :     R = Sum ( Ai X ( K*Bm )**Pi X 10**Qi )
     xd =  4.000000"    t =  0.375000"
     A1 =  1.790000   A2 =  1.760000   A3 =  2.040000    K =  0.064660
     P1 =         1   P2 =         3   P3 =         5   Q1 =         0   Q2 =        -1   Q3 =        -1

Modified exponential model  :
     Bm = Sum ( Ai X ( 1 - exp( - R/( 2*i*C ) ) ) )  +      Sum ( RKj X ( R - Rj0 ) )  +  Bm0
      C =   0.67084167E+00            Bmo=   0.00000000E+00            Nexp=  6        Nliner=  1
     AI =  -0.50805023E+03   0.49647764E+04  -0.18926937E+05   0.37107901E+05  -0.34728809E+05   0.12633887E+05
    Rj0 =         15.5000
    RKj =   0.15993937E+01
```

No	Rotation (radians) X 1/1000	Moment (kip-inch) Expri.	Moment (kip-inch) Poly.	Moment (kip-inch) M.Expo.	Connection stiffness (kip-inch) X 1000 Expri.	Connection stiffness (kip-inch) X 1000 Poly.	Connection stiffness (kip-inch) X 1000 M.Expo.
1	0.00	0.00	0.00	0.00	0.8544E+02	0.8640E+01	0.7619E+02
3	0.73	62.66	6.22	64.02	0.8545E+02	0.8131E+01	0.9273E+02
5	1.43	125.33	11.39	127.70	0.9400E+02	0.6506E+01	0.8718E+02
7	2.10	188.00	15.14	181.93	0.8045E+02	0.4785E+01	0.7499E+02
9	3.17	250.67	19.22	250.66	0.5875E+02	0.3069E+01	0.5432E+02
11	4.80	313.33	23.14	319.46	0.2848E+02	0.1912E+01	0.3212E+02
13	7.00	376.00	26.54	372.86	0.2286E+02	0.1268E+01	0.1869E+02
15	10.20	423.00	29.86	420.26	0.1250E+02	0.8620E+00	0.1201E+02
17	15.50	470.00	33.56	466.91	0.6829E+01	0.5749E+00	0.5986E+01
19	21.30	496.67	36.43	499.44	0.4597E+01	0.4283E+00	0.4233E+01
21	27.25	521.25	38.70	521.54	0.3689E+01	0.3430E+00	0.3439E+01
23	33.35	543.75	40.61	542.02	0.3689E+01	0.2868E+00	0.3280E+01
25	39.12	561.75	42.15	560.28	0.2477E+01	0.2494E+00	0.3020E+01
27	44.58	575.25	43.44	575.83	0.2477E+01	0.2228E+00	0.2680E+01
29	50.25	588.50	44.64	590.04	0.2204E+01	0.2009E+00	0.2338E+01
31	56.15	601.50	45.77	602.98	0.2204E+01	0.1827E+00	0.2061E+01
33	62.57	615.34	46.88	615.50	0.2115E+01	0.1667E+00	0.1859E+01
35	69.50	630.00	47.99	627.90	0.2115E+01	0.1525E+00	0.1731E+01

A.6 – 43

VI - 22

```
Connection type : Flush end-plate connections
           Mode : All bolted with column stiffener

Tested by  :  J.R.Ostrander (1970)                   Canada
Test  Id.  :  TEST 21

Column : W8x24                       Fasteners: A325- -3/4"D
Beam   : W12x27                            13/16" Oversize holes
Plate thickness : 1/2"               Material : A36
Stiffener thickness : 0.2500"             Fy =  42.40 ksi
                                          Fu =  71.10 ksi

                        Major parameters

     ct =  0.5000"    pt =  2.5000"    li =  7.0000"    pc =  2.5000"
     cc =  0.5000"    gt =  4.0000"    gc =  4.0000"
     tp =  0.5000"    nt = 2 X  1      nc = 2 X  1

Remark  1) 3 x 7 1/8 x 1/4 column web stiffeners used, fy = 41.4 ksi, fu = 57.0 ksi.
        2) fy = 58.0, fu = 78.8 ksi  for beam.
```

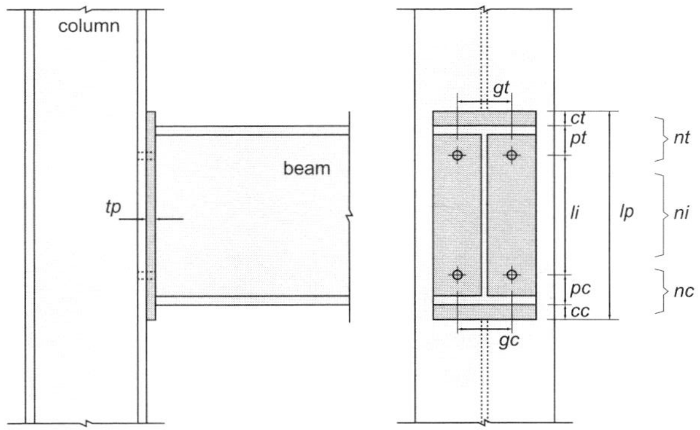

No	Moment (k-in)	Rotation (radians) X 1/1000	No	Moment (k-in)	Rotation (radians) X 1/1000
1	0.00	0.00	26	683.67	30.40
2	31.33	0.33	27	698.00	32.80
3	62.66	0.67	28	707.00	35.03
4	94.00	1.00	29	716.00	37.27
5	125.33	1.37	30	725.00	39.50
6	156.67	1.73	31	738.00	42.60
7	188.00	2.10	32	746.00	44.55
8	219.34	2.63	33	754.00	46.50
9	250.67	3.17	34	764.33	49.60
10	282.00	3.70	35	774.67	52.70
11	313.33	4.37	36	785.00	55.80
12	344.66	5.03	37	790.75	58.35
13	376.00	5.70	38	796.50	60.90
14	407.33	6.83	39	802.25	63.45
15	438.67	7.97	40	808.00	66.00
16	470.00	9.10			
17	491.50	10.30			
18	513.00	11.50			
19	539.00	13.25			
20	565.00	15.00			
21	587.50	17.50			
22	610.00	20.00			
23	632.50	22.80			
24	655.00	25.60			
25	669.33	28.00			

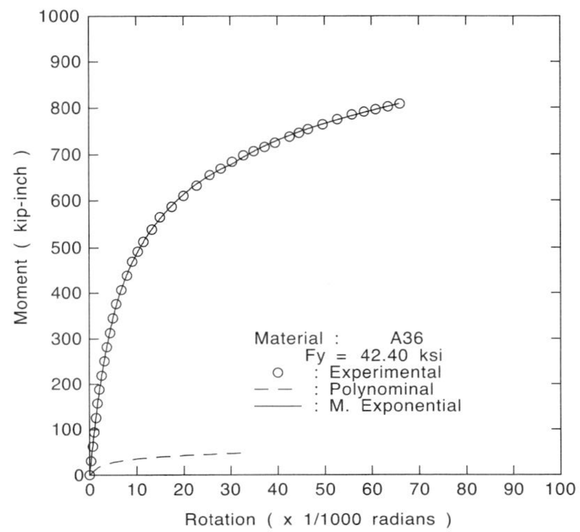

```
Moment-rotation prediction equations      ( R : X 1/1000 radians )

Frye and Morris polynominal model   :     R = Sum ( Ai X ( K*Bm )**Pi X 10**Qi )
     xd =  4.000000"   t =  0.500000"
     A1 =  1.790000   A2 =  1.760000   A3 =  2.040000   K =  0.054409
     P1 =         1   P2 =         3   P3 =         5   Q1 =        0   Q2 =       -1   Q3 =       -1

Modified exponential model  :
     Bm = Sum ( Ai X ( 1 - exp( - R/( 2*i*C ) ) ) )  +     Sum ( RKj X ( R - Rj0 ) )  +  Bm0
       C =   0.63332500E+00          Bmo=   0.00000000E+00          Nexp=  6        Nliner=  1
      AI =  -0.43381741E+02   0.35225060E+03  -0.26236511E+04   0.10889200E+05  -0.16348178E+05   0.85001317E+04
     Rj0 =           30.4000
     RKj =   0.23378485E+01
```

No	Rotation (radians) X 1/1000	Moment (kip-inch) Expri.	Poly.	M.Expo.	Connection stiffness (kip-inch) X 1000 Expri.	Poly.	M.Expo.
1	0.00	0.00	0.00	0.00	0.9401E+02	0.1027E+02	0.1007E+03
3	0.67	62.66	6.74	64.58	0.9399E+02	0.9779E+01	0.9264E+02
5	1.37	125.33	13.02	125.99	0.8547E+02	0.7951E+01	0.8272E+02
7	2.10	188.00	17.99	182.83	0.7457E+02	0.5687E+01	0.7240E+02
9	3.17	250.67	22.84	252.72	0.5875E+02	0.3647E+01	0.5903E+02
11	4.37	313.33	26.47	315.98	0.4700E+02	0.2527E+01	0.4692E+02
13	5.70	376.00	29.36	371.43	0.3983E+02	0.1879E+01	0.3679E+02
15	7.97	438.67	32.91	440.87	0.2765E+02	0.1316E+01	0.2548E+02
17	10.30	491.50	35.59	491.72	0.1792E+02	0.1015E+01	0.1871E+02
19	13.25	539.00	38.23	539.09	0.1486E+02	0.7933E+00	0.1389E+02
21	17.50	587.50	41.17	589.40	0.9000E+01	0.6099E+00	0.1017E+02
23	22.80	632.50	44.03	635.36	0.8036E+01	0.4785E+00	0.7334E+01
25	28.00	669.33	46.29	667.71	0.5973E+01	0.3979E+00	0.5182E+01
27	32.80	698.00	48.07	694.22	0.4967E+01	0.3459E+00	0.5927E+01
29	37.27	716.00	49.53	718.03	0.4030E+01	0.3093E+00	0.4789E+01
31	42.60	738.00	51.09	740.83	0.4137E+01	0.2753E+00	0.3827E+01
33	46.50	754.00	52.12	754.77	0.3805E+01	0.2553E+00	0.3347E+01
35	52.70	774.67	53.62	773.88	0.3334E+01	0.2293E+00	0.2864E+01
37	58.35	790.75	54.87	789.30	0.2254E+01	0.2100E+00	0.2620E+01
39	63.45	802.25	55.90	802.31	0.2254E+01	0.1956E+00	0.2495E+01

VI - 23

Connection type : Flush end-plate connections
Mode : All bolted with column stiffener

Tested by : J.R.Ostrander (1970) Canada
Test Id. : TEST 22

Column : W8x24
Beam : W12x27
Plate thickness : 5/8"
Stiffener thickness : 0.2500"

Fasteners: A325- -3/4"D
13/16" Oversize holes
Material : A36
Fy = 42.70 ksi
Fu = 71.20 ksi

Major parameters

ct = 0.5000" pt = 2.5000" li = 7.0000" pc = 2.5000"
cc = 0.5000" gt = 4.0000" gc = 4.0000"
tp = 0.6250" nt = 2 X 1 nc = 2 X 1

Remark 1) 3 x 7 1/8 x 1/4 column web stiffeners used, fy = 41.4 ksi, fu = 57.0 ksi.
2) fy = 58.0, fu = 78.8 ksi for beam.

No	Moment (k-in)	Rotation (radians) X 1/1000	No	Moment (k-in)	Rotation (radians) X 1/1000
1	0.00	0.00	26	650.80	20.68
2	31.33	0.17	27	662.60	22.66
3	62.66	0.33	28	674.40	24.64
4	94.00	0.50	29	686.20	26.62
5	125.33	0.63	30	698.00	28.60
6	156.67	0.77	31	708.00	30.50
7	188.00	0.90	32	718.00	32.40
8	219.34	1.13	33	728.00	34.30
9	250.67	1.37	34	738.00	36.20
10	282.00	1.60	35	744.00	37.97
11	313.33	2.00	36	750.00	39.73
12	344.66	2.40	37	756.00	41.50
13	376.00	2.80			
14	407.33	3.63			
15	438.67	4.47			
16	470.00	5.30			
17	490.50	6.25			
18	511.00	7.20			
19	537.50	8.65			
20	554.00	10.10			
21	584.00	11.80			
22	604.00	13.50			
23	615.66	15.23			
24	627.33	16.97			
25	639.00	18.70			

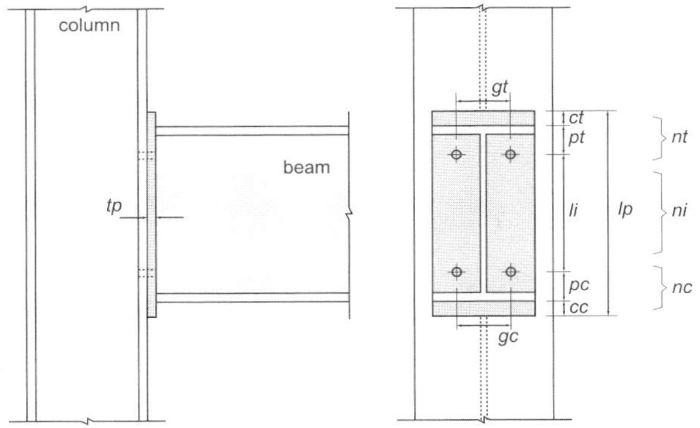

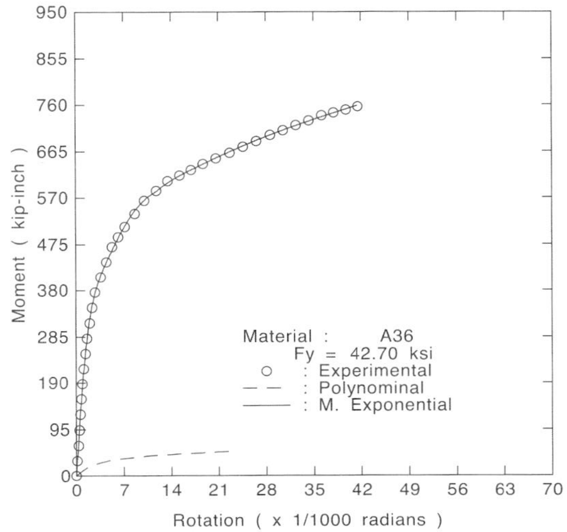

A.6 – 46

```
Moment-rotation prediction equations      ( R : X 1/1000 radians )

Frye and Morris polynominal model  :     R = Sum ( Ai X ( K*Bm )**Pi X 10**Qi )
     xd =  4.000000"   t =  0.625000"
     A1 =  1.790000   A2 =  1.760000   A3 =  2.040000    K =  0.047591
     P1 =        1    P2 =        3    P3 =        5    Q1 =        0    Q2 =      -1    Q3 =      -1

Modified exponential model  :
     Bm = Sum ( Ai X ( 1 - exp( - R/( 2*i*C ) ) ) )  +     Sum ( RKj X ( R - Rj0 ) )  +  Bm0
       C =   0.38750833E+00          Bmo=   0.00000000E+00          Nexp=  6        Nliner=  1
      AI =  -0.65185344E+03   0.68786191E+04  -0.24063807E+05   0.41849362E+05  -0.35541619E+05   0.12165406E+05
     Rj0 =          11.8000
     RKj =   0.41511887E+01
```

No	Rotation (radians) X 1/1000	Moment (kip-inch) Expri.	Moment (kip-inch) Poly.	Moment (kip-inch) M.Expo.	Connection stiffness (kip-inch) X 1000 Expri.	Connection stiffness (kip-inch) X 1000 Poly.	Connection stiffness (kip-inch) X 1000 M.Expo.
1	0.00	0.00	0.00	0.00	0.1880E+03	0.1174E+02	0.1907E+03
3	0.33	62.66	3.90	67.51	0.1880E+03	0.1161E+02	0.2071E+03
5	0.63	125.33	7.33	128.25	0.2350E+03	0.1124E+02	0.1951E+03
7	0.90	188.00	10.26	177.68	0.1984E+03	0.1065E+02	0.1748E+03
9	1.37	250.67	14.88	249.81	0.1343E+03	0.9090E+01	0.1343E+03
11	2.00	313.33	19.90	319.36	0.7833E+02	0.6812E+01	0.8799E+02
13	2.80	376.00	24.47	374.06	0.6512E+02	0.4793E+01	0.5283E+02
15	4.47	438.67	30.55	438.01	0.3760E+02	0.2815E+01	0.3088E+02
17	6.25	490.50	34.69	488.97	0.2157E+02	0.1944E+01	0.2654E+02
19	8.65	537.50	38.61	543.18	0.1828E+02	0.1383E+01	0.1807E+02
21	11.80	584.00	42.32	582.57	0.1176E+02	0.1015E+01	0.7906E+01
23	15.23	615.66	45.39	614.76	0.6731E+01	0.7941E+00	0.7630E+01
25	18.70	639.00	47.88	638.97	0.6371E+01	0.6558E+00	0.6589E+01
27	22.66	662.60	50.26	664.12	0.5960E+01	0.5501E+00	0.6118E+01
29	26.62	686.20	52.28	687.29	0.5960E+01	0.4759E+00	0.5571E+01
31	30.50	708.00	54.02	707.87	0.5263E+01	0.4216E+00	0.5057E+01
33	34.30	728.00	55.54	726.34	0.5263E+01	0.3802E+00	0.4683E+01
35	37.97	744.00	56.87	743.05	0.3396E+01	0.3479E+00	0.4450E+01
37	41.50	756.00	58.05	758.51	0.3396E+01	0.3220E+00	0.4315E+01

VI - 24

Connection type : Flush end-plate connections
Mode : All bolted with column stiffener

Tested by : J.R.Ostrander (1970) Canada
Test Id. : TEST 24

Column : W8x48
Beam : W12x27
Plate thickness : 5/8"
Stiffener thickness : 0.2500"

Fasteners: A325- -3/4"D
13/16" Oversize holes
Material : A36
Fy = 42.70 ksi
Fu = 71.20 ksi

Major parameters

ct = 0.5000" pt = 2.5000" li = 7.0000" pc = 2.5000"
cc = 0.5000" gt = 4.0000" gc = 4.0000"
tp = 0.6250" nt = 2 X 1 nc = 2 X 1

Remark 1) 4 x 6 3/4 x 1/4 column web stiffeners used, fy = 41.4 ksi, fu = 57.0 ksi.
2) fy = 58.0, fu = 78.8 ksi for beam, fy = 38.4, fu = 65.7 ksi for column.

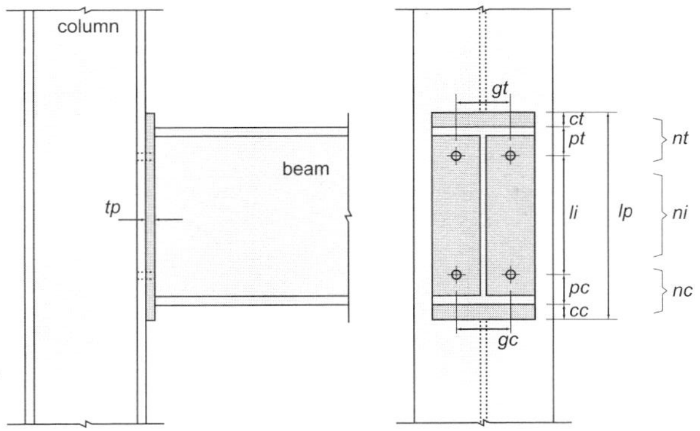

No	Moment (k-in)	Rotation (radians) X 1/1000	No	Moment (k-in)	Rotation (radians) X 1/1000
1	0.00	0.00	26	915.33	12.17
2	47.00	0.15	27	928.00	13.10
3	94.00	0.30	28	940.00	13.85
4	141.00	0.40	29	952.00	14.60
5	188.00	0.50	30	955.66	15.30
6	235.00	0.65	31	959.33	16.00
7	282.00	0.80	32	963.00	16.70
8	329.00	1.00	33	970.50	17.45
9	376.00	1.20	34	978.00	18.20
10	423.00	1.45	35	966.50	19.05
11	470.00	1.70	36	955.00	19.90
12	517.00	2.05			
13	564.00	2.40			
14	611.00	2.90			
15	658.00	3.40			
16	702.50	4.30			
17	747.00	5.20			
18	771.00	5.80			
19	795.00	6.40			
20	817.00	7.25			
21	839.00	8.10			
22	856.00	8.83			
23	873.00	9.57			
24	890.00	10.30			
25	902.67	11.23			

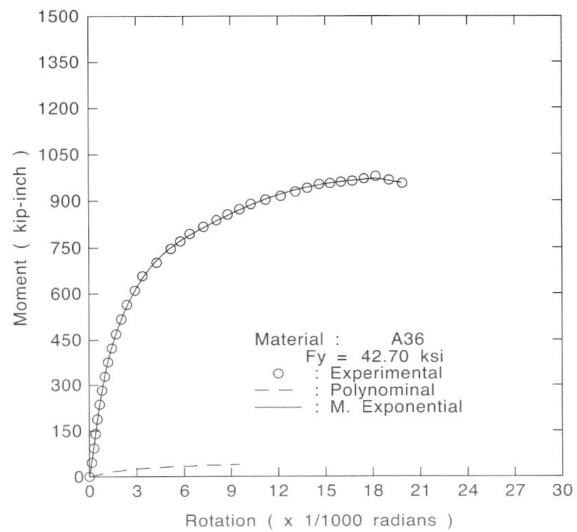

```
Moment-rotation prediction equations     ( R : X 1/1000 radians )

Frye and Morris polynominal model  :     R = Sum ( Ai X ( K*Bm )**Pi X 10**Qi )
     xd =  4.000000"    t =  0.625000"
     A1 =  1.790000    A2 =  1.760000    A3 =  2.040000     K =  0.047591
     P1 =         1    P2 =         3    P3 =         5    Q1 =         0    Q2 =        -1    Q3 =        -1

Modified exponential model  :
     Bm = Sum ( Ai X ( 1 - exp( - R/( 2*i*C ) ) ) )  +      Sum ( RKj X ( R - Rj0 ) )  +  Bm0
       C =   0.20333333E+00              Bmo=   0.00000000E+00           Nexp=  6          Nliner=  1
      AI =  -0.70087820E+03   0.78785044E+04  -0.36616334E+05   0.82394145E+05  -0.85734421E+05   0.33759283E+05
     Rj0 =          18.2000
     RKj =  -0.11006535E+02
```

No	Rotation (radians) X 1/1000	Moment (kip-inch) Expri.	Moment (kip-inch) Poly.	Moment (kip-inch) M.Expo.	Connection stiffness (kip-inch) X 1000 Expri.	Connection stiffness (kip-inch) X 1000 Poly.	Connection stiffness (kip-inch) X 1000 M.Expo.
1	0.00	0.00	0.00	0.00	0.3133E+03	0.1174E+02	0.2733E+03
3	0.30	94.00	3.51	104.18	0.4074E+03	0.1164E+02	0.3798E+03
5	0.50	188.00	5.82	178.82	0.4074E+03	0.1144E+02	0.3605E+03
7	0.80	282.00	9.18	277.68	0.2797E+03	0.1090E+02	0.2961E+03
9	1.20	376.00	13.32	379.15	0.2141E+03	0.9698E+01	0.2154E+03
11	1.70	470.00	17.70	469.56	0.1656E+03	0.7844E+01	0.1527E+03
13	2.40	564.00	22.39	559.65	0.1177E+03	0.5673E+01	0.1100E+03
15	3.40	658.00	27.05	651.77	0.7809E+02	0.3845E+01	0.7639E+02
17	5.20	747.00	32.44	751.27	0.4378E+02	0.2376E+01	0.3903E+02
19	6.40	795.00	34.98	791.26	0.3416E+02	0.1895E+01	0.2919E+02
21	8.10	839.00	37.82	836.50	0.2443E+02	0.1479E+01	0.2489E+02
23	9.57	873.00	39.81	871.26	0.2318E+02	0.1249E+01	0.2237E+02
25	11.23	902.67	41.73	905.22	0.1357E+02	0.1065E+01	0.1817E+02
27	13.10	928.00	43.57	934.09	0.1492E+02	0.9169E+00	0.1279E+02
29	14.60	952.00	44.87	950.33	0.1043E+02	0.8267E+00	0.8974E+01
31	16.00	959.33	45.99	960.83	0.5240E+01	0.7576E+00	0.6163E+01
33	17.45	970.50	47.05	968.13	0.1000E+02	0.6986E+00	0.4030E+01
35	19.05	966.50	48.12	963.86	-0.1353E+02	0.6441E+00	-0.8567E+01

VI - 25

Connection type : Flush end-plate connections
Mode : All bolted without column stiffener

Tested by : J.B.Davison et al (1987) United Kingdom
Test Id. : JT/11

Column : UC 152x152x23
Beam : UB 254x102x22
Plate thickness : 12 mm

Fasteners: 10.9- -M16
23/32" Oversize holes
Material : 43A
Fy = -- ksi
Fu = -- ksi

Major parameters

ct = 0.1969"	pt = 1.9685"	li = 5.9055"	pc = 2.1260"
cc = 0.2362"	pit= 1.9685"	pi = 0.0000"	pic= 3.9370"
gt = 2.9921"	gi = 2.9921"	gc = 2.9921"	
tp = 0.4724"	nt = 2 X 1	ni = 2 X 1	nc = 2 X 1

Remark 1) Suddenly failed at 0,01 radian and 407 kip-in
2) (A) Premature weld failure (B) Minor axis connection

No	Moment (k-in)	Rotation (radians) X 1/1000	No	Moment (k-in)	Rotation (radians) X 1/1000
1	0.00	0.00	26	379.80	6.25
2	18.44	0.20	27	387.17	6.73
3	36.87	0.33	28	390.86	7.21
4	55.31	0.52	29	394.54	7.69
5	73.75	0.71	30	396.38	8.17
6	92.18	0.81	31	398.23	8.65
7	110.62	1.00	32	401.92	9.13
8	129.06	1.10	33	405.60	9.62
9	147.49	1.19	34	409.30	10.10
10	165.93	1.33	35	413.00	10.58
11	184.37	1.48			
12	202.80	1.50			
13	221.24	1.72			
14	239.67	1.80			
15	258.11	1.92			
16	276.55	2.16			
17	294.98	2.40			
18	313.42	2.88			
19	322.64	3.17			
20	331.86	3.46			
21	342.92	3.85			
22	352.14	4.33			
23	361.36	4.81			
24	366.89	5.29			
25	372.42	5.77			

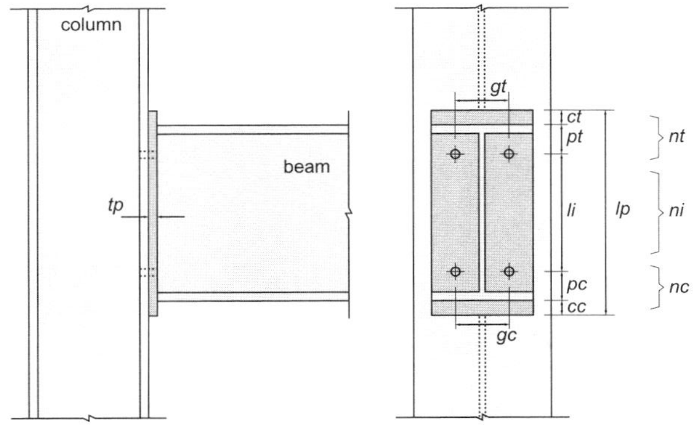

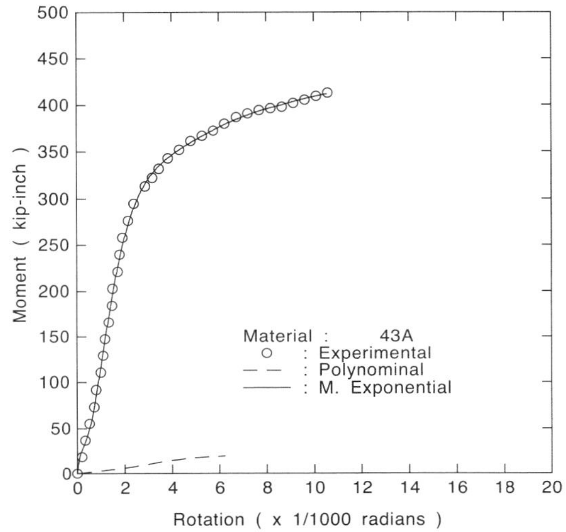

```
Moment-rotation prediction equations     ( R : X 1/1000 radians )

Frye and Morris polynominal model  :     R = Sum ( Ai X ( K*Bm )**Pi X 10**Qi )
     xd =  2.992100"    t =  0.472400"  tf =  0.629900"
     A1 =  1.830000   A2 = -1.040000   A3 =  6.380000     K =  0.194548
     P1 =         1    P2 =         3   P3 =         5    Q1 =         0    Q2 =       -1    Q3 =       -3

Modified exponential model  :
     Bm = Sum ( Ai X ( 1 - exp( - R/( 2*i*C ) ) ) )  +      Sum ( RKj X ( R - Rj0 ) )  +  Bm0
      C =  0.13871250E+00              Bmo=  0.00000000E+00          Nexp=  6         Nliner=  1
     AI =  0.47042166E+03  -0.28841378E+04  0.32594087E+04  0.45191372E+04  -0.10619619E+05  0.56672343E+04
    Rj0 =          8.6500
    RKj =  0.23321958E+01
```

No	Rotation (radians) X 1/1000	Moment (kip-inch) Expri.	Moment (kip-inch) Poly.	Moment (kip-inch) M.Expo.	Connection stiffness (kip-inch) X 1000 Expri.	Connection stiffness (kip-inch) X 1000 Poly.	Connection stiffness (kip-inch) X 1000 M.Expo.
1	0.00	0.00	0.00	0.00	0.9113E+02	0.2809E+01	0.2351E+03
3	0.33	36.87	0.94	34.76	0.1222E+03	0.2825E+01	0.6539E+02
5	0.71	73.75	2.02	72.16	0.1601E+03	0.2883E+01	0.1357E+03
7	1.00	110.62	2.87	116.99	0.1659E+03	0.2961E+01	0.1646E+03
9	1.19	147.49	3.42	147.56	0.1713E+03	0.3027E+01	0.1647E+03
11	1.48	184.37	4.33	193.53	0.8261E+03	0.3163E+01	0.1472E+03
13	1.72	221.24	5.10	226.30	0.2015E+03	0.3308E+01	0.1254E+03
15	1.92	258.11	5.79	249.70	0.1222E+03	0.3459E+01	0.1061E+03
17	2.40	294.98	7.56	290.78	0.6391E+02	0.3940E+01	0.6687E+02
19	3.17	322.64	10.95	326.68	0.3196E+02	0.4799E+01	0.3168E+02
21	3.85	342.92	14.03	343.42	0.2449E+02	0.4029E+01	0.2003E+02
23	4.81	361.36	16.99	359.80	0.1532E+02	0.2304E+01	0.1508E+02
25	5.77	372.42	18.78	373.22	0.1343E+02	0.1534E+01	0.1294E+02
27	6.73	387.17	20.05	384.53	0.1150E+02	0.1149E+01	0.1054E+02
29	7.69	394.54	21.04	393.42	0.5751E+01	0.9244E+00	0.7958E+01
31	8.65	398.23	21.86	399.93	0.5752E+01	0.7769E+00	0.7978E+01
33	9.62	405.60	22.55	406.69	0.7681E+01	0.6726E+00	0.6152E+01
35	10.58	413.00	23.16	411.92	0.7696E+01	0.5949E+00	0.4825E+01

VI - 26

Connection type : Flush end-plate connections
Mode : All bolted without column stiffener

Tested by : J.B.Davison et al (1987) United Kingdom
Test Id. : JT/11B

Column : UC 152x152x23
Beam : UB 254x102x22
Plate thickness : 12 mm

Fasteners: 10.9- -M16
23/32" Oversize holes
Material : 43A
Fy = -- ksi
Fu = -- ksi

Major parameters

ct = 0.1969"	pt = 1.9685"	li = 5.9055"	pc = 2.1260"
cc = 0.2362"	pit= 1.9685"	pi = 0.0000"	pic= 3.9370"
gt = 2.9921"	gi = 2.9921"	gc = 2.9921"	
tp = 0.4724"	nt = 2 X 1	ni = 2 X 1	nc = 2 X 1

Remark 1) The bolt stripping was the cause of failure
2) Minor axis connection

No	Mcment (k-in)	Rotation (radians) X 1/1000	No	Moment (k-in)	Rotation (radians) X 1/1000
1	0.00	0.00	26	387.17	3.85
2	18.44	0.05	27	395.77	4.49
3	36.87	0.10	28	404.38	5.13
4	55.31	0.14	29	412.98	5.77
5	73.75	0.19	30	418.51	6.25
6	92.18	0.24	31	424.04	6.73
7	110.62	0.29	32	430.19	7.21
8	129.06	0.34	33	436.33	7.69
9	147.49	0.38	34	442.48	8.17
10	165.93	0.43	35	443.95	8.65
11	184.37	0.48	36	445.43	9.13
12	202.80	0.53	37	446.90	9.62
13	221.24	0.58	38	446.90	10.10
14	239.67	0.62	39	446.90	10.58
15	258.11	0.67	40	442.97	11.07
16	276.55	0.72	41	439.03	11.56
17	294.98	0.79	42	435.10	12.05
18	309.73	0.96	43	425.27	12.52
19	320.79	1.25	44	415.43	12.99
20	331.86	1.54	45	405.60	13.46
21	342.93	1.73			
22	354.00	1.92			
23	368.73	2.40			
24	374.88	2.88			
25	381.02	3.37			

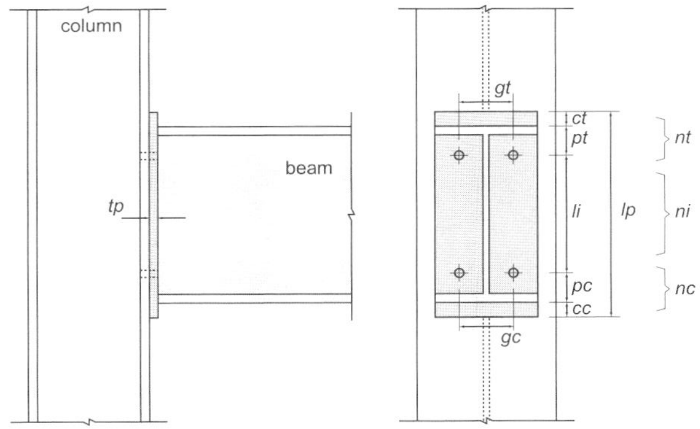

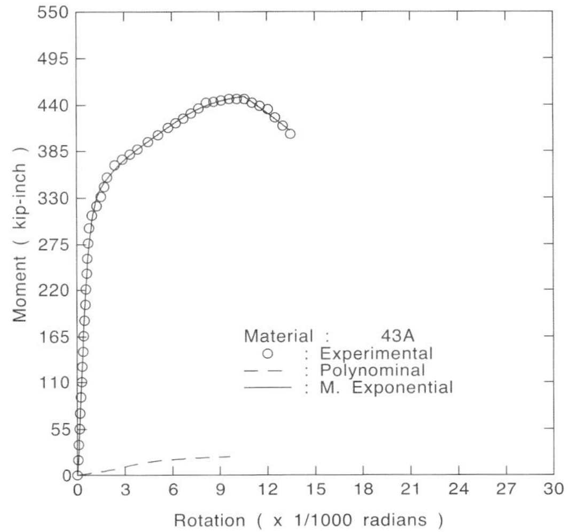

A.6 – 52

```
Moment-rotation prediction equations     ( R : X 1/1000 radians )

Frye and Morris polynominal model  :    R = Sum ( Ai X ( K*Bm )**Pi X 10**Qi )
     xd =  2.992100"    t =  0.472400"  tf =  0.629900"
     A1 =  1.830000   A2 = -1.040000   A3 =  6.380000    K =  0.194548
     P1 =        1    P2 =        3    P3 =        5    Q1 =        0    Q2 =       -1    Q3 =       -3

Modified exponential model  :
     Bm = Sum ( Ai X ( 1 - exp( - R/( 2*i*C ) ) ) )  +     Sum ( RKj X ( R - Rj0 ) )  +  Bm0
       C =   0.12420000E+00            Bmo=   0.00000000E+00          Nexp=  6        Nliner=  2
      AI =  -0.74614207E+03   0.55949559E+04  -0.19360171E+05   0.38588641E+05  -0.38080533E+05   0.14460155E+05
     Rj0 =           7.6900          10.5800
     RKj =  -0.55719545E+00  -0.15033936E+02
```

No	Rotation (radians) X 1/1000	Moment (kip-inch) Expri.	Moment (kip-inch) Poly.	Moment (kip-inch) M.Expo.	Connection stiffness (kip-inch) X 1000 Expri.	Connection stiffness (kip-inch) X 1000 Poly.	Connection stiffness (kip-inch) X 1000 M.Expo.
1	0.00	0.00	0.00	0.00	0.3833E+03	0.2809E+01	0.1571E+03
3	0.10	36.87	0.27	27.33	0.3837E+03	0.2810E+01	0.3802E+03
5	0.19	73.75	0.54	68.54	0.3837E+03	0.2814E+01	0.4598E+03
7	0.29	110.62	0.81	113.24	0.3833E+03	0.2821E+01	0.4601E+03
9	0.38	147.49	1.08	155.75	0.3833E+03	0.2830E+01	0.4202E+03
11	0.48	184.37	1.36	193.52	0.3837E+03	0.2842E+01	0.3635E+03
13	0.58	221.24	1.63	225.56	0.3833E+03	0.2857E+01	0.3034E+03
15	0.67	258.11	1.90	251.98	0.3837E+03	0.2875E+01	0.2469E+03
17	0.79	294.98	2.24	277.12	0.2183E+03	0.2901E+01	0.1881E+03
19	1.25	320.79	3.61	328.68	0.3836E+02	0.3052E+01	0.6031E+02
21	1.73	342.93	5.13	347.71	0.5757E+02	0.3315E+01	0.2760E+02
23	2.40	368.73	7.56	363.20	0.2170E+02	0.3940E+01	0.2038E+02
25	3.37	381.02	11.88	380.32	0.1279E+02	0.4788E+01	0.1550E+02
27	4.49	395.77	16.18	396.31	0.1342E+02	0.2750E+01	0.1371E+02
29	5.77	412.98	18.79	413.73	0.1232E+02	0.1533E+01	0.1324E+02
31	6.73	424.04	20.05	425.74	0.1214E+02	0.1149E+01	0.1153E+02
33	7.69	436.33	21.04	435.64	0.1278E+02	0.9244E+00	0.8465E+01
35	8.65	443.95	21.86	442.54	0.3067E+01	0.7769E+00	0.5926E+01
37	9.62	446.90	22.55	447.18	0.1533E+01	0.6726E+00	0.3810E+01
39	10.58	446.90	23.16	450.05	-0.3961E+01	0.5949E+00	0.2243E+01
41	11.56	439.03	23.71	436.96	-0.8008E+01	0.5335E+00	-0.1388E+02
43	12.52	425.27	24.21	423.26	-0.2091E+02	0.4850E+00	-0.1457E+02
45	13.46	405.60	24.64	409.35	-0.2091E+02	0.4466E+00	-0.1498E+02

VI - 27

Connection type : Flush end-plate connections
Mode : All bolted without column stiffener

Tested by : J.B.Davison et al (1987) United Kingdom
Test Id. : JT/12

Column : UC 152x152x23
Beam : UB 254x102x22
Plate thickness : 12 mm

Fasteners: 10.9- -M16
23/32" Oversize holes
Material : 43A
Fy = -- ksi
Fu = -- ksi

Major parameters

ct = 0.1969"	pt = 1.9685"	li = 5.9055"	pc = 2.1260"
cc = 0.2362"	pit= 1.9685"	pi = 0.0000"	pic= 3.9370"
gt = 2.9921"	gi = 2.9921"	gc = 2.9921"	
tp = 0.4724"	nt = 2 X 1	ni = 2 X 1	nc = 2 X 1

Remark 1) The bolt stripping was the failure
2) Minor axis connection

No	Moment (k-in)	Rotation (radians) X 1/1000
1	0.00	0.00
2	9.22	0.13
3	18.44	0.26
4	28.27	0.38
5	38.10	0.51
6	47.93	0.64
7	56.54	0.77
8	65.14	0.90
9	73.75	1.03
10	82.96	1.15
11	92.18	1.28
12	101.40	1.41
13	110.62	1.54
14	119.47	1.92
15	127.95	2.40
16	136.43	2.88
17	141.96	3.37
18	147.49	3.85
19	158.55	4.81
20	165.93	5.77
21	169.62	6.73
22	173.30	7.69
23	176.99	8.65
24	178.47	9.62
25	181.41	10.58

No	Moment (k-in)	Rotation (radians) X 1/1000
26	184.37	11.54
27	188.05	12.50
28	191.74	13.46
29	193.95	14.42
30	199.11	15.38
31	201.32	16.35
32	201.32	17.31
33	202.80	18.27
34	206.50	19.23
35	210.18	20.19
36	213.86	21.15
37	216.81	22.11
38	219.03	22.88
39	221.24	23.65

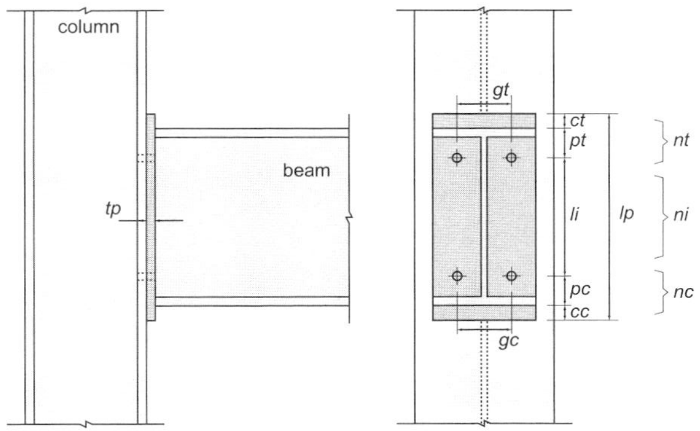

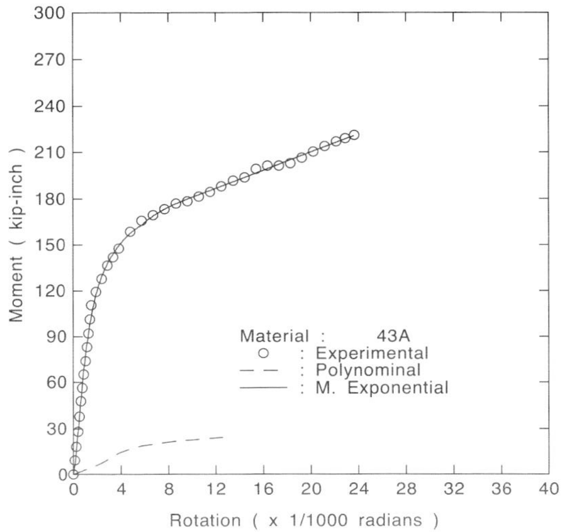

```
Moment-rotation prediction equations     ( R : X 1/1000 radians )

Frye and Morris polynominal model  :    R = Sum ( Ai X ( K*Bm )**Pi X 10**Qi )
     xd =  2.992100"    t =  0.472400"  tf =  0.629900"
     A1 =  1.830000    A2 = -1.040000   A3 =  6.380000    K =  0.194548
     P1 =         1    P2 =         3   P3 =         5   Q1 =         0    Q2 =        -1    Q3 =        -3

Modified exponential model  :
     Bm = Sum ( Ai X ( 1 - exp( - R/( 2*i*C ) ) ) )  +      Sum ( RKj X ( R - Rj0 ) )  +  Bm0
       C =   0.22913333E+00          Bmo=   0.00000000E+00          Nexp=  6        Nliner=  1
      AI =  -0.19432277E+03   0.12286440E+04  -0.32286758E+04   0.45044907E+04  -0.28806291E+04   0.73905667E+03
     Rj0 =           5.7700
     RKj =   0.29287909E+01
```

No	Rotation (radians) X 1/1000	Moment (kip-inch) Expri.	Moment (kip-inch) Poly.	Moment (kip-inch) M.Expo.	Connection stiffness (kip-inch) X 1000 Expri.	Connection stiffness (kip-inch) X 1000 Poly.	Connection stiffness (kip-inch) X 1000 M.Expo.
1	0.00	0.00	0.00	0.00	0.7190E+02	0.2809E+01	0.3698E+02
3	0.26	18.44	0.72	16.05	0.7430E+02	0.2818E+01	0.7952E+02
5	0.51	38.10	1.45	37.75	0.7670E+02	0.2847E+01	0.8596E+02
7	0.77	56.54	2.18	58.89	0.6711E+02	0.2896E+01	0.7758E+02
9	1.03	73.75	2.93	77.17	0.6951E+02	0.2968E+01	0.6473E+02
11	1.28	92.18	3.70	92.11	0.7191E+02	0.3066E+01	0.5205E+02
13	1.54	110.62	4.50	104.03	0.5968E+02	0.3194E+01	0.4128E+02
15	2.40	127.95	7.56	129.19	0.1764E+02	0.3940E+01	0.2031E+02
17	3.37	141.96	11.88	144.17	0.1150E+02	0.4788E+01	0.1221E+02
19	4.81	158.55	16.98	157.61	0.9588E+01	0.2309E+01	0.6886E+01
21	6.73	169.62	20.06	168.94	0.3833E+01	0.1148E+01	0.5336E+01
23	8.65	176.99	21.86	176.95	0.2686E+01	0.7768E+00	0.3350E+01
25	10.58	181.41	23.16	182.76	0.3068E+01	0.5949E+00	0.2840E+01
27	12.50	188.05	24.19	188.18	0.3835E+01	0.4864E+00	0.2830E+01
29	14.42	193.95	25.05	193.69	0.3825E+01	0.4140E+00	0.2900E+01
31	16.35	201.32	25.80	199.33	0.1140E+01	0.3618E+00	0.2943E+01
33	18.27	202.80	26.45	205.00	0.2696E+01	0.3225E+00	0.2957E+01
35	20.19	210.18	27.04	210.68	0.3833E+01	0.2916E+00	0.2955E+01
37	22.11	216.81	27.58	216.34	0.2964E+01	0.2664E+00	0.2948E+01
39	23.65	221.24	27.98	220.88	0.2877E+01	0.2498E+00	0.2943E+01

VI - 28

Connection type : Flush end-plate connections
Mode : All bolted with column stiffener

Tested by : J.Phillips & J.A.Packer (1981) Canada
Test Id. : BM1

Column : W200x100
Beam : W250x33
Plate thickness : 3/8"
Stiffener thickness : 0.2400"

Fasteners: A325- -7/8"D
7/8" Oversize holes
Material : --
Fy = 41.30 ksi
Fu = -- ksi

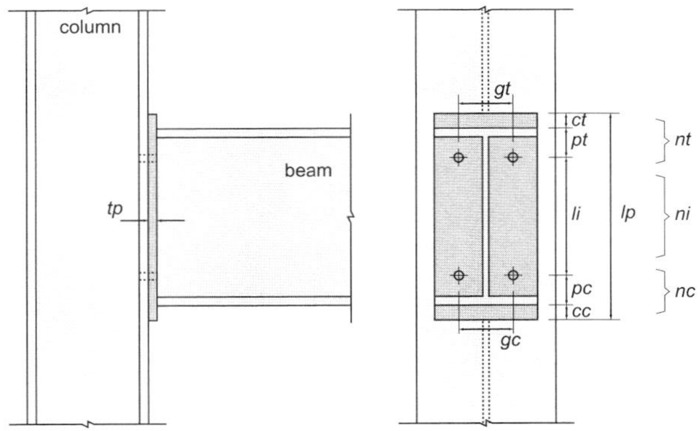

Major parameters

ct = 0.4269"	pt = 1.7385"	li = 6.6930"	pc = 1.7385"
cc = 0.4269"	pit= 2.3622"	pi = 0.0000"	pic= 4.3308"
gt = 3.1496"	gi = 3.1496"	gc = 3.1496"	
tp = 0.3750"	nt = 2 X 1	ni = 2 X 1	nc = 2 X 1

Remark 1) Slight increase in bolt force in top bolt row before separation
2) Bolt preloaded to aproximatly 39 kip

No	Moment (k-in)	Rotation (radians) X 1/1000
1	0.00	0.00
2	41.92	0.06
3	73.36	0.17
4	104.80	0.29
5	131.00	0.43
6	157.20	0.58
7	183.40	0.65
8	209.60	0.72
9	235.80	0.87
10	261.99	1.01
11	288.19	1.09
12	314.39	1.16
13	340.59	1.38
14	366.79	1.59
15	392.99	1.81
16	419.19	2.03
17	445.39	2.10
18	471.59	2.17
19	497.79	2.39
20	523.99	2.61
21	558.92	2.95
22	593.85	3.29
23	628.78	3.62
24	660.22	3.99
25	691.66	4.35
26	733.58	5.51
27	759.78	6.38
28	785.98	7.25
29	809.04	8.70
30	838.38	10.15
31	859.34	11.60
32	890.78	13.05
33	901.26	14.49
34	911.74	15.95
35	922.22	17.39
36	929.21	18.84
37	936.19	20.29
38	943.18	21.74
39	953.66	23.19
40	964.14	24.64
41	968.33	26.09
42	974.62	27.54
43	983.70	28.99
44	992.78	30.44
45	1001.86	31.89
46	1006.06	33.34
47	1012.34	34.79
48	1016.54	36.24
49	1018.63	37.69
50	1022.83	39.14
51	1027.02	40.59

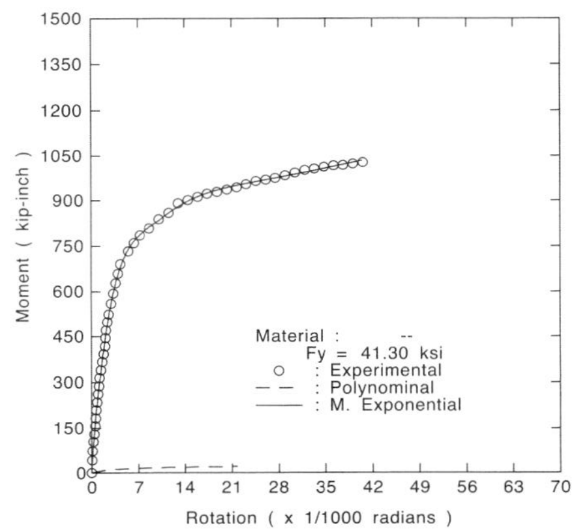

```
Moment-rotation prediction equations     ( R : X 1/1000 radians )

Frye and Morris polynominal model  :     R = Sum ( Ai X ( K*Bm )**Pi X 10**Qi )
     xd =  3.149600"    t =  0.375000"
     A1 =  1.790000   A2 =  1.760000   A3 =  2.040000    K =  0.114754
     P1 =         1   P2 =         3   P3 =         5   Q1 =         0    Q2 =        -1    Q3 =        -1

Modified exponential model  :
     Bm = Sum ( Ai X ( 1 - exp( - R/( 2*i*C ) ) ) )  +      Sum ( RKj X ( R - Rj0 ) )  +  Bm0
       C =  0.35275000E+00              Bmo=  0.00000000E+00          Nexp=  6        Nliner=  1
      AI =  0.68101436E+03  -0.72064730E+04  0.28228940E+05  -0.49000080E+05  0.40274519E+05  -0.12091767E+05
     Rj0 =          10.1500
     RKj =  0.48599904E+01
```

No	Rotation (radians) X 1/1000	Moment (kip-inch) Expri.	Poly.	M.Expo.	Connection stiffness (kip-inch) X 1000 Expri.	Poly.	M.Expo.
1	0.00	0.00	0.00	0.00	0.7228E+03	0.4868E+01	0.3926E+03
3	0.17	73.36	0.85	62.09	0.2711E+03	0.4855E+01	0.3254E+03
5	0.43	131.00	2.10	137.75	0.1807E+03	0.4777E+01	0.2604E+03
7	0.65	183.40	3.13	190.50	0.3614E+03	0.4647E+01	0.2272E+03
9	0.87	235.80	4.12	237.39	0.1807E+03	0.4448E+01	0.2054E+03
11	1.09	288.19	5.06	280.31	0.3616E+03	0.4181E+01	0.1904E+03
13	1.38	340.59	6.21	333.31	0.1205E+03	0.3754E+01	0.1758E+03
15	1.81	392.99	7.70	405.87	0.1205E+03	0.3087E+01	0.1583E+03
17	2.10	445.39	8.54	450.13	0.3614E+03	0.2693E+01	0.1470E+03
19	2.39	497.79	9.26	491.09	0.1205E+03	0.2362E+01	0.1355E+03
21	2.95	558.92	10.43	560.11	0.1033E+03	0.1877E+01	0.1130E+03
23	3.62	628.78	11.56	627.54	0.9532E+02	0.1482E+01	0.8688E+02
25	4.35	691.66	12.53	681.66	0.7470E+02	0.1204E+01	0.6341E+02
27	6.38	759.78	14.49	766.73	0.3013E+02	0.7889E+00	0.2692E+02
29	8.70	809.04	16.04	813.08	0.1808E+02	0.5703E+00	0.1579E+02
31	11.60	859.34	17.46	858.77	0.1807E+02	0.4280E+00	0.1594E+02
33	14.49	901.26	18.57	898.10	0.7230E+01	0.3452E+00	0.1121E+02
35	17.39	922.22	19.49	924.70	0.6026E+01	0.2908E+00	0.7411E+01
37	20.29	936.19	20.28	942.72	0.4820E+01	0.2523E+00	0.5278E+01
39	23.19	953.66	20.96	956.53	0.7235E+01	0.2234E+00	0.4417E+01
41	26.09	968.33	21.58	968.97	0.3615E+01	0.2009E+00	0.4235E+01
43	28.99	983.70	22.13	981.34	0.6262E+01	0.1829E+00	0.4323E+01
45	31.89	1001.86	22.65	994.10	0.4580E+01	0.1680E+00	0.4475E+01
47	34.79	1012.34	23.11	1007.27	0.3616E+01	0.1556E+00	0.4609E+01
49	37.69	1018.63	23.55	1020.78	0.2166E+01	0.1451E+00	0.4706E+01
51	40.59	1027.02	23.96	1034.54	0.2890E+01	0.1361E+00	0.4769E+01

VI - 29

Connection type : Flush end-plate connections
Mode : All bolted with column stiffener

Tested by : J.Phillips & J.A.Packer (198_) Canada
Test Id. : BM2

Column : W200x100	Fasteners: A325- -7/8"D
Beam : W250x33	7/8" Oversize holes
Plate thickness : 5/8"	Material : --
Stiffener thickness : 0.2400"	Fy = 43.50 ksi
	Fu = -- ksi

Major parameters

ct = 0.4269"	pt = 1.7385"	li = 6.6930"	pc = 1.7385"
cc = 0.4269"	pit= 2.3622"	pi = 0.0000"	pic= 4.3308"
gt = 3.1496"	gi = 3.1496"	gc = 3.1496"	
tp = 0.5250"	nt = 2 X 1	ni = 2 X 1	nc = 2 X 1

Remark 1) Slight increase in bolt force in top bolt row before sepration
2) Bolt preloaded upto aproximately 39 kip

No	Moment (k-in)	Rotation (radians) X 1/1000	No	Moment (k-in)	Rotation (radians) X 1/1000
1	0.00	0.00	26	1047.18	14.49
2	52.40	0.07	27	1068.94	15.95
3	104.80	0.14	28	1089.89	17.39
4	157.20	0.43	29	1100.37	18.84
5	209.59	0.58	30	1110.85	20.29
6	261.99	0.72	31	1121.33	21.74
7	314.39	0.78	32	1131.81	23.19
8	366.79	0.87	33	1136.00	24.64
9	419.19	1.01	34	1144.39	26.09
10	461.11	1.09	35	1152.77	27.54
11	503.03	1.16	36	1163.25	28.99
12	544.95	1.26	37	1173.73	30.44
13	586.87	1.35	38	1184.21	31.89
14	628.78	1.45	39	1194.69	33.34
15	681.18	2.03	40	1198.88	34.79
16	738.82	2.32	41	1205.17	36.24
17	796.46	2.61	42	1215.65	37.69
18	843.62	3.33			
19	890.78	4.06			
20	943.18	5.80			
21	964.14	7.25			
22	985.10	8.70			
23	995.58	10.15			
24	1016.53	11.60			
25	1037.50	13.05			

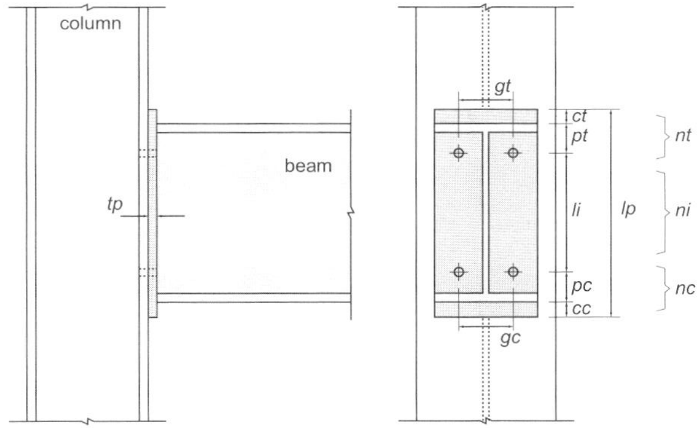

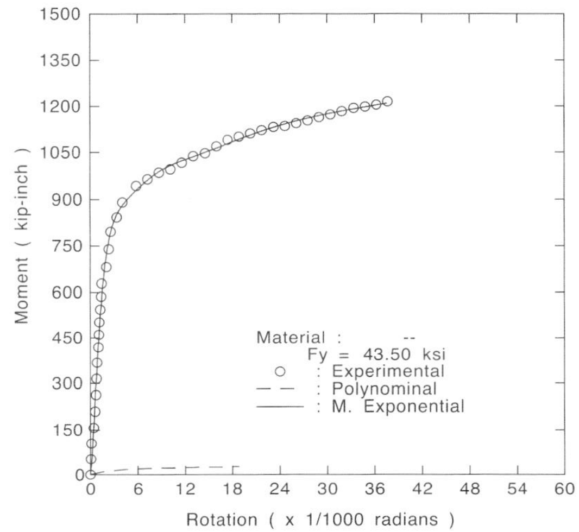

```
Moment-rotation prediction equations     ( R : X 1/1000 radians )

Frye and Morris polynominal model  :    R = Sum ( Ai X ( K*Bm )**Pi X 10**Qi )
     xd =  3.149600"    t =  0.625000"
     A1 =  1.790000   A2 =  1.760000   A3 =  2.040000    K =  0.084462
     P1 =        1    P2 =        3    P3 =        5    Q1 =        0    Q2 =       -1    Q3 =       -1

Modified exponential model  :
     Bm = Sum ( Ai X ( 1 - exp( - R/( 2*i*C ) ) ) )  +      Sum ( RKj X ( R - Rj0 ) )  +  Bm0
      C =   0.33218333E+00            Bmo=   0.00000000E+00          Nexp=  6        Nliner=  1
     AI =  -0.19399025E+04   0.16016222E+05  -0.53985243E+05   0.98418922E+05  -0.88368113E+05   0.30954931E+05
    Rj0 =            7.2500
    RKj =   0.37529494E+01
```

No	Rotation (radians) X 1/1000	Moment (kip-inch) Expri.	Poly.	M.Expo.	Connection stiffness (kip-inch) X 1000 Expri.	Poly.	M.Expo.
1	0.00	0.00	0.00	0.00	0.7227E+03	0.6614E+01	0.2459E+03
3	0.14	104.80	0.96	45.02	0.6143E+03	0.6601E+01	0.3651E+03
5	0.58	209.59	3.79	236.85	0.3615E+03	0.6383E+01	0.4703E+03
7	0.78	314.39	5.05	329.40	0.8102E+03	0.6167E+01	0.4502E+03
9	1.01	419.19	6.46	429.92	0.5065E+03	0.5811E+01	0.4037E+03
11	1.16	503.03	7.28	485.96	0.5163E+03	0.5544E+01	0.3693E+03
13	1.35	586.87	8.31	552.76	0.4337E+03	0.5151E+01	0.3218E+03
15	2.03	681.18	11.33	718.84	0.1627E+03	0.3783E+01	0.1780E+03
17	2.61	796.46	13.25	797.63	0.1607E+03	0.2921E+01	0.1008E+03
19	4.06	890.78	16.53	880.89	0.5479E+02	0.1769E+01	0.3404E+02
21	7.25	964.14	20.56	960.91	0.1446E+02	0.9355E+00	0.1863E+02
23	10.15	995.58	22.83	1008.07	0.1083E+02	0.6634E+00	0.1146E+02
25	13.05	1037.50	24.52	1036.33	0.1058E+02	0.5187E+00	0.9128E+01
27	15.95	1068.94	25.89	1063.84	0.1473E+02	0.4285E+00	0.9843E+01
29	18.84	1100.37	27.03	1092.48	0.7230E+01	0.3669E+00	0.9735E+01
31	21.74	1121.33	28.03	1119.23	0.7230E+01	0.3218E+00	0.8611E+01
33	24.64	1136.00	28.91	1142.14	0.4339E+01	0.2874E+00	0.7198E+01
35	27.54	1152.77	29.70	1161.16	0.6507E+01	0.2601E+00	0.5978E+01
37	30.44	1173.73	30.42	1177.13	0.7230E+01	0.2380E+00	0.5099E+01
39	33.34	1194.69	31.09	1191.03	0.5060E+01	0.2197E+00	0.4530E+01
41	36.24	1205.17	31.70	1203.62	0.5785E+01	0.2042E+00	0.4185E+01

VI - 30

Connection type : Flush end-plate connections
Mode : All bolted with column stiffener

Tested by : J.Phillips & J.A.Packer (1981) Canada
Test Id. : BM3

Column : W200x100
Beam : W250x33
Plate thickness : 3/4"
Stiffener thickness : 0.2400"

Fasteners: A325- -7/8"D
7/8" Oversize holes
Material : --
Fy = 41.30 ksi
Fu = -- ksi

Major parameters

ct = 0.4269"	pt = 1.7385"	li = 6.6930"	pc = 1.7385"
cc = 0.4269"	pit= 2.3622"	pi = 0.0000"	pic= 4.3308"
gt = 3.1496"	gi = 3.1496"	gc = 3.1496"	
tp = 0.7500"	nt = 2 X 1	ni = 2 X 1	nc = 2 X 1

Remark 1) (A) Bolts were preloaded (B) Failure occured due to bolt fracture
2) (A) Clear separation occured (3) End plate collapsed

No	Moment (k-in)	Rotation (radians) X 1/1000	No	Moment (k-in)	Rotation (radians) X 1/1000
1	0.00	0.00	26	1108.76	15.95
2	52.40	0.07	27	1123.43	17.39
3	104.80	0.14	28	1132.86	18.84
4	157.20	0.22	29	1142.29	20.29
5	209.59	0.29	30	1152.77	21.74
6	261.99	0.43	31	1169.54	23.19
7	314.39	0.58	32	1176.88	24.64
8	366.79	0.72	33	1184.21	26.09
9	419.19	0.87	34	1192.60	27.54
10	471.59	1.01	35	1200.98	28.99
11	523.99	1.16	36	1208.32	30.44
12	576.39	1.30	37	1215.65	31.89
13	628.78	1.45	38	1236.61	33.34
14	681.18	1.84	39	1247.10	34.79
15	733.58	2.22	40	1252.34	36.24
16	785.98	2.61	41	1257.57	37.69
17	838.38	3.19			
18	890.78	4.35			
19	943.18	5.80			
20	985.10	7.25			
21	1006.06	8.70			
22	1027.02	10.15			
23	1047.97	11.60			
24	1068.94	13.05			
25	1100.38	14.49			

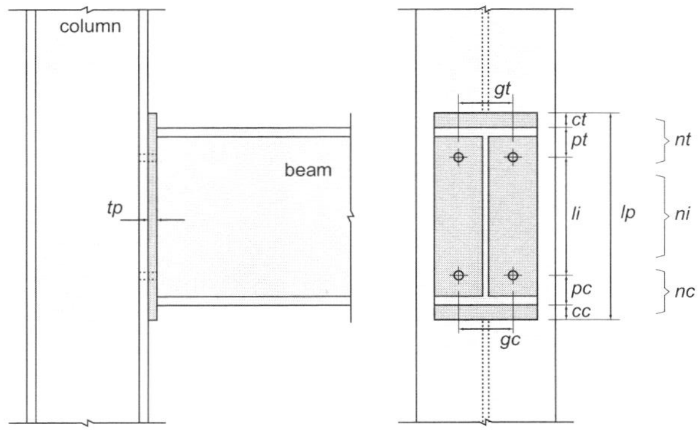

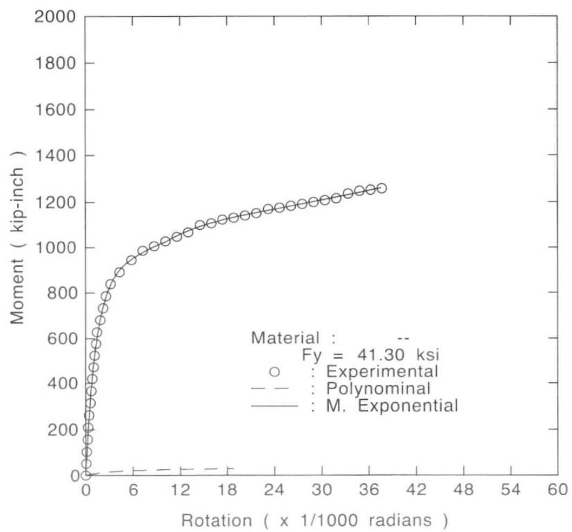

```
Moment-rotation prediction equations      ( R : X 1/1000 radians )

Frye and Morris polynominal model  :    R = Sum ( Ai X ( K*Bm )**Pi X 10**Qi )
     xd =  3.149600"    t =  0.750000"
     A1 =  1.790000   A2 =  1.760000   A3 =  2.040000    K =  0.075709
     P1 =        1    P2 =        3    P3 =        5    Q1 =        0    Q2 =       -1    Q3 =       -1

Modified exponential model  :
     Bm = Sum ( Ai X ( 1 - exp( - R/( 2*i*C ) ) ) )  +      Sum ( RKj X ( R - Rj0 ) )  +  Bm0
      C =   0.33218333E+00             Bmo=   0.00000000E+00             Nexp=  6         Nliner=  1
     AI =   0.24791008E+03  -0.18876920E+04   0.13200129E+05  -0.27637029E+05   0.24721130E+05  -0.75745429E+04
    Rj0 =          10.1500
    RKj =   0.69409585E+01
```

No	Rotation (radians) X 1/1000	Moment (kip-inch) Expri.	Moment (kip-inch) Poly.	Moment (kip-inch) M.Expo.	Connection stiffness (kip-inch) X 1000 Expri.	Connection stiffness (kip-inch) X 1000 Poly.	Connection stiffness (kip-inch) X 1000 M.Expo.
1	0.00	0.00	0.00	0.00	0.7227E+03	0.7379E+01	0.7175E+03
3	0.14	104.80	1.07	97.42	0.7232E+03	0.7365E+01	0.6287E+03
5	0.29	209.59	2.14	182.93	0.6023E+03	0.7320E+01	0.5536E+03
7	0.58	314.39	4.23	325.30	0.3614E+03	0.7121E+01	0.4347E+03
9	0.87	419.19	6.24	437.92	0.3615E+03	0.6742E+01	0.3464E+03
11	1.16	523.99	8.12	528.18	0.3616E+03	0.6185E+01	0.2793E+03
13	1.45	628.78	9.82	601.30	0.2998E+03	0.5518E+01	0.2273E+03
15	2.22	733.58	13.42	737.97	0.1356E+03	0.3862E+01	0.1354E+03
17	3.19	838.38	16.47	836.08	0.7531E+02	0.2600E+01	0.7467E+02
19	5.80	943.18	21.23	945.86	0.3254E+02	0.1325E+01	0.2503E+02
21	8.70	1006.06	24.31	1003.71	0.1446E+02	0.8644E+00	0.1648E+02
23	11.60	1047.97	26.47	1052.55	0.1446E+02	0.6487E+00	0.1716E+02
25	14.49	1100.38	28.15	1093.96	0.1374E+02	0.5232E+00	0.1172E+02
27	17.39	1123.43	29.54	1122.69	0.8316E+01	0.4408E+00	0.8466E+01
29	20.29	1142.29	30.73	1144.80	0.6869E+01	0.3824E+00	0.7022E+01
31	23.19	1169.54	31.77	1164.36	0.8312E+01	0.3386E+00	0.6575E+01
33	26.09	1184.21	32.71	1183.32	0.5423E+01	0.3046E+00	0.6546E+01
35	28.99	1200.98	33.55	1202.42	0.5423E+01	0.2773E+00	0.6641E+01
37	31.89	1215.65	34.33	1221.83	0.9760E+01	0.2546E+00	0.6744E+01
39	34.79	1247.10	35.03	1241.50	0.5424E+01	0.2359E+00	0.6822E+01
41	37.69	1257.57	35.69	1261.36	0.3612E+01	0.2200E+00	0.6873E+01

A.6 – 61

VI - 31

Connection type : Flush end-plate connections
Mode : All bolted with column stiffener

Tested by : J.Phillips & J.A.Packer (1981) Canada
Test Id. : BM4

Column : W200x100
Beam : W250x33
Plate thickness : 7/8"
Stiffener thickness : 0.2400"

Fasteners: A325- -7/8"D
7/8" Oversize holes
Material : --
Fy = 41.30 ksi
Fu = -- ksi

Major parameters

ct = 0.4269"	pt = 1.7385"	li = 6.6930"	pc = 1.7385"
cc = 0.4269"	pit= 2.3622"	pi = 0.0000"	pic= 4.3308"
gt = 3.1496"	gi = 3.1496"	gc = 3.1496"	
tp = 0.8750"	nt = 2 X 1	ni = 2 X 1	nc = 2 X 1

Remark 1) (A) Bolts were preloaded (B) Failure due to bolt fracture
2) (A) Clear separation occured (3) End plate collapsed

No	Moment (k-in)	Rotation (radians) X 1/1000	No	Moment (k-in)	Rotation (radians) X 1/1000
1	0.00	0.00	26	1131.81	13.05
2	52.40	0.07	27	1152.77	14.49
3	104.80	0.14	28	1168.49	15.94
4	157.20	0.22	29	1184.21	17.39
5	209.59	0.29	30	1199.93	18.84
6	261.99	0.36	31	1215.65	20.29
7	314.39	0.43	32	1219.84	21.74
8	366.79	0.51	33	1228.23	23.19
9	419.19	0.58	34	1236.61	24.64
10	471.59	0.72	35	1245.00	26.09
11	523.99	0.87	36	1256.53	27.54
12	576.39	1.01	37	1268.05	28.99
13	628.78	1.16	38	1278.53	30.44
14	681.18	1.30	39	1282.72	31.89
15	733.58	1.45	40	1289.01	33.34
16	778.99	1.93			
17	824.41	2.42			
18	869.82	2.90			
19	906.50	3.62			
20	943.18	4.35			
21	995.58	5.80			
22	1047.98	7.25			
23	1068.94	8.70			
24	1089.89	10.15			
25	1110.85	11.60			

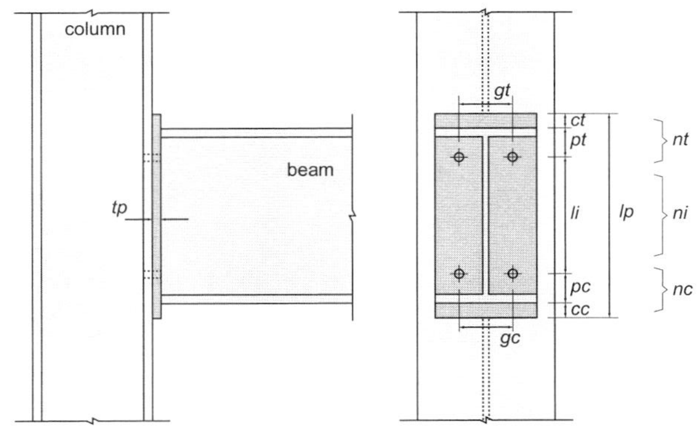

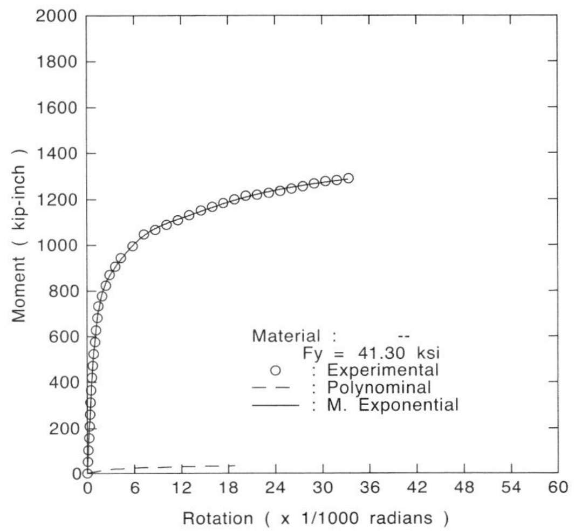

```
Moment-rotation prediction equations     ( R : X 1/1000 radians )

Frye and Morris polynominal model   :     R = Sum ( Ai X ( K*Bm )**Pi X 10**Qi )
     xd =  3.149600"   t =  0.875000"
     A1 =  1.790000   A2 =  1.760000   A3 =  2.040000    K =  0.069021
     P1 =         1   P2 =         3   P3 =         5   Q1 =         0   Q2 =        -1   Q3 =        -1

Modified exponential model  :
     Bm = Sum ( Ai X ( 1 - exp( - R/( 2*i*C ) ) ) )  +      Sum ( RKj X ( R - Rj0 ) )  +  Bm0
       C =   0.29595833E+00             Bmo=   0.00000000E+00          Nexp=  6        Nliner=  1
      AI =  -0.95921141E+03   0.13575756E+05  -0.51739303E+05   0.97766465E+05  -0.89051670E+05   0.31596992E+05
     Rj0 =           7.2500
     RKj =   0.37991494E+01
```

No	Rotation (radians) X 1/1000	Moment (kip-inch) Expri.	Moment (kip-inch) Poly.	Moment (kip-inch) M.Expo.	Connection stiffness (kip-inch) X 1000 Expri.	Connection stiffness (kip-inch) X 1000 Poly.	Connection stiffness (kip-inch) X 1000 M.Expo.
1	0.00	0.00	0.00	0.00	0.7227E+03	0.8094E+01	0.8104E+03
3	0.14	104.80	1.17	113.68	0.7232E+03	0.8078E+01	0.7549E+03
5	0.29	209.59	2.34	218.27	0.7227E+03	0.8029E+01	0.6874E+03
7	0.43	314.39	3.50	312.72	0.7232E+03	0.7942E+01	0.6149E+03
9	0.58	419.19	4.64	396.55	0.6023E+03	0.7811E+01	0.5425E+03
11	0.87	523.99	6.85	533.93	0.3615E+03	0.7396E+01	0.4087E+03
13	1.16	628.78	8.91	635.83	0.3616E+03	0.6784E+01	0.2986E+03
15	1.45	733.58	10.77	709.60	0.2996E+03	0.6053E+01	0.2145E+03
17	2.42	824.41	15.50	836.23	0.9399E+02	0.3883E+01	0.7715E+02
19	3.62	906.50	19.23	904.40	0.5061E+02	0.2463E+01	0.4714E+02
21	5.80	995.58	23.29	997.76	0.3614E+02	0.1454E+01	0.3653E+02
23	8.70	1068.94	26.66	1073.87	0.1446E+02	0.9482E+00	0.1800E+02
25	11.60	1110.85	29.03	1113.94	0.1446E+02	0.7116E+00	0.1182E+02
27	14.49	1152.77	30.88	1147.95	0.1266E+02	0.5740E+00	0.1177E+02
29	17.39	1184.21	32.41	1181.14	0.1085E+02	0.4836E+00	0.1088E+02
31	20.29	1215.65	33.71	1210.07	0.6868E+01	0.4194E+00	0.9001E+01
33	23.19	1228.23	34.85	1233.36	0.5786E+01	0.3714E+00	0.7126E+01
35	26.09	1245.00	35.87	1251.87	0.6869E+01	0.3341E+00	0.5736E+01
37	28.99	1268.05	36.80	1267.11	0.7589E+01	0.3041E+00	0.4855E+01
39	31.89	1282.72	37.65	1280.38	0.3614E+01	0.2793E+00	0.4349E+01

VI - 32

Connection type : Flush end-plate connecticns
Mode : All bolted with column stiffener

Tested by : J.Phillips & J.A.Packer (1981) Canada
Test Id. : BM5

Column : W200x100
Beam : W250x33
Plate thickness : 1"
Stiffener thickness : 0.2400"

Fasterers: A325- -7/8"D
7/8" Oversize holes
Material : --
Fy = 41.30 ksi
Fu = -- ksi

Major parameters

ct = 0.4269"	pt = 1.7385"	li = 6.6930"	pc = 1.7385"
cc = 0.4269"	pit= 2.3622"	pi = 0.0000"	pic= 4.3308"
gt = 3.1496"	gi = 3.1496"	gc = 3.1496"	
tp = 1.0000"	nt = 2 X 1	ni = 2 X 1	nc = 2 X 1

Remark 1) (A) Bolts were preloaded (B) Failure due bolt fracture
2) (A) Clear separation occured (B) End plate yielded

No	Moment (k-in)	Rotation (radians) X 1/1000	No	Moment (k-in)	Rotation (radians) X 1/1000
1	0.00	0.00	26	1137.05	7.97
2	52.40	0.05	27	1152.77	8.70
3	104.80	0.10	28	1163.25	9.42
4	157.20	0.14	29	1173.73	10.15
5	209.60	0.18	30	1184.21	10.87
6	262.00	0.24	31	1194.69	11.60
7	314.39	0.29	32	1199.93	12.32
8	366.79	0.38	33	1205.17	13.05
9	419.19	0.46	34	1217.05	14.02
10	471.59	0.60	35	1228.92	14.98
11	523.99	0.73	36	1240.80	15.95
12	576.39	0.80	37	1251.28	16.67
13	628.78	0.87	38	1261.76	17.39
14	681.18	1.01	39	1280.62	18.12
15	733.58	1.16	40	1299.49	18.84
16	785.98	1.45	41	1309.97	19.57
17	838.38	1.84	42	1320.45	20.29
18	890.78	2.22	43	1323.59	21.02
19	943.18	2.61	44	1326.74	21.74
20	1006.06	3.48	45	1328.84	22.47
21	1047.97	4.35	46	1330.93	23.19
22	1068.93	5.07	47	1341.41	24.35
23	1089.89	5.80			
24	1105.61	6.52			
25	1121.33	7.25			

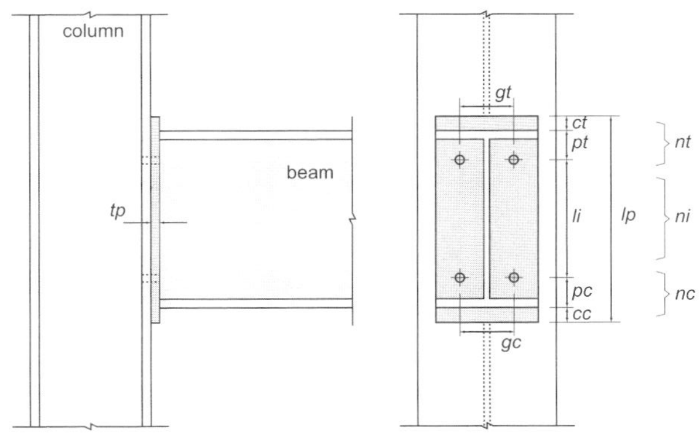

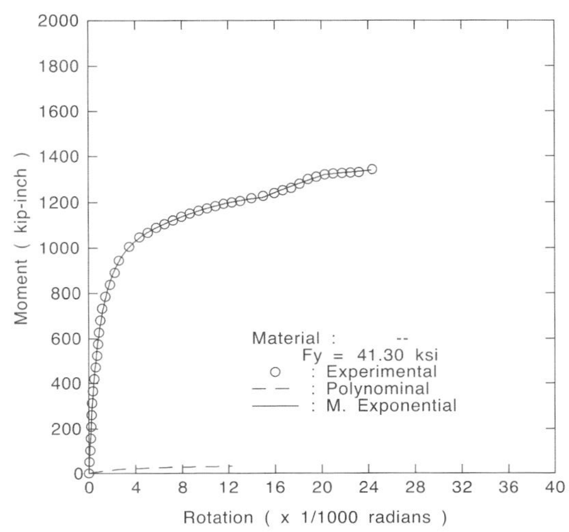

A.6 – 64

```
Moment-rotation prediction equations      ( R : X 1/1000 radians )

Frye and Morris polynominal model  :    R = Sum ( Ai X ( K*Bm )**Pi X 10**Qi )
     xd =  3.149600"   t =  1.000000"
     A1 =  1.790000   A2 =  1.760000   A3 =  2.040000    K =  0.063707
     P1 =         1   P2 =         3   P3 =         5   Q1 =         0   Q2 =       -1   Q3 =       -1

Modified exponential model  :
     Bm = Sum ( Ai X ( 1 - exp( - R/( 2*i*C ) ) ) )  +     Sum ( RKj X ( R - Rj0 ) )  +  Bm0
       C =   0.21541667E+00          Bmo=   0.00000000E+00          Nexp=  6        Nliner=  2
      AI =   0.25217080E+03   0.15532262E+04  -0.96654329E+04   0.28694211E+05  -0.35044795E+05   0.15450292E+05
     Rj0 =          14.9800          20.2900
     RKj =   0.15688663E+02  -0.11595936E+02
```

No	Rotation (radians) X 1/1000	Moment (kip-inch) Expri.	Moment (kip-inch) Poly.	Moment (kip-inch) M.Expo.	Connection stiffness (kip-inch) X 1000 Expri.	Connection stiffness (kip-inch) X 1000 Poly.	Connection stiffness (kip-inch) X 1000 M.Expo.
1	0.00	0.00	0.00	0.00	0.1048E+04	0.8769E+01	0.1269E+04
3	0.10	104.80	0.88	117.09	0.1156E+04	0.8761E+01	0.1079E+04
5	0.18	209.60	1.61	202.00	0.1134E+04	0.8741E+01	0.9461E+03
7	0.29	314.39	2.54	294.52	0.8423E+03	0.8699E+01	0.8059E+03
9	0.46	419.19	4.04	418.39	0.5213E+03	0.8580E+01	0.6281E+03
11	0.73	523.99	6.26	558.06	0.6180E+03	0.8262E+01	0.4453E+03
13	0.87	628.78	7.42	616.27	0.6134E+03	0.8013E+01	0.3761E+03
15	1.16	733.58	9.65	709.64	0.3013E+03	0.7350E+01	0.2756E+03
17	1.84	838.38	14.00	848.96	0.1356E+03	0.5497E+01	0.1537E+03
19	2.61	943.18	17.57	940.92	0.1161E+03	0.3872E+01	0.9185E+02
21	4.35	1047.97	22.57	1044.82	0.3769E+02	0.2168E+01	0.3874E+02
23	5.80	1089.89	25.23	1089.65	0.2530E+02	0.1575E+01	0.2551E+02
25	7.25	1121.33	27.25	1122.73	0.2169E+02	0.1241E+01	0.2078E+02
27	8.70	1152.77	28.89	1150.70	0.1807E+02	0.1027E+01	0.1785E+02
29	10.15	1173.73	30.26	1174.39	0.1446E+02	0.8796E+00	0.1479E+02
31	11.60	1194.69	31.46	1193.49	0.1084E+02	0.7710E+00	0.1157E+02
33	13.05	1205.17	32.51	1208.09	0.9399E+01	0.6875E+00	0.8576E+01
35	14.98	1228.92	33.75	1221.40	0.1231E+02	0.6027E+00	0.5378E+01
37	16.67	1251.28	34.73	1255.21	0.1447E+02	0.5446E+00	0.1909E+02
39	18.12	1280.62	35.49	1281.97	0.2604E+02	0.5041E+00	0.1792E+02
41	19.57	1309.97	36.19	1307.33	0.1446E+02	0.4696E+00	0.1712E+02
43	21.02	1323.59	36.85	1323.30	0.4339E+01	0.4398E+00	0.4994E+01
45	22.47	1328.84	37.47	1330.27	0.2892E+01	0.4140E+00	0.4652E+01
47	24.35	1341.41	38.22	1338.75	0.9050E+01	0.3851E+00	0.4389E+01

VI - 33

Connection type : Flush end-plate connections
Mode : All bolted without column stiffener

Tested by : C.Bernuzzi et al. (1996) Italy
Test Id. : FPC/M

Column : --
Beam : IPE300
Plate thickness : 12 mm

Fasteners: G8.8- -M20
7/8" Oversize holes
Material : --
Fy = 45.40 ksi
Fu = 66.57 ksi

Major parameters

ct = 0.0000" pt = 2.3622" li = 7.0866" pc = 2.3622"
cc = 0.0000" gt = 4.1339" gc = 4.1339"
tp = 0.4724" nt = 2 X 1 nc = 2 X 1

Remark 1)
2)

No	Moment (k-in)	Rotation (radians) X 1/1000	No	Moment (k-in)	Rotation (radians) X 1/1000
1	0.00	0.00	26	534.64	35.15
2	45.93	0.10	27	537.23	37.22
3	91.86	0.20	28	548.53	39.28
4	137.79	0.30	29	557.35	41.35
5	183.72	0.40	30	559.15	43.42
6	224.78	0.60	31	557.11	45.49
7	267.34	1.02	32	564.69	47.55
8	296.46	1.51	33	577.38	49.62
9	322.18	2.04	34	581.70	51.69
10	345.22	3.06	35	591.92	53.76
11	375.12	4.14	36	593.81	55.82
12	396.11	6.20	37	593.81	57.89
13	411.30	8.27	38	595.40	59.96
14	424.09	10.34	39	604.54	62.03
15	442.68	12.41	40	612.39	64.09
16	450.84	14.47	41	622.81	66.16
17	461.97	16.54	42	629.62	68.23
18	474.89	18.61	43	633.78	70.30
19	479.43	20.68	44	633.41	72.36
20	490.87	22.74	45	631.38	74.43
21	498.86	24.81	46	627.43	76.50
22	515.71	26.88			
23	519.22	28.95			
24	521.68	31.01			
25	522.12	33.08			

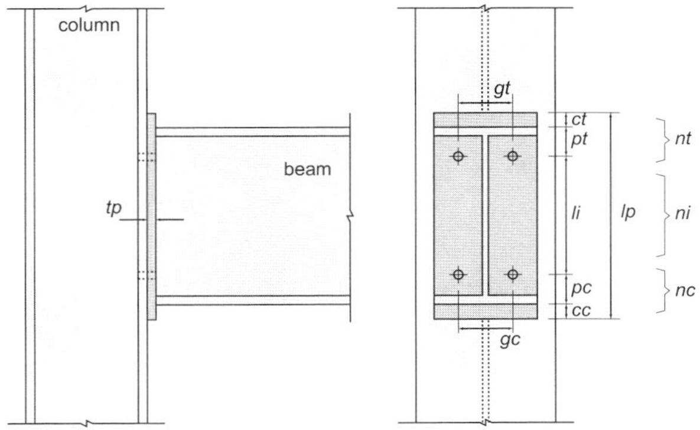

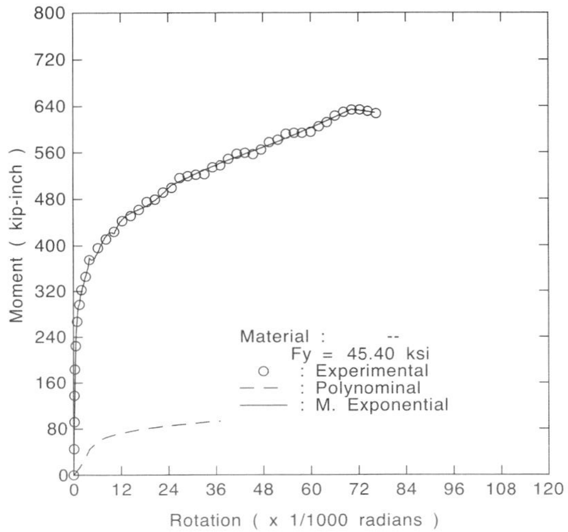

```
Moment-rotation prediction equations     ( R : X 1/1000 radians )

Frye and Morris polynominal model  :    R = Sum ( Ai X ( K*Bm )**Pi X 10**Qi )
     xd =  4.133900"   t =  0.472440"  tf =  0.787400"
     A1 =  1.830000   A2 = -1.040000   A3 =  6.380000     K =  0.064077
     P1 =        1     P2 =        3    P3 =        5     Q1 =        0     Q2 =       -1     Q3 =       -3

Modified exponential model  :
     Bm = Sum ( Ai X ( 1 - exp( - R/( 2*i*C ) ) ) )  +      Sum ( RKj X ( R - Rj0 ) )  +  Bm0
      C =   0.66250000E+00              Bmo=   0.00000000E+00              Nexp=  6           Nliner=  6
     AI =   0.34684645E+04  -0.20648853E+05   0.57536322E+05  -0.84280798E+05   0.63663012E+05  -0.19416606E+05
    Rj0 =           1.5100           3.0600           4.1400          10.3400          24.8100
                   70.3000
    RKj =   0.60594844E+02  -0.18343542E+02  -0.56504096E+02   0.21738200E+02  -0.41470415E+01
           -0.41682080E+01
```

No	Rotation (radians) X 1/1000	Moment (kip-inch) Expri.	Poly.	M.Expo.	Connection stiffness (kip-inch) X 1000 Expri.	Poly.	M.Expo.
1	0.00	0.00	0.00	0.00	0.4593E+03	0.8528E+01	0.5654E+03
3	0.20	91.86	1.71	97.84	0.4593E+03	0.8545E+01	0.4186E+03
5	0.40	183.72	3.42	169.45	0.3746E+03	0.8598E+01	0.3021E+03
7	1.02	267.34	8.86	278.81	0.8199E+02	0.9006E+01	0.8017E+02
9	2.04	322.18	18.81	318.65	0.3966E+02	0.1081E+02	0.2913E+02
11	4.14	375.12	45.90	378.48	0.2167E+02	0.1028E+02	0.4384E+02
13	8.27	411.30	65.42	416.34	0.6757E+01	0.2517E+01	0.8908E+01
15	12.41	442.68	73.32	445.21	0.6465E+01	0.1489E+01	0.7602E+01
17	16.54	461.97	78.53	461.96	0.5811E+01	0.1085E+01	0.2862E+01
19	20.68	479.43	82.53	478.13	0.3877E+01	0.8646E+00	0.5189E+01
21	24.81	498.86	85.79	503.83	0.6001E+01	0.7248E+00	0.6991E+01
23	28.95	519.22	88.58	516.78	0.1443E+01	0.6272E+00	0.3231E+01
25	33.08	522.12	91.01	529.69	0.3131E+01	0.5551E+00	0.2977E+01
27	37.22	537.23	93.19	541.34	0.3373E+01	0.4992E+00	0.2672E+01
29	41.35	557.35	95.16	552.01	0.2565E+01	0.4547E+00	0.2526E+01
31	45.49	557.11	96.98	562.45	0.1353E+01	0.4178E+00	0.2536E+01
33	49.62	577.38	98.64	573.12	0.4110E+01	0.3875E+00	0.2641E+01
35	53.76	591.92	100.19	584.34	0.2921E+01	0.3617E+00	0.2782E+01
37	57.89	593.81	101.63	596.12	0.3842E+00	0.3395E+00	0.2919E+01
39	62.03	604.54	103.00	608.45	0.4113E+01	0.3201E+00	0.3036E+01
41	66.16	622.81	104.27	621.19	0.4162E+01	0.3032E+00	0.3127E+01
43	70.30	633.78	105.51	634.28	0.9123E+00	0.2878E+00	0.3195E+01
45	74.43	631.38	106.66	630.37	-0.1444E+01	0.2744E+00	-0.9253E+00

VI - 34

Connection type : Flush end-plate connections
Mode : All bolted without column stiffener

Tested by : B.Bose et al. (1996) U.K.
Test Id. : 1

Column : 254x254 UC 73
Beam : 406x178 UB 60
Plate thickness : 12 mm

Fasteners: G8.8- -M20
7/8" Oversize holes
Material : S275
Fy = -- ksi
Fu = -- ksi

Major parameters

ct = 1.0551"	pt = 2.3622"	li = 11.2756"	pc = 2.3622"
cc = 1.0551"	pit= 3.5433"	pi = 0.0000"	pic= 7.7323"
gt = 3.5433"	gi = 3.5433"	gc = 3.5433"	
tp = 0.4724"	nt = 2 X 1	ni = 2 X 1	nc = 2 X 1

Remark 1) Failure mode: Thread stripping
2) Plate size: 460x200x12 mm

No	Moment (k-in)	Rotation (radians) X 1/1000
1	0.00	0.00
2	113.32	0.51
3	192.35	1.01
4	285.06	1.52
5	376.55	2.02
6	475.84	2.53
7	560.78	3.03
8	619.63	3.54
9	669.21	4.04
10	759.70	5.05
11	835.15	6.06
12	885.30	7.07
13	926.76	8.08
14	971.09	9.09
15	1004.98	10.11
16	1040.17	11.12
17	1063.11	12.13
18	1096.88	13.14
19	1130.04	14.15
20	1147.93	15.16
21	1167.44	16.17
22	1185.84	17.18
23	1203.54	18.19
24	1221.24	19.20
25	1230.09	20.21
26	1239.25	21.22
27	1255.43	22.23
28	1270.53	23.24
29	1283.19	24.25
30	1294.44	25.26
31	1309.73	26.27
32	1322.66	27.28
33	1333.26	28.29
34	1345.13	29.31
35	1357.80	30.32
36	1373.54	31.33
37	1388.45	32.34
38	1389.38	33.35
39	1391.71	34.36
40	1396.46	35.37
41	1398.23	36.38

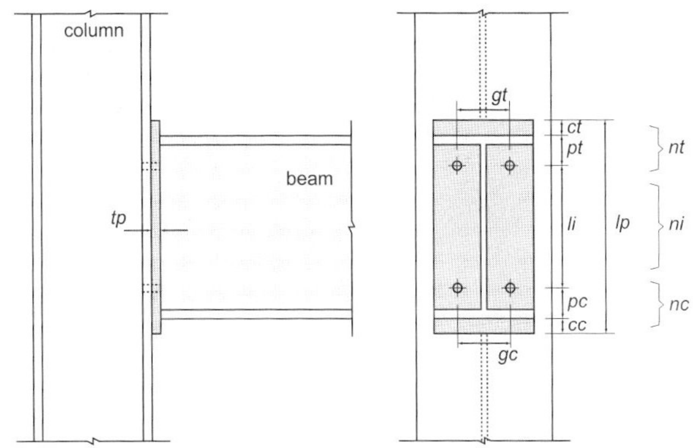

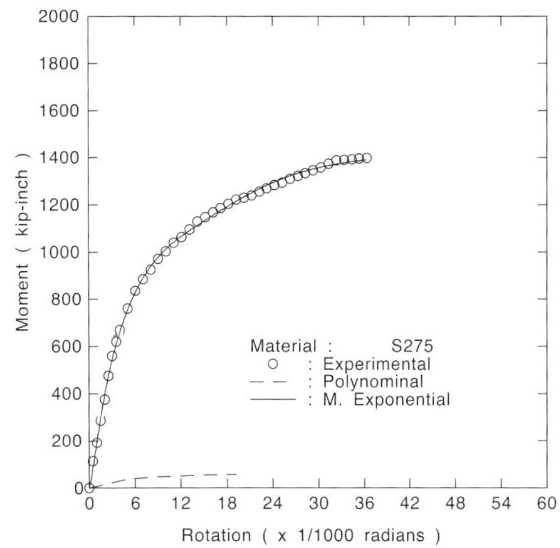

```
Moment-rotation prediction equations     ( R : X 1/1000 radians )

Frye and Morris polynominal model  :    R = Sum ( Ai X ( K*Bm )**Pi X 10**Qi )
     xd =  3.543300"    t =  0.472440"  tf =  0.787400"
     A1 =  1.830000    A2 = -1.040000   A3 =  6.380000    K =  0.092766
     P1 =        1     P2 =        3    P3 =        5     Q1 =       0     Q2 =      -1     Q3 =      -3

Modified exponential model  :
     Bm = Sum ( Ai X ( 1 - exp( - R/( 2*i*C ) ) ) )  +    Sum ( RKj X ( R - Rj0 ) )  + Bm0
      C =   0.43066667E+00           Bmo=   0.00000000E+00          Nexp=  6        Nliner=  0
     AI =  -0.80815368E+03   0.89774205E+04  -0.46877165E+05   0.11442230E+06  -0.12519034E+06   0.50886715E+05
```

No	Rotation (radians) X 1/1000	Moment (kip-inch) Expri.	Moment (kip-inch) Poly.	Moment (kip-inch) M.Expo.	Connection stiffness (kip-inch) X 1000 Expri.	Connection stiffness (kip-inch) X 1000 Poly.	Connection stiffness (kip-inch) X 1000 M.Expo.
1	0.00	0.00	0.00	0.00	0.2222E+03	0.5891E+01	0.1201E+03
3	1.01	192.35	6.06	194.46	0.1698E+03	0.6214E+01	0.2157E+03
5	2.02	376.55	12.84	394.24	0.1888E+03	0.7428E+01	0.1748E+03
7	3.03	560.78	21.54	547.65	0.1429E+03	0.9864E+01	0.1312E+03
9	4.04	669.21	30.98	664.15	0.9600E+02	0.7543E+01	0.1015E+03
11	6.06	835.15	40.28	829.88	0.6218E+02	0.2919E+01	0.6599E+02
13	8.08	926.76	44.86	939.38	0.4247E+02	0.1799E+01	0.4401E+02
15	10.11	1004.98	47.97	1013.80	0.3404E+02	0.1321E+01	0.3064E+02
17	12.13	1063.11	50.35	1068.12	0.2807E+02	0.1057E+01	0.2404E+02
19	14.15	1130.04	52.30	1113.56	0.2527E+02	0.8866E+00	0.2138E+02
21	16.17	1167.44	53.96	1155.51	0.1877E+02	0.7676E+00	0.2026E+02
23	18.19	1203.54	55.42	1195.46	0.1752E+02	0.6793E+00	0.1925E+02
25	20.21	1230.09	56.72	1232.96	0.8916E+01	0.6110E+00	0.1779E+02
27	22.23	1255.43	57.90	1267.00	0.1549E+02	0.5564E+00	0.1585E+02
29	24.25	1283.19	58.97	1296.81	0.1184E+02	0.5117E+00	0.1363E+02
31	26.27	1309.73	59.97	1322.05	0.1397E+02	0.4743E+00	0.1136E+02
33	28.29	1333.26	60.89	1342.78	0.1106E+02	0.4426E+00	0.9205E+01
35	30.32	1357.80	61.76	1359.47	0.1407E+02	0.4151E+00	0.7280E+01
37	32.34	1388.45	62.59	1372.48	0.7840E+01	0.3910E+00	0.5651E+01
39	34.36	1391.71	63.35	1382.49	0.3505E+01	0.3703E+00	0.4313E+01
41	36.38	1398.23	64.08	1390.08	0.1752E+01	0.3519E+00	0.3245E+01

A.6 – 69

VI - 35

Connection type : Flush end-plate connections
Mode : All bolted without column stiffener

Tested by : B.Bose et al. (1996) U.K.
Test Id. : 2

Column : 254x254 UC 89
Beam : 406x178 UB 60
Plate thickness : 12 mm

Fasteners: G8.8- -M20
7/8" Oversize holes
Material : S275
Fy = 58.74 ksi
Fu = 79.92 ksi

Major parameters

ct = 0.8583"	pt = 2.3622"	li = 11.2756"	pc = 2.3622"
cc = 0.8583"	pit= 3.5433"	pi = 0.0000"	pic= 7.7323"
gt = 3.5433"	gi = 3.5433"	gc = 3.5433"	
tp = 0.4724"	nt = 2 X 1	ni = 2 X 1	nc = 2 X 1

Remark 1) Failure mode: Thread stripping
2) Plate size: 450x200x12 mm

No	Moment (k-in)	Rotation (radians) X 1/1000	No	Moment (k-in)	Rotation (radians) X 1/1000
1	0.00	0.00	26	1095.93	21.06
2	109.21	0.50	27	1102.65	22.06
3	191.75	1.00	28	1106.19	23.07
4	279.40	1.50	29	1106.19	24.07
5	377.69	2.01	30	1107.10	25.07
6	474.43	2.51	31	1107.96	26.07
7	555.95	3.01	32	1107.96	27.08
8	607.08	3.51	33	1107.96	28.08
9	657.46	4.01	34	1107.96	29.08
10	721.00	5.01	35	1108.85	30.09
11	777.38	6.02	36	1106.19	31.09
12	812.02	7.02	37	1107.96	32.09
13	851.09	8.02	38	1109.06	33.20
14	878.54	9.03	39	1109.73	34.50
15	917.51	10.03	40	1109.73	35.60
16	974.25	11.03			
17	1001.01	12.03			
18	1017.70	13.04			
19	1024.19	14.04			
20	1035.40	15.04			
21	1041.97	16.05			
22	1044.25	17.05			
23	1053.75	18.05			
24	1061.95	19.05			
25	1063.63	20.06			

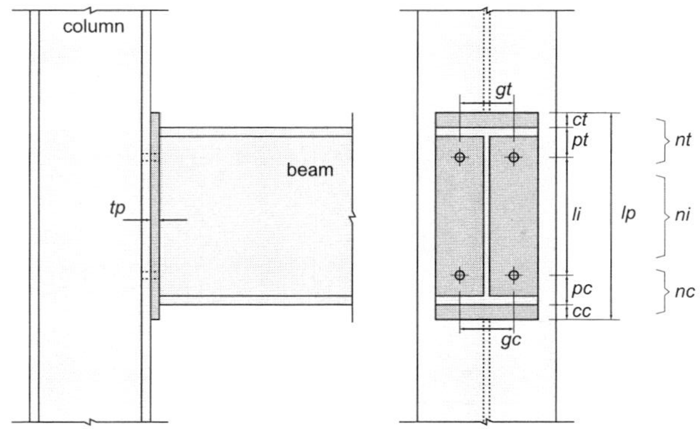

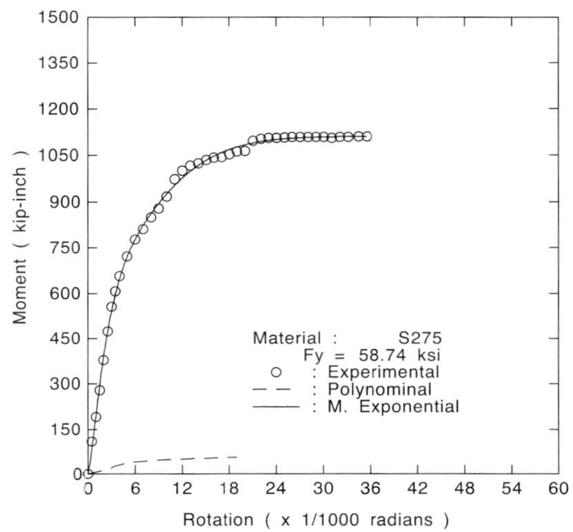

A.6 – 70

```
Moment-rotation prediction equations     ( R : X 1/1000 radians )

Frye and Morris polynominal model  :    R = Sum ( Ai X ( K*Bm )**Pi X 10**Qi )
     xd =  3.543300"   t =  0.472440"  tf =  0.787400"
     A1 =  1.830000   A2 = -1.040000   A3 =  6.380000    K =  0.092766
     P1 =         1   P2 =         3   P3 =         5   Q1 =         0    Q2 =        -1    Q3 =        -3

Modified exponential model  :
     Bm = Sum ( Ai X ( 1 - exp( - R/( 2*i*C ) ) ) )  +      Sum ( RKj X ( R - Rj0 ) )  +  Bm0
       C =   0.42166667E+00            Bmo=   0.00000000E+00          Nexp=  6        Nliner=  0
      AI =  -0.39958363E+03  -0.16562933E+03   0.76712170E+04  -0.17096199E+05   0.13671503E+05  -0.25708800E+04
```

No	Rotation (radians) X 1/1000	Moment (kip-inch) Expri.	Moment (kip-inch) Poly.	Moment (kip-inch) M.Expo.	Connection stiffness (kip-inch) X 1000 Expri.	Connection stiffness (kip-inch) X 1000 Poly.	Connection stiffness (kip-inch) X 1000 M.Expo.
1	0.00	0.00	0.00	0.00	0.2184E+03	0.5891E+01	0.1262E+03
3	1.00	191.75	5.99	192.57	0.1702E+03	0.6207E+01	0.2160E+03
5	2.01	377.69	12.77	393.43	0.1931E+03	0.7409E+01	0.1749E+03
7	3.01	555.95	21.34	542.08	0.1326E+03	0.9825E+01	0.1241E+03
9	4.01	657.46	30.75	646.39	0.8836E+02	0.7681E+01	0.8714E+02
11	6.02	777.38	40.16	777.91	0.4518E+02	0.2956E+01	0.4998E+02
13	8.02	851.09	44.75	862.04	0.3315E+02	0.1819E+01	0.3610E+02
15	10.03	917.51	47.86	926.62	0.4785E+02	0.1335E+01	0.2862E+02
17	12.03	1001.01	50.24	977.77	0.2167E+02	0.1067E+01	0.2264E+02
19	14.04	1024.19	52.20	1017.73	0.8850E+01	0.8943E+00	0.1724E+02
21	16.05	1041.97	53.87	1047.57	0.4382E+01	0.7737E+00	0.1259E+02
23	18.05	1053.75	55.32	1068.89	0.8850E+01	0.6847E+00	0.8878E+01
25	20.06	1063.63	56.63	1083.74	0.1706E+02	0.6155E+00	0.6044E+01
27	22.06	1102.65	57.80	1093.67	0.5123E+01	0.5606E+00	0.4006E+01
29	24.07	1106.19	58.88	1100.20	0.4515E+00	0.5153E+00	0.2581E+01
31	26.07	1107.96	59.87	1104.34	0.4356E+00	0.4777E+00	0.1626E+01
33	28.08	1107.96	60.80	1106.93	0.0000E+00	0.4456E+00	0.9968E+00
35	30.09	1108.85	61.67	1108.50	-0.8981E+00	0.4180E+00	0.5954E+00
37	32.09	1107.96	62.49	1109.42	0.1399E+01	0.3938E+00	0.3462E+00
39	34.50	1109.73	63.41	1110.03	0.2377E+00	0.3689E+00	0.1715E+00

VI - 36

Connection type : Flush end-plate connections
Mode : All bolted without column stiffener

Tested by : B.Bose et al. (1996) U.K.
Test Id. : 3

Column : 254x254 UC 89
Beam : 457x191 UB 74
Plate thickness : 12 mm

Fasteners: G8.8- -M20
7/8" Oversize holes
Material : S275
Fy = -- ksi
Fu = -- ksi

Major parameters

ct = 1.0433"	pt = 2.3622"	li = 13.2677"	pc = 2.3622"
cc = 1.0433"	pit= 3.5433"	pi = 0.0000"	pic= 9.7244"
gt = 3.5433"	gi = 3.5433"	gc = 3.5433"	
tp = 0.4724"	nt = 2 X 1	ni = 2 X 1	nc = 2 X 1

Remark 1) Failure mode: Thread stripping
2) Plate size: 510x200x12 mm

No	Moment (k-in)	Rotation (radians) X 1/1000	No	Moment (k-in)	Rotation (radians) X 1/1000
1	0.00	0.00	26	1382.61	13.02
2	105.03	0.36	27	1425.69	13.75
3	175.36	0.72	28	1467.46	14.47
4	259.89	0.96	29	1506.73	15.19
5	332.37	1.09	30	1531.49	15.92
6	395.39	1.45	31	1550.98	16.64
7	505.21	1.81	32	1571.23	17.36
8	597.98	1.94	33	1583.90	18.09
9	655.37	2.17	34	1594.33	18.81
10	700.86	2.53	35	1603.48	19.54
11	751.39	2.89	36	1612.38	20.26
12	824.63	3.26	37	1620.38	20.98
13	877.50	3.62	38	1628.41	21.71
14	959.58	4.34	39	1639.08	22.43
15	1029.23	5.06	40	1649.75	23.15
16	1067.23	5.79	41	1654.87	23.88
17	1097.61	6.51	42	1661.95	24.60
18	1115.83	7.24			
19	1148.63	7.96			
20	1178.58	8.68			
21	1202.83	9.41			
22	1234.93	10.13			
23	1263.40	10.85			
24	1308.07	11.58			
25	1339.82	12.30			

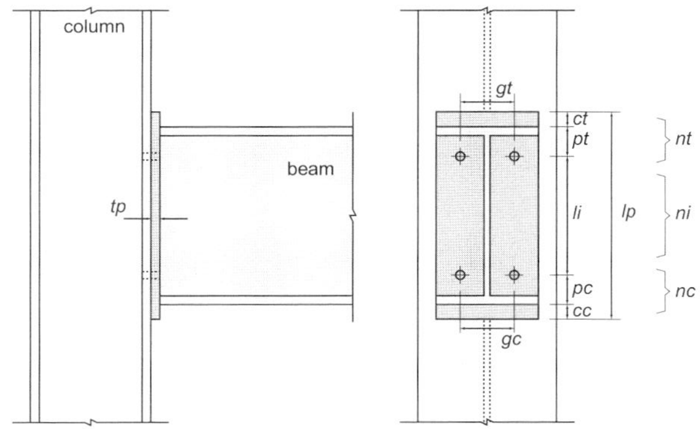

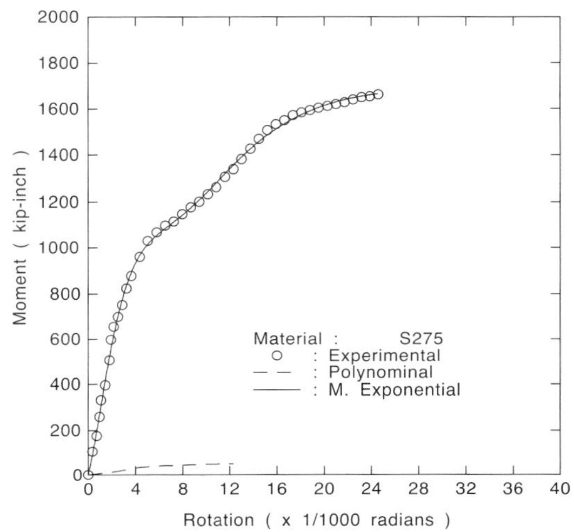

A.6 – 72

```
Moment-rotation prediction equations     ( R : X 1/1000 radians )

Frye and Morris polynominal model  :    R = Sum ( Ai X ( K*Bm )**Pi X 10**Qi )
     xd =  3.543300"    t =  0.472440"  tf =  0.787400"
     A1 =  1.830000   A2 = -1.040000   A3 =  6.380000    K =  0.092766
     P1 =        1     P2 =        3    P3 =        5    Q1 =        0    Q2 =       -1    Q3 =       -3

Modified exponential model  :
     Bm = Sum ( Ai X ( 1 - exp( - R/( 2*i*C ) ) ) ) +      Sum ( RKj X ( R - Rj0 ) )  + Bm0
       C =   0.29500000E+00           Bmo=   0.00000000E+00          Nexp=  6        Nliner=  0
      AI =   0.35804235E+03  -0.37647106E+04  -0.75364356E+04   0.69182450E+05  -0.11075605E+06   0.54209908E+05
```

No	Rotation (radians) X 1/1000	Moment (kip-inch) Expri.	Moment (kip-inch) Poly.	Moment (kip-inch) M.Expo.	Connection stiffness (kip-inch) X 1000 Expri.	Connection stiffness (kip-inch) X 1000 Poly.	Connection stiffness (kip-inch) X 1000 M.Expo.
1	0.00	0.00	0.00	0.00	0.2917E+03	0.5891E+01	0.2423E+03
3	0.72	175.36	4.28	195.75	0.2895E+03	0.6051E+01	0.2993E+03
5	1.09	332.37	6.55	308.88	0.4561E+03	0.6270E+01	0.3090E+03
7	1.81	505.21	11.32	523.91	0.6053E+03	0.7071E+01	0.2797E+03
9	2.17	655.37	13.98	619.52	0.2015E+03	0.7725E+01	0.2506E+03
11	2.89	751.39	20.18	776.54	0.1688E+03	0.9553E+01	0.1855E+03
13	3.62	877.50	27.40	889.77	0.1359E+03	0.9410E+01	0.1271E+03
15	5.06	1029.23	36.77	1016.39	0.7455E+02	0.4282E+01	0.5919E+02
17	6.51	1097.61	41.51	1084.40	0.3363E+02	0.2555E+01	0.4038E+02
19	7.96	1148.63	44.64	1143.56	0.4358E+02	0.1840E+01	0.4288E+02
21	9.41	1202.83	47.00	1210.24	0.3894E+02	0.1452E+01	0.4883E+02
23	10.85	1263.40	48.90	1282.93	0.5029E+02	0.1209E+01	0.5137E+02
25	12.30	1339.82	50.53	1356.55	0.5177E+02	0.1039E+01	0.4952E+02
27	13.75	1425.69	51.94	1425.00	0.5851E+02	0.9153E+00	0.4449E+02
29	15.19	1506.73	53.18	1484.43	0.4430E+02	0.8206E+00	0.3791E+02
31	16.64	1550.98	54.32	1534.33	0.2759E+02	0.7448E+00	0.3095E+02
33	18.09	1583.90	55.35	1574.43	0.1591E+02	0.6832E+00	0.2447E+02
35	19.54	1603.48	56.30	1605.71	0.1244E+02	0.6319E+00	0.1885E+02
37	20.98	1620.38	57.18	1629.42	0.1106E+02	0.5888E+00	0.1425E+02
39	22.43	1639.08	58.01	1647.30	0.1482E+02	0.5516E+00	0.1057E+02
41	23.88	1654.87	58.80	1660.47	0.8432E+01	0.5188E+00	0.7728E+01

VI - 37

Connection type : Flush end-plate connections
Mode : All bolted without column stiffener

Tested by : B.Bose et al. (1996) U.K.
Test Id. : 4

Column : 254x254 UC 89
Beam : 406x178 UB 60
Plate thickness : 15 mm

Fasteners: G8.8- -M24
33/32" Oversize holes
Material : S275
Fy = 44.96 ksi
Fu = 74.11 ksi

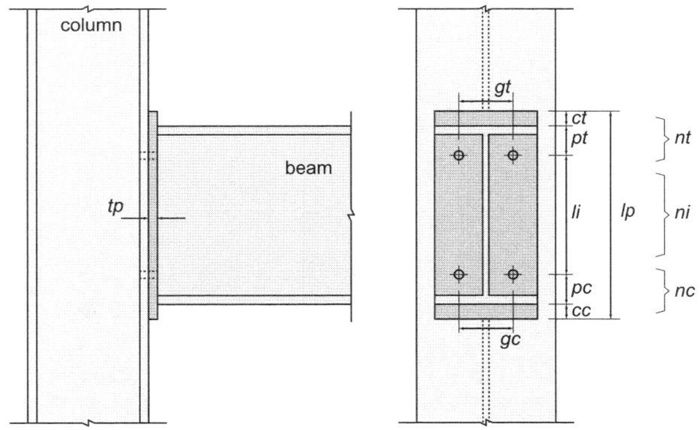

Major parameters

ct = 0.8583"	pt = 2.3622'	li = 11.2756"	pc = 2.3622"
cc = 0.8583"	pit= 3.5433'	pi = 0.0000"	pic= 7.7323"
gt = 3.5433"	gi = 3.5433'	gc = 3.5433"	
tp = 0.5906"	nt = 2 X 1	ni = 2 X 1	nc = 2 X 1

Remark 1) Failure mode: Thread stripping
2) Plate size: 450x200x15 mm

No	Moment (k-in)	Rotation (radians) X 1/1000	No	Moment (k-in)	Rotation (radians) X 1/1000	No	Moment (k-in)	Rotation (radians) X 1/1000
1	0.00	0.00	26	1733.26	21.04	51	2040.71	46.09
2	201.07	0.15	27	1749.77	22.04	52	2042.21	47.09
3	363.76	0.50	28	1766.48	23.05	53	2044.78	48.10
4	494.60	1.00	29	1782.35	24.05	54	2045.13	49.10
5	641.06	1.50	30	1803.08	25.05	55	2047.79	50.10
6	747.25	2.00	31	1818.29	26.05			
7	837.53	2.51	32	1834.92	27.05			
8	927.80	3.01	33	1850.85	28.06			
9	1035.66	4.01	34	1861.21	29.06			
10	1121.22	5.01	35	1878.56	30.06			
11	1193.99	6.01	36	1891.20	31.06			
12	1261.01	7.01	37	1902.38	32.06			
13	1316.06	8.02	38	1913.21	33.07			
14	1371.46	9.02	39	1924.77	34.07			
15	1409.13	10.02	40	1936.01	35.07			
16	1449.73	11.02	41	1946.81	36.07			
17	1484.05	12.02	42	1957.79	37.07			
18	1517.88	13.03	43	1967.68	38.08			
19	1551.32	14.03	44	1984.36	39.08			
20	1579.24	15.03	45	2001.58	40.08			
21	1608.07	16.03	46	2019.89	41.08			
22	1635.78	17.03	47	2026.06	42.08			
23	1664.24	18.04	48	2029.88	43.09			
24	1696.87	19.04	49	2035.62	44.09			
25	1713.68	20.04	50	2041.56	45.09			

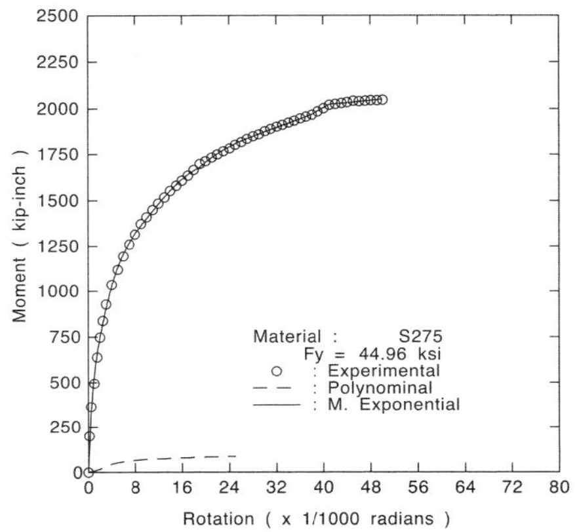

```
Moment-rotation prediction equations      ( R : X 1/1000 radians )

Frye and Morris polynominal model  :    R = Sum ( Ai X ( K*Bm )**Pi X 10**Qi )
     xd =  3.543300"    t =  0.590550"  tf =  0.944900"
     A1 =  1.830000    A2 = -1.040000   A3 =  6.380000    K =  0.064542
     P1 =         1    P2 =         3   P3 =         5   Q1 =         0    Q2 =        -1    Q3 =        -3

Modified exponential model  :
     Bm = Sum ( Ai X ( 1 - exp( - R/( 2*i*C ) ) ) )  +      Sum ( RKj X ( R - Rj0 ) )  +  Bm0
      C =  0.45500000E+00            Bmo=   0.00000000E+00          Nexp=  6         Nliner=  3
     AI =  0.25893084E+04  -0.22348681E+05  0.92080971E+05  -0.17326841E+06  0.14891669E+06  -0.46280357E+05
    Rj0 =         19.0400          38.0800          41.0800
    RKj =  0.14041878E+02   0.50849210E+01  -0.15276508E+02
```

No	Rotation (radians) X 1/1000	Moment (kip-inch) Expri.	Moment (kip-inch) Poly.	Moment (kip-inch) M.Expo.	Connection stiffness (kip-inch) X 1000 Expri.	Connection stiffness (kip-inch) X 1000 Poly.	Connection stiffness (kip-inch) X 1000 M.Expo.
1	0.00	0.00	0.00	0.00	0.1340E+04	0.8467E+01	0.9467E+03
3	0.50	363.76	4.25	345.61	0.3812E+03	0.8576E+01	0.4939E+03
5	1.50	641.06	13.22	655.85	0.2527E+03	0.9566E+01	0.2050E+03
7	2.51	837.53	24.06	829.36	0.1788E+03	0.1227E+02	0.1513E+03
9	4.01	1035.66	44.19	1028.92	0.9671E+02	0.1104E+02	0.1144E+03
11	6.01	1193.99	57.69	1208.37	0.6989E+02	0.4263E+01	0.6795E+02
13	8.02	1316.06	64.32	1318.48	0.5496E+02	0.2614E+01	0.4572E+02
15	10.02	1409.13	68.77	1403.02	0.3914E+02	0.1921E+01	0.4027E+02
17	12.02	1484.05	72.20	1481.53	0.3391E+02	0.1535E+01	0.3809E+02
19	14.03	1551.32	75.02	1554.12	0.3068E+02	0.1286E+01	0.3365E+02
21	16.03	1608.07	77.41	1614.91	0.2827E+02	0.1113E+01	0.2687E+02
23	18.04	1664.24	79.51	1661.29	0.3042E+02	0.9847E+00	0.1929E+02
25	20.04	1713.68	81.37	1706.83	0.1819E+02	0.8856E+00	0.2644E+02
27	22.04	1749.77	83.06	1753.92	0.1653E+02	0.8064E+00	0.2091E+02
29	24.05	1782.35	84.62	1791.68	0.1830E+02	0.7413E+00	0.1691E+02
31	26.05	1818.29	86.04	1822.69	0.1592E+02	0.6871E+00	0.1431E+02
33	28.06	1850.85	87.38	1849.77	0.1305E+02	0.6409E+00	0.1279E+02
35	30.06	1878.56	88.62	1874.50	0.1500E+02	0.6014E+00	0.1204E+02
37	32.06	1902.38	89.80	1898.28	0.1095E+02	0.5665E+00	0.1180E+02
39	34.07	1924.77	90.90	1922.03	0.1140E+02	0.5364E+00	0.1186E+02
41	36.07	1946.81	91.94	1945.96	0.1089E+02	0.5097E+00	0.1208E+02
43	38.08	1967.68	92.95	1970.53	0.1325E+02	0.4853E+00	0.1237E+02
45	40.08	2001.58	93.89	2005.73	0.1777E+02	0.4640E+00	0.1775E+02
47	42.08	2026.06	94.81	2026.23	0.4982E+01	0.4443E+00	0.2742E+01
49	44.09	2035.62	95.69	2031.98	0.5841E+01	0.4263E+00	0.2980E+01
51	46.09	2040.71	96.51	2038.15	0.3263E+00	0.4104E+00	0.3177E+01
53	48.10	2044.78	97.33	2044.70	0.1442E+01	0.3953E+00	0.3338E+01
55	50.10	2047.79	98.12	2051.51	0.2655E+01	0.3815E+00	0.3464E+01

```
Connection type : Flush end-plate connections
           Mode : All bolted without column stiffener

Tested by   :  B.Bose et al. (1996)                      U.K.
Test   Id.  :  5

Column : 254x254 UC 89                  Fasteners: G8.8- -M24
Beam   : 457x191 UB 74                       33/32"  Oversize holes
Plate thickness : 15 mm                 Material :  S275
                                             Fy =    --   ksi
                                             Fu =    --   ksi

                          Major parameters

     ct =  1.0433"    pt =  2.3622'    li = 13.2677"    pc =  2.3622"
     cc =  1.0433"    pit=  3.5433'    pi =  0.0000"    pic=  9.7244"
     gt =  3.5433"    gi =  3.5433'    gc =  3.5433"
     tp =  0.5906"    nt = 2 X  1      ni = 2 X  1      nc = 2 X  1

Remark  1) Failure mode: Thread stripping
        2) Plate size: 510x200x15 mm
```

No	Moment (k-in)	Rotation (radians) X 1/1000	No	Moment (k-in)	Rotation (radians) X 1/1000
1	0.00	0.00	26	2050.56	20.00
2	175.11	0.50	27	2076.92	21.00
3	324.18	1.00	28	2098.88	22.00
4	383.81	1.50	29	2125.81	23.00
5	444.32	2.00	30	2150.44	24.00
6	527.47	2.13	31	2179.55	25.00
7	610.62	2.25	32	2211.11	26.00
8	745.75	2.50	33	2243.65	27.00
9	881.93	3.00	34	2275.61	28.00
10	1031.86	4.00	35	2298.77	29.00
11	1159.80	5.00	36	2331.23	30.00
12	1276.83	6.00	37	2350.14	31.00
13	1391.82	7.00	38	2364.11	32.00
14	1486.47	8.00	39	2373.36	33.00
15	1566.93	9.00	40	2384.12	34.00
16	1637.37	10.00	41	2389.48	35.00
17	1687.77	11.00	42	2396.34	36.00
18	1746.44	12.00	43	2402.70	37.00
19	1800.46	13.00	44	2409.47	38.00
20	1873.74	14.00	45	2417.75	39.00
21	1927.77	15.00	46	2425.07	40.00
22	1960.75	16.00	47	2430.89	41.00
23	1984.32	17.00	48	2437.17	42.00
24	2009.28	18.00			
25	2030.57	19.00			

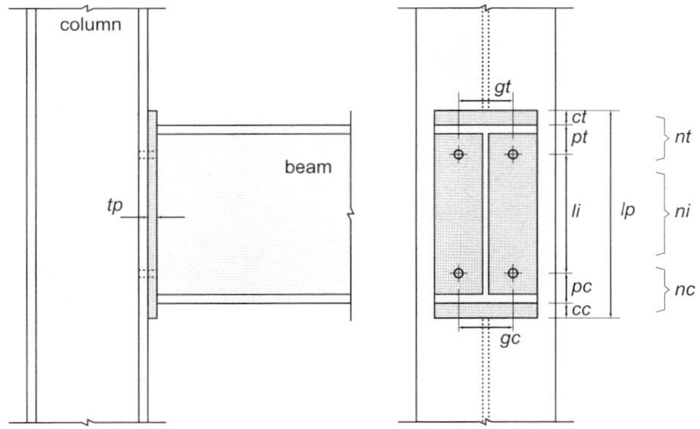

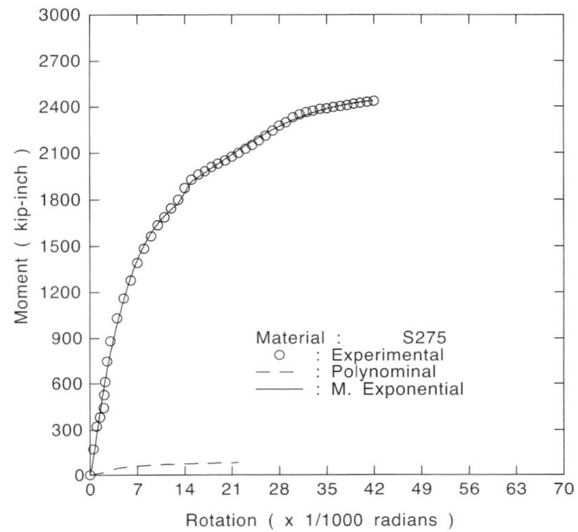

A.6 – 76

```
Moment-rotation prediction equations     ( R : X 1/1000 radians )

Frye and Morris polynominal model  :    R = Sum ( Ai X ( K*Bm )**Pi X 10**Qi )
     xd =  3.543300"    t =  0.590550"  tf =  0.944900"
     A1 =  1.830000   A2 = -1.040000   A3 =  6.380000    K =  0.064542
     P1 =        1    P2 =        3    P3 =        5    Q1 =        0    Q2 =       -1    Q3 =       -3

Modified exponential model  :
     Bm = Sum ( Ai X ( 1 - exp( - R/( 2*i*C ) ) ) )  +      Sum ( RKj X ( R - Rj0 ) )  +  Bm0
       C =   0.47500000E+00              Bmo=   0.00000000E+00          Nexp=  6         Nliner=  2
      AI =  -0.13705127E+04   0.19307321E+05  -0.10726486E+06   0.25854709E+06  -0.27553246E+06   0.10869833E+06
     Rj0 =          13.0000          15.0000
     RKj =   0.41691058E+02  -0.41541049E+02
```

No	Rotation (radians) X 1/1000	Moment (kip-inch) Expri.	Moment (kip-inch) Poly.	Moment (kip-inch) M.Expo.	Connection stiffness (kip-inch) X 1000 Expri.	Connection stiffness (kip-inch) X 1000 Poly.	Connection stiffness (kip-inch) X 1000 M.Expo.
1	0.00	0.00	0.00	0.00	0.3502E+03	0.8467E+01	0.1841E+03
3	1.00	324.18	8.62	271.49	0.2087E+03	0.8922E+01	0.3035E+03
5	2.00	444.32	18.25	558.24	0.5326E+03	0.1062E+02	0.2630E+03
7	2.25	610.62	20.97	622.32	0.6435E+03	0.1135E+02	0.2497E+03
9	3.00	881.93	30.53	795.79	0.2315E+03	0.1409E+02	0.2141E+03
11	5.00	1159.80	52.47	1155.21	0.1225E+03	0.6329E+01	0.1510E+03
13	7.00	1391.82	61.36	1410.89	0.1048E+03	0.3238E+01	0.1059E+03
15	9.00	1566.93	66.67	1584.54	0.7545E+02	0.2217E+01	0.6947E+02
17	11.00	1687.77	70.55	1697.02	0.5454E+02	0.1708E+01	0.4518E+02
19	13.00	1800.46	73.63	1773.31	0.6365E+02	0.1402E+01	0.3283E+02
21	15.00	1927.77	76.22	1917.11	0.4351E+02	0.1196E+01	0.7027E+02
23	17.00	1984.32	78.45	1973.87	0.2426E+02	0.1047E+01	0.2840E+02
25	19.00	2030.57	80.43	2031.30	0.2064E+02	0.9342E+00	0.2902E+02
27	21.00	2076.92	82.21	2089.54	0.2416E+02	0.8455E+00	0.2905E+02
29	23.00	2125.81	83.84	2146.75	0.2578E+02	0.7730E+00	0.2798E+02
31	25.00	2179.55	85.32	2200.76	0.3034E+02	0.7138E+00	0.2589E+02
33	27.00	2243.65	86.70	2249.83	0.3225E+02	0.6640E+00	0.2310E+02
35	29.00	2298.77	87.98	2292.96	0.2781E+02	0.6213E+00	0.1999E+02
37	31.00	2350.14	89.19	2329.79	0.1644E+02	0.5844E+00	0.1686E+02
39	33.00	2373.36	90.31	2360.50	0.1000E+02	0.5523E+00	0.1390E+02
41	35.00	2389.48	91.40	2385.62	0.6112E+01	0.5233E+00	0.1126E+02
43	37.00	2402.70	92.42	2405.80	0.6561E+01	0.4981E+00	0.8986E+01
45	39.00	2417.75	93.40	2421.81	0.7802E+01	0.4750E+00	0.7080E+01
47	41.00	2430.89	94.32	2434.36	0.6049E+01	0.4547E+00	0.5522E+01

VI - 39

Connection type : Flush end-plate connections
Mode : All bolted without column stiffener

Tested by : B.Bose et al. (1996) U.K.
Test Id. : 6

Column : 254x254 UC 89
Beam : 686x254 UB 125
Plate thickness : 15 mm

Fasteners: G8.8- -M24
33/32" Oversize holes
Material : S275
Fy = 44.96 ksi
Fu = 74.11 ksi

Major parameters

ct = 0.8268"	pt = 2.3622"	li = 21.9685"	pc = 2.3622"
cc = 0.8268"	pit= 3.5433"	pi = 0.0000"	pic= 18.4252"
gt = 3.5433"	gi = 3.5433"	gc = 3.5433"	
tp = 0.5906"	nt = 2 X 1	ni = 2 X 1	nc = 2 X 1

Remark 1) Failure mode: Bolt fracture
2) Plate size: 720x250x15 mm

No	Moment (k-in)	Rotation (radians) X 1/1000	No	Moment (k-in)	Rotation (radians) X 1/1000
1	0.00	0.00	26	3412.50	17.79
2	162.21	0.25	27	3436.64	18.78
3	324.43	0.49	28	3493.65	19.76
4	471.65	0.74	29	3513.16	20.75
5	618.87	0.99	30	3534.96	21.74
6	817.37	1.23	31	3568.18	22.73
7	1015.86	1.48	32	3599.03	23.72
8	1225.76	1.73	33	3632.06	24.70
9	1435.66	1.98	34	3659.44	25.69
10	1652.81	2.47	35	3696.44	26.68
11	1918.55	2.96	36	3720.05	27.67
12	2176.04	3.95	37	3730.49	28.66
13	2362.13	4.94	38	3743.05	29.65
14	2508.44	5.93	39	3748.94	30.63
15	2608.06	6.92	40	3751.89	31.62
16	2706.06	7.91	41	3753.98	32.70
17	2798.23	8.89	42	3764.60	34.00
18	2880.44	9.88			
19	2959.56	10.87			
20	3047.87	11.86			
21	3119.38	12.85			
22	3180.17	13.83			
23	3241.85	14.82			
24	3306.79	15.81			
25	3370.20	16.80			

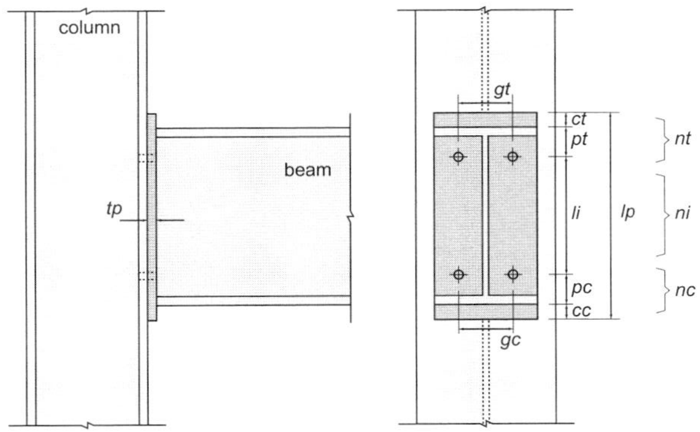

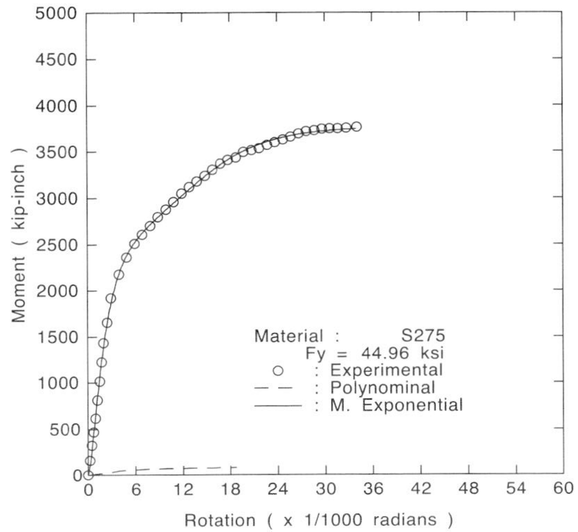

```
Moment-rotation prediction equations     ( R : X 1/1000 radians )

Frye and Morris polynominal model  :     R = Sum ( Ai X ( K*Bm )**Pi X 10**Qi )
     xd =  3.543300"    t =  0.590550"  tf =  0.944900"
     A1 =  1.830000   A2 = -1.040000   A3 =  6.380000    K =  0.064542
     P1 =        1     P2 =        3    P3 =        5    Q1 =        0     Q2 =       -1     Q3 =       -3

Modified exponential model  :
     Bm = Sum ( Ai X ( 1 - exp( - R/( 2*i*C ) ) ) )  +      Sum ( RKj X ( R - Rj0 ) )  +  Bm0
      C =  0.34583333E+00            Bmo=  0.00000000E+00          Nexp=  6        Nliner=  0
     AI = -0.19992753E+04  0.11965407E+05 -0.63853563E+05  0.18794999E+06 -0.23371635E+06  0.10341559E+06
```

No	Rotation (radians) X 1/1000	Moment (kip-inch) Expri.	Moment (kip-inch) Poly.	Moment (kip-inch) M.Expo.	Connection stiffness (kip-inch) X 1000 Expri.	Connection stiffness (kip-inch) X 1000 Poly.	Connection stiffness (kip-inch) X 1000 M.Expo.
1	0.00	0.00	0.00	0.00	0.6489E+03	0.8467E+01	0.2589E+03
3	0.49	324.43	4.17	271.68	0.6333E+03	0.8571E+01	0.7449E+03
5	0.99	618.87	8.53	671.64	0.7104E+03	0.8912E+01	0.8129E+03
7	1.48	1015.86	13.03	1052.36	0.8168E+03	0.9533E+01	0.7286E+03
9	1.98	1435.66	18.04	1385.88	0.7057E+03	0.1057E+02	0.6041E+03
11	2.96	1918.55	29.97	1866.67	0.4489E+03	0.1397E+02	0.3893E+03
13	4.94	2362.13	52.09	2380.13	0.1679E+03	0.6514E+01	0.1682E+03
15	6.92	2608.06	61.10	2629.01	0.9981E+02	0.3301E+01	0.9727E+02
17	8.89	2798.23	66.43	2797.98	0.8857E+02	0.2255E+01	0.7886E+02
19	10.87	2959.56	70.32	2949.93	0.8456E+02	0.1733E+01	0.7551E+02
21	12.85	3119.38	73.42	3097.18	0.6710E+02	0.1420E+01	0.7278E+02
23	14.82	3241.85	76.00	3235.07	0.6395E+02	0.1211E+01	0.6660E+02
25	16.80	3370.20	78.24	3358.14	0.5339E+02	0.1060E+01	0.5734E+02
27	18.78	3436.64	80.22	3461.29	0.4137E+02	0.9452E+00	0.4679E+02
29	20.75	3513.16	81.99	3543.26	0.2086E+02	0.8555E+00	0.3657E+02
31	22.73	3568.18	83.63	3606.47	0.3236E+02	0.7819E+00	0.2753E+02
33	24.70	3632.06	85.11	3653.18	0.3070E+02	0.7221E+00	0.2016E+02
35	26.68	3696.44	86.48	3687.13	0.3061E+02	0.6714E+00	0.1439E+02
37	28.66	3730.49	87.77	3711.14	0.1161E+02	0.6281E+00	0.1007E+02
39	30.63	3748.94	88.97	3727.74	0.4507E+01	0.5909E+00	0.6950E+01
41	32.70	3753.98	90.15	3739.58	0.4763E+01	0.5569E+00	0.4636E+01

VI - 40

Connection type : Flush end-plate connections
Mode : All bolted without column stiffener

Tested by : B.Bose et al. (1996) U.K.
Test Id. : 8

Column : 254x254 UC 73
Beam : 406x178 UB 60
Plate thickness : 12 mm

Fasteners: G8.8- -M20
7/8" Oversize holes
Material : S275
Fy = -- ksi
Fu = -- ksi

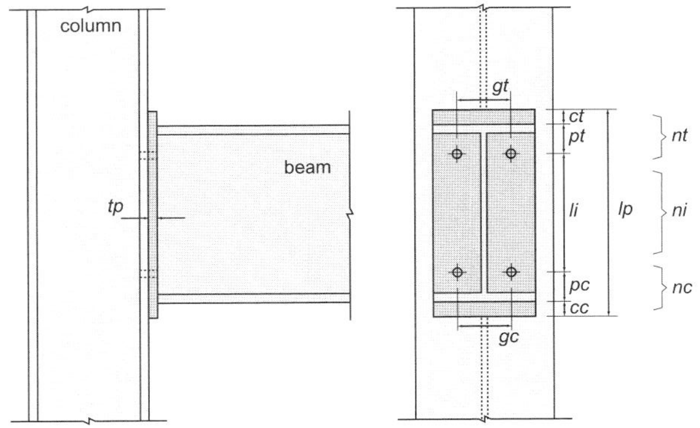

Major parameters

ct = 1.0551"	pt = 2.3622"	li = 11.2756"	pc = 2.3622"
cc = 1.0551"	pit= 3.5433"	pi = 4.1890"	pic= 3.5433"
gt = 3.5433"	gi = 3.5433"	gc = 3.5433"	
tp = 0.4724"	nt = 2 X 1	ni = 2 X 2	nc = 2 X 1

Remark 1) Failure mode: Column web buckling
2) Plate size: 460x200x12 mm

No	Moment (k-in)	Rotation (radians) X 1/1000
1	0.00	0.00
2	113.82	0.25
3	227.64	0.50
4	308.55	0.76
5	389.46	1.01
6	456.16	1.26
7	522.87	1.51
8	574.23	1.77
9	625.59	2.02
10	687.29	2.53
11	749.00	3.03
12	808.76	3.54
13	868.53	4.04
14	952.88	5.05
15	1022.71	6.06
16	1082.51	7.07
17	1133.15	8.07
18	1178.89	9.08
19	1223.45	10.09
20	1270.91	11.10
21	1313.74	12.11
22	1350.91	13.12
23	1375.93	14.13
24	1386.68	15.14
25	1390.93	16.15
26	1396.35	17.16
27	1397.94	18.17
28	1401.42	19.18
29	1404.23	20.19
30	1405.51	21.20
31	1406.30	22.21
32	1409.43	23.21
33	1411.14	24.22
34	1411.82	25.23
35	1412.39	26.24
36	1415.35	27.25
37	1416.53	28.26
38	1417.23	29.27
39	1417.91	30.28
40	1418.49	31.29
41	1417.24	32.30
42	1417.35	33.31
43	1418.75	34.32
44	1420.75	35.33
45	1423.88	36.34
46	1424.56	37.35
47	1425.72	38.35
48	1426.82	39.36
49	1427.43	40.37
50	1427.43	41.38
51	1427.95	42.39
52	1428.19	43.40
53	1428.27	44.41
54	1428.32	46.00

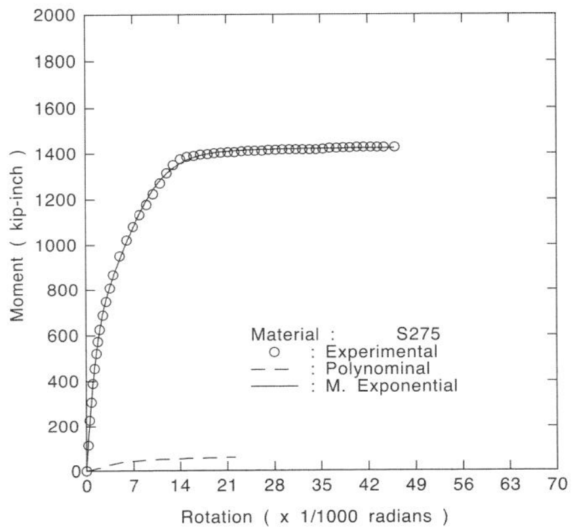

```
Moment-rotation prediction equations      ( R : X 1/1000 radians )

Frye and Morris polynominal model  :    R = Sum ( Ai X ( K*Bm )**Pi X 10**Qi )
     xd =  3.543300"    t =  0.472440"  tf =  0.787400"
     A1 =  1.830000    A2 = -1.040000   A3 =  6.380000     K =  0.092766
     P1 =         1    P2 =         3   P3 =         5    Q1 =         0    Q2 =        -1    Q3 =        -3

Modified exponential model  :
     Bm = Sum ( Ai X ( 1 - exp( - R/( 2*i*C ) ) ) )  +      Sum ( RKj X ( R - Rj0 ) )  +  Bm0
       C =   0.44583333E+00              Bmo=   0.00000000E+00         Nexp=  6        Nliner=  0
      AI =  -0.51576369E+03   0.77168836E+04  -0.23931671E+05   0.32556141E+05  -0.20288446E+05   0.58872802E+04
```

No	Rotation (radians) X 1/1000	Moment (kip-inch) Expri.	Moment (kip-inch) Poly.	Moment (kip-inch) M.Expo.	Connection stiffness (kip-inch) X 1000 Expri.	Connection stiffness (kip-inch) X 1000 Poly.	Connection stiffness (kip-inch) X 1000 M.Expo.
1	0.00	0.00	0.00	0.00	0.4553E+03	0.5891E+01	0.4800E+03
3	0.50	227.64	2.96	216.74	0.3846E+03	0.5967E+01	0.3862E+03
5	1.01	389.46	6.06	390.21	0.2952E+03	0.6214E+01	0.2965E+03
7	1.51	522.87	9.27	519.97	0.2329E+03	0.6667E+01	0.2257E+03
9	2.02	625.59	12.85	620.67	0.1777E+03	0.7428E+01	0.1723E+03
11	3.03	749.00	21.54	759.67	0.1203E+03	0.9864E+01	0.1112E+03
13	4.04	868.53	30.98	857.33	0.1076E+03	0.7543E+01	0.8606E+02
15	6.06	1022.71	40.28	1012.23	0.6418E+02	0.2919E+01	0.7025E+02
17	8.07	1133.15	44.84	1141.49	0.4797E+02	0.1802E+01	0.5764E+02
19	10.09	1223.45	47.94	1242.30	0.4556E+02	0.1325E+01	0.4204E+02
21	12.11	1313.74	50.33	1312.24	0.3960E+02	0.1059E+01	0.2772E+02
23	14.13	1375.93	52.28	1356.74	0.1771E+02	0.8880E+00	0.1699E+02
25	16.15	1390.93	53.95	1383.38	0.4785E+01	0.7686E+00	0.9930E+01
27	18.17	1397.94	55.41	1398.74	0.2513E+01	0.6801E+00	0.5663E+01
29	20.19	1404.23	56.71	1407.49	0.2022E+01	0.6116E+00	0.3241E+01
31	22.21	1406.30	57.89	1412.57	0.1963E+01	0.5569E+00	0.1923E+01
33	24.22	1411.14	58.96	1415.66	0.1186E+01	0.5123E+00	0.1223E+01
35	26.24	1412.39	59.96	1417.71	0.1743E+01	0.4748E+00	0.8440E+00
37	28.26	1416.53	60.88	1419.18	0.9334E+00	0.4430E+00	0.6276E+00
39	30.28	1417.91	61.75	1420.30	0.6214E+00	0.4156E+00	0.4907E+00
41	32.30	1417.24	62.57	1421.19	-0.5615E+00	0.3915E+00	0.3933E+00
43	34.32	1418.75	63.34	1421.90	0.1681E+01	0.3707E+00	0.3169E+00
45	36.34	1423.88	64.06	1422.48	0.1886E+01	0.3523E+00	0.2541E+00
47	38.35	1425.72	64.76	1422.93	0.1125E+01	0.3356E+00	0.2018E+00
49	40.37	1427.43	65.42	1423.30	0.3048E+00	0.3208E+00	0.1582E+00
51	42.39	1427.95	66.06	1423.58	0.3765E+00	0.3071E+00	0.1224E+00
53	44.41	1428.27	66.67	1423.79	0.5622E-01	0.2947E+00	0.9359E-01

A.6 – 81

VI - 41

Connection type : Flush end-plate connections
Mode : All bolted without column stiffener

Tested by : B.Bose et al. (1996) U.K.
Test Id. : 9

Column : 254x254 UC 89
Beam : 457x191 UB 74
Plate thickness : 12 mm

Fasteners: G8.8- -M20
7/8" Oversize holes
Material : S275
Fy = -- ksi
Fu = -- ksi

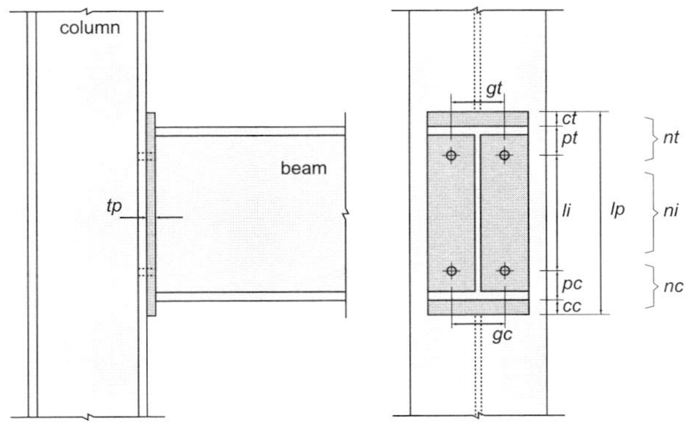

Major parameters

ct = 1.0433"	pt = 2.3622"	li = 13.2677"	pc = 2.3622"
cc = 1.0433"	pit= 3.5433"	pi = 6.1811"	pic= 3.5433"
gt = 3.5433"	gi = 3.5433"	gc = 3.5433"	
tp = 0.4724"	nt = 2 X 1	ni = 2 X 2	nc = 2 X 1

Remark 1) Failure mode: Column web buckling
2) Plate size: 510x200x12 mm

No	Moment (k-in)	Rotation (radians) X 1/1000	No	Moment (k-in)	Rotation (radians) X 1/1000	No	Moment (k-in)	Rotation (radians) X 1/1000
1	0.00	0.00	26	2C46.05	16.97	51	2457.52	41.94
2	96.12	0.25	27	2C62.74	17.97	52	2457.52	42.94
3	192.24	0.50	28	2C75.65	18.97	53	2458.57	43.93
4	272.02	0.75	29	2C92.47	19.97	54	2460.34	44.93
5	351.80	1.00	30	2118.34	20.97	55	2462.05	45.93
6	430.73	1.50	31	2157.62	21.97	56	2463.15	46.93
7	551.27	2.00	32	2200.98	22.97	57	2464.26	47.93
8	666.66	2.13	33	2248.37	23.96	58	2464.60	48.93
9	782.05	2.27	34	2290.78	24.96	59	2464.60	49.92
10	867.65	2.63	35	2315.58	25.96	60	2466.20	50.92
11	953.26	3.00	36	2333.10	26.96	61	2468.50	51.92
12	1049.92	3.50	37	2349.84	27.96	62	2469.03	52.92
13	1146.58	3.99	38	2369.05	28.96			
14	1267.57	4.99	39	2395.17	29.95			
15	1380.32	5.99	40	2417.62	30.95			
16	1492.10	6.99	41	2430.98	31.95			
17	1597.78	7.99	42	2438.50	32.95			
18	1671.23	8.99	43	2441.29	33.95			
19	1752.74	9.98	44	2443.92	34.95			
20	1820.50	10.98	45	2447.12	35.95			
21	1852.98	11.98	46	2449.70	36.94			
22	1897.80	12.98	47	2451.74	37.94			
23	1962.28	13.98	48	2454.31	38.94			
24	1997.47	14.98	49	2455.73	39.94			
25	2025.22	15.98	50	2456.84	40.94			

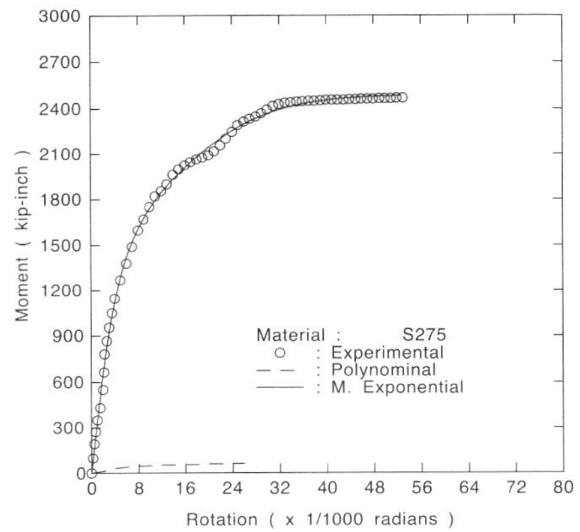

A.6 – 82

```
Moment-rotation prediction equations     ( R : X 1/1000 radians )

Frye and Morris polynominal model  :     R = Sum ( Ai X ( K*Bm )**Pi X 10**Qi )
     xd =  3.543300"    t =  0.472440"  tf =  0.787400"
     A1 =  1.830000    A2 = -1.040000   A3 =  6.380000    K =  0.092766
     P1 =         1    P2 =         3   P3 =         5   Q1 =         0    Q2 =        -1    Q3 =        -3

Modified exponential model  :
     Bm = Sum ( Ai X ( 1 - exp( - R/( 2*i*C ) ) ) )  +     Sum ( RKj X ( R - Rj0 ) )  +  Bm0
      C =   0.50350000E+00            Bmo=   0.00000000E+00             Nexp=  6         Nliner=  0
     AI =  -0.42431767E+03   0.34943135E+04  -0.25818052E+05   0.85532599E+05  -0.11222953E+06   0.51934121E+05
```

No	Rotation (radians) X 1/1000	Moment (kip-inch) Expri.	Moment (kip-inch) Poly.	Moment (kip-inch) M.Expo.	Connection stiffness (kip-inch) X 1000 Expri.	Connection stiffness (kip-inch) X 1000 Poly.	Connection stiffness (kip-inch) X 1000 M.Expo.
1	0.00	0.00	0.00	0.00	0.3845E+03	0.5891E+01	0.3076E+03
3	0.50	192.24	2.96	163.97	0.3518E+03	0.5967E+01	0.3403E+03
5	1.00	351.80	5.99	334.53	0.2654E+03	0.6207E+01	0.3379E+03
7	2.00	551.27	12.70	651.57	0.7542E+03	0.7391E+01	0.2913E+03
9	2.27	782.05	14.76	728.11	0.6600E+03	0.7944E+01	0.2757E+03
11	3.00	953.26	21.24	914.09	0.2152E+03	0.9805E+01	0.2343E+03
13	3.99	1146.58	30.59	1120.98	0.1722E+03	0.7774E+01	0.1853E+03
15	5.99	1380.32	40.07	1416.76	0.1123E+03	0.2985E+01	0.1166E+03
17	7.99	1597.78	44.69	1607.57	0.8957E+02	0.1829E+01	0.7802E+02
19	9.98	1752.74	47.79	1740.09	0.7509E+02	0.1343E+01	0.5736E+02
21	11.98	1852.98	50.19	1842.82	0.3865E+02	0.1072E+01	0.4655E+02
23	13.98	1962.28	52.15	1929.60	0.4984E+02	0.8985E+00	0.4077E+02
25	15.98	2025.22	53.82	2007.30	0.2438E+02	0.7773E+00	0.3712E+02
27	17.97	2062.74	55.27	2078.10	0.1480E+02	0.6878E+00	0.3406E+02
29	19.97	2092.47	56.57	2143.14	0.2134E+02	0.6183E+00	0.3094E+02
31	21.97	2157.62	57.75	2201.71	0.4132E+02	0.5628E+00	0.2760E+02
33	23.96	2248.37	58.83	2253.19	0.4515E+02	0.5176E+00	0.2413E+02
35	25.96	2315.58	59.82	2297.98	0.2116E+02	0.4796E+00	0.2067E+02
37	27.96	2349.84	60.75	2336.00	0.1798E+02	0.4474E+00	0.1738E+02
39	29.95	2395.17	61.61	2367.56	0.2443E+02	0.4198E+00	0.1439E+02
41	31.95	2430.98	62.44	2393.62	0.1044E+02	0.3954E+00	0.1172E+02
43	33.95	2441.29	63.20	2414.71	0.2712E+01	0.3744E+00	0.9428E+01
45	35.95	2447.12	63.92	2431.57	0.2898E+01	0.3557E+00	0.7495E+01
47	37.94	2451.74	64.62	2444.85	0.2305E+01	0.3388E+00	0.5905E+01
49	39.94	2455.73	65.28	2455.31	0.1268E+01	0.3239E+00	0.4606E+01
51	41.94	2457.52	65.92	2463.44	0.3400E+00	0.3100E+00	0.3564E+01
53	43.93	2458.57	66.53	2469.69	0.1412E+01	0.2976E+00	0.2742E+01
55	45.93	2462.05	67.10	2474.50	0.1406E+01	0.2864E+00	0.2094E+01
57	47.93	2464.26	67.67	2478.16	0.7251E+00	0.2759E+00	0.1591E+01
59	49.92	2464.60	68.22	2480.92	0.7940E+00	0.2662E+00	0.1204E+01
61	51.92	2468.50	68.73	2483.02	0.1414E+01	0.2575E+00	0.9057E+00

VI - 42

Connection type : Flush end-plate connections
Mode : All bolted without column stiffener

Tested by : B.Bose et al. (1996) U.K.
Test Id. : 10

Column : 254x254 UC 73
Beam : 406x178 UB 60
Plate thickness : 15 mm

Fasteners: G8.8- -M24
33/32" Oversize holes
Material : S275
Fy = -- ksi
Fu = -- ksi

Major parameters

ct = 1.0551"	pt = 2.3622"	li = 11.2756"	pc = 2.3622"
cc = 1.0551"	pit= 3.5433"	pi = 4.1890"	pic= 3.5433"
gt = 3.5433"	gi = 3.5433"	gc = 3.5433"	
tp = 0.5906"	nt = 2 X 1	ni = 2 X 2	nc = 2 X 1

Remark 1) Failure mode: Column web buckling
2) Plate size: 460x200x15 mm

No	Moment (k-in)	Rotation (radians) X 1/1000	No	Moment (k-in)	Rotation (radians) X 1/1000
1	0.00	0.00	26	1430.03	28.97
2	94.32	0.50	27	1434.47	30.97
3	174.82	1.00	28	1439.12	32.97
4	267.23	1.50	29	1441.09	34.97
5	374.34	2.00	30	1445.13	36.96
6	493.27	2.50	31	1448.92	38.96
7	581.38	3.00	32	1451.76	40.96
8	650.40	3.50	33	1455.80	42.96
9	719.42	4.00	34	1460.14	44.96
10	832.80	5.00	35	1464.37	46.95
11	943.10	5.99	36	1464.69	48.95
12	1029.56	6.99	37	1465.49	50.95
13	1111.45	7.99			
14	1179.95	8.99			
15	1239.82	9.99			
16	1293.05	10.99			
17	1353.46	11.99			
18	1396.62	12.99			
19	1401.90	14.99			
20	1406.41	16.98			
21	1410.26	18.98			
22	1415.35	20.98			
23	1419.52	22.98			
24	1422.47	24.98			
25	1425.54	26.97			

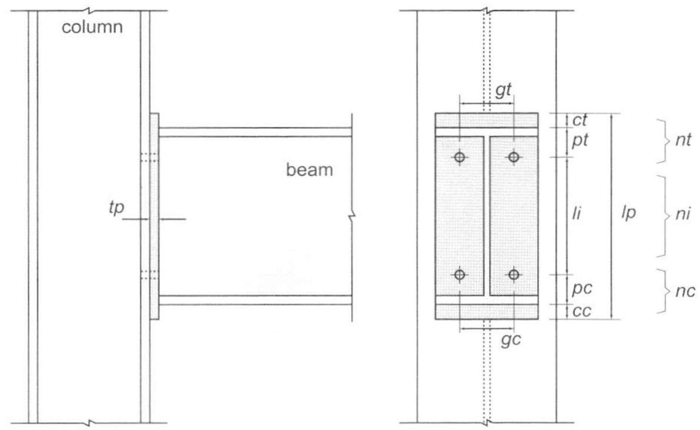

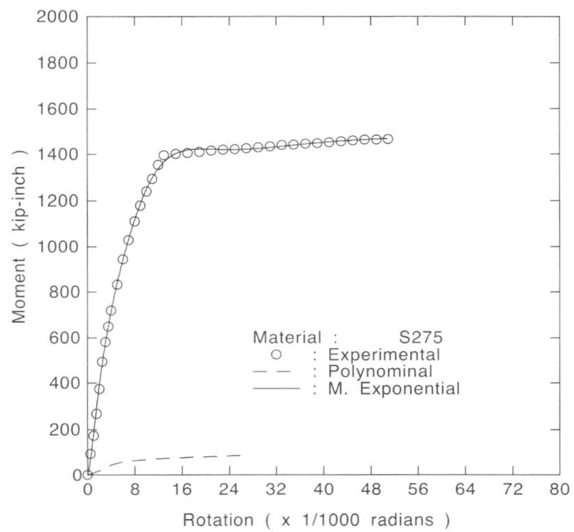

```
Moment-rotation prediction equations     ( R : X 1/1000 radians )

Frye and Morris polynominal model  :    R = Sum ( Ai X ( K*Bm )**Pi X 10**Qi )
     xd =  3.543300"   t =  0.590550"  tf =  0.944900"
     A1 =  1.830000   A2 = -1.040000   A3 =  6.380000    K =  0.064542
     P1 =         1   P2 =         3   P3 =         5   Q1 =         0   Q2 =        -1   Q3 =        -3

Modified exponential model  :
     Bm = Sum ( Ai X ( 1 - exp( - R/( 2*i*C ) ) ) )  +     Sum ( RKj X ( R - Rj0 ) )  +  Bm0
      C =   0.54958333E+00          Bmo=   0.00000000E+00          Nexp=  6        Nliner=  0
     AI =  -0.16201761E+04   0.17818589E+05  -0.75649324E+05   0.14312982E+06  -0.12202192E+06   0.39818771E+05
```

No	Rotation (radians) X 1/1000	Moment (kip-inch) Expri.	Poly.	M.Expo.	Connection stiffness (kip-inch) X 1000 Expri.	Poly.	M.Expo.
1	0.00	0.00	0.00	0.00	0.1886E+03	0.8467E+01	0.7934E+02
3	1.00	174.82	8.62	175.07	0.1729E+03	0.8922E+01	0.2204E+03
5	2.00	374.34	18.25	390.36	0.2260E+03	0.1062E+02	0.2005E+03
7	3.00	581.38	30.53	570.28	0.1571E+03	0.1409E+02	0.1602E+03
9	4.00	719.42	44.09	714.62	0.1298E+03	0.1111E+02	0.1307E+03
11	5.99	943.10	57.60	939.74	0.9900E+02	0.4291E+01	0.9946E+02
13	7.99	1111.45	64.24	1116.78	0.7519E+02	0.2629E+01	0.7761E+02
15	9.99	1239.82	68.71	1249.33	0.5655E+02	0.1928E+01	0.5489E+02
17	11.99	1353.46	72.15	1337.53	0.5179E+02	0.1539E+01	0.3393E+02
19	14.99	1401.90	76.21	1403.22	0.2453E+01	0.1196E+01	0.1201E+02
21	18.98	1410.26	80.41	1422.65	0.2234E+01	0.9351E+00	0.4306E+00
23	22.98	1419.52	83.81	1420.60	0.1779E+01	0.7744E+00	-0.5183E+00
25	26.97	1425.54	86.67	1421.77	0.1894E+01	0.6651E+00	0.1162E+01
27	30.97	1434.47	89.16	1429.16	0.2272E+01	0.5852E+00	0.2373E+01
29	34.97	1441.09	91.37	1439.48	0.1509E+01	0.5241E+00	0.2651E+01
31	38.96	1448.92	93.36	1449.56	0.1657E+01	0.4759E+00	0.2340E+01
33	42.96	1455.80	95.19	1457.89	0.2097E+01	0.4365E+00	0.1813E+01
35	46.95	1464.37	96.86	1464.07	0.1144E+01	0.4039E+00	0.1295E+01
37	50.95	1465.49	98.42	1468.37	0.3982E+00	0.3763E+00	0.8722E+00

A.6 – 85

VI - 43

Connection type : Flush end-plate connections
Mode : All bolted without column stiffener

Tested by : N.D.Brown & D.Anderson (2001) U.K.
Test Id. : TEST1

Column : 203x203 UC 52
Beam : 457x152 UB 52
Plate thickness : 15 mm

Fasteners: G8.8- -M20
7/8" Oversize holes
Material : S275
Fy = 31.18 ksi
Fu = 44.24 ksi

Major parameters

ct = 0.9842"	pt = 1.9685"	li = 13.7795"	pc = 1.9685"
cc = 0.9842"	pit= 10.2362"	pi = 0.0000"	pic= 3.5433"
gt = 3.5433"	gi = 3.5433"	gc = 3.5433"	
tp = 0.5906"	nt = 2 X 1	ni = 2 X 1	nc = 2 X 1

Remark 1) Plate size: 500x200x15 mm
2)

No	Moment (k-in)	Rotation (radians) X 1/1000
1	0.00	0.00
2	110.47	0.24
3	220.94	0.49
4	357.62	0.97
5	476.11	1.46
6	570.43	1.94
7	650.30	2.43
8	720.44	2.92
9	821.22	3.89
10	913.40	4.86
11	964.88	5.83
12	1035.45	6.81
13	1101.97	7.78
14	1156.34	8.75
15	1191.00	9.72
16	1207.42	10.70
17	1250.33	11.67
18	1284.26	12.64
19	1311.64	13.61
20	1331.11	14.59
21	1351.02	15.56
22	1363.92	16.53
23	1364.51	17.51
24	1385.19	18.48
25	1435.40	19.45

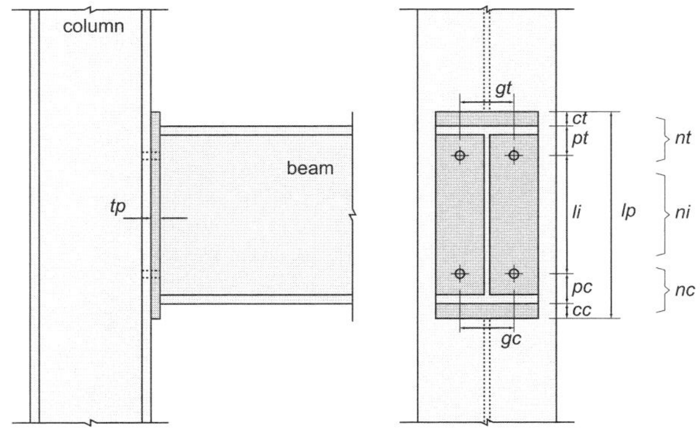

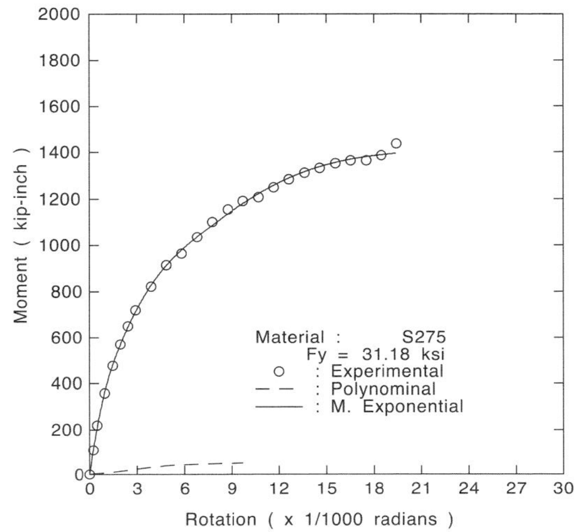

```
Moment-rotation prediction equations     ( R : X 1/1000 radians )

Frye and Morris polynominal model  :    R = Sum ( Ai X ( K*Bm )**Pi X 10**Qi )
     xd =  3.543300"   t =  0.590550"  tf =  0.787400"
     A1 =  1.830000   A2 = -1.040000   A3 =  6.380000    K =  0.084845
     P1 =        1    P2 =        3    P3 =        5    Q1 =        0    Q2 =       -1    Q3 =       -3

Modified exponential model  :
     Bm = Sum ( Ai X ( 1 - exp( - R/( 2*i*C ) ) ) )  +     Sum ( RKj X ( R - Rj0 ) )  +  Bm0
      C =   0.22208333E+00              Bmo=   0.00000000E+00         Nexp=  6        Nliner=  0
     AI =  -0.41193693E+03   0.80195156E+04  -0.45612999E+05   0.11460034E+06  -0.12888002E+06   0.53697676E+05
```

No	Rotation (radians) X 1/1000	Moment (kip-inch) Expri.	Moment (kip-inch) Poly.	Moment (kip-inch) M.Expo.	Connection stiffness (kip-inch) X 1000 Expri.	Connection stiffness (kip-inch) X 1000 Poly.	Connection stiffness (kip-inch) X 1000 M.Expo.
1	0.00	0.00	0.00	0.00	0.4603E+03	0.6441E+01	0.4890E+03
2	0.24	110.47	1.55	112.08	0.4513E+03	0.6459E+01	0.4397E+03
3	0.49	220.94	3.17	213.63	0.3881E+03	0.6520E+01	0.3725E+03
4	0.97	357.62	6.35	365.13	0.2635E+03	0.6765E+01	0.2664E+03
5	1.46	476.11	9.77	478.77	0.2189E+03	0.7228E+01	0.2040E+03
6	1.94	570.43	13.40	568.02	0.1799E+03	0.7964E+01	0.1709E+03
7	2.43	650.30	17.56	646.19	0.1531E+03	0.9105E+01	0.1493E+03
8	2.92	720.44	22.38	714.92	0.1300E+03	0.1052E+02	0.1316E+03
9	3.89	821.22	32.58	827.18	0.9946E+02	0.9005E+01	0.1007E+03
10	4.86	913.40	39.22	912.75	0.7405E+02	0.5155E+01	0.7721E+02
11	5.83	964.88	43.28	979.96	0.6250E+02	0.3443E+01	0.6278E+02
12	6.81	1035.45	46.20	1037.32	0.7028E+02	0.2580E+01	0.5516E+02
13	7.78	1101.97	48.44	1088.69	0.6231E+02	0.2082E+01	0.5107E+02
14	8.75	1156.34	50.29	1136.65	0.4589E+02	0.1754E+01	0.4783E+02
15	9.72	1191.00	51.87	1181.33	0.2629E+02	0.1520E+01	0.4417E+02
16	10.70	1207.42	53.27	1222.50	0.3056E+02	0.1344E+01	0.3972E+02
17	11.67	1250.33	54.50	1258.65	0.3960E+02	0.1209E+01	0.3475E+02
18	12.64	1284.26	55.62	1289.86	0.3160E+02	0.1101E+01	0.2960E+02
19	13.61	1311.64	56.65	1316.13	0.2407E+02	0.1012E+01	0.2461E+02
20	14.59	1331.11	57.60	1337.93	0.2020E+02	0.9373E+00	0.1998E+02
21	15.56	1351.02	58.48	1355.32	0.1691E+02	0.8743E+00	0.1596E+02
22	16.53	1363.92	59.30	1369.09	0.6982E+01	0.8200E+00	0.1254E+02
23	17.51	1364.51	60.08	1379.94	0.1101E+02	0.7724E+00	0.9699E+01
24	18.48	1385.19	60.81	1388.21	0.3654E+02	0.7309E+00	0.7429E+01
25	19.45	1435.40	61.50	1394.51	0.5177E+02	0.6941E+00	0.5631E+01

VI - 44

Connection type : Flush end-plate connections
Mode : All bolted without column stiffener

Tested by : A.W.Thomson & B.M.Broderick (2002) Ireland
Test Id. : J1-M

Column : 203x203 UC 86
Beam : 254x146 UB 37
Plate thickness : 10 mm

Fasteners: G8.8- -M20
7/8" Oversize holes
Material : G43
Fy = -- ksi
Fu = -- ksi

Major parameters

ct = 0.4724"	pt = 2.3622"	li = 5.3543"	pc = 2.3622"
cc = 0.4724"	gt = 3.5433"	gc = 3.5433"	
tp = 0.3937"	nt = 2 X 1	nc = 2 X 1	

Remark 1) Plate size: 280x200x10 mm
2)

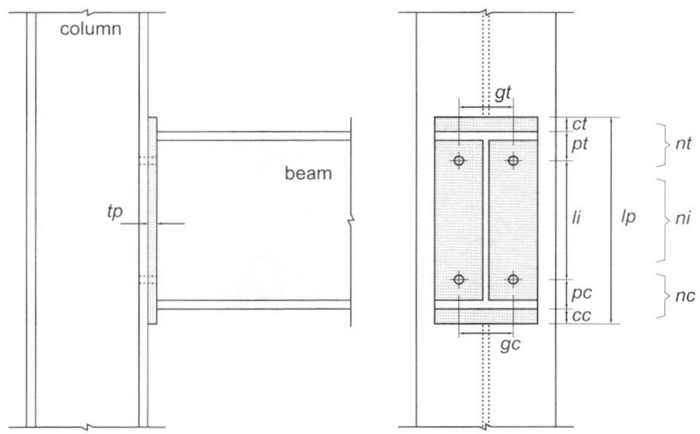

No	Moment (k-in)	Rotation (radians) X 1/1000	No	Moment (k-in)	Rotation (radians) X 1/1000
1	0.00	0.00	26	409.79	44.90
2	32.52	1.00	27	413.35	46.94
3	66.01	2.04	28	417.87	48.98
4	96.86	3.01	29	421.41	51.02
5	128.53	4.08	30	425.05	53.06
6	151.21	5.02	31	428.01	55.11
7	174.72	6.12	32	431.82	57.15
8	214.07	8.16	33	434.81	59.19
9	248.61	10.20	34	438.16	61.23
10	276.88	12.25	35	441.47	63.27
11	298.33	14.29	36	444.96	65.31
12	315.81	16.33			
13	329.13	18.37			
14	340.68	20.41			
15	349.51	22.45			
16	357.51	24.49			
17	364.17	26.53			
18	371.21	28.57			
19	377.07	30.61			
20	382.71	32.66			
21	387.31	34.70			
22	392.02	36.74			
23	396.55	38.78			
24	400.97	40.82			
25	405.51	42.86			

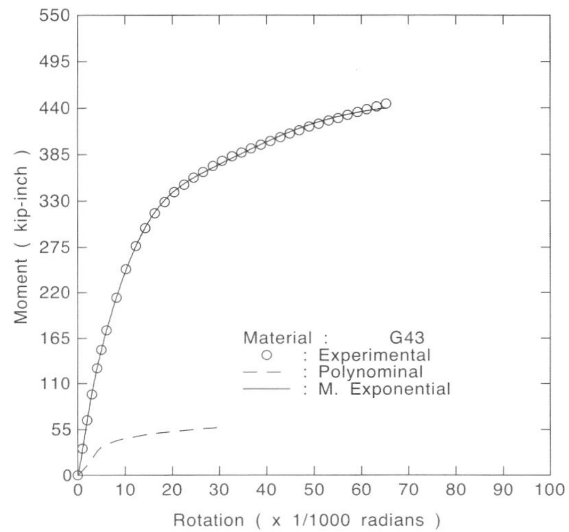

```
Moment-rotation prediction equations     ( R : X 1/1000 radians )

Frye and Morris polynominal model  :    R = Sum ( Ai X ( K*Bm )**Pi X 10**Qi )
     xd =  3.543300"   t =  0.393700"  tf =  0.787400"
     A1 =  1.830000   A2 = -1.040000   A3 =  6.380000    K =  0.099784
     P1 =        1    P2 =        3    P3 =        5    Q1 =        0    Q2 =       -1    Q3 =       -3

Modified exponential model  :
     Bm = Sum ( Ai X ( 1 - exp( - R/( 2*i*C ) ) ) )  +      Sum ( RKj X ( R - Rj0 ) )  +  Bm0
      C =   0.79425000E+00             Bmo=   0.00000000E+00            Nexp=  6       Nliner=  0
     AI =  -0.30313580E+03   0.42419714E+04  -0.21636895E+05   0.48671286E+05  -0.49466544E+05   0.18941659E+05
```

No	Rotation (radians) X 1/1000	Moment (kip-inch) Expri.	Moment (kip-inch) Poly.	Moment (kip-inch) M.Expo.	Connection stiffness (kip-inch) X 1000 Expri.	Connection stiffness (kip-inch) X 1000 Poly.	Connection stiffness (kip-inch) X 1000 M.Expo.
1	0.00	0.00	0.00	0.00	0.3252E+02	0.5476E+01	0.2330E+02
3	2.04	66.01	12.08	66.89	0.3200E+02	0.6939E+01	0.3400E+02
5	4.08	128.53	29.08	128.08	0.2668E+02	0.6842E+01	0.2599E+02
7	6.12	174.72	37.61	175.06	0.2064E+02	0.2663E+01	0.2064E+02
9	10.20	248.61	44.72	247.29	0.1537E+02	0.1213E+01	0.1514E+02
11	14.29	298.33	48.74	298.82	0.9540E+01	0.8152E+00	0.1007E+02
13	18.37	329.13	51.64	330.99	0.6096E+01	0.6252E+00	0.5987E+01
15	22.45	349.51	53.94	350.24	0.4125E+01	0.5123E+00	0.3747E+01
17	26.53	364.17	55.87	363.48	0.3359E+01	0.4369E+00	0.2909E+01
19	30.61	377.07	57.53	374.87	0.2811E+01	0.3826E+00	0.2730E+01
21	34.70	387.31	59.01	385.97	0.2282E+01	0.3414E+00	0.2699E+01
23	38.78	396.55	60.33	396.81	0.2195E+01	0.3091E+00	0.2591E+01
25	42.86	405.51	61.54	406.95	0.2160E+01	0.2829E+00	0.2361E+01
27	46.94	413.35	62.65	415.95	0.1980E+01	0.2613E+00	0.2044E+01
29	51.02	421.41	63.68	423.58	0.1760E+01	0.2431E+00	0.1694E+01
31	55.11	428.01	64.64	429.81	0.1657E+01	0.2275E+00	0.1354E+01
33	59.19	434.81	65.54	434.69	0.1555E+01	0.2140E+00	0.1050E+01
35	63.27	441.47	66.40	438.44	0.1665E+01	0.2020E+00	0.7954E+00

A.6 – 89

VI - 45

Connection type : Flush end-plate connections
Mode : All bolted without column stiffener

Tested by : A.W.Thomson & B.M.Broderick (2002) Ireland
Test Id. : J5-M

Column : 203x203 UC 52
Beam : 254x146 UB 37
Plate thickness : 12 mm

Fasteners: G8.8- -M20
7/8" Oversize holes
Material : G43
Fy = -- ksi
Fu = -- ksi

Major parameters

ct = 0.4724"	pt = 2.3622"	li = 5.3543"	pc = 2.3622"
cc = 0.4724"	gt = 3.5433"	gc = 3.5433"	
tp = 0.4724"	nt = 2 X 1	nc = 2 X 1	

Remark 1) Plate size: 280x200x12 mm
2)

No	Moment (k-in)	Rotation (radians) X 1/1000
1	0.00	0.00
2	32.87	0.99
3	65.50	1.98
4	95.48	2.97
5	122.11	3.97
6	145.63	4.96
7	166.43	5.95
8	204.44	7.93
9	238.23	9.92
10	266.71	11.90
11	289.56	13.88
12	307.57	15.86
13	322.27	17.85
14	333.44	19.83
15	342.69	21.81
16	350.49	23.80
17	357.81	25.78
18	364.25	27.76
19	370.18	29.75
20	375.66	31.73
21	380.38	33.71
22	384.92	35.69
23	389.53	37.68
24	394.07	39.66
25	398.36	41.64
26	402.46	43.63
27	406.38	45.61
28	410.30	47.59
29	414.10	49.58
30	417.85	51.56
31	421.56	53.54
32	425.17	55.52
33	428.58	57.51
34	431.47	59.49
35	434.42	61.47
36	437.83	63.46
37	441.24	65.44

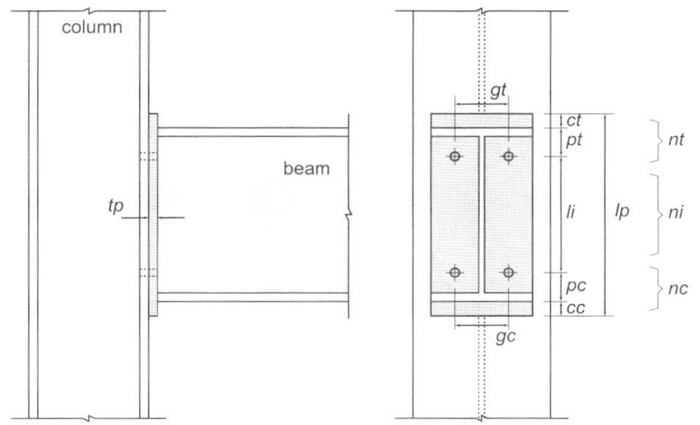

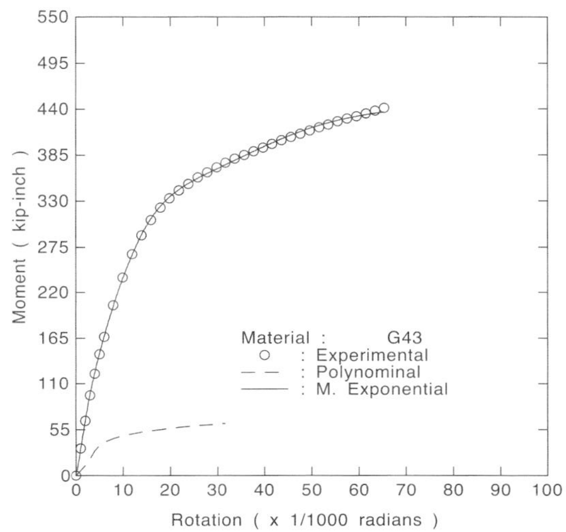

A.6 – 90

```
Moment-rotation prediction equations     ( R : X 1/1000 radians )

Frye and Morris polynominal model  :    R = Sum ( Ai X ( K*Bm )**Pi X 10**Qi )
     xd =  3.543300"    t =  0.472440"  tf =  0.787400"
     A1 =  1.830000   A2 = -1.040000   A3 =  6.380000    K =  0.092766
     P1 =         1    P2 =         3    P3 =         5    Q1 =         0    Q2 =        -1    Q3 =        -3

Modified exponential model  :
     Bm = Sum ( Ai X ( 1 - exp( - R/( 2*i*C ) ) ) )  +     Sum ( RKj X ( R - Rj0 ) )  +  Bm0
      C =   0.79283333E+00             Bmo=   0.00000000E+00          Nexp=  6       Nliner=  0
      AI =  -0.29781149E+03   0.45018332E+04  -0.23091778E+05   0.51609039E+05  -0.52099026E+05   0.19822493E+05
```

No	Rotation (radians) X 1/1000	Moment (kip-inch) Expri.	Moment (kip-inch) Poly.	Moment (kip-inch) M.Expo.	Connection stiffness (kip-inch) X 1000 Expri.	Connection stiffness (kip-inch) X 1000 Poly.	Connection stiffness (kip-inch) X 1000 M.Expo.
1	0.00	0.00	0.00	0.00	0.3320E+02	0.5891E+01	0.2652E+02
3	1.98	65.50	12.55	65.57	0.3162E+02	0.7354E+01	0.3289E+02
5	3.97	122.11	30.43	122.72	0.2519E+02	0.7871E+01	0.2480E+02
7	5.95	166.43	39.95	166.59	0.2041E+02	0.3025E+01	0.2010E+02
9	9.92	238.23	47.71	237.03	0.1568E+02	0.1354E+01	0.1564E+02
11	13.88	289.56	52.06	289.67	0.1032E+02	0.9056E+00	0.1084E+02
13	17.85	322.27	55.19	323.61	0.6510E+01	0.6925E+00	0.6511E+01
15	21.81	342.69	57.66	343.74	0.4297E+01	0.5668E+00	0.3959E+01
17	25.78	357.81	59.74	357.02	0.3475E+01	0.4828E+00	0.2924E+01
19	29.75	370.18	61.53	368.00	0.2875E+01	0.4224E+00	0.2689E+01
21	33.71	380.38	63.10	378.64	0.2337E+01	0.3769E+00	0.2688E+01
23	37.68	389.53	64.53	389.25	0.2306E+01	0.3411E+00	0.2641E+01
25	41.64	398.36	65.82	399.41	0.2113E+01	0.3122E+00	0.2469E+01
27	45.61	406.38	67.01	408.69	0.1980E+01	0.2882E+00	0.2191E+01
29	49.58	414.10	68.11	416.74	0.1902E+01	0.2680E+00	0.1858E+01
31	53.54	421.56	69.14	423.42	0.1848E+01	0.2508E+00	0.1517E+01
33	57.51	428.58	70.10	428.80	0.1585E+01	0.2359E+00	0.1201E+01
35	61.47	434.42	71.02	433.00	0.1602E+01	0.2227E+00	0.9266E+00
37	65.44	441.24	71.88	436.21	0.1722E+01	0.2113E+00	0.6996E+00

A.6 – 91

VI - 46

Connection type : Flush end-plate connections
Mode : All bolted without column stiffener

Tested by : A.W.Thomson & B.M.Broderick (2002) Ireland
Test Id. : J7-M

Column : 203x203 UC 86
Beam : 254x146 UB 37
Plate thickness : 12 mm

Fasteners: 10.9- -M20
7/8" Oversize holes
Material : G43
Fy = -- ksi
Fu = -- ksi

Major parameters

ct = 0.4724"	pt = 2.3622"	li = 5.3543"	pc = 2.3622"
cc = 0.4724"	gt = 3.5433"	gc = 3.5433"	
tp = 0.4724"	nt = 2 X 1	nc = 2 X 1	

Remark 1) Plate size: 280x200x12 mm
2)

No	Moment (k-in)	Rotation (radians) X 1/1000	No	Moment (k-in)	Rotation (radians) X 1/1000
1	0.00	0.00	26	490.55	42.51
2	50.76	1.00	27	494.75	44.53
3	103.56	2.02	28	498.49	46.55
4	140.40	2.99	29	501.70	48.58
5	174.11	4.05	30	504.56	50.60
6	200.47	4.98	31	507.82	52.63
7	229.60	6.07	32	511.13	54.65
8	251.72	6.98	33	514.20	56.67
9	275.29	8.10	34	516.85	58.70
10	313.55	10.12	35	519.95	60.72
11	343.62	12.14	36	523.37	62.75
12	366.19	14.17	37	526.90	64.77
13	383.66	16.19			
14	397.93	18.22			
15	410.09	20.24			
16	421.69	22.26			
17	432.70	24.29			
18	443.32	26.31			
19	452.10	28.34			
20	459.35	30.36			
21	466.51	32.39			
22	472.90	34.41			
23	478.42	36.43			
24	482.61	38.46			
25	486.61	40.48			

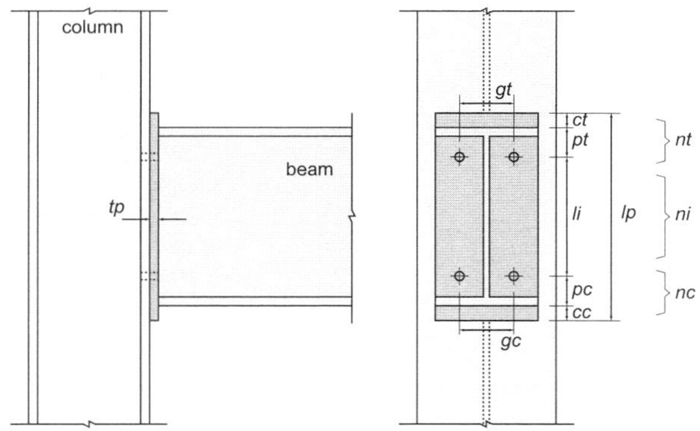

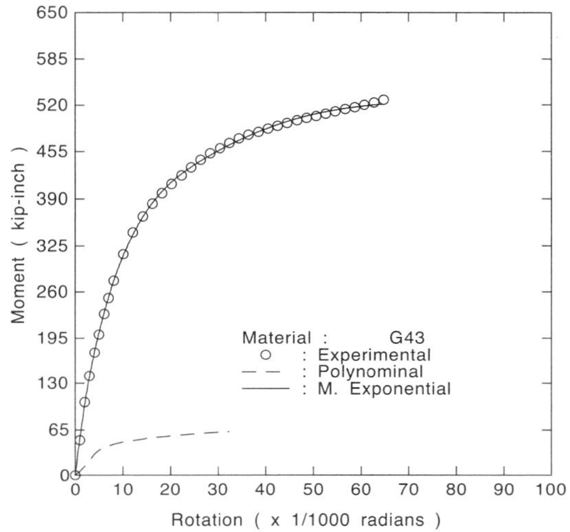

A.6 – 92

```
Moment-rotation prediction equations     ( R : X 1/1000 radians )

Frye and Morris polynominal model  :    R = Sum ( Ai X ( K*Bm )**Pi X 10**Qi )
     xd =  3.543300"   t =  0.472440"  tf =  0.787400"
     A1 =  1.830000   A2 = -1.040000   A3 =  6.380000    K =  0.092766
     P1 =        1    P2 =        3    P3 =        5    Q1 =        0    Q2 =       -1    Q3 =       -3

Modified exponential model  :
     Bm = Sum ( Ai X ( 1 - exp( - R/( 2*i*C ) ) ) )  +     Sum ( RKj X ( R - Rj0 ) )  +  Bm0
      C =   0.78975000E+00              Bmo=   0.00000000E+00          Nexp=  6         Nliner=  0
     AI =  -0.87709621E+02   0.19029057E+04  -0.11283939E+05   0.28502424E+05  -0.31601444E+05   0.13095276E+05
```

No	Rotation (radians) X 1/1000	Moment (kip-inch) Expri.	Moment (kip-inch) Poly.	Moment (kip-inch) M.Expo.	Connection stiffness (kip-inch) X 1000 Expri.	Connection stiffness (kip-inch) X 1000 Poly.	Connection stiffness (kip-inch) X 1000 M.Expo.
1	0.00	0.00	0.00	0.00	0.5076E+02	0.5891E+01	0.5717E+02
3	2.02	103.56	12.84	100.75	0.4470E+02	0.7427E+01	0.4248E+02
5	4.05	174.11	31.05	174.75	0.2996E+02	0.7498E+01	0.3133E+02
7	6.07	229.60	40.31	230.65	0.2541E+02	0.2910E+01	0.2453E+02
9	8.10	275.29	44.89	275.49	0.2029E+02	0.1792E+01	0.1987E+02
11	12.14	343.62	50.36	341.36	0.1301E+02	0.1055E+01	0.1313E+02
13	16.19	383.66	53.98	384.46	0.7839E+01	0.7666E+00	0.8505E+01
15	20.24	410.09	56.74	412.87	0.5881E+01	0.6101E+00	0.5806E+01
17	24.29	432.70	59.00	433.29	0.5341E+01	0.5108E+00	0.4444E+01
19	28.34	452.10	60.92	449.77	0.3956E+01	0.4418E+00	0.3760E+01
21	32.39	466.51	62.60	464.07	0.3344E+01	0.3908E+00	0.3318E+01
23	36.43	478.42	64.09	476.66	0.2400E+01	0.3515E+00	0.2916E+01
25	40.48	486.61	65.45	487.63	0.1960E+01	0.3200E+00	0.2496E+01
27	44.53	494.75	66.71	496.87	0.1965E+01	0.2941E+00	0.2069E+01
29	48.58	501.70	67.85	504.41	0.1498E+01	0.2726E+00	0.1663E+01
31	52.63	507.82	68.92	510.40	0.1624E+01	0.2544E+00	0.1300E+01
33	56.67	514.20	69.91	515.01	0.1412E+01	0.2388E+00	0.9930E+00
35	60.72	519.95	70.85	518.50	0.1609E+01	0.2251E+00	0.7426E+00
37	64.77	526.90	71.73	521.10	0.1750E+01	0.2132E+00	0.5458E+00

VI - 47

Connection type : Flush end-plate connections
Mode : All bolted without column stiffener

Tested by : A.M.G.Coelho & F.S.K.Bijlaard (2007) Portugal
Test Id. : F1EP_15_2

Column : HE300M
Beam : HE320A
Plate thickness : 15 mm

Fasteners: 12.9- -M24
33/32" Oversize holes
Material : S690
Fy = 112.26 ksi
Fu = 118.06 ksi

Major parameters

ct = 0.9842"	pt = 2.9528"	li = 6.2992"	pc = 2.9528"
cc = 0.9842"	gt = 5.9055"	gc = 5.9055"	
tp = 0.5807"	nt = 2 X 1	nc = 2 X 1	

Remark 1) Failure mode: Fracture of bolt at tension side
2) Plate size: 360x300x15 mm

No	Moment (k-in)	Rotation (radians) X 1/1000	No	Moment (k-in)	Rotation (radians) X 1/1000
1	0.00	0.00	26	1861.41	25.00
2	133.48	1.00	27	1882.55	26.00
3	260.62	2.00	28	1903.69	27.00
4	383.76	3.00	29	1924.69	28.00
5	502.53	4.00	30	1945.00	29.00
6	624.73	5.00	31	1965.31	30.00
7	746.61	6.00	32	1985.71	31.00
8	856.87	7.00	33	2006.73	32.00
9	965.78	8.00	34	2027.75	33.00
10	1051.56	9.00	35	2044.25	34.00
11	1137.19	10.00	36	2032.86	35.00
12	1218.19	11.00	37	1967.04	36.00
13	1299.19	12.00	38	1849.56	37.00
14	1363.34	13.00			
15	1427.31	14.00			
16	1483.37	15.00			
17	1535.69	16.00			
18	1586.10	17.00			
19	1626.46	18.00			
20	1666.81	19.00			
21	1704.25	20.00			
22	1740.18	21.00			
23	1776.11	22.00			
24	1804.79	23.00			
25	1833.48	24.00			

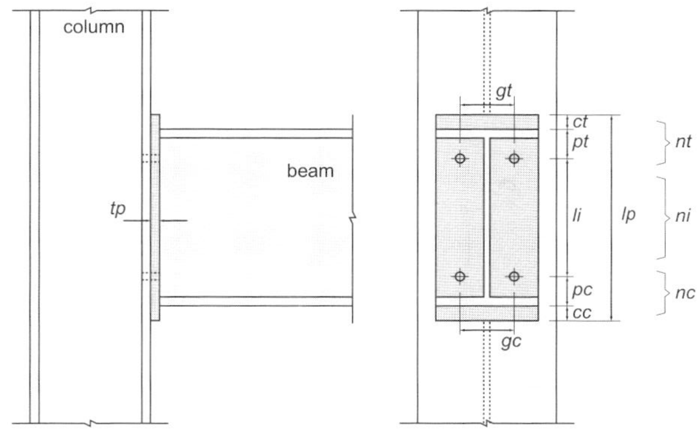

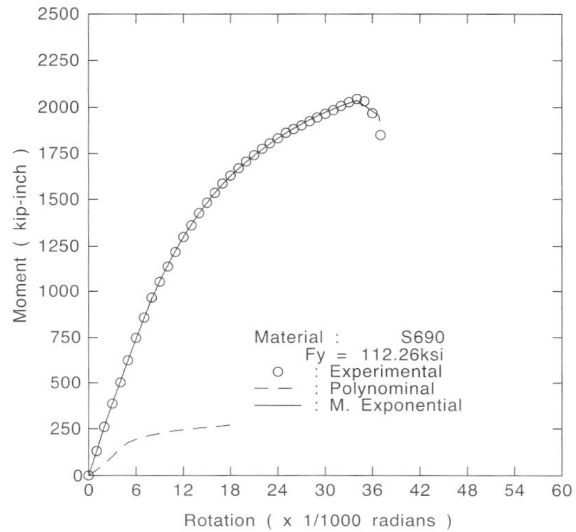

```
Moment-rotation prediction equations     ( R : X 1/1000 radians )

Frye and Morris polynominal model  :    R = Sum ( Ai X ( K*Bm )**Pi X 10**Qi )
     xd =  5.905500"    t =  0.580710"  tf =  0.944900"
     A1 =  1.830000   A2 = -1.040000   A3 =  6.380000    K =  0.019069
     P1 =         1   P2 =         3   P3 =         5   Q1 =         0   Q2 =        -1   Q3 =        -3

Modified exponential model  :
     Bm = Sum ( Ai X ( 1 - exp( - R/( 2*i*C ) ) ) )  +      Sum ( RKj X ( R - Rj0 ) )  +  Bm0
       C =   0.55833333E+00            Bmo=   0.00000000E+00         Nexp=  6        Nliner=  2
      AI =  -0.41201032E+03   0.85536224E+04  -0.51033994E+05   0.12760157E+06  -0.14576213E+06   0.63215842E+05
     Rj0 =          34.0000          36.9000
     RKj =  -0.41053624E+02  -0.99174263E+03
```

No	Rotation (radians) X 1/1000	Moment (kip-inch) Expri.	Moment (kip-inch) Poly.	Moment (kip-inch) M.Expo.	Connection stiffness (kip-inch) X 1000 Expri.	Connection stiffness (kip-inch) X 1000 Poly.	Connection stiffness (kip-inch) X 1000 M.Expo.
1	0.00	0.00	0.00	0.00	0.1335E+03	0.2866E+02	0.1231E+03
3	2.00	260.62	61.75	261.13	0.1251E+03	0.3595E+02	0.1268E+03
5	4.00	502.53	149.20	506.53	0.1205E+03	0.3760E+02	0.1200E+03
7	6.00	746.61	195.10	741.42	0.1161E+03	0.1448E+02	0.1139E+03
9	8.00	965.78	217.52	957.60	0.9735E+02	0.8882E+01	0.1013E+03
11	10.00	1137.19	232.64	1144.36	0.8331E+02	0.6519E+01	0.8532E+02
13	12.00	1299.19	244.26	1299.52	0.7257E+02	0.5205E+01	0.7022E+02
15	14.00	1427.31	253.77	1427.22	0.6002E+02	0.4364E+01	0.5801E+02
17	16.00	1535.69	261.88	1533.59	0.5137E+02	0.3776E+01	0.4882E+02
19	18.00	1626.46	268.98	1624.08	0.4035E+02	0.3340E+01	0.4200E+02
21	20.00	1704.25	275.31	1702.60	0.3668E+02	0.3003E+01	0.3672E+02
23	22.00	1776.11	281.04	1771.58	0.3231E+02	0.2734E+01	0.3237E+02
25	24.00	1833.48	286.28	1832.43	0.2831E+02	0.2514E+01	0.2854E+02
27	26.00	1882.55	291.12	1885.95	0.2114E+02	0.2330E+01	0.2502E+02
29	28.00	1924.69	295.61	1932.68	0.2066E+02	0.2174E+01	0.2175E+02
31	30.00	1965.31	299.82	1973.08	0.2035E+02	0.2039E+01	0.1870E+02
33	32.00	2006.73	303.84	2007.63	0.2102E+02	0.1921E+01	0.1590E+02
35	34.00	2044.25	307.55	2036.85	0.2552E+01	0.1819E+01	0.1337E+02
37	36.00	1967.04	311.07	1979.18	-0.9165E+02	0.1728E+01	-0.2993E+02

VI - 48

Connection type : Flush end-plate connections
Mode : All bolted without column stiffener

Tested by : A.M.G.Coelho & F.S.K.Bijlaard (2007) Portugal
Test Id. : F2EP_15_2

Column : HE300M
Beam : HE320A
Plate thickness : 15 mm

Fasteners: 12.9- -M24
33/32" Oversize holes
Material : S690
Fy = 112.26 ksi
Fu = 118.06 ksi

Major parameters

ct = 0.0000" pt = 2.9528" li = 6.2992" pc = 2.9528"
cc = 0.0000" gt = 5.9055" gc = 5.9055"
tp = 0.5764" nt = 2 X 1 nc = 2 X 1

Remark 1) Failure mode: Fracture of bolt at tension side
2) Plate size: 310x300x15 mm

No	Moment (k-in)	Rotation (radians) X 1/1000	No	Moment (k-in)	Rotation (radians) X 1/1000
1	0.00	0.00	26	1710.62	25.00
2	114.41	1.00	27	1737.21	26.00
3	228.89	2.00	28	1764.14	27.00
4	341.63	3.00	29	1791.07	28.00
5	448.69	4.00	30	1814.16	29.00
6	557.25	5.00	31	1836.28	30.00
7	665.53	6.00	32	1858.41	31.00
8	770.59	7.00	33	1880.53	32.00
9	864.45	8.00	34	1902.65	33.00
10	953.96	9.00	35	1895.85	34.00
11	1038.58	10.00	36	1848.02	35.00
12	1122.63	11.00	37	1782.61	36.00
13	1204.80	12.00	38	1723.14	37.00
14	1265.95	13.00	39	1672.57	38.00
15	1321.84	14.00	40	1619.47	39.00
16	1377.21	15.00			
17	1427.61	16.00			
18	1466.55	17.00			
19	1505.49	18.00			
20	1545.72	19.00			
21	1586.28	20.00			
22	1621.88	21.00			
23	1650.04	22.00			
24	1674.18	23.00			
25	1684.07	24.00			

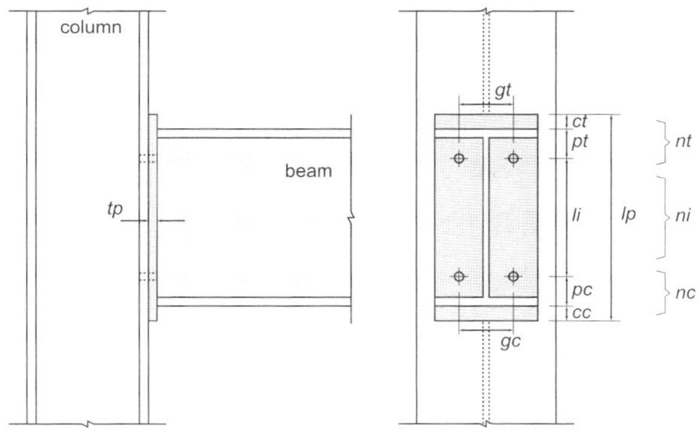

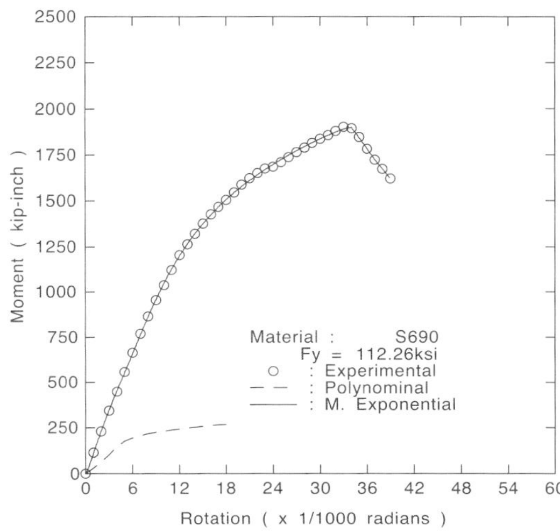

```
Moment-rotation prediction equations     ( R : X 1/1000 radians )

Frye and Morris polynominal model  :    R = Sum ( Ai X ( K*Bm )**Pi X 10**Qi )
     xd =  5.905500"   t =  0.576380"  tf =  0.944900"
     A1 =  1.830000   A2 = -1.040000   A3 =  6.380000    K =  0.019126
     P1 =        1     P2 =        3     P3 =        5     Q1 =        0     Q2 =       -1     Q3 =       -3

Modified exponential model  :
     Bm = Sum ( Ai X ( 1 - exp( - R/( 2*i*C ) ) ) )  +     Sum ( RKj X ( R - Rj0 ) )  +  Bm0
      C =   0.57500000E+00             Bmo=   0.00000000E+00         Nexp=  6        Nliner=  3
     AI =  -0.10501358E+04   0.16394160E+05  -0.84887052E+05   0.19297897E+06  -0.20362324E+06   0.82254211E+05
     Rj0 =         33.0000         34.0000         37.0000
     RKj =  -0.76868807E+01  -0.64215242E+02   0.84200095E+01
```

No	Rotation (radians) X 1/1000	Moment (kip-inch) Expri.	Moment (kip-inch) Poly.	Moment (kip-inch) M.Expo.	Connection stiffness (kip-inch) X 1000 Expri.	Connection stiffness (kip-inch) X 1000 Poly.	Connection stiffness (kip-inch) X 1000 M.Expo.
1	0.00	0.00	0.00	0.00	0.1144E+03	0.2857E+02	0.6990E+02
3	2.00	228.89	61.57	232.72	0.1136E+03	0.3584E+02	0.1188E+03
5	4.00	448.69	148.75	451.21	0.1078E+03	0.3749E+02	0.1041E+03
7	6.00	665.53	194.52	659.30	0.1067E+03	0.1443E+02	0.1042E+03
9	8.00	864.45	216.87	863.49	0.9169E+02	0.8856E+01	0.9838E+02
11	10.00	1038.58	231.95	1047.48	0.8433E+02	0.6499E+01	0.8476E+02
13	12.00	1204.80	243.53	1201.11	0.7166E+02	0.5190E+01	0.6896E+02
15	14.00	1321.84	253.01	1324.73	0.5563E+02	0.4351E+01	0.5519E+02
17	16.00	1427.61	261.10	1424.26	0.4467E+02	0.3765E+01	0.4493E+02
19	18.00	1505.49	268.17	1506.60	0.3959E+02	0.3330E+01	0.3788E+02
21	20.00	1586.28	274.48	1577.32	0.3808E+02	0.2994E+01	0.3313E+02
23	22.00	1650.04	280.20	1640.03	0.2615E+02	0.2726E+01	0.2975E+02
25	24.00	1684.07	285.42	1696.75	0.1822E+02	0.2506E+01	0.2704E+02
27	26.00	1737.21	290.25	1748.35	0.2676E+02	0.2323E+01	0.2459E+02
29	28.00	1791.07	294.73	1795.14	0.2501E+02	0.2167E+01	0.2220E+02
31	30.00	1836.28	298.93	1837.17	0.2212E+02	0.2033E+01	0.1982E+02
33	32.00	1880.53	302.93	1874.45	0.2212E+02	0.1915E+01	0.1747E+02
35	34.00	1895.85	306.63	1899.39	-0.2732E+02	0.1813E+01	0.7497E+01
37	36.00	1782.61	310.14	1783.77	-0.6244E+02	0.1723E+01	-0.5887E+02
39	38.00	1672.57	313.54	1672.43	-0.5183E+02	0.1641E+01	-0.5244E+02

A.6 – 97

VI - 49

Connection type : Flush end-plate connections
Mode : All bolted without column stiffener

Tested by : A.M.G.Coelho & F.S.K.Bijlaard (2007) Portugal
Test Id. : F1EP_10_2

Column : HE300M
Beam : HE320A
Plate thickness : 10 mm

Fasteners: 12.9- -M24
33/32" Oversize holes
Material : S690
Fy = 112.26 ksi
Fu = 118.06 ksi

Major parameters

ct = 0.9842" pt = 2.9528" li = 6.2992" pc = 2.9528"
cc = 0.9842" gt = 5.9055" gc = 5.9055"
tp = 0.3996" nt = 2 X 1 nc = 2 X 1

Remark 1) Failure mode: weld failure along inner welds zone
2) Plate size: 360x300x10 mm

No	Moment (k-in)	Rotation (radians) X 1/1000	No	Moment (k-in)	Rotation (radians) X 1/1000
1	0.00	0.00	26	1083.36	25.00
2	96.28	1.00	27	1099.40	26.00
3	192.57	2.00	28	1114.58	27.00
4	283.38	3.00	29	1127.78	28.00
5	361.72	4.00	30	1140.97	29.00
6	430.46	5.00	31	1154.17	30.00
7	499.42	6.00	32	1162.78	31.00
8	562.75	7.00	33	1170.78	32.00
9	616.24	8.00	34	1178.77	33.00
10	669.19	9.00	35	1191.02	34.00
11	712.45	10.00	36	1203.81	35.00
12	755.70	11.00	37	1217.91	36.00
13	792.28	12.00	38	1237.62	37.00
14	826.00	13.00	39	1252.05	38.00
15	857.78	14.00	40	1256.64	39.00
16	882.69	15.00	41	1182.25	40.00
17	907.59	16.00	42	1135.34	41.00
18	931.27	17.00	43	1105.28	42.00
19	952.23	18.00	44	1084.42	43.00
20	973.18	19.00	45	1071.73	44.00
21	993.88	20.00	46	1059.59	45.00
22	1012.35	21.00	47	1053.10	46.00
23	1030.81	22.00			
24	1049.28	23.00			
25	1067.33	24.00			

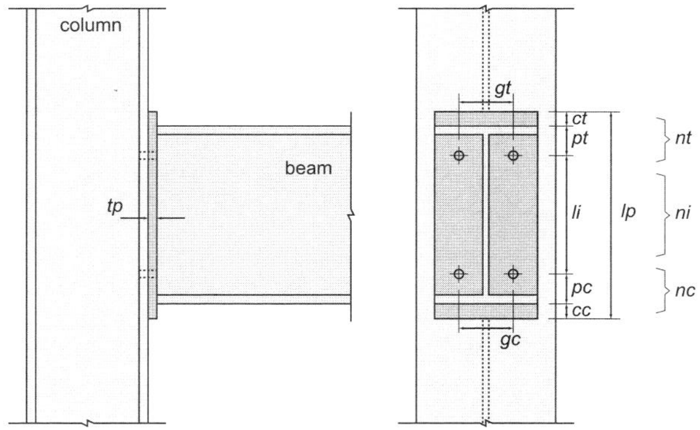

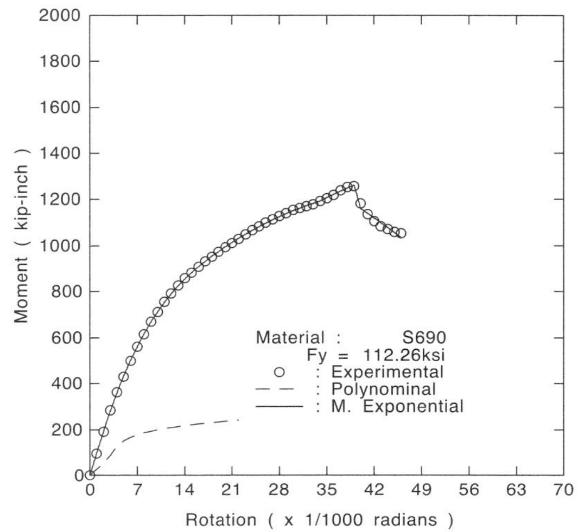

A.6 – 98

```
Moment-rotation prediction equations     ( R : X 1/1000 radians )

Frye and Morris polynominal model   :    R = Sum ( Ai X ( K*Bm )**Pi X 10**Qi )
     xd =  5.905500"    t =  0.399610"  tf =  0.944900"
     A1 =  1.830000   A2 = -1.040000   A3 =  6.380000     K =  0.022144
     P1 =         1   P2 =         3   P3 =         5    Q1 =         0    Q2 =        -1    Q3 =        -3

Modified exponential model  :
     Bm = Sum ( Ai X ( 1 - exp( - R/( 2*i*C ) ) ) )  +      Sum ( RKj X ( R - Rj0 ) )  +  Bm0
       C =  0.63333333E+00              Bmo=  0.00000000E+00          Nexp=  6        Nliner=  3
      AI = -0.11085592E+03  0.27644597E+04 -0.20699531E+05  0.59569591E+05 -0.73034263E+05  0.32806050E+05
     Rj0 =          34.0000         39.0000          40.0000
     RKj =  0.60548678E+01 -0.11310157E+03  0.82126267E+02
```

No	Rotation (radians) X 1/1000	Moment (kip-inch) Expri.	Moment (kip-inch) Poly.	Moment (kip-inch) M.Expo.	Connection stiffness (kip-inch) X 1000 Expri.	Connection stiffness (kip-inch) X 1000 Poly.	Connection stiffness (kip-inch) X 1000 M.Expo.
1	0.00	0.00	0.00	0.00	0.9628E+02	0.2468E+02	0.9849E+02
3	2.00	192.57	53.17	192.55	0.9355E+02	0.3096E+02	0.9073E+02
5	4.00	361.72	128.48	360.09	0.7354E+02	0.3238E+02	0.7696E+02
7	6.00	499.42	168.01	501.19	0.6614E+02	0.1247E+02	0.6432E+02
9	8.00	616.24	187.31	618.07	0.5322E+02	0.7649E+01	0.5278E+02
11	10.00	712.45	200.34	713.37	0.4325E+02	0.5614E+01	0.4284E+02
13	12.00	792.28	210.34	790.79	0.3515E+02	0.4483E+01	0.3493E+02
15	14.00	857.78	218.53	854.46	0.2834E+02	0.3758E+01	0.2904E+02
17	16.00	907.59	225.51	908.08	0.2429E+02	0.3252E+01	0.2482E+02
19	18.00	952.23	231.63	954.54	0.2095E+02	0.2877E+01	0.2180E+02
21	20.00	993.88	237.08	995.79	0.1958E+02	0.2586E+01	0.1954E+02
23	22.00	1030.81	242.01	1033.00	0.1846E+02	0.2355E+01	0.1772E+02
25	24.00	1067.33	246.52	1066.82	0.1704E+02	0.2165E+01	0.1612E+02
27	26.00	1099.40	250.69	1097.55	0.1561E+02	0.2006E+01	0.1462E+02
29	28.00	1127.78	254.56	1125.34	0.1320E+02	0.1872E+01	0.1317E+02
31	30.00	1154.17	258.19	1150.26	0.1091E+02	0.1756E+01	0.1176E+02
33	32.00	1170.78	261.65	1172.40	0.7991E+01	0.1654E+01	0.1039E+02
35	34.00	1191.02	264.84	1191.87	0.1252E+02	0.1566E+01	0.9090E+01
37	36.00	1217.91	267.87	1220.93	0.1691E+02	0.1488E+01	0.1393E+02
39	38.00	1252.05	270.81	1247.65	0.9506E+01	0.1417E+01	0.1281E+02
41	40.00	1182.25	273.55	1159.14	-0.6065E+02	0.1355E+01	-0.1013E+03
43	42.00	1105.28	276.23	1119.88	-0.2546E+02	0.1297E+01	-0.2007E+02
45	44.00	1071.73	278.80	1078.93	-0.1242E+02	0.1245E+01	-0.2086E+02
47	46.00	1053.10	281.20	1036.52	-0.6495E+01	0.1198E+01	-0.2154E+02

VI - 50

Connection type : Flush end-plate connections
Mode : All bolted without column stiffener

Tested by : A.M.G.Coelho & F.S.K.Bijlaard (2007) Portugal
Test Id. : F2EP_10_2

Column : HE300M
Beam : HE320A
Plate thickness : 10 mm

Fasteners: 12.9- -M24
33/32" Oversize holes
Material : S690
Fy = 112.26 ksi
Fu = 118.06 ksi

Major parameters

ct = 0.0000" pt = 2.9528" li = 6.2992" pc = 2.9528"
cc = 0.0000" gt = 5.9055" gc = 5.9055"
tp = 0.4035" nt = 2 X 1 nc = 2 X 1

Remark 1) Failure mode: weld failure along inner welds zone
2) Plate size: 310x300x10 mm

No	Moment (k-in)	Rotation (radians) X 1/1000	No	Moment (k-in)	Rotation (radians) X 1/1000
1	0.00	0.00	26	927.09	25.00
2	70.42	1.00	27	937.84	26.00
3	130.34	2.00	28	948.59	27.00
4	185.34	3.00	29	958.61	28.00
5	231.46	4.00	30	968.55	29.00
6	279.57	5.00	31	979.22	30.00
7	332.64	6.00	32	992.19	31.00
8	386.65	7.00	33	1005.16	32.00
9	455.00	8.00	34	1018.13	33.00
10	513.26	9.00	35	1027.35	34.00
11	565.49	10.00	36	1035.40	35.00
12	610.09	11.00	37	1018.66	36.00
13	650.08	12.00	38	1001.91	37.00
14	686.47	13.00	39	985.17	38.00
15	717.15	14.00	40	970.09	39.00
16	748.38	15.00	41	958.87	40.00
17	775.52	16.00	42	946.90	41.00
18	797.98	17.00			
19	817.37	18.00			
20	835.23	19.00			
21	853.09	20.00			
22	867.21	21.00			
23	881.22	22.00			
24	896.25	23.00			
25	911.79	24.00			

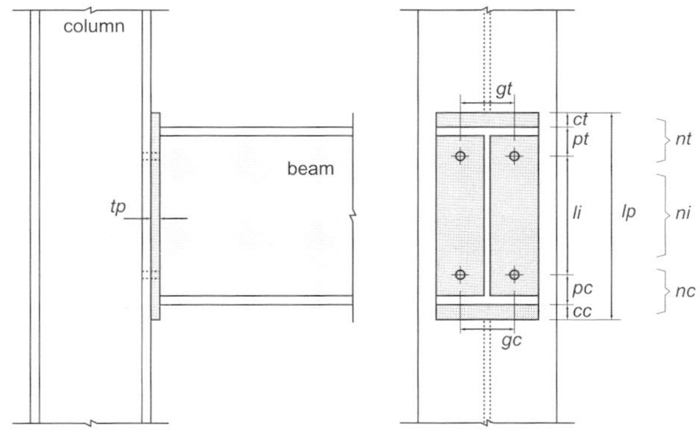

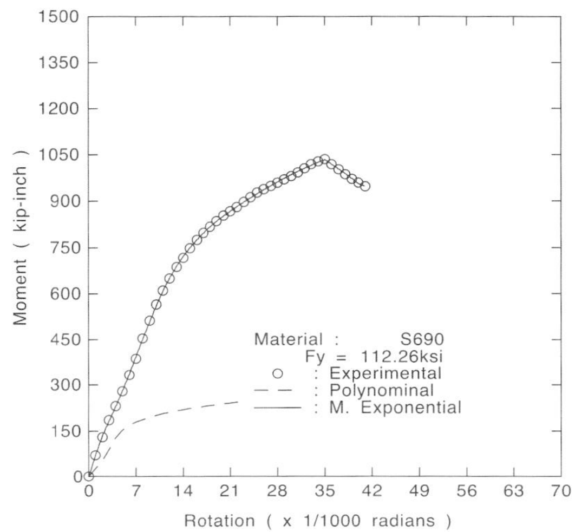

```
 Moment-rotation prediction equations     ( R : X 1/1000 radians )

Frye and Morris polynominal model  :    R = Sum ( Ai X ( K*Bm )**Pi X 10**Qi )
     xd =  5.905500"   t =  0.403540"  tf =  0.944900"
     A1 =  1.830000   A2 = -1.040000   A3 =  6.380000    K =  0.022057
     P1 =        1    P2 =        3    P3 =        5    Q1 =        0     Q2 =       -1     Q3 =       -3

Modified exponential model  :
     Bm = Sum ( Ai X ( 1 - exp( - R/( 2*i*C ) ) ) )  +     Sum ( RKj X ( R - Rj0 ) )  +  Bm0
      C =  0.59166667E+00             Bmo=  0.00000000E+00          Nexp=  6        Nliner=  2
     AI = -0.66528932E+03   0.12585539E+05  -0.65182088E+05   0.14332998E+06  -0.14494400E+06   0.56007733E+05
    Rj0 =          35.0000          40.0000
    RKj = -0.22919984E+02   0.88191835E+01
```

No	Rotation (radians) X 1/1000	Moment (kip-inch) Expri.	Moment (kip-inch) Poly.	Moment (kip-inch) M.Expo.	Connection stiffness (kip-inch) X 1000 Expri.	Connection stiffness (kip-inch) X 1000 Poly.	Connection stiffness (kip-inch) X 1000 M.Expo.
1	0.00	0.00	0.00	0.00	0.7042E+02	0.2477E+02	0.6628E+02
3	2.00	130.34	53.38	131.90	0.5746E+02	0.3108E+02	0.5421E+02
5	4.00	231.46	128.98	229.85	0.4711E+02	0.3251E+02	0.4876E+02
7	6.00	332.64	168.67	335.64	0.5354E+02	0.1251E+02	0.5678E+02
9	8.00	455.00	188.05	451.76	0.6331E+02	0.7679E+01	0.5769E+02
11	10.00	565.49	201.12	560.84	0.4842E+02	0.5636E+01	0.5044E+02
13	12.00	650.08	211.16	651.47	0.3819E+02	0.4500E+01	0.4008E+02
15	14.00	717.15	219.39	721.74	0.3095E+02	0.3773E+01	0.3053E+02
17	16.00	775.52	226.40	775.22	0.2480E+02	0.3265E+01	0.2340E+02
19	18.00	817.37	232.53	817.01	0.1862E+02	0.2888E+01	0.1876E+02
21	20.00	853.09	238.01	851.53	0.1599E+02	0.2596E+01	0.1602E+02
23	22.00	881.22	242.96	881.87	0.1452E+02	0.2364E+01	0.1446E+02
25	24.00	911.79	247.49	909.76	0.1542E+02	0.2173E+01	0.1349E+02
27	26.00	937.84	251.67	935.96	0.1075E+02	0.2014E+01	0.1273E+02
29	28.00	958.61	255.56	960.67	0.9980E+01	0.1879E+01	0.1197E+02
31	30.00	979.22	259.20	983.77	0.1182E+02	0.1763E+01	0.1111E+02
33	32.00	1005.16	262.67	1005.06	0.1297E+02	0.1660E+01	0.1016E+02
35	34.00	1027.35	265.88	1024.36	0.8636E+01	0.1572E+01	0.9134E+01
37	36.00	1018.66	268.92	1018.65	-0.1674E+02	0.1494E+01	-0.1484E+02
39	38.00	985.17	271.87	987.93	-0.1591E+02	0.1423E+01	-0.1588E+02
41	40.00	958.87	274.62	955.18	-0.1159E+02	0.1360E+01	-0.1686E+02

VI - 51

Connection type : Flush end-plate connections
Mode : All bolted without column stiffener

Tested by : P. Zoetemeijer (1981) Netherlands
Test Id. : 16

Column : --
Beam : HE300A
Plate thickness : 12 mm

Fasteners: G8.8- -M24
33/32" Oversize holes
Material : --
Fy = 42.70 ksi
Fu = -- ksi

Major parameters

ct = 0.0000" pt = 1.9134" li = 7.5906" pc = 1.9134"
cc = 0.0000" gt = 4.9331" gc = 4.9331"
tp = 0.4724" nt = 2 X 1 nc = 2 X 1

Remark 1) Plate size: 290x300x12 mm
2)

No	Moment (k-in)	Rotation (radians) X 1/1000	No	Moment (k-in)	Rotation (radians) X 1/1000
1	0.00	0.00	26	958.66	40.00
2	69.85	0.50	27	972.93	42.00
3	139.69	1.00	28	987.43	44.00
4	192.53	1.50	29	1002.04	46.00
5	245.36	2.00	30	1016.98	48.00
6	330.25	3.00	31	1031.92	50.00
7	399.31	4.00	32	1044.16	52.00
8	445.75	5.00	33	1056.23	54.00
9	492.18	6.00	34	1068.45	56.00
10	561.84	8.00	35	1080.76	58.00
11	613.72	10.00	36	1097.35	60.00
12	654.49	12.00			
13	682.80	14.00			
14	711.18	16.00			
15	740.10	18.00			
16	768.61	20.00			
17	793.82	22.00			
18	818.74	24.00			
19	839.03	26.00			
20	859.14	28.00			
21	877.17	30.00			
22	895.21	32.00			
23	913.34	34.00			
24	930.94	36.00			
25	944.80	38.00			

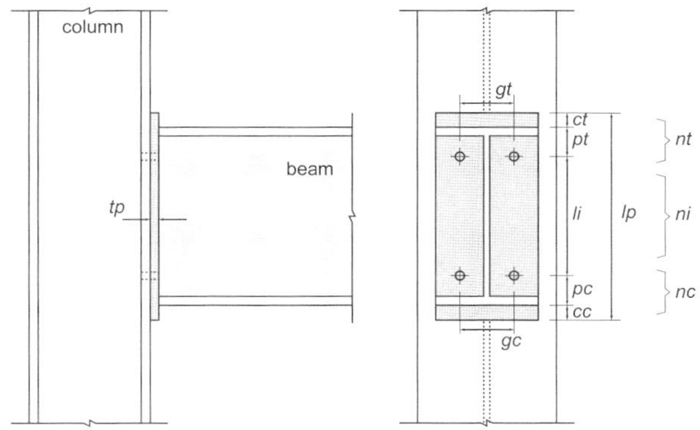

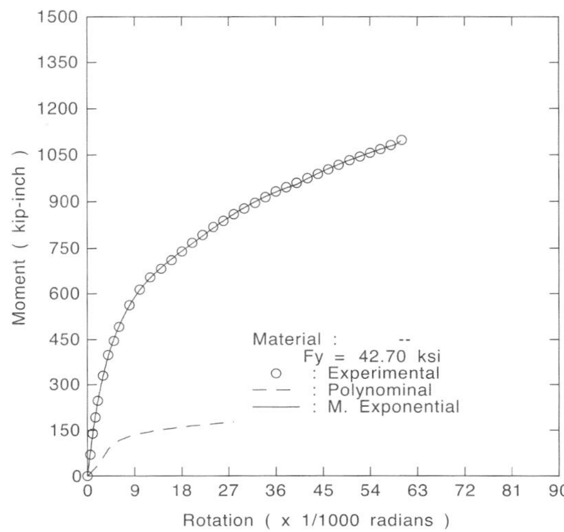

A.6 – 102

```
Moment-rotation prediction equations     ( R : X 1/1000 radians )

Frye and Morris polynominal model  :    R = Sum ( Ai X ( K*Bm )**Pi X 10**Qi )
     xd =  4.933070"   t =  0.472440"  tf =  0.944900"
     A1 =  1.830000   A2 = -1.040000   A3 =  6.380000    K =  0.031893
     P1 =         1   P2 =         3   P3 =         5   Q1 =         0   Q2 =        -1   Q3 =        -3

Modified exponential model  :
     Bm = Sum ( Ai X ( 1 - exp( - R/( 2*i*C ) ) ) )  +     Sum ( RKj X ( R - Rj0 ) )  +  Bm0
      C =   0.62500000E+00           Bmo=   0.00000000E+00         Nexp=  6        Nliner=  2
     AI =  -0.21093390E+03   0.39232179E+04  -0.23732492E+05   0.65002527E+05  -0.77257655E+05   0.33282644E+05
    Rj0 =          40.0000          59.0000
    RKj =   0.44132979E+01   0.80996573E+01
```

No	Rotation (radians) X 1/1000	Moment (kip-inch) Expri.	Moment (kip-inch) Poly.	Moment (kip-inch) M.Expo.	Connection stiffness (kip-inch) X 1000 Expri.	Connection stiffness (kip-inch) X 1000 Poly.	Connection stiffness (kip-inch) X 1000 M.Expo.
1	0.00	0.00	0.00	0.00	0.1397E+03	0.1713E+02	0.1488E+03
3	1.00	139.69	17.43	136.64	0.1227E+03	0.1806E+02	0.1229E+03
5	2.00	245.36	36.93	245.61	0.9874E+02	0.2150E+02	0.9573E+02
7	4.00	399.31	89.20	396.17	0.5775E+02	0.2248E+02	0.5880E+02
9	6.00	492.18	116.65	492.81	0.4257E+02	0.8655E+01	0.3987E+02
11	10.00	613.72	139.09	612.25	0.2316E+02	0.3898E+01	0.2245E+02
13	14.00	682.80	151.73	685.48	0.1417E+02	0.2609E+01	0.1530E+02
15	18.00	740.10	160.82	741.09	0.1436E+02	0.1997E+01	0.1299E+02
17	22.00	793.82	168.03	791.38	0.1253E+02	0.1635E+01	0.1221E+02
19	26.00	839.03	174.05	838.42	0.1010E+02	0.1393E+01	0.1122E+02
21	30.00	877.17	179.26	880.34	0.9017E+01	0.1219E+01	0.9661E+01
23	34.00	913.34	183.86	915.29	0.8932E+01	0.1088E+01	0.7792E+01
25	38.00	944.80	188.00	942.71	0.6930E+01	0.9846E+00	0.5943E+01
27	42.00	972.93	191.77	971.98	0.7192E+01	0.9011E+00	0.8746E+01
29	46.00	1002.04	195.23	1004.28	0.7389E+01	0.8321E+00	0.7460E+01
31	50.00	1031.92	198.44	1032.09	0.6794E+01	0.7739E+00	0.6494E+01
33	54.00	1056.23	201.43	1056.60	0.6073E+01	0.7243E+00	0.5802E+01
35	58.00	1080.76	204.24	1078.79	0.7224E+01	0.6812E+00	0.5322E+01

VI - 52

```
Connection type : Flush end-plate connections
           Mode : All bolted without column stiffener

Tested by   :  P. Zoetemeijer (1981)                      Netherlands
Test  Id.   :  17

Column : --                          Fasteners: G8.8- -M24
Beam   : HE300A                          33/32"  Oversize holes
Plate thickness : 16 mm              Material :  --
                                         Fy =  41.47 ksi
                                         Fu =   --   ksi

                          Major parameters

     ct =  0.0000"    pt =  1.9134"    li =  7.5906"    pc =  1.9134"
     cc =  0.0000"    gt =  4.9331"    gc =  4.9331"
     tp =  0.6299"    nt = 2 X  1      nc = 2 X  1

Remark  1) Plate size: 290x300x16 mm
        2)
```

No	Moment (k-in)	Rotation (radians) X 1/1000	No	Moment (k-in)	Rotation (radians) X 1/1000
1	0.00	0.00	26	958.81	21.93
2	118.09	0.50	27	977.75	22.93
3	222.31	1.00	28	996.69	23.92
4	305.35	1.50	29	1015.64	24.92
5	366.68	1.99	30	1034.58	25.92
6	415.88	2.49	31	1054.16	26.91
7	461.79	2.99	32	1074.04	27.91
8	517.47	3.99	33	1093.92	28.91
9	567.35	4.98	34	1113.80	29.90
10	599.32	5.98	35	1133.67	30.90
11	631.29	6.98	36	1151.45	31.90
12	661.36	7.97	37	1168.27	32.90
13	685.33	8.97	38	1185.10	33.89
14	709.31	9.97	39	1201.93	34.89
15	733.29	10.97	40	1218.76	35.89
16	757.26	11.96	41	1235.29	36.88
17	777.78	12.96	42	1251.69	37.88
18	798.24	13.96	43	1268.09	38.88
19	818.70	14.95	44	1284.49	39.87
20	839.15	15.95	45	1300.88	40.87
21	859.54	16.95			
22	879.48	17.94			
23	899.42	18.94			
24	919.36	19.94			
25	939.31	20.93			

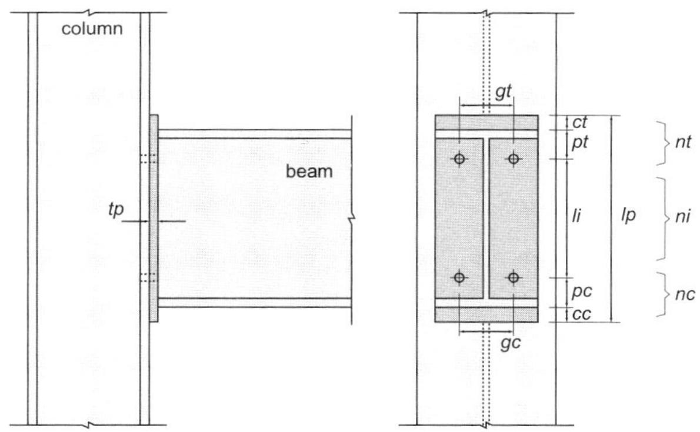

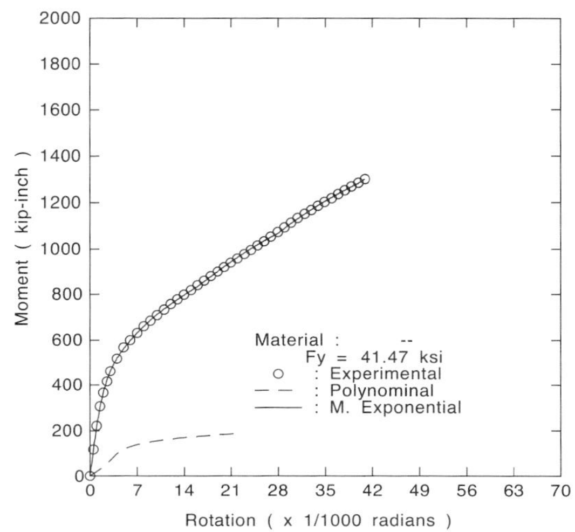

A.6 – 104

```
Moment-rotation prediction equations     ( R : X 1/1000 radians )

Frye and Morris polynominal model  :    R = Sum ( Ai X ( K*Bm )**Pi X 10**Qi )
     xd =  4.933070"    t =  0.629920"  tf =  0.944900"
     A1 =  1.830000   A2 = -1.040000   A3 =  6.380000    K =  0.028427
     P1 =         1   P2 =         3   P3 =         5   Q1 =         0   Q2 =        -1   Q3 =        -3

Modified exponential model  :
     Bm = Sum ( Ai X ( 1 - exp( - R/( 2*i*C ) ) ) )  +      Sum ( RKj X ( R - Rj0 ) )  +  Bm0
       C =   0.46558333E+00              Bmo=   0.00000000E+00          Nexp=  6          Nliner=  3
      AI =  -0.66084328E+03   0.97364735E+04  -0.48369591E+05   0.11559345E+06  -0.12694025E+06   0.51764115E+05
     Rj0 =          15.9500          27.9100          39.8700
     RKj =   0.33631869E+01   0.81289487E+01   0.52191432E+01
```

No	Rotation (radians) X 1/1000	Moment (kip-inch) Expri.	Moment (kip-inch) Poly.	Moment (kip-inch) M.Expo.	Connection stiffness (kip-inch) X 1000 Expri.	Connection stiffness (kip-inch) X 1000 Poly.	Connection stiffness (kip-inch) X 1000 M.Expo.
1	0.00	0.00	0.00	0.00	0.2362E+03	0.1922E+02	0.2383E+03
3	1.00	222.31	19.56	220.67	0.1873E+03	0.2026E+02	0.1876E+03
5	1.99	366.68	41.19	368.98	0.1119E+03	0.2406E+02	0.1156E+03
7	2.99	461.79	69.00	459.76	0.7978E+02	0.3193E+02	0.7081E+02
9	4.98	567.35	118.85	561.67	0.4123E+02	0.1451E+02	0.3962E+02
11	6.98	631.29	139.18	631.74	0.3117E+02	0.7388E+01	0.3147E+02
13	8.97	685.33	151.22	688.65	0.2398E+02	0.5056E+01	0.2588E+02
15	10.97	733.29	160.06	735.93	0.2410E+02	0.3891E+01	0.2174E+02
17	12.96	777.78	167.05	776.78	0.2049E+02	0.3193E+01	0.1963E+02
19	14.95	818.70	172.91	815.03	0.2056E+02	0.2724E+01	0.1897E+02
21	16.95	859.54	178.00	856.13	0.2026E+02	0.2384E+01	0.2214E+02
23	18.94	899.42	182.48	899.85	0.1994E+02	0.2128E+01	0.2173E+02
25	20.93	939.31	186.51	942.27	0.1982E+02	0.1926E+01	0.2081E+02
27	22.93	977.75	190.23	982.51	0.1904E+02	0.1760E+01	0.1936E+02
29	24.92	1015.64	193.59	1019.30	0.1894E+02	0.1626E+01	0.1757E+02
31	26.91	1054.16	196.71	1052.32	0.1983E+02	0.1512E+01	0.1561E+02
33	28.91	1093.92	199.63	1089.68	0.1998E+02	0.1415E+01	0.2177E+02
35	30.90	1133.67	202.36	1131.14	0.1883E+02	0.1331E+01	0.1993E+02
37	32.90	1168.27	204.93	1169.33	0.1691E+02	0.1257E+01	0.1829E+02
39	34.89	1201.93	207.39	1204.29	0.1683E+02	0.1192E+01	0.1688E+02
41	36.88	1235.29	209.69	1236.69	0.1655E+02	0.1134E+01	0.1571E+02
43	38.88	1268.09	211.93	1267.11	0.1648E+02	0.1081E+01	0.1475E+02
45	40.87	1300.88	214.01	1300.88	0.1640E+02	0.1035E+01	0.1920E+02

VI - 53

Connection type : Flush end-plate connections
Mode : All bolted without column stiffener

Tested by : P. Zoetemeijer (1981) Netherlands
Test Id. : 18

Column : --
Beam : HE300A
Plate thickness : 12 mm

Fasteners: G8.8- -M24
33/32" Oversize holes
Material : --
Fy = 42.70 ksi
Fu = -- ksi

Major parameters

ct = 0.0000"	pt = 2.7402"	li = 5.9370"	pc = 2.7402"
cc = 0.0000"	gt = 4.9331"	gc = 4.9331"	
tp = 0.4724"	nt = 2 X 1	nc = 2 X 1	

Remark 1) Failure mode: weld failure
2) Plate size: 290x300x12 mm

No	Moment (k-in)	Rotation (radians) X 1/1000	No	Moment (k-in)	Rotation (radians) X 1/1000
1	0.00	0.00	26	686.63	40.17
2	41.96	0.50	27	700.73	42.17
3	83.92	1.00	28	713.79	44.18
4	120.12	1.51	29	726.85	46.19
5	156.32	2.01	30	739.64	48.20
6	202.98	3.01	31	752.20	50.21
7	243.64	4.02	32	764.76	52.22
8	276.70	5.02	33	775.37	54.22
9	309.76	6.02	34	785.52	56.23
10	364.29	8.03	35	795.66	58.24
11	409.38	10.04			
12	441.22	12.05			
13	469.86	14.06			
14	489.16	16.07			
15	508.46	18.07			
16	525.38	20.08			
17	542.19	22.09			
18	559.01	24.10			
19	575.87	26.11			
20	592.86	28.12			
21	610.06	30.12			
22	627.22	32.13			
23	642.43	34.14			
24	657.63	36.15			
25	672.32	38.16			

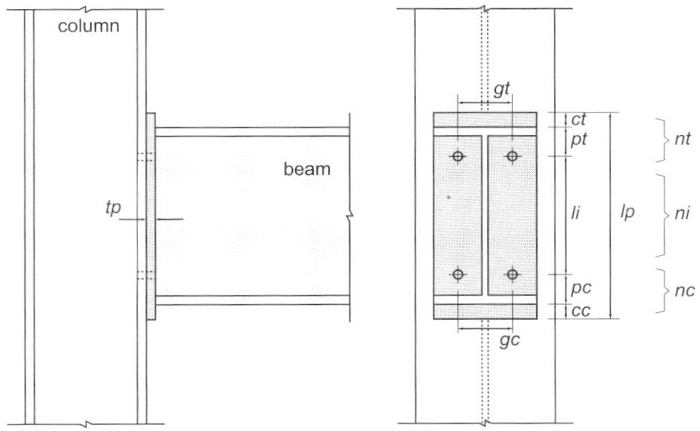

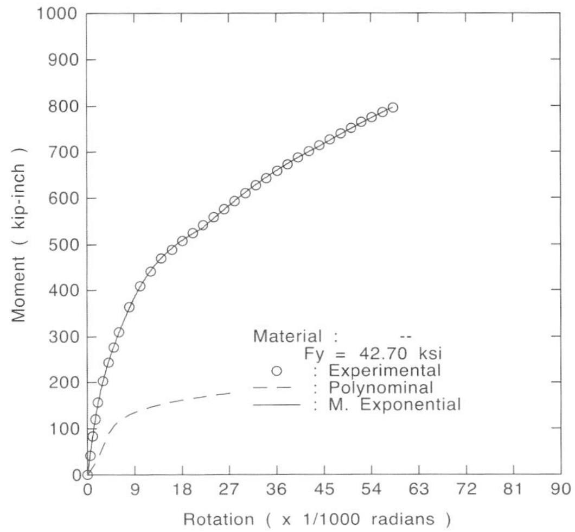

```
 Moment-rotation prediction equations     ( R : X 1/1000 radians )

Frye and Morris polynominal model  :    R = Sum ( Ai X ( K*Bm )**Pi X 10**Qi )
     xd =  4.933070"    t =  0.472440"  tf =  0.944900"
     A1 =  1.830000   A2 = -1.040000   A3 =  6.380000    K =  0.031893
     P1 =        1     P2 =        3    P3 =        5    Q1 =        0    Q2 =       -1    Q3 =       -3

Modified exponential model  :
     Bm = Sum ( Ai X ( 1 - exp( - R/( 2*i*C ) ) ) )  +      Sum ( RKj X ( R - Rj0 ) )  +  Bm0
      C =   0.61033333E+00              Bmo=   0.00000000E+00         Nexp=  6       Nliner=  2
     AI =  -0.29613181E+03   0.48959367E+04  -0.24522195E+05   0.55304642E+05  -0.57225641E+05   0.22489915E+05
    Rj0 =          22.0900          44.1800
    RKj =   0.38369532E+01   0.11555106E+01
```

No	Rotation (radians) X 1/1000	Moment (kip-inch) Expri.	Moment (kip-inch) Poly.	Moment (kip-inch) M.Expo.	Connection stiffness (kip-inch) X 1000 Expri.	Connection stiffness (kip-inch) X 1000 Poly.	Connection stiffness (kip-inch) X 1000 M.Expo.
1	0.00	0.00	0.00	0.00	0.8392E+02	0.1713E+02	0.8776E+02
3	1.00	83.92	17.43	84.09	0.7751E+02	0.1806E+02	0.7704E+02
5	2.01	156.32	37.14	152.52	0.6382E+02	0.2155E+02	0.5877E+02
7	4.02	243.64	89.66	244.78	0.3664E+02	0.2221E+02	0.3656E+02
9	6.02	309.76	116.82	308.91	0.3109E+02	0.8599E+01	0.2880E+02
11	10.04	409.38	139.25	407.15	0.1914E+02	0.3878E+01	0.2008E+02
13	14.06	469.86	151.89	470.93	0.1192E+02	0.2597E+01	0.1211E+02
15	18.07	508.46	160.96	509.57	0.9037E+01	0.1989E+01	0.7772E+01
17	22.09	542.19	168.18	536.96	0.8365E+01	0.1628E+01	0.6141E+01
19	26.11	575.87	174.21	575.57	0.8419E+01	0.1387E+01	0.9278E+01
21	30.12	610.06	179.41	611.47	0.8570E+01	0.1215E+01	0.8608E+01
23	34.14	642.43	184.02	644.47	0.7563E+01	0.1084E+01	0.7787E+01
25	38.16	672.32	188.16	674.02	0.7215E+01	0.9809E+00	0.6916E+01
27	42.17	700.73	191.92	700.11	0.6774E+01	0.8979E+00	0.6117E+01
29	46.19	726.85	195.39	725.64	0.6430E+01	0.8291E+00	0.6609E+01
31	50.21	752.20	198.60	751.13	0.6249E+01	0.7711E+00	0.6100E+01
33	54.22	775.37	201.59	774.81	0.5177E+01	0.7217E+00	0.5732E+01
35	58.24	795.66	204.40	797.31	0.5048E+01	0.6788E+00	0.5475E+01

A.6 – 107

VI - 54

Connection type : Flush end-plate connections
Mode : All bolted without column stiffener

Tested by : P. Zoetemeijer (1981) Netherlands
Test Id. : 19

Column : --
Beam : HE300A
Plate thickness : 16 mm

Fasteners: G8.8- -M24
33/32" Oversize holes
Material : --
Fy = 41.47 ksi
Fu = -- ksi

Major parameters

ct = 0.0000" pt = 2.7402' li = 5.9370" pc = 2.7402"
cc = 0.0000" gt = 4.9331' gc = 4.9331"
tp = 0.6299" nt = 2 X 1 nc = 2 X 1

Remark 1) Plate size: 290x300x16 mm
2)

No	Moment (k-in)	Rotation (radians) X 1/1000
1	0.00	0.00
2	65.65	0.50
3	131.29	1.00
4	218.48	2.00
5	297.94	3.00
6	367.36	3.99
7	428.40	4.99
8	485.58	5.99
9	535.07	6.99
10	584.56	7.99
11	645.93	9.99
12	693.34	11.98
13	719.47	13.98
14	739.10	15.98
15	756.78	17.98
16	773.25	19.97
17	789.22	21.97
18	804.80	23.97
19	820.20	25.96
20	836.41	27.96
21	853.20	29.96
22	868.74	31.96
23	882.95	33.95
24	896.99	35.95
25	910.50	37.95

Nc	Moment (k-in)	Rotation (radians) X 1/1000
26	924.00	39.95
27	937.32	41.94
28	950.65	43.94
29	962.01	45.94
30	972.06	47.94
31	982.10	49.93
32	993.07	51.93
33	1004.10	53.93
34	1015.13	55.92

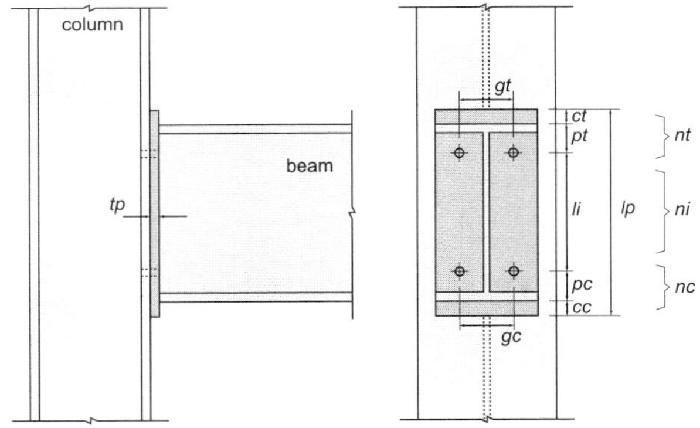

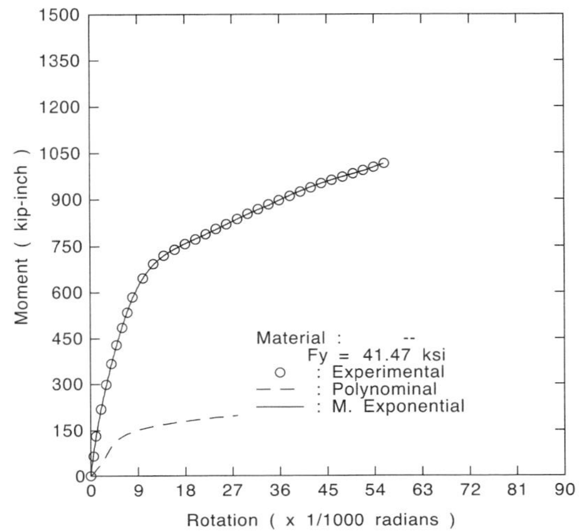

```
Moment-rotation prediction equations     ( R : X 1/1000 radians )

Frye and Morris polynominal model  :     R = Sum ( Ai X ( K*Bm )**Pi X 10**Qi )
     xd =  4.933070"   t =  0.629920"  tf =  0.944900"
     A1 =  1.830000   A2 = -1.040000   A3 =  6.380000    K =  0.028427
     P1 =        1    P2 =        3    P3 =        5    Q1 =        0    Q2 =       -1    Q3 =       -3

Modified exponential model  :
     Bm = Sum ( Ai X ( 1 - exp( - R/( 2*i*C ) ) ) ) +      Sum ( RKj X ( R - Rj0 ) )  +  Bm0
      C =   0.59100000E+00          Bmo=   0.00000000E+00          Nexp=  6         Nliner=  3
     AI =   0.11778337E+03  0.23850598E+04  -0.22190129E+05  0.62247707E+05  -0.70441921E+05   0.28811508E+05
    Rj0 =          23.9700          33.9500          51.9300
    RKj =   0.11001452E+01  0.21958472E+01   0.19766719E+01
```

No	Rotation (radians) X 1/1000	Moment (kip-inch) Expri.	Moment (kip-inch) Poly.	Moment (kip-inch) M.Expo.	Connection stiffness (kip-inch) X 1000 Expri.	Connection stiffness (kip-inch) X 1000 Poly.	Connection stiffness (kip-inch) X 1000 M.Expo.
1	0.00	0.00	0.00	0.00	0.1313E+03	0.1922E+02	0.1600E+03
3	1.00	131.29	19.56	127.56	0.1166E+03	0.2026E+02	0.1040E+03
5	3.00	297.94	69.27	296.78	0.7477E+02	0.3198E+02	0.7330E+02
7	4.99	428.40	119.00	430.08	0.5911E+02	0.1444E+02	0.6071E+02
9	6.99	535.07	139.25	537.56	0.4949E+02	0.7370E+01	0.4655E+02
11	9.99	645.93	156.01	646.33	0.2725E+02	0.4379E+01	0.2691E+02
13	13.98	719.47	170.18	720.06	0.1144E+02	0.2932E+01	0.1231E+02
15	17.98	756.78	180.39	758.39	0.8557E+01	0.2243E+01	0.7974E+01
17	21.97	789.22	188.47	788.67	0.7889E+01	0.1837E+01	0.7445E+01
19	25.96	820.20	195.22	820.32	0.7920E+01	0.1565E+01	0.8348E+01
21	29.96	853.20	201.07	852.38	0.8083E+01	0.1370E+01	0.7587E+01
23	33.95	882.95	206.23	880.39	0.7082E+01	0.1222E+01	0.6411E+01
25	37.95	910.50	210.87	912.26	0.6751E+01	0.1106E+01	0.7338E+01
27	41.94	937.32	215.09	939.22	0.6679E+01	0.1012E+01	0.6208E+01
29	45.94	962.01	218.98	962.16	0.5352E+01	0.9346E+00	0.5304E+01
31	49.93	982.10	222.58	981.92	0.5266E+01	0.8694E+00	0.4637E+01
33	53.93	1004.10	225.94	1003.42	0.5528E+01	0.8135E+00	0.6144E+01

A.6 – 109

VI - 55

Connection type : Flush end-plate connections
Mode : All bolted without column stiffener

Tested by : P. Zoetemeijer (1981) Netherlands
Test Id. : 31

Column : --
Beam : IPE400
Plate thickness : 12 mm

Fasteners: G8.8- -M24
33/32" Oversize holes
Material : --
Fy = 42.70 ksi
Fu = -- ksi

Major parameters

ct = 0.0000"	pt = 2.0118"	li = 11.7244"	pc = 2.0118"
cc = 0.0000"	gt = 3.8661"	gc = 3.8661"	
tp = 0.4724"	nt = 2 X 1	nc = 2 X 1	

Remark 1) Plate size: 400x180x12 mm
2)

No	Moment (k-in)	Rotation (radians) X 1/1000	No	Moment (k-in)	Rotation (radians) X 1/1000
1	0.00	0.00	26	960.22	16.14
2	53.50	0.25	27	979.84	17.15
3	106.99	0.50	28	999.45	18.16
4	154.37	0.76	29	1018.99	19.17
5	201.75	1.01	30	1038.48	20.18
6	243.70	1.26	31	1057.96	21.19
7	285.66	1.51	32	1077.45	22.20
8	332.97	1.77	33	1096.08	23.21
9	380.27	2.02	34	1114.69	24.22
10	428.29	2.27	35	1133.30	25.23
11	476.30	2.52	36	1151.92	26.23
12	544.39	3.03	37	1169.78	27.24
13	609.81	3.53	38	1187.39	28.25
14	651.24	4.04	39	1205.01	29.26
15	706.66	5.05	40	1222.62	30.27
16	743.23	6.05	41	1239.92	31.28
17	766.03	7.06	42	1256.79	32.29
18	788.83	8.07	43	1273.66	33.30
19	810.86	9.08	44	1290.53	34.31
20	832.60	10.09	45	1307.41	35.32
21	854.34	11.10	46	1323.20	36.32
22	875.98	12.11	47	1338.64	37.33
23	897.52	13.12	48	1354.08	38.34
24	919.06	14.13	49	1369.52	39.35
25	940.60	15.14	50	1384.96	40.36

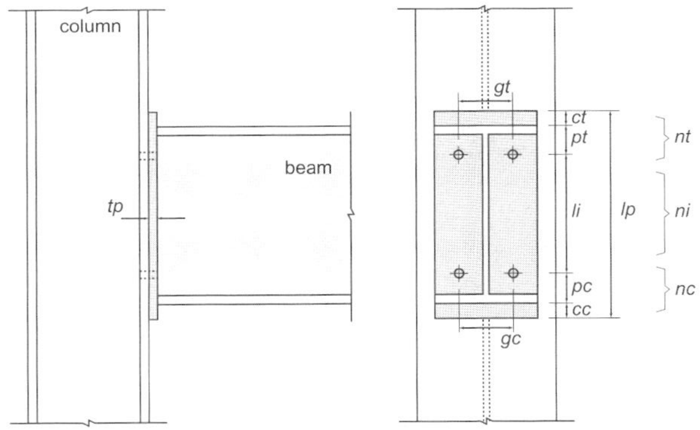

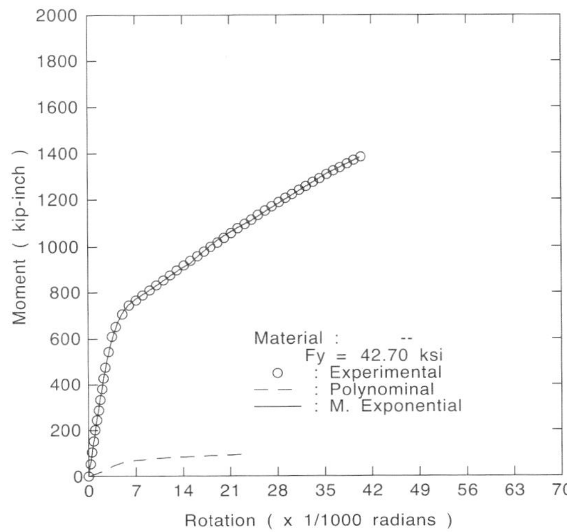

A.6 – 110

```
Moment-rotation prediction equations      ( R : X 1/1000 radians )

Frye and Morris polynominal model  :    R = Sum ( Ai X ( K*Bm )**Pi X 10**Qi )
     xd =  3.866140"    t =  0.472440"  tf =  0.944900"
     A1 =  1.830000    A2 = -1.040000   A3 =  6.380000     K =  0.057242
     P1 =         1    P2 =         3   P3 =         5    Q1 =         0    Q2 =        -1    Q3 =        -3

Modified exponential model  :
     Bm = Sum ( Ai X ( 1 - exp( - R/( 2*i*C ) ) ) )  +      Sum ( RKj X ( R - Rj0 ) )  +  Bm0
      C =  0.39883333E+00             Bmo=   0.00000000E+00           Nexp=  6        Nliner=  3
     AI =  0.75540356E+03  -0.13319273E+05   0.59856704E+05  -0.11108238E+06   0.93121857E+05  -0.28370613E+05
    Rj0 =         15.1400          24.2200          35.3200
    RKj =  0.14319803E+02   0.48786863E+01  -0.40293619E+01
```

No	Rotation (radians) X 1/1000	Moment (kip-inch) Expri.	Poly.	M.Expo.	Connection stiffness (kip-inch) X 1000 Expri.	Poly.	M.Expo.
1	0.00	0.00	0.00	0.00	0.2140E+03	0.9546E+01	0.2173E+03
3	0.50	106.99	4.79	100.49	0.1984E+03	0.9669E+01	0.1928E+03
5	1.01	201.75	9.81	198.44	0.1787E+03	0.1007E+02	0.1920E+03
7	1.51	285.66	15.02	293.78	0.1747E+03	0.1080E+02	0.1879E+03
9	2.02	380.27	20.82	386.54	0.1906E+03	0.1204E+02	0.1743E+03
11	2.52	476.30	27.27	468.73	0.1728E+03	0.1387E+02	0.1536E+03
13	3.53	609.81	43.00	599.15	0.1063E+03	0.1573E+02	0.1046E+03
15	5.05	706.66	59.52	710.87	0.4568E+02	0.6970E+01	0.4798E+02
17	7.06	766.03	69.41	772.46	0.2257E+02	0.3600E+01	0.2055E+02
19	9.08	810.86	75.37	810.90	0.2167E+02	0.2469E+01	0.1936E+02
21	11.10	854.34	79.73	852.35	0.2147E+02	0.1904E+01	0.2127E+02
23	13.12	897.52	83.21	894.27	0.2133E+02	0.1564E+01	0.1960E+02
25	15.14	940.60	86.12	929.63	0.2047E+02	0.1335E+01	0.1514E+02
27	17.15	979.84	88.63	983.67	0.1942E+02	0.1170E+01	0.2434E+02
29	19.17	1018.99	90.86	1028.15	0.1932E+02	0.1044E+01	0.1989E+02
31	21.19	1057.96	92.87	1064.86	0.1929E+02	0.9449E+00	0.1666E+02
33	23.21	1096.08	94.71	1096.28	0.1844E+02	0.8640E+00	0.1462E+02
35	25.23	1133.30	96.39	1129.48	0.1852E+02	0.7979E+00	0.1838E+02
37	27.24	1169.78	97.93	1165.86	0.1756E+02	0.7425E+00	0.1790E+02
39	29.26	1205.01	99.38	1201.86	0.1744E+02	0.6948E+00	0.1779E+02
41	31.28	1239.92	100.74	1237.87	0.1692E+02	0.6535E+00	0.1789E+02
43	33.30	1273.66	102.01	1274.20	0.1671E+02	0.6176E+00	0.1808E+02
45	35.32	1307.41	103.25	1310.95	0.1624E+02	0.5853E+00	0.1830E+02
47	37.33	1338.64	104.38	1339.83	0.1529E+02	0.5572E+00	0.1447E+02
49	39.35	1369.52	105.50	1369.23	0.1529E+02	0.5314E+00	0.1464E+02

A.6 – 111

VI - 56

Connection type : Flush end-plate connections
Mode : All bolted without column stiffener

Tested by : P. Zoetemeijer (1981) Netherlands
Test Id. : 34

Column : --
Beam : IPE400
Plate thickness : 16 mm

Fasteners: G8.8- -M24
33/32" Oversize holes
Material : --
Fy = 41.47 ksi
Fu = -- ksi

Major parameters

ct = 0.0000"	pt = 2.0118"	li = 11.7244"	pc = 2.0118"
cc = 0.0000"	gt = 3.8661"	gc = 3.8661"	
tp = 0.6299"	nt = 2 X 1	nc = 2 X 1	

Remark 1) Failure mode: bolt failure
2) Plate size: 400x180x16 mm

No	Moment (k-in)	Rotation (radians) X 1/1000	No	Moment (k-in)	Rotation (radians) X 1/1000
1	0.00	0.00	26	1577.67	19.23
2	101.78	0.12	27	1599.85	20.24
3	203.56	0.25	28	1622.03	21.25
4	305.35	0.37	29	1644.21	22.26
5	407.13	0.50	30	1664.63	23.28
6	508.23	0.76	31	1683.41	24.29
7	609.34	1.01	32	1702.19	25.30
8	744.48	1.50	33	1720.98	26.31
9	849.72	2.02	34	1739.76	27.32
10	982.49	3.04	35	1755.11	28.34
11	1059.14	4.05	36	1769.87	29.35
12	1124.71	5.06	37	1784.64	30.36
13	1179.35	6.07	38	1799.40	31.37
14	1221.55	7.08	39	1814.17	32.38
15	1259.37	8.10	40	1828.37	33.40
16	1295.38	9.11	41	1842.09	34.41
17	1326.98	10.12	42	1855.80	35.42
18	1358.58	11.13	43	1869.52	36.43
19	1390.18	12.14	44	1883.24	37.44
20	1421.59	13.16	45	1896.95	38.46
21	1452.25	14.17			
22	1480.58	15.18			
23	1505.33	16.19			
24	1530.09	17.20			
25	1554.84	18.22			

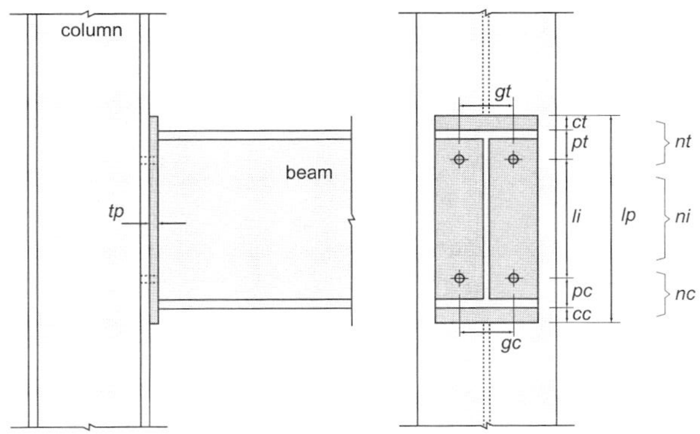

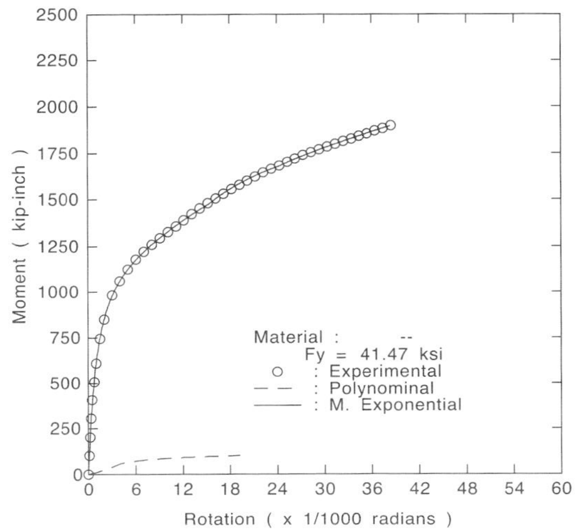

```
Moment-rotation prediction equations      ( R : X 1/1000 radians )

Frye and Morris polynominal model  :     R = Sum ( Ai X ( K*Bm )**Pi X 10**Qi )
     xd =  3.866140"    t =  0.629920"  tf =  0.944900"
     A1 =  1.830000    A2 = -1.040000   A3 =  6.380000    K =  0.051020
     P1 =         1    P2 =         3   P3 =         5   Q1 =         0    Q2 =       -1     Q3 =       -3

Modified exponential model  :
     Bm = Sum ( Ai X ( 1 - exp( - R/( 2*i*C ) ) ) )  +      Sum ( RKj X ( R - Rj0 ) )  +  Bm0
       C =  0.35050000E+00            Bmo=  0.00000000E+00            Nexp=  6        Nliner=  3
      AI =  0.47568100E+03  0.21146353E+04 -0.16791881E+05  0.53143284E+05 -0.69022382E+05  0.31728140E+05
     Rj0 =          15.1800         24.2900         35.4200
     RKj =  0.79692253E+01  0.47003792E+01 -0.23170323E+00
```

No	Rotation (radians) X 1/1000	Moment (kip-inch) Expri.	Poly.	M.Expo.	Connection stiffness (kip-inch) X 1000 Expri.	Poly.	M.Expo.
1	0.00	0.00	0.00	0.00	0.8482E+03	0.1071E+02	0.1006E+04
3	0.25	203.56	2.68	218.32	0.8169E+03	0.1074E+02	0.7546E+03
5	0.50	407.13	5.37	383.16	0.6516E+03	0.1085E+02	0.5741E+03
7	1.01	609.34	11.01	611.48	0.3610E+03	0.1130E+02	0.3460E+03
9	2.02	849.72	23.35	849.38	0.1780E+03	0.1350E+02	0.1611E+03
11	4.05	1059.14	56.47	1061.72	0.7040E+02	0.1363E+02	0.7181E+02
13	6.07	1179.35	73.29	1177.56	0.4794E+02	0.5291E+01	0.4668E+02
15	8.10	1259.37	81.63	1260.35	0.3636E+02	0.3259E+01	0.3643E+02
17	10.12	1326.98	87.24	1329.08	0.3129E+02	0.2399E+01	0.3213E+02
19	12.14	1390.18	91.56	1391.00	0.3104E+02	0.1919E+01	0.2918E+02
21	14.17	1452.25	95.12	1446.88	0.2920E+02	0.1609E+01	0.2576E+02
23	16.19	1505.33	98.14	1502.97	0.2451E+02	0.1394E+01	0.2971E+02
25	18.22	1554.84	100.80	1558.96	0.2343E+02	0.1233E+01	0.2547E+02
27	20.24	1599.85	103.16	1606.34	0.2196E+02	0.1109E+01	0.2151E+02
29	22.26	1644.21	105.30	1646.25	0.2099E+02	0.1010E+01	0.1810E+02
31	24.29	1683.41	107.27	1680.06	0.1860E+02	0.9288E+00	0.1533E+02
33	26.31	1720.98	109.07	1718.27	0.1860E+02	0.8611E+00	0.1790E+02
35	28.34	1755.11	110.76	1752.90	0.1483E+02	0.8034E+00	0.1631E+02
37	30.36	1784.64	112.33	1784.62	0.1462E+02	0.7539E+00	0.1516E+02
39	32.38	1814.17	113.83	1814.39	0.1427E+02	0.7102E+00	0.1436E+02
41	34.41	1842.09	115.22	1842.93	0.1358E+02	0.6725E+00	0.1380E+02
43	36.43	1869.52	116.54	1870.15	0.1358E+02	0.6391E+00	0.1318E+02
45	38.46	1896.95	117.82	1896.64	0.1345E+02	0.6086E+00	0.1293E+02

VI - 57

Connection type : Flush end-plate connections
Mode : All bolted without column stiffener

Tested by : P. Zoetemeijer (1981) Netherlands
Test Id. : 37

Column : --
Beam : IPE400
Plate thickness : 32 mm

Fasteners: G8.8- -M24
33/32" Oversize holes
Material : --
Fy = 34.78 ksi
Fu = -- ksi

Major parameters

ct = 0.0000"	pt = 2.1378'	li = 11.4724"	pc = 2.1378"
cc = 0.0000"	gt = 4.1181'	gc = 4.1181"	
tp = 1.2598"	nt = 2 X 1	nc = 2 X 1	

Remark 1) Failure mode: bolt failure
2) Plate size: 400x180x32 mm

No	Moment (k-in)	Rotation (radians) X 1/1000	No	Moment (k-in)	Rotation (radians) X 1/1000
1	0.00	0.00	26	1674.76	10.49
2	86.76	0.10	27	1667.27	10.99
3	173.51	0.19	28	1641.81	11.49
4	277.30	0.29	29	1616.32	11.99
5	389.55	0.39	30	1560.22	12.49
6	514.16	0.68	31	1503.40	12.99
7	622.69	1.00	32	1451.81	13.49
8	739.48	1.50	33	1421.73	13.99
9	845.50	2.00	34	1393.63	14.49
10	939.71	2.50			
11	1027.61	3.00			
12	1102.37	3.50			
13	1169.86	4.00			
14	1231.36	4.50			
15	1292.52	5.00			
16	1345.91	5.49			
17	1399.30	5.99			
18	1445.29	6.49			
19	1491.39	6.99			
20	1538.40	7.49			
21	1583.25	7.99			
22	1612.70	8.49			
23	1642.15	8.99			
24	1662.97	9.49			
25	1670.19	9.99			

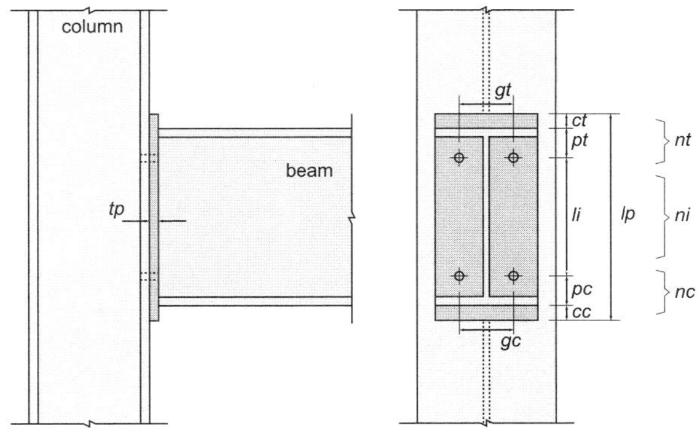

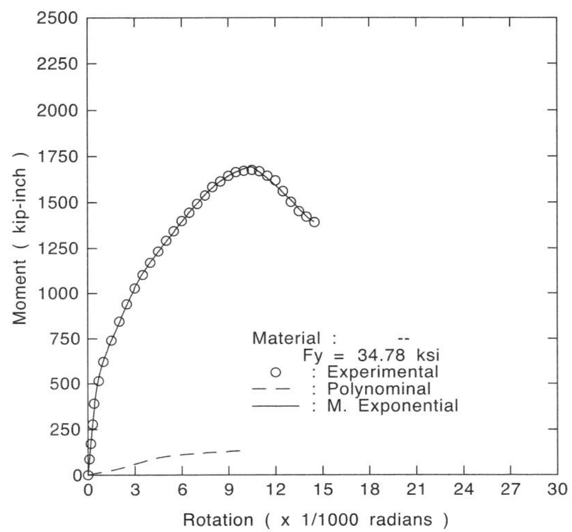

```
Moment-rotation prediction equations     ( R : X 1/1000 radians )

Frye and Morris polynominal model  :    R = Sum ( Ai X ( K*Bm )**Pi X 10**Qi )
     xd =  4.118110"   t =  1.259840"  tf =  0.944900"
     A1 =  1.830000   A2 = -1.040000   A3 =  6.380000     K =  0.033229
     P1 =         1   P2 =         3   P3 =         5    Q1 =         0    Q2 =        -1    Q3 =        -3

Modified exponential model  :
     Bm = Sum ( Ai X ( 1 - exp( - R/( 2*i*C ) ) ) )  +      Sum ( RKj X ( R - Rj0 ) )  +  Bm0
      C =   0.14575000E+00             Bmo=   0.00000000E+00          Nexp=  6         Nliner=  3
     AI =  -0.15519684E+04   0.23096530E+05  -0.11533863E+06   0.27113900E+06  -0.29501651E+06   0.11965718E+06
    Rj0 =           3.5000          10.4900          13.9900
    RKj =  -0.26258625E+02  -0.76071907E+02   0.35959488E+02
```

No	Rotation (radians) X 1/1000	Moment (kip-inch) Expri.	Moment (kip-inch) Poly.	Moment (kip-inch) M.Expo.	Connection stiffness (kip-inch) X 1000 Expri.	Connection stiffness (kip-inch) X 1000 Poly.	Connection stiffness (kip-inch) X 1000 M.Expo.
1	0.00	0.00	0.00	0.00	0.8676E+03	0.1644E+02	0.9413E+03
3	0.19	173.51	3.13	187.25	0.9990E+03	0.1648E+02	0.9595E+03
5	0.39	389.55	6.42	358.34	0.9448E+03	0.1657E+02	0.7381E+03
7	1.00	622.69	16.73	630.97	0.2980E+03	0.1733E+02	0.2572E+03
9	2.00	845.50	35.45	837.69	0.2002E+03	0.2063E+02	0.2009E+03
11	3.00	1027.61	59.30	1028.98	0.1627E+03	0.2737E+02	0.1743E+03
13	4.00	1169.86	85.63	1173.23	0.1290E+03	0.2157E+02	0.1184E+03
15	5.00	1292.52	101.93	1286.89	0.1156E+03	0.1229E+02	0.1111E+03
17	5.99	1399.30	111.88	1395.16	0.9938E+02	0.8333E+01	0.1066E+03
19	6.99	1491.39	119.10	1496.02	0.9311E+02	0.6310E+01	0.9350E+02
21	7.99	1583.25	124.76	1579.64	0.7431E+02	0.5108E+01	0.7292E+02
23	8.99	1642.15	129.45	1641.12	0.5027E+02	0.4312E+01	0.5008E+02
25	9.99	1670.19	133.46	1680.48	0.1178E+02	0.3746E+01	0.2919E+02
27	10.99	1667.27	136.99	1662.78	-0.3295E+02	0.3321E+01	-0.6386E+02
29	11.99	1616.32	140.14	1592.23	-0.8159E+02	0.2990E+01	-0.7657E+02
31	12.99	1503.40	142.99	1510.88	-0.1084E+03	0.2725E+01	-0.8557E+02
33	13.99	1421.73	145.60	1422.06	-0.5817E+02	0.2506E+01	-0.9167E+02

VI - 58

```
Connection type : Flush end-plate connections
           Mode : All bolted with column stiffener

Tested by  :  G. Shi et al (2007)                          China
Test  Id.  :  JD1

Column : H300x250x8x12              Fasteners: 10.9- -M20
Beam   : H300x200x8x12                      7/8" Oversize holes
Plate thickness : 20 mm             Material : Q345
Stiffener thickness : 0.3937"           Fy =  54.04 ksi
                                        Fu =  77.89 ksi
```

Major parameters

```
ct = 0.3937"   pt =  2.4409"   li =  6.9291"   pc =  2.4409"
cc = 0.3937"   pit=  3.4646"   pi =  0.0000"   pic=  3.4646"
gt = 4.2520"   gi =  4.2520"   gc =  4.2520"
tp = 0.7874"   nt = 2 X  1     ni = 2 X  1     nc = 2 X  1
```

Remark 1) Plate size: 320x200x20 mm
2) Column flange thickness is taken the same as the end-plate thickness

No	Moment (k-in)	Rotation (radians) X 1/1000
1	0.00	0.00
2	108.76	0.50
3	235.30	1.02
4	368.87	1.63
5	495.41	2.30
6	614.92	3.03
7	727.40	3.70
8	832.85	4.56
9	924.24	5.55
10	1008.60	7.09
11	1092.96	8.17
12	1184.35	9.41
13	1240.59	10.80
14	1296.83	12.65
15	1374.16	14.96
16	1409.31	17.89
17	1465.55	20.66
18	1507.73	23.44
19	1549.91	26.68
20	1585.06	29.76
21	1606.15	32.53
22	1627.24	35.31
23	1641.30	37.77
24	1641.30	40.39
25	1641.30	43.01

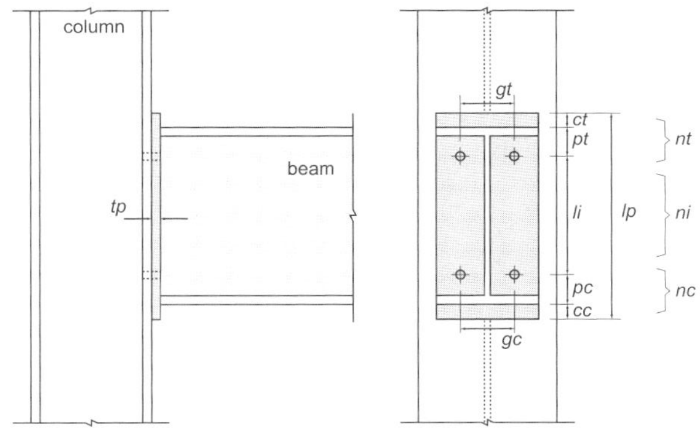

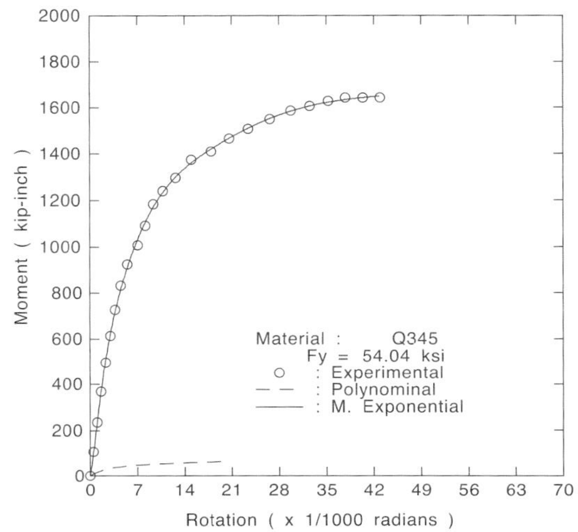

```
Moment-rotation prediction equations     ( R : X 1/1000 radians )

Frye and Morris polynominal model  :    R = Sum ( Ai X ( K*Bm )**Pi X 10**Qi )
     xd =  4.251970"   t =  0.787400"
     A1 =  1.790000   A2 =  1.760000   A3 =  2.040000    K =  0.035782
     P1 =        1    P2 =        3    P3 =        5    Q1 =        0    Q2 =       -1    Q3 =       -1

Modified exponential model  :
     Bm = Sum ( Ai X ( 1 - exp( - R/( 2*i*C ) ) ) )  +     Sum ( RKj X ( R - Rj0 ) )  +  Bm0
      C =   0.48341667E+00           Bmo=   0.00000000E+00          Nexp=  6        Nliner=  0
     AI =  -0.99975389E+03   0.10889211E+05  -0.50863481E+05   0.11460573E+06  -0.11960668E+06   0.47637440E+05
```

No	Rotation (radians) X 1/1000	Moment (kip-inch) Expri.	Moment (kip-inch) Poly.	Moment (kip-inch) M.Expo.	Connection stiffness (kip-inch) X 1000 Expri.	Connection stiffness (kip-inch) X 1000 Poly.	Connection stiffness (kip-inch) X 1000 M.Expo.
1	0.00	0.00	0.00	0.00	0.2175E+03	0.1561E+02	0.1655E+03
2	0.50	108.76	7.74	104.07	0.2302E+03	0.1522E+02	0.2357E+03
3	1.02	235.30	15.32	230.22	0.2321E+03	0.1370E+02	0.2425E+03
4	1.63	368.87	22.82	371.50	0.2046E+03	0.1077E+02	0.2177E+03
5	2.30	495.41	29.00	505.46	0.1768E+03	0.7890E+01	0.1823E+03
6	3.03	614.92	33.95	625.84	0.1659E+03	0.5833E+01	0.1490E+03
7	3.70	727.40	37.44	717.62	0.1481E+03	0.4640E+01	0.1261E+03
8	4.56	832.85	40.98	816.63	0.1085E+03	0.3658E+01	0.1055E+03
9	5.55	924.24	44.22	912.49	0.7763E+02	0.2941E+01	0.8913E+02
10	7.09	1008.60	48.18	1034.93	0.6849E+02	0.2259E+01	0.7074E+02
11	8.17	1092.96	50.44	1105.46	0.7606E+02	0.1950E+01	0.6008E+02
12	9.41	1184.35	52.68	1173.18	0.5803E+02	0.1689E+01	0.4942E+02
13	10.80	1240.59	54.87	1234.78	0.3614E+02	0.1472E+01	0.3960E+02
14	12.65	1296.83	57.39	1298.65	0.3177E+02	0.1261E+01	0.3011E+02
15	14.96	1374.16	60.08	1359.23	0.2401E+02	0.1074E+01	0.2308E+02
16	17.89	1409.31	62.97	1419.65	0.1627E+02	0.9086E+00	0.1872E+02
17	20.66	1465.55	65.32	1468.07	0.1774E+02	0.7959E+00	0.1636E+02
18	23.44	1507.73	67.41	1510.58	0.1418E+02	0.7096E+00	0.1421E+02
19	26.68	1549.91	69.58	1552.32	0.1220E+02	0.6317E+00	0.1154E+02
20	29.76	1585.06	71.43	1583.89	0.9412E+01	0.5731E+00	0.8989E+01
21	32.53	1606.15	72.95	1605.85	0.7600E+01	0.5298E+00	0.6908E+01
22	35.31	1627.24	74.37	1622.51	0.6594E+01	0.4930E+00	0.5138E+01
23	37.77	1641.30	75.55	1633.53	0.2948E+01	0.4648E+00	0.3867E+01
24	40.39	1641.30	76.73	1642.20	0.0000E+00	0.4385E+00	0.2802E+01
25	43.01	1641.30	77.85	1648.44	0.0000E+00	0.4152E+00	0.1996E+01

A.7 – 1

VII - 1

Connection type : Header plate connections
Mode : All bolted

Tested by : W.H.Sommer (1969) Canada
Test Id. : TEST 5

Column : --
Beam : W18X45
Plate thickness : 1/4"

Fasteners: A325-X-3/4"D
15/16" Oversize holes
Material : G40.12
Fy = 50.10 ksi
Fu = 76.50 ksi

Major parameters

lt = 1.5000"	ct = 1.5000"	p = 3.0000"	cc = 1.5000"
lc = 1.3750"	g = 4.0000"	lp = 15.0000"	tp = 0.2500"
tbw= 0.3350"			
n = 2 X 5			

Remark 1) Actual strength Fy = 50.0 ksi, Fu = 76.1 ksi for beam web.
2) No washers were used.

No	Moment (k-in)	Rotation (radians) X 1/1000	No	Moment (k-in)	Rotation (radians) X 1/1000
1	0.00	0.00	26	805.00	61.60
2	18.60	0.20	27	829.00	63.30
3	42.80	0.40	28	853.00	64.70
4	66.90	0.80	29	877.00	66.10
5	115.00	1.10	30	935.00	68.80
6	163.00	1.40	31	984.00	71.00
7	211.50	2.05	32	1022.00	72.90
8	260.00	2.70	33	1060.00	74.80
9	296.00	3.60	34	1095.00	76.90
10	332.00	4.50	35	1130.00	79.00
11	387.50	8.65	36	1175.00	82.95
12	443.00	12.80	37	1220.00	86.90
13	479.50	17.00	38	1245.00	89.95
14	516.00	21.20	39	1270.00	93.00
15	543.33	25.20	40	1285.00	96.05
16	570.66	29.20	41	1300.00	99.10
17	598.00	33.20			
18	626.66	36.87			
19	655.33	40.53			
20	684.00	44.20			
21	703.33	47.57			
22	722.66	50.93			
23	742.00	54.30			
24	767.00	57.60			
25	791.00	60.10			

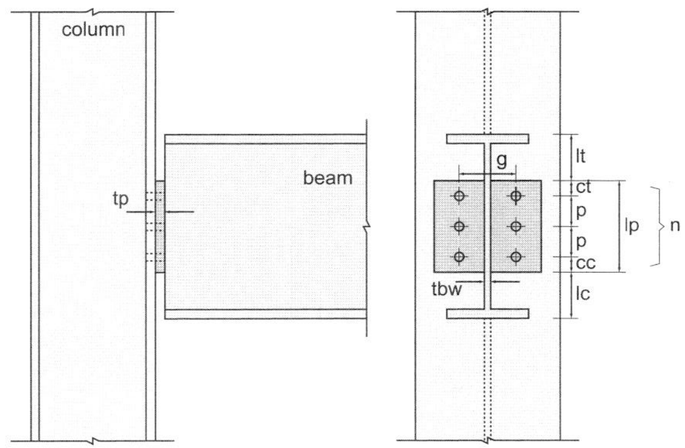

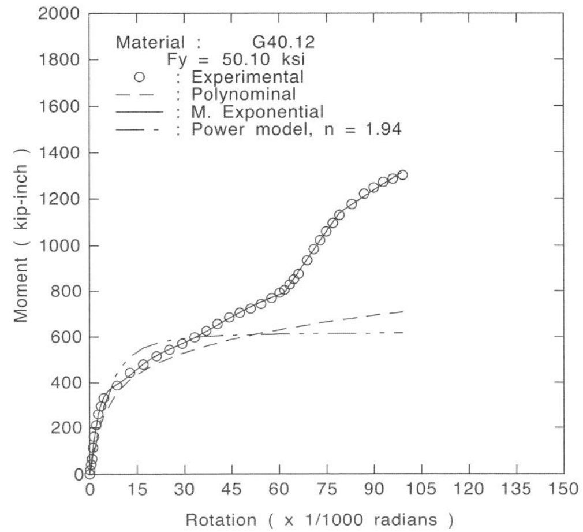

A.7 – 2

```
Moment-rotation prediction equations      ( R : X 1/1000 radians )

Frye and Morris polynominal model  :     R = Sum ( Ai X ( K*Bm )**Pi X 10**Qi )
     g =  4.000000"   d = 15.000000"  t  =  0.250000"   w =  0.335000"
    A1 =  5.100000   A2 =  6.200000   A3 =  2.400000    K =  0.287779
    P1 =        1    P2 =        3    P3 =        5    Q1 =       -2     Q2 =       -7     Q3 =      -10

Modified exponential model  :
     Bm = Sum ( Ai X ( 1 - exp( - R/( 2*i*C ) ) ) )  +      Sum ( RKj X ( R - Rj0 ) )  +  Bm0
      C =   0.87583333E+00            Bmo=   0.00000000E+00           Nexp=  6          Nliner=  3
     AI =  -0.10838473E+04   0.15618712E+05  -0.71985901E+05   0.15566904E+06  -0.15674286E+06   0.59310846E+05
    Rj0 =          12.8000          61.6000          79.0000
    RKj =   0.12055727E+01   0.16537299E+02  -0.96606226E+01

Power model  :    Bm = ( Rki X R ) / ( 1 + ( R/R0 )**rn )**( 1/rn )
     rn = 1.944      rki =   0.74770927E+02         rmu =   0.61799577E+03
```

No	Rotation (radians) X 1/1000	Moment (kip-inch) Expri.	Poly.	M.Expo.	P.Model.	Connection stiffness (kip-inch) X 1000 Expri.	Poly.	M.Expo.	P.Model.
1	0.00	0.00	0.00	0.00	0.00	0.9300E+02	0.6813E+02	0.1051E+03	0.7477E+02
3	0.40	42.80	27.25	43.24	29.87	0.1008E+03	0.6798E+02	0.1097E+03	0.7446E+02
5	1.10	115.00	74.46	118.52	81.42	0.1602E+03	0.6669E+02	0.1030E+03	0.7258E+02
7	2.05	211.50	135.68	206.36	148.28	0.7462E+02	0.6137E+02	0.8076E+02	0.6782E+02
9	3.60	296.00	218.69	301.77	245.20	0.4000E+02	0.4502E+02	0.4407E+02	0.5681E+02
11	8.65	387.50	356.85	392.84	442.35	0.1337E+02	0.1703E+02	0.8638E+01	0.2444E+02
13	17.00	479.50	452.52	481.58	551.83	0.8690E+01	0.8125E+01	0.1047E+02	0.6413E+01
15	25.20	543.33	506.82	546.30	584.46	0.6833E+01	0.5484E+01	0.6195E+01	0.2384E+01
17	33.20	598.00	545.14	596.88	597.71	0.7347E+01	0.4219E+01	0.6884E+01	0.1131E+01
19	40.53	655.33	573.29	651.34	604.01	0.7819E+01	0.3508E+01	0.7772E+01	0.6483E+00
21	47.57	703.33	596.20	705.23	607.66	0.5742E+01	0.3033E+01	0.7360E+01	0.4120E+00
23	54.30	742.00	615.43	751.12	609.96	0.6669E+01	0.2694E+01	0.6198E+01	0.2821E+00
25	60.10	791.00	630.48	783.72	611.38	0.9430E+01	0.2460E+01	0.5046E+01	0.2107E+00
27	63.30	829.00	638.11	827.01	612.00	0.1578E+02	0.2351E+01	0.2098E+02	0.1814E+00
29	66.10	877.00	644.63	885.08	612.48	0.1863E+02	0.2262E+01	0.2050E+02	0.1601E+00
31	71.00	984.00	655.33	983.72	613.19	0.2105E+02	0.2125E+01	0.1978E+02	0.1301E+00
33	74.80	1060.00	663.16	1058.00	613.65	0.1842E+02	0.2031E+01	0.1933E+02	0.1119E+00
35	79.00	1130.00	671.59	1138.30	614.08	0.1484E+02	0.1936E+01	0.1893E+02	0.9545E-01
37	86.90	1220.00	686.32	1209.25	614.74	0.9588E+01	0.1782E+01	0.8742E+01	0.7233E-01
39	93.00	1270.00	696.81	1261.76	615.14	0.6557E+01	0.1682E+01	0.8493E+01	0.5935E-01
41	99.10	1300.00	706.72	1313.04	615.47	0.4918E+01	0.1593E+01	0.8333E+01	0.4930E-01

A.7 – 3

VII - 2

Connection type : Header plate connections
Mode : All bolted

Tested by : W.H.Sommer (1969) Canada
Test Id. : TEST 6

Column : --
Beam : W24X76
Plate thickness : 1/4"

Fasteners: A325-X-3/4"D
15/16" Oversize holes
Material : G40.12
Fy = 49.30 ksi
Fu = 74.60 ksi

Major parameters

lt = 1.5000" ct = 1.5000" p = 3.0000" cc = 1.5000"
lc = 13.3750" g = 4.0000" lp = 9.0000" tp = 0.2500"
tbw= 0.4410"
n = 2 X 3

Remark 1) Actual strength Fy = 51.5 ksi, Fu = 77.9 ksi for beam web.
2) No washers were used.

No	Moment (k-in)	Rotation (radians) X 1/1000	No	Moment (k-in)	Rotation (radians) X 1/1000
1	0.00	0.00	26	768.00	27.40
2	26.20	2.50	27	817.00	28.20
3	40.70	4.50	28	866.00	29.00
4	55.20	6.50	29	904.50	30.15
5	69.65	8.35	30	943.00	31.30
6	84.10	10.20	31	991.50	33.10
7	103.40	12.47	32	1040.00	34.90
8	122.70	14.73	33	1085.00	36.80
9	142.00	17.00	34	1130.00	38.70
10	158.00	19.13	35	1150.00	40.65
11	174.00	21.27	36	1170.00	42.60
12	190.00	23.40	37	1215.00	44.35
13	211.00	23.70	38	1260.00	46.10
14	240.00	24.20	39	1290.00	48.13
15	283.00	24.80	40	1320.00	50.17
16	316.00	25.00	41	1350.00	52.20
17	364.00	25.20	42	1365.00	54.15
18	414.00	25.40	43	1380.00	56.10
19	456.50	25.55	44	1340.00	56.70
20	499.00	25.70	45	1294.00	57.02
21	538.00	25.85	46	1248.00	57.34
22	577.00	26.00	47	1202.00	57.66
23	625.00	26.20	48	1156.00	57.98
24	673.00	26.40	49	1110.00	58.30
25	720.50	26.90			

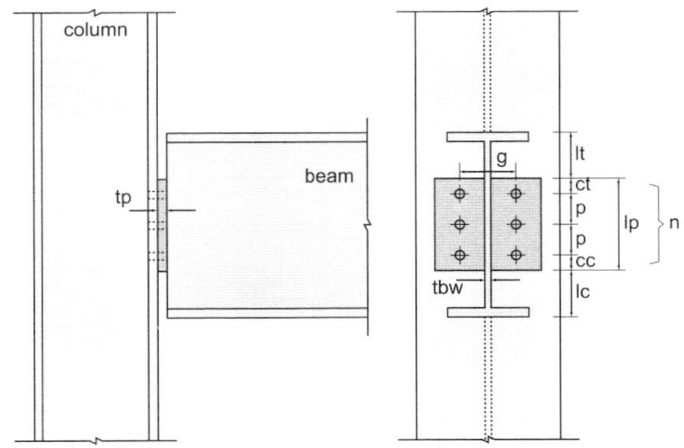

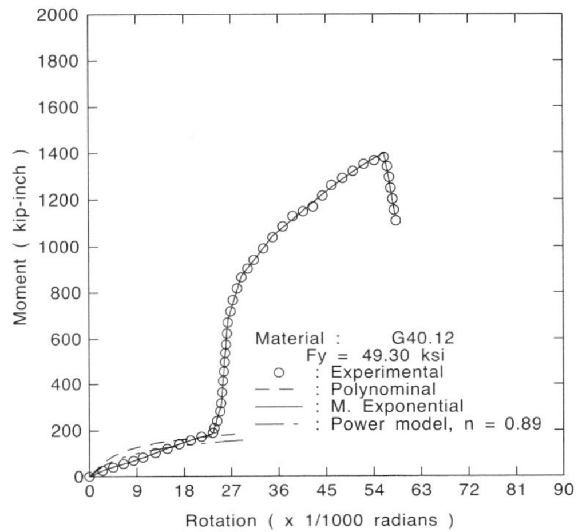

A.7 – 4

```
Moment-rotation prediction equations     ( R : X 1/1000 radians )

Frye and Morris polynominal model  :    R = Sum ( Ai X ( K*Bm )**Pi X 10**Qi )
      g =  4.000000"   d =  9.000000"  t  =  0.250000"   w =  0.441000"
     A1 =  5.100000   A2 =  6.200000   A3 =  2.400000    K =  0.812110
     P1 =        1    P2 =        3    P3 =        5    Q1 =       -2     Q2 =       -7     Q3 =      -10

Modified exponential model  :
     Bm = Sum ( Ai X ( 1 - exp( - R/( 2*i*C ) ) ) )  +      Sum ( RKj X ( R - Rj0 ) )  +  Bm0
       C =   0.11108333E+01                Bmo=   0.00000000E+00          Nexp=  6         Nliner=  7
      AI =  -0.50562273E+03   0.12738751E+05  -0.91006044E+05   0.26142970E+06  -0.32168317E+06   0.14105882E+06
     Rj0 =           0.0000          23.4000          25.0000          26.4000          29.0000
                    34.9000          56.1000
     RKj =  -0.20126044E+02   0.69600663E+02   0.19161748E+03  -0.19480933E+03  -0.44989307E+02
            -0.11569984E+02  -0.13772868E+03

Power model  :    Bm = ( Rki X R ) / ( 1 + ( R/R0 )**rn )**( 1/rn )
     rn = 0.890      rki =   0.24545764E+02         rmu =   0.22304995E+03
```

No	Rotation (radians) X 1/1000	Moment (kip-inch) Expri.	Poly.	M.Expo.	P.Model.	Connection stiffness (kip-inch) X 1000 Expri.	Poly.	M.Expo.	P.Model.
1	0.00	0.00	0.00	0.00	0.00	0.1048E+02	0.2414E+02	0.2656E+02	0.2455E+02
3	4.50	40.70	90.39	42.07	68.24	0.7251E+01	0.1285E+02	0.7150E+01	0.9879E+01
5	8.35	69.65	124.66	68.59	98.05	0.7810E+01	0.6279E+01	0.7528E+01	0.6092E+01
7	12.47	103.40	145.08	103.39	118.58	0.8515E+01	0.3998E+01	0.8978E+01	0.4091E+01
9	17.00	142.00	160.36	142.34	134.10	0.7992E+01	0.2879E+01	0.7902E+01	0.2873E+01
11	21.27	174.00	171.30	172.93	144.76	0.7500E+01	0.2296E+01	0.6572E+01	0.2174E+01
13	23.70	211.00	176.59	209.42	149.69	0.6550E+02	0.2063E+01	0.7593E+02	0.1887E+01
15	24.80	283.00	178.84	292.95	151.70	0.1417E+03	0.1973E+01	0.7596E+02	0.1776E+01
17	25.20	364.00	179.62	361.67	152.41	0.2450E+03	0.1942E+01	0.2676E+03	0.1739E+01
19	25.55	456.50	180.30	455.34	153.01	0.2833E+03	0.1917E+01	0.2676E+03	0.1707E+01
21	25.85	538.00	180.87	535.63	153.52	0.2600E+03	0.1895E+01	0.2677E+03	0.1680E+01
23	26.20	625.00	181.53	629.33	154.10	0.2400E+03	0.1871E+01	0.2677E+03	0.1649E+01
25	26.90	720.50	182.84	719.38	155.23	0.9500E+02	0.1823E+01	0.7304E+02	0.1591E+01
27	28.20	817.00	185.13	814.49	157.24	0.6125E+02	0.1744E+01	0.7330E+02	0.1491E+01
29	30.15	904.50	188.41	906.16	160.01	0.3347E+02	0.1637E+01	0.2879E+02	0.1358E+01
31	33.10	991.50	193.03	992.21	163.76	0.2695E+02	0.1499E+01	0.2953E+02	0.1189E+01
33	36.80	1085.00	198.30	1080.81	167.84	0.2368E+02	0.1359E+01	0.1858E+02	0.1020E+01
35	40.65	1150.00	203.30	1152.58	171.48	0.1026E+02	0.1240E+01	0.1857E+02	0.8801E+00
37	44.35	1215.00	207.74	1220.13	174.53	0.2571E+02	0.1144E+01	0.1783E+02	0.7718E+00
39	48.13	1290.00	211.92	1285.04	177.27	0.1475E+02	0.1063E+01	0.1638E+02	0.6809E+00
41	52.20	1350.00	216.08	1347.49	179.87	0.1115E+02	0.9881E+00	0.1424E+02	0.6004E+00
43	56.10	1380.00	219.81	1398.40	182.09	-0.4917E+02	0.9267E+00	0.1182E+02	0.5362E+00
45	57.02	1294.00	220.66	1282.29	182.57	-0.1438E+03	0.9134E+00	-0.1265E+03	0.5226E+00
47	57.66	1202.00	221.24	1201.19	182.91	-0.1438E+03	0.9044E+00	-0.1269E+03	0.5135E+00
49	58.30	1110.00	221.82	1119.82	183.23	-0.1438E+03	0.8955E+00	-0.1274E+03	0.5046E+00

A.7 – 5

VII - 3

Connection type : Header plate connections
Mode : All bolted

Tested by : W.H.Sommer (1969) Canada
Test Id. : TEST 7

Column : --
Beam : W24X76
Plate thickness : 1/4"

Fasteners: A325-X-3/4"D
15/16" Oversize holes
Material : G40.12
Fy = 49.30 ksi
Fu = 74.60 ksi

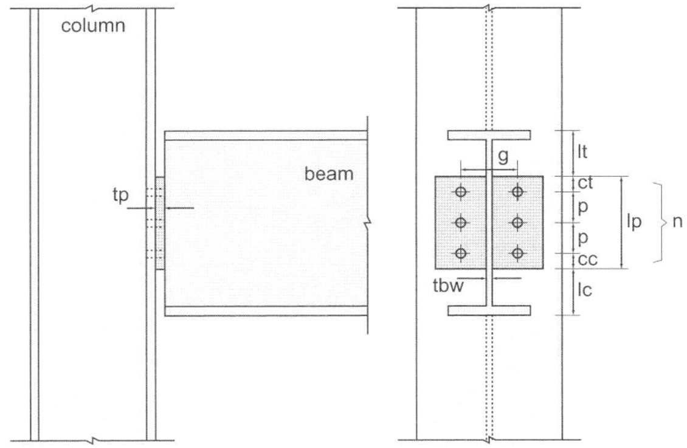

Major parameters

lt = 1.5000"	ct = 1.5000"	p = 3.0000"	cc = 1.5000"
lc = 10.3750"	g = 4.0000"	lp = 12.0000"	tp = 0.2500"
tbw= 0.4400"			
n = 2 X 4			

Remark 1) Actual strength Fy = 51.5 ksi, Fu = 77.9 ksi for beam web.
2) No washers were used.

No	Moment (k-in)	Rotation (radians) X 1/1000
1	0.00	0.00
2	26.30	0.40
3	55.40	1.60
4	84.50	2.80
5	113.75	3.45
6	143.00	4.10
7	177.00	5.40
8	211.00	6.70
9	235.25	8.90
10	259.50	11.10
11	283.75	13.30
12	308.00	15.50
13	322.50	17.30
14	337.00	19.10
15	361.00	20.50
16	385.00	21.50
17	424.00	23.10
18	473.00	23.50
19	506.50	23.70
20	540.00	23.90
21	574.00	24.10
22	608.00	24.30
23	656.50	24.55
24	705.00	24.80
25	753.50	25.05
26	802.00	25.30
27	850.66	25.57
28	899.33	25.83
29	948.00	26.10
30	994.00	26.60
31	1040.00	27.10
32	1090.00	28.33
33	1140.00	29.57
34	1190.00	30.80
35	1240.00	32.50
36	1290.00	34.20
37	1335.00	35.90
38	1380.00	37.60
39	1430.00	39.25
40	1480.00	40.90
41	1530.00	43.20
42	1580.00	45.50
43	1595.00	47.80
44	1610.00	50.10
45	1605.00	51.90
46	1600.00	53.70
47	1545.00	54.23
48	1490.00	54.75
49	1435.00	55.27
50	1380.00	55.80
51	1325.00	56.33
52	1270.00	56.85
53	1215.00	57.38
54	1160.00	57.90

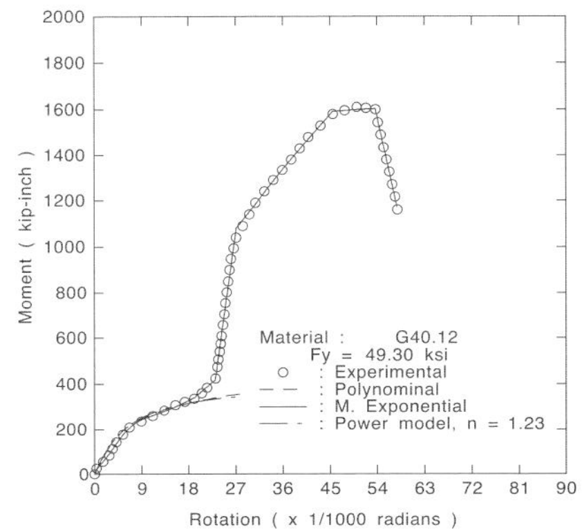

A.7 – 6

```
Moment-rotation prediction equations     ( R : X 1/1000 radians )

Frye and Morris polynominal model  :     R = Sum ( Ai X ( K*Bm )**Pi X 10**Qi )
      g =  4.000000"   d = 12.000000"  t  =  0.250000"    w =  0.440000"
     A1 =  5.100000   A2 =  6.200000   A3 =  2.400000     K =  0.419516
     P1 =         1   P2 =         3   P3 =         5    Q1 =        -2     Q2 =       -7     Q3 =      -10

Modified exponential model  :
     Bm = Sum ( Ai X ( 1 - exp( - R/( 2*i*C ) ) ) )  +      Sum ( RKj X ( R - Rj0 ) )  +  Bm0
       C =   0.58250000E+00               Bmo=   0.00000000E+00           Nexp=  6          Nliner=  6
      AI =   0.55769000E+03  -0.50981917E+04   0.11876062E+05  -0.18623086E+04  -0.17886657E+05   0.12939963E+05
     Rj0 =          19.1000          23.1000          27.1000          30.8000          45.5000
                    53.7000
     RKj =   0.74952921E+01   0.14766549E+03  -0.13535064E+03   0.41795480E+01  -0.23456179E+02
            -0.10685890E+03

Power model  :    Bm = ( Rki X R ) / ( 1 + ( R/R0 )**rn )**( 1/rn )
      rn = 1.225       rki =   0.56540501E+02         rmu =   0.39649904E+03
```

No	Rotation (radians) X 1/1000	Moment (kip-inch) Expri.	Poly.	M.Expo.	P.Model.	Connection stiffness (kip-inch) X 1000 Expri.	Poly.	M.Expo.	P.Model.
1	0.00	0.00	0.00	0.00	0.00	0.6576E+02	0.4674E+02	0.6956E+02	0.5654E+02
3	1.60	55.40	73.62	56.34	79.95	0.2425E+02	0.4425E+02	0.2578E+02	0.4294E+02
5	3.45	113.75	145.27	113.46	146.58	0.4500E+02	0.3201E+02	0.3446E+02	0.2994E+02
7	5.40	177.00	195.19	176.69	195.57	0.2615E+02	0.2030E+02	0.2829E+02	0.2098E+02
9	8.90	235.25	247.66	242.44	251.51	0.1102E+02	0.1130E+02	0.1119E+02	0.1208E+02
11	13.30	283.75	287.06	279.93	291.73	0.1102E+02	0.7216E+01	0.8398E+01	0.6874E+01
13	17.30	322.50	312.08	319.52	314.03	0.8053E+01	0.5474E+01	0.1128E+02	0.4511E+01
15	20.50	361.00	328.13	367.78	326.52	0.2114E+02	0.4610E+01	0.1951E+02	0.3373E+01
17	23.10	424.00	339.42	417.92	334.41	0.1029E+03	0.4096E+01	0.1891E+02	0.2727E+01
19	23.70	506.50	341.86	517.79	336.01	0.1675E+03	0.3993E+01	0.1663E+03	0.2603E+01
21	24.10	574.00	343.45	584.28	337.03	0.1700E+03	0.3928E+01	0.1661E+03	0.2525E+01
23	24.55	656.50	345.21	658.99	338.15	0.1940E+03	0.3858E+01	0.1659E+03	0.2441E+01
25	25.05	753.50	347.13	741.88	339.35	0.1940E+03	0.3783E+01	0.1657E+03	0.2352E+01
27	25.57	850.66	349.08	827.40	340.54	0.1825E+03	0.3708E+01	0.1654E+03	0.2265E+01
29	26.10	948.00	351.04	915.51	341.73	0.1510E+03	0.3635E+01	0.1651E+03	0.2180E+01
31	27.10	1040.00	354.64	1080.27	343.83	0.7716E+02	0.3505E+01	0.1645E+03	0.2033E+01
33	29.57	1140.00	362.94	1150.20	348.45	0.4054E+02	0.3225E+01	0.2759E+02	0.1725E+01
35	32.50	1240.00	371.91	1235.65	353.08	0.2941E+02	0.2953E+01	0.3004E+02	0.1439E+01
37	35.90	1335.00	381.49	1334.81	357.52	0.2647E+02	0.2691E+01	0.2835E+02	0.1186E+01
39	39.25	1430.00	390.22	1427.51	361.16	0.3030E+02	0.2476E+01	0.2705E+02	0.9946E+00
41	43.20	1530.00	399.50	1532.05	364.73	0.2174E+02	0.2270E+01	0.2595E+02	0.8213E+00
43	47.80	1595.00	409.47	1595.42	368.14	0.6522E+01	0.2071E+01	0.1669E+01	0.6694E+00
45	51.90	1605.00	417.71	1601.27	370.66	-0.2780E+01	0.1923E+01	0.1218E+01	0.5660E+00
47	54.23	1545.00	422.04	1547.78	371.92	-0.1048E+03	0.1850E+01	-0.1058E+03	0.5173E+00
49	55.27	1435.00	423.97	1436.64	372.45	-0.1048E+03	0.1818E+01	-0.1059E+03	0.4972E+00
51	56.33	1325.00	425.88	1325.44	372.97	-0.1048E+03	0.1788E+01	-0.1059E+03	0.4783E+00
53	57.38	1215.00	427.75	1214.18	373.46	-0.1048E+03	0.1759E+01	-0.1060E+03	0.4604E+00

A.7 – 7

VII - 4

Connection type : Header plate connections
Mode : All bolted

Tested by : W.H.Sommer (1969) Canada
Test Id. : TEST 8

Column : --
Beam : W24X76
Plate thickness : 1/4"

Fasteners: A325-X-3/4"D
15/16" Oversize holes
Material : G40.12
Fy = 50.50 ksi
Fu = 74.20 ksi

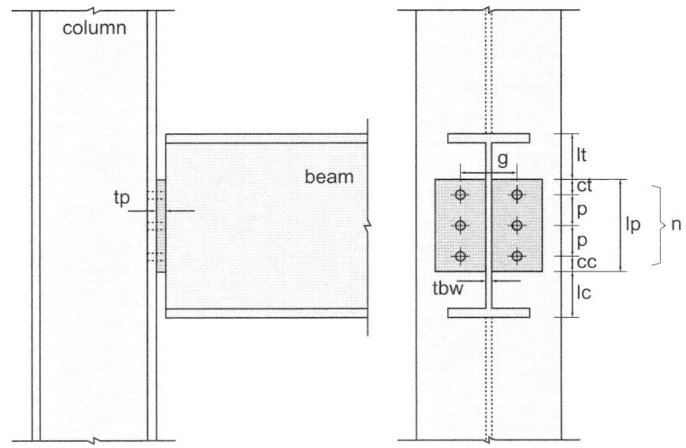

Major parameters

lt = 1.5000"	ct = 1.5000"	p = 3.0000"	cc = 1.5000"
lc = 7.3750"	g = 4.0000"	lp = 15.0000"	tp = 0.2500"
tbw= 0.4410"			
n = 2 X 5			

Remark 1) Actual strength Fy = 51.5 ksi, Fu = 77.9 ksi for beam web.
2) No washers were used.

No	Moment (k-in)	Rotation (radians) X 1/1000	No	Moment (k-in)	Rotation (radians) X 1/1000	No	Moment (k-in)	Rotation (radians) X 1/1000
1	0.00	0.00	26	625.00	25.90	51	1440.00	41.50
2	26.40	0.30	27	654.00	27.10	52	1440.00	43.50
3	44.70	0.50	28	678.50	27.50	53	1397.78	43.69
4	74.60	0.80	29	703.00	27.90	54	1355.55	43.88
5	106.00	1.10	30	739.25	28.18	55	1313.33	44.07
6	139.00	1.60	31	775.50	28.45	56	1271.11	44.26
7	172.00	2.10	32	811.75	28.73	57	1228.89	44.44
8	196.50	2.50	33	848.00	29.00	58	1186.66	44.63
9	221.00	2.90	34	879.67	29.23	59	1144.44	44.82
10	244.50	3.30	35	911.33	29.47	60	1102.22	45.01
11	268.00	3.70	36	943.00	29.70	61	1060.00	45.20
12	292.00	4.35	37	976.50	30.10	62	1020.00	47.40
13	316.00	5.00	38	1010.00	30.50	63	982.75	47.95
14	340.00	5.85	39	1043.34	30.70	64	945.50	48.50
15	364.00	6.70	40	1076.67	30.90	65	908.25	49.05
16	388.00	7.55	41	1110.00	31.10	66	871.00	49.60
17	412.00	8.40	42	1150.00	31.50			
18	448.00	10.35	43	1190.00	31.90			
19	484.00	12.30	44	1230.00	32.30			
20	511.50	14.40	45	1270.00	32.77			
21	539.00	16.50	46	1310.00	33.23			
22	557.00	18.50	47	1350.00	33.70			
23	580.50	20.85	48	1375.00	35.25			
24	604.00	23.20	49	1400.00	36.80			
25	614.50	24.55	50	1420.00	39.15			

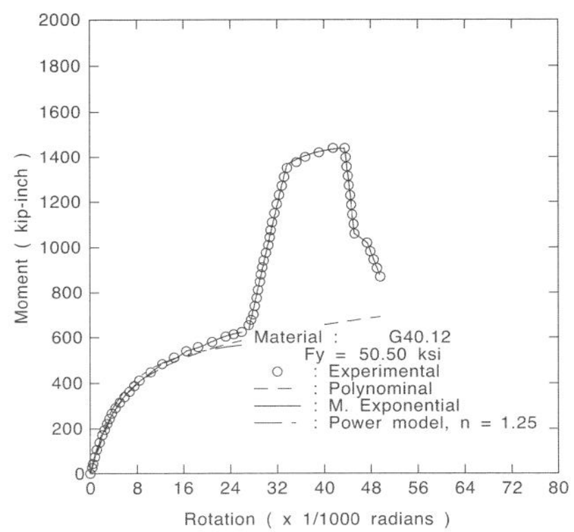

```
Moment-rotation prediction equations     ( R : X 1/1000 radians )

Frye and Morris polynominal model   :     R = Sum ( Ai X ( K*Bm )**Pi X 10**Qi )
     g =  4.000000"   d = 15.000000"  t  =  0.250000"   w =  0.441000"
    A1 =  5.100000   A2 =  6.200000   A3 =  2.400000    K =  0.250820
    P1 =         1   P2 =         3   P3 =         5   Q1 =        -2    Q2 =        -7    Q3 =       -10

Modified exponential model  :
    Bm = Sum ( Ai X ( 1 - exp( - R/( 2*i*C ) ) ) )  +      Sum ( RKj X ( R - Rj0 ) )  +  Bm0
     C =   0.48833333E+00            Bmo=    0.00000000E+00           Nexp=  6         Nliner=  8
    AI =  -0.57861081E+03   0.13810048E+05  -0.98653630E+05   0.27971429E+06  -0.33444256E+06   0.14160242E+06
   Rj0 =          12.3000          16.5000          25.9000          27.1000          33.7000
                  43.5000          45.2000          47.4000
   RKj =  -0.20913863E+02  -0.15824918E+02   0.36797387E+01   0.11635427E+03  -0.87999841E+02
          -0.22398412E+03   0.20618600E+03  -0.48344144E+02

Power model  :    Bm = ( Rki X R ) / ( 1 + ( R/R0 )**rn )**( 1/rn )
    rn = 1.254       rki =   0.10922676E+03         rmu =   0.63466434E+03
```

No	Rotation (radians) X 1/1000	Moment (kip-inch) Expri.	Moment (kip-inch) Poly.	Moment (kip-inch) M.Expo.	Moment (kip-inch) P.Model.	Connection stiffness (kip-inch) X 1000 Expri.	Connection stiffness (kip-inch) X 1000 Poly.	Connection stiffness (kip-inch) X 1000 M.Expo.	Connection stiffness (kip-inch) X 1000 P.Model.
1	0.00	0.00	0.00	0.00	0.00	0.8800E+02	0.7817E+02	0.8430E+02	0.1092E+03
3	0.50	44.70	39.07	45.86	52.68	0.9477E+02	0.7789E+02	0.9465E+02	0.1007E+03
5	1.10	106.00	85.43	100.62	109.46	0.9015E+02	0.7651E+02	0.8602E+02	0.8853E+02
7	2.10	172.00	159.18	175.69	188.51	0.6337E+02	0.6993E+02	0.6479E+02	0.7018E+02
9	2.90	221.00	211.57	222.46	239.73	0.6000E+02	0.6063E+02	0.5303E+02	0.5829E+02
11	3.70	268.00	255.96	261.65	282.38	0.5043E+02	0.5043E+02	0.4541E+02	0.4868E+02
13	5.00	316.00	312.29	314.73	337.58	0.3316E+02	0.3706E+02	0.3662E+02	0.3693E+02
15	6.70	364.00	365.33	368.47	390.91	0.2824E+02	0.2637E+02	0.2690E+02	0.2657E+02
17	8.40	412.00	404.47	408.02	429.93	0.2527E+02	0.2021E+02	0.2041E+02	0.1978E+02
19	12.30	484.00	467.53	488.45	488.00	0.1588E+02	0.1314E+02	0.2490E+02	0.1114E+02
21	16.50	539.00	514.48	541.25	524.51	0.1100E+02	0.9614E+01	0.2067E+02	0.6760E+01
23	20.85	580.50	551.50	581.50	548.29	0.9998E+01	0.7581E+01	0.1113E+02	0.4408E+01
25	24.55	614.50	577.34	615.89	562.26	0.7778E+01	0.6455E+01	0.6315E+01	0.3228E+01
27	27.10	654.00	593.02	628.87	569.73	0.5198E+02	0.5867E+01	0.3878E+01	0.2662E+01
29	27.90	703.00	597.69	724.22	571.80	0.1031E+03	0.5705E+01	0.1181E+03	0.2513E+01
31	28.45	775.50	600.81	788.79	573.16	0.1318E+03	0.5599E+01	0.1167E+03	0.2418E+01
33	29.00	848.00	603.87	852.56	574.46	0.1339E+03	0.5497E+01	0.1152E+03	0.2328E+01
35	29.47	911.33	606.43	906.05	575.53	0.1357E+03	0.5414E+01	0.1140E+03	0.2255E+01
37	30.10	976.50	609.84	977.71	576.93	0.8376E+02	0.5306E+01	0.1123E+03	0.2161E+01
39	30.70	1043.34	613.01	1044.64	578.20	0.1667E+03	0.5207E+01	0.1108E+03	0.2077E+01
41	31.10	1110.00	615.09	1088.76	579.02	0.1444E+03	0.5144E+01	0.1098E+03	0.2024E+01
43	31.90	1190.00	619.18	1175.79	580.60	0.1000E+03	0.5022E+01	0.1078E+03	0.1923E+01
45	32.77	1270.00	623.50	1268.34	582.22	0.8572E+02	0.4896E+01	0.1058E+03	0.1822E+01
47	33.70	1350.00	627.88	1366.08	583.87	0.6961E+02	0.4773E+01	0.1037E+03	0.1721E+01
49	36.80	1400.00	642.21	1405.07	588.75	0.1310E+02	0.4395E+01	0.9714E+01	0.1438E+01
51	41.50	1440.00	661.62	1434.60	594.72	0.3913E+01	0.3939E+01	0.3359E+01	0.1122E+01
53	43.69	1397.78	670.04	1397.34	597.05	-0.2235E+03	0.3758E+01	-0.2226E+03	0.1008E+01
55	44.07	1313.33	671.46	1313.17	597.43	-0.2235E+03	0.3729E+01	-0.2229E+03	0.9901E+00
57	44.44	1228.89	672.86	1228.91	597.80	-0.2236E+03	0.3700E+01	-0.2232E+03	0.9726E+00
59	44.82	1144.44	674.26	1144.52	598.16	-0.2235E+03	0.3672E+01	-0.2235E+03	0.9555E+00
61	45.20	1060.00	675.64	1060.03	598.52	-0.2073E+03	0.3644E+01	-0.2238E+03	0.9389E+00
63	47.95	982.75	685.40	982.82	600.95	-0.6773E+02	0.3455E+01	-0.6749E+02	0.8293E+00
65	49.05	908.25	689.18	908.31	601.84	-0.6773E+02	0.3385E+01	-0.6797E+02	0.7906E+00

VII - 5

Connection type : Header plate connections
Mode : All bolted

Tested by : W.H.Sommer (1969) Canada
Test Id. : TEST 9

Column : --
Beam : W24X76
Plate thickness : 1/4"

Fasteners: A325-X-3/4"D
15/16" Oversize holes
Material : G40.12
Fy = 47.40 ksi
Fu = 73.10 ksi

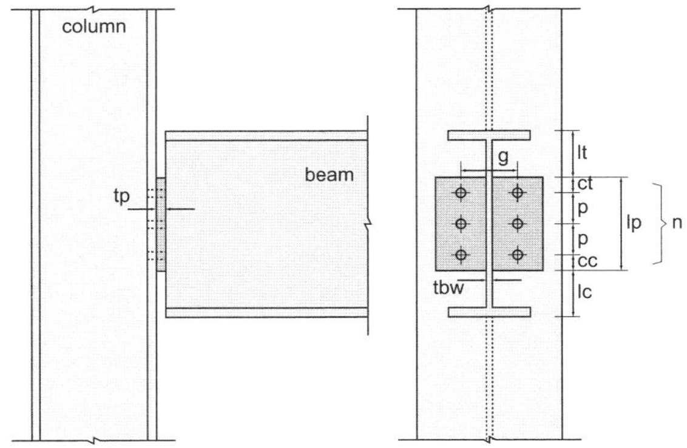

Major parameters

lt = 1.5000"	ct = 1.5000"	p = 3.0000"	cc = 1.5000"
lc = 4.3750"	g = 4.0000"	lp = 18.0000"	tp = 0.2500"
tbw= 0.4410"			
n = 2 X 6			

Remark 1) Actual strength Fy = 51.5 ksi, Fu = 77.9 ksi for beam web.
2) No washers were used.

No	Moment (k-in)	Rotation (radians) X 1/1000	No	Moment (k-in)	Rotation (radians) X 1/1000
1	0.00	0.00	26	909.00	21.67
2	26.60	0.20	27	938.00	23.90
3	55.70	0.40	28	954.50	25.20
4	95.40	0.60	29	971.00	26.50
5	134.00	0.90	30	1000.50	28.30
6	183.00	1.20	31	1030.00	30.10
7	216.50	1.45	32	1050.00	31.70
8	250.00	1.70	33	1070.00	33.30
9	284.50	2.05	34	1105.00	34.30
10	319.00	2.40	35	1140.00	35.30
11	357.00	2.90	36	1190.00	36.10
12	395.00	3.40	37	1240.00	36.90
13	429.00	3.90	38	1290.00	38.10
14	463.00	4.40	39	1340.00	39.30
15	497.50	5.00	40	1395.00	40.65
16	532.00	5.60	41	1450.00	42.00
17	570.50	6.40	42	1405.00	43.50
18	609.00	7.20	43	1360.00	45.00
19	658.50	8.75	44	1307.50	45.42
20	708.00	10.30	45	1255.00	45.85
21	754.00	12.20	46	1202.50	46.27
22	786.33	13.87	47	1150.00	46.70
23	818.67	15.53			
24	851.00	17.20			
25	880.00	19.43			

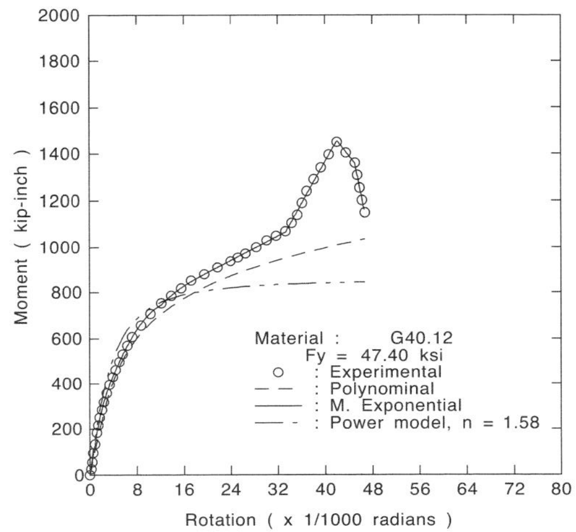

A.7 – 10

```
Moment-rotation prediction equations      ( R : X 1/1000 radians )

Frye and Morris polynominal model   :     R = Sum ( Ai X ( K*Bm )**Pi X 10**Qi )
      g =  4.000000"   d = 18.000000"  t  =  0.250000"    w =  0.441000"
     A1 =  5.100000   A2 =  6.200000   A3 =  2.400000     K =  0.164910
     P1 =        1     P2 =        3    P3 =        5    Q1 =       -2     Q2 =       -7     Q3 =      -10

Modified exponential model   :
     Bm = Sum ( Ai X ( 1 - exp( - R/( 2*i*C ) ) ) )  +      Sum ( RKj X ( R - Rj0 ) )  +  Bm0
      C =   0.43916667E+00             Bmo=   0.00000000E+00           Nexp=  6        Nliner=  4
     AI =  -0.60022391E+03   0.64949294E+04  -0.23021454E+05   0.36293523E+05  -0.26069541E+05   0.76979927E+04
     Rj0 =          12.2000         33.3000          42.0000          45.0000
     RKj =   0.13025953E+02   0.31434880E+02  -0.76785484E+02  -0.90697028E+02

Power model   :    Bm = ( Rki X R ) / ( 1 + ( R/R0 )**rn )**( 1/rn )
      rn = 1.578       rki =   0.18743302E+03          rmu =   0.85781483E+03
```

No	Rotation (radians) X 1/1000	Moment (kip-inch) Expri.	Poly.	M.Expo.	P.Model.	Connection stiffness (kip-inch) X 1000 Expri.	Poly.	M.Expo.	P.Model.
1	0.00	0.00	0.00	0.00	0.00	0.1330E+03	0.1189E+03	0.1319E+03	0.1874E+03
3	0.40	55.70	47.55	59.35	73.98	0.1720E+03	0.1186E+03	0.1579E+03	0.1811E+03
5	0.90	134.00	106.56	137.85	160.96	0.1460E+03	0.1173E+03	0.1518E+03	0.1661E+03
7	1.45	216.50	170.28	215.28	246.97	0.1340E+03	0.1140E+03	0.1288E+03	0.1464E+03
9	2.05	284.50	236.77	284.69	328.33	0.9857E+02	0.1071E+03	0.1034E+03	0.1250E+03
11	2.90	357.00	321.79	360.94	422.74	0.7601E+02	0.9222E+02	0.7833E+02	0.9805E+02
13	3.90	429.00	404.29	430.83	507.79	0.6799E+02	0.7308E+02	0.6343E+02	0.7327E+02
15	5.00	497.50	474.97	495.97	576.97	0.5750E+02	0.5637E+02	0.5580E+02	0.5368E+02
17	6.40	570.50	543.28	569.11	639.59	0.4813E+02	0.4233E+02	0.4862E+02	0.3705E+02
19	8.75	658.50	625.77	666.64	706.04	0.3194E+02	0.2931E+02	0.3391E+02	0.2134E+02
21	12.20	754.00	709.08	747.26	759.05	0.2165E+02	0.2016E+02	0.1424E+02	0.1092E+02
23	15.53	818.67	767.91	819.59	787.09	0.1940E+02	0.1557E+02	0.1757E+02	0.6433E+01
25	19.43	880.00	821.89	879.46	806.55	0.1298E+02	0.1237E+02	0.1401E+02	0.3846E+01
27	23.90	938.00	871.65	940.28	820.02	0.1280E+02	0.1008E+02	0.1346E+02	0.2354E+01
29	26.50	971.00	896.55	975.29	825.43	0.1424E+02	0.9118E+01	0.1347E+02	0.1835E+01
31	30.10	1030.00	927.41	1023.79	831.09	0.1433E+02	0.8074E+01	0.1346E+02	0.1345E+01
33	33.30	1070.00	952.04	1066.78	834.90	0.2635E+02	0.7342E+01	0.1340E+02	0.1049E+01
35	35.30	1140.00	966.52	1156.39	836.85	0.5028E+02	0.6948E+01	0.4478E+02	0.9078E+00
37	36.90	1240.00	977.28	1228.00	838.22	0.5417E+02	0.6672E+01	0.4473E+02	0.8132E+00
39	39.30	1340.00	992.78	1335.29	840.03	0.4123E+02	0.6297E+01	0.4467E+02	0.6951E+00
41	42.00	1450.00	1009.27	1455.82	841.76	0.7230E+01	0.5925E+01	0.4461E+02	0.5888E+00
43	45.00	1360.00	1026.49	1359.23	843.38	-0.1029E+03	0.5565E+01	-0.3222E+02	0.4953E+00
45	45.85	1255.00	1031.20	1254.74	843.79	-0.1235E+03	0.5471E+01	-0.1229E+03	0.4726E+00
47	46.70	1150.00	1035.83	1150.25	844.18	-0.1235E+03	0.5381E+01	-0.1229E+03	0.4513E+00

A.7 – 11

VII - 6

Connecticn type : Header plate connections
Mode : All bolted

Tested by : W.H.Sommer (1969) Canada
Test Id. : TEST 10

Column : --
Beam : W18X45
Plate thickness : 3/8"

Fasteners: A325-X-3/4"D
15/16" Oversize holes
Material : G40.12
Fy = 57.20 ksi
Fu = 75.70 ksi

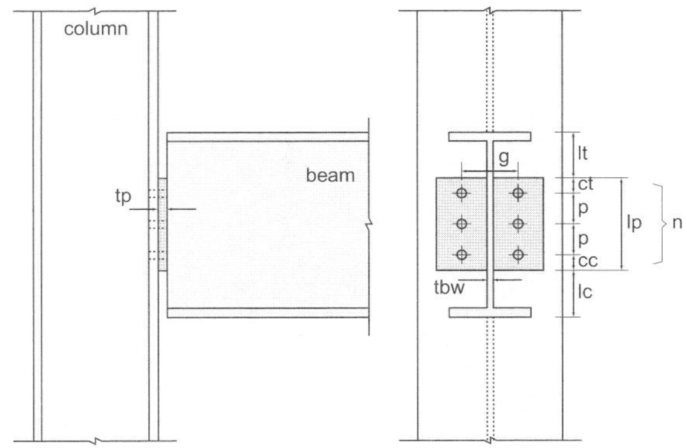

Major parameters

lt = 1.5000" ct = 1.5000" p = 3.0000" cc = 1.5000"
lc = 7.3750" g = 4.0000" lp = 9.0000" tp = 0.3750"
tbw= 0.3350"
n = 2 X 3

Remark 1) Actual strength Fy = 50.0 ksi, Fu = 76.1 ksi for beam web.
2) No washers were used.

No	Moment (k-in)	Rotation (radians) X 1/1000	No	Moment (k-in)	Rotation (radians) X 1/1000
1	0.00	0.00	26	796.67	31.13
2	24.30	0.60	27	853.33	31.57
3	63.00	1.60	28	910.00	32.00
4	87.20	2.20	29	960.00	32.45
5	111.00	2.90	30	1010.00	32.90
6	160.00	4.70	31	1055.00	33.90
7	184.00	5.70	32	1100.00	34.90
8	208.00	7.00	33	1150.00	36.70
9	232.00	8.60	34	1200.00	39.40
10	256.00	10.60	35	1225.00	41.05
11	268.50	12.10	36	1250.00	42.70
12	281.00	13.60	37	1275.00	44.55
13	293.00	15.90	38	1300.00	46.40
14	305.00	18.20	39	1302.00	49.06
15	317.00	21.00	40	1304.00	51.72
16	329.00	23.80	41	1306.00	54.38
17	358.00	25.10	42	1308.00	57.04
18	382.00	26.90	43	1310.00	59.70
19	426.00	28.20			
20	462.00	28.55			
21	498.00	28.90			
22	558.50	29.35			
23	619.00	29.80			
24	679.50	30.25			
25	740.00	30.70			

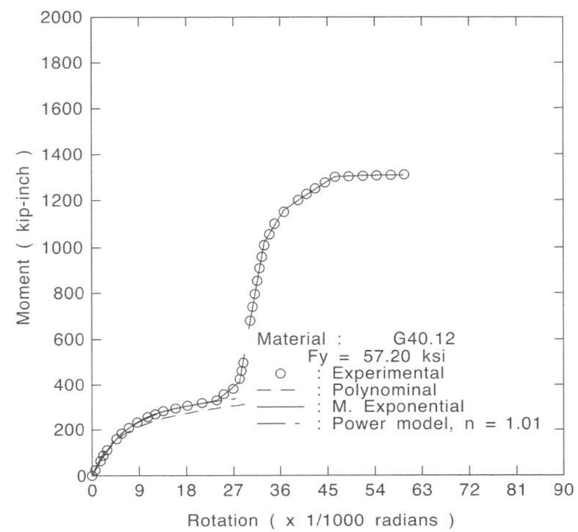

A.7 – 12

```
Moment-rotation prediction equations     ( R : X 1/1000 radians )

Frye and Morris polynominal model  :    R = Sum ( Ai X ( K*Bm )**Pi X 10**Qi )
     g =  4.000000"   d =  9.000000"  t  =  0.375000"    w =  0.335000"
     A1 =  5.100000   A2 =  6.200000   A3 =  2.400000     K =  0.487041
     P1 =        1    P2 =        3    P3 =        5     Q1 =       -2      Q2 =       -7      Q3 =      -10

Modified exponential model  :
     Bm = Sum ( Ai X ( 1 - exp( - R/( 2*i*C ) ) ) )  +      Sum ( RKj X ( R - Rj0 ) )  +  Bm0
      C =   0.64750000E+00             Bmo=   0.00000000E+00          Nexp=  6        Nliner=  5
     AI =  -0.15878415E+03   0.25934010E+04  -0.15980889E+05   0.41666718E+05  -0.46956736E+05   0.19263089E+05
     Rj0 =          23.8000          28.2000          32.9000          36.7000          46.4000
     RKj =   0.14957917E+02   0.10927286E+03  -0.91880402E+02  -0.20061223E+02  -0.12711962E+02

Power model  :     Bm = ( Rki X R ) / ( 1 + ( R/R0 )**rn )**( 1/rn )
     rn = 1.010       rki =   0.57334902E+02          rmu =   0.42638696E+03
```

No	Rotation (radians) X 1/1000	Moment (kip-inch) Expri.	Poly.	M.Expo.	P.Model.	Connection stiffness (kip-inch) X 1000 Expri.	Poly.	M.Expo.	P.Model.
1	0.00	0.00	0.00	0.00	0.00	0.4050E+02	0.4026E+02	0.3614E+02	0.5733E+02
3	1.60	63.00	63.41	64.18	75.85	0.3972E+02	0.3812E+02	0.3915E+02	0.3912E+02
5	2.90	111.00	108.96	110.62	120.35	0.3210E+02	0.3122E+02	0.3218E+02	0.2994E+02
7	5.70	184.00	173.24	183.16	186.31	0.2159E+02	0.1642E+02	0.2077E+02	0.1853E+02
9	8.60	232.00	210.39	232.41	230.28	0.1367E+02	0.1012E+02	0.1365E+02	0.1241E+02
11	12.10	268.50	239.40	269.41	265.88	0.8336E+01	0.6889E+01	0.7992E+01	0.8338E+01
13	15.90	293.00	261.93	293.31	292.38	0.5217E+01	0.5145E+01	0.5094E+01	0.5829E+01
15	21.00	317.00	284.60	316.77	316.73	0.4286E+01	0.3876E+01	0.4464E+01	0.3913E+01
17	25.10	358.00	299.13	354.87	330.74	0.1854E+02	0.3253E+01	0.1958E+02	0.2983E+01
19	28.20	426.00	308.66	415.52	339.18	0.8822E+02	0.2909E+01	0.1951E+02	0.2483E+01
21	28.90	498.00	310.69	505.66	340.88	0.1167E+03	0.2841E+01	0.1287E+03	0.2387E+01
23	29.80	619.00	313.22	621.49	342.98	0.1344E+03	0.2759E+01	0.1287E+03	0.2272E+01
25	30.70	740.00	315.69	737.24	344.97	0.1326E+03	0.2682E+01	0.1286E+03	0.2165E+01
27	31.57	853.33	318.00	848.62	346.81	0.1308E+03	0.2612E+01	0.1285E+03	0.2070E+01
29	32.45	960.00	320.29	962.04	348.60	0.1111E+03	0.2545E+01	0.1283E+03	0.1978E+01
31	33.90	1055.00	323.83	1056.08	351.36	0.4500E+02	0.2445E+01	0.3623E+02	0.1841E+01
33	36.70	1150.00	330.43	1156.85	356.19	0.2407E+02	0.2271E+01	0.3574E+02	0.1613E+01
35	41.05	1225.00	339.87	1223.31	362.57	0.1515E+02	0.2047E+01	0.1488E+02	0.1335E+01
37	44.55	1275.00	346.78	1274.34	366.92	0.1351E+02	0.1900E+01	0.1429E+02	0.1160E+01
39	49.06	1302.00	354.92	1303.52	371.73	0.7525E+00	0.1743E+01	0.9582E+00	0.9806E+00
41	54.38	1306.00	363.77	1307.11	376.50	0.7525E+00	0.1589E+01	0.4260E+00	0.8179E+00
43	59.70	1310.00	371.96	1308.38	380.50	0.7525E+00	0.1462E+01	0.7904E-01	0.6926E+00

VII - 7

Connection type : Header plate connections
Mode : All bolted

Tested by : W.H.Sommer (1969) Canada
Test Id. : TEST 11

Column : --
Beam : W18X45
Plate thickness : 3/8"

Fasteners: A325-X-3/4"D
15/16" Oversize holes
Material : G40.12
Fy = 57.20 ksi
Fu = 75.70 ksi

Major parameters

lt = 1.5000" ct = 1.5000" p = 3.0000" cc = 1.5000"
lc = 4.3750" g = 4.0000" lp = 12.0000" tp = 0.3750"
tbw= 0.3350"
n = 2 X 4

Remark 1) Actual strength Fy = 50.0 ksi, Fu = 76.1 ksi for beam web.
2) No washers were used.

No	Moment (k-in)	Rotation (radians) X 1/1000	No	Moment (k-in)	Rotation (radians) X 1/1000
1	0.00	0.00	26	705.00	46.90
2	18.90	0.10	27	774.00	47.40
3	69.20	0.50	28	822.00	48.10
4	81.30	0.60	29	879.00	49.20
5	147.00	1.20	30	928.00	49.90
6	194.00	1.70	31	986.00	50.60
7	232.50	2.20	32	1053.00	51.20
8	271.00	2.70	33	1120.00	51.80
9	319.00	3.60	34	1173.33	53.17
10	387.00	5.20	35	1226.67	54.53
11	426.00	6.80	36	1280.00	55.90
12	455.00	8.90	37	1343.34	57.57
13	484.00	11.00	38	1406.67	59.23
14	513.00	13.90	39	1470.00	60.90
15	534.00	16.85	40	1500.00	63.80
16	555.00	19.80	41	1533.33	66.90
17	572.50	22.55	42	1566.66	70.00
18	590.00	25.30	43	1600.00	73.10
19	602.67	27.90	44	1530.00	76.50
20	615.33	30.50			
21	628.00	33.10			
22	644.66	36.33			
23	661.33	39.57			
24	678.00	42.80			
25	691.50	44.85			

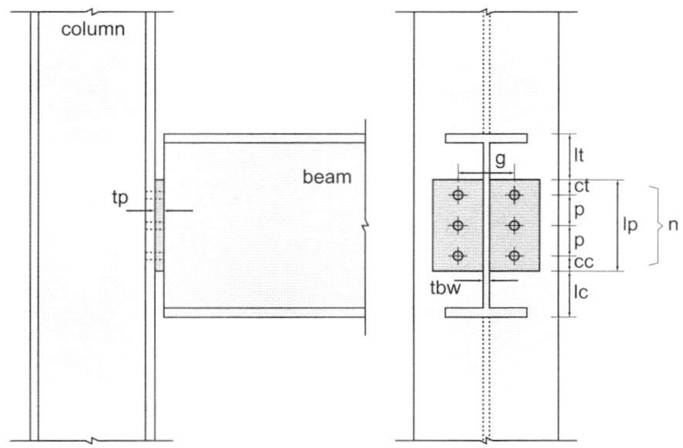

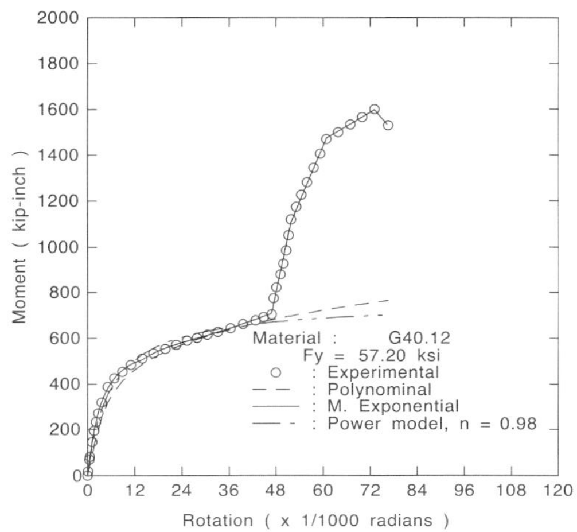

A.7 – 14

```
Moment-rotation prediction equations     ( R : X 1/1000 radians )

Frye and Morris polynominal model  :    R = Sum ( Ai X ( K*Bm )**Pi X 10**Qi )
     g =  4.000000"    d = 12.000000"  t  =  0.375000"   w =  0.335000"
    A1 =  5.100000    A2 =  6.200000   A3 =  2.400000    K =  0.251308
    P1 =         1    P2 =         3   P3 =         5   Q1 =        -2     Q2 =        -7     Q3 =       -10

Modified exponential model  :
    Bm = Sum ( Ai X ( 1 - exp( - R/( 2*i*C ) ) ) )  +      Sum ( RKj X ( R - Rj0 ) )  +  Bm0
      C =  0.66250000E+00               Bmo=  0.00000000E+00          Nexp=  6           Nliner=  5
     AI = -0.10210273E+03   0.13972597E+04  -0.26474650E+04  -0.43917684E+03   0.63086274E+04  -0.40463083E+04
    Rj0 =          13.9000          46.9000          51.8000          60.9000          73.1000
    RKj =  0.70187684E+01   0.74651519E+02  -0.40853784E+02  -0.30690343E+02  -0.30071595E+02

Power model  :     Bm = ( Rki X R ) / ( 1 + ( R/R0 )**rn )**( 1/rn )
      rn = 0.980        rki =   0.13132524E+03          rmu =   0.75802126E+03
```

No	Rotation (radians) X 1/1000	Moment (kip-inch) Expri.	Poly.	M.Expo.	P.Model.	Connection stiffness (kip-inch) X 1000 Expri.	Poly.	M.Expo.	P.Model.
1	0.00	0.00	0.00	0.00	0.00	0.1890E+03	0.7802E+02	0.1446E+03	0.1313E+03
3	0.50	69.20	38.96	67.66	60.09	0.1219E+03	0.7773E+02	0.1261E+03	0.1102E+03
5	1.20	147.00	92.88	147.44	129.26	0.1010E+03	0.7598E+02	0.1024E+03	0.8870E+02
7	2.20	232.50	165.80	235.37	206.72	0.7699E+02	0.6879E+02	0.7482E+02	0.6768E+02
9	3.60	319.00	250.43	320.25	287.31	0.4943E+02	0.5155E+02	0.4859E+02	0.4898E+02
11	6.80	426.00	367.21	424.24	404.37	0.1981E+02	0.2586E+02	0.2151E+02	0.2735E+02
13	11.00	484.00	448.54	485.51	490.77	0.1221E+02	0.1483E+02	0.9449E+01	0.1548E+02
15	16.85	534.00	516.83	535.66	558.19	0.7119E+01	0.9387E+01	0.8600E+01	0.8586E+01
17	22.55	572.50	562.80	574.10	597.41	0.6364E+01	0.7001E+01	0.5477E+01	0.5515E+01
19	27.90	602.67	596.49	601.20	622.30	0.4871E+01	0.5695E+01	0.4902E+01	0.3923E+01
21	33.10	628.00	623.77	627.34	640.00	0.4998E+01	0.4845E+01	0.5224E+01	0.2956E+01
23	39.57	661.33	652.57	663.18	656.43	0.5155E+01	0.4106E+01	0.5857E+01	0.2183E+01
25	44.85	691.50	673.04	695.30	666.77	0.6585E+01	0.3663E+01	0.6280E+01	0.1757E+01
27	47.40	774.00	682.31	748.85	671.04	0.1091E+03	0.3481E+01	0.8109E+02	0.1594E+01
29	49.20	879.00	688.35	894.90	673.82	0.6293E+02	0.3369E+01	0.8118E+02	0.1493E+01
31	50.60	986.00	693.04	1008.60	675.86	0.9837E+02	0.3284E+01	0.8125E+02	0.1421E+01
33	51.80	1120.00	696.96	1106.12	677.53	0.8950E+02	0.3216E+01	0.8129E+02	0.1363E+01
35	54.53	1226.67	705.47	1216.79	681.09	0.3903E+02	0.3073E+01	0.4053E+02	0.1244E+01
37	57.57	1343.34	714.67	1339.86	684.68	0.3800E+02	0.2927E+01	0.4061E+02	0.1129E+01
39	60.90	1470.00	724.08	1475.34	688.26	0.2791E+02	0.2786E+01	0.4067E+02	0.1021E+01
41	66.90	1533.33	740.12	1535.46	693.88	0.1075E+02	0.2565E+01	0.1005E+02	0.8611E+00
43	73.10	1600.00	755.41	1597.92	698.81	-0.4194E+01	0.2373E+01	0.1009E+02	0.7326E+00

A.7 – 15

VII - 8

Connection type : Header plate connections
Mode : All bolted

Tested by : W.H.Sommer (1969) Canada
Test Id. : TEST 12

Column : --
Beam : W18X45
Plate thickness : 3/8"

Fasteners: A325-X-3/4"D
15/16" Oversize holes
Material : G40.12
Fy = 51.60 ksi
Fu = 76.00 ksi

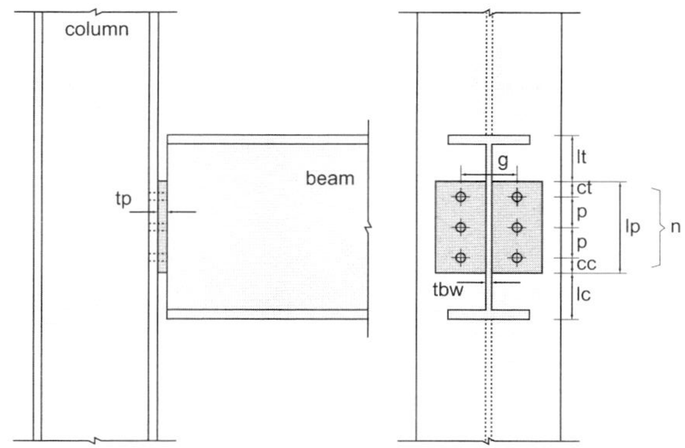

Major parameters

lt = 1.5000"   ct = 1.5000"   p = 3.0000"   cc = 1.5000"
lc = 1.3750"   g = 4.0000"   lp = 15.0000"   tp = 0.3750"
tbw= 0.3350"
n = 2 X 5

Remark 1) Actual strength Fy = 50.0 ksi, Fu = 76.1 ksi for beam web.
2) No washers were used.

No	Moment (k-in)	Rotation (radians) X 1/1000
1	0.00	0.00
2	18.90	0.20
3	45.50	0.30
4	69.65	0.45
5	93.80	0.60
6	116.90	0.75
7	140.00	0.90
8	164.00	1.10
9	188.00	1.30
10	227.00	1.70
11	266.00	2.10
12	302.00	2.45
13	338.00	2.80
14	369.50	3.20
15	401.00	3.60
16	434.00	4.03
17	467.00	4.47
18	500.00	4.90
19	541.34	5.63
20	582.67	6.37
21	624.00	7.10
22	654.00	7.80
23	684.00	8.50
24	724.00	9.95
25	764.00	11.40

No	Moment (k-in)	Rotation (radians) X 1/1000
26	798.67	13.57
27	833.33	15.73
28	868.00	17.90
29	891.50	20.40
30	915.00	22.90
31	938.50	25.40
32	962.00	27.90
33	974.00	29.93
34	986.00	31.95
35	998.00	33.98
36	1010.00	36.00
37	1025.00	38.65
38	1040.00	41.30
39	1030.00	43.53
40	1020.00	45.77
41	1010.00	48.00
42	973.75	49.45
43	937.50	50.90
44	901.25	52.35
45	865.00	53.80

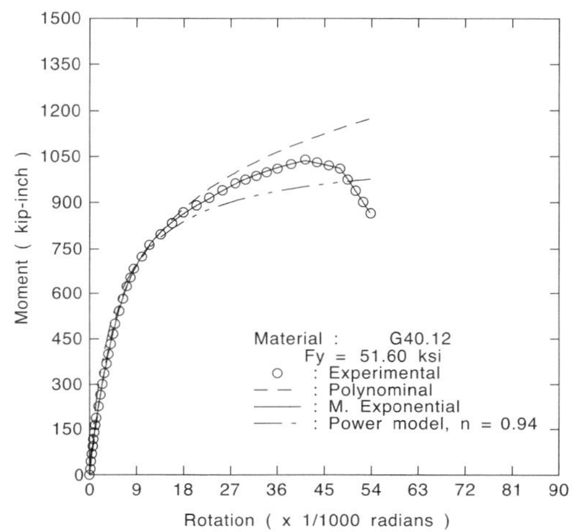

A.7 – 16

```
Moment-rotation prediction equations     ( R : X 1/1000 radians )

Frye and Morris polynominal model  :    R = Sum ( Ai X ( K*Bm )**Pi X 10**Qi )
      g =  4.000000"   d = 15.000000"  t  =  0.375000"   w =  0.335000"
     A1 =  5.100000   A2 =  6.200000  A3 =  2.400000    K =  0.150423
     P1 =         1   P2 =         3  P3 =         5   Q1 =        -2    Q2 =        -7    Q3 =       -10

Modified exponential model  :
     Bm = Sum ( Ai X ( 1 - exp( - R/( 2*i*C ) ) ) )  +      Sum ( RKj X ( R - Rj0 ) )  +  Bm0
      C =  0.49833333E+00           Bmo=  0.00000000E+00           Nexp=  6        Nliner=  3
     AI =  0.22527156E+02  0.18216926E+04 -0.14458024E+05  0.40049489E+05 -0.46335194E+05  0.19897490E+05
    Rj0 =         22.9000         41.3000         48.0000
    RKj =  0.25810032E+01 -0.72522892E+01 -0.20829964E+02

Power model  :    Bm = ( Rki X R ) / ( 1 + ( R/R0 )**rn )**( 1/rn )
     rn = 0.944      rki =   0.25235188E+03         rmu =   0.10684522E+04
```

No	Rotation (radians) X 1/1000	Moment (kip-inch) Expri.	Poly.	M.Expo.	P.Model.	Connection stiffness (kip-inch) X 1000 Expri.	Poly.	M.Expo.	P.Model.
1	0.00	0.00	0.00	0.00	0.00	0.9450E+02	0.1304E+03	0.1762E+03	0.2524E+03
3	0.30	45.50	39.10	49.78	69.63	0.2240E+03	0.1302E+03	0.1565E+03	0.2145E+03
5	0.60	93.80	78.12	94.31	129.61	0.1575E+03	0.1296E+03	0.1410E+03	0.1865E+03
7	0.90	140.00	116.83	134.73	182.11	0.1394E+03	0.1286E+03	0.1289E+03	0.1643E+03
9	1.30	188.00	167.84	183.67	242.90	0.1125E+03	0.1262E+03	0.1164E+03	0.1407E+03
11	2.10	266.00	265.42	269.46	341.07	0.1004E+03	0.1166E+03	0.9941E+02	0.1071E+03
13	2.80	338.00	342.56	335.30	408.66	0.9161E+02	0.1032E+03	0.8916E+02	0.8704E+02
15	3.60	401.00	418.29	402.75	471.30	0.7751E+02	0.8614E+02	0.7972E+02	0.7046E+02
17	4.47	467.00	485.65	467.87	526.43	0.7615E+02	0.6990E+02	0.7066E+02	0.5744E+02
19	5.63	541.34	557.27	543.63	585.74	0.5636E+02	0.5391E+02	0.5939E+02	0.4502E+02
21	7.10	624.00	626.14	621.21	643.51	0.4945E+02	0.4105E+02	0.4671E+02	0.3447E+02
23	8.50	684.00	677.77	679.29	686.67	0.3789E+02	0.3324E+02	0.3661E+02	0.2756E+02
25	11.40	764.00	758.99	762.86	752.31	0.2294E+02	0.2382E+02	0.2247E+02	0.1860E+02
27	15.73	833.33	845.27	836.99	816.08	0.1600E+02	0.1684E+02	0.1339E+02	0.1165E+02
29	20.40	891.50	913.84	889.87	860.54	0.9401E+01	0.1292E+02	0.9693E+01	0.7794E+01
31	25.40	938.50	971.68	937.71	893.18	0.9401E+01	0.1041E+02	0.9496E+01	0.5472E+01
33	29.93	974.00	1015.18	975.52	914.79	0.5926E+01	0.8901E+01	0.7279E+01	0.4168E+01
35	33.98	998.00	1049.11	1001.64	929.95	0.5926E+01	0.7899E+01	0.5692E+01	0.3362E+01
37	38.65	1025.00	1083.87	1025.01	944.02	0.5660E+01	0.7010E+01	0.4396E+01	0.2694E+01
39	43.53	1030.00	1116.27	1028.05	955.88	-0.4478E+01	0.6287E+01	-0.3693E+01	0.2191E+01
41	48.00	1010.00	1143.12	1010.48	964.86	-0.1692E+02	0.5755E+01	-0.4136E+01	0.1845E+01
43	50.90	937.50	1159.59	937.80	969.94	-0.2500E+02	0.5456E+01	-0.2514E+02	0.1663E+01
45	53.80	865.00	1174.92	864.70	974.53	-0.2500E+02	0.5194E+01	-0.2527E+02	0.1507E+01

VII - 9

Connection type : Header plate connections
Mode : All bolted

Tested by : W.H.Sommer (1969) Canada
Test Id. : TEST 13

Column : --
Beam : W24X76
Plate thickness : 3/8"

Fasteners: A325-X-3/4"D
15/16" Oversize holes
Material : G40.12
Fy = 57.20 ksi
Fu = 75.70 ksi

Major parameters

lt = 1.5000" ct = 1.5000" p = 3.0000" cc = 1.5000"
lc = 13.3750" g = 4.0000" lp = 9.0000" tp = 0.3750"
tbw= 0.4410"
n = 2 X 3

Remark 1) Actual strength Fy = 51.5 ksi, Fu = 77.9 ksi for beam web.
2) No washers were used.

No	Moment (k-in)	Rotation (radians) X 1/1000
1	0.00	0.00
2	37.10	1.40
3	74.20	2.80
4	122.00	4.80
5	170.00	7.30
6	218.00	9.60
7	250.00	11.77
8	282.00	13.93
9	314.00	16.10
10	341.75	18.88
11	369.50	21.65
12	397.25	24.43
13	425.00	27.20
14	482.00	27.50
15	566.00	27.65
16	650.00	27.80
17	746.00	28.20
18	842.00	28.40
19	914.00	28.55
20	986.00	28.70
21	1080.00	29.10
22	1155.00	29.25
23	1230.00	29.40
24	1290.00	29.60
25	1350.00	29.80
26	1410.00	30.05
27	1470.00	30.30
28	1530.00	30.60
29	1590.00	30.90
30	1650.00	31.35
31	1710.00	31.80
32	1770.00	32.80
33	1830.00	33.80
34	1890.00	35.15
35	1950.00	36.50
36	2010.00	38.60
37	2070.00	40.70
38	2130.00	43.45
39	2190.00	46.20
40	2213.33	48.53
41	2236.66	50.87
42	2260.00	53.20
43	2260.00	55.60

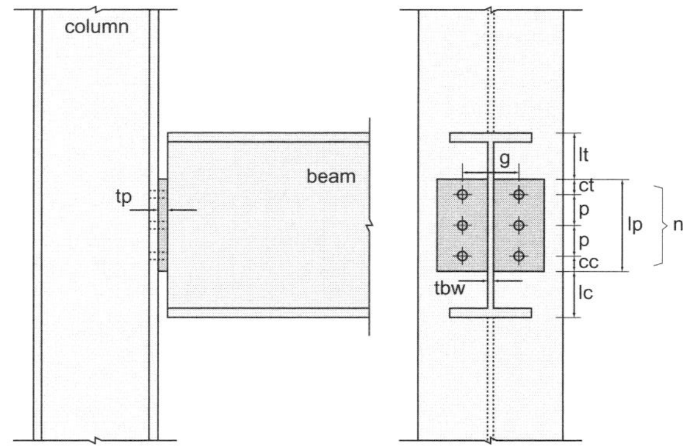

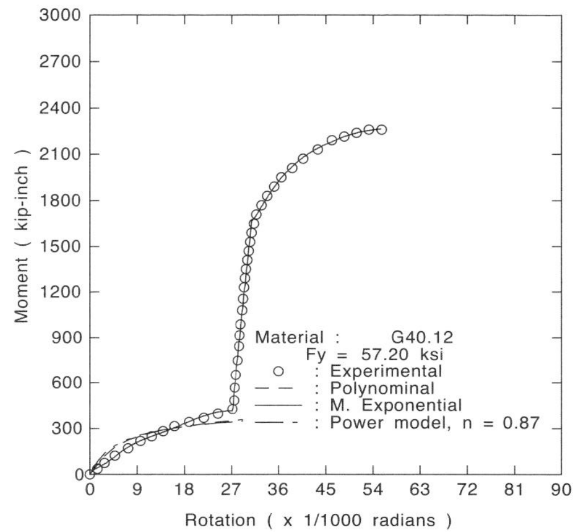

A.7 – 18

```
Moment-rotation prediction equations     ( R : X 1/1000 radians )

Frye and Morris polynominal model  :    R = Sum ( Ai X ( K*Bm )**Pi X 10**Qi )
      g =  4.000000"   d =  9.000000"  t  =  0.375000"   w =  0.441000"
     A1 =  5.100000   A2 =  6.200000   A3 =  2.400000    K =  0.424491
     P1 =         1   P2 =         3   P3 =         5   Q1 =        -2     Q2 =        -7     Q3 =       -10

Modified exponential model  :
     Bm = Sum ( Ai X ( 1 - exp( - R/( 2*i*C ) ) ) )  +      Sum ( RKj X ( R - Rj0 ) )  +  Bm0
      C =   0.81333333E+00             Bmo=   0.00000000E+00           Nexp=  6         Nliner=  7
     AI =   0.91988861E+03  -0.11767422E+05   0.37326682E+05  -0.11421234E+05  -0.76619509E+05   0.65573976E+05
    Rj0 =           0.0000          27.2000          27.5000          30.9000          31.8000
                   36.5000          46.2000
    RKj =  -0.84499326E+02   0.43863262E+03  -0.10032418E+03  -0.27934745E+03   0.17398218E+02
           -0.56205905E+01   0.43335297E+01

Power model  :    Bm = ( Rki X R ) / ( 1 + ( R/R0 )**rn )**( 1/rn )
     rn = 0.874      rki =   0.82841952E+02         rmu =   0.43699791E+03
```

No	Rotation (radians) X 1/1000	Moment (kip-inch) Expri.	Poly.	M.Expo.	P.Model.	Connection stiffness (kip-inch) X 1000 Expri.	Poly.	M.Expo.	P.Model.
1	0.00	0.00	0.00	0.00	0.00	0.2650E+02	0.4619E+02	0.1403E+03	0.8284E+02
3	2.80	74.20	121.46	68.83	138.00	0.2543E+02	0.3657E+02	0.2128E+02	0.3130E+02
5	7.30	170.00	224.74	178.92	230.02	0.2007E+02	0.1408E+02	0.2105E+02	0.1353E+02
7	11.77	250.00	272.02	248.75	275.72	0.1477E+02	0.8147E+01	0.1247E+02	0.7767E+01
9	16.10	314.00	301.69	304.70	303.12	0.1268E+02	0.5827E+01	0.1393E+02	0.5154E+01
11	21.65	369.50	329.38	380.96	326.32	0.1000E+02	0.4315E+01	0.1169E+02	0.3397E+01
13	27.20	425.00	350.75	413.14	342.23	0.1724E+03	0.3454E+01	-0.1611E+01	0.2421E+01
15	27.65	566.00	352.30	594.43	343.31	0.5600E+03	0.3400E+01	0.3353E+03	0.2362E+01
17	28.20	746.00	354.16	778.35	344.58	0.4000E+03	0.3337E+01	0.3335E+03	0.2292E+01
19	28.55	914.00	355.33	894.88	345.38	0.4800E+03	0.3297E+01	0.3324E+03	0.2249E+01
21	29.10	1080.00	357.13	1077.18	346.60	0.4277E+03	0.3238E+01	0.3305E+03	0.2185E+01
23	29.40	1230.00	358.10	1176.19	347.25	0.4143E+03	0.3206E+01	0.3295E+03	0.2151E+01
25	29.80	1350.00	359.38	1307.72	348.10	0.2733E+03	0.3165E+01	0.3282E+03	0.2107E+01
27	30.30	1470.00	360.96	1471.37	349.14	0.2218E+03	0.3116E+01	0.3264E+03	0.2053E+01
29	30.90	1590.00	362.82	1666.62	350.35	0.1733E+03	0.3058E+01	0.3244E+03	0.1992E+01
31	31.80	1710.00	365.55	1705.73	352.11	0.1106E+03	0.2976E+01	0.4189E+02	0.1906E+01
33	33.80	1830.00	371.31	1817.37	355.74	0.5338E+02	0.2812E+01	0.5237E+02	0.1733E+01
35	36.50	1950.00	378.66	1946.52	360.15	0.3823E+02	0.2617E+01	0.4339E+02	0.1535E+01
37	40.70	2070.00	389.05	2078.18	366.06	0.2565E+02	0.2369E+01	0.2528E+02	0.1291E+01
39	46.20	2190.00	401.34	2179.77	372.46	0.1542E+02	0.2111E+01	0.1232E+02	0.1051E+01
41	50.87	2236.66	410.77	2237.75	377.00	0.9999E+01	0.1936E+01	0.8613E+01	0.8979E+00
43	55.60	2260.00	419.58	2263.84	380.94	0.0000E+00	0.1788E+01	0.2751E+01	0.7750E+00

VII - 10

Connection type : Header plate connections
Mode : All bolted

Tested by : W.H.Sommer (1969) Canada
Test Id. : TEST 14

Column : --
Beam : W24X76
Plate thickness : 3/8"

Fasteners: A325-X-3/4"D
15/16" Oversize holes
Material : G40.12
Fy = 57.20 ksi
Fu = 75.70 ksi

Major parameters

lt = 1.5000"	ct = 1.5000"	p = 3.0000"	cc = 1.5000"
lc = 10.3750"	g = 4.0000"	lp = 12.0000"	tp = 0.3750"
tbw= 0.4400"			
n = 2 X 4			

Remark 1) Actual strength Fy = 51.5 ksi, Fu = 77.9 ksi for beam web.
2) No washers were used.

No	Moment (k-in)	Rotation (radians) X 1/1000
1	0.00	0.00
2	123.00	2.30
3	196.00	3.80
4	269.00	4.90
5	390.00	7.00
6	450.50	8.70
7	511.00	10.40
8	543.33	12.33
9	575.67	14.27
10	608.00	16.20
11	627.40	18.54
12	646.80	20.88
13	666.20	23.22
14	685.60	25.56
15	705.00	27.90
16	717.50	29.55
17	730.00	31.20
18	822.00	32.00
19	919.00	32.50
20	1044.50	33.00
21	1170.00	33.50
22	1265.00	33.85
23	1360.00	34.20
24	1480.00	34.67
25	1600.00	35.13

No	Moment (k-in)	Rotation (radians) X 1/1000
26	1720.00	35.60
27	1850.00	36.00
28	1930.00	36.40
29	2010.00	36.80
30	2085.00	37.15
31	2160.00	37.50
32	2255.00	38.65
33	2350.00	39.80
34	2450.00	42.10
35	2550.00	44.40
36	2607.50	47.12
37	2665.00	49.85
38	2722.50	52.58
39	2780.00	55.30

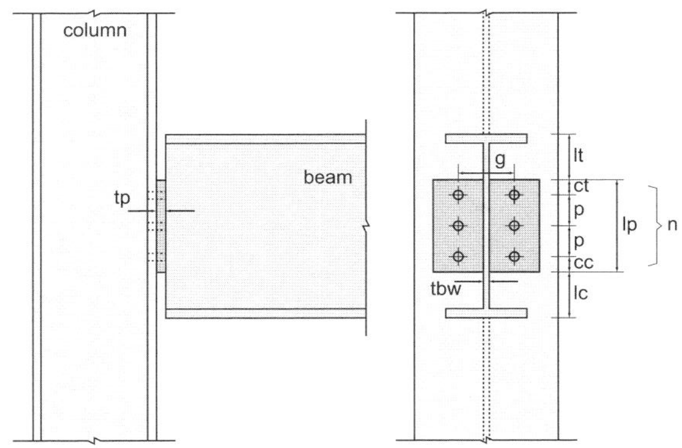

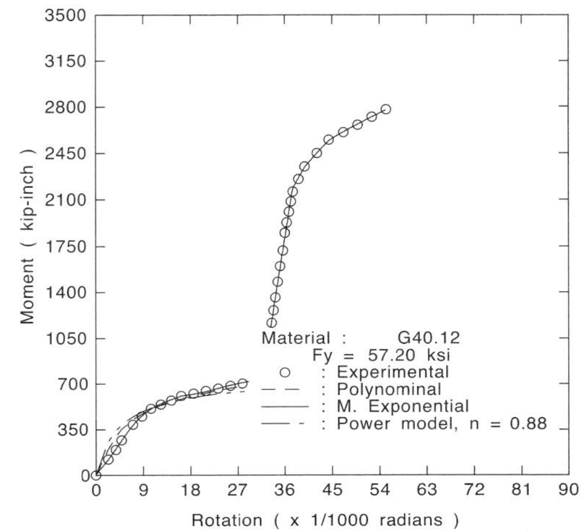

A.7 – 20

```
Moment-rotation prediction equations      ( R : X 1/1000 radians )

Frye and Morris polynominal model  :    R = Sum ( Ai X ( K*Bm )**Pi X 10**Qi )
      g =  4.000000"   d = 12.000000"  t  =  0.375000"   w =  0.440000"
     A1 =  5.100000   A2 =  6.200000   A3 =  2.400000    K =  0.219282
     P1 =         1   P2 =         3   P3 =         5   Q1 =        -2    Q2 =        -7    Q3 =       -10

Modified exponential model  :
     Bm = Sum ( Ai X ( 1 - exp( - R/( 2*i*C ) ) ) )  +      Sum ( RKj X ( R - Rj0 ) )  +  Bm0
       C =   0.10358333E+01              Bmo=   0.00000000E+00            Nexp=  6          Nliner=  6
      AI =   0.19194058E+04  -0.41285920E+05   0.24290916E+06  -0.58371541E+06   0.61282114E+06  -0.23245794E+06
     Rj0 =          16.2000          31.2000          32.0000          37.5000          39.8000
                    44.4000
     RKj =  -0.10158346E+02   0.96703066E+02   0.16058581E+03  -0.17233594E+03  -0.21320413E+02
            -0.15610711E+02

Power model  :    Bm = ( Rki X R ) / ( 1 + ( R/R0 )**rn )**( 1/rn )
      rn = 0.882       rki =   0.19082419E+03         rmu =   0.77679736E+03
```

No	Rotation (radians) X 1/1000	Moment (kip-inch) Expri.	Moment (kip-inch) Poly.	Moment (kip-inch) M.Expo.	Moment (kip-inch) P.Model.	Connection stiffness (kip-inch) X 1000 Expri.	Connection stiffness (kip-inch) X 1000 Poly.	Connection stiffness (kip-inch) X 1000 M.Expo.	Connection stiffness (kip-inch) X 1000 P.Model.
1	0.00	0.00	0.00	0.00	0.00	0.5348E+02	0.8942E+02	0.6688E+02	0.1908E+03
3	3.80	196.00	298.47	203.23	341.97	0.5888E+02	0.5631E+02	0.5900E+02	0.4636E+02
5	7.00	390.00	426.67	382.81	449.71	0.4544E+02	0.2864E+02	0.4831E+02	0.2458E+02
7	10.40	511.00	503.49	506.69	515.05	0.2676E+02	0.1810E+02	0.2564E+02	0.1506E+02
9	14.27	575.67	562.01	579.37	561.96	0.1673E+02	0.1281E+02	0.1499E+02	0.9788E+01
11	18.54	627.40	609.59	622.38	596.51	0.8290E+01	0.9758E+01	0.7283E+01	0.6688E+01
13	23.22	666.20	650.29	666.08	622.88	0.8290E+01	0.7796E+01	0.1052E+02	0.4749E+01
15	27.90	705.00	683.61	709.86	642.12	0.7871E+01	0.6527E+01	0.7038E+01	0.3560E+01
17	31.20	730.00	704.02	724.04	652.85	0.7993E+02	0.5870E+01	0.1240E+01	0.2976E+01
19	32.50	919.00	711.58	929.93	656.60	0.2225E+03	0.5646E+01	0.2558E+03	0.2785E+01
21	33.50	1170.00	717.19	1184.71	659.31	0.2630E+03	0.5487E+01	0.2537E+03	0.2651E+01
23	34.20	1360.00	721.01	1361.78	661.14	0.2653E+03	0.5382E+01	0.2522E+03	0.2563E+01
25	35.13	1600.00	726.00	1596.23	663.48	0.2572E+03	0.5248E+01	0.2502E+03	0.2453E+01
27	36.00	1850.00	730.53	1812.28	665.56	0.2625E+03	0.5130E+01	0.2484E+03	0.2357E+01
29	36.80	2010.00	734.42	2010.30	667.41	0.2076E+03	0.5031E+01	0.2467E+03	0.2273E+01
31	37.50	2160.00	737.93	2182.49	668.98	0.1836E+03	0.4944E+01	0.2453E+03	0.2204E+01
33	39.80	2350.00	749.14	2345.18	673.81	0.6956E+02	0.4677E+01	0.6861E+02	0.1997E+01
35	44.40	2550.00	769.44	2546.15	682.19	0.3324E+02	0.4236E+01	0.4055E+02	0.1664E+01
37	49.85	2665.00	791.34	2668.55	690.41	0.2110E+02	0.3816E+01	0.2061E+02	0.1368E+01
39	55.30	2780.00	811.18	2776.88	697.24	0.2110E+02	0.3478E+01	0.1961E+02	0.1147E+01

VII - 11

Connection type : Header plate connections
Mode : All bolted

Tested by : W.H.Sommer (1969) Canada
Test Id. : TEST 15

Column : --
Beam : W24X76
Plate thickness : 3/8"

Fasteners: A325-X-3/4"D
15/16" Oversize holes
Material : G40.12
Fy = 58.70 ksi
Fu = 75.10 ksi

Major parameters

lt = 1.5000"	ct = 1.5000"	p = 3.0000"	cc = 1.5000"
lc = 7.3750"	g = 5.5000"	lp = 15.0000"	tp = 0.3750"
tbw= 0.4410"			
n = 2 X 5			

Remark 1) Actual strength Fy = 51.5 ksi, Fu = 77.9 ksi for beam web.
2) No washers were used.

No	Moment (k-in)	Rotation (radians) X 1/1000	No	Moment (k-in)	Rotation (radians) X 1/1000
1	0.00	0.00	26	1476.67	46.23
2	75.40	0.60	27	1573.34	47.17
3	124.00	1.40	28	1670.00	48.10
4	220.00	2.70	29	1770.00	49.80
5	317.00	3.80	30	1870.00	51.50
6	389.50	4.95	31	1950.00	53.95
7	462.00	6.10	32	2030.00	56.40
8	534.50	9.15	33	2115.00	59.60
9	607.00	12.20	34	2200.00	62.80
10	636.20	15.42	35	2260.00	65.80
11	665.40	18.64	36	2320.00	68.80
12	694.60	21.86	37	2350.00	72.00
13	723.80	25.08	38	2380.00	75.20
14	753.00	28.30			
15	773.33	31.70			
16	793.66	35.10			
17	814.00	38.50			
18	849.00	40.80			
19	898.00	41.30			
20	964.00	42.00			
21	1030.00	42.70			
22	1110.00	43.40			
23	1190.00	44.10			
24	1285.00	44.70			
25	1380.00	45.30			

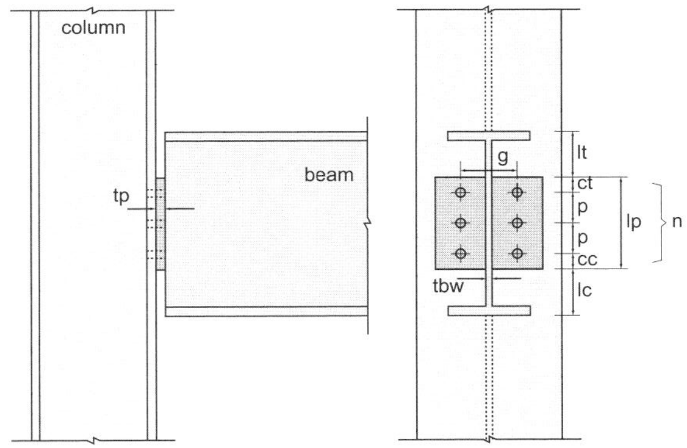

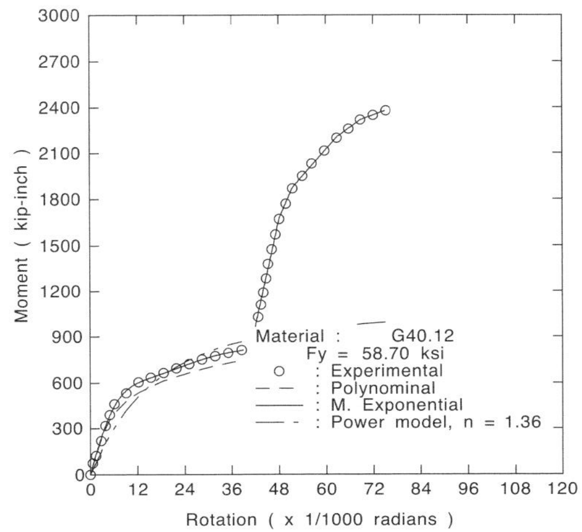

A.7 – 22

```
Moment-rotation prediction equations     ( R : X 1/1000 radians )

Frye and Morris polynominal model  :     R = Sum ( Ai X ( K*Bm )**Pi X 10**Qi )
     g =  5.500000"    d = 15.000000"  t  =  0.375000"    w =  0.441000"
     A1 =  5.100000    A2 =  6.200000   A3 =  2.400000    K =  0.218223
     P1 =        1     P2 =        3    P3 =        5    Q1 =       -2     Q2 =       -7     Q3 =      -10

Modified exponential model  :
     Bm = Sum ( Ai X ( 1 - exp( - R/( 2*i*C ) ) ) )  +      Sum ( RKj X ( R - Rj0 ) )  +  Bm0
       C =   0.77666667E+00             Bmo=   0.00000000E+00          Nexp=  6        Nliner=  6
      AI =   0.40147681E+03  -0.68148987E+04   0.28713136E+05  -0.46715239E+05   0.31939556E+05  -0.66837979E+04
     Rj0 =          38.5000          40.8000          48.1000          51.5000          62.8000
                    68.8000
     RKj =   0.45047140E+01   0.10968022E+03  -0.56700154E+02  -0.28634717E+02  -0.10003982E+02
           -0.92912071E+01

Power model  :    Bm = ( Rki X R ) / ( 1 + ( R/R0 )**rn )**( 1/rn )
     rn = 1.358      rki =   0.57728731E+02         rmu =   0.11044928E+04
```

No	Rotation (radians) X 1/1000	Moment (kip-inch) Expri.	Poly.	M.Expo.	P.Model.	Connection stiffness (kip-inch) X 1000 Expri.	Poly.	M.Expo.	P.Model.
1	0.00	0.00	0.00	0.00	0.00	0.1257E+03	0.8985E+02	0.1031E+03	0.5773E+02
3	1.40	124.00	124.36	128.85	79.15	0.6574E+02	0.8643E+02	0.8500E+02	0.5496E+02
5	3.80	317.00	300.09	314.21	202.95	0.7589E+02	0.5654E+02	0.6841E+02	0.4805E+02
7	6.10	462.00	400.76	447.02	305.67	0.5229E+02	0.3384E+02	0.4680E+02	0.4135E+02
9	12.20	607.00	535.82	602.68	511.71	0.1662E+02	0.1524E+02	0.1220E+02	0.2718E+02
11	18.64	665.40	613.55	664.10	654.20	0.9069E+01	0.9750E+01	0.9104E+01	0.1786E+02
13	25.08	723.80	667.48	724.06	749.60	0.9069E+01	0.7266E+01	0.9054E+01	0.1223E+02
15	31.70	773.33	710.37	776.44	817.78	0.5980E+01	0.5810E+01	0.6534E+01	0.8643E+01
17	38.50	814.00	746.39	810.75	867.98	0.1149E+02	0.4849E+01	0.3668E+01	0.6291E+01
19	41.30	898.00	759.55	887.15	884.56	0.9645E+02	0.4547E+01	0.1169E+03	0.5574E+01
21	42.70	1030.00	765.87	1050.57	892.14	0.1043E+03	0.4410E+01	0.1165E+03	0.5258E+01
23	44.10	1190.00	771.99	1213.47	899.29	0.1380E+03	0.4281E+01	0.1162E+03	0.4965E+01
25	45.30	1380.00	777.10	1352.74	905.11	0.1369E+03	0.4178E+01	0.1159E+03	0.4732E+01
27	47.17	1573.34	784.63	1568.79	913.62	0.1036E+03	0.4030E+01	0.1156E+03	0.4398E+01
29	49.80	1770.00	795.12	1776.16	924.64	0.5882E+02	0.3836E+01	0.5846E+02	0.3980E+01
31	53.95	1950.00	810.54	1947.66	939.97	0.3265E+02	0.3570E+01	0.2939E+02	0.3426E+01
33	59.60	2115.00	829.65	2112.70	957.57	0.2656E+02	0.3270E+01	0.2907E+02	0.2830E+01
35	65.80	2260.00	849.24	2262.35	973.50	0.2000E+02	0.2994E+01	0.1891E+02	0.2330E+01
37	72.00	2350.00	867.09	2349.62	986.71	0.9375E+01	0.2767E+01	0.9557E+01	0.1945E+01

VII - 12

Connection type : Header plate connections
Mode : All bolted

Tested by : W.H.Sommer (1969) Canada
Test Id. : TEST 16

Column : --
Beam : W24X76
Plate thickness : 3/8"

Fasteners: A325-X-3/4"D
15/16" Oversize holes
Material : G40.12
Fy = 61.30 ksi
Fu = 75.90 ksi

Major parameters

lt = 1.5000" ct = 1.5000" p = 3.0000" cc = 1.5000"
lc = 4.3750" g = 5.5000" lp = 18.0000" tp = 0.3750"
tbw= 0.4410"
n = 2 X 6

Remark 1) Actual strength Fy = 51.5 ksi, Fu = 77.9 ksi for beam web.
2) No washers were used.

No	Moment (k-in)	Rotation (radians) X 1/1000
1	0.00	0.00
2	110.00	0.55
3	220.00	1.10
4	317.00	1.80
5	389.50	2.35
6	462.00	2.90
7	583.00	4.30
8	704.00	5.70
9	776.50	7.35
10	849.00	9.00
11	912.67	12.43
12	976.34	15.87
13	1040.00	19.30
14	1077.50	22.60
15	1115.00	25.90
16	1152.50	29.20
17	1190.00	32.50
18	1220.00	36.10
19	1250.00	39.70
20	1280.00	43.30
21	1306.67	46.03
22	1333.33	48.77
23	1360.00	51.50
24	1410.00	53.30
25	1480.00	54.70

No	Moment (k-in)	Rotation (radians) X 1/1000
26	1570.00	56.20
27	1645.00	57.20
28	1720.00	58.20
29	1795.00	59.15
30	1870.00	60.10
31	1940.00	61.20
32	2010.00	62.30
33	2085.00	63.45
34	2160.00	64.60
35	2255.00	66.50
36	2350.00	68.40
37	2445.00	71.15
38	2540.00	73.90
39	2595.00	77.75
40	2650.00	81.60

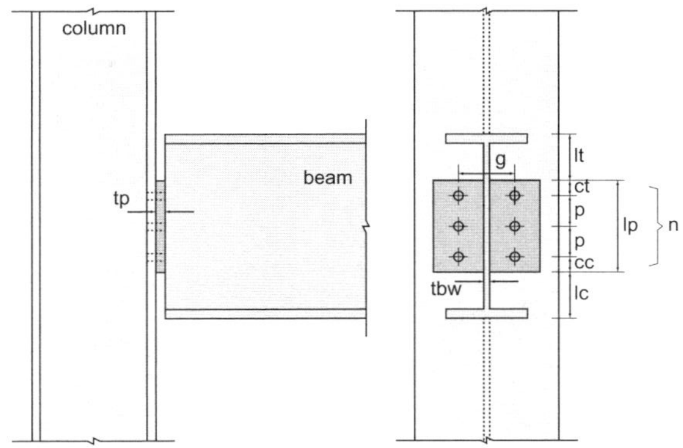

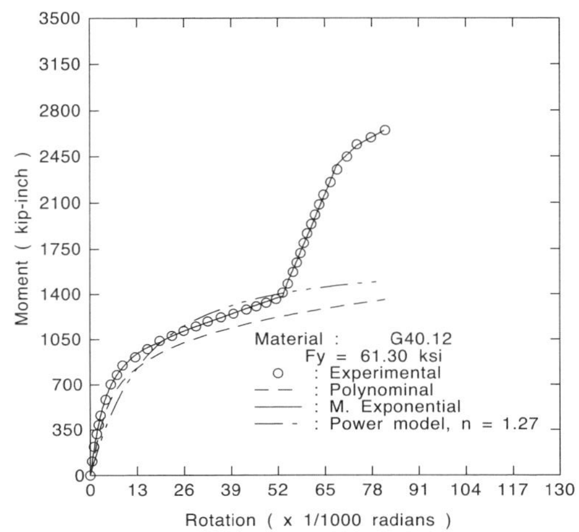

A.7 – 24

```
Moment-rotation prediction equations      ( R : X 1/1000 radians )

Frye and Morris polynominal model  :    R = Sum ( Ai X ( K*Bm )**Pi X 10**Qi )
    g =  5.500000"   d = 18.000000"  t  =  0.375000"   w =  0.441000"
   A1 =  5.100000   A2 =  6.200000   A3 =  2.400000    K =  0.143478
   P1 =         1   P2 =         3   P3 =         5   Q1 =        -2    Q2 =       -7    Q3 =      -10

Modified exponential model  :
   Bm = Sum ( Ai X ( 1 - exp( - R/( 2*i*C ) ) ) )  +     Sum ( RKj X ( R - Rj0 ) )  +  Bm0
     C =  0.81750000E+00          Bmo=  0.00000000E+00          Nexp=  6        Nliner=  5
    AI = -0.53478411E+03  0.71574712E+04 -0.29266073E+05  0.58584976E+05 -0.54163971E+05  0.19356713E+05
   Rj0 =          19.3000         53.3000         54.7000         68.4000         73.9000
   RKj =  0.79784484E+01  0.69763028E+02 -0.13527065E+02 -0.38029441E+02 -0.11188505E+02

Power model  :    Bm = ( Rki X R ) / ( 1 + ( R/R0 )**rn )**( 1/rn )
     rn = 1.269       rki =   0.97418188E+02         rmu =   0.16609164E+04
```

No	Rotation (radians) X 1/1000	Moment (kip-inch) Expri.	Poly.	M.Expo.	P.Model.	Connection stiffness (kip-inch) X 1000 Expri.	Poly.	M.Expo.	P.Model.
1	0.00	0.00	0.00	0.00	0.00	0.2000E+03	0.1367E+03	0.2007E+03	0.9742E+02
3	1.10	220.00	149.35	207.11	104.62	0.1730E+03	0.1338E+03	0.1723E+03	0.9226E+02
5	2.35	389.50	308.27	396.09	215.32	0.1318E+03	0.1177E+03	0.1303E+03	0.8477E+02
7	4.30	583.00	496.69	597.10	369.11	0.8643E+02	0.7626E+02	0.7993E+02	0.7311E+02
9	7.35	776.50	666.98	773.91	567.26	0.4394E+02	0.4132E+02	0.4232E+02	0.5743E+02
11	12.43	912.67	820.30	927.89	808.57	0.1854E+02	0.2270E+02	0.2179E+02	0.3894E+02
13	19.30	1040.00	942.82	1027.18	1020.87	0.1488E+02	0.1432E+02	0.8528E+01	0.2437E+02
15	25.90	1115.00	1024.13	1114.93	1153.44	0.1136E+02	0.1071E+02	0.1111E+02	0.1650E+02
17	32.50	1190.00	1087.43	1182.96	1245.34	0.9914E+01	0.8633E+01	0.9891E+01	0.1173E+02
19	39.70	1250.00	1143.96	1253.71	1317.12	0.8333E+01	0.7171E+01	0.9805E+01	0.8458E+01
21	46.03	1306.67	1186.37	1315.48	1364.27	0.9755E+01	0.6266E+01	0.9666E+01	0.6547E+01
23	51.50	1360.00	1218.91	1367.66	1396.64	0.2062E+02	0.5664E+01	0.9407E+01	0.5353E+01
25	54.70	1480.00	1236.80	1495.14	1412.85	0.5483E+02	0.5363E+01	0.7899E+02	0.4793E+01
27	57.20	1645.00	1249.75	1658.62	1424.35	0.7500E+02	0.5158E+01	0.6532E+02	0.4411E+01
29	59.15	1795.00	1259.73	1785.89	1432.68	0.7895E+02	0.5005E+01	0.6521E+02	0.4142E+01
31	61.20	1940.00	1269.92	1919.46	1440.91	0.6364E+02	0.4856E+01	0.6510E+02	0.3884E+01
33	63.45	2085.00	1280.75	2065.81	1449.35	0.6522E+02	0.4703E+01	0.6499E+02	0.3627E+01
35	66.50	2255.00	1294.66	2263.82	1459.93	0.5000E+02	0.4514E+01	0.6485E+02	0.3314E+01
37	71.15	2445.00	1314.93	2460.38	1474.36	0.3455E+02	0.4256E+01	0.2665E+02	0.2907E+01
39	77.75	2595.00	1341.95	2592.55	1491.95	0.1428E+02	0.3940E+01	0.1528E+02	0.2442E+01

VII - 13

Connection type : Header plate connections
Mode : All bolted

Tested by : W.H.Sommer (1969) Canada
Test Id. : TEST 17

Column : --
Beam : W24X76
Plate thickness : 1/4"

Fasteners: A325-X-3/4"D
15/16" Oversize holes
Material : G40.12
Fy = 49.30 ksi
Fu = 74.60 ksi

Major parameters

lt = 1.5000"	ct = 1.5000"	p = 3.0000"	cc = 1.5000"
lc = 10.3750"	g = 5.5000"	lp = 18.0000"	tp = 0.2500"
tbw= 0.4410"			
n = 2 X 4			

Remark 1) Actual strength Fy = 51.5 ksi, Fu = 77.9 ksi for beam web.
2) No washers were used.

No	Moment (k-in)	Rotation (radians) X 1/1000	No	Moment (k-in)	Rotation (radians) X 1/1000
1	0.00	0.00	26	653.00	31.80
2	37.20	2.00	27	689.25	33.17
3	74.40	4.10	28	725.50	34.55
4	98.70	5.15	29	761.75	35.92
5	123.00	6.20	30	798.00	37.30
6	147.00	9.10	31	830.33	39.13
7	171.00	12.00	32	862.66	40.97
8	183.67	14.30	33	895.00	42.80
9	196.34	16.60	34	927.00	44.70
10	209.00	18.90	35	959.00	46.60
11	223.50	21.55	36	991.00	48.50
12	238.00	24.20	37	1010.50	50.45
13	258.00	25.60	38	1030.00	52.40
14	296.00	27.10	39	1060.00	55.20
15	320.50	27.65	40	1075.00	56.75
16	345.00	28.20	41	1090.00	58.30
17	369.00	28.60	42	1051.00	58.67
18	393.00	29.00	43	1012.00	59.03
19	426.50	29.35	44	973.00	59.40
20	460.00	29.70	45	934.00	59.77
21	492.33	30.03	46	895.00	60.13
22	524.67	30.37	47	856.00	60.50
23	557.00	30.70			
24	589.00	31.07			
25	621.00	31.43			

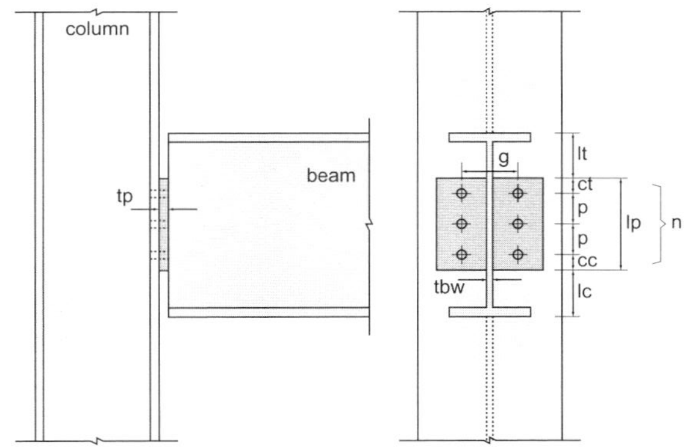

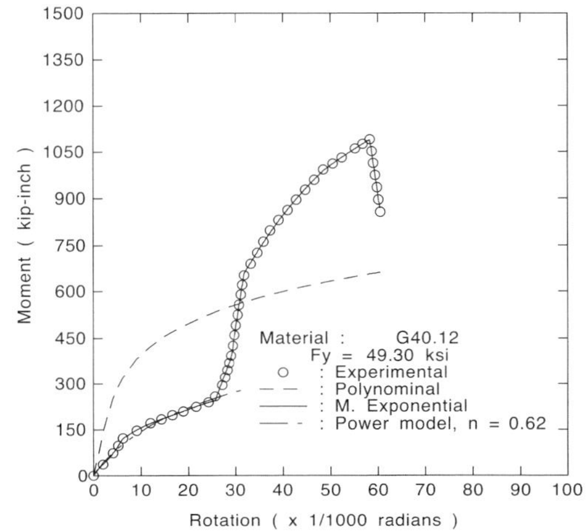

```
Moment-rotation prediction equations      ( R : X 1/1000 radians )

Frye and Morris polynominal model  :    R = Sum ( Ai X ( K*Bm )**Pi X 10**Qi )
      g =  5.500000"   d = 18.000000"  t  =  0.250000"   w =  0.441000"
     A1 =  5.100000   A2 =  6.200000  A3 =  2.400000    K =  0.274492
     P1 =        1    P2 =        3   P3 =        5    Q1 =       -2    Q2 =       -7    Q3 =      -10

Modified exponential model  :
     Bm = Sum ( Ai X ( 1 - exp( - R/( 2*i*C ) ) ) )  +      Sum ( RKj X ( R - Rj0 ) )  +  Bm0
       C =   0.10041667E+01           Bmo=   0.00000000E+00          Nexp=  6         Nliner=  6
      AI =   0.64660258E+03  -0.10511079E+05  0.54982919E+05  -0.12001572E+06  0.11227076E+06  -0.36641019E+05
     Rj0 =          6.2000          25.6000         29.0000          31.8000         48.5000
                   58.3000
     RKj =  -0.14695233E+02   0.38823165E+02  0.57270527E+02  -0.69213921E+02  -0.24490542E+01
            -0.11540635E+03

Power model  :    Bm = ( Rki X R ) / ( 1 + ( R/R0 )**rn )**( 1/rn )
      rn = 0.617       rki =   0.28864648E+02          rmu =   0.81562723E+03
```

No	Rotation (radians) X 1/1000	Moment (kip-inch) Expri.	Poly.	M.Expo.	P.Model.	Connection stiffness (kip-inch) X 1000 Expri.	Poly.	M.Expo.	P.Model.
1	0.00	0.00	0.00	0.00	0.00	0.1860E+02	0.7143E+02	0.3092E+02	0.2886E+02
3	4.10	74.40	251.46	74.62	76.98	0.2133E+02	0.4183E+02	0.2077E+02	0.1440E+02
5	6.20	123.00	321.25	122.54	104.67	0.1919E+02	0.2640E+02	0.2407E+02	0.1213E+02
7	12.00	171.00	423.53	170.07	163.44	0.6732E+01	0.1234E+02	0.6253E+01	0.8569E+01
9	16.60	196.34	470.99	196.05	198.92	0.5508E+01	0.8730E+01	0.5505E+01	0.6966E+01
11	21.55	223.50	508.71	223.97	230.31	0.5470E+01	0.6703E+01	0.5591E+01	0.5789E+01
13	25.60	258.00	533.62	244.37	252.26	0.1962E+02	0.5663E+01	0.4213E+01	0.5077E+01
15	27.65	320.50	544.83	331.35	262.35	0.4454E+02	0.5258E+01	0.4177E+02	0.4776E+01
17	28.60	369.00	549.79	370.71	266.83	0.6000E+02	0.5090E+01	0.4108E+02	0.4647E+01
19	29.35	426.50	553.58	421.34	270.28	0.9571E+02	0.4965E+01	0.9777E+02	0.4550E+01
21	30.03	492.33	556.95	487.96	273.36	0.9700E+02	0.4858E+01	0.9721E+02	0.4465E+01
23	30.70	557.00	560.18	552.58	276.31	0.9237E+02	0.4757E+01	0.9665E+02	0.4385E+01
25	31.43	621.00	563.65	623.23	279.49	0.8728E+02	0.4652E+01	0.9602E+02	0.4300E+01
27	33.17	689.25	571.41	693.97	286.81	0.2636E+02	0.4426E+01	0.2527E+02	0.4109E+01
29	35.92	761.75	583.14	760.07	297.73	0.2636E+02	0.4110E+01	0.2282E+02	0.3837E+01
31	39.13	830.33	595.81	828.88	309.58	0.1764E+02	0.3798E+01	0.2012E+02	0.3559E+01
33	42.80	895.00	609.17	897.61	322.12	0.1725E+02	0.3499E+01	0.1746E+02	0.3284E+01
35	46.60	959.00	621.96	959.59	334.11	0.1684E+02	0.3239E+01	0.1526E+02	0.3036E+01
37	50.45	1010.50	633.99	1010.26	345.37	0.1000E+02	0.3016E+01	0.1118E+02	0.2817E+01
39	55.20	1060.00	647.75	1059.98	358.18	0.1005E+02	0.2783E+01	0.9868E+01	0.2584E+01
41	58.30	1090.00	656.27	1089.70	365.98	-0.8416E+02	0.2649E+01	0.9340E+01	0.2449E+01
43	59.03	1012.00	658.21	1011.89	367.77	-0.1064E+03	0.2620E+01	-0.1062E+03	0.2419E+01
45	59.77	934.00	660.12	934.00	369.53	-0.1064E+03	0.2592E+01	-0.1062E+03	0.2389E+01
47	60.50	856.00	662.02	856.06	371.27	-0.1064E+03	0.2564E+01	-0.1063E+03	0.2361E+01

VII - 14

Connection type : Header plate connections
Mode : All bolted

Tested by : W.H.Sommer (1969) Canada
Test Id. : TEST 18

Column : --
Beam : W24X76
Plate thickness : 1/4"

Fasteners: A325-X-3/4"D
15/16" Oversize holes
Material : G40.12
Fy = 49.30 ksi
Fu = 74.60 ksi

Major parameters

lt = 1.5000" ct = 1.5000" p = 3.0000" cc = 1.5000"
lc = 7.3750" g = 5.5000" lp = 15.0000" tp = 0.2500"
tbw= 0.4410"
n = 2 X 5

Remark 1) Actual strength Fy = 51.5 ksi, Fu = 77.9 ksi for beam web.
2) No washers were used.

No	Moment (k-in)	Rotation (radians) X 1/1000	No	Moment (k-in)	Rotation (radians) X 1/1000
1	0.00	0.00	26	1290.00	56.30
2	74.40	2.80	27	1350.00	58.30
3	123.00	4.60	28	1410.00	60.75
4	171.00	6.70	29	1470.00	63.20
5	238.00	11.50	30	1530.00	65.30
6	277.00	15.90	31	1590.00	67.40
7	316.00	20.30	32	1650.00	70.60
8	340.00	23.70	33	1710.00	73.80
9	364.00	27.10	34	1725.00	77.67
10	393.00	29.60	35	1740.00	81.55
11	431.00	30.50	36	1755.00	85.42
12	460.00	31.10	37	1770.00	89.30
13	508.50	31.85	38	1730.00	94.00
14	557.00	32.60	39	1685.00	97.85
15	605.00	33.35	40	1640.00	101.70
16	653.00	34.10			
17	701.50	34.80			
18	750.00	35.50			
19	822.50	38.10			
20	895.00	40.70			
21	967.50	43.65			
22	1040.00	46.60			
23	1110.00	50.40			
24	1170.00	52.35			
25	1230.00	54.30			

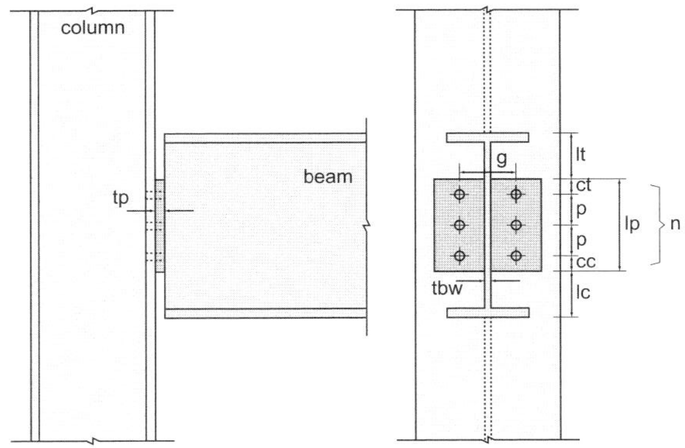

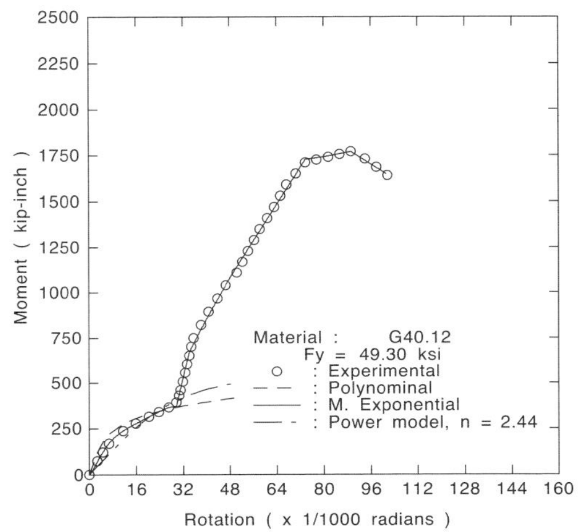

A.7 – 28

```
Moment-rotation prediction equations     ( R : X 1/1000 radians )

Frye and Mcrris polynominal model  :    R = Sum ( Ai X ( K*Bm )**Pi X 10**Qi )
      g =  5.500000"   d = 15.000000"  t  =  0.250000"   w =  0.441000"
     A1 =  5.100000   A2 =  6.200000   A3 =  2.400000    K =  0.417491
     P1 =         1   P2 =         3   P3 =         5   Q1 =        -2    Q2 =       -7    Q3 =      -10

Modified exponential model  :
     Bm = Sum ( Ai X ( 1 - exp( - R/( 2*i*C ) ) ) )  +      Sum ( RKj X ( R - Rj0 ) )  +  Bm0
      C =   0.15475000E+01            Bmo=   0.00000000E+00          Nexp=  6         Nliner=  5
     AI =  -0.69527770E+03   0.59813648E+04  -0.11571189E+05  -0.73209794E+04   0.34245539E+05  -0.20453469E+05
    Rj0 =          27.1000          29.6000          34.1000          73.8000          89.3000
    RKj =  -0.45031286E+01   0.67440657E+02  -0.33916964E+02  -0.23008050E+02  -0.13914523E+02

Power model  :     Bm = ( Rki X R ) / ( 1 + ( R/R0 )**rn )**( 1/rn )
     rn = 2.437       rki =   0.17104809E+02         rmu =   0.56640780E+03
```

No	Rotation (radians) X 1/1000	Moment (kip-inch) Expri.	Poly.	M.Expo.	P.Model.	Connection stiffness (kip-inch) X 1000 Expri.	Poly.	M.Expo.	P.Model.
1	0.00	0.00	0.00	0.00	0.00	0.2657E+02	0.4697E+02	0.1560E+02	0.1710E+02
3	4.60	123.00	178.29	124.01	78.42	0.2509E+02	0.2441E+02	0.2575E+02	0.1691E+02
5	11.50	238.00	274.34	237.64	190.88	0.1130E+02	0.8496E+01	0.1015E+02	0.1543E+02
7	20.30	316.00	328.84	313.57	311.45	0.7846E+01	0.4675E+01	0.8215E+01	0.1177E+02
9	27.10	364.00	356.28	365.46	380.91	0.9677E+01	0.3525E+01	0.6672E+01	0.8711E+01
11	30.50	431.00	367.64	431.09	408.18	0.4588E+02	0.3148E+01	0.6812E+02	0.7360E+01
13	31.85	508.50	371.85	522.60	417.79	0.6467E+02	0.3021E+01	0.6745E+02	0.6869E+01
15	33.35	605.00	376.23	623.20	427.70	0.6400E+02	0.2895E+01	0.6669E+02	0.6357E+01
17	34.80	701.50	380.38	695.62	436.58	0.6928E+02	0.2782E+01	0.3203E+02	0.5894E+01
19	38.10	822.50	389.12	798.54	454.44	0.2788E+02	0.2559E+01	0.3037E+02	0.4954E+01
21	43.65	967.50	402.46	960.15	478.28	0.2458E+02	0.2260E+01	0.2797E+02	0.3701E+01
23	50.40	1110.00	416.74	1141.59	499.37	0.2658E+02	0.1985E+01	0.2597E+02	0.2619E+01
25	54.30	1230.00	424.22	1241.43	508.66	0.3039E+02	0.1857E+01	0.2528E+02	0.2160E+01
27	58.30	1350.00	431.42	1341.68	516.52	0.2752E+02	0.1743E+01	0.2488E+02	0.1783E+01
29	63.20	1470.00	439.66	1463.07	524.33	0.2669E+02	0.1623E+01	0.2472E+02	0.1423E+01
31	67.40	1590.00	446.37	1567.00	529.78	0.2468E+02	0.1533E+01	0.2479E+02	0.1182E+01
33	73.80	1710.00	455.80	1726.70	536.40	0.1202E+02	0.1416E+01	0.2515E+02	0.9029E+00
35	81.55	1740.00	466.22	1745.59	542.42	0.3871E+01	0.1299E+01	0.2755E+01	0.6656E+00
37	89.30	1770.00	475.90	1769.49	546.91	-0.1724E+01	0.1201E+01	0.3409E+01	0.5012E+00
39	97.85	1685.00	485.84	1682.54	550.62	-0.1169E+02	0.1110E+01	-0.9852E+01	0.3747E+00

Connection type : Header plate connections
Mode : All bolted

Tested by : W.H.Sommer (1969) Canada
Test Id. : TEST 19

Column : --
Beam : W24X76
Plate thickness : 1/2"

Fasteners: A325-X-3/4"D
15/16" Oversize holes
Material : G40.12
Fy = 51.10 ksi
Fu = 76.10 ksi

Major parameters

lt = 1.5000" ct = 1.5000" p = 3.0000" cc = 1.5000"
lc = 10.3750" g = 5.5000" lp = 12.0000" tp = 0.5000"
tbw= 0.4410"
n = 2 X 4

Remark 1) Actual strength Fy = 51.5 ksi, Fu = 77.9 ksi for beam web.
2) No washers were used.

No	Moment (k-in)	Rotation (radians) X 1/1000	No	Moment (k-in)	Rotation (radians) X 1/1000
1	0.00	0.00	26	1430.00	50.70
2	124.00	2.20	27	1515.00	51.10
3	196.50	3.85	28	1600.00	51.50
4	269.00	5.50	29	1685.00	51.85
5	366.00	9.30	30	1770.00	52.20
6	439.00	12.00	31	1870.00	52.80
7	512.00	14.70	32	1970.00	53.40
8	536.25	18.20	33	2090.00	55.00
9	560.50	21.70	34	2175.00	56.70
10	584.75	25.20	35	2260.00	58.40
11	609.00	28.70	36	2323.33	61.40
12	633.00	31.90	37	2386.67	64.40
13	657.00	35.10	38	2450.00	67.40
14	669.00	37.60	39	2470.00	71.30
15	681.00	40.10	40	2490.00	75.20
16	693.50	42.90	41	2500.00	78.30
17	706.00	45.70	42	2510.00	81.40
18	752.00	46.90			
19	778.00	47.40			
20	851.00	47.90			
21	924.00	48.30			
22	997.00	48.70			
23	1118.50	49.30			
24	1240.00	49.90			
25	1335.00	50.30			

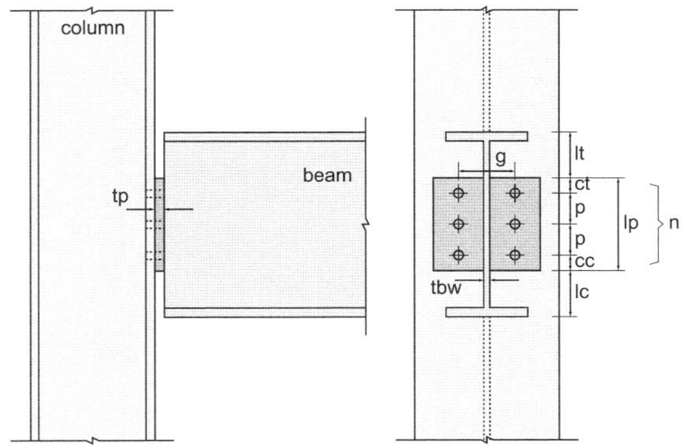

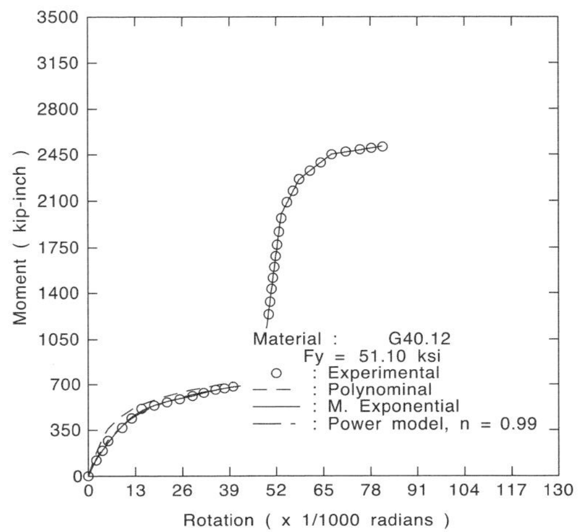

```
Moment-rotation prediction equations     ( R : X 1/1000 radians )

Frye and Morris polynominal model  :    R = Sum ( Ai X ( K*Bm )**Pi X 10**Qi )
     g =  5.500000"   d = 12.000000"  t  =  0.500000"   w =  0.441000"
    A1 =  5.100000   A2 =  6.200000   A3 =  2.400000    K =  0.230087
    P1 =         1   P2 =         3   P3 =         5   Q1 =       -2    Q2 =       -7    Q3 =      -10

Modified exponential model  :
    Bm = Sum ( Ai X ( 1 - exp( - R/( 2*i*C ) ) ) )  +      Sum ( RKj X ( R - Rj0 ) )  +  Bm0
      C =   0.12283333E+01            Bmo=   0.00000000E+00           Nexp=  6         Nliner=  6
     AI =   0.91769952E+03  -0.13475908E+05   0.70214852E+05  -0.16072370E+06   0.16648437E+06  -0.62974380E+05
    Rj0 =          28.7000          45.7000          47.9000          53.4000          58.4000
                   67.4000
    RKj =   0.33342312E+01   0.57812256E+02   0.15461777E+03  -0.15974602E+03  -0.30630493E+02
           -0.17234306E+02

Power model  :    Bm = ( Rki X R ) / ( 1 + ( R/R0 )**rn )**( 1/rn )
     rn = 0.989       rki =    0.73080426E+02          rmu =    0.88908172E+03
```

No	Rotation (radians) X 1/1000	Moment (kip-inch) Expri.	Poly.	M.Expo.	P.Model.	Connection stiffness (kip-inch) X 1000 Expri.	Poly.	M.Expo.	P.Model.
1	0.00	0.00	0.00	0.00	0.00	0.5636E+02	0.8522E+02	0.8341E+02	0.7308E+02
3	3.85	196.50	287.12	196.13	212.37	0.4394E+02	0.5302E+02	0.3824E+02	0.4177E+02
5	9.30	366.00	459.65	379.93	382.18	0.2641E+02	0.1958E+02	0.2794E+02	0.2326E+02
7	14.70	512.00	540.86	493.60	482.64	0.1828E+02	0.1182E+02	0.1490E+02	0.1489E+02
9	21.70	560.50	608.10	565.73	565.43	0.6928E+01	0.7941E+01	0.7326E+01	0.9400E+01
11	28.70	609.00	656.42	608.00	620.04	0.7227E+01	0.6054E+01	0.4981E+01	0.6476E+01
13	35.10	657.00	691.59	654.64	655.92	0.5984E+01	0.5010E+01	0.6150E+01	0.4853E+01
15	40.10	681.00	715.13	680.40	677.88	0.4641E+01	0.4431E+01	0.4141E+01	0.3976E+01
17	45.70	706.00	738.49	697.49	698.01	0.2818E+02	0.3933E+01	0.2021E+01	0.3249E+01
19	47.40	778.00	745.13	798.73	703.38	0.9900E+02	0.3804E+01	0.5928E+02	0.3068E+01
21	48.30	924.00	748.54	913.81	706.10	0.1825E+03	0.3740E+01	0.2136E+03	0.2978E+01
23	49.30	1118.50	752.26	1127.29	709.03	0.2025E+03	0.3672E+01	0.2133E+03	0.2883E+01
25	50.30	1335.00	755.92	1340.51	711.87	0.2375E+03	0.3606E+01	0.2131E+03	0.2792E+01
27	51.10	1515.00	758.79	1510.90	714.07	0.2125E+03	0.3555E+01	0.2129E+03	0.2723E+01
29	51.85	1685.00	761.45	1670.50	716.09	0.2429E+03	0.3509E+01	0.2127E+03	0.2660E+01
31	52.80	1870.00	764.77	1872.49	718.58	0.1667E+03	0.3452E+01	0.2125E+03	0.2583E+01
33	55.00	2090.00	772.09	2084.01	724.08	0.6288E+02	0.3331E+01	0.5240E+02	0.2418E+01
35	58.40	2260.00	783.26	2261.41	731.91	0.3955E+02	0.3156E+01	0.5199E+02	0.2193E+01
37	64.40	2386.67	801.26	2388.63	744.04	0.2111E+02	0.2897E+01	0.2113E+02	0.1865E+01
39	71.30	2470.00	820.36	2467.93	755.86	0.5128E+01	0.2649E+01	0.4171E+01	0.1572E+01
41	78.30	2500.00	838.30	2499.05	766.02	0.3226E+01	0.2440E+01	0.4753E+01	0.1340E+01

VII - 16

Connection type : Header plate connections
Mode : All bolted

Tested by : W.H.Sommer (1969) Canada
Test Id. : TEST 20

Column : --
Beam : W24X76
Plate thickness : 1/2"

Fasteners: A325-X-3/4"D
15/16" Oversize holes
Material : G40.12
Fy = 54.10 ksi
Fu = 77.50 ksi

Major parameters

lt = 1.5000"	ct = 1.5000"	p = 3.0000"	cc = 1.5000"
lc = 7.3750"	g = 5.5000"	lp = 15.0000"	tp = 0.5000"
tbw= 0.4410"			
n = 2 X 5			

Remark 1) Actual strength Fy = 51.5 ksi, Fu = 77.9 ksi for beam web.
2) No washers were used.

No	Moment (k-in)	Rotation (radians) X 1/1000	No	Moment (k-in)	Rotation (radians) X 1/1000
1	0.00	0.00	26	1180.00	48.90
2	61.50	0.45	27	1280.00	49.60
3	123.00	0.90	28	1380.00	50.30
4	220.00	1.50	29	1475.00	50.85
5	316.00	2.40	30	1570.00	51.40
6	388.50	3.10	31	1665.00	52.10
7	461.00	3.80	32	1760.00	52.80
8	533.50	4.95	33	1856.67	53.60
9	606.00	6.10	34	1953.33	54.40
10	678.00	8.10	35	2050.00	55.20
11	750.00	10.10	36	2115.00	56.30
12	798.33	13.13	37	2180.00	57.40
13	846.66	16.17	38	2202.50	60.10
14	895.00	19.20	39	2225.00	62.80
15	919.25	22.00	40	2247.50	65.50
16	943.50	24.80	41	2270.00	68.20
17	967.75	27.60			
18	992.00	30.40			
19	1016.00	33.80			
20	1040.00	37.20			
21	1053.34	39.70			
22	1066.67	42.20			
23	1080.00	44.70			
24	1100.00	47.50			
25	1130.00	48.40			

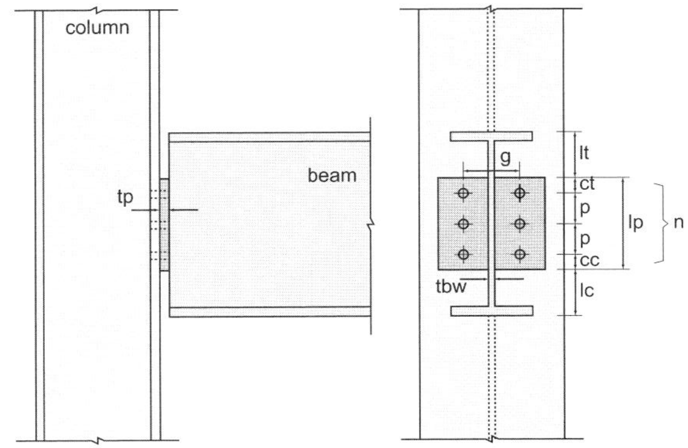

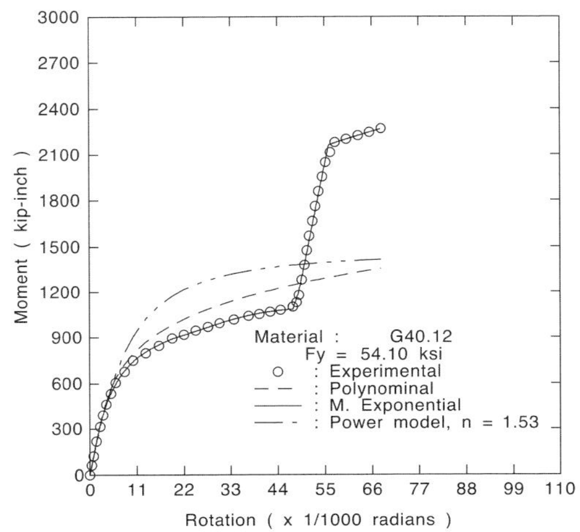

A.7 – 32

```
 Moment-rotation prediction equations     ( R : X 1/1000 radians )

Frye and Morris polynominal model  :     R = Sum ( Ai X ( K*Bm )**Pi X 10**Qi )
     g =  5.500000"   d = 15.000000"  t  =  0.500000"    w =  0.441000"
    A1 =  5.100000   A2 =  6.200000  A3 =  2.400000    K =  0.137721
    P1 =        1    P2 =        3   P3 =        5    Q1 =       -2     Q2 =       -7     Q3 =      -10

Modified exponential model  :
    Bm = Sum ( Ai X ( 1 - exp( - R/( 2*i*C ) ) ) )  +      Sum ( RKj X ( R - Rj0 ) )  +  Bm0
      C =   0.68083333E+00              Bmo=   0.00000000E+00          Nexp=  6          Nliner=  4
     AI =  -0.60325610E+03   0.77747393E+04  -0.42099725E+05   0.10714682E+06  -0.12084964E+06   0.49851587E+05
    Rj0 =          19.2000          47.5000          48.9000          56.0000
    RKj =  -0.29379514E+01   0.80848854E+02   0.54703670E+02  -0.12539968E+03

Power model  :     Bm = ( Rki X R ) / ( 1 + ( R/R0 )**rn )**( 1/rn )
     rn = 1.525      rki =   0.13683847E+03          rmu =   0.14707473E+04
```

No	Rotation (radians) X 1/1000	Moment (kip-inch) Expri.	Poly.	M.Expo.	P.Model.	Connection stiffness (kip-inch) X 1000 Expri.	Poly.	M.Expo.	P.Model.
1	0.00	0.00	0.00	0.00	0.00	0.1367E+03	0.1424E+03	0.1294E+03	0.1368E+03
3	0.90	123.00	127.60	126.51	121.35	0.1474E+03	0.1405E+03	0.1434E+03	0.1318E+03
5	2.40	316.00	327.28	321.52	308.23	0.1049E+03	0.1216E+03	0.1125E+03	0.1166E+03
7	3.80	461.00	475.28	455.78	460.21	0.8824E+02	0.8965E+02	0.8068E+02	0.1005E+03
9	6.10	606.00	635.02	600.25	662.84	0.5317E+02	0.5362E+02	0.4880E+02	0.7644E+02
11	10.10	750.00	792.97	741.88	904.41	0.2803E+02	0.2977E+02	0.2570E+02	0.4690E+02
13	16.17	846.66	931.17	852.06	1109.83	0.1593E+02	0.1788E+02	0.1338E+02	0.2397E+02
15	22.00	919.25	1019.87	915.27	1216.77	0.8661E+01	0.1309E+02	0.9015E+01	0.1389E+02
17	27.60	967.75	1085.33	966.60	1279.15	0.8661E+01	0.1050E+02	0.9165E+01	0.8888E+01
19	33.80	1016.00	1144.36	1019.71	1323.80	0.7059E+01	0.8670E+01	0.7664E+01	0.5810E+01
21	39.70	1053.34	1191.78	1057.64	1352.59	0.5333E+01	0.7470E+01	0.5114E+01	0.4086E+01
23	44.70	1080.00	1227.14	1077.62	1370.49	0.6187E+01	0.6704E+01	0.2915E+01	0.3130E+01
25	48.40	1130.00	1251.11	1158.50	1381.07	0.7618E+02	0.6239E+01	0.8236E+02	0.2611E+01
27	49.60	1280.00	1258.55	1295.38	1384.11	0.1429E+03	0.6102E+01	0.1367E+03	0.2468E+01
29	50.85	1475.00	1266.14	1465.96	1387.11	0.1727E+03	0.5966E+01	0.1363E+03	0.2331E+01
31	52.10	1665.00	1273.56	1636.07	1389.94	0.1357E+03	0.5837E+01	0.1359E+03	0.2203E+01
33	53.60	1856.67	1282.26	1839.63	1393.14	0.1208E+03	0.5690E+01	0.1355E+03	0.2063E+01
35	55.20	2050.00	1291.30	2056.14	1396.33	0.9484E+02	0.5542E+01	0.1351E+03	0.1926E+01
37	57.40	2180.00	1303.07	2177.36	1400.38	0.4440E+02	0.5356E+01	0.9281E+01	0.1758E+01
39	62.80	2225.00	1331.08	2225.08	1408.92	0.8333E+01	0.4944E+01	0.8455E+01	0.1423E+01
41	68.20	2270.00	1356.80	2269.23	1415.89	0.8333E+01	0.4599E+01	0.7940E+01	0.1170E+01

VII - 17

Connection type : Header plate connections
Mode : All bolted

Tested by : W.H.Sommer (1969) Canada
Test Id. : TEST 25

Column : --
Beam : W18X45
Plate thickness : 1/4"

Fasteners: A325-X-3/4"D
15/16" Oversize holes
Material : G40.12
Fy = 49.30 ksi
Fu = 74.60 ksi

Major parameters

lt = 1.5000" ct = 1.5000" p = 3.0000" cc = 1.5000"
lc = 4.3750" g = 4.0000" lp = 12.0000" tp = 0.2500"
tbw= 0.3350"
n = 2 X 4

Remark 1) Actual strength Fy = 50.0 ksi, Fu = 76.1 ksi for beam web.
2) No washers were used.

No	Moment (k-in)	Rotation (radians) X 1/1000
1	0.00	0.00
2	28.60	0.85
3	57.20	1.70
4	77.50	2.45
5	97.80	3.20
6	130.90	4.60
7	164.00	6.00
8	188.00	7.45
9	212.00	8.90
10	236.00	12.22
11	260.00	15.55
12	284.00	18.88
13	308.00	22.20
14	324.00	25.83
15	340.00	29.47
16	356.00	33.10
17	369.00	36.03
18	382.00	38.97
19	395.00	41.90
20	419.00	45.25
21	443.00	48.60
22	472.00	49.30
23	501.00	50.00
24	533.67	51.00
25	566.34	52.00
26	599.00	53.00
27	630.67	54.00
28	662.33	55.00
29	694.00	56.00
30	726.33	58.83
31	758.67	61.67
32	791.00	64.50
33	801.34	67.00
34	811.67	69.50
35	822.00	72.00
36	787.33	72.63
37	752.66	73.27
38	718.00	73.90

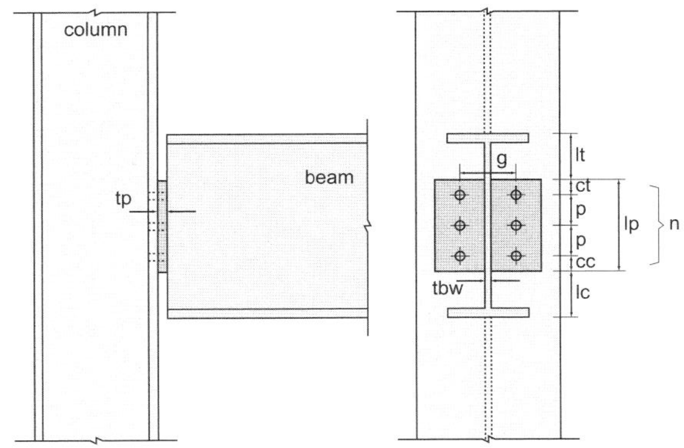

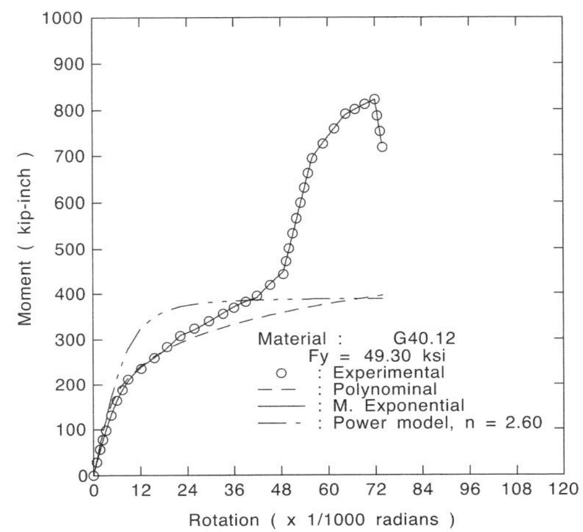

```
Moment-rotation prediction equations      ( R : X 1/1000 radians )

Frye and Morris polynominal model   :     R = Sum ( Ai X ( K*Bm )**Pi X 10**Qi )
       g =  4.000000"    d = 12.000000"  t  =  0.250000"    w =  0.335000"
      A1 =  5.100000    A2 =  6.200000   A3 =  2.400000     K =  0.480787
      P1 =         1    P2 =         3   P3 =         5    Q1 =        -2     Q2 =       -7     Q3 =      -10

Modified exponential model   :
      Bm = Sum ( Ai X ( 1 - exp( - R/( 2*i*C ) ) ) )  +      Sum ( RKj X ( R - Rj0 ) )  +  Bm0
       C =   0.82833333E+00                Bmo=   0.00000000E+00           Nexp=  6          Nliner=  6
      AI =   0.61227884E+02   0.54083011E+03  -0.14049795E+05   0.56916752E+05  -0.80472582E+05   0.37668683E+05
     Rj0 =          22.2000          41.9000          48.6000          56.0000          64.4000
                    72.0000
     RKj =  -0.65904242E+01   0.73764990E+01   0.27356308E+02  -0.20111445E+02  -0.53741093E+01
            -0.58531294E+02

Power model  :     Bm = ( Rki X R ) / ( 1 + ( R/R0 )**rn )**( 1/rn )
      rn = 2.597       rki =   0.38911183E+02          rmu =   0.38920165E+03
```

No	Rotation (radians) X 1/1000	Moment (kip-inch) Expri.	Poly.	M.Expo.	P.Model.	Connection stiffness (kip-inch) X 1000 Expri.	Poly.	M.Expo.	P.Model.
1	0.00	0.00	0.00	0.00	0.00	0.3365E+02	0.4078E+02	0.3692E+02	0.3891E+02
3	1.70	57.20	68.08	55.69	65.90	0.3015E+02	0.3826E+02	0.2994E+02	0.3838E+02
5	3.20	97.80	119.57	98.11	122.12	0.2587E+02	0.2961E+02	0.2665E+02	0.3628E+02
7	6.00	164.00	180.31	163.04	213.26	0.2016E+02	0.1567E+02	0.1947E+02	0.2809E+02
9	8.90	212.00	216.12	208.42	279.90	0.1372E+02	0.9859E+01	0.1214E+02	0.1809E+02
11	15.55	260.00	263.53	260.50	349.96	0.7218E+01	0.5332E+01	0.6069E+01	0.5428E+01
13	22.20	308.00	292.89	309.42	371.81	0.5873E+01	0.3716E+01	0.9117E+01	0.1875E+01
15	29.47	340.00	316.33	338.98	380.51	0.4404E+01	0.2826E+01	0.5033E+01	0.7358E+00
17	36.03	369.00	333.18	371.13	383.96	0.4432E+01	0.2340E+01	0.4370E+01	0.3686E+00
19	41.90	395.00	345.98	391.40	385.63	0.5707E+01	0.2037E+01	0.2424E+01	0.2176E+00
21	48.60	443.00	358.72	448.47	386.76	0.3550E+02	0.1781E+01	0.7255E+01	0.1290E+00
23	50.00	501.00	361.20	496.58	386.93	0.3782E+02	0.1735E+01	0.3411E+02	0.1166E+00
25	52.00	566.34	364.64	564.12	387.15	0.3267E+02	0.1675E+01	0.3344E+02	0.1015E+00
27	54.00	630.67	367.96	630.35	387.34	0.3166E+02	0.1619E+01	0.3281E+02	0.8876E-01
29	56.00	694.00	371.08	695.39	387.51	0.2638E+02	0.1569E+01	0.3223E+02	0.7800E-01
31	61.67	758.67	379.68	760.09	387.88	0.1141E+02	0.1439E+01	0.1078E+02	0.5533E-01
33	67.00	801.34	387.05	801.06	388.13	0.4133E+01	0.1338E+01	0.4499E+01	0.4115E-01
35	72.00	822.00	393.53	821.96	388.32	-0.4284E+02	0.1256E+01	0.3897E+01	0.3182E-01
37	73.27	752.66	395.11	752.68	388.36	-0.5474E+02	0.1237E+01	-0.5476E+02	0.2990E-01

VII - 18

Connection type : Header plate connections
Mode : All bolted

Tested by : W.H.Sommer (1969) Canada
Test Id. : TEST 26

Column : --
Beam : W18X45
Plate thickness : 1/4"

Fasteners: A325-X-3/4"D
15/16" Oversize holes
Material : G40.12
Fy = 49.30 ksi
Fu = 74.60 ksi

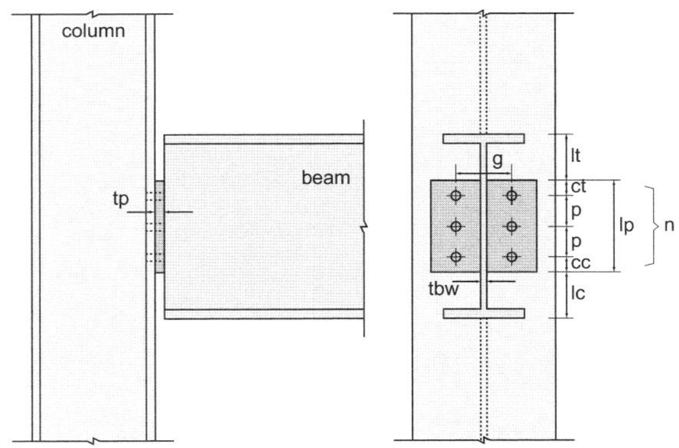

Major parameters

lt =	1.5000"	ct =	1.5000"	p =	3.0000"	cc =	1.5000"
lc =	7.3750"	g =	4.0000"	lp =	9.0000"	tp =	0.2500"
tbw=	0.3350"						
n =	2 X 3						

Remark 1) Actual strength Fy = 50.0 ksi, Fu = 76.1 ksi for beam web.
2) No washers were used.

No	Moment (k-in)	Rotation (radians) X 1/1000	No	Moment (k-in)	Rotation (radians) X 1/1000
1	0.00	0.00	26	463.00	29.60
2	16.95	0.60	27	485.50	31.43
3	33.90	1.20	28	508.00	33.27
4	50.85	1.85	29	530.50	35.10
5	67.80	2.50	30	553.00	36.93
6	91.40	4.80	31	575.50	38.77
7	115.00	7.10	32	598.00	40.60
8	124.60	9.92	33	612.00	42.65
9	134.20	12.74	34	626.00	44.70
10	143.80	15.56	35	639.00	47.73
11	153.40	18.38	36	652.00	50.77
12	163.00	21.20	37	665.00	53.80
13	188.00	23.90	38	661.00	56.37
14	205.67	24.67	39	657.00	58.93
15	223.33	25.43	40	653.00	61.50
16	241.00	26.20	41	631.17	61.67
17	265.00	26.60	42	609.34	61.83
18	289.00	27.00	43	587.50	62.00
19	311.33	27.30	44	565.67	62.17
20	333.67	27.60	45	543.83	62.33
21	356.00	27.90	46	522.00	62.50
22	377.40	28.24			
23	398.80	28.58			
24	420.20	28.92			
25	441.60	29.26			

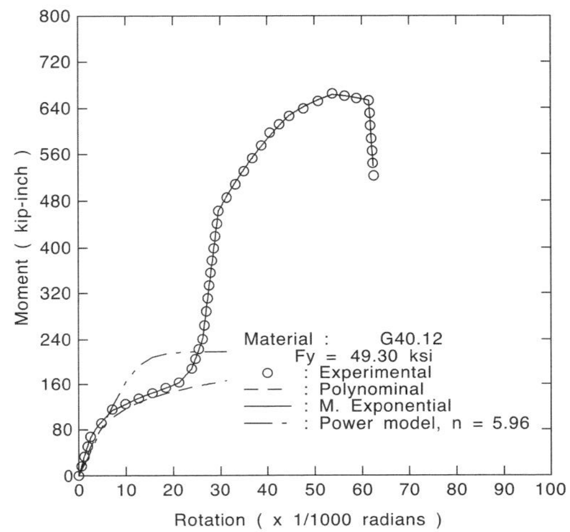

A.7 – 36

```
Moment-rotation prediction equations      ( R : X 1/1000 radians )

Frye and Morris polynominal model  :    R = Sum ( Ai X ( K*Bm )**Pi X 10**Qi )
      g =  4.000000"   d =  9.000000"  t  =  0.250000"   w =  0.335000"
     A1 =  5.100000   A2 =  6.200000  A3 =  2.400000    K =  0.931777
     P1 =         1   P2 =         3  P3 =         5   Q1 =       -2    Q2 =       -7     Q3 =      -10

Modified exponential model  :
     Bm = Sum ( Ai X ( 1 - exp( - R/( 2*i*C ) ) ) )  +      Sum ( RKj X ( R - Rj0 ) )  +  Bm0
      C =   0.67083333E+00              Bmo=   0.00000000E+00          Nexp=  6        Nliner=  8
     AI =  -0.39740207E+03   0.67064618E+04  -0.37889702E+05   0.95407775E+05  -0.10789154E+06   0.44587137E+05
    Rj0 =           7.1000          21.2000          23.9000          26.2000          29.6000
                   44.7000          53.8000          61.5000
    RKj =  -0.63984496E+01   0.52996691E+01   0.12943555E+02   0.42948403E+02  -0.52021011E+02
           -0.24977707E+01  -0.33018806E+01  -0.12902662E+03

Power model  :    Bm = ( Rki X R ) / ( 1 + ( R/R0 )**rn )**( 1/rn )
      rn = 5.958      rki =   0.16988119E+02         rmu =   0.21892593E+03
```

No	Rotation (radians) X 1/1000	Moment (kip-inch) Expri.	Poly.	M.Expo.	P.Model.	Connection stiffness (kip-inch) X 1000 Expri.	Poly.	M.Expo.	P.Model.
1	0.00	0.00	0.00	0.00	0.00	0.2825E+02	0.2104E+02	0.2292E+02	0.1699E+02
3	1.20	33.90	25.05	34.67	20.39	0.2721E+02	0.2049E+02	0.2883E+02	0.1699E+02
5	2.50	67.80	50.15	65.39	42.47	0.2259E+02	0.1766E+02	0.1823E+02	0.1699E+02
7	7.10	115.00	101.08	113.75	120.04	0.7180E+01	0.6628E+01	0.9321E+01	0.1644E+02
9	12.74	134.20	127.41	135.02	193.75	0.3404E+01	0.3401E+01	0.3586E+01	0.7864E+01
11	18.38	153.40	143.09	152.97	214.78	0.3404E+01	0.2316E+01	0.3219E+01	0.1257E+01
13	23.90	188.00	154.27	188.20	218.01	0.1999E+02	0.1784E+01	0.9641E+01	0.2244E+00
15	25.43	223.33	156.96	222.97	218.29	0.2304E+02	0.1677E+01	0.2275E+02	0.1469E+00
17	26.60	265.00	158.84	266.72	218.44	0.6000E+02	0.1608E+01	0.6573E+02	0.1080E+00
19	27.30	311.33	159.96	312.73	218.51	0.7445E+02	0.1568E+01	0.6571E+02	0.9035E-01
21	27.90	356.00	160.90	352.14	218.56	0.6905E+02	0.1535E+01	0.6567E+02	0.7779E-01
23	28.58	398.80	161.93	396.78	218.61	0.6294E+02	0.1500E+01	0.6560E+02	0.6589E-01
25	29.26	441.60	162.95	441.35	218.65	0.6294E+02	0.1467E+01	0.6550E+02	0.5602E-01
27	31.43	485.50	166.00	487.86	218.75	0.1227E+02	0.1372E+01	0.1300E+02	0.3413E-01
29	35.10	530.50	170.78	533.40	218.83	0.1227E+02	0.1237E+01	0.1177E+02	0.1588E-01
31	38.77	575.50	175.11	573.82	218.87	0.1227E+02	0.1129E+01	0.1026E+02	0.7966E-02
33	42.65	612.00	179.30	610.50	218.90	0.6831E+01	0.1034E+01	0.8647E+01	0.4102E-02
35	47.73	639.00	184.29	642.03	218.91	0.4286E+01	0.9338E+00	0.4311E+01	0.1875E-02
37	53.80	665.00	189.66	662.95	218.92	0.1120E+01	0.8388E+00	0.2691E+01	0.8157E-03
39	58.93	657.00	193.79	657.31	218.92	-0.1558E+01	0.7735E+00	-0.1526E+01	0.4327E-03
41	61.67	631.17	195.90	631.13	218.92	-0.1310E+03	0.7426E+00	-0.1309E+03	0.3156E-03
43	62.00	587.50	196.15	587.49	218.92	-0.1310E+03	0.7391E+00	-0.1309E+03	0.3040E-03
45	62.33	543.83	196.40	543.85	218.92	-0.1310E+03	0.7356E+00	-0.1310E+03	0.2929E-03

VII - 19

Connection type : Header plate connections
Mode : All bolted

Tested by : W.H.Sommer (1969) Canada
Test Id. : TEST 27

Column : --
Beam : W12X27
Plate thickness : 1/4"

Fasteners: A325-X-3/4"D
15/16" Oversize holes
Material : G40.12
Fy = 49.30 ksi
Fu = 74.60 ksi

Major parameters

lt = 1.5000" ct = 1.5000" p = 3.0000" cc = 1.5000"
lc = 1.5000" g = 4.0000" lp = 9.0000" tp = 0.2500"
tbw= 0.2390"
n = 2 X 3

Remark 1) Actual strength Fy = 48.4 ksi, Fu = 74.0 ksi for beam web.
2) No washers were used.

No	Moment (k-in)	Rotation (radians) X 1/1000	No	Moment (k-in)	Rotation (radians) X 1/1000
1	0.00	0.00	26	279.00	47.25
2	10.33	0.63	27	291.00	48.77
3	20.66	1.27	28	303.00	50.30
4	31.00	1.90	29	311.00	54.17
5	41.33	2.57	30	319.00	58.03
6	51.67	3.23	31	327.00	61.90
7	62.00	3.90	32	332.00	65.90
8	74.00	5.10	33	337.00	69.90
9	86.00	6.30	34	342.00	74.65
10	98.00	7.50	35	347.00	79.40
11	110.00	8.70	36	349.25	83.22
12	123.00	11.63	37	351.50	87.05
13	136.00	14.57	38	353.75	90.88
14	149.00	17.50	39	356.00	94.70
15	155.34	21.47	40	344.00	96.06
16	161.67	25.43	41	332.00	97.42
17	168.00	29.40	42	320.00	98.78
18	181.00	31.60	43	308.00	100.14
19	194.00	33.80	44	296.00	101.50
20	207.00	36.00			
21	219.00	38.05			
22	231.00	40.10			
23	243.00	42.15			
24	255.00	44.20			
25	267.00	45.73			

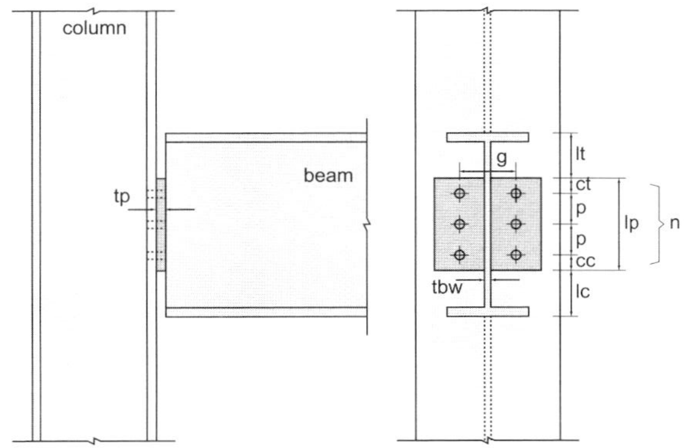

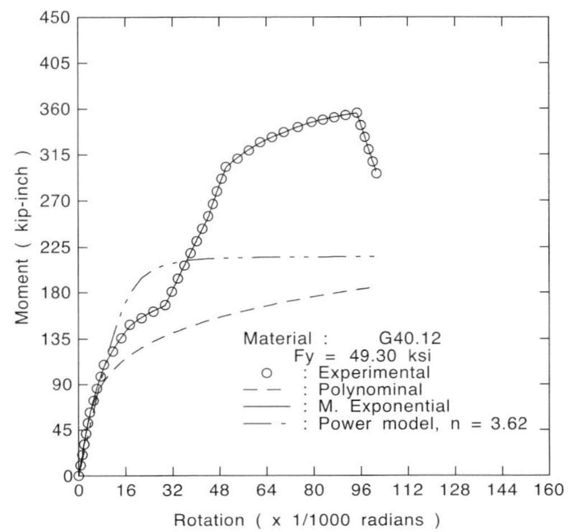

```
Moment-rotation prediction equations     ( R : X 1/1000 radians )

Frye and Morris polynominal model  :    R = Sum ( Ai X ( K*Bm )**Pi X 10**Qi )
      g =  4.000000"   d =  9.000000"  t  =  0.250000"   w =  0.239000"
     A1 =  5.100000   A2 =  6.200000  A3 =  2.400000    K =  1.103152
     P1 =        1    P2 =        3   P3 =        5    Q1 =       -2     Q2 =       -7     Q3 =      -10

Modified exponential model  :
     Bm = Sum ( Ai X ( 1 - exp( - R/( 2*i*C ) ) ) )  +      Sum ( RKj X ( R - Rj0 ) )  +  Bm0
      C =   0.10041583E+01             Bmo=   0.00000000E+00          Nexp=  6         Nliner=  7
     AI =  -0.13199049E+03   0.21143055E+04  -0.12495207E+05   0.33105212E+05  -0.38978570E+05   0.16736052E+05
    Rj0 =           8.7000          17.5000          29.4000          44.2000          50.3000
                   69.9000          94.7000
    RKj =  -0.19701924E+01  -0.86094578E+00   0.50023999E+01   0.25774885E+01  -0.49685056E+01
            0.36808222E+00  -0.91162514E+01

Power model  :    Bm = ( Rki X R ) / ( 1 + ( R/R0 )**rn )**( 1/rn )
      rn = 3.620      rki =   0.12471913E+02         rmu =   0.21550402E+03
```

No	Rotation (radians) X 1/1000	Moment (kip-inch) Expri.	Poly.	M.Expo.	P.Model.	Connection stiffness (kip-inch) X 1000 Expri.	Poly.	M.Expo.	P.Model.
1	0.00	0.00	0.00	0.00	0.00	0.1631E+02	0.1777E+02	0.1495E+02	0.1247E+02
3	1.27	20.66	22.31	20.74	15.80	0.1631E+02	0.1724E+02	0.1684E+02	0.1247E+02
5	2.57	41.33	43.35	41.76	32.00	0.1550E+02	0.1473E+02	0.1523E+02	0.1246E+02
7	3.90	62.00	60.44	60.50	48.58	0.1354E+02	0.1092E+02	0.1288E+02	0.1240E+02
9	6.30	86.00	80.59	87.20	78.02	0.1000E+02	0.6444E+01	0.9658E+01	0.1207E+02
11	8.70	110.00	93.32	108.09	106.13	0.8383E+01	0.4412E+01	0.7941E+01	0.1126E+02
13	14.57	136.00	112.48	136.80	161.28	0.4432E+01	0.2489E+01	0.4036E+01	0.7195E+01
15	21.47	155.34	126.45	155.98	194.27	0.1596E+01	0.1674E+01	0.1787E+01	0.2834E+01
17	29.40	168.00	137.79	167.19	207.54	0.4370E+01	0.1234E+01	0.1176E+01	0.8993E+00
19	33.80	194.00	142.86	194.07	210.54	0.5909E+01	0.1082E+01	0.6042E+01	0.5044E+00
21	38.05	219.00	147.22	219.43	212.21	0.5854E+01	0.9697E+00	0.5882E+01	0.3027E+00
23	42.15	243.00	151.01	243.11	213.20	0.5854E+01	0.8829E+00	0.5654E+01	0.1928E+00
25	45.73	267.00	154.07	266.80	213.78	0.7869E+01	0.8196E+00	0.7971E+01	0.1340E+00
27	48.77	291.00	156.49	290.73	214.13	0.7869E+01	0.7732E+00	0.7714E+01	0.1002E+00
29	54.17	311.00	160.45	311.79	214.56	0.2069E+01	0.7042E+00	0.2253E+01	0.6229E-01
31	61.90	327.00	165.58	326.49	214.92	0.1666E+01	0.6256E+00	0.1562E+01	0.3388E-01
33	69.90	337.00	170.32	336.51	215.13	0.1160E+01	0.5622E+00	0.9679E+00	0.1941E-01
35	79.40	347.00	175.37	346.65	215.27	0.7958E+00	0.5031E+00	0.8354E+00	0.1080E-01
37	87.05	351.50	179.07	351.97	215.33	0.5882E+00	0.4646E+00	0.5735E+00	0.7074E-02
39	94.70	356.00	182.50	355.66	215.38	-0.6355E+01	0.4321E+00	0.4044E+00	0.4798E-02
41	97.42	332.00	183.67	331.90	215.39	-0.8824E+01	0.4217E+00	-0.8755E+01	0.4210E-02
43	100.14	308.00	184.81	308.04	215.40	-0.8824E+01	0.4118E+00	-0.8792E+01	0.3708E-02

VII - 20

Connection type : Header plate connections
Mode : All bolted

Tested by : W.H.Sommer (1969) Canada
Test Id. : TEST 28

Column : --
Beam : W12X27
Plate thickness : 1/4"

Fasteners: A325-X-3/4"D
15/16" Oversize holes
Material : G40.12
Fy = 49.30 ksi
Fu = 74.60 ksi

Major parameters

lt = 1.5000"   ct = 1.5000"   p = 3.0000"   cc = 1.5000"
lc = 4.5000"   g = 4.0000"   lp = 6.0000"   tp = 0.2500"
tbw= 0.2390"
n = 2 X 2

Remark 1) Actual strength Fy = 48.4 ksi, Fu = 74.0 ksi for beam web.
2) No washers were used.

No	Moment (k-in)	Rotation (radians) X 1/1000	No	Moment (k-in)	Rotation (radians) X 1/1000
1	0.00	0.00	26	332.00	88.30
2	13.70	2.50	27	341.67	92.80
3	24.55	7.60	28	351.33	97.30
4	35.40	12.70	29	361.00	101.80
5	46.25	17.80	30	370.75	106.53
6	57.10	22.90	31	380.50	111.25
7	62.90	27.05	32	390.25	115.97
8	68.70	31.20	33	400.00	120.70
9	76.40	37.00	34	400.00	125.00
10	88.70	38.60			
11	101.00	40.20			
12	120.00	41.45			
13	139.00	42.70			
14	155.00	43.73			
15	171.00	44.77			
16	187.00	45.80			
17	203.33	48.83			
18	219.67	51.87			
19	236.00	54.90			
20	250.50	58.75			
21	265.00	62.60			
22	279.50	68.15			
23	294.00	73.70			
24	313.00	77.70			
25	322.50	83.00			

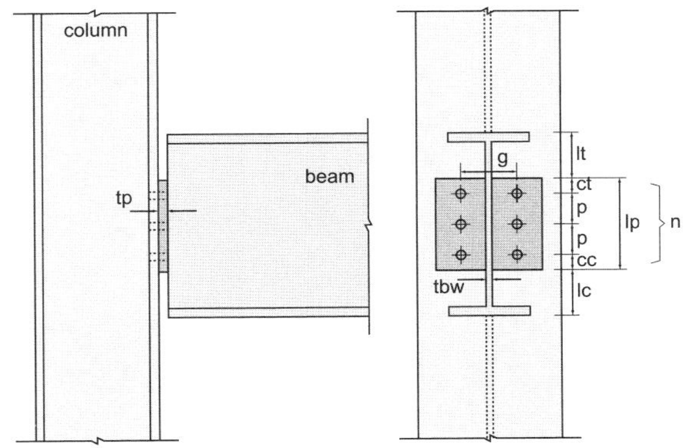

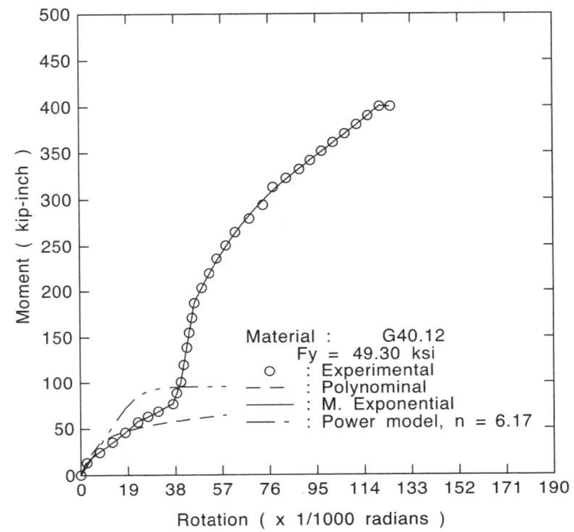

A.7 – 40

```
Moment-rotation prediction equations     ( R : X 1/1000 radians )

Frye and Morris polynominal model   :    R = Sum ( Ai X ( K*Bm )**Pi X 10**Qi )
      g =  4.000000"    d =  6.000000"  t  =  0.250000"    w =  0.239000"
     A1 =  5.100000    A2 =  6.200000   A3 =  2.400000     K =  2.803142
     P1 =         1    P2 =         3   P3 =         5    Q1 =        -2    Q2 =        -7    Q3 =      -10

Modified exponential model  :
     Bm = Sum  ( Ai X ( 1 - exp( - R/( 2*i*C ) ) ) )  +     Sum ( RKj X ( R - Rj0 ) )  +  Bm0
      C =    0.16666667E+01              Bmo=   0.00000000E+00           Nexp=  6         Nliner=  7
     AI =    0.14372681E+03  -0.25211128E+04   0.15935685E+05  -0.39886547E+05   0.41910023E+05  -0.15465960E+05
    Rj0 =           22.9000          37.0000          40.2000          45.8000          62.6000
                    77.7000         120.7000
    RKj =   -0.15481755E+01   0.60661482E+01   0.85276682E+01  -0.99331136E+01  -0.26941252E+00
            -0.22780713E+00  -0.24252112E+01

Power model  :     Bm = ( Rki X R ) / ( 1 + ( R/R0 )**rn )**( 1/rn )
     rn = 6.165        rki =    0.41288342E+01          rmu =    0.95779564E+02
```

No	Rotation (radians) X 1/1000	Moment (kip-inch) Expri.	Moment (kip-inch) Poly.	Moment (kip-inch) M.Expo.	Moment (kip-inch) P.Model.	Connection stiffness (kip-inch) X 1000 Expri.	Connection stiffness (kip-inch) X 1000 Poly.	Connection stiffness (kip-inch) X 1000 M.Expo.	Connection stiffness (kip-inch) X 1000 P.Model.
1	0.00	0.00	0.00	0.00	0.00	0.5482E+01	0.6995E+01	0.8332E+01	0.4129E+01
3	7.60	24.55	34.67	25.11	31.37	0.2127E+01	0.2031E+01	0.1920E+01	0.4124E+01
5	17.80	46.25	47.11	45.74	71.40	0.2127E+01	0.7957E+00	0.2216E+01	0.3356E+01
7	27.05	62.90	53.04	62.68	90.82	0.1398E+01	0.5259E+00	0.1284E+01	0.9383E+00
9	37.00	76.40	57.53	76.72	94.93	0.6313E+01	0.3916E+00	0.1381E+01	0.1366E+00
11	40.20	101.00	58.74	100.31	95.27	0.1190E+02	0.3628E+00	0.7283E+01	0.7730E-01
13	42.70	139.00	59.63	139.62	95.42	0.1535E+02	0.3431E+00	0.1563E+02	0.5076E-01
15	44.77	171.00	60.32	171.74	95.51	0.1548E+02	0.3289E+00	0.1545E+02	0.3642E-01
17	48.83	203.33	61.60	203.67	95.62	0.5385E+01	0.3041E+00	0.5126E+01	0.1970E-01
19	54.90	236.00	63.35	232.86	95.70	0.4672E+01	0.2738E+00	0.4495E+01	0.8563E-02
21	62.60	265.00	65.34	264.54	95.75	0.3293E+01	0.2438E+00	0.3759E+01	0.3354E-02
23	73.70	294.00	67.85	298.74	95.77	0.3854E+01	0.2113E+00	0.2740E+01	0.1043E-02
25	83.00	322.50	69.72	321.30	95.77	0.1793E+01	0.1905E+00	0.2184E+01	0.4454E-03
27	92.80	341.67	71.49	341.93	95.78	0.2148E+01	0.1731E+00	0.2057E+01	0.2002E-03
29	101.80	361.00	72.99	360.41	95.78	0.2107E+01	0.1598E+00	0.2063E+01	0.1032E-03
31	111.25	380.50	74.44	380.20	95.78	0.2063E+01	0.1482E+00	0.2132E+01	0.5461E-04
33	120.70	400.00	75.80	400.77	95.78	0.9832E+00	0.1382E+00	0.2223E+01	0.3045E-04

VII - 21

Connection type : Header plate connections
Mode : All bolted

Tested by : J.B.Davison et al (1987) United Kingdom
Test Id. : JT/14

Column : UC 152x152x23
Beam : UB 254x102x22
Plate thickness : 12 mm

Fasteners: 10.9- -M16
23/32" Oversize holes
Material : 43A
Fy = -- ksi
Fu = -- ksi

Major parameters

lt = -- " ct = 2.1654" p = 3.9370" cc = 2.3622"
lc = -- " g = 2.9920" lp = 10.4331" tp = 0.4724"
tbw= 0.2284"
n = 2 X 3

Remark 1) Flush end plate welded to the web of the beam
2)

No	Moment (k-in)	Rotation (radians) X 1/1000
1	0.00	0.00
2	12.29	0.09
3	24.58	0.19
4	36.87	0.28
5	49.16	0.32
6	61.45	0.35
7	73.75	0.38
8	86.04	0.51
9	98.33	0.64
10	110.62	0.77
11	122.91	0.83
12	135.20	0.90
13	147.49	0.96
14	159.78	1.04
15	172.07	1.12
16	184.37	1.19
17	196.66	1.44
18	208.95	1.68
19	221.24	1.92
20	235.99	2.40
21	250.74	2.88
22	261.80	3.37
23	272.86	3.85
24	283.92	4.81
25	293.14	5.29

No	Moment (k-in)	Rotation (radians) X 1/1000
26	302.36	5.77
27	311.58	6.25
28	320.80	6.73
29	331.86	7.69
30	341.07	8.65
31	350.29	9.62
32	354.00	10.58
33	357.70	11.54
34	361.67	12.50
35	365.64	13.46
36	368.73	14.42
37	368.73	15.20
38	367.62	15.77
39	366.52	16.35
40	354.00	17.31
41	350.30	18.27
42	346.61	19.23
43	345.13	20.18
44	345.13	21.16
45	331.86	21.72
46	317.11	22.11

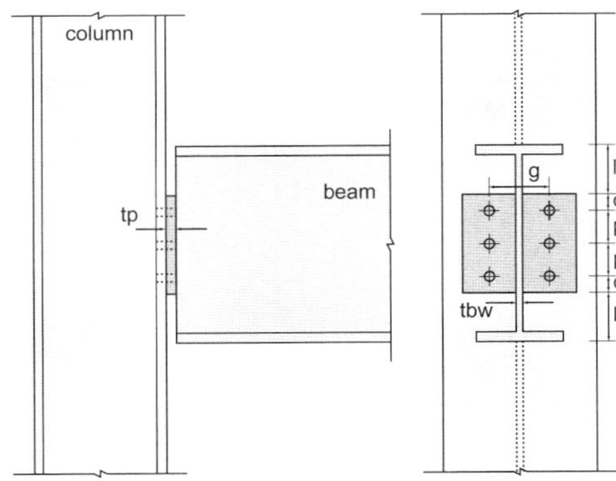

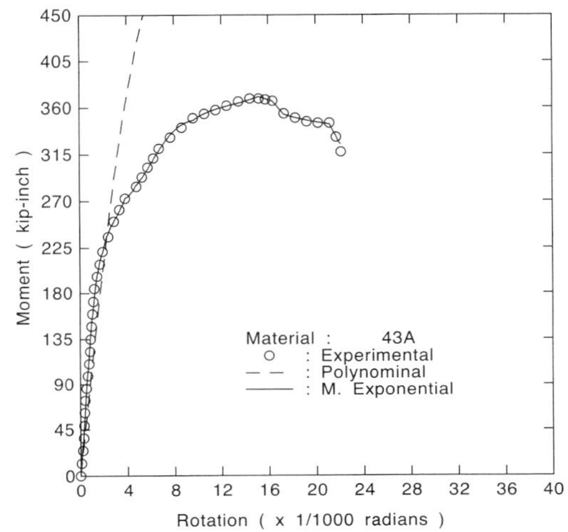

A.7 – 42

```
Moment-rotation prediction equations      ( R : X 1/1000 radians )

Frye and Morris polynominal model   :    R = Sum ( Ai X ( K*Bm )**Pi X 10**Qi )
      g =  2.992000"    d = 10.433100"  t  =  0.472400"    w =  0.228400"
     A1 =  5.100000    A2 =  6.200000   A3 =  2.400000     K =  0.182366
     P1 =         1    P2 =         3   P3 =         5    Q1 =        -2     Q2 =        -7     Q3 =       -10

Modified exponential model   :
     Bm = Sum ( Ai X ( 1 - exp( - R/( 2*i*C ) ) ) )  +      Sum ( RKj X ( R - Rj0 ) )  +  Bm0
      C =   0.20788000E+00              Bmo=  0.00000000E+00           Nexp=  6          Nliner= 10
     AI =   0.21831906E+03  -0.72401627E+04   0.41315164E+05  -0.89287323E+05   0.84289152E+05  -0.29110667E+05
    Rj0 =           0.0000           1.9200           3.8500           4.8100           9.6200
                   15.2000          16.3500          17.3100          21.1600          22.1100
    RKj =   0.12609394E+02   0.22735295E+02  -0.23973084E+02  -0.87482334E+01   0.65473095E+01
           -0.84818432E+01  -0.10676744E+02   0.95223851E+01  -0.20322708E+02  -0.17592468E+04
```

No	Rotation (radians) X 1/1000	Moment (kip-inch) Expri.	Moment (kip-inch) Poly.	Moment (kip-inch) M.Expo.	Connection stiffness (kip-inch) X 1000 Expri.	Connection stiffness (kip-inch) X 1000 Poly.	Connection stiffness (kip-inch) X 1000 M.Expo.
1	0.00	0.00	0.00	0.00	0.1302E+03	0.1075E+03	0.1303E+03
3	0.19	24.58	20.29	28.65	0.1303E+03	0.1075E+03	0.1592E+03
5	0.32	49.16	34.06	49.45	0.3631E+03	0.1074E+03	0.1644E+03
7	0.38	73.75	41.33	60.60	0.3078E+03	0.1073E+03	0.1651E+03
9	0.64	98.33	68.81	102.14	0.9587E+02	0.1068E+03	0.1562E+03
11	0.83	122.91	89.34	130.76	0.1917E+03	0.1063E+03	0.1405E+03
13	0.96	147.49	102.95	147.95	0.1772E+03	0.1058E+03	0.1275E+03
15	1.12	172.07	119.21	166.26	0.1598E+03	0.1052E+03	0.1104E+03
17	1.44	196.66	152.51	195.81	0.5046E+02	0.1032E+03	0.7442E+02
19	1.92	221.24	201.70	220.69	0.4380E+02	0.9842E+02	0.5314E+02
21	2.88	250.74	289.70	250.72	0.2685E+02	0.8367E+02	0.2013E+02
23	3.85	272.86	362.00	272.87	0.1918E+02	0.6696E+02	0.2929E+02
25	5.29	293.14	443.73	292.57	0.1918E+02	0.4783E+02	0.1641E+02
27	6.25	311.58	485.43	310.32	0.1918E+02	0.3937E+02	0.1915E+02
29	7.69	331.86	535.59	334.27	0.1054E+02	0.3081E+02	0.1286E+02
31	9.62	350.29	587.51	347.93	0.6721E+01	0.2379E+02	0.2004E+01
33	11.54	357.70	628.73	358.46	0.3989E+01	0.1939E+02	0.3449E+01
35	13.46	365.64	662.96	364.18	0.3671E+01	0.1639E+02	0.3029E+01
37	15.20	368.73	689.71	370.44	-0.1119E+01	0.1440E+02	-0.4183E+01
39	16.35	366.52	705.53	366.23	-0.6065E+01	0.1335E+02	-0.3185E+01
41	18.27	350.30	729.77	350.22	-0.3843E+01	0.1191E+02	-0.2877E+01
43	20.18	345.13	751.46	345.75	-0.7821E+00	0.1077E+02	-0.1862E+01
45	21.72	331.86	767.51	331.86	-0.3198E+02	0.1001E+02	-0.2166E+02

VII - 22

Connection type : Header plate connections
Mode : All bolted with column stiffener

Tested by : A.K.Aggarwal (1989) P. New Guinea
Test Id. : M 2

Column : UC 203x203x46
Beam : UB 203x133x25
Plate thickness : 16 mm

Fasteners: 10.9- -M20
7/8" Oversize holes
Material : --
Fy = 36.00 ksi
Fu = -- ksi

Major parameters

lt = 1.0472"	ct = 1.5748"	p = 2.7559"	cc = 1.5748"
lc = 1.0472"	g = 3.9370"	lp = 5.9055"	tp = 0.6299"
tbw= 0.2284"			
n = 2 X 2			

Remark 1) Failed with buckling of the compression flange of the beam
2) Plate welded after coping the beam web

No	Moment (k-in)	Rotation (radians) X 1/1000
1	0.00	0.00
2	18.75	0.51
3	37.51	1.03
4	56.26	1.54
5	75.02	2.05
6	93.77	2.97
7	112.52	3.90
8	131.28	5.03
9	150.03	6.15
10	168.79	6.67
11	187.54	7.18
12	206.29	7.90
13	225.05	8.62
14	243.80	9.44
15	262.56	10.26
16	281.31	12.31
17	300.10	14.36
18	307.58	17.44
19	315.07	20.51
20	322.57	23.93
21	330.07	27.35
22	337.57	30.77
23	354.45	35.90
24	371.33	41.03
25	384.45	46.15
26	397.58	51.28
27	408.84	54.36
28	408.84	57.95
29	408.84	61.54
30	408.84	64.96
31	408.84	68.38
32	408.84	71.80
33	408.84	76.93
34	408.84	82.05
35	408.84	85.47
36	408.84	88.89
37	408.84	92.31
38	416.34	97.44
39	423.84	102.56

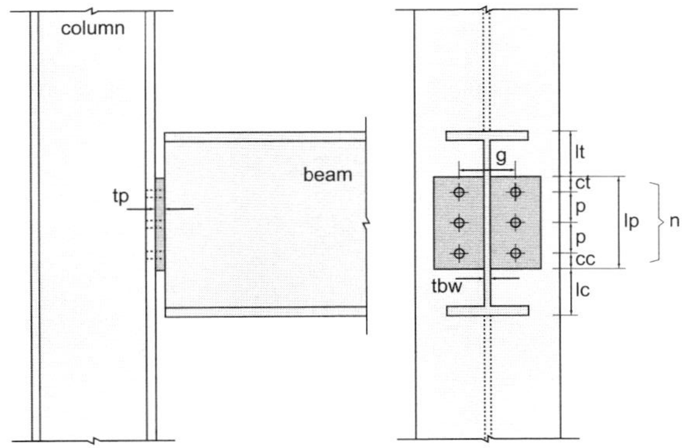

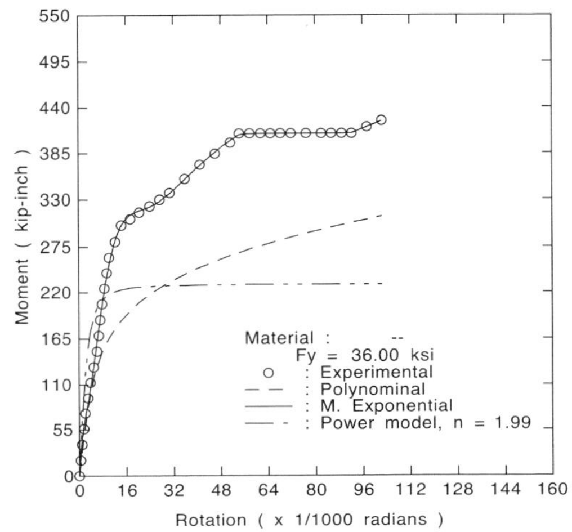

A.7 – 44

```
Moment-rotation prediction equations     ( R : X 1/1000 radians )

Frye and Morris polynominal model  :     R = Sum ( Ai X ( K*Bm )**Pi X 10**Qi )
     g =  3.937000"   d =  5.905500"  t  =  0.629900"    w =  0.228400"
    A1 =  5.100000   A2 =  6.200000   A3 =  2.400000     K =  0.660985
    P1 =         1   P2 =         3   P3 =         5    Q1 =        -2    Q2 =       -7     Q3 =      -10

Modified exponential model  :
    Bm = Sum ( Ai X ( 1 - exp( - R/( 2*i*C ) ) ) )  +      Sum ( RKj X ( R - Rj0 ) )  +  Bm0
     C =   0.98293333E+00              Bmo=   0.00000000E+00          Nexp=  6         Nliner=  4
    AI =   0.87302003E+03  -0.89256443E+04   0.31035357E+05  -0.49046437E+05   0.37062167E+05  -0.10708503E+05
   Rj0 =           0.0000          54.3600          92.3100         102.5600
   RKj =   0.20535697E+01  -0.18982765E+01   0.13504707E+01   0.21045359E+02

Power model  :    Bm = ( Rki X R ) / ( 1 + ( R/R0 )**rn )**( 1/rn )
    rn = 1.987       rki =    0.68458066E+02          rmu =    0.22954547E+03
```

No	Rotation (radians) X 1/1000	Moment (kip-inch) Expri.	Poly.	M.Expo.	P.Model.	Connection stiffness (kip-inch) X 1000 Expri.	Poly.	M.Expo.	P.Model.
1	0.00	0.00	0.00	0.00	0.00	0.3657E+02	0.2966E+02	0.6175E+02	0.6846E+02
3	1.03	37.51	30.26	44.56	67.08	0.3657E+02	0.2913E+02	0.2923E+02	0.5972E+02
5	2.05	75.02	59.11	68.43	119.55	0.3077E+02	0.2671E+02	0.1991E+02	0.4233E+02
7	3.90	112.52	100.85	106.22	173.61	0.1865E+02	0.1824E+02	0.2259E+02	0.1897E+02
9	6.15	150.03	132.91	162.35	201.20	0.3033E+02	0.1106E+02	0.2594E+02	0.7532E+01
11	7.18	187.54	143.23	188.60	207.65	0.3221E+02	0.9220E+01	0.2498E+02	0.5222E+01
13	8.62	225.05	155.11	222.34	213.63	0.2460E+02	0.7448E+01	0.2176E+02	0.3298E+01
15	10.26	262.56	166.15	254.12	217.95	0.1894E+02	0.6100E+01	0.1687E+02	0.2080E+01
17	14.36	300.10	186.84	299.72	223.38	0.6469E+01	0.4220E+01	0.6257E+01	0.8194E+00
19	20.51	315.07	208.30	317.09	226.45	0.2319E+01	0.2923E+01	0.1213E+01	0.2942E+00
21	27.35	330.07	225.59	328.12	227.78	0.2194E+01	0.2207E+01	0.2332E+01	0.1268E+00
23	35.90	354.45	242.12	353.85	228.51	0.3292E+01	0.1708E+01	0.3362E+01	0.5682E-01
25	46.15	384.45	257.68	386.32	228.92	0.2560E+01	0.1357E+01	0.2827E+01	0.2697E-01
27	54.36	408.84	268.02	407.09	229.09	0.1968E+01	0.1172E+01	0.2270E+01	0.1658E-01
29	61.54	408.84	275.97	408.66	229.19	-0.1416E-14	0.1049E+01	0.9806E-01	0.1146E-01
31	68.38	408.84	282.82	408.91	229.26	-0.4250E-14	0.9566E+00	-0.5334E-02	0.8373E-02
33	76.93	408.84	290.58	408.76	229.32	0.0000E+00	0.8631E+00	-0.1568E-01	0.5895E-02
35	85.47	408.84	297.68	408.77	229.36	-0.3008E-14	0.7873E+00	0.2305E-01	0.4306E-02
37	92.31	408.84	302.88	409.05	229.39	0.5852E+00	0.7368E+00	0.5862E-01	0.3423E-02
39	102.56	423.84	310.06	423.84	229.42	0.1463E+01	0.6735E+00	0.2250E+02	0.2500E-02

VII - 23

Connection type : Header plate connections
Mode : All bolted with column stiffener

Tested by : A.K.Aggarwal (1989) P. New Guinea
Test Id. : M 3

Column : UC 203x203x46
Beam : UB 203x133x25
Plate thickness : 12 mm

Fasteners: 10.9- -M20
7/8" Oversize holes
Material : --
Fy = 36.00 ksi
Fu = -- ksi

Major parameters

lt = 1.0472" ct = 1.5748" p = 2.7559" cc = 1.5748"
lc = 1.0472" g = 3.9370" lp = 5.9055" tp = 0.4724"
tbw= 0.2284"
n = 2 X 2

Remark 1) Failed at 7.194 kip shear
2) Failed by excessive deformation of end plate and lateral bending

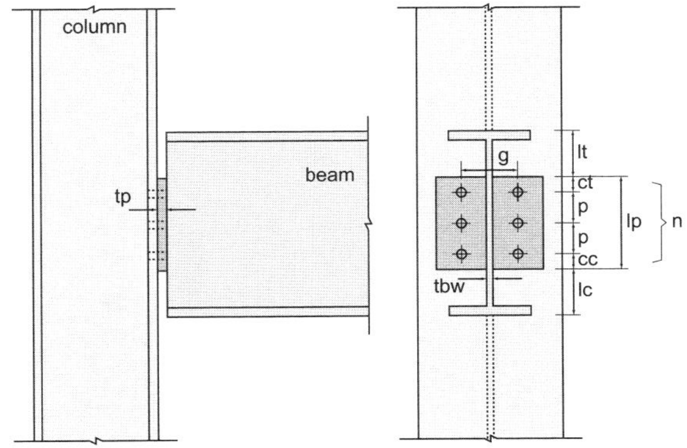

No	Moment (k-in)	Rotation (radians) X 1/1000	No	Moment (k-in)	Rotation (radians) X 1/1000
1	0.00	0.00	26	175.54	47.86
2	9.38	0.51	27	177.04	51.28
3	18.75	1.03	28	178.04	54.70
4	28.13	1.54	29	179.04	58.12
5	37.51	2.05	30	180.04	61.54
6	46.88	2.56	31	181.54	65.64
7	56.26	3.08	32	183.04	69.74
8	65.64	3.59	33	184.54	73.85
9	75.02	4.10	34	186.04	77.95
10	84.39	5.13	35	187.54	82.05
11	93.77	6.15	36	188.36	86.29
12	103.14	6.67	37	189.18	90.53
13	112.52	7.18	38	190.00	94.77
14	117.90	8.72			
15	123.28	10.26			
16	131.20	13.68			
17	139.11	17.09			
18	147.03	20.51			
19	154.28	23.93			
20	161.54	27.35			
21	168.79	30.77			
22	170.04	34.19			
23	171.29	37.61			
24	172.54	41.03			
25	174.04	44.45			

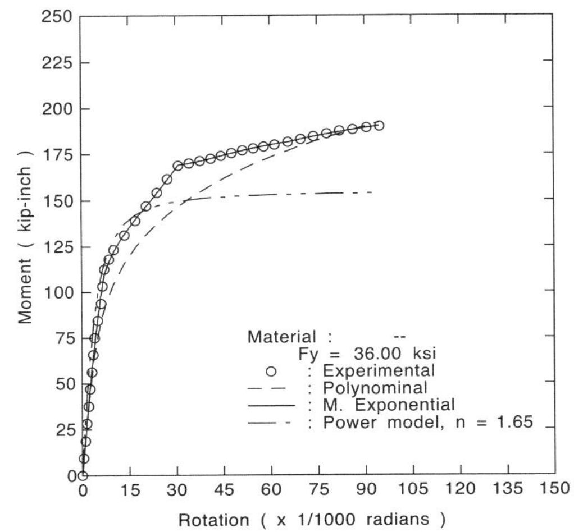

A.7 – 46

```
Moment-rotation prediction equations     ( R : X 1/1000 radians )

Frye and Morris polynominal model  :    R = Sum ( Ai X ( K*Bm )**Pi X 10**Qi )
      g =  3.937000"    d =  5.905500"  t  =  0.472400"    w =  0.228400"
     A1 =  5.100000    A2 =  6.200000   A3 =  2.400000     K =  1.047443
     P1 =          1   P2 =          3  P3 =          5   Q1 =        -2     Q2 =       -7     Q3 =      -10

Modified exponential model  :
     Bm = Sum ( Ai X ( 1 - exp( - R/( 2*i*C ) ) ) )  +      Sum ( RKj X ( R - Rj0 ) )  +  Bm0
       C =   0.91795000E+00               Bmo=   0.00000000E+00          Nexp=  6            Nliner=  4
      AI =  -0.75677134E+02   0.71852573E+03  -0.32783283E+04   0.79395648E+04  -0.86149578E+04   0.33903781E+04
     Rj0 =           0.0000           7.1800          30.7700          65.6400
     RKj =   0.73678812E+01  -0.52176950E+01  -0.21121374E+01   0.17739046E+00

Power model  :    Bm = ( Rki X R ) / ( 1 + ( R/R0 )**rn )**( 1/rn )
     rn = 1.651       rki =    0.28876162E+02          rmu =   0.15452831E+03
```

No	Rotation (radians) X 1/1000	Moment (kip-inch) Expri.	Poly.	M.Expo.	P.Model.	Connection stiffness (kip-inch) X 1000 Expri.	Poly.	M.Expo.	P.Model.
1	0.00	0.00	0.00	0.00	0.00	0.1829E+02	0.1872E+02	0.9680E+01	0.2888E+02
3	1.03	18.75	19.09	18.87	28.50	0.1828E+02	0.1838E+02	0.1906E+02	0.2608E+02
5	2.05	37.51	37.30	38.15	52.90	0.1829E+02	0.1686E+02	0.1827E+02	0.2140E+02
7	3.08	56.26	53.19	56.02	72.44	0.1828E+02	0.1397E+02	0.1650E+02	0.1680E+02
9	4.10	75.02	65.93	71.95	87.64	0.1524E+02	0.1095E+02	0.1459E+02	0.1299E+02
11	6.15	93.77	83.86	98.48	108.46	0.1523E+02	0.6979E+01	0.1149E+02	0.7801E+01
13	7.18	112.52	90.39	109.69	115.57	0.1458E+02	0.5819E+01	0.1041E+02	0.6133E+01
15	10.26	123.28	104.86	122.32	129.34	0.3130E+01	0.3849E+01	0.3292E+01	0.3210E+01
17	17.09	139.11	124.55	139.70	142.21	0.2316E+01	0.2219E+01	0.2149E+01	0.1066E+01
19	23.93	154.28	137.29	153.98	147.14	0.2121E+01	0.1584E+01	0.2109E+01	0.4781E+00
21	30.77	168.79	146.86	169.09	149.54	0.1244E+01	0.1245E+01	0.2321E+01	0.2563E+00
23	37.61	171.29	154.59	171.17	150.90	0.3655E+00	0.1032E+01	0.3807E+00	0.1542E+00
25	44.45	174.04	161.12	174.02	151.76	0.4390E+00	0.8860E+00	0.4333E+00	0.1005E+00
27	51.28	177.04	166.79	176.89	152.33	0.3658E+00	0.7789E+00	0.3965E+00	0.6950E-01
29	58.12	179.04	171.83	179.35	152.73	0.2924E+00	0.6966E+00	0.3210E+00	0.5022E-01
31	65.64	181.54	176.79	181.44	153.06	0.3656E+00	0.6257E+00	0.4122E+00	0.3657E-01
33	73.85	184.54	181.66	184.50	153.31	0.3656E+00	0.5644E+00	0.3373E+00	0.2688E-01
35	82.05	187.54	186.09	187.04	153.51	0.2809E+00	0.5150E+00	0.2864E+00	0.2040E-01
37	90.53	189.18	190.27	189.32	153.66	0.1934E+00	0.4731E+00	0.2543E+00	0.1576E-01

A.7 – 47

VII - 24

Connection type : Header plate connections
Mode : All bolted with column stiffener

Tested by : A.K.Aggarwal (1989) P. New Guinea
Test Id. : M 4

Column : UC 203x203x46
Beam : UB 203x133x25
Plate thickness : 12 mm

Fasteners: 10.9- -M20
7/8" Oversize holes
Material : --
Fy = 36.00 ksi
Fu = -- ksi

Major parameters

lt = 1.0472" ct = 1.5748" p = 2.7559" cc = 1.5748"
lc = 1.0472" g = 3.9370" lp = 5.9055" tp = 0.4724"
tbw= 0.2284"
n = 2 X 2

Remark 1) (a) Failed at 13.5 kip shear (b) Welded after coping the beam web
2) Failed by excessive deformation of the end plate and lateral bending

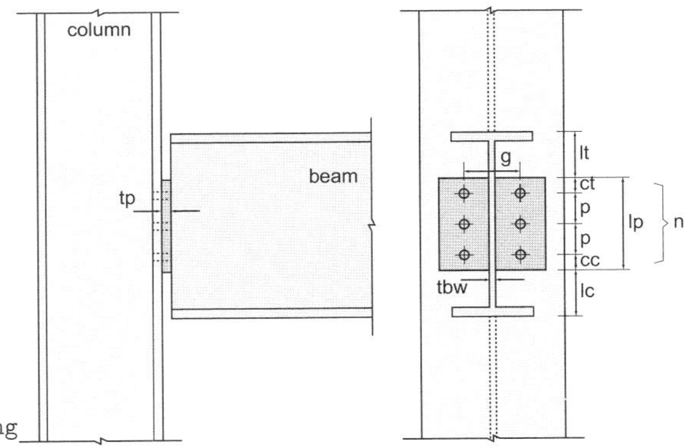

No	Moment (k-in)	Rotation (radians) X 1/1000
1	0.00	0.00
2	12.50	0.34
3	25.01	0.68
4	37.51	1.03
5	46.88	1.54
6	56.26	2.05
7	68.76	3.08
8	81.27	4.10
9	93.77	5.13
10	106.27	6.84
11	118.78	8.55
12	131.28	10.26
13	140.65	12.82
14	150.03	15.39
15	161.28	17.95
16	172.54	20.51
17	187.54	25.64
18	205.04	32.14
19	222.55	38.63
20	240.05	45.13
21	251.30	51.28
22	262.56	56.41
23	270.06	61.54
24	279.44	66.67
25	288.81	71.80

No	Moment (k-in)	Rotation (radians) X 1/1000
26	298.19	76.93
27	307.57	82.05
28	311.32	87.18
29	315.07	92.31
30	318.82	97.44
31	322.57	102.56
32	326.32	107.69
33	330.07	112.82
34	333.82	117.95
35	337.57	123.08
36	341.32	128.21
37	345.07	133.34
38	350.70	138.47
39	356.33	143.59
40	358.21	147.70
41	360.10	151.80

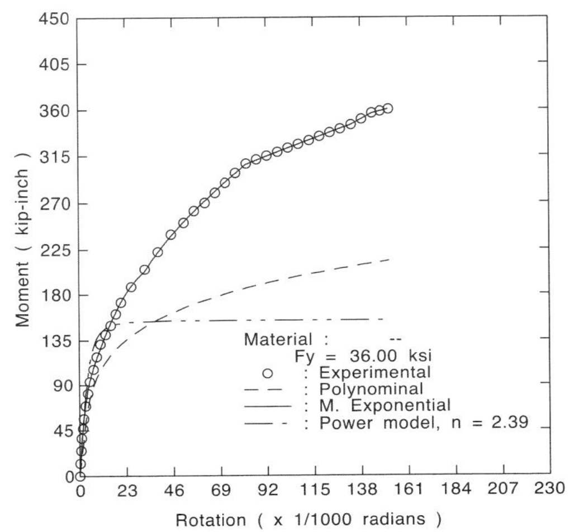

A.7 – 48

```
Moment-rotation prediction equations     ( R : X 1/1000 radians )

Frye and Morris polynominal model  :    R = Sum ( Ai X ( K*Bm )**Pi X 10**Qi )
      g =  3.937000"   d =  5.905500"  t  =  0.472400"   w =  0.228400"
     A1 =  5.100000   A2 =  6.200000  A3 =  2.400000    K =  1.047443
     P1 =         1   P2 =         3  P3 =         5   Q1 =        -2   Q2 =       -7   Q3 =      -10

Modified exponential model  :
     Bm = Sum ( Ai X ( 1 - exp( - R/( 2*i*C ) ) ) )  +      Sum ( RKj X ( R - Rj0 ) )  +  Bm0
      C =  0.13504750E+01              Bmo=  0.00000000E+00          Nexp=  6        Nliner=  4
     AI =  0.13428713E+03  -0.11162606E+03  -0.30844489E+03   0.75657836E+03  -0.61734321E+03   0.37671920E+03
    Rj0 =         32.1400          82.0500         133.3400         143.5900
    RKj =  0.15584608E+01  -0.83166561E+00   0.35830796E+00  -0.62533627E+00

Power model  :     Bm = ( Rki X R ) / ( 1 + ( R/R0 )**rn )**( 1/rn )
     rn = 2.390       rki =    0.28876162E+02          rmu =    0.15452831E+03
```

No	Rotation (radians) X 1/1000	Moment (kip-inch) Expri.	Moment (kip-inch) Poly.	Moment (kip-inch) M.Expo.	Moment (kip-inch) P.Model.	Connection stiffness (kip-inch) X 1000 Expri.	Connection stiffness (kip-inch) X 1000 Poly.	Connection stiffness (kip-inch) X 1000 M.Expo.	Connection stiffness (kip-inch) X 1000 P.Model.
1	0.00	0.00	0.00	0.00	0.00	0.3657E+02	0.1872E+02	0.3855E+02	0.2888E+02
3	0.68	25.01	12.77	23.29	19.69	0.3657E+02	0.1858E+02	0.2997E+02	0.2858E+02
5	1.54	46.88	28.39	45.38	43.51	0.1829E+02	0.1782E+02	0.2219E+02	0.2691E+02
7	3.08	68.76	53.20	72.25	80.49	0.1219E+02	0.1397E+02	0.1368E+02	0.2065E+02
9	5.13	93.77	75.92	94.16	113.12	0.1036E+02	0.8621E+01	0.8469E+01	0.1159E+02
11	8.55	118.78	97.58	117.40	137.29	0.7313E+01	0.4740E+01	0.5785E+01	0.3955E+01
13	12.82	140.65	113.58	140.00	147.16	0.3653E+01	0.3005E+01	0.4905E+01	0.1265E+01
15	17.95	161.28	126.40	163.00	151.08	0.4394E+01	0.2111E+01	0.4040E+01	0.4422E+00
17	25.64	187.54	139.91	188.70	153.02	0.2823E+01	0.1481E+01	0.2672E+01	0.1379E+00
19	38.63	222.55	155.64	222.42	153.96	0.2694E+01	0.1007E+01	0.2706E+01	0.3508E-01
21	51.28	251.30	166.80	251.93	154.24	0.2029E+01	0.7789E+00	0.2053E+01	0.1352E-01
23	61.54	270.06	174.15	271.63	154.34	0.1645E+01	0.6623E+00	0.1816E+01	0.7302E-02
25	71.80	288.81	180.50	289.58	154.40	0.1829E+01	0.5784E+00	0.1696E+01	0.4335E-02
27	82.05	307.57	186.09	306.62	154.43	0.1281E+01	0.5150E+00	0.8025E+00	0.2759E-02
29	92.31	315.07	191.10	314.66	154.46	0.7312E+00	0.4652E+00	0.7689E+00	0.1852E-02
31	102.56	322.57	195.66	322.45	154.47	0.7314E+00	0.4250E+00	0.7504E+00	0.1296E-02
33	112.82	330.07	199.89	330.08	154.48	0.7312E+00	0.3915E+00	0.7401E+00	0.9387E-03
35	123.08	337.57	203.75	337.64	154.49	0.7310E+00	0.3637E+00	0.7342E+00	0.6990E-03
37	133.34	345.07	207.36	345.16	154.50	0.9148E+00	0.3399E+00	0.7310E+00	0.5329E-03
39	143.59	356.33	210.73	356.31	154.50	0.7435E+00	0.3193E+00	0.4621E+00	0.4146E-03
41	151.80	360.10	213.27	360.10	154.51	0.4593E+00	0.3048E+00	0.4612E+00	0.3434E-03

VII - 25

Connection type : Header plate connections
Mode : All bolted with column stiffener

Tested by : A.K.Aggarwal (1939) P. New Guinea
Test Id. : M 5

Column : UC 203x203x46
Beam : UB 203x133x25
Plate thickness : 16 mm

Fasteners: 10.9- -M24
7/8" Oversize holes
Material : --
Fy = 36.00 ksi
Fu = -- ksi

Major parameters

lt = 1.0472"	ct = 1.5748"	p = 2.7559"	cc = 1.5748"
lc = 1.0472"	g = 3.3370"	lp = 5.9055"	tp = 0.6299"
tbw= 0.2284"			
n = 2 X 2			

Remark 1) Large deformation of the column flange and stiffener were observed
2) (a) Welded after coping the beam web (b) The comp. fla. buckled

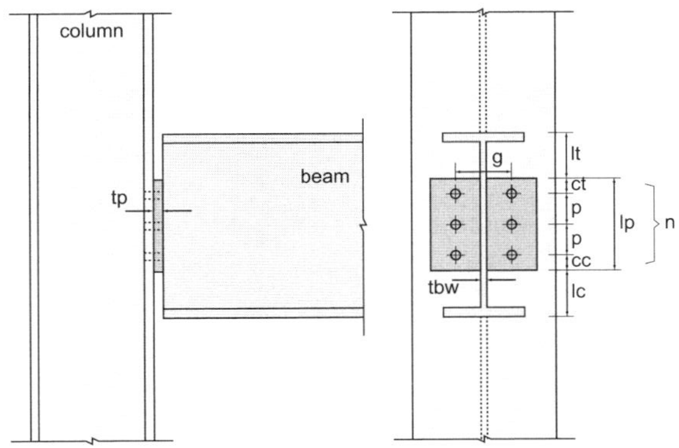

No	Moment (k-in)	Rotation (radians) X 1/1000
1	0.00	0.00
2	12.26	0.35
3	24.52	0.69
4	36.78	1.04
5	47.81	1.56
6	58.84	2.08
7	73.56	3.33
8	82.75	4.26
9	91.94	5.20
10	104.82	6.23
11	117.70	7.27
12	132.41	8.83
13	147.11	10.39
14	159.37	12.12
15	171.63	13.85
16	183.89	15.59
17	196.15	17.32
18	208.41	19.05
19	220.67	20.78
20	235.38	22.86
21	246.41	24.94
22	257.44	27.01
23	266.63	29.09
24	275.83	31.17
25	285.02	32.41
26	294.22	33.66
27	303.42	35.01
28	312.61	36.37
29	321.81	37.92
30	331.00	39.48
31	339.58	41.56
32	348.16	43.64
33	356.74	45.72

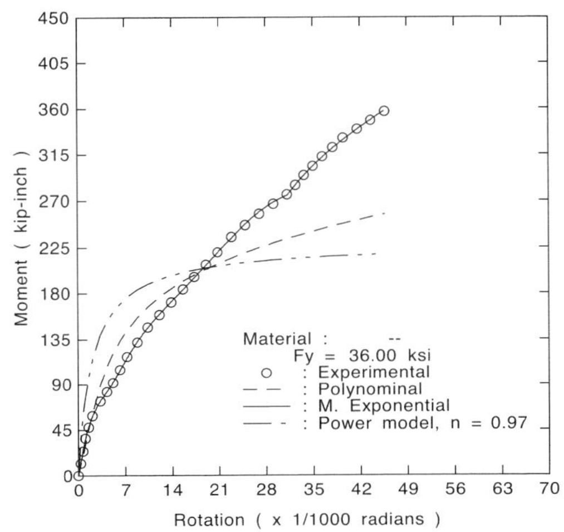

A.7 – 50

```
Moment-rotation prediction equations      ( R : X 1/1000 radians )

Frye and Morris polynominal model  :    R = Sum ( Ai X ( K*Bm )**Pi X 10**Qi )
      g =  3.937000"   d =  5.905500"  t  =  0.629900"   w =  0.228400"
     A1 =  5.100000   A2 =  6.200000   A3 =  2.400000    K =  0.660985
     P1 =         1   P2 =         3   P3 =         5   Q1 =        -2    Q2 =       -7    Q3 =      -10

Modified exponential model  :
     Bm = Sum ( Ai X ( 1 - exp( - R/( 2*i*C ) ) ) )  +      Sum ( RKj X ( R - Rj0 ) )  +  Bm0
      C =   0.46754167E+00              Bmo=   0.00000000E+00         Nexp=  6        Nliner=  4
     AI =  -0.27180865E+03   0.51427072E+04  -0.28874308E+05   0.70637209E+05  -0.77818232E+05   0.31573608E+05
    Rj0 =           3.3300          13.8500          22.8600          31.1700
    RKj =  -0.27143953E+01  -0.30702735E+00   0.76162437E+00   0.55709010E+01

Power model  :    Bm = ( Rki X R ) / ( 1 + ( R/R0 )**rn )**( 1/rn )
      rn = 0.966      rki =   0.11317907E+03           rmu =   0.22954547E+03
```

No	Rotation (radians) X 1/1000	Moment (kip-inch) Expri.	Poly.	M.Expo.	P.Model.	Connection stiffness (kip-inch) X 1000 Expri.	Poly.	M.Expo.	P.Model.
1	0.00	0.00	0.00	0.00	0.00	0.3540E+02	0.2966E+02	0.3497E+02	0.1132E+03
2	0.35	12.26	10.27	12.65	32.99	0.3540E+02	0.2961E+02	0.3684E+02	0.8065E+02
3	0.69	24.52	20.50	24.94	57.29	0.3540E+02	0.2944E+02	0.3364E+02	0.6107E+02
4	1.04	36.78	30.64	35.74	76.04	0.2973E+02	0.2911E+02	0.2860E+02	0.4802E+02
5	1.56	47.81	45.55	48.61	97.39	0.2123E+02	0.2819E+02	0.2117E+02	0.3520E+02
6	2.08	58.84	59.82	58.11	113.39	0.1846E+02	0.2662E+02	0.1579E+02	0.2696E+02
7	3.33	73.56	89.68	74.11	139.31	0.1068E+02	0.2091E+02	0.1153E+02	0.1604E+02
8	4.26	82.75	107.18	82.68	152.13	0.9832E+01	0.1671E+02	0.9741E+01	0.1171E+02
9	5.20	91.94	121.20	92.46	161.69	0.1105E+02	0.1346E+02	0.1110E+02	0.8940E+01
10	6.23	104.82	133.78	104.44	169.84	0.1240E+02	0.1089E+02	0.1177E+02	0.6881E+01
11	7.27	117.70	144.09	116.55	176.21	0.1121E+02	0.9080E+01	0.1140E+02	0.5463E+01
12	8.83	132.41	156.72	133.14	183.53	0.9435E+01	0.7234E+01	0.9782E+01	0.4040E+01
13	10.39	147.11	166.97	147.04	189.06	0.8320E+01	0.6011E+01	0.8133E+01	0.3111E+01
14	12.12	159.37	176.51	160.08	193.80	0.7080E+01	0.5066E+01	0.7082E+01	0.2412E+01
15	13.85	171.63	184.66	172.06	197.53	0.7080E+01	0.4385E+01	0.6554E+01	0.1926E+01
16	15.59	183.89	191.79	183.57	200.55	0.7080E+01	0.3871E+01	0.6784E+01	0.1574E+01
17	17.32	196.15	198.13	195.58	203.04	0.7080E+01	0.3471E+01	0.7071E+01	0.1311E+01
18	19.05	208.41	203.85	207.94	205.13	0.7080E+01	0.3150E+01	0.7158E+01	0.1109E+01
19	20.78	220.67	209.07	220.19	206.90	0.7075E+01	0.2886E+01	0.6941E+01	0.9507E+00
20	22.86	235.38	214.80	234.02	208.72	0.6190E+01	0.2626E+01	0.6290E+01	0.8017E+00
21	24.94	246.41	220.02	247.71	210.26	0.5312E+01	0.2412E+01	0.6090E+01	0.6856E+00
22	27.01	257.44	224.84	259.20	211.58	0.4868E+01	0.2233E+01	0.4963E+01	0.5930E+00
23	29.09	266.63	229.32	268.31	212.74	0.4425E+01	0.2081E+01	0.3802E+01	0.5181E+00
24	31.17	275.83	233.50	275.05	213.75	0.6276E+01	0.1949E+01	0.2700E+01	0.4566E+00
25	32.41	285.02	235.93	284.96	214.29	0.7386E+01	0.1877E+01	0.7663E+01	0.4249E+00
26	33.66	294.22	238.18	294.15	214.81	0.7104E+01	0.1813E+01	0.7103E+01	0.3964E+00
27	35.01	303.42	240.63	303.37	215.32	0.6799E+01	0.1747E+01	0.6553E+01	0.3686E+00
28	36.37	312.61	242.91	311.90	215.80	0.6381E+01	0.1688E+01	0.6065E+01	0.3437E+00
29	37.92	321.81	245.54	320.96	216.32	0.5900E+01	0.1622E+01	0.5575E+01	0.3179E+00
30	39.48	331.00	247.97	329.32	216.80	0.5141E+01	0.1564E+01	0.5160E+01	0.2950E+00
31	41.56	339.58	251.15	339.55	217.38	0.4129E+01	0.1493E+01	0.4708E+01	0.2681E+00
32	43.64	348.16	254.18	348.95	217.91	0.4129E+01	0.1428E+01	0.4357E+01	0.2448E+00
33	45.72	356.74	257.15	357.72	218.40	0.4129E+01	0.1367E+01	0.4087E+01	0.2243E+00

VII - 26

Connection type : Header plate connections
Mode : All bolted with column stiffener

Tested by : A.K.Aggarwal (1989) P. New Guinea
Test Id. : M 6

Column : UC 203x203x46
Beam : UB 203x133x25
Plate thickness : 16 mm

Fasteners: 10.9- -M16
7/8" Oversize holes
Material : --
Fy = 36.00 ksi
Fu = -- ksi

Major parameters

lt = 1.0472"	ct = 1.5748"	p = 2.7559"	cc = 1.5748"
lc = 1.0472"	g = 3.9370"	lp = 5.9055"	tp = 0.6299"
tbw= 0.2284"			
n = 2 X 2			

Remark 1) Large deformation at the column flange
2) Welded after coping the beam web

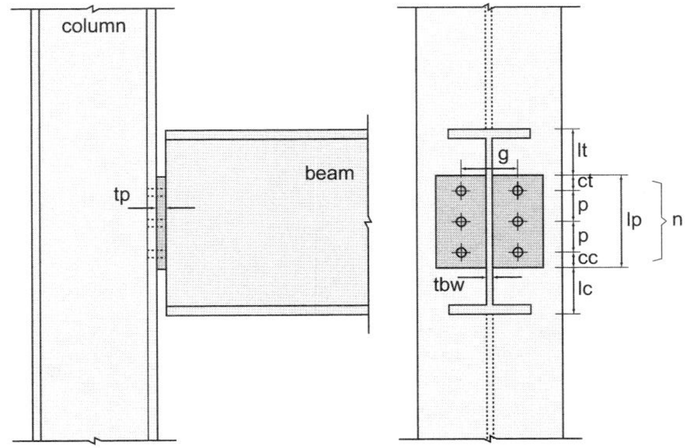

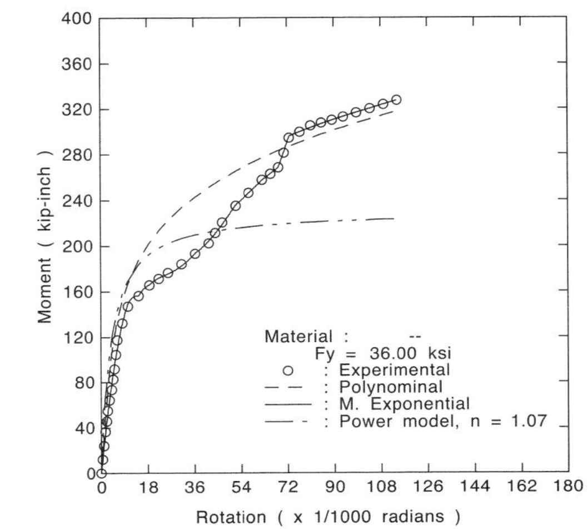

No	Moment (k-in)	Rotation (radians) X 1/1000	No	Moment (k-in)	Rotation (radians) X 1/1000
1	0.00	0.00	26	257.44	62.34
2	12.26	0.55	27	262.96	65.45
3	24.52	1.11	28	268.48	68.57
4	36.78	1.66	29	281.35	70.65
5	45.98	2.08	30	294.22	72.73
6	55.17	2.49	31	299.74	76.89
7	64.36	3.33	32	305.25	81.04
8	73.56	4.16	33	307.70	85.20
9	82.75	4.66	34	310.15	89.35
10	91.94	5.16	35	312.60	93.51
11	104.81	5.70	36	316.29	98.71
12	117.69	6.23	37	319.97	103.90
13	132.40	8.31	38	323.64	109.10
14	147.11	10.39	39	327.32	114.29
15	156.31	14.54			
16	165.50	18.70			
17	171.01	22.34			
18	176.53	25.98			
19	183.89	31.17			
20	193.09	36.37			
21	202.28	41.56			
22	211.47	44.16			
23	220.67	46.76			
24	235.38	51.95			
25	246.41	57.15			

A.7 – 52

```
Moment-rotation prediction equations      ( R : X 1/1000 radians )

Frye and Morris polynominal model  :     R = Sum ( Ai X ( K*Bm )**Pi X 10**Qi )
     g =  3.937000"    d =  5.905500"  t  =  0.629900"    w =  0.228400"
    A1 =  5.100000    A2 =  6.200000   A3 =  2.400000     K =  0.660985
    P1 =         1    P2 =         3   P3 =         5    Q1 =        -2      Q2 =        -7      Q3 =        -10

Modified exponential model  :
    Bm = Sum ( Ai X ( 1 - exp( - R/( 2*i*C ) ) ) )  +      Sum ( RKj X ( R - Rj0 ) )  +  Bm0
      C =   0.10909167E+01                Bmo=   0.00000000E+00           Nexp=  6          Nliner= 10
     AI =   0.74905472E+02  -0.25071099E+03  -0.35123669E+04   0.14964317E+05  -0.18209656E+05   0.70582005E+04
    Rj0 =          10.3900          14.5400          18.7000          36.3700          51.9500
                   62.3400          68.5700          70.6500          72.7300         103.9000
    RKj =  -0.19237258E+01   0.47297899E+01   0.13115126E+01  -0.89989999E+00  -0.14819585E+01
           -0.38362559E+00   0.44455713E+01   0.85670509E+00  -0.60817041E+01   0.46927896E-01

Power model  :     Bm = ( Rki X R ) / ( 1 + ( R/RO )**rn )**( 1/rn )
    rn = 1.068        rki =   0.54068636E+02          rmu =   0.22954547E+03
```

No	Rotation (radians) X 1/1000	Moment (kip-inch) Expri.	Poly.	M.Expo.	P.Model.	Connection stiffness (kip-inch) X 1000 Expri.	Poly.	M.Expo.	P.Model.
1	0.00	0.00	0.00	0.00	0.00	0.2213E+02	0.2966E+02	0.2488E+02	0.5407E+02
3	1.11	24.52	32.65	24.98	49.05	0.2213E+02	0.2902E+02	0.2067E+02	0.3576E+02
5	2.08	45.98	59.82	43.94	78.53	0.2210E+02	0.2662E+02	0.1859E+02	0.2578E+02
7	3.33	64.36	89.63	65.94	105.33	0.1106E+02	0.2092E+02	0.1677E+02	0.1790E+02
9	4.66	82.75	113.47	87.10	125.50	0.1838E+02	0.1522E+02	0.1503E+02	0.1281E+02
11	5.70	104.81	127.57	101.99	137.37	0.2389E+02	0.1211E+02	0.1361E+02	0.1018E+02
13	8.31	132.40	152.80	132.49	158.25	0.7079E+01	0.7765E+01	0.9634E+01	0.6243E+01
15	14.54	156.31	187.61	156.30	183.75	0.2213E+01	0.4163E+01	0.3996E+01	0.2673E+01
17	22.34	171.01	213.42	171.78	198.23	0.1515E+01	0.2686E+01	0.1182E+01	0.1287E+01
19	31.17	183.89	233.51	182.43	206.64	0.1594E+01	0.1949E+01	0.1679E+01	0.7043E+00
21	41.56	202.28	251.15	204.45	212.23	0.2947E+01	0.1493E+01	0.2490E+01	0.4105E+00
23	46.76	220.67	258.50	218.99	214.13	0.3302E+01	0.1341E+01	0.3061E+01	0.3277E+00
25	57.15	246.41	271.21	246.37	216.94	0.2123E+01	0.1121E+01	0.2098E+01	0.2223E+00
27	65.45	262.96	279.97	262.98	218.56	0.1772E+01	0.9940E+00	0.1771E+01	0.1705E+00
29	70.65	281.35	284.96	281.35	219.38	0.6187E+01	0.9296E+00	0.6182E+01	0.1467E+00
31	76.89	299.74	290.55	299.77	220.22	0.1327E+01	0.8635E+00	0.8905E+00	0.1241E+00
33	85.20	307.70	297.41	306.77	221.15	0.5894E+00	0.7900E+00	0.7970E+00	0.1013E+00
35	93.51	312.60	303.72	313.07	221.92	0.6428E+00	0.7291E+00	0.7204E+00	0.8414E-01
37	103.90	319.97	310.96	320.19	222.71	0.7080E+00	0.6661E+00	0.7018E+00	0.6816E-01
39	114.29	327.32	317.66	327.26	223.35	0.7080E+00	0.6136E+00	0.6632E+00	0.5630E-01

Index